Contents

About this Book

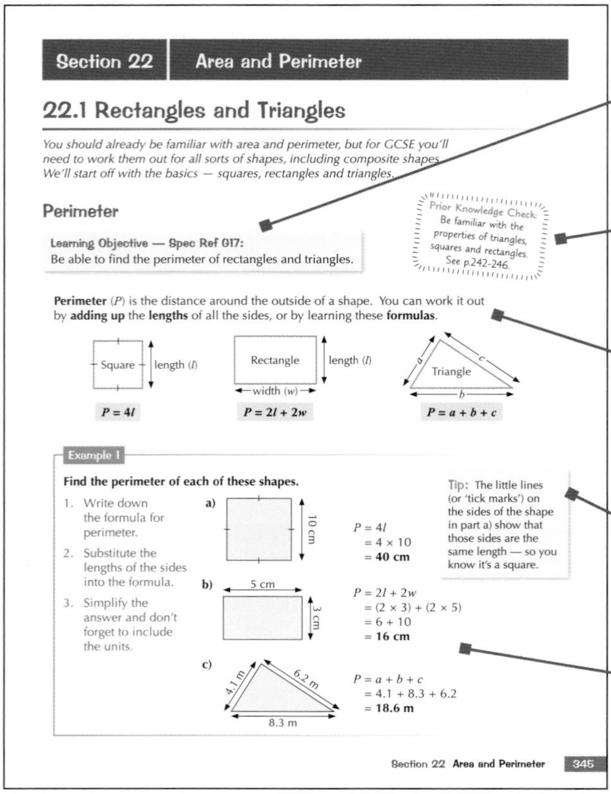

Section 22 Area and Perimeter

22.1 Rectangles and Triangles

You should already be familiar with area and perimeter, but for GCSE you'll need to work them out for all sorts of shapes, including composite shapes. We'll start off with the basics — squares, rectangles and triangles.

Perimeter

Prior Knowledge Check: Be familiar with the properties of triangles, squares and rectangles. See p.242-246.

Learning Objective — Spec Ref G17:
Be able to find the perimeter of rectangles and triangles.

Perimeter (*P*) is the distance around the outside of a shape. You can work it out by **adding up** the **lengths** of all the sides, or by learning these **formulas**.

$P = 4l$ $P = 2l + 2w$ $P = a + b + c$

Example 1

Find the perimeter of each of these shapes.

1. Write down the formula for perimeter.
2. Substitute the lengths of the sides into the formula.
3. Simplify the answer and don't forget to include the units.

a) 10 cm — $P = 4l$ $= 4 \times 10$ $= \mathbf{40\ cm}$

Tip: The little lines (or 'tick marks') on the sides of the shape in part a) show that those sides are the same length — so you know it's a square.

b) 5 cm, 3 cm — $P = 2l + 2w$ $= (2 \times 3) + (2 \times 5)$ $= 6 + 10$ $= \mathbf{16\ cm}$

c) 4.1 m, 6.2 m, 8.3 m — $P = a + b + c$ $= 4.1 + 8.3 + 6.2$ $= \mathbf{18.6\ m}$

Section 22 Area and Perimeter 345

Learning Objectives

Showing which bits of the specification are covered in each section.

Prior Knowledge Checks

Pointing you to the parts of the book that you should be familiar with before moving on to this topic.

Explanations

Clear explanations of every topic.

Tips

Lots of useful tips to help you get your head around the tricky bits.

Examples

Plenty of step-by-step worked examples.

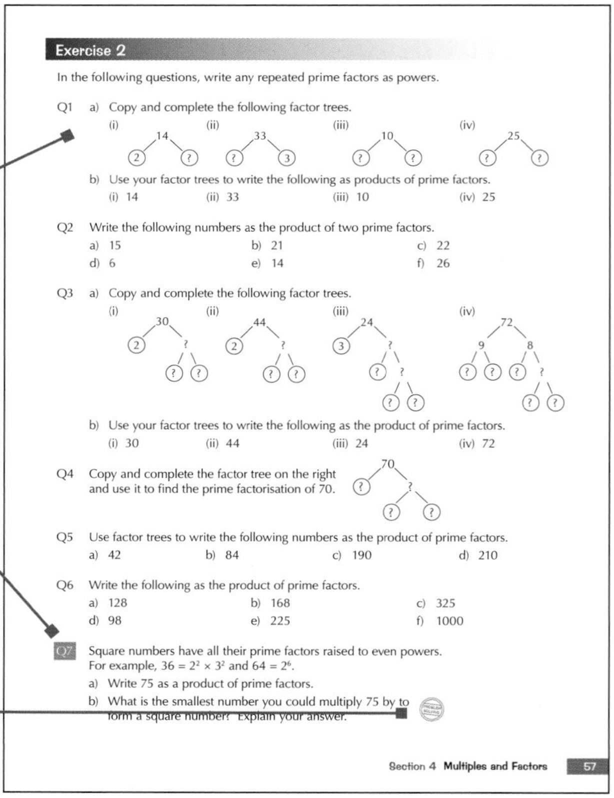

Exercise 2

In the following questions, write any repeated prime factors as powers.

Q1 a) Copy and complete the following factor trees.
(i) 14 (ii) 33 (iii) 10 (iv) 25
b) Use your factor trees to write the following as products of prime factors.
(i) 14 (ii) 33 (iii) 10 (iv) 25

Q2 Write the following numbers as the product of two prime factors.
a) 15 b) 21 c) 22
d) 6 e) 14 f) 26

Q3 a) Copy and complete the following factor trees.
(i) 30 (ii) 44 (iii) 24 (iv) 72
b) Use your factor trees to write the following as the product of prime factors.
(i) 30 (ii) 44 (iii) 24 (iv) 72

Q4 Copy and complete the factor tree on the right and use it to find the prime factorisation of 70.

Q5 Use factor trees to write the following numbers as the product of prime factors.
a) 42 b) 84 c) 190 d) 210

Q6 Write the following as the product of prime factors.
a) 128 b) 168 c) 325
d) 98 e) 225 f) 1000

Q7 Square numbers have all their prime factors raised to even powers.
For example, $36 = 2^2 \times 3^2$ and $64 = 2^6$.
a) Write 75 as a product of prime factors.
b) What is the smallest number you could multiply 75 by to form a square number? Explain your answer.

Section 4 Multiples and Factors 57

Exercises (with worked answers)

- Lots of practice for every topic, with fully worked answers at the back of the book.
- The more challenging questions are marked like this: Q1

Problem Solving

Problem solving questions involve skills such as combining different areas of maths or interpreting given information to identify what's being asked for. Questions that involve problem solving are marked with stamps.

PROBLEM SOLVING

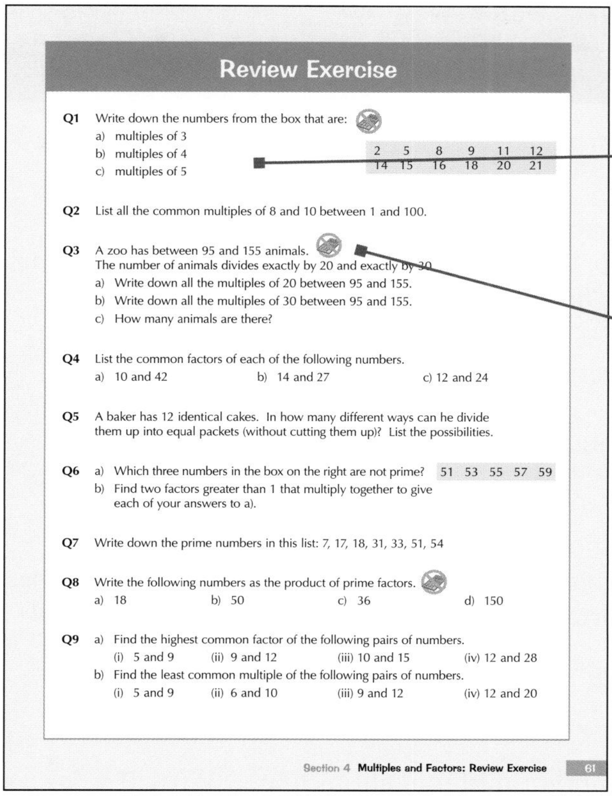

Review Exercise

Q1 Write down the numbers from the box that are:
a) multiples of 3
b) multiples of 4
c) multiples of 5

2	5	8	9	11	12
14	15	16	18	20	21

Q2 List all the common multiples of 8 and 10 between 1 and 100.

Q3 A zoo has between 95 and 155 animals.
The number of animals divides exactly by 20 and exactly by 30.
a) Write down all the multiples of 20 between 95 and 155.
b) Write down all the multiples of 30 between 95 and 155.
c) How many animals are there?

Q4 List the common factors of each of the following numbers.
a) 10 and 42 b) 14 and 27 c) 12 and 24

Q5 A baker has 12 identical cakes. In how many different ways can he divide them up into equal packets (without cutting them up)? List the possibilities.

Q6 a) Which three numbers in the box on the right are not prime? 51 53 55 57 59
b) Find two factors greater than 1 that multiply together to give each of your answers to a).

Q7 Write down the prime numbers in this list: 7, 17, 18, 31, 33, 51, 54

Q8 Write the following numbers as the product of prime factors.
a) 18 b) 50 c) 36 d) 150

Q9 a) Find the highest common factor of the following pairs of numbers.
(i) 5 and 9 (ii) 9 and 12 (iii) 10 and 15 (iv) 12 and 28
b) Find the least common multiple of the following pairs of numbers.
(i) 5 and 9 (ii) 6 and 10 (iii) 9 and 12 (iv) 12 and 20

Section 4 **Multiples and Factors: Review Exercise** 61

Review Exercises

Mixed questions covering the whole section, with fully worked answers.

Non-Calculator Questions

There are some methods you'll have to be able to do without a calculator. Stamps on certain questions let you know that you can't use a calculator for them.

Exam-Style Questions

Questions in the same style as the ones you'll get in the exam, with worked solutions and mark schemes.

Mixed Exam-Style Questions

At the end of the book is a set of exam questions covering a mixture of different topics from across the GCSE 9-1 course.

Glossary

All the definitions you need to know for the exam, plus other useful words.

Formula Page

Contains all the formulas that you need to know for your GCSE exams. You'll find it inside the back cover.

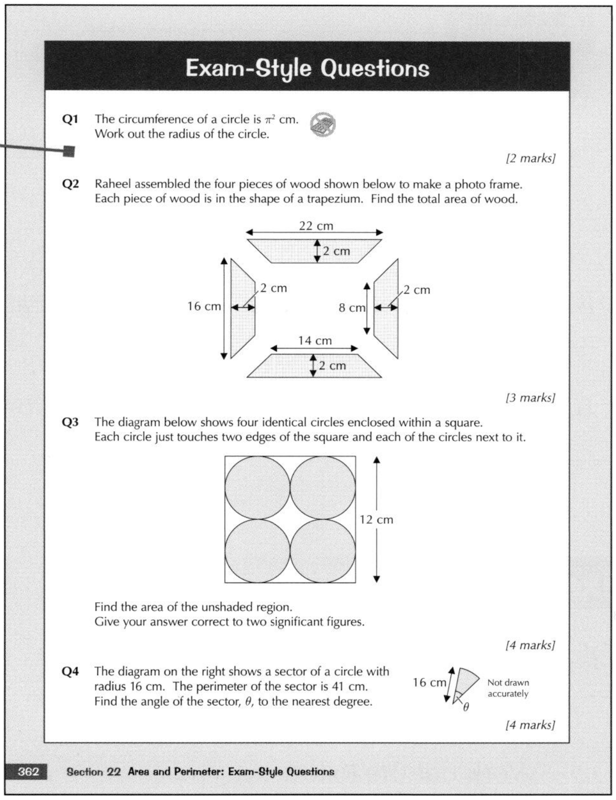

Exam-Style Questions

Q1 The circumference of a circle is π^2 cm.
Work out the radius of the circle.
[2 marks]

Q2 Raheel assembled the four pieces of wood shown below to make a photo frame.
Each piece of wood is in the shape of a trapezium. Find the total area of wood.

[3 marks]

Q3 The diagram below shows four identical circles enclosed within a square.
Each circle just touches two edges of the square and each of the circles next to it.

Find the area of the unshaded region.
Give your answer correct to two significant figures.
[4 marks]

Q4 The diagram on the right shows a sector of a circle with radius 16 cm. The perimeter of the sector is 41 cm.
Find the angle of the sector, θ, to the nearest degree.

[4 marks]

362 Section 22 **Area and Perimeter: Exam-Style Questions**

Section 1 Non-Calculator Arithmetic

1.1 Orders of Operations

What a nice, fun start to the book — no calculators allowed. It's all about adding, subtracting, multiplying and dividing, but the trick is knowing how to do it all in the right order.

Learning Objective — Spec Ref N3:
Know the correct order in which to apply operations, including dealing with brackets.

Operations in mathematics are things like addition, subtraction, multiplication and division. The **order** in which each of these is carried out within a calculation is **very important**.

Tip: The 'Other' operations are things like powers or roots — see Section 3.

BODMAS tells you the correct order to carry out these operations.

Brackets, **O**ther, **D**ivision, **M**ultiplication, **A**ddition, **S**ubtraction

If there are two or more **consecutive** divisions and/or multiplications (e.g. $3 \times 6 \div 9 \times 5$), they should be done in order, **from left to right**. The same goes for **addition and subtraction**.

Example 1

Work out 20 – 3 × (4 + 2).

1. There are **B**rackets, so do that operation first. ⟶ $20 - 3 \times (4 + 2) = 20 - 3 \times 6$
2. There are no '**O**ther' operations.
3. There is no **D**ivision.
4. There is **M**ultiplication, so do that operation next. ⟶ $20 - 3 \times 6 = 20 - 18$
5. There is no **A**ddition.
6. There is **S**ubtraction, so do that operation next. ⟶ $20 - 18 = \mathbf{2}$

Exercise 1

Q1 Work out the following.

a) $5 + 1 \times 3$ b) $11 - 2 \times 5$ c) $24 \div 4 + 2$ d) $18 - 10 \div 5$

Q2 Work out the following.

a) $6 \times (4 + 3)$ b) $11 - (2 + 3)$ c) $(8 - 7) \times (6 + 5)$ d) $56 \div (2 \times 4)$

Q3 Work out the following.

a) $2 \times (8 + 4) - 7$ b) $18 \div (9 - 12 \div 4)$ c) $100 \div (8 + 3 \times 4)$

d) $20 - (5 \times 3 + 2)$ e) $48 \div 3 - 7 \times 2$ f) $36 - (7 + 4 \times 4)$

Q4 Copy each of the following and insert brackets to make the calculation correct.

a) $9 \times 7 - 5 = 18$ b) $18 - 6 \div 3 = 4$ c) $5 + 2 \times 6 - 2 = 28$

d) $21 \div 4 + 3 = 3$ e) $13 - 5 \times 13 - 1 = 96$ f) $6 + 8 - 7 \times 5 = 35$

Q5 Copy each of the following and: (i) fill in the blanks with either +, –, × or ÷ ,
(ii) add any brackets needed to complete the calculation.

a) 16 □ 6 ÷ 3 = 14 b) 11 □ 3 + 5 = 38 c) 3 □ 6 □ 9 = 9

d) 8 □ 2 □ 6 = 10 e) 3 □ 7 □ 4 = 40 f) 14 □ 6 □ 8 = 1

Example 2

Work out $\frac{2 \times 4 + 12}{9 - 5}$.

1. You need to divide the top line by the bottom line.
2. Work out what's on the top and the bottom separately.

Top line: $2 \times 4 + 12$

1. No **B**rackets, '**O**ther' operations or **D**ivision
2. There is **M**ultiplication, so do that operation first. ⟶ $2 \times 4 = 8$
3. Then do the **A**ddition. ⟶ $8 + 12 = \mathbf{20}$

Bottom line: $9 - 5$

There is only **S**ubtraction here. ⟶ $9 - 5 = \mathbf{4}$

3. After putting the values for the top and bottom line back into the fraction, you get this expression. ⟶ $\frac{20}{4}$
4. It only has **D**ivision. ⟶ $20 \div 4 = \mathbf{5}$

Exercise 2

Q1 Work out the following.

a) $\frac{4 - 1 + 5}{2 \times 2}$ b) $\frac{6 + (11 - 8)}{7 - 5}$ c) $\frac{16}{4 \times (5 - 3)}$ d) $\frac{4 \times (7 + 5)}{6 + 3 \times 2}$

e) $\frac{12 \div (9 - 5)}{25 \div 5}$ f) $\frac{8 \times 2 \div 4}{5 - 6 + 7}$ g) $\frac{3 \times 3}{21 \div (12 - 5)}$ h) $\frac{36 \div (11 - 2)}{8 - 8 \div 2}$

1.2 Negative Numbers

You can add, subtract, multiply and divide with negative numbers too, but there are few things that might catch you out if you're not careful. Be sure to keep track of all those minus signs...

Adding and Subtracting Negative Numbers

Learning Objective — Spec Ref N2:
Add and subtract negative numbers.

Negative numbers are numbers which are **less than zero**.

You can count places on a **number line** to help with calculations involving negative numbers.

Example 1

Work out: **a) –1 + 4** **b) –2 – 3 + 1**

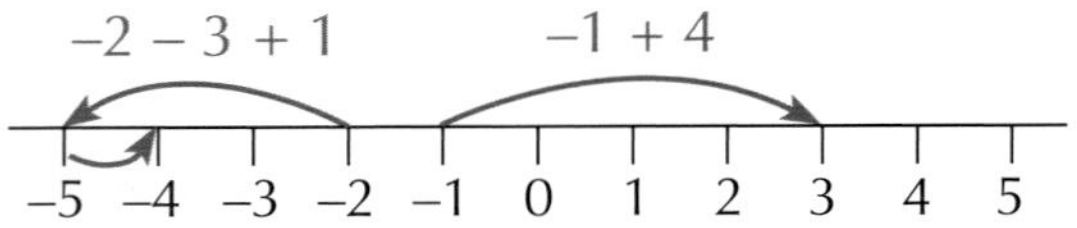

a) Start at –1 and count 4 places to the right. ⟶ –1 + 4 = **3**

b) Start at –2, count 3 places to the left, and then 1 place to the right. ⟶ –2 – 3 + 1 = **–4**

Tip: If you find it useful, draw your own number line when answering negative number questions.

Exercise 1

Q1 Work out the following.

a) –4 + 3 b) –1 + 5 c) –2 + 1
d) 6 – 17 e) –13 + 18 f) 11 – 19

Q2 Work out the following.

a) –3 + 2 – 1 b) –2 + 8 – 5 c) 8 – 5 – 3
d) –1 – 7 – 4 e) –9 + 13 + 11 f) 7 – 18 + 11

Q3 Copy the following calculations and fill in the blanks to make the equations correct.

a) –3 + □ = –1 b) 5 – □ = –10 c) □ + 7 = 4 d) □ – 2 = –6

The thermometer on the right shows temperature. Use it to answer **Questions 4-7**.

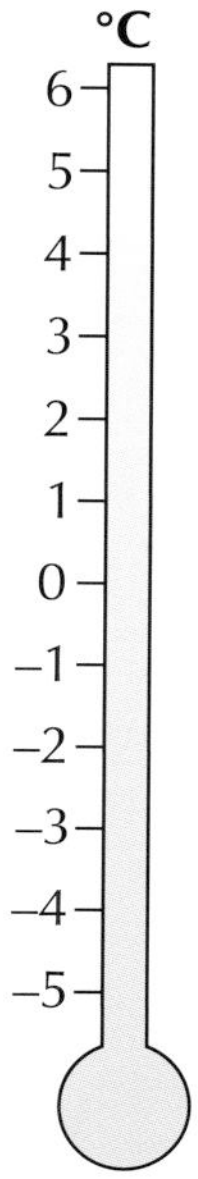

Q4 a) Find the temperature that is 3 °C lower than –1 °C.
b) Find the temperature that is 8 °C higher than –5 °C.

Q5 Find the difference between the following temperatures.
a) –4 °C and 3 °C
b) –2 °C and –5 °C

Q6 At midday the temperature was 6 °C. By midnight, the temperature had decreased by 7 °C. What was the temperature at midnight?

Q7 At 5 am, the temperature was –5 °C. At 11 pm, the temperature was –1 °C. What was the difference in temperature between these two times?

Rules for Adding and Subtracting Negative Numbers

When adding and subtracting negative numbers, there are two rules you need to know:

1. Adding a negative number is the same as subtracting a positive number. → **'+' next to '–' means subtract**
2. Subtracting a negative number is the same as adding a positive number. → **'–' next to '–' means add**

Example 2

Work out:

a) 1 – (–4) '–' next to '–' means add. 1 – (–4) = 1 + 4 = **5**

b) –5 + (–2) '+' next to '–' means subtract. –5 + (–2) = –5 – 2 = **–7**

Exercise 2

Q1 Work out the following.
a) 4 – (–2) b) –3 – (–5) c) –7 + (–2) d) –5 + (–5)
e) 9 + (– 2) f) –13 – (–3) g) –6 + (–3) h) 1 – (–12)

Q2 Work out the following.
a) –1 + (–4) – 1 b) 6 – (–2) + (–3) c) 7 + (–6) – (–8)
d) –8 + (–8) – (–12) e) 9 – (–13) + (–2) f) –3 + (–11) + (–6)

Q3 Find the difference between the following numbers by subtracting one from the other.
a) 6 and –5 b) –10 and –6 c) –8 and 4 d) –5 and –12

Multiplying and Dividing Negative Numbers

Learning Objective — Spec Ref N2:
Multiply and divide negative numbers.

When you multiply or divide two numbers which have the **same** sign, the answer is **positive** — e.g. $(-6) \times (-7) = 42$ and $(-40) \div (-10) = 4$.

When you multiply or divide two numbers with **opposite** signs, the answer is **negative** — e.g. $(-4) \times 6 = -24$ and $35 \div (-7) = -5$.

Tip: You can think of two minus signs cancelling each other out. If there's only one minus, it stays.

Example 3

Work out:

a) $\mathbf{24 \div (-6)}$

The signs are opposite (one positive, one negative), so the answer is negative. → $24 \div -6 = \mathbf{-4}$

b) $\mathbf{-5 \times (-8)}$

The signs are the same (both negative), so the answer is positive. → $-5 \times -8 = \mathbf{40}$

c) $\mathbf{-3 \times 7}$

The signs are different (one negative, one positive), so the answer is negative. → $-3 \times 7 = \mathbf{-21}$

Exercise 3

Q1 Work out the following.

a) $3 \times (-4)$ b) $2 \times (-8)$ c) $(-9) \times 6$

d) $(-15) \div 3$ e) $(-15) \div (-3)$ f) $12 \div (-4)$

Q2 Copy the following calculations and fill in the blanks.

a) $-3 \times \square = -6$ b) $-14 \div \square = -2$ c) $\square \times 4 = -16$ d) $\square \div (-2) = -5$

e) $-8 \times \square = -24$ f) $-18 \div \square = 3$ g) $\square \times (-3) = 36$ h) $\square \div 11 = -7$

Q3 Work out the following.

a) $2 \times 4 \div (-2)$ b) $(-5) \times (-6) \div 3$ c) $(-3) \times (-5) \times (-6)$

d) $(-63) \times (-2) \div (-9)$ e) $[55 \div (-11)] \times (-9)$ f) $[(-24) \div 8)] \div 3$

1.3 Whole Number Arithmetic

There's no getting around it — you need to be able to handle big numbers without a calculator.

Addition and Subtraction

Learning Objective — Spec Ref N2:
Add and subtract large numbers without using a calculator.

When adding or subtracting **large numbers**, write one number **above the other**, then add or subtract **one column at a time**.

It's important to make sure that the digits are lined up correctly, or you'll get the wrong answer.

Tip: The units in each number should be lined up, as should the tens, and the hundreds, etc.

Example 1

Work out 145 + 28.

1. Write one number above the other and make sure each column lines up, from right to left.
2. Add together the digits in each column, starting with the furthest right.
3. If the digits in a column add up to 10 or more, you'll need to 'carry' a 1. In the first column here 5 + 8 = 13, so you write 3 in the column you're working with, and carry 1 to the next — add that on when you add the next column (4 + 2 + 1 here).

$$\begin{array}{r} 1\,4\,5 \\ +\ \ 2\,8 \\ \hline {}_{1}3 \end{array} \qquad \begin{array}{r} 1\,4\,5 \\ +\ \ 2\,8 \\ \hline 7_{1}3 \end{array} \qquad \begin{array}{r} 1\,4\,5 \\ +\ \ 2\,8 \\ \hline \mathbf{1\,7_{1}3} \end{array}$$

Example 2

Work out 364 – 128.

1. Write the bigger number above the smaller one.
2. Subtract the digits in each column, starting with the furthest right.
3. If the top digit is smaller than the bottom digit, borrow 10 from the next column. To do this, subtract 1 from the top digit in that column, and write 1 in front of the digit in your current column. Doing this to the 1st column here gives 14 – 8 = 6.
4. Don't forget you've borrowed when you subtract the next column. In the 2nd column here, you get 5 – 2 = 3.

$$\begin{array}{r} 3\,\overset{5}{\cancel{6}}\,{}^{1}4 \\ -\,1\,2\,8 \\ \hline 6 \end{array} \qquad \begin{array}{r} 3\,\overset{5}{\cancel{6}}\,{}^{1}4 \\ -\,1\,2\,8 \\ \hline 3\,6 \end{array} \qquad \begin{array}{r} 3\,\overset{5}{\cancel{6}}\,{}^{1}4 \\ -\,1\,2\,8 \\ \hline \mathbf{2\,3\,6} \end{array}$$

Example 3

Tasmin has £305 in her bank account. She uses £146 to buy a television. How much money is left in her account?

1. You need to subtract 146 from 305, so set out the subtraction as usual.
2. In the column furthest to the right, 5 is less than 6 so you need to borrow from the middle column. But there's a 0 there, so you have to borrow from the column furthest to the left first.

$$\begin{array}{r} 3\,0\,5 \\ -\ \underline{1\,4\,6} \end{array} \longrightarrow \begin{array}{r} {}^{2}\cancel{3}\,{}^{1}0\,5 \\ -\ \underline{1\,4\,6} \end{array} \longrightarrow \begin{array}{r} {}^{2}\cancel{3}\,\overset{9}{\cancel{{}^{1}0}}\,{}^{1}5 \\ -\ \underline{1\,4\,6} \end{array}$$

3. Once the borrowing is sorted, subtract each column as usual.

$$\begin{array}{r} {}^{2}\cancel{3}\,\overset{9}{\cancel{{}^{1}0}}\,{}^{1}5 \\ -\ \underline{1\,4\,6} \\ 9 \end{array} \longrightarrow \begin{array}{r} {}^{2}\cancel{3}\,\overset{9}{\cancel{{}^{1}0}}\,{}^{1}5 \\ -\ \underline{1\,4\,6} \\ 5\,9 \end{array} \longrightarrow \begin{array}{r} {}^{2}\cancel{3}\,\overset{9}{\cancel{{}^{1}0}}\,{}^{1}5 \\ -\ \underline{1\,4\,6} \\ 1\,5\,9 \end{array}$$

So she has **£159** left.

Exercise 1

Q1 Copy and work out the following.

a) $\begin{array}{r} 2\,3 \\ +\ \underline{5\,6} \end{array}$ b) $\begin{array}{r} 1\,2\,2 \\ +\ \underline{9\,7} \end{array}$ c) $\begin{array}{r} 2\,4\,3 \\ +\ \underline{1\,7\,8} \end{array}$

d) $\begin{array}{r} 7\,3 \\ -\ \underline{2\,7} \end{array}$ e) $\begin{array}{r} 1\,8\,1 \\ -\ \underline{3\,5} \end{array}$ f) $\begin{array}{r} 2\,3\,3 \\ -\ \underline{1\,8\,7} \end{array}$

Q2 Work out the following.

a) 342 + 679 b) 604 + 288 c) 506 – 278

d) 2513 + 241 e) 2942 – 324 f) 4003 – 1235

Q3 Work out the following.

a) 41 + 112 + 213 b) 764 + 138 – 345 c) 123 + 478 + 215

d) 221 + 126 – 98 e) (498 – 137) + 556 f) (987 – 451) + 221

Q4 Lisa has 38 marbles, John has 52 marbles and Bina has 65 marbles.
How many marbles do they have altogether?

Q5 An art club has £2146 in the bank. It spends £224 on art supplies.
How much money does it have left?

Q6 The village of Great Missingham has a population of 1845.
The neighbouring village of Fentley has a population of 1257.
How many more people live in Great Missingham than Fentley?

Multiplication

Learning Objective — Spec Ref N2:
Multiply large numbers without using a calculator.

There are two methods for multiplying large numbers. Example 4 uses **long multiplication**, while Example 5 uses a **grid method.**

For both methods it's **very important** that the digits are **lined up** properly, starting on the right hand side.

Tip: If you're multiplying in an exam, just pick your favourite method.

Example 4

Work out 254 × 26.

1. Write one number above the other, making sure the columns line up.
2. Multiply each digit of 254 by 6, working from right to left. If the answer is 10 or more, carry the ten's digit. This is 254 × 6.
3. Now put a 0 in the right-hand column and multiply each digit of 254 by 2, carrying digits where necessary. This is 254 × 20.
4. Add the two rows together to find the final answer.

```
  2 5 4        2 5 4        2 5 4
×   2 6      ×   2 6      ×   2 6
      4          2 4      1 5 2 4
     2          3 2         3 2

  2 5 4        2 5 4        2 5 4        2 5 4
×   2 6      ×   2 6      ×   2 6      ×   2 6
1 5 2 4      1 5 2 4      1 5 2 4      1 5 2 4
  3 2          3 2          3 2          3 2
      0          8 0      1 0 8 0      5 0 8 0
                         1            1

  1 5 4
×   2 6
1 5 2 4
  3 2
5 0 8 0
 1
6 6 0 4
   1
```

So 254 × 26 = **6604**

Exercise 2

Use **long multiplication** to answer these questions.

Q1 Work out:

a) 66 × 72 b) 79 × 86 c) 83 × 81

d) 100 × 26 e) 114 × 30 f) 216 × 54

Q2 Jilly tells her friend that there are 46 days until Christmas. How many hours is this?

Q3 Dom's monthly electricity bill is £33. How much does he spend on electricity each year?

Q4 Work out the following.

a) 1347 × 20 b) 3669 × 21 c) 2623 × 42 d) 2578 × 36

Example 5

Work out 254 × 26.

1. Split each number up into units, tens, hundreds etc.
2. Put these around the outside of a grid.
3. Multiply the numbers at the edge of each box. To multiply by 200, multiply by 2, then by 100. To multiply by 50, multiply by 5, then by 10.
4. Add the numbers from the boxes to find the final answer.

	200	50	4
20	20 × 200 = 4000	20 × 50 = 1000	20 × 4 = 80
6	6 × 200 = 1200	6 × 50 = 300	6 × 4 = 24

$$\begin{array}{r} 4000 \\ 1000 \\ 80 \\ 1200 \\ 300 \\ +\quad 24 \\ \hline 6604 \end{array}$$

(carry: 1)

Tip: When adding, line up all the numbers on top of each other, from right to left.

So 254 × 26 = **6604**

Exercise 3

Q1 Work out the following using the grid method.

a) 13 × 12 b) 11 × 17 c) 52 × 10 d) 16 × 24

e) 367 × 62 f) 498 × 69 g) 511 × 55 h) 568 × 74

Q2 Sheila earns £243 per week. How much does she earn in 52 weeks?

Q3 The weight of a storage container is 14 tonnes. A cargo ship can carry 230 of these containers. How many tonnes can the cargo ship carry?

Q4 A box of matches contains 85 matches. There are 160 boxes in a carton. How many matches are there in a carton?

Q5 Hamza is making biscuits. Each biscuit uses 22 g of flour. Can he make 27 biscuits from a 500 g bag of flour? (PROBLEM SOLVING)

Q6 Work out the following.

a) 2271 × 25 b) 5624 × 42 c) 1211 × 34 d) 4526 × 19

Division

Learning Objective — Spec Ref N2:
Divide numbers without using a calculator.

There are different ways to divide numbers without using a calculator. Example 6 shows '**short division**' and Example 7 shows '**long division**'.

Example 6

Work out 6148 ÷ 4.

Write the division with the number you want to divide inside and the number you're dividing by to the left.

1. Look at the first digit, 6. 4 goes into 6 once with 2 left over, so write 1 above 6, keeping the columns in line. Then carry the remainder, 2, over to the next column.

$$\begin{array}{r} 1\phantom{\,{}^{2}1\,4\,8} \\ 4\overline{)6\,{}^{2}1\,4\,8} \end{array}$$

2. Move to the next column — 2 was carried over, so find the number of times 4 goes into 21, **not** 1. 4 goes into 21 five times with 1 left over, so write 5 above the line and carry the 1 over to the next column.

$$\begin{array}{r} 1\,5\phantom{\,{}^{1}4\,8} \\ 4\overline{)6\,{}^{2}1\,{}^{1}4\,8} \end{array}$$

3. In the next column, 4 goes into 14 three times with 2 left over, so write 3 above the line and carry the 2 to the next column.

$$\begin{array}{r} 1\,5\,3\phantom{\,{}^{2}8} \\ 4\overline{)6\,{}^{2}1\,{}^{1}4\,{}^{2}8} \end{array}$$

4. Finally, 4 goes into 28 seven times with nothing left over, so write 7 above the line.

$$\begin{array}{r} \mathbf{1\,5\,3\,7} \\ 4\overline{)6\,{}^{2}1\,{}^{1}4\,{}^{2}8} \end{array}$$

So 6148 ÷ 4 = **1537**

Long division is similar to short division, but uses **written subtraction** to find **remainders** (the "left over" numbers) instead of carrying digits across.

Example 7

Work out 385 ÷ 16.

1. 16 does not go into 3, so write 0 above the line and look at the first **two** digits instead, 38.

2. 16 × 2 = 32 but 16 × 3 = 48, so 16 goes into 38 twice. Write 2 above the line at the top, then write 32 underneath 38 and subtract. This gives a difference of 6.

$$\begin{array}{r} 0\,2 \\ 16\overline{)3\,8\,5} \\ -\underline{3\,2} \\ 6 \end{array}$$

3. Instead of carrying the remainder over to the next column, bring the digit in the next column down.

$$\begin{array}{r} 0\,2 \\ 16\overline{)3\,8\,5} \\ -\underline{3\,2}\downarrow \\ 6\,5 \end{array}$$

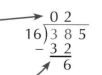

4. Look at how many 16's go into 65. 16 × 4 = 64, so write 4 above the line at the top, and write 64 underneath 65 and subtract.

5. When there are no more digits to bring down, the number on top gives you the whole part of the answer, and any number at the bottom is the remainder. To get an exact answer for the division, write the remainder as a fraction of the number you divided by.

$$\begin{array}{r} 0\,2\,4 \\ 16\overline{)3\,8\,5} \\ -\,3\,2 \\ \hline 6\,5 \\ -\,6\,4 \\ \hline 1 \end{array}$$

$385 \div 16 =$ **24 remainder 1**

$= \mathbf{24\frac{1}{16}}$

Exercise 4

Q1 Use short division to work out the following.

a) $7\overline{)357}$ b) $7\overline{)238}$ c) $9\overline{)603}$

d) $8\overline{)3616}$ e) $4\overline{)2240}$ f) $6\overline{)3732}$

Q2 Use short division to work out the following, giving the remainder as a fraction.

a) $361 \div 5$ b) $5213 \div 3$ c) $4198 \div 5$

d) $671 \div 9$ e) $2545 \div 6$ f) $2674 \div 8$

Q3 Use long division to work out the following.

a) $23\overline{)644}$ b) $19\overline{)608}$ c) $18\overline{)738}$

d) $17\overline{)816}$ e) $28\overline{)8680}$ f) $26\overline{)9152}$

Q4 Use long division to work out the following. Give any remainders as fractions.

a) $846 \div 32$ b) $945 \div 45$ c) $711 \div 21$

d) $6718 \div 25$ e) $8323 \div 31$ f) $8539 \div 44$

Q5 A postage stamp costs 41p. How many stamps could you buy for £5.00, and how much change would you receive?

Q6 672 people are divided into 24 equal groups. How many people are in each group?

Q7 An egg box can hold 6 eggs. How many egg boxes are needed to hold 1350 eggs?

Q8 22 washing machines cost £7634 in total. How much does each washing machine cost?

Q9 a) Mike earns £22 464 per year. How much does he earn per week? (Assume 1 year = 52 weeks.)

b) Andrea earns £24 636 per year. How much does she earn per month?

1.4 Decimals

Decimals are a way to show numbers that aren't whole numbers. They will come up a lot in maths, so make sure you understand what they mean and how they work before moving on.

Decimals

Learning Objective — Spec Ref N2:
Understand place values in decimals.

The digits that come after a **decimal point** show the **part** of a number that is **less than 1**. You can work out the **value** of each digit by looking at its **position** after the decimal point.

6	2	7	.	3	8	1
Hundreds	Tens	Units	Decimal Point	Tenths	Hundredths	Thousandths

Example 1

What is the value of each of the digits in 0.692?

1. The 6 is in the tenths column. Six tenths = $\frac{6}{10}$
2. The 9 is in the hundredths column. Nine hundredths = $\frac{9}{100}$
3. The 2 is in the thousandths column. Two thousandths = $\frac{2}{1000}$

Tip: You don't need to simplify these fractions any further.

Exercise 1

Q1 Write down the value of the 7 in 4.271.

Q2 Write down the value of the 3 in 6.382.

Q3 Write down the value of the 8 in 0.718.

Q4 Write down the value of the 1 in 9.0361.

Q5 Write down the value of the 2 in 2.37.

Q6 Write down the following as decimal numbers.
a) seven tenths
b) two hundredths
c) five thousandths
d) $\frac{7}{100}$
e) $\frac{1}{1000}$
f) $\frac{8}{10}$

Ordering Decimals

Learning Objective — Spec Ref N1:
Order and compare decimals.

You can order decimals by comparing the digits at each **place value**.

If you have a few numbers, it can help to arrange them **on top of each other**, lining up their decimal points. Then **compare** the digits in each column, working from **left to right**.

Tip: Be careful with negative numbers — remember they're always smaller than positive ones.

Example 2

Put the decimals 8.092, 8.2, 8.09 and 8.9 in order of size, from smallest to largest.

1. Write the numbers on top of each other and fill in any gaps with 0's. Then compare the digits in each column of the list.

8.092	8.092	8.090
8.200 →	8.090 →	8.092
8.090	8.200	8.200
8.900	8.900	8.900

2. The 1st column doesn't help. In the 2nd, 9 is bigger than 2, and 2 is bigger than 0. So make a new list with 8.9 at the bottom (as it's biggest), and 8.2 next up.
3. Then look at the 3rd column in the new list. The two 9's don't help, so move onto the 4th column. 2 is bigger than 0, so swap 8.092 and 8.09 to get the final order.

The correct order is **8.09, 8.092, 8.2, 8.9**

Example 3

Put the decimals –1.02, 2.1, 0.12 and –1.2 in order, from lowest to highest.

Compare the negative and positive numbers separately, then combine. With negatives, a larger digit means that number comes lower in the list.

Comparing the tenths shows –1.2 is less than –1.02

Comparing the units shows 0.12 is less than 2.1

The correct order is: **–1.2, –1.02, 0.12, 2.1**

Exercise 2

Q1 Write down the larger number in each of the following pairs.

a) 0.3, 0.31 b) 0.09, 0.009 c) 0.427, 0.472 d) 18.07, 17.08

Q2 Put each of the following lists of decimals in order of size, from lowest to highest.

a) 0.02, 0.2, 0.15 b) 0.6, 6.1, –0.6, –6 c) 4.05, 5.04, 5.4, 4.5

d) 1.05, 1.5, –1.5, 1.55 e) 0.61, 0.51, 0.16, 0.15 f) 0.9, –0.05, –0.09, –0.095

1.5 Adding and Subtracting Decimals

Performing operations with decimals is very similar to the methods you know for whole numbers. The main difference is that it's really important that the decimal point is kept in the right place.

Learning Objective — Spec Ref N2:
Add and subtract decimals without using a calculator.

To add and subtract **decimals**, arrange them in columns like you would for whole numbers — just make sure you line up the **decimal points**. You might have to **add in 0's** to fill in any gaps.

Example 1

Work out 4.53 + 1.6

1. Write one number above the other, lining up the decimal points.
2. Fill in any gaps with 0's.
3. Add the digits one column at a time.
4. Carry digits just like when adding whole numbers.

$$\begin{array}{r} 4\,.\,5\,3 \\ +\;1\,.\,6\,0 \\ \hline .\;\;3 \end{array} \qquad \begin{array}{r} 4\,.\,5\,3 \\ +\;1\,.\,6\,0 \\ \hline {}_{1}.\,1\,3 \end{array} \qquad \begin{array}{r} 4\,.\,5\,3 \\ +\;1\,.\,6\,0 \\ \hline 6_{1}.\,1\,3 \end{array}$$

So 4.53 + 1.6 = **6.13**

Example 2

Work out 8.5 – 3.07

1. Write one number above the other, lining up the decimal points.
2. Fill in any gaps with 0's.
3. Subtract the digits one column at a time.
4. Borrow digits just like when subtracting whole numbers.

$$\begin{array}{r} 8\,.\,\overset{4}{\not5}\,{}^{1}0 \\ -\;3\,.\,0\,7 \\ \hline .\;\;3 \end{array} \qquad \begin{array}{r} 8\,.\,\overset{4}{\not5}\,{}^{1}0 \\ -\;3\,.\,0\,7 \\ \hline .\,4\,3 \end{array} \qquad \begin{array}{r} 8\,.\,\overset{4}{\not5}\,{}^{1}0 \\ -\;3\,.\,0\,7 \\ \hline 5\,.\,4\,3 \end{array}$$

So 8.5 – 3.07 = **5.43**

Exercise 1

Q1 Copy and work out the following.

a) $\begin{array}{r} 5\,.\,1 \\ +\;1\,.\,8 \\ \hline \end{array}$ b) $\begin{array}{r} 6\,.\,3 \\ +\;5\,.\,4 \\ \hline \end{array}$ c) $\begin{array}{r} 5\,.\,7 \\ +\;1\,2\,.\,6 \\ \hline \end{array}$ d) $\begin{array}{r} 4\,.\,8 \\ +\;5\,.\,3 \\ \hline \end{array}$

Q2 Copy and work out the following.

a) $\begin{array}{r} 5.6 \\ -\ 0.3 \\ \hline \end{array}$ b) $\begin{array}{r} 9.9 \\ -\ 4.2 \\ \hline \end{array}$ c) $\begin{array}{r} 5.3 \\ -\ 2.8 \\ \hline \end{array}$ d) $\begin{array}{r} 8.5 \\ -\ 1.9 \\ \hline \end{array}$

Q3 Work out the following.

a) 10.83 + 7.4 b) 0.029 + 1.8 c) 91.7 + 0.492 d) 6.474 + 0.92
e) 7.89 + 4.789 f) 0.888 + 1.02 g) 0.02 + 0.991 h) 3.41 + 22.169

Q4 Work out the following.

a) 24.63 – 7.5 b) 6.78 – 5.6 c) 73.46 – 8.5 d) 9.915 – 3.7
e) 3.52 – 0.126 f) 9.1 – 7.02 g) 11.2 – 1.89 h) 1.1 – 0.839

Q5 Work out the following.

a) 3.81 + 9.54 b) 2.81 – 0.16 c) 2.75 + 9.45 d) 8.67 + 0.95
e) 3.8 – 0.59 f) 15.1 – 0.08 g) 6.25 + 5.6 h) 1.97 + 21.7

Example 3

Work out 4 – 0.91

1. Write any whole numbers as decimals by adding 0's after the decimal point, then line up the decimal points.

$$\begin{array}{r} 4.00 \\ -\ 0.91 \\ \hline . \end{array}$$

2. Subtract the digits one column at a time. As you can't borrow from 0, borrow from the 4 instead.

$$\begin{array}{r} \overset{3}{\cancel{4}}.\overset{9}{\cancel{0}}\overset{1}{0} \\ -\ 0.91 \\ \hline .9 \end{array} \qquad \begin{array}{r} \overset{3}{\cancel{4}}.\overset{9}{\cancel{0}}\overset{1}{0} \\ -\ 0.91 \\ \hline 3.09 \end{array}$$

So 4 – 0.91 = **3.09**

Tip: Put the decimal point in the answer straight away, directly below the others.

Exercise 2

Q1 Work out the following.

a) 6 – 5.1 b) 23 – 18.51 c) 12 – 5.028 d) 13 – 6.453

Q2 Work out the following.

a) 2 + 1.8 b) 3.7 + 6 c) 12.7 + 7.34 d) 9.49 + 13
e) 38 + 6.92 f) 5 – 0.8 g) 9 – 4.2 h) 24 – 5.7

Q3 Work out the following.

a) 6.474 + 0.92 + 3 b) 2.39 + 8 + 0.26 c) 12.24 + 4 + 1.2
d) 2 + 4.123 + 1.86 e) 16.8 + 4.17 + 2 f) 2.64 + 13 + 1.012

Example 4

A rope is 8 m long. 6.46 m of the rope is cut off. How much of the original rope is left?

1. Turn the question into a calculation. You want to find 8 – 6.46, so write the 8 as a decimal by adding 0's, then line up the decimal points.

$$\begin{array}{r} 8.00 \\ -\ 6.46 \\ \hline . \end{array}$$

2. Subtract the digits one column at a time. As you can't borrow from 0, borrow from the 8 instead.

$$\begin{array}{r} 8.00 \\ -\ 6.46 \\ \hline . \end{array} \quad \begin{array}{r} \overset{7}{\cancel{8}}.\overset{9}{\cancel{\overset{1}{0}}}\overset{1}{0} \\ -\ 6.46 \\ \hline .\ \ 4 \end{array} \quad \begin{array}{r} \overset{7}{\cancel{8}}.\overset{9}{\cancel{\overset{1}{0}}}\overset{1}{0} \\ -\ 6.46 \\ \hline .54 \end{array} \quad \begin{array}{r} \overset{7}{\cancel{8}}.\overset{9}{\cancel{\overset{1}{0}}}\overset{1}{0} \\ -\ 6.46 \\ \hline 1.54 \end{array}$$

3. Remember to make sure you are answering the question. You are looking for a length in metres.

8 – 6.46 = 1.54, so **1.54 m** of rope is left.

Tip: Always read the question carefully to check whether it's asking you to perform addition or subtraction.

Exercise 3

Q1 Malcolm travels 2.3 km to the shops, then a further 4.6 km to his aunt's house.
How far does he travel in total?

Q2 Sunita buys a hat for £18.50 and a bag for £31.
How much does she spend altogether?

Q3 A plank of wood is 4 m long.
A 2.75 m long piece is cut from the plank.
What length of wood is left?

Q4 Joan goes out for a meal.
The bill for the meal comes to £66.50.
She uses a voucher which entitles her to £15 off her meal.
How much does she have left to pay?

Q5 Ashkan spends £71.42 at the supermarket.
His receipt says that he has saved £11.79 on special offers.
How much would he have spent if there had been no special offers?

Q6 On his first run, Ted sprints 100 m in 15.32 seconds.
On his second run, he is 0.47 seconds quicker.
How long did he take on his second run?

1.6 Multiplying and Dividing Decimals

Multiplying or dividing decimals is a little trickier than addition or subtraction. Rather than keeping the decimal point in the same place, it can move depending on the numbers being used.

Multiplying and Dividing by 10, 100, 1000

Learning Objective — Spec Ref N2:
Multiply and divide decimals by 10, 100, 1000.

When multiplying a number by 10, 100 or 1000, each digit in the number moves to the **left**.

× 10 — each digit moves **one place** to the left.
× 100 — each digit moves **two places** to the left.
× 1000 — each digit moves **three places** to the left.

Tip: The rule continues for bigger powers of 10 — e.g. × 10 000 means each digit moves 4 places to the left, etc.

Example 1

Work out: a) 0.478 × 100 b) 1.35 × 1000

a) Move each digit two places to the left. 0.478 × 100 = **47.8**

b) Move each digit three places to the left. Fill in any gaps with 0's. 1.35 × 1000 = **1350**

When dividing a number by 10, 100 or 1000, each digit in the number moves to the **right**.

÷ 10 — each digit moves **one place** to the right.
÷ 100 — each digit moves **two places** to the right.
÷ 1000 — each digit moves **three places** to the right.

Tip: You can use the same idea for other powers of 10 — e.g. ÷ 10 000 means each digit moves 4 places to the right, etc.

Example 2

Work out: a) 923.1 ÷ 100 b) 51.4 ÷ 1000

a) Move each digit two places to the right. 923.1 ÷ 100 = **9.231**

b) Move each digit three places to the right. Fill in any gaps after the decimal point with 0's. 51.4 ÷ 1000 = **0.0514**

Exercise 1

Q1 Work out the following.

a) 0.92×10 b) 1.41×100 c) 0.23×1000

d) 14.6×100 e) 0.019×1000 f) 13.04×10

Q2 Work out the following.

a) $861.5 \div 100$ b) $381.7 \div 10$ c) $549.1 \div 1000$

d) $6.3 \div 10$ e) $5.1 \div 1000$ f) $0.94 \div 100$

Multiplying Decimals

Learning Objectives — Spec Ref N2:

- Multiplying two decimals without using a calculator.
- Multiply a decimal and a whole number without using a calculator.

Prior Knowledge Check: Be able to multiply whole numbers — see p.9.

If you know the result of a multiplication involving **whole numbers**, you can use it to find the result of related **decimal** multiplications.

You do this by considering how to turn the whole numbers into the decimals you're interested in, by **multiplying or dividing** by 10, 100 or 1000.

Example 3

Given that 167 × 486 = 81 162, work out 1.67 × 0.486

1. Work out how to turn 167 into 1.67. — $167 \div 100 = 1.67$
2. Work out how to turn 486 into 0.486. — $486 \div 1000 = 0.486$
3. Apply both operations to the answer of the original calculation. — So $1.67 \times 0.486 = 81\,162 \div 100 \div 1000$ $= \mathbf{0.81162}$

Exercise 2

Q1 You are given that $132 \times 238 = 31\,416$.
Use this information to work out the following.

a) 13.2×238 b) 1.32×23.8 c) 1.32×0.238 d) 0.132×0.238

Q2 You are given that $401 \times 119 = 47\,719$.
Use this information to work out the following.

a) 40.1×11.9 b) 4.01×1.19 c) 0.401×1.19 d) 0.401×0.119

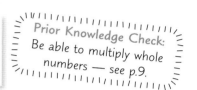

When multiplying two decimals from scratch:

- **Ignore** the decimal point and multiply them as **whole numbers**.
- Then add in the decimal point in the **right place** at the end.

To work out where the decimal point should go, **add** together the number of digits **after the decimal points** in the two numbers you're multiplying — this will tell you the number of digits after the decimal point in your answer.

Tip: What you're actually doing here is multiplying by powers of 10 and then dividing by them again.

Example 4

Work out 0.32 × 0.6

1. Ignore the decimal points and treat it as a whole number multiplication.
2. There are three digits in total after the decimal points in the question (two in 0.32 and one in 0.6), so the answer should have three digits after the decimal point.

$$\begin{array}{r} 3\,2 \\ \times \quad 6 \\ \hline {}_{1}2 \end{array} \qquad \begin{array}{r} 3\,2 \\ \times \quad 6 \\ \hline 1\,9_{1}2 \end{array}$$

So 0.32 × 0.6 = **0.192**

Example 5

Work out 0.57 × 2.4

1. Ignore the decimal points and treat it as a whole number multiplication.

$$\begin{array}{r} 5\,7 \\ \times\ 2\,4 \\ \hline {}_{2}8 \end{array} \qquad \begin{array}{r} 5\,7 \\ \times\ 2\,4 \\ \hline 2\,2_{2}8 \end{array}$$

2. There are three digits in total after the decimal points in the question (two in 0.57, one in 2.4), so the answer should have three digits after the decimal point.

$$\begin{array}{r} 5\,7 \\ \times\ 2\,4 \\ \hline 2\,2_{2}8 \\ 0 \end{array} \qquad \begin{array}{r} 5\,7 \\ \times\ 2\,4 \\ \hline 2\,2_{2}8 \\ {}_{1}4\,0 \end{array} \qquad \begin{array}{r} 5\,7 \\ \times\ 2\,4 \\ \hline 2\,2_{2}8 \\ 1\,1_{1}4\,0 \end{array} \qquad \begin{array}{r} 5\,7 \\ \times\ 2\,4 \\ \hline 2\,2_{2}8 \\ 1\,1_{1}4\,0 \\ \hline 1\,3\,6\,8 \end{array}$$

So 0.57 × 2.4 = **1.368**

Exercise 3

Q1 Work out the following.

a) 6.7 × 8　　b) 0.65 × 9　　c) 0.9 × 0.8　　d) 0.6 × 0.3

e) 0.01 × 0.6　　f) 0.61 × 0.7　　g) 0.33 × 0.02　　h) 0.007 × 0.006

Q2 Work out the following.

a) 6.3 × 2.1　　b) 1.4 × 2.3　　c) 2.4 × 1.8

d) 8.6 × 6.9　　e) 0.16 × 3.3　　f) 5.1 × 0.23

Example 6

Work out 1.36 × 200

1. 1.36 × 200 = 1.36 × 2 × 100, so break the multiplication down into two stages.
2. First calculate 1.36 × 2 by ignoring the decimal point and adding it in at the end.
3. There are two digits after the decimal point in the question (1.36), so the answer will have two digits after the decimal place.
4. Then multiply by 100 by moving each digit two places to the left.

$$\begin{array}{r} 1\,3\,6 \\ \times \quad 2 \\ \hline {}_1 2 \end{array} \qquad \begin{array}{r} 1\,3\,6 \\ \times \quad 2 \\ \hline 7\,{}_1 2 \end{array} \qquad \begin{array}{r} 1\,3\,6 \\ \times \quad 2 \\ \hline 2\,7\,{}_1 2 \end{array}$$

136 × 2 = 272
so 1.36 × 2 = 2.72

1.36 × 200 = 2.72 × 100
= **272**

Exercise 4

Q1 Work out the following.

a) 3.1 × 40 b) 0.7 × 600 c) 0.061 × 2000 d) 11.06 × 80
e) 12.1 × 30 f) 1.007 × 400 g) 101.8 × 60 h) 0.903 × 5000

Dividing Decimals

Learning Objective — Spec Ref N2:
Divide decimals without using a calculator.

Prior Knowledge Check: Be able to divide whole numbers — see p.11.

When dividing decimals, it's important that you **set up** the division properly — make sure the **decimal point** in the answer is **directly above** the decimal point inside the division.

Example 7

Work out 5.16 ÷ 8

1. Position the decimal point in the answer directly above the one inside the division.
2. Divide as you would with whole numbers.
3. You may need to add zeros to the decimal you're dividing.

$$8\overline{)5.{}^{5}16}\ \text{ gives } 0.$$

$$8\overline{)5.{}^{5}1{}^{3}6}\ \text{ gives } 0.6 \qquad 8\overline{)5.{}^{5}1{}^{3}6{}^{4}0}\ \text{ gives } 0.64 \qquad 8\overline{)5.{}^{5}1{}^{3}6{}^{4}0}\ \text{ gives } 0.645$$

So 5.16 ÷ 8 = **0.645**

Tip: You could also use long division.

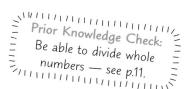

Exercise 5

Q1 Copy and work out the following.

a) $2\overline{)5.4}$ b) $3\overline{)9.6}$ c) $6\overline{)9.24}$ d) $5\overline{)2.65}$

e) $7\overline{)7.21}$ f) $9\overline{)4.05}$ g) $7\overline{)98.7}$ h) $4\overline{)0.924}$

Q2 Work out the following.

a) $8.52 \div 4$ b) $112.8 \div 4$ c) $1.02 \div 3$ d) $5.62 \div 8$

e) $0.052 \div 5$ f) $12.06 \div 8$ g) $3.061 \div 5$ h) $0.0612 \div 6$

When the number you're **dividing by** is a decimal, convert the calculation into a division by a whole number by **multiplying** both numbers by a **power of 10**. Then divide as usual.

Example 8

Work out 8.16 ÷ 0.2

1. Multiply both numbers by 10, so you are dividing by a whole number.
2. Divide as usual.

$$\times 10 \quad 8.16 \div 0.2 \quad \times 10$$
$$= 81.6 \div 2$$

$$\begin{array}{r} 4\ .\ \ \\ 2\overline{)81.6} \end{array} \qquad \begin{array}{r} 40.\ \ \\ 2\overline{)81.{}^{1}6} \end{array} \qquad \begin{array}{r} 40.8 \\ 2\overline{)81.{}^{1}6} \end{array}$$

So $8.16 \div 0.2 =$ **40.8**

Tip: Only the number that you're dividing by needs to be a whole number.

Exercise 6

Q1 Consider the calculation 6.4 ÷ 0.4.

a) Multiply both numbers by 10 to form an equivalent calculation which involves dividing by a whole number.

b) Use your answer to work out 6.4 ÷ 0.4.

Q2 Consider the calculation 0.384 ÷ 0.12.

a) Multiply both numbers by 100 to form an equivalent calculation which involves dividing by a whole number.

b) Use your answer to work out 0.384 ÷ 0.12.

Q3 Consider the calculation 3.8 ÷ 0.008.

a) Write down an equivalent calculation which involves dividing by a whole number.

b) Use your answer to work out 3.8 ÷ 0.008.

Q4 Work out the following.

a) $6.4 \div 0.2$ b) $3.54 \div 0.4$ c) $0.624 \div 0.3$

Q5 Work out the following.

a) $22.56 \div 0.03$ b) $0.257 \div 0.05$ c) $0.039 \div 0.06$

Q6 Work out the following.

a) $0.081 \div 0.009$ b) $0.008 \div 0.4$ c) $1.44 \div 1.2$

When you are dividing by a whole number that's a **multiple** of a **power of 10**, start by **splitting** up that number. E.g. if you were dividing by 4000, you'd write this as 4×1000. Then you can start **dividing**:

- First divide by the **smaller** number — so in the example above you'd divide by 4 first.
- Then divide by the correct **power of 10** to get the final answer — so you'd divide by 1000 in the example.

Tip: Always split the number using the biggest power of 10 that you can. E.g. $4000 = 4 \times 1000$ instead of using $4000 = 40 \times 100$.

Example 9

Work out 96.6 ÷ 300.

1. Write 300 as something multiplied by a power of 10. — $300 = 3 \times 100$
2. First work out 96.6 ÷ 3. — $96.6 \div 3 = 3\overline{)9\ 6.6}$ = 3 2.2
3. Then divide by 100 to get the final answer. — $32.2 \div 100 =$ **0.322**

Exercise 7

Q1 Consider the calculation $7.3 \div 50$.

a) Work out $7.3 \div 5$.

b) Use your answer to work out $7.3 \div 50$.

Q2 Consider the calculation $2.41 \div 400$.

a) Work out $2.41 \div 4$.

b) Use your answer to work out $2.41 \div 400$.

Q3 Work out the following.

a) $6.08 \div 40$ b) $5.74 \div 700$ c) $25.47 \div 900$ d) $13.722 \div 3000$

Review Exercise

Q1 Work out the following.

a) $6 + 4 \times 2$ b) $48 \div (4 + 2)$ c) $(13 - 5) \times 12$

Q2 Work out the following.

a) $-5 + 8$ b) $6 - (-2)$ c) $-8 \times (-5)$ d) $54 \div (-9)$

Q3 Work out the following.

a) $256 + 312$ b) $841 - 346$ c) $1632 + 421$ d) $2830 - 394$

Q4 Laura receives £14 pocket money each week.
How much money does she get in 52 weeks?

Q5 462 pupils are going on a school trip. A coach can seat 54 children. PROBLEM SOLVING
How many coaches will be needed for the trip?

Q6 Write down the value of the 6 in 0.956.

Q7 Put each of the following lists of decimals in order of size, from lowest to highest.

a) $-0.01, 0.1, -0.09, -0.1$ b) $-0.5, -0.45, -0.55, -5$ c) $-7, -7.1, -7.07, 0.007$

Q8 Work out the following.

a) $6.78 - 5.6$ b) $1.6 + 4.35$ c) $0.78 + 1.3$ d) $4.32 - 2.17$

Q9 Work out the following.

a) 7.8×1000 b) 0.006×100 c) $25.9 \div 10$ d) $901.5 \div 100$

Q10 You are given that $221 \times 168 = 37\ 128$. Work out the following.

a) 2.21×1.68 b) 0.221×1.68 c) 221×0.168

Q11 1 litre is equal to 1.76 pints. How many pints are there in 5 litres?

Q12 A 2.72 m ribbon is cut into pieces of length 0.08 m. How many pieces will there be?

Exam-Style Questions

Q1 In New York, the evening temperature was –4 °C.
During the night, the temperature dropped by 7 °C.
What was the lowest temperature that night?

[1 mark]

Q2 Work out the following calculation.
$12 + 1.3 + 0.25$

[1 mark]

Q3 You are given that $539 \times 28 = 15092$.
Use the above result to work out the value of:

a) 539×14

[1 mark]

b) 5390×0.28

[1 mark]

c) $1\,509\,200 \div 53.9$

[2 marks]

Q4 Rani has to work out $5 + 3 \times 4$. She gives the answer 32.
Explain what she has done wrong and give the correct answer.

[2 marks]

Q5 It costs £35.55 to buy nine identical books.
How much would it cost to buy seven of these books?

[2 marks]

Q6 Put one pair of brackets into each of these calculations so that the answer is correct.

a) $3 \times 2 - 4 \div 2 = -3$

[1 mark]

b) $8 + 6 \div 5 \times 10 = 92$

[1 mark]

2.1 Rounding — Whole Numbers

Numbers can be approximated by (or rounded to) whole numbers. As you'll see later, these approximations can be used to find estimates and to check answers to difficult calculations.

Learning Objective — Spec Ref N15:
Round numbers to the nearest whole number, ten, hundred etc.

Prior Knowledge Check: You'll need to know about decimal numbers (p.13).

Numbers can be **rounded** to make them easier to work with.
For example, a number like 5468.9 could be rounded to:

- the nearest **whole number** (= 5469)
- the nearest **ten** (= 5470)
- the nearest **hundred** (= 5500)
- the nearest **thousand** (= 5000)

To round any number to a **whole number** use the following method:

1. **Identify** the position of the '**last digit**' that you want to keep.
   E.g. if you're rounding to the nearest ten, then this last digit is in the tens place.
2. Look at the next digit to the **right** — called **the decider**.
3. If the decider is **5 or more**, then **round up** the last digit.
   If the decider is **4 or less**, then **leave** the last digit as it is.
4. All digits to the right of the last digit should be **zero**, and there should be **no digits** after the decimal point.

Tip: When you're rounding up a digit, just add 1 to it.

Example 1

Round 18.6 to the nearest ten.

1. The 'last digit' is in the tens place. — 18.6 — the 'last digit' is 1.
2. The decider is the next digit to the right. — 18.6 — the decider is 8.
3. The decider is more than 5, so round up.
4. All digits after the 'last digit' should be zero.

18.6 rounds up to **20**

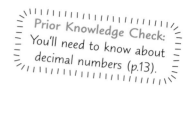

Things are a little trickier if you need to **round up** and the last digit is 9.
You need to replace the 9 with a 0 and **carry 1 to the left** (by adding a 1 to the digit to its left).

This is easier to understand by looking at an example:

Example 2

Round 39 742 to the nearest thousand.

1. The 'last digit' is in the thousands place. The decider is the next digit to the right. — 3⑨742 — the 'last digit' is 9.
2. The decider is more than 5, so round up. The last digit is a 9, so replace with 0 and carry the 1. — 39⑦42 — the decider is 7. → ~~3~~0 000 (4 written above the 3)
3. Replace the decider and all the digits to its right with 0s. — 39 742 rounds up to **40 000**

Exercise 1

Q1 Write down all the numbers from the box that round to 14 to the nearest whole number.

14.1	14.9	14.02	15.499	14.5	15.01
13.7	14.09	13.3	13.4999	14.4999	13.901

Q2 Round the following to the nearest whole number.
a) 9.7 b) 8.4 c) 12.2 d) 39.8

Q3 Round the following to the nearest hundred.
a) 158 b) 596 c) 650 d) 4714

Q4 Round the following to the nearest thousand.
a) 2536 b) 8516 c) 7218 d) 9500

Q5 Round the numbers on these calculator displays to the nearest: (i) whole number (ii) ten
a) 18.2 b) 16.479 c) 201.50001 d) 1.4999999

Q6 Matilda says Italy has an area of 301 225 km². Round this figure to the nearest thousand.

Q7 At its closest, Jupiter is about 390 682 810 miles from Earth.
Write this distance to the nearest million miles.

2.2 Rounding — Decimal Places

Decimal places are the digits that come after a decimal point. They show parts of numbers — tenths, hundredths, thousandths etc. You can round to decimal places, just like with whole numbers.

Learning Objective — Spec Ref N15:
Round numbers to a given number of decimal places.

Prior Knowledge Check:
You'll need to know about decimal numbers (p.13).

You can round to different numbers of **decimal places (d.p.)**.

For example, a number like 8.9471 could be rounded:

- to **one** decimal place (= 8.9)
- to **two** decimal places (= 8.95)
- to **three** decimal places (= 8.947)

Tip: 1 decimal place is the same as 'the nearest tenth', 2 decimal places is the same as 'the nearest hundredth' and so on.

The method for rounding to a given number of **decimal places** is basically the same as the method on page 26:

1. **Identify** the position of the '**last digit**' that you want to keep.
   E.g. when rounding to 2 d.p., it's the digit in the hundredths place.
2. Look at the next digit to the **right** (**the decider**).
3. If the decider is **5 or more**, then **round up** the last digit.
   If the decider is **4 or less**, then **leave** the last digit as it is.
4. There must be **no more digits** after the last digit (not even zeros).

Example 1

a) Round 4.7623 to one decimal place.

1. The 'last digit' is in the first d.p. (the tenths place). — 4.7623 — the 'last digit' is 7.
2. The decider is the next digit to the right. — 4.7623 — the decider is 6.
3. The decider is more than 5, so round up the last digit and remove all other digits. — 4.7623 rounds up to **4.8**

b) Round 4.7623 to three decimal places.

1. The 'last digit' is in the third d.p. (the thousandths place). — 4.7623 — the 'last digit' is 2.
2. The decider is the next digit to the right. — 4.7623 — the decider is 3.
3. The decider is less than 5, so leave the last digit as it is and remove all the other digits. — 4.7623 rounds down to **4.762**

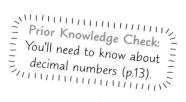

Just like with whole numbers, things get a little tricky when you need to round up and the **last digit is a 9**. Just like on page 27, you need to **carry a 1** to the left.

After rounding up, your last digit will be a **zero**. You still need to **remove** all other digits after the last digit (even if they are more zeros).

Example 2

Round 3.896 to two decimal places.

1. The 'last digit' is in the second d.p. (the hundredths place). — 3.896 — the 'last digit' is 9.
   The decider is the next digit to the right. — 3.896 — the decider is 6.
2. The decider is more than 5, so round up.
   The last digit is 9, so replace with 0 and carry the 1. → 3.90 (8 crossed out and replaced by 9)
3. Remove all the other digits after the last digit. — 3.896 rounds up to **3.90**

Exercise 1

Q1 Write down all the numbers from the box that round to 0.4 to one decimal place.

0.41	0.45	0.347	0.47204	0.335	0.405
0.35	0.4295	0.5216	0.4124	0.4671	0.307

Q2 Round the following numbers to one decimal place.
a) 0.23 b) 0.678 c) 2.6893 d) 0.9324

Q3 Round the following numbers to two decimal places.
a) 4.567 b) 0.0424 c) 6.2971 d) 0.35273

Q4 Round the following numbers to three decimal places.
a) 0.96734 b) 0.25471 c) 2.43658 d) 6.532561

Q5 Round these numbers to the number of decimal places (d.p.) specified.
a) 0.19745 — to 2 d.p. b) 0.68361 — to 1 d.p.
c) 5.73174 — to 3 d.p. d) 0.000635 — to 3 d.p.

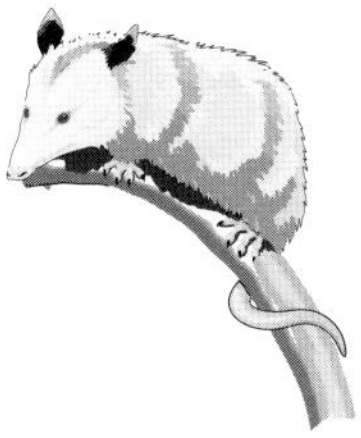
Not shown actual size.

Q6 The mass of the field vole on the right is 0.0384 kg.
Round this mass to two decimal places.

Q7 Usain measures his height to be 1.7 m to the nearest 10 cm.
Is it possible for Usain's exact height to be 1.76 m? Explain your answer.

2.3 Rounding — Significant Figures

One more type of rounding now. It's a biggie, but not too tough. Once you know how to identify significant figures, you can round just like rounding to whole numbers or decimal places.

Learning Objective — Spec Ref N15:
Round whole numbers and decimals to a given number of significant figures.

The **first significant figure** in a number is the first digit that **isn't 0**.

All the digits that follow the first significant figure are also **significant figures**, regardless of whether or not they're **zeros**. So, 307 has **three** significant figures — 3, 0 and 7.

You can round to different numbers of **significant figures** (s.f.).
For example, 217 304 could be rounded:

- to **one** significant figure (= 200 000)
- to **two** significant figures (= 220 000)
- to **three** significant figures (= 217 000)

The method for significant figures is **identical** to the one for decimal places — except it can be a bit harder to locate the **last digit**. There are a few key things to note:

- The '**last digit**' is the significant figure you're **rounding to**. E.g. rounding to 2 s.f. means the 'last digit' is the **second** s.f.
- After rounding, fill in all the places up to the decimal point with 0s. (You may need to include a 0 after the decimal point to make up the correct number of significant figures — e.g. 32.0)

Tip: If the last s.f. is **after** the decimal point, don't add any extra zeros to the end.

Example 1

a) Round 56 291 to two significant figures.

1. The 'last digit' is the second s.f. and the decider is the next digit to the right.

   56291 — the 'last digit' is 6.
   56291 — the decider is 2.

2. The decider is less than 5, so leave the last digit as it is and fill in all the places up to the decimal point with zeros.

   56 291 rounds down to **56 000**

b) Round 6.597 to three significant figures.

1. The decider is more than 5, so round up. The last digit is a 9, so you'll need to carry the 1.

   6.597 — the 'last digit' is 9.
   6.597 — the decider is 7.

2. Keep the zero so you've got the correct number of significant figures.

   $6.\overset{6}{\not{5}}0$

3. The last significant figure is after the decimal point, so don't add any more zeros.

   6.597 rounds up to **6.60**

Exercise 1

Q1 Round the following numbers to: (i) 1 s.f. (ii) 2 s.f. (iii) 3 s.f.

a) 7036 b) 6551 c) 7067 d) 2649

Q2 Round the following numbers to the number of significant figures (s.f.) indicated.

a) 45.89 — to 1 s.f. b) 5689.6 — to 3 s.f. c) 6.497 — to 3 s.f.
d) 360.8 — to 2 s.f. e) 6527 — to 2 s.f. f) 756 557 — to 3 s.f.
g) 46.745 — to 3 s.f. h) 376.25 — to 2 s.f. i) 79 477 — to 2 s.f.

Q3 The speed of sound is approximately 1236 km/h.
Round this speed to two significant figures.

Rounding Decimals to Significant Figures

Zeros at the start of a decimal do **not** count as significant figures.
The first significant figure is the first **non-zero digit**.

For example, 0.001520 has **four** significant figures — 1, 5, 2 and 0.

You can round decimals to significant figures using the **same method** as on the previous page.

Tip: Remember — **all** the digits following the first s.f. are also significant figures, regardless of whether or not they're zeros.

Example 2

Round 0.06826 to one significant figure.

1. The 'last digit' is the first significant figure and the decider is the next digit to the right. — 0.06826 — the 'last digit' is 6. 0.06826 — the decider is 8.
2. The decider is more than 5, so round up. — 0.06826 rounds up to **0.07**
3. Remove any digits to the right, and do not add any more zeros.

Exercise 2

Q1 Round the following numbers to: (i) 1 s.f. (ii) 2 s.f. (iii) 3 s.f.

a) 0.003753 b) 0.02644 c) 0.0001792
d) 0.03970 e) 0.5635 f) 0.0007049

Q2 Round the following numbers to the number of significant figures (s.f.) indicated.

a) 0.004567 — to 1 s.f. b) 0.1962 — to 2 s.f. c) 0.0043862 — to 3 s.f.
d) 0.006204 — to 1 s.f. e) 0.009557 — to 2 s.f. f) 0.00060384 — to 3 s.f.

Q3 The density of the hydrogen gas in a balloon is 0.0899 kg/m^3.
Round this density to one significant figure.

2.4 Estimating Answers

Rounded numbers can be used in calculations to estimate answers.
By comparing an exact calculation with an estimate, you can see if it looks 'about right'.

Learning Objectives — Spec Ref N14:
- Find estimates using approximate values.
- Check answers to calculations using estimates.

Using rounded numbers in a **calculation** gives an **estimate** of the actual answer. By **simplifying** in this way, you get an **approximate value**. To find an estimate of a calculation:

- Round every number to one significant figure.
- Work out the answer using the rounded values.

Tip: The symbol '≈' is used when estimating. It means 'is approximately equal to'.

Example 1

By rounding each number to one significant figure, estimate $\frac{78.43 \times 6.24}{19.76}$.

1. Round each number to 1 s.f. $78.43 = 80$ (1 s.f.), $6.24 = 6$ (1 s.f.), $19.76 = 20$ (1 s.f.)
2. Replace the numbers in the calculation with the rounded values. $\frac{78.43 \times 6.24}{19.76} \approx \frac{80 \times 6}{20}$
3. Work out the estimate. $\frac{80 \times 6}{20} = \frac{480}{20} = \mathbf{24}$

An estimate to a calculation will probably be **different** to the actual answer.
You can usually figure out if your answer will be an **overestimate** or an **underestimate**:

- **Addition** or **multiplication** — if both numbers are rounded **up** you'll get an **overestimate** and if both numbers are rounded **down** you'll get an **underestimate**.
- **Subtraction** or **division** — you'll get an **overestimate** if you rounded the **1st number up** and the **2nd number down**, and an **underestimate** if you rounded the **1st number down** and the **2nd number up**.

Example 2

By rounding each number to 1 s.f., estimate 42.6 ÷ 7.8.
Is your answer an overestimate or an underestimate?

1. Round each number to 1 s.f. to find an estimate. $42.6 \div 7.8 \approx 40 \div 8 = \mathbf{5}$
2. The calculation is a division — the 1st number was rounded down and the 2nd number was rounded up. **underestimate.**

Exercise 1

Q1 By rounding each number to one significant figure, estimate the following.

a) 437 + 175 b) 310 + 876 c) 784 – 279

d) 0.516 – 0.322 e) 184 + 722 f) 838 – 121

Q2 By rounding each number to one significant figure:
(i) estimate the calculation (ii) say whether this is an overestimate or an underestimate.

a) 23 + 43 b) 59×5.7 c) 40.4 + 5.1

d) 18×79 e) 276 + 19 f) 587×8.81

Q3 Use rounding to choose the correct answer (A, B or C) for each of the following calculations.

a) 1.76×6.3	A: 1.328	B: 5.788	C: 11.088
b) 582×2.1	A: 119.52	B: 1222.2	C: 4545.2
c) $\frac{57.5 \times 3.78}{16.1}$	A: 1.65	B: 6.3	C: 13.5

Q4 By rounding each number to one significant figure:
(i) estimate the calculation (ii) say whether this is an overestimate or an underestimate.

a) $\frac{8.9}{3.1}$ b) 33 – 17 c) $\frac{43}{18}$

d) 37.3 – 5.2 e) 112 – 68 f) $\frac{9.98}{2.14}$

Q5 By rounding each number to one significant figure, estimate the following.

a) $\frac{68.8 + 27.3}{23.7}$ b) $\frac{5.6 \times 9.68}{5.14}$ c) $\frac{\sqrt{38.6 + 56.3}}{1.678}$

Q6 The decimal points have been missed out from each of the answers to these calculations. Use rounding to find an approximate answer in each case, and then decide where the decimal point should be.

a) $18.5 \times 3.2 = 592$ b) $\frac{325.26}{5.2} = 6255$

c) $\frac{19.8 \times 27.4}{3.3} = 1644$ d) $\frac{\sqrt{48.4 \times 8.1}}{4.8} = 4125$

Q7 It costs £4.70 to buy the toys needed for one 'Child's Party Bag'. If 21 children attend a party, estimate how much it would cost to buy all the toys for the party bags.

Q8 A smoothie factory operates for 62 hours per week.
It makes 324 litres of smoothie per hour.
Each litre of smoothie contains 14 strawberries.

a) Estimate the number of strawberries used each week.

b) Is your answer to a) an overestimate or an underestimate? Give a reason for your answer.

Use Estimates to Check Answers

Even though an **estimated** answer **isn't exact**, it can still be useful. You can use an estimate to **check** your calculations — in other words, to see if your actual answer looks 'about right'.

Example 3

Hiromi uses a calculator to work out $\frac{27\,891 \times 628}{18}$ and gets the answer 973 086.

Use approximations to decide whether Hiromi's answer is sensible.

1. Round each number to 1 s.f.

   27 891 = 30 000 (1 s.f.),
   628 = 600 (1 s.f.) and
   18 = 20 (1 s.f.)

2. Replace the numbers in the calculation with the rounded values and work out the estimate.

   $$\frac{27\,891 \times 628}{18} \approx \frac{30\,000 \times 600}{20}$$
   $$= \frac{18\,000\,000}{20} = 900\,000$$

3. Compare the estimate to the calculated answer given in the question. Decide if the numbers seem close to each other.

   The approximate answer is close to the calculated answer, so **Hiromi's answer is sensible.**

Exercise 2

Q1 For each of the following:
  (i) Use your calculator to work out the value of the calculation and write down your full calculator display.
  (ii) Use approximations to check if your answer to part (i) is sensible.

  a) 112.62×268.9 b) $\frac{52.668 \times \sqrt{104.04}}{3.78}$ c) $5.39^2 \times \sqrt[3]{1012} \div 2.36$

Q2 Sam has done the calculation 56.2 × 34.7 on his calculator. The calculator display is shown on the right. Alex says Sam must have pressed a wrong button at some point.

17686.14

  a) By rounding each number in the calculation to 1 s.f., estimate 56.2 × 34.7.
  b) Do you think Sam pressed a wrong button at some point? Explain your answer.

Q3 A company posts an update saying "We had a great day today, selling 987 items at a price of £27.85 each, so we made over £32 000 in total." Use approximations to show that the company is wrong.

Q4 Karen earns £6.85 per hour. One week she works for 42 hours and is paid £287.70. Use approximations to see if it looks like Karen has been paid correctly or not.

2.5 Rounding Errors

Rounding leaves some uncertainty between the actual number and the rounded number. Error intervals show you the range of values that a rounded number could actually be.

Error Intervals for Rounded Numbers

Learning Objectives — Spec Ref N15/N16:
- Interpret rounded numbers.
- Use inequality notation to describe error intervals.

Prior Knowledge Check: Be able to use the inequality notation '$<$' and '$\leq$' (p.151)

When a number has been rounded to a given **rounding unit**, the actual number could be up to **half a unit** bigger or smaller than the rounded number. For example, if a number has been rounded to 1 d.p., the **rounding unit** is **0.1** so the actual value is anything up to **0.05 either side**.

Once you know these **maximum** and **minimum values** of a rounded number, you can show the **error interval** using **inequalities**:

- The actual value is **greater than or equal to** the **minimum**.
- The actual value is **strictly less than** the **maximum**.
  (If it was exactly equal to the maximum value, it would round up to the next unit.)

The error interval is the interval of **possible values** the actual number can be, from the **smallest possible number** that rounds **up** to the given number to the **largest possible number** that rounds **down** to the given number.

Example 1

m = 20 when rounded to 1 s.f. Find the error interval for m.

1. The first significant figure of 20 lies in the tens column. So the rounding unit is 10. Half the rounding unit is $10 \div 2 = 5$.

   Minimum value of m is $20 - 5 = 15$.
   Maximum value of m is $20 + 5 = 25$.

2. The actual value of m can be greater than or equal to 15 but is strictly less than 25.

   The error interval is $\mathbf{15 \leq m < 25}$.

Exercise 1

Q1 Find the minimum and maximum values of the following rounded values.

a) 80 rounded to the nearest whole number

b) 400 rounded to 1 significant figure.

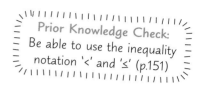

Q2 Each of the following values have been rounded as shown.
Write down the error interval for each value.

a) $a = 60$ to the nearest ten
b) $b = 9$ to the nearest whole number
c) $c = 500$ to the nearest hundred
d) $d = 15\ 000$ to the nearest thousand

Q3 Write down the error interval for each of the following values, which have been rounded as shown.

a) $a = 7.6$ to 1 d.p.
b) $b = 0.3$ to 1 s.f.
c) $c = 2.55$ to 2 d.p.
d) $d = 50$ to 1 s.f.
e) $e = 109.9$ to 1 d.p.
f) $f = 540$ to 2 s.f.

Q4 The length, l cm, of a piece of rope is 76 cm correct to the nearest centimetre.
Write down the error interval for l.

Q5 Alasdair is canoeing down a river and has travelled 10 km to the nearest 100 m.
Write down the error interval for d, the actual distance in km he has travelled.

Q6 The number of sweets in a jar is 670, to the nearest 10.
What is the maximum possible number of sweets in the jar?

Error Intervals for Truncated Numbers

Learning Objectives — Spec Ref N15:
- Be able to truncate numbers.
- Use inequality notation to describe error intervals.

You **truncate** a number by chopping off decimal places — e.g. 77.889 truncated to 1 d.p. is 77.8.
The **actual value** of a truncated number can be up to a **whole rounding unit bigger but no smaller**.

Example 2

$x = 62.1$ truncated to 1 decimal place.
Find the error interval for x.

The rounding unit is 0.1. x has been truncated, so it can be up to 0.1 bigger, but no smaller.

$$62.1 \le x < 62.2$$

Tip: Just like before, it's a strict inequality (<) on the right.

Exercise 2

Q1 Truncate the following values to the given number of decimal places.

a) 1.354 to 1 d.p.
b) 19.133 to 2 d.p.
c) 103.67183 to 3 d.p.

Q2 Find the error interval for the following numbers, which have been truncated as shown.

a) $x = 1.3$ to 1 d.p.
b) $y = 5.13$ to 2 d.p.
c) $z = 7.731$ to 3 d.p.

Review Exercise

Q1 Round the following to: (i) the nearest ten
(ii) the nearest hundred
(iii) the nearest thousand

a) 6724 b) 25 361 c) 8499.3 d) 3822.8

Q2 Round the following to: (i) 1 d.p. (ii) 2 d.p. (iii) 3 d.p.

a) 2.6893 b) 0.3249 c) 5.6023 d) 0.0525

Q3 Round the following numbers to the number of significant figures (s.f.) indicated.

a) 4589 — to 1 s.f. b) 56 986 — to 3 s.f. c) 6.792 — to 2 s.f.
d) 360.8 — to 1 s.f. e) 6527 — to 2 s.f. f) 756 557 — to 3 s.f.

Q4 Jade buys four items costing £1.35, £8.52, £14.09 and £17.93.
Estimate how much she spent by rounding each price to the nearest pound.

Q5 By rounding each number to 1 s.f., estimate each of the following calculations.

a) $\frac{64.4 \times 5.6}{17 \times 9.5}$ b) $\frac{310.33 \times 2.68}{316.39 \times 0.82}$ c) $\frac{13.7 \times 5.2}{12.3 \div 3.9}$

Q6 a) Pens cost 32 pence each. Estimate the cost of 14 pens.
Give your answer in pounds.

b) Is your answer to a) an underestimate or an overestimate?
Give a reason for your answer.

Q7 a) Work out the value of $24.37 \div \sqrt{3.9}$.
Write down all the numbers on our calculator.

b) Use an estimate to decide whether your answer to part a) is sensible.

Q8 Write down the error interval for the following rounded numbers.

a) $a = 50$ to 1 s.f. b) $b = 5690$ to 3 s.f. c) $c = 7$ to 1 s.f.
d) $d = 360$ to 2 s.f. e) $e = 6500$ to 2 s.f. f) $f = 757\,000$ to 3 s.f.

Q9 Given that the following values have been truncated to 2 d.p.,
find the error interval for each value.

a) $s = 6.57$ b) $t = 25.71$ c) $w = 13.29$

Exam-Style Questions

Q1 Write the number 45.768 to 1 decimal place.

[1 mark]

Q2 a) Work out $\frac{3.5^4}{\sqrt{0.007}}$. Write down your entire calculator display.

[2 marks]

b) Write your answer to part a):

(i) to 3 decimal places

[1 mark]

(ii) to 3 significant figures

[1 mark]

Q3 $x = \frac{628}{\sqrt{97} + 9.6}$

a) By rounding each number to 1 significant figure, estimate the value of x.

[2 marks]

b) Explain why your answer in part a) is an underestimate of the actual value of x.

[1 mark]

Q4 The audience at a rock concert, r, was 7300, correct to the nearest hundred.
Write down the error interval for r.

[2 marks]

Q5 The weight of a dog, w is 8 kg, rounded to the nearest kg.

a) Write down the error interval for w.

[2 marks]

b) Another dog weighs 6 kg, rounded to the nearest kg.
Kayla says "the total weight of the two dogs must be at least 14 kg."
Give an example to show that Kayla is wrong.

[2 marks]

3.1 Squares, Cubes and Roots

Section 3 is all about powers — where numbers are multiplied by themselves a number of times. Squares and cubes are important types of powers, so make sure you get your head around them.

Squares and Cubes

Learning Objectives — Spec Ref N6:
- Recognise square numbers and cube numbers.
- Evaluate square numbers and cube numbers.

Prior Knowledge Check:
Know how to multiply and divide by negative numbers — see p.6.

Squares show that a number is multiplied by itself.
For example, $4 \times 4 = 4^2 = 16$ is the **square** of 4, and 16 is called a **square number**.

If you square a number, you always get a **positive** answer because you're multiplying the **same signs** together.

The first 10 square numbers are shown in this table.

1^2	2^2	3^2	4^2	5^2	6^2	7^2	8^2	9^2	10^2
1	4	9	16	25	36	49	64	81	100

Cubes show that a number is multiplied by itself and then by itself again.
For example $4 \times 4 \times 4 = 4^3 = 64$ is the **cube** of 4, and 64 is called a **cube number**.

Cubing a number can give a **positive or negative** answer since there are **three multiples** of the same sign — the answer will be the same sign as whatever the **original** number was.

The first 5 cube numbers are shown in this table.

1^3	2^3	3^3	4^3	5^3
1	8	27	64	125

Example 1

Find: **a)** 5^2 — $5^2 = 5 \times 5 = \mathbf{25}$

b) $(-5)^2$ — $(-5)^2 = -5 \times -5 = \mathbf{25}$

c) 5^3 — $5^3 = 5 \times 5 \times 5 = \mathbf{125}$

d) $(-5)^3$ — $(-5)^3 = -5 \times -5 \times -5 = \mathbf{-125}$

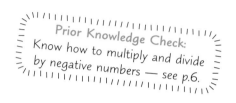

Exercise 1

Q1 Evaluate the following. Use a calculator where necessary.

a) 6^2 b) 12^2 c) 15^2 d) 20^2

e) $(-4)^2$ f) 0.3^2 g) 0.6^2 h) $(-0.2)^2$

Q2 Evaluate the following. Use a calculator where necessary.

a) 3^3 b) 6^3 c) 11^3 d) 20^3

e) $(-3)^3$ f) $(-10)^3$ g) 0.4^3 h) $(-0.5)^3$

Square Roots

Learning Objective — Spec Ref N6/N7:
Evaluate square roots.

Finding a **square root** is the **opposite** of squaring.
A square root of x is a number that **multiplies with itself** to give x.
Every **positive** number has **two** square roots — one **positive** ($\sqrt{x}$) and one **negative** ($-\sqrt{x}$).

For example, the positive square root of 2 is $\sqrt{2}$.
And the negative square root of 2 is $-\sqrt{2}$.

Negative numbers **don't have** square roots.

$$\sqrt{x} \times \sqrt{x} = (\sqrt{x})^2 = x$$
$$(-\sqrt{x}) \times (-\sqrt{x}) = (-\sqrt{x})^2 = x$$

Example 2

Find both square roots of 16.

1. $4^2 = 16$, so the positive square root is 4. $\sqrt{16} = \mathbf{4}$
2. There's also the negative square root. Remember, $(-4) \times (-4) = 16$ too. $-\sqrt{16} = \mathbf{-4}$

Tip: Make sure you know your square numbers — then it'll be easy to write down their square roots.

Exercise 2

Q1 Find the positive and negative square roots of each of these numbers.

a) 1 b) 4 c) 9 d) 16

e) 25 f) 36 g) 64 h) 100

Q2 Calculate the following.

a) $\sqrt{49}$ b) $-\sqrt{49}$ c) $\sqrt{81}$ d) $-\sqrt{81}$
e) $\sqrt{121}$ f) $\sqrt{169}$ g) $-\sqrt{144}$ h) $\sqrt{400}$

Q3 Find both square roots of the following numbers. Use a calculator where necessary.

a) 64 b) 121 c) 10 000 d) 196

Q4 Calculate the following.

a) $\sqrt{9}+\sqrt{16}$ b) $\sqrt{25}-\sqrt{4}$ c) $\sqrt{100}-\sqrt{49}$

Cube Roots

Learning Objective — Spec Ref N6/N7:
Evaluate cube roots.

The **cube root** of a number x is the number that when **multiplied by itself** and then **by itself again** gives x. Finding a cube root is the **opposite of cubing**.

Every number has **exactly one** cube root.
The symbol $\sqrt[3]{}$ is used for cube roots.

$$\sqrt[3]{x}\times\sqrt[3]{x}\times\sqrt[3]{x}=(\sqrt[3]{x})^3=x$$

Example 3

Find the cube root of: a) 64 b) –64

a) $4^3 = 64$, so the cube root of 64 is 4. $\sqrt[3]{64} = \mathbf{4}$

b) $(-4)^3 = -64$, so the cube root of –64 is –4. $\sqrt[3]{-64} = \mathbf{-4}$

Tip: Just like square numbers, knowing your cube numbers makes finding cube roots much easier.

Exercise 3

Q1 Copy and complete the table.

x	1	8	27	1000	–1	–8	–27	–1000
$\sqrt[3]{x}$				10				

Q2 Find the following cube roots. Use a calculator where necessary.

a) $\sqrt[3]{64}$ b) $\sqrt[3]{125}$ c) $\sqrt[3]{1331}$ d) $\sqrt[3]{-64}$
e) $\sqrt[3]{-125}$ f) $\sqrt[3]{512}$ g) $\sqrt[3]{216}$ h) $\sqrt[3]{-729}$

Q3 Calculate:

a) $\sqrt[3]{15-7}$ b) $\sqrt[3]{39+5^2}$ c) $\sqrt[3]{4^2-43}$

3.2 Indices

Indices (or powers) are a useful shorthand that allow you to write repeated multiplications with just two symbols — a base and an index. Squares and cubes are two simple examples of indices.

Indices

Tip: Indices is the plural of index.

Learning Objective — Spec Ref A1/A4:
Work with numbers in index notation.

Powers show something that is being multiplied by itself. Powers are usually written using '**index notation**' — involving a **base** and an **index**.

base → 2^3 ← index

For example, $2 \times 2 \times 2 \times 2 = 2^4$ — this is four 2's multiplied together, and is read as "**2 to the power 4**".
And $5 \times 5 \times 5 \times 5 \times 5 \times 5 = 5^6$ — this is six 5's multiplied together, and is read as "**5 to the power 6**".

Example 1

a) Rewrite 3 × 3 × 3 × 3 × 3 using index notation.

There are five 3's multiplied together. $3 \times 3 \times 3 \times 3 \times 3 = \mathbf{3^5}$

b) Rewrite 100 000 using powers of 10.

Multiply 10 by 10 until you reach 100 000. $100\,000 = 10 \times 10 \times 10 \times 10 \times 10 = \mathbf{10^5}$

c) Evaluate 5^4.

There are four 5's multiplied together. $5^4 = 5 \times 5 \times 5 \times 5 = \mathbf{625}$

Exercise 1

Q1 Write the following using index notation.

a) 3×3
b) $2 \times 2 \times 2$
c) $7 \times 7 \times 7 \times 7 \times 7$
d) $9 \times 9 \times 9 \times 9 \times 9 \times 9$
e) $12 \times 12 \times 12 \times 12$
f) $17 \times 17 \times 17$

Q2 a) Use a calculator to evaluate these powers of 10.

(i) 10^5
(ii) 10^7
(iii) 10^8
(iv) 10^9

b) Copy and complete the following sentences. (Here, n is a positive whole number.)

(i) *"10^{15} can be written as a '1' followed by ________ zeros."*
(ii) *"10^n can be written as a '1' followed by ________ zeros."*

Q3 Rewrite the following as powers of 10.

a) 100 b) 1000 c) 10 000 d) 1 million

Q4 Use a calculator to evaluate these powers.

a) 2^4 b) 2^5 c) 3^4 d) 4^6

e) 6^4 f) 17^3 g) 5^5 h) 3^5

Q5 Evaluate the following using a calculator.
(Remember to work out powers **before** carrying out any addition or subtraction.)

a) $3^4 + 2^3$ b) $2^6 + 3^5$ c) $3^7 - 4^2$ d) $10^3 - 6^4$

Q6 Evaluate the following using a calculator.

a) $8^7 \div 4^6$ b) $10^4 \times 10^3$ c) $2^4 \times 2^2$ d) $3^4 \div 5^4$

Q7 Evaluate the following using a calculator.

a) $(5 - 2)^3$ b) $(2^2)^2$ c) $(3^2)^2$ d) $(8 - 5)^4$

e) $(7 + 3)^5$ f) $6^4 - 7^2$ g) $2 + 10^4$ h) $(150 - 50)^6$

Example 2

a) Rewrite $a \times a \times a \times a \times a \times a$ using index notation.

There are 6 a's multiplied together. $a \times a \times a \times a \times a \times a = \boldsymbol{a^6}$

b) Rewrite $b \times b \times b \times b \times c \times c$ using index notation.

There are 4 b's multiplied together, and 2 c's. $b \times b \times b \times b \times c \times c = b^4 \times c^2 = \boldsymbol{b^4c^2}$

Tip: Indices work in exactly the same way for letters as they do for numbers.

Exercise 2

Q1 Rewrite the following using index notation.

a) $h \times h \times h \times h$ b) $t \times t \times t \times t \times t$ c) $s \times s \times s \times s \times s \times s \times s$

Q2 Rewrite the following using index notation.

a) $a \times a \times b \times b \times b$ b) $k \times k \times k \times k \times f \times f \times f$ c) $m \times m \times m \times m \times n \times n$

d) $s \times s \times s \times s \times t$ e) $w \times w \times w \times v \times v \times v$ f) $p \times p \times q \times q \times q \times q \times q$

Q3 Evaluate these powers using a calculator, given that $x = 2$ and $y = 5$.

a) x^2y^2 b) x^3y^2 c) x^2y^3

d) x^5y^2 e) x^4y^3 f) x^4y^4

3.3 Laws of Indices

The laws of indices are important rules that help you work with expressions involving powers.

Working with Indices

Learning Objective — Spec Ref N7/A4:
Use the laws of indices to simplify expressions.

You can use the laws of indices to multiply and divide powers with the same base.

1) To **multiply** two powers with the same base, **add** the indices: $a^m \times a^n = a^{m+n}$
2) To **divide** two powers with the same base, **subtract** the indices: $a^m \div a^n = a^{m-n}$
3) To **raise** one power to another power, **multiply** the indices: $(a^m)^n = a^{m \times n}$
4) To raise a **fraction** to a power, apply the power to the **numerator** and **denominator** (see p.63): $\left(\frac{a}{b}\right)^n = \frac{a^n}{b^n}$

There are two other important index facts you need to know.

1) Anything to the **power 1** is **itself**: $a^1 = a$
2) Anything to the **power 0** is **1**: $a^0 = 1$

Example 1

Simplify the following, leaving the answers in index form.

a) $3^8 \times 3^5$ — This is multiplication, so add the indices. $3^8 \times 3^5 = 3^{8+5} = \mathbf{3^{13}}$

b) $\frac{10^8}{10^5}$ — This is division ($\frac{10^8}{10^5} = 10^8 \div 10^5$), so subtract the indices. $10^8 \div 10^5 = 10^{8-5} = \mathbf{10^3}$

c) $(2^7)^2$ — For one power raised to another power, multiply the indices. $(2^7)^2 = 2^{7 \times 2} = \mathbf{2^{14}}$

d) $a^7 \times a^3$ — The laws of indices work in exactly the same way with variables. $a^7 \times a^3 = a^{7+3} = \boldsymbol{a^{10}}$

Exercise 1

Q1 Simplify these expressions. Leave your answers in index form.

a) $3^2 \times 3^6$ b) $10^7 \times 10^3$ c) $4^7 \times 4^4$ d) 7×7^6

Q2 Simplify these expressions. Leave your answers in index form.

a) $6^7 \div 6^4$ b) $8^6 \div 8^3$ c) $5^7 \div 5^2$ d) $6^8 \div 6^6$

Q3 Simplify these expressions. Leave your answers in index form.

a) $(4^3)^3$ b) $(11^2)^5$ c) $(100^3)^{23}$

d) $\frac{2^8}{2^5}$ e) $\left(\frac{2^7}{5}\right)^3$ f) $\left(\frac{4^6}{4^3}\right)^2$

Q4 Simplify these expressions. Leave your answers in index form.

a) $4^5 \times 4^{11}$ b) $12^7 \div 12^3$ c) $8^2 \times 8^9$ d) $(6^8)^4$

e) $(3^{12})^4$ f) $7^{11} \div 7^6$ g) $4^{15} \div 4^7$ h) $(11^0)^9$

Q5 For each of the following, find the number that should replace the square.

a) $s^9 \times s^{\blacksquare} = s^{14}$ b) $t^5 \div t^{\blacksquare} = t^3$ c) $r^7 \times r^{\blacksquare} = r^{13}$ d) $(p^7)^{\blacksquare} = p^{49}$

e) $k^{\blacksquare} \div k^5 = 1$ f) $a^{\blacksquare} \times a^7 = a^{15}$ g) $m^5 \div m^{-4} = m^{\blacksquare}$ h) $(q^5)^{\blacksquare} = q^{25}$

Negative Indices

Prior Knowledge Check: Know how to find the reciprocal of a number — see p.80.

You can evaluate powers that have a **negative index** by taking the **reciprocal** of the base and making the index **positive**.

The reciprocal is found by turning the base **upside down**.

E.g. the reciprocal of a is $\frac{1}{a}$ and the reciprocal of $\frac{a}{b}$ is $\frac{b}{a}$

$$a^{-n} = \frac{1}{a^n}$$

$$\left(\frac{a}{b}\right)^{-m} = \left(\frac{b}{a}\right)^m = \frac{b^m}{a^m}$$

The **laws of indices** work in exactly the **same way** for negative indices.
It's often easier to leave the index negative when simplifying expressions.

Example 2

Evaluate 5^{-3}

1. Take the reciprocal of the base and make the index positive.
2. Evaluate the denominator.

$$5^{-3} = \frac{1}{5^3} = \frac{1}{5 \times 5 \times 5} = \mathbf{\frac{1}{125}}$$

Exercise 2

Q1 Evaluate the following. Give fractions in their simplest form.

a) 8^{-2} b) 2^{-3} c) 5^{-2} d) 3^{-3}

e) 2^{-4} f) $\left(\frac{1}{9}\right)^{-2}$ g) $\left(\frac{4}{5}\right)^{-2}$ h) $\left(\frac{2}{6}\right)^{-3}$

Q2 Simplify these expressions. Leave your answers in index form.

a) $j^{-13} \div j^7$ b) $(n^7)^{-3}$ c) $p^{-8} \times p^{-6}$ d) $y^8 \div y^{-2}$

e) $(k^{-3})^6$ f) $\frac{b^5}{b^9}$ g) $d^{-7} \times d^2$ h) $\frac{x^{60}}{x^{-8}}$

Q3 For each of the following, find the number that should replace the square.

a) $l^{\blacksquare} \times l^{-8} = l^3$ b) $b^{\blacksquare} \div b^7 = b^{-10}$ c) $c^{-15} \times c^{\blacksquare} = c^8$ d) $(y^{-4})^{\blacksquare} = y^{16}$

Calculating with Indices

Learning Objective — Spec Ref N7/A4:
Use the laws of indices to work with complex expressions.

Prior Knowledge Check: Be able to use the laws of indices — see page 44.

Complicated expressions, made up of many terms with indices, can often be **simplified** to a single term by using the **laws of indices**.

Example 3

Simplify each expression without using a calculator. Leave your answers in index form.

a) $\mathbf{2^6 \times 2^8 \div 2^2}$

Apply the index laws to each step. $2^6 \times 2^8 \div 2^2 = 2^{6+8-2} = \mathbf{2^{12}}$

b) $\mathbf{\dfrac{2^3 \times 2^5}{2^8 \div 2^6}}$

1. Work out the top and bottom lines of the fraction separately.
2. Then you can do the final division.

$$\frac{2^3 \times 2^5}{2^8 \div 2^6} = \frac{2^{3+5}}{2^{8-6}} = \frac{2^8}{2^2}$$
$$= 2^{8-2} = \mathbf{2^6}$$

Tip: These questions can look scary at first, but once you break them into steps they really aren't that bad.

Exercise 3

In Questions 1 and 2, simplify each expression, leaving your answers in index form.

Q1 a) $3^2 \times 3^5 \times 3^7$ b) $(8^6)^2 \times 8^5$ c) $(12^8 \div 12^4)^3$ d) $(4^3)^6 \div 4^{16}$

Q2 a) $\dfrac{3^4 \times 3^5}{3^6}$ b) $\dfrac{8^{25} \div 8^2}{8^6 \times 8^{10}}$ c) $\dfrac{(7^5)^7 \div 7^{12}}{7^5 \times 7^9}$ d) $\dfrac{(5^{10} \div 5^8)^4}{5^4 \div 5^2}$

Q3 Which of the expressions in the box below are equal to 1?

$\dfrac{4^4 \div 4^3}{4}$ $\dfrac{7^{16}}{7^8 \times 7^2}$ $\dfrac{3^8 - 3^7}{3}$ $\dfrac{5^5 \times 5^9}{(5^2)^7}$ $\dfrac{(9^2)^2 - 9^0}{9^3}$

Q4 Simplify these expressions. Leave your answers in index form.

a) $\left(\dfrac{2^{-5} \times 2^7}{2^3}\right)^5$ b) $\left(\dfrac{7^3}{7}\right)^3 \times 7^{-2}$ c) $\dfrac{9^{-3} \times 9^{15}}{(9^{-3})^{-2}}$ d) $\left(\dfrac{3^{-8} \times 3^{12}}{3^2}\right)^{-6}$

Q5 Simplify each of the following expressions.

a) $a^6 \times a^5 \div a^4$ b) $(p^5 \div p^3)^6$ c) $\dfrac{(t^6 \div t^3)^4}{t^9 \div t^4}$ d) $\dfrac{(c^{-4})^3}{c^{-8} \div c^4}$

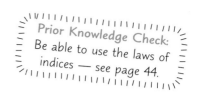

3.4 Standard Form

Standard form allows you to write very large or very small numbers without having to write lots of zeros. E.g. 12 000 000 000 can be written as 1.2×10^{10} and 0.000 000 345 as 3.45×10^{-7}.

Standard Form

Learning Objective — Spec Ref N9:
Write and interpret numbers in standard form.

Prior Knowledge Check: Know how to multiply and divide decimals by 10, 100 and 1000 — see p.18.

In **standard form** (or standard index form), numbers are written like this:

$$A \times 10^n$$

A can be **any number** between 1 and 10 (but not 10 itself)

n can be **any integer** (whole number).

There are three vital things you need to know about standard form:

- The **front number**, A, must always be **between 1 and 10** (i.e. $1 \le A < 10$).
- The power of 10, n, is how far the **decimal point moves**.
- n is **positive** for **BIG** numbers and **negative** for **SMALL** numbers.

Tip: It's handy to think of the decimal point moving, but it's actually the digits that shift around it.

Example 1

Write these numbers in standard form: a) 360 000 b) 0.000036

a) 1. Move the decimal point until 360 000 becomes 3.6. The decimal point has moved 5 places.

2. The number is big so n must be +5.

$360000.0 = \mathbf{3.6 \times 10^5}$

b) 1. As in part a), the decimal point moves 5 places to make the number 3.6.

2. The number is small so n must be –5.

$0.000036 = \mathbf{3.6 \times 10^{-5}}$

Tip: Once you have an answer, check that it satisfies the three vital things described above.

Exercise 1

Q1 Write the following numbers in standard form.

a) 250 b) 7340 c) 48 000 d) 5 900 000

Q2 Write the following numbers in standard form.

a) 0.375 b) 0.0067 c) 0.000078 d) 0.07070

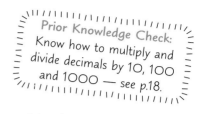

Example 2

Write the following standard form numbers as ordinary numbers.

a) 3.5×10^3
1. The power is positive so the number will be big.
2. The decimal point moves 3 places. $3.5 \times 10^3 = \mathbf{3500.0}$

b) 4.67×10^{-5}
1. The power is negative so the number will be small.
2. The decimal point moves 5 places. $4.67 \times 10^{-5} = \mathbf{0.0000467}$

Exercise 2

Q1 Write the following out as ordinary numbers.

a) 3×10^6 b) 9.4×10^4 c) 1.989×10^8 d) 7.20×10^0

e) 3.56×10^{-6} f) 4.23×10^{-2} g) 8.88×10^{-5} h) 1.9×10^{-8}

Calculations in Standard Form

Learning Objectives — Spec Ref N9:
- Multiply and divide numbers in standard form.
- Add and subtract numbers in standard form.

Prior Knowledge Check: Know how to use the laws of indices — see p.44.

To **multiply** or **divide** numbers in standard form:

- **Rearrange** the calculations so that the **front numbers** are **together**. E.g. rewrite $(4 \times 10^6) \times (8 \times 10^2)$ as $(4 \times 8) \times (10^6 \times 10^2)$.
- **Multiply or divide** the front numbers and use the **laws of indices** (p.44) to multiply or divide the powers of 10.
- Make sure your answer is still in **standard form** — if not, use the method from p.47.

Example 3

Calculate $(2.4 \times 10^7) \times (5.2 \times 10^3)$. Give your answer in standard form.

1. Rearrange to put the front numbers and powers of 10 together. $(2.4 \times 5.2) \times (10^7 \times 10^3)$
2. Multiply the front numbers and use the laws of indices. $= 12.48 \times 10^{7+3}$
3. 12.48 isn't between 1 and 10 so this isn't in standard form. Convert 12.48 to standard form. $= 12.48 \times 10^{10}$ $= 1.248 \times 10 \times 10^{10}$
4. Add the indices again to get the answer in standard form. $= \mathbf{1.248 \times 10^{11}}$

Example 4

Calculate $(9.6 \times 10^7) \div (1.2 \times 10^4)$. Give your answer in standard form.

1. Rewrite as a fraction.
2. Separate the front numbers and powers of 10.
3. Simplify the two fractions.

$$\frac{9.6 \times 10^7}{1.2 \times 10^4} = \frac{9.6}{1.2} \times \frac{10^7}{10^4}$$

$$= 8 \times 10^{7-4} = \mathbf{8 \times 10^3}$$

Exercise 3

Give your answers to these questions in standard form.

Q1 a) $(3 \times 10^7) \times (2 \times 10^4)$ b) $(4 \times 10^9) \times (2 \times 10^{-4})$ c) $(6 \times 10^5) \times (1.4 \times 10^2)$

Q2 a) $(9 \times 10^6) \div (3 \times 10^4)$ b) $(1.8 \times 10^{-4}) \div (0.9 \times 10^8)$ c) $(8.1 \times 10^{-1}) \div (9 \times 10^{-3})$

To **add** or **subtract** numbers in standard form:

- Check the **powers of 10** are the **same** in both terms.
- **Add or subtract** the front numbers.
- Make sure your answer is still in **standard form** at the end.

Example 5

Calculate $(3.7 \times 10^3) + (8.2 \times 10^3)$ without using a calculator. Give your answer in standard form.

1. The powers of 10 match so add the numbers.
2. The answer isn't in standard form (because 11.9 is bigger than 10), so convert it to standard form.

$(3.7 + 8.2) \times 10^3$

$= 11.9 \times 10^3$

$= 1.19 \times 10 \times 10^3$

$= \mathbf{1.19 \times 10^4}$

Tip: For subtraction questions, use the same method, but subtract the front numbers instead.

Exercise 4

Give your answers to these questions in standard form.

Q1 a) $(5 \times 10^3) + (3 \times 10^3)$ b) $(6.4 \times 10^2) + (3.2 \times 10^2)$ c) $(6.9 \times 10^{-4}) + (3.8 \times 10^{-4})$

Q2 a) $(4.5 \times 10^{-2}) - (3.3 \times 10^{-2})$ b) $(1.8 \times 10^4) - (1.2 \times 10^4)$ c) $(6.4 \times 10^2) - (6.3 \times 10^2)$

Review Exercise

Q1 Evaluate the following. Use a calculator where necessary.

a) $(-5)^3$ b) 0.5^3 c) $(-0.3)^3$ d) $(-12)^3$

e) 0.1^2 f) $(-0.4)^2$ g) $((-2)^2)^3$ h) $((-2)^3)^2$

Q2 Evaluate the following using a calculator.

a) $3^2 - 2^3$ b) $5^2 - 6^2$ c) 3×2^8 d) 8×5^4

Q3 Evaluate the following. Use a calculator where necessary.

a) $-\sqrt{36}$ b) $\sqrt{361}$ c) $\sqrt[3]{-343}$ d) $-\sqrt{10^2 - 19}$

Q4 Rewrite the following using index notation.

a) $k \times k \times l \times l \times k \times k$ b) $z \times y \times y \times y$ c) $m \times n \times m \times n \times m$

Q5 Simplify these expressions. Leave your answers in index form.

a) $a^6 \times a^4$ b) $15^{12} \div 15^{-14}$ c) $(45^2)^{-9}$ d) $\frac{20^{222}}{20^{210}}$

Q6 Evaluate the following. Give any fractions in their simplest forms.

a) 8^{-1} b) 4^{-2} c) $\left(\frac{1}{3}\right)^{-3}$ d) $\left(\frac{4}{6}\right)^{-2}$

Q7 Simplify the following. Leave your answer in index form.

a) $\frac{4^4 \times 4^6}{4^8 \times 4}$ b) $\frac{(5^5 \times 5^5)^2}{5^8 \div 5^3}$ c) $\left(\frac{2^5 \times 2^5}{2^3}\right)^4$

Q8 Write the following numbers in standard form.

a) 330 b) 2 750 000 c) 0.0025 d) 0.0005002

Q9 Write the following out as ordinary numbers.

a) 4×10^2 b) 8.8×10^5 c) 6.69×10^{-1} d) 7.05×10^{-6}

Q10 Do these questions. Give your answers in standard form.

a) $(7 \times 10^5) \times (1.3 \times 10^2)$ b) $(8.8 \times 10^3) \div (4 \times 10^8)$

c) $(1.9 \times 10^6) + (9.1 \times 10^6)$ d) $(5.9 \times 10^{-8}) - (3.4 \times 10^{-8})$

Exam-Style Questions

Q1 Find the value of $(4.2 - 0.81)^2 + \sqrt{289}$

[1 mark]

Q2 Simplify the following expressions:

a) $2x^3 \times 4x^4$

[1 mark]

b) $(3y^2)^4$

[2 marks]

c) $5z^0$

[1 mark]

Q3 a) Find the value of n when $4^n = 64$

[1 mark]

b) Find the value of k when $2^2 \times 3^k \times 5 = 540$.

[3 marks]

Q4 Mustafa has attempted to write the numbers 650 million and 0.000234 in standard form. His working is shown below.
For each attempt, explain a mistake that Mustafa has made.

a) 650 million = 650 000 000 = 0.65×10^9

[1 mark]

b) $0.000234 = 2.34 \times 10^4$

[1 mark]

Q5 Here is some information about Mercury and Venus. PROBLEM SOLVING

Mercury	
Number of Earth days to orbit Sun	88
Distance travelled in orbit of Sun	3.6×10^8 km

Venus	
Number of Earth days to orbit Sun	225
Distance travelled in orbit of Sun	6.8×10^8 km

Calculate which planet travels further in one Earth day, and state by how much.
Give your answer in standard form to a suitable degree of accuracy.

[5 marks]

4.1 Finding Multiples and Factors

To tackle multiples and factors, you need to know your times tables — that's all there is to it.

Multiples

Learning Objective — Spec Ref N4:
Identify and find multiples and common multiples.

A **multiple** of a number is one that is in its **times table**.
E.g. the multiples of 2 are 2, 4, 6, 8, 10... and the multiples of 5 are 5, 10, 15, 20, 25...

A **common multiple** of two (or more) numbers is a multiple of both (or all) of those numbers.
10 is in both lists above, so 10 is a common multiple of 2 and 5.
Both 2 and 5 divide into 10 exactly.

Example 1

a) **List the multiples of 5 between 23 and 43.**

Starting at 20, the times table of 5 is: 20, 25, 30, 35, 40, 45...
So the multiples between 23 and 43 are: **25, 30, 35, 40**

b) **Which of the numbers in the box below are:**

24	7	28	35	39

(i) **multiples of 3?**

3 divides into 24 and 39 exactly — so 24 and 39 are multiples of 3. But 3 doesn't divide exactly into 7, 28 or 35 — these aren't multiples of 3.

(i) **24 and 39**

(ii) **multiples of 5?**

The only number in the box that 5 divides into exactly is 35.

(ii) **35**

(iii) **common multiples of 4 and 7?**

The multiples of 4 are 24 and 28, while the multiples of 7 are 7, 28 and 35. So the only common multiple is 28.

(iii) **28**

Exercise 1

Q1 List the first five multiples of: a) 4 b) 10 c) 3 d) 6 e) 7

Q2 a) List the multiples of 8 between 10 and 20.
b) List the multiples of 9 between 20 and 50.

Q3 Write down the numbers from the box that are:
a) multiples of 10
b) multiples of 15
c) common multiples of 10 and 15

5	10	15	20	25	30	35
40	45	50	55	60	65	70
75	80	85	90	95	100	105

Q4 List all the common multiples of 5 and 6 between 1 and 40.

Q5 List the common multiples of 3 and 4 between 19 and 35.

Factors

Learning Objective — Spec Ref N4:
Identify and find factors and common factors.

The **factors** of a number are the numbers that divide into it exactly. E.g. the factors of 8 are 1, 2, 4 and 8 — all these numbers divide into 8 exactly.

A **common factor** of two (or more) numbers is a factor of both (or all) those numbers. E.g. the factors of 12 are 1, 2, 3, 4, 6 and 12, so the common factors of 8 and 12 are 1, 2 and 4.

Tip: Any two factors that multiply to give the number are called a **factor pair.** For example, 3 and 6 are a factor pair of 18.

Example 2

Write down all the factors of 18.

1. Check if 1, 2, 3, 4... divide into the number. Write down each number that divides in exactly, and also its 'factor partner' in a multiplication.
2. Stop checking when you reach a number already in an earlier multiplication, or when a factor is repeated in a multiplication.
3. List all the numbers in your multiplications.

1 × 18 = 18 — so 1 and 18 are factors
2 × 9 = 18 — so 2 and 9 are factors
3 × 6 = 18 — so 3 and 6 are factors
4 × – = 18 — so 4 is not a factor
5 × – = 18 — so 5 is not a factor
6 × 3 = 18 — 6 and 3 are repeated so stop.

So the factors of 18 are **1, 2, 3, 6, 9, 18.**

Example 3

Find the common factors of 6 and 20.

1. Find the factors of 6 and 20.

$1 \times 6 = 6$ — so 1 and 6 are factors
$2 \times 3 = 6$ — so 2 and 3 are factors
$3 \times 2 = 6$ — 3 and 2 are repeated so stop.

So the factors of 6 are: **1, 2, 3, 6**

$1 \times 20 = 20$ — so 1 and 20 are factors
$2 \times 10 = 20$ — so 2 and 10 are factors
$3 \times - = 20$ — so 3 is not a factor
$4 \times 5 = 20$ — so 4 and 5 are factors
$5 \times 4 = 20$ — 5 and 4 are repeated so stop.

So the factors of 20 are: **1, 2, 4, 5, 10, 20**

2. The common factors are the numbers that appear in both lists.
So the common factors of 6 and 20 are **1 and 2**.

Exercise 2

Q1 List the numbers from the box on the right that are factors of: 2 3 5 6 12 15

a) 6 b) 24 c) 30
d) 36 e) 20 f) 45

Q2 List all the factors of the following numbers.

a) 10 b) 4 c) 13
d) 25 e) 24 f) 35

Q3 a) List all the factors of 15.
b) List all the factors of 21.
c) List the common factors of 15 and 21.

Q4 List the common factors of the following pairs of numbers.

a) 15, 20 b) 12, 15 c) 30, 45
d) 50, 90 e) 25, 50 f) 21, 22

Q5 a) Which number is a factor of all other numbers?
b) Which two numbers are factors of all even numbers?
c) Which two numbers must be factors of all numbers whose last digit is 5?

Q6 List the common factors of the following sets of numbers.

a) 15, 20, 25 b) 12, 18, 20
c) 30, 45, 50 d) 15, 16, 17

4.2 Prime Numbers

There's a key definition coming up — prime numbers. Once you know what they are (and how to find them), you can move on to writing numbers as products of prime factors.

Prime Numbers

Learning Objective — Spec Ref N4:
Identify and find prime numbers.

A **prime number** is a number that has no other factors except **itself** and **1**.

In other words, the only numbers that **divide exactly** into a prime number are itself and 1.

Here are a few things to note about prime numbers:

- 1 is **not** classed as a prime number — this is a common mistake.
- 2 is the only **even** prime number — so any other even number is **not** prime.
- Prime numbers end in **1**, **3**, **7** or **9** (2 and 5 are the only exceptions to this rule). But **not all** numbers ending in 1, 3, 7 or 9 are prime (e.g. $27 = 3 \times 9$, so it isn't prime).

Example 1

Which of the numbers in the box on the right are prime? 16 17 18 19 20

1. Look for factors of each of the numbers.
2. If you can find factors other than 1 and the number itself, then the number isn't prime.
3. If there are no factors other than 1 and the number itself, then the number is prime.

$16 = 2 \times 8$, and so 16 isn't prime
$18 = 3 \times 6$, and so 18 isn't prime
$20 = 4 \times 5$, and so 20 isn't prime

17 has no factors other than 1 and 17.
19 has no factors other than 1 and 19.

So the prime numbers are **17 and 19.**

Exercise 1

Q1 a) Which three numbers in the box on the right are not prime? 31 33 35 37 39

b) Find two factors greater than 1 that multiply together to give each of your answers to a).

Q2 Write down the prime numbers in this list: 5, 15, 22, 34, 47, 51, 59

Q3 Find all the prime numbers between 20 and 30.

Q4 a) For each of the following, find a factor greater than 1 but less than the number itself:
(i) 4 (ii) 14 (iii) 34 (iv) 74

b) Explain why any number with last digit 4 cannot be prime.

Q5 Without doing any calculations, explain how you can tell that none of the following numbers in the list below are prime.

20	30	40	50	70	90	110	130

Writing a Number as a Product of Primes

Learning Objectives — Spec Ref N4:
- Understand the unique factorisation theorem.
- Find the prime factorisation of a number.
- Write a prime factorisation using product notation.

Prior Knowledge Check:
Be able to find factors, recognise prime numbers and use index notation. See p.42 and p.53.

Whole numbers which are **not** prime can be broken down into **prime factors**. The product of these prime factors is the original number — the group of prime factors is known as the **prime factorisation** of the number. For example, the prime factorisation of 28 is $2 \times 2 \times 7$.

The prime factorisation of every number is **unique** — each number only has **one** prime factorisation, and no two numbers can have the **same one**.

To find a prime factorisation, you can use a **factor tree**. A factor tree breaks a number into factors then breaks these factors into smaller factors, and keeps going until all of the factors are prime.

If the prime factorisation has **repeated factors**, you can write it using **index notation** (i.e. as a product of powers). So the prime factorisation of 28 can be written as $2^2 \times 7$.

Example 2

Write 12 as the product of prime factors.

Make a factor tree.

1. First find any two factors whose product is 12. $12 = 2 \times 6$, so create two branches with these factors and circle any factors that are prime.
2. Repeat step 1 for any factors you didn't circle ($6 = 2 \times 3$).
3. Stop when all the branches end in a circle. The product of all the circled primes is the number you started with.

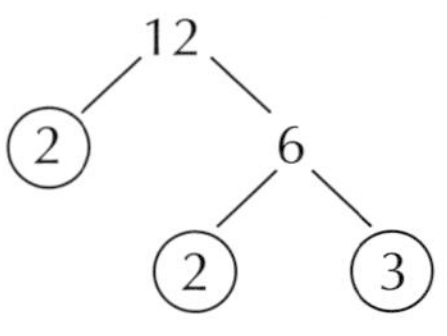

$12 = 2 \times 2 \times 3$, or $\mathbf{12 = 2^2 \times 3}$

Tip: You could have started with any two factors of 12 on the first branches — e.g. 3 and 4. The rest of the tree would look a bit different, but the prime factorisation would be exactly the same.

Exercise 2

In the following questions, write any repeated prime factors as powers.

Q1 a) Copy and complete the following factor trees.

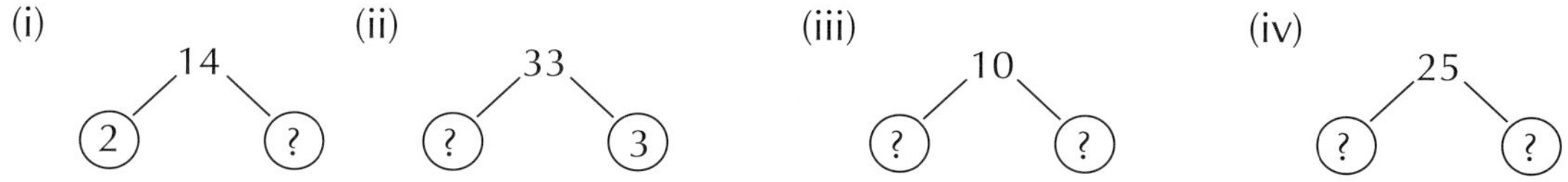

b) Use your factor trees to write the following as products of prime factors.

(i) 14 (ii) 33 (iii) 10 (iv) 25

Q2 Write the following numbers as the product of two prime factors.

a) 15 b) 21 c) 22

d) 6 e) 14 f) 26

Q3 a) Copy and complete the following factor trees.

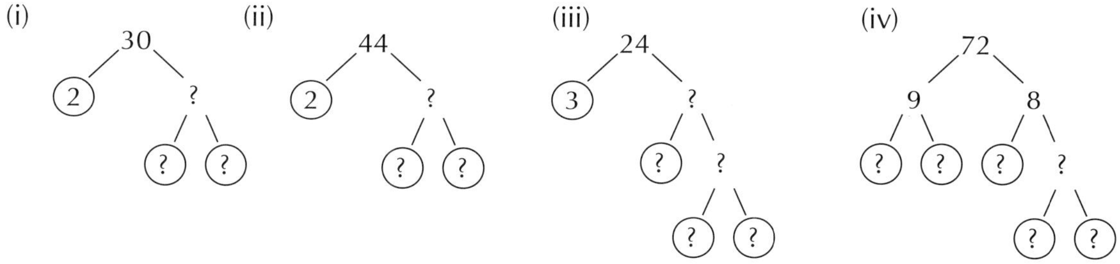

b) Use your factor trees to write the following as the product of prime factors.

(i) 30 (ii) 44 (iii) 24 (iv) 72

Q4 Copy and complete the factor tree on the right and use it to find the prime factorisation of 70.

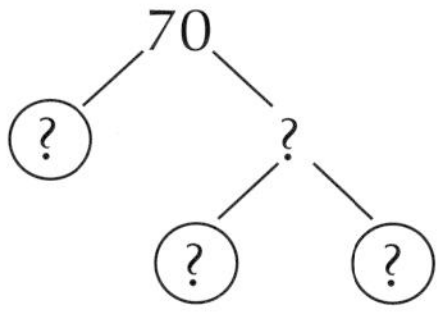

Q5 Use factor trees to write the following numbers as the product of prime factors.

a) 42 b) 84 c) 190 d) 210

Q6 Write the following as the product of prime factors.

a) 128 b) 168 c) 325

d) 98 e) 225 f) 1000

Q7 Square numbers have all their prime factors raised to even powers. For example, $36 = 2^2 \times 3^2$ and $64 = 2^6$.

a) Write 75 as a product of prime factors.

b) What is the smallest number you could multiply 75 by to form a square number? Explain your answer. PROBLEM SOLVING

4.3 LCM and HCF

Now that you've got your head around multiples and factors, it's time to start looking at the 'lowest common multiple' and 'highest common factor' — or LCM and HCF for short.

Lowest Common Multiple (LCM)

Learning Objectives — Spec Ref N4:
- Understand the term 'lowest common multiple'.
- Be able to find the lowest common multiple of a set of numbers.

Prior Knowledge Check: Be able to work with multiples. See p.52.

If you're given two (or more) numbers, then you can find their **lowest common multiple** (LCM).

As the name suggests, the LCM is basically the smallest number that is a multiple of both (or all) of the numbers. In other words:

The LCM is the **smallest** number that will **divide** by **all** the numbers in a list.

Tip: The LCM is the smallest number in the times table of all the numbers in the list.

If you're given a set of numbers, you can find their LCM by **listing multiples** of each number, then identifying the **first one** that appears in **every list**.

Example 1

Find the lowest common multiple (LCM) of 4, 6 and 8.

1. Find the multiples of 4, 6 and 8.
2. The LCM is the smallest number that appears in all three lists.

Multiples of 4 are: 4, 8, 12, 16, 20, (24), 28...
Multiples of 6 are: 6, 12, 18, (24), 30, 36...
Multiples of 8 are: 8, 16, (24), 32, 40, 48...

So the LCM of 4, 6 and 8 is **24**.

Exercise 1

Q1 a) Find the first 5 multiples of 9 and the first 5 multiples of 12.
b) Hence find the LCM of 9 and 12.

Q2 a) Find the first 10 multiples of 5 and the first 10 multiples of 7.
b) Hence find the LCM of 5 and 7.

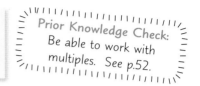

Q3 Find the LCM of each of the following pairs of numbers.

a) 3 and 4
b) 6 and 8
c) 2 and 10
d) 6 and 7
e) 10 and 15
f) 15 and 20

Q4 Find the LCM of each of the sets of numbers below.

a) 3, 6, 8
b) 3, 5, 6
c) 4, 9, 12

Example 2

Jane and Alec are running around a small circular track.
It takes Jane 10 seconds to run one lap and Alec 12 seconds.

If they both start from the same point on the track at exactly the same time, how long will it be before they next cross the start line together?

1. Find the number of seconds after which Jane will cross the start line.

Jane crosses the start line after the following numbers of seconds: 10, 20, 30, 40, 50, 60, 70...

2. Find the number of seconds after which Alec will cross the start line.

Alec crosses the start line after the following numbers of seconds: 12, 24, 36, 48, 60, 72...

3. Find the smallest number that is in both lists

So they will next cross the start line together after **60 seconds**.

Exercise 2

Q1 Laurence and Naima are cycling around a course. They leave the start-line at the same time and each do 10 laps. It takes Laurence 8 minutes to do one lap and Naima 12 minutes.

a) After how many minutes does Laurence pass the start-line? Write down all possible answers.
b) After how many minutes does Naima pass the start-line? Write down all possible answers.
c) When will they first pass the start-line together?

Q2 Mike visits Oscar every 4 days, while Narinda visits Oscar every 5 days. If they both visited today, how many days will it be before they visit on the same day again? PROBLEM SOLVING

Q3 Jill divides a pile of sweets into 5 equal piles. Kay then divides the same sweets into 7 equal piles. What is the smallest number of sweets there could be? PROBLEM SOLVING

Q4 A garden centre has between 95 and 205 potted plants. They can be arranged exactly in rows of 25 and exactly in rows of 30. How many plants are there? PROBLEM SOLVING

Highest Common Factor (HCF)

Learning Objectives — Spec Ref N4:
- Understand the term 'highest common factor'.
- Be able to find the highest common factor of a set of numbers.

Prior Knowledge Check: Be able to find the factors of a number. See p.53.

You may also be asked to find the **highest common factor** (HCF) of a list of numbers. The HCF is just the largest value that is a factor of all the numbers in the list. In other words:

HCF is the **largest** number that will **divide into all** the numbers in your list.

To find the HCF of a set of numbers, list the factors of each number, then pick the biggest one that's in every list.

Example 3

Find the highest common factor of 12 and 15.

1. Find the factors of both numbers. The common factors are the numbers in both lists.

   The factors of 12 are: (1) 2, (3) 4, 6, 12

   The factors of 15 are: (1) (3) 5, 15

2. The HCF is the biggest number that appears in both lists.

   So the highest common factor of 12 and 15 is **3**.

Exercise 3

Q1 a) Find the common factors of 12 and 20.
b) Hence find the highest common factor (HCF) of 12 and 20.

Q2 a) List the common factors of 20 and 30.
b) Use your list to find the HCF of 20 and 30.

Q3 Find the HCF of the following pairs of numbers.
a) 8 and 12 b) 24 and 32 c) 18 and 24
d) 36 and 60 e) 14 and 15 f) 12 and 36

Q4 Write down the following:
a) the HCF of 11 and 12 b) the HCF of 21 and 22

Q5 Find the HCF of the following sets of numbers.
a) 6, 8, 16 b) 12, 15, 18 c) 24, 30, 36 d) 18, 36, 72

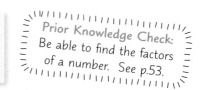

Review Exercise

Q1 Write down the numbers from the box that are:

a) multiples of 3
b) multiples of 4
c) multiples of 5

2	5	8	9	11	12
14	15	16	18	20	21

Q2 List all the common multiples of 8 and 10 between 1 and 100.

Q3 A zoo has between 95 and 155 animals.
The number of animals divides exactly by 20 and exactly by 30.

a) Write down all the multiples of 20 between 95 and 155.
b) Write down all the multiples of 30 between 95 and 155.
c) How many animals are there?

Q4 List the common factors of each of the following numbers.

a) 10 and 42 b) 14 and 27 c) 12 and 24

Q5 A baker has 12 identical cakes. In how many different ways can he divide them up into equal packets (without cutting them up)? List the possibilities.

Q6 a) Which three numbers in the box on the right are not prime?

51	53	55	57	59

b) Find two factors greater than 1 that multiply together to give each of your answers to a).

Q7 Write down the prime numbers in this list: 7, 17, 18, 31, 33, 51, 54

Q8 Write the following numbers as the product of prime factors.

a) 18 b) 50 c) 36 d) 150

Q9 a) Find the highest common factor of the following pairs of numbers.

(i) 5 and 9 (ii) 9 and 12 (iii) 10 and 15 (iv) 12 and 28

b) Find the least common multiple of the following pairs of numbers.

(i) 5 and 9 (ii) 6 and 10 (iii) 9 and 12 (iv) 12 and 20

Exam-Style Questions

Q1 Here is a list of numbers.

8 11 15 17 21 28

From the list, write down a number that is:

a) a multiple of both 2 and 7

[1 mark]

b) a factor of 32

[1 mark]

c) 1 more than a prime number.

[1 mark]

Q2 a) Write down all the factors of 50, giving your answer in ascending order.

[1 mark]

b) Find the smallest prime number that is bigger than 50.

[1 mark]

Q3 Find the HCF of 48 and 60.

[2 marks]

Q4 Write 380 as a product of its prime factors.

[2 marks]

Q5 The lowest common multiple of two numbers is 60.
The highest common factor of the same two numbers is 4.
Neither of the numbers is 4 or 60. What are the numbers?

[2 marks]

Q6 For this question, you are given that $525 = \sqrt{275\,625}$.

a) Express 525 as a product of prime factors.

[2 marks]

b) Use your answer to part a) to write 275 625 as a product of prime factors.

[1 mark]

5.1 Equivalent Fractions

Fractions are a way of writing one number divided by another. As different divisions can give the same answer (e.g $1 \div 3$ is the same as $2 \div 6$), different fractions can have the same value.

Learning Objective — Spec Ref N3:
Be able to find equivalent fractions.

The **fraction** $\frac{a}{b}$ is another way of writing the **division** $a \div b$.

- The number **above the line** (a) is called the **numerator**.
- The number **below the line** (b) is called the **denominator**.

There are lots of **different ways** to write the **same amount** using fractions — these are known as **equivalent fractions**.

For example, one half is the same as two quarters, three sixths, etc. You can see this on the diagrams below:

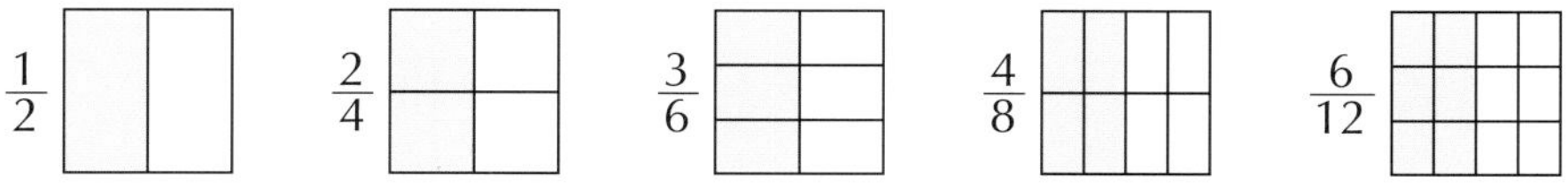

Tip: Notice that the same proportion of the shape is shaded in each diagram.

To find an equivalent fraction, **multiply** or **divide** both the numerator and the denominator by the **same thing** — for example:

- $\frac{3}{9}$ and $\frac{1}{3}$ are equivalent fractions because $\frac{3}{9} = \frac{3 \div 3}{9 \div 3} = \frac{1}{3}$.
- $\frac{1}{4}$ and $\frac{2}{8}$ are equivalent fractions because $\frac{1}{4} = \frac{1 \times 2}{4 \times 2} = \frac{2}{8}$.

Example 1

Find the value that needs to replace the square: $\frac{1}{5} = \frac{\square}{20}$

1. Find what you need to multiply by to get from one denominator to the other.

$$\frac{1}{5} = \frac{\square}{20} \quad (5 \xrightarrow{\times 4} 20)$$

2. Multiply the numerator by the same number.

$$\frac{1}{5} = \frac{\mathbf{4}}{20} \quad (1 \xrightarrow{\times 4} 4)$$

Example 2

Find the value of b if $\frac{12}{30} = \frac{4}{b}$.

1. Find what you need to divide by to get from one numerator to the other. $\frac{12}{30} = \frac{4}{b}$ (numerator $\div 3$)
2. Divide the denominator by the same number. $\frac{12}{30} = \frac{4}{10}$ (denominator $\div 3$) So $b = 10$.

Exercise 1

Q1 Replace the stars to make fractions equivalent to $\frac{1}{3}$. Use the circles to help.

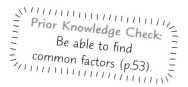

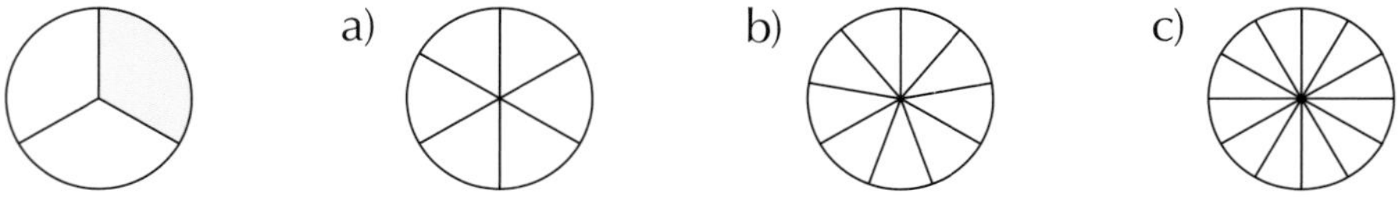

a) b) c)

$\frac{1}{3} = \frac{☆}{6} = \frac{☆}{9} = \frac{☆}{12}$

Q2 Find the values of the letters in the following fractions.

a) $\frac{1}{5} = \frac{a}{10}$ b) $\frac{1}{4} = \frac{b}{12}$ c) $\frac{1}{5} = \frac{5}{c}$ d) $\frac{1}{6} = \frac{3}{d}$

e) $\frac{e}{9} = \frac{15}{27}$ f) $\frac{f}{51} = \frac{9}{17}$ g) $\frac{11}{g} = \frac{55}{80}$ h) $\frac{1}{h} = \frac{11}{121}$

Simplifying Fractions

Learning Objectives — Spec Ref N3/R3:
- Be able to write fractions in their simplest form.
- Be able to write one quantity as a fraction of another.

Prior Knowledge Check: Be able to find common factors (p.53).

Simplifying a fraction means finding an equivalent fraction with the **smallest** possible whole numbers. This is also known as 'expressing a fraction in its **lowest terms**'.

To simplify a fraction:

- Look for a **common factor** of the numerator and denominator.
- **Divide both** the numerator and denominator by this factor.
- **Repeat** this process with the new fraction. When you can find no common factor (except 1) the fraction is in its **simplest form**.

Tip: You can get to the simplest form in one step if you spot the **highest common factor** right away.

Example 3

Express $\frac{24}{30}$ as a fraction in its lowest terms.

1. Divide the numerator and denominator by any common factor. Here, 3 is a common factor of 24 and 30. → $\frac{24}{30} = \frac{8}{10}$ (÷ 3)
2. Repeat this until the numerator and denominator have no more common factors. Here, 2 is a common factor of 8 and 10. → $\frac{8}{10} = \mathbf{\frac{4}{5}}$ (÷ 2)
3. 4 and 5 have no common factors, so this fraction is in its lowest terms.

Tip: Dividing 24 and 30 by their highest common factor (= 6) will get the answer in a single step.

To express one number **as a fraction** of another:

- Write the **first number** as the **numerator**
- Write the **second number** as the **denominator**.
- **Simplify** and put the fraction in its **lowest terms**.

Example 4

At a restaurant, 24 out of 36 customers said they liked their food. Write this as a fraction in its simplest form.

1. Write the 24 as the numerator and 36 as the denominator.
2. Simplify by dividing 24 and 36 by a common factor — here it's 12.
3. 2 and 3 have no common factor, so this fraction is in its lowest terms.

$\frac{24}{36} = \mathbf{\frac{2}{3}}$ (÷ 12)

Exercise 2

Q1 Write the following fractions in their lowest terms.

a) $\frac{3}{9}$ b) $\frac{5}{20}$ c) $\frac{8}{16}$ d) $\frac{4}{32}$

e) $\frac{9}{45}$ f) $\frac{15}{36}$ g) $\frac{24}{64}$ h) $\frac{30}{40}$

Q2 Simplify these fractions. Then state which fraction is not equivalent to the other two.

a) $\frac{6}{18}, \frac{4}{20}, \frac{9}{27}$ b) $\frac{6}{8}, \frac{9}{15}, \frac{15}{25}$ c) $\frac{4}{18}, \frac{6}{33}, \frac{10}{45}$

Q3 There are 300 animals on a farm. 50 of them are cows, 70 are pigs and the rest are sheep. What is each type of animal as a fraction of the total number of animals? Write each of your answers in its simplest form.

5.2 Mixed Numbers

All fractions you've seen so far have been smaller than 1. You can write numbers bigger than 1 as fractions in two different ways — either as mixed numbers or as improper fractions.

Learning Objective — Spec Ref N2:
Be able to convert between mixed numbers and improper fractions.

Mixed numbers and improper fractions are two ways of writing numbers greater than 1 that **aren't whole numbers**.

- A **mixed number** is a quantity that has a **whole part** and a **fraction part**. For example, $2\frac{1}{2}$ (two-and-a-half) is a mixed number.
- An **improper fraction** is a fraction where the numerator is **bigger** than the denominator. For example, $\frac{5}{2}$ is an improper fraction.

Converting from Mixed Numbers to Improper Fractions

To **convert** a mixed number to an **improper fraction**:

- Find a **fraction** that is **equivalent** to the **whole part** and has the **same denominator** as the **fraction part**.
- Then **add** this to the **fraction part** of the mixed number.

Example 1

Write the mixed number $4\frac{3}{5}$ as an improper fraction.

1. Find the fraction which is equivalent to 4 and has 5 as the denominator.

$$4 = \frac{4}{1} \overset{\times 5}{=} \frac{20}{5}$$

2. Combine the two fractions into one improper fraction by adding the numerators.

$$\text{So } 4\frac{3}{5} = \frac{20}{5} + \frac{3}{5} = \frac{20+3}{5} = \mathbf{\frac{23}{5}}$$

Exercise 1

Q1 Use the diagrams to form improper fractions equivalent to the whole numbers.

a) $2 = \frac{☆}{3}$

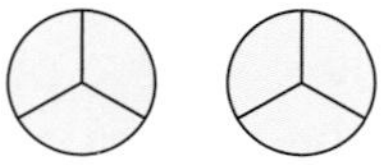

Clue: *How many thirds are there in 2?*

b) $3 = \frac{☆}{4}$

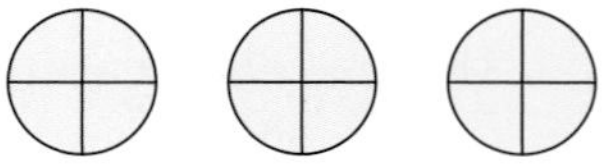

Clue: *How many quarters are there in 3?*

Q2 a) Find the value of a if $4 = \frac{a}{3}$.

b) Use your answer to write the mixed number $4\frac{1}{3}$ as an improper fraction.

Q3 Find the values of the letters to write the following mixed numbers as improper fractions.

a) $1\frac{1}{3} = \frac{a}{3}$ b) $1\frac{2}{7} = \frac{b}{7}$ c) $2\frac{1}{2} = \frac{c}{2}$

Q4 Write the following mixed numbers as improper fractions.

a) $1\frac{4}{5}$ b) $1\frac{5}{12}$ c) $2\frac{9}{10}$

d) $5\frac{3}{10}$ e) $4\frac{3}{4}$ f) $9\frac{5}{6}$

Converting from Improper Fractions to Mixed Numbers

To **convert** an improper fraction to a **mixed number**:

- **Divide** the numerator by the denominator — the answer gives you the **whole part.**
- The **remainder** gives the **numerator** of the **fraction part** — the **denominator** stays the same.

Example 2

Write the improper fraction $\frac{13}{5}$ as a mixed number.

1. Divide the numerator by the denominator. $13 \div 5 = 2$ remainder 3
2. The whole part of the result (2) is the whole part of your mixed number.
   The remainder (3) tells you the fraction part.

So $\frac{13}{5} = 2 + \frac{3}{5} = \mathbf{2\frac{3}{5}}$

Exercise 2

Q1 a) Work out $11 \div 7$, giving your answer as a whole number and a remainder.

b) Use your answer to write $\frac{11}{7}$ as a mixed number in its simplest terms.

Q2 Write the following improper fractions as mixed numbers.

a) $\frac{5}{3}$ b) $\frac{9}{5}$ c) $\frac{13}{11}$

d) $\frac{9}{4}$ e) $\frac{20}{9}$ f) $\frac{11}{3}$

Q3 a) Write '12 out of 5' as an improper fraction.

b) Convert the improper fraction into a mixed number.

5.3 Ordering Fractions

It can be difficult to figure out whether one fraction is bigger or smaller than another just by looking at them. The trick is to find equivalent fractions that have the same denominator.

Finding a Common Denominator

Learning Objective — Spec Ref N4:
Be able to write fractions over a common denominator.

Prior Knowledge Check: Be able to find the lowest common multiple of a set of numbers (p.58).

To put two fractions over a **common denominator**:

- Find the **lowest common multiple** (LCM) of the denominators of the fractions.
- Write each fraction as an **equivalent fraction** with this number as its denominator.

Example 1

Rewrite the following pairs of fractions so they have a common denominator.

a) $\mathbf{\frac{1}{2}}$ **and** $\mathbf{\frac{1}{8}}$

Use 8 as the common denominator, since 8 is a multiple of 2 and 8.

So find a fraction equivalent to $\frac{1}{2}$ with 8 as the denominator.

$$\frac{1}{2} \overset{\times 4}{=} \frac{4}{8}$$

So the fractions are equivalent to $\mathbf{\frac{4}{8}}$ **and** $\mathbf{\frac{1}{8}}$.

b) $\mathbf{\frac{5}{6}}$ **and** $\mathbf{\frac{3}{8}}$

24 is a multiple of both 6 and 8, so use this as the common denominator.

Rewrite the two fractions so that they have 24 as the denominator.

$$\frac{5}{6} \overset{\times 4}{=} \mathbf{\frac{20}{24}} \qquad \frac{3}{8} \overset{\times 3}{=} \mathbf{\frac{9}{24}}$$

Exercise 1

Q1 Rewrite the following pairs of fractions so that they have a common denominator.

a) $\frac{1}{3}, \frac{1}{6}$ b) $\frac{1}{5}, \frac{3}{10}$ c) $\frac{1}{4}, \frac{5}{16}$ d) $\frac{2}{5}, \frac{7}{20}$

e) $\frac{2}{9}, \frac{1}{3}$ f) $\frac{2}{3}, \frac{3}{4}$ g) $\frac{5}{6}, \frac{1}{7}$ h) $\frac{2}{9}, \frac{1}{2}$

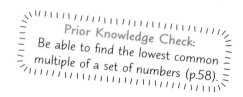

Q2 Rewrite the following groups of fractions so that they have a common denominator.

a) $\frac{3}{4}, \frac{5}{8}, \frac{7}{12}$ b) $\frac{1}{5}, \frac{7}{10}, \frac{9}{20}$ c) $\frac{1}{7}, \frac{4}{21}, \frac{5}{14}$

d) $\frac{1}{2}, \frac{3}{8}, \frac{2}{3}$ e) $\frac{2}{5}, \frac{5}{12}, \frac{11}{30}$ f) $\frac{1}{8}, \frac{7}{20}, \frac{3}{5}$

Ordering Fractions

Learning Objective — Spec Ref N1:
Be able to order and compare positive and negative fractions.

To put fractions in **order**:

- Put all the fractions over a **common denominator**.
- **Order** the fractions by ordering their **numerators**. The **smaller** the numerator, the **smaller** the fraction.

Tip: You can use the '<' (less than) and '>' (more than) symbols to show how two fractions compare.

Example 2

Rewrite $\frac{5}{6}$ and $\frac{7}{8}$ so that they have a common denominator and say which is larger.

1. The lowest common multiple of 6 and 8 is 24, so find equivalent fractions that have a denominator of 24.

$$\frac{5}{6} \overset{\times 4}{=} \frac{20}{24} \qquad \frac{7}{8} \overset{\times 3}{=} \frac{21}{24}$$

2. Compare the numerators: $20 < 21$.

$\frac{20}{24} < \frac{21}{24}$, so $\frac{21}{24} = \mathbf{\frac{7}{8}}$ **is larger**.

Example 3

Put the fractions $\frac{1}{2}$, $\frac{3}{8}$ and $\frac{3}{4}$ in order from smallest to largest.

1. The LCM of 2, 8 and 4 is 8, so find equivalent fractions that have a denominator of 8.

$$\frac{1}{2} \overset{\times 4}{=} \frac{4}{8} \qquad \frac{3}{4} \overset{\times 2}{=} \frac{6}{8}$$

So the fractions are equivalent to $\frac{4}{8}$, $\frac{3}{8}$ and $\frac{6}{8}$.

2. Use the numerators to put the fractions in order.

From smallest to largest, these are: $\frac{3}{8}, \frac{4}{8}, \frac{6}{8}$.

So in order, the original fractions are $\mathbf{\frac{3}{8}, \frac{1}{2}, \frac{3}{4}}$.

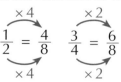

Exercise 2

Q1 Find the larger fraction in each pair below.

a) $\frac{1}{4}, \frac{5}{8}$ b) $\frac{3}{5}, \frac{7}{10}$ c) $\frac{4}{7}, \frac{9}{14}$ d) $\frac{11}{18}, \frac{2}{3}$

e) $\frac{5}{6}, \frac{3}{4}$ f) $\frac{2}{3}, \frac{3}{5}$ g) $\frac{2}{3}, \frac{3}{4}$ h) $\frac{7}{10}, \frac{3}{4}$

In **Questions 2 and 3**, put each of the sets of fractions in order, from smallest to largest.

Q2 a) $\frac{1}{2}, \frac{5}{8}, \frac{7}{16}$ b) $\frac{2}{5}, \frac{3}{10}, \frac{7}{20}$ c) $\frac{3}{4}, \frac{7}{12}, \frac{5}{8}$

Q3 a) $\frac{7}{8}, \frac{5}{6}, \frac{13}{16}$ b) $\frac{4}{15}, \frac{7}{27}, \frac{13}{45}$ c) $\frac{5}{16}, \frac{7}{20}, \frac{9}{25}$

Negative Fractions

Just like whole numbers and decimals, fractions can be **negative**.

- The **minus sign** is placed **in front** of a negative fraction.
- **Negative fractions** can be **ordered** with other fractions (both positive and negative). Put all fractions over a common denominator, as before, and treat the **numerator** of a negative fraction as a **negative number**.
  For example, $-\frac{2}{3}$ is **less than** $\frac{1}{3}$ because –2 is **less than** 1.

Tip: Mixed numbers can also be negative. To order these, first convert them to improper fractions.

Example 4

Put the fractions $-\frac{5}{6}, \frac{1}{2}$ and $-\frac{2}{3}$ in order from smallest to largest.

1. The lowest common multiple of 6, 2 and 3 is 6. Find equivalent fractions with a denominator of 6.

$$\frac{1}{2} \xrightarrow{\times 3} \frac{3}{6} \qquad -\frac{2}{3} \xrightarrow{\times 2} -\frac{4}{6}$$

2. Compare the numerators: –5 < –4 < 3.

$$-\frac{5}{6} < -\frac{4}{6} < \frac{3}{6}, \quad \text{so } \mathbf{-\frac{5}{6} < -\frac{2}{3} < \frac{1}{2}}$$

Exercise 3

Q1 a) Rewrite the fractions $-\frac{4}{15}, \frac{2}{9}$ and $-\frac{1}{3}$ so that they have a common denominator.

b) Use your answer to write the fractions $-\frac{4}{15}, \frac{2}{9}$ and $-\frac{1}{3}$ in order (smallest to largest).

Q2 Write down the smallest fraction in each list below:

a) $-\frac{4}{9}, -\frac{5}{12}$ b) $-\frac{9}{10}, \frac{4}{5}$ c) $-\frac{1}{4}, \frac{5}{2}\ -\frac{3}{11}$

5.4 Adding and Subtracting Fractions

You can add and subtract fractions just like whole and decimal numbers. The result is often an improper fraction, so you might be asked to convert it to a mixed number. You can also add and subtract mixed numbers, as well as combinations of mixed numbers and fractions.

Learning Objective — Spec Ref N2:
Be able to add and subtract fractions.

Prior Knowledge Check: Be able to find common denominators — see p.68.

To **add or subtract** fractions:

- Put the fractions over a **common denominator**.
- Then add or subtract the **numerators**.
- **Simplify** the resulting fraction if necessary.

Tip: If your answer has a negative numerator, move the minus sign in front of the fraction.

Example 1

Work out $\frac{2}{5} + \frac{2}{3}$. Give you answer as a mixed number.

1. Rewrite the fractions with a common denominator of 15.
2. Rewrite the addition using your new fractions. Then add or subtract the numerators.
3. Convert the improper fraction to a mixed number.

$\frac{2}{5} = \frac{6}{15}$ (× 3) $\qquad \frac{2}{3} = \frac{10}{15}$ (× 5)

So $\frac{2}{5} + \frac{2}{3} = \frac{6}{15} + \frac{10}{15}$

$= \frac{6+10}{15}$

$= \frac{16}{15} = \mathbf{1\frac{1}{15}}$

Exercise 1

Q1 Work out the following. Give your answers in their simplest form.

a) $\frac{1}{3} + \frac{1}{3}$ b) $\frac{4}{5} - \frac{2}{5}$ c) $\frac{5}{11} - \frac{3}{11}$ d) $\frac{1}{10} + \frac{3}{10}$

Q2 Work out the following. Give your answers as mixed numbers in their simplest form.

a) $\frac{5}{8} + \frac{7}{8}$ b) $\frac{3}{4} + \frac{3}{4}$ c) $\frac{8}{15} + \frac{13}{15} - \frac{2}{15}$ d) $\frac{17}{20} + \frac{19}{20} - \frac{7}{20}$

Q3 a) Rewrite the fractions $\frac{1}{2}$ and $\frac{1}{4}$ with a common denominator.

b) Work out $\frac{1}{2} + \frac{1}{4}$.

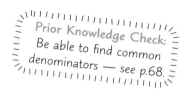

Q4 Work out the following:

a) $\frac{3}{5} + \frac{1}{10}$ b) $\frac{1}{4} + \frac{3}{8}$ c) $\frac{4}{9} - \frac{1}{3}$ d) $\frac{3}{4} - \frac{3}{8}$

Q5 Work out the following.
Give any answer that is greater than 1 as a mixed number in its simplest form.

a) $\frac{1}{9} + \frac{5}{9} + \frac{11}{18}$ b) $\frac{3}{4} + \frac{1}{8} - \frac{7}{16}$ c) $\frac{6}{7} + \frac{1}{14} - \frac{1}{2}$ d) $\frac{1}{4} + \frac{2}{3} + \frac{5}{6}$

If you're given a **wordy question** then you'll have to figure out what to do with the **fractions**.

Example 2

In a maths exam $\frac{1}{2}$ of the questions are on number topics, $\frac{1}{3}$ of the questions are on algebra, and the rest are on geometry. What fraction of the questions are geometry questions?

1. The fractions of number questions, algebra questions and geometry questions must add up to 1.

   Fraction of geometry questions $= 1 - \frac{1}{2} - \frac{1}{3}$

2. Put the fractions over a common denominator. (Remember, anything divided by itself is 1, so you can write $1 = \frac{6}{6}$.)

   $1 = \frac{6}{6}$ $\frac{1}{2} \overset{\times 3}{=} \frac{3}{6}$ $\frac{1}{3} \overset{\times 2}{=} \frac{2}{6}$

3. Subtract to find the fraction of geometry questions.

   Fraction of geometry questions $= \frac{6}{6} - \frac{3}{6} - \frac{2}{6} = \frac{6-3-2}{6} = \mathbf{\frac{1}{6}}$

Exercise 2

Q1 $\frac{4}{9}$ of the pupils in one class are boys. What fraction of the class are girls?

Q2 Jake, Amar and Olga are sharing a cake. Jake eats $\frac{2}{7}$ of the cake, Amar eats $\frac{3}{7}$ and Olga eats the rest. What fraction of the cake does Olga eat?

Q3 $\frac{1}{5}$ of the flowers in a garden are roses. $\frac{3}{10}$ of the flowers are tulips. What fraction of the flowers are neither roses nor tulips?

Q4 A bag contains a mixture of sweets. $\frac{2}{5}$ of the sweets are chocolates, $\frac{1}{4}$ are toffees and the rest are mints. What fraction of the sweets are mints?

Q5 In a school survey, $\frac{1}{2}$ of the pupils said they walk to school. $\frac{1}{5}$ said they catch the bus. The rest arrive by car. What fraction come to school by car?

Adding and Subtracting Mixed Numbers

Learning Objective — Spec Ref N2:
Be able to add and subtract mixed numbers.

Prior Knowledge Check: Be able to convert between improper fractions and mixed numbers (p.66-67).

There are two methods for adding and subtracting **mixed numbers**.

Method 1 — Use Improper Fractions

Change the mixed numbers into **improper fractions** (see p.66), then add and subtract by getting them over a **common denominator** — like you did on page 71.

Example 3

Work out $1\frac{1}{3} + 2\frac{5}{6}$ by converting each term to an improper fraction.

1. Write the mixed numbers as improper fractions. $1\frac{1}{3} = \frac{3}{3} + \frac{1}{3} = \frac{4}{3}$ and $2\frac{5}{6} = \frac{12}{6} + \frac{5}{6} = \frac{17}{6}$
2. Rewrite the improper fractions with a common denominator of 6. $\frac{4}{3} \overset{\times 2}{=} \frac{8}{6}$ and $\frac{17}{6}$
3. Add the numerators. $1\frac{1}{3} + 2\frac{5}{6} = \frac{8}{6} + \frac{17}{6} = \frac{25}{6}$
4. Give your answer as a mixed number in its simplest form. $25 \div 6 = 4$ remainder 1, so $\frac{25}{6} = \mathbf{4\frac{1}{6}}$

Tip: If one denominator is a multiple of the other, you only need to rewrite one fraction.

Exercise 3

Q1 a) Write the mixed number $1\frac{4}{5}$ as an improper fraction.
b) Use part a) to work out $1\frac{4}{5} + \frac{3}{5}$, giving your answer as a mixed number.

Q2 a) Write the following as improper fractions: (i) $2\frac{1}{5}$ (ii) $1\frac{3}{5}$
b) Use your answers to work out $2\frac{1}{5} - 1\frac{3}{5}$.

Q3 By writing each term as a proper or improper fraction, work out the following. Give your answers as mixed numbers in their simplest form, where appropriate.

a) $1\frac{2}{3} + \frac{1}{3}$ b) $2\frac{3}{8} + \frac{7}{8}$ c) $1\frac{5}{6} + 1\frac{1}{6}$

d) $4\frac{5}{12} - 2\frac{11}{12}$ e) $5\frac{7}{11} + \frac{5}{11}$ f) $3\frac{2}{5} - \frac{4}{5}$

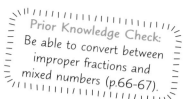

Method 2 — Separate the Whole and Fraction Parts

Add/subtract the whole parts and the fraction parts **separately**, and then **add** the results. This method is particularly useful when the **whole** parts or the **denominators** are **big**.

Example 4

Work out $5\frac{1}{4} + 3\frac{1}{2}$ by adding whole parts and fraction parts separately.

1. Add the whole parts. $5 + 3 = 8$
2. Add the fraction parts in the usual way. $\frac{1}{4} + \frac{1}{2} = \frac{1}{4} + \frac{2}{4} = \frac{3}{4}$
3. Add the results. $8 + \frac{3}{4} = \mathbf{8\frac{3}{4}}$

In a **subtraction**, you **subtract** the whole and fraction parts **separately**, but you still **add** the results at the end. Note: subtracting the fractions might give you a **negative** fraction — e.g. subtracting rather than adding in Example 4 would give:
$5\frac{1}{4} - 3\frac{1}{2} = (5 - 3) + \left(\frac{1}{4} - \frac{1}{2}\right) = 2 + \left(-\frac{1}{4}\right) = 2 - \frac{1}{4} = 1\frac{3}{4}$.

In an **addition**, adding the fraction parts might result in an **improper fraction**.

Example 5

Work out $6\frac{2}{3} + 1\frac{3}{4}$ by adding whole parts and fraction parts separately.

1. Add the whole parts. $6 + 1 = 7$
2. Add the fraction parts in the usual way. The result is an improper fraction, so convert it to a mixed number. $\frac{2}{3} + \frac{3}{4} = \frac{8}{12} + \frac{9}{12} = \frac{17}{12} = 1\frac{5}{12}$
3. Add the results. $7 + 1\frac{5}{12} = \mathbf{8\frac{5}{12}}$

Exercise 4

Q1 By separating into whole and fraction parts, work out the following.

a) $4\frac{1}{9} + 3\frac{4}{9}$ b) $3\frac{1}{5} + 2\frac{3}{7}$ c) $2\frac{5}{8} + 6\frac{2}{3}$

Q2 a) Work out $\frac{2}{5} - \frac{4}{5}$.

b) Use your answer to work out $2\frac{2}{5} - 1\frac{4}{5}$.

Q3 By separating into whole and fraction parts, work out the following.

a) $3\frac{3}{4} - \frac{5}{7}$ b) $2\frac{1}{4} - 1\frac{6}{7}$ c) $5\frac{2}{5} - 3\frac{7}{9}$

5.5 Multiplying and Dividing Fractions

Being able to multiply by fractions is really useful — it lets you find fractions of amounts. Once you've got multiplying sorted, you'll find dividing by fractions to be no harder —to divide, all you need to do is flip the fraction upside-down and multiply instead.

Multiplying by Unit Fractions

Learning Objective — Spec Ref N2/N12:
Be able to multiply whole numbers by unit fractions.

Prior Knowledge Check:
Be able to simplify fractions (p.64) and convert between improper fractions and mixed numbers (p.66).

A **unit fraction** has 1 as its numerator — for example, these are all unit fractions: $\frac{1}{2}, \frac{1}{3}, \frac{1}{5}, \frac{1}{10}$.
Multiplying by a **unit fraction** is the same as **dividing** by a **whole number**.

For example, multiplying by $\frac{1}{2}$ is the same as dividing by 2.

Example 1

Work out: a) $12 \times \frac{1}{3}$ b) $10 \times -\frac{1}{4}$

a) Multiplying by $\frac{1}{3}$ is the same as dividing by 3. Remember $12 \div 3$ can be written as $\frac{12}{3}$.

$12 \times \frac{1}{3} = \frac{12}{3} = \mathbf{4}$

b) Multiplying by $-\frac{1}{4}$ is the same as dividing by –4. $10 \div (-4)$ can be written as $-\frac{10}{4}$.

$10 \times -\frac{1}{4} = -\frac{10}{4} = -2\frac{2}{4} = \mathbf{-2\frac{1}{2}}$

Tip: Have a look back to page 6 to remind yourself how to multiply and divide with negatives.

Exercise 1

Q1 Work out the following.

a) $8 \times \frac{1}{4}$ b) $10 \times \frac{1}{5}$ c) $15 \times \frac{1}{5}$ d) $45 \times \frac{1}{3}$

Q2 Work out the following. Write your answers as mixed numbers.

a) $18 \times \frac{1}{4}$ b) $15 \times \frac{1}{6}$ c) $17 \times \frac{1}{2}$ d) $25 \times \frac{1}{10}$

Q3 Work out the following. Write your answers as mixed numbers.

a) $48 \times -\frac{1}{6}$ b) $80 \times -\frac{1}{10}$ c) $25 \times -\frac{1}{6}$ d) $40 \times -\frac{1}{3}$

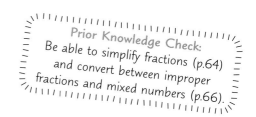

Multiplying Whole Numbers by Fractions

Learning Objectives — Spec Ref N2/N12:
- Be able to multiply whole numbers by fractions.
- Be able to find fractions of amounts.

Prior Knowledge Check:
Be able to identify common factors (p.53).

To multiply a **whole number** by a fraction:
- **Multiply** by the **numerator**.
- **Divide** by the **denominator**.

You can do these steps in either order. If **multiplying** first is going to give you a **big** number, and the **denominator** is a **factor** of the **whole** number, **dividing** first might be better.

Example 2

Work out these multiplications:

a) $15 \times \frac{2}{5}$

You need to multiply 15 by 2, and then divide by 5. (Or you can divide by 5, and then multiply by 2 — you'll get the same answer.)

$$15 \times \frac{2}{5} = \frac{15 \times 2}{5} = \frac{30}{5} = \mathbf{6}$$

b) $5 \times \frac{3}{4}$

You need to multiply 5 by 3, and then divide by 4. (Or you can divide by 4, and then multiply by 3.)

$$5 \times \frac{3}{4} = \frac{5 \times 3}{4} = \frac{15}{4} = \mathbf{3\frac{3}{4}}$$

If the whole number and the fraction's denominator have a **common factor**, then you can **cancel** it before carrying out the rest of the calculation.

Example 3

Work out $35 \times \frac{7}{25}$. Give your answer as a mixed number in its simplest form.

1. 25 and 35 are both multiples of 5. 25 = 5 × 5 and 35 = 7 × 5.
2. Cancel the 5s...
3. ... then multiply and divide as usual.

$$35 \times \frac{7}{25} = \frac{35 \times 7}{25}$$
$$= \frac{7 \times 5 \times 7}{5 \times 5}$$
$$= \frac{7 \times \not{5} \times 7}{\not{5} \times 5}$$
$$= \frac{7 \times 7}{5} = \frac{49}{5} = \mathbf{9\frac{4}{5}}$$

Tip: If you didn't cancel first, you'd get $\frac{245}{25}$, which then simplifies to $9\frac{4}{5}$.

Exercise 2

Q1 Work out the following. Write your answers as simply as possible.

a) $12 \times \frac{2}{3}$ b) $28 \times \frac{3}{4}$ c) $15 \times \frac{4}{5}$ d) $48 \times \frac{3}{8}$

e) $60 \times \frac{5}{12}$ f) $32 \times \frac{7}{16}$ g) $100 \times -\frac{7}{25}$ h) $-96 \times -\frac{7}{12}$

Q2 Work out the following. Give your answers as mixed numbers in their simplest form.

a) $15 \times \frac{3}{4}$ b) $22 \times \frac{2}{5}$ c) $7 \times -\frac{3}{11}$ d) $6 \times -\frac{5}{8}$

Q3 Calculate $45 \times \frac{5}{18}$. Give your answer as a mixed number in its simplest form.

You can find **fractions of amounts** by **multiplying** the amount by the fraction in question.

Example 4

Find $\frac{3}{4}$ of 18. Give your answer as a mixed number in its simplest form.

1. You can replace 'of' with a multiplication sign. $\frac{3}{4}$ of $18 = \frac{3}{4} \times 18$
2. Since 18 isn't a multiple of 4, do 3 × 18 and write this on the top of the fraction. $\frac{3}{4} \times 18 = \frac{3 \times 18}{4} = \frac{54}{4}$
3. Simplify the fraction and convert to a mixed number. $\frac{54}{4} = \frac{27}{2} = \frac{26+1}{2} = \mathbf{13\frac{1}{2}}$

Exercise 3

Q1 Find the following.

a) $\frac{3}{4}$ of 36 b) $\frac{2}{3}$ of 33 c) $\frac{3}{8}$ of 64 d) $\frac{5}{12}$ of 72

e) $\frac{5}{6}$ of 18 f) $\frac{3}{5}$ of 15 g) $\frac{5}{6}$ of 33 h) $\frac{7}{12}$ of 45

Q2 Find the following.

a) $\frac{3}{2}$ of 18 b) $\frac{5}{3}$ of 21 c) $\frac{11}{9}$ of 72

Q3 Out of a class of 27 students, $\frac{2}{3}$ prefer rounders to cricket.
How many students prefer rounders?

Multiplying Fractions and Mixed Numbers

Learning Objectives — Spec Ref N2/N12:
- Be able to multiply fractions together.
- Be able to multiply mixed numbers together.

To multiply two or more **fractions**, multiply their numerators and denominators **separately**.

Example 5

Work out: a) $\frac{2}{3}\times\frac{4}{5}$ b) $\frac{3}{4}\times\frac{2}{9}$

1. Multiply the numerators together to get the new numerator, and the denominators together to get the new denominator.
2. Simplify your answer as much as possible.

a) $\frac{2}{3}\times\frac{4}{5}=\frac{2\times4}{4\times5}=\mathbf{\frac{8}{15}}$

b) $\frac{3}{4}\times\frac{2}{9}=\frac{3\times2}{4\times9}=\frac{6}{36}=\mathbf{\frac{1}{6}}$

To make calculations easier, you can **cancel** any **factors** that appear in the numerator and denominator of **either** fraction. It helps to cancel all **common factors** as early as possible.

Example 6

Work out $\frac{7}{25}\times\frac{15}{16}$.

1. 25 and 15 share a common factor (5), so cancel that first.
2. Multiply the numerators together and the denominators together.

$$\frac{7}{25}\times\frac{15}{16}=\frac{7}{5\times\cancel{5}}\times\frac{3\times\cancel{5}}{16}$$
$$=\frac{7}{5}\times\frac{3}{16}$$
$$=\frac{7\times3}{5\times16}=\mathbf{\frac{21}{80}}$$

Tip: If you didn't cancel the 5, you'd get $\frac{7\times15}{25\times16}=\frac{105}{400}$. This still simplifies to $\frac{21}{80}$, but it's much harder to work with.

Exercise 4

Q1 Work out the following.

a) $\frac{1}{6}\times\frac{1}{3}$ b) $\frac{2}{5}\times\frac{1}{3}$ c) $\frac{3}{4}\times\frac{1}{7}$ d) $\frac{1}{5}\times\frac{3}{5}$

e) $\frac{5}{6}\times\frac{1}{4}$ f) $\frac{4}{5}\times\frac{2}{7}$ g) $-\frac{2}{7}\times\frac{5}{7}$ h) $-\frac{3}{8}\times-\frac{7}{10}$

Q2 Work out the following. Give your answers in their lowest terms.

a) $\frac{1}{4} \times \frac{2}{3}$ b) $\frac{3}{5} \times \frac{1}{6}$ c) $\frac{5}{6} \times \frac{2}{15}$ d) $\frac{5}{12} \times \frac{3}{4}$

e) $\frac{4}{42} \times \frac{18}{8}$ f) $\frac{22}{5} \times \frac{15}{77}$ g) $-\frac{6}{7} \times \frac{7}{8}$ h) $\frac{7}{10} \times -\frac{5}{14}$

Q3 Calculate $\frac{3}{5}$ of $\frac{15}{8}$. Give your answer as a mixed number in its simplest form.

To multiply by **mixed numbers**, change them into **improper fractions** first.

Example 7

Work out $4\frac{1}{2} \times 3\frac{3}{5}$.

1. Write the mixed numbers as improper fractions.
   $4\frac{1}{2} = \frac{8}{2} + \frac{1}{2} = \frac{9}{2}$ and $3\frac{3}{5} = \frac{15}{5} + \frac{3}{5} = \frac{18}{5}$
2. Multiply the two fractions.
   So $4\frac{1}{2} \times 3\frac{3}{5} = \frac{9}{2} \times \frac{18}{5} = \frac{9 \times 18}{2 \times 5}$
3. Write your answer as a mixed number in its simplest form.
   $= \frac{162}{10} = 16\frac{2}{10} = \mathbf{16\frac{1}{5}}$.
   (or $\frac{162}{10} = \frac{81 \times \cancel{2}}{5 \times \cancel{2}} = \frac{81}{5} = \mathbf{16\frac{1}{5}}$)

Exercise 5

Q1 Work out the following.

a) $1\frac{1}{2} \times \frac{1}{3}$ b) $2\frac{1}{5} \times \frac{3}{4}$ c) $1\frac{5}{6} \times \frac{2}{3}$

d) $3\frac{3}{4} \times \frac{2}{5}$ e) $2\frac{1}{7} \times \frac{2}{9}$ f) $2\frac{4}{9} \times \frac{3}{8}$

Q2 a) Write the mixed number $3\frac{2}{5}$ as an improper fraction.

b) Write the mixed number $1\frac{1}{2}$ as an improper fraction.

c) Use your answer to work out $3\frac{2}{5} \times 1\frac{1}{2}$.

Q3 Work out the following.

a) $1\frac{1}{5} \times 1\frac{1}{4}$ b) $2\frac{2}{5} \times 1\frac{2}{3}$ c) $1\frac{3}{5} \times 1\frac{3}{4}$

d) $2\frac{3}{7} \times 3\frac{1}{6}$ e) $3\frac{4}{9} \times 1\frac{7}{8}$ f) $2\frac{6}{7} \times 2\frac{1}{9}$

Dividing by Fractions

Learning Objectives — Spec Ref N2/N12:
- Be able to find the reciprocal of a number or fraction.
- Be able to divide by fractions and mixed numbers.

- The **reciprocal** of a number is just **1 ÷ that number.** The reciprocal of a **whole number** is always a **unit fraction**. For example, the reciprocal of 4 is $\frac{1}{4}$.
- To find the reciprocal of a **fraction**, just **swap** its numerator and denominator (in other words, **flip it upside-down**). For example, the reciprocal of $\frac{3}{5}$ is $\frac{5}{3}$.
- To find the reciprocal of a **mixed number**, convert it to an **improper fraction** first.

Tip: The reciprocal of a negative number is also negative.

Example 8

Find the reciprocal of each number, giving your answer as a fraction in its lowest terms.

a) 3 **b)** $\frac{4}{5}$ **c)** $2\frac{5}{6}$

a) The reciprocal of 3 is 1 ÷ 3. So write this as a fraction.
The reciprocal of 3 is $\mathbf{\frac{1}{3}}$.

b) To find the reciprocal, just swap the numerator and the denominator.
The reciprocal of $\frac{4}{5}$ is $\mathbf{\frac{5}{4}}$.

c) 1. Convert to an improper fraction.
$2\frac{5}{6} = \frac{12}{6} + \frac{5}{6} = \frac{17}{6}$

2. Flip the fraction upside-down.
The reciprocal of $2\frac{5}{6}$ is $\mathbf{\frac{6}{17}}$.

Exercise 6

Q1 Find the reciprocal of each of the following.
a) 5 b) 12 c) 9 d) 27

Q2 Find the reciprocal of each of the following.
a) $\frac{1}{3}$ b) $-\frac{1}{7}$ c) $\frac{4}{5}$ d) $-\frac{5}{8}$

Q3 Find the reciprocal of each of the following. Start by writing each as an improper fraction.
a) $1\frac{3}{5}$ b) $2\frac{1}{7}$ c) $1\frac{4}{9}$ d) $-2\frac{3}{4}$

Dividing by Fractions

Dividing by a number is the same as **multiplying** by its **reciprocal**.

So, to divide a number by a **fraction**:

- Calculate the **reciprocal** of the fraction.
- **Multiply** the number by this reciprocal.

> **Tip:** Once you've got a multiplication, the method is exactly the same as on p.78.

Example 9

Work out: a) $\frac{1}{2} \div \frac{2}{3}$ b) $\frac{4}{7} \div 2$

a) Multiply $\frac{1}{2}$ by the reciprocal of $\frac{2}{3}$. $\frac{1}{2} \div \frac{2}{3} = \frac{1}{2} \times \frac{3}{2} = \frac{1\times 3}{2\times 2} = \mathbf{\frac{3}{4}}$

b) Multiply $\frac{4}{7}$ by the reciprocal of 2. $\frac{4}{7} \div 2 = \frac{4}{7} \times \frac{1}{2} = \frac{4\times 1}{7\times 2} = \frac{4}{14} = \mathbf{\frac{2}{7}}$

Example 10

Find $2\frac{2}{3} \div 1\frac{1}{5}$ as a mixed number in its lowest terms.

1. Write both numbers as improper fractions. $2\frac{2}{3} = \frac{6}{3} + \frac{2}{3} = \frac{8}{3}$ and $1\frac{1}{5} = \frac{5}{5} + \frac{1}{5} = \frac{6}{5}$
2. Flip the $\frac{6}{5}$ upside down and change the ÷ into a ×. $2\frac{2}{3} \div 1\frac{1}{5} = \frac{8}{3} \div \frac{6}{5}$ $= \frac{8}{3} \times \frac{5}{6} = \frac{\cancel{2}\times 4}{3} \times \frac{5}{\cancel{2}\times 3} = \frac{4\times 5}{3\times 3}$
3. Convert the improper fraction into a mixed number. $= \frac{20}{9} = \mathbf{2\frac{2}{9}}$

Exercise 7

Q1 Work out the following.

a) $\frac{1}{5} \div \frac{2}{3}$ b) $\frac{1}{6} \div \frac{2}{5}$ c) $2 \div \frac{2}{3}$ d) $\frac{2}{3} \div 4$

e) $\frac{5}{12} \div \frac{4}{5}$ f) $\frac{15}{7} \div 6$ g) $-\frac{3}{8} \div \frac{2}{5}$ h) $-\frac{7}{10} \div \frac{5}{7}$

Q2 By first rewriting any mixed numbers as improper fractions, work out the following.

a) $1\frac{1}{2} \div 4$ b) $3\frac{1}{3} \div 6$ c) $5 \div 1\frac{3}{5}$ d) $2 \div 1\frac{7}{8}$

Q3 By first rewriting any mixed numbers as improper fractions, work out the following.

a) $1\frac{1}{3} \div \frac{2}{5}$ b) $2\frac{1}{2} \div \frac{1}{3}$ c) $\frac{3}{4} \div 2\frac{1}{3}$ d) $\frac{4}{7} \div 3\frac{1}{2}$

e) $1\frac{1}{4} \div 1\frac{1}{5}$ f) $2\frac{2}{3} \div 1\frac{1}{4}$ g) $4\frac{5}{6} \div 2\frac{1}{3}$ h) $3\frac{2}{3} \div \left(-2\frac{1}{10}\right)$

5.6 Fractions and Decimals

Fractions and decimals are different ways of writing numbers. You'll need to know how to write fractions as decimals, with and without the help of a calculator.

Converting Fractions to Decimals using a Calculator

Learning Objective — Spec Ref N10:
Be able to convert fractions to decimals using a calculator.

You can use a **calculator** to convert a fraction to a decimal — just **divide** the numerator by the denominator. Make sure you write down **all the digits** that you see on your calculator display.

Example 1

Use a calculator to convert the following fractions to decimals.

a) $\frac{5}{16}$ Divide the numerator by the denominator. $\frac{5}{16} = 5 \div 16 = \mathbf{0.3125}$

b) $1\frac{2}{5}$ You can convert to an improper fraction. $1\frac{2}{5} = \frac{7}{5} = 7 \div 5 = \mathbf{1.4}$

Or, you can work out the fraction part, then add on the whole part. $\frac{2}{5} = 2 \div 5 = 0.4$, so $1\frac{2}{5} = 1 + 0.4 = \mathbf{1.4}$

Tip: Your calculator might have a button to go between fractions and decimals, but you should still learn these methods.

Exercise 1

Q1 Use a calculator to convert the following fractions to decimals.

a) $\frac{5}{8}$ b) $\frac{7}{20}$ c) $\frac{7}{16}$ d) $\frac{5}{32}$

e) $\frac{7}{40}$ f) $\frac{23}{50}$ g) $\frac{176}{200}$ h) $\frac{329}{500}$

Q2 Use a calculator to convert the following mixed numbers to decimals.

a) $1\frac{3}{8}$ b) $2\frac{1}{8}$ c) $6\frac{7}{20}$ d) $2\frac{37}{100}$

e) $4\frac{719}{1000}$ f) $5\frac{19}{25}$ g) $7\frac{11}{32}$ h) $8\frac{7}{16}$

Converting Fractions to Decimals Without Using a Calculator

Learning Objective — Spec Ref N10:
Convert fractions to decimals without using a calculator.

Prior Knowledge Check:
Be able to multiply with decimals (p.19) and use short division to get decimal answers (p.21).

There are a couple of ways to **convert** fractions to **decimals** if you don't have a calculator.

Method 1 — Find an Equivalent Fraction

You can find an **equivalent fraction** with a **denominator** of **10, 100** or **1000** and then use one of the following conversions:

$$\frac{1}{10} = 0.1, \qquad \frac{1}{100} = 0.01, \qquad \frac{1}{1000} = 0.001$$

Tip: These conversions are just dividing 1 by 10, 100 and 1000 (see p.18).

Example 2

Write $\frac{7}{10}$ as a decimal.

1. Write the fraction as a multiple of $\frac{1}{10}$.
2. Use the conversion $\frac{1}{10} = 0.1$

$\frac{7}{10} = 7 \times \frac{1}{10}$

$= 7 \times 0.1 = \mathbf{0.7}$

Example 3

Write the following fractions as decimals.

a) $\frac{4}{5}$

1. Multiply top and bottom to find an equivalent fraction with a denominator of 10.
2. Then rewrite as a decimal.

$\frac{4}{5} \overset{\times 2}{=} \frac{8}{10}$ (top and bottom × 2)

$\frac{8}{10} = 8 \times \frac{1}{10} = 8 \times 0.1 = \mathbf{0.8}$

b) $\frac{123}{300}$

1. Divide top and bottom to find an equivalent fraction with a denominator of 100.
2. Then rewrite as a decimal.

$\frac{123}{300} \overset{\div 3}{=} \frac{41}{100}$ (top and bottom ÷ 3)

$\frac{41}{100} = 41 \times \frac{1}{100} = 41 \times 0.01 = \mathbf{0.41}$

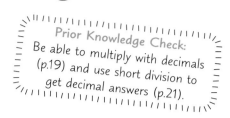

Exercise 2

Q1 Write the following fractions as decimals.

a) $\frac{9}{10}$ b) $\frac{2}{10}$ c) $\frac{3}{10}$ d) $\frac{8}{10}$

e) $\frac{91}{100}$ f) $\frac{42}{100}$ g) $\frac{99}{100}$ h) $\frac{8}{100}$

i) $\frac{7}{1000}$ j) $\frac{201}{1000}$ k) $\frac{41}{1000}$ l) $\frac{27}{1000}$

Q2 a) Find a fraction which is equivalent to $\frac{8}{25}$ and which has a denominator of 100.

b) Use your answer to write $\frac{8}{25}$ as a decimal.

Q3 Write the following fractions as decimals.

a) $\frac{3}{5}$ b) $\frac{9}{30}$ c) $\frac{45}{50}$ d) $\frac{22}{25}$

e) $\frac{96}{300}$ f) $\frac{33}{250}$ g) $\frac{103}{200}$ h) $\frac{306}{3000}$

Q4 Write the following numbers as decimals.

a) $\frac{11}{10}$ b) $\frac{14}{5}$ c) $5\frac{7}{25}$ d) $3\frac{11}{200}$

Method 2 — Using Short Division

You can divide the numerator by the denominator using **short division**.
You'll have to keep putting **zeros** after the decimal point until the division is completed.

Example 4

Write $\frac{1}{8}$ as a decimal.

You need to work out 1 ÷ 8.

1. 8 doesn't go into 1, so the answer will begin with "0.". Write the decimal point in the answer directly above the decimal point in 1.0. → $\begin{array}{r} 0. \\ 8\overline{)1.{}^{1}0} \end{array}$
2. 8 goes into 10 once, with remainder 2. → $\begin{array}{r} 0.\,1 \\ 8\overline{)1.{}^{1}0{}^{2}0} \end{array}$
3. 8 goes into 20 twice, with remainder 4. → $\begin{array}{r} 0.\,1\;2 \\ 8\overline{)1.{}^{1}0{}^{2}0{}^{4}0} \end{array}$
4. 8 goes into 40 exactly 5 times. → $\begin{array}{r} 0.\,1\;2\;5 \\ 8\overline{)1.{}^{1}0{}^{2}0{}^{4}0} \end{array}$

So $\frac{1}{8}$ = **0.125**

Tip: Write the 1 as 1.0 to start with, then add zeros as needed.

Exercise 3

Q1 a) Write $\frac{3}{8}$ as a decimal by finding 3 ÷ 8.

b) Write $\frac{5}{16}$ as a decimal by finding 5 ÷ 16.

Q2 Write the following fractions as decimals by using short division.

a) $\frac{1}{4}$ b) $\frac{3}{4}$ c) $\frac{1}{20}$ d) $\frac{1}{40}$

e) $\frac{1}{16}$ f) $\frac{7}{8}$ g) $\frac{7}{40}$ h) $\frac{13}{80}$

Converting Decimals to Fractions

Learning Objective — Spec Ref N10:
Convert decimals to fractions without using a calculator.

Prior Knowledge Check: Be able to simplify fractions (p.64).

You can quickly convert a decimal to a fraction by using a **denominator** of **10, 100, 1000** or another **power of 10**.

Converting Decimals that are Smaller Than 1

To convert a decimal that is **smaller than 1** to a fraction:

- Write the bit after the decimal point as the **numerator**.
- Use a power of 10 as the **denominator** with the same **number of zeros** as the number of decimal places.

Tip: A decimal smaller than 1 has only a 0 before the decimal point.

Example 5

Write the following decimals as fractions in their simplest form.

a) 0.24 There are two decimal places, so the denominator should have two zeros — it's 100. Remember to simplify your answer.

$$0.24 = \frac{24}{100} = \mathbf{\frac{6}{25}}$$ (÷4 top and bottom)

b) 0.025 There are three decimal places, so the denominator should have three zeros — it's 1000. Remember to simplify your answer.

$$0.025 = \frac{25}{1000} = \mathbf{\frac{1}{40}}$$ (÷25 top and bottom)

Exercise 4

Q1 Convert the following decimals to fractions. Give your answers in their simplest form.

a) 0.7 b) 0.9 c) 0.1 d) 0.4

Q2 Convert the following decimals to fractions.

a) 0.93 b) 0.07 c) 0.23 d) 0.47

Q3 Convert the following decimals to fractions. Give your answers in their simplest form.

a) 0.004 b) 0.801 c) 0.983 d) 0.098

Q4 Convert 0.1002 to a fraction, giving your answer in its simplest form.

Converting Decimals that are Bigger Than 1

To convert a decimal that is **bigger than 1** to a fraction:

- Write the number as the **numerator**, removing the decimal point.
- Use a power of 10 as the **denominator** with the same **number of zeros** as the number of decimal places.

Tip: A decimal bigger than 1 has a non-zero digit before the decimal point.

Example 6

Write the following decimals as improper fractions in their simplest form.

a) 3.5 There's just 1 decimal place, so the denominator should have one zero — it's 10. Remember to simplify your answer.

$$3.5 = \frac{35}{10} \overset{\div 5}{=} \mathbf{\frac{7}{2}}$$

b) 1.28 There are two decimal places, so the denominator should have two zeros — it's 100. Remember to simplify your answer.

$$1.28 = \frac{128}{100} \overset{\div 4}{=} \mathbf{\frac{32}{25}}$$

Exercise 5

Q1 Convert the following decimals to fractions. Give your answers in their simplest form.

a) 1.2 b) 3.4 c) 4.7 d) 8.4

Q2 Convert the following decimals to fractions. Give your answers in their simplest form.

a) 3.02 b) 1.55 c) 2.05 d) 18.2

Review Exercise

Q1 Find the values of the letters in the following equivalent fractions.

a) $\frac{3}{4} = \frac{a}{16}$ b) $\frac{7}{12} = \frac{35}{b}$ c) $\frac{c}{3} = \frac{10}{15}$

Q2 a) Which of the following is a common factor of 45 and 75? 9 25 2 5

b) Write $\frac{45}{75}$ in its lowest terms.

Q3 Express each of the following as a fraction in its simplest terms.

a) 4 as a fraction of 24 b) 12 as a fraction of 66

Q4 a) Write $7\frac{2}{3}$ as an improper fraction.

b) Write $\frac{26}{4}$ as a mixed number in its simplest form.

Q5 Rewrite the following pairs of fractions so they have a common denominator, and say which of each pair is the smaller of the two fractions.

a) $\frac{2}{7}, \frac{5}{28}$ b) $\frac{1}{8}, \frac{1}{6}$ c) $\frac{2}{5}, \frac{4}{9}$

Q6 Work out the following. Giving your answers in their simplest form.

a) $\frac{1}{5} + \frac{1}{3}$ b) $\frac{9}{10} - \frac{5}{6}$ c) $6\frac{3}{8} - \frac{7}{8}$

Q7 a) Write the mixed number $1\frac{3}{7}$ as an improper fraction.

b) Use your answer from a) to work out $1\frac{3}{7} \times \frac{2}{3}$.

Q8 a) What is the reciprocal of $\frac{3}{8}$?

b) Calculate $\frac{1}{6} \div \frac{3}{8}$. Give your answer in its simplest form.

Q9 Use a calculator to express $\frac{23}{40}$ as a decimal.

Q10 a) Convert $\frac{12}{25}$ to a decimal.

b) Convert 0.35 to a fraction. Give your answer in its simplest form.

Exam-Style Questions

Q1 Work out:

a) $\frac{5}{12}$ of 78 cm

[2 marks]

b) $15 \div 2\frac{1}{2}$

[2 marks]

Q2 Write 3.125 as a mixed number in its simplest form.

[2 marks]

Q3 Ruby buys some meat and uses $\frac{3}{4}$ of it to make a cottage pie.
She then uses $\frac{3}{5}$ of the remaining meat to make ravioli.
Work out the fraction of the meat Ruby has left over.
Give your answer in its lowest terms.

[3 marks]

Q4 Sam makes some dumplings. Each dumpling uses $5\frac{1}{3}$ g of butter.
Sam has 120 g of butter. Work out the number of dumplings Sam can make.

[3 marks]

Q5 A supermarket has boxes of peppers which are either green, red or yellow.
In one box, $\frac{2}{5}$ of the peppers are green, there are 24 red peppers,
and there are four times as many red as there are yellow peppers.

a) Work out the total number of peppers in the box.

[3 marks]

b) Write the fraction of red peppers in the box as a decimal.

[2 marks]

6.1 Ratios

Like fractions and percentages, ratios are a way of showing proportion. They tell you how the size of one thing relates to the size of another thing, or sometimes to several other things. They crop up in lots of real-life situations, such as on map scales and in recipes.

Simplifying Ratios

Learning Objective — Spec Ref R4:
Write ratios in their simplest form.

Prior Knowledge Check:
Be able to find common factors of numbers. See p.53.

Ratios are used to compare quantities. If you have an amount a of one thing and an amount b of another thing, then the **ratio of a to b** is written $\boldsymbol{a:b}$.

- For example, here the ratio of circles to squares is **6 : 4**.

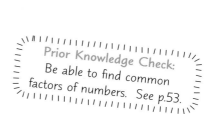

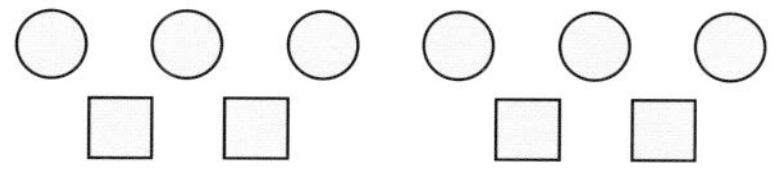

You can **simplify** ratios by dividing each number by a **common factor**, just like you do with fractions.

- The ratio 6 : 4 simplifies to **3 : 2** by dividing both sides by 2.

A ratio is in its **simplest form** when all parts are whole numbers, but they have no common factor that is greater than 1.

Tip: A ratio $a:b:c$, with three quantities, can be simplified by dividing all parts by a common factor.

Example 1

There are 15 fiction books and 10 non-fiction books on a shelf.

a) Write down the ratio of fiction books to non-fiction books.

Write down the ratio with the quantities in the order that you're asked to give them.

fiction : non-fiction
15 : 10

b) Write this ratio in its simplest form.

1. Divide both sides by the same number — a common factor of the two sides.
2. Stop when you can't divide any further and leave whole numbers on each side.

5 is a common factor.

$$15:10 \xrightarrow{\div 5} 3:2$$

The simplest form is **3 : 2**.

Tip: You can get to the simplest form of a ratio in one step by dividing both sides by their highest common factor.

Exercise 1

Q1 Write down the ratio of stars to triangles.

Q2 On a farm there are 15 pigs and 23 cows. Write down the ratio of cows to pigs.

Q3 Write down each of the following ratios in its simplest form.
a) 2 : 8   b) 5 : 15   c) 40 : 10   d) 4 : 6

Q4 Write down each of the following ratios in its simplest form.
a) 6 : 2 : 4   b) 15 : 12 : 3   c) 14 : 10 : 2   d) 24 : 12 : 20

Q5 A floor is made up of 24 black tiles and 8 white tiles.
Find the ratio of black tiles to white tiles in its simplest form.

Q6 At a party there are 36 girls and 27 boys. What is the ratio of girls to boys?
Give your answer in its simplest form.

Q7 A school has 595 pupils and 170 computers.
Write down the ratio of computers to pupils in its simplest form.

Q8 Paul and Soraya share a bag of 42 sweets. Paul has 16 sweets and Soraya has the rest.
Find the ratio of Soraya's sweets to Paul's sweets, giving your answer in its simplest form.

Ratios with Different Units

When you have a ratio involving **different units**, you must **convert** all quantities to the **same unit** before simplifying.

You don't need to include the units in the ratio after **simplifying**.

Tip: You'll need to know your unit conversions — see p.263 for these.

Example 2

Write the ratio 1 m : 40 cm in its simplest form.

1. Rewrite the ratio so that the units are the same. — 1 m = 100 cm, so 1 m : 40 cm is the same as 100 cm : 40 cm.
2. Remove the units altogether. → 100 : 40
3. Simplify as usual by dividing each side by the highest common factor, 20. → ÷20 on each side gives 5 : 2

The simplest form is **5 : 2**.

Exercise 2

Q1 Write these ratios in their simplest form.

a) 10p : £1 b) 20 mm : 4 cm c) 10 g : 1 kg

d) 2 weeks : 7 days e) 40p : £1 f) 30 cm : 2 m

Q2 Give the following ratios in their simplest form.

a) 1 m : 150 mm b) 8 cm : 1.1 m c) 9 g : 0.3 kg

d) 2.5 hours : 20 mins e) £1.25 : 75p f) 65 m : 1.3 km

Q3 A jug of orange squash is made using 50 ml of orange concentrate and 1 litre of water. Find the ratio of concentrate to water in its simplest form.

Q4 Alexsy runs in two cross-country races. He runs the first in 45 minutes and the second in 3½ hours. Find the ratio of his first race time to his second race time in its simplest form.

Q5 The icing for some cupcakes is made by mixing 1.6 kg of icing sugar with 640 g of butter. Find the ratio of butter to icing sugar. Give your answer in its simplest form.

Writing Ratios in the Form 1 : n

Learning Objective — Spec Ref R4:
Convert ratios to the form $1:n$.

Practical problems can often involve finding 'how much of b is needed for **every one** of a'.

In these situations, it's useful to have the ratio in the form $\mathbf{1:n}$. To write a ratio $a:b$ in this form, just **divide both sides** by $\boldsymbol{a}$.

$$a:b \longrightarrow 1:\frac{b}{a}$$

Tip: b doesn't have to be a factor of a. You can leave $\frac{b}{a}$ as a fraction or as a decimal.

Example 3

Nigel makes his favourite smoothie by mixing half a litre of blueberry juice with 100 millilitres of plain yoghurt. How much yoghurt does he use for every millilitre of juice?

1. Write down the ratio of juice to yoghurt and remove the units.
   1 litre = 1000 ml, so 0.5 litres : 100 ml is the same as 500 ml : 100 ml.
2. You want the juice side (the left-hand side) to equal 1, so divide both sides by 500.
   500 : 100 → (÷500, ÷500) → 1 : 0.2
3. Use the simplified ratio to answer the question.
   So he uses **0.2 ml** of yoghurt for every 1 ml of juice.

Exercise 3

Q1 Write down the following ratios in the form $1:n$.

a) 2 : 6 b) 7 : 35 c) 6 : 24 d) 30 : 120

e) 2 : 7 f) 4 : 26 g) 8 : 26 h) 2 : 1

Q2 Write down each of these ratios in the form $1:n$.

a) 10 mm : 5 cm b) 12p : £6 c) 30 mins : 2 hours d) 500 g : 20 kg

Q3 In a pond there are 7 frogs and 56 fish. Find the ratio of frogs to fish in the form $1:n$.

Q4 On a garage forecourt there are 15 red cars and 45 silver cars.
Find the ratio of red cars to silver cars in the form $1:n$.

Q5 A recipe uses 125 ml of chocolate syrup and 2.5 litres of milk.
Write the ratio of chocolate syrup to milk in the form $1:n$.

Q6 Two towns 4.8 km apart are shown on a map 12 cm apart.
Find the ratio of the map distance to the true distance in the form $1:n$.

Ratios and Fractions

Learning Objectives — Spec Ref N11/R8:
- Use ratios to find fractions.
- Use fractions to find ratios.

Prior Knowledge Check:
Be able to simplify, add and subtract fractions, and write one quantity as a fraction of another. See p.64 and p.71.

If you have a ratio $a:b$ (or $a:b:c$), you can find the **fraction** of one part out of the total:

- First find the **total** number of **parts** in the ratio, i.e. $a + b$ (or $a + b + c$).
- Make a **fraction** with the **part** you're interested in as the fraction's **numerator**, and the **total** number of parts as the fraction's **denominator** — then **simplify** if necessary.

Example 4

A box of doughnuts contains jam doughnuts and chocolate doughnuts in the ratio 3 : 5. What fraction of the doughnuts are chocolate flavoured?

1. Add the numbers to find the total number of parts. — $3 + 5 = 8$ parts altogether.
2. Write a fraction with the number of parts that are chocolate as the numerator and the total number of parts as the denominator. — $\frac{5}{8}$ of the doughnuts are chocolate flavoured.

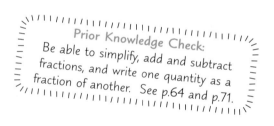

Exercise 4

Q1 A tiled floor has blue and white tiles in the ratio 1 : 3. What fraction of the tiles are blue?

Q2 A necklace has yellow beads and red beads in the ratio 3 : 2.
What fraction of the beads are red?

Q3 A bag contains 21 balls. The balls are blue, green and red in the ratio 1 : 2 : 4. PROBLEM SOLVING
Decide whether each of these statements is true or false.

a) $\frac{1}{7}$ of the balls are blue.
b) 3 of the balls are blue.
c) $\frac{2}{4}$ of the balls are red or green.
d) The ratio of blue to red balls is 1 : 4.
e) There are more red balls than green balls.

When quantities are given as fractions, you can put them into ratio form.

- Express each quantity as a fraction with a **common denominator**.
- Then the **numerators** of each fraction go into the ratio.

Example 5

All of Ishan's DVDs are either horror films or comedies.
If $\frac{2}{7}$ of his DVDs are horror films, find the ratio of comedies to horror films.

1. Write each part as a fraction over a common denominator.
2. Write the numerators in ratio form in the order you were asked for.

$\frac{2}{7}$ are horror films,
so $1 - \frac{2}{7} = \frac{5}{7}$ are comedies.

Ratio of comedies to horror is **5 : 2**.

Exercise 5

Q1 Aiden has a bag of red and green Jelly Babies. $\frac{1}{3}$ of the Jelly Babies are red.
Write down the ratio of red to green Jelly Babies.

Q2 $\frac{3}{19}$ of the members of a chess club are left-handed.
Write down the ratio of right-handed club members to left-handed club members.

Q3 During one day at a pizza restaurant, $\frac{3}{8}$ of the pizzas ordered were pepperoni, $\frac{1}{2}$ were goat's cheese and the rest were spicy chicken.
Write down the ratio of pepperoni to goat's cheese to spicy chicken.

6.2 Using Ratios

Now that you're familiar with ratios, it's time to see how they can be used to solve different types of problems. These aren't too tricky as long as you understand what the ratio tells you, so look back at the previous few pages if you need a little reminder.

Part : Part Ratios

Learning Objective — Spec Ref R5:
Be able to use ratios to find unknown amounts.

A **part : part** ratio $a:b$ describes the **relationship** between two **quantities** — for every $\boldsymbol{a}$ lots of the first quantity, there are $\boldsymbol{b}$ lots of the second quantity.

If you know the **ratio** of one quantity to another, and you know **how big one** of them is, you can use this information to find the size of the **other**.

Example 1

The ratio of men to women in an office is 3 : 4.
If there are 9 men in the office, how many women are there?

1. Write down what you know and what you need to find out. $3:4 = 9:?$
2. Work out what you have to multiply the left-hand side of the ratio by to get from 3 to 9. Then multiply the right-hand side by the same number.

$$\times 3 \quad 3:4 \quad \times 3$$
$$9:12$$

So there are **12 women**.

Exercise 1

Q1 For a class of pupils, the ratio of blue eyes to brown eyes is 2 : 3.
If 8 pupils have blue eyes, how many pupils have brown eyes?

Q2 The ages of a father and son are in the ratio 8 : 3. If the father is 48, how old is the son?

Q3 The ratio of red to yellow sweets in a bag is 3 : 4.
If the bag contains 12 yellow sweets, how many red sweets are there?

Q4 In a wood there are oak trees and beech trees in the ratio 2 : 9.
If there are 42 oak trees, how many beech trees are there?

Q5 In a supermarket the ratio of apples to bananas is 5 : 3.
If there are 450 bananas, how many apples are there?

Q6 A recipe uses sugar and butter in the ratio 2 : 1.
How much butter would be needed with 100 g of sugar?

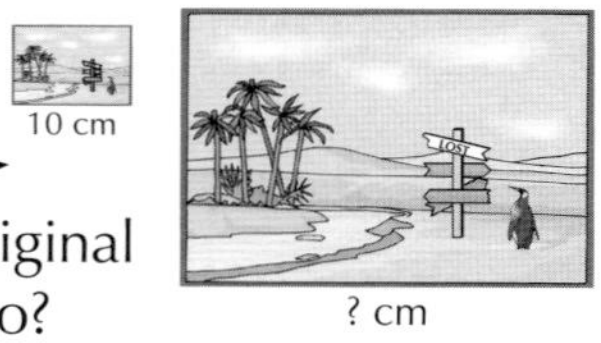

Q7 A photo with a width of 10 cm is enlarged so that the ratio of the original and enlarged photos' widths is 2 : 7. How wide is the enlarged photo?

Q8 The ratio of children to adults in a swimming pool must be 5 : 1 or less. If there are 32 children, how many adults must there be? PROBLEM SOLVING

Q9 A TV-show producer is selecting a studio audience. He wants the ratio of under 30s to those aged 30 or over to be at least 8 : 1. If 100 under 30s are selected, find the maximum number of people aged 30 or over. PROBLEM SOLVING

A **part** : **part** : **part** ratio $a : b : c$ can be used in the same way.

- It contains three **part** : **part** ratios — these are $a : b$, $b : c$ and $a : c$.
- Given a part : part : part ratio and **one quantity**, you can work out the **other two** quantities.

Tip: You won't always need every bit of the ratio, as in the example below.

Example 2

A cereal contains raisins, nuts and oats in the ratio 2 : 3 : 5. If a box of the cereal contains 200 g of raisins and 300 g of nuts, find how many grams of oats it contains.

1. Write the ratio of one of the quantities you know to the one you want to find.
2. Find the number that both sides of the ratio have to be multiplied by. Remember to give the mass in grams.

The ratio of nuts to oats is 3 : 5.

3 : 5
= 300 : ? ⟶ × 100 (3 : 5 → 300 : 500) × 100

So it contains **500 g oats**.

Exercise 2

Q1 A fruit punch is made by mixing pineapple juice, orange juice, and lemonade in the ratio 1 : 3 : 6. If 500 ml of pineapple juice is used, how much orange juice is needed?

Q2 Mai, Lizzy and Dave have heights in the ratio 31 : 33 : 37.
Mai is 155 cm tall. How tall are Lizzy and Dave?

Q3 Max, Molly and Hasan are at a bus stop. The number of minutes they have waited can be represented by the ratio 3 : 7 : 2. Molly has been waiting for 1 hour and 10 minutes. Calculate how long Max and Hasan have been waiting for.

Part : Whole Ratios

Learning Objective — Spec Ref R5:
Be able to change between part : part and part : whole ratios.

A **part : whole** ratio describes how **one** quantity relates to the **whole** (or **total**).

- To convert a **part : part** ratio $a:b$ into a **part : whole** ratio, choose which part you want to keep (a or b) and find the **total** $a + b$. Then the part : whole ratio is $\boldsymbol{a : a + b}$ (or $\boldsymbol{b : a + b}$).
- To convert a **part : whole** ratio $a:w$ into a **part : part** ratio, you subtract the **known part**, a, from the **whole**, w, to find the **unknown** part $\boldsymbol{w - a}$. Then the part : part ratio is $\boldsymbol{a : w - a}$.

Example 3

A recipe uses white flour and brown flour in a ratio of 2 : 1.
What is the ratio of white flour used to the total amount of flour used?

1. Find the total number of parts. — There are $2 + 1 = 3$ parts altogether.
2. So 2 parts of white flour is used for every 3 parts in total. — So white flour : total flour is **2 : 3**.

Example 4

A shop sells only green and brown garden sheds. The ratio of green sheds sold to the total number of sheds sold is 5 : 7. What is the ratio of green sheds sold to brown sheds sold?

Subtract to find the number of brown sheds sold for every 5 green sheds, then write as a ratio. — $7 - 5 = 2$ brown sheds per 5 green sheds. So green : brown is **5 : 2**.

Exercise 3

Q1 A biscuit tin only contains digestives and bourbons. The ratio of digestives to the total number of biscuits is 4 : 7. What is the ratio of digestives to bourbons?

Q2 When Dan makes a cup of tea, the ratio of water to milk is 15 : 2. What is the ratio of water to the total amount of liquid in the cup?

Q3 A rock album is sold as a digital download and on a CD. The ratio of digital downloads to total albums sold is 53 : 99. What is the ratio of digital downloads to CDs sold?

Q4 A bouquet of flowers contains only roses, tulips and daisies in the ratio 3 : 5 : 9. What is the ratio of roses to the total number of flowers? PROBLEM SOLVING

6.3 Dividing in a Given Ratio

Ratios can be used to share things out unequally. The ratio tells you how many shares there are and what the size of each share should be. You can use both part : part and part : whole ratios for sharing, but you might have to come up with the ratio yourself from information in the question.

Learning Objectives — Spec Ref R5/R6:
- Divide quantities in a given part : part or part : whole ratio.
- Be able to write a ratio given a multiplicative relationship between two quantities.

Part : part ratios can be used to divide an amount into two or more **shares**. The numbers in the ratio show how many parts of the whole each share gets. To share an amount in a given ratio:

- **Add up** all the quantities to get the **total** number of parts, i.e. $a + b$ or $a + b + c$.
- **Divide** the amount to be shared by the total number of parts. This tells you the amount that **one part** is worth.
- **Multiply** the amount for one part by each quantity in the ratio. This gives you the amount in **each share**.

Tip: After you've calculated the shares, check they add up to the total amount.

Example 1

Divide £54 in the ratio 4 : 5.

1. Add the numbers to find the total number of parts. — $4 + 5 = 9$ parts altogether
2. Work out the amount for one part. — 9 parts = £54
   So 1 part = £54 ÷ 9 = £6
3. Then multiply the amount for one part by each quantity in the ratio. — £6 × 4 = £24
   £6 × 5 = £30
   So the shares are **£24** and **£30**.
   (Check: £24 + £30 = £54 ✔)

Example 2

A drink is made using apple juice, blackcurrant juice and lemonade in the ratio 3 : 2 : 5. How much lemonade is needed to make 5 litres of the drink?

1. Add the numbers to find the total number of parts. — $3 + 2 + 5 = 10$ parts altogether
2. Work out the amount for one part. — 10 parts = 5 litres
   So 1 part = 0.5 litres
3. Then multiply the amount for one part by the quantity for lemonade in the ratio. — 0.5 litres × 5
   = **2.5 litres of lemonade**

Exercise 1

Q1 Divide £48 in the following ratios.

a) 2 : 1 b) 1 : 3 c) 5 : 1 d) 7 : 5

Q2 Share 90 kg in these ratios.

a) 4 : 1 b) 7 : 2 c) 8 : 7 d) 12 : 18

Q3 Divide 72 cm in the following ratios.

a) 2 : 3 : 1 b) 2 : 2 : 5 c) 5 : 3 : 4 d) 7 : 6 : 5

Q4 Kat and Lincoln share 30 cupcakes in the ratio 3 : 2. How many do they each get?

Q5 Three friends win £6000 between them. They decide to share the money in the ratio 3 : 5 : 4. Calculate the amounts they each receive.

Q6 The length and width of a rectangle are in the ratio 5 : 1. If the perimeter of the rectangle is 72 cm, calculate the length and the width of the rectangle.

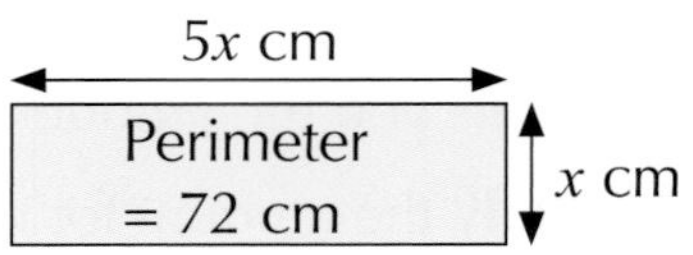

If you're given a **part : whole** ratio, you can just **divide** the amount by the total number of parts (the whole). You may need to work out the **unknown part** of the ratio, as you did on p.96.

Example 3

A box of 45 chocolates contains dark and milk chocolates. The ratio of milk chocolates to the total number of chocolates is 7 : 9. How many of each chocolate are in the box?

1. Work out the unknown part. — There are 9 – 7 = 2 parts dark chocolate.
2. Work out the amount for one part. The ratio 2 : 9 is part : whole, so there are 9 parts in total. — 9 parts = 45 chocolates, so 1 part = 45 ÷ 9 = 5 chocolates
3. Multiply the amount for one part by the number of parts for each share. — 5 × 7 = **35 milk chocolates**; 5 × 2 = **10 dark chocolates**

Exercise 2

Q1 A farmer has a flock of 60 sheep that he splits into two parts. Write down the size of each part when the flock is divided according to the following part : whole ratios.

a) 1 : 2 b) 2 : 3 c) 4 : 5 d) 7 : 12

Q2 Orange paint is made by mixing yellow and red paint.
The ratio of yellow paint to the total volume of paint is 4 : 7.
How much of each colour is needed to make 42 litres of orange paint?

Q3 In a school of 600 pupils, the ratio of right-handed pupils to the total number of students is 7 : 8. How many left-handed pupils are there?

Multiplicative Relationships

Instead of a ratio, you might be given a **multiplicative relationship** between two or more quantities. To divide an amount into shares, you will first need to turn this relationship into a **ratio**.

For example, if you're told there are five times as many sparrows as pigeons, first write down the ratio of sparrows to pigeons — it's 5 : 1.

Tip: These can be wordy questions. But once you've picked out the numbers, you'll find they're the usual ratio problems.

Example 4

A reptile house contains only snakes and lizards, with twice as many snakes as lizards. If there's a total of 24 reptiles, how many snakes and how many lizards are there?

1. Work out the ratio from the information you're given.
   There are 2 snakes for every lizard, so the ratio of snakes to lizards is 2 : 1.
2. Work out the amount for one part.
   There are 2 + 1 = 3 parts in total, so 1 part = 24 ÷ 3 = 8 reptiles.
3. Multiply the amount for one part by each quantity in the ratio.
   8 × 2 = **16 snakes**
   8 × 1 = **8 lizards**

Exercise 3

PROBLEM SOLVING

Q1 A piggy bank contains only 5p and 20p coins. There are three times as many 5p coins as 20p coins, and 32 coins in total. How many 20p coins does the piggy bank contain?

Q2 There are 30 passengers on a bus. There are four times as many passengers not on the phone as there are on the phone. How many passengers are not on their phones?

Q3 Daniel gets one-and-a-half times as much profit from a business as his partner Elsa. How much of a £5700 profit does Daniel get?

Q4 For a fruit salad, Celia uses twice the weight of raspberries as grapes, and three times the weight of strawberries as grapes. How much of each fruit did Celia use in a 450 g fruit salad?

Q5 Nicky, Jacinta and Samir share a bag of 35 sweets so that Nicky gets half as much as Jacinta, and Samir gets half as much as Nicky. How many sweets do each of them get?

Review Exercise

Q1 There are 17 boys and 14 girls in a class. Write down the ratio of girls to boys.

Q2 Write down each of the following ratios in its simplest form.

a) 24 : 6 b) 2 cm : 8 mm c) 6 : 3 : 15 d) 0.03 kg : 10 g : 25 g

Q3 An animal sanctuary has 120 animals, 40 of which are donkeys.
Write the ratio of donkeys to other animals in the form 1 : n.

Q4 In Amy's sock drawer there are spotty, stripy and plain socks in the ratio 5 : 1 : 4.
What fraction of Amy's socks are stripy?

Q5 $\frac{5}{12}$ of the children at a school eat school dinners and the rest bring a packed lunch.
What is the ratio of children with a packed lunch to those who eat school dinners?

Q6 The ratio of green to red peppers in a risotto is 2 : 5. A restaurant uses 20 red peppers to make some risotto. How many green peppers does it use?

Q7 Meera and Sabrina share a holiday job. They split the money they make in the ratio 7 : 6 (Meera : Sabrina). If Sabrina gets £48, how much does Meera get?

Q8 Share 56 m in these ratios.

a) 1 : 7 b) 4 : 4 c) 10 : 4 d) 22 : 6

Q9 Gemma, Alisha and Omar have a combined height of 496 cm.
If their heights are in the ratio 19 : 20 : 23, how tall are they?

Q10 Lauren and Cara receive £1200 in total from their Grandad. The ratio of the money given to Lauren to the total amount is 2 : 3. How much money did Cara get?

Q11 A patch of garden contains daisy, dandelion and thistle plants. There are twice as many dandelions as daisies, and three times as many thistles as dandelions. PROBLEM SOLVING

a) What is the ratio of daisies to dandelions to thistles?

b) If there are 54 plants in total, how many thistles are there?

Exam-Style Questions

Q1 a) Explain why the ratio 30 : 48 can be written as 5 : 8.

[1 mark]

b) Use part a) to write the ratio 30 : 48 in the form 1 : n, where n is a decimal.

[2 marks]

Q2 A football squad of 24 players has three times as many right-footed players as it does left-footed players.

a) Write down the ratio of right-footed players to left-footed players.

[1 mark]

b) How many players are right-footed?

[2 marks]

c) What fraction of players are left-footed?

[2 marks]

Q3 A badminton club has 330 members. 240 are adults and the rest are children. Two thirds of the children are boys and the rest are girls.

a) Show that the ratio of adults : boys : girls is 8 : 2 : 1

[2 marks]

A year ago, the club had 200 adult members but the ratio adults : boys : girls was the same as it is now.

b) Work out the number of boys in the club a year ago.

[2 marks]

Q4 The weight of flour in two bags is in the ratio 4 : 5.

a) Write this ratio in the form 1 : n.

[1 mark]

Alan says "If I add the same amount of flour to each bag, then the two bags will still be in the same ratio."

b) Show that Alan is incorrect.

[2 marks]

c) Explain what Alan could have done to the amount of flour in each bag to keep the ratio the same.

[1 mark]

7.1 Percentages

'Per cent' means 'out of 100'. Writing an amount as a percentage means writing it as a number out of 100. Percentages are written using the % symbol and are useful for showing proportions.

Writing One Number as a Percentage of Another

Learning Objectives — Spec Ref R9:
- Be able to write one number as a percentage of another without a calculator.
- Be able to write one number as a percentage of another using a calculator.

Prior Knowledge Check: Be able to find equivalent fractions — see p.63.

Here's how to write one number as a **percentage** of another number without a calculator.
- Make a **fraction** by writing the first number over the second number.
- Change that fraction into an **equivalent fraction** that has a denominator of 100.
- The **numerator** of the equivalent fraction will then tell you the percentage.

Example 1

Out of 100 cars in a car park, 38 are red. What percentage of the cars are red?

1. Write the amount as a fraction. $\frac{38}{100}$
2. The amount is already written 'out of 100'.
3. Write the amount as a percentage. So **38%** of the cars are red.

Example 2

Express 15 as a percentage of 50.

1. Write the amount as a fraction. $\frac{15}{50}$
2. Write an equivalent fraction which is 'out of 100' by multiplying the top and bottom by the same number. $\frac{15}{50} = \frac{30}{100}$ (× 2 top and bottom)
3. The percentage is the numerator of the equivalent fraction. So 15 is **30%** of 50.

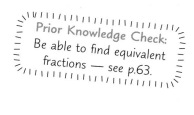

Exercise 1

Q1 Each grid below is made up of 100 small squares.
Find the percentage of each grid that is shaded.

a) 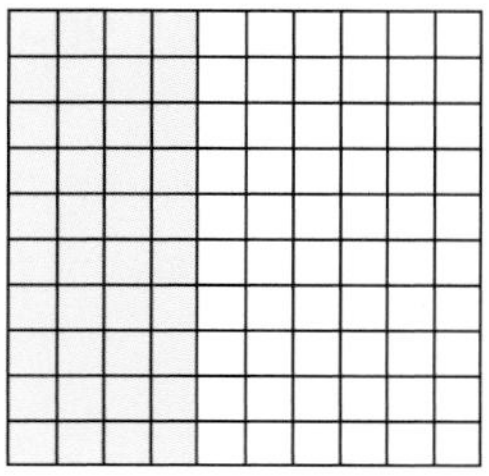

b) 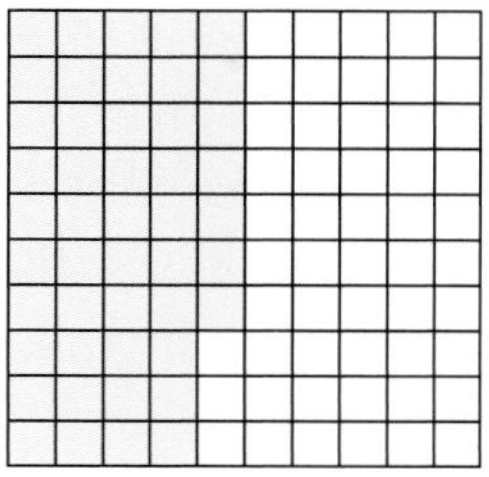

c) 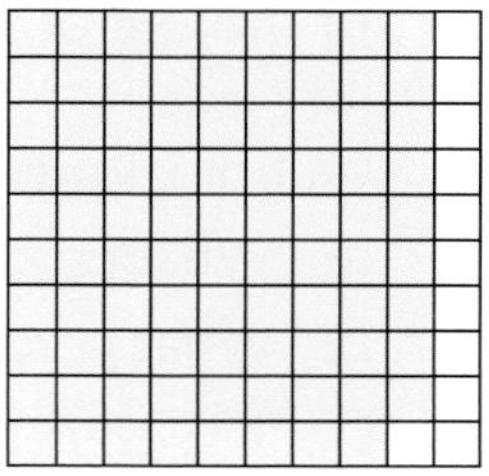

Q2 Write each of the following amounts as a percentage.

a) 13 out of 100 b) 27 out of 100 c) 76 out of 100 d) 243 out of 100

Q3 A football team scored 100 goals in one season.
13 of these goals were penalties.
What percentage of the goals scored were **not** penalties?

Q4 There are 300 coloured counters in a bag. 45 of these counters are green.

a) Write the number of green counters as a fraction of the total number of counters.

b) Find the fraction equivalent to your answer to part a) which has 100 as the denominator.

c) Hence express the amount of green counters as a percentage.

Q5 Write each of the following amounts as a percentage.

a) 11 out of 25 b) 33 out of 50 c) 3 out of 20 d) 21 out of 10

e) 12 out of 200 f) 99 out of 300 g) 600 out of 400 h) 890 out of 1000

You might need to use **more than one step** to find the equivalent fraction.

Example 3

Express 45 as a percentage of 180.

1. Write the amount as a fraction.
2. It may take more than one step to write the fraction out of 100.
3. Write the amount as a percentage.

$$\frac{45}{180} \overset{\div 9}{=} \frac{5}{20} \overset{\times 5}{=} \frac{25}{100}$$

So 45 is **25%** of 180.

Tip: When finding percentages that require more than one step, try to first get the denominator equal to a factor of 100. E.g. 4, 5, 10, 20, etc.

Exercise 2

Q1 Write each of the following amounts as a percentage.

a) 8 out of 32
b) 36 out of 60
c) 24 out of 40
d) 48 out of 120
e) 34 out of 170
f) 42 out of 35

Q2 Out of 24 pupils in a class, 18 walk to school.

a) What percentage of the class walk to school?
b) What percentage of the class do not walk to school?

Q3 There are 55 chocolates in a tin. 33 of the chocolates are milk chocolate. The rest are dark chocolate. What percentage are dark chocolate?

Q4 39 out of 65 people in a book club have blonde hair. What percentage do not have blonde hair?

Sometimes it's **hard** to find an equivalent fraction that has a denominator of 100. When this happens, it is easier to **divide** the top number by the bottom number and then **multiply** the result **by 100**. You can **use a calculator** if the division is tricky.

Example 4

Express 333 as a percentage of 360.

1. Write the amount as a fraction. $\frac{333}{360}$
2. It's not easy to rewrite this as a fraction out of 100. So divide using a calculator instead. $333 \div 360 = 0.925$
3. Multiply by 100% to write as a percentage. $0.925 \times 100\% =$ **92.5%**

Tip: Here we're converting from a fraction to a decimal, then to a percentage — see p.108.

Example 5

The original price of a car was £6500. During a sale, Jennifer bought the car for £4550. What percentage of the original price did she pay?

1. Write the amount as a fraction. $\frac{£4550}{£6500}$
2. Divide using a calculator. $4550 \div 6500 = 0.7$
3. Multiply by 100% to write as a percentage. $0.7 \times 100\% =$ **70%**

Exercise 3

Q1 Write each of the following amounts as a percentage.

a) 15 out of 24 b) 77 out of 275 c) 61 out of 500 d) 1512 out of 375

Q2 A school has 875 pupils. 525 are boys. What is this as a percentage?

Q3 Express £252 as a percentage of £560.

Q4 171 out of 180 raffle tickets were sold for a summer fete.
What percentage of the tickets were sold?

Q5 a) Express 31.36 as a percentage of 32.

b) Express £117.30 as a percentage of £782.

Q6 The jackpot for a lottery was £10 250. Caitlin won £1896.25.
What percentage of the total jackpot did she win?

Q7 Curtis receives £5.60 pocket money per week from his parents, and £2.40 pocket money per week from his grandparents.
What percentage of his total pocket money comes from his grandparents?

Finding a Percentage without a Calculator

Learning Objective — Spec Ref N12:
Be able to find a percentage of an amount without a calculator.

You can find some percentages without a calculator using the following rules.

- **50%** $= \frac{1}{2}$, so find 50% of something by **dividing by 2** (or multiplying by $\frac{1}{2}$).
- **25%** $= \frac{1}{4}$, so find 25% of something by **dividing by 4** (or multiplying by $\frac{1}{4}$).
- **10%** $= \frac{1}{10}$, so find 10% of something by **dividing by 10** (or multiplying by $\frac{1}{10}$).
- **5%** $= \frac{1}{20}$, so find 5% of something by **dividing by 20** (or by dividing 10% of something by 2).
- **1%** $= \frac{1}{100}$, so find 1% of something by **dividing by 100** (or multiplying by $\frac{1}{100}$).

To find other percentages, add up combinations of the percentages above,
e.g. 65% = 50% + 10% + 5%.

Example 6

Find 75% of 44.

1. First find 25% by dividing by 4. 25% of $44 = 44 \div 4 = 11$
2. $75\% = 3 \times 25\%$, so multiply by 3. So 75% of $44 = 3 \times 11 =$ **33**

Tip: You could also find 75% by adding together 50%, 20% and 5%.

Example 7

Find 35% of 70.

1. First find 10% of 70 by dividing by 10. 10% of $70 = 70 \div 10 = 7$
2. $30\% = 3 \times 10\%$, so multiply 7 by 3 to find 30%. 30% of $70 = 3 \times 7 = 21$
3. $5\% = 10\% \div 2$, so divide 7 by 2 to find 5%. 5% of $70 = 7 \div 2 = 3.5$
4. $35\% = 30\% + 5\%$, so add the two amounts to find 35%. So 35% of $70 = 21 + 3.5$ $=$ **24.5**

Exercise 4

Q1 Find each of the following.

a) 50% of 24 b) 50% of 15 c) 25% of 36

d) 25% of 120 e) 10% of 90 f) 10% of 270

Q2 a) Find 25% of 48.

b) Use your answer to find 75% of 48.

Q3 a) Find 10% of 120.

b) Use your answer to find the following.

(i) 5% of 120 (ii) 20% of 120 (iii) 25% of 120

Q4 Find each of the following.

a) 75% of 12 b) 125% of 20 c) 5% of 260

d) 31% of 200 e) 110% of 70 f) 46% of 500

Q5 What is 83% of £30?

Q6 A wooden plank is 9 m long. 55% of the plank is cut off.
What length of wood has been cut off?

Q7 Shima has £1400 in her savings account. She gives 95% of her savings to charity.
How much does she give to charity?

Finding a Percentage with a Calculator

Learning Objective — Spec Ref N12:
Be able to find a percentage of an amount using a calculator.

To find a percentage of an amount, first **divide** the percentage by **100%** to change the percentage into a **decimal**. Then multiply the amount by the decimal.

Example 8

Find 67% of 138.

1. Divide by 100% to turn 67% into a decimal. — $67\% \div 100\% = 0.67$
2. Multiply 138 by the decimal. — $0.67 \times 138 =$ **92.46**

Exercise 5

Q1 Find each of the following.

a) 17% of 200 b) 9% of 11 c) 3% of 210

d) 158% of 615 e) 59% of 713 f) 282% of 823

Q2 What is 12% of 68 kg?

Q3 Jeff is on a journey of 385 km.
So far, he has completed 31% of his journey.
How far has he travelled?

Q4 125 people work in an office. 52% of the office workers are men.
How many workers is this?

Q5 The cost of an adult's ticket for a theme park is £42.
A child's ticket is 68% of the price of an adult's ticket.
How much does a child's ticket cost?

Q6 Which is larger, 22% of £57 or 46% of £28? By how much?

Q7 A jug can hold 2.4 litres of water. It is 34% full. PROBLEM SOLVING
How much more water will fit in the jug?

7.2 Percentages, Fractions and Decimals

Percentages, fractions and decimals are three different ways of showing a proportion of something. It's really important that you are able to convert between all three of them.

Converting between Percentages, Fractions and Decimals

Learning Objective — Spec Ref R9:
Be able to convert between percentages, fractions and decimals.

Prior Knowledge Check:
Be able to convert between fractions and decimals. See p.82-86.

You can switch between percentages, fractions and decimals in the following ways.

Fraction —Divide→ **Decimal** —× 100%→ **Percentage**

Fraction ←Use place value (tenths, hundredths, etc.)— **Decimal** ←÷ 100%— **Percentage**

You can convert directly from a fraction to a percentage by finding an **equivalent fraction** with a **denominator of 100** — then the **numerator** gives the percentage.

You can also convert directly from a percentage to a fraction by writing the **percentage as the numerator** and **100 as the denominator** — then **simplify** the fraction.

Example 1

Write 24% as: a) a decimal b) a fraction in its simplest terms.

a) Divide by 100% to write as a decimal. $24\% \div 100\% = \mathbf{0.24}$

b) The final digit of 0.24 (the '4') is in the hundredths column, so write 0.24 as 24 hundredths. Then simplify by dividing the top and bottom by 4.

$$0.24 = \frac{24}{100} = \mathbf{\frac{6}{25}} \quad (\div 4 \text{ top and bottom})$$

Tip: For a reminder about simplifying fractions, have a look at page 64.

Example 2

Write $\frac{3}{8}$ as: a) a decimal b) a percentage.

a) Calculate 3 ÷ 8 using short division. The answer will be a decimal.

$$8\overline{)3.{}^{3}0{}^{6}0{}^{4}0} = 0.375 \qquad \text{So } \frac{3}{8} = \mathbf{0.375}$$

b) Multiply the decimal by 100% to find the percentage. $0.375 \times 100\% = \mathbf{37.5\%}$

Exercise 1

Q1 Each grid below is made up of 100 small squares.
Find the proportion of each grid that is shaded as:
(i) a percentage, (ii) a decimal, and (iii) a fraction in its simplest terms.

a) 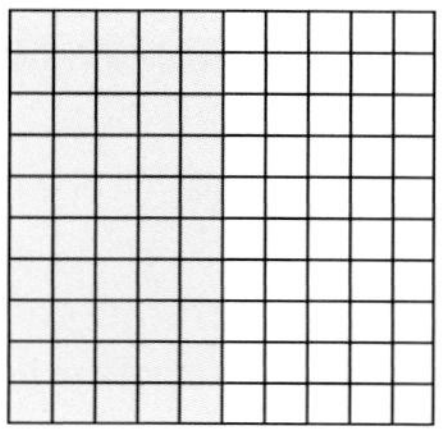

b) 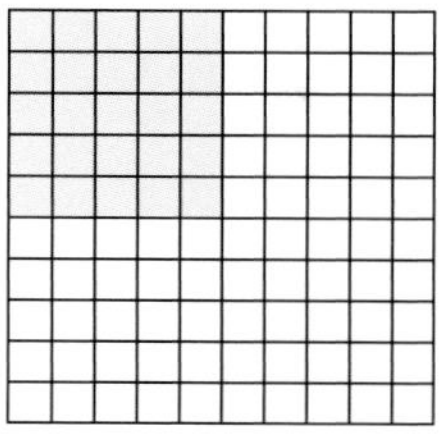

c) 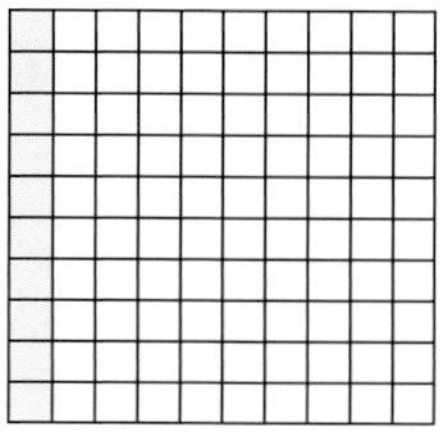

Q2 a) Find the fraction equivalent to $\frac{3}{20}$ which has 100 as the denominator.
b) Write $\frac{3}{20}$ as (i) a percentage, and (ii) a decimal.

Q3 Write each of the following percentages as
(i) a decimal, and (ii) a fraction in its simplest terms.
a) 30% b) 5% c) 13% d) 96%

Q4 Write each of the following fractions as (i) a decimal, and (ii) a percentage.
a) $\frac{79}{100}$ b) $\frac{2}{5}$ c) $\frac{4}{25}$ d) $\frac{7}{8}$

Q5 Write each of the following decimals as
(i) a percentage, and (ii) a fraction in its simplest terms.
a) 0.35 b) 0.86 c) 1.2 d) 0.125

Q6 Raj answers 76% of the questions in a test correctly.
Write this as a decimal.

Q7 $\frac{3}{5}$ of the pupils in a class are right-handed.
What percentage of the class are right-handed?

Comparing Percentages, Fractions and Decimals

Learning Objective — Spec Ref R9:
Be able to compare and order percentages, fractions and decimals.

As percentages, fractions and decimals are all used to represent proportions, it's useful to be able to **compare** them — you'll have to **convert** them all into the **same form** first.

Example 3

Put $\frac{1}{4}$, 24% and 0.244 in order, from smallest to largest.

Write the amounts in the same form.
In this example, they'll all be written as decimals.

1. Calculate 1 ÷ 4 to write $\frac{1}{4}$ as a decimal. — $\frac{1}{4} = 0.25$
2. Calculate 24% ÷ 100% to write 24% as a decimal. — 24% ÷ 100% = 0.24
3. Put the decimals in order, from smallest to largest. — 0.24, 0.244, 0.25
4. Rewrite in their original forms. — **24%, 0.244, $\frac{1}{4}$**

Tip: It's best to write the amounts as either all decimals or all percentages. Fractions are a bit trickier as you need to make sure the denominators are the same before comparing them.

Exercise 2

Q1 For each of the following pairs, write down which is larger.

a) 0.35, 32%
b) 0.4, 4%
c) 0.09, 90%
d) 0.2, $\frac{21}{100}$
e) 0.6, $\frac{7}{10}$
f) 0.55, $\frac{3}{5}$

Q2 Put the numbers in each of the following lists in order, from smallest to largest.

a) 0.42, 25%, $\frac{2}{5}$
b) 0.505, 45%, $\frac{1}{2}$
c) 0.37, 38%, $\frac{4}{10}$
d) 22%, 0.2, $\frac{6}{25}$
e) 0.13, 12.5%, $\frac{3}{20}$
f) 0.25, 23%, $\frac{9}{40}$

Q3 Two shops are having a sale. PROBLEM SOLVING
Shop A is offering $\frac{1}{25}$ off all items, while Shop B is offering 5% off all items.
Which shop is reducing its prices by the greater percentage?

Q4 Margaret is buying a car. She needs to pay $\frac{2}{5}$ of the total cost as a deposit. PROBLEM SOLVING
Her parents give her 35% of the cost of the car to help her buy it.
Is this enough to pay the deposit?

Q5 In a season, Team X won 14 out of the 20 matches they played. PROBLEM SOLVING
Team Y won 60% of their matches.
Which team had the highest proportion of wins?

Q6 Oliver and Jen each try flicking a set of counters into a box. PROBLEM SOLVING
Oliver gets 65% of the counters into the box. Jen gets $\frac{11}{20}$ of the counters into the box.
Who got more counters into the box?

Percentage, Fraction and Decimal Problems

Learning Objective — Spec Ref R9:
Be able to solve problems involving percentages, fractions and decimals.

You might need to convert between percentages, fractions and decimals before you're able to solve a problem. The first step will usually be **converting** them to the **same form**. Once you've converted them there's still work to do — don't forget to answer the question.

Example 4

$\frac{1}{4}$ of pupils in a school bring a packed lunch, 65% have school dinners, and the rest go home for lunch. What percentage of pupils go home for lunch?

1. Write $\frac{1}{4}$ as a percentage by writing it as a fraction out of 100.

   $\frac{1}{4} = \frac{25}{100}$ (× 25 on top and bottom) So $\frac{1}{4} = 25\%$

2. Find the percentage that don't go home for lunch by adding the percentages for 'packed lunches' and 'school dinners'.

   25% + 65% = 90%

3. Subtract this from 100% to find the percentage who do go home for lunch.

   100% – 90% = **10%**

Tip: Think about your final answer when deciding how to convert. If you need to find a percentage, it's usually best to convert everything to percentages.

Exercise 3

Q1 $\frac{3}{4}$ of the people at a concert arrived by train, 0.05 walked, and the rest came by car. What percentage came by car?

Q2 $\frac{3}{5}$ of the footballs in a bag are white, 20% are black, and the rest are blue. What percentage of footballs are blue?

Q3 Beverley eats 0.3 of a pie, Victoria eats $\frac{1}{10}$, Patrick eats 20%, and Gus eats the rest. What percentage of the pie does Gus eat?

Q4 Ainslie is keeping a record of the birds in his garden. Of the birds he has seen this month, $\frac{3}{8}$ were sparrows, 41.5% were blackbirds, and the rest were robins. What percentage were robins?

7.3 Percentage Increase and Decrease

Percentages are often used to describe a change in an amount. The change could be an increase or a decrease, so make sure you know which one you're trying to find before tackling a question.

Calculating Amounts After a Percentage Increase or Decrease

Learning Objective — Spec Ref R9/N12:
Be able to find new amounts after a percentage increase or decrease.

Prior Knowledge Check:
Be able to find percentages with and without a calculator — see p.105-107.

To **increase** an amount by a percentage without using a calculator:

- Calculate the percentage.
- **Add it** to the original amount.

To **decrease** an amount by a percentage, **subtract** the percentage from the original amount.

Example 1

Increase 450 by 15% without using your calculator.

1. Find 10% and 5% of 450. — 10% of $450 = 450 \div 10 = 45$
   5% of $450 = 45 \div 2 = 22.5$
2. Add these to find 15% of 450. — So 15% of $450 = 45 + 22.5 = 67.5$
3. Add this to the original amount. — $450 + 67.5 =$ **517.5**

Exercise 1

Q1 a) Find 50% of 360.
b) Increase 360 by 50%.

Q2 a) Find 30% of 120.
b) Increase 120 by 30%.

Q3 a) Find 10% of 160.
b) Decrease 160 by 10%.

Q4 a) Find 20% of 84.
b) Decrease 84 by 20%.

Q5 Increase each of the following amounts by the percentage given.
a) 90 by 10% b) 11 by 80% c) 140 by 45%

Q6 Decrease each of the following amounts by the percentage given.
a) 24 by 25% b) 55 by 70% c) 150 by 55%

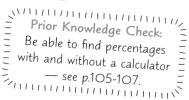

Using a Multiplier

If you're using a calculator, you can find a percentage increase or decrease in one go using a **multiplier**. To find the multiplier:

- **Convert** the percentage to a decimal
- **Add it to 1** for a percentage increase, **or subtract it from 1** for a percentage decrease.

Then **multiply the amount** in the question by the multiplier to find the answer.

Tip: The method on the previous page also works with a calculator — use the method you prefer.

Example 2

Increase 425 by 18%.

1. Convert the percentage to a decimal. — $18\% \div 100\% = 0.18$
2. It's a percentage increase, so add to 1 to find the multiplier. — Multiplier $= 1 + 0.18$ $= 1.18$
3. Multiply 425 by the multiplier. — $425 \times 1.18 =$ **501.5**

Tip: A percentage increase has a multiplier greater than 1. A percentage decease has a multiplier less than 1.

Example 3

Decrease 326 by 12%.

1. Convert the percentage to a decimal. — $12\% \div 100\% = 0.12$
2. It's a percentage decrease, so subtract from 1 to find the multiplier. — Multiplier $= 1 - 0.12 = 0.88$
3. Multiply 326 by the multiplier. — $326 \times 0.88 =$ **286.88**

Exercise 2

Q1 Increase each of the following amounts by the percentage given.

a) 490 by 11% b) 101 by 16% c) 55 by 37%
d) 2523 by 67% e) 1036 by 23% f) 36 500 by 32%

Q2 Decrease each of the following amounts by the percentage given.

a) 77 by 8% b) 36 by 21% c) 101 by 43%
d) 8612 by 39% e) 75 250 by 4% f) 116 000 by 18%

Q3 a) Increase £89.50 by 62%. b) Decrease 58 kg by 19%.

Simple Interest

Learning Objective — Spec Ref R9:
Be able to solve problems involving simple interest.

Interest is a percentage of money that is added on to an amount over a period of time. **Simple interest** is when a certain percentage of the **original amount** is paid at regular intervals (usually once per year). Since simple interest is always based on the original amount, and the original amount **never changes**, the interest paid is the **same** every time.

Example 4

Fabian deposits £150 into an account which pays 5% simple interest per year. Without using a calculator, work out how much will be in the account after one year.

1. Find 5% of £150. — 10% of £150 = £15, so 5% of £150 = £15 ÷ 2 = £7.50
2. Add this to £150. — £150 + £7.50 = **£157.50**

Exercise 3

Q1 Leroy deposits £230 into an account which pays 5% simple interest per year. How much will be in the account after one year?

Q2 Kimberley deposits £890 into an account which pays 3% simple interest per year. How much will be in the account after three years?

Percentage Increase and Decrease Problems

Learning Objective — Spec Ref R9:
Be able to solve problems involving percentage increase and decrease.

You might come across **wordy problems** that involve percentage increase or decrease.

- Read the question **carefully** and work out if it's an **increase** or **decrease** problem (or both).
- Calculate the result of the increase or decrease using the **methods** you've learnt.

Tip: Check that you've answered the question fully — some problems have more than one step.

Example 5

Shop A sells a type of oven for £300 and then increases its prices by 5%.
Shop B sells the same type of oven for £290 and increases its prices by 10%.
Without using a calculator, work out which shop now has the cheaper price.

1. Calculate the new price for each shop.

	Shop A	Shop B
	10% of 300 = 30 So 5% of 300 = 15	10% of 290 = 29
	£300 + £15 = £315	£290 + £29 = £319

2. Compare the new prices to see which is cheaper.

£315 < £319 ⇒ **Shop A** is cheaper

Exercise 4

Q1 Damelza's salary of £24 500 is increased by 15%.
What is her new salary?

Q2 A kettle originally costing £42 is reduced by 75% in a sale.
What is the sale price of the kettle?

Q3 20% VAT is added to the basic price of a TV to give the selling price.
The basic price of the TV is £485. What is the selling price?

Q4 A couple go out for a meal in a restaurant.
The total cost of the meal is £63, but the couple have a voucher for 13% off the bill.
How much do the couple pay?

Q5 David's height increased by 20% between the ages of 6 and 10.
He was 50 inches tall when he was 6. How tall was he when he was 10?

Q6 At the start of a journey, Natalie's car had 8 gallons of fuel, and Jason's car had 12 gallons of fuel.
During the journey, Natalie used 25% of her fuel, and Jason used 40% of his fuel.
Who had more fuel left at the end of the journey? By how much?

Q7 Last year, Elsie's gas bill was £480 and her electricity bill was £612.
This year, her gas bill has increased by 2% and her electricity bill has decreased by 4%.
Which is now the more expensive bill? By how much?

Q8 Alison earns £31 000 per year. One year, she gets a pay rise of 3%.
The following year, she gets a pay cut of 2%. How much does she now earn?

Finding a Change as a Percentage

Learning Objective — Spec Ref R9:
Be able to express the change in an amount as a percentage.

To find a percentage increase or decrease:

- Calculate the **difference** between the new amount and the original amount.
- Find this as a percentage of the **original amount**.

Example 6

In a sale, the cost of a CD is reduced from £15 to £9. Find the percentage decrease.

1. Find the difference between the new cost and the original cost. £15 – £9 = £6
2. Write the difference as a fraction of the original cost...
3. ...then write the fraction out of 100 to find the percentage decrease.

$$\frac{6}{15} \overset{\div 3}{=} \frac{2}{5} \overset{\times 20}{=} \frac{40}{100} = 40\%$$

So it is a **40%** decrease.

Tip: Remember to always divide by the original cost.

Exercise 5

Q1 Find the percentage increase when:

a) a price of £10 is increased to £12.
b) a price of £20 is increased to £22.
c) a price of £140 is increased to £161.
d) a price of £120 is increased to £174.

Q2 Find the percentage decrease when:

a) a price of £10 is decreased to £8.
b) a price of £25 is decreased to £22.
c) a price of £80 is decreased to £64.
d) a price of £150 is decreased to £138

Q3 The number of people working for a company increases from 50 to 72.

a) Find the difference in the number of people working for the company.
b) Write your answer to part a) as a fraction of the original amount.
c) Find the percentage increase in the number of people working for the company.

Q4 The price of a local newspaper increases from 80p to £1.
Find the percentage increase.

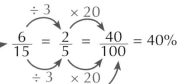

Q5 In a sale, the price of a toaster is reduced from £50 to £30.
Find the percentage reduction.

Q6 Percy is on a healthy eating plan.
His weight drops from 90 kg to 72 kg.

a) Find the amount of weight Percy has lost.
b) Find the percentage decrease in Percy's weight.

When you are allowed to **use a calculator** to find a change as a percentage, use the same method but find the percentage using the calculator technique shown on page 104.

Example 7

**The price of a holiday increases from £320 to £364.80.
Find the percentage increase.**

1. Find the difference between the new cost and the original cost. — £364.80 – £320 = £44.80
2. Divide the difference by the original amount using a calculator. — $\frac{44.80}{320} = 44.80 \div 320 = 0.14$
3. Multiply by 100 to find the percentage increase. — $0.14 \times 100 = 14$
So it is a **14%** increase.

Exercise 6

Q1 A shop owner buys a pair of trainers for £52 and sells them for £70.20.
What is her percentage profit?

Q2 The height of a sunflower increases from 1.3 m to 2.08 m over the course of summer.
What is the percentage increase in its height?

Q3 During a season, the average attendance for a local sports team's matches was 11 350.
The following season, the average attendance was 11 123.
Find the percentage decrease.

Q4 A car is bought for £12 950.
Three years later, it is sold for £8806.
After another three years, it is sold again for £4403.

a) Find the percentage decrease in the car's price over the first three years.
b) Find the percentage decrease in the car's price over the next three years.
c) Find the percentage decrease in the car's price over the whole six years.

Finding the Original Value

Learning Objective — Spec Ref R9:
Be able to find the original amount after a percentage increase or decrease.

If you know the **new amount** after a percentage increase or decrease, you can find the **original value** using the following steps:

- Write the new amount as a **percentage** of the original value.
- **Divide** the new amount by the percentage to find **1%** of the original value.
- **Multiply by 100** to find the original value (100%).

Example 8

The value of a painting increases by 12% to £1680.
What was the painting worth before it increased in value?

1. A 12% increase means the new amount is 112% of the original value.
2. Divide by 112 to find 1% of the original value.
3. Multiply by 100 to get the original value.

112% = £1680

÷ 112 ÷ 112

1% = £15

× 100 × 100

100% = **£1500**

Tip: If it's a percentage increase, add the percentage to 100% to get the percentage of the original amount. For a percentage decrease, subtract from 100%.

Exercise 7

Q1 A fridge costs £200 after a 50% reduction.
Calculate the original price of the fridge.

Q2 Andy buys a top hat that has been reduced in a sale by 35%.
If the sale price is £13.00, find the original price.

Q3 The owner of a furniture shop increased the price of a bookcase by 26%.
If the bookcase is now priced at £819, what was the price before the increase?

Q4 An estate agent reduced the price of a house by 4%.
If the house is now for sale at £192 000, what was its original price?

Q5 In the past year, the number of frogs living in a pond has increased by 10% to 528, and the number of newts living there has increased by 15% to 621.
How many frogs and how many newts lived in the pond a year ago?

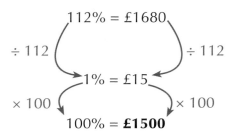

7.4 Compound Growth and Decay

Compound percentage changes involve repeating a percentage increase or decrease. At each step, the percentage change is applied to the current amount, rather than the original amount.

Compound Growth

Learning Objective — Spec Ref R16:
Be able to solve compound growth problems.

Compound growth is when a **percentage increase** is **repeatedly** applied to a quantity. The percentage increase is applied to the **current value** at each step. Compound growth applied to money is known as **compound interest**. E.g. if £1000 is increased by 10% each year:

- After 1 year: £1000 × 1.1 = £1100
- After 2 years: £1100 × 1.1 = £1210
- After 3 years: £1210 × 1.1 = £1331

Example 1

Mr Zupnik invests £600 in a bank account at 5% per annum compound interest. How much money will he have in the bank after two years?

1. Find the multiplier for the percentage increase. — 5% ÷ 100% = 0.05
   So the multiplier = 1 + 0.05 = 1.05
2. 'Per annum' means 'per year'.
   Multiply to find the amount after one year. — £600 × 1.05 = £630
3. Multiply again to find the amount after two years. — £630 × 1.05 = **£661.50**

Exercise 1

Q1 A bank pays 3% per annum compound interest.
Calculate how much interest these accounts will earn.
Give your answer to the nearest penny.

a) £250 invested for 1 year.
b) £45 invested for 3 years.
c) £1500 invested for 2 years.

Q2 Josephine invests £3500 in a bank account that pays annual compound interest of 4.5%.
How much money (to the nearest penny) will she have in the bank after:

a) 2 years, b) 3 years?

Compound Decay

Learning Objective — Spec Ref R16:
Be able to solve compound decay problems.

Compound decay is the opposite of compound growth — it's when a **percentage decrease** is **repeatedly** applied to a quantity, and acts on the **current amount** each time.

When applied to money, compound decay is known as **depreciation**.

Example 2

Mrs Jones buys a laptop for £360. It depreciates at a rate of 35% per annum. How much will the laptop be worth after three years? Give your answer to 2 d.p.

1. Find the multiplier for the percentage decrease.
   35% ÷ 100% = 0.35
   So the multiplier = 1 – 0.35
   = 0.65
2. Multiply to find the amount after one year.
   £360 × 0.65 = £234
3. Multiply again to find the amount after two years.
   £234 × 0.65 = £152.10
4. Multiply one last time to find the amount after three years.
   £152.10 × 0.65 = £98.865
   = **£98.87 (2 d.p.)**

Tip: After the first step, remember to multiply the **current** amount not the original amount.

Exercise 2

Q1 Mr Quasar buys a television for £320.
Its value depreciates at a rate of 15% per year.
What is its value after: a) 2 years? b) 3 years?

Q2 Alan is visiting Las Vegas and has taken $1000 spending money.
He spends his money at a rate of 5% a day.
How much money will Alan have after 2 days?

Q3 A painting is bought for £12 000.
The value of the painting depreciates at a rate of 6% per year for the first two years, and then by 17% in the third year. What is the value of the painting three years after it was bought? Give your answer to the nearest penny.

The Formula

Learning Objective — Spec Ref R16:
Use the formula for compound growth and decay.

The **formula** for compound growth and decay is: $P_n = P_0 \times (\text{multiplier})^n$

where P_n = amount after n hours/days/years,
P_0 = initial amount,
n = number of hours/days/years

Example 3

Calculate the total compound interest paid on £500 over 6 years if it's paid at a rate of 4% per year.

1. Use the formula for compound growth: $P_n = P_0 \times (\text{multiplier})^n$
2. Plug in the numbers... $P_0 = £500$, multiplier = 1.04, $n = 6$
   $P_6 = £500 \times (1.04)^6 = £632.6595...$
3. Subtract the initial figure from the final value to find the interest paid. Interest = $P_6 - P_0$
   = £632.6595... – £500
   = £132.6595... = **£132.66**

Tip: Always check the context before giving your final answer. Here it makes sense to round to 2 d.p. as it's money, but sometimes only a whole number fits the context.

Exercise 3

Q1 Calculate the amount of money (to the nearest penny) you would have if you invested:
a) £1000 at an annual rate of 4% for 8 years.
b) £600 at an annual rate of 5.2% for 7 years.

Q2 Calculate the value (to the nearest £100) of a house after six years if it was bought for £650 000 and house prices are falling at a rate of 2% per annum.

Q3 A colony of ants has set up home in Mr Murphy's shed. PROBLEM SOLVING
At 9 am on Monday there were 250 ants.
If the colony of ants grows at a rate of 6% per day, how many ants will there be at 9 am on Saturday?

Q4 Mr Butterworth is on a diet and is losing weight at a rate of 2% of his total body weight every week. PROBLEM SOLVING
If he weighs 110 kg when he starts his diet, what will his body weight be at the end of 8 weeks? Give your answer to one decimal place.

Review Exercise

Q1 A chess club has 25 members. 12 of these members are female.
Express the number of female members of the club as a percentage.

Q2 A coat normally costs £160. In a sale, the coat's price was reduced by 35%.
How much cheaper is the sale price than the normal price?

Q3 Each grid on the right is made up of 25 small squares.
Find the proportion of each grid that is shaded as
(i) a fraction in its simplest terms,
(ii) a percentage,
(iii) a decimal.

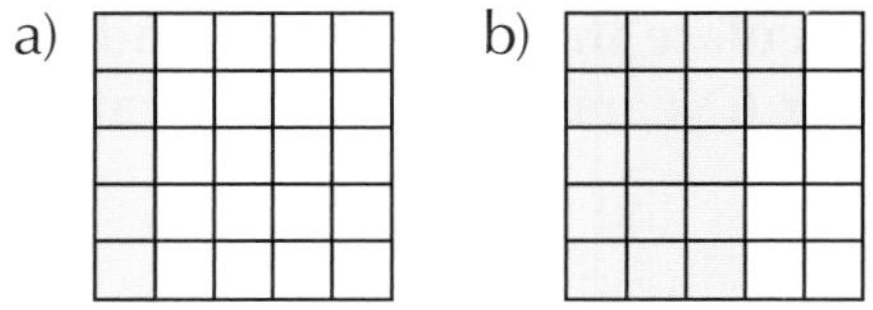

Q4 Write $\frac{31}{500}$, 0.061 and 6% in order, from smallest to largest.

Q5 A pot of yoghurt normally contains 450 g.
A special offer pot contains 55% extra free.
How many grams of yoghurt does the special offer pot contain?

Q6 Two shops had a sale on a suit that had previously cost £92.
Shop A had 70% off the original price, while Shop B had $\frac{7}{8}$ off.
Which shop had the lower price, and by how much?

Q7 Tony deposits £3250 into a savings account that earns 8.5% simple interest per year.
How much will he have at the end of the year?

Q8 In an experiment, the mass of a chemical drops from 75 g to 69 g.
Find the percentage decrease.

Q9 Dave loves a bargain and buys a feather boa which has been reduced in price by 70%.
If the sale price is £2.85, what was the original price of the boa?

Q10 The population of an island is 12 500. Each year, the population increases by 8%.
What is the population of the island after 5 years?

Exam-Style Questions

Q1 Gemma's phone contract gives her 750 MB of data each month.
One month, she buys 15% more data.
How much data does she have this month?

[2 marks]

Q2 Increase 3.7 metres by 24%.
Give your answer correct to the nearest centimetre.

[3 marks]

Q3 In 2016 a football club was worth £375 million.
In 2017 the football club was worth £450 million.
Show that the percentage increase in the value of the football club from 2016 to 2017 was 20%.

[2 marks]

Q4 A large box in a school gym's equipment cupboard contains 20 basketballs and 12 footballs.
The school receives eight new basketballs which are added to the box.
Work out the percentage of balls in the box which are now basketballs.

[2 marks]

Q5 Claire buys a sports car for £68 000.
It will depreciate at 20% per annum for the first two years and at 15% per annum for the next three years.
What is the value of the sports car after 5 years?

[4 marks]

Q6 30% of the animals in a wildlife park are lemurs.
40% of the lemurs are ring-tailed lemurs.
What percentage of the animals in the wildlife park are ring-tailed lemurs?

[2 marks]

Section 8 Algebraic Expressions

8.1 Simplifying Expressions

Algebraic expressions involve variables (letters that represent numbers) and don't contain an equals sign — e.g. x, $3a + b$, $y^2 + z^2$. You can simplify them to avoid writing the same thing over and over.

Collecting Like Terms

Learning Objective — Spec Ref A1/A3/A4:
Be able to simplify algebraic expressions by collecting like terms.

These are all algebraic expressions:

a $6b$ xyz $a + b$ $x^2 + y^2 + z^2$

Tip: a is the same as $1a$, and $6b$ is the same as $6 \times b$.

Terms are the individual parts of an expression and **include** the plus or minus signs.
E.g. in the expression $2x + 6 - 3xy$, the terms are $2x$, 6 and $-3xy$.

Expressions can sometimes be **simplified** by collecting **like terms**.
'Like terms' are terms that contain **exactly the same** combination of letters.
For example xy, $-3xy$ and $2yx$ are all like terms, since the only letters are xy.

Example 1

Simplify this expression by collecting like terms: $4a + 3b - a - 7b$

1. First write the like terms next to each other. Take the sign with the term when you move it.
2. $4a$ and $-a$ are like terms as they just contain a. Together these give $4a - a = 3a$.
3. $3b$ and $-7b$ are like terms as they just contain b. Together these give $3b - 7b = -4b$.

$4a + 3b - a - 7b = 4a - a + 3b - 7b$
$= 3a + 3b - 7b$
$= \mathbf{3a - 4b}$

Example 2

Simplify the expression $x + 4y + 4 - 2x - y - 3$ by collecting like terms.

1. Terms involving no letters at all are like terms.
2. Collect them together in the same way as terms containing letters.

$x + 4y + 4 - 2x - y - 3$
$= x - 2x + 4y - y + 4 - 3$
$= \mathbf{-x + 3y + 1}$

Exercise 1

Q1 Simplify these expressions by collecting like terms.

a) $2x + 3x + x$ b) $7p - 2p + 3p - 4p$ c) $c + c + c + d + d$

d) $a + b + a - a + b$ e) $5a - 2a + 5b + 2b$ f) $4b + 8c - b - 5c$

Q2 Simplify these expressions by collecting like terms.

a) $2c + 4 + c + 7$ b) $3x + 6 - 6x - 4$ c) $5y + 12 - 9y - 7$

d) $-m + 5 - 8m + 16$ e) $-4x - 2 + 7x + 12$ f) $13a + 8 + 8a + 2$

Q3 Simplify these expressions.

a) $x + 7 + 4x + y + 5$ b) $a + 2b + b - 8 - 5a$

c) $13a - 5b + 8a + 12b + 7$ d) $8p + 6q + 14 - 6r - 4p - 14r - 2q - 23$

Example 3

Simplify the expression $x + 3x^2 + yx + 7 + 4x^2 + 2xy - 3$ by collecting like terms.

There are four sets of like terms:
(i) terms involving just x
(ii) terms involving x^2
(iii) terms involving xy (or yx, which means the same)
(iv) terms involving just numbers

$$x + 3x^2 + yx + 7 + 4x^2 + 2xy - 3$$
$$= x + (3x^2 + 4x^2) + (yx + 2xy) + (7 - 3)$$

Collect the different sets together separately.

$$= \mathbf{x + 7x^2 + 3xy + 4}$$

Exercise 2

Q1 Simplify the following expressions by collecting like terms.

a) $x^2 + 3x + 2 + 2x + 3$ b) $x^2 + 4x + 1 + 3x - 3$ c) $x^2 + 4x + x^2 + 2x + 4$

d) $x^2 + 2x^2 + 4x - 3x$ e) $p^2 - 5p + 2p^2 + 3p$ f) $3p^2 + 6q + p^2 - 4q + 3p^2$

g) $8 + 6p^2 - 5 + pq + p^2$ h) $4p + 5q - pq + p^2 - 7q$ i) $a^2 + 7b + 2a^2 + 5ab - 3b$

Q2 Simplify the following expressions by collecting like terms.

a) $ab + cd - xy + 3ab - 2cd + 3yx + 2x^2$ b) $pq + 3pq + p^2 - 2qp + q^2$

Q3 By collecting number terms and root terms, simplify the following.

a) $7 + 3\sqrt{3} + 6 - 2\sqrt{3}$ b) $-2 - 13\sqrt{7} - 7 + 3\sqrt{7}$ c) $11 - 7\sqrt{5} - 11 - 8\sqrt{5}$

Multiplying and Dividing Letters

Learning Objectives — Spec Ref A1/A4:
- Use and interpret algebraic notation.
- Use the laws of indices when multiplying and dividing variables.

Prior Knowledge Check: Be able to use the laws of indices — see p.44.

When you've got numbers and variables **multiplied** or **divided** by each other, you should deal with the numbers and each letter **separately**.
You can use the **laws of indices** to write each term as **simply** as possible.

The most useful laws of indices are: $a^m \times a^n = a^{m+n}$ $\quad a^m \div a^n = a^{m-n}$ $\quad (a^m)^n = a^{m \times n}$

There are some more rules that might help you with these questions:
- abc means $a \times b \times c$ — the × symbols are left out to make it clearer.
- $ab^2 = a \times b \times b$ — only the b is squared.
- $(ab)^2 = ab \times ab = a \times a \times b \times b = a^2b^2$ — the whole bracket is squared.
- $(-a)^2 = (-a) \times (-a) = (-1) \times (-1) \times a \times a = a^2$ — squaring a negative makes it positive.
- $2ab \times 3a = 2 \times 3 \times a \times a \times b = 6a^2b$ — multiply the numbers and variables separately.
- $\frac{a}{b} = a \div b$ — use the power rules from p.44 if dividing by powers of the same letter.
- $6a^5 \div 3a^3 = (6 \div 3)(a^5 \div a^3) = 2a^2$ — divide the numbers and variables separately.

Example 4

Simplify these expressions: **a)** $b \times b \times b \times b$ **b)** $4a \times 5b$ **c)** $6a^2 \div 3a$

1. If the same letter is multiplied by itself, write it as a power.
2. Multiply numbers and letters separately.

a) $b \times b \times b \times b = \mathbf{b^4}$

b) $4a \times 5b = 4 \times 5 \times a \times b = \mathbf{20ab}$

c) $6a^2 \div 3a = (6 \div 3) \times (a^2 \div a) = \mathbf{2a}$

Exercise 3

Q1 Simplify the following expressions.

a) $a \times a \times a$ b) $2a \times 3a$ c) $-8p \times 2q$ d) $3a \times 7a$

e) $5x \times 3y$ f) $m \times m \times -m \times m$ g) $12a \times 4b$ h) $6p \times 8p$

Q2 Simplify the following expressions.

a) $a \times ab$ b) $4m^3 \div m$ c) $(r^2)^3$ d) $3(st)^2$

e) $4a^2 \times 5a$ f) $-9s^4 \div -3s^3$ g) $4a^4b^3 \times 2ab^2$ h) $\frac{6y^2}{3y}$

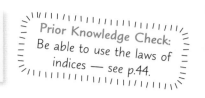

8.2 Expanding Brackets

When brackets show up, you'll often want to expand them to get rid of them. Single brackets are pretty straightforward, but with two sets of brackets there's a lot to keep track of.

a(b + c)

Learning Objective — Spec Ref A4:
Be able to expand a single term multiplied by a bracket.

You can **expand** (or remove) brackets by multiplying everything **inside** the brackets by the letter or number **in front**.
Remember that $a(b + c) = a \times (b + c) = (a \times b) + (a \times c)$.

$$a(b + c) = ab + ac$$
$$a(b - c) = ab - ac$$

Example 1

Expand the brackets in these expressions: **a)** $3(a + 2)$ **b)** $8(2n - 3)$

a) Multiply both a and 2 by 3. $3(a + 2) = (3 \times a) + (3 \times 2) = \mathbf{3a + 6}$

b) Multiply both $2n$ and 3 by 8. $8(2n - 3) = (8 \times 2n) - (8 \times 3) = \mathbf{16n - 24}$

Example 2

Expand the brackets in these expressions: **a)** $m(n + 7)$ **b)** $a(a - 4)$

a) Multiply both n and 7 by m. $m(n + 7) = (m \times n) + (m \times 7) = \mathbf{mn + 7m}$

b) Multiply both a and 4 by a. $a(a - 4) = (a \times a) - (a \times 4) = \mathbf{a^2 - 4a}$

Exercise 1

Q1 Expand the brackets in these expressions.

a) $2(a + 5)$ b) $4(b + 3)$ c) $5(d + 7)$ d) $3(p + 4)$
e) $3(5 + p)$ f) $7(6 + g)$ g) $5(3 - y)$ h) $8(a - b)$

Q2 Expand the brackets in these expressions.

a) $x(y + 5)$ b) $p(q + 2)$ c) $x(8 - x)$ d) $a(b - 12)$
e) $3(2p + 4)$ f) $5(4t - 8)$ g) $3(u + 8v)$ h) $7(5n - 6m)$

Example 3

Expand the brackets in this expression: $-b(4 - 2b)$

1. Multiply 4 and $2b$ by $-b$.
2. Be careful with the signs — multiplying by a negative will change the signs of the terms inside the brackets.

$$-b(4 - 2b) = (-b \times 4) - (-b \times 2b)$$
$$= (-4b) - (-2b^2)$$
$$= \mathbf{-4b + 2b^2}$$

Exercise 2

Q1 Expand the brackets in the following expressions.

a) $-(q + 2)$ b) $-(x + 7)$ c) $-8(7 - w)$ d) $-5(5 - x)$

e) $-v(v + 4)$ f) $-v(v - 5)$ g) $-x(12 - x)$ h) $-y(4 + y)$

Q2 Expand the brackets in the following expressions.

a) $-6(5g - 3)$ b) $-7(4v + 8)$ c) $-2(5 + 4m)$ d) $-5(10 - 8v)$

e) $-5(2 + 3n)$ f) $-4z(8 - 2z)$ g) $-2(6b - 3)$ h) $-4y(2y + 6)$

When doing **calculations** involving brackets, expand all brackets **first**.

Example 4

Simplify the following expressions: a) $2(a + 5) + 3(a + 2)$ b) $3(x + 2) - 5(2x + 1)$

a) Multiply out both brackets. Then collect like terms.

$$2(a + 5) + 3(a + 2) = (2a + 10) + (3a + 6)$$
$$= 2a + 10 + 3a + 6$$
$$= \mathbf{5a + 16}$$

b) Multiply out the individual brackets. The minus sign before the second set of brackets reverses the sign of each term inside those brackets.

$$3(x + 2) - 5(2x + 1) = (3x + 6) - (10x + 5)$$
$$= 3x + 6 - 10x - 5$$
$$= \mathbf{-7x + 1}$$

Exercise 3

Q1 Simplify the following expressions.

a) $2(z + 3) + 4(z + 2)$ b) $3(c + 1) + 5(c + 7)$ c) $4(u + 6) + 8(u + 5)$

d) $7(t - 3) + 2(t + 12)$ e) $8(m - 2) + 9(m + 5)$ f) $5(p - 3) - (p + 6)$

g) $2(j - 5) - (j - 3)$ h) $5(y - 4) - (y - 2)$ i) $5(3c - 6) - (c - 3)$

Q2 Simplify the following expressions.

a) $5(2q + 5) - 2(q - 2)$ b) $2(3c - 8) - 8(c + 4)$ c) $5(q - 2) - 3(q - 4)$

Q3 Simplify the following expressions.

a) $2(-z + 2) + 3z(3z + 6)$ b) $4p(3p + 5) - 3(p + 1)$ c) $9b(2b + 5) + 4b(6b + 6)$

(a + b)(c + d)

Learning Objective — Spec Ref A4:
Be able to expand two brackets multiplied together.

When expanding **pairs** of brackets, multiply each term in the left bracket by each term in the right bracket. If each bracket contains two terms (brackets like this are called **binomials**), you can use **FOIL** to keep track of which terms you need to multiply:

F IRST — multiply the first term from each bracket
O UTSIDE — multiply the terms on the outside
I NSIDE — multiply the terms on the inside
L AST — multiply the last term from each bracket

$$(a + b)(c + d) = ac + ad + bc + bd$$

You'll always get four terms after multiplying binomials together this way.
Sometimes you can then **simplify** by collecting like terms — see p.124.

Example 5

Expand the brackets in the following expression: $(q + 4)(p + 3)$

1. Multiply each term in the left bracket by each term in the right bracket using FOIL.
2. It's a good idea to write the letters in each term of the answer in alphabetical order.

$(q + 4)(p + 3) = qp + 3q + 4p + 12$

$= \mathbf{pq + 3q + 4p + 12}$

Example 6

Expand the following expression: $(2x + 1)(3x - 2)$

1. Expand the brackets using FOIL — be careful when multiplying terms with coefficients.
2. Simplify by collecting like terms.

$(2x + 1)(3x - 2) = 6x^2 - 4x + 3x - 2$

$= \mathbf{6x^2 - x - 2}$

Exercise 4

Q1 Expand the brackets in the following expressions.

a) $(a + 2)(b + 3)$ b) $(j + 4)(k - 5)$ c) $(x - 4)(y - 1)$ d) $(x + 6)(y + 2)$

e) $(9 - a)(b - 3)$ f) $(t - 5)(s + 3)$ g) $(5x + 4)(3 - y)$ h) $(3a - 1)(2b + 2)$

Q2 Expand the brackets in the following expressions.

a) $(x + 8)(x + 3)$ b) $(b + 2)(b - 4)$ c) $(a - 1)(a + 2)$ d) $(d + 7)(d + 6)$

e) $(z - 12)(z + 9)$ f) $(c + 5)(3 - c)$ g) $(3y - 8)(6 - y)$ h) $(2x + 2)(2x + 3)$

$(a + b)^2$

Learning Objective — Spec Ref A4:
Be able to expand the square of a bracket.

If you have to expand **squared brackets** write them out as **two sets** of brackets — e.g. $(a + b)^2 = (a + b)(a + b)$
Then you can expand using **FOIL** — like on p.129.

Tip: Remember that you cannot just square each term inside the brackets to expand them. E.g. $(a + b)^2 \neq a^2 + b^2$.

Example 7

Expand the brackets in the following expression: $3(x - 2)^2$

1. Write out $(x - 2)^2$ as two sets of brackets.
2. Use FOIL to expand the brackets — leave the '3 ×' alone for now.
3. Collect like terms, then multiply each term in the brackets by 3.

$3(x - 2)^2 = 3 \times (x - 2)(x - 2)$

$= 3 \times (x^2 - 2x - 2x + 4)$

$= 3 \times (x^2 - 4x + 4)$

$= \mathbf{3x^2 - 12x + 12}$

Exercise 5

Q1 Expand the brackets in the following expressions.

a) $(x + 1)^2$ b) $(x + 4)^2$ c) $(x + 5)^2$ d) $(x - 2)^2$

e) $(x - 3)^2$ f) $(x - 7)^2$ g) $3(x + 3)^2$ h) $2(x - 6)^2$

Q2 Expand the brackets in the following expressions.

a) $(5x + 2)^2$ b) $(2x + 6)^2$ c) $(3x - 1)^2$

8.3 Factorising

Factorising an expression means adding brackets in where there weren't any before.
It's called factorising because you need to look for common factors of all the different terms.

Taking Out Common Factors

Learning Objective — Spec Ref A4:
Be able to factorise expressions by taking out common factors.

Prior Knowledge Check: Be able to find the highest common factor of a pair of numbers — see p.60.

Factorising is the opposite of expanding brackets.
You look for a **common factor** of all the terms in an expression and 'take it outside' the brackets.
These common factors could be **numbers**, **variables** or **both**.

To factorise an expression **fully**, you need to find the **highest common factor** of all the terms.

Tip: You can check you have factorised correctly by expanding the brackets again.

Example 1

Factorise the expression $12x - 18y$.

1. 6 is the highest common factor of $12x$ and $18y$. So 6 goes outside the brackets. $12x - 18y = 6(\quad - \quad)$
2. Divide each term by the common factor, and write the results inside the brackets.
   $12x \div 6 = 2x$
   and $18y \div 6 = 3y$
   So $12x - 18y = \mathbf{6(2x - 3y)}$

Tip: If you used 2 or 3 as the factor instead of 6, you would get an expression that wasn't fully factorised, e.g. $2(6x - 9y)$.

When factorising **variables**, the highest common factor is the **highest power** of the variable that will go into every term.

You'll need to remember how to **divide** two powers: $a^m \div a^n = \frac{a^m}{a^n} = a^{m-n}$

Example 2

Factorise $3x^2 + 2x$.

1. x is the only common factor of $3x^2$ and $3x$. So x goes outside the brackets. $3x^2 + 2x = x(\quad + \quad)$
2. Divide each term by the common factor, and write the results inside the brackets.
   $3x^2 \div x = 3x$ and $2x \div x = 2$
   So $3x^2 + 2x = \mathbf{x(3x + 2)}$

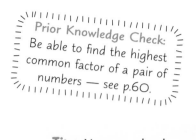

Exercise 1

Q1 Factorise the following expressions.

a) $2a + 10$ b) $3b + 12$ c) $20c + 15$ d) $18 + 12x$

e) $8c + 12f$ f) $25d + 35e$ g) $12x + 16y$ h) $3x + 9y$

Q2 Factorise the following expressions.

a) $3a^2 + 7a$ b) $4b^2 + 19b$ c) $2x^2 + 9x$ d) $7y + 15y^2$

e) $4x^2 - 9x$ f) $21q^2 - 16q$ g) $15y - 7y^2$ h) $27z^2 + 11z$

Sometimes the highest common factor of all the terms might have both **numbers and variables** in it.

It's usually best to work out the highest common factor for the numbers and for the letters **separately**.

Tip: You have factorised fully when the terms left inside the brackets have no common factors.

Example 3

Factorise the expression $15x^2 - 10xy$.

1. The HCF of 15 and 10 is 5, and the HCF of x^2 and xy is x.
2. The HCF of $15x^2$ and $10xy$ is $5x$. So $5x$ goes outside the brackets $\quad 15x^2 - 10xy = 5x(\quad - \quad)$
3. Divide each term by the common factor. $\quad 15x^2 \div 5x = 3x$ and $10xy \div 5x = 2y$
4. Write the results inside the brackets. $\quad 15x^2 - 10xy = \mathbf{5x(3x - 2y)}$

Tip: $15x^2$ doesn't have a y in it, so neither will the HCF.

Exercise 2

Q1 The expression $4xy^2 + 8x^2y$ contains two terms.

a) What is the highest numerical common factor of both terms?

b) What is the highest power of x that is common to both terms?

c) What is the highest power of y that is common to both terms?

d) Factorise the expression.

Q2 Factorise these expressions.

a) $5a^2 + 5a$ b) $4b + 8b^2$ c) $6c^2 - 9c$ d) $12d - 16d^2$

Q3 Factorise these expressions.

a) $10c^2 - 5cd$ b) $20x^2 - 10xy$ c) $9x^2 + 6xy$ d) $12x^2 + 8xy$

e) $6a^2 + 9ab$ f) $12pq - 8p^2$ g) $8a^2 + 6ab^2$ h) $24x^2y - 16xy^2$

Factorising Quadratics

Learning Objective — Spec Ref A4:
Be able to factorise a quadratic expression.

A **quadratic expression** is an expression where the highest power of the variable (e.g. x) is **2**.
They have the form $\boldsymbol{x^2 + bx + c}$ where b and c are numbers.

You can **factorise** some quadratics into the form $\boldsymbol{(x + d)(x + e)}$, but it's usually **not clear** what the values of d and e are.

Tip: If a quadratic can be factorised, only one combination of d and e will work.

Here's the method to find them:

- Write out the brackets as $(x \quad)(x \quad)$ — don't put the signs or numbers in yet.
- Find pairs of numbers that **multiply to give c** — this is just like finding **factors** (see p.53). You can **ignore the sign** of c for now.
- Choose the pair of numbers that also **add** or **subtract** to give $\boldsymbol{b}$ (ignoring signs here too).
- Write one number in each bracket, then fill in the + or – signs so that b and c work out with the **correct signs**. Check they're right by expanding the brackets to get back to the original expression.

If c is **positive**, then the two brackets will have the same sign (both + or both –), and if c is **negative** then the signs will be different (one + and one –).

Example 4

Factorise: a) $x^2 + 7x + 10$ b) $x^2 + 2x - 15$

a)
1. Find all the pairs of numbers that multiply to give 10 .

   1×10 or 2×5
2. Find the pair that add/subtract to give 7.

   $1 + 10 = 11$, $10 - 1 = 9$
   $2 + 5 = 7$ (circled), $5 - 2 = 3$
3. You need +2 and +5 to give +7, so both brackets should have + signs.

   So $x^2 + 7x + 10 = \boldsymbol{(x + 2)(x + 5)}$
4. Check your answer by expanding the brackets.

   $(x + 2)(x + 5) = x^2 + 2x + 5x + 10$
   $= x^2 + 7x + 10$

b)
1. Find all the pairs of numbers that multiply to give 15.

   1×15 or 3×5
2. Find the pair that add/subtract to give 2.

   $1 + 15 = 16$, $15 - 1 = 14$
   $3 + 5 = 8$, $5 - 3 = 2$ (circled)
3. You need +5 and –3 to give +2, so put a + with the 5 and a – with the 3.

   So $x^2 + 2x - 15 = \boldsymbol{(x + 5)(x - 3)}$
4. Check your answer by expanding the brackets.

   $(x + 5)(x - 3) = x^2 - 3x + 5x - 15$
   $= x^2 + 2x - 15$

Exercise 3

Q1 Factorise each of the following expressions.

a) $x^2 + 7x + 6$ b) $x^2 + 7x + 12$ c) $x^2 + 8x + 7$ d) $x^2 + 6x + 9$

e) $x^2 + 6x + 8$ f) $y^2 + 8y + 15$ g) $z^2 + 9z + 14$ h) $v^2 + 11v + 24$

Q2 Factorise each of the following expressions.

a) $x^2 + 2x - 3$ b) $x^2 - 6x + 8$ c) $x^2 - 2x - 8$ d) $x^2 - 5x + 4$

e) $x^2 - 3x - 10$ f) $x^2 + 2x - 8$ g) $r^2 + 6r - 27$ h) $u^2 - 15u + 54$

Difference of Two Squares

Learning Objective — Spec Ref A4:
Be able to factorise a difference of two squares.

Prior Knowledge Check:
Recognise square numbers and find square roots — p.39-40.

Some quadratic expressions have **no middle term** e.g. $x^2 - 49$.
They're of the form 'one thing squared' take away 'another thing squared'.

When you **factorise** these, you get two brackets that are **the same**, except that one has a **+ sign** and one has a **– sign**. For example, $x^2 - 49$ factorises to $(x + 7)(x - 7)$.

The general rule is $\mathbf{a^2 - b^2 = (a + b)(a - b)}$ — this is known as the **difference of two squares**.

Example 5

Factorise: a) $x^2 - 16$ b) $x^2 - 7$

a) Write the expression in the form $a^2 - b^2$, then use the rule given above.

$x^2 - 16 = x^2 - 4^2 = \mathbf{(x + 4)(x - 4)}$

b) 7 isn't a square number, but you can write 7 as $(\sqrt{7})^2$ so that you can use the formula.

$x^2 - 7 = x^2 - (\sqrt{7})^2$
$= \mathbf{(x + \sqrt{7})(x - \sqrt{7})}$

Exercise 4

Q1 Factorise each of the following expressions.

a) $x^2 - 25$ b) $x^2 - 9$ c) $x^2 - 4$ d) $x^2 - 36$

e) $x^2 - 81$ f) $x^2 - 64$ g) $b^2 - 121$ h) $t^2 - 144$

Q2 Factorise each of the following expressions.

a) $x^2 - 5$ b) $x^2 - 3$ c) $x^2 - 11$ d) $x^2 - y^2$

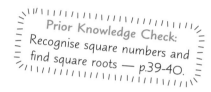

Review Exercise

Q1 Simplify these expressions by collecting like terms.

a) $4s + s + 2s + 7s$

b) $5m - 2m + 8m - 6m$

c) $x + y + x + y + x - y$

d) $16p + 4q + 4 - 2p + 3q - 8$

e) $5s + 7t^2 - 3s^2 + 2s + 2t^2 - s$

f) $6b^2 + 7b + 9 - 4b^2 + 5b - 2$

Q2 Simplify the following expressions.

a) $a \times ab$

b) $4a^2 \div 2a$

c) $2p \times 7q^2$

d) $12e^3 \div 4e$

e) $-3i^2 \times 8i^3$

f) $15d^4 \div 5d^2$

Q3 Expand the brackets in these expressions.

a) $4(x + 8)$

b) $6(5 - r)$

c) $-2(7 + y)$

d) $8(h - 2)$

e) $h(h + 3)$

f) $-4n(n + 2)$

g) $4w(u - 7)$

h) $-2x(12 - v)$

Q4 Simplify the following expressions.

a) $4(c + 3) + 6(c + 2)$

b) $5(u + 4) + 3(u + 8)$

c) $5(b - 6) + 7(b + 4)$

d) $2(c - 6) - (c + 5)$

e) $5(q - 3) - (q + 1)$

f) $2(j - 5) - (j - 3)$

Q5 Expand and simplify the brackets in the following expressions.

a) $(2 + x)(8 + y)$

b) $(x - 3)(x - 5)$

c) $3(j - 2)(k + 4)$

d) $(n + 5)(m - 4)$

e) $(3 + r)(s + 4)$

f) $(z - 8)^2$

Q6 Factorise these expressions.

a) $2x + 4y$

b) $8x + 24$

c) $4y - 6y^2$

d) $20y + 12xy$

e) $15ab - 10a^2$

f) $60x + 144y$

g) $28r + 40r^2s$

h) $4ab - 8a^2b$

i) $14m^2n - 35mn^2$

Q7 Factorise each of the following expressions.

a) $y^2 + 10y + 21$

b) $x^2 - 4x - 5$

c) $t^2 - 8t + 16$

d) $x^2 - x - 12$

e) $x^2 + 5x - 6$

f) $x^2 - 4x - 45$

Q8 Factorise each of the following expressions.

a) $x^2 - 100$

b) $y^2 - 36$

c) $y^2 - 121$

Exam-Style Questions

Q1 Complete this table.

Expression	Simplified expression
$a + a + a$	
$3a^2 + 2a^2$	
$4a^2 + 6a - a^2$	

[3 marks]

Q2 Fully simplify the following expressions:

a) $v + v + w - w - w - w - w$

[1 mark]

b) $5 \times 7 \times x \times x$

[1 mark]

c) $6y - 3yz - 8y + 10yz$

[1 mark]

Q3 Expand and simplify:

a) $x(x^2 - 4y) + 9xy$

[2 marks]

b) $(2x - 7)^2$

[2 marks]

Q4 Factorise the following expressions:

a) $7c + 56$

[1 mark]

b) $d^2 + 5d - 2de$

[2 marks]

Q5 a) Factorise $y^2 + 12y + 32$

[2 marks]

b) Factorise $x^2 - 169$

[2 marks]

9.1 Solving Equations

Solving an equation means finding the value of an unknown letter that makes both sides equal. For example, the solution to $2x + 3 = 11$ is $x = 4$ — because if $x = 4$, both sides equal 11.

Basic Equations

Learning Objective — Spec Ref A17:
Be able to solve algebraic equations.

Prior Knowledge Check:
Be able to simplify algebraic expressions — see p.124.

You can solve some simpler equations using the '**common sense**' approach.

E.g. Start with the equation: $9 + x = 12$
This means: $9 + \text{something} = 12$
And you know that: $9 + 3 = 12$
So the 'something' must equal 3, which means $\boldsymbol{x = 3}$.

For harder equations, the 'common sense' approach isn't very useful. Instead, you'll need to solve most equations by **rearranging** them, until you end up with '$x =$ ' on one side.

When you're rearranging, remember these rules:

- Whatever you do to **one side** of the equation you **must** do to the **other side**.
- To get rid of something, do the **opposite**:
  + and – are opposites.
  × and ÷ are opposites.
- Keep rearranging the equation until x is **on its own**.

Tip: Remember, $2x$ means $2 \times x$. So to rearrange something like $2x = 12$, you do the opposite of × which is ÷. So ÷ both sides by 2 to get $x = 6$.

Example 1

Solve the equation $x + 8 = 15$.

1. You need to get rid of the '+ 8' to get x on its own.
   $x + 8 = 15$
2. The opposite of + is –, so subtract 8 from both sides.
   $x + 8 - 8 = 15 - 8$
   $x = 15 - 8$
   $\boldsymbol{x = 7}$
3. Check your answer by putting it in the original equation and checking both sides are the same.
   $x + 8 = 7 + 8 = 15$ ✔

Tip: It's really important to check your answer at the end. It's very easy to make a silly mistake when rearranging.

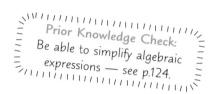

When there is a **minus sign** in front of the x, get rid of it by **adding x** to both sides.

Example 2

Solve the equation $15 - x = 7$.

1. Add x to both sides of the equation
2. Now that x is positive, you can solve the equation as usual.

$$15 - x = 7$$
$$15 = x + 7$$
$$15 - 7 = x$$
$$8 = x \Rightarrow \mathbf{x = 8}$$

Tip: It doesn't matter which side the x ends up on, as long as it's on its own.

Exercise 1

Q1 Get x alone on one side of the equation to solve these equations.

a) $x + 9 = 12$ b) $x + 5 = 16$ c) $x - 2 = 14$

d) $x - 7 = -19$ e) $-2 = 7 + x$ f) $32 = x - 17$

Q2 Solve these equations. Start by adding x to both sides.

a) $12 - x = 9$ b) $4 - x = 2$ c) $2 - x = 7$

d) $19 = 14 - x$ e) $14 = 8 - x$ f) $7 = 5 - x$

Q3 Solve the following equations.

a) $x + 7 = 12$ b) $5 - x = 21$ c) $16 = x + 10$

d) $x - 8 = 14$ e) $12 - x = 23$ f) $35 = 31 - x$

Q4 Solve the following equations.

a) $x - 7.3 = 1.6$ b) $6.03 - x = 0.58$ c) $3.47 = 7.18 + x$ d) $5.2 = 2.8 - x$

If one side of the equation has something like $5x$ or $-4x$, you **divide** to get x on its own.
If it's something like $\frac{x}{7}$ or $-\frac{x}{3}$, you **multiply**.

Example 3

Solve the equation $\frac{x}{5} = 3$.

1. Multiply both sides by 5.
2. Remember: $\frac{x}{5} \times 5 = x$

$$\frac{x}{5} = 3 \longrightarrow \frac{x}{5} \times 5 = 3 \times 5$$
$$\mathbf{x = 15}$$

Sometimes you'll need to multiply or divide by a **negative** number — make sure you use the rules for negative numbers given on p.6.

Example 4

Solve the equation $-6x = 9$.

1. Divide both sides by –6. $-6x = 9 \longrightarrow -6x \div (-6) = 9 \div (-6)$
2. Remember: $-6x \div (-6) = x$ $\mathbf{x = -1.5}$

Exercise 2

Q1 Solve the following equations.

a) $\frac{x}{3} = 2$ b) $\frac{x}{6} = -3$ c) $\frac{x}{3} = 0.4$ d) $\frac{x}{11} = -0.5$

e) $8x = 24$ f) $4.5x = 81$ g) $5x = -20$ h) $3.5x = -7$

Q2 Solve these equations by multiplying or dividing both sides by a negative number.

a) $-7x = -56$ b) $-9x = 108$ c) $-4.5x = -2.7$

d) $-\frac{x}{4} = 3$ e) $-\frac{x}{5} = 6$ f) $-\frac{x}{10} = 1.1$

Sometimes the x term is being **both** multiplied and divided.
Treat the multiplication and division **separately**, and rearrange the equation one step at a time.

Example 5

Solve the equation $\frac{3x}{4} = 6$.

1. Multiply both sides by 4. $\frac{3x}{4} = 6 \Rightarrow 3x = 6 \times 4 = 24$
2. Then divide both sides by 3. $\Rightarrow x = 24 \div 3 \Rightarrow \mathbf{x = 8}$

Tip: You could have divided by 3 first and then multiplied by 4.

Exercise 3

Q1 Solve the following equations.

a) $\frac{4x}{3} = 12$ b) $\frac{2x}{5} = 6$ c) $\frac{6x}{7} = 12$ d) $\frac{5x}{6} = 3$

e) $\frac{7x}{5} = 1.4$ f) $\frac{2x}{1.5} = -0.2$ g) $\frac{3x}{0.1} = -0.6$ h) $\frac{11x}{2.2} = 6.3$

Two-Step Equations

Learning Objective — Spec Ref A17:
Be able to solve equations that involve more than one step.

Prior Knowledge Check: Be able to solve basic equations — see p.137.

'Two-step equations' are equations that look something like $3x - 2 = 10$ or $15 - 4x = 3$. They need to be solved in **two stages**, which you need to do in the **right order**:

- Addition or subtraction **first**
- Multiplication or division **second**.

Following these rules avoids dealing with too many **fractions** and makes solving much **easier**.

Example 6

Solve the equation $2x + 3 = 11$.

1. $2x + 3$ means: (i) multiply your value of x by 2, (ii) add 3 to it. To get x on its own, 'undo' these steps, but in the opposite order. $2x + 3 = 11$
2. First, subtract 3. $2x = 11 - 3 = 8$
3. Then divide by 2 to find the answer. $x = 8 \div 2 \Rightarrow \boldsymbol{x = 4}$
4. Check your answer. $2 \times 4 + 3 = 8 + 3 = 11$ ✔

Exercise 4

Q1 Solve the following two-step equations.

a) $8x + 10 = 66$ b) $10x + 15 = 115$ c) $12x + 9 = 105$
d) $1.5x + 3 = 93$ e) $4x + 12 = -8$ f) $2x + 9 = -2$

Q2 Solve the following equations.

a) $16x - 6 = 10$ b) $15x - 8 = 22$ c) $14x - 17 = 25$
d) $2.6x - 7 = -59$ e) $18x - 6 = -60$ f) $20x - 12 = -132$

Q3 The expression $\frac{x}{2} - 1$ means 'divide x by 2, then subtract 1'.

'Undo' these two steps in the opposite order to solve this equation: $\frac{x}{2} - 1 = 3$

Q4 Solve the following equations.

a) $\frac{x}{2} + 1 = 7$ b) $\frac{x}{6} + 4 = 16$ c) $\frac{x}{10} - 3 = -1$
d) $\frac{x}{4} - 5 = -9$ e) $\frac{x}{2} - 1 = 3.5$ f) $\frac{x}{7} - 8 = -11$

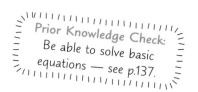

Q5 a) Write down the equation you get if you add $5x$ to both sides of the equation $20 - 5x = 10$.
b) Solve your equation to find x.

Q6 Solve the following equations.
a) $12 - 4x = 8$ b) $47 - 9x = 11$ c) $8 - 7x = 22$ d) $17 - 10x = 107$

Harder Equations

Learning Objectives — Spec Ref A17:
- Be able to solve equations involving brackets.
- Be able to solve equations where the unknown appears twice.

Some equations involve **brackets** — you need to **get rid** of the brackets before you can solve the equation. There are **two ways** to do this:

- **expand** the brackets
- **divide** both sides by the number in front of the brackets.

Tip: See p.127-130 for how to expand brackets.

You can use whichever method you find **easier** as both will give you the same answer. Both methods are shown in the example below.

Example 7

Solve the equation $8(x + 2) = 36$.

Either:

1. Multiply out the brackets. $8x + 16 = 36$
2. Subtract 16. $8x = 36 - 16 = 20$
3. And then divide by 8. $x = 20 \div 8 = 2.5$

$\mathbf{x = 2.5}$

Or:

1. $8(x + 2)$ means: "add 2 to x, then multiply by 8". $8(x + 2) = 36$
2. So to find x, first divide by 8. $x + 2 = 36 \div 8 = 4.5$
3. And then subtract 2. $x = 4.5 - 2 = 2.5$

$\mathbf{x = 2.5}$

Exercise 5

Q1 Solve the following equations.

a) $7(x + 4) = 63$ b) $8(x + 4) = 88$ c) $11(x + 3) = 132$

d) $16(x - 3) = -80$ e) $13(x - 4) = -91$ f) $14(x - 2) = -98$

Q2 Solve the following equations.

a) $315 = 21(6 - x)$ b) $12.5(x - 4) = 75$ c) $36 = 7.2(2 - x)$

Some equations will have an x term on **both sides** of the equation.
To tackle these, first rearrange the equation so you have all the ***x*-terms** on **one side** and all the **numbers** on the **other side**. You can then solve them in the usual way.

Example 8

Solve the equation $5x + 6 = 2x + 18$.

1. First subtract $2x$ from both sides, which leaves x terms on only one side of the equation.

$$5x + 6 = 2x + 18$$
$$5x + 6 - 2x = 2x + 18 - 2x$$
$$3x + 6 = 18$$

2. Now you can solve the equation. Subtract 6 from both sides... ...and then divide by 3.

$$3x = 18 - 6 = 12$$
$$x = 12 \div 3 \Rightarrow \mathbf{x = 4}$$

3. Plug your answer back into the equation to check — you should get the same number on both sides.

Check: $(5 \times 4) + 6 = 26$
$(2 \times 4) + 18 = 26$ ✔

Exercise 6

Q1 Solve these equations.

a) $6x - 4 = 2x + 16$ b) $17x + 2 = 7x - 8$ c) $9x - 26 = 5x - 14$

d) $8x + 4 = 2x + 40$ e) $15x - 8 = 4x + 47$ f) $21x - 5 = 5x + 11$

Q2 Solve the following equations.

a) $13x - 35 = 45 - 3x$ b) $20x + 18 = 54 - 16x$ c) $17x - 9 = 57 - 5x$

d) $82 - 8x = 10 - 6x$ e) $33 - 7x = -12 - 2x$ f) $4x - 15 = 147 - 14x$

Q3 Solve the following equations.

a) $4x - 3 = 0.5 - 3x$ b) $10x - 18 = 10.2 + 4x$

c) $4x - 8.6 = 48.1 - 5x$ d) $-x + 1 = 28 + 2x$

You might get brackets on **both sides** of the equation.
Expand both brackets before doing any other rearranging.

Example 9

Solve the equation $4(x + 2) = 2(x + 6)$.

1. Multiply out both of the brackets.
2. Now you can solve the equation in the usual way.
3. Remember to check your answer.

$4(x + 2) = 2(x + 6)$
$4x + 8 = 2x + 12$
$2x + 8 = 12$
$2x = 4 \Rightarrow \boldsymbol{x = 2}$
$4 \times (2 + 2) = 16$ and $2 \times (2 + 6) = 16$ ✔

Exercise 7

Q1 Solve the following equations by first multiplying out the brackets.

a) $3(x + 2) = x + 14$ b) $9(x - 1) = x + 15$ c) $6(x + 2) = 3x + 48$
d) $8(x - 8) = 2(x - 2)$ e) $4(4 - x) = 2(x - 1)$ f) $20(x - 2) = 5(x + 1)$

Q2 Solve the following equations.

a) $5(x - 5) = 2(x - 14)$ b) $2(x - 2) = 5(x - 8)$ c) $4(x - 2) = 6(x + 3)$
d) $6(x - 1.5) = 2(x - 3.5)$ e) $9(x - 3.3) = -6(x + 1.7)$ f) $-4(x - 3) = 8(0.7 - x)$

Q3 Solve the following equations.

a) $7(3x + 2) = 5(9x - 0.08)$ b) $7(2x + \frac{1}{7}) = 14(3x - 0.5)$
c) $10(x - 2) = -2(\frac{4}{3} + 7x)$ d) $4(3x - 3) = -2(\frac{76}{9} + 5x)$

Most of the time it's better to get rid of a **fraction** before starting to rearrange anything else. You do this by multiplying **both sides** of the equation by the **denominator** of the fraction.

Example 10

Solve the equation $\frac{x}{3} = 7 - 2x$.

1. Multiply both sides by 3.
2. Then solve in the normal way.

$\frac{x}{3} = 7 - 2x \Rightarrow x = 3(7 - 2x)$
$x = 21 - 6x$
$7x = 21 \Rightarrow \boldsymbol{x = 3}$

Tip: If you tried to rearrange by adding $2x$ first you would have $2x + \frac{x}{3}$, which makes things messier.

Exercise 8

Q1 Solve the following equations. Start by multiplying both sides by a number.

a) $\frac{x}{4} = 1 - x$ b) $\frac{x}{3} = 8 - x$ c) $\frac{x}{5} = 11 - 2x$

d) $\frac{x}{4} = 10 - x$ e) $\frac{x}{5} = x + 4$ f) $\frac{x}{5} = -22 - 2x$

Q2 Solve the following equations.

a) $\frac{x}{3} = 2(x - 5)$ b) $\frac{x}{2} = 4(x - 7)$ c) $\frac{x}{5} = 2(x + 9)$

d) $\frac{x}{2} = 2(x + 5)$ e) $\frac{x}{4} = -2(x + 18)$ f) $\frac{x}{4} = 3(x - 55)$

If there's a **fraction** on **both sides** of the equation, there's a trick known as **cross-multiplying** to get rid of both fractions at the **same time**.

You need to multiply the **numerator** of each fraction by the **denominator** of the other.

$\frac{a}{b} = \frac{c}{d}$ becomes $a \times d = c \times b$

Example 11

Solve the equation $\frac{x-2}{2} = \frac{6-x}{6}$.

1. Cross-multiply — multiply the top of each fraction by the bottom of the other.
2. Then solve in the normal way. Expand the brackets...
   ... get x terms on one side...
   ... and divide by 8.

$$\frac{x-2}{2} = \frac{6-x}{6}$$

$$6(x - 2) = 2(6 - x)$$

$$6x - 12 = 12 - 2x$$

$$8x = 24$$

$$\mathbf{x = 3}$$

Tip: Step 1 is the same as multiplying both sides by 2 and then multiplying both sides by 6.

Exercise 9

Q1 Solve the following equations.

a) $\frac{x+4}{2} = \frac{x+10}{3}$ b) $\frac{x+2}{2} = \frac{x+4}{6}$ c) $\frac{x+3}{4} = \frac{x+9}{7}$

d) $\frac{x-2}{3} = \frac{x+4}{5}$ e) $\frac{x-3}{4} = \frac{x+2}{8}$ f) $\frac{x-6}{5} = \frac{x+3}{8}$

Q2 Solve the following equations.

a) $\frac{x-6}{2} = \frac{8-2x}{4}$ b) $\frac{x-9}{2} = \frac{2-3x}{4}$ c) $\frac{x-12}{6} = \frac{4-2x}{3}$

9.2 Forming Your Own Equations

Some problems in the real world can be solved by forming an equation and then solving it. You'll need to be able to write an equation using the information from a wordy problem.

Word Problems

Learning Objectives — Spec Ref A21:
- Be able to set up an algebraic equation for a given situation.
- Know how to interpret the solution of an algebraic equation in context.

Sometimes you'll need to **write** your own equation based on a **description** of a situation.

- Read the question **very carefully** so you don't **miss** any information.
- Call the **unknown** quantity in the situation 'x'.
- Make an equation in **terms of x** by using the information in the **question**.
  For example, if someone has **3 more** than the unknown, this would be $x + 3$,
  if there is **twice as much**, this would be $2x$, etc.
- **Simplify** your equation as much as possible, then **solve** it using the usual methods.

Example 1

I think of a number, double it, and add 3. The result equals 17.
What is the number I thought of?

1. You don't know what the number is yet. — Call the number x.
2. Doubling x gives $2x$. Then adding 3 gives $2x + 3$. The result is 17. → Then: $2x + 3 = 17$
3. Solve the equation in the normal way. → $2x = 17 - 3 = 14$ $\Rightarrow$ $\mathbf{x = 7}$

Tip: Remember to check your answer.

Exercise 1

PROBLEM SOLVING

Q1 Which number did I think of in each situation below?
"I think of a number, and then..."

a) ...I add 5 to it. The result equals 12.

b) ...I multiply it by 2, and then subtract 5. The result equals 15.

c) ...I divide it by 4, and then add 10. The result equals 14.

Example 2

One day a furniture store sells 3 times as many leather sofas as it does fabric sofas. On this day it sells a total of 28 sofas. How many leather sofas were sold?

1. Call the number of fabric sofas sold x. Then the number of leather sofas sold is $3x$.
2. 28 sofas are sold in total, so use this to write an equation. $x + 3x = 28$, $4x = 28$
3. Solve the equation to find x. $x = 28 \div 4 = 7$
4. Multiply by 3 to find the number of leather sofas sold. $3x = 3 \times 7 = 21$
   21 leather sofas were sold.

Exercise 2

PROBLEM SOLVING

Q1 A bride and groom each invite an equal number of guests to their wedding. All of the groom's guests are able to come, but 8 of the bride's guests can't come. If a total of 60 guests can attend, how many guests did the groom invite?

Q2 One day a bed shop sells a total of 54 beds. 10 fewer king-size beds were sold than single beds, and 16 more double beds were sold than single beds. How many single beds were sold?

Q3 Rufina makes and sells 20 fruit scones and 10 cheese scones. The ingredients cost £y in total. She sells each fruit scone for £x and each cheese scone for 10p more than a fruit scone.

a) Write an expression in terms of x and y for the profit, in £, that she makes.

b) If she spent £6 on ingredients and made a profit of £10, what was the selling price of a fruit scone?

Shape Problems

Learning Objective — Spec Ref A21:
Be able to form and solve equations using known properties of shapes.

You might have to use the **properties of shapes** to form an equation, which you can then use to find side lengths or angles. You'll need to use your knowledge of:

- **Angle rules** of shapes, e.g. sum of angles in a triangle = 180°.
- The formulas for **areas** of different shapes.
- Finding the **perimeter** of different shapes.

Tip: These are covered in Section 15 and Section 22.

Example 3

The perimeter of this rectangle is 78 cm. Write an equation involving x to show this. Then solve your equation to find x.

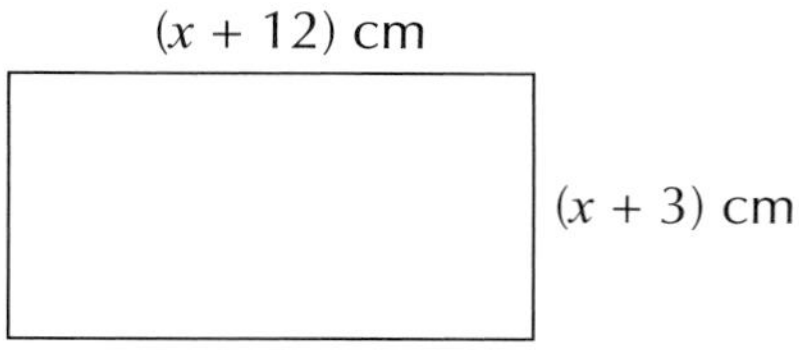

1. Find the perimeter by adding together the lengths of all the sides. This must equal 78.

   Perimeter $= (x + 12) + (x + 3) + (x + 12) + (x + 3)$
   $= 4x + 30$
   So $4x + 30 = 78$.

2. Solve the equation in the normal way.

   $4x = 78 - 30 = 48$
   $x = 48 \div 4$
   $\boldsymbol{x = 12}$

Tip: Check your answer works in the context — e.g. length must be positive.

Example 4

Use the triangle to write an equation involving x. Solve your equation to find x.

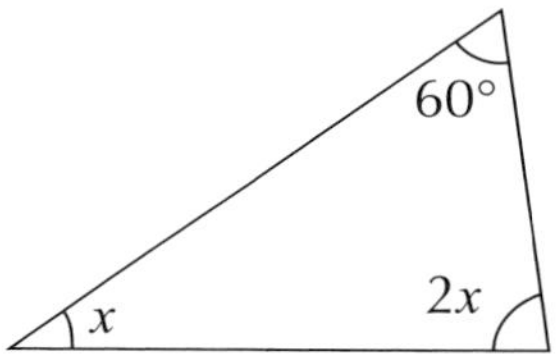

1. The angles in a triangle always add up to 180°.
   $x + 2x + 60° = 180°$
2. Simplify your equation.
   $3x + 60° = 180°$
3. Solve the equation in the normal way.
   $3x = 180° - 60° = 120°$
   $x = 120° \div 3$
   $\boldsymbol{x = 40°}$

Exercise 3

Q1 For the triangle on the right:

a) Write an equation involving x. (All the angles are measured in degrees.)

b) Solve your equation to find x.

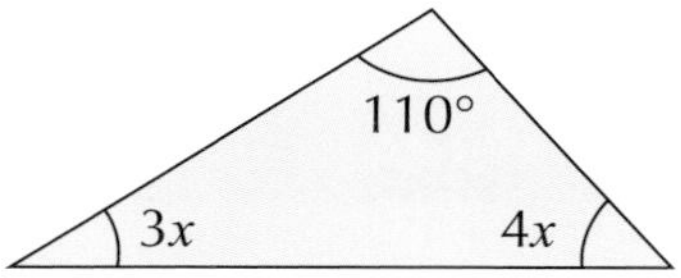

Q2 For each shape below:

(i) Write an equation involving x.

(ii) Solve your equation to find x.

a)

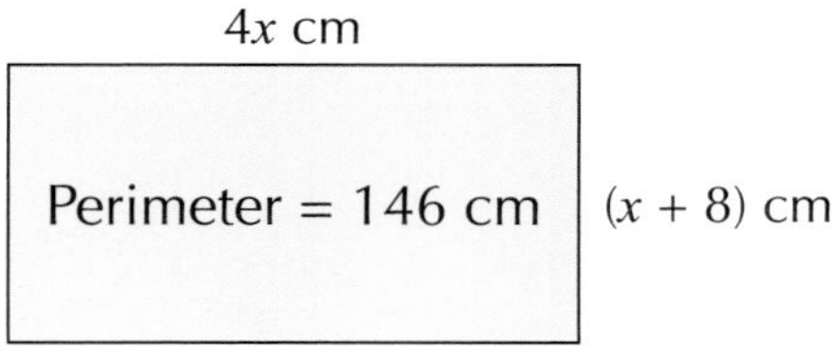

b)

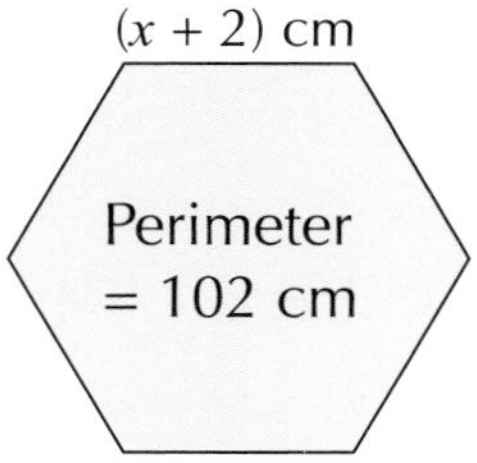

9.3 Identities

Identities look like equations with an extra line, but there's an important difference — equations are only true for a particular value (or values) of x, but identities are always true, no matter what.

Learning Objective — Spec Ref A3/A6:
Understand the difference between equations and identities.

An **equation** is a way of showing that two expressions are equal for some particular values of an unknown.

Identities are like equations, but are **always true**, for **any value** of the unknown. Identities have the symbol '≡' instead of '='.

E.g. $x - 1 = 2$ is an equation — it's only true when $x = 3$.
$x + 1 \equiv 1 + x$ is an identity — it's always true, whatever the value of x.

In identity questions, you should **rearrange** the expressions on either side **separately** to see if they're the **same**.

You **don't** need to take things to the other side, like you would if you were solving an equation.

Example 1

Show that $(x + 6)(x - 3) \equiv x(x + 3) - 18$

1. Expand and simplify both sides of the identity separately.

LHS: $(x + 6)(x - 3) = x^2 - 3x + 6x - 18$
$= x^2 + 3x - 18$

RHS: $x(x + 3) - 18 = x^2 + 3x - 18$

2. Both sides give the same expression.

So $(x + 6)(x - 3) \equiv x(x + 3) - 18$

Tip: LHS means "left-hand side" and RHS is "right-hand side".

Exercise 1

Q1 For each of the following, state whether or not you could replace the box with the symbol '≡'.

a) $4x \square 10$

b) $x^2 + 2x + 1 \square 0$

c) $-x^2 + 3 \square 3 - x^2$

d) $2(x + 1) \square x - 1$

e) $3(x + 2) - x \square 2(x + 3)$

f) $3(2 - 3x) + 2 \square 8x$

Q2 Show that:

a) $(x + 4)^2 - 4 \equiv (x + 6)(x + 2)$

b) $5(x + 2) + (x^2 - 4) \equiv (x + 4)(x + 1) + 2$

9.4 Proof

Proof questions can be confusing — it's not always obvious where to start and where you want to end up. Don't worry though, there are some handy tricks you can use to help you to get started.

Learning Objective — Spec Ref A6:
Be able to show that mathematical statements are true or false.

Proof is all about showing that something is **true or false**.

You can use these **facts** to make proof questions much easier:

- Any **even number** can be written as $\mathbf{2n}$ — i.e. as "2 × an integer".
- Any **odd number** can be written as $\mathbf{2n + 1}$ — i.e. as "(2 × an integer) + 1".
- **Consecutive numbers** can be written as $\mathbf{n, (n + 1), (n + 2)}$, etc.
- **Consecutive even** numbers are written as $\mathbf{2n, (2n + 2), (2n + 4)}$, etc. and **consecutive odd** numbers as $\mathbf{(2n + 1), (2n + 3), (2n + 5)}$, etc. **Nonconsecutive numbers** need different letters, such as $\mathbf{a, b, c}$ etc.
- The **sum**, **difference** and **product** of integers is **always** an integer.

Tip: n is a common letter used for integers (whole numbers), but you can use any letter you like.

Example 1

Show that the product of two odd numbers is odd.

1. Take two odd numbers. — $2a + 1$ and $2b + 1$, where a and b are integers.
2. Multiply them together and rearrange into the form $2n + 1$ (where n is an integer). —

$$(2a + 1)(2b + 1) = 4ab + 2a + 2b + 1$$
$$= 2(2ab + a + b) + 1 = 2n + 1$$

where $n = (2ab + a + b)$ is an integer

So the **product of two odd numbers is odd.**

Exercise 1

Q1 Show that the product of an even number and an odd number is always even.

Q2 Show that the sum of two consecutive square numbers is always odd.

To show that a statement is **false**, you can just find an **example** where it **doesn't work**. This is called a **counter example**.

There are usually lots of counter examples you could give, but you only need to find **one**. Make sure you show clearly **why** the statement doesn't work for your example.

Example 2

Show that the following statement is false by finding a counter example:
"The difference between two consecutive square numbers is always prime."

Try consecutive square numbers until you find a pair that doesn't work:

1 and 4 — difference = 3 (prime)
4 and 9 — difference = 5 (prime)
9 and 16 — difference = 7 (prime)
16 and 25 — difference = 9 (NOT prime), so the **statement is false.**

Tip: You don't have to go through loads of examples if you can spot one that's wrong straightaway.

Exercise 2

Show that the statements in Q1 and Q2 are false by finding a suitable counter example for each.

Q1 "The sum of three consecutive integers is always bigger than each individual number."

Q2 "The difference between any two prime numbers is always an even number."

To show that something is a **multiple** of a particular number, you need to write it as that number multiplied by an integer. You'll often need to **factorise** to show this.

E.g. if you wanted to show that something was a multiple of 4, you'd need to be able to write it as $\mathbf{4n}$, where n is some integer. If you **can't** write it in this form then it is **not** a multiple of 4.

Example 3

Show that the sum of three consecutive integers is a multiple of 3.

1. Take three consecutive integers.
   n, $n + 1$ and $n + 2$, where n is an integer.
2. Add them together and factorise into the form 3 × integer.
   $n + n + 1 + n + 2 = 3n + 3 = 3(n + 1)$
   where $n + 1$ is an integer
   So the sum of three consecutive numbers is a **multiple of 3.**

Exercise 3

Q1 Show that the sum of two consecutive odd integers is a multiple of 4.

Q2 Let $x = 2(y + 5) + 4(y + 1) - 2$, where y is an integer. Show that x is a multiple of 6.

9.5 Inequalities

Inequality symbols can be used to compare numbers or to compare algebraic expressions. They can also be used to describe the range of values that a variable can take (e.g. $x < 4$).

Inequalities

Learning Objectives — Spec Ref A3/A22:

- Understand and be able to use the four inequality symbols.
- Know how to represent inequalities on a number line.

Prior Knowledge Check: Be able to solve linear equations — see pages 137-144.

You write inequalities using these symbols:

> greater than **< less than** **≥ greater than or equal to** **≤ less than or equal to**

Greater than (>) and less than (<) are called **strict inequalities**.

You can show inequalities **visually** on a **number line**, using a **circle** to show the **boundary value** and an **arrow** to show which values are part of the solution.

Example 1

Show the following inequalities on a number line: a) $x > 1$ b) $x \leq 1$

a) The empty circle shows 1 is not included, as $x = 1$ does not make $x > 1$ true.

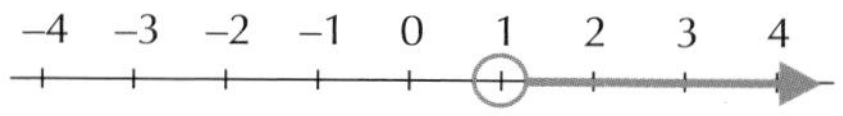

b) The solid circle shows 1 is included, as $x = 1$ does make $x \leq 1$ true.

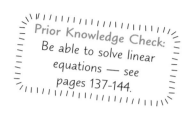

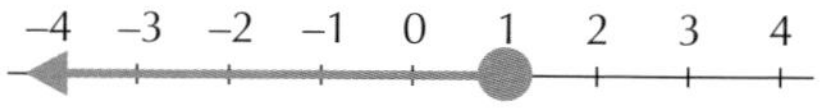

Exercise 1

Q1 Insert > or < in each of the boxes to complete the following inequalities.

a) 6 ☐ 1 b) 2 ☐ 8 c) –1 ☐ –3 d) –7 ☐ 1

Q2 Describe in words what is meant by the following inequalities.

a) $x \geq 1$ b) $x < 7$ c) $x > -4$ d) $x \leq 9$

Q3 Write each of the following as an inequality.

a) x is greater than 4 b) x is less than or equal to 12 c) x is less than 3

Q4 Show the following inequalities on a number line.

a) $x \geq 12$ b) $x < 22$ c) $x > -6$ d) $x \leq -3$

Algebra with Inequalities

Learning Objective — Spec Ref A3/A22:
Be able to solve inequalities using algebra.

The rules for solving inequalities are **very similar** to the rules for solving equations. The **solution** to an inequality will usually be an inequality with x on one side and a **number** on the other.

Tip: You have to be careful if you need to multiply or divide by a negative number — see next page.

Example 2

Solve the inequalities: a) $x + 4 < 8$ b) $2 \leq x - 7$
Show your solutions on a number line.

1. You need to get x on its own on one side. But you must always do the same to both sides of an inequality.

a) $x + 4 < 8$

$x + 4 - 4 < 8 - 4$

$\boldsymbol{x < 4}$

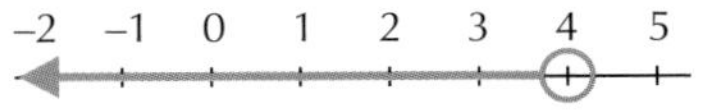

2. If the question uses < or >, so will your answer. If the question uses ≤ or ≥, so will your answer.

b) $2 \leq x - 7$

$2 + 7 \leq x - 7 + 7$

$9 \leq x \Rightarrow \boldsymbol{x \geq 9}$

Exercise 2

Q1 Solve the following inequalities. Show each of your solutions on a number line.
a) $x + 9 > 14$ b) $x + 3 \leq 12$ c) $x - 5 < -3$ d) $x + 1 \leq -1$

Q2 Solve the following inequalities.
a) $x - 9 > 8$ b) $x + 7 < 17$ c) $x + 12 < -18$ d) $x - 8 \leq -3$

Q3 This question is about the inequality $6 > x$.
a) Describe in words what is meant by this inequality.
b) Rewrite the inequality by completing the following: x ☐ 6
c) Show the solution to this inequality on a number line.

Q4 Rewrite each of these inequalities with x on the left-hand side.
a) $12 \geq x$ b) $4 < x$ c) $15 \leq x$ d) $14 > x$

Q5 Solve the following inequalities. Show each of your solutions on a number line.
a) $18 < x + 2$ b) $12 \leq x - 4$ c) $1 > x - 17$ d) $31 \geq x + 30$

If you need to **multiply or divide** to solve an inequality, you can solve it the same way as you would an equation — as long as you are multiplying or dividing by a **positive** number.

Example 3

Solve the following inequalities: a) $4x < 12$ b) $\frac{x}{3} \geq 5$

1. Do the same to both sides of an inequality.
2. You are only multiplying or dividing by positive numbers so you can just solve as usual.

a)

$$4x < 12$$
$$4x \div 4 < 12 \div 4$$
$$\mathbf{x < 3}$$

b)

$$\frac{x}{3} \geq 5$$
$$\frac{x}{3} \times 3 \geq 5 \times 3$$
$$\mathbf{x \geq 15}$$

If you're **multiplying** or **dividing** by a **negative number** then the inequality sign **'flips over'**.

- '<' turns into '>' and '>' turns into '<'.
- '≤' turns into '≥' and '≥' turns into '≤'.

Tip: This is the most important rule to remember when it comes to inequalities.

Example 4

Solve the following inequalities: a) $-3x < 9$ b) $-\frac{x}{4} \geq 8$

1. Do the same to both sides of an inequality.
2. Multiplying or dividing by a negative number changes the direction of the inequality sign.

a)

$$-3x < 9$$
$$-3x \div -3 > 9 \div -3$$
$$\mathbf{x > -3}$$

b)

$$-\frac{x}{4} \geq 8$$
$$-\frac{x}{4} \times -4 \leq 8 \times -4$$
$$\mathbf{x \leq -32}$$

Exercise 3

Q1 Solve the following inequalities.

a) $3x \geq 9$ b) $5x < -25$ c) $2x > 8$ d) $7x \leq 21$

Q2 Solve the following inequalities.

a) $\frac{x}{2} \geq 3$ b) $\frac{x}{5} < 2$ c) $\frac{x}{5.5} < 1.2$ d) $\frac{x}{2.5} > -3.2$

Q3 Solve the following inequalities.

a) $-4x < -16$ b) $-9x > -72$ c) $-11x \leq 33$ d) $-2x < 45$

Q4 Solve the following inequalities.

a) $-\frac{x}{3} < 8$ b) $-\frac{x}{5} \leq -4$ c) $-\frac{x}{1.1} \geq 10$ d) $-\frac{x}{0.2} > -2.1$

You can solve harder inequalities in the **same way** as you solved harder equations on page 141.

Example 5

Solve the following inequality: $-\frac{x+4}{2} \leq 3$

1. First multiply by –2 and switch the inequality sign.
2. Then subtract 4.

$$-\frac{x+4}{2} \leq 3$$
$$x + 4 \geq 3 \times -2$$
$$x + 4 \geq -6$$
$$x \geq -6 - 4$$
$$\mathbf{x \geq -10}$$

Tip: Remember to switch the inequality sign if you multiply or divide by a negative number.

Exercise 4

Solve each of the inequalities in Questions 1-3.

Q1 a) $7x - 12 > 65$ b) $2x + 16 \geq -8$ c) $-8x - 4.2 < 12.6$ d) $4x + 2.6 \leq 28.6$

Q2 a) $\frac{x+2}{3} < 1$ b) $\frac{x-8}{2} > 7$ c) $\frac{x+4}{5} \geq 2$ d) $-\frac{x-6}{4} \leq 0.5$

Q3 a) $\frac{x}{4} - 2.5 \geq 1$ b) $\frac{x}{2} + 5.5 > 7$ c) $-\frac{x}{8} - 3.1 < -1$ d) $\frac{x}{3.2} + 1.3 \leq 5$

Compound Inequalities

Learning Objectives — Spec Ref A3/A22:
- Be able to solve compound inequalities.
- Know how to represent compound inequalities on a number line.

A **compound inequality** combines multiple inequalities into one. For example, $3 < x \leq 9$ means that $x > 3$ **and** $x \leq 9$ — so if x is an integer (whole number), the solutions are 4, 5, 6, 7, 8 and 9.

To solve a compound inequality, you can just **split it up** into two **simple inequalities** and solve each one separately. Then **combine** your solutions back into one inequality at the **end**.

Just like for simple inequalities, you can give a solution using a **number line** The only difference is that this time the solution will be shown by **two circles** with a line between them. E.g. If the solution is $-2 < x \leq 1$, then on a number line it would be:

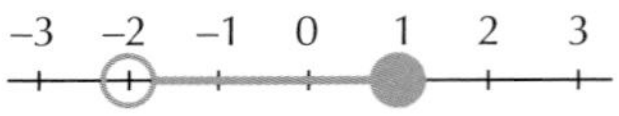

Example 6

Show the inequality $2 \le x < 4$ on a number line.

1. Write down the two separate inequalities. — $2 \le x < 4$ means $2 \le x$ and $x < 4$.
2. Find the number x is greater than... ...and the number x is less than. — $2 \le x$ is the same as $x \ge 2$, so $x \ge 2$ and $x < 4$.
3. Draw the number line. Here, x is between 2 and 4 (including 2, but not including 4).

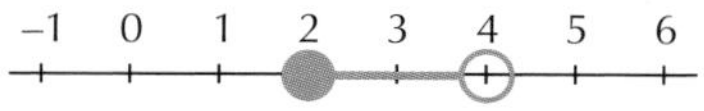

Example 7

Solve the inequality $-1 < 2x + 2 < 4$. Show your solution on a number line.

1. Write down the two separate inequalities. — $-1 < 2x + 2$ and $2x + 2 < 4$
2. Solve the inequalities separately, then write the solution as a compound inequality.

① $-1 < 2x + 2$
$-3 < 2x$
$-1.5 < x$

② $2x + 2 < 4$
$2x < 2$
$x < 1$

So $\mathbf{-1.5 < x < 1}$.

3. Draw the number line.

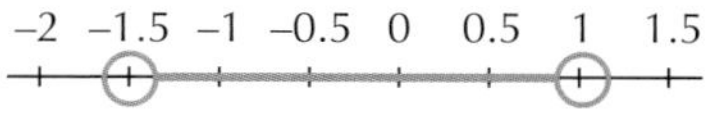

Exercise 5

Q1 List the **integers** which satisfy the following inequalities.
a) $2 < x \le 4$ b) $-5 \le x < 1$ c) $6 \le x \le 13$

Q2 Show the following inequalities on a number line.
a) $1 < x \le 6$ b) $-1 \le x < 8$ c) $-2.4 \le x < 1.6$

Q3 Solve the following inequalities. Show each of your solutions on a number line.
a) $7 < x + 3 \le 15$ b) $2 \le x - 4 \le 12$ c) $-5.6 < x - 6.8 < 12.9$

Q4 Solve the following inequalities.
a) $32 < 2x \le 42$ b) $-24 < 8x \le 40$ c) $27 < 4.5x \le 72$

Q5 Solve the following inequalities.
a) $17 < 6x + 5 < 29$ b) $8 < 3x - 4 \le 26$
c) $-42 \le 7x + 7 < 91$ d) $9 \le 1.5x + 3 \le 9.9$

9.6 Simultaneous Equations

Simultaneous equations are a pair of equations, which contain two unknowns (e.g. x and y). The solution to these equations will be a pair of values for x and y, that make both equations true.

Solving Simultaneous Equations

Learning Objective — Spec Ref A19:
Be able to solve two equations simultaneously.

You can solve simultaneous equations using the **elimination method**, where you **add** the two equations together (or **subtract** one from the other) so that **one variable** is eliminated.

This works if either x or y has the **same coefficient** in **both equations** — you **eliminate** the variable with **matching** coefficients:

- If the matching coefficients have the **same sign** (both + or both –), **subtract** one equation from the other.
- If the coefficients have **opposite signs** (one + and one –), **add** the two equations.

Tip: A coefficient is just the number in front of a variable.

Example 1

Solve the simultaneous equations: **(1)** $\boldsymbol{x + y = 11}$ **(2)** $\boldsymbol{x - 3y = 7}$

1. Subtract equation (2) from equation (1) to eliminate x.
2. Solve the equation for y.
3. Put $y = 1$ into one of the original equations and solve for x.
4. Use the other equation to check the answer.

$$\begin{array}{r} x + y = 11 \\ -\,(x - 3y = 7) \\ \hline 4y = 4 \\ y = 1 \end{array}$$

Plug into (1): $x + 1 = 11 \Rightarrow x = 10$

Check in (2): $x - 3y = 10 - 3 \times 1 = 7$ ✔
So $\boldsymbol{x = 10, y = 1}$

Tip: Start off by looking for matching coefficients — here the coefficient of x is 1 in both equations, so you eliminate x.

Exercise 1

Solve each of the following pairs of simultaneous equations.

Q1 a) $x + 3y = 10$, $x + y = 6$ b) $x + 3y = 13$, $x - y = 5$ c) $x + 2y = 6$, $x + y = 2$

d) $2x - y = 7$, $4x + y = 23$ e) $3x - 2y = 16$, $2x + 2y = 14$ f) $2x + 4y = 16$, $3x + 4y = 24$

If **neither** variable has the same coefficient, you'll have to **multiply** one or both equations to make one set of **coefficients match**.
For example, if you had the equations $2x + y = 4$ and $3x + 2y = 7$, you'd multiply the first equation by 2 to make the y-coefficients match.

You then add or subtract using the **elimination method** to find the solutions.

Example 2

Solve the simultaneous equations: **(1) $5x - 4y = 13$**
(2) $2x + 6y = -10$

1. Multiply equation (1) by 3 and equation (2) by 2 to get $-12y$ in one and $+12y$ in the other.
2. Add the resulting equations to eliminate y.
3. Solve the equation for x.
4. Put $x = 1$ into one of the equations and solve for y.
5. Use the other equation to check.

$$\begin{aligned} 3 \times (1):&\quad 15x - 12y = 39 \\ 2 \times (2):&\ + \ 4x + 12y = -20 \\ &\quad 19x = 19 \\ &\quad x = 1 \end{aligned}$$

Plug into (1): $5(1) - 4y = 13$
$-4y = 8 \Rightarrow y = -2$

Check in (2): $2x + 6y = 2(1) + 6(-2) = -10$ ✔
So $\boldsymbol{x = 1,\ y = -2}$

Exercise 2

Q1
a) $3x + 2y = 16$
$2x + y = 9$

b) $4x + 3y = 16$
$5x - y = 1$

c) $5x - 3y = 12$
$2x - y = 5$

d) $2e + 5f = 16$
$3e - 2f = 5$

e) $3d - 2e = 8$
$5d - 3e = 14$

f) $5k + 3l = 4$
$3k + 2l = 3$

Forming Simultaneous Equations

Learning Objective — Spec Ref A21:
Be able to form a set of simultaneous equations using real-life information.

Prior Knowledge Check:
Be able to form equations from word problems — see page 145.

Some **word problems** contain **two unknowns**. You'll need to use the **information** in the question to work out what the unknowns are (they could be the number of two different items, or the cost of the items, etc.). Then **form** your own **simultaneous equations** from the information given to you.

Tip: Read the question carefully to make sure you are forming both of the equations correctly.

Follow these rules to form your own simultaneous equations:

- **Assign** a letter to each of your unknowns.
- Form the two simultaneous equations using the **information in the question.**
- **Simplify** both equations as much as possible (this will make **solving** them easier).
- You may need to **rearrange** one (or both) of the equations into the form $ax + by = c$.

Example 3

Sue buys 4 dining chairs and 1 table for £142. Ken buys 6 of the same chairs and 2 of the same tables for £254. What is the price of one chair? What is the price of one table?

1. Choose some variables.

   Let the cost of one chair be £c and the cost of one table be £t.

2. Write the question as two simultaneous equations — one equation about what Sue bought and one about what Ken bought.

   $$(1)\quad 4c + t = 142$$
   $$(2)\quad 6c + 2t = 254$$

3. Multiply equation (1) by 2 to give the same coefficients of t and label it (3). Subtract equation (2) from (3)

   $$2 \times (1) = (3):\quad 8c + 2t = 284$$
   $$(2):\quad -(6c + 2t = 254)$$
   $$2c = 30$$

4. Solve the resulting equation for c.

   $$c = 15$$

5. Put $c = 15$ into equation (1) and solve for t.

   $4 \times 15 + t = 142$

   $60 + t = 142 \Rightarrow t = 82$

6. Use equation (2) to check the answer.

   $6c + 2t = 6 \times 15 + 2 \times 82 = 254$ ✔

7. Write the answer in terms of the original question.

   The chairs cost £15 each.
   The tables cost £82 each.

Tip: Always give your answer in the context of the question.

Exercise 3

PROBLEM SOLVING

Q1 The sum of two numbers, x and y, is 58. The difference between x and y is 22. Given that x is greater than y, use simultaneous equations to find both numbers.

Q2 Mountain bikes and road bikes are sold in a shop.
One day the shop sells 1 mountain bike and 2 road bikes, for a total of £350.
The price of a mountain bike is £50 more than the price of a road bike.
Calculate the price of each bike.

Q3 A grandfather with 7 grandchildren bought 6 sherbet dips and 1 Supa-Choc bar for £1.70 last week.
He bought 3 sherbet dips and 4 Supa-Choc bars for £2.60 the week before.
Calculate the price of each item.

9.7 Solving Quadratic Equations

You've already seen quadratic expressions of the form $x^2 + bx + c$.
When such an expression is set equal to zero, this forms a quadratic equation.

Learning Objective — Spec Ref A18:
Be able to solve quadratic equations using factorisation.

Prior Knowledge Check:
Be able to factorise quadratics — see p.133.

Quadratic equations are equations of the form $\mathbf{x^2 + bx + c = 0}$, where b and c are constants.
The way you **solve** them depends on the **values** of $\mathbf{b}$ and $\mathbf{c}$.
If $\mathbf{b = 0}$, there is no x term, so you can just **rearrange** and take the **square root**.

Example 1

Solve the equation $\mathbf{x^2 - 16 = 0}$.

1. Rearrange the equation to get x^2 on its own. $x^2 = 16$
2. Square root both sides. $x = \pm 4$
3. Write down the positive and negative values for x. $\mathbf{x = 4}$ or $\mathbf{x = -4}$

Tip: Remember to take the positive and negative square roots.

Exercise 1

Q1 Solve the following equations, giving exact solutions.

a) $x^2 - 4 = 0$ b) $x^2 - 1 = 0$ c) $x^2 - 25 = 0$ d) $x^2 - 64 = 0$
e) $x^2 - 121 = 0$ f) $x^2 - 100 = 0$ g) $x^2 - 2 = 0$ h) $x^2 - 7 = 0$

When $\mathbf{c = 0}$ the quadratic has the form $\mathbf{x^2 + bx = 0}$, which can be factorised to give: $\mathbf{x(x + b) = 0}$.
If two numbers multiply to make 0, then one of them must be 0.
So $\mathbf{x = 0}$ **or** $\mathbf{x + b = 0}$, which means $\mathbf{x = 0}$ and $\mathbf{x = -b}$ are the solutions of the quadratic.

Example 2

Solve the equation $\mathbf{x^2 - 7x = 0}$.

1. Factorise the left-hand side. $x(x - 7) = 0$
2. Set each factor equal to zero. $x = 0$ or $x - 7 = 0$
3. Solve to find the two possible values for x. $\mathbf{x = 0}$ or $\mathbf{x = 7}$

Tip: If you get an x on its own outside the brackets, then you know one of the solutions is $x = 0$.

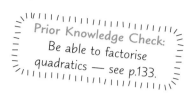

Exercise 2

Q1 Find the possible values of x for each of the following.

a) $x(x + 8) = 0$ b) $x(x - 5) = 0$ c) $x(x + 6) = 0$

d) $x(3 - x) = 0$ e) $x(4 - x) = 0$ f) $x(-2 - x) = 0$

Q2 Solve the following equations.

a) $x^2 + 6x = 0$ b) $x^2 - 6x = 0$ c) $x^2 - 24x = 0$

d) $x^2 + 5x = 0$ e) $x - x^2 = 0$ f) $12x - x^2 = 0$

If b and c are **non-zero**, then the quadratic equation has the standard form $\mathbf{x^2 + bx + c = 0}$. You'll need to **factorise** the quadratic into **two brackets** to get an equation in the form $\mathbf{(x + m)(x + n) = 0}$ (where m and n are integers that can be positive or negative). Then either $x + m = 0$ or $x + n = 0$, which you can **solve** to find the two possible ***x*-values**.

Example 3

Solve the equation $x^2 - 3x + 2 = 0$.

1. Factorise the left-hand side. $(x - 1)(x - 2) = 0$
2. Put each factor equal to zero. $x - 1 = 0$ or $x - 2 = 0$
3. Solve to find the two possible values for x. $\mathbf{x = 1}$ or $\mathbf{x = 2}$

Exercise 3

Q1 Find the possible values of x for each of the following.

a) $(x - 5)(x - 1) = 0$ b) $(x + 2)(x + 6) = 0$ c) $(x - 9)(x + 7) = 0$

Q2 a) Factorise the following expressions.

(i) $x^2 + 7x + 10$ (ii) $x^2 + 9x + 20$

(iii) $x^2 + 13x + 36$ (iv) $x^2 + 2x - 24$

b) Use your answers to part (a) to solve the following equations.

(i) $x^2 + 7x + 10 = 0$ (ii) $x^2 + 9x + 20 = 0$

(iii) $x^2 + 13x + 36 = 0$ (iv) $x^2 + 2x - 24 = 0$

Q3 Solve the following equations by factorising.

a) $x^2 + 2x + 1 = 0$ b) $x^2 - 7x + 12 = 0$ c) $x^2 + 4x + 4 = 0$

d) $x^2 - 4x + 4 = 0$ e) $x^2 + 2x - 15 = 0$ f) $x^2 - 4x - 21 = 0$

Review Exercise

Q1 Solve the following equations.

a) $\frac{x+8}{3} = 4$ b) $9 + 5x = 54$

c) $72 = 4.5(8 + 2x)$ d) $7(x - 3) = 3(x - 6)$

Q2 An electrician charges £x for each hour worked plus a £35 call-out charge. PROBLEM SOLVING
She does a job lasting 4 hours for which her total bill is £170.
How much does the electrician charge per hour?

Q3 a) Use the triangle on the right to write an equation involving x.

b) Use your equation to find the sizes of the triangle's angles.

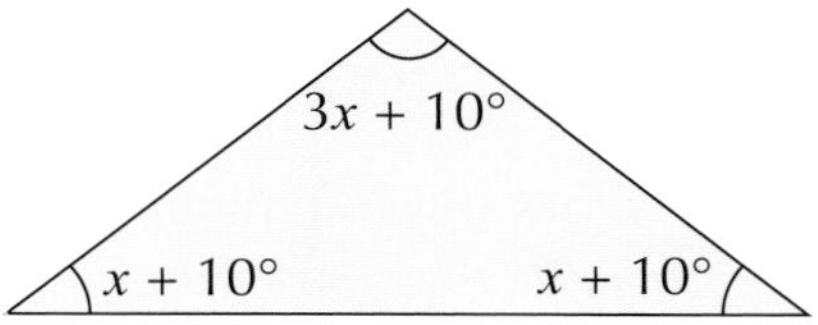

Q4 Show that: $x(x - 1) + 2(x - 3) \equiv (x + 3)(x - 2)$

Q5 Use a counter example to show that the following statement is false:
"The sum of two square numbers is never a square number."

Q6 Show that the sum of an even number and an odd number is always odd.

Q7 Solve the following inequalities. Show each of your solutions on a number line.

a) $x + 2.7 \geq 6.2$ b) $-\frac{x}{9} < 7$ c) $-5 \leq 12x + 7 < 43$

Q8 Solve the following simultaneous equations.

a) $m - 3n = 7$
$5m + 4n = -3$

b) $4u + 7v = 15$
$5u - 2v = 8$

Q9 Solve the following equations.

a) $x^2 - 36 = 0$ b) $x^2 - 81 = 0$ c) $x^2 - 3 = 0$

Q10 Solve the following equations.

a) $x^2 - 4x = 0$ b) $x^2 + 9x + 18 = 0$ c) $x^2 + 9x - 22 = 0$

Exam-Style Questions

Q1 Solve:

a) $2x + 5 = 19$

[1 mark]

b) $3(2x - 1) = 2(x + 4)$

[3 marks]

Q2 $7 \leq x < 15$ is a compound inequality, where x is an integer

a) Write down the maximum value of x.

[1 mark]

b) Work out the minimum value of x^2.

[1 mark]

Q3 Anish is x years old. Bethany is three times as old as Anish.
Cate is two years younger than Anish. The sum of all their ages is 58 years.
By forming and solving an equation, find the value of x.

[3 marks]

Q4 Bill buys 2 chews and 3 lollies and pays 84p.
Aisha buys 3 chews and 1 lolly and pays 63p.
How much will Rita pay when she buys 7 chews and 5 lollies?

[4 marks]

Q5 The diagram shows a rectangle with area 10 cm^2.
The length is 3 cm more than the width.

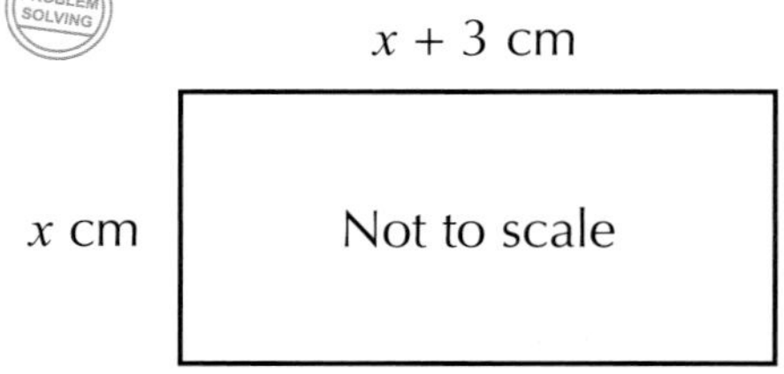

a) Show that $x^2 + 3x - 10 = 0$.

[2 marks]

b) Solve $x^2 + 3x - 10 = 0$ and write down the length and width of the rectangle.

[4 marks]

Section 10 Formulas and Functions

10.1 Formulas

A formula is a mathematical set of instructions for working something out. For example, $s = 4t + 3$ is a formula for working out the value of s — it tells you how to find s, assuming you know the value of t. You can use formulas to help solve real-life problems mathematically.

Writing Formulas

Learning Objective — Spec Ref A3/A5:
Use given information to write a formula.

Prior Knowledge Check:
Be able to simplify algebraic expressions. See p.124.

Formulas show **relationships** between different **quantities**, which are represented by **letters**.

For example, the formula for the area of a rectangle, $A = l \times w$, shows how the area (A) is connected to the length (l) and the width (w).

There are **two parts** to every formula:

- The **letter** before the equals sign is called the **subject** of the formula.
- The part after the equals sign is an **algebraic expression** which tells you the **value** of the subject.

So to **write** a formula, you need to be able to write an **expression** to go on the **right-hand side**. That means picking out the **key bits** of maths from the given **information**. These questions can be quite wordy, so take note of the important stuff as you read along.

Example 1

Write an expression for the number of marbles I have in each case below.

a) I have a bag containing m marbles. I then lose 8 marbles.

1. "I have a bag containing m marbles" — so use the letter m to represent the number of marbles in the bag. m
2. "I then lose 8 marbles" — so you need to subtract 8. $\mathbf{m - 8}$

b) My brother has 12 marbles, which is m more marbles than I own.

1. "My brother has 12 marbles..." — jot this down. 12
2. "...which is m more marbles than I own" — this means I have m fewer marbles than my brother, so subtract m. $\mathbf{12 - m}$

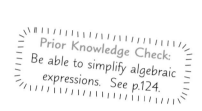

Example 2

My friend has m marbles. I had twice as many marbles as my friend, but then I lost 6 of them. Write an expression for how many marbles I have now.

1. "My friend has m marbles" — use the letter m. m
2. "I had twice as many marbles" — so multiply m by 2. $2m$
3. "...then I lost 6 of them" — so subtract 6. $\mathbf{2m - 6}$

Tip: '$2m$' is a short way of writing '$2 \times m$' — you don't need to write the × sign.

Exercise 1

Q1 Write an algebraic expression for each of the quantities asked for below.

a) I have c carrots. Su has 6 more carrots than me. How many carrots does Su have?

b) Daisy has p plants. Iris has 8 fewer plants. How many plants does Iris have?

Q2 Claudia owns f films. Barry owns twice as many films as Claudia.

a) How many films does Barry own?

b) How many films do Claudia and Barry own in total?

c) How many films would they own in total if they each gave away 3 of their films?

Q3 I have b flower bulbs. To find the number of flowers that should grow from them, multiply the number of bulbs by 3 and then add 5. How many flowers should I expect to grow?

Q4 Alf has £18. He then works in a shop for h hours. For each hour he works, he is paid £8. How much money (in pounds) does Alf have now?

Q5 I'm thinking of a number, x, and a smaller number, y. Write an expression for:

a) The square of the smaller number minus the cube of the larger number.

b) The square root of the sum of both numbers.

All that practice at writing expressions will help you out with **writing formulas**.

To write a **formula** from given information you need to:

- **Identify** the **subject** of the formula — for example, if you're asked to write a formula for the cost (C) of..., then $\boldsymbol{C}$ is the subject.
- Use the **information** to write an **expression** that will give you the **value** of the **subject**.
- Write out the formula by putting the subject **equal to** the expression.

Example 3

The cost (C) of hiring a bike is £5 per hour plus a fixed cost of £25.
Write a formula for the cost in pounds of hiring a bike for h hours.

1. Identify the subject of the formula. — The subject is C, the total cost of hiring a bike.
2. You need an expression for the total cost. Multiply the number of hours (h) by the cost per hour (£5). Then add on the fixed cost of £25. — Cost (in £) for h hours = $h \times 5 = 5h$. Total cost (in £) is $5h + 25$
3. A formula for C must start with 'C = ...'. — So $\mathbf{C = 5h + 25}$

Exercise 2

Q1 It costs £3 per hour to park a car.
Write a formula for the cost in pounds (C) to park for h hours.

Q2 It takes 2 minutes to drive 1 km.
Write a formula for the time taken in minutes (T) to drive k km.

Q3 Tom gets paid w pounds for each hour he works in his local shop.
Write a formula for the total amount he gets paid in pounds (P) if he works for 8 hours.

Q4 Kojo runs r km, but Ellie runs 5 km less.
Write a formula for the distance in km (d) that Ellie runs.

Q5 Write a formula for the cost in pounds (C) of hiring a minibus for n hours if it costs £5.50 for each hour plus a fixed charge of £F.

Q6 The instructions for cooking a goose are to cook for 50 minutes per kg, plus 25 minutes.
Write a formula to find the time taken (t minutes) to cook a goose weighing n kg.

Q7 Write a formula for the cost in pounds (C) of having t trees cut down if it costs p pounds per tree plus a fixed amount of £25.

Q8 To hire a boat costs a £10 fixed fee plus 22 pence per mile for each mile covered.
Write a formula for finding the cost in pounds (C) of hiring a boat and covering m miles.

Q9 To hire a bouncy castle costs a £125 fixed fee plus 80 pence per minute it is used.
Write a formula for the cost in pounds (C) of hiring a bouncy castle for h hours.

Substituting into Formulas

Learning Objective — Spec Ref A2:
Substitute values into a given formula.

Prior Knowledge Check:
Be able to use BODMAS. See p.2.

You can **evaluate** a formula by replacing the **letters** in the formula with actual **values**. This is called **substitution**. Here's the method to follow:

- Write out the formula.
- Write it out again, substituting numbers for letters.
- Work out the calculation — using the correct order of operations (BODMAS).

Example 4

Use the formula $v = u + 5a$ to find v if $u = 3$ and $a = -18$.

1. Write out the formula. $v = u + 5a$
2. Replace each letter with its value. $v = 3 + 5 \times (-18)$
3. Work out the calculation step by step — you do multiplication before addition. $v = 3 + (-90)$, $v = \mathbf{-87}$

Tip: Be careful substituting in negative numbers. You can use brackets so the minus sign doesn't get lost.

A formula can include all sorts of operations, such as **powers** and **brackets**, which can make things a bit trickier. Stick to the **method** above and you'll be able to evaluate these too.

Example 5

The sum of the first n square numbers is $1^2 + 2^2 + 3^2 + 4^2 + \ldots + n^2$.
This sum (S) is given by the formula $S = \frac{n}{6}(2n^2 + 3n + 1)$.
Use the formula to find S when $n = 10$.

1. Write out the formula. $S = \frac{n}{6}(2n^2 + 3n + 1)$
2. Replace n with 10. $= \frac{10}{6}(2 \times 10^2 + 3 \times 10 + 1)$
3. Work out the calculation using BODMAS:
   Evaluate the brackets first.
   Inside the brackets: (i) Find 10^2 first. $= \frac{10}{6}(2 \times 100 + 3 \times 10 + 1)$
   (ii) Then do the multiplications. $= \frac{10}{6}(200 + 30 + 1)$
   (iii) Then do the additions. $= \frac{10}{6} \times 231$

   Now evaluate the rest. $= \mathbf{385}$

Exercise 3

Q1 Find the value of y in each of the following given that $x = 7$.

a) $y = x + 4$ b) $y = x - 3$ c) $y = 12 - x$ d) $y = 6x$

Q2 Find the value of y in each of the following given that $m = -3$.

a) $y = m - 8$ b) $y = 3m^2$ c) $y = -4 + m$ d) $y = \frac{12}{m}$

Q3 If $m = 4$ and $n = 3$, then find the value of p in each of the following.

a) $p = mn$ b) $p = m^2$ c) $p = m - n^2$ d) $p = \frac{3m}{n}$

Q4 If $x = -4$ and $y = -3$, then find the value of z in each of the following.

a) $z = x + 2$ b) $z = y - 1$ c) $z = -x + 2y$ d) $z = 6x - y$

Q5 Use the formula $S = \frac{1}{2}n(n + 1)$ to find S when:

a) $n = 10$ b) $n = 100$ c) $n = 1000$ d) $n = 5000$

Q6 Use the formula $s = ut + \frac{1}{2}at^2$ to find s if:

a) $u = 7$, $a = 2$ and $t = 4$ b) $u = 24$, $a = 11$ and $t = 13$

c) $u = -11$, $a = -9.81$ and $t = 12.2$ d) $u = 66.6$, $a = -1.64$ and $t = 14.2$

Wordy problems work in the same way — you write out the formula and substitute in the values. If a question involves **units**, don't forgot to include them in your **answer**.

Example 6

The temperature in degrees Fahrenheit (f) is given by the formula $f = 1.8c + 32$, where c is the temperature in degrees Celsius. Convert the following temperatures to degrees Fahrenheit: a) –17 °C b) 37.4 °C

1. Write out the formula. **a)** $f = 1.8c + 32$ **b)** $f = 1.8c + 32$
2. Identify the number to substitute for c. $c = -17$ $c = 37.4$
3. Replace c with its value, and do the calculation. $f = 1.8 \times (-17) + 32 = -30.6 + 32 = 1.4$ $f = 1.8 \times 37.4 + 32 = 67.32 + 32 = 99.32$
4. Give your answer with units. So –17 °C = **1.4 °F** So 37.4 °C = **99.32 °F**

Example 7

The taxi fare, £T, for a journey is calculated using the formula $T = 0.8m + 2.5$, where m is the distance of the journey in miles. How much is the fare for a 5 mile journey?

1. Write out the formula. — $T = 0.8m + 2.5$
2. The journey is 5 miles so $m = 5$. — $T = 0.8 \times 5 + 2.5$
   $= 6.5$
3. Give your answer in the context of the question. — So the fare is **£6.50**

Exercise 4

Q1 Use the formula $A = \frac{1}{2}bh$ to find the area (A, in m^2) of a triangle with base b and height h, if:

a) $b = 4$ m, $h = 6$ m b) $b = 2$ m, $h = 3$ m c) $b = 0.4$ m, $h = 1.8$ m

Q2 The formula for working out the speed (s, in metres per second) of a moving object is $s = \frac{d}{t}$, where d is the distance travelled (in metres) and t is the time taken (in seconds). Find the speed (in metres per second) of each of the following. Give your answers to 3 significant figures.

a) a runner who travels 800 metres in 110 seconds

b) a cheetah that travels 400 metres in 14 seconds

Q3 The length, c, of the longest side of a right-angled triangle is given by the formula $c = \sqrt{a^2 + b^2}$, where a and b are the lengths of the other two sides. Use the formula to work out the length of the longest side of the triangle drawn on the right.

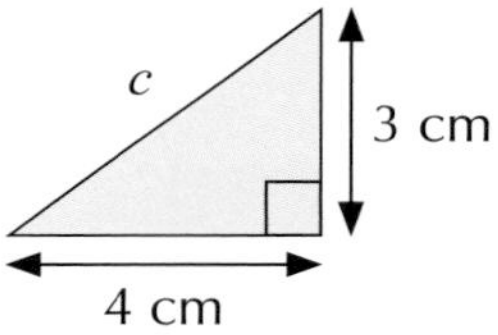

Q4 The area (A, in cm^2) of this circle is given by the formula $A = \pi r^2$. Find A for the values of r below. Give your answers to 1 decimal place.

a) $r = 5$ cm b) $r = 3.5$ cm

c) $r = 11.1$ cm d) $r = 6.4$ cm

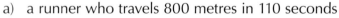

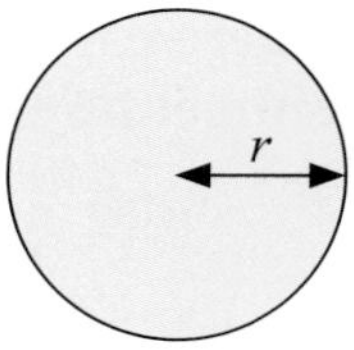

Q5 The number of seconds (T) taken for a pendulum to swing forwards and then backwards once is given by the formula $T = 2\pi\sqrt{\frac{l}{10}}$, where l is the length of the pendulum in metres. Calculate (to 1 decimal place) how long it will take a pendulum to swing backwards and forwards once if:

a) $l = 1$ metre b) $l = 0.5$ metres c) $l = 16$ metres

Rearranging Formulas

Learning Objective — Spec Ref A5:
Rearrange formulas to change the subject.

Prior Knowledge Check: Be able to solve equations. See p.137-144.

Rearranging formulas means making a different letter the **subject**. For example, getting a formula beginning '$y = ...$' from '$x = 3y + 2$'.

You're aiming to get the new subject **on its own** on one side of the '=' sign. The method is just like solving an equation — you carry out **opposite operations** until you get the letter on its own, making sure to always do the **same thing** on **both sides**.

Tip: – is the opposite of +, and ÷ is the opposite of ×.

Example 8

Make x the subject of the following formulas.

a) $y = x + 4$

1. Write down the original formula. You want to get x on its own. $y = x + 4$
2. The opposite of + 4 is – 4, so subtract 4 from both sides. $y - 4 = x + 4 - 4$; $y - 4 = x$
3. x is on its own, but you need to write the formula with x on the left-hand side. $\mathbf{x = y - 4}$

b) $y = 6x$

1. Write down the original formula. You want to get x on its own. $y = 6x$
2. The opposite of × 6 is ÷ 6, so divide both sides by 6. $\frac{y}{6} = \frac{6x}{6} = x$
3. Write the formula with x on the left-hand side. $\mathbf{x = \frac{y}{6}}$

If the subject is **squared**, you'll need to take **square roots** — remember, there's a **negative root** as well as a positive root, so you'll need a **± sign**.

Example 9

Make z the subject of the formula $w = z^2$.

1. Write down the original formula. You want to get z on its own. $w = z^2$
2. z is squared, so take the square root of each side. $\pm\sqrt{w} = z$
3. Write the formula with z on the left-hand side. $\mathbf{z = \pm\sqrt{w}}$

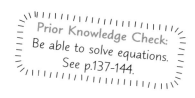

Exercise 5

Q1 Make x the subject of the following formulas. All your answers should begin '$x =$'.

a) $y = x + 2$ b) $b = x - 5$ c) $z = 7 + x$

Q2 Make x the subject of the following formulas.

a) $z = 4x$ b) $p = 17x$ c) $r = 4.2x$

Q3 Make x the subject of the following formulas.

a) $y = \frac{x}{8}$ b) $z = \frac{x}{17}$ c) $t = \frac{x}{8.6}$

Q4 Make x the subject of the following formulas.

a) $abc = 2x$ b) $t = xy$ c) $uv + y = 4.2x$

Q5 Make s the subject of the following formulas.

a) $m = \frac{4}{5}s$ b) $r = -16s$ c) $p = -14.2s$

d) $a = \frac{5}{4}s$ e) $b = \sqrt{s}$ f) $c = s^2$

Two-Step Rearrangements

For trickier formulas you need to carry out **two steps** to change the **subject**. For example, you might need to **add** a number to both sides in the **first step**, then **multiply** both sides by a number in the **second step**.

You'll need to decide which order to do the steps — do what makes the maths easier.

Example 10

Rearrange $a = 3b + 4$ to make b the subject of the formula.

$3b + 4$ means 'take your value of b and then (i) multiply by 3 and (ii) add 4'. To get b on its own, you need to undo these steps, but in the opposite order.

1. Write down the original formula. $a = 3b + 4$
2. The opposite of $+ 4$ is $- 4$, so subtract 4 from both sides. $a - 4 = 3b + 4 - 4$
   $a - 4 = 3b$
3. The opposite of $\times 3$ is $\div 3$, so divide both sides by 3. $\frac{a-4}{3} = \frac{3b}{3} = b$
4. Write the formula with b on the left-hand side. $\boldsymbol{b = \frac{a-4}{3}}$

Tip: Be careful when dividing the left-hand side here — make sure you divide everything by 3.

Example 11

Make r the subject of $s = \frac{r-1}{2}$

$\frac{r-1}{2}$ means 'take your value of r and then (i) subtract 1 and (ii) divide by 2'.
To get r on its own, you need to undo these steps, but in the opposite order.

1. Write down the original formula. $s = \frac{r-1}{2}$
2. The opposite of ÷ 2 is × 2, so multiply both sides by 2. $2s = r - 1$
3. The opposite of – 1 is + 1, so add 1 to both sides. $2s + 1 = r$
4. Write the formula with r on the left-hand side. $\mathbf{r = 2s + 1}$

Exercise 6

Q1 Make x the subject of the following formulas.
a) $y = 5x + 3$ b) $z = 8x - 2$ c) $p = 15x + 18$

Q2 Make y the subject of the following formulas.
a) $z = \frac{y+4}{3}$ b) $x = \frac{7+y}{4}$ c) $s = \frac{y-2}{9}$

Q3 Make x the subject of the following formulas.
a) $u = 4(x - 2)$ b) $v = 8(x + 4)$ c) $w = 3(x - 4)$

Q4 Make y the subject of the following formulas.
a) $p + 3 = 4y - 2$ b) $q + 7 = 9y + 11$ c) $r - 5 = 21y - 9$

Q5 The surface area (A) of the shape on the right is given approximately by the formula $A = 21.5d^2$.
a) Rearrange the formula to make d the subject.
b) Find d if:
(i) $A = 344\text{ cm}^2$ (ii) $A = 134.375\text{ cm}^2$

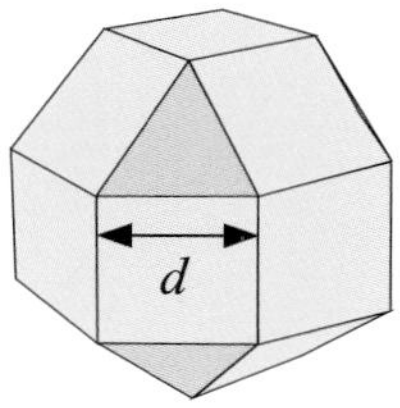

Q6 The perimeter (P) of the shape on the right is given by the formula $P = 2(2x + y)$.
a) Rearrange the formula to make x the subject.
b) Find x if:
(i) $P = 14$ and $y = 3$ (ii) $P = 32$ and $y = 4$

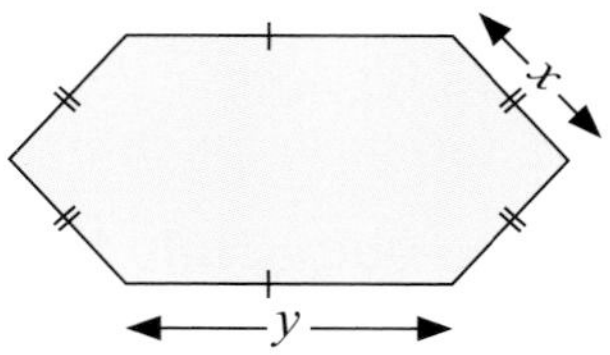

Q7 Make x the subject of these formulas:
a) $x + y = 6x$ b) $y = \sqrt{x} + 2$ c) $\frac{3 + xy + x}{x + 1} = y$

10.2 Functions

A function is a mathematical instruction that turns one number into a new number. Diagrams called 'function machines' are used to show how a function works, step-by-step.

Learning Objective — Spec Ref A7:
Be able to use functions to find inputs and outputs.

A **function** takes a number (the **input**), does a calculation with it, then gives back a new number (the **output**). For example, the calculation could be 'multiply by 4 then add 5', or 'subtract 3 then divide by 8'.

A **function machine** is a **diagram** that breaks a function down into **steps**. To find the **output** from a given **input**, just follow the steps and see what comes out.

For example, this **function machine** represents the function 'multiply by 4 then add 5':

input → [× 4] → [+ 5] → **output**

There are two steps to this machine: the first is '**multiply** the **input** by **4**' and the second is '**add 5** to the **number** you get **from step 1** to give the **output**'.

Example 1

This function machine represents the function 'multiply by 3 then add 4'.

x → [**× 3**] → [**+ 4**] → y

Find the value of y when $x = 3$.

Put $x = 3$ into the machine and work forwards through the steps.

$$3 \xrightarrow{\times 3} 9 \xrightarrow{+4} 13$$

So $y = \mathbf{13}$.

You can use a function machine **in reverse** to **find the input** from a **given output**:

- Put the given output **into the end** of the function machine.
- Work backwards by doing the **opposite operations** in **reverse order**.

Example 2

Here is a function machine:

$$x \longrightarrow \boxed{\times 3} \longrightarrow \boxed{+ 4} \longrightarrow y$$

Find the value of x when $y = 37$.

Put $y = 37$ into the end of the machine and work backwards using opposite operations.

1. The opposite of + 4 is – 4.
2. The opposite of × 3 is ÷ 3.

$$37 \xrightarrow{-4} 33 \xrightarrow{\div 3} 11$$

So $\boldsymbol{x = 11}$

Tip: Check your answer by putting it back into the machine:
11 × 3 = 33
33 + 4 = **37** ✓

Exercise 1

Q1 The function machine below represents the function 'divide by 5 then add 7'.

$$x \longrightarrow \boxed{\div 5} \longrightarrow \boxed{+ 7} \longrightarrow y$$

Find the value of y for the following values of x.

a) $x = 20$ b) $x = 35$ c) $x = 45$ d) $x = -10$

Q2 The function machine below represents the function 'subtract 3 then multiply by 6'.

$$x \longrightarrow \boxed{- 3} \longrightarrow \boxed{\times 6} \longrightarrow y$$

a) What is the value of y when $x = 11$? b) Given that $y = 72$, find the value of x.

Q3 The equation $y = 7x - 2$ is represented by the function machine below.

$$x \longrightarrow \boxed{\times 7} \longrightarrow \boxed{- 2} \longrightarrow y$$

a) Given that $x = -1$, find the value of y. b) When $y = 19$, find the value of x.

Q4 Jared wants to write the equation $y = 4x + 1$ as a function machine.

a) Copy and complete this function machine to show Jared's equation.

$$x \longrightarrow \boxed{} \longrightarrow \boxed{} \longrightarrow y$$

b) Use the function machine to find the value of x when $y = 17$.

Q5 For the function machine below, show there's a value where the input is equal to the output.

$$x \longrightarrow \boxed{\times 6} \longrightarrow \boxed{- 10} \longrightarrow y$$

Review Exercise

Q1 Chloe gets 45 fewer free minutes on her mobile phone each month than Wassim.

a) If Wassim gets w free minutes, write a formula for the number of free minutes (c) Chloe gets.

b) Find c when $w = 125$.

Q2 The number of matchsticks (m) needed to make h hexagons as shown is given by the formula $m = 5h + 1$.

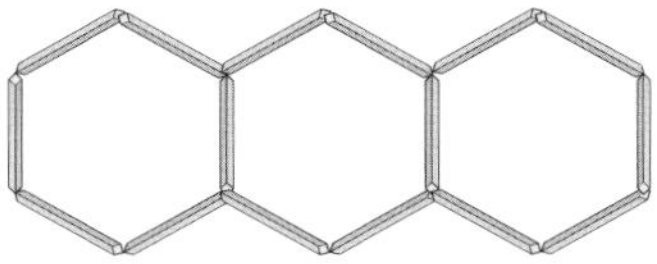

a) Rearrange this formula to make h the subject.

b) Use your formula to find h when $m = 36$.

Q3 An isosceles triangle has one angle of size x and two angles of size y.
The formula for y in terms of x is $y = \frac{180° - x}{2}$.

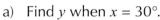

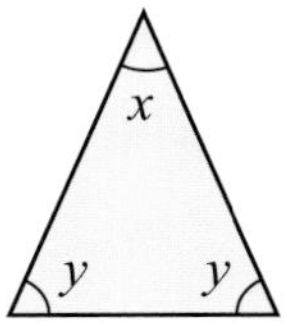

a) Find y when $x = 30°$.

b) Rearrange the formula to make x the subject.

Q4 The formula for calculating the cost (C, in £) of gas is represented by the following function machine, where n is the number of units used.

n ⟶ [× 0.06] ⟶ [+ 7.5] ⟶ C

a) Find the cost if 275 units of gas were used.

b) How many units of gas were used if the gas bill is £40.50?

Q5 Debi is decorating her bathroom walls with black tiles and white tiles. She uses 3 white tiles for every black tile, plus an extra 50 white tiles.

a) Copy and complete the function machine below to represent the number of white tiles used, w, given the number of black tiles used, b.

b ⟶ [   ] ⟶ [+ 50] ⟶ w

b) If Debi uses 200 black tiles, how many white tiles does she use?

c) The function machine below shows the total number of tiles used, t, given the number of black tiles used, b.

b ⟶ [× 4] ⟶ [+ 50] ⟶ t

If Debi uses 530 tiles in total, how many black tiles does she use?

Exam-Style Questions

Q1 Make b the subject of the formula $y = a + \frac{b}{x}$.

[2 marks]

Q2 Look at the function machine below.

input → -4 → $\div 25$ → output

a) Work out the output when the input is 17, giving your answer as a decimal.

[1 mark]

b) Work out the input when the output is 5.

[2 marks]

c) Find an expression in terms of x for the input when the output is $3x$, giving your answer in its simplest form.

[1 mark]

Q3 To book a swimming pool for a party, there is a fixed charge of £30, plus a fee of £1.25 for each person who attends.

a) Write a formula to calculate the hire cost (C, in £) for n people.

[1 mark]

b) Calculate the hire cost when 32 people attend.

[1 mark]

c) Rearrange your formula to make n the subject.

[2 marks]

d) If the total cost of hiring the pool was £80, how many people attended the party?

[1 mark]

Q4 Fill in the missing operation to complete the function machine below.

7 → ☐ → -4 → 10

[2 marks]

11.1 Term to Term Rules

A sequence is a list of numbers or shapes which follows a particular rule. Each number or shape in the sequence is called a term. Term to term rules tell you how to go from one term to the next.

Number Sequences

Learning Objectives — Spec Ref A23:
- Find rules for simple number sequences.
- Use the rule for a sequence to find terms in the sequence.

There are different types of **number sequences** that follow different rules to get from one term to the next. The first ones you should know about are **arithmetic** and **geometric** sequences.

Arithmetic Sequences

Arithmetic sequences are sequences where you **add** (or **subtract)** the **same number** each time to get from one term to the next. The number you add is called the **common difference**.

2, 5, 8, 11, 14
+3 +3 +3 +3

The rule is 'add 3 to the previous term'.

27, 21, 15, 9, 3
−6 −6 −6 −6

The rule is 'subtract 6 from the previous term'.

Geometric Sequences

Geometric sequences are sequences where you find the next term by **multiplying** (or **dividing)** by the **same number** each time. The number you multiply by is called the **common ratio**.

2, 10, 50, 250, 1250
×5 ×5 ×5 ×5

The rule is 'multiply the previous term by 5'.

48, 24, 12, 6, 3
÷2 ÷2 ÷2 ÷2

The rule is 'divide the previous term by 2'.

Example 1

The first term of a sequence is 3. The rule for finding the next term in the sequence is 'add 4 to the previous term'. Write down the first 5 terms of the sequence.

1. The first term is 3, so write this down.
2. Add 4 each time to find the terms.

3, 7, 11, 15, 19
+4 +4 +4 +4

Example 2

Consider the sequence 2, 6, 18, 54...
a) Explain the rule for finding the next term in the sequence.
b) Write down the next three terms in the sequence.

1. This time the rule for finding the next term involves multiplication.

$$2 \xrightarrow{\times 3} 6 \xrightarrow{\times 3} 18 \xrightarrow{\times 3} 54...$$

2. Multiply each term by 3 to get the next term.

a) Multiply the previous term by 3.

b) 54, **162**, **486**, **1458**. (× 3, × 3, × 3)

Exercise 1

Q1 The first term of a sequence is 5. The rule for finding the next term in the sequence is 'add 4 to the previous term'. Write down the first 5 terms of the sequence.

Q2 The first term of a sequence is 2. The rule for finding the next term in the sequence is 'multiply the previous term by 2'. Write down the first 5 terms of the sequence.

Q3 Write down the first 5 terms of the sequence with:
- a) first term = 100; further terms generated by the rule 'subtract 6 from the previous term'.
- b) first term = 40; further terms generated by the rule 'divide the previous term by 2'.
- c) first term = 11; further terms generated by the rule 'multiply the previous term by –2'.

Q4 The first four terms of a sequence are 3, 6, 9, 12.
- a) Write down what you add to each term in the sequence to find the next term.
- b) Write down the next three terms in the sequence.

Q5 For the following sequences: (i) explain the rule for finding the next term in the sequence.
(ii) find the next three terms in the sequence.

a) 3, 5, 7, 9... b) 4, 12, 36, 108... c) 16, 8, 4, 2... d) 5, 3, 1, –1...
e) 1, 1.5, 2, 2.5... f) –1, –3, –9, –27... g) 1000, 100, 10, 1... h) 1, –2, 4, –8...

Q6 The first four terms of a sequence are 4, 10, 16, 22.
- a) Explain the rule for finding the next term in this sequence.
- b) Use your rule to find: (i) the 5th term (ii) the 6th term (iii) the 8th term

Q7 Copy the following sequences and fill in the blanks.

a) 9, 5, □, –3, –7, –11
b) –72, □, –18, –9, □, –2.25
c) □, 0.8, 3.2, 12.8, □, 204.8
d) –63, –55, □, □, □, –23

Special Number Sequences

Learning Objective — Spec Ref A24:
Recognise and use sequences of square, cube and triangular numbers, and quadratic and Fibonacci-type sequences.

Quadratic Sequences

In a **quadratic** sequence, the **difference** between terms **changes** by the **same amount** each time.

1, 2, 4, 7, 11, 16, 22
+ 1 + 2 + 3 + 4 + 5 + 6
+ 1 + 1 + 1 + 1 + 1

Fibonacci-Type Sequences

The **Fibonacci sequence** is a special sequence that starts 1, 1, 2, 3, 5... and follows the rule '**add together** the **previous two terms**'. A **Fibonacci-type sequence** is a sequence that follows the same rule, but starts with a different number.

2, 2, 4, 6, 10, 16, 26
2 + 2 2 + 4 4 + 6 6 + 10 10 + 16

Sequences of Triangular, Square and Cube Numbers

Some sequences are harder to spot by their term to term rules. Common ones to watch out for are the **square numbers**: 1, 4, 9... (i.e. 1^2, 2^2, 3^2...), the **cube** numbers: 1, 8, 27 (i.e. 1^3, 2^3, 3^3...), and **triangular numbers** (shown below):

To make the sequence of triangular numbers, **start at 1** and then **add 2**, then **3**, then **4**, then **5** etc. to each new term.

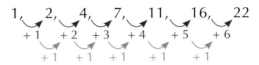

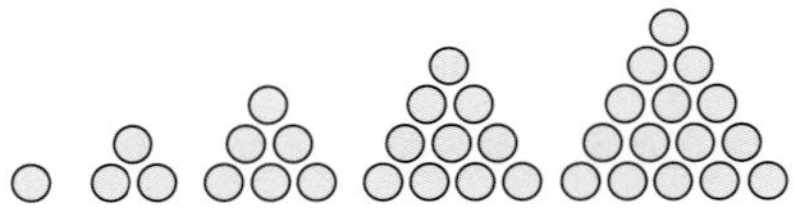

1, 3, 6, 10, 15
+ 2 + 3 + 4 + 5

Example 3

Find the next three terms in the sequence 4, 5, 7, 10...

1. Try finding the difference between neighbouring terms.
2. Here, the difference is increasing by 1 each time.
3. Use this to find the next three terms in the sequence. Start with 10. Then add 4. Then add 5. Then add 6.

4 5 7 10
+ 1 + 2 + 3

10 **14** **19** **25**
+ 4 + 5 + 6

Exercise 2

Q1 Consider the sequence 7, 8, 10, 13, 17...

a) Find the difference between each term and the next for the first five terms.

b) Find the next three terms in the sequence.

Q2 For each sequence below, find:
(i) the difference between each term and the next for the first five terms,
(ii) the next three terms in the sequence.

a) 5, 7, 11, 17, 25...

b) 20, 18, 15, 11, 6...

c) 3, 4, 6, 9, 13...

d) 1, 2, 0, 3, –1...

Q3 The sequence 1, 1, 2, 3, 5... is known as the Fibonacci sequence. Each term in the sequence is found by adding together the previous two terms.
Find the next three terms in the sequence.

Shape Sequences

Learning Objective — Spec Ref A23:
Find rules for patterns of shapes and use them to find patterns in a sequence.

Patterns of **shapes** can form **sequences**. Just like number sequences, you need to find the **rule** to get from one pattern to the next — then use that rule to find more patterns in the sequence.

Example 4

The matchstick shapes on the right form the first three patterns in a sequence.

Draw the fourth and fifth patterns in the sequence.

1. First work out how to get from one pattern to the next. You have to add 3 matchsticks to add an extra square to the previous pattern.

2. The fourth pattern is the next one in the sequence, so add 3 matchsticks to the third pattern to get the fourth...

3. Add another 3 matchsticks to that to get the fifth.

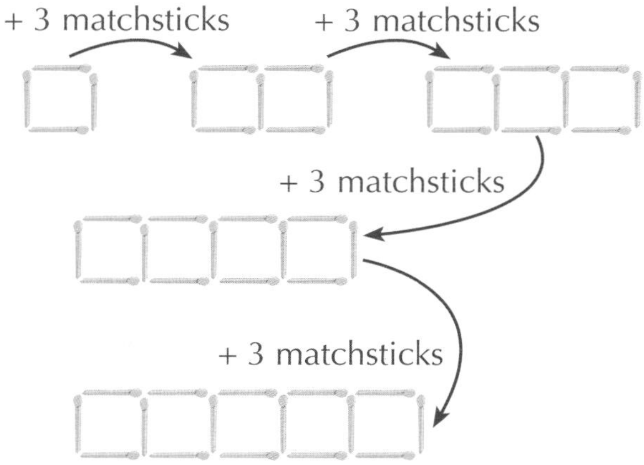

Example 5

The black and white circles on the right form the first 3 patterns in a sequence.
Find the number of black circles and the number of white circles in the 9th pattern.

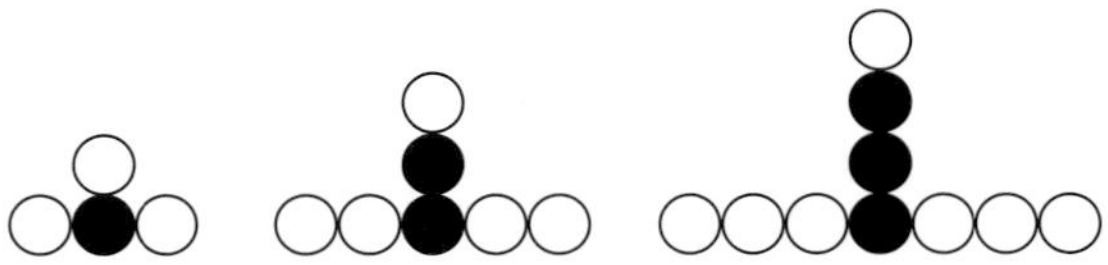

1. Write the numbers of each colour of circle in each pattern as two number sequences.

   Black circles: 1, 2, 3
   White circles: 3, 5, 7

2. Find the term to term rules for each sequence.

   Rule for black circles: +1 each time.
   Rule for white circles: +2 each time.

3. Starting at the 3rd term, apply these rules 6 times to get the 9th term.
   Black: add on 6 lots of 1 to the 3rd term.
   White: add on 6 lots of 2 to the 3rd term.

   $3 + (6 \times 1) =$ **9 black circles**
   $7 + (6 \times 2) =$ **19 white circles**

Exercise 3

Q1 The first three patterns of several sequences are shown below.
For each of the sequences: (i) explain the rule for making the next pattern,
(ii) draw the fourth and fifth patterns in the sequence,
(iii) find the number of matches needed for the sixth pattern.

a) 

b) 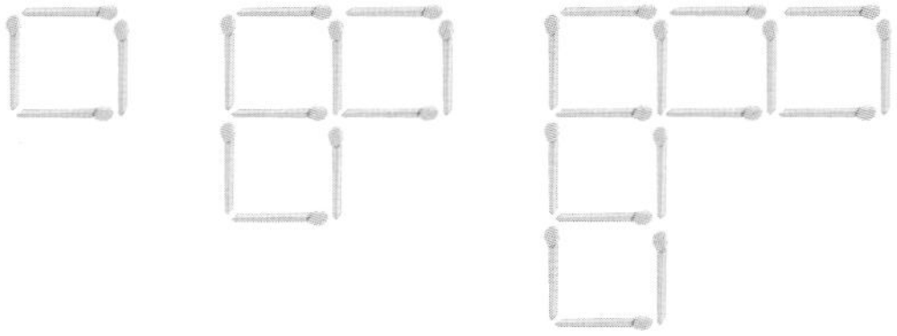

c) 

Q2 Below is a sequence of triangles made up of different numbers of circles.
The numbers of circles in each triangle form a sequence called the 'triangle numbers'.
For example, the first three triangle numbers are 1, 3 and 6.

a) Draw the next three triangles in the sequence, and write down the corresponding triangle numbers.
b) Explain the rule for generating the next triangle number in the sequence.
c) Find the 7th triangle number.

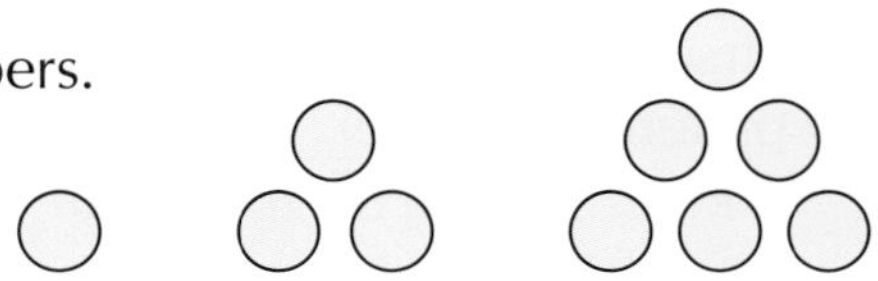

Q3 For each of the sequences below:
(i) Draw the next three patterns in the sequence.
(ii) Explain the rules for finding the number of circles of each colour in the next pattern.
(iii) Work out how many white circles there are in the 7th pattern.
(iv) Work out how many black circles there are in the 10th pattern.

a) 

b) 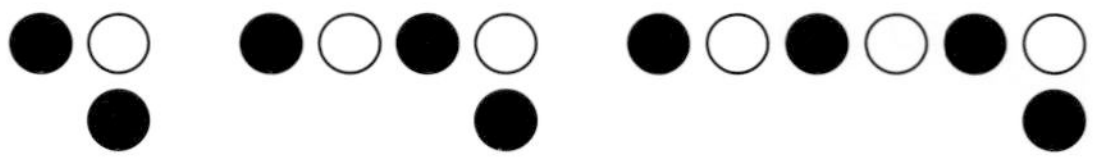

11.2 Position to Term Rules

Using a term to term rule is fine for the first few terms, but it can be a chore if you want the hundredth term. That's where using a position to term rule is helpful — instead of defining a term by the terms that come before, you define it by its position in the sequence.

Learning Objective — Spec Ref A23:
Use the nth term to find terms in a sequence.

Prior Knowledge Check:
Be able to substitute values into a given formula (p.166) and solve equations (p.137-144).

You can work out the **value** of a term in a sequence by using its **position** (n) in the sequence — the 1st term has $n = 1$, the 2nd term has $n = 2$, the 10th term has $n = 10$, and so on.

The **nth term formula** tells you what to do to n to get the value of that term.
So if the nth term was $5n + 2$, you'd **multiply** n by 5 then **add** 2 to get the term.

Example 1

The nth term of a sequence is $2n - 1$. Find the first four terms of the sequence.

1. To find the 1st, 2nd, 3rd and 4th terms of the sequence, substitute the values $n = 1$, $n = 2$, $n = 3$ and $n = 4$ into the formula.

$(2 \times 1) - 1 = 1$

$(2 \times 2) - 1 = 3$

$(2 \times 3) - 1 = 5$

$(2 \times 4) - 1 = 7$

2. Write the terms in order to form the sequence.

So the first four terms are **1, 3, 5, 7**.

Exercise 1

Q1 The nth term of a sequence is $20 - 2n$. Find the value of:
a) the 1st term b) the 2nd term c) the 3rd term d) the 4th term

Q2 Find the first four terms of a sequence if the nth term is given by:
a) $n + 5$ b) $3n + 2$ c) $10n - 8$ d) $12 - n$
e) $-7 - 3n$ f) $3 - 4n$ g) $2n^2$ h) $2n^2 + 3$

Q3 The nth term of a sequence is $2n + 20$. Find the value of:
a) the 5th term b) the 10th term c) the 20th term d) the 100th term

Q4 The nth term of a sequence is $100 - 3n$. Find the value of:
a) the 3rd term b) the 10th term c) the 30th term d) the 40th term

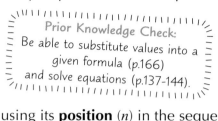

If you know the value of a term in a sequence, but not its position, set up and solve an **equation** using the *n*th term rule to find the position, n.

Example 2

The *n*th term of a sequence is $4n + 5$. Which term has the value 41?

1. Make the *n*th term equal to 41.
2. Solve the equation to find n.

$$4n + 5 = 41$$
$$4n = 36$$
$$n = 9$$

So the **9th term** is 41.

Example 3

The *n*th term of a sequence is $2n - 7$.
Which is the first term in this sequence to have a value greater than 50?

1. Find the value of n which would give a value of 50. As before, set up and solve an equation for n.
2. The sequence is increasing, so the first term that will give a value over 50 will be the next whole number value of n.
3. Check your answer by working out some terms in the sequence.

$$2n - 7 = 50$$
$$2n = 57$$
$$n = 28.5$$

So the first term with a value greater than 50 is the **29th term**.

Check:
28th term = $(2 \times 28) - 7 = 49$ (<50)
29th term = $(2 \times 29) - 7 = 51$ (>50) ✔

Exercise 2

Q1 The *n*th term of a sequence is $2n + 6$. Which term has the value 20?

Q2 The *n*th term of a sequence is $17 - 2n$. Find which terms have the following values.
a) 3 b) 9 c) –7

Q3 The *n*th term of a sequence is $n^2 + 1$. Find which terms have the following values.
a) 5 b) 50 c) 82

Q4 The *n*th term of a sequence is $4n - 10$.
Which term in the sequence is the last to have a value less than 75?

Q5 The nth term of a sequence is $6n + 2$.
Which term in the sequence is the first to have a value greater than 40?

Q6 The formulas for the number of matches in the nth pattern of the following 'matchstick sequences' are shown below. For each of the sequences:
(i) find the number of matches needed to make the 6th pattern,
(ii) find the value of n for the last pattern you could make with 100 matches.

a) 

Number of matches in nth pattern = $2n + 1$

b) 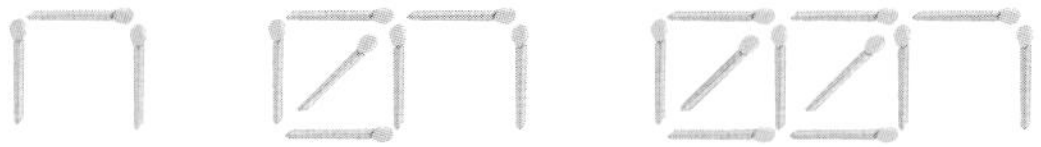

Number of matches in nth pattern = $4n - 1$

You can use the nth term to **check** if a given value is a term in that sequence. As before, set up and solve an equation to find the value of n for a given (suspected) term. The term is **only** part of the sequence if n is a **whole number**. If not, you know the value is **not** part of the sequence.

Example 4

A sequence has nth term $3n + 2$. Is 37 a term in the sequence?

1. Make an equation by setting the formula for the nth term equal to 37.
2. Then solve your equation to find n.
3. Since n is not a whole number, 37 is not a term in the sequence.

$3n + 2 = 37$

$3n = 35$

$n = 11.666...$

So 37 **is not** a term in the sequence.

Exercise 3

Q1 A sequence has nth term $2n + 1$. Show that 54 is not a term in this sequence.

Q2 Show that 80 is a term in the sequence with nth term equal to $3n - 1$.

Q3 A sequence has nth term $21 - 2n$. Show that -1 is a term in this sequence, and write down the corresponding value of n to show its position.

Q4 A sequence has nth term equal to $17 + 3n$.
Determine whether each of the following is a term in this sequence.
a) 52 b) 98 c) 248 d) 996

Q5 A sequence has nth term equal to $4n - 9$. Determine whether each of the following is a term in this sequence. For those that are, state the corresponding value of n.
a) 43 b) 71 c) 138 d) 879

11.3 Finding a Position to Term Rule

You know how to find the terms in a sequence from the nth term rule, but you also need to be able to do it in reverse — working out the rule yourself from some terms in a sequence.

Learning Objective — Spec Ref A25:
Be able to write formulas for the nth term in a sequence.

Prior Knowledge Check:
Be able to substitute values into a given formula (p.166) and solve equations (p.137-144).

You can find the nth term of an arithmetic sequence by looking at its terms:

- Find the **coefficient** of n by finding the **common difference** — the bit you add or subtract to get from one term to the next.
- For example, if the common difference was 2, the ***n*th term rule** would be '$2n$ + something'.
- Find the 'something' by comparing **multiples** of the common difference with the terms in the sequence, and seeing what you need to **add** and **subtract**.

Tip: The coefficient is just the number next to n — so the coefficient of $3n$ is 3, and of $-4n$ is -4.

This might be easier to understand by looking at an example:

Example 1

a) Find the *n*th term of the sequence 5, 7, 9, 11...

1. Find the difference between each term and the next.
2. Here, the terms increase by 2 each time, so the coefficient of n is 2.

5 → 7 → 9 → 11 (+ 2, + 2, + 2)

3. Work out what you need to add or subtract to get from $2n$ to the term in the sequence.
4. To get to each term from $2n$, you need to add 3. This tells you that '+ 3' is in the nth term rule.

$2n$:	2	4	6	8
	↓+ 3	↓+ 3	↓+ 3	↓+ 3
Term:	5	7	9	11

5. Combine the two parts to find the nth term rule.

So the nth term is $\mathbf{2n + 3}$

6. Check your formula by using it to find a term you know.

Check: 2nd term ($n = 2$) is $(2 \times 2) + 3 = 7$ ✔

b) Find the 50th term in the sequence.

1. You worked out the nth rule formula in part a), so now you can use it to find the 50th term.

The nth term is $\mathbf{2n + 3}$

2. Substitute $n = 50$ into $2n + 3$.

$(2 \times 50) + 3 = 100 + 3 = \mathbf{103}$

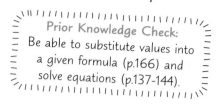

Example 2

Find the nth term of the sequence 45, 42, 39, 36...

1. Find the difference between each term and the next.
2. Here, the terms decrease by 3 each time, so the coefficient of n is –3.
3. Work out what you need to add or subtract to get from $-3n$ to the term in the sequence.
4. To get to each term from $-3n$, you need to add 48. This tells you that '+ 48' is in the nth term rule.
5. Combine the two parts to find the nth term rule.
6. Check your formula by using it to find a term you know.

45 → 42 → 39 → 36
(– 3, – 3, – 3)

$-3n$:	–3	–6	–9	–12
	↓+ 48	↓+ 48	↓+ 48	↓+ 48
Term:	45	42	39	36

So the nth term is $\mathbf{-3n + 48}$

Check: 2nd term ($n = 2$) is
$(-3 \times 2) + 48 = 42$ ✔

Exercise 1

Q1 The first 4 terms of a sequence are: 9, 13, 17, 21
   a) Find the difference between each term and the next.
   b) Find the formula for the nth term of the sequence, and use a term to check it.

Q2 Find the formula for the nth term of each of the following sequences.
   a) 7, 13, 19, 25...   b) 6, 16, 26, 36...   c) 41, 81, 121, 161...
   d) –1, 1, 3, 5...   e) –9, –5, –1, 3...   f) –45, –26, –7, 12...

Q3 Find the formula for the nth term of each of the following sequences.
   a) 10, 8, 6, 4...   b) 40, 37, 34, 31...   c) 78, 69, 60, 51...
   d) 4, –1, –6, –11...   e) –10, –25, –40, –55...   f) –39, –51, –63, –75...

Q4 Find the number of matchsticks in the nth pattern of the following sequences.
   a)
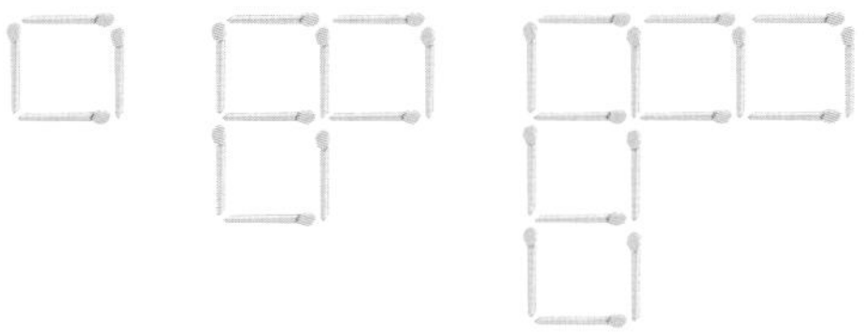
   b)
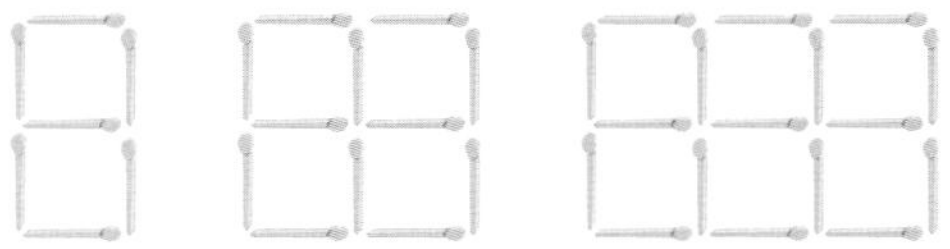

Q5 a) Find the nth term of the sequence 8, 11, 14, 17...   PROBLEM SOLVING
   b) Use your answer to part a) to write down the nth term of the sequence 9, 12, 15, 18...

Review Exercise

Q1 For the following sequences: (i) explain the rule for finding the next term,
(ii) write down the next three terms in the sequence.

a) 4, 7, 10, 13... b) 192, 96, 48, 24... c) 0, –4, –8, –12...
d) –4, –2, 0, 2... e) 11, 8, 5, 2... f) 2, –6, 18, –54...

Q2 A sequence begins 7, 9, 13, 19, 27...

a) Find the difference between the neighbouring terms for the first five terms.
b) Find the next three terms in the sequence.

Q3 For the following sequence of circles:

a) Draw the next three patterns in the sequence.
b) Explain the rule for generating the number of circles in the next pattern.
c) Work out how many circles there are in the 8th pattern.

Q4 A sequence has the nth term $-3 - 2n$. Write down the first four terms of the sequence.

Q5 Each of the following gives the nth term for a different sequence.
For each sequence, find: (i) the 5th term (ii) the 10th term (iii) the 100th term

a) $4n + 12$ b) $30 - 3n$ c) $100n - 8$ d) $n(n - 1)$

Q6 The nth term of a sequence is $50 - 6n$. Find which terms have the following values.

a) 2 b) 8 c) 14 d) 26

Q7 An expression for the nth term of a sequence is $7n - 3$. Find:

a) the position of the last term in the sequence that is smaller than 100,
b) the position of the first term in the sequence that is greater than 205.

Q8 a) Find the nth term of the sequence whose first four terms are 12, 18, 24, 30...
b) Is 86 a term in this sequence?

Q9 Find an expression in terms of n for the number of dots in the nth term of the sequence on the right.

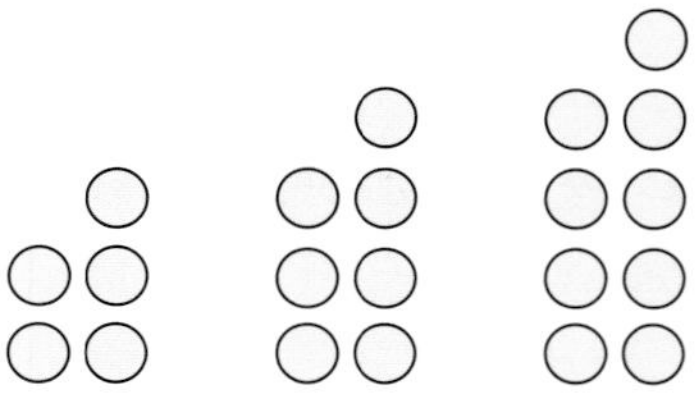

Exam-Style Questions

Q1 A quadratic sequence starts 3, 5, 9, 15... Find the next term in the sequence.

[2 marks]

Q2 In a chemistry lesson, Yuki uses a set of balls and rods to make models of molecules. His first model uses 5 balls and 4 rods, and the models form the following sequence:

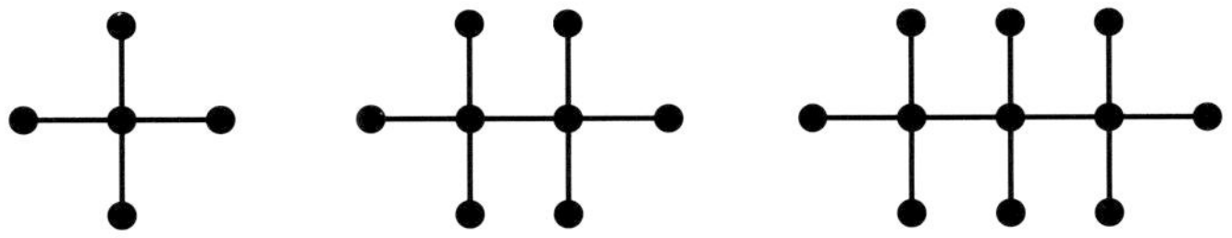

Design 1 Design 2 Design 3

a) Draw design number 4.

[1 mark]

b) How many rods will be needed to make design number 8?

[2 marks]

c) Design n will require $3n + 2$ balls. Write down a simplified expression in terms of n for the number of rods needed for design n.

[1 mark]

Q3 a) Write down the next term in this sequence: 2, 6, 10, 14...

[1 mark]

b) The nth term of this sequence is $4n - 2$. Work out the 210th term.

[2 marks]

c) Jane says that 74 is a term in this sequence. Is Jane correct? Show how you decide.

[2 marks]

Q4 Find the nth term of this sequence: –11, –2, 7, 16...

[2 marks]

Section 12 Graphs and Equations

12.1 Coordinates

Axes, coordinates and quadrants are used to help pin-point a location. Just remember that in a pair of coordinates, the horizontal (x) one comes first, followed by the vertical (y) one. And make sure you don't get tripped up by the negative numbers in the different quadrants.

Coordinates and Quadrants

Learning Objective — Spec Ref A8:
Be able to identify, specify and plot points using their coordinates in all four quadrants.

A grid can be split into **four** different **quadrants** (regions) using a line going **left-to-right** (the ***x*-axis**) and a line going **top-to-bottom** (the ***y*-axis**).

Coordinates describe the **position** of a point. They are written in **pairs** inside brackets with:

- The ***x*-coordinate** (left or right) **first**.
- The ***y*-coordinate** (up or down) **second**.

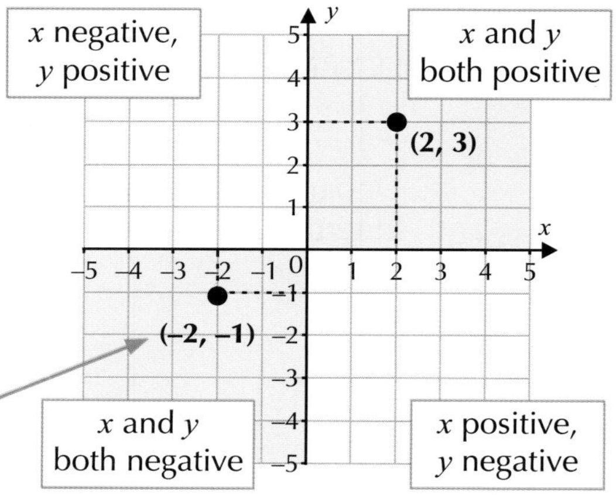

The **origin** is the point **(0, 0)** — this is where the *x*- and *y*-axis intersect. If you go **left** from the origin the ***x*-coordinate** becomes **negative**, and if you **down** from the origin the ***y*-coordinate** becomes **negative**.

For example, the point **(–2, –1)** is positioned **2 to the left** on the ***x*-axis** and **1 down** on the ***y*-axis**, but the point **(2, 3)** is positioned **2 to the right** on the ***x*-axis** and **3 up** on the ***y*-axis**.

Example 1

Write down the coordinates of the vertices of the triangle *ABC* shown below.

1. Follow the grid line up or down from each point and read off the *x*-coordinate.
2. Follow the grid line left or right from each point and read off the *y*-coordinate.
3. Write the coordinates in brackets with the *x*-coordinate first.

   The coordinates are ***A*(–4, –3)**, ***B*(–2, 4)**, ***C*(3, –2)**.

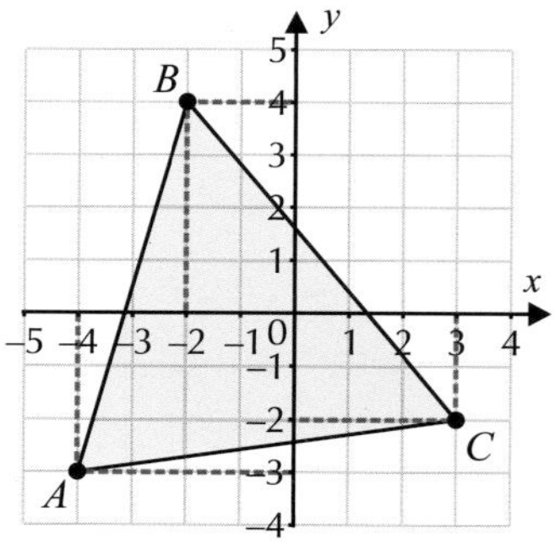

Exercise 1

Q1 The four quadrants are labelled on the grid on the right.
Write down which quadrants each of the following points lies in.

a) (2, 2) b) (1, –1) c) (–2, –1) d) (–1, 1)

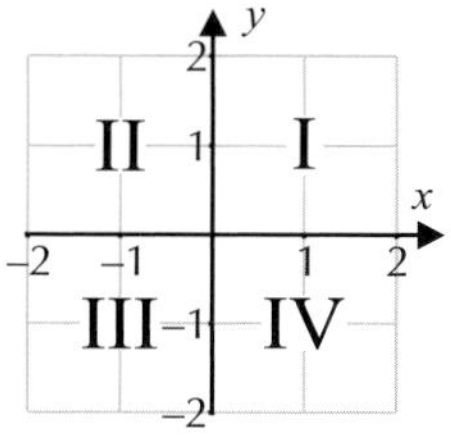

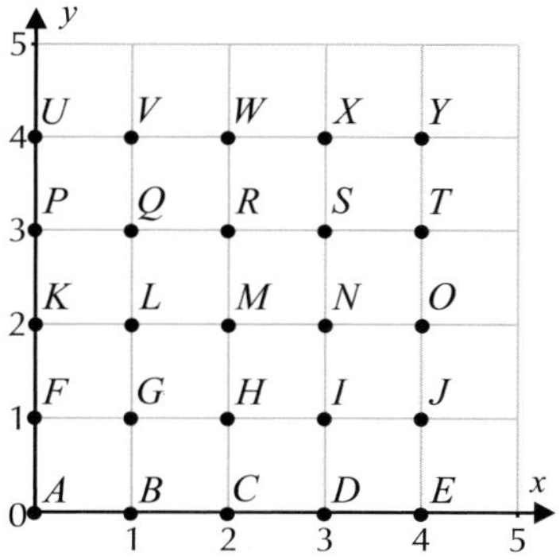

Q2 Use the grid on the left to answer this question.

a) Write down the coordinates of the following points.
(i) *A* (ii) *M* (iii) *Q* (iv) *U* (v) *Y*

b) Write down the sentence given by the letters with the following coordinates.

(3, 4) - (2, 0) (4, 2) (4, 2) (2, 3) (3, 0) (3, 1) (3, 2) (0, 0) (4, 3) (4, 0) (2, 0) (4, 2) (2, 2) (4, 0) (3, 3) (0, 1) (3, 1) (2, 3) (3, 3) (4, 3)

Q3 Use the grid on the right for this question.

a) The following sets of coordinates spell out the names of shapes. Write down the name of each shape.
(i) (2, –4) (–4, –4) (–5, –6) (–4, 5) (–2, –4) (–5, 3)
(ii) (2, 5) (–4, 1) (–2, –4) (2, 5) (4, 1) (–5, 3)
(iii) (2, 5) (–5, –6) (–2, 5) (3, –2) (–4, 1) (4, 5)

b) Write down the sets of coordinates which spell out the names of the following shapes.
(i) kite (ii) sphere (iii) triangle

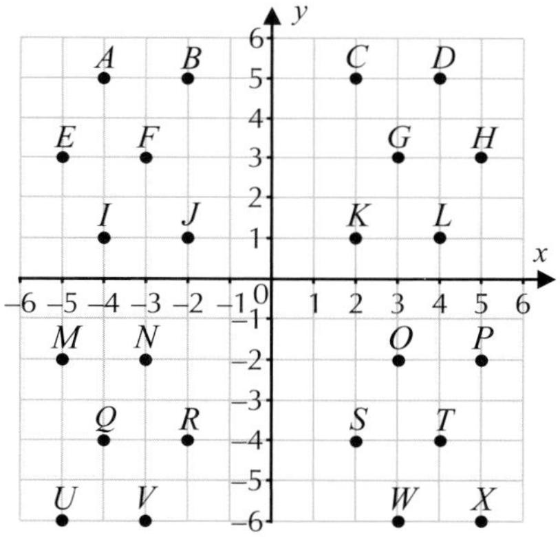

Plotting a point means **marking** its **position** on a grid. To plot a point, find its ***x*-coordinate** and then read up or down to the correct ***y*-coordinate** — draw a **dot** to mark this position.

Example 2

Draw the shape *WXYZ* with vertices *W*(–4, 5), *X*(3, 2), *Y*(4, –3) and *Z*(–5, –2).

1. Some of the coordinates are negative, so your axes need to go below zero.
2. Read across the horizontal axis for the *x*-coordinates.
3. Read up and down the vertical axis for the *y*-coordinates.
4. Plot the points and connect them to draw the shape. In a shape like *WXYZ*, connect adjacent letters (*WX*, *XY* and *YZ*) and also the first and last (*WZ*).

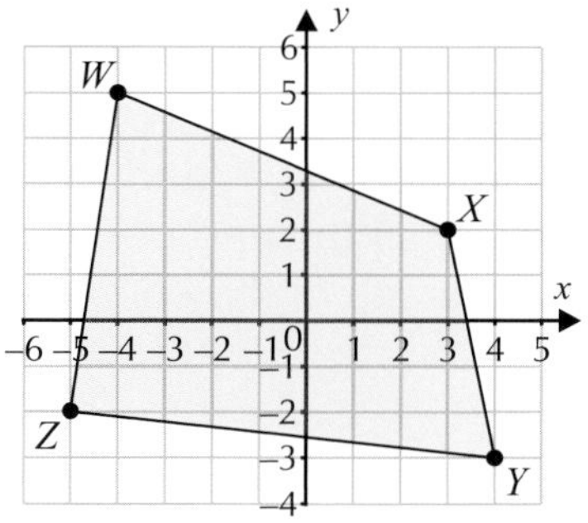

Exercise 2

Q1 On separate copies of the grid on the right, draw the shapes whose vertices are given by the following sets of coordinates.

a) $A(2, 1)$, $B(2, 4)$, $C(5, 4)$, $D(5, 1)$

b) $E(1, 1)$, $F(6, 3)$, $G(3, 5)$

c) $H(1, 3)$, $I(3, 5)$, $J(6, 3)$, $K(3, 1)$

d) $L(2, 6)$, $M(5, 5)$, $N(5, 3)$, $O(2, 1)$

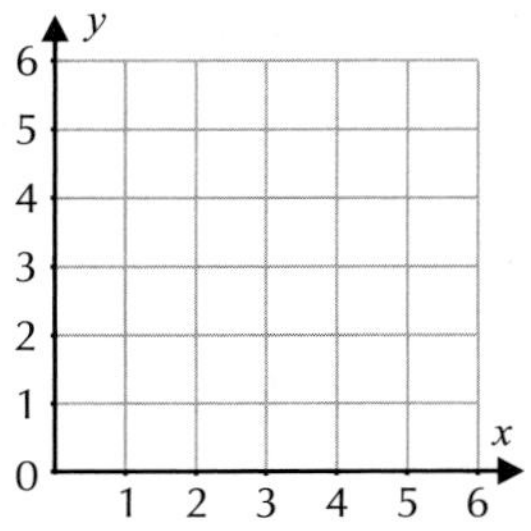

Q2 Draw a coordinate grid ranging from –4 to 4 on the x-axis and y-axis. Plot the following points on your grid. Join the points you have plotted.

$A(4, 3)$ $B(1, -3)$ $C(-3, -3)$ $D(-2, 1)$

Q3 Draw a coordinate grid ranging from –5 to 5 on the x-axis and y-axis. Plot the following points on your grid. Join the points you have plotted.

$A(0, 5)$ $B(1, 1)$ $C(5, 1)$ $D(2, -1)$ $E(3, -4)$

$F(0, -2)$ $G(-3, -4)$ $H(-2, -1)$ $I(-5, 1)$ $J(-1, 1)$

Midpoints of Line Segments

Learning Objective — Spec Ref A8:
Be able to find the coordinates of the midpoint of a line segment.

A **line segment** is part of a line, lying between two **end points**.

The **midpoint** of a line segment is **halfway** between the end points. If (x_1, y_1) and (x_2, y_2) are the coordinates of the **end points**, you can use the following **formula** to find the coordinates of the **midpoint**.

Tip: Line segments have end points, unlike lines which go on forever in both directions.

$$\textbf{Midpoint} = \left(\frac{x_1 + x_2}{2}, \frac{y_1 + y_2}{2}\right)$$

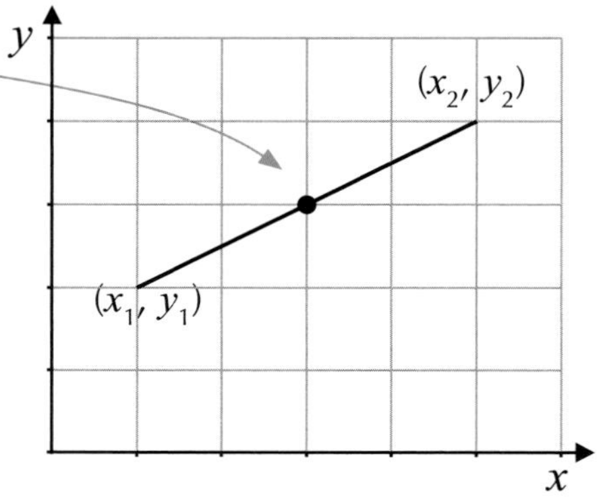

The midpoint's x- and y-coordinates equal the **average** of the end points' **x-coordinates** and the average of the end points' **y-coordinates**.

Example 3

Find the midpoint of the line segment *AB*, shown on the right.

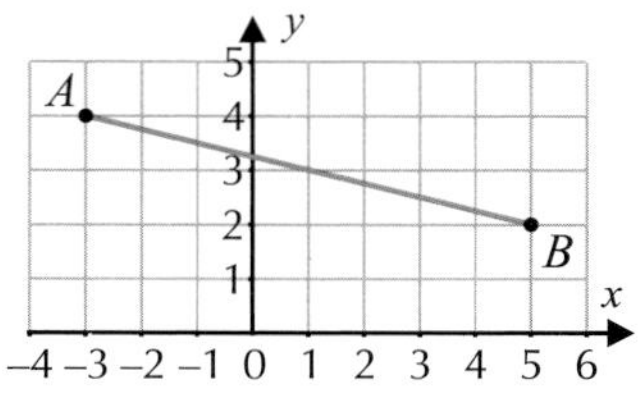

1. Write down the coordinates of the end points *A* and *B*.
   $A(-3, 4)$ and $B(5, 2)$
2. Write down the formula for the midpoint.
   $$\text{Midpoint} = \left(\frac{x_1 + x_2}{2}, \frac{y_1 + y_2}{2}\right)$$
3. Put in the *x*- and *y*-coordinates of *A* and *B*.
   $$= \left(\frac{-3 + 5}{2}, \frac{4 + 2}{2}\right)$$
   $$= \mathbf{(1, 3)}$$

Exercise 3

Q1 A line segment *XY* is shown on the grid on the right.

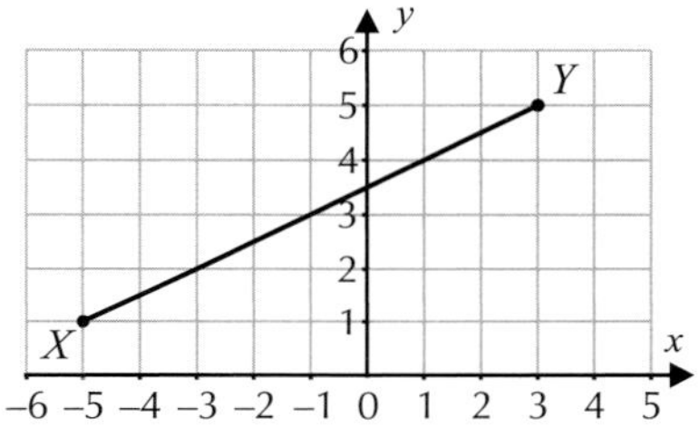

a) Write down the coordinates of the points *X* and *Y*.

b) Use the coordinates of *X* and *Y* to calculate the coordinates of *M*, the midpoint of *XY*.

c) Plot the point *M* on a copy of the diagram. Use a ruler to check that the distances *XM* and *MY* are equal.

Q2 The points *P* and *Q* have coordinates $P(1, 0)$ and $Q(3, 5)$.

a) Find the coordinates of *M*, the midpoint of *PQ*.

b) Check your answer:

(i) Plot *P* and *Q* on a square grid. Join the points to form the line segment *PQ*.

(ii) Plot the point *M*. Check that *M* is the midpoint of *PQ*.

Q3 Find the coordinates of the midpoint of the line segment *AB*, where *A* and *B* have coordinates:

a) $A(1, 1)$, $B(3, 5)$ b) $A(0, 1)$, $B(6, 3)$

c) $A(0, -4)$, $B(-5, 1)$ d) $A(-2, 0)$, $B(1, -8)$

Q4 Use the diagram on the right to find the midpoints of the following line segments.

a) *AF* b) *AC* c) *DF*

d) *BE* e) *BF* f) *CE*

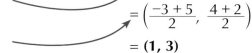

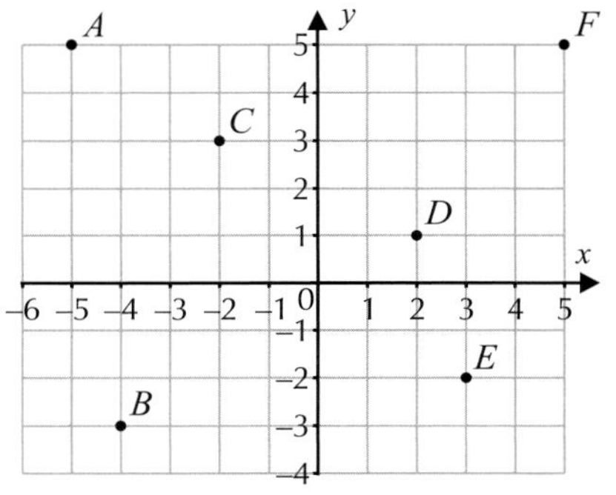

12.2 Horizontal and Vertical Graphs

This is a gentle introduction to straight-line graphs — horizontal lines are parallel to the x-axis (they go across the page) and vertical lines are parallel to the y-axis (they go down the page).

Learning Objective — Spec Ref A9:
Recognise and draw horizontal and vertical lines on a set of axes.

All **horizontal** lines have the equation $\boldsymbol{y = a}$ (where a is a number), as every point on the same horizontal line has the same y-coordinate.

All **vertical** lines have the equation $\boldsymbol{x = b}$ (where b is a number), as every point on the same vertical line has the same x-coordinate.

Tip: $y = a$ intersects the y-axis at $(0, a)$. $x = b$ intersects the x-axis at $(b, 0)$.

The equation of the **x-axis** is $\boldsymbol{y = 0}$ and the equation of the **y-axis** is $\boldsymbol{x = 0}$.

Example 1

Write down the equations of the lines marked *A*-*D*.

1. Every point on the line marked A has y-coordinate 1. — A is the line $\boldsymbol{y = 1}$
2. Every point on the line marked B has x-coordinate 2. — B is the line $\boldsymbol{x = 2}$
3. Every point on the line marked C has y-coordinate –4. — C is the line $\boldsymbol{y = -4}$
4. Every point on the line marked D has x-coordinate –2. — D is the line $\boldsymbol{x = -2}$

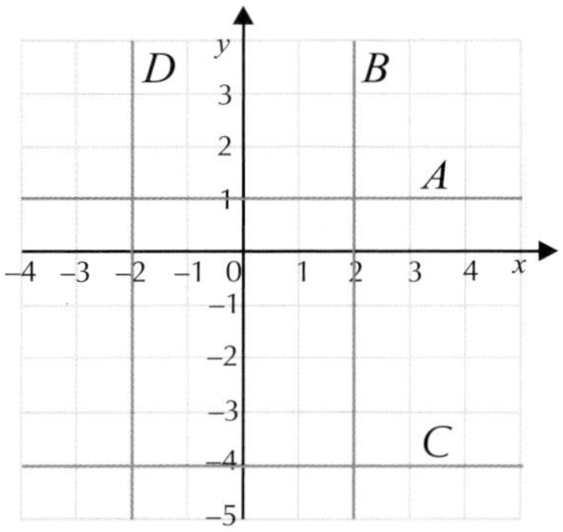

To **draw** the line $y = a$, draw a **horizontal** line that passes through a on the **y-axis**.

Example 2

Plot the graphs of the following equations.

a) $\boldsymbol{y = 3}$ The graph with equation $y = 3$ is a horizontal line through 3 on the y-axis.

b) $\boldsymbol{y = -1}$ The graph with equation $y = -1$ is a horizontal line through –1 on the y-axis.

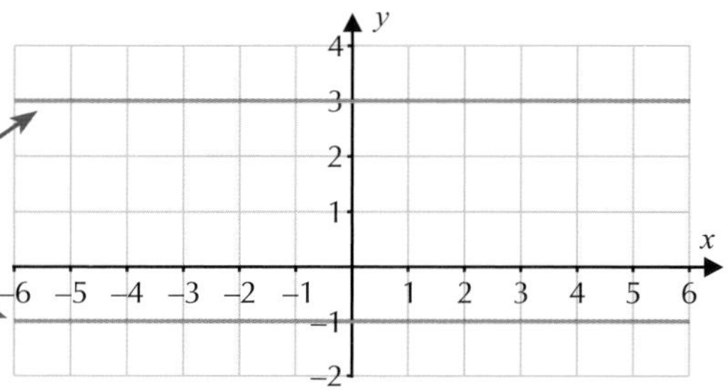

To **draw** the line $x = b$, draw a **vertical** line that passes through b on the **x-axis**.

Example 3

Plot the graphs of the following equations.

a) $x = -4$ The graph with equation $x = -4$ is a vertical line through –4 on the x-axis.

b) $x = 2$ The graph with equation $x = 2$ is a vertical line through 2 on the x-axis.

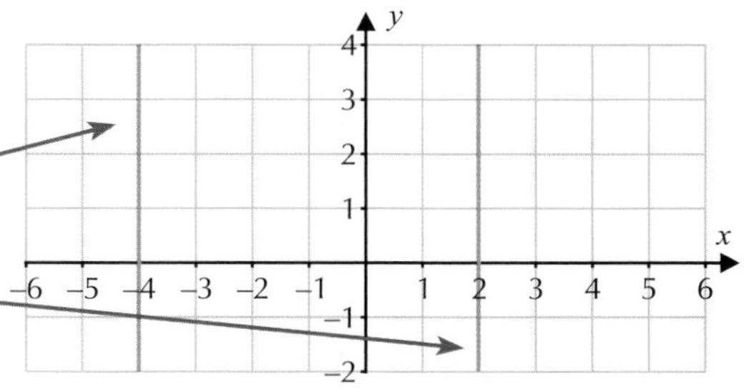

Exercise 1

Q1 Write down the equations of each of the lines labelled A to E on this diagram.

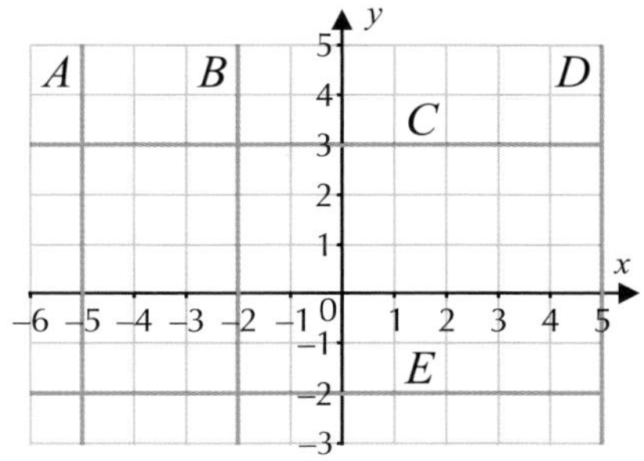

Q2 Draw a set of coordinate axes and plot the graphs with the following equations.

a) $y = 3$ b) $y = -6$ c) $y = -1$

d) $x = 2$ e) $x = 4$ f) $x = -4$

Q3 a) What is the y-coordinate of every point on the x-axis?

b) Write down the equation of the x-axis.

Q4 Draw a set of coordinate axes.

a) Plot the horizontal line which passes through the point (2, –2).

b) What is the equation of this line?

Q5 Draw a set of coordinate axes.

a) Plot the vertical line which passes through the point (1, –3).

b) What is the equation of this line?

Q6 Write down the equation of the line which is parallel to the x-axis, and which passes through the point (4, 8).

Q7 Write down the equation of the line which is parallel to the y-axis, and which passes through the point (–2, –6).

Q8 Draw a set of coordinate axes.

a) Plot the points P(3, –1) and Q(–1, –1).

b) What is the equation of the line that contains points P and Q?

12.3 Other Straight-Line Graphs

Horizontal and vertical lines are certainly nice, but life isn't always so straightforward. You need to do a bit more preparation work when you're plotting other straight lines.

Drawing Straight-Line Graphs

Learning Objective — Spec Ref A9/A12:
Be able to draw straight-line graphs on a set of axes.

Prior Knowledge Check: Be able to rearrange linear equations (p.137).

The equation of a straight line which **isn't** horizontal or vertical contains **both x and y** (e.g. $y = 2x + 4$). If an equation **only** contains x and y terms (e.g. $y = 5x$), then the line passes through the **origin (0, 0)**.

There are a couple of different methods you can use to draw these straight-line graphs:

- Make a **table of values** — find the values of y for different values of x, plot the points and join with a straight line. You only have to plot two points to be able to sketch the graph, but it's often useful to plot more than two, in case one of the points you plot is incorrect.
- Find the **value of x** when **$y = 0$** and the **value of y** when **$x = 0$**. Plot these two points and join with a straight line. Both points should lie on the axes. **Extend** your line to cover the range of x-values required (usually specified in the question).

Example 1

a) Complete the table to show the value of $y = 2x + 1$ for values of x from 0 to 5.

x	0	1	2	3	4	5
y						

Use the equation $y = 2x + 1$ to find the y-value corresponding to each value of x.

x	0	1	2	3	4	5
y	$2 \times 0 + 1 = \mathbf{1}$	$2 \times 1 + 1 = \mathbf{3}$	$2 \times 2 + 1 = \mathbf{5}$	$2 \times 3 + 1 = \mathbf{7}$	$2 \times 4 + 1 = \mathbf{9}$	$2 \times 5 + 1 = \mathbf{11}$

b) Plot the points from the table, and hence draw the graph of $y = 2x + 1$ for x from 0 to 5.

1. Use your table to find the coordinates to plot — just read off the x- and y-values from each column.

   The points to plot are (0, 1), (1, 3), (2, 5), (3, 7), (4, 9) and (5, 11).

2. Plot each point on the grid, then join them up with a straight line.

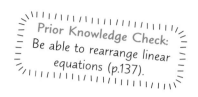

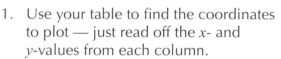

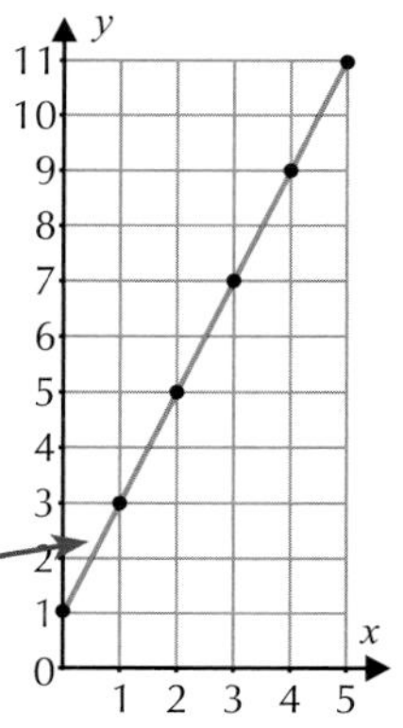

Tip: Choose sensible values for x, e.g. whole numbers are usually easier to plot.

Example 2

Draw the graph of $y + 2x = 4$ for $-1 \le x \le 3$.

1. Put $x = 0$ into the equation to find the value of y — this is where it crosses the y-axis.

   When $x = 0$, $y + 2(0) = 4$, so $y = 4$.

2. Put $y = 0$ into the equation to find the value of x — this is where it crosses the x-axis.

   When $y = 0$, $0 + 2x = 4$, so $2x = 4 \Rightarrow x = 2$.
   So the graph crosses the axes at (0, 4) and (2, 0).

3. Mark the points (0, 4) and (2, 0) on your graph and draw a straight line passing through them. Make sure you extend it to cover the whole range of x-values asked for in the question.

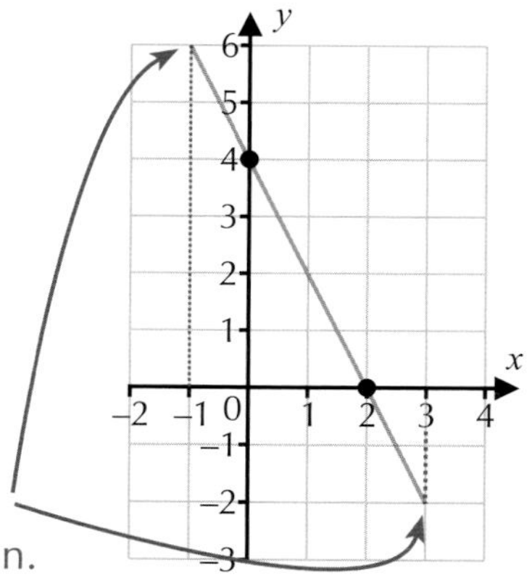

Exercise 1

Q1 a) Copy and complete the table below to show the value of $y = x + 2$ for values of x from 0 to 5.

x	0	1	2	3	4	5
y	2	3				
Coordinates	(0, 2)					

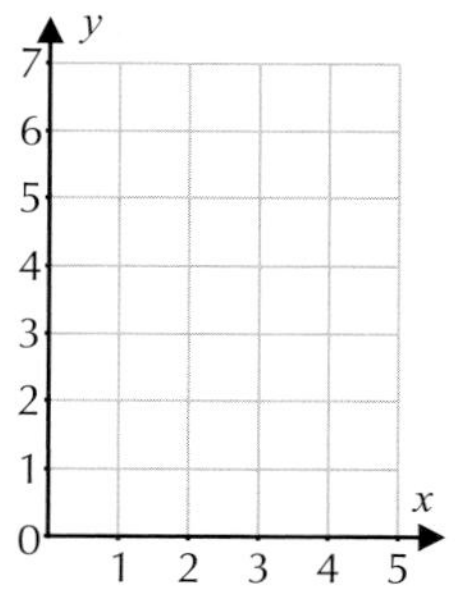

b) Copy the grid on the right and plot the coordinates from your table.

c) Join up the points to draw the graph of $y = x + 2$ for values of x from 0 to 5.

Q2 a) Make y the subject of the equation $2y - 2x = -8$ (so it's in the form '$y = ...$').

b) Copy and complete the table below to show the value of y for values of x from 0 to 5.

x	0	1	2	3	4	5
y	−4					
Coordinates						

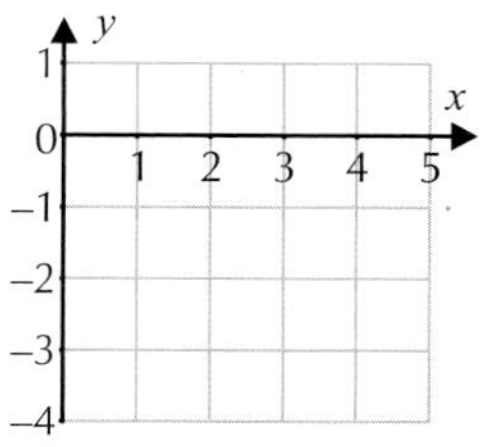

c) Copy the grid on the right and plot the coordinates from your table.

d) Join up the points to draw the graph for values of x from 0 to 5.

Q3 Draw the graphs of the following equations by first finding the value of y when $x = 0$, and the value of x when $y = 0$.

a) $y = x + 7$ for x from −7 to 0

b) $y = 3x - 6$ for x from 0 to 4

c) $y + 2x = 8$ for x from 0 to 5

d) $y + 5x = 7.5$ for x from 0 to 2

Solving Equations Graphically

Learning Objective — Spec Ref A17:
Be able to solve linear equations graphically.

You can use the graph of $y = ax + b$ to **solve** the equation $ax + b = 0$.

- Find where this graph **crosses** the ***x*-axis** — this is the point where $y = ax + b = 0$.
- Read off the ***x*-coordinate** of this point — this is the **solution** to the linear equation.

Tip: Check your answer by substituting the value of x back into the equation.

Example 3

Use the graph on the right to solve the equation $2x - 1 = 0$.

1. Find where the graph crosses the x-axis. — The graph crosses the x-axis at $(\frac{1}{2}, 0)$.
2. Read off the x-coordinate. — So the solution is $x = \frac{1}{2}$.
3. Check your answer. — Check: $2 \times \frac{1}{2} - 1 = 1 - 1 = 0$ ✔

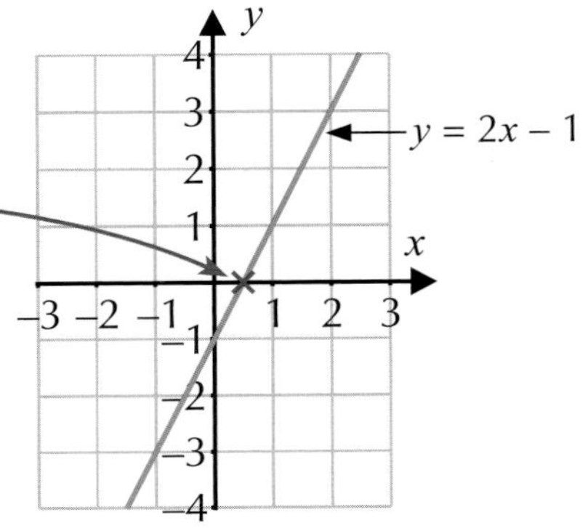

Exercise 2

Q1 The graph of $y = 2x - 4$ is shown below.

a) Write down the coordinates of the point where the graph crosses the x-axis.

b) Hence, solve the equation $2x - 4 = 0$.

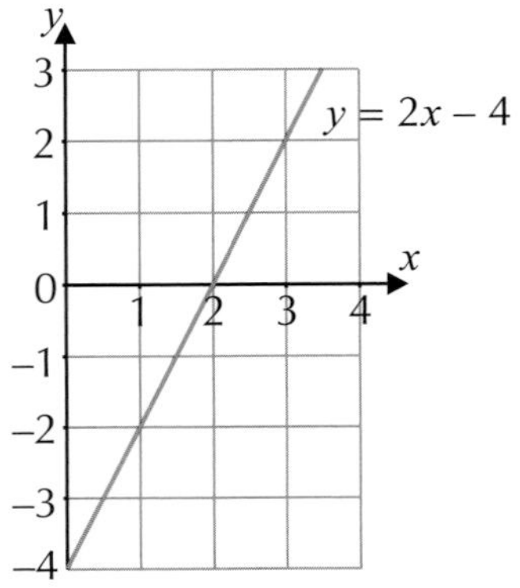

Q2 Use the graphs on the grid below to solve the following equations:

a) $0.5x - 2 = 0$

b) $-2x + 1 = 0$

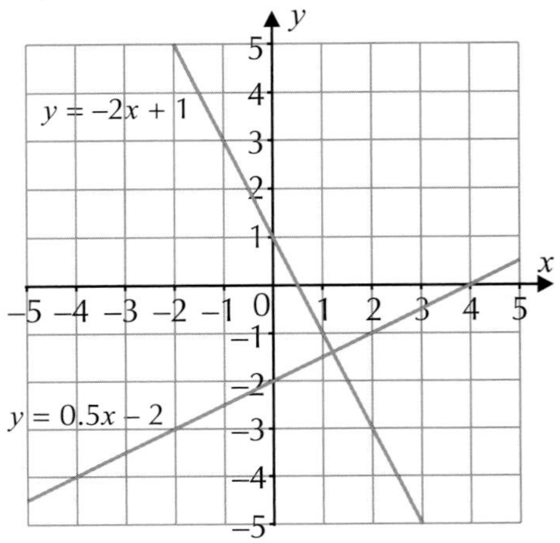

If you're given the graph of $y = ax + b$, you can actually **solve** the equation $ax + b = c$ for **any number** c. On the same axes, draw the graph of $y = c$. Now find the **point** where the two lines **intersect** — the **x-coordinate** of this point is the **solution**.

You can solve any equation that can be **rearranged** into the form '$ax + b = c$' in this way.

Example 4

The graph of $y = 2x - 1$ is shown below.

a) Use the graph to solve the equation $2x - 1 = 4$.

1. Draw the graph of $y = 4$. — $y = 4$ crosses $y = 2x - 1$
2. Find the point where the two lines cross, and read off the x-coordinate. — at $(2\frac{1}{2}, 4)$. So the solution is $x = 2\frac{1}{2}$.
3. Check your answer. — $2 \times 2\frac{1}{2} - 1 = 5 - 1 = 4$ ✔

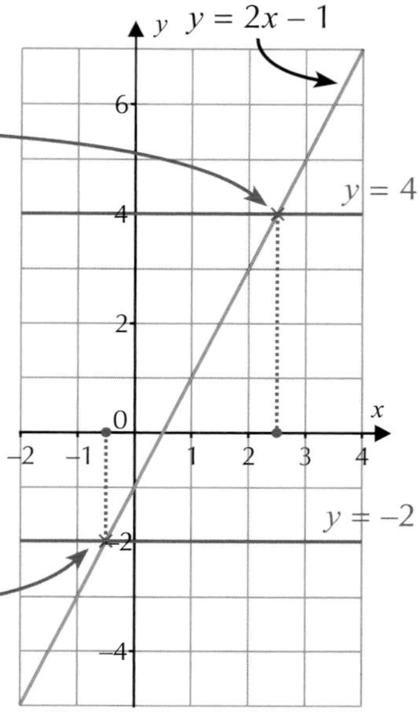

b) Use the graph to solve the equation $2x + 1 = 0$.

1. Subtract 2 from both sides to get '$2x - 1$' on the left. — $2x + 1 - 2 = 0 - 2$, $2x - 1 = -2$
2. Draw the graph of $y = -2$. — $y = -2$ crosses $y = 2x - 1$
3. Find the point where the two lines cross, and read off the x-coordinate. — at $(-\frac{1}{2}, -2)$. So the solution is $x = -\frac{1}{2}$.
4. Check your answer. — $2 \times \left(-\frac{1}{2}\right) + 1 = -1 + 1 = 0$ ✔

Exercise 3

Use the diagram on the right for Questions 1-3.

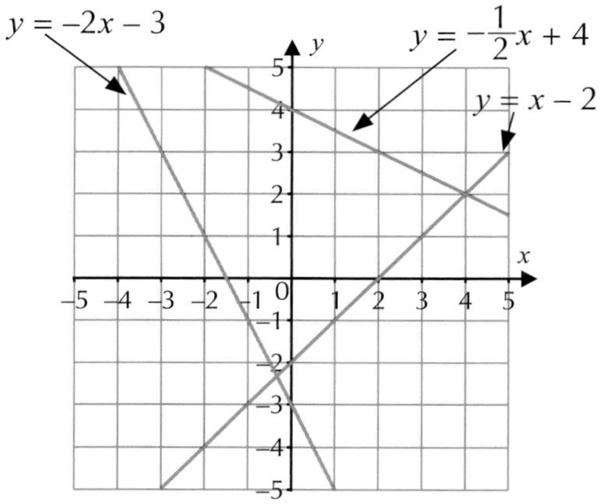

Q1 Use the graphs to solve the following equations.

a) $-2x - 3 = 1$ b) $x - 2 = -4$

c) $-2x - 3 = -5$ d) $-\frac{1}{2}x + 4 = 2\frac{1}{2}$

Q2 a) Rearrange the equation $-2x + 2 = 1$ to get '$-2x - 3$' on the left-hand side.

b) Hence, use the graph of $y = -2x - 3$ to solve the equation $-2x + 2 = 1$.

Q3 Use the graphs to solve the following equations.

a) $x - 1 = 2$ b) $-2x - 5 = 0$ c) $-\frac{1}{2}x - 1 = -2$

Solving Simultaneous Equations Graphically

Learning Objective — Spec Ref A19:
Be able to solve simultaneous equations graphically.

You can solve **simultaneous equations** by plotting the graphs of both equations and finding the point where they **intersect**.

Tip: You plot straight lines like before, with two on the same grid.

Example 5

Solve the following simultaneous equations graphically: $y = 8 - x$ and $y = 2x + 2$.

1. Both equations are straight lines. Make a table of x- and y-values and plot them.

$y = 8 - x$

x	0	4	8
y	8	4	0

$y = 2x + 2$

x	–1	0	1
y	0	2	4

2. Draw a straight line through each set of points.

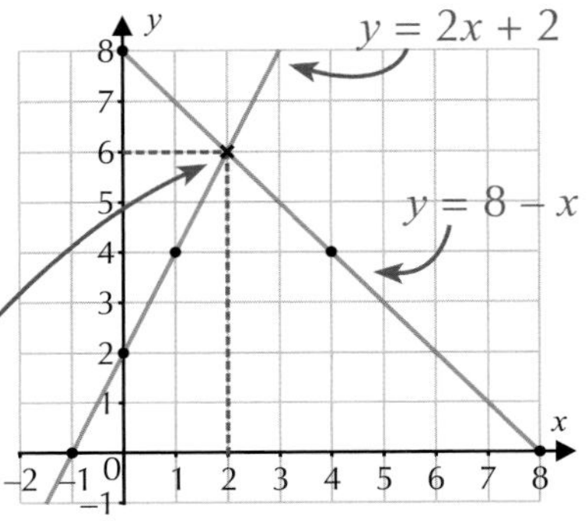

3. Find the x- and y-values of the point where the graphs intersect.

The graphs cross at (2, 6). So the solution is **$x = 2$ and $y = 6$.**

4. Check your answer.

$8 - 2 = 6$ ✔ and $2 \times 2 + 2 = 4 + 2 = 6$ ✔

Exercise 4

Q1 The diagram on the right shows the graphs of $y = x + 3$ and $y = -x + 7$.

Use the diagram to find the solution to the simultaneous equations $y = x + 3$ and $y = -x + 7$.

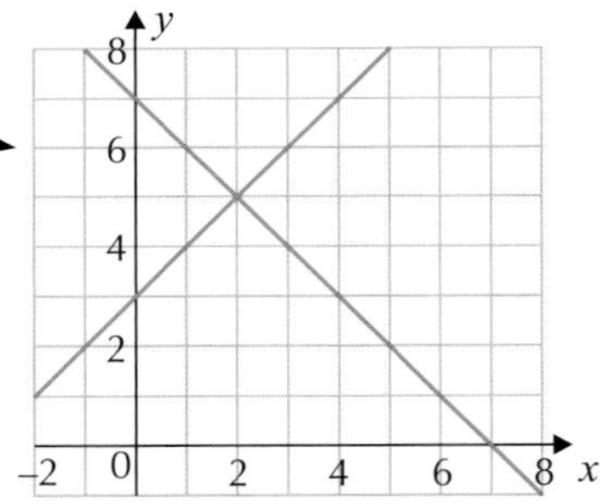

Q2 a) Draw the graph of $y = 5 - x$.

b) On the same axes, draw the graph of $y = x - 3$.

c) Use your graphs to solve the simultaneous equations $y = 5 - x$ and $y = x - 3$.

Q3 a) Draw the graphs of $y = 2 - 2x$ and $y = x + 5$ on the same axes.

b) Use your graphs to solve the equation $2 - 2x = x + 5$.

Q4 a) Draw the graphs of $y = x + 3$ and $y = x - 2$.

b) Explain how this shows that the simultaneous equations $y = x + 3$ and $y = x - 2$ have no solutions.

12.4 Gradients

The gradient of a straight-line tells you how steep it is — the bigger the number, the steeper the line. Gradients come in handy for lots of things — for example, in real life, you'll see gradients on road signs to describe the steepness of hills and slopes.

Learning Objective — Spec Ref A10:
Be able to find the gradient of a straight line.

To find the **gradient** of a line, divide the '**vertical distance**' (the change in the y-coordinates) between **two points** on the line by the '**horizontal distance**' (the change in the x-coordinates) between those points.

$$\textbf{Gradient} = \frac{\textbf{Vertical distance}}{\textbf{Horizontal distance}} = \frac{\textbf{Change in } y}{\textbf{Change in } x} = \frac{y_2 - y_1}{x_2 - x_1}$$

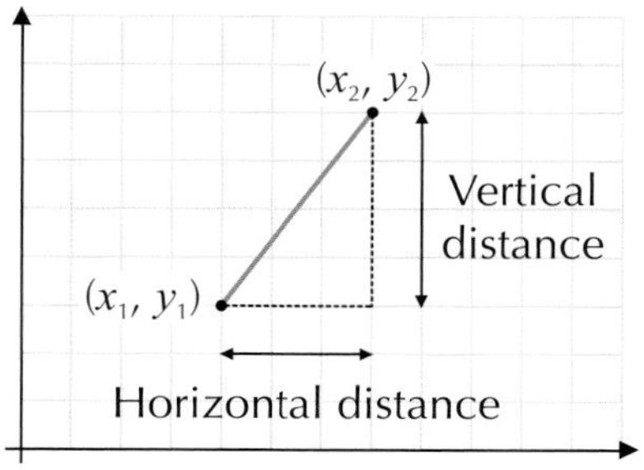

Make sure you subtract the x-coordinates and y-coordinates in the **same order** — i.e. if you do $y_2 - y_1$ on the numerator, you must do $x_2 - x_1$ on the denominator.

The **sign** of the gradient is important:

- A line sloping **upwards** from left to right has a **positive gradient.**
- A line sloping **downwards** from left to right has a **negative gradient.**

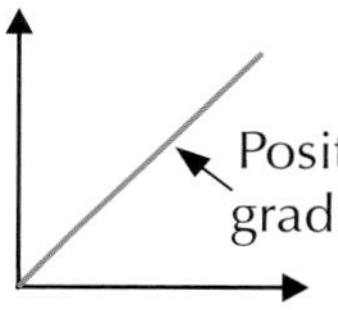

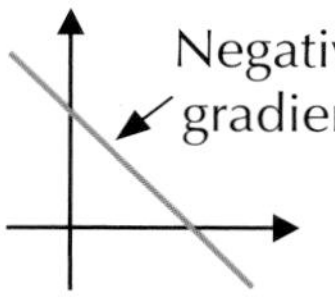

Gradients can be given in different ways — e.g. if the gradient is $\frac{1}{4}$, it has a **ratio** of **1:4**, a **percentage** of **25%** and you can say that the **slope** is **"1 in 4"**. Generally, your answer should be a **fraction.**

Road signs use %

Example 1

Find the gradient of the line containing the points $P(-3, 1)$ and $Q(4, 5)$.

1. Call the coordinates of P (x_1, y_1). Call the coordinates of Q (x_2, y_2).
2. Use the formula to calculate the gradient.
3. The line slopes upwards from left to right, so you should get a positive answer.

$$\text{Gradient} = \frac{y_2 - y_1}{x_2 - x_1} = \frac{5 - 1}{4 - (-3)} = \frac{4}{7}$$

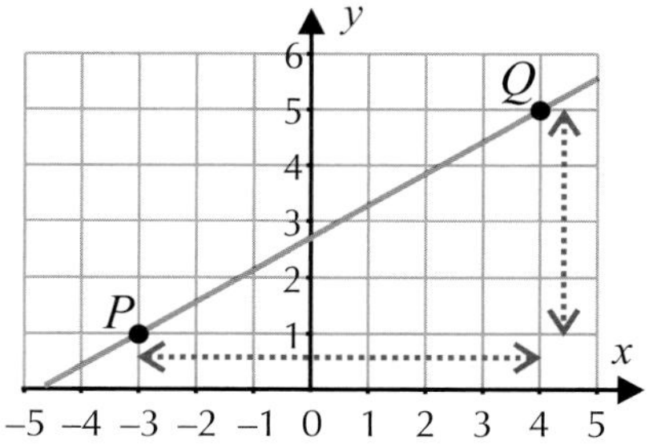

Exercise 1

Q1 The grid spacing in this question is 1 unit.

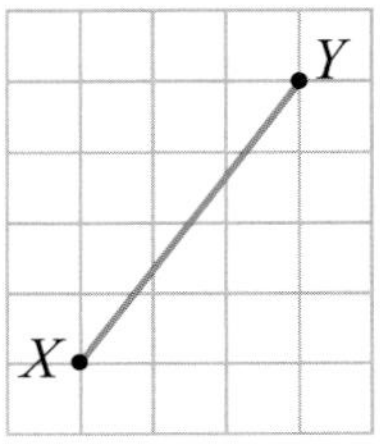

a) Write down the vertical distance between the points X and Y.

b) Write down the horizontal distance between X and Y.

c) State whether the gradient is positive or negative.

d) Calculate the gradient of the line segment XY.

Q2 Find the gradient of each of the following line segments. The grid spacing is 1 unit.

a) 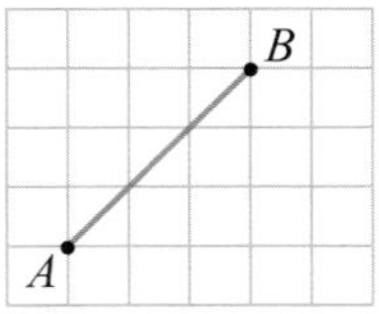

b) 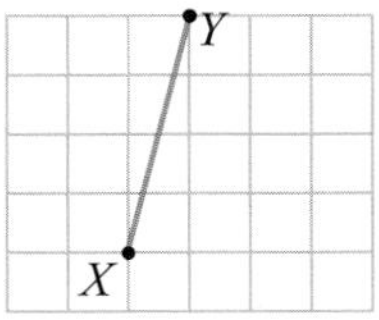

c) 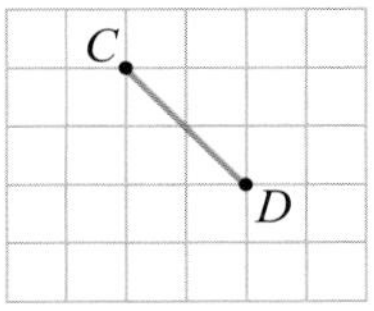

d) 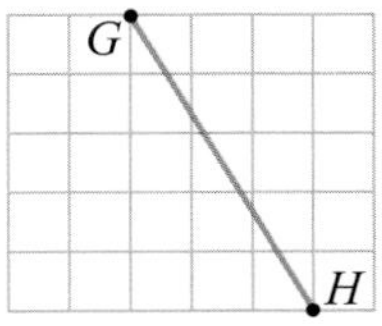

Q3 Use the points shown to find the gradient of each of the following lines.

a) 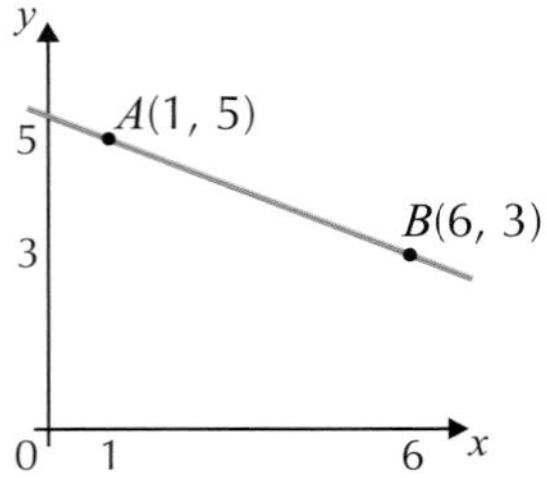

b) 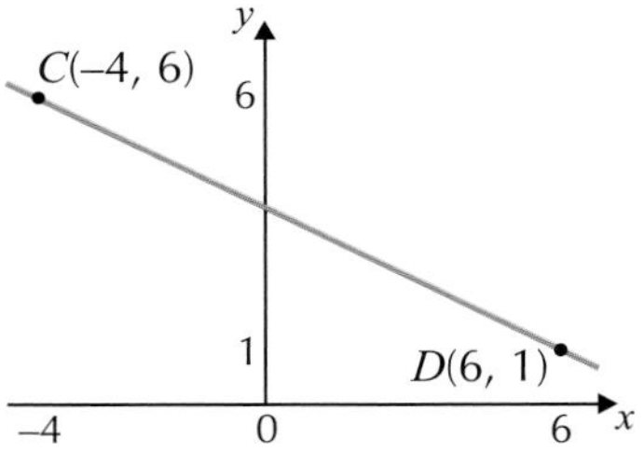

c) 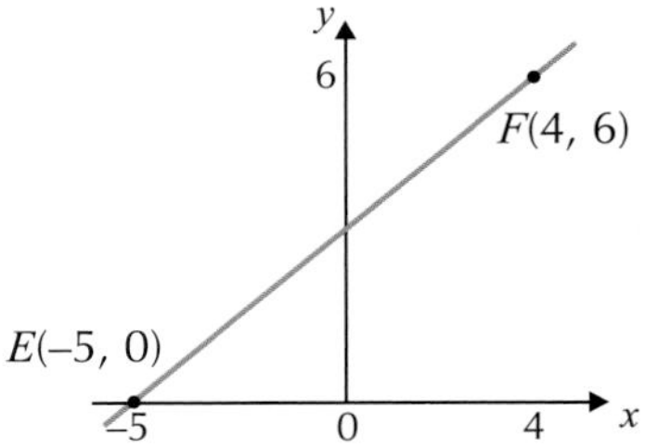

Q4 For each line shown below:

(i) Use the axes to find the coordinates of each of the marked points.

(ii) Find the gradient of each of the lines.

a) 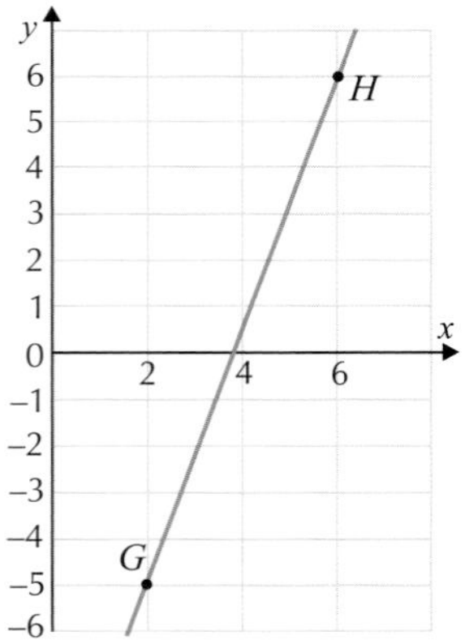

b) 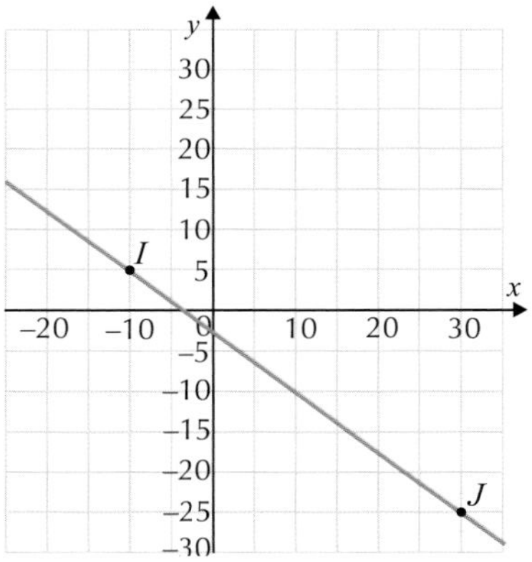

c) 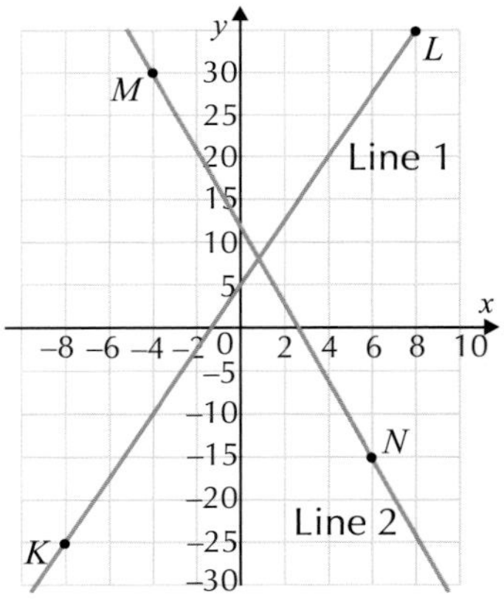

Q5 Find the missing coordinates, marked by letters, in each of the following.

a) A line with a gradient of 1 that goes through points (3, 5) and (7, a).

b) A line with a gradient of 2 that goes through points (1, 1) and (b, 5).

c) A line with a gradient of $\frac{1}{2}$ that goes through points (0, 0) and (8, c).

12.5 Equations of Straight-Line Graphs

The equation of a straight line depends on two things — the gradient of the line, and the point where it crosses the y-axis (the y-intercept). If you know these you can write down the equation.

y = mx + c

Learning Objective — Spec Ref A10:
Identify and interpret gradients and the y-intercept in equations of straight lines.

Prior Knowledge Check: Be able to find the gradient of a line. See p.199.

The **equation** of a **straight line** can be written in the form $\boldsymbol{y = mx + c}$,
where m and c are numbers — for example, $y = 3x + 5$ ($m = 3$, $c = 5$).

When written in this form:

- m is the **gradient** of the line,
- c tells you the **y-intercept** — the point where the line crosses the y-axis.

Make sure you don't mix up m and c when you get something like $y = 5 + 2x$.
Remember, m is the number in front of the x and c is the number on its own.

Watch out for **minus signs** too — both m and c can be negative (e.g. $y = -2x - 5$).

Example 1

Write down the gradient and the coordinates of the y-intercept of $y = 2x + 1$.

The equation is in the form $y = mx + c$, so you just need to read the values for the gradient and y-intercept from the equation.

$y = (2)x + (1)$ — m is 2, c is 1

The question asks for the coordinates of the y-intercept, so don't forget the x-coordinate.

gradient = **2**
y-intercept = **(0, 1)**

Tip: The x-coordinate of the y-intercept will always be 0.

Exercise 1

Q1 Write down the equation of the straight line with gradient = 2 and y-intercept = (0, 3).

Q2 Find the gradient and the coordinates of the y-intercept for each of the following graphs.

a) $y = 3x + 2$ b) $y = 2x - 4$ c) $y = 5x - 11$

d) $y = -3x + 7$ e) $y = 4x$ f) $y = \frac{1}{2}x - 1$

g) $y = x - 6$ h) $y = -x + 5$ i) $y = 3 - 6x$

Q3 Match the graphs to the correct equation from the box.

$y = x + 2$	$y = \frac{7}{3}x - 1$	$y = -x + 6$	$y = 3x$	$y = -\frac{1}{3}x + 4$	$y = \frac{1}{3}x + 2$

A

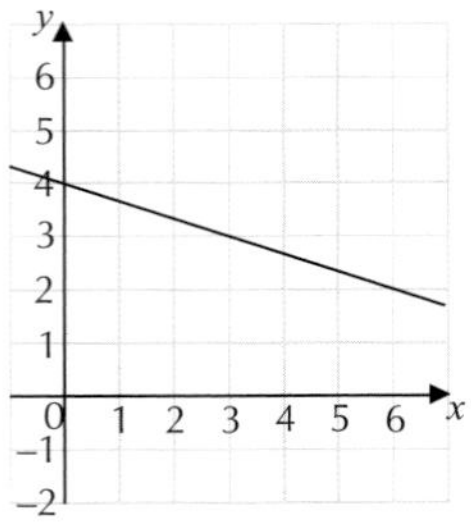

B

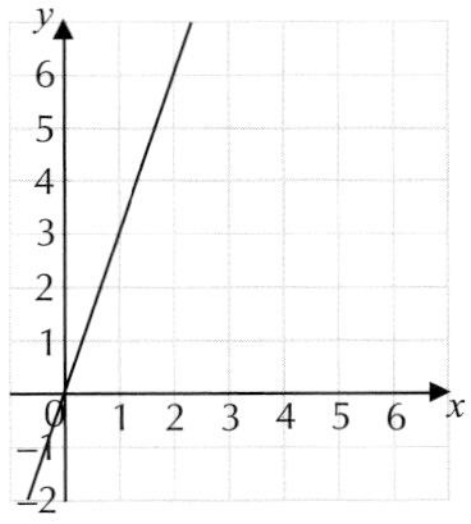

C

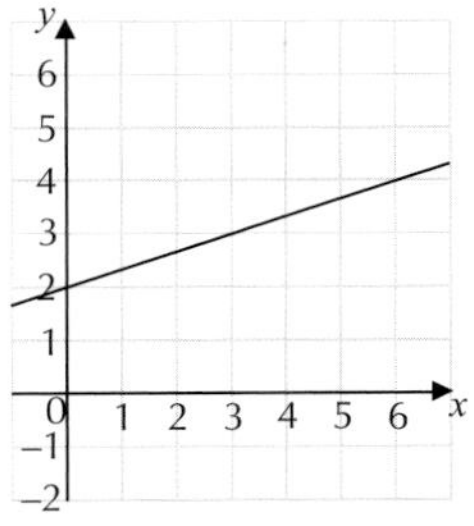

D

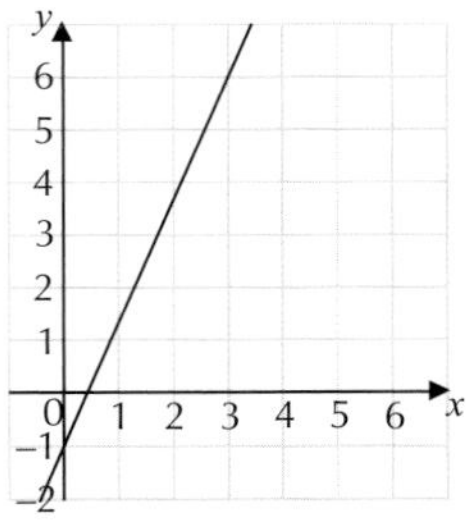

E

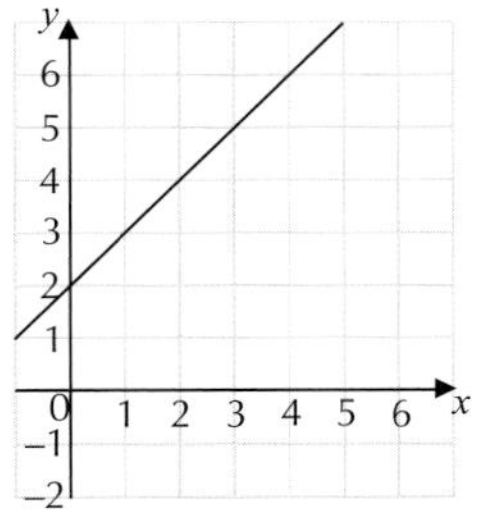

F

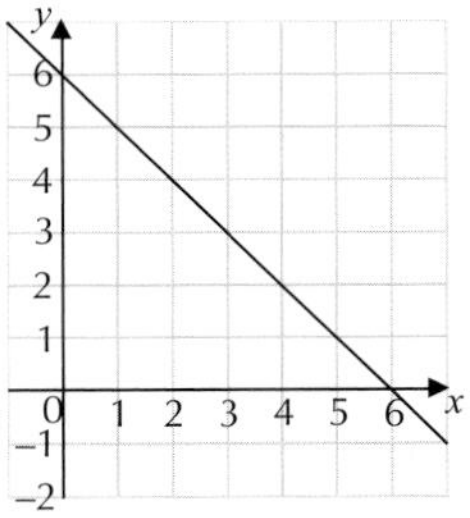

If you're given an equation that isn't in $y = mx + c$ form, **rearrange** it into this form. For example, $y - 3x = 2$ can be rearranged to $y = 3x + 2$ (with $m = 3$ and $c = 2$).

Example 2

Find the gradient and the coordinates of the *y*-intercept of $2x + 3y = 12$.

1. Rearrange the equation into the form $y = mx + c$.

$$2x + 3y = 12 \quad (-\,2x)$$
$$3y = -2x + 12 \quad (\div\, 3)$$
$$y = -\frac{2}{3}x + 4$$

($m = -\frac{2}{3}$, $c = 4$)

2. Write down the values for the gradient and *y*-intercept.

gradient = $\mathbf{-\frac{2}{3}}$
y-intercept = **(0, 4)**

Tip: To check your answer, substitute the *x*- and *y*-values of the *y*-intercept into the original equation:
$2 \times 0 + 3 \times 4 = 12$ ✔

Exercise 2

Q1 Find the gradient and the coordinates of the *y*-intercept for each of the following graphs.

a) $3y = 9 - 3x$

b) $y - 5 = 7x$

c) $3x + y = 1$

d) $3y - 6x = 15$

e) $4y - 6x + 8 = 0$

f) $6x - 3y + 1 = 0$

Parallel Lines

Learning Objective — Spec Ref A9:
Know that parallel lines have the same gradient.

Lines that are **parallel** have the **same gradient** — so their equations (in $y = mx + c$ form) all have the same value of $\boldsymbol{m}$.

For example, the lines $y = \mathbf{3}x$, $y = \mathbf{3}x - 2$ and $y = \mathbf{3}x + 4$ are all parallel.

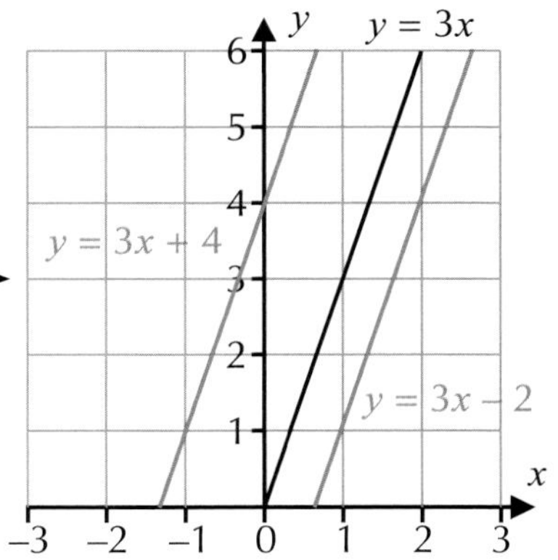

To **check** if two lines are parallel, **rearrange** their equations so that they're both in $y = mx + c$ form, then **compare** the values of m.

Example 3

Which of the following lines is parallel to the line $2x + y = 5$?
A: $y = 3 - 2x$ B: $x + y = 5$ C: $y - 2x = 6$

1. Rearrange the equation into the form $y = mx + c$ to find its gradient.

$2x + y = 5$
$y = 5 - 2x$
$y = -2x + 5$, so the gradient $(m) = -2$.

2. Rearrange the other equations in the same way. Any that have $m = -2$ will be parallel to $2x + y = 5$.

A:	B:	C:
$y = 3 - 2x$	$x + y = 5$	$y - 2x = 6$
$y = -2x + 3$	$y = -x + 5$	$y = 2x + 6$
$m = -2$	$m = -1$	$m = 2$

So **line A** is parallel to $2x + y = 5$.

Exercise 3

Q1 Write down the equations of three lines that are parallel to:

a) $y = 5x - 1$ b) $x + y = 7$

Q2 Which of the following are the equations of lines parallel to: a) $y = 2x - 1$ b) $2x - 3y = 0$

A: $y - 2x = 4$ B: $2y = 2x + 5$ C: $2x - y = 2$

D: $2x + y + 7 = 0$ E: $3y + 2x = 2$ F: $6x - 9y = -2$

Q3 Which of the lines listed below are parallel to the line shown in the diagram?

A: $y + 3x = 2$ B: $3y = 7 - x$ C: $y = 4 - 3x$

D: $x - 3y = 8$ E: $y = 3 - \frac{1}{3}x$ F: $6y = -2x$

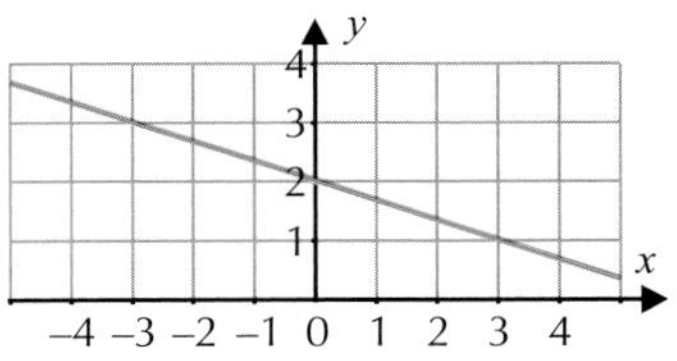

Finding the Equation of a Line

Learning Objectives — Spec Ref A9:

- Find the equation of a straight line given its gradient and a point on the line.
- Find the equation of a straight line given the coordinates of two points on the line.

You can find the equation of a line using its **gradient** and **one point** on the line.

- First, substitute the values of the **gradient** (m) and the **coordinates** of the known point (x, y) into $y = mx + c$. You'll be left with an equation where c is the only unknown.
- **Solve** this equation to find the value of c.
- Finally, put your values of m and c into $y = mx + c$ to give the **equation of the line**.

Tip: You might need to find the gradient from the equation of a parallel line — look back at p.203 for this.

Example 4

Find the equation of the line which passes through the point (5, 8) and has a gradient of 3.

1. Write down the equation $y = mx + c$, then substitute the values of the gradient m and the known point (x, y).

   $y = mx + c$
   $8 = 3 \times 5 + c$
   $8 = 15 + c$

2. Solve the equation to find the value of c.

   $8 - 15 = 15 + c - 15$
   $-7 = c$

3. Put your values of m and c into $y = mx + c$.

   So the equation of the line is $\mathbf{y = 3x - 7}$.

Tip: It's a good idea to check the known point works in your equation:
$3 \times 5 - 7 = 15 - 7 = 8$ ✔

Exercise 4

Q1 Find the equations of the following lines based on the information given.

a) gradient = 8, passes through (0, 2)
b) gradient = –1, passes through (0, 7)
c) gradient = 3, passes through (1, 10)
d) gradient = $\frac{1}{2}$, passes through (4, –5)
e) gradient = –7, passes through (2, –4)
f) gradient = 5, passes through (–3, –7)

Q2 For each of the following, find the equation of the line which is parallel to the given line and passes through the given point. Give your answers in the form $y = mx + c$.

a) $y = 3x + 2$, (0, 5)
b) $y = 5x - 3$, (1, –4)
c) $y = 2x + 1$, (1, 6)
d) $y = \frac{1}{2}x + 3$, (6, –7)
e) $2y = 6x + 3$, (–3, 4)
f) $x + y = 4$, (8, 8)

If you only know **two points** on the line, you can calculate the **gradient** of the line using the method on p.199. Then follow the method from the previous page (using either of the two points) to find the **equation of the line**.

Example 5

Find the equation of the straight line that passes through the points $A(-3, -4)$ and $B(-1, 2)$.

1. Write down the equation for a straight line. $y = mx + c$
2. Find the gradient (m) of the line. $m = \frac{y_2 - y_1}{x_2 - x_1} = \frac{2-(-4)}{-1-(-3)} = \frac{6}{2} = 3$
3. Substitute the value for the gradient and the x and y values for one of the points into $y = mx + c$, then solve to find c.
   At point B, $x = -1$ and $y = 2$.
   $2 = 3 \times (-1) + c$
   $2 = -3 + c$
   $c = 5$
4. Put your values of m and c into $y = mx + c$. So the equation of the line is $\boldsymbol{y = 3x + 5}$.

Exercise 5

Q1 Find the equations of the lines passing through the following points.

a) (3, 7) and (5, 11) b) (5, 1) and (2, –5) c) (4, 1) and (–3, –6)

d) (–2, 1) and (1, 7) e) (2, 8) and (–1, –1) f) (–3, 2) and (–2, 5)

Q2 Find the equations of the lines A to H shown below.
Write all your answers in the form $y = mx + c$.

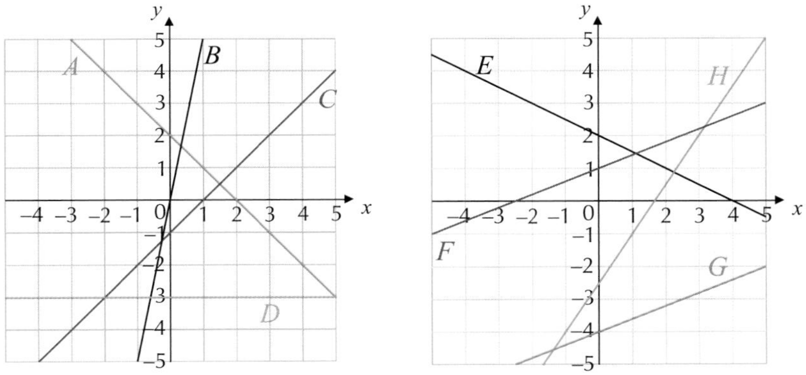

Q3 Find the equation of the line shown in the diagram on the right.

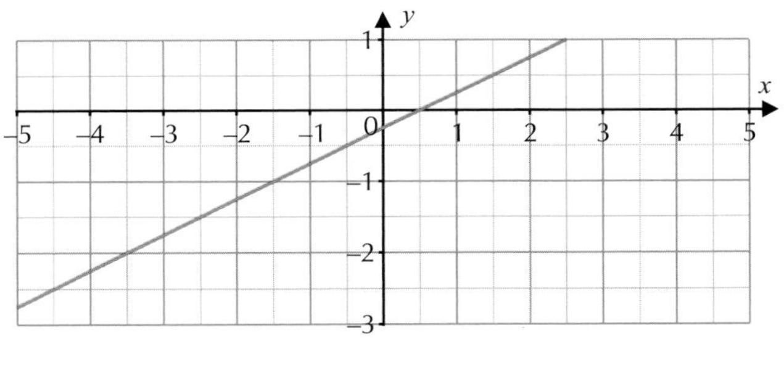

12.6 Quadratic Graphs

Quadratic expressions and equations were covered in Section 8 and Section 9. In a quadratic function $y = ax^2 + bx + c$, values of x and y are connected using a quadratic expression. You can plot graphs of such functions using a table of values like you did for straight lines.

Drawing Quadratic Graphs

Learning Objective — Spec Ref A12:
Be able to draw the graphs of quadratic functions.

Quadratic functions always involve an x^2 **term** (but no higher powers of x, such as x^3, x^4...). The graph of a quadratic function is always a **curve** called a **parabola**.

- If the coefficient of x^2 is **positive** (e.g. $3x^2 + x - 2$), the parabola is **u-shaped**.
- If the coefficient of x^2 is **negative** (e.g. $-3x^2 - x + 2$), the parabola is **n-shaped**.

Making a **table** can help with **drawing** quadratic graphs:

- In the **first row** write the ***x*-values** that you're going to plot.
- Make a **row** for each **term** in the quadratic and find the value of that term for each x-value.
- The **last row** in the table is the **quadratic itself** — because you have calculated each term, you can just **add** the values in that column.
- Now you can **plot** the graph using the values in the **first and last row** of each column. Join the points with a **smooth curve** — if it's not smooth, then you've got a point wrong.

Tip: The range of x-values will be given in the question.

Example 1

a) Complete the table to find the value of $y = x^2 - 3$ for values of x from –3 to 3.

b) Draw the graph of $y = x^2 - 3$ using the values of x and $y = x^2 - 3$ in your table.

x	–3	–2	–1	0	1	2	3
x^2							
–3							
$x^2 - 3$							

1. Fill in the table one row at a time.

x	–3	–2	–1	0	1	2	3
x^2	9	4	1	0	1	4	9
–3	–3	–3	–3	–3	–3	–3	–3
$x^2 - 3$	6	1	–2	–3	–2	1	6

2. Now plot each x-value from the first row against the corresponding y-value ($= x^2 - 3$) from the third row, and join the points with a smooth curve.

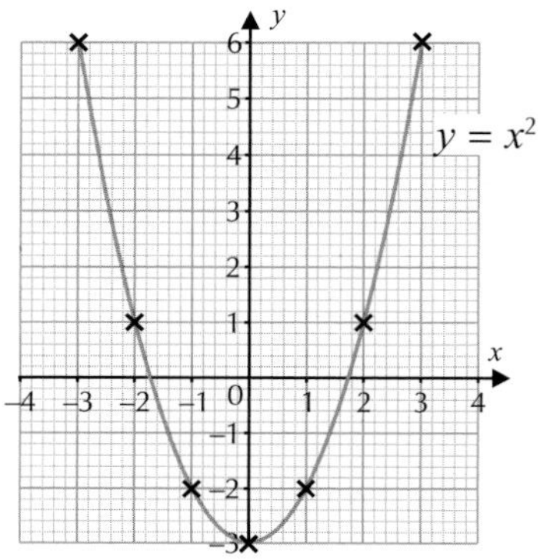

Exercise 1

Q1 a) Copy and complete the table below to find the value of $y = x^2 + 2$ for values of x from –3 to 3.

x	–3	–2	–1	0	1	2	3
x^2	9		1			4	
2	2		2			2	
$x^2 + 2$	11		3			6	

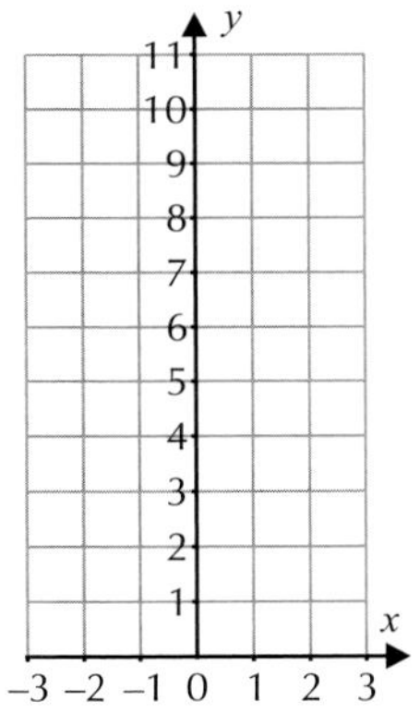

b) Copy the grid on the right and plot the points from your table.

c) Join up the points to draw the graph of $y = x^2 + 2$.

Q2 a) Copy and complete the table to show the value of $y = x^2 - 1$ for values of x from –3 to 3.

b) Draw a set of axes with x-values from –3 to 3 and y-values from –1 to 8. Draw the graph of $y = x^2 - 1$ on your axes.

x	–3	–2	–1	0	1	2	3
x^2		4					9
–1		–1					–1
$x^2 - 1$		3					8

Q3 For each of the following equations:
(i) Complete a table to show the value of y for values of x from –3 to 3.
(ii) Draw a graph of the equation on suitable axes.

a) $y = x^2 + 3$ b) $y = 5 - x^2$

Example 2

a) Complete the table to find the value of $y = x^2 - x$ for values of x from –3 to 3.

b) Draw the graph of $y = x^2 - x$ for values of x from –3 to 3.

x	–3	–2	–1	0	1	2	3
x^2		4			1		
$-x$		2			–1		
$x^2 - x$		6			0		

1. Fill in the table one row at a time. Find the entry for the '$x^2 - x$' row by adding the entries in the 'x^2' and '$-x$' rows.

x	–3	–2	–1	0	1	2	3
x^2	9	4	1	0	1	4	9
$-x$	3	2	1	0	–1	–2	–3
$x^2 - x$	12	6	2	0	0	2	6

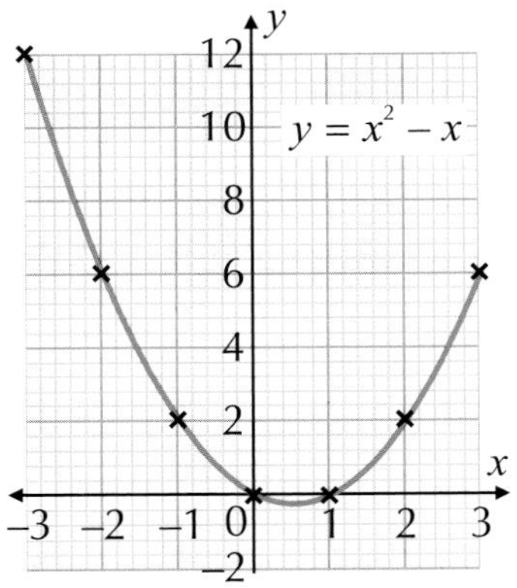

2. Now plot each x-value from the first row against the corresponding y-value ($= x^2 - x$) from the fourth row. Join the points with a smooth curve.

Exercise 2

Q1 a) Copy and complete the table to show the value of $y = x^2 - 2x$ for values of x from –3 to 3.

x	–3	–2	–1	0	1	2	3
x^2		4					9
$-2x$		4					–6
$x^2 - 2x$		8					3

b) Draw a set of axes with x-values from –3 to 3 and y-values from –1 to 15. Draw the graph of $y = x^2 - 2x$ on your axes.

Q2 For each of the following quadratic equations:
(i) Complete a table to show the value of y for values of x from –3 to 3.
(ii) Draw a graph of the equation on suitable axes.

a) $y = x^2 + 3x$ b) $y = x^2 - 4x$

Example 3

Draw the graph of $y = x^2 + 3x - 2$ for values of x from –3 to 3.

1. Make a table — include separate rows for x^2, $3x$ and -2. The last row should add the three entries above.

x	–3	–2	–1	0	1	2	3
x^2	9	4	1	0	1	4	9
$3x$	–9	–6	–3	0	3	6	9
-2	–2	–2	–2	–2	–2	–2	–2
$x^2 + 3x - 2$	–2	–4	–4	–2	2	8	16

2. Now plot each x-value from the first row against the corresponding y-value ($= x^2 + 3x - 2$) from the final row. Join the points with a smooth curve.

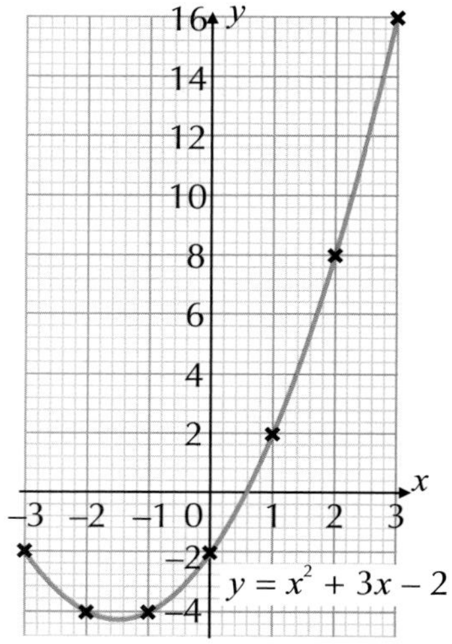

Exercise 3

Q1 For each of the following quadratic equations:
(i) Complete a table to show the value of y for values of x from –3 to 3.
(ii) Draw a graph of the equation on suitable axes.

a) $y = x^2 + 2x + 5$ b) $y = -x^2 - x - 1$

Q2 a) Complete a table to show the value of $y = 2x^2 + 3x - 7$ for values of x from –4 to 4.
b) Draw a graph of $y = 2x^2 + 3x - 7$ on suitable axes.

Solving Quadratic Equations Graphically

Learning Objective — Spec Ref A11/A18:
Be able to solve quadratic equations graphically.

You can use the **graph** of $y = ax^2 + bx + c$ to estimate solutions to the **quadratic equation** $ax^2 + bx + c = 0$ — the method is:

- Find the **points** where the x-axis **intersects** the quadratic graph — this is where $y = 0$. It can happen **once**, **twice** or **not at all**.
- **Read off** the **x-coordinates** of these points — they will be **approximate solutions** to the quadratic equation.

Tip: Often, you can only estimate because it's hard to read off the exact value — read the scale carefully and just be as accurate as you can.

Example 4

This graph shows the curve $y = x^2 - 2x - 1$. Use the graph to estimate both solutions (or roots) to the equation $x^2 - 2x - 1 = 0$. Give your answers to 1 decimal place.

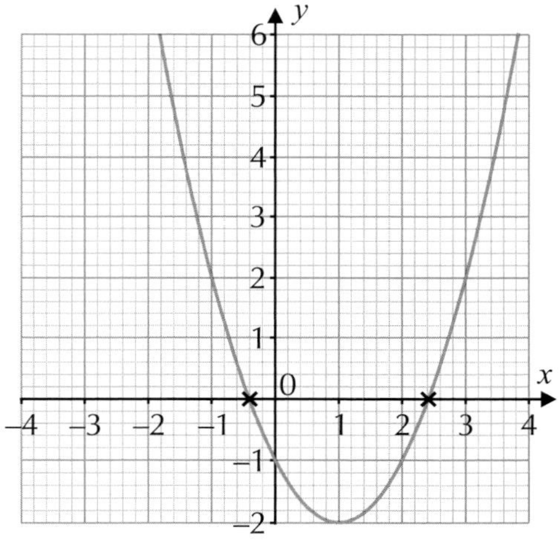

1. The left hand side of the equation $x^2 - 2x - 1 = 0$ is represented by the curve $y = x^2 - 2x - 1$ and the right-hand side is represented by the line $y = 0$ (the x-axis).
2. So, the estimated solutions are found where the curve crosses the x-axis. Read off the x-values:

$\mathbf{x = -0.4}$ and $\mathbf{x = 2.4}$

Exercise 4

Q1 Using the graphs shown below, find the solutions to the following equations.

a) $x^2 + 4x = 0$

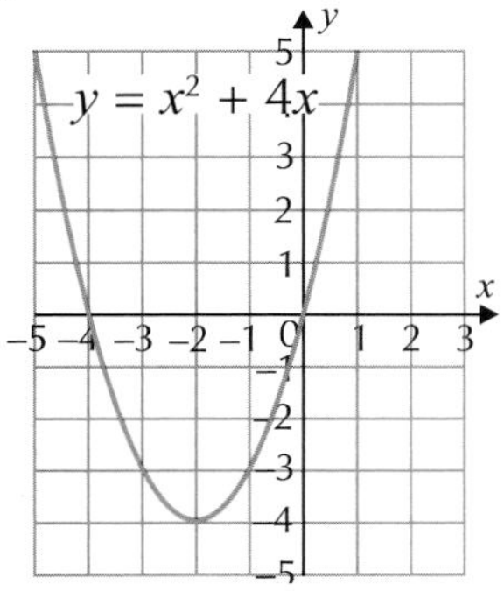

b) $x^2 - 2x - 3 = 0$

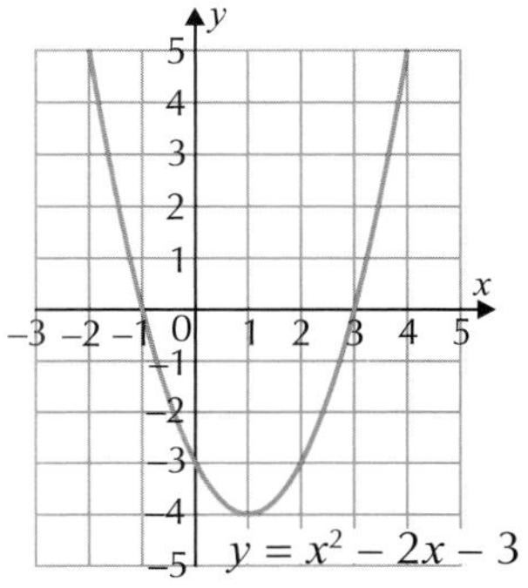

c) $-x^2 + 4 = 0$

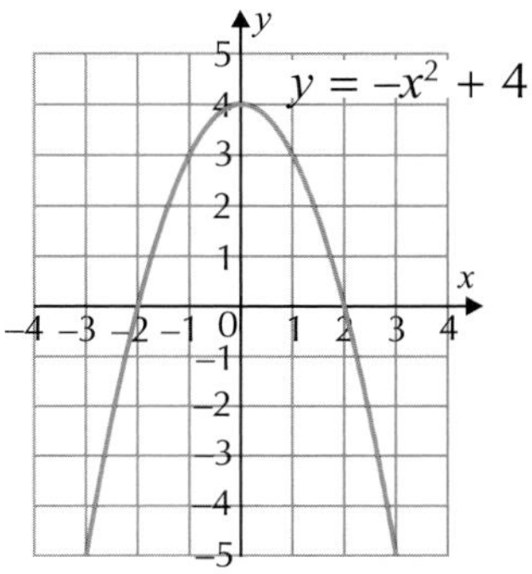

Q2 a) Draw the graph of $y = x^2 - 5x + 3$ for values of x from 0 to 5.
b) Use your graph to estimate the solutions to the equation $x^2 - 5x + 3 = 0$. Give your answers to 1 decimal place.

Turning Points of Quadratics

Learning Objective — Spec Ref A11:
Be able to identify the turning points of quadratic curves.

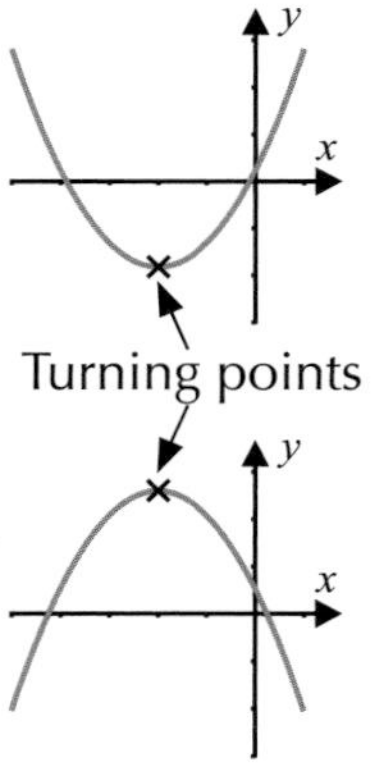

The **turning point** of a quadratic graph is at the **minimum** y-value if the curve is **u-shaped** or at the **maximum** y-value if the curve is **n-shaped**.

The **coordinates** of the turning point can be found as follows.

- The **x-coordinate** of the turning point is always **halfway** between any two points on the curve with the **same y-value:** (x_1, y) and (x_2, y). The points where the curve **intersects the x-axis** are often used, as they both have $y = 0$.
- The turning point has x-coordinate $\frac{x_1 + x_2}{2}$. Put this value of x into the **equation** of the graph to find the **y-coordinate** of the turning point.

Example 5

Find the turning point of the graph of $y = x^2 + x - 6$.

1. Find two points on the curve with the same y-value.

   The curve crosses the x-axis at (–3, 0) and (2, 0).

2. The x-coordinate of the turning point is halfway between these two x-values.

   x-coordinate = (–3 + 2) ÷ 2 = –0.5

3. Put the x-coordinate back into the equation to find the y-coordinate.

   y-coordinate = $(-0.5)^2 + (-0.5) - 6 = -6.25$
   So the turning point has coordinates **(–0.5, –6.25).**

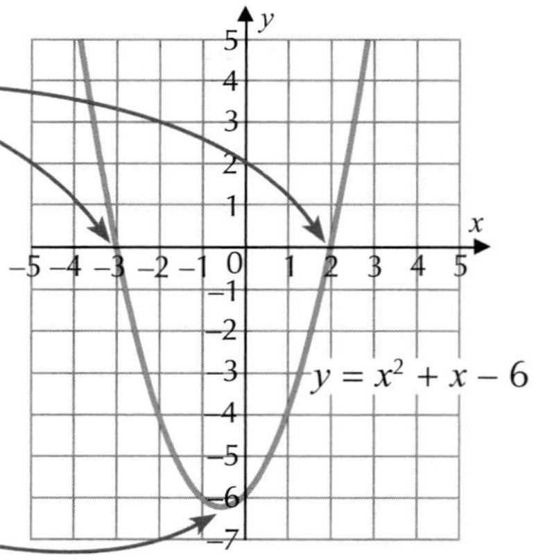

Exercise 5

Q1 For each of the following graphs, decide whether the turning point is a maximum or minimum.

a) $y = x^2 + 2$ b) $y = 7 - x^2$ c) $y = x^2 + 2x$

d) $y = 3x^2 - 10x + 7$ e) $y = -x^2 + 9x - 5$ f) $y = -4x^2 + 6x + 7$

Q2 Find the coordinates of the turning points of these graphs.

a)

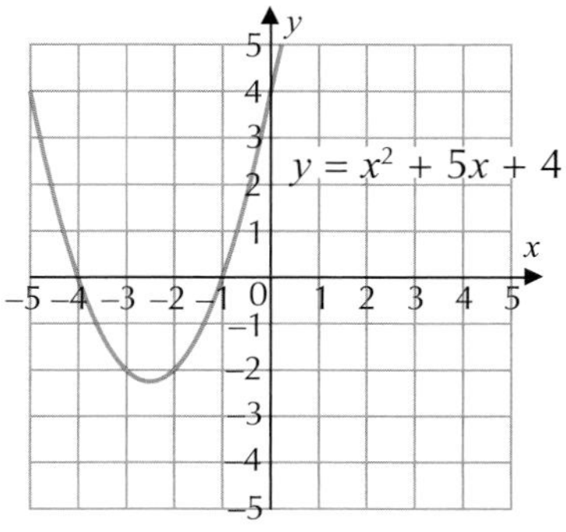

b)

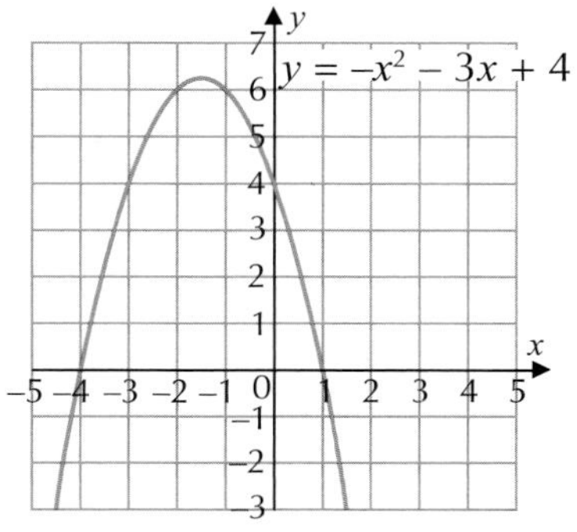

Q3 Find the turning points of these equations given the two points on the curve.

a) $y = x^2 - 6x$ goes through points (–1, 7) and (7, 7).

b) $y = x^2 - 2x + 3$ goes through points (0, 3) and (2, 3).

c) $y = 8x - x^2$ goes through points (2, 12) and (6, 12).

You can also find the **turning point** if you only have the **equation** of the quadratic. Make a **table** as if you were going to plot the graph and look for **two** x-values that give the **same** y-value. Once you have these x-values, you can follow the method from the previous page.

Tip: You don't need to draw the graph — the table is just to make spotting the x-values easier.

Example 6

Find the turning point of the graph of $y = x^2 - 3x + 7$.

1. Draw a table of values and look for two x-values that give the same y-value.

x	−2	−1	0	1	2
y	17	11	7	5	5

2. The x-coordinate of the turning point is halfway between these two x-values.

$x = 1$ and $x = 2$ both give $y = 5$

x-coordinate = $(1 + 2) \div 2 = 1.5$

3. Put the x-coordinate back into the equation to find the y-coordinate.

y-coordinate = $1.5^2 - (3 \times 1.5) + 7 = 4.75$
So the turning point = **(1.5, 4.75)**.

Exercise 6

Q1 By drawing tables for x values in the range $-3 \leq x \leq 3$, find the turning points of each of these curves.

a) $y = x^2 - 2x$
b) $y = x^2 + 3x + 11$
c) $y = x^2 + 5x - 9$
d) $y = 3x^2 + 12x - 8$
e) $y = -2x^2 - 2x + 9$
f) $y = -6x^2 + 18x - 3$

12.7 Harder Graphs

In this topic you'll learn about cubic and reciprocal functions — cubics have an x^3 term and reciprocals involve dividing by x. The graphs of these functions have their own distinctive shape.

Cubic Graphs

Learning Objective — Spec Ref A12:
Recognise and be able to sketch the graphs of cubic functions.

Cubic functions have x^3 as the **highest power** of x.
Cubic graphs all have the same basic shape — a curve with a '**wiggle**' in the middle.

- If the coefficient of x^3 is **positive**, the curve goes **up** from the **bottom left**.

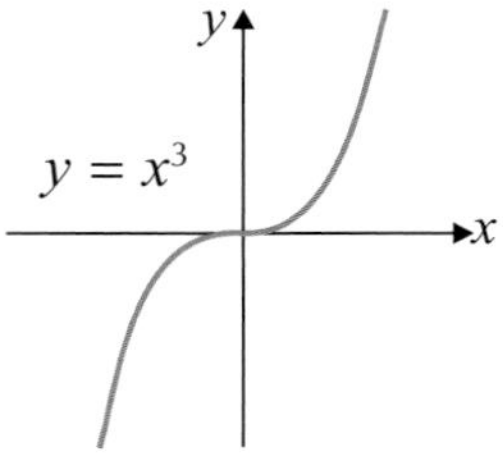

- If the coefficient of x^3 is **negative**, the curve goes **down** from the **top left**.

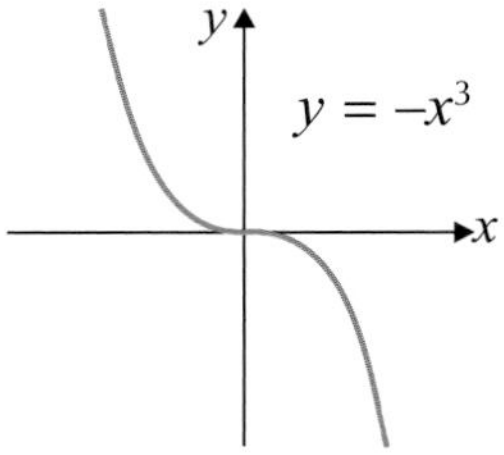

To **draw the graph** of a cubic function $y = ax^3 + bx^2 + cx + d$, find y-values using x-values in a **table**, then **plot** the coordinates and draw a **smooth curve** through the points.

Tip: This is like the method for plotting quadratics on p.206.

Example 1

Draw the graph of $y = x^3 + 1$.

1. Draw a table with rows for each term to help you work out the y-value for each value of x.

x	–3	–2	–1	0	1	2
x^3	–27	–8	–1	0	1	8
+1	1	1	1	1	1	1
$x^3 + 1$	–26	–7	0	1	2	9

2. Plot the coordinates and join the points with a smooth curve — use a pencil, as it may take a few tries to get right.

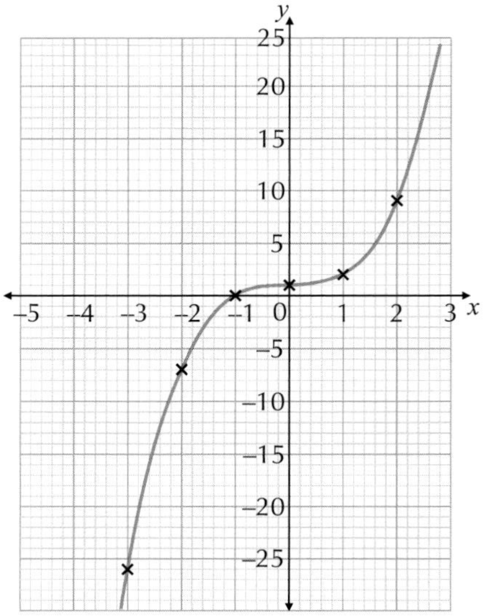

Exercise 1

Q1 Copy and complete each table and draw graphs for the following cubic equations.

a) $y = x^3 + 5$

x	–3	–2	–1	0	1	2	3
x^3	–27					8	
$x^3 + 5$	–22					13	

b) $y = 5 - x^3$

x	–3	–2	–1	0	1	2	3
5		5					5
$-x^3$		8					–27
$5 - x^3$		13					–22

c) $y = -4 - x^3$

x	–3	–2	–1	0	1	2	3
–4	–4				–4		
$-x^3$	27				–1		
$-4 - x^3$	23				–5		

Q2 Draw the graph of $y = x^3 + 3$ for values of x between –3 and 3.

Q3 Draw the graph of $y = x^3 - 3x + 7$ for values of x between –4 and 4.

Reciprocal Graphs

Learning Objective — Spec Ref A12:
Recognise and be able to sketch the graphs of reciprocal functions.

The equations of basic **reciprocal** graphs have the form: $y = \frac{A}{x}$ where A is constant

A reciprocal graph appears as **two curves**,
which are **symmetrical** about the lines $y = x$ and $y = -x$ and don't touch either the x- or y-axis.

When A is **positive** the curves are in the **top-right** and **bottom-left** quadrants.

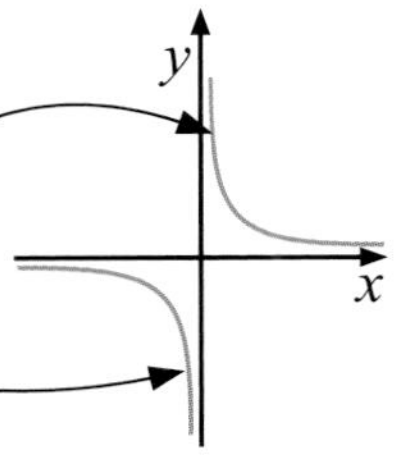

When A is **negative** the curves are in the **top-left** and **bottom-right** quadrants.

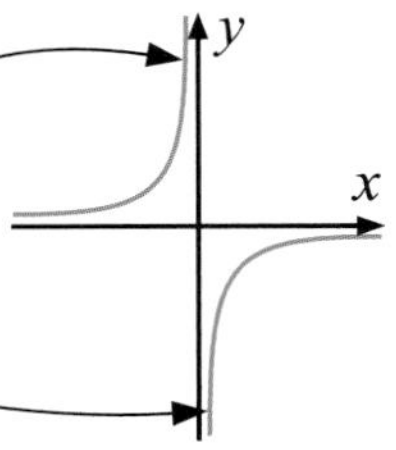

To **draw the graph** of a reciprocal function, make a table of x- and y-values — use both **positive** and **negative** values of x, and include **decimal values** between **–1** and **1**.
Plot the coordinates from the table and draw **two smooth curves** through the points.

Example 2

Draw the graph of $y = \frac{1}{x}$ for values of x between –5 and 5.

1. Draw a table of values for the equation.

x	–5	–4	–2	–1	–0.5	–0.2
$\frac{1}{x}$	–0.2	–0.25	–0.5	–1	–2	–5

x	0.2	0.5	1	2	4	5
$\frac{1}{x}$	5	2	1	0.5	0.25	0.2

2. Plot the coordinates and join the points to make two smooth curves.

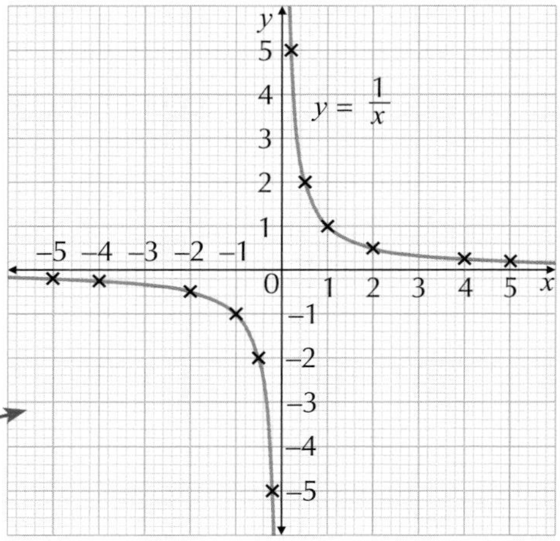

Exercise 2

Q1 Which of these equations would produce a reciprocal graph?

A: $y = \frac{x}{3}$	B: $y = \frac{9}{x}$	C: $y = \frac{6}{-x}$	D: $9 = \frac{y}{x}$
E: $xy = 5$	F: $3 = \frac{x}{y}$	G: $y = \frac{8}{2x}$	H: $y = \frac{2}{x} + x$

Q2 Copy and complete each table and draw graphs for the following reciprocal functions.

a) $y = \frac{2}{x}$

x	–5	–4	–2	–1	–0.5	–0.1	0.1	0.5	1	2	4	5
$\frac{2}{x}$		–0.5		–2						1		

b) $y = -\frac{1}{x}$

x	–5	–4	–2	–1	–0.5	–0.1	0.1	0.5	1	2	4	5
$-\frac{1}{x}$		0.25		1								

c) $y = \frac{3}{x}$

x	–5	–4	–2	–1	–0.5	–0.1	0.1	0.5	1	2	4	5
$\frac{3}{x}$		–0.75		–3								

Review Exercise

Q1 a) Plot the points $A(0, 4)$, $B(2, 6)$, $C(4, 4)$ and $D(2, 0)$ on a set of axes.

b) Join the points. What kind of shape have you made?

Q2 The points A and B have coordinates $A(2, 5)$ and $B(-4, 1)$.
Find the coordinates of the midpoint of the line segment AB.

Q3 a) Write down the coordinates of each of the points A to E.

b) Find the coordinates of the midpoint of AB.

c) Find the coordinates of the midpoint of BC.

d) Write down the equation of the line that passes through points A and E.

e) Find the gradient of the line BC.

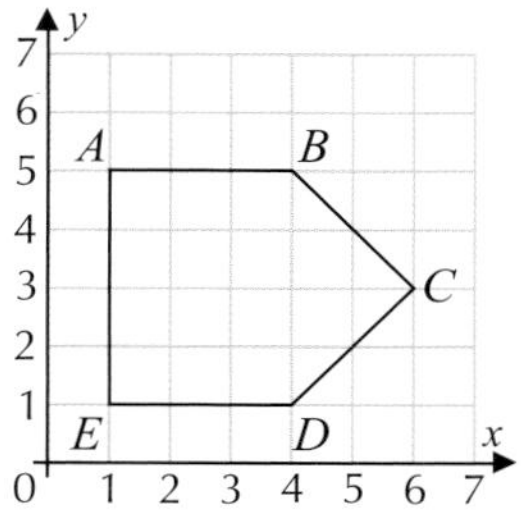

Q4 Draw each of the following lines on a set of axes.

a) $x = 1$ b) $y = -3$ c) $x = -1$ d) $y = 5$

Q5 Find the gradient of each of the line segments shown below.

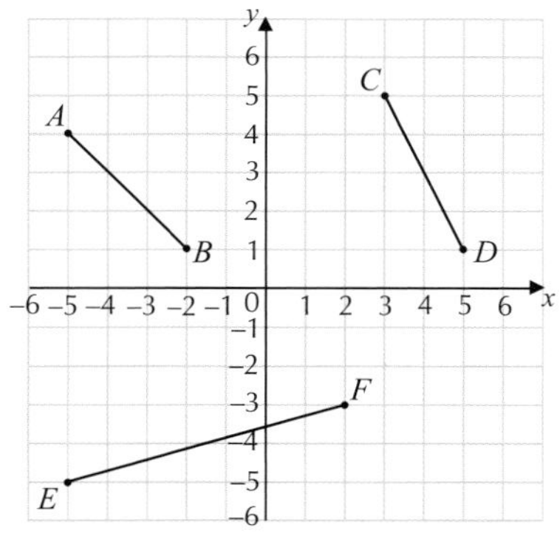

Q6 Match the graphs shown below with the correct equation from the list:

$y = x$ $x = -5$ $y = x^2$ $y = -4$

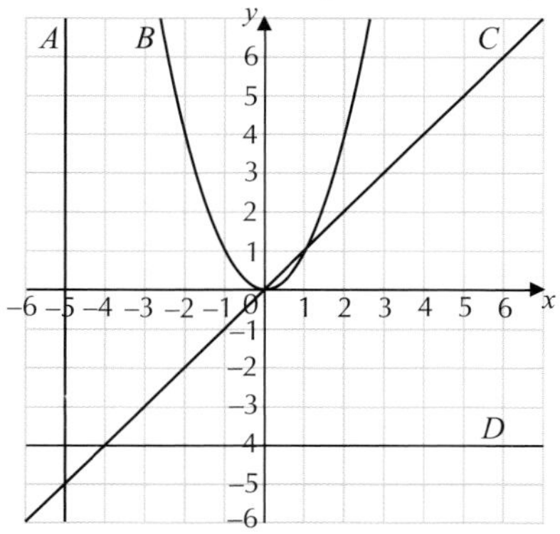

Q7 a) Copy and complete the table to show the value of $y = 2x + 2$ for values of x from –3 to 3.

b) Plot the graph of $y = 2x + 2$.

c) Use your graph solve the equation $2x + 2 = 5$.

x	–3	–2	–1	0	1	2	3
y							
Coordinates							

Q8 a) On the same axes, draw the graphs of $y = 2x + 3$ and $y = 4x + 2$.

b) Use your graphs to solve the simultaneous equations $y = 2x + 3$ and $y = 4x + 2$.

Q9 Find the gradient of the line passing through the following points.

a) (0, 0) and (3, 15) b) (–4, –5) and (–1, 4)

c) (–7, 2) and (–4, –1) d) (5, –2) and (2, 10)

Q10 For each of the following equations, write down:

(i) the gradient of the line

(ii) the coordinates of the line's y-intercept

a) $y = 2x + 3$ b) $y = 3 - x$ c) $y = -\frac{2}{3}x - 1$

Q11 a) Draw the line $y = 6 - 3x$ for values of x from –1 to 3.

b) Find the equation of the line parallel to this one that passes through (2, –4).

Q12 Sketch the general shape of the following:

a) A quadratic graph where the number in front of the x^2 term is positive.

b) A cubic graph where the number in front of the x^3 term is negative.

c) A reciprocal graph of the form $y = \frac{A}{x}$ where A is positive.

Q13 The graph on the right shows the curve $y = 3 - x - x^2$. Use the graph to estimate the solutions to the equation $3 - x - x^2 = 0$. Give your answers to 1 decimal place.

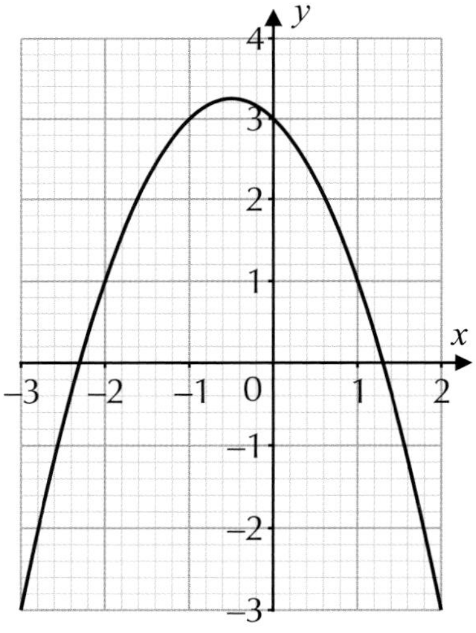

Q14 a) Copy and complete the table below to show the values of $y = x^2 + 5x$ for values of x from –4 to 2.

x	–4	–3	–2	–1	0	1	2
x^2			4				
$5x$			–10				
$x^2 + 5x$			–6				

b) Draw the graph of $y = x^2 + 5x$ for values of x from –4 to 2.

c) Find the coordinates of the turning point of your graph.

Exam-Style Questions

Q1 The diagram on the right shows a 1 cm^2 grid.
Point B has coordinates that are whole numbers and point B is 5 cm from point A.
On the grid, mark two points where point B could be.

[2 marks]

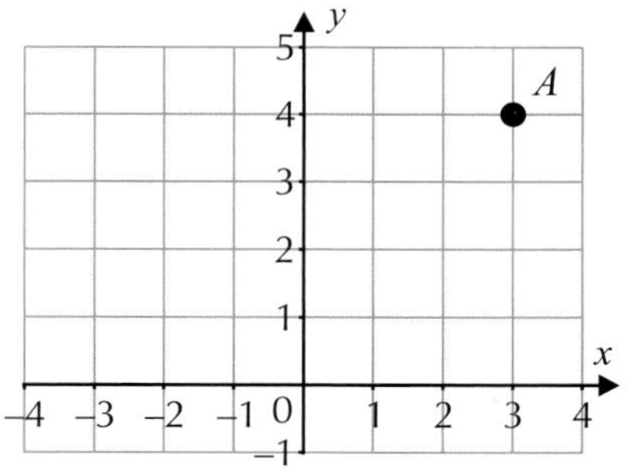

Q2 Point A has coordinates (–2, 1). Point B has coordinates (3, k).
The gradient of the line AB is –1.5.
Work out the value of k.

[3 marks]

Q3 The diagram on the right shows a line drawn on a pair of axes.

a) Write down the coordinates of the two points A and B.

[2 marks]

b) Complete the equation for this line:
$y = 5$..................

[2 marks]

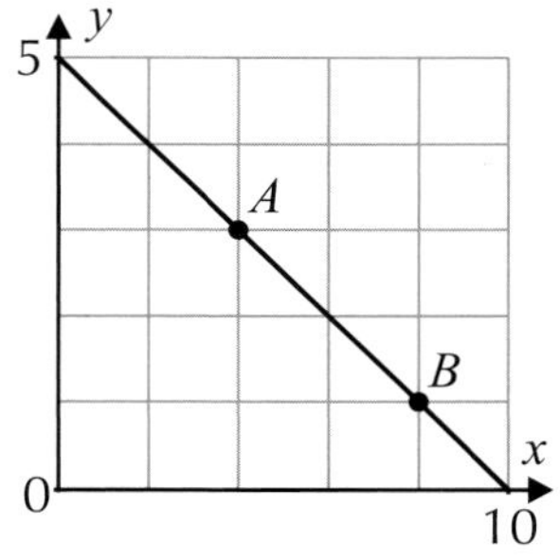

Q4 The line BC is shown on the grid below.

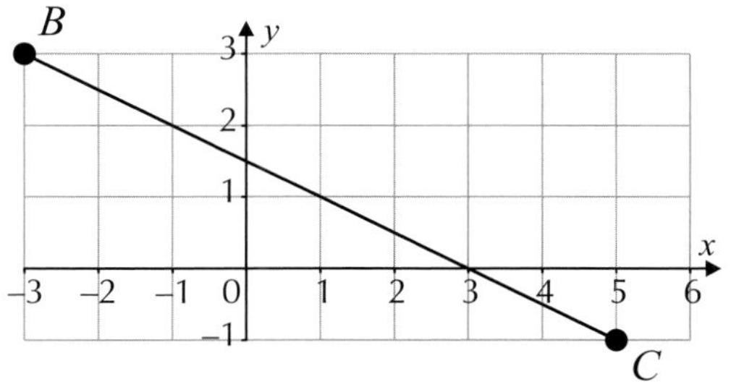

a) Write down the coordinates of where the line segment BC intersects the y-axis.

[1 mark]

b) Work out the coordinates of the midpoint of the line segment BC.

[1 mark]

Q5 The graph of a quadratic $y = ax^2 + bx + c$ is shown on the right for values of x from –3 to 5.

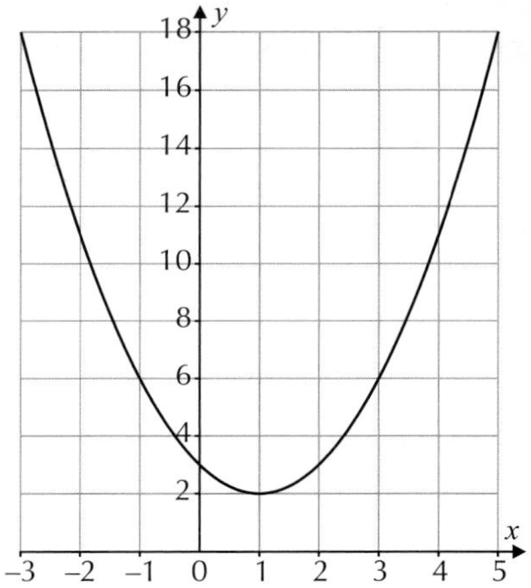

a) Write down the coordinates of the turning point of the graph.

[1 mark]

b) Use the graph to write down the value of y for $x = 3$.

[1 mark]

c) Use the graph to find the negative value of x for which $y = 6$.

[1 mark]

Q6 This is part of the graph of $y = 2x^2 - 4x - 5$.

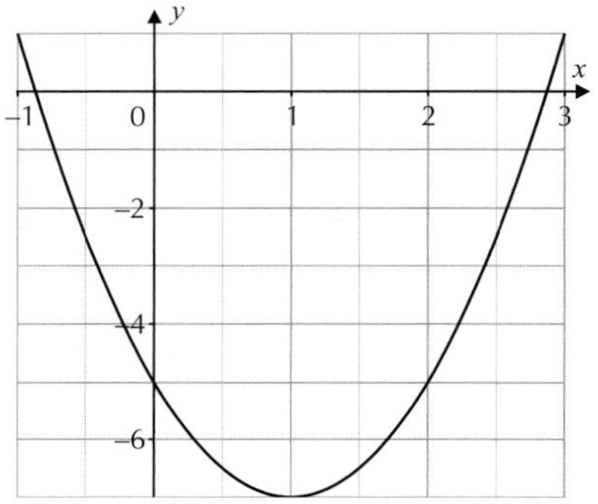

a) Write down the equation of the line of symmetry of the graph.

[1 mark]

b) Use the graph to estimate the solutions of $2x^2 - 4x - 5 = 0$ to 1 d.p.

[2 marks]

Q7 Mike has sketched the graph of $y = \frac{1}{x}$. Write down two things that are wrong with his sketch.

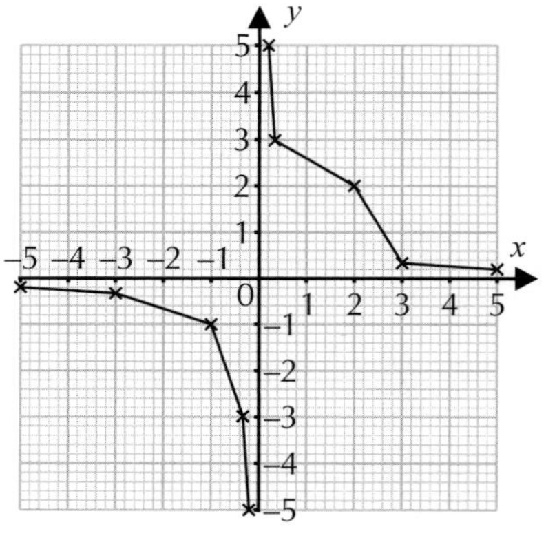

[2 marks]

Section 13 Real-Life Graphs

13.1 Interpreting Real-Life Graphs

Sometimes, graphs show something more interesting than just how y changes with x. They can be used to illustrate motion (such as in distance-time graphs), unit conversions (such as changing between temperatures in °C and °F), and many other connections between real-life quantities.

Linear Graphs

Learning Objective — Spec Ref A14:
Understand and interpret straight-line graphs that represent real-life situations.

Prior Knowledge Check:
Be familiar with gradients (p.199).

Straight-line graphs can be used to show how one thing **changes** in relation to another when the **rate of change** is **fixed** — for example, when **converting** between different **units**.

To **read off** values from a graph:

- Draw a **straight line** from **one axis** to the **graph**.
- Draw **another** straight line at a **right angle** to the first from the **graph** to the **other axis**.
- **Read off** the **value** from this axis, including any **units**.

Tip: To convert units, start from the axis that shows the units you're converting from, and finish on the axis showing those you're converting to.

Example 1

The graph shown can be used to convert between temperatures in degrees Celsius (°C) and degrees Fahrenheit (°F).

a) Convert 10 °C to °F.

Draw a line up from 10 °C until you reach the graph. Then draw a line across to find the amount in °F.

10 °C = **50 °F**

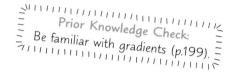

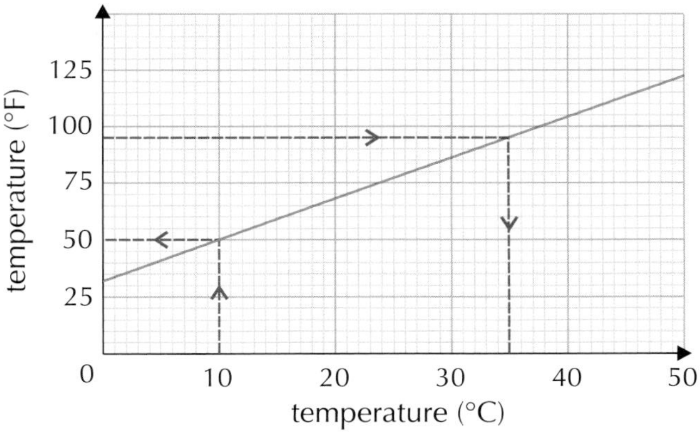

b) Convert 95 °F to °C.

Draw a line across from 95 °F until you reach the graph. Then draw a line down to find the amount in °C.

95 °F = **35 °C**

Exercise 1

The graph on the right can be used to convert between pounds (£) and euros (€). Use the graph to answer these questions.

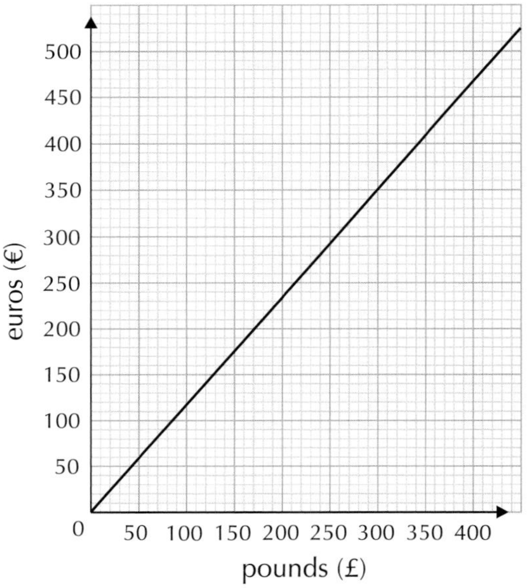

Q1 Use the graph to convert the following amounts from pounds to euros.

a) £50 b) £250 c) £110

Q2 Use the graph to convert the following amounts from euros to pounds.

a) €50 b) €200 c) €360

Q3 A dress costs €130. How much is this in pounds?

Q4 a) A TV costs £420 in the UK. How much is this in euros?

b) The price of the TV in France is €470. How much is this in pounds?

c) In which country is the TV cheaper?

When **describing** straight-line graphs, look at the following **features** and **interpret** them in the **context** of the graph:

- The **direction** of the graph — i.e. is the variable on the vertical axis **increasing** or **decreasing** as the other increases?
- The **gradient** (steepness) — this shows the **rate of change** of one variable with the other. The **steeper** the line, the **faster** the quantity on the **vertical axis** is changing compared to the quantity on the **horizontal axis**.

Tip: Graphs are sometimes made up of a few straight lines with different gradients — you'll need to work out what's happening in each section.

Example 2

The graph shows the cost of parking in a multistorey car park. Describe how the cost changes for different parking durations.

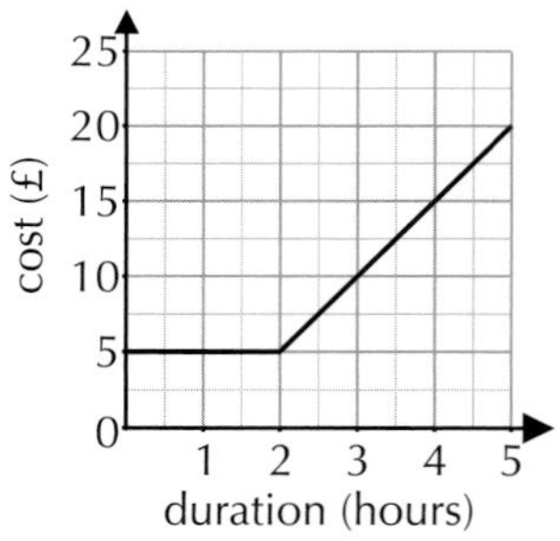

1. The graph is flat initially — it has a constant value of £5 between 0 and 2 hours.
2. After 2 hours, the cost increases as the duration increases, with a constant gradient.

The cost of parking is **fixed** at £5 for up to 2 hours. From 2 hours onwards, the cost **increases at a constant rate** of £5 per hour.

Exercise 2

The graph on the right shows the cost of a fine for returning a library book late. Use the graph to answer **Questions 1-3**.

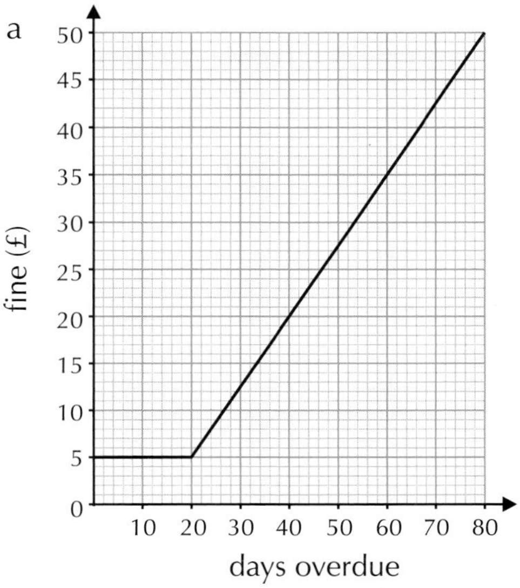

Q1 There is a basic fine, which is fixed up to a certain number of days.

a) What is the cost of this basic fine?

b) What is the longest time a book can be overdue and still receive the basic fine?

Q2 Estimate the fine for a book that is 50 days overdue.

Q3 Tamar receives a fine of £38.
How many days was her book overdue?

Q4 Katherine (K), Lemar (L) and Morag (M) climbed a mountain with a summit that is 1 km above sea level. The graphs below show their progress over 4 hours.

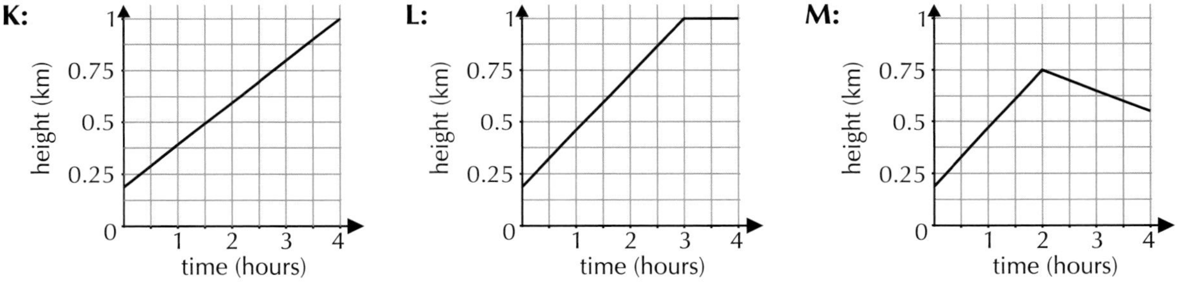

a) Which climbers reached the summit? Explain your answer.

b) Which climber reached the summit first?

Non-linear Graphs

Learning Objective — Spec Ref A14:
Understand and interpret curves that represent real-life situations.

Tip: A non-linear graph is a graph that isn't a straight line.

Graphs describing **real-life situations** can be almost **any shape**. Whatever the shape, the gradient (steepness of the graph) still shows the **rate of change** — how fast the y-axis quantity is changing compared to the x-axis quantity.

When describing a **non-linear graph**, look at the **direction** and **gradient** like you did for linear graphs on p.220 — and look out for any **change** in **direction** or **gradient**.

Use the **method** you saw on p.219 to **read off** values from a non-linear graph.

Example 3

The graph shows the temperature of an oven as it heats up.

a) Describe how the temperature of the oven changes during the first 10 minutes shown on the graph.

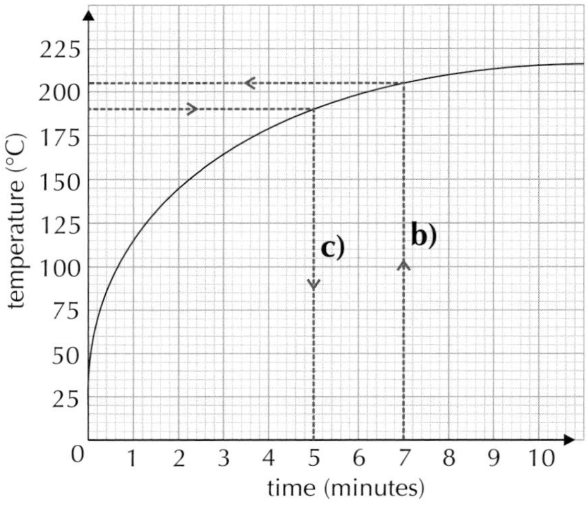

1. Look at the direction of the graph — temperature is increasing with time.
2. The gradient of the graph is steep initially and quite flat towards the end.
3. The graph doesn't change direction (it keeps increasing), but the gradient decreases over time.
4. Relate these features to the context:

The temperature of the oven **rises** for the entire 10 minutes. This rise is **rapid at first** but then **becomes slower** as the oven heats up.

b) What is the temperature of the oven after 7 minutes?

Follow the grid upwards from 7 minutes until you reach the curve, then read across to find the temperature.

After 7 minutes the oven is at **205 °C**.

c) How long does it take for the temperature to reach 190 °C?

Follow the grid across from 190 °C until you reach the curve, then read down to find the number of minutes.

It takes **5 minutes** for the oven to reach 190 °C.

Exercise 3

Q1 Each statement below describes one of the graphs to the right. Match each statement with the correct graph.

a) The temperature rose quickly, and then fell again gradually.

b) The number of people who needed hospital treatment stayed at the same level all year.

c) The cost of gold went up more and more quickly.

d) The temperature fell overnight, but then climbed quickly again the next morning.

K:

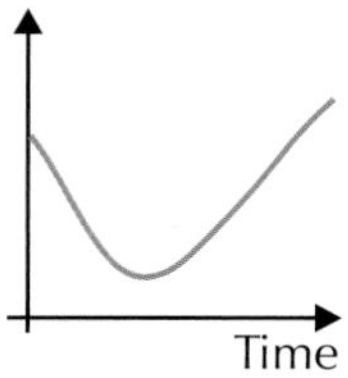

L:

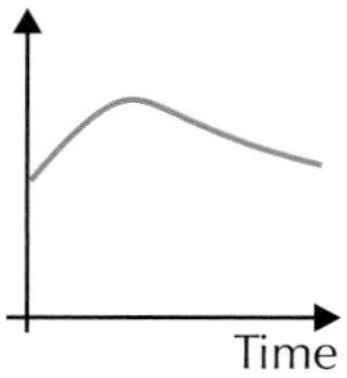

M:

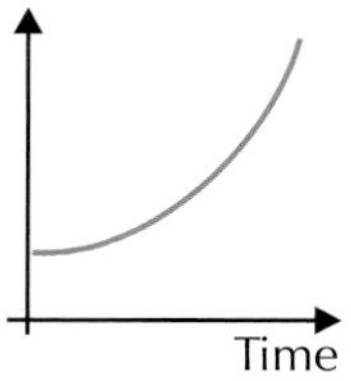

N:

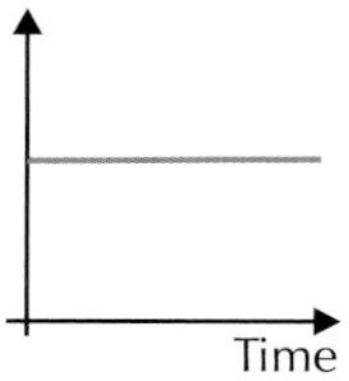

Q2 The graph shows the depth of water in a harbour one day between the times of 08:00 and 20:00.

a) Describe how the depth of water changed over this time period.

b) At approximately what time was the depth of water the greatest?

c) What was the minimum depth of water during this period?

d) At approximately what times was the water 3 m deep?

e) Mike's boat floats when the depth of the water is 1.6 m or over. Estimate the amount of time that his boat was not floating during this period.

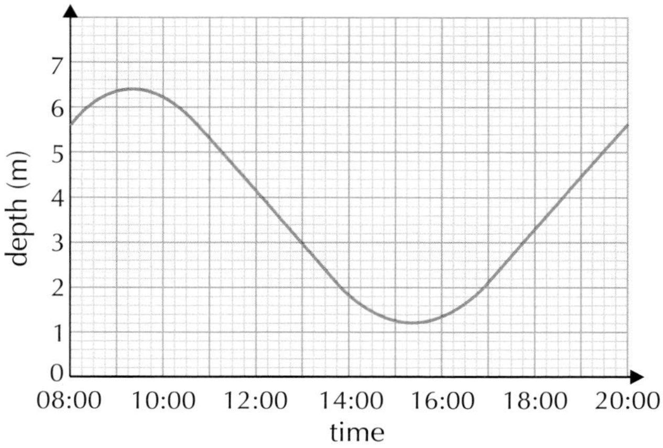

Q3 The graph on the left shows the temperature in two ovens as they warm up.

a) Which oven reaches 100 °C more quickly?

b) Which oven reaches a higher maximum temperature?

c) Estimate how long it takes Oven 2 to reach its maximum temperature.

d) (i) After how long are the two ovens at the same temperature?

(ii) What temperature do they both reach at this time?

e) What do the gradients of the lines represent?

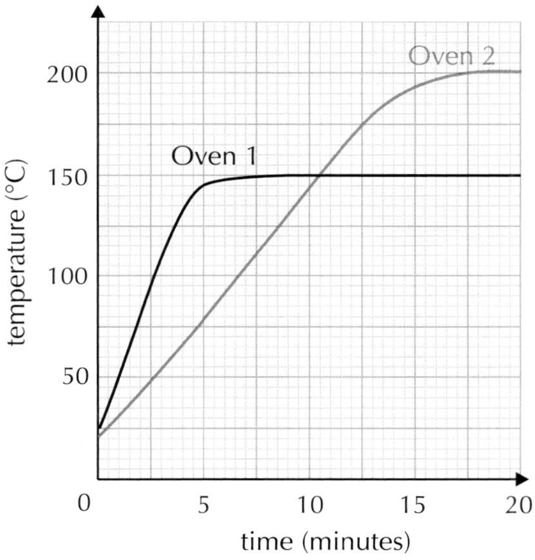

Q4 a) Vase P is 30 cm tall. The depth of water in Vase P as it is filled up is shown on the graph below. By how much does the depth of water in P increase:

(i) between 0 and 5 seconds?

(ii) between 10 and 15 seconds?

(iii) between 25 and 30 seconds?

b) Vase Q is also 30 cm tall. The depth of water in Vase Q is also shown on the graph. By how much does the depth of water in Q increase:

(i) between 0 and 5 seconds?

(ii) between 10 and 15 seconds?

(iii) between 25 and 30 seconds?

c) Describe the difference between how the depth of water increases in Vase P and Vase Q.

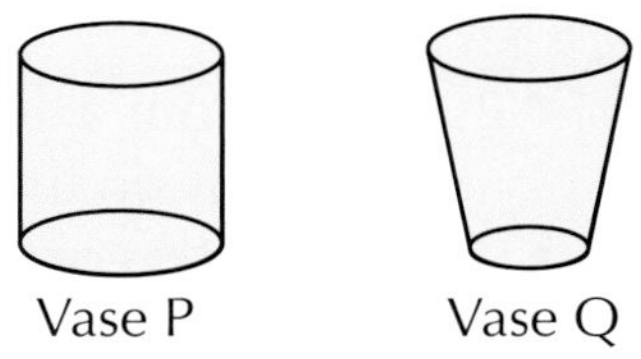

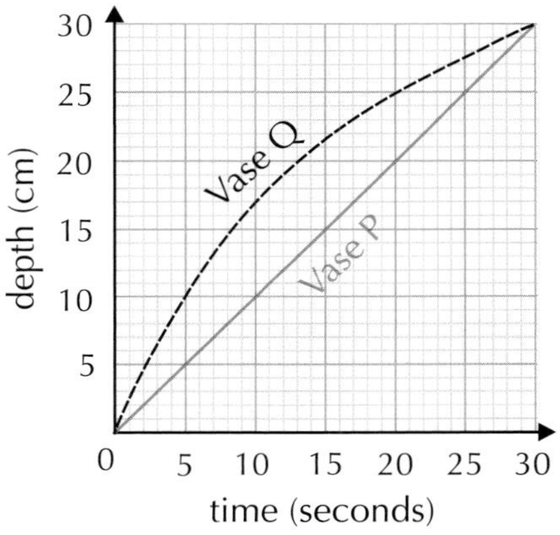

13.2 Drawing Real-Life Graphs

Now you know how to interpret real-life graphs, it's time to have a go at drawing them. You'll need a ruler to draw straight lines and a steady hand to draw smooth curves.

Learning Objective — Spec Ref A14:
Draw graphs that represent real-life situations.

Prior Knowledge Check: Be able to draw axes and plot coordinates (p.188).

To **draw** a graph representing a **real-life situation** you need to:

- Fill a **table of values** of points to plot. For a **straight line**, you'll need at least **three** points. But if the graph is a **curve**, you'll need **more than three** points.
- Draw and label your **axes** — the variable that **depends** on the other usually goes on the **vertical axis**.
- Then **plot** the points from the table as **coordinates** and join them by drawing a **straight line** or a **smooth curve** through them.

Tip: Make sure your axes are long enough to fit on all the points from your table.

Example 1

A plumber charges customers a standard fee of £40, plus £30 per hour for all work carried out.

a) Draw a graph to show how the plumber's fee varies with the amount of time the job takes.

1. Make a table showing the fee for different numbers of hours.
   A 1-hour job costs £40 + £30 = £70.
   A 2-hour job costs £40 + (2 × £30) = £100.
   A 3-hour job costs £40 + (3 × £30) = £130, etc.

Time (hours)	1	2	3	4	5
Fee (£)	70	100	130	160	190

2. Draw your axes. The cost of the job depends on the time it takes to complete, so the fee goes on the vertical axis and time on the horizontal axis. Make sure you label them and choose a scale that makes the graph easy to read.
3. Plot the values and join the points to draw the graph. For each extra hour, the fee increases by £30, so this is a straight-line graph.

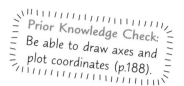

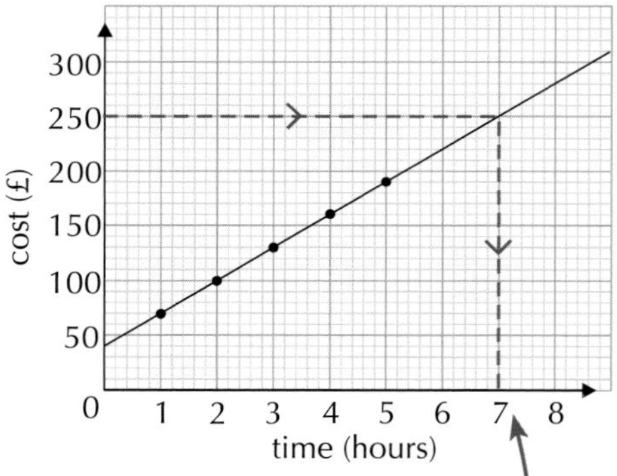

b) Use the graph to estimate the amount of time taken to do a job costing £250.

Follow the grid across from £250, then read downwards to find the correct time.

The job would have taken **7 hours**.

Example 2

The temperature of a cup of tea was measured at regular intervals as it cooled down. Draw a graph to illustrate how the temperature of the tea changes over time.

Time (minutes)	1	2	3	4	5	6
Temperature (°C)	80	78	73	62	50	43

1. Draw the axes — the temperature goes on the vertical axis and time on the horizontal axis. The temperature values are between 43 °C and 80 °C, so you can cut some of the vertical axis from the bottom of the scale.
2. Plot the values in the table as coordinates.
3. Draw a smooth curve through the points. There shouldn't be any sudden changes of direction or sharp kinks.

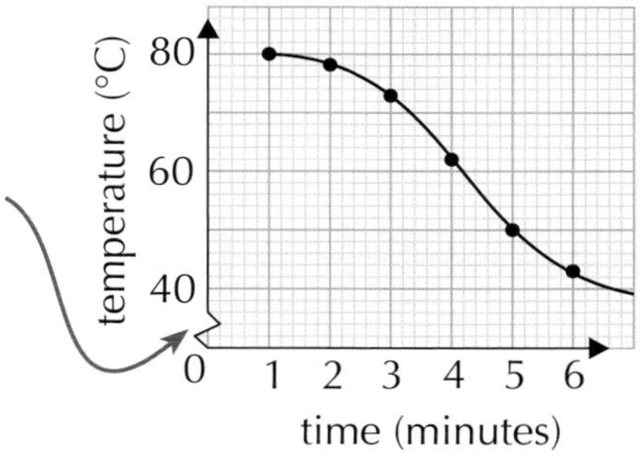

Exercise 1

Q1 The instructions for cooking different weights of chicken are as follows:

'Cook for 35 mins per kg, plus an extra 25 minutes.'

a) Copy and complete the table to show the cooking times for chickens of different weights.

Weight (kg)	1	2	3	4	5
Time (minutes)					

b) Draw a set of axes on a sheet of graph paper. Plot the time on the vertical axis and the weight on the horizontal axis. Then plot the values from your table to draw a graph showing the cooking times for different weights of chicken.

c) A chicken cooks in 110 minutes. What is the weight of the chicken?

Q2 This table shows how the fuel efficiency of a car in miles per gallon (mpg) varies with the speed of the car in miles per hour (mph).

Speed (mph)	55	60	65	70	75	80
Fuel Efficiency (mpg)	32.3	30.7	28.9	27.0	24.9	22.7

a) Plot the points from the table on a pair of axes and join them with a smooth curve.

b) Use your graph to predict the fuel efficiency of the car when it is travelling at 73 mph.

Review Exercise

Q1 The graph on the right can be used to convert between kilometres per hour (km/h) and miles per hour (mph).

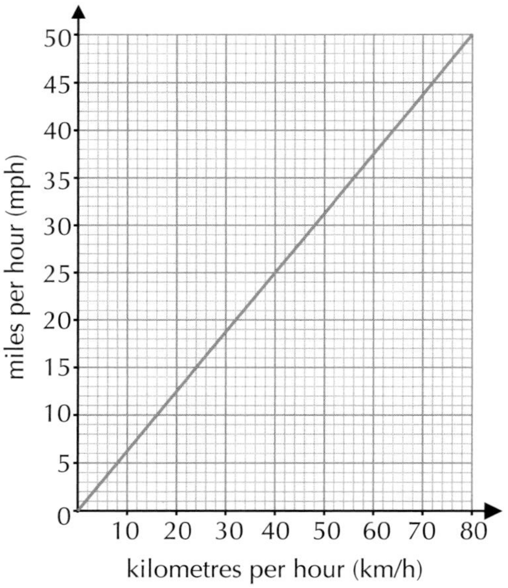

a) Convert 38 km/h into miles per hour.

b) Convert 27 mph into km/h.

c) The speed limit on a particular road is 30 mph. A driver travels at 52 km/h. By how many miles per hour is the driver breaking the speed limit?

d) The maximum speed limit in the UK is 70 mph. The maximum speed limit in Spain is 120 km/h. Which country has the higher speed limit, and by how much?

Q2 A scientist is conducting an experiment. The graph on the right shows the temperature of the experiment after t seconds.

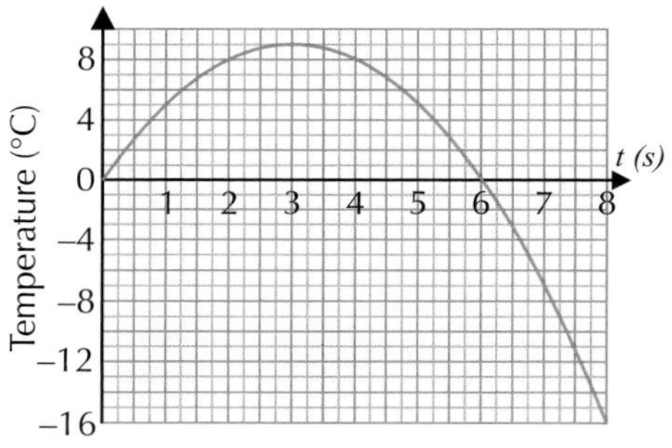

a) Give a brief description of how the temperature changes in the first 8 seconds of the experiment.

b) State the maximum temperature that it reaches.

c) A temperature of 8 °C was recorded twice. At what times was the temperature 8 °C?

Q3 Helena is a baby girl. A health visitor records the weight of Helena every two months. The measurements are shown in the table below.

Age (months)	0	2	4	6	8	10	12	14	16
Weight (kg)	3.2	4.6	5.9	7.0	7.9	8.7	9.3	9.8	10.2

a) Draw a graph to show this information. Plot age on the horizontal axis and weight on the vertical axis. Join your points with a smooth curve.

b) Keira is 9 months old and has a weight of 9.1 kg. Use your graph to estimate how much heavier Keira is than Helena was at 9 months old.

Exam-Style Questions

Q1 The graph below shows how the depth of a lake varies, measured between a point on beach A and a point on beach B on the opposite side of the lake.

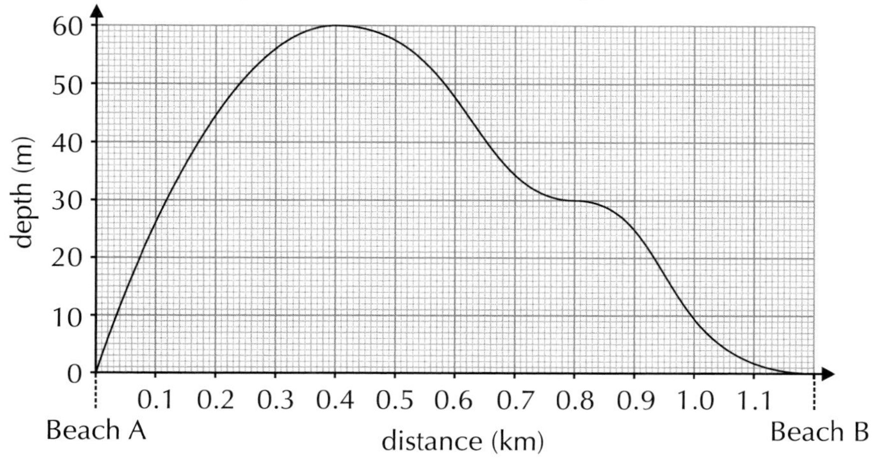

a) What is the maximum depth of the lake between the two beaches?

[1 mark]

b) Herons are long-legged birds that fish in shallow water. On which beach would you most likely find herons? Explain your answer.

[2 marks]

Q2 A cafe buys coffee beans from suppliers in Kenya and Peru.
The Kenyan supplier charges a rate of £2.50 per kilogram.
The Peruvian supplier charges a rate of £1 per kilogram, plus a fixed fee of £25.

a) Complete the following table of values.

Weight (kg)	10	20	30	40	50
Cost (£) of Kenyan coffee	25	50			
Cost (£) of Peruvian coffee	35				

[2 marks]

b) Using your table, draw a graph showing how the cost of Kenyan coffee beans varies with the weight of the beans purchased.

[2 marks]

c) On the same axes as part a), draw the graph for Peruvian coffee beans.

[2 marks]

d) Estimate the weight at which the two suppliers charge the same amount.

[1 mark]

Section 14 Proportion

14.1 Direct Proportion

Proportion is about how things change in relation to each other.
If two things are in direct proportion, they increase and decrease together.

Learning Objective — Spec Ref R7/R10:
Understand and use direct proportion.

Saying two things are in **direct proportion** means the **ratio** between them is always the **same**. For example, if one item costs £2, two items will cost £4, three items will cost £6, etc. — the ratio is always 1 item : £2. Quantities that are in direct proportion **increase** or **decrease** by the same **scale factor** — e.g. if you double the number of items, the cost doubles as well.

You can solve direct proportion problems using this method:

- **Divide** to find out how much of one thing you get for every **one** of the other — e.g. the price per item.
- Then **multiply** to find the **scaled** amount — e.g. the cost of 8 items.

Just remember: **DIVIDE FOR ONE, THEN MULTIPLY FOR ALL**

Example 1

If 8 chocolate bars cost £6, calculate the cost of 10 chocolate bars.

1. Divide the cost of 8 bars by 8 to find the cost of 1 bar.
2. Then multiply by 10 to find the cost of 10 chocolate bars.

8 bars cost £6
1 bar costs £6 ÷ 8 = £0.75

10 bars cost £0.75 × 10 = **£7.50**

Example 2

Oliver has 30 euros (€) left over from a holiday in France. If the exchange rate is £1 = €1.14, how many pounds can he exchange his euros for?

1. Divide both sides of the equation "£1 = €1.14" by 1.14 to find the number of pounds per euro.
2. Then multiply this by the number of euros Oliver has.

€1.14 = £1
€1 = £1 ÷ 1.14 = £0.877...

€30 = 30 × £0.877...
= **£26.32**
(to the nearest penny)

Tip: You're given the ratio of euros to pounds, but you need to find the ratio of pounds to euros.

Exercise 1

Q1 If 1 pair of jeans costs £35, find the cost of the following:
a) 2 pairs of jeans b) 5 pairs of jeans c) 20 pairs of jeans

Q2 The cost of 8 identical books is £36. What is the cost of 12 of these books?

Q3 If 1 DVD costs £7.50, work out how many DVDs you can buy for: a) £22.50 b) £60

Q4 A car uses 35 litres of petrol to travel 250 km.
a) How far, to the nearest km, can the car travel on 50 litres of petrol?
b) How many litres of petrol would the car use to travel 400 km?

Q5 Grace buys 11 pens for £12.32 and 6 note pads for £5.88. PROBLEM SOLVING
How much would she pay altogether for 8 pens and 5 note pads?

You can use the proportion method to work out which of a set of options is the **best value for money**. Just use the 'divide for one' part to get the **cost per item** (or amount of the item per penny), and then **compare** them.

Example 3

A supermarket sells bread buns in packs of 6, 12 and 24.
The pack of 6 costs 78p, the pack of 12 costs £1.44 and the pack of 24 costs £3.00.
Which pack size represents the best value for money?

1. For each pack divide the cost by the pack size to get the price per bread bun.
2. The pack with the best value for money is the one with the lowest cost per bread bun.

6 pack: 78p ÷ 6 = 13p per bread bun
12 pack: £1.44 ÷ 12 = 12p per bread bun
24 pack: £3.00 ÷ 24 = 12.5p per bread bun
The **12 pack** is the best value for money.

Exercise 2

Q1 A greengrocer's sells apples individually, in packs of 6, or in bags of 10. They cost 30p each individually, the pack of 6 costs £1.50, and the bag of 10 costs £2.40.
Which of these options represents the best value for money?

Q2 A coffee shop sells coffee in small, medium and large cups. A small coffee is 240 ml and costs £2, a medium coffee is 350 ml and costs £3, and a large coffee is 470 ml and costs £4. Which size of cup represents the best value for money?

Q3 Sausages are sold in packs of 6, 8 or 10. The pack of 6 costs £2.18, the pack of 8 costs £2.80, and the pack of 10 costs £3.46. Which pack represents the best value for money?

Graphs of Direct Proportion

Learning Objective — Spec Ref R10:

Recognise, interpret and be able to draw the graphs of two directly proportional quantities.

Prior Knowledge Check:
Be able to draw and interpret straight lines on a graph. See p.194, p.219-220.

Two quantities are **directly proportional** if, when you plot them on a graph, you get a **straight line** through the **origin**.

The **general equation** for two quantities in direct proportion is **$y = \mathrm{A}x$** where y is directly proportional to x and A is just a number.

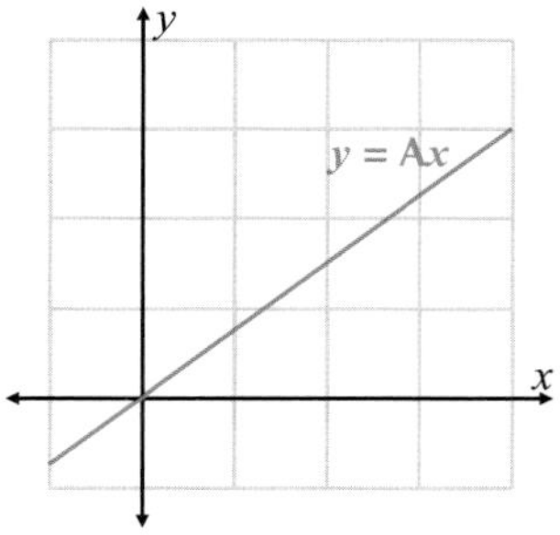

Example 4

***y* is directly proportional to *x*. When $x = 10$, $y = 25$.**

Sketch the graph of this direct proportion.

1. A sketch only requires the rough shape of the graph with the important points labelled.
2. As x and y are directly proportional, the graph is a straight line through the origin. You also know that it goes through the point (10, 25), so mark these two points and draw a straight line through them.

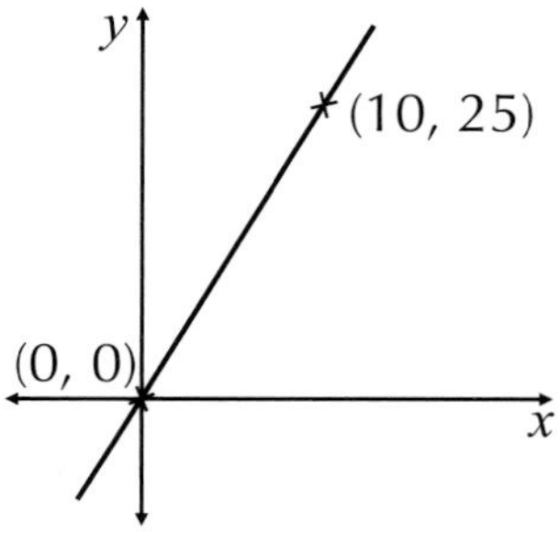

Example 5

***y* is directly proportional to *x*. Fill in the gaps in the table below.**

x	3	5	10	12	
y		25			100

1. From the table you can see that when $x = 5$, $y = 25$. You can use the 'divide for one, multiply for all' method to find the missing values.
2. Divide x and y by 5 to find the value of y when $x = 1$. When $x = 1$, $y = 25 \div 5 = 5$.
3. Now multiply each value of x in the table by 5 to get the missing values of y, and divide the value of y by 5 to get the missing value of x.

x	3	5	10	12	$100 \div 5 =$ **20**
y	$3 \times 5 =$ **15**	25	$10 \times 5 =$ **50**	$12 \times 5 =$ **60**	100

Exercise 3

Q1 Which of these graphs shows that y is directly proportional to x? Explain your answer.

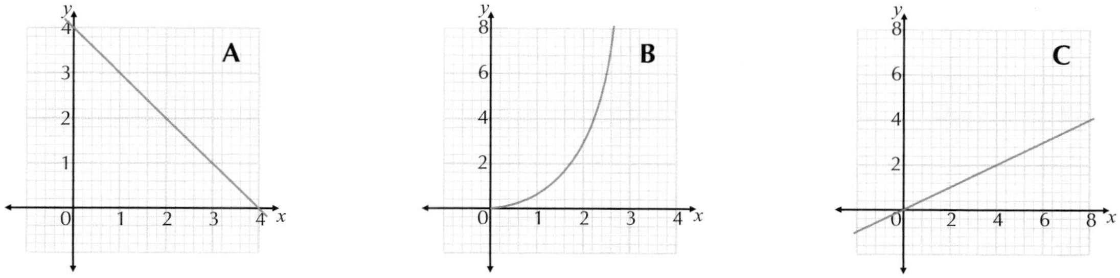

Q2 Sketch the graphs of the following proportions:

a) y is directly proportional to x. When $x = 20$, $y = 8$.

b) s is directly proportional to t. When $t = 80$, $s = 15$.

Q3 In each of the following tables, y is directly proportional to x.

(i) Copy and complete each table. (ii) Draw the graph of each direct proportion.

a)

x	8	12
y	2	

b)

x	2	7	
y		21	30

c)

x	–2	0	3	
y	–10			30

Example 6

y is directly proportional to x. When $y = 42$, $x = 6$.

a) Write an equation in the form $y = Ax$ to represent this direct proportion.

1. Put the values of x and y into the equation $y = Ax$ and rearrange it to find A.

$y = Ax$
$42 = A \times 6$
$A = 42 \div 6 = 7$

2. Put your value of A into the equation $y = Ax$.

$\mathbf{y = 7x}$

b) Calculate the value of y when $x = 51$.

Substitute $x = 51$ into your equation from part a).

$y = 7 \times 51 = \mathbf{357}$

Exercise 4

Q1 j is directly proportional to h. When $j = 15$, $h = 5$.

a) Write an equation to represent this direct proportion.

b) Work out the value of j when $h = 40$.

c) Sketch the graph of this direct proportion.

Q2 p is directly proportional to q. When $p = 14$, $q = 4$.

a) Accurately draw the graph of this direct proportion.

b) Use your graph to find the value of p when $q = 8$.

14.2 Inverse Proportion

If two things are inversely proportional, one thing gets smaller as the other gets bigger.

Learning Objective — Spec Ref R7/R10:
Understand and use inverse proportion.

If the two quantities are **inversely proportional**, then one quantity decreases as another increases. For example, **speed** and **time** are inversely proportional, so if you run faster (speed **increases**), it will take less time to finish a race (time **decreases**).

Solving inverse proportion problems is the **opposite** of solving direct proportion problems, so you need to switch the method from p.228 around.

For inverse proportion problems:

MULTIPLY FOR ONE, THEN DIVIDE FOR ALL

Example 1

Four people take five hours to dig a hole.
How long would it take ten people to dig the same sized hole at the same rate?

1. Multiply the time by 4 (the number of people) to find the time it would take one person.
   4 people take 5 hours
   1 person will take
   $5 \times 4 = 20$ hours
2. Divide by 10 (the new number of people) to find the time for 10 people.
   10 people will take
   $20 \div 10 =$ **2 hours.**

Tip: Check your answer makes sense — there are more people digging, so it should take less time.

Exercise 1

Q1 It takes three people two hours to paint a wall.
How long would it take six people to paint the same wall at the same rate?

Q2 A journey takes two hours when travelling at an average speed of 30 mph.
How long would the same journey take when travelling at an average speed of 45 mph?

Q3 Four chefs can prepare a meal in 20 minutes. They hire an extra chef. How long will it take five chefs to prepare the same meal, working at the same rate as the original four?

Q4 It will take five builders 62 days to complete a particular project.

a) At this rate, how long would the project take if there were only two builders?

b) If the project needed completing in under 40 days, what is the minimum number of builders that would be required?

When **two quantities change**, split the calculation up into **stages** and change one quantity at a time. Decide if the amount should go **up** or **down** at each stage of the calculation.

Example 2

It takes 5 examiners 4 hours to mark 125 exam papers.
How long would it take 8 examiners to mark 200 papers at the same rate?

1. Split the calculation up into stages — first change the number of examiners, then the number of papers.
2. Find how long it would take 1 examiner to mark 125 papers (multiply by 5), then find how long it would take 8 examiners (divide by 8).

   To mark 125 papers:
   5 examiners take 4 hours
   1 examiner takes $4 \times 5 = 20$ hours
   8 examiners take $20 \div 8 = 2.5$ hours

3. Find how long it would take 8 examiners to mark 1 paper (divide by 125), then how long it would take them to mark 200 papers (multiply by 200).

   For 8 examiners:
   1 paper takes $2.5 \div 125 = 0.02$ hours
   200 papers take $0.02 \times 200 =$ **4 hours.**

Exercise 2

Q1 It takes 3 hours for two people to clean six identical rooms.
How long would it take five people to clean 20 of the same rooms at the same rate?

Q2 It takes 144 minutes for two bakers to bake 72 identical buns. How long would it take for five bakers to bake 90 of these buns if they each work at the same rate?

Q3 Fourteen people can paint 35 identical plates in two hours.
At this rate, how long would it take 20 people to paint 60 identical plates?

Graphs of Inverse Proportion

Prior Knowledge Check: Recognising and sketching reciprocal graphs. See p.213.

Learning Objective — Spec Ref R10/R13:
Recognise, interpret and be able to draw the graphs of two inversely proportional quantities.

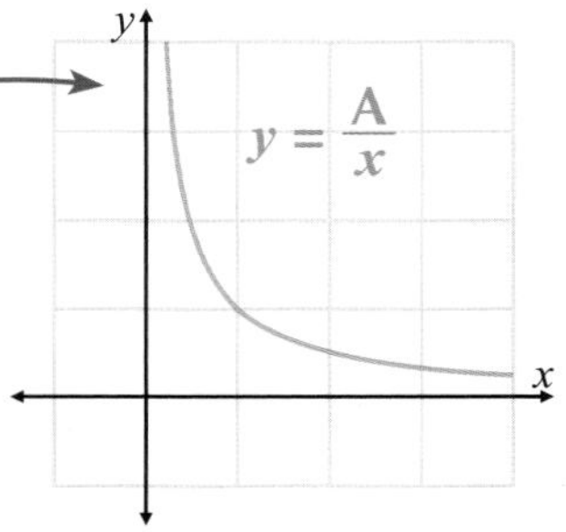

When two quantities are **inversely proportional**, plotting them on a graph looks like this. This is a **reciprocal** graph.

The **general equation** for an inverse proportion is: $y = \frac{A}{x}$ **or** $xy = A$

Saying 'y is **inversely proportional** to x' is the same as saying 'y is **proportional** to $\frac{1}{x}$' — the two statements are **equivalent**.

Example 3

y is inversely proportional to x. Fill in the gaps in the table below.

x	1		5	10
y		100		20

1. From the table you can see that when $x = 10$, $y = 20$. Using this you can work out A using the equation $xy = A$.

When $x = 10$ and $y = 20$
$A = 10 \times 20 = 200$

2. Once you know A, you can substitute in either x or y to find the other. E.g. when $x = 5$, $5y = 200$ so $y = 40$

x	1	$x \times 100 = 200$ $x = 200 \div 100 = \mathbf{2}$	5	10
y	$y = \mathbf{200}$	100	$5 \times y = 200$ $y = 200 \div 5 = \mathbf{40}$	20

Exercise 3

Q1 Which of these graphs shows that y is inversely proportional to x? Explain your answer.

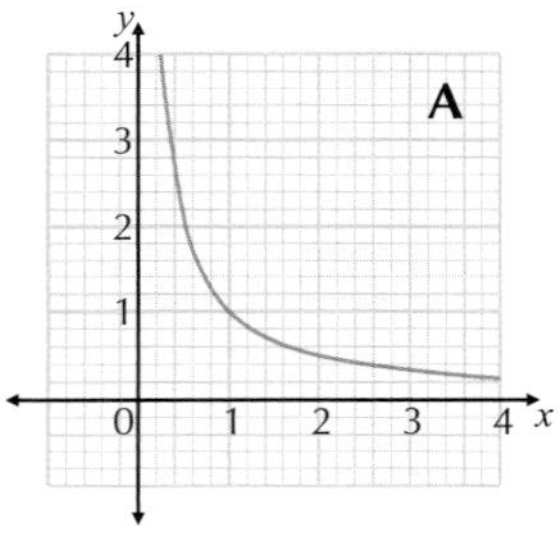

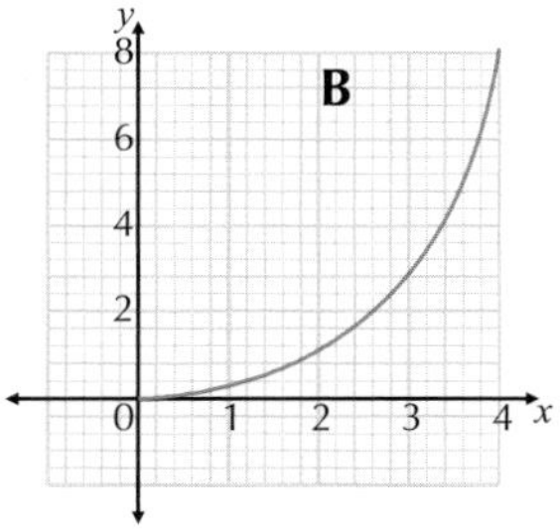

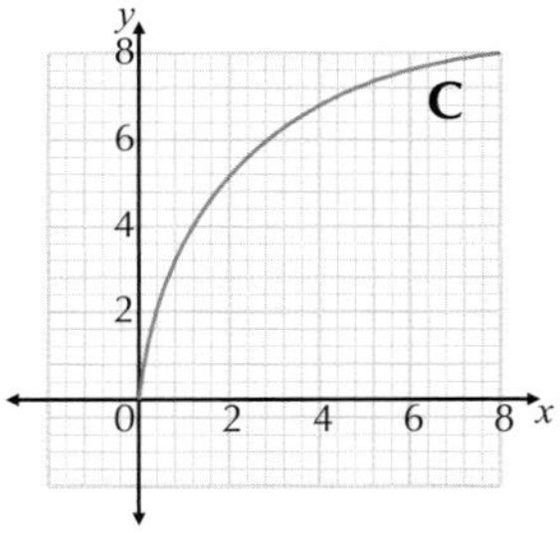

Q2 Which of the equations below show that y is inversely proportional to x?

$y = \frac{4}{x}$ $\quad y = \frac{7}{x} + 3$ $\quad \frac{y}{x} = 8$ $\quad yx = 9$ $\quad y = 3 - \frac{1}{x}$

Q3 In each of the following tables, y is inversely proportional to x. Use this information to copy and complete each table.

a)

x	12	6
y	15	

b)

x	2	4	8	10
y			20	

c)

x		4	20	100
y	320		4	

Q4 p is inversely proportional to q.

a) Copy and complete the following table:

q	0.2	0.4	0.5	1	2	4	5	10
p					1			

b) Use the table to draw the graph of this inverse proportion.

Review Exercise

Q1 It costs £20 to put 12.5 litres of petrol in a car.
How much will it cost for a full tank of petrol, if the tank holds 60 litres?

Q2 Emma changed £500 into rand (R) before going on holiday to South Africa.
The exchange rate at the time was £1 = R 17.5.

a) How many rand did she get for her £500?

On holiday Emma spent R 6200. When she got home, she changed the rand she had left over back into pounds. The exchange rate had changed to £1 = R 16.9.

b) How much money, in pounds, did she get back?

Q3 It takes a runner 8 minutes to run 5 laps of a 400 m track. How many minutes will it take the runner, running at the same speed, to complete 9 laps of a 200 m track?

Q4 A shop sells chocolates in different sized boxes. A box of 8 chocolates costs £3.50, a box of 12 chocolates costs £4.70, and a box of 20 chocolates costs £8.15.
Which of these represents the best value for money?

Q5 b is directly proportional to a.

a	2	4	7	10
b	6			

a) Copy and complete the table.

b) Use the table to draw an accurate graph of this direct proportion.

Q6 y is directly proportional to x and when $y = 32$, $x = 4$.

a) Find y when $x = 8$.

b) Find x when $y = 48$.

c) Find an equation in the form $y = Ax$ to represent this direct proportion.

Q7 8 chickens will lay 20 eggs in 3 days. The chickens always lay eggs at the same rate.

a) How many days will it take 12 chickens to lay 20 eggs?

b) How many chickens would be required if a farmer needed 30 eggs each day?

Q8 In each of the following tables, y is inversely proportional to x.
Use this information to copy and complete the following tables.

a)

x		7
y	25.2	9

b)

x		3	6	30
y	180		15	

Exam-Style Questions

Q1 Kitchen rolls each have 100 sheets per roll and cost £1.20. The manufacturer reduces the number of sheets on a roll to 90 and the price to £1.10.

Is the new roll better value for money than the original? Show your working.

[2 marks]

Q2 Hermain uses this recipe to make muffins.

Makes 12 muffins:

360 g	self-raising flour
200 g	caster sugar
250 ml	milk
125 ml	vegetable oil
4	medium eggs

a) Work out how much self-raising flour she will need to make 20 muffins.

[2 marks]

b) Sketch a graph to represent the relationship between the number of muffins Hermain makes and the amount of caster sugar she needs.

[2 marks]

Q3 Niall is cleaning the carpets in a hotel. The foyer area has 258 m^2 of carpet which takes him 24 minutes to clean. The bar area has 68.8 m^2 of carpet.

Work out how long it will take Niall to clean the carpet in the bar area if he works at the same rate, giving your answer in minutes and seconds.

[3 marks]

Q4 At an election, a council employed 840 staff, who counted 45 802 votes in 96 minutes. A second council employed 560 staff to count half as many votes.

a) Estimate how long it took the staff of the second council to count their votes.

[3 marks]

b) What assumption have you made in calculating your answer to part a)?

[1 mark]

Section 15 Angles and 2D Shapes

15.1 Basic Angle Properties

Angles along a straight line or around a single point always add up to a particular number.
You can use these facts to find the size of an unknown angle by forming and solving an equation.

Learning Objectives — Spec Ref G3:
- Find angles that lie on a straight line.
- Find angles that form a right angle.
- Find angles around a point.

Prior Knowledge Check:
Be able to write and solve equations — see p137-147.

Angles at a point **on a straight line** add up to **180°**.

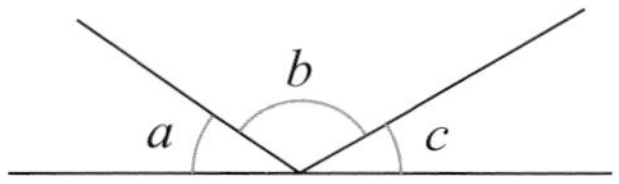

$a + b + c = 180°$

Perpendicular lines meet at 90° to form a **right angle** (shown by the little **square**).

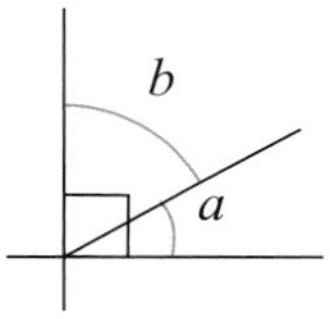

Angles within a **right angle** add up to **90°**.

$a + b = 90°$

Example 1

Find the size of angle *a* in the diagram on the right.

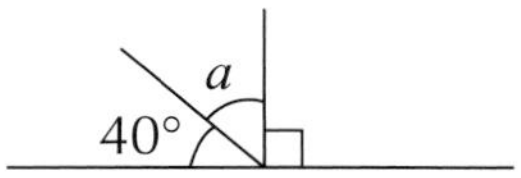

1. Angles on a straight line add up to 180°. Use this to form an equation in terms of *a*. $40° + a + 90° = 180°$
2. Simplify the equation. $a + 130° = 180°$
3. Solve the equation to find *a*. $a = 180° - 130°$

$\mathbf{a = 50°}$

Tip: Most of the time angles aren't drawn accurately, so you can't measure them with a protractor.

Exercise 1

Q1 Find the value of each letter in the diagrams below. None of the angles are drawn accurately.

a)

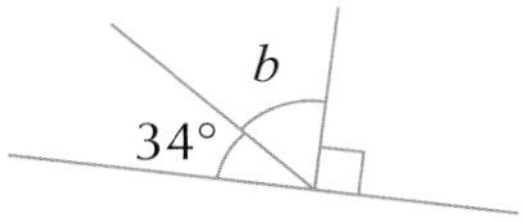

b)

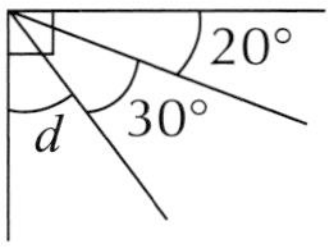

c)

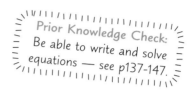

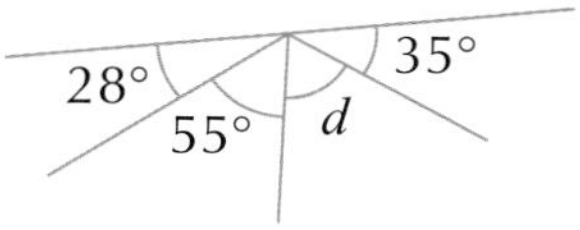

Q2 Find the value of each letter in the diagrams below.
None of the angles are drawn accurately.

a) 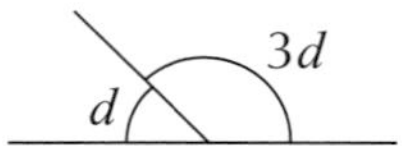

b) 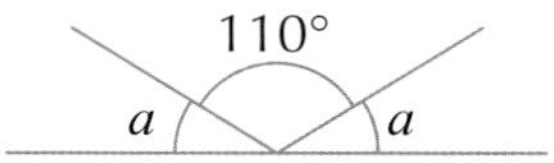

c) 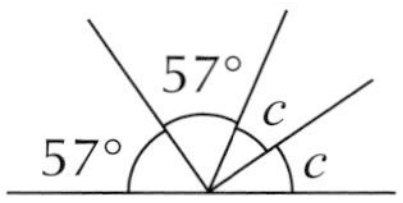

Angles around a **point** add up to **360°**.

You can use this fact to find **unknown** angles by writing and **solving** equations, like in the example shown below.

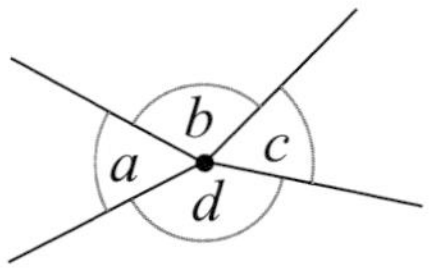

$\boldsymbol{a + b + c + d = 360°}$

Example 2

Find the value of *a* in the diagram on the right.

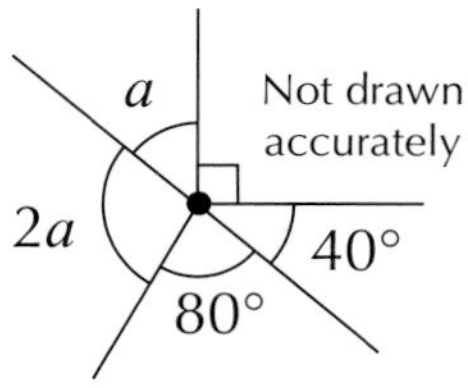

1. The angles are around a point so they must add up to 360°. Use this to form an equation. $a + 2a + 80° + 40° + 90° = 360°$
2. Simplify the equation. $3a + 210° = 360°$
   $3a = 150°$
3. Then solve the equation to find a. $a = 150° \div 3$
   $\boldsymbol{a = 50°}$

Tip: Remember to write the degree symbol in your answer.

Exercise 2

Q1 Find the value of each letter in the diagrams below. None of the angles are drawn accurately.

a) 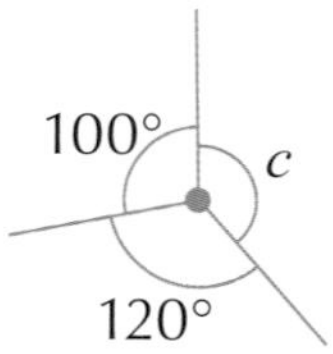

b) 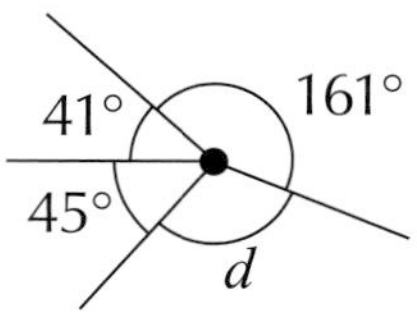

c) 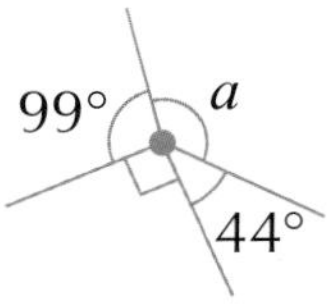

Q2 Find the value of each letter in the diagrams below. None of the angles are drawn accurately.

a) 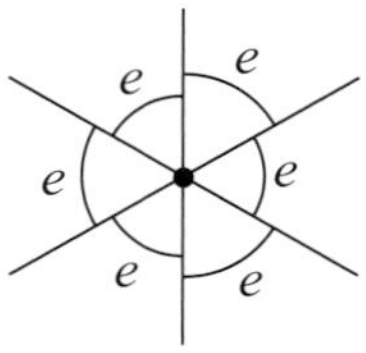

b) 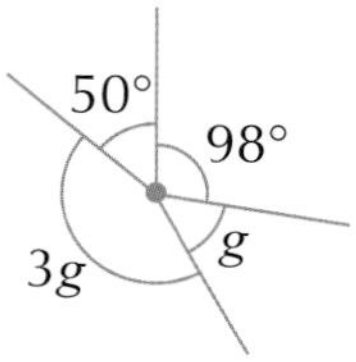

c) 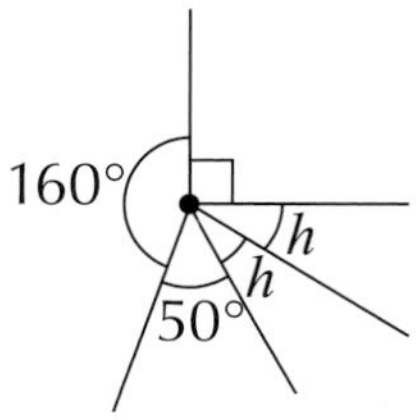

15.2 Parallel and Intersecting Lines

There are a few different rules which connect the angles created where two lines intersect, and where a line crosses two parallel lines. Let's start with the one where just two lines cross...

Vertically Opposite Angles

Learning Objective — Spec Ref G3:
Find vertically opposite angles.

When any **two lines intersect**, it produces **two pairs** of angles.
The angles opposite one another are known as **vertically opposite angles**, and vertically opposite angles are always **equal**.

In the diagram on the right, the two angles **labelled a** are **vertically opposite**, so they are **equal**. Similarly, the two angles **labelled b** are vertically opposite, so they are equal as well.

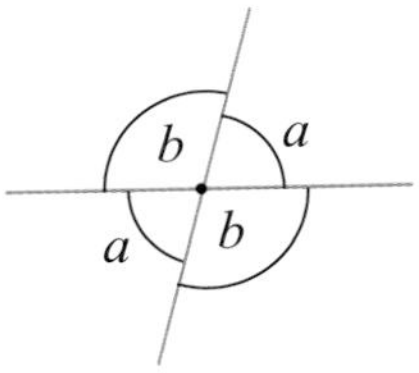

Also, because the two **distinct** angles (i.e. a and b) form a straight line, they add up to **180°** — so $a + b = 180°$.

$a + b = 180°$

Example 1

Find the values of a, b and c shown in the diagram below.

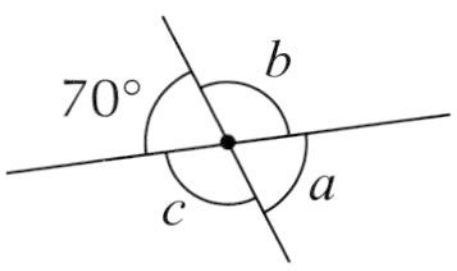

1. a and 70° are vertically opposite angles, so they are equal. **$a = 70°$**
2. b and the angle marked 70° lie on a straight line, so they add up to 180°. $70° + b = 180°$, $b = 180° - 70°$, **$b = 110°$**
3. c and b are vertically opposite angles, so they are equal. $c = b =$ **110°**

Exercise 1

Q1 Find the missing angles marked by letters. The angles aren't drawn accurately.

a)

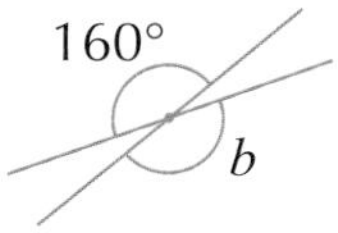

b)

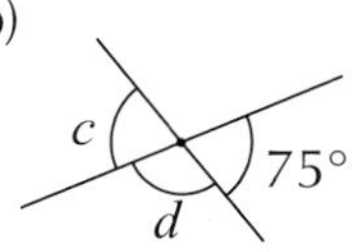

c)

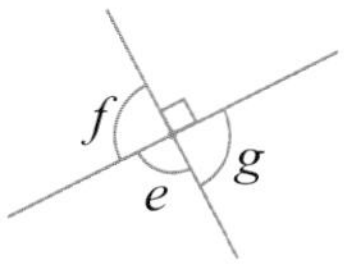

Q2 Find the value of a in the diagram shown on the right.

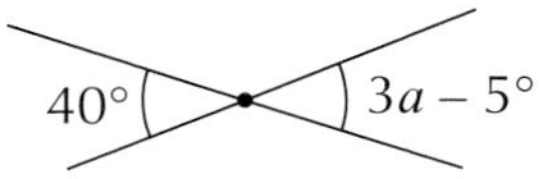

Alternate Angles

Learning Objective — Spec Ref G3:
Find alternate angles.

When a straight line crosses two **parallel** lines, it forms two pairs of **alternate angles** in a sort of **Z-shape** — as shown in the diagrams below. Alternate angles are always **equal**.

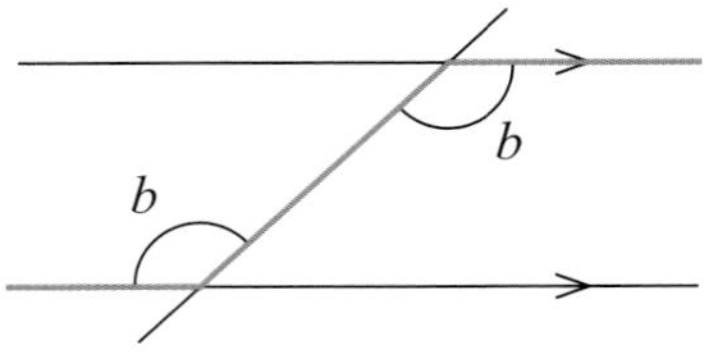

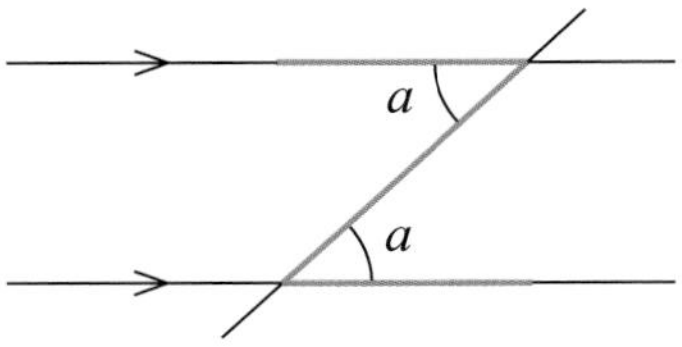

Tip: The small arrows in the diagram show that those two lines are parallel.

Example 2

Find the values of *a*, *b* and *c* in the diagram below.

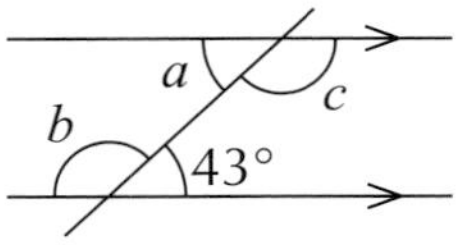

1. *a* and 43° are alternate angles, so they are equal. — $\mathbf{a = 43°}$
2. *b* and the angle marked 43° lie on a straight line, so they add up to 180°. — $43° + b = 180°$, $b = 180° - 43°$, $\mathbf{b = 137°}$
3. *c* and *b* are alternate angles, so they are equal. — $c = b = \mathbf{137°}$

Exercise 2

Q1 Find the missing angles marked by letters.

a)

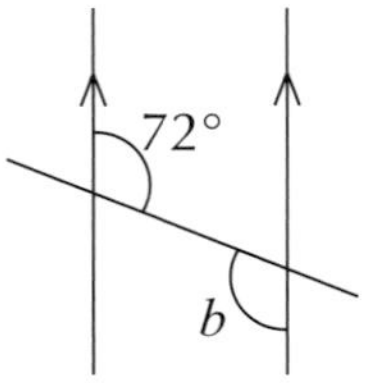

b)

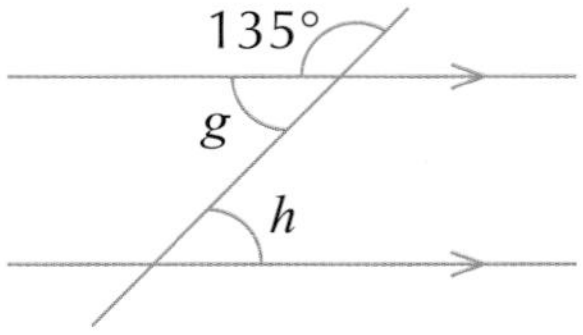

c)

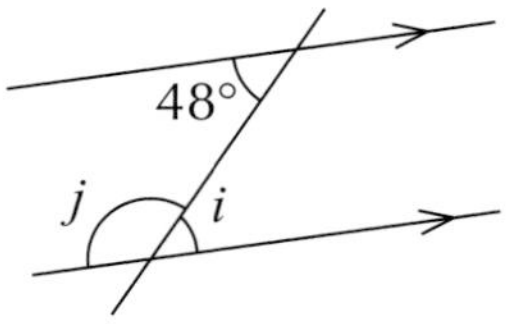

Q2 The diagram on the right shows a staircase between two parallel floors of a building. The staircase makes an angle of 42° with the lower floor.

a) Write down the angle that the staircase makes with the upper floor, marked *x* on the diagram.

b) Give a reason for your answer.

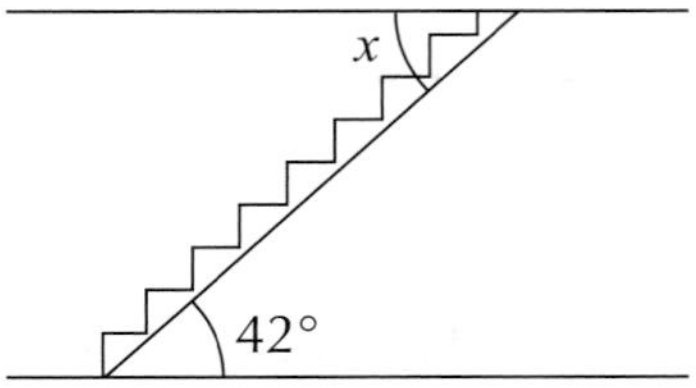

Corresponding Angles and Allied Angles

Learning Objective — Spec Ref G3:
Find corresponding and allied angles.

Corresponding angles form an **F-shape**. Corresponding angles are always **equal**.

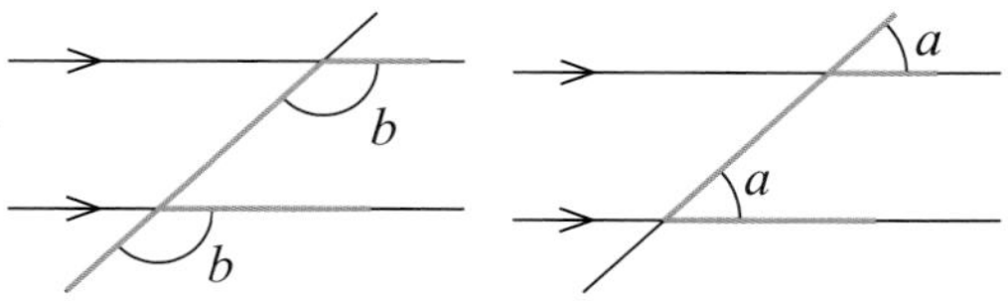

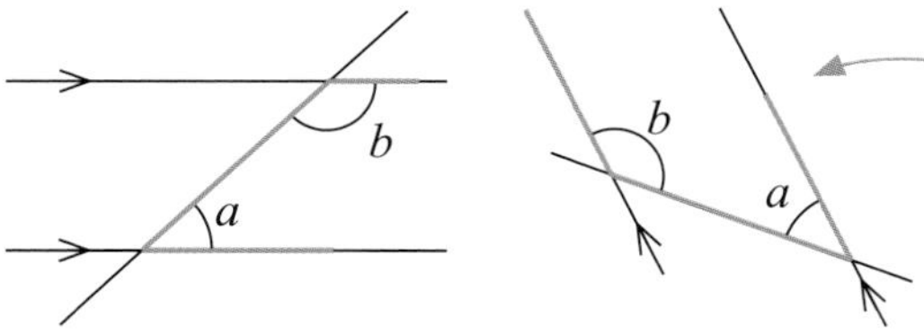

Allied angles form a **C-** or **U-shape**. They always **add up to 180°**.

$a + b = 180°$

Tip: State which rules you're using when solving geometry problems — e.g. say "because these are allied angles" or "as the angles all lie on a straight line". Make sure you use the proper terms — don't describe them as "angles in a Z-shape".

Example 3

Find the values of *a*, *b* and *c* shown in the diagram below.

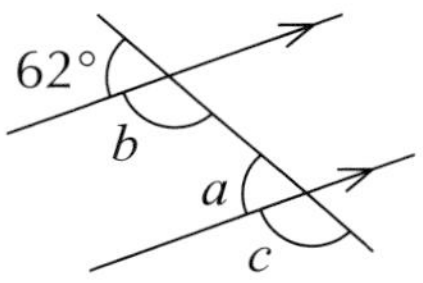

1. a and 62° are corresponding angles, so they are equal. — $a = 62°$
2. b and the angle marked 62° lie on a straight line, so they add up to 180°. (Or you could say a and b are allied angles, so they add up to 180°.) — $62° + b = 180°$, $b = 180° - 62°$, $b = 118°$
3. c and b are corresponding angles, so they're equal. — $c = b = 118°$

Exercise 3

Q1 Find the missing angles marked by letters.

a)

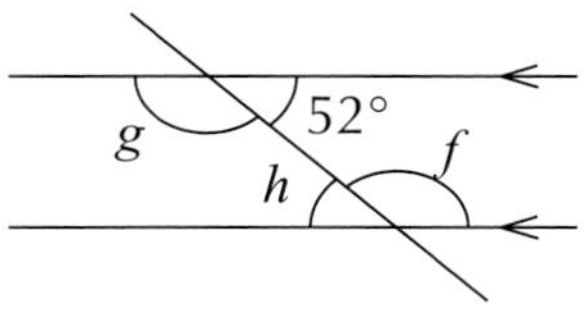

b)

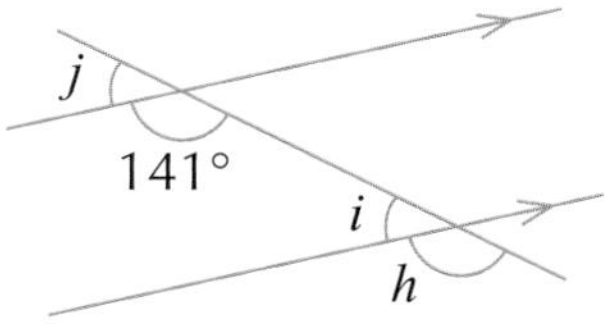

c)

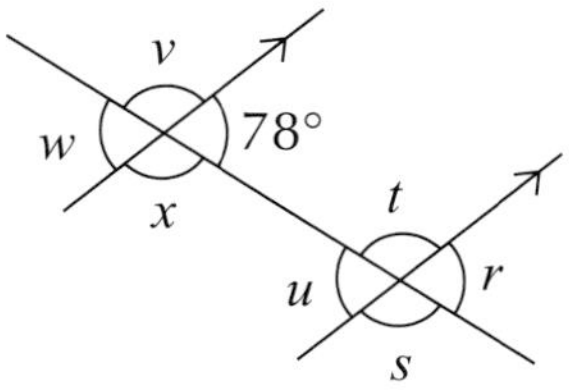

Q2 Two wooden posts stand vertically on sloped ground. The first post makes an angle of 99° with the downward slope, as shown. Find the angle that the second post makes with the upward slope, labelled y on the diagram.

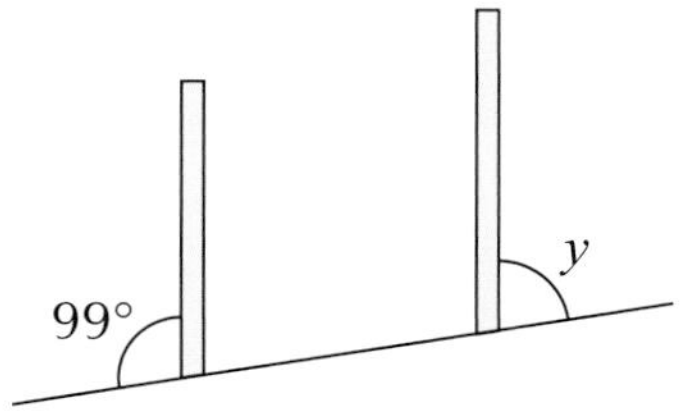

15.3 Triangles

Triangles are perhaps the simplest of the 2D shapes because they only have three sides. But there are different types depending on the lengths of their sides and the sizes of their angles.

Learning Objectives — Spec Ref G3/G4:
- Know the properties of different types of triangles.
- Know and be able to prove that the angles in a triangle sum to 180°.
- Be able to find missing angles in triangles.

There are **different** types of triangles that you need to know about. Make sure you know the defining features of each type.

Tip: 'Tick marks' are used to show sides are the same length.

An **equilateral** triangle has 3 equal sides and 3 equal angles (each of 60°).

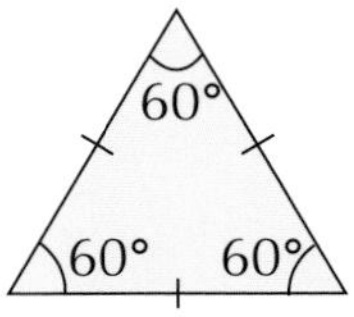

An **isosceles** triangle has 2 equal sides and 2 equal angles.

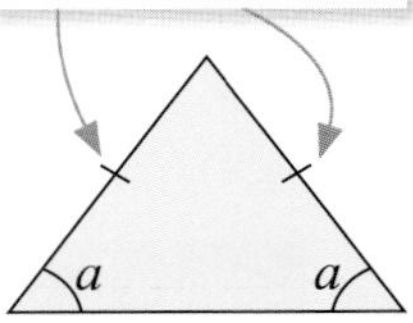

The sides and angles of a **scalene** triangle are all different.

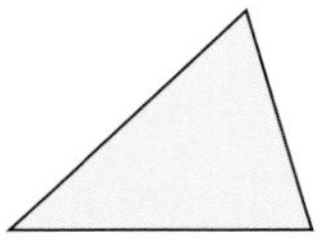

A **right-angled** triangle has 1 right angle (90°).

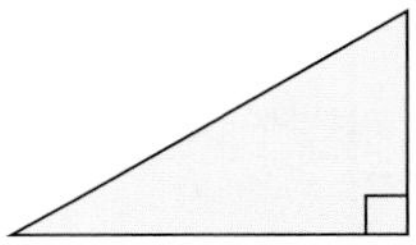

You might occasionally see triangles described by the size of their angles:

An **acute-angled** triangle has 3 acute angles (less than 90°).

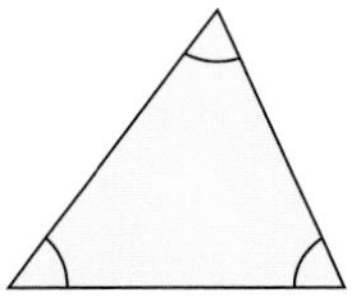

An **obtuse-angled** triangle has 1 obtuse angle (between 90° and 180°).

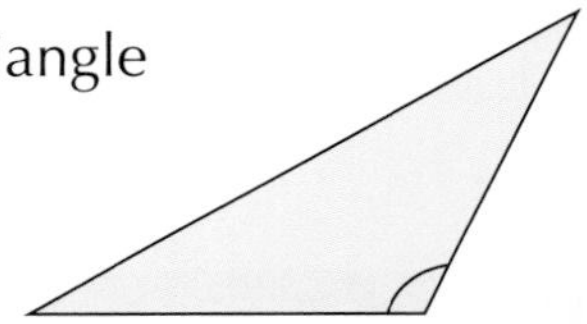

Exercise 1

Q1 Describe each of these triangles using the above definitions.

a)

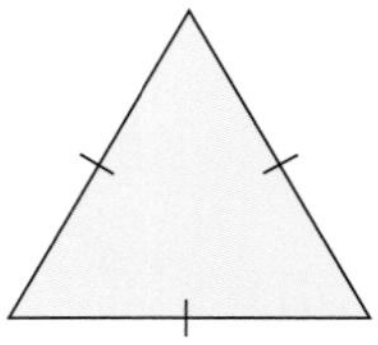

b)

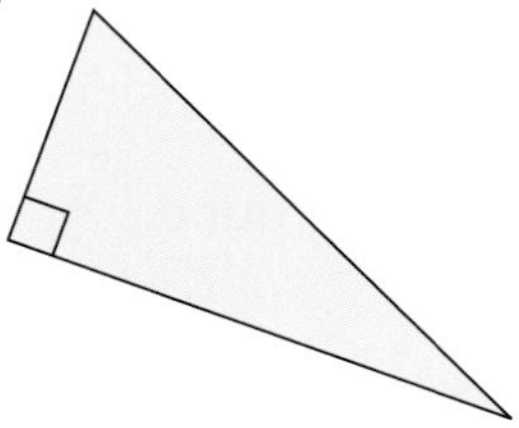

c)

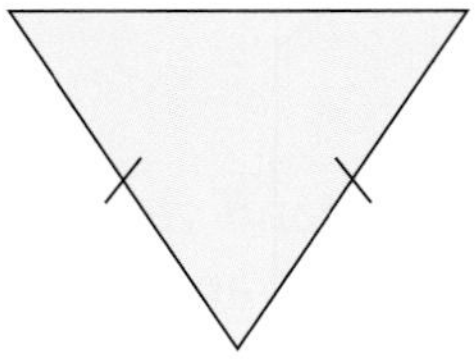

d)

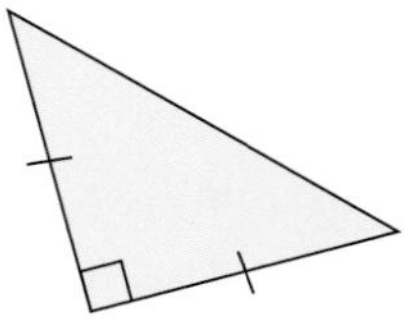

For **any** triangle, the angles inside (*a*, *b* and *c*) **add up to 180°**. You can use this to set up an **equation** that you can **solve** to find missing angles.

$$a + b + c = 180°$$

To **prove** this rule:

- Draw **parallel lines** at the top and base of the triangle, as shown on the right.
- Then use the fact that **alternate** angles are equal (see page 240) to work out the angles that lie on a **straight line with *b***.

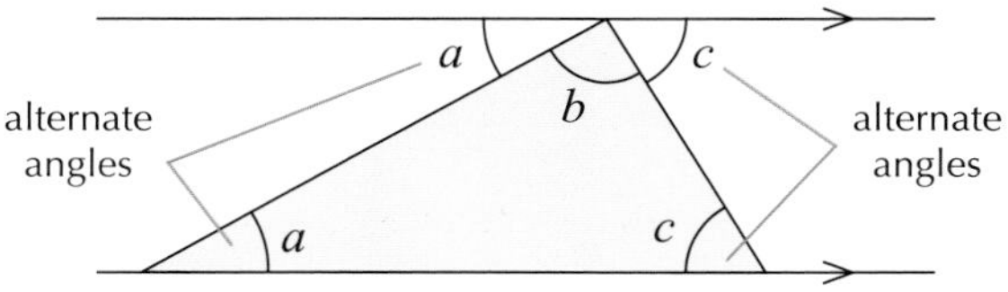

- **Angles on a straight line** add up to **180°** (see p.237) — so $a + b + c = 180°$.

Example 1

Find the value of *x* in the triangle shown below.

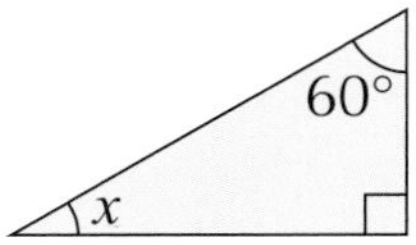

1. The angles in a triangle add up to 180°. Use this to form an equation in terms of *x*.
   $x + 60° + 90° = 180°$
   $x + 150° = 180°$
2. Solve your equation to find *x*.
   $x = 180° - 150° = \mathbf{30°}$

Exercise 2

Q1 Find the missing angles marked with letters. Diagrams have not been drawn accurately.

a)

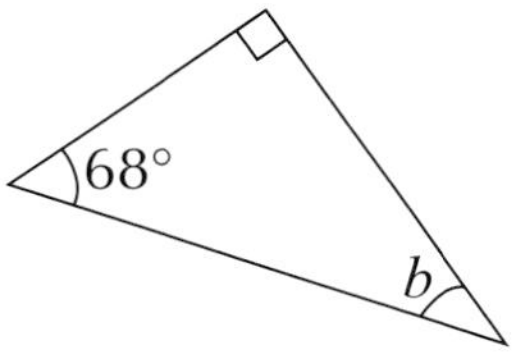

b)

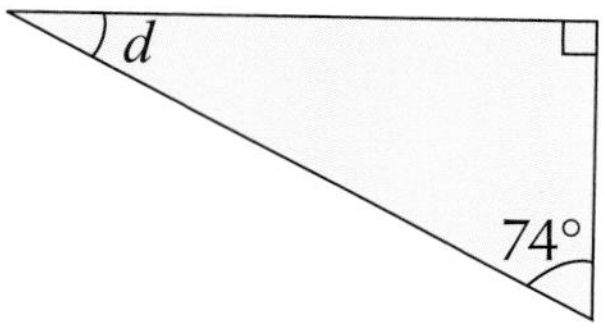

c)

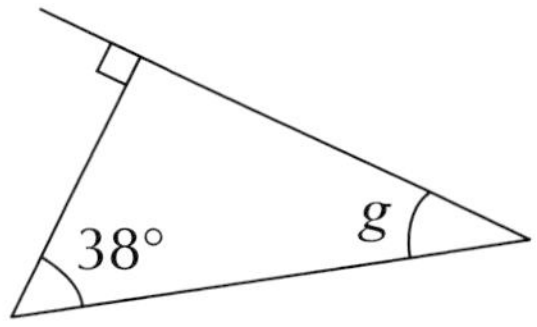

Q2 Look at the diagram on the right. Explain why a triangle with the angles shown in the diagram **cannot** exist.

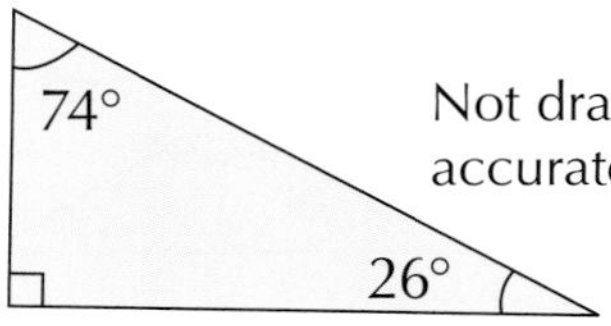

Not drawn accurately

Q3 Find the missing angles marked with letters. Diagrams have not been drawn accurately.

a)

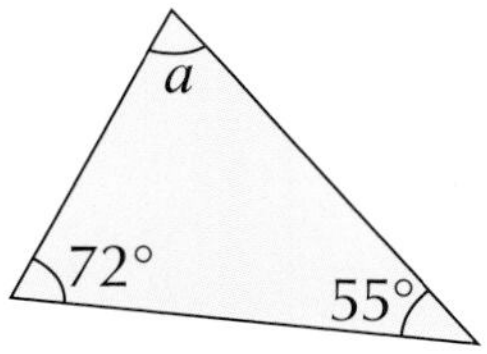

b)

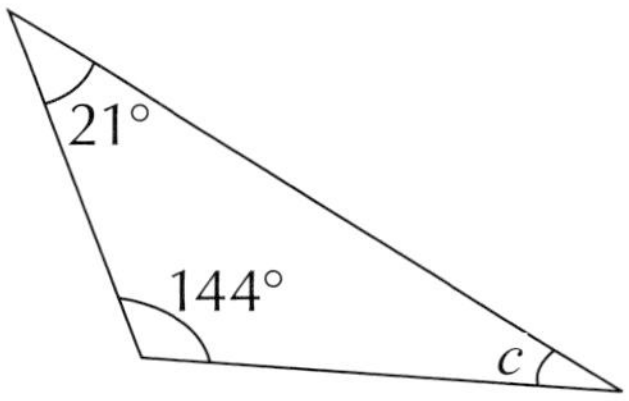

c)

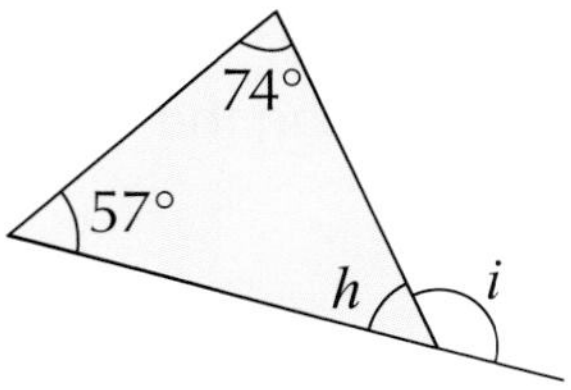

Example 2

a) Find the value of x in the triangle below.

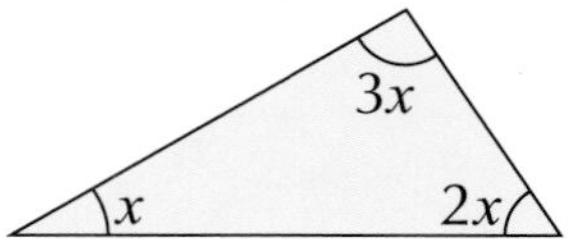

1. The angles in the triangle must add up to 180°, so set up an equation in x.
   $x + 2x + 3x = 180°$
   $6x = 180°$
2. Solve your equation to find x.
   $x = 180° \div 6 = \mathbf{30°}$

b) Find the value of y in the isosceles triangle below.

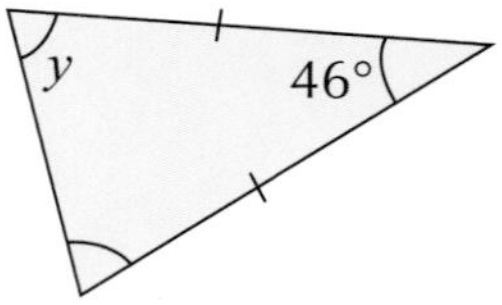

1. The triangle is isosceles so the unmarked angle must be equal to y.
2. All three angles must sum to 180°, so form an equation in y. → $y + y + 46° = 180°$
   $2y = 134°$
3. Solve your equation to find y.
   $y = 134° \div 2 = \mathbf{67°}$

Exercise 3

The angles in this exercise have not been drawn accurately, so don't try to measure them.

Q1 Find the values of the letters shown in the following diagrams.

a)

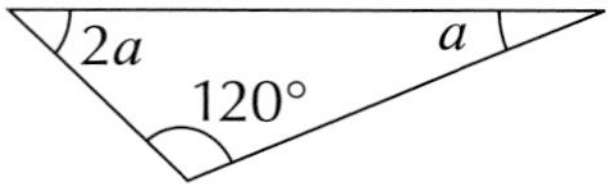

b)

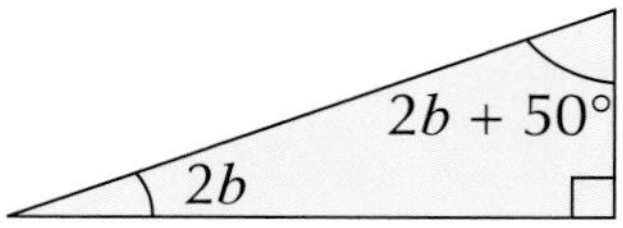

c)

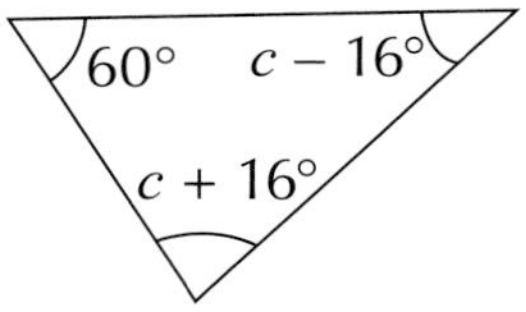

Q2 Find the values of the letters in each of these diagrams.

a)

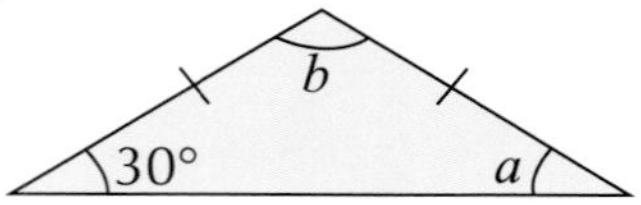

b)

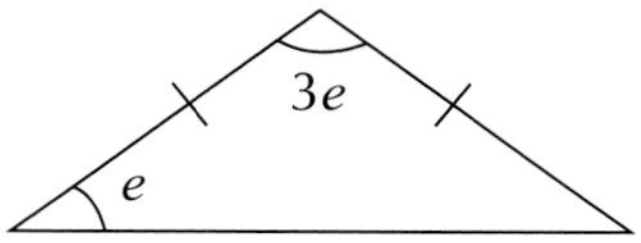

c)

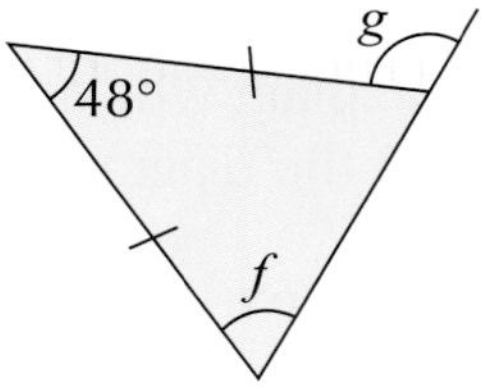

Q3 a) Find the value of x.
b) Find the value of y.

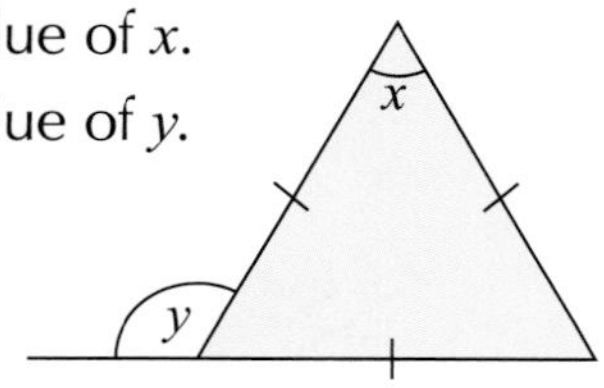

Q4 a) Find the value of p.
b) Find the value of q.

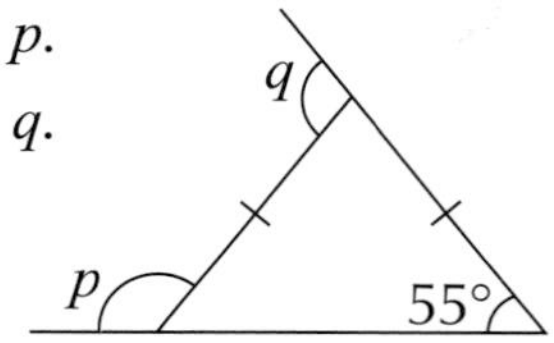

15.4 Quadrilaterals

Quadrilaterals are shapes with four sides. You're probably familiar with squares and rectangles, two of the simplest quadrilaterals, but the next few pages will introduce you to lots of other types.

Quadrilaterals

Learning Objectives — Spec Ref G3:
- Know that the angles in a quadrilateral add up to 360°.
- Be able to find missing angles in quadrilaterals.

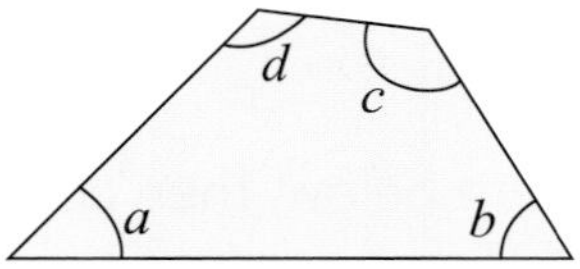

$a + b + c + d = 360°$

The **angles** in a quadrilateral always **add up to 360°**.

You can find **unknown** angles by forming an equation and then **rearranging**.

Tip: There's a method on page 250 which you can use to show that the angles add up to 360°.

Example 1

Find the missing angle x in this quadrilateral.

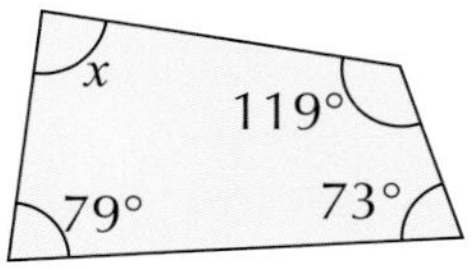

1. The angles in a quadrilateral add up to 360°, so write an equation involving x.
2. Then solve your equation to find the value of x.

$79° + 73° + 119° + x = 360°$

$271° + x = 360°$

$x = 360° - 271° = $ **89°**

Exercise 1

Q1 Find the size of the angles marked by letters in the following quadrilaterals. (They're not drawn accurately, so don't try to measure them.)

a) 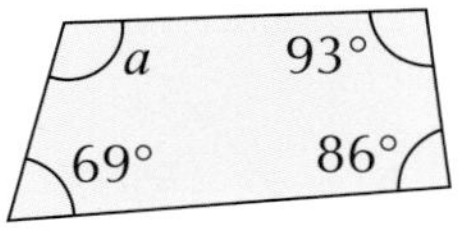

b) 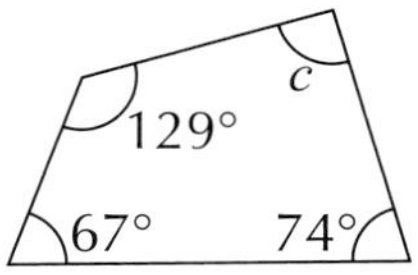

c) 

Q2 Find the values of the letters in the quadrilaterals below.

a) 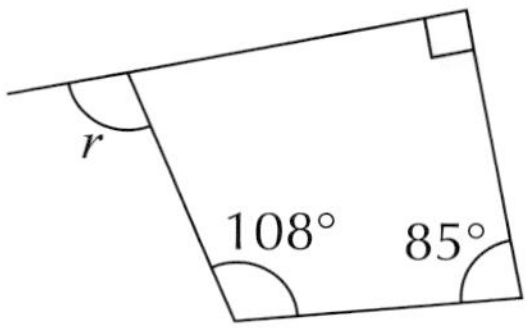

b) 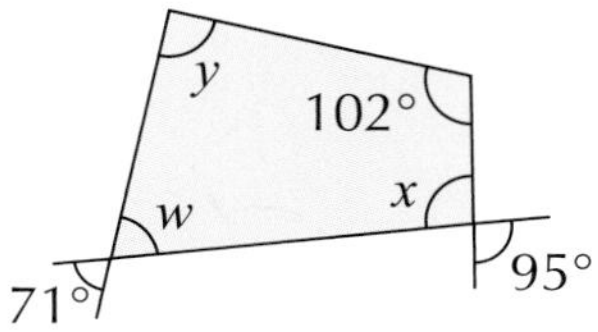

Squares and Rectangles

Learning Objective — Spec Ref G4:
Know the properties of squares and rectangles.

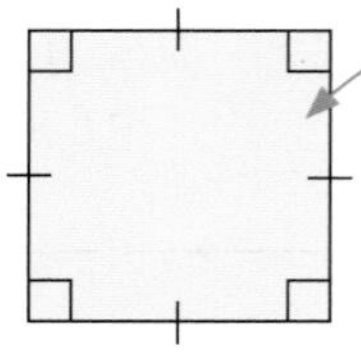

A **square** is a quadrilateral with 4 equal sides and 4 angles of 90°.

A **rectangle** is a quadrilateral with 4 angles of 90° and opposite sides of the same length.

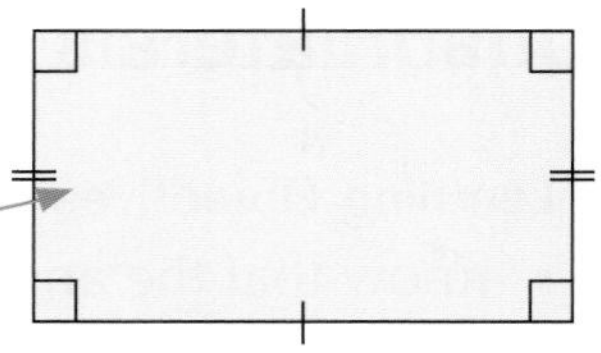

Exercise 2

Q1 Copy the diagram below, then add two more points to form a square. Join the points to complete the square.

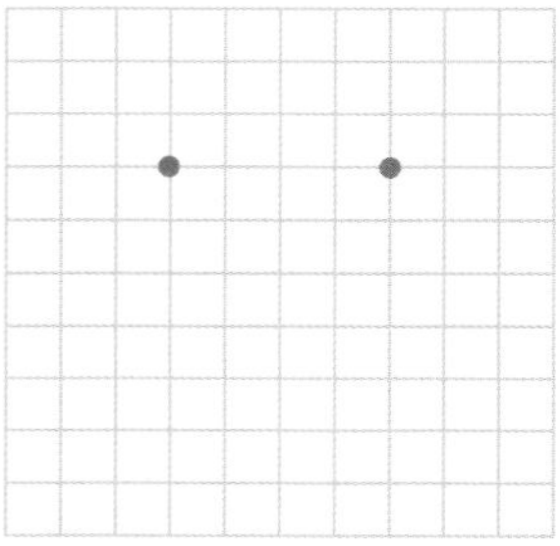

Q2 Copy the diagram below, then add two more points to form a rectangle. Join the points to complete the rectangle.

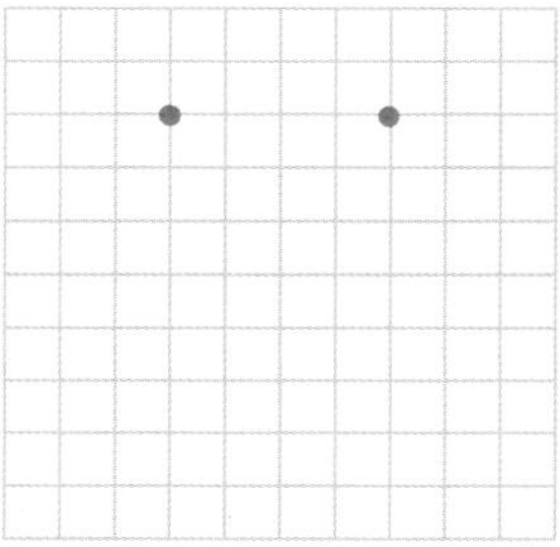

Q3 a) Measure the length of the diagonals on the square on the right. What do you notice?

b) At what angle do the two diagonals cross?

c) By measuring diagonals, determine which of the following are squares.

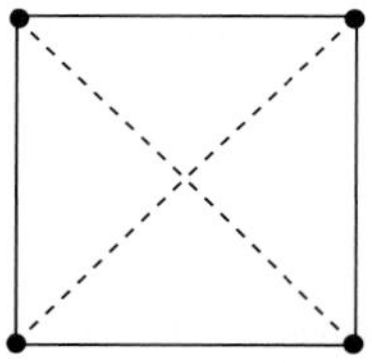

(i)

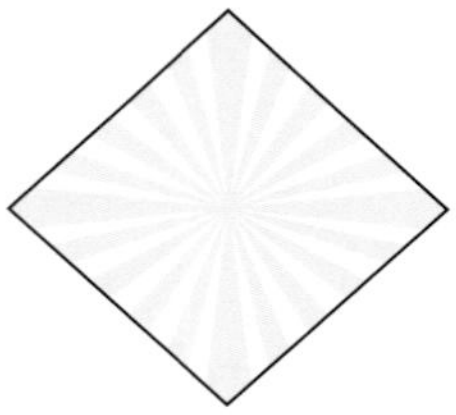

(ii)

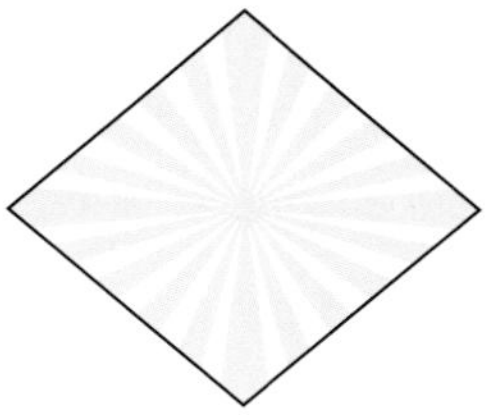

(iii)

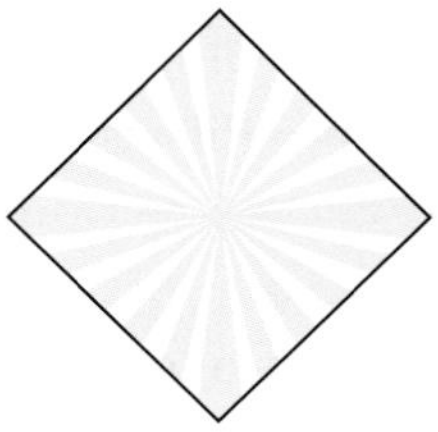

Q4 Look at this diagram of a rectangle. Explain why one of the side lengths must be **incorrect**.

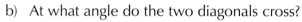

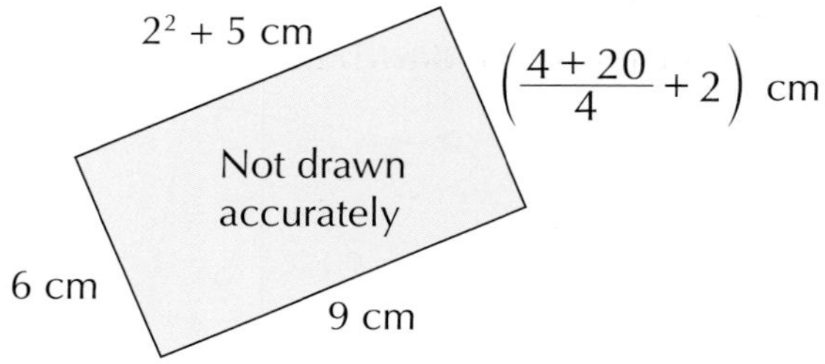

Parallelograms and Rhombuses

Learning Objective — Spec Ref G4:
Know the properties of parallelograms and rhombuses.

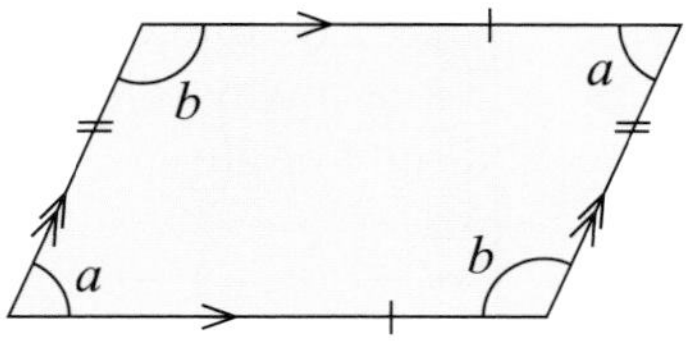

A **parallelogram** is a quadrilateral with 2 pairs of equal, parallel sides.

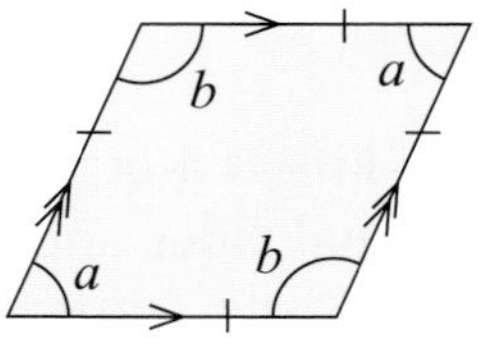

A **rhombus** is a parallelogram where all the sides are the same length. The **diagonals** of a rhombus **bisect** the angles (i.e. cut them in half) and cross at a **right angle**.

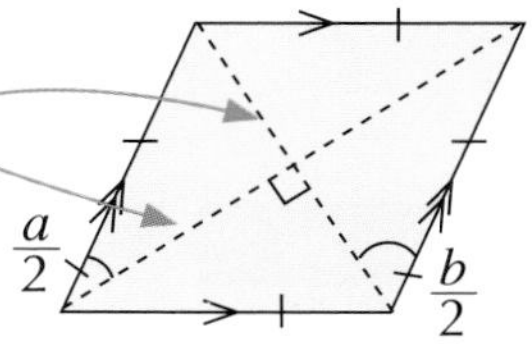

Opposite angles in parallelograms and rhombuses are **equal**, and **neighbouring angles** always add up to **180°**: $\boldsymbol{a + b = 180°}$
This is because the **parallel lines** that make up the sides of these shapes mean a and b are **allied angles** (page 241).

Example 2

Find the size of the angles marked with letters in this rhombus.

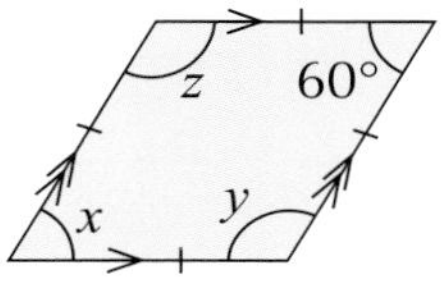

1. Opposite angles in a rhombus are equal. Use this fact to find angle x. $\quad \boldsymbol{x = 60°}$
2. Neighbouring angles in a rhombus add up to 180°. Use this fact to find angle y. $\quad 60° + y = 180°$, $y = 180° - 60°$, $\boldsymbol{y = 120°}$
3. Opposite angles in a rhombus are equal, so z is the same size as y. $\quad \boldsymbol{z = 120°}$

Tip: You could have found z first, using the fact that z and 60° are neighbouring angles.

Exercise 3

Q1 a) Copy the diagram on the right. Add one more point and join the points to form a rhombus.

b) On a new grid, plot four points to form a parallelogram.

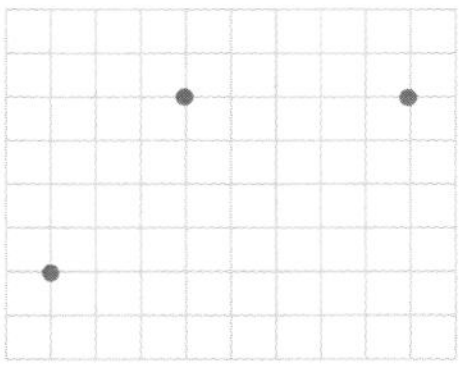

Q2 Calculate the size of the angles marked by letters in these quadrilaterals. (They're not drawn accurately, so don't try to measure them.)

a)

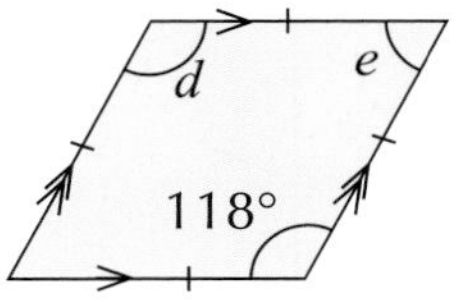

b)

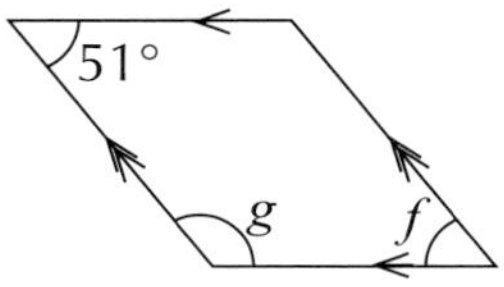

c)

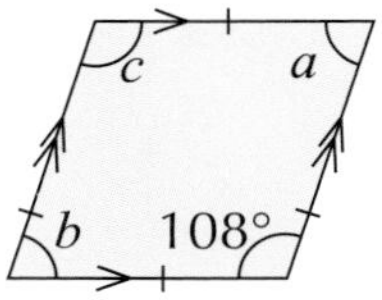

Kites

Learning Objective — Spec Ref G4:
Know the properties of kites.

A **kite** is a quadrilateral with **2 pairs** of **equal sides** and **1 pair** of **equal angles** in opposite corners, as shown on the diagram. (The other pair of angles aren't generally equal.)

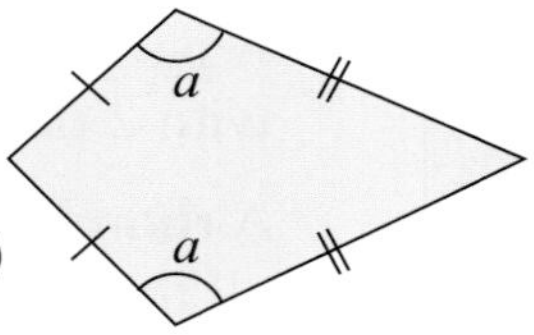

Tip: The diagonals of a kite always cross at a right angle.

Example 3

Find the size of the angles marked with letters in the kite below.

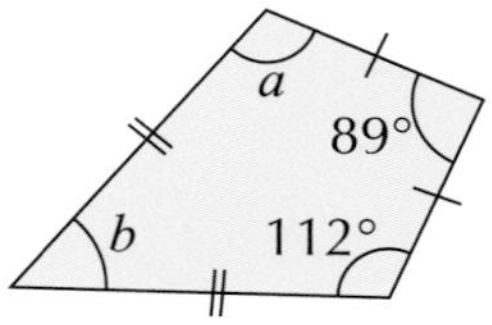

1. a and the 112° angle must be equal. — $\boldsymbol{a = 112°}$
2. A kite is a quadrilateral, so its angles add up to 360°. Use this to write an equation in terms of b. — $112° + 89° + 112° + b = 360°$

   $313° + b = 360°$

   $\boldsymbol{b} = 360° - 313° = \mathbf{47°}$

Exercise 4

Q1 Which letter goes in each box to complete the sentences about this kite?

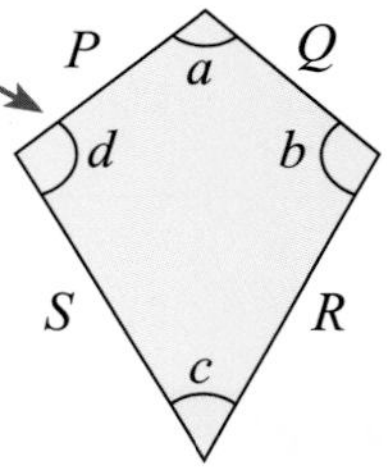

a) Angle b is the same size as angle ☐

b) The length of side P is the same as the length of side ☐

c) The length of side R is the same as the length of side ☐

Q2 Find the size of the angles marked by letters in these kites. (They're not drawn accurately, so don't try to measure them.)

a)

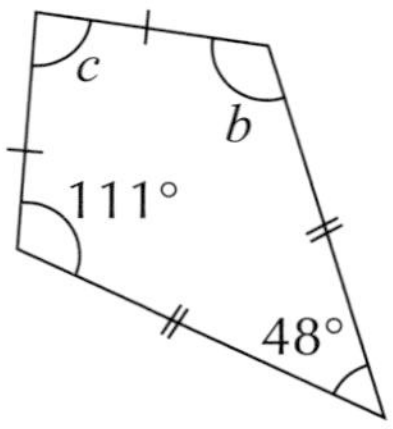

b)

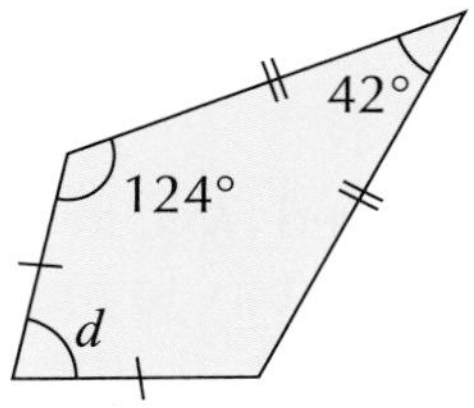

c)

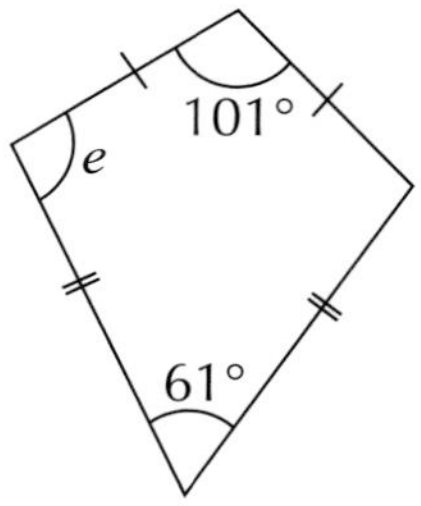

Trapeziums

Learning Objective — Spec Ref G4:
Know the properties of trapeziums and isosceles trapeziums.

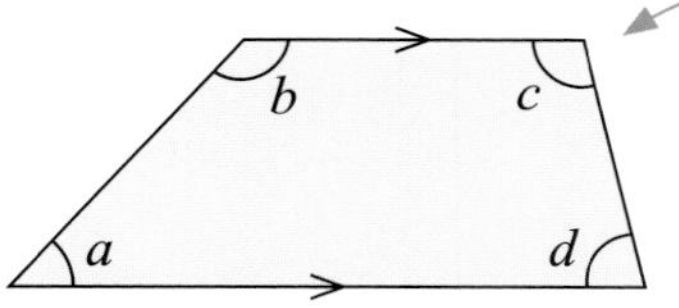

$a + b = 180°$ **$c + d = 180°$**

A **trapezium** is a quadrilateral with 1 pair of parallel sides.

An **isosceles trapezium** is a trapezium with 2 pairs of equal angles and 2 sides of the same length.

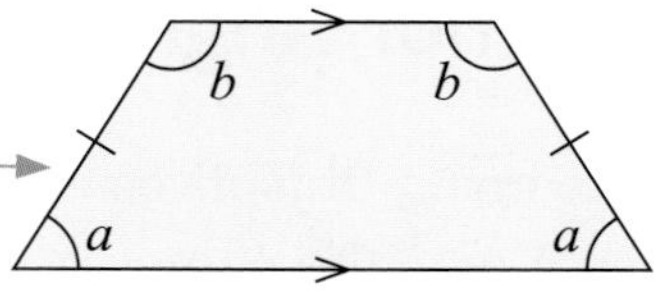

$a + b = 180°$

Because of the **allied angles** created by the parallel sides, pairs of angles **add up to 180°** as shown.

Example 4

Find the size of the angles marked with letters in the trapezium below.

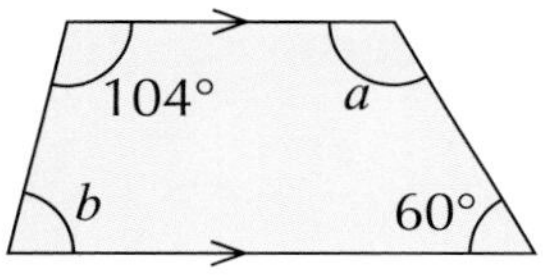

1. a and the 60° angle must add up to 180°. $a + 60° = 180°$
   $\boldsymbol{a} = 180° - 60° = \mathbf{120°}$
2. Similarly, b and the 104° angle must add up to 180°. $b + 104° = 180°$
   $\boldsymbol{b} = 180° - 104° = \mathbf{76°}$

Exercise 5

Q1 Choose the correct word from each pair to complete the following sentences.

a) A trapezium is a quadrilateral with one pair of (**parallel / equal**) sides.
b) An isosceles trapezium has (**one pair / two pairs**) of equal angles.
c) An isosceles trapezium has (**one pair / two pairs**) of parallel sides.
d) An isosceles trapezium has (**one pair / two pairs**) of equal sides.
e) The angles in a trapezium add up to (**180° / 360°**).

Q2 Find the size of the angles marked by letters in these trapeziums.
(They're not drawn accurately, so don't try to measure them.)

a)
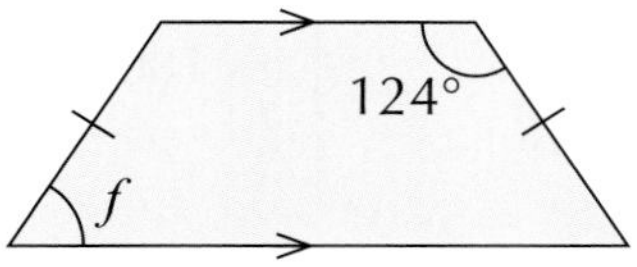

b)
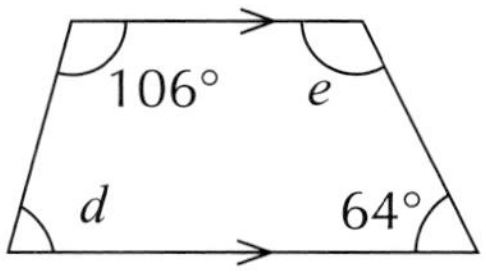

c)
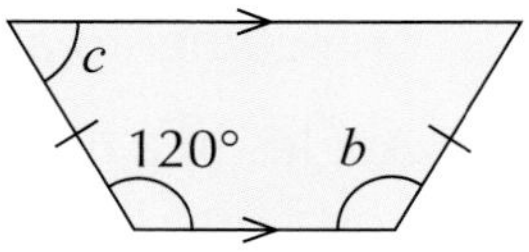

15.5 Interior and Exterior Angles

You've met lots of polygons before — they're just 2D shapes with straight sides. These pages will show you how to work out the number of sides and the sizes of angles in polygon problems.

Interior Angles

Learning Objectives — Spec Ref G1/G3:
- Know the names of different types of polygons.
- Be able to find the sum of the interior angles in a polygon by splitting it into triangles.

A polygon is a **2D shape** whose sides are all **straight**.

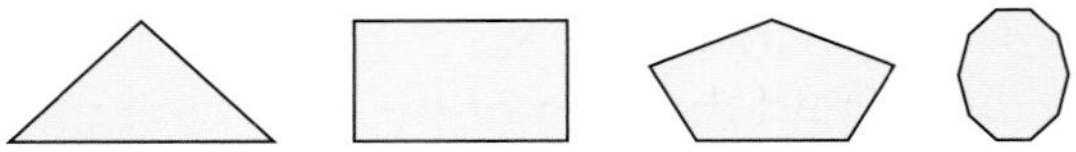

The **triangles** and **quadrilaterals** on the previous pages were three- and four-sided polygons. The box on the right shows the names of some other polygons — their names depend on the **number of sides** they have.

pentagon = **5** sides	**oct**agon = **8** sides
hexagon = **6** sides	**non**agon = **9** sides
heptagon = **7** sides	**dec**agon = **10** sides

A **regular** polygon has sides of **equal length** and angles that are all **equal**.
An **equilateral triangle** is a regular triangle and a **square** is a regular quadrilateral.

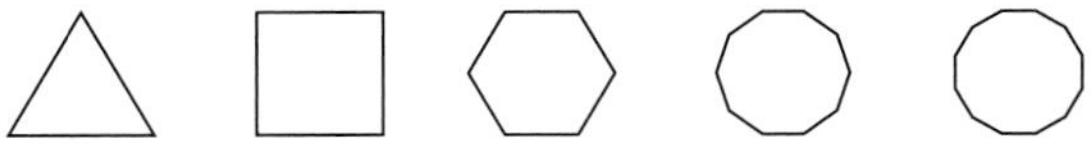

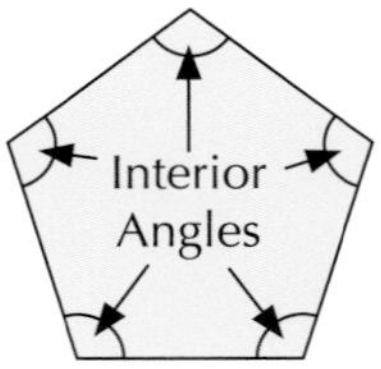

The **interior angles** of a polygon are the angles inside each vertex (corner). The interior angles of a regular pentagon are shown on the left.

You can find the **sum of the interior angles** by splitting the polygon into **triangles** and then using the rule that angles in a triangle add up to **180°** — you'll see how this works in the example below.

Example 1

Find the sum of the interior angles of a pentagon.

1. First draw any pentagon — it doesn't have to be regular.
2. Split the pentagon into triangles by drawing lines from one corner to all the others.
3. Angles in a triangle add up to 180°, and there are 3 triangles.

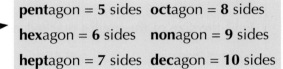

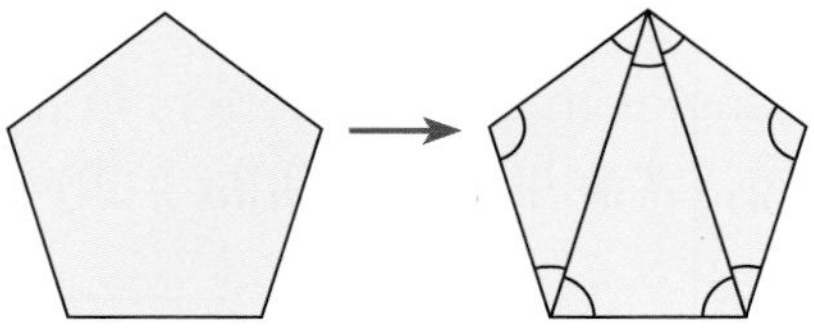

The sum of the interior angles of a pentagon is 3 × 180° = **540°**

Tip: Quadrilaterals can be split into two triangles. This explains why the sum of their angles is 360° (see page 245) — it's 180° × 2.

Example 2

A pentagon has four interior angles of 100°. Find the size of the fifth angle.

1. The interior angles of a pentagon add up to 540°. Use this fact to write an equation for the size of the missing angle, x.

   $100° + 100° + 100° + 100° + x = 540°$
   $400° + x = 540°$

2. Solve your equation to find x.

   $x = 540° - 400° =$ **140°**

Exercise 1

Q1 By dividing each shape into triangles, find the sum of the interior angles of the shapes on the right.

a) 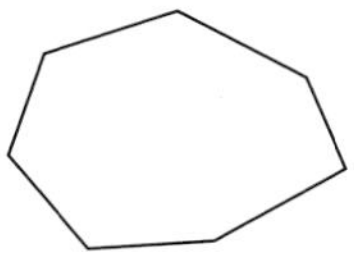

b) 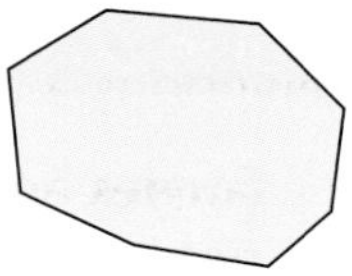

Q2 Sketch the polygon shown below.

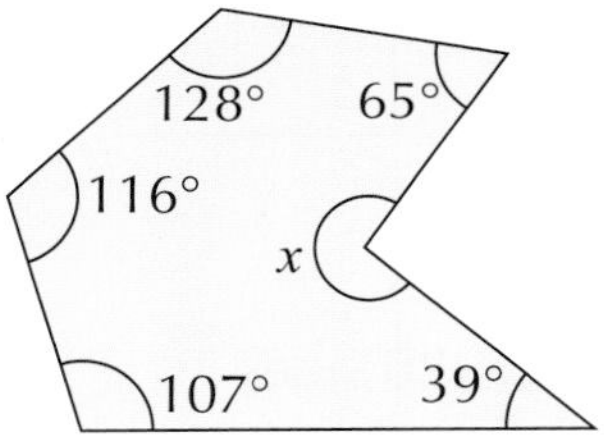

a) What happens when you try to divide the polygon into triangles by drawing a line from the 65° angle to each of the other corners?

b) Join the corner marked x to each of the other corners to split the shape into triangles.

c) Find the sum of the interior angles of the shape.

d) Find the size of the angle marked x.

The Formula for the Sum of the Interior Angles

Learning Objective — Spec Ref G3:
Know and be able to use the formula for the sum of the interior angles in a polygon.

You can always divide a polygon of n sides into **n – 2 triangles**.
Using this and the fact that the sum of the angles in a triangle is **180°**, you can create a **formula** for the sum of the interior angles:

$$S = (n - 2) \times 180°$$

where S is the **sum** of a polygon's **interior angles** and n is the **number of sides**.

This formula works for both **regular** and **irregular** polygons. Once you've found the value of the **sum of the interior angles**, you can use it to find the size of **individual angles** in the shape — as shown in the example on the next page.

Example 3

Find the size of each of the interior angles of a regular hexagon.

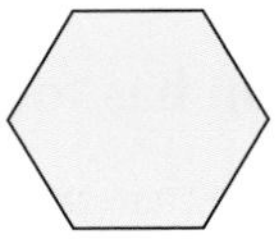

1. Use the formula to find the sum of the interior angles of a hexagon. A hexagon has 6 sides so $n = 6$.

   $S = (6 - 2) \times 180°$
   $= 4 \times 180° = 720°$

2. All the angles in the hexagon are equal, so divide the sum by the number of angles.

   So each interior angle of a regular hexagon is $720° \div 6 =$ **120°**

Tip: The angles are equal because the polygon is **regular**.

Exercise 2

Q1 For each of the shapes below, determine whether or not it is a polygon, and give reasons why.

a) 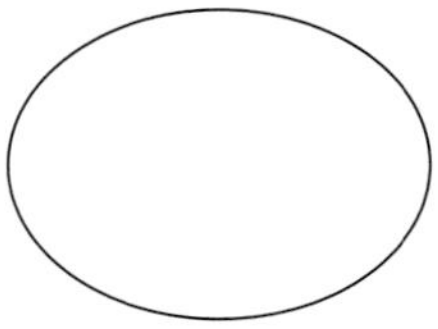

b) 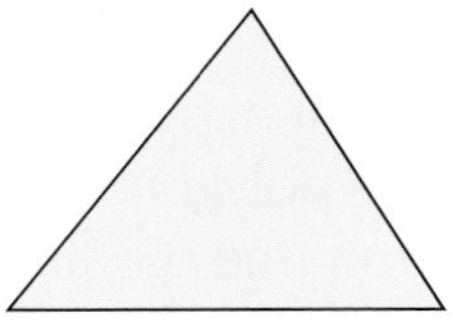

c) 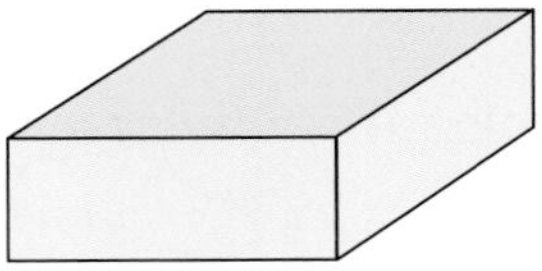

Q2 Find the sum of the interior angles of a polygon with:

a) 10 sides b) 12 sides c) 20 sides

Q3 A regular heptagon is shown on the right.

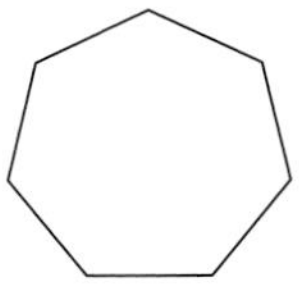

a) Find the sum of the interior angles of a heptagon.

b) Find the size of each of the interior angles of a regular heptagon to 2 d.p.

Q4 Find the size of each of the interior angles in the following shapes.

a) Regular octagon (8 sides)

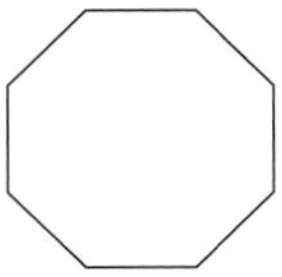

b) Regular nonagon (9 sides)

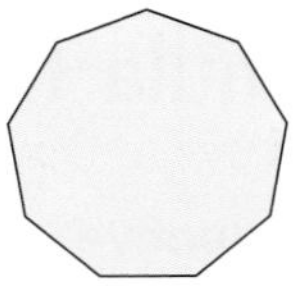

Q5 The shapes below are not drawn accurately. For each of the shapes:

(i) Find the sum of the interior angles (ii) Find the size of the missing angle.

a) 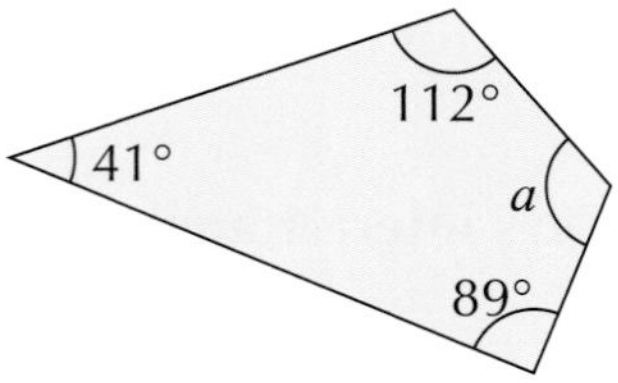

b) 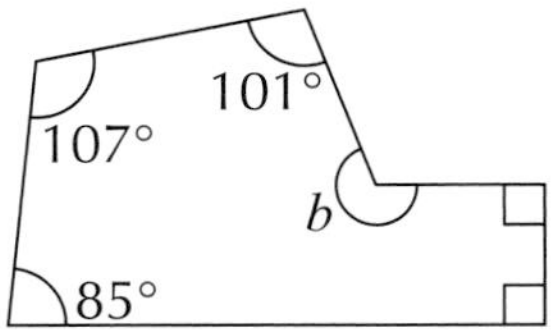

c) 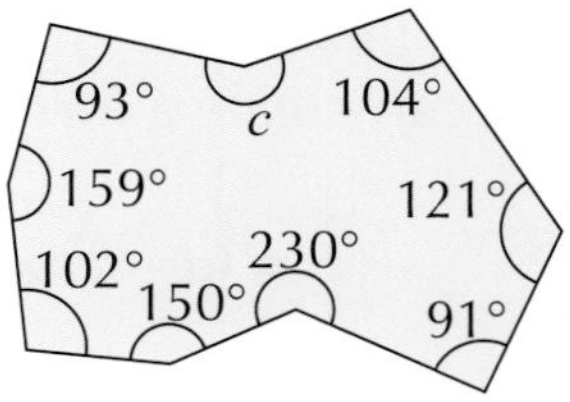

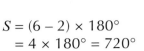

Exterior Angles

Learning Objective — Spec Ref G3:
Find interior and exterior angles of regular and irregular polygons.

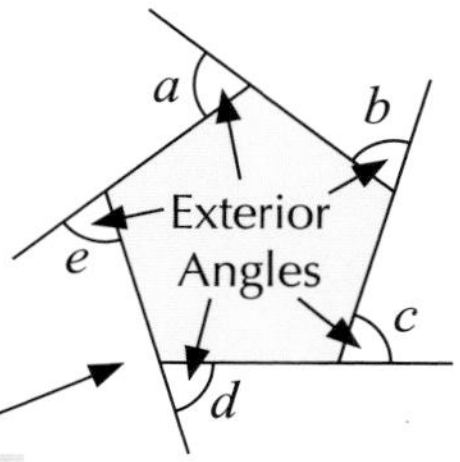

An **exterior angle** of a polygon is an angle between a **side** and a **line** that extends out from one of the **neighbouring sides**. For example, the exterior angles of a regular pentagon are shown on the right.

For any polygon, the exterior angles always **add up to 360°**. In the case of the pentagon, this is: $\boldsymbol{a + b + c + d + e = 360°}$

Since the exterior angle and the neighbouring interior angle lie on a straight line, they must **add up to 180°** (see page 237). So you can use this formula to find an **interior angle** given the exterior angle:

Interior angle = 180° – Exterior angle

For a **regular** polygon, the interior angles are all equal, and this means the **exterior angles are all equal** too. Then the formula for the size of an exterior angle for a regular n-sided polygon is:

Exterior angle = $\boldsymbol{\frac{360°}{n}}$

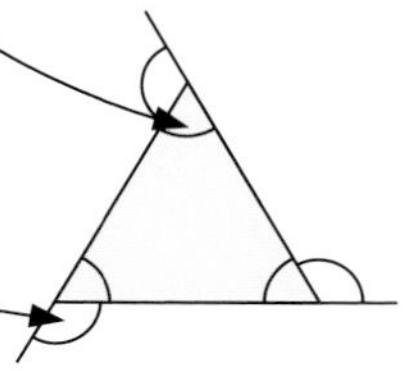

Example 4

Find the size of each of the exterior angles of a regular hexagon.

A hexagon has 6 sides.
The hexagon is regular so put $n = 6$ into the exterior angle formula.

$360° \div 6 = 60°$

So each exterior angle is **60°**.

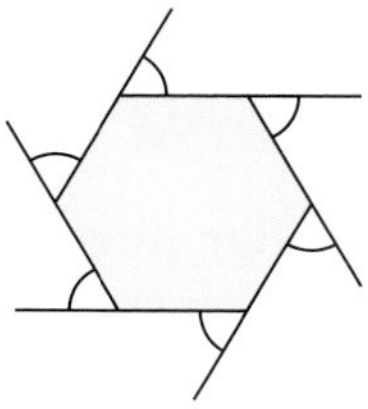

Example 5

Prove that the exterior angle of a triangle is equal to the sum of the two non-adjacent interior angles.

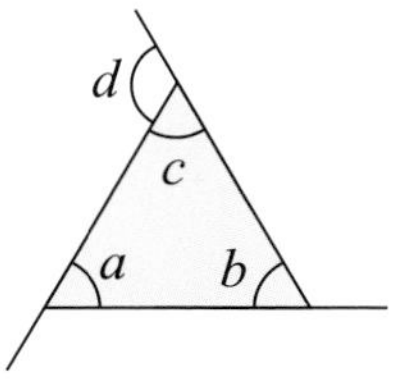

1. You need to show that $a + b = d$.
2. The exterior angle and neighbouring interior angle add up to 180°. $d + c = 180° \Rightarrow c = 180° - d$
3. The angles in the triangle also add up to 180°. $a + b + c = 180°$
4. Substitute in the expression for c, then rearrange to get the result. $a + b + (180° - d) = 180°$
   $\boldsymbol{a + b = d}$

Tip: Writing what you are trying to show is useful, as it tells you what to aim for.

Exercise 3

Q1 The diagram on the right shows a regular pentagon with the exterior angles marked on.

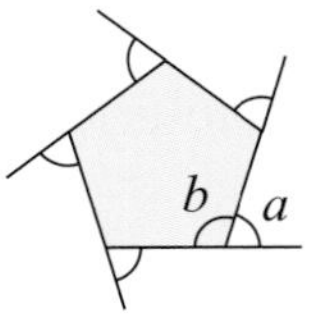

a) Find a, the size of each of the exterior angles of the pentagon.

b) Hence find b, the size of each of the interior angles of the pentagon.

Q2 a) Find the size of each of the exterior angles (to 2 d.p.) of the following polygons.

(i) regular heptagon (ii) regular octagon (iii) regular nonagon

b) Use your answers to part a) to find the size of each of the interior angles of these polygons.

Q3 Find the size of the angles marked by letters in these diagrams.

a)

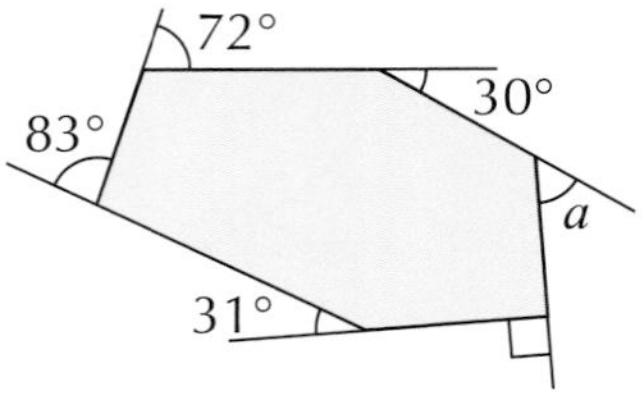

Diagrams not drawn accurately.

b)

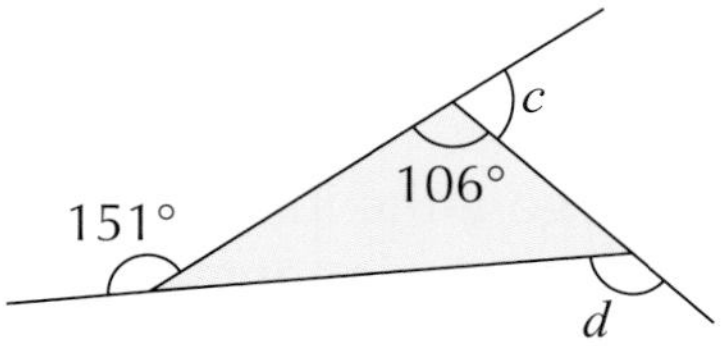

Q4 Find the size of the unknown exterior angle in a shape whose other exterior angles are:

a) 100°, 68°, 84° and 55° b) 30°, 68°, 45°, 52°, 75° and 50°

Example 6

A regular polygon has exterior angles of 30°. How many sides does the polygon have?

Tip: A 12-sided polygon is called a dodecagon.

1. It's a regular polygon so put 30° into the exterior angle formula. $30° = \frac{360°}{n}$
2. Solve the equation for n. $n = 360° \div 30° = 12$

So the regular polygon has **12 sides**.

Exercise 4

Q1 The exterior angles of some regular polygons are given below. For each exterior angle, find:

(i) the number of sides the polygon has,

(ii) the size of each of the polygon's interior angles,

(iii) the sum of the polygon's interior angles.

a) 90° b) 40° c) 6° d) 4°

15.6 Symmetry

Symmetry is when a shape can be reflected or rotated and still look exactly the same afterwards. E.g. rotating a square by 90° puts each vertex in a different place but the square looks the same.

Learning Objectives — Spec Ref G1:
- Recognise lines of symmetry and rotational symmetry of 2D shapes.
- Be able to find the sum of the interior angles in a polygon by splitting it into triangles.

A line of **symmetry** on a shape is a **mirror line**, where you can fold the shape so that both halves match up **exactly**. Each side of the line of symmetry is a **reflection** of the other.

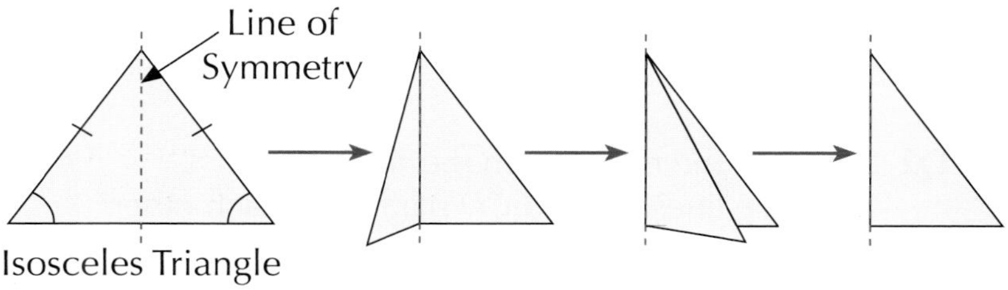

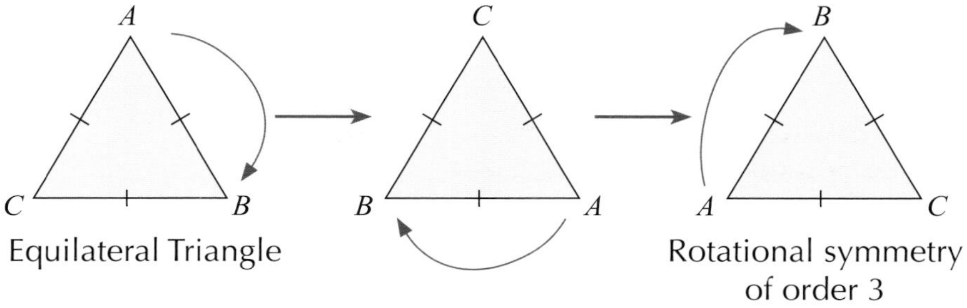

The order of **rotational symmetry** of a shape is the number of positions you can **rotate** (turn) the shape into so that it looks **exactly** the same.

A shape that only looks the same **once** every complete turn has rotational symmetry of **order 1** (or no rotational symmetry).

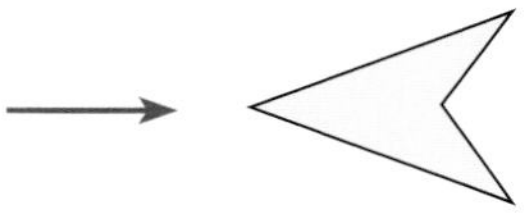

Exercise 1

Q1 Sketch each of the shapes below, then draw on any lines of symmetry. State the number of lines of symmetry you have drawn for each shape.

a) rectangle b) rhombus c) isosceles trapezium

d) regular pentagon e) regular hexagon f) regular heptagon

Q2 Find the order of rotational symmetry of each of the following shapes.

a)

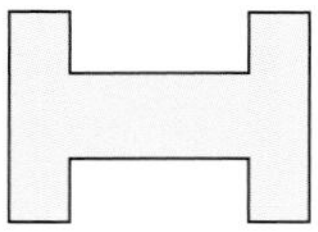

b)

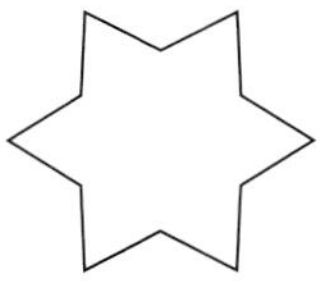

c)

d)

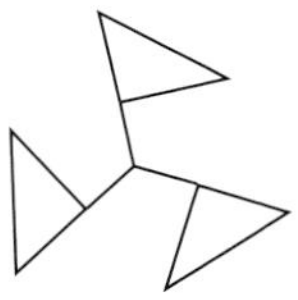

Review Exercise

The angles in this exercise have not been drawn accurately, so don't try to measure them.

Q1 Find the value of each letter.

a)
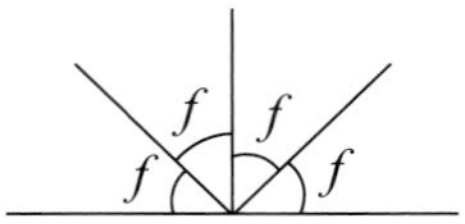

b)
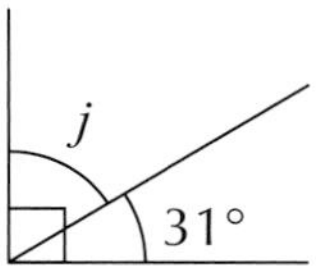

c)
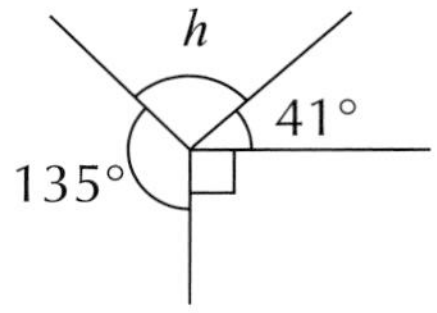

Q2 Look at the diagram on the right.
Are the lines *AB* and *CD* parallel to each other?
Explain your answer.

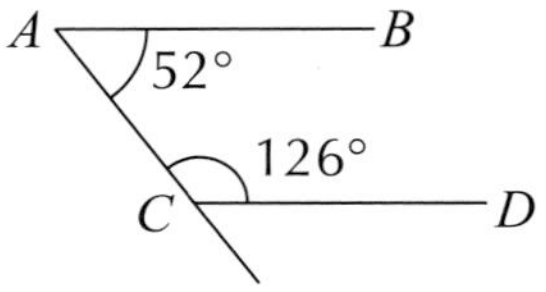

Q3 Find the value of each letter.

a)
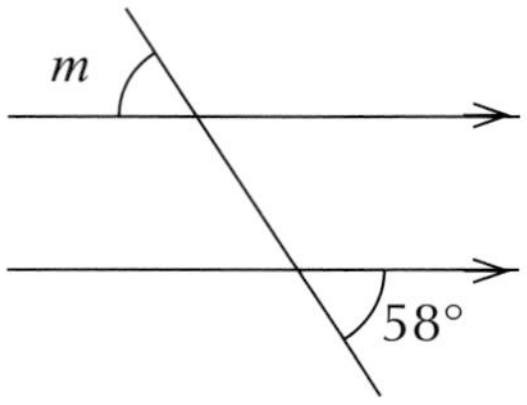

b)
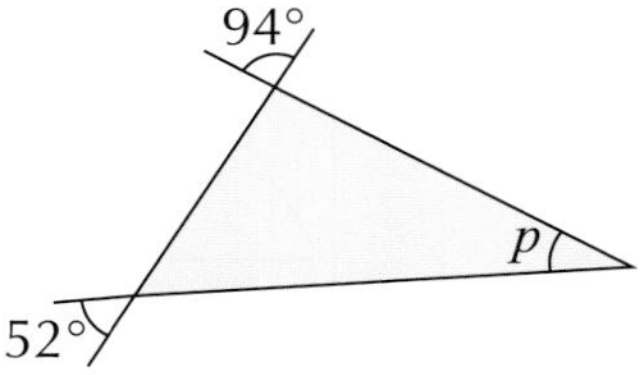

c)
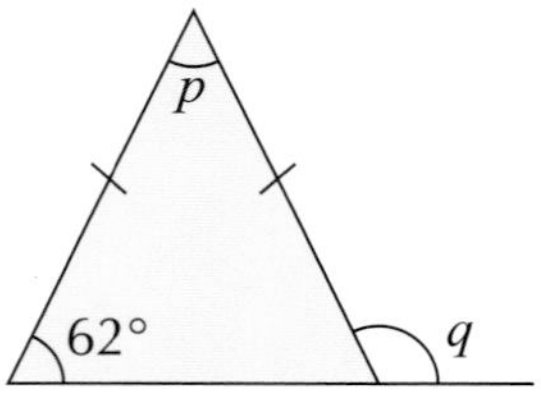

d)
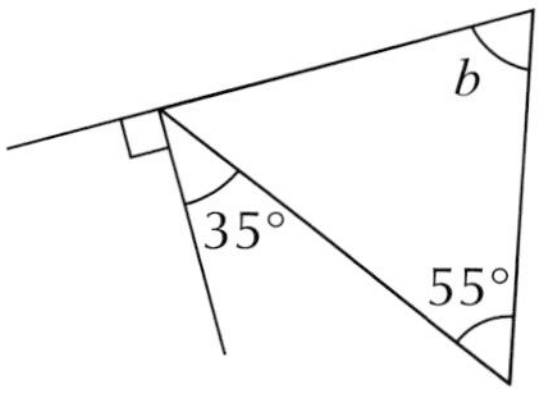

e)
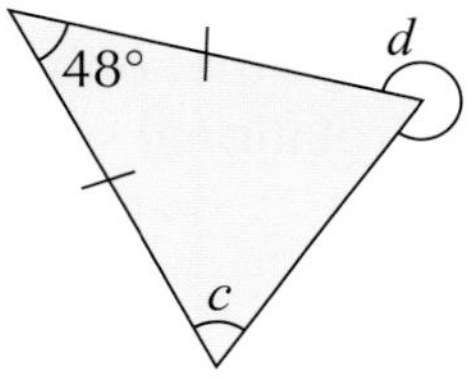

Q4 Match one name from the box to each of the quadrilaterals below.

kite	rectangle	trapezium	parallelogram

a)
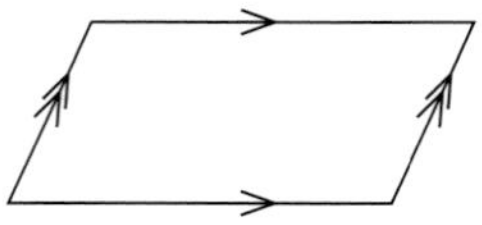

b)
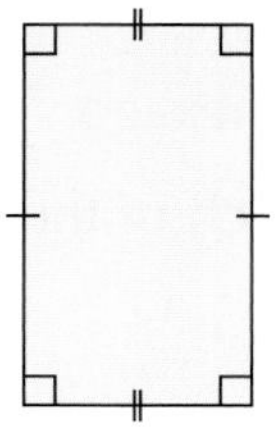

c)
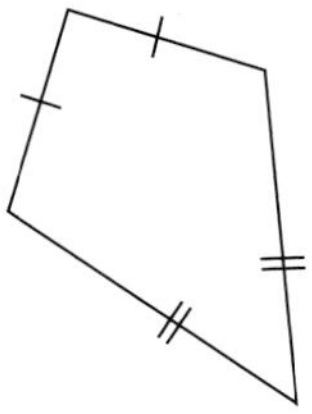

d)
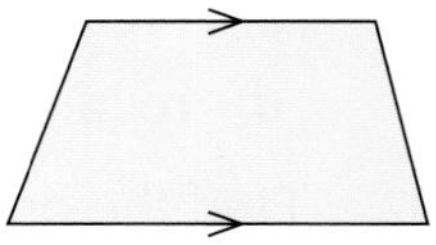

Q5 A quadrilateral has 40°, 83° and 99° interior angles.
Find the size of the fourth angle.

Q6 Find the value of each letter.

a)

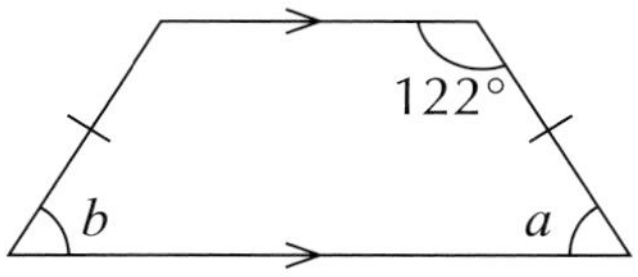

b)

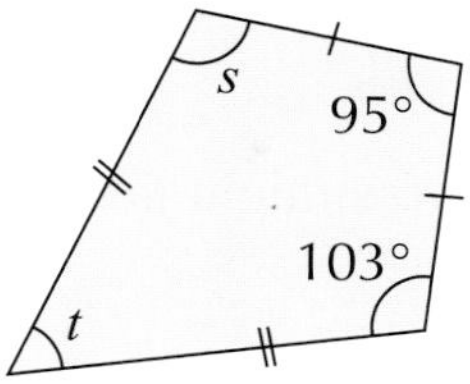

c)

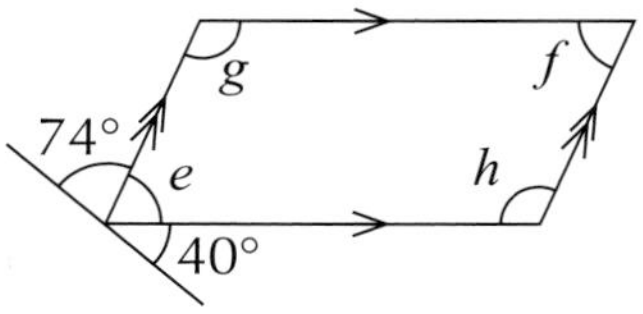

Q7 The diagram on the right shows a kite and a square.

a) Write down the value of a.

b) Use your answer to find the size of angles b and c.

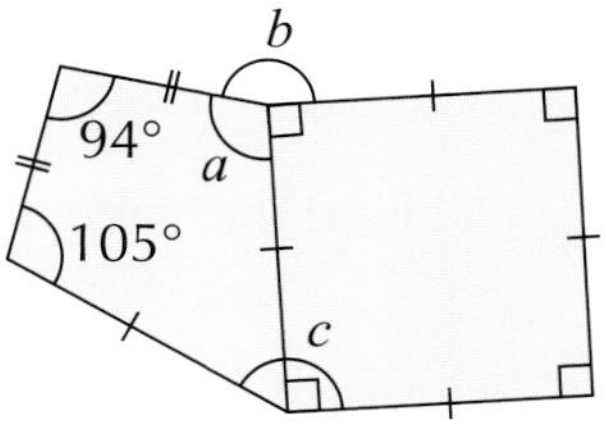

Q8

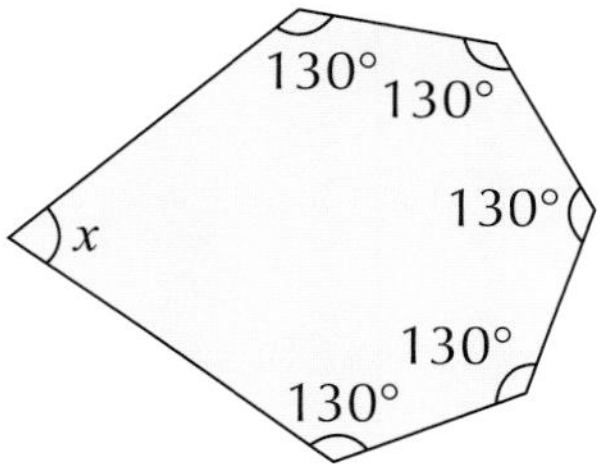

The shape shown on the left has 6 sides.

a) Calculate the sum of the interior angles of a hexagon.

b) Use your answer to find angle x.

Q9 A regular polygon has exterior angles of 45°.

a) How many sides does the polygon have?
What is the name of this kind of polygon?

b) Sketch the polygon.

c) What is the size of each of the polygon's interior angles?

d) What is the sum of the polygon's interior angles?

Q10 Decide which of the shapes below matches each of the descriptions.

A

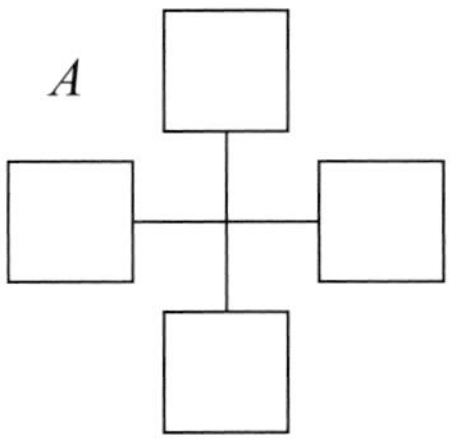

B

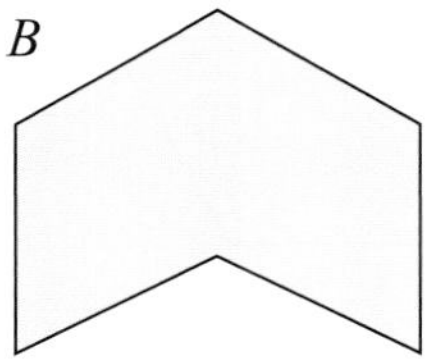

C

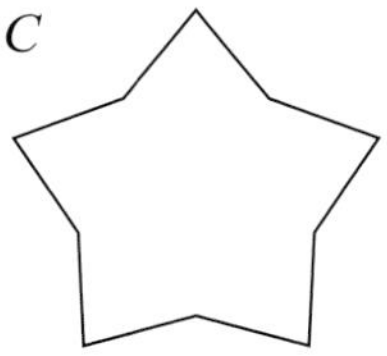

D

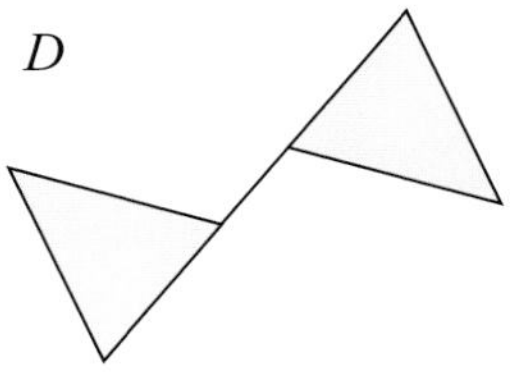

a) 1 line of symmetry, no rotational symmetry.

b) 4 lines of symmetry, rotational symmetry of order 4.

c) No lines of symmetry, rotational symmetry of order 2.

d) 5 lines of symmetry, rotational symmetry of order 5.

Exam-Style Questions

Q1 The diagram below shows two straight lines meeting at a point.

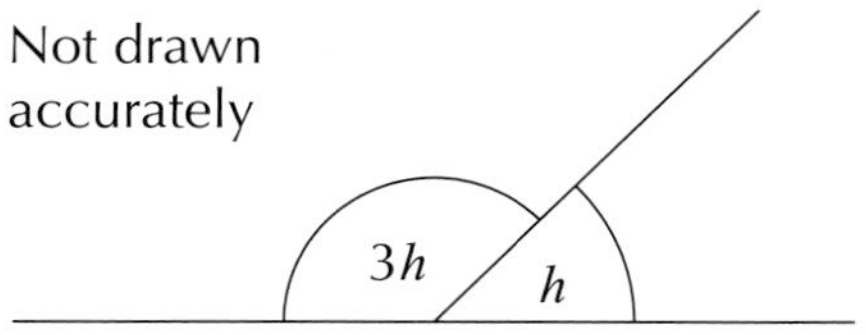

Find the value of h.

[2 marks]

Q2 The diagram below shows a quadrilateral.

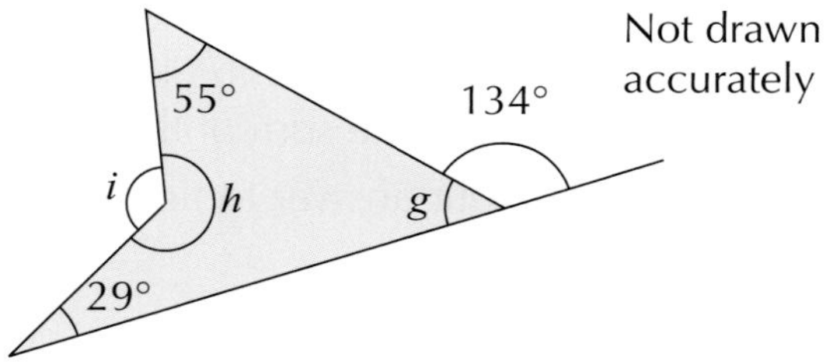

Find the values of the letters.

[3 marks]

Q3 The diagram shows a parallelogram.
$ABCD$ is a straight line. EBF is a straight line.
Angle BFG is 56°.

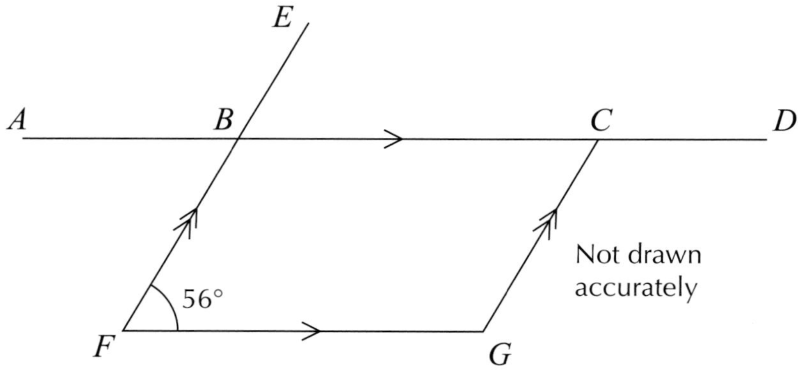

a) Write down the size of angle EBC.

[1 mark]

b) Find the size of angle DCG. Give a reason for each step of your working.

[2 marks]

Q4 The diagram below shows a seven-sided shape.

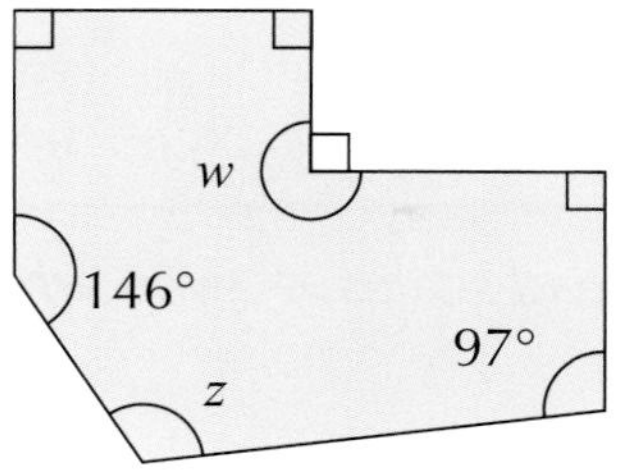

Not drawn accurately

Find the value of angles w and z.

[3 marks]

Q5 The diagram below shows a quadrilateral.

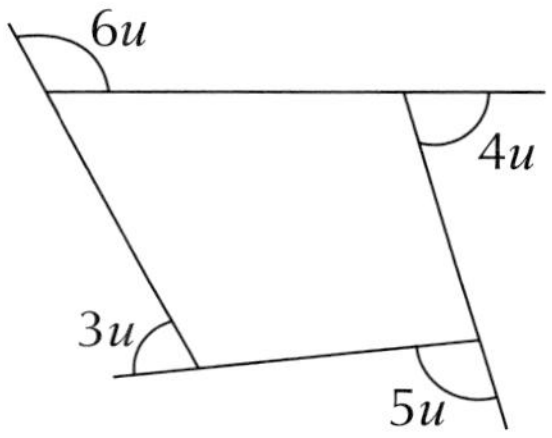

Find the value of u.

[2 marks]

Q6 The diagram below shows an equilateral triangle ABC and an isosceles triangle ACE where $AC = EC$.
The triangles have a common side, AC, and EAD is a straight line.
Angle $ECB = 38°$.

Not drawn accurately

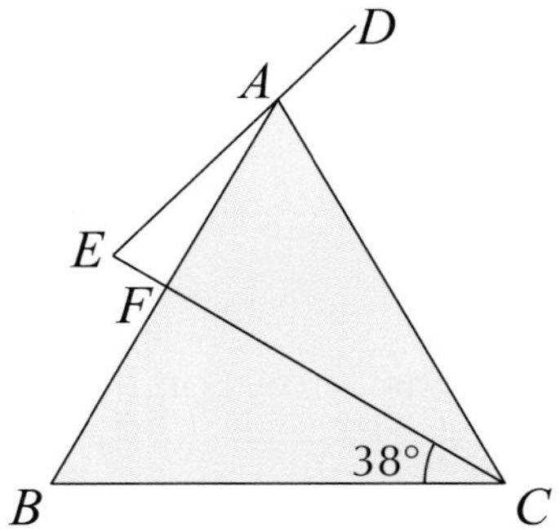

Prove that angle $DAC = 101°$.

[3 marks]

16.1 Reading Scales

You always use the same method to read scales — even when they use different units.

Reading Scales

Learning Objective — Spec Ref G14:
Be able to read values off scales.

The **main values** on a scale are **labelled**, but the smaller values are not. To find the **smaller values**, you need to work out what each **gap** between the smaller marks represents.

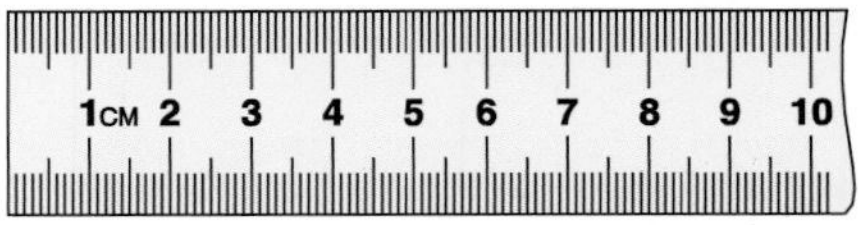

The **gap** between each pair of labelled marks on a scale is always the **same amount** — on the ruler above, the gap between each big mark is 1 cm. To work out what each small gap represents, **divide** the distance between two big marks by the **number of small gaps**. There are 10 gaps between each big mark, so each small gap is 1 cm ÷ 10 = 0.1 cm, or **1 mm**.

Example 1

Write down the volume of liquid in the beaker on the right.

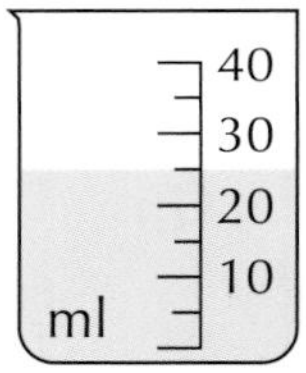

1. Find the interval between the numbered marks. $40 - 30 = 10$ ml
2. Divide this interval by the number of smaller divisions between each number to find what each division represents. Here, there are 2 divisions between each number. $10 \div 2 = 5$ ml
3. The amount in the beaker is 20 ml, plus the amount represented by one small division. $20 + 5 =$ **25 ml**

Exercise 1

Q1 Write down the lengths shown by the arrows on the ruler, giving your answers in cm.

a)

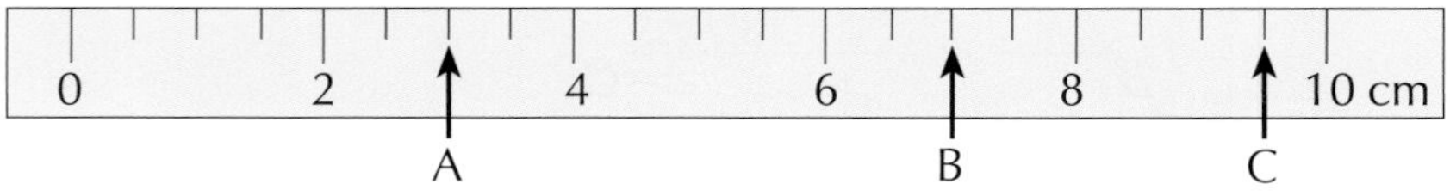

b)

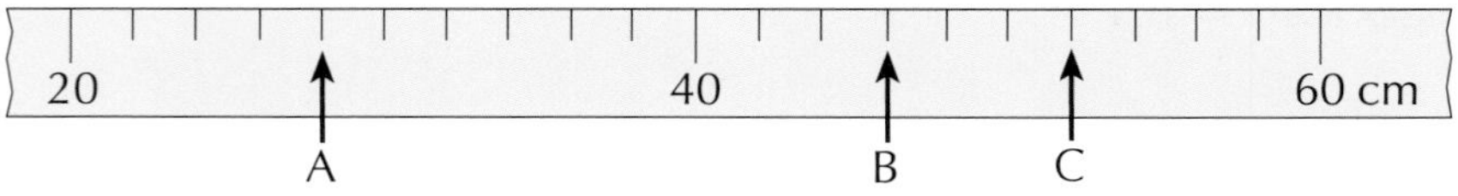

Q2 Use the fact that 1 cm = 10 mm to write down the length of each of the bugs in mm.

a)

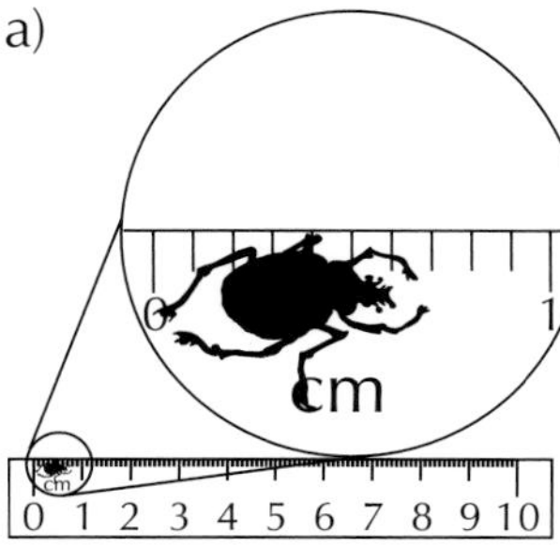

b)

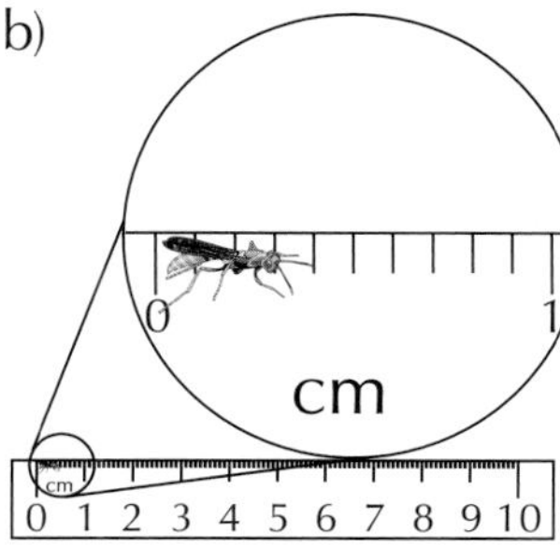

c)

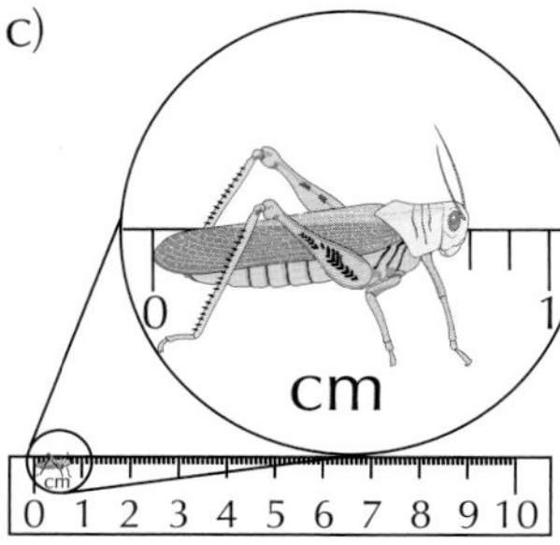

Q3 Write down the mass shown by the arrow on each scale.

a)

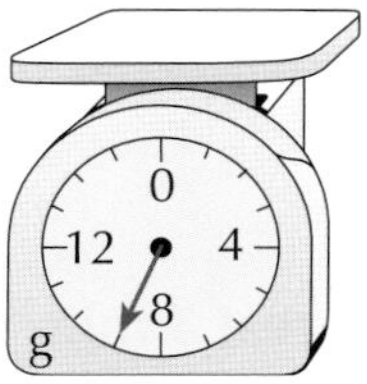

b)

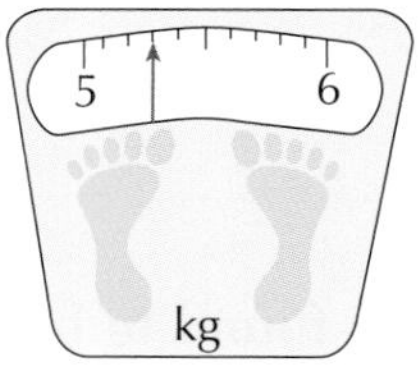

c)

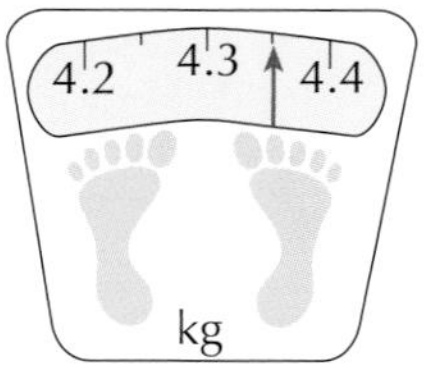

Q4 Write down:

a) the weight of the bananas,

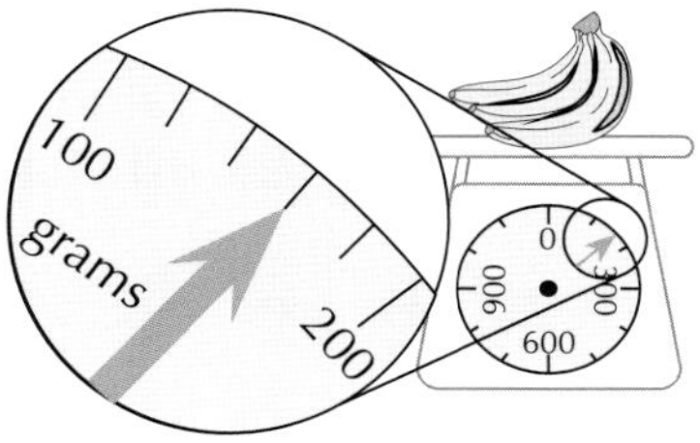

b) the volume of liquid in the bottle.

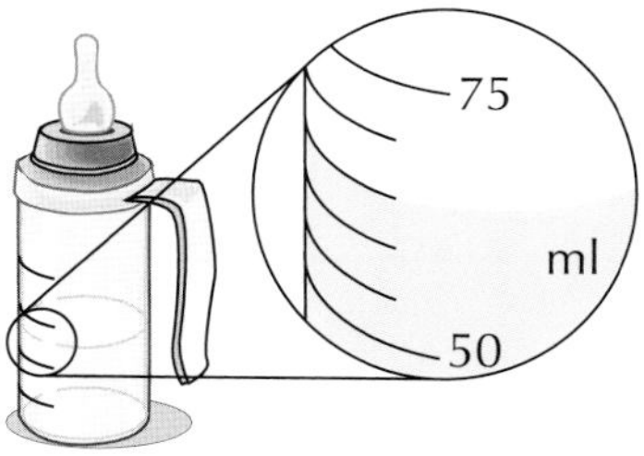

Inaccuracy of Measurements

Learning Objectives — Spec Ref N15/N16:

- Understand that measurements may be inaccurate due to rounding errors.
- Find the maximum and minimum possible values of a measurement.

Prior Knowledge Check:
Be able to find rounding errors. See p35.

The **accuracy** of a measurement depends on the device being used to make that measurement.

For example, this ruler measures to the nearest millimetre — so both these bugs seem to be 9 mm long. But using a more **precise** piece of equipment, you can see the first bug is actually slightly shorter than the second.

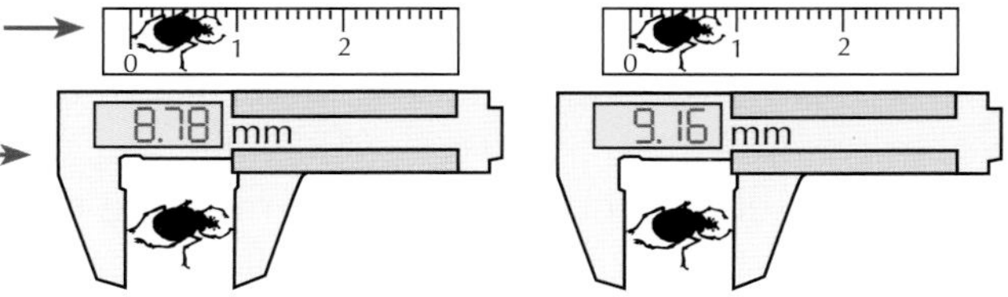

When you measure to the nearest millimetre, anything from 8.5 mm up to 9.5 mm would round to 9 mm. We say that 8.5 mm is the **minimum value** and 9.5 mm is the **maximum value**.
In general, the max. and min. values for measurements are '**+ half a unit**' and '**– half a unit**'.

Example 2

The length of a pen is measured and found to be 15 cm, to the nearest 1 cm. Write down the maximum and minimum length the pen could be.

1. The length has been rounded to the nearest cm. So half a unit is 1 cm ÷ 2 = 0.5 cm.
2. Find the minimum value by subtracting half a unit. — Minimum value: 15 – 0.5 = **14.5 cm**
3. Find the maximum value by adding half a unit. — Maximum value: 15 + 0.5 = **15.5 cm**

Tip: The biggest value that would round down to 15 cm is 15.4999..., but we call 15.5 cm the maximum because it's easier to write.

Example 3

The volume of a teacup is measured and found to be 135 ml, to the nearest 5 ml. Write down the maximum and minimum possible volume of the teacup.

1. The volume has been rounded to the nearest 5 ml. So half a unit is 5 ml ÷ 2 = 2.5 ml.
2. Find the minimum value by subtracting half a unit.
3. Find the maximum value by adding half a unit.

Minimum value: 135 – 2.5 = **132.5 ml**

Maximum value: 135 + 2.5 = **137.5 ml**

Exercise 2

Q1 Find the maximum and minimum values for each of these measurements.
- a) A length of 10 cm, which has been measured to the nearest 1 cm.
- b) A volume of 18 litres, which has been measured to the nearest 1 litre.
- c) A volume of 65 litres, which has been measured to the nearest 5 litres.
- d) A length of 20 m, which has been measured to the nearest 2 m.

Q2 A pipette contains 5.7 cm^3 of liquid, measured to the nearest 0.1 cm^3. What is the greatest amount of liquid that could be in the pipette?

Q3 Elliot's house has a front door frame which is 96.5 cm wide. He wants to buy a table which is given as 95 cm wide, to the nearest 2 cm. Can he be sure that the table will fit through his door frame? PROBLEM SOLVING

Q4 A set of kitchen scales can measure masses correct to the nearest 10 g. Janek weighs out 100 g of flour. Work out the maximum and minimum possible mass of the flour.

Q5 Will and Kyle measure their heights. They are both 170 cm tall to the nearest 10 cm. What is the maximum possible difference between their heights?

16.2 Converting Units — Length, Mass and Volume

Metric units are a system of units based on powers of 10. E.g., a centimetre is 10 millimetres, a metre is 100 centimetres etc. This topic covers the metric units for length, mass and volume.

Learning Objective — Spec Ref N13/R1/G14:
Convert between different metric units for length, mass and volume.

Prior Knowledge Check: Be able to multiply and divide by 10, 100, 1000, etc. See p18.

The **metric units** for length, mass and volume are as follows.

- **Length** is measured in **millimetres** (mm), **centimetres** (cm), **metres** (m) and **kilometres** (km).
- **Mass** is measured in **milligrams** (mg), **grams** (g), **kilograms** (kg) and **tonnes**.
- **Volume** is measured in **millilitres** (ml), **litres** (l) and **cubic centimetres** (cm^3).

To **convert** between different units of length, mass or volume, you **multiply** or **divide** by a **conversion factor**. The most commonly used conversion factors are shown below.

Length	Mass	Volume
1 cm = 10 mm	1 g = 1000 mg	1 litre (l) = 1000 ml
1 m = 100 cm	1 kg = 1000 g	1 ml = 1 cm^3
1 km = 1000 m	1 tonne = 1000 kg	

When converting from small units to **bigger units** (e.g. cm to m), you **divide** by the conversion factor. When converting from big units to **smaller units**, you **multiply** by the conversion factor.

Always **check** your answers to make sure they seem **reasonable** — e.g. if you converted an elephant's height from m to cm and got 0.025 cm, you'd know you'd gone wrong. You'd expect **more** small units than big units — there are 100 cm in 1 m, so you'd expect there to be more centimetres than metres.

Example 1

a) Convert 2.5 m into cm.

1. There are 100 cm in 1 m, so the conversion factor is 100. — 1 m = 100 cm
2. cm are smaller than m, so you need to multiply by the conversion factor to change m into cm. — So 2.5 m = 2.5 × 100 = **250 cm**

b) Convert 2500 m into km.

1. There are 1000 m in 1 km, so the conversion factor is 1000. — 1 km = 1000 m
2. km are bigger than m, so you need to divide by the conversion factor to change m into km. — 2500 m = 2500 ÷ 1000 = **2.5 km**

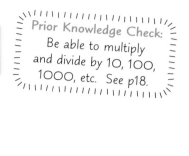

Exercise 1

For Questions 1-3, convert each measurement into the units given.

Q1 a) 2 cm into mm b) 15 ml into cm^3 c) 2.3 tonnes into kg

Q2 a) 3400 m into km b) 50 cm into m c) 246 kg into tonnes

Q3 a) 3 kg into g b) 379 mm into cm c) 22.3 mg into g

Q4 a) Convert 1.2 kg to grams.
b) Use your answer to find how many 30 g servings there are in a 1.2 kg box of cereal.

Q5 Convert each measurement into the units given.
a) 0.6 tonnes into g b) 62 m into mm c) 302 300 mg into kg

To **compare** measurements, or to add or subtract them, they need to be in the **same units**.

Example 2

Find the total of 0.2 tonnes, 31.8 kg and 1700 g. Give your answer in kg.

1. Convert tonnes to kg. kg are smaller than tonnes, so multiply by the conversion factor. $0.2 \text{ tonnes} = 0.2 \times 1000 = 200 \text{ kg}$
2. Convert g to kg. kg are bigger than g, so divide by the conversion factor. $1700 \text{ g} = 1700 \div 1000 = 1.7 \text{ kg}$
3. The masses are all now in kg, so they can be added together. $200 \text{ kg} + 31.8 \text{ kg} + 1.7 \text{ kg} =$ **233.5 kg**

Exercise 2

Q1 Complete each calculation.
a) 3200 ml + 75.3 litres = ☐ litres b) 681 cm + 51.2 m = ☐ cm
c) 3 kg + 375 g + 0.2 kg = ☐ kg d) 100 cm + 0.35 m + 12.6 m = ☐ cm
e) 4000 g + 200 kg + 1 tonne = ☐ kg f) 300 cm^3 + 0.7 litres + 250 ml = ☐ ml

Q2 How many metres further is a journey of 3.4 km than a journey of 1800 m?

Q3 A recycling van collects 3200 g of paper, 15 kg of aluminium, 0.72 tonnes of glass and 3.2 kg of cardboard. What is the mass of all the recycling in kg?

Q4 Amirah, Trevor and Elsie get into a cable car while skiing. Amirah weighs 55.2 kg, Trevor weighs 78.1 kg and Elsie weighs 65.9 kg. Each person's equipment weighs 9000 g. The cable car is unsafe when carrying a mass of over a quarter of a tonne. Will Amirah, Trevor and Elsie be safe?

Example 3

Seo-yun is making orange squash. She uses 240 ml of cordial and 3.5 litres of water.

a) How much orange squash does she make? Give your answer in cm³.

1. Convert litres to ml. ml are smaller than litres, so you need to multiply by the conversion factor.
   1 litre = 1000 ml
   3.5 litres = 3.5 × 1000
   = 3500 ml
2. Add 240 ml to 3500 ml.
   3500 ml + 240 ml = 3740 ml
3. 1 cm³ = 1 ml, so swap ml for cm³.
   3740 ml = **3740 cm³**

b) She pours the orange squash into glasses that each hold 0.25 litres. What is the minimum number of glasses needed to hold all the squash?

1. Convert ml to litres. Litres are bigger than ml, so you need to divide by the conversion factor.
   1000 ml = 1 litre
   3740 ml = 3740 ÷ 1000
   = 3.74 litres
2. Divide 3.74 by 0.25 to find how many glasses she will need.
   3.74 ÷ 0.25 = 14.96
3. You need to round up to a whole number.
   Seo-yun will need **15 glasses.**

Exercise 3

PROBLEM SOLVING

Q1 Callum buys 2 tonnes of topsoil for his gardening business. If he uses 250 kg each day, how long will his supply last?

Q2 Milly runs a 1500 m fun run, a 100 m sprint and a 13.2 km race. How many km has she run in total?

Q3 A café puts two slices of ham into every ham sandwich. A slice of ham weighs 10 g. If the café expects to sell 500 ham sandwiches, how many 1.5 kg packs of ham do they need to order?

Q4 A reservoir contains 600 000 litres of water. A stream flowing into it adds 750 000 ml of water a day.

a) If no water was removed from the reservoir, how many litres would the volume of water increase by each day?

b) The reservoir can only hold 800 000 litres of water. How many days will it take for the reservoir to start overflowing?

Q5 A recipe for lasagne needs 0.7 kg of minced beef, 400 g of tomato sauce, 300 g of cheese sauce, 0.2 kg of pasta sheets and 2500 mg of herbs and spices.

a) How many kg do these ingredients weigh?

b) If 0.2 kg of ingredients will feed one person, how many people does this recipe feed?

16.3 Converting Units — Area and Volume

Converting between units for area and volume is similar to what you've already seen with length and mass, but now you've got more dimensions to deal with...

Area

Learning Objective — Spec Ref R1/G14:
Convert between different metric units for area.

Prior Knowledge Check: Be able to find areas of rectangles. See page 346.

The area of a shape is found by multiplying **two lengths** — so area is measured in **units squared** (e.g. m^2, cm^2, mm^2).

To convert between different units of area, **multiply** or **divide** by the **square** of the 'length' conversion factor. For example, the **conversion factor** between cm and mm is 10. This means the conversion factor between cm^2 and mm^2 is $10^2 = 10 \times 10 = 100$.

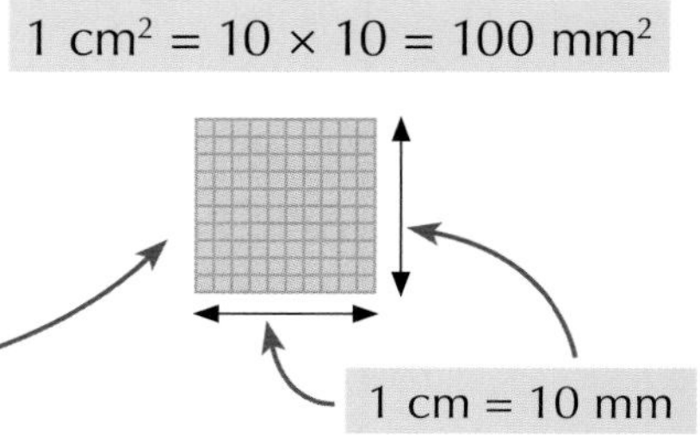

Here are some commonly used conversion factors for area.

Area
$1\ cm^2 = 10^2\ mm^2 = 100\ mm^2$
$1\ m^2 = 100^2\ cm^2 = 10\ 000\ cm^2$
$1\ km^2 = 1000^2\ m^2 = 1\ 000\ 000\ m^2$

Tip: You're dividing or multiplying by the conversion factor twice — once for each dimension.

As on p.263, **divide** by the conversion factor when converting from small units to **bigger units** and **multiply** by the conversion factor when converting from big units to **smaller units**.

Example 1

Find the area of the rectangle on the right in mm^2 by:
a) first converting its length and width to mm

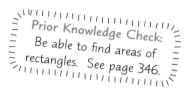

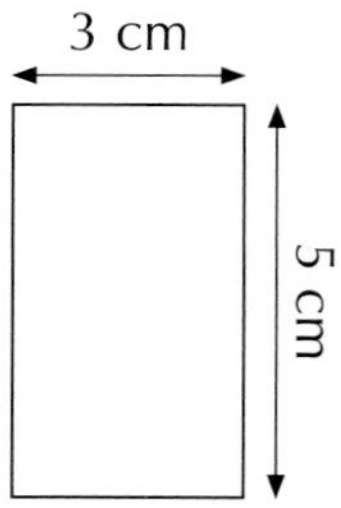

1. Convert the sides to mm. mm are smaller than cm, so multiply by the conversion factor. $5\ cm = 5 \times 10 = 50\ mm$ $3\ cm = 3 \times 10 = 30\ mm$
2. Multiply to get the area. $50 \times 30 =$ **$1500\ mm^2$**

b) first finding the area in cm^2 and then using the appropriate conversion factor.

1. Find the area in cm^2. $5 \times 3 = 15\ cm^2$
2. Find the conversion factor from cm^2 to mm^2. $1\ cm = 10\ mm$, so $1\ cm^2 = 10 \times 10 = 100\ mm^2$
3. Use the conversion factor to convert $15\ cm^2$ to mm^2. $15\ cm^2 = 15 \times 100 =$ **$1500\ mm^2$**

Tip: mm are smaller than cm, so multiply by the conversion factor for the area.

Example 2

Convert an area of 600 cm² to m².

1. Work out the conversion factor from cm² to m². — 1 m = 100 cm, so 1 m^2 = 100 × 100 = 10 000 cm^2
2. m are bigger than cm, so divide by the conversion factor. — 600 cm^2 = 600 ÷ 10 000 = **0.06 m^2**

Exercise 1

Q1 Work out the area (in m^2) of a rectangular farm measuring 2 km by 3 km by:

a) first converting its length and width to m

b) first finding the area in km^2 and then using the appropriate conversion factor.

Q2 Convert each of the following areas into the units given.

a) 26 cm^2 into mm^2 b) 1.05 m^2 into cm^2 c) 1.2 m^2 into cm^2

d) 1750 cm^2 into m^2 e) 8500 mm^2 into cm^2 f) 27 cm^2 into m^2

Q3 One bottle of weedkiller can treat an area of 16 m^2. How many bottles of this weedkiller should Ali buy to treat her lawn measuring 990 cm by 430 cm?

Q4 Sandeesh wants to carpet two rectangular rooms. One of the rooms measures 1.7 m by 3 m, the other is 670 cm by 420 cm. How many square metres of carpet will she need?

Volume

Prior Knowledge Check: Be able to find volumes of cuboids. See p.375.

Learning Objective — Spec Ref R1/G14:
Convert between different metric units for volume.

1 cm^3 = 10 × 10 × 10 = 1000 mm^3

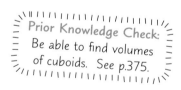

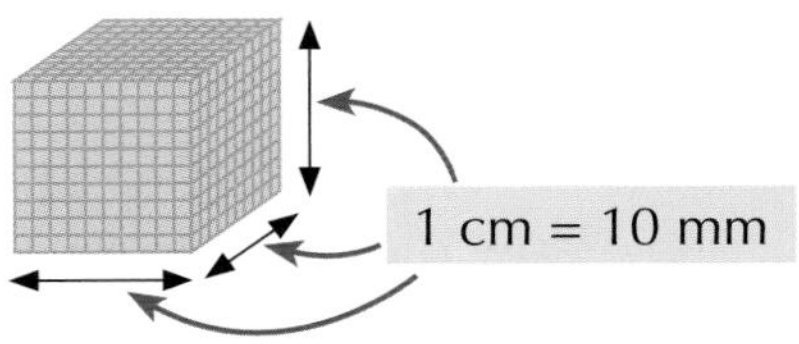

The volume of a shape is found by multiplying **three lengths**, so it is measured in **units cubed** (e.g. m^3, cm^3, mm^3).

To convert between different units of volume, **multiply** or **divide** by the cube of the 'length' conversion factor.

For example, the **conversion factor** between cm and mm is 10. This means the conversion factor between cm^3 and mm^3 is 10^3 = 10 × 10 × 10 = 1000.

The table on the right shows some commonly used conversion factors for volume.

Volume
1 cm^3 = 10^3 mm^3 = 1000 mm^3
1 m^3 = 100^3 cm^3 = 1 000 000 cm^3
1 km^3 = 1000^3 m^3 = 1 × 10^9 m^3

Example 3

Convert a volume of 382 000 cm³ to m³.

1. Work out the conversion factor from cm³ to m³.
   1 m = 100 cm, so
   1 m³ = 100³ = 100 × 100 × 100
   = 1 000 000 cm³
2. m are bigger than cm, so divide by the conversion factor.
   So 382 000 cm³ = 382 000 ÷ 1 000 000 = **0.382 m³**

Exercise 2

Q1 Work out the volume (in m³) of the swimming pool shown by:
a) first converting its length, width and depth to m
b) first finding the volume in cm³ and then using the appropriate conversion factor.

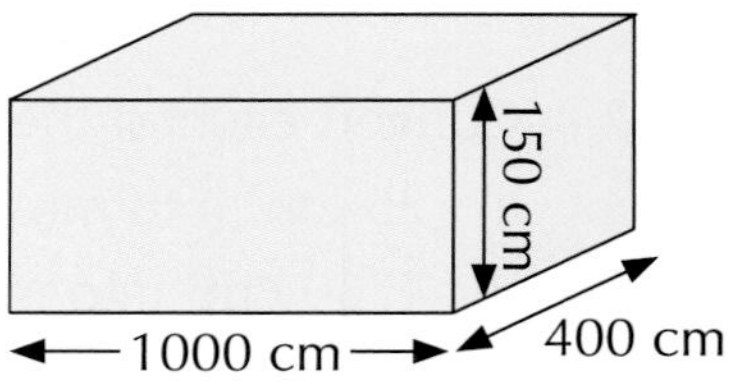

Q2 Convert each of these volumes into the units given.
a) 0.001 km³ into m³
b) 17.6 m³ into cm³
c) 1.2 km³ into m³
d) 16 000 mm³ into cm³
e) 150 m³ into km³
f) 35.9 cm³ into m³

Q3 A brand of coffee powder is sold in cuboid packets with dimensions 20 cm by 25 cm by 10 cm. A volume of 0.003 m³ of coffee powder has already been used from one packet. What volume (in m³) of coffee powder is left, assuming the packet was initially full?

Q4 A box is 10.5 cm long by 5.3 cm wide by 8.67 cm tall. Find the volume of the box in mm³.

Q5 What is the maximum number of these solid plastic bricks that can be made out of 1 m³ of plastic?

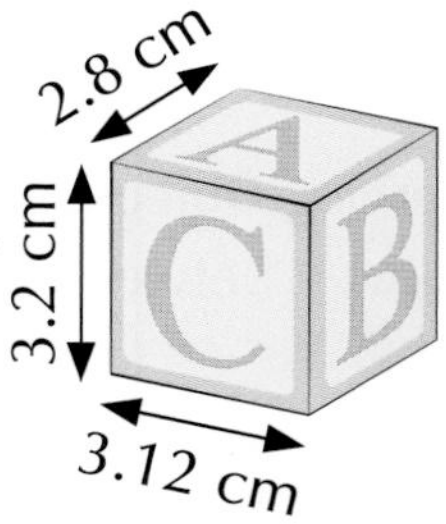

Q6 A swimming pool is 3 m deep and has a base with area 375 m².
a) Work out the volume of the swimming pool in m³.
b) Find the volume of the pool in cm³.
c) How many litres of water can the pool hold?

Q7 a) Convert 0.56 m³ to cm³.
b) Use your answer to a) to convert 0.56 m³ into mm³.

16.4 Metric and Imperial Units

The conversion factors for imperial units aren't as nice as the metric ones, but you can still use the same method to switch to different imperial units or between metric and imperial units.

Learning Objective — Spec Ref N13/R1/G14:
Convert between metric and imperial units.

The **imperial units** for length, mass and volume are as follows:

- **Length** is measured in **inches** (in), **feet**, **yards** and **miles**. There are **12 inches** in 1 foot, **3 feet** in 1 yard and **1760 yards** in 1 mile.
- **Mass** is measured in **ounces** (oz), **pounds** (lb) and **stones**. There are **16 ounces** in 1 pound and **14 pounds** in a stone.
- **Volume** is measured in **pints** and **gallons**. There are **8 pints** in 1 gallon.

Tip: As 1 foot = 12 inches, the conversion factor from feet to inches is 12.

There are **approximate conversion factors** to switch between metric and imperial units — e.g. there are approximately 2.5 cm in 1 inch. The symbol '≈' means 'approximately equal to'.

To **convert** between metric and imperial units, just **multiply** or **divide** by the conversion factors below.

Length	Mass	Volume
1 inch ≈ 2.5 cm	1 ounce ≈ 28 g	1 litre ≈ 1.76 pints
1 foot ≈ 30 cm	1 pound ≈ 450 g	1 gallon ≈ 4.5 litres
1 yard ≈ 90 cm	1 stone ≈ 6400 g	
1 mile ≈ 1.6 km	1 kg ≈ 2.2 pounds	

Tip: You'll be given metric to imperial conversion factors in an exam if they're needed. But don't be surprised if the numbers are slightly different to these.

Example 1

a) Convert 15 miles into km.

1. 1 mile ≈ 1.6 km so the conversion factor is 1.6. — 1 mile ≈ 1.6 km
2. A km is smaller than a mile, so multiply by the conversion factor. — So 15 miles ≈ 15 × 1.6 = **24 km**

b) Convert 10 km into miles.

1. Again, the conversion factor is 1.6. — 1 mile ≈ 1.6 km
2. A mile is bigger than a km, so divide by the conversion factor. — So 10 km ≈ 10 ÷ 1.6 = **6.25 miles**

Exercise 1

In these questions, use the conversion factors given on the previous page.

Q1 Convert these measurements into different imperial units:

a) 2 feet into inches
b) 0.5 gallons into pints
c) 9 gallons into pints
d) 56 pounds into stone
e) 60 inches into feet
f) 32 pints into gallons

Q2 Convert these measurements from imperial units to metric units.

a) 4 inches into cm
b) 3 ounces into g
c) 10 stone into g
d) 5 yards into cm
e) 25 miles into km
f) 6 feet into cm

Q3 Convert these measurements from metric units to imperial units.

a) 8 km into miles
b) 100 pints into litres
c) 12 800 g into stone
d) 25 cm into inches
e) 56 g into ounces
f) 16.5 pounds into kg

Q4 A running track is 400 m long. How many laps of the track make one mile? PROBLEM SOLVING

Q5 a) Convert 18 yards into cm.

b) Use your answer to part a) to convert 18 yards into metres.

Q6 Convert each measurement into the units given.

a) 11 feet into m
b) 1 stone into kg
c) 16 pints into ml

Often when you're converting between metric and imperial units, you'll need to convert the imperial units **to** and **from** a **mixture of big and small units** (e.g. feet and inches):

- To write small imperial units as a **mixture of big and small units** (e.g. inches to feet and inches), you **divide** by the conversion factor and keep the **remainder** in the smaller units.
- To write a **mixture** of big and small units in **smaller units** (e.g. feet and inches to inches), you **multiply** the big unit by the conversion factor and **add** the remaining small units.

Example 2

Convert 6 pounds and 4 ounces into kilograms.

1. Write the whole mass in the same unit. Ounces are smaller than pounds, so multiply by the conversion factor.

   1 pound = 16 ounces,
   6 pounds = $6 \times 16 = 96$ ounces

   So 6 pounds and 4 ounces = 96 + 4
   = 100 ounces

2. Convert this from ounces into grams. Grams are smaller than ounces, so multiply by the conversion factor.

   1 ounce ≈ 28 grams,
   100 ounces ≈ $100 \times 28 = 2800$ g

3. Then convert the result from grams to kilograms.

   2800 g = $2800 \div 1000$ = **2.8 kg**

Example 3

Convert 160 cm into feet and inches.

1. First, convert 160 cm into inches. Inches are bigger than cm, so divide by the conversion factor.
   1 inch ≈ 2.5 cm,
   160 cm ≈ 160 ÷ 2.5 = 64 inches
2. Convert 64 inches into feet and inches. Feet are bigger than inches, so divide by the conversion factor.
   1 foot = 12 inches,
   64 ÷ 12 = 5 remainder 4
3. Keep the remainder in inches.
   So 160 cm ≈ **5 feet 4 inches**

Exercise 2

In these questions, use the conversion factors given on page 269.

Q1 Convert each of the following measurements into the units given.

a) 3 ft 7 in to inches b) 12 ft 5 in to inches c) 5 lb 2 oz to ounces
d) 280 in to feet and inches e) 72 oz to lb and oz f) 200 oz to lb and oz

Q2 For each of the following:
(i) Convert the mass into ounces. (ii) Write this in pounds and ounces.

a) 1904 g b) 840 g c) 4.9 kg d) 0.98 kg

Q3 Convert each of the following into feet and inches.

a) 50 cm b) 105 cm c) 2 m d) 3.4 m

Q4 A ride at a theme park states you must be 140 cm or over to ride.
Maddie is 4 foot 5 inches and Lily is 4 foot 9 inches.
Can they both go on the ride?

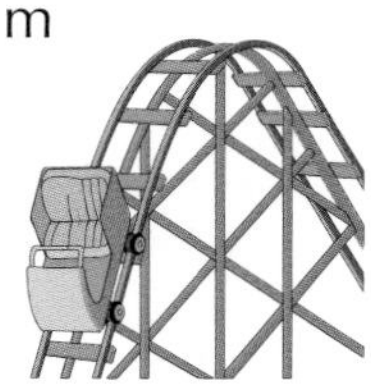

Q5 Jamie and Oliver are cooking. They need 1 pound and 12 ounces of tofu for their recipe. They see a 750 g packet of tofu in the supermarket. Will this be enough? PROBLEM SOLVING

Q6 The weights of 8 people getting into a lift are shown below.

7 stone 2 pounds	11 stone 4 pounds	16 stone	15 stone 4 pounds
10 stone 3 pounds	12 stone	8 stone 9 pounds	13 stone 1 pound

The lift has a weight limit of 0.8 tonnes.
Will the total weight of the 8 people exceed the limit? PROBLEM SOLVING

16.5 Estimating in Real Life

When something is tricky to measure properly, you can look at how big it is compared to something else, and use that to estimate its size.

Learning Objective — Spec Ref N14:
Estimate the size of real-life objects by comparison.

You can **estimate** how big something is by comparing it with something you already know the size of. For example, the average height of a man is approximately 1.8 m, so something half as tall as an average-height man would be just under a metre tall.

An estimate does not have to be completely accurate, but make sure your answer seems **realistic** and is given in **suitable units**.

Example 1

Estimate the height of this lamp post.

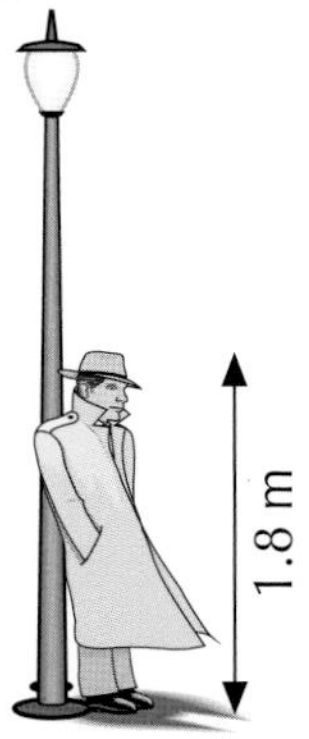

1. Estimate the height of the man. — Average height of a man ≈ 1.8 m.
2. Compare the height of the lamp post and the man. — The lamp post is roughly twice the height of the man.
3. Estimate the height of the lamp post by multiplying the height of the man by 2. — Height of the lamp post ≈ 2 × 1.8 m = **3.6 m**

Example 2

Give sensible units for measuring the height of a room.

When you give a measurement, use units that mean the value isn't too big or too small.

Most rooms are taller than an average person, but not that much taller — roughly 2 to 3 metres. So it makes sense to measure the height of a room in metres (or possibly cm).

A sensible unit is **metres**.

You wouldn't use km or mm to measure the height of a room — the number in km would be very small (0.002 km), and the number in mm would be very big (2000 mm).

Tip: It's usually easier to estimate in real life using metric units, but feet (or feet and inches) would be an acceptable answer in imperial units.

Exercise 1

Q1 For each of the following, suggest a sensible unit of measurement, using:

(i) metric units (ii) imperial units

a) the length of a pencil
b) the mass of a tomato
c) the height of a house
d) the length of an ant
e) the mass of a bus
f) the weight of a baby
g) the distance from Birmingham to Manchester

Q2 Estimate each of the following, using sensible metric units.

a) the height of your bedroom
b) the height of a football goal
c) the arm span of an average man
d) the volume of a mug

For Questions 3-5 give your answer in sensible units.

Q3 Estimate:

a) the height of this house,
b) the height of this elephant.

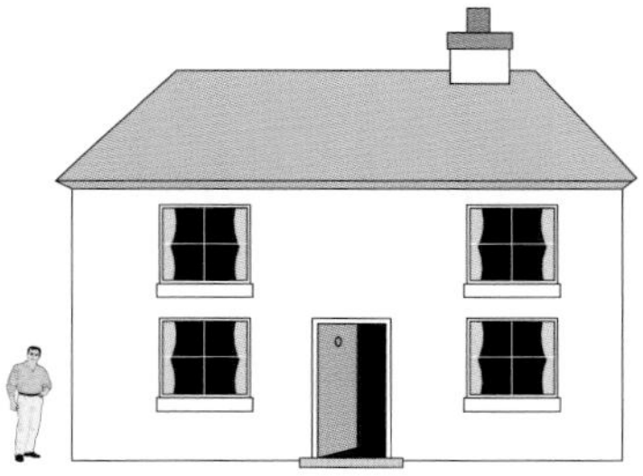

Q4 Estimate the length and height of the bus.

Q5 Estimate:

a) the height and length of this dinosaur by comparing it with a chicken,
b) the height of this rhino by comparing it with a domestic cat.

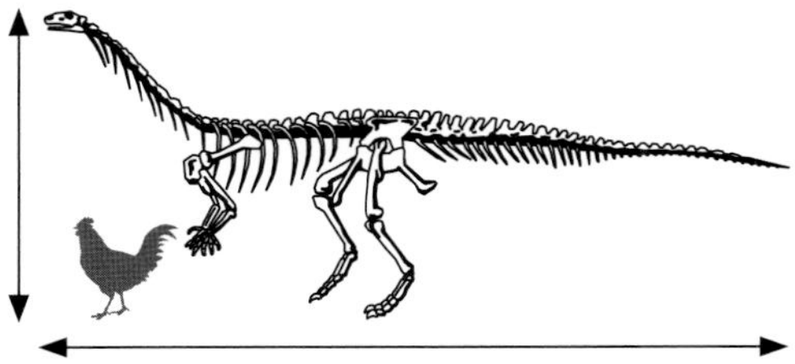

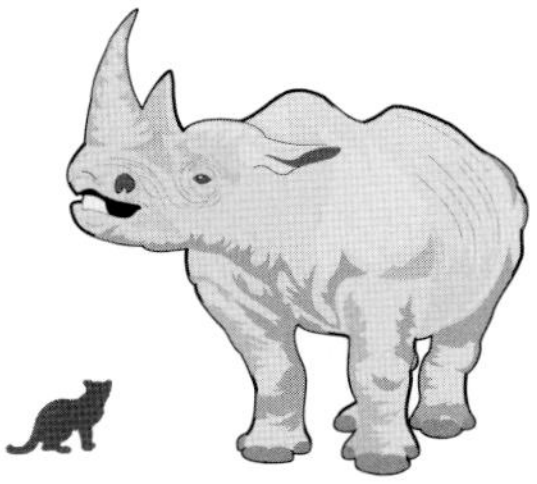

Review Exercise

Q1 Write down the volume of liquid shown in each of these containers.

a) 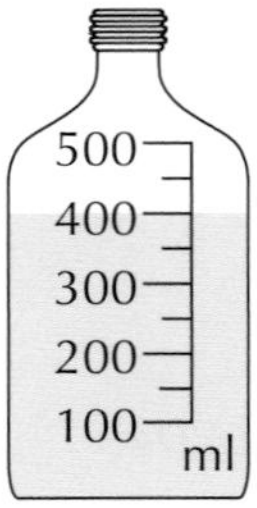

b) 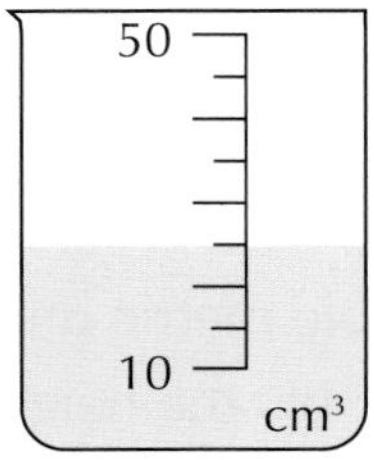

c) 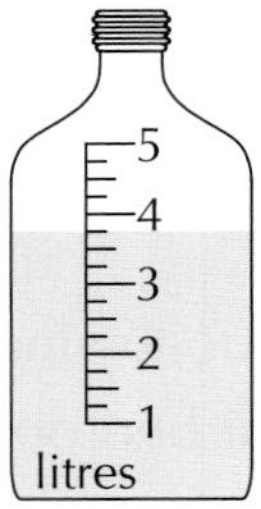

d) 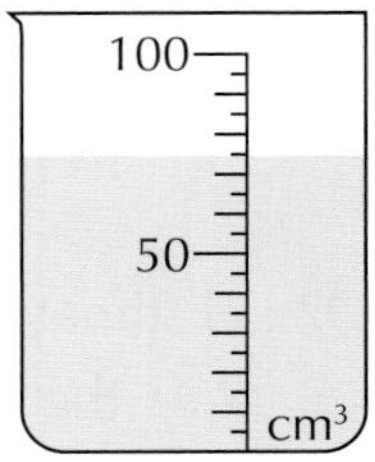

Q2 A snake is measured to be 10.6 m long to the nearest 0.2 m.
Write down the maximum and minimum possible length of the snake.

Q3 Jane is having a party for 30 guests. Her glasses have a capacity of 400 ml each. If she wants everyone to have a full glass of juice, how many 2 litre bottles of juice should she buy?

Q4 A recipe uses 450 g of flour, 0.2 kg of margarine, 300 g of fruit and 0.1 kg of sugar. How much more than 1 kg is the total mass of these ingredients? Give your answer in kg.

Q5 a) Calculate the area (in mm^2) of a sticker measuring 20 mm by 40 mm.
b) Convert this area to cm^2 by using the appropriate conversion factor.

Q6 A carton of apple juice measures 7 cm by 50 mm by 0.17 m, and is completely full.
a) Find the volume of the carton in cm^3.
b) Hati fills 7 glasses, each of which has a volume of 20 cm^3, with juice. How much juice (in cm^3) does she use?
c) How much juice (in mm^3) will she have left in the carton?

Q7 Convert these measurements from imperial units to metric.
a) 3 ounces into g
b) 2 gallons into litres
c) 10 stone into g
d) 5 yards into cm
e) 2.7 inches into cm
f) 4.5 pints into litres

Q8 Estimate each of the following, using sensible units:
a) the length of a family car
b) the diameter of a football
c) the height of a doorway
d) the volume of a lunch box carton of juice

Exam-Style Questions

Q1 a) Change 625 cm into metres.

[1 mark]

b) Change 3.94 kg into grams.

[1 mark]

Q2 The diagram on the right shows an adult of average height stood on level ground next to a double-decker bus. Use the diagram to estimate the height of the bus, giving your answer in an appropriate metric unit.

[2 marks]

Q3 Jack buys 1.2 kg of flour. How many pizzas can he make if each pizza needs 300 g of flour?

[1 mark]

Q4 Grace has made 7.35 litres of fruit punch to serve at a party. She serves it by filling glasses which hold exactly 250 ml.

a) How many full glasses will she be able to serve?

[2 marks]

b) How much punch will she have left over once she has filled as many glasses as possible? Write your answer in litres.

[1 mark]

Q5 1 stone ≈ 6400 g, 1 pound ≈ 450 g, 1 ounce ≈ 28 g.
An airline allows each passenger to travel with 18 kg of luggage.
Iona has one bag that weighs 10 pounds 2 ounces, and a second bag that weighs 2 stone. Can she take both bags with her on the flight?

PROBLEM SOLVING

[3 marks]

Q6 Lotte's car travels 55 miles per gallon of petrol. Her car contains 45 litres of petrol.

a) How many miles can she travel?
Use the approximation 1 gallon ≈ 4.5 litres.

[2 marks]

b) If petrol costs £5 per gallon, how much did the petrol in her car cost?

[1 mark]

17.1 Speed, Distance and Time

The speed of an object is the distance it travels divided by the time it takes to travel the distance. Speed is an example of a compound measure. Compound measures are a combination of two or more other measurements — in this case, distance and time.

Finding an Object's Speed

Learning Objective — Spec Ref N13/R1/R11:
Be able to calculate an object's speed given distance and time.

Prior Knowledge Check: Be able to convert between different units. See Section 16.

Distance, time and (average) speed are connected by the **formula**:

$$\textbf{Speed} = \frac{\textbf{Distance}}{\textbf{Time}}$$

The formula gives the **average speed** of the journey — the actual speed is likely to change throughout. The units for speed are **distance per unit time**, e.g. **km per hour** or **metres per second**.

Tip: Other examples of compound measures include rates of pay (e.g. £ per hour) and prices per unit mass/volume (e.g. £ per kg or pence per ml).

Example 1

A car travels 150 km in 3 hours. What is the average speed of the car?

1. Substitute the distance and time into the formula for speed.
2. The units of speed are a combination of the units of distance (here, km) and time (here, hours).

$$\text{Speed} = \frac{\text{Distance}}{\text{Time}} = \frac{150\text{ km}}{3\text{ hours}} = \mathbf{50\ km/h}.$$

Example 2

A car travels 81 km in 45 minutes. What is the average speed of the car in km/h?

1. Convert the time to hours.

$$45\text{ minutes} = 45 \div 60 = 0.75\text{ hours}$$

2. Substitute the distance and time into the formula. The units of speed are a combination of the units of distance (km) and time (hours).

$$\text{Speed} = \frac{\text{Distance}}{\text{Time}} = \frac{81\text{ km}}{0.75\text{ hours}} = \mathbf{108\ km/h}$$

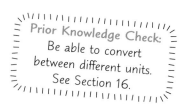

Exercise 1

Q1 Calculate the average speed of each of the following journeys.

a) distance = 30 km, time = 2 hours
b) distance = 60 km, time = 3 hours
c) distance = 150 miles, time = 5 hours
d) distance = 140 miles, time = 4 hours

Q2 Find the average speed of the following.

a) a car travelling 80 km in 2 hours
b) a snail crawling 50 cm in 500 seconds
c) a plane flying 1800 miles in 3 hours
d) a lift travelling 100 m in 80 seconds

Q3 Find the average speed of the following in km/h.

a) a tractor moving 10 km in 30 minutes
b) a train travelling 30 000 m in 2.5 hours
c) a river flowing 2.25 km in 45 minutes
d) a balloon rising 700 m in 3 minutes

Tip: Convert the units to kilometres and hours first.

Finding Distance or Time

Learning Objective — Spec Ref N13/R1/R11:
Know and be able to use the formula linking speed, distance and time to find the distance of a journey or the time a journey took.

You can **rearrange** the speed formula to give formulas for **distance** and **time**:

$$\textbf{Distance} = \textbf{Speed} \times \textbf{Time}$$

$$\textbf{Time} = \frac{\textbf{Distance}}{\textbf{Speed}}$$

You can use the **formula triangle** given below to help you remember all three of the formulas. To use the formula triangle, **cover up** the measurement that you want to find and **write down** the two measurements that are left.

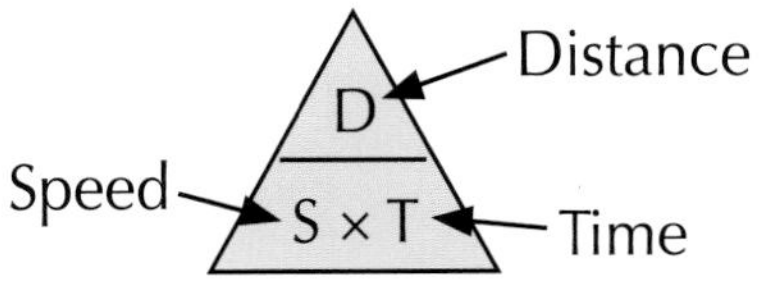

- To find **speed**, cover up **S** to leave $\frac{\mathbf{D}}{\mathbf{T}}$
- To find **distance**, cover up **D** to leave **S × T**
- To find **time**, cover up **T** to leave $\frac{\mathbf{D}}{\mathbf{S}}$

Remember to always **check the units** before doing a calculation. If you have **speed** in **mph** and **time** in **minutes**, then doing speed × time won't give you the correct answer.

Example 3

A man runs for 30 minutes with an average speed of 12 km/h. How far does he run?

1. The speed is in km/h, so convert the time into hours. — 30 minutes = 0.5 hours
2. Write down the formula for distance. — Distance = Speed × Time
3. Substitute the speed and time into the formula. — 12 km/h × 0.5 hours = **6 km.**
4. The speed is in km/h, so the distance will be in km.

Example 4

A train travels 60 miles at a speed of 100 mph. How many minutes will the journey take?

1. Write down the formula for time. — $\text{Time} = \frac{\text{Distance}}{\text{Speed}}$
2. Substitute the distance and speed into the formula. — $\frac{60\text{ miles}}{100\text{ mph}} = 0.6\text{ hours}$
3. Convert your answer into minutes. — 0.6 hours = 0.6 × 60 minutes = **36 minutes**

Exercise 2

Q1 For each of the following, use the speed and time given to calculate the distance travelled.
a) speed = 20 km/h, time = 2 hours
b) speed = 10 m/s, time = 50 seconds
c) speed = 3 km/h, time = 24 hours
d) speed = 70 mph, time = 2.5 hours

Q2 For each of the following, use the speed and distance given to calculate the time taken.
a) speed = 2 km/h, distance = 4 km
b) speed = 15 m/s, distance = 45 m
c) speed = 60 mph, distance = 150 miles
d) speed = 24 km/h, distance = 6 km

Q3 Find the distance travelled by a bus moving at 30 mph for 4 hours.

Q4 A dart is thrown with a speed of 15 m/s. It hits a dartboard 2.4 m away.
For how long is the dart in the air?

Q5 A flight to Spain takes 2 hours. The plane travels at an average speed of 490 mph.
How far does the plane travel?

Q6 A train travels at 56 km/h for 5.6 km. How many minutes does the journey take?

Q7 A girl skates at an average speed of 7.5 mph. How far does she skate in 15 minutes?

Converting Between Units of Speed

Learning Objective — Spec Ref N13/R1/R11:
Be able to convert between units of speed.

To **convert** between units of speed, you can convert the **individual units** that make it up. For example, to convert **km/h to m/s** you would need to do **two separate conversions:**

- First convert **km** to **metres** (to give m per hour),
- Then convert **hours** to **seconds** (to give m/s).

Tip: It doesn't matter which unit you convert first, but always check your conversions to make sure your answers are sensible.

Example 5

A swallow is flying at a speed of 9 m/s. What is its speed in km/h?

1. Work out the conversion factor for m to km. There'll be fewer km/s than m/s, so divide.

   1 km = 1000 m
   9 m/s = (9 ÷ 1000) km/s = 0.009 km/s

2. Work out the conversion factor for s to h. There'll be more km/h than km/s, so multiply.

   1 hour = 60 min = (60 × 60) s
   0.009 km/s = (0.009 × 60 × 60) km/h
   = **32.4 km/h**

Example 6

Using the conversion 1 mile ≈ 1.6 km, what is 8 km/h in mph?

Use the conversion factor for km to miles. There'll be fewer miles per hour than km/h, so divide.

1 mile ≈ 1.6 km

8 km/h ≈ $\frac{8}{1.6}$ = **5 mph**

Exercise 3

Q1 Convert each of the speeds below into the given units.
a) 50 m/s into km/h b) 72 km/h into m/s c) 26.5 m/s into km/h

Q2 James is cycling at 18 km/h. What is his speed in m/s?

Q3 Using the conversion 1 mile ≈ 1.6 km, convert each of these speeds into the given units.
a) 54 km/h to mph b) 25 mph to km/h c) 94.4 km/h to mph

Q4 A slug is moving at 0.5 cm/s. What is its speed in metres per minute (m/min)?

Q5 A train travels at 40 m/s. What is the train's speed in mph?

17.2 Density, Mass and Volume

Density is another compound measure — it's a combination of mass and volume.

Learning Objective — Spec Ref N13/R1/R11:
Know and be able to use the formula linking density, mass and volume.

Prior Knowledge Check: Be able to convert between different units of volume. See p267.

Density is the **mass per unit volume** of a substance and is usually measured in kg/m³ or g/cm³. Different substances have **different densities** — for example, gold has a **higher density** than ice. The **formula** that connects density, mass and volume is:

$$\textbf{Density} = \frac{\textbf{Mass}}{\textbf{Volume}}$$

You can rearrange the formula to find mass and volume.

$$\textbf{Volume} = \frac{\textbf{Mass}}{\textbf{Density}} \qquad \textbf{Mass} = \textbf{Density} \times \textbf{Volume}$$

The **formula triangle** on the right gives a summary of all the formulas. Remember to **check the units** of the measurements that you're putting into a formula — that way you'll know the units of the measurement that comes out.

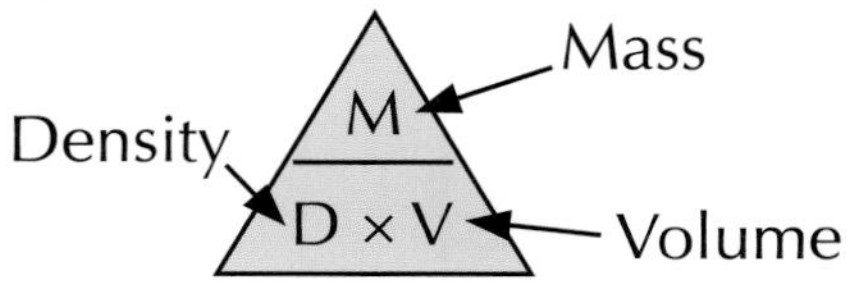

Example 1

A 20 g block of wood has a volume of 50 cm³.

a) Calculate the density of the wooden block in g/cm³.

1. Substitute the mass and volume into the formula for density.
2. The units of density are a combination of the units of mass (g) and volume (cm³).

$$\text{Density} = \frac{\text{Mass}}{\text{Volume}} = \frac{20\text{ g}}{50\text{ cm}^3} = \mathbf{0.4\ g/cm^3}$$

b) What is its density in kg/m³?

1. Work out the conversion factor for g to kg. There'll be fewer kg/cm³ than g/cm³, so divide.

$$1\text{ kg} = 1000\text{ g}$$
$$0.4\text{ g/cm}^3 = (0.4 \div 1000)\text{ kg/cm}^3 = 0.0004\text{ kg/cm}^3$$

2. Work out the conversion factor for cm³ to m³. There'll be more kg/m³ than kg/cm³, so multiply.

$$1\text{ m}^3 = (100 \times 100 \times 100)\text{ cm}^3 = 1\,000\,000\text{ cm}^3$$
$$0.0004\text{ kg/cm}^3 = (0.0004 \times 1\,000\,000)\text{ kg/m}^3 = \mathbf{400\ kg/m^3}$$

Example 2

The mass of water in a bathtub is 180 kg. Water has a density of 1000 kg/m³. What is the volume of water in the bathtub?

1. Write down the formula for volume.

$$\text{Volume} = \frac{\text{Mass}}{\text{Density}}$$

2. Check units — the mass is in kg and the density is in kg/m³, so your answer will be in m³.

Mass = 180 kg, Density = 1000 kg/m³

3. Substitute the numbers into the formula.

$$\text{Volume} = \frac{180\text{ kg}}{1000\text{ kg/m}^3} = \mathbf{0.18\ m^3}$$

Exercise 1

Q1 For each of the following, use the mass and volume to calculate the density in kg/m³.

a) mass = 20 kg, volume = 5 m³

b) mass = 300 kg, volume = 5 m³

c) mass = 1000 kg, volume = 4 m³

d) mass = 63 000 kg, volume = 700 m³

Q2 A 1840 kg concrete block has a volume of 0.8 m³.
Calculate the density of the concrete block.

Q3 a) Write down the formula to calculate volume from density and mass.

b) Use your answer to calculate the volume for each of the following.

(i) density = 8 kg/m³, mass = 40 kg

(ii) density = 15 kg/m³, mass = 750 kg

(iii) density = 240 kg/m³, mass = 4800 kg

Q4 A limestone statue has a volume of 0.4 m³.
The limestone has a density of 2600 kg/m³.

a) Write down the formula to calculate a mass from a volume and a density.

b) Calculate the mass of the statue.

Q5 A cricket ball has a mass of 0.15 kg and a volume of 200 cm³.
Calculate the density of the cricket ball in kg/m³.

Q6 A paperweight has a volume of 8 cm³ and a density of 11 500 kg/m³.
Calculate the mass of the paperweight.

Q7 A roll of aluminium foil has a density of 2.7 g/cm³.
What is the density of aluminium in kg/m³?

17.3 Pressure, Force and Area

Like speed and density, pressure is a compound measure — it's a combination of force and area.

Learning Objective — Spec Ref N13/R1/R11:
Know and be able to use the formula linking pressure, force and area.

Prior Knowledge Check:
Be familiar with areas of shapes. See Section 22.

Pressure is the **force** exerted by an object **per unit area**. Pressure is usually measured in **N/m²** (also known as pascals, **Pa**), or **N/cm²**, where **N** is **newtons**, the unit of force.

The **formula** that connects pressure, force and area is:

$$\textbf{Pressure} = \frac{\textbf{Force}}{\textbf{Area}}$$

You can rearrange the formula to find **force** and **area**:

$$\textbf{Force} = \textbf{Pressure} \times \textbf{Area} \qquad \textbf{Area} = \frac{\textbf{Force}}{\textbf{Pressure}}$$

The **formula triangle** on the right gives a summary of these formulas. Don't forget to **check your units** throughout so that you know the units of your final measurement.

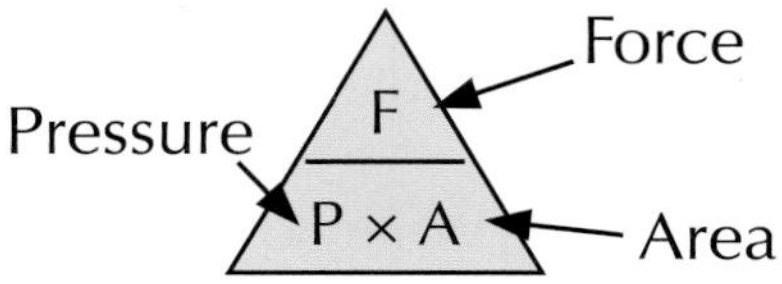

Example 1

An object is resting with its base on horizontal ground. The area of the object's base is 20 cm² and the object weighs 60 N.

a) What pressure is the object exerting on the ground?

1. Write down the formula for pressure.

$$\text{Pressure} = \frac{\text{Force}}{\text{Area}}$$

2. Check units — you have force in N and area in cm², so your answer will be in N/cm².

Force = 60 N
Area = 20 cm²

3. Substitute the numbers into the formula.

$$\text{Pressure} = \frac{60\ \text{N}}{20\ \text{cm}^2} = \mathbf{3\ N/cm^2}$$

b) What is the pressure in N/m²?

1. Work out the conversion factor for N/cm² to N/m².

$1\ \text{m}^2 = (100 \times 100)\ \text{cm}^2$
$= 10\,000\ \text{cm}^2$

2. There will be more N/m² than N/cm², so multiply.

$3\ \text{N/cm}^2 = 3 \times 10\,000$
$= \mathbf{30\,000\ N/m^2}$

Example 2

A laptop with a base of 0.07 m² is resting on a desk and exerting a pressure of 330 N/m². What is the weight of the laptop?

1. Write down the formula for force. — Force = Pressure × Area
2. Check units — you have area in m^2 and pressure in N/m^2, so your answer will be in N. — Pressure = 330 N/m^2, Area = 0.07 m^2

   Tip: Weight is the force of an object on a surface due to gravity.

3. Substitute the numbers into the formula. — Force = 330 N/m^2 × 0.07 m^2 = **23.1 N**

Exercise 1

Q1 For each of the following, use the area and force to calculate the pressure.

a) Area = 3 cm^2, Force = 27 N
b) Area = 5 m^2, Force = 125 N
c) Area = 4 m^2, Force = 4800 N
d) Area = 80 cm^2, Force = 640 N

Q2 Calculate the area in each of the following situations.

a) Pressure = 6 N/cm^2, Force = 36 N
b) Pressure = 120 N/m^2, Force = 840 N
c) Pressure = 2.5 N/cm^2, Force = 10 N
d) Pressure = 180 N/m^2, Force = 540 N

Q3 Calculate the force in each of the following situations.

a) Pressure = 300 N/m^2, Area = 5 m
b) Pressure = 36 N/cm^2, Area = 30 cm^2

Q4 A cube has edges of length 10 cm. When it rests on horizontal ground it exerts a pressure of 0.02 N/cm^2. What is its weight? PROBLEM SOLVING

Q5 A cube of metal with a volume of 512 cm^3 is resting with one of its faces on horizontal ground. The cube has a weight of 1792 N.

a) What is the area of each face of the cube?
b) What pressure is the cube exerting on the ground?

Q6 A square-based pyramid is resting with its square face on horizontal ground. The pyramid has a weight of 45 N and its square face has a side length of 3 cm. What pressure is the pyramid exerting on the ground? PROBLEM SOLVING

Give your answer in: a) N/cm^2 b) N/m^2

17.4 Distance-Time Graphs

Distance-time graphs are used to show the journey of an object over a given period of time.

Representing a Journey on a Distance-Time Graph

Learning Objective — Spec Ref A14:
Draw and interpret distance-time graphs.

Prior Knowledge Check:
Be able to read off graphs.
See p219-222.

A **distance-time** graph shows how far an object has travelled in a particular period of time.

- When the graph is **going up**, the object is **moving away**.
- When the graph is **going down**, the object is **coming back**.
- A **straight** line shows the object is moving at a **constant speed**.
- A **horizontal** line means the object is **stationary**.

Tip: Be careful when reading distance-time graphs — they show distance from a point, not always total distance travelled.

Example 1

Danny cycles 5 miles in 20 minutes. He stops and rests for 10 minutes, then returns to his starting point in 30 minutes. Copy the axes on the right and use them to draw a graph of Danny's journey.

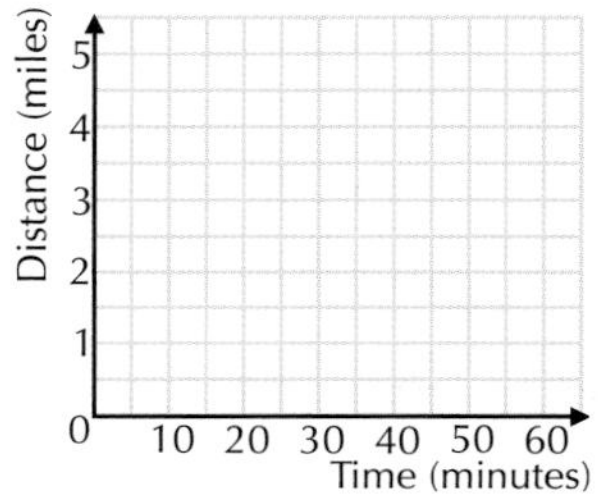

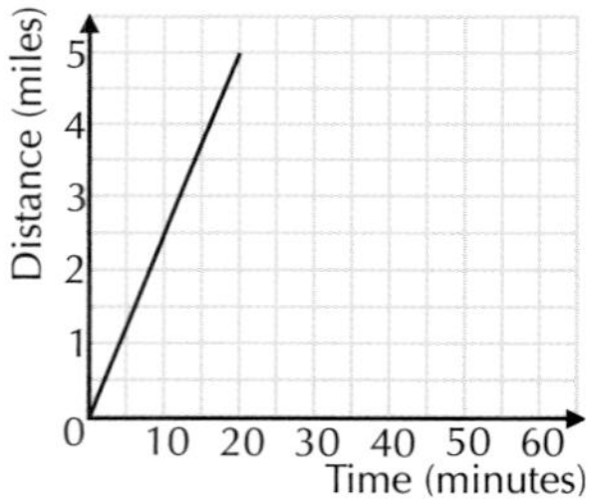

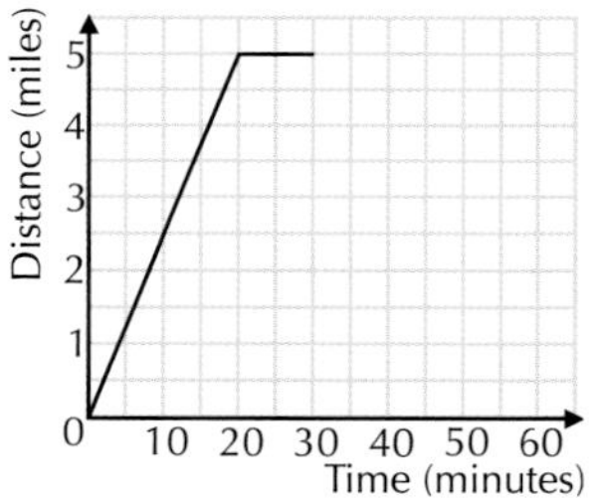

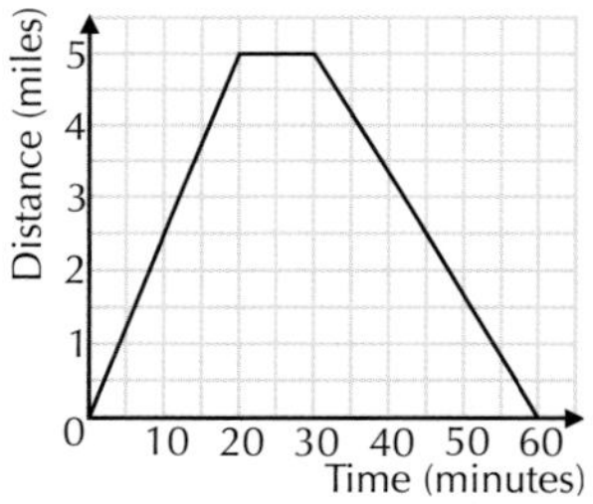

1. The first part of the graph shows 5 miles being covered in 20 minutes. It's a straight line from the origin to the point representing 20 minutes on the x-axis and 5 miles on the y-axis.
2. The second part of the graph shows no distance covered during the next 10 minutes. It's a horizontal line from 20 minutes to 30 minutes (on the x-axis).
3. The final part of the graph shows the 5-mile return journey taking 30 minutes. It's a straight line back to a distance of 0 miles (on the y-axis) and to a time of 60 minutes (on the x-axis).

Exercise 1

Q1 Adi is a keen cyclist. The following points describe the different stages of one of her bike rides.
- Adi cycles 30 km in 2 hours.
- She stops and rests for half an hour.
- She then cycles a further 40 km in 2.5 hours.

Copy the coordinate grid on the right.

a) Draw a straight line on the grid to represent the first stage of Adi's journey.

b) The second stage of her ride is represented by a horizontal line. Draw this part of the graph.

c) (i) After the third stage of her ride, how far is Adi from her original starting point?

(ii) Draw the straight line representing the third stage of Adi's ride.

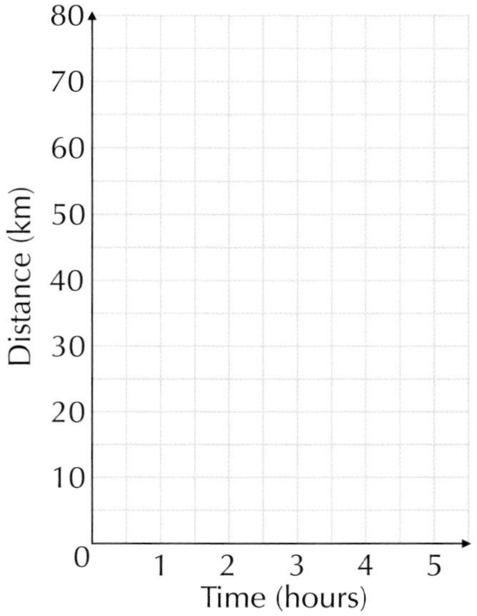

Draw a graph representing each of the journeys described in **Questions 2-3**.

Q2 Yemi drives 50 km in 1 hour, stops at a service station for half an hour, then drives a further 30 km in half an hour.

Q3 Sandy walks 2 km to the bus stop in 20 minutes. She waits for 10 minutes for a bus to arrive. She then travels a further 5 km on the bus in 15 minutes.

Q4 The graph on the right shows a family's car journey. The family left home at 8:00 am.

a) How far had the family travelled when they stopped?

b) How long did the family stay at their destination before setting off home?

c) (i) What time did they start the journey back home?

(ii) How long did the journey home take?

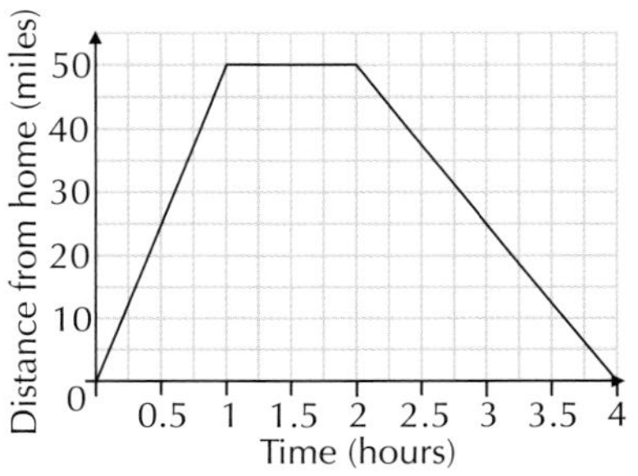

Q5 Describe the journey that is represented by each of these distance-time graphs.

a)

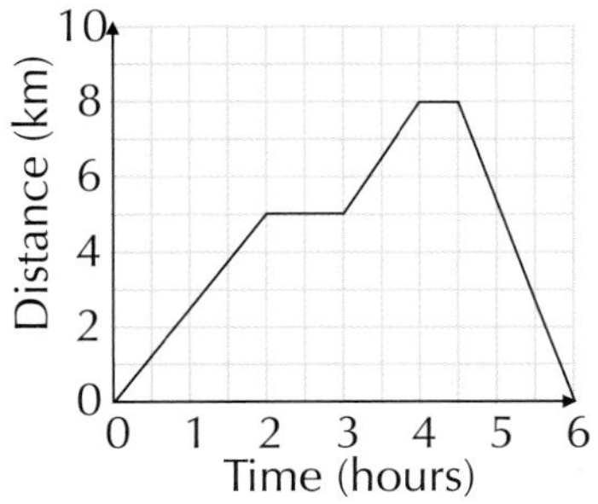

b)

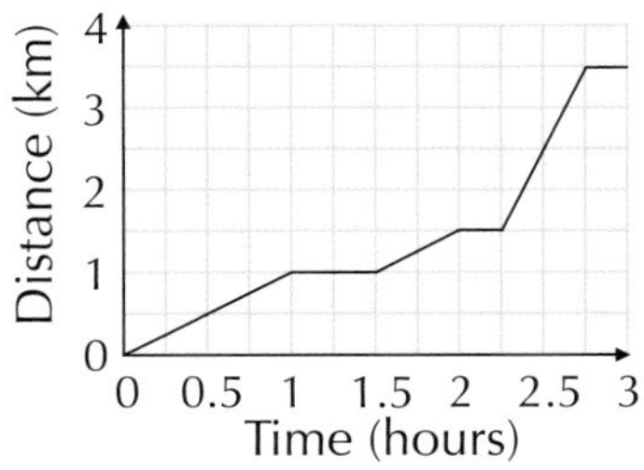

c)

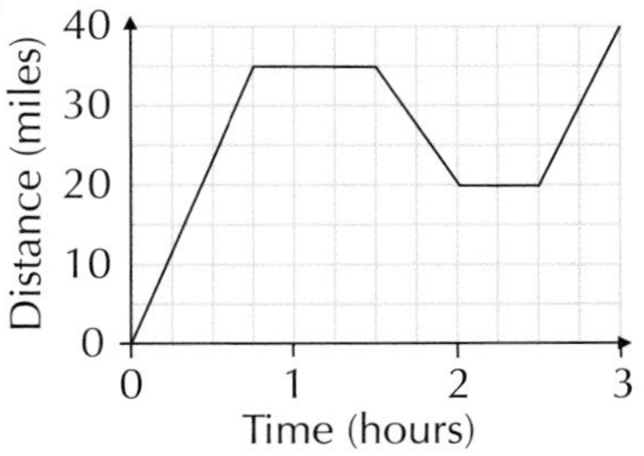

Finding Speed from a Distance-Time Graph

Learning Objective — Spec Ref A14/R14:
Be able to find speed from a distance-time graph.

The **gradient** of a distance-time graph shows the **speed** of the object.

- The steeper the graph, the faster the object is moving.
- Remember, a horizontal line means the object is stationary.

You can work out the speed at any stage of the journey by dividing the **distance travelled** by the **time taken**.

To calculate the **average speed** across the whole journey, divide the **total distance travelled** by the **total time taken**.

Tip: If the graph goes up and down, add up the distance travelled at each stage to find the total distance travelled.

Example 2

The graph below represents a train journey from Clumpton Station to Hillybrook Station.

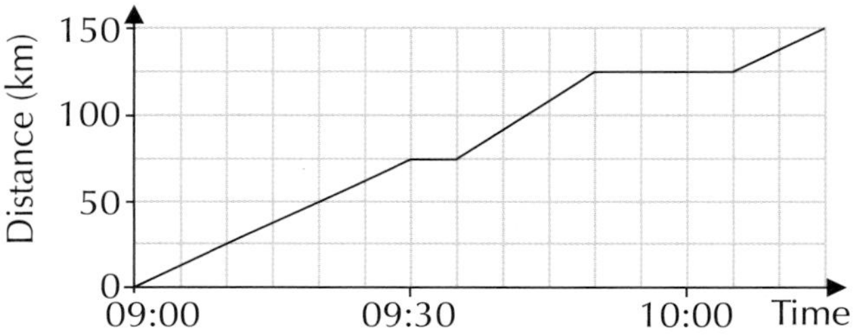

a) Find the speed of the train (in km/h) at 9:15.

1. The speed is constant from 9:00 until 9:30, so the speed at 9:15 is equal to the average speed over this time.
2. Use the graph to find the distance travelled in this time.
3. Substitute the distance and time into the formula for speed.

Distance = 75 km

Time = 30 mins = 0.5 hours

$$\text{Speed} = \frac{\text{Distance}}{\text{Time}} = \frac{75\text{ km}}{0.5\text{ hours}} = \mathbf{150\text{ km/h}}$$

b) Find the average speed of the train (in km/h) for this journey.

1. Work out the total distance travelled and the total time taken.
2. Substitute the distance and time into the formula for speed.

Distance = 150 – 0 = 150 km

Time = 09:00 to 10:15 = 1.25 hours

$$\text{Speed} = \frac{\text{Distance}}{\text{Time}} = \frac{150\text{ km}}{1.25\text{ hours}} = \mathbf{120\text{ km/h}}$$

Exercise 2

Q1 Find the speed of the object represented by each of the following distance-time graphs.

a) 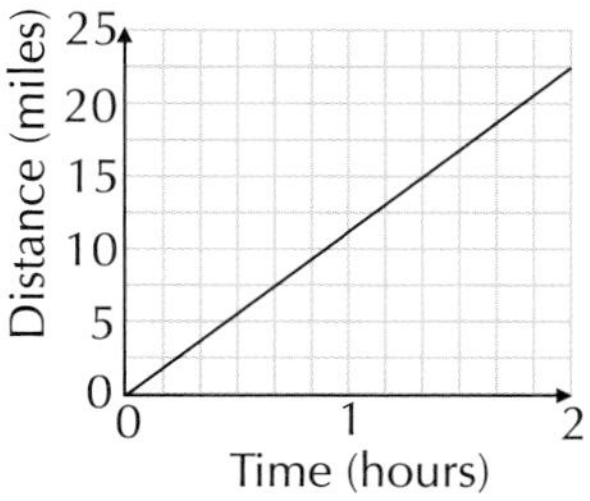

b) 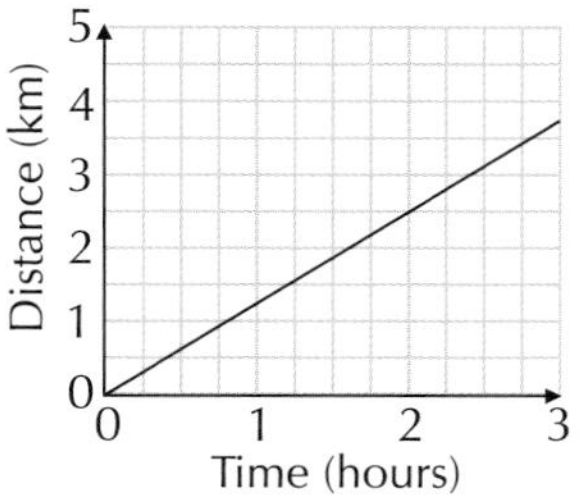

c) 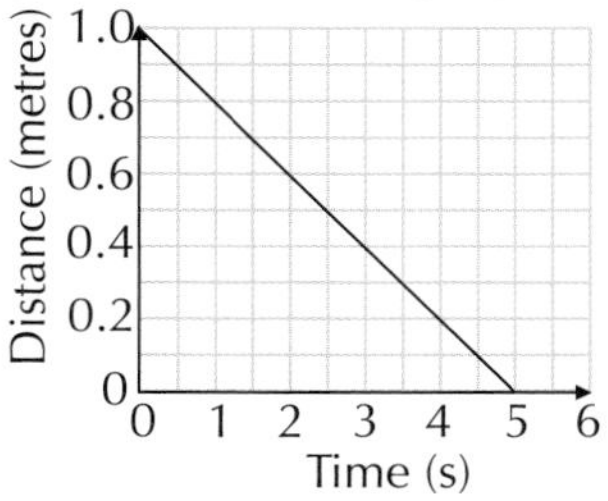

Q2 This graph shows Gil's journey to work. His journey consisted of two stages of travelling, separated by a break of 30 minutes.

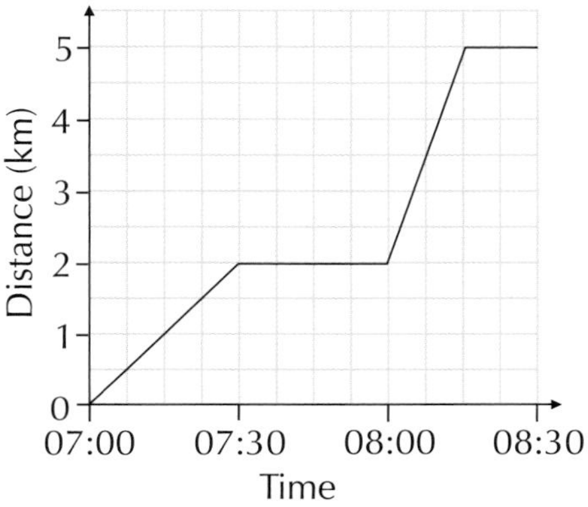

a) Without carrying out any calculations, state which of the journey's two stages was at a higher speed. Explain your answer.

b) (i) How far did he travel in stage 1 of his journey (the first 30 minutes)?

(ii) What was his speed (in km/h) for the first stage?

c) What was his speed (in km/h) during the second stage of his journey?

Q3 The graph shows Sophie's trek.
She stopped for a rest at 12:00 and for lunch at 13:00.

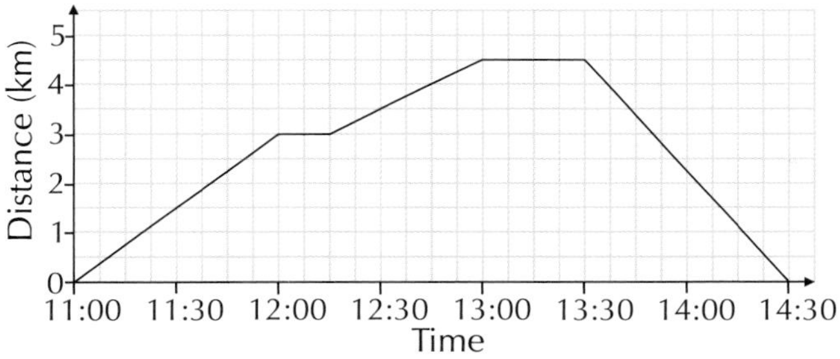

a) Without carrying out any calculations, state the times between which she was walking fastest. Explain your answer.

b) (i) For how long did she walk before she first stopped for a rest?

(ii) How far did she walk during this time?

(iii) What was her average speed during this first part of her walk?

c) What was Sophie's average speed after her rest but before she stopped for lunch?

d) What was her average speed after lunch?

Q4 Draw distance-time graphs to show the following journeys.

a) Ashna catches a train at 10:00, then travels 200 miles at an average speed of 100 mph.

b) Jasper sets off at 09:00 and drives 120 miles at an average speed of 40 mph.
After 1 hour at his destination, he drives back to his starting point at an average speed of 60 mph.

Review Exercise

Q1 A football is passed between two players 10 m apart. The ball travels for 2.5 seconds. Find the average speed of the ball in m/s.

Q2 A yacht travels 13.5 km in 15 minutes.

a) What is its average speed in km/h? b) What is this speed in m/s?

Q3 An object has a mass of 2500 kg and a volume of 50 m^3.

a) What is the object's density in kg/m^3?

b) What is its density in g/cm^3?

Q4 For each of the following, calculate the missing measure.

a) Pressure = ? N/m^2, Area = 4 m^2, Force = 4800 N

b) Pressure = 180 N/m^2, Area = ? m^2, Force = 540 N

c) Pressure = 28 N/cm^2, Area = 14 cm^2, Force = ? N

Q5 Helena is going on a journey. The following points describe her journey.

- Helena sets off from home and travels 40 km by bus, as shown on the graph.
- She gets off the bus and then waits half an hour for a train.
- She gets on the train and travels for a further 1 hour 30 minutes, as shown on the graph.
- She spends 30 minutes at her destination.
- She then returns home by taxi at an average speed of 60 km/h.

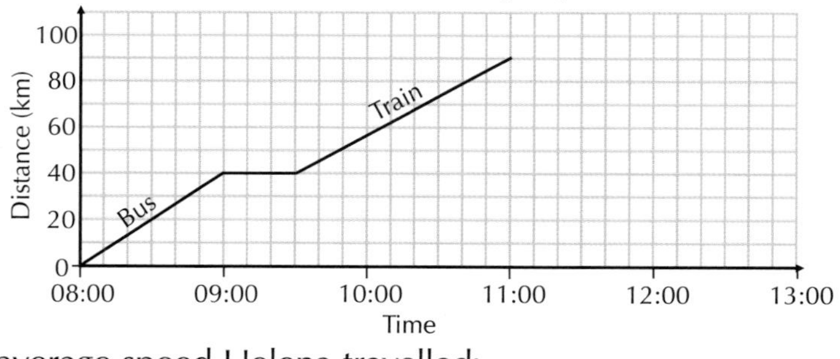

a) Find the average speed Helena travelled:

(i) by bus. (ii) by train.

b) Copy the above graph, and extend the line to show Helena's 30 minutes at her destination.

c) Find the time it took Helena to travel home by taxi.

d) Show Helena's taxi ride home on your graph.

Exam-Style Questions

Q1 Tash goes for a run. Her run is shown on the distance-time graph to the right.

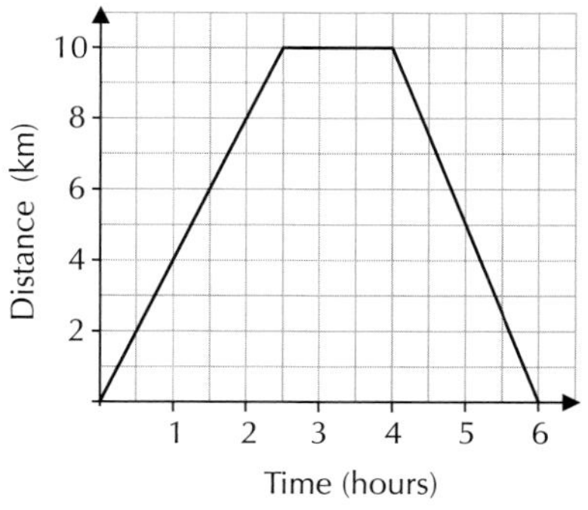

a) Describe the three stages of her run.

[3 marks]

b) Explain what the gradient of the graph represents.

[1 mark]

c) Calculate her speed on her return journey in km/h.

[2 marks]

Q2 Will drives his vintage bus to a bus rally 60 miles away. His average speed for the journey is 24 mph and he arrives at 11:10 am. Work out what time he set off.

[2 marks]

Q3 The density of steel is approximately 8 g/cm^3. A solid steel cube has a side length of 5 cm. Find the mass of the cube in kilograms.

[2 marks]

Q4 A skip exerts a pressure of 625 N/m^2 over a ground area of 40 000 cm^2. Calculate the force in N exerted by the skip on the ground.

[3 marks]

Q5 Charlotte is a runner. She runs at a pace of 1 km in 4 minutes 45 seconds. Calculate the time that Charlotte will take to run 800 m at this pace.

[3 marks]

Q6 36 cm^3 of copper and 4 cm^3 of tin are melted down and mixed to make a bronze medal. The densities of copper and tin are 9 g/cm^3 and 7 g/cm^3 respectively. Work out the mass of the medal.

[2 marks]

Section 18 Scale Drawings and Bearings

18.1 Scale Drawings

It's important for things like maps to show distances accurately, but scaled to a useful size. Using scales is all about ratios and converting between different units of length.

Scales with Units

Learning Objective — Spec Ref R2/G15:
Interpret scale drawings and maps where the units are given.

Prior Knowledge Check:
Use ratio notation and apply ratios to problems with real contexts (Section 6). Convert between metric units of length (p.263).

A **scale** tells you the **relationship** between distances on a **map** or **plan** and distances in **real life**. For example, a map scale of 1 cm : 10 m means that 1 cm on the **map** represents a **real-life** distance of 10 m.

- To work out what a distance on the map **represents in real life**, **multiply** the **map distance** by the number on the right-hand side of the **map scale** ratio. So for a map with a scale of 1 cm : 10 m, a map distance of 2.5 cm would be 2.5 × 10 = 25 m in real life.
- To convert **real-life distances** to **map distances**, you **divide** the real-life distance by the map scale number. So 500 m in real life would be represented by 500 ÷ 10 = 50 cm on a map with scale 1 cm : 10 m.

Be careful with units — the units of **map distances** should match the **left-hand side** of the **scale** ratio, and **real-life distances** should have the same units as the **right-hand side**.

Tip: You might need to convert the units of a measurement before you start calculating.

Example 1

A plan of a garden is drawn to a scale of 1 cm : 5 m.

a) The distance between two trees is measured on the plan as 3 cm. What is the actual distance between the trees?

The scale is 1 cm : 5 m, so to convert from cm on the map to m in real life, multiply by 5.

1 cm represents 5 m, so
3 cm represents 3 × 5 = **15 m**

b) The actual distance between the garden shed and the pond is measured as 250 cm. What would the distance between the shed and the pond be on the plan?

1. The RHS of the scale is in m, so convert 250 cm to m. 250 cm = 2.5 m
2. Then divide by 5 to find the map distance. 2.5 m is shown as 2.5 ÷ 5 = **0.5 cm**

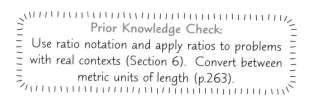

Example 2

The distance between two villages is 12 km. This is represented on a map by a distance of 24 cm. Express the scale of the map in the form 1 cm : n km.

1. Write down the distances you are given as a scale. — 24 cm : 12 km
2. Divide both sides by 24 to find the scale in the form 1 cm : n km. — (24 ÷ 24) cm : (12 ÷ 24) km
   1 cm : 0.5 km

Exercise 1

Q1 A map scale is given as 1 cm : 2 km.

a) Convert the following distances on the map to actual distances.
(i) 4 cm (ii) 22 cm (iii) 0.5 cm (iv) 0.25 cm

b) Convert the following actual distances to distances on the map.
(i) 10 km (ii) 14 km (iii) 7 km (iv) 1.4 km

Q2 An atlas uses a scale of 1 cm : 100 km. Find the actual distances represented by:
a) 7 cm b) 1.5 cm c) 6.22 cm d) 43 mm

Q3 The scale on a map of Europe is 1 cm : 50 km. Find the distance used on the map to represent the following actual distances.
a) 150 km b) 600 km c) 1000 km d) 15 000 m

Q4 The distance from Madrid to Malaga is shown on a map as 11 cm.
The actual distance is 440 km. Express the scale of the map in the form 1 cm : n km.

Q5 The distance from Thenford to Syresham is 12 km. This is shown on a map as 4 cm.

a) Express the scale of the map in the form 1 cm : n km.

b) The same map shows the distance from Chacombe to Badby as 7 cm.
What is the actual distance between these two villages?

Q6 You are asked to draw up the plans for a building using the scale 1 cm : 0.5 m.
Find the lengths you should draw on the plan to represent the following actual distances.
a) 4 m b) 18 m c) 21 m d) 1180 cm

Q7 To the right is the plan for a kitchen surface.
Measure the appropriate lengths on the plan to find the actual dimensions of the following:
a) the sink area b) the hob area

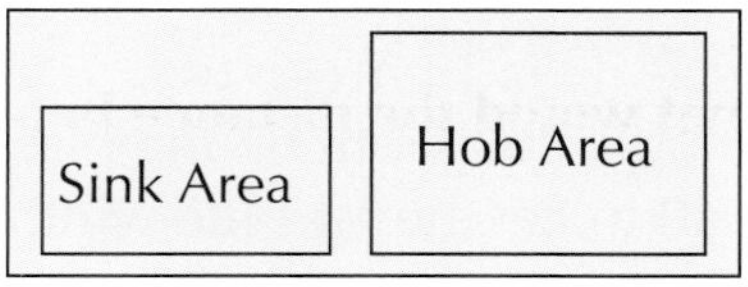

1 mm : 3 cm

Scales without Units

Learning Objective — Spec Ref R2/G15:
Interpret scale drawings and maps where no units are given.

A scale **without units** (e.g. 1 : 100) means you can use **any units** as long as they're the **same on both sides**. So 1 : 100 could be 1 cm : 100 cm, 1 mm : 100 mm, etc. It's often best to pick the units that **match** the measurement you're given in the problem.

Example 3

A map uses a scale of 1 : 200.
What is the actual distance between two points which appear 35 cm apart on the map?

1. Write the scale down using centimetres, to match the distance given in the question. — 1 cm : 200 cm
2. Multiply the map distance by the number on the right-hand side of the scale. — 35 cm represents 35 × 200 = 7000 cm.
3. Give your answer using sensible units. — 1 m = 100 cm, so the actual distance is: 7000 ÷ 100 = **70 m**.

Scales **without units** sometimes have a **big number** on the right-hand side. Once you've written such a scale **with units**, it often makes sense to **convert** one side to **different units**, so you end up with a smaller number in the ratio. E.g. you can write the ratio 1 : 100 as 1 cm : 100 cm, and then use the conversion 1 m = 100 cm to write this as 1 cm : 1 m.

Example 4

The plan of the grounds of a stately home has a scale of 1 : 4000.

a) Represent this scale in the form 1 cm : *n* m.

1. Write the scale down using centimetres, to match the left-hand side of 1 cm : *n* m. — 1 cm : 4000 cm
2. Convert the right-hand side to metres by dividing by 100. — 1 cm : (4000 ÷ 100) m

1 cm : 40 m

Tip: You could also do it starting with m:
1 m : 4000 m
100 cm : 4000 m
1 cm : 40 m

b) On the plan, one of the lakes is 11.5 cm wide. Calculate the actual width of the lake.

The scale is 1 cm : 40 m. Multiply the width on the plan by 40 to get the real-life width.

1 cm represents 40 m, so
11.5 cm represents 11.5 × 40 = **460 m**

Exercise 2

Q1 A map scale is given as 1 : 350.

a) Convert the following distances on the map to actual distances in m.

(i) 2 cm (ii) 7 cm (iii) 9.9 cm (iv) 25.7 cm

b) Convert the following actual distances to distances on the map.

(i) 700 cm (ii) 875 cm (iii) 945 cm (iv) 1.47 m

Q2 A plan uses the scale 1 : 75. Find the actual distances, in m, represented by:

a) 4 cm b) 11 cm c) 7.9 cm d) 17.2 cm

Q3 Convert these actual distances to the lengths they will appear on a map with scale 1 : 1000. Give your answers in cm.

a) 300 m b) 1400 m c) 120 m d) 0.43 km

Q4 Toy furniture is made to a scale of 1 : 40. Find the dimensions of the actual furniture, in m, when the toys have the following measurements.

a) Width of bed: 3.5 cm b) Length of table: 3.2 cm c) Height of chair: 2.4 cm

Q5 A road of length 6.7 km is to be drawn on a map. The scale of the map is 1 : 250 000. How long will the road be on the map? Give your answer in cm.

Q6 A model railway uses a scale of 1 : 500. Use the actual measurements given below to find measurements for the model in cm.

a) Length of carriage: 20 m b) Height of coal tower: 30 m

c) Length of footbridge: 100 m d) Height of signal box: 6 m

Q7 On the plans for a house, 3 cm represents the length of a garden with actual length 18 m.

a) Find the map scale in the form 1 cm : d m.

b) On the plan, the width of the garden is 1.2 cm. What is its actual width in metres?

c) The lounge has a length of 4.5 m. What is the corresponding length on the plan?

Q8 A path of length 4.5 km is shown on a map as a line of length 3 cm.

a) Express the scale in the form 1 cm : n km.

b) Express the scale in the form 1 : k.

Q9 The plan to the right shows the main tourist attractions of a major city.

Find the actual distances, in km, between:

a) the museum and the cathedral

b) the art gallery and the theatre

c) the cathedral and the park

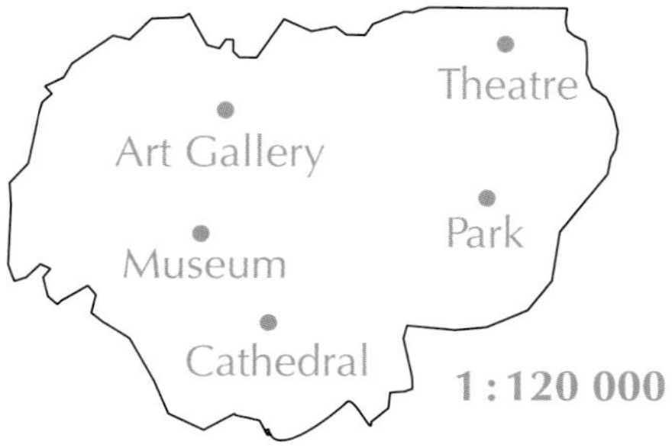

Constructing Scale Drawings

Learning Objective — Spec Ref R2/G15:
Construct accurate plans using a given scale.

If you know the **measurements** of something, you can draw an **accurate plan** using a given scale. **Convert** all the distances to lengths on the plan, then use your **ruler** to draw lines of the required lengths.

Tip: Use a sharp pencil to keep things nice and precise.

Example 5

The diagram shows a rough sketch of a garden.
Use the scale 1 : 400 to draw an accurate plan of the garden.

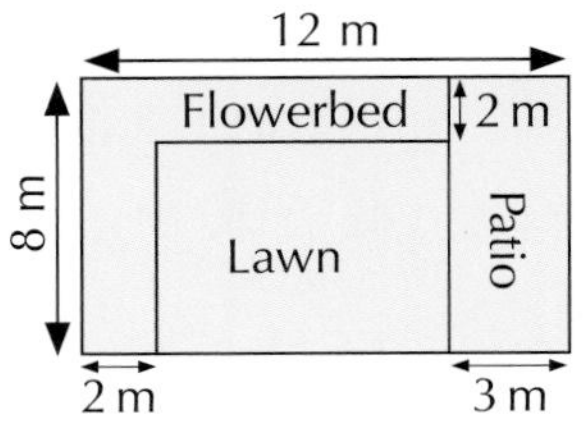

1. Write down the scale in cm. — 1 cm : 400 cm
2. Change the right-hand side to metres by dividing by 100. — 1 cm : 4 m
3. Use the scale to work out the lengths on the plan, then use these lengths to draw your plan. — 4 m is shown as 1 cm, so:
   12 m is shown as 12 ÷ 4 = 3 cm
   8 m is shown as 8 ÷ 4 = 2 cm
   2 m is shown as 2 ÷ 4 = 0.5 cm
   3 m is shown as 3 ÷ 4 = 0.75 cm

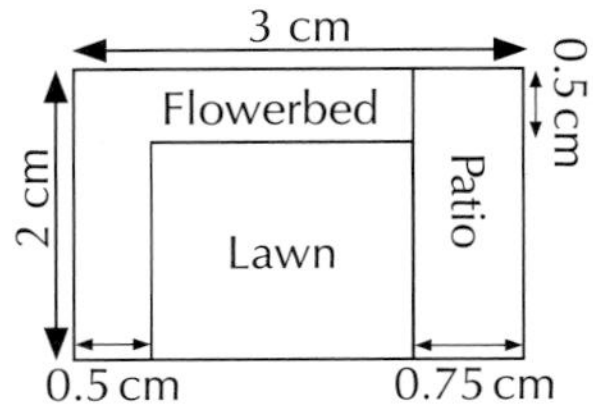

Exercise 3

Q1 A sketch of the floor plan for a symmetrical squash court is shown below. Use the scale 1 : 50 to draw an accurate plan of the court.

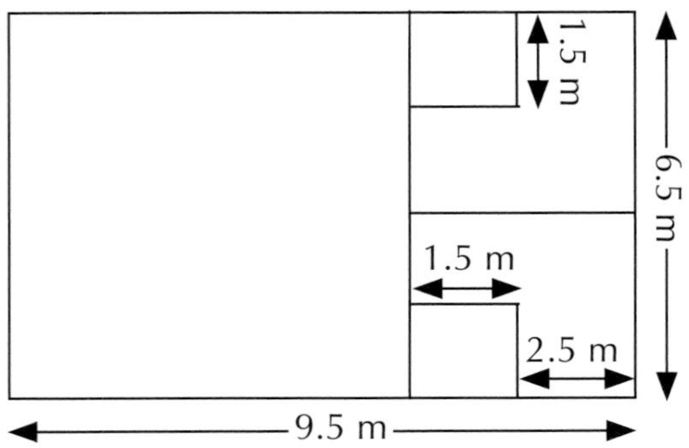

Q2 Below is a sketch of a park lake.

a) Draw an accurate plan of the lake using the scale 1 cm : 3 m.

b) There is a duck house at the intersection of *AC* and *BD*. Use your plan to find the actual distance from the duck house to point *B*.

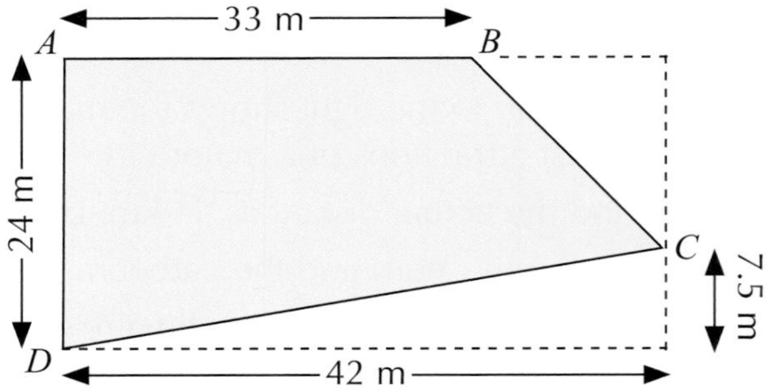

18.2 Bearings

Bearings are used in navigation to describe which direction something is in, relative to a north line.

Learning Objective — Spec Ref G15:
Understand and use bearings.

Prior Knowledge Check:
Use properties of angles and lines. See p237-241.

A **bearing** tells you the direction of one point from another. Bearings are given as **three-figure angles**, measured **clockwise** from the **north line**. For example, a bearing that's 60° clockwise from north will be written as 060°. To find a bearing:

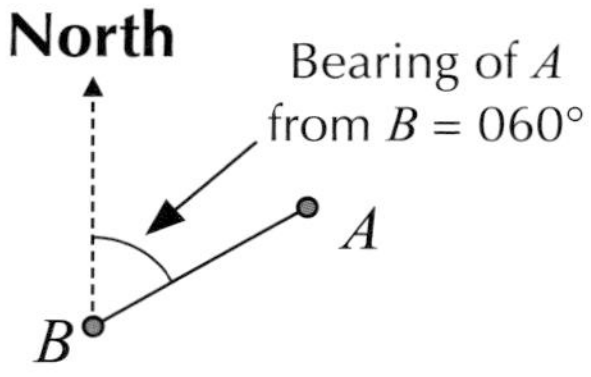

- Find the point you are going '**from**'.
- Draw the **north line** at this point.
- Then go **clockwise** to find the angle you want — this is the **bearing**.

As well as the information you're given, you might have to use the **properties of angles** (angles **around a point**, on a **straight line**, and on **parallel lines** — see Section 15) to find a bearing.

Example 1

a) Find the bearing of B from A.

b) Find the bearing of C from A.

1. Draw a north line at the point you're going 'from' — here it's A.
2. Find the clockwise angle from the north line.
3. Give the bearing as a three-figure number.

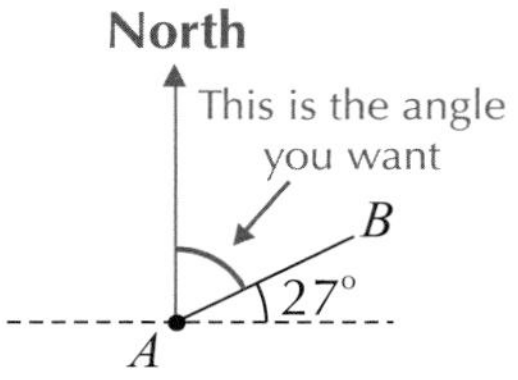

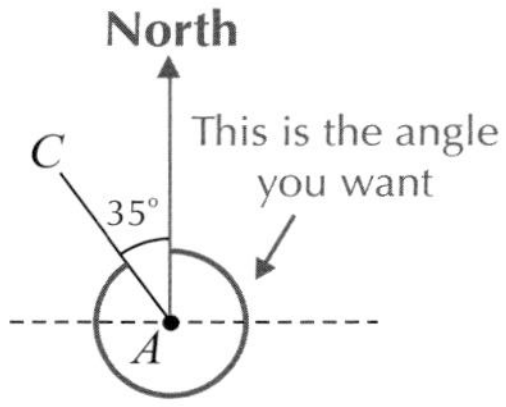

a) $90° - 27° = 63°$
So the bearing of B from A is **063°**.

b) $360° - 35° = 325°$
So the bearing of C from A is **325°**.

Exercise 1

Q1 Write these compass directions as bearings.

a) East b) Northeast c) South d) Northwest

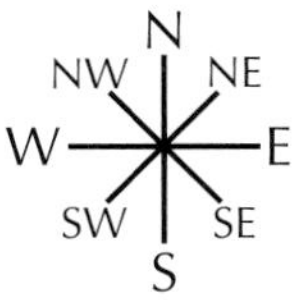

Q2 Find the bearing of B from A in the following.

a)
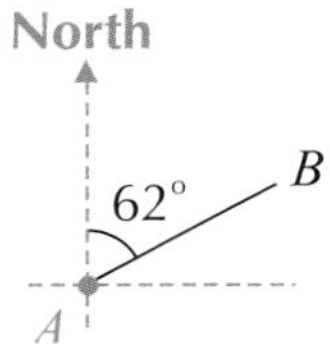

b)
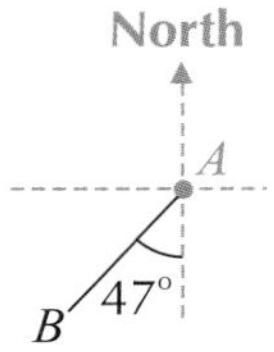

c)
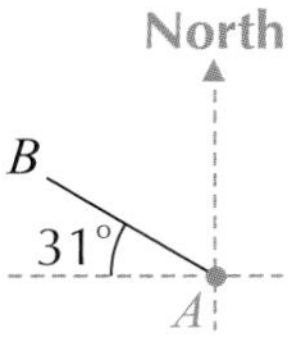

d)
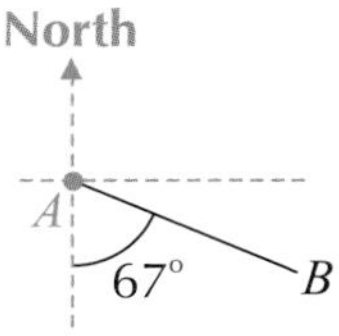

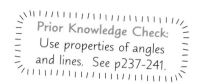

Q3 A ship travels in a direction which is 25° north of due east. Write this as a bearing.

Q4 Find the angle θ in each of the following using the information given.

a)

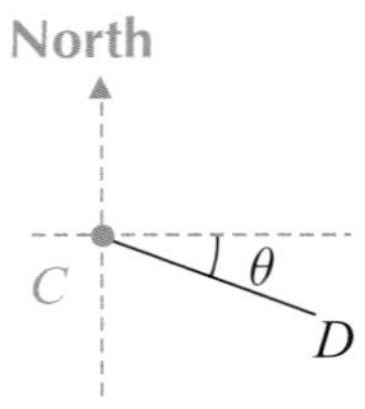

Bearing of D from C is 111°

b)

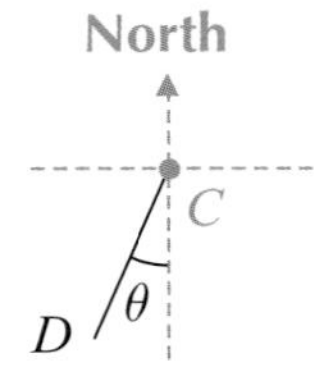

Bearing of D from C is 203°

c)

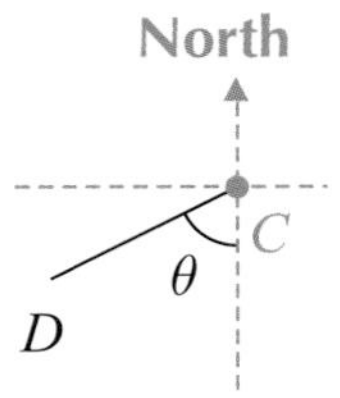

Bearing of D from C is 243°

d)

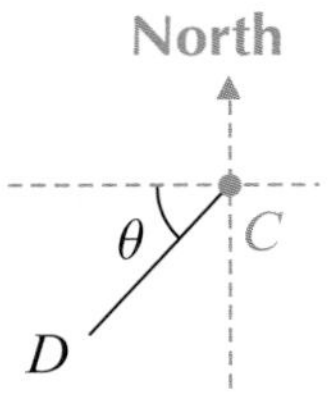

Bearing of D from C is 222°

Q5 Mark a point O and draw in a north line. Use a protractor to help you draw the points a) to d) with the following bearings from O.

a) 040° b) 321° c) 163° d) 263°

Since **north lines** are always **parallel** to each other, you can use the properties of **alternate**, **allied** and **corresponding** angles (see p.239-241) to work out the bearing of A from B, when **given** the bearing of B from A.

Tip: If you're not given a diagram, always start by doing a sketch yourself.

You might notice that the bearing of A from B is always either **180° more** or **180° less** than the bearing of B from A.

- If the given bearing is **less than 180°**, then you can **add** 180° to get the other bearing.
- If the bearing is **more than 180°**, then **subtract** 180° to find the bearing you want.

Example 2

The bearing of X from Y is 244°. Find the bearing of Y from X.

1. Draw a diagram showing what you know. Label the angle you're trying to find.

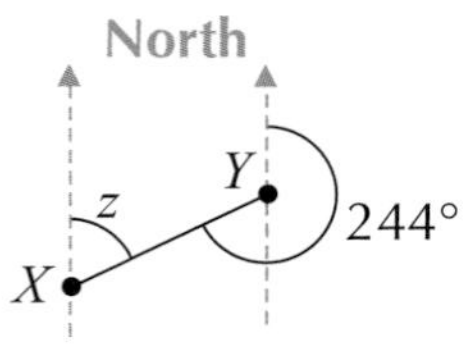

2. Find the alternate angle to the one you're looking for.
244° – 180° = 64°

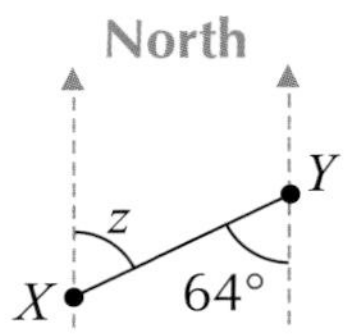

3. Alternate angles are equal, so these two are the same. Make sure you give your answer as a three-figure bearing.

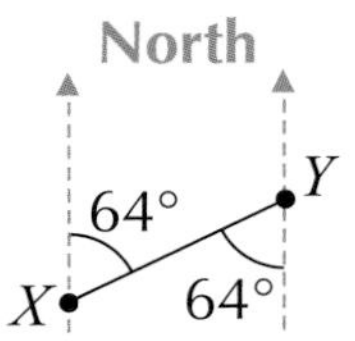

The bearing of Y from X is **064°**.

OR

1. Draw a diagram showing what you know.
2. 244° is greater than 180°, so subtract 180° to find the bearing.

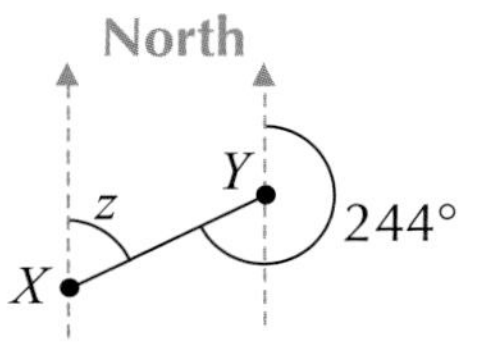

244° – 180° = 64°

So the bearing of Y from X is **064°**.

Exercise 2

Q1 a) The bearing of *B* from *A* is 218°. Find the bearing of *A* from *B*.

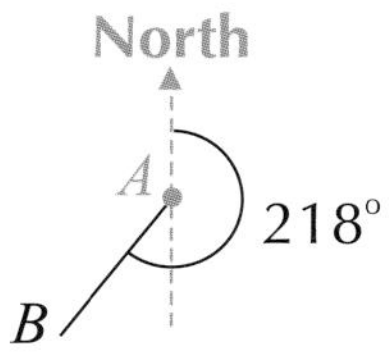

b) The bearing of *D* from *C* is 125°. Find the bearing of *C* from *D*.

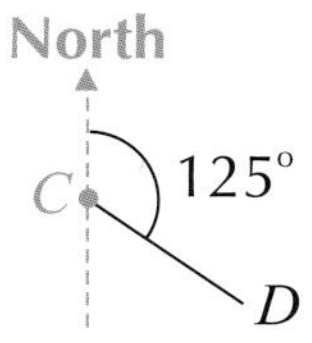

c) The bearing of *F* from *E* is 310°. Find the bearing of *E* from *F*.

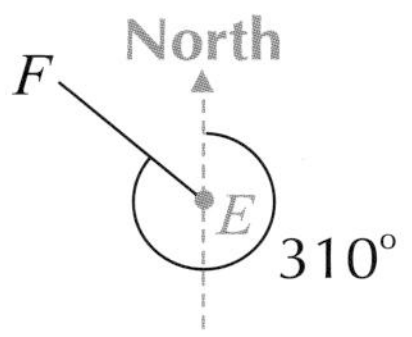

Q2 Find the bearing of *N* from *M* given that the bearing of *M* from *N* is:

a) 200° b) 330° c) 117° d) 015°

Q3 a) Measure the angle θ in the diagram.
b) Write down the bearing of *R* from *S*.

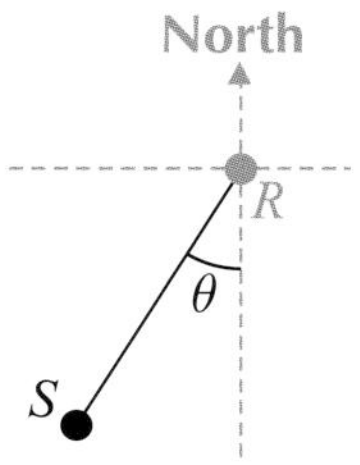

Q4 The point *Z* lies southeast of the point *Y*.
a) Write down the bearing of *Z* from *Y*.
b) Find the bearing of *Y* from *Z*.

Q5 Find the bearing of:

a) *B* from *A* b) *C* from *A* c) *E* from *A*
d) *A* from *B* e) *A* from *C* f) *A* from *D*

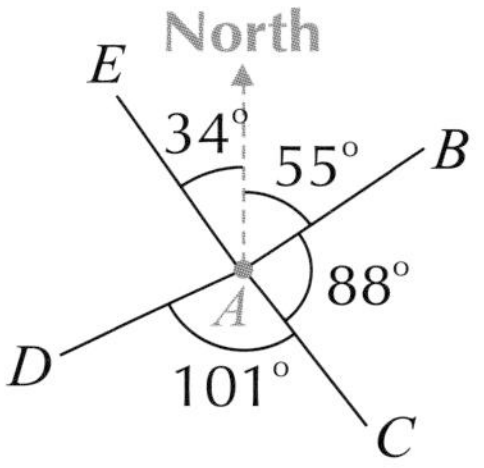

Bearings can also be used to draw **scale diagrams** (see method on p.294).
Use your **protractor** to draw any bearings and make sure any **north lines** are vertical.

Example 3

The points *P* and *Q* are a distance of 75 km apart. *Q* lies on a bearing of 055° from *P*. Use the scale 1 cm : 25 km to draw an accurate scale diagram of *P* and *Q*.

1. Draw *P* and a north line. Use your protractor to measure and draw the required angle clockwise from North.

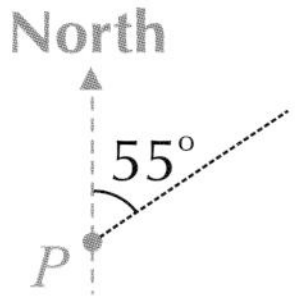

2. Use the scale to work out the distance between the two points.

25 km is shown by 1 cm, so 75 km is shown by 75 ÷ 25 = **3 cm**.

3. Draw *Q* the correct distance and direction from *P*.

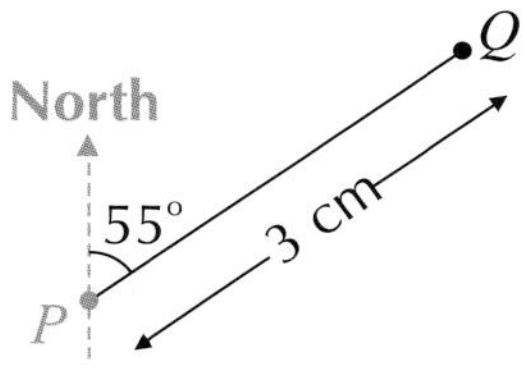

Exercise 3

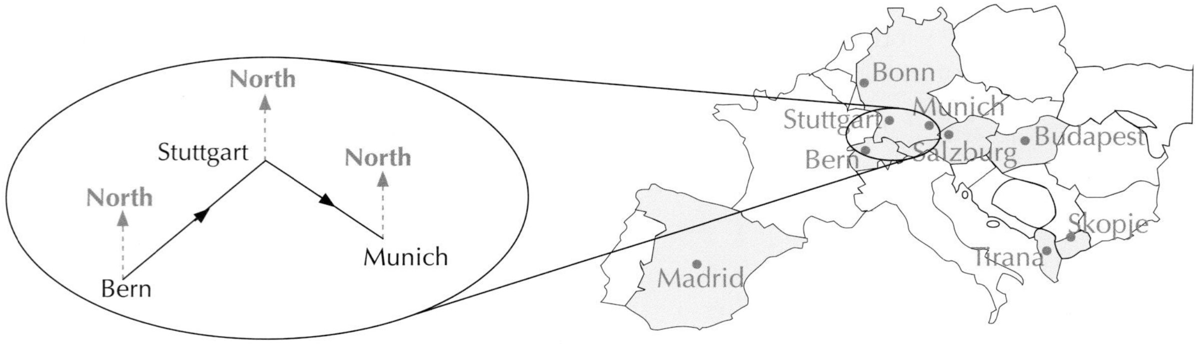

Q1 Above is a rough map of part of Europe.
A pilot flies from Bern to Stuttgart, then on to Munich.
This is shown on the enlarged part of the map, which uses a scale of 1 cm : 100 km.

a) Find the distance and bearings of the following stages of the journey.
   (i) Bern to Stuttgart (ii) Stuttgart to Munich

b) The pilot returns directly from Munich to Bern.
   Find the actual distance travelled in this stage.

Q2 Skopje is 150 km from Tirana, on a bearing of 048°.
Draw an accurate scale diagram of Skopje and Tirana using the scale 1 cm : 30 km.

Q3 Salzburg lies 540 km from Bonn, on a bearing of 125°.
Draw an accurate scale diagram of the two locations using the scale 1 cm : 90 km.

Q4 A pilot flies 2000 km from Budapest to Madrid, on a bearing of 242°.
Draw an accurate scale diagram of the journey using the scale 1 : 100 000 000.

Q5 Use the scale 1 : 22 000 000 to draw an accurate scale diagram of a 880 km journey from Budapest to Bern, on a bearing of 263°.

Q6 The scale drawing on the right shows three more European cities.
The scale of the diagram is 1 : 10 000 000.

a) Use the diagram to find the following distances.
   (i) PQ
   (ii) QR
   (iii) PR

b) Use a protractor to find the following bearings.
   (i) Q from P
   (ii) R from Q
   (iii) P from R

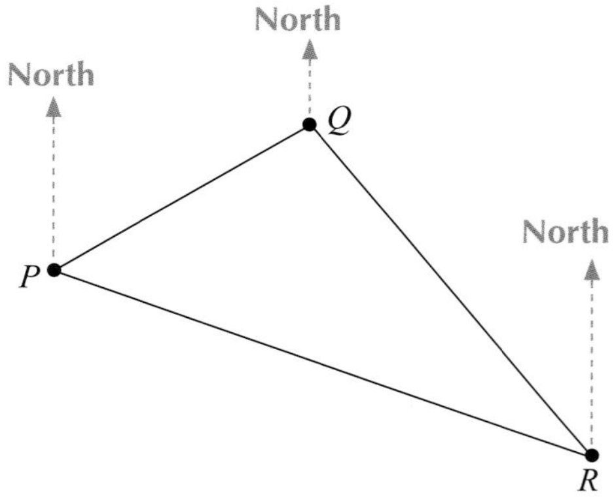

Review Exercise

Q1 The floor plan of a house is drawn to a scale of 1 cm : 2 m.
Find the actual dimensions of the rooms if they are shown on the plan as:

a) 2.7 cm by 1.5 cm
b) 3.2 cm by 2.2 cm
c) 1.85 cm by 1.4 cm
d) 0.9 cm by 1.35 cm

Q2 The plan to the right represents the distances between A, B and C.
The actual distance AB is 150 m.

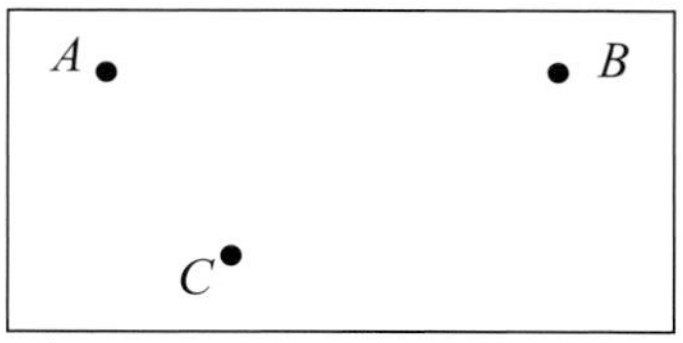

a) Measure AB on the plan to find the scale in the form 1 cm : n m.
b) Use your scale to find:
(i) the distance AC
(ii) the distance BC

Q3 Draw an accurate plan of the kitchen shown to the right using the scale 1 : 20.

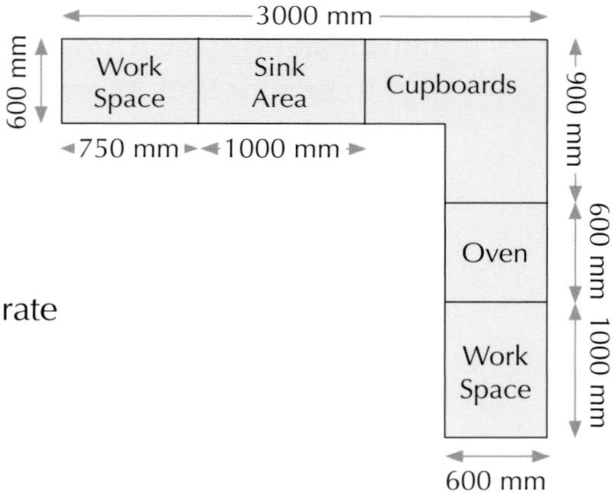

Q4 A jogger runs 230 m on a bearing of 020°, then 390 m on a bearing of 110°.

a) Use the scale 1 cm : 100 m to draw an accurate scale diagram of the two stages of his run.
b) Find the angle the jogger changes direction by after the first stage of his run.

Q5 Town A is 14 km north of Town B, which is 14 km east of Town C.
Find the following bearings.

a) Town B from Town A
b) Town C from Town B
c) Town C from Town A

Q6 Leicester is 100 km south of Doncaster. King's Lynn is 100 km east of Leicester.

a) Sketch the layout of the three locations.
b) Find the bearing of King's Lynn from Leicester.
c) Draw a north line through Doncaster. Find the bearings from Doncaster of:
(i) Leicester
(ii) King's Lynn
d) Draw a north line through King's Lynn. Find the bearings from King's Lynn of:
(i) Leicester
(ii) Doncaster

Exam-Style Questions

Q1 On a 1 : 25 000 scale map, the length of a reservoir is 3.8 cm. Work out the real life length of the reservoir, giving your answer in metres.

[2 marks]

Q2 The diagram on the right shows an accurate scale plan of a walk. The actual distance from The Knott to Hartsop Village is 12 km.

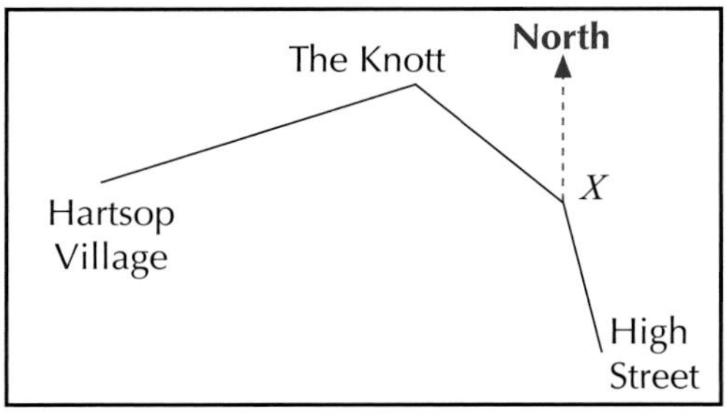

a) Find the scale of the plan in the form 1 cm : n km.

[2 marks]

b) Point X lies between High Street and The Knott. There are two paths leading away from X. What bearing should a walker take at X to ensure they take the correct path to The Knott?

[1 mark]

Q3 Part of a city is laid out as a rectangle, divided into 12 identical squares, as shown on the diagram on the right. 3 locations in this city ware labelled as the points A, B and C.

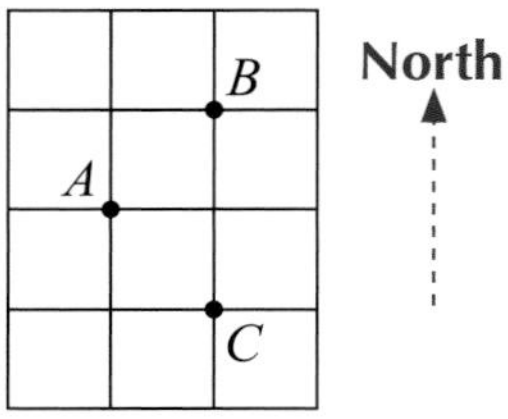

a) Write down the three figure bearing of B from A.

[1 mark]

b) Work out the three figure bearing of A from C.

[1 mark]

Q4 A boat sails on a bearing of 055° for 2000 m.
It then changes course and sails on a bearing of 100° for 1500 m.

a) Draw a scale diagram of the boat's journey. Use the scale 1 cm : 500 m.

[3 marks]

The boat returns directly to its starting point. Use your scale diagram to find:

b) the direct return distance,

[1 mark]

c) the bearing of the return journey.

[1 mark]

19.1 Pythagoras' Theorem

Pythagoras' theorem can be applied to all right-angled triangles. You use it to find the length of one side if you know the lengths of the other two sides.

Learning Objective — Spec Ref G6/G20:
Use Pythagoras' theorem to find missing lengths in right-angled triangles.

Prior Knowledge Check:
Be able to square numbers and find square roots. See p.39-41.

The lengths of the sides in a **right-angled triangle** always follow the rule:

$$h^2 = a^2 + b^2$$

h is the **hypotenuse** — this is the **longest** side, which is always **opposite** the right angle. a and b are the **shorter** sides.

This is **Pythagoras' theorem** and it is used to find lengths in right-angled triangles.

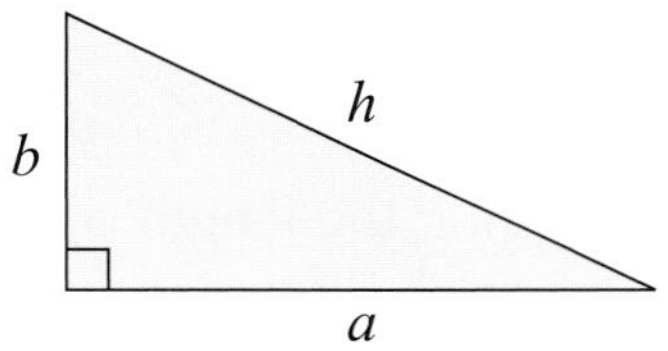

Finding the Hypotenuse

To find the length of the hypotenuse in a right-angled triangle:

- **Substitute** the values of a and b into the formula given above.
- **Add** together the squared lengths, a^2 and b^2, to get h^2.
- Take the **square root** of h^2 to find the hypotenuse, h.

Example 1

Find the length x on the triangle shown. Give your answer to 2 decimal places.

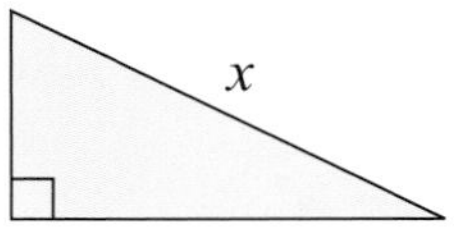

1. x is the hypotenuse, so substitute x for h in the formula and replace a and b with 7 and 4.
   $h^2 = a^2 + b^2$
   $x^2 = 7^2 + 4^2$
2. Add the squared lengths together to get x^2.
   $x^2 = 49 + 16 = 65$
3. Take the square root to find x. Don't forget to round your answer and use the correct units.
   $x = \sqrt{65} = 8.0622...$
   So $x =$ **8.06 cm** (2 d.p.)

Tip: If the question had asked for an exact length, you'd leave the square root in your answer — so the exact length of x is $\sqrt{65}$.

Exercise 1

Q1 Find the length of the hypotenuse in each of the triangles below.

a) 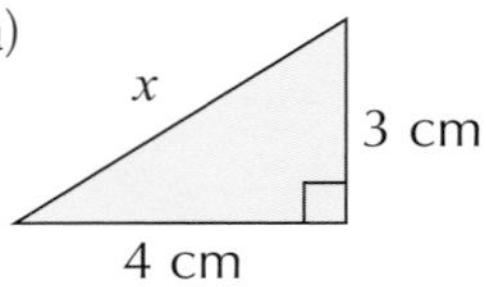

b) 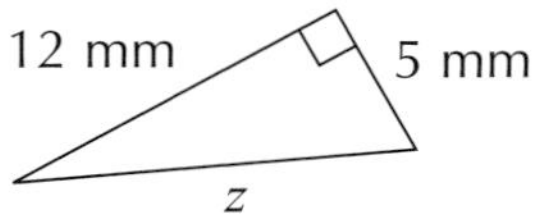

c) 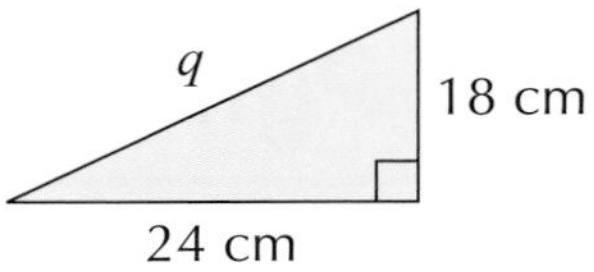

Q2 Find the length of the hypotenuse in each of the triangles below. Give your answers to 2 decimal places.

a) 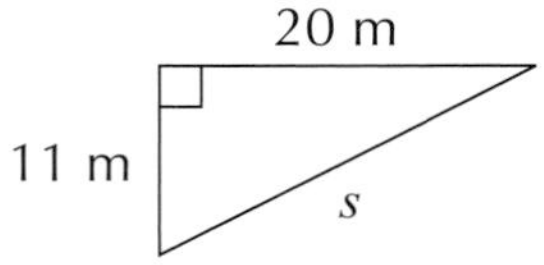

b) 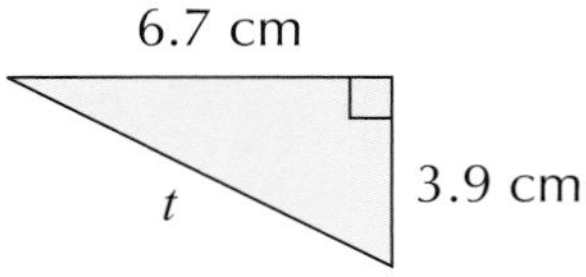

c) 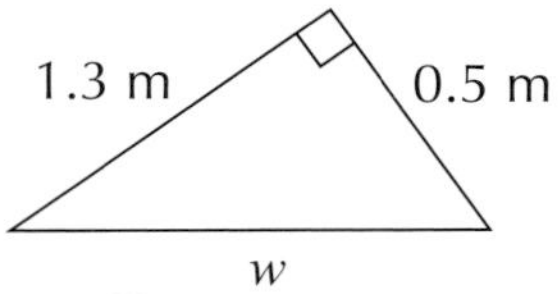

Q3 Find the exact length of the hypotenuse in each of the triangles below.

a) 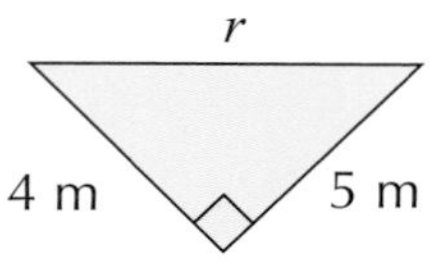

b) 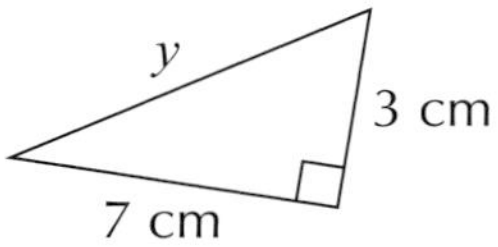

c) 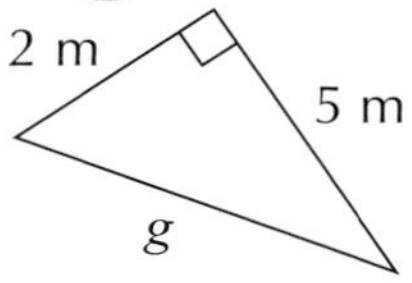

Q4 Find the length of the longest side of a right-angled triangle if the other sides are 8.7 cm and 6.1 cm in length. Give your answer to 2 decimal places.

Q5 Find the length of the hypotenuse of a right-angled triangle if the shorter sides have the following lengths. Give your answers to 2 decimal places.

a) $a = 5$ cm, $b = 7$ cm b) $a = 4$ cm, $b = 11$ cm c) $a = 6.3$ mm, $b = 1.9$ mm

Q6 In triangle XYZ, angle XYZ is 90°, XY is 21 cm and YZ is 32 cm. Find the exact length XZ.

Finding a Shorter Side

You can also use Pythagoras' theorem to find one of the **shorter sides** if you know the hypotenuse and the other side. To do this, **substitute** in the values and then **rearrange** the formula to make the unknown length the subject.

Example 2

Find the length *a* on the triangle shown. Give your answer to 2 decimal places.

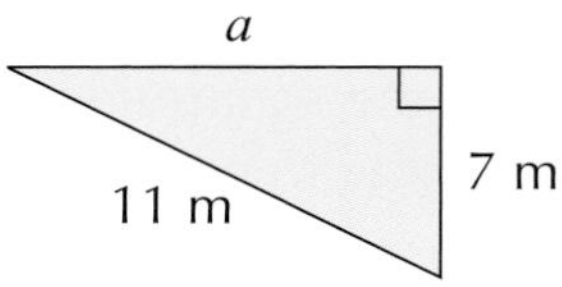

1. a is one of the shorter sides, so use Pythagoras' formula with $h = 11$ and $b = 7$.

$h^2 = a^2 + b^2$
$11^2 = a^2 + 7^2$

2. Rearrange to find a^2.

$a^2 = 11^2 - 7^2$
$a^2 = 121 - 49 = 72$

3. Take the square root, round your answer and use the correct units.

$a = \sqrt{72} = 8.4852...$
So a = **8.49 m** (2 d.p.)

Exercise 2

Q1 Find the lengths of the missing sides in these triangles.

a) 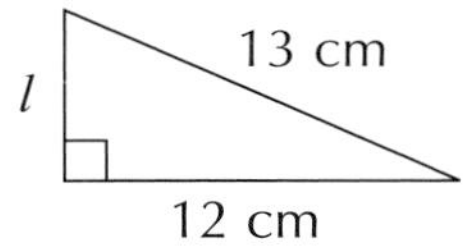

b) 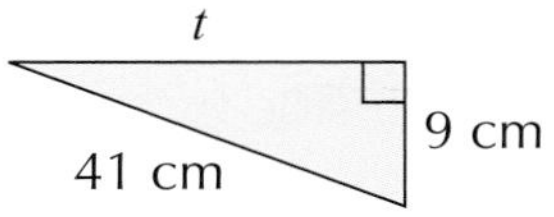

c) 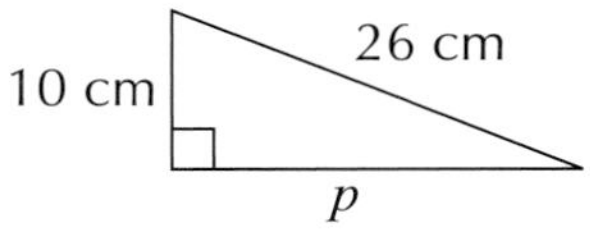

Q2 Find the length of the missing side in each of the triangles below.
Give your answers to 2 decimal places.

a) 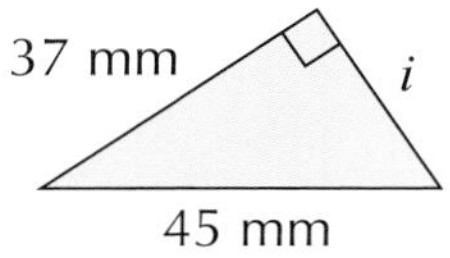

b) 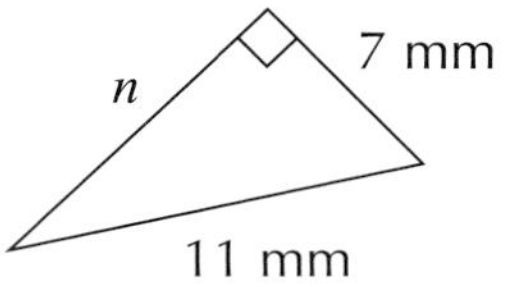

c) 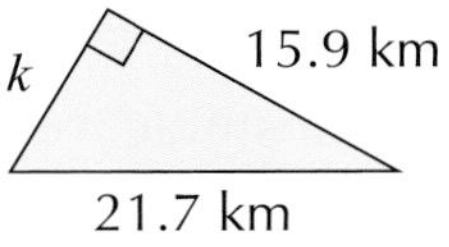

Q3 Find the exact length of the missing side in each of the triangles below.

a) 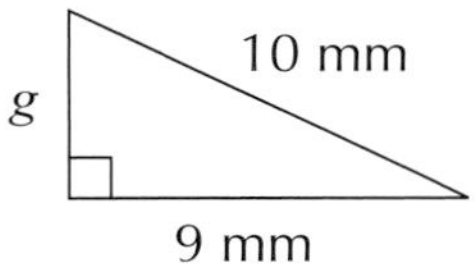

b) 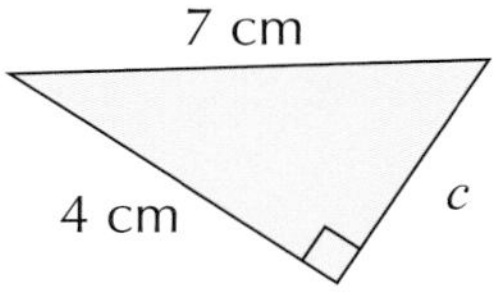

c) 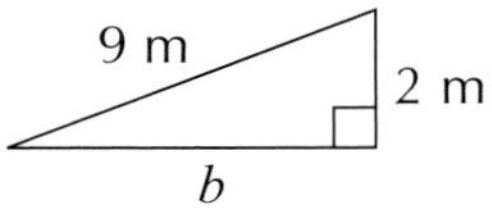

Q4 Find the length b in the right-angled triangle with the values of a and h given below.
Give your answers to 2 decimal places.

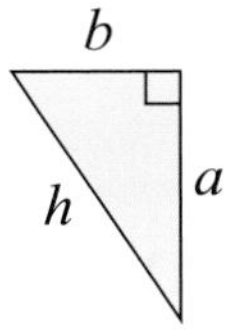

a) $a = 1$ cm, $h = 8$ cm b) $a = 6$ m, $h = 13$ m

c) $a = 4.1$ mm, $h = 11.3$ mm d) $a = 17.7$ cm, $h = 22.9$ cm

Q5 The lengths of the longest and shortest sides of a right-angled triangle are 17.3 cm and 6.6 cm. Find the length of the third side to 2 decimal places.

Using Pythagoras' Theorem

You can use Pythagoras' theorem to **say** whether a triangle is **right-angled** or **not**. Label the sides a, b and c, where c is the longest side. If the sides **satisfy** the rule $a^2 + b^2 = c^2$, then the triangle **is** right-angled. If they don't satisfy the rule, the triangle is not right-angled.

For example, consider a triangle with sides of length 3 cm, 6 cm and 7 cm. $3^2 + 6^2 = 9 + 36 = 45$, but $7^2 = 49$. $45 \neq 49$, so the triangle is **not right-angled**.

Tip: The symbol ≠ means 'is not equal to'.

Pythagoras' theorem can be used in lots of other situations too — you just have to look for ways to **create** a right-angled triangle. For example:

- **Splitting** an **equilateral** or **isosceles triangle** in half can create two identical right-angled triangles.

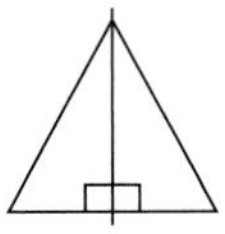

- The **straight line** between two pairs of **coordinates** forms the hypotenuse of a right-angled triangle with shorter sides equal to the difference in the x-coordinates and the difference in the y-coordinates (see Example 3 on the next page).

Example 3

Find the exact distance between points A and B on the grid.

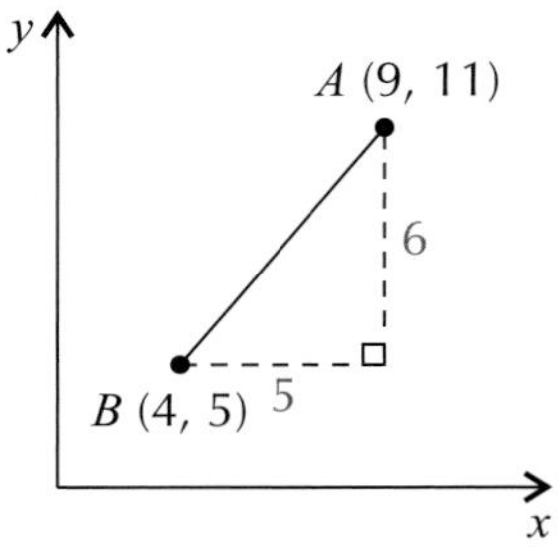

1. Create a right-angled triangle with hypotenuse AB.
2. Find the length of the horizontal side by working out the difference in the x-coordinates.
   Difference in x-coordinates: $9 - 4 = 5$
3. Find the length of the vertical side by working out the difference in the y-coordinates.
   Difference in y-coordinates: $11 - 5 = 6$
4. Substitute the values into the formula.
   $AB^2 = 5^2 + 6^2 = 25 + 36 = 61$
5. Take the square root to find the distance. Leave your answer as a square root.
   $AB = \sqrt{\mathbf{61}}$

Tip: Remember — 'exact distance' means you don't work out the value of the square root.

Pythagoras' theorem can also be applied to **real life situations**. The formula is used in the **same way**, you just have to link your answer back to the **context** of the situation. If you're struggling to work out what's going on, **sketch a diagram** to help you picture it.

Example 4

A TV has a height of 40 cm and width of w cm.
Its diagonal measures 82 cm. Will the TV fit in a box 75 cm wide?

1. The height, width and diagonal form a right-angled triangle, so substitute these values into the formula.
   $h^2 = a^2 + b^2$
   $82^2 = 40^2 + w^2$
2. Rearrange to find w^2.
   $w^2 = 82^2 - 40^2 = 6724 - 1600$
   $= 5124$
3. Take the square root to find w.
   $w = \sqrt{5124} = 71.58$ cm (2 d.p.)
4. Use your result to draw a conclusion — make sure you actually answer the question.
   $71.58 < 75$ so **yes, the TV will fit in the box.**

Exercise 3

Q1 Use Pythagoras' theorem to decide whether or not the following triangles are right-angled.

a) $a = 9$ cm, $b = 12$ cm, $h = 15$ cm

b) $a = 8$ mm, $b = 7$ mm, $h = 11$ mm

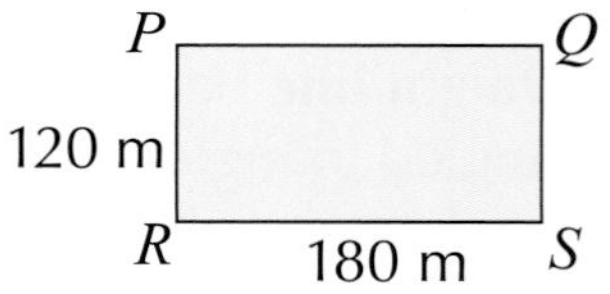

Q2 Find the distance PS in the rectangle on the right.
Give your answer to 2 decimal places.

Q3 I run 540 m south and then 970 m east. What is my final distance from my starting point? Give your answer to the nearest metre.

In Questions 4-13, give your answers to 2 decimal places (unless told otherwise).

Q4 XYZ is an isosceles triangle. M is the mid-point of YZ. Find the length of XY.

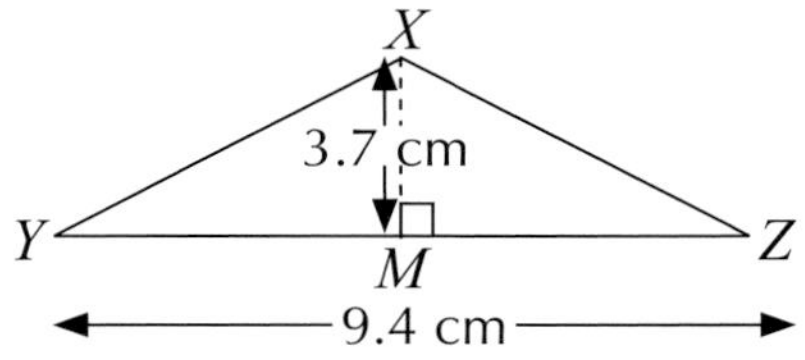

Q5 The end of a ladder of length 3.3 m is placed on the ground 0.8 m from the base of a wall. When leant against the wall, how high up the wall does the ladder reach?

Q6 A slice of toast is in the shape of a right-angled triangle. The hypotenuse of the triangle has length 14 cm, and a shorter side is of length 11 cm. Find the length of the third side.

Q7 The triangle JKL is drawn inside a circle centred on O, as shown. JK and KL have lengths 4.9 cm and 6.8 cm respectively.

a) Find the length of JL.

b) Find the radius of the circle.

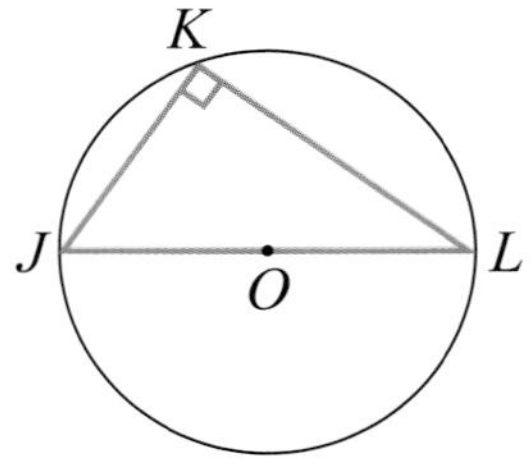

Q8 Find the exact length of each line segment AB.

a)

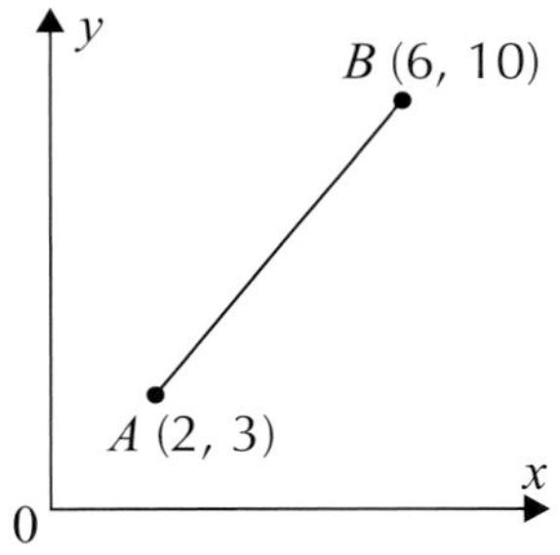

b)

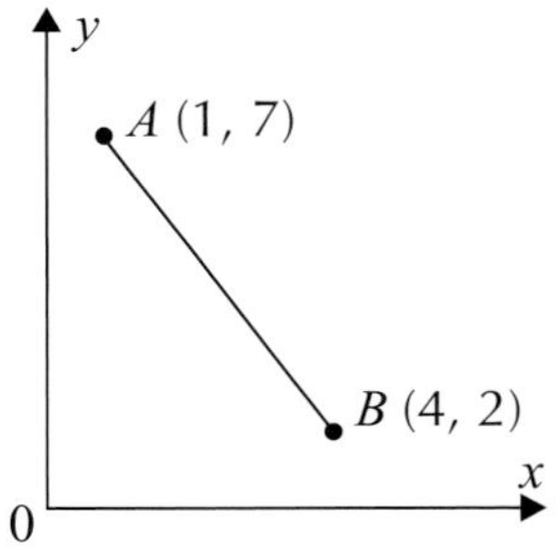

Q9 Find the radius of the base of the cone shown on the right.

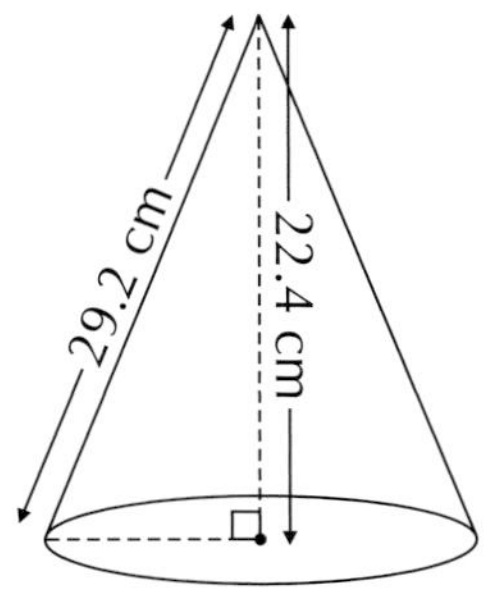

Q10 A kite gets stuck at the top of a tree. The kite's 15 m string is taut, and its other end is held on the ground, 8.5 m from the base of the tree. Find the height of the tree.

Q11 Newtown is 88 km northwest of Oldtown. Bigton is 142 km from Newtown, and lies northeast of Oldtown. What is the distance from Bigton to Oldtown? Give your answer to the nearest kilometre.

Q12 A spaghetti jar is in the shape of a cylinder. The jar has radius 6 cm and height 28 cm. What is the length of the longest stick of dried spaghetti that will fit inside the jar?

Q13 What is the height of the kite on the right?

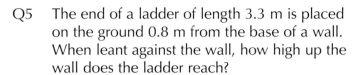

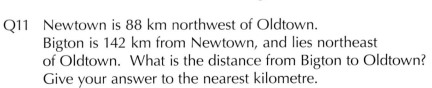

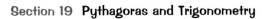

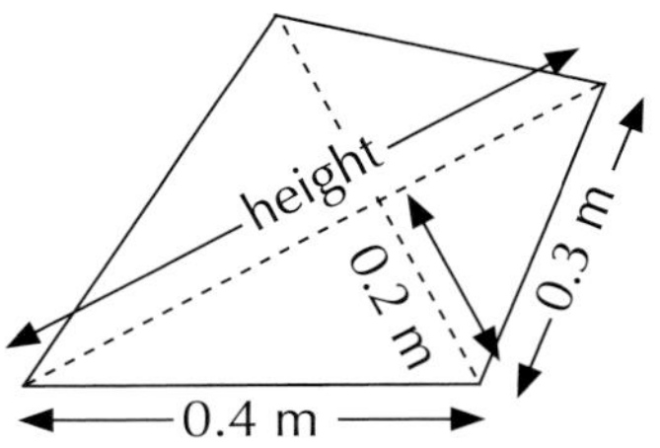

19.2 Trigonometry — Sin, Cos and Tan

Trigonometry allows you to find missing sides and angles in triangles. For right-angled triangles, you'll need to know the sine (sin), cosine (cos) and tangent (tan) formulas.

The Three Formulas

Learning Objective — Spec Ref G20:
Use trigonometry to find missing lengths and angles in right-angled triangles.

Trigonometry can be used to find lengths or angles in **right-angled triangles**. For a given angle x (as shown on the diagram on the right):

- The side opposite the right angle is the **hypotenuse**.
- The side opposite the given angle is the **opposite**.
- The side between the given angle and the right-angle is the **adjacent**.

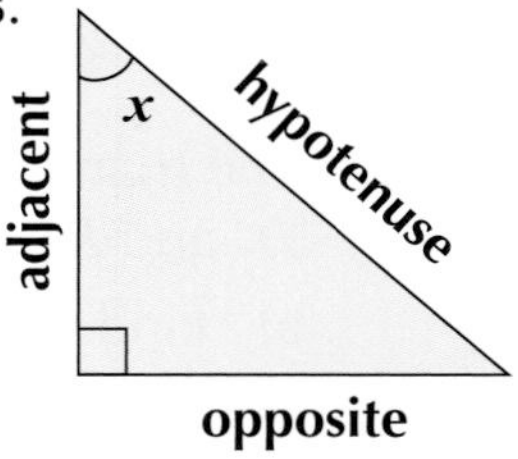

The three sides of a right-angled triangle are linked by the following formulas:

$$\sin x = \frac{\text{opp}}{\text{hyp}}, \quad \cos x = \frac{\text{adj}}{\text{hyp}}, \quad \tan x = \frac{\text{opp}}{\text{adj}}$$

If you're given **one angle** and **one side**, you can use trigonometry to find an **unknown side length**.

Start by looking at the side you've been **given** and the side you **want to find** — this will tell you which formula to use. For example, if you had the **hypotenuse** and wanted to find the **adjacent**, you'd use the **cos** formula as it contains these two sides. **Substitute** the values you know into the formula and **rearrange** it to find the length you want.

Tip: Remember 'SOH CAH TOA' to help you decide which formula you need to use — and always label the sides of your triangle that you're interested in.

Example 1

Find the length of side y. Give your answer correct to 3 significant figures.

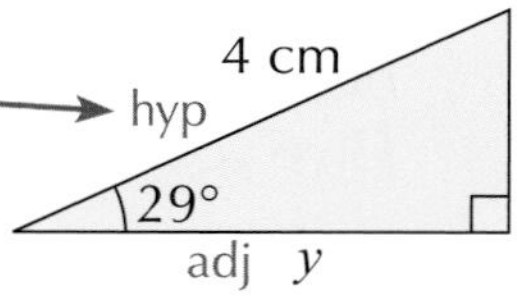

1. Start by labelling the two sides.
2. You can see from the labels that you're given the hypotenuse and asked to find the adjacent, so use the formula for cos x. $\cos x = \frac{\text{adj}}{\text{hyp}}$
3. Put the numbers into the formula... $\cos 29° = \frac{y}{4}$
4. ...and rearrange to find y. $4 \cos 29° = y$
5. Input '4 cos 29' into your calculator and press '=' to find the value of y. $y = 3.4984... =$ **3.50 cm** (3 s.f.)

Example 2

Find the length of side *w*. Give your answer correct to 3 significant figures.

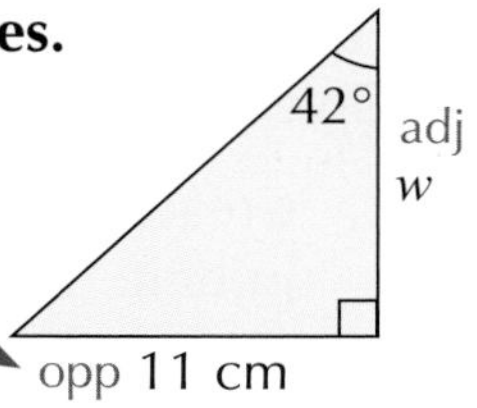

1. Start by labelling the two sides.
2. You can see from the labels that you're given the opposite and asked to find the adjacent, so use the formula for tan *x*. $\tan x = \frac{\text{opp}}{\text{adj}}$
3. Put the numbers into the formula... $\tan 42° = \frac{11}{w}$
4. ...and rearrange to find *w*. $w \tan 42° = 11$
5. Input '11 ÷ tan 42' into your calculator and press '=' to find the value of *w*. $w = \frac{11}{\tan 42°} = 12.216...$
   $= \mathbf{12.2\ cm}$ (3 s.f.)

Tip: Be careful — *w* is on the bottom of the fraction, so the rearrangement is slightly different.

Exercise 1

Q1 Label each of the triangles below with letters to show the hypotenuse (H), opposite (O) and adjacent (A) sides, in relation to the labelled angle.

a) 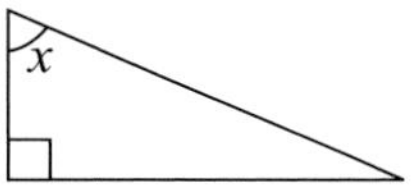

b) 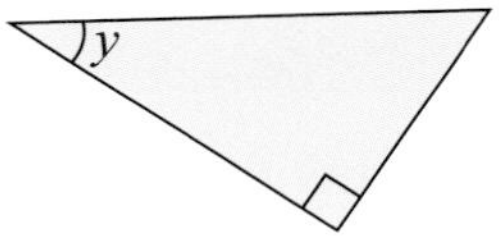

c) 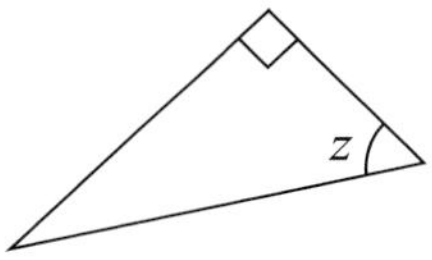

Q2 Find the lengths of the missing sides marked with letters.
Give your answers to 3 significant figures.

a) 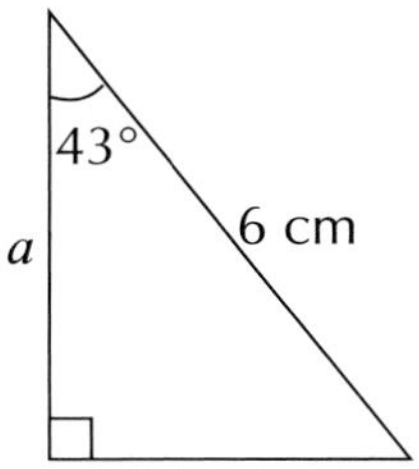

b) 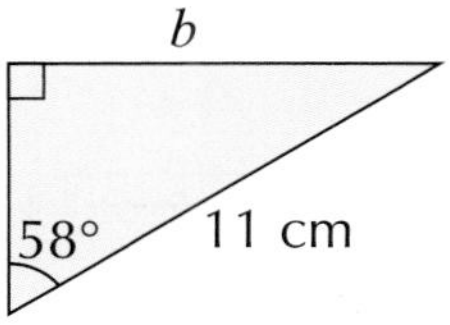

c) 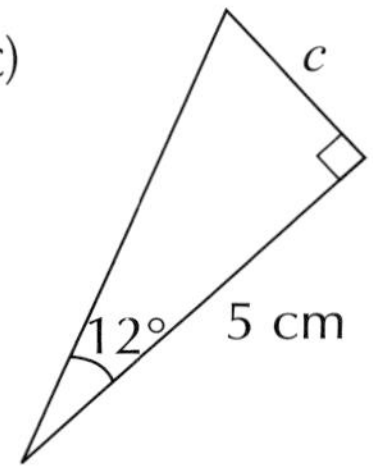

d) 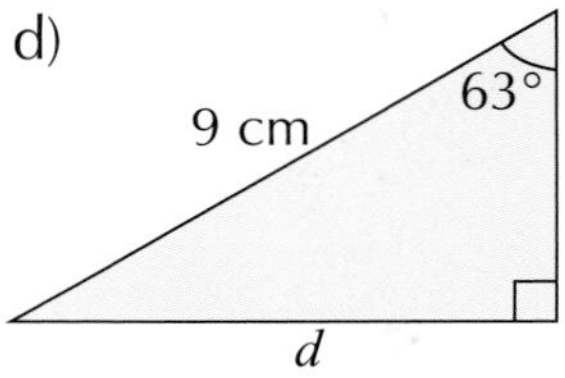

Q3 Find the lengths of the missing sides marked with letters.
Give your answers to 3 significant figures.

a) 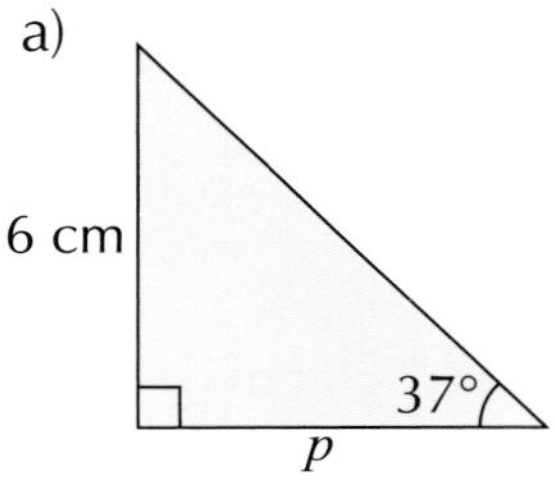

b) 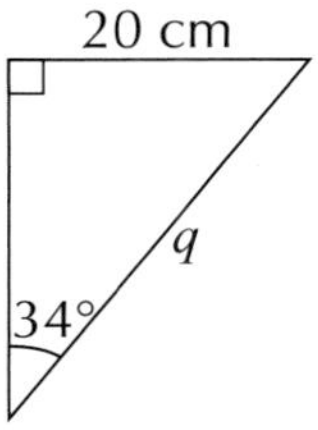

c) 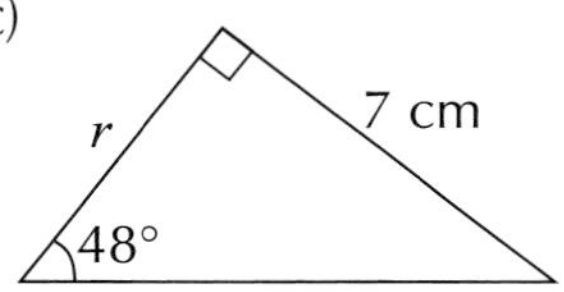

d) 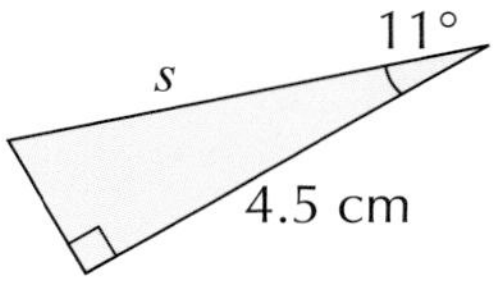

Example 3

Find the missing side length of the isosceles triangle shown. Give your answer correct to 3 significant figures.

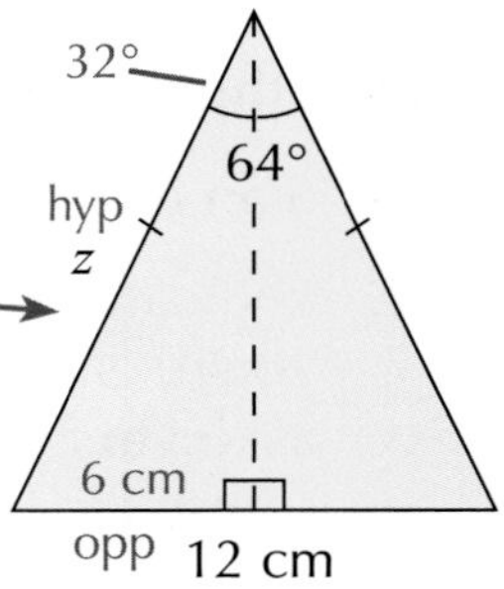

1. Create a right-angled triangle by splitting the triangle in half. Label the two sides of your new triangle.
2. Divide the angle by 2 to find the angle in your right-angled triangle. $64 \div 2 = 32°$
   Divide the length of the base by 2 to find the base length of your right-angled triangle. $12 \div 2 = 6$ cm
3. Now you have the opposite and you need to find the hypotenuse, so use the formula for sin x. $\sin x = \frac{\text{opp}}{\text{hyp}}$
4. Put the numbers into the formula... $\sin 32° = \frac{6}{z}$
5. ...and rearrange to find z. $z \sin 32° = 6$
6. Input '6 ÷ sin 32' into your calculator and press '=' to find the value of z. $z = \frac{6}{\sin 32°}$
   $= 11.322... =$ **11.3 cm** (3 s.f.)

Exercise 2

Q1 Find the length of the side labelled with a letter in each of the triangles below. Give your answers to 3 significant figures.

a)

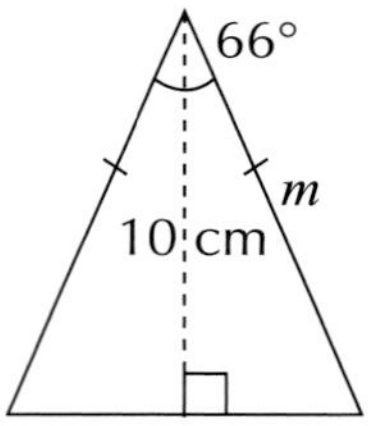

b)

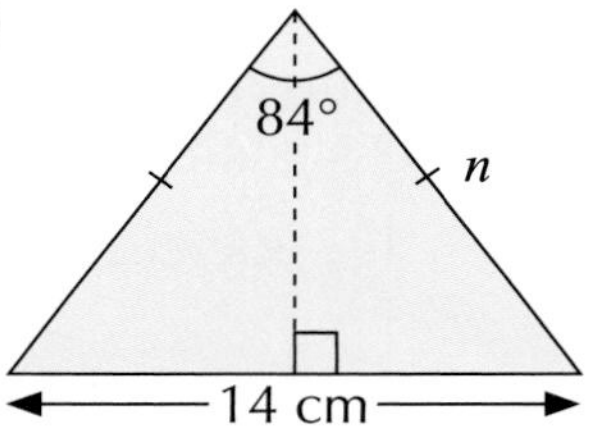

Q2 Find the length of the side labelled with a letter in each of the triangles below. Give your answers to 3 significant figures.

a)

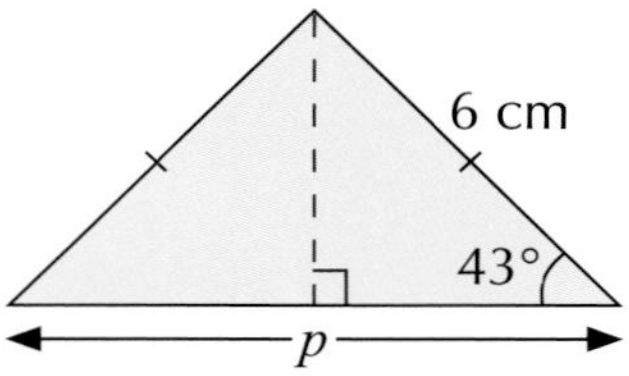

b)

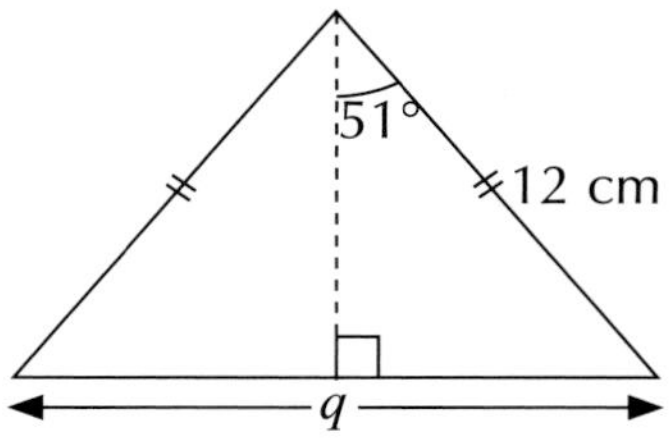

Q3 An isosceles triangle has two sides measuring 22 cm, with an angle of 98° between them. What is the length of the third side of the triangle? Give your answer to 3 significant figures.

You can also use the trig formulas to find an **angle** if you know two side lengths. You have to use the **inverse functions** of sin, cos and tan (written **sin⁻¹**, **cos⁻¹** and **tan⁻¹**), which return an **angle**.

To find an angle, work out which formula you need from the sides you're given as before, then **substitute** in the known values — this will give you a **fraction** on the right-hand side, e.g. $\sin x = \frac{1}{2}$. Take the **inverse trig function** of the fraction to get the angle — so here you'd do $x = \sin^{-1}\left(\frac{1}{2}\right) = 30°$.

Tip: Inverse trig functions are usually found on a calculator by pressing 'shift' or '2nd' before pressing sin, cos or tan.

Example 4

Find the size of angle x. Give your answer correct to 1 decimal place.

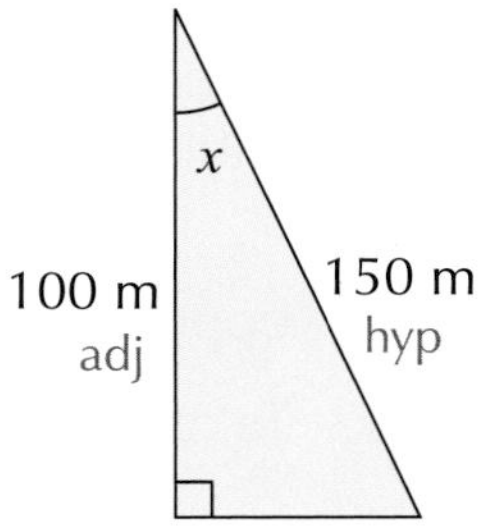

1. Start by labelling the sides of the triangle.
2. You're given the adjacent and the hypotenuse, so use the formula for cos x. $\cos x = \frac{\text{adj}}{\text{hyp}} = \frac{100}{150}$
3. Take the inverse of cos to find the angle. $x = \cos^{-1}\left(\frac{100}{150}\right)$
4. Input 'cos⁻¹(100 ÷ 150)' into your calculator and press '=' to find the value of x. $x = 48.189... =$ **48.2°** (1 d.p.)

Example 5

Find the size of angle x. Give your answer correct to 1 decimal place.

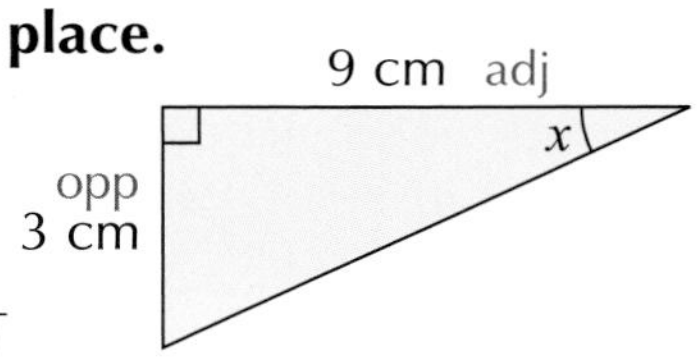

1. Start by labelling the sides of the triangle.
2. You're given the opposite and the adjacent, so use the formula for tan x. $\tan x = \frac{\text{opp}}{\text{adj}} = \frac{3}{9}$
3. Take the inverse of tan to find the angle. $x = \tan^{-1}\left(\frac{3}{9}\right)$
4. Use your calculator to find the value of x. $x = 18.434... =$ **18.4°** (1 d.p.)

Trigonometry can be used to work out angles of **elevation** and **depression**.

The **angle of elevation** is the angle between a **horizontal line** and the **line of sight** of an observer at the same level **looking up**. E.g. the angle made when looking up at a hovering helicopter.

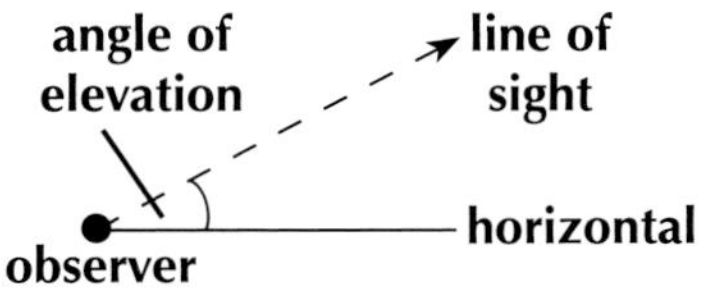

The **angle of depression** is the angle between a **horizontal line** and the **line of sight** of an observer at the same level **looking down**. E.g. the angle made when looking down from a window.

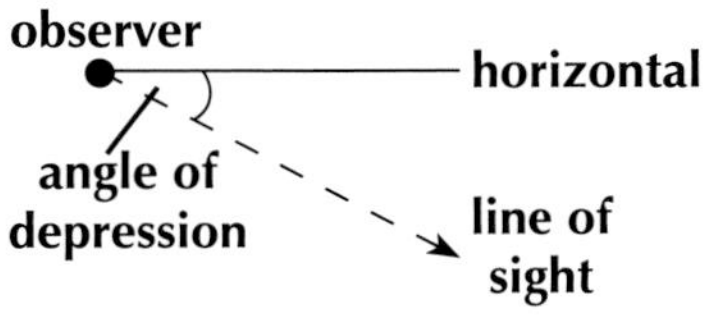

For problems that ask for an angle of elevation or depression, use the information given to **draw** a right-angled triangle and use the formulas in the **same way** as usual. Remember to relate your answer to the **original context** of the problem.

Example 6

Liz holds one end of a 7 m paper chain out of her window. Phil stands in the garden below holding the other end so it's taut. Phil's end of the paper chain is 6 m vertically below Liz's end. Find the size of the angle of depression from Liz to Phil. Give your answer to 1 decimal place.

1. Use the information to draw a right-angled triangle — the angle of depression is the angle below the horizontal. Label the sides of your triangle.

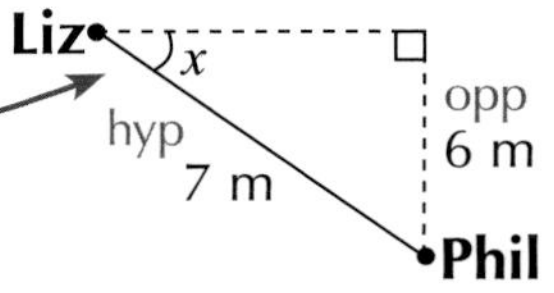

2. You're given the hypotenuse and the opposite, so use the formula for sin x.
   $\sin x = \frac{\text{opp}}{\text{hyp}} = \frac{6}{7}$
3. Take the inverse of sin to find the angle.
   $x = \sin^{-1}\left(\frac{6}{7}\right)$
4. Use your calculator to find the value of x.
   $x = 58.997... =$ **59.0°** (1 d.p.)

Exercise 3

Q1 Find the sizes of the missing angles marked with letters. Give your answers to 1 d.p.

a)
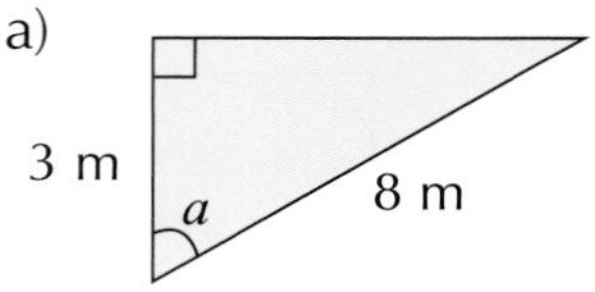

b)
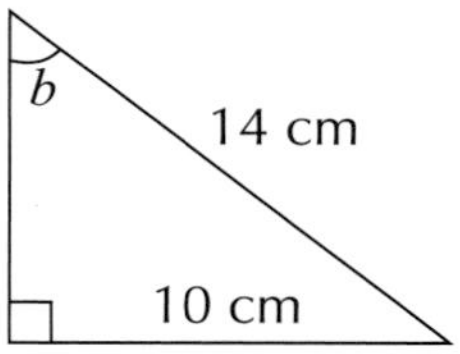

c)
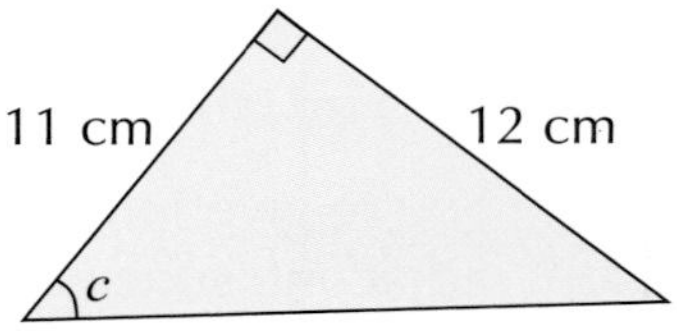

Q2 Find the sizes of the missing angles marked with letters, giving your answers to 1 d.p.

a)
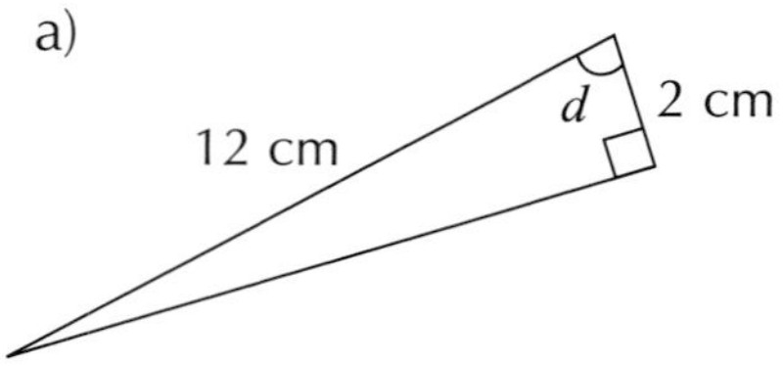

b)
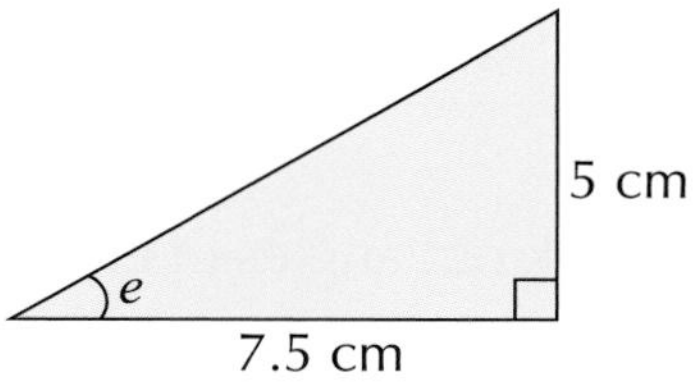

c)
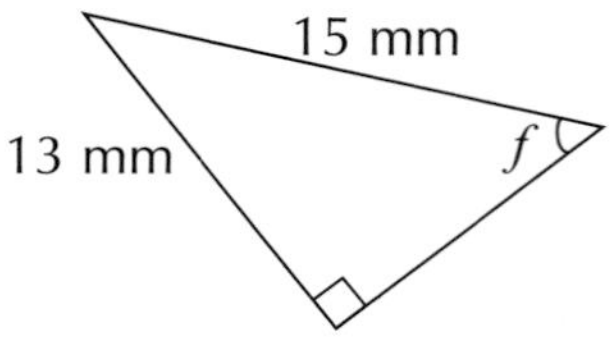

Q3 Maha is building slides for an adventure playground.

a) The first slide she builds has an 8 m high vertical ladder and a slide of length 24 m. Find m, the slide's angle of elevation, to 1 d.p.

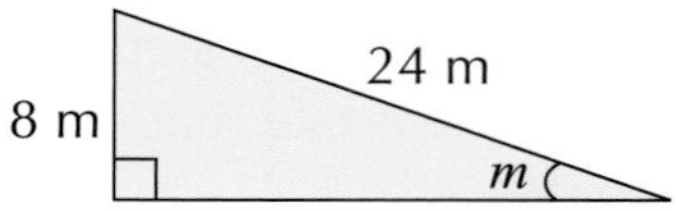

b) A second slide has a 4 m vertical ladder, and the base of the slide reaches the ground 5.5 m from the base of the ladder as shown. Find q, the angle of depression at the top of the this slide, to 1 d.p.

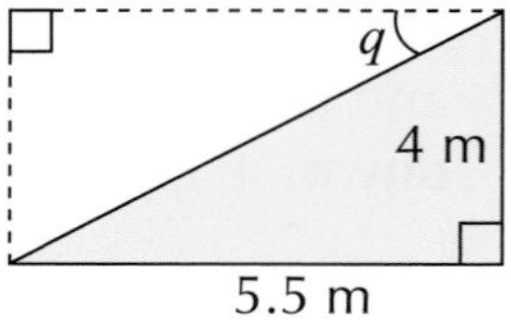

Common Trig Values

Learning Objective — Spec Ref G21:
Know and be able to find trig values for common angles.

The sin, cos and tan of some angles have **exact values** as shown below.
You need to know these for trigonometry questions where you're not allowed a calculator.

$\sin 0° = 0$	$\sin 30° = \frac{1}{2}$	$\sin 45° = \frac{\sqrt{2}}{2}$	$\sin 60° = \frac{\sqrt{3}}{2}$	$\sin 90° = 1$
$\cos 0° = 1$	$\cos 30° = \frac{\sqrt{3}}{2}$	$\cos 45° = \frac{\sqrt{2}}{2}$	$\cos 60° = \frac{1}{2}$	$\cos 90° = 0$
$\tan 0° = 0$	$\tan 30° = \frac{\sqrt{3}}{3}$	$\tan 45° = 1$	$\tan 60° = \sqrt{3}$	

Example 7

Without using a calculator, find the exact length of side *y* in the triangle below.

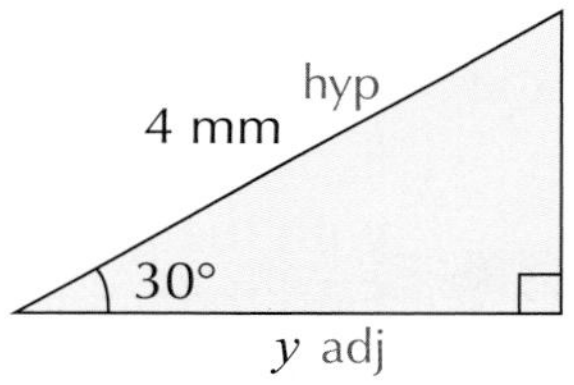

1. Label the triangle.
2. You're given the hypotenuse and want to find the adjacent, so use the formula for cos *x*. $\cos x = \frac{\text{adj}}{\text{hyp}}$
3. Substitute in the values you know and rearrange the formula to make *y* the subject. $\cos 30° = \frac{y}{4}$
4. Replace cos 30° with its exact value. $y = 4 \times \cos 30° = 4 \times \frac{\sqrt{3}}{2} = \mathbf{2\sqrt{3}\ mm}$

Exercise 4

Q1 Find the size of the angles marked with letters.

a)

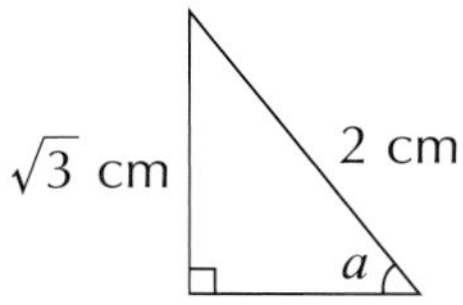

b)

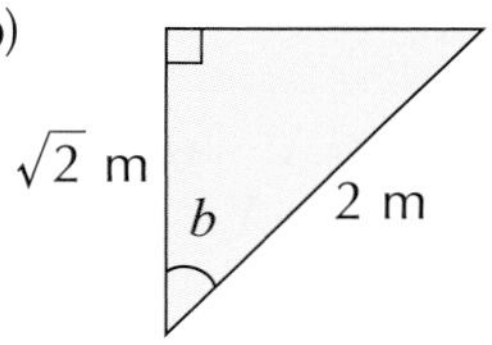

c)

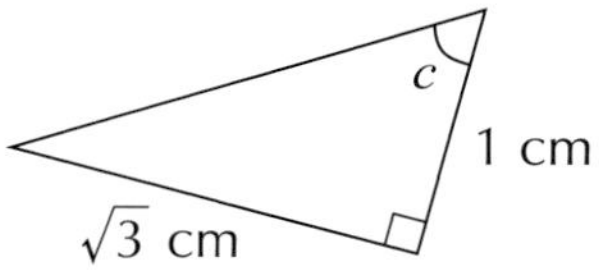

Q2 Find the exact length of the sides marked with letters.

a)

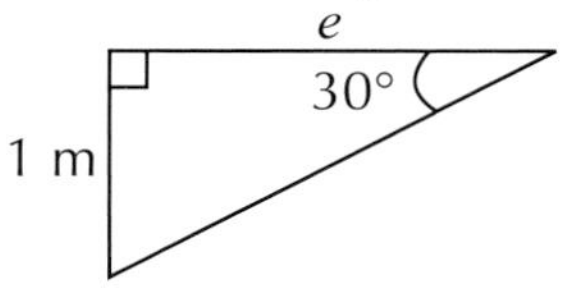

b)

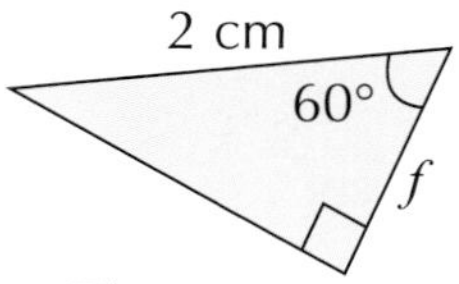

c)

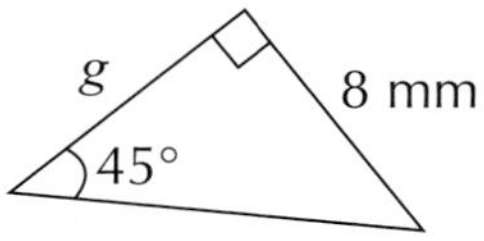

Q3 Show that: $\tan 45° + \sin 60° = \frac{2 + \sqrt{3}}{2}$

Q4 Triangle *DEF* is isosceles. $DF = 7\sqrt{2}$ cm and angle $DEF = 90°$.
What is the exact length of side *DE*?

Review Exercise

Q1 The points P and R have coordinates $P(1, 3)$ and $R(7, 8)$.
Find the length of the line segment PR, correct to 1 d.p.

Q2 A pilot flies 150 km on a bearing of 090°, then 270 km on a bearing of 180°.
Find the direct distance from his start point to his end point. Give your answer to the nearest kilometre.

Q3 Find the lengths of the sides marked with letters in each of the following.
Give your answers to 3 significant figures.

a)
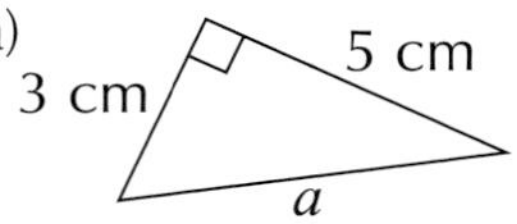

b)
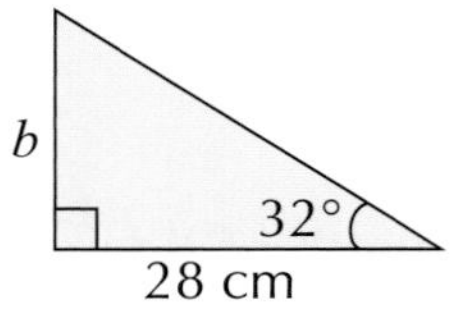

c)
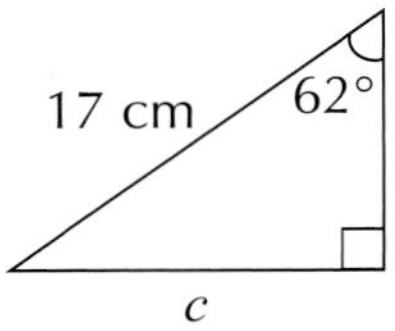

d)
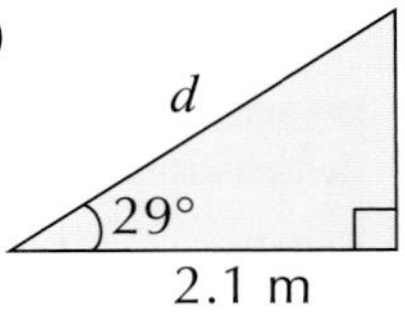

Q4 A ladder is leaning against the side of a tower. The base of the ladder is placed 8 m away from the bottom of the tower, making an angle of 68° with the ground. It reaches a window h m above the ground.
Find the value of h, giving your answer to 1 decimal place.

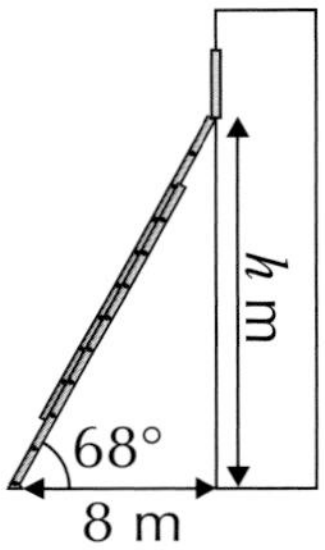

Q5 A kite with taut string of length 5.8 m is flying in the air.
The kite is 4.1 m vertically higher than the other end of the string.
Find the angle between the string and the horizontal, to 1 d.p.

Q6 The shape shown is made up of two right-angled triangles.
Find the value of x. Give your answer to 1 d.p.

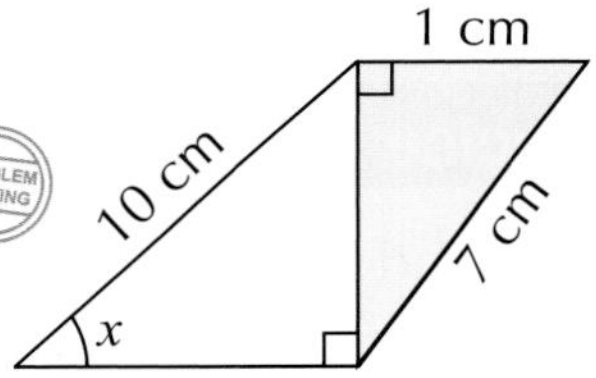

Q7 Town W is 25 km due south of Town X. Town Y is 42 km due east of Town W.
Find the bearing of Town Y from Town X, to 1 d.p.

Q8 Find the exact lengths of the sides marked with letters in each of the following.

a)
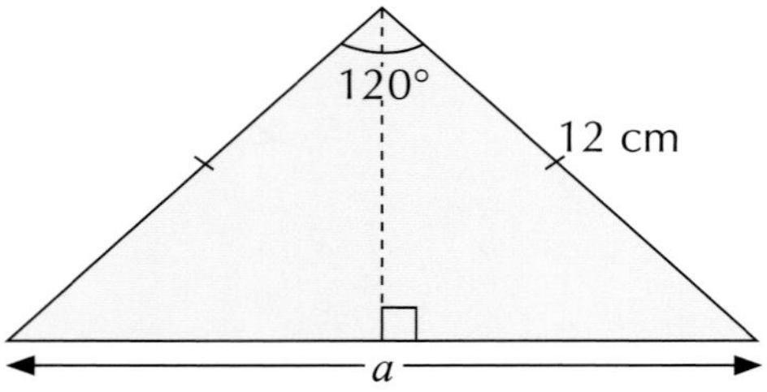

b)
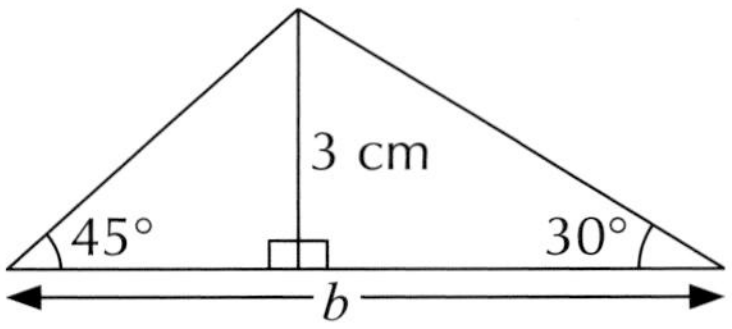

Exam-Style Questions

Q1 A pilot flies 860 km from Lyon to Prague.
She then flies 426 km south to Ljubljana, which is directly east of Lyon. Find the distance from Ljubljana to Lyon to the nearest kilometre.

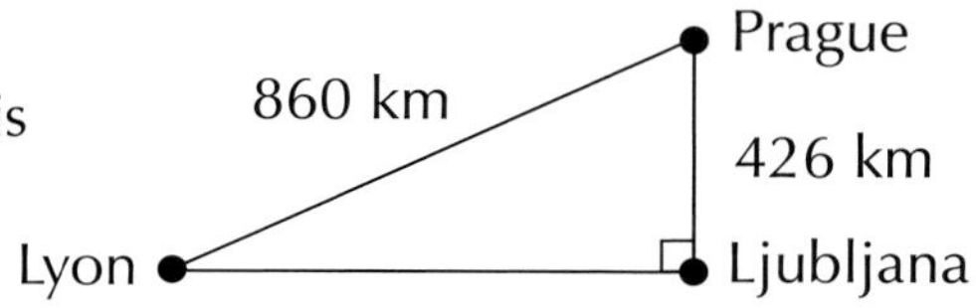

[2 marks]

Q2 Calculate the length of the shortest side of the rectangle below.
Give your answer to 2 decimal places.

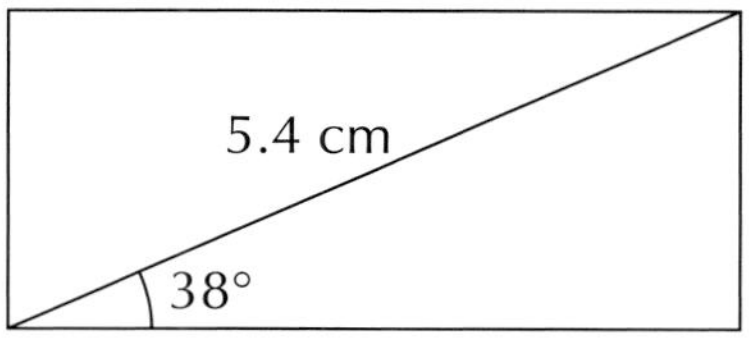

[2 marks]

Q3 *ABC* is a right-angled triangle.
Find the value of sin *ACB*,
giving your answer as a fraction.

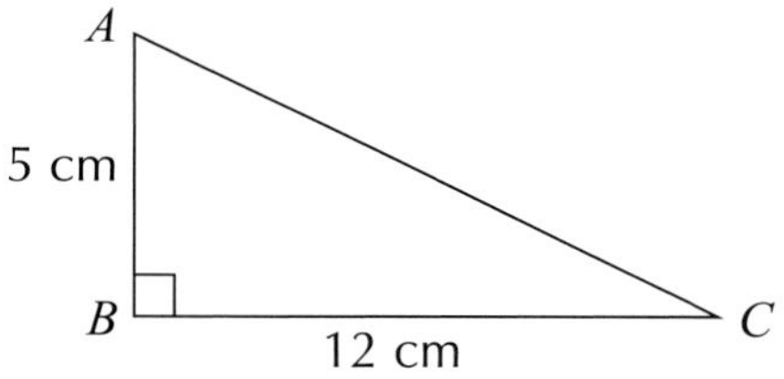

[3 marks]

Q4

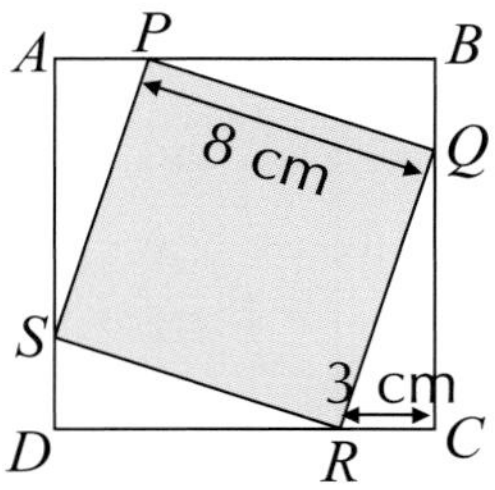

The square *PQRS* is drawn inside another square *ABCD*, as shown in the diagram on the left.
Find the side length of *ABCD* to 1 decimal place.

[3 marks]

Q5 Find the exact length of diagonal *AC* of the rhombus on the right.

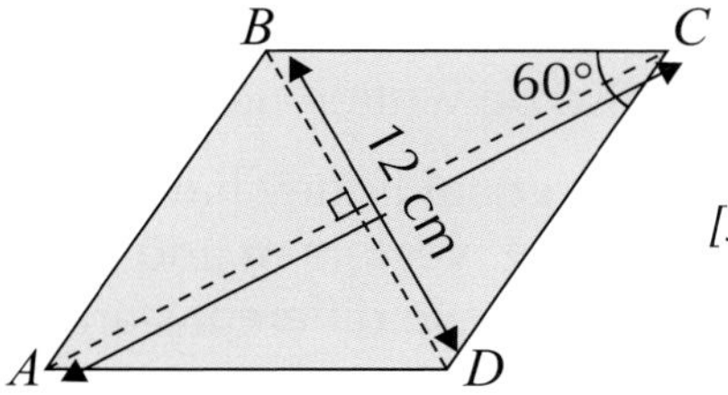

[3 marks]

20.1 Column Vectors

Time for something completely different now — vectors. You can think of them as straight lines from one point in space to another.

Vectors and Scalars

Learning Objectives — Spec Ref G25:
- Understand and use vector notation.
- Be able to multiply vectors by scalars.

A **vector** has **magnitude** (size or length) and **direction**.

There are various ways to represent a vector — it can be:

- written as a **column vector** (positive numbers mean right or up, negative numbers mean left or down).

$\begin{pmatrix} 3 \\ 2 \end{pmatrix}$

x-component (**horizontal**): 3 units right

y-component (**vertical**): 2 units up

- shown on a diagram by an **arrow**.

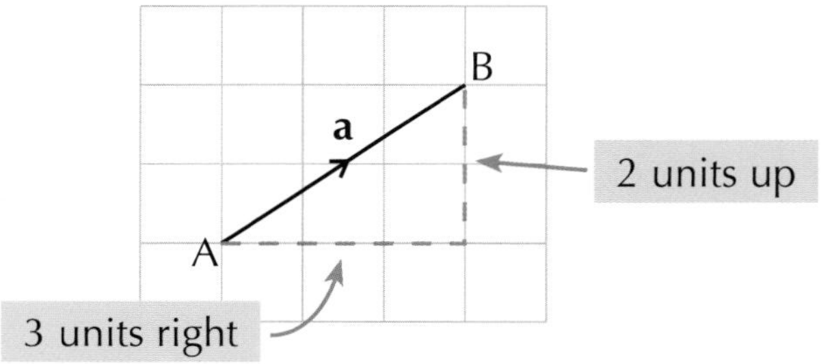

Tip: These two vectors are exactly the same — they're just shown in different ways.

Vectors can be written using their **end points**, so $\overrightarrow{AB}$ means the vector **from A to B** and $\overrightarrow{BA}$ means the vector **from B to A**. $\overrightarrow{AB}$ and $\overrightarrow{BA}$ are **different** vectors — they have the **same magnitude** but **different directions**.

Vectors can also be written using **bold** letters (**a**) or **underlined** letters ($\underline{a}$ or $\underset{\sim}{a}$).

Two vectors are **equal** if they have the **same magnitude** and **direction**. E.g. vectors **m** and **n** are equal vectors, even though they have different start and end points.

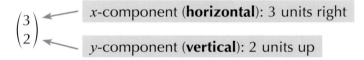

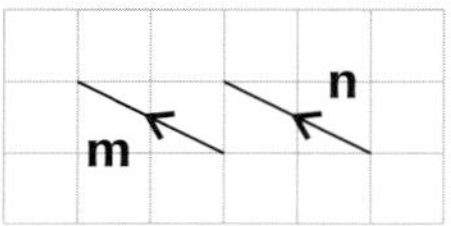

Example 1

a) Draw the vector m = $\begin{pmatrix} -3 \\ 2 \end{pmatrix}$.

1. The positive and negative directions in column vectors are the same as they are for coordinates.
2. So –3 in the x-component means '3 units left', and 2 in the y-component means '2 units up'.
3. Make sure that you label the vector **m**, and add an arrow to show its direction.

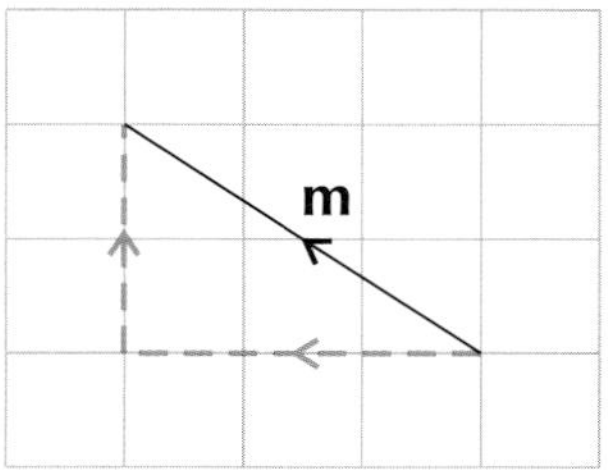

Tip: It doesn't matter where the vector is drawn on the grid. You can choose any starting point — as long as the end point is 3 units to the left and 2 units up.

b) The vector $\overrightarrow{PQ}$ is drawn on the right. Write $\overrightarrow{PQ}$ as a column vector.

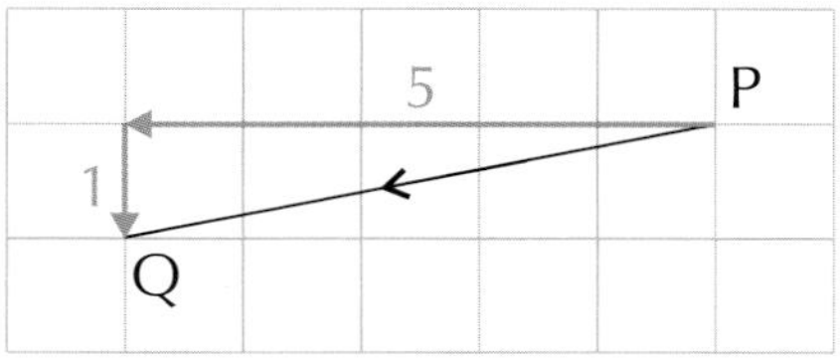

1. Start at P and count the number of units in the horizontal and vertical directions.
2. Write the number of horizontal units on the top and the number of vertical units on the bottom. Remember that if the direction is left or down, you write the negative.

$$\overrightarrow{PQ} = \begin{pmatrix} -5 \\ -1 \end{pmatrix}$$

Exercise 1

Q1 Draw arrows to represent the following vectors.

a) $\begin{pmatrix} 1 \\ 4 \end{pmatrix}$ b) $\begin{pmatrix} 3 \\ 5 \end{pmatrix}$ c) $\begin{pmatrix} -2 \\ 4 \end{pmatrix}$ d) $\begin{pmatrix} 0 \\ 5 \end{pmatrix}$

e) $\begin{pmatrix} -3 \\ -5 \end{pmatrix}$ f) $\begin{pmatrix} 3 \\ 0 \end{pmatrix}$ g) $\begin{pmatrix} -3 \\ -3 \end{pmatrix}$ h) $\begin{pmatrix} 0 \\ -3 \end{pmatrix}$

Q2 Write down the column vectors represented by these arrows:

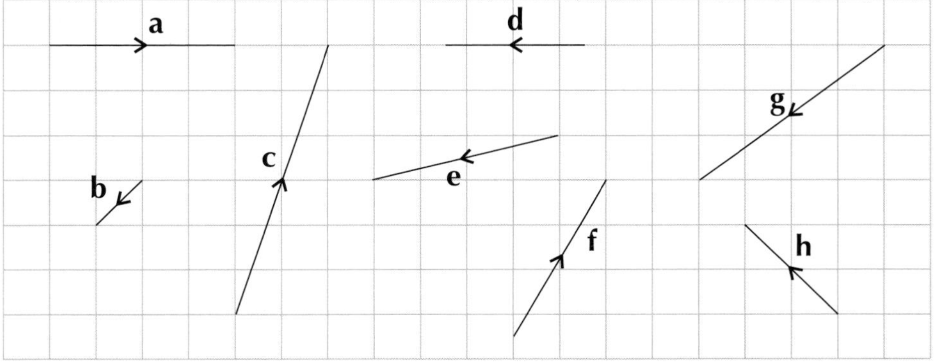

A **scalar** is just a number (like all the ones you're used to working with). Unlike vectors, scalars have magnitude (size) but **no direction**. A vector can be **multiplied** by a scalar to give another vector. To do the multiplication, you multiply each of the vector's **components** by the scalar.

The resulting vector is **parallel** to the original vector.
If the scalar is **negative**, the direction of the vector is **reversed**.

Example 2

If vector p = $\begin{pmatrix} 2 \\ -3 \end{pmatrix}$, write the following as column vectors: a) 2p b) $\frac{1}{2}$p c) –p

Multiply a vector by a scalar by multiplying the x-component and the y-component separately.

a) $2\mathbf{p} = \begin{pmatrix} 2 \times 2 \\ 2 \times -3 \end{pmatrix} = \begin{pmatrix} \mathbf{4} \\ \mathbf{-6} \end{pmatrix}$

b) $\frac{1}{2}\mathbf{p} = \begin{pmatrix} \frac{1}{2} \times 2 \\ \frac{1}{2} \times -3 \end{pmatrix} = \begin{pmatrix} \mathbf{1} \\ \mathbf{-\frac{3}{2}} \end{pmatrix}$

c) $-\mathbf{p} = \begin{pmatrix} -1 \times 2 \\ -1 \times -3 \end{pmatrix} = \begin{pmatrix} \mathbf{-2} \\ \mathbf{3} \end{pmatrix}$

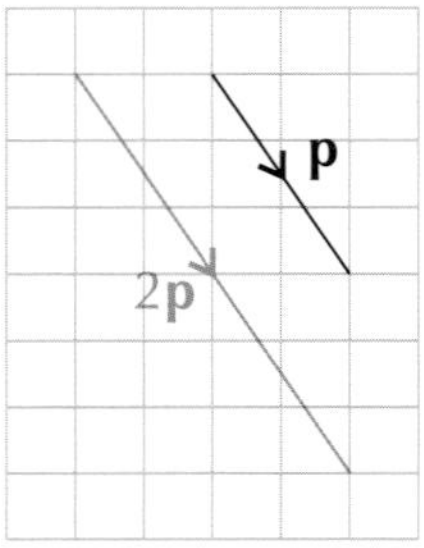

The direction has stayed the same but the magnitude has doubled.

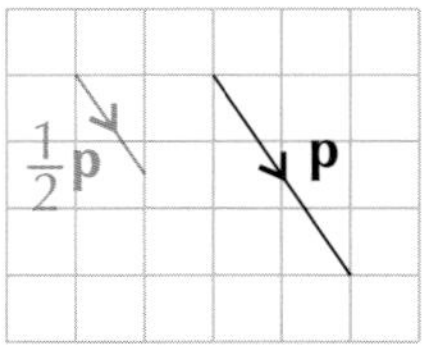

The direction has stayed the same but the magnitude has halved.

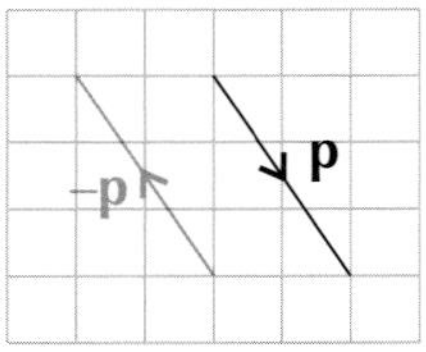

The direction has reversed but the magnitude has stayed the same.

Exercise 2

Q1 If $\mathbf{q} = \begin{pmatrix} -1 \\ 3 \end{pmatrix}$, find and draw the following vectors.

a) 3**q** b) 5**q** c) $\frac{3}{2}$**q** d) –2**q**

Q2

$\mathbf{a} = \begin{pmatrix} 4 \\ -2 \end{pmatrix}$	$\mathbf{b} = \begin{pmatrix} -1 \\ 4 \end{pmatrix}$	$\mathbf{c} = \begin{pmatrix} 3 \\ 12 \end{pmatrix}$	$\mathbf{d} = \begin{pmatrix} 8 \\ -4 \end{pmatrix}$
$\mathbf{e} = \begin{pmatrix} 1 \\ 4 \end{pmatrix}$	$\mathbf{f} = \begin{pmatrix} 0 \\ 3 \end{pmatrix}$	$\mathbf{g} = \begin{pmatrix} 3 \\ -12 \end{pmatrix}$	$\mathbf{h} = \begin{pmatrix} 6 \\ 0 \end{pmatrix}$

From the list of vectors above:

a) Which vector is equal to 2**a**?

b) Which vector is equal to –3**b**?

c) Which vector is the same length as **e**?

d) Which vector is parallel to **c**?

Adding and Subtracting Vectors

Learning Objective — Spec Ref G25:
Be able to add and subtract vectors.

To **add** or **subtract** column vectors, you add or subtract the x-components and y-components separately. The sum of two or more vectors is called the **resultant vector.**

Vectors can also be added by drawing them in a chain, nose-to-tail. The resultant vector goes in a **straight line** from the **start** to the **end** of the chain of vectors.

When you add two vectors, it doesn't matter which comes first, i.e. **a** + **b** = **b** + **a**. Be careful when subtracting though — just like with ordinary numbers **a** – **b** = –**b** + **a**, not **b** – **a**.

Example 3

Look at these three vectors: $\mathbf{p} = \begin{pmatrix} 3 \\ 1 \end{pmatrix}$ $\mathbf{q} = \begin{pmatrix} -2 \\ 0 \end{pmatrix}$ $\mathbf{r} = \begin{pmatrix} 1 \\ -3 \end{pmatrix}$

Work out the following and draw the resultant vectors:

a) p + q **b) r – q** **c) 2q + p – r**

Add and subtract the x-components and the y-components separately.

a) $\begin{pmatrix} 3 \\ 1 \end{pmatrix} + \begin{pmatrix} -2 \\ 0 \end{pmatrix} = \begin{pmatrix} 3 + (-2) \\ 1 + 0 \end{pmatrix} = \begin{pmatrix} \mathbf{1} \\ \mathbf{1} \end{pmatrix}$ →

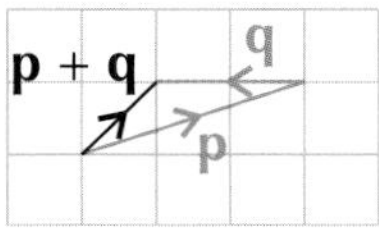

b) $\begin{pmatrix} 1 \\ -3 \end{pmatrix} - \begin{pmatrix} -2 \\ 0 \end{pmatrix} = \begin{pmatrix} 1 - (-2) \\ -3 - 0 \end{pmatrix} = \begin{pmatrix} \mathbf{3} \\ \mathbf{-3} \end{pmatrix}$ →

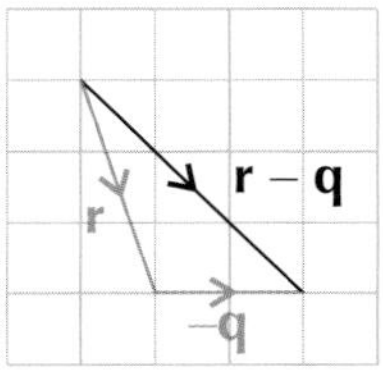

c) $2\begin{pmatrix} -2 \\ 0 \end{pmatrix} + \begin{pmatrix} 3 \\ 1 \end{pmatrix} - \begin{pmatrix} 1 \\ -3 \end{pmatrix} = \begin{pmatrix} 2 \times (-2) + 3 - 1 \\ 2 \times 0 + 1 - (-3) \end{pmatrix} = \begin{pmatrix} \mathbf{-2} \\ \mathbf{4} \end{pmatrix}$ →

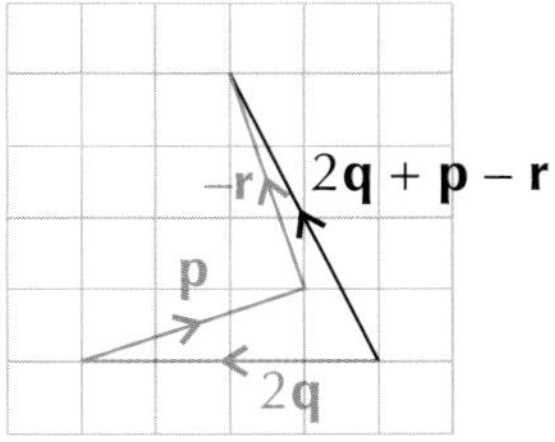

Exercise 3

Q1 Write the answers to the following calculations as column vectors. For each expression, draw arrows to represent the two given vectors and the resultant vector.

a) $\begin{pmatrix} 5 \\ 2 \end{pmatrix} + \begin{pmatrix} 3 \\ 4 \end{pmatrix}$ b) $\begin{pmatrix} 4 \\ -1 \end{pmatrix} + \begin{pmatrix} 1 \\ 6 \end{pmatrix}$ c) $\begin{pmatrix} 2 \\ -1 \end{pmatrix} - \begin{pmatrix} -2 \\ 2 \end{pmatrix}$ d) $\begin{pmatrix} -3 \\ 0 \end{pmatrix} - \begin{pmatrix} 6 \\ 2 \end{pmatrix}$

Q2 If $\mathbf{a} = \begin{pmatrix} 2 \\ 3 \end{pmatrix}$, $\mathbf{b} = \begin{pmatrix} 0 \\ -2 \end{pmatrix}$ and $\mathbf{c} = \begin{pmatrix} -1 \\ 4 \end{pmatrix}$, work out:

a) **b** + **c** b) **c** – **a** c) 2**c** + **a** d) 3**a** + **b**
e) **a** – 2**c** f) **a** + **b** – **c** g) 5**b** + 4**c** h) 4**a** – **b** + 3**c**

Q3 $\mathbf{u} = \begin{pmatrix} 6 \\ -2 \end{pmatrix}$, $\mathbf{v} = \begin{pmatrix} -2 \\ 3 \end{pmatrix}$ and $\mathbf{w} = \begin{pmatrix} 1 \\ 2 \end{pmatrix}$

a) Work out $\mathbf{u} + 2\mathbf{v}$.

b) Draw the vectors $\mathbf{u} + 2\mathbf{v}$ and $\mathbf{w}$.

c) What do you notice about the directions of the vectors $\mathbf{u} + 2\mathbf{v}$ and $\mathbf{w}$?

You can also write a vector as a **sum** of given vectors.

Example 4

Describe the following vectors in terms of a and b.

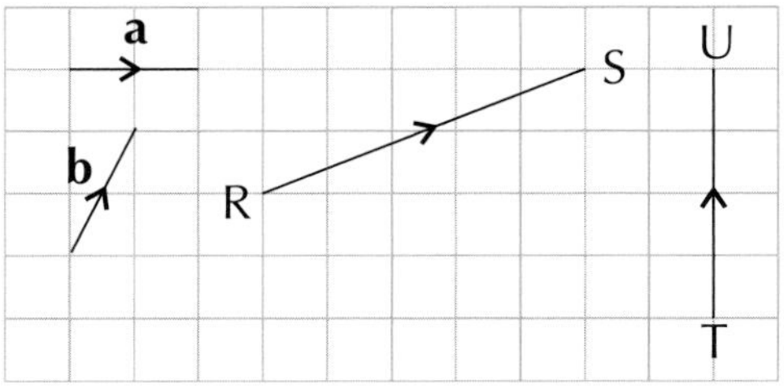

a) $\overrightarrow{\mathbf{RS}}$

1. Find a route from R to S using just vectors **a** and **b**.

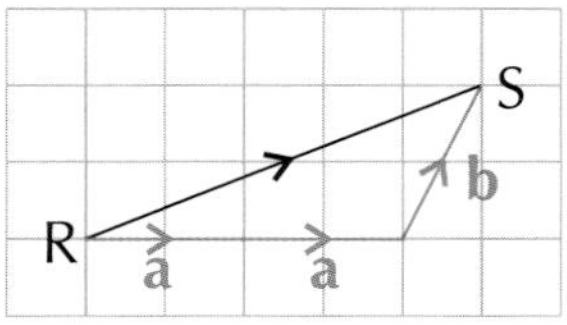

2. So two lots of vector **a** and then vector **b** takes you from R to S.

$\overrightarrow{RS} = \mathbf{2a + b}$

3. You can check the answer by doing the vector addition — from the diagram, $\overrightarrow{RS} = \begin{pmatrix} 5 \\ 2 \end{pmatrix}$, $\mathbf{a} = \begin{pmatrix} 2 \\ 0 \end{pmatrix}$ and $\mathbf{b} = \begin{pmatrix} 1 \\ 2 \end{pmatrix}$.

Check: $2\mathbf{a} + \mathbf{b} = 2\begin{pmatrix} 2 \\ 0 \end{pmatrix} + \begin{pmatrix} 1 \\ 2 \end{pmatrix}$

$= \begin{pmatrix} 4 \\ 0 \end{pmatrix} + \begin{pmatrix} 1 \\ 2 \end{pmatrix} = \begin{pmatrix} 5 \\ 2 \end{pmatrix} = \overrightarrow{RS}$ ✓

b) $\overrightarrow{\mathbf{TU}}$

1. The reverse of **a** and then two lots of **b** takes you from T to U.

2. Remember to check the answer by doing the vector addition.

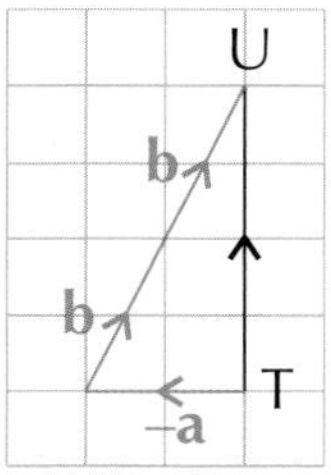

$\overrightarrow{TU} = \mathbf{-a + 2b}$

Check: $-\mathbf{a} + 2\mathbf{b} = -\begin{pmatrix} 2 \\ 0 \end{pmatrix} + 2\begin{pmatrix} 1 \\ 2 \end{pmatrix}$

$= \begin{pmatrix} -2 \\ 0 \end{pmatrix} + \begin{pmatrix} 2 \\ 4 \end{pmatrix} = \begin{pmatrix} 0 \\ 4 \end{pmatrix} = \overrightarrow{TU}$ ✓

Exercise 4

Q1

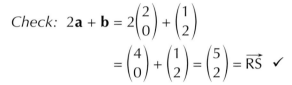

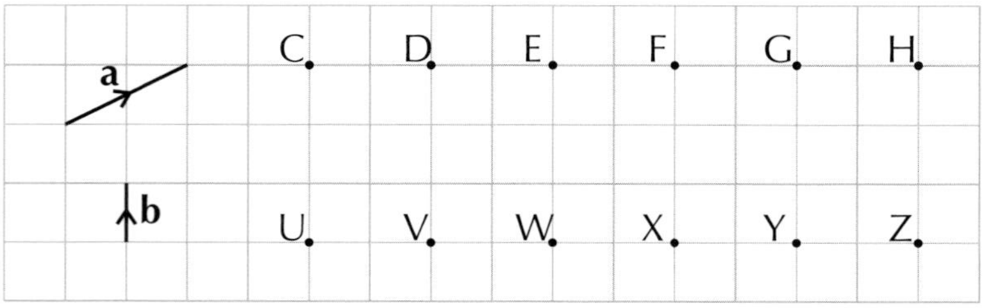

Write the following vectors in terms of **a** and **b**.

a) $\overrightarrow{WH}$ b) $\overrightarrow{ZH}$ c) $\overrightarrow{FX}$ d) $\overrightarrow{UD}$

e) $\overrightarrow{DU}$ f) $\overrightarrow{FZ}$ g) $\overrightarrow{CZ}$ h) $\overrightarrow{DW}$

20.2 Vector Geometry

Vectors can be used in geometry to describe the lines that make up shapes — you can then use the vectors along with properties of the shape to solve problems.

Learning Objective — Spec Ref G25:
Use vectors to solve geometry problems.

Prior Knowledge Check:
Understand and use vector notation (p.314) and shape properties (Section 15).

The first step to any vector geometry problem is working out what information you already **know** and what you need to **find**. Always work with a **diagram** — if you're not given one, **draw your own**. Then it'll be easier to see which vectors you need to add, subtract or multiply to get to the answer.

Example 1

In triangle OAB, $\overrightarrow{OA}$ = a and $\overrightarrow{OB}$ = b.
Write down, in terms of a and b: a) $\overrightarrow{AO}$ b) $\overrightarrow{AB}$ c) $\overrightarrow{BA}$

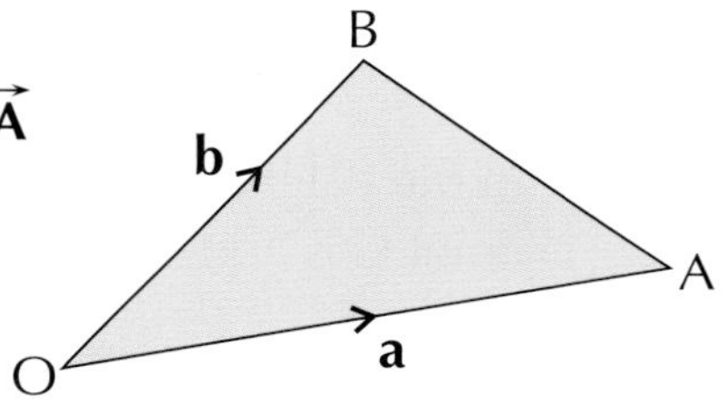

a) To get from A to O, you go backwards along the vector **a**, so you need the reverse vector of **a**. $\overrightarrow{AO} = -\mathbf{a}$

b) To get from A to B, you go backwards along **a** and then along **b**. $\overrightarrow{AB} = -\mathbf{a} + \mathbf{b}$

c) To get from B to A, you go backwards along **b** and then along **a**. $\overrightarrow{BA} = -\mathbf{b} + \mathbf{a}$

Example 2

WXYZ is a parallelogram. $\overrightarrow{WX}$ = u and $\overrightarrow{WZ}$ = v.
M is the midpoint of WY.
Write down, in terms of u and v: a) $\overrightarrow{WY}$ b) $\overrightarrow{WM}$

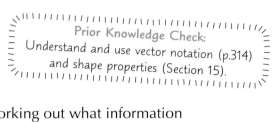

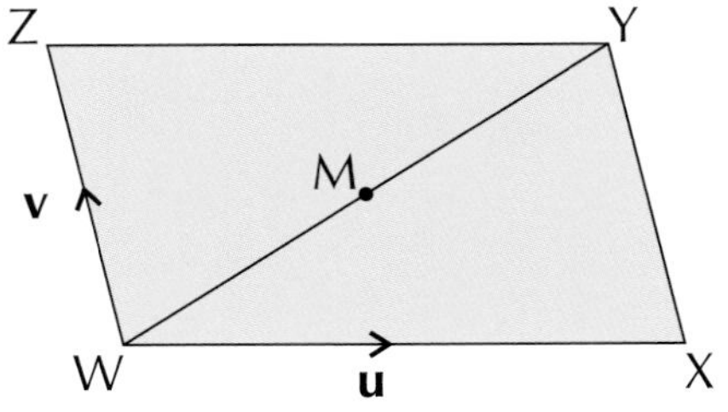

a) 1. To get from W to Y, go from W to X and then from X to Y. $\overrightarrow{WY} = \overrightarrow{WX} + \overrightarrow{XY}$

2. To get from W to X, you go along **u**. $\overrightarrow{WX} = \mathbf{u}$

3. The line XY isn't labelled with a vector — but the shape is a parallelogram, so XY and WZ are parallel and the same length. So the vectors between X and Y and between W and Z are the same. $\overrightarrow{XY} = \overrightarrow{WZ} = \mathbf{v}$

So $\overrightarrow{WY} = \overrightarrow{WX} + \overrightarrow{XY}$
$= \mathbf{u} + \mathbf{v}$

Tip: You could also go from W to Z and then from Z to Y. The vector $\overrightarrow{ZY} = \overrightarrow{WX}$, so you end up with the answer of **u** + **v**.

b) 1. W to M is the same direction as W to Y, but it's half the distance.

2. Halve the vector $\overrightarrow{WY}$ by multiplying the answer from part a) by $\frac{1}{2}$.

$$\overrightarrow{WM} = \frac{1}{2}\overrightarrow{WY} = \frac{1}{2}(\mathbf{u} + \mathbf{v}) = \frac{1}{2}\mathbf{u} + \frac{1}{2}\mathbf{v}$$

Exercise 1

Q1 ABCD is a trapezium. $\overrightarrow{AB} = 4\mathbf{p}$, $\overrightarrow{AD} = \mathbf{q}$ and $\overrightarrow{DC} = \mathbf{p}$. Write the following vectors in terms of **p** and **q**:

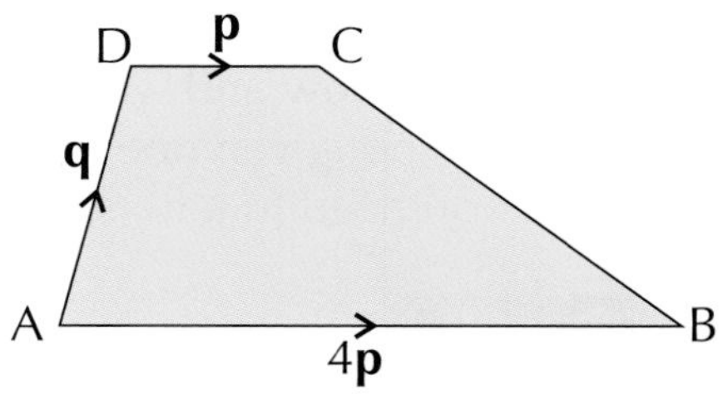

a) $\overrightarrow{BA}$ b) $\overrightarrow{CD}$

c) $\overrightarrow{AC}$ d) $\overrightarrow{CA}$

e) $\overrightarrow{CB}$ f) $\overrightarrow{BD}$

Q2 In the diagram on the right, $\overrightarrow{OA} = \mathbf{a}$ and $\overrightarrow{OB} = \mathbf{b}$. Point C is added such that $\overrightarrow{OC} = \mathbf{a} + \mathbf{b}$.

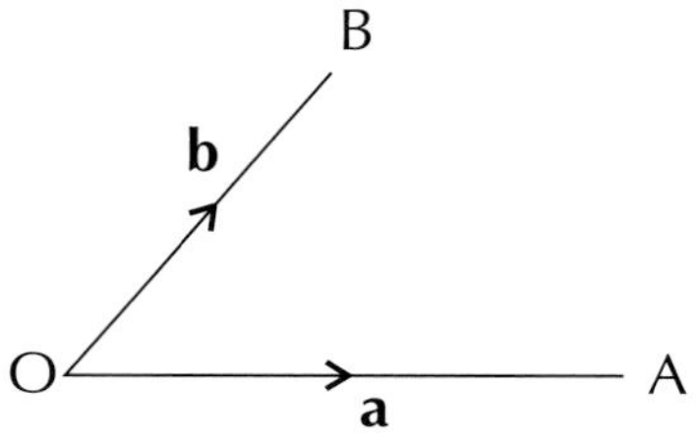

a) Write down, in terms of **a** and **b**:
(i) $\overrightarrow{CO}$ (ii) $\overrightarrow{AB}$

b) What type of shape is OACB?

Q3 In the triangle OBD, $\overrightarrow{OA} = \mathbf{a}$ is $\frac{1}{4}$ of the length of $\overrightarrow{OB}$ and $\overrightarrow{OC} = \mathbf{c}$ is $\frac{1}{3}$ of the length of $\overrightarrow{OD}$.

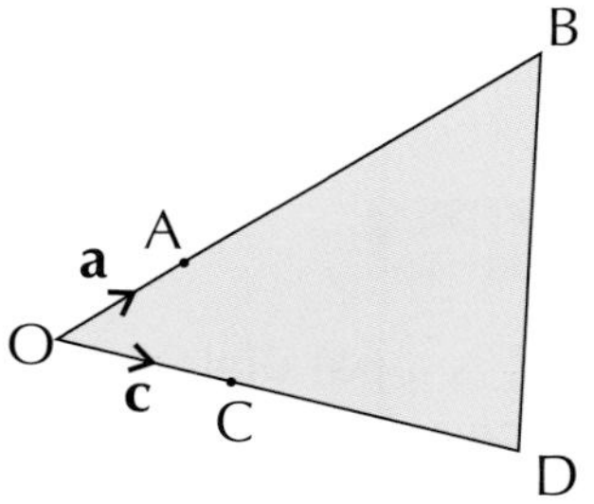

Write down, in terms of **a** and **c**:

a) $\overrightarrow{OB}$ b) $\overrightarrow{OD}$ c) $\overrightarrow{AB}$

d) $\overrightarrow{CD}$ e) $\overrightarrow{DB}$ f) $\overrightarrow{BD}$

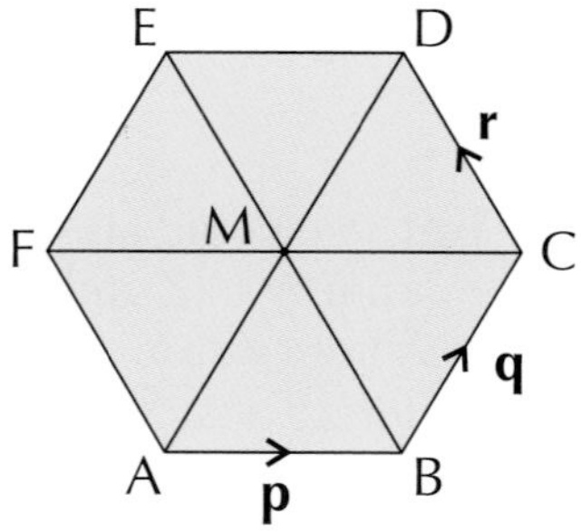

Q4 ABCDEF is a regular hexagon.
M is the centre of the hexagon.
$\overrightarrow{AB} = \mathbf{p}$ and $\overrightarrow{BC} = \mathbf{q}$ and $\overrightarrow{CD} = \mathbf{r}$.

The following vectors can all be written, in terms of **p**, **q** and **r**, in multiple different ways. For each vector, write down two ways.

a) $\overrightarrow{ED}$ b) $\overrightarrow{FE}$ c) $\overrightarrow{AF}$

d) $\overrightarrow{EF}$ e) $\overrightarrow{MF}$ f) $\overrightarrow{EM}$

Review Exercise

Q1 a) Match the following column vectors to the correct vectors in the diagram below.

$\begin{pmatrix} 1 \\ 4 \end{pmatrix}$ $\begin{pmatrix} -2 \\ -4 \end{pmatrix}$

$\begin{pmatrix} -4 \\ -1 \end{pmatrix}$ $\begin{pmatrix} 4 \\ -2 \end{pmatrix}$

$\begin{pmatrix} 0 \\ -5 \end{pmatrix}$ $\begin{pmatrix} 3 \\ 4 \end{pmatrix}$

$\begin{pmatrix} 4 \\ -1 \end{pmatrix}$ $\begin{pmatrix} 1 \\ -4 \end{pmatrix}$

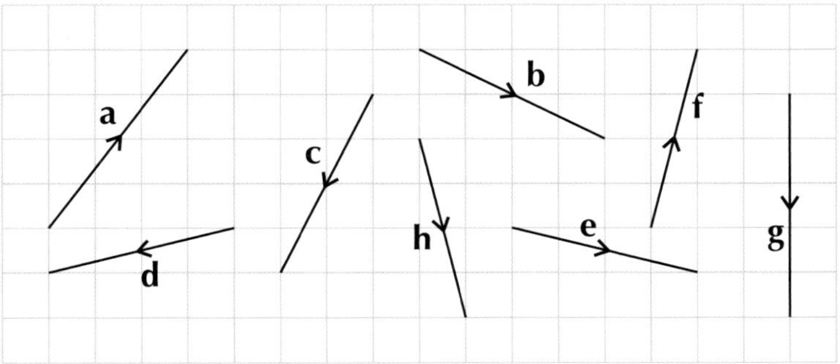

b) Which vector has the same length as **c**?

c) Four of the vectors have the same length. Which vectors are these?

Q2 If $\mathbf{p} = \begin{pmatrix} 4 \\ -3 \end{pmatrix}$, $\mathbf{q} = \begin{pmatrix} 0 \\ 2 \end{pmatrix}$ and $\mathbf{r} = \begin{pmatrix} -1 \\ 5 \end{pmatrix}$, find:

a) $3\mathbf{p}$ b) $2\mathbf{q} + \mathbf{r}$ c) $\mathbf{r} - 2\mathbf{p}$ d) $\mathbf{p} + 5\mathbf{r} - 3\mathbf{q}$

Q3 $\mathbf{a} = \begin{pmatrix} 6 \\ -4 \end{pmatrix}$ $\mathbf{b} = \begin{pmatrix} -2 \\ 3 \end{pmatrix}$ $\mathbf{c} = \begin{pmatrix} 1 \\ 2 \end{pmatrix}$ $\mathbf{d} = \begin{pmatrix} 4 \\ -2 \end{pmatrix}$ $\mathbf{e} = \begin{pmatrix} 3 \\ 6 \end{pmatrix}$ $\mathbf{f} = \begin{pmatrix} 4 \\ -6 \end{pmatrix}$

a) Draw the vector **a** + **b** and write down the resultant vector as a column vector.

b) Draw the vector **e** – **d** and write down the resultant vector as a column vector.

c) Which two of the above vectors can be added to give the resultant vector $\begin{pmatrix} 10 \\ -10 \end{pmatrix}$?

d) Which two of the above vectors can be added to give the resultant vector $\begin{pmatrix} 1 \\ 9 \end{pmatrix}$?

e) Which vector above is parallel to **c**?

f) Which vector above is parallel to **b**?

Q4 PQRS is a trapezium. $\overrightarrow{PQ} = \mathbf{m}$ and $\overrightarrow{SP} = \mathbf{n}$.
The side SR is five times as long as the side PQ.

a) Write down $\overrightarrow{SR}$ in terms of **m** and **n**.

b) Find $\overrightarrow{QR}$ in terms of **m** and **n**.

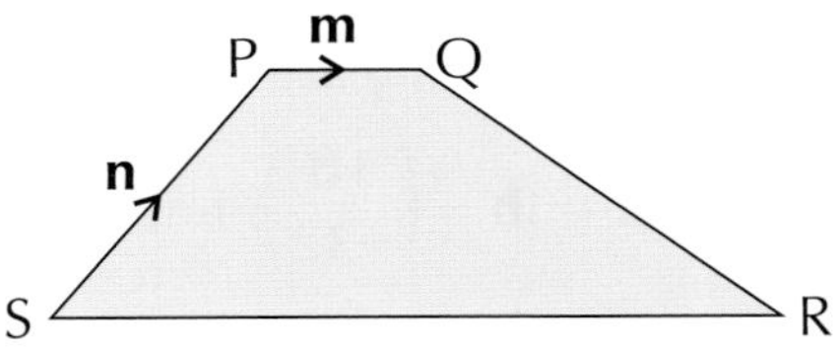

Exam-Style Questions

Q1 The triangle ABC is shown below. $\overrightarrow{AB} = \mathbf{u}$ and $\overrightarrow{CB} = \mathbf{v}$.
Write the vector $\overrightarrow{AC}$ in terms of **u** and **v**.

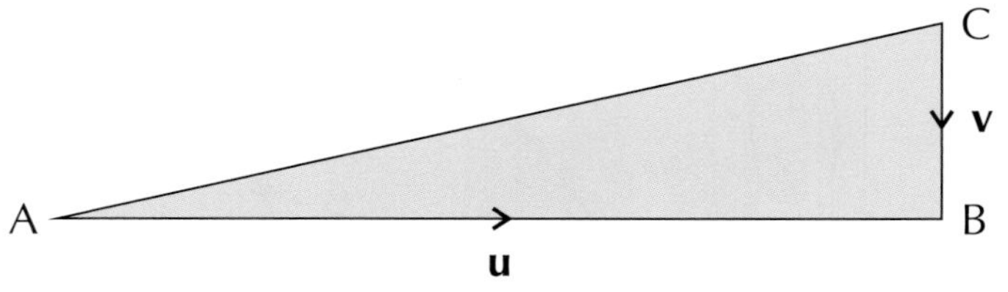

[2 marks]

Q2 $\mathbf{m} = \begin{pmatrix} 5 \\ 0 \end{pmatrix}$ $\mathbf{n} = \begin{pmatrix} -3 \\ -1 \end{pmatrix}$

Write $\mathbf{m} + 4\mathbf{n}$ as a column vector.

[2 marks]

Q3 The vectors **a**, **b**, **c** and **d** are drawn on the square grid below.
Write the vectors **c** and **d** each in terms of **a** and **b**.

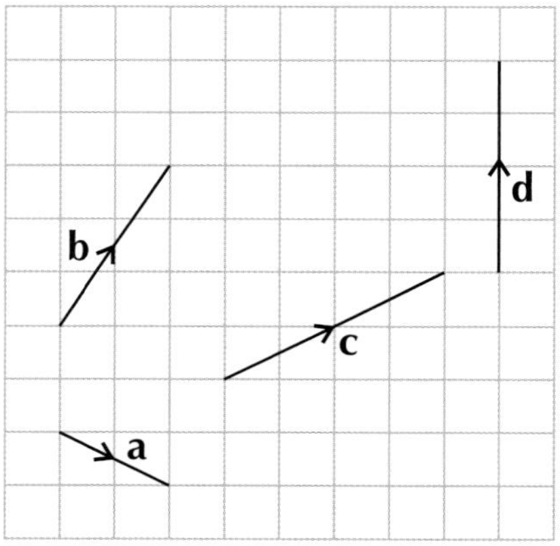

[2 marks]

Q4 **a** and **b** are column vectors such that $\mathbf{a} = \begin{pmatrix} 5p \\ 4q \end{pmatrix}$ and $\mathbf{b} = \begin{pmatrix} 4 \\ 0 \end{pmatrix}$,
where p and q are integers.

If $2\mathbf{a} - 3\mathbf{b} = \begin{pmatrix} 18 \\ -32 \end{pmatrix}$, find the values of p and q.

[4 marks]

Section 21 Constructions

21.1 Circles

Circles come up a lot in Maths... they'll crop up with loci later in this section, and with area and perimeter in Section 22 First, you need to know what the different parts of a circle are called.

Radius and Diameter

Learning Objective — Spec Ref G9:
Be familiar with the diameter and radius of a circle.

Radius (r): a line from the centre of a circle to the edge.
The circle's centre is the same distance from all points on the edge.

Diameter (d): a line from one side of a circle to the other through the centre.

The diameter is **twice** the radius: $d = 2r$

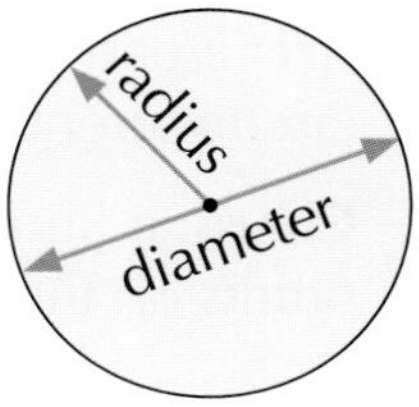

Example 1

a) A circle has radius 4 cm. What is its diameter?

1. Use the formula $d = 2r$. — $d = 2r$
2. Remember to use the correct units. — $d = 2 \times 4 =$ **8 cm**

b) A circle has diameter 7 m. What is its radius?

1. Rearrange the formula $d = 2r$. — $r = \frac{d}{2}$
2. Remember to use the correct units. — So $r = 7 \div 2 =$ **3.5 m**

Exercise 1

Q1 Find the diameter of a circle with radius:
a) 4 cm b) 6 cm c) 30 mm d) 4.2 mm

Q2 Find the radius of a circle with diameter:
a) 2 cm b) 12 cm c) 13 m d) 0.02 cm

Q3 Use compasses to draw a circle with:
a) a radius of 4 cm b) a diameter of 80 mm c) a diameter of 6 cm

More Parts of Circles

Learning Objective — Spec Ref G9:
Be able to identify the different parts of a circle.

There are a few more parts of a circle that you need to know.

Circumference: the distance around the outside of a circle.

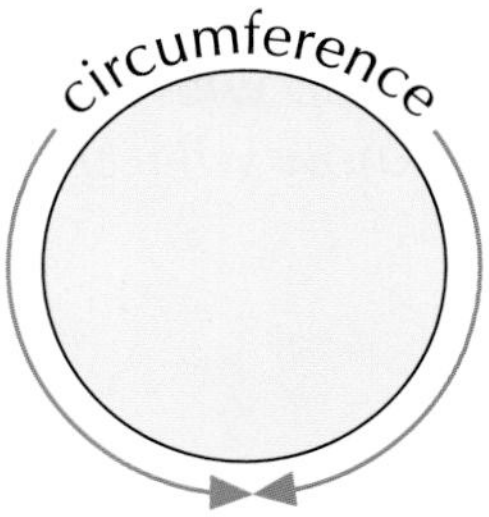

Tip: Make sure you are completely happy with these terms — you will need to use them frequently.

Tangent: a straight line that just touches the circle.

Arc: a part of the circumference.

Chord: a line between two points on the edge of the circle.

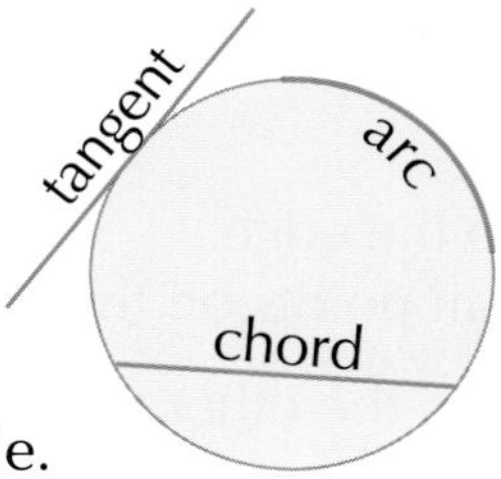

Sector: an area of a circle like a "slice of pie".

Segment: an area of a circle between an arc and a chord.

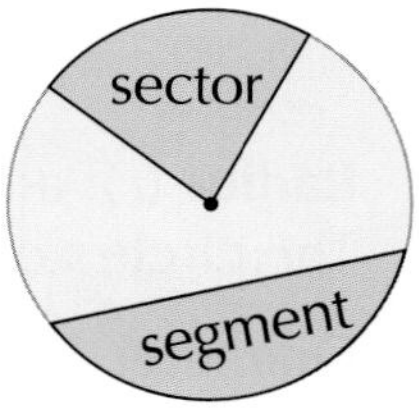

Exercise 2

Q1 Name the feature highlighted in each diagram below.

a)

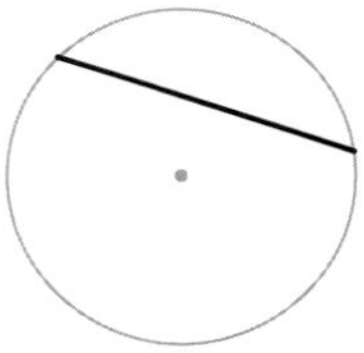

b)

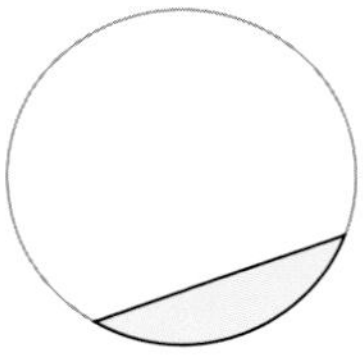

c)

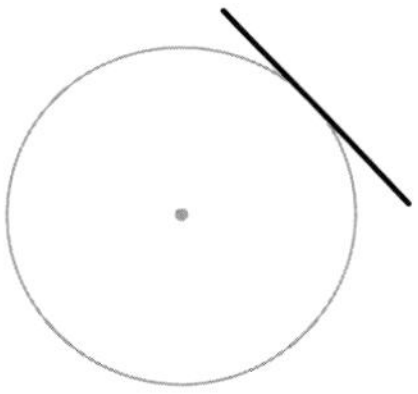

d)

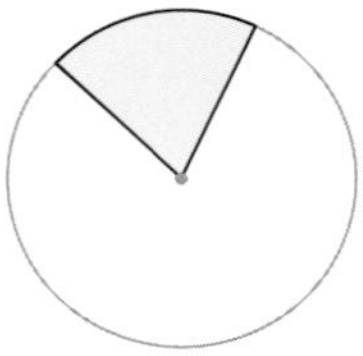

e)

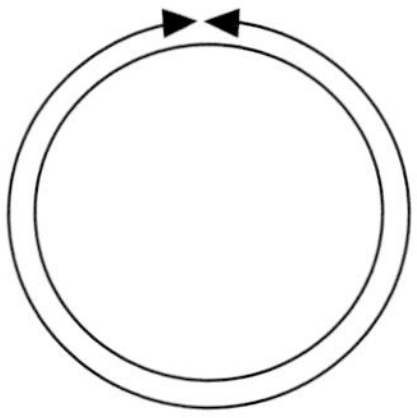

f)

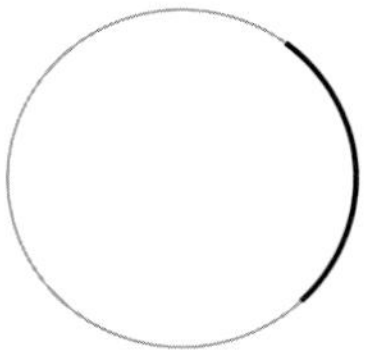

Q2 Draw a circle. Then draw and label the following features.

a) a chord b) an arc c) a tangent d) a diameter

Q3 Draw a circle. Then draw and label the following features.

a) a sector b) a segment c) a radius

21.2 Lines, Angles and Triangles

Over the next few pages you'll learn how to "construct" triangles — i.e. draw them accurately using just a pencil, a ruler, a pair of compasses and a protractor. Before you start constructing triangles, you need to make sure you can use a ruler, protractor and pair of compasses correctly.

Learning Objectives — Spec Ref G1/G15:

- Be able to measures lines and angles accurately.
- Be able to draw triangles accurately using a ruler and a protractor.

For this topic, you'll need to know how to use:

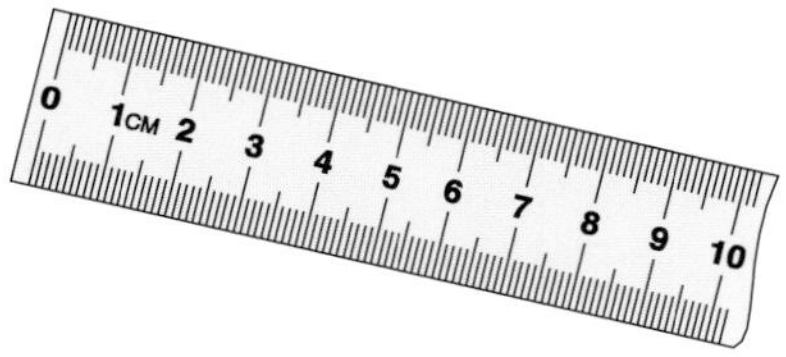

A **ruler** to measure **lengths**.

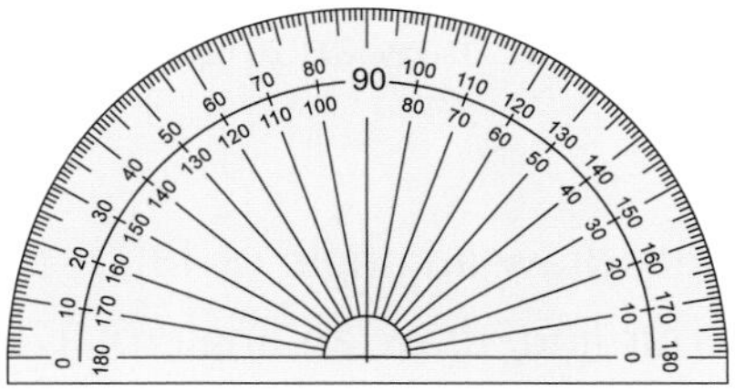

A **protractor** to measure **angles**.

Tip: A protractor has two scales. Always count up in steps of 10° from your start line (at 0°) to make sure you're using the correct scale.

Example 1

Measure the size of angle *a*.

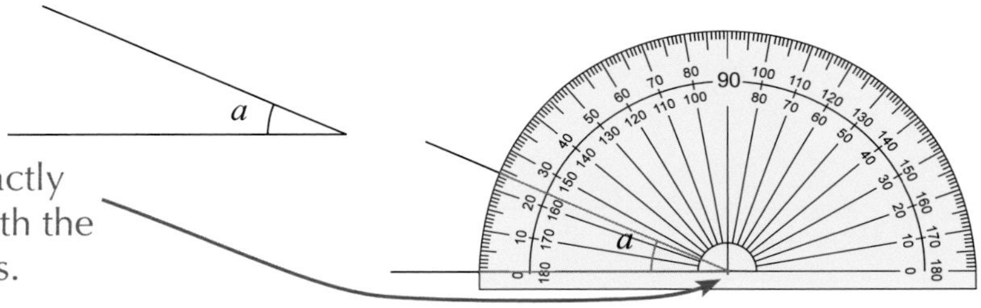

1. Put the protractor's cross exactly where the two lines meet with the 0° line along one of the lines.
2. Count up in steps of 10°, then use the smaller lines on the scale to find the final answer.
3. The answer is 24°, not 156°, because you counted up to 24° from 0°.

$a = 24°$

It can be **difficult** to describe which angle you are talking about if there is more than one angle at that point.

In a diagram whose vertices have been labelled with letters, each angle can be described using three letters.

E.g. In the diagram, angle $ABD = 72°$.

- The **middle** letter is where the angle is.
- The other two letters tell you which two lines enclose the angle.

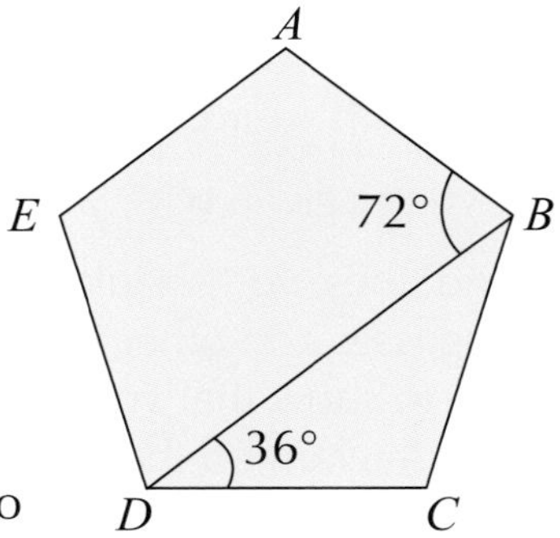

Tip: Always include the word '**angle**' before the three letters or you'll be talking about the **triangle** made from those three points, not the angle. E.g. angle $BDC = 36°$, but BDC is a triangle.

Exercise 1

Q1 In each diagram below, measure the size of the angle and the length of the lines.

a)

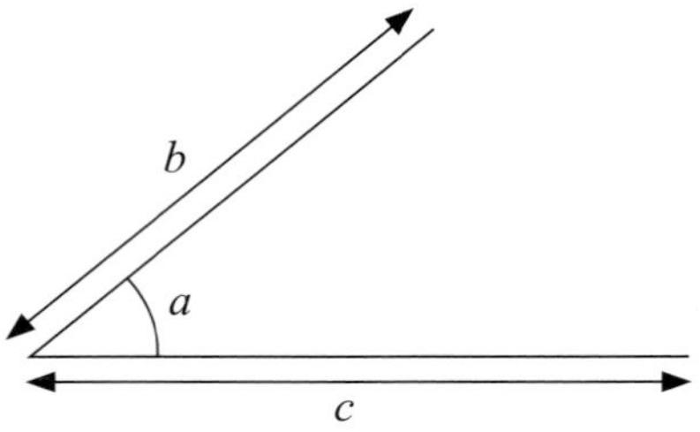

b)

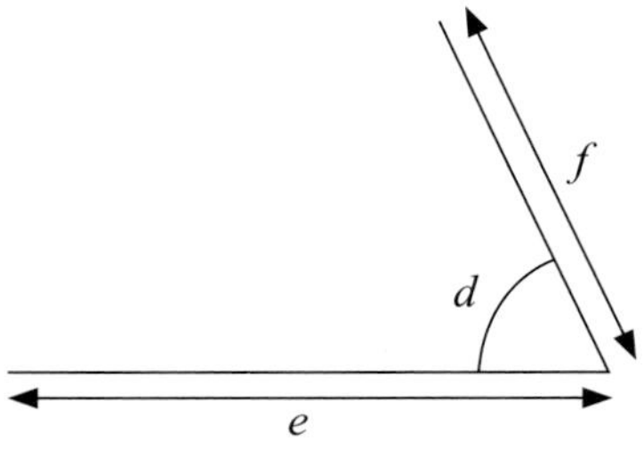

Q2 Draw a line 6 cm long.

a) On one end of your line, draw an angle of 70°.

b) On the other end of your line, draw an angle of 30°.

Q3 Draw a line 8.5 cm long.

a) On one end of your line, draw an angle of 130°.

b) On the other end of your line, draw an angle of 170°.

Q4 Measure all the angles in the triangles below.
Make sure your answers add up to 180°.

a)

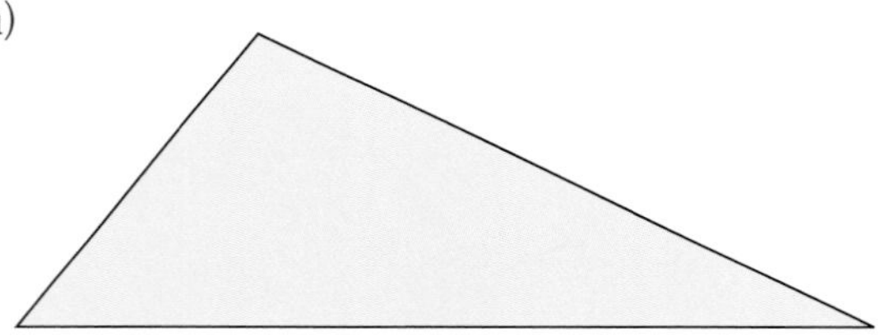

b)

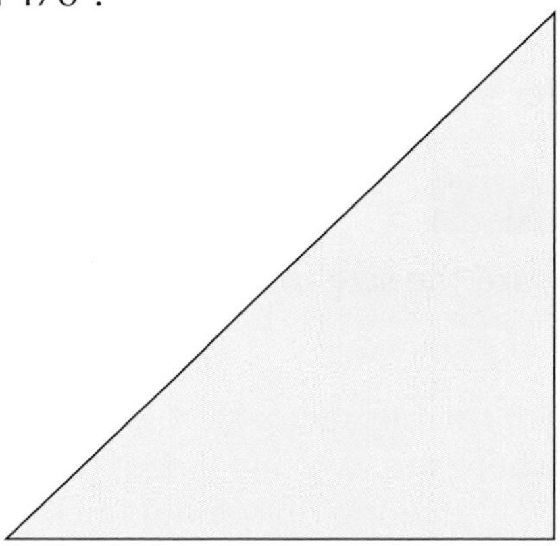

Constructing a triangle

To construct a triangle, you need **three** pieces of information about it. These could be **lengths** of the sides or the **angles** between them.

- If you're given a **side length**, use your **ruler** to measure the length and draw an arc with your **compasses** if necessary.
- Measure any **angles** that you're given with your **protractor**.

When constructing shapes and lines, you should **always** leave the **arcs from your compasses** and other **construction lines** on your finished drawing to show that you've used the right method — like in the diagram on the right.

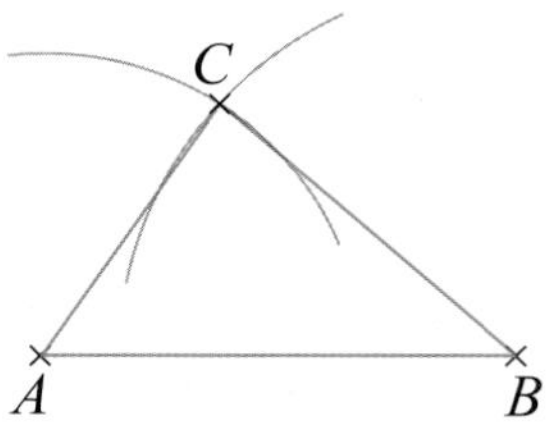

Example 2

Draw triangle *ABC*, where side *AB* is 4 cm, angle *BAC* is 55°, and angle *ABC* is 35°.

1. Draw and label the side you know the length of.

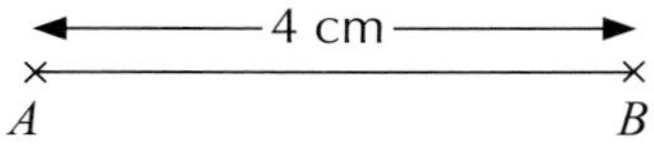

2. Measure an angle of 55° at *A* with a protractor. Mark the angle with a dot.

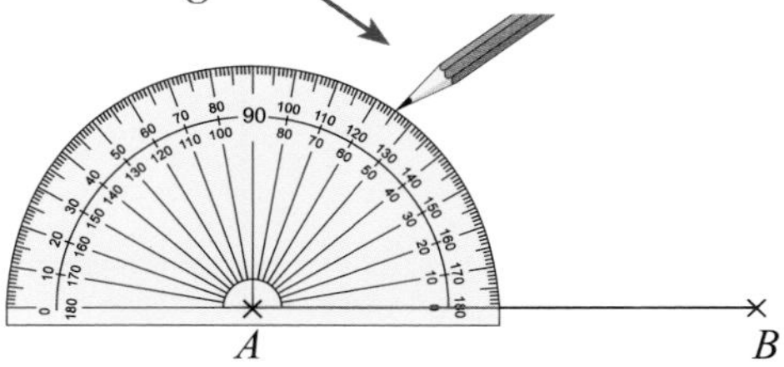

3. Draw a faint line from *A* through the dot.

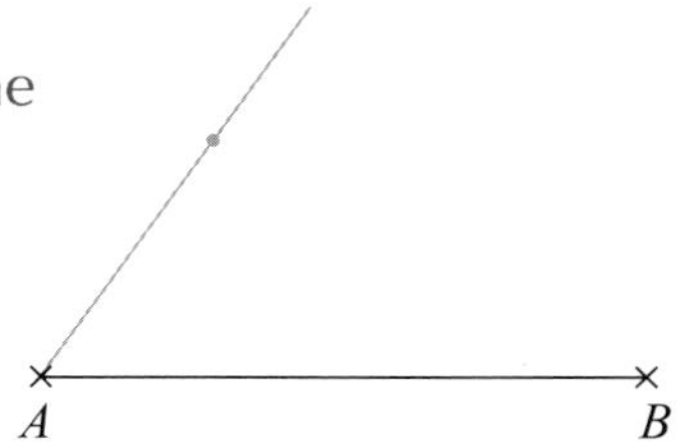

4. Draw the second angle in the same way, and complete the triangle.

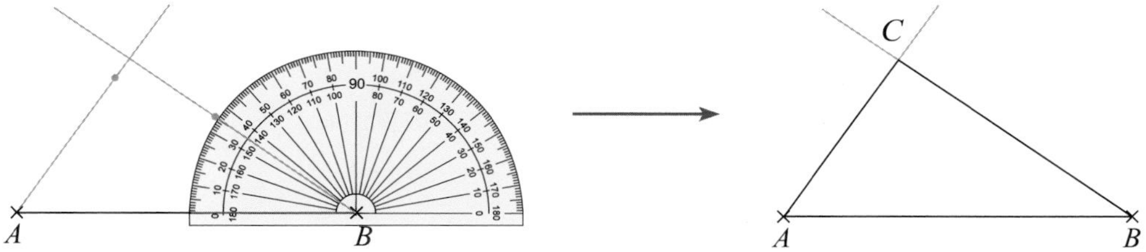

Exercise 2

Q1 Draw the following triangles accurately, then measure the lengths marked *l*.

a) 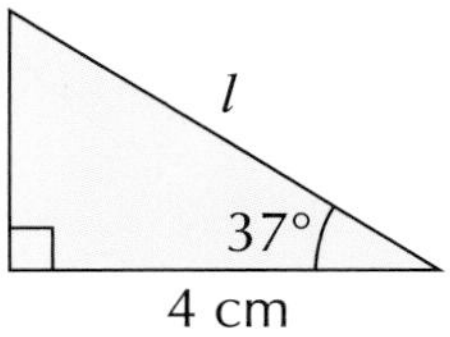

b) 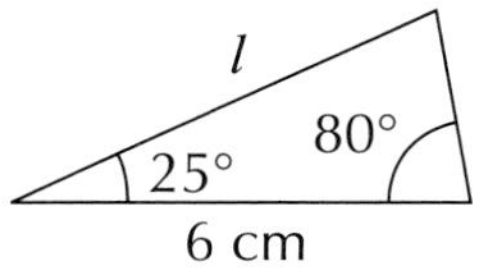

c) 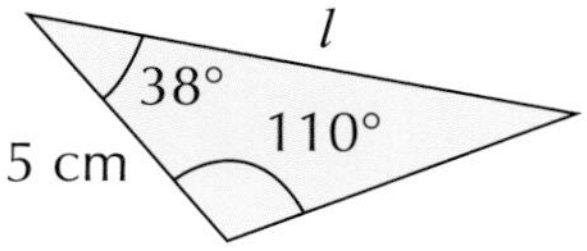

Q2 a) Draw each of the triangles *ABC* described below.

(i) *AB* = 4 cm, angle *BAC* = 55°, angle *CBA* = 35°.

(ii) *AB* = 8 cm, angle *BAC* = 22°, angle *CBA* = 107°.

(iii) *AB* = 6.5 cm, angle *BAC* = 65°, angle *CBA* = 30°.

b) Measure the length of side *BC* in each of your triangles in part a).

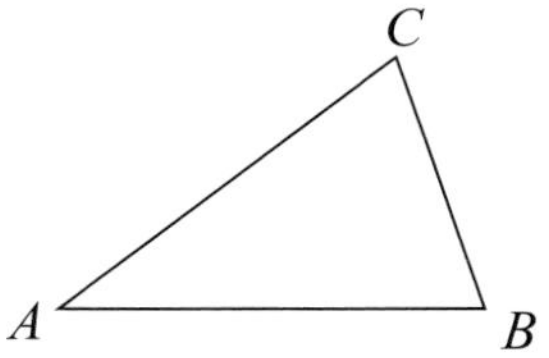

Q3 Alec is standing 5 km directly west of Brenda. At noon, Alec starts to walk northeast, while Brenda starts to walk at the same speed northwest.

By carefully drawing their paths on a scale drawing using 1 cm to represent 1 km, find how far from their starting points they eventually meet.

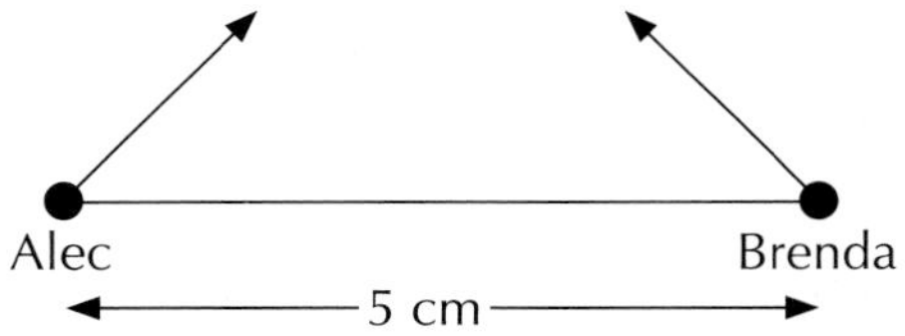

Example 3

Draw triangle *ABC*, where *AB* is 4 cm, *BC* is 4 cm, and angle *ABC* is 25°.

1. Draw and label a side you know the length of.

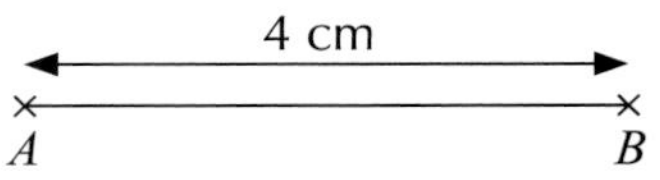

2. Measure an angle of 25° at *B* with a protractor. Mark the angle with a dot.

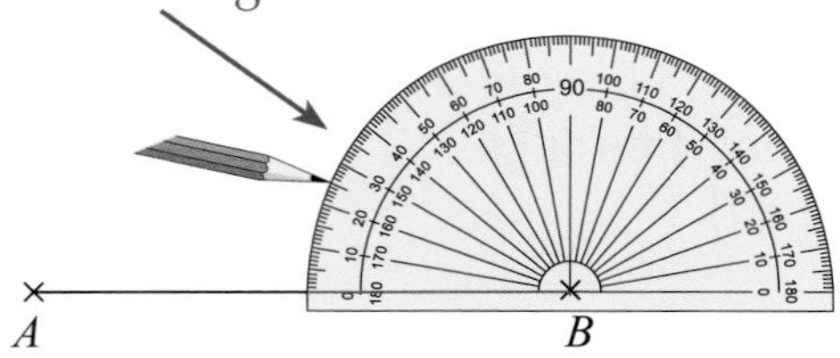

3. Draw a faint line from *B* through the dot.

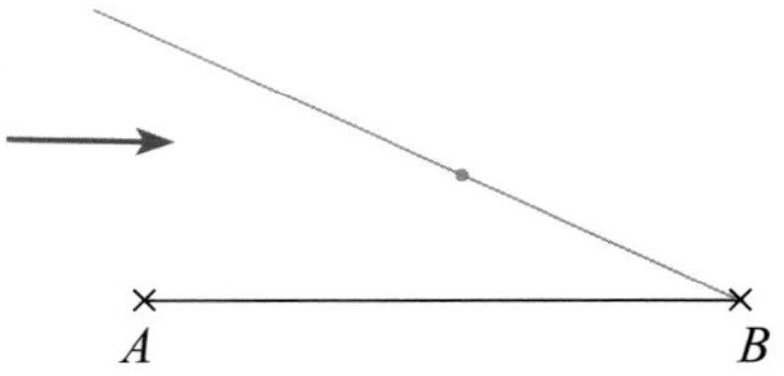

4. Point *C* is 4 cm along this line.
5. Complete the triangle by drawing the line *AC*.

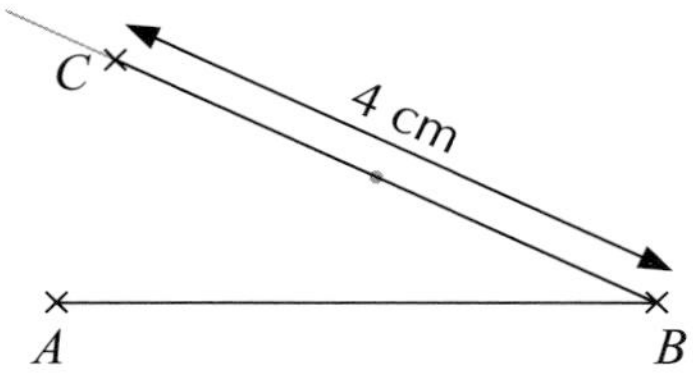

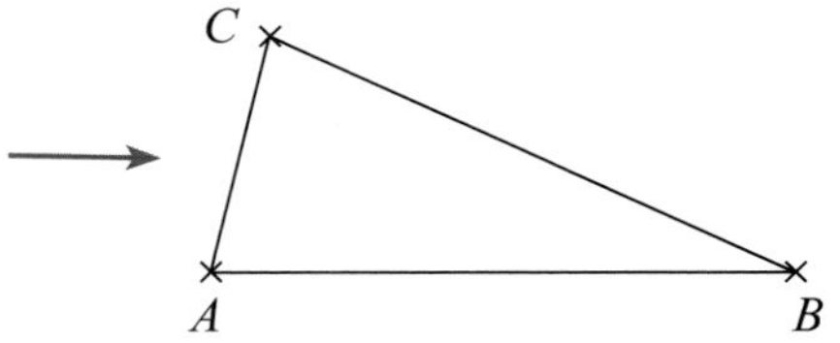

Exercise 3

Q1 Draw the following triangles accurately, then measure the lengths marked *l*.

a)

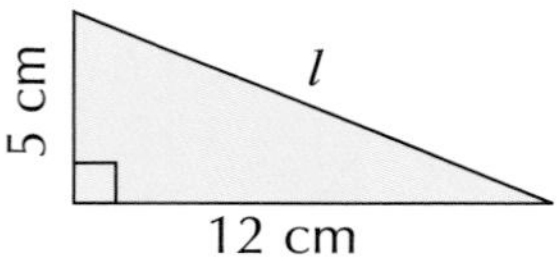

b)

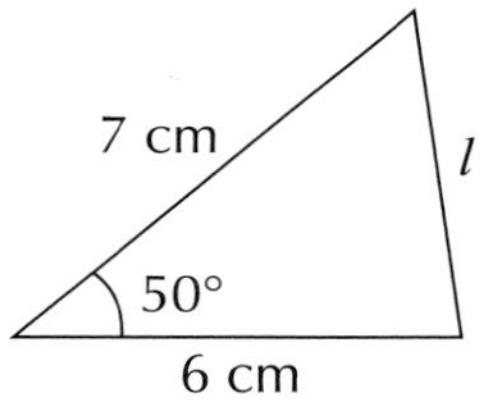

c)

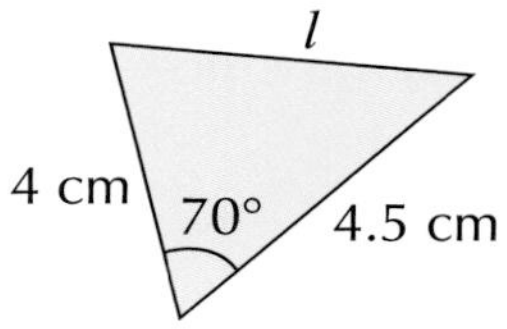

Q2 a) Draw each of the triangles *ABC* described below.

(i) *AB* = 6 cm, *BC* = 7 cm, angle *ABC* = 40°.

(ii) *AB* = 4 cm, *BC* = 3 cm, angle *ABC* = 110°.

(iii) *AB* = 65 mm, *BC* = 53 mm, angle *ABC* = 20°.

(iv) *AB* = 45 mm, *BC* = 45 mm, angle *ABC* = 45°.

b) Measure the length of side *AC* in each of your triangles in part a).

Q3 Draw an isosceles triangle with an angle of 50° between its two 5 cm long sides.

Q4 a) Draw a rhombus with sides measuring 6 cm and two angles of 40°.

b) Draw a rhombus with sides measuring 4.5 cm and two angles of 110°.

Example 4

Draw triangle *ABC*, where *AB* is 3 cm, *BC* is 2.5 cm, and *AC* is 2 cm.

1. Draw and label one of the sides.

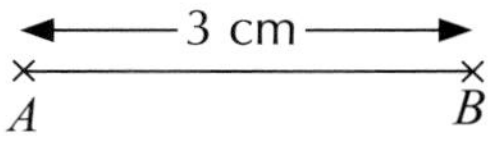

2. Set your compasses to 2.5 cm.
   Draw an arc 2.5 cm from *B*.

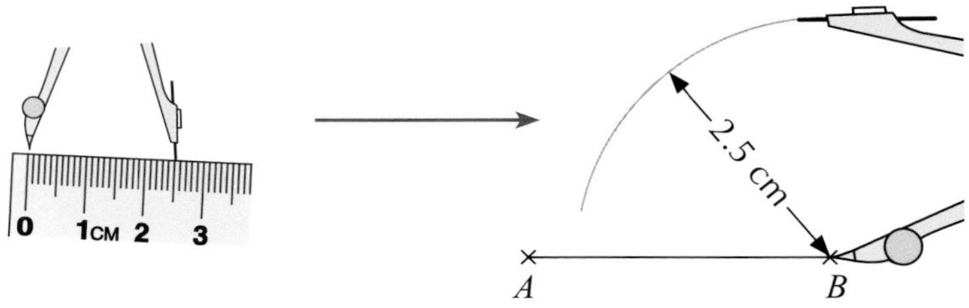

3. Now set your compasses to 2 cm.
   Draw an arc 2 cm from *A*.

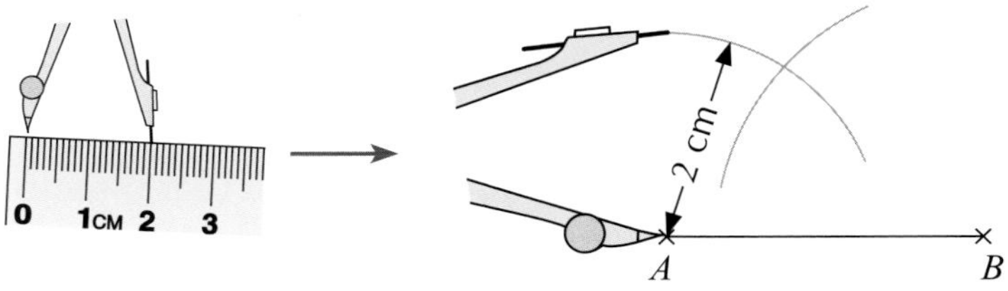

4. *C* is where your arcs cross.

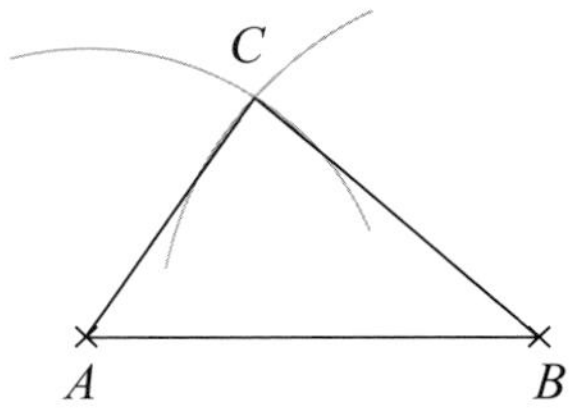

Tip: Make sure that your compasses don't change length while you are drawing — most have a screw that you can tighten.

Exercise 4

Q1 These triangles are not drawn accurately.
Draw them accurately using the measurements given.

a)

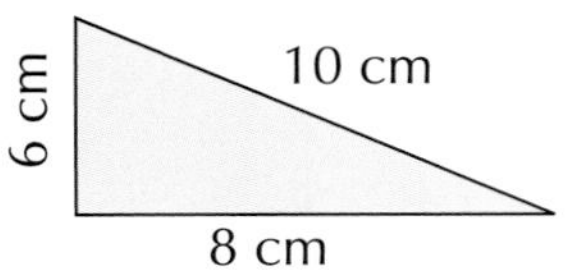

b)

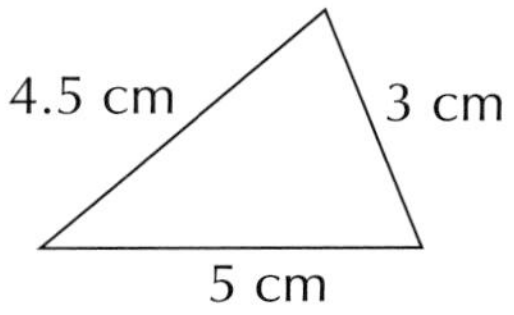

c)

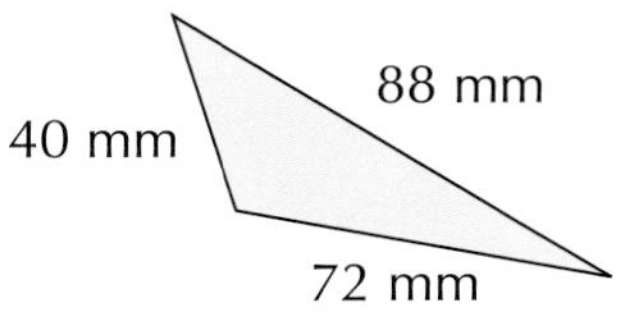

Q2 Draw each of the triangles *ABC* described below.

a) *AB* is 5 cm, *BC* is 6 cm, *AC* is 7 cm.

b) *AB* is 4 cm, *BC* is 7 cm, *AC* is 9 cm.

c) *AB* is 4.6 cm, *BC* is 5.4 cm, *AC* is 8.4 cm.

21.3 More Constructions

There are many other constructions that you can make using your pencil, ruler and compasses. In this topic you'll see them being used to draw perpendicular lines, split angles in half, and more.

Constructing a Perpendicular Bisector

Learning Objective — Spec Ref G1/G2:
Be able to construct a perpendicular bisector.

The **perpendicular bisector** of a line AB is at **right angles** to the line, and cuts it in **half**.

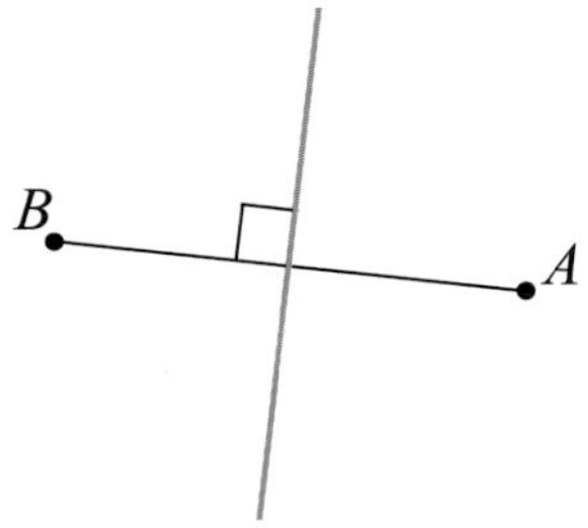

All points on the perpendicular bisector are **equally far** from both A and B (this is important when drawing loci — see page 339).

You can use this fact to draw perpendicular bisectors **without measuring** the distances — just use your compasses to find two points that are the **same distance** from A and B, then the perpendicular bisector will **pass through both** of these points.

Example 1

Draw a line AB which is 3 cm long and construct its perpendicular bisector.

1. Draw AB.

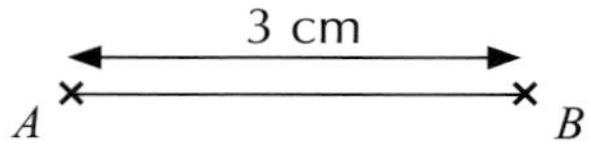

2. Place the compass point at A, with the radius set at more than half of the length AB. Draw two arcs as shown.
3. Keep the radius the same and put the compass point at B. Draw two more arcs.
4. Use a ruler to draw a straight line through the points where the arcs meet. This is the perpendicular bisector.

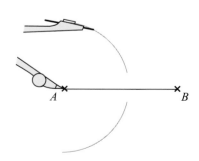

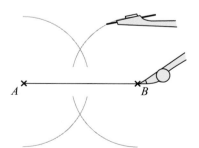

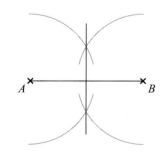

Exercise 1

Q1 Draw a horizontal line PQ 5 cm long.
Construct its perpendicular bisector using a ruler and compasses only.

Q2 Draw a vertical line XY 9 cm long.
Construct its perpendicular bisector using a ruler and compasses only.

Q3 Draw a line AB 7 cm long.
Construct its perpendicular bisector using a ruler and compasses only.

Q4 a) Draw a line AB 6 cm long.
Construct the perpendicular bisector of AB.
b) Draw the rhombus $ACBD$ with diagonals 6 cm and 8 cm.

Q5 a) Draw a circle with radius 5 cm, and draw any two chords.
Label your chords AB and CD.
b) Construct the perpendicular bisector of chord AB.
c) Construct the perpendicular bisector of chord CD.
d) Where do the two perpendicular bisectors meet?

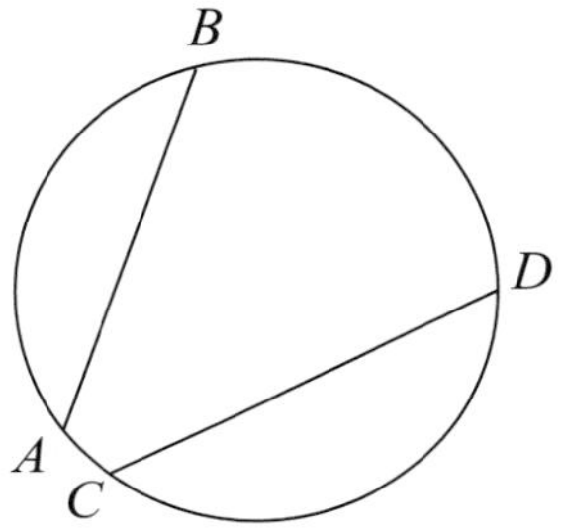

Constructing an Angle Bisector

Learning Objective — Spec Ref G1/G2:
Be able to construct an angle bisector.

An **angle bisector** is the line that cuts an angle in half.
All points on the angle bisector are the **same distance** from each of the two lines that enclose the angle (this is also useful for drawing loci — see page 339).

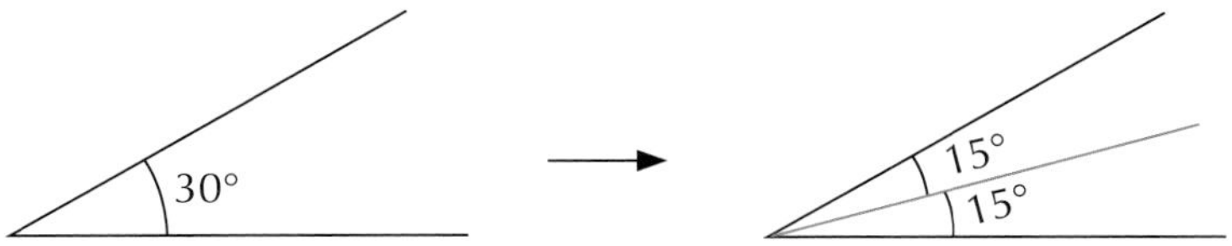

Just like the perpendicular bisector, you can use your ruler and compasses to construct an angle bisector **without measuring** the angles with a protractor. The **method** for constructing an angle bisector is shown in the next example.

Tip: After constructing an angle bisector you know the two new angles — they're both half the original angle.

Example 2

Draw an angle of 60° using a protractor, then construct the angle bisector using only a ruler and compasses.

1. Place the point of the compasses on the angle... ...and draw arcs crossing both lines... ...using the same radius.

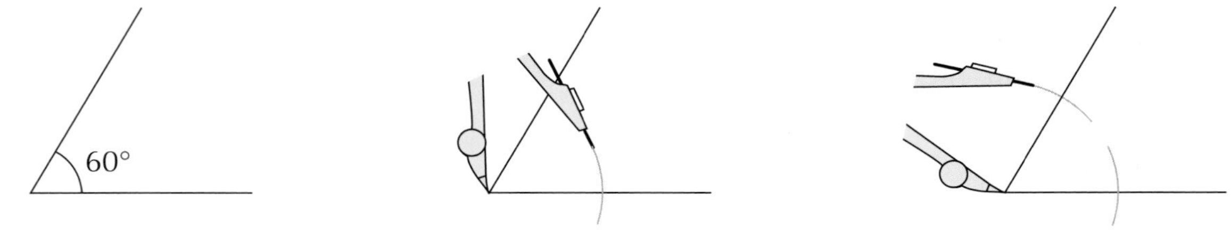

2. Now place the point of the compasses where your arcs cross the lines and, from each point, draw a new arc (using the same radius).

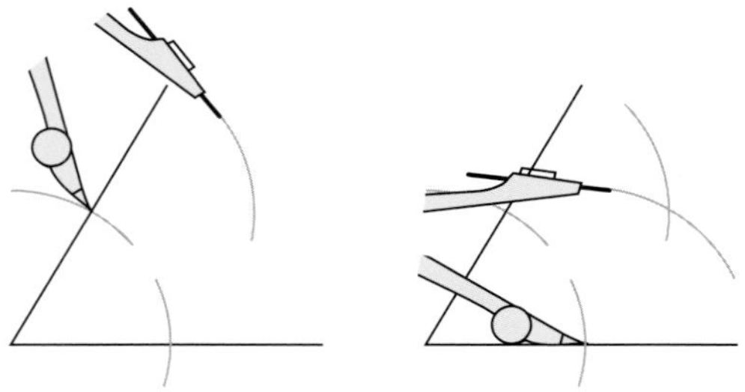

3. Draw the angle bisector through the point where the arcs cross.

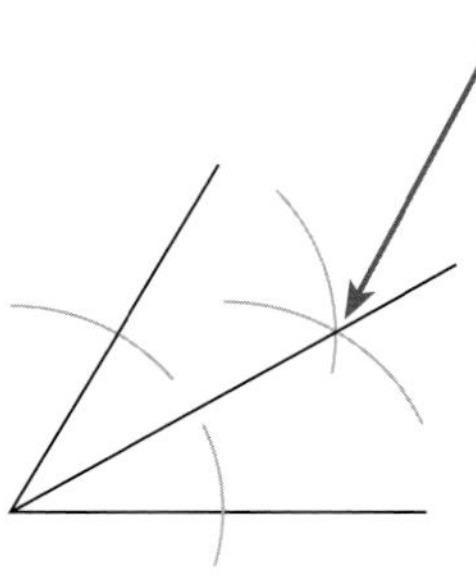

Tip: You can check your answer by measuring the angles with a protractor — they should both be 30°.

Exercise 2

Q1 Draw the following angles using a protractor.
For each angle, construct the angle bisector using a ruler and compasses.

a) 100° b) 44° c) 70° d) 65°

Check each of your angle bisectors with a protractor.

Q2 a) Draw any triangle. Use a ruler to make sure all the sides are straight.

b) Construct the bisectors of each of the angles.
What do you notice about these bisectors?

Q3 a) Use a protractor to draw an angle ABC of 110°, with $AB = BC = 5$ cm.
Construct the bisector of angle ABC.

b) Mark point D on your drawing, where D is the point on the angle bisector with $BD = 8$ cm. What kind of quadrilateral is $ABCD$?

Constructing a Perpendicular from a Point to a Line

Learning Objective — Spec Ref G1/G2:
Be able to construct a perpendicular from a point to a line.

The **perpendicular** from a point to a line is the **shortest path** between them.
It should **pass through** the point, and meet the line at **90°**.

Constructing one of these is similar to constructing a **perpendicular bisector**, except you need to work **backwards** — first, you find two points **on the line**, then you use them to find the perpendicular.

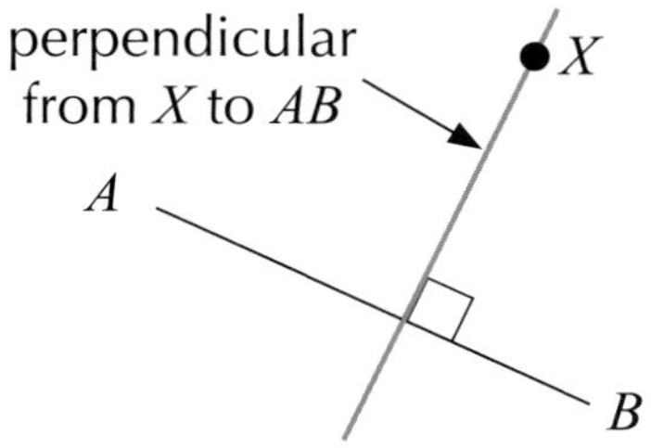

Example 3

Construct the perpendicular from the point X to the line AB using only a ruler and a pair of compasses.

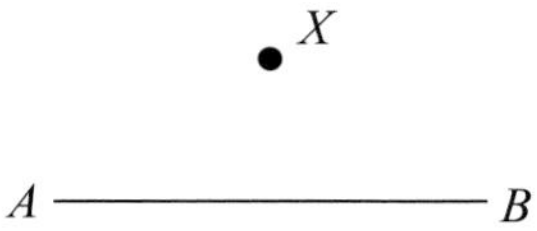

1. Draw an arc centred on X cutting the line twice.

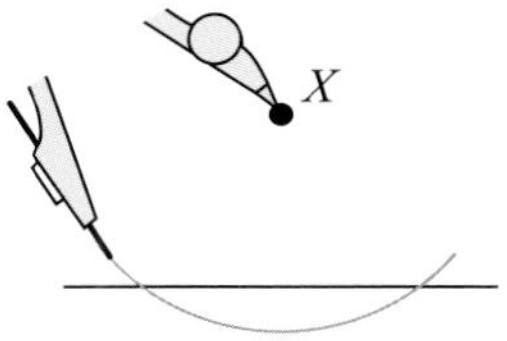

Tip: You may need to extend the line with a ruler to make the arc cut the line twice.

2. Draw an arc centred on one of the points where your arc meets the line.

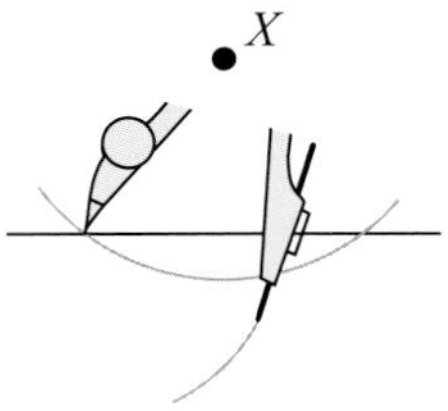

3. Do the same for the other point, keeping the radius the same.

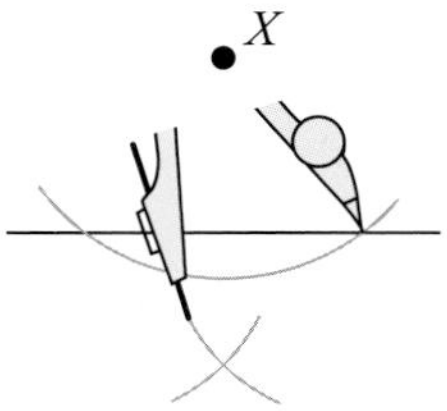

4. Draw the perpendicular to where the arcs cross.

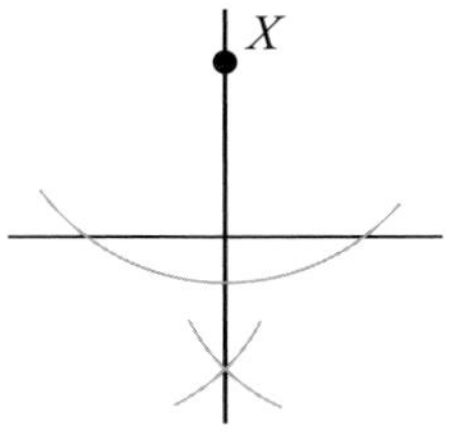

Exercise 3

Q1 Use a ruler to draw a triangle like the one on the right.
Construct the perpendicular from X to the line YZ.

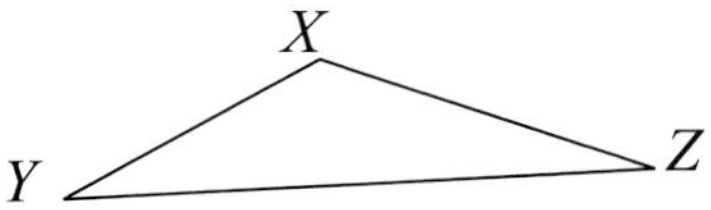

Q2 a) On squared paper draw axes with x-values from 0 to 10 and y-values from 0 to 10.
b) Plot the points $A(1, 2)$, $B(9, 1)$ and $C(6, 8)$.
c) Construct the perpendicular from point C to the line AB.

Q3 Draw three points not on a straight line.
Label your points *P*, *Q* and *R*.
Draw a long straight line passing through your points *P* and *Q*.
Construct the perpendicular from *R* to this line.

Q4 a) Draw any triangle. Use a ruler to make sure all the sides are straight.
b) Construct a perpendicular from each of the triangle's corners to the opposite side.
What do you notice about these lines?

Q5 a) Construct triangle *DEF*, where *DE* = 5 cm, *DF* = 6 cm and angle *FDE* = 55°.
b) Construct the perpendicular from *F* to *DE*.
Label the point where the perpendicular meets *DE* as point *G*.
c) Measure the length *FG*.
Use your result to work out the area of the triangle to 1 decimal place.

Constructing an Angle of 60°

Learning Objective — Spec Ref G1/G2:
Be able to construct an angle of 60°.

To construct a **60° angle** you can take advantage of the fact that all the interior angles of an **equilateral triangle** are 60°.

Start by drawing a line, then set your compasses to match its **length** and draw an arc from **each end** of the line. The point where the arcs cross will form an **equilateral triangle** with the **end points** of the line.

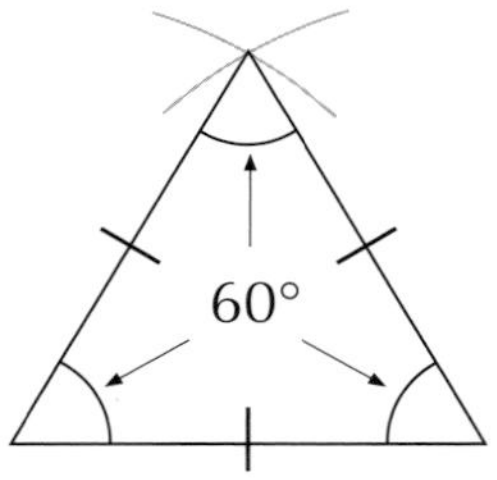

Example 4

Draw a line *AB* and construct an angle of 60° at *A*.

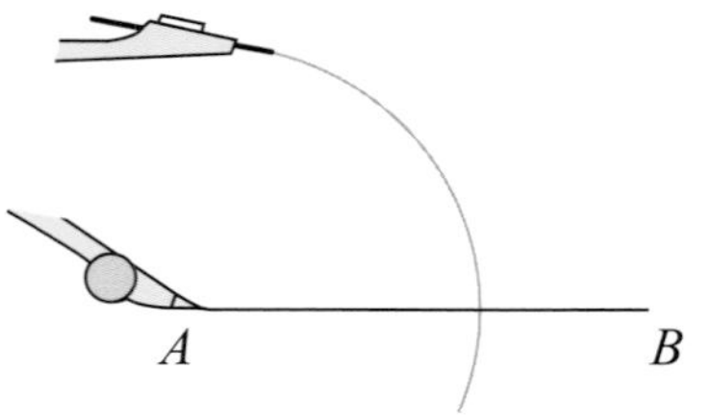

1. Place the compass point on *A* and draw a long arc that crosses the line *AB*.

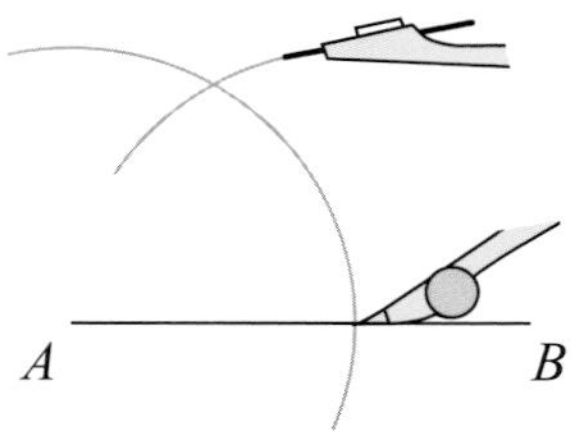

2. Place the compass point where the arc meets the line, and draw another arc of the same radius.

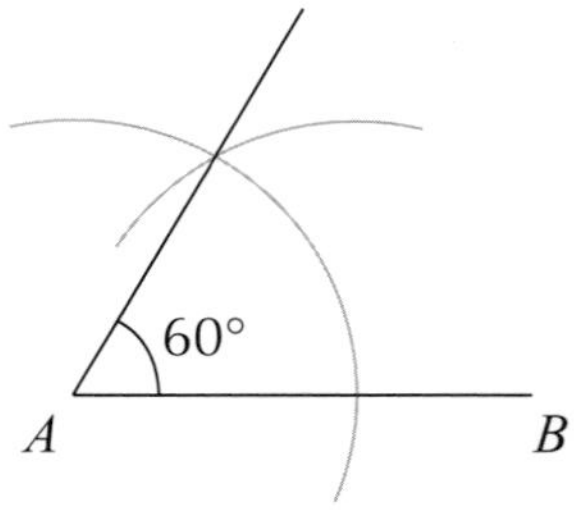

3. Draw a straight line through *A* and the point where your arcs cross. The angle will be 60°.

Exercise 4

Q1 Draw a line AB measuring 5 cm. Construct an angle of 60° at A.

Q2 a) Draw a line measuring 6 cm. Construct an angle of 60° at each end of the line. Join your lines to form a triangle.

b) By measuring the lengths of the sides, check that your triangle is equilateral.

Constructing an Angle of 30°

Learning Objective — Spec Ref G1/G2:
Be able to construct an angle of 30°.

To construct a **30° angle**, first construct a 60° angle.
You can then construct the **angle bisector** to get **two** angles of 30°.

Tip: You form the angle bisector in exactly the same way as usual — see p.331.

Example 5

Draw a line AB and construct an angle of 30° at A.

1. Construct an angle of 60° at A.

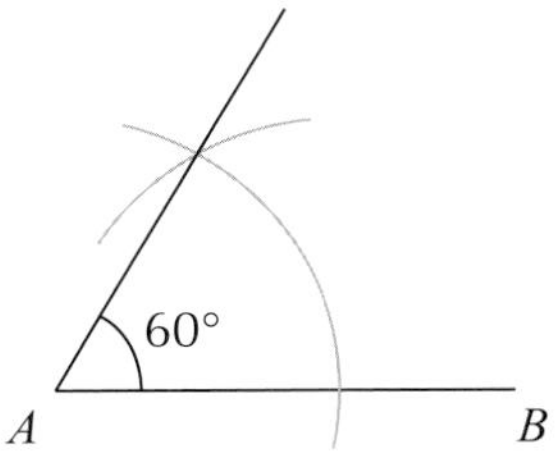

2. Now bisect this angle. You can use the arc from step 1 — place the point of the compasses where your arc crosses the lines and, from each point, draw a new arc (using the same radius).

3. Finally, draw a straight line through A and the point where your arcs cross to get a 30° angle.

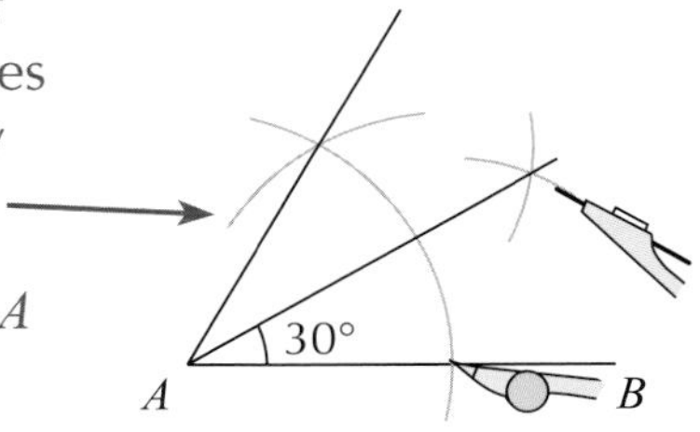

Exercise 5

Q1 Draw a line AB measuring 6 cm. Construct an angle of 30° at A.

Q2 a) Construct the triangle ABC where $AB = 7$ cm, angle $CAB = 60°$ and angle $CBA = 30°$.

b) Check that angle ACB is a right angle using a protractor.

Q3 Construct an isosceles triangle PQR where $PQ = 8$ cm and the angles RPQ and RQP are both 30°.

Constructing an Angle of 90°

Learning Objective — Spec Ref G1/G2:
Be able to construct an angle of 90°.

Constructing a **90° angle** is a bit like constructing a **perpendicular from a point to a line** (see p.333) except now the point is **on the line**.

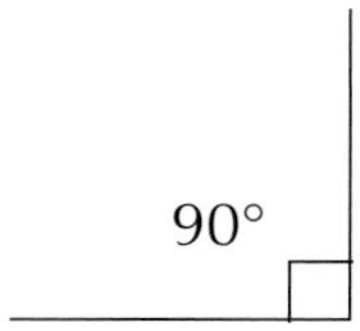

Tip: Remember that a small square shows an angle is 90°.

Example 6

Construct an angle of 90°.

1. Draw a straight line, and mark the point where you want to form the right angle.

2. Draw arcs of the same radius on either side of your point.

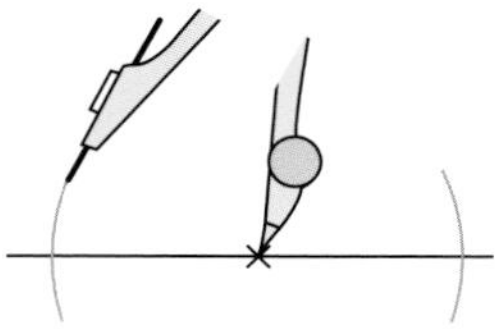

Tip: The size of the radius doesn't matter as long as it's the same for both arcs.

3. Increase the radius of your compasses, and draw two arcs of the same radius — one arc centred on each of the intersections.

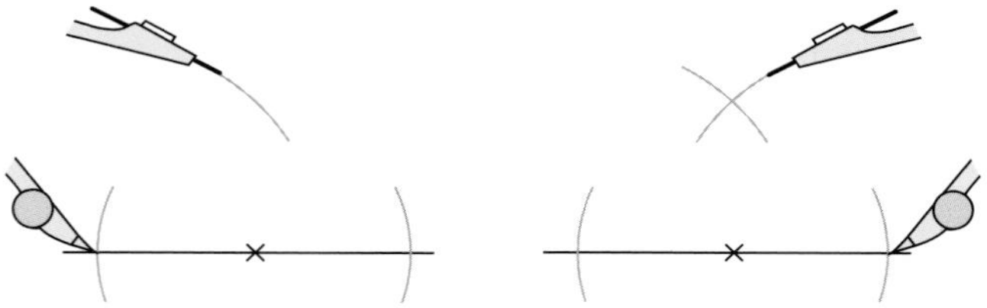

4. Draw a straight line to complete the right angle.

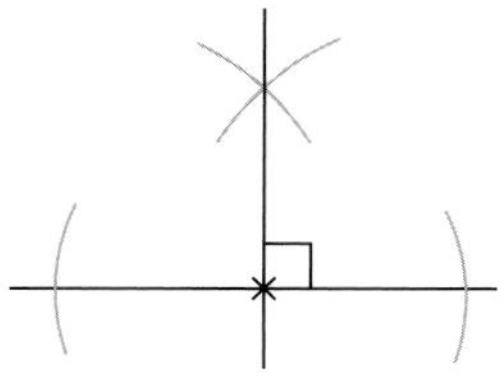

Exercise 6

Q1 a) Draw a straight line, and mark a point roughly halfway along it. Label the point X.
b) Construct a right angle at X using only a ruler and compasses.

Q2 Using a ruler and compasses only, construct a rectangle with sides of length 5 cm and 7 cm.

Q3 Using a ruler and compasses only, construct a square with sides of length 6 cm.

Constructing an Angle of 45°

Learning Objective — Spec Ref G1/G2:
Be able to construct an angle of 45°.

Tip: Using a sharp pencil will improve the accuracy of the angles in your constructions.

If you need to construct an angle of 45°, just construct the **angle bisector** (p.331) of the 90° angle.

Example 7

Construct an angle of 45°.

1. Construct an angle of 90° (see previous page).

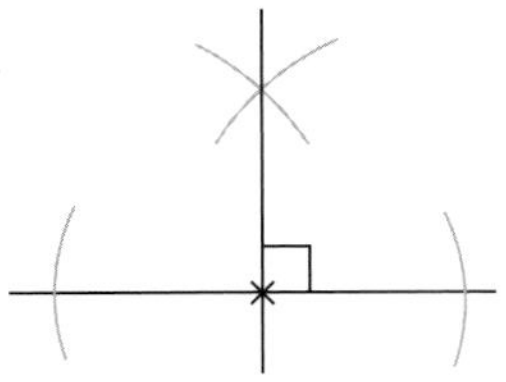

2. Form the angle bisector to make an angle of 45° using the same method as on page 331.

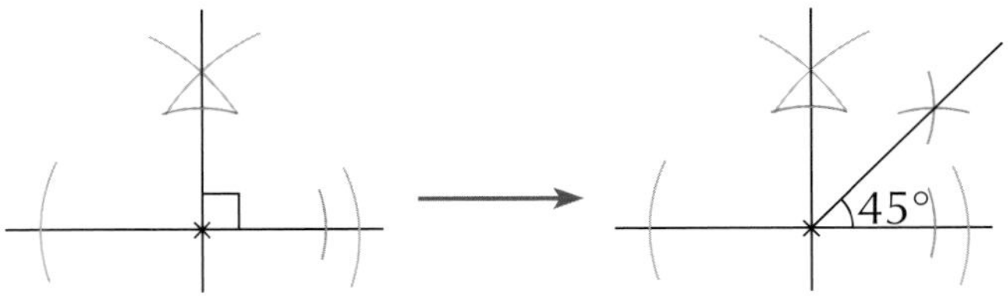

Exercise 7

Q1 a) Draw a straight line, and mark a point roughly halfway along it. Label the point X.

b) Construct an angle of 45° at X using only a ruler and compasses.

Q2 Construct an isosceles triangle ABC where $AB = 8$ cm and the angles CAB and CBA are both 45°.

Constructing Parallel Lines

Learning Objective — Spec Ref G1/G2:
Be able to construct parallel lines.

To construct a line that is **parallel** to another, passing **through a given point**, the first step is to construct the **perpendicular from the point to the line** (see p.333).

Once you've done that, you just need to construct a **right angle** at the point using the method on the previous page.

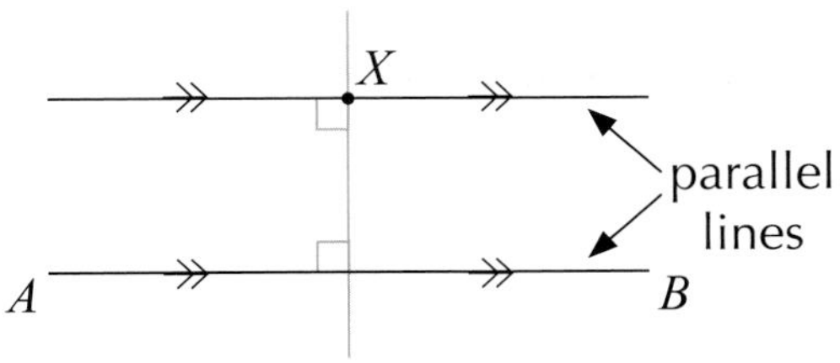

Example 8

Construct a line parallel to *AB* through the point *P*.

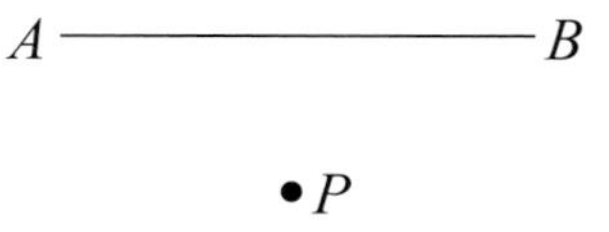

1. Construct the line perpendicular to *AB* passing through *P* (p.330).

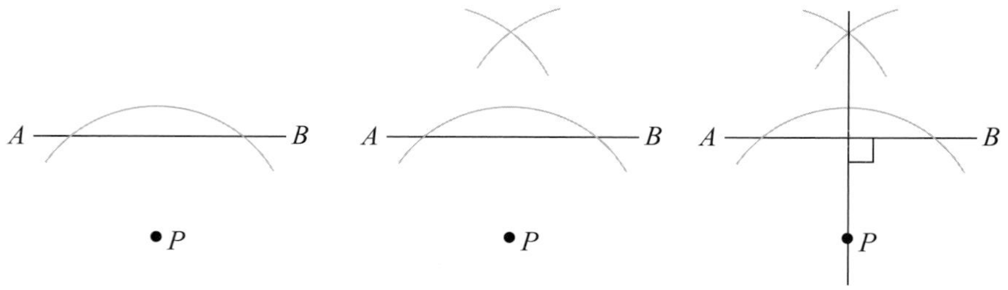

2. Construct a right angle (p.336) to this line at point *P*. This will be parallel to *AB*.

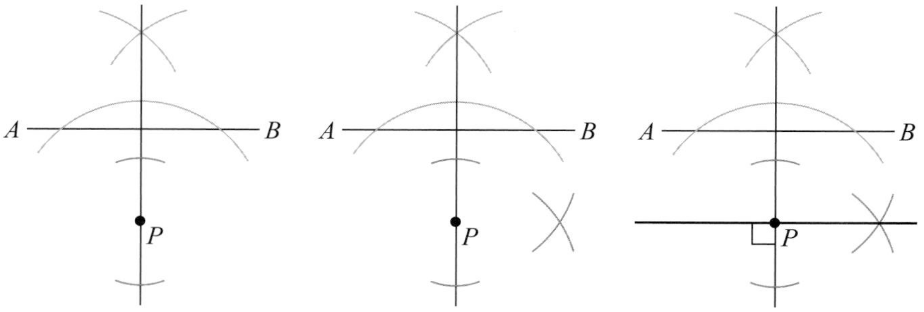

Tip: You could also construct the parallel line through *P* by constructing a right angle anywhere on line *AB*, then constructing the perpendicular from *P* to this new line.

Exercise 8

Q1 Draw a line *AB*, and mark a point *P* approximately 4 cm from your line. Construct a line parallel to *AB* through the point *P*.

Q2 a) On squared paper draw axes with *x*-values from 0 to 10 and *y*-values from 0 to 10.

b) Plot the points *A*(5, 2), *B*(10, 4) and *C*(1, 6).

c) Construct a line parallel to *BC* that passes through the point *A*.

Q3 Draw two straight lines that cross each other at a single point. By adding two parallel lines, construct a parallelogram.

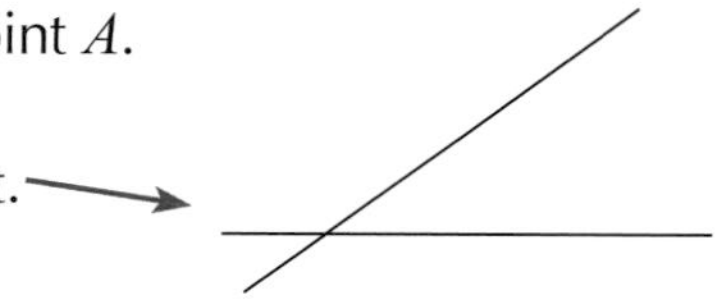

Q4 Draw a line *AB* 10 cm long.

a) Construct angles of 60° at points *A* and *B*, as shown.

b) Mark point *C* on the line 3 cm from *B*.

c) By drawing a line parallel to *AB*, complete the trapezium *ABCD*.

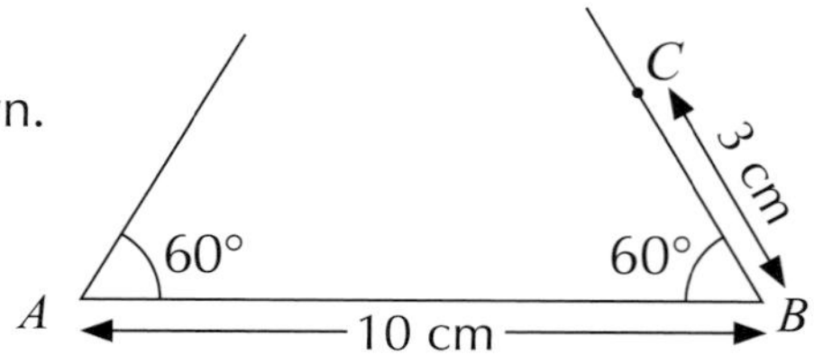

21.4 Loci

A common use for constructions is to find all the points that are a given distance from something or the same distance from two things. These sets of points are called loci (pronounced low-kai).

Learning Objectives — Spec Ref G2:
- Construct the locus of points that are a given distance from a point or a line.
- Construct the locus of points that are equidistant from two points or two lines.
- Solve problems involving loci.

A **locus** (plural **loci**) is a **set of points** which satisfy a particular condition.

The types of loci you need to know are:

- The set of points that are a **fixed distance away** from a point, a line or another kind of shape.
- The set of points that are **equidistant** (i.e. the **same distance**) from two points or two lines.

Tip: Make sure you're comfortable using your ruler and compasses before drawing loci.

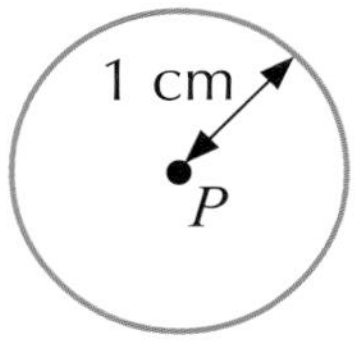

The locus of points that are a fixed distance, e.g. 1 cm, from a **point** P is a **circle** with radius 1 cm centred on P.

To construct this, set your **compasses** to the given distance and draw a circle around the point.

The locus of points that are a fixed distance from a **line** AB is a 'sausage shape'.

To construct this, use your compasses to draw the ends, which are **semicircles**, then join them up with your ruler.

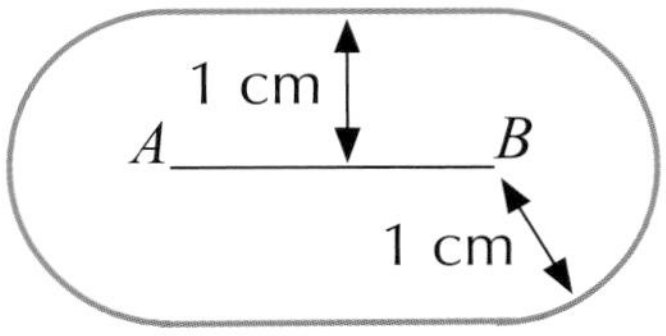

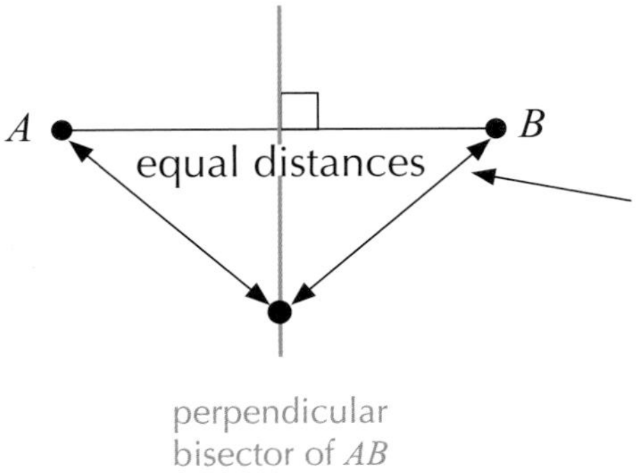

The locus of points equidistant from **two points** A and B is the **perpendicular bisector** of AB.

Tip: There's help with constructing perpendicular bisectors on page 330.

The locus of points equidistant from **two lines** is their **angle bisector** (see page 331).

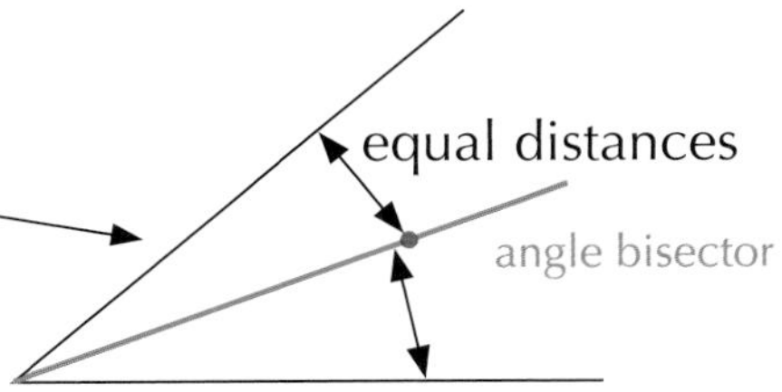

Example 1

The line AB is 2.6 cm long.
Construct the locus of points that are 5 mm from AB.

1. Draw the line AB using a ruler.
2. Set your compasses to 5 mm and draw arcs around each end of the line. Make sure each arc is slightly more than a semicircle.
3. Using your ruler, join the tops and bottoms of the arcs with straight lines.
4. Mark the locus of points, leaving your construction lines on the diagram. Remember to use your compasses again to draw an accurate diagram.

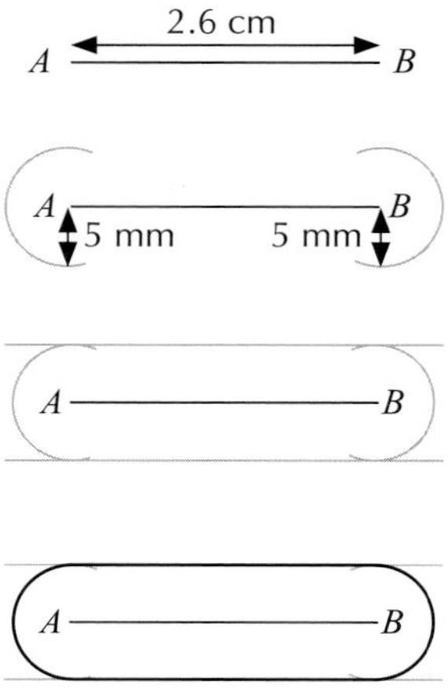

Exercise 1

Q1 Draw a 7 cm long line AB.
Construct the locus of all the points 2 cm from the line.

Q2 a) Mark a point X on your page.
Draw the locus of all points which are 3 cm from X.
b) Shade the locus of all points that are less than 3 cm from X.

Q3 Mark two points A and B on your page 6 cm apart.
Construct the locus of all points which are equidistant from A and B.

Q4 Draw two lines that meet at an angle of 50°.
Construct the locus of all points which are equidistant from the two lines.

Q5 Draw a line AB 6 cm long.
Draw the locus of all points which are 3 cm from AB.

Q6 a) Draw axes on squared paper with x- and y-values from 0 to 10.
Plot the points $P(2, 7)$ and $Q(10, 3)$.
b) Construct the locus of points which are equidistant from P and Q.

You might have to draw **more than one locus** to find the region that satisfies **multiple conditions**.

Example 2

Shade the locus of points inside rectangle *ABCD* that are:
- **more than 2 cm from point *A*,**
- **less than 1 cm from side *CD*.**

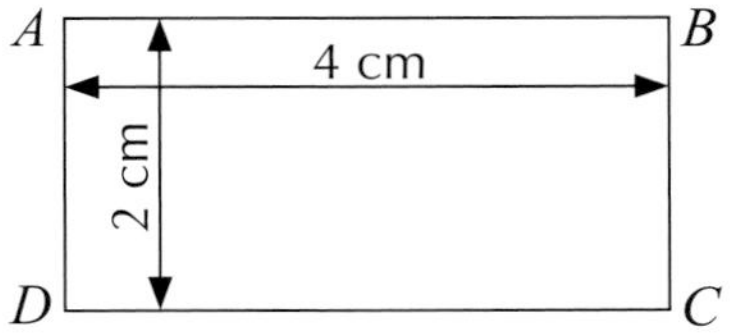

1. Use your compasses to construct an arc of radius 2 cm around *A*. The first condition means that points on or within this arc are excluded.

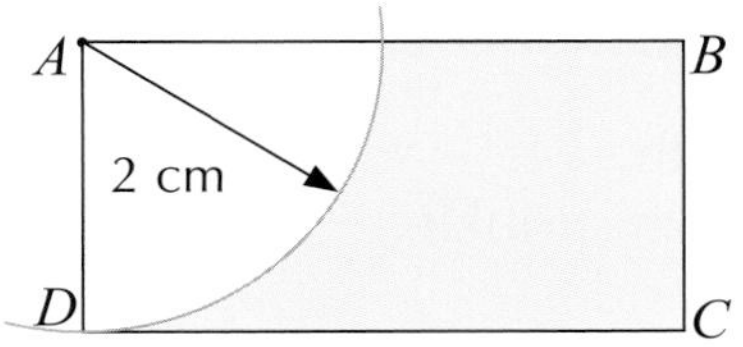

2. Using a ruler, draw a line 1 cm from *CD*. The second condition means that points that are on or above this line are also excluded.

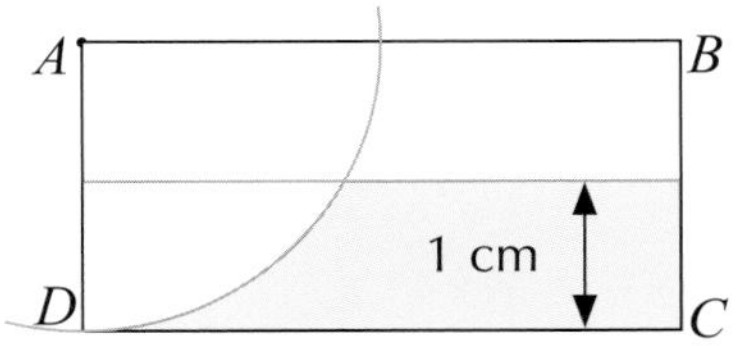

Tip: Read the question carefully to check whether it's asking for less than, more than, or equal to a distance.

Exercise 2

Q1 a) Mark points *P* and *Q* on your page 5 cm apart.
b) Draw the locus of points which are 3 cm from *P*.
c) Draw the locus of points which are 4 cm from *Q*.
d) Show clearly which points are both 3 cm from *P* and 4 cm from *Q*.

Q2 Draw the 5 cm square *ABCD* as shown on the right.
Shade the locus of points inside the square *ABCD* that are:
- less than 3 cm from point *B* and
- more than 4 cm from side *AD*

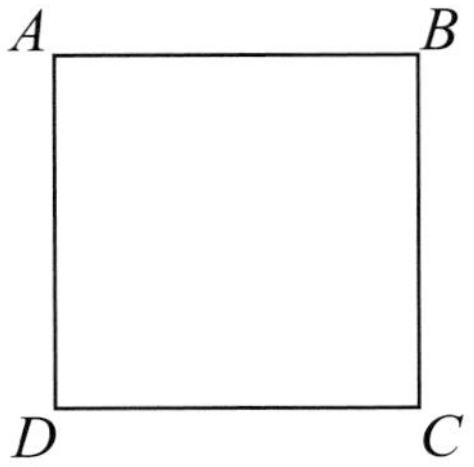

Q3 a) Construct a triangle with sides 4 cm, 5 cm and 6 cm.
b) Draw the locus of all points which are exactly 1 cm from any of the triangle's sides.

Q4 a) Construct an isosceles triangle *DEF* with *DE* = *EF* = 5 cm and *DF* = 3 cm.
b) Draw the locus of points which are equidistant from *D* and *F* and less than 2 cm from *E*.

Loci can also be used to solve **real-life problems**, particularly on **scale diagrams** (see p.290).

Example 3

The diagram shows a plan of a greenhouse, drawn at a scale of 1 cm to 4 m. The greenhouse has a path through the middle modelled by a straight line, and four sprinklers, shown as dots on the plan.

The sprinklers can water plants within a 4 m radius. The gardener can water anything up to 6 m away from the path using a hosepipe. Shade the area on the diagram that can be watered.

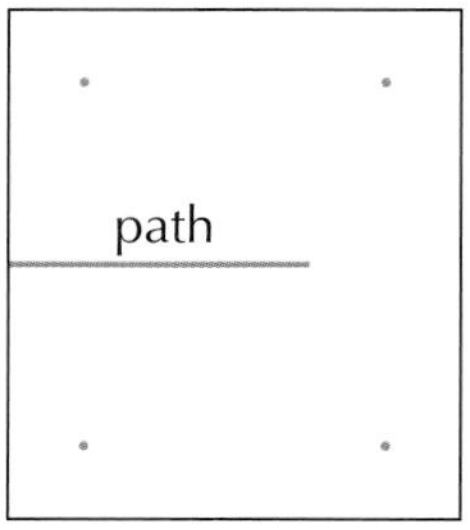

1. 1 cm = 4 m. So draw arcs of radius 1 cm around each sprinkler, and shade the region inside.

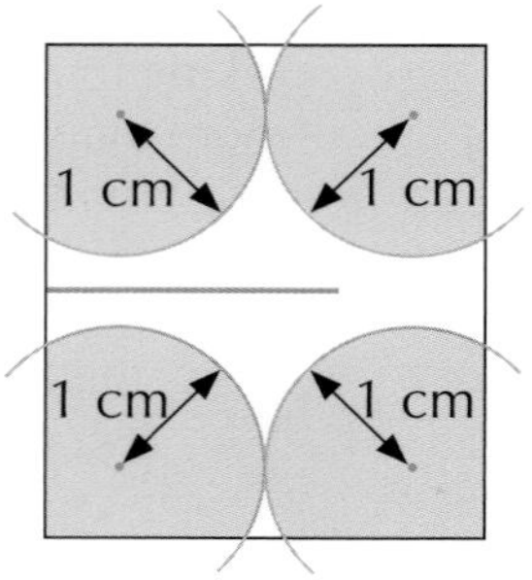

2. 6 m in real life is 6 ÷ 4 = 1.5 cm on the diagram. Construct the locus of points that are 1.5 cm away from the path and shade the region inside.

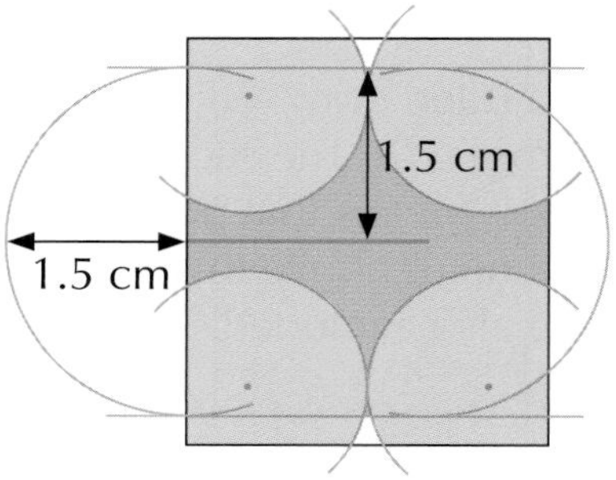

Exercise 3

PROBLEM SOLVING

Q1 A ship sails so that it is always the same distance from a port P and a lighthouse L. The lighthouse and the port are 3 km apart.

a) Draw a scale diagram showing the port and lighthouse. Use a scale of 1 cm:1 km.

b) Show the path of the ship on your diagram.

Q2 Two camels set off at the same time from towns A and B, located 50 miles apart in the desert.

a) Draw a scale diagram showing towns A and B. Use a scale of 1 cm : 10 miles.

b) If a camel can walk up to 40 miles in a day, show on your diagram the region where the camels could possibly meet each other after walking for one day.

Q3 A walled rectangular yard has length 4 m and width 2 m. A dog is secured by a lead of length 1 m to a post in the corner of the yard.

a) Show on an accurate scale drawing the area in which the dog can move. Use the scale 1 cm : 1 m.

b) The post is replaced with a 3 m rail mounted horizontally along one of the long walls, with one end in the corner. If the end of the lead attached to the rail is free to slide, show the area in which the dog can move.

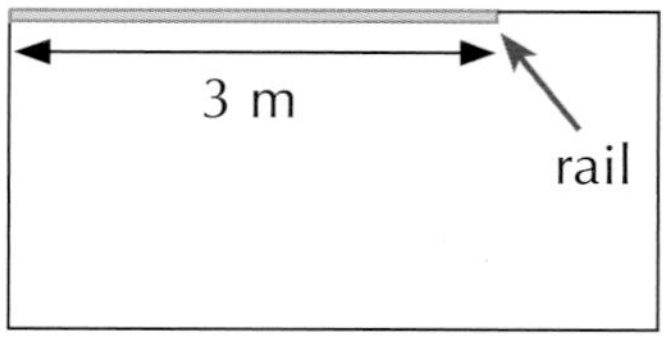

Review Exercise

Q1 What part of the circle does each label show in the diagrams below?

a)

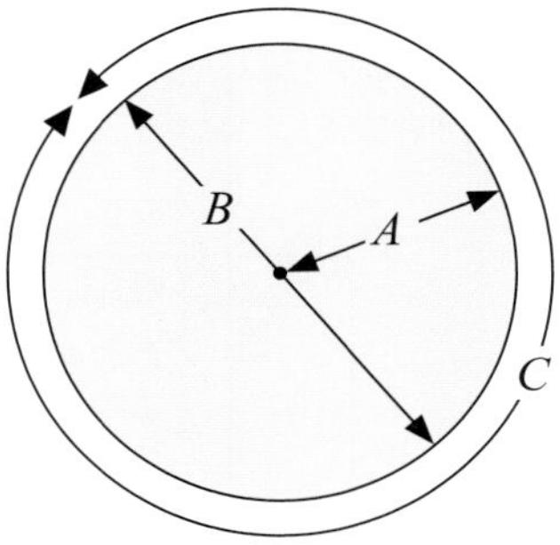

b)

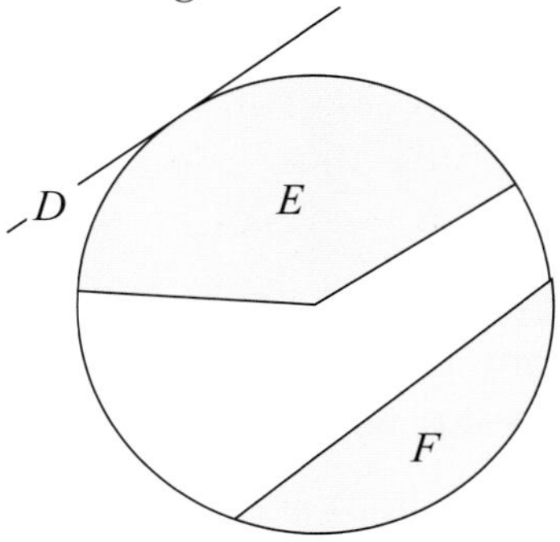

Q2 In the diagram to the right, measure the size of the angle and the length of the lines.

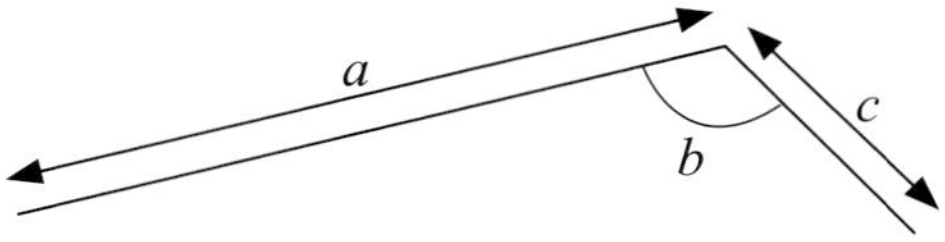

Q3 These triangles are not drawn accurately.
Draw them accurately using the measurements given.

a)

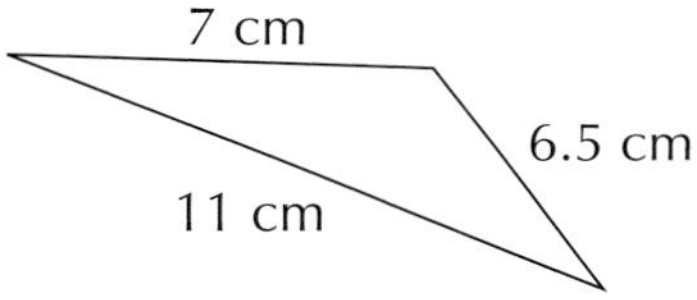

b)

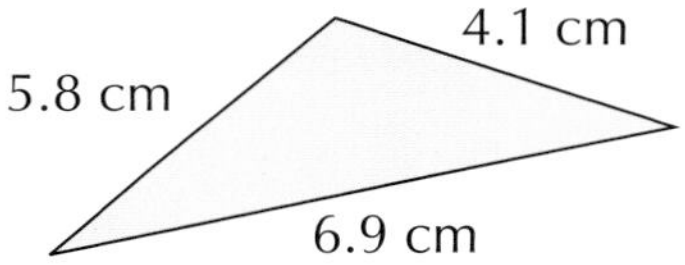

Q4 a) Construct an equilateral triangle *DEF* with sides of length 5.8 cm.

b) Construct a line that is parallel to side *DE* and passes through point *F*.

Q5 a) Using only a ruler and a pair of compasses, construct the triangle *ABC* with *AB* = 7.4 cm, angle *CAB* = 60° and angle *ABC* = 45°.

b) Calculate the size of angle *ACB*.
Check the angle in your drawing using a protractor.

Q6 Two walls of a field meet at an angle of 80°.
A bonfire has to be the same distance from each wall and 3 m from the corner. Copy the diagram on the right, then use a ruler and pair of compasses to show the position of the fire.

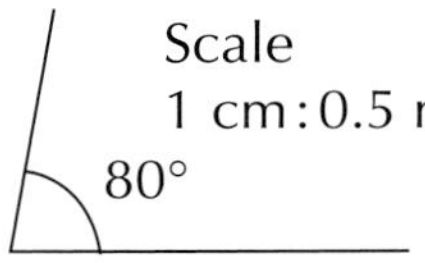

Scale
1 cm : 0.5 m

Exam-Style Questions

Q1 Construct an isosceles triangle with two sides of length 7.1 cm and an angle of 22° between them.

[3 marks]

Q2 A swimmer is in the sea at point *S*.
They want to get to the beach by swimming the shortest possible distance.

Copy the diagram and construct a line from *S* to the beach to show the route they should take.

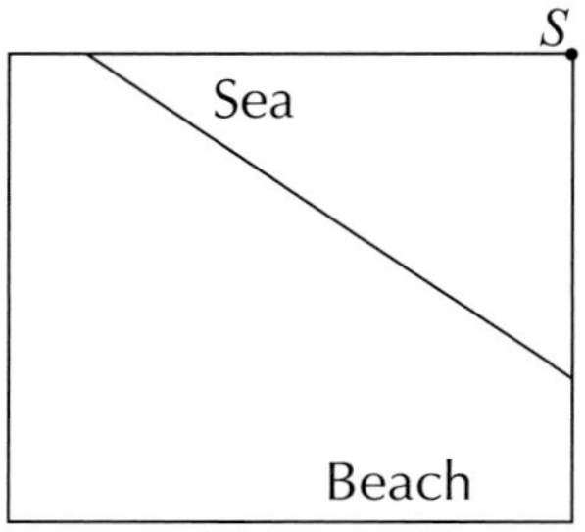

[3 marks]

Q3 Draw a line *AB* 8 cm long.

a) Construct an angle of 60° at *A*.

[2 marks]

b) Complete the construction of a rhombus *ABCD* with sides of length 8 cm.

[3 marks]

Q4 The council is planning to build a new road to the north of a motorway.
The road must be:

- parallel to the motorway
- 400 m from the motorway

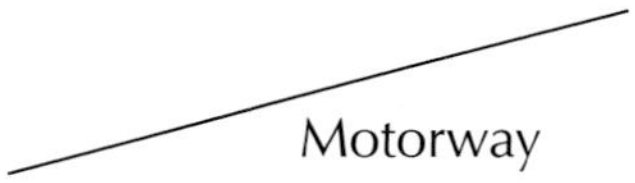

By copying the diagram on the right and using a scale of 1 cm : 100 m, construct a diagram showing the path of the new road.

[3 marks]

Q5 Some students are doing a treasure hunt.
They know the treasure is:

- located in a square region *ABCD*, which measures 10 m × 10 m
- the same distance from *AB* as from *AD*
- 7 m from corner *C*.

Draw a scale diagram to show the location of the treasure. Use a scale of 1 cm : 1 m.

[4 marks]

22.1 Rectangles and Triangles

You should already be familiar with area and perimeter, but for GCSE you'll need to work them out for all sorts of shapes, including composite shapes. We'll start off with the basics — squares, rectangles and triangles.

Perimeter

Learning Objective — Spec Ref G17:
Be able to find the perimeter of rectangles and triangles.

Prior Knowledge Check: Be familiar with the properties of triangles, squares and rectangles. See p.242-246.

Perimeter (P) is the distance around the outside of a shape. You can work it out by **adding up** the **lengths** of all the sides, or by learning these **formulas**.

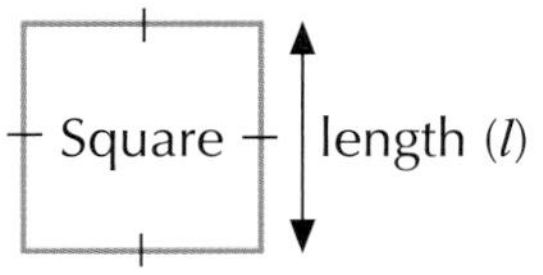

$P = 4l$

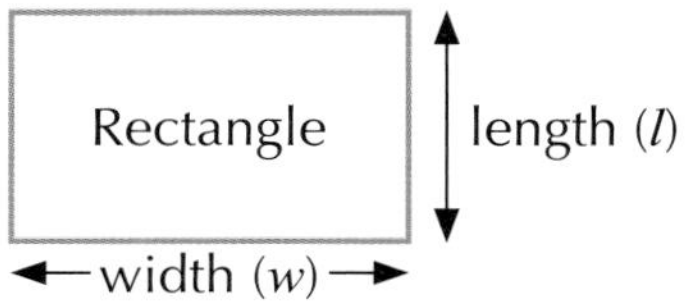

$P = 2l + 2w$

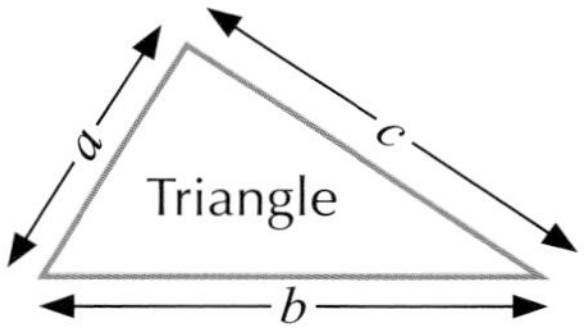

$P = a + b + c$

Example 1

Find the perimeter of each of these shapes.

1. Write down the formula for perimeter.
2. Substitute the lengths of the sides into the formula.
3. Simplify the answer and don't forget to include the units.

a)

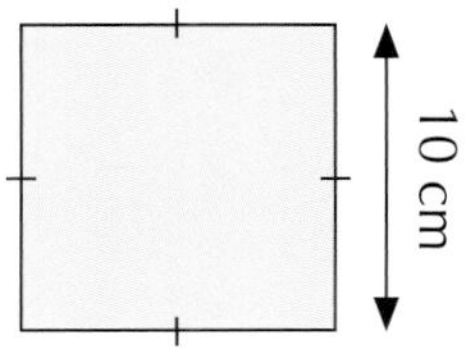

$P = 4l$
$= 4 \times 10$
$=$ **40 cm**

Tip: The little lines (or 'tick marks') on the sides of the shape in part a) show that those sides are the same length — so you know it's a square.

b)

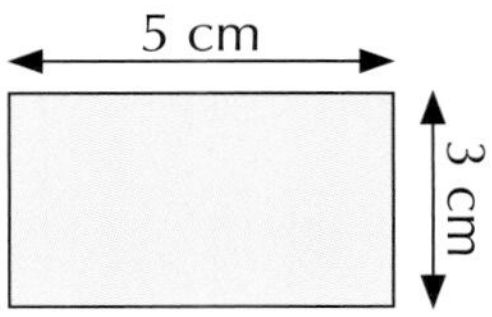

$P = 2l + 2w$
$= (2 \times 3) + (2 \times 5)$
$= 6 + 10$
$=$ **16 cm**

c)

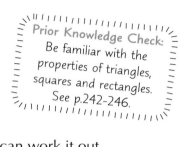

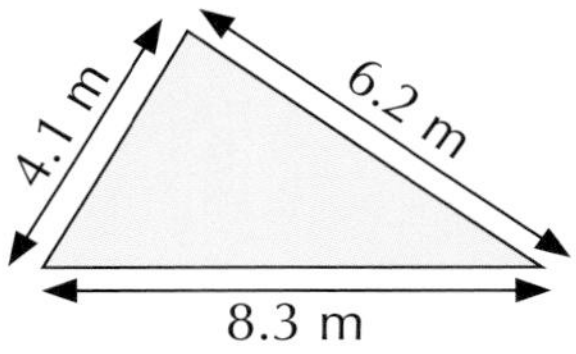

$P = a + b + c$
$= 4.1 + 8.3 + 6.2$
$=$ **18.6 m**

Exercise 1

Q1 Find the perimeters of the shapes below.

a)
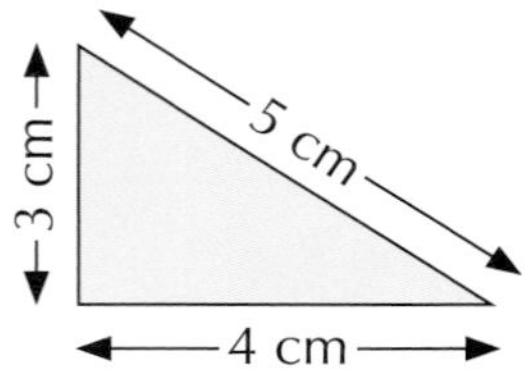

b)
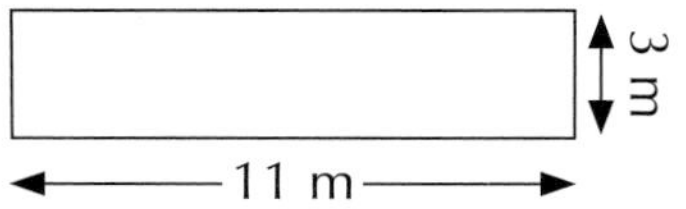

c)
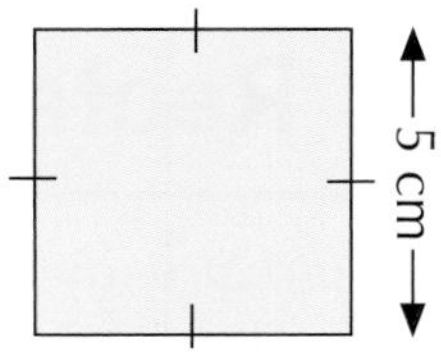

d)
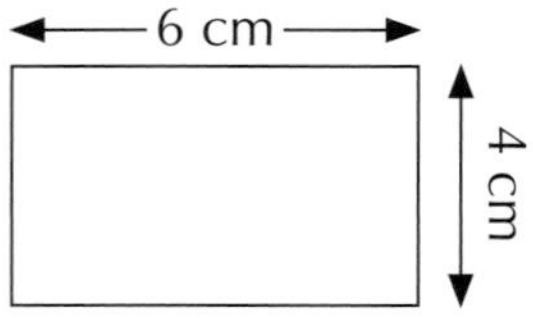

e)
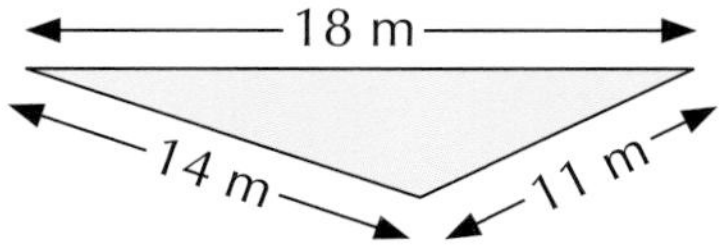

f)
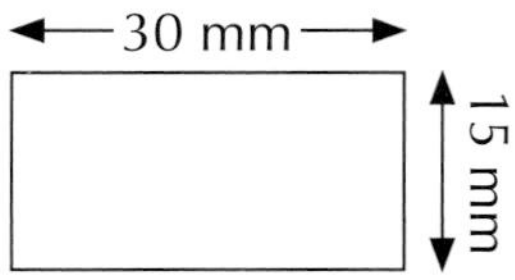

Q2 Find the perimeter of each shape described below.

a) A square with sides of length 4 cm

b) A triangle where two sides measure 5 cm and one side measures 7 cm

c) A rectangle of width 6 m and length 8 m

Q3 The police need to cordon off and then search a rectangular crime scene measuring 2.1 m by 2.8 m. What is the perimeter of the crime scene?

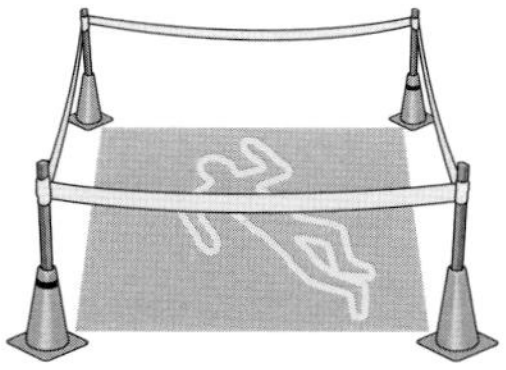

Area

Learning Objective — Spec Ref G16:
Be able to find the area of rectangles and triangles.

Area (A) is the amount of space inside a shape. It's measured in '**units squared**' — e.g. if the shape has sides in cm then the area will be in cm². Use these **formulas** to work it out:

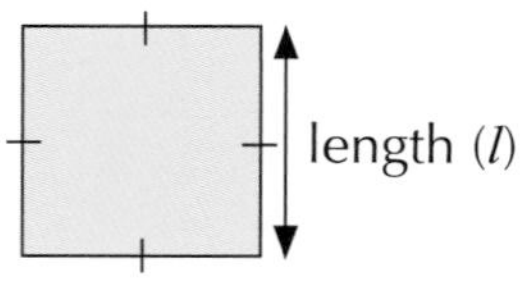

Area = (side length)²

$A = l^2$

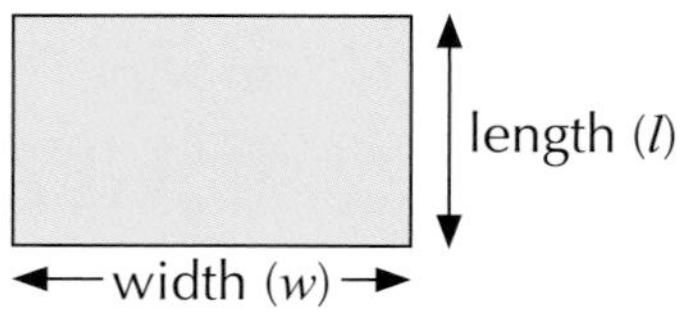

Area = length × width

$A = lw$

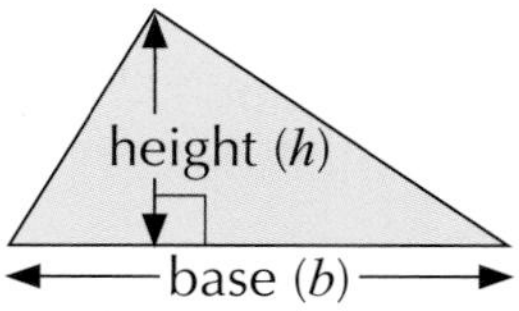

Area = $\frac{1}{2}$ × base × perpendicular height

$A = \frac{1}{2}bh$

Example 2

Find the area of each of these shapes.

a)

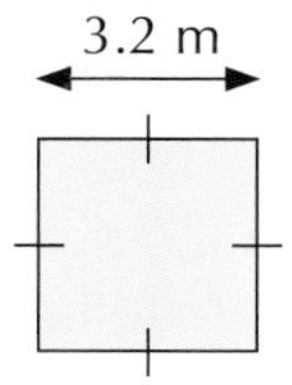

b)

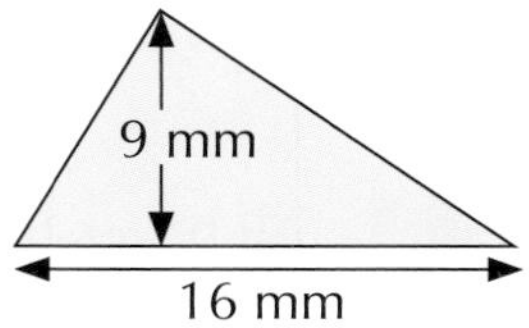

1. Write down the formula for area.
2. Substitute the lengths into the formula.
3. It's an area, so use 'squared' units.

a) $A = l^2$

$= 3.2^2$

$=$ **10.24 m²**

b) $A = \frac{1}{2}bh$

$= \frac{1}{2} \times 16 \times 9$

$=$ **72 mm²**

Exercise 2

Q1 Find the areas of the shapes below.

a)

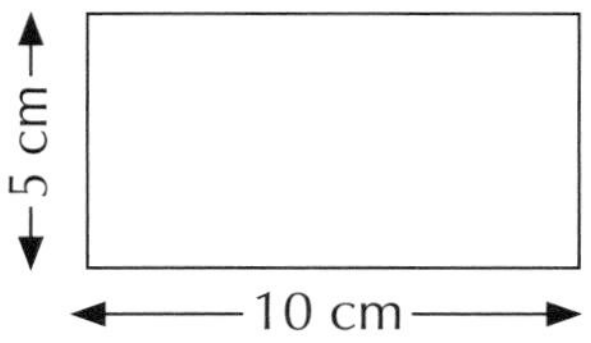

b)

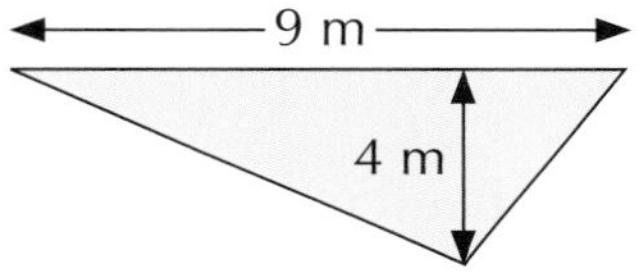

c)

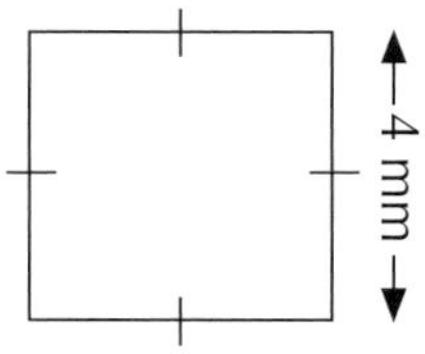

d)

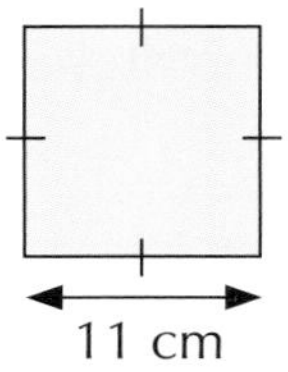

e)

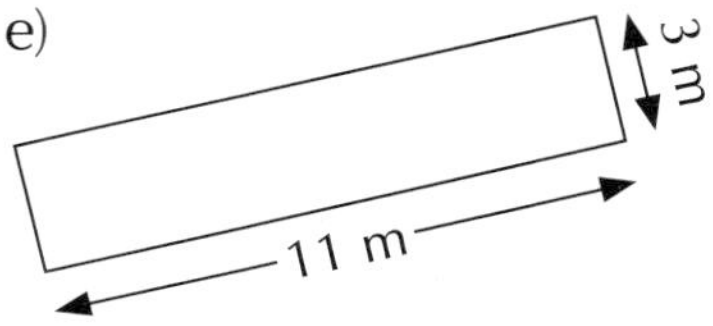

f)

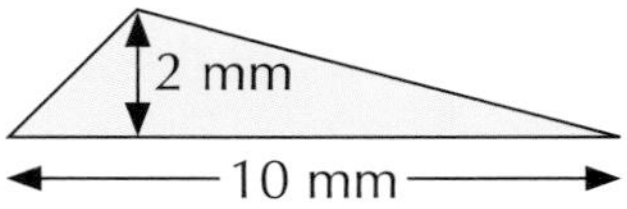

Q2 Find the area of each shape described below.

a) A rectangle 23 mm long and 15 mm wide

b) A square with 17 m sides

c) A triangle with a base of 4 cm and a height of 12.5 cm

Q3 Barb has a rectangular lawn 23.5 m long by 17.3 m wide. She is going to mow the lawn. What area will Barb have to mow (to the nearest m²)?

Q4 A rectangular floor measures 9 m by 7.5 m.
It is to be tiled using square tiles with sides of length 0.5 m.

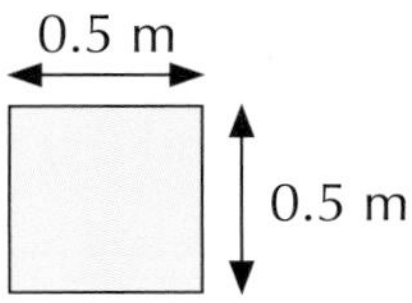

a) What is the area of the floor?

b) What is the area of one of the tiles?

c) How many tiles will be needed to cover the floor?

Q5 A rectangular garden 24 m long and 5.4 m wide is to be re-turfed. Turf is bought in rolls that are 60 cm wide and 8 m long. How many rolls of turf are needed to cover the garden? PROBLEM SOLVING

Q6 Ali bakes the cake shown on the right. PROBLEM SOLVING If the top and the four sides are to be iced, what area of icing will be needed?

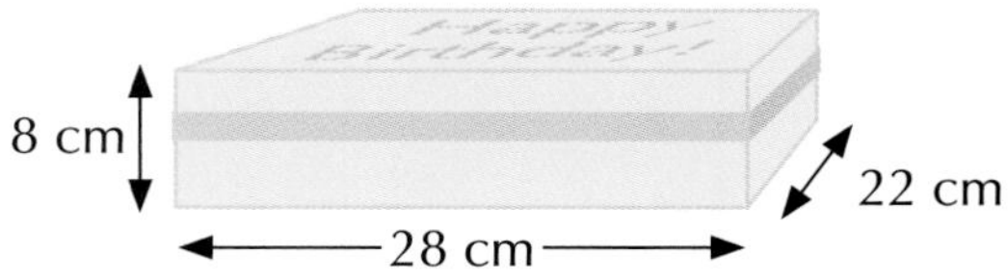

Composite Shapes involving Rectangles and Triangles

Learning Objective — Spec Ref G16/G17:
Be able to find the perimeter and area of composite shapes made from rectangles and triangles.

A **composite shape** is one that has been made up from two or more basic shapes.

To find the **perimeter** of a composite shape, use the lengths you're given to find any that are **missing** and then **add up** the lengths of all of the sides.

To find the **area** of a composite shape, **split it up** into shapes that you recognise. Then work out the area of each of these shapes separately and **add them** up at the end.

Example 3

Find the perimeter of the shape below.

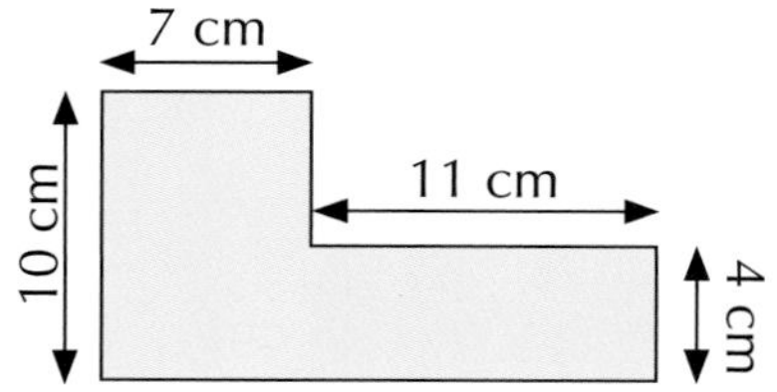

Tip: To find a missing side length, look at the lengths of the sides that are parallel to it.

1. Label the missing sides, and find their lengths.

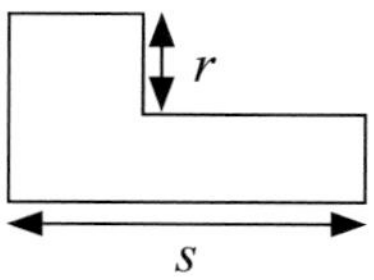

$r = 10 - 4 = 6$ cm
$s = 7 + 11 = 18$ cm

2. Add up the lengths of all the sides to find the perimeter.

Perimeter $= 7 + 6 + 11 + 4 + 18 + 10$
$=$ **56 cm**

Example 4

Find the area of the shape from Example 3.

1. Split the shape into rectangles A and B, and find their areas.

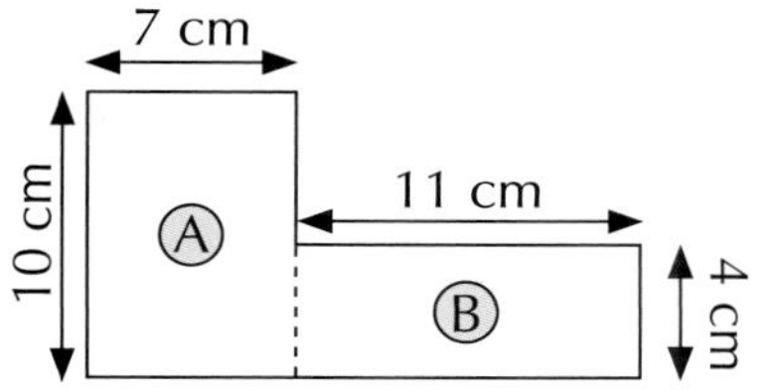

Area of rectangle A = length × width
= 10 × 7
= 70 cm^2

Area of rectangle B = length × width
= 4 × 11
= 44 cm^2

2. Add these up to find the total area.

Total area of shape = 70 + 44 = **114 cm^2**

Exercise 3

Q1 Find the area of each shape below.

a) 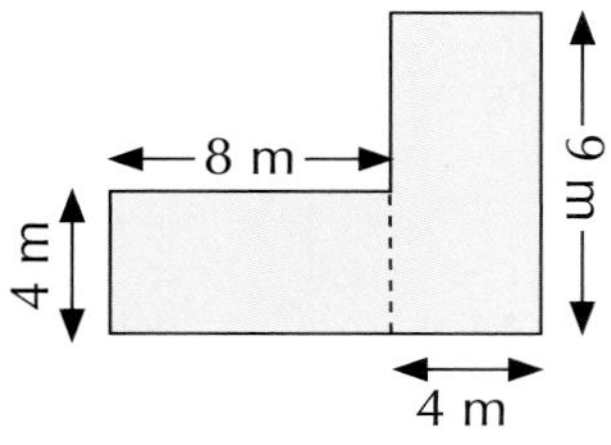

b) 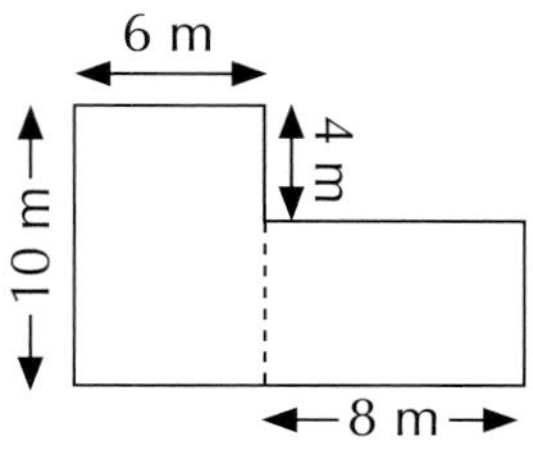

c) 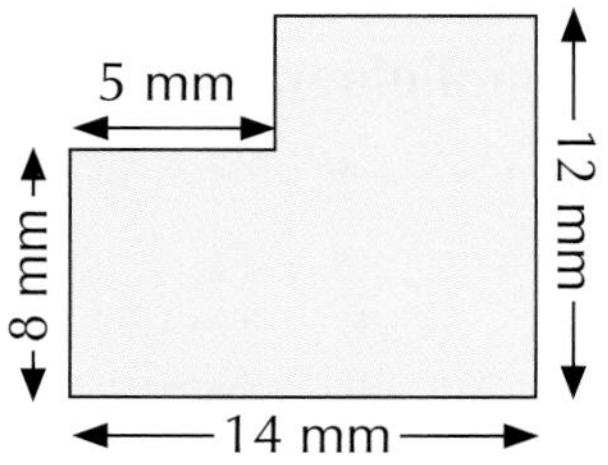

Q2 For each shape below, find: (i) its perimeter, (ii) its area.

a) 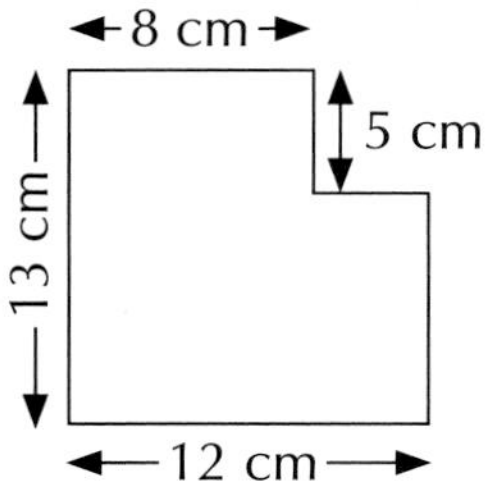

b) 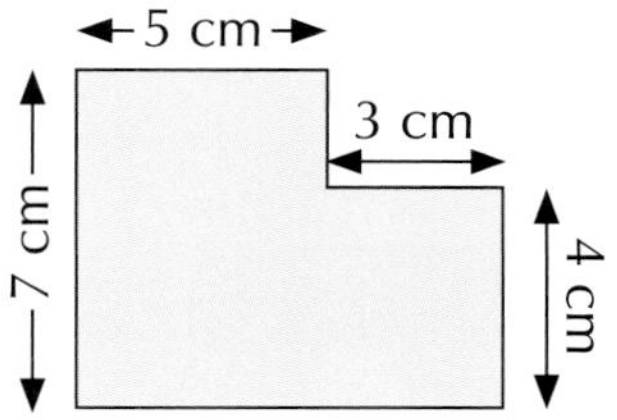

c) 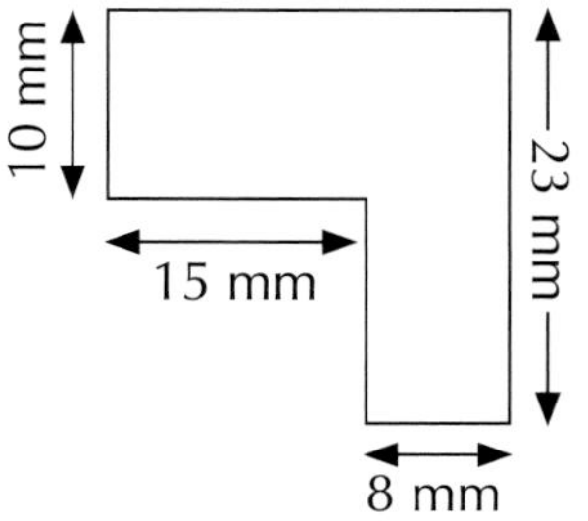

Q3 Find the area of the shapes below by splitting them into a rectangle and a triangle.

a) 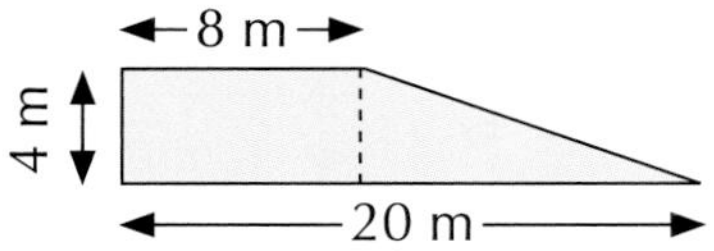

b) 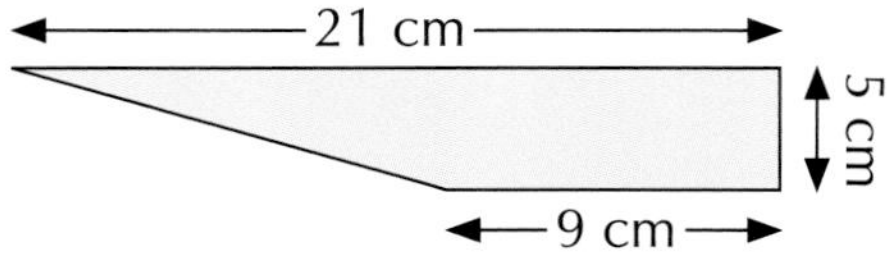

22.2 Other Quadrilaterals

Now that the simple shapes are out of the way, it's time to move on to the perimeter and area of some more complicated shapes — parallelograms and trapeziums.

Parallelograms

Learning Objective — Spec Ref G16/G17:
Be able to find the perimeter and area of parallelograms.

Prior Knowledge Check: Be familiar with the properties of parallelograms. See p.247.

Parallelograms have **two pairs** of **parallel** sides, with opposite sides being the same length. Parallel sides are shown with **matching arrows**.

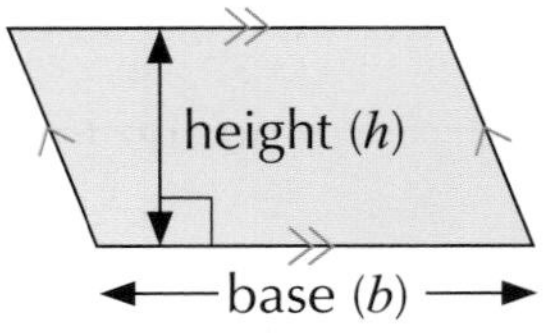

The **area** of a parallelogram is given by the formula: $A = bh$

Here, h is the **perpendicular height** — it's measured at right angles to the base.

Example 1

Find the area of the parallelogram on the right.

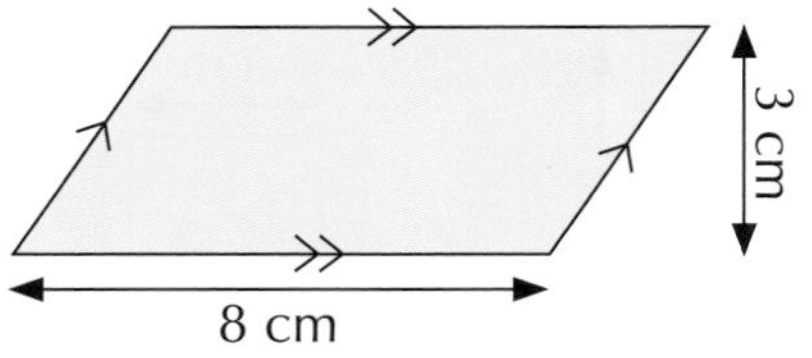

1. Write down the formula for the area of a parallelogram. $A = bh$
2. Substitute in the values for b and h. $= 8 \times 3$
3. Use 'squared' units — here, it's cm^2. $= \mathbf{24\ cm^2}$

Tip: Remember that h is the perpendicular height, not the length of the second set of parallel sides.

Exercise 1

Q1 Find the area of each parallelogram below.

a)
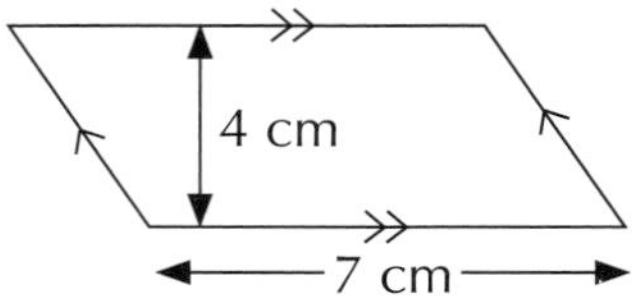

b)
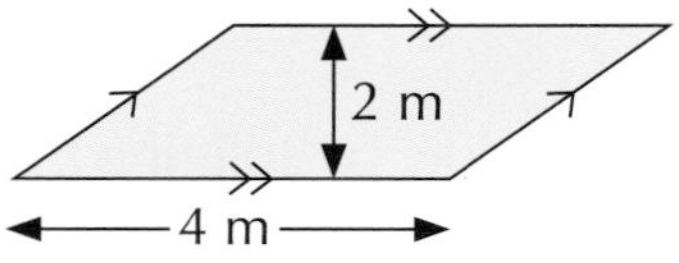

c)
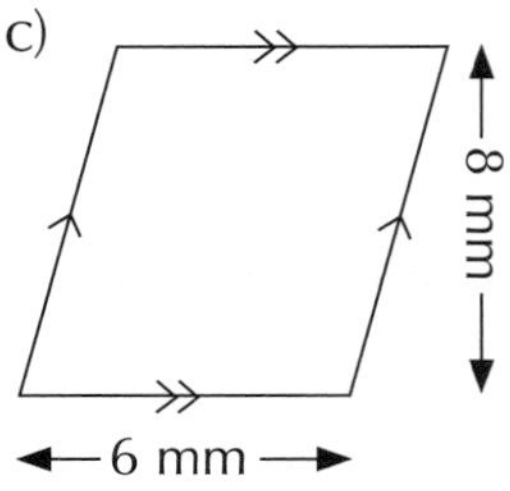

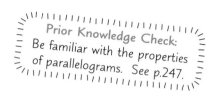

Q2 For each of these parallelograms, find: (i) the area, (ii) the perimeter.

a)

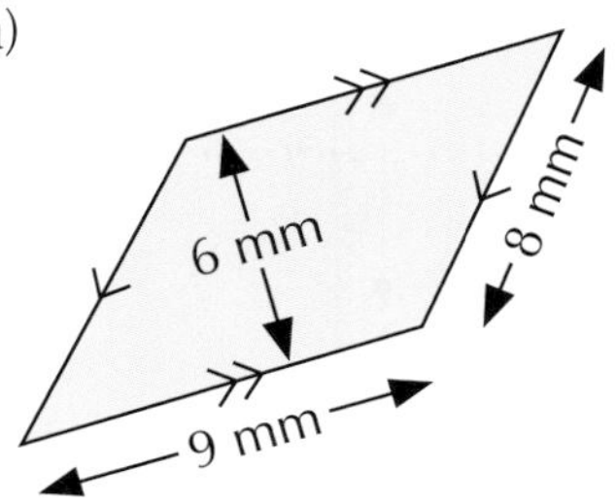

b)

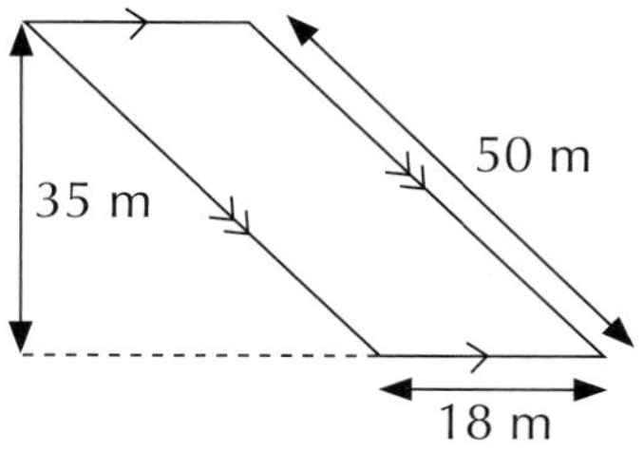

c)

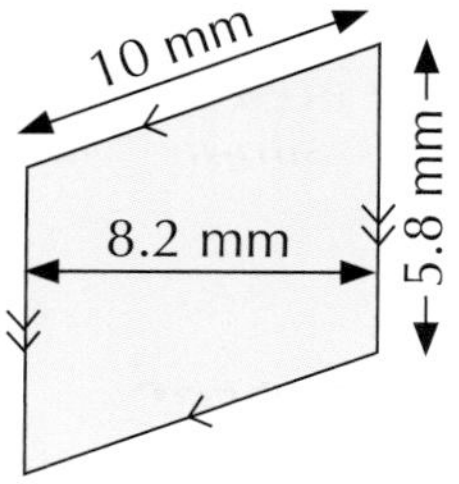

Q3 The logo of a company that designs rockets consists of two identical parallelograms, as shown on the right. Find the total area of the logo.

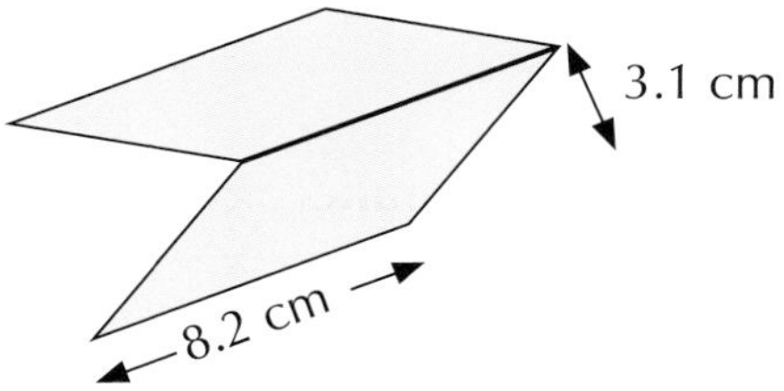

Q4 The picture on the right shows part of a tiled wall. All the tiles are identical parallelograms. Find the area of one tile.

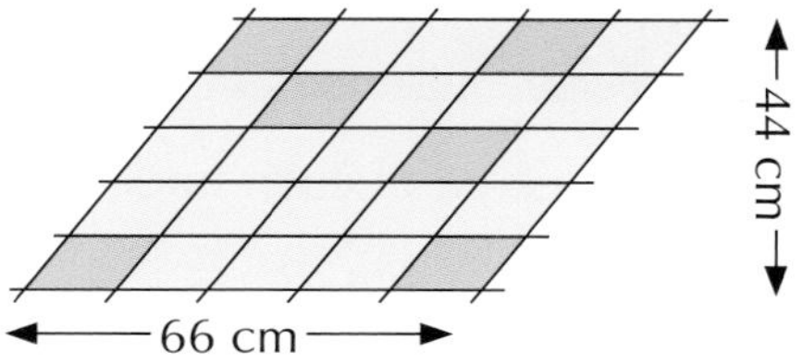

Q5 Without doing any calculations, explain which of the shapes below has the larger area.

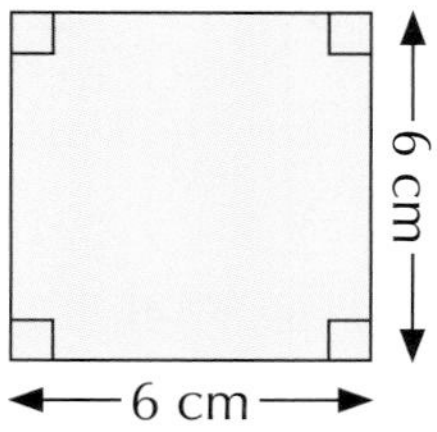

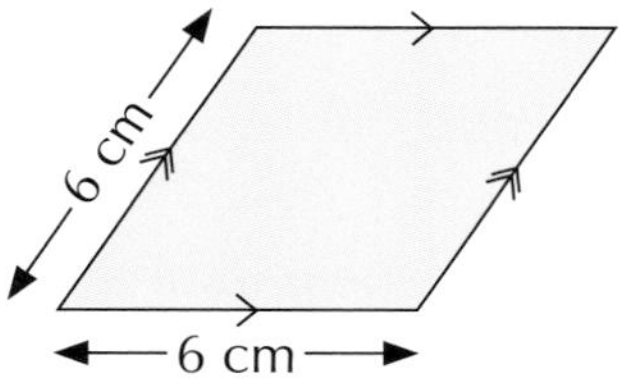

Trapeziums

Learning Objective — Spec Ref G16/G17:
Be able to find the perimeter and area of trapeziums.

Prior Knowledge Check: Be familiar with the properties of trapeziums. See p.249.

The shape on the right is a **trapezium**.
It has **one pair** of **parallel** sides.

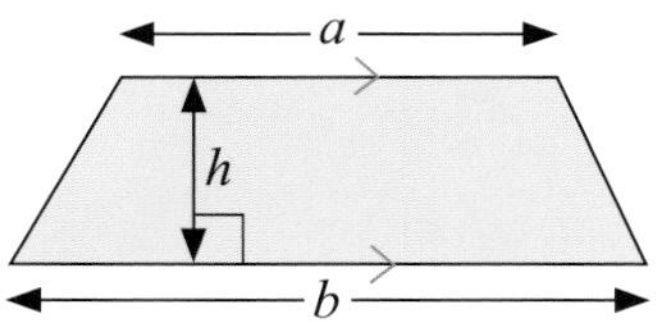

The **area** of a trapezium is given by this formula: $A = \frac{1}{2}(a + b) \times h$

Remember, a and b are the parallel sides, and h is the **perpendicular** height.

Example 2

Find the area of the trapezium on the right.

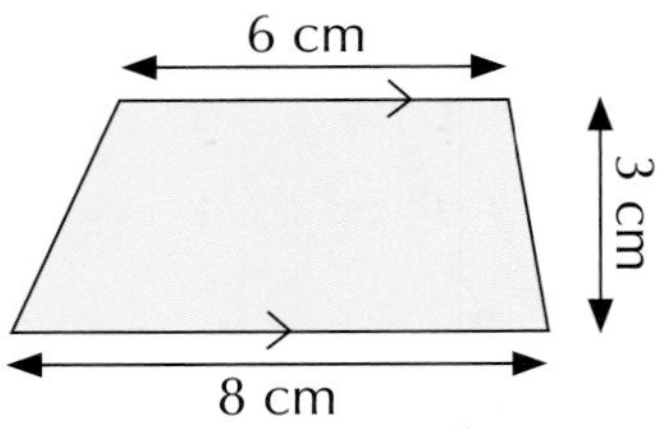

1. Write down the formula for the area of a trapezium. $A = \frac{1}{2}(a + b) \times h$
2. Substitute in the values for a, b and h. $= \frac{1}{2}(6 + 8) \times 3$

$= \frac{1}{2} \times 14 \times 3$

3. Use 'squared' units — here, it's cm². $= 7 \times 3 =$ **21 cm²**

Exercise 2

Q1 Find the area of each trapezium below.

a)

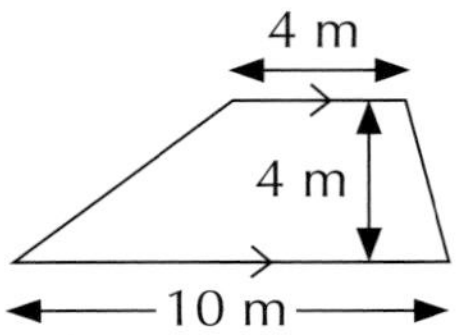

b)

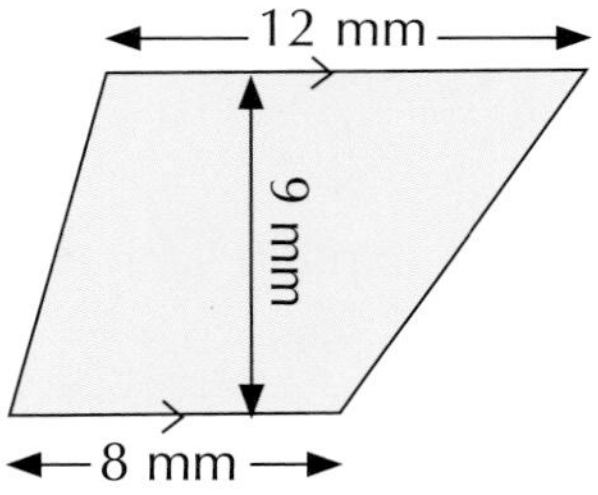

c)

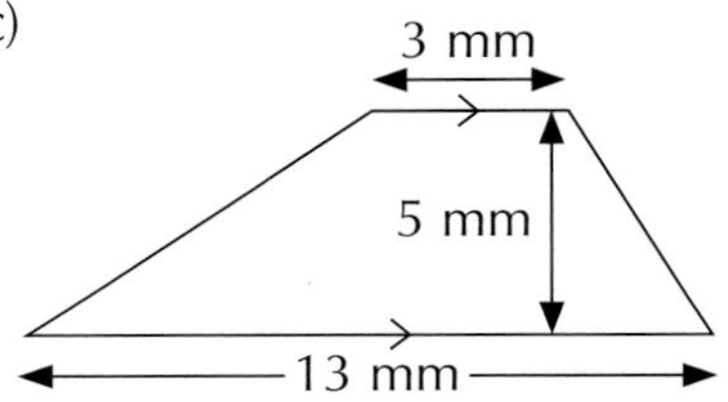

Q2 The picture below shows the end of a barn. Find: a) its total area, b) its perimeter.

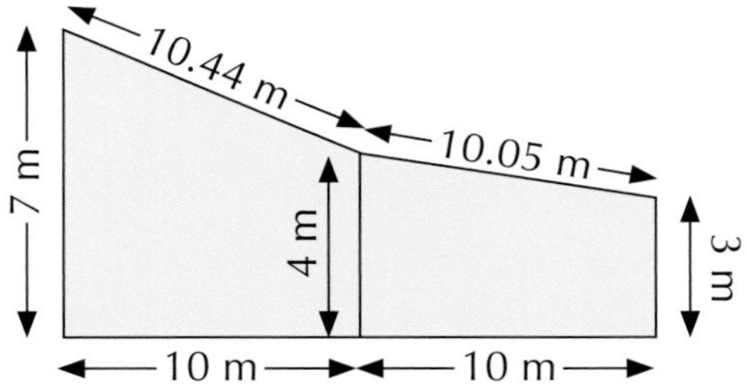

Q3 What can you say about the area of these two shapes? Explain your answer. PROBLEM SOLVING

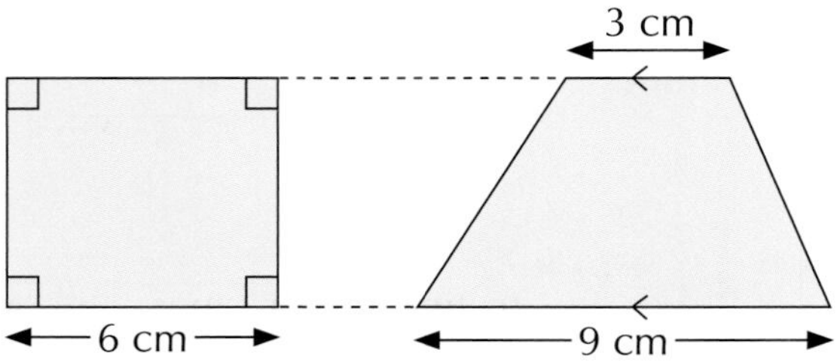

Composite Shapes involving Parallelograms and Trapeziums

Learning Objective — Spec Ref G16/G17:
Be able to find the perimeter and area of composite shapes made from parallelograms and trapeziums.

Composite shapes made up of parallelograms and trapeziums can be treated in the same way as shapes made of rectangles and triangles. When finding perimeters, remember that **opposite sides** on a parallelogram are the **same length** — this can be helpful for finding missing lengths.

Sometimes, instead of two shapes being joined together to make a bigger shape, a **big shape** can have a smaller shape **cut out** of it. In this case, **subtract** the area of the **smaller** shape from the area of the bigger one.

Example 3

For the composite shape on the right, find:
a) its area, b) its perimeter.

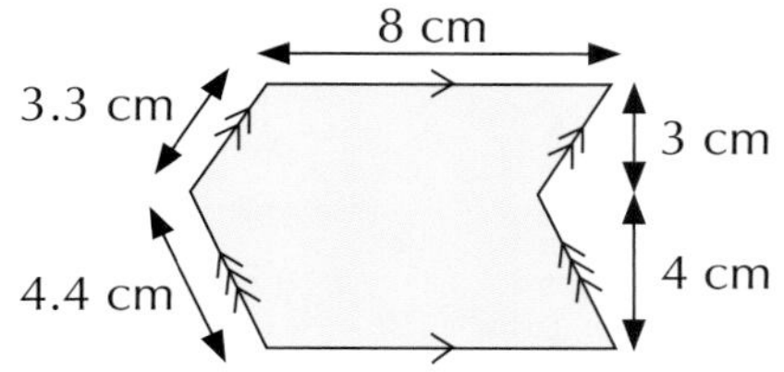

a) 1. Split the shape into two parallelograms, A and B, and find their areas.

2. Add these areas to find the total area.

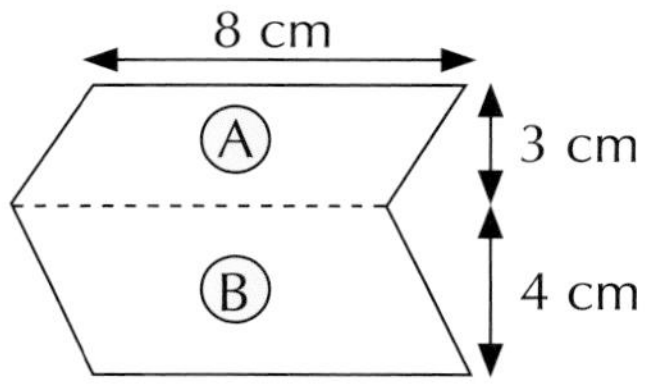

Area of A = $bh = 8 \times 3 = 24$ cm²
Area of B = $bh = 8 \times 4 = 32$ cm²
Total area = 24 + 32
= **56 cm²**

b) The perimeter is the distance around the outside of the shape.

$P = 8 + 3.3 + 4.4 + 8 + 4.4 + 3.3$
= **31.4 cm**

Example 4

Find the area of the shape on the right.

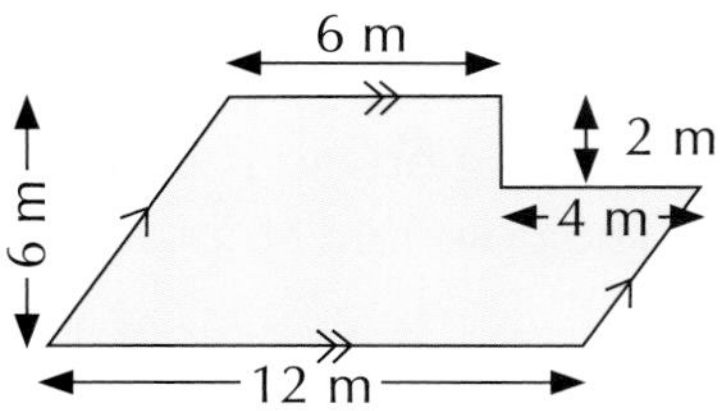

1. Form a parallelogram (A) by adding a trapezium (B) to the corner of the original shape.

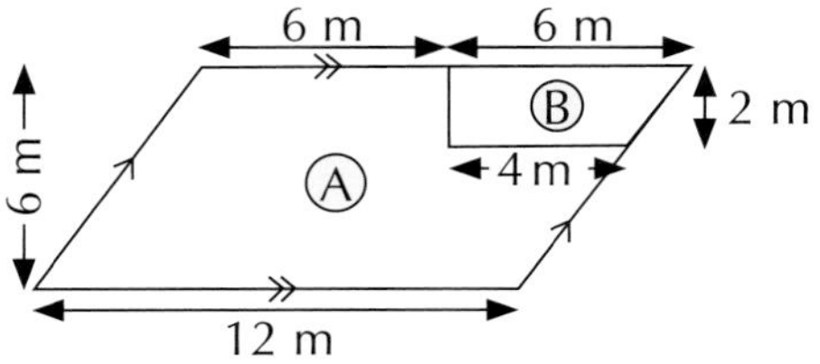

2. Find the areas of parallelogram A and trapezium B.

Area of A = bh
= $12 \times 6 = 72$ m²

Area of B = $\frac{1}{2}(a + b) \times h$
= $\frac{1}{2}(6 + 4) \times 2 = 10$ m²

3. Subtract B from A to find the original shape's area.

Area = 72 – 10
= **62 m²**

Exercise 3

Q1 Find each shaded area below.

a)

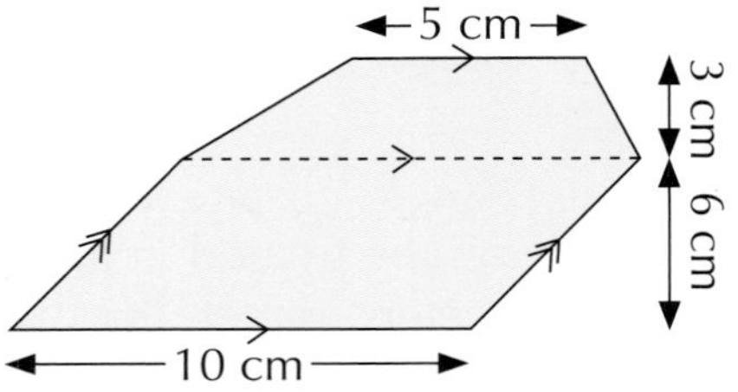

b)

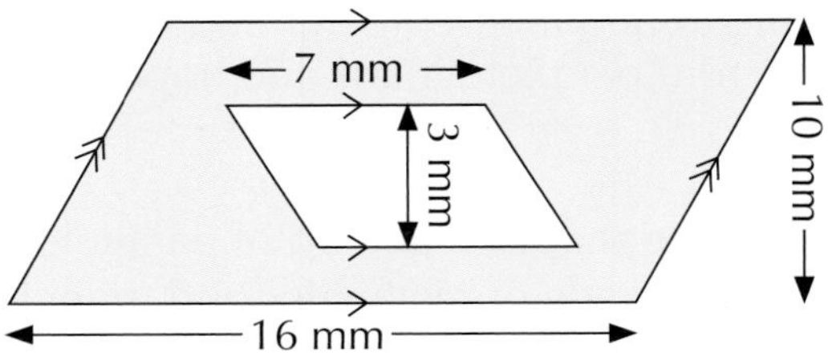

Q2 For each shape below, find: (i) the area, (ii) the perimeter.
The dotted lines show lines of symmetry.

a)

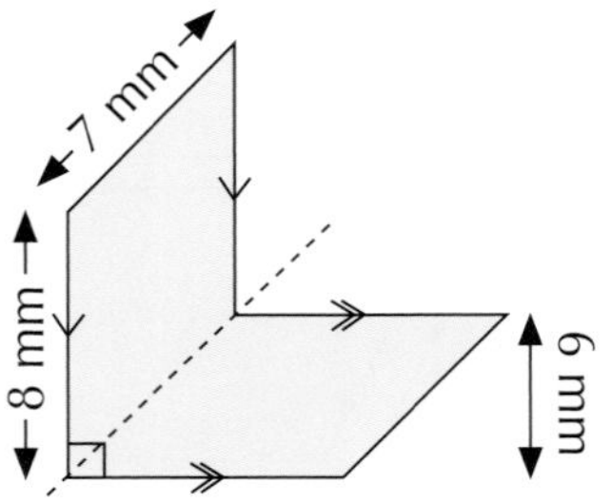

b)

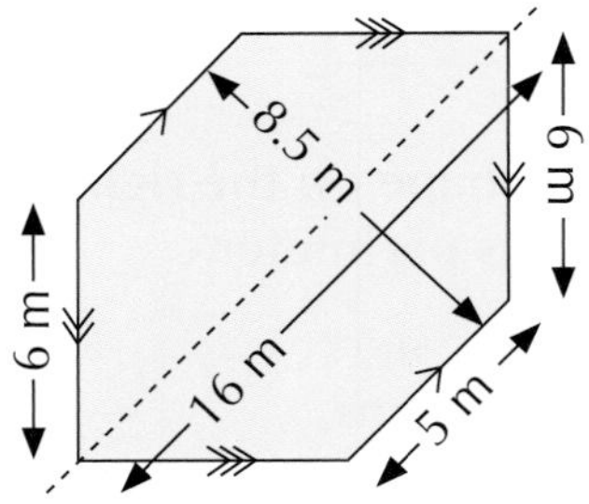

c)

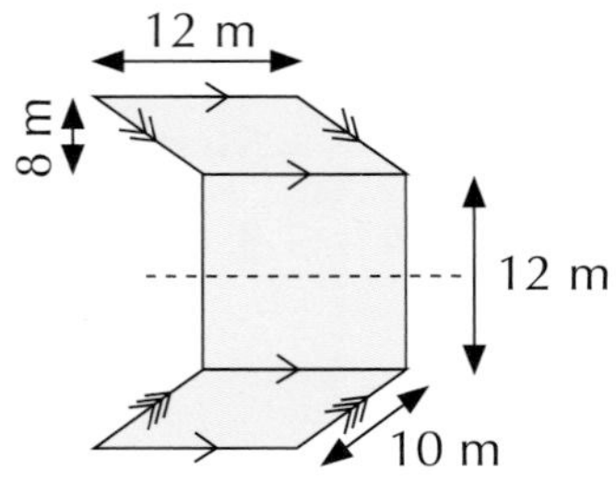

d)

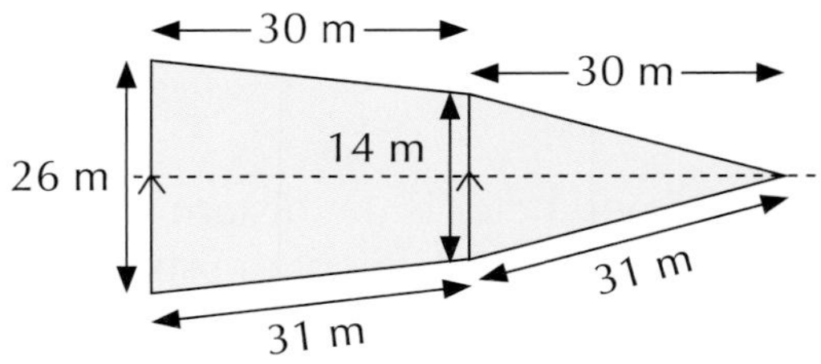

Q3 The flag below is in the shape of a trapezium.
The coloured strips along the top and bottom edges are parallelograms.

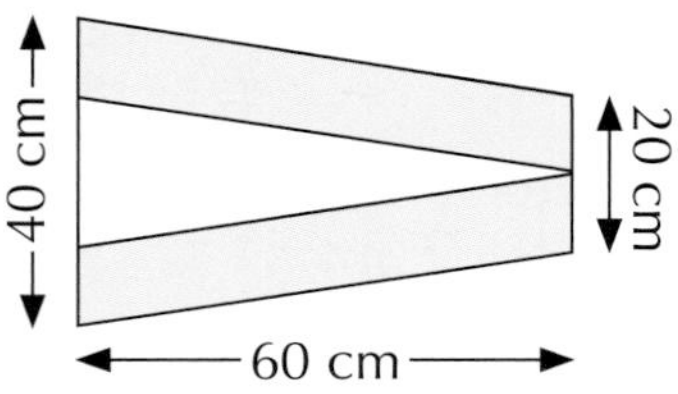

a) Find the total area of the flag.

b) Find the total area of the coloured strips.

22.3 Circumference of a Circle

There are more formulas you need to know on the next few pages, this time featuring your good friend π. Make sure you know where the π button lives on your calculator — you'll need it a lot for circle questions.

Learning Objective — Spec Ref G17:
Be able to find the circumference of a circle.

Prior Knowledge Check:
Recognise the radius, diameter and circumference of a circle. See p.323.

The **circumference** of a circle is the distance all the way around its edge. You can find the circumference (C) of a circle if you know its **diameter** (d) or its **radius** (r) using this formula: $\boldsymbol{C = \pi d = 2\pi r}$

Remember that the diameter of a circle is always **twice as long** as the radius: $\boldsymbol{d = 2r}$

You might be asked to give an **exact answer** or to give your answer **in terms of π**. This means that you **shouldn't** use your calculator to evaluate π — so instead of 31.415... you would write your answer as 10π.

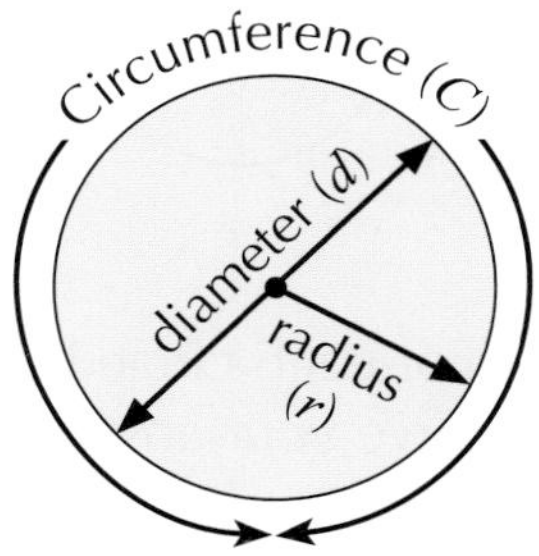

Example 1

Find the circumference of the circle shown below. Give your answer to one decimal place.

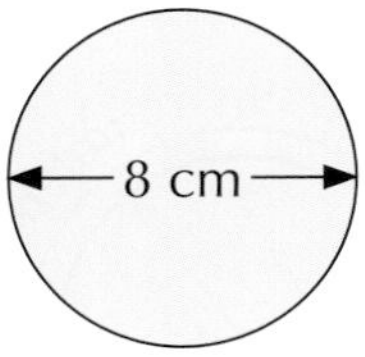

1. Write down the diameter. $d = 8$ cm
2. Use the formula $C = \pi d$. $C = \pi d$
   $= \pi \times 8$
3. Round your answer and use the correct units.
   $= 25.1327...$
   $=$ **25.1 cm** (1 d.p.)

Tip: If your calculator doesn't have a π button, then use the value 3.142.

Example 2

Find the circumference of a circle which has radius 6 m. Give your answer in terms of π.

1. Write down the radius. $r = 6$ m
2. Use the formula $C = 2\pi r$. $C = 2\pi r$
   $= 2 \times \pi \times 6$
3. Simplify your answer and use the correct units.
   $=$ **12π m**

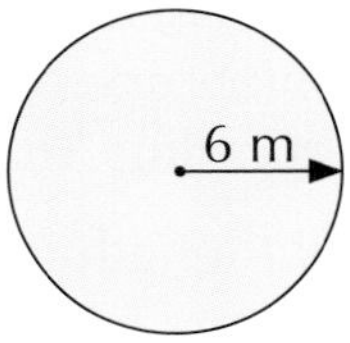

Exercise 1

Q1 Find the circumference of each circle. Give your answers to 1 decimal place.

a) 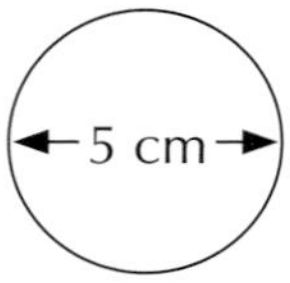

b) 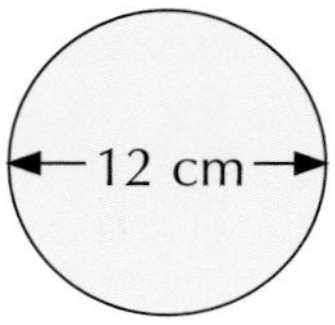

c) 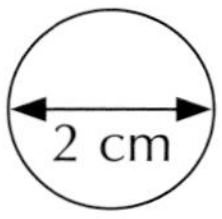

d) 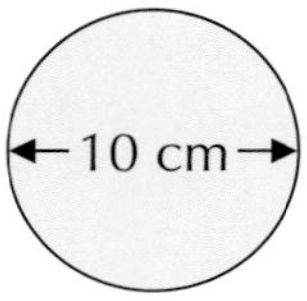

Q2 Find the exact circumference of each of these circles.

a) 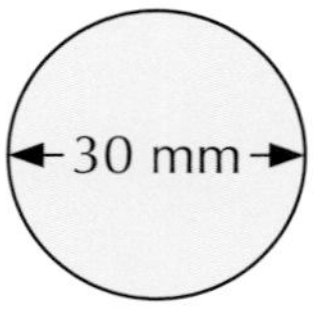

b) 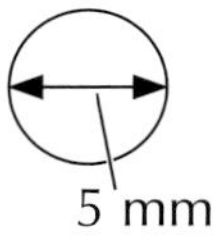

c) 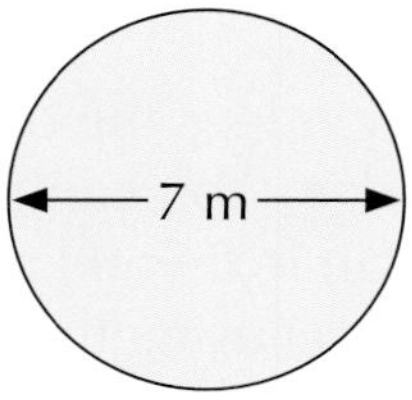

d) 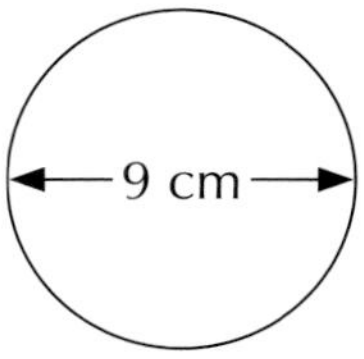

Q3 For each circle below: (i) write down the diameter, (ii) find its circumference.
Give your answers to 1 decimal place.

a) 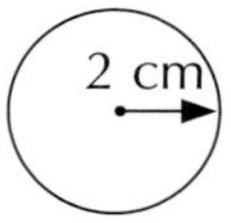

b) 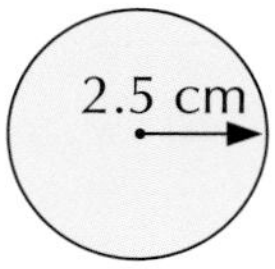

c) 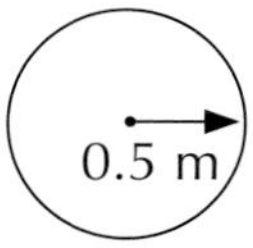

d) 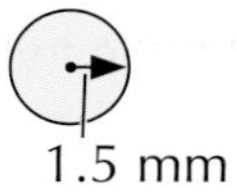

Q4 Find the exact circumference of the circles with the diameter (d) or radius (r) given below.

a) $d = 4$ cm b) $d = 8$ mm c) $r = 14$ km d) $r = 0.1$ km

A **semicircle** is **half** of a circle. You might need to find the perimeter of semicircles or **composite shapes** which include parts of circles. You can make use of the formula for the **circumference** of a circle to find the length of any **curved sides** — e.g. the curved part of a semicircle has a length of $\frac{1}{2} \times C$, where C is the circumference of the **full circle**.

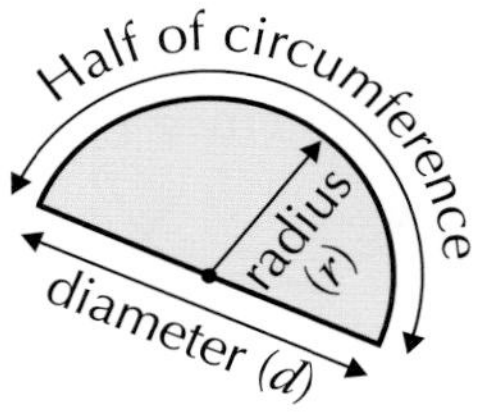

Example 3

Find the perimeter of the semicircle below, correct to one decimal place.

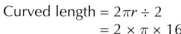

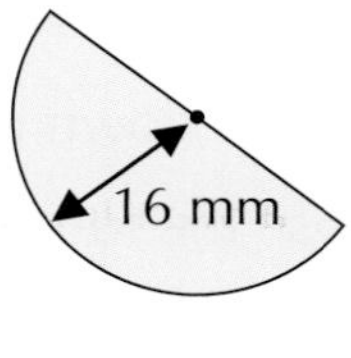

1. The curved length is half the circumference of a circle with radius 16 mm.
   Curved length $= 2\pi r \div 2$
   $= 2 \times \pi \times 16 \div 2$
   $= 50.265...$ mm
2. The straight length is the diameter of the full circle.
   Straight length $= d = 2r = 2 \times 16$
   $= 32$ mm
3. Add up the two lengths to find the perimeter.
   Total length = curved length + straight length
   $= 50.265... + 32 =$ **82.3 mm** (1 d.p.)

Example 4

The shape on the right consists of a semicircle on top of a rectangle. Find the perimeter of the shape to one decimal place.

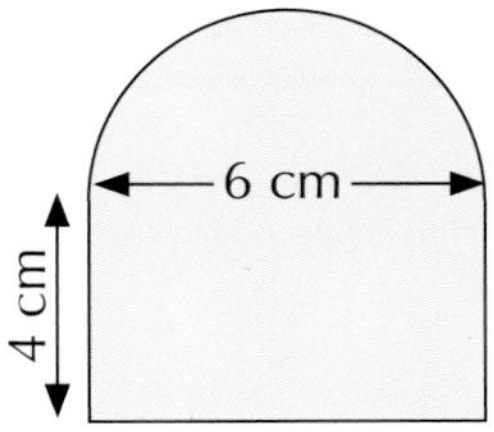

1. Find the curved length. This is half the circumference of a circle with diameter 6 cm.

   Curved length $= \pi d \div 2$
   $= \pi \times 6 \div 2$
   $= 9.424...$ cm

2. Find the total length of the straight sides.

   Total of straight sides $= 4 + 6 + 4$
   $= 14$ cm

3. Add the two parts.

   Total length = curved length + straight length
   $= 9.424... + 14$
   $=$ **23.4 cm** (1 d.p.)

Exercise 2

Q1 Find the perimeter of each shape below. Give your answers to 1 decimal place.

a) 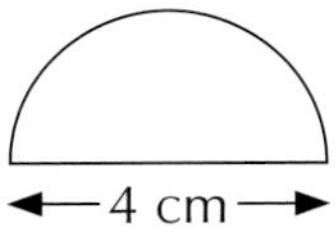

b) 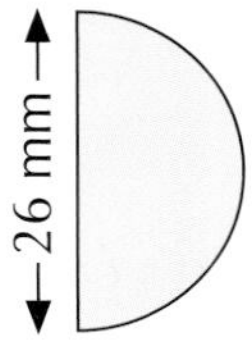

c) 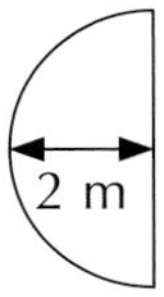

d) 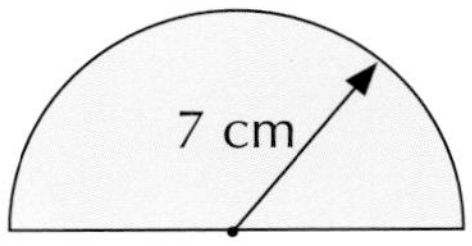

Q2 Find the perimeter of each shape below. Give your answers to 1 decimal place.

a) 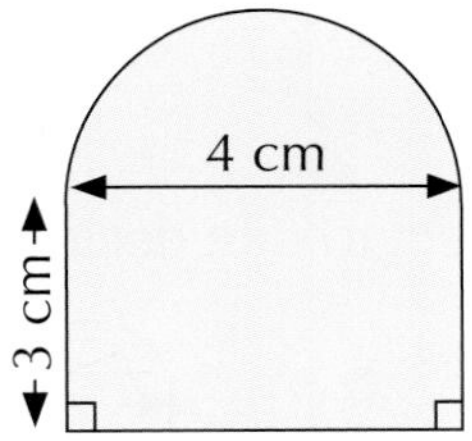

b) 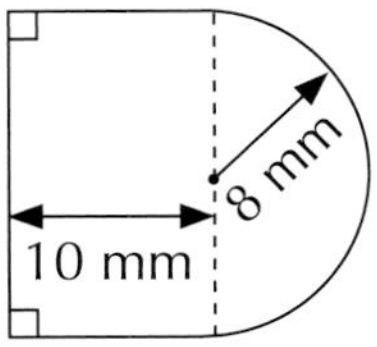

c) 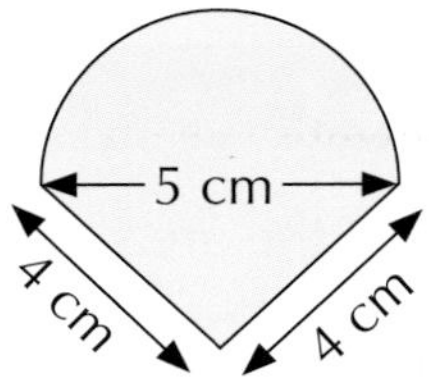

d) 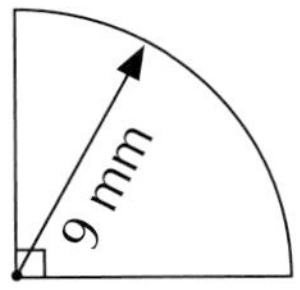

e) 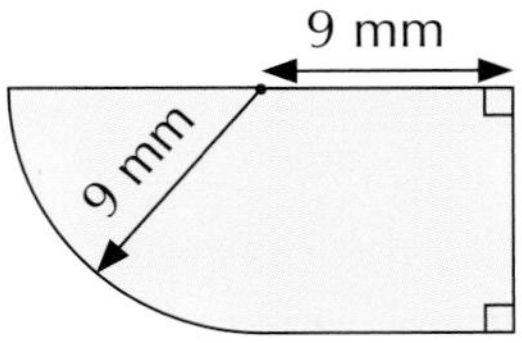

f) 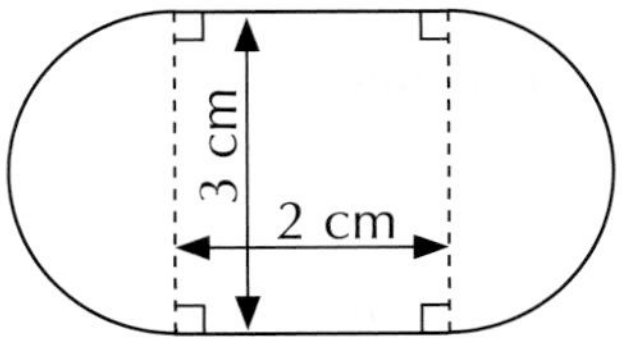

Q3 Alec used wooden fencing to build a semicircular sheep pen, with a wall forming the straight side. If the radius of the semicircle is 16 m, how many metres of fencing did Alec have to use (to 1 d.p.)?

Q4 A running track consists of two semicircles of radius 80 m, joined together by two straight sections, each with length 100 m. Calculate the length of the running track to the nearest metre.

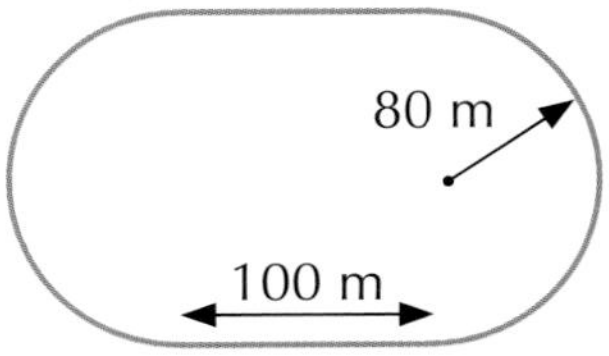

22.4 Area of a Circle

Another page, another formula. This one also involves circles and π, so make sure that you don't get it confused with the circumference formulas from earlier on.

Learning Objective — Spec Ref G17:
Be able to find the area of a circle.

The **area** (A) of a circle is given by this formula: $A = \pi r^2$

If you're given the **diameter** of the circle, make sure you **divide by 2** to get the radius before you use the formula.

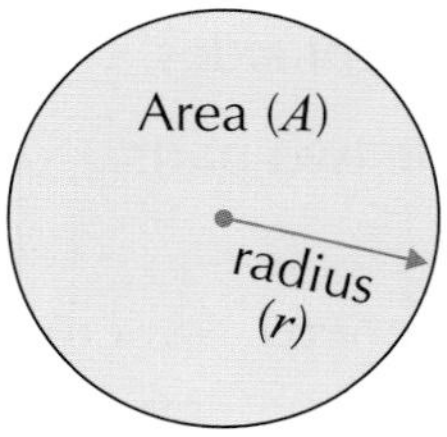

Example 1

Find the area of the circle below. Give your answer to one decimal place.

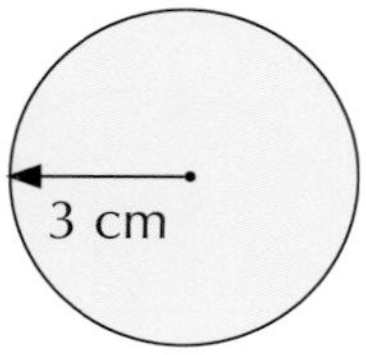

1. Write down the radius. $r = 3$ cm
2. Use the formula $A = \pi r^2$. (Remember: $r^2 = r \times r$)
   $A = \pi r^2$
   $= \pi \times 3 \times 3$
   $= 28.274...$
3. Round your answer. And use 'squared' units — here, it's cm². $A =$ **28.3 cm²** (1 d.p.)

Exercise 1

Q1 Find the area of each of these circles. Give your answers correct to one decimal place.

a) 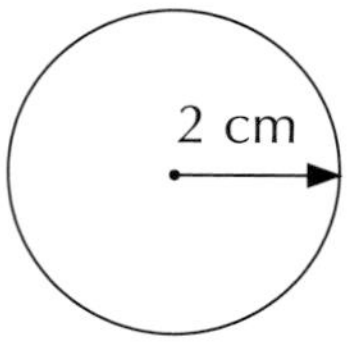

b) 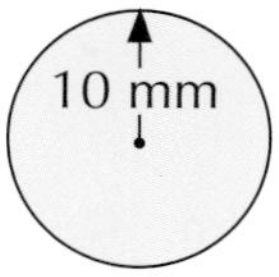

c) 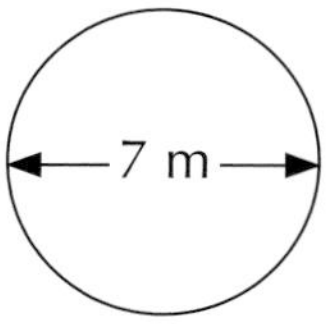

d) 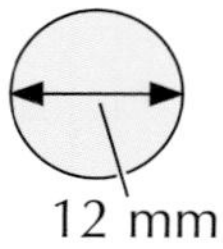

Q2 Find the areas of the circles with the diameter (d) or radius (r) given below. Give your answers in terms of π.

a) $r = 7$ mm b) $r = 5$ cm c) $d = 18$ km d) $d = 3$ m

Q3 What is the area of a circular reservoir of radius 0.82 km? Give your answer to two decimal places.

Q4 Which of the two shapes on the right has the greater area?

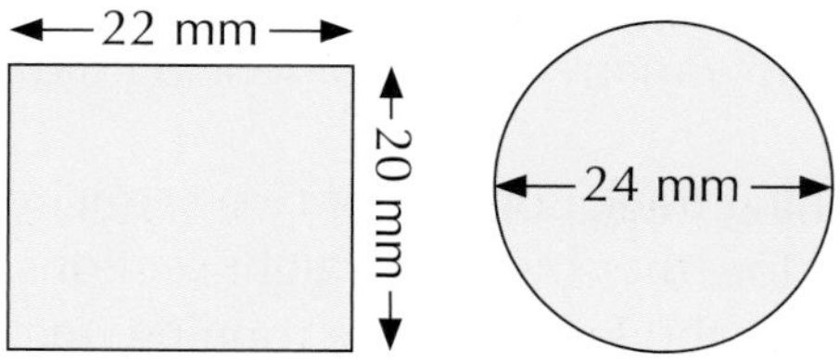

Find the area of **composite shapes** involving parts of circles by **splitting** the shape into ones that you recognise. The area of a **semicircle** is just **half** the area of the full circle.

Example 2

Find the area of the shape on the right. Give your answer to one decimal place.

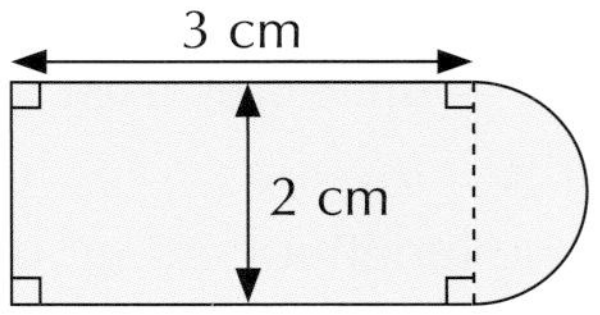

1. Divide the shape into a rectangle and a semicircle.
2. Find the area of the rectangle. Area of rectangle = $l \times w = 2 \times 3 = 6$ cm^2
3. Find the area of the semicircle. The diameter is 2 cm, so the radius is $2 \div 2 = 1$ cm. Area of semicircle = (area of circle of radius 1 cm) $\div 2$
$= \pi r^2 \div 2 = \pi \times 1^2 \div 2$
$= 1.5707...$ cm^2
4. Add up the two areas. Total area = area of rectangle + area of semicircle
$= 6 + 1.5707...$
$=$ **7.6 cm^2** (1 d.p.)

Exercise 2

Q1 Find the area of each shape below. Give your answers to one decimal place.

a)

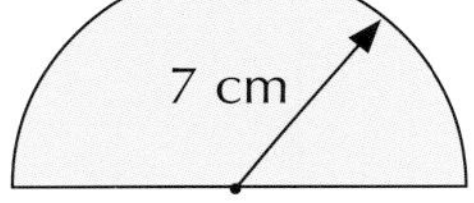

b)

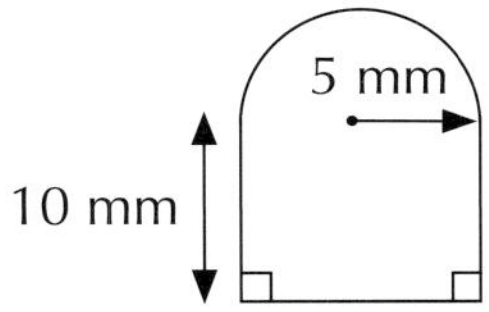

c)

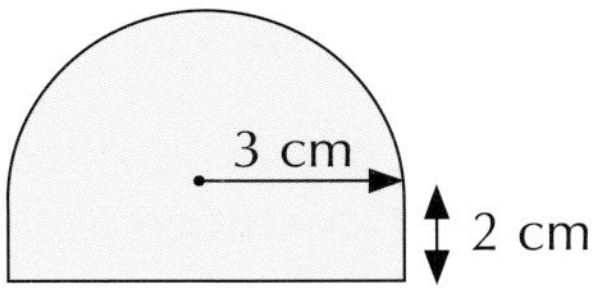

d)

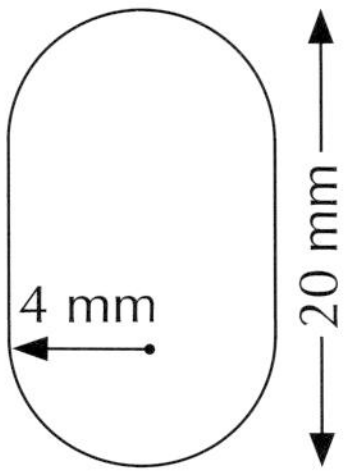

e)

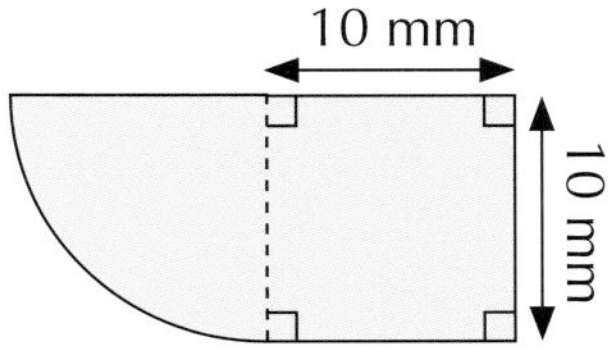

f)

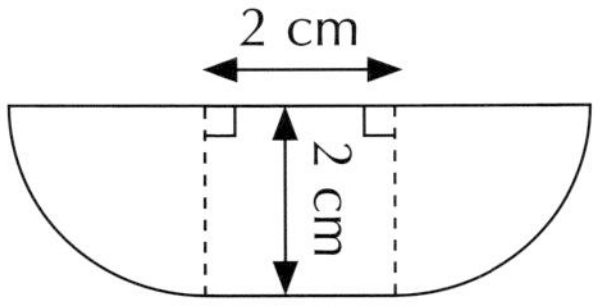

Q2 A church window is in the shape of a rectangle 1 m wide and 2 m high, with a semicircle on top, as shown on the right. What is its area to two decimal places?

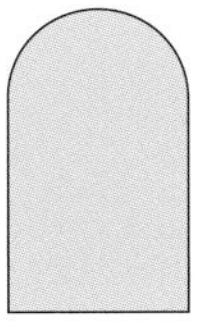

Q3 A circular pond of diameter 4 m is surrounded by a path 1 m wide, as shown in the diagram. What is the area of the path to the nearest square metre?

PROBLEM SOLVING

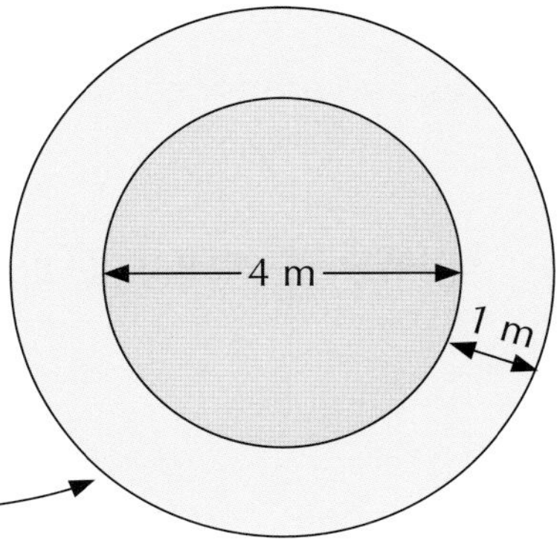

22.5 Arcs and Sectors of Circles

A couple more definitions and two more formulas to learn, and then that's circles all done.

Learning Objectives — Spec Ref G18:
- Be able to find the area of a sector of a circle.
- Be able to find the length of an arc of a circle.

Prior Knowledge Check: Recognise arcs and sectors of circles. See p.324.

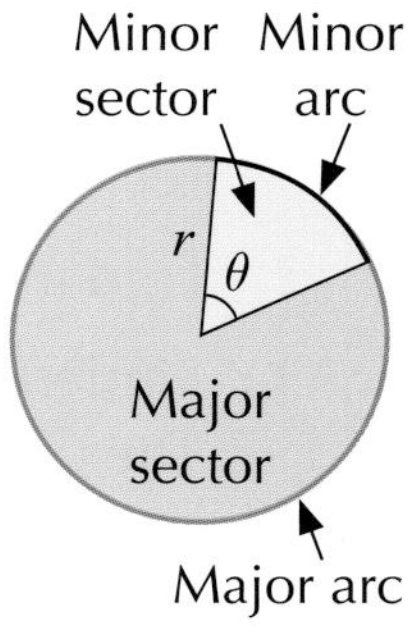

You saw on p.324 that a **sector** is a 'slice' of a circle and an **arc** is a part of the circumference. If the angle in a sector is **less than 180°** then it's called a **minor** sector, and if the angle is **more than 180°** then it's called a **major** sector.

You can use the following **formulas** to find an arc length and a sector area:

Length of arc = $\frac{\theta}{360°} \times$ circumference of circle
$= \boldsymbol{\frac{\theta}{360°} \times 2\pi r}$

Area of sector = $\frac{\theta}{360°} \times$ area of circle
$= \boldsymbol{\frac{\theta}{360°} \times \pi r^2}$

Example 1

For the circle on the right, calculate the exact length of the minor arc and the exact area of the minor sector.

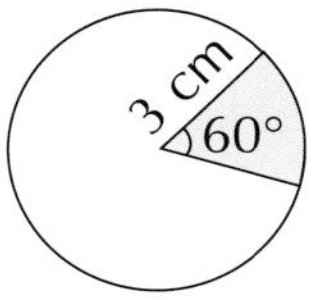

Tip: 60° < 180°, so the shaded region is the minor sector.

1. Put $\theta = 60°$ and $r = 3$ cm into the formulas.
2. For exact solutions, leave the answers in terms of π.

Length of arc $= \frac{60°}{360°} \times (2 \times \pi \times 3)$
$= \frac{1}{6} \times 6\pi = \boldsymbol{\pi}$ **cm**

Area of sector $= \frac{60°}{360°} \times \pi \times 3^2$
$= \frac{1}{6} \times 9\pi = \boldsymbol{\frac{3}{2}\pi}$ **cm²**

Exercise 1

Q1 For the circles below, find the exact minor arc length and exact minor sector area.

a) 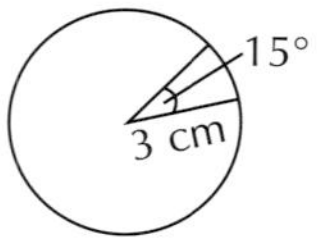

b) 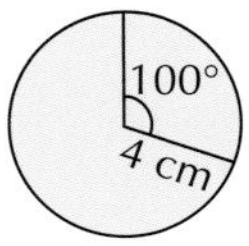

c) 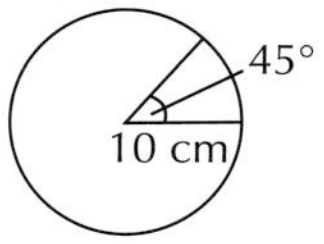

d) 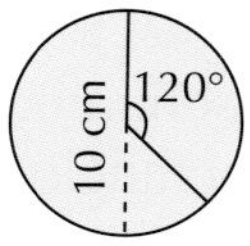

Q2 For the circles below, find the major sector area and major arc length to 2 decimal places.

a) 

b) 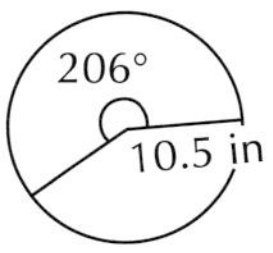

c) 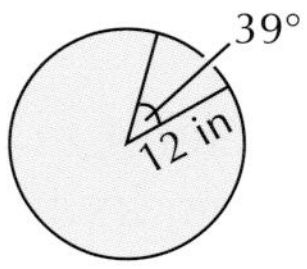

d) 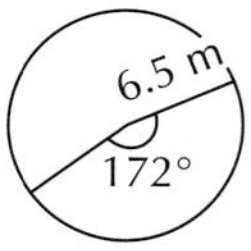

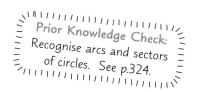

Review Exercise

Q1 For each shape below, find: (i) its perimeter, (ii) its area.

a)
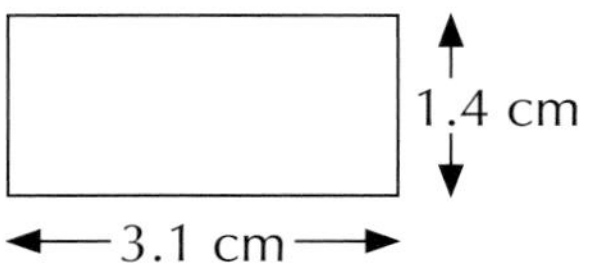

b)
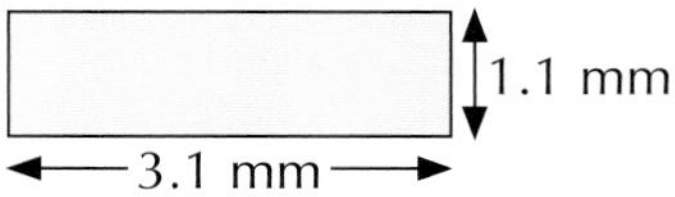

c)
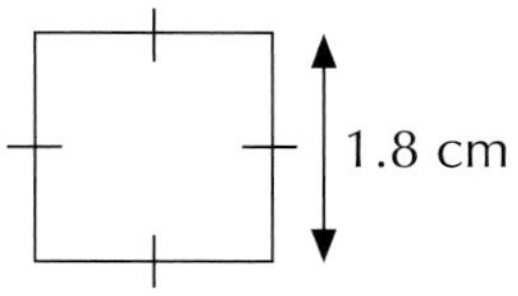

Q2 For each of the triangles below, find: (i) its perimeter, (ii) its area.

a)
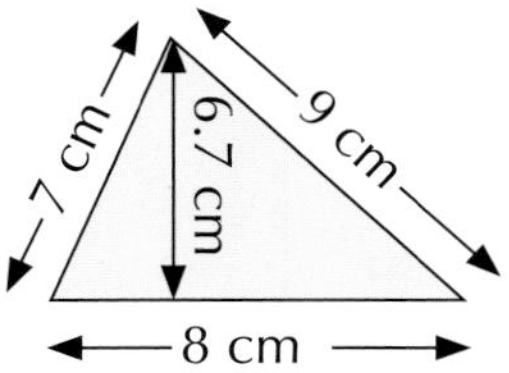

b)
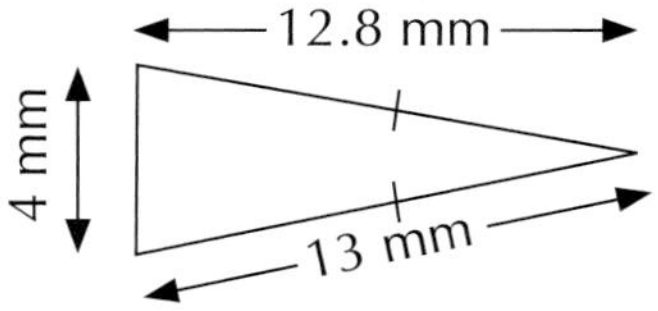

c)
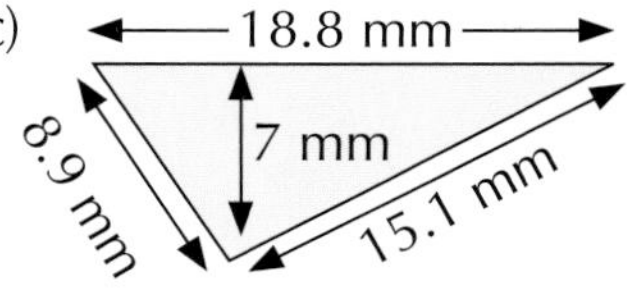

Q3 Find the area of each shape below.

a)
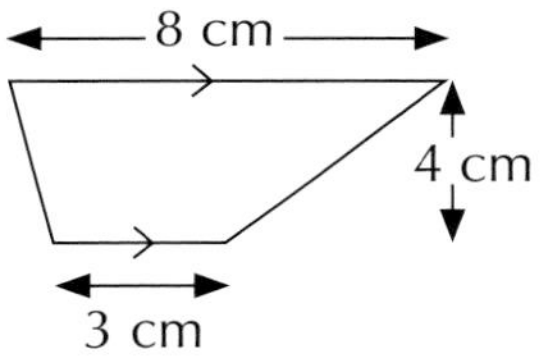

b)
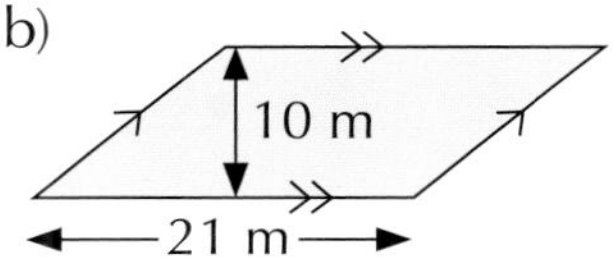

c)
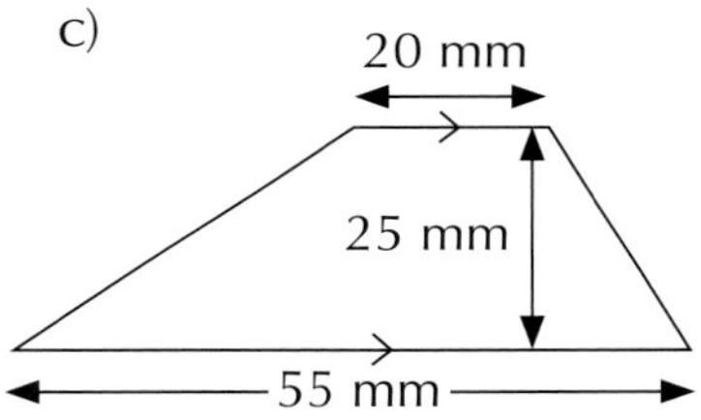

Q4 Find, to 1 decimal place, the shaded area in each of the following shapes.

a)
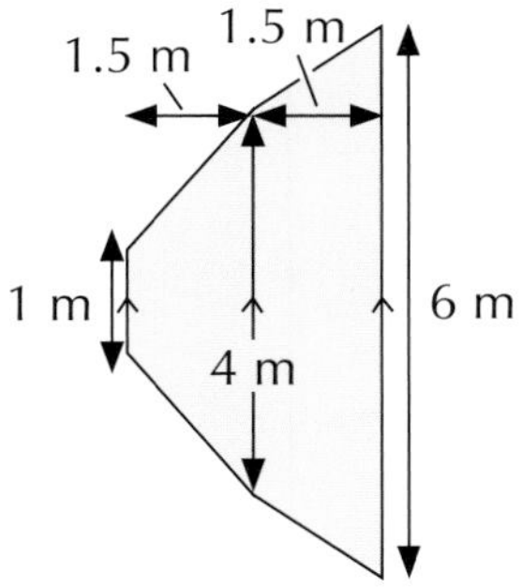

b)
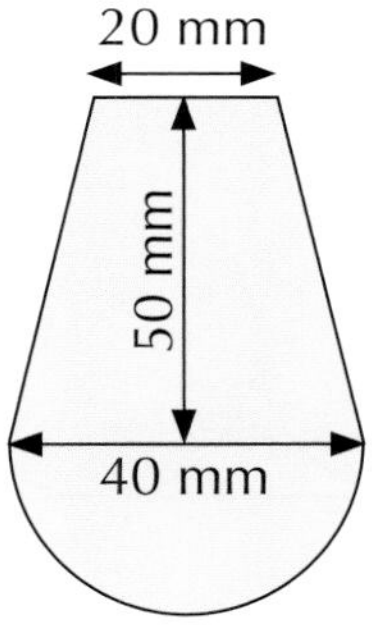

c)
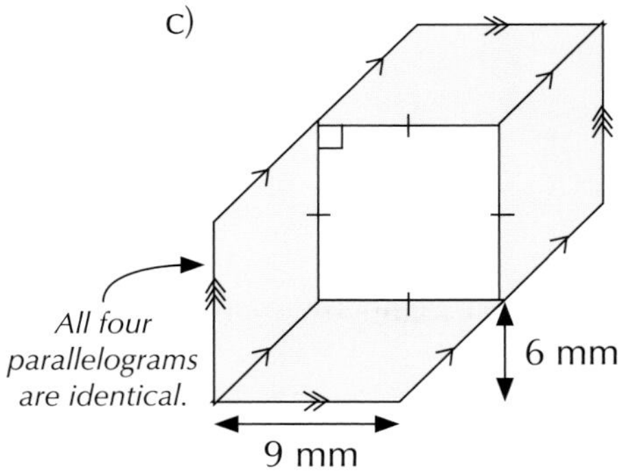

Q5 a) A 2p coin has a radius of 1.3 cm. What is its circumference to two decimal places?

b) A 5p coin has a diameter of 1.8 cm.
Which has the greater area: a 2p coin or two 5p coins?

Q6 Find the exact minor arc length and sector area for a circle with diameter 10 m and major sector angle 320°.

Exam-Style Questions

Q1 The circumference of a circle is π^2 cm.
Work out the radius of the circle.

[2 marks]

Q2 Raheel assembled the four pieces of wood shown below to make a photo frame.
Each piece of wood is in the shape of a trapezium. Find the total area of wood.

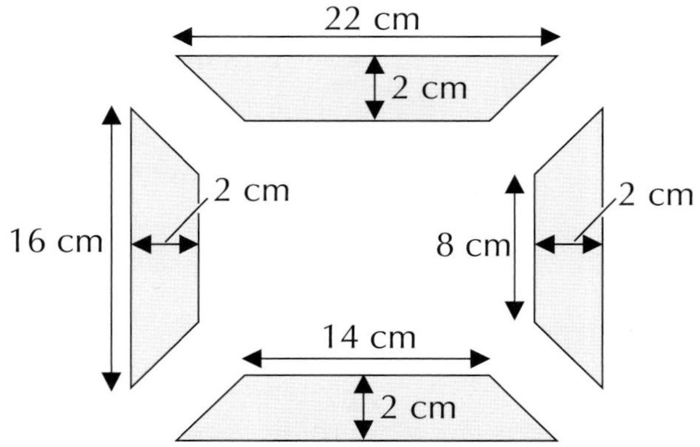

[3 marks]

Q3 The diagram below shows four identical circles enclosed within a square.
Each circle just touches two edges of the square and each of the circles next to it.

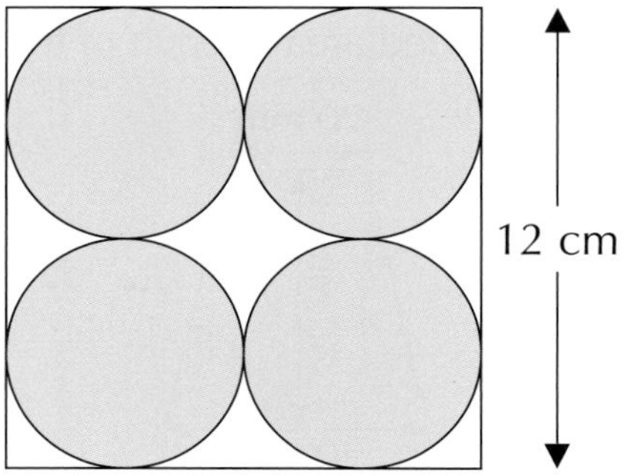

Find the area of the unshaded region.
Give your answer correct to two significant figures.

[4 marks]

Q4 The diagram on the right shows a sector of a circle with radius 16 cm. The perimeter of the sector is 41 cm.
Find the angle of the sector, θ, to the nearest degree.

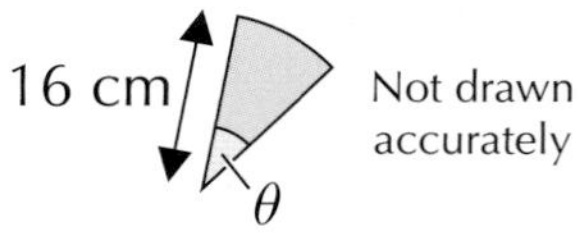

[4 marks]

Section 23 3D Shapes

23.1 Properties of 3D Shapes

3D (three-dimensional) shapes are objects that fill the real world. They have length, width and depth, unlike 2D shapes which are flat. They take all sorts of different forms, so let's start off by going through the ones you'll see most often in maths questions.

Different Solids

Learning Objective — Spec Ref G12:
Identify the faces, vertices and edges of 3D shapes.

Tip: 'Solids' is just another word for 3D shapes.

These are the most **common** solids that you need to be able to recognise.

Cube

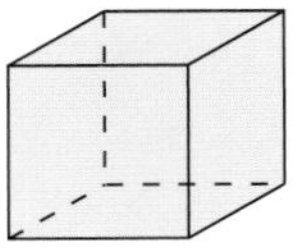

Cuboid

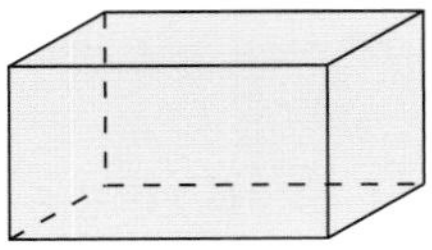

Cylinder

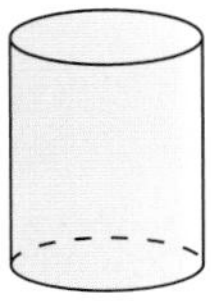

Triangular prism

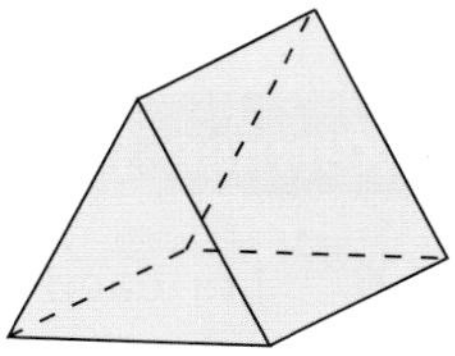

Tetrahedron (Triangular-based pyramid)

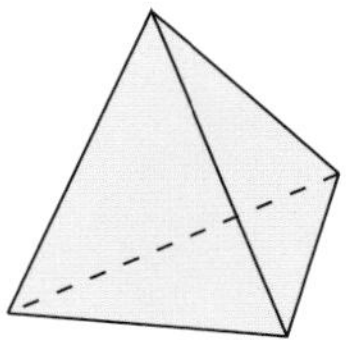

Square-based pyramid

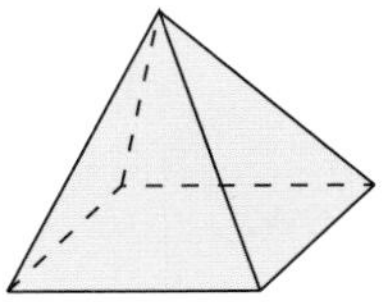

Cone

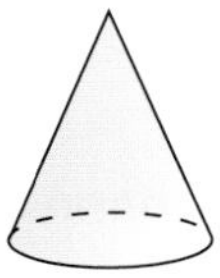

Sphere

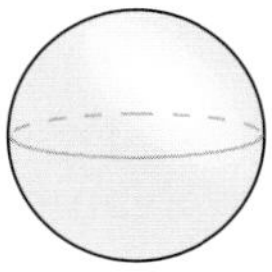

A **prism** is a 3D shape which has a **constant cross-section**.

This means that if you **slice** the shape anywhere along its length **parallel** to the faces at the end of the length, the new face you produce is **exactly the same** as those faces at the end.

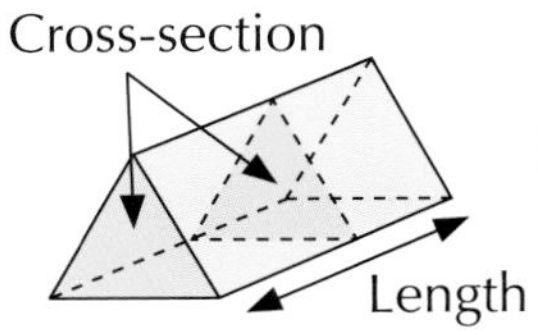

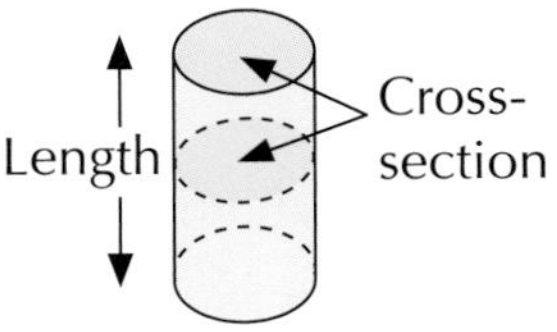

For example, a **triangular prism** has a **triangle** as its constant cross-section and a **cylinder** has a **circle**.

You need to know how to **describe** the different **parts** of a 3D shape.

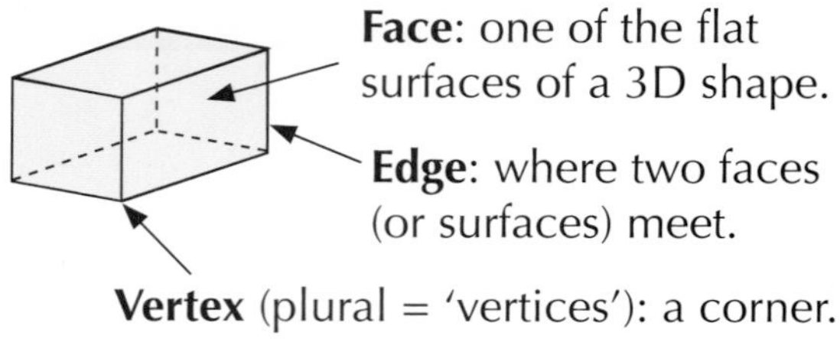

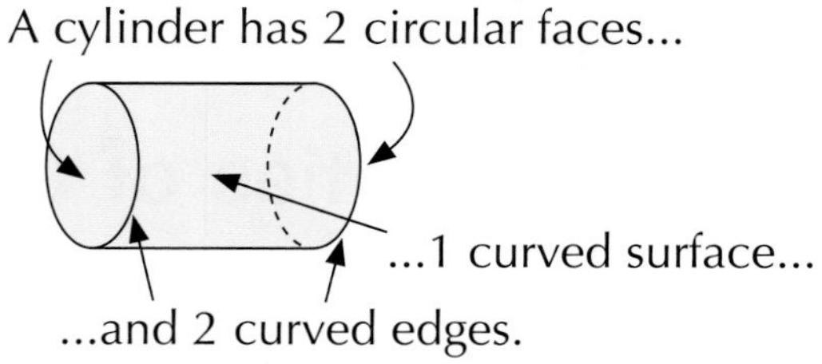

Example 1

How many faces, vertices and edges does a cube have?

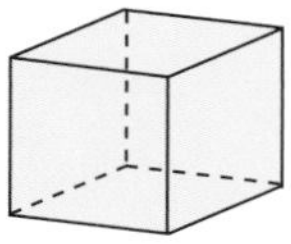

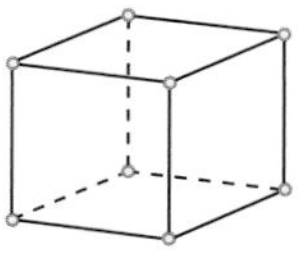

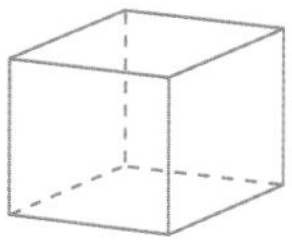

1. Count the number of flat surfaces.
2. Count the number of corners.
3. Count how many places two faces meet.

So a cube has **6 faces**, **8 vertices** and **12 edges**.

Exercise 1

Q1 For each of the solids below:

(i) give the name of the solid,

(ii) write down how many faces, how many vertices and how many edges it has.

a) 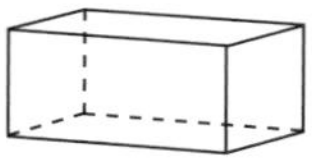 b) 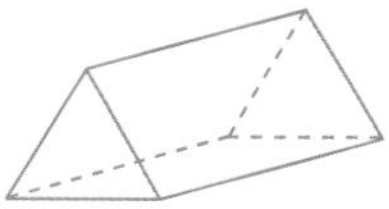 c) 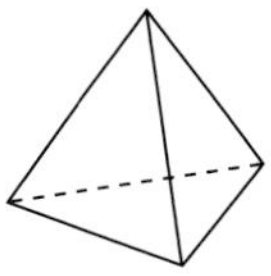 d) 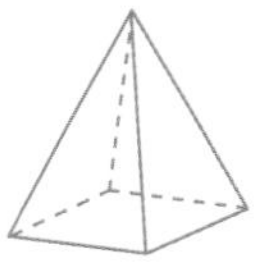

Q2 Name the 3D shapes described below.

a) 6 identical faces, 8 vertices and 12 edges

b) 2 parallel triangular faces, 3 rectangular faces

c) 4 triangular faces, 6 edges

d) 1 square face, 4 identical triangular faces

e) 2 circular faces, 1 curved surface and 2 curved edges

Q3 Which of the following 3D shapes are prisms?

A 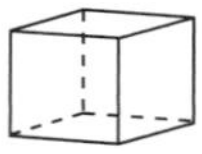 *B* 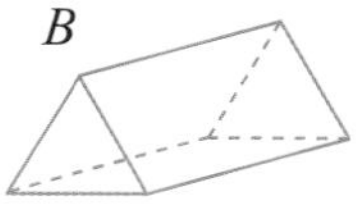 *C* 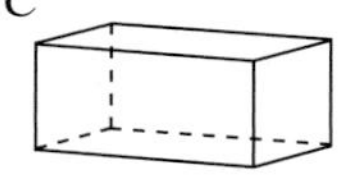 *D* 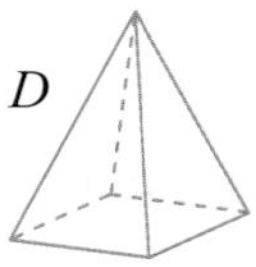 *E* 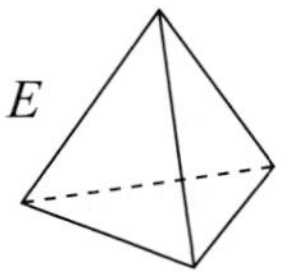

Nets

Learning Objective — Spec Ref G12:
Be able to use and draw nets of 3D shapes.

A **net** of a 3D shape is a **2D shape** (2D means two-dimensional, or '**flat**') that can be **folded** to make the 3D shape. For example, the nets of a cube and of a cylinder are shown below.

Net of a cube

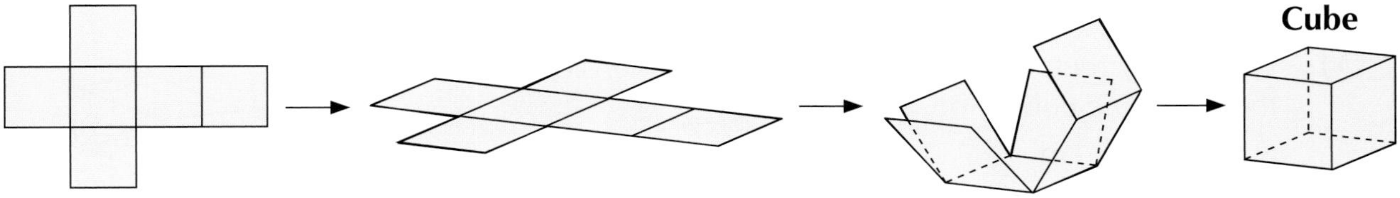

Net of a cylinder

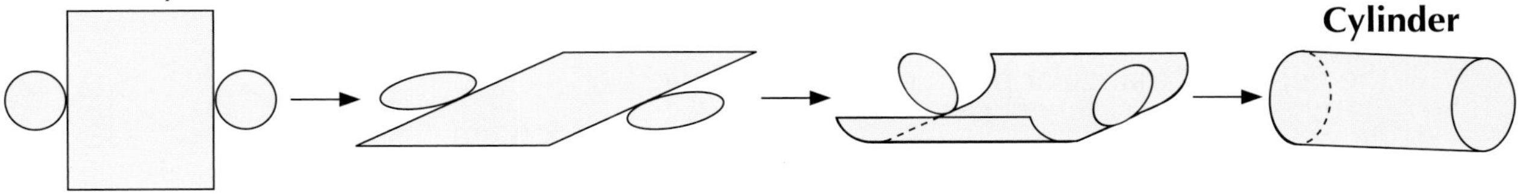

Example 2

How many triangles and how many rectangles are there in the net of a triangular prism?

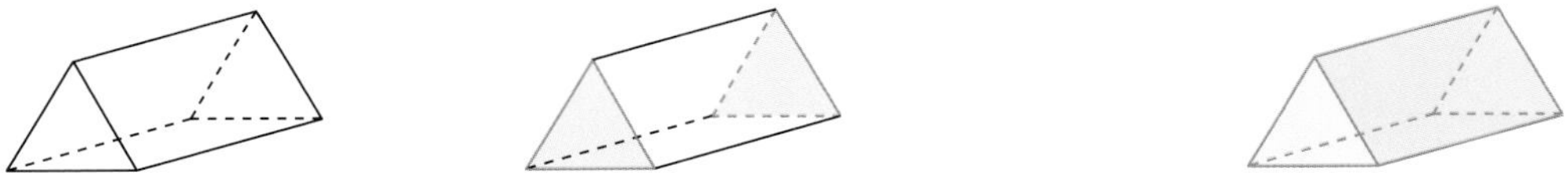

1. Sketch a triangular prism.
2. Count the triangular faces — this will be the number of triangles in the net.
3. Count the rectangular faces — this will be the number of rectangles in the net.

So the net will have **2 triangles** and **3 rectangles**.

Exercise 2

Q1 State how many (i) squares and (ii) triangles there will be in the nets of the following.

a) tetrahedron b) cube c) square-based pyramid

Q2 How many faces will the 3D shape with this net have?

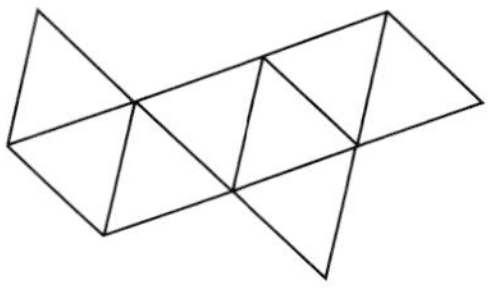

Q3 Name and sketch the 3D shapes with the following nets.
Mark the edge lengths on your sketches.

a)

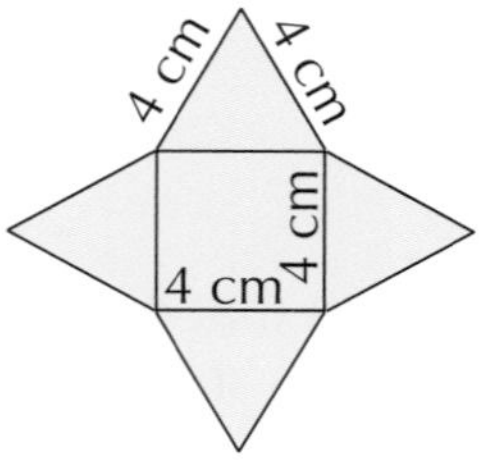

b)

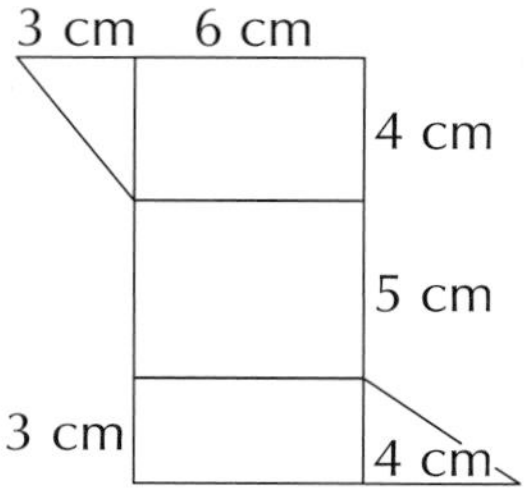

c)

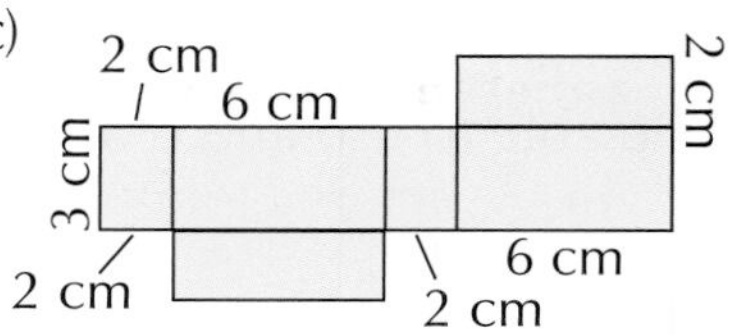

Q4 How many rectangles with the following dimensions will the net of the cuboid on the right have?

a) 2 cm × 3 cm b) 2 cm × 4 cm

c) 3 cm × 4 cm d) 2 cm × 2 cm

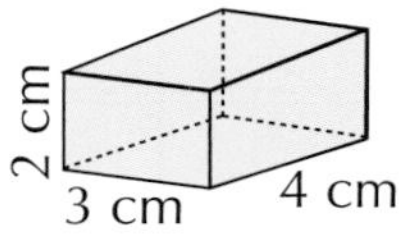

Q5 How many rectangles with the following dimensions will the net of the triangular prism on the right have?

a) 4 cm × 6 cm b) 5 cm × 6 cm

c) 3 cm × 6 cm d) 3 cm × 4 cm

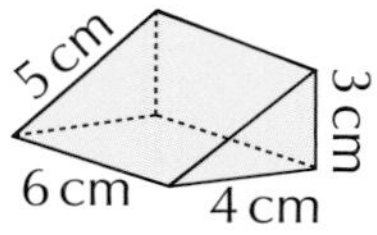

Q6 Which of the nets *A*, *B* or *C* is the net of the triangular prism shown below?

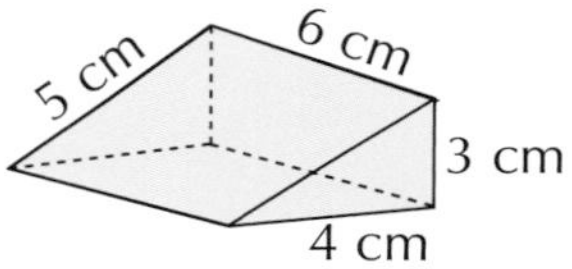

A

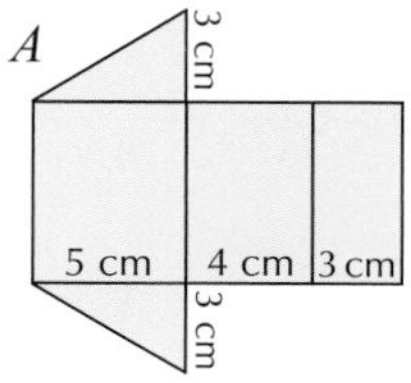

B

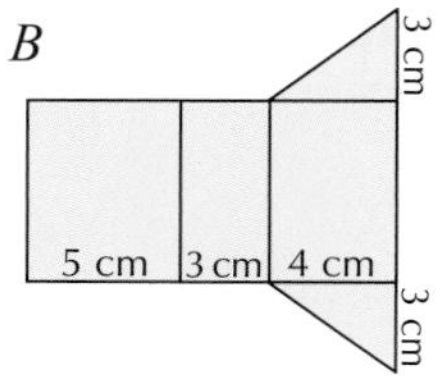

C

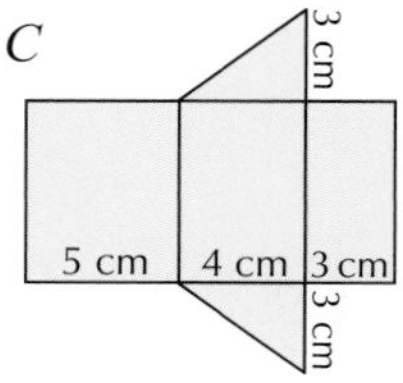

To draw the net of a 3D shape, imagine '**unfolding**' the shape so that **all its faces** are laid out **flat**. If you're not given a diagram of the 3D shape, it'll help if you make a quick **sketch** of it first.

Tip: Don't forget any hidden faces, such as the shape's base.

Example 3

Draw a net for a cuboid with dimensions 2 cm × 2 cm × 3 cm. Label the net with its dimensions.

1. Sketch the cuboid.

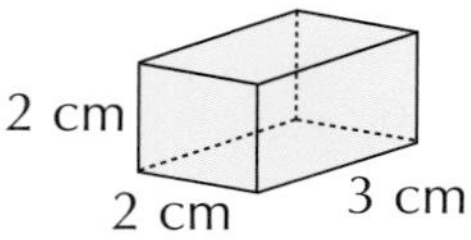

2. Draw it 'unfolded' — there could be several ways of doing this.

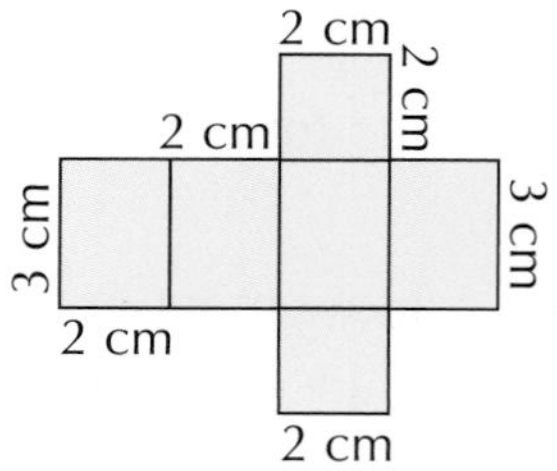

Example 4

Draw a net for the cylinder shown on the right. Label the net with its dimensions.

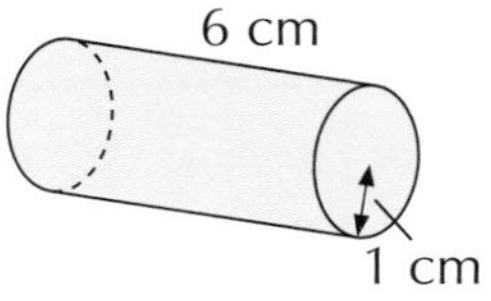

1. The tube can be 'unfolded' to give a rectangle. Its width will be 6 cm but you need to calculate its length, l.
2. The length is the same as the circumference of the circular ends of the cylinder, so use $C = 2\pi r$ (p.355).

   $l = 2\pi r$
   $= 2 \times \pi \times 1$
   $= 6.28$ cm (2 d.p.)

3. Draw and label the rectangle and add on the circular ends to the sides.

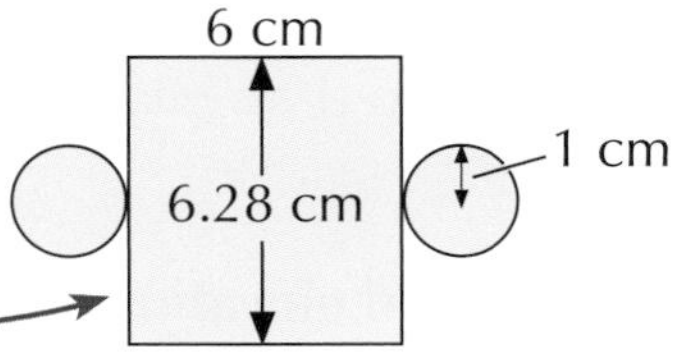

Exercise 3

Q1 Copy and complete the unfinished nets of the objects shown below.

a)

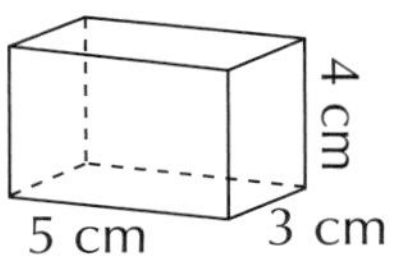

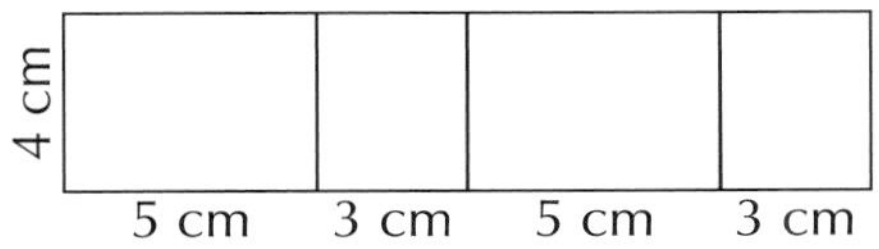

b)

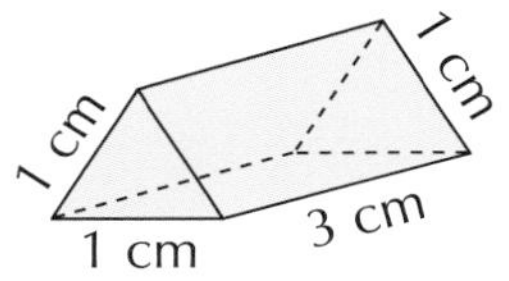

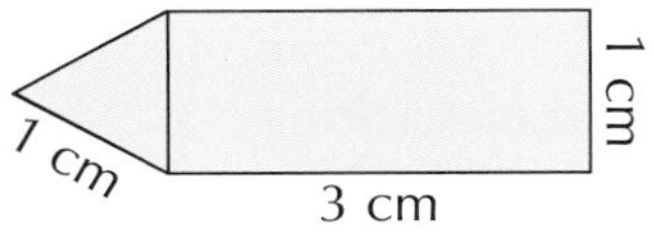

Q2 Draw a net of each of the following objects. Label each net with its dimensions.

a)

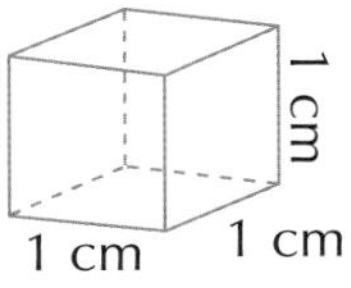

b)

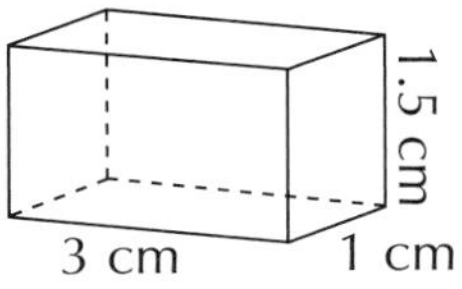

c)

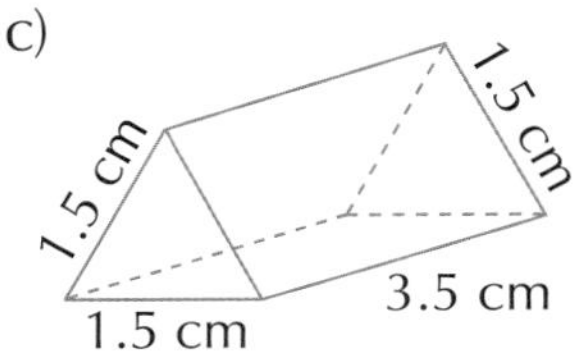

d)

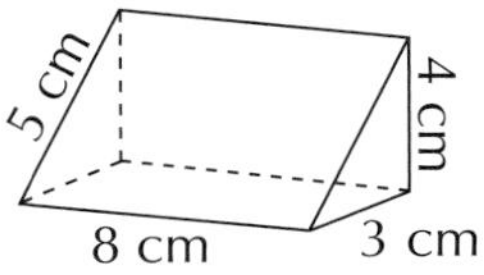

e)

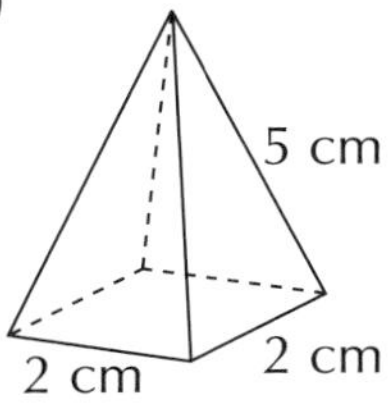

f)

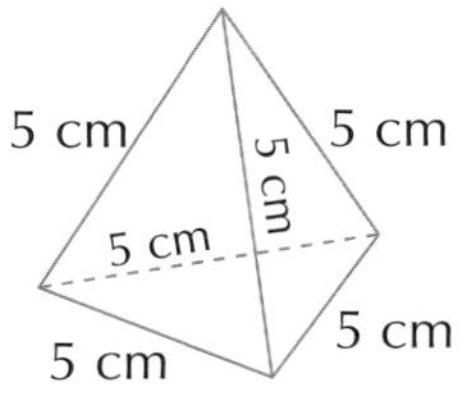

g)

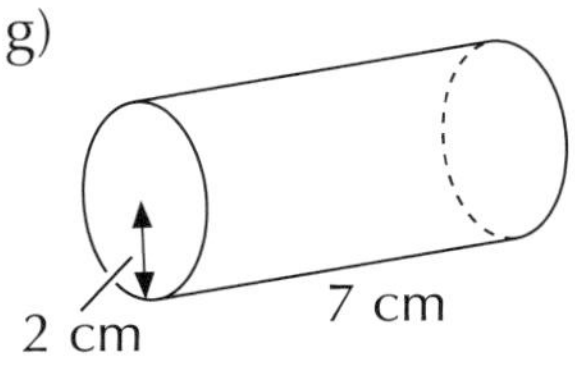

h)

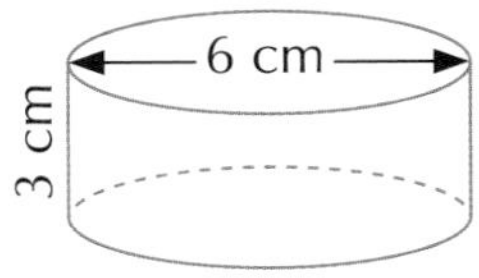

Q3 Draw a net of each of the following objects. Label each net with its dimensions.

a) a cube with 2 cm edges

b) a 1.5 cm × 2 cm × 2.5 cm cuboid

c) a tetrahedron with 3.5 cm edges

d) a cylinder of length 4 cm and radius 2.5 cm

23.2 Plans and Elevations

It can be difficult to draw 3D shapes accurately on paper — sometimes it's clearer to draw 2D plans and elevations of the 3D shape to show what it looks like from different sides.

Learning Objective — Spec Ref G13:
Be able to draw plans and elevations of 3D shapes.

Plans and **elevations**, also known as **projections**, are **2D representations** of **3D objects** viewed from particular directions. There are **three** different projections:

- **Plan** — the 2D view looking **vertically downwards** on the 3D object.
- **Front elevation** — the 2D view looking **horizontally** from the **front** of the 3D object.
- **Side elevation** — the 2D view looking **horizontally** from the **side** of the 3D object.

Tip: The directions for the front and side elevations are usually indicated by arrows.

Example 1

Draw the plan and elevations of the shape on the right from the directions indicated.

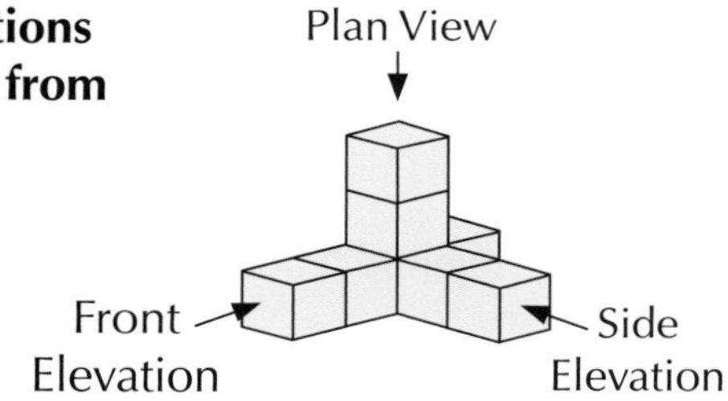

Tip: For the elevations, imagine you're standing in front and to the side of the shape.

1. Viewed from above, the shape has 6 squares arranged in a sideways T-shape. You can't tell that there are 2 cubes on top of the base layer.

Plan View

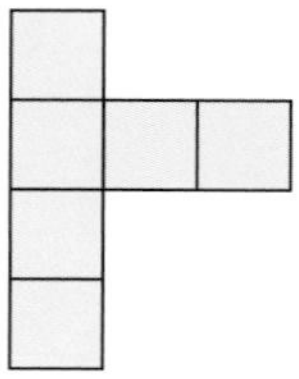

2. Viewed from the front, the shape has 5 squares in an L-shape. You can't see the change in depth.

Front Elevation

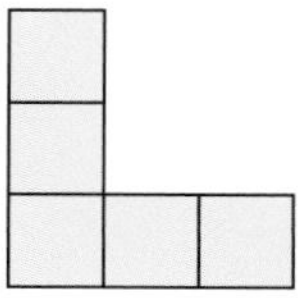

3. Viewed from the side, the shape has 6 squares in an upside-down T-shape. Again, you can't see a change in depth from this elevation.

Side Elevation

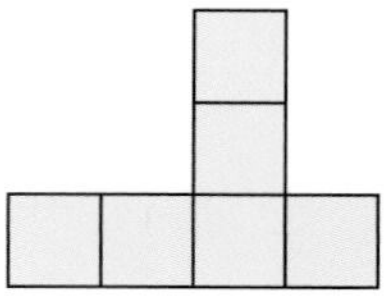

Exercise 1

Q1 Below are the front and side elevations of the given objects.
Draw the plan view for each.

a)

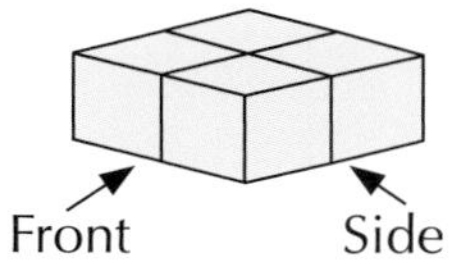

Front Elevation 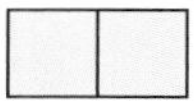

Side Elevation 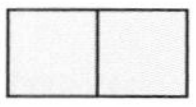

b)

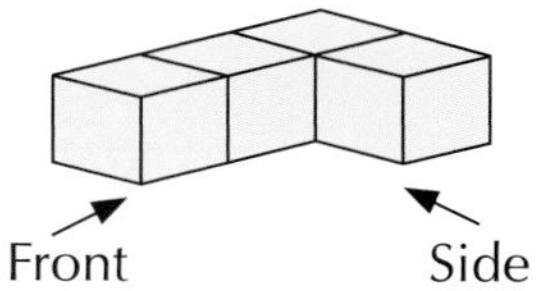

Front Elevation 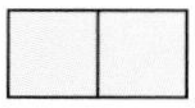

Side Elevation 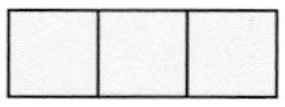

Q2 Below are the plan view and front elevation of the given objects.
Draw the side elevation for each.

a)

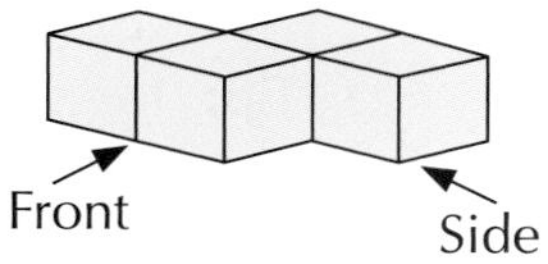

Plan View 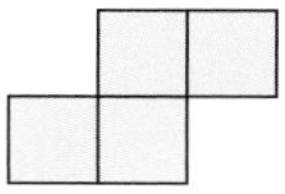

Front Elevation 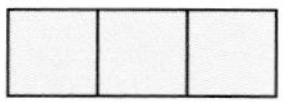

b)

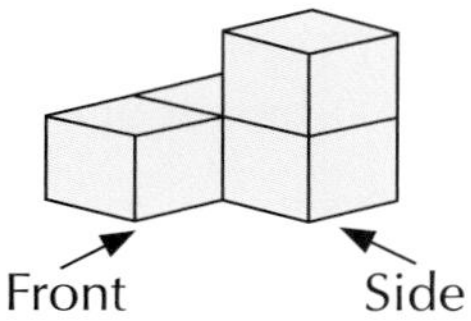

Plan View 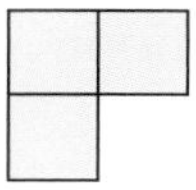

Front Elevation 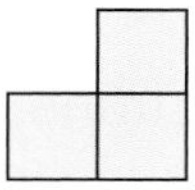

Q3 For each of the following, draw:

(i) the plan view,

(ii) the front and side elevations, using the directions shown in a).

a)

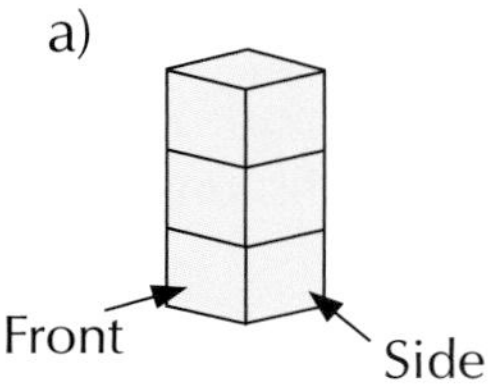

b)

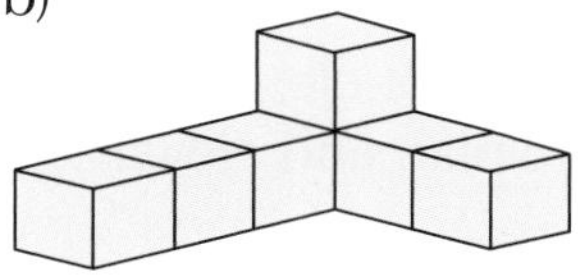

c)

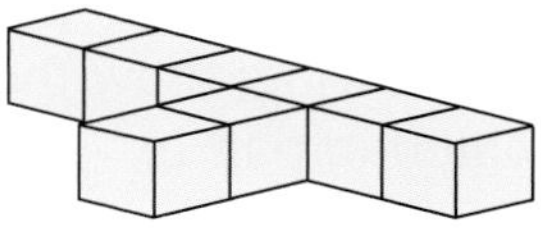

d)

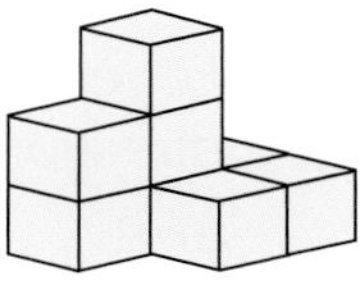

Q4 Which of the solid objects *A*, *B* or *C* below corresponds to the plan and elevations shown?

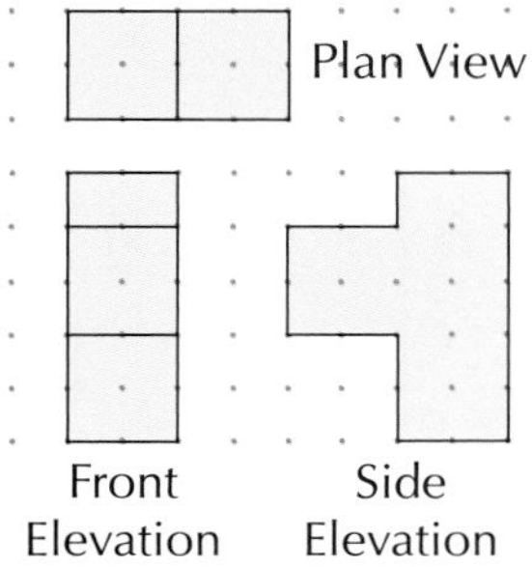

A

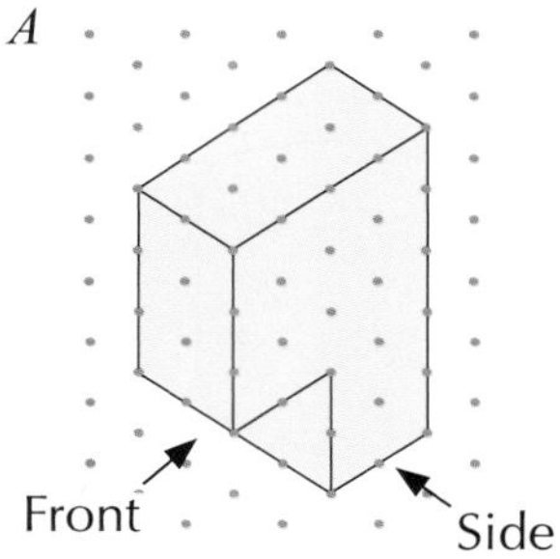

B

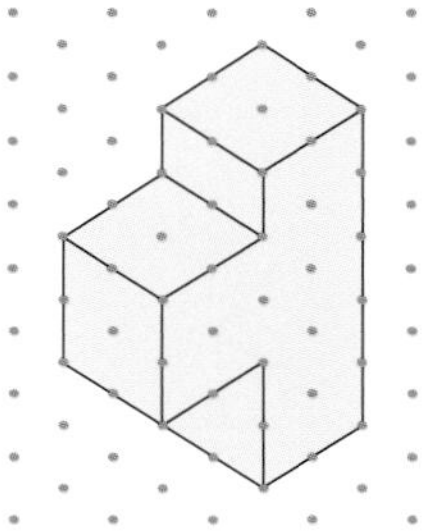

C

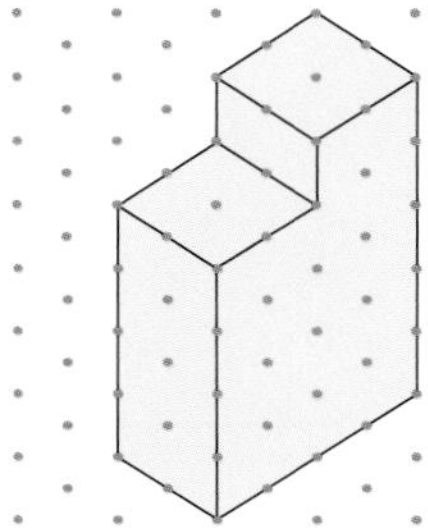

Projections are trickier when the 3D shape has **sloped** or **curved** faces, but it helps to think about how the shape would look if it was **squashed flat** from the direction you're given. A **sloping** face is always projected as the **same type** of shape (e.g. a sloping triangular face will be projected onto a triangle) — but its **dimensions** will be decided by the dimensions of **other faces** of the shape. The example below shows how this works.

Example 2

Draw the plan and elevations of this triangular prism. Label the plan and elevations with their dimensions.

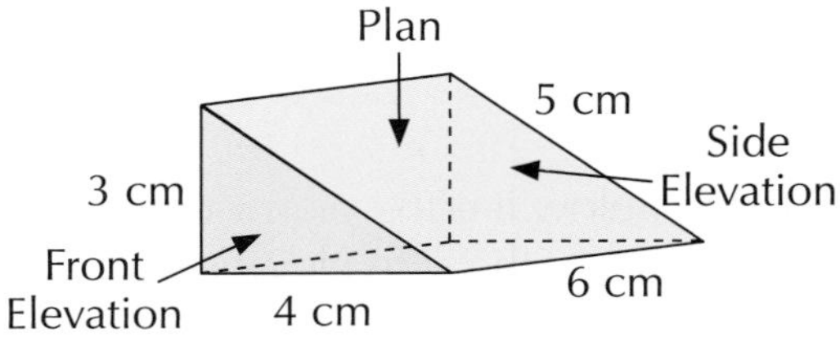

1. Viewed from above, the triangular prism appears as a 6 cm × 4 cm rectangle.

Plan:

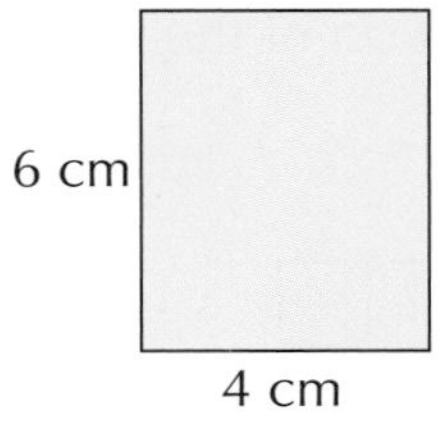

2. Viewed from the side, the triangular prism appears as a 6 cm × 3 cm rectangle.

Side elevation:

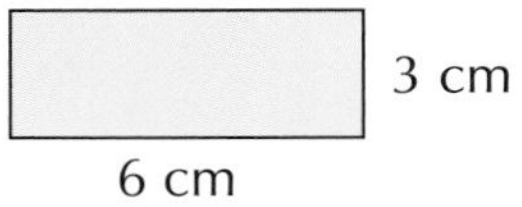

3. Viewed from the front, the triangular prism appears as a triangle with side lengths 3 cm, 4 cm and 5 cm.

Front elevation:

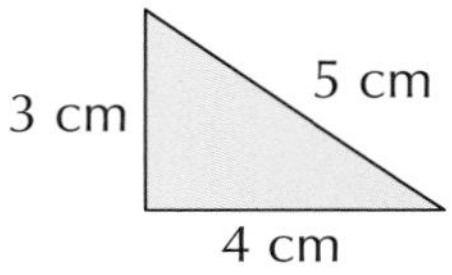

Tip: The width of the plan is decided by the width of the prism's base, and the height of the side elevation is decided by the height of the prism.

Exercise 2

Q1 For each of the following, draw the plan, front elevation and side elevation from the directions indicated in part a). Label the plan and elevations with their dimensions.

a)

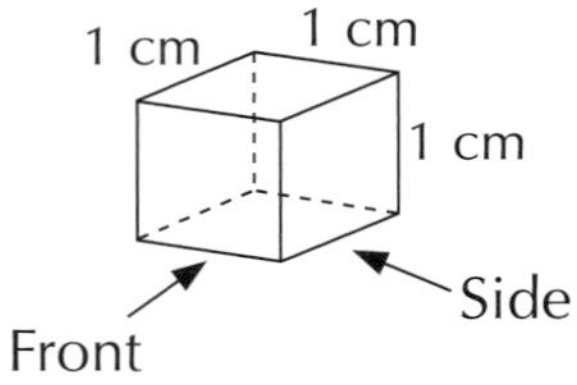

b)

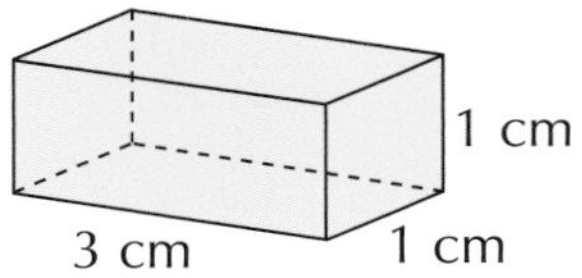

c)

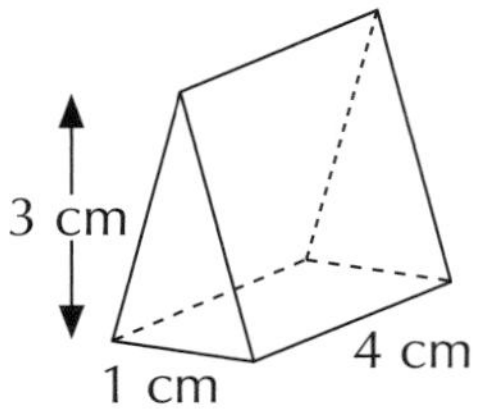

d)

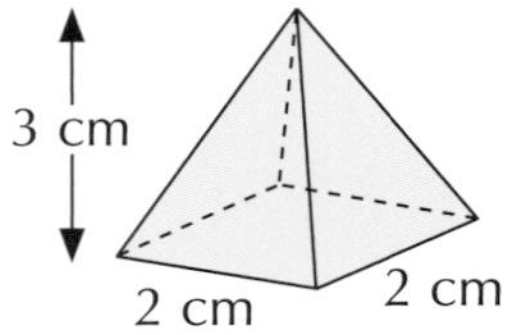

e)

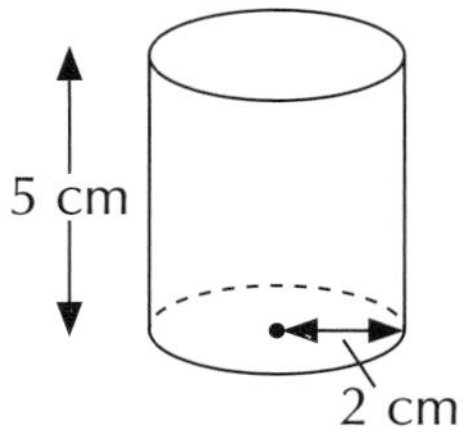

f)

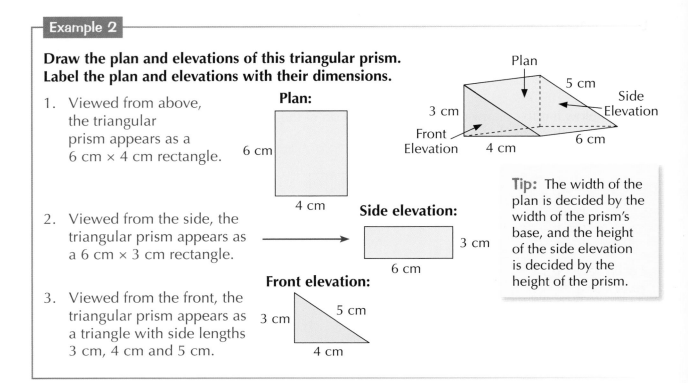

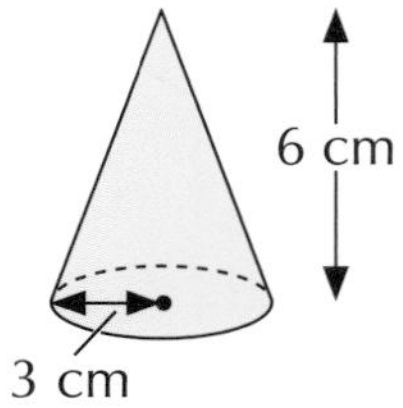

23.3 Isometric Drawings

Pictures of 3D shapes drawn on a grid of dots or lines arranged in a pattern of equilateral triangles are called isometric drawings.

Learning Objective — Spec Ref G13:
Be able to draw 3D shapes on isometric paper.

The grid of **dots** and grid of **lines** on the right are examples of **isometric paper**.

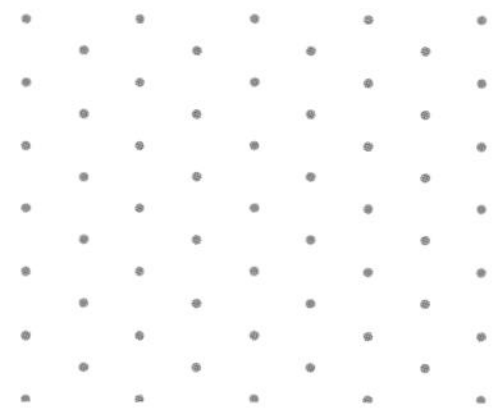

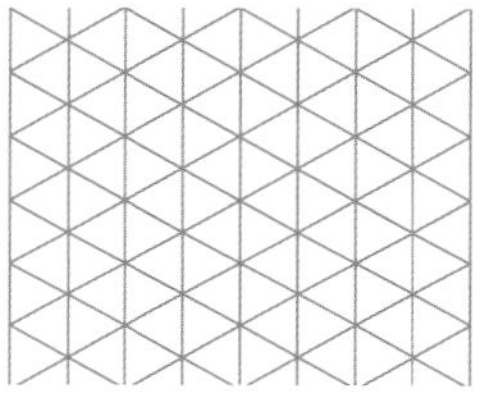

To draw a 3D shape on isometric paper, you have to use the following rules:

- **Vertical lines** on the shape are shown by **vertical lines** on the isometric paper.
- **Horizontal lines** on the shape are shown by **diagonal lines** on the isometric paper.

Each space between the dots in the vertical or diagonal directions represents **one unit of length** (e.g. 1 cm or 1 m).

Example 1

Draw the triangular prism shown on the right on isometric paper.

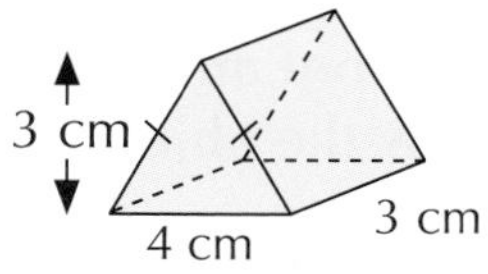

1. Join the dots with vertical lines for the vertical lines in the 3D shape and with diagonal lines for the horizontal lines.
2. Build up the drawing as shown below, using the dots to show the dimensions.
3. The triangle has a vertical height of 3 cm, so draw a vertical line 3 spaces long.
4. It has a horizontal width of 4 cm, so draw a diagonal line 4 spaces long.
5. The depth of the object is 3 cm so draw a diagonal line 3 spaces long.

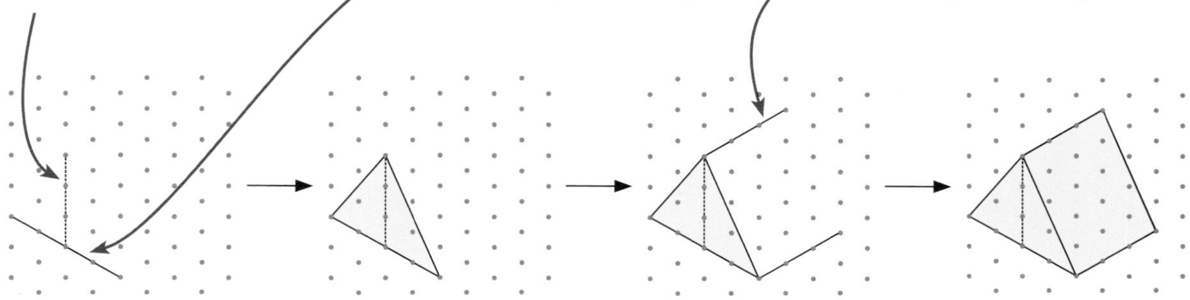

Exercise 1

Q1 Draw the following cuboids and prisms on isometric paper.

a) 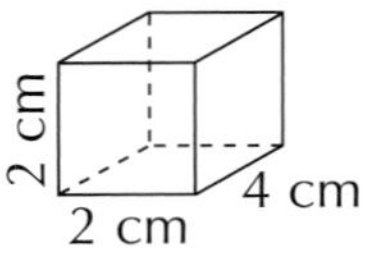

b) 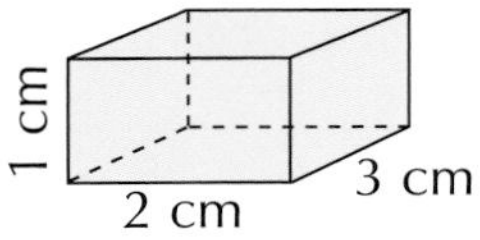

c) 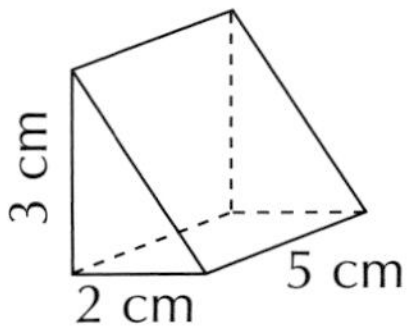

d) 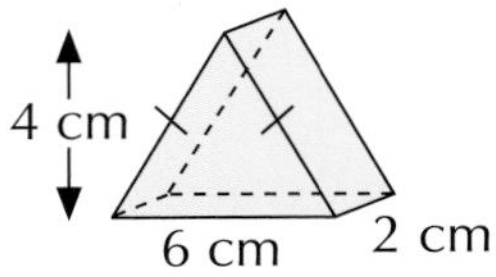

Q2 Draw the following prisms on isometric paper.

a) 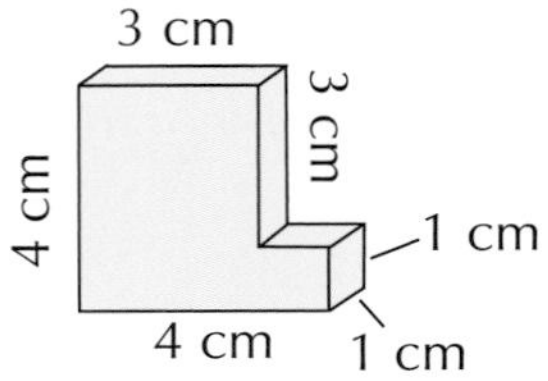

b) 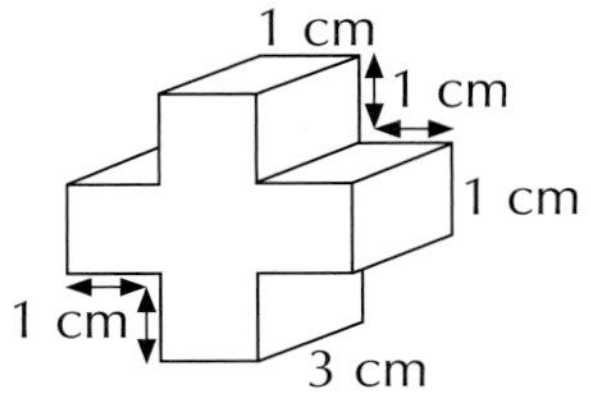

c) 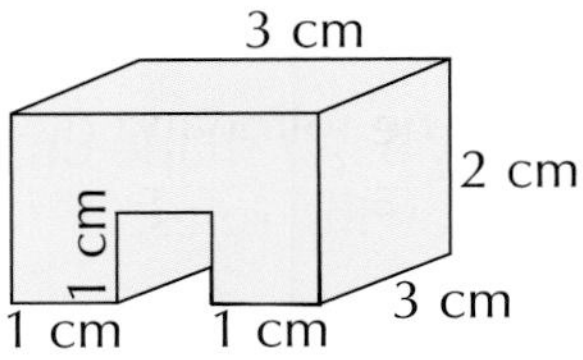

You can also use **projections** (see p.368) to draw 3D shapes on isometric paper. It's often helpful to **picture** or **sketch the shape** first, then use the dimensions to draw it **accurately**.

Example 2

The diagram on the right shows the plan, and front and side elevations of a 3D object. Draw the object on isometric paper.

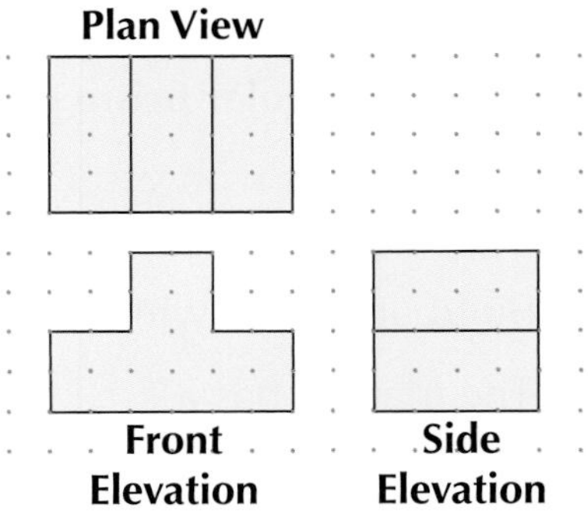

1. Try to picture the shape. It's a prism with the front elevation as its cross-section, which you can see since the plan view and side elevation have the same height all the way across.
2. Draw the cross-section on the isometric paper first, using the dots for the dimensions.
3. Use the side elevation and plan view to complete the drawing.

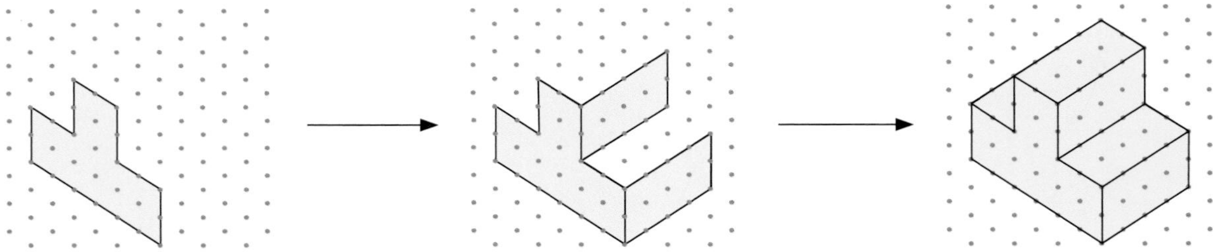

Exercise 2

Q1 The following diagrams show the constant cross-section of prisms of length 3 cm. Draw each prism on isometric paper.

a) 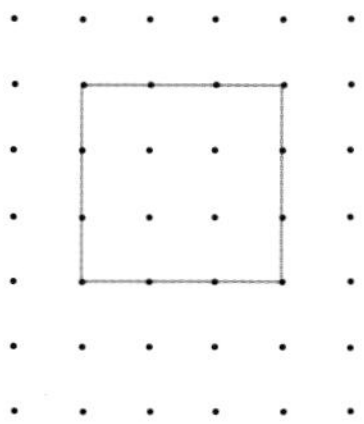

b) 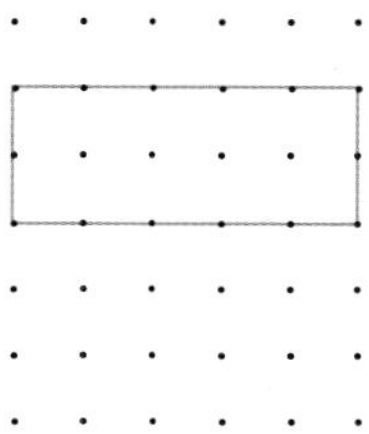

c) 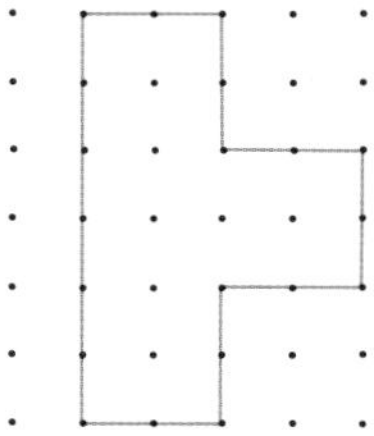

Q2 The following diagrams show the constant cross-section of prisms of length 2 cm. Draw each prism on isometric paper.

a) 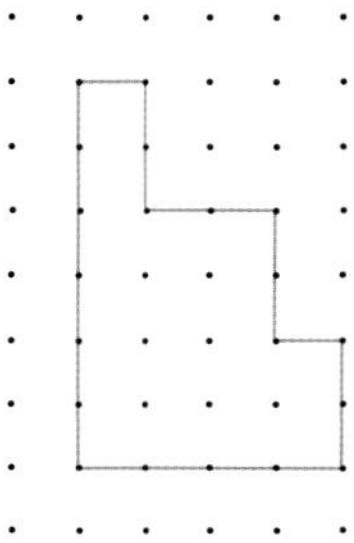

b) 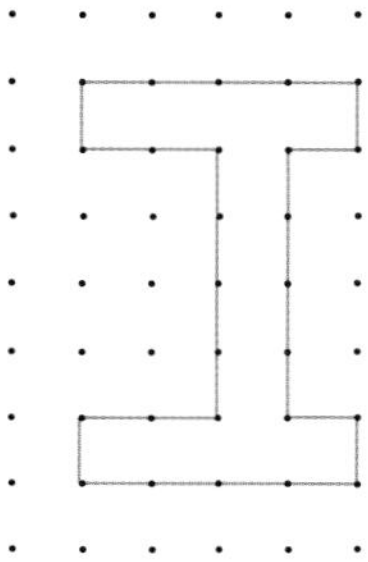

c) 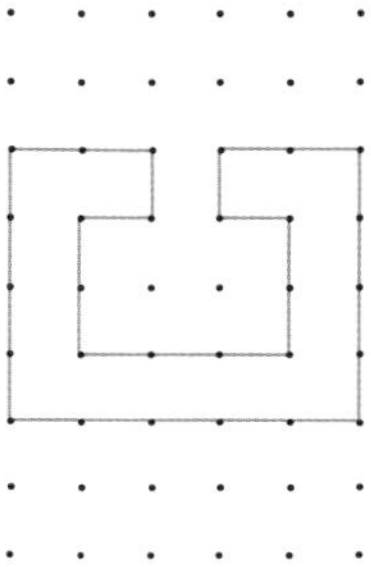

Q3 The following diagrams show the plan, and front and side elevations of different 3D objects. For each of the objects:

(i) use the projections to sketch the object and label the dimensions,

(ii) draw each object on isometric paper.

a) 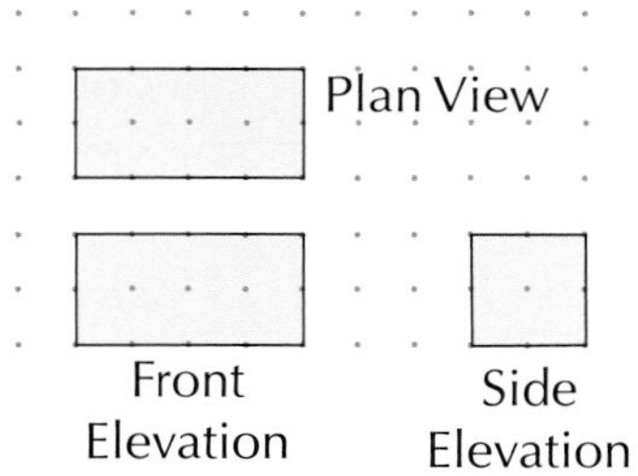

b) 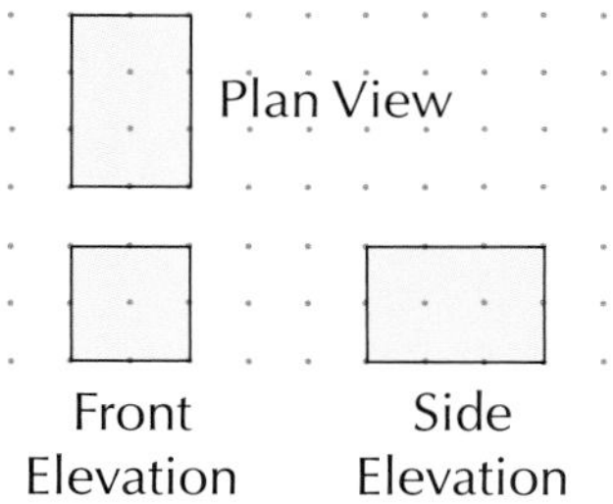

c) 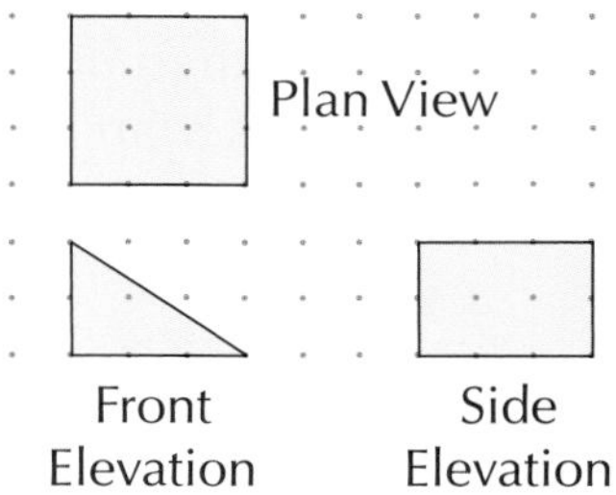

d) 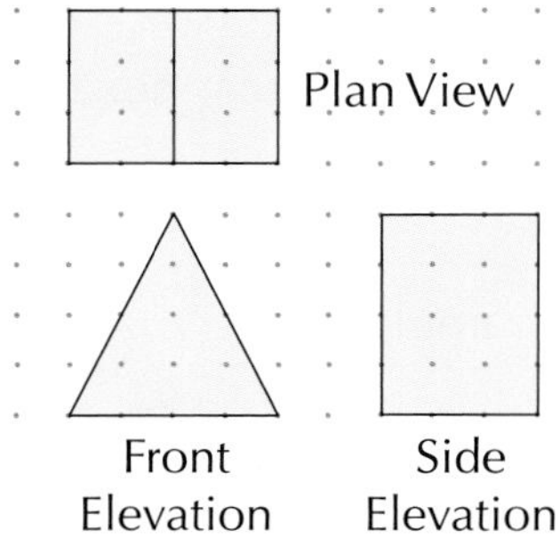

e) 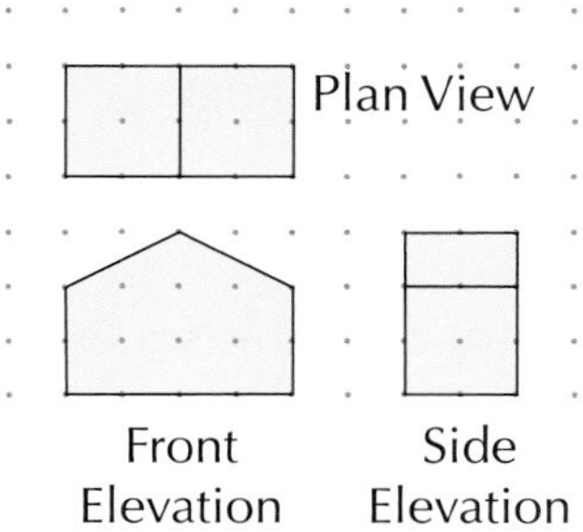

f) 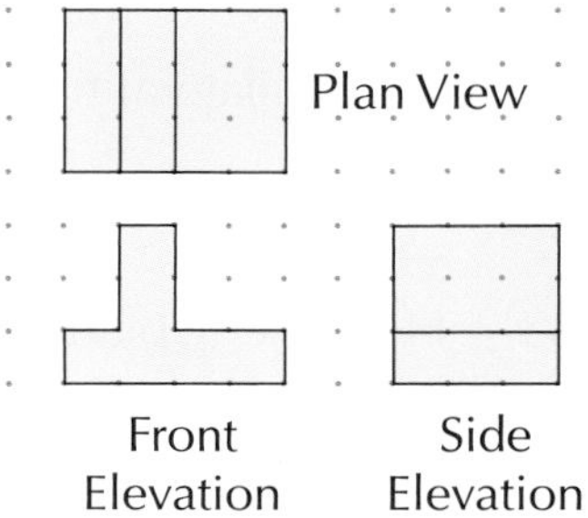

23.4 Symmetry of 3D Shapes

While 2D objects have lines of symmetry, 3D objects have planes of symmetry.

Learning Objective — Spec Ref G12:
Be able to draw the planes of symmetry of a 3D shape.

Prior Knowledge Check:
Be able to find lines of symmetry in 2D shapes — see p.255.

A **plane of symmetry** cuts a solid into **two identical halves**. To picture where they are, imagine **slicing** through the solid with a knife. The two pieces that you create should be **exactly the same** and **mirror images** of one another.

To find planes of symmetry, think about the **lines of symmetry** of the faces. You might be able to cut along these to create a plane of symmetry. For **prisms**, you should first look at the face that shows the **cross-section**, while for **pyramids**, you should look at the **base**.

Example 1

Draw the planes of symmetry of this isosceles triangular prism.

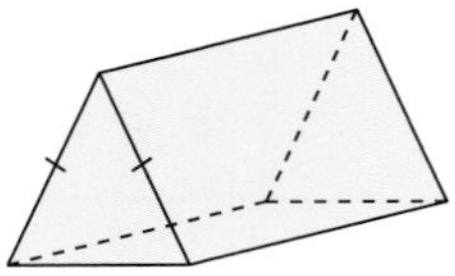

1. Imagine cutting through the solid to produce two halves.
2. First, look at the lines of symmetry of the cross-section. There's only one, and cutting along here splits the prism into two new prisms with right-angled triangles as their cross-section.

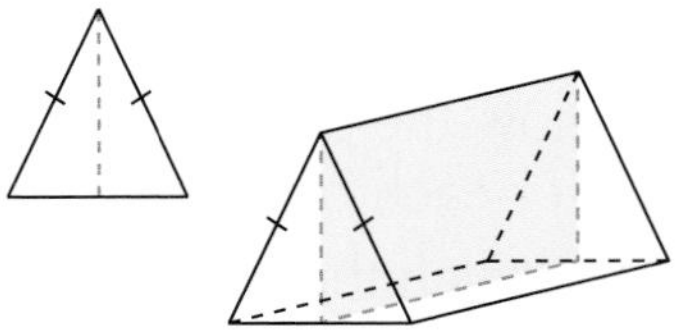

3. You can also cut halfway along the length to produce two smaller isosceles triangular prisms.

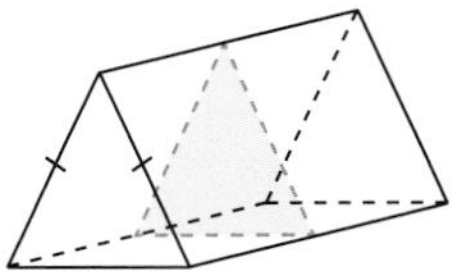

Tip: Prisms usually have one more plane of symmetry than the number of lines of symmetry of the cross-section. The only exception is a cube which has nine planes of symmetry.

Exercise 1

Q1 Draw two of the planes of symmetry of the prisms with the following cross-sections.
a) equilateral triangle b) rectangle c) regular hexagon

Q2 Draw all nine of the planes of symmetry of a cube.

Q3 How many planes of symmetry does this square-based pyramid have?

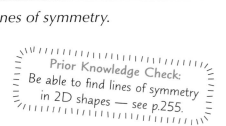

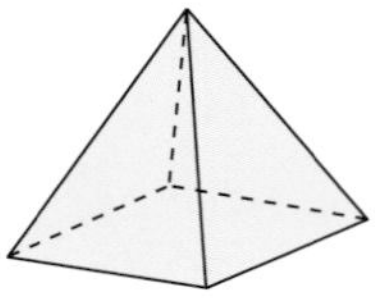

23.5 Volume

The volume of a 3D shape is the amount of space inside the shape — it's measured in cubic units, e.g. cm³ or m³.

Cubes and Cuboids

Learning Objective — Spec Ref G16:
Be able to find the volume of cubes and cuboids.

A **cuboid** is a 3D shape with **six rectangular faces** (see p.363).
The **volume** of a cuboid is found using the following formula:

Volume = Length × Width × Height

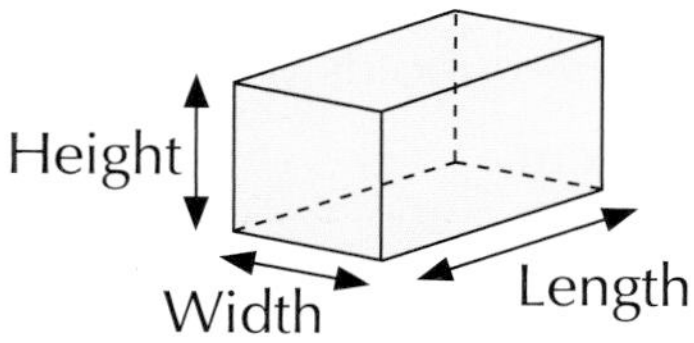

A **cube** is a special type of cuboid — all six of the faces are **squares**.
This means all of the edges have the **same length**, so the volume is given by:

Volume = Length × Length × Length = (Length)³

Tip: The units for volume are cubed because you're multiplying three dimensions — length, width, height. Make sure all three are in the same units.

Example 1

Find the volume of the cuboid shown in the diagram.

1. Write down the formula for volume. Volume = length × width × height
2. Substitute the values into the formula. Volume = 6 cm × 4 cm × 8 cm
3. Calculate the volume — don't forget the units. = **192 cm³**

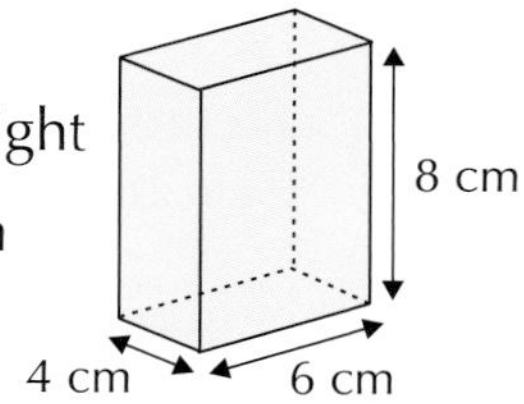

Exercise 1

Q1 The following shapes are made up of cubes with edges of length 1 cm.
Find the volumes of the shapes.

a) 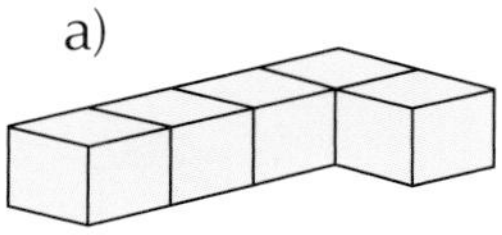

b) 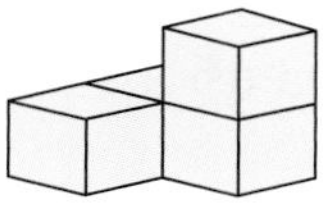

c) 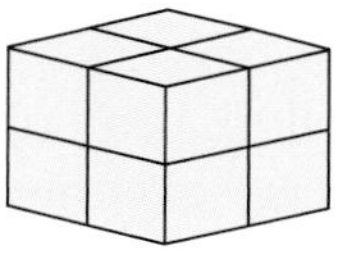

d) 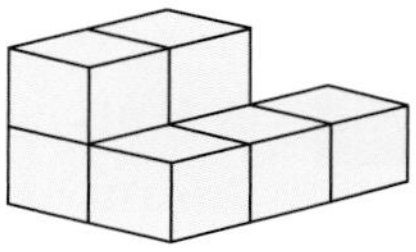

Q2 Find the volumes of the following cuboids.

a)

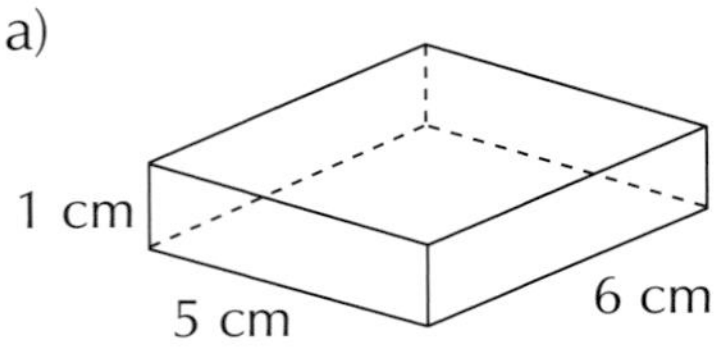

b)

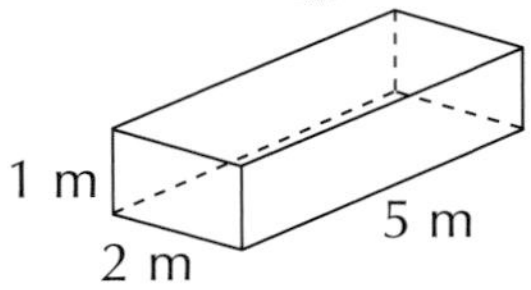

c)

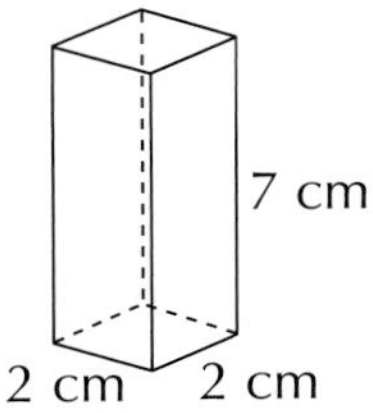

d)

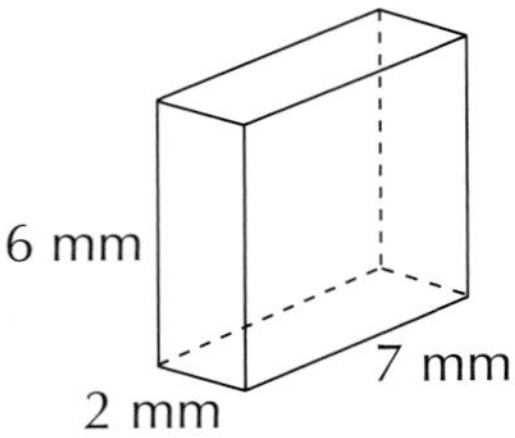

e)

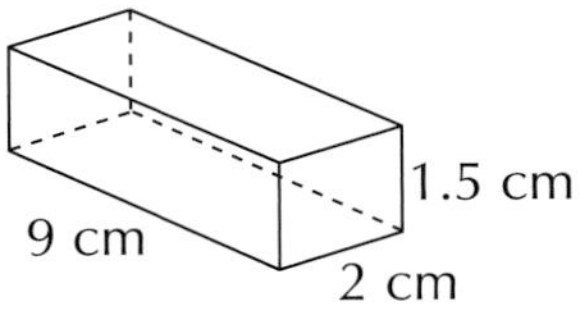

f)

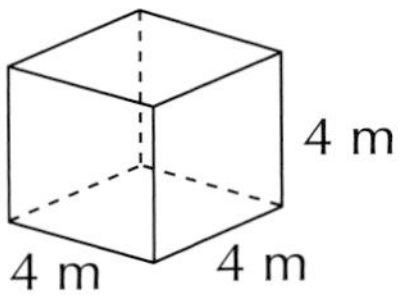

Q3 Find the volumes of the cuboids with the following dimensions.

a) 3 cm × 2 cm × 4 cm
b) 5 m × 2 m × 7 m
c) 20 cm × 10 cm × 8 cm
d) 2.5 cm × 3.0 cm × 4.2 cm
e) 18 m × 14 m × 3 m
f) 1.8 mm × 3.2 mm × 6.1 mm

Q4 a) Estimate the volume of a cube whose edges are 3.2 mm long by rounding the measurements to 1 s.f.

b) Is this an overestimate or an underestimate?

Q5 Will 3.5 m^3 of sand fit in a cuboid-shaped box with dimensions 1.7 m × 1.8 m × 0.9 m?

Q6 A cereal box is 9 cm long, 23 cm wide and 32 cm high. The box is half full of cereal. By rounding measurements to 1 s.f., estimate the volume of cereal in the box.

Q7 Split the following 3D objects into cuboids to find their volumes.

a)

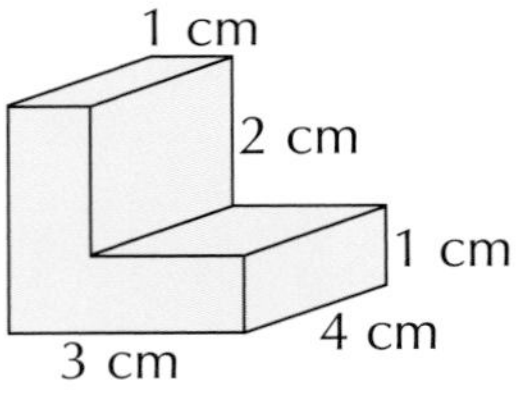

b)

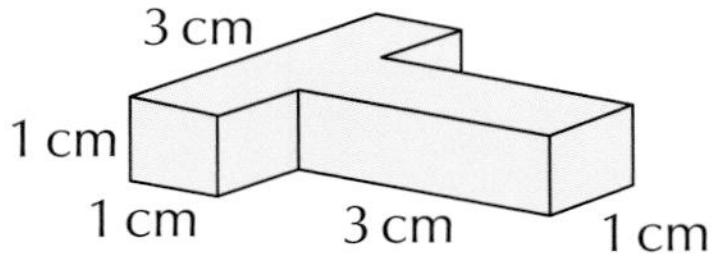

Q8 A matchbox is 5 cm long and 3 cm wide and has a volume of 18 cm^3. What is its height?

Q9 A bath can be modelled as a cuboid with dimensions 1.5 m × 0.5 m × 0.6 m, as shown.

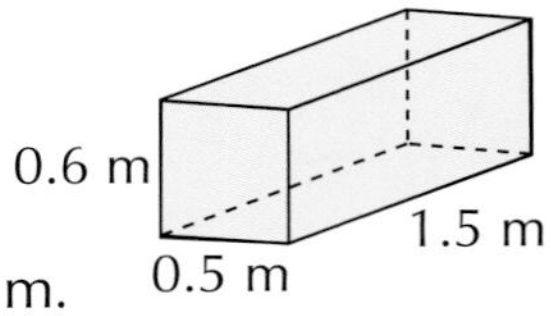

a) What is the maximum volume of water that the bath will hold?

b) Find the volume of water needed to fill the bath to a height of 0.3 m.

c) Find the height of the water if the volume of water in the bath is 0.3 m^3.

Prisms

Learning Objective — Spec Ref G16:
Be able to find the volume of prisms, including cylinders.

Prior Knowledge Check:
Be able to find the area of a triangle (p.346) and of a circle (p.358).

A **prism** is a 3D shape which has a **constant cross-section** (see p.363).
The **volume** of a prism is given by the following formula:

Volume = Area of Cross-Section × Length

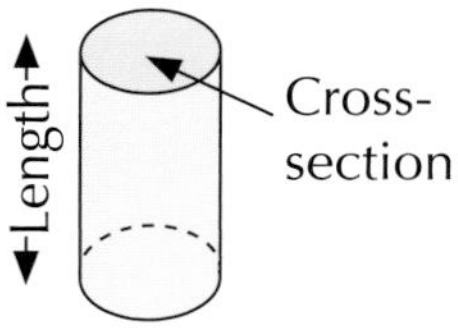

The area of the cross-section for a **triangular prism** is the area of a triangle ($\frac{1}{2}$ **× base × height**), and for a **cylinder** it's the area of a circle (πr^2).

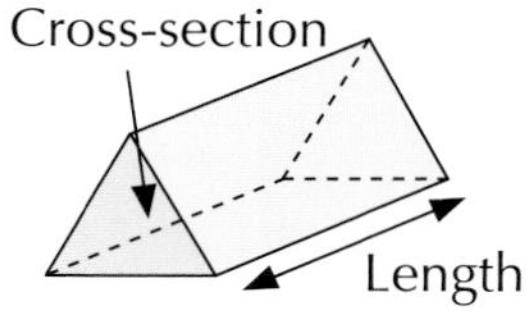

Example 2

Find the volume of each of the shapes shown on the right.

a)

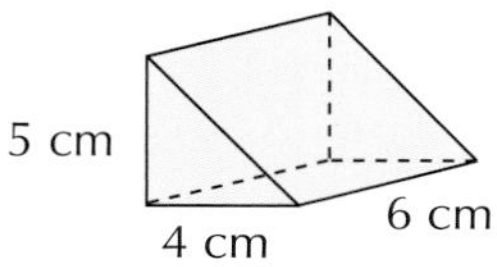

b)

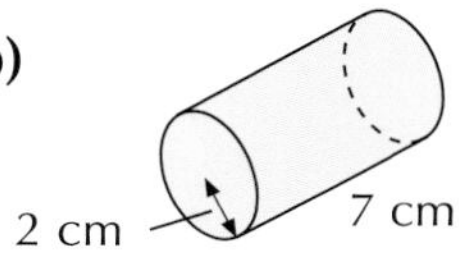

a) 1. Work out the area of the cross-section. Here it's a triangle, so calculate the area of the triangle.

Area of cross-section = area of triangle
$= \frac{1}{2} \times \text{base} \times \text{height} = \frac{1}{2} \times 4 \times 5 = 10 \text{ cm}^2$

2. Multiply the cross-sectional area by the length of the prism.

Volume = area of cross-section × length
$= 10 \times 6 =$ **60 cm³**

b) 1. Work out the area of the cross-section. Here it's a circle, so use area $= \pi r^2$.

Area of cross-section = area of circle
$= \pi r^2 = \pi \times 2^2 = 4\pi \text{ cm}^2$

2. Multiply the cross-sectional area by the length of the prism.

Volume = area of cross-section × length
$= 4\pi \times 7 =$ **88.0 cm³** (1 d.p.)

Example 3

Find the volume of the shape shown on the right.

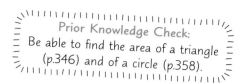

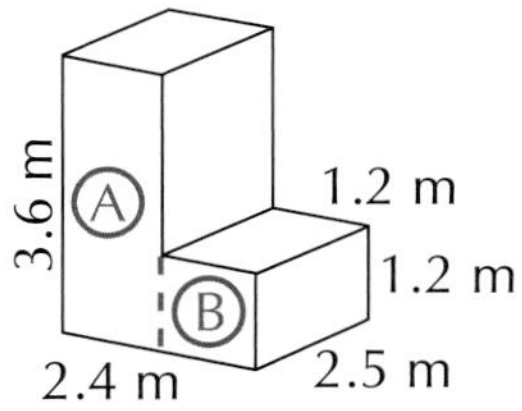

1. To work out the area of the cross-section, split it into 2 separate rectangles, as shown.

Area of rectangle A $= 1.2 \times 3.6$
$= 4.32 \text{ m}^2$
Area of rectangle B $= 1.2 \times 1.2$
$= 1.44 \text{ m}^2$

2. Add the two areas together.

Area of cross-section $= 4.32 + 1.44 = 5.76 \text{ m}^2$

3. Multiply the cross-sectional area by the length of the prism.

Volume = area of cross-section × length
$= 5.76 \times 2.5 =$ **14.4 m³**

Exercise 2

Q1 Find the volumes of the prisms with the following cross-sectional areas and lengths.

a) area = 2 cm^2, length = 3 cm
b) area = 6 cm^2, length = 9 cm
c) area = 1.5 m^2, length = 6 m
d) area = 3.5 cm^2, length = 3.5 cm
e) area = 9.25 mm^2, length = 1.75 mm
f) area = 11.6 mm^2, length = 9.1 mm

Q2 Find the volumes of the following prisms.

a)

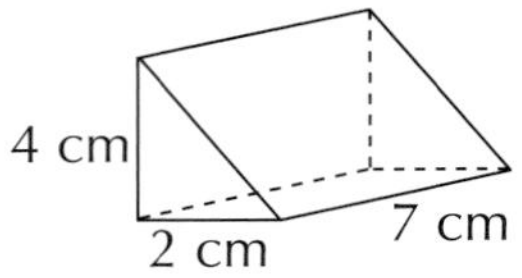

b)

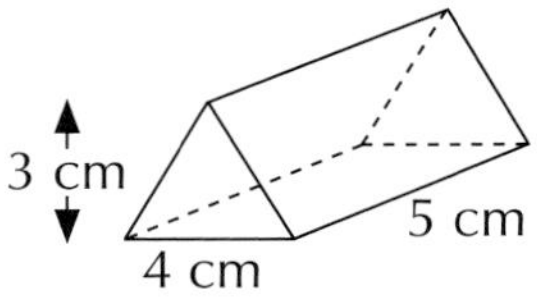

c)

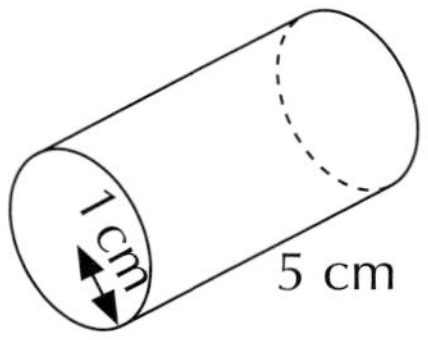

Q3 The following diagrams show the cross-sections of prisms of the length shown. Find the volume of each prism. The grid spacing is 1 cm.

a)

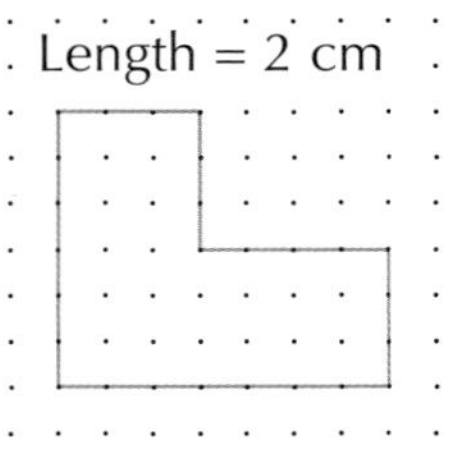

b)

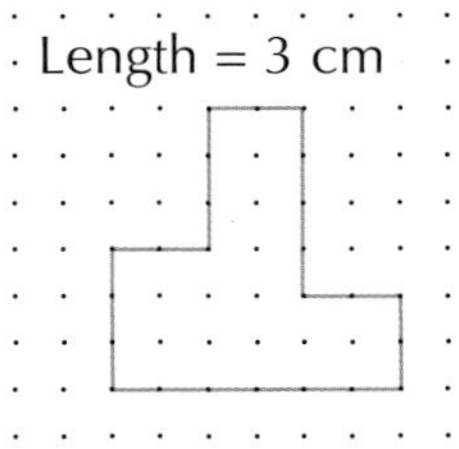

c)

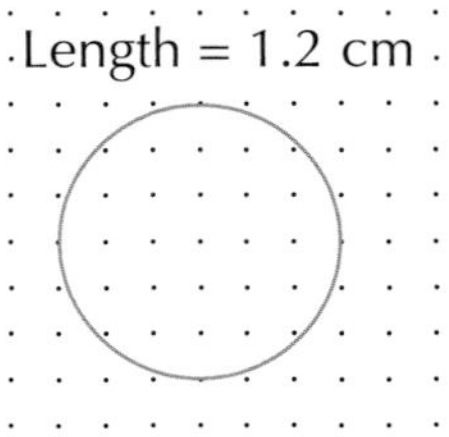

Q4 Find the volumes of the following prisms.

a) a triangular prism of base 13 cm, vertical height 12 cm and length 8 cm
b) a triangular prism of base 4.2 m, vertical height 1.3 m and length 3.1 m
c) a cylinder of radius 4 m and length 18 m
d) a prism with parallelogram cross-section of base 3 m and vertical height 4.2 m, and length 1.5 m

Q5 By first calculating their cross-sectional areas, find the volumes of the following prisms.

a)

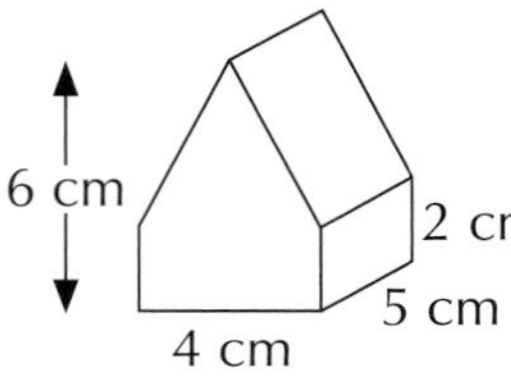

b)

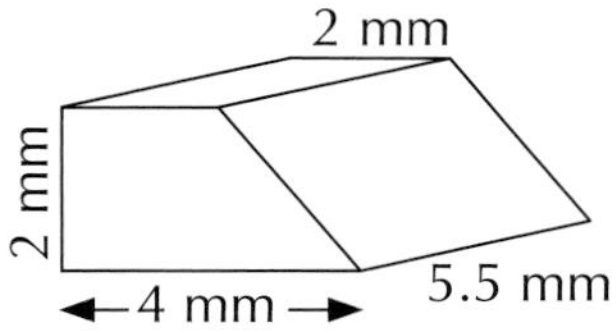

c)

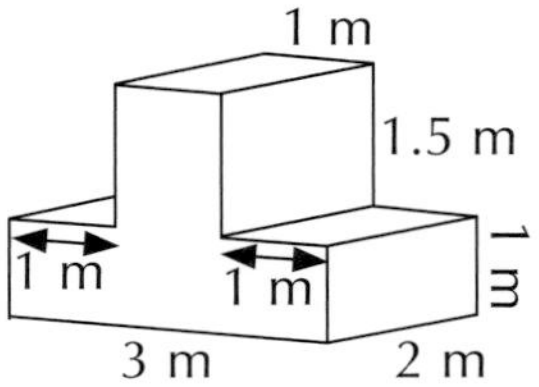

Q6 The following diagrams show prisms drawn on isometric paper with grid spacing 1 cm. Find the volume of each prism.

a)

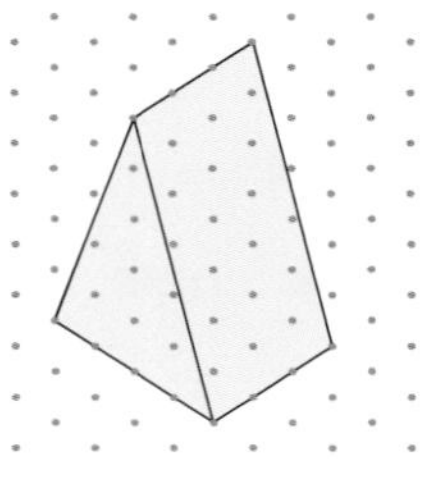

b)

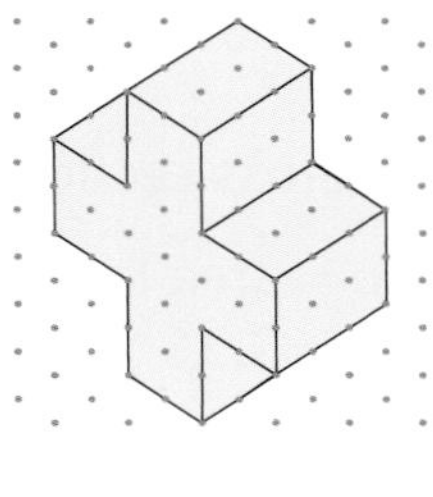

c)

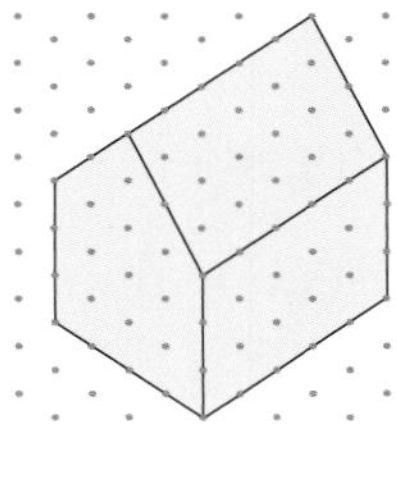

d)

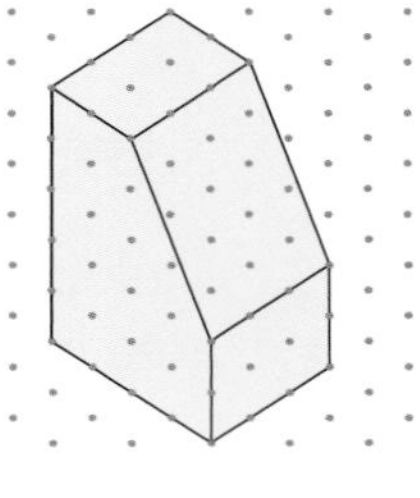

Spheres

Learning Objective — Spec Ref G17:
Be able to find the volume of spheres.

A **sphere** has one curved face, no vertices and no edges.
To find the volume of a sphere, you need to know its **radius** — the distance from the **centre** of the sphere to **any point on its surface**.
For a sphere with radius r, the formula for **volume** is as follows:

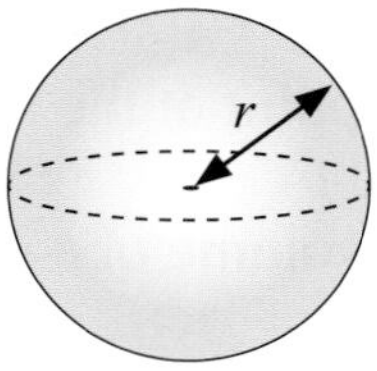

$$\textbf{Volume} = \frac{4}{3}\pi r^3$$

Example 4

Find the exact volume of a sphere with radius 6 cm.

1. Substitute $r = 6$ into the formula for the volume and work it through.
2. 'Exact' means you should leave your answer in terms of π.

$\text{Volume} = \frac{4}{3}\pi r^3 = \frac{4}{3} \times \pi \times 6^3$
$= \mathbf{288\pi\ cm^3}$

Example 5

A sphere has a volume of 2304π cm³. What is its radius?

1. Substitute the volume given into the formula for the volume of a sphere. $\text{Volume} = \frac{4}{3}\pi r^3$, so $2304\pi = \frac{4}{3}\pi r^3$
2. Divide by $\frac{4}{3}\pi$ to find r^3. $r^3 = 2304\pi \div \left(\frac{4}{3}\pi\right) = 1728\ \text{cm}^3$
3. Cube root to find r. $r = \sqrt[3]{1728} = \mathbf{12\ cm}$

Tip: The π's cancel.

Exercise 3

Q1 For each of the following, find the exact volume of the sphere with the given radius, r.
a) $r = 3$ cm b) $r = 2$ m c) $r = 5$ mm

Q2 For each of the following, find the volume of the sphere with the given radius, r. Give your answers to 2 d.p.
a) $r = 4$ cm b) $r = 8$ m c) $r = 9.6$ mm d) $r = 15.7$ m

Q3 Find the radius of the sphere with volume 24 429 cm³. Give your answer correct to 1 d.p.

Cones and Frustums

Learning Objective — Spec Ref G17:
Be able to find the volume of cones and frustums.

A **cone** is a 3D shape that has a **circular base** and goes up to a **point** at the top. To find the **volume** of a cone, you need to know its base **radius** $\boldsymbol{r}$ and **perpendicular height** $\boldsymbol{h}$. The **volume** of a cone is given by:

$$\textbf{Volume} = \frac{1}{3}\pi r^2 h$$

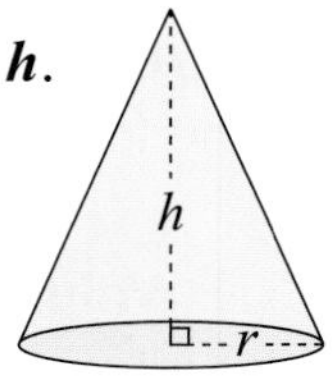

Tip: The perpendicular height goes from the centre of the base to the point.

Example 6

Find the volume of the cone shown on the right. Give your answer to 2 d.p.

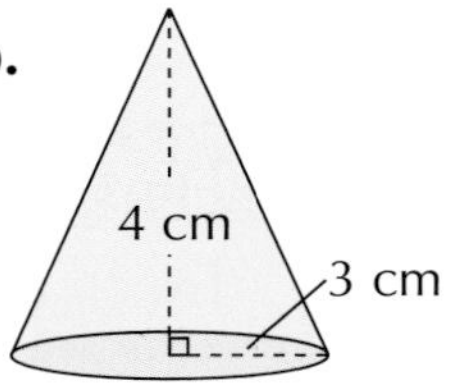

Substitute $r = 3$ and $h = 4$ into the formula for volume.

Volume $= \frac{1}{3}\pi r^2 h = \frac{1}{3} \times \pi \times 3^2 \times 4$
$= 37.699... =$ **37.70 cm³** (2 d.p.)

Exercise 4

Q1 Find the exact volume of the cones with the given properties.
a) base radius = 5 m, height = 12 m
b) base radius = 7 cm, height = 24 cm
c) base radius = 15 m, height = 8 m
d) base radius = 30 mm, height = 5.5 mm

Q2 Find the volume of the cones with the given properties. Give your answers to 2 d.p.
a) base radius = 4 cm, height = 10 cm
b) base radius = 14 mm, height = 32 mm
c) base radius = 7 m, height = 25 m
d) base radius = 3.6 cm, height = 7.2 cm

Frustums

A **frustum** of a cone is the 3D shape left once you **chop** the top bit off a cone **parallel** to its circular base. The smaller, removed cone is always **similar** to the larger, original cone — see p.414 for more on similarity.

To find the **volume** of a frustum, find the volume of the **original cone** and take away the volume of the **removed cone** (the top bit).

Volume of frustum = **Volume of original cone** − **Volume of removed cone**

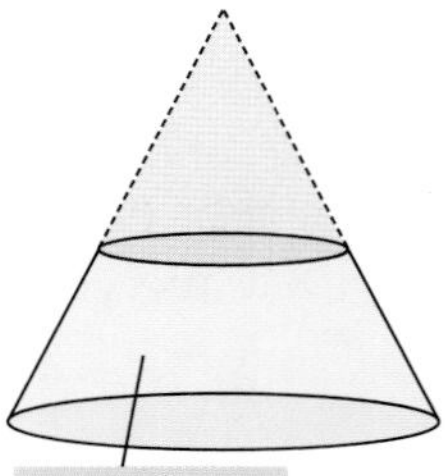

Frustum

Example 7

Find the exact volume of the frustum shown on the right.

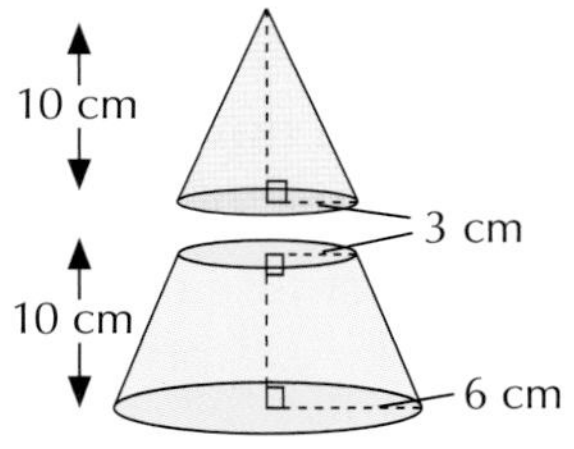

1. Find the volume of the original cone.

$$\text{Volume of original cone} = \frac{1}{3}\pi \times 6^2 \times (10 + 10)$$
$$= \frac{1}{3}\pi \times 36 \times 20 = 240\pi \text{ cm}^3$$

2. Find the volume of the removed cone.

$$\text{Volume of removed cone} = \frac{1}{3}\pi \times 3^2 \times 10 = \frac{1}{3}\pi \times 9 \times 10 = 30\pi \text{ cm}^3$$

3. Subtract the volume of the removed cone from the volume of the original cone.

Volume of frustum = vol. of original cone – vol. of removed cone
$= 240\pi - 30\pi =$ **210π cm³**

Exercise 5

Q1 A cone of height 4 cm and base radius 3 cm is removed from a cone of height 16 cm and base radius 12 cm, as shown in the sketch on the right.

a) Find the exact volume of the original cone.
b) Find the exact volume of the removed cone.
c) Hence find the exact volume of the frustum.

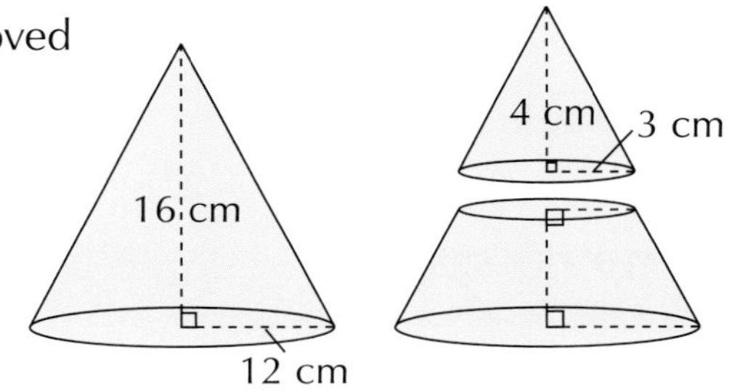

Q2

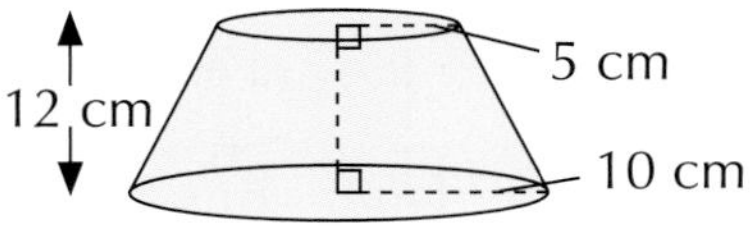

The frustum on the left is what remains after the top 12 cm is removed from a 24 cm tall cone. What is the volume of the frustum? Give your answer to 2 d.p.

Pyramids

Learning Objective — Spec Ref G17:
Be able to find the volume of pyramids.

A **pyramid** is a 3D shape that has a **polygon base** (see page 250) which rises to a **point**.

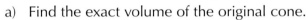

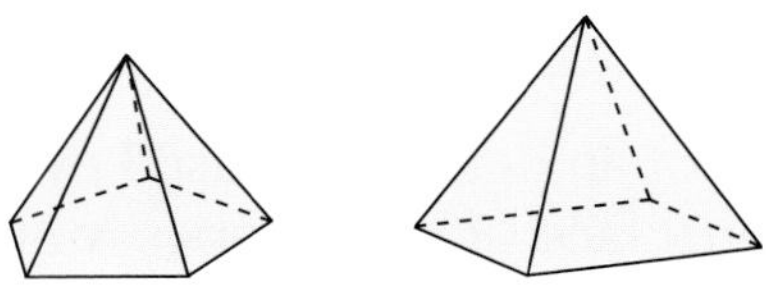

Tip: A cone is a bit like a pyramid with a circular base.

The **volume** of a pyramid can be found by using this formula:

$$\textbf{Volume} = \frac{1}{3} \times \textbf{base area} \times \textbf{height}$$

Example 8

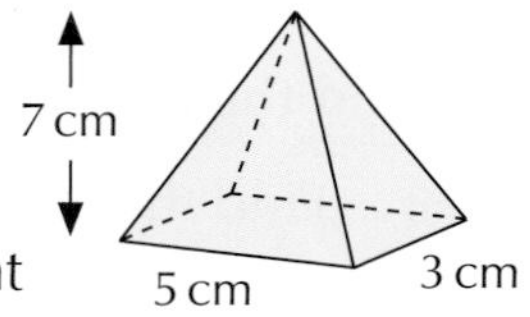

Find the volume of the rectangular-based pyramid shown.

1. Write down the formula for volume of a pyramid. $\text{Volume} = \frac{1}{3} \times \text{base area} \times \text{height}$
2. Substitute the values into the formula — since the base is a rectangle, base area = length × width. $= \frac{1}{3} \times (5 \times 3) \times 7$

   $= \mathbf{35\ cm^3}$

Example 9

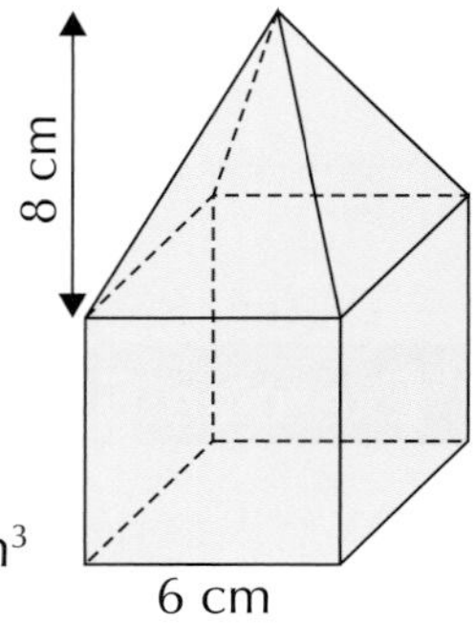

The 3D shape on the right is made up of a square-based pyramid on top of a cube, with dimensions as shown. Find the volume of the shape.

1. Find the volume of the cube (see p.375).

   Volume of cube = $(\text{length})^3 = 6^3 = 216\ \text{cm}^3$
2. Find the volume of the pyramid — the base is a square, so its area is $(\text{side length})^2$.

   Volume of pyramid = $\frac{1}{3} \times \text{base area} \times \text{height} = \frac{1}{3} \times 6^2 \times 8 = 96\ \text{cm}^3$
3. Add the two volumes together to find the total volume.

   Total volume = 216 + 96 = **312 cm³**

Exercise 6

Q1 Find the volumes of triangular-based pyramids with the following dimensions.
a) base area = 18 cm^2, height = 10 cm b) base area = 8 m^2, height = 3 m

Q2 Find the volume of a rectangular-based pyramid of height 10 cm with base dimensions 4 cm × 9 cm.

Q3 A hexagon-based pyramid is 15 cm tall. The area of its base is 18 cm^2. Calculate its volume.

Q4 Just to be controversial, Pharaoh Tim has decided he wants a pentagon-based pyramid. The area of the pentagon is 27 m^2 and the height of the pyramid is 12 m. What is the volume of Tim's pyramid?

Q5 A 3D shape is made up of an octagonal-based pyramid on top of an octagonal prism. The area of the octagonal face of both shapes is 24 cm^2, the height of the prism is 10 cm and the height of the pyramid is also 10 cm. What is the volume of the shape?

Q6 A square-based pyramid has height 13.5 cm and volume 288 cm^3. Find the side length of its base.

Rates of Flow

Learning Objective — Spec Ref G17:
Know how to calculate rates of flow into and out of 3D shapes.

Prior Knowledge Check:
Be able to convert between different units. See Section 16.

The **rate of flow** tells you **how quickly** a liquid is moving **into**, **out of** or **through** a certain **space**. To work out the rate of flow, you need to know the **total volume** of the space and the **time** it would take to **completely fill** (or empty) the space. The volume **divided** by the time gives the rate of flow.

$$\textbf{Rate of Flow} = \frac{\textbf{Volume of container}}{\textbf{Total time taken}}$$

Rates of flow are measured in **volume per unit time**, e.g. litres per second or m^3 per hour (sometimes written as litres/s or m^3/hr).

Example 10

Water flows into this cuboid-shaped tank at a rate of 150 cm^3 per second. How long does it take to fill the tank?

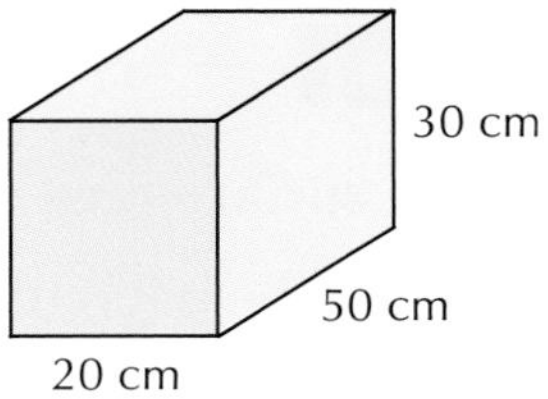

1. Calculate the volume of the cuboid.

   Volume of cuboid = length × width × height
   = 50 × 20 × 30 = 30 000 cm^3

2. Rearrange the formula and divide the volume by the rate of flow to get the time taken.

   $$\text{Time taken to fill tank} = \frac{\text{Volume of container}}{\text{Rate of flow}} = \frac{30\,000}{150} = \textbf{200 seconds}$$

Exercise 7

Q1 Work out the average rates of flow when containers with these volumes are completely filled in the given time:

a) Volume = 600 cm^3, Time = 15 seconds

b) Volume = 7200 litres, Time = 40 minutes

c) Volume = 385 m^3, Time = 5 minutes

d) Volume = 150 litres, Time = 8 hours

Q2 A cube with a side length of 30 cm is filled with sand at a rate of 15 cm^3 per second. How many minutes does it take to fill the cube?

Q3 Grain is being poured into the empty cylindrical tank shown on the right. The grain is flowing at a rate of 12 m^3 every minute. How long will it take to fill the tank? Give your answer to the nearest minute.

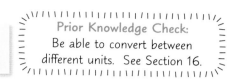

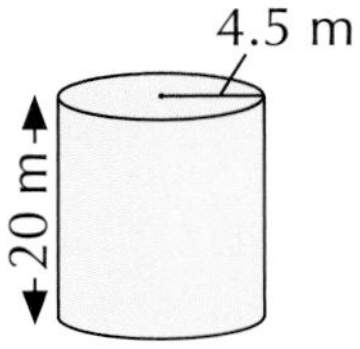

23.6 Surface Area

The surface area of a 3D shape is the sum of the areas of each face — it's measured in square units, e.g. cm² or m².

Cubes, Cuboids, Prisms and Pyramids

Learning Objective — Spec Ref G17:
Be able to find the surface area of cubes, cuboids, prisms and pyramids.

Prior Knowledge Check:
Be able to find the areas of squares, rectangles and triangles (see p.345).

The **surface area** of a 3D shape is the **total area** of all of the **faces** of the shape added together. To find the surface area of a **cube**, **cuboid**, **prism** or **pyramid**, you should sketch the **net** of the shape first to make sure that you include **all** of the faces of the shape.

Tip: See p.365 for more on nets.

Example 1

Find the surface area of the cuboid shown.

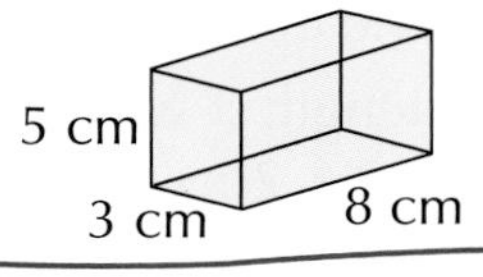

1. Sketch the net of the cuboid.
2. Find the area of each face.

2 faces of area $8 \times 5 = 40$ cm²
2 faces of area $8 \times 3 = 24$ cm²
2 faces of area $5 \times 3 = 15$ cm²

3. Add together the areas of the faces.

So the total surface area is
$(2 \times 40) + (2 \times 24) + (2 \times 15)$

4. Give your answer with correct units.

$= 80 + 48 + 30 =$ **158 cm²**

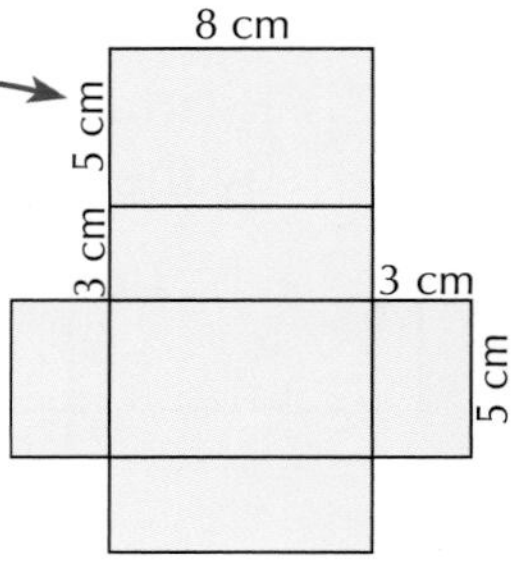

Example 2

Find this prism's surface area by considering its net.

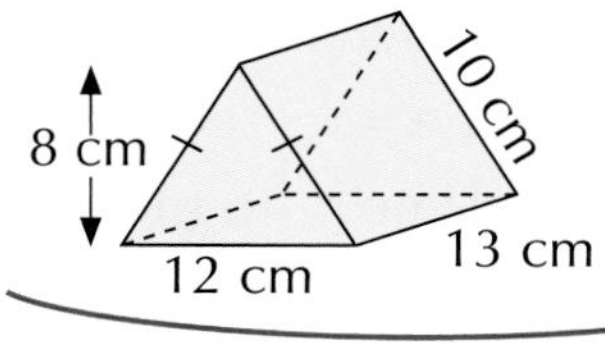

1. Draw the net of the triangular prism.

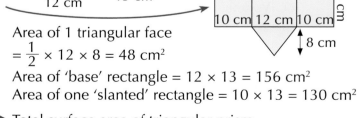

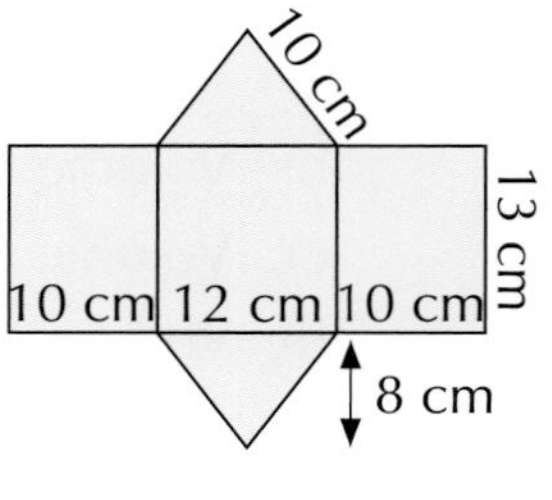

2. Find the area of each face.

Area of 1 triangular face
$= \frac{1}{2} \times 12 \times 8 = 48$ cm²
Area of 'base' rectangle $= 12 \times 13 = 156$ cm²
Area of one 'slanted' rectangle $= 10 \times 13 = 130$ cm²

3. Add the different areas to find the total surface area. Make sure you include the correct number of each face shape.

Total surface area of triangular prism

4. Give your answer with correct units.

$= (2 \times 48) + 156 + (2 \times 130) =$ **512 cm²**

Exercise 1

Q1 A cube has 6 faces, each of area 2 cm². Find the total surface area of the cube.

Q2 a) Find the area of one face of the cube shown.

b) Find the total surface area of the cube.

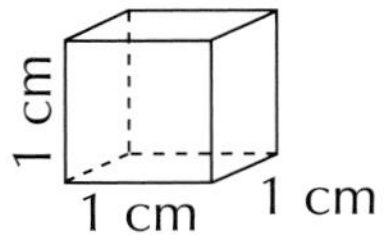

Q3 a) Find the area of one face of a cube with edges of length 2 m.

b) Find the total surface area of the cube.

Q4 a) Find the area of face *A* of the cuboid shown.

b) Find the area of face *B*.

c) Find the area of face *C*.

d) Find the total surface area of the cuboid.

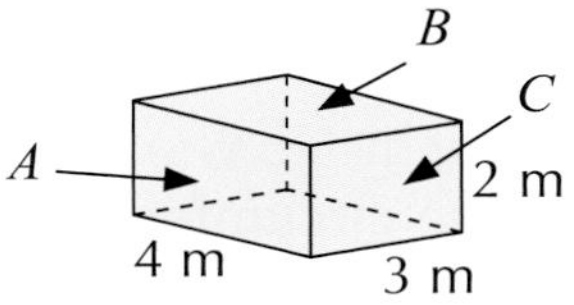

Q5 Find the surface area of the following cubes and cuboids.

a)

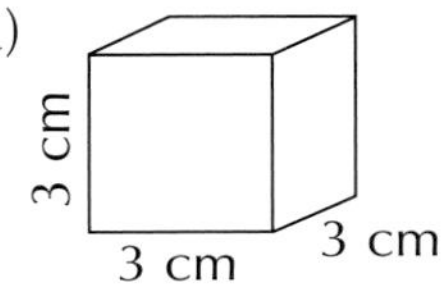

b)

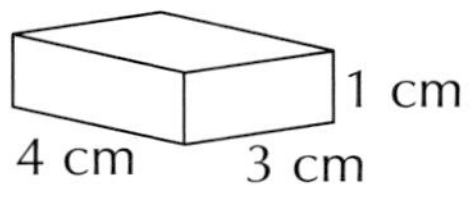

c)

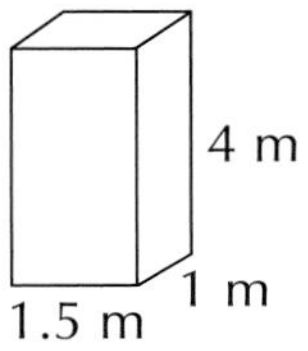

d)

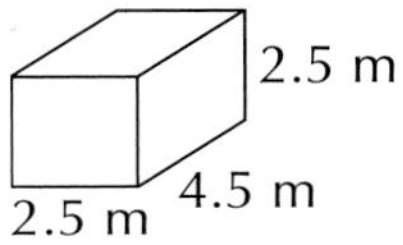

Q6 By considering their nets, find the surface area of these shapes.

a)

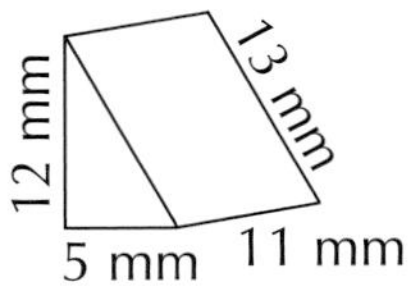

b)

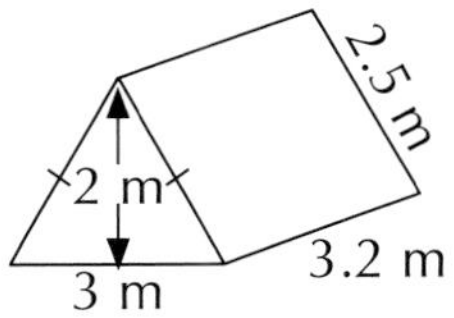

c)

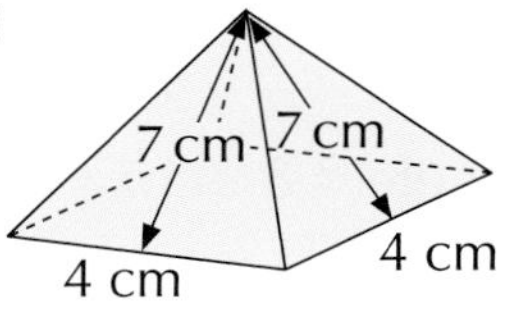

d)

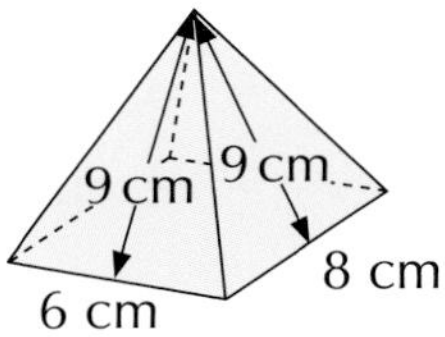

Q7 Find the surface area of the following shapes.

a) a cube with edges of length 5 m

b) a cube with edges of length 6 mm

c) a 1.5 m × 2 m × 6 m cuboid

d) a 7.5 m × 0.5 m × 8 m cuboid

e) a prism of length 2.5 m whose cross-section is an isosceles triangle of height 4 m, slant edge 5 m and base 6 m

Q8 The local scouts are waterproofing their tent, shown on the right.

a) Find the surface area of the outside of the tent.

b) How many tins of waterproofing spray will they need to buy to cover the outside of their tent if each tin covers an area of 4 m²?

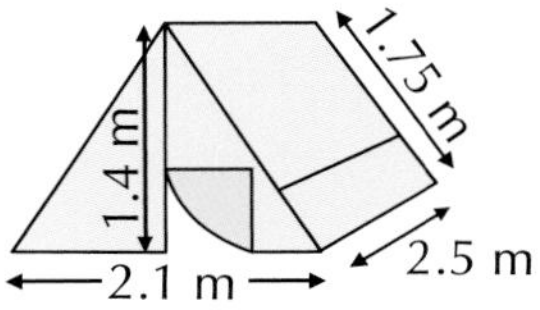

Cylinders

Learning Objective — Spec Ref G17:
Be able to find the surface area of cylinders.

Prior Knowledge Check:
Be able to find the circumference and area of a circle (see p.355 and p.358).

Find the surface area of a **cylinder** using the same method as on p.384 — **sketch its net** and work out the **area** of each **face**. It's just a bit trickier to work out the areas you need.

You saw on p.365 that the **net** of a cylinder has **two circular faces** and one **rectangular face**. The **width** of the rectangle will be the **height** of the cylinder, but you'll have to work out the **length** yourself. The rectangular face has to wrap all around the circular faces, so its length is equal to the **circumference** of the circle — use the formula $\boldsymbol{C = 2\pi r}$.

Tip: You use the formula $A = \pi r^2$ to find the area of the circular faces.

Example 3

Find this cylinder's surface area by considering its net.

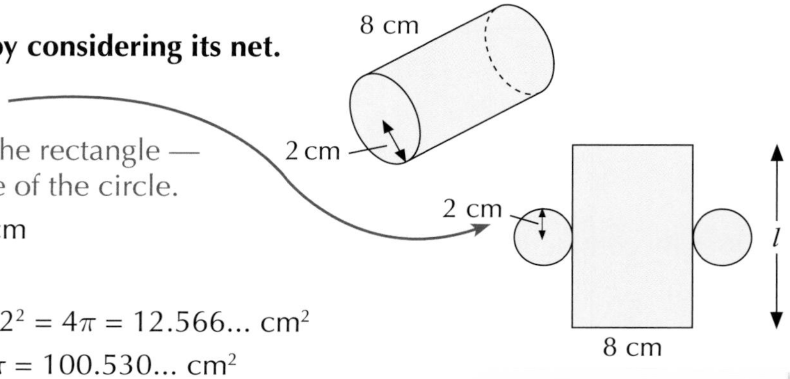

1. Draw the net of the cylinder.
2. Find the missing length, l, of the rectangle — it's equal to the circumference of the circle.

   $l = 2\pi r = 2 \times \pi \times 2 = 4\pi$ cm
3. Find the area of each face.

   Area of circle $= \pi r^2 = \pi \times 2^2 = 4\pi = 12.566...$ cm^2

   Area of rectangle $= 8 \times 4\pi = 100.530...$ cm^2
4. Add together the individual areas.

   So total surface area $= (2 \times 12.566...) + 100.530...$

   $=$ **125.7 cm²** (to 1 d.p.)

Tip: Don't forget to use the correct units in your answer.

Exercise 2

Round any inexact answers to 2 decimal places unless told otherwise.

Q1 a) Draw the net of the cylinder shown on the right.
b) Find the area of each surface of the cylinder.
c) Find the surface area of the cylinder, correct to 1 decimal place.

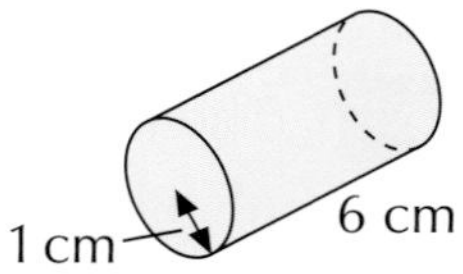

Q2 a) Draw the net of the cylinder shown on the right.
b) Find the area of each surface of the cylinder.
c) Find the surface area of the cylinder, correct to 1 decimal place.

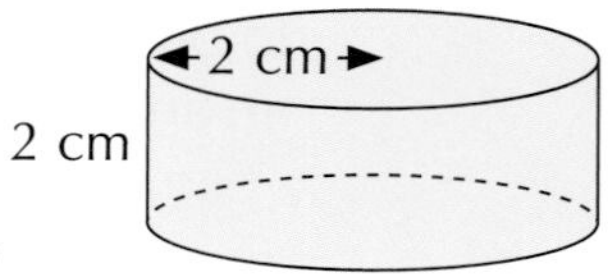

Q3 Find the surface area of the following cylinders.

a)

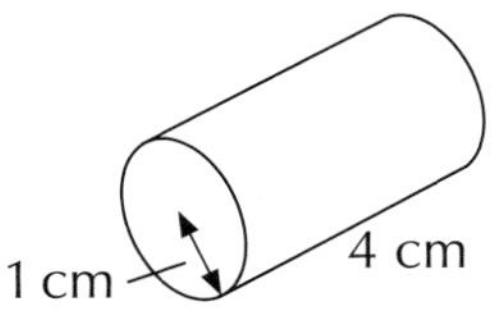

b)

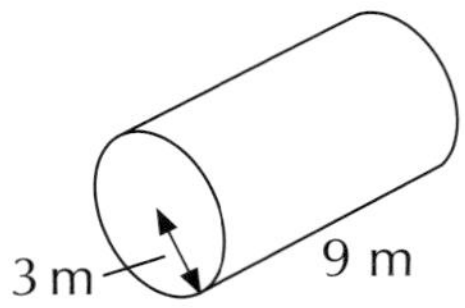

c)

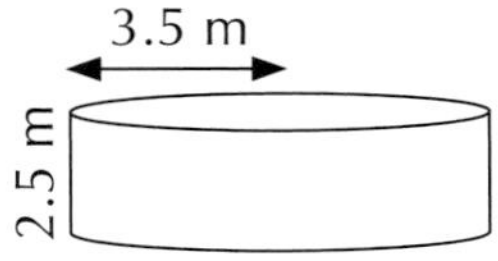

d)

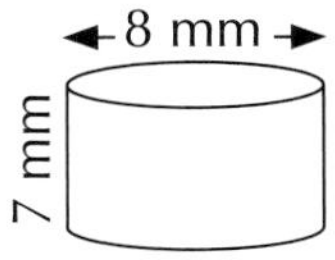

e)

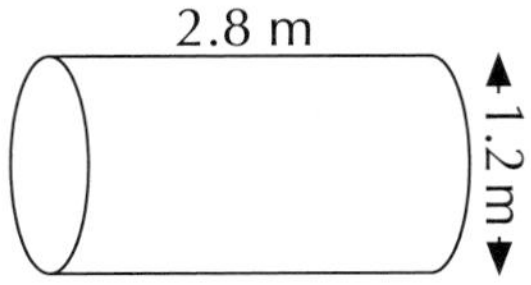

f)

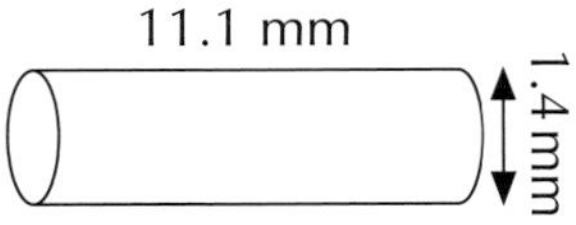

Q4 Find the surface area of the cylinders with the following dimensions.
Give your answers correct to 1 decimal place.

a) radius = 2 m, length = 7 m

b) radius = 7.5 mm, length = 2.5 mm

c) radius = 12.2 cm, length = 9.9 cm

d) diameter = 22.1 m, length = 11.1 m

Q5 Maeve is painting her cylindrical gas tank. The tank has radius 0.8 m and length 3 m. She uses tins of paint which cover an area of 14 m^2.

a) What is the surface area of the tank?

b) How many tins of paint will Maeve need?

Q6 A cylindrical metal pipe has radius 2.2 m and length 7.1 m. The ends of the pipe are open.

a) Find the curved surface area of the outside of the pipe.

b) A system of pipes consists of 9 of the pipes described above.
What area of metal is required to build the system of pipes?
Give your answer correct to 1 decimal place.

Q7 The diagram shows a cylindrical bin which is closed at one end and open at the other. Find the total surface area of the outside of the bin.

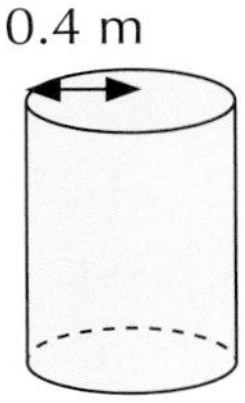

Spheres and Cones

Learning Objective — Spec Ref G17:
Be able to find the surface area of spheres and cones.

For spheres and cones, you **don't** need to sketch the net — just use the **formulas** below.

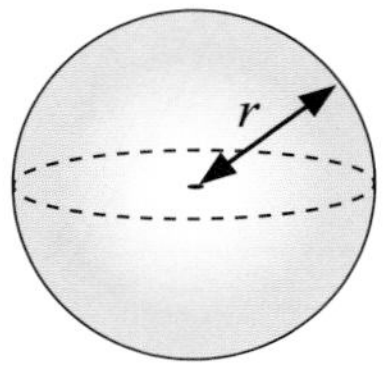

Surface area of a sphere = $4\pi r^2$

Where r is the radius of the sphere.

Surface area of a cone = $\pi rl + \pi r^2$

Where r is the radius at the base of the cone and l is the slant height — the distance from the edge of the base to the point.

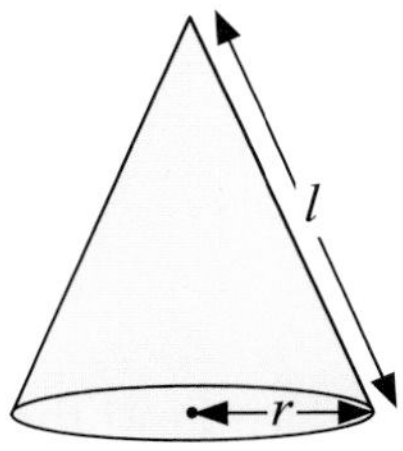

Example 4

Find the exact surface area of:

a) a sphere with radius 3 cm

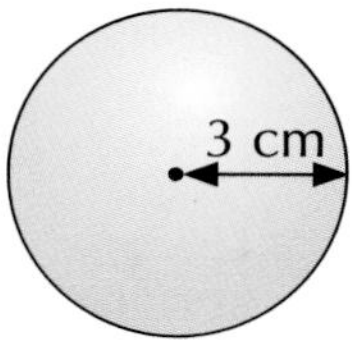

Put $r = 3$ into the formula.
Leave your answer in terms of π
(as the question asks for an exact value).

Surface area $= 4\pi r^2$
$= 4 \times \pi \times 3^2 =$ **36π cm²**

b) a cone with base radius 4 cm and a slant height of 10 cm

Put $r = 4$ and $l = 10$ into the formula
— leave your answer in terms of π.

Surface area $= \pi rl + \pi r^2$
$= (\pi \times 4 \times 10) + (\pi \times 4^2)$
$= 40\pi + 16\pi =$ **56π cm²**

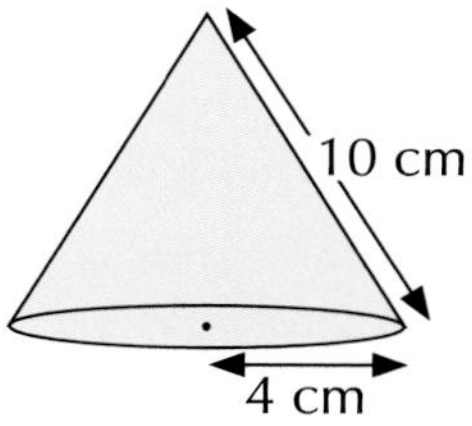

Tip: For the surface area of a cone, make sure you use the slant height (l), not the perpendicular height.

Exercise 3

Q1 For each of the following, find the exact surface area of the sphere with the given radius, r.

a) $r = 2$ cm b) $r = 5$ cm c) $r = 7$ m
d) $r = 9$ mm e) $r = 10.5$ m f) $r = 3.25$ mm

Q2 For each of the following, find the surface area of the sphere with the given radius, r. Give your answers to 2 d.p.

a) $r = 2.5$ cm b) $r = 7.1$ m c) $r = 11.2$ mm
d) $r = 19.3$ m e) $r = 4.88$ cm f) $r = 6.35$ mm

Q3 Find the exact surface area of the cones with the given properties.

a) $r = 5$ m, $l = 13$ m b) $r = 7$ cm, $l = 25$ cm c) $r = 15$ m, $l = 17$ m

Q4 Find the surface area of the cones with the given properties. Give your answers to 2 d.p.

a) $r = 12$ cm, $l = 30$ cm b) $r = 1.5$ m, $l = 4.5$ m c) $r = 2.8$ mm, $l = 8.2$ mm

Q5 Find the surface area of each of the following, giving your answers to 2 d.p.

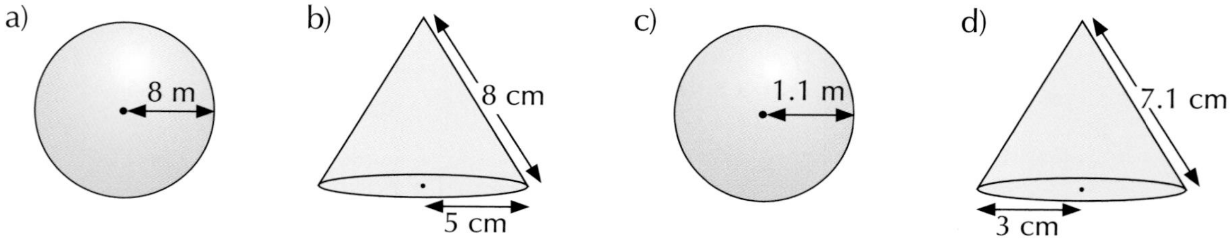

Q6 Find the radius of the sphere with surface area 265.9 cm². Give your answer to 1 d.p.

Review Exercise

Where appropriate, give your answers to these questions to 2 decimal places.

Q1 a) Sketch the 3D shape with the net shown.

b) Use isometric paper to accurately draw the shape.

c) Write down how many faces, vertices and edges the shape has.

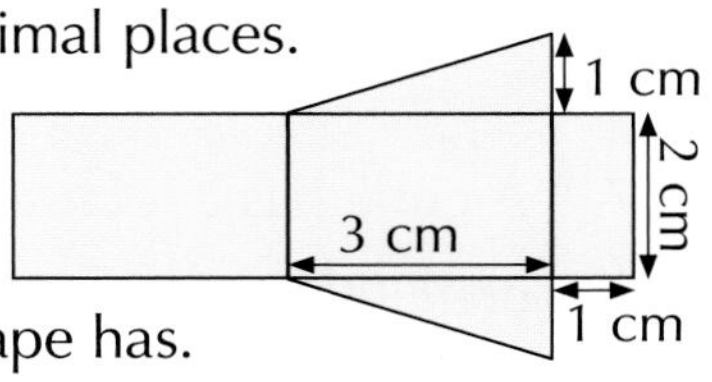

Q2 a) Draw an accurate, full-sized net of the triangular prism shown.

b) Measure the appropriate length on your drawing to find the vertical height, h, of the prism.

c) Draw the plan, and the front and side elevations of the prism from the directions shown.

d) How many planes of symmetry does the prism have?

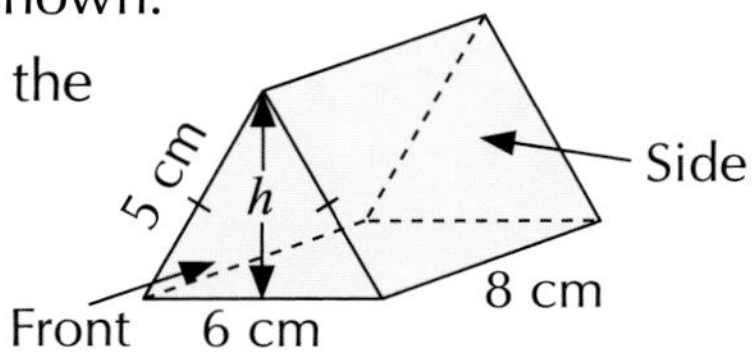

Q3 For each of the shapes P and Q on the right, complete the following.

a) Draw a net of the shape.

b) Draw the plan, and front and side elevations of the shape from the directions indicated.

c) Draw the shape on isometric paper.

d) Calculate the shape's volume.

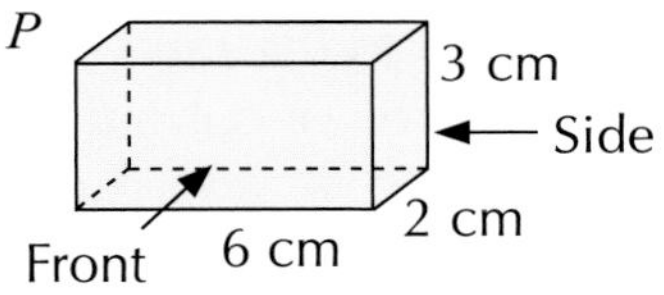

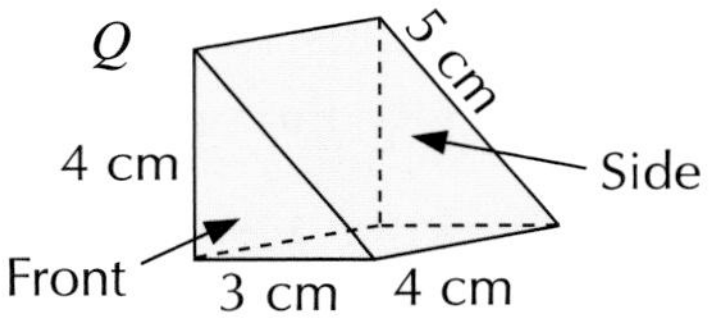

Q4 24 cubes with edges of length 8 cm fit exactly in a tray of depth 8 cm, as shown.

a) Find the volume of one cube.

b) Find the volume of the tray.

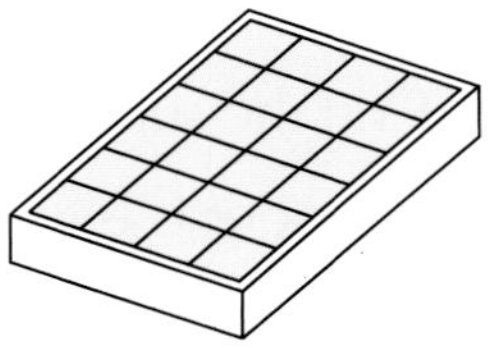

Q5 Beans are sold in cylindrical tins of diameter 7.4 cm and height 11 cm.

a) Find the volume of one tin.

The tins are stored in boxes which hold 12 tins in three rows of 4, as shown.

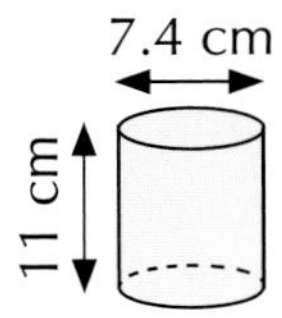

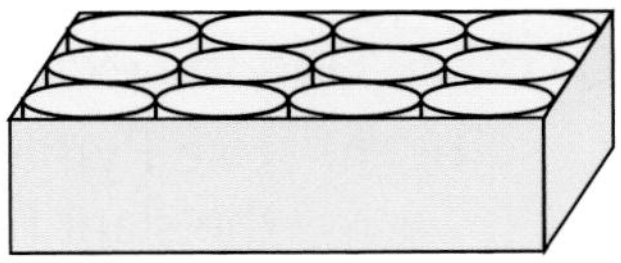

b) Find the dimensions of the box.

c) Find the volume of the box.

d) Calculate the volume of the box that is not taken up by tins when it is fully packed.

Give your answer correct to 1 decimal place.

Q6 The shape on the right is made by joining together the bases of two identical square-based pyramids. Each pyramid has height 9 cm and base side length 7 cm. Find the volume of this shape.

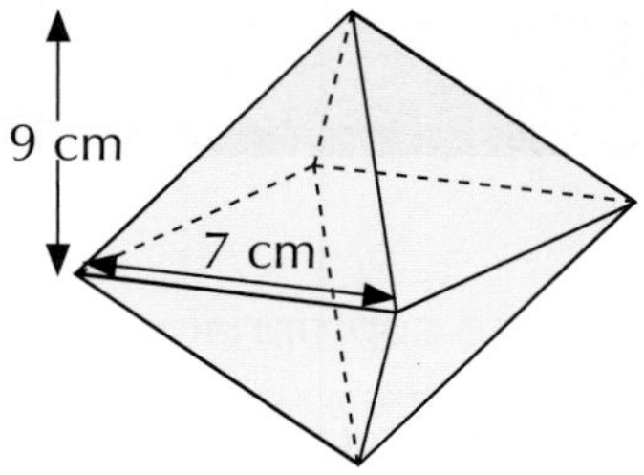

Q7 A cube has a total surface area of 54 cm².

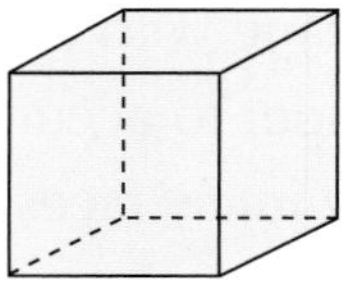

a) Find the area of one face of the cube.

b) Find the length of the edges of the cube.

c) Find the cube's volume.

d) How long will it take to fill the cube with water flowing at a rate of 2.5 cm³/s?

e) Draw the cube on isometric paper.

Q8 Two cuboids, A and B, are shown on the right.

a) Find the ratio of the volume of cuboid A to the volume of cuboid B. Give your answer in its simplest form.

b) Find the ratio of the surface area of cuboid A to the surface area of cuboid B. Give your answer in its simplest form.

Cuboid A

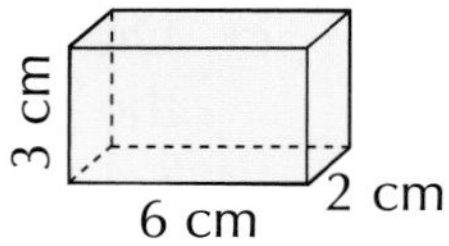

Cuboid B

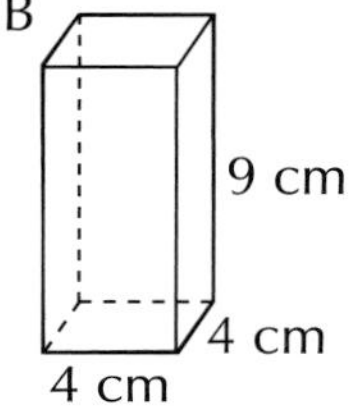

Q9 a) Draw an accurate net of a cylinder of length 3 cm and radius 1.5 cm. Label your diagram with the net's dimensions.

b) Use the net to find the cylinder's surface area.

Q10 A cylinder of diameter 25 mm has volume 7854 mm³.

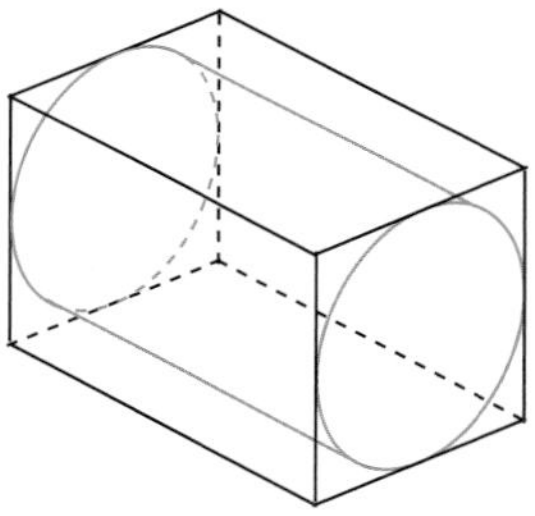

a) Find the length of the cylinder.

b) Find the surface area of the cylinder to the nearest mm².

The cylinder is enclosed in a cuboid, as shown.

c) Find the volume of the cuboid to the nearest mm³.

d) Find the volume of the empty space between the cylinder and the cuboid to the nearest mm³.

Q11 a) Find the exact volume of the cone shown on the right.

b) (i) Use Pythagoras' theorem $h^2 = a^2 + b^2$ to find h, the slant height of the cone.

(ii) Find the exact surface area of the cone.

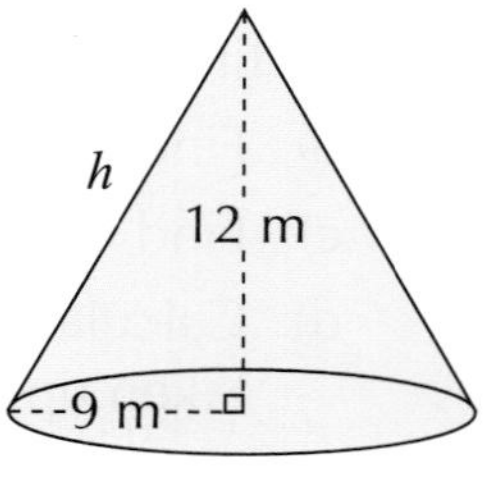

The cone is cut parallel to the circular face and a cone of height 8 m and base radius 6 m is removed to leave a frustum.

c) Find the exact volume of the frustum.

Exam-Style Questions

Q1 The diagram on the right shows three solid shapes, A, B and C.

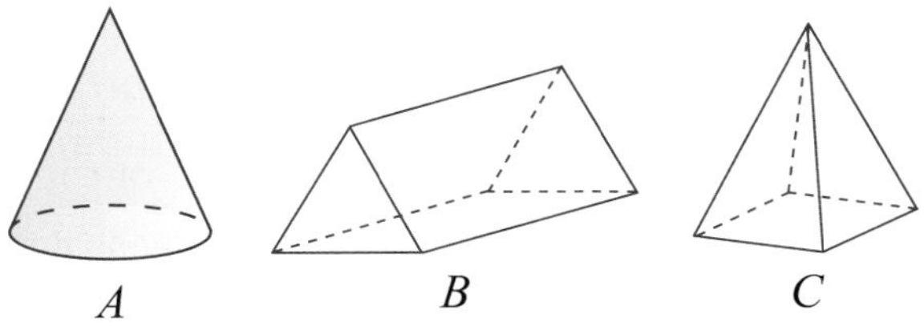

a) Write down the name of each shape.

[2 marks]

b) For shape B, write down:

(i) the number of vertices,

[1 mark]

(ii) the number of faces.

[1 mark]

Q2 The net of a cylinder is drawn on a 1 cm^2 grid, as shown on the right.

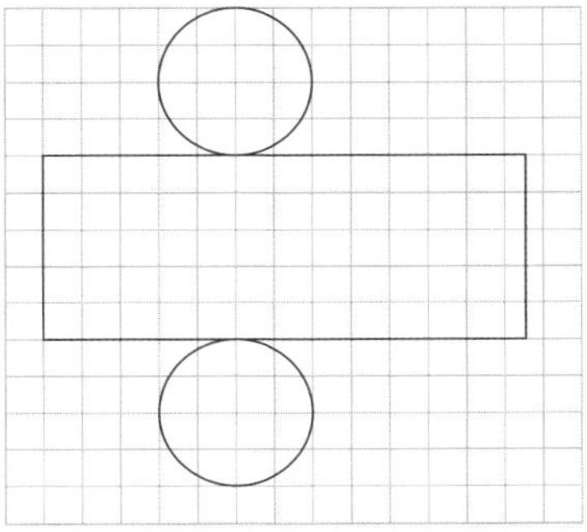

a) Write down the height of the cylinder in cm.

[1 mark]

b) Hence find the exact surface area of the cylinder.

[3 marks]

Q3 The diagram on the right shows a cuboid with a volume of 300 cm^3.
Each edge is a whole number of centimetres long.
The width of the cuboid is double the height.
The area of the shaded face is 30 cm^2.
Find the dimensions of the cuboid.

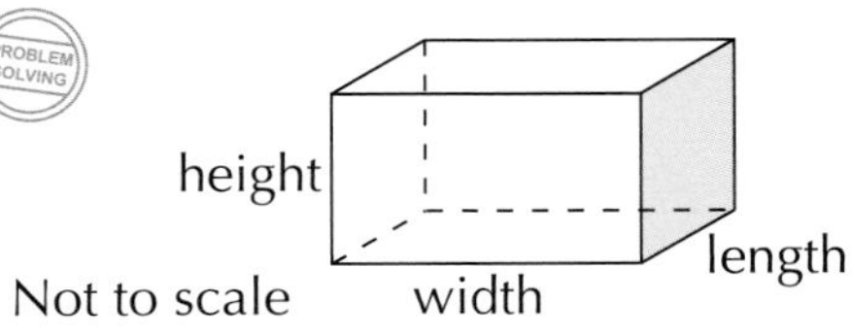

[3 marks]

Q4 Sadiq is making jam for the summer fete. His jam pan is cylindrical with radius 20 cm. The jam in his pan is 18 cm deep. Jam jars are cylindrical with radius 3 cm and can be filled to a depth of 10 cm.
How many jars can Sadiq fill with the amount of jam he has made?

[3 marks]

24.1 Reflections

Transformations can be used to move and resize shapes on a coordinate grid.
In this section, you'll meet four different transformations — the first of which is reflection.

Learning Objectives — Spec Ref G7:
- Reflect a shape in a given line.
- Describe the reflection that transforms a shape.

Prior Knowledge Check:
Be familiar with straight-line graphs and coordinates (see Section 12).

To **reflect** a shape, you draw its mirror image.

- First reflect the **vertices** of the shape in the line of symmetry (also known as the **mirror line**). The **reflected points** (called the **image points**) should be the **same distance** from the line as the original points but on the **other side** of it.
- Then **join up** the image points to create the reflected shape. The reflected shape will be the **same size** and **shape** as the original — so the shapes are **congruent** (see p.411).
- A reflection in the **y-axis** will send a point (x, y) to $\mathbf{(-x, y)}$.
- A reflection in the **x-axis** will send a point (x, y) to $\mathbf{(x, -y)}$.

To **describe** a reflection, you just need to find the **equation** of the mirror line.

Tip: Mirror lines of the form $x = a$ are vertical and mirror lines of the form $y = a$ are horizontal.

Example 1

Reflect the shape below in the mirror line.

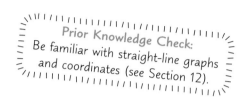

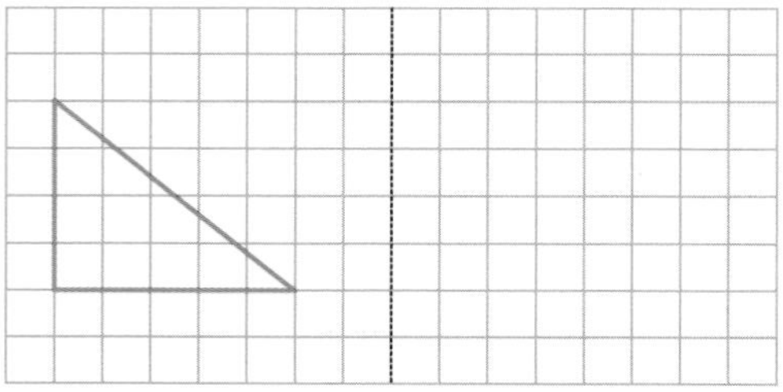

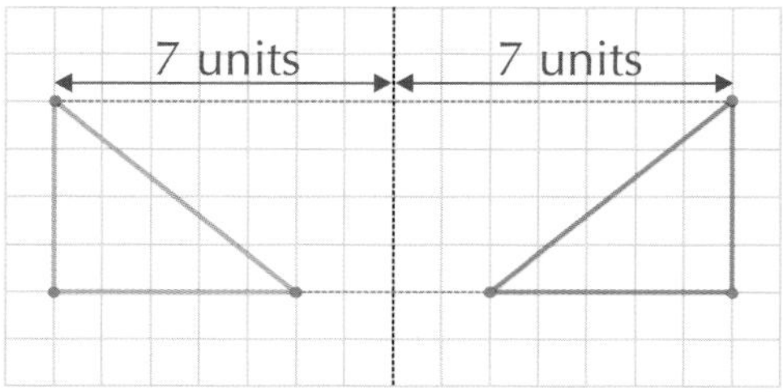

1. Reflect one corner at a time.
2. The image of each point is the same distance from the mirror line as the original.
3. Join up the corners to create the image.

Exercise 1

Q1 Copy the diagrams below, and reflect each of the shapes in the mirror line.

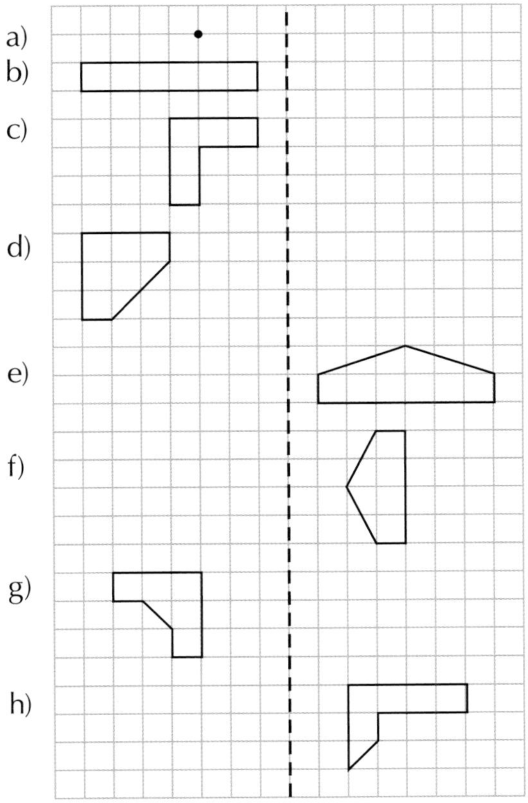

Q2 Copy the diagrams below, and reflect each of the shapes in the mirror line.

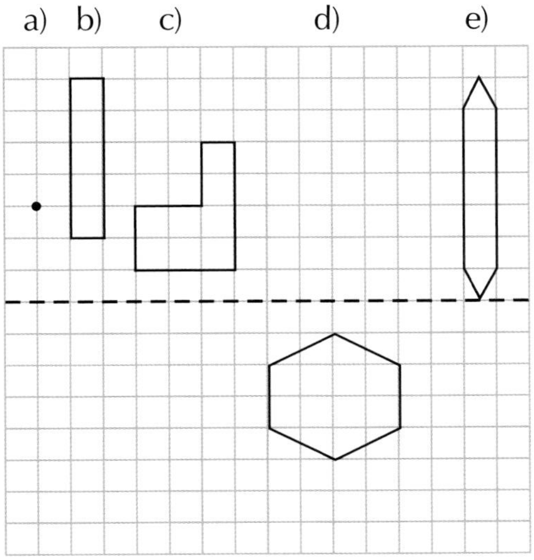

Q3 Copy the diagram.
Reflect the shape in the mirror line.

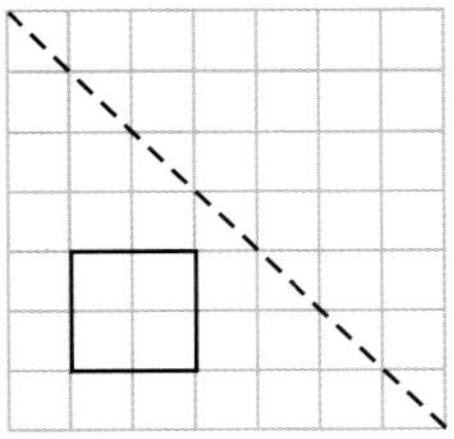

Example 2

a) Reflect the shape *ABCDE* in the *y*-axis.

b) Label the image points A_1, B_1, C_1, D_1 and E_1 with their coordinates.

1. Here the *y*-axis is the mirror line.
2. Each image point should be the same distance from the *y*-axis as the original point. E.g. *C* is 2 units to the left of the *y*-axis so its image C_1 should be 2 units to the right of the *y*-axis.
3. Write down the coordinates of each of the image points. Each point (x, y) becomes $(-x, y)$.

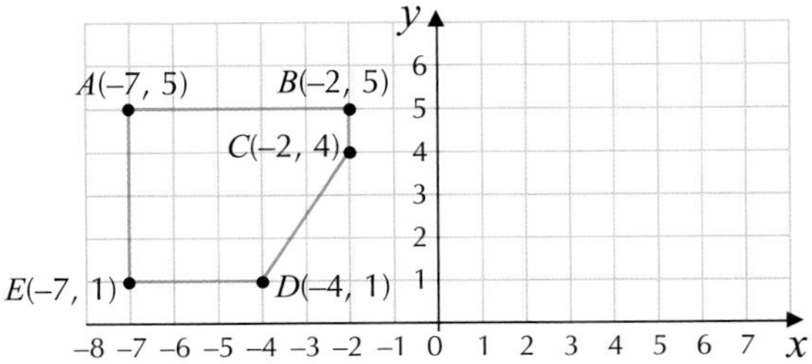

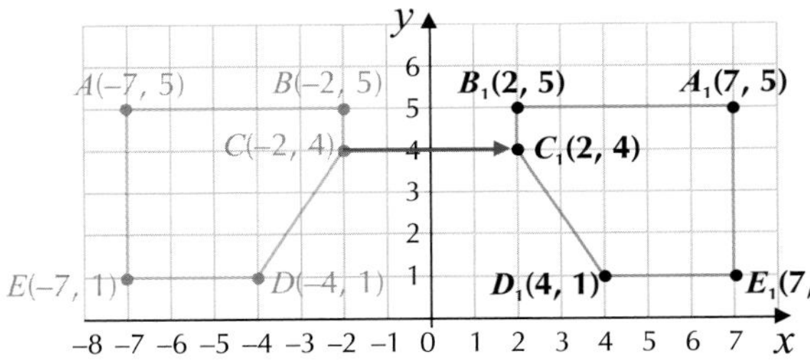

Example 3

Reflect the shape below in the line $x = 5$.

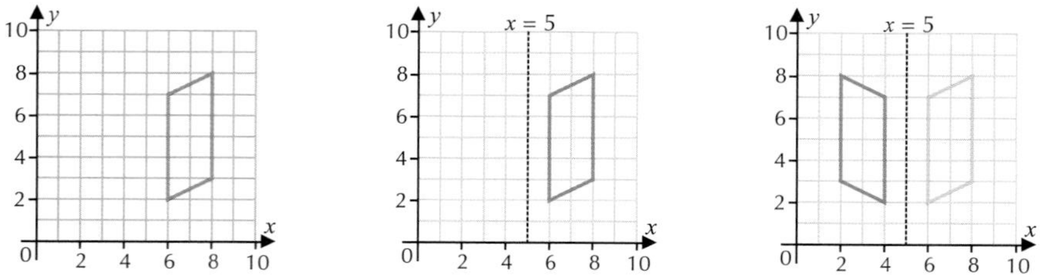

1. Draw in the mirror line, $x = 5$.
2. Reflect the shape in the line as usual, one point at a time.

Tip: The image of each point should be the same distance from the line $x = 5$ as the original point.

Exercise 2

Q1 a) Copy the diagram on the right, and reflect the shape in the y-axis.

b) Label the image points A_1, B_1, C_1, D_1 and E_1 with their coordinates.

c) Describe a rule connecting the coordinates of A, B, C, D and E and the coordinates of A_1, B_1, C_1, D_1 and E_1.

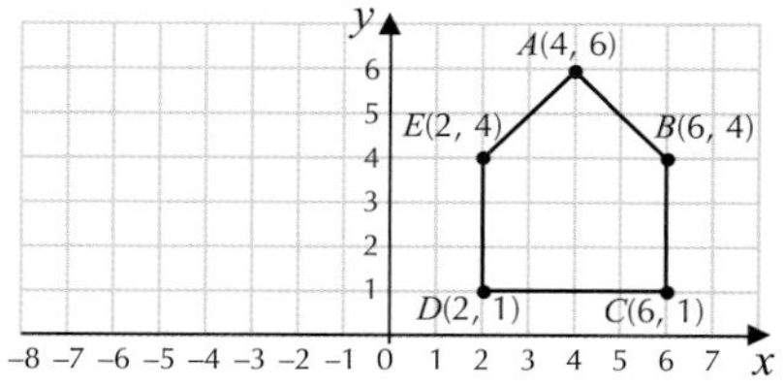

Q2 Copy each of the diagrams below, and reflect the shapes in:

a) the y-axis,

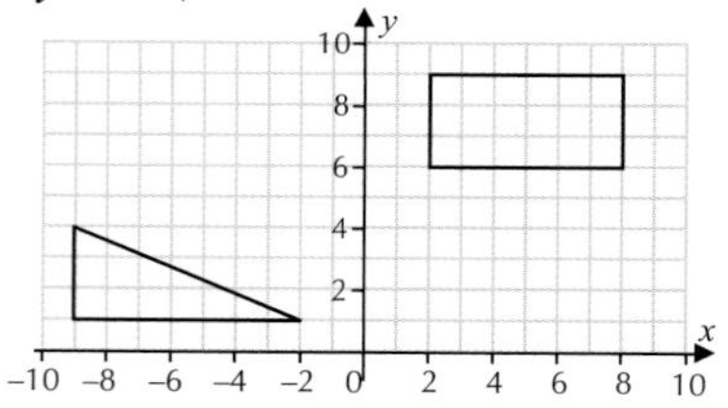

b) the x-axis.

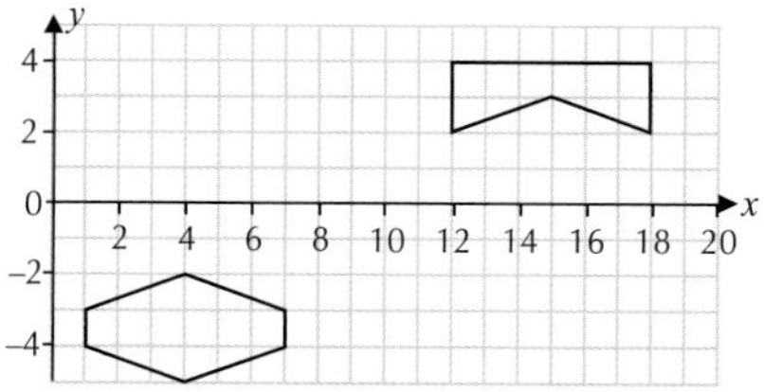

Q3 a) The following points are reflected in the x-axis. Find the coordinates of the image.

(i) (1, 2) (ii) (3, 0) (iii) (–2, 4) (iv) (–1, –3)

b) The following points are reflected in the y-axis. Find the coordinates of the image.

(i) (4, 5) (ii) (7, 2) (iii) (–1, 3) (iv) (–3, –1)

Q4 Copy the diagram shown on the right.

a) Reflect shape A in the line $x = -4$. Label the image A_1.

b) Reflect shape B in the line $y = 4$. Label the image B_1.

c) Reflect shape C in the line $y = 5$. Label the image C_1.

d) Reflect shape D in the line $x = 5$. Label the image D_1.

e) Shape C_1 is the reflection of shape B in which mirror line?

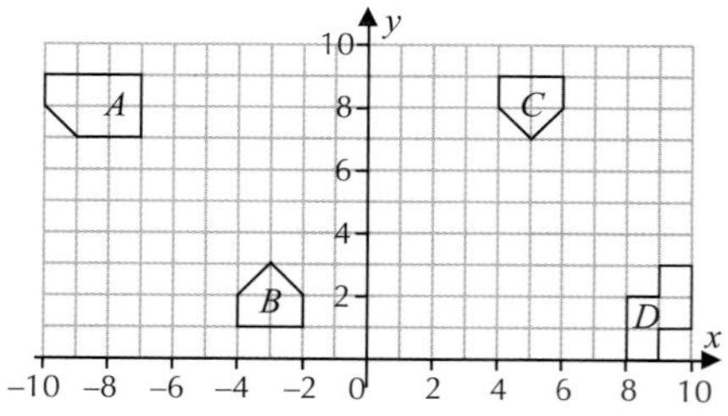

The mirror line isn't always one of the axes or a line parallel to an axis — it could be a **diagonal** line such as $y = x$ or $y = -x$.

(i) A reflection in $y = x$ sends (x, y) to (y, x). (ii) A reflection in $y = -x$ sends (x, y) to $(-y, -x)$.

Follow the same method as before, **reflecting each vertex** and then joining up the **image points**.

Make sure the image points are the **same distance** from the mirror line as the original ones — you measure this distance **perpendicular** to the mirror line.

Example 4

Reflect the shape below in the line $y = x$.

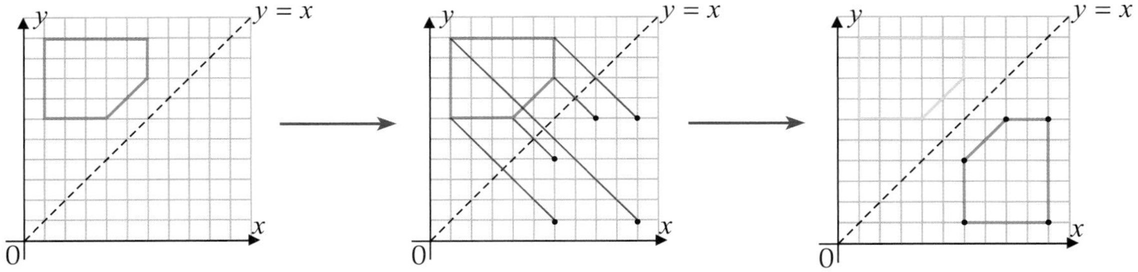

Reflect each corner in the mirror line then join up the vertices. Each image point should be the same perpendicular distance from the line $y = x$ as the original point.

Tip: Perpendicular distance is measured at right angles to the mirror line.

Exercise 3

Q1 a) Copy the diagram below, and reflect the shape $ABCD$ in the line $y = x$.

b) Describe a rule connecting the coordinates of a point and the coordinates of its image after reflection in the line $y = x$.

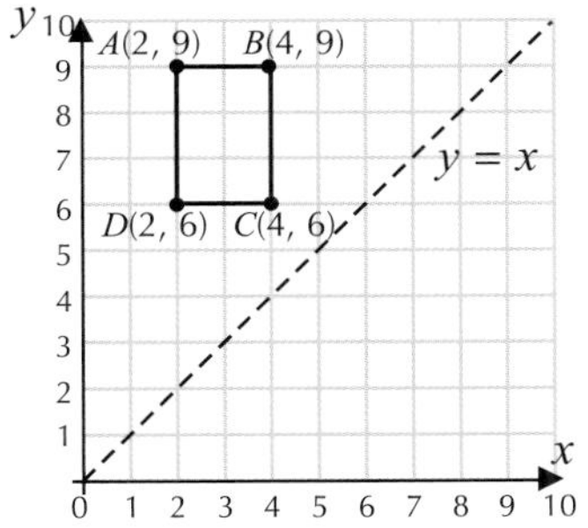

Q2 Copy the diagram below, and reflect the shapes in the line $y = -x$.

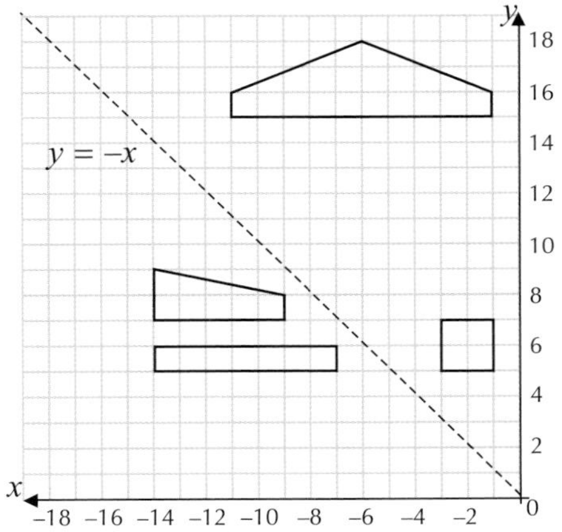

Q3 Find the coordinates of the following points after they have been reflected in:

a) the line $y = x$, b) the line $y = -x$.

(i) (1, 2) (ii) (3, 0) (iii) (–2, 4) (iv) (–1, –3)

24.2 Rotations

The second transformation to learn about is rotation. Rotations spin objects around a fixed point.

Learning Objectives — Spec Ref G7:
- Rotate a shape around a given point.
- Describe the rotation that transforms a shape.

Prior Knowledge Check:
Be familiar with coordinates (see p.188).

When an object is **rotated** about a point, its size and shape stay the same — so the new shape is **congruent** to the original shape. Also, the **distance** of each vertex from the centre of rotation doesn't change. To describe a rotation, you need to give **three** pieces of information:

(i) the **centre** of rotation (ii) the **direction** of rotation (iii) the **angle** of rotation

The **centre** of rotation can be **any point** — e.g. (5, 1) or the origin (0, 0).
The **direction** of rotation will be either **clockwise** or **anticlockwise** and the **angle** might be given in **degrees** or as a **fraction of a turn** (e.g. 90° or a quarter-turn).

Example 1

Rotate the shape below 180° about point *P*.

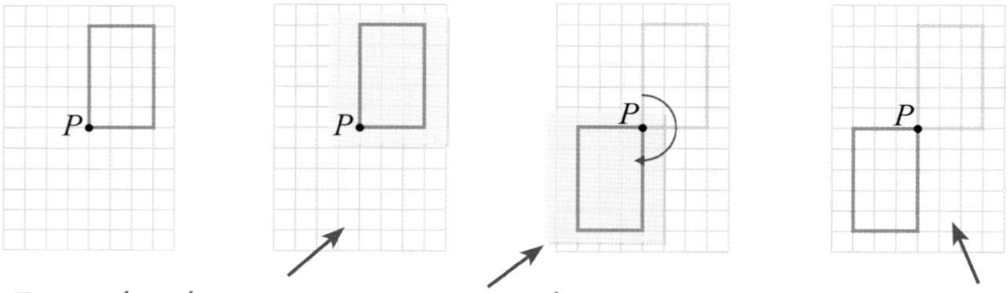

1. Draw the shape on a piece of tracing paper. (Or imagine a drawing of it.)
2. Rotate the tracing paper half a turn about *P*, the centre of rotation. ('About *P*' means *P* doesn't move.)
3. Draw the image in its new position.

Tip: A rotation of 180° gives the same result clockwise or anticlockwise, so you can rotate in either direction.

Exercise 1

Q1 Copy the diagrams below, then rotate the shapes 180° about *P*.

a)

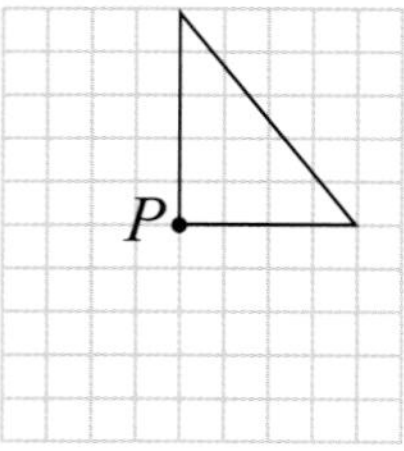

b)

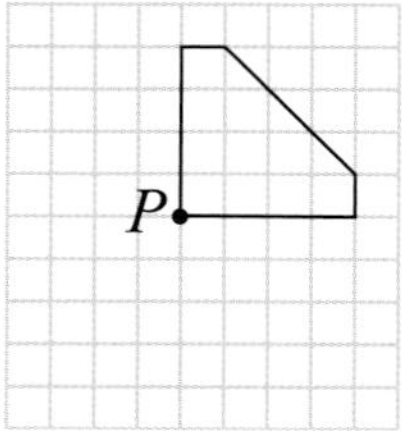

Q2 Copy the diagrams below, then rotate the shapes 90° clockwise about *P*.

a)

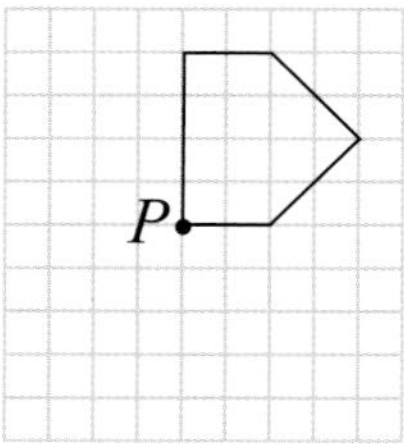

b)

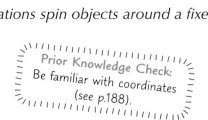

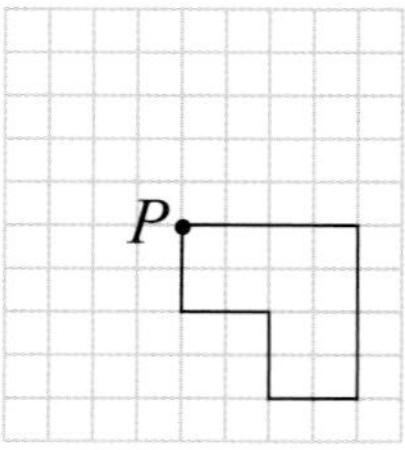

Example 2

Rotate the shape below 90° clockwise about point *P*.

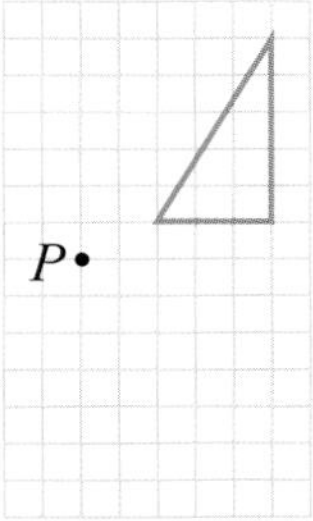

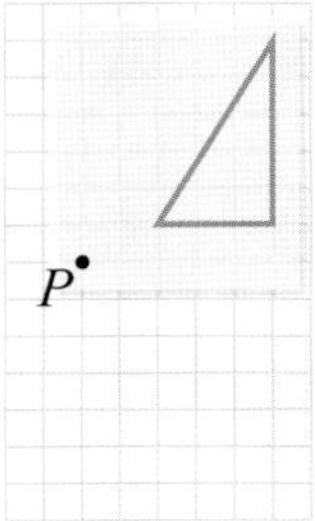

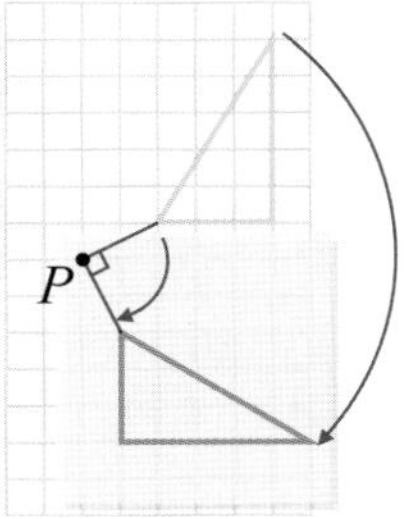

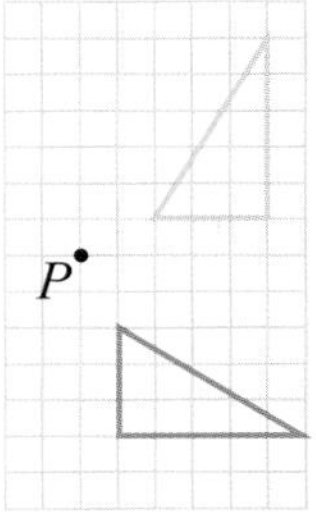

1. Here the centre of rotation is not a point on the shape.
2. Draw the shape on a piece of tracing paper. (Or imagine a drawing of it.)
3. Rotate the tracing paper 90° clockwise about *P*.
4. Draw the image.

Exercise 2

Q1 Copy the diagrams below, then rotate the shapes 90° anticlockwise about *P*.

a)

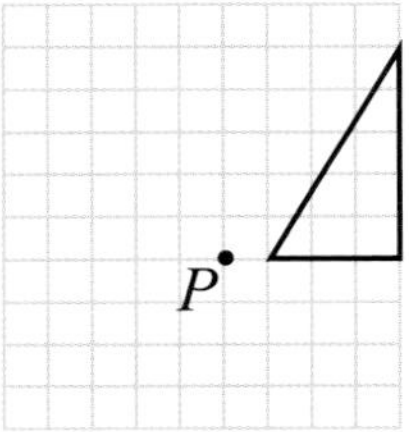

b)

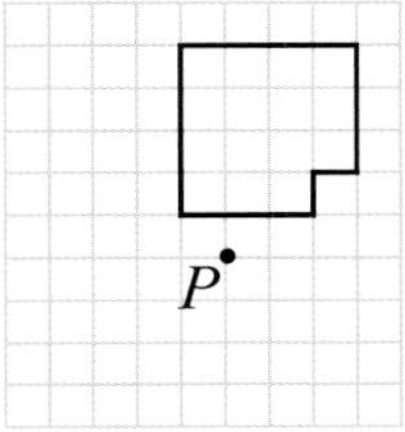

c)

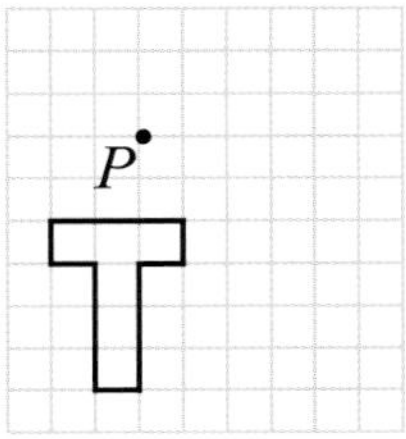

d)

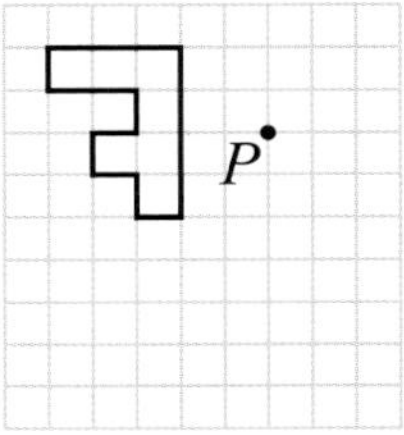

Q2

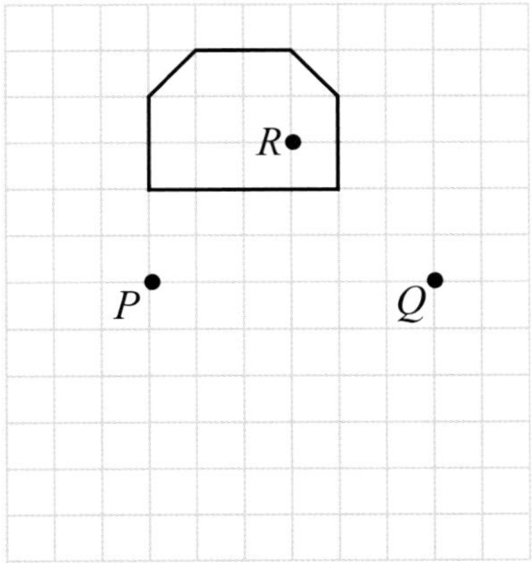

On separate copies of the diagram above, rotate the shape as follows.

a) 90° clockwise about *P*

b) 270° clockwise about *Q*

c) 180° about *R*

Q3

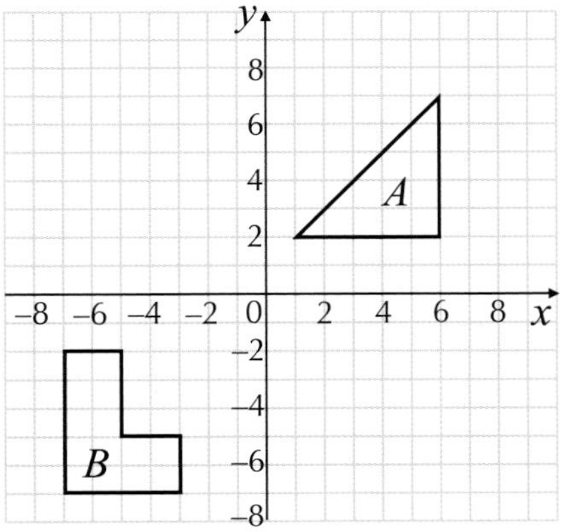

Copy the diagram above, then complete the following.

a) Rotate *A* 90° clockwise about the origin.

b) Rotate *B* 270° anticlockwise about the origin.

Q4 Copy the diagram on the right, then complete the following.

a) Rotate A 90° clockwise about (–8, 5).

b) Rotate B 90° anticlockwise about (6, 4).

c) Rotate C 90° clockwise about (8, –4).

d) Rotate D 180° about (–2, –5).

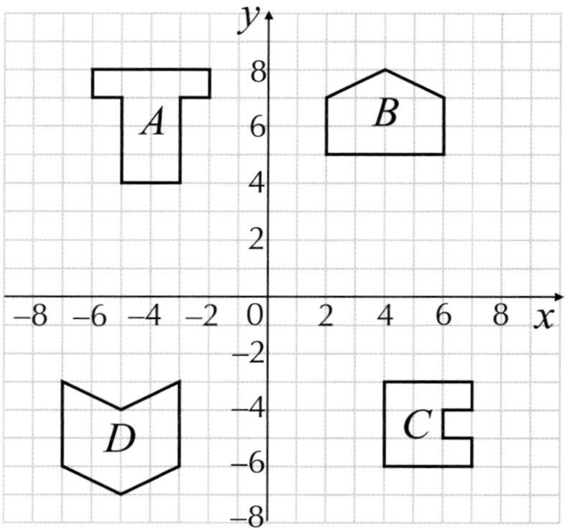

Q5 The triangle ABC has vertices $A(-2, 1)$, $B(-2, 6)$ and $C(4, 1)$.

a) Draw the triangle on a pair of axes.

b) Rotate the triangle 90° anticlockwise about (5, 4). Label the image $A_1B_1C_1$.

c) Write down the coordinates of A_1, B_1 and C_1.

Example 3

Describe fully the rotation that transforms shape A to shape B.

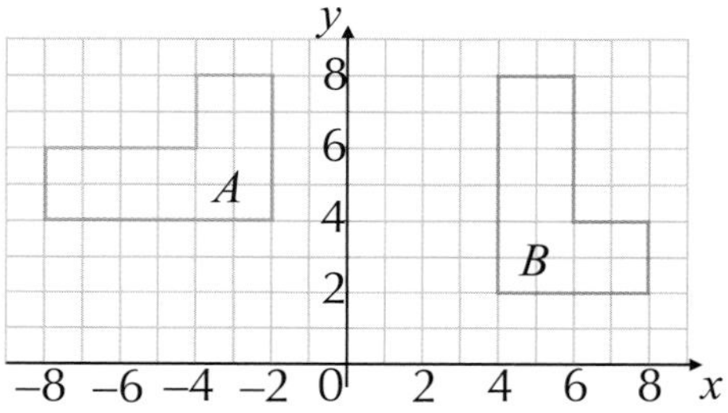

1. The shape looks like it has been rotated clockwise by 90°.
2. Trace shape A using tracing paper. Put your pencil on different possible centres of rotation and turn the tracing paper 90° clockwise until you find a centre that takes shape A onto shape B.

Tip: Remember:
90° = a ¼-turn,
180° = a ½-turn,
270° = a ¾-turn.

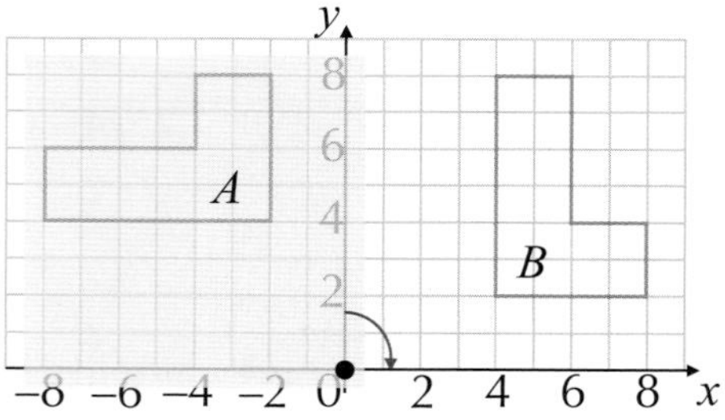

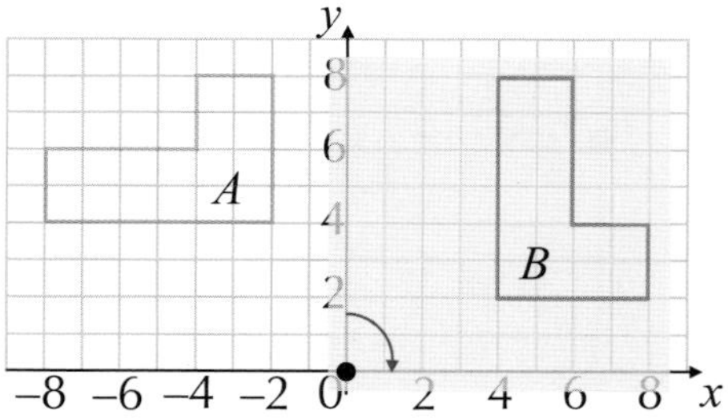

3. To fully describe the rotation, you need to write down the centre, direction and angle of rotation.

So A is transformed to B by a rotation of **90° clockwise** about **the origin (0, 0)**.

Exercise 3

Q1 Shapes *A*, *B*, *C* and *D* are shown on the grid below.

a) Describe fully the rotation that transforms shape *A* to shape *B*.

b) Describe fully the rotation that transforms shape *C* to shape *D*

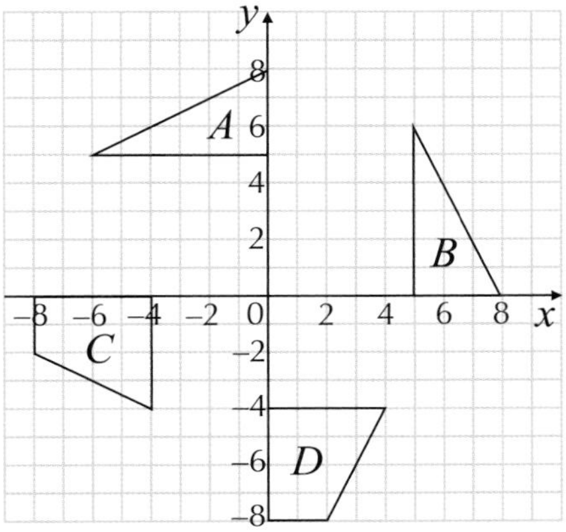

Q2 Shapes *E*, *F*, *G* and *H* are shown on the grid below.

a) Describe fully the rotation that transforms shape *E* to shape *F*.

b) Describe fully the rotation that transforms shape *G* to shape *H*.

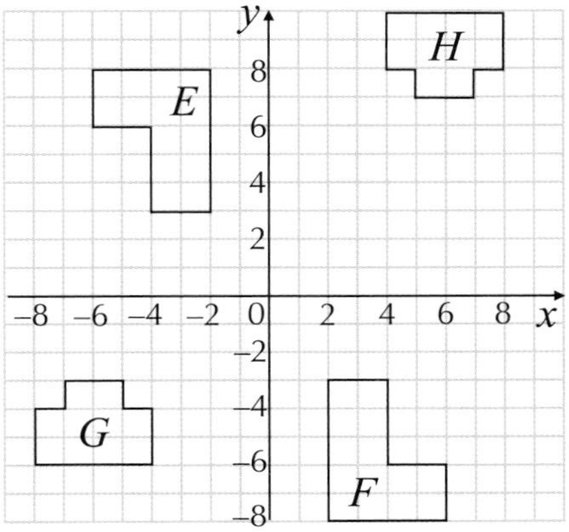

Q3 Shapes *I*, *J*, *K* and *L* are shown on the grid below.

a) Describe fully the rotation that transforms shape *I* to shape *J*.

b) Describe fully the rotation that transforms shape *K* to shape *L*.

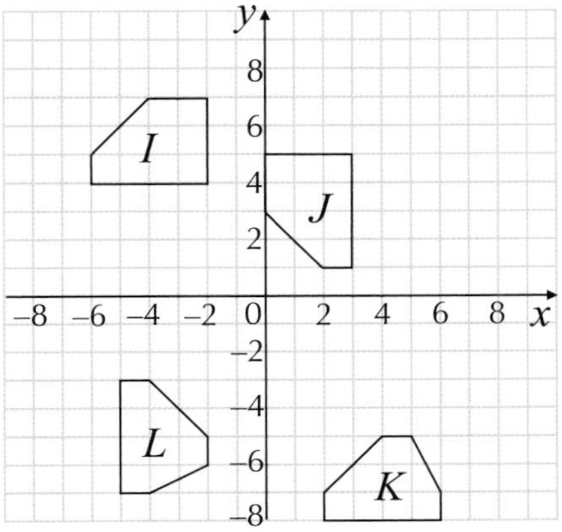

Q4 Shapes *M*, *N* and *P* are shown on the grid below.

a) Describe fully the rotation that transforms shape *M* to shape *N*.

b) Describe fully the rotation that transforms shape *M* to shape *P*.

c) Describe fully the rotation that transforms shape *N* to shape *P*.

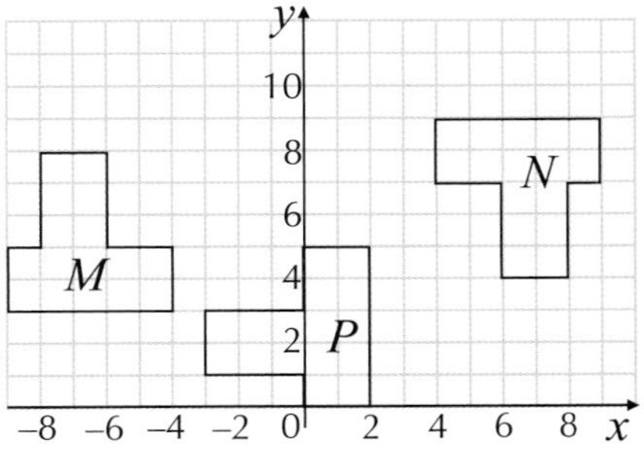

24.3 Translations

Translations are simple — they just slide a shape up/down and left/right.

Learning Objectives — Spec Ref G7/G24:
- Translate a shape on a coordinate grid.
- Find the vector that describes a translation.

Prior Knowledge Check: Be familiar with column vectors (see p.314).

To **translate** an object, you need to know the **distance** that it moves and in which **direction** — these are broken down into **horizontal** and **vertical** movements.

- For horizontal movements, **positive** numbers move the shape **right** and **negative** numbers move it **left**.
- For vertical movements, **positive** numbers move the shape **up** and **negative** numbers move it **down**.

Tip: Start by working out where the corners of a shape move to under a translation — then join them to complete the shape.

You can use **column vectors** to represent this information. For example:

$\begin{pmatrix} 3 \\ -2 \end{pmatrix}$

- 3 — the object moves 3 units to the right (**positive *x*-direction**)
- -2 — the object moves 2 units down (**negative *y*-direction**)

When an object is translated, its **size** and **shape** stay the **same**.
Only its **position** changes — so the new shape is **congruent** to the original shape.

Example 1

Translate the shape below by the vector $\begin{pmatrix} 5 \\ -3 \end{pmatrix}$.

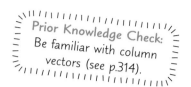

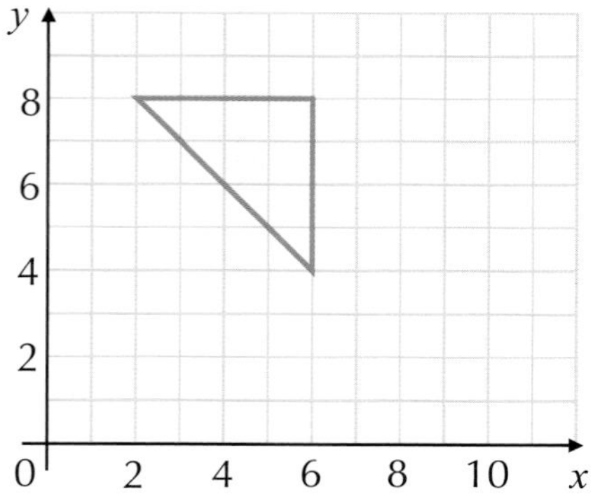

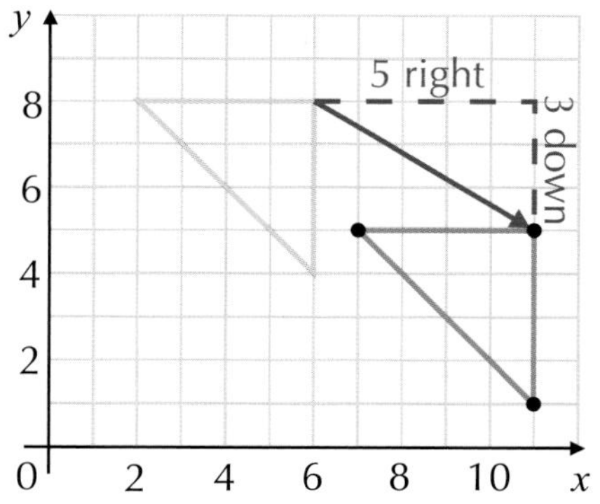

1. $\begin{pmatrix} 5 \\ -3 \end{pmatrix}$ is a translation of: i) 5 units to the right ii) 3 units down.
2. Translate each vertex, then join them up to create the translated shape.

Example 2

Find the coordinates of the point (2, –5) after it has been translated by $\begin{pmatrix} 6 \\ -4 \end{pmatrix}$.

The x-coordinate increases by 6, while the y-coordinate decreases by 4.

$(2 + 6, -5 - 4) = \mathbf{(8, -9)}$

Exercise 1

Q1 Write down in words the translations described by the following vectors.

a) $\begin{pmatrix} 1 \\ 1 \end{pmatrix}$ b) $\begin{pmatrix} 2 \\ 0 \end{pmatrix}$ c) $\begin{pmatrix} -2 \\ 6 \end{pmatrix}$ d) $\begin{pmatrix} -3 \\ -2 \end{pmatrix}$

Q2 Copy the diagram below, then translate each shape by the vector written next to it.

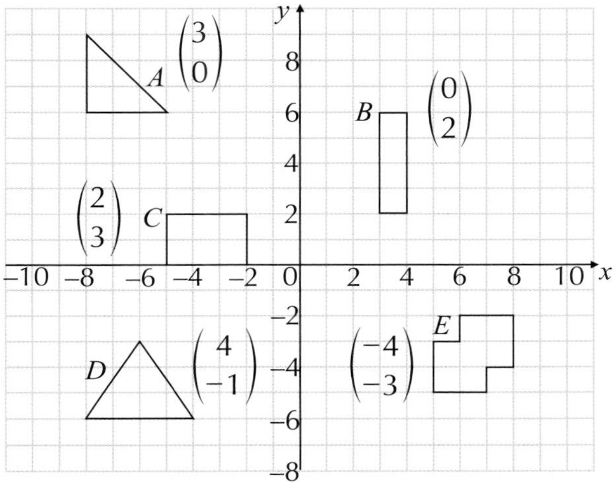

Q3 Copy the diagram on the right, then:

a) Translate the triangle ABC by the vector $\begin{pmatrix} -10 \\ -1 \end{pmatrix}$. Label the image $A_1B_1C_1$.

b) Label A_1, B_1 and C_1 with their coordinates.

c) Describe a rule connecting the coordinates of A, B and C and the coordinates of A_1, B_1 and C_1.

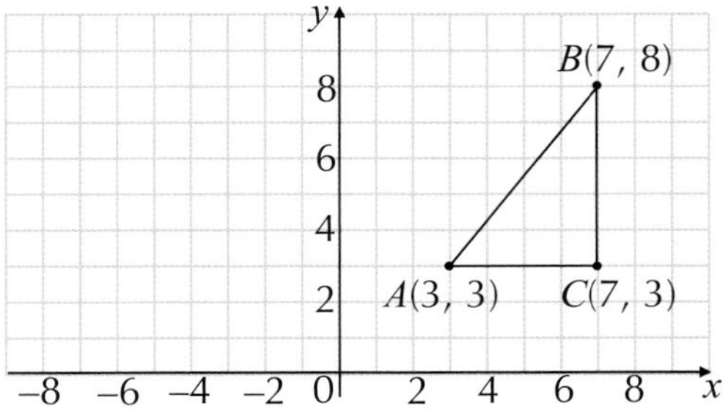

Q4 Find the coordinates of the image of the point (3, –4) after it has been translated by:

a) $\begin{pmatrix} 0 \\ 1 \end{pmatrix}$ b) $\begin{pmatrix} 3 \\ 0 \end{pmatrix}$ c) $\begin{pmatrix} 4 \\ -2 \end{pmatrix}$ d) $\begin{pmatrix} -1 \\ -5 \end{pmatrix}$

Q5 The triangle DEF has corners $D(1, 1)$, $E(3, -2)$ and $F(4, 0)$. After the translation $\begin{pmatrix} -2 \\ 2 \end{pmatrix}$, the image of DEF is $D_1E_1F_1$. Find the coordinates of D_1, E_1 and F_1.

Example 3

Describe the transformation that maps shape *A* onto shape *B*.

1. The shape hasn't been rotated or reflected and hasn't changed size, so this must be a translation.
2. Choose a pair of corresponding vertices on the two shapes and count how many units horizontally and vertically the point on *A* has moved to become the point on *B*.
3. Write the translation as a vector.

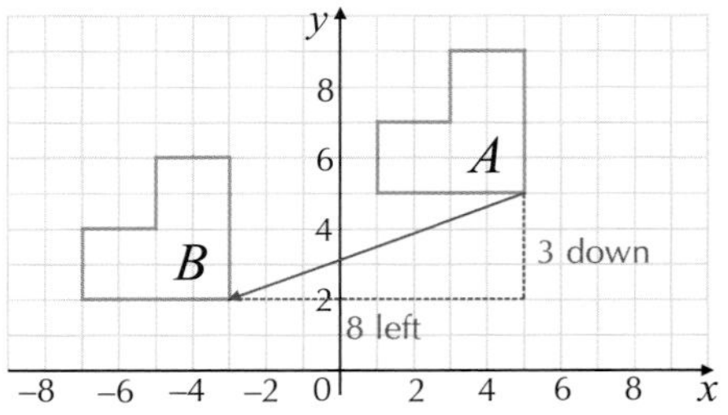

A has moved 8 units to the left and 3 units down so the transformation is a **translation described by the vector** $\begin{pmatrix} -8 \\ -3 \end{pmatrix}$.

Exercise 2

Q1 Write the following translations in vector form.

a) 1 unit to the right, 2 units up
b) 1 unit to the right, 2 units down
c) 3 units down
d) 4 units to the left, 3 units down
e) 6 units to the right, 7 units up
f) 6 units up

Q2

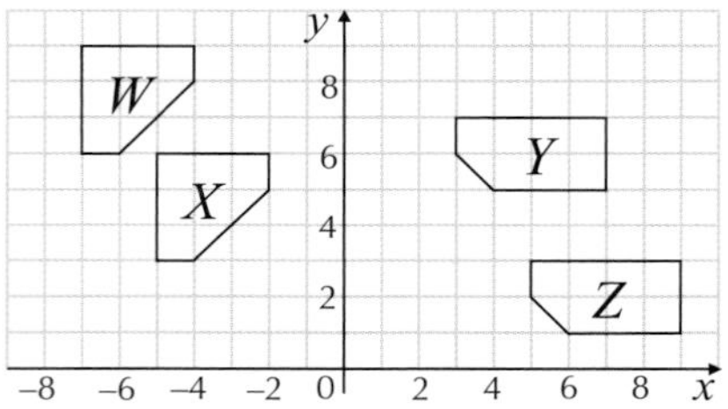

a) Write down in words the translation that maps *W* onto *X*.
b) Write the vector describing the translation that maps *W* onto *X*.
c) Describe fully, using a vector, the translation that maps *Y* onto *Z*.

Q3

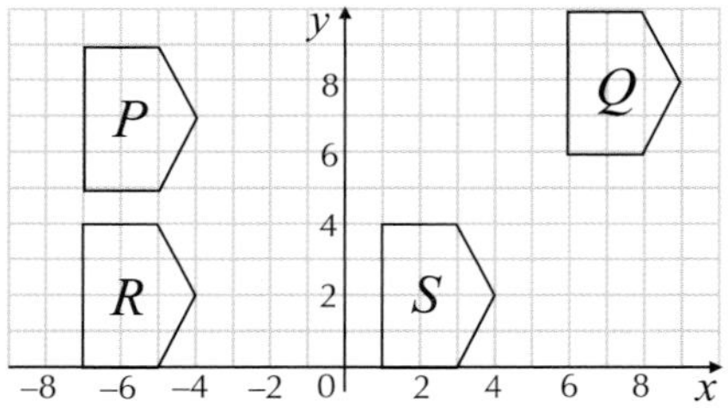

Give the translation vector that maps:

a) P to R
b) R to S
c) *P* to *Q*
d) S to R
e) Q to R
f) *S* to *P*

Q4 The triangle *DEF* has vertices *D*(–3, –2), *E*(1, –1) and *F*(0, 2).
The triangle *GHI* has vertices *G*(0, 2), *H*(4, 3) and *I*(3, 6).

a) Sketch triangles *DEF* and *GHI*.
b) Give the translation vector that maps *DEF* onto *GHI*.

Q5 The triangle *JKL* has vertices *J*(1, 0), *K*(–2, 4) and *L*(–4, 7). The triangle *MNP* has vertices *M*(0, 2), *N*(–3, 6) and *P*(–5, 9). Give the translation vector that maps:

a) *JKL* onto *MNP*,
b) *MNP* onto *JKL*.

Sometimes, you might be told a **translation vector**, and be shown the **image** of a shape after the translation. To find the **original shape** you'll need to work **backwards** and do the opposite translation, represented by the original vector with the **signs** of both components **changed**.

Example 4

Shape B is the image of shape A after the translation $\begin{pmatrix} 7 \\ -3 \end{pmatrix}$. Draw shape A.

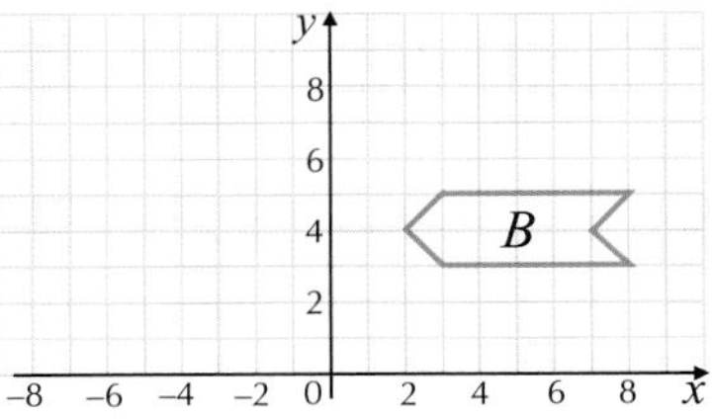

1. Find the vector that maps B onto A — it's the negative of the vector that maps A to B.

   A to B is given by $\begin{pmatrix} 7 \\ -3 \end{pmatrix}$.

   So B to A must be given by $\begin{pmatrix} -7 \\ 3 \end{pmatrix}$.

2. So shape A must be 7 units to the left and 3 units up from shape B.

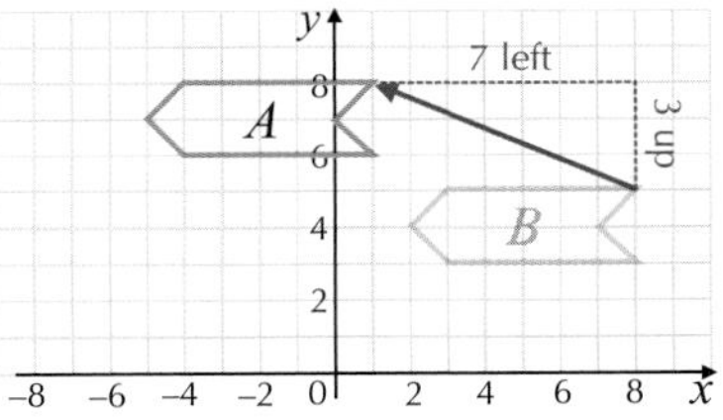

Exercise 3

Q1 This question is about the diagram on the right.

a) Give the translation vector that maps X onto Y.

b) Give the translation vector that maps Y onto X.

c) What do you notice about your answers to a) and b)?

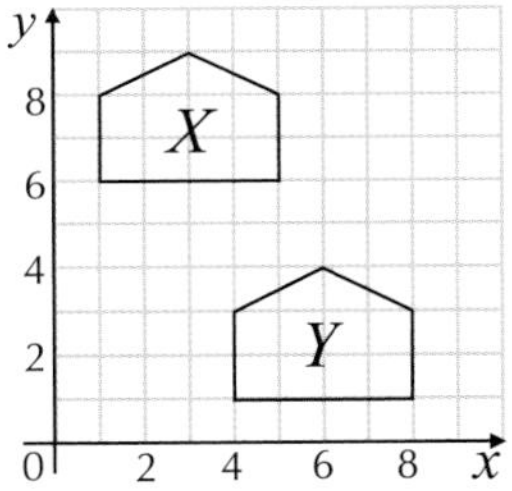

Q2 Shape Z is the image of shape W after the translation $\begin{pmatrix} 1 \\ -4 \end{pmatrix}$. Write down the translation that maps Z onto W.

Q3

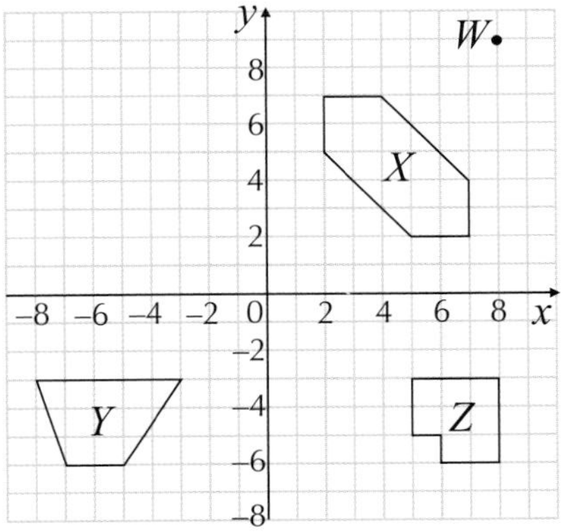

Copy the diagram on the left.

a) W is the image of A after the translation $\begin{pmatrix} 0 \\ 2 \end{pmatrix}$. Draw A.

b) Draw B, given that X is the image of B after the translation $\begin{pmatrix} 5 \\ 0 \end{pmatrix}$.

c) Y is the image of C after the translation $\begin{pmatrix} -1 \\ -5 \end{pmatrix}$. Draw C.

d) Draw D, given that Z is the image of D after the translation $\begin{pmatrix} 3 \\ 2 \end{pmatrix}$.

Q4 The triangle PQR has vertices $P(-1, 0)$, $Q(-4, 4)$ and $R(3, 2)$. PQR is the image of the triangle DEF after the translation $\begin{pmatrix} -1 \\ 4 \end{pmatrix}$. Find the coordinates of D, E and F.

24.4 Enlargements

Enlargements can be a little bit trickier than the other transformations. They are described by two properties — a scale factor (a number) and a centre of enlargement (given using coordinates).

Scale Factors

Learning Objective — Spec Ref G7:
Enlarge a shape using a positive or fractional scale factor.

When an object is **enlarged**, its shape stays the same, but its **size changes**. The image of a shape after an enlargement is **similar** to the original shape (see page 414 for more on similar shapes).

The **scale factor** of an enlargement tells you how many **times longer** the sides of the new shape are **compared** to the old shape. For example, enlarging by a **scale factor of 2** makes each side **twice as long**.

Scale factors can also be **fractions**. If the scale factor is a **fraction** between –1 and 1, then the enlargement will **shrink** the shape. E.g. enlarging by a **scale factor of $\frac{1}{2}$** makes each side **half as long**.

Tip: To draw an enlargement, start by multiplying each side length of the original shape by the scale factor. Then draw the new shape using those lengths.

Example 1

Enlarge the shape on the grid by scale factor 2.

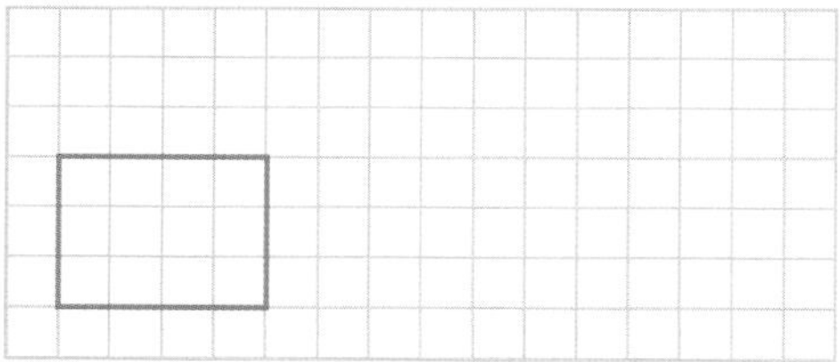

1. Find the dimensions of the shape you're enlarging.

   The original shape is 4 units wide and 3 units tall.

2. Multiply these by the scale factor to find the dimensions of the image.

   The enlargement will be
   4 × 2 = 8 units wide and
   3 × 2 = 6 units tall.

3. Draw the image using the dimensions you've found.

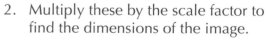

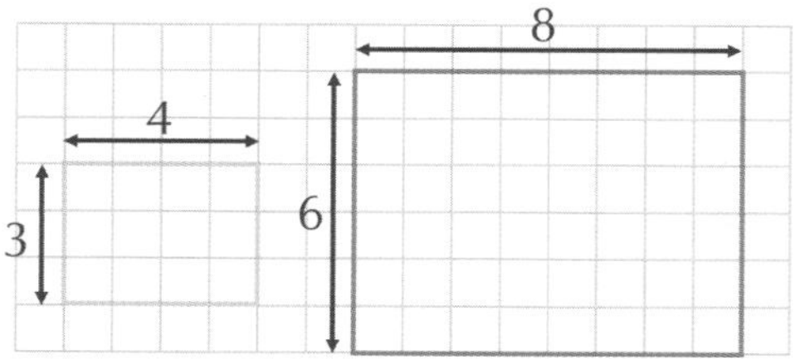

Example 2

Enlarge the shape on the grid by scale factor $\frac{1}{2}$.

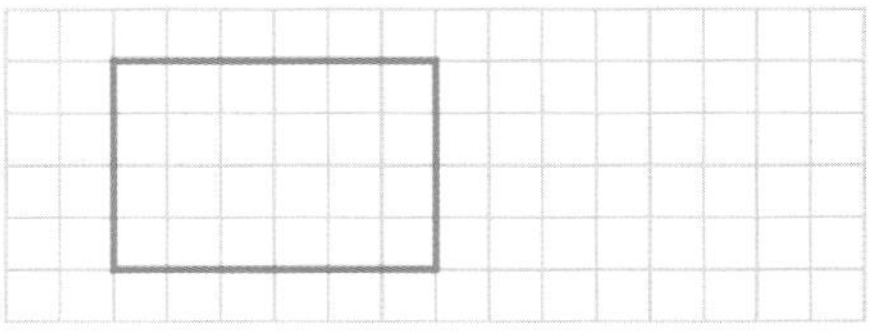

1. Multiply the dimensions of the shape by the scale factor — i.e. multiply by $\frac{1}{2}$ (or divide by 2).

   $6 \times \frac{1}{2} = 3$ and $4 \times \frac{1}{2} = 2$.

2. Draw the image using the new dimensions — this time, the image will be smaller than the original.

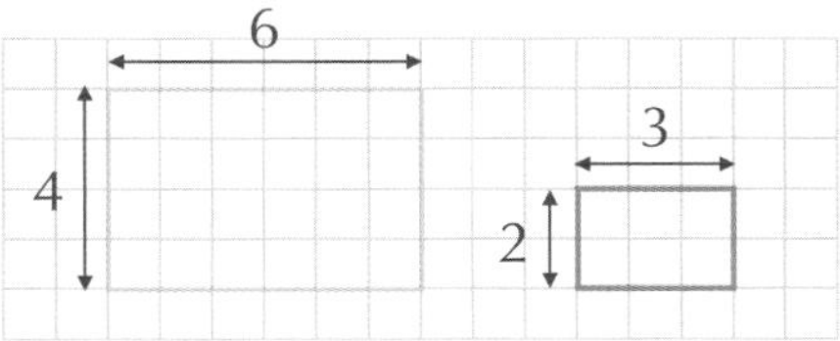

Exercise 1

Q1 Copy the diagram below, then enlarge the shapes by scale factor 2.

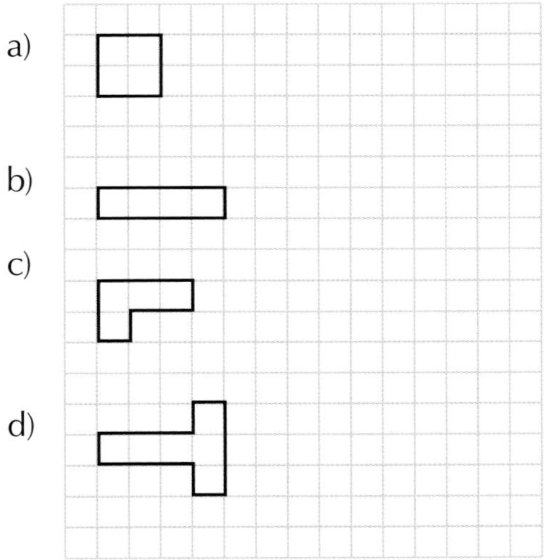

Q2 Copy the diagram below, then enlarge the shapes by scale factor 3.

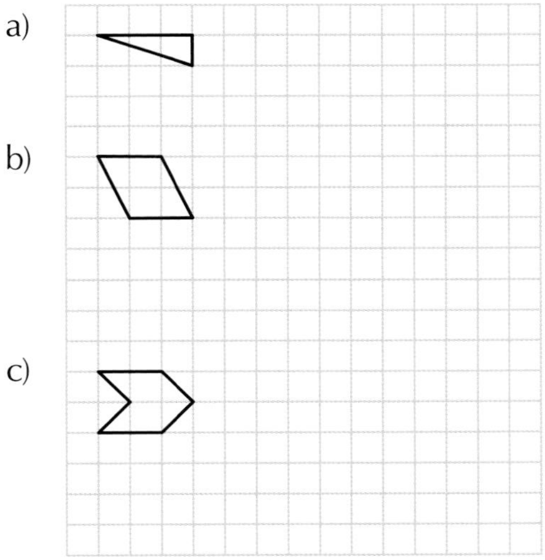

Q3 Sketch the following shapes after they have been enlarged by scale factor 5.

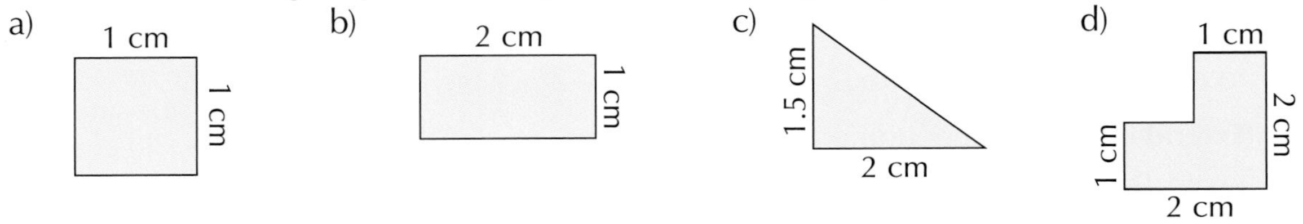

Q4 Copy the diagram below, then enlarge the shapes by scale factor $\frac{1}{2}$.

a)

b)

c)

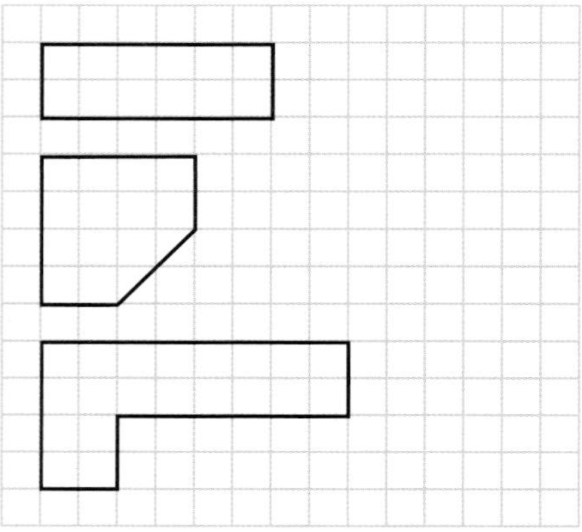

Q5 Copy the diagram below, then enlarge the shapes by scale factor $\frac{1}{3}$.

a)

b)

c)

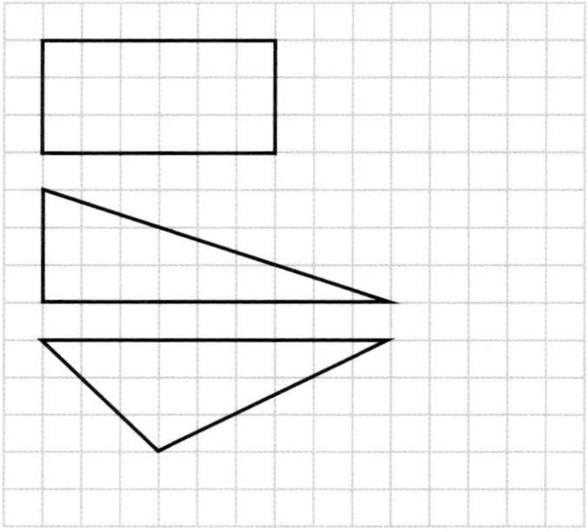

Q6 A square with sides of length 16 cm is enlarged by scale factor $\frac{1}{4}$. How long are the square's sides after the enlargement?

Q7 A 15 cm × 35 cm rectangle is enlarged by scale factor $\frac{1}{5}$. What are the dimensions of the rectangle after the enlargement?

Centres of Enlargement

Learning Objectives — Spec Ref G7:

- Enlarge a shape on a coordinate grid with a given centre of enlargement.
- Describe an enlargement that transforms a shape.

The **centre of enlargement** tells you where the enlargement is measured from. As well as telling you how much bigger the sides are, the **scale factor** tells you **how much further** the points on the new shape are from the **centre of enlargement** than the points on the old shape. For example, enlarging by a **scale factor of 3** makes each point **three times as far** from the centre of enlargement.

To enlarge a shape when you know the centre of enlargement:

- **Draw lines** from the centre of enlargement to **each vertex** of the shape.
- **Extend** each line depending on the scale factor (e.g. if the scale factor is 3 the line needs to be 3 times as long). Mark the vertices of the new shape at the ends of these extended lines.
- **Join up** the new vertices to create the enlarged shape.

Tip: Check your new shape is the correct size once you're done — the side lengths should equal those of the original shape multiplied by the scale factor.

Example 3

Enlarge the shape below by scale factor 2 with centre of enlargement (2, 2).

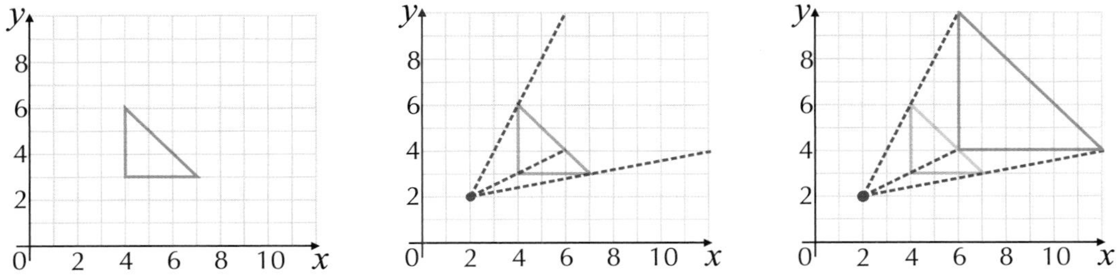

1. Draw a line from (2, 2) through each corner of the shape. Continue each line until it is twice as far away from (2, 2) as the original corner.
2. Join up the ends of the lines to create the image.

Exercise 2

Q1 The triangle *PQR* is shown on the left. $P_1Q_1R_1$ is the image of *PQR* after it has been enlarged by scale factor 2 with centre of enlargement (2, 2).

a) Find the distance of each of the following corners from (2, 2).
(i) *P* (ii) *Q* (iii) *R*

b) What will the distance of the following corners be from (2, 2)?
(i) P_1 (ii) Q_1 (iii) R_1

c) Copy the diagram and draw a line from the point (2, 2) through *P* and *Q*.
Mark the points P_1, Q_1 and R_1 on your diagram.
Join up the points P_1, Q_1 and R_1 to draw $P_1Q_1R_1$.

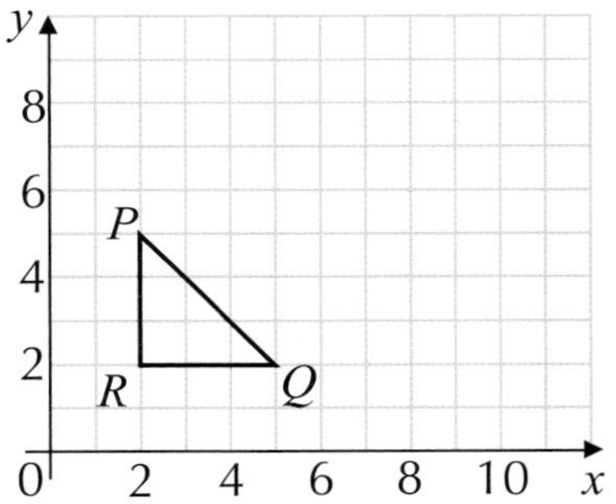

Q2 Copy the diagram on the right.

a) Enlarge *A* by scale factor 2 with centre of enlargement (–8, 9).

b) Enlarge *B* by scale factor 2 with centre of enlargement (9, 9).

c) Enlarge *C* by scale factor 3 with centre of enlargement (9, –8).

d) Enlarge *D* by scale factor 4 with centre of enlargement (–8, –8).

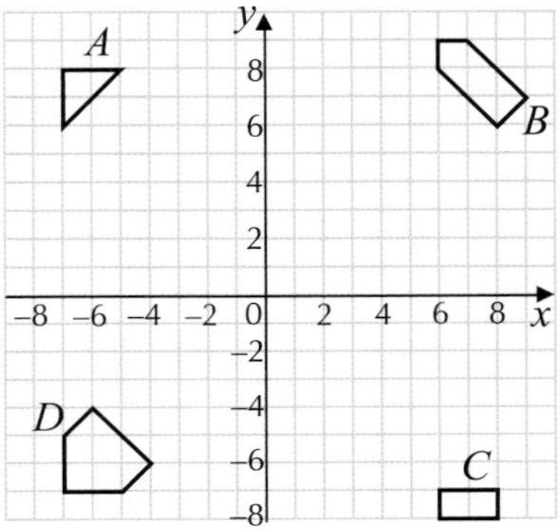

Q3 The triangle *PQR* has corners at *P*(1, 1), *Q*(1, 4) and *R*(4, 2).

a) Draw *PQR* on a pair of axes.

b) Enlarge *PQR* by scale factor 2 with centre of enlargement (–1, 1).

If the **scale factor** is a fraction, each point on the new shape will be **closer** to the **centre of enlargement**.
To enlarge by a fractional scale factor:

- First **draw lines** from the **centre of enlargement** to **each vertex**, just like before.
- Then mark the vertices of the **new shape** a **fraction** of the way along these lines (depending on the **scale factor**). E.g. a scale factor of $\frac{1}{4}$ means the new vertices are a quarter of the way along the lines.
- **Join up** the new vertices to form the new shape.

Tip: When a shape is enlarged, all the angles stay the same — always check your shape looks right.

Example 4

Enlarge the shape below by scale factor $\frac{1}{2}$ with centre of enlargement (2, 7).

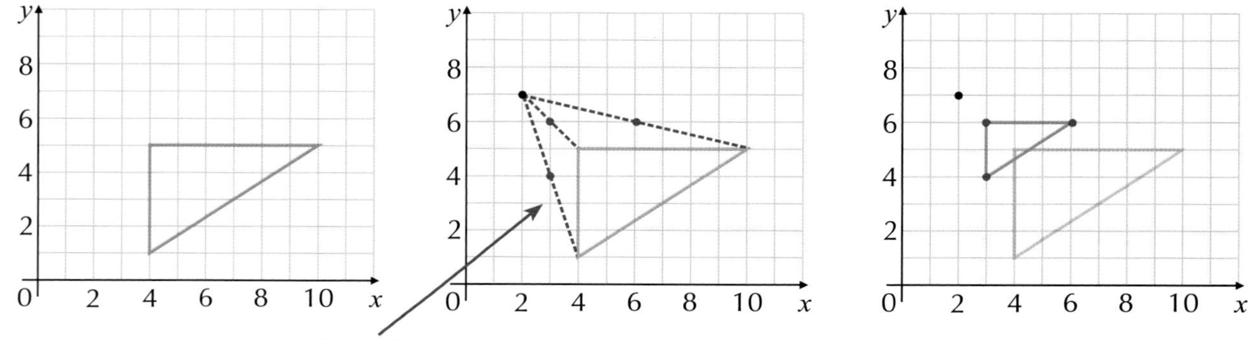

1. Draw lines from (2, 7) to each corner. The image points will lie $\frac{1}{2}$ as far from the centre of enlargement as the original corners.
2. Join up the image points.

Exercise 3

Q1 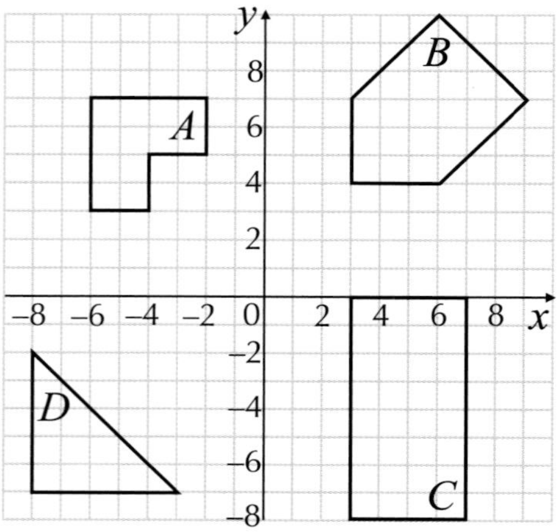

Copy the diagram on the left.

a) Enlarge A by scale factor $\frac{1}{2}$ with centre of enlargement (–8, 9).

b) Enlarge B by scale factor $\frac{1}{3}$ with centre of enlargement (0, 1).

c) Enlarge C by scale factor $\frac{1}{4}$ with centre of enlargement (–1, –8).

d) Enlarge D by scale factor $\frac{1}{5}$ with centre of enlargement (2, 3).

Q2 Copy the diagram on the right.

a) Enlarge A by scale factor $\frac{1}{2}$ with centre of enlargement (–4, 5).

b) Enlarge B by scale factor $\frac{1}{3}$ with centre of enlargement (7, 7).

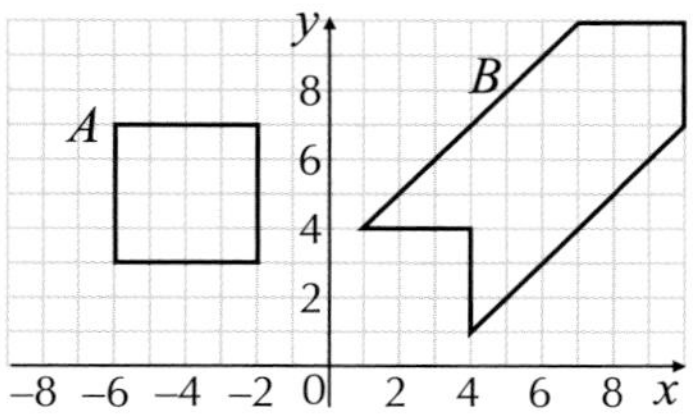

To describe an enlargement, you need to give the **scale factor** and the **centre of enlargement**. To find the **scale factor**, take the length of **any side** on the new shape and the length of its corresponding side on the old shape and use the **formula**:

$$\textbf{scale factor} = \frac{\textbf{new length}}{\textbf{old length}}$$

Tip: You can check the scale factor using a different pair of corresponding sides.

To find the **centre of enlargement**, **draw** and **extend lines** that go through corresponding vertices of both shapes and see where they all **intersect**.

Example 5

Describe the enlargement that maps shape X onto shape Y.

1. Pick any side of the shape, and see how many times bigger this side is on the image than the original. This is the scale factor.
2. Draw a line from each corner on the image through the corresponding corner on the original shape.
3. The point where these lines meet is the centre of enlargement.

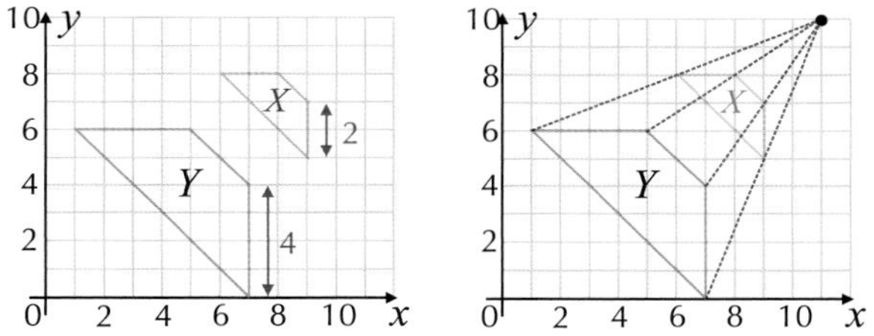

The scale factor = 4 ÷ 2 = 2.
The lines through the vertices intersect at (11, 10).
Enlargement by **scale factor 2**, **centre (11, 10)**.

Exercise 4

Q1 For each of the following, describe the enlargement that maps shape A onto shape B.

a)

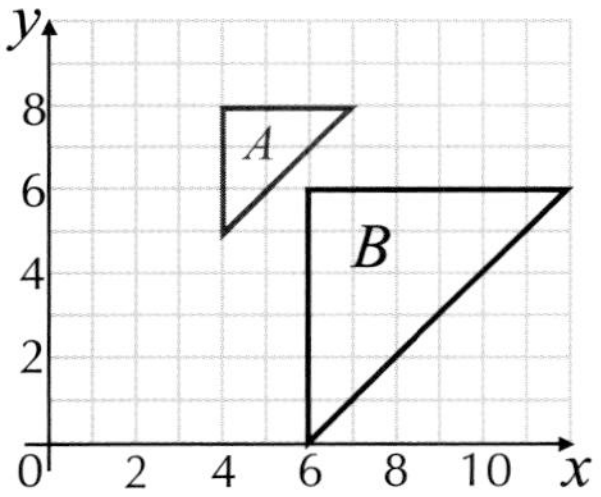

b)

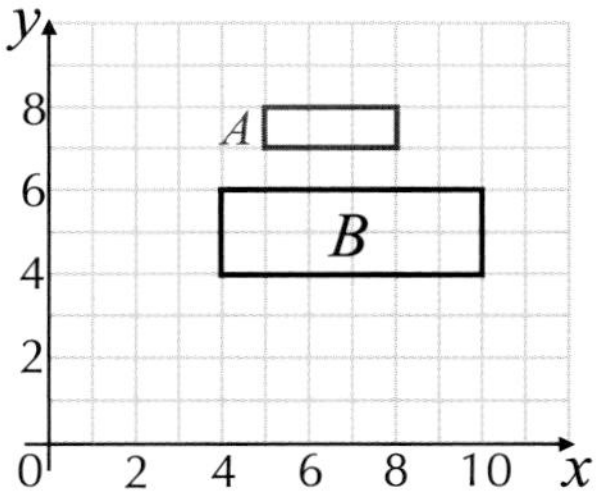

c)

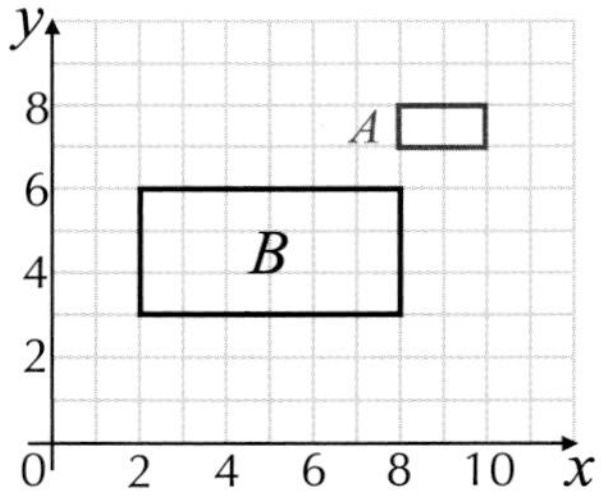

Q2 For each of the following, describe the enlargement that maps shape B onto shape A.

a) 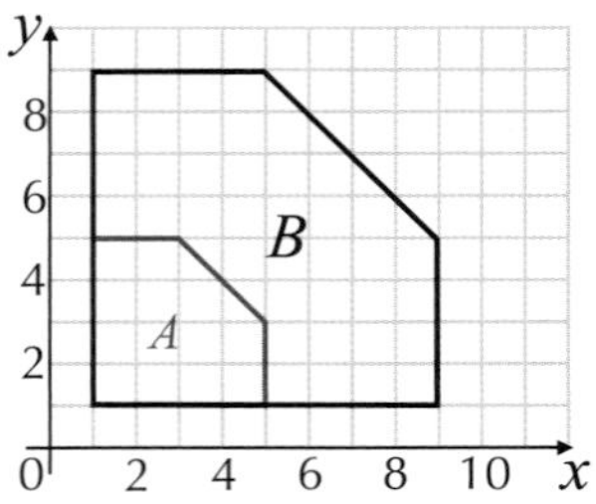

b) 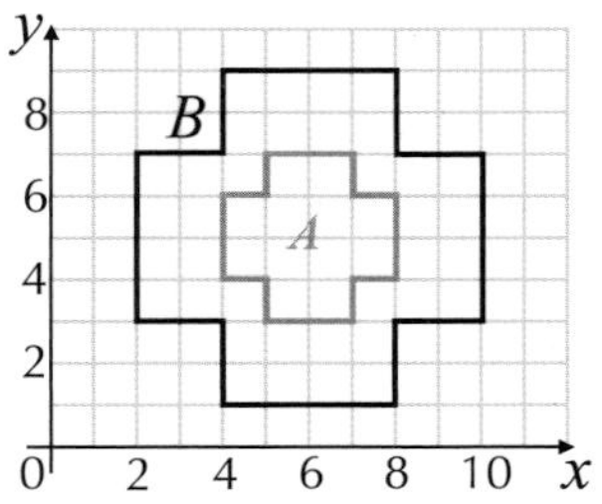

c) 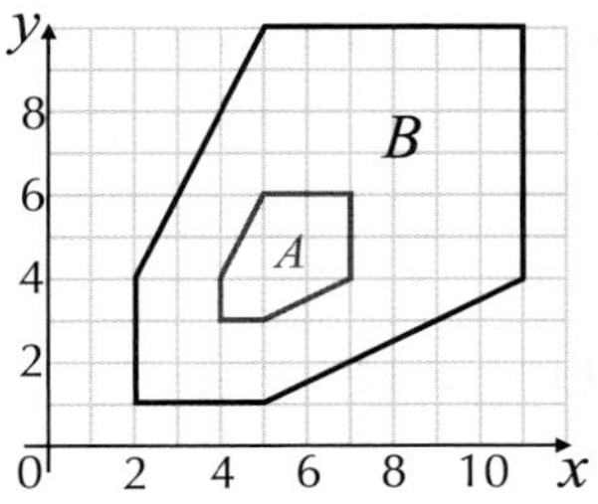

Enlargements and Perimeters

Learning Objective — Spec Ref G19:
Know how an enlargement affects the perimeter of a shape.

You know that in an **enlargement** the side lengths of a shape are multiplied by the scale factor. This means that the **perimeter** is also **multiplied by the scale factor.**

Example 6

Find the perimeter of this triangle after it has been enlarged by scale factor 4.

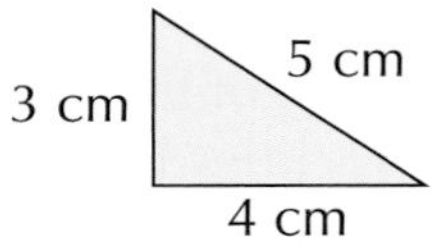

1. Find the perimeter of the original triangle.

Perimeter = 3 + 4 + 5 = 12 cm

2. The perimeter of the enlarged triangle will be 12 + 16 + 20 = 48 cm. This is the same as the original perimeter multiplied by the scale factor.

Perimeter after enlargement
= 12 cm × 4 = **48 cm**

Exercise 5

Q1 The rectangle $WXYZ$ has vertices $W(1, 1)$, $X(1, 2)$, $Y(4, 2)$ and $Z(4, 1)$.

a) Draw $WXYZ$ on a pair of axes, and find the perimeter of $WXYZ$.

b) Enlarge $WXYZ$ by scale factor 2 with centre of enlargement (0, 0). Label the image $W_1X_1Y_1Z_1$.

c) Find the perimeter of $W_1X_1Y_1Z_1$.

Q2 The shapes below are enlarged by the given scale factor. Find the perimeter of each image.

a) a square with sides 3 cm; scale factor 3

b) a 2 m × 8 m rectangle; scale factor 5

24.5 Congruence and Similarity

Two shapes are congruent if they are exactly the same shape and size.
Two shapes are similar if they are exactly the same shape but different sizes.

Congruence

Learning Objective — Spec Ref G6:
Be able to identify congruent shapes.

If two shapes are congruent, all the **side lengths** and **angles** on one shape are **identical** to the side lengths and angles on the other shape.

A **translated**, **rotated** or **reflected** shape is always **congruent** to the original shape.

Tip: You can use tracing paper to check whether two shapes are identical.

Example 1

Which two of the shapes *A*, *B* and *C* are congruent?

Compare each shape to see if they are the same shape and size. Tracing paper may help with this.

1. *A* and *B* are different sizes.
2. *A* and *C* are different sizes.
3. *B* and *C* are the same size and shape, just in different orientations.

The two congruent shapes are ***B* and *C***.

Exercise 1

Q1 Write down the letters of the congruent pairs of shapes shown in the box below. For example, *A* is congruent to *G*.

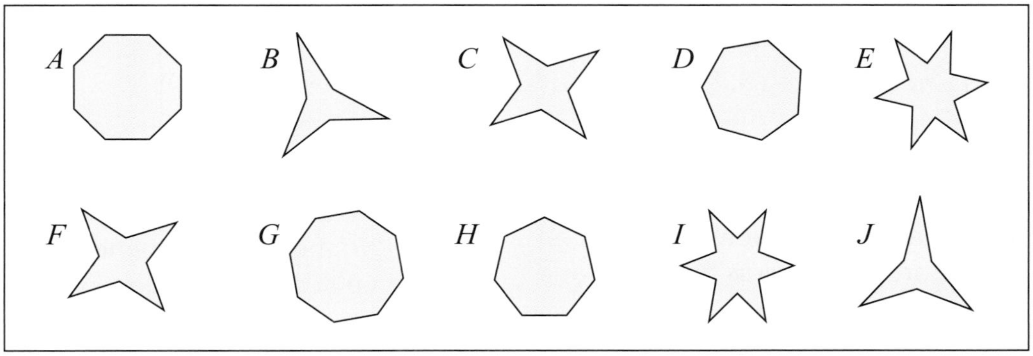

Q2 For each of the following, decide which shape is not congruent to the others.

a) 

b) 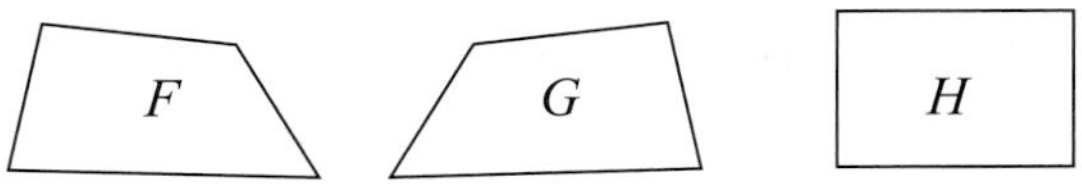

c) 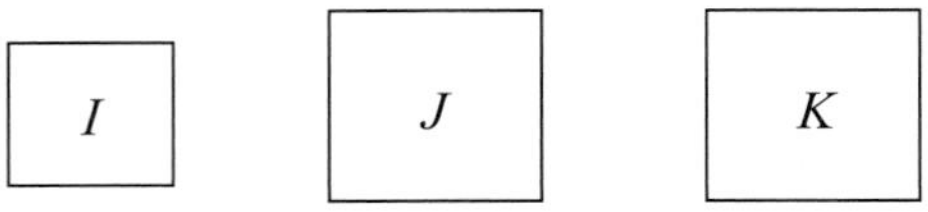

d) 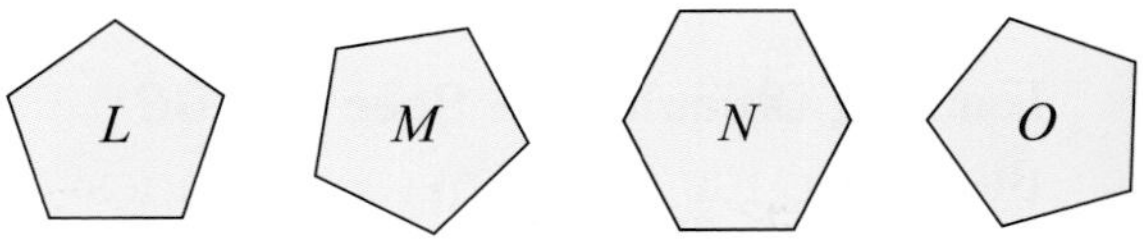

e) 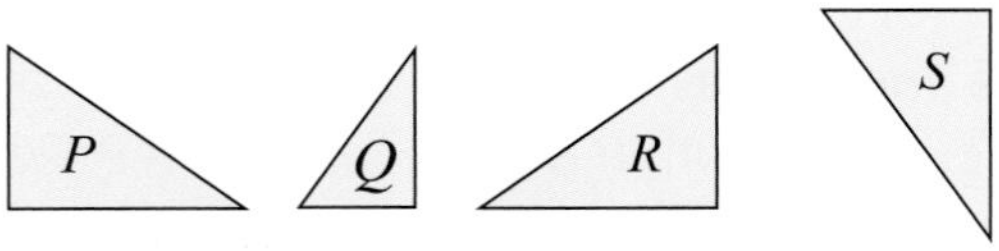

f) 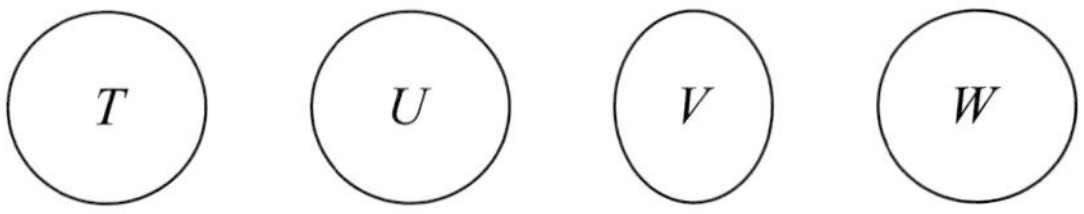

Congruent Triangles

Learning Objective — Spec Ref G5/G6:
Know the congruence conditions for triangles.

To show that two **triangles** are **congruent**, you **don't** need to know all the sides and angles — you just need to show that they satisfy any **one** of the following '**congruence conditions**'. (If any of these conditions are true, trigonometry can be used to show that the other sides and angles in the triangles must also be the same.)

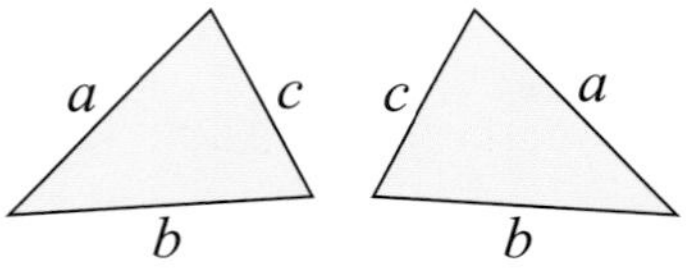

Side, Side, Side:
The three sides on one triangle are the same as the three sides on the other triangle.

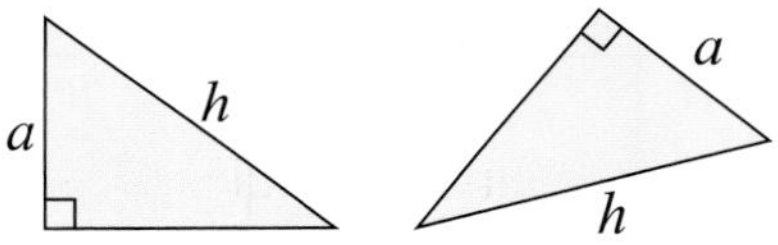

Right angle, Hypotenuse, Side:
Both triangles have a right angle, both triangles have the same hypotenuse and one other side is the same.

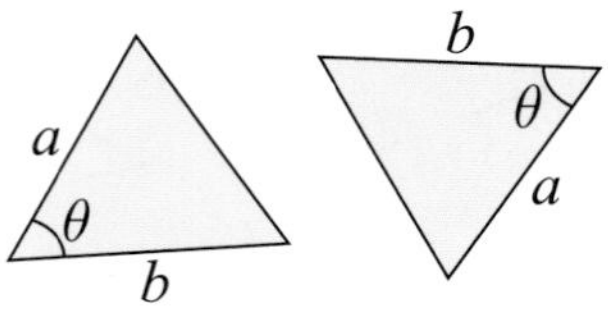

Side, Angle, Side:
Two sides and the angle between them on one triangle are the same as two sides and the angle between them on the other triangle.

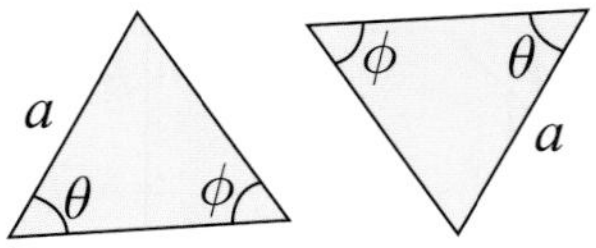

Angle, Angle, Side:
Two angles and any side on one triangle are the same as two angles and the corresponding side on the other triangle.

Be very careful when using 'Side, Angle, Side' and 'Angle, Angle, Side'. For SAS, you have to have the **correct combination** of sides and angles. For AAS, you need to make sure you're comparing **corresponding sides** on the two triangles.

If it isn't obvious which of the conditions apply, you might have to work out some side lengths or angles yourself. **Pythagoras' theorem**, the **properties of isosceles triangles** and the **sum of angles in a triangle** will come in handy — have a look at Sections 15 and 19 for a reminder.

Example 2

Are these two triangles congruent? Give a reason for your answer.

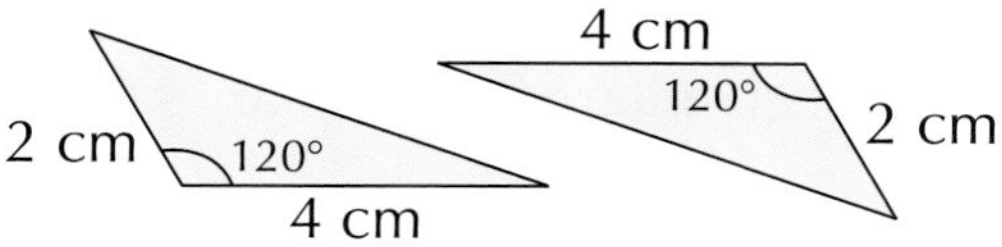

Look to see if the triangles satisfy any of the conditions for congruency. The orientation of the triangles doesn't matter.

Two of the sides and the angle between them are the same on both triangles.

Condition SAS holds, so the triangles **are congruent**.

Tip: The 'congruence conditions' are often abbreviated to SSS, RHS, SAS and AAS.

Exercise 2

Q1 The following pairs of triangles are congruent. Find the values marked with letters.

a)

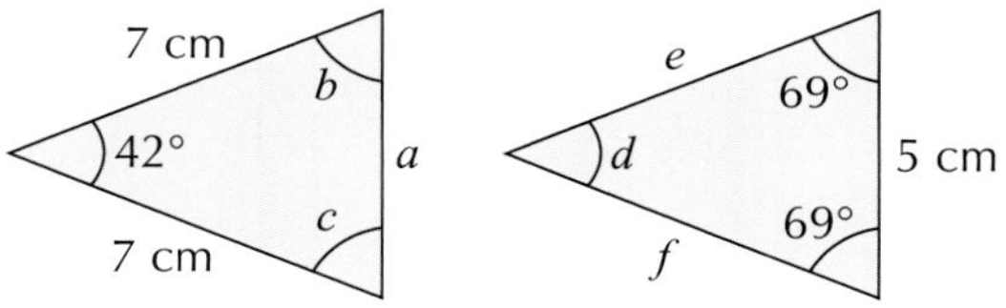

b)

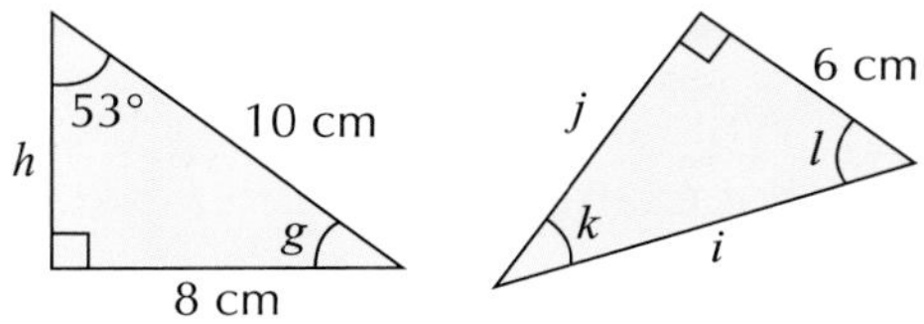

Q2 For each pair of triangles, write down whether the triangles are congruent and explain why.

a)

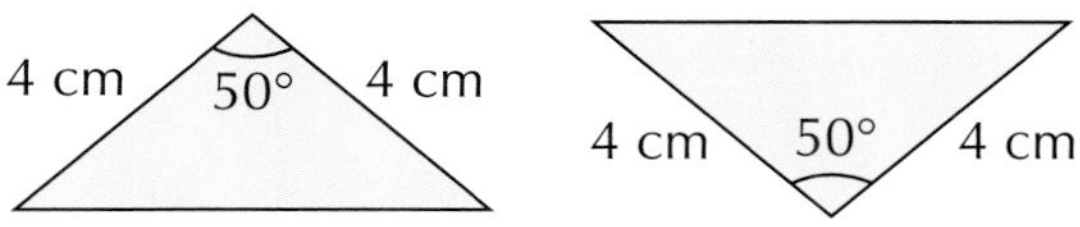

b)

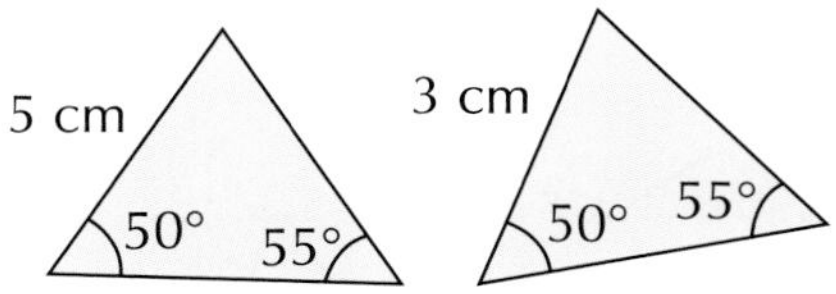

c)

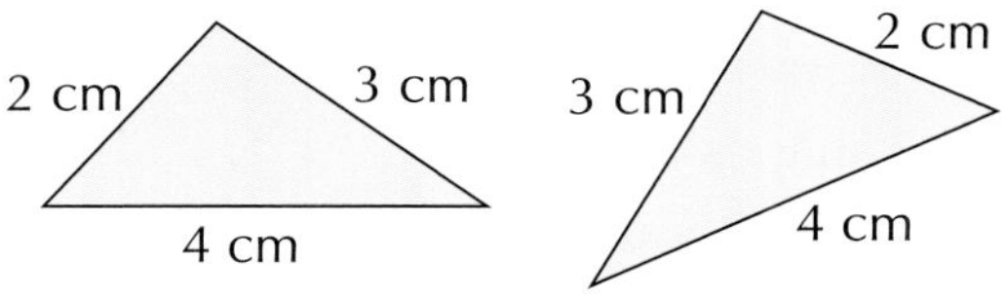

d)

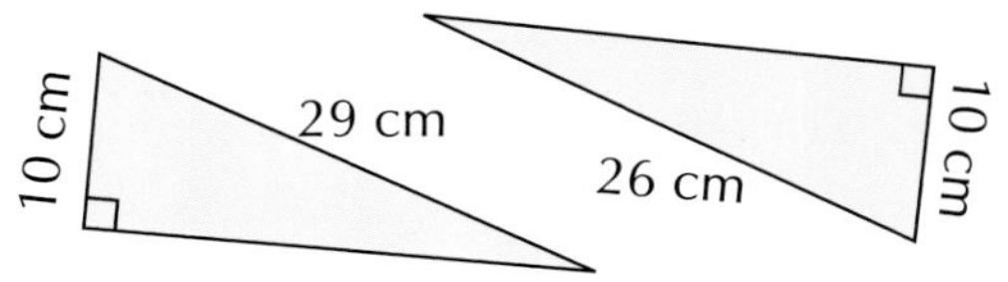

Q3 Show that each pair of triangles below is congruent.

a)

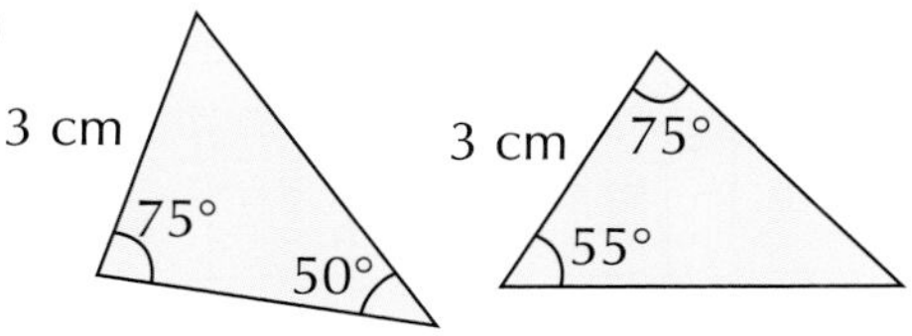

b)

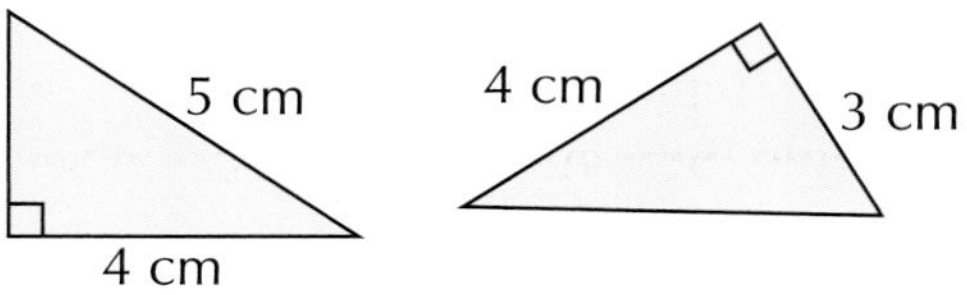

Similarity

Learning Objective — Spec Ref G6:
Be able to identify similar shapes.

Prior Knowledge Check:
Be familiar with enlargements
— see p.404-409.

Similar shapes are the same shape but can be different sizes. They have the **same angles** as each other, but the **side lengths** can be **different** — they are enlarged by the same **scale factor**. This means that the image of a shape after it has been **enlarged** is **similar** to the original shape.

Tip: Regular polygons with the same number of sides are always similar.

Example 3

Which two of the shapes *P*, *Q* and *R* are similar?

1. Compare *P* and *R* — they are different shapes.
2. Compare *Q* and *R* — they are different shapes.
3. Compare *P* and *Q* — they are the same shape, but different sizes.

The two similar shapes are ***P* and *Q***.

Exercise 3

Q1 Write down the letters of the similar shapes shown in the box below. For example, *A*, *C* and *F* are similar.

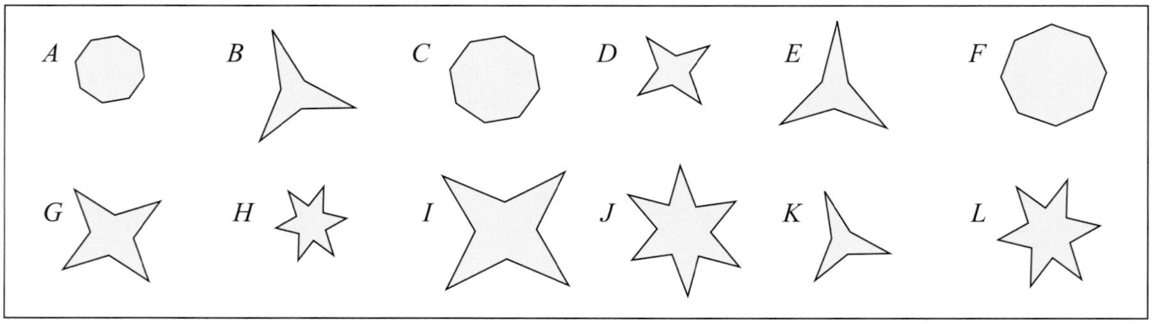

Q2 For each of the following, decide which two shapes are similar.

a)

b)

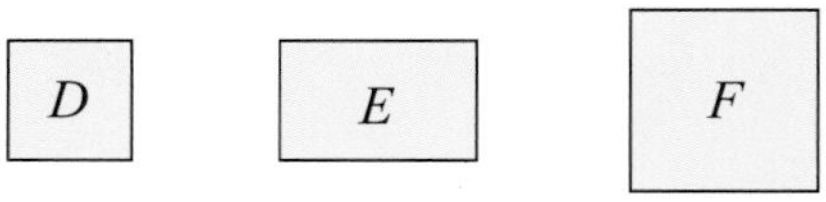

c)

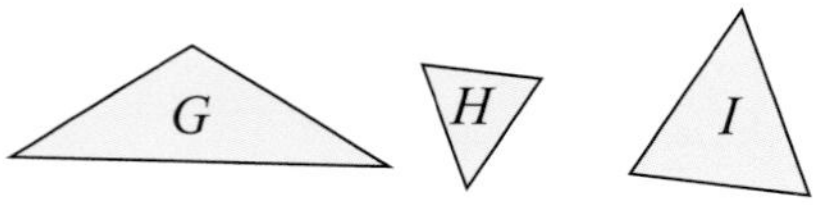

d)

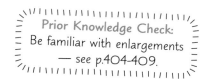

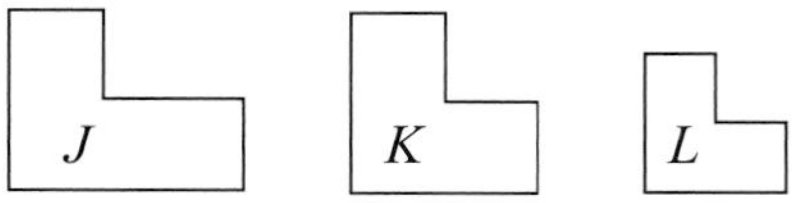

Similar Triangles

Learning Objectives — Spec Ref G6:
- Be able to prove that two triangles are similar.
- Find unknown lengths and angles in similar triangles.

As with congruent triangles, you **don't** need to know **all** the side lengths and angles to show similarity for triangles. Two **triangles** are **similar** if they satisfy any **one** of the following **conditions**:

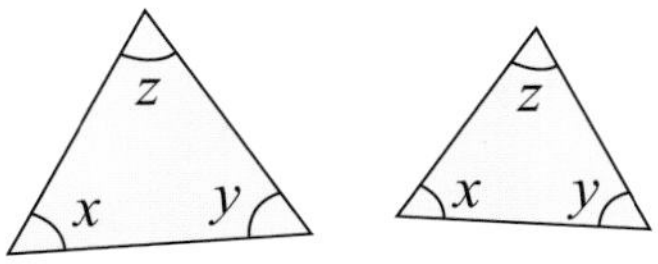

All the **angles** in one triangle are the **same** as the angles in the other triangle.

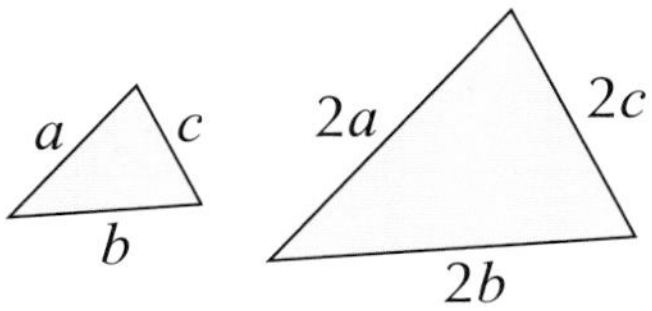

All corresponding **sides** of the two triangles are in the **same ratio**.

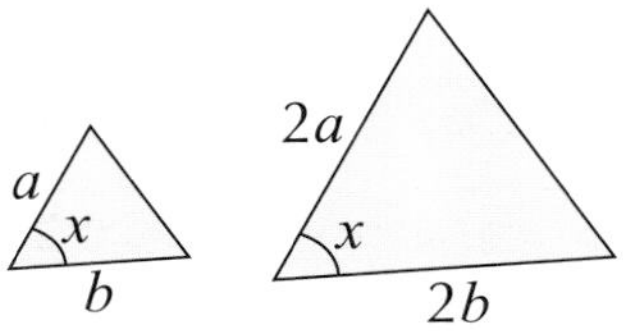

Two sides of the triangles are in the **same ratio** and the **angle between** them is the **same** for both triangles.

Tip: "In the same ratio" means you multiply each side on one shape by the same number to get the other shape.

Example 4

Are these two triangles similar? Give a reason for your answer.

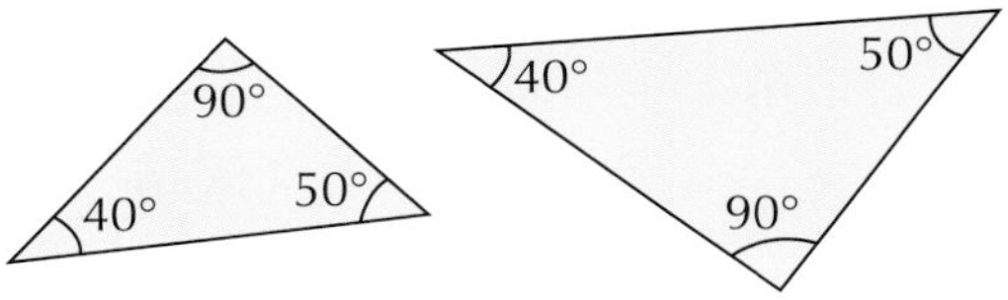

Look to see whether the triangles satisfy any of the conditions for similarity. The orientation of the triangles doesn't matter.

The angles in one triangle are the same as the angles in the other triangle.

So the triangles **are similar**.

Example 5

Explain why these two triangles are similar.

Look to see which condition for similarity the triangles satisfy. Again, the orientation of the triangles doesn't matter.

All the sides in the first triangle are twice the length of those in the second triangle.
So **corresponding sides are in the same ratio**.

Exercise 4

Q1 Explain why each of the following pairs of triangles are similar.

a)

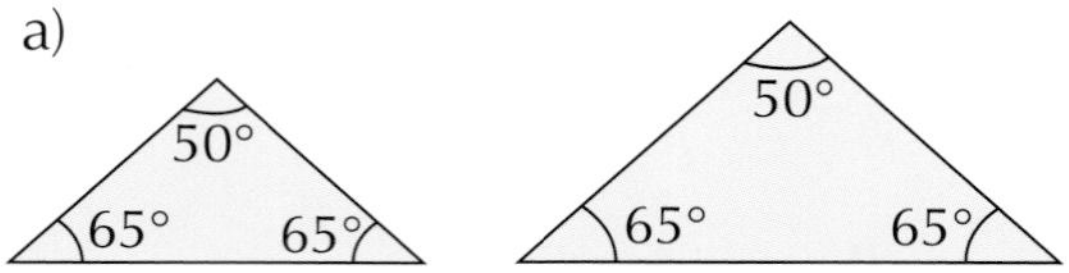

b)

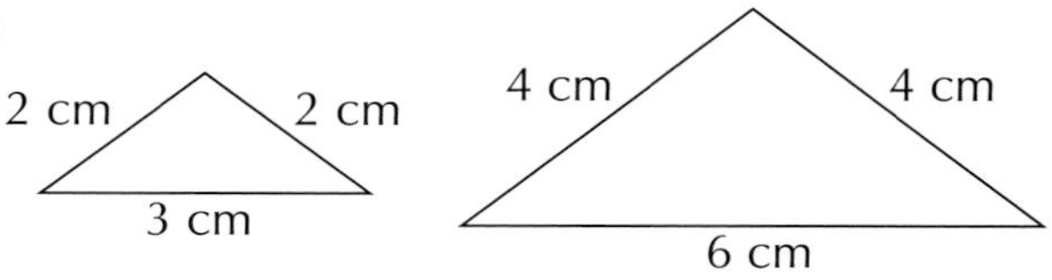

c)

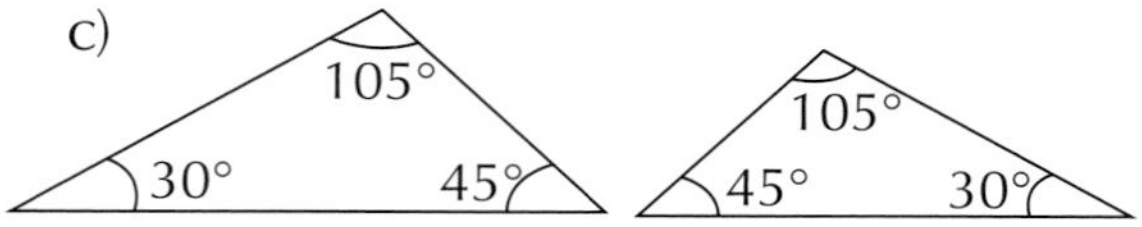

d)

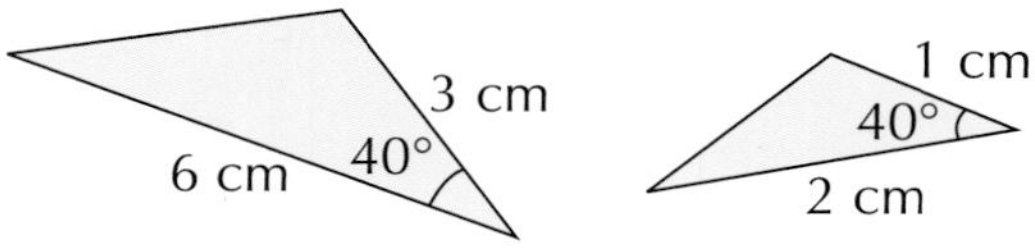

Q2 Decide whether each of the following pairs of triangles are similar.

a)

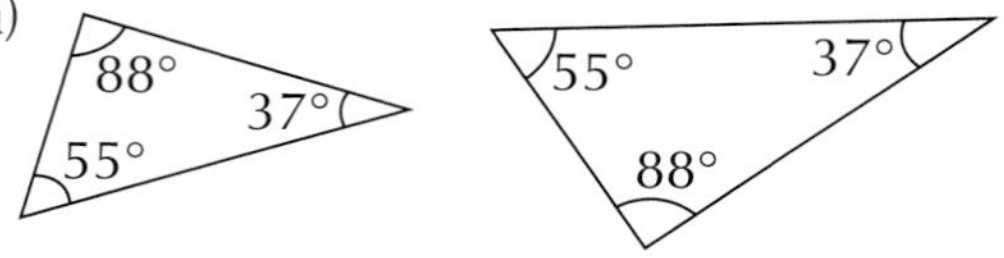

b)

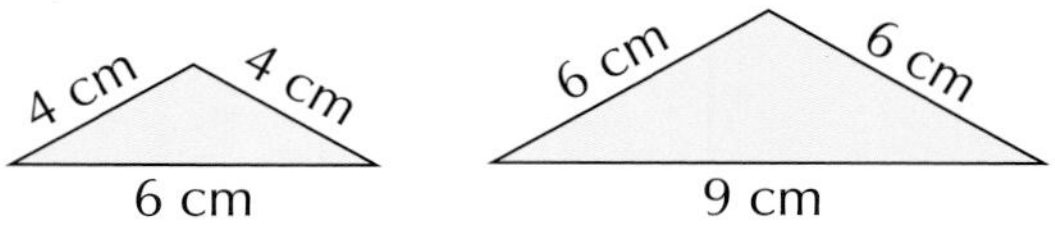

c)

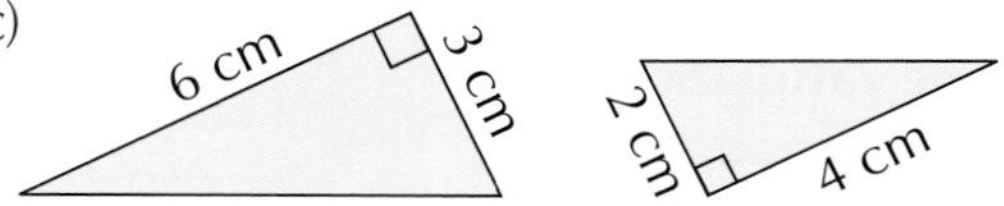

d)

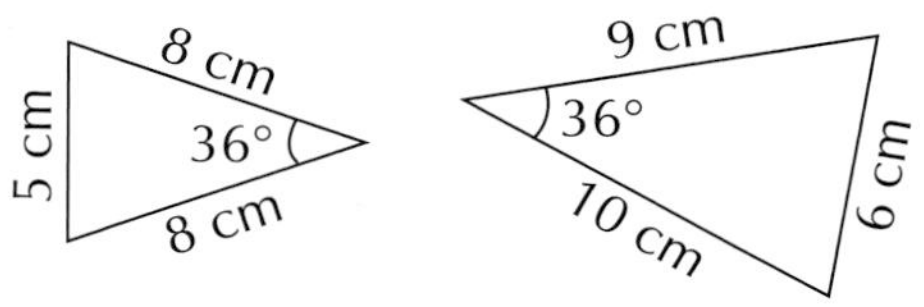

e)

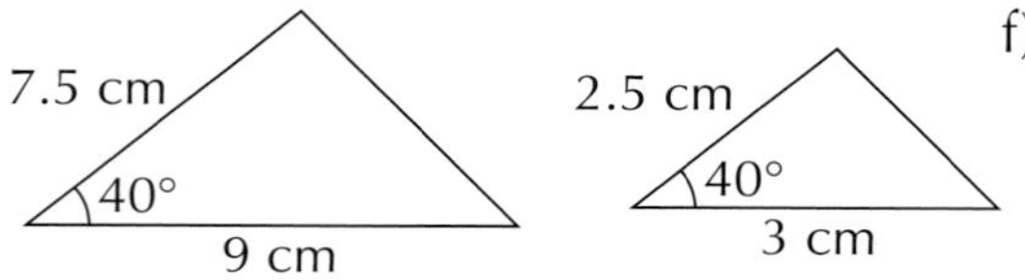

f)

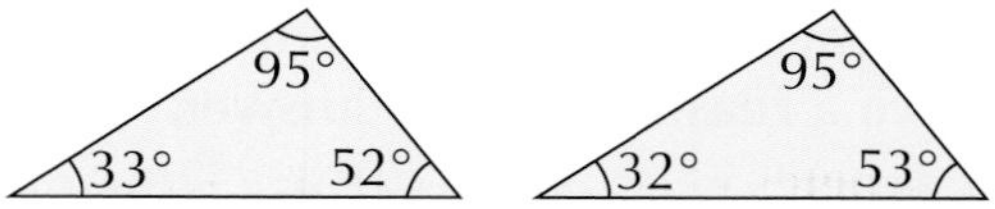

If you **know** that two shapes are **similar**, you can find **missing sides** and **angles** by identifying **corresponding sides and angles** — just remember that similar shapes have **all angles** the **same**, and **all corresponding sides** of the two shapes are in the **same ratio**.

To find a **missing side** using the corresponding side on the other shape, you'll need to work out the **scale factor** that connects the two shapes. If you've got similar shapes A and B, identify a pair of **corresponding** sides that you know the lengths of, and then use the **formula**:

$$\textbf{scale factor} = \frac{\textbf{side length of shape } B}{\textbf{side length of shape } A}$$

Tip: Make sure you use corresponding sides — you can't just pick any two sides.

To find a missing side on shape B, **multiply** the corresponding side on shape A by the **scale factor**.

Example 6

Triangles *ABC* and *DEF* are similar. Find the lengths of *DE* and *AC*.

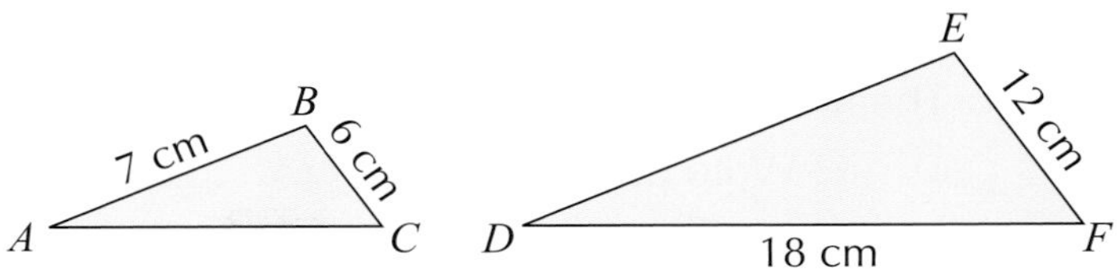

1. Identify a pair of corresponding side lengths.
2. Use the formula to calculate the scale factor.
3. Use the scale factor to find the missing side lengths.

Side *BC* corresponds to *EF*.

Scale factor $= \frac{EF}{BC} = \frac{12}{6} = 2$

So $DE = 2 \times AB$

$= 2 \times 7 =$ **14 cm**

$DF = 2 \times AC$ So $AC = DF \div 2$

$= 18 \div 2 =$ **9 cm**

A **question** might not tell you directly the value of the side or angle corresponding to the one you need to find. So you might need to use other rules too, such as the **sum of angles** in a triangle (p.243), the properties of **isosceles triangles** (p.242) and **Pythagoras' theorem** (p.301).

You might also need to use **vertically opposite angles** (p.239), and the angle rules you know for **parallel lines** (p.240-241).

Example 7

Triangles *PQR* and *UVW* are similar. Find angle *x*.

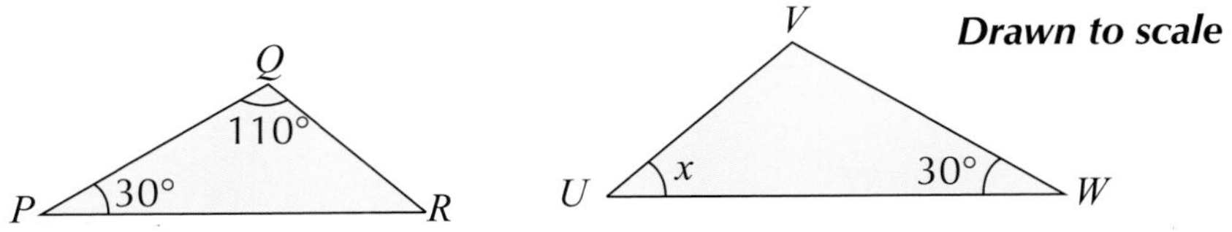

1. Find the missing angle in triangle *PQR*.
2. Work out which angles in the triangles correspond to each other.
3. Use the properties of similar triangles to determine the required angle.

Angle $QRP = 180° - (110° + 30°)$
$= 40°$

The diagrams are drawn to scale, so the obtuse angle *PQR* corresponds to angle *UVW*.

Angle *QPR* corresponds to angle *VWU*, so angle *QRP* must correspond to angle *VUW*.

This means $\boldsymbol{x = 40°}$.

Exercise 5

Q1 The triangles in each pair below are similar.

a) (i) Write down the value of x.

(ii) Find the value of y.

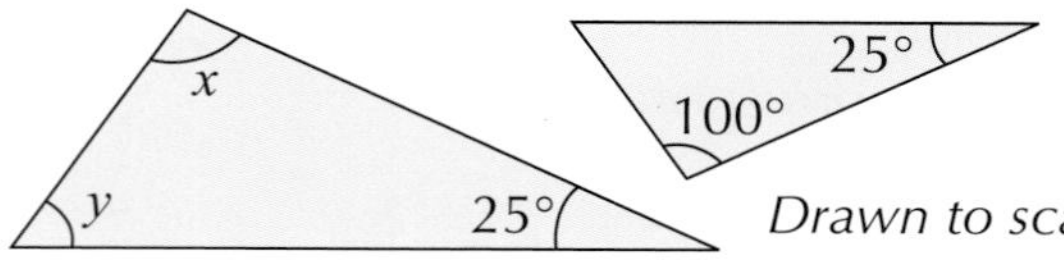

Drawn to scale

b) Find the value of z.

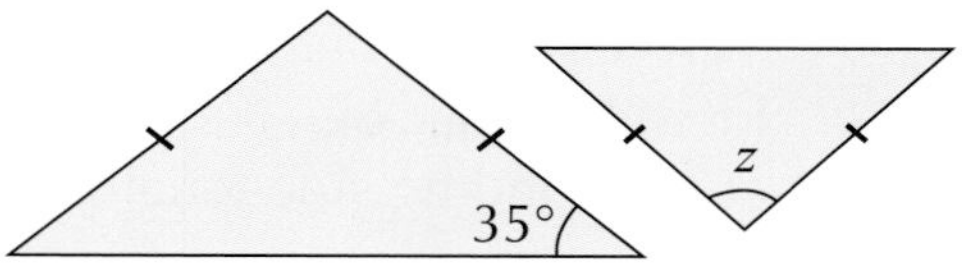

In **Questions 2-5**, the lines marked with arrows are parallel.

Q2 The diagram shows two similar triangles, ABC and ADE.

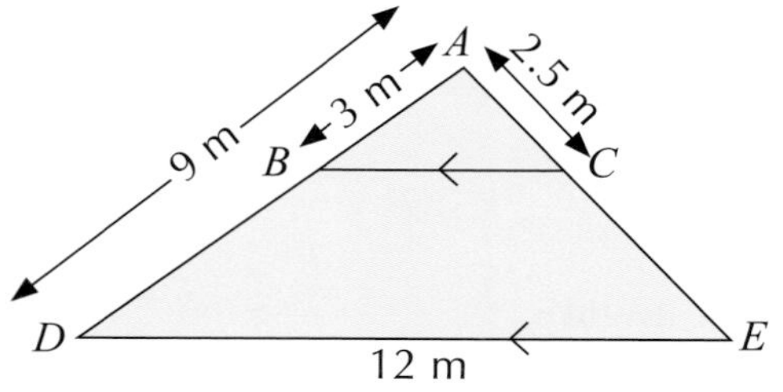

a) State which angle in triangle ABC corresponds to:

(i) angle ADE

(ii) angle AED

(iii) angle DAE

b) Find the scale factor connecting triangles ABC and ADE.

c) Find the length of: (i) AE (ii) BC

Q3 The diagram below shows two similar triangles, QRU and QST.

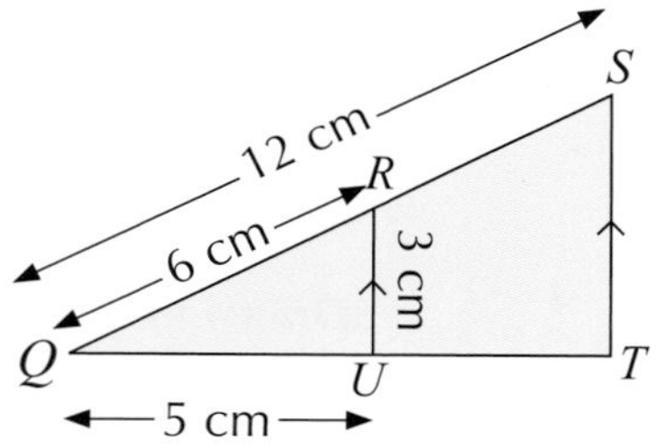

a) Find the length of ST.

b) Find the length of QT.

c) Find the length of UT.

Q4 The diagram below shows two similar triangles, STW and SUV.

a) Find the length of TW.

b) Find the length of SU.

c) Find the length of TU.

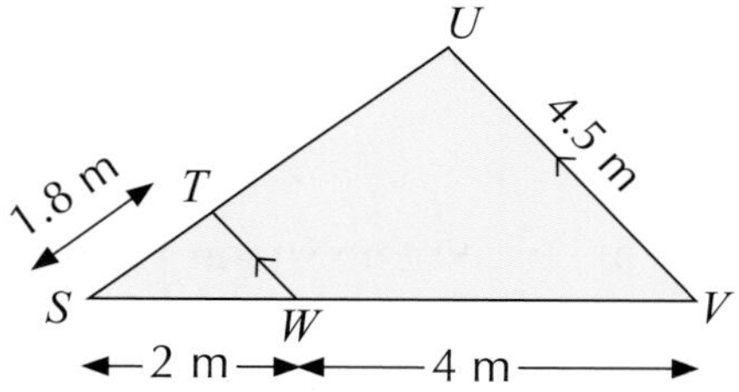

Q5 Triangles XRQ and XYZ are similar. RXY and QXZ are straight lines.

a) State the size of angle QXR. Explain your answer.

b) Use alternate angles to state which angle in triangle XRQ is equal to angle XYZ.

c) Find the scale factor connecting triangles XRQ and XYZ.

d) Use your answers to find the length of:

(i) YZ (ii) XQ

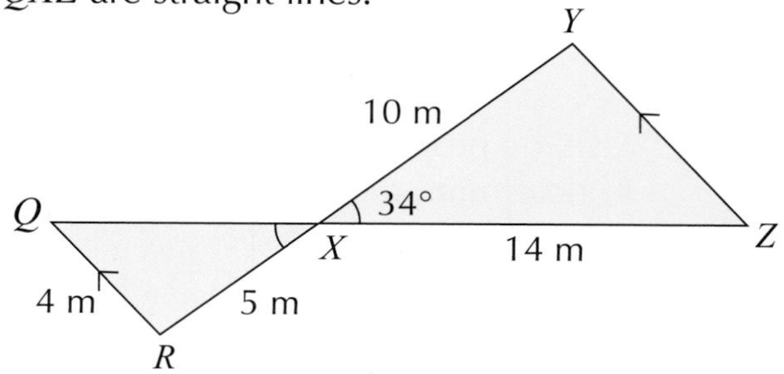

Review Exercise

Q1 a) Copy the diagram on the right.

(i) Reflect shape A in the line $y = -1$. Label the new image B.

(ii) Rotate B 90° clockwise about the origin. Label the final image C.

b) Copy the diagram again.

(i) Rotate shape A 90° clockwise about the origin. Label the new image D.

(ii) Reflect this new image in the line $y = -1$. Label the final image E.

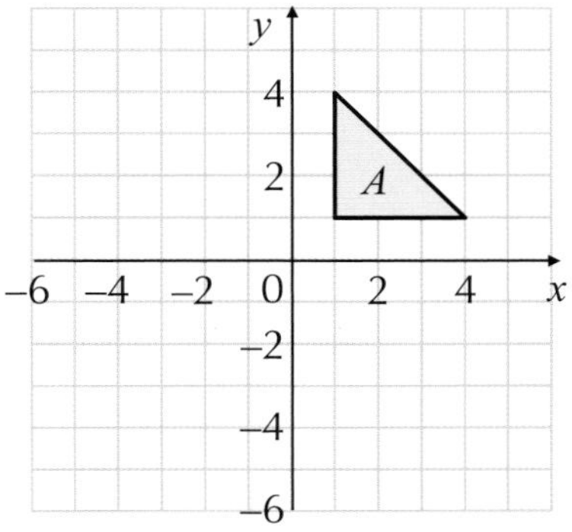

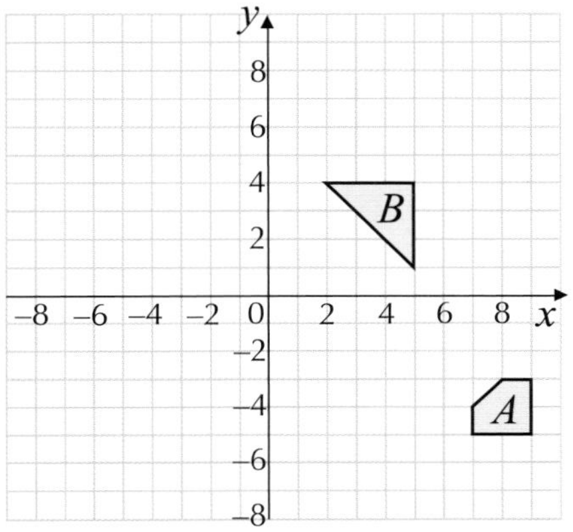

Q2 Copy the diagram on the left.

a) (i) Translate A by $\begin{pmatrix} 0 \\ -2 \end{pmatrix}$. Label the new image A_1.

(ii) Enlarge A_1 by scale factor 3 with centre of enlargement (9, –7). Label the final image A_2.

b) (i) Translate B by $\begin{pmatrix} -2 \\ -1 \end{pmatrix}$. Label the new image B_1.

(ii) Reflect B_1 in the y-axis. Label the final image B_2.

Q3 For each of parts a), b), c) and d), start by making a new copy of the diagram below.

a) (i) Rotate shape P 180° about (4, 5). Label the image P_1.

(ii) Translate the image by $\begin{pmatrix} 2 \\ 2 \end{pmatrix}$. Label the final image P_2.

b) Rotate shape P 180° about (5, 6). Label the final image P_3.

c) (i) Rotate shape Q 180° about (4, 5). Label the image Q_1.

(ii) Translate the image by $\begin{pmatrix} 2 \\ 2 \end{pmatrix}$. Label the final image Q_2.

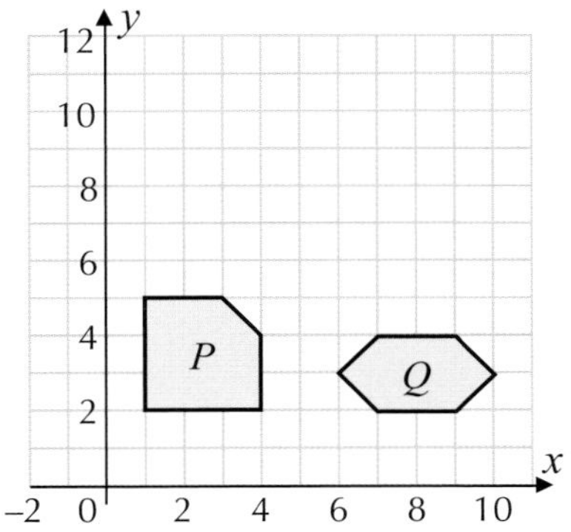

d) Rotate shape Q 180° about (5, 6). Label the final image Q_3.

e) What do you notice about images P_2 and P_3 and about images Q_2 and Q_3?

Q4 Triangle *PQR* has corners at *P*(0, 0), *Q*(4, 2) and *R*(4, 0).

a) Draw *PQR* on a pair of axes.

(i) Rotate *PQR* 90° clockwise about the point (4, 2). Label the image $P_1Q_1R_1$.

(ii) Enlarge $P_1Q_1R_1$ using a scale factor of 1.5 and centre of enlargement (2, 6). Label the image $P_2Q_2R_2$.

b) Draw *PQR* on a new pair of axes.

(i) Enlarge the original shape *PQR* using a scale factor of 1.5 and centre of enlargement (2, 6). Label the image $P_3Q_3R_3$.

(ii) Rotate $P_3Q_3R_3$ 90° clockwise about the point (4, 2). Label the image $P_4Q_4R_4$.

Q5 By considering triangle *STU* with corners at *S*(1, 1), *T*(3, 1) and *U*(1, 2), find the single transformation equivalent to a rotation of 180° about the origin, followed by a translation by $\begin{pmatrix} 6 \\ 2 \end{pmatrix}$.

Q6 Explain whether each of the following pairs of shapes are congruent, similar or neither.

a)

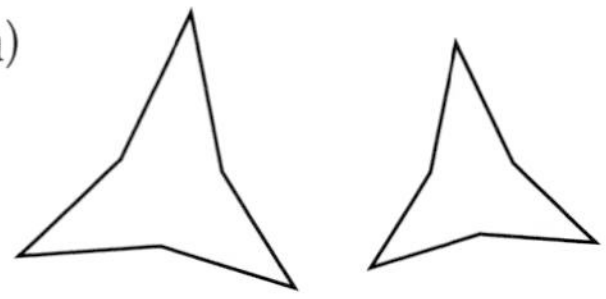

b)

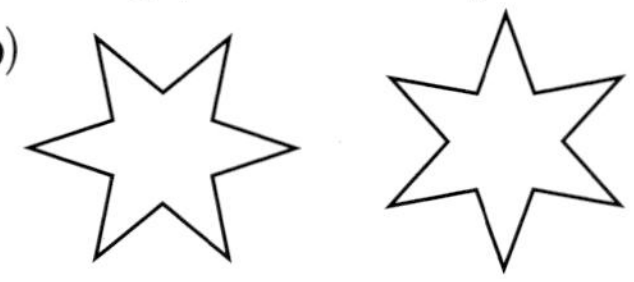

c)

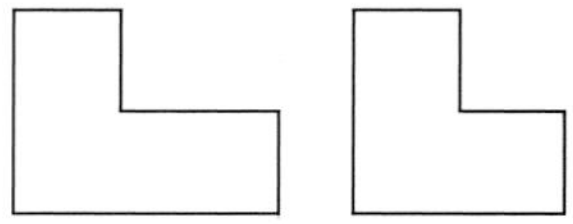

Q7 For each of the following, decide whether the two triangles are congruent, similar or neither. In each case, explain your answer.

a)

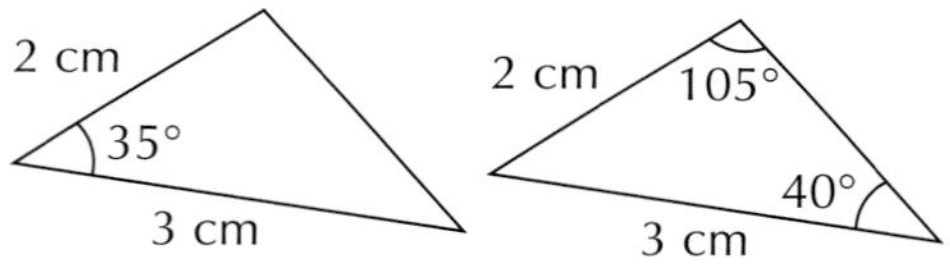

b)

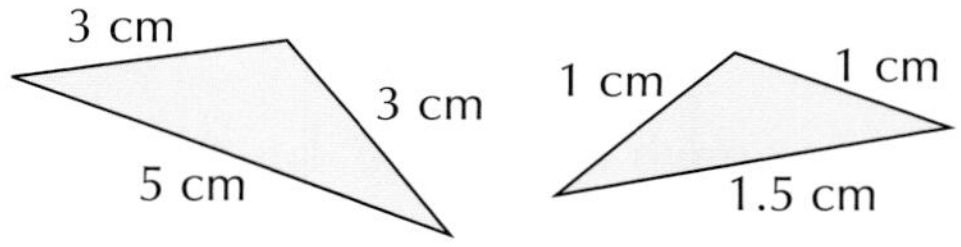

c)

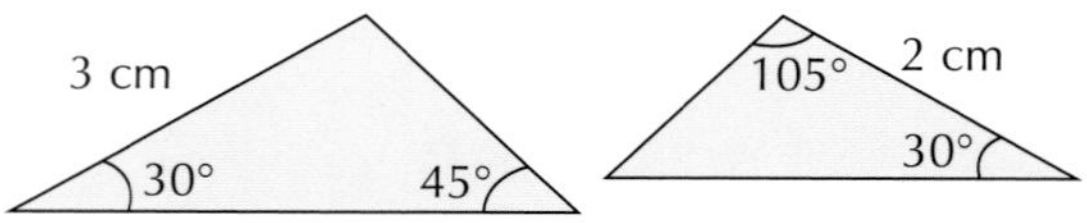

d)

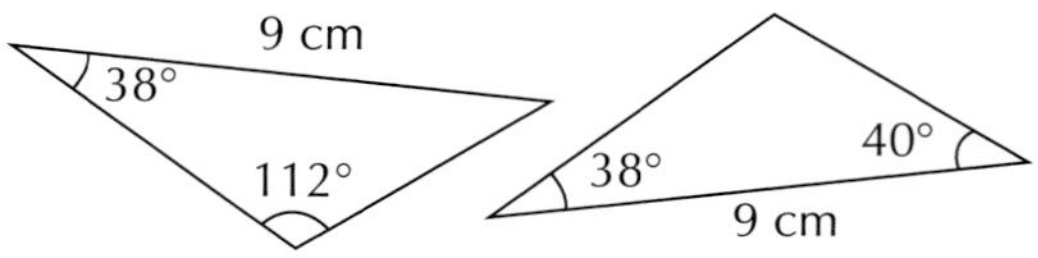

Q8 *UVW* and *WXY* are triangles.
UWY and *VWX* are straight lines.
Find the missing angles and lengths marked *a*-*f* in the diagram.

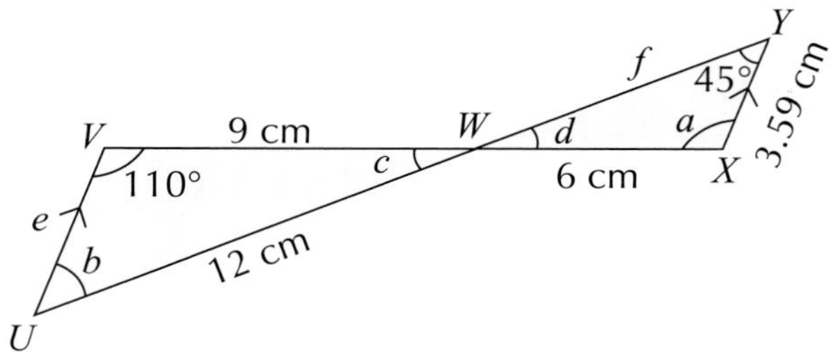

Exam-Style Questions

Q1 In the diagram, shape A was produced by reflecting shape B in the line $y = x$.

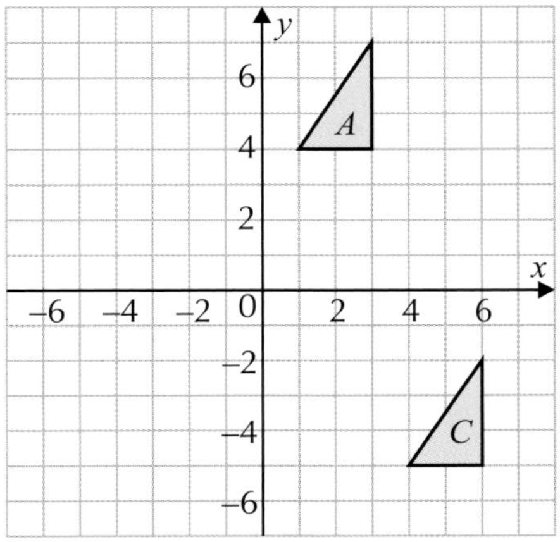

a) Draw shape B on the grid.

[2 marks]

Shape C can be mapped onto shape A by a translation of $\begin{pmatrix} p \\ q \end{pmatrix}$.

b) Write down the values of p and q.

[1 mark]

Q2 Shape P is transformed by a translation of $\begin{pmatrix} 1 \\ -4 \end{pmatrix}$ to produce an image Q.

Shape Q is then transformed by an enlargement of scale factor 2 with centre of enlargement (–2, 3) to give shape R.
Shape P is shown on the diagram below.

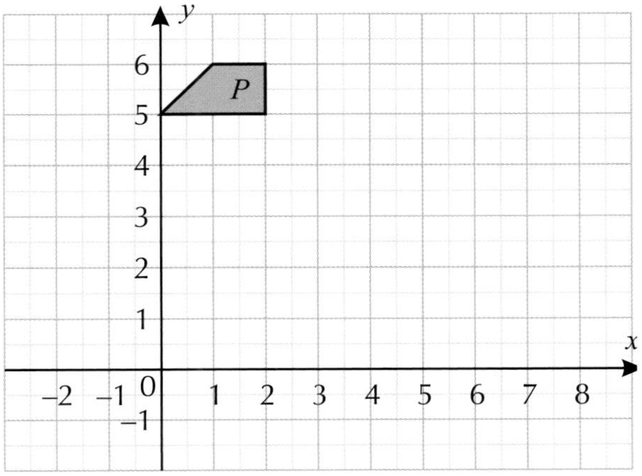

Copy the diagram and draw shape R. Show all your working.

[3 marks]

Q3 The diagram below shows two similar triangles, ABC and ADE.

a) Find length BC.

[2 marks]

b) Find angle ACB.

[1 mark]

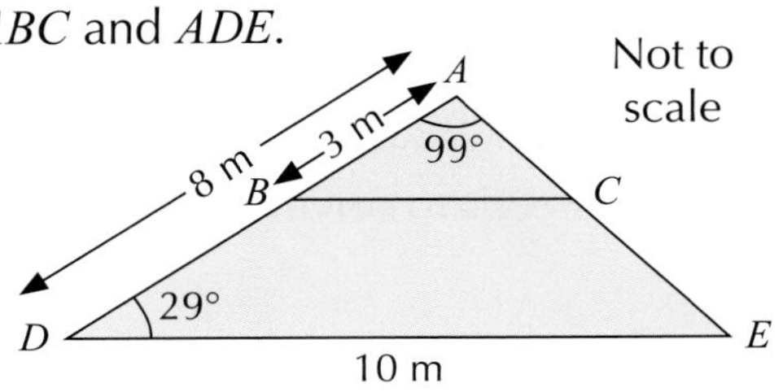

25.1 Using Different Types of Data

To test a hypothesis or conduct an investigation, you first need to plan what data you need, how you're going to collect it and how you're going to analyse it. Data comes in different types depending on who first gathered it and what sort of values/format it takes.

Primary and Secondary Data

Learning Objective — Spec Ref S1:
Know the difference between primary and secondary data.

Primary data is data you **collect yourself**, e.g. by doing a survey or experiment. It's good to use primary data since you can **control** what information is gathered, and how **accurate** it is.

Secondary data is data that has been **collected by someone else**. You can get secondary data from things like newspapers or the internet. You're relying on someone else to collect the data — but secondary data is often **cheap** and **quick** to obtain. This is especially useful if you need a **large** amount of information, or you need data that would be hard for you to **gather** yourself.

Example 1

Marya wants to know if there's a link between area and population in 10 African countries.

a) **What pieces of data does Marya need to find?**
This is the information she needs to answer her question. → The **areas and populations** of the 10 countries she's interested in.

b) **Where could Marya get the data from?** → E.g. she could find it on the **internet**.

c) **Is Marya's data primary or secondary data?**
The data has been collected by someone else, so... → It's **secondary** data.

Example 2

Jon wants to test whether a six-sided dice has an equal chance of landing on any side. Explain how he could collect data for the test. Will his data be primary or secondary data?

1. Jon needs to do an experiment. He could roll the dice lots of times and record how many times it lands on each side using a tally chart.
2. He collects the data himself, so... His data will be **primary** data.

Exercise 1

For **Questions 1-3**, decide if the data is primary or secondary.

Q1 Megan and Jane want to find out the types of music that students like.

a) Megan plans to send a questionnaire to students from her college.

b) Jane plans to use the results of an online music survey on the college's website.

Q2 Faheem is going to use the data on election results that appeared in his local newspaper.

Q3 Nancy is going to time how long it takes her friends to run 100 metres.

For **Questions 4-7**: a) Describe what data is needed and give a suitable method for collecting it.
b) Say whether the data will be primary or secondary data.

Q4 Nikita wants to know what the girls in her class think about school dinners.

Q5 Anne wants to compare the daily rainfall in London and Manchester last August.

Q6 Skylar wants to see if the boys in his class can throw a ball further than the girls.

Q7 Jim wants to find out how the temperature in his garden at 10 am each morning compares with the temperature recorded by the Met Office for his local area.

Discrete and Continuous Data

Learning Objectives — Spec Ref S1:
- Know the difference between qualitative and quantitative data.
- Know the difference between discrete and continuous data.

Data can be either **qualitative** or **quantitative**. **Qualitative** data is **descriptive**, meaning it uses **words** instead of numbers — e.g. favourite flavours of ice cream ('vanilla', 'chocolate', 'strawberry', etc.) or opinions ('happy', 'neutral' etc.).

Quantitative data is **numerical** — e.g. the heights of players in a football team or pupils' scores in a Science test. Quantitative data can be split up into two further types:

- **Discrete** data can only take **certain values**. For example, the number of goals scored by a football team — the data must be whole numbers, the team can't score half a goal.
- **Continuous** data can take **any value** in a given **range** — e.g. lengths, heights or weights.

How you **process** and **analyse** your data will depend on the data type. Some of the different types of **graphs** and **diagrams** in Section 26 would be **inappropriate** for showing certain data types — e.g. you shouldn't put qualitative data onto a line graph, but you can make a bar chart or pie chart out of it.

Example 3

Is the following data qualitative, discrete quantitative or continuous quantitative?

a) The hometowns of 100 people.

This data is in the form of words, so... It's **qualitative** data.

b) The weights of the bags of potatoes on sale in a greengrocer's.

This data is numerical and isn't restricted to certain specific values. It's **continuous quantitative** data.

c) The number of students in each class at a school.

This data is numerical and can be counted, so... It's **discrete quantitative** data.

Exercise 2

Q1 Say whether the following data is qualitative, discrete quantitative or continuous quantitative.

a) The number of words in your favourite song.

b) Your favourite food.

c) The numbers of pets in 20 households.

d) The sizes of the crowds at 10 rugby matches.

e) The heights of 100 tomato plants.

f) The time it takes Matt to walk to school.

g) The nationalities of the people in a park.

h) The lengths of 30 worms.

i) The distances of planets from the Sun.

j) The hair colours of 50 people.

Q2 Janka wants to find out if tea or coffee is more popular in a café. She decides to stand outside the café one morning, and ask people as they leave which drink they ordered.

a) Say whether this data is primary data or secondary data.

b) Say whether this data is qualitative, discrete quantitative or continuous quantitative?

Q3 Gemma thinks there is a link between the average number of chocolate bars eaten each week by pupils in her class and how fast they can run 100 metres.

a) Describe two sets of data Gemma should collect to investigate this link.

b) Describe suitable methods for collecting the data.

c) Say whether each set of data is discrete quantitative or continuous quantitative.

d) Say whether each set of data is primary data or secondary data.

25.2 Data Collection

As part of an investigation, you'll probably collect lots of data, so you need a way of recording and organising it. Your collection method should fit the type of data you're trying to collect.

Data Collection Sheets

Learning Objective — Spec Ref S1/S2:
Design and use data collection sheets.

Prior Knowledge Check:
Be able to use inequality signs. See p.151.

Tally charts and **frequency tables** can be used to record qualitative or quantitative data — the only thing that changes is the name of the **categories** or **groups**. There's more about these types of diagrams on p.435.

For **qualitative** data, the **names** of each category are used (e.g. if you're asking about favourite colours, then the categories might be 'red', 'green, 'blue', ...). It's useful to have an **'Other'** category so that all the possible options are included without having to list every category.

Example 1

A restaurant manager asks 50 customers to choose their favourite way to eat potatoes from the following four options: boiled, mashed, baked and roast.
Design a tally chart that could be used to record the data.

Column for data names...

...with a row for every possible answer

Potato type	Tally	Frequency
Boiled		
Mashed		
Baked		
Roast		

Tally column with plenty of space to record the marks

Frequency column for adding up the tally marks

Tip: The tally column lets you record data as you collect it, e.g. as you ask people or count something.

Exercise 1

Q1 Give two things that are wrong with the tally chart on the right for recording the colours of cars in a car park.

Car colour	Frequency
Red	
Black	
Blue	
Silver	

Q2 Design a tally chart that could be used to record the answers to each of the questions below.

a) How many times a week do you go to the supermarket?

b) Where did you go for your last holiday?

For **discrete quantitative** data, you can use **individual** values (e.g. 0, 1, 2) if the range is small or **groups** of values if the range is large (e.g. 0-4, 5-9, 10-14). However, you lose some **accuracy** by grouping the data values as you no longer know the actual data values.

For **continuous quantitative** data, you have to group the data. The groups of values need to cover every possible option, so you write the groups using **inequalities**.

Example 2

A group of students take a test. The tests are marked and given a score out of 50. Design a tally chart to record the students' scores. (Assume that all scores are whole numbers.)

1. This is discrete data, but a large range of scores are possible, so group them into a sensible number of classes.
2. Make sure the classes cover all possible scores between 0 and 50.
3. Make sure the classes don't overlap — each score should only be able to go into one class.

Test Score	Tally	Frequency
0 – 10		
11 – 20		
21 – 30		
31 – 40		
41 – 50		

Example 3

An assault course is designed to take about 10 minutes to complete.
Design a tally chart that could be used to record the times taken by a group of people.

1. The data is continuous, so use inequalities for the classes.
2. Make sure there are no gaps between classes and that classes don't overlap. Using '$t \le$' to end a class and '$< t$' to start the next class, means all values can go into exactly one class.
3. Leave the last class open-ended to cover all possible times.

Time (t mins)	Tally	Frequency
$0 < t \le 5$		
$5 < t \le 10$		
$10 < t \le 15$		
$15 < t \le 20$		
$t > 20$		

Exercise 2

Q1 Give two things that are wrong with these tally charts, and design an improved version.

a) Chart for recording the number of people watching a band play at each venue on their tour (max. venue capacity = 20 000).

No. of people	Tally	Frequency
0 – 5000		
6000 – 10 000		
11 000 – 15 000		

b) Chart for recording the weights of pumpkins.

Weight (w kg)	Tally	Frequency
$w \le 3$		
$3 \le w \le 3.5$		
$3.5 \le w \le 4$		

Q2 Write down five classes that could be used in a tally chart to record the sets of data below.
a) The heights of 50 plants, which range from 5 cm to 27 cm.
b) The number of quiz questions, out of 20, answered correctly by some quiz teams.
c) The weights of 30 bags of apples, where each bag should weigh roughly 200 g.
d) The volumes of 50 cups of tea as they're served in a café. Each cup can hold 300 ml.

Q3 Design a tally chart that could be used to record the following data.
a) The average lengths of time 100 people spend watching TV each week.
b) The numbers of pairs of socks owned by 50 students.
c) The lengths of 20 people's feet.
d) The distances that 30 people travel to get to work and back each day.

Two-Way Tables

Learning Objective — Spec Ref S1/S2:
Design two-way tables.

A **two-way table** is a data collection sheet that allows you to record **two** different pieces of information about the **same subject** at once — e.g. for each person, you could collect data on their gender as well as hair colour in one table. One piece of information is covered by the **rows** of the table and the other by the **columns**.

Example 4

Raymond is investigating how fast the players at a tennis club can serve. He's interested in whether being right-handed or left-handed has any effect on average speed.

Design a two-way table he could use to record the data he needs.

One variable goes down the side and the other goes across the top

Use inequalities for continuous data

	Serve average speed (s mph)				
	$s < 40$	$40 \leq s < 50$	$50 \leq s < 60$	$60 \leq s < 70$	$70 \leq s$
Right-handed					
Left-handed					

Rows and columns should cover every possible answer

Make sure there are no gaps and classes don't overlap

Space for tally marks or frequencies

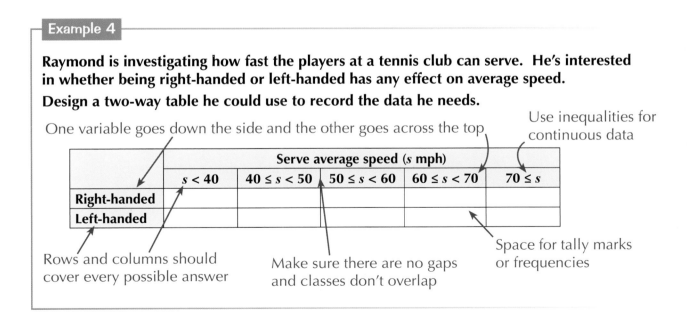

Exercise 3

Q1 The table below has been designed to record the hair colour and age (in whole years) of 100 adults.

a) Give three criticisms of the table.

b) Design an improved version of the table.

	Age (in whole years)					
	0 – 15	15 – 30	30 – 45	45 – 60	60 – 75	75 or older
Blonde						
Light brown						
Dark brown						

Q2 Design a two-way table that could be used to record the data for each of the following:

a) The favourite season of the year of 50 men and 50 women.

b) The type of music adults and children prefer listening to out of pop, classical and rock.

c) The average length of time spent doing homework each evening by pupils in each of the school years 7-11. Assume no one spends more than an average of 4 hours an evening on homework.

d) The total number of books people read last year and their favourite genre out of horror, romance, sci-fi and fantasy.

Q3 For each investigation below, design a two-way table for recording the data.

a) Hoi Wan is going to ask 50 adults if they prefer cats or dogs. She wants to find out if it's true that men prefer dogs and women prefer cats.

b) Nathan wants to find out whether children watch more TV on average each day than adults.

c) Felicity is investigating whether pupils in different school years use different methods of transport to get to school.

d) Matthew is going to ask people how tall they are and how many portions of fruit they eat on average each day.

Q4 Olivia is going to ask some students at her school about their favourite subject. She wants to find out what the most popular subject is in each year group.

a) Identify two reasons why the table below is unsuitable to record her results.

b) Design an improved version of the table.

	Favourite Subject				
	English	Maths	P.E.	Science	Music
Years 7-9					
Years 10-11					

25.3 Sampling and Bias

It might not be practical or possible to collect data on every individual that you're interested in. If you wanted to know about the wingspan of all the birds in a forest, you wouldn't be able to find and measure every one. Instead, you'll need to collect data from a smaller group — but you have to pick it carefully.

Populations and Samples

Learning Objectives — Spec Ref S1:
- Know the difference between a population and a sample.
- Understand the reasons for using a population and for using a sample.

When you're collecting data, the **whole group** of people or things you want to find out about is called the **population**. But populations can be very big and so it is often **impractical** to collect data from **every member**. Usually, it's **quicker**, **cheaper** and **easier** to collect data from a **sample** of the population, rather than the whole thing.

A sample is a **smaller group** taken from the population. Different samples will give different results — the **bigger** the sample, the more reliable it should be. It's also important that your sample **fairly represents** the population, so you can apply any **conclusions** to the whole population.

Example 1

Michael and Tina have written a questionnaire to find out what students at their college think about public transport. Michael gives the questionnaire to 10 students and Tina gives the questionnaire to 50 students. Michael concludes that 50% of students are happy with public transport, whereas Tina concludes that only 30% are happy.

a) Why have Michael and Tina only given the questionnaire to some of the students?

Think about the advantages of sampling...

E.g. there are fewer copies to print, so **print costs will be lower**.
Also, it will take them **much less time** to collect and analyse the results.

b) Whose results are likely to be more reliable? Explain your answer.

Think about the size of the sample...

Tina's results are likely to be more reliable because she has used a **bigger sample**.

Exercise 1

Q1 Identify the population and the sample in the following scenario:
A university wants to find out how its students feel about plans to build a new library. It randomly distributes a survey to 800 of its students.

Q2 Jenny wants to know how long it takes the pupils in Year 7 to type out a poem. She plans to time a sample of 30 out of the 216 Year 7 pupils.
Give two advantages for Jenny of using a sample.

Q3 Patrick is testing a box of matches to see how long they burn for. He does this by lighting 10 of the matches. Explain why Patrick doesn't test all the matches.

Q4 An audience of 1000 people are watching a musical. Khadija wants to know what they think about it and plans to interview 5 people from the audience afterwards. What's wrong with Khadija's plan?

Q5 Melissa and Gareth are doing an experiment to see if a coin is fair. They each toss the coin 100 times and record the number of heads. 52% of Melissa's tosses are heads and 47% of Gareth's tosses are heads. Explain why Melissa and Gareth get different results.

Q6 Alfie and Keith want to find out what the most popular flavour of ice cream is. Alfie asks 30 people and finds that 'chocolate' is the most popular. Keith asks 15 people and finds that it's 'strawberry'. Based on this information, what would you say is the most popular flavour? Explain your answer.

Q7 Jack, Nikhil and Daisy bake a batch of 200 cupcakes to sell on their market stall. Jack thinks they should taste one cake to check the quality is OK. Nikhil thinks they should taste 10 cakes and Daisy thinks they should taste 50 cakes.
Say who you agree with and explain why.

Bias and Fairness

Learning Objectives — Spec Ref S1:
- Consider the bias or fairness of a sample.
- Know how to take a random sample.

Tip: When deciding if a sample is biased, see if you can think of any groups that would be excluded.

When you're choosing a sample, it's important to make sure that your selection process is **fair** so that results are likely to reflect the **whole population**. A sample is **biased** if some members of the population are **more likely** to be included than others.

To spot bias, think about **when**, **where** and **how** the sample was taken, and **how many** members are in it — a **small** sample is also likely to be biased.

Example 2

Perry plans to sample 50 shoppers at a supermarket to see what they think about the supermarket's decision to close earlier in the evenings. He decides to stand outside the supermarket on a Wednesday morning and ask the first 50 people he sees leaving the shop.

a) Explain why this will give him a biased sample.

Think about when he selects the sample...

E.g. Perry is only asking people who are free to shop on a Wednesday morning, so he's excluding all the shoppers who are at work or school. He's also excluding people who only shop in the evening, who will probably be more affected by the earlier closing.

b) Suggest how Perry could choose a fairer sample.

Think about how he could include more groups of shoppers...

E.g. Perry should ask people on different days of the week and at different times of day.

Exercise 2

Q1 Explain why the following methods of selecting a sample will result in a biased sample.

a) Kelechi wants to know what people at her college think about a particular film, so she asks 10 of her friends for their opinions.

b) A school cook wants to know whether the pupils want all school dinners to be vegetarian. He asks all the members of the school's animal rights group.

c) A library needs to reduce its opening hours. The librarian asks 20 people on a Monday morning whether the library should close on a Monday or a Friday.

d) A market research company wants to find out about people's working hours. They select 100 home telephone numbers and call them at 2 pm one afternoon.

A **fair** way to select a sample is at **random** — every member of the population has an **equal chance** of being selected. To take a random sample:

- Make a **list** of **every member** of the population and assign everyone a **number**.
- Use e.g. a computer or calculator to create a list of **random numbers**.
- **Match** the random numbers to the assigned numbers of the population.

Exercise 3

Q1 A head teacher wants to know what pupils in Year 7 think about after-school clubs. Explain how he could select a random sample of 50 Year 7 pupils.

Q2 The manager of a health club wants to survey a random sample of 40 female members. Explain how she could do this.

Using a Sample to Make Estimates About the Population

Learning Objective — Spec Ref S1:
Know how to infer properties of a population using a sample.

You can use a fair sample to make **estimates** about the **population**.
E.g. if **a third of the people in a sample** have a certain characteristic, you can estimate that **a third of the population** will also have this characteristic.

Example 3

There are 10 000 residents in Sheila's town. She wants to find out how satisfied they are with the street furniture in the town, so she fairly samples 1000 residents. The results are in the table.

Opinion	No. of residents
Very dissatisfied	250
Quite dissatisfied	150
Don't care	216
Quite satisfied	204
Very satisfied	180

a) Estimate how many residents in the town are either quite satisfied or very satisfied with the street furniture.

b) Could you use this sample to estimate how many people in the country are satisfied with the street furniture in their town?

a) 1. Work out the proportion of satisfied residents from the sample.

There are 180 + 204 = 384 residents satisfied in the sample, so the proportion of residents satisfied in the sample is $\frac{384}{1000} = 0.384$.

2. Your estimate is the same proportion in the population.

The number of people satisfied in the town is estimated to be $0.384 \times 10\,000 =$ **3840**.

b) Think about whether the sample is still fair or representative.

The sample is **not representative** of the country — it is **biased** in favour of people in one particular town. You **can't use it** to estimate the opinion of the country.

Exercise 4

Q1 A castle has 450 rooms. The king of the castle believes that many of them have damp and so hires a dirty rascal to remove it. The king doesn't have time to inspect all the rooms for damp, so he chooses 20 rooms at random to inspect. Given that seven of these rooms have damp, estimate how many of the rooms in the entire castle have damp.

Q2 A journalist interviews a sample of 50 out of 650 politicians to get their views on three policies X, Y and Z. The results are in the table on the right.

	X	Y	Z
In Favour	34	4	13
Against	13	45	14
Neutral	3	1	23

a) Use the table to estimate how many of the 650 politicians:

(i) are in favour of policy X, (ii) are neutral about policy Y, (iii) are against policy Z.

b) What assumption has been made in using this data to make these estimations?

Review Exercise

Q1 Steven wants to investigate the most common favourite colour amongst his classmates.

a) Describe the data he needs and suggest a way that he could collect it.

b) Say whether Steven's data is primary or secondary data.

Q2 Is the following data qualitative, continuous quantitative or discrete quantitative?

a) The amount of time it takes people to travel to school.

b) The number of siblings people have.

c) People's favourite film.

d) The prices of different pairs of shoes in a shoe shop.

Q3 Lars wants to know how many days there were in the month that people were born. Design a tally chart that could be used to record his data.

Q4 Pearl makes a tally chart to record how many times people went to the cinema last year. Identify two things that are wrong with her tally chart and design an improved version.

No. of cinema trips	Tally	Frequency
1 – 10		
10 – 20		
20 – 30		
30 – 40		
40 or more		

Q5 Ronaldo wants to do an investigation to test his theory that children spend more time outside each week than adults. Design a data collection sheet for his investigation.

Q6 A supermarket chain wants to know what people in a town think of their plan to build a supermarket there. They hire some researchers to interview a sample of 500 people. Give one reason why they wouldn't want to interview everyone in the town.

Q7 Jasper and Connie want to find out which football team is most popular at their school. Jasper asks a random selection of 30 pupils and Connie asks the first 5 pupils she sees at lunchtime. Give two reasons why Jasper's sample will be more representative of the whole school.

Q8 A factory makes hundreds of the same component each day. PROBLEM SOLVING
Describe how a representative sample of 50 components could be tested each day to make sure the machinery is working properly.

Q9 Sadie is making doughnuts for 50 party guests. She asks 12 party guests at random whether they prefer doughnuts with jam or with sprinkles. Four of them prefer jam. Based on her sample, how many party guests would you expect to prefer jam?

Exam-Style Questions

Q1 Mehran claims:

"Left-handed people can throw farther than right-handed people."

He tests his claim by measuring how far left- and right-handed people throw a ball.

a) Are the distances Mehran measures discrete or continuous data?

[1 mark]

b) Design a suitable data collection sheet for Mehran's investigation.

[2 marks]

Q2 Jill wants to know what the most popular chocolate bar is in her year of 150 students. She decides to select a sample of 10 students from her class to ask about their favourite chocolate bars. The results are shown in the tally chart below.

Chocolate bar	Cocoa Cream	Elephant Bar	Choco Crunch	Martian Milk	Other
Tally	II	IIII	III		I
Frequency	2	4	3	0	1

Jill concludes that the 'Elephant Bar' is the most popular chocolate bar with students in her year. Make two comments on the accuracy of Jill's conclusion.

[2 marks]

Q3 Tahani is investigating how long it takes 200 people to complete a crossword puzzle. She designs a tally chart to record her results:

Time (t minutes)	**Tally**	**Frequency**
$0 < t \leq 10$		
$10 \leq t \leq 20$		
$20 \leq t \leq 30$		

a) Give two problems with her table.

[2 marks]

b) Tahani takes a random sample of 40 of the people and records their times. She finds that 12 people in the sample take between 10 and 20 minutes to complete the crossword. How many of the 200 participants can she expect to complete the crossword puzzle in between 10 and 20 minutes? Write down any assumptions you have made.

[3 marks]

Section 26 Analysing Data

26.1 Tally Charts and Frequency Tables

You saw how to design tables in Section 25 — now you need to use them to analyse data.

Learning Objective — Spec Ref S2:
Complete and interpret tally charts and frequency tables.

Prior Knowledge Check:
Be familiar with tally charts and frequency tables. See Section 25.

The **frequency** of a data value is the **number of times it occurs** in a data set.

Frequency tables can show frequencies of **individual** data values or of **groups** of values. **Grouping** data makes it **easier to analyse**, but the **exact values** are lost.

Example 1

A mathematics test is marked out of 40. Here are the marks for 20 students.

10	18	28	38	40	40	29	11	13	16
18	20	31	40	27	25	22	40	9	34

a) Complete the grouped frequency table to show the marks.

1. Add a tally mark (|) in the Tally column for every test score. Write the tally marks in groups of 5, with the fifth tally mark across the others to make a '5-bar gate'.
2. Find each frequency by counting the tally marks.

Mark	Tally	Frequency
0-10	||	**2**
11-20	𝍸 |	**6**
21-30	𝍸	**5**
31-40	𝍸 ||	**7**

b) The pass mark for this test was 21 out of 40. How many students passed the test?

Add the frequencies for the bottom two rows. 5 + 7 = **12**

Exercise 1

Q1 Sharla asks a group of people what type of music they like most. Here are their responses.

Pop	R&B	Rock	Indie	Pop	R&B	Indie	Pop	Jazz	Pop
Pop	Rock	Pop	Pop	Jazz	Pop	Pop	Classical	Indie	Rock

a) Copy and complete the frequency table on the right.

b) Use your table to answer the following.

(i) What is the most popular type of music?

(ii) How many more people like pop the most compared to rock?

Type of Music	Tally	Frequency
Classical		
Indie		
Pop		
Rock		
Other		

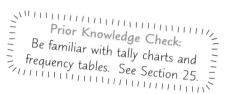

26.2 Averages and Range

Finding averages and the range is a useful way of summarising a data set — these values give you a good idea of the data set as a whole without needing to see every piece of data within it.

Averages: Mean, Median and Mode

Learning Objective — Spec Ref S4:
Find the mode, median and mean of a set of data.

An **average** is a way of **representing** a whole set of data using a **single value** — it is the **central** or **typical** value within the data set. The **mode**, **median** and **mean** are three common averages that are used. The most **suitable** one to use depends on the data.

The Mode

Mode (or modal value) is the most common value

The mode is the value that appears in the data set **more often** than any other. It's easy to find and **isn't distorted** by **extreme** values (called **outliers**). However, there could be **no mode** (or more than one) and it might **not represent the data well.**

Tip: The mode can be used with qualitative data.

Example 1

Find the mode of these numbers: 3 7 4 8 3

The mode is the most common number. Mode = **3**

Exercise 1

Q1 Find the mode of the following sets of data.

a) 8, 5, 3, 8, 4   b) 6, 9, 2, 7, 7, 6, 5, 9, 6   c) 16, 8, 12, 13, 13, 8, 8, 17

Q2 The test scores of nine students are given below. Find the modal score.

34	67	86	58	51	52	71	65	58

Q3 Find the mode for the following data:

red, yellow, red, black, orange, purple, red, green, black

Q4 The data below shows how long in days it took for 10 tomato seeds to germinate.

3, 6, 17, 4, 3, 6, 6, 5, 4, 3

Is the mode a suitable measure of average for this data?

The Median

Median = middle value once the values have been put in order from smallest to largest

If there is an **even number of values**, there will be **two numbers** in the middle. In this case the median will be **halfway** between the two middle numbers.

The median **isn't distorted** by **outliers**, but it might not be a good representation of the data as it **doesn't take into account the value of every piece of data**.

Example 2

a) Find the median of these numbers: **3 7 4 8 3**

1. Put the numbers in order first.
2. There's an odd number of values, so there will be just one value in the middle.

From smallest to largest:
3 3 4 7 8

The median is the 3rd value, so the median = **4**

Tip: There should be an equal number of values either side of the median position, i.e. 3 3 (4) 7 8.

b) The number 5 is added to the list. Find the new median.

1. Write the numbers in order.
2. There's an even number of values so the median will be halfway between the two middle numbers. Find it by adding them together and dividing by 2.

From smallest to largest:
3 3 4 5 7 8

The median is halfway between the 3rd and 4th values, so the median = $\frac{4+5}{2}$ = **4.5**

Exercise 2

Q1 Find the median of the following sets of data.
a) 8, 5, 3, 8, 4 b) 6, 9, 2, 7, 7, 6, 5, 9, 6 c) 16, 18, 12, 13, 17, 8, 8, 17

Q2 Find the median of the following sets of data.
a) 1.5, 2.7, 3.8, 4.8, 5.6 b) 15, 14, 22, 17 c) 3, 3, 3, 3, 3, 3, 3, 4

Q3 The times (to the nearest second) of nine athletes running the 400 m hurdles are:

78	78	84	81	90	79	84	78	95

Find the median time.

Q4 The median of the following data set is 16.7, but one value is missing. What is the smallest that the missing value could be? PROBLEM SOLVING

16.5	16.9	15.8	14.3	18.9	?

The Mean

Mean = the total of all the values ÷ the number of values

The **mean** is usually the most **representative** average as it **uses all of the data values**. However, it can be **distorted** by **outliers**.

To find the mean, **add up** all the values in the data set and **divide** by the number of values.

Example 3

Find the mean of these numbers: 3 7 4 8 3

1. Find the total of the values. — The total is $3 + 7 + 4 + 8 + 3 = 25$
2. Divide the total by the number of values. — So the mean $= 25 \div 5 = \mathbf{5}$

Exercise 3

Q1 Find the mean of the following sets of data.
a) 8, 5, 3, 8, 4
b) 6, 9, 2, 7, 7, 6, 5, 9, 6
c) 16, 18, 12, 13, 13, 8, 8, 17

Q2 Find the mean of the following sets of data.
a) 1.5, 2.7, 3.8, 4.8, 5.6
b) 15.85, 16.96, 22.04, 17.45
c) 3, 3, 3, 3, 3, 3, 3, 4

Q3 The test scores of nine students are given below. Find the mean score.

34	67	86	58	51	52	71	65	58

Q4 The times (to the nearest second) of nine athletes running the 400 m hurdles are:

78	78	84	81	90	79	84	78	95

Find the mean time.

Q5 Find the mean, median and mode for each of the following two sets of data.
a) 2, 3, 2, 1, 3, 2, 8, 5
b) 2, 3, 3, 1, 5, 3, 4, 3

Q6 If the mean of the following data set is 7, find the missing value. PROBLEM SOLVING

6	5	8	8	5	?

Q7 If the mean of the following data set is 16.3, find the missing value. PROBLEM SOLVING

16.6	16.9	15.8	14.3	18.9	?

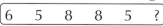

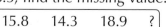

The Range

Learning Objective — Spec Ref S4:
Find the range of a set of data.

The **range** is the difference between the **largest value** and the **smallest value** in a data set — it tells you how **spread out** the values are.

Range = largest value – smallest value

Data sets with a **small range** are more **consistent** than those with a **large range** — this means there is less variation in the values.

The range can be a **misleading** measure of spread for data sets that contain **outliers** (extreme values). Most of the data could be much **closer together** than the value of the range suggests.

Example 4

Find the range of these numbers: **3 7 4 8 3**

Subtract the lowest value from the highest. Range = 8 – 3 = **5**

Exercise 4

Q1 Find the range of the following sets of data.

a) 8, 5, 3, 8, 4 b) 6, 9, 2, 7, 7, 6, 5, 9, 6 c) 16, 8, 12, 13, 13, 8, 8, 17

Q2 Find the range of the following sets of data.

a) 1.5, 2.7, 3.8, 4.8, 5.6 b) 15.85, 16.96, 22.04, 17.45 c) 3, 3, 3, 3, 3, 3, 3, 4

Q3 Hayley takes the temperature (in °C) nine times over the year.
The results are below. Find the range of the temperatures.

15	13	–2	8	3	–1	0	22	10

Q4 The profits made by a number of lemonade stands last Tuesday are given below.
A negative number means the lemonade stand made a loss. Find the range of the data.

£30	£12	£1	–£4	–£40	£8.50

Q5 The range of the following data set is 6: 6, 5, 8, 8, 5, ? PROBLEM SOLVING
What are the two possible values for the missing number?

Q6 The number of skips Simone managed when skimming stones on a lake are given below.

3, 6, 17, 4, 3, 6, 6, 5, 4, 3

a) What is the range of the data?

b) Do you think the range is a good measure of the spread of this data?

Finding Averages and the Range from a Frequency Table

Learning Objective — Spec Ref S2/S4:
Find the mode, median, mean and range for data in a frequency table.

Large data sets are often put into **frequency tables** to make them easier to deal with.

The data looks different, but you can find the averages and range as usual:

- The **mode** is the value that appears the most — the one with the **highest** frequency.
- The **median** is the middle value (or halfway between two middle values).
  Find its position as normal, then count through the table to see which value it takes.
- The **range** is the **difference** between the **largest** and the **smallest** values given in the table.

Example 5

This frequency table shows the number of mobile phones owned by a group of people.

Number of mobile phones	0	1	2	3
Frequency	4	10	4	2

a) Find the modal number of mobile phones owned.

Most people own 1 mobile phone since it has the highest frequency (10). So the mode is 1. Modal number = **1 mobile phone**

b) What is the median number of mobile phones owned?

1. Find the number of data values and position of the median.
   Total frequency = 4 + 10 + 4 + 2 = 20,
   so the median is halfway between the 10th and 11th values.
2. Count through the 'Frequency' row to find the 10th and 11th values. The raw data is 0, 0, 0, 0, 1, 1, 1, 1, 1, 1, 1, 1, 1, 1...
   10th value = 1 and 11th value = 1,
   so median = $\frac{1+1}{2}$ = **1 mobile phone**

Tip: For larger data sets, it's easier to use $(n + 1) \div 2$ to find the median position, where n is the total frequency.
E.g. if $n = 20$,
$(20 + 1) \div 2 = 10.5$

c) Find the range for this data.

Subtract the fewest number of phones from the greatest number of phones: 3 – 0 = **3**

Finding the **mean** from a frequency table is a bit trickier. To work out the total of all the values, it's best to add a third row (or column) to the table. Take a look at the example below...

Example 6

This frequency table shows the number of mobile phones owned by a group of people.

Number of mobile phones	0	1	2	3
Frequency	4	10	4	2

a) How many people were in the group altogether?

This is the total of the frequencies. Total number of people = 4 + 10 + 4 + 2 = **20**

b) What is the total number of mobile phones owned by this group of people?

1. First multiply each number of mobile phones by its frequency — add a third row to record your values.

Number of mobile phones	0	1	2	3
Frequency	4	10	4	2
Phones × frequency	0	10	8	6

2. Then add the results together. Total number of mobile phones = 0 + 10 + 8 + 6 = **24**

c) Find the mean number of mobile phones owned by each person.

Divide the total number of phones by the number of people in the group.

$$\text{Mean} = \frac{\text{Total number of phones}}{\text{Total number of people}}$$

$$= \frac{24}{20} = \textbf{1.2 mobile phones}$$

Tip: Make sure you remember to divide by the total frequency (i.e. 20) and not the number of groups.

Exercise 5

Q1 This frequency table shows the number of pets the students in a class have.

Number of pets	0	1	2	3	4
Frequency	5	10	5	5	2

a) What is the modal number of pets owned?

b) What is the median number of pets owned?

c) How many students had:
(i) no pets? (ii) 1 pet? (iii) 2 pets? (iv) 3 pets? (v) 4 pets?

d) Use your answer to c) to find the total number of pets owned by the students.

e) How many students are in the class altogether?

f) Use your answers to d) and e) to find the mean number of pets owned by each student.

Q2 The table shows the number of people living in each of 30 houses.

Number of people	1	2	3	4	5
Frequency	7	9	1	10	3

a) Write down the modal number of people living in a house.

b) Find the median number of people living in a house.

c) Calculate the mean number of people living in a house.

d) Work out the range of the data.

Q3 During June, a student wrote down the temperature in his garden in degrees Celsius (°C) every day at noon, as shown in the table.

Temperature (°C)	Frequency
16	4
17	9
18	2
19	5
20	4
21	6

a) Find the median noon temperature.

b) Find the mean noon temperature (correct to 1 decimal place).

c) The average noon temperature in June in the UK is approximately 18.5 °C. What does this suggest about these results?

Finding Averages from a Grouped Frequency Table

Learning Objective — Spec Ref S2/S4:
Find the mode, median, mean and range for data in a grouped frequency table.

If you're given data in the form of a **grouped frequency table**, then you don't know the **exact** data values — so you can't find exact values for the averages or range. You can only **identify** the **modal group** and **group containing the median**, and **estimate** the **mean** and the **range**.

- The **modal group** is the group (sometimes called a **class**) that has the **highest frequency**.
- Find the **group containing the median** by working out the **position** of the median in the usual way. Then use the **group frequencies** to identify which group the median falls into.
- To estimate the mean, you need to find the midpoint of each group — this is used as an **estimate** for all the data values within the group. To find the **midpoint** of a group, add the **lower** and **upper** bounds together and **divide by 2**. The method is then the same as for non-grouped tables.
- The **estimated range** is found by **subtracting** the **lower bound** of the smallest group from the **upper bound** of the largest group — this gives you the **largest possible range** for the data set.

Example 7

This grouped frequency table shows the number of hours that 20 people spent exercising during one week.

Hours, h	$0 \le h < 4$	$4 \le h < 8$	$8 \le h < 12$	$12 \le h < 16$	$16 \le h < 20$	$20 \le h < 24$
Frequency	2	3	6	4	4	1

a) **Write down the modal group.**

This is the one with the highest frequency.

Modal group is $\mathbf{8 \le h < 12}$

b) **Which group contains the median?**

Find the position of the median and then count through the groups until you reach this position.

There are 20 values, so $n = 20$.
$(n + 1) \div 2 = (20 + 1) \div 2 = 10.5$
So the median lies halfway between the 10th and 11th values, so it's in the group $\mathbf{8 \le h < 12}$.

c) **Find an estimate for the mean.**

You don't know how long each person spent exercising, so you assume that each of their times is in the middle of their group. So work out the midpoints:

Hours, h	$0 \le h < 4$	$4 \le h < 8$	$8 \le h < 12$	$12 \le h < 16$	$16 \le h < 20$	$20 \le h < 24$
Frequency	2	3	6	4	4	1
Midpoint of group	$\frac{0+4}{2} = 2$	$\frac{4+8}{2} = 6$	$\frac{8+12}{2} = 10$	$\frac{12+16}{2} = 14$	$\frac{16+20}{2} = 18$	$\frac{20+24}{2} = 22$

Now find the mean as before, using the midpoints instead of the actual data values.

Freq. × midpoint	$2 \times 2 = 4$	$3 \times 6 = 18$	$6 \times 10 = 60$	$4 \times 14 = 56$	$4 \times 18 = 72$	$1 \times 22 = 22$

Add the results together. Total number of hours = $4 + 18 + 60 + 56 + 72 + 22 = 232$

Divide the total number of hours by the total frequency. Mean = $\frac{232}{20}$ = **11.6 hours**

d) Estimate the range.

Subtract the lower bound of the smallest group from the upper bound of the largest group.

Range = $24 - 0$ = **24 hours**

Exercise 6

Q1 The test results of 25 people are shown in this grouped frequency table.

Marks scored	1-5	6-10	11-15	16-20
Frequency	4	5	7	9

a) Write down the modal group.

b) Which group contains the median?

c) Find an estimate for the mean.

d) Estimate the range of marks.

Q2 The table on the right shows some information about the weights of some tangerines in a supermarket.

Weight in grams, w	Frequency
$0 \le w < 20$	1
$20 \le w < 40$	6
$40 \le w < 60$	9
$60 \le w < 80$	24

a) Find an estimate for the mean.

b) Find an estimate for the range of weights.

c) Explain why your answer to part b) might be very different from the actual range.

Q3 Troy collected some information about the number of hours (to the nearest whole hour) students spent watching television over a week.

Time in hours	Frequency
0-5	3
6-10	8
11-15	11
16-20	4

a) Write down the modal group.

b) Which group contains the median?

c) Find an estimate for the mean to 1 decimal place.

Q4 This table shows information about the heights of 200 people.

Height (h) in m	Frequency
$1.50 \le h < 1.60$	27
$1.60 \le h < 1.70$	92
$1.70 \le h < 1.80$	63
$1.80 \le h < 1.90$	18

a) Write down the modal group.

b) Which group contains the median?

c) Find an estimate for the mean to 3 significant figures.

d) Find an estimate for the range.

26.3 Two-Way Tables

Two-way tables are really clever, they show two variables (pieces of information) at the same time.

Learning Objective — Spec Ref S2:
Complete and interpret data in two-way tables.

Prior Knowledge Check:
Be able to write one number as a percentage of another. See p.102.

Two-way tables are used to show the frequencies for **two different variables** — e.g. the hair colour and eye colour of school pupils.

- **Rows** represent the **categories** for one variable (e.g. 'brown hair', 'blonde hair', etc.).
- **Columns** represent the **categories** for the other variable (e.g. 'blue eyes', 'brown eyes', etc.).
- Each **cell** then shows the number of items in a particular row AND a particular column — e.g. the number of pupils with brown hair AND blue eyes.
- **Row** and **column totals** show the **total** number of items in each **category** and the bottom-right cell shows the **overall** total.

To interpret two-way tables you may need to find missing values or work out percentages. You can find percentages of the **whole group** or percentages within a **category**.

Tip: Be careful with which totals you use. You need the overall total if you're looking at the whole group, or the row/column totals if you're looking within a category.

Example 1

This table shows how students in a class travel to school.

	Walk	Bus	Car	Total
Boys	8	7		19
Girls	6		2	
Total		12		

a) Complete the table.

Add entries to find row/column totals. Subtract from row/column totals to find other entries.

	Walk	Bus	Car	Total
Boys	8	7	19 – 8 – 7 = **4**	19
Girls	6	12 – 7 = **5**	2	6 + 5 + 2 = **13**
Total	8 + 6 = **14**	12	4 + 2 = **6**	19 + 13 = **32**

b) How many girls take the bus to school?

Find the entry in the Bus column and the Girls row.

5 girls take the bus

c) How many students walk to school?

This is the total of the Walk column.

14 students walk

d) What percentage of students take the bus to school?

Divide the total of the Bus column by the overall total, and multiply by 100 to get the percentage.

12 students take the bus and there are 32 students in total, so:
12 ÷ 32 = 0.375, 0.375 × 100 = **37.5%**

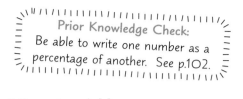

Exercise 1

Q1 Copy and complete these two-way tables, showing how groups of students travel to school.

a)

	Walk	Bus	Car	Total
Boys	9	5	3	
Girls	5	8	6	
Total				

b)

	Walk	Bus	Car	Total
Boys			8	
Girls	11		4	30
Total	21			60

Q2 This two-way table shows how many male and female language students from a university's languages department went to study in Germany, France or Spain.

	Germany	France	Spain	Total
Male	16	7	12	
Female	18	4	16	
Total				

a) Copy the table and fill in the entries in the final row and final column.

b) How many female students went to France?

c) How many male students went to Germany?

d) How many students in total went to Spain?

e) How many female students went to either Germany, France or Spain to study?

f) What was the total number of students from this department that went to either Germany, France or Spain to study?

Q3 This two-way table gives information about some students' favourite type of snack.

	Chocolate	Crisps	Jellies	Total
Male	3			
Female		7	2	
Total	6	12		30

a) Copy and complete the two-way table.

b) How many students preferred crisps?

c) How many females preferred jellies?

d) What percentage of males preferred chocolate?

Q4 This two-way table gives information about the colours of the vehicles in a car park.

	Red	Black	Blue	White	Total
Cars	8	7	4		22
Vans		2	1	10	
Motorbikes	2	1		2	
Total	12		6		

a) Copy and complete the two-way table.

b) How many motorbikes were blue?

c) How many vans were there?

d) What percentage of vehicles were: (i) cars? (ii) vans? (iii) red?

26.4 Bar Charts and Pictograms

Bar charts show how many items fall into different categories.
A pictogram uses symbols to represent a certain number of items.

Bar Charts

Learning Objective — Spec Ref S2:
Display and interpret data in bar charts.

Bar charts show the **number** (or frequency) of items in **different categories**. They're used for **qualitative** or **discrete quantitative data** (see p.423). Each bar represents a different category — the bars **shouldn't touch** because the categories are **distinct**.

Tip: Bars can be replaced by lines, and drawn either vertically or horizontally.

You can easily read the **mode** (see p.436) from a bar chart — it's the category with the highest frequency, so it's shown by the **tallest bar**.

If the data is numerical, you can find (or estimate) the mean, median and range too.

Example 1

The eye colour of 50 students is shown in the frequency table on the right. Draw a bar chart to display this information.

Eye colour	Blue	Brown	Green	Other
Frequency	15	19	10	6

1. Draw a bar of equal width for each eye colour, with height equal to the frequency, and a space between each bar.
2. Label your axes with 'Eye colour' on the horizontal axis and 'Frequency' on the vertical axis.

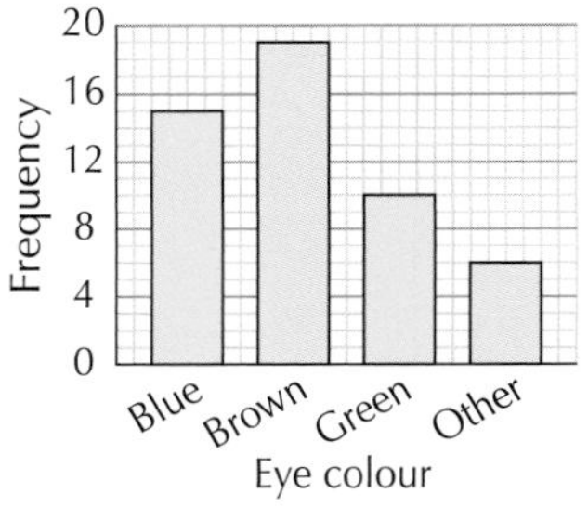

Exercise 1

Q1 Arthur asked some students in the canteen which meal they had bought at lunchtime. His results are in the table below. Draw a bar chart to show this information.

Meal	Pie & Chips	Pasta	Baked potato	Baguette	Salad
Frequency	11	3	7	10	4

Q2 Some people at a bus stop were asked which bus they were waiting for. Their responses are shown in the frequency table below. Draw a bar chart to show this information.

Bus	Number 7	Number 23	Coastlander	X94	12a
Frequency	6	5	2	3	2

Q3 One morning a coffee shop recorded the first 100 drinks that were ordered. Their results are in the frequency table below. Draw a bar chart to show this information

Drink	Espresso	Latte	Cappuccino	Mocha	Tea	Other
Frequency	23	19	12	15	22	9

Multiple sets of data can be displayed on the **same bar chart** — e.g. data for boys and girls or children and adults.

- **Dual bar charts** have two bars per category — one for each data set.
- **Composite bar charts** have single bars split into different sections for each data set.

Example 2

Manpreet and Jack recorded how many TV programmes they watched each day for a week. Their results are shown in the table on the right. Draw a dual bar chart to display this information.

Day	**M**	**T**	**W**	**T**	**F**	**S**	**S**
No. watched by Manpreet	1	2	4	2	3	7	4
No. watched by Jack	2	1	0	2	3	4	0

1. Each day should have two bars — one for Manpreet and one for Jack.
2. The height of each bar is the frequency.
3. Mark up the bars for Manpreet and Jack in different ways (e.g. shade one and not the other) — and make sure to include a key showing which is which.

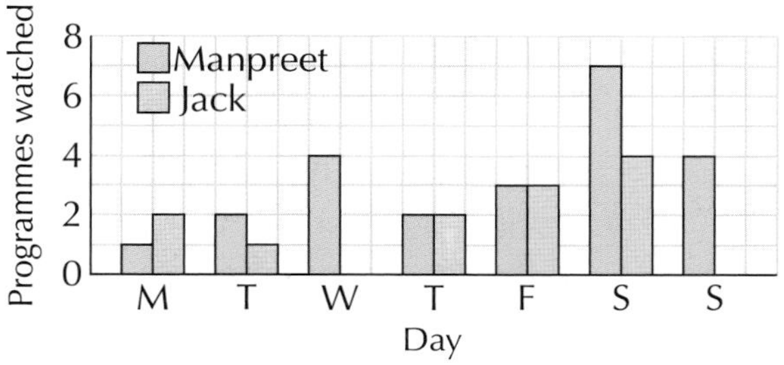

Example 3

The number of cars sold at two car showrooms over a week is shown in the table below. Draw a composite bar chart to display this information.

Day	**M**	**T**	**W**	**T**	**F**	**S**	**S**
No. sold at Showroom 1	3	1	4	2	2	6	3
No. sold at Showroom 2	2	1	1	0	2	5	1

1. Start with Showroom 1 — the height of the orange bar is the frequency for Showroom 1.
2. Add grey bars onto the top of the orange ones. The height of each grey bar should be the frequency for Showroom 2. Make sure you include a key to show which colour is for which showroom.
3. The total height of the composite bar (the orange and grey bars combined) is the total frequency for both showrooms.

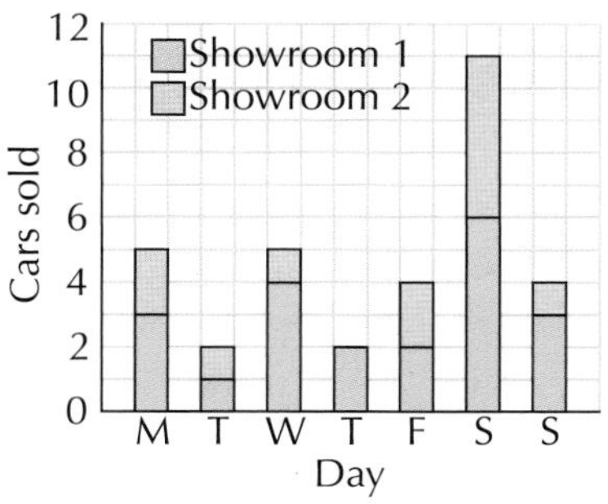

Exercise 2

Q1 The eye colour of 50 students is shown in the table.

Eye colour	Blue	Brown	Green	Other
No. of males	8	7	5	4
No. of females	7	12	5	2

a) Draw a dual bar chart to display the data.

b) Draw a composite bar chart to display the data.

c) Which chart is best for comparing males and females?

d) What is the modal eye colour?

e) Which chart was easier to use to find the mode?

Q2 This composite bar chart shows the ages in whole years of the members of a small gym.

a) How many members are in the age range 26-35 years?

b) How many members does the gym have in total?

c) What is the modal age range for the female members?

d) How many more men aged 26-35 use the gym than women aged 26-35?

e) Which age range has the greatest difference in numbers of men and women?

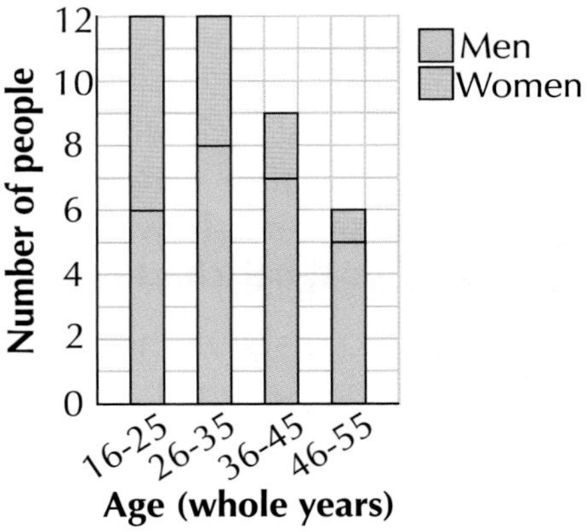

Pictograms

Learning Objective — Spec Ref S2:
Display and interpret data in pictograms.

Pictograms show the number (or frequency) of items using **symbols**. Every pictogram has a key telling you what one symbol represents.

One symbol usually represents more than one item, so a fraction of the symbol is used to show fewer items than the whole symbol represents.

For example:

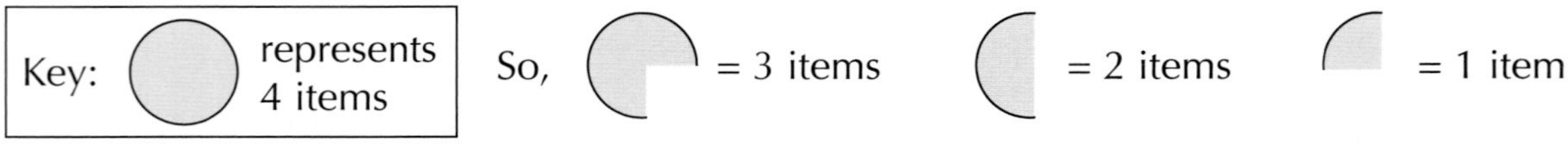

Key:
represents 4 items
So,
= 3 items
= 2 items
= 1 item

Example 4

This table shows the number of TVs in four secondary schools.

School	Cool School	Great Hall	St. Jimmy's	Parker's Park
TVs	20	10	8	13

Using the symbol ▭ to represent 4 TVs, draw a pictogram to show this information.

1. Work out how many symbols you need to draw for each school. For example, for Great Hall:

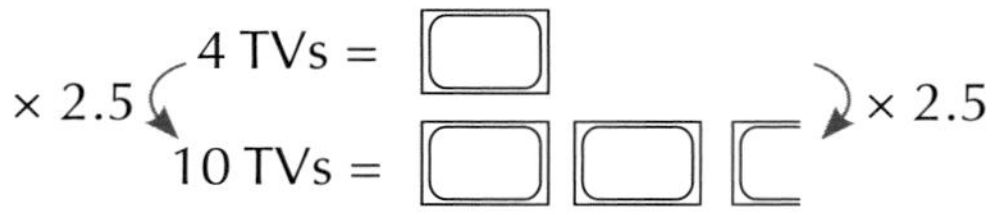

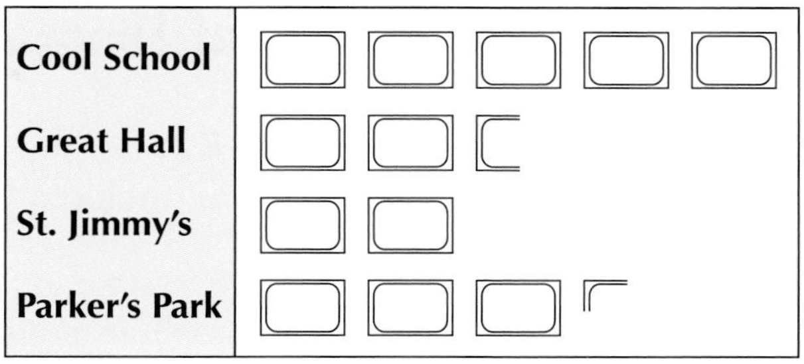

2. Include a key to show what one symbol represents. ⟶ Key: ▭ represents 4 TVs.

Exercise 3

Q1 Members of a hockey club were asked which activity they would like to do on a team day out. The results are in this frequency table.

Activity	Theme Park	Bowling	Cinema	Boat Trip
Frequency	14	10	5	3

Draw a pictogram to show this information. Show 2 club members using the symbol: ◯

Q2 The incomplete pictogram on the right shows the number of chocolate bars a shop sold on Monday, Tuesday and Wednesday.

a) How many chocolate bars were sold on:

(i) Tuesday? (ii) Wednesday?

b) Copy the pictogram, and use the information below to complete it.

- On Thursday the shop sold 20 bars.
- On Friday they sold 50 bars.
- On Saturday they sold 65 bars.
- On Sunday the shop was closed.

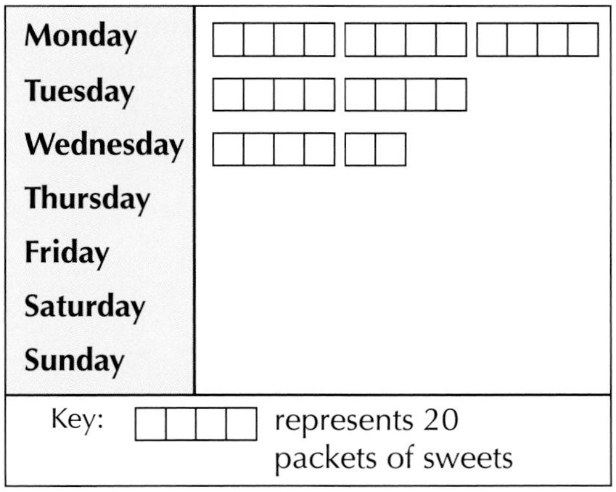

Q3 This pictogram shows the number of letters that were delivered to each of the 6 houses in a street one week.

a) How many letters were delivered to Number 4?

b) Which house received fewest letters during the week?

c) How many letters in total were delivered to the 6 houses?

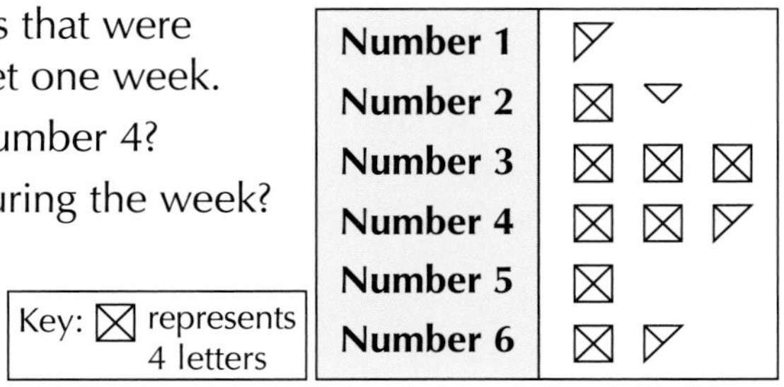

26.5 Stem and Leaf Diagrams

Stem and leaf diagrams are used to display sets of discrete data. They are useful for showing the spread of data visually, but still showing each data value.

Building Stem and Leaf Diagrams

Learning Objective — Spec Ref S2:
Construct stem and leaf diagrams, including back-to-back diagrams.

In **stem and leaf diagrams**, data values are split up into '**stems**' (their first digit(s)) and '**leaves**' (the remaining digit). So for the value **25**, the stem would be **2** and the leaf would be **5**.
The leaves are then **ordered** numerically.

Stem and leaf diagrams always have a **key** — e.g. '2 | 5 means 25'.

Tip: Decimals and three-figure numbers can be shown using different keys, e.g. 0 | 3 = 0.3 or 20 | 4 = 204.

Example 1

Here are the marks scored by pupils in a class test.

56, 52, 82, 65, 76, 82, 57, 63, 69, 73, 58, 81, 73, 52, 73, 71, 67, 59, 63

Use this data to build an ordered stem and leaf diagram.

1. Write down the 'stems' — here use the first digit of the marks. The smallest first digit is 5 and the largest is 8, so use 5, 6, 7 and 8.

2. Next, make a 'leaf' for each data value by adding the second digit to the correct stem.

5	6	2	7	8	2	9
6	5	3	9	7	3	
7	6	3	3	3	1	
8	2	2	1			

3. Put the leaves in each row in order — from lowest to highest.

5	**2**	**2**	**6**	**7**	**8**	**9**
6	**3**	**3**	**5**	**7**	**9**	
7	**1**	**3**	**3**	**3**	**6**	
8	**1**	**2**	**2**			

4. Always include a key. → Key: 5 | 2 means 52

Back-to-back stem and leaf diagrams can be used to display **two sets of data** next to one another — e.g. they might show data for different genders or age groups.

In these diagrams, the stem is in the **centre** and the leaves from the two data sets are placed on **either side** of the stem. This means that one data set has to be read '**backwards**' (the **smaller leaves** are closest to the centre) — so the **key** is really important.

Example 2

The times taken (in minutes) by 14 girls and 14 boys to complete a puzzle were recorded. Using the results below, construct an ordered back-to-back stem and leaf diagram.

Boys: 17, 27, 14, 33, 32, 5, 14, 6, 6, 19, 29, 9, 38, 7

Girls: 32, 31, 2, 25, 23, 28, 38, 29, 37, 34, 2, 4, 25, 28

1. Write down the 'stem' — use the 'tens' digit. For single-digit numbers, this is 0.
2. Make the 'leaves' for one data set by adding the second digit of each value to the correct stem.
3. Repeat step 2 for the other data set, but write the digits on the other side of the stem.
4. Put the leaves in each row in order — from lowest to highest as you read outwards from the stem.
5. Remember to include a key.

Girls		Boys
4 2 2	0	5 6 6 9 7
	1	7 4 4 9
8 5 9 8 3 5	2	7 9
4 7 8 1 2	3	3 2 8

Girls		**Boys**
4 2 2	**0**	**5 6 6 7 9**
	1	**4 4 7 9**
9 8 8 5 5 3	**2**	**7 9**
8 7 4 2 1	**3**	**2 3 8**

Key: 2 | 7 for boys means 27 minutes
3 | 2 for girls means 23 minutes

Exercise 1

Q1 Use the data set below to make an ordered stem and leaf diagram.

41, 48, 51, 54, 59, 65, 65, 69, 74, 80, 86, 89

Q2 Only some of the values from the data sets below have been added to the ordered stem and leaf diagrams. Copy and complete the diagrams by adding the rest of the data values.

a) Key: 3 | 1 means 3.1

3	1
4	0 4
5	1 3 4
6	0
7	1

~~3.1~~	~~4.0~~	~~4.4~~	~~5.3~~
5.7	5.9	~~6.0~~	7.7
3.4	4.9	~~5.4~~	~~5.1~~
6.1	5.7	~~7.1~~	4.4

b) Key: 20 | 1 means 201

20	3 5
30	1
40	2 2
50	0
60	1 6

~~203~~	~~205~~	~~301~~	~~402~~
409	~~500~~	~~606~~	608
304	409	404	501
403	~~601~~	503	~~402~~

Q3 The amount of rainfall (in cm) over Morecambe Bay was recorded every week for 16 weeks. Use the data below to make an ordered stem and leaf diagram.

0.0	3.8	3.6	0.1	2.7	0.6	0.3	1.1
2.0	1.3	0.0	1.6	4.1	0.0	2.5	3.1

Q4 Use these two data sets to draw an ordered back-to-back stem and leaf diagram.

18, 8, 38, 29, 1, 28, 33, 24, 12, 37, 32

27, 25, 19, 15, 22, 18, 13, 23, 22, 32, 13

Using Stem and Leaf Diagrams

Learning Objective — Spec Ref S2/S4:
Interpret data presented in stem and leaf diagrams.

Prior Knowledge Check:
Be able to find the mode, median, and range. See p.436-439.

Once your data is in a **stem and leaf diagram**, you can easily find the **mode**, **median** and **range**, and you can compare two data sets from a back-to-back diagram. You can also quickly see how many values fall between certain limits — e.g. 'between 2 and 8'.

Example 3

The stem and leaf diagram on the right shows the number of seconds 19 children took to open a jar of jam. Use it to find the mode, median and range of the data.

0	6 8 8
1	0 2 4 4 4 5 5 7
2	0 4 5 6 6 7
3	1 3

Key: 2 | 1 means 21 seconds

1. Find the mode by looking for the number that repeats most often in one of the rows — here, there are three 4's in the second row. Use the key to work out what it represents.
2. There are 19 data values, so the median is the 10th value. Count along the 'leaves' to find it, starting with the first row.
3. Find the range by subtracting the first number from the last.

Mode = **14 seconds**

Median = **15 seconds**

Range = 33 – 6
= **27 seconds**

Exercise 2

Q1 Use the stem and leaf diagram on the right to find:
a) the mode of the data
b) the median of the data
c) the range of the data

0	5 8
1	2 3 7 9
2	0 1 3 9
3	2 2 2

Key: 2 | 1 means 21

Q2 The times in seconds that 16 people took to run a 100 m race are shown in the box.
a) Use these times to create an ordered stem and leaf diagram.
b) Find the mode, median and range of the times.
c) To qualify for the final, a contestant had to finish in under 13 seconds. How many of those who took part qualified?

10.2	13.1	13.9	14.2
17.3	11.7	11.4	12.9
15.4	13.6	13.9	10.6
12.8	13.9	12.4	13.3

Q3 The heart rates in beats per minute (bpm) of 15 people at rest and after exercise are shown on the right.
a) Find the median of each data set.
b) What conclusion can you draw from your answer to part a)?

At rest		After exercise
8 7 6 4 3 2 2	6	5 8 8 9
9 8 6 3 2 2	7	4 5 7 7 8
4 1	8	5 6 7
	9	1 3 7

Key: 6 | 5 after exercise means 65 bpm
2 | 6 at rest means 62 bpm

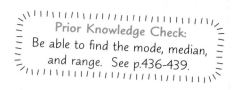

26.6 Pie Charts

In a pie chart, the size of the angle of each sector represents the frequency of a data value.

Drawing Pie Charts

Learning Objective — Spec Ref S2:
Construct pie charts.

Prior Knowledge Check:
Be familiar with angles.
See Sections 15 and 21.

Pie charts show how data is divided into categories, but they show the **proportion** in each category, rather than the actual frequency. The sizes of the **angles** of the sectors are **proportional** to the **frequencies**.

To **draw** a pie chart:

- Find the **total frequency** by **adding** up the **frequencies** of each category.
- **Divide 360°** (the full circle) by the **total frequency** to find the **angle** which represents a **frequency of 1**.
- **Multiply** this value by the **frequency** of each **category** to find the **angle** of the **sector**.

Tip: The sector angles should add up to 360°.
You can use this to check your working.

Example 1

Kamali asked everyone in her class to name their favourite colour. The frequency table on the right shows her results.

Colour	Red	Green	Blue	Pink
Frequency	12	7	5	6

Draw a pie chart to show her results.

1. Calculate the total frequency — the total number of people in Kamali's class.

Total frequency = 12 + 7 + 5 + 6 = 30

2. Divide 360° by the total frequency to find the number of degrees needed to represent each person.

Each person is represented by
360° ÷ 30 = 12°

3. Multiply each frequency by the number of degrees for one person.
This tells you the angle for each colour.
(Check the angles add up to 360°.
Here: 144° + 84° + 60° + 72° = 360°)

Colour	Red	Green	Blue	Pink
Frequency	12	7	5	6
Angle	144°	84°	60°	72°

4. Draw a pie chart — the sizes of the sectors are the angles you've just calculated.

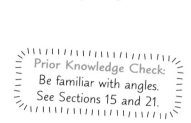

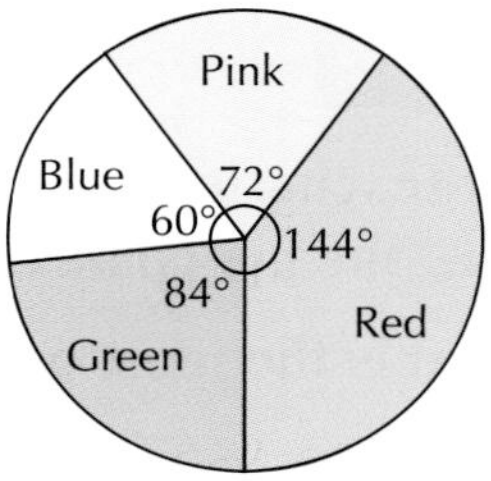

Exercise 1

Q1 Albert recorded the colours of cars that passed his school. His results are shown in the table on the right.

a) Find the total number of cars Albert recorded.

b) Find how many degrees represent one car.

c) Calculate the angle needed to represent each colour.

d) Draw a pie chart to illustrate this data.

Colour	Frequency
Black	25
Silver	17
Red	8
Other	10

Q2 Becky asked her friends which football team they support. Their answers are shown in the table below.

Team	Carlisle United	Kendal Town	Millom Reds	Bristol
Frequency	13	9	8	6

a) Find the total frequency.

b) Calculate the angle needed to represent each of the four football teams.

c) Draw a pie chart showing Becky's results.

Q3 Vicky asked people entering a sports centre what activity they were going to do.

- 33 were going to play squash
- 21 were going swimming
- 52 were going to use the gym
- 14 had come to play table tennis

Draw a pie chart to show this data.

Interpreting Pie Charts

Learning Objective — Spec Ref S2:
Interpret data presented in pie charts.

Prior Knowledge Check:
Be familiar with proportions.
See Section 14.

Pie charts let you **compare** the frequencies of different items, and say which ones occurred more often, and which ones occurred less often. However, pie charts **don't show the frequencies** themselves.

To work out the frequencies from a pie chart, you need to know the sizes of the angles of the sectors.

Tip: You can only compare frequencies in two different pie charts if you know the total number of items that each chart represents.

Example 2

A headteacher carries out a survey to find out how pupils travel to school. The pie chart on the right shows the results of the survey.

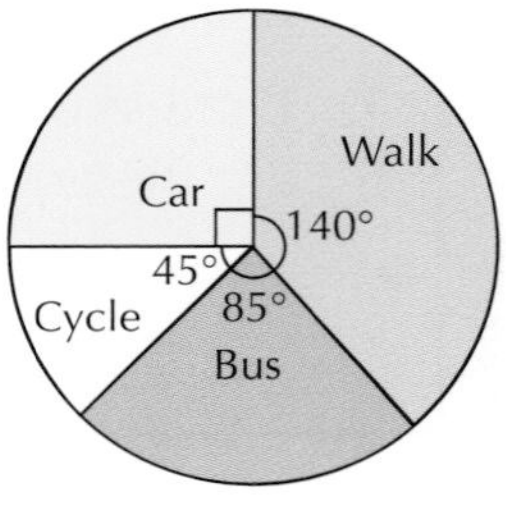

a) What is the most popular way to travel to school?

This is the sector with the largest angle. **Walking**

b) Which method of transport is twice as common as cycling?

Cycling is represented by a sector with an angle of 45°, so look for a sector with an angle of 45° × 2 = 90°. **Travelling by car**

c) 280 pupils walk to school. How many pupils took part in the survey altogether?

1. Work out how many pupils are represented by 1°. — 140° represents 280 pupils. So 1° represents 280 ÷ 140 = 2 pupils.
2. Use this to work out how many pupils the whole pie chart represents. — This means 360° represents 360 × 2 = **720 pupils**.

d) How many pupils cycle to school?

1. You know what 1° represents. — 1° represents 2 pupils.
2. Multiply this by the 'Cycle' sector angle to find how many pupils cycle to school. — So 45° represents 2 × 45 = **90 pupils**.

Exercise 2

Q1 Match pie charts *P* and *Q* to the correct data set.

P

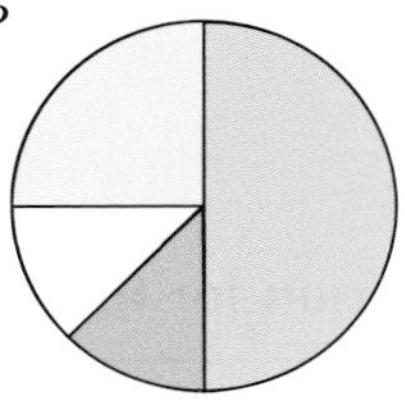

Q

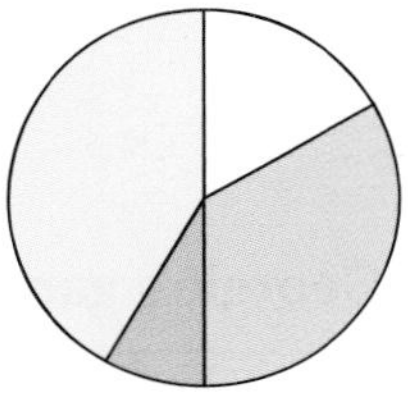

Data Set 1	A	B	C	D
Frequency	24	12	6	6

Data Set 2	A	B	C	D
Frequency	25	20	10	5

Q2 Keemia asked pupils in her school to name their favourite type of pizza. The pie chart on the right shows the results.

a) Which was the most popular type of pizza?

b) What fraction of the pupils said cheese and tomato was their favourite type of pizza?

c) 17 pupils said that vegetable was their favourite type of pizza. Calculate the total number of pupils that Keemia asked.

d) Use your answer to part c) to calculate the number of pupils that said spicy chicken was their favourite type of pizza.

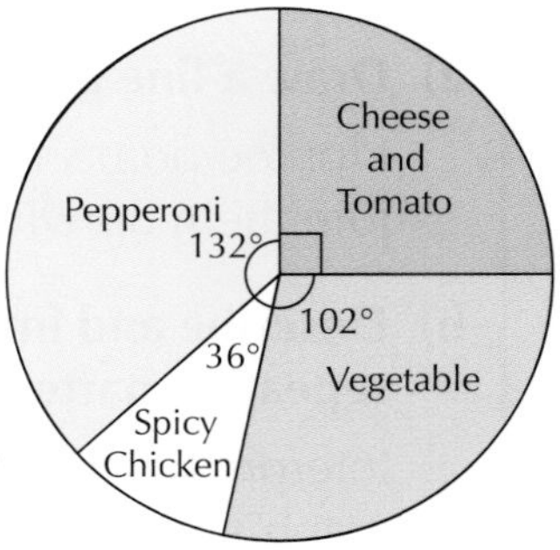

Q3 Tom and Brooke record the number of homework tasks they are set in their Maths, English and Science lessons during one term.

Their data is displayed in the pie charts on the right.

Tom

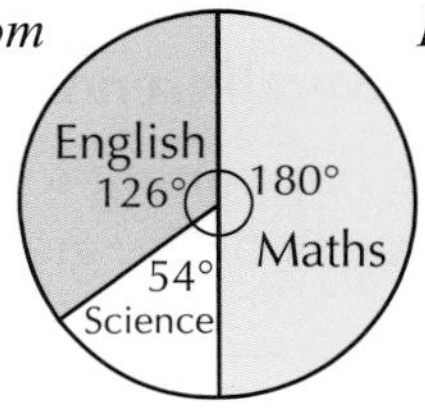

Brooke

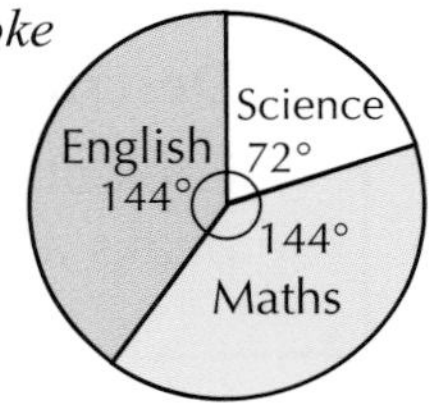

a) Brooke says: "I have a smaller proportion of English homework than Tom." Is she correct?

b) Do the pie charts tell you who spent more time on their English homework? Explain your answer.

26.7 Time Series

Some line graphs can be used to show how things change over time — these are known as time series graphs. They're useful for spotting repeating patterns or trends in data.

Learning Objective — Spec Ref S2:
Display and interpret time series on a line graph.

A **time series** is a set of data collected at **regular intervals** over time — e.g. every day for a week, every month for a year or over several years.
Time series graphs are **line graphs**, with 'time' along the bottom and a scale for the data values up the side. You plot the points and **join them with straight lines**.

You can use time series to find **trends** in the data over time. Look for:

- **Seasonality** — a basic pattern that is **repeated** regularly over time. E.g. the average monthly temperatures will follow a similar pattern year on year.
- An **overall trend** — where the data values **generally** get **smaller** or **larger** over time (ignoring seasonal patterns). E.g. the price of a weekly shop may steadily increase or decrease.

Tip: Seasonality doesn't have to match the actual seasons — e.g. tide levels show seasonality over the course of a day.

Example 1

The data below shows the total rainfall in Georgie's garden each season for three years.

Season	Spr	Sum	Aut	Win	Spr	Sum	Aut	Win	Spr	Sum	Aut	Win
Rainfall (mm)	225	200	275	375	250	225	300	425	300	250	325	475

a) **Draw a line graph to display this data.**

Plot the points from the table and join them up with straight lines.

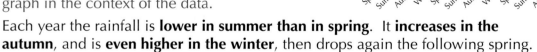

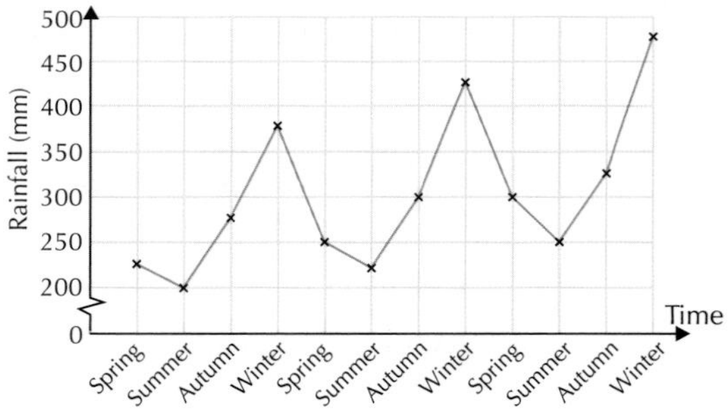

b) **Describe and interpret the repeating pattern in the data.**

Interpret what you see on the graph in the context of the data.

Each year the rainfall is **lower in summer than in spring**. It **increases in the autumn**, and is **even higher in the winter**, then drops again the following spring.

c) **Describe the overall trend in the data.**

Look at what's happening to the peaks and troughs of the graph.

There is an **upward trend** in the amount of rainfall.

Exercise 1

Q1 This table shows the amount of gas used (in m^3) by a family over a period of three years.

Season	Spring	Summer	Autumn	Winter
2012	202	80	170	298
2013	196	76	161	283
2014	183	69	149	259

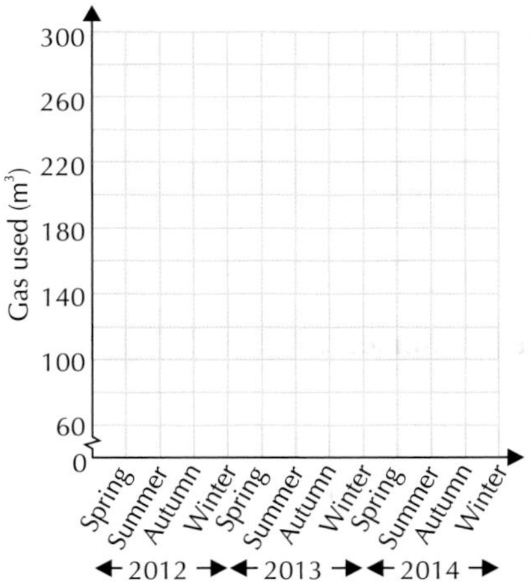

a) Copy the axes on the right, and draw a graph to show how the amount of gas used varied over the three years.

b) Describe and interpret the repeating pattern seen on the graph.

c) Describe the overall trend seen in the data.

Q2 This table shows the highest temperature, in °C, recorded in London and Sydney each month during one year.

Month	Jan	Feb	Mar	Apr	May	Jun	Jul	Aug	Sep	Oct	Nov	Dec
London	7	8	11	13	17	20	23	22	19	15	11	8
Sydney	26	27	25	22	20	17	16	18	19	22	24	25

a) Draw a line graph to show the highest monthly temperature in London during the year.

b) On the same axes, draw a line graph to show the highest monthly temperature in Sydney during the year.

c) Describe the shapes of your two lines. What does this tell you about the way the temperature changes in the two cities during the year?

Q3 The line graph on the right shows the number of pairs of sunglasses sold by a shop over four years.

a) Describe and interpret the repeating pattern seen in the graph.

b) Describe the overall trend shown by the graph.

c) Give a reason why the graph might be misleading.

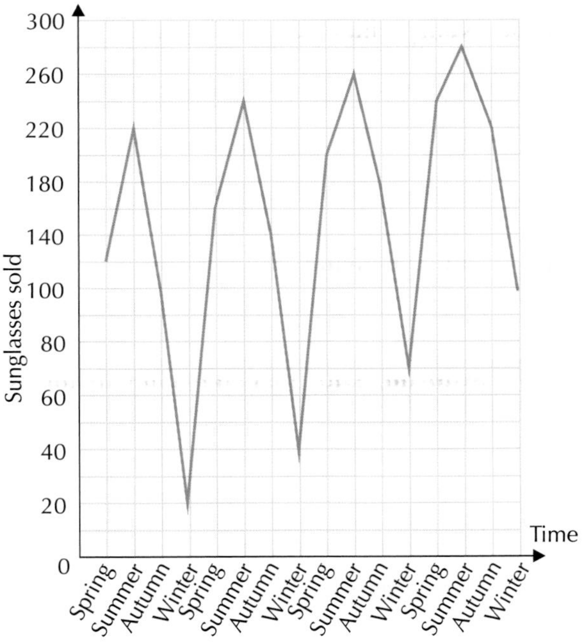

26.8 Scatter Graphs

Scatter graphs show two variables plotted against each other. They are used to show how closely two variables are related to one another — this is known as correlation.

Drawing Scatter Graphs

Learning Objective — Spec Ref S6:
Draw and interpret scatter graphs.

A **scatter graph** shows **two variables** plotted against each other, e.g. height and weight or temperature and BBQ sales. To **draw** a scatter graph:

- Decide **which variable** should go on **which axis** — the one that you think **depends** on the other should go on the **vertical** axis.
- **Plot** your data as points (x, y), where x is the variable on the horizontal axis and y is the variable on the vertical axis.

Example 1

Dougal measured the height and shoe size of 10 people.

Height (cm)	165	159	173	186	176	172	181	169	179	194
Shoe size	6	5	8	9	8.5	7	8	6	8	11

a) Use his results to plot a scatter graph of shoe size against height in cm.

1. Draw and label your axes. Here, height has been plotted along the horizontal axis and shoe size up the vertical axis.
2. Plot each point carefully (don't join them up).

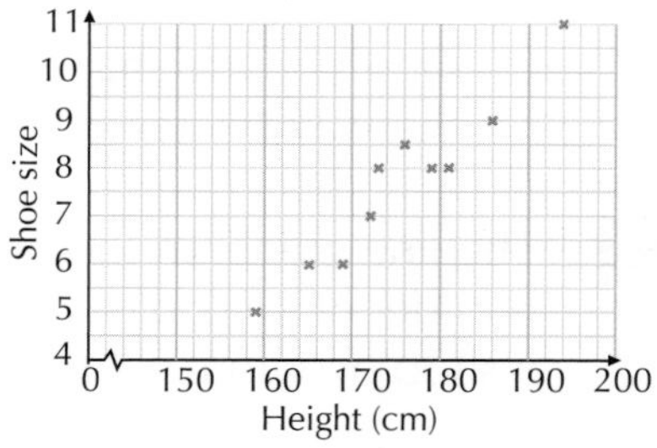

b) What is the shoe size of the smallest person?

Look at the first point on the 'height' scale and read off the value on the 'shoe size' scale.

Smallest height is 159 cm, and the corresponding shoe size is **5**

Exercise 1

Q1 The outside temperature and the number of ice creams sold in a cafe were recorded for six days.

Temperature (°C)	28	25	26	21	23	29
Ice creams sold	30	22	27	5	13	33

Copy the axes on the right, then use the data from the table to plot a scatter graph.

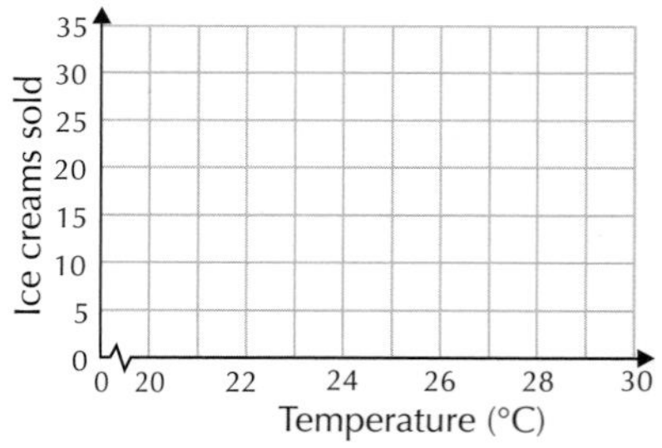

Q2 Ten children of different ages were asked how many baby teeth they still had. Use the results below to plot a scatter graph.

Age (years)	5	6	8	7	9	7	10	6	8	9
Baby teeth	20	17	11	15	7	17	5	19	13	8

Correlation and Lines of Best Fit

Learning Objectives — Spec Ref S6:
- Recognise and describe correlation.
- Draw a line of best fit and use it to estimate and predict data values.
- Be able to recognise outliers.

If two variables are **related** to each other then they are **correlated**.
Variables can be **positively correlated**, **negatively correlated** or **not correlated** at all.

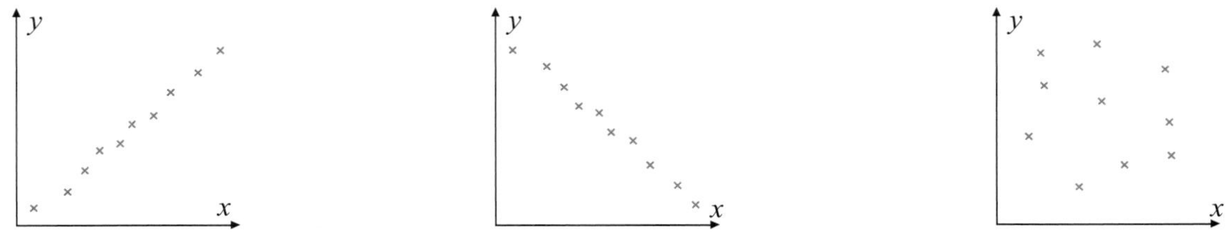

Positive correlation — both variables **increase and decrease together**. The **points** on the scatter graph will look like a line sloping **upward** from left to right.

Negative correlation — as one variable **increases**, the other **decreases**. The **points** on the graph will look like a line sloping **downward** from left to right.

No correlation — there is **no linear relationship** between the variables. The **points** on the graph will look **randomly scattered**.

You can describe the **strength** of the correlation as well. The **closer** the points are to forming a **straight line**, the **stronger** the correlation.

- If most of your points are close to a **straight line**, then you have **strong correlation**.
- If your points are spread **loosely around** a straight line, then you have **moderate correlation**.
- If your points **don't line up** nicely but you can still see that there is a **relationship** between the two variables, then you have **weak correlation**.

Tip: If two variables are correlated it doesn't necessarily mean that one causes the other. There could be a third factor affecting both, or it could just be a coincidence.

Example 2

Describe the strength and type of correlation shown by each of the scatter graphs.

a)

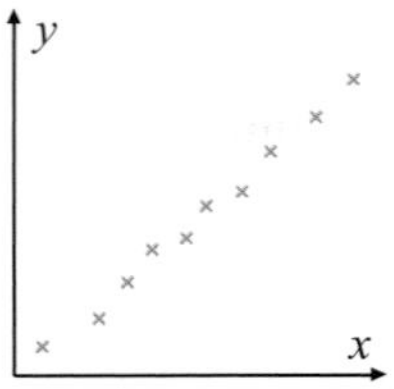

The points form an upward slope fairly close to a straight line, so this is...

Strong positive correlation

b)

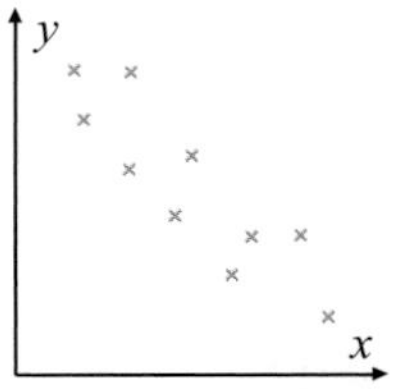

The points form a downward slope loosely around a straight line, so this is...

Moderate negative correlation

c)

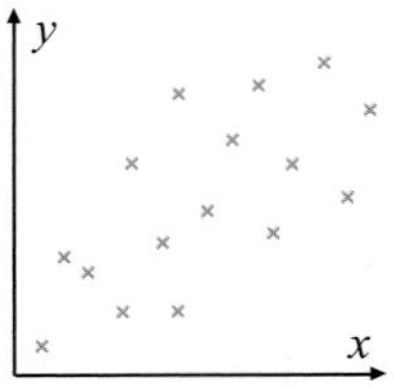

The points form an upward slope, but do not lie close to a straight line, so this is...

Weak positive correlation

Exercise 2

Q1 Describe the strength and type of correlation seen in the scatter graphs below.

a)

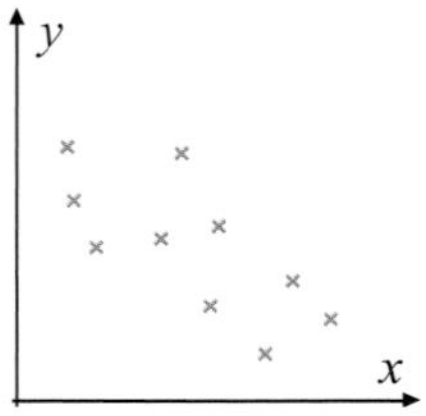

b)

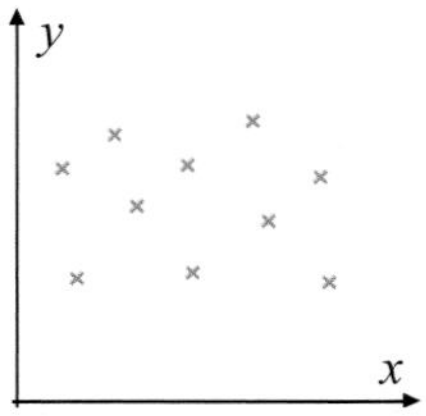

c)

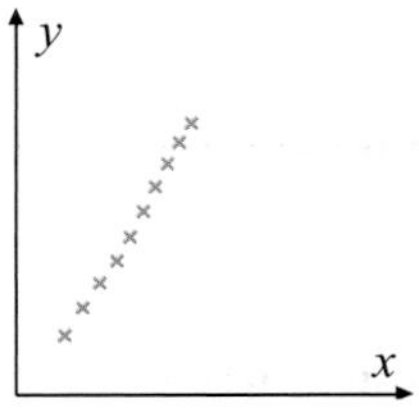

Q2 Are the following pairs of variables likely to show positive correlation, negative correlation or no correlation? Explain your answers.

a) Outside temperature and ice cream sales

b) Outside temperature and hot chocolate sales

c) Outside temperature and bread sales

d) Age of a child and his or her height

e) Speed limit in a street and the average speed of cars as they drive down that street

Q3 Jacob measured the wind speed and boat speed at different times during a sail. The results are shown by the scatter graph to the right.

Describe the relationship between wind speed and boat speed.

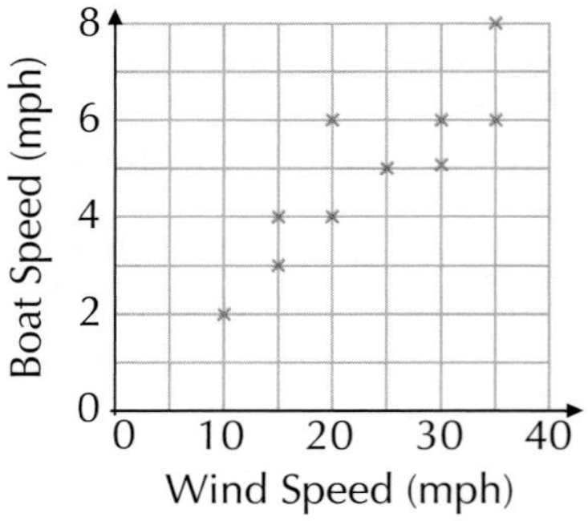

If two variables are correlated, then you can draw a **line of best fit** on their scatter graph. This is a **straight line** that passes through the **middle of the points** with a roughly **equal number** on either side.

Outliers are points that **don't fit** the general pattern of the rest of the data. They can **move** your line of best fit away from other values, so are usually **ignored** when drawing your line. Outliers can sometimes be caused by **errors** in the data, but not always — they can just be **unusually high** or **low values**.

You can use a line of best fit to **predict values** for one variable when you know the value of the other. All you have to do is **draw a line** from the value you're given to the line of best fit, then **read off** the value for the variable on the other axis.

- Predicting values **within** the range of data you have is known as **interpolation**, and should be **fairly reliable**.
- Predicting values **outside** the range of the data is known as **extrapolation**. This can be **unreliable** because you don't know that the pattern continues outside the data range.

Example 3

The scatter graph on the right shows the marks a class of pupils achieved in a Maths test plotted against the marks they achieved in an English test.

a) Draw a line of best fit on the graph.

b) Jimmy was ill on the day of the Maths test. If he scored 75 in his English test, predict what his Maths mark would have been.

c) Elena was ill on the day of the English test. If she scored 35 on her Maths test, predict what her English result would have been.

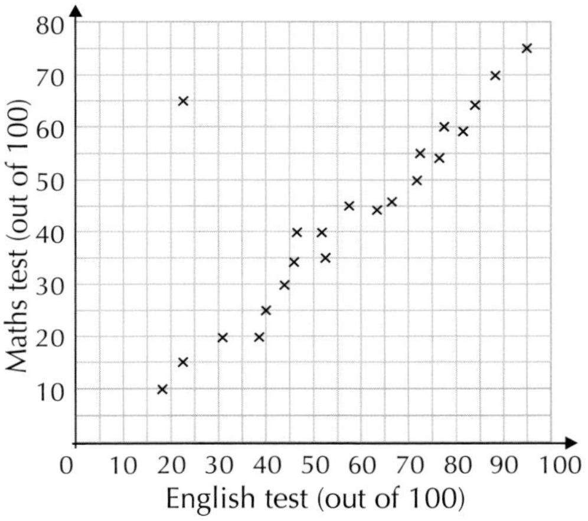

This point is an outlier, so ignore it when you draw your line.

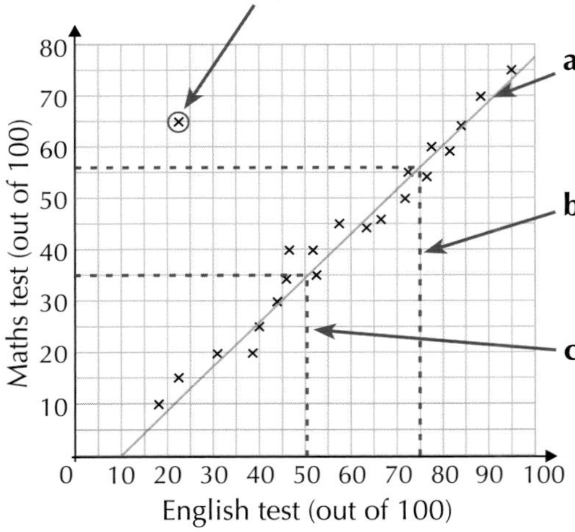

a) Draw a straight line through the middle of the points — have approximately the same number of points on either side of the line.

b) Draw a line up from 75 on the 'English' axis to the line of best fit, then across to the 'Maths' axis.

Predicted Maths mark for Jimmy = **56**

c) Draw a line across from 35 on the 'Maths' axis to the line of best fit, and then down.

Predicted English mark for Elena = **51**

Exercise 3

Q1 Could a line of best fit be drawn on each of these graphs? Explain your answers.

a)

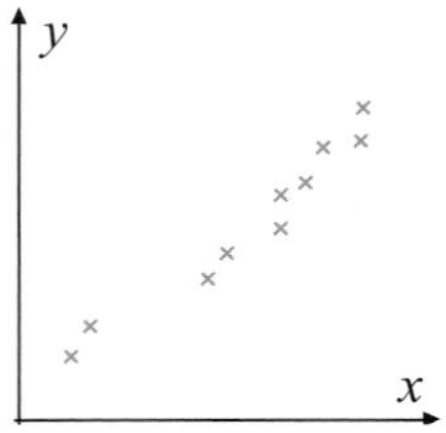

b)

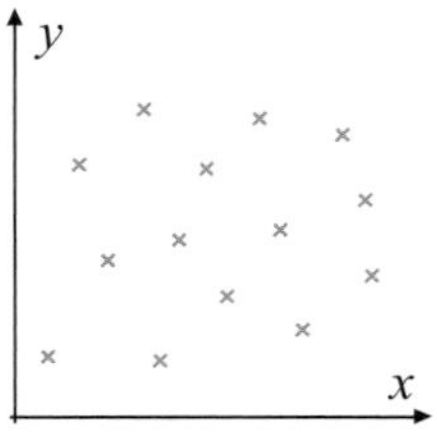

c)

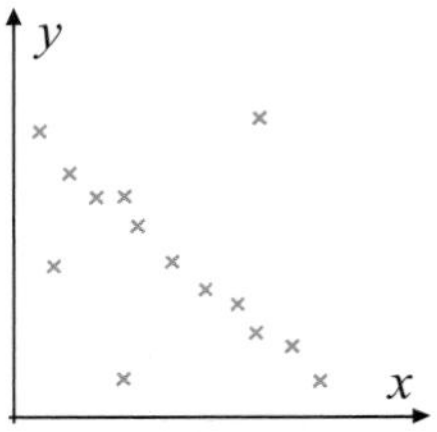

Q2 Pupils in a class were asked how many hours at the weekend they spent doing homework and watching television. The results are shown on this scatter graph.

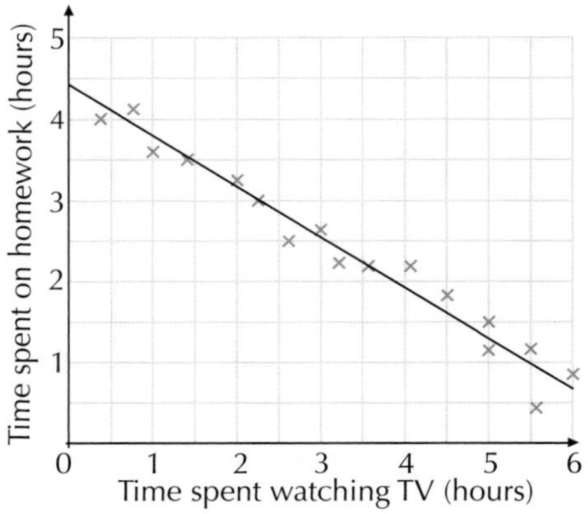

a) Describe the correlation between time spent watching TV and time spent doing homework.

b) Use the line of best fit to predict how long a pupil spends on homework if they watch 5.5 hours of TV at the weekend.

c) Amelia did 3.5 hours of homework at the weekend. Predict how many hours she spent watching TV.

Q3 The graph on the right shows the height of various types of tree plotted against the width of their trunks.

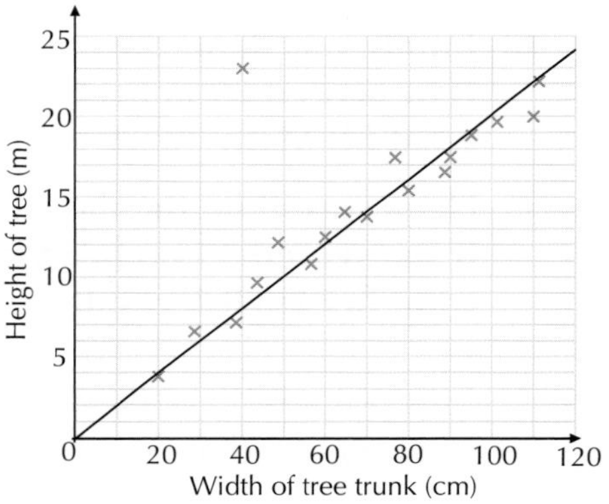

a) Describe the correlation between the width of the trunks and the height of the trees.

b) Give the width and the height of the tree that appears as an outlier on the graph.

c) Use the graph to predict the width of a tree's trunk if it is 13 m tall.

d) Explain whether the prediction that you made in part c) is reliable.

Q4 Anton wants to buy a particular model of car. The table below shows the cost of several of these cars that are for sale, as well as their mileage.

Mileage	5000	20 000	10 000	12 000	5000	25 000	27 000
Cost (£)	3500	2000	3000	2500	3900	1000	500

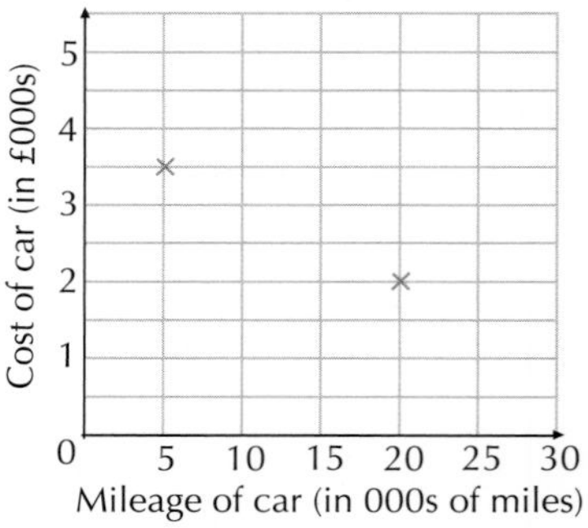

a) The first two points have already been plotted on the scatter graph on the right. Copy and complete the graph.

b) Draw a line of best fit through your points.

c) Thelma has seen a car of this model with a mileage of 50 000. She plans to use the trend shown by the graph to predict the cost of this car. Comment on the reliability of this estimate.

Review Exercise

Q1 Eight students sat two maths tests. Their scores out of 20 are shown below.

Test 1: 19, 14, 8, 17, 17, 20, 6, 13

Test 2: 7, 14, 16, 7, 10, 8, 10, 12

a) Calculate the mean score for each test.

b) Find the range for each test.

c) Use your results from parts a) and b) to compare the distributions of the scores in the two tests.

Q2 Susie buys a bag of 5 apples. The weights of the apples are: PROBLEM SOLVING

57 g, 60 g, 69 g, 72 g, 75 g

After Susie has eaten two of the apples, the mean weight of the three that remain is 62 g. What were the weights of the two apples that she ate?

Q3 This table shows the number of goals scored one week by 18 teams in the premier division.

Number of goals	0	1	2	3	4	5
Number of teams	1	3	4	5	3	2

a) Find the median number of goals.

b) Find the mean number of goals. Give your answer to 1 decimal place.

c) Write down the mode.

d) The mean, median and modal numbers of goals scored in the same week of the previous year were all 2. How do these results compare?

Q4 This table shows the heights (in metres) of 20 people.

a) Complete the frequency column.

b) Write down the modal group.

c) Which group contains the median?

d) Find an estimate for the mean.

e) Estimate the range.

Height in metres, h	**Tally**	**Frequency**
$1.5 < h \le 1.6$	II	
$1.6 < h \le 1.7$	𝍸 IIII	
$1.7 < h \le 1.8$	𝍸 II	
$1.8 < h \le 1.9$	II	

Q5 The vertical line graph shows information on the number of washing machines sold by an electrical shop each day for 40 days.

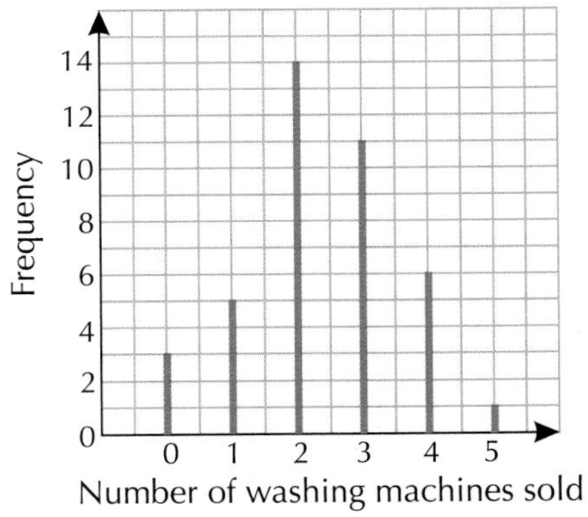

a) What is the median number of washing machines sold each day?

b) What is the modal number sold each day?

c) Calculate the mean number of machines sold each day.

d) Find the range of the number of machines sold each day.

Q6 The children at a youth club were asked to name their favourite flavour of ice cream.
The dual bar chart on the right shows the results.

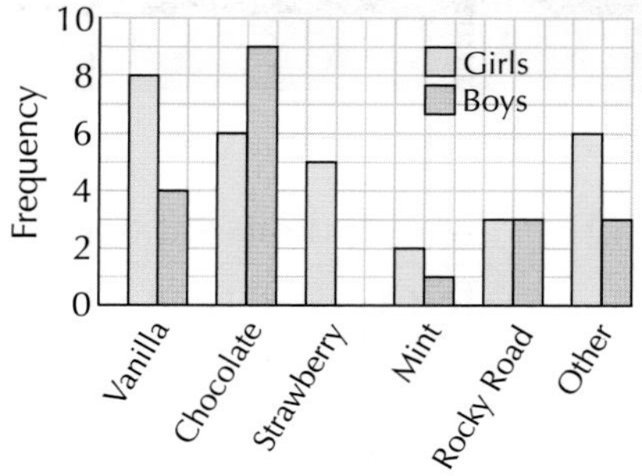

a) How many girls were asked altogether?

b) How many more boys than girls chose chocolate?

c) What is the modal flavour for the girls?

The pie chart on the right shows the same data for the girls.

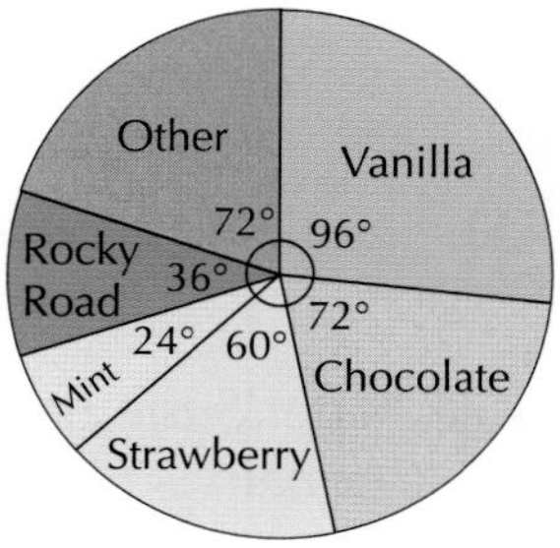

d) Work out the number of degrees used to represent one girl.

e) Draw a pie chart to show the data for the boys.

f) Rocky road was chosen by the same number of girls and boys. Explain why the sectors representing rocky road in each pie chart are different sizes.

Q7 The data below shows the ages of people queuing in a post office at 10 am and at 3 pm.

10 am:	65	48	51	27	29	35	58	51	54	60	59
3 pm:	15	23	32	31	35	22	24	18	27		

a) Draw a back-to-back stem and leaf diagram to show the data.

b) Find and compare the median age of the people queuing at each time.

Q8 The outside temperature and the number of drinks sold by two vending machines were recorded over a 10-day period. The results are shown in this table.

Temperature (°C)	14	29	23	19	22	31	33	18	27	21
Drinks sold from Machine 1	6	24	16	13	15	28	31	13	22	14
Drinks sold from Machine 2	7	25	18	15	17	32	35	14	24	17

For each machine:

a) Draw a scatter graph of drinks sold against temperature.

b) Draw a line of best fit.

c) Predict the number of drinks that would be sold if the temperature was 25 °C.

d) Explain why it might not be appropriate to use your lines of best fit to estimate the number of drinks sold from each machine if the outside temperature was 3 °C.

e) Lucas says, "An increase in temperature causes more vending machine drinks to be sold." Do you agree with Lucas's statement? Explain your answer.

Exam-Style Questions

Q1 Two groups of students in Mr Green's class took the same test. Both groups contained 5 students who each scored a mark that is a whole number.

Mr Green found the median and range of marks for both groups:

Group	Median	Range
A	50	3
B	48	3

Show that the student who scored the lowest mark could be in group A.

[2 marks]

Q2 The number of eggs laid by a farmer's hens on each day in one week were:

22 34 6 28 5 31 29

a) Work out the median for this data.

[1 mark]

b) Explain why the mean would not be a good measure of the average number of eggs laid per day.

[1 mark]

Q3 The amount of time (t minutes) spent on a social media website in one day was recorded for some students. The results are shown in the frequency table.

Time (t minutes)	Frequency
$0 \le t < 40$	3
$40 \le t < 80$	4
$80 \le t < 120$	7
$120 \le t < 160$	15
$160 \le t < 200$	11

a) Write down the modal class.

[1 mark]

b) Calculate an estimate for the mean time spent.

[3 marks]

c) Explain why your answer to b) is only an estimate.

[1 mark]

Q4 Leah draws a pictogram to show how many books people read in a year. She shows 14 books like this:

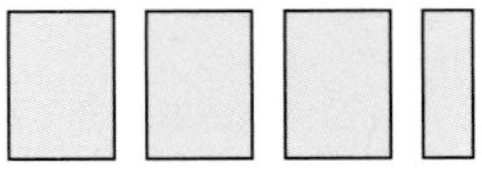

Draw how Leah would show 7 books.

[2 marks]

Q5 Over the last year, some students at a school received gold, silver or bronze Duke of Edinburgh awards. Of the awards received, $\frac{1}{3}$ were silver. Five times as many bronze awards were received than gold awards.

a) Show this information as a pie chart.

[3 marks]

18 students received silver awards

b) Work out the number of students who received bronze awards.

[2 marks]

Q6 At the end of each month for a year, Midas recorded the price of gold per gram to the nearest £0.10. He displayed the data he recorded in the time series graph below.

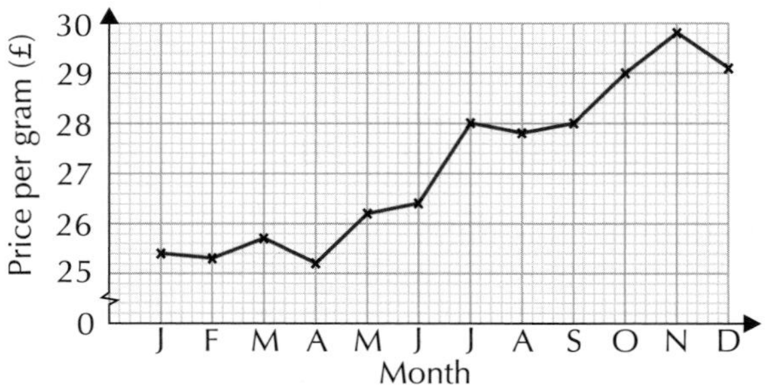

a) Describe the overall trend in gold prices over the course of the year.

[1 mark]

b) Midas owns a 12 kg gold bar. According to the graph, what is the maximum amount of money he could have made from selling the gold bar this year?

[2 marks]

Q7 The scatter diagram shows the percentage obtained in a geography exam and the number of lunchtime revision classes attended by 6 students.

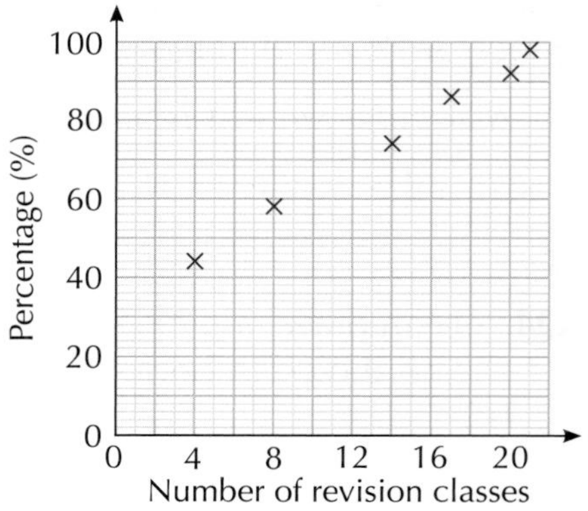

a) Describe the correlation between the percentage obtained and the number of revision classes attended.

[1 mark]

b) Duane was due to sit the exam but was absent. He says:

"I didn't attend any revision classes but I can see from the diagram that I would have got over 20%."

Is Duane correct? Refer to the diagram to explain your answer.

[1 mark]

27.1 Probability — The Basics

Probability is about how likely an event is to happen. For example, if you flip a coin it's just as likely to land on heads as it is tails. You would say the likelihood of each outcome is equal.

Likelihood

Learning Objective — Spec Ref P3:
Understand that probability measures how likely an event is to happen.

The **probability** of any event happening is between **impossible** (definitely won't happen) and **certain** (definitely will happen). Events can be put on a **probability scale** like this:

Example 1

One of these four cards is chosen at random. → 1 2 3 4

Mark each of statements A-D in the correct place on the scale below.

A — the card is 4 or less — All the cards are 4 or less.

B — the card is an odd number — Half the cards are odd numbers.

C — the card is a 1 — There's one 1, so it's not impossible, but it's less than 50/50.

D — the card is a 5 — None of the cards is a 5.

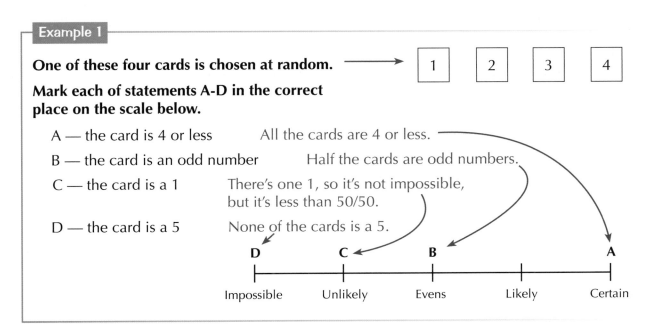

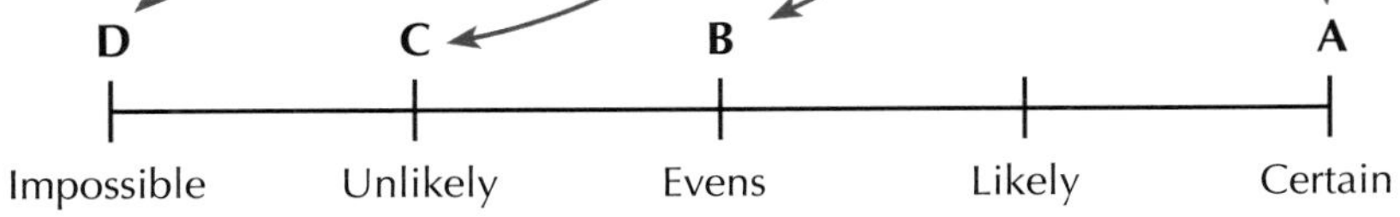

Exercise 1

Q1 Choose from the words *impossible, unlikely, evens, likely* and *certain* to describe:

a) Tossing a coin and getting tails.

b) A person growing to be 10 metres tall.

c) Rolling an even number on a fair dice.

d) Rolling 1 or more on a fair dice.

e) Picking a red card from a pack of 52.

f) Spinning 2 on a fair spinner labelled 1-4.

Q2 One card is picked at random from eight cards numbered 1 to 8.
Make a copy of this probability scale and add arrows to show the probability of each of the events described below. Each arrow should match one of the words.

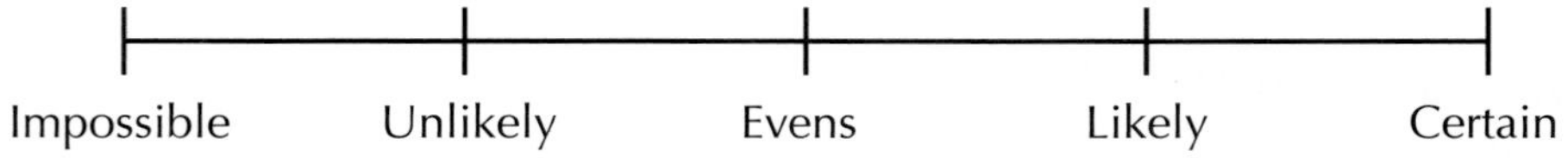

a) The card is less than 9.

b) The card is an odd number.

c) The card is greater than 2.

d) The card is greater than or equal to 7.

e) The card is 6 or less.

f) The card is a zero.

Q3 These eight cards are placed face down on a table and one is selected at random.

P	A	R	A	L	L	E	L

a) Which letter is twice as likely to be on the selected card as the letter P?

b) Which letter is three times as likely to be on the selected card as the letter E?

c) The three arrows below show the probability of selecting each of the letters P, A, R, L and E. Match each letter to one of the arrows. (You can use each arrow for more than one letter.)

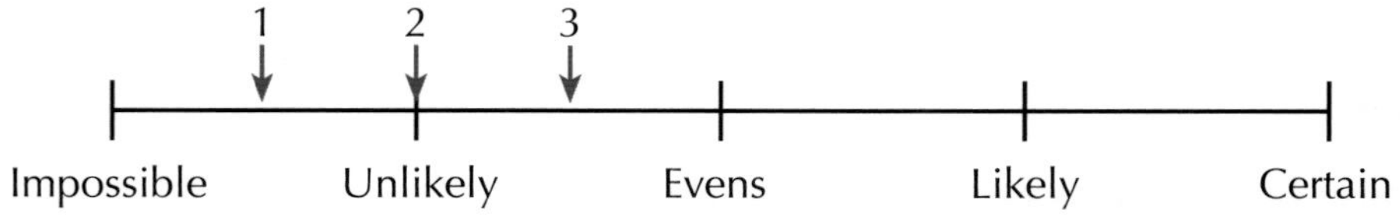

Writing Probabilities as Numbers

Learning Objective — Spec Ref P3:
Be able to write probabilities as fractions, decimals and percentages.

Prior Knowledge Check:
Be able to convert between fractions, decimals and percentages. See p.108.

All probabilities can be written as a **number between 0 and 1**.
An event that's **impossible** has a probability of **0**
and an event that's **certain** has a probability of **1**.

So, using **fractions**, the probability scale looks like this.

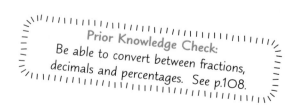

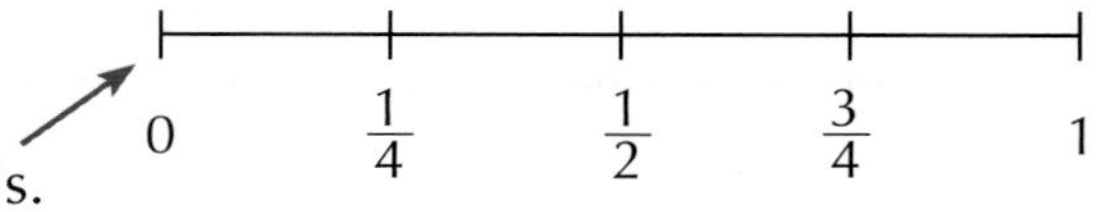

Probabilities can also be written as **decimals** and **percentages**.
For instance, if the probability of something happening is 1 out of 4, it can be written as $\frac{1}{4}$, or as the decimal 0.25, or the percentage 25%.

Example 2

A normal, fair, six-sided dice is rolled. On the scale below, the probabilities of three possible results are labelled by A, B and C.

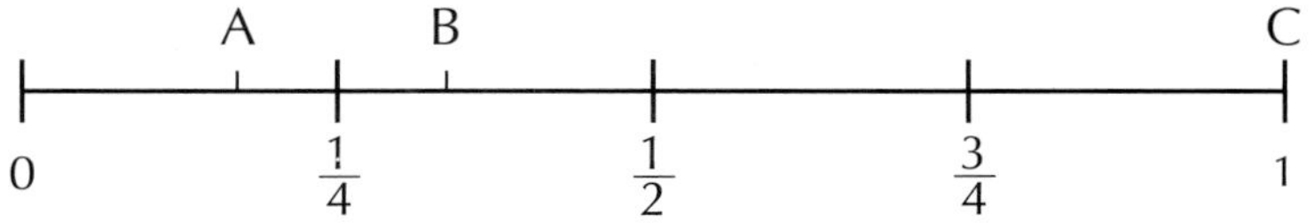

Match each of the results below to the correct letter on the scale.

a) 6 or less is rolled

All possible rolls are 6 or less, so this is certain. **C**

b) 5 is rolled

There's one 5 but six numbers altogether. So the probability of rolling 5 is a sixth of the way along the scale. **A**

c) 1 or 2 is rolled

There are two possibilities for this, meaning it's twice as likely as rolling a 1. This means the probability is twice as big. **B**

Tip: If the dice wasn't fair, the probabilities of rolling each number wouldn't be equal.

Exercise 2

Q1 Match each letter on this scale to the correct probability.

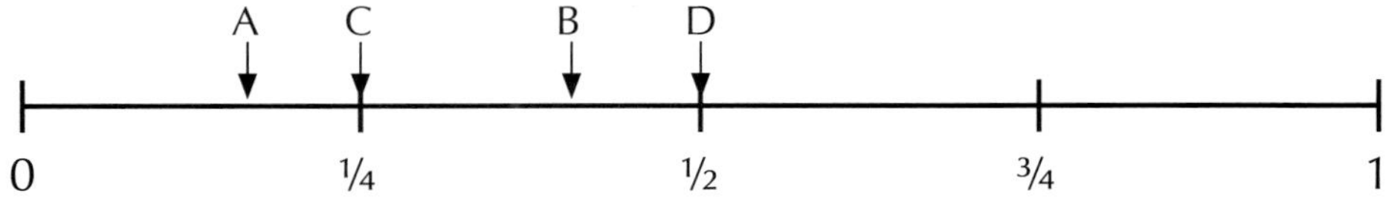

a) 0.5 b) 25% c) $\frac{1}{6}$ d) 0.4

Q2 Match each letter on this probability scale to one of the events below.

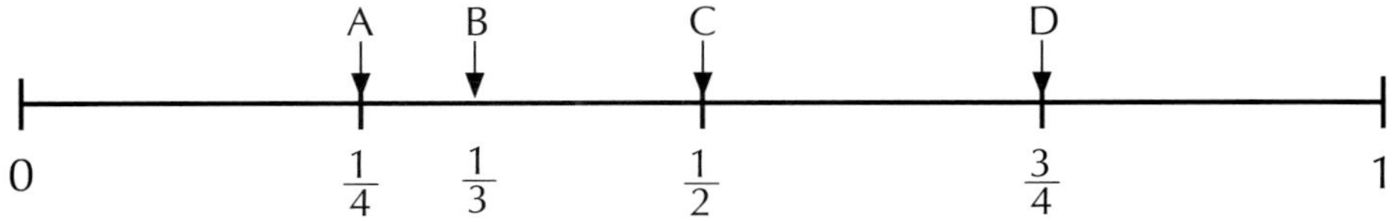

a) Rolling an odd number on a normal fair dice.
b) Selecting a card at random from a standard pack of 52 cards and getting a diamond.
c) Selecting a card at random from a standard pack of 52 cards and not getting a diamond.
d) Spinning blue on a fair, three-sided spinner with 1 blue section and 2 orange sections.

27.2 Calculating Probabilities

If you know (or can work out) the total number of possible outcomes of an activity, you can find the probability of any given event, assuming the outcomes are equally likely.

Learning Objective — Spec Ref P2/P3:
Find the probability of a given event.

Prior Knowledge Check:
Be able to simplify fractions.
See Section 5.

In **probability**, an **outcome** is the result of an activity (e.g. 'getting tails on a coin flip' or 'rolling 3 on a dice') and an **event** is an outcome or a set of outcomes that you find a **probability** for.

The probability of something happening depends on the **total number of possible outcomes**. When all possible outcomes are **equally likely**, you can find the probabilities using the formula:

$$\textbf{Probability of event} = \frac{\textbf{number of ways the event can happen}}{\textbf{total number of possible outcomes}}$$

Tip: You should always simplify fractions if you can.

The probability of **event A** happening is often written as **P(A)**.

Example 1

A box contains 20 coloured counters numbered 1 to 20.
11 of the counters are blue, 7 of the counters are purple and 2 of the counters are yellow.
If one counter is selected at random, work out the following probabilities:

a) The counter is the number 12.

1. Find the total number of outcomes. There are 20 different counters which can be picked. — Total outcomes = 20
2. Count the number of ways that you could get a 12. Only one counter has the number 12. — There's 1 way that the counter is 12.
3. Put the numbers into the formula. $P(12) = \frac{\text{number of ways the event can happen}}{\text{total number of possible outcomes}} = \mathbf{\frac{1}{20}}$

b) The counter is yellow.

1. Count the number of outcomes that are yellow. There are 2 yellow counters. — There are 2 ways of getting yellow.
2. Put the numbers into the formula. (You already know there are 20 outcomes in total.) $P(\text{yellow}) = \frac{2}{20} = \mathbf{\frac{1}{10}}$

c) The counter is either blue or yellow.

1. Count how many counters are either blue or yellow. — 11 blue + 2 yellow = 13 counters
2. Put the numbers into the formula. (You already know there are 20 outcomes in total.) $P(\text{blue or yellow}) = \mathbf{\frac{13}{20}}$

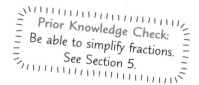

Exercise 1

Q1 State the total number of possible outcomes in each of the following situations.

a) A coin is tossed.

b) One card is selected from a pack of 52.

c) A ten-sided dice is rolled.

d) A fair spinner with 8 sections is spun.

e) One day of the week is chosen at random.

f) One day of the year is chosen at random.

Q2 This fair spinner is spun once.

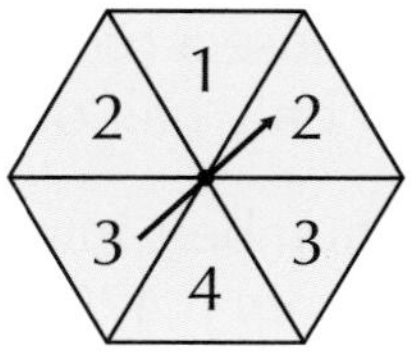

a) What is the total number of possible outcomes?

b) What is the probability of spinning a 1?

c) What is the probability of spinning a 3?

Q3 Calculate the probability of rolling a fair, six-sided dice and getting each of the following:

a) 6 b) 2 c) 7

d) 4 or 5 e) a multiple of 3 f) a factor of 6

Q4 Nine cards numbered 1 to 9 are face down on a table. If one of the cards is selected at random, find the probability of selecting each of the following:

a) card 4 b) an even number c) a number less than 6

Q5 A fair spinner has 12 equal sections. 5 are yellow, 3 are green and the rest are purple. Find the probability that the spinner lands on the following colours:

a) green b) purple c) yellow or green d) not green

Q6 A bag contains some coloured balls — 2 black, 4 blue, 2 green, 3 red, 2 yellow, 1 orange, 1 brown and 1 purple. If a ball is selected at random, find the probabilities that it will be:

a) green b) red c) orange d) black

e) blue or green f) red, green or brown g) not purple h) white

Q7 A standard pack of 52 playing cards is shuffled and one card is selected at random. Find the probability of selecting each of the following:

a) a club b) an ace c) a red card

d) the two of hearts e) not a spade f) a 4 or a 5

Q8 For each of the following, draw a copy of the spinner on the right and number the sections so that the spinner fits the rules.

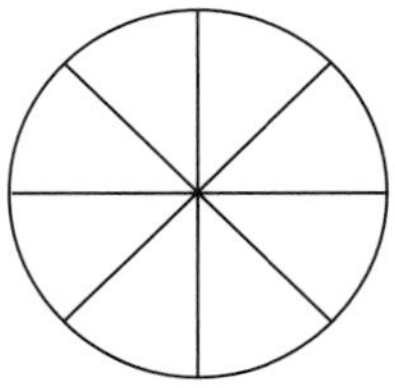

a) The probability of getting 2 is $\frac{3}{8}$.

b) The probability of getting 3 is $\frac{1}{2}$.

c) The probability of getting 5 and the probability of getting 6 are both $\frac{1}{4}$.

Example 2

Maya has 4 cats, 2 dogs and 10 chickens. Half of her cats, dogs and chickens are brown. She decides to give a treat to one animal at random.

Find the probability that she chooses a brown cat.

1. Find the total number of outcomes. — 4 cats + 2 dogs + 10 chickens = 16 outcomes
2. Count the number of ways Maya could choose a brown cat. She has 4 cats, but only half of them are brown. — Number of brown cats = 4 ÷ 2 = 2
3. Put the numbers into the formula from p.470. — P(brown cat) = $\frac{2}{16} = \mathbf{\frac{1}{8}}$

Exercise 2

Q1 Lewis was born in 2004 (a leap year). A friend randomly tries to guess Lewis's exact birthday. What is the probability that he guesses:

a) the correct month? b) the exact date?

Q2 At a summer fair, 100 raffle tickets are sold. Each ticket is bought by a different person. The winning number is drawn at random.

a) What is the probability that the first person to buy a ticket wins the first prize drawn?

b) What is the probability that the last person to buy a ticket wins the first prize drawn?

Q3 Tove has 20 different pairs of socks and has picked 1 sock at random. If she then picks another sock at random from the remaining socks, what is the probability that the 2 socks make a pair?

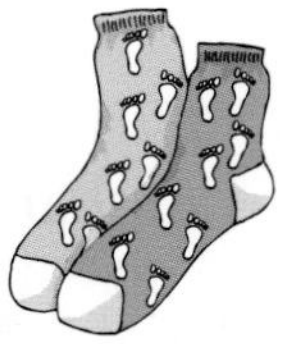

Q4 A box of chocolates contains 8 caramels, 6 truffles and 4 pralines. Half of each type of chocolate are fully coated in milk chocolate and half are fully coated in white chocolate. All the chocolates are individually wrapped in identical paper.

Chelsea selects a chocolate at random. She doesn't like pralines or white chocolate.

a) What is the probability that she gets a white-chocolate-coated praline?

b) (i) How many chocolates are neither praline nor coated in white chocolate?

(ii) What is the probability that Chelsea gets a chocolate she likes?

Q5 Explain why the following statement is false.

"When a football team plays a match there are 3 possible outcomes for the team — win, draw and lose. So the probability that they win the match is always $\frac{1}{3}$."

Mutually Exclusive Events

Learning Objective — Spec Ref P4:
Know that the probabilities of mutually exclusive events add up to 1.

Events that **can't happen at the same time** are called **mutually exclusive**. For example, a coin landing on **heads** and a coin landing on **tails** are **mutually exclusive events** — if the coin lands on heads, it can't also land on tails in the same coin flip.

The probabilities of **mutually exclusive events** covering **all possible outcomes** always **add up to 1**.

Tip: Be careful — this only works for mutually exclusive events.

For any event, there are only **two possibilities** — it either **happens** or it **doesn't happen**. These two facts can be written as the **rule**:

Probability something doesn't happen = 1 – Probability it does happen

Example 3

Jon gets the train to work every day. The probability that his train is late on any day is 0.05.

a) Explain whether the events 'train is late' and 'train is not late' are mutually exclusive.

The train can't be both late and not late, so the events **are mutually exclusive**.

b) Work out the probability that Jon's train is not late.

The train is either late or not late, so you can use the rule above.

P(not late) = 1 – P(late) = 1 – 0.05 = **0.95**

Exercise 3

Q1 A six-sided dice, numbered 1-6, is rolled. Here are three possible events:

A — a 5 is rolled, B — a number less than 3 is rolled, C — an odd number is rolled

Say whether the following pairs of events are mutually exclusive:

a) A and B b) A and C c) B and C

Q2 Charlie chooses one month of the year at random. Here are four possible events:

A — the name begins with J B — the name begins with M
C — the name ends in Y D — it's one of the first four months

Say whether the following pairs of events are mutually exclusive:

a) A and B b) A and C c) A and D d) B and C

Q3 The probability that Aasir's school bus is late is 0.2. What's the probability that it's not late?

Q4 The probability that it will snow in a particular Canadian town on any particular day in February is $\frac{5}{8}$. What is the probability that it won't snow there on 6th February?

Q5 In a class, the probability that a randomly selected pupil is a boy is 0.45, and the probability that a pupil has blond hair is 0.2.
Find the probability that a randomly selected pupil:

a) is not a boy

b) doesn't have blond hair

Activities with mutually exclusive events can have **more than two outcomes**. For example, rolling a dice has six possible outcomes, and each outcome is a **mutually exclusive event** — you can't roll both a 1 and a 3 in the same dice roll.

You can use the fact that the **probabilities** of mutually exclusive events **add up to 1** in problems with more than two outcomes. To find the probability that one thing happens, just **subtract** the probabilities of all the other things happening from **1**.

Example 4

A bag contains red, green, blue and white counters. The table below shows the probabilities of randomly selecting a red, green or white counter.

What is the probability of selecting a blue counter?

Colour	Red	Green	Blue	White
Probability	0.2	0.1		0.5

1. These are mutually exclusive events, so the probabilities must add up to 1.
2. This means the probability of selecting blue is 1 minus the probability of selecting red, green or white.

$$\text{P(blue)} = 1 - (0.2 + 0.1 + 0.5)$$
$$= 1 - 0.8 = \mathbf{0.2}$$

Exercise 4

Q1 This table shows the probabilities of getting one of the five possible colours on a spinner.

a) Find the missing probability.

b) Find the probability of spinning 'not pink'.

Colour	Red	Blue	Green	Pink	Black
Probability		0.2	0.1	0.1	0.3

Q2 A bag contains some equal-sized discs. The discs are either yellow, orange or red. If Jack takes out one disc without looking, the probability that it's yellow is $\frac{1}{4}$ and the probability that it's orange is $\frac{3}{8}$. Work out the probability that Jack takes a red disc.

Q3 One counter is selected at random from a box containing blue, green and red counters. The probability that it's a blue counter is 0.5 and the probability that it's a green counter is 0.4. If there are 4 red counters in the box, how many counters are there altogether?

27.3 Listing Outcomes

One of the trickiest bits of probability is figuring out what all the possible outcomes are, especially if you've got more than one activity. But don't worry, this topic tells you how to do it all.

Learning Objectives — Spec Ref P6/P7:
- List outcomes of two or more events.
- Use sample space diagrams to list outcomes of more complicated situations.
- Use sample space diagrams to find probabilities.

When **two things** are happening at once, e.g. a coin toss and a dice roll, it's much easier to work out probabilities if you first **list all the possible outcomes** in a systematic way so you don't miss any. For example, a coin toss has two outcomes (heads 'H' and tails 'T'), and a dice roll has 6 outcomes ('1', '2', '3', '4', '5' and '6'), so if both happen together, there will be **12 possible outcomes**.

You can record all the possible outcomes using a **sample space diagram** (also called a **possibility diagram**). A sample space diagram can take the form of a **list** of outcomes — and it can be useful to put those in a **table**. This sets them out in an **ordered** and **logical** way.

Tip: In a simple table, each row is a mutually exclusive outcome.

Example 1

Anna has three tickets for a theme park. She chooses two friends at random to go with her. She chooses one girl from Bea, Claire and Daisy, and one boy from Ethan and Fatik.

a) List all the possible combinations of friends she could choose.

1. Make a sample space diagram — a simple table works here. Have 1 column for the girls and 1 column for the boys.
2. Write in the first girl and then fill in all the possibilities for the boys.
3. Repeat for the other 2 girls.
4. Each row of the table is a possible outcome.

Girls	Boys
Bea	Ethan
Bea	Fatik
Claire	Ethan
Claire	Fatik
Daisy	Ethan
Daisy	Fatik

b) What is the probability that Anna chooses Claire to go with her?

1. Count the number of rows that Claire's name appears in.
   There are 2 rows with outcomes that include Claire.
2. The total number of rows is the total number of outcomes.
   There are 6 rows in total.
3. Divide the two numbers to find the probability.
   P(Claire) = $\frac{2}{6} = \mathbf{\frac{1}{3}}$

Exercise 1

Q1 Use a sample space diagram to list all the possible outcomes when:

a) two coins are tossed b) a standard six-sided dice is rolled and a coin is tossed

Q2 A bag contains two balls — one green, one blue. A ball is picked at random and replaced. Then a second ball is picked at random. List all the possible combinations of colours.

Q3 A burger bar offers the meal deal shown on the right.

Choose 1 burger and 1 drink	
Burgers	***Drinks***
Hamburger	Cola
Cheeseburger	Lemonade
Veggie burger	Coffee

a) List all the different combinations available.

Jana picks one combination at random.

b) What is the probability she chooses a veggie burger and cola?

c) What is the probability she chooses a cheeseburger?

Q4 The fair spinner on the right is spun twice.

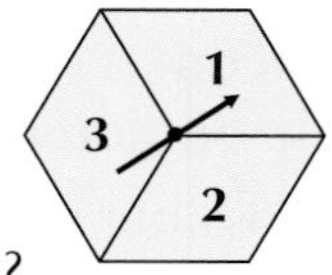

a) Copy and complete the table to list all the possible combinations of scores.

1st spin	2nd spin
1	1
1	2
⋮	⋮

b) What is the probability of spinning 3 on each spin?

c) What is the probability of getting a total of 4 or more over the two spins?

Q5 A fair coin is tossed three times.

a) List all the possible outcomes in a sample space diagram.

b) Work out the probability of getting: (i) three tails (ii) one head and two tails

When two activities have **lots of possible outcomes**, you need more than just a simple table to show all the different possibilities. A table with the outcomes of one activity **along the top** and the outcomes of another activity **down the side** is a good way of doing this.

Example 2

A white four-sided dice and a blue four-sided dice, both numbered 1-4, are rolled together.

a) Draw a sample space diagram to show all the possible total scores.

1. Put the outcomes for one dice across the top and those for the second dice down the side.
2. Fill in each square with the score for the row and the score for the column added together.

		White dice			
		1	**2**	**3**	**4**
Blue dice	**1**	2	3	4	5
	2	3	4	5	6
	3	4	5	6	7
	4	5	6	7	8

b) If both dice are fair, what is the probability of scoring a total of 4?

Count how many times a total of 4 appears in the table and then divide by the total number of outcomes.

4 appears 3 times in the table, and there are 16 outcomes in total, so P(total of 4) = $\mathbf{\frac{3}{16}}$

Exercise 2

Q1 A coin is tossed and a six-sided dice is rolled.
Copy and complete the sample space diagram below to show all the possible outcomes.

	1	2	3	4	5	6
H	H1	H2				
T						T6

Q2 Two fair, six-sided dice are rolled. A sample space diagram for the results is below.

a) Copy and complete the table to show all the possible total scores.

b) How many possible outcomes are there?

c) Find the probability of each of the following total scores.

(i) 6 (ii) less than 8

(iii) more than 8 (iv) an even number

	1	2	3	4	5	6
1						
2						
3						
4						
5						
6						

Q3 A bag contains 3 balls — 1 blue, 1 green and 1 yellow.
A second bag contains 4 balls — 1 blue, 2 green and 1 yellow.
One ball is taken at random from each bag.

a) Copy and complete this sample space diagram to show all the possible outcomes.

b) Use the table to find the probability of selecting:

(i) 2 blue balls (ii) 2 green balls

(iii) 2 balls the same colour (iv) at least 1 yellow ball

	B	G	G	Y
B				
G				
Y				

Q4 Tom rolls a fair, six-sided dice and spins a fair spinner with four sections — A, B, C and D.

a) Draw a sample space diagram to show all the possible outcomes.

b) Use your diagram to find the probability that Tom gets each of the following.

(i) C and 5 (ii) B and less than 3 (iii) A or B and more than 4

Q5 Hayley and Asha are playing a game.
In each round they both spin a fair spinner with five sections labelled 1 to 5.

a) Copy and complete this sample space diagram to show all the possible outcomes for their spins.

b) What is the probability that Hayley gets a higher score than Asha in a round?

		Asha's score (2nd number)				
		1	**2**	**3**	**4**	**5**
Hayley's score (1st number)	**1**	1, 1	1, 2			
	2	2, 1				
	3	3, 1				
	4					
	5					

27.4 Probability from Experiments

So far this section has been about theoretical probabilities, but in reality it's difficult to be sure if outcomes are equally likely. The best way to find out is to run some experiments...

Estimating Probabilities

Learning Objective — Spec Ref P1/P5:
Estimate the probability of an event using relative frequency.

You can **estimate** probabilities using the **results** of an experiment or what you know has already happened. Your estimate is called a **relative frequency** (or an **experimental probability**). You can work out relative frequency using this formula:

$$\text{Relative Frequency} = \frac{\textbf{number of times the result has happened}}{\textbf{number of times the experiment has been carried out}}$$

The **more times** you do the experiment, the **more accurate** the estimate should be — i.e. the experimental probability gets **closer** to the theoretical probability. E.g. if you tossed a fair coin **10 times**, you might get 7 heads, so you'd end up with a relative frequency of **0.7**. But if you tossed the coin **100 times**, the relative frequency for heads should end up much **closer to 0.5**.

Example 1

A biased dice is rolled 100 times. Here are the results.

Score	1	2	3	4	5	6
Frequency	11	14	27	15	17	16

a) Estimate the probability of rolling a 1.

1. Look at the table to find the number of times 1 was rolled. — 1 was rolled 11 times.
2. Divide by the total number of rolls. — $P(1) = \frac{\mathbf{11}}{\mathbf{100}}$

b) Estimate the probability of rolling a 3.

1. Look at the table to find the number of times 3 was rolled. — 3 was rolled 27 times.
2. Divide by the total number of rolls. — $P(3) = \frac{\mathbf{27}}{\mathbf{100}}$

Exercise 1

Q1 Kano wants to estimate the probability that a drawing pin lands with its point up when dropped. He drops the drawing pin 50 times and finds that it lands point-up 17 times.

a) Use these results to estimate the probability the drawing pin lands with its point up.

b) Estimate the probability that the drawing pin doesn't land with its point up.

Q2 A spinner with four sections is spun 100 times. The results are shown in the table on the right.

Colour	Red	Green	Yellow	Blue
Frequency	49	34	8	9

a) Find the relative frequency of each colour.

b) Sam uses these relative frequencies to estimate the probability of spinning each colour. How could these estimates be made more accurate?

Q3 Stacy rolls a six-sided dice 50 times and 2 comes up 13 times. Jamal rolls the same dice 100 times and 2 comes up 18 times.

a) Use Stacy's results to estimate the probability of rolling a 2 on this dice.

b) Use Jamal's results to estimate the probability of rolling a 2 on this dice.

c) Whose estimate should be more accurate and why?

Q4 Suki wants to know how likely it is that her school bus will arrive on time. She keeps a record for a month and finds that the bus is on time 20 times out of 24 days.

Estimate the probability that the next time Suki gets the bus it will arrive on time.

Q5 George has burnt 12 of the last 20 cakes he's baked. Estimate the probability that the next cake he bakes won't be burnt.

Q6 Lilia wants to estimate the probability that the football team she supports will win a match. Describe how she could do this.

Expected Frequency

Learning Objective — Spec Ref P2/P3:
Find the expected frequency of an event.

If you know (or have an estimate of) the **probability** of an event, you can work out how many times you would **expect** the event to happen in a given number of experiments using the equation:

Expected frequency = number of times the experiment is done × probability of the event happening

Tip: This isn't always going to happen in reality — it's an estimate of how many times you think it will occur.

Example 2

The probability of a biased dice landing on 1 is 0.3.
How many times would you expect to roll a 1 if you rolled the dice 50 times?

Multiply the number of rolls by the probability of rolling a 1.

$50 \times 0.3 =$ **15 times**

Exercise 2

Q1 The probability that a biased dice lands on 4 is 0.75.
How many times would you expect the dice to land on 4 if it's rolled:

a) 20 times? b) 60 times? c) 100 times? d) 1000 times?

Q2 A fair, six-sided dice is rolled 120 times. How many times would you expect to roll:

a) a 5? b) a 6? c) an even number? d) higher than 1?

Q3 The spinner on the right has 3 equal sections.
How many times would you expect to spin 'penguin' in:

a) 60 spins? b) 300 spins? c) 480 spins?

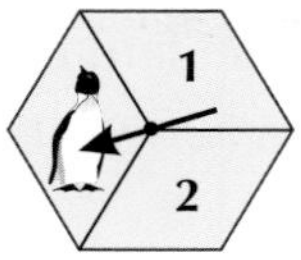

Q4 60% of the people who buy bread from a bakery choose brown bread. If 20 people buy bread there tomorrow, how many of them would you expect to choose brown bread?

Frequency Trees

Learning Objective — Spec Ref P1:
Use frequency trees to display outcomes and find probabilities.

When an experiment has **two or more steps**, you can record the results in a **frequency tree**. The **branches** of a frequency tree show the **different possible outcomes** of each event. The **number** at the end of each branch shows **how many times** that event or **combination** of events happened.

Tip: The numbers at the end of a set of branches should always add up to the number at the start of the branches.

Example 3

James asks 200 people of various ages to roll a dice. There were 120 people under the age of 20 and 16 of them rolled a six. Of the people aged 20 and over, 18 rolled a six.

a) Fill in the frequency tree to show this information.

1. Fill in the bits of the tree given in the question first.
2. Work out the remaining numbers and fill them in.

200 – 120 = 80 people are aged 20 and over
120 – 16 = 104 people under 20 didn't roll a six
80 – 18 = 62 people aged 20 and over didn't roll a six

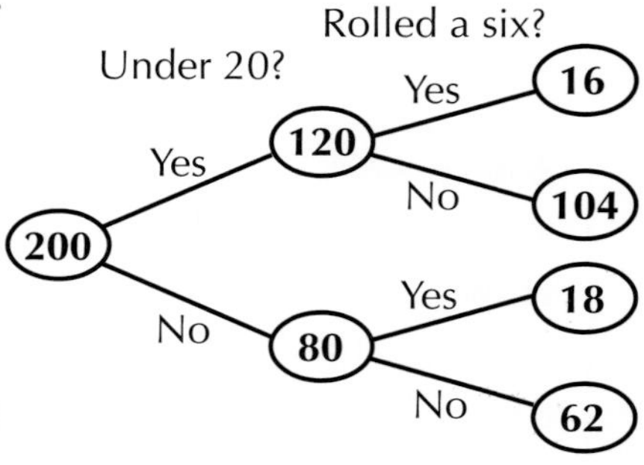

b) What is the probability that a randomly selected roll was a six rolled by someone under 20?

1. Follow the branches of the tree to find the frequency of people under 20 who rolled a six. — 16 people under 20 rolled a six
2. Divide the frequency by the total number of people. — probability = $\frac{16}{200} = \mathbf{\frac{2}{25}}$

Exercise 3

Q1 Copy and complete the frequency trees below:

a)

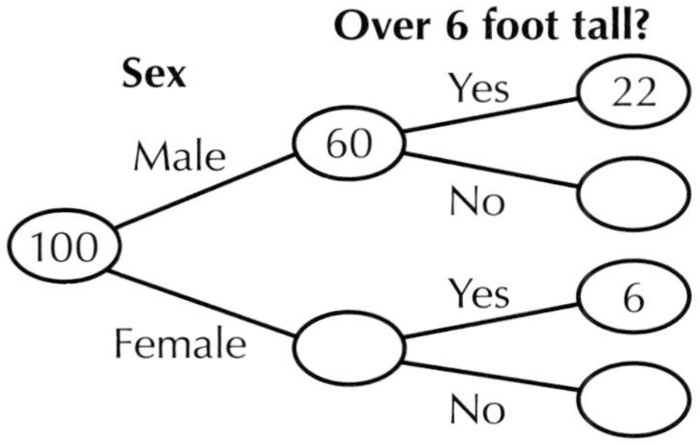

b)

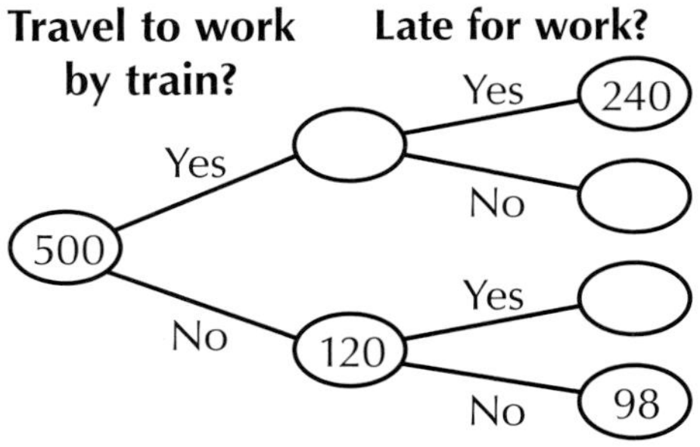

Q2 720 people took an eye test. 640 said their vision was fine.
180 of the people who said their vision was fine failed the eye test.
30 of the people who said their vision wasn't fine passed the eye test.

a) Draw and complete a frequency tree to represent this information.

b) A person who said their vision wasn't fine is selected at random. What is the probability that they failed the eye test?

c) Use your frequency tree to estimate the probability that a randomly selected person says their vision is fine and passes the eye test. Give your answer as a fraction in its simplest form.

Fair or Biased?

Learning Objective — Spec Ref P2:
Use experimental data to decide if outcomes are fair or biased.

Prior Knowledge Check: Be able to find relative frequencies. See p.478.

Things like dice and spinners are **fair** if they are **equally likely** to land on any side or section. If they're more likely to give some outcomes than others, they're called **biased**. To decide whether something is fair or biased, you need to do an **experiment**. Then you can **compare** the experimental results with what you would expect in theory (the **theoretical probability**).

For example, if you flip a coin, you expect it to land on heads about half the time — so if you flipped it 100 times, you'd expect **about** 50 heads. If you actually got 80 heads, you could be pretty sure the coin was **biased**.

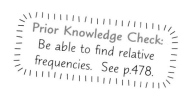

Example 4

Amir thinks his dice is biased. He rolls it 60 times and records the results shown in the table on the right.

Score	1	2	3	4	5	6
Frequency	12	3	9	10	14	12

a) Work out the relative frequencies of each score.

1. For each, work out frequency ÷ total number of rolls.
2. Write the probabilities as decimals (to 2 d.p. where necessary) so they're easier to compare.

$P(1) = \frac{12}{60} = \mathbf{0.20}$ $P(2) = \frac{3}{60} = \mathbf{0.05}$

$P(3) = \frac{9}{60} = \mathbf{0.15}$ $P(4) = \frac{10}{60} = \mathbf{0.17}$

$P(5) = \frac{14}{60} = \mathbf{0.23}$ $P(6) = \frac{12}{60} = \mathbf{0.20}$

b) Do you think the dice is fair or biased? Explain your answer.

Compare these probabilities to the theoretical probability of $\frac{1}{6} = 0.166...$ for each score.

The relative frequency of a score of 2 is very different from the theoretical probability, so the experiment suggests that the dice is **biased**.

Exercise 4

Q1 A spinner has four sections: blue, green, white and pink. The table shows the results of 100 spins.

Colour	Blue	Green	White	Pink
Frequency	22	21	18	39

a) Work out the relative frequencies of the four colours. Write your answers as decimals.

b) What are the theoretical probabilities of getting each of the colours, assuming the spinner is fair? Write your answers as decimals.

c) Explain whether you think the spinner is fair or biased.

Q2 A six-sided dice is rolled 120 times and 4 comes up 32 times.

a) How many times would you expect 4 to come up on a fair dice in 120 rolls?

b) Use your answer to part a) to explain whether you think the dice is fair or biased.

c) Explain how to find a more accurate estimate for the probability of rolling a 4 on this dice.

Q3 Three friends each toss the same coin and record how many heads they get. The table shows their results.

	Amy	Steve	Hal
Number of tosses	20	60	100
Number of heads	12	33	49
Relative frequency			

a) Copy and complete the table.

b) Explain whose results are likely to be the most accurate.

c) Explain whether you think the coin is fair or biased.

Q4 Information on 800 UK residents' hair and eye colour was collected. The data was recorded in the two-way table shown on the right.

		Has brown hair?	
		Yes	No
Has brown eyes?	Yes	220	105
	No	335	140

a) Show the information from this two-way table on a frequency tree.

b) Find the relative frequency of someone having brown hair and brown eyes.

c) If you collected the same data on another group of 2000 UK residents, how many would you estimate have brown hair and brown eyes?

27.5 The AND / OR Rules

The AND rule lets you work out the probability that one event AND another event will both happen. However, it only works when the events don't affect each other.

The AND Rule

Learning Objective — Spec Ref P8:
Use the AND rule to find the probability of two independent events both occurring.

Prior Knowledge Check:
Be able to multiply fractions and decimals. See p.19 and p.78.

Two events are said to be **independent** if one of them happening has **no effect** on the probability of the other happening.

For example, imagine drawing **two cards** at random from a standard deck of playing cards. If you **put the first card back** and shuffle the deck, then it **doesn't affect** the second card drawn, so the events are **independent** (but if you **don't** put the first card back, then the probabilities for the second draw **depend** on the result of the first).

If A and B are **independent** events, you can use the **AND rule**: the probability of **both A and B** happening is equal to the probability of A happening **multiplied** by the probability of B happening.
This can be written as:

P(A and B) = P(A) × P(B)

Tip: The AND rule doesn't just apply to two events. If there are more than two independent events you multiply together all of the probabilities.

Example 1

The probability of a biased dice landing on 6 is 0.2. The dice is rolled twice. What is the probability that two sixes are rolled?

The first roll has no effect on the second roll, so the events '6 on first roll' and '6 on second roll' are independent. This means you can use the AND rule.

P(6 on 1st roll and 6 on 2nd roll)
= P(6 on 1st) × P(6 on 2nd)
= 0.2 × 0.2 = **0.04**

Exercise 1

Q1 A fair coin is tossed and a fair, standard dice is rolled.
Find the probability the results are:

a) a head and a 6
b) a head and an odd number
c) a tail and a square number
d) a tail and a prime number
e) a head and a multiple of 3
f) a tail and a factor of 5

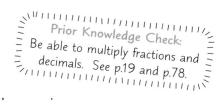

Q2 10% of the pupils in a school are left-handed and 15% wear glasses. Assuming that the hand they write with and glasses wearing are independent, find the probability that a pupil picked at random:

a) is right-handed

b) doesn't wear glasses

c) wears glasses and is left-handed

d) doesn't wear glasses and is right-handed

Q3 A bag contains ten coloured balls. Five of them are red and three of them are blue. A ball is taken from the bag at random, then replaced. A second ball is then selected at random. Find the probability that:

a) both balls are red

b) both balls are blue

c) neither ball is blue

The OR Rules

Learning Objective — Spec Ref P8:
Use the OR rules to find the probabilities of mutually exclusive events and non-mutually exclusive events.

Two events are **mutually exclusive** if they **can't both happen** at the same time (see p.473) — for example, rolling a 2 and a 3 on the same dice roll.

The **OR rule** (or **addition rule**) says that if events are **mutually exclusive**, then the probability that one of the events happens is the **sum** of the probabilities of each event happening:

P(A or B) = P(A) + P(B)

Tip: Always check whether the events are mutually exclusive or not before doing your calculations.

Example 2

A bag contains red, yellow and blue counters. The probabilities of randomly selecting each colour are shown in the table opposite.

Red	Yellow	Blue
0.3	0.5	0.2

Find the probability that a randomly selected counter is red or blue.

The counter can't be both red and blue, so the events 'counter is red' and 'counter is blue' are mutually exclusive.

P(red or blue) = P(red) + P(blue)
= 0.3 + 0.2 = **0.5**

When two events are **not mutually exclusive** (e.g. rolling an even number and a number less than 3 on the same dice roll), you need a different version of the **OR rule**. To find the probability of **at least one event happening**, you **add** the probabilities of the **individual events** as before, but then **subtract** the probability that **both events happen**:

P(A or B) = P(A) + P(B) – P(A and B)

Example 3

The fair spinner shown on the right is spun once.
What is the probability that it lands on a 2 or a shaded sector?

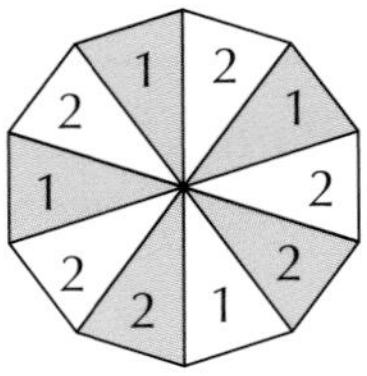

The spinner can land on both 2 and a shaded sector at the same time so they are not mutually exclusive.

This means you use the OR rule
P(A or B) = P(A) + P(B) – P(A and B)

$$P(2 \text{ or shaded}) = P(2) + P(\text{shaded}) - P(2 \text{ and shaded})$$
$$= \frac{6}{10} + \frac{5}{10} - \frac{2}{10} = \frac{9}{10} = \mathbf{0.9}$$

Exercise 2

Q1 A bag contains some coloured balls. The probability of randomly selecting a pink ball is 0.5, selecting a red ball is 0.4 and selecting an orange ball is 0.1. Find the probability that a randomly selected ball will be:

a) pink or orange b) pink or red c) red or orange

Q2 Chocolates in a box are wrapped in four different colours of foil — gold, silver, red or blue. The table shows the probabilities of randomly picking a chocolate wrapped in each colour.

Colour	Gold	Silver	Red	Blue
Probability	0.4	0.26	0.14	0.2

Find the probability of picking a chocolate wrapped in:

a) red or gold foil b) silver or red foil
c) gold or blue foil d) silver or gold foil

Q3 On sports day, pupils are split into three equal-sized teams — the Eagles, the Falcons and the Ospreys.
What is the probability that a pupil picked at random belongs to:

a) the Eagles? b) the Eagles or the Falcons? c) the Falcons or the Ospreys?

Q4 A fair 20-sided dice, numbered 1-20, is rolled. What is the probability that it lands on:

a) a multiple of 3 or an odd number? b) a factor of 20 or a multiple of 4?

Q5 A card is randomly chosen from a standard pack of 52 playing cards.

a) What's the probability that it's a red suit or a queen?
b) What's the probability that it's a club or a picture card?

Q6 Jane is told that in a class of 30 pupils, 4 wear glasses and 10 have blond hair.
She says that the probability that a pupil picked at random from the class will have blond hair or wear glasses is $\frac{10}{30} + \frac{4}{30} = \frac{14}{30}$. Do you agree with Jane? Explain your answer.

27.6 Tree Diagrams

Tree diagrams can be used to show all the possible results from an experiment — they're really useful for working out probabilities of combinations of events.

Learning Objective — Spec Ref P6/P8:
Use tree diagrams to represent events and to find probabilities.

Tree diagrams are similar to **frequency trees** (see p.480), except they show the **probabilities** of the events rather than the frequency from an experiment.

The really useful thing about tree diagrams is that you can find the probability of specific **results** (e.g. P(A happens and B doesn't happen)) by **multiplying along the branches** that you follow to get to that result.

Since branches from the same point show all the outcomes of a single activity, their probabilities should **add up to 1**.

Tip: Tree diagrams can get crowded if they show lots of events. Always make sure you leave enough space to draw them.

Example 1

A fair coin is tossed twice.

a) Draw a tree diagram showing all the possible results for the two tosses.

1. Draw a set of branches for the first toss. You need 1 branch for each of the 2 results.
2. Draw a set of branches for the second toss. Again, there are two possible results.
3. Write on the probability for each branch — here it's 0.5 for each possible outcome.

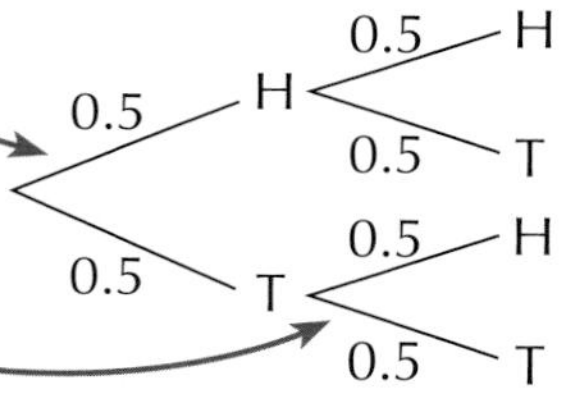

b) Find the probability of getting two heads.

You need to use the AND rule so multiply along the branches for a head AND a head.

P(2H) = 0.5 × 0.5 = **0.25**

c) Find the probability of getting heads and tails in any order.

1. Multiply along the branches for a head AND a tail. P(H then T) = 0.5 × 0.5 = 0.25
2. Multiply along the branches for a tail AND a head. P(T then H) = 0.5 × 0.5 = 0.25
3. Both these results give heads and tails. You need to use the OR rule to find the probability of HT OR TH, so add the probabilities. P(1H and 1T) = 0.25 + 0.25 = **0.5**

Exercise 1

Q1 Copy and complete the following tree diagrams.

a) A fair spinner has five equal sections — three are red and two are blue. The spinner is spun twice.

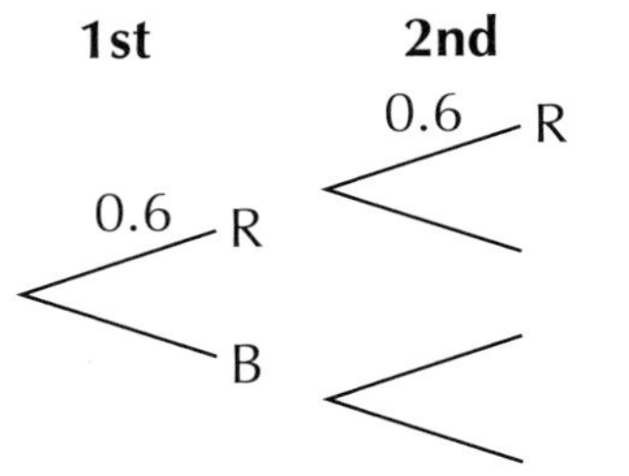

b) A bag contains ten coloured balls — five red, three blue and two green. A ball is selected at random and replaced, then a second ball is selected at random.

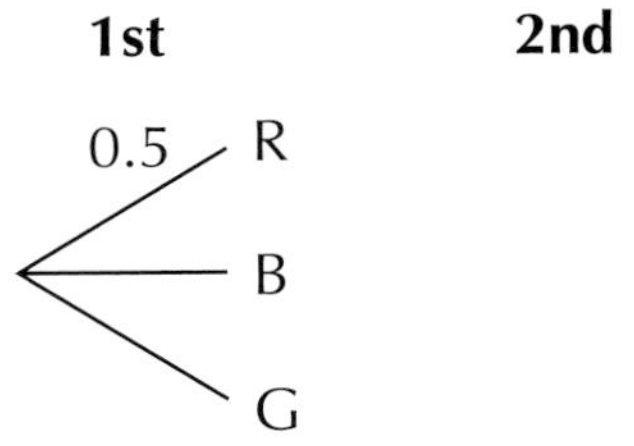

Q2 For each of these situations, draw a tree diagram showing all the possible results. Write the probability on each branch.

a) A biased coin lands on heads with a probability of 0.65. The coin is tossed twice.

b) The probability a football team wins is 0.7, draws is 0.1 and loses is 0.2. The team plays two matches.

Q3 Freddie and James play two games of pool. The probability that Freddie beats James is 0.8.

a) Draw a tree diagram to show all the possible results for the two games.

b) Find the probability that Freddie wins both games.

Q4 Ifrath owns 12 DVDs, four of which are comedies. Jesse owns 20 DVDs, eight of which are comedies. They each select one of their DVDs at random to watch over the weekend.

a) Draw a tree diagram showing the probabilities of each choice being 'comedy' or 'not a comedy'.

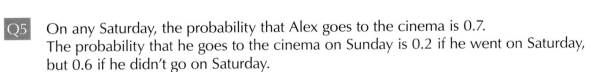

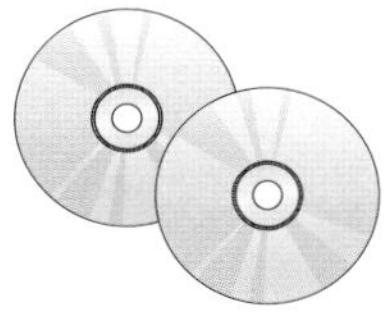

b) Find the probability that neither of them chooses a comedy.

c) Find the probability that at least one of them chooses a comedy.

Q5 On any Saturday, the probability that Alex goes to the cinema is 0.7. The probability that he goes to the cinema on Sunday is 0.2 if he went on Saturday, but 0.6 if he didn't go on Saturday.

a) Draw a tree diagram showing the probabilities of Alex going and not going to the cinema on Saturday and Sunday.

b) What is the probability that Alex goes to the cinema on both days?

c) What is the probability that Alex goes to the cinema on exactly one of the two days?

Q6 In a class there are 16 boys and 14 girls. They all put their names into a box. One name is selected at random and **not** replaced, then another name is selected at random.

a) Draw a tree diagram to show the probabilities of each name selected being a boy or a girl.

b) What is the probability that two boys are picked?

c) What is the probability that exactly one girl is picked?

27.7 Sets and Venn Diagrams

A set is a collection of things — it can be considered as an object in its own right. Venn diagrams can be used to display sets. They're great at showing the overlap between sets.

Sets

Learning Objectives — Spec Ref P6:
- Understand and use set notation.
- Be able to list the elements of a set.

Prior Knowledge Check: Be familiar with multiples and factors. See Section 4.

A **set** is a group of items or numbers. Sets are written in pairs of **curly brackets { }**. Each item in a set is called an **element** or **member** of the set. You can describe a set by **listing every element** in that set (e.g. {2, 4, 6}) or by **giving a rule** that all elements follow (e.g. {all red objects}).

Here's the **set notation** you need to become familiar with:

A = {...}	A is the **set** of ...
$x \in$ **A**	x is an **element** of A
n(A)	the **number of elements** in A
ξ	the **universal** set (all the elements that you need to consider)
A ∪ B	the **union** of sets A and B (all elements in either set)
A ∩ B	the **intersection** of sets A and B (all elements in both sets)
A′	the **complement** of set A (all elements not in set A)

Example 1

Given that ξ = {positive integers less than 20}, list the elements of the following sets:

a) A = {multiples of 2}

A consists of the multiples of 2 in the universal set.

A = {2, 4, 6, 8, 10, 12, 14, 16, 18}

b) B = {x : x is a factor of 40}

B consists of the factors of 40 in the universal set.

B = {1, 2, 4, 5, 8, 10}

Tip: The colon means 'such that'.

c) A ∪ B

List all the elements from either set. If a number appears in both sets, it should only be listed once.

A ∪ B = {1, 2, 4, 5, 6, 8, 10, 12, 14, 16, 18}

d) A ∩ B

List only those elements that appear in both sets.

A ∩ B = {2, 4, 8, 10}

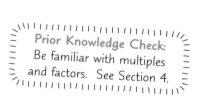

Exercise 1

Q1 a) List the elements of the following sets.

(i) A = {even numbers between 11 and 25}
(ii) B = {prime numbers less than 30}
(iii) C = {x : x is a square number < 200}
(iv) D = {factors of 30}

b) Write down: (i) n(A) (ii) n(B) (iii) n(C) (iv) n(D)

Q2 A = {1, 2, 4, 5, 7, 8}, B = {2, 4, 6, 8}. Find the following: a) $A \cup B$ b) $A \cap B$

Q3 ξ = {positive integers less than 10}, A = {1, 3, 7, 9}, B = {2, 3, 5, 6, 8, 9}. Find the following:

a) n(A) b) n(B) c) A′ d) $A \cup B$
e) B′ f) $A \cap B$ g) $n(A \cup B)$ h) $A' \cup B$

Venn Diagrams

Learning Objective — Spec Ref P6:
Use Venn diagrams to represent sets and to find probabilities.

Venn diagrams use **circles** to represent sets — the **space inside** the circle represents everything in the set. Each **circle** is labelled with a **letter** — this tells you **which set** the circle represents.

The **numbers inside a circle** can tell you either the **number of members** of that set or **actual elements** of the set.

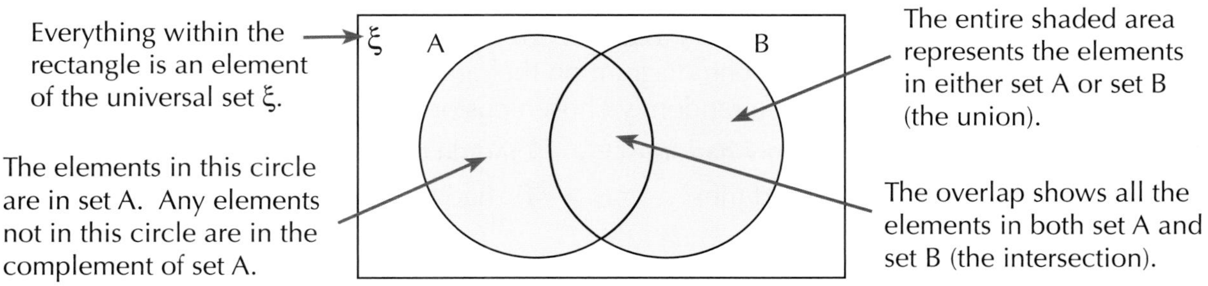

Everything within the rectangle is an element of the universal set ξ.
The elements in this circle are in set A. Any elements not in this circle are in the complement of set A.
ξ
A
B
The entire shaded area represents the elements in either set A or set B (the union).
The overlap shows all the elements in both set A and set B (the intersection).

Finding probabilities from Venn diagrams

If you know the **number of elements** in each set in a Venn diagram, you can work out **probabilities** from the Venn diagram. To find the probability of randomly selecting a member of a **particular set** from the **universal set**, you use a similar **formula** to the one for **calculating probabilities** on p.470. The **number of elements** in a set is **divided by the total number of elements** in the universal set:

$$\textbf{Probability of an element being in the set} = \frac{\textbf{number of elements in the set}}{\textbf{total number of elements in the universal set}}$$

Example 2

a) Given that ξ = {positive integers less than or equal to 12}, draw a Venn diagram for the sets A = {x : x is a multiple of 3} and B = {x : x is a factor of 30}.

1. Write out sets A and B:
   A = {3, 6, 9, 12} B = {1, 2, 3, 5, 6, 10}
2. A ∩ B = {3, 6} so these elements go in the overlap (intersection) of the two circles.
3. The elements in the circle for A that aren't in the overlap are the elements of A that aren't in B (and similarly for B).
4. The elements of ξ that aren't in either set go outside the circles.

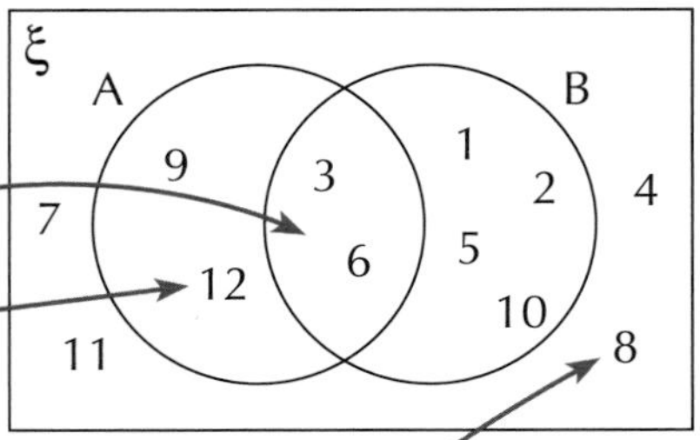

b) What is the probability that a randomly selected element of ξ is in set A?

Divide the number of elements of set A by the total number of elements in ξ.

$$\frac{n(A)}{n(\xi)} = \frac{4}{12} = \mathbf{\frac{1}{3}}$$

Exercise 2

Q1 Draw Venn diagrams to show the following sets, where ξ = {positive integers ≤ 10}.

a) A = {1, 3, 5} B = {1, 3, 7} b) A = {2, 3, 4, 5} B = {1, 3, 5, 7, 9}

c) A = {2, 6, 10} B = {1, 3, 6, 9} d) A = {2, 3, 4, 5, 6} B = {1, 3, 7}

Q2 At a fast food restaurant there are two options for side dishes — fries or salad. The number of customers (out of a total of 200) who chose each side dish is recorded in the Venn diagram on the right.
What is the probability that a randomly chosen customer:

a) only had salad? b) only had fries? c) didn't have a side?

d) had salad? e) had fries? f) had fries and salad?

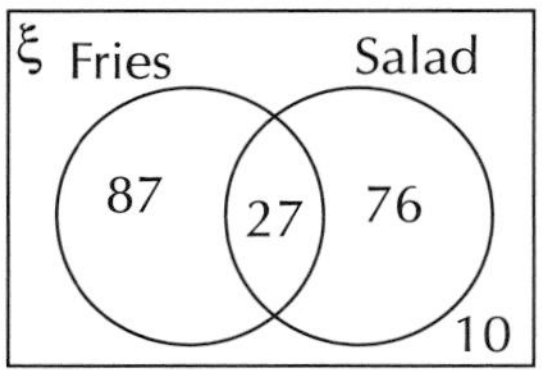

Q3 a) Given that ξ = {positive integers ≤ 10}, draw a Venn diagram for the sets A = {x : x is an odd number} and B = {x : x is a prime number}.

b) What is the probability that a randomly chosen element of ξ is:

(i) in B? (ii) in A ∩ B? (iii) in A ∩ B′?

Q4 In a class of 30 pupils, 23 like Maths (M), 18 like English (E) and 15 like Maths and English.

a) Draw a Venn diagram to show this information.

b) What is the probability that a randomly selected pupil:

(i) likes Maths and English? (ii) only likes Maths?

(iii) only likes English? (iv) doesn't like Maths or English?

Review Exercise

Q1 In a fruit bowl, there are 4 apples, 6 bananas, 5 pears and 5 oranges.
Bianca picks out one piece of fruit at random. On the scale, indicate with an arrow:

a) How likely she is to pick an apple or banana.

b) How likely it is that she does not pick an orange.

c) How likely it is that she picks a lemon.

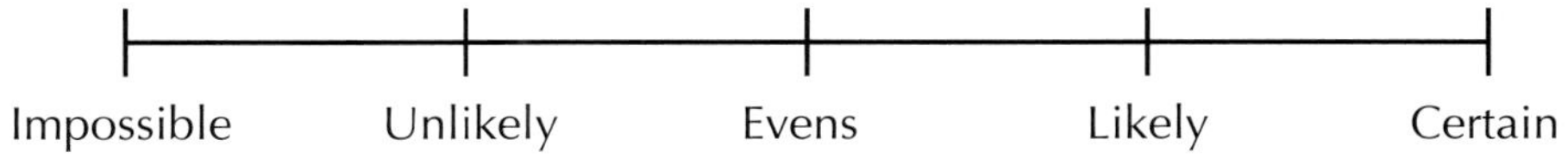

Q2 A fair, standard dice is rolled.

a) What is the probability of rolling an even number?

b) What is the probability of rolling a multiple of 3?

c) Explain why your answers to parts a) and b) don't add up to 1.

Q3 Derek and Eileen have played golf against each other 15 times.
Derek has won 8 times, and there have been no draws.

a) Estimate the probability that Derek will win the next time they play.

b) Estimate the probability that Eileen will win the next time they play.

Q4 Miley has a bag of sweets. There are 4 blue sweets, 5 white sweets and 11 red sweets. She selects one at random, puts it back, then selects another at random.

a) Are these events independent?

b) Find the probability that Miley selects:

(i) two blue sweets (ii) a white sweet followed by a red sweet

Q5 A box contains ten coloured marbles — five blue, four white and one red.
Two marbles are picked at random and are not replaced.

a) Draw a tree diagram showing all the possible results.

b) Work out the probability that:

(i) both are blue (ii) at least one is blue (iii) they are different

Q6 An ice cream van sells ice creams and ice lollies.
The Venn diagram on the right shows the sales for an afternoon.

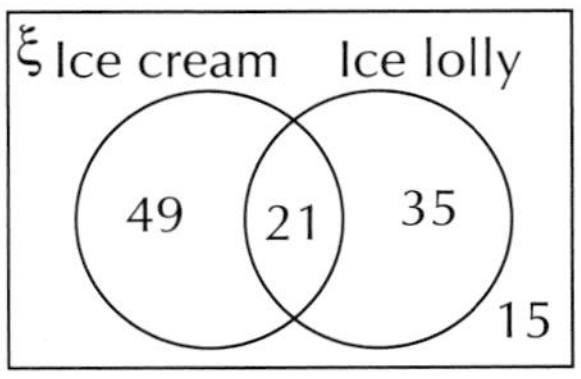

a) How many people bought an ice cream?

b) What is the probability that a randomly selected customer only bought an ice lolly?

Exam-Style Questions

Q1 In Box A are three cards, numbered 1, 2 and 3. In Box B are two cards, numbered 7 and 8. Leila picks one card from each box at random and puts them down in a random order to make a 2-digit number.

Find the probability that she makes an even number.

[3 marks]

Q2 Tania drives to and from work every weekday. There is a set of traffic lights on her route. The probability that she will have to stop at the lights on any given journey is 0.7.

a) Copy and complete the tree diagram below.

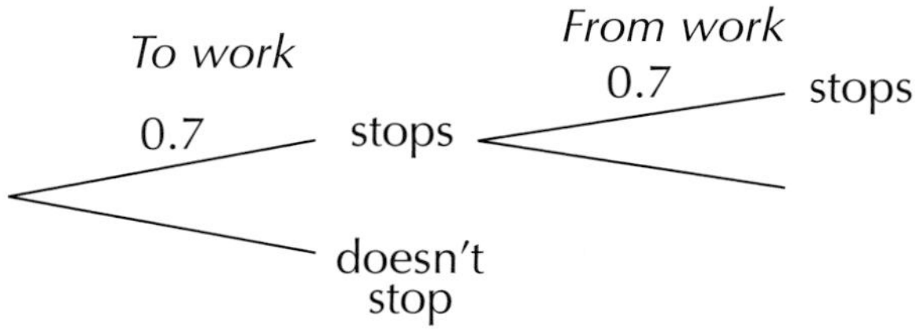

[2 marks]

b) Calculate the probability that on any given day she will stop at the traffic lights exactly once.

[2 marks]

Q3 Ian asked 7 boys and 9 girls in a Year 11 class at his school whether they went to the cinema last week. 5 boys had been to the cinema and 2 boys hadn't. Two thirds of the girls had been to the cinema and the rest of the girls hadn't.

a) What is the probability that a randomly selected pupil from the ones who were asked went to the cinema last week?

[3 marks]

Ian says $\frac{5}{7}$ of Year 11 boys at his school went to the cinema last week.

b) Give one reason why Ian's statement may be unreliable.

[1 mark]

Q4 30 members of a youth club go to an outdoor activity centre to take part in activities. 17 **only** do basketball (B), 10 do archery (A) and 6 do both.

a) Draw a Venn diagram to show the information given.

[2 marks]

b) Find the value of $n(A \cup B)$.

[1 mark]

Mixed Exam-Style Questions

PROBLEM SOLVING

Q1 A cinema is showing a film twice every evening.
The first showing starts at 18:40 and finishes at 20:15.
The second showing starts at 20:30.

a) Work out the time that the second showing finishes.

[2 marks]

Tickets to see the film are the same price for all customers.
Robert buys 5 tickets for him and his family to see the film together.
He pays with one £20 note and three £10 notes. He receives £1.25 change.

b) Work out the cost of a cinema ticket.

[3 marks]

Q2 Here is a list of numbers: 3 5 12 18 25

a) Write down the even number which has 9 as a factor.

[1 mark]

b) Write down the square number.

[1 mark]

c) Write down the two numbers which have a product of 60.

[1 mark]

d) Four of the numbers are added. The total is 58.
Which number was not included in this sum?

[1 mark]

e) The difference between 2 of the numbers is found. The answer is odd.
Work out the largest value that this difference could be.

[1 mark]

Q3 5 shapes are shown on the grid.

a) Write down the letters of the two shapes that are congruent.

[1 mark]

b) Write down the letters of the two shapes which are similar but not congruent.

[1 mark]

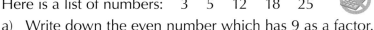

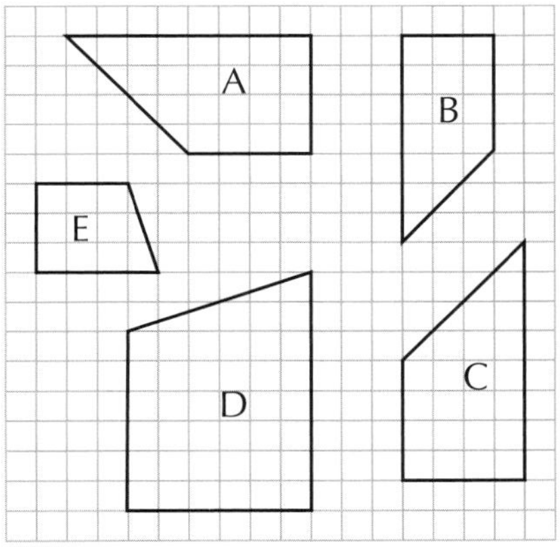

Q4 Ali has some 2p and 5p coins in his pocket.
He buys a fruit bag for 40p and uses all of his coins.
Show that there are five possible way that he could do this.

[2 marks]

Q5 The table shows the populations of China and Italy in 2017, both correct to 4 significant figures.

Country	Population
Italy	5.936×10^7
China	1.410×10^9

Find the ratio of the larger population to the smaller population in the form $n:1$.

[3 marks]

Q6 Every Friday, a school canteen offers a choice of a mini pizza or pasta. The number of Year 11 students who had pizza or pasta over three consecutive Fridays is shown on the dual bar chart below.

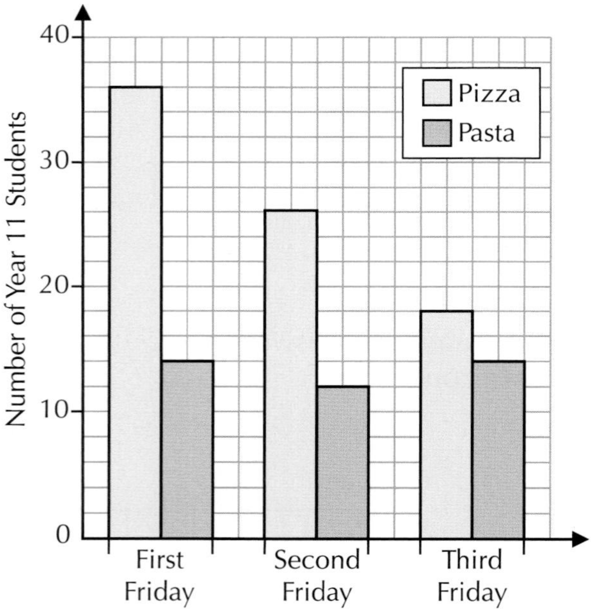

a) What percentage of students who ate in the canteen on the first Friday chose pasta?

[2 marks]

b) What fraction of students who ate in the canteen on the second Friday chose pizza? Give your answer in its simplest form.

[2 marks]

c) Describe a trend in the data over the course of the three Fridays.

[1 mark]

Q7 Uma is paid £9.46 per standard hour of work. She gets double this rate for overtime.

Last week, Uma's standard hours were 7.5 hours on Monday, Tuesday, Wednesday and Thursday and 7 hours on Friday. On Saturday she worked some hours of overtime.

She received £406.78 in total for the hours she worked, including overtime.
Work out how many hours of overtime Uma did last week. Show your working.

[3 marks]

Q8 A motorboat crosses a lake in a straight line from point A to point B, as shown on the map below. The average speed of the boat is 12 km/h. Work out how long it will take for the boat to cross the lake.

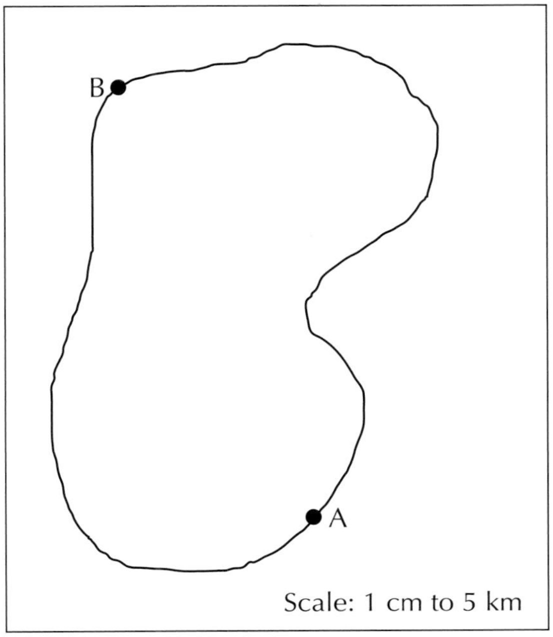

[3 marks]

Q9 For each of the following, work out what single-digit positive integer should go in the box to make the statement true:

a) $\frac{3}{4} = \frac{6}{\square}$

[1 mark]

b) $\frac{3}{4} < \frac{\square}{11}$

[1 mark]

Q10 A brand of breakfast cereal is sold in two different sizes.
The "family" size box contains 750 g.
The "standard" box normally contains 500 g, but as a special offer, these boxes currently contain an extra 25% of cereal.

The "family" box costs £2.25. The special offer "standard" boxes cost £2.
Work out which size of box is the better value for money.
You must show your working.

[4 marks]

Q11 The diagram shows an equilateral triangle.
The length of each side is a whole number of centimetres.

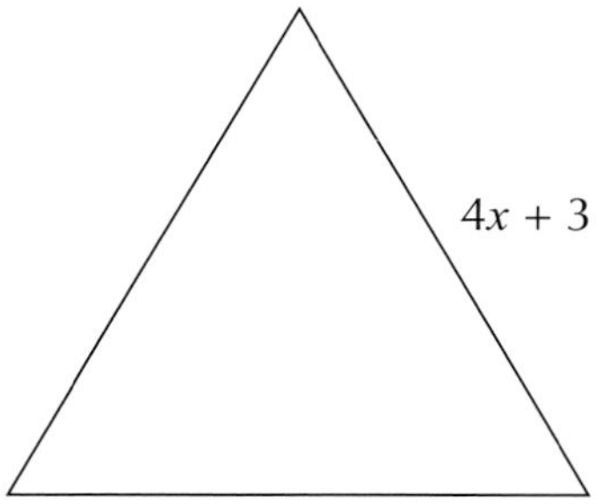

The perimeter of the diagram is greater than 480 cm.
Write an inequality, in terms of x, and solve it to find the lower limit for x.

[4 marks]

Q12 A shape is made when two identical squares overlap at the mid-point of their sides.

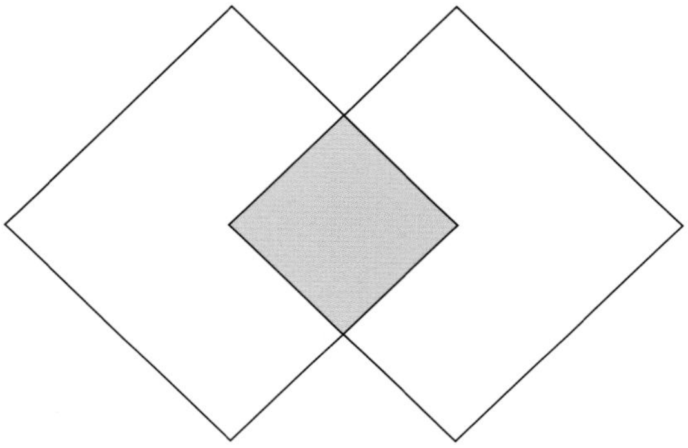

a) What fraction of the shape is shaded?

[1 mark]

b) The area of the shaded shape is 25 cm^2.
What is the length of one side of a large square?

[2 marks]

Q13 Muneera has a bowl of grapes.
8 of the grapes are green, 12 are red and the rest are black.
When Muneera picks a grape from the bowl at random,
the probability that she picks a green grape is $\frac{1}{3}$.
How many black grapes are there in the bowl?

[3 marks]

Q14 Jeremy is looking to buy a new computer on the internet.

A UK seller offers the model he wants with free delivery for £420.

The same model is also available from an American seller priced at 432 US dollars ($). Delivery from America will cost $30 and there will also be a 20% import tax on both the cost of the computer and the delivery charge.

The exchange rate is $1 = £0.75. Work out which is the better deal.

[3 marks]

Q15 The diagram shows a parallelogram with the lengths of three of its sides given in centimetres in terms of x. Show that the perimeter of the parallelogram is 23.2 cm.

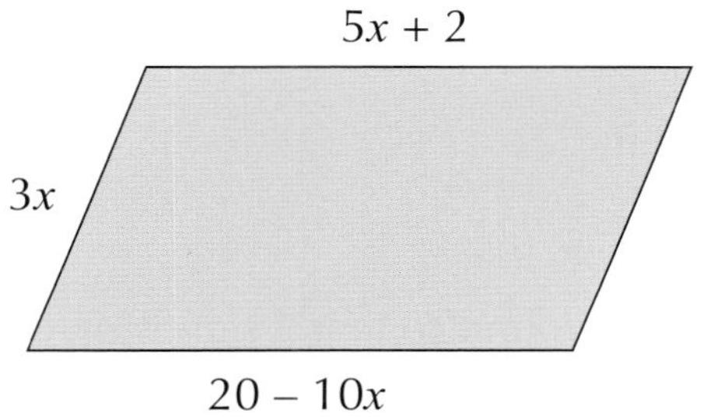

Not drawn accurately

[4 marks]

Q16 The lines $x = -1$, $y = 1$ and $y = 2x + 7$ have been sketched on the axes below. They intersect at points A, B and C.

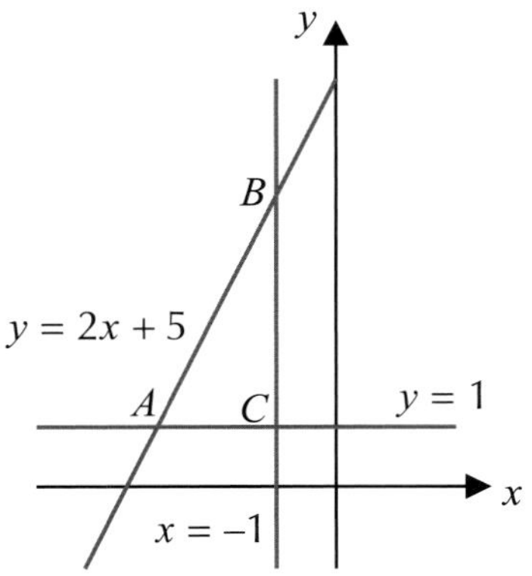

a) Find the coordinates of A, B and C.

[5 marks]

b) Show that the length of AB is $\sqrt{5}$.

[4 marks]

Q17 Three adjacent interior angles of a pentagon are right angles. The other two angles are in the ratio 4 : 11. Work out the size of the largest interior angle in the pentagon.

[4 marks]

Q18 A sealed water tank is partially filled with water.
It is in the shape of a cuboid with dimensions as shown on the diagram.
It stands on one of its smallest faces on level ground and the depth of water is 1.75 m.

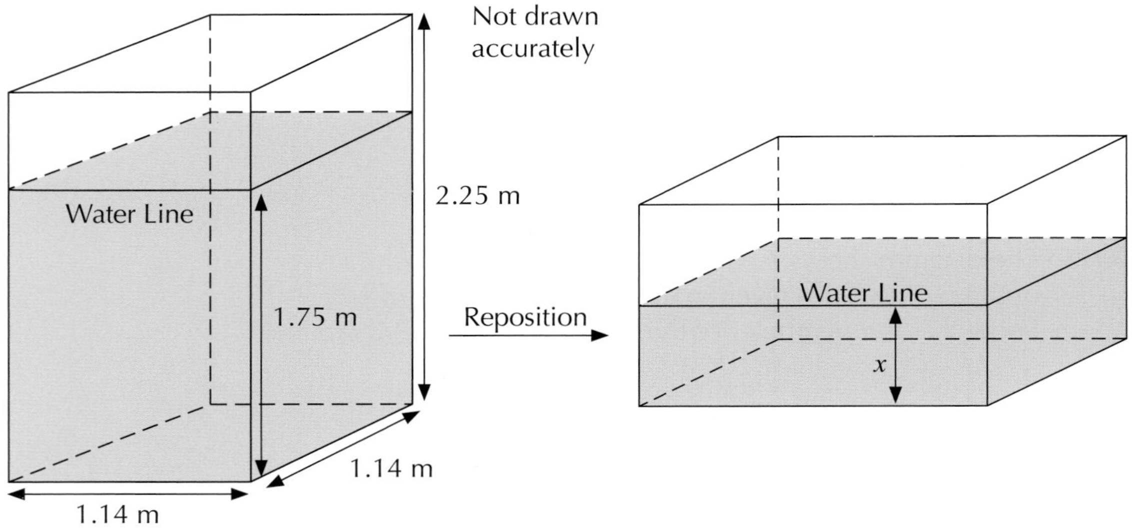

The tank is repositioned on level ground so that it stands on one of its largest faces.
By forming and solving an equation, work out x, the new depth of the water.
Give your answer in metres to an appropriate degree of accuracy.

[3 marks]

Q19 The diagram shows a regular hexagon with a right-angled triangle drawn on one side.
PQ is a diagonal, which divides the hexagon into two identical trapeziums.

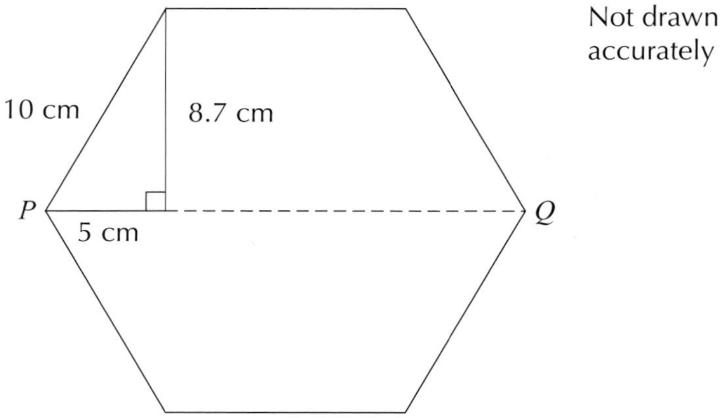

Find the area of the hexagon.

[4 marks]

Answers

Section 1 — Non-Calculator Arithmetic

1.1 Order of Operations

Page 2   Exercise 1

1 a) $5 + 1 \times 3 = 5 + 3 = \mathbf{8}$
   b) $11 - 2 \times 5 = 11 - 10 = \mathbf{1}$
   c) $24 \div 4 + 2 = 6 + 2 = \mathbf{8}$
   d) $18 - 10 \div 5 = 18 - 2 = \mathbf{16}$

2 a) $6 \times (4 + 3) = 6 \times 7 = \mathbf{42}$
   b) $11 - (2 + 3) = 11 - 5 = \mathbf{6}$
   c) $(8 - 7) \times (6 + 5) = 1 \times 11 = \mathbf{11}$
   d) $56 \div (2 \times 4) = 56 \div 8 = \mathbf{7}$

3 a) $2 \times (8 + 4) - 7 = 2 \times 12 - 7 = 24 - 7 = \mathbf{17}$
   b) $18 \div (9 - 12 \div 4) = 18 \div (9 - 3)$
      $= 18 \div 6 = \mathbf{3}$
   c) $100 \div (8 + 3 \times 4) = 100 \div (8 + 12)$
      $= 100 \div 20 = \mathbf{5}$
   d) $20 - (5 \times 3 + 2) = 20 - (15 + 2)$
      $= 20 - 17 = \mathbf{3}$
   e) $48 \div 3 - 7 \times 2 = 16 - 14 = \mathbf{2}$
   f) $36 - (7 + 4 \times 4) = 36 - (7 + 16)$
      $= 36 - 23 = \mathbf{13}$

4 a) $9 \times (7 - 5) = 18$
      Don't panic if you can't spot the answer to
      questions like this straight away — try brackets
      in different positions and see what works.
   b) $(18 - 6) \div 3 = 4$
   c) $(5 + 2) \times (6 - 2) = 28$
   d) $21 \div (4 + 3) = 3$
   e) $(13 - 5) \times (13 - 1) = 96$
   f) $(6 + 8 - 7) \times 5 = 35$

5 a) $16 - 6 \div 3 = 14$
   b) $11 \times 3 + 5 = 38$
   c) $3 \times 6 - 9 = 9$
   d) $8 \div 2 + 6 = 10$ or $8 \times 2 - 6 = 10$
   e) $(3 + 7) \times 4 = 40$
   f) $14 \div (6 + 8) = 1$

Page 3   Exercise 2

1 a) Top line: $4 - 1 + 5 = 3 + 5 = 8$
      Bottom line: $2 \times 2 = 4$
      Top line $\div$ bottom line $= 8 \div 4 = \mathbf{2}$
   b) Top line: $6 + (11 - 8) = 6 + 3 = 9$
      Bottom line: $7 - 5 = 2$
      Top line $\div$ bottom line $= 9 \div 2 = \dfrac{\mathbf{9}}{\mathbf{2}}$

Using the same method for c)-h):

   c) **2**   d) **4**   e) $\dfrac{\mathbf{3}}{\mathbf{5}}$
   f) $\dfrac{\mathbf{2}}{\mathbf{3}}$ g) **3**   h) **1**

1.2 Negative Numbers

Page 4   Exercise 1

1 a) $-4 + 3 = \mathbf{-1}$ b) $-1 + 5 = \mathbf{4}$
   c) $-2 + 1 = \mathbf{-1}$ d) $6 - 17 = \mathbf{-11}$
   e) $-13 + 18 = \mathbf{5}$ f) $11 - 19 = \mathbf{-8}$

2 a) $-3 + 2 - 1 = \mathbf{-2}$
   b) $-2 + 8 - 5 = \mathbf{1}$
   c) $8 - 5 - 3 = \mathbf{0}$
   d) $-1 - 7 - 4 = \mathbf{-12}$
   e) $-9 + 13 + 11 = \mathbf{15}$
   f) $7 - 18 + 11 = \mathbf{0}$

3 a) $-3 + 2 = \mathbf{-1}$ b) $5 - 15 = \mathbf{-10}$
   c) $-3 + 7 = \mathbf{4}$ d) $-4 - 2 = \mathbf{-6}$

4 a) Start at -1°C and count 3 places down
      to get $\mathbf{-4° C}$.
   b) Start at -5°C and count 8 places up
      to get to $\mathbf{3° C}$.

5 a) Start at 3°C and count how many
      places down it is to -4° C. The number of
      places is 7 so the difference is $\mathbf{7° C}$.
   b) Start at -2°C and count how many
      places down it is to -5° C. The number of
      places is 3 so the difference is $\mathbf{3° C}$.

6 Start at 6° C and count down 7 places
   to get to $\mathbf{-1° C}$.

7 Start at -5°C and count how many places
   up it is to -1° C. The number of places is 4
   so the difference is $\mathbf{4° C}$.

Page 5   Exercise 2

1 a) $4 - (-2) = \mathbf{6}$ b) $-3 - (-5) = \mathbf{2}$
   c) $-7 + (-2) = \mathbf{-9}$ d) $-5 + (-5) = \mathbf{-10}$
   e) $9 + (-2) = \mathbf{7}$ f) $-13 - (-3) = \mathbf{-10}$
   g) $-6 + (-3) = \mathbf{-9}$ h) $1 - (-12) = \mathbf{13}$

2 a) $-1 + (-4) - 1 = \mathbf{-6}$
   b) $6 - (-2) + (-3) = \mathbf{5}$
   c) $7 + (-6) - (-8) = \mathbf{9}$
   d) $-8 + (-8) - (-12) = \mathbf{-4}$
   e) $9 - (-13) + (-2) = \mathbf{20}$
   f) $-3 + (-11) + (-6) = \mathbf{-20}$

3 a) $6 - (-5) = \mathbf{11}$ b) $-6 - (-10) = \mathbf{4}$
   c) $4 - (-8) = \mathbf{12}$ d) $-5 - (-12) = \mathbf{7}$

Page 6   Exercise 3

1 a) $3 \times (-4) = \mathbf{-12}$ b) $2 \times (-8) = \mathbf{-16}$
   c) $(-9) \times 6 = \mathbf{-54}$ d) $(-15) \div 3 = \mathbf{-5}$
   e) $(-15) \div (-3) = \mathbf{5}$ f) $12 \div (-4) = \mathbf{-3}$

2 a) $-3 \times \mathbf{2} = -6$ b) $-14 \div \mathbf{7} = -2$
   c) $\mathbf{-4} \times 4 = -16$ d) $\mathbf{10} \div (-2) = -5$
   e) $-8 \times 3 = \mathbf{-24}$ f) $-18 \div (-6) = \mathbf{3}$
   g) $-12 \times (-3) = \mathbf{36}$ h) $-77 \div 11 = \mathbf{-7}$

3 a) $2 \times 4 \times (-2) = 8 \times (-2) = \mathbf{-4}$
   b) $(-5) \times (-6) \div 3 = 30 \div 3 = \mathbf{10}$
   c) $(-3) \times (-5) \times (-6) = 15 \times (-6) = \mathbf{-90}$
   d) $(-63) \times (-2) \div (-9) = 126 \div (-9) = \mathbf{-14}$
   e) $[55 \div (-11)] \times (-9) = -5 \times (-9) = \mathbf{45}$
   f) $[(-24) \div 8)] \div 3 = -3 \div 3 = \mathbf{-1}$

1.3 Whole Number Arithmetic

Page 8   Exercise 1

1 a) $\begin{array}{r} 2\,3 \\ +\ 5\,6 \\ \hline 7\,9 \end{array}$ b) $\begin{array}{r} 1\,2\,2 \\ +\ \ 9\,7 \\ \hline 2_{\,1}1\,9 \end{array}$

   c) $\begin{array}{r} 2\,4\,3 \\ +1\,7\,8 \\ \hline 4_{\,1}2_{\,1}1 \end{array}$ d) $\begin{array}{r} {}^{6}\!\nabla^{1}3 \\ -\ \ 2\,7 \\ \hline 4\,6 \end{array}$

   e) $\begin{array}{r} {}^{7}\!8^{1}1 \\ -\ \ 3\,5 \\ \hline 1\,4\,6 \end{array}$ f) $\begin{array}{r} 1\,{}^{12}\!2^{1}3 \\ -1\,8\,7 \\ \hline 4\,6 \end{array}$

2 a) $\begin{array}{r} 3\,4\,2 \\ +6\,7\,9 \\ \hline 1_{\,1}0_{\,1}2_{\,1}1 \end{array}$ b) $\begin{array}{r} 6\,0\,4 \\ +2\,8\,8 \\ \hline 8\,9_{\,1}2 \end{array}$

   c) $\begin{array}{r} 4\,{}^{9}\!5^{0}6 \\ -2\,7\,8 \\ \hline 2\,2\,8 \end{array}$ d) $\begin{array}{r} 2\,5\,1\,3 \\ +\ \ 2\,4\,1 \\ \hline 2\,7\,5\,4 \end{array}$

   e) $\begin{array}{r} 2\,9\,{}^{3}\!4^{1}2 \\ -\ \ 3\,2\,4 \\ \hline 2\,6\,1\,8 \end{array}$ f) $\begin{array}{r} 3\,{}^{9}\!4^{9}0^{1}3 \\ -1\,2\,3\,5 \\ \hline 2\,7\,6\,8 \end{array}$

3 a) Add the first two numbers: $\begin{array}{r} {}^{1}{}^{1}2 \\ +\ \ 4\,1 \\ \hline 1\,5\,3 \end{array}$

   Then add on 213: $\begin{array}{r} 1\,5\,3 \\ +\ 2\,1\,3 \\ \hline 3\,6\,6 \end{array}$

   So $41 + 112 + 213 = \mathbf{366}$
   You could do part a) all in one go by writing the
   three numbers on top of each other and adding.
   You couldn't do the same for b) though, as that's
   an addition followed by a subtraction.

   b) Add the first two numbers: $\begin{array}{r} 7\,6\,4 \\ +\ 1\,3\,8 \\ \hline 9\,0\,2 \end{array}$

   The subtract 345: $\begin{array}{r} {}^{8}\!9^{9}0^{1}2 \\ -\ 3\,4\,5 \\ \hline 5\,5\,7 \end{array}$

   Using the same method for c)-f):
   c) **816**   d) **249**
   e) **917**   f) **757**

4   $\begin{array}{r} 3\,8 \\ 5\,2 \\ +\ 6\,5 \\ \hline 1\,5_{\,1}5 \end{array}$

5   $\begin{array}{r} £\,{}^{1}2^{1}1\,4\,6 \\ -\ \ 2\,2\,4 \\ \hline £\,1\,9\,2\,2 \end{array}$

6   $\begin{array}{r} 1\,{}^{7}\!8^{3}4^{1}5 \\ -1\,2\,5\,7 \\ \hline 5\,8\,8 \end{array}$

Page 9   Exercise 2

1 a) $\begin{array}{r} 6\,6 \\ \times\ 7\,2 \\ \hline 1\,3_{\,1}2 \\ 4\,6_{\,4}2\,0 \\ \hline 4\,7\,5\,2 \end{array}$ b) $\begin{array}{r} 7\,9 \\ \times\ 8\,6 \\ \hline 4\,7_{\,5}4 \\ 6\,3_{\,7}2\,0 \\ \hline 6\,7\,9\,4 \end{array}$

   Using the same method for c)-f):
   c) **6723**   d) **2600**
   e) **3420**   f) **11 664**

2   $\begin{array}{r} 4\,6 \\ \times\ 2\,4 \\ \hline 1\,8_{\,2}4 \\ 9_{\,1}2\,0 \\ \hline 1\,1_{\,1}0\,4 \end{array}$

3   $\begin{array}{r} 3\,3 \\ \times\ 1\,2 \\ \hline 6\,6 \\ 3\,3\,0 \\ \hline 3\,9\,6 \end{array}$

4 a) $\begin{array}{r} 1\,3\,4\,7 \\ \times\ \ \ 2\,0 \\ \hline 0\,0\,0\,0 \\ 2\,6\,9_{\,1}4\,0 \\ \hline 2\,6\,9\,4\,0 \end{array}$ b) $\begin{array}{r} 3\,6\,6\,9 \\ \times\ \ \ 2\,1 \\ \hline 3\,6\,6\,9 \\ 7\,3_{\,1}3_{\,1}8\,0 \\ \hline 7\,7_{\,1}0_{\,1}4\,9 \end{array}$

   c) $\begin{array}{r} 2\,6\,2\,3 \\ \times\ \ \ 4\,2 \\ \hline 5_{\,1}2\,4\,6 \\ 1\,0_{\,2}4\,9\,2\,0 \\ \hline 1\,1_{\,1}0\,1\,6\,6 \end{array}$ d) $\begin{array}{r} 2\,5\,7\,8 \\ \times\ \ \ 3\,6 \\ \hline 1\,5_{\,4}4_{\,6}4_{\,8}8 \\ 7_{\,1}7_{\,3}3_{\,4}4\,0 \\ \hline 9_{\,1}2\,8_{\,1}0\,8 \end{array}$

Page 10 Exercise 3

1 a)

	10	3
10	10 × 10 = 100	10 × 3 = 30
2	2 × 10 = 20	2 × 3 = 6

So 13 × 12 = 100 + 30 + 20 + 6 = **156**

b)

	10	1
10	10 × 10 = 100	10 × 1 = 10
7	7 × 10 = 70	7 × 1 = 7

So 11 × 17 = 100 + 10 + 70 + 7 = **187**

Using the same method for c)-h):

c) 520 **d) 384** **e) 22 754**
f) 34 362 **g) 28 105** **h) 42 032**

2

	200	40	3
50	200 × 50 = 10 000	40 × 50 = 2000	3 × 50 = 150
2	200 × 2 = 400	40 × 2 = 80	3 × 2 = 6

So £243 × 52
= 10 000 + 2000 + 150 + 400 + 80 + 6
= **£12 636**

3

	200	30
10	200 × 10 = 2000	30 × 10 = 300
4	200 × 4 = 800	30 × 4 = 120

So 230 × 14 = 2000 + 300 + 800 + 120
= **3220**

4

	100	60
80	100 × 80 = 8000	60 × 80 = 4800
5	100 × 5 = 500	60 × 5 = 300

So 160 × 85 = 8000 + 4800 + 500 + 300
= **13 600**

5

	20	2
20	20 × 20 = 400	2 × 20 = 40
7	20 × 7 = 140	2 × 7 = 14

27 biscuits requires 22 × 27
= 400 + 40 + 140 + 14 = 594 g of flour,
so **Hamza cannot make 27 biscuits.**

6 a)

	2000	200	70	1
20	2000 × 20 = 40 000	200 × 20 = 4000	70 × 20 = 1400	1 × 20 = 20
5	2000 × 5 = 10 000	200 × 5 = 1000	70 × 5 = 350	1 × 5 = 5

So 2271 × 25
= 40 000 + 4000 + 1400 + 20
+ 10 000 + 1000 + 350 + 5 = **56 775**

Double check you've added all the numbers from the grid — it's easy to miss one when you're dealing with big numbers.

b)

	5000	600	20	4
40	5000 × 40 = 200 000	600 × 40 = 24 000	20 × 40 = 800	4 × 40 = 160
2	5000 × 2 = 10 000	600 × 2 = 1200	20 × 2 = 40	4 × 2 = 8

So 5624 × 42
= 200 000 + 24 000 + 800 + 160
+ 10 000 + 1200 + 40 + 8 = **236 208**

Using the same method for c) and d):

c) 41 174 **d) 85 994**

Page 12 Exercise 4

1 a) 0 5 1
7)3 5 7
So 357 ÷ 7 = **51**

b) 0 3 4
7)2 3 ^{2}8
So 238 ÷ 7 = **34**

Using the same method for c)-f):

c) 67 **d) 452**
e) 560 **f) 622**

2 a) 0 7 2
5)3 6 ^{1}1 remainder 1
So 361 ÷ 5 = **$72\frac{1}{5}$**

b) 1 7 3 7
3)5 ^{2}2 ^{1}1 ^{2}3 remainder 2
So 5213 ÷ 3 = **$1737\frac{2}{3}$**

Using the same method for c)-f):

c) $839\frac{3}{5}$ **d) $74\frac{5}{9}$**
e) $424\frac{1}{6}$ **f) $334\frac{1}{4}$**

3 a) 2 8
23)6 4 4
−4 6
1 8 4
−1 8 4
0
So 644 ÷ 23 = **28**

b) 3 2
19)6 0 8
−5 7
3 8
− 3 8
0
So 608 ÷ 19 = **32**

Using the same method for c)-f):

c) 41 **d) 48**
e) 310 **f) 352**

4 a) 2 6
32)8 4 6
− 6 4
2 0 6
− 1 9 2
1 4
So 846 ÷ 32 = **$26\frac{14}{32}$** = **$26\frac{7}{16}$**

b) 2 1
45)9 4 5
−9 0
4 5
− 4 5
0
So 945 ÷ 45 = **21**

Using the same method for c)-f):

c) $33\frac{18}{21}$ = $33\frac{6}{7}$ **d) $268\frac{18}{25}$**
e) $268\frac{15}{31}$ **f) $194\frac{3}{44}$**

5 £5.00 = 500p
1 2
41)5 0 0
−4 1
9 0
− 8 2
8
So 500 ÷ 41 = 12 remainder 8
So you can buy **12 stamps** and you would get **8p change.**

Make sure you interpret what the remainder means in the situation in the question.

6 2 8
24)6 7 2
−4 8
1 9 2
−1 9 2
0
Each group will have **28 people**.

7 2 2 5
6)1 3 ^{1}5 ^{3}0
You need **225 boxes**.

8 3 4 7
22)7 6 3 4
−6 6
1 0 3
− 8 8
1 5 4
− 1 5 4
0
Each washing machine costs **£347**.

9 a) 4 3 2
52)2 2 4 6 4
−2 0 8
1 6 6
−1 5 6
1 0 4
− 1 0 4
0
Mike earns **£432 a week**.

b) 0 2 0 5 3
12)2 4 6 ^{6}3 ^{3}6
Andrea earns **£2053 a month**.

1.4 Decimals

Page 13 Exercise 1

1 The 7 is in the hundredths column so has value $\frac{7}{100}$.

2 The 3 is in the tenths column so has value $\frac{3}{10}$.

3 The 8 is in the thousandths column so has value $\frac{8}{1000}$.

4 The 1 is in the ten thousandths column so has value $\frac{1}{10000}$.

5 The 2 is in the units column so has value **2**.

6 a) Seven tenths = **0.7**

b) Two hundredths = **0.02**

c) Five thousandths = **0.005**

d) $\frac{7}{100}$ = seven hundredths = **0.07**

e) $\frac{1}{1000}$ = one thousandth = **0.001**

f) $\frac{8}{10}$ = eight tenths = **0.8**

Page 14 Exercise 2

1 a) 0.3 and 0.31 can be written as 0.30 and 0.31. Columns are equal until the hundredths column where 1 is bigger than 0, so the larger number is **0.31**

b) 0.09 and 0.009 can be written as 0.090 and 0.009. Columns are equal until the hundredth column where 9 is bigger than 0, so the larger number is **0.09**

c) Compare 0.427 and 0.472. Columns are equal until the hundredth column where 7 is bigger than 2, so the larger number is **0.472**

d) Compare 18.07 and 17.08. Columns are equal until the units column where 8 is bigger than 7, so the larger number is **18.07**

2 a) Write the numbers as 0.02, 0.20 and 0.15. Columns are equal until the tenths column where 2 is the largest, 1 is the next largest and then 0 is the smallest, so the correct order is **0.02, 0.15, 0.2**

b) Positive numbers are always bigger than the negative numbers.
Sorting the positive numbers 0.6 and 6.1: In the units column 6 is bigger than 0, so 6.1 is bigger than 0.6.
Sorting the negative numbers –0.6 and –6: In the units column 6 is bigger than 0, so –6 is lower than –0.6.
Remember — with negative numbers, the number with the larger digit in the column you're comparing is the lower number.
So the correct order is **–6, –0.6, 0.6, 6.1**

Using the same method for c)-f):
c) 4.05, 4.5, 5.04, 5.4
d) –1.5, 1.05, 1.5, 1.55
e) 0.15, 0.16, 0.51, 0.61
f) –0.095, –0.09, –0.05, 0.9

1.5 Adding and Subtracting Decimals

Page 15 Exercise 1

1 a) 5.1 + 1.8 = **6.9** **b)** 6.3 + 5.4 = **11.7**
c) 5.7 + 12.6 = **18.3** **d)** 4.8 + 5.3 = **10.1**

2 a) 5.6 – 0.3 = **5.3** **b)** 9.9 – 4.2 = **5.7**
c) 5.3 – 2.8 = **2.5** **d)** 8.5 – 1.9 = **6.6**

3 a) 10.83 + 7.40 = **18.23** **b)** 0.029 + 1.800 = **1.829**
c) 91.700 + 0.492 = **92.192**
Make sure you get those decimal points lined up — add zeros after the decimal point if your numbers aren't the same length.

Using the same method for d)-h):
d) 7.394 **e) 12.679** **f) 1.908**
g) 1.011 **h) 25.579**

4 a) 24.63 – 7.50 = **17.13** **b)** 6.78 – 5.60 = **1.18**
c) 73.46 – 8.50 = **64.96**

Using the same method for d)-h):
d) 6.215 **e) 3.394** **f) 2.08**
g) 9.31 **h) 0.261**

5 a) 3.81 + 9.54 = **13.35** **b)** 2.81 – 0.16 = **2.65**
c) 2.75 + 9.45 = **12.20**

Using the same method for d)-h):
d) 9.62 **e) 3.21** **f) 15.02**
g) 11.85 **h) 23.67**

Page 16 Exercise 2

1 a) 6.0 – 5.1 = **0.9** **b)** 23.00 – 18.51 = **4.49**
c) 12.000 – 5.028 = **6.972** **d)** 13.000 – 6.453 = **6.547**

2 a) 2.0 + 1.8 = **3.8** **b)** 3.7 + 6.0 = **9.7**
c) 12.70 + 7.34 = **20.04**

Using the same method for d)-h):
d) 22.49 **e) 44.92** **f) 4.2**
g) 4.8 **h) 18.3**

3 a) 6.474 + 0.920 = 7.394
So 6.474 + 0.92 + 3 = 7.394 + 3
7.394 + 3.000 = 10.394
So 6.474 + 0.92 + 3 = **10.394**

b) 2.39 + 8.00 = 10.39
So 2.39 + 8 + 0.26 = 10.39 + 0.26
10.39 + 0.26 = 10.65
So 2.39 + 8 + 0.26 =**10.65**

Using the same method for c)-f):
c) 17.44 **d) 7.983**
e) 22.97 **f) 16.652**

Page 17 Exercise 3

1 2.3 + 4.6 = 6.9
⇒ he travels a distance of **6.9 km**.
Remember to give your answer in the context of the question.

2 18.50 + 31.00 = 49.50 ⇒ she spends a total of **£49.50**.

3 4.00 – 2.75 = 1.25 ⇒ **1.25 m** of wood is left.

4 66.50 – 15.00 = 51.50 ⇒ she has **£51.50** left to pay.

5 71.42 + 11.79 = 83.21 ⇒ he would have spent **£83.21**.

6 15.32 – 0.47 = 14.85 ⇒ his second run took **14.85 seconds**.

1.6 Multiplying and Dividing Decimals

Page 19 Exercise 1

1 a) Multiplying by 10 moves the decimal point one place to the right, so 0.92 × 10 = **9.2**
b) Multiplying by 100 moves the decimal point two places to the right, so 1.41 × 100 = **141**
c) Multiplying by 1000 moves the decimal point three places to the right, so 0.23 × 1000 = **230**
Using the same methods for d)-f):
d) 1460 **e) 19** **f) 130.4**

2 a) Dividing by 100 moves the decimal point two places to the left, so 861.5 ÷ 100 = **8.615**
b) Dividing by 10 moves the decimal point one place to the left, so 381.7 ÷ 10 = **38.17**
c) Dividing by 1000 moves the decimal point three places to the left, so 549.1 ÷ 1000 = **0.5491**
Using the same methods for d)-f):
d) 0.63 **e) 0.0051** **f) 0.0094**

Page 19 Exercise 2

1 a) 132 ÷ 10 = 13.2
So 13.2 × 238 = 132 × 238 ÷ 10
= 31 416 ÷ 10 = **3141.6**
b) 132 ÷ 100 = 1.32 and 238 ÷ 10 = 23.8
So 1.32 × 23.8 = 132 × 238 ÷ 100 ÷ 10
= 31 416 ÷ 1000 = **31.416**
c) 132 ÷ 100 = 1.32 and 238 ÷ 1000 = 0.238
So 1.32 × 0.238 = 132 × 238 ÷ 100 ÷ 1000
= 31 416 ÷ 100 000 = **0.31416**
d) 132 ÷ 1000 = 0.132 and 238 ÷ 1000 = 0.238, so
0.132 × 0.238 = 132 × 238 ÷ 1000 ÷ 1000
= 31 416 ÷ 1 000 000 = **0.031416**

2 a) 401 ÷ 10 = 40.1 and 119 ÷ 10 = 11.9
So 40.1 × 11.9 = 401 × 119 ÷ 10 ÷ 10
= 47 719 ÷ 100 = **477.19**
b) 401 ÷ 100 = 4.01 and 119 ÷ 100 = 1.19
So 4.01 × 1.19 = 401 × 119 ÷ 100 ÷ 100
= 47 719 ÷ 10 000
= **4.7719**
c) 401 ÷ 1000 = 0.401 and 119 ÷ 100 = 1.19
So 0.401 × 1.19 = 401 × 119 ÷ 1000 ÷ 100
= 47 719 ÷ 100 000
= **0.47719**
d) 401 ÷ 1000 = 0.401 and 119 ÷ 1000 = 0.119
So 0.401 × 0.119
= 401 × 119 ÷ 1000 ÷ 1000
= 47 719 ÷ 1 000 000 = **0.047719**

Page 20 Exercise 3

1 a) 67 × 8 = 536
6.7 has one digit after the decimal point and 8 has none, so the answer should have one digit after the decimal point.
6.7 × 8 = **53.6**
b) 65 × 9 = 585
0.65 has two digits after the decimal point and 9 has none, so the answer should have two digits after the decimal point.
0.65 × 9 = **5.85**
c) 9 × 8 = 72
0.9 and 0.8 both have one digit after the decimal point, so the answer should have two digits after the decimal point.
0.9 × 0.8 = **0.72**

d) 6 × 3 = 18
0.6 and 0.3 both have one digit after the decimal point, so the answer should have two digits after the decimal point.
0.6 × 0.3 = **0.18**

e) 1 × 6 = 6
0.01 has two digits after the decimal point and 0.6 has one digit after the decimal point. So the answer should have three digits after the decimal point.
0.01 × 0.6 = **0.006**
Don't ignore 0's after the decimal point — here 0.01 has two digits after the decimal point, not one. (NB this only applies if the 0's come before other digits — 0.10 only really has one digit after the decimal point, as you can just knock the 0 off the end.)

f)
$$\begin{array}{r} 6\,1 \\ \times\ \ 7 \\ \hline 4\,2\,7 \end{array}$$
0.61 has two digits after the decimal point and 0.7 has one digit after the decimal point. So the answer should have three digits after the decimal point.
0.61 × 0.7 = **0.427**

g) 33 × 2 = 66
0.33 has two digits after the decimal point and 0.02 has two digits after the decimal point. So the answer should have four digits after the decimal point.
0.33 × 0.02 = **0.0066**

h) 7 × 6 = 42
0.007 and 0.006 both have three digits after the decimal point, so the answer should have six digits after the decimal point.
0.007 × 0.006 = **0.000042**

2 a)
$$\begin{array}{r} 6\,3 \\ \times\ 2\,1 \\ \hline 6\,3 \\ 1\,2\,6\,0 \\ \hline 1\,3_{1}2\,3 \end{array}$$
6.3 has one digit after the decimal point, and 2.1 has one digit after the decimal point, so the answer should have two digits after the decimal point.
6.3 × 2.1 = **13.23**

b)
$$\begin{array}{r} 1\,4 \\ \times\ 2\,3 \\ \hline 4_{1}2 \\ 2\,8\,0 \\ \hline 3_{1}2\,2 \end{array}$$
1.4 has one digit after the decimal point, and 2.3 has one digit after the decimal point, so the answer should have two digits after the decimal point.
1.4 × 2.3 = **3.22**

Using the same method for c)-f):
c) 4.32 **d) 59.34**
e) 0.528 **f) 1.173**

Page 21 Exercise 4

1 a) 3.1 × 40 = 3.1 × 4 × 10
$$\begin{array}{r} 3\,1 \\ \times\ \ 4 \\ \hline 1\,2\,4 \end{array}$$
3.1 has one digit after the decimal point and 4 has none, so the answer should have one digit after the decimal point.
3.1 × 4 = 12.4
So 3.1 × 40 = 12.4 × 10 = **124**
If you're doing a multiplication or division, always see if you can split it into a power of 10 and a smaller number.

b) 0.7 × 600 = 0.7 × 6 × 100
7 × 6 = 42
0.7 has one digit after the decimal point and 6 has none, so the answer should have one digit after the decimal point.
0.7 × 6 = 4.2
So 0.7 × 600 = 4.2 × 100 = **420**

c) 0.061 × 2000 = 0.061 × 2 × 1000
$$\begin{array}{r} 6\,1 \\ \times\ \ 2 \\ \hline 1\,2\,2 \end{array}$$
0.061 has three digits after the decimal point and 2 has none, so the answer should have three digits after the decimal point.
So 0.061 × 2 = 0.122
So 0.061 × 2000 = 0.122 × 1000 = **122**

d) 11.06 × 80 = 11.06 × 8 × 10
$$\begin{array}{r} 1\,1\,0\,6 \\ \times\ \ \ \ 8 \\ \hline 8\,8\,4_{4}8 \end{array}$$
11.06 has two digits after the decimal point and 8 has none, so the answer should have two digits after the decimal point.
11.06 × 8 = 88.48
So 11.06 × 80 = 88.48 × 10 = **884.8**

Using the same method for e)-h):
e) 363 **f) 402.8**
g) 6108 **h) 4515**

Page 22 Exercise 5

1 a) $\begin{array}{r} 2.7 \\ 2\overline{)5.^{1}4} \end{array}$ so 5.4 ÷ 2 = **2.7**

b) $\begin{array}{r} 3.2 \\ 3\overline{)9.6} \end{array}$ so 9.6 ÷ 3 = **3.2**

c) $\begin{array}{r} 1.5\,4 \\ 6\overline{)9.^{3}2\,^{2}4} \end{array}$ so 9.24 ÷ 6 = **1.54**

d) $\begin{array}{r} 0.5\,3 \\ 5\overline{)2.^{2}6\,^{1}5} \end{array}$ so 2.65 ÷ 5 = **0.53**

Using the same method for e)-h):
e) 1.03 **f) 0.45**
g) 14.1 **h) 0.231**

2 a) $\begin{array}{r} 2.1\,3 \\ 4\overline{)8.5\,^{1}2} \end{array}$ so 8.52 ÷ 4 = **2.13**

b) $\begin{array}{r} 0\,2\,8.2 \\ 4\overline{)1\,^{1}1\,^{3}2.8} \end{array}$ so 112.8 ÷ 4 = **28.2**

c) $\begin{array}{r} 0.3\,4 \\ 3\overline{)1.^{1}0\,^{1}2} \end{array}$ so 1.02 ÷ 3 = **0.34**

d) $\begin{array}{r} 0.7\,0\,2\,5 \\ 8\overline{)5.^{5}6\,2\,^{2}0\,^{4}0} \end{array}$ so 5.62 ÷ 8 = **0.7025**

Using the same method for e)-h):
e) 0.0104 **f) 1.5075**
g) 0.6122 **h) 0.0102**

Page 22 Exercise 6

1 a) 6.4 × 10 = 64 and 0.4 × 10 = 4,
so the calculation becomes **64 ÷ 4**

b) $\begin{array}{r} 1\,6 \\ 4\overline{)6\,^{2}4} \end{array}$ so 6.4 ÷ 0.4 = **16**

2 a) 0.384 × 100 = 38.4 and 0.12 × 100 = 12,
so the calculation becomes **38.4 ÷ 12**

b) $\begin{array}{r} 0\,3.2 \\ 12\overline{)3\,8.^{2}4} \end{array}$ so 0.384 ÷ 0.12 = **3.2**

3 a) 3.8 × 1000 = 3800 and 0.008 × 1000 = 8,
so the calculation becomes **3800 ÷ 8**

b) $\begin{array}{r} 0\,4\,7\,5 \\ 8\overline{)3\,8\,^{6}0\,^{4}0} \end{array}$ so 3.8 ÷ 0.008 = **475**

4 a) 6.4 × 10 = 64 and 0.2 × 10 = 2,
so the calculation becomes 64 ÷ 2.
$\begin{array}{r} 3\,2 \\ 2\overline{)6\,4} \end{array}$ so 6.4 ÷ 0.2 = **32**
Make sure the number you are dividing by is a whole number.

b) 3.54 × 10 = 35.4 and 0.4 × 10 = 4,
so the calculation becomes 35.4 ÷ 4.
$\begin{array}{r} 0\,8.8\,5 \\ 4\overline{)3\,^{3}5.^{3}4\,^{2}0} \end{array}$ so 3.54 ÷ 0.4 = **8.85**

c) 0.624 × 10 = 6.24 and 0.3 × 10 = 3,
so the calculation becomes 6.24 ÷ 3.
$\begin{array}{r} 2.0\,8 \\ 3\overline{)6.2\,^{2}4} \end{array}$ so 0.624 ÷ 0.3 = **2.08**

5 a) 22.56 × 100 = 2256 and 0.03 × 100 = 3,
so the calculation becomes 2256 ÷ 3.
$\begin{array}{r} 0\,7\,5\,2 \\ 3\overline{)2\,2\,^{1}5\,6} \end{array}$ so 22.56 ÷ 0.03 = **752**

b) 0.257 × 100 = 25.7 and 0.05 × 100 = 5,
so the calculation becomes 25.7 ÷ 5.
$\begin{array}{r} 0\,5.1\,4 \\ 5\overline{)2\,^{2}5.7\,^{2}0} \end{array}$ so 0.257 ÷ 0.05 = **5.14**

c) 0.039 × 100 = 3.9 and 0.06 × 100 = 6,
so the calculation becomes 3.9 ÷ 6.
$\begin{array}{r} 0.\,6\,5 \\ 6\overline{)3.^{3}9\,^{3}0} \end{array}$ so 0.039 ÷ 0.06 = **0.65**

6 a) 0.081 × 1000 = 81 and 0.009 × 1000 = 9,
so the calculation becomes 81 ÷ 9 = **9**

b) 0.008 × 10 = 0.08 and 0.4 × 10 = 4,
so the calculation becomes 0.08 ÷ 4
$\begin{array}{r} 0.0\,2 \\ 4\overline{)0.0\,8} \end{array}$ so 0.008 ÷ 0.4 = **0.02**

c) 1.44 × 10 = 14.4 and 1.2 × 10 = 12,
so the calculation becomes 14.4 ÷ 12
$\begin{array}{r} 0\,1.\,2 \\ 12\overline{)1\,4.^{2}4} \end{array}$ so 1.44 ÷ 1.2 = **1.2**

Page 23 Exercise 7

1 a) $\begin{array}{r} 1.\,4\,6 \\ 5\overline{)7.^{2}3\,^{3}0} \end{array}$ so 7.3 ÷ 5 = **1.46**

b) 7.3 ÷ 50 = (7.3 ÷ 5) ÷ 10
= 1.46 ÷ 10 = **0.146**

2 a) $\begin{array}{r} 0.6\,0\,2\,5 \\ 4\overline{)2.4\,^{1}0\,^{2}0} \end{array}$ so 2.41 ÷ 4 = **0.6025**

b) 2.41 ÷ 400 = (2.41 ÷ 4) ÷ 100
= 0.6025 ÷ 100 = **0.006025**

3 a) 40 = 4 × 10, so first divide by 4 and then divide by 10.
$\begin{array}{r} 1.5\,2 \\ 4\overline{)6.0\,8} \end{array}$ so 6.08 ÷ 4 = 1.52
6.08 ÷ 40 = (6.08 ÷ 4) ÷ 10
= 1.52 ÷ 10 = **0.152**

b) 700 = 7 × 100, so first divide by 7 and then divide by 100.
$\begin{array}{r} 0.\,8\,2 \\ 7\overline{)5.^{5}7\,^{1}4} \end{array}$ so 5.74 ÷ 7 = 0.82
5.74 ÷ 700 = (5.74 ÷ 7) ÷ 100
= 0.82 ÷ 100 = **0.0082**

c) 900 = 9 × 100, so first divide by 9 and then divide by 100.
$\begin{array}{r} 0\,2.\,8\,3 \\ 9\overline{)2\,4.^{7}4\,^{2}7} \end{array}$ so 25.47 ÷ 9 = 2.83
25.47 ÷ 900 = (25.47 ÷ 9) ÷ 100
= 2.83 ÷ 100 = **0.0283**

d) 3000 = 3 × 1000, so first divide by 3 and then divide by 1000.
$\begin{array}{r} 0\,4.\,5\,7\,4 \\ 3\overline{)1\,^{1}3.^{1}7\,^{2}2\,^{1}2} \end{array}$ so 13.722 ÷ 3 = 4.574
13.722 ÷ 3000 = (13.722 ÷ 3) ÷ 1000
= 4.574 ÷ 100 = **0.004574**

Page 24 Review Exercise

1 a) $6 + 4 \times 2 = 6 + 8 =$ **14**

b) $48 \div (4 + 2) = 48 \div 6 =$ **8**

c) $(13 - 5) \times 12 = 8 \times 12 =$ **96**

2 a) $-5 + 8 =$ **3**   b) $6 - (-2) = 6 + 2 =$ **8**

c) $-8 \times -5 =$ **40**   d) $54 \div (-9) =$ **−6**

3 a)
```
  2 5 6
+ 3 1 2
-------
  5 6 8
```
b)
```
 7 8³4¹1
- 3 4 6
-------
  4 9 5
```
c)
```
  1 6 3 2
+   4 2 1
---------
  2 0 5 3
   1
```
d)
```
   7¹2 1
  2 8 3 0
-   3 9 4
---------
  2 4 3 6
```

4
```
   1 4
 × 5 2
------
   2 8
  7 0 0
   2
------
  7 2 8
```
so £14 × 52 = **£728**

5
```
     0 0 8
54 ) 4 6 2
    −4 3 2
    ------
       3 0
```
So 462 ÷ 54 = 8 remainder 30, which means **9 coaches are needed.**

As there is a remainder, 8 coaches wouldn't be enough for everyone on the trip, so you have to round up to 9.

6 The 6 is in the thousands column so it has value $\frac{6}{1000}$.

7 a) **−0.1, −0.09, −0.01, 0.1**

b) **−5, −0.55, −0.5, −0.45**

c) **−7.1, −7.07, −7, 0.007**

8 a)
```
  6 . 7 8
− 5 . 6 0
---------
  1 . 1 8
```
b)
```
  1 . 6 0
+ 4 . 3 5
---------
  5 . 9 5
```
c)
```
  0 . 7 8
+ 1 . 3 0
---------
  2 . 0 8
  1
```
d)
```
  4 .²3¹2
− 2 . 1 7
---------
  2 . 1 5
```

9 a) $7.8 \times 1000 =$ **7800**

b) $0.006 \times 100 =$ **0.6**

c) $25.9 \div 10 =$ **2.59**

d) $901.5 \div 100 =$ **9.015**

10 a) $221 \div 100 = 2.21$ and $168 \div 100 = 1.68$

So $2.21 \times 1.68 = 221 \times 168 \div 100 \div 100$
$= 37\,128 \div 10\,000$
$=$ **3.7128**

b) $221 \div 1000 = 0.221$ and $168 \div 100 = 1.68$,

So $0.221 \times 1.68 = 221 \times 168 \div 1000 \div 100$
$= 37\,128 \div 100\,000$
$=$ **0.37128**

c) $168 \div 1000 = 0.168$,

So $221 \times 0.168 = 221 \times 168 \div 1000$
$= 37\,128 \div 1000 =$ **37.128**

11
```
  1 7 6
×     5
-------
  8 8 0
   3 3
```
there are two digits after the decimal point in 1.76 so 1.76 × 5 = **8.8 pints**

12 $2.72 \times 100 = 272$ and $0.08 \times 100 = 8$, so the calculation becomes **272 ÷ 8.**
```
   0 3 4
8 ) 2 ²7 ³2
```
so $2.72 \div 0.08 =$ **34 pieces**

Page 25 Exam-Style Questions

1 $-4° - 7° =$ **−11°** *[1 mark]*

2
```
  1 2 . 0
+   1 . 3
---------
  1 3 . 3

  1 3 . 3 0
+   0 . 2 5
-----------
  1 3 . 5 5
```
So $12 + 1.3 + 0.25 =$ **13.55** *[1 mark]*

3 a) $539 \times 14 = 539 \times 28 \div 2$
$= 15\,092 \div 2 =$ **7546** *[1 mark]*

b) $539 \times 10 = 5390$ and $28 \div 100 = 0.28$
So $5390 \times 0.28 = 539 \times 28 \times 10 \div 100$
$= 15\,092 \div 10$
$=$ **1509.2** *[1 mark]*

c) Rearrange the equation given:
$15\,092 \div 539 = 28.$ *[1 mark]*
$15\,092 \times 100 = 1\,509\,200$
and $539 \div 10 = 53.9$
So $1\,509\,200 \div 53.9$
$= (15\,092 \times 100) \div (539 \div 10)$
$= 15\,092 \div 539 \times 100 \times 10$
$= 28 \times 1000 =$ **28 000** *[1 mark]*

Careful here — dividing by (539 ÷ 10) is the same as dividing by 539 and then multiplying by 10.

4 E.g. She has performed the addition before the multiplication, to get $5 + 3 \times 4 = 8 \times 4 = 32$, which doesn't follow the rules of BODMAS. *[1 mark]*

She should have done the multiplication first, to get the correct answer of $5 + 3 \times 4 = 5 + 12 =$ **17**. *[1 mark]*

5
```
    0 3 . 9 5
9 ) 3 ³5 . ⁸5 ⁴5
```
so one book costs £35.55 ÷ 9 = £3.95 *[1 mark]*
```
   3 9 5
 ×     7
--------
  2 7 6 5
   2 6 3
```
so 7 books cost £3.95 × 7 = **£27.65** *[1 mark]*

6 a) $3 \times (2 - 4) \div 2 = -3$ *[1 mark]*

b) $(8 + 6 \div 5) \times 10 = 92$ *[1 mark]*

Section 2 — Approximations

2.1 Rounding — Whole Numbers

Page 27 Exercise 1

1 The last digit is in the units place and the decider is the first number after the decimal point.

For 14.1 the decider is 1 so round down to 14.
For 14.9 the decider is 9 so round up to 15.
For 14.02 the decider is 0 so round down to 14.

Use the same method to round the remaining numbers. So the following numbers from the box all round to 14:
14.1, 14.02, 13.7, 14.09, 14.4999, 13.901

2 The last digit is in the units place and the decider is the first number after the decimal point.

a) The decider is 7 so round up to **10**.

b) The decider is 4 so round down to **8**.

c) The decider is 2 so round down to **12**.

d) The decider is 8 so round up to **40**.

3 The last digit is in the hundreds place and the decider is in the tens place.

a) The decider is 5 so round up to **200**.

b) The decider is 9 so round up to **600**.

c) The decider is 5 so round up to **700**.

d) The decider is 1 so round down to **4700**.

4 The last digit is in the thousands place and the decider is in the hundreds place.

a) The decider is 5 so round up to **3000**.

b) The decider is 5 so round up to **9000**.

c) The decider is 2 so round down to **7000**.

d) The decider is 5 so round up to **10 000**.

5 a) (i) The last digit is 8 and the decider is 2 so round down to **18**.

(ii) The last digit is 1 and the decider is 8 so round up to **20**.

b) (i) The last digit is 6 and the decider is 4 so round down to **16**

(ii) The last digit is 1 and the decider is 6 so round up to **20**

Using the same method for c)-d):

c) (i) **202**   (ii) **200**

d) (i) **1**   (ii) **0**

6 The last digit is 1 and the decider is 2, so round down to **301 000 km²**.

7 The last (millions) digit is 0 for 39<u>0</u> 682 810. The decider is 6, so round up to **391 000 000 miles**.

2.2 Rounding — Decimal Places

Page 29 Exercise 1

1 The last digit is in the first d.p. and the decider is in the second d.p.

For 0.41, the decider is 1 so round down to 0.4.
For 0.45, the decider is 5 so round up to 0.5.
For 0.347, the decider is 4 so round down to 0.3.

Use the same method to round the remaining numbers. The following numbers from the box all round to 0.4:
0.41, 0.405, 0.35, 0.4295, 0.4124

2 The last digit is in the first d.p. and the decider is in the second d.p.

a) The decider is 3 so round down to **0.2**.

b) The decider is 7 so round up to **0.7**.

c) The decider is 8 so round up to **2.7**.

d) The decider is 3 so round down to **0.9**.

3 The last digit is in the second d.p. and the decider is in the third d.p.

a) The decider is 7 so round up to **4.57**.

b) The decider is 2 so round down to **0.04**.

c) The decider is 7 so round up to **6.30**.

The answer is 6.30 not 6.3 as you're rounding to 2 d.p.

d) The decider is 2 so round down to **0.35**.

4 The last digit is in the third d.p. and the decider is in the fourth d.p.

a) The decider is 3 so round down to **0.967**.

b) The decider is 7 so round up to **0.255**.

c) The decider is 5 so round up to **2.437**.

d) The decider is 5 so round up to **6.533**.

5 a) The last digit is 9 and the decider is 7, so round up to **0.20**.

b) The last digit is 6 and the decider is 8, so round up to **0.7**.

c) The last digit is 1 and the decider is 7, so round up to **5.732**.

d) The last digit is 0 and the decider is 6, so round up to **0.001**.

6 The last digit is 3 and the decider is 8 so round up to **0.04 kg.**

Remember to give the units in your answer — here it's 'kg'.

7 10 cm = 0.1 m so the last digit is in the first d.p. Suppose Usain's exact height was 1.76 m. The last digit is 7 and the decider is 6 so his height would round up to 1.8 m to the nearest 10 cm. So **no**, his exact height could not be 1.76 m.

2.3 Rounding — Significant Figures

Page 31 Exercise 1

1 a) (i) The last digit is the first s.f., which is 7. The decider is 0 so round down to **7000.**
(ii) The last digit is the second s.f., which is 0. The decider is 3 so round down to **7000.**
(iii) The last digit is the third s.f., which is 3. The decider is 6 so round up to **7040**.
b) (i) The last digit is the first s.f., which is 6. The decider is 5 so round up to **7000**.
(ii) The last digit is the second s.f., which is 5. The decider is 5 so round up to **6600**.
(iii) The last digit is the third s.f., which is 5. The decider is 1 so round down to **6550**.
Use the same method for c)-d):
c) (i) **7000** (ii) **7100** (iii) **7070**
d) (i) **3000** (ii) **2600** (iii) **2650**

2 a) The last digit is 4 and the decider is 5, so round up to **50**.
b) The last digit is 8 and the decider is 9, so round up to **5690**.
c) The last digit is 9 and the decider is 7, so round up to **6.50**.
Use the same method for d)-i):
d) **360** e) **6500** f) **757 000**
g) **46.7** h) **380** i) **79 000**

3 The last digit is 2 and the decider is 3, so round down to **1200 km/h.**

Page 31 Exercise 2

1 a) (i) The last digit is the first s.f., which is 3. The decider is 7 so round up to **0.004**.
(ii) The last digit is the second s.f., which is 7. The decider is 5 so round up to **0.0038**.
(iii) The last digit is the third s.f., which is 5. The decider is 3 so round down to **0.00375**.
b) (i) The last digit is the first s.f., which is 2. The decider is 6 so round up to **0.03**.
(ii) The last digit is the second s.f., which is 6. The decider is 4 so round down to **0.026**.
(iii) The last digit is the third s.f., which is 4. The decider is 4 so round down to **0.0264**.
Use the same method for c)-f):
c) (i) **0.0002** (ii) **0.00018**
(iii) **0.000179**
d) (i) **0.04** (ii) **0.040**
(iii) **0.0397**
e) (i) **0.6** (ii) **0.56**
(iii) **0.564**
f) (i) **0.0007** (ii) **0.00070**
(iii) **0.000705**

2 a) The decider is 5 so round up to **0.005**.
b) The decider is 6 so round up to **0.20**.
c) The decider is 6 so round up to **0.00439**.
Use the same method for d)-f):
d) **0.006** e) **0.0096** f) **0.000604**

3 The last digit is 8 and the decider is 9, so round up to **0.09 kg/m³**.

2.4 Estimating Answers

Page 33 Exercise 1

1 a) $437 + 175 \approx 400 + 200 =$ **600**
b) $310 + 876 \approx 300 + 900 =$ **1200**
c) $784 - 279 \approx 800 - 300 =$ **500**
Using the same method for d)-f):
d) **0.2** e) **900** f) **700**

2 a) (i) $23 + 43 \approx 20 + 40 =$ **60**
(ii) Both numbers were rounded down so it will be an **underestimate**.
b) (i) $59 \times 5.7 \approx 60 \times 6 =$ **360**
(ii) Both numbers were rounded up so it will be an **overestimate**.
Using the same method for c)-f):
c) (i) **45** (ii) **Underestimate**
d) (i) **1600** (ii) **Overestimate**
e) (i) **320** (ii) **Overestimate**
f) (i) **5400** (ii) **Overestimate**

3 a) $1.76 \times 6.3 \approx 2 \times 6 = 12$.
So correct answer is **C**.
b) $582 \times 2.1 \approx 600 \times 2 = 1200$.
So correct answer is **B**.
c) $\frac{57.5 \times 3.78}{16.1} \approx \frac{60 \times 4}{20} = \frac{240}{20} = 12$.
So correct answer is **C**.

4 a) (i) $\frac{8.9}{3.1} \approx \frac{9}{3} = \mathbf{3}$
(ii) First number was rounded up and the second number was rounded down so it will be an **overestimate**.
b) (i) $33 - 17 \approx 30 - 20 =$ **10**
(ii) First number was rounded down and the second number was rounded up so it will be an **underestimate**.
Using the same method for c)-f):
c) (i) **2** (ii) **Underestimate**
d) (i) **35** (ii) **Overestimate**
e) (i) **30** (ii) **Underestimate**
f) (i) **5** (ii) **Overestimate**

5 a) $\frac{68.8 + 27.3}{23.7} \approx \frac{70 + 30}{20} = \frac{100}{20} = \mathbf{5}$
b) $\frac{5.6 \times 9.68}{5.14} \approx \frac{6 \times 10}{5} = \frac{60}{5} = \mathbf{12}$
c) $\frac{\sqrt{38.6 + 56.3}}{1.678} \approx \frac{\sqrt{40 + 60}}{2} = \frac{\sqrt{100}}{2} = \mathbf{5}$

6 a) $18.5 \times 3.2 \approx 20 \times 3 = 60$, so actual answer is **59.2**.
b) $\frac{325.26}{5.2} \approx \frac{300}{5} = 60$, so actual answer is **62.55**.
c) $\frac{19.8 \times 27.4}{3.3} \approx \frac{20 \times 30}{3} = 200$, so actual answer is **164.4**.
d) $\frac{\sqrt{48.4 \times 8.1}}{4.8} \approx \frac{\sqrt{50 \times 8}}{5} = \frac{\sqrt{400}}{5} = 4$, so actual answer is **4.125**.

7 £4.70 × 21 ≈ £5 × 20 = **£100**

8 a) $62 \times 324 \times 14 \approx 60 \times 300 \times 10$ = **180 000 strawberries**
b) All numbers were rounded down so it will be an **underestimate**.

Page 34 Exercise 2

1 a) (i) **30283.518**
(ii) $112.62 \times 268.9 \approx 100 \times 300 =$ **30 000**
b) (i) **142.12**
(ii) $\frac{52.668 \times \sqrt{104.04}}{3.78} \approx \frac{50 \times \sqrt{100}}{4} = \frac{500}{4}$ = **125**
c) (i) **123.5925705**
(ii) $5.39^2 \times \sqrt[3]{1012} \div 2.36$
$\approx 5^2 \times \sqrt[3]{1000} \div 2$
$= 25 \times 10 \div 2 =$ **125**

2 a) $56.2 \times 34.7 \approx 60 \times 30 =$ **1800**
b) **Yes**, as Sam's answer is approximately 10 times bigger than the estimate.

3 987 × £27.85 ≈ 1000 × £30 = £30 000
Both values have been rounded up, so the approximation is an overestimate. So the company must be wrong as their figure is even higher.

4 £6.85 × 42 ≈ 7 × 40 = £280, so it looks like Karen has been paid correctly.

2.5 Rounding Errors

Page 35 Exercise 1

1 a) Rounding unit is 1 and half the rounding unit is 0.5.
Maximum value is 80 + 0.5 = **80.5**
Minimum value is 80 – 0.5 = **79.5**
b) Rounding unit is 100 and half the rounding unit is 50.
Maximum value is 400 + 50 = **450**
Minimum value is 400 – 50 = **350**

2 a) Rounding unit is 10 and half the rounding unit is 5.
Maximum value is 60 + 5 = 65
Minimum value is 60 – 5 = 55
So error interval is $\mathbf{55 \le a < 65}$
Remember: the actual value is greater than or equal to the minimum value, but strictly less than the maximum value.
b) Rounding unit is 1 and half the rounding unit is 0.5.
Maximum value is 9 + 0.5 = 9.5
Minimum value is 9 – 0.5 = 8.5
So error interval is $\mathbf{8.5 \le b < 9.5}$
c) Rounding unit is 100 and half the rounding unit is 50.
Maximum value is 500 + 50 = 550
Minimum value is 500 – 50 = 450
So error interval is $\mathbf{450 \le c < 550}$
d) Rounding unit is 1000 and half the rounding unit is 500.
Maximum value is 15 000 + 500 = 15 500
Minimum value is 15 000 – 500 = 14 500
So error interval is $\mathbf{14\,500 \le d < 15\,500}$

3 a) Rounding unit is 0.1 and half the rounding unit is 0.05.
Maximum value is 7.6 + 0.05 = 7.65
Minimum value is 7.6 – 0.05 = 7.55
So error interval is $\mathbf{7.55 \le a < 7.65}$
b) Rounding unit is 0.1 and half the rounding unit is 0.05.
Maximum value is 0.3 + 0.05 = 0.35
Minimum value is 0.3 – 0.05 = 0.25
So error interval is $\mathbf{0.25 \le b < 0.35}$
c) Rounding unit is 0.01 and half the rounding unit is 0.005.
Maximum value is 2.55 + 0.005 = 2.555
Minimum value is 2.55 – 0.005 = 2.545
So error interval is $\mathbf{2.545 \le c < 2.555}$
Using the same method for d)-f):
d) $\mathbf{45 \le d < 55}$
e) $\mathbf{109.85 \le e < 109.95}$
f) $\mathbf{535 \le f < 545}$

4 The rounding unit is 1 and half the rounding unit is 0.5.
Maximum length is 76 + 0.5 = 76.5
Minimum length is 76 – 0.5 = 75.5.
So the error interval is **75.5 cm** $\le l <$ **76.5 cm**.

5 1 km = 1000 m so 100 m = 0.1 km.
The rounding unit is 0.1 and half the rounding unit is 0.05.
Maximum distance is 10 km + 0.05 = 10.05 km
Minimum distance is 10 km – 0.05 = 9.95 km
So the error interval is
9.95 km ≤ *d* < 10.05 km.

6 The rounding unit is 10 and half the rounding unit is 5.
Maximum number of sweets is 670 + 5 = 675.
But this would round up to 680, so maximum number of sweets is **674**.
The maximum number of sweets in the jar is 674.999999..., but the number of sweets must be a whole number, so the answer is 674.

Page 36 Exercise 2

1 **a)** To truncate 1.354 to 1 d.p. delete all the digits after the first decimal place.
So it's **1.3**.
Use the same method for b)-c):
b) 19.13 **c) 103.671**

2 **a)** The rounding unit is 0.1.
So the error interval is **1.3 ≤ *x* < 1.4**
b) The rounding unit is 0.01.
So the error interval is **5.13 ≤ *y* < 5.14**
c) The rounding unit is 0.001.
So the error interval is **7.731 ≤ *z* < 7.732**
Remember — the actual value of a truncated number can be up to a whole rounding unit bigger, but no smaller.

Page 37 Review Exercise

1 **a)** **(i)** The decider is 4, so round down to **6720**.
(ii) The decider is 2, so round down to **6700**.
(iii) The decider is 7, so round up to **7000**.
b) **(i)** The decider is 1, so round down to **25 360**.
(ii) The decider is 6, so round up to **25 400**.
(iii) The decider is 3, so round down to **25 000**.
Use the same method for c)-d):
c) (i) 8500 **(ii) 8500** **(iii) 8000**
d) (i) 3820 **(ii) 3800** **(iii) 4000**

2 **a)** **(i)** The last digit is 6 and the decider is 8 so round up to **2.7**.
(ii) The last digit is 8 and the decider is 9, so round up to **2.69**.
(iii) The last digit is 9 and the decider is 3, so round down to **2.689**.
b) **(i)** The last digit is 3 and the decider is 2, so round down to **0.3**.
(ii) The last digit is 2 and the decider is 4, so round down to **0.32**.
(iii) The last digit is 4 and the decider is 9, so round up to **0.325**.
Use the same method for c)-d):
c) (i) 5.6 **(ii) 5.60** **(iii) 5.602**
d) (i) 0.1 **(ii) 0.05** **(iii) 0.053**

3 **a)** The decider is 5, so round up to **5000**.
b) The decider is 8, so round up to **57 000**.
c) The decider is 9, so round up to **6.8**.
d) The decider is 6, so round up to **400**.
e) The decider is 2, so round down to **6500**.
f) The decider is 5, so round up to **757 000**.

4 £1.35 + £8.52 + £14.09 + £17.93
≈ £1 + £9 + £14 + £18 = **£42**.

5 **a)** $\frac{64.4 \times 5.6}{17 \times 9.5} \approx \frac{60 \times 6}{20 \times 10} = \frac{360}{200} = \mathbf{1.8}$
b) $\frac{310.33 \times 2.68}{316.39 \times 0.82} \approx \frac{300 \times 3}{300 \times 1} = \frac{900}{300} = \mathbf{3}$
c) $\frac{13.7 \times 5.2}{12.3 \div 3.9} \approx \frac{10 \times 5}{10 \div 4} = \frac{50}{2.5} = \mathbf{20}$

6 **a)** Round each number to 1 s.f. to estimate.
32p × 14 ≈ 30p × 10 = 300p = **£3**.
b) Both numbers have been rounded down, so it's an **underestimate**.

7 **a)** Use a calculator to find $24.37 \div \sqrt{3.9}$ = **12.3402...**
b) $24.37 \div \sqrt{3.9} \approx 20 \div \sqrt{4} = 20 \div 2 = 10$.
Yes, the answer to a) is sensible.

8 **a)** The rounding unit is 10, so half the rounding unit is 5.
Maximum value is 50 + 5 = 55
Minimum value is 50 – 5 = 45
So the error interval is **45 ≤ *a* < 55**
b) The rounding unit is 10, so half the rounding unit is 5.
Maximum value is 5690 + 5 = 5695
Minimum value is 5690 – 5 = 5685
So the error interval is **5685 ≤ *b* < 5695**
c) The rounding unit is 1, so half the rounding unit is 0.5.
Maximum value is 7 + 0.5 = 7.5
Minimum value is 7 – 0.5 = 6.5
So the error interval is **6.5 ≤ *c* < 7.5**
Use the same method for d)-f):
d) 355 ≤ *d* < 365
e) 6450 ≤ *e* < 6550
f) 756 500 ≤ *f* < 757 500

9 For all parts, the rounding unit is 0.01. The actual value can be a whole rounding unit bigger than the truncated number, but no smaller.
a) 6.57 ≤ *s* < 6.58
b) 25.71 ≤ *t* < 25.72
c) 13.29 ≤ *w* < 13.30

Page 38 Exam-Style Questions

1 The last digit is 7 and the decider is 6, so round up to **45.8** *[1 mark]*.

2 **a)** $\frac{3.5^4}{\sqrt{0.007}} = \frac{150.0625}{0.08366...}$ *[1 mark]*
= **1793.589932** *[1 mark]*
Your answer may be different depending on how many digits your calculator can display.
b) **(i)** The decider is 9, so round up to **1793.590** *[1 mark]*
The final '0' is required to earn the mark.
(ii) The decider is 3, so round down to **1790** *[1 mark]*

3 **a)** $\frac{628}{\sqrt{97} + 9.6} \approx \frac{600}{\sqrt{100} + 10} = \frac{600}{10 + 10} = \frac{600}{20}$
= **30**
[2 marks available — 1 mark for correctly rounding all numbers to 1 s.f., 1 mark for correct answer]
b) The calculation is a division where the numerator has been rounded down and the denominator has been rounded up, so it's an **underestimate**. *[1 mark]*
Both numbers in the denominator (97 and 9.6) have been rounded up, so overall their sum has also been rounded up.

4 The rounding unit is 100 and half the rounding unit is 50.
Maximum value is 7300 + 50 = 7350
Minimum value is 7300 – 50 = 7250
So the error interval is **7250 ≤ *r* < 7350**.
[2 marks available — 1 mark for correct numbers, and 1 mark for correct inequality signs]

5 **a)** The rounding unit is 1 and half the rounding unit is 0.5.
Maximum value is 8 + 0.5 = 8.5
Minimum value is 8 – 0.5 = 7.5
So the error interval is **7.5 kg ≤ *r* < 8.5 kg**
[2 marks available — 1 mark for correct numbers, and 1 mark for correct inequality signs]
b) If the dogs weigh 7.5 kg and 5.5 kg, then their weight would round to 8 kg and 6 kg to the nearest kg.
But the total would be 7.5 kg + 5.5 kg = 13 kg, which is less than 14 kg.
[2 marks available — 1 mark for each weight rounded correctly, 1 mark for showing sum of the weights is less than 14 kg.]

Section 3 — Powers and Roots

3.1 Squares, Cubes and Roots

Page 40 Exercise 1

1 **a)** $6^2 = 6 \times 6 = \mathbf{36}$
Using the same method for b)-d):
b) 144 **c) 225** **d) 400**
e) $(-4)^2 = -4 \times -4 = \mathbf{16}$
Using the same method for f)-h):
f) 0.09 **g) 0.36** **h) 0.04**

2 **a)** $3^3 = 3 \times 3 \times 3 = \mathbf{27}$
This is one of the standard cubes that is worth remembering.
Using the same method for b)-d):
b) 216 **c) 1331** **d) 8000**
e) $(-3)^3 = -3 \times -3 \times -3 = \mathbf{-27}$
Using the same method for f)-h):
f) –1000 **g) 0.064** **h) –0.125**

Page 40 Exercise 2

1 **a) 1, –1** **b) 2, –2** **c) 3, –3** **d) 4, –4**
e) 5, –5 **f) 6, –6** **g) 8, –8** **h) 10, –10**

2 **a) 7** **b) –7** **c) 9** **d) –9**
e) 11 **f) 13** **g) –12** **h) 20**

3 **a) 8, –8** **b) 11, –11** **c) 100, –100**
d) 14, –14

4 **a)** $\sqrt{9} + \sqrt{16} = 3 + 4 = \mathbf{7}$
b) $\sqrt{25} - \sqrt{4} = 5 - 2 = \mathbf{3}$
c) $\sqrt{100} - \sqrt{49} = 10 - 7 = \mathbf{3}$

Page 41 Exercise 3

1

x	1	8	27	1000	–1	–8	–27	–1000
$\sqrt[3]{x}$	**1**	**2**	**3**	**10**	**–1**	**–2**	**–3**	**–10**

2 **a) 4** **b) 5** **c) 11** **d) –4**
e) –5 **f) 8** **g) 6** **h) –9**

3 **a)** $\sqrt[3]{15 - 7} = \sqrt[3]{8} = \mathbf{2}$
b) $\sqrt[3]{(39 + 5^2)} = \sqrt[3]{39 + 25} = \sqrt[3]{64} = \mathbf{4}$
c) $\sqrt[3]{4^2 - 43} = \sqrt[3]{16 - 43} = \sqrt[3]{-27} = \mathbf{-3}$

3.2 Indices

Page 42 Exercise 1

1 **a)** $\mathbf{3^2}$ **b)** $\mathbf{2^3}$ **c)** $\mathbf{7^5}$
d) $\mathbf{9^6}$ **e)** $\mathbf{12^4}$ **f)** $\mathbf{17^3}$

2 **a)** **(i) 100 000** **(ii) 10 000 000**
(iii) 100 000 000 **(iv) 1 000 000 000**
b) **(i)** 10^{15} can be written as a '1' followed by **15** zeros.
(ii) 10^n can be written as a '1' followed by ***n*** zeros.
The rule in part (ii) is a key one to remember — it makes it really easy to work out powers of 10.

3 a) $\mathbf{10^2}$ b) $\mathbf{10^3}$ c) $\mathbf{10^4}$ d) $\mathbf{10^6}$

4 a) **16** b) **32** c) **81** d) **4096**
e) **1296** f) **4913** g) **3125** h) **243**

5 a) $3^4 + 2^3 = 81 + 8 = \mathbf{89}$
b) $2^6 + 3^5 = 64 + 243 = \mathbf{307}$
c) $3^7 - 4^2 = 2187 - 16 = \mathbf{2171}$
d) $10^3 - 6^4 = 1000 - 1296 = \mathbf{-296}$

6 a) $8^7 \div 4^6 = 2\,097\,152 \div 4096 = \mathbf{512}$
b) $10^4 \times 10^3 = 10\,000 \times 1000 = \mathbf{10\,000\,000}$
c) $2^4 \times 2^2 = 16 \times 4 = \mathbf{64}$
d) $3^4 \div 5^4 = 81 \div 625 = \mathbf{0.1296}$

7 a) $(5 - 2)^3 = 3^3 = \mathbf{27}$
b) $(2^2)^2 = 4^2 = \mathbf{16}$
c) $(3^2)^2 = 9^2 = \mathbf{81}$
d) $(8 - 5)^4 = 3^4 = \mathbf{81}$
e) $(7 + 3)^5 = 10^5 = \mathbf{100\,000}$
f) $6^4 - 7^2 = 1296 - 49 = \mathbf{1247}$
g) $2 + 10^4 = 2 + 10\,000 = \mathbf{10\,002}$
h) $(150 - 50)^6 = 100^6 = \mathbf{1\,000\,000\,000\,000}$

Page 43 Exercise 2

1 a) $\mathbf{h^4}$ b) $\mathbf{t^5}$ c) $\mathbf{s^7}$

2 a) $\mathbf{a^2b^3}$ b) $\mathbf{k^4f^3}$ c) $\mathbf{m^4n^2}$
d) $\mathbf{s^4t}$ e) $\mathbf{w^3v^3}$ f) $\mathbf{p^2q^5}$

3 a) $2^2 \times 5^2 = 4 \times 25 = \mathbf{100}$
b) $2^3 \times 5^2 = 8 \times 25 = \mathbf{200}$
c) $2^2 \times 5^3 = 4 \times 125 = \mathbf{500}$
d) $2^5 \times 5^2 = 32 \times 25 = \mathbf{800}$
e) $2^4 \times 5^3 = 16 \times 125 = \mathbf{2000}$
f) $2^4 \times 5^4 = 16 \times 625 = \mathbf{10\,000}$

3.3 Laws of Indices

Page 44 Exercise 1

1 a) $3^2 \times 3^6 = 3^{2+6} = \mathbf{3^8}$
b) $10^7 \times 10^3 = 10^{7+3} = \mathbf{10^{10}}$
c) $4^7 \times 4^4 = 4^{7+4} = \mathbf{4^{11}}$
d) $7 \times 7^6 = 7^{1+6} = \mathbf{7^7}$

2 a) $6^7 \div 6^4 = 6^{7-4} = \mathbf{6^3}$
b) $8^6 \div 8^3 = 8^{6-3} = \mathbf{8^3}$
c) $5^7 \div 5^2 = 5^{7-2} = \mathbf{5^5}$
d) $6^8 \div 6^6 = 6^{8-6} = \mathbf{6^2}$

3 a) $(4^3)^3 = 4^{3\times3} = \mathbf{4^9}$
b) $(11^2)^5 = 11^{2\times5} = \mathbf{11^{10}}$
c) $(100^3)^{23} = 100^{3\times23} = \mathbf{100^{69}}$
d) $\frac{2^8}{2^5} = 2^8 \div 2^5 = \mathbf{2^3}$
e) $\left(\frac{2^7}{5}\right)^3 = \mathbf{\frac{2^{21}}{5^3}}$
f) $\left(\frac{4^6}{4^3}\right)^2 = (4^{6-3})^2 = (4^3)^2 = \mathbf{4^6}$

4 a) $4^5 \times 4^{11} = 4^{5+11} = \mathbf{4^{16}}$
b) $12^7 \div 12^3 = 12^{7-3} = \mathbf{12^4}$
c) $8^2 \times 8^9 = 8^{2+9} = \mathbf{8^{11}}$
d) $(6^8)^4 = 6^{8\times4} = \mathbf{6^{32}}$
e) $(3^{12})^4 = 3^{12\times4} = \mathbf{3^{48}}$
f) $7^{11} \div 7^6 = 7^{11-6} = \mathbf{7^5}$
g) $4^{15} \div 4^7 = 4^{15-7} = \mathbf{4^8}$
h) $(11^0)^9 = 11^{0\times9} = 11^0 = \mathbf{1}$
For this final part you could also write $(11^0)^9 = (1)^9 = 1$.

5 a) $9 + ■ = 14 \Rightarrow ■ = 5$
b) $5 - ■ = 3 \Rightarrow ■ = 2$
c) $7 + ■ = 13 \Rightarrow ■ = 6$
d) $7 \times ■ = 49 \Rightarrow ■ = 7$
e) $■ - 5 = 0 \Rightarrow ■ = 5$
f) $■ + 7 = 15 \Rightarrow ■ = 8$
g) $5 - (-4) = 9 \Rightarrow ■ = 9$
h) $5 \times ■ = 25 \Rightarrow ■ = 5$

Page 45 Exercise 2

1 a) $8^{-2} = \frac{1}{8^2} = \mathbf{\frac{1}{64}}$
b) $2^{-3} = \frac{1}{2^3} = \mathbf{\frac{1}{8}}$
c) $5^{-2} = \frac{1}{5^2} = \mathbf{\frac{1}{25}}$
d) $3^{-3} = \frac{1}{3^3} = \mathbf{\frac{1}{27}}$
e) $2^{-4} = \frac{1}{2^4} = \mathbf{\frac{1}{16}}$
f) $\left(\frac{1}{9}\right)^{-2} = \left(\frac{9}{1}\right)^2 = 9^2 = \mathbf{81}$
g) $\left(\frac{4}{5}\right)^{-2} = \left(\frac{5}{4}\right)^2 = \frac{5^2}{4^2} = \mathbf{\frac{25}{16}}$
h) $\left(\frac{2}{6}\right)^{-3} = \left(\frac{6}{2}\right)^3 = 3^3 = \mathbf{27}$

2 a) $j^{-13} \div j^7 = j^{-13-7} = \mathbf{j^{-20}}$
b) $(n^7)^{-3} = n^{7\times-3} = \mathbf{n^{-21}}$
c) $p^{-8} \times p^{-6} = p^{-8+(-6)} = \mathbf{p^{-14}}$
d) $y^8 \div y^{-2} = y^{8-(-2)} = \mathbf{y^{10}}$
e) $(k^{-3})^6 = k^{-3\times6} = \mathbf{k^{-18}}$
f) $\frac{b^5}{b^9} = b^{5-9} = \mathbf{b^{-4}}$
g) $d^{-7} \times d^2 = d^{-7+2} = \mathbf{d^{-5}}$
h) $\frac{x^{60}}{x^{-8}} = x^{60} \div x^{-8} = x^{60-(-8)} = \mathbf{x^{68}}$

3 a) $■ + (-8) = 3 \Rightarrow ■ = \mathbf{11}$
b) $■ - 7 = -10 \Rightarrow ■ = \mathbf{-3}$
c) $-15 + ■ = 8 \Rightarrow ■ = \mathbf{23}$
d) $-4 \times ■ = 16 \Rightarrow ■ = \mathbf{-4}$

Page 46 Exercise 3

1 a) $3^2 \times 3^5 \times 3^7 = 3^{2+5+7} = \mathbf{3^{14}}$
b) $(8^6)^2 \times 8^5 = 8^{6\times2+5} = 8^{12+5} = \mathbf{8^{17}}$
c) $(12^8 \div 12^4)^3 = 12^{(8-4)\times3} = 12^{4\times3} = \mathbf{12^{12}}$
d) $(4^3)^6 \div 4^{16} = 4^{3\times6-16} = 4^{18-16} = \mathbf{4^2}$
Remember to read the question carefully to see if it wants the answers in index form.

2 a) $\frac{3^4\times3^5}{3^6} = \frac{3^9}{3^6} = 3^{9-6} = \mathbf{3^3}$
b) $\frac{8^{25}\div8^2}{8^6\times8^{10}} = \frac{8^{25-2}}{8^{6+10}} = \frac{8^{23}}{8^{16}} = 8^{23-16} = \mathbf{8^7}$
c) $\frac{(7^5)^7\div7^{12}}{7^5\times7^9} = \frac{7^{5\times7}\div7^{12}}{7^{5+9}} = \frac{7^{35-12}}{7^{14}} = \frac{7^{23}}{7^{14}}$
$= 7^{23-14} = \mathbf{7^9}$
d) $\frac{(5^{10}\div5^8)^4}{5^4\div5^2} = \frac{(5^{10-8})^4}{5^{4-2}} = \frac{5^{2\times4}}{5^2} = 5^{8-2} = \mathbf{5^6}$

3 $\frac{4^4\div4^3}{4} = \frac{4^{4-3}}{4} = \frac{4}{4} = \mathbf{1}$
$\frac{7^{16}}{7^8\times7^2} = \frac{7^{16}}{7^{8+2}} = \frac{7^{16}}{7^{10}} = 7^6 \neq 1$
$\frac{3^8-3^7}{3} = 3^7 - 3^6 \neq 1$
$\frac{5^5\times5^9}{(5^2)^7} = \frac{5^{5+9}}{5^{2\times7}} = \frac{5^{14}}{5^{14}} = \mathbf{1}$
$\frac{(9^2)^2-9^0}{9^3} = \frac{(9^2)^2-9^0}{9^3} = \frac{9^4-1}{9^3} = 9 - 9^{-3} \neq 1$
So $\mathbf{\frac{4^4\div4^3}{4}}$ and $\mathbf{\frac{5^5\times5^9}{(5^2)^7}}$ are equal to 1.

4 a) $\left(\frac{2^{-5}\times2^7}{2^3}\right)^5 = \left(\frac{2^{-5+7}}{2^3}\right)^5 = \left(\frac{2^2}{2^3}\right)^5 = (2^{-1})^5$
$= 2^{-1\times5} = \mathbf{2^{-5}}$
b) $\left(\frac{7^3}{7}\right)^3 \times 7^{-2} = (7^2)^3 \times 7^{-2} = 7^6 \times 7^{-2} = \mathbf{7^4}$
c) $\frac{9^{-3}\times9^{15}}{(9^{-3})^{-2}} = \frac{9^{12}}{9^6} = \mathbf{9^6}$
d) $\left(\frac{3^{-8}\times3^{12}}{3^2}\right)^{-6} = \left(\frac{3^4}{3^2}\right)^{-6} = (3^2)^{-6} = \mathbf{3^{-12}}$

5 a) $a^6 \times a^5 \div a^4 = a^{6+5-4} = \mathbf{a^7}$
b) $(p^5 \div p^3)^6 = (p^{5-3})^6 = (p^2)^6 = \mathbf{p^{12}}$
c) $\frac{(t^6\div t^3)^4}{t^9\div t^4} = \frac{(t^{6-3})^4}{t^{9-4}} = \frac{(t^3)^4}{t^5} = \frac{t^{12}}{t^5} = \mathbf{t^7}$
d) $\frac{(c^{-4})^3}{c^{-8}\div c^4} = \frac{c^{-12}}{c^{-8-4}} = \frac{c^{-12}}{c^{-12}} = c^0 = \mathbf{1}$

3.4 Standard Form

Page 47 Exercise 1

1 a) $\mathbf{2.5 \times 10^2}$ b) $\mathbf{7.34 \times 10^3}$
c) $\mathbf{4.8 \times 10^4}$ d) $\mathbf{5.9 \times 10^6}$

2 a) $\mathbf{3.75 \times 10^{-1}}$ b) $\mathbf{6.7 \times 10^{-3}}$
c) $\mathbf{7.8 \times 10^{-5}}$ d) $\mathbf{7.07 \times 10^{-2}}$

Page 48 Exercise 2

1 a) **3 000 000** b) **94 000**
c) **198 900 000** d) **7.2**
e) **0.00000356** f) **0.0423**
g) **0.0000888** h) **0.000000019**

Page 49 Exercise 3

1 a) $(3 \times 10^7) \times (2 \times 10^4)$
$= (3 \times 2) \times (10^7 \times 10^4)$
$= 6 \times 10^{7+4} = \mathbf{6 \times 10^{11}}$
b) $(4 \times 10^9) \times (2 \times 10^{-4})$
$= (4 \times 2) \times (10^9 \times 10^{-4})$
$= 8 \times 10^{9-4} = \mathbf{8 \times 10^5}$
c) $(6 \times 10^5) \times (1.4 \times 10^2)$
$= (6 \times 1.4) \times (10^5 \times 10^2)$
$= 8.4 \times 10^{5+2} = \mathbf{8.4 \times 10^7}$

2 a) $(9 \times 10^6) \div (3 \times 10^4)$
$= \frac{9\times10^6}{3\times10^4} = \frac{9}{3}\times\frac{10^6}{10^4} = 3 \times 10^{6-4} = \mathbf{3 \times 10^2}$
b) $(1.8 \times 10^{-4}) \div (0.9 \times 10^8)$
$= \frac{1.8\times10^{-4}}{0.9\times10^8} = \frac{1.8}{0.9}\times\frac{10^{-4}}{10^8} = 2 \times 10^{-4-8}$
$= \mathbf{2 \times 10^{-12}}$
c) $(8.1 \times 10^{-1}) \div (9 \times 10^{-3})$
$= \frac{8.1\times10^{-1}}{9\times10^{-3}} = \frac{8.1}{9}\times\frac{10^{-1}}{10^{-3}} = 0.9 \times 10^{-1-(-3)}$
$= 0.9 \times 10^2$
$= 9 \times 10^{-1} \times 10^2 = \mathbf{9 \times 10^1}$
Make sure your final answer is actually in standard form, with the front number between 1 and 10, otherwise you will lose marks.

Page 49 Exercise 4

1 a) $(5 \times 10^3) + (3 \times 10^3)$
$= (5 + 3) \times 10^3 = \mathbf{8 \times 10^3}$
b) $(6.4 \times 10^2) + (3.2 \times 10^2)$
$= (6.4 + 3.2) \times 10^2 = \mathbf{9.6 \times 10^2}$
c) $(6.9 \times 10^{-4}) + (3.8 \times 10^{-4})$
$= (6.9 + 3.8) \times 10^{-4}$
$= 10.7 \times 10^{-4}$
$= 1.07 \times 10 \times 10^{-4} = \mathbf{1.07 \times 10^{-3}}$

2 a) $(4.5 \times 10^{-2}) - (3.3 \times 10^{-2})$
$= (4.5 - 3.3) \times 10^{-2} = \mathbf{1.2 \times 10^{-2}}$
b) $(1.8 \times 10^4) - (1.2 \times 10^4)$
$= (1.8 - 1.2) \times 10^4$
$= 0.6 \times 10^4$
$= 6 \times 10^{-1} \times 10^4 = \mathbf{6 \times 10^3}$
c) $(6.4 \times 10^2) - (6.3 \times 10^2)$
$= (6.4 - 6.3) \times 10^2$
$= 0.1 \times 10^2$
$= 1 \times 10^{-1} \times 10^2 = \mathbf{1 \times 10^1}$

Page 50 Review Exercise

1 a) $(-5)^3 = (-5 \times -5 \times -5) = \mathbf{-125}$
b) $(0.5)^3 = (0.5 \times 0.5 \times 0.5) = \mathbf{0.125}$
c) $(-0.3)^3 = (-0.3 \times -0.3 \times -0.3) = \mathbf{-0.027}$
d) $(-12)^3 = (-12 \times -12 \times -12) = \mathbf{-1728}$
e) $0.1^2 = 0.1 \times 0.1 = \mathbf{0.01}$
f) $(-0.4)^2 = -0.4 \times -0.4 = \mathbf{0.16}$
g) $((-2)^2)^3 = (-2 \times -2)^3 = 4^3 = \mathbf{64}$
h) $((-2)^3)^2 = (-2 \times -2 \times -2)^2 = (-8)^2 = \mathbf{64}$

2 a) $3^2 - 2^3 = 9 - 8 = \mathbf{1}$
b) $5^2 - 6^2 = 25 - 36 = \mathbf{-11}$
c) $3 \times 2^8 = 3 \times 256 = \mathbf{768}$
d) $8 \times 5^4 = 8 \times 625 = \mathbf{5000}$
3 a) **–6** b) **19** c) **–7**
d) $-\sqrt{10^2 - 19} = -\sqrt{100 - 19} = -\sqrt{81} = \mathbf{-9}$
4 a) $\boldsymbol{k^4l^2}$ b) $\boldsymbol{zy^3}$ c) $\boldsymbol{m^3n^2}$
5 a) $a^6 \times a^4 = a^{6+4} = \boldsymbol{a^{10}}$
b) $15^{12} \div 15^{-14} = 15^{12-(-14)} = \mathbf{15^{26}}$
c) $(45^2)^{-9} = 45^{2 \times (-9)} = \mathbf{45^{-18}}$
d) $\dfrac{20^{222}}{20^{210}} = 20^{222-210} = \mathbf{20^{12}}$
6 a) $8^{-1} = \mathbf{\frac{1}{8}}$
b) $4^{-2} = \left(\frac{1}{4}\right)^2 = \frac{1}{4^2} = \mathbf{\frac{1}{16}}$
c) $\left(\frac{1}{3}\right)^{-3} = \left(\frac{3}{1}\right)^3 = 3^3 = \mathbf{27}$
d) $\left(\frac{4}{6}\right)^{-2} = \left(\frac{6}{4}\right)^2 = \left(\frac{3}{2}\right)^2 = \frac{3^2}{2^2} = \mathbf{\frac{9}{4}}$
When you're evaluating quantities with negative indices, it's often helpful to rewrite them using positive indices.
7 a) $\dfrac{4^4 \times 4^6}{4^8 \times 4} = \dfrac{4^{10}}{4^9} = \mathbf{4}$
b) $\dfrac{(5^5 \times 5^5)^2}{5^8 \div 5^3} = \dfrac{(5^{10})^2}{5^5} = \dfrac{5^{20}}{5^5} = \mathbf{5^{15}}$
c) $\left(\dfrac{2^5 \times 2^5}{2^3}\right)^4 = \left(\dfrac{2^{10}}{2^3}\right)^4 = (2^7)^4 = \mathbf{2^{28}}$
8 a) $\mathbf{3.3 \times 10^2}$ b) $\mathbf{2.75 \times 10^6}$
c) $\mathbf{2.5 \times 10^{-3}}$ d) $\mathbf{5.002 \times 10^{-4}}$
9 a) **400** b) **880 000**
c) **0.669** d) **0.00000705**
10 a) $(7 \times 10^5) \times (1.3 \times 10^2)$
$= (7 \times 1.3) \times (10^5 \times 10^2)$
$= (9.1) \times 10^{5+2} = \mathbf{9.1 \times 10^7}$
b) $(8.8 \times 10^3) \div (4 \times 10^8)$
$= \dfrac{8.8 \times 10^3}{4 \times 10^8} = \dfrac{8.8}{4} \times \dfrac{10^3}{10^8} = \mathbf{2.2 \times 10^{-5}}$
c) $(1.9 \times 10^6) + (9.1 \times 10^6)$
$= (1.9 + 9.1) \times 10^6$
$= 11 \times 10^6$
$= 1.1 \times 10 \times 10^6 = \mathbf{1.1 \times 10^7}$
d) $(5.9 \times 10^{-8}) - (3.4 \times 10^{-8})$
$= (5.9 - 3.4) \times 10^{-8} = \mathbf{2.5 \times 10^{-8}}$

Page 51 Exam-Style Questions

1 $(4.2 - 0.81)^2 + \sqrt{289} = 3.39^2 + 17 = \mathbf{28.4921}$
[1 mark]
2 a) $2x^3 \times 4x^4 = (2 \times 4) \times (x^3 \times x^4) = 8x^{3+4} = \mathbf{8x^7}$
[1 mark]
b) $(3y^2)^4 = 3^4 \times (y^2)^4 = 81y^{2 \times 4} = \mathbf{81y^8}$
[2 marks available – 1 mark for the 81 and one mark for the y^8]
c) $5z^0 = 5 \times 1 = \mathbf{5}$ *[1 mark]*
3 a) $4^3 = 64$ so $\boldsymbol{n} = \mathbf{3}$ *[1 mark]*
It's important to be able to recognise the first few cube numbers.
b) $2^2 \times 3^k \times 5 = 540$
$4 \times 3^k \times 5 = 540$
$20 \times 3^k = 540$ *[1 mark]*
$3^k = 27$ *[1 mark]*
Therefore $\boldsymbol{k} = \mathbf{3}$ *[1 mark]*
4 a) **0.65 is not between 1 and 10, so the number is not in correct standard form.** *[1 mark]*
b) $0.000234 = 2.34 \times 10^{-4}$, so the **power of 10 should be –4 rather than 4.** *[1 mark]*

5 In one Earth day, Mercury travels:
3.6×10^8 km $\div 88 = 3.6 \div 88 \times 10^8$ km
$= 0.040909 \times 10^8$ km (to 5 s.f) *[1 mark]*
$= 4.0909 \times 10^6$ km (to 5 s.f)
In one Earth day, Venus travels:
6.8×10^8 km $\div 225 = 6.8 \div 225 \times 10^8$ km
$= 0.030222 \times 10^8$ km (to 5 s.f) *[1 mark]*
$= 3.0222 \times 10^6$ km (to 5 s.f)
So Mercury travels further than Venus in one Earth day. *[1 mark]*
The difference is
4.0909×10^6 km – 3.0222×10^6 km
$= \mathbf{1.07 \times 10^6}$ **km (to 3 s.f.)**
[1 mark for subtracting the two distances, 1 mark for correct answer in standard form]
You could leave the distances per day as multiples of 10^8, and just convert to correct standard form at the end.

Section 4 — Multiples and Factors

4.1 Finding Multiples and Factors

Page 53 Exercise 1

1 a) **4, 8, 12, 16, 20**
b) **10, 20, 30, 40, 50**
c) **3, 6, 9, 12, 15**
d) **6, 12, 18, 24, 30**
e) **7, 14, 21, 28, 35**
2 a) **16** b) **27, 36, 45**
3 a) **10, 20, 30, 40, 50, 60, 70, 80, 90, 100**
b) **15, 30, 45, 60, 75, 90, 105**
c) **30, 60, 90**
4 Multiples of 5: 5, 10, 15, 20, 25, 30, 35
Multiples of 6: 6, 12, 18, 24, 30, 36
Common multiple: **30**
5 Multiples of 3: 21, 24, 27, 30, 33
Multiples of 4: 20, 24, 28, 32
Common multiple: **24**

Page 54 Exercise 2

1 a) 2 × 3 = 6 so both 2 and 3 are factors.
5 × – = 6 so 5 is not a factor.
6 × 1 = 6 so 6 is a factor.
The other numbers cannot be factors as they are greater than 6.
Factors are **2, 3, 6**
b) 2 × 12 = 24 so both 2 and 12 are factors.
3 × 8 = 24 so 3 is factor.
5 × – = 24 so 5 is not a factor.
6 × 4 = 24 so 6 is a factor.
15 × – = 24 so 15 is not a factor.
Factors are **2, 3, 6, 12**
Using the same method for c)-f):
c) **2, 3, 5, 6, 15** d) **2, 3, 6, 12**
e) **2, 5** f) **3, 5, 15**
2 a) 1 × 10 = 10 so both 1 and 10 are factors.
2 × 5 = 10 so both 2 and 5 are factors.
3 × – = 10 so 3 is not a factor.
4 × – = 10 so 4 is not a factor.
Stop as 5 has already been used.
So factors are **1, 2, 5, 10**
b) 1 × 4 = 4 so both 1 and 4 are factors.
2 × 2 = 4 so 2 is a factor.
3 × – = 4 so 3 is not a factor.
Stop as 4 has already been used.
So factors are **1, 2, 4**
Using the same method for c)-f):
c) **1, 13** d) **1, 5, 25**
e) **1, 2, 3, 4, 6, 8, 12, 24** f) **1, 5, 7, 35**
3 a) 1 × 15 = 15
2 × – = 15
3 × 5 = 15
4 × – = 15
So the factors of 15 are **1, 3, 5, 15**
b) 1 × 21 = 21
2 × – = 21
3 × 7 = 21
4 × – = 21
5 × – = 21
6 × – = 21
So the factors of 21 are **1, 3, 7, 21**
c) Common factors appear on both lists, so the common factors of 15 and 21 are **1 and 3.**
4 a) Factors of 15: 1, 3, 5, 15
Factors of 20: 1, 2, 4, 5, 10, 20
Common factors: **1, 5**
b) Factors of 12: 1, 2, 3, 4, 6, 12
Factors of 15: 1, 3, 5, 15
Common factors: **1, 3**
Using the same method for c)-f):
c) **1, 3, 5, 15** d) **1, 2, 5, 10**
e) **1, 5, 25** f) **1**
5 a) **1** b) **1, 2** c) **1, 5**
6 a) Factors of 15: 1, 3, 5, 15
Factors of 20: 1, 2, 4, 5, 10, 20
Factors of 25: 1, 5, 25
Common factors: **1, 5**
Using the same method for b)-d):
b) **1, 2** c) **1, 5** d) **1**

4.2 Prime Numbers

Page 55 Exercise 1

1 a) 33 = 11 × 3 so 33 is not prime.
35 = 5 × 7 so 35 is not prime.
39 = 3 × 13 so 39 is not prime.
So the numbers that are not prime are **33, 35, 39**
b) Using part a)
Two factors of 33 are **3 and 11**
Two factors of 35 are **5 and 7**
Two factors of 39 are **3 and 13**
2 15 = 3 × 5 so 15 is not prime.
22 = 2 × 11 so 22 is not prime.
34 = 2 × 17 so 34 is not prime.
51 = 3 × 17 so 51 is not prime.
So primes are **5, 47, 59**
3 **23, 29**
4 a) (i) 4 = 2 × 2 so a factor of 4 is **2**.
(ii) 14 = 2 × 7 so a factor of 14 is **2**.
7 is also a factor.
(iii) 34 = 2 × 17 so a factor of 34 is **2**.
17 is also a factor.
(iv) 74 = 2 × 37 so a factor of 74 is **2**.
37 is also a factor.
b) Because 2 will always be a factor.
5 E.g. Because they all end in zero so 10 will always be a factor.
You could also have said that all the numbers in the list are even, so they must have 2 as a factor.

Page 57 Exercise 2

1 a) (i)
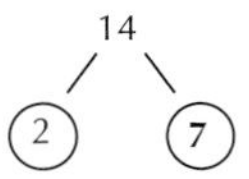

(ii)
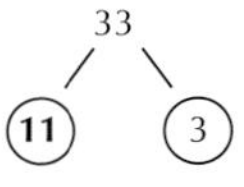

(iii)
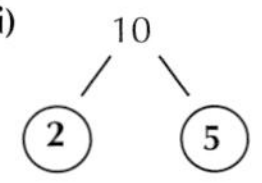

(iv)
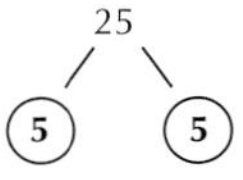

b) (i) 14 = **2 × 7** (ii) 33 = **3 × 11**
(iii) 10 = **2 × 5** (iv) 26 = **2 × 13**
2 a) 15 = **3 × 5** b) 21 = **3 × 7**
c) 22 = **2 × 11** d) 6 = **2 × 3**
e) 14 = **2 × 7** f) 26 = **2 × 13**

3 a) (i)

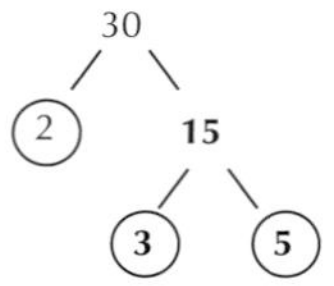

(ii)

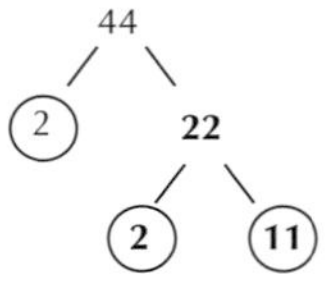

(iii)

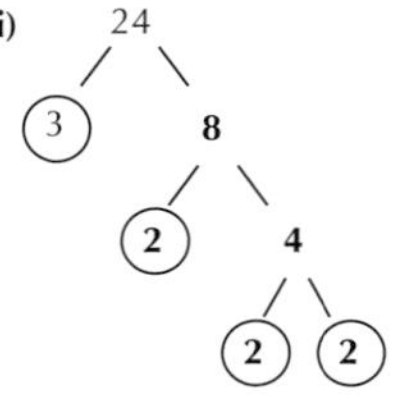

(iv)

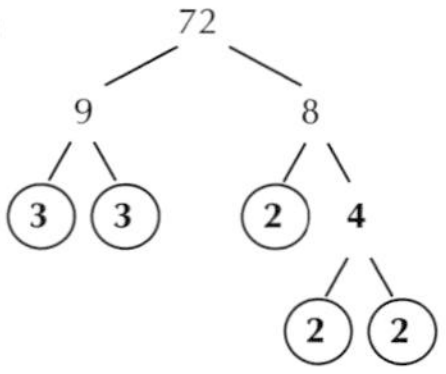

b) (i) $30 = \mathbf{2 \times 3 \times 5}$ (ii) $44 = \mathbf{2^2 \times 11}$
(iii) $24 = \mathbf{2^3 \times 3}$ (iv) $72 = \mathbf{2^3 \times 3^2}$

4 E.g.

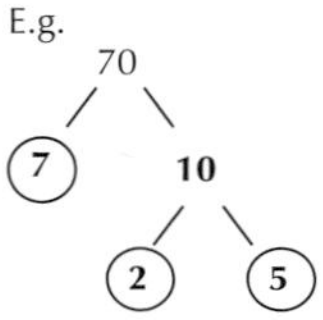

$70 = \mathbf{2 \times 5 \times 7}$

You might have a different looking factor tree (e.g. you might have split 70 into 2 × 35). But your final prime factorisation of 70 should always turn out to be 2 × 5 × 7.

5 a) E.g.

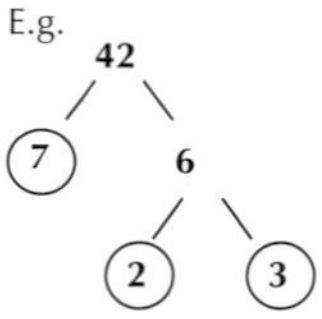

$42 = \mathbf{2 \times 3 \times 7}$

Using the same method for b)-d):

b) $84 = \mathbf{2^2 \times 3 \times 7}$

c) $190 = \mathbf{2 \times 5 \times 19}$

d) $210 = \mathbf{2 \times 3 \times 5 \times 7}$

6 a) $128 = \mathbf{2^7}$ b) $168 = \mathbf{2^3 \times 3 \times 7}$
c) $325 = \mathbf{5^2 \times 13}$ d) $98 = \mathbf{2 \times 7^2}$
e) $225 = \mathbf{3^2 \times 5^2}$ f) $1000 = \mathbf{2^3 \times 5^3}$

7 a) $75 = 3 \times 5 \times 5 = \mathbf{3 \times 5^2}$

b) **3** — this is the smallest number you can multiply by that gives even powers of all the prime factors, giving $3^2 \times 5^2 = 15^2$.

4.3 LCM and HCF

Page 58 Exercise 1

1 a) First five multiples of 9: **9, 18, 27, 36, 45**
First five multiples of 12: **12, 24, 36, 48, 60**

b) The LCM is the smallest number in both lists: **36**

2 a) First ten multiples of 5: **5, 10, 15, 20, 25, 30, 35, 40, 45, 50**
First ten multiples of 7: **7, 14, 21, 28, 35, 42, 49, 56, 63, 70**

b) The LCM is the smallest number in both lists: **35**

3 a) Multiples of 3: 3, 6, 9, 12, 15...
Multiples of 4: 4, 8, 12, 16...
LCM = smallest number in both lists = **12**

b) Multiples of 6: 6, 12, 18, 24, 30...
Multiples of 8: 8, 16, 24, 32...
LCM = smallest number in both lists = **24**

Using the same method for c)-f):

c) **10** d) **42** e) **30** f) **60**

4 a) Multiples of 3:
3, 6, 9, 12, 15, 18, 21, 24, 27...
Multiples of 6: 6, 12, 18, 24, 30...
Multiples of 8: 8, 16, 24, 32...
LCM = smallest number in all lists = **24**

Using the same method for b) and c):

b) **30** c) **36**

Page 59 Exercise 2

1 a) **8, 16, 24, 32, 40, 48, 56, 64, 72, 80**

b) **12, 24, 36, 48, 60, 72, 84, 96, 108, 120**

c) **After 24 minutes**

2 Mike visits Oscar on day: 4, 8, 12, 16, 20...
Narinda visits Oscar on day: 5, 10, 15, 20...
So they will both visit Oscar on the same day after **20 days**.

3 Number of sweets must be a multiple of 5. Number of sweets must also be a multiple of 7.
Multiples of 5: 5, 10, 15, 20, 25, 30, 35, 40...
Multiples of 7: 7, 14, 21, 28, 35...
The smallest possible number of sweets is the LCM of 5 and 7, which is **35**.

4 Number of plants must be a multiple of 25. Number of plants must also be a multiple of 30. So you need to find a common multiple of 25 and 30. Between 95 and 205, the multiples of 25 are: 100, 125, 150, 175, 200 and the multiples of 30 are: 120, 150, 180. So there must be **150 plants**.

Page 60 Exercise 3

1 a) Factors of 12: 1, 2, 3, 4, 6, 12
Factors of 20: 1, 2, 4, 5, 10, 20
Common factors of 12 and 20: **1, 2, 4**

b) **4**

2 a) Factors of 20: 1, 2, 4, 5, 10, 20
Factors of 30: 1, 2, 3, 5, 6, 10, 15, 30
Common factors of 20 and 30: **1, 2, 5, 10**

b) **10**

3 a) Factors of 8: 1, 2, 4, 8
Factors of 12: 1, 2, 3, 4, 6, 12
The HCF is the biggest number in both lists: **4**.

b) Factors of 24: 1, 2, 3, 4, 6, 8, 12, 24
Factors of 32: 1, 2, 4, 8, 16, 32
The HCF is the biggest number in both lists: **8**.

Using the same method for c)-f):

c) **6** d) **12**
e) **1** f) **12**

4 a) Factors of 11: 1, 11
Factors of 12: 1, 2, 3, 4, 6, 12
HCF is the biggest number in both lists: **1**

b) Factors of 21: 1, 3, 7, 21
Factors of 22: 1, 2, 11, 22
HCF is the biggest number in both lists: **1**

5 a) Factors of 6: 1, 2, 3, 6
Factors of 8: 1, 2, 4, 8
Factors of 12: 1, 2, 3, 4, 6, 12
HCF is the biggest number in all three lists: **2**

b) Factors of 12: 1, 2, 3, 4, 6, 12
Factors of 15: 1, 3, 5, 15
Factors of 18: 1, 2, 3, 6, 9, 18
HCF is the biggest number in all three lists: **3**

Using the same method for c) and d):

c) **6** d) **18**

Page 61 Review Exercise

1 a) Multiples of 3: **9, 12, 15, 18, 21**

b) Multiples of 4: **8, 12, 16, 20**

c) Multiples of 5: **5, 15, 20**

2 Multiples of 8: 8, 16, 24, 32, 40, 48, 56, 64, 72, 80, 88, 96
Multiples of 10: 10, 20, 30, 40, 50, 60, 70, 80, 90, 100
Common multiples of 8 and 10: **40, 80**

3 a) Multiples of 20 between 95 and 155: **100, 120, 140**

b) Multiples of 30 between 95 and 155: **120, 150**

c) Number of animals is the only common multiple of 20 and 30 between 95 and 155: **120**

4 a) Factors of 10: 1, 2, 5, 10
Factors of 42: 1, 2, 3, 6, 7, 21, 28, 42
Common factors: **1, 2**

b) Factors of 14: 1, 2, 7, 14
Factors of 27: 1, 3, 9, 27
Common factors: **1**

c) Factors of 12: 1, 2, 3, 4, 6, 12
Factors of 24: 1, 2, 3, 4, 6, 8, 12, 24
Common factors: **1, 2, 3, 4, 6, 12**

5 The baker could have:
1 packet of 12 cakes, 2 packets of 6 cakes
3 packets of 4 cakes, 4 packets of 3 cakes
6 packets of 2 cakes, 12 packets of 1 cake
So there are **6 ways** the baker could divide up the cakes.

6 a) $51 = 3 \times 17$ so 51 is not prime.
$55 = 5 \times 11$ so 55 is not prime.
$57 = 3 \times 19$ so 57 is not prime.
So numbers which are not prime are **51, 55, 57.**

b) From a):
Two factors of 51 are **3 and 17**.
Two factors of 55 are **5 and 11**.
Two factors of 57 are **3 and 19**.

7 $18 = 3 \times 6$ so 18 is not prime.
$33 = 3 \times 11$ so 33 is not prime.
$51 = 3 \times 17$ so 51 is not prime.
$54 = 6 \times 9$ so 54 is not prime.
So prime numbers are **7, 17, 31**.

8 a)

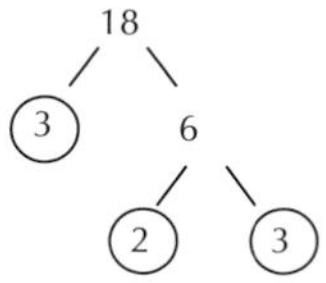

$18 = \mathbf{2 \times 3^2}$

Using the same method for b)-d):

b) $50 = \mathbf{2 \times 5^2}$

c) $36 = \mathbf{2^2 \times 3^2}$

d) $150 = \mathbf{2 \times 3 \times 5^2}$

9 a) (i) Factors of 5: 1, 5
Factors of 9: 1, 3, 9
HCF is the biggest number in both lists: **1**.

Using the same method for (ii)-(iv):

(ii) **3** (iii) **5** (iv) **4**

b) **(i)** Multiples of 5: 5, 10, 15, 20, 25, 30, 35, 40, 45, 50...
Multiples of 9: 9, 18, 27, 36, 45...
LCM is the smallest number in both lists: **45**.

Using the same method for (ii)-(iv):

(ii) **30** **(iii)** **36** **(iv)** **60**

Page 62 Exam-Style Questions

1 **a)** 2 × 14 = 28 and 4 × 7 = 28, so **28** is a multiple of 4 and 7 *[1 mark]*.

b) 8 × 4 = 32, so **8** is a factor of 32 *[1 mark]*.

c) **8** is 1 more than 7, which is a prime number *[1 mark]*.

2 **a)** Factors of 50: **1, 2, 5, 10, 25, 50** *[1 mark]*

b) 51 is divisible by 3 so not prime.
52 is divisible by 2 so not prime.
53 is the smallest prime number that is bigger than 50 *[1 mark]*.

3 Factors of 48: 1, 2, 3, 4, 6, 8, 12, 16, 24, 48

Factors of 60: 1, 2, 3, 4, 5, 6, 10, 12, 15, 20, 30, 60

HCF is biggest number in both lists: **12**

[2 marks available — 1 mark for correctly listing factors of 48 and 60, 1 mark for HCF]

4

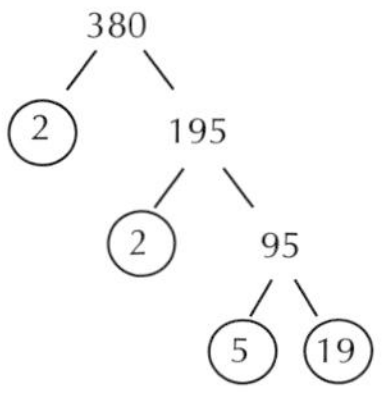

$380 = \mathbf{2^2 \times 5 \times 19}$

[2 marks available — 1 mark for at least two correct divisions by a prime number, 1 mark for correct prime factorisation]

You don't have to use a factor tree to find the prime factorisation, but you do need to write down some working showing how you got your final answer.

5 As the LCM is 60, the two numbers must be factors of 60: 1, 2, 3, 4, 5, 6, 10, 12, 15, 20, 30, 60. But as the HCF is 4, both numbers must also be multiples of 4, which leaves 4, 12, 20 and 60 *[1 mark for correct reasoning]*. Since the numbers are not 4 and 60, they must be **12** *[1 mark]* and **20** *[1 mark]*.

6 **a)**

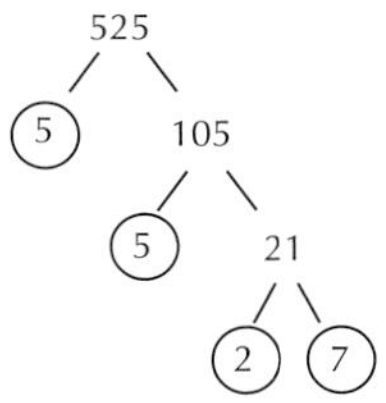

$525 = \mathbf{3 \times 5^2 \times 7}$

[2 marks available — 1 mark for at least two correct divisions by a prime number, 1 mark for correct prime factorisation]

b) $275\,625 = 525^2 = (3 \times 5^2 \times 7)^2$
$= \mathbf{3^2 \times 5^4 \times 7^2}$ *[1 mark]*

Section 5 — Fractions

5.1 Equivalent Fractions

Page 64 Exercise 1

1 **a)** $\frac{1}{3} = \frac{\star}{6}$, $\frac{1}{3} = \frac{1\times 2}{3\times 2} = \frac{2}{6}$, so $\star$ = **2**

b) $\frac{1}{3} = \frac{\star}{9}$, $\frac{1}{3} = \frac{1\times 3}{3\times 3} = \frac{3}{9}$, so $\star$ = **3**

c) $\frac{1}{3} = \frac{\star}{12}$, $\frac{1}{3} = \frac{1\times 4}{3\times 4} = \frac{4}{12}$, so $\star$ = **4**

2 **a)** $\frac{1}{5} = \frac{1\times 2}{5\times 2} = \frac{2}{10}$, so $\boldsymbol{a = 2}$

b) $\frac{1}{4} = \frac{1\times 3}{4\times 3} = \frac{3}{12}$, so $\boldsymbol{b = 3}$

Using the same method for d)-h):

c) $\boldsymbol{c = 25}$ **d)** $\boldsymbol{d = 18}$ **e)** $\boldsymbol{e = 5}$

f) $\boldsymbol{f = 27}$ **g)** $\boldsymbol{g = 16}$ **h)** $\boldsymbol{h = 11}$

Page 65 Exercise 2

1 **a)** $\frac{3}{9} = \frac{3\div 3}{9\div 3} = \mathbf{\frac{1}{3}}$ **b)** $\frac{5}{20} = \frac{5\div 5}{20\div 5} = \mathbf{\frac{1}{4}}$

Using the same method for c)-h):

c) $\mathbf{\frac{1}{2}}$ **d)** $\mathbf{\frac{1}{8}}$ **e)** $\mathbf{\frac{1}{5}}$

f) $\mathbf{\frac{5}{12}}$ **g)** $\mathbf{\frac{3}{8}}$ **h)** $\mathbf{\frac{3}{4}}$

2 **a)** $\frac{6}{18} = \frac{6\div 6}{18\div 6} = \frac{1}{3}$, $\frac{4}{20} = \frac{4\div 4}{20\div 4} = \frac{1}{5}$,
$\frac{9}{27} = \frac{9\div 9}{27\div 9} = \frac{1}{3}$

So the fraction that is not equivalent to the other two is $\mathbf{\frac{4}{20}}$.

b) $\frac{6}{8} = \frac{6\div 2}{8\div 2} = \frac{3}{4}$, $\frac{9}{15} = \frac{9\div 3}{15\div 3} = \frac{3}{5}$,
$\frac{15}{25} = \frac{15\div 5}{25\div 5} = \frac{3}{5}$

So the fraction that is not equivalent to the other two is $\mathbf{\frac{6}{8}}$.

c) $\frac{4}{18} = \frac{4\div 2}{18\div 2} = \frac{2}{9}$, $\frac{6}{33} = \frac{6\div 3}{33\div 3} = \frac{2}{11}$,
$\frac{10}{45} = \frac{10\div 5}{45\div 5} = \frac{2}{9}$

So the fraction that is not equivalent to the other two is $\mathbf{\frac{6}{33}}$.

The answers for Q2 always simplify by dividing by the highest common factor. You might have simplified some of the fractions using more than one step — that's fine as long as you got to the correct final answer.

3 cows: $\frac{\text{cows}}{\text{animals}} = \frac{50}{300} = \frac{50\div 50}{300\div 50} = \mathbf{\frac{1}{6}}$

pigs: $\frac{\text{pigs}}{\text{animals}} = \frac{70}{300} = \frac{70\div 10}{300\div 10} = \mathbf{\frac{7}{30}}$

300 – 50 – 70 = 180 sheep, so:

sheep: $\frac{\text{sheep}}{\text{animals}} = \frac{180}{300} = \frac{180\div 60}{300\div 60} = \mathbf{\frac{3}{5}}$

5.2 Mixed Numbers

Page 66 Exercise 1

1 **a)** $2 = \frac{2}{1} = \frac{2\times 3}{1\times 3} = \frac{6}{3}$, so $\star$ = **6**

b) $3 = \frac{3}{1} = \frac{3\times 4}{1\times 4} = \frac{12}{4}$, so $\star$ = **12**

2 **a)** $4 = \frac{a}{3}$, $4 = \frac{4}{1} = \frac{4\times 3}{1\times 3} = \frac{12}{3}$, so $\boldsymbol{a = 12}$

b) $4\frac{1}{3} = 4 + \frac{1}{3} = \frac{12}{3} + \frac{1}{3} = \frac{12+1}{3} = \mathbf{\frac{13}{3}}$

3 **a)** $1\frac{1}{3} = \frac{3}{3} + \frac{1}{3} = \frac{4}{3}$, so $\boldsymbol{a = 4}$

b) $1\frac{2}{7} = \frac{7}{7} + \frac{2}{7} = \frac{9}{7}$, so $\boldsymbol{b = 9}$

c) $2\frac{1}{2} = \frac{4}{2} + \frac{1}{2} = \frac{5}{2}$, so $\boldsymbol{c = 5}$

4 **a)** $1\frac{4}{5} = \frac{5}{5} + \frac{4}{5} = \mathbf{\frac{9}{5}}$

b) $1\frac{5}{12} = \frac{12}{12} + \frac{5}{12} = \mathbf{\frac{17}{12}}$

Using the same method for c)-f):

c) $\mathbf{\frac{29}{10}}$ **d)** $\mathbf{\frac{53}{10}}$

e) $\mathbf{\frac{19}{4}}$ **f)** $\mathbf{\frac{59}{6}}$

Page 67 Exercise 2

1 **a)** 11 ÷ 7 = **1 remainder 4**

b) $\frac{11}{7} = \mathbf{1\frac{4}{7}}$

2 **a)** 5 ÷ 3 = 1 remainder 2, so $\frac{5}{3} = \mathbf{1\frac{2}{3}}$

b) 9 ÷ 5 = 1 remainder 4, so $\frac{9}{5} = \mathbf{1\frac{4}{5}}$

Using the same method for c)-f):

c) $\mathbf{1\frac{2}{11}}$ **d)** $\mathbf{2\frac{1}{4}}$ **e)** $\mathbf{2\frac{2}{9}}$ **f)** $\mathbf{3\frac{2}{3}}$

3 **a)** $\mathbf{\frac{12}{5}}$

b) 12 ÷ 5 = 2 remainder 2, so $\frac{12}{5} = \mathbf{2\frac{2}{5}}$

5.3 Ordering Fractions

Page 68 Exercise 1

1 **a)** E.g. $\frac{1}{3} = \frac{1\times 2}{3\times 2} = \mathbf{\frac{2}{6}}$, $\frac{1}{6} = \mathbf{\frac{1}{6}}$

b) E.g. $\frac{1}{5} = \frac{1\times 2}{5\times 2} = \mathbf{\frac{2}{10}}$, $\frac{3}{10} = \mathbf{\frac{3}{10}}$

Using the same method for c)-h):

c) $\mathbf{\frac{4}{16}, \frac{5}{16}}$ **d)** $\mathbf{\frac{8}{20}, \frac{7}{20}}$ **e)** $\mathbf{\frac{2}{9}, \frac{3}{9}}$

f) $\mathbf{\frac{8}{12}, \frac{9}{12}}$ **g)** $\mathbf{\frac{35}{42}, \frac{6}{42}}$ **h)** $\mathbf{\frac{4}{18}, \frac{9}{18}}$

2 **a)** The lowest common multiple of 4, 8 and 12 is 24.

$\frac{3}{4} = \frac{3\times 6}{4\times 6} = \mathbf{\frac{18}{24}}$, $\frac{5}{8} = \frac{5\times 3}{8\times 3} = \mathbf{\frac{15}{24}}$,
$\frac{7}{12} = \frac{7\times 2}{12\times 2} = \mathbf{\frac{14}{24}}$

b) The lowest common multiple of 5, 10 and 20 is 20.

$\frac{1}{5} = \frac{1\times 4}{5\times 4} = \mathbf{\frac{4}{20}}$, $\frac{7}{10} = \frac{7\times 2}{10\times 2} = \mathbf{\frac{14}{20}}$,
$\frac{9}{20} = \mathbf{\frac{9}{20}}$

Using the same method for c)-f):

c) $\mathbf{\frac{6}{42}, \frac{8}{42}, \frac{15}{42}}$ **d)** $\mathbf{\frac{12}{24}, \frac{9}{24}, \frac{16}{24}}$

e) $\mathbf{\frac{24}{60}, \frac{25}{60}, \frac{22}{60}}$ **f)** $\mathbf{\frac{5}{40}, \frac{14}{40}, \frac{24}{40}}$

Page 70 Exercise 2

1 **a)** $\frac{1}{4} = \frac{1\times 2}{4\times 2} = \frac{2}{8}$, $\frac{2}{8} < \frac{5}{8}$, so $\mathbf{\frac{5}{8}}$ **is larger.**

b) $\frac{3}{5} = \frac{3\times 2}{5\times 2} = \frac{6}{10}$, $\frac{6}{10} < \frac{7}{10}$, so $\mathbf{\frac{7}{10}}$ **is larger.**

Using the same method for c)-h):

c) $\mathbf{\frac{9}{14}}$ **d)** $\mathbf{\frac{2}{3}}$ **e)** $\mathbf{\frac{5}{6}}$

f) $\mathbf{\frac{2}{3}}$ **g)** $\mathbf{\frac{3}{4}}$ **h)** $\mathbf{\frac{3}{4}}$

2 **a)** E.g. $\frac{1}{2} = \frac{1\times 8}{2\times 8} = \frac{8}{16}$, $\frac{5}{8} = \frac{5\times 2}{8\times 2} = \frac{10}{16}$, $\frac{7}{16}$

$\frac{7}{16} < \frac{8}{16} < \frac{10}{16}$,

so order from smallest to largest is

$\mathbf{\frac{7}{16}, \frac{1}{2}, \frac{5}{8}}$.

b) E.g. $\frac{2}{5} = \frac{2\times 4}{5\times 4} = \frac{8}{20}$,

$\frac{3}{10} = \frac{3\times 2}{10\times 2} = \frac{6}{20}$, $\frac{7}{20}$

$\frac{6}{20} < \frac{7}{20} < \frac{8}{20}$,

so order from smallest to largest is

$\mathbf{\frac{3}{10}, \frac{7}{20}, \frac{2}{5}}$.

c) E.g. $\frac{3}{4} = \frac{3\times 6}{4\times 6} = \frac{18}{24}$, $\frac{7}{12} = \frac{7\times 2}{12\times 2} = \frac{14}{24}$,

$\frac{5}{8} = \frac{5\times 3}{8\times 3} = \frac{15}{24}$

$\frac{14}{24} < \frac{15}{24} < \frac{18}{24}$,

so order from smallest to largest is

$\mathbf{\frac{7}{12}, \frac{5}{8}, \frac{3}{4}}$.

You could have used different common denominators for any of the parts in Q2 — the ones used here are the lowest common denominators.

3 **a)** E.g. $\frac{7}{8} = \frac{7\times 6}{8\times 6} = \frac{42}{48}$, $\frac{5}{6} = \frac{5\times 8}{6\times 8} = \frac{40}{48}$,

$\frac{13}{16} = \frac{13\times 3}{16\times 3} = \frac{39}{48}$

$\frac{39}{48} < \frac{40}{48} < \frac{42}{48}$,

so order from smallest to largest is

$\mathbf{\frac{13}{16}, \frac{5}{6}, \frac{7}{8}}$.

b) E.g. $\frac{4}{15} = \frac{4\times9}{15\times9} = \frac{36}{135}$,

$\frac{7}{27} = \frac{7\times5}{27\times5} = \frac{35}{135}$, $\frac{13}{45} = \frac{13\times3}{45\times3} = \frac{39}{135}$.

$\frac{35}{135} < \frac{36}{135} < \frac{39}{135}$,

so order from smallest to largest is

$\mathbf{\frac{7}{27}}$, $\mathbf{\frac{4}{15}}$, $\mathbf{\frac{13}{45}}$.

c) E.g. $\frac{5}{16} = \frac{5\times25}{16\times25} = \frac{125}{400}$,

$\frac{7}{20} = \frac{7\times20}{20\times20} = \frac{140}{400}$, $\frac{9}{25} = \frac{9\times16}{25\times16} = \frac{144}{400}$

$\frac{125}{400} < \frac{140}{400} < \frac{144}{400}$,

so order from smallest to largest is

$\mathbf{\frac{5}{16}}$, $\mathbf{\frac{7}{20}}$, $\mathbf{\frac{9}{25}}$.

Page 70 Exercise 3

1 a) E.g. $-\frac{4}{15} = -\frac{4\times3}{15\times3} = -\mathbf{\frac{12}{45}}$,

$\frac{2}{9} = \frac{2\times5}{9\times5} = \mathbf{\frac{10}{45}}$, $-\frac{1}{3} = -\frac{1\times15}{3\times15} = -\mathbf{\frac{15}{45}}$

b) $-\frac{15}{45} < -\frac{12}{45} < \frac{10}{45}$

so $\mathbf{-\frac{1}{3} < -\frac{4}{15} < \frac{2}{9}}$

2 a) E.g. $-\frac{4}{9} = -\frac{4\times4}{9\times4} = -\frac{16}{36}$,

$-\frac{5}{12} = -\frac{5\times3}{12\times3} = -\frac{15}{36}$

$-\frac{16}{36} < -\frac{15}{36}$, so $\mathbf{-\frac{4}{9}}$ **is smaller.**

b) E.g. $\frac{4}{5} = \frac{4\times2}{5\times2} = \frac{8}{10}$

$-\frac{9}{10} < \frac{8}{10}$, so $\mathbf{-\frac{9}{10}}$ **is smaller.**

c) E.g. $-\frac{1}{4} = -\frac{1\times11}{4\times11} = -\frac{11}{44}$,

$\frac{5}{2} = \frac{5\times22}{2\times22} = \frac{110}{44}$,

$-\frac{3}{11} = -\frac{3\times4}{11\times4} = -\frac{12}{44}$

$-\frac{12}{44} < -\frac{11}{44} < \frac{110}{44}$, so $\mathbf{-\frac{3}{11}}$ **is smaller.**

5.4 Adding and Subtracting Fractions

Page 71 Exercise 1

1 a) $\frac{1}{3} + \frac{1}{3} = \frac{1+1}{3} = \mathbf{\frac{2}{3}}$

b) $\frac{4}{5} - \frac{2}{5} = \frac{4-2}{5} = \mathbf{\frac{2}{5}}$

c) $\frac{5}{11} - \frac{3}{11} = \frac{5-3}{11} = \mathbf{\frac{2}{11}}$

d) $\frac{1}{10} + \frac{3}{10} = \frac{1+3}{10} = \frac{4}{10} = \mathbf{\frac{2}{5}}$

Don't forget to simplify your answer in part d).

2 a) $\frac{5}{8} + \frac{7}{8} = \frac{5+7}{8} = \frac{12}{8} = 1\frac{4}{8} = \mathbf{1\frac{1}{2}}$

b) $\frac{3}{4} + \frac{3}{4} = \frac{3+3}{4} = \frac{6}{4} = 1\frac{2}{4} = \mathbf{1\frac{1}{2}}$

c) $\frac{8}{15} + \frac{13}{15} - \frac{2}{15} = \frac{8+13-2}{15} = \frac{19}{15} = \mathbf{1\frac{4}{15}}$

d) $\frac{17}{20} + \frac{19}{20} - \frac{7}{20} = \frac{17+19-7}{20} = \frac{29}{20} = \mathbf{1\frac{9}{20}}$

3 a) $\frac{1}{2} = \frac{1\times2}{2\times2} = \mathbf{\frac{2}{4}}$ and $\mathbf{\frac{1}{4}}$

b) $\frac{1}{2} + \frac{1}{4} = \frac{2}{4} + \frac{1}{4} = \frac{2+1}{4} = \mathbf{\frac{3}{4}}$

4 a) $\frac{3}{5} + \frac{1}{10} = \frac{6}{10} + \frac{1}{10} = \frac{6+1}{10} = \mathbf{\frac{7}{10}}$

b) $\frac{1}{4} + \frac{3}{8} = \frac{2}{8} + \frac{3}{8} = \frac{2+3}{8} = \mathbf{\frac{5}{8}}$

c) $\frac{4}{9} - \frac{1}{3} = \frac{4}{9} - \frac{3}{9} = \frac{4-3}{9} = \mathbf{\frac{1}{9}}$

d) $\frac{3}{4} - \frac{3}{8} = \frac{6}{8} - \frac{3}{8} = \frac{6-3}{8} = \mathbf{\frac{3}{8}}$

5 a) $\frac{1}{9} + \frac{5}{9} + \frac{11}{18} = \frac{2}{18} + \frac{10}{18} + \frac{11}{18}$

$= \frac{2+10+11}{18} = \frac{23}{18} = \mathbf{1\frac{5}{18}}$

b) $\frac{3}{4} + \frac{1}{8} - \frac{7}{16} = \frac{12}{16} + \frac{2}{16} - \frac{7}{16}$

$= \frac{12+2-7}{16} = \mathbf{\frac{7}{16}}$

Using the same method for c) and d):

c) $\mathbf{\frac{3}{7}}$ **d)** $\mathbf{1\frac{3}{4}}$

Page 72 Exercise 2

1 The fraction of boys and fraction of girls must add up to 1.

So the fraction of girls is

$1 - \frac{4}{9} = \frac{9}{9} - \frac{4}{9} = \frac{9-4}{9} = \mathbf{\frac{5}{9}}$.

2 The fractions of cake must add up to 1.

So Olga eats $1 - \frac{2}{7} - \frac{3}{7} = \frac{7}{7} - \frac{2}{7} - \frac{3}{7}$

$= \frac{7-2-3}{7} = \mathbf{\frac{2}{7}}$ of the cake.

3 The fractions of flowers must add up to 1. So the fraction that is neither a rose nor a tulip is $1 - \frac{1}{5} - \frac{3}{10} = \frac{10}{10} - \frac{2}{10} - \frac{3}{10}$

$= \frac{10-2-3}{10} = \frac{5}{10} = \mathbf{\frac{1}{2}}$.

4 The fractions of sweets must add up to 1. So the fraction of sweets that are mints is

$1 - \frac{2}{5} - \frac{1}{4} = \frac{20}{20} - \frac{8}{20} - \frac{5}{20} = \frac{20-8-5}{20} = \mathbf{\frac{7}{20}}$.

5 The fractions of pupils must add up to 1. So the fraction of pupils that come to school by car is

$1 - \frac{1}{2} - \frac{1}{5} = \frac{10}{10} - \frac{5}{10} - \frac{2}{10} = \frac{10-5-2}{10} = \mathbf{\frac{3}{10}}$.

Page 73 Exercise 3

1 a) $1\frac{4}{5} = \frac{5}{5} + \frac{4}{5} = \frac{5+4}{5} = \mathbf{\frac{9}{5}}$

b) $1\frac{4}{5} + \frac{3}{5} = \frac{9}{5} + \frac{3}{5} = \frac{12}{5}$

$12 \div 5 = 2$ remainder 2, so $\frac{12}{5} = \mathbf{2\frac{2}{5}}$

2 a) (i) $2\frac{1}{5} = \frac{10}{5} + \frac{1}{5} = \mathbf{\frac{11}{5}}$

(ii) $1\frac{3}{5} = \frac{5}{5} + \frac{3}{5} = \mathbf{\frac{8}{5}}$

b) $2\frac{1}{5} - 1\frac{3}{5} = \frac{11}{5} - \frac{8}{5} = \frac{11-8}{5} = \mathbf{\frac{3}{5}}$

3 a) $1\frac{2}{3} = \frac{3}{3} + \frac{2}{3} = \frac{5}{3}$,

so $1\frac{2}{3} + \frac{1}{3} = \frac{5}{3} + \frac{1}{3} = \frac{6}{3} = \mathbf{2}$

b) $2\frac{3}{8} = \frac{16}{8} + \frac{3}{8} = \frac{19}{8}$,

so $2\frac{3}{8} + \frac{7}{8} = \frac{19}{8} + \frac{7}{8} = \frac{26}{8}$

$26 \div 8 = 3$ remainder 2, so $\frac{26}{8} = 3\frac{2}{8} = \mathbf{3\frac{1}{4}}$

Using the same method for c)-f):

c) 3 **d)** $\mathbf{1\frac{1}{2}}$

e) $\mathbf{6\frac{1}{11}}$ **f)** $\mathbf{2\frac{3}{5}}$

Page 74 Exercise 4

1 a) $4 + 3 = 7$ and $\frac{1}{9} + \frac{4}{9} = \frac{5}{9}$,

so $4\frac{1}{9} + 3\frac{4}{9} = \mathbf{7\frac{5}{9}}$

b) $3 + 2 = 5$ and $\frac{1}{5} + \frac{3}{7} = \frac{7}{35} + \frac{15}{35} = \frac{22}{25}$,

so $3\frac{1}{5} + 2\frac{3}{7} = 5 + \frac{22}{35} = \mathbf{5\frac{22}{35}}$

c) $2 + 6 = 8$ and

$\frac{5}{8} + \frac{2}{3} = \frac{15}{24} + \frac{16}{24} = \frac{31}{24} = 1\frac{7}{24}$,

so $2\frac{5}{8} + 6\frac{2}{3} = 8 + 1\frac{7}{24} = \mathbf{9\frac{7}{24}}$

2 a) $\frac{2}{5} - \frac{4}{5} = \frac{2-4}{5} = \mathbf{-\frac{2}{5}}$

b) $2\frac{2}{5} - 1\frac{4}{5} = (2-1) + (\frac{2}{5} - \frac{4}{5})$

$= 1 - \frac{2}{5} = \frac{5}{5} - \frac{2}{5} = \mathbf{\frac{3}{5}}$

3 a) $3 - 0 = 3$ and $\frac{3}{4} - \frac{5}{7} = \frac{21}{28} - \frac{20}{28} = \frac{1}{28}$,

so $3\frac{3}{4} - \frac{5}{7} = 3 + \frac{1}{28} = \mathbf{3\frac{1}{28}}$

b) $2 - 1 = 1$ and $\frac{1}{4} - \frac{6}{7} = \frac{7}{28} - \frac{24}{28} = -\frac{17}{28}$,

so $2\frac{1}{4} - 1\frac{6}{7} = 1 - \frac{17}{28} = \frac{28}{28} - \frac{17}{28} = \mathbf{\frac{11}{28}}$

c) $5 - 3 = 2$ and $\frac{2}{5} - \frac{7}{9} = \frac{18}{45} - \frac{35}{45} = -\frac{17}{45}$,

so $5\frac{2}{5} - 3\frac{7}{9} = 2 - \frac{17}{45} = \frac{90}{45} - \frac{17}{45}$

$= \frac{73}{45} = \mathbf{1\frac{28}{45}}$

Watch your signs in b) and c) — subtracting the fraction parts gives you a negative fraction, so you end up subtracting a fraction from the whole number when you combine the parts.

5.5 Multiplying and Dividing Fractions

Page 75 Exercise 1

1 a) $8 \times \frac{1}{4} = \frac{8}{4} = \mathbf{2}$ **b)** $10 \times \frac{1}{5} = \frac{10}{5} = \mathbf{2}$

c) $15 \times \frac{1}{5} = \frac{15}{5} = \mathbf{3}$ **d)** $45 \times \frac{1}{3} = \frac{45}{3} = \mathbf{15}$

2 a) $18 \times \frac{1}{4} = \frac{18}{4} = 4\frac{2}{4} = \mathbf{4\frac{1}{2}}$

b) $15 \times \frac{1}{6} = \frac{15}{6} = 2\frac{3}{6} = \mathbf{2\frac{1}{2}}$

Using the same method for c) and d):

c) $\mathbf{8\frac{1}{2}}$ **d)** $\mathbf{2\frac{1}{2}}$

3 a) $48 \times -\frac{1}{6} = -\frac{48}{6} = \mathbf{-8}$

b) $80 \times -\frac{1}{10} = -\frac{80}{10} = \mathbf{-8}$

c) $25 \times -\frac{1}{6} = -\frac{25}{6} = \mathbf{-4\frac{1}{6}}$

d) $40 \times -\frac{1}{3} = -\frac{40}{3} = \mathbf{-13\frac{1}{3}}$

Page 77 Exercise 2

1 a) E.g. multiply by the numerator, $12 \times 2 = 24$, then divide by the denominator, $24 \div 3 = \mathbf{8}$.

b) E.g. divide by the denominator, $28 \div 4 = 7$, then multiply by the numerator, $7 \times 3 = \mathbf{21}$.

Remember you can multiply and divide in either order — just do whatever works best for the numbers you have.

Using the same method for c)-h):

c) 12 **d) 18** **e) 25**

f) 14 **g) −28** **h) 56**

2 a) $15 \times \frac{3}{4} = \frac{15\times3}{4} = \frac{45}{4} = \mathbf{11\frac{1}{4}}$

b) $22 \times \frac{2}{5} = \frac{22\times2}{5} = \frac{44}{5} = \mathbf{8\frac{4}{5}}$

c) $7 \times -\frac{3}{11} = -\frac{7\times3}{11} = -\frac{21}{11} = \mathbf{-1\frac{10}{11}}$

d) $6 \times -\frac{5}{8} = -\frac{6\times5}{8} = -\frac{30}{8} = -3\frac{6}{8} = \mathbf{-3\frac{3}{4}}$

3 $45 \times \frac{5}{18} = \frac{45\times5}{18} = \frac{5\times\cancel{9}\times5}{2\times\cancel{9}} = \frac{25}{2} = \mathbf{12\frac{1}{2}}$

Page 77 Exercise 3

1 a) $\frac{3}{4}$ of $36 = \frac{3}{4} \times 36 = 3 \times \frac{36}{4} = 3\times9 = \mathbf{27}$

b) $\frac{2}{3}$ of $33 = \frac{2}{3} \times 33 = 2 \times \frac{33}{3} = 2\times11 = \mathbf{22}$

Using the same method for c)-h):

c) 24 **d) 30** **e) 15**

f) 9 **g)** $\mathbf{27\frac{1}{2}}$ **h)** $\mathbf{26\frac{1}{4}}$

2 a) $\frac{3}{2}$ of $18 = \frac{3}{2} \times 18 = 3 \times \frac{18}{2} = 3\times9 = \mathbf{27}$

b) $\frac{5}{3}$ of $21 = \frac{5}{3} \times 21 = 5 \times \frac{21}{3} = 5\times7 = \mathbf{35}$

c) $\frac{11}{9}$ of $72 = 11 \times \frac{72}{9} = 11\times8 = \mathbf{88}$

3 $\frac{2}{3}$ of $27 = \frac{2}{3} \times 27 = 2 \times \frac{27}{3} = 2 \times 9$

= **18** students prefer rounders

Page 78 Exercise 4

1 a) $\frac{1}{6} \times \frac{1}{3} = \frac{1\times1}{6\times3} = \mathbf{\frac{1}{18}}$

b) $\frac{2}{5} \times \frac{1}{3} = \frac{2\times1}{5\times3} = \mathbf{\frac{2}{15}}$

Using the same method for c)-h):

c) $\mathbf{\frac{3}{28}}$ d) $\mathbf{\frac{3}{25}}$ e) $\mathbf{\frac{5}{24}}$

f) $\mathbf{\frac{8}{35}}$ g) $\mathbf{-\frac{10}{49}}$ h) $\mathbf{\frac{21}{80}}$

In h) the two minus signs cancel out (negative × negative = positive).

2 a) $\frac{1}{4} \times \frac{2}{3} = \frac{1 \times 2}{4 \times 3} = \frac{2}{12} = \mathbf{\frac{1}{6}}$

b) $\frac{3}{5} \times \frac{1}{6} = \frac{3 \times 1}{5 \times 6} = \frac{3}{30} = \mathbf{\frac{1}{10}}$

c) $\frac{5}{6} \times \frac{2}{15} = \frac{\cancel{5}}{\cancel{2} \times 3} \times \frac{\cancel{2}}{3 \times \cancel{5}} = \frac{1}{3} \times \frac{1}{3} = \frac{1 \times 1}{3 \times 3} = \mathbf{\frac{1}{9}}$

d) $\frac{5}{12} \times \frac{3}{4} = \frac{5}{\cancel{3} \times 4} \times \frac{\cancel{3}}{4} = \frac{5}{4 \times 4} = \mathbf{\frac{5}{16}}$

Using the same method for e)-h):

e) $\mathbf{\frac{3}{14}}$ f) $\mathbf{\frac{6}{7}}$

g) $\mathbf{-\frac{3}{4}}$ h) $\mathbf{-\frac{1}{4}}$

3 $\frac{3}{5}$ of $\frac{15}{8} = \frac{3}{5} \times \frac{15}{8} = \frac{3}{\cancel{5}} \times \frac{3 \times \cancel{5}}{8} = \frac{3 \times 3}{8} = \frac{9}{8} = \mathbf{1\frac{1}{8}}$

Page 79 Exercise 5

1 a) $1\frac{1}{2} = \frac{2}{2} + \frac{1}{2} = \frac{3}{2}$, so $1\frac{1}{2} \times \frac{1}{3} = \frac{\cancel{3}}{2} \times \frac{1}{\cancel{3}} = \mathbf{\frac{1}{2}}$

b) $2\frac{1}{5} = \frac{10}{5} + \frac{1}{5} = \frac{11}{5}$, so $2\frac{1}{5} \times \frac{3}{4} = \frac{11}{5} \times \frac{3}{4} = \frac{11 \times 3}{5 \times 4} = \frac{33}{20} = \mathbf{1\frac{13}{20}}$

Using the same method for c)-f):

c) $\mathbf{1\frac{2}{9}}$ d) $\mathbf{1\frac{1}{2}}$

e) $\mathbf{\frac{10}{21}}$ f) $\mathbf{\frac{11}{12}}$

2 a) $3\frac{2}{5} = \frac{15}{5} + \frac{2}{5} = \mathbf{\frac{17}{5}}$

b) $1\frac{1}{2} = \frac{2}{2} + \frac{1}{2} = \mathbf{\frac{3}{2}}$

c) $3\frac{2}{5} \times 1\frac{1}{2} = \frac{17}{5} \times \frac{3}{2} = \frac{51}{10} = \mathbf{5\frac{1}{10}}$

3 a) $1\frac{1}{5} = \frac{5}{5} + \frac{1}{5} = \frac{6}{5}$ and $1\frac{1}{4} = \frac{4}{4} + \frac{1}{4} = \frac{5}{4}$, so $1\frac{1}{5} \times 1\frac{1}{4} = \frac{6}{\cancel{5}} \times \frac{\cancel{5}}{4} = \frac{6}{4} = 1\frac{2}{4} = \mathbf{1\frac{1}{2}}$

b) $2\frac{2}{5} = \frac{10}{5} + \frac{2}{5} = \frac{12}{5}$ and $1\frac{2}{3} = \frac{3}{3} + \frac{2}{3} = \frac{5}{3}$, so $2\frac{2}{5} \times 1\frac{2}{3} = \frac{12}{\cancel{5}} \times \frac{\cancel{5}}{3} = \frac{12}{3} = \mathbf{4}$

Using the same method for c)-f):

c) $\mathbf{2\frac{4}{5}}$ d) $\mathbf{7\frac{29}{42}}$

e) $\mathbf{6\frac{11}{24}}$ f) $\mathbf{6\frac{2}{63}}$

Page 80 Exercise 6

1 a) Reciprocal of 5 is $1 \div 5 = \mathbf{\frac{1}{5}}$.

b) Reciprocal of 12 is $1 \div 12 = \mathbf{\frac{1}{12}}$.

c) Reciprocal of 9 is $1 \div 9 = \mathbf{\frac{1}{9}}$.

d) Reciprocal of 27 is $1 \div 27 = \mathbf{\frac{1}{27}}$.

2 a) Reciprocal of $\frac{1}{3} = \frac{3}{1} = \mathbf{3}$.

b) Reciprocal of $-\frac{1}{7} = -\frac{7}{1} = \mathbf{-7}$.

c) Reciprocal of $\frac{4}{5}$ is $\mathbf{\frac{5}{4}}$.

d) Reciprocal of $-\frac{5}{8}$ is $\mathbf{-\frac{8}{5}}$.

Don't forget to keep the minus sign when you find the reciprocal of a negative number.

3 a) $1\frac{3}{5} = \frac{5}{5} + \frac{3}{5} = \frac{8}{5}$, so reciprocal of $1\frac{3}{5}$ is $\mathbf{\frac{5}{8}}$.

b) $2\frac{1}{7} = \frac{14}{7} + \frac{1}{7} = \frac{15}{7}$, so reciprocal of $2\frac{1}{7}$ is $\mathbf{\frac{7}{15}}$.

c) $1\frac{4}{9} = \frac{9}{9} + \frac{4}{9} = \frac{13}{9}$, so reciprocal of $1\frac{4}{9}$ is $\mathbf{\frac{9}{13}}$.

d) $-2\frac{3}{4} = -\left(\frac{8}{4} + \frac{3}{4}\right) = -\frac{11}{4}$, so reciprocal of $-2\frac{3}{4}$ is $\mathbf{-\frac{4}{11}}$.

Page 81 Exercise 7

1 a) $\frac{1}{5} \div \frac{2}{3} = \frac{1}{5} \times \frac{3}{2} = \frac{1 \times 3}{5 \times 2} = \mathbf{\frac{3}{10}}$

b) $\frac{1}{6} \div \frac{2}{5} = \frac{1}{6} \times \frac{5}{2} = \frac{1 \times 5}{6 \times 2} = \mathbf{\frac{5}{12}}$

Using the same method for c)-h):

c) **3** d) $\mathbf{\frac{1}{6}}$ e) $\mathbf{\frac{25}{48}}$

f) $\mathbf{\frac{5}{14}}$ g) $\mathbf{-\frac{15}{16}}$ h) $\mathbf{-\frac{49}{50}}$

2 a) $1\frac{1}{2} = \frac{2}{2} + \frac{1}{2} = \frac{3}{2}$, so $1\frac{1}{2} \div 4 = \frac{3}{2} \div 4 = \frac{3}{2} \times \frac{1}{4} = \frac{3 \times 1}{2 \times 4} = \mathbf{\frac{3}{8}}$

b) $3\frac{1}{3} = \frac{9}{3} + \frac{1}{3} = \frac{10}{3}$, so $3\frac{1}{3} \div 6 = \frac{10}{3} \times \frac{1}{6} = \frac{10 \times 1}{3 \times 6} = \frac{10}{18} = \mathbf{\frac{5}{9}}$

c) $1\frac{3}{5} = \frac{5}{5} + \frac{3}{5} = \frac{8}{5}$, so $5 \div 1\frac{3}{5} = 5 \div \frac{8}{5} = 5 \times \frac{5}{8} = \frac{25}{8} = \mathbf{3\frac{1}{8}}$

d) $1\frac{7}{8} = \frac{8}{8} + \frac{7}{8} = \frac{15}{8}$, so $2 \div 1\frac{7}{8} = 2 \div \frac{15}{8} = 2 \times \frac{8}{15} = \frac{2 \times 8}{15} = \frac{16}{15} = \mathbf{1\frac{1}{15}}$

3 a) $1\frac{1}{3} = \frac{3}{3} + \frac{1}{3} = \frac{4}{3}$, so $1\frac{1}{3} \div \frac{2}{5} = \frac{4}{3} \times \frac{5}{2} = \frac{4 \times 5}{3 \times 2} = \frac{20}{6} = 3\frac{2}{6} = \mathbf{3\frac{1}{3}}$

b) $2\frac{1}{2} = \frac{4}{2} + \frac{1}{2} = \frac{5}{2}$, so $2\frac{1}{2} \div \frac{1}{3} = \frac{5}{2} \times \frac{3}{1} = \frac{15}{2} = \mathbf{7\frac{1}{2}}$

c) $2\frac{1}{3} = \frac{6}{3} + \frac{1}{3} = \frac{7}{3}$, so $\frac{3}{4} \div 2\frac{1}{3} = \frac{3}{4} \times \frac{3}{7} = \mathbf{\frac{9}{28}}$

d) $3\frac{1}{2} = \frac{6}{2} + \frac{1}{2} = \frac{7}{2}$, so $\frac{4}{7} \div 3\frac{1}{2} = \frac{4}{7} \times \frac{2}{7} = \mathbf{\frac{8}{49}}$

e) $1\frac{1}{4} = \frac{4}{4} + \frac{1}{4} = \frac{5}{4}$ and $1\frac{1}{5} = \frac{5}{5} + \frac{1}{5} = \frac{6}{5}$, so $1\frac{1}{4} \div 1\frac{1}{5} = \frac{5}{4} \times \frac{5}{6} = \frac{25}{24} = \mathbf{1\frac{1}{24}}$

f) $2\frac{2}{3} = \frac{6}{3} + \frac{2}{3} = \frac{8}{3}$ and $1\frac{1}{4} = \frac{4}{4} + \frac{1}{4} = \frac{5}{4}$, so $2\frac{2}{3} \div 1\frac{1}{4} = \frac{8}{3} \times \frac{4}{5} = \frac{32}{15} = \mathbf{2\frac{2}{15}}$

g) $4\frac{5}{6} = \frac{24}{6} + \frac{5}{6} = \frac{29}{6}$ and $2\frac{1}{3} = \frac{6}{3} + \frac{1}{3} = \frac{7}{3}$, so $4\frac{5}{6} \div 2\frac{1}{3} = \frac{29}{6} \times \frac{3}{7} = \frac{29}{\cancel{3} \times 2} \times \frac{\cancel{3}}{7} = \frac{29}{14} = \mathbf{2\frac{1}{14}}$

h) $3\frac{2}{3} = \frac{9}{3} + \frac{2}{3} = \frac{11}{3}$ and $-2\frac{1}{10} = -\left(\frac{20}{10} + \frac{1}{10}\right) = -\frac{21}{10}$, so $3\frac{2}{3} \div \left(-2\frac{1}{10}\right) = \frac{11}{3} \times \left(-\frac{10}{21}\right) = -\frac{110}{63} = \mathbf{-1\frac{47}{63}}$

5.6 Fractions and Decimals

Page 82 Exercise 1

1 a) $\frac{5}{8} = 5 \div 8 = \mathbf{0.625}$

b) $\frac{7}{20} = 7 \div 20 = \mathbf{0.35}$

No trick questions here — just pop the numbers into your calculator.

Using the same method for c)-h):

c) **0.4375** d) **0.15625** e) **0.175**

f) **0.46** g) **0.88** h) **0.658**

2 a) $\frac{3}{8} = 3 \div 8 = 0.375$, so $1\frac{3}{8} = 1 + 0.375 = \mathbf{1.375}$ (or $1\frac{3}{8} = \frac{8}{8} + \frac{3}{8} = \frac{11}{8} = 11 \div 8 = \mathbf{1.375}$)

b) $\frac{1}{8} = 1 \div 8 = 0.125$, so $2\frac{1}{8} = 2 + 0.125 = \mathbf{2.125}$ (or $2\frac{1}{8} = \frac{16}{8} + \frac{1}{8} = \frac{17}{8} = 17 \div 8 = \mathbf{2.125}$)

Using the same method for c)-h):

c) **6.35** d) **2.37** e) **4.719**

f) **5.76** g) **7.34375** h) **8.4375**

Page 84 Exercise 2

1 a) $\frac{9}{10} = 9 \times \frac{1}{10} = 9 \times 0.1 = \mathbf{0.9}$

Using the same method for b)-d):

b) **0.2** c) **0.3** d) **0.8**

e) $\frac{91}{100} = 91 \times \frac{1}{100} = 91 \times 0.01 = \mathbf{0.91}$

Using the same method for f)-h):

f) **0.42** g) **0.99** h) **0.08**

i) $\frac{7}{1000} = 7 \times \frac{1}{1000} = 7 \times 0.001 = \mathbf{0.007}$

Using the same method for j)-l):

j) **0.201** k) **0.041** l) **0.027**

2 a) $\frac{8}{25} = \frac{8 \times 4}{25 \times 4} = \mathbf{\frac{32}{100}}$

b) $\frac{32}{100} = 32 \times \frac{1}{100} = 32 \times 0.01 = \mathbf{0.32}$

3 a) $\frac{3}{5} = \frac{3 \times 2}{5 \times 2} = \frac{6}{10} = 6 \times \frac{1}{10} = 6 \times 0.1 = \mathbf{0.6}$

b) $\frac{9}{30} = \frac{9 \div 3}{30 \div 3} = \frac{3}{10} = 3 \times \frac{1}{10} = 3 \times 0.1 = \mathbf{0.3}$

c) $\frac{45}{50} = \frac{45 \times 2}{50 \times 2} = \frac{90}{100} = 90 \times \frac{1}{100} = 90 \times 0.01 = \mathbf{0.9}$

d) $\frac{22}{25} = \frac{22 \times 4}{25 \times 4} = \frac{88}{100} = 88 \times \frac{1}{100} = 88 \times 0.01 = \mathbf{0.88}$

e) $\frac{96}{300} = \frac{96 \div 3}{300 \div 3} = \frac{32}{100} = 32 \times \frac{1}{100} = 32 \times 0.01 = \mathbf{0.32}$

f) $\frac{33}{250} = \frac{33 \times 4}{250 \times 4} = \frac{132}{1000} = 132 \times \frac{1}{1000} = 132 \times 0.001 = \mathbf{0.132}$

g) $\frac{103}{200} = \frac{103 \times 5}{200 \times 5} = \frac{515}{1000} = 515 \times \frac{1}{1000} = 515 \times 0.001 = \mathbf{0.515}$

h) $\frac{306}{3000} = \frac{306 \div 3}{3000 \div 3} = \frac{102}{1000} = 102 \times \frac{1}{1000} = 102 \times 0.001 = \mathbf{0.102}$

4 a) E.g. $\frac{11}{10} = 11 \times 0.1 = \mathbf{1.1}$

b) E.g. $\frac{14}{5} = \frac{14 \times 2}{5 \times 2} = \frac{28}{10} = 28 \times 0.1 = \mathbf{2.8}$

c) E.g. $5\frac{7}{25} = 5 + \frac{7}{25} = 5 + \frac{7 \times 4}{25 \times 4} = 5 + \frac{28}{100} = 5 + 28 \times 0.01 = 5 + 0.28 = \mathbf{5.28}$

d) E.g. $3\frac{11}{200} = 3 + \frac{11}{200} = 3 + \frac{11 \times 5}{200 \times 5} = 3 + \frac{55}{1000} = 3 + 55 \times 0.001 = 3 + 0.055 = \mathbf{3.055}$

To convert mixed numbers to decimals you can either separate the whole and fraction parts, as in parts c) and d) above, or you can convert them to improper fractions.

Page 85 Exercise 3

1 a) $\begin{array}{r}0.\,3\,7\,5\\ 8\overline{)3.{}^{3}0{}^{6}0{}^{4}0}\end{array}$, so $\frac{3}{8} = 3 \div 8 = \mathbf{0.375}$

b) $\begin{array}{r}0.\,3\,1\,2\,5\\ 16\overline{)5.{}^{5}0{}^{2}0{}^{4}0{}^{8}0}\end{array}$, so $\frac{5}{16} = 5 \div 16 = \mathbf{0.3125}$

2 a) $\begin{array}{r}0.\,2\,5\\ 4\overline{)1.{}^{1}0{}^{2}0}\end{array}$, so $\frac{1}{4} = 1 \div 4 = \mathbf{0.25}$

b) $\begin{array}{r}0.\,7\,5\\ 4\overline{)3.{}^{3}0{}^{2}0}\end{array}$, so $\frac{3}{4} = 3 \div 4 = \mathbf{0.75}$

c) $\begin{array}{r}0.\,0\,5\\ 20\overline{)1.{}^{1}0{}^{10}0}\end{array}$, so $\frac{1}{20} = 1 \div 20 = \mathbf{0.05}$

Using the same method for d)-h):

d) **0.025** e) **0.0625** f) **0.875**

g) **0.175** h) **0.1625**

Page 86 Exercise 4

1 a) 0.7 has one decimal place, so the denominator is 10.
$0.7 = \mathbf{\frac{7}{10}}$

b) 0.9 has one decimal place, so the denominator is 10.
$0.9 = \mathbf{\frac{9}{10}}$

Using the same method for c) and d):
c) $\mathbf{\frac{1}{10}}$ d) $\mathbf{\frac{2}{5}}$

2 a) 0.93 has two decimal places, so the denominator is 100.
$0.93 = \mathbf{\frac{93}{100}}$

b) 0.07 has two decimal places, so the denominator is 100.
$0.07 = \mathbf{\frac{7}{100}}$

Using the same method for c) and d):
c) $\mathbf{\frac{23}{100}}$ d) $\mathbf{\frac{47}{100}}$

3 a) 0.004 has 3 decimal places, so the denominator is 1000.
$0.004 = \frac{4}{1000} = \mathbf{\frac{1}{250}}$

b) 0.801 has 3 decimal places, so the denominator is 1000.
$0.801 = \mathbf{\frac{801}{1000}}$

Using the same method for c) and d):
c) $\mathbf{\frac{983}{1000}}$ d) $\mathbf{\frac{49}{500}}$

4 0.1002 has 4 decimal places, so the denominator is 10 000.
$0.1002 = \frac{1002}{10\,000} = \mathbf{\frac{501}{5000}}$

Page 86 Exercise 5

1 a) 1.2 has one decimal place, so the denominator is 10.
$1.2 = \frac{12}{10} = \frac{12 \div 2}{10 \div 2} = \mathbf{\frac{6}{5}}$

b) 3.4 has one decimal place, so the denominator is 10.
$3.4 = \frac{34}{10} = \frac{34 \div 2}{10 \div 2} = \mathbf{\frac{17}{5}}$

Using the same method for c) and d):
c) $\mathbf{\frac{47}{10}}$ d) $\mathbf{\frac{42}{5}}$

2 a) 3.02 has two decimal places, so the denominator is 100.
So $3.02 = \frac{302}{100} = \frac{302 \div 2}{100 \div 2} = \mathbf{\frac{151}{50}}$

b) 1.55 has two decimal places, so the denominator is 100.
So $1.55 = \frac{155}{100} = \frac{155 \div 5}{100 \div 5} = \mathbf{\frac{31}{20}}$

c) 2.05 has two decimal places, so the denominator is 100.
So $2.05 = \frac{205}{100} = \frac{205 \div 5}{100 \div 5} = \mathbf{\frac{41}{20}}$

d) 18.2 has one decimal place, so the denominator is 10.
So 18.2 = $= \frac{182}{10} = \frac{182 \div 2}{10 \div 2} = \mathbf{\frac{91}{5}}$

Careful with d) — although there are 3 digits, there's only one number after the decimal point, so the denominator's 10.

Page 87 Review Exercise

1 a) $\frac{3}{4} = \frac{3 \times 4}{4 \times 4} = \frac{12}{16}$, so $\mathbf{a = 12}$.

b) $\frac{7}{12} = \frac{7 \times 5}{12 \times 5} = \frac{35}{60}$, so $\mathbf{b = 60}$.

c) $\frac{10}{15} = \frac{10 \div 5}{15 \div 5} = \frac{2}{3}$, so $\mathbf{c = 2}$.

2 a) 45 = 9 × 5 and 75 = 15 × 5, so **5 is the common factor.**

b) E.g. $\frac{45}{75} = \frac{45 \div 5}{75 \div 5} = \frac{9}{15} = \frac{9 \div 3}{15 \div 3} = \mathbf{\frac{3}{5}}$.

3 a) 4 out of 24 = $\frac{4}{24} = \frac{4 \div 4}{24 \div 4} = \mathbf{\frac{1}{6}}$.

b) 12 out of 66 = $\frac{12}{66} = \frac{12 \div 6}{66 \div 6} = \mathbf{\frac{2}{11}}$.

4 a) $7\frac{2}{3} = \frac{21}{3} + \frac{2}{3} = \mathbf{\frac{23}{3}}$

b) 26 ÷ 4 = 6 remainder 2,
so $\frac{26}{4} = 6\frac{2}{4} = \mathbf{6\frac{1}{2}}$

5 a) $\frac{2}{7} = \frac{2 \times 4}{7 \times 4} = \mathbf{\frac{8}{28}}$, $\frac{5}{28} = \mathbf{\frac{5}{28}}$,
so $\mathbf{\frac{5}{28}}$ is the smaller fraction.

b) $\frac{1}{8} = \frac{1 \times 3}{8 \times 3} = \mathbf{\frac{3}{24}}$, $\frac{1}{6} = \frac{1 \times 4}{6 \times 4} = \mathbf{\frac{4}{24}}$,
so $\mathbf{\frac{1}{8}}$ is the smaller fraction.

c) $\frac{2}{5} = \frac{2 \times 9}{5 \times 9} = \mathbf{\frac{18}{45}}$, $\frac{4}{9} = \frac{4 \times 5}{9 \times 5} = \mathbf{\frac{20}{45}}$,
so $\mathbf{\frac{2}{5}}$ is the smaller fraction.

6 a) $\frac{1}{5} + \frac{1}{3} = \frac{3}{15} + \frac{5}{15} = \frac{3+5}{15} = \mathbf{\frac{8}{15}}$

b) $\frac{9}{10} - \frac{5}{6} = \frac{27}{30} - \frac{25}{30} = \frac{27-25}{30} = \frac{2}{30} = \mathbf{\frac{1}{15}}$

c) $6\frac{3}{8} = \frac{48}{8} + \frac{3}{8} = \frac{51}{8}$,
so $6\frac{3}{8} - \frac{7}{8} = \frac{51}{8} - \frac{7}{8} = \frac{44}{8} = \frac{(5 \times 8) + 4}{8}$
$= 5\frac{4}{8} = \mathbf{5\frac{1}{2}}$

7 a) $1\frac{3}{7} = \frac{7}{7} + \frac{3}{7} = \mathbf{\frac{10}{7}}$

b) $1\frac{3}{7} \times \frac{2}{3} = \frac{10}{7} \times \frac{2}{3} = \frac{10 \times 2}{7 \times 3} = \mathbf{\frac{20}{21}}$

8 a) $\mathbf{\frac{8}{3}}$

b) $\frac{1}{6} \div \frac{3}{8} = \frac{1}{6} \times \frac{8}{3} = \frac{1 \times 8}{6 \times 3} = \frac{8}{18} = \mathbf{\frac{4}{9}}$

9 $23 \div 40 = \mathbf{0.575}$

10 a) $\frac{12}{25} = \frac{12 \times 4}{25 \times 4} = \frac{48}{100} = 48 \times \frac{1}{100}$
$= 48 \times 0.01 = \mathbf{0.48}$

b) 0.35 has two decimal places, so the denominator is 100.
$0.35 = \frac{35}{100} = \frac{35 \div 5}{100 \div 5} = \mathbf{\frac{7}{20}}$

Page 88 Exam-Style Questions

1 a) $\frac{5}{12}$ of $78 = \frac{5}{12} \times 78 = \frac{5 \times 78}{12}$ *[1 mark]*
$= \frac{5 \times \cancel{2} \times \cancel{3} \times 13}{2 \times \cancel{2} \times \cancel{3}} = \frac{65}{2} = \mathbf{32\frac{1}{2}}$ **cm**
(or **32.5 cm**) *[1 mark]*

b) $2\frac{1}{2} = \frac{4}{2} + \frac{1}{2} = \frac{5}{2}$ *[1 mark]*
$15 \div 2\frac{1}{2} = 15 \div \frac{5}{2}$
$= 15 \times \frac{2}{5} = \frac{15 \times 2}{5} = \frac{30}{5} = \mathbf{6}$ *[1 mark]*

2 $3.125 = 3 + 0.125 = 3 + \frac{125}{1000}$ *[1 mark]*
$= 3 + \frac{125 \div 25}{1000 \div 25} = 3 + \frac{5}{40} = 3 + \frac{1}{8} = \mathbf{3\frac{1}{8}}$
[1 mark]

3 $\frac{3}{4}$ is used, so $1 - \frac{3}{4} = \frac{1}{4}$ is left over.
$\frac{3}{5}$ is then used, so $1 - \frac{3}{5} = \frac{2}{5}$ of $\frac{1}{4}$ is left over *[1 mark]*.
$\frac{2}{5}$ of $\frac{1}{4} = \frac{2}{5} \times \frac{1}{4} = \frac{2 \times 1}{5 \times 4} = \frac{2}{20}$ *[1 mark]*
$= \mathbf{\frac{1}{10}}$ *[1 mark]*

4 Divide the total butter by the butter per dumpling:
$120 \div 5\frac{1}{3} = 120 \div \frac{16}{3} = 120 \times \frac{3}{16}$ *[1 mark]*
$= \frac{120 \times 3}{16} = \frac{30 \times 3}{4} = \frac{15 \times 3}{2} = \frac{45}{2}$ *[1 mark]*
= 22.5
So Sam can make **22 dumplings** *[1 mark]*.

5 a) $\frac{2}{5}$ of the box are green, so $1 - \frac{2}{5} = \frac{3}{5}$ of the box are red or yellow *[1 mark]*.
There are 24 red and $\frac{24}{4} = 6$ yellow,
so 24 + 6 = 30 peppers equal $\frac{3}{5}$ of the box *[1 mark]*.
So $\frac{1}{5}$ of the box is 30 ÷ 3 = 10 peppers, which means the box has 5 × 10
= **50 peppers** *[1 mark]*.
Part a) might look tricky — but once you've found number of peppers in a fifth of the box, you can just scale up to find the number in the full box.

b) $\frac{24}{50} = \frac{24 \times 2}{50 \times 2} = \frac{48}{100}$ *[1 mark]*
$= 48 \times \frac{1}{100} = 48 \times 0.01 = \mathbf{0.48}$ *[1 mark]*

Section 6 — Ratios

6.1 Ratios

Page 90 Exercise 1

1 There are 4 stars and 3 triangles.
So the ratio of stars to triangles is **4:3**.

2 The ratio cows : pigs is **23:15**.
Make sure you write the ratio in the correct order. The question asks for the ratio of cows to pigs, so write down the 23 cows first.

3 a) The highest common factor of 2 and 8 is 2. 2 ÷ 2 = 1 and 8 ÷ 2 = 4, so the simplest form of 2 : 8 is **1:4**.

b) The highest common factor of 5 and 15 is 5. 5 ÷ 5 = 1 and 15 ÷ 5 = 3, so the simplest form of 5 : 15 is **1:3**.

Use the same method for c) and d):
c) (÷ 10) **4:1** d) (÷ 2) **2:3**

4 a) The highest common factor of 6, 2 and 4 is 2. 6 ÷ 2 = 3, 2 ÷ 2 = 1 and 4 ÷ 2 = 2.
So the simplest form is **3:1:2**.

b) The highest common factor of 15, 12 and 3 is 3. 15 ÷ 3 = 5, 12 ÷ 3 = 4 and 3 ÷ 3 is 1.
So the simplest form is **5:4:1**.

Use the same method for c) and d):
c) (÷ 2) **7:5:1** d) (÷ 4) **6:3:5**

5 black tiles : white tiles = 24 : 8. 8 is highest common factor of 24 and 8. 24 ÷ 8 = 3 and 8 ÷ 8 = 1, so the simplest form is **3:1**.

6 girls : boys = 36 : 27. 9 is the highest common factor of 36 and 27. 36 ÷ 9 = 4 and 27 ÷ 9 = 3, so the simplest form is **4:3.**

7 computers : pupils = 170 : 595
Divide both sides by 5:
170 ÷ 5 : 595 ÷ 5 = 34 : 119
Then divide both sides by 17:
34 ÷ 17 : 119 ÷ 17 = **2:7**
You could get straight to the final answer by dividing by the highest common factor, which is 85.

8 Soraya has 42 – 16 = 26 sweets.
The ratio of Soraya's sweets : Paul's sweets is 26 : 16. Divide both sides by 2 to get **13:8.**

Page 91 Exercise 2

1 a) £1 = 100p, so 10p : £1 = 10p : 100p.
The highest common factor of 10 and 100 is 10. Divide both sides by 10 to get **1:10.**

b) 1 cm = 10 mm, so 20 mm : 4 cm = 20 mm : 40 mm.
The highest common factor of 20 and 40 is 20. Divide both sides by 20 to get **1:2.**

c) 1 kg = 1000 g, so 10 g : 1 kg = 10 g : 1000 g.
The highest common factor of 10 and 1000 is 10. Divide both sides by 10 to get **1:100.**

d) 1 week = 7 days,
so 2 weeks : 7 days = 14 days : 7 days.
The highest common factor of 14 and 7 is 7. Divide both sides by 7 to get **2:1**.

e) £1 = 100p, so 40p : £1 = 40p : 100p.
The highest common factor of 40 and 100 is 20. Divide both sides by 20 to get **2:5**.

f) 1 m = 100 cm,
so 30 cm : 2 m = 30 cm : 200 cm.
The highest common factor of 30 and 200 is 10. Divide both sides by 10 to get **3:20**.

2 a) 1 m = 1000 mm,
so 1 m : 150 mm = 1000 mm : 150 mm.
The highest common factor of 1000 and 150 is 50. Divide both sides by 50 to get the ratio **20:3**.

b) 1 m = 100 cm, so
8 cm : 1.1 m = 8 cm : 110 cm.
The highest common factor of 8 and 110 is 2. Divide both sides by 2 to get **4:55**.

c) 1 kg = 1000 g, so 9 g : 0.3 kg = 9 g : 300 g.
The highest common factor of 9 and 300 is 3. Divide both sides by 3 to get **3:100**.

d) 1 hr = 60 mins,
so 2.5 hrs : 20 mins = 150 mins : 20 mins.
The highest common factor of 150 and 20 is 10. Divide both sides by 10 to get **15:2**.

e) £1 = 100p, so £1.25 : 75p = 125p : 75p.
The highest common factor of 125 and 75 is 25. Divide both sides by 25 to get **5:3**.

f) 1 km = 1000 m,
so 65 m : 1.3 km = 65 m : 1300 km.
The highest common factor of 65 and 1300 is 65. Divide both sides by 65 to get **1:20**.

3 1 litre = 1000 ml,
so 50 ml : 1 litre = 50 ml : 1000 ml.
The highest common factor of 50 and 1000 is 50. 50 ÷ 50 = 1 and 1000 ÷ 50 = 20, so the ratio in its simplest form is **1:20**.

4 1 hour = 60 minutes, so 45 minutes : 3.5 hours = 45 minutes : 210 minutes. The highest common factor of 45 and 210 is 15.
45 ÷ 15 = 3 and 210 ÷ 15 = 14,
so the ratio in its simplest form is **3:14**.

5 1 kg = 1000 g. The ratio of butter : icing sugar is 640 g : 1.6 kg = 640 g : 1600 g.
640 : 1600 = 640 ÷ 10 : 1600 ÷ 10 = 64 : 160
64 : 160 = 64 ÷ 16 : 160 ÷ 16 = 4 : 10 = **2:5**.

Page 92 Exercise 3

1 a) $2:6 = 1:\frac{6}{2} = \mathbf{1:3}$

b) $7:35 = 1:\frac{35}{7} = \mathbf{1:5}$
Use the same method for c)-h):

c) (÷ 6) **1:4** **d)** (÷ 30) **1:4**
e) (÷ 2) **1:3.5** **f)** (÷ 4) **1:6.5**
g) (÷ 8) **1:3.25** **h)** (÷ 2) **1:0.5**

2 a) 1 cm = 10 mm,
so 10 mm : 5 cm = 10 mm : 50 mm.
Divide both sides by 10 to get **1:5**.

b) £1 = 100p, so 12p : £6 = 12p : 600p.
Divide both sides by 12 to get **1:50**.

c) 1 hour = 60 mins,
so 30 mins : 2 hours = 30 mins : 120 mins.
Divide both sides by 30 to get **1:4**.

d) 1 kg = 1000 g,
so 500 g : 20 kg = 500 g : 20 000 g.
Divide both sides by 500 to get **1:40**.

3 The ratio of frogs to fish is
$7:56 = 1:\frac{56}{7} = \mathbf{1:8}$.

4 The ratio of red to silver cars is
$15:45 = 1:\frac{45}{15} = \mathbf{1:3}$.

5 1 litre = 1000 ml,
so 125 ml : 2.5 litres = 125 ml : 2500 ml.
The ratio of syrup to milk is
$125:2500 = 1:\frac{2500}{125} = \mathbf{1:20}$.

6 1 km = 1000 m and 1 m = 100 cm,
so 4.8 km = 4800 m = 480 000 cm.
So 12 cm : 4.8 km = 12 cm : 480 000 cm.
The ratio of map distance to true distance is $12:480\,000 = 1:\frac{480\,000}{12} = \mathbf{1:40\,000}$.
The question asks for the ratio of map distance to true distance, so remember to give your answer in the correct order.

Page 93 Exercise 4

1 The ratio blue : white = 1 : 3 has 1 + 3 = 4 parts, with 1 part blue. So the fraction of blue tiles is $\mathbf{\frac{1}{4}}$.

2 The ratio yellow : red = 3 : 2 has 3 + 2 = 5 parts, with 2 parts red. So the fraction of red beads is $\mathbf{\frac{2}{5}}$.

3 a) **True**. The total number of parts is 1 + 2 + 4 = 7. The blue part of the ratio is 1, so $\frac{1}{7}$ of the balls are blue.

b) **True**. As a) is true, you know the bag contains $\frac{1}{7} \times 21 = 3$ blue balls.

c) **False**. From a) $\frac{1}{7}$ balls are blue, so $1 - \frac{1}{7} = \frac{6}{7}$ are red or green.

d) **True**. The ratio blue : red = 1 : 4 is contained in the given ratio — it's found by ignoring the green part.

e) **True**. The ratio says there are 4 red balls for every 2 green, so there must be more red balls than green balls.

Page 93 Exercise 5

1 If $\frac{1}{3}$ of the Jelly Babies are red,
then $1 - \frac{1}{3} = \frac{2}{3}$ are green.
The two fractions have a common denominator (3), so the ratio of red to green is **1:2**.

2 If $\frac{3}{19}$ members are left-handed, then $1 - \frac{3}{19} = \frac{16}{19}$ are right-handed. The two fractions have a common denominator (19), so the ratio of right-handed to left-handed members is **16:3**.

3 The fraction of pepperoni pizzas is $\frac{3}{8}$.
The fraction of goat's cheese pizzas is
$\frac{1}{2} = \frac{1 \times 4}{2 \times 4} = \frac{4}{8}$.
So the fraction of spicy chicken pizzas is
$1 - \frac{3}{8} - \frac{4}{8} = \frac{1}{8}$.
These fractions have a common denominator (8), so the ratio is **3:4:1**.

6.2 Using Ratios

Page 94 Exercise 1

1 blue eyes : brown eyes = 2 : 3 = 8 : ?
Multiply both sides by 8 ÷ 2 = 4, so the ratio becomes 8 : 12.
So there are **12 brown eyed pupils**.

2 father's age : son's age = 8 : 3 = 48 : ?
Multiply both sides by 48 ÷ 8 = 6, so the ratio becomes 48 : 18. So the son's age is **18**.

3 red : yellow = 3 : 4 = ? : 12.
Multiply both sides by 12 ÷ 4 = 3, so the ratio becomes 9 : 12. So there are **9 red sweets** in the bag.

4 oak : beech = 2 : 9 = 42 : ?
Multiply both sides by 42 ÷ 2 = 21, so the ratio becomes 42 : 189. So there are **189 beech trees**.

5 apples : bananas = 5 : 3 = ? : 450
Multiply both sides by 450 ÷ 3 = 150, so the ratio becomes 750 : 450.
So there are **750 apples**.

6 sugar : butter = 2 : 1 = 100 : ?
Multiply both sides by 100 ÷ 2 = 50, so the ratio becomes 100 : 50. So **50 g** of butter is needed.
If there are units in this type of question, don't forget to include them in your answer.

7 small photo : large photo = 2 : 7 = 10 : ?
Multiply both sides by 10 ÷ 2 = 5, so the ratio becomes 2 : 35.
So the enlarged photo **35 cm** wide.

8 The ratio children : adults is 5 : 1, so there must be at least 1 adult for every 5 children.
If there are 32 children, then there must be at least $\frac{32}{5} = 6.4$ adults — you need a whole number, so the answer is **7 adults** (round up, as 6 adults is too few).

9 The ratio 'under 30s' : '30 or over' is at least 8 : 1, so there must be at least 8 'under 30s' for every '30 or over'. If there are 100 'under 30s', at most $\frac{100}{8} = 12.5$ can be '30 or over' — you need a whole number, so the answer is **12** (round down, as 13 is too many).

Page 95 Exercise 2

1 pineapple juice : orange juice = 1 : 3 = 500 : ?
Multiply both sides by 500, so the ratio becomes 500 : 1500. So **1500 ml** of orange juice are needed.

2 Mai : Lizzy : Dave = 31 : 33 : 37 = 155 : ? : ?
Multiply all parts by 155 ÷ 31 = 5, so the ratio becomes 155 : 165 : 185. So Lizzy is **165 cm** tall and Dave is **185 cm** tall.

3 1 hour and 10 minutes = 60 + 10 minutes = 70 minutes.
Max : Molly : Hasan = 3 : 7 : 2 = ? : 70 : ?
Multiply all parts by 70 ÷ 7 = 10, so the ratio becomes 30 : 70 : 20. So Max has been waiting **30 minutes** and Hasan has been waiting **20 minutes**.

Page 96 Exercise 3

1 7 – 4 = 3 parts are bourbons, so the ratio of digestives : bourbons = **4:3**.

2 There are 15 + 2 = 17 liquid parts in total, of which 15 parts are water. So the ratio water : total is **15:17**.

3 There are 99 parts in total, of which 53 parts are digital downloads. So 99 – 53 = 46 parts are CDs. So the ratio digital downloads : CDs sold is **53:46**.

4 There are 3 + 5 + 9 = 17 parts in total, of which 3 are roses. So the ratio roses : total is **3:17**.

6.3 Dividing in a Given Ratio

Page 98 Exercise 1

1 a) There are 2 + 1 = 3 parts in total.
3 parts = £48, so 1 part = 48 ÷ 3 = £16.
The shares are 2 × 16 = **£32** and **£16**.
Check your answer makes sense by seeing if the shares add up to the total. £32 + £16 = £48, so the answer looks good.

b) There are 1 + 3 = 4 parts in total.
4 parts = £48, so 1 part = 48 ÷ 4 = £12.
The shares are **£12** and 3 × 12 = **£36**.
Use the same method for c) and d):
c) **£40** and **£8** **d)** **£28** and **£20**

2 a) There are 4 + 1 = 5 parts in total.
5 parts = 90 kg, so 1 part = 18 kg.
The shares are 4 × 18 = **72 kg** and **18 kg**.

b) There are 7 + 2 = 9 parts in total.
9 parts = 90 kg, so 1 part = 10 kg.
The shares are 7 × 10 = **70 kg** and
2 × 10 = **20 kg**.

Use the same method for c) and d):

c) 48 kg and **42 kg** **d) 36 kg** and **54 kg**

3 a) There are 2 + 3 + 1 = 6 parts in total.
6 parts = 72 cm, so 1 part = 12 cm. The shares are 2 × 12 = **24 cm**, 3 × 12 = **36 cm** and **12 cm**.

b) There are 2 + 2 + 5 = 9 parts in total.
9 parts = 72 cm, so 1 part = 8 cm.
The shares are 2 × 8 = **16 cm**, **16 cm** and 5 × 8 = **40 cm**.

Use the same method for c) and d):

c) 30 cm, **18 cm** and **24 cm**

d) 28 cm, **24 cm** and **20 cm**

4 The total number of parts is 3 + 2 = 5.
5 parts = 30 cupcakes, so 1 part = 6 cupcakes.
The shares are 3 × 6 = **18** for Kat and 2 × 6 = **12** for Lincoln.

5 The total number of parts is 3 + 5 + 4 = 12.
12 parts = £6000, so 1 part = £500.
The shares are 3 × £500 = **£1500**,
5 × £500 = **£2500** and 4 × 500 = **£2000**.

6 The perimeter is 72 cm, so the length + width is half of this: 72 ÷ 2 = 36 cm. You need to share 36 cm in the ratio 5 : 1.
Number of parts = 5 + 1 = 6.
1 part = 36 ÷ 6 = 6 cm.
So length = 5 × 6 = **30 cm** and width = **6 cm**.

Page 98 Exercise 2

1 a) The first share is 1 part, the second share is 2 – 1 = 1 part.
2 parts = 60 sheep, so
1 part = 60 ÷ 2 = 30 sheep.
The flock is divided into **30 sheep** and **30 sheep**.

b) The first share is 2 parts, the second share is 3 – 2 = 1 part.
3 parts = 60 sheep, so
1 part = 60 ÷ 3 = 20 sheep.
The flock is divided into 2 × 20 = **40 sheep** and **20 sheep**.

Use the same method for c) and d):

c) 48 sheep and **12 sheep**

d) 35 sheep and **25 sheep**

2 The ratio yellow paint : total is 4 : 7, so the red part is 7 – 4 = 3.
7 parts = 42 litres, so 1 part = 6 litres.
The shares are 4 × 6 = **24 litres of yellow** and 3 × 6 = **18 litres of red**.

3 The ratio right-handed : total is 7 : 8, so the left-handed part is 8 – 7 = 1.
8 parts = 600 pupils, so 1 part = 75 pupils.
The left-handed share is **75 pupils**.

Page 99 Exercise 3

1 The ratio of 5p coins to 20p coins is 3 : 1.
The total number of parts is 3 + 1 = 4.
4 parts = 32 coins, so
1 part = 32 ÷ 4 = 8 coins. 20p coins make up 1 part, so there are **8** of them.

2 The ratio of passengers not on the phone to on the phone is 4 : 1. The total number of parts is 4 + 1 = 5. 5 parts = 30 passengers, so 1 part = 6 passengers. So there are 4 × 6 = **24 passengers not on their phone**.

3 The ratio Elsa : Daniel is 1 : 1.5. The total number of parts is 1 + 1.5 = 2.5.
2.5 parts = £5700, so 1 part = £2280.
So the profit that goes to Daniel is 1.5 × 2280 = **£3420**.

4 The ratio grapes : raspberries : strawberries is 1 : 2 : 3. The total number of parts is 1 + 2 + 3 = 6. 6 parts = 450 g, so 1 part = 75 g.
The shares are grapes: 1 × 75 = **75 g**, raspberries: 2 × 75 = **150 g** and strawberries: 3 × 75 = **225 g**.

5 The ratio Jacinta : Nicky : Samir is $1 : \frac{1}{2} : \frac{1}{4}$
= 4 : 2 : 1. The total number of parts is 4 + 2 + 1 = 7. 7 parts = 35 sweets, so 1 part = 5 sweets.
The shares are Jacinta: 4 × 5 = **20 sweets**, Nicky: 2 × 5 = **10 sweets** and Samir: **5 sweets**.

Page 100 Review Exercise

1 It's girls : boys so 14 comes first, the ratio is **14 : 17**.

2 a) The highest common factor of 24 and 6 is 6. Divide both sides by 6, so the ratio becomes **4 : 1**.

b) 1 cm = 10 mm,
so 2 cm : 8 mm = 20 mm : 8 mm.
The highest common factor of 20 and 8 is 4. Divide both sides by 4, so the ratio becomes **5 : 2**.

c) The highest common factor of 6, 3 and 15 is 3. Divide all parts by 3, so the ratio becomes **2 : 1 : 5**.

d) 1 kg = 1000 g, so 0.03 kg = 30 g, so the ratio is 30 g : 10 g : 25 g.
The highest common factor of 30, 10 and 25 is 5. Divide all parts by 5, so the ratio becomes **6 : 2 : 5**.

3 There are 120 – 40 = 80 other animals.
So the ratio of donkeys : other animals = 40 : 80. Divide both sides by 40 to get the answer **1 : 2**.

4 The total number of parts is 5 + 1 + 4 = 10.
The stripy part is 1. So as a fraction it's $\mathbf{\frac{1}{10}}$.

5 As $\frac{5}{12}$ students eat school dinners,
$1 - \frac{5}{12} = \frac{7}{12}$ students have a packed lunch.
The fractions already have a common denominator (12), so put the numerators into a ratio. It's packed lunch : school diners, so the answer is **7 : 5**.

6 green : red = 2 : 5 = ? : 20.
As 20 = 5 × 4, the multiplier is 4, so there are 2 × 4 = **8 green peppers**.

7 Meera : Sabrina = 7 : 6 = ? : 48. As 48 = 8 × 6 the multiplier is 8, so Meera gets 8 × 7 = **£56**.

8 a) The total number of parts is 1 + 7 = 8.
8 parts = 56 m, so 1 part = 7 m.
The shares are **7 m** and 7 × 7 = **49 m**.

b) The total number of parts is 4 + 4 = 8.
8 parts = 56 m, so 1 part = 7 m.
The shares are both 4 × 7 = **28 m**.

Use the same methods for c) and d):

c) 40 m and **16 m** **d) 44 m** and **12 m**

9 The total is 19 + 20 + 23 = 62.
62 parts = 496, so 1 part = 496 ÷ 62 = 8.
The shares are: **Gemma**: 19 × 8 = **152 cm**, **Alisha**: 20 × 8 = **160 cm** and **Omar**: 23 × 8 = **184 cm**.

10 Cara gets 3 – 2 = 1 part.
3 parts = £1200 so 1 part = 1200 ÷ 3 = £400.
So Cara gets **£400**.

11 a) daisies : dandelions = 1 : 2 and dandelions : thistles = 1 : 3 = 2 : 6, so daisies : dandelions : thistles = **1 : 2 : 6**

b) The total number of parts is 1 + 2 + 6 = 9.
9 parts = 54 plants, so
1 part = 54 ÷ 9 = 6 plants
Number of thistles = 6 × 6 = **36**

Page 101 Exam-Style Questions

1 a) 30 : 48 = 5 : 8 by dividing both sides by 6 *[1 mark]*

b) $30 : 48 = 5 : 8 = 1 : \frac{8}{5}$ *[1 mark]*
$= 1 : 1\frac{3}{5}$ = **1 : 1.6** *[1 mark]*

2 a) For every left-footed player there are 3 right-footed players, so the ratio right-footed : left-footed is **3 : 1** *[1 mark]*

b) The total number of parts is 3 + 1 = 4.
4 parts = 24 players, so 1 part = 6 players *[1 mark]*.
The number of right-footed players is 6 × 3 = **18** *[1 mark]*.

c) E.g. there are 4 parts in total and left-footed players make up 1 part *[1 mark]*,
so $\mathbf{\frac{1}{4}}$ of the players are left-footed *[1 mark]*.

3 a) adults : children is 240 : 90 *[1 mark]*
so adults : boys : girls is 240 : 60 : 30 = **8 : 2 : 1** *[1 mark — show all working]*
You could simplify the adult : children ratio to 8 : 3 first and then split up the 3 to give 8 : 2 : 1.

b) 8 parts = 200, so 1 part = 200 ÷ 8 *[1 mark]* = 25, so the number of boys = 2 × 25 = 50 *[1 mark]*.
Or you could do 200 ÷ 240 × 60 = 50.

4 a) Divide both sides by 4 to get the ratio 1 : 1.25 *[1 mark]*.

b) E.g. if the two bags contained 400 g and 500 g of flour, then their weights would be in the ratio 400 : 500 = 4 : 5 = 1 : 1.25.
If 100 g is added to each bag, the ratio becomes 500 : 600 = 1 : 600 ÷ 500 = 1 : 1.2.
So the ratio is different from 1 : 1.25.
[2 marks — 1 mark for example weight in correct ratio, 1 mark for calculating correct ratio after adding same weight to each bag]

c) Alan could multiply or divide the weight in each bag by the same amount. *[1 mark]*

Section 7 — Percentages

7.1 Percentages

Page 103 Exercise 1

1 a) 40 squares out of 100 are shaded, which is **40%**.

b) 47 squares out of 100 are shaded, which is **47%**.

c) 89 squares out of 100 are shaded, which is **89%**.

2 a) 13% **b) 27%** **c) 76%** **d) 243%**

3 13 out of 100 goals were penalties, so 100 – 13 = 87 goals were not penalties.
87 out of 100 goals is **87%**.
If something is out of 100, you can write down the percentage straight away.

4 a) $\mathbf{\frac{45}{300}}$

b) Divide the numerator and denominator by 3.
$\frac{45}{300} = \mathbf{\frac{15}{100}}$

c) 15%

5 a) $\frac{11}{25} = \frac{44}{100}$ = **44%**

b) $\frac{33}{50} = \frac{66}{100}$ = **66%**

c) $\frac{3}{20} = \frac{15}{100}$ = **15%**

d) $\frac{21}{10} = \frac{210}{100}$ = **210%**

e) $\frac{12}{200} = \frac{6}{100} =$ **6%**

f) $\frac{99}{300} = \frac{33}{100} =$ **33%**

g) $\frac{600}{400} = \frac{150}{100} =$ **150%**

h) $\frac{890}{1000} = \frac{89}{100} =$ **89%**

Page 104 Exercise 2

1 a) $\frac{8}{32} = \frac{1}{4} = \frac{25}{100} =$ **25%**

b) $\frac{36}{60} = \frac{6}{10} = \frac{60}{100} =$ **60%**

c) $\frac{24}{40} = \frac{6}{10} = \frac{60}{100} =$ **60%**

d) $\frac{48}{120} = \frac{4}{10} = \frac{40}{100} =$ **40%**

e) $\frac{34}{170} = \frac{2}{10} = \frac{20}{100} =$ **20%**

f) $\frac{42}{35} = \frac{6}{5} = \frac{120}{100} =$ **120%**

2 a) $\frac{18}{24} = \frac{3}{4} = \frac{75}{100} =$ **75%**

b) 100% – 75% = **25%**

3 55 – 33 = 22
$\frac{22}{55} = \frac{2}{5} = \frac{40}{100} =$ **40%**

4 65 – 39 = 26
$\frac{26}{65} = \frac{2}{5} = \frac{40}{100} =$ **40%**

Page 105 Exercise 3

1 a) (15 ÷ 24) × 100% = **62.5%**
b) (77 ÷ 275) × 100% = **28%**
c) (61 ÷ 500) × 100% = **12.2%**
d) (1512 ÷ 375) × 100% = **403.2%**

2 (525 ÷ 875) × 100% = **60%**

3 (£252 ÷ £560) × 100% = **45%**

4 (171 ÷ 180) × 100% = **95%**

5 a) (31.36 ÷ 32) × 100% = **98%**
b) (£117.30 ÷ £782) × 100% = **15%**

6 (£1896.25 ÷ £10 250) × 100% = **18.5%**

7 Total pocket money = £5.60 + £2.40 = £8
(£2.40 ÷ £8) × 100% = **30%**

Page 106 Exercise 4

1 a) 50% of 24 = 24 ÷ 2 = **12**
b) 50% of 15 = 15 ÷ 2 = **7.5**
c) 25% of 36 = 36 ÷ 4 = **9**
d) 25% of 120 = 120 ÷ 4 = **30**
e) 10% of 90 = 90 ÷ 10 = **9**
f) 10% of 270 = 270 ÷ 10 = **27**

2 a) 25% of 48 = 48 ÷ 4 = **12**
b) 75% = 25% × 3 = 12 × 3 = **36**

3 a) 10% of 120 = 120 ÷ 10 = **12**
b) (i) 5% = 10% ÷ 2 = 12 ÷ 2 = **6**
(ii) 20% = 10% × 2 = 12 × 2 = **24**
(iii) 25% = 20% + 5% = 24 + 6 = **30**

4 a) 25% of 12 = 12 ÷ 4 = 3
75% = 25% × 3 = 3 × 3 = **9**
b) 25% of 20 = 20 ÷ 4 = 5
125% = 25% × 5 = 5 × 5 = **25**
c) 10% of 260 = 260 ÷ 10 = 26
5% = 10% ÷ 2 = 26 ÷ 2 = **13**
d) 10% of 200 = 200 ÷ 10 = 20
1% = 10% ÷ 10 = 20 ÷ 10 = 2
30% = 10% × 3 = 20 × 3 = 60
31% = 30% + 1% = 60 + 2 = **62**
e) 10% of 70 = 70 ÷ 10 = 7
110% = 100% + 10% = 70 + 7 = **77**
f) 10% of 500 = 500 ÷ 10 = 50
40% = 10% × 4 = 50 × 4 = 200
1% = 10% ÷ 10 = 50 ÷ 10 = 5
6% = 1% × 6 = 5 × 6 = 30
46% = 40% + 6% = 200 + 30 = **230**

5 10% of £30 = £30 ÷ 10 = £3
80% of £30 = 10% × 8 = £3 × 8 = £24
1% of £30 = 10% ÷ 10 = £3 ÷ 10 = £0.30
3% of £30 = 1% × 3 = £0.30 × 3 = £0.90
83% of £30 = 80% + 3%
= £24 + £0.90 = **£24.90**

6 50% of 9 m = 9 m ÷ 2 = 4.5 m
10% of 9 m = 9 m ÷ 10 = 0.9 m
5% of 9 m = 10% ÷ 2 = 0.9 m ÷ 2 = 0.45 m
55% of 9 m = 50% + 5%
= 4.5 m + 0.45 m = **4.95 m**

7 10% of £1400 = £1400 ÷ 10 = £140
5% of £1400 = 10% ÷ 2 = £140 ÷ 2 = £70
90% of £1400 = 10% × 9 = £140 × 9
= £1260
95% of £1400 = 90% + 5% = £1260 + £70
= **£1330**
You could also subtract 5% from 100%.

Page 107 Exercise 5

1 a) 17% ÷ 100% = 0.17
0.17 × 200 = **34**
b) 9% ÷ 100% = 0.09
0.09 × 11 = **0.99**
Using the same method for c)-f)
c) **6.3** d) **971.7**
e) **420.67** f) **2320.86**

2 12% ÷ 100% = 0.12
0.12 × 68 = **8.16 kg**

3 31% ÷ 100% = 0.31
0.31 × 385 = **119.35 km**

4 52% ÷ 100% = 0.52
0.52 × 125 = **65 men**

5 68% ÷ 100% = 0.68
0.68 × £42 = **£28.56**

6 22% of £57: 22% ÷ 100% = 0.22
0.22 × £57 = £12.54
46% of £28: 46% ÷ 100% = 0.46
0.46 × £28 = £12.88
£12.88 > £12.54, so **46% of £28 is larger** than 22% of £57.
The difference is £12.88 – £12.54 = **£0.34**

7 34% ÷ 100% = 0.34
0.34 × 2.4 = 0.816 litres
It can hold another 2.4 – 0.816 = **1.584 litres.**

7.2 Percentages, Fractions and Decimals

Page 109 Exercise 1

1 a) 50 out of 100 squares are shaded.
(i) **50%** (ii) **0.5** (iii) $\mathbf{\frac{1}{2}}$
b) 25 out of 100 squares are shaded.
(i) **25%** (ii) **0.25** (iii) $\mathbf{\frac{1}{4}}$
c) 10 out of 100 squares are shaded.
(i) **10%** (ii) **0.1** (iii) $\mathbf{\frac{1}{10}}$

2 a) $\frac{3}{20} = \frac{3 \times 5}{20 \times 5} = \mathbf{\frac{15}{100}}$
b) (i) **15%** (ii) 15% ÷ 100% = **0.15**

3 a) (i) 30% ÷ 100% = **0.3**
(ii) $0.3 = \mathbf{\frac{3}{10}}$
b) (i) 5% ÷ 100% = **0.05**
(ii) $0.05 = \frac{5}{100} = \mathbf{\frac{1}{20}}$
Using the same method for c) and d):
c) (i) **0.13** (ii) $\mathbf{\frac{13}{100}}$
d) (i) **0.96** (ii) $\mathbf{\frac{24}{25}}$

4 a) (i) 79 ÷ 100 = **0.79**
(ii) 0.79 × 100% = **79%**
b) (i) $\frac{2}{5} = \frac{4}{10} =$ **0.4**
(ii) 0.4 × 100% = **40%**
c) (i) $\frac{4}{25} = \frac{16}{100} =$ **0.16**
(ii) 0.16 × 100% = **16%**
d) (i) $7 \div 8 = \begin{array}{r} 0.\,8\,7\,5 \\ 8\overline{)7\,.{}^{7}0\,{}^{6}0\,{}^{4}0} \end{array} =$ **0.875**
(ii) 0.875 × 100% = **87.5%**

5 a) (i) 0.35 × 100% = **35%**
(ii) $0.35 = \frac{35}{100} = \mathbf{\frac{7}{20}}$
b) (i) 0.86 × 100% = **86%**
(ii) $0.86 = \frac{86}{100} = \mathbf{\frac{43}{50}}$
c) (i) 1.2 × 100% = **120%**
(ii) $1.2 = \frac{12}{10} = \mathbf{\frac{6}{5}}$
d) (i) 0.125 × 100% = **12.5%**
(ii) $0.125 = \frac{125}{1000} = \mathbf{\frac{1}{8}}$

6 76% ÷ 100% = **0.76**

7 $\frac{3}{5} = \frac{60}{100} =$ **60%**

Page 110 Exercise 2

1 a) 0.35 × 100% = 35%
35% > 32% so **0.35** is larger.
You could have changed 32% into a decimal instead.
b) 0.4 × 100% = 40%
40% > 4% so **0.4** is larger.
c) 0.09 × 100% = 9%
90% > 9% so **90%** is larger.
d) $0.2 = \frac{20}{100}$
$\frac{21}{100} > \frac{20}{100}$ so $\mathbf{\frac{21}{100}}$ is larger.
e) $0.6 = \frac{6}{10}$
$\frac{7}{10} > \frac{6}{10}$ so $\mathbf{\frac{7}{10}}$ is larger.
f) $0.55 = \frac{55}{100}$
$\frac{3}{5} = \frac{60}{100}$
$\frac{60}{100} > \frac{55}{100}$ so $\mathbf{\frac{3}{5}}$ is larger.

2 a) 25% ÷ 100% = 0.25
$\frac{2}{5} = \frac{4}{10} = 0.4$
0.25 < 0.4 < 0.42, so ordered list is
25%, $\mathbf{\frac{2}{5}}$, **0.42**
b) 45% ÷ 100% = 0.45
$\frac{1}{2} = 0.5$
0.45 < 0.5 < 0.505, so ordered list is
45%, $\mathbf{\frac{1}{2}}$, **0.505**
Using the same method for c)-f):
c) **0.37**, **38%**, $\mathbf{\frac{4}{10}}$
d) **0.2**, **22%**, $\mathbf{\frac{6}{25}}$
e) **12.5%**, **0.13**, $\mathbf{\frac{3}{20}}$
f) $\mathbf{\frac{9}{40}}$, **23%**, **0.25**

3 $\frac{1}{25} = \frac{4}{100} = 4\%$
5% > 4% so **Shop B** is reducing their prices by the greater percentage.

4 $\frac{2}{5} = \frac{40}{100} = 40\%$
40% > 35%, so 35% is **not enough** to pay the deposit.

5 $\frac{14}{20} = \frac{7}{10} = \frac{70}{100} = 70\%$
70% > 60%, so **Team X** had a higher proportion of wins.

6 $\frac{11}{20} = \frac{55}{100} = 55\%$
55% < 65%, so **Oliver** got more counters into the box.

Page 111 Exercise 3

1 $\frac{3}{4} = \frac{75}{100}$ = 75% arrived by train.
0.05 × 100% = 5% walked.
People that came by car = 100% – 75% – 5% = **20%**

2 $\frac{3}{5} = \frac{60}{100}$ = 60% are white footballs.
20% are black footballs.
Blue footballs = 100% – 60% – 20% = **20%**

3 Beverley eats 0.3 × 100% = 30% of the pie.
Victoria eats = $\frac{1}{10} = \frac{10}{100}$ = 10%.
Patrick eats 20%.
So Gus eats 100% – 30% – 10% – 20% = **40%**.

4 $\frac{3}{8}$ = 3 ÷ 8 = $8\overline{)3.3^{3}0^{6}0^{4}0}$ with quotient 0.375 = 0.375
0.375 × 100% = 37.5% were sparrows and 41.5% were blackbirds, so
100% – 37.5% – 41.5% = **21%** were robins.
Look back at section 1 for help with dividing without a calculator.

7.3 Percentage Increase and Decrease

Page 112 Exercise 1

1 a) 50% of 360 = 360 ÷ 2 = **180**
b) 360 + 180 = **540**

2 a) 10% of 120 = 120 ÷ 10 = 12
30% of 120 = 10% × 3 = 12 × 3 = **36**
b) 120 + 36 = **156**

3 a) 10% of 160 = 160 ÷ 10 = **16**
b) 160 – 16 = **144**

4 a) 10% of 84 = 84 ÷ 10 = 8.4
20% of 84 = 10% × 2 = 8.4 × 2 = **16.8**
b) 84 – 16.8 = **67.2**

5 a) 10% of 90 = 90 ÷ 10 = 9
90 + 9 = **99**
b) 10% of 11 = 11 ÷ 10 = 1.1
80% of 11 = 10% × 8 = 1.1 × 8 = 8.8
11 + 8.8 = **19.8**
c) 10% of 140 = 140 ÷ 10 = 14
5% of 140 = 10% ÷ 2 = 14 ÷ 2 = 7
40% of 140 = 10% × 4 = 14 × 4 = 56
45% of 140 = 40% + 5% = 56 + 7 = 63
140 + 63 = **203**

6 a) 25% of 24 = 24 ÷ 4 = 6
24 – 6 = **18**
b) 10% of 55 = 55 ÷ 10 = 5.5
70% of 55 = 10% × 7 = 5.5 × 7 = 38.5
55 – 38.5 = **16.5**
c) 10% of 150 = 150 ÷ 10 = 15
5% of 150 = 10% ÷ 2 = 15 ÷ 2 = 7.5
50% of 150 = 150 ÷ 2 = 75
55% of 150 = 50% + 5% = 75 + 7.5 = 82.5
150 – 82.5 = **67.5**

Page 113 Exercise 2

1 a) 11% ÷ 100% = 0.11
Multiplier = 0.11 + 1 = 1.11
490 × 1.11 = **543.9**
b) 16% ÷ 100% = 0.16
Multiplier = 0.16 + 1 = 1.16
101 × 1.16 = **117.16**
Using the same method for c)-f):
c) **75.35** d) **4213.41**
e) **1274.28** f) **48 180**

2 a) 8% ÷ 100% = 0.08
Multiplier = 1 – 0.08 = 0.92
77 × 0.92 = **70.84**
b) 21% ÷ 100% = 0.21
Multiplier = 1 – 0.21 = 0.79
36 × 0.79 = **28.44**
Using the same method for c)-f):
c) **57.57** d) **5253.32**
e) **72 240** f) **95 120**

3 a) 62% ÷ 100% = 0.62
Multiplier = 1 + 0.62 = 1.62
£89.50 × 1.62 = **£144.99**
Remember to include the units given in the question.
b) 19% ÷ 100% = 0.19
Multiplier = 1 – 0.19 = 0.81
58 kg × 0.81 = **46.98 kg**

Page 114 Exercise 3

1 10% of £230 = £230 ÷ 10 = £23
5% of £230 = £23 ÷ 2 = £11.50
Total after 1 year = £230 + £11.50 = **£241.50**

2 1% of £890 = £890 ÷ 100 = £8.90
3% of £890 = 1% × 3 = £8.90 × 3 = £26.70
After three years Kimberley has
£890 + 3 × £26.70 = **£970.10**

Page 115 Exercise 4

1 10% of £24 500 = £24 500 ÷ 10 = £2450
5% of £24 500 = £2450 ÷ 2 = £1225
15% of £24 500 = £2450 + £1225 = £3675
New salary = £24 500 + £3675 = **£28 175**

2 25% of £42 = £42 ÷ 4 = £10.50
75% of £42 = £10.50 × 3 = £31.50
Sale price = £42 – £31.50 = **£10.50**

3 10% of £485 = £485 ÷ 10 = £48.50
20% of £485 = £48.50 × 2 = £97
Selling price = £485 + £97 = **£582**

4 13% ÷ 100% = 0.13
Multiplier = 1 – 0.13 = 0.87
Total = £63 × 0.87 = **£54.81**

5 20% ÷ 100% = 0.2
Multiplier = 1 + 0.2 = 1.2
Height at 10 years old = 50 × 1.2 = **60 inches**

6 25% of 8 = 8 ÷ 4 = 2
Natalie's remaining fuel = 8 – 2 = 6 gallons
10% of 12 = 12 ÷ 10 = 1.2
40% of 12 = 1.2 × 4 = 4.8
Jason's remaining fuel = 12 – 4.8 = 7.2 gallons
7.2 > 6 and 7.2 – 6 = 1.2
So **Jason had 1.2 gallons more fuel** than Natalie at the end of the journey.

7 2% ÷ 100% = 0.02
Multiplier = 1 + 0.02 = 1.02
Gas bill = £480 × 1.02 = £489.60
4% ÷ 100% = 0.04
Multiplier = 1 – 0.04 = 0.96
Electricity bill = £612 × 0.96 = £587.52
£587.52 > £489.60 and £587.52 – £489.60 = £97.92, so **electricity bill** is still more expensive, by **£97.92**.
In questions 7 and 8, one step of the question is a % increase and the other is a % decrease, so you need to be really careful with your multipliers.

8 3% ÷ 100% = 0.03
Multiplier = 1 + 0.03 = 1.03
Salary after 3% increase = £31 000 × 1.03 = £31 930
2% ÷ 100% = 0.02
Multiplier = 1 – 0.02 = 0.98
New salary after 2% decrease = £31 930 × 0.98 = **£31 291.40**

Page 116 Exercise 5

1 a) £12 – £10 = £2
$\frac{2}{10} = \frac{20}{100}$ = 20% ⇒ **20% increase**
b) £22 – £20 = £2
$\frac{2}{20} = \frac{10}{100}$ = 10% ⇒ **10% increase**
c) £161 – £140 = £21
$\frac{21}{140} = \frac{3}{20} = \frac{15}{100}$ = 15%
⇒ **15% increase**
d) £174 – £120 = £54
$\frac{54}{120} = \frac{9}{20} = \frac{45}{100}$ = 45%
⇒ **45% increase**

2 a) £10 – £8 = £2
$\frac{2}{10} = \frac{20}{100}$ = 20% ⇒ **20% decrease**
b) £25 – £22 = £3
$\frac{3}{25} = \frac{12}{100}$ = 12% ⇒ **12% decrease**
c) £80 – £64 = £16
$\frac{16}{80} = \frac{2}{10} = \frac{20}{100}$ = 20% ⇒ **20% decrease**
d) £150 – £138 = £12
$\frac{12}{150} = \frac{4}{50} = \frac{8}{100}$ = 8% ⇒ **8% decrease**

3 a) 72 – 50 = **22**
b) $\mathbf{\frac{22}{50}}$
c) $\frac{22}{50} = \frac{44}{100}$ = 44% ⇒ **44% increase**

4 100p – 80p = 20p
$\frac{20}{80} = \frac{1}{4} = \frac{25}{100}$ = 25% ⇒ **25% increase**

5 £50 – £30 = £20
$\frac{20}{50} = \frac{40}{100}$ = 40% ⇒ **40% reduction**

6 a) 90 kg – 72 kg = **18 kg**
b) $\frac{18}{90} = \frac{2}{10} = \frac{20}{100}$ = 20% ⇒ **20% decrease**

Page 117 Exercise 6

1 £70.20 – £52 = £18.20
(18.20 ÷ 52) × 100% = 35% ⇒ **35% profit**

2 2.08 m – 1.3 m = 0.78 m
(0.78 ÷ 1.3) × 100% = 60% ⇒ **60% increase**

3 11 350 – 11 123 = 227
(227 ÷ 11 350) × 100% = 2%
⇒ **2% decrease**

4 a) £12 950 – £8806 = £4144
(4144 ÷ 12 950) × 100% = 32%
⇒ **32% decrease** in first 3 years.
b) £8806 – £4403 = £4403
(4403 ÷ 8806) × 100% = 50%
⇒ **50% decrease** in the next 3 years.
c) £12 950 – £4403 = £8547
(8547 ÷ 12 950) × 100% = 66%
⇒ **66% decrease** over 6 years.

Page 118 Exercise 7

1 50% of original price = £200
100% of original price = £200 × 2 = **£400**

2 100% – 35% = 65%,
so 65% of original price = £13.00
1% of original price = £13.00 ÷ 65 = £0.20
Original price = £0.20 × 100 = **£20**

3 100% + 26% = 126%,
so 126% of original price = £819
1% of original price = £819 ÷ 126 = £6.50
Original price = £6.50 × 100 = **£650**

4 100% – 4% = 96%,
so 96% of original price = £192 000
1% of original price = £192 000 ÷ 96 = £2000
Original price = £2000 × 100 = **£200 000**

5 100% + 10% = 110%
110% of number of frogs last year = 528
Number of frogs last year = (528 ÷ 110) × 100 = **480**
100% + 15% = 115%
115% of number of newts last year = 621
Number of newts last year = (621 ÷ 115) × 100 = **540**

7.4 Compound Growth and Decay

Page 119 Exercise 1

1 3% ÷ 100% = 0.03
Multiplier = 1 + 0.03 = 1.03

a) £250 × 1.03 = £257.50
Interest earned = £257.50 – £250 = **£7.50**
You need to subtract the original amount to find the interest earned.

b) First year: £45 × 1.03 = £46.35
Second year: £46.35 × 1.03 = £47.7405
Third year: £47.7405 × 1.03 = £49.172715
Interest earned = £49.172715 – £45
= £4.172715 = **£4.17** (nearest penny)

c) First year: £1500 × 1.03 = £1545
Second year: £1545 × 1.03 = £1591.35
Interest earned = £1591.35 – £1500
= **£91.35**

2 a) 4.5% ÷ 100% = 0.045
Multiplier = 1 + 0.045 = 1.045
First year: £3500 × 1.045 = £3657.50
Second year: £3657.50 × 1.045
= £3822.0875 = **£3822.09** (nearest penny)

b) Third year: £3822.0875 × 1.045
= £3994.081438
= **£3994.08** (nearest penny)

Page 120 Exercise 2

1 a) 15% ÷ 100% = 0.15
Multiplier = 1 – 0.15 = 0.85
First year: £320 × 0.85 = £272
Second year: £272 × 0.85 = **£231.20**

b) Third year: £231.20 × 0.85 = **£196.52**

2 5% ÷ 100% = 0.05
Multiplier = 1 – 0.05 = 0.95
Day 1: $1000 × 0.95 = $950
Day 2: $950 × 0.95 = **$902.50**

3 6% ÷ 100% = 0.06
Multiplier for first two years = 1 – 0.06 = 0.94
First year: £12 000 × 0.94 = £11 280
Second year: £11 280 × 0.94 = £10 603.20
17% ÷ 100% = 0.17
Multiplier for third year = 1 – 0.17 = 0.83
Third year: £10 603.20 × 0.83 = £8800.656
= **£8800.66** (nearest penny)

Page 121 Exercise 3

1 a) 4% ÷ 100% = 0.04
Multiplier = 1 + 0.04 = 1.04
Final amount = £1000 × $(1.04)^8$
= £1368.56905
= **£1368.57** (nearest penny)

b) 5.2% ÷ 100% = 0.052
Multiplier = 1 + 0.052 = 1.052
Final amount = £600 × $(1.052)^7$
= £855.5815862
= **£855.58** (nearest penny)

2 2% ÷ 100% = 0.02
Multiplier = 1 – 0.02 = 0.98
Final value = £650 000 × $(0.98)^6$
= £575 797.5476 = **£575 800** (nearest £100)

3 6% ÷ 100% = 0.06
Multiplier = 1 + 0.06 = 1.06
n = number of days = 5
Number of ants = 250 × $(1.06)^5$
= 334.556... = **334 ants**
Check to see if you should round your final answer based on the context. Here, you can't have '0.55...' of an ant, so you need to round down.

4 2% ÷ 100% = 0.02
Multiplier = 1 – 0.02 = 0.98
Final body weight = 110 kg × $(0.98)^8$
= 93.583... kg = **93.6 kg** (1 d.p.)

Page 122 Review Exercise

1 $\frac{12}{25} = \frac{48}{100}$ = **48%**

2 10% of £160 = £160 ÷ 10 = £16
5% of £160 = £16 ÷ 2 = £8
30% of £160 = 10% × 3 = £16 × 3 = £48
35% of £160 = 30% + 5% = £48 + £8 = **£56**

3 a) There are 5 shaded squares.
(i) $\frac{5}{25} = \mathbf{\frac{1}{5}}$
(ii) $\frac{5}{25} = \frac{20}{100}$ = **20%**
(iii) 20% ÷ 100% = **0.2**

b) There are 17 shaded squares.
(i) $\mathbf{\frac{17}{25}}$
(ii) $\frac{17}{25} = \frac{68}{100}$ = **68%**
(iii) 68% ÷ 100% = **0.68**

4 $\frac{31}{500}$ = 0.062, 6% ÷ 100% = 0.06,
so ordered list from smallest to largest is
6%, 0.061, $\mathbf{\frac{31}{500}}$

5 50% of 450 g = 450 g ÷ 2 = 225 g
10% of 450 g = 450 g ÷ 10 = 45 g
5% of 450 g = 45 g ÷ 2 = 22.5 g
55% of 450 g = 225 g + 22.5 g = 247.5 g
450 g + 247.5 g = **697.5 g**

6 Shop A: 70% ÷ 100% = 0.7
Multiplier = 1 – 0.7 = 0.3
Sale price = £92 × 0.3 = £27.60
Shop B: $\frac{7}{8}$ = 0.875
Multiplier = 1 – 0.875 = 0.125
Sale price = £92 × 0.125 = £11.50
£27.60 – £11.50 = £16.10
So **Shop B** had the lower price by **£16.10**
You could also have found the price in Shop B by calculating $\frac{7}{8}$ using fraction methods *(see page 105).*

7 8.5% ÷ 100% = 0.085
Multiplier = 1 + 0.085 = 1.085
Amount at end of the year
= £3250 × 1.085 = **£3526.25**

8 75 g – 69 g = 6 g
$\frac{6}{75} = \frac{2}{25} = \frac{8}{100}$ = **8%**

9 100% – 70% = 30%
So 30% of original price = £2.85
1% of original price = £2.85 ÷ 30 = £0.095
Original price = 100 × £0.095 = **£9.50**

10 8% ÷ 100% = 0.08
Multiplier = 1 + 0.08 = 1.08
Using the formula, the population after 5 years is given by: 12 500 × $(1.08)^5$ = 18 366.600...
So population will be **18 366**.

Page 123 Exam-Style Questions

1 10% of 750 MB = 750 MB ÷ 10 = 75 MB
5% of 750 MB = 10% ÷ 2 = 75 MB ÷ 2
= 37.5 MB
15% = 75 MB + 37.5 MB = 112.5 MB
[1 mark]
750MB + 112.5 MB = **862.5 MB** *[1 mark]*

2 24% ÷ 100% = 0.24
Multiplier = 1 + 0.24 = 1.24 *[1 mark]*
3.7 m × 1.24 = 4.588 m *[1 mark]*
= **4.59 m** (nearest centimetre)
[1 mark]

3 £450 million – £375 million
= £75 million *[1 mark]*
$\frac{75}{375}$ × 100% = **20%** *[1 mark]*

4 Total number of balls in the box
= 20 + 12 + 8 = 40
Total number of basketballs = 20 + 8 = 28
Percentage of balls that are basketballs
= $\frac{28}{40} = \frac{7}{10} = \frac{70}{100}$ = **70%**
[2 marks available — 1 mark for calculating both totals, 1 mark for correct percentage]

5 20% ÷ 100% = 0.2
Multiplier for first two years = 1 – 0.2 = 0.8
First year: £68 000 × 0.8 = £54 400
Second year: £54 400 × 0.8 = £43 520
15% ÷ 100% = 0.15
Multiplier for following three years = 1 – 0.15 = 0.85
Third year: £43 520 × 0.85 = £36 992
Fourth year: £36 992 × 0.85 = £31 443.20
Fifth year: £31 443.20 × 0.85 = **£26 726.72**
[4 marks available — 1 mark for each correct multiplier, 1 mark for correct answer after two years, 1 mark for correct final answer]

6 If 30% are lemurs and 40% of lemurs are ring-tailed lemurs then you need to find 40% of 30%.
10% of 30% = 3%
40% of 30% = 4 × 3% = **12%**
[2 marks available — 1 mark for correct method of finding 40% of 30%, 1 mark for the correct answer]
There's more than one correct method — you could work out $\frac{40}{100} \times \frac{30}{100}$, then simplify and read off the numerator.

Section 8 — Algebraic Expressions

8.1 Simplifying Expression

Page 125 Exercise 1

1 a) $2x + 3x + x = \mathbf{6x}$
b) $7p - 2p + 3p - 4p = \mathbf{4p}$
c) $c + c + c + d + d = \mathbf{3c + 2d}$
d) $a + b + a - a + b = (a + a - a) + (b + b)$
$= \mathbf{a + 2b}$
e) $5a - 2a + 5b + 2b = (5a - 2a) + (5b + 2b)$
$= \mathbf{3a + 7b}$
f) $4b + 8c - b - 5c = (4b - b) + (8c - 5c)$
$= \mathbf{3b + 3c}$

2 a) $2c + 4 + c + 7 = (2c + c) + (4 + 7) = \mathbf{3c + 11}$
b) $3x + 6 - 6x - 4 = (3x - 6x) + (6 - 4)$
$= \mathbf{-3x + 2}$
Using the same method for c)-f):
c) $\mathbf{-4y + 5}$ **d)** $\mathbf{-9m + 21}$
e) $\mathbf{3x + 10}$ **f)** $\mathbf{21a + 10}$

3 a) $x + 7 + 4x + y + 5 = (x + 4x) + y + (7 + 5)$
$= \mathbf{5x + y + 12}$
Using the same method for b)-d):
b) $\mathbf{-4a + 3b - 8}$ **c)** $\mathbf{21a + 7b + 7}$
d) $\mathbf{4p + 4q - 20r - 9}$

Page 125 Exercise 2

1 a) $x^2 + 3x + 2 + 2x + 3 = \mathbf{x^2 + 5x + 5}$
b) $x^2 + 4x + 1 + 3x - 3 = \mathbf{x^2 + 7x - 2}$
Using the same method for c)-f):
c) $\mathbf{2x^2 + 6x + 4}$ **d)** $\mathbf{3x^2 + x}$
e) $\mathbf{3p^2 - 2p}$ **f)** $\mathbf{7p^2 + 2q}$
g) $8 + 6p^2 - 5 + pq + p^2$
$= (6p^2 + p^2) + pq + (8 - 5)$
$= \mathbf{7p^2 + pq + 3}$
The extra step of grouping 'like terms' makes it less likely you'll make an error.
h) $4p + 5q - pq + p^2 - 7q$
$= p^2 + 4p - pq + (5q - 7q)$
$= \mathbf{p^2 + 4p - 2q - pq}$
i) $a^2 + 7b + 2a^2 + 5ab - 3b$
$= (a^2 + 2a^2) + (7b - 3b) + 5ab$
$= \mathbf{3a^2 + 4b + 5ab}$

2 a) $ab + cd - xy + 3ab - 2cd + 3yx + 2x^2$
$= (ab + 3ab) + (cd - 2cd) + (3xy - xy) + 2x^2$
$= \mathbf{4ab - cd + 2xy + 2x^2}$

b) $pq + 3pq + p^2 - 2qp + q^2$
$= p^2 + (pq + 3pq - 2pq) + q^2$
$= \mathbf{p^2 + 2pq + q^2}$

3 a) $7 + 3\sqrt{3} + 6 - 2\sqrt{3}$
$= (7 + 6) + (3\sqrt{3} - 2\sqrt{3})$
$= \mathbf{13 + \sqrt{3}}$

b) $-2 - 13\sqrt{7} - 7 + 3\sqrt{7}$
$= (-2 - 7) + (3\sqrt{7} - 13\sqrt{7})$
$= \mathbf{-9 - 10\sqrt{7}}$

c) $11 - 7\sqrt{5} - 11 - 8\sqrt{5}$
$= (11 - 11) + (-7\sqrt{5} - 8\sqrt{5})$
$= \mathbf{-15\sqrt{5}}$

Page 126 Exercise 3

1 a) $a \times a \times a = \mathbf{a^3}$

b) $2a \times 3a = 2 \times 3 \times a \times a = \mathbf{6a^2}$

c) $-8p \times 2q = -8 \times 2 \times p \times q = \mathbf{-16pq}$

Using the same method for d)-h):

d) $\mathbf{21a^2}$ e) $\mathbf{15xy}$ f) $\mathbf{-m^4}$
g) $\mathbf{48ab}$ h) $\mathbf{48p^2}$

2 a) $a \times ab = a \times a \times b = \mathbf{a^2b}$

b) $4m^3 \div m = (4 \div 1) \times (m^3 \div m) = \mathbf{4m^2}$

c) $(r^2)^3 = r^{2 \times 3} = \mathbf{r^6}$

Using the same method for d)-h):

d) $\mathbf{3s^2t^2}$ e) $\mathbf{20a^3}$ f) $\mathbf{3s}$
g) $\mathbf{8a^5b^5}$ h) $\mathbf{2y}$

8.2 Expanding Brackets

Page 127 Exercise 1

1 a) $2(a + 5) = (2 \times a) + (2 \times 5) = \mathbf{2a + 10}$

b) $4(b + 3) = (4 \times b) + (4 \times 3) = \mathbf{4b + 12}$

c) $5(d + 7) = (5 \times d) + (5 \times 7) = \mathbf{5d + 35}$

Using the same method for d)-h):

d) $\mathbf{3p + 12}$ e) $\mathbf{15 + 3p}$ f) $\mathbf{42 + 7g}$
g) $\mathbf{15 - 5y}$ h) $\mathbf{8a - 8b}$

2 a) $x(y + 5) = (x \times y) + (x \times 5) = \mathbf{xy + 5x}$

b) $p(q + 2) = (p \times q) + (p \times 2) = \mathbf{pq + 2p}$

c) $x(8 - x) = (x \times 8) + (x \times -x) = \mathbf{8x - x^2}$
Be careful when expanding a bracket with a negative term inside it.

d) $a(b - 12) = (a \times b) + (a \times -12) = \mathbf{ab - 12a}$

e) $3(2p + 4) = (3 \times 2p) + (3 \times 4) = \mathbf{6p + 12}$

f) $5(4t - 8) = (5 \times 4t) + (5 \times -8) = \mathbf{20t - 40}$

g) $3(u + 8v) = (3 \times u) + (3 \times 8v) = \mathbf{3u + 24v}$

h) $7(5n - 6m) = (7 \times 5n) + (7 \times -6m)$
$= \mathbf{35n - 42m}$

Page 128 Exercise 2

1 a) $-(q + 2) = (-1 \times q) + (-1 \times 2) = \mathbf{-q - 2}$

b) $-(x + 7) = (-1 \times x) + (-1 \times 7) = \mathbf{-x - 7}$

Using the same method for c)-h):

c) $\mathbf{-56 + 8w}$ d) $\mathbf{-25 + 5x}$ e) $\mathbf{-v^2 - 4v}$
f) $\mathbf{-v^2 + 5v}$ g) $\mathbf{-12x + x^2}$ h) $\mathbf{-4y - y^2}$

2 a) $-6(5g - 3) = (-6 \times 5g) - (-6 \times 3)$
$= -30g - (-18)$
$= \mathbf{-30g + 18}$

b) $-7(4v + 8) = (-7 \times 4v) + (-7 \times 8)$
$= -28v + (-56)$
$= \mathbf{-28v - 56}$

Using the same method for c)-h):

c) $\mathbf{-10 - 8m}$ d) $\mathbf{-50 + 40v}$
e) $\mathbf{-10 - 15n}$ f) $\mathbf{-32z + 8z^2}$
g) $\mathbf{-12b + 6}$ h) $\mathbf{-8y^2 - 24y}$

Page 128 Exercise 3

1 a) $2(z + 3) + 4(z + 2) = 2z + 6 + 4z + 8$
$= \mathbf{6z + 14}$

b) $3(c + 1) + 5(c + 7) = 3c + 3 + 5c + 35$
$= \mathbf{8c + 38}$

c) $4(u + 6) + 8(u + 5) = 4u + 24 + 8u + 40$
$= \mathbf{12u + 64}$

Using the same method for d)-i):

d) $\mathbf{9t + 3}$ e) $\mathbf{17m + 29}$ f) $\mathbf{4p - 21}$
g) $\mathbf{j - 7}$ h) $\mathbf{4y - 18}$ i) $\mathbf{14c - 27}$

2 a) $5(2q + 5) - 2(q - 2) = 10q + 25 - (2q - 4)$
$= 10q - 2q + 25 + 4$
$= \mathbf{8q + 29}$

b) $2(3c - 8) - 8(c + 4) = 6c - 16 - (8c + 32)$
$= 6c - 8c - 16 - 32$
$= \mathbf{-2c - 48}$

c) $5(q - 2) - 3(q - 4) = 5q - 10 - (3q - 12)$
$= 5q - 3q - 10 + 12$
$= \mathbf{2q + 2}$

3 a) $2(-z + 2) + 3z(3z + 6) = -2z + 4 + 9z^2 + 18z$
$= \mathbf{9z^2 + 16z + 4}$

b) $4p(3p + 5) - 3(p + 1)$
$= 12p^2 + 20p - (3p + 3)$
$= 12p^2 + 20p - 3p - 3$
$= \mathbf{12p^2 + 17p - 3}$

c) $9b(2b + 5) + 4b(6b + 6)$
$= 18b^2 + 45b + 24b^2 + 24b$
$= \mathbf{42b^2 + 69b}$

Page 130 Exercise 4

1 a) $(a + 2)(b + 3) = \mathbf{ab + 3a + 2b + 6}$

b) $(j + 4)(k - 5) = \mathbf{jk - 5j + 4k - 20}$

Using the same method for c)-h):

c) $\mathbf{xy - x - 4y + 4}$ d) $\mathbf{xy + 2x + 6y + 12}$
e) $\mathbf{9b - 27 - ab + 3a}$ f) $\mathbf{st + 3t - 5s - 15}$
g) $\mathbf{15x - 5xy + 12 - 4y}$
h) $\mathbf{6ab + 6a - 2b - 2}$

2 a) $(x + 8)(x + 3) = x^2 + 3x + 8x + 24$
$= \mathbf{x^2 + 11x + 24}$

b) $(b + 2)(b - 4) = b^2 - 4b + 2b - 8$
$= \mathbf{b^2 - 2b - 8}$

c) $(a - 1)(a + 2) = a^2 + 2a - a - 2$
$= \mathbf{a^2 + a - 2}$

Using the same method for d)-f):

d) $\mathbf{d^2 + 13d + 42}$ e) $\mathbf{z^2 - 3z - 108}$
f) $\mathbf{-c^2 - 2c + 15}$

g) $(3y - 8)(6 - y) = 18y - 3y^2 - 48 + 8y$
$= \mathbf{-3y^2 + 26y - 48}$

h) $(2x + 2)(2x + 3) = 4x^2 + 6x + 4x + 6$
$= \mathbf{4x^2 + 10x + 6}$

Page 130 Exercise 5

1 a) $(x + 1)^2 = (x + 1)(x + 1) = x^2 + x + x + 1$
$= \mathbf{x^2 + 2x + 1}$

b) $(x + 4)^2 = (x + 4)(x + 4) = x^2 + 4x + 4x + 16$
$= \mathbf{x^2 + 8x + 16}$

Using the same method for c)-f):

c) $\mathbf{x^2 + 10x + 25}$ d) $\mathbf{x^2 - 4x + 4}$
e) $\mathbf{x^2 - 6x + 9}$ f) $\mathbf{x^2 - 14x + 49}$

g) $3(x + 3)^2 = 3 \times (x + 3)(x + 3)$
$= 3 \times (x^2 + 3x + 3x + 9)$
$= 3 \times (x^2 + 6x + 9)$
$= \mathbf{3x^2 + 18x + 27}$

h) $2(x - 6)^2 = 2 \times (x - 6)(x - 6)$
$= 2 \times (x^2 - 6x - 6x + 36)$
$= 2 \times (x^2 - 12x + 36)$
$= \mathbf{2x^2 - 24x + 72}$

2 a) $(5x + 2)^2 = (5x + 2)(5x + 2)$
$= (25x^2 + 10x + 10x + 4)$
$= \mathbf{25x^2 + 20x + 4}$

b) $(2x + 6)^2 = (2x + 6)(2x + 6)$
$= (4x^2 + 12x + 12x + 36)$
$= 4x^2 + 24x + 36$

c) $(3x - 1)^2 = (3x - 1)(3x - 1)$
$= (9x^2 - 3x - 3x + 1)$
$= \mathbf{9x^2 - 6x + 1}$

8.3 Factorising

Page 132 Exercise 1

1 a) $2a + 10 = (2 \times a) + (2 \times 5) = \mathbf{2(a + 5)}$
Remember to check your factorisation by expanding afterwards.

b) $3b + 12 = (3 \times b) + (3 \times 4) = \mathbf{3(b + 4)}$

Using the same method for c)-h):

c) $\mathbf{5(4c + 3)}$ d) $\mathbf{6(3 + 2x)}$
e) $\mathbf{4(2c + 3f)}$ f) $\mathbf{5(5d + 7e)}$
g) $\mathbf{4(3x + 4y)}$ h) $\mathbf{3(x + 3y)}$

2 a) $3a^2 + 7a = (a \times 3a) + (a \times 7) = \mathbf{a(3a + 7)}$

b) $4b^2 + 19b = (b \times 4b) + (b \times 19) = \mathbf{b(4b + 19)}$

Using the same method for c)-h):

c) $\mathbf{x(2x + 9)}$ d) $\mathbf{y(7 + 15y)}$
e) $\mathbf{x(4x - 9)}$ f) $\mathbf{q(21q - 16)}$
g) $\mathbf{y(15 - 7y)}$ h) $\mathbf{z(27z + 11)}$

Page 132 Exercise 2

1 a) **4** b) $\mathbf{x}$ c) $\mathbf{y}$

d) The HCF of $4xy^2$ and $8x^2y$ is $4xy$.
$4xy^2 + 8x^2y = (y \times 4xy) + (2x \times 4xy)$
$= \mathbf{4xy(y + 2x)}$

2 a) The HCF of 5 and 5 is 5.
The HCF of a^2 and a is a.
So the HCF of $5a^2$ and $5a$ is $5a$.
$5a^2 + 5a = (5a \times a) + (5a \times 1)$
$= \mathbf{5a(a + 1)}$

Using the same method for b)-d):

b) $\mathbf{4b(1 + 2b)}$ c) $\mathbf{3c(2c - 3)}$
d) $\mathbf{4d(3 - 4d)}$

3 a) The HCF of 10 and 5 is 5.
The HCF of c^2 and c is c.
There is no d in $10c^2$ so d won't come outside the brackets.
So the HCF of $10c^2$ and $5cd$ is $5c$.
$10c^2 - 5cd = (5c \times 2c) - (5c \times d)$
$= \mathbf{5c(2c - d)}$

b) The HCF of 20 and 10 is 10.
The HCF of x^2 and x is x.
There is no y in $20x^2$ so y won't come outside the brackets.
So the HCF of $20x^2$ and $10xy$ is $10x$.
$20x^2 - 10xy = (10x \times 2x) - (10x \times y)$
$= \mathbf{10x(2x - y)}$

Using the same method for c)-h):

c) $\mathbf{3x(3x + 2y)}$ d) $\mathbf{4x(3x + 2y)}$
e) $\mathbf{3a(2a + 3b)}$ f) $\mathbf{4p(3q - 2p)}$
g) $\mathbf{2a(4a + 3b^2)}$ h) $\mathbf{8xy(3x - 2y)}$

Page 134 Exercise 3

In this exercise the order of the brackets in the factorisations does not matter.

1 a) Find pairs of numbers that multiply to give 6: 1×6, 2×3
To make +7, you need to do +1 + 6, so:
$x^2 + 7x + 6 = \mathbf{(x + 1)(x + 6)}$
If you can spot the 2 numbers you need, you can write down the answer straight away.

b) Find pairs of numbers that multiply to give 12: 1×12, 2×6, 3×4
To make +7, you need to do +3 + 4, so:
$x^2 + 7x + 12 = \mathbf{(x + 3)(x + 4)}$

Using the same method for c)-h):

c) $\mathbf{(x + 1)(x + 7)}$ d) $\mathbf{(x + 3)^2}$
$(x + 3)(x + 3) = (x + 3)^2$
e) $\mathbf{(x + 2)(x + 4)}$ f) $\mathbf{(y + 3)(y + 5)}$
g) $\mathbf{(z + 2)(z + 7)}$ h) $\mathbf{(v + 3)(v + 8)}$

2 a) Find pairs of numbers that multiply to give 3: 1 × 3
To make +2, you need to do –1 + 3, so:
$x^2 + 2x - 3 = \mathbf{(x - 1)(x + 3)}$

b) Find pairs of numbers that multiply to give 8: 1 × 8, 2 × 4
To make –6, you need to do –2 – 4, so:
$x^2 - 6x + 8 = \mathbf{(x - 2)(x - 4)}$

Using the same method for c)-h):

c) $\mathbf{(x - 4)(x + 2)}$ **d)** $\mathbf{(x - 4)(x - 1)}$
e) $\mathbf{(x - 5)(x + 2)}$ **f)** $\mathbf{(x + 4)(x - 2)}$
g) $\mathbf{(r - 3)(r + 9)}$ **h)** $\mathbf{(u - 6)(u - 9)}$

Page 134 Exercise 4

1 a) $25 = 5^2$, so:
$x^2 - 25 = x^2 - 5^2 = \mathbf{(x + 5)(x - 5)}$

b) $9 = 3^2$, so:
$x^2 - 9 = x^2 - 3^2 = \mathbf{(x + 3)(x - 3)}$

Using the same method for c)-h):

c) $\mathbf{(x + 2)(x - 2)}$ **d)** $\mathbf{(x + 6)(x - 6)}$
e) $\mathbf{(x + 9)(x - 9)}$ **f)** $\mathbf{(x + 8)(x - 8)}$
g) $\mathbf{(b + 11)(b - 11)}$ **h)** $\mathbf{(t + 12)(t - 12)}$

2 a) $x^2 - 5 = x^2 - \sqrt{5}^2$
$= \mathbf{(x + \sqrt{5})(x - \sqrt{5})}$

Using the same method for b)-c):

b) $\mathbf{(x + \sqrt{3})(x - \sqrt{3})}$
c) $\mathbf{(x + \sqrt{11})(x - \sqrt{11})}$
d) $\mathbf{(x + y)(x - y)}$

Page 135 Review Exercise

1 a) $\mathbf{14s}$ **b)** $\mathbf{5m}$

c) $x + y + x + y + x - y = (x + x + x) + (y + y - y)$
$= \mathbf{3x + y}$

d) $16p + 4q + 4 - 2p + 3q - 8$
$= (16p - 2p) + (4q + 3q) + (4 - 8)$
$= \mathbf{14p + 7q - 4}$

Using the same method for e)-f):

e) $\mathbf{6s + 9t^2 - 3s^2}$ **f)** $\mathbf{2b^2 + 12b + 7}$

2 a) $a \times ab = a \times a \times b = \mathbf{a^2b}$

b) $4a^2 \div 2a = (4 \div 2) \times (a^2 \div a) = \mathbf{2a}$

Using the same method for c)-f):

c) $\mathbf{14pq^2}$ **d)** $\mathbf{3e^2}$ **e)** $\mathbf{-24i^5}$ **f)** $\mathbf{3d^2}$

3 a) $4(x + 8) = (4 \times x) + (4 \times 8) = \mathbf{4x + 32}$
b) $6(5 - r) = (6 \times 5) + (6 \times r) = \mathbf{30 - 6r}$
c) $-2(7 + y) = (-2 \times 7) + (-2 \times y) = \mathbf{-14 - 2y}$
d) $8(h - 2) = (8 \times h) - (8 \times 2) = \mathbf{8h - 16}$
e) $h(h + 3) = (h \times h) + (h \times 3) = \mathbf{h^2 + 3h}$
f) $-4n(n + 2) = (-4n \times n) + (-4n \times 2)$
$= \mathbf{-4n^2 - 8n}$
g) $4w(u - 7) = (4w \times u) - (4w \times 7) = \mathbf{4uw - 28w}$
h) $-2x(12 - v) = (-2x \times 12) - (-2x \times v)$
$= \mathbf{-24x + 2xv}$

4 a) $4(c + 3) + 6(c + 2) = 4c + 12 + 6c + 12$
$= (4c + 6c) + (12 + 12) = \mathbf{10c + 24}$
b) $5(u + 4) + 3(u + 8) = 5u + 20 + 3u + 24$
$= (5u + 3u) + (20 + 24) = \mathbf{8u + 44}$

Using the same method for c)-f):

c) $\mathbf{12b - 2}$ **d)** $\mathbf{c - 17}$
e) $\mathbf{4q - 16}$ **f)** $\mathbf{j - 7}$

5 a) $(2 + x)(8 + y) = \mathbf{16 + 2y + 8x + xy}$
b) $(x - 3)(x - 5) = x^2 - 5x - 3x + 15$
$= \mathbf{x^2 - 8x + 15}$
c) $3(j - 2)(k + 4) = 3 \times (jk + 4j - 2k - 8)$
$= \mathbf{3jk + 12j - 6k - 24}$
d) $(n + 5)(m - 4) = \mathbf{nm + 5m - 4n - 20}$
e) $(3 + r)(s + 4) = \mathbf{3s + 12 + rs + 4r}$
f) $(z - 8)^2 = (z - 8)(z - 8)$
$= z^2 - 8z - 8z + 64$
$= \mathbf{z^2 - 16z + 64}$

6 a) The HCF of $2x$ and $4y$ is 2.
$2x + 4y = (2 \times x) + (2 \times 2y) = \mathbf{2(x + 2y)}$
b) The HCF of $8x$ and 24 is 8.
$8x + 24 = (8 \times x) + (8 \times 3) = \mathbf{8(x + 3)}$
c) The HCF of $4y$ and $6y^2$ is $2y$.
$4y - 6y^2 = (2y \times 2) - (2y \times 3y) = \mathbf{2y(2 - 3y)}$

Using the same method for d)-i):

d) $\mathbf{4y(5 + 3x)}$ **e)** $\mathbf{5a(3b - 2a)}$
f) $\mathbf{12(5x + 12y)}$ **g)** $\mathbf{4r(7 + 10rs)}$
h) $\mathbf{4ab(1 - 2a)}$ **i)** $\mathbf{7mn(2m - 5n)}$

7 a) Find pairs of numbers that multiply to give 21: 3 × 7
To make +10, you need to do 3 + 7, so:
$y^2 + 10y + 21 = \mathbf{(y + 3)(y + 7)}$
b) Find pairs of numbers that multiply to give 5: 1 × 5
To make –4, you need to do –5 + 1, so:
$x^2 - 4x - 5 = \mathbf{(x - 5)(x + 1)}$

Using the same method for c)-f):

c) $\mathbf{(t - 4)^2}$ **d)** $\mathbf{(x - 4)(x + 3)}$
e) $\mathbf{(x + 6)(x - 1)}$ **f)** $\mathbf{(x - 9)(x + 5)}$

8 a) $100 = 10^2$, so:
$x^2 - 100 = x^2 - 10^2 = \mathbf{(x + 10)(x - 10)}$
b) $36 = 6^2$, so:
$y^2 - 36 = y^2 - 6^2 = \mathbf{(y - 6)(y + 6)}$
c) $y^2 - 121 = y^2 - 11^2$
$= \mathbf{(y + 11)(y - 11)}$

Page 136 Exam-Style Questions

1

Expression	Simplified expression
$a + a + a$	$\mathbf{3a}$
$3a^2 + 2a^2$	$\mathbf{5a^2}$
$4a^2 + 6a - a^2$	$\mathbf{3a^2 + 6a}$

[3 marks available — 1 mark for each correct answer in the table]

2 a) $v + v + w - w - w - w - w = 2v + w - 4w$
$= \mathbf{2v - 3w}$ *[1 mark]*
b) $5 \times 7 \times x \times x = \mathbf{35x^2}$ *[1 mark]*
c) $6y - 3yz - 8y + 10yz = 6y - 8y + 10yz - 3yz$
$= \mathbf{-2y + 7yz}$ *[1 mark]*

3 a) $x(x^2 - 4y) + 9xy = x^3 - 4xy + 9xy$ *[1 mark]*
$= \mathbf{x^3 + 5xy}$ *[1 mark]*
b) $(2x - 7)^2 = (2x - 7)(2x - 7)$
$= 4x^2 - 14x - 14x + 49$ *[1 mark]*
$= \mathbf{4x^2 - 28x + 49}$ *[1 mark]*

4 a) $7c + 56 = (7 \times c) + (7 \times 8)$
$= \mathbf{7(c + 8)}$ *[1 mark]*
b) $d^2 + 5d - 2de = (d \times d) + (d \times 5) - (d \times 2e)$
$= \mathbf{d(d + 5 - 2e)}$
[2 marks available – 1 mark for taking out a common factor of d, 1 mark for a fully correct expression]

5 a) Find pairs of numbers that multiply to give 32: 1 × 32, 2 × 16, 4 × 8
To make 12, you need to do 4 + 8.
So $y^2 + 12y + 32 = \mathbf{(y + 4)(y + 8)}$
[2 marks available — 1 mark for the correct numbers in the brackets, 1 mark for the correct signs]
b) $169 = 13^2$ so you can use difference of two squares.
$x^2 - 169 = \mathbf{(x + 13)(x - 13)}$
[2 marks available — 1 mark for using the fact that 13 is the square root of 169, 1 mark for fully correct factorisation]

Section 9 — Equations, Identities and Inequalities

9.1 Solving Equations

Page 138 Exercise 1

1 a) $x + 9 = 12$
$x = 12 - 9 \Rightarrow \mathbf{x = 3}$
b) $x + 5 = 16$
$x = 16 - 5 \Rightarrow \mathbf{x = 11}$
c) $x - 2 = 14$
$x = 14 + 2 \Rightarrow \mathbf{x = 16}$
d) $x - 7 = -19$
$x = -19 + 7 \Rightarrow \mathbf{x = -12}$
e) $-2 = 7 + x$
$-2 - 7 = x \Rightarrow \mathbf{x = -9}$
f) $32 = x - 17$
$32 + 17 = x \Rightarrow \mathbf{x = 49}$

2 a) $12 - x = 9$
$12 = 9 + x$
$12 - 9 = x \Rightarrow \mathbf{x = 3}$
b) $4 - x = 2$
$4 = 2 + x$
$4 - 2 = x \Rightarrow \mathbf{x = 2}$
c) $2 - x = 7$
$2 = 7 + x$
$2 - 7 = x \Rightarrow \mathbf{x = -5}$
d) $19 = 14 - x$
$19 + x = 14$
$x = 14 - 19 \Rightarrow \mathbf{x = -5}$
e) $14 = 8 - x$
$14 + x = 8$
$x = 8 - 14 \Rightarrow \mathbf{x = -6}$
f) $7 = 5 - x$
$7 + x = 5$
$x = 5 - 7 \Rightarrow \mathbf{x = -2}$

3 a) $x + 7 = 12$
$x = 12 - 7 \Rightarrow \mathbf{x = 5}$
b) $5 - x = 21$
$5 = 21 + x$
$5 - 21 = x \Rightarrow \mathbf{x = -16}$
c) $16 = x + 10$
$16 - 10 = x \Rightarrow \mathbf{x = 6}$
d) $x - 8 = 14$
$x = 14 + 8 \Rightarrow \mathbf{x = 22}$
e) $12 - x = 23$
$12 = 23 + x$
$12 - 23 = x \Rightarrow \mathbf{x = -11}$
f) $35 = 31 - x$
$35 + x = 31$
$x = 31 - 35 \Rightarrow \mathbf{x = -4}$

4 a) $x - 7.3 = 1.6$
$x = 1.6 + 7.3 \Rightarrow \mathbf{x = 8.9}$
b) $6.03 - x = 0.58$
$6.03 = 0.58 + x$
$6.03 - 0.58 = x \Rightarrow \mathbf{x = 5.45}$
c) $3.47 = 7.18 + x$
$3.47 - 7.18 = x \Rightarrow \mathbf{x = -3.71}$
d) $5.2 = 2.8 - x$
$5.2 + x = 2.8$
$x = 2.8 - 5.2 \Rightarrow \mathbf{x = -2.4}$

Page 139 Exercise 2

1 a) $\frac{x}{3} = 2 \Rightarrow \frac{x}{3} \times 3 = 2 \times 3 \Rightarrow \mathbf{x = 6}$
b) $\frac{x}{6} = -3 \Rightarrow \frac{x}{6} \times 6 = -3 \times 6 \Rightarrow \mathbf{x = -18}$
c) $\frac{x}{3} = 0.4 \Rightarrow \frac{x}{3} \times 3 = 0.4 \times 3 \Rightarrow \mathbf{x = 1.2}$
d) $\frac{x}{11} = -0.5 \Rightarrow \frac{x}{11} \times 11 = -0.5 \times 11$
$\Rightarrow \mathbf{x = -5.5}$
e) $8x = 24 \Rightarrow 8x \div 8 = 24 \div 8 \Rightarrow \mathbf{x = 3}$
f) $4.5x = 81 \Rightarrow 4.5x \div 4.5 = 81 \div 4.5$
$\Rightarrow \mathbf{x = 18}$

g) $5x = -20 \Rightarrow 5x \div 5 = -20 \div 5 \Rightarrow \boldsymbol{x = -4}$

h) $3.5x = -7 \Rightarrow 3.5x \div 3.5 = -7 \div 3.5$
$\Rightarrow \boldsymbol{x = -2}$

2 a) $-7x = -56 \Rightarrow -7x \div -7 = -56 \div -7$
$\Rightarrow \boldsymbol{x = 8}$
You could add 7x to both sides and then rearrange, but it would take more steps.

b) $-9x = 108 \Rightarrow -9x \div -9 = 108 \div -9$
$\Rightarrow \boldsymbol{x = -12}$

c) $-4.5x = -2.7 \Rightarrow -4.5x \div -4.5$
$= -2.7 \div -4.5 \Rightarrow \boldsymbol{x = 0.6}$

d) $-\frac{x}{4} = 3 \Rightarrow -\frac{x}{4} \times -4 = 3 \times -4 \Rightarrow \boldsymbol{x = -12}$

e) $-\frac{x}{5} = 6 \Rightarrow -\frac{x}{5} \times -5 = 6 \times -5 \Rightarrow \boldsymbol{x = -30}$

f) $-\frac{x}{10} = 1.1 \Rightarrow -\frac{x}{10} \times -10 = 1.1 \times -10$
$\Rightarrow \boldsymbol{x = -11}$

Page 139 Exercise 3

1 a) $\frac{4x}{3} = 12 \Rightarrow 4x = 12 \times 3 = 36$
$\Rightarrow x = 36 \div 4 \Rightarrow \boldsymbol{x = 9}$

b) $\frac{2x}{5} = 6 \Rightarrow 2x = 6 \times 5 = 30 \Rightarrow x = 30 \div 2$
$\Rightarrow \boldsymbol{x = 15}$

Using the same method for c)-h):

c) $\boldsymbol{x = 14}$ **d)** $\boldsymbol{x = 3.6}$ **e)** $\boldsymbol{x = 1}$
f) $\boldsymbol{x = -0.15}$ **g)** $\boldsymbol{x = -0.02}$ **h)** $\boldsymbol{x = 1.26}$

Page 140 Exercise 4

1 a) $8x + 10 = 66$
$8x = 66 - 10 = 56$
$x = 56 \div 8 \Rightarrow \boldsymbol{x = 7}$

b) $10x + 15 = 115$
$10x = 115 - 15 = 100$
$x = 100 \div 10 \Rightarrow \boldsymbol{x = 10}$

Using the same method for c)-f):

c) $\boldsymbol{x = 8}$ **d)** $\boldsymbol{x = 60}$
e) $\boldsymbol{x = -5}$ **f)** $\boldsymbol{x = -5.5}$

2 a) $16x - 6 = 10$
$16x = 10 + 6 = 16$
$x = 16 \div 16 \Rightarrow \boldsymbol{x = 1}$

b) $15x - 8 = 22$
$15x = 22 + 8 = 30$
$x = 30 \div 15 = 2 \Rightarrow \boldsymbol{x = 2}$

Using the same method for c)-f):

c) $\boldsymbol{x = 3}$ **d)** $\boldsymbol{x = -20}$
e) $\boldsymbol{x = -3}$ **f)** $\boldsymbol{x = -6}$

3 $\frac{x}{2} - 1 = 3$
$\frac{x}{2} - 1 + 1 = 3 + 1 \Rightarrow \frac{x}{2} = 4$
$\frac{x}{2} \times 2 = 4 \times 2 \Rightarrow \boldsymbol{x = 8}$ is the solution.
'Undo' means doing the opposite. Adding 1 is the opposite of subtracting 1, and multiplying by 2 is the opposite of dividing by 2.

4 a) $\frac{x}{2} + 1 = 7$
$\frac{x}{2} = 7 - 1 = 6$
$x = 6 \times 2 \Rightarrow \boldsymbol{x = 12}$

b) $\frac{x}{6} + 4 = 16$
$\frac{x}{6} = 16 - 4 = 12$
$x = 12 \times 6 \Rightarrow \boldsymbol{x = 72}$

Using the same method for c)-f):

c) $\boldsymbol{x = 20}$ **d)** $\boldsymbol{x = -16}$
e) $\boldsymbol{x = 9}$ **f)** $\boldsymbol{x = -21}$

5 a) $20 - 5x = 10$
$20 - 5x + 5x = 10 + 5x$
$\boldsymbol{20 = 10 + 5x}$

b) $20 = 10 + 5x$
$20 - 10 = 5x \Rightarrow 5x = 10$
$x = 10 \div 5 \Rightarrow \boldsymbol{x = 2}$

6 a) $12 - 4x = 8$
$12 = 8 + 4x$
$12 - 8 = 4x \Rightarrow 4x = 4$
$x = 4 \div 4 \Rightarrow \boldsymbol{x = 1}$

b) $47 - 9x = 11$
$47 = 11 + 9x$
$47 - 11 = 9x \Rightarrow 9x = 36$
$x = 36 \div 9 \Rightarrow \boldsymbol{x = 4}$

Using the same method for c) and d):

c) $\boldsymbol{x = -2}$ **d)** $\boldsymbol{x = -9}$

Page 142 Exercise 5

1 a) $7(x + 4) = 63$
$7x + 28 = 63$
$7x = 63 - 28 = 35$
$x = 35 \div 7 \Rightarrow \boldsymbol{x = 5}$
Alternatively, you could start by dividing both sides by 7, and then rearrange.

b) $8(x + 4) = 88$
$8x + 32 = 88$
$8x = 88 - 32 = 56$
$x = 56 \div 8 \Rightarrow \boldsymbol{x = 7}$

Using the same method for c)-f):

c) $\boldsymbol{x = 9}$ **d)** $\boldsymbol{x = -2}$
e) $\boldsymbol{x = -3}$ **f)** $\boldsymbol{x = -5}$

2 a) $315 = 21(6 - x)$
$315 \div 21 = 6 - x \Rightarrow 15 = 6 - x$
$\Rightarrow x = 6 - 15 \Rightarrow \boldsymbol{x = -9}$
Dividing first keeps the numbers small, but you could start by expanding the brackets if you wanted to.

b) $12.5(x - 4) = 75$
$x - 4 = 75 \div 12.5 \Rightarrow x - 4 = 6$
$\Rightarrow x = 6 + 4 \Rightarrow \boldsymbol{x = 10}$

c) $36 = 7.2(2 - x)$
$36 \div 7.2 = 2 - x \Rightarrow 5 = 2 - x$
$\Rightarrow x = 2 - 5 \Rightarrow \boldsymbol{x = -3}$

Page 142 Exercise 6

1 a) $6x - 4 = 2x + 16$
$6x - 2x - 4 = 16 \Rightarrow 4x - 4 = 16$
$4x = 16 + 4 = 20$
$x = 20 \div 4 \Rightarrow \boldsymbol{x = 5}$

b) $17x + 2 = 7x - 8$
$17x - 7x + 2 = -8 \Rightarrow 10x + 2 = -8$
$10x = -8 - 2 = -10$
$x = -10 \div 10 \Rightarrow \boldsymbol{x = -1}$

Using the same method for c)-f):

c) $\boldsymbol{x = 3}$ **d)** $\boldsymbol{x = 6}$
e) $\boldsymbol{x = 5}$ **f)** $\boldsymbol{x = 1}$

2 a) $13x - 35 = 45 - 3x$
$13x + 3x - 35 = 45 \Rightarrow 16x - 35 = 45$
$16x = 45 + 35 = 80$
$x = 80 \div 16 = 5 \Rightarrow \boldsymbol{x = 5}$

b) $20x + 18 = 54 - 16x$
$20x + 16x + 18 = 54 \Rightarrow 36x + 18 = 54$
$36x = 54 - 18 = 36$
$x = 36 \div 36 \Rightarrow \boldsymbol{x = 1}$

Using the same method for c)-f):

c) $\boldsymbol{x = 3}$ **d)** $\boldsymbol{x = 36}$
e) $\boldsymbol{x = 9}$ **f)** $\boldsymbol{x = 9}$

3 a) $4x - 3 = 0.5 - 3x$
$4x + 3x - 3 = 0.5 \Rightarrow 7x - 3 = 0.5$
$7x = 0.5 + 3 = 3.5$
$x = 3.5 \div 7 \Rightarrow \boldsymbol{x = \frac{1}{2}}$

b) $10x - 18 = 10.2 + 4x$
$10x - 4x - 18 = 10.2 \Rightarrow 6x - 18 = 10.2$
$6x = 10.2 + 18 = 28.2$
$x = 28.2 \div 6 \Rightarrow \boldsymbol{x = 4.7}$

c) $4x - 8.6 = 48.1 - 5x$
$4x + 5x - 8.6 = 48.1 \Rightarrow 9x - 8.6 = 48.1$
$9x = 48.1 + 8.6 = 56.7$
$x = 56.7 \div 9 \Rightarrow \boldsymbol{x = 6.3}$

d) $-x + 1 = 28 + 2x$
$1 = 28 + 2x + x \Rightarrow 1 = 28 + 3x$
$1 - 28 = 3x \Rightarrow 3x = -27$
$x = -27 \div 3 \Rightarrow \boldsymbol{x = -9}$

Page 143 Exercise 7

1 a) $3(x + 2) = x + 14$
$3x + 6 = x + 14$
$3x - x + 6 = 14 \Rightarrow 2x + 6 = 14$
$2x = 14 - 6 = 8$
$x = 8 \div 2 \Rightarrow \boldsymbol{x = 4}$

b) $9(x - 1) = x + 15$
$9x - 9 = x + 15$
$9x - x - 9 = 15 \Rightarrow 8x - 9 = 15$
$8x = 15 + 9 = 24$
$x = 24 \div 8 \Rightarrow \boldsymbol{x = 3}$

c) $6(x + 2) = 3x + 48$
$6x + 12 = 3x + 48$
$6x - 3x + 12 = 48 \Rightarrow 3x + 12 = 48$
$3x = 48 - 12 = 36$
$x = 36 \div 3 \Rightarrow \boldsymbol{x = 12}$

d) $8(x - 8) = 2(x - 2)$
$8x - 64 = 2x - 4$
$8x - 2x - 64 = -4 \Rightarrow 6x - 64 = -4$
$6x = -4 + 64 = 60$
$x = 60 \div 6 \Rightarrow \boldsymbol{x = 10}$

e) $4(4 - x) = 2(x - 1)$
$16 - 4x = 2x - 2$
$16 - 4x - 2x = -2 \Rightarrow -6x + 16 = -2$
$-6x = -2 - 16 = -18$
$x = -18 \div -6 \Rightarrow \boldsymbol{x = 3}$

f) $20(x - 2) = 5(x + 1)$
$20x - 40 = 5x + 5$
$20x - 5x - 40 = 5 \Rightarrow 15x - 40 = 5$
$15x = 5 + 40 = 45$
$x = 45 \div 15 \Rightarrow \boldsymbol{x = 3}$

2 a) $5(x - 5) = 2(x - 14)$
$5x - 25 = 2x - 28$
$5x - 2x - 25 = -28 \Rightarrow 3x - 25 = -28$
$3x = -28 + 25 = -3$
$x = -3 \div 3 \Rightarrow \boldsymbol{x = -1}$

b) $2(x - 2) = 5(x - 8)$
$2x - 4 = 5x - 40$
$-4 = 5x - 2x - 40 \Rightarrow -4 = 3x - 40$
$\Rightarrow -4 + 40 = 3x \Rightarrow 36 = 3x$
$x = 36 \div 3 \Rightarrow \boldsymbol{x = 12}$
You could have done 2x − 5x but you would have a negative multiple of x, which would make rearranging things slightly trickier.

c) $4(x - 2) = 6(x + 3)$
$4x - 8 = 6x + 18$
$-8 = 6x - 4x + 18 \Rightarrow -8 = 2x + 18$
$\Rightarrow -8 - 18 = 2x \Rightarrow -26 = 2x$
$x = -26 \div 2 \Rightarrow \boldsymbol{x = -13}$

d) $6(x - 1.5) = 2(x - 3.5)$
$6x - 9 = 2x - 7$
$6x - 2x - 9 = -7 \Rightarrow 4x - 9 = -7$
$4x = -7 + 9 = 2$
$x = 2 \div 4 \Rightarrow \boldsymbol{x = 0.5}$

e) $9(x - 3.3) = -6(x + 1.7)$
$9x - 29.7 = -6x - 10.2$
$9x + 6x - 29.7 = -10.2$
$\Rightarrow 15x - 29.7 = -10.2$
$15x = -10.2 + 29.7 = 19.5$
$x = 19.5 \div 15 \Rightarrow \boldsymbol{x = 1.3}$

f) $-4(x - 3) = 8(0.7 - x)$
$-4x + 12 = 5.6 - 8x$
$-4x + 8x + 12 = 5.6 \Rightarrow 4x + 12 = 5.6$
$4x = 5.6 - 12 = -6.4$
$x = -6.4 \div 4 \Rightarrow \boldsymbol{x = -1.6}$

3 a) $7(3x + 2) = 5(9x - 0.08)$
$21x + 14 = 45x - 0.4$
$45x - 21x - 0.4 = 14 \Rightarrow 24x - 0.4 = 14$
$24x = 14 + 0.4 = 14.4$
$x = 14.4 \div 24 \Rightarrow \boldsymbol{x = 0.6}$

b) $7(2x + \frac{1}{7}) = 14(3x - 0.5)$

$14x + 1 = 42x - 7$
$1 = 42x - 14x - 7 \Rightarrow 1 = 28x - 7$
$\Rightarrow 28x = 1 + 7 = 8$
$x = 8 \div 28 \Rightarrow \mathbf{x = \frac{2}{7}}$

c) $10(x - 2) = -2(\frac{4}{3} + 7x)$
$10x - 20 = -\frac{8}{3} - 14x$
$10x = -14x + 20 - \frac{8}{3} \Rightarrow 10x = -14x + \frac{52}{3}$
$24x = \frac{52}{3}$
$x = \frac{52}{72} = \mathbf{\frac{13}{18}}$

d) $4(3x - 3) = -2(\frac{76}{9} + 5x)$
$12x - 12 = -\frac{152}{9} - 10x$
$12x + 10x - 12 = -\frac{152}{9}$
$22x - 12 = -\frac{152}{9}$
$22x = -\frac{152}{9} + 12 = -\frac{44}{9}$
$x = -\frac{44}{9} \div 22 \Rightarrow \mathbf{x = -\frac{2}{9}}$

Page 144 Exercise 8

1 **a)** $\frac{x}{4} = 1 - x$
$x = 4(1 - x) = 4 - 4x$
$x + 4x = 4 \Rightarrow 5x = 4 \Rightarrow x = \mathbf{\frac{4}{5}}$ or **0.8**

b) $\frac{x}{3} = 8 - x$
$x = 3(8 - x) = 24 - 3x$
$x + 3x = 24 \Rightarrow 4x = 24$
$x = 24 \div 4 \Rightarrow \mathbf{x = 6}$

Using the same method for c)-f):

c) $\mathbf{x = 5}$ **d)** $\mathbf{x = 8}$
e) $\mathbf{x = -5}$ **f)** $\mathbf{x = -10}$

2 **a)** $\frac{x}{3} = 2(x - 5)$
$x = 3 \times 2(x - 5) = 6(x - 5) = 6x - 30$
$x + 30 = 6x$
$30 = 6x - x \Rightarrow 5x = 30$
$x = 30 \div 5 \Rightarrow \mathbf{x = 6}$

b) $\frac{x}{2} = 4(x - 7)$
$x = 2 \times 4(x - 7) = 8(x - 7) = 8x - 56$
$x + 56 = 8x$
$56 = 8x - x \Rightarrow 56 = 7x$
$x = 56 \div 7 \Rightarrow \mathbf{x = 8}$

Using the same method for c)-f):

c) $\mathbf{x = -10}$ **d)** $\mathbf{x = -\frac{20}{3}}$
e) $\mathbf{x = -16}$ **f)** $\mathbf{x = 60}$

Page 144 Exercise 9

1 **a)** $\frac{x+4}{2} = \frac{x+10}{3}$
$3(x + 4) = 2(x + 10)$
$3x + 12 = 2x + 20$
$3x - 2x + 12 = 20$
$x + 12 = 20$
$x = 20 - 12 \Rightarrow \mathbf{x = 8}$

b) $\frac{x+2}{2} = \frac{x+4}{6}$
$6(x + 2) = 2(x + 4)$
$6x + 12 = 2x + 8$
$6x - 2x + 12 = 8$
$4x + 12 = 8$
$4x = 8 - 12 = -4$
$x = -4 \div 4 \Rightarrow \mathbf{x = -1}$

Using the same method for c)-f):

c) $\mathbf{x = 5}$ **d)** $\mathbf{x = 11}$
e) $\mathbf{x = 8}$ **f)** $\mathbf{x = 21}$

2 **a)** $\frac{x-6}{2} = \frac{8-2x}{4}$
$4(x - 6) = 2(8 - 2x)$
$4x - 24 = 16 - 4x$
$4x + 4x - 24 = 16 \Rightarrow 8x - 24 = 16$
$8x = 16 + 24 = 40$
$x = 40 \div 8 \Rightarrow \mathbf{x = 5}$

b) $\frac{x-9}{2} = \frac{2-3x}{4}$
$4(x - 9) = 2(2 - 3x)$
$4x - 36 = 4 - 6x$
$4x + 6x - 36 = 4 \Rightarrow 10x - 36 = 4$
$10x = 4 + 36 = 40$
$x = 40 \div 10 \Rightarrow \mathbf{x = 4}$

c) $\frac{x-12}{6} = \frac{4-2x}{3}$
$3(x - 12) = 6(4 - 2x)$
$3x - 36 = 24 - 12x$
$3x + 12x - 36 = 24 \Rightarrow 15x - 36 = 24$
$15x = 24 + 36 = 60$
$x = 60 \div 15 \Rightarrow \mathbf{x = 4}$

9.2 Forming Your Own Equations

Page 145 Exercise 1

1 **a)** Call the number x.
Add 5 $\Rightarrow x + 5$
Result equals 12, so:
$x + 5 = 12 \Rightarrow x = 12 - 5 = 7$
The number they were thinking of was **7**.

b) Call the number x.
Multiply by 2 $\Rightarrow 2x$
Subtract 5 $\Rightarrow 2x - 5$
Result equals 15, so:
$2x - 5 = 15$
$2x = 15 + 5 = 20$
$x = 20 \div 2 = 10$
The number they were thinking of was **10**.

c) Call the number x.
Divide by 4 $\Rightarrow \frac{x}{4}$
Add 10 $\Rightarrow \frac{x}{4} + 10$
Result equals 14, so:
$\frac{x}{4} + 10 = 14$
$\frac{x}{4} = 14 - 10 = 4$
$x = 4 \times 4 = 16$
The number they were thinking of was **16**.

Page 146 Exercise 2

1 Call the number of people that the bride and the groom each invited x.
The total number of people invited was $2x$.
8 people couldn't come so $2x - 8$ could attend.
So $2x - 8 = 60$
$2x = 60 + 8 = 68$
$x = 68 \div 2 = 34$
So the groom invited **34 guests**.

2 Call the number of single beds x.
Number of king-size beds = $x - 10$
Number of double beds = $x + 16$.
So $x + (x - 10) + (x + 16) = x + x + x + 16 - 10$
$= 3x + 6 = 54$
$3x = 54 - 6 = 48$
$x = 48 \div 3 = 16$
So **16 single beds** were sold that day.

3 **a)** A fruit scone sells for £x and a cheese scone sells for £x + 0.1, so the total amount in £ that she sells the scones for is:
$20x + 10(x + 0.1) = 20x + 10x + 1$
$= 30x + 1$
Subtract the cost of the ingredients to find the profit in £:
profit = $\mathbf{30x + 1 - y}$.

b) Substitute profit = 10 and y = 6 into the expression from a):
$10 = 30x + 1 - 6 \Rightarrow 10 = 30x - 5$
$10 + 5 = 30x \Rightarrow 15 = 30x$
$x = 30 \div 15 = 0.5$
So a fruit scone costs **50p**.

Page 147 Exercise 3

1 **a)** Sum of angles in a triangle = 180°.
This is a rule for all triangles that you need to know.
So 110° + $3x$ + $4x$ = 180°
$\Rightarrow$ **110° + 7x = 180°**

b) 110° + $7x$ = 180°
$7x$ = 180° – 110° = 70°
x = 70° ÷ 7 $\Rightarrow$ **x = 10°**

2 **a) (i)** $4x + (x + 8) + 4x + (x + 8) = 146$
$4x + x + 8 + 4x + x + 8 = 146$
So $\mathbf{10x + 16 = 146}$

(ii) $10x + 16 = 146$
$10x = 146 - 16 = 130$
$x = 130 \div 10 \Rightarrow$ **x = 13 cm**

b) (i) The hexagon is regular, so all sides are the same length:
$(x + 2) + (x + 2) + (x + 2) + (x + 2) + (x + 2) + (x + 2) = 102$
$6(x + 2) = 102$
$\mathbf{6x + 12 = 102}$

(ii) $6x + 12 = 102$
$6x = 102 - 12 = 90$
$x = 90 \div 6 \Rightarrow$ **x = 15 cm**

9.3 Identities

Page 148 Exercise 1

1 **a)** $4x = 10$ only when $x = 2.5$.
So **no** the '≡' symbol can't be used.

b) When $x = 1$,
$x^2 + 2x + 1 = 1 + 2 + 1 = 4 \neq 0$.
So **no**, the '≡' symbol can't be used.
If you can find one value of x that makes the identity false, then it isn't an identity.

c) Rearranging the order of $-x^2 + 3$ gives $3 - x^2$, which is the same as the right-hand side. So **yes**, the '≡' symbol can be used.

d) $2(x + 1) = 2x + 2$ which is different to the right-hand side.
So **no**, the '≡' symbol can't be used.

e) Expanding the left-hand side gives
$3(x + 2) - x = 3x + 6 - x = 2x + 6$
Expanding the right-hand side gives
$2(x + 3) = 2x + 6$, which is the same as the left-hand side.
So **yes**, the '≡' symbol can be used.

f) Expanding the left-hand side gives
$3(2 - 3x) + 2 = 6 - 9x + 2 = 4 - 9x$
$4 - 9x$ is different to $8x$ so **no**, the '≡' symbol can't be used.

2 **a)** Expanding the left-hand side gives
$(x + 4)^2 - 4 = (x + 4)(x + 4) - 4$
$= x^2 + 4x + 4x + 16 - 4$
$= \mathbf{x^2 + 8x + 12}$
Expanding the right-hand side gives
$(x + 6)(x + 2) = x^2 + 2x + 6x + 12$
$= \mathbf{x^2 + 8x + 12}$
The two sides are the same so the identity is true.

b) Expanding the left-hand side gives
$5(x + 2) + (x^2 - 4) = 5x + 10 + x^2 - 4$
$= \mathbf{x^2 + 5x + 6}$
Expanding the right-hand side gives
$(x + 4)(x + 1) + 2 = x^2 + x + 4x + 4 + 2$
$= \mathbf{x^2 + 5x + 6}$
The two sides are the same so the identity is true.

9.4 Proof

Page 149 Exercise 1

1 Take an even number and an odd number — $2a$ and $(2b + 1)$. Their product is
$2a(2b + 1) = 4ab + 2a$
$= 2(2ab + a)$
$= \mathbf{2n}$, where $n = 2ab + a$
So the product of an even number and an odd number is **even**.

2 Take two consecutive square numbers — a^2 and $(a + 1)^2$.
Their sum is $a^2 + (a + 1)^2 = a^2 + (a + 1)(a + 1)$
$= a^2 + a^2 + a + a + 1$
$= 2a^2 + 2a + 1$
$= 2(a^2 + a) + 1$
$= \mathbf{2n + 1}$,
where $n = a^2 + a$.
So the sum of two consecutive square numbers is **odd**.

Page 150 Exercise 2

1 E.g. $(-3) + (-2) + (-1) = -6$ which is **less** than each individual number.

2 E.g. 2 and 3 are both prime, but $3 - 2 = 1$ which is **odd**.

Page 150 Exercise 3

1 Take $2n + 1$ and $2n + 3$ as consecutive odd integers.
Then $(2n + 1) + (2n + 3) = 4n + 4 = \mathbf{4(n + 1)}$ where $n + 1$ is an integer.
So the sum of two consecutive odd integers is a **multiple of 4**.

2 Expanding gives: $x = 2(y + 5) + 4(y + 1) - 2$
$= 2y + 10 + 4y + 4 - 2$
$= 6y + 12$
$= 6(y + 2)$
$= \mathbf{6n}$, where $n = y + 2$.
As n is an integer, x is a **multiple of 6**.

9.5 Inequalities

Page 151 Exercise 1

1 a) $\mathbf{6 > 1}$ b) $\mathbf{2 < 8}$ c) $\mathbf{-1 > -3}$ d) $\mathbf{-7 < 1}$

2 a) **x is greater than or equal to 1.**
b) **x is less than 7.**
c) **x is greater than –4.**
d) **x is less than or equal to 9.**

3 a) $\mathbf{x > 4}$ b) $\mathbf{x \le 12}$ c) $\mathbf{x < 3}$

4 a) 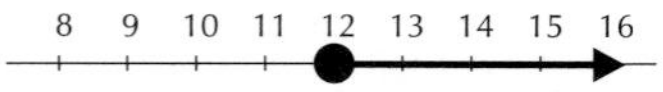

Remember, it's a black circle if x can take that value, or a white circle if it cannot take that value.

b) 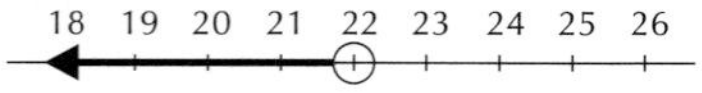

c) 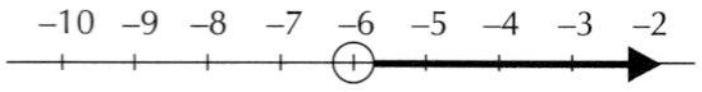

d) 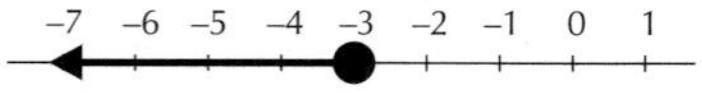

Page 152 Exercise 2

1 a) $x + 9 > 14$
$x > 14 - 9$
$\mathbf{x > 5}$

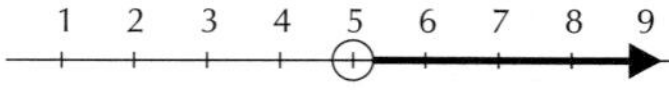

b) $x + 3 \le 12$
$x \le 12 - 3$
$\mathbf{x \le 9}$

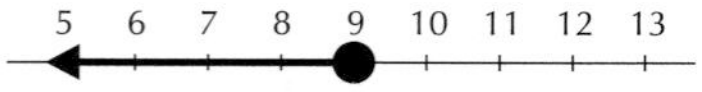

c) $x - 5 < -3$
$x < -3 + 5$
$\mathbf{x < 2}$

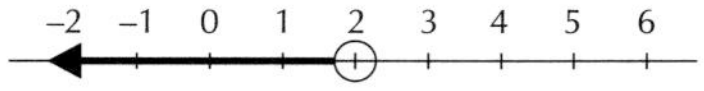

d) $x + 1 \le -1$
$x \le -1 - 1$
$\mathbf{x \le -2}$

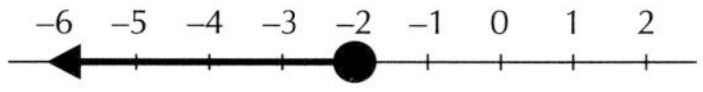

2 a) $x - 9 > 8$
$x > 8 + 9$
$\mathbf{x > 17}$

b) $x + 7 < 17$
$x < 17 - 7$
$\mathbf{x < 10}$

c) $x + 12 < -18$
$x < -18 - 12$
$\mathbf{x < -30}$

d) $x - 8 \le -3$
$x \le -3 + 8$
$\mathbf{x \le 5}$

3 a) **6 is greater than x** or **x is less than 6.**
b) $\mathbf{x < 6}$
c) 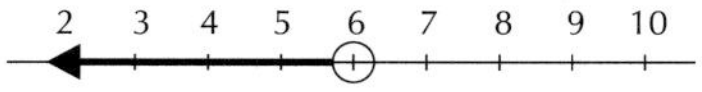

4 a) $\mathbf{x \le 12}$
If 12 if greater than or equal to x, then x is less than or equal to 12.
b) $\mathbf{x > 4}$ c) $\mathbf{x \ge 15}$ d) $\mathbf{x < 14}$

5 a) $18 < x + 2$
$18 - 2 < x$
$16 < x$
$\mathbf{x > 16}$

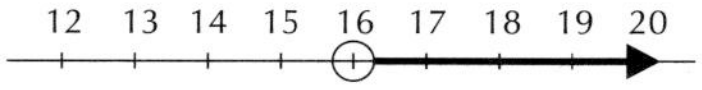

b) $12 \le x - 4$
$12 + 4 \le x$
$16 \le x$
$\mathbf{x \ge 16}$

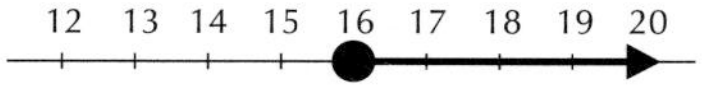

c) $1 > x - 17$
$1 + 17 > x$
$18 > x$
$\mathbf{x < 18}$

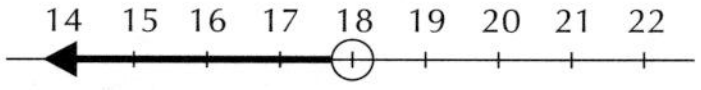

d) $31 \ge x + 30$
$31 - 30 \ge x$
$1 \ge x$
$\mathbf{x \le 1}$

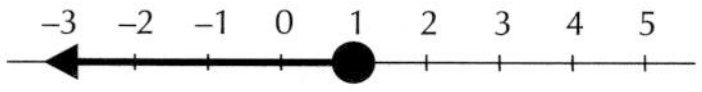

Page 153 Exercise 3

1 a) $3x \ge 9 \Rightarrow x \ge 9 \div 3 \Rightarrow \mathbf{x \ge 3}$
b) $5x < -25 \Rightarrow x < -25 \div 5 \Rightarrow \mathbf{x < -5}$
c) $2x > 8 \Rightarrow x > 8 \div 2 \Rightarrow \mathbf{x > 4}$
d) $7x \le 21 \Rightarrow x \le 21 \div 7 \Rightarrow \mathbf{x \le 3}$

2 a) $\frac{x}{2} \ge 3 \Rightarrow x \ge 3 \times 2 \Rightarrow \mathbf{x \ge 6}$
b) $\frac{x}{5} < 2 \Rightarrow x < 2 \times 5 \Rightarrow \mathbf{x < 10}$
c) $\frac{x}{5.5} < 1.2 \Rightarrow x < 1.2 \times 5.5 \Rightarrow \mathbf{x < 6.6}$
d) $\frac{x}{2.5} > -3.2 \Rightarrow x > -3.2 \times 2.5 \Rightarrow \mathbf{x > -8}$

3 a) $-4x < -16 \Rightarrow x > -16 \div (-4) \Rightarrow \mathbf{x > 4}$
Remember to swap the inequality sign when dividing or multiplying by a negative number.
b) $-9x > -72 \Rightarrow x < -72 \div (-9) \Rightarrow \mathbf{x < 8}$
c) $-11x \le 33 \Rightarrow x \ge 33 \div (-11) \Rightarrow \mathbf{x \ge -3}$
d) $-2x < 45 \Rightarrow x > 45 \div (-2) \Rightarrow \mathbf{x > -22.5}$

4 a) $-\frac{x}{3} < 8 \Rightarrow x > 8 \times (-3) \Rightarrow \mathbf{x > -24}$
b) $-\frac{x}{5} \le -4 \Rightarrow x \ge -4 \times (-5) \Rightarrow \mathbf{x \ge 20}$
c) $-\frac{x}{1.1} \ge 10 \Rightarrow x \le 10 \times (-1.1) \Rightarrow \mathbf{x \le -11}$
d) $-\frac{x}{0.2} > -2.1 \Rightarrow x < (-2.1) \times (-0.2)$
$\Rightarrow \mathbf{x < 0.42}$

Page 154 Exercise 4

1 a) $7x - 12 > 65$
$7x > 65 + 12 \Rightarrow 7x > 77$
$x > 77 \div 7 \Rightarrow \mathbf{x > 11}$
b) $2x + 16 \ge -8$
$2x \ge -8 - 16 \Rightarrow 2x \ge -24$
$x \ge -24 \div 2 \Rightarrow \mathbf{x \ge -12}$
c) $-8x - 4.2 < 12.6$
$-8x < 12.6 + 4.2 \Rightarrow -8x < 16.8$
$x > 16.8 \div (-8) \Rightarrow \mathbf{x > -2.1}$
d) $4x + 2.6 \le 28.6$
$4x \le 28.6 - 2.6 \Rightarrow 4x \le 26$
$x \le 26 \div 4 \Rightarrow \mathbf{x \le 6.5}$

2 a) $\frac{x+2}{3} < 1$
$x + 2 < 1 \times 3 \Rightarrow x + 2 < 3$
$x < 3 - 2 \Rightarrow \mathbf{x < 1}$
b) $\frac{x-8}{2} > 7$
$x - 8 > 7 \times 2 \Rightarrow x - 8 > 14$
$x > 14 + 8 \Rightarrow \mathbf{x > 22}$
c) $\frac{x+4}{5} \ge 2$
$x + 4 \ge 2 \times 5 \Rightarrow x + 4 \ge 10$
$x \ge 10 - 4 \Rightarrow \mathbf{x \ge 6}$
d) $-\frac{x-6}{4} \le 0.5$
$x - 6 \ge 0.5 \times -4 \Rightarrow x - 6 \ge -2$
$x \ge -2 + 6 \Rightarrow \mathbf{x \ge 4}$

3 a) $\frac{x}{4} - 2.5 \ge 1$
$\frac{x}{4} \ge 1 + 2.5 \Rightarrow \frac{x}{4} \ge 3.5$
$x \ge 3.5 \times 4 \Rightarrow \mathbf{x \ge 14}$
b) $\frac{x}{2} + 5.5 > 7$
$\frac{x}{2} > 7 - 5.5 \Rightarrow \frac{x}{2} > 1.5$
$x > 1.5 \times 2 \Rightarrow \mathbf{x > 3}$
c) $-\frac{x}{8} - 3.1 < -1$
$-\frac{x}{8} < -1 + 3.1 \Rightarrow -\frac{x}{8} < 2.1$
$x > 2.1 \times -8 \Rightarrow \mathbf{x > -16.8}$
d) $\frac{x}{3.2} + 1.3 \le 5$
$\frac{x}{3.2} \le 5 - 1.3 \Rightarrow \frac{x}{3.2} \le 3.7$
$x \le 3.7 \times 3.2 \Rightarrow \mathbf{x \le 11.84}$

Page 155 Exercise 5

1 a) **3, 4**
Each number must satisfy both parts of the inequality.
b) **–5, –4, –3, –2, –1, 0**
c) **6, 7, 8, 9, 10, 11, 12, 13**

2 a) 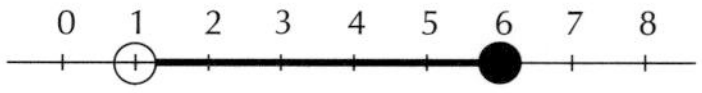

b) 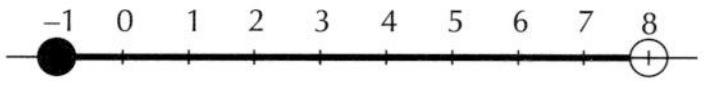

c) 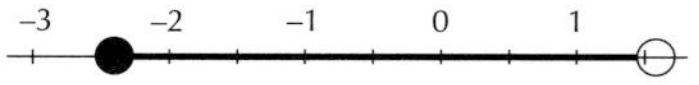

3 a) $7 < x + 3 \le 15$
1st inequality: $7 < x + 3$
$7 - 3 < x \Rightarrow 4 < x$
2nd inequality: $x + 3 \le 15$
$x \le 15 - 3 \Rightarrow x \le 12$
So combining the two solved inequalities gives $\mathbf{4 < x \le 12}$.

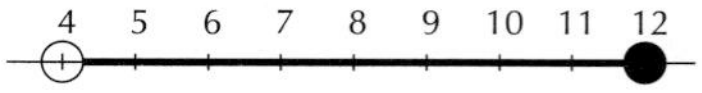

b) $2 \le x - 4 \le 12$
1st inequality: $2 \le x - 4$
$2 + 4 \le x \Rightarrow 6 \le x$
2nd inequality: $x - 4 \le 12$
$x \le 12 + 4 \Rightarrow x \le 16$
So combining the two solved inequalities gives $\mathbf{6 \le x \le 16}$.

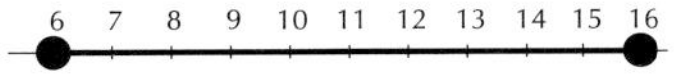

c) $-5.6 < x - 6.8 < 12.9$
1st inequality: $-5.6 < x - 6.8$
$-5.6 + 6.8 < x \Rightarrow 1.2 < x$
2nd inequality: $x - 6.8 < 12.9$
$x < 12.9 + 6.8 \Rightarrow x < 19.7$
So combining the two solved inequalities gives $\mathbf{1.2 < x < 19.7}$.

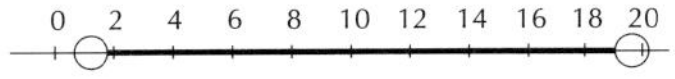

4 a) $32 < 2x \le 42$
1st inequality: $32 < 2x$
$32 \div 2 < x \Rightarrow 16 < x$
2nd inequality: $2x \le 42$
$x \le 42 \div 2 \Rightarrow x \le 21$
So combining the two solved inequalities gives $\mathbf{16 < x \le 21}$.

b) $-24 < 8x \le 40$
1st inequality: $-24 < 8x$
$-24 \div 8 < x \Rightarrow -3 < x$
2nd inequality: $8x \le 40$
$x \le 40 \div 8 \Rightarrow x \le 5$
So combining the two solved inequalities gives $\mathbf{-3 < x \le 5}$

c) $27 < 4.5x \le 72$
1st inequality: $27 < 4.5x$
$27 \div 4.5 < x \Rightarrow 6 < x$
2nd inequality: $4.5x \le 72$
$x \le 72 \div 4.5 \Rightarrow x \le 16$
So combining the two solved inequalities gives $\mathbf{6 < x \le 16}$

5 a) $17 < 6x + 5 < 29$
1st inequality: $17 < 6x + 5$
$17 - 5 < 6x \Rightarrow 12 < 6x$
$12 \div 6 < x \Rightarrow 2 < x$
2nd inequality: $6x + 5 < 29$
$6x < 29 - 5 \Rightarrow 6x < 24$
$x < 24 \div 6 \Rightarrow x < 4$
So combining the two solved inequalities gives $\mathbf{2 < x < 4}$.

b) $8 < 3x - 4 \le 26$
1st inequality: $8 < 3x - 4$
$8 + 4 < 3x \Rightarrow 12 < 3x$
$12 \div 3 < x \Rightarrow 4 < x$
2nd inequality: $3x - 4 \le 26$
$3x \le 26 + 4 \Rightarrow 3x \le 30$
$x \le 30 \div 3 \Rightarrow x \le 10$
So combining the two solved inequalities gives $\mathbf{4 < x \le 10}$

c) $-42 \le 7x + 7 < 91$
1st inequality: $-42 \le 7x + 7$
$-42 - 7 \le 7x \Rightarrow -49 \le 7x$
$-49 \div 7 \le x \Rightarrow -7 \le x$
2nd inequality: $7x + 7 < 91$
$7x < 91 - 7 \Rightarrow 7x < 84$
$x < 84 \div 7 \Rightarrow x < 12$
So combining the two solved inequalities gives $\mathbf{-7 \le x < 12}$

d) $9 \le 1.5x + 3 \le 9.9$
1st inequality: $9 \le 1.5x + 3$
$9 - 3 \le 1.5x \Rightarrow 6 \le 1.5x$
$6 \div 1.5 \le x \Rightarrow 4 \le x$
2nd inequality: $1.5x + 3 \le 9.9$
$1.5x \le 9.9 - 3$
$\Rightarrow 1.5x \le 6.9$
$x \le 6.9 \div 1.5 \Rightarrow x \le 4.6$
So combining the two solved inequalities gives $\mathbf{4 \le x \le 4.6}$

9.6 Simultaneous Equations

Page 156 Exercise 1

1 a) $x + 3y = 10$
$\underline{-(x + y = 6)}$
$2y = 4$
$\mathbf{y = 2}$
$x + 3 \times 2 = 10 \Rightarrow x = 10 - 6 \Rightarrow \mathbf{x = 4}$
Don't forget to put your values into the other equation to check them.

b) $x + 3y = 13$
$\underline{-(x - y = 5)}$
$4y = 8$
$\mathbf{y = 2}$
$x + 3 \times 2 = 13 \Rightarrow x = 13 - 6 \Rightarrow \mathbf{x = 7}$

c) $x + 2y = 6$
$\underline{-(x + y = 2)}$
$\mathbf{y = 4}$
$x + 2 \times 4 = 6 \Rightarrow x = 6 - 8 \Rightarrow \mathbf{x = -2}$

d) $2x - y = 7$
$\underline{+(4x + y = 23)}$
$6x = 30$
$\mathbf{x = 5}$
$2 \times 5 - y = 7 \Rightarrow 10 - 7 = y \Rightarrow \mathbf{y = 3}$

e) $3x - 2y = 16$
$\underline{+(2x + 2y = 14)}$
$5x = 30$
$\mathbf{x = 6}$
$3 \times 6 - 2y = 16 \Rightarrow 18 - 16 = 2y$
$\Rightarrow 2y = 2 \Rightarrow \mathbf{y = 1}$

f) $2x + 4y = 16$
$\underline{-(3x + 4y = 24)}$
$-x = -8$
$\mathbf{x = 8}$
$2 \times 8 + 4y = 16 \Rightarrow 4y = 16 - 16 = 0$
$\Rightarrow \mathbf{y = 0}$

Page 157 Exercise 2

1 a) (1) $3x + 2y = 16$,
(2) $2x + y = 9$
(1) $3x + 2y = 16$
(2) × 2: $\underline{-(4x + 2y = 18)}$
$-x = -2$
$\mathbf{x = 2}$
$2 \times 2 + y = 9 \Rightarrow 4 + y = 9 \Rightarrow y = 9 - 4$
$\Rightarrow \mathbf{y = 5}$

b) (1) $4x + 3y = 16$,
(2) $5x - y = 1$
(1) $4x + 3y = 16$
(2) × 3: $\underline{+(15x - 3y = 3)}$
$19x = 19$
$\mathbf{x = 1}$
$5 \times 1 - y = 1 \Rightarrow 5 - 1 = y \Rightarrow \mathbf{y = 4}$

c) (1) $5x - 3y = 12$,
(2) $2x - y = 5$
(2) × 3: $6x - 3y = 15$
(1) $\underline{-(5x - 3y = 12)}$
$\mathbf{x = 3}$
$2 \times 3 - y = 5 \Rightarrow 6 - 5 = y \Rightarrow \mathbf{y = 1}$

d) (1) $2e + 5f = 16$
(2) $3e - 2f = 5$
(1) × 2: $4e + 10f = 32$
(2) × 5: $\underline{+(15e - 10f = 25)}$
$19e = 57$
$\mathbf{e = 3}$
$3 \times 3 - 2f = 5 \Rightarrow 9 - 5 = 2f \Rightarrow 4 = 2f$
$\Rightarrow \mathbf{f = 2}$
There's more than one way to solve equations like this — here you could have started by multiplying (1) by 3 and (2) by 2 instead, and then subtracting the equations.

e) (1) $3d - 2e = 8$,
(2) $5d - 3e = 14$
(1) × 3: $(9d - 6e = 24)$
(2) × 2: $\underline{-(10d - 6e = 28)}$
$-d = -4$
$\mathbf{d = 4}$
$12 - 2e = 8 \Rightarrow 12 - 8 = 2e \Rightarrow 4 = 2e$
$\Rightarrow \mathbf{e = 2}$

f) (1) $5k + 3l = 4$,
(2) $3k + 2l = 3$
(2) × 3: $9k + 6l = 9$
(1) × 2: $\underline{-(10k + 6l = 8)}$
$-k = 1$
$\mathbf{k = -1}$
$3 \times (-1) + 2l = 3 \Rightarrow 2l = 3 + 3 = 6$
$\Rightarrow \mathbf{l = 3}$

Page 158 Exercise 3

1 If the sum of x and y is 58, then $x + y = 58$.
If the difference between x and y is 22, then $x - y = 22$.
Solving the two simultaneous equations gives:
$x + y = 58$
$\underline{+(x - y = 22)}$
$2x = 80$
$\mathbf{x = 40}$
$40 + y = 58 \Rightarrow \mathbf{y = 18}$

2 Let M be the price of a mountain bike and R be the cost of a road bike.
Then considering the total price of the bikes sold gives:
$M + 2R = 350$
Considering the difference in price gives:
$M = R + 50 \Rightarrow M - R = 50$.
Solve this pair of simultaneous equations:
$M + 2R = 350$
$\underline{-(M - R = 50)}$
$3R = 300$
$R = 100$
$M - 100 = 50 \Rightarrow M = 150$
So the price of a road bike is **£100** and the price of a mountain bike is **£150**.

3 Let d = price of sherbet dip, c = price of chocolate bar
Considering the total price last week gives:
$6d + c = 1.7$
Considering the total price the week before gives: $3d + 4c = 2.6$
(1) $6d + c = 1.7$
(2) $3d + 4c = 2.6$
(1) × 4: $24d + 4c = 6.8$
(2): $\underline{-(3d + 4c = 2.6)}$
$21d = 4.2$
$d = 0.2$
$6 \times 0.2 + c = 1.7$
$1.2 + c = 1.7 \Rightarrow c = 1.7 - 1.2 = 0.5$
So a sherbet dip costs **20p** and a chocolate bar costs **50p**.

9.7 Solving Quadratic Equations

Page 159 Exercise 1

1 a) $x^2 - 4 = 0 \Rightarrow x^2 = 4$
$\mathbf{x = -2, x = 2}$
There are always two solutions when taking the square root.

b) $x^2 - 1 = 0 \Rightarrow x^2 = 1$
$\mathbf{x = -1, x = 1}$
Using the same method for c)-h):
c) $\mathbf{x = -5, x = 5}$
d) $\mathbf{x = -8, x = 8}$
e) $\mathbf{x = -11, x = 11}$
f) $\mathbf{x = -10, x = 10}$
g) $\mathbf{x = -\sqrt{2}, x = \sqrt{2}}$
h) $\mathbf{x = -\sqrt{7}, x = \sqrt{7}}$

Page 160 Exercise 2

1 a) $x(x + 8) = 0 \Rightarrow x = 0$ or $x + 8 = 0$
So $\mathbf{x = 0}$ or $\mathbf{x = -8}$

b) $x(x - 5) = 0 \Rightarrow x = 0$ or $x - 5 = 0$
So $\mathbf{x = 0}$ or $\mathbf{x = 5}$

Using the same method for c)-f):

c) $\mathbf{x = 0}$ or $\mathbf{x = -6}$ **d)** $\mathbf{x = 0}$ or $\mathbf{x = 3}$

e) $\mathbf{x = 0}$ or $\mathbf{x = 4}$ **f)** $\mathbf{x = 0}$ or $\mathbf{x = -2}$

2 a) $x^2 + 6x = 0$
$x(x + 6) = 0 \Rightarrow x = 0$ or $x + 6 = 0$
So $\mathbf{x = 0}$ or $\mathbf{x = -6}$

b) $x^2 - 6x = 0$
$x(x - 6) = 0 \Rightarrow x = 0$ or $x - 6 = 0$
So $\mathbf{x = 0}$ or $\mathbf{x = 6}$

c) $x^2 - 24x = 0$
$x(x - 24) = 0 \Rightarrow x = 0$ or $x - 24 = 0$
So $\mathbf{x = 0}$ or $\mathbf{x = 24}$

d) $x^2 + 5x = 0$
$x(x + 5) = 0 \Rightarrow x = 0$ or $x + 5 = 0$
So $\mathbf{x = 0}$ or $\mathbf{x = -5}$

e) $x - x^2 = 0$
$x(1 - x) = 0 \Rightarrow x = 0$ or $1 - x = 0$
So $\mathbf{x = 0}$ or $\mathbf{x = 1}$

f) $12x - x^2 = 0$
$x(12 - x) = 0 \Rightarrow x = 0$ or $12 - x = 0$
So $\mathbf{x = 0}$ or $\mathbf{x = 12}$

Page 160 Exercise 3

1 a) $(x - 5)(x - 1) = 0 \Rightarrow x - 5 = 0$ or $x - 1 = 0$
So $\mathbf{x = 5}$ or $\mathbf{x = 1}$

b) $(x + 2)(x + 6) = 0 \Rightarrow x + 2 = 0$ or $x + 6 = 0$
So $\mathbf{x = -2}$ or $\mathbf{x = -6}$

c) $(x - 9)(x + 7) = 0 \Rightarrow x - 9 = 0$ or $x + 7 = 0$
So $\mathbf{x = 9}$ or $\mathbf{x = -7}$

2 a) (i) $\mathbf{(x + 2)(x + 5)}$ **(ii)** $\mathbf{(x + 4)(x + 5)}$
(iii) $\mathbf{(x + 4)(x + 9)}$ **(iv)** $\mathbf{(x + 6)(x - 4)}$

b) (i) $x^2 + 7x + 10 = 0 \Rightarrow (x + 2)(x + 5) = 0$
$\Rightarrow x + 2 = 0$
or $x + 5 = 0$
So $\mathbf{x = -2}$ or $\mathbf{x = -5}$

(ii) $x^2 + 9x + 20 = 0 \Rightarrow (x + 4)(x + 5) = 0$
$\Rightarrow x + 4 = 0$
or $x + 5 = 0$
So $\mathbf{x = -4}$ or $\mathbf{x = -5}$

Using the same method for (iii) and (iv):
(iii) $\mathbf{x = -4}$ or $\mathbf{x = -9}$
(iv) $\mathbf{x = -6}$ or $\mathbf{x = 4}$

3 a) $x^2 + 2x + 1 = 0 \Rightarrow (x + 1)(x + 1) = 0$
$\Rightarrow (x + 1)^2 = 0$
$\Rightarrow x + 1 = 0$
So $\mathbf{x = -1}$

b) $x^2 - 7x + 12 = 0 \Rightarrow (x - 3)(x - 4) = 0$
$\Rightarrow x - 3 = 0$ or $x - 4 = 0$
So $\mathbf{x = 3}$ or $\mathbf{x = 4}$

c) $x^2 + 4x + 4 = 0 \Rightarrow (x + 2)(x + 2) = 0$
$\Rightarrow (x + 2)^2 = 0$
$\Rightarrow x + 2 = 0$
So $\mathbf{x = -2}$

Using the same method for d)-f):
d) $\mathbf{x = 2}$ **e)** $\mathbf{x = -5}$ or $\mathbf{x = 3}$
f) $\mathbf{x = 7}$ or $\mathbf{x = -3}$

Page 161 Review Exercise

1 a) $\frac{x+8}{3} = 4$
$x + 8 = 4 \times 3 \Rightarrow x + 8 = 12$
$x = 12 - 8 \Rightarrow \mathbf{x = 4}$

b) $9 + 5x = 54$
$5x = 54 - 9 \Rightarrow 5x = 45$
$x = 45 \div 5 \Rightarrow \mathbf{x = 9}$

c) $72 = 4.5(8 + 2x)$
$72 \div 4.5 = 8 + 2x \Rightarrow 16 = 8 + 2x$
$16 - 8 = 2x \Rightarrow 8 = 2x$
$x = 8 \div 2 \Rightarrow \mathbf{x = 4}$

d) $7(x - 3) = 3(x - 6)$
$7x - 21 = 3x - 18$
$7x - 3x - 21 = -18 \Rightarrow 4x - 21 = -18$
$4x = -18 + 21 \Rightarrow 4x = 3$
$\mathbf{x = \frac{3}{4}}$ or **0.75**

2 Four hours at £x an hour cost £$4x$.
Adding the call-out charge gives £$4x + 35$.
So $4x + 35 = 170$.
$4x = 170 - 35 \Rightarrow 4x = 135$
$x = 135 \div 4 \Rightarrow x = 33.75$
So the electrician charges **£33.75** per hour.

3 a) Sum of angles in a triangle = 180°, so:
$(3x + 10°) + (x + 10°) + (x + 10°) = 180°$
$3x + x + x + 10° + 10° + 10° = 180°$
$\Rightarrow \mathbf{5x + 30° = 180°}$

b) $5x + 30° = 180°$
$5x = 180° - 30° = 150°$
$x = 150° \div 5 \Rightarrow x = 30°$
Bottom left angle = 30° + 10° = **40°**
Bottom right angle = 30° + 10° = **40°**
Top angle = 3 × 30° + 10° = **100°**
Check these angles work for a triangle:
40° + 40° + 100° = 180°.

4 Expanding the left-hand side gives
$x(x - 1) + 2(x - 3) = x^2 - x + 2x - 6 = \mathbf{x^2 + x - 6}$
Expanding the right-hand side gives
$(x + 3)(x - 2) = x^2 - 2x + 3x - 6 = \mathbf{x^2 + x - 6}$
The two sides are the same
so the identity is **true.**

5 E.g. $3^2 + 4^2 = 9 + 16 = 25 = 5^2$
so the statement is **false**.

6 Take an even number and an odd number — $2a$ and $(2b + 1)$. Their sum is
$2a + 2b + 1 = 2(a + b) + 1$
$= \mathbf{2n + 1}$, where $n = a + b$
So the sum of an even number and an odd number is **odd**.

7 a) $x + 2.7 \geq 6.2$
$x \geq 6.2 - 2.7$
$\mathbf{x \geq 3.5}$

b) $-\frac{x}{9} < 7$
$x > 7 \times -9$
$\mathbf{x > -63}$

c) $-5 \leq 12x + 7 < 43$
1st inequality: $-5 \leq 12x + 7$
$-5 - 7 \leq 12x \Rightarrow -12 \leq 12x$
$-12 \div 12 \leq x \Rightarrow -1 \leq x$
2nd inequality: $12x + 7 < 43$
$12x < 43 - 7 \Rightarrow 12x < 36$
$x < 36 \div 12 \Rightarrow x < 3$
So combining the two solved inequalities gives $\mathbf{-1 \leq x < 3}$.

8 a) (1) $m - 3n = 7$,
(2) $5m + 4n = -3$

(1) × 5: $5m - 15n = 35$
(2) : $-\underline{(5m + 4n = -3)}$
$-19n = 38$
$n = (38 \div -19)$
$\mathbf{n = -2}$
$m - 3 \times (-2) = 7 \Rightarrow m + 6 = 7 \Rightarrow \mathbf{m = 1}$

b) (1) $4u + 7v = 15$,
(2) $5u - 2v = 8$

(1) × 2: $8u + 14v = 30$
(2) × 7: $+\underline{(35u - 14v = 56)}$
$43u = 86$
$\mathbf{u = 2}$
$5 \times 2 - 2v = 8 \Rightarrow 10 - 8 = 2v \Rightarrow 2 = 2v$
$\Rightarrow \mathbf{v = 1}$
You could also have started by multiplying (1) by 5 and (2) by 4.

9 a) $x^2 - 36 = 0$
$x^2 = 36$
$\mathbf{x = -6, x = 6}$

b) $x^2 - 81 = 0$
$x^2 = 81$
$\mathbf{x = -9, x = 9}$

c) $x^2 - 3 = 0$
$x^2 = 3$
$\mathbf{x = -\sqrt{3}, x = \sqrt{3}}$

10 a) $x^2 - 4x = 0 \Rightarrow x(x - 4) = 0$
$\Rightarrow x = 0$ or $x - 4 = 0$
So $\mathbf{x = 0}$ or $\mathbf{x = 4}$

b) $x^2 + 9x + 18 = 0 \Rightarrow (x + 3)(x + 6) = 0$
$\Rightarrow x + 3 = 0$ or $x + 6 = 0$
So $\mathbf{x = -3}$ or $\mathbf{x = -6}$

c) $x^2 + 9x - 22 = 0 \Rightarrow (x + 11)(x - 2) = 0$
$\Rightarrow x + 11 = 0$ or $x - 2 = 0$
So $\mathbf{x = -11}$ or $\mathbf{x = 2}$

Page 162 Exam-Style Questions

1 a) $2x + 5 = 19$
$2x = 19 - 5 \Rightarrow 2x = 14$
$x = 14 \div 2 \Rightarrow \mathbf{x = 7}$ *[1 mark]*

b) $3(2x - 1) = 2(x + 4)$
$6x - 3 = 2x + 8$ *[1 mark]*
$6x - 2x - 3 = 8 \Rightarrow 4x - 3 = 8$
$4x = 8 + 3 \Rightarrow 4x = 11$ *[1 mark]*
$x = 11 \div 4 \Rightarrow \mathbf{x = 2.75}$ *[1 mark]*

2 a) **14** *[1 mark]*

b) Minimum value of $x = 7$, so minimum value of x^2 is $7^2 = \mathbf{49}$ *[1 mark]*

3 Bethany is 3 times as old as Anish, so Bethany's age = $3x$.
Cate is two years younger than Anish, so Cate's age = $x - 2$.
[1 mark for both expressions]
The sum of their ages is $x + 3x + (x - 2)$
$= 5x - 2$.
So $5x - 2 = 58$ *[1 mark]*
$\Rightarrow 5x = 58 + 2 = 60 \Rightarrow \mathbf{x = 12}$ *[1 mark]*

4 Let x be the cost of a chew and let y be the cost of a lolly.
Bill buys 2 chews ($2x$) and 3 lollies ($3y$) and pays 84p.
So $2x + 3y = 84$
Aisha buys 3 chews ($3x$) and 1 lolly (y) and pays 63p.
So $3x + y = 63$
Solving:
(1) $2x + 3y = 84$
(2) $3x + y = 63$
(2) × 3: $9x + 3y = 189$
(1) $-\underline{(2x + 3y = 84)}$
$7x = 105$
$x = 105 \div 7 \Rightarrow x = 15$
$3 \times 15 + y = 63$
$45 + y = 63$
$y = 63 - 45 \Rightarrow y = 18$
For 7 chews and 5 lollies:
$7 \times (15) + 5 \times (18) = 195$p or £1.95
[4 marks available — 1 mark for forming both equations, 1 mark for combining the two equations to eliminate a variable, 1 mark for correctly finding both variables, 1 mark for the correct answer]

5 a) The area of the rectangle is given by
$x(x + 3) = x^2 + 3x$ *[1 mark]*
From the question the area of the rectangle is 10 cm², so $x^2 + 3x = 10$.
Rearranging gives $\mathbf{x^2 + 3x - 10 = 0}$ *[1 mark]*

b) $x^2 + 3x - 10 = 0 \Rightarrow (x + 5)(x - 2) = 0$
[1 mark]
So $x = -5$ or $x = 2$ *[1 mark]*
A distance can't be negative, so $x = 2$.
Length = $x + 3 = 2 + 3 =$ **5 cm** *[1 mark]*
Width = $x =$ **2 cm** *[1 mark]*

Section 10 — Formulas and Functions

10.1 Formulas

Page 164 Exercise 1

1 a) "I have c carrots" — start with the letter c. "Su has 6 more carrots than me" — so add 6. The expression is $\boldsymbol{c+6}$.

b) "Daisy has p plants" — start with the letter p. "Iris has 8 fewer plants" — so subtract 8. The expression is $\boldsymbol{p-8}$

2 a) "Claudia has f films" — start with the letter f. "Barry owns twice as many..." — so multiply by 2. $2 \times f = $ **$2f$ films**.

b) The total is $f + 2f = $ **$3f$ films**.

c) Subtract 3 for both Claudia and Barry from $3f$. $3f - 3 - 3 = $ **$3f - 6$ films**.

3 Start with b bulbs. Multiply by 3 to get $3b$. Then add 5 to get: **$3b + 5$ flowers**.

4 Alf earns £8 per hour, so multiply h (number of hours) by 8 to get $8h$. Then add the £18 that Alf started with to get: $\boldsymbol{8h+18}$.

5 a) Square the smaller number, y, to get y^2. Cube the larger number, x, to get x^3. Now subtract x^3 from y^2 to get the expression: $\boldsymbol{y^2 - x^3}$.

b) Add the numbers to get $x + y$, then take the square root to get the expression: $\boldsymbol{\sqrt{x+y}}$.
Make sure you put the square root sign over all of 'x + y'.

Page 165 Exercise 2

1 The subject is cost, C, so the formula begins '$C =$'. It costs £3 per hour, so multiply h (number of hours) by 3 to get the formula $\boldsymbol{C = 3h}$.

2 The subject is time, T, so the formula begins '$T =$'. It takes 2 minutes per km, so multiply k (number of km) by 2 to get the formula $\boldsymbol{T = 2k}$.

3 The subject is amount paid, P, so the formula begins '$P =$'. Tom is paid w pounds per hour and works for 8 hours, so multiply w by 8 to get the formula $\boldsymbol{P = 8w}$.

4 The subject is Ellie's distance, d, so the formula begins '$d =$'. Kojo runs r km and Ellie runs 5 km less, so subtract 5 from r to get the formula $\boldsymbol{d = r - 5}$.

5 The subject is total cost, C, so the formula begins '$C =$'. It's £5.50 per hour so multiply n (number of hours) by 5.5 to get $5.5n$. Add on the fixed cost, £F, to get the formula $\boldsymbol{C = F + 5.5n}$.

6 The subject is time, t, so the formula begins '$t =$'. It takes 50 minutes per kg and the goose weighs n kg, so multiply 50 by n to get $50n$. Add on the fixed time of 25 minutes to get the formula $\boldsymbol{t = 50n + 25}$.

7 The subject is total cost, C, so the formula begins '$C =$'. It costs £p per tree and t trees are cut down, so multiply p by t to get pt. Add on the fixed cost of £25 to get the formula $\boldsymbol{C = pt + 25}$.

8 The subject is cost, C, so the formula begins '$C =$'. It costs 22p = £0.22 per mile and m miles are covered, so multiply 0.22 by m to get $0.22m$. Add on the fixed cost of £10 to get the formula $\boldsymbol{C = 0.22m + 10}$.
Watch out for where the units need converting. The question asks for the answer in pounds, so convert pence to pounds here.

9 The subject is cost, C, so the formula begins '$C =$'. It costs 80p per minute, but C is a cost in pounds and h is a number of hours, so work out the cost in pounds per hour. 80p = £0.80 and 1 hour = 60 minutes, so it costs £0.80 × 60 = £48 per hour. So the cost of h hours is $48 \times h = 48h$. Add on the fixed cost of £125 to get the formula $\boldsymbol{C = 48h + 125}$.

Page 167 Exercise 3

1 a) $x = 7$, so $y = x + 4 = 7 + 4 = \mathbf{11}$

b) $x = 7$, so $y = x - 3 = 7 - 3 = \mathbf{4}$

Using the same method for c)-d):

c) 5 **d) 42**

2 a) $m = -3$, so $y = m - 8 = -3 - 8 = \mathbf{-11}$

b) $m = -3$, so $y = 3m^2 = 3 \times (-3)^2 = 3 \times 9 = \mathbf{27}$

Using the same method for c)-d):

c) –7 **d) –4**

3 a) $m = 4$ and $n = 3$, so $p = mn = 4 \times 3 = \mathbf{12}$

b) $m = 4$, so $p = m^2 = 4^2 = \mathbf{16}$

c) $m = 4$ and $n = 3$, so $p = m - n^2 = 4 - 3^2 = 4 - 9 = \mathbf{-5}$

d) $m = 4$ and $n = 3$, so $p = \frac{3m}{n} = \frac{\cancel{3} \times 4}{\cancel{3}} = \mathbf{4}$

4 a) $x = -4$, so $z = x + 2 = -4 + 2 = \mathbf{-2}$

b) $y = -3$, so $z = y - 1 = -3 - 1 = \mathbf{-4}$

c) $x = -4$ and $y = -3$, so
$z = -x + 2y = -(-4) + 2 \times (-3) = 4 - 6 = \mathbf{-2}$

d) $x = -4$ and $y = -3$, so
$z = 6x - y = 6 \times (-4) + 3 = -24 + 3 = \mathbf{-21}$

5 a) $n = 10$, so $S = \frac{1}{2}n(n+1)$
$= \frac{1}{2} \times 10 \times (10 + 1) = 5 \times 11 = \mathbf{55}$

b) $n = 100$, so $S = \frac{1}{2}n(n+1)$
$= \frac{1}{2} \times 100 \times (100 + 1)$
$= 50 \times 101 = \mathbf{5050}$

Using the same method for c)-d):

c) 500 500 **d) 12 502 500**

6 a) $u = 7$, $a = 2$ and $t = 4$, so:
$s = ut + \frac{1}{2}at^2 = 7 \times 4 + \frac{1}{2} \times 2 \times 4^2$
$= 28 + 16 = \mathbf{44}$

b) $u = 24$, $a = 11$ and $t = 13$, so:
$s = ut + \frac{1}{2}at^2 = 24 \times 13 + \frac{1}{2} \times 11 \times 13^2$
$= 312 + 929.5 = \mathbf{1241.5}$

Using the same method for c)-d):

c) –864.2602 **d) 780.3752**

Page 168 Exercise 4

1 a) $b = 4$ and $h = 6$, so $A = \frac{1}{2}bh$
$= \frac{1}{2} \times 4 \times 6 = \mathbf{12\ m^2}$
Remember to give your answer with units — here it's m².

b) $b = 2$ and $h = 3$, so $A = \frac{1}{2}bh$
$= \frac{1}{2} \times 2 \times 3 = \mathbf{3\ m^2}$

c) $b = 0.4$ and $h = 1.8$, so $A = \frac{1}{2}bh$
$= \frac{1}{2} \times 0.4 \times 1.8 = \mathbf{0.36\ m^2}$

2 a) The runner travels 800 metres, so $d = 800$, and the time taken is 110 seconds, so $t = 110$.
$s = \frac{d}{t} = \frac{800}{110} = $ **7.27 m/s (to 3 s.f.)**

b) The cheetah travels 400 metres, so $d = 400$, and the time taken is 14 seconds, so $t = 14$.
$s = \frac{d}{t} = \frac{400}{14} = $ **28.6 m/s (to 3 s.f.)**

3 $a = 3$ and $b = 4$, so:
$c = \sqrt{a^2 + b^2} = \sqrt{3^2 + 4^2} = \sqrt{9 + 16} = \sqrt{25} = $ **5 cm**
You could have used a = 4 and b = 3 instead.

4 a) $r = 5$ cm, so:
$A = \pi r^2 = \pi \times 5^2 = \pi \times 25 = 78.539...$
$=$ **78.5 cm² (to 1 d.p.)**

b) $r = 3.5$ cm, so:
$A = \pi \times 3.5^2 = \pi \times 12.25 = 38.484...$
$=$ **38.5 cm² (to 1 d.p.)**

Using the same method for c)-d):

c) 387.1 cm² (to 1 d.p.)

d) 128.7 cm² (to 1 d.p.)

5 a) $l = 1$, so:
$T = 2\pi\sqrt{\frac{l}{10}} = 2 \times \pi \times \sqrt{\frac{1}{10}} = 1.986...$
$=$ **2.0 s (to 1 d.p.)**

b) $l = 0.5$, so:
$T = 2\pi\sqrt{\frac{l}{10}} = 2 \times \pi \times \sqrt{\frac{0.5}{10}} = 1.404...$
$=$ **1.4 s (to 1 d.p.)**

c) $l = 16$, so:
$T = 2\pi\sqrt{\frac{l}{10}} = 2 \times \pi \times \sqrt{\frac{16}{10}} = 7.947...$
$=$ **7.9 s (to 1 d.p.)**

Page 170 Exercise 5

1 a) The opposite of + 2 is – 2, so subtract 2 from both sides:
$y = x + 2$, so $y - 2 = x$. Rewrite as $\boldsymbol{x = y - 2}$.

b) The opposite of – 5 is + 5, so add 5 to both sides: $b = x - 5$, so $b + 5 = x$.
Rewrite as $\boldsymbol{x = b + 5}$.

Using the same method for c):

c) $\boldsymbol{x = z - 7}$

2 a) The opposite of × 4 is ÷ 4, so divide both sides by 4:
$z = 4x$, so $\frac{z}{4} = x$. Rewrite as $\boldsymbol{x = \frac{z}{4}}$.

b) The opposite of × 17 is ÷ 17, so divide both sides by 17:
$p = 17x$, so $\frac{p}{17} = x$. Rewrite as $\boldsymbol{x = \frac{p}{17}}$.

Using the same method for c):

c) $\boldsymbol{x = \frac{r}{4.2}}$

3 a) The opposite of ÷ 8 is × 8, so multiply both sides by 8:
$y = \frac{x}{8}$, so $8y = x$. Rewrite as $\boldsymbol{x = 8y}$.

b) The opposite of ÷ 17 is × 17, so multiply both sides by 17:
$z = \frac{x}{17}$, so $17z = x$. Rewrite as $\boldsymbol{x = 17z}$.

Using the same method for c):

c) $\boldsymbol{x = 8.6t}$

4 a) The opposite of × 2 is ÷ 2, so divide both sides by 2:
$abc = 2x$, so $\frac{abc}{2} = x$. Rewrite as $\boldsymbol{x = \frac{abc}{2}}$.

b) The opposite of × y is ÷ y, so divide both sides by y:
$t = xy$, so $\frac{t}{y} = x$. Rewrite as $\boldsymbol{x = \frac{t}{y}}$.

Using the same method for c):

c) $\boldsymbol{x = \frac{uv + y}{4.2}}$

5 a) The opposite of $\times \frac{4}{5}$ is $\div \frac{4}{5}$.
$\div \frac{4}{5}$ is the same as $\times \frac{5}{4}$.
So multiply both sides by $\frac{5}{4}$:
$m = \frac{4}{5}s$, so $\frac{5}{4}m = s$. Rewrite as $\boldsymbol{s = \frac{5}{4}m}$.

b) The opposite of ×(–16) is ÷(–16), so divide both sides by –16:
$r = -16s$, so $\frac{r}{-16} = s$. Rewrite as $\boldsymbol{s = -\frac{r}{16}}$.

c) The opposite of ×(–14.2) is ÷(–14.2), so divide both sides by –14.2:
$p = -14.2s$, so $\frac{p}{-14.2} = s$.
Rewrite as $\boldsymbol{s = -\frac{p}{14.2}}$.

d) The opposite of $\times \frac{5}{4}$ is $\div \frac{5}{4}$. $\div \frac{5}{4}$ is the same as $\times \frac{4}{5}$.
So multiply both sides by $\frac{4}{5}$:
$a = \frac{5}{4}s$, so $\frac{4}{5}a = s$. Rewrite as $\boldsymbol{s = \frac{4}{5}a}$.

e) s is under a square root, so square both sides:
$b = \sqrt{s}$, so $b^2 = s$. Rewrite as $\boldsymbol{s = b^2}$.

f) s is squared, so take square roots:
$c = s^2$, so $\pm\sqrt{c} = s$. Rewrite as $\boldsymbol{s = \pm\sqrt{c}}$.

Page 171 Exercise 6

1 a) $y = 5x + 3$
Subtract 3 from both sides: $y - 3 = 5x$
Divide both sides by 5: $\frac{y-3}{5} = x$,
so $\boldsymbol{x = \frac{y-3}{5}}$

b) $z = 8x - 2$
Add 2 to both sides: $z + 2 = 8x$
Divide both sides by 8: $\frac{z+2}{8} = x$,
so $\boldsymbol{x = \frac{z+2}{8}}$

c) $p = 15x + 18$
Subtract 18 from both sides: $p - 18 = 15x$
Divide both sides by 15: $\frac{p-18}{15} = x$,
so $\boldsymbol{x = \frac{p-18}{15}}$

2 a) $z = \frac{y+4}{3}$
Multiply both sides by 3: $3z = y + 4$
Subtract 4 from both sides: $3z - 4 = y$,
so $\boldsymbol{y = 3z - 4}$

b) $x = \frac{7+y}{4}$
Multiply both sides by 4: $4x = 7 + y$
Subtract 7 from both sides: $4x - 7 = y$,
so $\boldsymbol{y = 4x - 7}$

c) $s = \frac{y-2}{9}$
Multiply both sides by 9: $9s = y - 2$
Add 2 to both sides: $9s + 2 = y$,
so $\boldsymbol{y = 9s + 2}$

3 a) $u = 4(x - 2)$
Divide both sides by 4: $\frac{u}{4} = x - 2$
Add 2 to both sides: $\frac{u}{4} + 2 = x$,
so $\boldsymbol{x = \frac{u}{4} + 2}$ or $\boldsymbol{x = \frac{u+8}{4}}$

b) $v = 8(x + 4)$
Divide both sides by 8: $\frac{v}{8} = x + 4$
Subtract 4 from both sides: $\frac{v}{8} - 4 = x$,
so $\boldsymbol{x = \frac{v}{8} - 4}$ or $\boldsymbol{x = \frac{v-32}{8}}$

c) $w = 3(x - 4)$
Divide both sides by 3: $\frac{w}{3} = x - 4$
Add 4 to both sides: $\frac{w}{3} + 4 = x$,
so $\boldsymbol{x = \frac{w}{3} + 4}$ or $\boldsymbol{x = \frac{w+12}{3}}$

4 a) $p + 3 = 4y - 2$
Add 2 to both sides: $p + 3 + 2 = 4y$,
so $p + 5 = 4y$
Divide both sides by 4: $\frac{p+5}{4} = y$,
so $\boldsymbol{y = \frac{p+5}{4}}$

b) $q + 7 = 9y + 11$
Subtract 11 from both sides:
$q + 7 - 11 = 9y$, so $q - 4 = 9y$
Divide both sides by 9: $\frac{q-4}{9} = y$,
so $\boldsymbol{y = \frac{q-4}{9}}$

c) $r - 5 = 21y - 9$
Add 9 to both sides: $r - 5 + 9 = 21y$,
so $r + 4 = 21y$
Divide both sides by 21: $\frac{r+4}{21} = y$,
so $\boldsymbol{y = \frac{r+4}{21}}$

5 a) $A = 21.5d^2$
Divide both sides by 21.5: $\frac{A}{21.5} = d^2$
d is squared, so take the square root of each side:
$\sqrt{\frac{A}{21.5}} = d$, so $\boldsymbol{d = \sqrt{\frac{A}{21.5}}}$
The positive square root is used because d is a length.

b) (i) $A = 344$, so $d = \sqrt{\frac{344}{21.5}} = \sqrt{16} =$ **4 cm**

(ii) $A = 134.375$, so
$d = \sqrt{\frac{134.375}{21.5}} = \sqrt{6.25} =$ **2.5 cm**

6 a) $P = 2(2x + y)$
Divide both sides by 2: $\frac{P}{2} = 2x + y$
Subtract y from both sides: $\frac{P}{2} - y = 2x$
Divide both sides by 2: $\frac{P}{4} - \frac{y}{2} = x$,
so $\boldsymbol{x = \frac{P}{4} - \frac{y}{2}}$ or $\boldsymbol{x = \frac{P-2y}{4}}$

b) (i) $P = 14$ and $y = 3$,
so $x = \frac{P}{4} - \frac{y}{2} = \frac{14}{4} - \frac{3}{2} = \mathbf{2}$

(ii) $P = 32$ and $y = 4$,
so $x = \frac{P}{4} - \frac{y}{2} = \frac{32}{4} - \frac{4}{2} = \mathbf{6}$

7 a) $x + y = 6x$
Subtract x from both sides: $y = 5x$
Divide both sides by 5: $\frac{y}{5} = x$, so $\boldsymbol{x = \frac{y}{5}}$

b) $y = \sqrt{x} + 2$
Subtract 2 from both sides: $y - 2 = \sqrt{x}$
x is under a square root, so square both sides:
$(y - 2)^2 = x$, so $\boldsymbol{x = (y - 2)^2}$

c) $\frac{3 + xy + x}{x + 1} = y$
Multiply both sides by $(x + 1)$:
$3 + xy + x = y(x + 1)$
Multiply out the brackets: $3 + xy + x$
$= xy + y$
Subtract xy from both sides: $3 + x = y$
Subtract 3 from both sides: $\boldsymbol{x = y - 3}$

10.2 Functions

Page 173 Exercise 1

1 a) $x = 20$, divide by 5 to get $20 \div 5 = 4$.
Add 7 to get $4 + 7 = 11$. So $\boldsymbol{y = 11}$.
Using the same method for b)-d):
b) $\boldsymbol{y = 14}$ c) $\boldsymbol{y = 16}$ d) $\boldsymbol{y = 5}$

2 a) $x = 11$, subtract 3 to get $11 - 3 = 8$.
Multiply by 6 to get $8 \times 6 = 48$. So $\boldsymbol{y = 48}$.

b) Work backwards using opposite operations.
$y = 72$, divide by 6 to get $72 \div 6 = 12$.
Add 3 to get $12 + 3 = 15$. So $\boldsymbol{x = 15}$.

3 a) $x = -1$, multiply by 7 to get $-1 \times 7 = -7$.
Subtract 2 to get $-7 - 2 = -9$. So $\boldsymbol{y = -9}$.

b) Work backwards using opposite operations.
$y = 19$, add 2 to get $19 + 2 = 21$.
Divide by 7 to get $21 \div 7 = 3$. So $\boldsymbol{x = 3}$.

4 a) x is multiplied by 4, then 1 is added to get y. So the function machine is:
$\boldsymbol{x}$ → [**× 4**] → [**+ 1**] → $\boldsymbol{y}$

b) Work backwards using opposite operations.
$y = 17$, subtract 1 to get $17 - 1 = 16$.
Divide by 4 to get $16 \div 4 = 4$. So $\boldsymbol{x = 4}$.

5 The output is equal to the input when $\boldsymbol{x = 2}$,
as $2 \times 6 = 12$ and $12 - 10 = 2 = x$

Page 174 Review Exercise

1 a) The subject is c, the number of free minutes Chloe gets, so the formula begins '$c =$'.
Wassim gets w free minutes and Chloe gets 45 fewer minutes, so subtract 45 from w.
The formula is $\boldsymbol{c = w - 45}$.

b) Substitute $w = 125$ into $c = w - 45$ to find $c = 125 - 45 =$ **80**.

2 a) $m = 5h + 1$
Subtract 1 from both sides: $m - 1 = 5h$
Divide both sides by 5: $\frac{m-1}{5} = h$,
so $\boldsymbol{h = \frac{m-1}{5}}$

b) Substitute $m = 36$ into $h = \frac{m-1}{5}$ to find
$h = \frac{36-1}{5} = \mathbf{7}$.

3 a) Substitute $x = 30°$ into $y = \frac{180° - x}{2}$ to find
$y = \frac{180° - 30°}{2} = \mathbf{75°}$.

b) $y = \frac{180° - x}{2}$
Multiply both sides by 2: $2y = 180° - x$
Add x to both sides: $2y + x = 180°$
Subtract $2y$ from both sides: $\boldsymbol{x = 180° - 2y}$

4 a) $n = 275$ units, multiply by 0.06 to get
$275 \times 0.06 = 16.5$.
Next add 7.5 to get $16.5 + 7.5 = 24$.
So $C =$ **£24**.

b) Work backwards using opposite operations.
$C = £40.50$, subtract 7.5 to get
$40.5 - 7.5 = 33$.
Next divide by 0.06 to get $33 \div 0.06 = 550$.
So **550 units** were used.

5 a) Use the letter b for the number of black tiles. '3 white tiles for every black tile'
— so multiply b by 3.
'plus an extra 50 white tiles'
— so add 50 to get w.
So the completed function machine is:
b → [**× 3**] → [+ 50] → w

b) Put $b = 200$ into the machine.
Multiply by 3 to get $200 \times 3 = 600$,
add 50 to get $600 + 50 = 650$.
So **650 white tiles** are used.

c) Work backwards using opposite operations.
$t = 530$ total tiles, subtract 50 to get
$530 - 50 = 480$.
Next divide by 4 to get $b = 480 \div 4 = 120$.
So **120 black tiles** are used.

Page 175 Exam-Style Questions

1 $y = a + \frac{b}{x}$
Subtract a from both sides: $y - a = \frac{b}{x}$ *[1 mark]*
Multiply both sides by x: $xy - xa = b$,
so $\boldsymbol{b = xy - xa}$ *[1 mark]*

2 a) $17 - 4 = 13$, $13 \div 25 =$ **0.52** *[1 mark]*

b) $5 \times 25 = 125$ *[1 mark]*,
$125 + 4 =$ **129** *[1 mark]*

c) $3x \times 25 + 4 = \mathbf{75x + 4}$ *[1 mark]*

3 a) Multiply n (number of people) by £1.25 (cost per person), then add on the fixed fee of £30. $\boldsymbol{C = 1.25n + 30}$ *[1 mark]*

b) Substitute $n = 32$ into $C = 1.25n + 30$.
$C = 1.25 \times 32 + 30 =$ **£70** *[1 mark]*

c) $C = 1.25n + 30$, so $C - 30 = 1.25n$ *[1 mark]*
and $\frac{C-30}{1.25} = n$, so $\boldsymbol{n = \frac{C-30}{1.25}}$ *[1 mark]*

d) Substitute $C = £80$ into $n = \frac{C-30}{1.25}$ to get
$n = \frac{80-30}{1.25} =$ **40 people** *[1 mark]*

4 Work backwards using opposite operations:
$10 + 4 = 14$ *[1 mark]*, so the first step needs to turn 7 into 14. $14 = 7 \times 2$, so the missing operation is **× 2** *[1 mark]*.
You could also do ÷ 0.5 or + 7.

Section 11 — Sequences

11.1 Term to Term Rules

Page 177 Exercise 1

1 1st term: **5**
2nd term: 5 + 4 = **9**
3rd term: 9 + 4 = **13**
4th term: 13 + 4 = **17**
5th term: 17 + 4 = **21**

2 1st term: **2**
2nd term: 2 × 2 = **4**
3rd term: 4 × 2 = **8**
4th term: 8 × 2 = **16**
5th term: 16 × 2 = **32**

3 **a)** 1st term: **100**
2nd term: 100 – 6 = **94**
3rd term: 94 – 6 = **88**
4th term: 88 – 6 = **82**
5th term: 82 – 6 = **76**

b) 1st term: **40**
2nd term: 40 ÷ 2 = **20**
3rd term: 20 ÷ 2 = **10**
4th term: 10 ÷ 2 = **5**
5th term: 5 ÷ 2 = **2.5**

c) 1st term: **11**
2nd term: 11 × –2 = **–22**
3rd term: –22 × –2 = **44**
4th term: 44 × –2 = **–88**
5th term: –88 × –2 = **176**

4 **a)** The difference between the terms is:
6 – 3 = 3, 9 – 6 = 3,
so the rule is '**add 3 to the previous term**'.

b) 5th term: 12 + 3 = **15**
6th term: 15 + 3 = **18**
7th term: 18 + 3 = **21**

5 **a) (i)** The difference between the terms is 5 – 3 = 2, 7 – 5 = 2, so the rule is '**add 2 to the previous term**'

(ii) 5th term: 9 + 2 = **11**
6th term: 11 + 2 = **13**
7th term: 13 + 2 = **15**

b) (i) The ratio between the terms is 12 ÷ 4 = 3, 36 ÷ 12 = 3, so the rule is '**multiply the previous term by 3**'

(ii) 5th term: 108 × 3 = **324**
6th term: 324 × 3 = **972**
7th term: 972 × 3 = **2916**

Using the same method for c)-h):

c) (i) The rule is '**divide the previous term by 2**'.
(ii) The next three terms are **1**, **0.5**, **0.25**.

d) (i) The rule is '**subtract 2 from the previous term**'.
(ii) The next three terms are **–3**, **–5**, **–7**.

e) (i) The rule is '**add 0.5 to the previous term**'.
(ii) The next three terms are **3**, **3.5**, **4**.

f) (i) The rule is '**multiply the previous term by 3**'.
(ii) The next three terms are **–81**, **–243**, **–729**.

g) (i) The rule is '**divide the previous term by 10**'.
(ii) The next three terms are **0.1**, **0.01**, **0.001**.

h) (i) The rule is '**multiply the previous term by –2**'.
(ii) The next three terms are **16**, **–32**, **64**.

6 **a)** The difference between the terms is 10 – 4 = 6, 16 – 10 = 6, so the rule is '**add 6 to the previous term**'.

b) (i) The 5th term is 22 + 6 = **28**
(ii) The 6th term is 28 + 6 = **34**
(iii) The 8th term is 34 + 6 + 6
= 40 + 6 = **46**

7 **a)** The terms are decreasing by 4 each time (5 – 9 = 4, – 7 – (–3) = –4) so the missing number is 5 – 4 = **1**.

b) The terms are being divided by 2 each time (–18 ÷ –9 = 2), so the missing terms are –72 ÷ 2 = **–36** and –9 ÷ 2 = **–4.5**.

c) The terms are being multiplied by 4 each time (3.2 ÷ 0.8 = 4), so the missing terms are 0.8 ÷ 4 = **0.2** and 204.8 ÷ 4 = **51.2**.

d) The terms are increasing by 8 each time (–55 – (–63) = 8), so the missing terms are –55 + 8 = **–47**, –47 + 8 = **–39** and –39 + 8 = **–31**.

Page 179 Exercise 2

1 **a)** 8 – 7 = 1   10 – 8 = 2
13 – 10 = 3   17 – 13 = 4
So the differences are **+1**, **+2**, **+3**, **+4**.

b) The rule is 'increase the amount added each time by 1':
6th term: 17 + 5 = **22**
7th term: 22 + 6 = **28**
8th term: 28 + 7 = **35**

2 **a) (i)** 7 – 5 = 2   11 – 7 = 4
17 – 11 = 6   25 – 17 = 8
So the differences are **+2**, **+4**, **+6**, **+8**.

(ii) The rule is 'increase the amount added each time by 2':
6th term: 25 + 10 = **35**
7th term: 35 + 12 = **47**
8th term: 47 + 14 = **61**

b) (i) 18 – 20 = –2   15 – 18 = –3
11 – 15 = –4   6 – 11 = –5
So the differences are **–2**, **–3**, **–4**, **–5**.

(ii) The rule is 'increase the amount subtracted each time by 1':
6th term: 6 – 6 = **0**
7th term: 0 – 7 = **–7**
8th term: –7 – 8 = **–15**

c) (i) 4 – 3 = 1   6 – 4 = 2
9 – 6 = 3   13 – 9 = 4
So the differences are **+1**, **+2**, **+3**, **+4**.

(ii) The rule is 'increase the amount added each time by 1':
6th term: 13 + 5 = **18**
7th term: 18 + 6 = **24**
8th term: 24 + 7 = **31**

d) (i) 2 – 1 = 1   0 – 2 = –2
3 – 0 = 3   –1 – 3 = –4
So the differences are **1**, **–2**, **3**, **–4**.

(ii) The rule is 'alternate between adding and subtracting and increase the amount each time by 1':
6th term: –1 + 5 = **4**
7th term: 4 – 6 = **–2**
8th term: –2 + 7 = **5**

3 Add together the two previous terms to get the next one:
6th term: 4th term + 5th term = 3 + 5 = **8**
7th term: 5th term + 6th term = 5 + 8 = **13**
8th term: 6th term + 7th term = 8 + 13 = **21**

Page 180 Exercise 3

1 **a) (i)** E.g. **Add 2 matches** to the right hand side to form another equilateral triangle.

(ii)

(iii) There are 11 matches in the 5th pattern (from part (ii)) and 2 matches added each time, so the 6th pattern has 11 + 2 = **13 matches**.

b) (i) E.g. **Add 3 matches** to form another square to the right of the top right square **and 3 matches** to form another square below the bottom left square.

(ii)

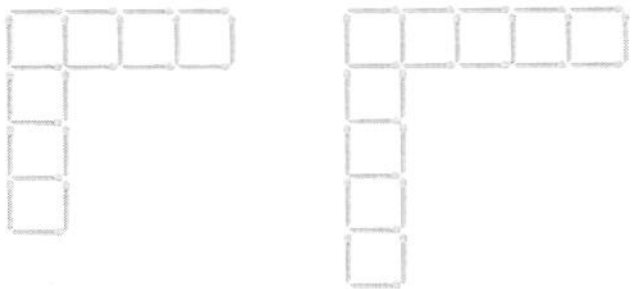

(iii) There are 28 matches in the 5th pattern (from part (ii)) and 3 + 3 matches added each time, so the 6th pattern has 28 + 3 + 3 = **34 matches**

c) (i) E.g. **Add 3 matches** to the left hand side to form a new square.

(ii)

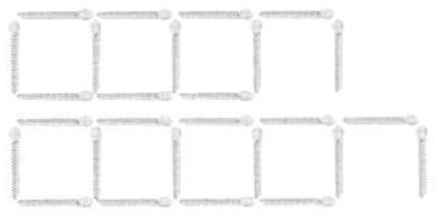

(iii) There are 15 matches in the 5th pattern (from part (ii)) and 3 matches added each time, so the 6th pattern has 15 + 3 = **18 matches**.

2 **a)**

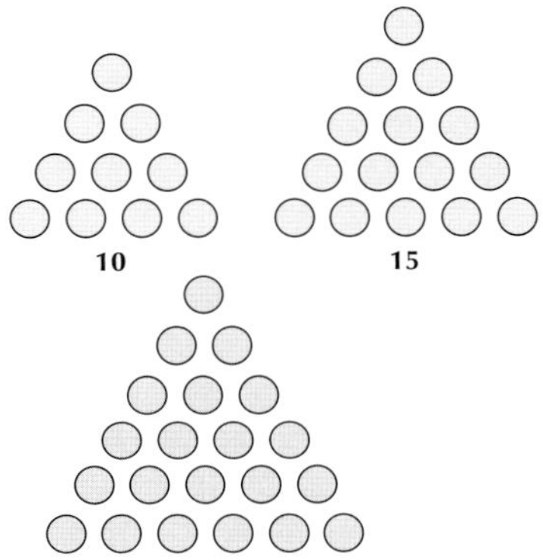

b) E.g. **Add a row of circles** to the bottom of the triangle, **with one more circle** in the new row than in the bottom row of the previous shape in the pattern.

c) The sequence of the number of dots is 1, 3, 6, 10, 15, 21...
The rule is 'increase the amount added each time by 1' and the difference between the 5th and 6th terms is 21 – 15 = 6, so the 7th term is 21 + 7 = **28**.

3 **a) (i)**

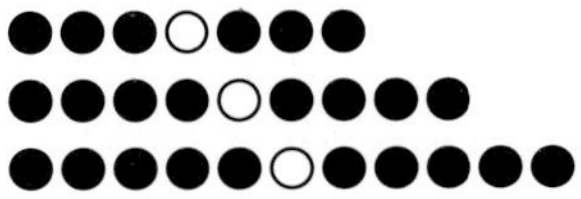

(ii) The number of white circles in each pattern is **always 1**.
The number of black circles in each pattern **increases by 2** each time.

(iii) There is always 1 white circle, so there will be **1 white circle** in the 7th pattern.

(iv) There are 10 black circles in the 6th pattern (from part (i)) and 2 added each time, so add on 4 lots of 2 to get the 10th pattern:
10 + (4 × 2) = **18 black circles**.

b) (i) 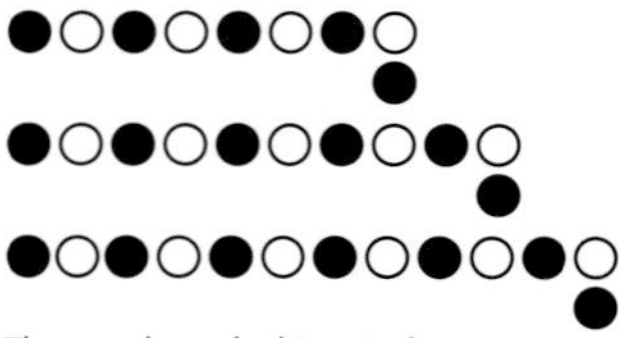

(ii) The number of white circles **increases by 1** each time.
The number of black circles **increases by 1** each time.

(iii) There are 6 white circles in the 6th pattern (from part (i)) and 1 added each time, so in the 7th pattern there are:
$6 + 1 =$ **7 white circles.**

(iv) There are 7 black circles in the 6th pattern (from part (i)) and 1 added each time, so add on 4 lots of 1 to get the 10th pattern:
$7 + (4 \times 1) =$ **11 black circles.**

11.2 Position to Term Rules

Page 181 Exercise 1

1 a) Substitute $n = 1$ into $20 - 2n$ to find the first term: $20 - 2 \times 1 =$ **18**

b) Substitute $n = 2$ into $20 - 2n$ to find the second term: $20 - 2 \times 2 =$ **16**

c) Substitute $n = 3$ into $20 - 2n$ to find the third term: $20 - 2 \times 2 =$ **14**

d) Substitute $n = 4$ into $20 - 2n$ to find the fourth term: $20 - 2 \times 2 =$ **12**

2 a) 1st term ($n = 1$): $1 + 5 =$ **6**
2nd term ($n = 2$): $2 + 5 =$ **7**
3rd term ($n = 3$): $3 + 5 =$ **8**
4th term ($n = 4$): $4 + 5 =$ **9**

b) 1st term ($n = 1$): $3 \times 1 + 2 =$ **5**
2nd term ($n = 2$): $3 \times 2 + 2 =$ **8**
3rd term ($n = 3$): $3 \times 3 + 2 =$ **11**
4th term ($n = 4$): $3 \times 4 + 2 =$ **14**

Using the same method for c)-h):

c) 2, 12, 22, 32

d) 11, 10, 9, 8

e) –10, –13, –16, –19

f) –1, –5, –9, –13

g) 2, 8, 18, 32

h) 5, 11, 21, 35

3 a) 5th term: $2 \times 5 + 20 =$ **30**

b) 10th term: $2 \times 10 + 20 =$ **40**

c) 20th term: $2 \times 20 + 20 =$ **60**

d) 100th term: $2 \times 100 + 20 =$ **220**

4 a) 5th term: $100 - 3 \times 3 =$ **91**

b) 10th term: $100 - 3 \times 10 =$ **70**

c) 20th term: $100 - 3 \times 30 =$ **10**

d) 100th term: $100 - 3 \times 40 =$ **–20**

Page 182 Exercise 2

1 Solve the equation $2n + 6 = 20 \Rightarrow 2n = 14$
$\Rightarrow$ $\boldsymbol{n = 7}$

2 a) Solve the equation $17 - 2n = 3 \Rightarrow 2n = 14$
$\Rightarrow$ $\boldsymbol{n = 7}$

b) Solve the equation $17 - 2n = 9 \Rightarrow 2n = 8$
$\Rightarrow$ $\boldsymbol{n = 4}$

c) Solve the equation $17 - 2n = -7$
$\Rightarrow 2n = 24 \Rightarrow$ $\boldsymbol{n = 12}$

3 a) Solve the equation $n^2 + 1 = 5 \Rightarrow n^2 = 4$
$\Rightarrow$ $\boldsymbol{n = 2}$

b) Solve the equation $n^2 + 1 = 50 \Rightarrow n^2 = 49$
$\Rightarrow$ $\boldsymbol{n = 7}$

c) Solve the equation $n^2 + 1 = 82 \Rightarrow n^2 = 81$
$\Rightarrow$ $\boldsymbol{n = 9}$

You can discount the negative solution to $n^2 = a$ here as n can only be positive.

4 Solve the equation $4n - 10 = 75 \Rightarrow 4n = 85$
$\Rightarrow n = 21.25$
Since $4n - 10$ is increasing, the last term with a value less than 75 will be the last whole number before 21.25, so $\boldsymbol{n = 21}$.

5 Solve the equation $6n + 2 = 40 \Rightarrow 6n = 38$
$\Rightarrow n = 6.333...$
Since $6n + 2$ is increasing, the first term with a value greater than 40 will be the next whole number after 6.333, so $\boldsymbol{n = 7}$.

6 a) (i) Substitute $n = 6$ into $2n + 1$:
$2 \times 6 + 1 =$ **13 matches**

(ii) Solve the equation $2n + 1 = 100$
$\Rightarrow 2n = 99 \Rightarrow n = 49.5$
You need to find the last pattern that uses fewer than 100 matches, so $\boldsymbol{n = 49}$.

b) (i) Substitute $n = 6$ into $4n - 1$:
$4 \times 6 - 1 =$ **23 matches**

(ii) Solve the equation $4n - 1 = 100$
$\Rightarrow 4n = 101 \Rightarrow n = 25.25$
You need to find the last pattern that uses fewer than 100 matches, so $\boldsymbol{n = 25}$.

Page 183 Exercise 3

1 Solve the equation $2n + 1 = 54 \Rightarrow 2n = 53$
$\Rightarrow n = 26.5$
n is not a whole number so **54 is not a term in the sequence.**

2 Solve the equation $3n - 1 = 80 \Rightarrow 3n = 81$
$\Rightarrow n = 27$
n is a whole number so **80 is a term in the sequence.**

3 Solve the equation $21 - 2n = -1 \Rightarrow 2n = 22$
$\Rightarrow n = 11$
n is a whole number so **–1 is in the sequence and has position $\boldsymbol{n = 11}$.**

4 a) Solve the equation $17 + 3n = 52$
$\Rightarrow 3n = 35 \Rightarrow n = 11.66...$
n is not a whole number so **52 is not a term in the sequence.**

b) Solve the equation $17 + 3n = 98$
$\Rightarrow 3n = 81 \Rightarrow n = 27$
n is a whole number so **98 is a term in the sequence.**

c) Solve the equation $17 + 3n = 248$
$\Rightarrow 3n = 231 \Rightarrow n = 77$
n is a whole number so **248 is a term in the sequence.**

d) Solve the equation $17 + 3n = 996$
$\Rightarrow 3n = 979 \Rightarrow n = 326.33...$
n is not a whole number so **996 is not a term in the sequence.**

5 a) Solve the equation $4n - 9 = 43 \Rightarrow 4n = 52$
$\Rightarrow n = 13$
n is a whole number so **52 is a term in the sequence with position $\boldsymbol{n = 13}$.**

b) Solve the equation $4n - 9 = 71 \Rightarrow 4n = 80$
$\Rightarrow n = 20$
n is a whole number so **71 is a term in the sequence with position $\boldsymbol{n = 20}$.**

c) Solve the equation $4n - 9 = 138$
$\Rightarrow 4n = 147 \Rightarrow n = 36.75$
n is not a whole number so **138 is not a term in the sequence.**

d) Solve the equation $4n - 9 = 879$
$\Rightarrow 4n = 888 \Rightarrow n = 222$
n is a whole number so **879 is a term in the sequence with position $\boldsymbol{n = 222}$.**

11.3 Finding a Position to Term Rule

Page 185 Exercise 1

1 a) $13 - 9 = 4$ $17 - 13 = 4$ $21 - 17 = 4$
The difference between each term and the next is **+4**.

b) Common difference is +4, so compare the sequence with $4n$:

$4n$:	4	8	12	16
	↓+5	↓+5	↓+5	↓+5
Term:	9	13	17	21

You need to add 5, so the nth term is $\boldsymbol{4n + 5}$.
Check your rule using e.g. $n = 2$:
$(4 \times 2) + 5 = 8 + 5 =$ **13** ✔

2 a) $13 - 7 = 6$ $19 - 13 = 6$ $25 - 19 = 6$
The difference between each term and the next is +6, so compare the sequence with $6n$:

$6n$:	6	12	18	24
	↓+1	↓+1	↓+1	↓+1
Term:	7	13	19	25

You need to add 1, so the nth term is $\boldsymbol{6n + 1}$.

b) $16 - 6 = 10$ $26 - 16 = 10$
$36 - 26 = 10$
The difference between each term and the next is +10, so compare the sequence with $10n$:

$10n$:	10	20	30	40
	↓–4	↓–4	↓–4	↓–4
Term:	6	16	26	36

You need to subtract 4, so the nth term is $\boldsymbol{10n - 4}$.

Using the same method for c)-f):

c) $\boldsymbol{40n + 1}$ **d) $\boldsymbol{2n - 3}$**

e) $\boldsymbol{4n - 13}$ **f) $\boldsymbol{19n - 64}$**

3 a) $8 - 10 = -2$ $6 - 8 = -2$ $4 - 6 = -2$
The difference between each term and the next is –2, so compare the sequence with $-2n$:

$-2n$:	–2	–4	–6	–8
	↓+12	↓+12	↓+12	↓+12
Term:	10	8	6	4

You need to add 12, so the nth term is $\boldsymbol{-2n + 12}$.

b) $37 - 40 = -3$ $34 - 37 = -3$
$31 - 34 = -3$
The difference between each term and the next is –3, so compare the sequence with $-3n$:

$-3n$:	–3	–6	–9	–12
	↓+43	↓+43	↓+43	↓+43
Term:	40	37	34	31

You need to add 43, so the nth term is $\boldsymbol{-3n + 43}$.

Using the same method for c)-f):

c) $\boldsymbol{-9n + 87}$

d) $\boldsymbol{-5n + 9}$

e) $\boldsymbol{-15n + 5}$

f) $\boldsymbol{-12n - 27}$

4 a) The sequence of the number of matchsticks is 4, 10, 16...
$10 - 4 = 6$ $16 - 10 = 6$
The difference between each term and the next is +6, so compare the sequence with $6n$:

$6n$:	6	12	18
	↓–2	↓–2	↓–2
Term:	4	10	16

You need to subtract 2, so the nth term is $\boldsymbol{6n - 2}$.

b) The sequence of the number of matchsticks is 7, 12, 17...
$12 - 7 = 5$ $\quad$ $17 - 12 = 5$
The difference between each term and the next is +5, so compare the sequence with $5n$:

$5n$:	5	10	15
	↓+2	↓+2	↓+2
Term:	7	12	17

You need to add 2, so the nth term is $\mathbf{5n + 2}$.

5 a) $11 - 8 = 3$ $\quad$ $14 - 11 = 3$
$17 - 14 = 3$
The difference between each term and the next is +3, so compare the sequence with $3n$:

$3n$:	3	6	9	12
	↓+5	↓+5	↓+5	↓+5
Term:	8	11	14	17

You need to add 5, so the nth term is $\mathbf{3n + 5}$.

b) Each term in the sequence is 1 greater than the corresponding term in the sequence from part a), so these terms can be found using the formula $(3n + 5) + 1 = \mathbf{3n + 6}$.

Page 186 Review Exercise

1 a) (i) The rule is '**add 3 to the previous term**'
(ii) 5th term: $13 + 3 = \mathbf{16}$
6th term: $16 + 3 = \mathbf{19}$
7th term: $19 + 3 = \mathbf{22}$

b) (i) The rule is '**divide the previous term by 2**'
(ii) 5th term: $24 \div 2 = \mathbf{12}$
6th term: $12 \div 2 = \mathbf{6}$
7th term: $6 \div 2 = \mathbf{3}$

c) (i) The rule is '**subtract 4 from the previous term**'
(ii) 5th term: $-12 - 4 = \mathbf{-16}$
6th term: $-16 - 4 = \mathbf{-20}$
7th term: $-20 - 4 = \mathbf{-24}$

d) (i) The rule is '**add 2 to the previous term**'
(ii) 5th term: $2 + 2 = \mathbf{4}$
6th term: $4 + 2 = \mathbf{6}$
7th term: $6 + 2 = \mathbf{8}$

e) (i) The rule is '**subtract 3 from the previous term**'
(ii) 5th term: $2 - 3 = \mathbf{-1}$
6th term: $-1 - 3 = \mathbf{-4}$
7th term: $-4 - 3 = \mathbf{-7}$

f) (i) The rule is '**multiply the previous term by –3**'
(ii) 5th term: $-54 \times -3 = \mathbf{162}$
6th term: $162 \times -3 = \mathbf{-486}$
7th term: $-486 \times -3 = \mathbf{1458}$

2 a) $9 - 7 = 2$ $\quad$ $13 - 9 = 4$
$19 - 13 = 6$ $\quad$ $27 - 19 = 8$
The differences between terms are **+2, +4, +6, +8**

b) The rule is 'increase the amount added each time by 2':
6th term: $27 + 10 = \mathbf{37}$
7th term: $37 + 12 = \mathbf{49}$
8th term: $49 + 14 = \mathbf{63}$

3 a)

b) Add two more circles on the right of the previous pattern.

c) There are 12 circles in the 6th pattern (from part a)), so in the 7th pattern there will be $12 + 2 = 14$ circles and in the 8th pattern there will be $14 + 2 = \mathbf{16}$ **circles.**

4 1st term ($n = 1$): $-3 - 2 \times 1 = \mathbf{-5}$
2nd term ($n = 2$): $-3 - 2 \times 2 = \mathbf{-7}$
3rd term ($n = 3$): $-3 - 2 \times 3 = \mathbf{-9}$
4th term ($n = 4$): $-3 - 2 \times 4 = \mathbf{-11}$

5 a) (i) 5th term: $4 \times 5 + 12 = \mathbf{32}$
(ii) 10th term: $4 \times 10 + 12 = \mathbf{52}$
(iii) 100th term: $4 \times 100 + 12 = \mathbf{412}$

b) (i) 5th term: $30 - 3 \times 5 = \mathbf{15}$
(ii) 10th term: $30 - 3 \times 10 = \mathbf{0}$
(iii) 100th term: $30 - 3 \times 100 = \mathbf{-270}$

Using the method for c)-d):
c) (i) 492 $\quad$ **(ii) 992** $\quad$ **(iii) 9992**
d) (i) 20 $\quad$ **(ii) 90** $\quad$ **(iii) 9900**

6 a) Solve the equation $50 - 6n = 2 \Rightarrow 6n = 48$ $\Rightarrow \mathbf{n = 8}$.

b) Solve the equation $50 - 6n = 8 \Rightarrow 6n = 42$ $\Rightarrow \mathbf{n = 7}$.

c) Solve the equation $50 - 6n = 14$ $\Rightarrow 6n = 36 \Rightarrow \mathbf{n = 6}$.

d) Solve the equation $50 - 6n = 26$ $\Rightarrow 6n = 24 \Rightarrow \mathbf{n = 4}$.

7 a) Solve the equation $7n - 3 = 100$ $\Rightarrow 7n = 103 \Rightarrow n = 14.714...$
You want the last whole number smaller than 14.714...
so the last term with value smaller than 100 is the **14th term**.

b) Solve the equation $7n - 3 = 205$ $\Rightarrow 7n = 208 \Rightarrow n = 29.714...$
You want the next whole number greater than 29.714...
so the first term with value greater than 205 is the **30th term**.

8 a) $18 - 12 = 6$ $\quad$ $24 - 18 = 6$
$30 - 24 = 6$
The difference between each term and the next is +6, so compare the sequence with $6n$:

$6n$:	6	12	18	24
	↓+6	↓+6	↓+6	↓+6
Term:	12	18	24	30

You need to add 6, so the nth term is $\mathbf{6n + 6}$.

b) Solve the equation $6n + 6 = 86 \Rightarrow 6n = 80$ $\Rightarrow n = 13.33...$
n is not a whole number, so **86 is not a term in the sequence.**

9 The sequence of the number of dots is 5, 7, 9...
$7 - 5 = 2$ $\quad$ $9 - 7 = 2$
The difference between each term and the next is +2, so compare the sequence with $2n$:

$2n$:	2	4	6
	↓+3	↓+3	↓+3
Term:	5	7	9

You need to add 3, so the nth term is $\mathbf{2n + 3}$.

Page 187 Exam-Style Questions

1 $5 - 3 = 2$ $\quad$ $9 - 5 = 4$ $\quad$ $15 - 9 = 6$
The rule is 'increase the amount added each time by 2'. *[1 mark]*
so the 5th term is $15 + 8 = \mathbf{23}$ *[1 mark]*

2 a)

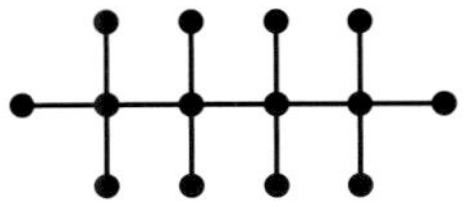

Design 4 *[1 mark]*

b) The number of rods forms the sequence 4, 7, 10, 13...
To get the next term of the sequence, you add 3 to the previous term *[1 mark]*.
So the 5th term is 16, 6th term is 19, 7th term is 22, and the **8th term is 25**. *[1 mark]*

c) The number of rods is one less than the number of balls, so the expression for the number of rods is:
$(3n + 2) - 1 = 3n + 1$ *[1 mark]*

3 a) $6 - 2 = 4$ $\quad$ $10 - 6 = 4$ $\quad$ $14 - 10 = 4$
The rule is 'add 4 to the previous term', so the next term is $14 + 4 = \mathbf{18}$ *[1 mark]*

b) Substitute $n = 210$ into $4n - 2$:
$4 \times 210 - 2 = \mathbf{838}$
[2 marks available — 1 mark for the expression 4 × 210 – 2, 1 mark for the correct answer]

c) $4n - 2 = 74 \Rightarrow 4n = 76 \Rightarrow n = 19$.
Yes, Jane is correct, because n is a whole number.
[2 marks available — 1 mark for the expression 4n – 2 = 74, 1 mark for correct answer]

4 $-2 - (-11) = 9$ $\quad$ $7 - (-2) = 9$ $\quad$ $16 - 7 = 9$
Common difference is +9, so compare the sequence with $9n$: *[1 mark]*

$9n$:	9	18	27	36
	↓–20	↓–20	↓–20	↓–20
Term:	–11	–2	7	16

You need to subtract 20, so the nth term is $\mathbf{9n - 20}$. *[1 mark]*

Section 12 — Graphs and Equations

12.1 Coordinates

Page 189 Exercise 1

1

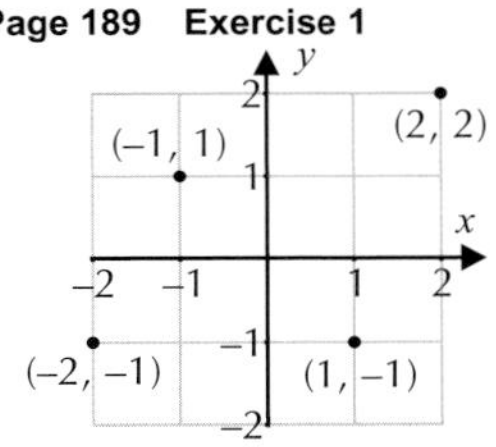

a) (2, 2) lies in quadrant **I**.
b) (1, –1) lies in quadrant **IV**.
c) (–2, –1) lies in quadrant **III**.
d) (–1, 1) lies in quadrant **II**.

2 a) (i) (0, 0) $\quad$ **(ii) (2, 2)**
(iii) (1, 3) $\quad$ **(iv) (0, 4)**
(v) (4, 4)

b) X-COORDINATE COMES FIRST

3 a) (i) SQUARE $\quad$ **(ii) CIRCLE**
(iii) CUBOID

b) (i) (2, 1) (–4, 1) (4, –4) (–5, 3)
(ii) (2, –4) (5, –2) (5, 3) (–5, 3) (–2, –4) (–5, 3)
(iii) (4, –4) (–2, –4) (–4, 1) (–4, 5) (–3, –2) (3, 3) (4, 1) (–5, 3)

Page 190 Exercise 2

1 a)

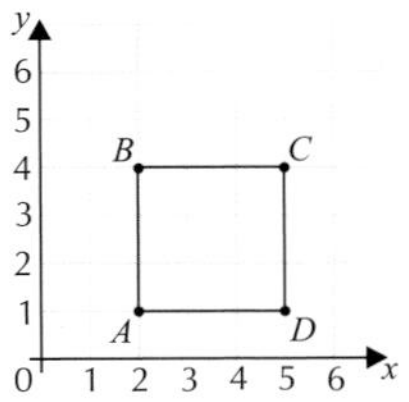

b)

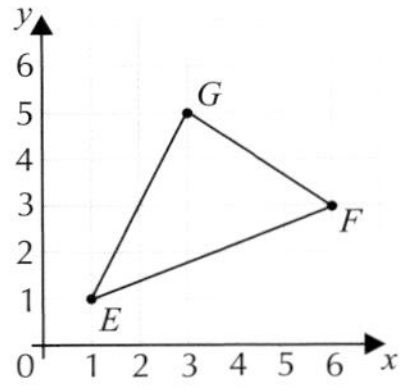

c) 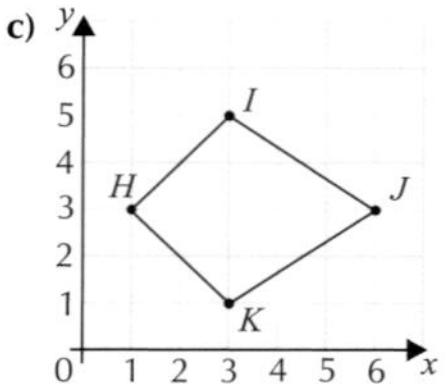

d) 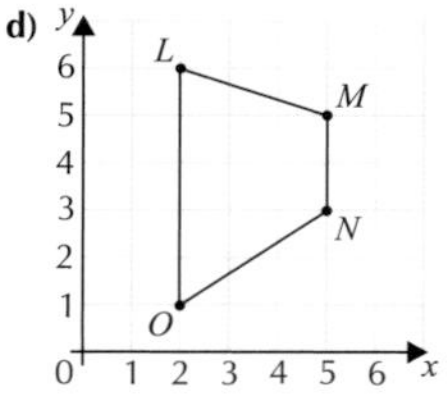

When joining points to make shapes, connect A to B, B to C, etc.

2 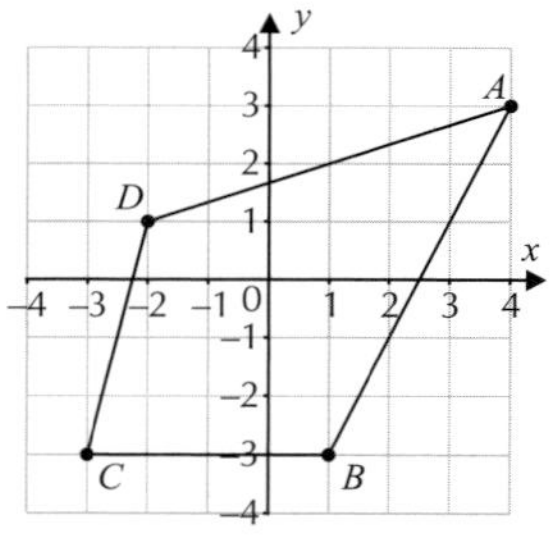

3 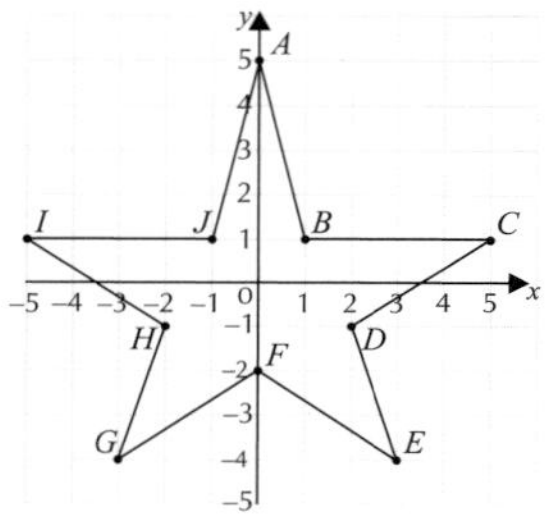

Page 191 Exercise 3

1 a) X: **(–5, 1)**, Y: **(3, 5)**

b) M: $\left(\frac{-5+3}{2}, \frac{1+5}{2}\right) = \left(\frac{-2}{2}, \frac{6}{2}\right) =$ **(–1, 3)**

c) 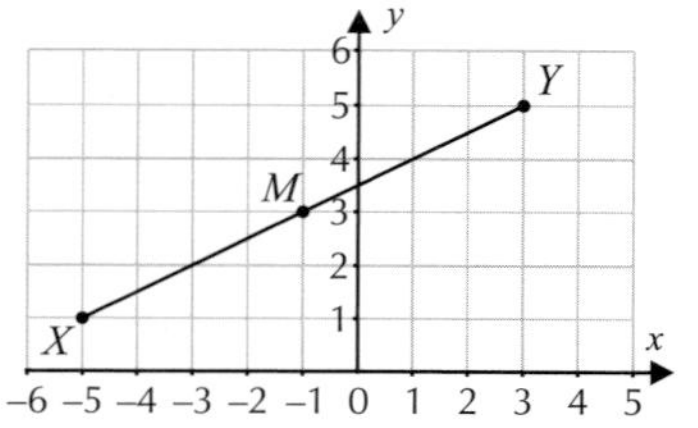

Check to see if the midpoint you have calculated looks about right. It should be halfway along the line segment from X to Y.

2 a) M: $\left(\frac{1+3}{2}, \frac{0+5}{2}\right) = \left(\frac{4}{2}, \frac{5}{2}\right) =$ **(2, 2.5)**

b) (i), (ii)

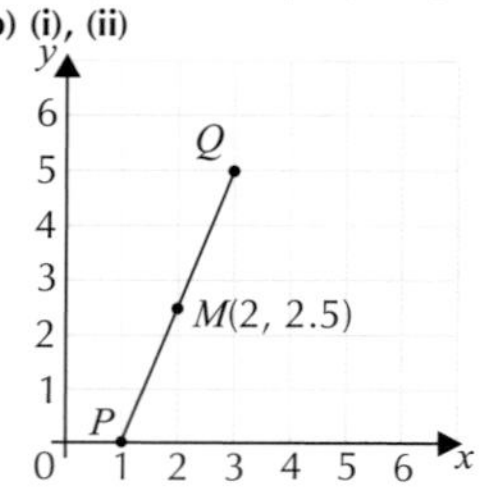

3 a) $\left(\frac{1+3}{2}, \frac{1+5}{2}\right) = \left(\frac{4}{2}, \frac{6}{2}\right) =$ **(2, 3)**

b) $\left(\frac{0+6}{2}, \frac{1+3}{2}\right) = \left(\frac{6}{2}, \frac{4}{2}\right) =$ **(3, 2)**

c) $\left(\frac{0+(-5)}{2}, \frac{-4+1}{2}\right) = \left(\frac{-5}{2}, \frac{-3}{2}\right)$

= **(–2.5, –1.5)**

d) $\left(\frac{-2+1}{2}, \frac{0+(-8)}{2}\right) = \left(\frac{-1}{2}, \frac{-8}{2}\right) =$ **(–0.5, –4)**

4 a) A is (–5, 5) and F is (5, 5), so midpoint of AF

$= \left(\frac{-5+5}{2}, \frac{5+5}{2}\right) = \left(\frac{0}{2}, \frac{10}{2}\right) =$ **(0, 5)**

b) A is (–5, 5) and C is (–2, 3), so midpoint of AC

$= \left(\frac{-5+-2}{2}, \frac{5+3}{2}\right) = \left(\frac{-7}{2}, \frac{8}{2}\right) =$ **(–3.5, 4)**

Using the same method for c)-f):

c) **(3.5, 3)** d) **(–0.5, –2.5)**

e) **(0.5, 1)** f) **(0.5, 0.5)**

12.2 Horizontal and Vertical Graphs

Page 193 Exercise 1

1 Every point on A has x-coordinate = –5, so it has equation x **= –5**.
Every point on B has x-coordinate = –2, so it has equation x **= –2**.
Every point on C has y-coordinate = 3, so it has equation y **= 3**.
Every point on D has x-coordinate = 5, so it has equation x **= 5**.
Every point on E has y-coordinate = –2, so it has equation y **= –2**.

2 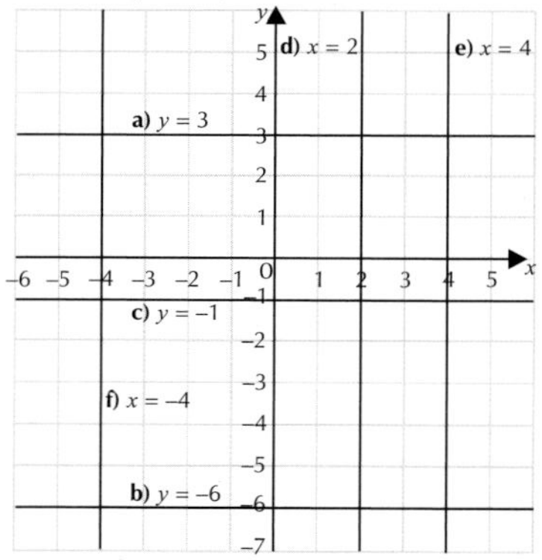

3 a) Every point on the x-axis has y-coordinate = **0**.

b) So the x-axis has equation y **= 0**.

Don't get mixed up — y = 0 on the x-axis, and x = 0 on the y-axis.

4 a) 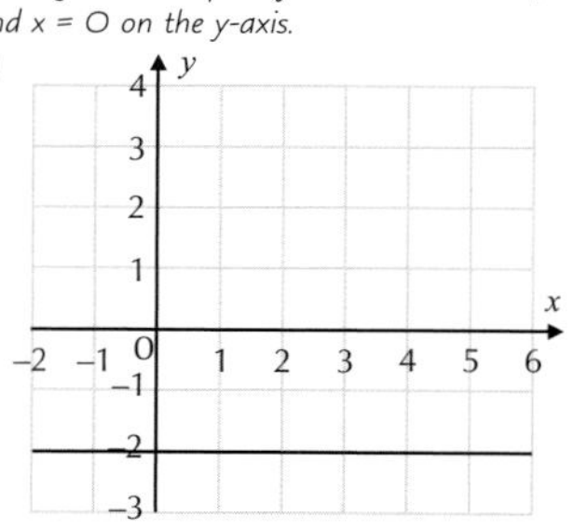

b) y **= –2**

5 a) 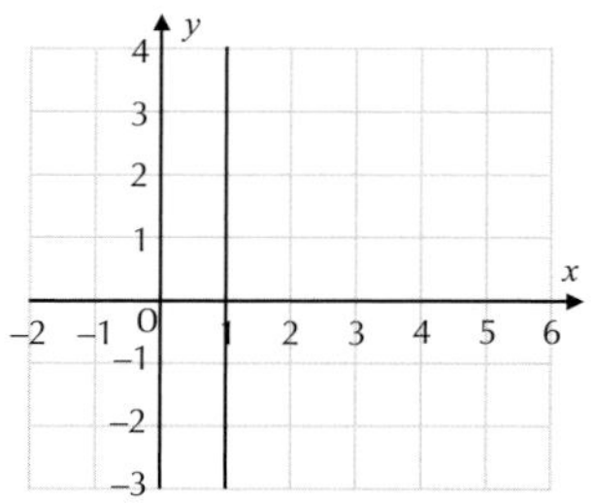

b) x **= 1**

6 'Parallel to the x-axis' means the line is horizontal.
The point (4, 8) has y-coordinate = 8, so the line has equation y **= 8**.

7 'Parallel to the y-axis' means the line is vertical.
The point (–2, –6) has x-coordinate = –2, so the line has equation x **= –2**.

8 a) 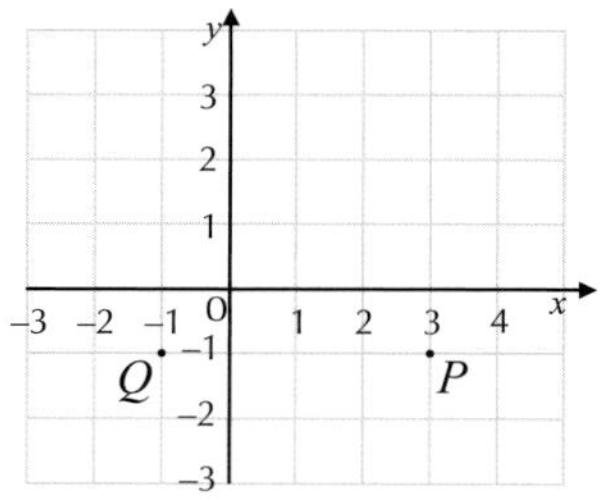

b) The line through Q and P is horizontal, so it has equation y **= –1**.

12.3 Other Straight-Line Graphs

Page 195 Exercise 1

1 a) When $x = 2$, $y = 2 + 2 = 4$.
When $x = 3$, $y = 3 + 2 = 5$.
When $x = 4$, $y = 4 + 2 = 6$.
When $x = 5$, $y = 5 + 2 = 7$.

x	0	1	2	3	4	5
y	2	3	**4**	**5**	**6**	**7**
Coords	(0, 2)	**(1, 3)**	**(2, 4)**	**(3, 5)**	**(4, 6)**	**(5, 7)**

b), c) 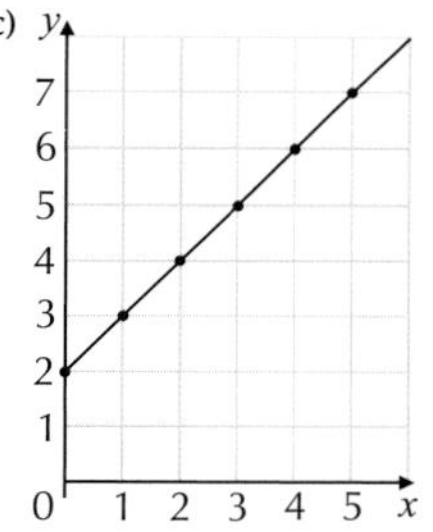

2 a) $2y - 2x = -8 \Rightarrow 2y = 2x - 8 \Rightarrow$ $\mathbf{y = x - 4}$.

b) When $x = 1$, $y = 1 - 4 = -3$.
When $x = 2$, $y = 2 - 4 = -2$.
When $x = 3$, $y = 3 - 4 = -1$.
When $x = 4$, $y = 4 - 4 = 0$.
When $x = 5$, $y = 5 - 4 = 1$.

x	0	1	2	3	4	5
y	–4	**–3**	**–2**	**–1**	**0**	**1**
Coords	**(0, –4)**	**(1, –3)**	**(2, –2)**	**(3, –1)**	**(4, 0)**	**(5, 1)**

c), d)

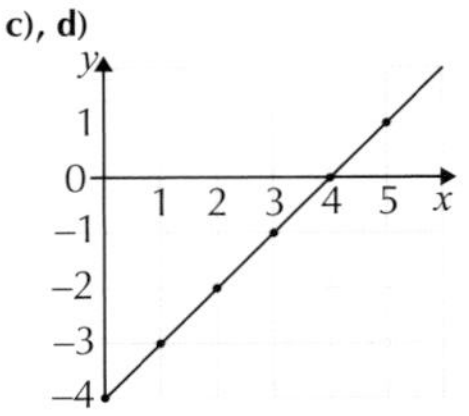

3 a) When $x = 0$, $y = 0 + 7 = 7$.
When $y = 0$, $0 = x + 7 \Rightarrow x = -7$.
You can check you have calculated the x- and y-intercepts correctly by putting the x- and y-values back into the equation.
Plot the points (0, 7) and (–7, 0) and draw a straight line between them.

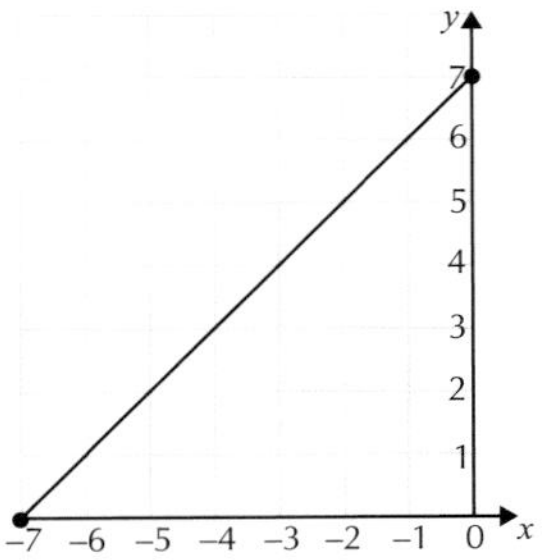

b) When $x = 0$, $y = 3 \times 0 - 6 = -6$.
When $y = 0$, $0 = 3x - 6 \Rightarrow x = 6 \div 3 = 2$.
Plot the points (0, –6) and (2, 0) and draw a straight line passing through them. Extend the line to cover the range of x from 0 to 4.

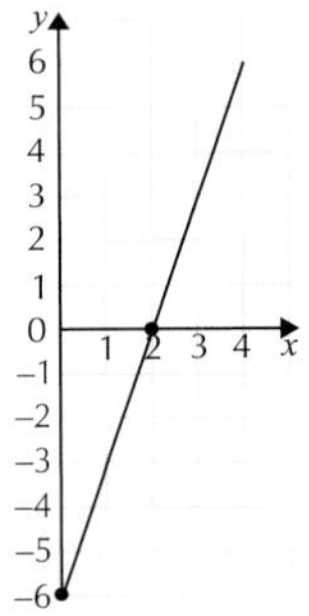

c) When $x = 0$, $y + 2 \times 0 = 8$.
When $y = 0$, $0 + 2x = 8 \Rightarrow x = 8 \div 2 = 4$.
Plot the points (0, 8) and (4, 0) and draw a straight line passing through them. Extend the line to cover the range of x from 0 to 5.

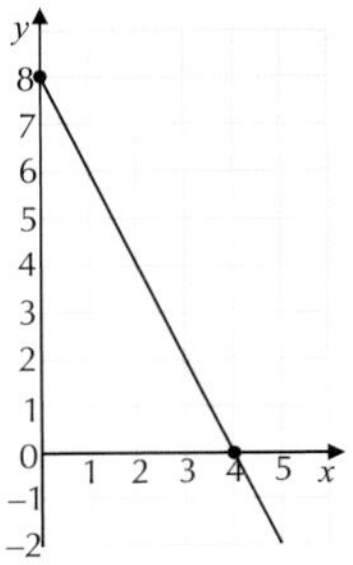

d) When $x = 0$, $y + 5 \times 0 = 7.5$.
When $y = 0$, $0 + 5x = 7.5$
$\Rightarrow x = 7.5 \div 5 = 1.5$.
Plot the points (0, 7.5) and (1.5, 0) and draw a straight line passing through them. Extend the line to cover the range of x from 0 to 2.

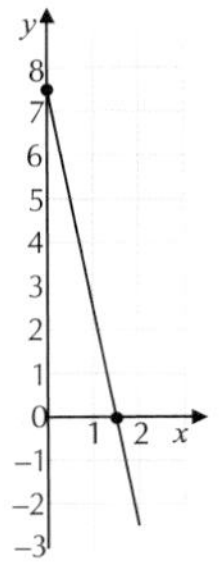

Page 196 Exercise 2

1 a) (2, 0)

b) Read off the x-coordinate: $\boldsymbol{x = 2}$

2 a) The graph of $y = 0.5x - 2$ crosses the x-axis at (4, 0). So the solution to the equation $0.5x - 2 = 0$ is $\boldsymbol{x = 4}$.

b) The graph of $y = -2x + 1$ crosses the x-axis at (0.5, 0). So the solution to the equation $-2x + 1 = 0$ is $\boldsymbol{x = 0.5}$.

Page 197 Exercise 3

1 a) Draw the line $y = 1$ and find the intersection point with the graph of $y = -2x - 3$.

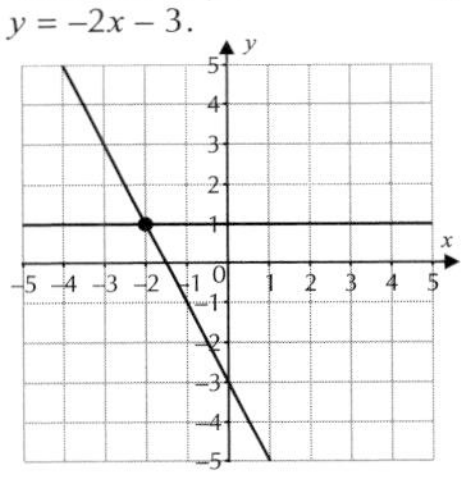

Read off the x-coordinate: $\boldsymbol{x = -2}$.
Check your answer by putting the x-value back into the equation.

b) Draw the line $y = -4$ and find the intersection point with the graph of $y = x - 2$.

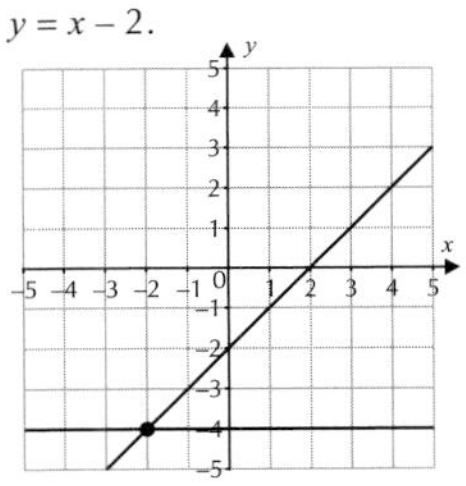

Read off the x-coordinate: $\boldsymbol{x = -2}$.

c) Draw the line $y = -5$ and find the intersection point with the graph of $y = -2x - 3$.

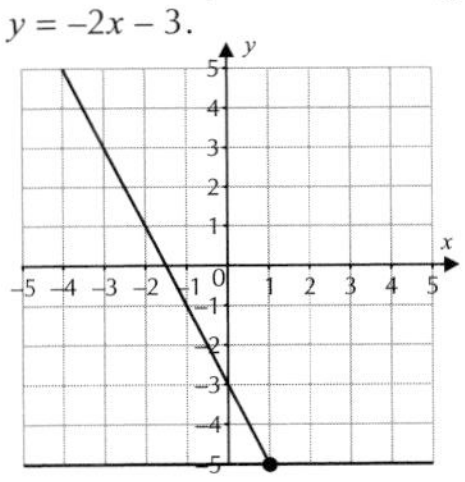

Read off the x-coordinate: $\boldsymbol{x = 1}$.

d) Draw the line $y = 2\frac{1}{2}$ and find the intersection point with the graph of $y = -\frac{1}{2}x + 4$.

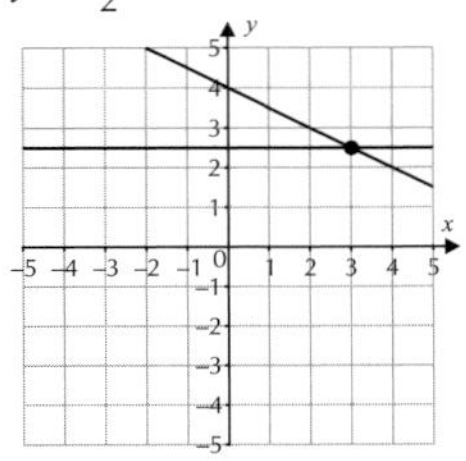

Read off the x-coordinate: $\boldsymbol{x = 3}$.

2 a) $-2x + 2 = 1 \Rightarrow -2x + 2 - 5 = 1 - 5$
$\boldsymbol{\Rightarrow -2x - 3 = -4}$

b) Draw the line $y = -4$ and find the intersection point with the graph of $y = -2x - 3$.

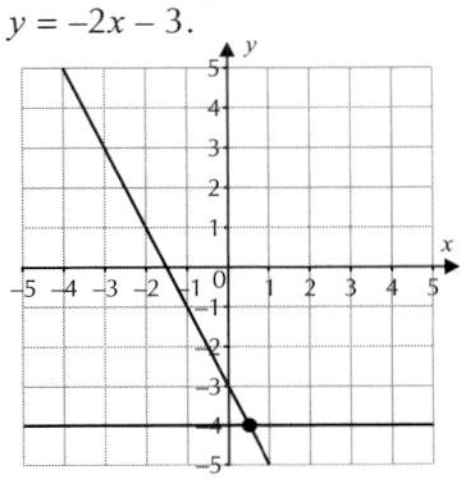

Read off the x-coordinate: $\boldsymbol{x = 0.5}$.

3 a) $x - 1 = 2 \Rightarrow x - 1 - 1 = 2 - 1 \Rightarrow x - 2 = 1$.
The step above transforms the equation so its left-hand side matches the right-hand side of the one the graph equations, y = x − 2.
Draw the line $y = 1$ and find the intersection point with the graph of $y = x - 2$.

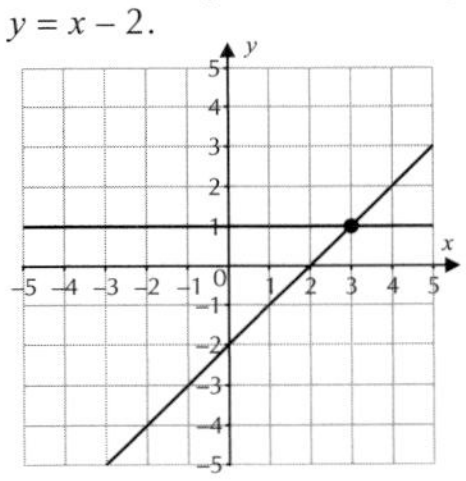

Read off the x-coordinate: $\boldsymbol{x = 3}$.

b) $-2x - 5 = 0 \Rightarrow -2x - 5 + 2 = 0 + 2$
$\Rightarrow -2x - 3 = 2$.
The step above transforms the equation so its left-hand side matches the right-hand side of the graph equation y = −2x − 3.
Draw the line $y = 2$ and find the intersection point with the graph of $y = -2x - 3$.

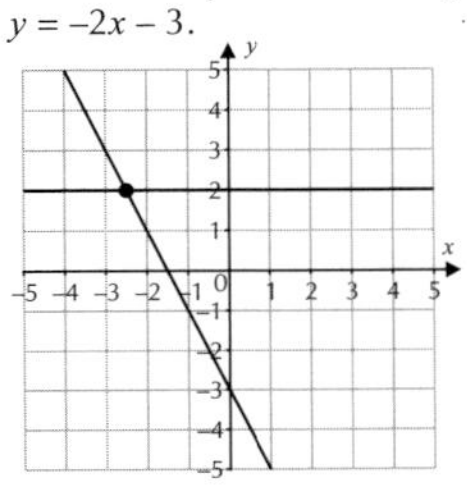

Read off the x-coordinate: $\boldsymbol{x = -2.5}$.

c) $-\frac{1}{2}x - 1 = -2 \Rightarrow -\frac{1}{2}x - 1 + 5 = -2 + 5$

$\Rightarrow -\frac{1}{2}x + 4 = 3.$

The step above transforms the equation so its left-hand side matches the right-hand side of the graph equation $y = -\frac{1}{2}x + 4$.

Draw the line $y = 3$ and find the intersection point with the graph of $y = -\frac{1}{2}x + 4$.

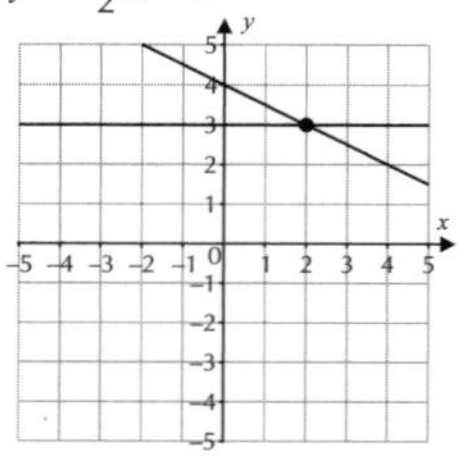

Read off the x-coordinate: $\mathbf{x = 2}$.

Page 198 Exercise 4

1 The lines intersect at the point (2, 5). So the solution is $\mathbf{x = 2}$ and $\mathbf{y = 5}$.
Check your answer by putting the x- and y-values into both equations.

2 a), b) Draw the graphs of $y = 5 - x$ and $y = x - 3$ using the methods on p.194.

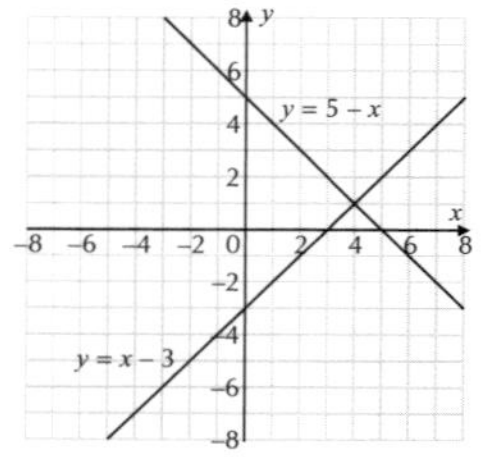

c) The lines intersect at the point (4, 1). So the solution is $\mathbf{x = 4}$ and $\mathbf{y = 1}$.

3 a) Draw the graphs of $y = 2 - 2x$ and $y = x + 5$ using the methods on p.194.

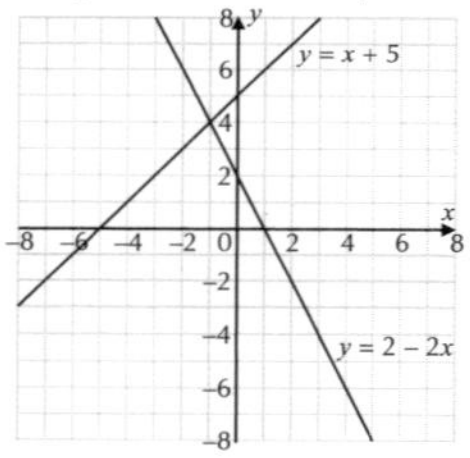

b) At the point where the lines cross, the values of y are the same, and hence the values of the right-hand side of the two equations are also the same.
So the solution to $2 - 2x = x + 5$ is at the point (–1, 4): $\mathbf{x = -1}$ and $\mathbf{y = 4}$.

4 a) Draw the graphs of $y = x + 3$ and $y = x - 2$ using the methods on p.194.

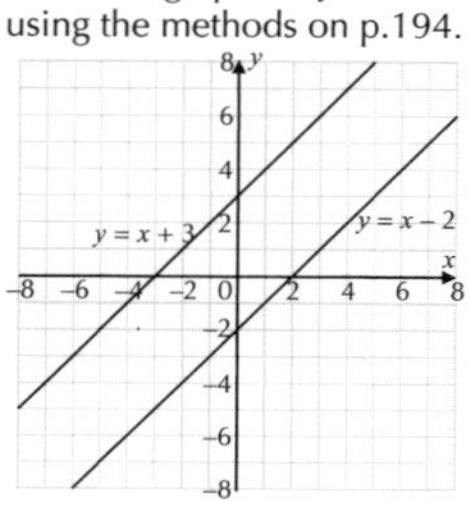

b) E.g. The lines are **parallel** and so **do not intersect**. This means there are **no points** where **both** the $\mathbf{x}$- and the $\mathbf{y}$**-values** are the **same** for **both lines**, and so there are **no solutions** to the corresponding simultaneous equations.

12.4 Gradients

Page 200 Exercise 1

1 a) **4** b) **3**

c) The line slopes upwards from left to right, so the gradient is **positive.**

d) gradient $= \frac{\text{change in } y}{\text{change in } x} = \mathbf{\frac{4}{3}}$

2 a) The line slopes upwards from left to right, so the gradient is positive.

gradient $= \frac{\text{change in } y}{\text{change in } x} = \frac{3}{3} = \mathbf{1}$

b) The line slopes upwards from left to right, so the gradient is positive.

gradient $= \frac{\text{change in } y}{\text{change in } x} = \frac{4}{1} = \mathbf{4}$

c) The line slopes downwards from left to right, so the gradient is negative.

gradient $= \frac{\text{change in } y}{\text{change in } x} = -\frac{2}{2} = \mathbf{-1}$

d) The line slopes downwards from left to right, so the gradient is negative.

gradient $= \frac{\text{change in } y}{\text{change in } x} = \mathbf{-\frac{5}{3}}$

3 a) gradient $= \frac{\text{change in } y}{\text{change in } x} = \frac{3-5}{6-1} = \frac{-2}{5} = \mathbf{-\frac{2}{5}}$

b) gradient $= \frac{\text{change in } y}{\text{change in } x} = \frac{1-6}{6-(-4)} = \frac{-5}{10} = \mathbf{-\frac{1}{2}}$

c) gradient $= \frac{\text{change in } y}{\text{change in } x} = \frac{6-0}{4-(-5)} = \frac{6}{9} = \mathbf{\frac{2}{3}}$

As a quick check of your answer, look at the direction of the line to see if the gradient should be positive or negative.

4 a) (i) $\mathbf{G}$**: (2, –5),** $\mathbf{H}$**: (6, 6)**

(ii) $\frac{6-(-5)}{6-2} = \mathbf{\frac{11}{4}}$

b) (i) $\mathbf{I}$**: (–10, 5),** $\mathbf{J}$**: (30, –25)**

(ii) $\frac{-25-5}{30-(-10)} = \frac{-30}{40} = \mathbf{-\frac{3}{4}}$

c) (i) **Line 1:** $\mathbf{K}$**: (–8, –25),** $\mathbf{L}$**: (8, 35)**
Line 2: $\mathbf{M}$**: (–4, 30),** $\mathbf{N}$**: (6, –15)**

(ii) **Line 1:** $\frac{35-(-25)}{8-(-8)} = \frac{60}{16} = \mathbf{\frac{15}{4}}$

Line 2: $\frac{-15-30}{6-(-4)} = \frac{-45}{10} = \mathbf{-\frac{9}{2}}$

5 a) gradient $= \frac{\text{change in } y}{\text{change in } x} \Rightarrow 1 = \frac{a-5}{7-3} = \frac{a-5}{4}$

$\Rightarrow 4 = a - 5$

$\Rightarrow \mathbf{a = 9}$

b) gradient $= \frac{\text{change in } y}{\text{change in } x} \Rightarrow 2 = \frac{5-1}{b-1} = \frac{4}{b-1}$

$\Rightarrow b - 1 = \frac{4}{2}$

$\Rightarrow \mathbf{b = 3}$

c) gradient $= \frac{\text{change in } y}{\text{change in } x} \Rightarrow \frac{1}{2} = \frac{c-0}{8-0} = \frac{c}{8}$

$\Rightarrow c = \frac{8}{2} \Rightarrow \mathbf{c = 4}$

12.5 Equations of Straight-Line Graphs

Page 201 Exercise 1

1 $m = 2$ and $c = 3$, so the equation is $\mathbf{y = 2x + 3}$.

2 For a line of the form $y = mx + c$, the gradient is m and the y-intercept has coordinates $(0, c)$.

a) gradient = **3**, y-intercept = **(0, 2)**
b) gradient = **2**, y-intercept = **(0, –4)**
c) gradient = **5**, y-intercept = **(0, –11)**
d) gradient = **–3**, y-intercept = **(0, 7)**
e) gradient = **4**, y-intercept = **(0, 0)**
f) gradient = $\mathbf{\frac{1}{2}}$, y-intercept = **(0, –1)**
g) gradient = **1**, y-intercept = **(0, –6)**
h) gradient = **–1**, y-intercept = **(0, 5)**
i) gradient = **–6**, y-intercept = **(0, 3)**

3 Start by looking at the y-intercept of the graphs and equations. If the two lines have the same y-intercept, look at the gradient.

$\mathbf{A}$: $\mathbf{y = -\frac{1}{3}x + 4}$ $\mathbf{B}$: $\mathbf{y = 3x}$

$\mathbf{C}$: $\mathbf{y = \frac{1}{3}x + 2}$ $\mathbf{D}$: $\mathbf{y = \frac{7}{3}x - 1}$

$\mathbf{E}$: $\mathbf{y = x + 2}$ $\mathbf{F}$: $\mathbf{y = -x + 6}$

Page 202 Exercise 2

1 Rearrange each equation into the form $y = mx + c$, then m is the gradient and $(0, c)$ is the y-intercept.

a) $3y = 9 - 3x \Rightarrow y = -x + 3$
gradient = **–1**, y-intercept = **(0, 3)**

b) $y - 5 = 7x \Rightarrow y = 7x + 5$
gradient = **7**, y-intercept = **(0, 5)**

c) $3x + y = 1 \Rightarrow y = -3x + 1$
gradient = **–3**, y-intercept = **(0, 1)**

d) $3y - 6x = 15 \Rightarrow 3y = 6x + 15$
$\Rightarrow y = 2x + 5$
gradient = **2**, y-intercept = **(0, 5)**

e) $4y - 6x + 8 = 0 \Rightarrow 4y = 6x - 8$
$\Rightarrow y = \frac{3}{2}x - 2$
gradient = $\mathbf{\frac{3}{2}}$, y-intercept = **(0, –2)**

f) $6x - 3y + 1 = 0 \Rightarrow 6x + 1 = 3y$
$\Rightarrow y = 2x + \frac{1}{3}$
gradient = **2**, y-intercept = $\mathbf{(0, \frac{1}{3})}$

Page 203 Exercise 3

1 a) Lines parallel to $y = 5x - 1$ have a gradient of 5, e.g. $\mathbf{y = 5x + 1}$**,** $\mathbf{y = 5x + 2}$**,** $\mathbf{y = 5x + 3}$
You can have $y = 5x + c$ for any value of c except –1 here.

b) $x + y = 7$, so $y = -x + 7$ and parallel lines have a gradient of –1, e.g. $\mathbf{y = -x + 6}$**,** $\mathbf{y = -x + 5}$**,** $\mathbf{y = -x + 3}$
You can have $y = -x + c$ for any value of c except 7 here.

2 Rearranging the lines A-F into $y = mx + c$ form:

A: $y = 2x + 4$ B: $y = x + 2.5$
C: $y = 2x - 2$ D: $y = -2x - 7$
E: $y = -\frac{2}{3}x + \frac{2}{3}$ F: $y = \frac{2}{3}x + \frac{2}{9}$

a) $y = 2x - 1$ has a gradient of 2, so lines $\mathbf{A}$ and $\mathbf{C}$ are parallel to it.

b) $2x - 3y = 0 \Rightarrow y = \frac{2}{3}x$. This line has a gradient of $\frac{2}{3}$, so line $\mathbf{F}$ is parallel to it.

3 (3, 1) and (–3, 3) are two points on the line, so the line has gradient $\frac{3-1}{(-3)-3} = \frac{2}{-6} = -\frac{1}{3}$.

You can pick any two points on the line to work out the gradient, but it's best to pick whole numbers that are nice and easy to work with.

Rearrange the lines A-F into $y = mx + c$ form:

A: $y = -3x + 2$ B: $y = -\frac{1}{3}x + \frac{7}{3}$

C: $y = -3x + 4$ D: $y = \frac{1}{3}x - \frac{8}{3}$

E: $y = -\frac{1}{3}x + 3$ F: $y = -\frac{1}{3}x$

So the lines ***B***, ***E*** and ***F*** are parallel to the line on the diagram.

Page 204 Exercise 4

1 For parts a) and b) you're given the gradient (m) and the y-intercept in the form (0, c), so put these values into $y = mx + c$.

a) $\mathbf{y = 8x + 2}$ **b)** $\mathbf{y = -x + 7}$

For the rest of the question, use the gradient and the given point to find the value of c.

c) Gradient = 3, so 10 = 3 × 1 + c, so c = 7. So the equation of the line is $\mathbf{y = 3x + 7}$.

d) Gradient = $\frac{1}{2}$, so $-5 = \frac{1}{2} \times 4 + c$, so $c = -7$. So the equation of the line is $\mathbf{y = \frac{1}{2}x - 7}$.

e) Gradient = –7, so –4 = –7 × 2 + c, so c = 10. So the equation of the line is $\mathbf{y = -7x + 10}$.

f) Gradient = 5, so –7 = 5 × (–3) + c, so c = 8. So the equation of the line is $\mathbf{y = 5x + 8}$.

2 For each part, find the gradient m from the original equation. Then use that and the given point to find the value of c in the equation $y = mx + c$ of the parallel line.

a) Gradient = 3, so $5 = 3 \times 0 + c \Rightarrow c = 5$ So the line has equation $\mathbf{y = 3x + 5}$.

b) Gradient = 5, so $-4 = 5 \times 1 + c \Rightarrow c = -9$ So the line has equation $\mathbf{y = 5x - 9}$.

c) Gradient = 2, so $6 = 2 \times 1 + c \Rightarrow c = 4$ So the line has equation $\mathbf{y = 2x + 4}$

d) Gradient = $\frac{1}{2}$, so $-7 = \frac{1}{2} \times 6 + c \Rightarrow c = -10$ So the line has equation $\mathbf{y = \frac{1}{2}x - 10}$

e) $2y = 6x + 3 \Rightarrow y = 3x + \frac{3}{2}$
Gradient = 3, so for the parallel line:
$4 = 3 \times (-3) + c \Rightarrow c = 13$
So the parallel line has equation $\mathbf{y = 3x + 13}$

f) $x + y = 4 \Rightarrow y = -x + 4$
Gradient = –1, so for the parallel line:
$8 = -1 \times 8 + c \Rightarrow c = 16$
So the parallel line has equation $\mathbf{y = -x + 16}$

Page 205 Exercise 5

1 a) Gradient = $\frac{11-7}{5-3} = \frac{4}{2} = 2$,
so 7 = 2(3) + c and c = 1.
So the equation of the line is $\mathbf{y = 2x + 1}$.

b) Gradient = $\frac{(-5)-1}{2-5} = \frac{-6}{-3} = 2$,
so 1 = 2(5) + c and c = –9.
So the equation of the line is $\mathbf{y = 2x - 9}$.

Using the same method for c)-f):

c) $\mathbf{y = x - 3}$ **d)** $\mathbf{y = 2x + 5}$

e) $\mathbf{y = 3x + 2}$ **f)** $\mathbf{y = 3x + 11}$

2 (–2, 4) and (2, 0) are two points on line A.
Gradient of line $A = \frac{4-0}{(-2)-2} = \frac{4}{-4} = -1$,
y-intercept = 2, so its equation is $\mathbf{y = -x + 2}$.
Gradient of line $B = \frac{5-0}{1-0} = \frac{5}{1} = 5$.
It goes through the origin so y-intercept = 0, so its equation is $\mathbf{y = 5x}$.
Using the same method for lines C-H:

line C: $\mathbf{y = x - 1}$ line D: $\mathbf{y = -3}$

line E: $\mathbf{y = -\frac{1}{2}x + 2}$ line F: $\mathbf{y = \frac{2}{5}x + 1}$

line G: $\mathbf{y = \frac{2}{5}x - 4}$ line H: $\mathbf{y = \frac{3}{2}x - \frac{5}{2}}$

3 For Q3, you need to pick your own two points on the line to find the gradient — choose points that lie on the grid lines so they're accurate.
E.g. the line passes through (–2.5, –1.5) and (1.5, 0.5),
so gradient = $\frac{0.5-(-1.5)}{1.5-(-2.5)} = \frac{2}{4} = \frac{1}{2} = 0.5$
so 0.5 = 0.5 × 1.5 + c and c = –0.25.
The equation of the line is $\mathbf{y = 0.5x - 0.25}$.

12.6 Quadratic Graphs

Page 207 Exercise 1

1 a)

x	–3	–2	–1	0	1	2	3
x^2	9	4	1	0	1	4	9
2	2	2	2	2	2	2	2
$x^2 + 2$	11	6	3	2	3	6	11

Don't get put off by the constant row (the 2 row here) — just fill in every entry in that row with the value given in the first column.

b), c)

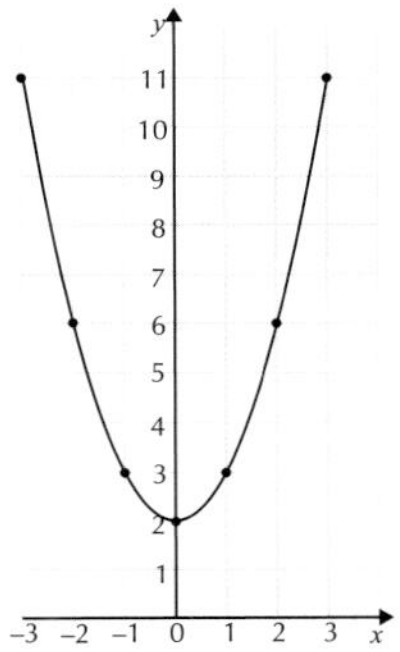

2 a)

x	–3	–2	–1	0	1	2	3
x^2	9	4	1	0	1	4	9
–1	–1	–1	–1	–1	–1	–1	–1
$x^2 - 1$	8	3	0	–1	0	3	8

b)

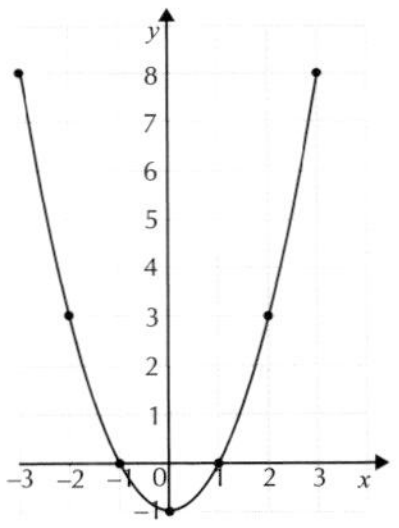

3 a) (i) E.g.

x	–3	–2	–1	0	1	2	3
x^2	9	4	1	0	1	4	9
3	3	3	3	3	3	3	3
$x^2 + 3$	12	7	4	3	4	7	12

(ii)

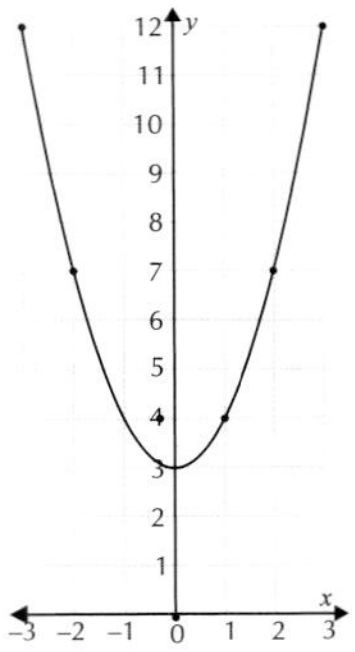

b) (i) E.g.

x	–3	–2	–1	0	1	2	3
5	5	5	5	5	5	5	5
$-x^2$	–9	–4	–1	0	–1	–4	–9
$5 - x^2$	–4	1	4	5	4	1	–4

(ii)

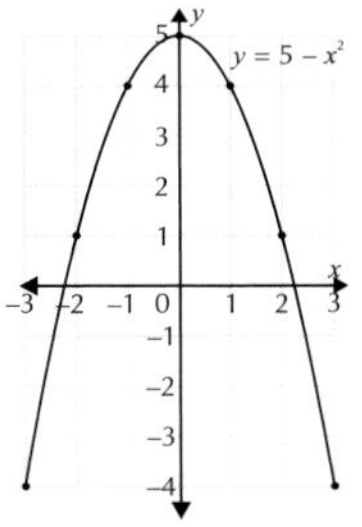

Remember, if the coefficient of x^2 is negative (–1 here), you get a n-shaped graph.

Page 208 Exercise 2

1 a)

x	–3	–2	–1	0	1	2	3
x^2	9	4	1	0	1	4	9
$-2x$	6	4	2	0	–2	–4	–6
$x^2 - 2x$	15	8	3	0	–1	0	3

b)

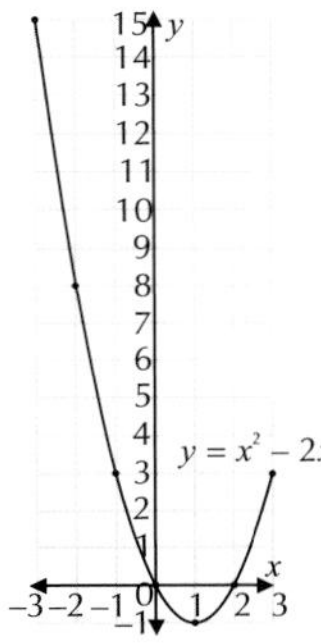

2 a) (i)

x	–3	–2	–1	0	1	2	3
x^2	9	4	1	0	1	4	9
$3x$	–9	–6	–3	0	3	6	9
$x^2 + 3x$	0	–2	–2	0	4	10	18

(ii)

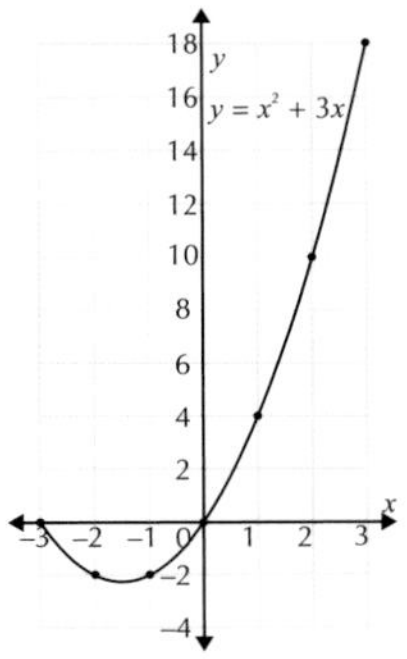

b) (i)

x	–3	–2	–1	0	1	2	3
x^2	9	4	1	0	1	4	9
$-4x$	12	8	4	0	–4	–8	–12
$x^2 - 4x$	21	12	5	0	–3	–4	–3

(ii)

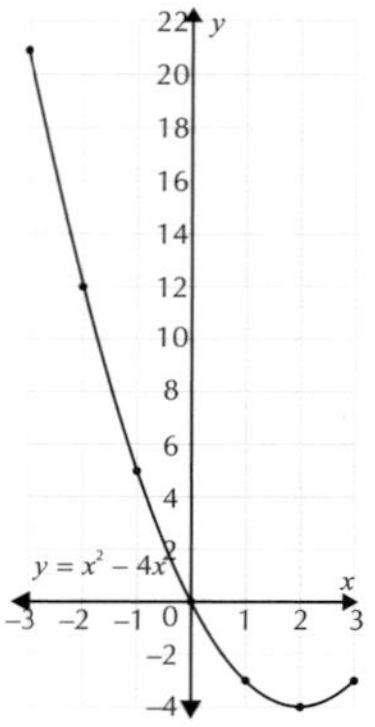

Page 208 Exercise 3

1 a) (i)

x	–3	–2	–1	0	1	2	3
x^2	9	4	1	0	1	4	9
$2x$	–6	–4	–2	0	2	4	6
5	5	5	5	5	5	5	5
$x^2 + 2x + 5$	8	5	4	5	8	13	20

(ii)

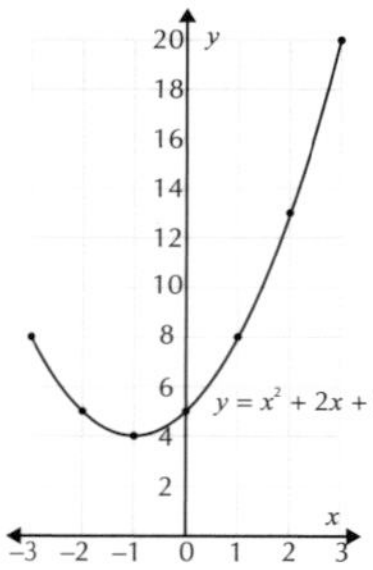

b) (i)

x	–3	–2	–1	0	1	2	3
$-x^2$	–9	–4	–1	0	–1	–4	–9
$-x$	3	2	1	0	–1	–2	–3
–1	–1	–1	–1	–1	–1	–1	–1
$-x^2 - x - 1$	–7	–3	–1	–1	–3	–7	–13

(ii)

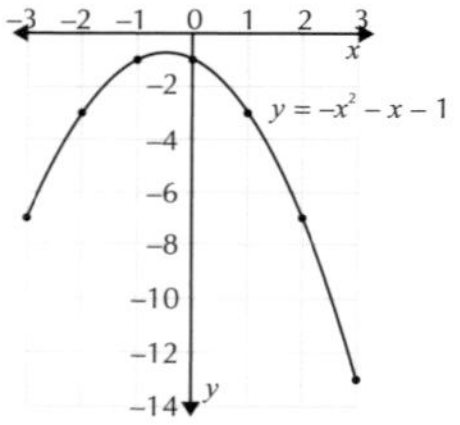

2 a)

x	–4	–3	–2	–1	0	1	2	3	4
$2x^2$	32	18	8	2	0	2	8	18	32
$3x$	–12	–9	–6	–3	0	3	6	9	12
–7	–7	–7	–7	–7	–7	–7	–7	–7	–7
$2x^2 + 3x - 7$	13	2	–5	–8	–7	–2	7	20	37

b)

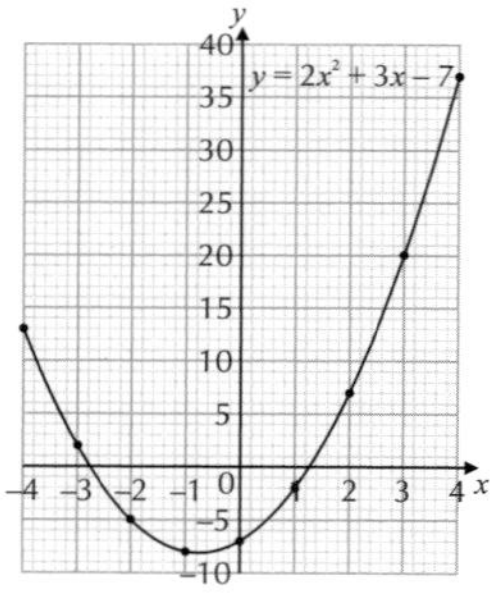

Page 209 Exercise 4

1 a) **$x = -4$ and $x = 0$**

The solutions are the points where the graph crosses the x-axis.

b) **$x = -1$** and **$x = 3$** c) **$x = -2$** and **$x = 2$**

2 a)

x	0	1	2	3	4	5
x^2	0	1	4	9	16	25
$-5x$	0	–5	–10	–15	–20	–25
3	3	3	3	3	3	3
$x^2 - 5x + 3$	3	–1	–3	–3	–1	3

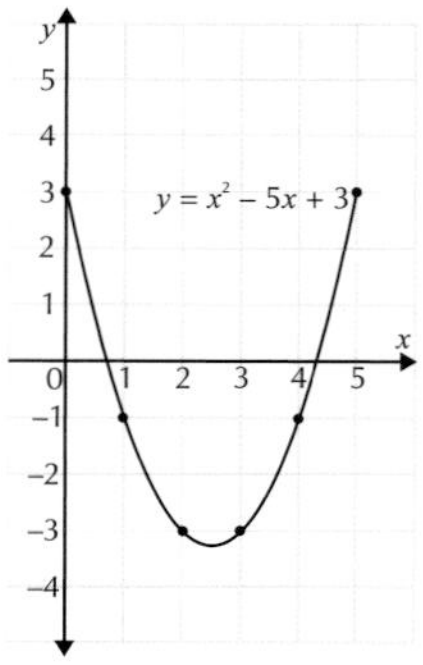

b) **$x = 0.7$ and $x = 4.3$**
(accept 0.6 to 0.8 and 4.2 to 4.4)

Page 210 Exercise 5

1 a) **minimum**

The turning point is a minimum if the x^2 term is positive.

b) **maximum**

The turning point is a maximum if the x^2 term is negative.

c) **minimum** d) **minimum**

e) **maximum** f) **maximum**

2 a) E.g. $y = 0$ at $x_1 = -4$ and $x_2 = -1$,
so turning point is at $x = \frac{-4+(-1)}{2} = -2.5$.
When $x = -2.5$,
$y = (-2.5)^2 + (5 \times -2.5) + 4 = -2.25$.
The turning point is **(–2.5, –2.25)**.

b) E.g. $y = 0$ at $x_1 = -4$ and $x_2 = 1$, so turning point is at $x = \frac{-4+1}{2} = -1.5$.
When $x = -1.5$,
$y = -(-1.5)^2 - (3 \times -1.5) + 4 = 6.25$.
The turning point is **(–1.5, 6.25)**.

3 For Q3 you don't know the points where the curve intersects the x-axis — but it doesn't matter, because you're given two other points with the same y-value. So just use those to find the x-value of the turning point.

a) $y = 7$ at $x_1 = -1$ and $x_2 = 7$, so turning point is at $x = \frac{-1+7}{2} = 3$.
When $x = 3$, $y = (3)^2 - (6 \times 3) = -9$.
The turning point is **(3, –9)**.

b) $y = 3$ at $x_1 = 0$ and $x_2 = 2$, so turning point is at $x = \frac{0+2}{2} = 1$.
When $x = 1$, $y = (1)^2 - (2 \times 1) + 3 = 2$.
The turning point is **(1, 2)**.

c) $y = 12$ at $x_1 = 2$ and $x_2 = 6$, so turning point is at $x = \frac{2+6}{2} = 4$.
When $x = 4$, $y = (8 \times 4) - (4)^2 = 16$.
The turning point is **(4, 16)**.

Page 211 Exercise 6

1 For each part a)-f), make a table of values and look for two x-values that give the same y-value. The average of these gives the x-coordinate of the turning point, and the y-coordinate can be found using either the table or the equation.

a)

x	–3	–2	–1	0	1	2	3
$x^2 - 2x$	15	8	3	0	–1	0	3

$x = 0$ and $x = 2$ have the same y-value of $y = 0$, so the x-coordinate of the turning point is $x = \frac{0+2}{2} = 1$. So using the table, the turning point is **(1, –1)**.

b)

x	–3	–2	–1	0	1	2	3
$x^2 + 3x + 11$	11	9	9	11	15	21	29

$x = -2$ and $x = -1$ have the same y-value of $y = 9$, so the x-coordinate of the turning point is $x = \frac{-2+(-1)}{2} = -1.5$.
The y-coordinate of the turning point is $y = (-1.5)^2 + (3 \times -1.5) + 11 = 8.75$ so the turning point is **(–1.5, 8.75)**

Using the same method for c)-f):

c) **(–2.5, –15.25)** d) **(–2, –20)**

e) **(–0.5, 9.5)** f) **(1.5, 10.5)**

12.7 Harder Graphs

Page 213 Exercise 1

1 a)

x	–3	–2	–1	0	1	2	3
x^3	–27	–8	–1	0	1	8	27
$x^3 + 5$	–22	–3	4	5	6	13	32

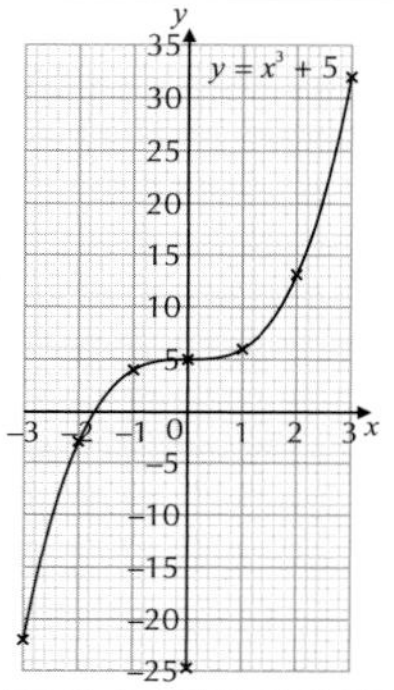

b)

x	–3	–2	–1	0	1	2	3
5	5	5	5	5	5	5	5
$-x^3$	27	8	1	0	–1	–8	–27
$5 - x^3$	32	13	6	5	4	–3	–22

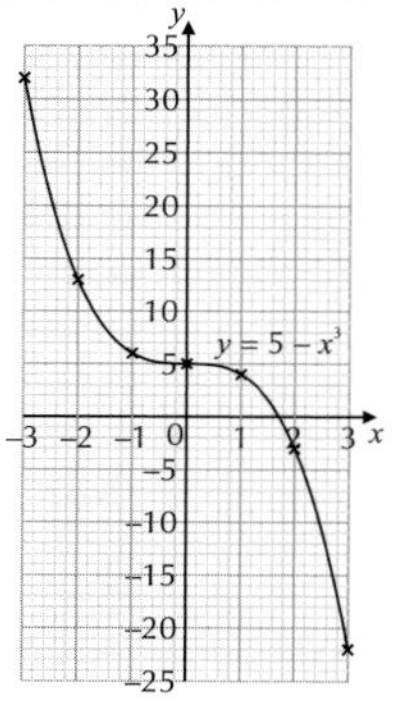

Remember, when the coefficient of x^3 is negative, it goes downwards from left to right — by looking at the coefficient you can do a quick check that your curve is the right shape.

c)

x	−3	−2	−1	0	1	2	3
−4	−4	−4	−4	−4	−4	−4	−4
$-x^3$	27	8	1	0	−1	−8	−27
$-4-x^3$	23	4	−3	−4	−5	−12	−31

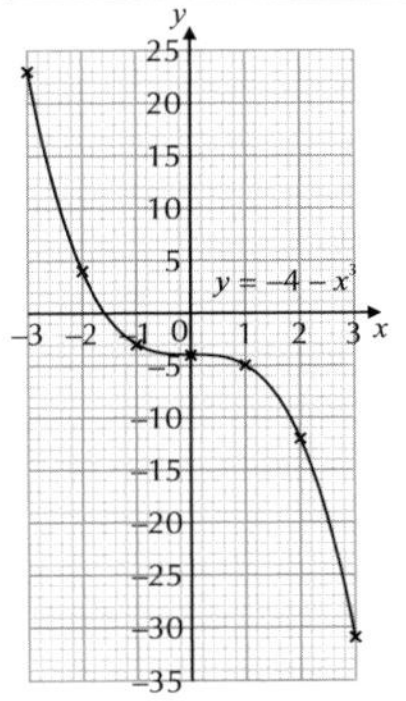

2

x	−3	−2	−1	0	1	2	3
x^3	−27	−8	−1	0	1	8	27
3	3	3	3	3	3	3	3
$x^3 + 3$	−24	−5	2	3	4	11	30

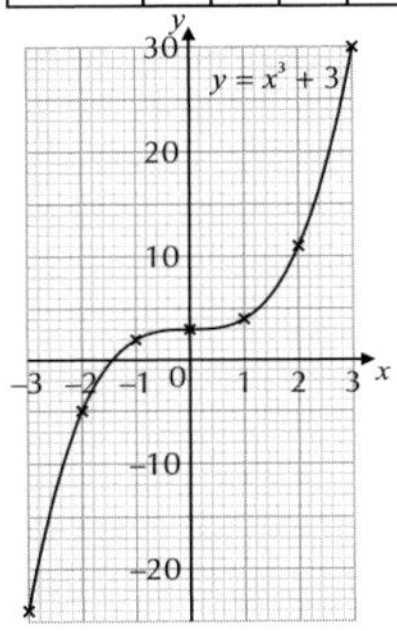

3

x	−4	−3	−2	−1	0	1	2	3	4
x^3	−64	−27	−8	−1	0	1	8	27	64
$-3x$	12	9	6	3	0	−3	−6	−9	−12
7	7	7	7	7	7	7	7	7	7
$x^3 + 3$	−45	−11	5	9	7	5	9	25	59

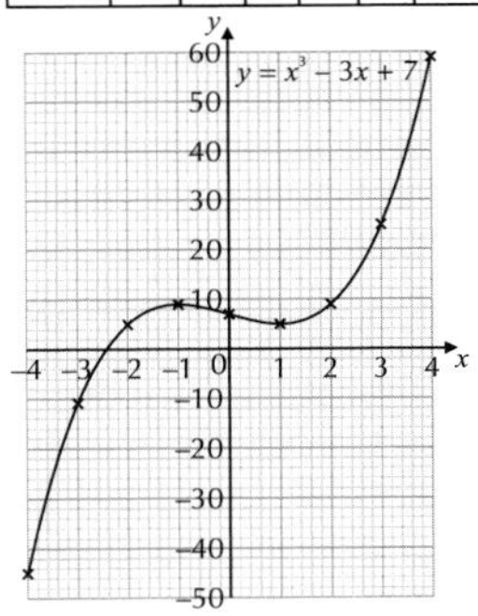

Page 214 Exercise 2

1 A basic reciprocal equation has the form $y = \frac{A}{x}$. The following equations are either in or can be rearranged into this form: ***B*, *C*, *E* and *G***

2 a)

x	−5	−4	−2	−1	−0.5	−0.1
$\frac{2}{x}$	**−0.4**	−0.5	**−1**	−2	**−4**	**−20**

x	0.1	0.5	1	2	4	5
$\frac{2}{x}$	**20**	**4**	**2**	**1**	**0.5**	**0.4**

Lots goes on between x = −1 and x = 1 in a reciprocal graph, so decimal values are included in the table to make sure nothing's missed.

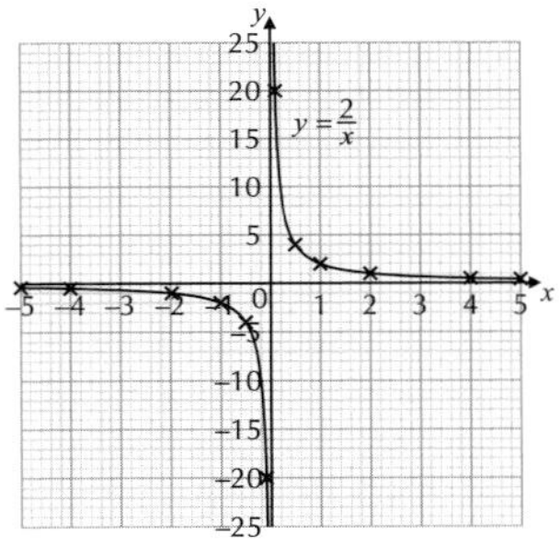

b)

x	−5	−4	−2	−1	−0.5	−0.1
$-\frac{1}{x}$	**0.2**	0.25	**0.5**	1	**2**	**10**

x	0.1	0.5	1	2	4	5
$-\frac{1}{x}$	**−10**	**−2**	**−1**	**−0.5**	**−0.25**	**−0.2**

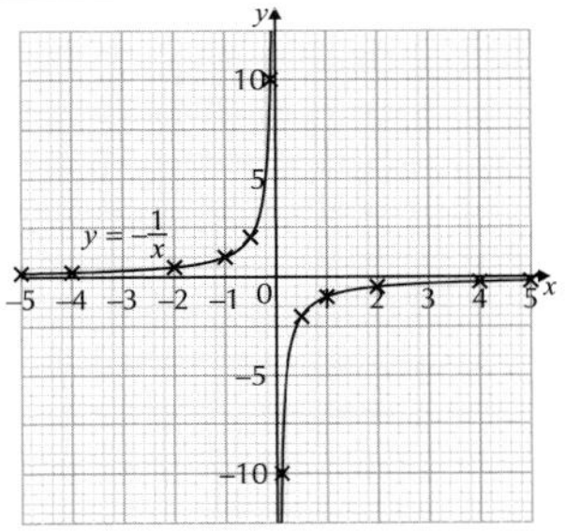

c)

x	−5	−4	−2	−1	−0.5	−0.1
$\frac{3}{x}$	**−0.6**	−0.75	**−1.5**	−3	**−6**	**−30**

x	0.1	0.5	1	2	4	5
$\frac{3}{x}$	**30**	**6**	**3**	**1.5**	**0.75**	**0.6**

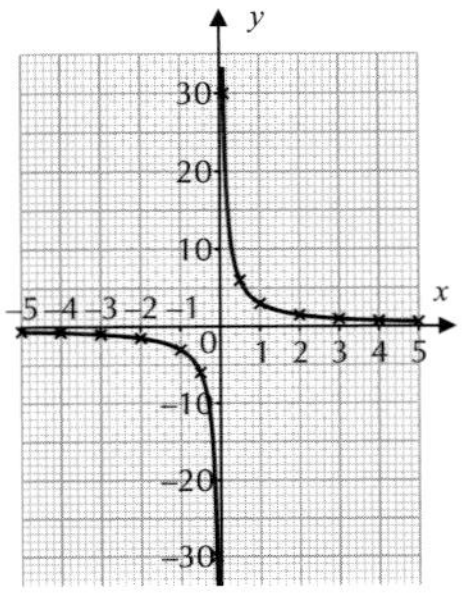

Page 215 Review Exercise

1 a), b)

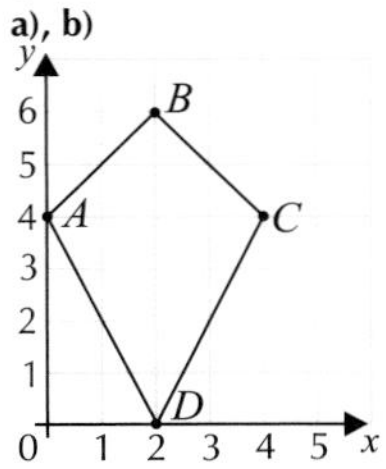

this is a kite.

2 Midpoint $= \left(\frac{2+(-4)}{2}, \frac{5+1}{2}\right) = \left(\frac{-2}{2}, \frac{6}{2}\right) =$ **(−1, 3)**

3 a) ***A*: (1, 5) *B*: (4, 5) *C*: (6, 3)**
***D*: (4, 1) *E*: (1, 1)**

b) Midpoint $= \left(\frac{1+4}{2}, \frac{5+5}{2}\right) = \left(\frac{5}{2}, \frac{10}{2}\right) =$ **(2.5, 5)**

c) Midpoint $= \left(\frac{4+6}{2}, \frac{5+3}{2}\right) = \left(\frac{10}{2}, \frac{8}{2}\right) =$ **(5, 4)**

d) $\mathbf{x = 1}$
Remember, vertical lines have an equation of the form x = a.

e) gradient $= \frac{3-5}{6-4} = \frac{-2}{2} = \mathbf{-1}$.

4 a)

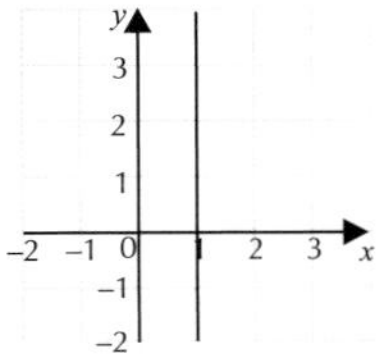

b)

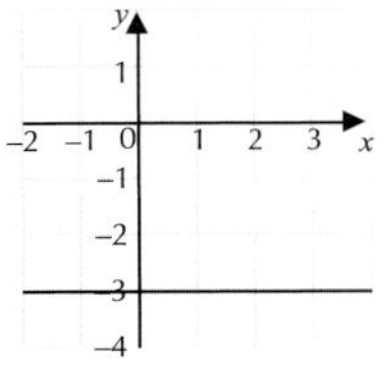

c)

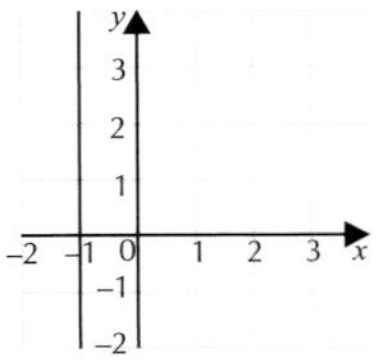

d)

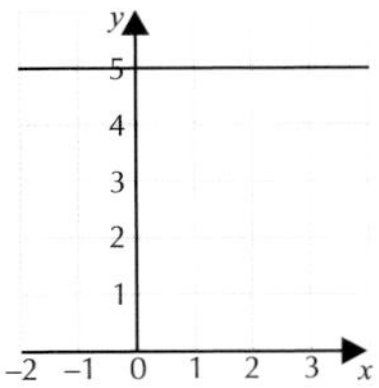

5 ***AB*:** gradient $= \frac{4-1}{-5-(-2)} = \frac{3}{-3} = \mathbf{-1}$

***CD*:** gradient $= \frac{5-1}{3-5} = \frac{4}{-2} = \mathbf{-2}$

***EF*:** gradient $= \frac{-3-(-5)}{2-(-5)} = \mathbf{\frac{2}{7}}$

6 ***A*: $x = -5$** — it's a vertical line through (−5, 0).
***B*: $y = x^2$** — it's a u-shaped parabola.
***C*: $y = x$** — it's a straight line through (0, 0) with a gradient of 1.
***D*: $y = -4$** — it's a horizontal line through (0, −4).
Look for the main features of each graph — e.g. is it straight or curved? What are the x- and y-intercepts?

7 a)

x	–3	–2	–1	0	1	2	3
y	–4	–2	0	2	4	6	8
	(–3, –4)	**(–2, –2)**	**(–1, 0)**	**(0, 2)**	**(1, 4)**	**(2, 6)**	**(3, 8)**

b)

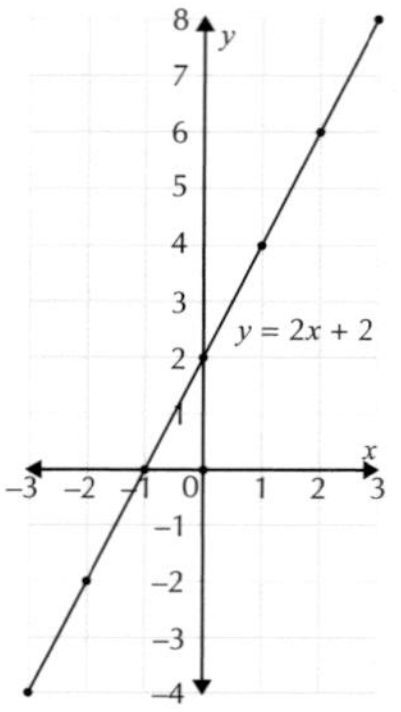

c) On the same axes, draw the line $y = 5$. Find its intersection with the graph of $y = 2x + 2$ and read off the x-coordinate to get the solution $\boldsymbol{x = 1.5}$.

8 a) Draw the graphs of $y = 2x + 3$ and $y = 4x + 2$ using the methods on p.194.

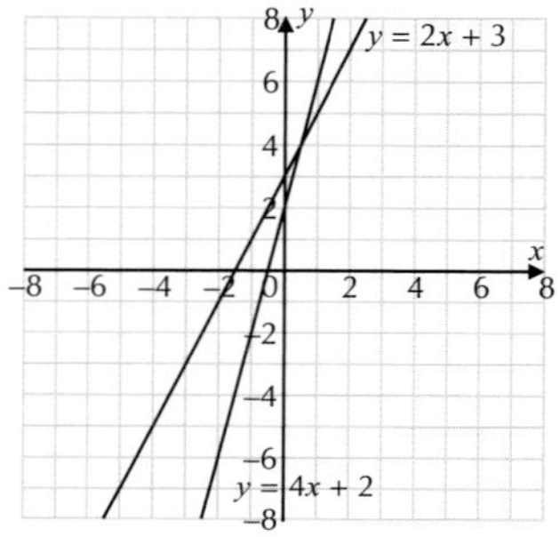

b) The lines intersect at the point ($\frac{1}{2}$, 4). So the solution is $\boldsymbol{x = \frac{1}{2}}$ and $\boldsymbol{y = 4}$.

9 a) gradient $= \frac{15-0}{3-0} = \frac{15}{3} = \mathbf{5}$

b) gradient $= \frac{4-(-5)}{-1-(-4)} = \frac{9}{3} = \mathbf{3}$

c) gradient $= \frac{-1-2}{-4-(-7)} = \frac{-3}{3} = \mathbf{-1}$

d) gradient $= \frac{10-(-2)}{2-5} = \frac{12}{-3} = \mathbf{-4}$

10 Compare each equation with $y = mx + c$. m is the gradient and $(0, c)$ is the y-intercept.

a) (i) 2 **(ii) (0, 3)**
b) (i) –1 **(ii) (0, 3)**
c) (i) $\boldsymbol{-\frac{2}{3}}$ **(ii) (0, –1)**

11 a) Draw the graph of $y = 6 - 3x$ using the methods on p.194.

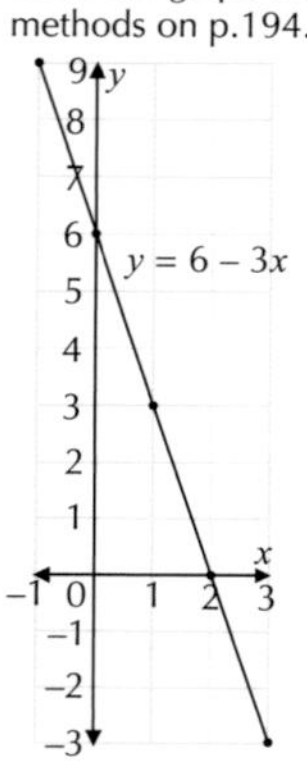

b) gradient = –3, so plugging this and the point (2, –4) into $y = mx + c$ gives $-4 = -3 \times 2 + c \Rightarrow c = 2$. So the equation of the line is $\boldsymbol{y = 2 - 3x}$.

12 a) E.g.

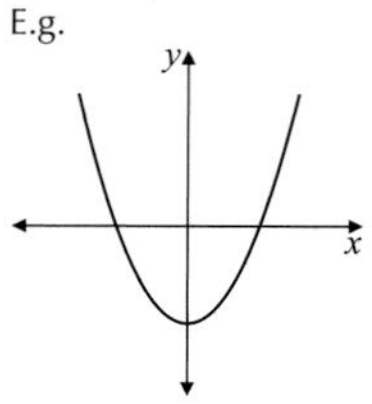

b) E.g.

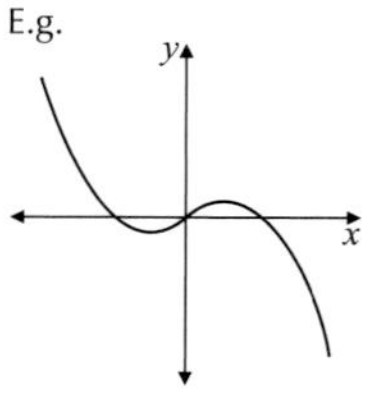

c)

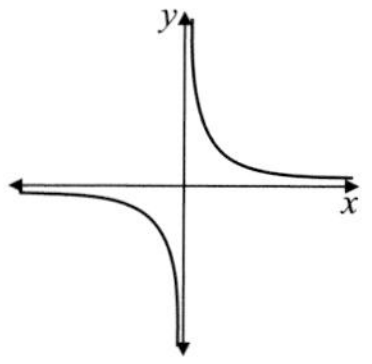

13 Find where the curve intersects the x-axis, and read off the x-coordinates to get the solutions $\boldsymbol{x = -2.3, x = 1.3}$

14 a)

x	–4	–3	–2	–1	0	1	2
x^2	**16**	**9**	4	**1**	**0**	**1**	**4**
$5x$	**–20**	**–15**	–10	**–5**	**0**	**5**	**10**
$x^2 + 5x$	**–4**	**–6**	–6	**–4**	**0**	**6**	**14**

b)

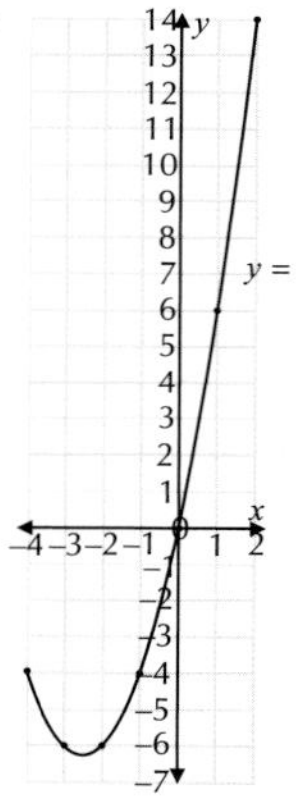

c) E.g. $y = -6$ at $x_1 = -3$ and $x_2 = -2$, so the turning point is at

$$x = \frac{-3+(-2)}{2} = \frac{-5}{2} = -2.5.$$

When $x = -2.5$, $y = (-2.5)^2 + (5 \times -2.5) = -6.25$.

The turning point is **(–2.5, –6.25)**.

You could also have used $x = -4$ and $x = -1$ as your two values, since $y = -4$ at both of those points — you'd still have come out with the same answer at the end.

Page 217 Exam-Style Questions

1

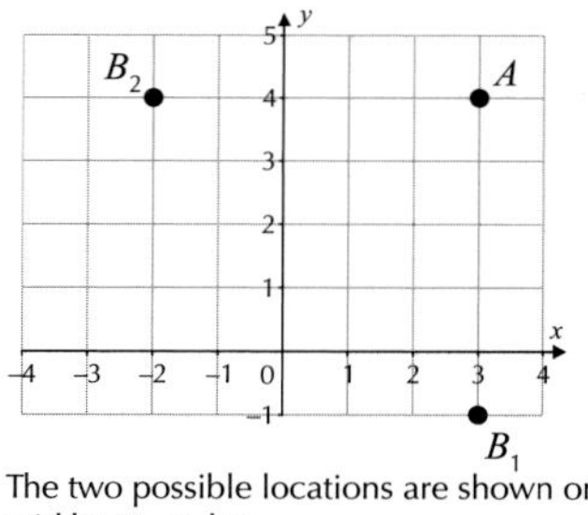

The two possible locations are shown on the grid by $\boldsymbol{B_1}$ and $\boldsymbol{B_2}$.
[2 marks available — 1 mark for each correct location]

2 Difference in x-values between A and B is $3 - (-2) = 5$.
Difference in y-values between A and B is $k - 1$.
Gradient $= \frac{k-1}{5}$ *[1 mark]*
So $\frac{k-1}{5} = -1.5$ *[1 mark]*
$k - 1 = -1.5 \times 5 \Rightarrow k - 1 = -7.5$
$k = -7.5 + 1 \Rightarrow \boldsymbol{k = -6.5}$ *[1 mark]*

3 a) $\boldsymbol{A = (4, 3)}$ *[1 mark]*
$\boldsymbol{B = (8, 1)}$ *[1 mark]*

b) Difference in x values between A and B is $8 - 4 = 4$.
Difference in y values between A and B is $1 - 3 = -2$
So gradient $= \frac{-2}{4} = -\frac{1}{2}$ *[1 mark]*
Equation of line is $\boldsymbol{y = 5 - \frac{1}{2}x}$ *[1 mark]*
The right-hand side of the equation begins with 5 because that's the value of the y-intercept on the graph.

4 a) (0, 1.5) *[1 mark]*

b) Midpoint $= \left(\frac{-3+5}{2}, \frac{3+(-1)}{2}\right) = \left(\frac{2}{2}, \frac{2}{2}\right)$
= **(1, 1)** *[1 mark]*

5 a) (1, 2) *[1 mark]*

b) Go up from 3 on the x-axis until you reach the curve, then go across to the y-axis. This gives $\boldsymbol{y = 6}$ *[1 mark]*.

c) Go left from 6 on the y-axis until you reach the curve, then go down to the x-axis. This gives $\boldsymbol{x = -1}$ *[1 mark]*.

6 a) $\boldsymbol{x = 1}$ *[1 mark]*

b) Find where the curve intersects the x-axis and read off the x-coordinates to get $\boldsymbol{x = -0.9}$ and $\boldsymbol{x = 2.9}$.
[2 marks available — 1 mark for each correct x-coordinate. Accept x = –0.8 and x = 2.8 as alternative answers.]

7 The curve has not been drawn smoothly *[1 mark]*.
The point (2, 2) should not lie on this graph of $y = \frac{1}{x}$ *[1 mark]*.

Section 13 — Real-Life Graphs

13.1 Interpreting Real-Life Graphs

Page 220 Exercise 1

1 a) Read up from £50 on the horizontal axis and then across to the vertical axis to get an answer of **€60**.
b) Read up from £250 on the horizontal axis and then across to the vertical axis to get an answer of **€290**.
c) Read up from £110 on the horizontal axis and then across to the vertical axis to get an answer of **€130**.

2 a) Read across from €50 on the vertical axis and then down to the horizontal axis to get an answer of **£40**.
b) Read across from €200 on the vertical axis and then down to the horizontal axis to get an answer of **£170**.
c) Read across from €360 on the vertical axis and then down to the horizontal axis to get an answer of **£310**.

3 Read across from €130 on the vertical axis and then down to the horizontal axis to get an answer of **£110**.

4 a) Read up from £420 on the horizontal axis and then across to the vertical axis to get an answer of **€490**.
b) Read across from €470 on the vertical axis and then down to the horizontal axis to get an answer of **£400**.
c) €470 = £400 is less than £420, so it's cheaper in **France**.
Make sure you're comparing numbers that have the same units.

Page 221 Exercise 2

1 a) The basic fine is shown by the horizontal line on the graph. This intersects the vertical axis at 5, so the basic fine is **£5**.
b) Read down from the right-hand end of the horizontal line.
The longest time that receives the basic fine is **20 days**.

2 Go up from 50 days until you reach the graph, then go across to the horizontal axis.
The fine for 50 days overdue is **£27.50**.
Any answer between £27 and £28 is OK.

3 Go across from £38 on the vertical axis until you reach the graph, then go down to the horizontal axis.
This shows the book was overdue by **64** days.

4 a) Look at whether the graphs for each climber reach 1 km:
Katherine reached a **height of 1 km** after **4 hours** and **Lemar** reached **1 km** after **3 hours**. **Morag's greatest height** was **0.75 km**, so she **didn't** reach the summit. So the climbers that **reached the summit** were **Katherine and Lemar**.
b) **Lemar** reached the summit in less time than Katherine, so he was first.

Page 222 Exercise 3

1 a) Look for a graph that increases with a steep gradient and then decreases gently. This matches graph **L**.
b) Look for a graph that neither rises nor falls (i.e. is horizontal). This matches graph **N**.
c) Look for a graph that increases and has a gradient which gets more and more steep. This matches graph **M**.
d) Look for a graph that decreases initially but then rises steeply. This matches graph **K**.

2

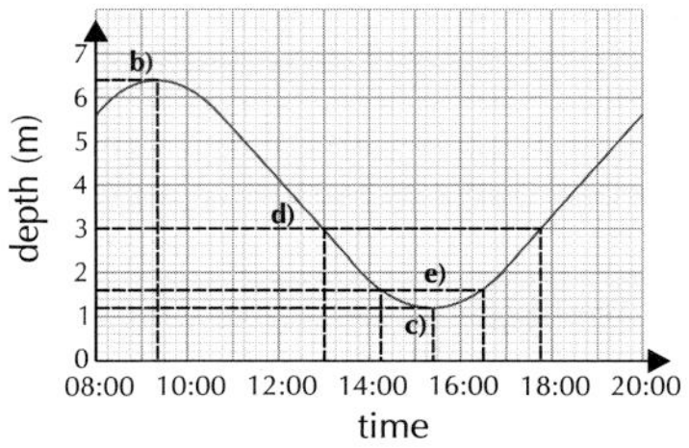

a) E.g. The water got **deeper** for about **an hour** (since the graph **increases**), then **shallower** for about **6 hours** (since the graph **decreases**). Finally it got **deeper** for about **5 hours** (since the graph **increases** again).
Make sure you describe each feature of the graph in the context of the question.
b) The greatest depth corresponds to the highest point on the graph. This occurs at about **09:20** (accept 09:15-09:30).
c) The minimum depth corresponds to the lowest point on the graph. At this point, the depth is **1.2 m**.
d) Read across from 3 m. The graph is at this depth twice, once at approx. **13:00** and again at approx. **17:45**.
e) The depth of the water is below 1.6 m between about 14:15 and 16:30. So his boat is not floating for **2h 15m** (135 mins). (Accept answers between 2 hours 10 minutes and 2 hours 20 minutes.)

3

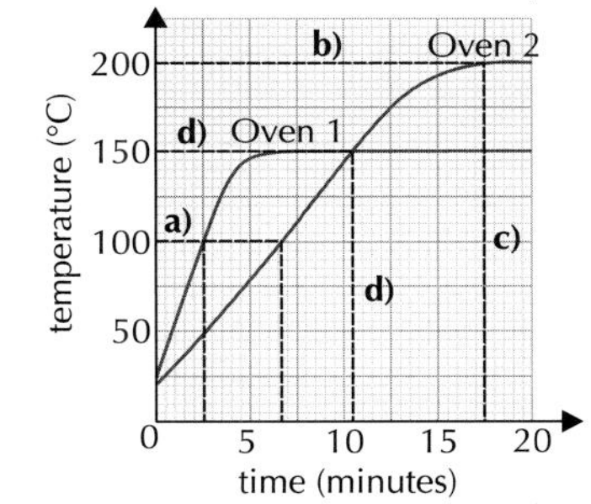

a) Reading across from 100 °C on the vertical axis, the graph for Oven 1 shows it's at this temperature after about 2.5 minutes, while the graph for Oven 2 shows it's at this temperature after about 6.5 minutes.
So **Oven 1** reaches 100 °C more quickly.
b) The maximum point on the graph for Oven 1 is at 150 °C, while the maximum point on the graph for Oven 2 is at about 200 °C. So **Oven 2** reaches a higher maximum temperature.
c) Reading down from the highest point on the Oven 2 graph gives a time of about **18 minutes** (accept 17.5-18.5 minutes).
d) The ovens are the same temperature where the lines intersect. This point occurs at:
(i) **10.5 minutes**
(ii) **150 °C**
e) E.g. The gradients show the change in the temperature of each oven per unit of time — this is the rate at which the temperature of each oven changes.

4 a) (i) At 0 s the depth is 0 cm, at 5 s the depth is 5 cm. The increase in depth is 5 cm – 0 cm = **5 cm**.
(ii) At 10 s the depth is 10 cm, at 15 s the depth is 15 cm. The increase in depth is 15 cm – 10 cm = **5 cm**.
(iii) At 25 s the depth is 25 cm, at 30 s the depth is 30 cm. The increase in depth is 30 cm – 25 cm = **5 cm**.
b) (i) At 0 s the depth is 0 cm, at 5 s the depth is 10 cm. The increase in depth is 10 cm – 0 cm = **10 cm**.
(ii) At 10 s the depth is 17 cm, at 15 s the depth is 21.5 cm. The increase in depth is 21.5 cm – 17 cm = **4.5 cm**.
(iii) At 25 s the depth is 27.5 cm, at 30 s the depth is 30 cm. The increase in depth is 30 cm – 27.5 cm = **2.5 cm**.
c) The depth of water in Vase P increases steadily while the depth of water in Vase Q increases quickly at first, and then more slowly as the vase fills up. When the vases are nearly empty, the depth of water in Vase Q increases more quickly than the depth of water in Vase P, but as the vases fill up the depth of water in Vase Q increases at a slower rate than the depth of water in vase P.

13.2 Drawing Real-Life Graphs

Page 225 Exercise 1

1 a) You would cook a chicken weighing 1 kg for (35 × 1) + 25 = 60 minutes.
You would cook a chicken weighing 2 kg for (35 × 2) + 25 = 95 minutes.
Using the same method for the remaining weights, the table can be completed as follows.

Weight (kg)	1	2	3	4	5
Time (minutes)	**60**	**95**	**130**	**165**	**200**

b) Plot each pair of values on suitable axes and join with a straight line:

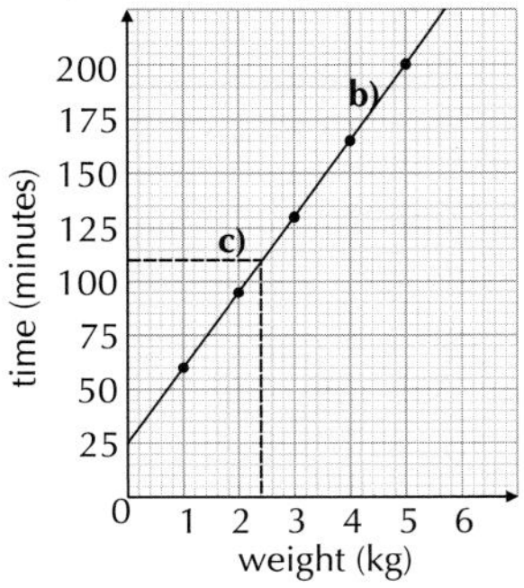

c) Reading across from 110 minutes on the vertical axis and then down to the horizontal axis, the weight is **2.4 kg**. (Accept 2.3-2.5 kg)

2 a) Fuel efficiency depends on speed, so fuel efficiency should go on the vertical axis and speed on the horizontal axis.
Plot each pair of values from the table and join them with a smooth curve:

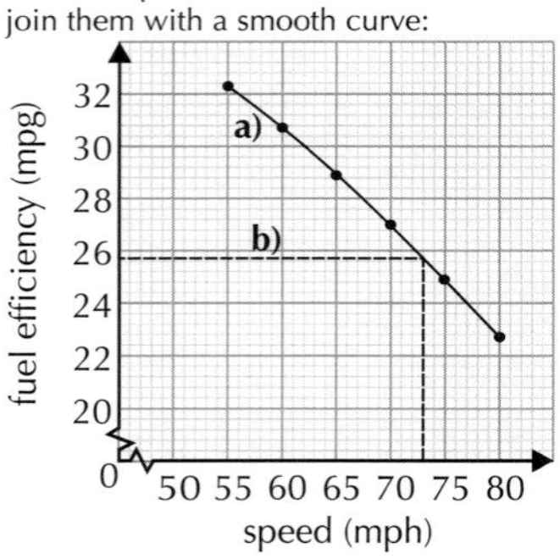

You don't need speeds lower than 55 mph or fuel efficiencies lower than 22.7 mpg so you can remove parts of both axes (shown by the squiggles).

b) Reading up from 73 mph on the horizontal axis and then across to the vertical axis, the fuel efficiency is about **25.7 mpg**.
(Accept 25.5-25.9 mpg)

Page 226 Review Exercise

1 a) Read up from 38 km/h on the horizontal axis and then across to the vertical axis to get an answer of **23.5 mph**.
(Accept 23-24 mph)

b) Read across from 27 mph on the vertical axis and then down to the horizontal axis to get an answer of **43 km/h**.
(Accept 42-44 km/h)

c) Read up from 52 km/h on the horizontal axis and then across to the vertical axis to find that 52 km/h ≈ 32.5 mph. So the driver is about 32.5 – 30 = **2.5 mph** over the speed limit. (Accept 2-3 mph)

d) Read up from 60 km/h on the horizontal axis and then across to the vertical axis to find 60 km/h ≈ 37.5 mph. Double this to find that 120 km/h ≈ 75 mph, which is greater than 70 mph. So the speed limit is greater in **Spain** by about 75 – 70 = **5 mph**. (Accept 4-6 mph)
Alternatively, you could convert the UK speed limit into km/h. Using the graph, 35 mph ≈ 56 km/h. Double that to find that 70 mph ≈ 112 km/h. So the speed limit is greater in Spain by about 8 km/h.

2 a) The graph initially increases and then decreases so the temperature **rises for the first 3 seconds** and then **decreases for the remaining 5 seconds**. Over the first 3 seconds, the gradient of the graph decreases to zero, so the temperature **rises at a slower rate over time** until it **stops rising**. In the final 5 seconds, the gradient gets steeper and steeper so the temperature **falls at a more rapid rate** as time goes on.

b) The graph is at its highest when $t = 3$. Reading across to the vertical axis, this is at a temperature of **9 °C**.

c) Reading across from 8 °C on the vertical axis and down to the horizontal axis, the times are **t = 2 seconds** and **t = 4 seconds**.

3 a) Plot each pair of values from the table on suitable axes and join them with a smooth curve:

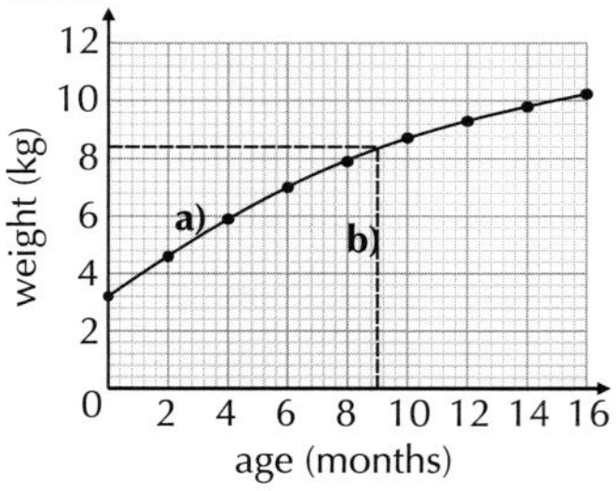

b) Reading up from 9 months on the horizontal axis and across to the vertical axis, the weight is about 8.4 kg.
So Keira is about 9.1 – 8.4 = **0.7 kg** heavier.
(Accept 0.5-0.9 kg)

Page 227 Exam-Style Questions

1 a) 60 m *[1 mark]*
Find the highest point on the graph and read across to the vertical axis.

b) Beach B *[1 mark]*
E.g. the graph **flattens out** towards Beach B so the **depth** of the water is **lower** there than near Beach A *[1 mark]*.

2 a)

Weight (kg)	10	20	30	40	50
Cost (£) of Kenyan coffee	25	50	**75**	**100**	**125**
Cost (£) of Peruvian coffee	35	**45**	**55**	**65**	**75**

[2 marks available — 1 mark for each correct row of the table]

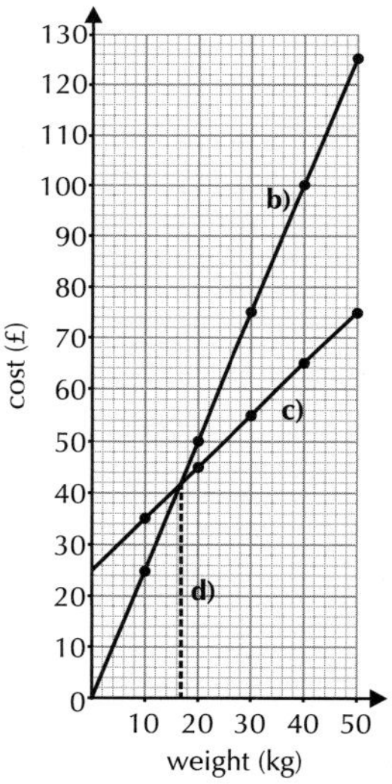

b) Plot each pair of values on suitable axes and join with a straight line (see above).
[2 marks available — 1 mark for plotting points, 1 mark for joining with a straight line]

c) Plot each pair of values and join with a straight line (see above).
[2 marks available — 1 mark for plotting points, 1 mark for joining with a straight line]

d) Find the intersection point of the two lines, then draw a line down to the vertical axis to get an estimate of **16.5 kg**.
[1 mark available — answers in the range 16.0-17.0 kg are acceptable]

Section 14 — Proportion

14.1 Direct Proportion

Page 229 Exercise 1

1 a) 1 pair of jeans costs £35, so 2 pairs cost £35 × 2 = **£70**

b) 5 pairs of jeans cost £35 × 5 = **£175**

c) 20 pairs of jeans cost £35 × 20 = **£700**

2 8 books cost £36, so 1 book costs £36 ÷ 8 = £4.50.
12 books cost £4.50 × 12 = **£54.**

3 a) Each DVD costs £7.50, so with £22.50 you can buy £22.50 ÷ £7.50 = **3 DVDs.**

b) With £60, you can buy £60 ÷ £7.50 = **8 DVDs.**
You could also divide 1 DVD by 7.50 to find the number of DVDs per pound, then multiply by 22.5 and 60.

4 a) On 1 litre of petrol, the car travels 250 ÷ 35 = 7.142... km
On 50 litres of petrol, the car travels 50 × 7.142... = 357.142... km
= **357 km (to the nearest km)**

b) Travelling 1 km uses 35 ÷ 250 = 0.14 litres of petrol. So travelling 400 km uses 400 × 0.14 = **56 litres**
Don't round any numbers while you're doing your working or you might get the wrong answer.

5 11 pens cost £12.32 ⇒ 1 pen costs £12.32 ÷ 11 = £1.12.
8 pens cost 8 × £1.12 = £8.96.
6 note pads cost £5.88
⇒ 1 note pad costs £5.88 ÷ 6 = £0.98.
Then 5 note pads cost 5 × £0.98 = £4.90.
In total 8 pens and 5 note pads cost £8.96 + £4.90 = **£13.86**

Page 229 Exercise 2

1 Pack of 6: £1.50 ÷ 6 = £0.25 or 25p per apple
Bag of 10: £2.40 ÷ 10 = £0.24 or 24p per apple
Individually: 30p
So the **bag of 10 apples** represents the best value for money.

2 Find how many ml you get per £1 for each size of coffee cup:
Small cup: 240 ÷ 2 = 120 ml per £1
Medium cup: 350 ÷ 3 = 116.666... ml per £1
Large cup: 470 ÷ 4 = 117.5 ml per £1
So the **small cup** represents the best value for money.
You could also find the price per ml and see which is cheapest.

3 6 pack: £2.18 = 218p ÷ 6
= 36.333...p per sausage
8 pack: £2.80 = 280p ÷ 8 = 35p per sausage
10 pack: £3.46 = 346p ÷ 10
= 34.6p per sausage
So the **pack of 10 sausages** represents the best value for money.

Page 231 Exercise 3

1 Graph C — it is a straight line through the origin.

Graph A does not go through the origin and Graph B isn't a straight line, so neither of them can be directly proportional to x.

2 a)

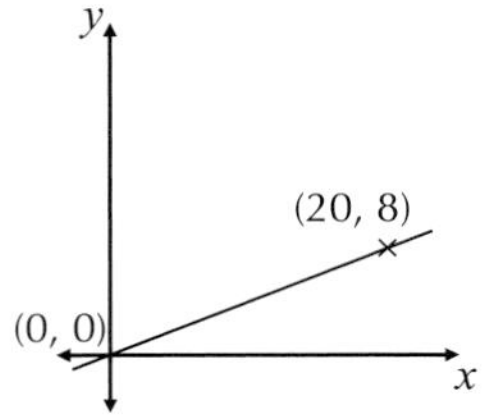

b)

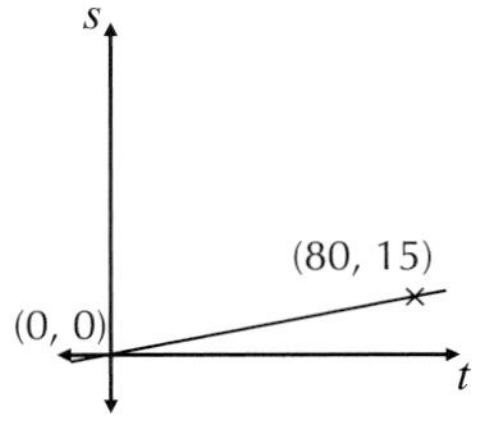

3 a) (i) Use the 'divide for one, multiply for all' method:
When $x = 1$, $y = 2 \div 8 = 0.25$,
so when $x = 12$, $y = 12 \times 0.25 = 3$.

x	8	12
y	2	**3**

(ii) x and y are directly proportional, so the graph is a straight line through the origin. Plot the points from your table and draw a line through them:

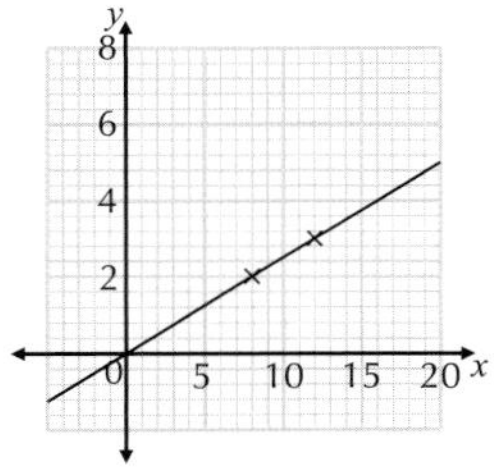

b) (i) When $x = 1$, $y = 21 \div 7 = 3$,
so when $x = 2$, $y = 2 \times 3 = 6$,
when $y = 30$, $x = 30 \div 3 = 10$.

x	2	7	**10**
y	**6**	21	30

(ii) Plot your points and draw a straight line through them:

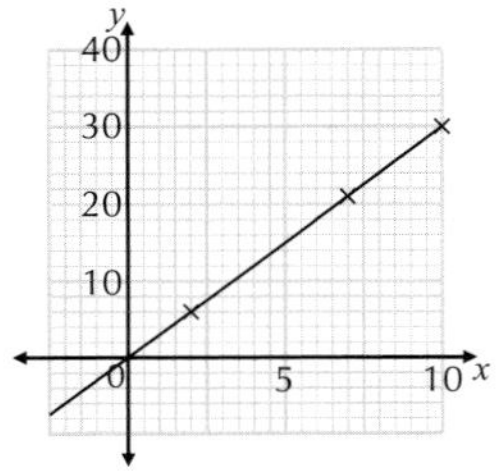

c) (i) When $x = 1$, $y = -10 \div -2 = 5$,
so when $x = 0$, $y = 0 \times 5 = 0$,
when $x = 3$, $y = 3 \times 5 = 15$,
when $y = 30$, $x = 30 \div 5 = 6$.

x	−2	0	3	**6**
y	−10	**0**	**15**	30

(ii) Plot your points and draw a straight line through them:

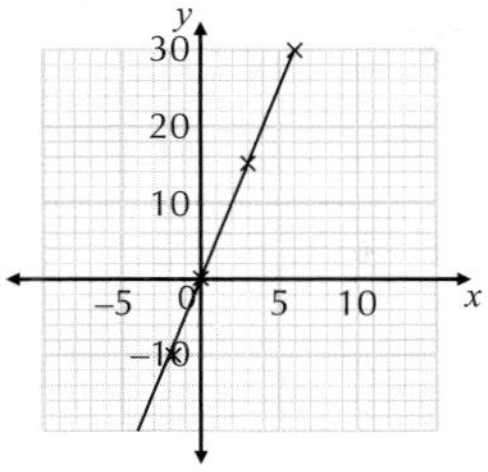

Page 231 Exercise 4

1 a) $j = Ah$
$15 = A \times 5$, so $A = 15 \div 5 = 3 \Rightarrow$ $\boldsymbol{j = 3h}$

b) When $h = 40$, $j = 3 \times 40 =$ **120**

c)

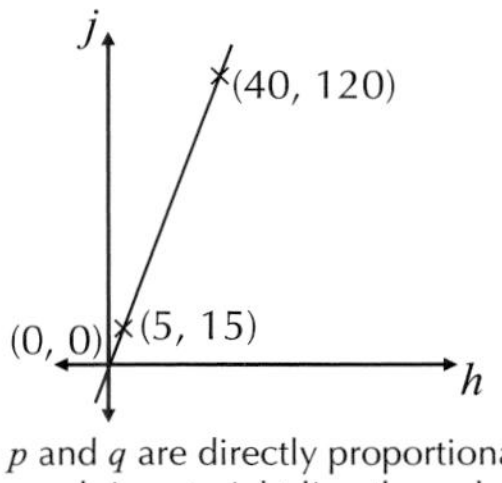

2 a) p and q are directly proportional, so the graph is a straight line through the origin. It also goes through the point (4, 14).

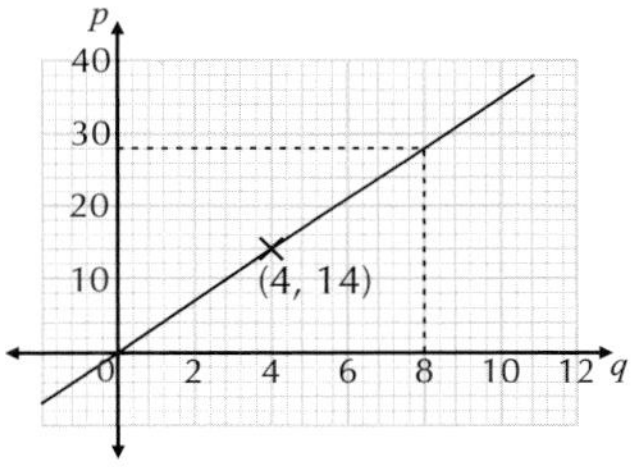

b) Read off the graph — go up from $q = 8$ to the line, then across to the axis for p (see dashed lines on graph above).
When $q = 8$, $\boldsymbol{p = 28}$.

14.2 Inverse Proportion

Page 232 Exercise 1

1 3 people take 2 hours to paint the wall
1 person would take $3 \times 2 = 6$ hours
6 people would take $6 \div 6 =$ **1 hour**

2 At 30 mph, the journey takes 2 hours
At 1 mph, it would take $2 \times 30 = 60$ hours
At 45 mph, it would take
$60 \div 45 =$ **1.333... hours**
= **1 hour and 20 minutes**

3 4 chefs take 20 minutes
1 chef would take $4 \times 20 = 80$ minutes
5 chefs would take $80 \div 5 =$ **16 minutes**

4 a) 5 builders take 62 days
1 builder would take $5 \times 62 = 310$ days
2 builders would take $310 \div 2 =$ **155 days**

b) 1 builder would take 310 days
To do it in 1 day, you would need
$310 \times 1 = 310$ builders
To do it in 40 days, you would need
$310 \div 40 = 7.75$ builders
So **8 builders** would be needed.

Page 233 Exercise 2

1 2 people take 3 hours to clean 6 rooms
First change the number of people:
1 person would take
$3 \times 2 = 6$ hours to clean 6 rooms
5 people would take
$6 \div 5 = 1.2$ hours to clean 6 rooms
Then change the number of rooms:
5 people would take
$1.2 \div 6 = 0.2$ hours to clean 1 room,
so cleaning 20 rooms would take 5 people
$0.2 \times 20 =$ **4 hours**

2 2 bakers take 144 minutes to bake 72 buns
First change the number of bakers:
1 baker would take
$144 \times 2 = 288$ minutes to bake 72 buns
5 bakers would take
$288 \div 5 = 57.6$ minutes to bake 72 buns
Then change the number of buns:
5 bakers would take
$57.6 \div 72 = 0.8$ minutes to bake 1 bun,
so baking 90 buns would take 5 bakers
$0.8 \times 90 =$ **72 minutes**

3 14 people take 2 hours to paint 35 plates
First change the number of people:
1 person would take
$2 \times 14 = 28$ hours to paint 35 plates
20 people would take
$28 \div 20 = 1.4$ hours to paint 35 plates
Then change the number of plates:
20 people would take
$1.4 \div 35 = 0.04$ hours to paint 1 plate,
so painting 60 plates would take
$0.04 \times 60 =$ **2.4 hours = 2 hours 24 minutes**

Page 234 Exercise 3

1 Graph A — y decreases as x increases

2 Look for equations in the form $y = \frac{A}{x}$ or $xy = A$
— $\boldsymbol{y = \frac{4}{x}}$ and $\boldsymbol{yx = 9}$ show inverse proportion.

3 a) When $x = 12$, and $y = 15$,
$A = xy = 12 \times 15 = 180$.
When $x = 6$, $y = 180 \div 6 = 30$.

x	12	6
y	15	**30**

b) When $x = 8$ and, $y = 20$,
$A = xy = 8 \times 20 = 160$,
so when $x = 2$, $y = 160 \div 2 = 80$,
when $x = 4$, $y = 160 \div 4 = 40$,
when $x = 10$, $y = 160 \div 10 = 16$.

x	2	4	8	10
y	**80**	**40**	20	**16**

c) When $x = 20$, and $y = 4$,
$A = xy = 20 \times 4 = 80$,
so when $x = 4$, $y = 80 \div 4 = 20$,
when $x = 100$, $y = 80 \div 100 = 0.8$,
when $y = 320$, $x = 80 \div 320 = 0.25$.

x	**0.25**	4	20	100
y	320	**20**	4	**0.8**

4 a) When $q = 2$, and $p = 1$, $A = pq = 2 \times 1 = 2$,
so when $q = 0.2$, $p = 2 \div 0.2 = 10$,
when $q = 0.4$, $p = 2 \div 0.4 = 5$.
Continue for each value of q to get:

q	0.2	0.4	0.5	1	2	4	5	10
p	**10**	**5**	**4**	**2**	1	**0.5**	**0.4**	**0.2**

b) Plot each point from the table in part a). p and q are inversely proportional so you will get a reciprocal graph.

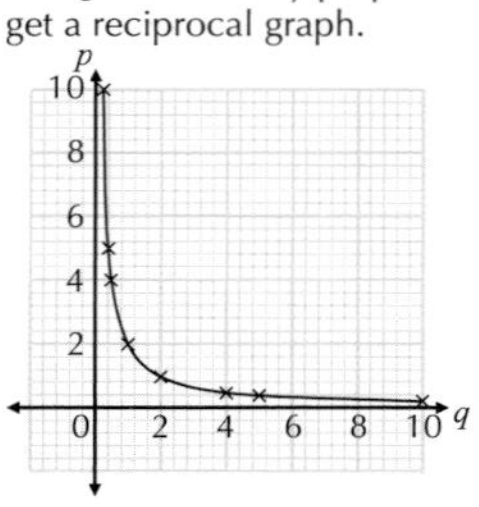

Page 235 Review Exercise

1 12.5 litres cost £20
1 litre costs 20 ÷ 12.5 = £1.60.
60 litres cost 60 × £1.60 = **£96**

2 a) £1 = R17.5 ⇒ £500 = 500 × 17.5
= **R8750**

b) Emma has 8750 – 6200 = R2550 left over from her holiday. The exchange rate is now £1 = R16.9, so
R1 = £1 ÷ 16.9 = £0.0591...
⇒ R2550 = 2550 × £0.0591... =
£150.887...
= **£150.89** (to the nearest penny)

3 5 laps of a 400 m track takes 8 minutes.
First change the length of the track:
1 lap of the 400 m track takes
8 ÷ 5 = 1.6 minutes
1 lap of a 200 m track would take
1.6 ÷ 2 = 0.8 minutes.
Then change the number of laps:
9 laps of the 200 m track would take
9 × 0.8 = **7.2 minutes**
= **7 mins, 12 secs**
(Alternative method:
5 × 400 m = 2000 m takes 8 minutes, so
1 m takes 8 ÷ 2000 = 0.004 minutes
9 × 200 m = 1800 m takes
0.004 × 1800 = **7.2 minutes**)

4 Box of 8: £3.50 ÷ 8 = 43.75p per chocolate
Box of 12: £4.70 ÷ 12
= 39.166...p per chocolate
Box of 20: £8.15 ÷ 20 = 40.75p per chocolate
So the best value for money is **the box of 12 chocolates**.

5 a) When $a = 1$, $b = 6 \div 2 = 3$,
so when $a = 4$, $b = 4 \times 3 = 12$,
when $a = 7$, $b = 7 \times 3 = 21$,
when $a = 10$, $b = 10 \times 3 = 30$.

a	2	4	7	10
b	6	**12**	**21**	**30**

b) Plot your points and draw a straight line through them and the origin:

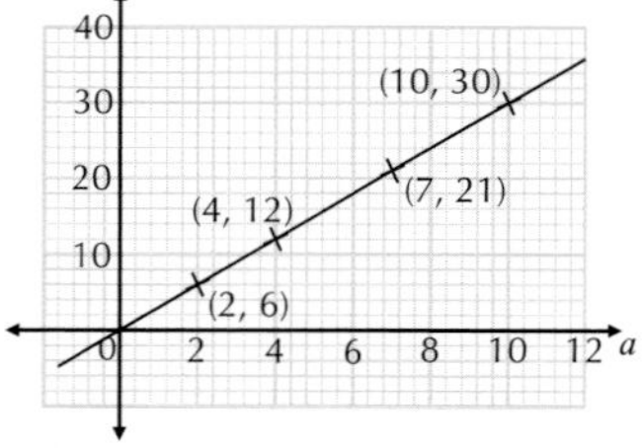

6 a) When $x = 4 \div 4 = 1$, $y = 32 \div 4 = 8$,
so when $x = 1 \times 8 = 8$, $y = 8 \times 8 =$ **64**

b) When $x = 1$, $y = 8$,
so when $y = 2$, $x = 2 \div 8 =$ **2.5**

c) $y = Ax$.
$A = 32 \div 4 = 8$, so $\boldsymbol{y = 8x}$

7 a) 8 chickens lay 20 eggs in 3 days
1 chicken lays 20 eggs in 8 × 3 = 24 days
12 chickens lay 20 eggs in 24 ÷ 12
= **2 days**

b) 12 chickens lay 20 eggs in 2 days, so they lay 10 eggs in 1 day, and 10 × 3 = 30 eggs in 3 days.
So to get 30 eggs in 1 day, you would need 3 × 12 = **36 chickens**

8 a) When $x = 7$, and $y = 9$, $A = xy = 7 \times 9 = 63$,
so when $y = 25.2$, $x = 63 \div 25.2 = 2.5$.

x	**2.5**	7
y	25.2	9

b) When $x = 6$, and $y = 15$,
$A = xy = 6 \times 15 = 90$,
so when $x = 3$, $y = 90 \div 3 = 30$,
when $x = 30$, $y = 90 \div 30 = 3$,
when $y = 180$, $x = 90 \div 180 = 0.5$.

x	**0.5**	3	6	30
y	180	**30**	15	**3**

Page 236 Exam-Style Questions

1 Original roll: 1 sheet costs £1.20 ÷ 100 = £0.012
New roll: 1 sheet costs £1.10 ÷ 90 = £0.01222...
The new roll **is not** better value for money.
[2 marks for finding both costs per sheet and correct answer, or 1 mark for finding one correct cost per sheet]
You could also find the number of sheets bought per £1 spent.

2 a) 12 muffins: 360 g
1 muffin: 360 ÷ 12 = 30 g *[1 mark]*
20 muffins: 20 × 30 g = **600 g** *[1 mark]*

b)

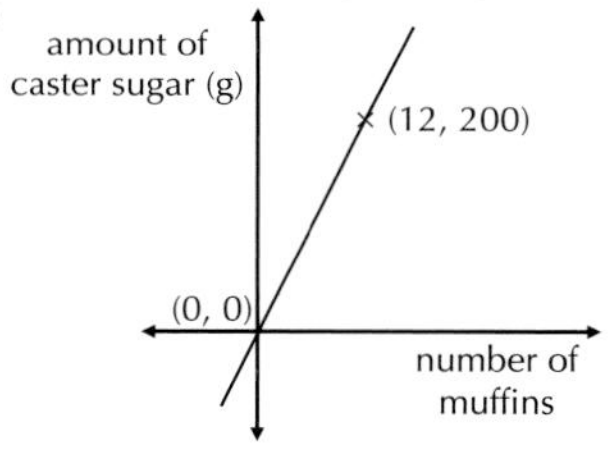

[2 marks available — 1 mark for drawing correctly labelled axes and a straight line through the origin, 1 mark for showing the point (12, 200)]

3 Cleaning 1 m² takes Niall 24 ÷ 258
= 0.093... minutes *[1 mark]*
⇒ 68.8 m² will take 68.8 × 0.093...
= 6.4 minutes *[1 mark]*
Convert 0.4 minutes into seconds:
0.4 × 60 = 24 seconds,
so it will take Niall **6 minutes 24 seconds** *[1 mark]*.
Alternatively, Niall cleans 258 ÷ 24 = 10.75 m² in one minute, so cleaning 68.8 m² would take 68.8 ÷ 10.75 = 6.4 minutes = 6 minutes 24 seconds.

4 a) It takes 96 minutes for 840 people to count 45 802 votes. Changing the number of votes:
840 people would take 96 ÷ 2 = 48 minutes to count half as many votes *[1 mark]*.
Changing the number of people:
1 person would take
840 × 48 = 40 320 minutes *[1 mark]*
560 people would take
40 320 ÷ 560 = **72 minutes** *[1 mark]*

b) The assumption is that the staff of the second council count votes at the **same rate** as the staff of the first council *[1 mark]*.

Section 15 — Angles and 2D Shapes

15.1 Basic Angle Properties

Page 237 Exercise 1

1 a) $34° + b + 90° = 180°$
$b = 180° - 90° - 34°$ ⇒ $\boldsymbol{b = 56°}$

b) $d + 30° + 20° = 90°$
$d = 90° - 30° - 20°$ ⇒ $\boldsymbol{d = 40°}$

c) $28° + 55° + d + 35° = 180°$
$d = 180° - 28° - 55° - 35°$ ⇒ $\boldsymbol{d = 62°}$

2 a) $d + 3d = 180°$ ⇒ $4d = 180°$
⇒ $d = 180° \div 4$ ⇒ $\boldsymbol{d = 45°}$

b) $a + 110° + a = 180°$ ⇒ $2a + 110° = 180°$
⇒ $2a = 180° - 110°$ ⇒ $2a = 70°$
⇒ $a = 70° \div 2$ ⇒ $\boldsymbol{a = 35°}$

c) $57° + 57° + c + c = 180°$
⇒ $114° + 2c = 180°$
⇒ $2c = 180° - 114°$ ⇒ $2c = 66°$
⇒ $c = 66° \div 2$ ⇒ $\boldsymbol{c = 33°}$

Page 238 Exercise 2

1 a) $100° + c + 120° = 360°$
$c = 360° - 120° - 100°$ ⇒ $\boldsymbol{c = 140°}$

b) $41° + 161° + d + 45° = 360°$
$d = 360° - 41° - 161° - 45°$ ⇒ $\boldsymbol{d = 113°}$

c) $99° + a + 44° + 90° = 360°$
$a = 360° - 99° - 44° - 90°$ ⇒ $\boldsymbol{a = 127°}$

2 a) $e + e + e + e + e + e = 360°$ ⇒ $6e = 360°$
⇒ $e = 360° \div 6$ ⇒ $\boldsymbol{e = 60°}$

b) $50° + 98° + g + 3g = 360°$
⇒ $148° + 4g = 360°$
⇒ $4g = 360° - 148°$ ⇒ $4g = 212°$
⇒ $g = 212° \div 4$ ⇒ $\boldsymbol{g = 53°}$

c) $160° + 90° + h + h + 50° = 360°$
⇒ $300° + 2h = 360°$
$2h = 360° - 300°$ ⇒ $2h = 60°$
⇒ $h = 60° \div 2$ ⇒ $\boldsymbol{h = 30°}$

15.2 Parallel and Intersecting Lines

Page 239 Exercise 1

1 a) 160° and b and vertically opposite so
$\boldsymbol{b = 160°}$

b) 75° and c are vertically opposite so $\boldsymbol{c = 75°}$
75° and d are on a straight line so
$75° + d = 180°$ ⇒ $d = 180° - 75°$
⇒ $\boldsymbol{d = 105°}$

c) e and the right angle are vertically opposite so $\boldsymbol{e = 90°}$
f and the right angle are on a straight line so $f + 90° = 180°$ ⇒ $f = 180° - 90°$
⇒ $\boldsymbol{f = 90°}$
f and g are vertically opposite so $\boldsymbol{g = 90°}$
You could also have used the fact that g and the right angle are on a straight line so
$g + 90° = 180°$ ⇒ $g = 90°$

2 The 40° angle and the angle labelled $3a - 5°$ are vertically opposite, so
$40° = 3a - 5°$
$40° + 5° = 3a$ ⇒ $45° = 3a$
⇒ $a = 45° \div 3$ ⇒ $\boldsymbol{a = 15°}$

Page 240 Exercise 2

1 a) 72° and b are alternate angles so $\boldsymbol{b = 72°}$

b) 135° and g are on a straight line so
$135° + g = 180°$
$g = 180° - 135°$ ⇒ $\boldsymbol{g = 45°}$
g and h are alternate angles so $\boldsymbol{h = g = 45°}$.

c) 48° and i are alternate angles so $\boldsymbol{i = 48°}$
i and j are on a straight line so
$i + j = 180°$ ⇒ $48° + j = 180°$
⇒ $j = 180° - 48°$ ⇒ $\boldsymbol{j = 132°}$

2 a) **42°**

b) x and the angle of 42° are **alternate angles** as the floors are **parallel** and the stairs are a **straight line** crossing them.

Try not to let the context of the question put you off — you just need to use the same rules as in question 1.

Page 241 Exercise 3

1 a) 52° and f are allied angles so
$52° + f = 180° \Rightarrow f = 180° - 52°$
$\Rightarrow$ $\boldsymbol{f = 128°}$
f and g are alternate angles so $\boldsymbol{g = 128°}$
52° and h are alternate angles so $\boldsymbol{h = 52°}$

There are often a few different methods you can use for questions like this — for example, you could use the fact that g and 52° lie on a straight line to work out g, and the fact that g and h are allied to work out h.

b) 141° and h are corresponding angles so $\boldsymbol{h = 141°}$
141° and i are allied angles so
$i + 141° = 180° \Rightarrow i = 180° - 141°$
$\Rightarrow$ $\boldsymbol{i = 39°}$
i and j are corresponding angles so $\boldsymbol{j = 39°}$

c) 78° and v are on a straight line so
$78° + v = 180° \Rightarrow v = 180° - 78°$
$\Rightarrow$ $\boldsymbol{v = 102°}$
78° and w are vertically opposite angles so $\boldsymbol{w = 78°}$
x and v are vertically opposite so $\boldsymbol{x = 102°}$
78° and r are corresponding angles so $\boldsymbol{r = 78°}$
x and s are corresponding angles so $\boldsymbol{s = 102°}$
v and t are corresponding angles so $\boldsymbol{t = 102°}$
w and u are corresponding angles so $\boldsymbol{u = 78°}$

2 The angle made by the second post and the downward slope is 99°, since this and the 99° angle shown in the diagram are corresponding angles. Then y and 99° lie on a straight line so
$y + 99° = 180° \Rightarrow y = 180° - 99° \Rightarrow$ $\boldsymbol{y = 81°}$

15.3 Triangles

Page 242 Exercise 1

1 a) **Equilateral triangle; acute-angled**

b) **Right-angled triangle**

This triangle is also a scalene triangle as all sides are different lengths.

c) **Isosceles triangle; acute-angled**

d) **Isosceles triangle; right-angled**

Make sure you look at every part of the triangle — in d) you need to spot the tick marks as well as the right angle.

Page 243 Exercise 2

1 a) $68° + 90° + b = 180° \Rightarrow 158° + b = 180°$
$\Rightarrow b = 180° - 158° \Rightarrow$ $\boldsymbol{b = 22°}$

b) $d + 90° + 74° = 180° \Rightarrow d + 164° = 180°$
$\Rightarrow d = 180° - 164° \Rightarrow$ $\boldsymbol{d = 16°}$

c) Call the unlabelled angle in the triangle h.
h and the right angle lie on a straight line, so $h + 90° = 180° \Rightarrow h = 180° - 90°$
$\Rightarrow h = 90°$
So as the angles in a triangle add up to 180°,
$90° + g + 38° = 180° \Rightarrow g + 128° = 180°$
$\Rightarrow g = 180° - 128° \Rightarrow$ $\boldsymbol{g = 52°}$

2 The angles in a triangle must add up to **180°**, but the angles in this triangle add up to **74° + 26° + 90° = 190°**, so such a triangle **can't exist.**

3 a) $a + 55° + 72° = 180° \Rightarrow a + 127° = 180°$
$\Rightarrow a = 180° - 127° \Rightarrow$ $\boldsymbol{a = 53°}$

b) $21° + c + 144° = 180° \Rightarrow c + 165° = 180°$
$\Rightarrow c = 180° - 165° \Rightarrow$ $\boldsymbol{c = 15°}$

c) $57° + 74° + h = 180° \Rightarrow h + 131° = 180°$
$h = 180° - 131° \Rightarrow$ $\boldsymbol{h = 49°}$
i and h lie on a straight line so $i + h = 180°$
$i + 49° = 180° \Rightarrow i = 180° - 49°$
$\Rightarrow$ $\boldsymbol{i = 131°}$

Page 244 Exercise 3

1 a) $2a + a + 120° = 180° \Rightarrow 3a + 120° = 180°$
$\Rightarrow 3a = 180° - 120° \Rightarrow 3a = 60°$
$\Rightarrow a = 60° \div 3 \Rightarrow$ $\boldsymbol{a = 20°}$

b) $2b + (2b + 50°) + 90° = 180°$
$\Rightarrow 4b + 140° = 180°$
$\Rightarrow 4b = 180° - 140° \Rightarrow 4b = 40°$
$\Rightarrow b = 40° \div 4 \Rightarrow$ $\boldsymbol{b = 10°}$

c) $60° + (c - 16°) + (c + 16°) = 180°$
$\Rightarrow 2c + 60° + 16° - 16° = 180°$
$\Rightarrow 2c + 60° = 180°$
$\Rightarrow 2c = 180° - 60° \Rightarrow 2c = 120°$
$\Rightarrow c = 120° \div 2 \Rightarrow$ $\boldsymbol{c = 60°}$

2 a) The triangle is isosceles, so $\boldsymbol{a = 30°}$.
As all the angles in a triangle add up to 180°, $30° + b + 30° = 180°$
$\Rightarrow b + 60° = 180° \Rightarrow b = 180° - 60°$
$\Rightarrow$ $\boldsymbol{b = 120°}$

b) As the triangle is isosceles, the non-labelled angle and angle e are equal.
As all the angles in a triangle add up to 180°,
$e + 3e + e = 180° \Rightarrow 5e = 180°$
$\Rightarrow e = 180° \div 5 \Rightarrow$ $\boldsymbol{e = 36°}$

c) Call the unlabelled angle in the triangle h.
The triangle is isosceles, so h and angle f are equal.
As all the angles in a triangle add up to 180°, $48° + f + f = 180° \Rightarrow 2f = 180° - 48°$
$\Rightarrow 2f = 132° \Rightarrow f = 132° \div 2 \Rightarrow$ $\boldsymbol{f = 66°}$
and $h = 66°$
Since h and g lie on a straight line:
$66° + g = 180° \Rightarrow g = 180° - 66°$
$\Rightarrow$ $\boldsymbol{g = 114°}$

3 a) The triangle is an equilateral triangle, so all the angles are $60° \Rightarrow$ $\boldsymbol{x = 60°}$

You could also work this out by calling each angle x. Then the sum of angles in a triangle gives 3x = 180° and so x = 60°.

b) y lies on a straight line with one of the angles in the triangle: $60° + y = 180°$
$\Rightarrow y = 180° - 60° \Rightarrow$ $\boldsymbol{y = 120°}$

4 a) Call the bottom left angle in the triangle l.
The triangle is isosceles, so $l = 55°$.
As l lies on a straight line with p, then
$p + 55° = 180° \Rightarrow p = 180° - 55°$
$\Rightarrow$ $\boldsymbol{p = 125°}$

b) Call the top angle in the triangle t.
$t = 180° - (55° + 55°)$
$\Rightarrow t = 180° - 110° \Rightarrow t = 70°$
t and q lie on a straight line so
$70° + q = 180° \Rightarrow q = 180° - 70°$
$\Rightarrow$ $\boldsymbol{q = 110°}$

15.4 Quadrilaterals

Page 245 Exercise 1

1 a) $a + 93° + 86° + 69° = 360°$
$a = 360° - 93° - 86° - 69° \Rightarrow$ $\boldsymbol{a = 112°}$

b) $129° + c + 74° + 67° = 360°$
$c = 360° - 129° - 74° - 67° \Rightarrow$ $\boldsymbol{c = 90°}$

c) $72° + 112° + d + 106° = 360°$
$d = 360° - 72° - 112° - 106° \Rightarrow$ $\boldsymbol{d = 70°}$

2 a) The missing angle in the quadrilateral is $360° - 90° - 108° - 85° = 77°$.
Then this angle and r lie on a straight line so $r + 77° = 180° \Rightarrow$ $\boldsymbol{r = 103°}$

b) w and 71° are vertically opposite angles so $\boldsymbol{w = 71°}$.
x and 95° are vertically opposite angles so $\boldsymbol{x = 95°}$.
$71° + 95° + 102° + y = 360° \Rightarrow$ $\boldsymbol{y = 92°}$

Page 246 Exercise 2

1

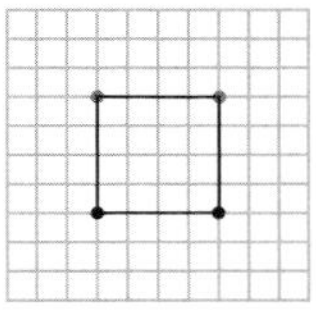

2 **E.g.**

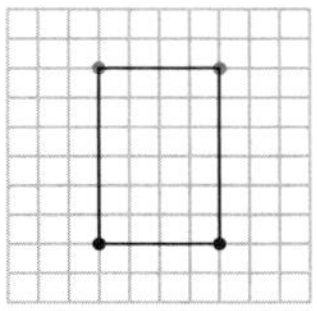

There are lots of different rectangles you could have drawn — make sure it has 4 right angles and 2 pairs of equal sides.

3 a) **They are the same length.**

b) **90°**

c) The only shape with diagonals of equal length is shape **iii)**, so that is the only square.

All squares have diagonals of equal length that meet at 90°.

4 In a rectangle, opposite side lengths are equal.
$2^2 + 5 = 4 + 5 = 9$ so the top and bottom sides are equal.
$\frac{4 + 20}{4} + 2 = \frac{24}{4} + 2 = 6 + 2 = 8 \neq 6$
so the left and right sides are not equal.
Therefore, either 6 cm or $\frac{4+20}{4} + 2$ cm must be **incorrect**.

Page 247 Exercise 3

1 a)

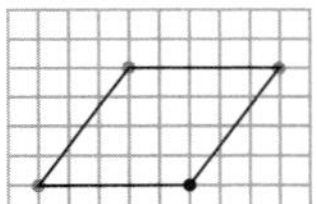

b) **E.g.**

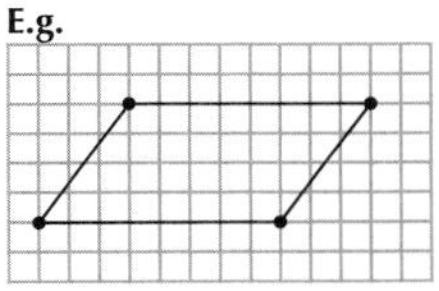

2 a) d and 118° are opposite angles so $d = 118°$
118° and e are neighbouring angles so
$118° + e = 180° \Rightarrow e = 180° - 118°$
$\Rightarrow$ $\boldsymbol{e = 62°}$

b) f and 51° are opposite angles so $\boldsymbol{f = 51°}$
51° and g are neighbouring angles so
$51° + g = 180° \Rightarrow g = 180° - 51°$
$\Rightarrow$ $\boldsymbol{g = 129°}$

c) 108° and a are neighbouring angles so
$108° + a = 180° \Rightarrow a = 180° - 108°$
$\Rightarrow$ $\boldsymbol{a = 72°}$
b and a are opposite angles so $\boldsymbol{b = 72°}$
c and 108° are opposite angles so $\boldsymbol{c = 108°}$

Page 248 Exercise 4

1 a) d b) Q c) S

2 a) 111° and b are equal angles so $\mathbf{b = 111°}$
$c + 111° + 48° + 111° = 360$
$\Rightarrow c + 270° = 360°$
$\Rightarrow c = 360° - 270° \Rightarrow \mathbf{c = 90°}$

b) The missing angle inside the kite and the angle marked 124° are equal angles, so the missing angle is also 124°.
Angles in a quadrilateral add up to 360° so
$42° + 124° + d + 124° = 360°$
$\Rightarrow 290° + d = 360°$
$\Rightarrow d = 360° - 290° \Rightarrow \mathbf{d = 70°}$

c) The missing angle inside the kite and the angle e are equal angles, so the missing angle equals e.
Angles in a quadrilateral add up to 360° so
$e + 101° + e + 61° = 360°$
$\Rightarrow 2e + 162° = 360°$
$2e = 360° - 162° \Rightarrow 2e = 198°$
$\Rightarrow e = 198° \div 2 \Rightarrow \mathbf{e = 99°}$

Page 249 Exercise 5

1 a) **parallel** b) **two pairs**
c) **one pair** d) **one pair** e) **360°**

2 a) This is an isosceles trapezium so both the base angles equal f.
$f + 124° = 180° \Rightarrow f = 180° - 124°$
$\Rightarrow \mathbf{f = 56°}$

b) $d + 106° = 180° \Rightarrow d = 180° - 106°$
$\Rightarrow \mathbf{d = 74°}$
$e + 64° = 180° \Rightarrow e = 180° - 64°$
$\Rightarrow \mathbf{e = 116°}$

c) This is an isosceles trapezium so $\mathbf{b = 120°}$
$c + 120° = 180° \Rightarrow c = 180° - 120°$
$\Rightarrow \mathbf{c = 60°}$

15.5 Interior and Exterior Angles

Page 251 Exercise 1

1 a) E.g.

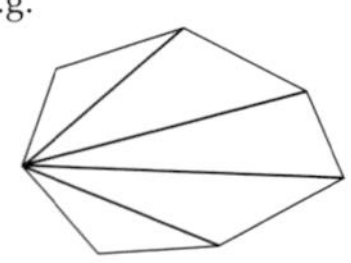

The shape can be divided into 5 triangles and each triangle's angles add up to 180°. So the sum of the interior angles is
$5 \times 180° = \mathbf{900°}$.

b) E.g.

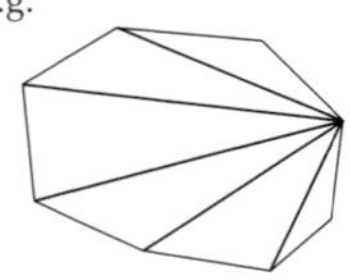

The shape can be divided into 6 triangles and each triangle's angles add up to 180°. So the sum of the interior angles is
$6 \times 180° = \mathbf{1080°}$.

2 a) **One triangle is partly outside the original shape.**

b)

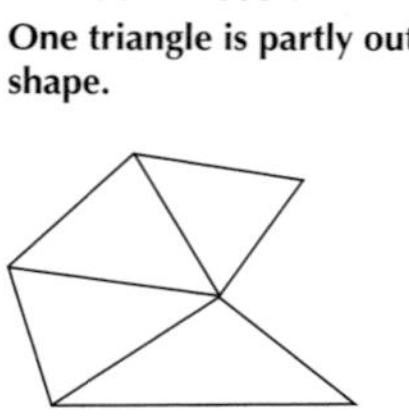

c) The shape can be divided into 4 triangles and each triangle's angles add up to 180°. So the sum of the interior angles is
$4 \times 180° = \mathbf{720°}$.

d) The sum of all the interior angles is 720° from part c).
$116° + 128° + 65° + x + 39° + 107° = 720°$
$x + 455° = 720° \Rightarrow \mathbf{x = 265°}$

Page 252 Exercise 2

1 a) **No** — as it has 1 curved edge and all polygons only have straight edges.
b) **Yes** — it's a triangle.
c) **No** — as it's a 3D shape and all polygons are 2D.

2 Using the formula for the sum of a polygon's interior angles:
a) $n = 10$ so $S = (10 - 2) \times 180°$
$= 8 \times 180° = \mathbf{1440°}$
b) $n = 12$ so $S = (12 - 2) \times 180°$
$= 10 \times 180° = \mathbf{1800°}$
c) $n = 20$ so $S = (20 - 2) \times 180°$
$= 18 \times 180° = \mathbf{3240°}$

3 a) Using the formula for the sum of a polygon's interior angles with $n = 7$,
$S = (7 - 2) \times 180° = 5 \times 180° = \mathbf{900°}$
b) There are seven interior angles which are all equal due to the shape being regular.
So one interior angle $= \frac{900°}{7} = 128.571...°$
$= \mathbf{128.57°}$ (2 d.p.)

4 a) Using the formula for the sum of a polygon's interior angles with $n = 8$,
$S = (8 - 2) \times 180° = 6 \times 180° = 1080°$
There are eight interior angles which are all equal due to the shape being regular. So one interior angle $= \frac{1080°}{8} = \mathbf{135°}$
b) Using the formula for the sum of a polygon's interior angles with $n = 9$,
$S = (9 - 2) \times 180° = 7 \times 180° = 1260°$
There are nine interior angles which are all equal due to the shape being regular. So one interior angle $= \frac{1260°}{9} = \mathbf{140°}$

5 a) (i) Using the formula for the sum of a polygon's interior angles with $n = 4$,
$S = (4 - 2) \times 180° = \mathbf{360°}$
This is a quadrilateral so you already know that the angles have to sum to 360°.
(ii) $41° + 112° + 89° + a = 360°$
$a = 360° - 41° - 112° - 89°$
$\Rightarrow \mathbf{a = 118°}$

b) (i) Using the formula for the sum of a polygon's interior angles with $n = 6$,
$S = (6 - 2) \times 180° = \mathbf{720°}$
(ii) $107° + 101° + b + 90° + 90° + 85° = 720°$
$b = 720° - 107° - 101°$
$- 90° - 90° - 85° \Rightarrow \mathbf{b = 247°}$

c) (i) Using the formula for the sum of a polygon's interior angles with $n = 9$,
$S = (9 - 2) \times 180° = \mathbf{1260°}$
(ii) $93° + c + 104° + 121° + 91° + 230° + 150° + 102° + 159° = 1260°$
$\Rightarrow c = 1260° - 93° - 104° - 121° - 91° - 230° - 150° - 102° - 159°$
$\Rightarrow \mathbf{c = 210°}$
Don't let the large number of angles put you off — the method is exactly the same as in parts a) and b).

Page 254 Exercise 3

1 a) $a = \frac{360°}{5} \Rightarrow \mathbf{a = 72°}$
b) $b + a = 180° \Rightarrow b + 72° = 180°$
$\Rightarrow b = 180° - 72° \Rightarrow \mathbf{b = 108°}$

2 a) (i) Exterior angle $= \frac{360°}{7} = 51.428...$
$= \mathbf{51.43°}$ (2 d.p.)
(ii) Exterior angle $= \frac{360°}{8} = \mathbf{45°}$
(iii) Exterior angle $= \frac{360°}{9} = \mathbf{40°}$

b) (i) Interior angle + 51.43° = 180°
Interior angle = 180° − 51.43°
= **128.57°** (2 d.p)
(ii) Interior angle + 45° = 180°
Interior angle = 180° − 45° = **135°**
(iii) Interior angle + 40° = 180°
Interior angle = 180° − 40° = **140°**

3 a) The exterior angles add up to 360°, so
$90° + 31° + 83° + 72° + 30° + a = 360°$
$a = 360° - 90° - 31° - 83° - 72° - 30°$
$\Rightarrow \mathbf{a = 54°}$
b) c and the angle marked 106° are on a straight line, so $c = 180° - 106° \Rightarrow \mathbf{c = 74°}$
The exterior angles add up to 360°, so
$151° + 74° + d = 360°$
$d = 360° - 151° - 74° \Rightarrow \mathbf{d = 135°}$

4 a) Call the unknown exterior angle x.
The exterior angles add up to 360°, so
$100° + 68° + 84° + 55° + x = 360°$
$x = 360° - 100° - 68° - 84° - 55° = \mathbf{53°}$
b) Call the unknown exterior angle x.
The exterior angles add up to 360°, so
$30° + 68° + 45° + 52° + 75° + 50° + x = 360°$
$x = 360° - 30° - 68° - 45° - 52° - 75° - 50° = \mathbf{40°}$

Page 254 Exercise 4

1 a) (i) $360° \div n = 90° \Rightarrow n = 360° \div 90° = \mathbf{4}$
(ii) $180° - 90° = \mathbf{90°}$
(iii) $4 \times 90° = \mathbf{360°}$
b) (i) $360° \div n = 40° \Rightarrow n = 360° \div 40° = \mathbf{9}$
(ii) $180° - 40° = \mathbf{140°}$
(iii) $9 \times 140° = \mathbf{1260°}$
c) (i) $360° \div n = 6° \Rightarrow n = 360° \div 6° = \mathbf{60}$
(ii) $180° - 6° = \mathbf{174°}$
(iii) $60 \times 174° = \mathbf{10\,440°}$
d) (i) $360° \div n = 4° \Rightarrow n = 360° \div 4° = \mathbf{90}$
(ii) $180° - 4° = \mathbf{176°}$
(iii) $90 \times 176° = \mathbf{15\,840°}$

15.6 Symmetry

Page 255 Exercise 1

1 a)

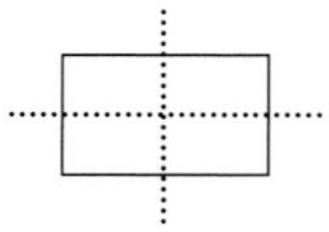

2 lines of symmetry

b)

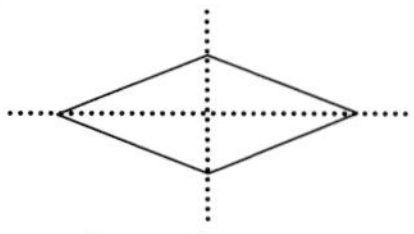

2 lines of symmetry

c)

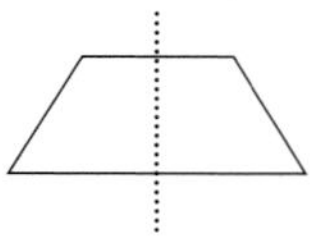

1 line of symmetry

d)

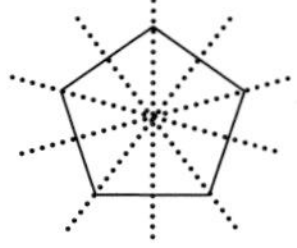

5 lines of symmetry

e)

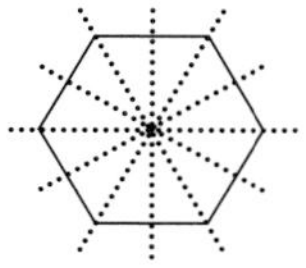

6 lines of symmetry

f) 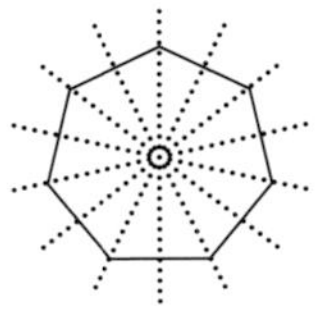

7 lines of symmetry

2 a) 2 b) 6 c) 1 d) 3

Page 256 Review Exercise

1 a) $f + f + f + f = 180° \Rightarrow 4f = 180°$
$f = 180° \div 4 \Rightarrow$ **$f = 45°$**

b) $j + 31° = 90°$
$j = 90° - 31° \Rightarrow$ **$j = 59°$**

c) $h + 41° + 90° + 135° = 360°$
$\Rightarrow h + 266° = 360° \Rightarrow h = 360° - 266°$
$\Rightarrow$ **$h = 94°$**

2 If the lines AB and CD were parallel then the angles BAC and DCA would be allied. Allied angles add up to 180°, but $52° + 126° = 178°$, so the lines are **not parallel**.
For this type of question, assume that the answer is yes and then see if the rules hold.

3 a) The angle that is vertically opposite to the angle labelled 58° is also corresponding with angle m, so **$m = 58°$**

b) The top angle in the triangle is vertically opposite to 94° so it equals 94°.
Similarly the bottom-left angle is vertically opposite to 52° so it equals 52°.
Angles in a triangle sum to 180° so
$94° + p + 52° = 180° \Rightarrow p + 146° = 180°$
$\Rightarrow p = 180° - 146° \Rightarrow$ **$p = 34°$**

c) The triangle is an isosceles triangle so the bottom-right angle is equal to 62°.
Angles in a triangle sum to 180° so
$62° + p + 62° = 180° \Rightarrow p + 124° = 180°$
$\Rightarrow p = 180° - 124° \Rightarrow$ **$p = 56°$**
The bottom-right angle and q lie on a straight line so they must add up to 180°.
$62° + q = 180° \Rightarrow q = 180° - 62°$
$\Rightarrow$ **$q = 118°$**

d) Call the top-left angle in the triangle l.
The 90° angle, 35° angle and l lie on a straight line, so they must add up to 180°.
$90° + 35° + l = 180° \Rightarrow 125° + l = 180°$
$\Rightarrow l = 180° - 125° \Rightarrow l = 55°$
Angles in a triangle add up to 180°, so
$55° + b + 55° = 180° \Rightarrow b + 110° = 180°$
$\Rightarrow b = 180° - 110° \Rightarrow$ **$b = 70°$**

e) The triangle is an isosceles triangle, so the top-right angle is equal to c.
Angles in a triangle add up to 180°, so
$48° + c + c = 180° \Rightarrow 48° + 2c = 180°$
$\Rightarrow 2c = 180° - 48° = 132°$
$\Rightarrow c = 132° \div 2 \Rightarrow$ **$c = 66°$**
As the top-right angle is equal to c, it must equal 66°.
Angles around a point add up to 360°, so
$66° + d = 360° \Rightarrow d = 360° - 66°$
$\Rightarrow$ **$d = 294°$**

4 a) **Parallelogram** b) **Rectangle**
c) **Kite** d) **Trapezium**

5 Angles in a quadrilateral add up to 360°, so
$40° + 83° + 99° +$ (fourth angle) $= 360°$
$222° +$ (fourth angle) $= 360°$
fourth angle $= 360° - 222°$
fourth angle **= 138°**

6 a) a and the angle labelled 122° add up to 180°.
$a + 122° = 180° \Rightarrow a = 180° - 122°$
$\Rightarrow$ **$a = 58°$**
As this is an isosceles trapezium,
$b = a = 58°$

b) Using the symmetry of the kite, **$s = 103°$**.
Angles in a quadrilateral add up to 360°, so
$103° + 95° + 103° + t = 360°$
$\Rightarrow 301° + t = 360°$
$\Rightarrow t = 360° - 301° \Rightarrow$ **$t = 59°$**

c) The angles 74°, e and 40° lie on a straight line, so $74° + e + 40° = 180°$
$\Rightarrow e + 114° = 180°$
$\Rightarrow e = 180° - 114° \Rightarrow$ **$e = 66°$**
Opposite angles in a parallelogram are equal, so **$f = 66°$**
Neighbouring angles always add up to 180°, so $66° + g = 180° \Rightarrow g = 180° - 66°$
$\Rightarrow$ **$g = 114°$**
As g and h are opposite, **$h = 114°$** as well.

7 a) Using the symmetry of the kite, **$a = 105°$**.

b) Angles around a point add up to 360°, so
$105° + b + 90° = 360° \Rightarrow b + 195° = 360°$
$\Rightarrow b = 360° - 195° \Rightarrow$ **$b = 165°$**
Call the bottom angle in the kite m.
c is made up of a right angle and m.
As angles in a quadrilateral add up to 360°,
$94° + 105° + m + 105° = 360°$
$\Rightarrow m + 304° = 360°$
$\Rightarrow m = 360° - 304° \Rightarrow m = 56°$
So $c = 56° + 90° \Rightarrow$ **$c = 146°$**

8 a) Using the formula,
$S = (6 - 2) \times 180° = 4 \times 180° =$ **720°**

b) So summing up the internal angles must give 720°.
$x + 130° + 130° + 130° + 130° + 130° = 720°$
$x + 650° = 720° \Rightarrow x = 720° - 650°$
$\Rightarrow$ **$x = 70°$**

9 a) Since the polygon is regular you can use the formula, $360° \div n = 45°$
$\Rightarrow n = 360° \div 45° =$ **8**
This is an **octagon**.

b) 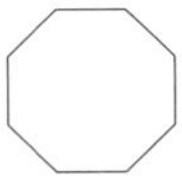

c) Interior angle $= 180° - 45° =$ **135°**

d) $8 \times 135° =$ **1080°**
You could also use the formula to get $(8 - 2) \times 180° = 1080°$.

10 a) *B* b) *A* c) *D* d) *C*

Page 258 Exam-Style Questions

1 Angles on a straight line add up to 180°, so
$3h + h = 180°$ *[1 mark]*
$4h = 180° \Rightarrow h = 180° \div 4 \Rightarrow$ **$h = 45°$**
[1 mark]

2 g and the angle labelled 134° lie on a straight line, so $g + 134° = 180°$
$\Rightarrow$ **g** $= 180° - 134°$ **= 46°** *[1 mark]*
Angles in a quadrilateral add up to 360°, so
$55° + 46° + 29° + h = 360°$
$\Rightarrow 130° + h = 360°$
$\Rightarrow$ **h** $= 360° - 130°$ **= 230°** *[1 mark]*
Angles around a point add up to 360°, so
$230° + i = 360°$
$\Rightarrow$ **i** $= 360° - 230°$ **= 130°** *[1 mark]*

3 a) **56°** *[1 mark]*
This is because angle EBC and angle BFG are corresponding.

b) E.g. As angle EBC and angle FBC lie on a straight line, they must add up to 180°, so 56° + (angle FBC) = 180°
angle FBC = 180° – 56° = 124° *[1 mark]*
Angle FBC and angle DCG are corresponding, so **angle DCG = 124°** *[1 mark]*.
You'll still get the marks if you used a different method, as long as you explain each step and get the correct answer. For example, you could have used the fact that angles BFG and FGC are allied, so FGC = 180° – 56° = 124°. Then FGC and DCG are alternate angles, meaning DCG is also 124°.

4 Angles around a point add up to 360°, so
$w + 90° = 360°$
$w = 360° - 90° \Rightarrow$ **$w = 270°$** *[1 mark]*
Using the formula for interior angles,
$S = (7 - 2) \times 180° = 5 \times 180° = 900°$ *[1 mark]*
So $90° + 90° + 270° + 90° + 97° + z + 146° = 900°$
$z + 783° = 900°$
$z = 900° - 783° \Rightarrow$ **$z = 117°$** *[1 mark]*

5 Exterior angles always add up to 360°, so
$6u + 4u + 5u + 3u = 360° \Rightarrow 18u = 360°$
[1 mark]
$\Rightarrow u = 360° \div 18 \Rightarrow$ **$u = 20°$** *[1 mark]*

6 As triangle ABC is equilateral, angle BCA is 60° and so angle ACE = 60° – 38° = 22° *[1 mark]*.
Triangle ACE is an isosceles so angle EAC = angle AEC
Angles in a triangle add up to 180°, so
2 × angle EAC + 22° = 180°
2 × angle EAC = 180° – 22° = 158°
angle EAC = 158° ÷ 2 = 79° *[1 mark]*.
Angle EAC and angle DAC lie on a straight line so they must add up to 180°.
79° + (angle DAC) = 180°
angle DAC = 180° – 79°, so **angle DAC = 101°**
[1 mark for a fully correct proof with all reasons included]

Section 16 — Units, Measuring and Estimating

16.1 Reading Scales

Page 260 Exercise 1

1 a) There are four divisions between the labelled marks, and the pairs of labelled marks have a difference of 4 – 2 = 2, so each division is worth 2 ÷ 4 = 0.5 cm.
A is two divisions along from 2, so 2 + (2 × 0.5) = **3 cm**.
B is two divisions along from 6, so 6 + (2 × 0.5) = **7 cm**.
C is three divisions along from 8, so 8 + (3 × 0.5) = **9.5 cm**.

b) There are ten divisions between the labelled marks, and the pairs of labelled marks have a difference of 40 – 20 = 20, so each division is worth 20 ÷ 10 = 2 cm.
A is four divisions along from 20, so 20 + (4 × 2) = **28 cm**.
B is three divisions along from 40, so 40 + (3 × 2) = **46 cm**.
C is six divisions along from 40, so 40 + (6 × 2) = **52 cm**.

2 Each division is one tenth of a centimetre, i.e. 1 millimetre, so:

a) The bug is 7 divisions long.
7×1 mm = **7 mm**

b) The bug is 4 divisions long.
4×1 mm = **4 mm**

c) The bug is $10 - 2 = 8$ divisions long.
8×1 mm = **8 mm**
You can't see the divisions to count them because the bug is in the way, but there are two divisions left before it reaches the 1 cm mark.

3 a) There are four divisions between the labelled marks, and the pairs of labelled marks have a difference of $12 - 8 = 4$, so each division is worth $4 \div 4 = 1$ g.
The arrow is pointing at the mark one division more than 8, so the mass shown is $8 + 1 =$ **9 g**.

b) There are ten divisions between the labelled marks, and the pairs of labelled marks have a difference of $6 - 5 = 1$, so each division is worth $1 \div 10 = 0.1$ kg.
The arrow is pointing at the mark three divisions more than 5, so the mass shown is $5 + (3 \times 0.1) =$ **5.3 kg**.

c) There are two divisions between the labelled marks, and the pairs of labelled marks have a difference of $4.4 - 4.3 = 0.1$, so each division is worth $0.1 \div 2 = 0.05$ kg.
The arrow is pointing at the mark one division more than 4.3, so the mass shown is $4.3 + 0.05 =$ **4.35 kg**.

4 There are five divisions between the labelled marks, and the pairs of labelled marks have a difference of $200 - 100 = 100$, so each division is worth $100 \div 5 = 20$ g.
The arrow is pointing at the mark three divisions more than 100, so the mass shown is $100 + (3 \times 20) =$ **160 g**.

5 There are five divisions between the labelled marks, and the pairs of labelled marks have a difference of $75 - 50 = 25$, so each division is worth $25 \div 5 = 5$ ml.
The liquid reaches the mark that is three divisions more than 50, so the volume shown is $50 + (3 \times 5) =$ **65 ml**.

Page 262 Exercise 2

1 The general maximum and minimum values are '+ half a unit' and '– half a unit'.

a) This is measured to the nearest cm, so the maximum value is
$10 + 0.5 =$ **10.5 cm**,
and the minimum value is
$10 - 0.5 =$ **9.5 cm**.
Remember — the 'real' maximum value is 10.499999.... cm, but we use 10.5 to make it easier to write down.

b) This is measured to the nearest litre, so the maximum value is $18 + 0.5 =$ **18.5 litres**, and the minimum value is
$18 - 0.5 =$ **17.5 litres**.

c) This is measured to the nearest 5 litres, so the maximum value is $65 + 2.5$
= **67.5 litres**,
and the minimum value is $65 - 2.5$
= **62.5 litres**.

d) This is measured to the nearest 2 m, so the maximum value is $20 + 1 =$ **21 m**,
and the minimum value is $20 - 1$
= **19 m**.

2 The volume is measured to the nearest 0.1 cm^3, so the maximum value is $0.1 \div 2$ = 0.05 cm^3 more: $5.7 + 0.05 =$ **5.75 cm^3**

3 To be sure that the table will fit through the door, Elliot needs to check that the maximum possible width of the table is smaller than the width of the door frame. The table is measured to the nearest 2 cm, so the maximum width of the table is $95 + 1 = 96$ cm.
96 cm < 96.5 cm, so **yes, Elliot can be sure that the table will fit.**

4 The scales are correct to 10 g, so the maximum and minimum values are
$100 + (10 \div 2) =$ **105 g** and
$100 - (10 \div 2) =$ **95 g**.

5 The maximum possible difference happens when one of their heights is the maximum value and the other is the minimum value.
The maximum value is $170 + (10 \div 2) =$ 175 cm and the minimum value is $170 - (10 \div 2) = 165$ cm, so the greatest possible difference is 175 cm – 165 cm = **10 cm**.

16.2 Converting Units — Length, Mass and Volume

Page 264 Exercise 1

1 a) The conversion factor for cm to mm is 10, and mm are smaller than cm so you need to multiply.
2 cm = 2×10 = **20 mm**.

b) The conversion factor for ml to cm^3 is 1, so 15 ml = **15 cm^3**.

c) The conversion factor for tonnes to kg is 1000, and kg are smaller than tonnes so you need to multiply.
2.3 tonnes = 2.3×1000 = **2300 kg**

2 a) The conversion factor for m to km is 1000, and km are bigger than m so you need to divide.
3400 m = $3400 \div 1000$ = **3.4 km**.

b) The conversion factor for cm to m is 100, and m are bigger than cm so you need to divide.
50 cm = $50 \div 100$ = **0.5 m**.

c) The conversion factor for kg to tonnes is 1000, and tonnes are bigger than kg so you need to divide.
246 kg = $246 \div 1000$ = **0.246 tonnes**.

3 a) The conversion factor for kg to g is 1000, and g are smaller than kg so you need to multiply.
3 kg = 3×1000 = **3000 g**.

b) The conversion factor for mm to cm is 10, and cm are bigger than mm so you need to divide.
379 mm = $379 \div 10$ = **37.9 cm**.

c) The conversion factor for mg into g is 1000, and g are bigger than mg so you need to divide.
22.3 mg = $22.3 \div 1000$ = **0.0223 g**.

4 a) The conversion factor for kg into g is 1000, and grams are smaller than kg so you need to multiply.
1.2 kg = 1.2×1000 = **1200 g**.

b) Part a) gives the weight in g, so
$1200 \div 30$ = **40 servings**

5 a) Do the conversion for tonnes into g in two parts. The conversion factor for tonnes into kg is 1000:
0.6 tonnes = 0.6×1000 = 600 kg.
The conversion factor for kg into g is 1000:
600 kg = 600×1000 = 600 000 g.
So 0.6 tonnes = **600 000 g**.

b) Do the conversion for m into mm in two parts. The conversion factor for m into cm is 100: 62 m = 62×100 = 6200 cm.
The conversion factor for cm into mm is 10:
6200 cm = 6200×10 = 62 000 mm.
So 62 m = **62 000 mm**.

c) Do the conversion for mg into kg in two parts. The conversion factor for mg into g is 1000:
302 300 mg = $302\,300 \div 1000$ = 302.3 g.
The conversion factor for g into kg is 1000:
302.3 g = $302.3 \div 1000$ = 0.3023 kg.
So 302 300 mg = **0.3023 kg**.

Page 264 Exercise 2

1 a) Convert 3200 ml into litres:
$3200 \div 1000 = 3.2$ litres.
Then $3.2 + 75.3$ = **78.5 litres**

b) Convert 51.2 m into cm:
$51.2 \times 100 = 5120$ cm
Then $681 + 5120$ = **5801 cm**

Using the same method for c)-f):

c) **3.575 kg** d) **1395 cm**
e) **1204 kg** f) **1250 ml**

2 Convert km into m:
3.4 km = 3.4×1000 = 3400 m.
Then the difference is $3400 - 1800$
= **1600 m**

3 Convert all of the masses into kg:
3200 g = $3200 \div 1000$ = 3.2 kg
0.72 tonnes = 0.72×1000 = 720 kg
Then add the masses together:
3.2 kg + 15 kg + 720 kg + 3.2 kg = **741.4 kg**

4 Convert the weight of one person's equipment into kg:
9000 g = $9000 \div 1000$ = 9 kg.
There are 3 people so the three lots of equipment weigh $9 \times 3 = 27$ kg.
1 tonne = 1×1000 = 1000 kg, so a quarter of a tonne is 250 kg.
The combined mass of Amirah, Trevor, Elsie and their equipment is
$55.2 + 78.1 + 65.9 + 27 = 226.2$ kg,
so **they will be safe**.

Page 265 Exercise 3

1 Convert 2 tonnes into kg:
2 tonnes = 2×1000 = 2000 kg
He uses 250 kg each day, so it will last
$2000 \div 250$ = **8 days**

2 Convert the distances in m into km:
1500 m = $1500 \div 1000$ = 1.5 km
100 m = $100 \div 1000$ = 0.1 km
So the total of the runs is $1.5 + 0.1 + 13.2$
= **14.8 km**

3 The cafe uses two slices of ham per sandwich, so they use 2×10 g = 20 g per sandwich.
The total weight of the ham they need is
20 g × 500 = 10 000 g.
Convert this into kg: $10\,000 \div 1000 = 10$ kg.
So the café needs $10 \div 1.5 = 6.666...$
= **7 packs of ham**

4 a) Convert 750 000 ml into litres:
$750\,000 \div 1000$ = **750 litres**

b) The reservoir will overflow when
800 000 – 600 000 = 200 000 litres of water has been added.
750 litres are added each day, so it will take $200\,000 \div 750 = 266.66...$ = **267 days** (to the nearest day)

5 a) Convert all of the weights in kg:
400 g = $400 \div 1000$ = 0.4 kg
300 g = $300 \div 1000$ = 0.3 kg
2500 mg = $2500 \div 1000$ = 2.5 g, and
2.5 g = $2.5 \div 1000$ = 0.0025 kg
Adding all of the weights in kg together:
$0.7 + 0.4 + 0.3 + 0.2 + 0.0025$ = **1.6025 kg**

b) $1.6025 \div 0.2 = 8.0125$, so the recipe will feed **8 people**.

16.3 Converting Units — Area and Volume

Page 267 Exercise 1

1 a) 2 km = 2 × 1000 = 2000 m and 3 km = 3 × 1000 = 3000 m So area = 2000 × 3000 = **6 000 000 m²**

b) Area in km² = 2 × 3 = 6 km². The conversion factor for km² to m² is 1000² = 1 000 000, and m are smaller than km so you need to multiply. 6 km² = 6 × 1 000 000 = **6 000 000 m²**

2 a) The conversion factor for cm² to mm² is 10² = 100, and mm are smaller than cm so you need to multiply. 26 cm² = 26 × 100 = **2600 mm²**.

Using the same method for b) and c):

b) 10 500 cm² **c) 12 000 cm²**

d) The conversion factor for cm² to m² is 100² = 10 000, and m are bigger than cm so you need to divide. 1750 cm² = 1750 ÷ 10 000 = **0.175 m²**.

Using the same method for e) and f):

e) 85 cm² **f) 0.0027 m²**

3 Find the area of the lawn in cm²: 990 × 430 = 425 700 cm² Convert this into m²: 425 700 cm² = 425 700 ÷ 100² = 425 700 ÷ 10 000 = 42.57 m² So Ali needs 42.57 ÷ 16 = 2.660... bottles, so she should buy **3 bottles of weedkiller**.

4 Convert the measurements from cm to m: 670 cm = 670 ÷ 100 = 6.7 m, and 420 cm = 420 ÷ 100 = 4.2 m The area of the rooms are 1.7 × 3 = 5.1 m² and 6.7 × 4.2 = 28.14 m². The total area is 5.1 + 28.14 = **33.24 m²**

Page 268 Exercise 2

1 a) 1000 cm = 1000 ÷ 100 = 10 m, 400 cm = 400 ÷ 100 = 4 m and 150 cm = 150 ÷ 100 = 1.5 m, so the volume in m³ is: 10 × 4 × 1.5 = **60 m³**

b) The volume in cm³ is 1000 × 400 × 150 = 60 000 000 cm³ The conversion factor from cm³ to m³ is 100³ = 1 000 000, and m are bigger than cm so you need to divide. 60 000 000 cm³ = 60 000 000 ÷ 1 000 000 = **60 m³**

2 a) The conversion factor for km³ to m³ is 1000³, and m are smaller than km so you need to multiply. 0.001 km³ = 0.001 × 1000³ = **1 000 000 m³**.

Using the same method for b) and c):

b) 17 600 000 cm³

c) 1 200 000 000 m³

d) The conversion factor for mm³ to cm³ is 10³ and cm are bigger than mm so you need to divide. 16 000 mm³ = 16 000 ÷ 10³ = 16 000 ÷ 1000 = **16 cm³**

Using the same method for e) and f):

e) 0.00000015 km³ **f) 0.0000359 m³**

3 Volume in cm³ is: 20 × 25 × 10 = 5000 cm³. The conversion factor for cm³ to m³ is 100³ 5000 cm³ = 5000 ÷ 100³ = 0.005 m³. So the volume of coffee powder left over is 0.005 – 0.003 = **0.002 m³**

You could also have converted the dimensions of the packet into m first.

4 The volume in cm³ is: 10.5 × 5.3 × 8.67 = 482.4855 cm³ The conversion factor for cm³ to mm³ is 10³ 482.4855 cm³ = 482.4855 × 10³ = **482 485.5 mm³**

5 Volume of one brick in cm³ is: 3.2 × 3.12 × 2.8 = 27.9552 cm³. 1 m³ = 100³ cm³. 100³ ÷ 27.9552 = 35 771.5..., so **35 771 complete bricks can be made out of 1 m³ of plastic.**

There's not enough plastic to make 35 772 bricks, so you have to round your answer down to 35 771.

6 a) The volume is 3 × 375 = **1125 m³**

b) Convert m³ to cm³: The conversion factor for cm³ to m³ is 100³ 1125 m³ = 1125 × 100³ = **1 125 000 000 cm³**

c) 1 125 000 000 cm³ = 1 125 000 000 ml 1 125 000 000 ml = 1 125 000 000 ÷ 1000 = **1 125 000 litres**

7 a) The conversion factor for cm³ to m³ is 100³ 0.56 m³ = 0.56 × 100³ = **560 000 cm³**

b) The conversion factor for cm³ to mm³ is 10³ 560 000 cm³ = 560 000 × 10³ = **560 000 000 mm³**

16.4 Metric and Imperial Units

Page 270 Exercise 1

1 a) The conversion factor for feet to inches is 12, and inches are smaller than feet so you need to multiply. 2 feet = 2 × 12 = **24 inches**

Using the same method for b) and c):

b) 4 pints **c) 72 pints**

d) The conversion factor for pounds to stone is 14, and stone are bigger than pounds so you need to divide. 56 pounds = 56 ÷ 14 = **4 stone**

Using the same method for e) and f):

e) 5 feet **f) 4 gallons**

2 a) The conversion factor for inches to cm is 2.5, and cm are smaller than inches so you need to multiply. 4 inches ≈ 4 × 2.5 = **10 cm**

b) The conversion factor for ounces to g is 28, and g are smaller than ounces so you need to multiply. 3 ounces ≈ 3 × 28 = **84 g**

c) The conversion factor for stone to g is 6400, and g are smaller than stone so you need to multiply. 10 stone ≈ 10 × 6400 = **64 000 g**

d) The conversion factor for cm to yards is 90 and cm are smaller than yards so you need to multiply. 5 yards ≈ 5 × 90 = **450 cm**

e) The conversion factor for miles to km is 1.6, and km are smaller than miles so you need to multiply. 25 miles ≈ 25 × 1.6 = **40 km**

f) The conversion factor for feet to cm is 30, and cm are smaller than feet so you need to multiply. 6 feet ≈ 6 × 30 = **180 cm**

3 a) The conversion factor for km to miles is 1.6, and miles are bigger than km so you need to divide. 8 km ≈ 8 ÷ 1.6 = **5 miles**

b) The conversion factor for pints to litres is 1.76 and litres are larger than pints so you need to divide. 100 pints ≈ 100 ÷ 1.76 = 56.818... = **57 litres** (to the nearest litre)

c) The conversion factor for g to stones is 6400, and stones are bigger than g so you need to divide. 12 800 g ≈ 12 800 ÷ 6400 = **2 stones**

d) The conversion factor for cm to inches is 2.5 and inches are larger than cm so you need to divide. 25 cm ≈ 25 ÷ 2.5 = **10 inches**

e) The conversion factor for g to ounces is 28, and ounces are bigger than g so you need to divide. 56 g ≈ 56 ÷ 28 = **2 ounces**

f) The conversion factor for pounds to kg is 2.2 and kg are bigger than pounds so you need to divide. 16.5 pounds ≈ 16.5 ÷ 2.2 = **7.5 kg**

4 1 mile ≈ 1.6 km = 1600 m, so 1 mile is 1600 ÷ 400 = **4 laps**

5 a) The conversion factor for yards to cm is 90, and cm are smaller than yards, so you need to multiply. 18 yards ≈ 18 × 90 = **1620 cm**

b) The conversion factor for cm to m is 100, and m are bigger than cm so you need to divide. 1620 cm = 1620 ÷ 100 = **16.2 m**

6 a) The conversion factor for feet to cm is 30, and cm are smaller than feet so you need to multiply. 11 feet = 11 × 30 = 330 cm 330 cm = 330 ÷ 100 = **3.3 m**

b) The conversion factor for stones to g is 6400 and g are smaller than stones so you need to multiply. 1 stone ≈ 1 × 6400 = 6400 g 6400 g = 6400 ÷1000 = **6.4 kg**

c) The conversion factor for pints to litres is 1.76 and litres are bigger than pints so you need to divide. 16 pints ≈ 16 ÷ 1.76 = 9.090... litres 9.090... litres = 9.090... × 1000 = **9091 ml** (to the nearest ml)

Page 271 Exercise 2

1 a) 3 ft 7 in = (3 × 12) + 7 = **43 inches**

b) 12 ft 5 in = (12 × 12) + 5 = **149 inches**

c) 5 lb 2 oz = (5 × 16) + 2 = **82 ounces**

d) 280 in = 280 ÷ 12 = 23 remainder 4 = **23 feet 4 inches**

e) 72 oz = 72 ÷ 16 = 4 remainder 8 = **4 pounds 8 ounces**

f) 200 oz = 200 ÷ 16 = 12 remainder 8 = **12 pounds 8 ounces**

2 a) (i) 1904 g ≈ 1904 ÷ 28 = **68 ounces**

(ii) There are 16 ounces in 1 pound. 68 ÷ 16 = 4 remainder 4, so 1904 g ≈ **4 pounds 4 ounces**

b) (i) 840 g = 840 ÷ 28 = **30 ounces**

(ii) There are 16 ounces in 1 pound. 30 ÷ 16 = 1 remainder 14, so 840 g ≈ **1 pound 14 ounces**

Using the same method for c) and d):

c) (i) 175 ounces

(ii) 10 pounds 15 ounces

d) (i) 35 ounces

(ii) 2 pounds 3 ounces

For c) and d), you have to convert kg to g first.

3 a) The conversion factor for cm to inches is 2.5. 50 cm ≈ 50 ÷ 2.5 = 20 inches. 1 foot = 12 inches, and 20 ÷ 12 = 1 remainder 8, so 50 cm ≈ **1 foot 8 inches**

Using the same method for b)-d):

b) 3 feet 6 inches **c) 6 feet 8 inches**

d) 11 feet 4 inches

In parts c) and d) you need to convert from metres to cm first.

4 Maddie: 4 foot 5 inches = (4 × 12) + 5 = 53 inches
53 inches ≈ 53 × 2.5 = 132.5 cm
Lily: 4 foot 9 inches = (4 × 12) + 9 = 57 inches
57 inches ≈ 57 × 2.5 = 142.5 cm
No, only Lily is tall enough to go on the ride.

5 1 pound 12 oz = (1 × 16) + 12 = 28 oz.
28 oz ≈ 28 × 28 = 784 g,
so **no, it will not be enough.**

6 7 stone 2 pounds = (7 × 14) + 2 = 100 pounds
11 stone 4 pounds = (11 × 14) + 4
= 158 pounds
16 stone = 16 × 14 = 224 pounds
15 stone 4 pounds = (15 × 14) + 4
= 214 pounds
10 stone 3 pounds = (10 × 14) + 3
= 143 pounds
12 stone = 12 × 14 = 168 pounds
8 stone 9 pounds = (8 × 14) + 9
= 121 pounds
13 stone 1 pounds = (13 × 14) + 1
= 183 pounds
The total sum of the weights is 1311 pounds.
The conversion factor for pounds to grams is 450, so 1311 pounds ≈ 1311 × 450
= 589 950 g
589 950 g = 589 950 ÷ 1000 = 589.95 kg
589.95 kg = 589.95 ÷ 1000 = 0.58995 tonnes
No, the total weight will not exceed the limit.

16.5 Estimating in Real Life

Page 273 Exercise 1

1 a) (i) **centimetres (or millimetres)**
(ii) **inches**
b) (i) **grams** (ii) **ounces**
c) (i) **metres** (ii) **yards (or feet)**
d) (i) **millimetres**
(ii) **inches (as a fraction)**
e) (i) **tonnes** (ii) **stone**
f) (i) **kg** (ii) **pounds**
g) (i) **km** (ii) **miles**

2 a) E.g **2.5 m** b) E.g. **2.5 m**
c) Between **1.5 and 2 m** d) E.g. **350 ml**

3 a) The height of the house is roughly 4 times the height of the man. Average height of man ≈ 1.8 m,
so height of house ≈ 1.8 × 4 = 7.2 m.
So the height of the house is between **6 m and 8 m**
b) Average height of a woman is 1.6 m, so the elephant is between **2 m and 2.5 m.**

4 Estimating the man's height to be 1.8 metres, the length of the bus is between **10 m and 12 m**, and the height of bus is between **3 m and 4 m.**

5 The dinosaur is about 3 times as tall as the chicken and 7 times as long. Estimating the height and length of a chicken at around 30 cm, the dinosaur will be **about 0.9 m tall and 2.1 m long.**

6 The rhino is about 6 times as tall as the cat. Estimating the height of an average cat at around 25 cm, the rhino will be **around 1.5 m tall.**

Page 274 Review Exercise

1 a) **400 ml**
b) There are 8 divisions between the labelled marks, and the difference between the labelled marks is 50 – 10 = 40 cm^3.
So each division represents
40 ÷ 8 = 5 cm^3.
The liquid is at the level three divisions above 10, so 10 + (3 × 5) = **25 cm^3**
c) There are 4 divisions between the labelled marks, and the difference between the labelled marks is 4 – 3 = 1.
So each division represents
1 ÷ 4 = 0.25 litres
The liquid is at the level three divisions above 3, so 3 + (3 × 0.25) = **3.75 litres**
d) There are 10 divisions between the labelled marks, and the difference between the labelled marks is 100 – 50 = 50.
So each division represents
50 ÷ 10 = 5 cm^3
The liquid is at the level five divisions above 50, so 50 + (5 × 5) = **75 cm^3**

2 Maximum value is 10.6 + (0.2 ÷ 2)
= 10.6 + 0.1 = **10.7 m**
Minimum value is 10.6 – (0.2 ÷ 2)
= 10.6 – 0.1 = **10.5 m**

3 30 × 400 ml = 12 000 ml.
Conversion factor for ml to litres is 1000, and litres are bigger than ml so divide:
12 000 ml = 12 000 ÷ 1000 = 12 litres
12 ÷ 2 = 6, so she needs **6 bottles of juice.**

4 Convert all weights to kg. Conversion factor for g to kg is 1000.
450 g = 450 ÷ 1000 = 0.45 kg,
300 g = 300 ÷ 1000 = 0.3 kg
The sum of the weights of all the ingredients is: 0.45 + 0.2 + 0.3 + 0.1 = 1.05 kg,
so the mass is **0.05 kg more.**

5 a) 20 × 40 = **800 mm^2**
b) The conversion factor from mm^2 to cm^2 is 10^2 = 100, so 800 mm^2 = 800 ÷ 100
= **8 cm^2**

6 a) Convert all measurements to cm:
50 mm = 50 ÷ 10 = 5 cm,
0.17 m = 0.17 × 100 = 17 cm
Volume = 7 × 5 × 17 = **595 cm^3**
b) 7 × 20 = **140 cm^3**
c) 595 – 140 = 455 cm^3
455 cm^3 = 455 × 10^3 = **455 000 mm^3**

7 a) 3 ounces ≈ 3 × 28 = **84 g**
b) 2 gallons ≈ 2 × 4.5 = **9 litres**
Using the same method for c)-f):
c) **64 000 g** d) **450 cm**
e) **6.75 cm** f) **2.565 litres**

8 a) E.g. **4.5 m**
b) E.g. **22 cm**
c) E.g. **2 m**
d) E.g **150 ml**

Page 275 Exam-Style Questions

1 a) 625 cm = 625 ÷ 100 = **6.25 m** *[1 mark]*
b) 3.94 kg = 3.94 × 1000 = **3940 g** *[1 mark]*

2 The average height for a woman is around 1.5 m to 1.7 m. The bus is about 2.5 times the height of the woman, so an approximate height for the bus would be **3.75 m – 4.25 m.**
[2 marks available — 1 mark for an appropriate estimate for the average height of a woman, 1 mark for using this to estimate the height of the bus]

3 1.2 kg = 1.2 × 1000 = 1200 g
1200 ÷ 300 = **4 pizzas** *[1 mark]*

4 a) 7.35 litres = 7.35 × 1000 = 7350 ml
[1 mark]
7350 ÷ 250 = 29.4, so she will be able to serve **29 glasses** *[1 mark]*
b) 250 ml × 29 = 7250 ml = 7.25 litres
So there is 7.35 – 7.25 = **0.1 litres left**
[1 mark]

5 10 pounds ≈ 10 × 450 = 4500 g and
2 ounces ≈ 2 × 28 = 56 g,
so the first bag weighs 4500 + 56 = 4556 g.
2 stone ≈ 2 × 6400 = 12 800 g
The total weight of her bags is 4556 + 12 800
= 17 356 g = 17.356 kg
17.356 kg < 18 kg, so **she can take both bags on her flight.**
[3 marks available — 1 mark for converting the first bag's weight units, 1 mark for converting the second bag's weight units, 1 mark for finding their total and stating the conclusion]

6 a) 45 litres ≈ 45 ÷ 4.5 = 10 gallons
10 gallons × 55 miles/gallon = **550 miles**
[2 marks available — 1 mark for conversion between litres and gallons, 1 mark for correct answer]
b) 45 litres ≈ 10 gallons (from a))
10 gallons × £5 = **£50** *[1 mark]*

Section 17 – Speed, Density and Pressure

17.1 Speed, Distance and Time

Page 277 Exercise 1

1 a) speed = $\frac{\text{distance}}{\text{time}} = \frac{30}{2}$ = **15 km/h**
Using the same method for b)-d):
b) **20 km/h** c) **30 mph** d) **35 mph**

2 a) speed = $\frac{\text{distance}}{\text{time}} = \frac{80}{2}$ = **40 km/h**
b) speed = $\frac{\text{distance}}{\text{time}} = \frac{50}{500}$ = **0.1 cm/s**
Using the same method for c) and d):
c) **600 mph** d) **1.25 m/s**

3 a) 1 hour = 60 minutes
30 min = 0.5 hours
speed = $\frac{10}{0.5}$ = **20 km/h**
b) 1 km = 1000 m
30 000 m = $\frac{30\,000}{1000}$ = 30 km
speed = $\frac{30}{2.5}$ = **12 km/h**
Using the same method for c) and d):
c) **3 km/h** d) **14 km/h**

Page 278 Exercise 2

1 a) distance = speed × time = 20 × 2 = **40 km**
Using the same method for b)-d):
b) **500 m** c) **72 km** d) **175 miles**

2 a) time = $\frac{\text{distance}}{\text{speed}} = \frac{4}{2}$ = **2 hours**
Using the same method for b)-d):
b) **3 s** c) **2.5 hours**
d) **0.25 hours (or 15 minutes)**

3 distance = speed × time = 30 × 4 = **120 miles**

4 time = $\frac{\text{distance}}{\text{speed}} = \frac{2.4}{15}$ = **0.16 s**

5 distance = 490 × 2 = **980 miles**

6 time = $\frac{\text{distance}}{\text{speed}} = \frac{5.6}{56}$ = 0.1 h
0.1 × 60 = **6 minutes**

7 15 min = $\frac{15}{60}$ = 0.25 h
distance = speed × time = 7.5 × 0.25
= **1.875 miles**

Page 279 Exercise 3

1 a) 1 km = 1000 m

$50 \text{ m/s} = \frac{50}{1000} = 0.05 \text{ km/s}$

1 h = (60 × 60) s = 3600 s

0.05 × 3600 = **180 km/h**

b) 1 km = 1000 m

72 km/h = 72 × 1000 = 72 000 m/h

1 h = (60 × 60) s = 3600 s

$72\,000 \text{ m/h} = \frac{72\,000}{3600} = \mathbf{20 \text{ m/s}}$

Using the same method for c):

c) **95.4 km/h**

2 1 km = 1000 m

18 km/h = 18 × 1000 = 18 000 m/h

1 h = (60 × 60) s = 3600 s

$18\,000 \text{ m/h} = \frac{18\,000}{3600} = \mathbf{5 \text{ m/s}}$

3 a) 1 mile ≈ 1.6 km

$54 \text{ km/h} \approx \frac{54}{1.6} = \mathbf{33.75 \text{ mph}}$

Remember — ≈ means 'is approximately equal to'.

b) 1 mile ≈ 1.6 km

25 mph ≈ 25 × 1.6 = **40 km/h**

c) 1 mile ≈ 1.6 km

$94.4 \text{ km/h} \approx \frac{94.4}{1.6} = \mathbf{59 \text{ mph}}$

4 100 cm = 1 m

$0.5 \text{ cm/s} = \frac{0.5}{100} = 0.005 \text{ m/s}$

60 seconds = 1 minute

0.005 m/s = 0.005 × 60 = **0.3 m/min**

5 1000 m = 1 km

$40 \text{ m/s} = \frac{40}{1000} = 0.04 \text{ km/s}$

1 h = 60 × 60 = 3600 s

0.04 km/s = 0.04 × 3600 = 144 km/h

1 mile ≈ 1.6 km

$144 \text{ km/h} \approx \frac{144}{1.6} = \mathbf{90 \text{ mph}}$

17.2 Density, Mass and Volume

Page 281 Exercise 1

1 a) $\text{density} = \frac{\text{mass}}{\text{volume}} = \frac{20}{5} = \mathbf{4 \text{ kg/m}^3}$

Using the same method for b)-d):

b) **60 kg/m³** c) **250 kg/m³** d) **90 kg/m³**

2 $\text{density} = \frac{\text{mass}}{\text{volume}} = \frac{1840}{0.8} = \mathbf{2300 \text{ kg/m}^3}$

3 a) $\mathbf{volume = \frac{mass}{density}}$

b) (i) $\text{volume} = \frac{40}{8} = \mathbf{5 \text{ m}^3}$

(ii) $\text{volume} = \frac{750}{15} = \mathbf{50 \text{ m}^3}$

(iii) $\text{volume} = \frac{4\,800}{240} = \mathbf{20 \text{ m}^3}$

4 a) **mass = density × volume**

b) mass = 2600 × 0.4 = **1040 kg**

5 1 m³ = (100 × 100 × 100) cm³

= 1 000 000 cm³

$200 \text{ cm}^3 = \frac{200}{1\,000\,000} = 0.0002 \text{ m}^3$

$\text{density} = \frac{\text{mass}}{\text{volume}} = \frac{0.15}{0.0002} = \mathbf{750 \text{ kg/m}^3}$

6 1 m³ = 1 000 000 cm³

$11\,500 \text{ kg/m}^3 = \frac{11500}{1\,000\,000} = 0.0115 \text{ kg/cm}^3$

mass = density × volume

= 0.0115 × 8 = **0.092 kg or 92 g**

You could also do this by converting the volume of the paperweight to m³ and finding the density.

7 1 m³ = 1 000 000 cm³

2.7 g/cm³ = 2.7 × 1 000 000 = 2 700 000 g/m³

1 kg = 1000 g

$2\,700\,000 \text{ g/m}^3 = \frac{2\,700\,000}{1000} = \mathbf{2700 \text{ kg/m}^3}$

17.3 Pressure, Force and Area

Page 283 Exercise 1

1 a) $\text{pressure} = \frac{\text{force}}{\text{area}} = \frac{27}{3} = \mathbf{9} \text{ N/cm}^2$

Using the same method for b)-d):

b) **25 N/m²** c) **1200 N/m²** d) **8 N/cm²**

2 a) $\text{area} = \frac{\text{force}}{\text{pressure}} = \frac{36}{6} = \mathbf{6 \text{ cm}^2}$

Using the same method for b)-d):

b) **7 m²** c) **4 cm²** d) **3 m²**

3 a) force = pressure × area = 300 × 5 = **1500 N**

b) force = pressure × area = 36 × 30 = **1080 N**

4 Area of cube face = 10 × 10 = 100 cm²

Force = pressure × area = 0.02 × 100 = **2 N**

5 a) Cube side length = $\sqrt[3]{512}$ = 8 cm

Area of cube face = 8 × 8 = **64 cm²**

b) $\text{pressure} = \frac{\text{force}}{\text{area}} = \frac{1792}{64} = \mathbf{28 \text{ N/cm}^2}$

6 a) Area of square face = 3 × 3 = 9 cm²

$\text{pressure} = \frac{\text{force}}{\text{area}} = \frac{45}{9} = \mathbf{5 \text{ N/cm}^2}$

b) E.g. 1 m² = (100 × 100) cm² = 10 000 cm²

5 N/m² = 5 × 10 000 = **50 000 N/m²**

You could also have found the area of the square face in m² (9 cm = 0.09 m) and used the formula for pressure again.

17.4 Distance-Time Graphs

Page 285 Exercise 1

1 a)

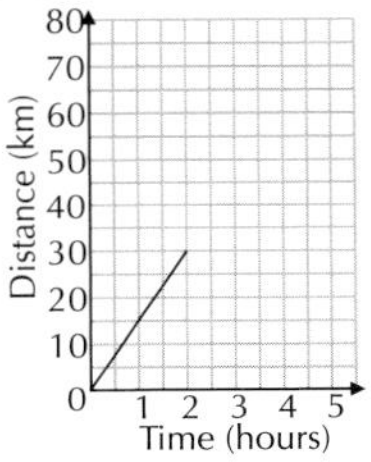

b)

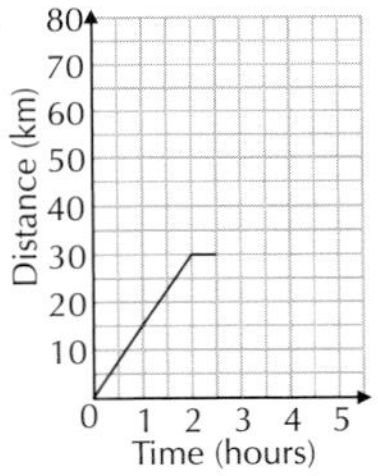

c) (i) 30 + 40 = **70 km**

(ii)

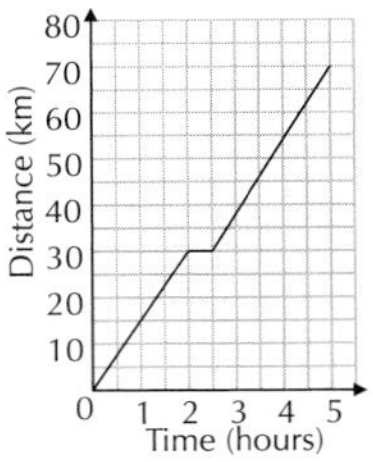

2

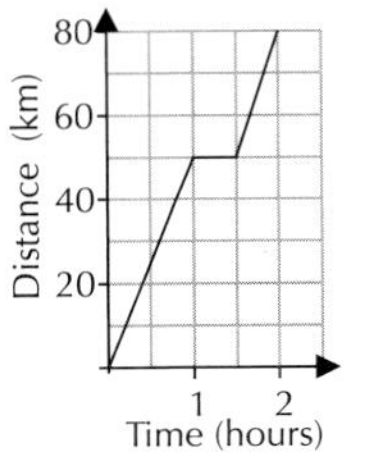

3

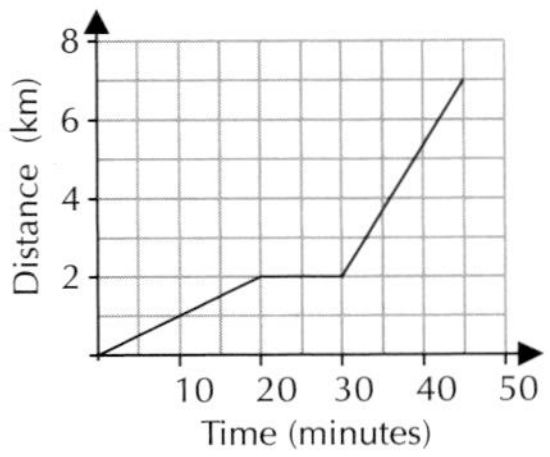

4 a) They stopped at the horizontal part of the graph, so, reading the y-axis at the point where they stopped, they travelled for **50 miles**.

b) The graph is horizontal between 1 hour and 2 hours, so they stayed at their destination for 2 – 1 = **1 hour**.

c) (i) The graph starts to decline at 2 hours, which shows they started the journey back at 8:00 am + 2 hours = **10:00 am**.

(ii) They set off after 2 hours and got back after 4 hours, so the journey home took 4 – 2 = **2 hours**.

5 a) The object travels 5 km in 2 hours. It stops for 1 hour. It then continues its journey for another hour, during which it travels a further 3 km. It then stops for 30 minutes. Finally, it travels back to its starting point in 1.5 hours.

b) The object travels 1 km in 1 hour. It then stops for 30 minutes. It then travels 0.5 km in 30 minutes before stopping again for 15 minutes. It then travels 2 km in 30 minutes and then stops for 15 minutes.

c) The object travels 35 miles in 45 minutes. It then stops for 45 minutes before it travels back towards its starting point, for 15 miles in 30 minutes. It stops for 30 minutes, and then travels away from its starting point again for 30 minutes, in which it travels for 20 miles.

Page 287 Exercise 2

1 a) distance travelled = 22.5 – 0 = 22.5 miles

time taken = 2 – 0 = 2 hours

$\text{speed} = \frac{\text{distance}}{\text{time}} = \frac{22.5}{2} = \mathbf{11.25 \text{ mph}}$

b) distance travelled = 3.75 – 0 = 3.75 km

time taken = 3 – 0 = 3 hours

$\text{speed} = \frac{\text{distance}}{\text{time}} = \frac{3.75}{3} = \mathbf{1.25 \text{ km/h}}$

c) distance travelled = 0 – 1 = –1 m

time taken = 5 – 0 = 5 s

$\text{speed} = \frac{\text{distance}}{\text{time}} = \frac{-1}{5} = -0.2$, so the speed is **0.2 m/s**

The gradient is negative, which means the object is moving back towards you — but an object can't have a negative speed so you take the positive value.

2 a) **The second stage** of the journey was at a higher speed since the gradient is steeper for that part of the graph.

b) (i) Reading off the y-axis, he travelled **2 km**.

(ii) Time taken = 30 min = 0.5 hours

$\text{speed} = \frac{\text{distance}}{\text{time}} = \frac{2}{0.5} = \mathbf{4 \text{ km/h}}$

c) distance travelled = 5 – 2 = 3 km

time taken = 15 min = 0.25 hours

$\text{speed} = \frac{\text{distance}}{\text{time}} = \frac{3}{0.25} = \mathbf{12 \text{ km/h}}$

3 a) She was walking fastest **between 13:30 and 14:30** since between these times the gradient is steepest.

b) (i) 12:00 – 11:00 = **1 hour**

(ii) Reading from the y-axis on the graph she had travelled **3 km** before she stopped for a rest.

(iii) speed = $\frac{\text{distance}}{\text{time}} = \frac{3}{1} =$ **3 km/h**

c) Time taken = 13:00 – 12:15 = 45 min = 0.75 hours.

Distance travelled = 4.5 – 3 = 1.5 km.

Speed = $\frac{\text{distance}}{\text{time}} = \frac{1.5}{0.75} =$ **2 km/h**

d) Time taken = 14:30 – 13:30 = 1 hour

Distance travelled = 4.5 – 0 = 4.5 km

Speed = $\frac{\text{distance}}{\text{time}} = \frac{4.5}{1} =$ **4.5 km/h**

4 a) Time = $\frac{\text{distance}}{\text{speed}} = \frac{200}{100} = 2$ hours.

Ashna starts travelling at 10:00 so she will stop at 12:00 having travelled 200 miles. This gives the graph:

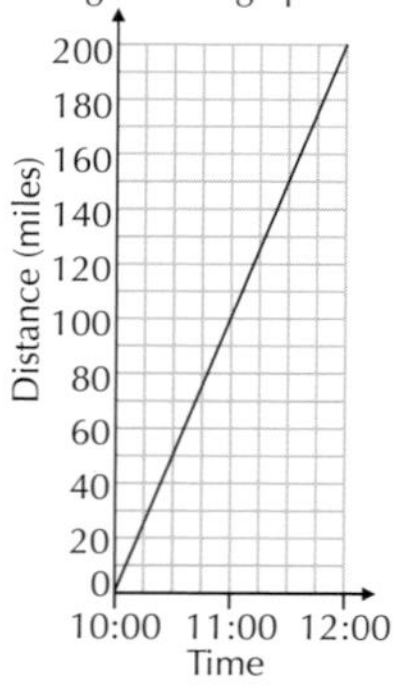

b) For the first leg of Jasper's journey, time = $\frac{\text{distance}}{\text{speed}} = \frac{120}{40} = 3$ hours. He sets of 9:00 so he will reach his destination at 12:00. After 1 hour, it will be 13:00 when he sets of back and it will take $\frac{120}{60} = 2$ hours, so he will arrive back at 15:00. This gives the graph:

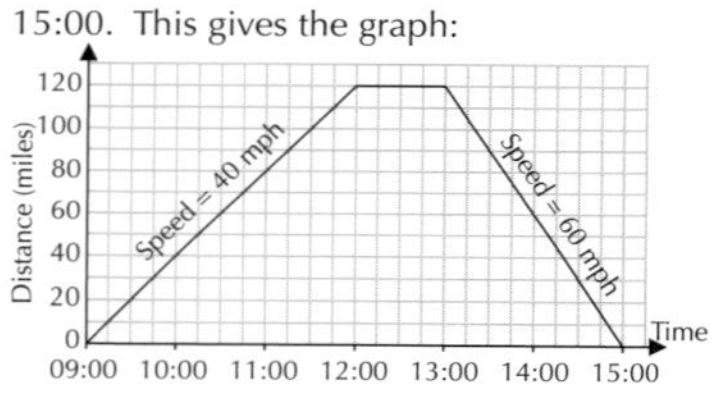

Page 288 Review Exercise

1 speed = $\frac{\text{distance}}{\text{time}} = \frac{10}{2.5} =$ **4 m/s**

2 a) 15 minutes = 0.25 hours

$\frac{13.5}{0.25} =$ **54 km/h**

b) 1 km = 1000 m

54 km/h = 54 × 1000 = 54 000 m/h

1 hour = (60 × 60) s = 3600 s

54 000 m/h = $\frac{54\,000}{3600} =$ **15 m/s**

3 a) density = $\frac{\text{mass}}{\text{volume}} = \frac{2500}{50} =$ **50 kg/m³**

b) 1 kg = 1000 g

50 kg/m³ = 50 000 g/m³

1 m³ = (100 × 100 × 100) cm³ = 1 000 000 cm³

50 000 g/m³ = $\frac{50\,000}{1\,000\,000} =$ **0.05 g/cm³**

4 a) pressure = $\frac{\text{force}}{\text{area}} = \frac{4800}{4} =$ **1200 N/m²**

b) area = $\frac{\text{force}}{\text{pressure}} = \frac{540}{180} =$ **3 m²**

c) force = pressure × area = 28 × 14 = **392 N**

5 a) i) Helena was on the bus between 8:00 and 9:00 so time = 1 hour, distance = 40 km

speed = $\frac{\text{distance}}{\text{time}} = \frac{40}{1} =$ **40 km/h**

ii) Helena was on the train for 1.5 hours, distance = 90 – 40 = 50 km

speed = $\frac{\text{distance}}{\text{time}} = \frac{50}{1.5} =$ **33.3 km/h** (to 1 d.p.)

b)

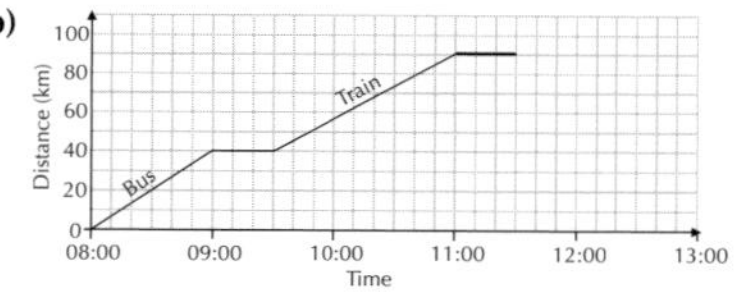

c) Helena is 90 km from home and she travels at 60 km/h

time = $\frac{\text{distance}}{\text{speed}} = \frac{90}{60} =$ **1.5 hours**

d)

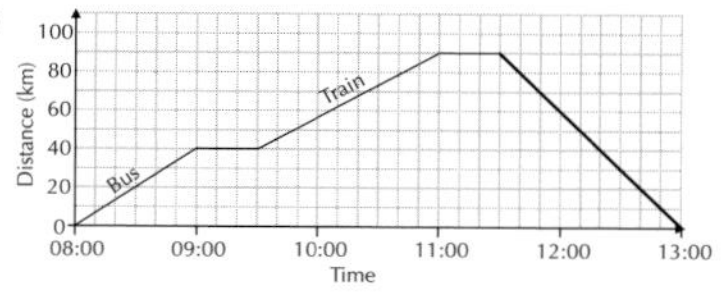

Page 289 Exam-Style Questions

1 a) Tash runs 10 km in 2.5 hours. *[1 mark]* She stops to rest for 1.5 hours. *[1 mark]* She then returns to her starting point in 2 hours. *[1 mark]*

b) The gradient of the graph represents Tash's speed *[1 mark]*

c) Find the total distance and time and then use the formula speed = $\frac{\text{distance}}{\text{time}}$ to calculate the speed.

distance = 0 – 10 = –10 km

time = 6 – 4 = 2 hours

speed = $\frac{-10}{2} =$ **5 km/h**

[2 marks available — 1 mark for finding the distance travelled and time taken and 1 mark for finding the speed in km/h]

The gradient is negative so the object is moving back towards you — but an object can't have a negative speed so you take the positive value.

2 time = $\frac{\text{distance}}{\text{speed}} = \frac{60}{24}$

= 2.5 hours (2 hours 30 minutes) *[1 mark]*

11:10 – 2.5 hours = **8:40 am** *[1 mark]*

3 Volume of cube = 5 × 5 × 5 = 125 cm³ *[1 mark]*

mass = density × volume = 8 × 125 = 1000 g = **1 kg** *[1 mark]*

4 1 m² = (100 × 100) cm² = 10 000 cm² *[1 mark]*

40 000 cm² = $\frac{40\,000}{10\,000} = 4$ m² *[1 mark]*

force = pressure × area = 625 × 4 = **2500 N** *[1 mark]*

5 1 km = 1000 m

4 min 45 s = (4 × 60) + 45 = 285 s,

speed = $\frac{\text{distance}}{\text{time}} = \frac{1000}{285}$

= 3.508... m/s *[1 mark]*

time = $\frac{\text{distance}}{\text{speed}}$

$\frac{800}{3.508...} = 228$ s *[1 marks]*

= **3 minutes 48 seconds** *[1 mark]*

6 density = $\frac{\text{mass}}{\text{volume}}$

⇒ mass = density × volume

Mass of copper = 9 × 36 = 324 g

Mass of tin = 4 × 7 = 28 g *[1 mark for both]*

Total mass = 324 + 28 = **352 g** *[1 mark]*

Section 18 – Scale Drawings and Bearings

18.1 Scale Drawings

Page 291 Exercise 1

1 a) The scale is 1 cm : 2 km, so to convert from cm on the map to km in real life, multiply by 2.

(i) 4 × 2 = **8 km**

(ii) 22 × 2 = **44 km**

(iii) 0.5 × 2 = **1 km**

(iv) 0.25 × 2 = **0.5 km**

b) The scale is 1 cm : 2 km, so to convert from km in real life to cm on the map, divide by 2.

(i) 10 ÷ 2 = **5 cm**

(ii) 14 ÷ 2 = **7 cm**

(iii) 7 ÷ 2 = **3.5 cm**

(iv) 1.4 ÷ 2 = **0.7 cm**

2 The scale is 1 cm : 100 km, so to convert from cm on the map to km in real life, multiply by 100.

a) 7 × 100 = **700 km**

b) 1.5 × 100 = **150 km**

c) 6.22 × 100 = **622 km**

d) 1 cm = 10 mm, so 43 mm = 43 ÷ 10 = 4.3 cm

4.3 × 100 = **430 km**

3 The scale is 1 cm : 50 km, so to convert from the real distance in km to cm on the map, divide by 50.

a) 150 ÷ 50 = **3 cm**

b) 600 ÷ 50 = **12 cm**

c) 1000 ÷ 50 = **20 cm**

d) 1 km = 1000 m, so 15 000 m = 15 000 ÷ 1000 = 15 km

15 ÷ 50 = **0.3 cm**

4 The distances give a scale of 11 cm : 440 km. Divide both sides by 11 to get 1 cm on the LHS: **1 cm : 40 km**

5 a) The measurements give a scale of 4 cm : 12 km.

Divide both sides by 4 to get 1 cm on the LHS: **1 cm : 3 km**

b) The scale is 1 cm : 3 km, so to convert from cm on the map to km in real life, multiply by 3.

7 × 3 = **21 km**

6 The scale is 1 cm : 0.5 m, so to convert from m in real life to cm on the diagram, divide by 0.5.

Dividing by 0.5 is the same as multiplying by 2.

a) 4 ÷ 0.5 = **8 cm**

b) 18 ÷ 0.5 = **36 cm**

c) 21 ÷ 0.5 = **42 cm**

d) 1 m = 100 cm, so 1180 cm = 1180 ÷ 100 = 11.8 m

11.8 ÷ 0.5 = **23.6 cm**

7 The scale is 1 mm:3 cm. The left-hand side is in mm, so the measurements you make should be in mm too. Multiply by 3 to find the real-life measurements in cm.

a) The measurements from the diagram for the sink are:
Width = 19 mm, Height = 10 mm, so multiplying by 3 gives:
Width = 19 × 3 = **57 cm**,
Height = 10 × 3 = **30 cm**

b) The measurements from the diagram for the hob are:
Width = 22 mm, Height = 15 mm, so multiplying by 3 gives:
Width = 22 × 3 = **66 cm**,
Height = 115 × 3 = **45 cm**

Page 292 Exercise 2

1 a) The map distances are in cm, so write the scale as 1 cm:350 cm. To convert from distances on the map to real life, multiply by 350. Then convert to m.

(i) 2 × 350 = 700 cm
100 cm = 1 m, so 700 cm = **7 m**

(ii) 7 × 350 = 2450 cm
2450 ÷ 100 = **24.5 m**

Using the same method for (iii)-(iv):

(iii) **34.65 m** (iv) **89.95 m**

b) To convert from distances in real life to distances on the map, divide by 350.

(i) 700 ÷ 350 = **2 cm**

(ii) 875 ÷ 350 = **2.5 cm**

(iii) 945 ÷ 350 = **2.7 cm**

(iv) 1.47 m = 1.47 × 100 = 147 cm
147 ÷ 350 = **0.42 cm**

2 The map distances are in cm, so write the scale as 1 cm:75 cm. To convert from distances on the plan to real life, multiply by 75. Then convert to m.

a) 4 × 75 = 300 cm = **3 m**

b) 11 × 75 = 825 cm = **8.25 m**

c) 7.9 × 75 = 592.5 cm = **5.925 m**

d) 17.2 × 75 = 1290 cm = **12.9 m**

3 The actual distances are in m, so write the scale as 1:1000 = 1 m:1000 m. To convert from distances in real life to distances on the map, divide by 1000. Then convert to cm.

a) 300 ÷ 1000 = 0.3 m = **30 cm**

b) 1400 ÷ 1000 = 1.4 m = **140 cm**

c) 120 ÷ 1000 = 0.12 m = **12 cm**

d) 1 km = 1000 m, so 0.43 km
= 0.43 × 1000 = 430 m
430 ÷ 1000 = 0.43 m = **43 cm**
You could write the scale as 1 cm:1000 cm = 1 cm:10 m. Then divide by 10 to convert the real-life distances in m directly to cm.

4 The toy measurements are in cm, so write the scale as 1 cm:40 cm. To convert from the toy measurements to actual furniture measurements, multiply by 40. Then convert to m.

a) 3.5 × 40 = 140 cm = **1.4 m**

b) 3.2 × 40 = 128 cm = **1.28 m**

c) 2.4 × 40 = 96 cm = **0.96 m**

5 You've got a big number on the RHS of the scale, so write the scale in cm, and then convert the units on the RHS.
1:250 000 = 1 cm:250 000 cm
= 1 cm:2500 m
= 1 cm:2.5 km

To convert from distances in real life to distances on the map, divide by 2.5:
6.7 ÷ 2.5 = **2.68 cm**
Starting with cm on both sides of the scale and then converting the units on the RHS makes the calculation a lot nicer — if you used 1 km:250 000 km you'd get the map distance as 0.0000268 km, which you'd then need to convert to cm.

6 The actual distances are in m, so write the scale as 1 m:500 m.
To convert from actual measurements to model measurements, divide by 500. Then convert to cm.

a) 20 ÷ 500 = 0.04 m = **4 cm**

b) 30 ÷ 500 = 0.06 m = **6 cm**

c) 100 ÷ 500 = 0.2 m = **20 cm**

d) 6 ÷ 500 = 0.012 m = **1.2 cm**
You could write the scale as 1 cm:500 cm = 1 cm:5 m and then divide by 5 to convert the actual measurements directly into model measurements in cm.

7 a) The measurements give a scale of 3 cm:18 m.
Divide both sides by 3 to get 1 cm on the LHS: **1 cm:6 m**

b) Using the scale from a),
actual width = 1.2 × 6 = **7.2 m**

c) Using the scale from a),
plan length = 4.5 ÷ 6 = **0.75 cm**

8 a) The measurements give a scale of 3 cm:4.5 km.
Divide both sides by 3 to get 1 cm on the LHS: **1 cm:1.5 km**

b) Convert the RHS of the ratio to cm so both sides have the same units, and then remove the units:
1 km = 1000 m = (100 × 1000) cm
= 100 000 cm
1.5 km = 1.5 × 100 000 = 150 000 cm
So the scale is 1 cm:150 000 cm
= **1:150 000**

9 1:120 000 = 1 cm:120 000 cm
= 1 cm:1200 m
= 1 cm:1.2 km

To convert cm on the map to km in real life, multiply by 1.2.

a) The museum to the cathedral measures 1 cm
1 × 1.2 = **1.2 km**
You may have a slightly different measurement — any answer between 1.08 km and 1.32 km is fine.

b) Art gallery to theatre measures 2.5 cm
2.5 × 1.2 = **3 km**
Any answer between 2.88 km and 3.12 km is fine.

c) Cathedral to park measures 2.2 cm
2.2 × 1.2 = **2.64 km**
Any answer between 2.52 km and 2.76 km is fine.

Page 294 Exercise 3

For this exercise, use a ruler to make sure that your drawings have the correct measurements.

1 1:50 = 1 cm:50 cm = 1 cm:0.5 m
So to convert from metres in real life to cm on the plan, divide by 0.5:
1.5 ÷ 0.5 = 3 cm 6.5 ÷ 0.5 = 13 cm
9.5 ÷ 0.5 = 19 cm 2.5 ÷ 0.5 = 5 cm

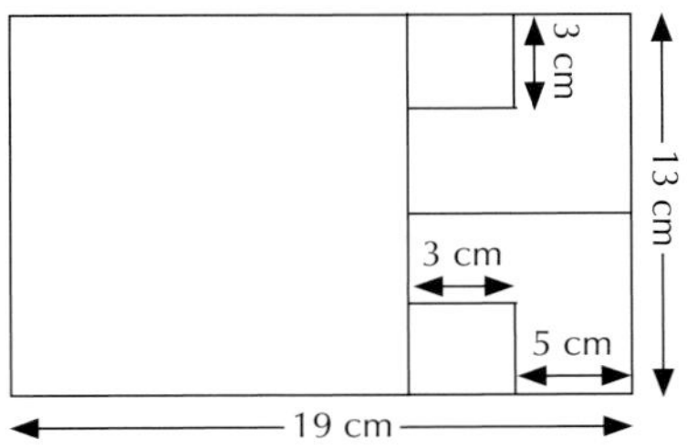

2 a) The scale is 1 cm:3 m, so to convert from metres in real life to cm on the plan, divide by 3:
33 ÷ 3 = 11 cm, 24 ÷ 3 = 8 cm
42 ÷ 3 = 14 cm, 7.5 ÷ 3 = 2.5 cm

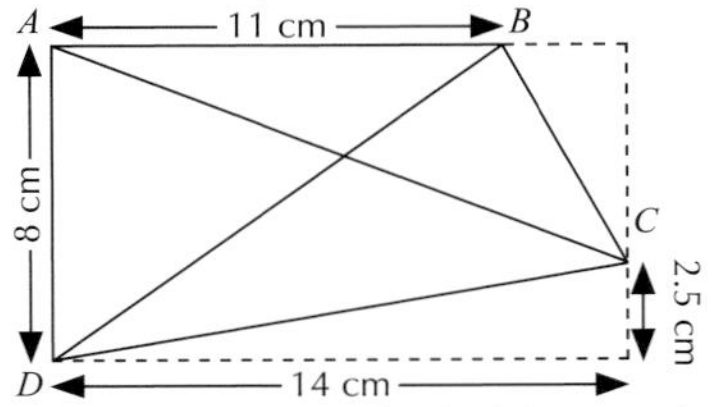

b) The distance between the duck house and *B* measures about 48 mm = 4.8 cm. To convert from cm on the plan to metres in real life, multiply by 3:
4.8 × 3 = **14.4 m**
You might have a slightly different measurement for the distance between the duck house and B — any answer between 13.8 m and 14.7 m is fine.

18.2 Bearings

Page 295 Exercise 1

1 a) East is 90° clockwise from North so the bearing is **090°**.

b) Northeast is 45° clockwise from North so the bearing is **045°**.

c) South is 180° clockwise from North so the bearing is **180°**.

d) Northwest is 45° anti-clockwise from North so the bearing is 360°– 45° = **315°**.
Remember — the bearing is always the clockwise angle.

2 a) **062°** b) 47° + 180° = **227°**
c) 31° + 270° = **301°** d) 180° – 67° = **113°**

3 East is 90° clockwise from North, so the bearing is 90° – 25° = **065°**

4 a) 111° – 90° = **21°** b) 203° – 180° = **23°**
c) 243° – 180° = **63°** d) 270° – 222° = **48°**

5

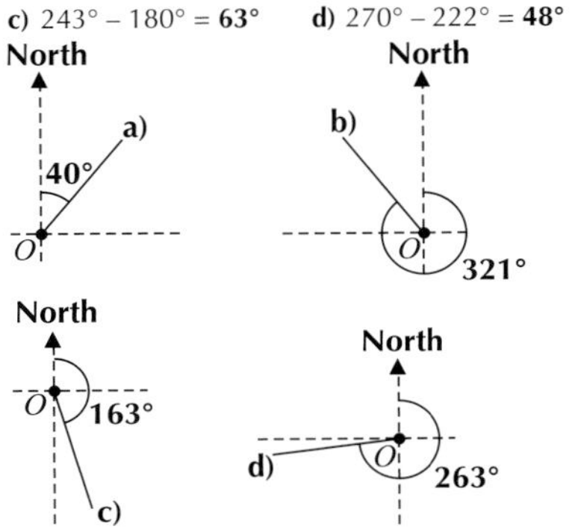

Page 297 Exercise 2

1 a) The angle you need to find is labelled z in the diagram below.

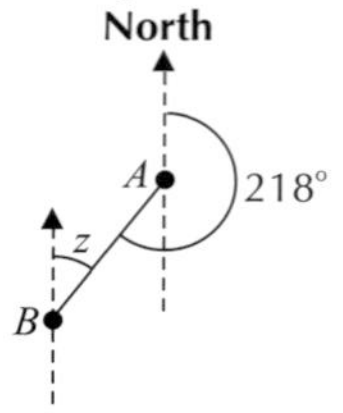

Using alternate angles, z is equal to $218° - 180° = 38°$.
So the bearing is **038°**.

b) The angle you need to find is labelled z in the diagram below.

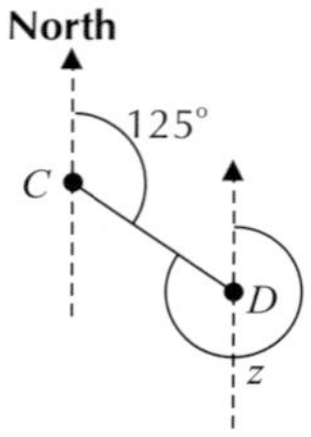

Using alternate angles, z is equal to $125° + 180° = 305°$.
So the bearing is **305°**.

c) The angle you need to find is labelled z in the diagram below.

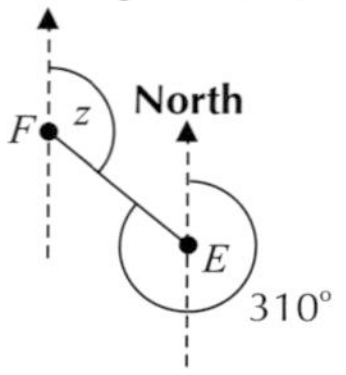

Using alternate angles, z is equal to $310° - 180° = 130°$.
So the bearing is **130°**.

2 a) $200° - 180° =$ **020°**

b) $330° - 180° =$ **150°**

c) $117° + 180° =$ **297°**

d) $15° + 180° =$ **195°**

If you're unsure if you should add or subtract 180°, draw a diagram to check.

3 a) **32°**

An answer between 31° and 33° is okay.

b) Using alternate angles, the angle clockwise from a north line at S to the line RS is also 32°. So the bearing is **032°**.

4 a) Bearing of Z from $Y = 90° + 45° =$ **135°**

b) The bearing of Z from Y is less than 180°, so add 180° to it to find the bearing of Y from Z:
$135° + 180° =$ **315°**

5 a) **055°**

b) $55° + 88° =$ **143°**

c) $360° - 34° =$ **326°**

d) $55° + 180° =$ **235°**

The bearing of B from A is 055°, as you found in part a). This is less than 180°, so you add it to 180° to find the bearing of A from B.

e) $143° + 180° =$ **323°**

143° is the bearing of C from A, which you worked out in part b) — this is less than 180°, so you add it to 180° to find the bearing of A from C.

f) Bearing of D from A
$= 55° + 88° + 101° = 244°$
So bearing of A from D is
$244° - 180° =$ **064°**.

Since the bearing of D from A is greater than 180°, you subtract 180° from it to find the bearing of A from D.

Page 298 Exercise 3

1 a) The scale is 1 cm : 100 km, so to convert from distances on the map to real life, multiply by 100.

(i) Bern and Stuttgart are 2.5 cm apart on the diagram.
$2.5 \times 100 =$ **250 km**

You may have a slightly different measurement — any answer between 240 km and 260 km is fine.

The angle from Bern to Stuttgart, measured clockwise from North, is 50°.
So the bearing is **050°**.

Any answer between 049° and 051° is fine.

(ii) Stuttgart and Munich are 1.9 cm apart on the diagram.
$1.9 \times 100 =$ **190 km**

Any answer between 180 km and 200 km is fine.

The angle from Stuttgart to Munich, measured clockwise from North, measures 125° so the bearing is **125°**.

Any answer between 124° and 126° is fine.

b) Bern and Munich are 3.5 cm apart on the diagram.
$3.5 \times 100 =$ **350 km**

You may have a slightly different measurement — any answer between 340 km and 360 km is fine.

2 The scale is 1 cm : 30 km so to convert from real-life distance in km to cm on the diagram, divide by 30.
$150 \div 30 = 5$, so the distance should be 5 cm on the diagram.

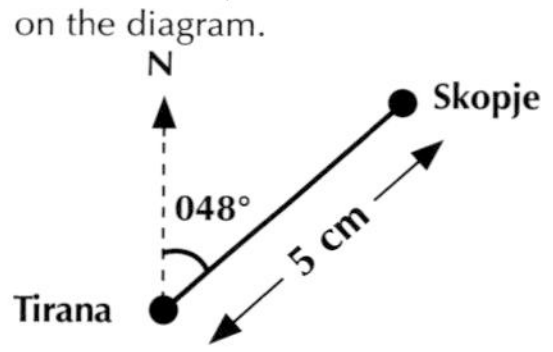

3 The scale is 1 cm : 90 km so to convert from real-life distance in km to cm on the diagram, divide by 90.
$540 \div 90 = 6$, so the distance should be 6 cm on the diagram.

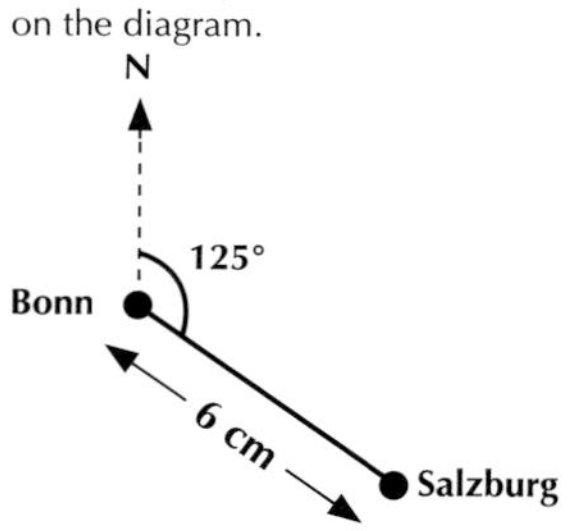

4 1 : 100 000 000 = 1 cm : 100 000 000 cm
= 1 cm : 1 000 000 m
= 1 cm : 1000 km

So to convert from distances in real life to distances on the map, divide by 1000.
$2000 \div 1000 = 2$, so the distance should be 2 cm on the diagram.

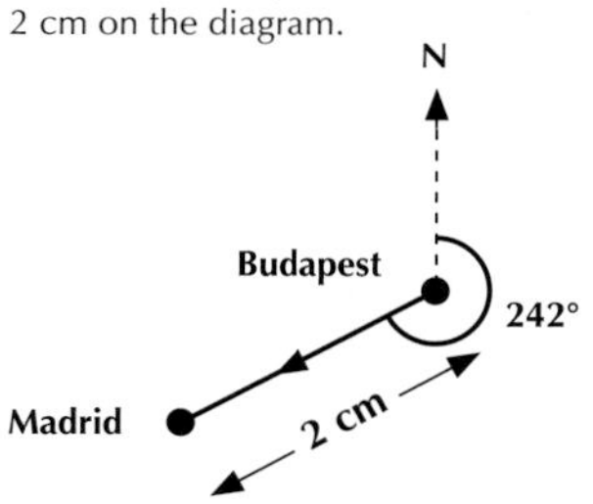

5 1 : 22 000 000 = 1 cm : 22 000 000 cm
= 1 cm : 220 000 m
= 1 cm : 220 km

So to convert from distances in real life to distances on the map, divide by 220.
$880 \div 220 = 4$, so the distance should be 4 cm on the diagram.

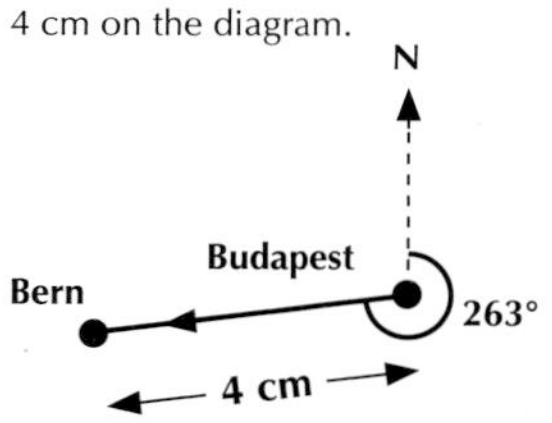

6 a) 1 : 10 000 000 = 1 cm : 10 000 000 cm
= 1 cm : 100 000 m
= 1 cm : 100 km

(i) Distance PQ on the diagram measures 3 cm. The scale is 1 cm : 100 km so to convert from distances on the plan to real life, multiply by 100.
$3 \times 100 =$ **300 km**

You may have a slightly different measurement — any answer between 290 km and 310 km is fine

Using the same method for (ii) and (iii):

(ii) 450 km

Any answer between 440 km and 460 km is fine.

(iii) 580 km

Any answer between 570 km and 590 km is fine.

b) (i) **060°**

You may have a slightly different measurement — any answer between 059° and 061° is fine.

(ii) **140°**

Any answer between 139° and 141° is fine.

(iii) **290°**

Any answer between 289° and 291° is fine.

Page 299 Review Exercise

1 The scale is 1 cm:2 m, so to convert from cm on the map to m in real life, multiply by 2.

a) 2.7 × 2 = **5.4 m** by 1.5 × 2 = **3 m**

b) 3.2 × 2 = **6.4 m** by 2.2 × 2 = **4.4 m**

c) 1.85 × 2 = **3.7 m** by 1.4 × 2 = **2.8 m**

d) 0.9 × 2 = **1.8 m** by 1.35 × 2 = **2.7 m**

2 a) The distance AB measures 3 cm, so the scale is 3 cm:150 m.
Divide by 3 to get 1 cm on the left-hand side: **1 cm:50 m**.

b) (i) The distance AC measures 1.5 cm. Multiply by 50 to find the real-life distance:
1.5 × 50 = **75 m**

(ii) The distance BC measures 2.5 cm. Multiply by 50 to get the real-life distance:
2.5 × 50 = **125 m**

3 The scale is 1:20, so to convert measurements in real life to measurements on the plan, divide by 20.
3000 mm ÷ 20 = 150 mm = 15 cm
900 mm ÷ 20 = 45 mm = 4.5 cm
600 mm ÷ 20 = 30 mm = 3 cm
1000 mm ÷ 20 = 50 mm = 5 cm
750 mm ÷ 20 = 37.5 mm = 3.75 cm

These lengths produce the plan below. Use a ruler to check your drawing has the correct measurements.

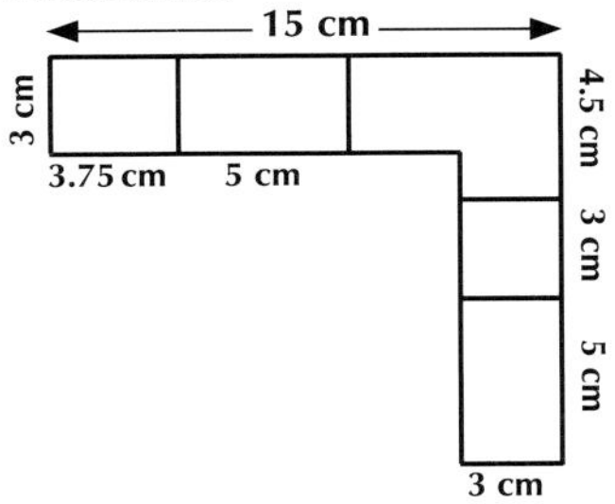

4 a) The scale is 1 cm:100 m so to convert from the real-life distances to the scale diagram distances, divide by 100.
230 ÷ 100 = 2.3 cm
390 ÷ 100 = 3.9 cm
This will give the diagram:

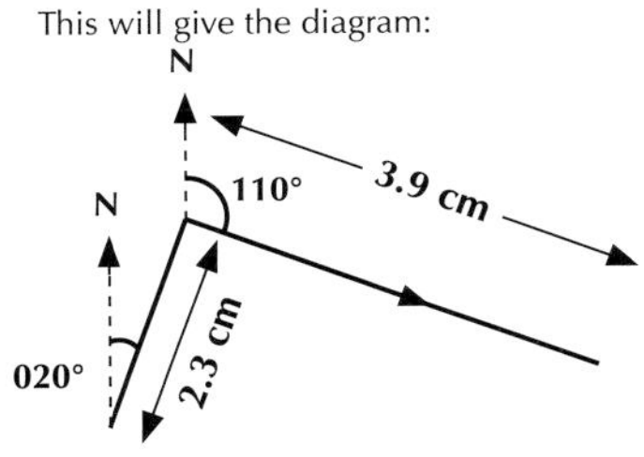

b) 110° – 20° = **90°**

5 a) Town B is south of town A so the bearing = **180°**

b) Town C is west of town B so the bearing = **270°**

c) Town C is southwest of town A so the bearing = **225°**
Draw a sketch if you're struggling to picture the towns.

6 a)

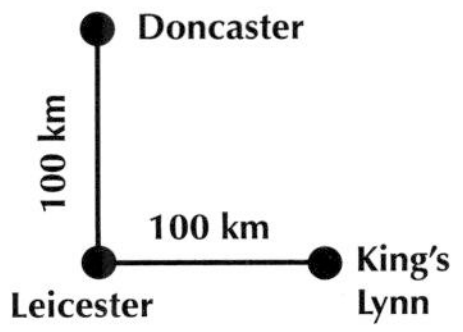

b) King's Lynn is due east of Leicester so the bearing is **090°**

c) (i) Leicester is due south of Doncaster so the bearing is **180°**

(ii) King's Lynn is due south east of Doncaster so, 90° + 45° = **135°**

d) (i) 90° + 180° = **270°**

(ii) 135° + 180° = **315°**

Page 300 Exam-Style Questions

1 1:25 000 = 1 cm:25 000 cm = 1 cm:250 m *[1 mark]*
So multiply by 250 to convert to the real-life distance: 3.8 × 250 = **950 m** *[1 mark]*

2 a) The distance from The Knott to Hartsop Village on the map is 3 cm, so the scale is 3 cm:12 km *[1 mark]*.
Dividing both sides by 3 gives **1 cm:4 km** *[1 mark]*.

b) The angle between North and the line between X and The Knott is 50°, measured anticlockwise.
So the bearing is 360° – 50° = **310°** *[1 mark]*.

3 a) The clockwise angle between a north line at A and the line AB is 45°, since the grid is made of squares.
So the bearing is **045°** *[1 mark]*.

b) The clockwise angle between a north line at C and the line CA is 360° – 45° (or 270° + 45°) = **315°** *[1 mark]*.

4 a) The scale is 1 cm:500 m so to convert from real-life distances to distances on the diagram, divide by 500.
2000 ÷ 500 = 4 cm, 1500 ÷ 500 = 3 cm *[1 mark]*
This will give the diagram:

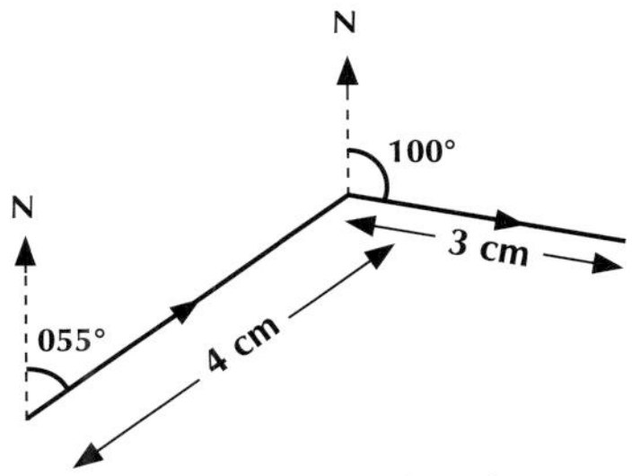

[2 marks — 1 mark for two lines drawn to the correct lengths, 1 mark for the lines drawn at the correct angles and with the bearings labelled.]

b) The distance measures 6.5 cm.
To convert cm on the map to m in real life, multiply by 500.
6.5 × 500 = **3250 m (or 3.25 km)** *[1 mark]*
You may have a slightly different measurement — any answer between 3200 m and 3300 m is okay.

c) **255°** *[1 mark]*
Any answer between 253° and 257° is fine.

Section 19 — Pythagoras and Trigonometry

19.1 Pythagoras' Theorem

Page 302 Exercise 1

1 a) $x^2 = 3^2 + 4^2 = 9 + 16 = 25$
$x = \sqrt{25} = \mathbf{5\ cm}$

b) $z^2 = 12^2 + 5^2 = 144 + 25 = 169$
$z = \sqrt{169} = \mathbf{13\ mm}$

c) $q^2 = 24^2 + 18^2 = 576 + 324 = 900$
$q = \sqrt{900} = \mathbf{30\ cm}$

Using the same method for Q2-Q6:

2 a) **22.83 m** (2 d.p.) b) **7.75 cm** (2 d.p.)

c) **1.39 m** (2 d.p.)

3 a) $\sqrt{41}$ **m** b) $\sqrt{58}$ **cm**

c) $\sqrt{29}$ **m**
Remember to leave the square roots in your answers as the question asks for the exact length.

4 **10.63 cm** (2 d.p.)

5 a) **8.60 cm** (2 d.p.) b) **11.70 cm** (2 d.p.)

c) **6.58 mm** (2 d.p.)

6 $\sqrt{1465}$ **cm**
Sketch the triangle if you're struggling to figure out which side goes where. You'll find that XZ is opposite the right angle, so it's the hypotenuse.

Page 303 Exercise 2

1 a) $13^2 = l^2 + 12^2$
$l^2 = 13^2 - 12^2$
$l^2 = 169 - 144 = 25$
$l = \sqrt{25} = \mathbf{5\ cm}$

b) $41^2 = t^2 + 9^2$
$t^2 = 41^2 - 9^2$
$t^2 = 1681 - 81 = 1600$
$t = \sqrt{1600} = \mathbf{40\ cm}$

c) $26^2 = p^2 + 10^2$
$p^2 = 26^2 - 10^2$
$p^2 = 676 - 100 = 576$
$p = \sqrt{576} = \mathbf{24\ cm}$

Using the same method for Q2-Q5:

2 a) **25.61 mm** (2 d.p.) b) **8.49 mm** (2 d.p.)

c) **14.77 km** (2 d.p.)

3 a) $\sqrt{19}$ **mm** b) $\sqrt{33}$ **cm**

c) $\sqrt{77}$ **m**

4 a) **7.94 cm** (2 d.p.) b) **11.53 m** (2 d.p.)

c) **10.53 mm** (2 d.p.) d) **14.53 cm** (2 d.p.)

5 **15.99 cm** (2 d.p.)

Page 304 Exercise 3

1 a) $9^2 + 12^2 = 81 + 144 = 225$
$15^2 = 225 = 9^2 + 12^2$ so the triangle **is right-angled.**

b) $8^2 + 7^2 = 64 + 49 = 113$
$11^2 = 121 \neq 8^2 + 7^2$ so the triangle is **not right-angled.**

2 $PS^2 = PR^2 + RS^2 = 120^2 + 180^2 = 46\ 800$
$PS = \sqrt{46\ 800} = \mathbf{216.33\ m}$

3 $\text{distance}^2 = 540^2 + 970^2 = 1\ 232\ 500$
$\text{distance} = \sqrt{1\ 232\ 500} = 1110.180... = \mathbf{1110\ m}$ (to the nearest m)
The distances run south and east form the two shorter sides of a right-angled triangle — so the final distance from the starting point is the hypotenuse of that triangle.

4 XYM is a right-angled triangle with hypotenuse XY.
Length $YM = YZ \div 2 = 9.4 \div 2 = 4.7$ cm
So $XY^2 = XM^2 + YM^2 = 3.7^2 + 4.7^2$
$= 13.69 + 22.09 = 35.78$
$XY = \sqrt{35.78} = 5.981... = \mathbf{5.98\ cm}$ (2 d.p.)

5 The ladder is the hypotenuse of a right-angled triangle, so $3.3^2 = 0.8^2 + (\text{height})^2$
$(\text{height})^2 = 3.3^2 - 0.8^2$
$(\text{height})^2 = 10.89 - 0.64 = 10.25$
height $= \sqrt{10.25} = 3.201... =$ **3.20 m** (2 d.p.)

6 Call the missing side of the piece of toast t.
$14^2 = 11^2 + t^2$
$t^2 = 14^2 - 11^2$
$t^2 = 196 - 121 = 75$
$t = \sqrt{75} = 8.660... =$ **8.66 cm** (2 d.p.)

7 a) $JL^2 = JK^2 + KL^2 = 4.9^2 + 6.8^2$
$= 24.01 + 46.24 = 70.25$
$JL = \sqrt{70.25} = 8.381... =$ **8.38 cm** (2 d.p.)

b) JL is the diameter of the circle,
so radius $= JL \div 2 = 8.381... \div 2 = 4.190...$
$=$ **4.19 cm** (2 d.p.)

8 a) Difference in x-coordinates $= 6 - 2 = 4$
Difference in y-coordinates $= 10 - 3 = 7$
$AB^2 = 4^2 + 7^2 = 16 + 49 = 65$
$AB = \sqrt{65}$ **units**

b) Difference in x-coordinates $= 4 - 1 = 3$
Difference in y-coordinates $= 7 - 2 = 5$
$AB^2 = 3^2 + 5^2 = 9 + 25 = 34$
$AB = \sqrt{34}$ **units**

9 $29.2^2 = 22.4^2 + (\text{radius})^2$
$(\text{radius})^2 = 29.2^2 - 22.4^2$
$(\text{radius})^2 = 852.64 - 501.76 = 350.88$
radius $= \sqrt{350.88} = 18.731...$
$=$ **18.73 cm** (2 d.p.)

10 The kite's string forms the hypotenuse of a right-angled triangle with base length equal to the distance between the end of the string and the tree.
$15^2 = 8.5^2 + (\text{height})^2$
$(\text{height})^2 = 15^2 - 8.5^2$
$(\text{height})^2 = 225 - 72.25 = 152.75$
height $= \sqrt{152.75} = 12.359...$
$=$ **12.36 m** (2 d.p.)

11 The lines NW and NE from the same point form a right angle. Use this to sketch a diagram:

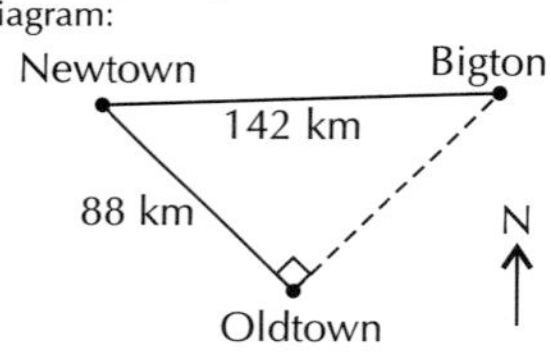

$NB^2 = ON^2 + OB^2$
$142^2 = 88^2 + OB^2$
$OB^2 = 142^2 - 88^2$
$OB^2 = 20\,164 - 7744 = 12\,420$
$OB = \sqrt{12\,420} = 111.445... =$ **111 km** (to the nearest km)
It really helps to draw a diagram for this one — it's a bit tricky to get your head around which town is where.

12 The longest stick of spaghetti that will fit in the jar will be the hypotenuse of a right-angled triangle with sides that are the height (28 cm) and full width (6 × 2 = 12 cm) of the jar.
$h^2 = 28^2 + 12^2 = 784 + 144 = 928$
$h = \sqrt{928} = 30.463... =$ **30.46 cm** (2 d.p.)
Careful here — you need to double the radius of the jar to get the side of the right-angled triangle you need.

13 The height of the kite is the sum of the missing sides of the two right-angled triangles. One triangle has hypotenuse 0.3 m and one side measuring 0.2 m, so the missing side of that triangle (call it A) is:
$0.3^2 = 0.2^2 + A^2$
$A^2 = 0.3^2 - 0.2^2$
$A^2 = 0.09 - 0.04 = 0.05$
$A = \sqrt{0.05} = 0.223...$
The other triangle has hypotenuse 0.4 m and one side measuring 0.2 m, so the missing side of that triangle (call it B) is:
$0.4^2 = 0.2^2 + B^2$
$B^2 = 0.4^2 - 0.2^2$
$B^2 = 0.16 - 0.04 = 0.12$
$B = \sqrt{0.12} = 0.346...$
So the height of the kite is:
$0.223... + 0.346... = 0.570... =$ **0.57 m** (2 d.p.)

19.2 Trigonometry — Sin, Cos and Tan

Page 307 Exercise 1

1 a)

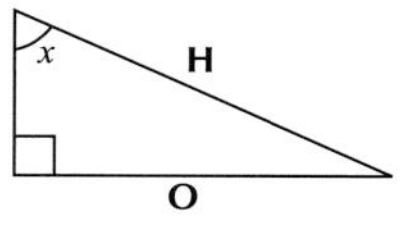

b)

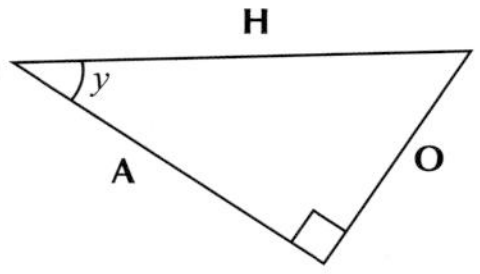

c)

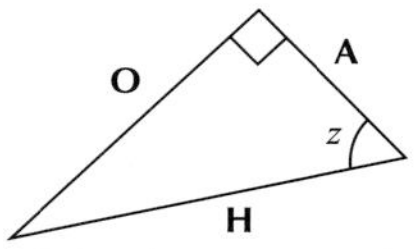

For all the trigonometry questions in this section, make sure you label the triangle first (draw your own sketch if you need to).

2 a) You know the hypotenuse and want to find the adjacent, so use the formula for cos x:
$\cos 43° = \frac{a}{6}$, so $a = 6 \cos 43° = 4.388...$
$=$ **4.39 cm** (3 s.f.)

b) You know the hypotenuse and want to find the opposite, so use the formula for sin x:
$\sin 58° = \frac{b}{11}$, so $b = 11 \sin 58° = 9.328...$
$=$ **9.33 cm** (3 s.f.)

c) You know the adjacent and want to find the opposite, so use the formula for tan x:
$\tan 12° = \frac{c}{5}$, so $c = 5 \tan 12° = 1.062...$
$=$ **1.06 cm** (3 s.f.)

d) You know the hypotenuse and want to find the opposite, so use the formula for sin x:
$\sin 63° = \frac{d}{9}$, so $d = 9 \sin 63° = 8.019...$
$=$ **8.02 cm** (3 s.f.)

3 a) You know the opposite and want to find the adjacent, so use the formula for tan x:
$\tan 37° = \frac{6}{p}$, so $p = \frac{6}{\tan 37°} = 7.962...$
$=$ **7.96 cm** (3 s.f.)

b) You know the opposite and want to find the hypotenuse, so use the formula for sin x:
$\sin 34° = \frac{20}{q}$, so $q = \frac{20}{\sin 34°} = 35.765...$
$=$ **35.8 cm** (3 s.f.)

c) You know the opposite and want to find the adjacent, so use the formula for tan x:
$\tan 48° = \frac{7}{r}$, so $r = \frac{7}{\tan 48°} = 6.302...$
$=$ **6.30 cm** (3 s.f.)

d) You know the adjacent and want to find the hypotenuse, so use the formula for cos x:
$\cos 11° = \frac{4.5}{s}$, so $s = \frac{4.5}{\cos 11°} = 4.584...$
$=$ **4.58 cm** (3 s.f.)

Page 308 Exercise 2

1 a) To find m, consider the right-angled triangle with angle $66° \div 2 = 33°$, and adjacent 10 cm. m is the hypotenuse, so use the formula for cos x:
$\cos 33° = \frac{10}{m}$, so $m = \frac{10}{\cos 33°} = 11.923...$
$=$ **11.9 cm** (3 s.f.)

b) To find n, consider the right-angled triangle with angle $84° \div 2 = 42°$, and opposite $14 \div 2 = 7$ cm. n is the hypotenuse, so use the formula for sin x:
$\sin 42° = \frac{7}{n}$, so $n = \frac{7}{\sin 42°} = 10.461...$
$=$ **10.5 cm** (3 s.f.)

2 a) p is double the length of the adjacent side of the right-angled triangle with angle 43° and hypotenuse 6 cm.
Use the formula for cos x:
$\cos 43° = \frac{\text{adj}}{6}$, so adj $= 6 \cos 43°$
$= 4.388...$
So $p = 4.388... \times 2 = 8.776...$
$=$ **8.78 cm** (3 s.f.)

b) q is double the length of the opposite side of the right-angled triangle with angle 51° and hypotenuse 12 cm.
Use the formula for sin x:
$\sin 51° = \frac{\text{opp}}{12}$, so opp $= 12 \sin 51°$
$= 9.325...$
So $q = 9.325... \times 2 = 18.651...$
$=$ **18.7 cm** (3 s.f.)

3 The third side is double the length of the opposite side of the right-angled triangle with angle $98° \div 2 = 49°$ and hypotenuse 22 cm.
Use the formula for sin x:
$\sin 49° = \frac{\text{opp}}{22}$, so opp $= 22 \sin 49°$
$= 16.603...$
So the length of the third side is
$16.603... \times 2 = 33.207... =$ **33.2 cm** (3 s.f.)
Sketch a diagram for this question if you're struggling to figure out which side goes where.

Page 310 Exercise 3

1 a) You know the adj and hyp, so use the formula for cos x:
$\cos a = \frac{3}{8}$, so $a = \cos^{-1}\left(\frac{3}{8}\right) = 67.975...$
$=$ **68.0°** (1 d.p.)

b) You know the opp and hyp, so use the formula for sin x:
$\sin b = \frac{10}{14}$, so $b = \sin^{-1}\left(\frac{10}{14}\right) = 45.584...$
$=$ **45.6°** (1 d.p.)

c) You know the opp and adj, so use the formula for tan x:
$\tan c = \frac{12}{11}$, so $c = \tan^{-1}\left(\frac{12}{11}\right) = 47.489...$
$=$ **47.5°** (1 d.p.)

2 a) You know the adj and hyp, so use the formula for cos x:

$\cos d = \frac{2}{12}$, so $d = \cos^{-1}\left(\frac{2}{12}\right) = 80.405...$
$= \mathbf{80.4°}$ (1 d.p.)

b) You know the opp and adj so use the formula for tan x:

$\tan e = \frac{5}{7.5}$, so $e = \tan^{-1}\left(\frac{5}{7.5}\right) = 33.690...$
$= \mathbf{33.7°}$ (1 d.p.)

c) You know the opp and hyp so use the formula for sin x:

$\sin f = \frac{13}{15}$, so $f = \sin^{-1}\left(\frac{13}{15}\right) = 60.073...$
$= \mathbf{60.1°}$ (1 d.p.)

3 a) You know the opp and hyp, so use the formula for sin x:

$\sin m = \frac{8}{24}$, so $m = \sin^{-1}\left(\frac{8}{24}\right) = 19.471...$

So the angle of elevation of the slide is **19.5°** (1 d.p.).

b) Using alternate angles, q is the same as the angle opposite the ladder. You know the opp and adj so use the formula for tan x:

$\tan q = \frac{4}{5.5}$, so $q = \tan^{-1}\left(\frac{4}{5.5}\right) = 36.027...$

So the angle of depression from the top of the slide is **36.0°** (1 d.p.).

Page 311 Exercise 4

Use the exact values from the green box on p.311 to find the values of sin, cos and tan of 30°, 45°, 60° and 90° in this exercise.

1 a) The sides given are the opposite and hypotenuse,

so $\sin a = \frac{\sqrt{3}}{2}$, which means $\boldsymbol{a = 60°}$.

b) The sides given are the adjacent and hypotenuse,

so $\cos b = \frac{\sqrt{2}}{2}$, which means $\boldsymbol{b = 45°}$.

c) The sides given are the opposite and adjacent,

so $\tan c = \frac{\sqrt{3}}{1} = \sqrt{3}$, which means $\boldsymbol{c = 60°}$.

2 a) You know the opposite and want to find the adjacent, so use the formula for tan x:

$\tan 30° = \frac{1}{e}$, which means $\frac{1}{\sqrt{3}} = \frac{1}{e}$ so

$\boldsymbol{e = \sqrt{3}}$ **m**

b) You know the hypotenuse and want to find the adjacent, so use the formula for cos x:

$\cos 60° = \frac{f}{2}$, which means $\frac{1}{2} = \frac{f}{2}$ so

$\boldsymbol{f = 1}$ **cm**

c) You know the opposite and want to find the adjacent, so use the formula for tan x:

$\tan 45° = \frac{8}{g}$, which means $1 = \frac{8}{g}$ so

$\boldsymbol{g = 8}$ **mm**

3 $\tan 45° + \sin 60° =$

$1 + \frac{\sqrt{3}}{2} = \frac{2}{2} + \frac{\sqrt{3}}{2} = \boldsymbol{\frac{2 + \sqrt{3}}{2}}$

4 DEF is isosceles, and as angle DEF is 90°, the other two angles must both be 45°. DF is opposite the right angle, so is the hypotenuse. DE is opposite one of the 45° angles, so use the formula for sin x:

$\sin 45° = \frac{DE}{7\sqrt{2}}$, so $\frac{1}{\sqrt{2}} = \frac{DE}{7\sqrt{2}}$, which means

$\boldsymbol{DE = 7}$ **cm**

You could have used the cos formula here — since both the other angles are 45°, DE is also adjacent to a 45° angle. cos 45° and sin 45° are the same, so you'd have ended up with the same answer. You might want to draw a sketch for this question.

Page 312 Review Exercise

1 Difference in x-coordinates = 7 – 1 = 6
Difference in y-coordinates = 8 – 3 = 5
$PR^2 = 6^2 + 5^2 = 36 + 25 = 61$
$PR = \sqrt{61} = 7.810... =$ **7.8 units**

2 Draw a sketch:

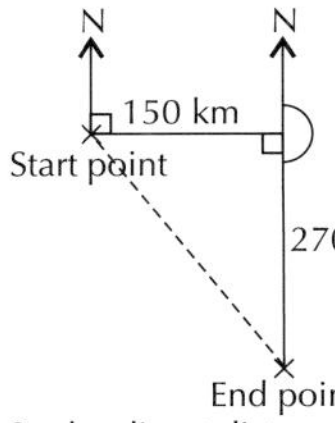

So the direct distance is the hypotenuse of the triangle with shorter sides measuring 150 km and 270 km.
$(\text{distance})^2 = 150^2 + 270^2$
$= 22\,500 + 72\,900 = 95\,400$
distance $= \sqrt{95\,400} = 308.868...$
$=$ **309 km** (to the nearest km)
See p295 for more on bearings.

3 a) $a^2 = 3^2 + 5^2 = 9 + 25 = 34$
$a = \sqrt{34} = 5.830... =$ **5.83 cm** (3 s.f.)

b) b is the opposite and you know the adjacent, so use the formula for tan x:

$\tan 32° = \frac{b}{28}$, so $b = 28 \tan 32° = 17.496...$
$=$ **17.5 cm** (3 s.f.)

c) c is the opposite and you know the hypotenuse, so use the formula for sin x:
$\sin 62° = \frac{c}{17}$, so $c = 17 \sin 62° = 15.010...$
$=$ **15.0 cm** (3 s.f.)

d) d is the hypotenuse and you know the adjacent, so use the formula for cos x:

$\cos 29° = \frac{2.1}{d}$, so $d = \frac{2.1}{\cos 29°} = 2.401...$
$=$ **2.40 m** (3 s.f.)

4 The ladder forms the hypotenuse of a right-angled triangle.
The 8 m side is the adjacent, and the height is the opposite, so use the formula for tan x:
$\tan 68° = \frac{h}{8}$, so $h = 8 \tan 68° = 19.800...$
$=$ **19.8 m** (3 s.f.)

5 The string forms the hypotenuse, and you also know the opposite, so use the formula for sin x:

$\sin x = \frac{4.1}{5.8}$, so $x = \sin^{-1}\left(\frac{4.1}{5.8}\right) = 44.982...$
$=$ **45.0°** (1 d.p.)

Draw a sketch here if you need to.

6 Call the side common to both triangles p. Use Pythagoras' theorem to find p:
$7^2 = 1^2 + p^2$, so $p^2 = 7^2 - 1^2 = 48$, so $p = \sqrt{48}$.
You now have the hypotenuse and opposite of the left-hand triangle, so use the formula for sin x:

$\sin x = \frac{\sqrt{48}}{10}$, so $x = \sin^{-1}\left(\frac{\sqrt{48}}{10}\right) = 43.853...$
$=$ **43.9°** (1 d.p.)

7 Draw a sketch:

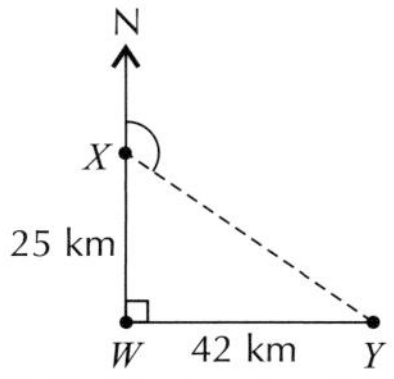

To find the bearing, first find angle WXY. You know the opposite and the adjacent, so use the formula for tan x:

$\tan WXY = \frac{42}{25}$, so

$WXY = \tan^{-1}\left(\frac{42}{25}\right) = 59.237...$

So the bearing of Y from X is:
$180° - 59.237...° = 120.762...°$
$=$ **120.8°** (1 d.p.)

8 a) To find a, first find the base of the right-angled triangle formed by splitting the isosceles triangle in two. This creates a right-angled triangle with angle 60° and hypotenuse 12 cm. You want to find the opposite, so use the formula for sin x:

$\sin 60° = \frac{\text{opp}}{12}$, so $\frac{\sqrt{3}}{2} = \frac{\text{opp}}{12}$,

which means opp $= \frac{12\sqrt{3}}{2} = 6\sqrt{3}$.

$a = 2 \times \text{opp} = 2 \times 6\sqrt{3} = \boldsymbol{12\sqrt{3}}$ **cm**

b) Split length b into two different bits, b_1 and b_2, where b_1 is the base of the left-hand triangle and b_2 is the base of the right-hand triangle.
The left-hand triangle has angles of 45° and 90°, so the missing angle must also be 45°, which means the triangle is isosceles. b_1 is the same as the 3 cm side, so $b_1 = 3$ cm.
b_2 is the adjacent in the right-hand triangle, and the opposite is 3 cm, so use the formula for tan x:

$\tan 30° = \frac{3}{b_2}$, which means $\frac{1}{\sqrt{3}} = \frac{3}{b_2}$,

so $b_2 = 3\sqrt{3}$ cm

$b = b_1 + b_2 = \boldsymbol{3 + 3\sqrt{3}}$ **cm**

You could have also used the value of tan 45° to find the value of b_1.

Page 313 Exam-Style Questions

1 $860^2 = 426^2 + (\text{distance})^2$
$(\text{distance})^2 = 860^2 - 426^2$
$(\text{distance})^2 = 739\,600 - 181\,476 = 558\,124$
distance $= \sqrt{558\,124} = 747.076... =$ **747 km**
(to the nearest km)
[2 marks available — 1 mark for using Pythagoras' theorem, 1 mark for the correct answer]

2 You know the hypotenuse and want to find the opposite, so use the formula for sin x:
$\sin 38° = \frac{\text{opp}}{5.4}$, so opp $= 5.4 \sin 38° = 3.324...$
$=$ **3.32 cm** (2 d.p.)

[2 marks available — 1 mark for using the correct trigonometric formula, 1 mark for the correct answer]

3 To use the formula for sin ACB, you need to find the hypotenuse AC. Find the length of side AC using Pythagoras' theorem:
$AC^2 = AB^2 + BC^2 = 5^2 + 12^2 = 25 + 144 = 169$
$AC = \sqrt{169} = 13$ cm
AB is the opposite and AC is the hypotenuse, so $\sin ACB = \mathbf{\frac{5}{13}}$.
[3 marks available — 1 mark for using Pythagoras' theorem to find AC, 1 mark for the correct value of AC, 1 mark for the correct value of sin ACB]

4 Find length QC using Pythagoras' theorem:
$QR^2 = RC^2 + QC^2$
$8^2 = 3^2 + QC^2$
$QC^2 = 8^2 - 3^2$
$QC^2 = 64 - 9 = 55$, so $QC = \sqrt{55}$ cm
$BC = BQ + QC = 3 + \sqrt{55} = 10.416...$
= **10.4 cm** (1 d.p.)
[3 marks available — 1 mark for using Pythagoras' theorem, 1 mark for the correct length of QC (or equivalent side), 1 mark for the correct answer]

5 Half of diagonal AC is the adjacent side of a right-angled triangle with opposite 12 ÷ 2 = 6 cm and angle 60° ÷ 2 = 30°, so use the formula for tan x:
$\tan 30° = \frac{6}{\text{adj}}$, which means $\frac{1}{\sqrt{3}} = \frac{6}{\text{adj}}$,
so adj = $6\sqrt{3}$ cm
AC is double this length, so
$AC = 6\sqrt{3} \times 2 = \mathbf{12\sqrt{3}}$ **cm**
[3 marks available — 1 mark for creating a right-angled triangle, 1 mark for using trigonometry to find a side, 1 mark for the correct answer]

Section 20 — Vectors

20.1 Column Vectors

Page 315 Exercise 1

1

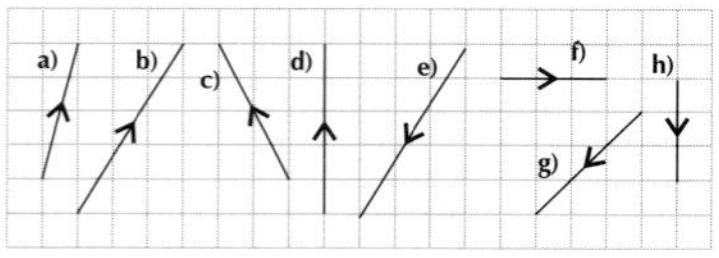

2 $\mathbf{a} = \begin{pmatrix} \mathbf{4} \\ \mathbf{0} \end{pmatrix}$ $\mathbf{b} = \begin{pmatrix} \mathbf{-1} \\ \mathbf{-1} \end{pmatrix}$ $\mathbf{c} = \begin{pmatrix} \mathbf{2} \\ \mathbf{6} \end{pmatrix}$

$\mathbf{d} = \begin{pmatrix} \mathbf{-3} \\ \mathbf{0} \end{pmatrix}$ $\mathbf{e} = \begin{pmatrix} \mathbf{-4} \\ \mathbf{-1} \end{pmatrix}$ $\mathbf{f} = \begin{pmatrix} \mathbf{2} \\ \mathbf{3.5} \end{pmatrix}$

$\mathbf{g} = \begin{pmatrix} \mathbf{-4} \\ \mathbf{-3} \end{pmatrix}$ $\mathbf{h} = \begin{pmatrix} \mathbf{-2} \\ \mathbf{2} \end{pmatrix}$

Page 316 Exercise 2

1 a) $3\mathbf{q} = 3\begin{pmatrix} -1 \\ 3 \end{pmatrix} = \begin{pmatrix} 3 \times (-1) \\ 3 \times 3 \end{pmatrix} = \begin{pmatrix} \mathbf{-3} \\ \mathbf{9} \end{pmatrix}$

b) $5\mathbf{q} = 5\begin{pmatrix} -1 \\ 3 \end{pmatrix} = \begin{pmatrix} 5 \times (-1) \\ 5 \times 3 \end{pmatrix} = \begin{pmatrix} \mathbf{-5} \\ \mathbf{15} \end{pmatrix}$

c) $\frac{3}{2}\mathbf{q} = \frac{3}{2}\begin{pmatrix} -1 \\ 3 \end{pmatrix} = \begin{pmatrix} \frac{3}{2} \times (-1) \\ \frac{3}{2} \times 3 \end{pmatrix} = \begin{pmatrix} \mathbf{-1.5} \\ \mathbf{4.5} \end{pmatrix}$

d) $-2\mathbf{q} = -2\begin{pmatrix} -1 \\ 3 \end{pmatrix} = \begin{pmatrix} -2 \times (-1) \\ -2 \times 3 \end{pmatrix} = \begin{pmatrix} \mathbf{2} \\ \mathbf{-6} \end{pmatrix}$

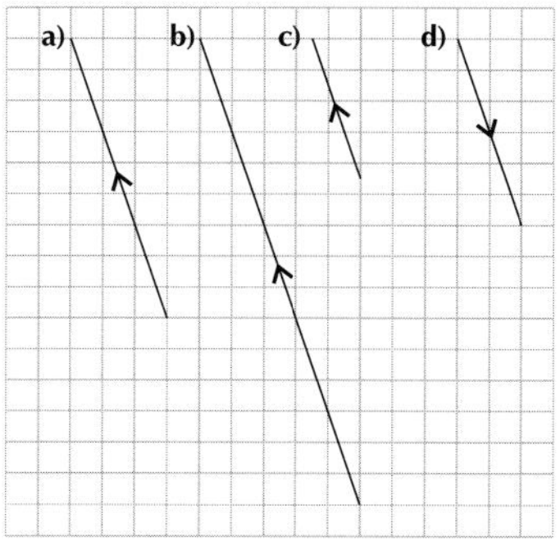

2 a) $2\mathbf{a} = 2\begin{pmatrix} 4 \\ -2 \end{pmatrix} = \begin{pmatrix} 2 \times 4 \\ 2 \times (-2) \end{pmatrix} = \begin{pmatrix} 8 \\ -4 \end{pmatrix} = \mathbf{d}$

b) $-3\mathbf{b} = -3\begin{pmatrix} -1 \\ 4 \end{pmatrix} = \begin{pmatrix} -3 \times (-1) \\ -3 \times 4 \end{pmatrix} = \begin{pmatrix} 3 \\ -12 \end{pmatrix} = \mathbf{g}$

c) Look at the x- and y-components of **e** — the x-component represents a distance of 1 horizontal unit and the y-component represents a distance of 4 vertical units. The same is true in the vector **b**. So **b** is the same length as **e**.

d) Parallel vectors are multiples of one another.
$\mathbf{c} = \begin{pmatrix} 3 \\ 12 \end{pmatrix} = \begin{pmatrix} 3 \times 1 \\ 3 \times 4 \end{pmatrix} = 3\begin{pmatrix} 1 \\ 4 \end{pmatrix} = 3\mathbf{e}$.
So **e** is parallel to **c**.

Page 317 Exercise 3

1 a) $\begin{pmatrix} 5 \\ 2 \end{pmatrix} + \begin{pmatrix} 3 \\ 4 \end{pmatrix} = \begin{pmatrix} 5+3 \\ 2+4 \end{pmatrix} = \begin{pmatrix} \mathbf{8} \\ \mathbf{6} \end{pmatrix}$

b) $\begin{pmatrix} 4 \\ -1 \end{pmatrix} + \begin{pmatrix} 1 \\ 6 \end{pmatrix} = \begin{pmatrix} 4+1 \\ -1+6 \end{pmatrix} = \begin{pmatrix} \mathbf{5} \\ \mathbf{5} \end{pmatrix}$

c) $\begin{pmatrix} 2 \\ -1 \end{pmatrix} - \begin{pmatrix} -2 \\ 2 \end{pmatrix} = \begin{pmatrix} 2-(-2) \\ -1-2 \end{pmatrix} = \begin{pmatrix} \mathbf{4} \\ \mathbf{-3} \end{pmatrix}$

d) $\begin{pmatrix} -3 \\ 0 \end{pmatrix} - \begin{pmatrix} 6 \\ 2 \end{pmatrix} = \begin{pmatrix} -3-6 \\ 0-2 \end{pmatrix} = \begin{pmatrix} \mathbf{-9} \\ \mathbf{-2} \end{pmatrix}$

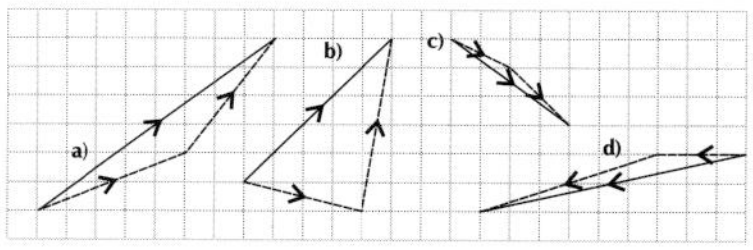

2 a) $\mathbf{b} + \mathbf{c} = \begin{pmatrix} 0 \\ -2 \end{pmatrix} + \begin{pmatrix} -1 \\ 4 \end{pmatrix} = \begin{pmatrix} 0+(-1) \\ -2+4 \end{pmatrix} = \begin{pmatrix} \mathbf{-1} \\ \mathbf{2} \end{pmatrix}$

b) $\mathbf{c} - \mathbf{a} = \begin{pmatrix} -1 \\ 4 \end{pmatrix} - \begin{pmatrix} 2 \\ 3 \end{pmatrix} = \begin{pmatrix} -1-2 \\ 4-3 \end{pmatrix} = \begin{pmatrix} \mathbf{-3} \\ \mathbf{1} \end{pmatrix}$

c) $2\mathbf{c} + \mathbf{a} = 2\begin{pmatrix} -1 \\ 4 \end{pmatrix} + \begin{pmatrix} 2 \\ 3 \end{pmatrix} = \begin{pmatrix} 2 \times (-1) + 2 \\ 2 \times 4 + 3 \end{pmatrix}$
$= \begin{pmatrix} \mathbf{0} \\ \mathbf{11} \end{pmatrix}$

Using the same method for d)-h):

d) $\begin{pmatrix} \mathbf{6} \\ \mathbf{7} \end{pmatrix}$ e) $\begin{pmatrix} \mathbf{4} \\ \mathbf{-5} \end{pmatrix}$ f) $\begin{pmatrix} \mathbf{3} \\ \mathbf{-3} \end{pmatrix}$ g) $\begin{pmatrix} \mathbf{-4} \\ \mathbf{6} \end{pmatrix}$

h) $\begin{pmatrix} \mathbf{5} \\ \mathbf{26} \end{pmatrix}$

Don't be put off by there being three vectors in parts f), g) and h). You still answer them in the same way by adding all the x-components and all the y-components separately.

3 a) $\mathbf{u} + 2\mathbf{v} = \begin{pmatrix} 6 \\ -2 \end{pmatrix} + 2\begin{pmatrix} -2 \\ 3 \end{pmatrix} = \begin{pmatrix} 6 + 2 \times (-2) \\ -2 + 2 \times 3 \end{pmatrix}$
$= \begin{pmatrix} \mathbf{2} \\ \mathbf{4} \end{pmatrix}$

b)

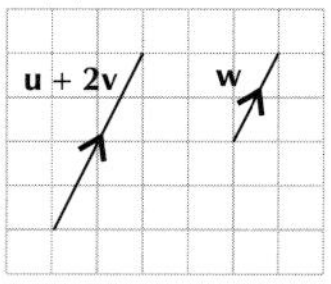

c) The two vectors are **parallel** because $\mathbf{u} + 2\mathbf{v} = 2\mathbf{w}$.

Page 318 Exercise 4

1 a)

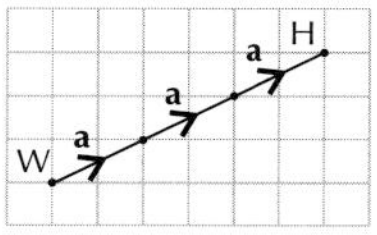

so $\overrightarrow{WH} = \mathbf{3a}$

b)

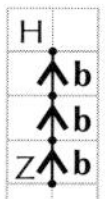

so $\overrightarrow{ZH} = \mathbf{3b}$

c)

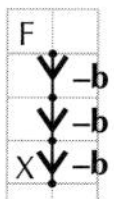

so $\overrightarrow{FX} = \mathbf{-3b}$

d)

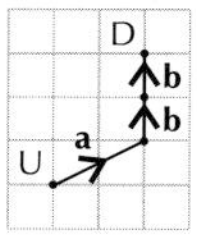

so $\overrightarrow{UD} = \mathbf{a + 2b}$

Using the same method for e)-h):

e) **–a – 2b** f) **2a – 5b**

g) **5a – 8b** h) **a – 4b**

*Be careful with the direction of the vectors. For instance, the vector **b** is pointing up but in part c) you're asked for the vector from F to X — this is going downwards. So you have to use the reverse of **b**, which is **–b**.*

20.2 Vector Geometry

Page 320 Exercise 1

1 a) $\overrightarrow{BA} = -\overrightarrow{AB} = \mathbf{-4p}$

b) $\overrightarrow{CD} = -\overrightarrow{DC} = \mathbf{-p}$

c) $\overrightarrow{AC} = \overrightarrow{AD} + \overrightarrow{DC} = \mathbf{q + p}$

d) $\overrightarrow{CA} = -\overrightarrow{AC} = -(\mathbf{q} + \mathbf{p}) = \mathbf{-q - p}$

e) $\overrightarrow{CB} = \overrightarrow{CD} + \overrightarrow{DA} + \overrightarrow{AB} = \overrightarrow{CD} - \overrightarrow{AD} + \overrightarrow{AB}$
$= -\mathbf{p} - \mathbf{q} + 4\mathbf{p} = \mathbf{3p - q}$

f) $\overrightarrow{BD} = \overrightarrow{BA} + \overrightarrow{AD} = \mathbf{-4p + q}$

2 Add point C onto the diagram.

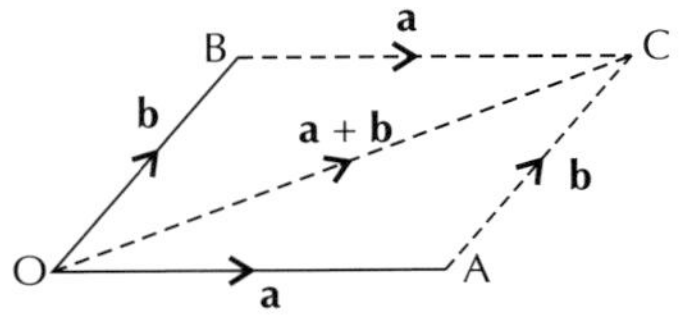

a) (i) $\overrightarrow{CO} = -\overrightarrow{OC} = -(\mathbf{a} + \mathbf{b}) = \mathbf{-a - b}$

(ii) $\overrightarrow{AB} = \overrightarrow{AO} + \overrightarrow{OB} = -\overrightarrow{OA} + \overrightarrow{OB} = \mathbf{-a + b}$

b) The sides OA and BC are both represented by vector **a**, so they're parallel, and the sides OB and AC are both represented by the vector **b**, so they're parallel. So the shape is a **parallelogram**.

3 a) $\overrightarrow{OB}$ is in the same direction as $\overrightarrow{OA}$ and it is 4 times as long, so $\overrightarrow{OB} = 4\overrightarrow{OA} = \mathbf{4a}$.

b) $\overrightarrow{OD}$ is in the same direction as $\overrightarrow{OC}$ and it is 3 times as long, so $\overrightarrow{OD} = 3\overrightarrow{OC} = \mathbf{3c}$.

c) $\overrightarrow{AB} = -\overrightarrow{OA} + \overrightarrow{OB} = -\mathbf{a} + 4\mathbf{a} = \mathbf{3a}$

d) $\overrightarrow{CD} = -\overrightarrow{OC} + \overrightarrow{OD} = -\mathbf{c} + 3\mathbf{c} = \mathbf{2c}$

e) $\overrightarrow{DB} = \overrightarrow{DO} + \overrightarrow{OB} = -\overrightarrow{OD} + \overrightarrow{OB} = \mathbf{-3c + 4a}$

f) $\overrightarrow{BD} = -\overrightarrow{DB} = -(-3\mathbf{c} + 4\mathbf{a}) = \mathbf{3c - 4a}$

4 a) E.g. $\overrightarrow{ED} = \mathbf{p}$ since ED and AB are parallel and the same length.
$\overrightarrow{ED} = \overrightarrow{EM} + \overrightarrow{MD} = \mathbf{-r + q}$ since EM and DC are parallel and the same length, and MD and BC are parallel and the same length.

Using the same method for b)-f):

b) E.g. **q** and **p + r**
c) E.g. **r** and **q – p**
d) E.g. **–q** and **–r – p**
e) E.g. **–p** and **r – q**
f) E.g. **–r** and **p – q**

Page 321 Review Exercise

1 a) $\mathbf{a} = \begin{pmatrix} 3 \\ 4 \end{pmatrix}$ $\mathbf{b} = \begin{pmatrix} 4 \\ -2 \end{pmatrix}$ $\mathbf{c} = \begin{pmatrix} -2 \\ -4 \end{pmatrix}$

$\mathbf{d} = \begin{pmatrix} -4 \\ -1 \end{pmatrix}$ $\mathbf{e} = \begin{pmatrix} 4 \\ -1 \end{pmatrix}$ $\mathbf{f} = \begin{pmatrix} 1 \\ 4 \end{pmatrix}$

$\mathbf{g} = \begin{pmatrix} 0 \\ -5 \end{pmatrix}$ $\mathbf{h} = \begin{pmatrix} 1 \\ -4 \end{pmatrix}$

b) **b**
c) **d**, **e**, **f**, **h**

2 a) $3\mathbf{p} = 3\begin{pmatrix} 4 \\ -3 \end{pmatrix} = \begin{pmatrix} 3 \times 4 \\ 3 \times (-3) \end{pmatrix} = \begin{pmatrix} \mathbf{12} \\ \mathbf{-9} \end{pmatrix}$

b) $2\mathbf{q} + \mathbf{r} = 2\begin{pmatrix} 0 \\ 2 \end{pmatrix} + \begin{pmatrix} -1 \\ 5 \end{pmatrix} = \begin{pmatrix} 2 \times 0 + (-1) \\ 2 \times 2 + 5 \end{pmatrix}$
$= \begin{pmatrix} \mathbf{-1} \\ \mathbf{9} \end{pmatrix}$

c) $\mathbf{r} - 2\mathbf{p} = \begin{pmatrix} -1 \\ 5 \end{pmatrix} - 2\begin{pmatrix} 4 \\ -3 \end{pmatrix} = \begin{pmatrix} -1 - 2 \times 4 \\ 5 - 2 \times (-3) \end{pmatrix}$
$= \begin{pmatrix} \mathbf{-9} \\ \mathbf{11} \end{pmatrix}$

d) $\mathbf{p} + 5\mathbf{r} - 3\mathbf{q} = \begin{pmatrix} 4 \\ -3 \end{pmatrix} + 5\begin{pmatrix} -1 \\ 5 \end{pmatrix} - 3\begin{pmatrix} 0 \\ 2 \end{pmatrix}$
$= \begin{pmatrix} 4 + 5 \times (-1) - 3 \times 0 \\ -3 + 5 \times 5 - 3 \times 2 \end{pmatrix} = \begin{pmatrix} \mathbf{-1} \\ \mathbf{16} \end{pmatrix}$

3 a)

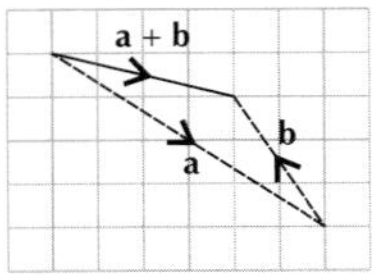

$\mathbf{a} + \mathbf{b} = \begin{pmatrix} \mathbf{4} \\ \mathbf{-1} \end{pmatrix}$

b)

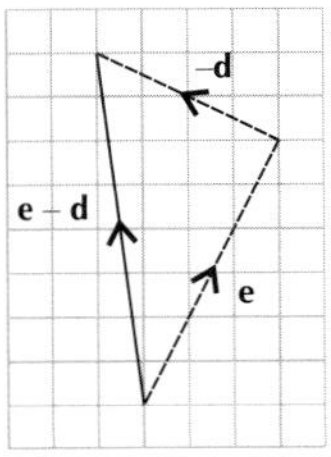

$\mathbf{e} - \mathbf{d} = \begin{pmatrix} \mathbf{-1} \\ \mathbf{8} \end{pmatrix}$

c) **a and f**
d) **b and e**

Look at the x- and y-components separately. E.g. in part c), you need the x-components to add to 10 and the y-components to add to –10.

e) $\mathbf{e} = \begin{pmatrix} 3 \\ 6 \end{pmatrix} = \begin{pmatrix} 3 \times 1 \\ 3 \times 2 \end{pmatrix} = 3\begin{pmatrix} 1 \\ 2 \end{pmatrix} = 3\mathbf{c}$, so **e** is parallel to **c**.

f) $\mathbf{f} = \begin{pmatrix} 4 \\ -6 \end{pmatrix} = \begin{pmatrix} -2 \times (-2) \\ -2 \times 3 \end{pmatrix} = -2\begin{pmatrix} -2 \\ 3 \end{pmatrix} = -2\mathbf{b}$, so **f** is parallel to **b**.

4 a) $\overrightarrow{SR} = 5 \times \overrightarrow{PQ} = \mathbf{5m}$

b) $\overrightarrow{QR} = \overrightarrow{QP} + \overrightarrow{PS} + \overrightarrow{SR} = -\overrightarrow{PQ} - \overrightarrow{SP} + \overrightarrow{SR}$
$= -\mathbf{m} - \mathbf{n} + 5\mathbf{m}$
$= \mathbf{4m - n}$

Page 322 Exam-Style Questions

1 $\overrightarrow{AC} = \overrightarrow{AB} + \overrightarrow{BC} = \overrightarrow{AB} - \overrightarrow{CB}$ *[1 mark]*
$= \mathbf{u - v}$ *[1 mark]*

2 $\mathbf{m} + 4\mathbf{n} = \begin{pmatrix} 5 \\ 0 \end{pmatrix} + 4\begin{pmatrix} -3 \\ -1 \end{pmatrix}$ *[1 mark]*

$= \begin{pmatrix} 5 + 4 \times (-3) \\ 0 + 4 \times (-1) \end{pmatrix} = \begin{pmatrix} \mathbf{-7} \\ \mathbf{-4} \end{pmatrix}$ *[1 mark]*

3

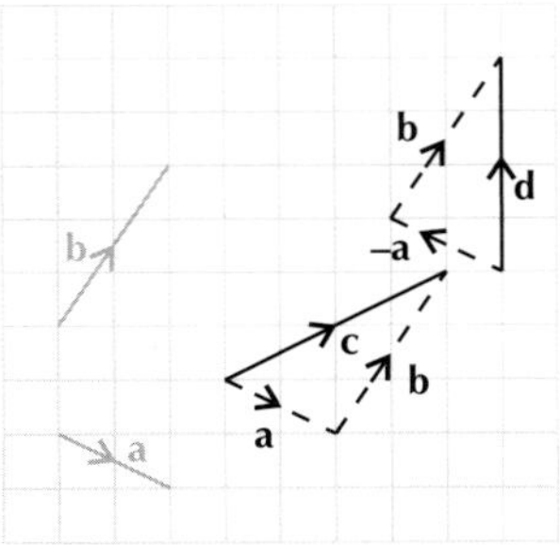

c = a + b *[1 mark]* **d = –a + b** *[1 mark]*

4 $2\mathbf{a} - 3\mathbf{b} = 2\begin{pmatrix} 5p \\ 4q \end{pmatrix} - 3\begin{pmatrix} 4 \\ 0 \end{pmatrix}$

$= \begin{pmatrix} (2 \times 5p) - (3 \times 4) \\ (2 \times 4q) - (3 \times 0) \end{pmatrix}$ *[1 mark]*

$= \begin{pmatrix} 10p - 12 \\ 8q \end{pmatrix}$ *[1 mark]*

$2\mathbf{a} - 3\mathbf{b} = \begin{pmatrix} 10p - 12 \\ 8q \end{pmatrix} = \begin{pmatrix} 18 \\ -32 \end{pmatrix}$, so:

$10p - 12 = 18 \Rightarrow 10p = 30$
$\Rightarrow \boldsymbol{p = 3}$ *[1 mark]*

$8q = -32 \Rightarrow q = -32 \div 8 \Rightarrow \boldsymbol{q = -4}$ *[1 mark]*

Section 21 — Constructions

21.1 Circles

Page 323 Exercise 1

1 a) $d = 2r = 2 \times 4$ cm $=$ **8 cm**

b) $d = 2r = 2 \times 6$ cm $=$ **12 cm**

c) $d = 2r = 2 \times 30$ mm $=$ **60 mm**

d) $d = 2r = 2 \times 4.2$ mm $=$ **8.4 mm**

2 a) $d = 2r \Rightarrow r = \frac{d}{2} = \frac{2\text{ cm}}{2} = \mathbf{1\ cm}$

b) $d = 2r \Rightarrow r = \frac{d}{2} = \frac{12\text{ cm}}{2} = \mathbf{6\ cm}$

c) $d = 2r \Rightarrow r = \frac{d}{2} = \frac{13\text{ m}}{2} = \mathbf{6.5\ m}$

d) $d = 2r \Rightarrow r = \frac{d}{2} = \frac{0.02\text{ cm}}{2} = \mathbf{0.01\ cm}$

3 a)

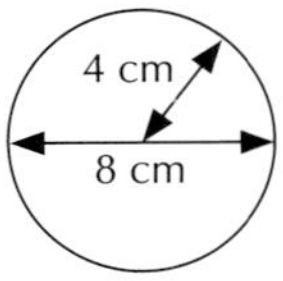

b)

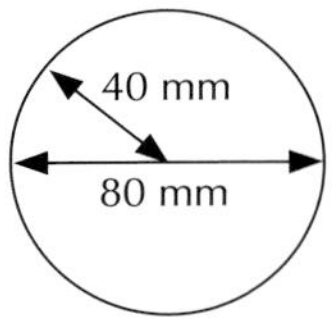

Remember that 80 mm is 8 cm.

c)

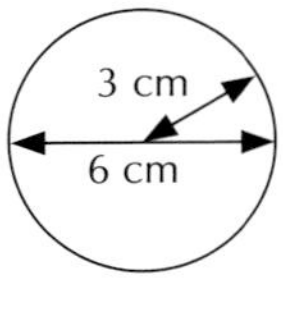

Page 324 Exercise 2

1 a) **chord** b) **segment**
c) **tangent** d) **sector**
e) **circumference** f) **arc**

2 E.g.

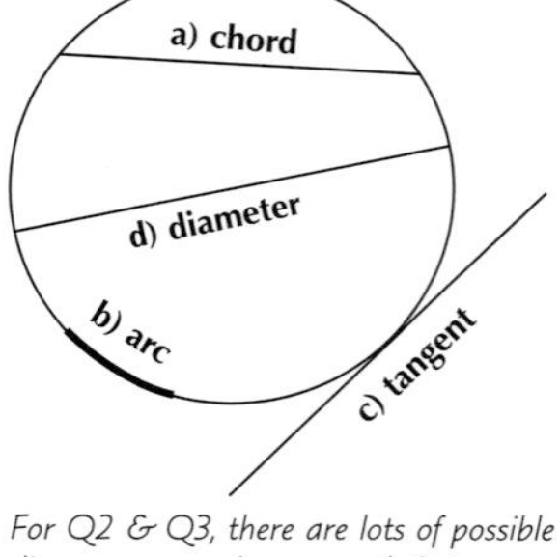

For Q2 & Q3, there are lots of possible correct diagrams — make sure each feature you've drawn has the properties described on p.323 & 324.

3 E.g.

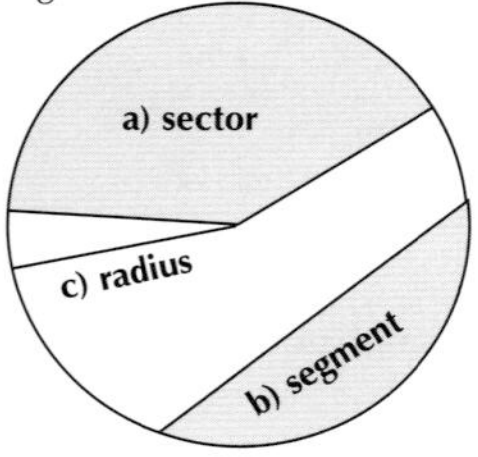

21.2 Lines, Angles and Triangles

Page 326 Exercise 1

1 a) a **= 40°** b **= 3.9 cm - 4 cm**
c **= 4.9 cm - 5 cm**
b) d **= 65°** e **= 4.4 cm - 4.5 cm**
f **= 2.9 cm - 3 cm**

2 **a) and b)**

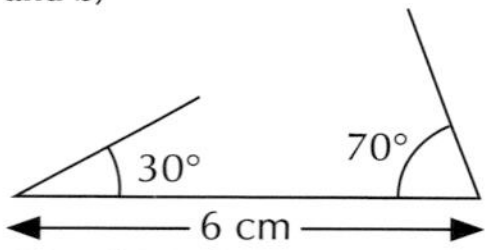

In Q2 and Q3, it's okay if you've drawn the angles at the other ends of the line.

3 **a) and b)**

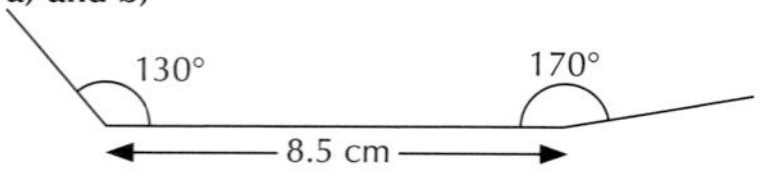

4 a)

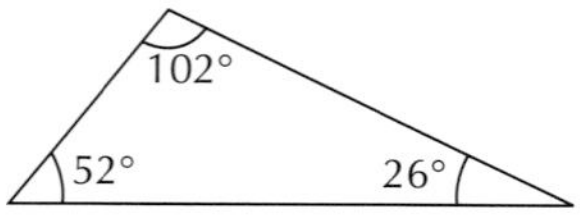

b)

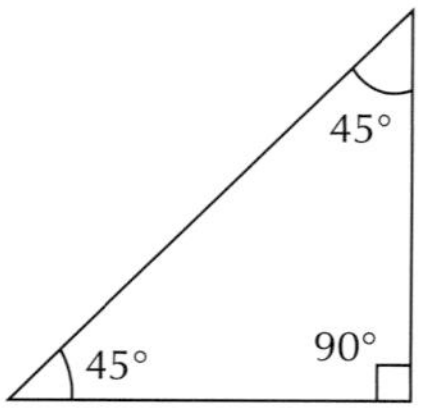

Page 327 Exercise 2

In this exercise, use a ruler and a protractor to make sure that your diagrams have the measurements shown (given rounded to the nearest mm or degree). Allow 1 mm or 1° either side for all answers — including written answers.

1 a) Draw a 4 cm line, using a ruler to measure the correct length. Measure an angle of 90° at the end using a protractor, mark it with a dot and draw a line from the end of the first line through the dot. Repeat at the other end of the line with an angle of 37°. Mark the point where these two lines intersect and join it to each end of the 4 cm line. Measuring side l ⇒ **l = 5.0 cm**

Draw the triangles in parts b)-c) using the same method as part a).

b) **l = 6.1 cm** c) **l = 8.9 cm**

Using the same method as Q1:

2 a) (i)

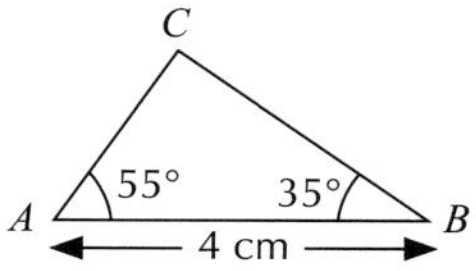

(ii)

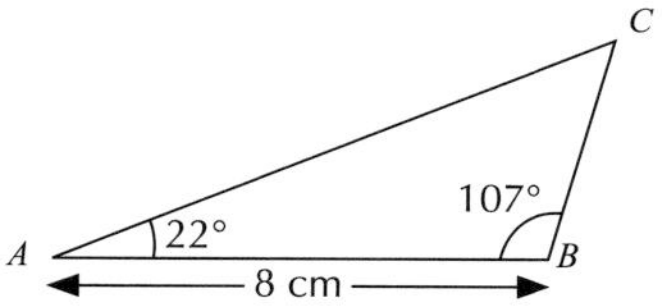

(iii)

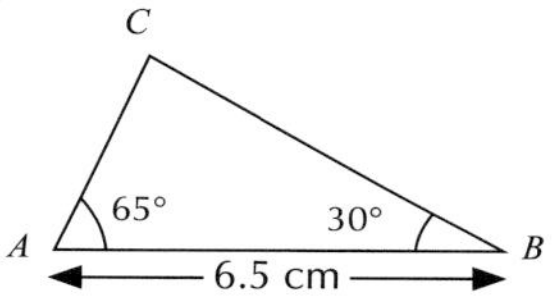

b) (i) **3.3 cm** (ii) **3.9 cm** (iii) **5.9 cm**

3 Using a protractor, draw a line with a 45° angle (measured anticlockwise from the 5 cm line) at Alec and another line with an angle of 45° (measured clockwise from the 5 cm line) at Brenda. Mark the intersection of these two lines.

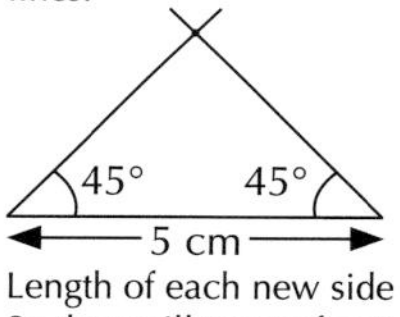

Length of each new side is 3.5 cm.
So they will meet after **3.5 km**.

Page 328 Exercise 3

In this exercise, use a ruler and a protractor to make sure that your diagrams have the measurements shown (given rounded to the nearest mm or degree). Allow 1 mm or 1° either side for all answers — including written answers.

1 a) Draw a 12 cm line, using a ruler to measure the correct length. Measure an angle of 90° at the end using a protractor and mark it with a dot. Draw a line from the end of the 12 cm line through the dot. Measure 5 cm along, mark this point, and join it to the other end of the 12 cm line. Measuring side l ⇒ **l = 13 cm**

Draw the triangles in parts b)-c) using the same method as part a).

b) **l = 5.6 cm** c) **l = 4.9 cm**

2 a) (i) Draw a 6 cm line, using a ruler to measure the correct length. Measure an angle of 40° at the end using a protractor and mark it with a dot. Draw a line from the end of the 6 cm line through the dot. Measure 7 cm along, mark this point, and join it to the other end of the 6 cm line.

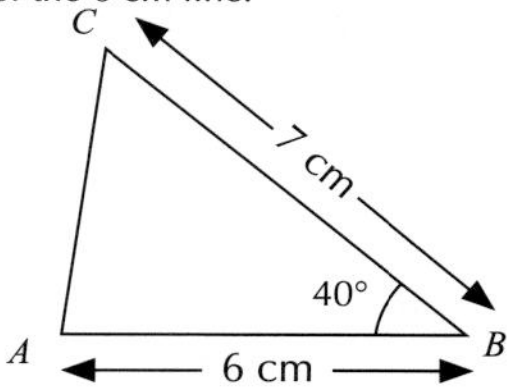

Draw the triangles in parts (ii)-(iv) using the same method as part a).

(ii)

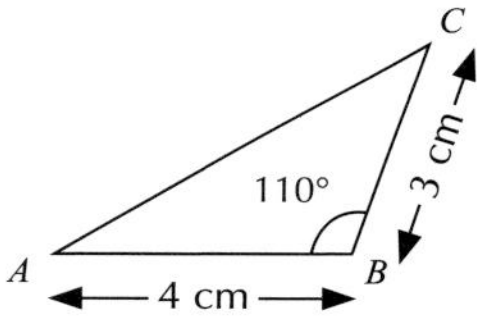

(iii)

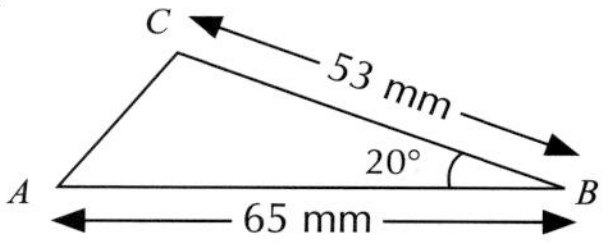

(iv)

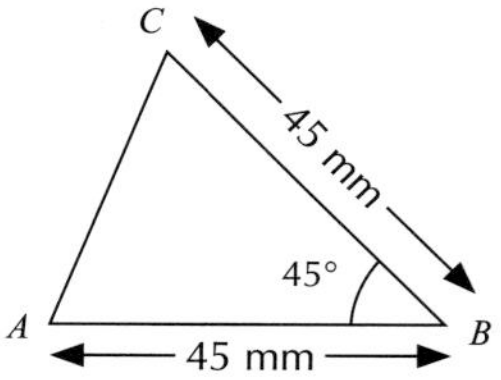

b) (i) **4.5 cm** (ii) **5.8 cm**
(iii) **24 mm** (iv) **34 mm**

3 Draw a 5 cm line, using a ruler to measure the correct length. Measure an angle of 50° at the end of the line using a protractor and mark it with a dot. Draw a line from the end of the 5 cm line through the dot. Then measure 5 cm along, mark this point, and join it to the other end of the original 5 cm line.

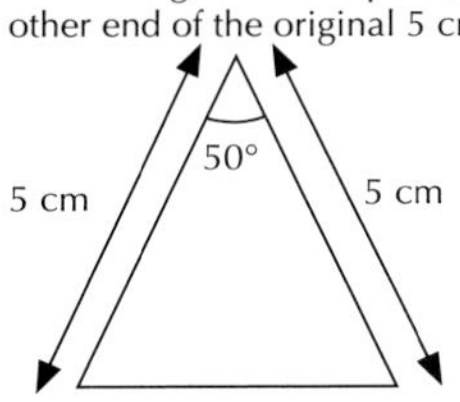

4 a) Angles in a quadrilateral add up to 360° and opposite angles in a rhombus are equal, so find the missing angle (x):
$40° + x + 40° + x = 360°$
$\Rightarrow 2x + 80° = 360°$
$2x = 280° \Rightarrow x = 140°$
Draw a 6 cm line, using a ruler to measure the correct length. Measure an angle of 40° at the end of the line using a protractor and mark it with a dot. Draw a line from the end of the 6 cm line through the dot. Then measure 6 cm along and mark this point. Measure an angle of 140° at this point and mark it with a dot.
Draw a line from the end of the line through the dot. Then measure 6 cm along and mark this point. Connect this point to the end of the original 6 cm line to complete the rhombus.

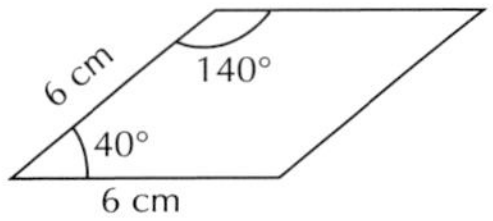

Using the same method for b):

b)

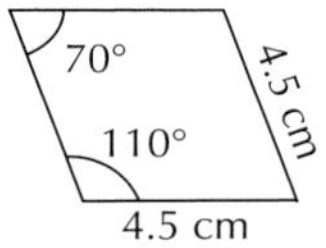

You could have also drawn the rhombus so that the 70° angle was in the bottom left corner.

Page 329 Exercise 4

In this exercise, use a ruler and a protractor to make sure that your diagrams have the measurements shown (given rounded to the nearest mm or degree). Allow 1 mm or 1° either side for all answers — including written answers.

1 a) E.g. Draw the 8 cm line, using a ruler to measure the correct length. Then, set your compasses to 10 cm, and draw an arc from the right-hand end. Next, set your compasses to 6 cm, and draw an arc from the left-hand end. Mark the point where these arcs cross and join it to the ends of the 8 cm line.

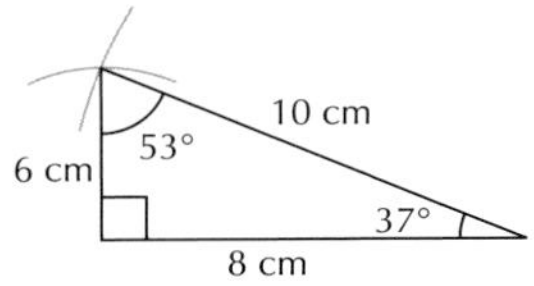

b) E.g. First, draw the 5 cm side. Next, set your compasses to 4.5 cm and draw an arc from the left-hand end of the 5 cm line. Then, set your compasses to 3 cm and draw an arc from the other end. Mark the point where these arcs cross and join it to the ends of the 5 cm line.

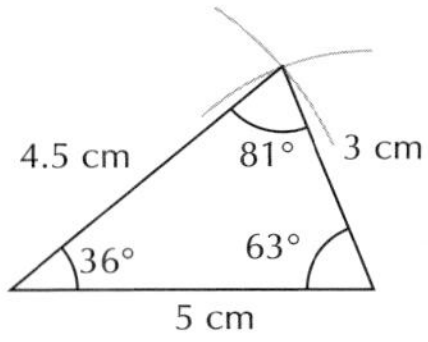

c) E.g. First, draw the 72 mm (= 7.2 cm) side. Next, set your compasses to 40 mm (= 4 cm) and draw an arc from the left-hand end of the 72 mm line. Then, set your compasses to 88 mm (= 8.8 cm) and draw an arc from the other end. Mark the point where these arcs cross and join it to the ends of the 72 mm line.

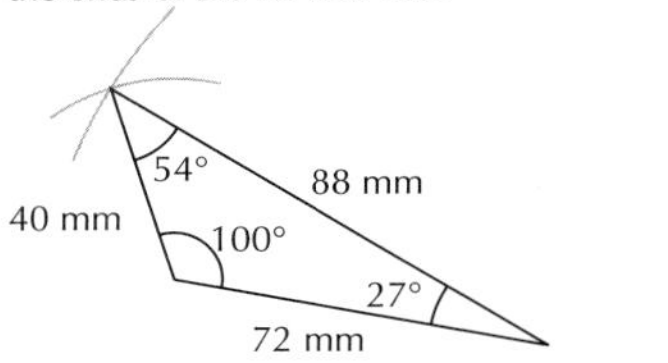

The angles don't appear to add up to 180° as they have been rounded to the nearest degree.

Using the same method for Q2:

2 a) E.g.

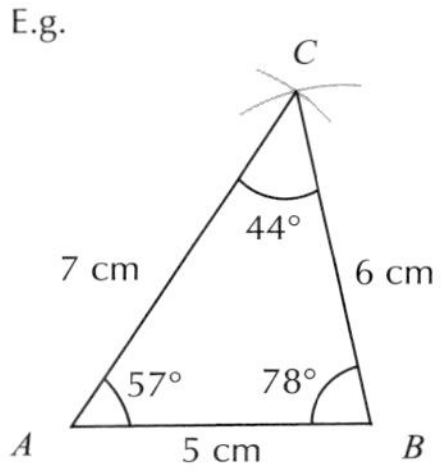

The angles don't appear to add up to 180° as they have been rounded to the nearest degree.

b) E.g.

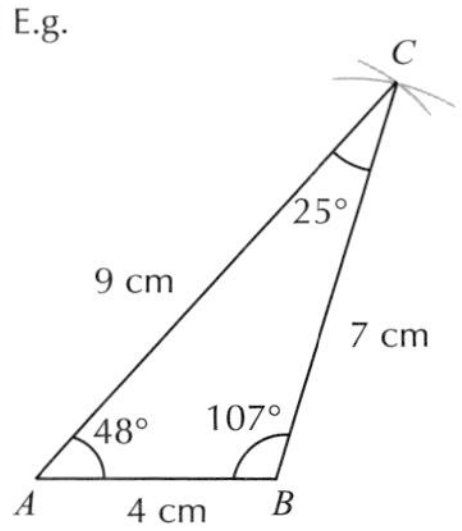

c) E.g.

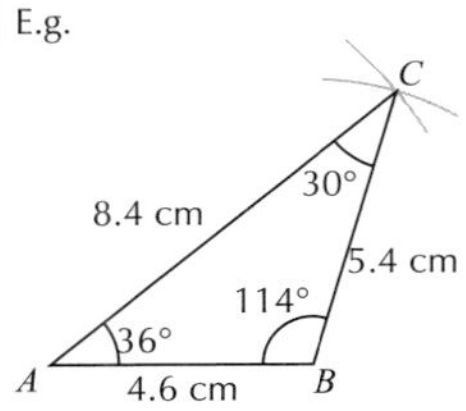

21.3 More Constructions

Page 331 Exercise 1

1 First, draw a horizontal line that is 5 cm long. Next, set your compasses so that they are more than 2.5 cm apart, and draw two arcs from P (one above PQ and one below) and two arcs from Q. Then, draw the line that passes through the points where the arcs cross.

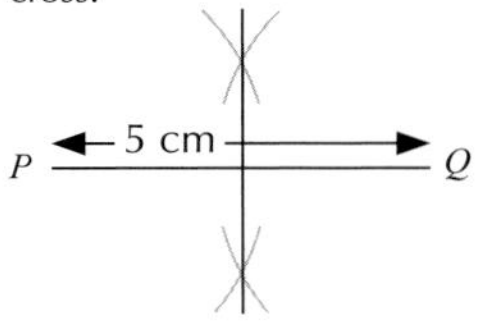

Leave the construction marks on the diagram to show your method.

Using the same method for Q2-4:

2

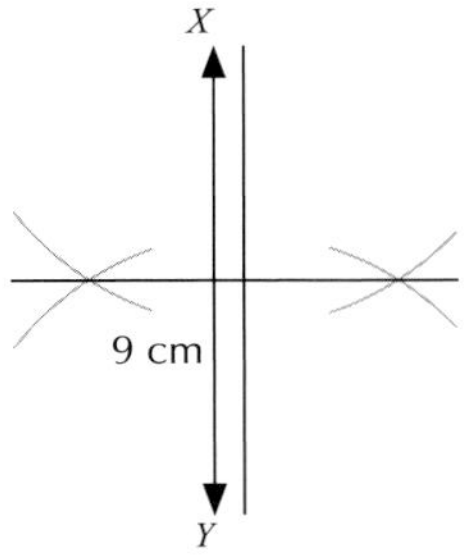

3

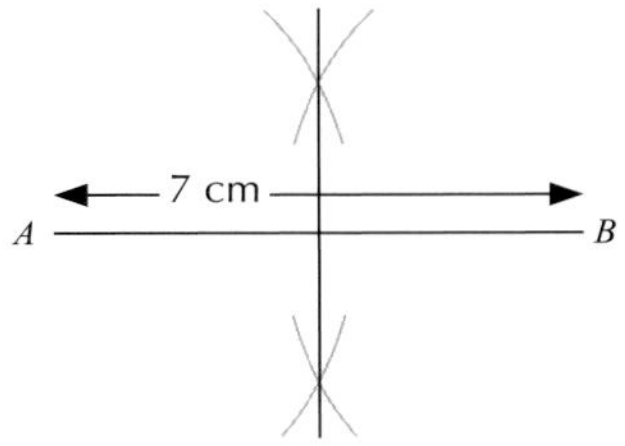

4 a)

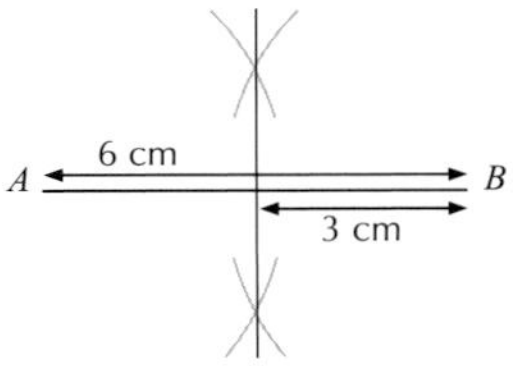

b) Measure 4 cm along the perpendicular bisector each way from the 6 cm line, and mark these points C and D. Since they lie the same distance along the bisector, AC, BC, AD and BD will all be the same length, so you can form a rhombus by joining the points $ACBD$.

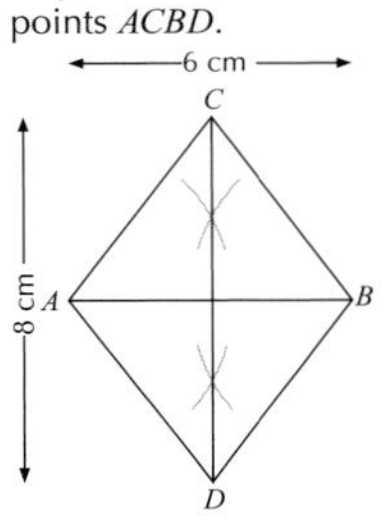

Remember, the diagonals of a rhombus cross at right angles — see page 247.

5 a) Set your compasses to 5 cm to draw a circle with radius 5 cm. Then draw your chords. E.g.

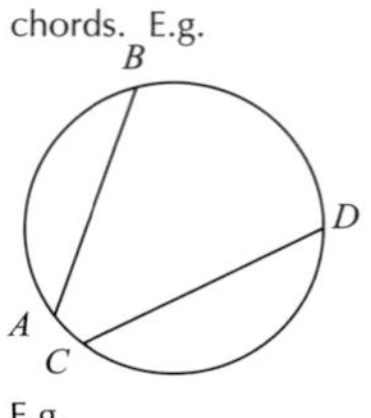

b) E.g.

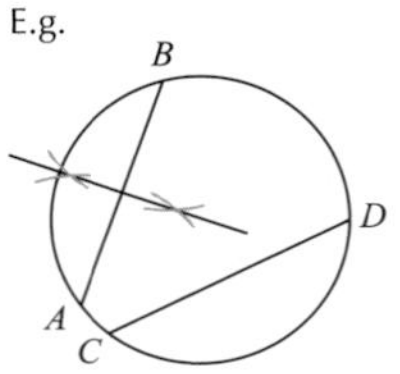

Construct the perpendicular bisectors using the same method used in question 1.

c) E.g.

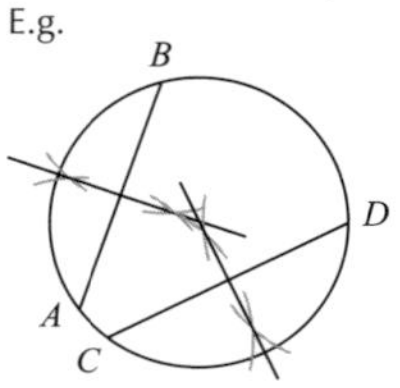

d) They meet at the circle's centre.

Page 332 Exercise 2

In this exercise, mark your answers by using a protractor to measure the angles and checking that they match the diagrams given. Allow 1° either side.

1 a) Draw the 100° angle using your protractor. Next, place the point of your compasses on the angle and draw arcs crossing both lines. Then, put the point of your compasses on the points where these arcs cross the lines and draw two more arcs using the same radius. Finally, draw a line through the point where these arcs cross.

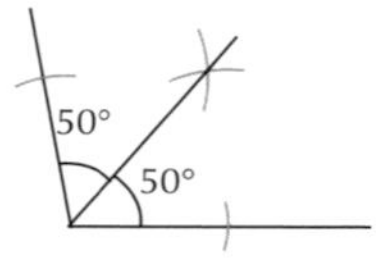

Using the same method for b)-d) and Q2-3:

b)

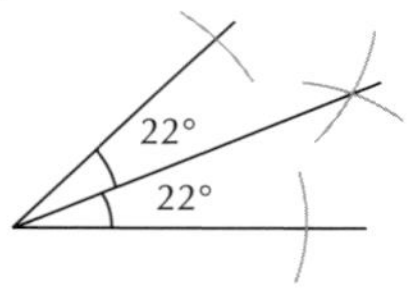

c)

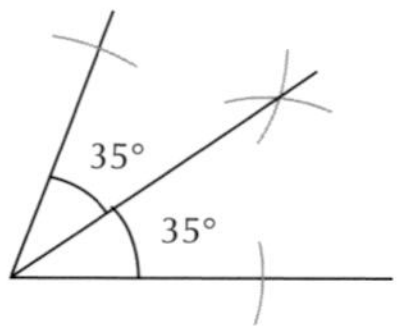

d)

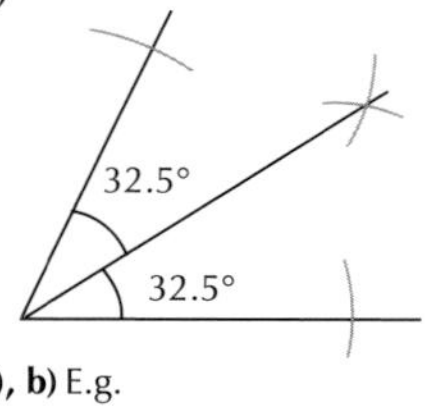

2 a), b) E.g.

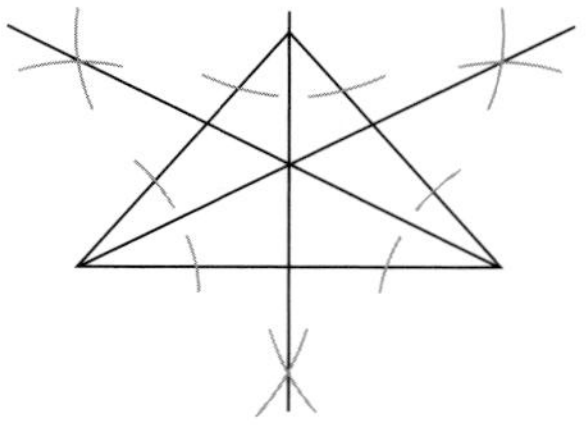

The bisectors all **intersect at a single point**.
This is true for all triangles.

3 a)

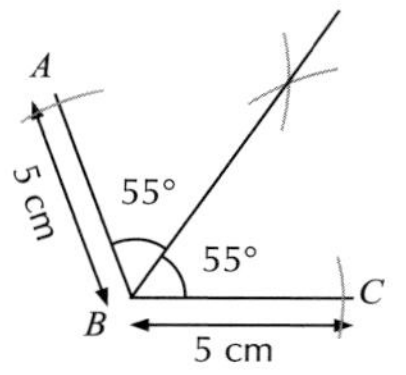

b) Use your ruler to measure 8 cm along the angle bisector:

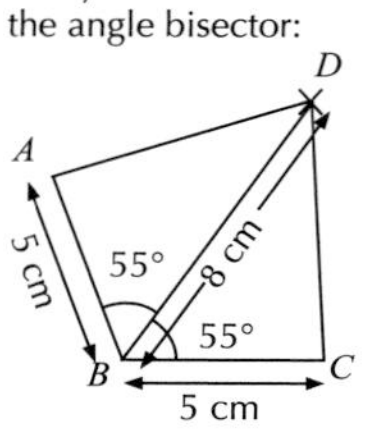

The shape is symmetrical along the line *BD*, so *AD* and *CD* are the same length, meaning *ABCD* is a **kite**.

Page 333 Exercise 3

In this exercise, mark your answers by using a protractor to check that all perpendiculars make an angle of 90° with the line.

1 Draw the triangle, then draw an arc centred on *X* that crosses the line *YZ* twice (you may need to extend the line so that you get two crossing points). Then, draw two arcs of the same radius centred on each of these crossing points. Finally, draw a line from *X* through the point where these two arcs intersect.
E.g.

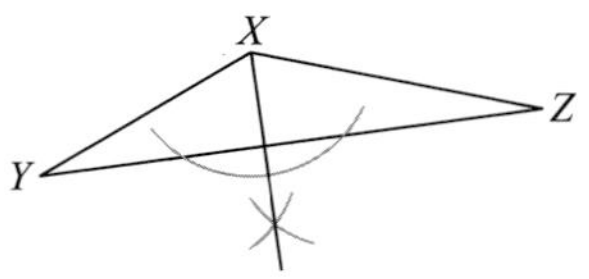

Using the same method for Q2-3:

2 a)-c)

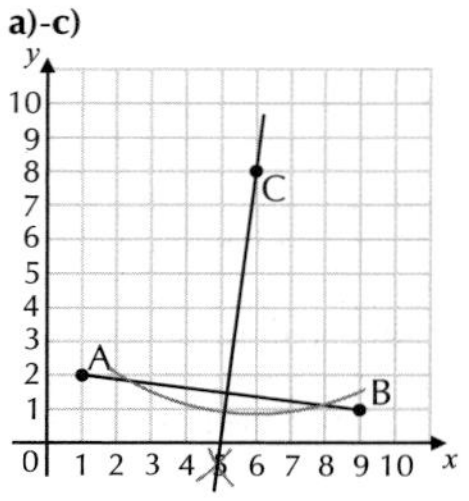

3 E.g.

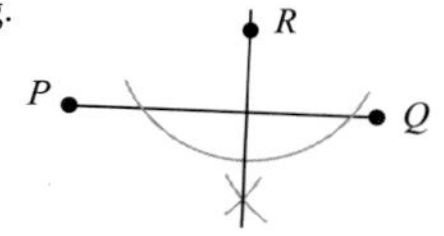

or

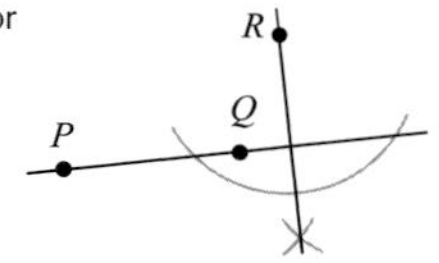

Your diagram might look different depending on where you placed P, Q and R.

4 a), b) E.g.

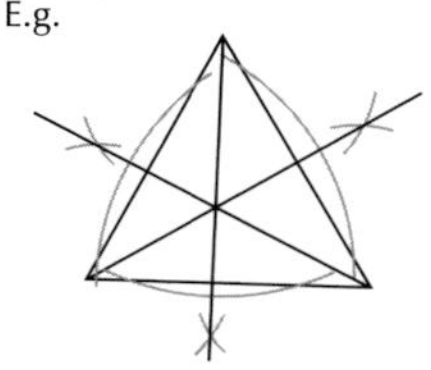

The perpendiculars **intersect at a single point.**
This is true for any triangle.

5 a) E.g. Draw *DE* by using your ruler to measure 5 cm.
Next, measure an angle of 55° at *D*, marking it with a dot.
Draw a line from *D* through this dot.
Finally, measure 6 cm from *D* with your ruler and mark *F*, joining it to *E* to complete the triangle.

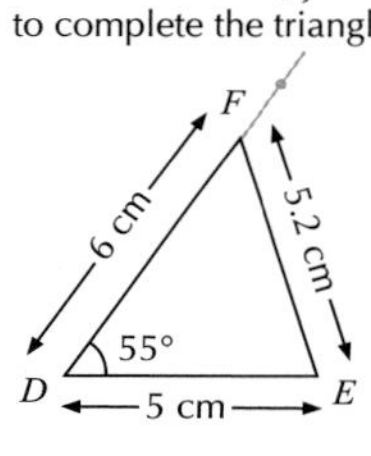

b) Using the same method as in Q1:

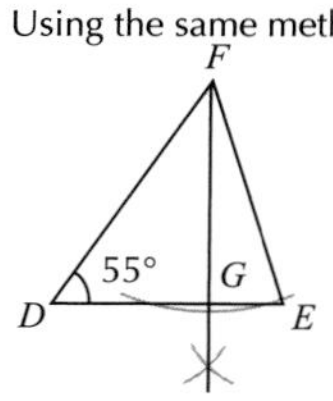

c) Using your ruler, *FG* = **4.9 cm**.
Accept 4.8-5.0 cm.

Area of a triangle = $\frac{1}{2}$ × base × height
= $\frac{1}{2}$ × 5 × 4.9 = **12.3 cm²** (1 d.p.)

Page 335 Exercise 4

In this exercise, mark your answers by using a ruler and protractor to measure the lengths and angles and checking that they match the diagrams given.

1 Use your ruler to measure a 5 cm line and label it AB. Then, place your compass point at A and draw a long arc that crosses AB. Next, draw an arc of the same radius centred on the point where the first arc crosses AB. Finally, draw a line from A through the point where these arcs cross.

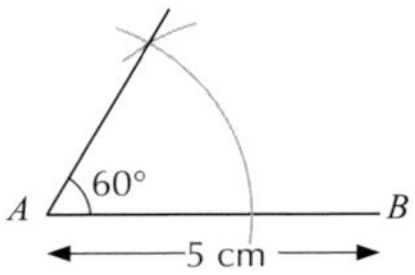

2 a), b) Use the same method as in Q1 to construct the 60° angle(s).

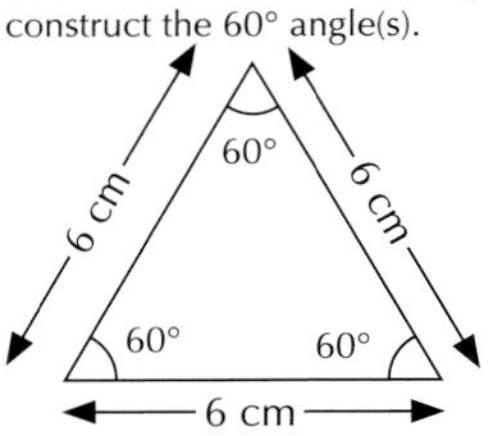

Page 335 Exercise 5

1 Use your ruler to measure a line 6 cm long, and construct a 60° angle at A (see p.334). Then, use your compasses to draw two arcs of the same radius about the points where the long arc from the 60° construction crosses each line. Finally, draw a line from A through the point where these arcs cross.

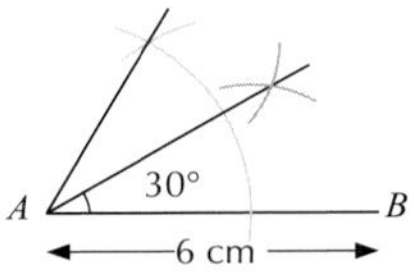

2 a) E.g. Use your ruler to measure a line 7 cm long and label it AB. Next, set your compasses to 7 cm and draw arcs from both ends of the line. Join A and B to the point where these lines cross to form an equilateral triangle. Then, bisect the 60° angle at B using the method from p.331. Mark the point where the angle bisector crosses the 60° line from A as point C, then join C to A and B to complete the triangle.

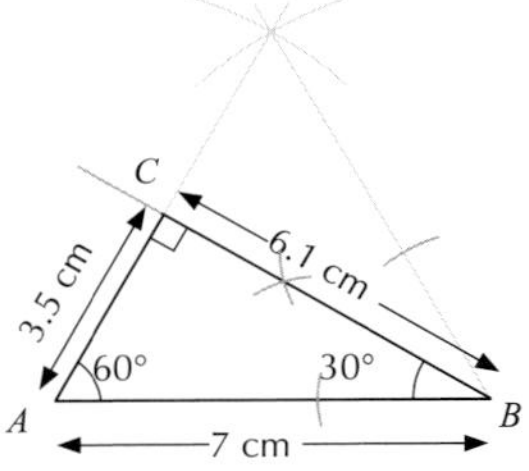

b) Check with protractor.

Using the same method for Q3:

3 Bisect both base angles.

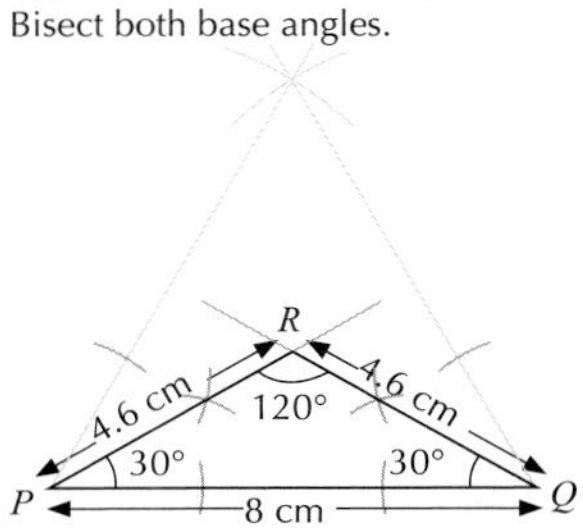

Page 336 Exercise 6

In this exercise, mark your answers by using a ruler and protractor to measure the lengths and angles and checking that they match the diagrams given.

1 a)

b) Place your compass point at X and draw two arcs that cross the line either side of X. Then, increase the radius of your compasses and draw an arc from each intersection. Finally, draw a line through X and the point where these arcs cross.

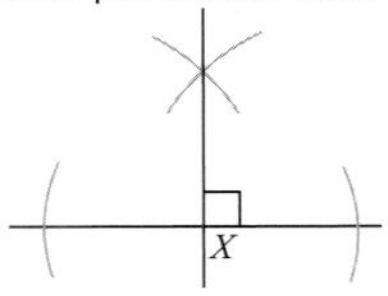

2 E.g. Draw the 7 cm line, mark each end of the line, then extend the line at each end. Next, construct 90° angles at each end of the 7 cm line using the same method as in Q1. Measure 5 cm along each line and mark these points. Finally, join these points to complete the rectangle.

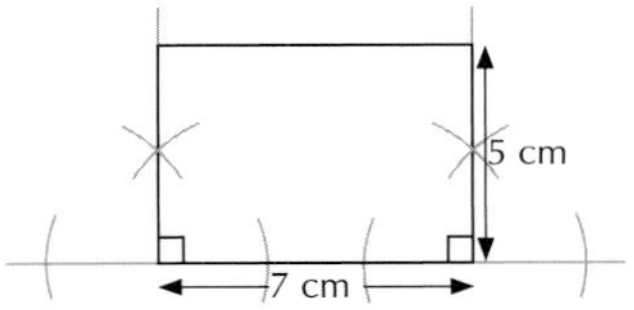

3 E.g. Draw the 6 cm line, mark each end of the line, then extend the line at each end. Next, construct 90° angles at each end of the 6 cm line using the same method as in Q1. Measure 6 cm along each line and mark these points. Finally, join these points to complete the square.

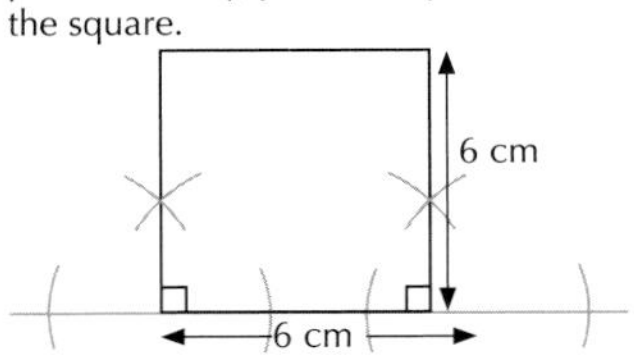

Page 337 Exercise 7

1 a)

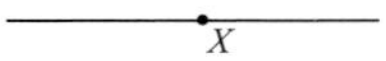

b) First construct a 90° angle (see page 336) and then place your compass point at X and draw two arcs — one through the original line and one through the perpendicular line.
Then, use your compasses to draw two arcs of the same radius about the intersections. Finally, draw a line from X through the point where these arcs cross.

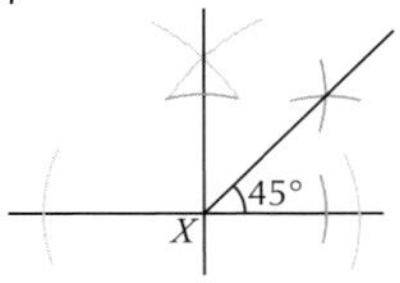

2 Draw a line 8 cm long, label the ends A and B, and extend it at each end. Then, construct 45° angles at A and B using the same method as in Q1. Label the point where the 45° lines intersect as C, then join it to A and B to complete the triangle.

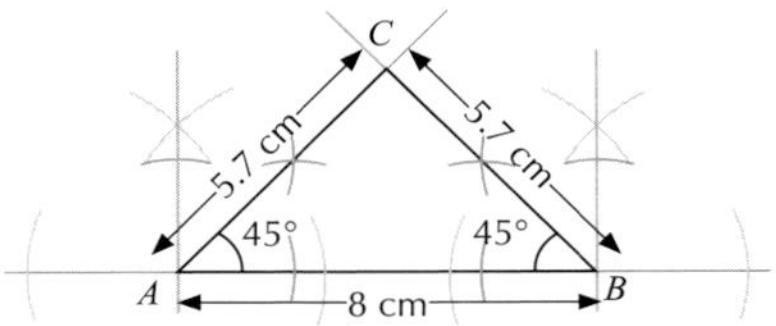

Check your measurements are within 0.1 cm or 1°.

Page 338 Exercise 8

In this exercise, mark your answers by using a protractor to measure any right angles and checking that they are all 90°.

1 Place your compass point at P and draw an arc that crosses AB twice, then draw an arc from each intersection.
Draw a line through P from the point where these arcs cross.
Next, draw two arcs of the same radius from P that cross this new line, and draw an arc from each intersection. Finally, draw another line through P and the point where these arcs cross.
E.g.

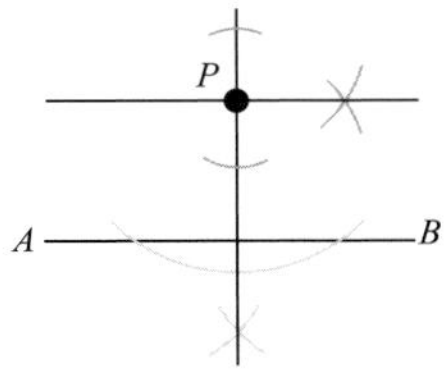

2 a)-c) Using the same method as in Q1.

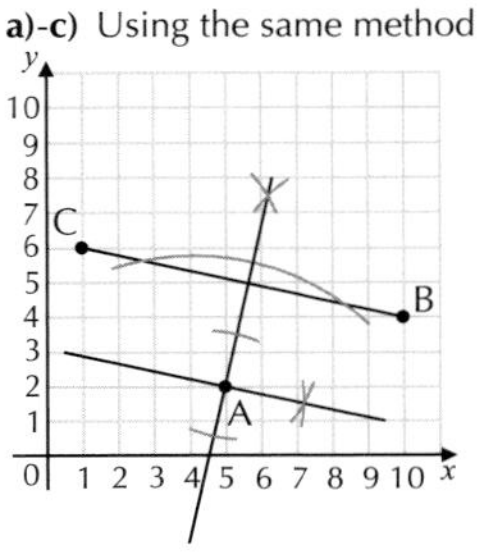

3 First, draw the two lines and draw a point not on the lines. Using the same method as for Q1, construct a line parallel to one of the original lines through this point. Then, draw a new point (or use the same one) and construct a line parallel to the other original line through this point. The shape enclosed by these four lines will be a parallelogram. E.g.

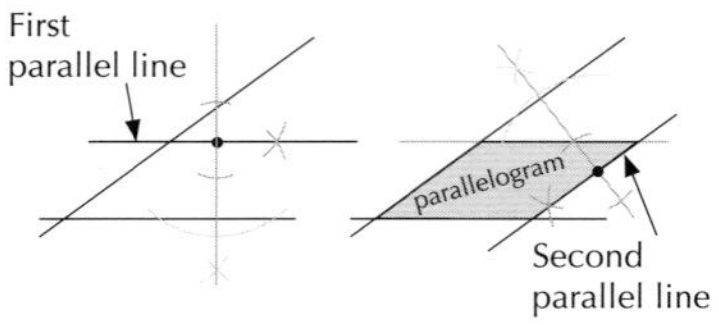

4 a) E.g. First, draw line AB by measuring 10 cm with your ruler. Next, construct a 60° angle at A and B using the method from page 334.

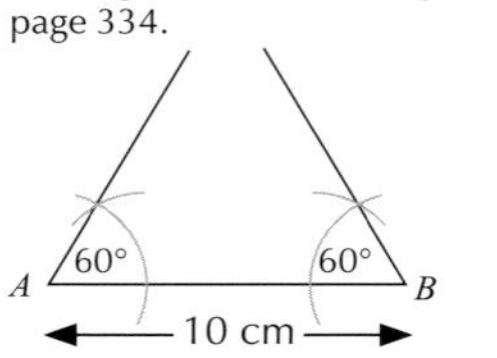

b) Draw a line 3 cm along the 60° line from B, marking C at the end.

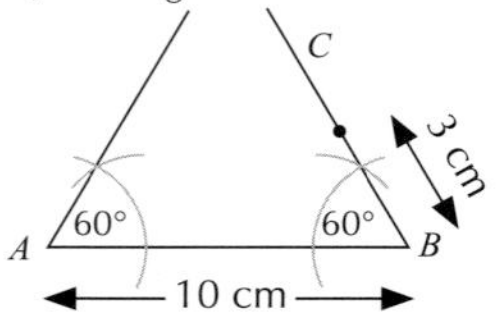

c) Finally, use the method from Q1 to construct a line parallel to AB that passes through C — the point where this line and the 60° line from A intersect is point D.

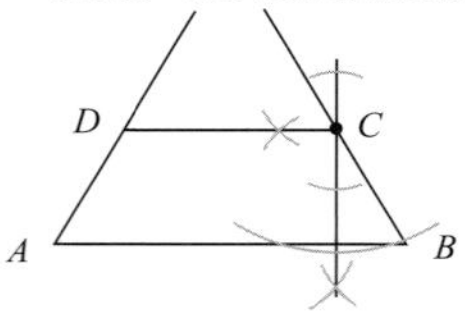

21.4 Loci

Page 340 Exercise 1

For this exercise, use a ruler to check that your drawings have the measurements shown.

1 Use your ruler to draw a 7 cm line. Next, set your compasses to 2 cm and draw a long arc around each end of the line. Finally, use your ruler to join the tops and the bottoms of each arc.

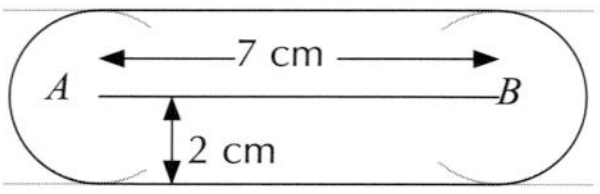

2 a) Set your compasses to 3 cm, place the point on X and draw a **circle**.

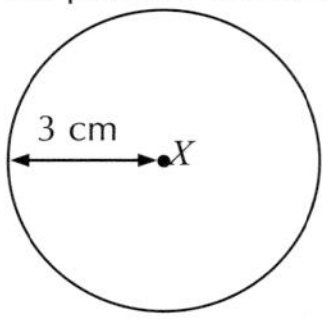

b) Shade the **inside** of the circle.

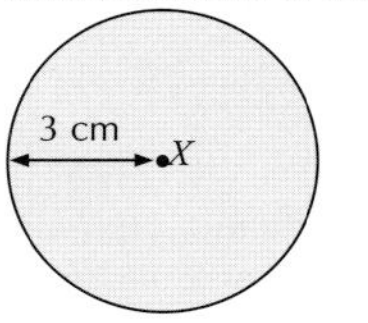

3 Use your ruler to draw two points 6 cm apart, then construct the **perpendicular bisector** of these points using the method from page 330.

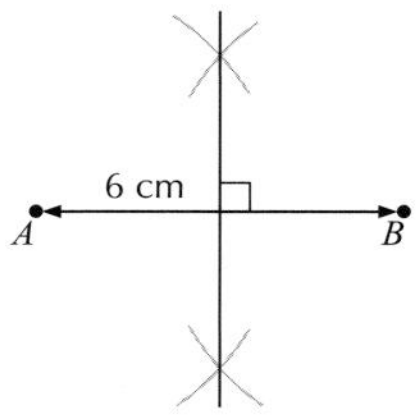

4 Use your protractor to draw two lines that meet at an angle of 50°, then construct the **angle bisector** of these lines using the method from page 331.

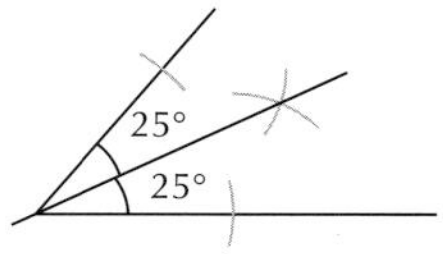

5 Using the same method as Q1:

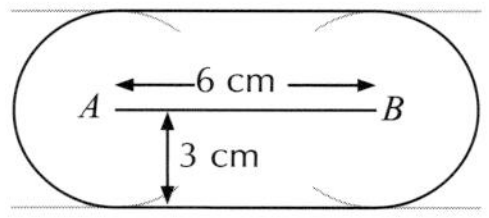

6 a), b) Using the same method as Q3:

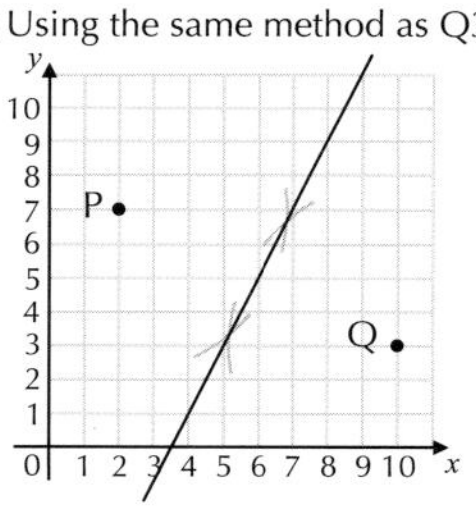

Page 341 Exercise 2

1 a)-d) Draw two points P and Q that are 5 cm apart. Draw a circle of radius 3 cm around P, and a circle of radius 4 cm around Q:

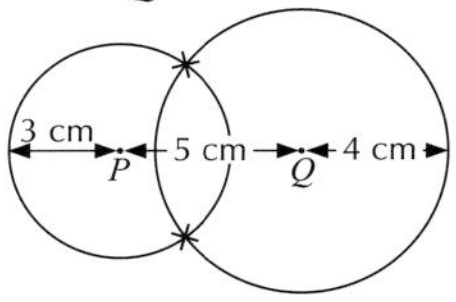

The points marked with crosses are both 3 cm from P and 4 cm from Q.

2 Set your compasses to 3 cm, place them at point B and draw an arc that is 3 cm from point B. Then using a ruler draw a line which is 4 cm from AD. Finally shade the required locus as showed below.

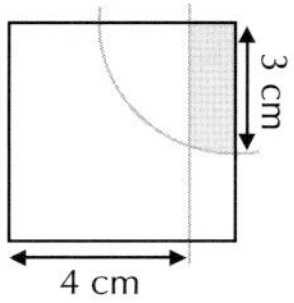

3 a) Using the method from page 329:

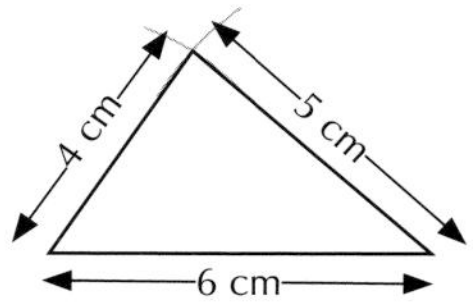

b) Set your compasses to 1 cm, and draw circles around each corner of the triangle. Then, draw two lines from each circle to each of the others, as shown in grey below. Finally, mark the required locus as shown below (it has two parts — one inside the original triangle and one outside).

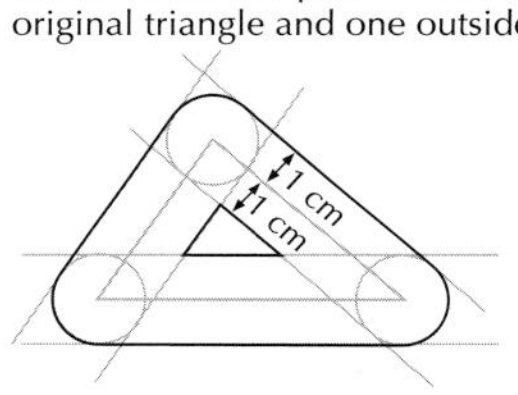

4 a) Using the method from page 329:

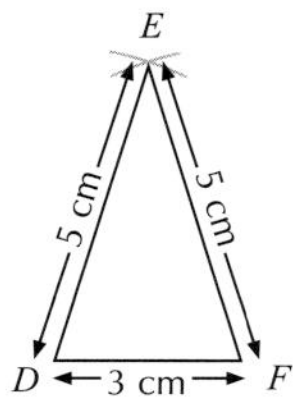

b) E.g. First, use your compasses to draw a circle of radius 2 cm around E. Next, construct the perpendicular bisector of DF using the method from page 330. The required locus is the set of points that are both on this line and inside the circle:

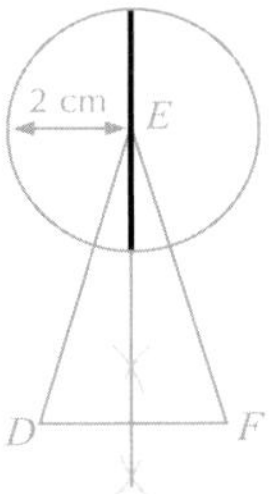

Page 342 Exercise 3

1 a)-b) The scale is 1 cm : 1 km, so 3 km in real life will be 3 cm on the diagram. Draw P and L 3 cm apart, then construct the perpendicular bisector using the method from page 330.

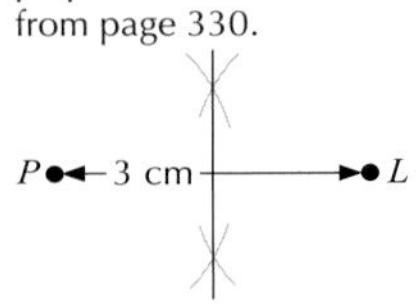

Remember to make all your measurements in relation to the scale given in the question.

2 a), b) The scale is 1 cm : 10 miles, so 50 miles will be 50 ÷ 10 = 5 cm on the diagram. Draw A and B 5 cm apart.
40 miles will be 40 ÷ 10 = 4 cm on the diagram, so draw circles of radius 4 cm around each point.
The area where these circles overlap is the region where the camels could possibly meet.

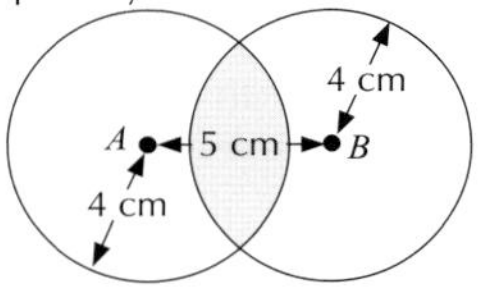

3 a) The scale is 1 cm : 1 m so the length of the yard on the diagram is 4 cm, the width is 2 cm and the dog's lead is 1 cm. Use your ruler to draw a 2 cm by 4 cm rectangle. Then, set your compasses to 1 cm and draw an arc from one of the corners.
The required area is the inside of this quarter-circle:

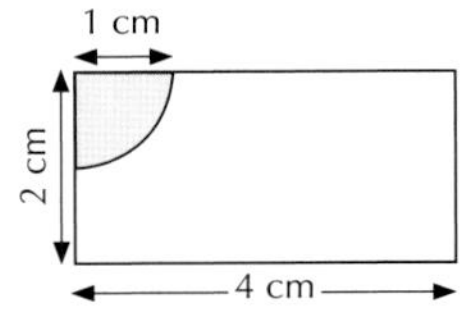

b) The rail will be 3 cm on the diagram. Use your ruler to measure 3 cm from the corner, then use your compasses to draw an arc of radius 1 cm around this point. Then, draw a horizontal line 1 cm from the top of the rectangle. Finally, shade the required area as shown below:

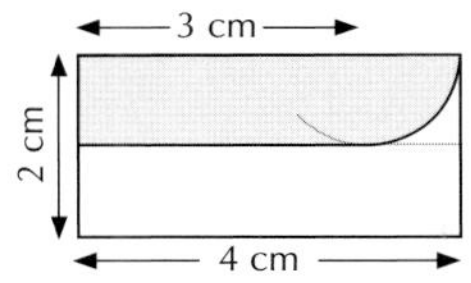

Page 343 Review Exercise

1 a) A: **radius** B: **diameter**
C: **circumference**

b) D: **tangent** E: **sector**
F: **segment**

2 a **= 4.8 cm**
You may have a slightly different measurement — any answer between 4.7 cm and 4.9 cm is okay.
b **= 120°**
Any answer between 118° and 122° is okay.
c **= 1.8 cm**
Any answer between 1.7 cm and 1.9 cm is okay.

3 a) E.g. Draw the 11 cm line, using a ruler to measure the correct length. Then, set your compasses to 6.5 cm, and draw an arc from the right-hand end. Next set your compasses to 7 cm, and draw an arc from the left-hand end. Mark the point where these arcs cross and join it to the ends of the 11 cm line. If drawn correctly, the angles inside the triangle will be:

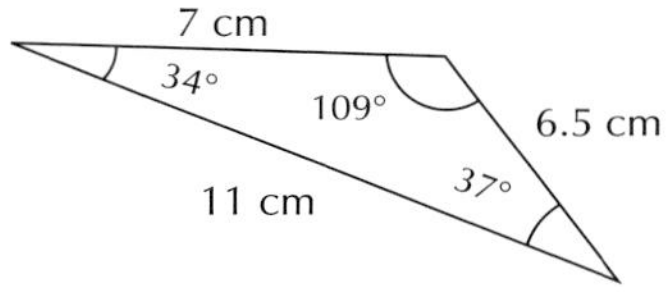

Using the same method for part b), the angles inside the triangle will be:

b)

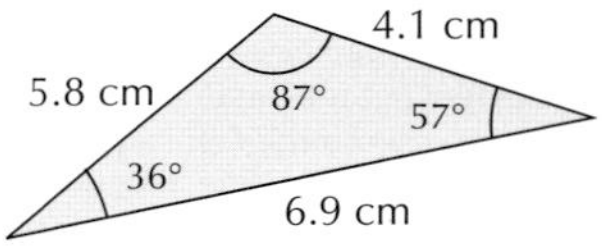

4 a) E.g. Use your ruler to draw line DE 5.8 cm long. Then, set your compasses to 5.8 cm and draw arcs from D and E. Mark point F where these arcs intersect. Then connect F to D and E using a ruler.

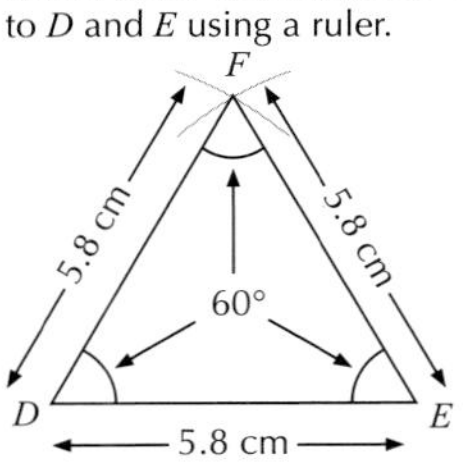

b) Using the method from page 337:

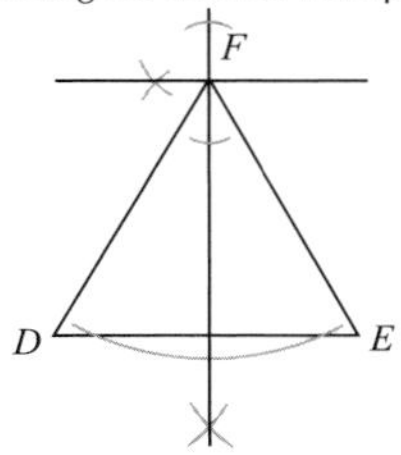

5 a) E.g. Use your ruler to draw line AB 7.4 cm long. Then, construct a 60° angle at A using the method from page 334.
Next, construct a 90° angle at B and bisect it, as shown on page 331. Mark the point where these two lines cross as C and join it to A and B.

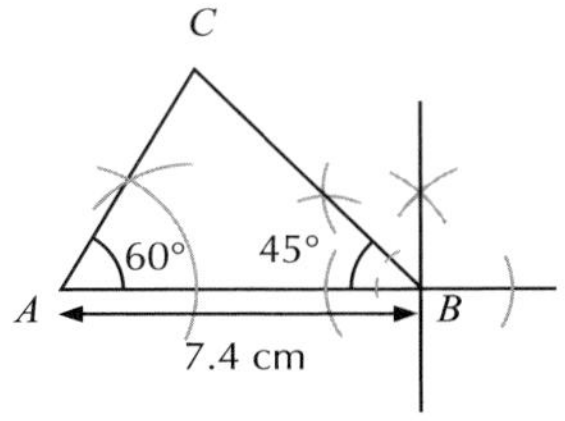

b) Angles in a triangle add up to 180°, so angle ACB + 45° + 60° = 180°
angle ACB = 180° – 45° – 60° = **75°**
Triangle rules are covered on page 242.

6 Copy the diagram, then construct the angle bisector using the method from page 331. The scale is 1 cm : 0.5 m, so 3 m in real life will be 6 cm on the diagram. Use your ruler to measure 6 cm along the angle bisector and mark the position of the bonfire.

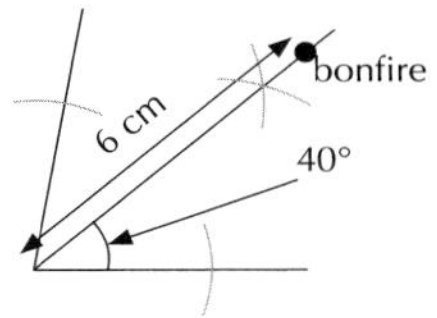

Page 344 Exam-Style Questions

1 Draw a 7.1 cm line, using a ruler to measure the correct length. Measure an angle of 22° at the end of the line using a protractor and mark it with a dot. Draw a line from the end of the 7.1 cm line through the dot. Then measure 7.1 cm along, mark this point, and join it to the other end of the original line.

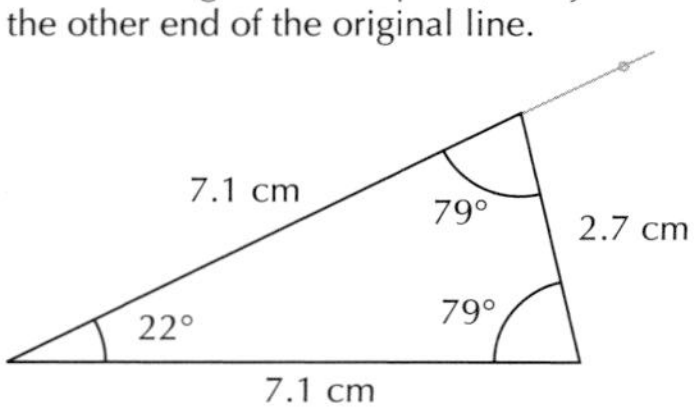

[3 marks available — 1 mark for accurately drawing one line, 1 mark for a correctly drawn angle of 22°, 1 mark for a fully correct triangle]

You can check you have the right triangle by measuring the other angles and making sure they are they same as those in the diagram above.

2 Construct a perpendicular line from S to the beach.

The shortest distance from a point to a line is always the perpendicular distance.

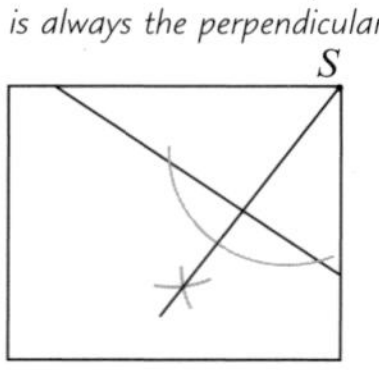

[3 marks available – 1 mark for the 2 arcs on the beach line (or a single arc crossing the line twice), 1 mark for the pair of intersecting arcs and 1 mark for the line from point S to the beach (or beyond) – the drawn line and the beach line must meet at angles between 89° and 91° inclusive]

3 a) Use your ruler to draw line AB 8 cm long. Set your compasses to 8 cm and draw arcs from A and B. Draw a line from A to where these arcs cross.

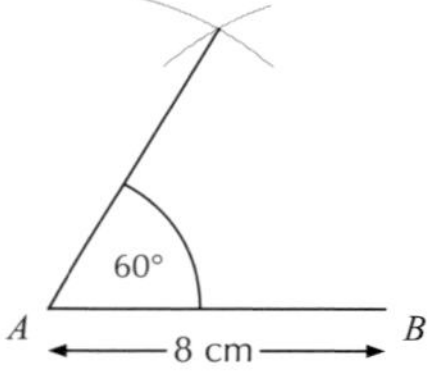

[2 marks available — 1 mark for two intersecting construction arcs and 1 mark for a fully correct construction of 60° angle]

b) E.g. Measure 8 cm along the 60° line and mark this point D.

This will be the point where the arcs crossed if you used the same method as above.

Set your compasses to 8 cm and draw arcs from B and D. Mark the point where these cross as point C and join it to B and D to complete the rhombus.

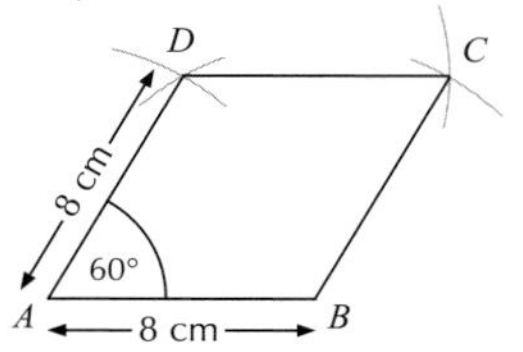

[3 marks available — 1 mark for accurately drawing one 8 cm line, 1 mark for two intersecting construction arcs and 1 mark for a fully correct rhombus]

4 The scale is 1 cm : 100 m so on the diagram the road is 4 cm above the motorway. First, construct a perpendicular line to the motorway using your compasses using the method shown on page 333. Measure 4 cm from the motorway along this line using a ruler, and mark this point with a dot. Construct a parallel line to the motorway at this dot using the method shown on page 337.

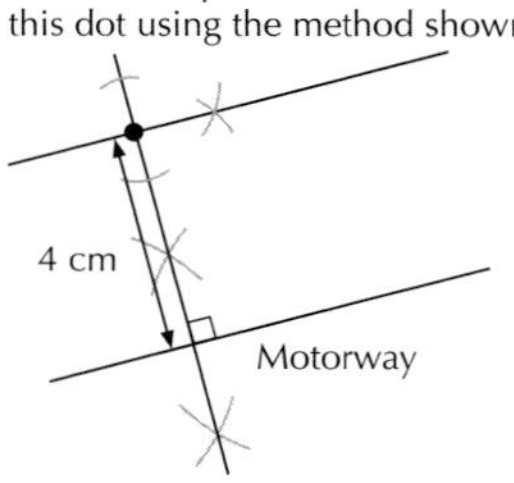

[3 marks available — 1 mark for at least one pair of two intersecting construction arcs, 1 mark for a correct perpendicular constructed to the motorway, 1 mark for a parallel line constructed 4 cm from the motorway]

5 The scale is 1 cm : 1 m so on the diagram, region $ABCD$ is 10 cm × 10 cm and the treasure is 7 cm from corner C.
E.g. First, use your ruler to draw a 10 cm by 10 cm square using a ruler and a protractor. Since the treasure is the same distance from AB and AD, the treasure must lie on the angle bisector of the angle between these two lines. Since this is a square, the angle bisector is AC, so join A and C with a straight line. Next, set your compasses to 7 cm and draw an arc around C. The treasure is at the point where the arc and angle bisector intersect.

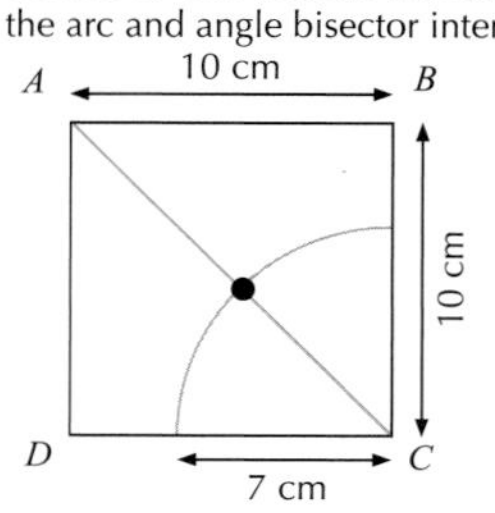

(location of treasure shown by the dot)

[4 marks available — 1 mark for correctly drawing a 10 cm × 10 cm square, 1 mark for finding the angle bisector of the angle between AB and AD, 1 mark for an arc of radius 7 cm from C and 1 mark for a fully correct answer with position of treasure clearly indicated]

Section 22 — Area and Perimeter

22.1 Rectangles and Triangles

Page 346 Exercise 1

1 a) $P = a + b + c = 3 + 4 + 5 =$ **12 cm**

b) $P = 2l + 2w = (2 \times 3) + (2 \times 11)$
$= 6 + 22 =$ **28 m**

c) $P = 4l = 4 \times 5 =$ **20 cm**

d) $P = 2l + 2w = (2 \times 4) + (2 \times 6)$
$= 8 + 12 =$ **20 cm**

e) $P = a + b + c = 14 + 11 + 18 =$ **43 m**

f) $P = 2l + 2w = (2 \times 15) + (2 \times 30)$
$= 30 + 60 =$ **90 mm**

2 a) $P = 4l = 4 \times 4 =$ **16 cm**

b) $P = a + b + c = 5 + 5 + 7 =$ **17 cm**

c) $P = 2l + 2w = (2 \times 8) + (2 \times 6)$
$= 16 + 12 =$ **28 m**

3 $P = (2 \times 2.1) + (2 \times 2.8) = 4.2 + 5.6 =$ **9.8 m**

Page 347 Exercise 2

1 a) $A = lw = 5 \times 10 =$ **50 cm²**

b) $A = \frac{1}{2}bh = \frac{1}{2} \times 9 \times 4 =$ **18 m²**

c) $A = l^2 = 4^2 =$ **16 mm²**

d) $A = l^2 = 11^2 =$ **121 cm²**

e) $A = lw = 3 \times 11 =$ **33 m²**

f) $A = \frac{1}{2}bh = \frac{1}{2} \times 10 \times 2 =$ **10 mm²**

2 a) $A = lw = 23 \times 15 =$ **345 mm²**

b) $A = l^2 = 17^2 =$ **289 m²**

c) $A = \frac{1}{2}bh = \frac{1}{2} \times 4 \times 12.5 =$ **25 cm²**

3 The lawn is a rectangle, so its area is:
$A = 23.5 \times 17.3 = 406.55\ m^2 =$ **407 m²** (to the nearest m²)

4 a) $A_{floor} = 9 \times 7.5 =$ **67.5 m²**

b) $A_{tile} = 0.5^2 =$ **0.25 m²**

c) $67.5 \div 0.25 =$ **270 tiles**

5 Area of the garden $= 24 \times 5.4 = 129.6\ m^2$
60 cm $= 60 \div 100 = 0.6$ m, so the area covered by one roll of turf is $8 \times 0.6 = 4.8\ m^2$.
So $129.6 \div 4.8 =$ **27 rolls** are needed.

6 The area of icing is made up from rectangles.

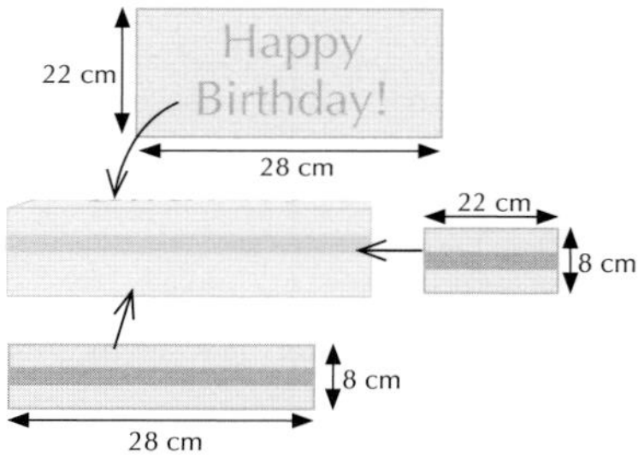

Area of top $= 22 \times 28 = 616\ cm^2$
Area of front / back $= 8 \times 28 = 224\ cm^2$
Area of sides $= 8 \times 22 = 176\ cm^2$
So area of icing needed
$= 616 + (2 \times 224) + (2 \times 176)$
$=$ **1416 cm²**

Page 349 Exercise 3

1 a) Split the shape into two rectangles along the dashed line.

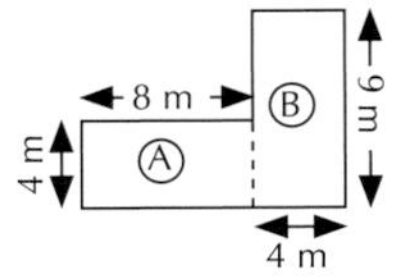

Area of A = 4 × 8 = 32 m^2
Area of B = 9 × 4 = 36 m^2
So total area = 32 + 36 = **68 m^2**

b)

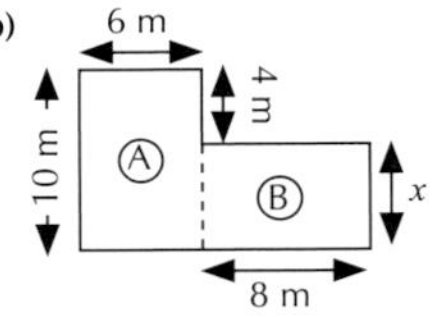

Missing side length x = 10 – 4 = 6 m
Area of A = 10 × 6 = 60 m^2
Area of B = 6 × 8 = 48 m^2
So total area = 60 + 48 = **108 m^2**

c)

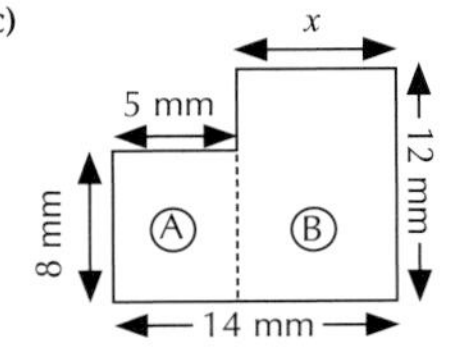

Missing side length x = 14 – 5 = 9 mm
Area of A = 8 × 5 = 40 mm^2
Area of B = 12 × 9 = 108 mm^2
So total area = 40 + 108 = **148 mm^2**
Alternatively, you could have split the shape horizontally to make one rectangle with dimensions 8 × 14 and one with dimensions 4 × 9.

2 a) Missing side lengths are 12 – 8 = 4 cm and 13 – 5 = 8 cm.
(i) P = 8 + 5 + 4 + 8 + 12 + 13 = **50 cm**
(ii) Splitting the shape vertically into two rectangles:
A = (13 × 8) + (8 × 4) = **136 cm^2**

b) Missing side lengths are 7 – 4 = 3 cm and 5 + 3 = 8 cm.
(i) P = 5 + 3 + 3 + 4 + 8 + 7 = **30 cm**
(ii) Splitting the shape vertically into two rectangles:
A = (7 × 5) + (4 × 3) = **47 cm^2**

c) Missing side lengths are 15 + 8 = 23 mm and 23 – 10 = 13 mm.
(i) P = 23 + 23 + 8 + 13 + 15 + 10 = **92 mm**
(ii) Splitting the shape vertically into two rectangles:
A = (10 × 15) + (23 × 8) = **334 mm^2**

3 a) Area of rectangle = 4 × 8 = 32 m^2
Base of triangle = 20 – 8 = 12 m,
so area of triangle = $\frac{1}{2}$ × 12 × 4 = 24 m^2
Total area = 32 + 24 = **56 m^2**

b) Area of rectangle = 5 × 9 = 45 cm^2
Base of triangle = 21 – 9 = 12 cm,
so area of triangle = $\frac{1}{2}$ × 12 × 5 = 30 cm^2
Total area = 45 + 30 = **75 cm^2**
The shapes in this question are trapeziums, so you could also have used the formula from page 351 to find their areas.

22.2 Other Quadrilaterals

Page 350 Exercise 1

1 a) $A = bh$ = 7 × 4 = **28 cm^2**
b) $A = bh$ = 4 × 2 = **8 m^2**
c) $A = bh$ = 6 × 8 = **48 mm^2**

2 a) (i) A = 9 × 6 = **54 mm^2**
(ii) P = (2 × 8) + (2 × 9) = **34 mm**
The lengths of opposite sides in a parallelogram are equal.
b) (i) A = 18 × 35 = **630 m^2**
(ii) P = (2 × 50) + (2 × 18) = **136 m**
c) (i) A = 5.8 × 8.2 = **47.56 mm^2**
(ii) P = (2 × 5.8) + (2 × 10) = **31.6 mm**

3 Area of one parallelogram = 8.2 × 3.1
= 25.42 cm^2
So area of logo = 2 × 25.42 = **50.84 cm^2**

4 Area of the part of the wall = 66 × 44
= 2904 cm^2
There are 25 tiles on the wall, so the area of one tile is 2904 ÷ 25 = **116.16 cm^2**.

5 **The square has the larger area.** The vertical height of the rhombus (a special type of parallelogram) is less than 6 cm, so its area will be less than the area of the square.

Page 352 Exercise 2

1 a) $A = \frac{1}{2}(a + b) \times h = \frac{1}{2}(4 + 10) \times 4$
$= \frac{1}{2} \times 14 \times 4$ = **28 m^2**
b) $A = \frac{1}{2}(a + b) \times h = \frac{1}{2}(12 + 8) \times 9$
$= \frac{1}{2} \times 20 \times 9$ = **90 mm^2**
c) $A = \frac{1}{2}(a + b) \times h = \frac{1}{2}(3 + 13) \times 5$
$= \frac{1}{2} \times 16 \times 5$ = **40 mm^2**

2 a) Find the area of the two trapeziums separately.
Area of left trapezium = $\frac{1}{2}(7 + 4) \times 10$
= 55 m^2
Area of right trapezium = $\frac{1}{2}(4 + 3) \times 10$
= 35 m^2
So the area of the end of the barn is 55 + 35 = **90 m^2**.
b) P = 10 + 10 + 7 + 10.44 + 10.05 + 3
= **50.49 m**

3 **They have the same area, since they have the same vertical height, and ½(9 + 3) = 6.**

Page 354 Exercise 3

1 a) Area of parallelogram = 10 × 6 = 60 cm^2
Area of trapezium = $\frac{1}{2}(5 + 10) \times 3$
= 22.5 cm^2
So the total area of the shaded region is 60 + 22.5 = **82.5 cm^2**.
b) Area of large parallelogram = 16 × 10
= 160 mm^2
Area of small parallelogram = 7 × 3
= 21 mm^2
So the total area of the shaded region is 160 – 21 = **139 mm^2**.

2 a) Split the shape along the dashed line to create two identical parallelograms.
(i) Area of one parallelogram = 8 × 6
= 48 mm^2
Area of shape = 2 × 48 = **96 mm^2**
(ii) P = 7 + 8 + 8 + 7+ 8 + 8 = **46 mm**

b) Split the shape along the dashed line to create two identical trapeziums.
(i) Height of one trapezium = 8.5 ÷ 2
= 4.25 m
Area of one trapezium
$= \frac{1}{2}(5 + 16) \times 4.25$ = 44.625 m^2
Area of shape = 2 × 44.625 = **89.25 m^2**
(ii) P = 6 + 5 + 6 + 6 + 5 + 6 = **34 m**

c) The shape is already split into two identical parallelograms and a square.
(i) Area of one parallelogram = 12 × 8
= 96 m^2
Area of square = 12^2 = 144 m^2
Area of shape = (2 × 96) + 144
= **336 m^2**
(ii) P = 12 + 10 + 12 + 10 + 12
+ 10 + 12 + 10 = **88 m**

d) The shape is already split into a trapezium and a triangle.
(i) Area of trapezium = $\frac{1}{2}(14 + 26) \times 30$
= 600 m^2
Area of triangle = $\frac{1}{2} \times 14 \times 30$
= 210 m^2
Area of shape = 600 + 210 = **810 m^2**
(ii) P = 26 + 31 + 31 + 31 + 31 = **150 m**

3 a) $A = \frac{1}{2}(20 + 40) \times 60$ = **1800 cm^2**
b) Base of one parallelogram = 20 ÷ 2 = 10 cm
Area of one parallelogram = 10 × 60
= 600 cm^2
Total area of coloured strips = 2 × 600
= **1200 cm^2**

22.3 Circumference of a Circle

Page 356 Exercise 1

1 a) $C = \pi d = \pi \times 5$ = 15.707......
= **15.7 cm** (1 d.p.)
b) $C = \pi d = \pi \times 12$ = 37.699...
= **37.7 cm** (1 d.p.)
c) $C = \pi d = \pi \times 2$ = 6.283... = **6.3 cm** (1 d.p.)
d) $C = \pi d = \pi \times 10$ = 31.415...
= **31.4 cm** (1 d.p.)

2 a) $C = \pi d = \pi \times 30$ = **30π mm**
Using the same method for b)-d):
b) **5π mm** c) **7π m**
d) **9π cm**

3 a) (i) $d = 2r$ = **4 cm**
(ii) $C = \pi d = \pi \times 4$ = 12.566...
= **12.6 cm** (1 d.p.)
b) (i) $d = 2r$ = **5 cm**
(ii) $C = \pi d = \pi \times 5$ = 15.707...
= **15.7 cm** (1 d.p.)
c) (i) $d = 2r$ = **1 m**
(ii) $C = \pi d = \pi \times 1$ = 3.141...
= **3.1 m** (1 d.p.)
d) (i) $d = 2r$ = **3 mm**
(ii) $C = \pi d = \pi \times 3$ = 9.424...
= **9.4 mm** (1 d.p.)

4 a) $C = \pi d$ = **4π cm**
b) $C = \pi d$ = **8π mm**
c) $C = 2\pi r = 2 \times \pi \times 14$ = **28π km**
d) $C = 2\pi r = 2 \times \pi \times 0.1$ = **0.2π km** or **$\frac{\pi}{5}$ km**

Page 357 Exercise 2

1 a) Circumference of full circle = $\pi d = \pi \times 4$
= 12.566... cm
Curved part of semicircle = 12.566... ÷ 2
= 6.283... cm
Perimeter = 6.283... + 4 = 10.283...
= **10.3 cm** (1 d.p.)

b) Circumference of full circle = $\pi d = \pi \times 26$
= 81.681... mm
Curved part of semicircle = 81.681... ÷ 2
= 40.840... mm
Perimeter = 40.840... + 26 = 66.840...
= **66.8 mm** (1 d.p.)

c) Circumference of full circle = $2\pi r$
= $2 \times \pi \times 2$ = 12.566... m
Curved part of semicircle = 12.566... ÷ 2
= 6.283... m
Perimeter = 6.283... + 4 = 10.283...
= **10.3 m** (1 d.p.)
The straight edge of the semicircle is the diameter of the circle, which is twice the radius (i.e. 2 × 2 = 4 m).

d) Circumference of full circle = $2\pi r$
= $2 \times \pi \times 7$ = 43.982... cm
Curved part of semicircle = 43.982... ÷ 2
= 21.991... cm
Perimeter = 21.991... + 14 = 35.991...
= **36.0 cm** (1 d.p.)

2 a) Curved side = $(\pi \times 4) \div 2$
= 6.283... cm
Perimeter = 3 + 4 + 3 + 6.283... = 16.283...
= **16.3 cm** (1 d.p.)

b) Curved side = $(2 \times \pi \times 8) \div 2$
= 25.132... mm
Perimeter = 10 + 16 + 10 + 25.132...
= 61.132... = **61.1 mm** (1 d.p.)
The vertical straight edge of the shape has the same length as the diameter of the full circle: 2 × 8 = 16 mm.

c) Curved side = $(\pi \times 5) \div 2$ = 7.853... cm
Perimeter = 4 + 4 + 7.853... = 15.853...
= **15.9 cm** (1 d.p.)

d) Curved side = $(2 \times \pi \times 9) \div 4$
= 14.137... mm
Perimeter = 9 + 9 + 14.137... = 32.137...
= **32.1 mm** (1 d.p.)
The shape is a quarter-circle, so you have to divide the circumference of the full circle by 4 to find the length of the curved side.

e) Curved side = $(2 \times \pi \times 9) \div 4$
= 14.137... mm
Perimeter = 9 + 9 + 9 + 9 + 14.137...
= 50.137... = **50.1 mm** (1 d.p.)

f) One curved side = $(\pi \times 3) \div 2$
= 4.712... cm
Perimeter = 2 + 4.712... + 2 + 4.712...
= 13.424... = **13.4 cm** (1 d.p.)

3 Circumference of full circle = $2 \times \pi \times 16$
= 100.530... m
So amount of fencing needed is
100.530... ÷ 2 = 50.265... = **50.3 m** (1 d.p.)

4 Length of one curved section
= $(2 \times \pi \times 80) \div 2$ = 251.327... m
Total length
= 251.327... + 100 + 251.327... + 100
= 702.654... = **703 m** (to the nearest metre)

22.4 Area of a Circle

Page 358 Exercise 1

1 a) $A = \pi r^2 = \pi \times 2^2$ = 12.566...
= **12.6 cm²** (1 d.p.)

b) $A = \pi r^2 = \pi \times 10^2$ = 314.159...
= **314.2 mm²** (1 d.p.)

c) d = 7 m, so r = 7 ÷ 2 = 3.5 m
$A = \pi r^2 = \pi \times 3.5^2$ = 38.484...
= **38.5 m²** (1 d.p.)

d) d = 12 mm, so r = 12 ÷ 2 = 6 mm
$A = \pi r^2 = \pi \times 6^2$ = 113.097...
= **113.1 mm²** (1 d.p.)

2 a) $A = \pi r^2 = \pi \times 7^2 = \mathbf{49\pi}$ **mm²**

b) $A = \pi r^2 = \pi \times 5^2 = \mathbf{25\pi}$ **cm²**

c) d = 18 km, so r = 18 ÷ 2 = 9 km
$A = \pi r^2 = \pi \times 9^2 = \mathbf{81\pi}$ **km²**

d) d = 3 m, so r = 3 ÷ 2 = 1.5 m
$A = \pi r^2 = \pi \times 1.5^2 = \mathbf{2.25\pi}$ **m²** or $\mathbf{\frac{9\pi}{4}}$ **m²**

3 Area of reservoir = $\pi r^2 = \pi \times 0.82^2$ = 2.112...
= **2.11 km²** (2 d.p.)

4 Area of rectangle = 20 × 22 = 440 mm²
Area of circle = $\pi \times 12^2$ = 452.389... mm²
So the **circle** has the greater area.
Remember to use the radius of the circle to find the area. The diameter of this circle is 24 mm, so the radius is 24 ÷ 2 = 12 mm.

Page 359 Exercise 2

1 a) Area of full circle = $\pi r^2 = \pi \times 7^2$
= 153.938... cm²
Area of semicircle = 153.938... ÷ 2
= 76.969... = **77.0 cm²** (1 d.p.)

b) Split the shape into a semicircle and a square.
Area of semicircle = $(\pi \times 5^2) \div 2$
= 39.269... mm²
Area of square = 10^2 = 100 mm²
Total area of shape = 39.269... + 100
= 139.269... mm² = **139.3 mm²** (1 d.p.)
The radius of the semicircle is 5 mm, so the diameter is 2 × 5 = 10 mm. This diameter is also the top side of the quadrilateral, so the width of the quadrilateral is 10 mm, making it a square.

c) Split the shape into a semicircle and a rectangle.
Area of semicircle = $(\pi \times 3^2) \div 2$
= 14.137... cm²
Area of rectangle = 2 × 6 = 12 cm²
Total area of shape = 14.137... + 12
= 26.137... = **26.1 cm²** (1 d.p.)

d) Split the shape into two identical semicircles and a rectangle.
Area of one semicircle = $(\pi \times 4^2) \div 2$
= 25.132... mm²
Length of rectangle = 20 – 4 – 4 = 12 mm
Area of rectangle = 12 × 8 = 96 mm²
Total area of shape = (2 × 25.132...) + 96
= 146.265... = **146.3 mm²** (1 d.p.)
You might have noticed that if you put the two semicircles together, then you get the full circle with radius 4 mm. So you could have just used the area of the circle and the rectangle, instead of using the area of two semicircles.

e) Split the shape into a quarter-circle and a square.
Area of quarter-circle = $(\pi \times 10^2) \div 4$
= 78.539... mm²
Area of square = 10^2 = 100 mm²
Total area of shape = 78.539... + 100
= 178.539... = **178.5 mm²** (1 d.p.)
The area of a quarter-circle is one quarter of the area of the full circle, so you divide by 4 to find it.

f) Split the shape into two quarter-circles and a square.
Area of one quarter-circle = $(\pi \times 2^2) \div 4$
= 3.141... cm²
Area of square = 2^2 = 4 cm²
Total area of shape = (2 × 3.141...) + 4
= 10.283... = **10.3 cm²** (1 d.p.)

2 Label the diagram with the length and width of the rectangle, and the radius of the semicircle.

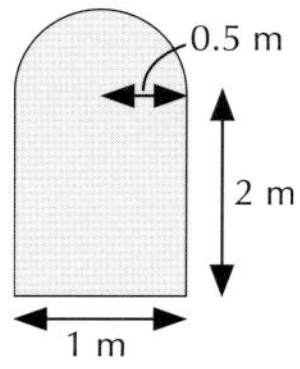

Area of semicircle = $(\pi \times 0.5^2) \div 2$
= 0.392... m²
Area of rectangle = 2 × 1 = 2 m²
Total area of church window
= 0.392... + 2
= 2.392... = **2.39 m²** (2 d.p.)

3 The pond is a circle of radius 2 m. Including the path creates a circle with radius 3 m. To find the area of the path, subtract the area of the pond from the total area.
Area including path = $\pi \times 3^2$ = 28.274... m²
Area of pond = $\pi \times 2^2$ = 12.566... m²
So area of path = 28.274... – 12.566...
= 15.707... = **16 m²** (to the nearest m²)

22.5 Arcs and Sectors of Circles

Page 360 Exercise 1

1 a) Arc length = $\frac{\theta}{360°} \times 2\pi r$
= $\frac{15°}{360°} \times 2 \times \pi \times 3 = \frac{1}{24} \times 6\pi = \mathbf{\frac{\pi}{4}}$ **cm**
Sector area = $\frac{\theta}{360°} \times \pi r^2$
= $\frac{15°}{360°} \times \pi \times 3^2 = \frac{1}{24} \times \pi \times 9 = \mathbf{\frac{3\pi}{8}}$ **cm²**

b) Arc length = $\frac{100°}{360°} \times 2 \times \pi \times 4 = \mathbf{\frac{20\pi}{9}}$ **cm**
Sector area = $\frac{100°}{360°} \times \pi \times 4^2 = \mathbf{\frac{40\pi}{9}}$ **cm²**

c) Arc length = $\frac{45°}{360°} \times 2 \times \pi \times 10 = \mathbf{\frac{5\pi}{2}}$ **cm**
Sector area = $\frac{45°}{360°} \times \pi \times 10^2 = \mathbf{\frac{25\pi}{2}}$ **cm²**

d) Diameter = 10 cm, so radius = 10 ÷ 2
= 5 cm
Arc length = $\frac{120°}{360°} \times 2 \times \pi \times 5 = \mathbf{\frac{10\pi}{3}}$ **cm**
Sector area = $\frac{120°}{360°} \times \pi \times 5^2 = \mathbf{\frac{25\pi}{3}}$ **cm²**

2 a) Sector area = $\frac{311°}{360°} \times \pi \times 5.5^2$
= 82.098... cm²
= **82.10 cm²** (2 d.p.)
Arc length = $\frac{311°}{360°} \times 2 \times \pi \times 5.5$
= 29.853... = **29.85 cm** (2 d.p.)

b) Sector area = $\frac{206°}{360°} \times \pi \times 10.5^2$
= 198.195... = **198.20 in²** (2 d.p.)
Arc length = $\frac{206°}{360°} \times 2 \times \pi \times 10.5$
= 37.751... = **37.75 in** (2 d.p.)

c) $39° < 180°$, so 39° is the angle of the minor sector. The angle of the major sector is $360° - 39° = 321°$.
Sector area $= \frac{321°}{360°} \times \pi \times 12^2$
$= 403.380... =$ **403.38 in²** (2 d.p.)
Arc length $= \frac{321°}{360°} \times 2 \times \pi \times 12$
$= 67.230... =$ **67.23 in** (2 d.p.)

d) The angle of the major sector is $360° - 172° = 188°$.
Sector area $= \frac{188°}{360°} \times \pi \times 6.5^2$
$= 69.315... =$ **69.32 m²** (2 d.p.)
Arc length $= \frac{188°}{360°} \times 2 \times \pi \times 6.5$
$= 21.327... =$ **21.33 m** (2 d.p.)

Page 361 Review Exercise

1 a) (i) $P = (2 \times 1.4) + (2 \times 3.1) =$ **9 cm**
(ii) $A = 1.4 \times 3.1 =$ **4.34 cm²**
b) (i) $P = (2 \times 1.1) + (2 \times 3.1) =$ **8.4 mm**
(ii) $A = 1.1 \times 3.1 =$ **3.41 mm²**
c) (i) $P = 4 \times 1.8 =$ **7.2 cm**
(ii) $A = 1.8^2 =$ **3.24 cm²**

2 a) (i) $P = 7 + 8 + 9 =$ **24 cm**
(ii) $A = \frac{1}{2} \times 8 \times 6.7 =$ **26.8 cm²**
b) (i) $P = 4 + 13 + 13 =$ **30 mm**
The tick marks on the triangle show that two of the sides are equal — so the triangle is isosceles (see p.242).
(ii) $A = \frac{1}{2} \times 4 \times 12.8 =$ **25.6 mm²**
c) (i) $P = 8.9 + 15.1 + 18.8 =$ **42.8 mm**
(ii) $A = \frac{1}{2} \times 18.8 \times 7 =$ **65.8 mm²**

3 a) $A = \frac{1}{2}(3 + 8) \times 4 =$ **22 cm²**
b) $A = 21 \times 10 =$ **210 m²**
c) $A = \frac{1}{2}(20 + 55) \times 25 =$ **937.5 mm²**

4 a) Split the shape into two trapeziums.
$A_1 = \frac{1}{2}(1 + 4) \times 1.5 = 3.75$ m²
$A_2 = \frac{1}{2}(4 + 6) \times 1.5 = 7.5$ m²
So total shaded area $= 3.75 + 7.5 = 11.25$
$=$ **11.3 m²** (1 d.p.)
b) Split the shape into a trapezium and a semicircle.
Area of trapezium $= \frac{1}{2}(20 + 40) \times 50$
$= 1500$ mm²
Area of semicircle $= (\pi \times 20^2) \div 2$
$= 628.318...$ mm²
So total shaded area $= 1500 + 628.318...$
$= 2128.318... =$
2128.3 mm² (1 d.p.)
c) Area of one parallelogram $= 9 \times 6$
$= 54$ mm²
So total area of shaded region $= 4 \times 54$
$=$ **216.0 mm²**

5 a) Circumference $= 2 \times \pi \times 1.3 = 8.168...$
$=$ **8.17 cm** (2 d.p.)
b) Area of 2p coin $= \pi \times 1.3^2 = 5.309...$ cm²
Radius of 5p coin $= 1.8 \div 2 = 0.9$ cm
Area of 5p coin $= \pi \times 0.9^2 = 2.544...$ cm²
So area of two 5p coins $= 2 \times 2.544...$
$= 5.089...$ cm²
So the **2p coin** has the greater area.

6 The angle of the major sector is 320°, so the angle of the minor sector is $360° - 320°$ $= 40°$. The radius is $10 \div 2 = 5$ m.
Arc length $= \frac{40°}{360°} \times 2 \times \pi \times 5 = \frac{1}{9} \times 10 \times \pi$
$=$ $\mathbf{\frac{10\pi}{9}}$ **m**
Sector area $= \frac{40°}{360°} \times \pi \times 5^2 = \frac{1}{9} \times \pi \times 25$
$=$ $\mathbf{\frac{25\pi}{9}}$ **m²**

Page 362 Exam-Style Questions

1 Circumference $= 2\pi r = \pi^2$ *[1 mark]*
So $r = \frac{\pi^2}{2\pi} =$ $\mathbf{\frac{\pi}{2}}$ **cm** *[1 mark]*

2 Let the top and bottom pieces of the frame have area A_1 and the two sides have area A_2:
$A_1 = \frac{1}{2}(14 + 22) \times 2 = 36$ cm²
$A_2 = \frac{1}{2}(8 + 16) \times 2 = 24$ cm²
[1 mark for a correct method to find the area of one of the trapeziums]
So total area $= (2 \times 36) + (2 \times 24)$ *[1 mark]*
$=$ **120 cm²** *[1 mark]*

3 Diameter of a circle $= 12 \div 2 = 6$ cm
Radius of a circle $= 6 \div 2 = 3$ cm
Area of one circle $= \pi \times 3^2 = 28.274...$ cm² *[1 mark]*
Area of all four circles $= 4 \times 28.274...$
$= 113.097...$ cm²
Area of square $= 12^2 = 144$ cm² *[1 mark]*
So area of unshaded region $= 144 - 113.097...$ *[1 mark]* $= 30.902... =$ **31 cm²** (2 s.f.) *[1 mark]*

4 Perimeter $= 16 + 16 +$ Arc length $= 41$ cm
$\Rightarrow$ Arc length $= 41 - 16 - 16 = 9$ cm
Arc length $= \frac{\theta}{360°} \times 2 \times \pi \times 16 = 9$
$\Rightarrow \frac{\theta}{360°} \times 32\pi = 9 \Rightarrow \frac{\theta}{360°} = 9 \div 32\pi$
$= 0.0895...$
So $\theta = 0.0895... \times 360° = 32.228... =$ **32°** (to the nearest degree)
[4 marks available — 1 mark for the correct arc length, 1 mark for using the formula for the arc length, 1 mark for attempting to rearrange to find θ, 1 mark for the correct final answer]

Section 23 — 3D Shapes

23.1 Properties of 3D Shapes

Page 364 Exercise 1

1 The faces are the flat surfaces, the vertices are the corners and the edges are the places where the faces meet.
a) (i) **Cuboid**
(ii) Faces:
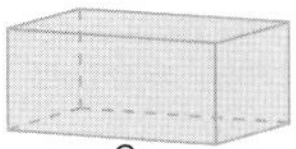
Vertices:
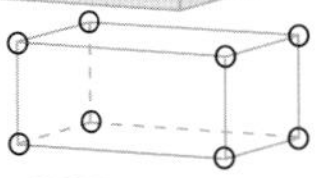
Edges:
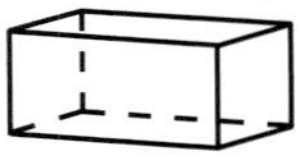
So there are **6 faces, 8 vertices, 12 edges**.

b) (i) **Triangular prism**
(ii) Faces:
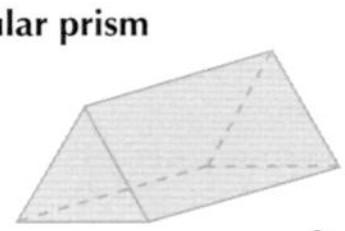
Vertices:
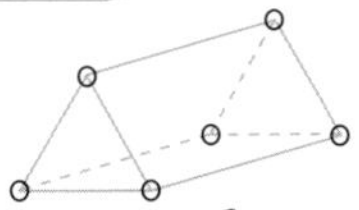
Edges:

So there are **5 faces, 6 vertices, 9 edges**.
Using the same method for c)-d):
c) (i) **Tetrahedron (or triangular based pyramid)**
(ii) **4 faces, 4 vertices, 6 edges**
d) (i) **Square-based pyramid**
(ii) **5 faces, 5 vertices, 8 edges**

2 a) **Cube** b) **Triangular prism**
c) **Tetrahedron (or triangular-based pyramid)**
d) **Square-based pyramid** e) **Cylinder**

3 Prisms have a constant cross-section. Imagine slicing each of the shapes along its length — if it's a prism, the new face will be the same as the face parallel to your slice: ***A*, *B*** and ***C*** are prisms.
A cube is a special type of prism — the cross-sections along its length, width and height are all the same.

Page 365 Exercise 2

1 a)
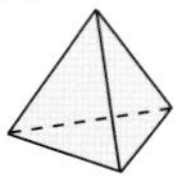
(i) **0 squares** (ii) **4 triangles**
b)
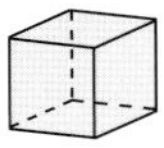
(i) **6 squares** (ii) **0 triangles**
c)
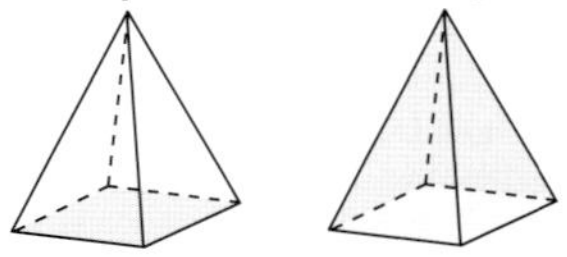
(i) **1 square** (ii) **4 triangles**

2 The 3D shape will have the same number of faces as the number of shapes that make up the net. There are 8 triangles in the net, so there will be **8 faces**.

3 Imagine folding up the nets to make them into solids.
a) **Square-based Pyramid**
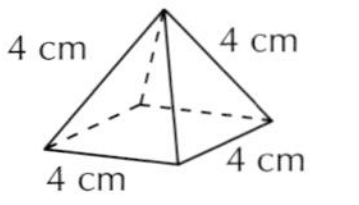

b) **Triangular Prism**
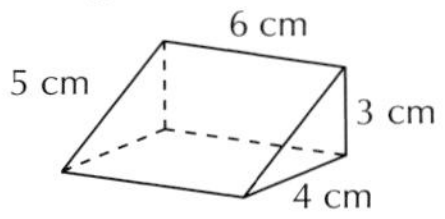

c) **Cuboid**
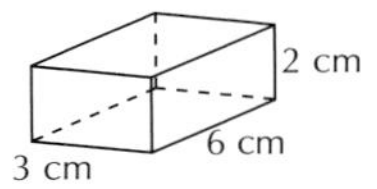

4 The base and top rectangles are 3 cm × 4 cm. The two smaller vertical rectangles are 2 cm × 3 cm.
The two larger vertical rectangles are 2 cm × 4 cm. This gives:

a) 2 **b)** 2 **c)** 2 **d)** 0

5 The base of the prism is 4 cm × 6 cm. The vertical rectangle at the back is 3 cm × 6 cm.
The sloping rectangle is 5 cm × 6 cm. This gives:

a) 1 **b)** 1 **c)** 1 **d)** 0

6 **Net *C*** because it's the only net where the edges match correctly when it's folded.

Page 367 Exercise 3

1 **a)** You need to add two 5 cm × 3 cm rectangles.
E.g.

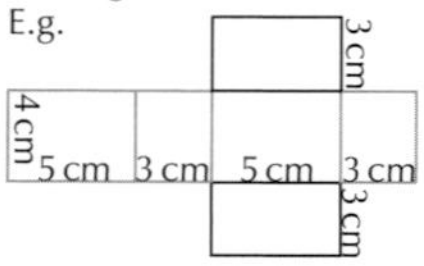

There's more than one possible net for any 3D shape, so your answers to this exercise might look different from the diagrams given. If your net could be folded up to make the given 3D shape then it's correct.

b) You need to add two 1 cm × 3 cm rectangles and one triangle with sides of length 1 cm.
E.g.

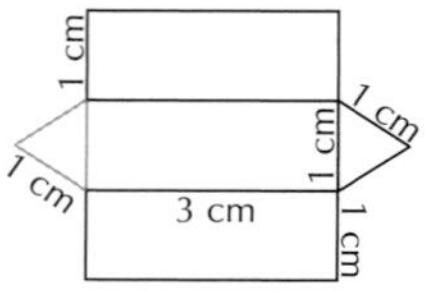

2 **a)** There should be six faces, each with 1 cm sides.
E.g.

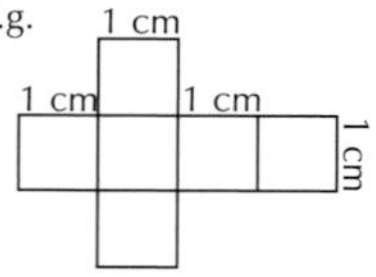

b) There should be two 3 cm × 1 cm rectangles, two 1 cm × 1.5 cm rectangles and two 3 cm × 1.5 cm rectangles. E.g.

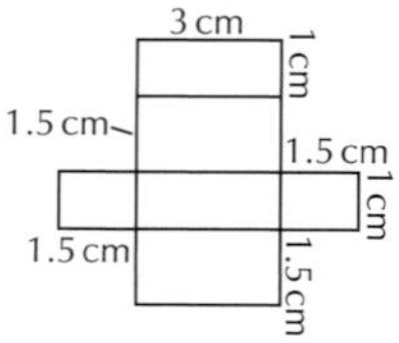

c) There should be two triangles and three rectangles with dimensions as shown. E.g.

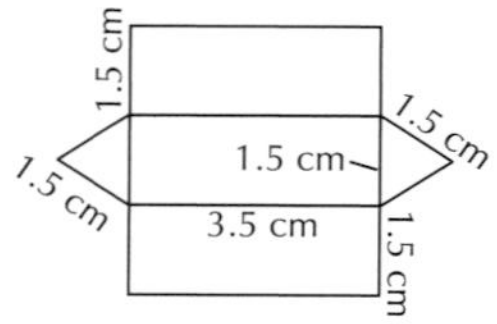

d) There should be three rectangles and two triangles, with dimensions as shown. E.g.

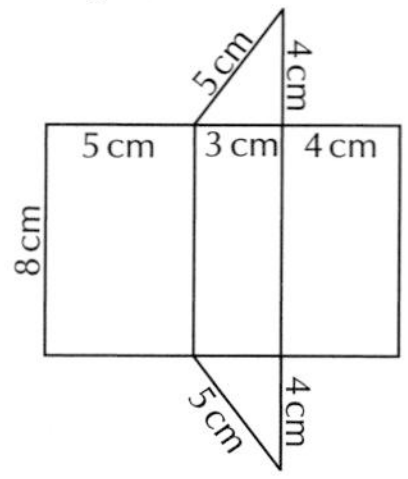

e) There should be a 2 cm × 2 cm square and four triangles with dimensions as shown.
E.g.

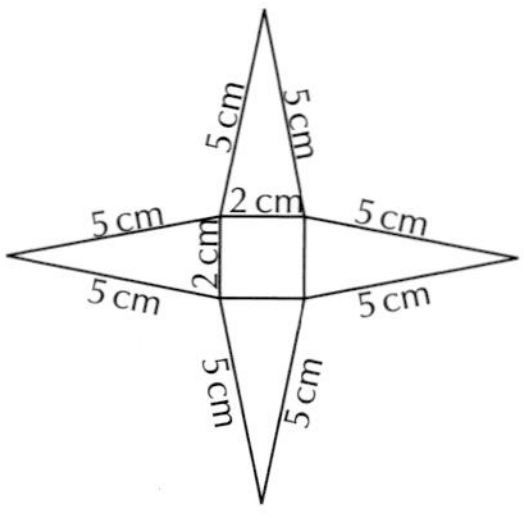

f) There should be four triangles with sides of 5 cm. E.g.

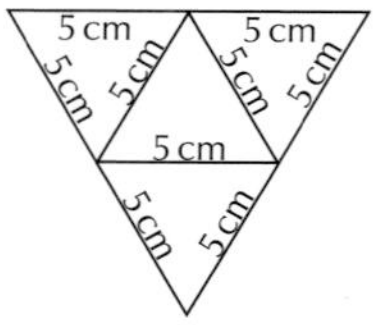

g) There should be one rectangular face of width 7 cm and length $2\pi r = 2 \times \pi \times 2$ = 12.57 cm (2 d.p.), and two circular faces with radius 2 cm attached to either side of the rectangle as shown. E.g.

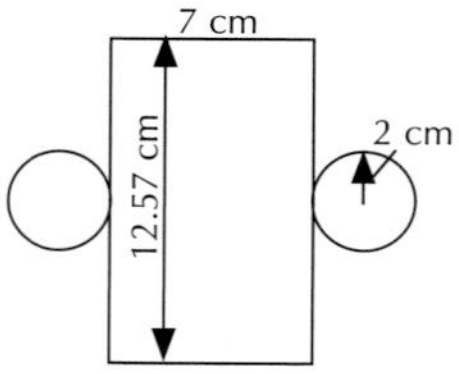

h) There should be one rectangular face of width 3 cm and length $\pi d = \pi \times 6$ = 18.85 cm (2 d.p.), and two circular faces with diameter 6 cm attached to either side of the rectangle as shown. E.g.

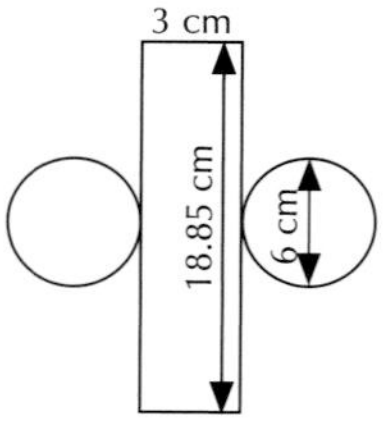

Using the same methods for Q3:

3 **a)** E.g.

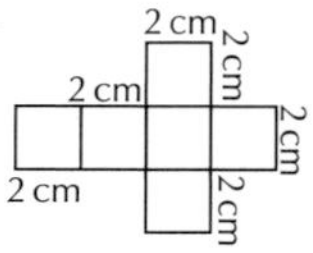

b) E.g.

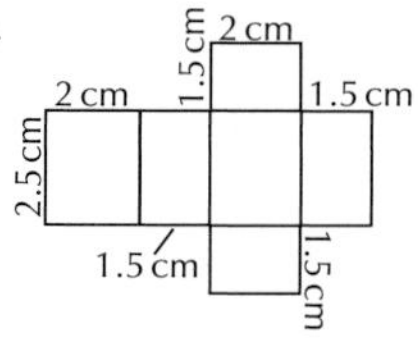

c) E.g.

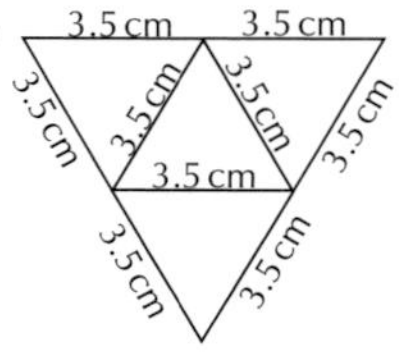

d) E.g.

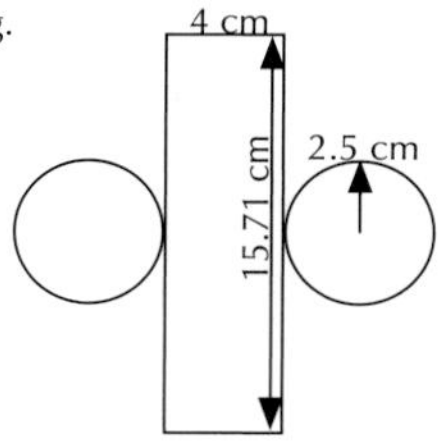

23.2 Plans and Elevations

Page 369 Exercise 1

1 The plan is the view of the shape from above.

a) From above, the shape looks like a 2 × 2 square:

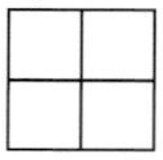

To picture the plan for a 3D shape, imagine you're hovering directly above the shape, looking down on it.

b) From above, the shape looks like an upside down L-shape 3 high and 2 across:

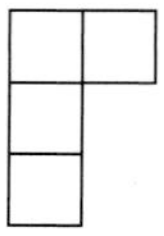

2 The side elevation is the view of the shape from the direction indicated 'Side'.

a) From the side, the shape looks like a 2 × 1 rectangle:

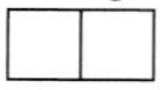

b) From the side, the shape looks like a backwards L-shape 2 high and 2 across:

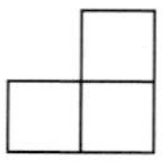

3 a) (i) The plan view is the view from above, so only one square is seen:

(ii) The front elevation and side elevation both show three squares on top of each other:

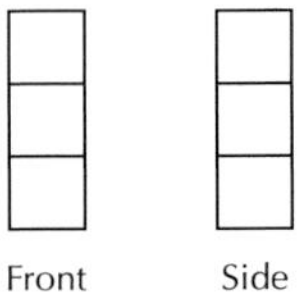

b) (i) Looking vertically from above, you can see an upside down L-shape:

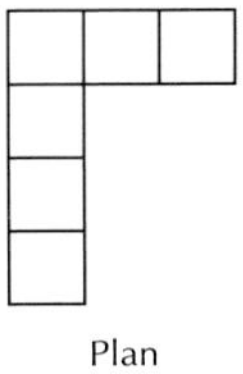

(ii) The views from the front and the side are also L-shapes:

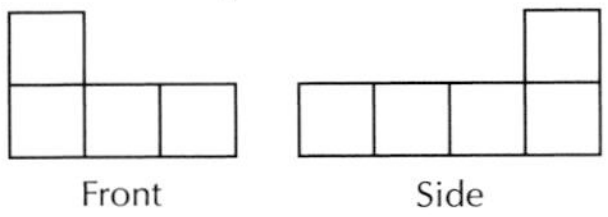

Check the "front" and "side" directions on the 3D diagram, to make sure you're labelling your elevations correctly.

Using the same method for c)-d):

c) (i)

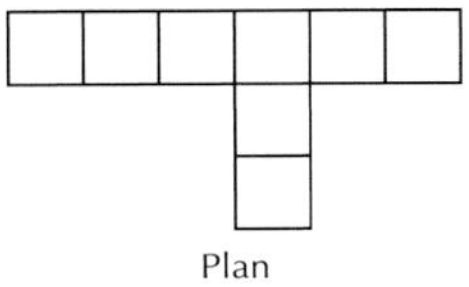

(ii)

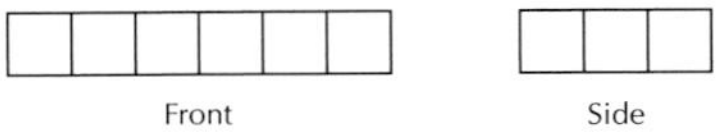

d) (i)

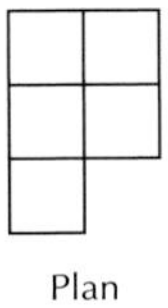

(ii)

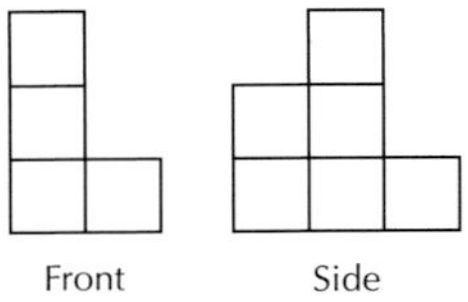

4 ***B*** — the side elevation is a sideways T-shape, which only matches shape *B*. The plan view matches *B* because the vertical line down the centre indicates the step on the top of the shape. The front elevation matches *B* because the two horizontal lines indicate where the piece sticks out at the front of the shape.

Page 370 Exercise 2

1 a) As the shape is a cube, all elevations will be the same:

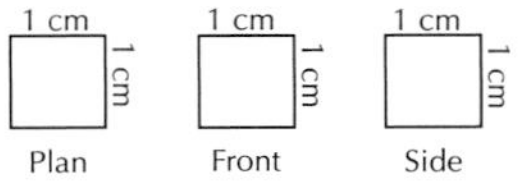

b) The plan and front elevation will be rectangles. The side elevation will be a square:

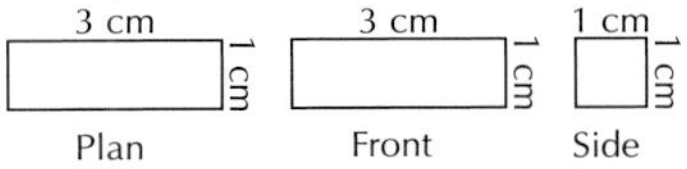

c) The plan will be a rectangle divided down the middle, the front elevation will be a triangle and the side elevation will be a rectangle.

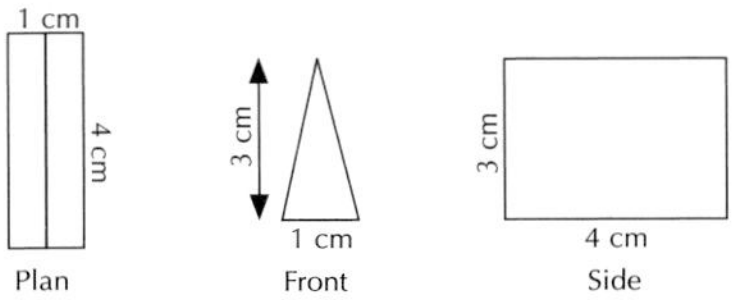

Take care with sloped surfaces — make sure you have chosen the correct dimensions from the diagram. The line down the centre of the plan indicates where the two sloping sides of the 3D shape meet.

d) The plan will be a square split into four triangles. The front and side elevations will be triangles.

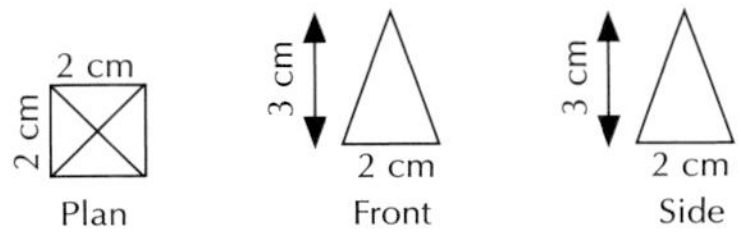

The diagonal lines on the plan indicate where the four sloping sides of the pyramid meet.

e) The plan will be a circle, radius 2 cm. The front and side elevations will be rectangles with width equal to the diameter of the circular face.

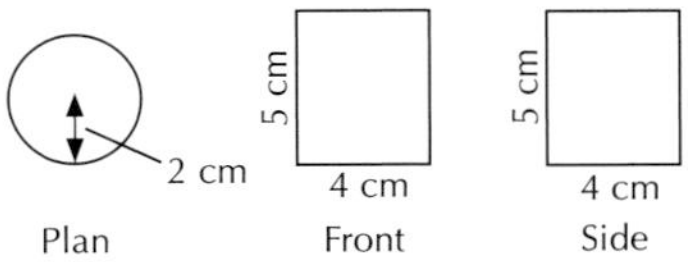

If you ignore the curvature of the cylinder, from the front or side it looks like a rectangle whose width is the diameter of the cylinder's base (2 × 2 cm = 4 cm) — that gives you the front and side elevations.

f) The plan will be a circle, radius 3 cm. The front and side elevations will be triangles.

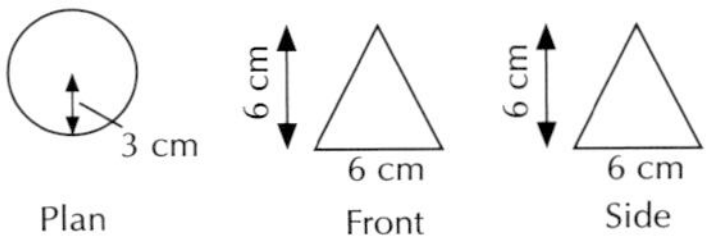

This time when you ignore the curvature, you get a triangle who's base length is the diameter of the base of the cone.

23.3 Isometric Drawings

Page 372 Exercise 1

1 Each space between the dots in the vertical and diagonal directions is equal to 1 cm.

a)

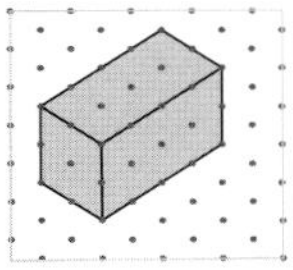

b)

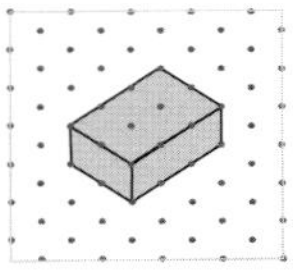

c)

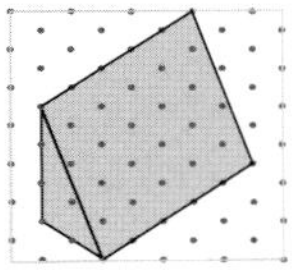

d)

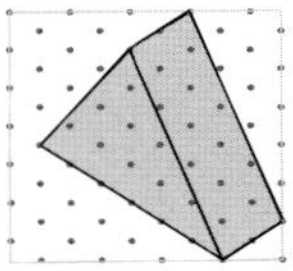

Draw the 3D shape's horizontal and vertical edges first, and then connect them up to form the sloped edges.

2 Each space between the dots in the vertical and diagonal directions is equal to 1 cm.

a)

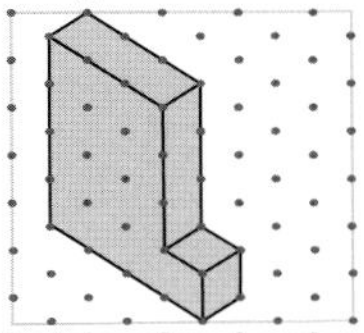

b)

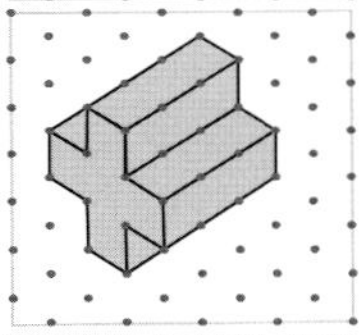

c)

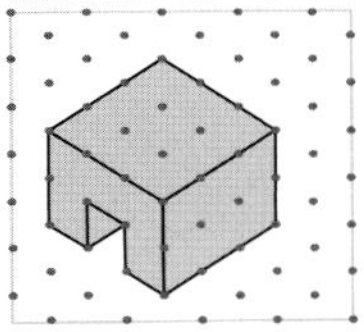

Page 373 Exercise 2

1 Draw the cross-sections given, then extend them by 3 dots diagonally to create the prisms.

a)

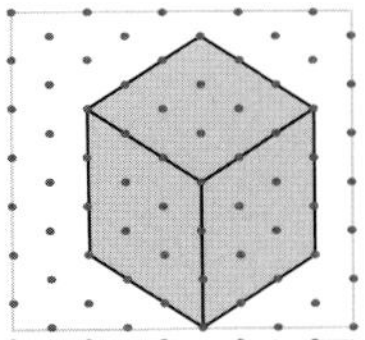

b)

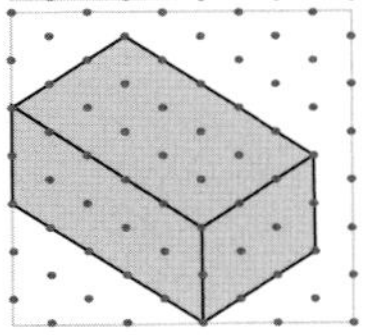

c)

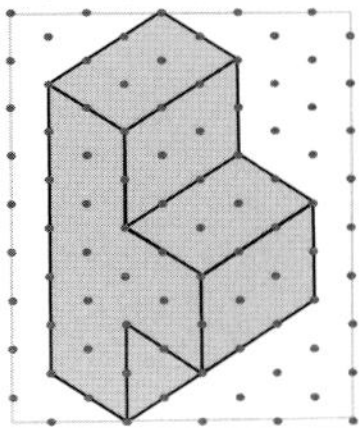

2 Draw the cross-sections given, then extend them by 2 dots diagonally to create the prisms.

a)
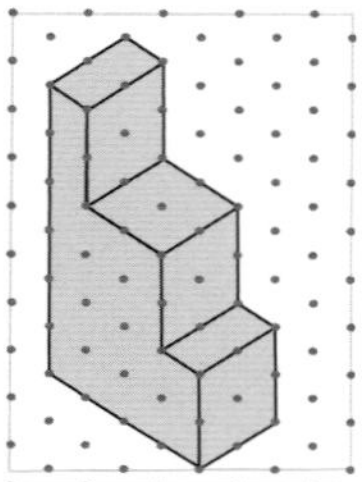

b)
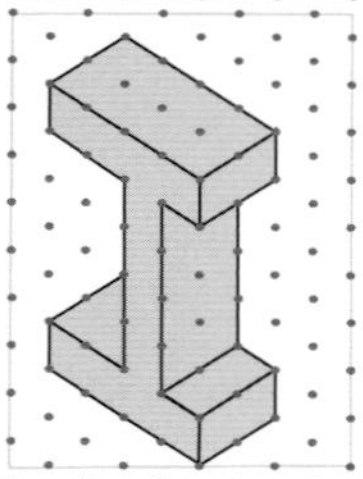

c)
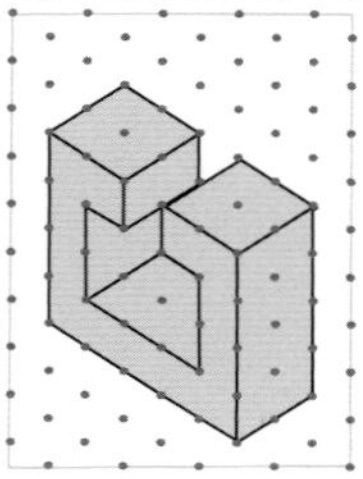

3 a) (i)
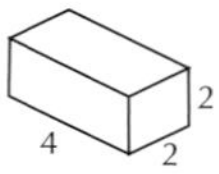

(ii)
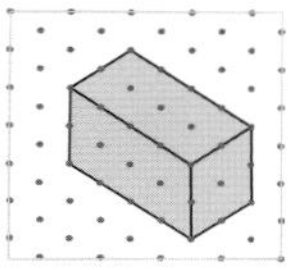

b) (i)
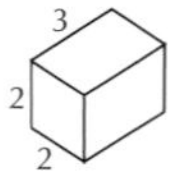

(ii)
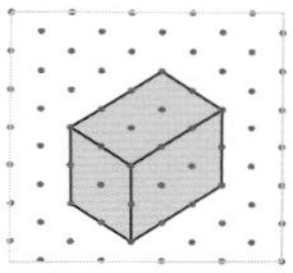

c) (i)
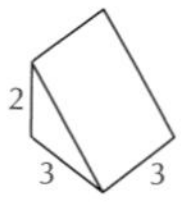

(ii)
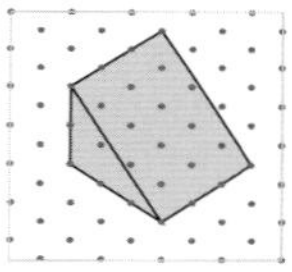

d) (i)
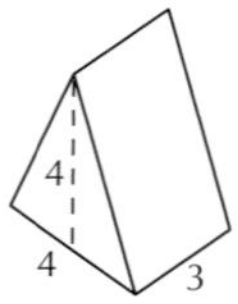

(ii)
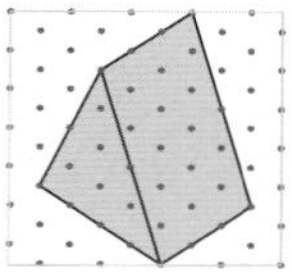

e) (i)
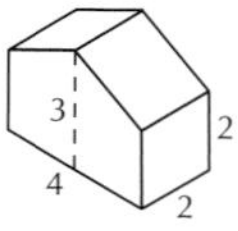

(ii)
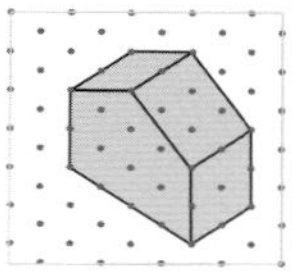

f) (i)
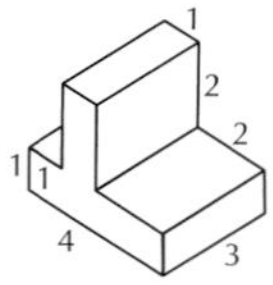

(ii)
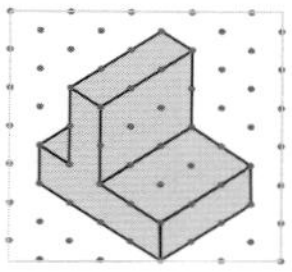

23.4 Symmetry of 3D Shapes

Page 374 Exercise 1

1 a) Any two from:

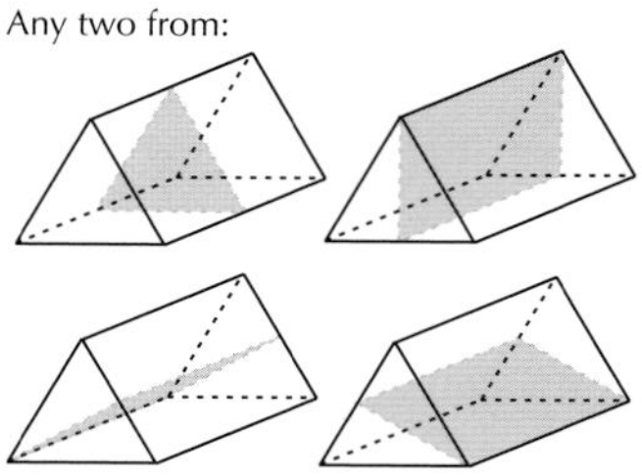

Remember, the number of planes of symmetry in a prism is usually one more than the number of lines of symmetry in the cross-section.

b) Any two from:

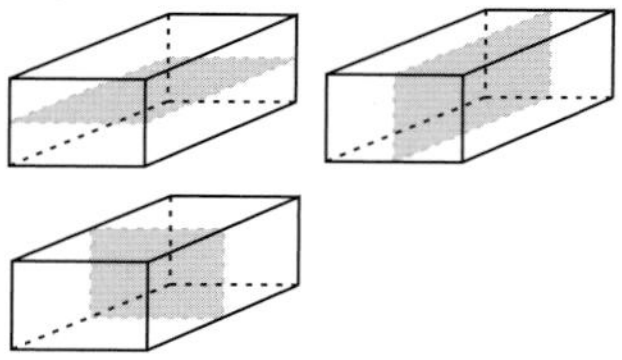

c) Any two from:

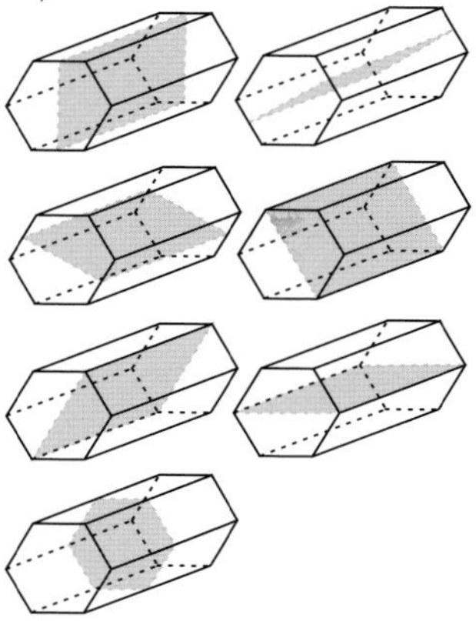

2 The planes of symmetry correspond to the lines of symmetry of the faces of the cube:

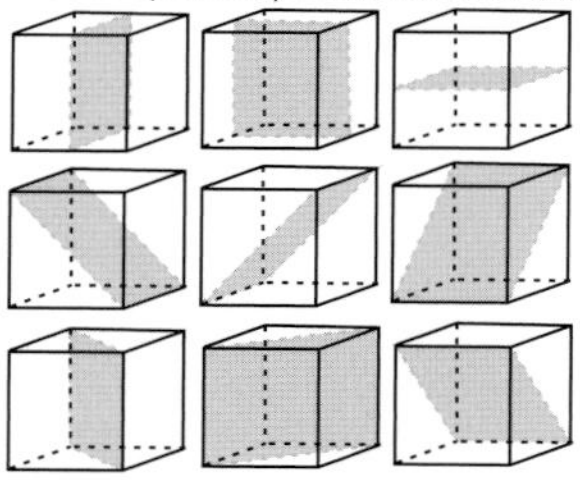

3 The base of the pyramid has 4 lines of symmetry, which correspond to **4 planes of symmetry** in the 3D shape.

23.5 Volume

Page 375 Exercise 1

1 Each cube has a length of 1 cm, so a volume of $1 \times 1 \times 1 = 1$ cm³. To find the volume of the solids, count the number of cubes.

a) **5 cm³** b) **4 cm³**

c) **8 cm³** d) **7 cm³**

Don't forget the hidden cubes you can't see.

2 a) Volume = length × width × height
= 6 × 5 × 1 = **30 cm³**

b) Volume = length × width × height
= 5 × 2 × 1 = **10 m³**

Using the same method for c)-f):

c) **28 cm³** d) **84 mm³**

e) **27 cm³** f) **64 m³**

3 a) Volume = 3 × 2 × 4 = **24 cm³**

b) Volume = 5 × 2 × 7 = **70 m³**

Using the same method for c)-f):

c) **1600 cm³** d) **31.5 cm³**

e) **756 m³** f) **35.136 mm³**

4 a) Length = 3 mm to 1 s.f.
So volume ≈ 3 × 3 × 3 = 27 mm³

b) The lengths were rounded down from 3.2, so this is an underestimate.
Actual volume = 3.2³ = 32.768 mm³

5 Volume of box = 1.7 × 1.8 × 0.9 = 2.754 m³
2.754 m³ < 3.5 m³, so **the sand will not fit.**

6 The dimensions of the box are 9 cm, 20 cm and 30 cm to 1 s.f.
Volume of box ≈ 9 × 20 × 30 = 5400 cm³
Box is half full, so volume of cereal ≈ 5400 ÷ 2 = **2700 cm³**.

7 a) Split the solid into two cuboids, e.g.

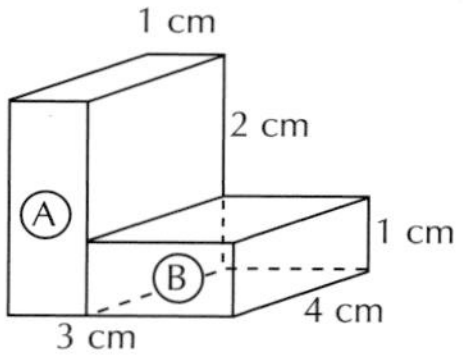

Volume of A = 4 × 3 × 1 = 12 cm³
Volume of B = 4 × 2 × 1 = 8 cm³
So volume of the solid = 12 + 8 = **20 cm³**

b)
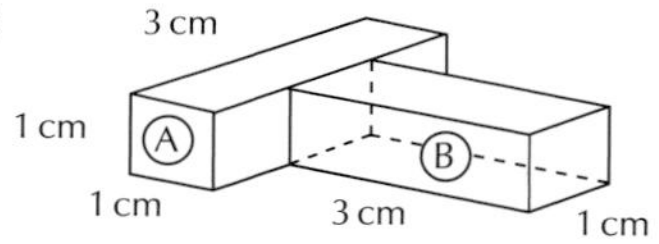

Volume of A = 3 × 1 × 1 = 3 cm³
Volume of B = 1 × 3 × 1 = 3 cm³
So volume of the solid = 3 + 3 = **6 cm³**
Always try to split the shapes into as few pieces as possible, so you don't have loads of calculations to do.

8 Volume = 5 cm × 3 cm × height
= 15 × height = 18 cm³

So height = $\frac{18}{15} = \frac{6}{5} = 1\frac{1}{5}$ = **1.2 cm**

9 a) Volume of bath = 1.5 × 0.5 × 0.6 = **0.45 m³**

b) 1.5 × 0.5 × 0.3 = **0.225 m³**

c) Volume = 1.5 × 0.5 × height
= 0.75 × height = 0.3 m³
So height = 0.3 ÷ 0.75 = **0.4 m**

Page 378 Exercise 2

1 a) Volume = cross-sectional area × length
$= 2 \times 3 =$ **6 cm³**

b) Volume = cross-sectional area × length
$= 6 \times 9 =$ **54 cm³**

c) Volume = cross-sectional area × length
$= 1.5 \times 6 =$ **9 m³**

Using the same method for d)-f):

d) 12.25 cm³ **e) 16.1875 mm³**

f) 105.56 mm³

2 a) Cross-sectional area $= \frac{1}{2} \times 2 \times 4 = 4$ cm²
So volume = 4 × 7 = **28 cm³**

b) Cross-sectional area $= \frac{1}{2} \times 4 \times 3 = 6$ cm²
So volume = 6 × 5 = **30 cm³**

c) Cross-sectional area $= \pi r^2 = \pi \times 1^2$
= 3.141... cm²
So volume = 3.141... × 5
= 15.707... = **15.71 cm³** (2 d.p.)

3 a) Split the cross-section into two rectangles to find its area.

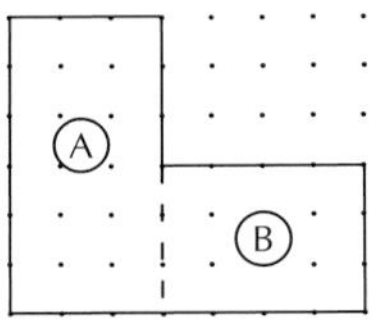

Area of A = 3 × 6 = 18 cm²
Area of B = 4 × 3 = 12 cm²
So area of cross-section = 18 + 12
= 30 cm²
Volume = 30 × 2 = **60 cm³**

b) Split the cross-section into three rectangles to find its area.

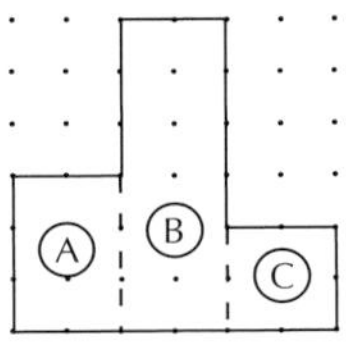

Area of A = 2 × 3 = 6 cm²
Area of B = 2 × 6 = 12 cm²
Area of C = 2 × 2 = 4 cm²
So area of cross-section = 6 + 12 + 4
= 22 cm²
Volume = 22 × 3 = **66 cm³**

c) Diameter = 6 cm, so radius = 6 ÷ 2 = 3 cm
Cross-sectional area $= \pi r^2 = \pi \times 3^2$
= 28.274... cm²
Volume = 28.274... × 1.2
= **33.93 cm³** (2 d.p.)

4 a) Area of cross-section $= \frac{1}{2} \times 13 \times 12$
= 78 cm²
Volume = 78 × 8 = **624 cm³**

b) Area of cross-section $= \frac{1}{2} \times 4.2 \times 1.3$
= 2.73 m²
Volume = 2.73 × 3.1 = **8.463 m³**

c) Area of cross-section $= \pi \times 4^2$
= 50.265... m²
Volume = 50.265... × 18
= **904.78 m³** (2 d.p.)

d) Area of cross-section = 3 × 4.2 = 12.6 m²
Volume = 12.6 × 1.5 = **18.9 m³**

5 a) The cross-section can be split into a rectangle and a triangle.

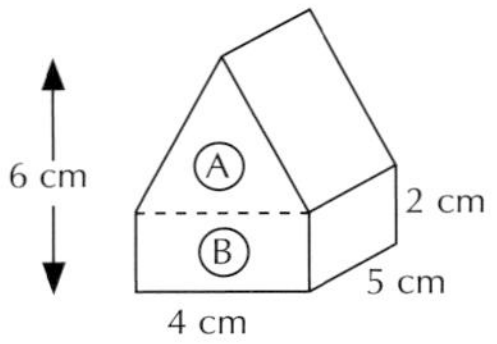

Area of A $= \frac{1}{2} \times 4 \times 4 = 8$ cm²
Area of B = 4 × 2 = 8 cm²
So area of cross-section = 8 + 8 = 16 cm²
Volume = 16 × 5 = **80 cm³**

b) The cross-section is a trapezium.

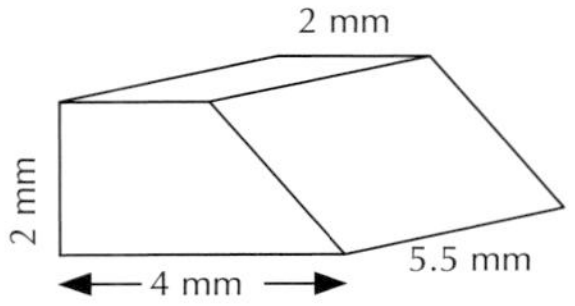

Area of cross-section $= \frac{1}{2} \times (2 + 4) \times 2$
$= \frac{1}{2} \times 6 \times 2 = 6$ mm²
Volume = 6 × 5.5 = **33 mm³**
You could also work out the cross-sectional area by splitting the shape into a triangle and a square.

c) The cross-section can be split into two squares and a rectangle.

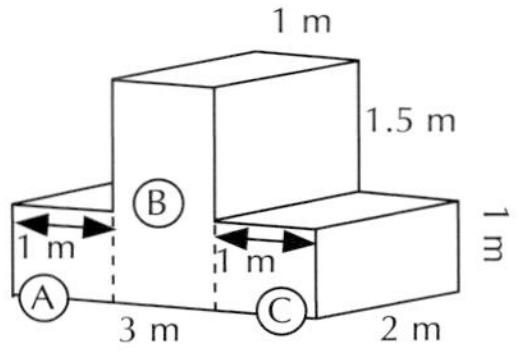

Area of A = 1² = 1 m²
Area of B = 1 × 2.5 = 2.5 m²
Area of C = 1² = 1 m²
So area of cross-section = 1 + 2.5 + 1
= 4.5 m²
Volume = 4.5 × 2 = **9 m³**
A and C are identical so have the same area — always look out for symmetry in your shapes to reduce the number of calculations you need to do.

6 a) Area of cross-section $= \frac{1}{2} \times 4 \times 5 = 10$ cm²
Volume = 10 × 3 = **30 cm³**

b) Area of cross-section = 2 × 2² + (6 × 2)
= 8 + 12 = 20 cm²
Volume = 20 × 3 = **60 cm³**

c) Area of cross-section
$= (4 \times 3) + (\frac{1}{2} \times 4 \times 2)$
= 12 + 4 = 16 cm²
Volume = 16 × 5 = **80 cm³**

d) Area of cross-section
$= (4 \times 2) + (2 \times 3) + (\frac{1}{2} \times 2 \times 3)$
= 8 + 6 + 3 = 17 cm²
Volume = 17 × 3 = **51 cm³**

Page 379 Exercise 3

1 a) Volume $= \frac{4}{3}\pi r^3 = \frac{4}{3}\pi \times 3^3 = 4\pi \times 9$
$=$ **36π cm³**
'Exact' means leave your answer in terms of π, and use fractions rather than decimals where needed.

b) Volume $= \frac{4}{3}\pi r^3 = \frac{4}{3}\pi \times 2^3$
$= \frac{4}{3}\pi \times 8 =$ $\mathbf{\frac{32}{3}\pi}$ **m³** or $\mathbf{10\frac{2}{3}\pi}$ **m³**

c) Volume $= \frac{4}{3}\pi r^3 = \frac{4}{3}\pi \times 5^3 = \frac{4}{3}\pi \times 125$
$=$ $\mathbf{\frac{500}{3}\pi}$ **mm³** or $\mathbf{166\frac{2}{3}\pi}$ **mm³**

2 a) Volume $= \frac{4}{3} \times \pi \times 4^3 = 268.082...$
= **268.08 cm³** (2 d.p.)

b) Volume $= \frac{4}{3} \times \pi \times 8^3 = 2144.660...$
= **2144.66 m³** (2 d.p.)

c) Volume $= \frac{4}{3} \times \pi \times 9.6^3 = 3705.973...$
= **3705.97 mm³** (2 d.p.)

d) Volume $= \frac{4}{3} \times \pi \times 15.7^3 = 16\,210.169...$
= **16 210.17 m³** (2 d.p.)

3 Volume $= \frac{4}{3}\pi r^3 = 24\,429$
So $r^3 = 24\,429 \div \frac{4}{3}\pi = 5831.9941...$
$\Rightarrow r = \sqrt[3]{5831.9941...} = 17.999...$
= **18.0 cm** (1 d.p.)

Page 380 Exercise 4

1 a) Volume $= \frac{1}{3}\pi r^2 h = \frac{1}{3}\pi \times 5^2 \times 12$
$= \pi \times 25 \times 4 =$ **100π m³**

b) Volume $= \frac{1}{3}\pi r^2 h = \frac{1}{3}\pi \times 7^2 \times 24$
$= \pi \times 49 \times 8 =$ **392π cm³**

c) Volume $= \frac{1}{3}\pi r^2 h = \frac{1}{3}\pi \times 15^2 \times 8$
$= \frac{1}{3}\pi \times 225 \times 8 = \pi \times 75 \times 8 =$ **600π m³**

d) Volume $= \frac{1}{3}\pi r^2 h = \frac{1}{3}\pi \times 30^2 \times 5.5$
$= \frac{1}{3}\pi \times 900 \times 5.5$
$= \pi \times 300 \times 5.5 =$ **1650π mm³**

2 a) Volume $= \frac{1}{3} \times \pi \times 4^2 \times 10$
= 167.551... = **167.55 cm³** (2 d.p.)

b) Volume $= \frac{1}{3} \times \pi \times 14^2 \times 32$
= 6568.023... = **6568.02 mm³** (2 d.p.)

c) Volume $= \frac{1}{3} \times \pi \times 7^2 \times 25$
= 1282.817... = **1282.82 m³** (2 d.p.)

d) Volume $= \frac{1}{3} \times \pi \times 3.6^2 \times 7.2$
= 97.716... = **97.72 cm³** (2 d.p.)

Page 381 Exercise 5

1 a) Volume of original cone
$= \frac{1}{3}\pi \times 12^2 \times 16 =$ **768π cm³**

b) Volume of removed cone $= \frac{1}{3}\pi \times 3^2 \times 4$
= **12π cm³**

c) Volume of frustum = 768π – 12π
= **756π cm³**

2 Volume of original cone $= \frac{1}{3}\pi \times 10^2 \times 24$
= 2513.274... cm³
Volume of removed cone $= \frac{1}{3}\pi \times 5^2 \times 12$
= 314.159... cm³
Volume of frustum = 2513.274... – 314.159...
= **2199.11 cm³**

Page 382 Exercise 6

1 a) Volume = $\frac{1}{3}$ × base area × height
$= \frac{1}{3} \times 18 \times 10 =$ **60 cm³**

b) Volume = $\frac{1}{3}$ × base area × height
$= \frac{1}{3} \times 8 \times 3 =$ **8 m³**

2 Area of base = 4 × 9 = 36 cm²
Volume = $\frac{1}{3} \times 36 \times 10 =$ **120 cm³**

3 Volume = $\frac{1}{3} \times 18 \times 15 =$ **90 cm³**

4 Volume = $\frac{1}{3} \times 27 \times 12 =$ **108 m³**

5 Find the volume of the pyramid and the prism separately.
Volume of pyramid = $\frac{1}{3} \times 24 \times 10 = 80$ cm³
Volume of prism = 24 × 10 = 240 cm³
So area of solid = 80 + 240 = **320 cm³**

6 Volume = $\frac{1}{3}$ × base area × 13.5
= base area × 4.5 = 288 cm³
So base area = 288 ÷ 4.5 = 64 cm²
Area of square = (length)² = 64 cm²
So side length = $\sqrt{64}$ = **8 cm**

Page 383 Exercise 7

1 a) Rate of flow = volume ÷ time
= 600 ÷ 15 = **40 cm³ per second**

b) Rate of flow = volume ÷ time
= 7200 ÷ 40 = **180 litres per minute**

c) Rate of flow = volume ÷ time
= 385 ÷ 5 = **77 m³ per minute**

d) Rate of flow = volume ÷ time
= 150 ÷ 8 = **18.75 litres per hour**

2 Volume of cube = 30³ = 27 000 cm³
Time = volume ÷ rate of flow
= 27 000 ÷ 15 = 1800 seconds
Convert to minutes: 1800 ÷ 60 = **30 minutes**

3 Area of cross-section = $\pi \times 4.5^2$
= 63.617... m²
Volume of tank = 63.617... × 20
= 1272.345... m³
Time = volume ÷ rate of flow
= 1272.345... ÷ 12 = 106.028...
= **106 minutes** (to the nearest minute)

23.6 Surface Area

Page 385 Exercise 1

For all the questions in this exercise, sketch the net of the shape to help you find its surface area.

1 The cube has 6 faces with area 2 cm², so its surface area is 6 × 2 = **12 cm²**

2 a) Area of 1 face = 1 × 1 = **1 cm²**

b) Surface area of cube = 6 × 1 = **6 cm²**

3 a) Area of 1 face = 2 × 2 = **4 m²**

b) Surface area of cube = 6 × 4 = **24 m²**

4 a) Area of face A = 4 × 2 = **8 m²**

b) Area of face B = 4 × 3 = **12 m²**

c) Area of face C = 3 × 2 = **6 m²**

d) Surface area of cuboid
= (2 × 8) + (2 × 12) + (2 × 6) = **52 m²**
There are two of each size of rectangle in the net.

5 a) Area of 1 face = 3 × 3 = 9 cm²
Surface area of cube = 6 × 9 = **54 cm²**

b) Surface area
= 2(3 × 1) + 2(4 × 1) + 2(4 × 3) = **38 cm²**

c) Surface area
= 2(1.5 × 4) + 2(1 × 4) + 2(1.5 × 1) = **23 m²**

d) Surface area
= 4(2.5 × 4.5) + 2(2.5 × 2.5) = **57.5 m²**

6 a) Area of triangular face = $\frac{1}{2} \times 5 \times 12$
= 30 mm²
Area of rectangular base = 5 × 11
= 55 mm²
Area of vertical rectangular face
= 12 × 11 = 132 mm²
Area of sloping rectangular face
= 13 × 11 = 143 mm²
Surface area = (2 × 30) + 55 + 132 + 143
= **390 mm²**
Remember, there are two triangular faces in a triangular prism.

b) Area of triangular face = $\frac{1}{2} \times 3 \times 2 = 3$ m²
Area of rectangular base = 3 × 3.2 = 9.6 m²
Area of sloping rectangular face
= 2.5 × 3.2 = 8 m²
Surface area = (2 × 3) + 9.6 + (2 × 8)
= **31.6 m²**

c) Area of square base = 4 × 4 = 16 cm²
Area of each triangular face
= $\frac{1}{2} \times 4 \times 7 = 14$ cm²
Surface area = 16 + (4 × 14) = **72 cm²**

d) Area of rectangular base = 6 × 8 = 48 cm²
Area of smaller triangular face
= $\frac{1}{2} \times 6 \times 9 = 27$ cm²
Area of larger triangular face
= $\frac{1}{2} \times 8 \times 9 = 36$ cm²
Surface area = 48 + (2 × 27) + (2 × 36)
= **174 cm²**

7 a) Surface area = 6(5 × 5) = **150 m²**

b) Surface area = 6(6 × 6) = **216 mm²**

c) Surface area
= 2(1.5 × 2) + 2(1.5 × 6) + 2(2 × 6) = **48 m²**

d) Surface area
= 2(7.5 × 0.5) + 2(7.5 × 8) + 2(0.5 × 8)
= **135.5 m²**

e) Area of triangular face = $\frac{1}{2} \times 6 \times 4 = 12$ m²
Area of rectangular base = 6 × 2.5 = 15 m²
Area of sloping rectangular face
= 5 × 2.5 = 12.5 m²
Surface area = (2 × 12) + 15 + (2 × 12.5)
= **64 m²**

8 a) Area of triangular face = $\frac{1}{2} \times 2.1 \times 1.4$
= 1.47 m²
Area of sloping rectangular face
= 2.5 × 1.75 = 4.375 m²
Surface area = (2 × 1.47) + (2 × 4.375)
= **11.69 m²**
You don't need to include the base here, as you're only interested in the outside of the tent.

b) 11.69 ÷ 4 = 2.9225, so they will need to buy **3 tins** to cover the outside of the tent.
2 tins isn't enough, so you need to round up.

Page 386 Exercise 2

1 a)

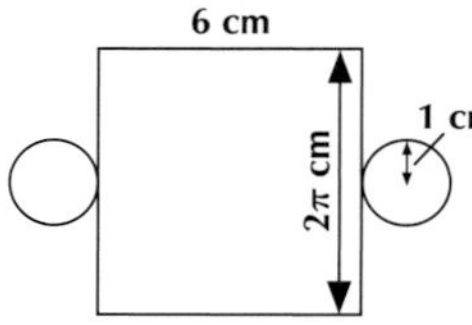

b) Area of each circular face = $\pi \times 1^2$
= 3.141... = **3.14 cm²**
Area of rectangular face = 6 × 2π
= 37.699... = **37.70 cm²** (both to 2 d.p.)

c) Surface area = (2 × 3.141...) + 37.699...
= 43.982... = **44.0 cm²** (1 d.p.)

2 a)

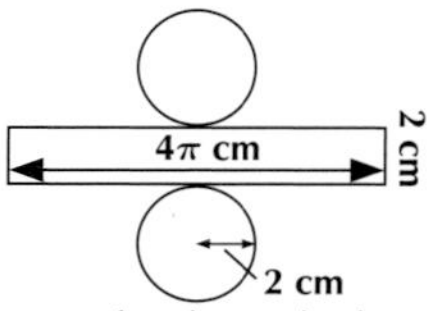

b) Area of each circular face = $\pi \times 2^2$
= 12.566... = **12.57 cm²**
Area of rectangular face = 2 × 4π
= 25.132... = **25.13 cm²** (both to 2 d.p.)

c) Surface area = (2 × 12.566...) + 25.132...
= 50.265... = **50.3 cm²** (1 d.p.)

For the remaining questions in this exercise, sketch the net first to help you find the areas.

3 a) Area of each circular face = $\pi \times 1^2$
= 3.141... cm²
Area of rectangular face = 4 × 2π × 1
= 25.132... cm²
Surface area = (2 × 3.141...) + 25.132...
= 31.415... = **31.42 cm²** (2 d.p.)

b) Area of each circular face = $\pi \times 3^2$
= 28.274... m²
Area of rectangular face = 9 × 2π × 3
= 169.646... m²
Surface area = (2 × 28.274...) + 169.646...
= ... 226.194...
= **226.19 m²** (2 d.p.)

Using the same method for c):

c) **131.95 m²** (2 d.p.)

d) Radius = diameter ÷ 2 = 8 ÷ 2 = 4 mm
Area of each circular face = $\pi \times 4^2$
= 50.265... mm²
Area of rectangular face = 7 × 8π
= 175.929... mm²
Surface area = (2 × 50.265...) + 175.929...
= 276.459... = **276.46 mm²** (2 d.p.)

Using the same method for e) and f):

e) **12.82 m²** f) **51.90 mm²**

4 a) Area of each circular face = $\pi \times 2^2$
= 12.566... m²
Area of rectangular face = 7 × 4π
= 87.964... m²
Surface area = (2 × 12.566...) + 87.964...
= 113.097... = **113.1 m²** (1 d.p.)

Using the same method for b)-c):

b) **471.2 mm²** (1 d.p.)

c) **1694.1 cm²** (1 d.p.)

d) Radius = diameter ÷ 2 = 22.1 ÷ 2
= 11.05 m
Area of each circular face = $\pi \times 11.05^2$
= 383.596... m²
Area of rectangular face = 11.1 × 22.1π
= 770.664... m²
Surface area = (2 × 383.596...) + 770.664...
= 1537.856... = **1537.9 m²** (1 d.p.)

5 a) Area of each circular face = $\pi \times 0.8^2$
= 2.010... m^2
Area of rectangular face = $3 \times 1.6\pi$
= 15.079... m^2
Surface area = (2 × 2.010...) + 15.079...
= 19.100... = **19.10 m²** (2 d.p.)

b) 1 tin covers 14 m^2 and 2 tins cover 28 m^2, so Maeve will need **2 tins** of paint.
You could do 19.100... ÷ 14 = 1.364... and round up to 2, but it's much easier to just use common sense here.

6 a) Curved surface area = $7.1 \times 4.4\pi$
= 98.143... = **98.14 m²** (2 d.p.)

b) Area of metal = 98.143... × 9
= 883.290... = **883.3 m²** (1 d.p.)

7 Area of circular face = $\pi \times 0.4^2$ = 0.502... m^2
Area of rectangular face = $1.1 \times 0.8\pi$
= 2.764... m^2
Surface area = 0.502... + 2.764... = 3.267...
= **3.27 m²** (2 d.p.)

Page 388 Exercise 3

1 a) Surface area = $4 \times \pi \times 2^2$ = **16π cm²**

b) Surface area = $4 \times \pi \times 5^2$ = **100π cm²**

Using the same method for c)-f):

c) **196π m²** d) **324π mm²**

e) **441π m²** f) **42.25π mm²**

2 a) Surface area = $4 \times \pi \times 2.5^2$ = 78.539...
= **78.54 cm²** (2 d.p.)

b) Surface area = $4 \times \pi \times 7.1^2$ = 633.470...
= **633.47 m²** (2 d.p.)

Using the same method for c)-f):

c) **1576.33 mm²** (2 d.p.)

d) **4680.85 m²** (2 d.p.)

e) **299.26 cm²** (2 d.p.)

f) **506.71 mm²** (2 d.p.)

3 a) Surface area = $(\pi \times 5 \times 13) + (\pi \times 5^2)$
= $65\pi + 25\pi$ = **90π m²**

b) Surface area = $(\pi \times 7 \times 25) + (\pi \times 7^2)$
= $175\pi + 49\pi$ = **224π cm²**

c) Surface area = $(\pi \times 15 \times 17) + (\pi \times 15^2)$
= $255\pi + 225\pi$ = **480π m²**

4 a) Surface area = $(\pi \times 12 \times 30) + (\pi \times 12^2)$
= 1583.362... = **1583.36 cm²** (2 d.p.)

b) Surface area = $(\pi \times 1.5 \times 4.5) + (\pi \times 1.5^2)$
= 28.274... = **28.27 m²** (2 d.p.)

c) Surface area = $(\pi \times 2.8 \times 8.2) + (\pi \times 2.8^2)$
= 96.761... = **96.76 mm²** (2 d.p.)

5 a) Surface area = $4 \times \pi \times 8^2$ = 804.247...
= **804.25 m²** (2 d.p.)

b) Surface area = $(\pi \times 5 \times 8) + (\pi \times 5^2)$
= 204.203... = **204.20 cm²** (2 d.p.)

c) Surface area = $4 \times \pi \times 1.1^2$ = 15.205...
= **15.21 m²** (2 d.p.)

d) Surface area = $(\pi \times 3 \times 7.1) + (\pi \times 3^2)$
= 95.190... = **95.19 cm²** (2 d.p.)

6 Surface area = $4\pi r^2$, so $265.9 = 4\pi r^2$
$r^2 = 265.9 \div 4\pi = 21.159...$
$r = \sqrt{21.159...} = 4.599...$ = **4.6 cm** (1 d.p.)

Page 389 Review Exercise

1 a)

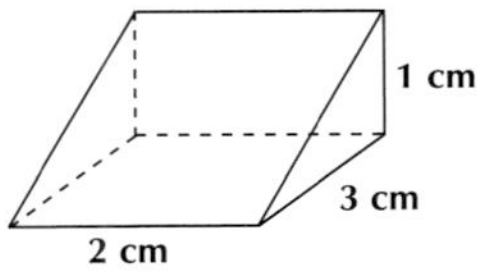

b)

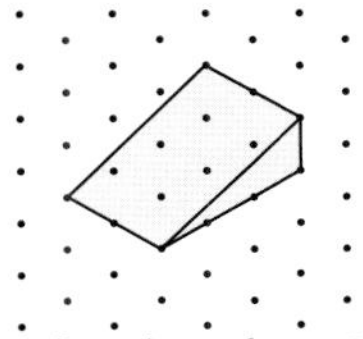

c) The shape has **5 faces**, **6 vertices** and **9 edges**.

2 a) E.g.

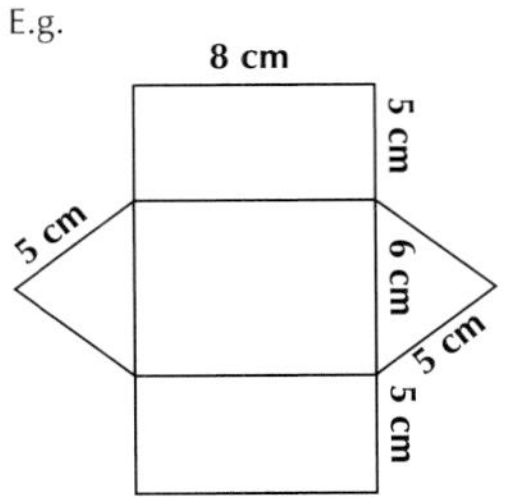

b) h = **4 cm**

c)

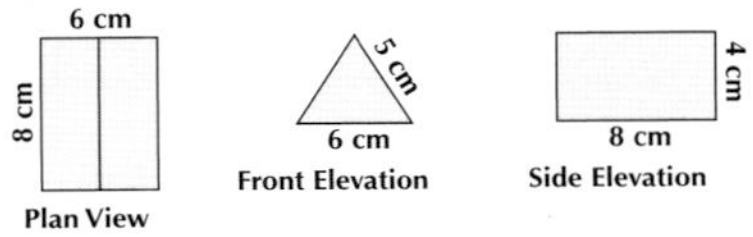

d) The prism has **2** planes of symmetry.
The cross-section is an isosceles triangle so it has one line of symmetry — that gives the prism one plane of symmetry. The other plane of symmetry cuts the prism in two halfway along its length.

3 a) E.g.

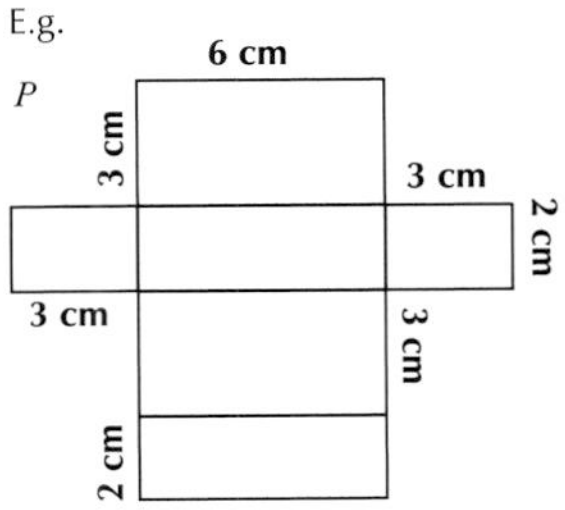

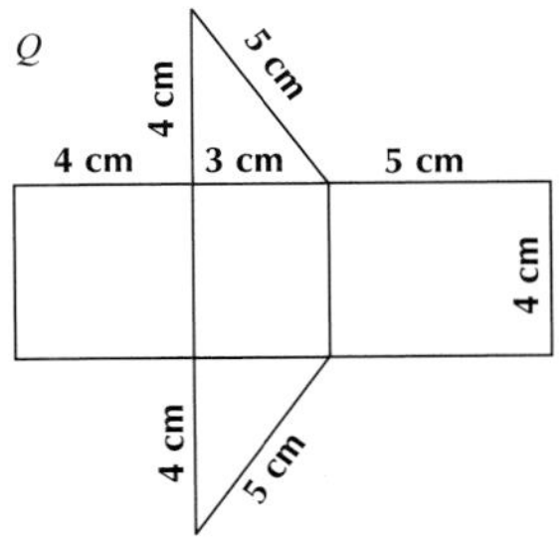

b) P

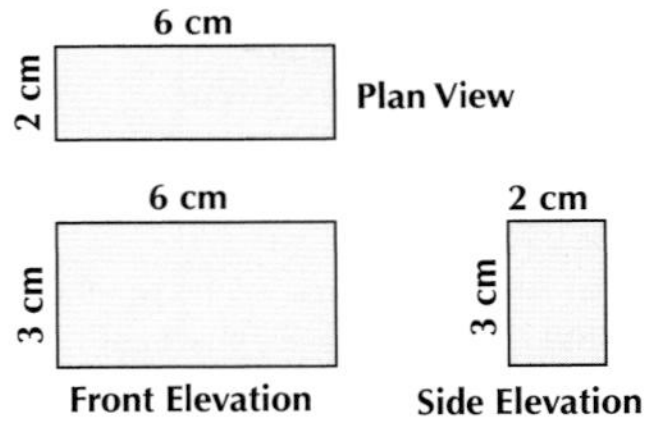

Q

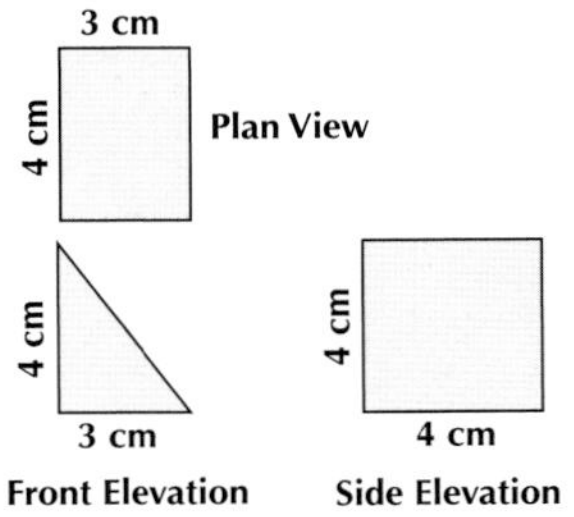

c) P

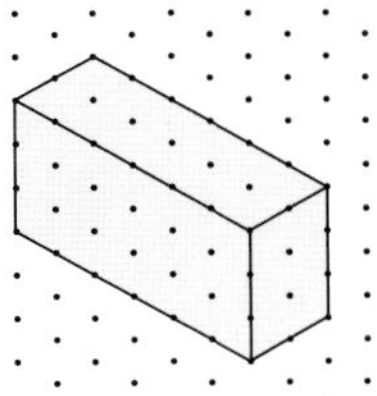

Q

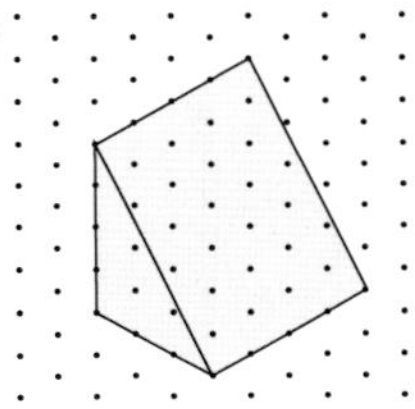

d) P: Volume = 6 × 2 × 3 = **36 cm³**
Q: Volume = $(\frac{1}{2} \times 3 \times 4) \times 4$ = **24 cm³**

4 a) Volume of one cube = $(\text{length})^3 = 8^3$
= **512 cm³**

b) Volume of tray = 24 × 512 = **12 288 cm³**

5 a) Radius = diameter ÷ 2 = 7.4 ÷ 2 = 3.7 cm
Cross-sectional area = $\pi r^2 = \pi \times 3.7^2$
= 43.008... cm^2
So volume = 43.008... × 11 = 473.092...
= **473.09 cm³** (2 d.p.)

b) 4 tins fit into the width of the box, so w = 7.4 × 4 = 29.6 cm
3 tins fit into the length of the box, so l = 7.4 × 3 = 22.2 cm
The height of the box is the height of 1 tin, so h = 11 cm
So the box has dimensions
29.6 cm × 22.2 cm × 11 cm

c) Volume = 29.6 × 22.2 × 11
= **7228.32 cm³**

d) Volume not taken up by tins
= 7228.32 – (473.092... × 12)
= 1551.210... = **1551.2 cm³** (1 d.p.)

6 Volume of 1 pyramid:
area of base = $7^2 = 49$ cm²
Volume = $\frac{1}{3} \times 49 \times 9 = 147$ cm³
So volume of whole shape = 147 × 2
= **294 cm³**

7 a) Area of one face = 54 ÷ 6 = **9 cm²**
b) Length of edge = $\sqrt{9}$ = **3 cm**
c) Volume = (length)³ = 3³ = **27 cm³**
d) Time = volume ÷ rate of flow
= 27 ÷ 2.5 = **10.8 seconds**
e)

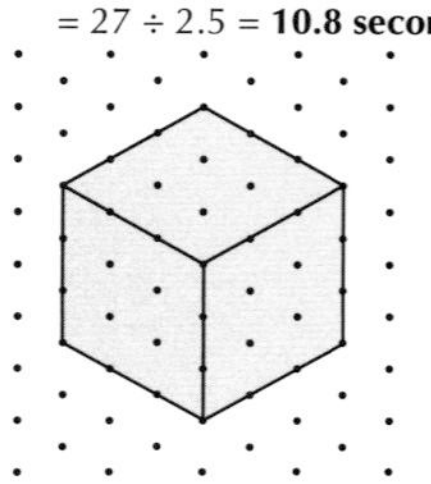

8 a) Volume of cuboid A = 6 × 2 × 3 = 36 cm³
Volume of cuboid B = 4 × 4 × 9 = 144 cm³
So ratio of volumes A : B = 36 : 144 = **1 : 4**
Have a look at Section 6 if you need a reminder about ratios.
b) Surface area of cuboid A
= 2(3 × 6) + 2(2 × 3) + 2(2 × 6) = 72 cm²
Surface area of cuboid B
= 2(4 × 4) + 4(4 × 9) = 176 cm²
So ratio of surface areas A : B = 72 : 176
= **9 : 22**

9 a)

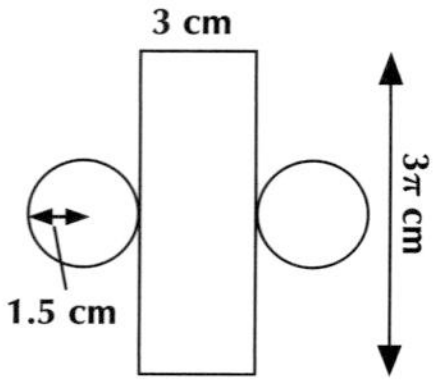

b) Area of each circular face = $\pi \times 1.5^2$
= 7.068... cm²
Area of rectangular face = 3 × 3π
= 28.274... cm²
Surface area = (2 × 7.068...) + 28.274...
= 42.411... = **42.41 cm²** (2 d.p.)

10 a) Radius = diameter ÷ 2 = 25 ÷ 2 = 12.5 mm
Volume = πr^2 × length,
so 7854 = $\pi \times 12.5^2$ × length
7854 = 156.25π × length
length = 7854 ÷ 156.25π
= 16.000... = **16.00 mm**
b) Area of each circular face = $\pi \times 12.5^2$
= 490.873... cm²
Area of rectangular face = 16.000... × 25π
= 1256.637... cm²
Surface area
= (2 × 490.873...) + 1256.637...
= 2238.384...
= **2238 mm²** (nearest mm²)
c) Length of cuboid = length of cylinder
= 16.000...
Width of cuboid = height of cuboid
= diameter of cylinder = 25 mm
So volume of cuboid = 16.000... × 25 × 25
= 10 000.023...
= **10 000 mm³** (nearest mm³)
d) Volume of empty space
= 10 000.023... – 7854
= 2146.023... = **2146 mm³** (nearest mm³)

11 a) Volume = $\frac{1}{3} \times \pi \times 9^2 \times 12$ = **324π m³**
b) (i) $h^2 = 9^2 + 12^2 = 81 + 144 = 225$
$h = \sqrt{225}$ = **15 m**
(ii) Surface area = (π × 9 × 15) + (π × 9²)
= **216π m²**
c) Volume of removed cone = $\frac{1}{3} \times \pi \times 6^2 \times 8$
= 96π m³
Volume of frustum = 324π – 96π
= **228π m³**

Page 391 Exam-Style Questions

1 a) **Cone**
Triangular prism
Square-based pyramid
[2 marks if all three shapes are named correctly, otherwise 1 mark if one or two shapes are named correctly]
b) (i) Number of vertices = **6** *[1 mark]*
(ii) Number of faces = **5** *[1 mark]*

2 a) Height = **5 cm** *[1 mark]*
Count the number of squares up one side of the rectangle that isn't attached to the circular faces.
b) Radius = 2 cm
Area of each circular face = π × 2²
= 4π cm² *[1 mark]*
Area of rectangular face = 5 × 4π
= 20π cm² *[1 mark]*
Surface area = (2 × 4π) + 20π
= **28π cm²** *[1 mark]*

3 Volume of cuboid
= area of shaded face × width, so
width = volume ÷ area of shaded face
= 300 ÷ 30 = **10 cm** *[1 mark]*.
Width = 2 × height, so height = width ÷ 2
= 10 ÷ 2 = **5 cm** *[1 mark]*.
Area of the shaded face = length × height, so
length = area of shaded face ÷ height
= 30 ÷ 5 = **6 cm** *[1 mark]*.

4 Volume of jam made: (π × 20²) × 18
= 7200π cm³ *[1 mark]*
Volume of one jar: (π × 3²) × 10 = 90π cm³ *[1 mark]*
So Sadiq can fill 7200π ÷ 90π = **80 jars** with jam *[1 mark]*.
Leaving the volumes in terms of π makes the numbers easier to deal with — but you'll still get the marks if you wrote them as decimals.

Section 24 — Transformations

24.1 Reflections

Page 393 Exercise 1

For each question in this exercise, reflect the vertices of the shape in the line of symmetry, and then join them up to form the shape. For each vertex, count how many squares it is from the mirror line. The image point is the same number of squares to the opposite side of the mirror line.

1 a) b) c) d) e) f) g) h)

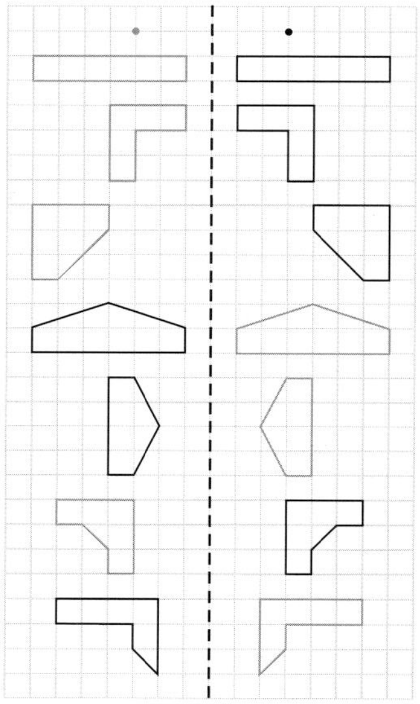

2 a) b) c) d) e)

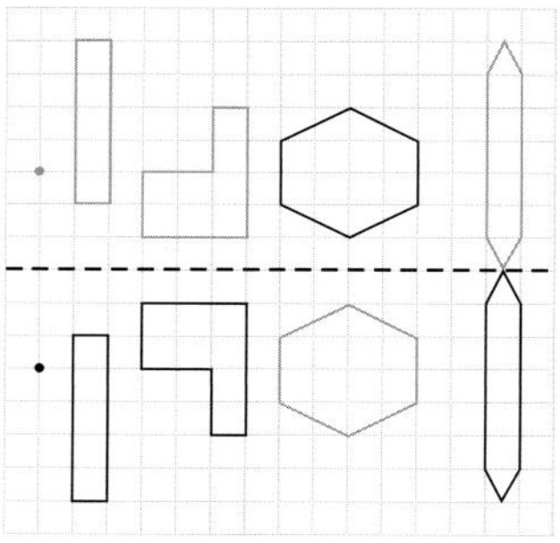

3

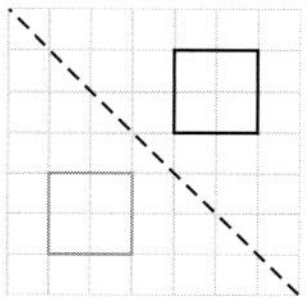

Page 394 Exercise 2

For Qs 1, 2 and 4, reflect the vertices of the shape in the line of symmetry, and then join them up to form the shape. For each vertex, count how many squares it is from the mirror line. The image point is the same number of squares to the opposite side of the mirror line.

1 a)

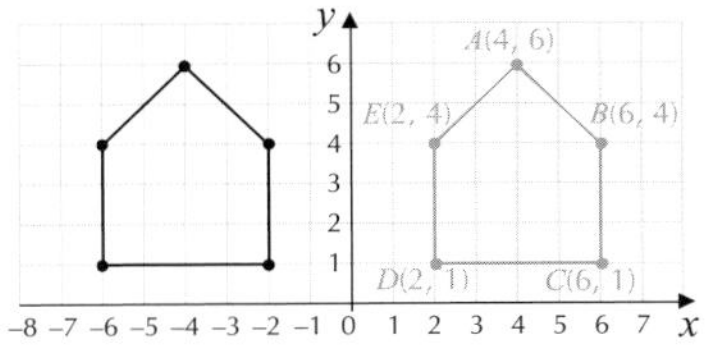

b)

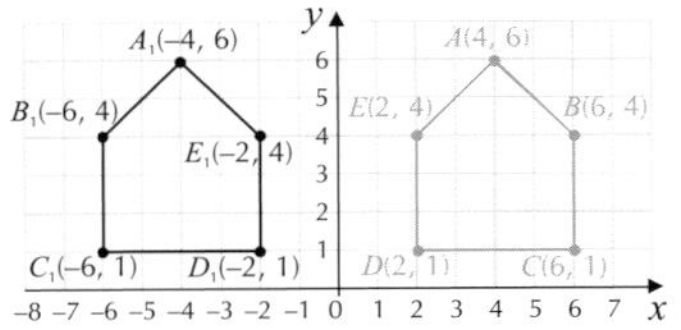

c) The x-coordinates become negative in the reflected shape, and the y-coordinates stay the same.

2 a)

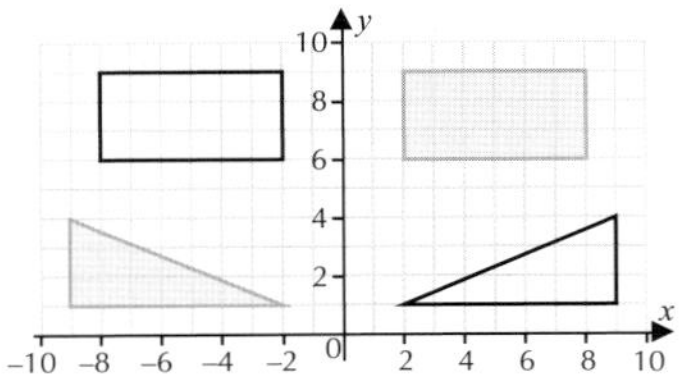

b)

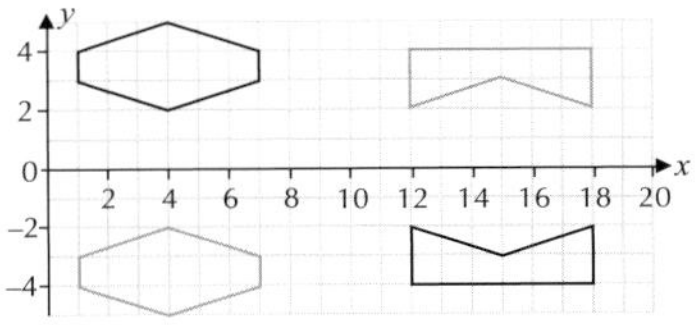

3 a) The point (x, y) is sent to $(x, -y)$ under a reflection in the x-axis, so the new coordinates are:

(i) (1, –2) **(ii) (3, 0)**
(iii) (–2, –4) **(iv) (–1, 3)**

b) The point (x, y) is sent to $(-x, y)$ under a reflection in the y-axis, so the new coordinates are:

(i) (–4, 5) **(ii) (–7, 2)**
(iii) (1, 3) **(iv) (3, –1)**

4 a)-d) Start by drawing the mirror line. Then reflect the vertices and join up the image points — see below.

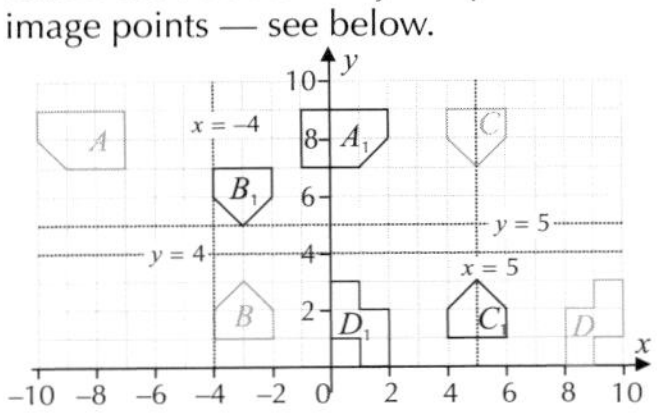

e) Pick a pair of corresponding vertices on shapes B and C_1, e.g. (–3, 3) on B and (5, 3) on C_1. They are 8 horizontal units apart and the mirror line must be the same distance from each (8 ÷ 2 = 4 units). So the mirror line is $x = 1$.

Page 395 Exercise 3

1 a) Each image point is the same perpendicular distance from the mirror line as the vertex but on the other side of it.

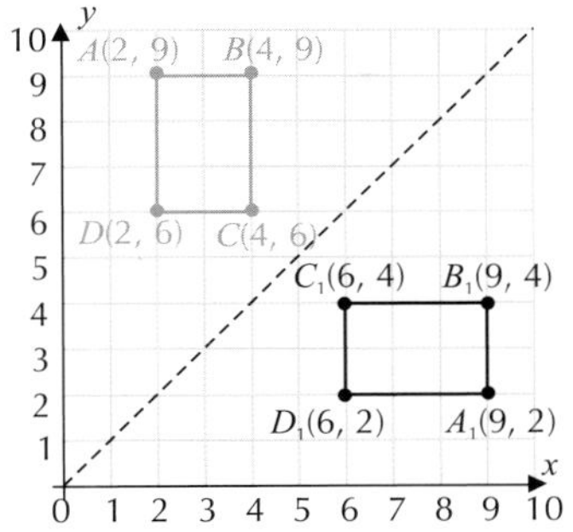

b) E.g. **The x-coordinate becomes the y-coordinate, and the y-coordinate becomes the x-coordinate.**

2 Each image point is the same perpendicular distance from the mirror line as the vertex but on the other side of it.

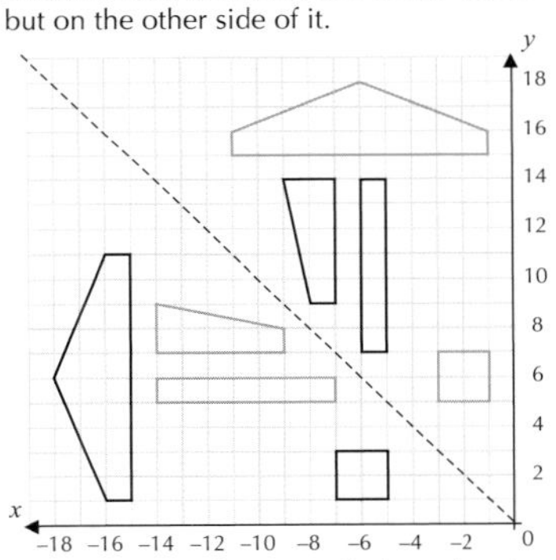

Make sure you reflect perpendicular to the mirror line. Turn your book so that the mirror line is horizontal or vertical if it helps.

3 a) The point (x, y) is sent to (y, x) under a reflection in $y = x$.

(i) (2, 1) **(ii) (0, 3)**
(iii) (4, –2) **(iv) (–3, –1)**

b) The point (x, y) is sent to $(-y, -x)$ under a reflection in $y = -x$.

(i) (–2, –1) **(ii) (0, –3)**
(iii) (–4, 2) **(iv) (3, 1)**

24.2 Rotations

Page 396 Exercise 1

For the rotations in this exercise, use tracing paper to draw the shapes then put your pencil on the centre of rotation and rotate the tracing paper by the required angle in the direction given. Draw the rotated shape in its new position.

1 a)

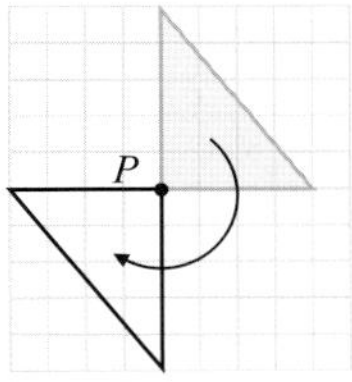

b)

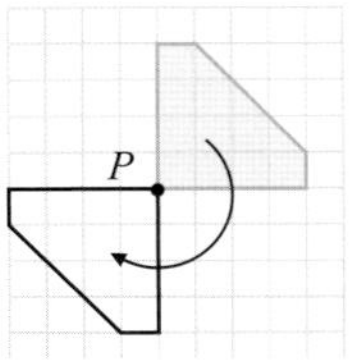

Since these rotations are by an angle of 180°, you could go clockwise or anticlockwise.

2 a)

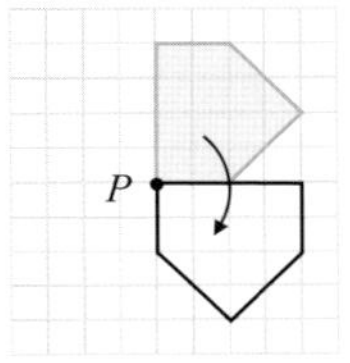

b)

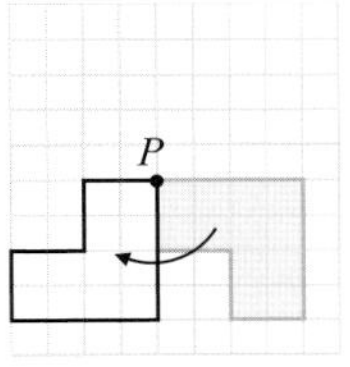

Page 397 Exercise 2

For the rotations in this exercise, use tracing paper to draw the shapes then put your pencil on the centre of rotation and rotate the tracing paper by the required angle in the direction given. Draw the rotated shape in its new position.

1 a)

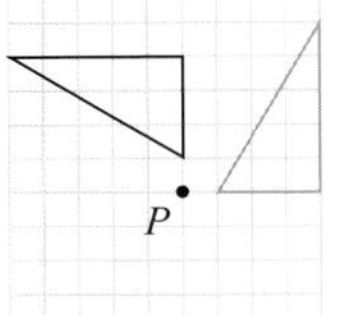

b)

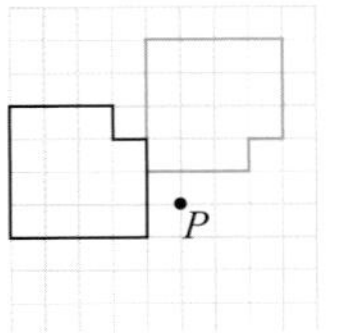

c)

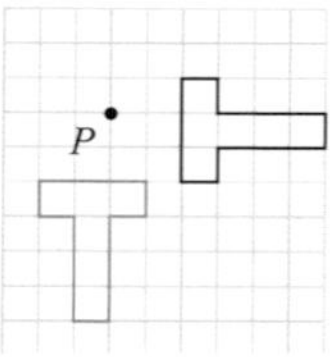

d)

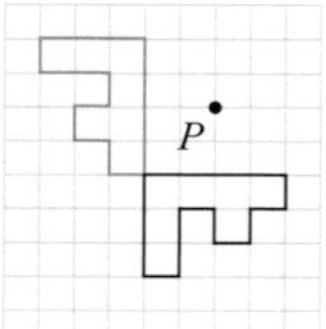

2 a)

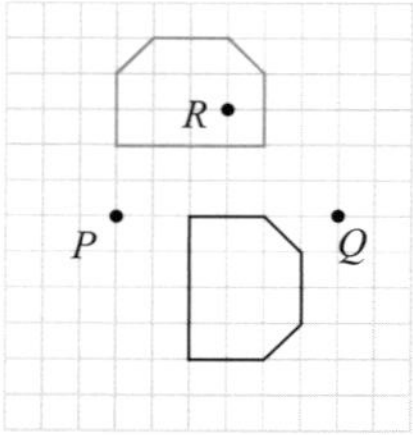

b)

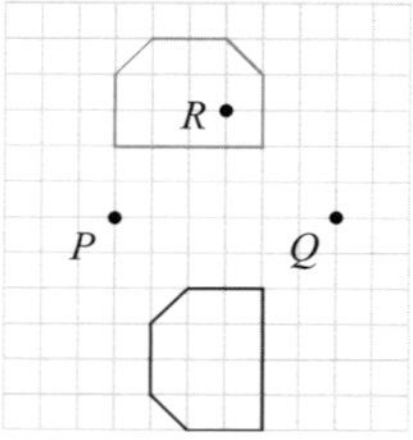

A 270° clockwise rotation is the same as a 90° anticlockwise rotation.

c)

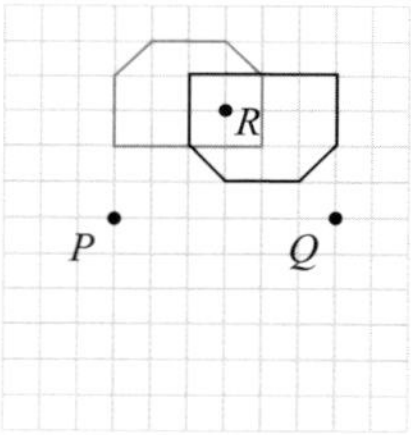

3 a), b)

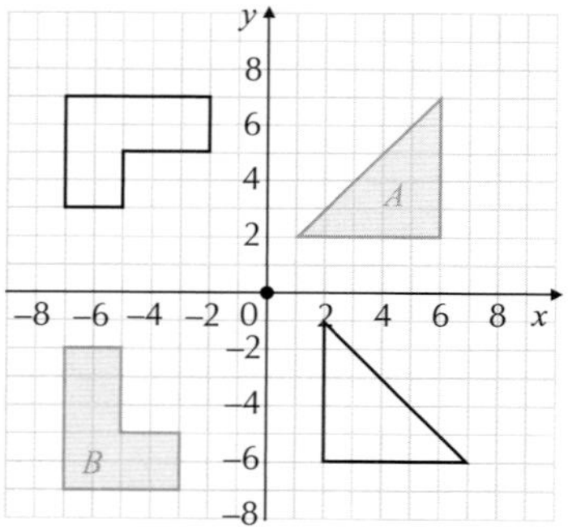

4 a), b), c), d)

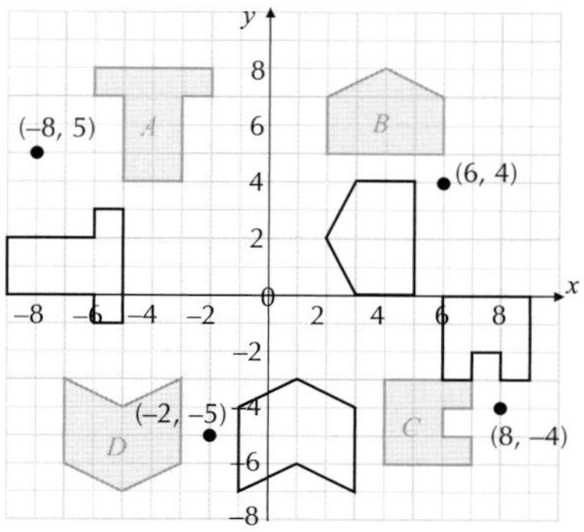

5 a), b)

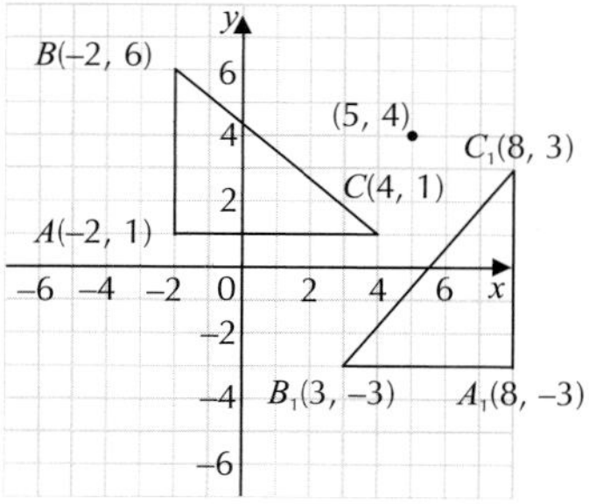

c) A_1**(8, –3),** B_1**(3, –3),** C_1**(8, 3).**

Page 399 Exercise 3

To describe the transformations, draw the shape being rotated on tracing paper. Then place your pencil at different centres of rotation and rotate the tracing paper until you find a point and rotation that takes it on top of the new shape.

1 a) **Rotation of 90° clockwise about the origin.**
Alternatively, this is a rotation 270° anticlockwise about (0, 0).

b) **Rotation of 90° anticlockwise about the origin.**
Alternatively, this is a rotation 270° clockwise about (0, 0).

2 a) **Rotation of 180° about the origin.**

b) **Rotation of 180° about (0, 2).**

3 a) **Rotation of 90° anticlockwise about (0, 7).**

b) **Rotation of 90° clockwise about (1, –2).**

4 a) **Rotation of 180° about (0, 6).**

b) **Rotation of 90° anticlockwise about (–2, 7).**

c) **Rotation of 90° clockwise about (1, 8).**

24.3 Translations

Page 401 Exercise 1

1 a) **1 to the right, 1 up**

b) **2 to the right**

c) **2 to the left, 6 up**

d) **3 to the left, 2 down**

2 Shape *A* moves 3 units to the right.
Shape *B* moves 2 units up.
Shape *C* moves 2 units to the right and 3 units up.
Shape *D* moves 4 units to the right and 1 unit down.
Shape *E* moves 4 units to the left and 3 units down.

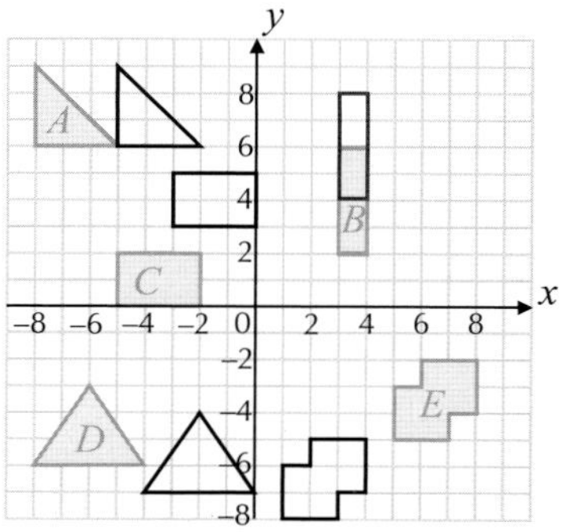

For each shape, start by working out where the corners of the shape move to and then join them to form the shape.

3 a), b) Translate each of *A*, *B* and *C* 10 units to the left and 1 unit down. Then join them up to create $A_1B_1C_1$.

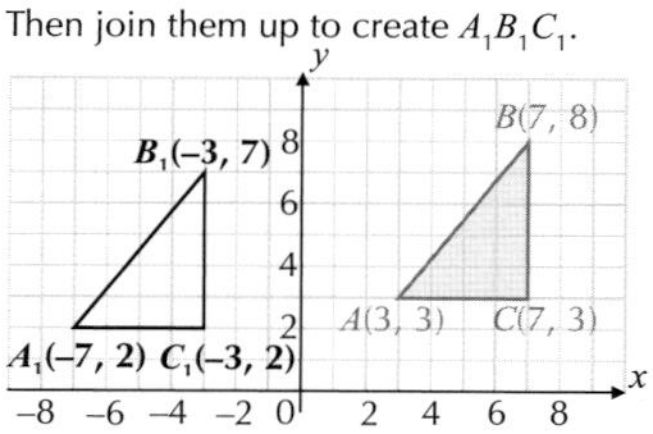

c) To get from the original point to its image, **subtract 10** from the ***x*-coordinate**, and **subtract 1** from the ***y*-coordinate.**
You could also write this as (x − 10, y − 1).

4 a) 3 + 0 = 3, –4 + 1 = –3, so the coordinates are **(3, –3)**

b) 3 + 3 = 6, –4 + 0 = –4, so the coordinates are **(6, –4)**

c) 3 + 4 = 7, –4 – 2 = –6, so the coordinates are **(7, –6)**

d) 3 – 1 = 2, –4 – 5 = –9, so the coordinates are **(2, –9)**

5 D: 1 + (–2) = –1, 1 + 2 = 3, so the new coordinates are D_1**(–1, 3)**
E: 3 + (–2) = 1, –2 + 2 = 0, so the new coordinates are E_1**(1, 0)**
F: 4 + (–2) = 2, 0 + 2 = 2, so the new coordinates are F_1**(2, 2)**

Page 402 Exercise 2

1 a) The translation is 1 unit in the positive *x* direction and 2 units in the positive *y* direction, so the vector is $\begin{pmatrix} \mathbf{1} \\ \mathbf{2} \end{pmatrix}$.

b) The translation is 1 unit in the positive *x* direction and 2 units in the negative *y* direction, so the vector is $\begin{pmatrix} \mathbf{1} \\ \mathbf{-2} \end{pmatrix}$.

c) The translation is 3 units in the negative *y* direction only, so the vector is $\begin{pmatrix} \mathbf{0} \\ \mathbf{-3} \end{pmatrix}$.

d) The translation is 4 units in the negative x direction and 3 units in the negative y direction, so the vector is $\begin{pmatrix}-4\\-3\end{pmatrix}$.

e) The translation is 6 units in the positive x direction and 7 units in the positive y direction, so the vector is $\begin{pmatrix}6\\7\end{pmatrix}$.

f) The translation is 6 units in the positive y direction only, so the vector is $\begin{pmatrix}0\\6\end{pmatrix}$.

2 a) 2 units to the right, 3 units down

b) The translation is 2 units in the positive x direction and 3 units in the negative y direction, so the vector is $\begin{pmatrix}2\\-3\end{pmatrix}$.

c) Z is 2 units in the positive x direction from Y and 4 units in the negative y direction, so the translation is $\begin{pmatrix}2\\-4\end{pmatrix}$.

3 a) The image R is 5 units in the negative y direction from P so the translation is described by the vector $\begin{pmatrix}0\\-5\end{pmatrix}$.

b) The image S is 8 units in the positive x direction from R so the translation is described by the vector $\begin{pmatrix}8\\0\end{pmatrix}$.

Using the same method for c)-f):

c) $\begin{pmatrix}13\\1\end{pmatrix}$ **d)** $\begin{pmatrix}-8\\0\end{pmatrix}$

e) $\begin{pmatrix}-13\\-6\end{pmatrix}$ **f)** $\begin{pmatrix}-8\\5\end{pmatrix}$

4 a) Drawing the points and connecting them up gives:

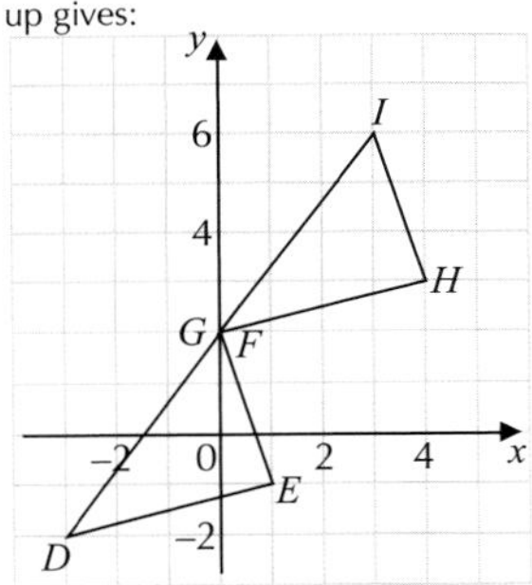

b) GHI is 3 units to the right fom DEF and 4 units up, so the vector is $\begin{pmatrix}3\\4\end{pmatrix}$.

5 a) To get from J to M, K to N and L to P, you subtract 1 from the x-coordinate and add 2 to the y-coordinate. So the vector describing the translation is $\begin{pmatrix}-1\\2\end{pmatrix}$.

b) MNP to JKL is opposite to the translation in a), so the vector is $\begin{pmatrix}1\\-2\end{pmatrix}$.

Page 403 Exercise 3

1 a) Y is 3 units to the right and 5 units down from X so the vector describing the translation is $\begin{pmatrix}3\\-5\end{pmatrix}$.

b) X is 3 units to the left and 5 units up from Y so the vector describing the translation is $\begin{pmatrix}-3\\5\end{pmatrix}$.

c) The x and y components have changed signs.

2 The vector describing the translation from Z onto W is the negative of the vector $\begin{pmatrix}1\\-4\end{pmatrix}$, i.e. $\begin{pmatrix}-1\\4\end{pmatrix}$.

3 The negatives of the vectors given in each part are:

a) $\begin{pmatrix}0\\-2\end{pmatrix}$ **b)** $\begin{pmatrix}-5\\0\end{pmatrix}$ **c)** $\begin{pmatrix}1\\5\end{pmatrix}$ **d)** $\begin{pmatrix}-3\\-2\end{pmatrix}$

Translate the each shape by the relevant vector above:

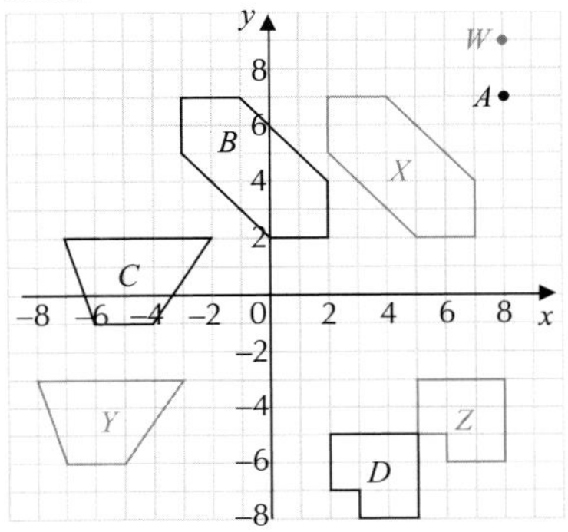

4 The vector describing the translation from PQR onto DEF is the negative of the vector $\begin{pmatrix}-1\\4\end{pmatrix}$, i.e. $\begin{pmatrix}1\\-4\end{pmatrix}$. Applying this vector to the coordinates of PQR gives:
D: $-1 + 1 = 0$, $0 - 4 = -4$, so $\boldsymbol{D = (0, -4)}$,
E: $-4 + 1 = -3$, $4 - 4 = 0$, so $\boldsymbol{E = (-3, 0)}$,
F: $3 + 1 = 4$, $2 - 4 = -2$, so $\boldsymbol{F = (4, -2)}$.

24.4 Enlargements

Page 405 Exercise 1

1 The scale factor is 2, so multiply each of the side lengths by 2.

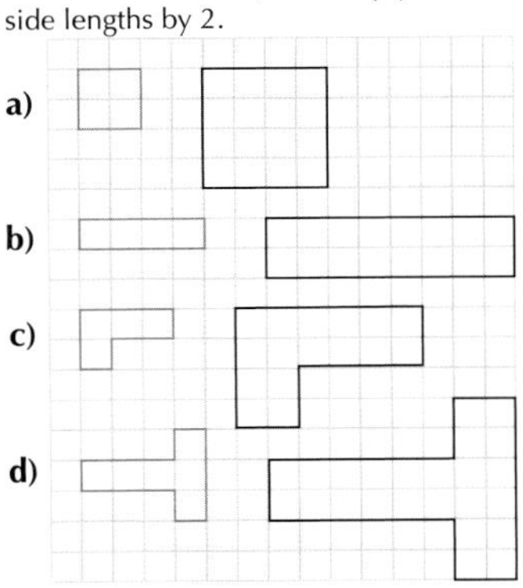

2 The scale factor is 3, so multiply each of the side lengths by 3.

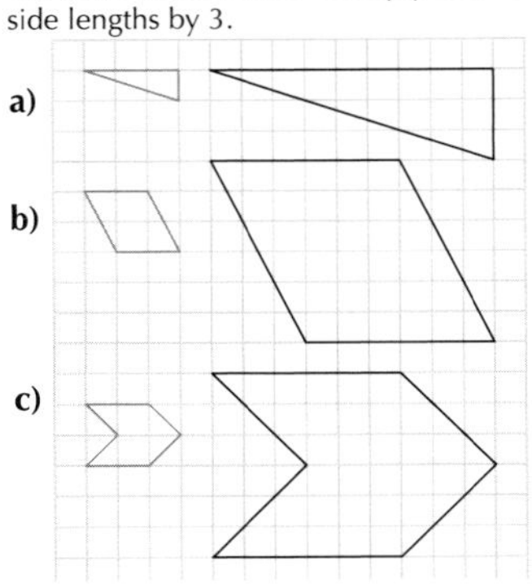

3 The scale factor is 5, so multiply each of the side lengths by 5.

a)

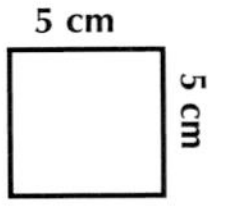

Your diagrams don't have to be full size, but they do need to be labelled correctly.

b)

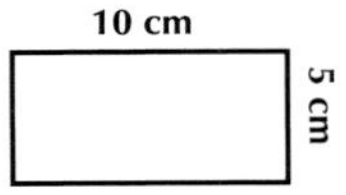

c)

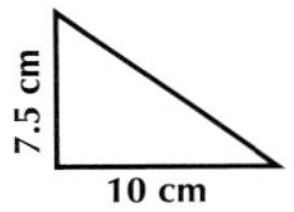

d)

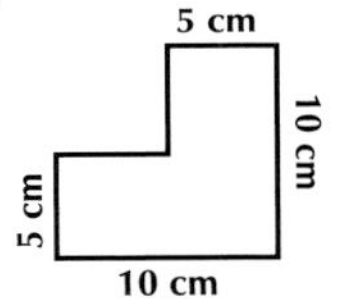

4 The scale factor is $\frac{1}{2}$, so multiply each of the side lengths by $\frac{1}{2}$.

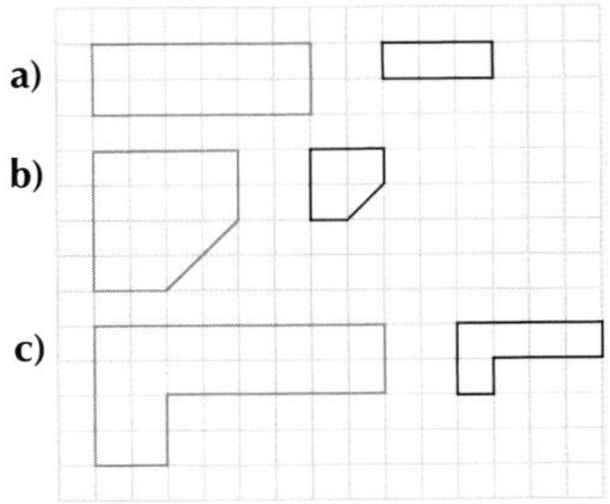

5 The scale factor is $\frac{1}{3}$, so multiply each of the side lengths by $\frac{1}{3}$.

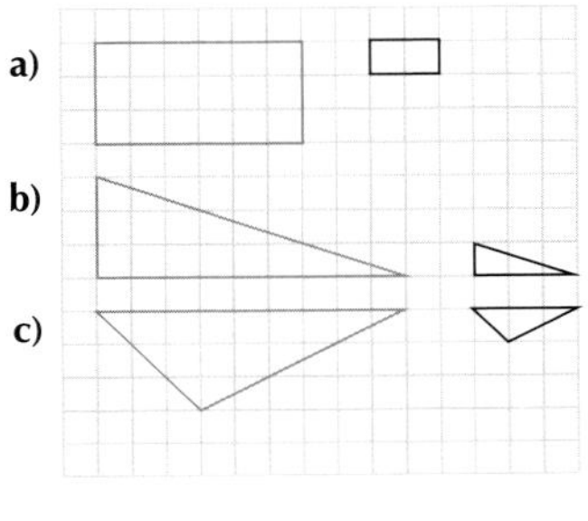

6 The shape is a square so all the sides will be the same length. The scale factor is $\frac{1}{4}$ so the new sides will be:

$16 \times \frac{1}{4} =$ **4 cm**

7 The scale factor is $\frac{1}{5}$ so the new side lengths will be:

$15 \times \frac{1}{5} =$ **3 cm** and $35 \times \frac{1}{5} =$ **7 cm**

Page 407 Exercise 2

1 a) (i) **3 units** (ii) **3 units** (iii) **0 units**

b) (i) $3 \times 2 =$ **6 units**

(ii) $3 \times 2 =$ **6 units**

(iii) $0 \times 2 =$ **0 units**

c)

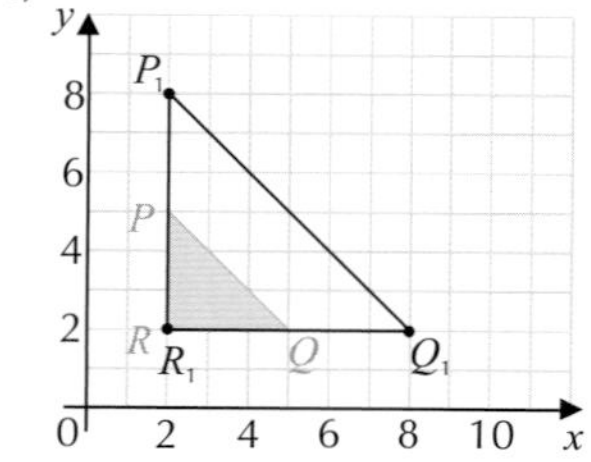

For the enlargements in Q2 and Q3, draw lines from the centre of enlargement to each vertex of the shape. Extend these lines depending on the scale factor (e.g. for a scale factor of 3 the lines should be 3 times as long). Mark the new vertices at the end of these lines and join them up to create the new shape.

2 a), b), c), d)

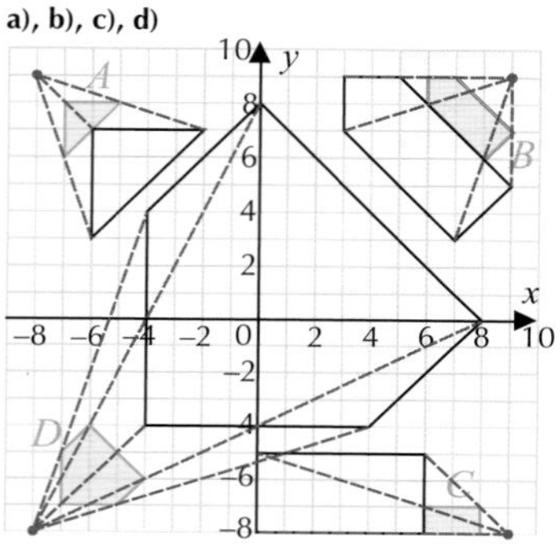

3 a), b)

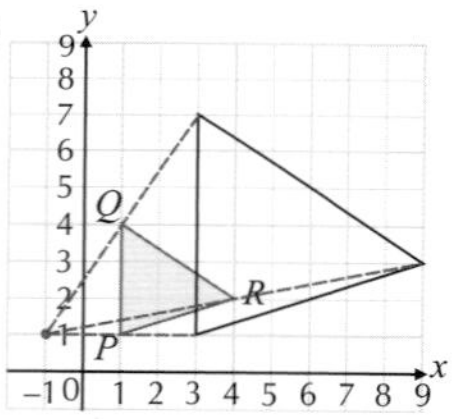

Page 408 Exercise 3

For each enlargement in this exercise, draw lines from the centre of enlargement to each vertex of the shape. Mark the new vertices a fraction of the way along these lines (e.g. for a scale factor of $\frac{1}{4}$ the vertices should be a quarter of the way along the lines from the centre of enlargement). Join the vertices up to create the new shape.

1 a), b)

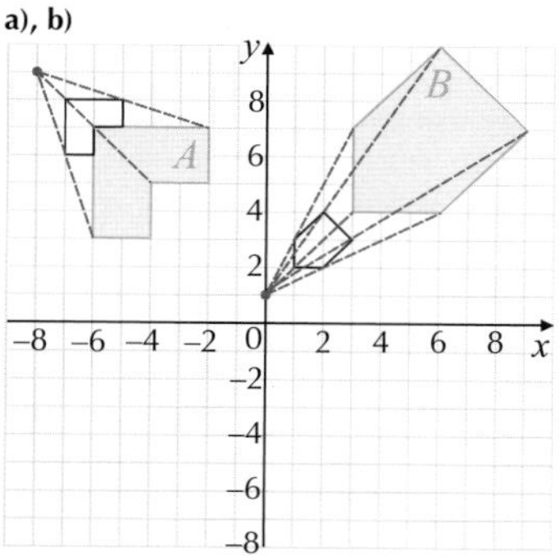

c), d)

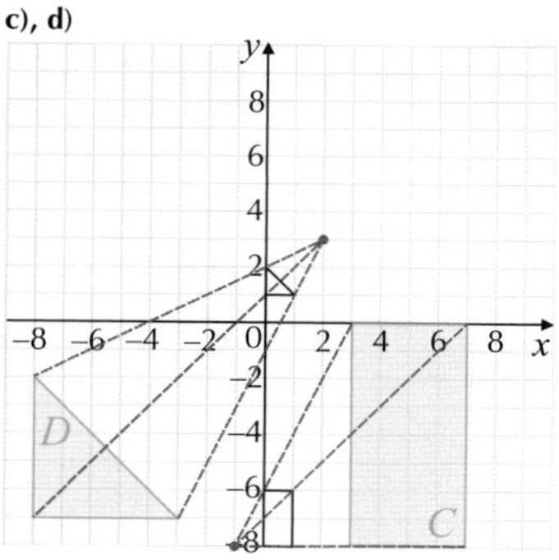

It's fine if you've drawn all the Q1 enlargements on the same grid.

2 a), b)

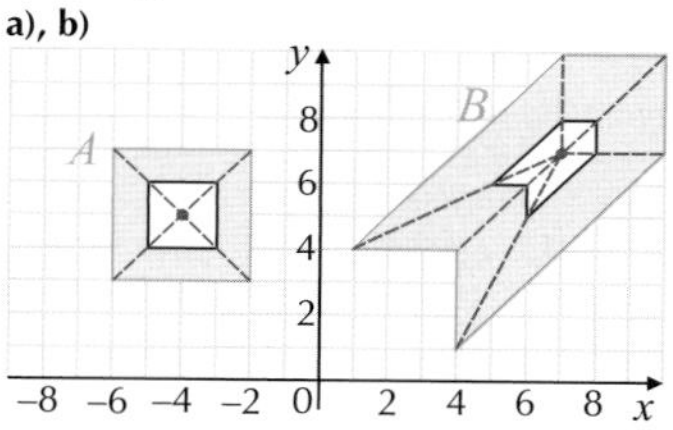

Page 409 Exercise 4

1 a) Measure corresponding sides of shapes A and B — e.g. the vertical side of A is 3 units and of B is 6 units. So the scale factor is $6 \div 3 = 2$. Draw lines through corresponding vertices of shapes A and B. These intersect at the point (2, 10). So the transformation is an enlargement, **scale factor 2, centre (2, 10)**.

b) E.g. vertical side of A is 1 unit and of B is 2 units. So the scale factor is $2 \div 1 = 2$. Draw lines through corresponding vertices of shapes A and B. These intersect at the point (6, 10). So the transformation is an enlargement, **scale factor 2, centre (6, 10)**.

c) E.g. vertical side of A is 1 unit and of B is 3 units. So the scale factor is $3 \div 1 = 3$. Draw lines through corresponding vertices of shapes A and B. These intersect at the point (11, 9). So the transformation is an enlargement, **scale factor 3, centre (11, 9)**.

2 a) Measure corresponding sides of shapes A and B — e.g. the bottom horizontal side of shape A is 4 units and of B is 8 units. The new shape is A, so the scale factor is $\frac{4}{8} = \frac{1}{2}$. Draw lines through corresponding vertices of shapes A and B. These intersect at the point (1, 1). So the transformation is an enlargement, **scale factor $\frac{1}{2}$, centre (1, 1)**.

b) E.g. the bottom horizontal side of shape A is 2 units and of B is 4 units. The new shape is A, so the scale factor is $\frac{2}{4} = \frac{1}{2}$. Draw lines through corresponding vertices of shapes A and B. These intersect at the point (6, 5). So the transformation is an enlargement, **scale factor $\frac{1}{2}$, centre (6, 5)**.

c) E.g. the bottom horizontal side of shape A is 1 unit and of B is 3 units. The new shape is A, so the scale factor is $\frac{1}{3}$. Draw lines through corresponding vertices of shapes A and B. These intersect at the point (5, 4). So the transformation is an enlargement, **scale factor $\frac{1}{3}$, centre (5, 4)**.

Page 410 Exercise 5

1 a)

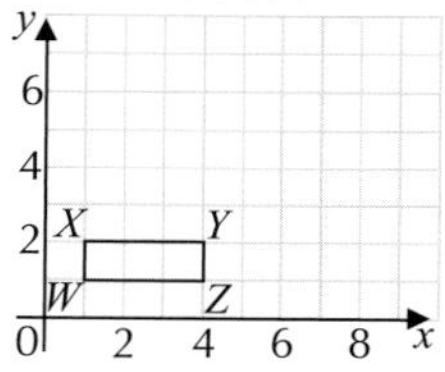

Perimeter = 3 + 1 + 3 + 1 = **8 units**

b) Draw lines from the origin to each vertex of the shape. The scale factor is 2 so extend these lines so they are twice as long. Mark the new vertices at the ends of these lines and join them up to create the new shape:

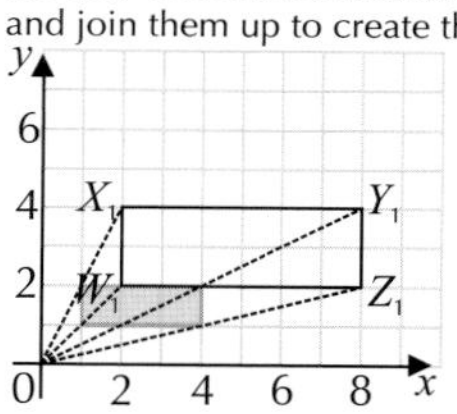

c) Perimeter of $W_1X_1Y_1Z_1$ is the old perimeter multiplied by the scale factor:

$8 \times 2 =$ **16 units**

You can check your answer by working out the perimeter from your diagram.

2 a) A square with sides 3 cm has a perimeter of $3 \times 4 = 12$ cm. Enlarging by scale factor 3 gives a perimeter of $12 \times 3 =$ **36 cm**

b) The rectangle has a perimeter of: $(2 \times 2) + (2 \times 8) = 20$ m. Enlarging by a scale factor of 5 gives a perimeter of $20 \times 5 =$ **100 m**

24.5 Congruence and Similarity

Page 411 Exercise 1

1 Shapes are congruent if they are the same size and shape, so:
A is congruent to G, B is congruent to J, C is congruent to F, D is congruent to H, E is congruent to I.

2 Shapes are congruent if they are the same size and shape, so the odd ones out are:

a) D **b)** H **c)** I
d) N **e)** Q **f)** V

Page 413 Exercise 2

1 **a)** The triangles are congruent, so match the sides and angles:
a = 5 cm, b = 69°, c = 69°, d = 42°, e = 7 cm, f = 7 cm

b) $g = 180° - 90° - 53° =$ **37°**
The triangles are congruent, so match the remaining sides and angles:
h = 6 cm, i = 10 cm, j = 8 cm, k = 37°, l = 53°

2 You need to show whether or not one of the 'congruence conditions' holds:

a) Yes. Two sides and the angle between on one triangle are the same as two sides and the angle between on the other triangle, so condition **SAS** holds.

b) No. The two angles shown are the same, but the corresponding sides are not equal, so **AAS** doesn't hold.

c) Yes, all the sides are the same in both triangles so **SSS** holds.

d) No. Their hypotenuses are different, so **RHS** doesn't hold.

3 You need to show whether or not one of the 'congruence conditions' holds:

a) E.g. Missing angle in first triangle: 180° – 75° – 50° = 55°. Two angles and one corresponding side are the same in both triangles so condition **AAS** holds and **the triangles are congruent**.

b) Using Pythagoras' theorem the hypotenuse of the triangle on the right is 5 cm. Both triangles have a right angle, the same hypotenuse and another side the same, so condition **RHS holds** and **the triangles are congruent**.

Page 414 Exercise 3

1 Shapes are similar if they have the same shape, but they can be different sizes so:
A, C and F are similar
B, E and K are similar
D, G and I are similar
H, J and L are similar

2 Shapes are similar if they have the same shape, but they can be different sizes, so the pairs of similar shapes are:

a) A **and** B **b)** D **and** F
c) H **and** I **d)** K **and** L

Page 416 Exercise 4

1 For each part of the question find the condition for similarity which applies:

a) All the angles in one triangle are the same as the angles in the other triangle.

b) All corresponding sides of the two triangles are in the same ratio, since each side in the 2nd triangle is twice as long as the corresponding side in the 1st.

c) All the angles in one triangle are the same as the angles in the other triangle.
The triangles are in a different orientation but they are still similar — one is an enlarged reflection of the other.

d) Two sides of the triangles are in the same ratio (the sides in the second triangle are a third of the length of those in the first) **and the angle between them is the same in both triangles.**

2 To prove if triangles are similar, you need to show whether or not one of the 'similarity conditions' holds:

a) Similar — all of the angles are the same in the two triangles.

b) Similar — all corresponding sides are in the same ratio, since each side in the second triangle is 1.5 times as long as the corresponding side in the first.

c) Similar — two sides are in the same ratio (the sides in the second triangle are two thirds of the length of those in the first) and there's a right angle between them in both triangles.

d) Not similar — the sides either side of the given angle are the same in one triangle and different in the other.

e) Similar — two sides are in the same ratio (the sides in the second triangle are a third of the length of those in the first) and there's a 40° angle between them in both triangles.

f) Not similar — only one angle matches.

Page 418 Exercise 5

1 **a) (i)** The angle is the same as the corresponding angle in the other triangle, so x = **100°**.
The diagrams are drawn to scale, and there's only one obtuse angle, so x must be 100°.

(ii) All angles in a triangle add up to 180° so $y = 180° - 100° - 25° =$ **55°**

b) The triangles are isosceles, so both base angles in the first triangle must be 35°. Since the triangles are similar, both angles at the top of the second triangle must also be 35°. So $z = 180° - 35° - 35° =$ **110°**.

2 **a)** The triangles are similar so the corresponding angles are:
(i) angle ABC
(ii) angle ACB
(iii) angle BAC

b) Scale factor = $AD \div AB = 9 \div 3 =$ **3**

c) (i) $AE = AC \times 3 = 2.5$ m $\times 3 =$ **7.5 m**
(ii) $DE = BC \times 3$ so $BC = DE \div 3$ = 12 m ÷ 3 = **4 m**

3 Scale factor = $QS \div QR$ = 12 cm ÷ 6 cm = **2**

a) $ST = RU \times 2$ = 3 cm × 2 = **6 cm**
b) $QT = QU \times 2$ = 5 cm × 2 = **10 cm**
c) $UT = QT - QU$ = 10 cm – 5 cm = **5 cm**

4 Use lengths SW and SV to find the scale factor.
$SV = SW + WV = 2 + 4 = 6$ m,
so the scale factor = $SV \div SW = 6 \div 2 = 3$

a) $TW = UV \div 3 = 4.5 \div 3 =$ **1.5 m**
b) $SU = ST \times 3 = 1.8 \times 3 =$ **5.4 m**
c) $TU = SU - ST = 5.4 - 1.8 =$ **3.6 m**

5 **a)** Angle QXR = **34°**
Angle QXR is vertically opposite angle ZXY so they are equal.

b) angle XYZ is alternate to **angle XRQ**

c) Scale factor = $XY \div RX = 10 \div 5 =$ **2**

d) (i) $YZ = QR \times 2 = 4 \times 2 =$ **8 m**
(ii) $XQ = XZ \div 2 = 14 \div 2 =$ **7 m**

Page 419 Review Exercise

1 **a) (i)** Start by drawing the mirror line at $y = -1$. Then reflect the vertices and join up the image points to form image B — see below.
Remember, the vertices should be the same perpendicular distance from the mirror line.

(ii) Use tracing paper to draw the shape, then put your pencil on the origin and rotate the tracing paper by 90° clockwise. Draw the rotated shape C in its new position — see below.

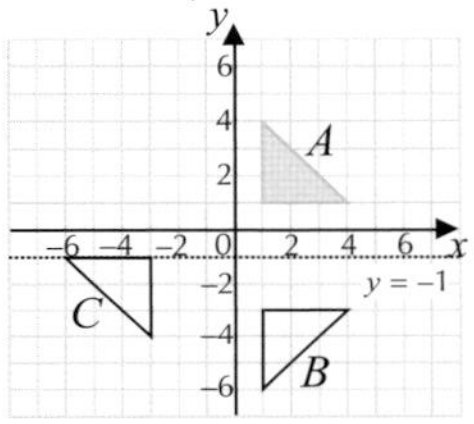

b) (i) Use tracing paper to draw the shape, then put your pencil on the origin and rotate the tracing paper by 90° clockwise. Draw the rotated shape D in its new position — see below.

(ii) Draw the mirror line at $y = -1$. Then reflect the vertices and join up the image points to form image E — see below.

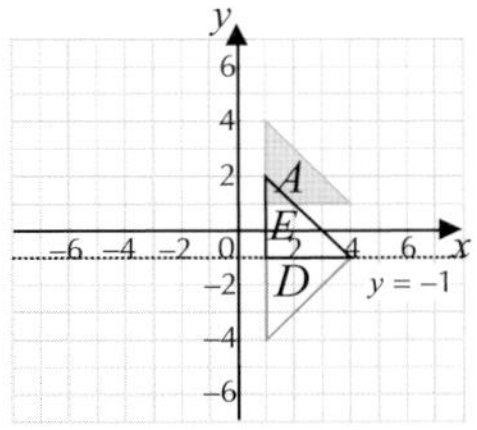

Notice that shapes C and E aren't in the same place even though you started with the same shape (A) and did the same transformations. This shows that it matters in which order you do the transformations.

2 **a) (i)** A_1 should be drawn 2 units directly down from A — see below.
Translate the vertices and then join them up to form the image.

(ii) Draw lines from (9, –7) through the vertices of A_1. The scale factor is 3, so extend the lines so that they are 3 times as long. Mark the new vertices at the ends of these lines and then join them up to make the new shape A_2 — see below.
You can check your answer by checking the sides of A_2 are 3 times as long as those of A and A_1.

b) (i) B_1 should be 2 units to the left of B and 1 unit down — see below.

(ii) Reflect the vertices and join up the image points to form image B_2 — see below.

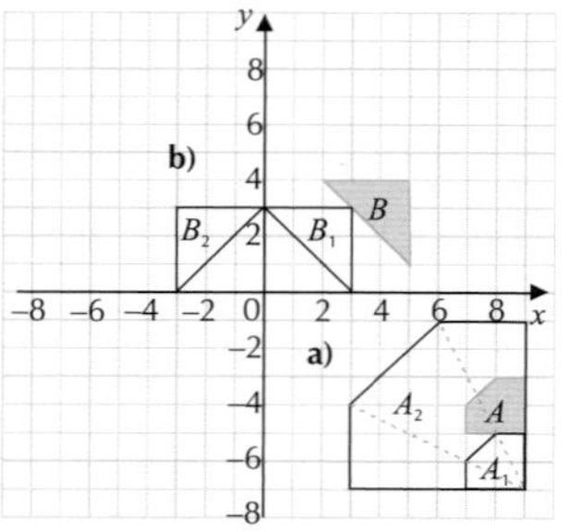

3 a) (i) Use tracing paper to draw shape P, then put your pencil on the point (4, 5) and rotate the tracing paper by 180°. Draw the rotated shape P_1 in its new position — see below.

(ii) P_2 should be 2 units to the right and 2 units up from P_1 — see below.

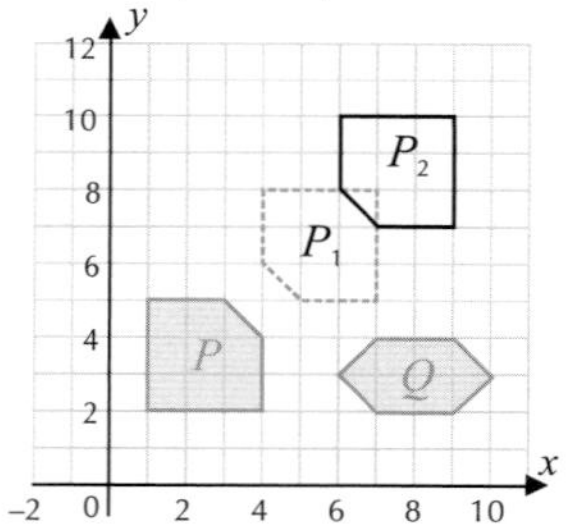

b) Place the tracing paper drawing from part a) back over P. Put your pencil on the point (5, 6) and rotate the tracing paper by 180°. Draw the rotated shape P_3 in its new position.

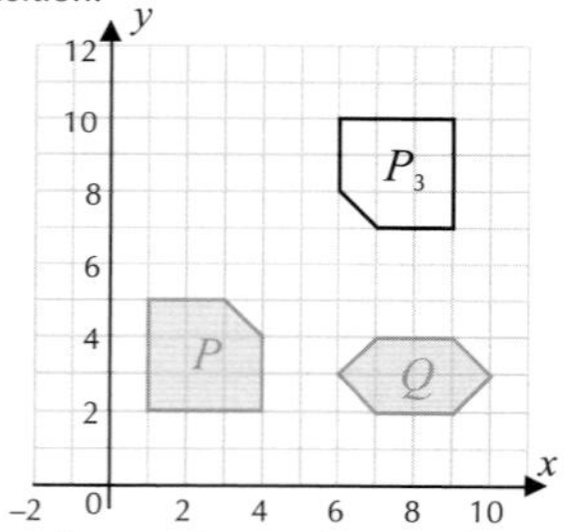

Since these rotations are by an angle of 180°, you could go clockwise or anticlockwise.

c) (i) Use tracing paper to draw shape Q, then put your pencil on the point (4, 5) and rotate the tracing paper by 180°. Draw the rotated shape Q_1 in its new position — see below.

(ii) Q_2 should be 2 units to the right and 2 units up from Q_1 — see below.

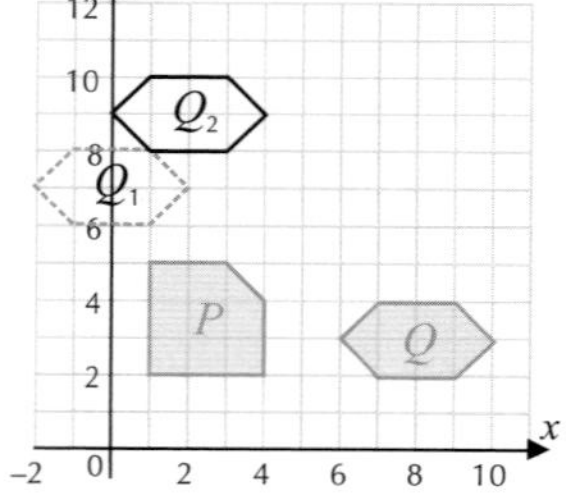

d) Place the tracing paper drawing from part c) back over Q. Put your pencil on the point (5, 6) and rotate the tracing paper by 180°. Draw the rotated shape Q_3 in its new position.

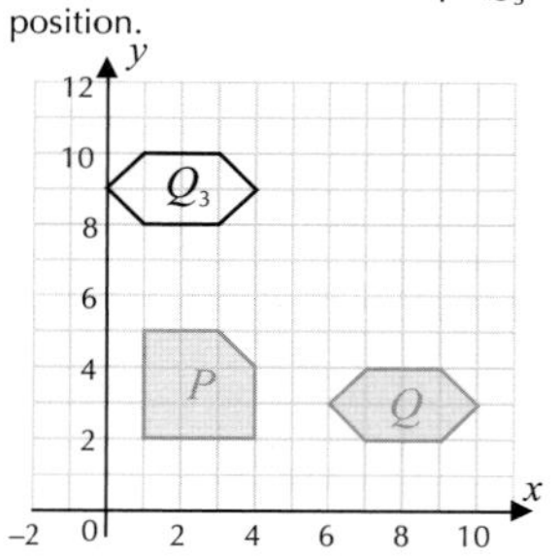

e) P_2 and P_3 are identical and in exactly the same position, and Q_2 and Q_3 are identical and in exactly the same position.

4 a) (i) Use tracing paper to draw the shape, then put your pencil on the point (4, 2) and rotate the tracing paper by 90° clockwise. Draw the rotated shape in its new position.

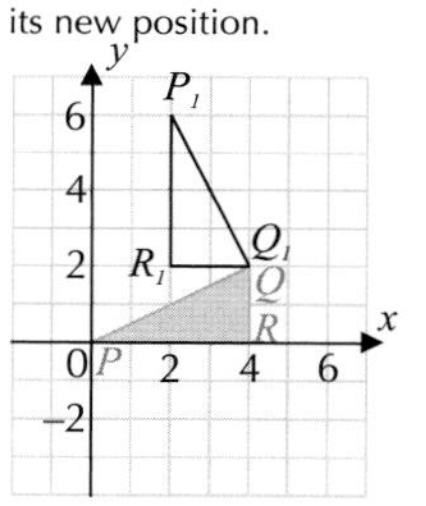

(ii) Draw lines from (2, 6) through the vertices of $P_1Q_1R_1$. The scale factor is 1.5, so extend the lines so that they are one and a half times as long. Mark the new vertices at the ends of these lines and then join them up to make the new shape $P_2Q_2R_2$.

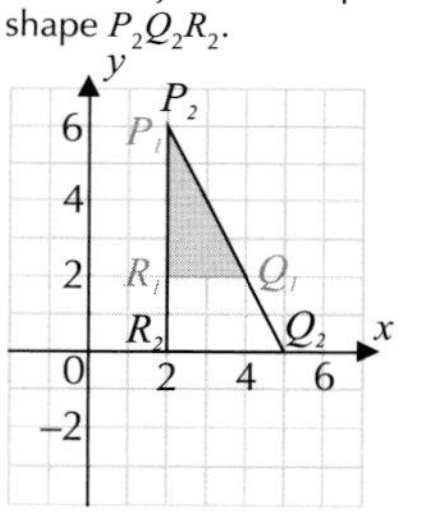

You can check your answer for the enlargement by dividing the side lengths of the image by the corresponding side lengths of the original shape, which should be equal to the scale factor.

b) (i) Draw lines from (2, 6) through the vertices of PQR. The scale factor is 1.5, so extend the lines so that they are one and a half times as long. Mark the new vertices at the ends of these lines and then join them up to make the new shape $P_3Q_3R_3$.

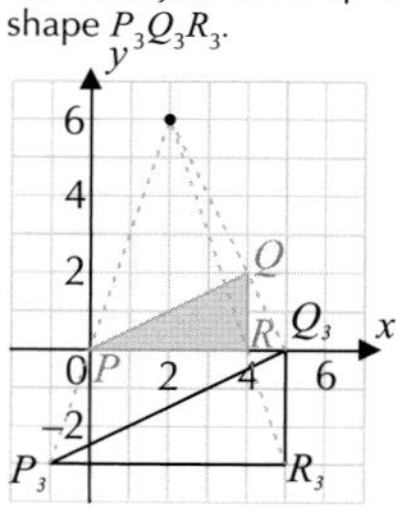

(ii) Use tracing paper to draw the shape, then put your pencil on the point (4, 2) and rotate the tracing paper by 90° clockwise. Draw the rotated shape in its new position.

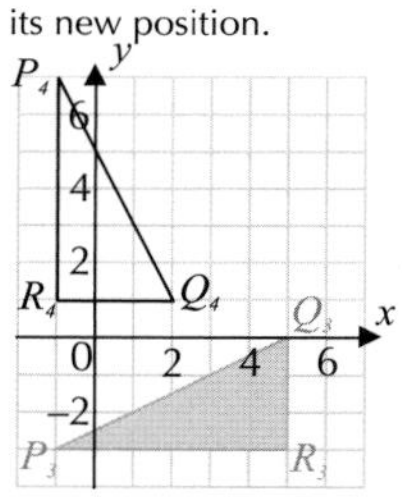

5 First draw the triangle onto a coordinate grid. Then find the image of the 180° rotation about the origin using tracing paper. Find the second image after the translation by moving the first image 6 units to the right and 2 units up. Then by comparing STU with the final image, you can see the final image is equivalent to a **rotation of 180° about (3, 1)**.

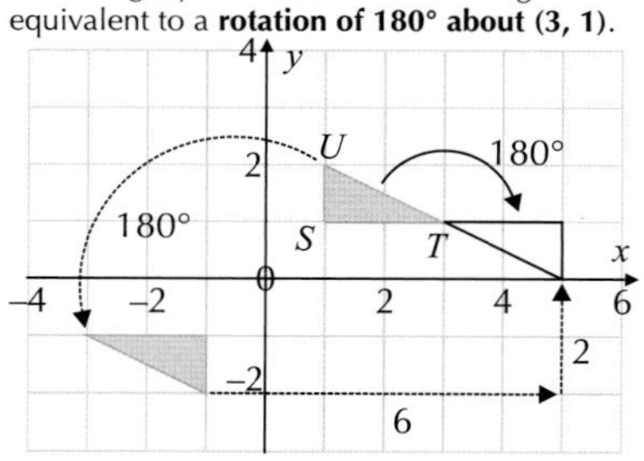

The rotations are 180° so you can rotate clockwise or anticlockwise.

6 a) The shapes are the same shape but different sizes so they are **similar**.

b) The shapes are the same shape and size so they are **congruent**.
Remember that orientation does not affect whether a shape is congruent or similar.

c) The shapes are different so are **neither congruent nor similar**.

7 a) The missing angle on right-hand triangle is $180° - 105° - 40° = 35°$, so two sides and the angle between on one triangle are the same as two sides and the angle between on the other triangle. Condition **SAS** holds so the triangles are **congruent**.

b) The 3 cm sides on the first triangle are 3 times the length of the 1 cm sides on the second. But the 5 cm side is not 3 times as long as the corresponding side on the second triangle ($1.5 \times 3 = 4.5$ cm $\neq 5$ cm). The lengths are not in the same ratio, so the triangles are **neither** similar nor congruent.

c) The missing angle on the left-hand triangle is $180° - 30° - 45° = 105°$.
The missing angle on the right-hand triangle is $180° - 105° - 30° = 45°$.
All the angles on one triangle are the same as the angles on the other triangle but corresponding sides are not the same length so the triangles are **similar**.

d) The missing angle on the left-hand triangle is $180° - 38° - 112° = 30°$, so there's no 40° angle in this triangle. The angles in one triangle are not all the same as the angles in the other triangle, so the triangles are **neither** similar nor congruent.

8 Using alternate angles a = **110°** and b = **45°**,
All angles in a triangle sum to 180° so
c = 180° – 45° – 110° = **25°**.
Using vertically opposite angles d = **25°**.
All of the angles in the two triangles are the same, so they are similar. The 9 cm and 6 cm sides are corresponding sides, so the scale factor is given by 9 ÷ 6 = 1.5.
e = 3.59 × 1.5 = **5.385 cm,**
f = 12 ÷ 1.5 = **8 cm**

Page 421 Exam-Style Questions

1 a) Draw in the mirror line $y = x$, reflect each point of A and then join them up to form the image B.

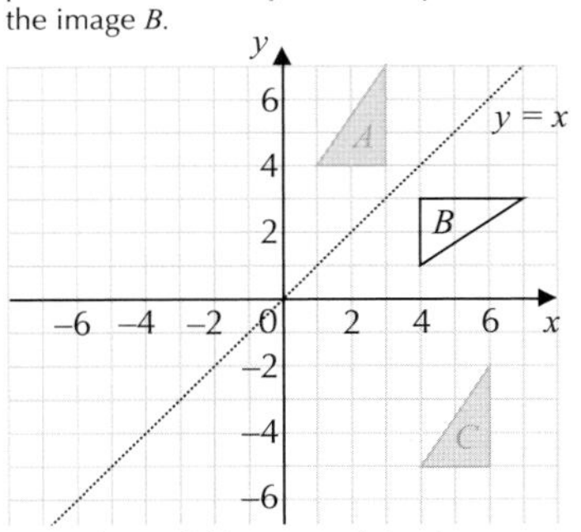

[2 marks available – 1 mark for the correct line $y = x$ used, 1 mark for shape B in the correct position.]
The vertices of the image should be the same perpendicular distance from the mirror line, but on the opposite side to the original shape.

b) Shape C needs to move 3 units left and 9 up so the translation vector would be $\begin{pmatrix}-3\\9\end{pmatrix}$.
So p = **–3** and q = **9**. *[1 mark]*

2 First, translate shape P by $\begin{pmatrix}1\\-4\end{pmatrix}$, so P moves 1 unit to the right and 4 units down to produce shape Q. Then enlarge shape Q by scale factor 2 with centre of enlargement (–2, 3) to produce shape R. Draw lines from (–2, 3) to each vertex of the shape. Extend these lines to twice as long. Mark the new vertices at the end of these lines and join them up to create the new shape R.

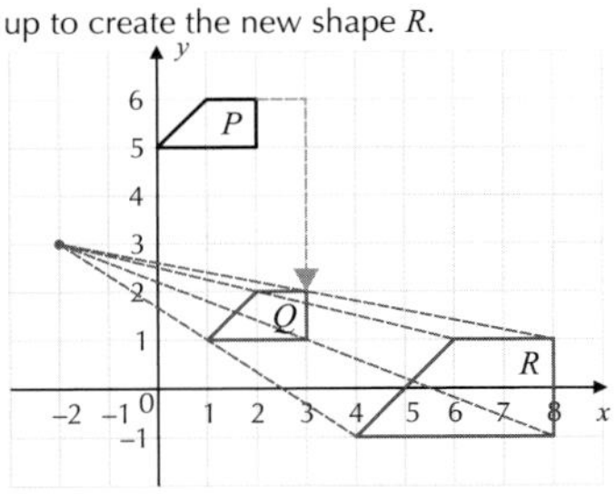

[3 marks available — 1 mark for translating P to the correct position to get Q, 1 mark for attempting to enlarge Q by scale factor 2 centred on (–2, 3), 1 mark for R drawn in the correct position]
You can check your answer for the enlargement by dividing the side lengths of the image by the corresponding side lengths of the original shape, which should be equal to the scale factor.

3 a) Scale factor = 8 ÷ 3 = $\frac{8}{3}$ *[1 mark]*
BC = 10 ÷ $\frac{8}{3}$ = **3.75 m** *[1 mark]*

b) The triangles are similar, so
angle ACB = angle AED = 180° – 29° – 99° = **52°** *[1 mark]*

Section 25 — Collecting Data

25.1 Using Different Types of Data

Page 423 Exercise 1

1 a) Megan collects the data herself, so it is **primary**.
b) Jane gets her data from the internet, so it is **secondary**.

2 Faheem gets his data from a newspaper, so it is **secondary**.

3 Nancy collects her data herself, so it is **primary**.

4 a) Data needed — **girls' answers to some questions about school dinners.**
Method of collecting — e.g. **Nikita could ask all the girls in her class to fill in a questionnaire.**
b) **Primary data**

5 a) Data needed — **daily rainfall figures for London and Manchester last August.**
Method of collecting — e.g. **Anne could look for rainfall figures on the internet.**
b) **Secondary data**

6 a) Data needed — **the distance an identical ball can be thrown by the boys and girls in his class.**
Method of collecting — e.g. **Skylar could ask everyone in the class to throw the same ball as far as they can. He could measure the distances and record them in a table, along with whether each thrower was male or female.**
b) **Primary data**

7 a) Data needed — **one set of data consists of the temperature readings in Jim's garden taken at 10 am each day. The other set consists of the Met Office's temperatures recorded for Jim's area at the same time.**
Method of collecting — e.g. **Jim could collect the data from his garden by taking readings from a thermometer. He can get the Met Office temperatures from their website. He should record both temperatures for each day in a table.**
b) Data collected in Jim's garden is **primary data**. The Met Office data is **secondary data**.

Page 424 Exercise 2

1 a) The number of words is numerical and it can only be whole values so this is **discrete quantitative**.
b) Foods are descriptive so this is **qualitative**.
c) The number of pets is numerical and it can only take whole values so this is **discrete quantitative**.
d) The size of a crowd is numerical and it can only take whole values so this is **discrete quantitative**.
e) Heights are numerical and they can take any value greater than zero so this is **continuous quantitative**.
f) Times are numerical and they can take any value greater than zero so this is **continuous quantitative**.
g) Nationalities are descriptive so this is **qualitative**.
h) Lengths are numerical and they can take any value greater than zero so this is **continuous quantitative**.
i) Distances are numerical and they can take any value greater than zero so this is **continuous quantitative**.
j) Colours are descriptive so this is **qualitative**.

2 a) Janka collects the data herself, so it is **primary data**.
b) Drinks are descriptive, so this is **qualitative**.

3 a) One set of data is the **average number of chocolate bars eaten each week by each pupil**. The other set of data is the **time it takes these pupils to run 100 metres**.
b) Gemma could **ask each pupil how many chocolate bars they eat on average each week**. She could **time how long it takes each pupil to run 100 m** and **record the data in a table, along with the chocolate bar data**.
c) The number of chocolate bars is numerical and can only take whole values so it is **discrete quantitative data**.
The running time is numerical and can take any value greater than zero so it is **continuous quantitative data**.
d) Both sets of data are **primary data**.

25.2 Data Collection

Page 425 Exercise 1

1 There is **no 'tally' column**.
There could be a car in the car park that isn't one of the colours listed — **the rows do not cover every possibility**.

2 a) E.g.

Visits	Tally	Frequency
0		
1		
2		
3		
More than 3		

b) E.g.

Destination	Tally	Frequency
UK		
Spain		
USA		
Asia		
Other		

Make sure your groups cover all possible options, e.g. by having an 'other' or 'more than...' row. Also, none of your groups should overlap — e.g. if you had a row for New York and for USA, then these would overlap (if you went to New York, then you also went to the USA).

Page 426 Exercise 2

1 a) **There are gaps between the data classes. The rows do not cover every possibility.**
E.g.

No. of people	Tally	Frequency
0 - 5000		
5001 - 10 000		
10 001 - 15 000		
15 001 - 20 000		

b) **The data classes overlap. The rows do not cover every possibility.**
E.g.

Weight (w kg)	Tally	Frequency
$w \le 3$		
$3 < w \le 3.5$		
$3.5 < w \le 4$		
$w > 4$		

2 **a)** The heights are continuous so you need classes with inequalities to cover values ranging between 5 and 27.
E.g. $5 \le h < 10$ $10 \le h < 15$
$15 \le h < 20$ $20 \le h < 25$
$25 \le h \le 27$

b) The number of quiz questions answered correctly is discrete. All whole numbers between 0 and 20 are possible so you need classes with grouped values.
E.g. **0-4 5-8 9-12 13-16 17-20**

c) The weights are continuous so you need classes with inequalities covering a sensible range around 200 g.
E.g. $w \le 180$ $180 < w \le 190$
$190 < w \le 200$ $200 < w \le 210$
$w > 210$

d) The volumes are continuous so you need classes with inequalities covering a sensible range up to 300 ml.
E.g. $v \le 260$ $260 < v \le 270$
$270 < v \le 280$ $280 < v \le 290$
$290 < v \le 300$

3 **a)** E.g.

Time (t hrs)	Tally	Frequency
$t \le 5$		
$5 < t \le 10$		
$10 < t \le 20$		
$20 < t \le 40$		
$t > 40$		

b) E.g.

No. of pairs	Tally	Frequency
0 - 4		
5 - 8		
9 - 12		
13 - 16		
17 or more		

c) E.g.

Length (s cm)	Tally	Frequency
$s \le 15$		
$15 < s \le 20$		
$20 < s \le 25$		
$25 < s \le 30$		
$s > 30$		

d) E.g.

Distance (d km)	Tally	Frequency
$d \le 5$		
$5 < d \le 10$		
$10 < d \le 20$		
$20 < d \le 40$		
$d > 40$		

Page 428 Exercise 3

1 **a) There are not enough hair colour data classes.**
The age data classes overlap.
The data is for adults so doesn't need a 0-15 data class.

b) E.g.

Hair Colour	Age (in whole years)				
	18-30	31-45	46-60	61-75	76 or more
Blonde					
Light brown					
Dark brown					
Ginger					
Grey					
Other					

2 **a)** E.g.

Season	Gender	
	Male	Female
Spring		
Summer		
Autumn		
Winter		

b) E.g.

Music	Age Group	
	Adult	Child
Pop		
Classical		
Rock		

c) E.g.

Time spent (t hours)	School Year				
	7	8	9	10	11
$t \le 1$					
$1 < t \le 2$					
$2 < t \le 3$					
$3 < t \le 4$					

d) E.g.

Total books read last year	Favourite Genre			
	Horror	Romance	Sci-Fi	Fantasy
0-10				
11-20				
21-30				
More than 30				

3 **a)** E.g.

Cats or dogs	Gender	
	Male	Female
Cats		
Dogs		

b) E.g.

TV time (t hours)	Age Group	
	Adult	Child
$t \le 1$		
$1 < t \le 2$		
$2 < t \le 3$		
$3 < t \le 4$		
$t > 4$		

c) E.g.

Transport	School Year				
	7	8	9	10	11
Walking					
Bus					
Car					
Bicycle					
Train					
Other					

d) E.g.

Height (h cm)	No. of fruit portions eaten			
	0-1	2-3	4-5	6 or more
$h \le 120$				
$120 < h \le 140$				
$140 < h \le 160$				
$160 < h \le 180$				
$h > 180$				

4 **a) The favourite subject data classes don't cover every option.**
Olivia wants to know the most popular subject in each year group so the year group data classes shouldn't be grouped.

b) E.g.

Year	Favourite Subject					
	English	Maths	P.E.	Science	Music	Other
Year 7						
Year 8						
Year 9						
Year 10						
Year 11						

25.3 Sampling and Bias

Page 430 Exercise 1

1 The population is all the students of that university, and the sample is the students who complete the survey.

2 E.g. It would take too long to time all 216 pupils and a smaller sample would create less disruption in the school routine.

3 E.g. It would take a long time to test all the matches, and Patrick wouldn't have any matches left to use at the end of it.

4 E.g. Khadija's sample is too small so her results could be very unrepresentative of the whole audience.

5 E.g. There is a random element to the results of tossing a coin, so different samples will usually give different results.

6 Alfie's sample was bigger than Keith's and would be expected to be more accurate, so 'chocolate' is more likely to be the most popular flavour.

7 E.g. Nikhil's idea is best, as only tasting one cake would not be reliable, and tasting 50 out of the 200 would take too long and use up a quarter of their cakes.

Page 431 Exercise 2

1 a) Kelechi's friends could have similar opinions about the film.

b) Animal rights activists are more likely to want vegetarian food, so the sample will be biased towards vegetarian food.

c) People using the library on a Monday are unlikely to want it to close on a Monday, so the sample will be biased away from a Monday closure.

d) People at work probably won't be able to answer the phone in the afternoon, so the only replies they will get will be from people who don't work, home workers, and people with the day off.

Page 431 Exercise 3

1 E.g. The teacher could make a list of all the Year 7 pupils and assign each pupil a number. He could then generate 50 random numbers with a computer or calculator, and match the numbers to the pupils to create the sample.

2 E.g. The manager could assign each of the female members on her database a number, generate 40 random numbers with a computer or calculator, and match the numbers to the members to create the sample.

Page 432 Exercise 4

1 $7 \div 20 = 0.35$ of the sampled rooms have damp, so you could expect $0.35 \times 450 = 157.5$ $\approx$ **158 rooms** in the castle to have damp.
The calculated answer is a decimal, so you need to round your answer to the nearest whole number.

2 a) (i) $34 \div 50 = 0.68$ so you could expect $0.68 \times 650 =$ **442 politicians** to be in favour of policy X.

(ii) $1 \div 50 = 0.02$ so you could expect $0.02 \times 650 =$ **13 politicians** to be neutral about policy Y.

(iii) $14 \div 50 = 0.28$ so you could expect $0.28 \times 650 =$ **182 politicians** to be against policy Z.

b) This assumes that the sample of 50 politicians is **representative of the population of 650 politicians**.

Page 433 Review Exercise

1 a) E.g. The data he needs is his classmates' favourite colours. He could use a tally chart to record his classmates' answers.

b) He collects the data himself so it is primary data.

2 a) The time taken to travel to school is numerical and can take any value greater than zero so it is continuous quantitative.

b) The number of siblings is numerical and can only take certain values so it is discrete quantitative.

c) Films are descriptive so it is qualitative.

d) Shoe prices are numerical and only take certain values so it is discrete quantitative.

3 E.g.

Days in birth month	Tally	Frequency
28		
29		
30		
31		

4 **There is no row for someone who went to the cinema 0 times.**
The data classes overlap.
E.g.

No. of trips	Tally	Frequency
0 - 10		
11 - 20		
21 - 30		
31 - 40		
41 or more		

5 E.g.

Average time spent outside per week (t hours)	Age Group	
	Adult	Child
$t \leq 3$		
$3 < t \leq 6$		
$6 < t \leq 9$		
$9 < t \leq 12$		
$t > 12$		

6 E.g. It would be **expensive/time-consuming/impractical** to interview everyone in the town.

7 Jasper's sample will be a much higher percentage of the total number of pupils, and Connie's sample is not random.
For example, it could be a group of friends who may all support the same football team.

8 E.g. If the factory runs for 24 hours a day, test one freshly made component roughly every 30 minutes (or at 50 random times throughout the day).

9 $4 \div 12 = 0.333...$ of the sample prefer jam, so you could expect that $0.333... \times 50 = 16.666... \approx$ **17 party guests** would prefer jam doughnuts.

Page 434 Exam-Style Questions

1 a) Distances can take any value greater than zero so the data is **continuous data** *[1 mark]*.

b) E.g.

Distance (s metres)	Left- or Right-handed	
	Left-handed	Right-handed
$s \leq 5$		
$5 < s \leq 10$		
$10 < s \leq 15$		
$15 < s \leq 20$		
$t > 20$		

[2 marks available — 1 mark for using a two-way table, 1 mark for choosing appropriate data classes]

2 E.g. Jill wants to find out about her entire year but the students in the sample are all from the same class so their views might not accurately represent the whole year *[1 mark]*.
Jill's sample is quite small — there are only 10 students in the sample to represent the views of 150 students, so the results could be unreliable *[1 mark]*.

3 a) The data classes don't cover every possibility — there's no row for taking longer than 30 minutes *[1 mark]*.
The data classes overlap *[1 mark]*.

b) $12 \div 40 = 0.3$ of the sample take between 10 and 20 minutes *[1 mark]*, so Tahani can expect $0.3 \times 200 = 60$ participants to take between 10 and 20 minutes to complete the puzzle *[1 mark]*. It is assumed that the sample is representative of the whole population of 200 participants *[1 mark]*.

Section 26 — Analysing Data

26.1 Tally Charts and Frequency Tables

Page 435 Exercise 1

1 a) Write a tally mark for every response and then count the marks to find the frequency:

Type of Music	Tally	Frequency
Classical	|	1
Indie	|||	3
Pop	卌 ||||	9
Rock	|||	3
Other	||||	4

b) (i) **Pop** music is the type with the highest frequency.

(ii) The frequency for pop is 9 and for rock is 3. So the difference is $9 - 3 =$ **6**.

26.2 Averages and Range

Page 436 Exercise 1

In questions 1-3, you need to count up how many times the different data values appear. The mode (or modal value) is the value that appears most.

1 a) **8** b) **6** c) **8**

2 **58**

3 **red**

4 E.g. There are two modes for this data, 3 and 6, so the mode does not give a good indication of the average value of the data.

Page 437 Exercise 2

1 a) The numbers in order are: 3, 4, 5, 8, 8. The middle value is the 3rd value, so the median is **5**.

b) The numbers in order are: 2, 5, 6, 6, 6, 7, 7, 9, 9. The middle value is the 5th value, so the median is **6**.

c) The numbers in order are: 8, 8, 12, 13, 16, 17, 17, 18. The middle values are the 4th and 5th values, so the median is halfway between 13 and 16: $(13 + 16) \div 2 =$ **14.5**.
You might find it easier to find the position of the median using the formula $(n + 1) \div 2$, where n is the number of values.

Using the same method for Q2-3:

2 a) **3.8** b) **16** c) **3**

3 **81 seconds**

4 The known data values in order are 14.3, 15.8, 16.5, 16.9, 18.9. There are six values in total, so 16.7 (the median) must be between the third and fourth values. So the missing value must be greater than 16.5, i.e. 16.5 is the third value. 16.7 is halfway between 16.5 and 16.9, so the missing value must be at least **16.9**.

Page 438 Exercise 3

1 a) Total $= 8 + 5 + 3 + 8 + 4 = 28$
Mean $= 28 \div 5 =$ **5.6**

b) Total $= 6 + 9 + 2 + 7 + 7 + 6 + 5 + 9 + 6 = 57$
Mean $= 57 \div 9 = 6.333...$ $=$ **6.33** (2 d.p.)

c) Total $= 16 + 18 + 12 + 13 + 13 + 8 + 8 + 17 = 105$
Mean $= 105 \div 8 =$ **13.125**

Using the same method for Q2-4:

2 **a)** **3.68** **b)** **18.075** **c)** **3.125**

3 60.222... = **60.2 marks** (1 d.p.)

4 **83 seconds**

5 **a)** Put the data in order: 1, 2, 2, 2, 3, 3, 5, 8
Total = 1 + 2 + 2 + 2 + 3 + 3 + 5 + 8 = 26
Mean = 26 ÷ 8 = **3.25**
The middle values are the 4th and 5th values, so the **median** is halfway between 2 and 3: (2 + 3) ÷ 2 = **2.5**
2 appears more than any other number so the **mode** = **2**

b) Put the data in order: 1, 2, 3, 3, 3, 3, 4, 5
Total = 1 + 2 + 3 + 3 + 3 + 3 + 4 + 5 = 24
Mean = 24 ÷ 8 = **3**
The middle values are the 4th and 5th values, so the **median** is halfway between 3 and 3, so **3**.
3 appears more than any other number so the **mode** = **3**

6 Including the missing value there are 6 values, so the total of the values
= the mean × number of values = 7 × 6
= 42. To find the missing value, subtract the known values from the total:
42 – (6 + 5 + 8 + 8 + 5) = 42 – 32 = **10**

7 Including the missing value there are 6 values, so the total of the values
= the mean × number of values = 16.3 × 6
= 97.8. To find the missing value, subtract the known values from the total:
97.8 – (16.6 + 16.9 +15.8 + 14.3 + 18.9)
= 97.8 – 82.5 = **15.3**

Page 439 Exercise 4

1 **a)** Highest value = 8, lowest value = 3, so the range = 8 – 3 = **5**

b) Highest value = 9, lowest value = 2, so the range = 9 – 2 = **7**

c) Highest value = 17, lowest value = 8, so the range = 17 – 8 = **9**

Using the same method for Q2:

2 **a)** **4.1** **b)** **6.19** **c)** **1**

3 22 – (–2) = 22 + 2 = **24 °C**

4 £30 – (–£40) = £30 + £40 = **£70**

5 The range is 6, so the two possible values are the highest known number minus 6, or the lowest known number plus 6.
Missing number = 8 – 6 = **2**, or 5 + 6 = **11**

6 **a)** Highest value = 17, lowest value = 3, so the range = 17 – 3 = **14 skips**

b) E.g. Most of the data is much closer together than the range suggests, so the range is not a good measure of spread. The range is affected by the data value 17, which is an outlier.

Page 441 Exercise 5

1 **a)** The frequency is highest for 1 pet, so mode = **1 pet**.

b) The total frequency is 5 + 10 + 5 + 5 + 2 = 27, so the median position is (27 + 1) ÷ 2 = 14. The 14th value is 1, so the median = **1 pet**.

c) Find the frequency for each group:
(i) 5 **(ii) 10** **(iii) 5**
(iv) 5 **(v) 2**

d) The total number of pets
= (0 × 5) + (1 × 10) + (2 × 5) + (3 × 5) + (4 × 2)
= 0 + 10 + 10 + 15 + 8 = **43**

e) The total number of students
= 5 + 10 + 5 + 5 + 2 = **27**

f) Mean = 43 ÷ 27 = 1.592...
= **1.6 pets** (1 d.p.)

2 **a)** The frequency is highest for 4 people, so mode = **4 people**.

b) There are 30 data values, so the position of the median is (30 + 1) ÷ 2 = 15.5.
The median is halfway between the 15th value = 2 and 16th value = 2, so median = **2 people.**

c) Total number of people
= (1 × 7) + (2 × 9) + (3 × 1) + (4 × 10) + (5 × 3) = 83
Mean = 83 ÷ 30 = 2.766...
= **2.8 people** (1 d.p.)

d) Smallest value = 1, largest value = 5, so the range is 5 – 1 = **4**.

3 **a)** Total frequency = 4 + 9 + 2 + 5 + 4 + 6 = 30
There are 30 data values, so the median position is (30 + 1) ÷ 2 = 15.5. The median is halfway between the 15th value = 18 and 16th value = 19,
so median = (18 + 19) ÷ 2 = **18.5 °C**.

b) Total of temperatures
= (16 × 4) + (17 × 9) + (18 × 2) + (19 × 5) + (20 × 4) + (21 × 6)
= 64 + 153 + 36 + 95 + 80 + 126
= 554
Mean = 554 ÷ 30 = 18.466...
= **18.5 °C** (1 d.p.)

c) E.g. The student's garden has experienced some fairly typical June temperatures for the UK.

Page 443 Exercise 6

1 **a)** The group with the highest frequency is 16-20 so the modal group is **16-20**.

b) The total frequency = 4 + 5 + 7 + 9 = 25, so the position of the median is (25 + 1) ÷ 2 = 13. The 13th value is in the 11-15 group so the group containing the median = **11-15**.

c)

Marks scored	1-5	6-10	11-15	16-20
Frequency	4	5	7	9
Midpoint	$\frac{1+5}{2}$ = 3	$\frac{6+10}{2}$ = 8	$\frac{11+15}{2}$ = 13	$\frac{16+20}{2}$ = 18
Freq. × Midpoint	4 × 3 = 12	5 × 8 = 40	7 × 13 = 91	9 × 18 = 162

Total frequency = 4 + 5 + 7 + 9 = 25
Total no. of marks = 12 + 40 + 91 + 162 = 305
Mean = 305 ÷ 25 = **12.2**

d) Lower bound of smallest group = 1
Upper bound of largest group = 20
Range = 20 – 1 = **19**

2 **a)**

Weight (w)	Freq.	Midpoint	Freq. × Midpoint
$0 \le w < 20$	1	(0 + 20) ÷ 2 = 10	1 × 10 = 10
$20 \le w < 40$	6	(20 + 40) ÷ 2 = 30	6 × 30 = 180
$40 \le w < 60$	9	(40 + 60) ÷ 2 = 50	9 × 50 = 450
$60 \le w < 80$	24	(60 + 80) ÷ 2 = 70	24 × 70 = 1680

Total frequency = 1 + 6 + 9 + 24 = 40
Total weight = 10 + 180 + 450 + 1680
= 2320 grams
Mean = 2320 ÷ 40 = **58 grams**

b) Lower bound of smallest group = 0
Upper bound of largest group = 80
Range = 80 – 0 = **80 grams**

c) E.g. You don't know where the data values lie within the groups. The one tangerine in the 0 ≤ w < 20 g group could weigh 19 g, in which case the actual range would be no greater than 61 g.

3 **a)** The group 11-15 has the highest frequency, so the modal group is **11-15**.

b) There are 3 + 8 + 11 + 4 = 26 data values, so the position of the median is (26 + 1) ÷ 2 = 13.5. The median is halfway between the 13th and 14th values. Both values are in the group 11-15, so the median is in **11-15**.

c)

Time (hours)	Freq.	Midpoint	Freq. × Midpoint
0-5	3	(0 + 5) ÷ 2 = 2.5	3 × 2.5 = 7.5
6-10	8	(6 + 10) ÷ 2 = 8	8 × 8 = 64
11-15	11	(11 + 15) ÷ 2 = 13	11 × 13 = 143
16-20	4	(16 + 20) ÷ 2 = 18	4 × 18 = 72

Total frequency = 3 + 8 + 11 + 4 = 26
Total time = 7.5 + 64 + 143 + 72
= 286.5 hours
Mean = 286.5 ÷ 26 = **11.0 hours** (1 d.p.)

4 **a)** The group $1.60 \le h < 1.70$ has the highest frequency, so the modal group is **$1.60 \le h < 1.70$**.

b) There are 200 data values, so the position of the median is (200 + 1) ÷ 2 = 100.5. The median is halfway between the 100th and 101st values. Both values are in the group $1.60 \le h < 1.70$, so the median is in **$1.60 \le h < 1.70$**.

c)

Height (h) in m	Freq.	Midpoint	Freq. × Midpoint
$1.50 \le h < 1.60$	27	(1.5 + 1.6) ÷ 2 = 1.55	27 × 1.55 = 41.85
$1.60 \le h < 1.70$	92	(1.6 + 1.7) ÷ 2 = 1.65	92 × 1.65 = 151.8
$1.70 \le h < 1.80$	63	(1.7 + 1.8) ÷ 2 = 1.75	63 × 1.75 = 110.25
$1.80 \le h < 1.90$	18	(1.8 + 1.9) ÷ 2 = 1.85	18 × 1.85 = 33.3

Total frequency = 27 + 92 + 63 + 18
= 200
Total height = 41.85 + 151.8 + 110.25 + 33.3 = 337.2 m
Mean = 337.2 ÷ 200 = **1.69 m** (3 s.f.)

d) Lower bound of smallest group = 1.50 m
Upper bound of largest group = 1.90 m
Range = 1.90 – 1.50 = **0.4 m**

26.3 Two-Way Tables

Page 445 Exercise 1

1 **a)**

	Walk	Bus	Car	Total
Boys	9	5	3	9 + 5 + 3 = **17**
Girls	5	8	6	5 + 8 + 6 = **19**
Total	9 + 5 = **14**	5 + 8 = **13**	3 + 6 = **9**	17 + 19 = **36**

b)

	Walk	Bus	Car	Total
Boys	21 – 11 = **10**	27 – 15 = **12**	8	60 – 30 = **30**
Girls	11	30 – 11 – 4 = **15**	4	30
Total	21	60 – 21 – 12 = **27**	8 + 4 = **12**	60

It's okay if you've used different calculations to work out the correct missing values.

2 **a)** Adding the entries for each column and row gives:

	Germany	France	Spain	Total
Male	16	7	12	**35**
Female	18	4	16	**38**
Total	**34**	**11**	**28**	**73**

b) The entry in the female row and France column is **4**.
c) The entry in the male row and Germany column is **16**.
d) The total of the Spain column is **28**.
e) The total of the female row is **38**.
f) The total number of students (in the bottom right cell) is **73**.

3 **a)**

	Choc.	Crisps	Jellies	Total
Male	3	12 – 7 = **5**	12 – 2 = **10**	3 + 5 + 10 = **18**
Female	6 – 3 = **3**	7	2	30 – 18 = **12**
Total	6	12	30 – 6 – 12 = **12**	30

b) The total of the crisps column is **12.**
c) The entry in the female row and jelly column is **2**.
d) Number of males who prefer chocolate = 3
Total number of males = 18
Percentage = (3 ÷ 18) × 100 = 16.666...
= **16.7%** (1 d.p.)

4 **a)** Using the same method for completing the table as in Q1-3:

	Red	Black	Blue	White	Total
Cars	8	7	4	**3**	22
Vans	**2**	2	1	10	**15**
Motorbikes	2	1	**1**	2	**6**
Total	12	**10**	6	**15**	**43**

b) The entry in the motorbikes row and blue column is **1**.
c) The total of the vans row is **15**.
d) (i) Total number of cars = 22
Total number of vehicles = 43
(22 ÷ 43) × 100 = 51.162...
= **51.2%** (3 s.f.)
(ii) Total number of vans = 15
Total number of vehicles = 43
(15 ÷ 43) × 100 = 34.883...
= **34.9%** (3 s.f.)
(iii) Total number of red vehicles = 12
Total number of vehicles = 43
(12 ÷ 43) × 100 = 27.906...
= **27.9%** (3 s.f.)

26.4 Bar Charts and Pictograms

Page 446 Exercise 1

1

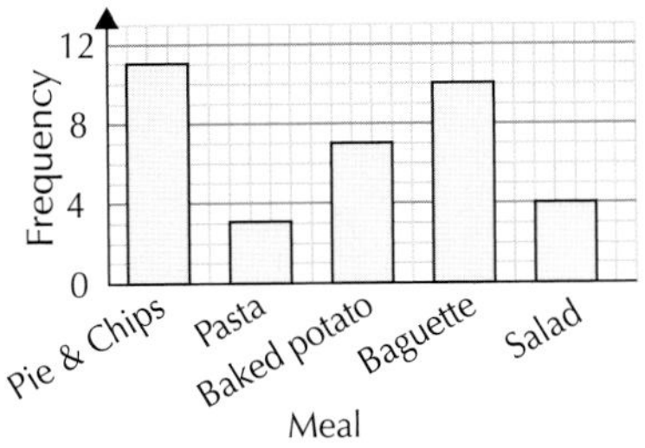

Remember to label your axes and to leave gaps between the bars.

2

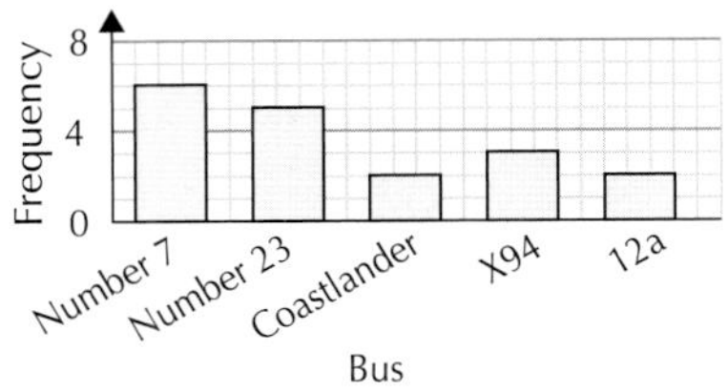

3

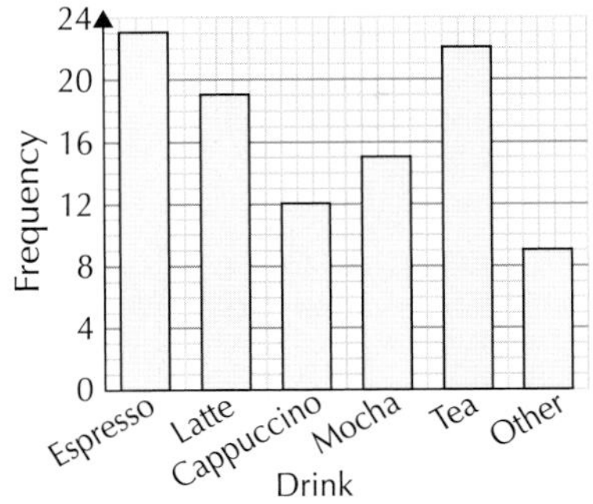

Page 448 Exercise 2

1 **a)**

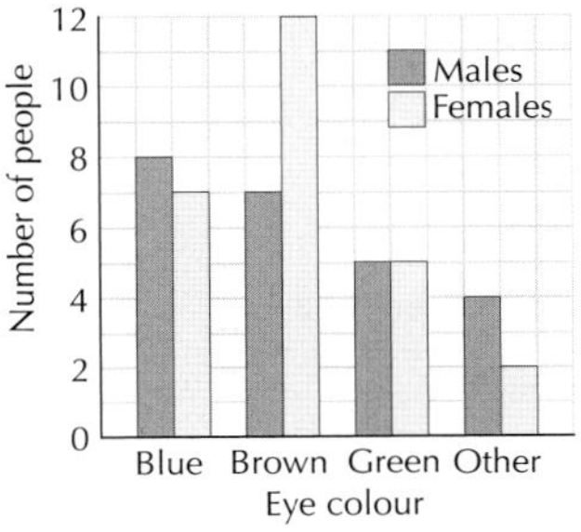

b)

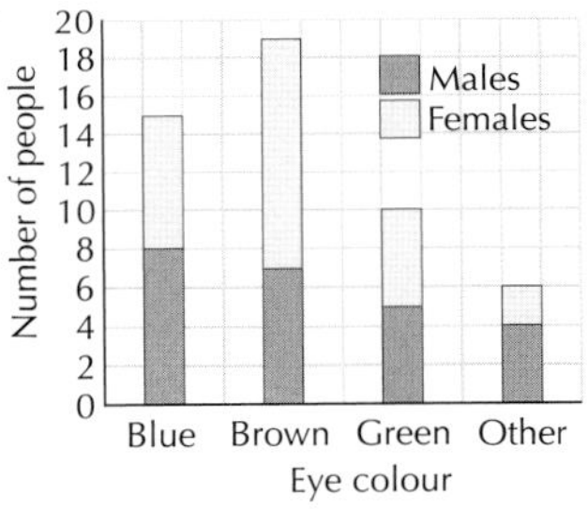

c) The **dual bar chart** is the best for comparing male and female eye colours — it is easy to see the difference in the heights of the pair of bars for each eye colour.
d) The most common eye colour is **brown**.
e) The **composite bar chart** is easier to use to find the modal eye colour, as you can just look for the tallest bar.

2 **a)** The height of the composite bar for the 26-35 age group is **12**.
b) 12 + 12 + 9 + 6 = **39**
c) The mode is the most common, so look for the tallest bar for women: **16-25 yrs**.
d) 8 men aged 26-35 use the gym.
4 women aged 26-35 use the gym.
So 8 – 4 = **4** more men aged 26-35 use the gym than women aged 26-35.
e) 16-25: Difference = 6 – 6 = 0
26-35: Difference = 4 (from part d) above)
36-45: Difference = 7 – 2 = 5
46-55: Difference = 5 – 1 = 4
So the age group with the greatest difference is **36-45 yrs**.

Page 449 Exercise 3

1

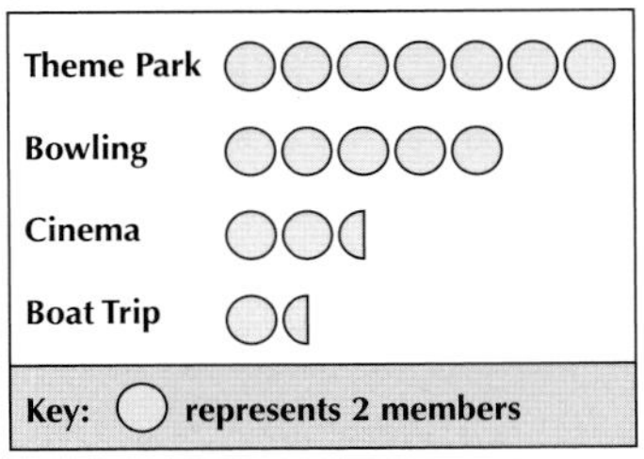

Use half a symbol to represent 1 member.

2 **a) (i)** In the Tuesday row there are 2 complete symbols, so there were 20 × 2 = **40** packets of sweets sold.
(ii) In the Wednesday row there are one and a half symbols, so there were 20 × 1.5 = **30** packets of sweets sold.

b)

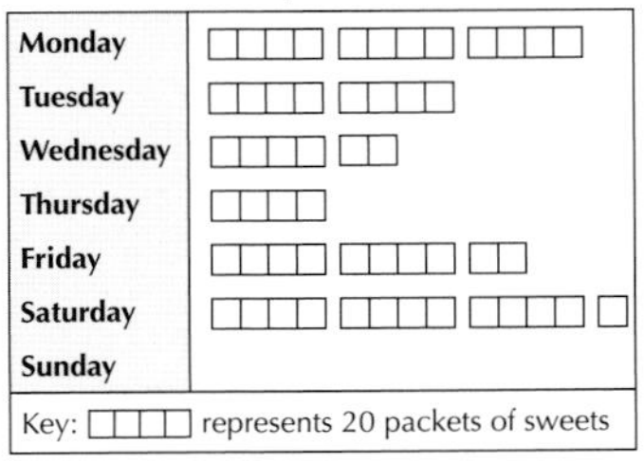

3 **a)** In the number 4 row there are 2 and a half symbols, so there were 4 × 2.5 = **10** letters delivered.
b) Number 1 has the fewest symbols in its row.
c) In the whole pictogram there are 8 whole symbols, 3 half symbols and one quarter symbol, so total letters
$= (8 \times 4) + (3 \times \frac{4}{2}) + (1 \times \frac{4}{4}) = 32 + 6 + 1$
= **39**

26.5 Stem and Leaf Diagrams

Page 451 Exercise 1

1

4	1 8	**Key: 4 | 1 means 41**
5	1 4 9	
6	5 5 9	
7	4	
8	0 6 9	

Don't forget to order the leaves.

2 **a)**

3	1 4	**Key: 3 | 1 means 3.1**
4	0 4 4 9	
5	1 3 4 7 7 9	
6	0 1	
7	1 7	

b)

20	3 5	**Key: 20 | 1 means 201**
30	1 4	
40	2 2 3 4 9 9	
50	0 1 3	
60	1 6 8	

3

0	0 0 0 1 3 6	
1	1 3 6	
2	0 5 7	
3	1 6 8	**Key: 0 | 1 means 0.1 cm**
4	1	

4 E.g.

Set 2		Set 1
	0	1 8
9 8 5 3 3	1	2 8
7 5 3 2 2	2	4 8 9
2	3	2 3 7 8

Key: 1 | 2 for Set 1 means 12
3 | 1 for Set 2 means 13

You might have put Set 1 on the left and Set 2 on the right.

Page 452 Exercise 2

1 a) Mode = most common value = **32**

b) There are 13 values so the median is the 7th value. Median = **20**

Remember, the median position for n values is (n + 1) ÷ 2.

c) Range = 32 – 5 = **27**

2 a)

10	2 6
11	4 7
12	4 8 9
13	1 3 6 9 9 9
14	2
15	4
16	
17	3

b) **Mode** = most common value = **13.9 seconds**
There are 16 values so the median lies halfway between the 8th value = 13.1 and the 9th value = 13.3.
Median = (13.1 + 13.3) ÷ 2 = **13.2 seconds**.
Range = 17.3 – 10.2 = **7.1 seconds.**

c) **7** contestants finished in under 13 seconds.

3 a) There are 15 data values, so the median is the 8th value.
Median for 'at rest' = **72 bpm**
Median for 'after exercise' = **77 bpm**

b) The higher median for the 'after exercise' data suggests that, on average, **people's heart rate was faster after they'd exercised.**

26.6 Pie Charts

Page 454 Exercise 1

1 a) 25 + 17 + 8 + 10 = **60**

b) 360° ÷ 60 = **6°**

c) **Black** = 25 × 6° = **150°**,
Silver = 17 × 6° = **102°**,
Red = 8 × 6° = **48°**,
Other = 10 × 6° = **60°**

d) Use the angles from part c) to draw and label the pie chart:

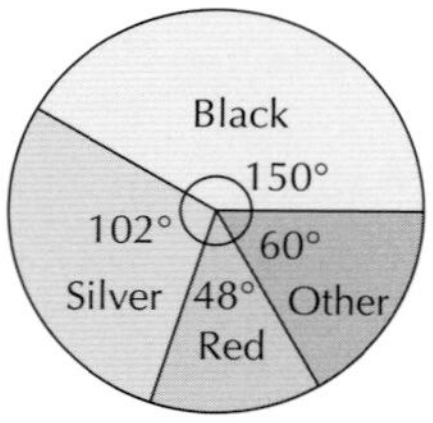

2 a) 13 + 9 + 8 + 6 = **36**

b) Degrees for 1 friend = 360° ÷ 36 = 10°, so:
Carlisle = 13 × 10° = **130°**,
Kendal = 9 × 10° = **90°**,
Millom = 8 × 10 = **80°**,
Bristol = 6 × 10° = **60°**

c) Use the angles from part b) to draw and label the pie chart:

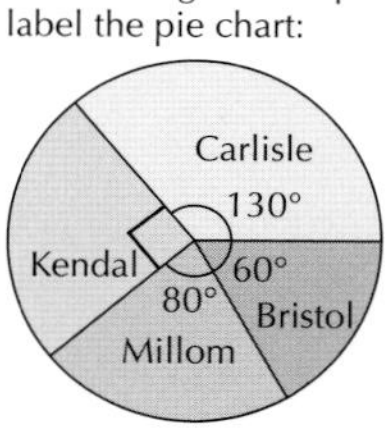

3 Total frequency = 33 + 52 + 21 + 14 = 120
Angle for one person = 360° ÷ 120 = 3°
Squash = 33 × 3° = 99°
Gym = 52 × 3° = 156°
Swimming = 21 × 3° = 63°
Table Tennis = 14 × 3° = 42°

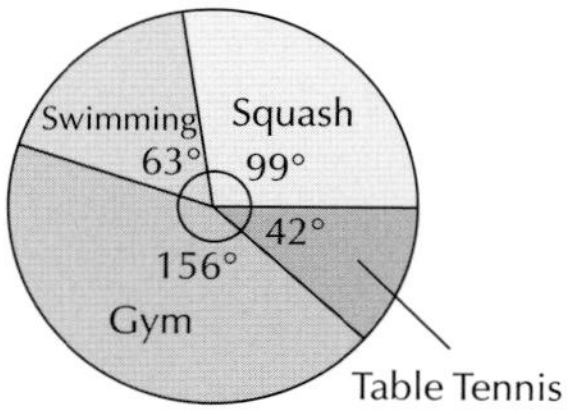

Page 455 Exercise 2

1 By comparing the proportions in the pie charts with the proportions of the frequency in the tables: **Set 1 = *P*** and **Set 2 = *Q***

2 a) The sector with the largest angle is **pepperoni**.

b) The angle for cheese and tomato is a right angle so is 90°. $\frac{90°}{360°} = \mathbf{\frac{1}{4}}$

c) 102° represents 17 pupils:
Degrees for one pupil = 102° ÷ 17 = 6°
Total pupils = 360° ÷ 6° = **60 pupils**

d) Fraction of pupils who like spicy chicken $= \frac{36°}{360°} = \frac{1}{10}$, then $60 \times \frac{1}{10} =$ **6 pupils**.

3 a) **No** — The **angle** for English for Brooke is 144° which is **larger** than Tom's which is 126°.

b) **No** — The pie charts **do not show times**, just the **proportion** of homework tasks set for each subject.

26.7 Time Series

Page 457 Exercise 1

1 a)

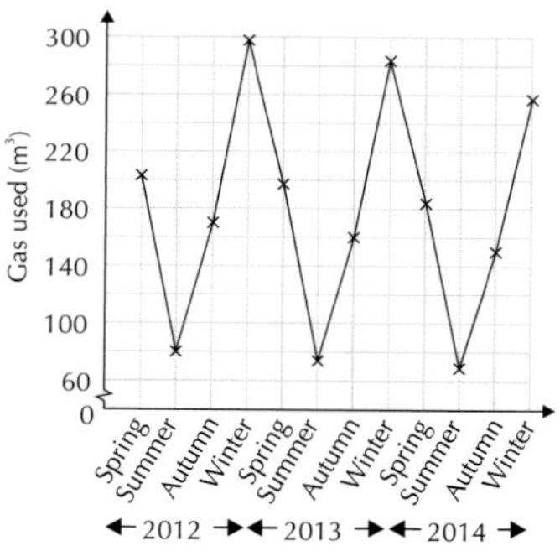

b) E.g. The amount of gas they use rises through autumn and winter. It then decreases in spring and drops further in summer, before rising again in autumn.

c) E.g. Overall there is a slight downward trend in the amount of gas they use.

2

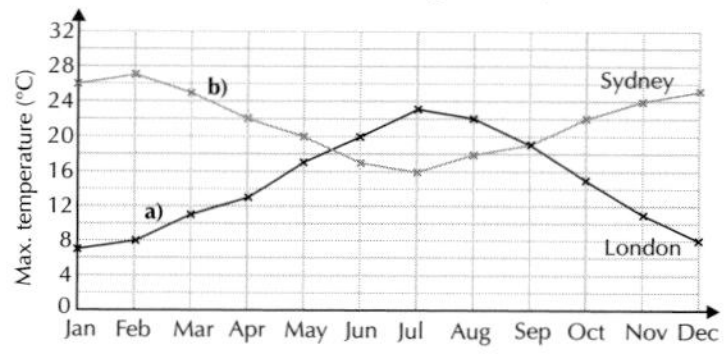

c) E.g. The London graph peaks in July, when the Sydney graph is at its lowest. The Sydney graph peaks in February, when the London graph is very low. Warmer temperatures in one city correspond to lower temperatures in the other city.

3 a) E.g. The number of sunglasses sold increases from spring to summer, then falls in autumn and again in winter, before rising again the next spring.

b) E.g. There is an overall upward trend in the number of sunglasses sold.

c) E.g. The scale on the vertical axis is inconsistent. At the lower end of the axis, the numbers on the scale increase in increments of 20 and each grid square represents 10 pairs of sunglasses. But after 100 it increases in increments of 40 and each square represents 20 pairs of sunglasses. This makes it hard to interpret the graph.

26.8 Scatter Graphs

Page 458 Exercise 1

1 Carefully plot the data points — the first point is (28, 30), the second is (25, 22), etc.

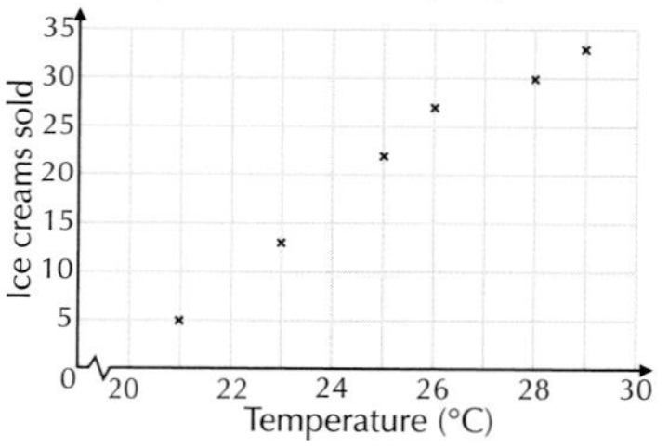

You can use the 'zig zag' on the 'Temperature' axis to miss out the numbers between 0 and 20.

2 The number of baby teeth will depend on age, so plot the number of baby teeth on the vertical axis. Then plot the data points: (5, 20), (6, 17), etc.

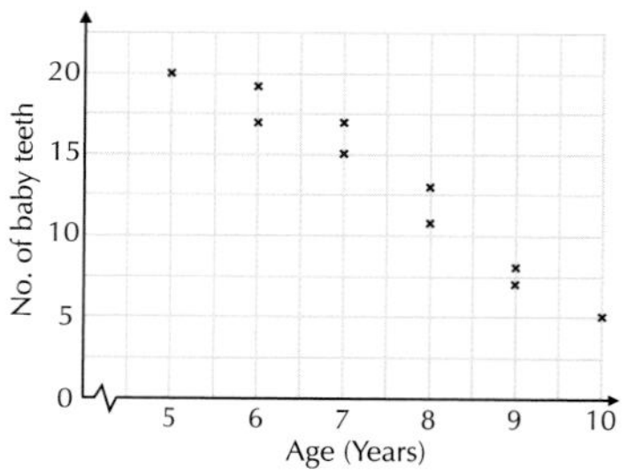

Page 460 Exercise 2

1 a) **Weak negative correlation** — the points form a downward slope but do not lie close to a straight line.

b) **No correlation** — the points don't seem to form a line in any direction.

c) **Strong positive correlation** — the points form an upward slope fairly close to a straight line.

2 a) **Positive correlation** — as it gets warmer, people will probably be more likely to buy ice cream.

b) **Negative correlation** — as it gets warmer, people will probably be less likely to buy hot chocolate.

c) **No correlation** — there probably isn't any connection between how hot it is and how much bread people buy.

d) **Positive correlation** — as a child grows, he or she gets taller.

e) **Positive correlation** — most people will probably drive more slowly in an area with a low speed limit.

3 There is **moderate positive correlation** between the wind speed and boat speed OR **the boat speed increased as the wind speed increased.**

Page 462 Exercise 3

1 a) **Yes**, the points will lie close to a straight line.

b) **No**, the points are randomly scattered.

c) **Yes**, most of the points will lie close to a straight line. There are 3 outliers which don't match the general pattern and so can be ignored.

2 a) The points form a downward slope fairly close to a straight line so there is **strong negative correlation** OR **the more time spent watching TV, the less time was spent on homework.**

b) Find 5.5 hours on the 'Time spent watching TV' axis. Draw a vertical line up to the line of best fit and then read off the value on the 'Time spent on homework' axis, which is **1 hour**.

c) Find 3.5 hours on the 'Time spent on homework' axis and draw a horizontal line across to the line of best fit. Then read off the value on the 'Time spent watching TV' axis, which is **1.5 hours**.

3 a) The points form an upward slope fairly close to a straight line so there is **strong positive correlation** OR **the wider the trunk the taller the tree.**

b) Width = **40 cm**, height = **23 m**

c) Draw a line from 13 m on the vertical axis to the line of best fit and read off the value on the horizontal axis, which is **65 cm** (allow 64-66 cm).

d) 13 m lies within the data range for the height of the tree, which means the prediction is from interpolation, so it should be **reliable**.

4 a), b)

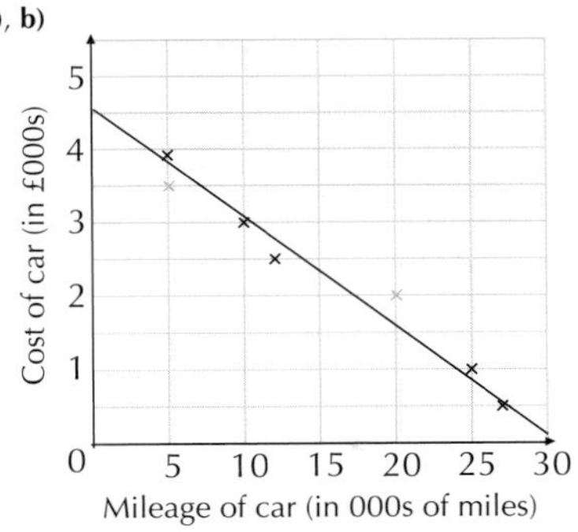

c) E.g. 50 000 miles lies outside the range of the data so the data would need to be extrapolated and her estimate would be **unreliable**. If the line of best fit was continued to 50 000 miles, you would get a negative cost.

Page 463 Review Exercise

1 a) Test 1: Total frequency
= 19 + 14 + 8 + 17 + 17 + 20 + 6 + 13
= 114
Mean = 114 ÷ 8 = **14.25**
Test 2: Total frequency
= 7 + 14 + 16 + 7 + 10 + 8 + 10 + 12
= 84
Mean = 84 ÷ 8 = **10.5**

b) Test 1: Range = 20 – 6 = **14**
Test 2: Range = 16 – 7 = **9**

c) E.g. The mean score was higher in Test 1, so the score achieved was generally higher in that test. The range of scores was lower in Test 2, so there was less variation in the scores in Test 2.

2 Total weight of original 5 apples
= 57 + 60 + 69 + 72 + 75 = 333 g
Total weight of remaining three = 62 × 3
= 186 g
Weight of two apples eaten = 333 – 186
= 147 g, so the apples eaten must be the ones which weigh **72 g** and **75 g**.

3 a) Total frequency = 1 + 3 + 4 + 5 + 3 + 2
= 18
Median will be the (18 + 1) ÷ 2
= 9.5th value, so the median lies halfway between the 9th value = 3 and 10th value = 3. So the median is **3 goals**.

b) Total goals = (0 × 1) + (1 × 3) + (2 × 4)
+ (3 × 5) + (4 × 3) + (5 × 2) = 48
Mean = 48 ÷ 18 = 2.66... = **2.7 goals** (1 d.p.)

c) Mode = most common value = **3 goals**

d) The values of all three averages were lower in the previous year, so **more goals were scored on average than in the previous year.**

4 a)

Height in metres, h	Tally	Frequency
$1.5 < h \leq 1.6$	||	2
$1.6 < h \leq 1.7$	~~||||~~ ||||	9
$1.7 < h \leq 1.8$	~~||||~~ ||	7
$1.8 < h \leq 1.9$	||	2

b) Modal group has the highest frequency so is **$1.6 < h \leq 1.7$**.

c) Total frequency is 20, so the position of the median is (20 + 1) ÷ 2 = 10.5, so the median lies between the 10th and 11th values. Both are in the $1.6 < h \leq 1.7$ group, so median group = **$1.6 < h \leq 1.7$**.

d) Total frequency = 20

Height in metres, h	Tally	Freq.	Freq. × Midpoint
$1.5 < h \leq 1.6$	||	2	2 × 1.55 = 3.1
$1.6 < h \leq 1.7$	~~||||~~ ||||	9	9 × 1.65 = 14.85
$1.7 < h \leq 1.8$	~~||||~~ ||	7	7 × 1.75 = 12.25
$1.8 < h \leq 1.9$	||	2	2 × 1.85 = 3.7

Estimate of total height
= 3.1 + 14.85 + 12.25 + 3.7 = 33.9
Mean = 33.9 ÷ 20 = **1.695 m**

e) Range = upper bound of largest group
– lower bound of smallest group
= 1.9 – 1.5 = **0.4 m**

5 a) There are 40 values so the median position is (40 + 1) ÷ 2 = 20.5.
The median lies between the 20th and 21st values. Both values are in the 2 washing machines group so the median = **2 washing machines.**

b) Mode = most common value
= **2 washing machines**

c) Total machines sold
= (0 × 3) + (1 × 5) + (2 × 14) + (3 × 11)
+ (4 × 6) + (5 × 1) = 95
Mean = 95 ÷ 40 = **2.375 washing machines**

d) Range = highest value – lowest value
= 5 – 0 = **5**

6 a) Total frequency of girls:
8 + 6 + 5 + 2 + 3 + 6 = **30**

b) 9 boys chose chocolate and 6 girls chose chocolate, so 9 – 6 = **3** more boys than girls chose chocolate.

c) The highest bar for girls is vanilla, so **vanilla** is the modal flavour for girls.

d) E.g. 360° ÷ 30 girls = **12° per girl**
Alternatively, you could have used any of the categories. E.g., using vanilla: 96° ÷ 8 = 12°.

e) Total frequency of boys:
4 + 9 + 1 + 3 + 3 = 20
360° ÷ 20 = 18° per boy
Vanilla: 4 × 18° = 72°
Chocolate: 9 × 18° = 162°
Strawberry: 0 × 18° = 0°
Mint: 1 × 18° = 18°
Rocky road: 3 × 18 = 54°
Other = 3 × 18° = 54°

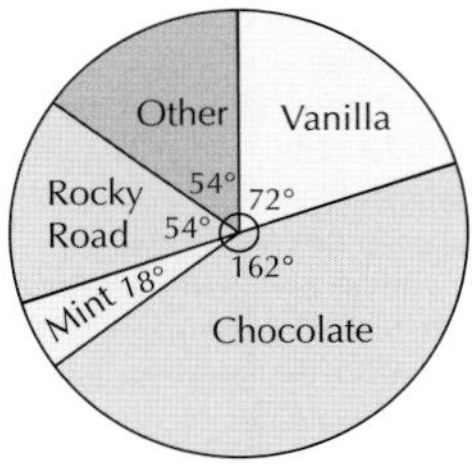

f) There are **fewer boys than girls** — so even though the number who chose rocky road is the same, the proportion of boys is larger, so the sector on the pie chart is larger.

7 a) E.g.

10 am		3 pm
	1	5 8
9 7	2	2 3 4 7
5	3	1 2 5
8	4	
9 8 4 1 1	5	
5 0	6	

Key: 7 | 2 for 10 am means 27
1 | 5 for 3 pm means 15

b) There are 11 people queuing at 10 am so the median is in the (11 + 1) ÷ 2 = 6th position. That person has an age of **51**. There are 9 people queueing at 3 pm so the median is in the (9 + 1) ÷ 2 = 5th position. That person has an age of **24**. So the median is **much lower for people queuing at 3 pm**.

8 a), b) The number of drinks sold is dependent on the temperature, so the number of drinks sold should go on the vertical axis.

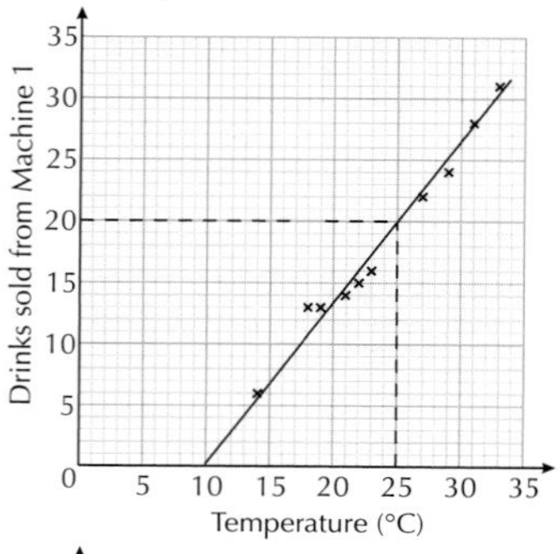

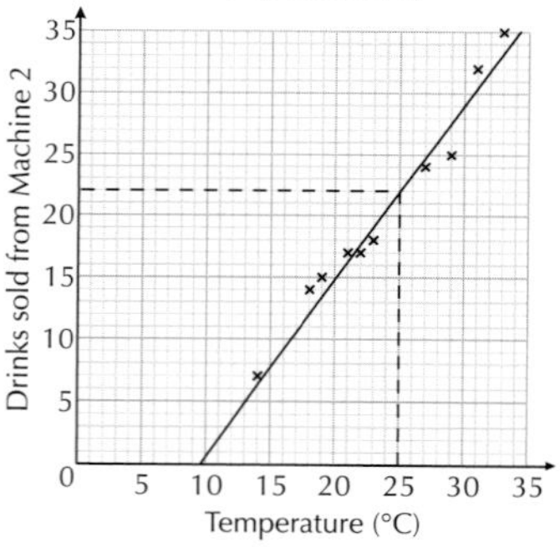

c) Draw a line from 25°C to the line of best fit on both graphs (see graphs above) and read off the number of drinks sold. The predicted number of drinks sold from Machine 1 is **20**. The predicted number of drinks sold from Machine 2 is **22**.
Your readings might be slightly different, depending on how you've drawn your line of best fit. We'll allow answers within 21 - 23 drinks.

d) 3 °C is well below the temperature range for this data — you'd need to **extrapolate** to find your answer, which is **unreliable**. In this case, it would give you a negative number of drinks sold which is impossible.

e) No — just because the two things are correlated does not mean that one thing causes the other. They both could be influenced by a third variable.

Page 465 Exam-Style Questions

1 The middle number must be 50 for A, but 50 could also be the highest number if there are 3 marks of 50. Similarly for group B, the lowest mark could be 48 if there are three marks of 48. So the lowest mark in group A could be 50 – 3 = 47 *[1 mark]* and the lowest mark in group B could be 48 *[1 mark]*, so the student who scored the lowest mark could be in group A.
You'd also pick up the marks if you wrote out a list of data values that work — e.g. A: 47, 48, 50, 50, 50 and B: 48, 48, 48, 50, 51.

2 a) There are 7 values so the median is the 4th value. The values in order are:
5, 6, 22, 28, 29, 31, 34, so the median is **28 eggs**. *[1 mark]*

b) E.g. There are two extreme values (5 and 6), which distort the value of the mean OR the data is not evenly spread out so the mean will not be a good measure of the centre of the data *[1 mark]*.

3 a) The group with the highest frequency is $120 \le t < 160$, so the modal class is $\mathbf{120 \le t < 160}$. *[1 mark]*

b) Use the midpoints of the groups to estimate the mean:

Time (t minutes)	Freq.	Midpoint	Freq. × Midpoint
$0 \le t < 40$	3	(0 + 40) ÷ 2 = 20	3 × 20 = 60
$40 \le t < 80$	4	(40 + 80) ÷ 2 = 60	4 × 60 = 240
$80 \le t < 120$	7	(80 + 120) ÷ 2 = 100	7 × 100 = 700
$120 \le t < 160$	15	(120 + 160) ÷ 2 = 140	15 × 140 = 2100
$160 \le t < 200$	11	(160 + 200) ÷ 2 = 180	11 × 180 = 1980

Total Frequency = 3 + 4 + 7 + 15 + 11 = 40
Total Time = 60 + 240 + 700 + 2100 + 1980 = 5080 *[1 mark]*
Mean = 5080 ÷ 40 *[1 mark]*
= **127 minutes** *[1 mark]*

c) E.g. The method assumes that all the values in a group are equal to the midpoint of the group (or are evenly spread across the group) OR you do not know the exact times/original values since the data is grouped *[1 mark]*.

4 14 books is shown by 3.5 shapes, so one shape represents 14 ÷ 3.5 = 4 books and $\frac{1}{4}$ of a shape represents 1 book.
7 books = 4 books + 3 books, so that's 1 full shape and $\frac{3}{4}$ of a shape.

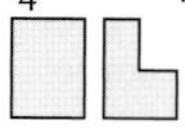

[2 marks available — 1 mark for finding that one shape represents 4 books, 1 mark for a correct diagram]

5 a) Silver requires a sector with angle = 360° ÷ 3 = 120°.
The remaining 360° – 120° = 240° needs to be split in the ratio 5 : 1. 6 parts = 240°, so 1 part = 240° ÷ 6 = 40°.
Bronze needs an angle of 5 × 40° = 200° and gold needs an angle of 40°.

[3 marks for a fully correct and labelled pie chart, otherwise 2 marks for all three angles calculated correctly, or one mark for finding the angle for silver]

b) Silver = 120° so 120° represents 18 students. 1° represents 18 ÷ 120 = 0.15 students *[1 mark]*, so Bronze awards were received by 200 × 0.15 = **30 students** *[1 mark]*.

6 a) The price of gold per gram showed an **upward trend** — there was a general **increase over time**.
[1 mark for the correct observation]

b) The highest price per gram was £29.80 in November. 12 kg = 12 × 1000 = 12 000 g. So Midas could have sold his gold bar for 29.8 × 12 000 = **£357 600**
[2 marks available — 1 mark for finding the highest price per gram, 1 mark for the correct answer]

7 a) The marks show a **strong positive correlation** with the number of revision classes. *[1 mark for the correct observation]*

b) No — to predict what a student would get if they didn't attend any revision classes requires extrapolation and this is unreliable.
[1 mark for a correct explanation of why Duane is incorrect]

Section 27 — Probability

27.1 Probability — The Basics

Page 468 Exercise 1

1 a) There are 2 possible outcomes that are equally likely — heads or tails. So tossing a coin and getting tails is **evens**.

b) No person can grow to 10 metres tall, so it's **impossible**.

c) Half of the numbers on a dice are even, so it's **evens**.

d) All numbers on a normal dice are 1 or more, so it's **certain**.

e) Half of the cards in a pack of 52 are red, so it's **evens**.

f) There are 4 possible outcomes — 1, 2, 3 and 4.
1 out of the 4 outcomes is 2, so it's **unlikely**.

2 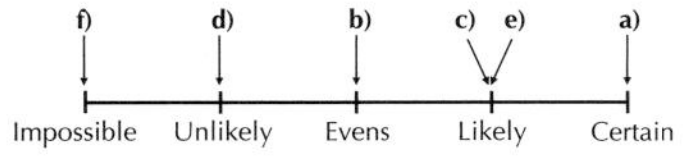

a) All the cards are less than 9, so this is **certain**.

b) Half of the numbers are odd, so this is **evens**.

c) All but two of the cards are greater than 2, so this is **likely**.

d) All but two of the cards are not greater than or equal to 7, so this is **unlikely**.

e) All but two of the cards are 6 or less, so this is **likely**.

f) None of the cards are zero, so this is **impossible**.

3 a) There's one P and two A's, so **A** is twice as likely to be selected as P.

b) There's one E and three L's, so **L** is three times as likely to be selected as E.

c) Count the number of times the letter appears. The more times a letter appears, the more likely it is to be chosen, so it will correspond to a higher-numbered arrow (as the arrows show an increase in likelihood as the numbers increase).

P: **1** A: **2** R: **1** L: **3** E: **1**

All of the probabilities are less likely than 'evens' — there are 8 letters in total and the maximum times one letter appears is 3 (for an even chance, there'd have to be 4 out of the 8 cards the same).

Page 469 Exercise 2

1 a) $0.5 = \frac{1}{2}$ so the correct probability is **D**.

b) $25\% = \frac{1}{4}$ so the correct probability is **C**.

c) $\frac{1}{6}$ is less than $\frac{1}{4}$ so the correct probability is **A**.

d) $0.4 = \frac{2}{5}$, which is less than $\frac{1}{2}$ but greater than $\frac{1}{4}$, so the correct probability is **B**.

2 a) Half of the numbers on a dice are odd, which means the probability is $\frac{1}{2}$, so the letter is **C**.

b) A quarter of the cards in a pack are diamonds, which means the probability is $\frac{1}{4}$, so the letter is **A**.

c) One quarter of the cards are diamonds, which means three-quarters of the cards are not diamonds, so the probability of not getting a diamond is $\frac{3}{4}$, so the letter is **D**.

d) 1 of the 3 sections is blue, which means the probability of spinning blue is $\frac{1}{3}$, so the letter is **B**.

27.2 Calculating Probabilities

Page 471 Exercise 1

1 a) You can either get heads or tails, so there are **2 outcomes**.

b) There are 52 cards, so there are **52 outcomes**.

Using the same method for c)-f):

c) 10 **d) 8** **e) 7**

f) 365 (or 366 for a leap year)

2 a) There are 6 sections, so there are **6 outcomes**.

b) There's 1 section with a 1, so $P(1) = \mathbf{\frac{1}{6}}$.

c) There are 2 sections with a 3, so $P(3) = \frac{2}{6} = \mathbf{\frac{1}{3}}$.

3 There are 6 possible outcomes: 1, 2, 3, 4, 5 and 6. So:

a) There's one 6, so $P(6) = \mathbf{\frac{1}{6}}$.

b) There's one 2, so $P(2) = \mathbf{\frac{1}{6}}$.

c) There is no 7, so P(7) = **0**.

d) Two outcomes will give a 4 or 5, so $P(4 \text{ or } 5) = \frac{2}{6} = \mathbf{\frac{1}{3}}$.

e) 3 and 6 are multiples of 3, so $P(\text{multiple of } 3) = \frac{2}{6} = \mathbf{\frac{1}{3}}$.

f) 1, 2, 3 and 6 are factors of 6, so $P(\text{factor of } 6) = \frac{4}{6} = \mathbf{\frac{2}{3}}$.

Remember to always simplify your fractions if possible.

4 a) One card is a 4, so $P(4) = \mathbf{\frac{1}{9}}$.

b) 4 of the 9 cards (2, 4, 6, and 8) are even, so $P(\text{even}) = \mathbf{\frac{4}{9}}$.

c) 5 of the cards are less than 6, so $P(\text{less than } 6) = \mathbf{\frac{5}{9}}$.

5 a) 3 of the 12 sections are green, so $P(\text{green}) = \frac{3}{12} = \mathbf{\frac{1}{4}}$.

b) 12 – 5 – 3 = 4, so there are 4 purple sections. So $P(\text{purple}) = \frac{4}{12} = \mathbf{\frac{1}{3}}$.

c) 5 + 3 = 8, so $P(\text{yellow or green}) = \frac{8}{12} = \mathbf{\frac{2}{3}}$.

d) 3 sections are green, so 12 – 3 = 9 sections are not green. $P(\text{not green}) = \frac{9}{12} = \mathbf{\frac{3}{4}}$.

6 First, find the total number of balls in the bag: 2 + 4 + 2 + 3 + 2 + 1 + 1 + 1 = 16 balls.

a) There are 2 green balls, so $P(\text{green}) = \frac{2}{16} = \mathbf{\frac{1}{8}}$.

b) There are 3 red balls, so $P(\text{red}) = \mathbf{\frac{3}{16}}$.

c) There is 1 orange ball, so $P(\text{orange}) = \mathbf{\frac{1}{16}}$.

d) There are 2 black balls, so $P(\text{black}) = \frac{2}{16} = \mathbf{\frac{1}{8}}$.

e) There are 4 blue balls and 2 green balls, so the number of outcomes = 4 + 2 = 6, and $P(\text{blue or green}) = \frac{6}{16} = \mathbf{\frac{3}{8}}$.

f) There are 3 red balls, 2 green balls and 1 brown ball, so the number of outcomes = 3 + 2 + 1 = 6, and $P(\text{red, green or brown}) = \frac{6}{16} = \mathbf{\frac{3}{8}}$.

g) There is only 1 purple ball, so there are 16 – 1 = 15 balls which are not purple. $P(\text{not purple}) = \mathbf{\frac{15}{16}}$.

h) There are no white balls so P(white) = **0**.

7 a) 13 cards are clubs, so $P(\text{clubs}) = \frac{13}{52} = \mathbf{\frac{1}{4}}$.

b) There are 4 aces in a pack, so $P(\text{ace}) = \frac{4}{52} = \mathbf{\frac{1}{13}}$.

c) 26 cards are red, so $P(\text{red}) = \frac{26}{52} = \mathbf{\frac{1}{2}}$.

d) Only one card is the 2 of hearts, so $P(2 \text{ of hearts}) = \mathbf{\frac{1}{52}}$.

e) 13 cards are spades, so 52 – 13 = 39 cards are not spades. $P(\text{not spades}) = \frac{39}{52} = \mathbf{\frac{3}{4}}$.

f) There are four 4s and four 5s in the pack. 4 + 4 = 8, so $P(4 \text{ or } 5) = \frac{8}{52} = \mathbf{\frac{2}{13}}$.

8 a) 3 of the 8 sections need to be a 2:

E.g. 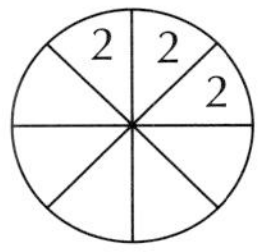

b) Half of 8 is 4, so 4 of the 8 sections need to be a 3:

E.g. 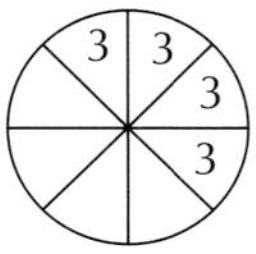

c) $\frac{1}{4}$ of 8 is 2, so there needs to be a 5 in two sections and a 6 in two sections:

E.g. 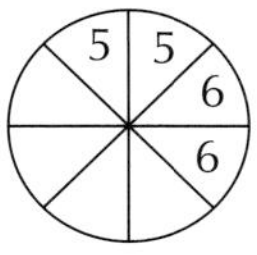

Page 472 Exercise 2

1 a) There are 12 months in a year, so $P(\text{correct month}) = \mathbf{\frac{1}{12}}$

b) There are 366 days in a leap year, so $P(\text{correct date}) = \mathbf{\frac{1}{366}}$

2 Each ticket is equally likely to be chosen, so:

a) $P(\text{first person wins}) = \mathbf{\frac{1}{100}}$

b) $P(\text{last person wins}) = \mathbf{\frac{1}{100}}$

3 20 pairs of socks = 40 socks. Tove picks 1 sock, so there are now 40 – 1 = 39 socks to pick from. So the probability of picking a second sock that makes a pair is $\mathbf{\frac{1}{39}}$.

4 a) There are 8 + 6 + 4 = 18 chocolates in the box. There are 4 pralines, half of which are white chocolate. This means there are 2 white-chocolate-coated pralines.

So P(white-chocolate-coated praline) $= \frac{2}{18} = \mathbf{\frac{1}{9}}$.

b) (i) 4 chocolates are praline, so 18 – 4 = 14 chocolates are not praline. Of these 14, you know half of them are white chocolate and half are milk chocolate. 14 ÷ 2 = 7, so there are **7 chocolates** that are neither praline nor white chocolate.

(ii) Chelsea will like 7 chocolates, so the probability that Chelsea gets a chocolate she likes is $\mathbf{\frac{7}{18}}$.

5 It doesn't take into account how good the team is or how good their opponents are. This means the **three outcomes are not equally likely**.

Page 473 Exercise 3

1 a) A 5 and a number less than 3 can't be rolled at the same time, so these events **are mutually exclusive**.

b) 5 is an odd number, so if a 5 was rolled, an odd number would also be rolled. This means these two events are **not mutually exclusive**.

c) If a number less than 3 is rolled it will be a 1 or a 2. 1 is an odd number, so these events are **not mutually exclusive**.

2 a) If the name of a month begins with J, it can't also begin with M, so these events **are mutually exclusive**.

b) January and July both begin with J and end in Y, so these events are **not mutually exclusive**.

c) The first four months are January, February, March and April. January begins with J, so these events are **not mutually exclusive**.

d) The month of May begins with M and ends with Y, so these events are **not mutually exclusive**.

It's a good idea to write out all the months of the year so you don't miss any.

3 Use the rule P(event happening) = 1 – P(event not happening).
So, P(not late) = 1 – P(late) = 1 – 0.2 = **0.8**

An event happening and the event not happening are always mutually exclusive events — things can't both happen and not happen.

4 P(no snow) = 1 – P(snow) = $1 - \frac{5}{8} = \mathbf{\frac{3}{8}}$

5 **a)** P(not boy) = 1 – P(boy) = 1 – 0.45 = **0.55**

b) P(not blond) = 1 – P(blond) = 1 – 0.2 = **0.8**

Page 474 Exercise 4

1 **a)** P(red) = 1 – (0.2 + 0.1 + 0.1 + 0.3) = 1 – 0.7 = **0.3**

b) P(not pink) = 1 – P(pink) = 1 – 0.1 = **0.9**

2 P(red) = $1 - (\frac{1}{4} + \frac{3}{8}) = 1 - \frac{5}{8} = \mathbf{\frac{3}{8}}$

3 P(red) = 1 – (0.5 + 0.4) = 1 – 0.9 = 0.1

There are 4 red counters, so P(red) = $\frac{4}{\text{total}}$ = 0.1,
so total = 4 ÷ 0.1 = **40 counters**.

27.3 Listing Outcomes

Page 475 Exercise 1

1 **a)**

1st coin	2nd coin
Heads	Heads
Heads	Tails
Tails	Heads
Tails	Tails

b)

Dice	Coin
1	Heads
1	Tails
2	Heads
2	Tails
3	Heads
3	Tails
4	Heads
4	Tails
5	Heads
5	Tails
6	Heads
6	Tails

Writing your table in order makes it much easier to spot if you miss out an outcome.

2

1st ball	2nd ball
Green	Green
Green	Blue
Blue	Green
Blue	Blue

3 **a)**

Burger	Drink
Hamburger	Cola
Hamburger	Lemonade
Hamburger	Coffee
Cheeseburger	Cola
Cheeseburger	Lemonade
Cheeseburger	Coffee
Veggie burger	Cola
Veggie burger	Lemonade
Veggie burger	Coffee

b) There are 9 possible combinations, and only 1 way to choose a veggie burger and cola, so P(veggie burger and cola) = $\mathbf{\frac{1}{9}}$.

c) There are 3 ways she could choose a cheeseburger,
so P(cheeseburger) = $\frac{3}{9} = \mathbf{\frac{1}{3}}$.

4 **a)**

1st spin	2nd spin
1	1
1	2
1	3
2	1
2	2
2	3
3	1
3	2
3	3

b) There are 9 possible outcomes but only 1 way to spin a 3 for both spins, so the probability is $\mathbf{\frac{1}{9}}$.

c) There are 6 ways of getting a total of 4 or more:
1 and 3, 2 and 2, 2 and 3, 3 and 1, 3 and 2, 3 and 3.
So the probability is $\frac{6}{9} = \mathbf{\frac{2}{3}}$.

5 **a)**

1st toss	2nd toss	3rd toss
Heads	Heads	Heads
Heads	Heads	Tails
Heads	Tails	Heads
Heads	Tails	Tails
Tails	Heads	Heads
Tails	Heads	Tails
Tails	Tails	Heads
Tails	Tails	Tails

b) (i) There are 8 possible outcomes and 1 way of getting three tails, so the probability is $\mathbf{\frac{1}{8}}$.

(ii) There are three ways of getting one head and two tails:
Heads, Tails, Tails; Tails, Heads, Tails; Tails, Tails, Heads
So the probability is $\mathbf{\frac{3}{8}}$.

Page 477 Exercise 2

1

	1	2	3	4	5	6
H	H1	H2	H3	H4	H5	H6
T	T1	T2	T3	T4	T5	T6

2 **a)**

	1	2	3	4	5	6
1	2	3	4	5	6	7
2	3	4	5	6	7	8
3	4	5	6	7	8	9
4	5	6	7	8	9	10
5	6	7	8	9	10	11
6	7	8	9	10	11	12

b) 6 × 6 = **36 outcomes**

c) (i) There are 5 ways to score 6:
1 and 5, 2 and 4, 3 and 3, 4 and 2, 5 and 1.
So P(6) = $\mathbf{\frac{5}{36}}$.

(ii) P(less than 8) = probability of scoring 2, 3, 4, 5, 6 or 7. There are 6 ways to score 7, 5 ways to score 6, 4 ways to score 5, 3 ways to score 4, 2 ways to score 3 and 1 way to score 2. So there are 6 + 5 + 4 + 3 + 2 + 1 = 21 ways to score less than 8.
P(less than 8) = $\frac{21}{36} = \mathbf{\frac{7}{12}}$.

(iii) P(more than 8) = probability of scoring 9, 10 , 11 or 12. There are 4 ways to score 9, 3 ways to score 10, 2 ways to score 11 and 1 way to score 12. So there are 4 + 3 + 2 + 1 = 10 ways to score more than 8.
P(more than 8) = $\frac{10}{36} = \mathbf{\frac{5}{18}}$.

(iv) 18 of the 36 options are even, so P(even) = $\frac{18}{36} = \mathbf{\frac{1}{2}}$.

3 **a)**

	B	G	G	Y
B	BB	BG	BG	BY
G	GB	GG	GG	GY
Y	YB	YG	YG	YY

b) There are 3 × 4 = 12 possible outcomes.

(i) There's only 1 way to get 2 blue balls, so P(2 blue balls) = $\mathbf{\frac{1}{12}}$.

(ii) There are 2 ways to get 2 green balls, so P(2 green balls) = $\frac{2}{12} = \mathbf{\frac{1}{6}}$.

(iii) There are 4 ways to get 2 balls of the same colour (BB, GG, GG, YY), so P(2 balls same colour) = $\frac{4}{12} = \mathbf{\frac{1}{3}}$.

(iv) There are 6 ways to get at least 1 yellow ball (YB, YG, YG, YY, BY, GY), so P(at least 1 yellow) = $\frac{6}{12} = \mathbf{\frac{1}{2}}$.

4 **a)**

	1	2	3	4	5	6
A	A1	A2	A3	A4	A5	A6
B	B1	B2	B3	B4	B5	B6
C	C1	C2	C3	C4	C5	C6
D	D1	D2	D3	D4	D5	D6

b) There are 4 × 6 = 24 possible outcomes.

(i) There's only 1 way to get C and 5, so P(C and 5) = $\mathbf{\frac{1}{24}}$.

(ii) There are 2 ways to get B and less than 3: B1 and B2,
so P(B and less than 3) = $\frac{2}{24} = \mathbf{\frac{1}{12}}$.

(iii) There are 4 ways to get A or B and more than 4:
A5, A6, B5, B6. So P(A or B and more than 4) = $\frac{4}{24} = \mathbf{\frac{1}{6}}$.

5 a)

		Asha's score (2nd number)				
		1	**2**	**3**	**4**	**5**
Hayley's score (1st number)	**1**	1, 1	1, 2	**1, 3**	**1, 4**	**1, 5**
	2	2, 1	**2, 2**	**2, 3**	**2, 4**	**2, 5**
	3	3, 1	**3, 2**	**3, 3**	**3, 4**	**3, 5**
	4	**4, 1**	**4, 2**	**4, 3**	**4, 4**	**4, 5**
	5	**5, 1**	**5, 2**	**5, 3**	**5, 4**	**5, 5**

b) There are 5 × 5 = 25 possible outcomes. If Hayley spins a 5, there are four outcomes where she beats Asha (if Asha rolls a 4, 3, 2 or 1). If she spins a 4, there are three outcomes where she beats Asha. If she spins a 3, there are two outcomes and if she spins a 2 then there is one outcome. If Hayley spins a 1 there are no outcomes where she beats Asha. So there are 4 + 3 + 2 + 1 = 10 outcomes where Hayley can get a higher score than Asha. So P(Hayley wins) = $\frac{10}{25} = \mathbf{\frac{2}{5}}$.

The outcomes where Hayley wins are found in the bottom-left corner of the table — you can just count them directly from there.

27.4 Probability from Experiments

Page 478 Exercise 1

1 a) The experiment was carried out 50 times, and the pin landed point up 17 times. So the probability is $\mathbf{\frac{17}{50}}$.

b) P(not point up) = 1 – P(point up) $= 1 - \frac{17}{50} = \mathbf{\frac{33}{50}}$.

2 a) Red came up 49 times out of 100, so its relative frequency is $\mathbf{\frac{49}{100}}$.
Green came up 34 times, so its relative frequency is $\frac{34}{100} = \mathbf{\frac{17}{50}}$.
Yellow came up 8 times, so its relative frequency is $\frac{8}{100} = \mathbf{\frac{2}{25}}$.
Blue came up 9 times, so its relative frequency is $\mathbf{\frac{9}{100}}$.

b) The more times an experiment is done, the more accurate the estimates should be. So Sam could perform the experiment more times.

3 a) Stacy rolled a two 13 times out of 50, so the estimated probability is $\mathbf{\frac{13}{50}}$.

b) Jamal rolled a two 18 times out of 100, so the estimated probability is $\frac{18}{100} = \mathbf{\frac{9}{50}}$.

c) **Jamal's estimate** should be more accurate as he has performed the experiment more times.

4 The bus is on time 20 out of 24 days, so the estimated probability that the bus is on time is $\frac{20}{24} = \mathbf{\frac{5}{6}}$.

5 The estimated probability that George will burn the next cake he bakes is $\frac{12}{20}$, so the estimated probability that he won't burn it is $1 - \frac{12}{20} = \frac{8}{20} = \mathbf{\frac{2}{5}}$.

6 E.g. Lilia could look at recent results of her team's matches against a similar level of opposition to find the estimated probability that her team will win. To do this, she could count the number of wins and divide it by the total number of matches in those records.

Page 480 Exercise 2

1 a) 20 × 0.75 = **15 times**

b) 60 × 0.75 = **45 times**

c) 100 × 0.75 = **75 times**

d) 1000 × 0.75 = **750 times**

2 a) P(5) = $\frac{1}{6}$, 120 × $\frac{1}{6}$ = **20 times**

b) P(6) = $\frac{1}{6}$, 120 × $\frac{1}{6}$ = **20 times**

c) P(even) = $\frac{3}{6} = \frac{1}{2}$, 120 × $\frac{1}{2}$ = **60 times**

d) P(higher than 1) = $\frac{5}{6}$, 120 × $\frac{5}{6}$ = **100 times**

3 There are 3 sections so the probability of spinning a penguin is $\frac{1}{3}$:

a) 60 × $\frac{1}{3}$ = **20 times**

b) 300 × $\frac{1}{3}$ = **100 times**

c) 480 × $\frac{1}{3}$ = **160 times**

4 60% = 0.6 and 0.6 × 20 = 12, so you would expect **12 people** to buy brown bread.
You could also have converted 60% to a fraction instead of a decimal and used that to find the number.

Page 481 Exercise 3

1 a)

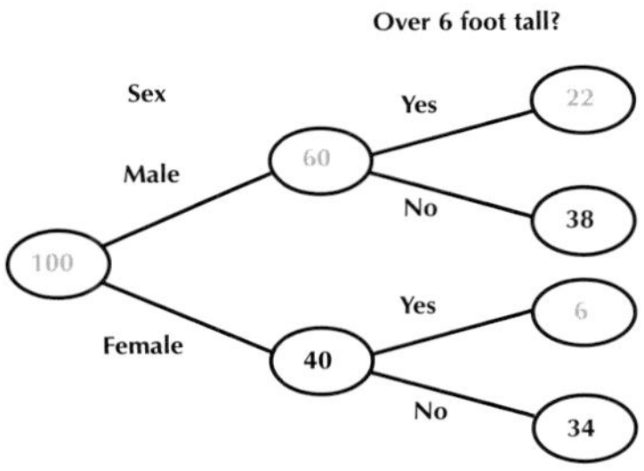

Add up all the numbers at the end to make sure they equal the number at the start. 22 + 38 + 6 + 34 = 100.

b)

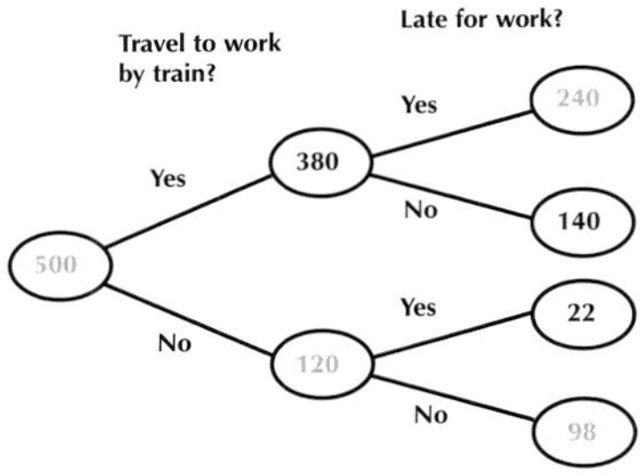

240 + 140 + 22 + 98 = 500, so this is correct.

2 a) E.g.

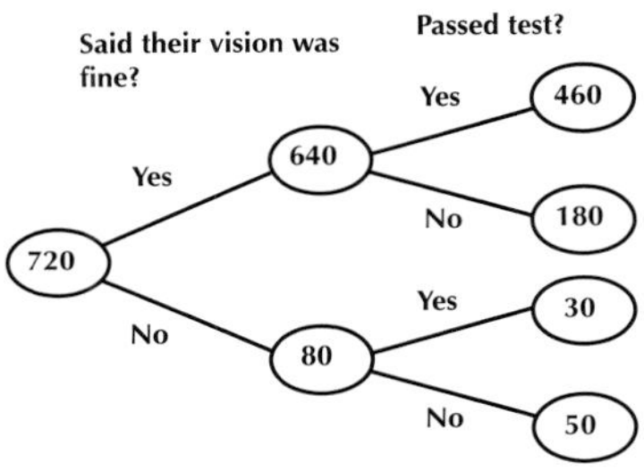

b) From the tree in part a) you know 80 drivers said their vision wasn't fine. Of these 80 drivers, 50 failed the test. So the probability is $\frac{50}{80} = \mathbf{\frac{5}{8}}$.

c) 460 drivers said that their vision was fine and passed the eye test. So the probability is $\frac{460}{720} = \mathbf{\frac{23}{36}}$.

Page 482 Exercise 4

1 a) Blue is $\frac{22}{100}$ = **0.22**, green is $\frac{21}{100}$ = **0.21**, white is $\frac{18}{100}$ = **0.18**, pink is $\frac{39}{100}$ = **0.39**.

b) There are 4 possible outcomes of equal probability, so blue is **0.25**, green is **0.25**, white is **0.25**, and pink is **0.25**.

c) The spinner seems to be biased as the relative frequency for pink (0.39) is a long way off the theoretical probability.

2 a) P(4) = $\frac{1}{6}$, so expected frequency of rolling a 4 is 120 × $\frac{1}{6}$ = **20 times**.

b) The dice may be biased as the number 4 comes up much more than expected — 32 times rather than 20 times.

c) Roll the dice more times.

3 a)

	Amy	Steve	Hal
No. of tosses	20	60	100
No. of heads	12	33	49
Relative frequency	**0.6**	**0.55**	**0.49**

b) **Hal's results** should be the most accurate because he performed the experiment the greatest number of times.

c) The coin seems fair as the theoretical probability of heads for a fair coin is 0.5 and Hal's result is very close to that (0.49).
You could also show that, as the same coin is used, the three friends's results can be combined to give a relative frequency of $\frac{94}{180}$ = 0.522...

4 a) E.g.

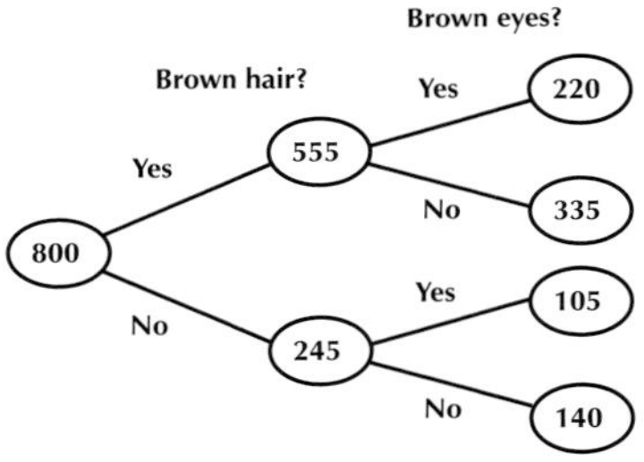

The frequency tree could be drawn the other way round, with eyes first and then hair.

b) 220 residents have brown hair and brown eyes, so the relative frequency is $\frac{220}{800} = \frac{11}{40}$ = **0.275**.

c) Estimated frequency = 2000 × 0.275 = **550 residents**

27.5 The AND / OR Rules

Page 483 Exercise 1

1 **a)** P(a head and a 6) = P(head) × P(6)
$= \frac{1}{2} \times \frac{1}{6} = \mathbf{\frac{1}{12}}$

b) P(a head and an odd number)
= P(head) × P(odd number)
$= \frac{1}{2} \times \frac{3}{6} = \frac{1}{2} \times \frac{1}{2} = \mathbf{\frac{1}{4}}$

c) There are 2 square numbers on a dice (1 and 4), so P(a tail and a square number)
= P(tail) × P(square number)
$= \frac{1}{2} \times \frac{2}{6} = \frac{1}{2} \times \frac{1}{3} = \mathbf{\frac{1}{6}}$

d) There are 3 prime numbers on a dice (2, 3 and 5), so P(a tail and a prime number)
= P(tail) × P(prime number)
$= \frac{1}{2} \times \frac{3}{6} = \frac{1}{2} \times \frac{1}{2} = \mathbf{\frac{1}{4}}$

e) There are 2 multiples of 3 on a dice (3 and 6), so P(a head and a multiple of 3)
= P(head) × P(multiple of 3)
$= \frac{1}{2} \times \frac{2}{6} = \frac{1}{2} \times \frac{1}{3} = \mathbf{\frac{1}{6}}$

f) There are 2 factors of 5 on a dice (1 and 5), so P(a tail and a factor of 5)
= P(tail) × P(factor of 5)
$= \frac{1}{2} \times \frac{2}{6} = \frac{1}{2} \times \frac{1}{3} = \mathbf{\frac{1}{6}}$

2 **a)** 10% = 0.1, so P(right-handed)
= 1 – P(left-handed) = 1 – 0.1 = **0.9**

b) 15% = 0.15, so P(no glasses)
= 1 – P(glasses) = 1 – 0.15 = **0.85**

c) P(left-handed) = 0.1, P(glasses) = 0.15
P(glasses and left-handed)
= P(glasses) × P(left-handed)
= 0.15 × 0.1 = **0.015**
Be careful with the decimal places here.

d) P(no glasses) = 0.85,
P(right-handed) = 0.9
P(no glasses and right-handed)
= P(no glasses) × P(right-handed)
= 0.85 × 0.9 = **0.765**

3 **a)** 5 out of the 10 balls are red, so P(red)
$= \frac{5}{10} = \frac{1}{2}$, so
P(red and red) = P(red) × P(red)
$= \frac{1}{2} \times \frac{1}{2} = \mathbf{\frac{1}{4}}$ or **0.25**

b) P(blue) $= \frac{3}{10}$, so P(blue and blue)
= P(blue) × P(blue)
$= \frac{3}{10} \times \frac{3}{10} = \mathbf{\frac{9}{100}} = \mathbf{0.09}$

c) P(not blue) = 1 – P(blue) $= 1 - \frac{3}{10} = \frac{7}{10}$
P(neither ball is blue)
= P(not blue) × P(not blue)
$= \frac{7}{10} \times \frac{7}{10} = \mathbf{\frac{49}{100}} = 0.49$

Page 485 Exercise 2

1 The events in this question are mutually exclusive — balls can't be two different colours.

a) P(pink or orange) = 0.5 + 0.1 = **0.6**
b) P(pink or red) = 0.5 + 0.4 = **0.9**
c) P(red or orange) = 0.4 + 0.1 = **0.5**

2 The events in this question are mutually exclusive — chocolate cannot be wrapped in two colours of foil.

a) P(red or gold) = 0.14 + 0.4 = **0.54**
b) P(silver or red) = 0.26 + 0.14 = **0.4**
c) P(gold or blue) = 0.4 + 0.2 = **0.6**
d) P(silver or gold) = 0.26 + 0.4 = **0.66**

3 The events in this question are mutually exclusive — a student can't be in two different teams.

a) P(Eagle) $= \mathbf{\frac{1}{3}}$
b) P(Eagle or Falcon) $= \frac{1}{3} + \frac{1}{3} = \mathbf{\frac{2}{3}}$
c) P(Falcon or Osprey) $= \frac{1}{3} + \frac{1}{3} = \mathbf{\frac{2}{3}}$

4 **a)** The events are not mutually exclusive — the dice can land on both an odd number and a multiple of 3.
P(multiple of 3) = P(3, 6, 9, 12, 15 or 18)
$= \frac{6}{20}$
P(odd) = P(1, 3, 5, 7, 9, 11, 13, 15, 17 or 19) $= \frac{10}{20}$
P(multiple of 3 and odd)
= P(3, 9 or 15) $= \frac{3}{20}$
So P(multiple of 3 or odd) $= \frac{6}{20} + \frac{10}{20} - \frac{3}{20}$
$= \mathbf{\frac{13}{20}}$

b) The events are not mutually exclusive — the dice can land on a factor of 20 which is also a multiple of 4.
P(factor of 20) = P(1, 2, 4, 5, 10 or 20)
$= \frac{6}{20}$
P(multiple of 4) = P(4, 8, 12, 16 or 20)
$= \frac{5}{20}$
P(factor of 20 and multiple of 4)
= P(4 or 20) $= \frac{2}{20}$
So P(factor of 20 or multiple of 4)
$= \frac{6}{20} + \frac{5}{20} - \frac{2}{20} = \mathbf{\frac{9}{20}}$

5 **a)** The events are not mutually exclusive — a card can be both a red suit and a queen.
P(red suit) = P(heart or diamond) $= \frac{26}{52}$
P(queen) = P(Q♠, Q♥, Q♣ or Q♦) $= \frac{4}{52}$
P(red suit and queen) = P(Q♥ or Q♦) $= \frac{2}{52}$
So P(red suit or queen) $= \frac{26}{52} + \frac{4}{52} - \frac{2}{52} = \frac{28}{52} = \mathbf{\frac{7}{13}}$

b) The events are not mutually exclusive — a card can be both a club and a picture card.
P(club) $= \frac{13}{52}$
P(picture card) = P(Jack, Queen or King)
$= \frac{12}{52}$
P(club and picture card) = P(J♣, Q♣ or K♣) $= \frac{3}{52}$
So P(club or picture card) $= \frac{13}{52} + \frac{12}{52} - \frac{3}{52}$
$= \frac{22}{52} = \mathbf{\frac{11}{26}}$
A pack of cards is a popular context in Maths questions — so it's a good idea to familiarise yourself with the different suits, colours, picture cards etc. (and how many there are of each).

6 No, having glasses and having blond hair are not mutually exclusive (some people could have both). This means that the probability of picking a pupil with blond hair or glasses isn't just the sum of the separate probabilities (unless you know that none of the pupils both wear glasses and have blond hair).

27.6 Tree Diagrams

Page 487 Exercise 1

1 **a)** P(R) $= \frac{3}{5} = 0.6$, P(B) $= \frac{2}{5} = 0.4$

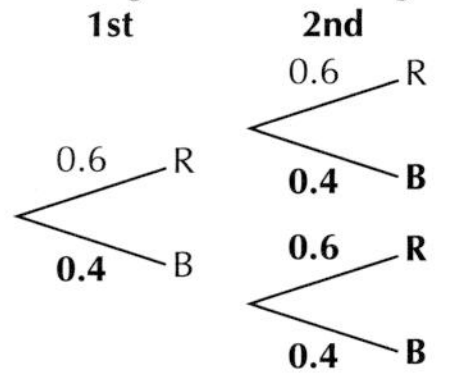

b) P(R) $= \frac{5}{10} = 0.5$, P(B) $= \frac{3}{10} = 0.3$,
P(G) $= \frac{2}{10} = 0.2$

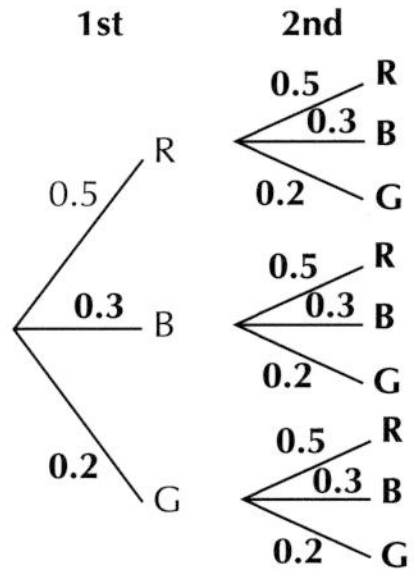

2 **a)** P(H) = 0.65, P(T) = 1 – 0.65 = 0.35

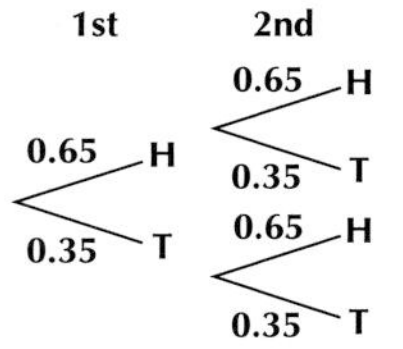

b)

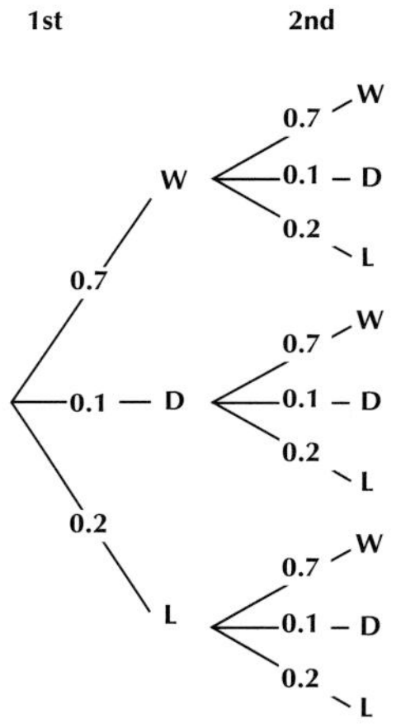

3 **a)** P(Freddie wins) = 0.8,
P(James wins) = 1 – 0.8 = 0.2

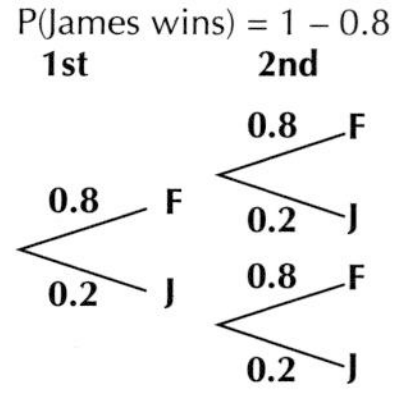

b) P(F, F) = 0.8 × 0.8 = **0.64**

4 a) P(Ifrath chooses a comedy) = $\frac{4}{12} = \frac{1}{3}$
P(Ifrath doesn't choose a comedy) = $1 - \frac{1}{3}$
$= \frac{2}{3}$
P(Jesse chooses a comedy) = $\frac{8}{20} = \frac{2}{5}$
P(Jesse doesn't choose a comedy) = $1 - \frac{2}{5}$
$= \frac{3}{5}$

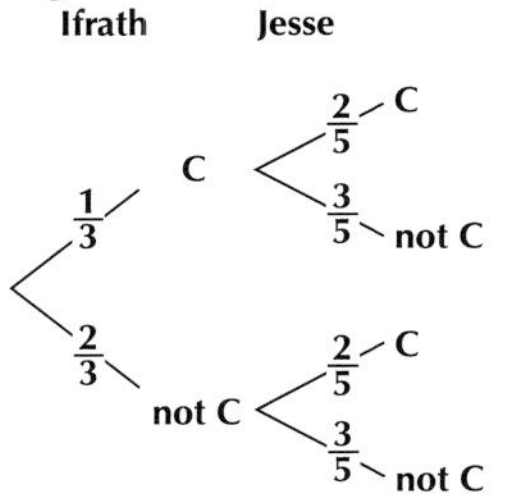

b) P(not C, not C) = $\frac{2}{3} \times \frac{3}{5} = \frac{6}{15} = \mathbf{\frac{2}{5}}$

c) P(at least one comedy) = 1 – P(not C, not C)
$= 1 - \frac{2}{5} = \mathbf{\frac{3}{5}}$

5 a)

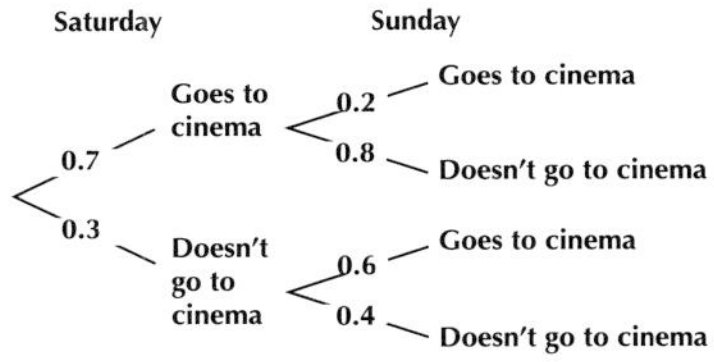

b) P(cinema, cinema) = 0.7 × 0.2 = **0.14**

c) P(cinema, not cinema)
+ P(not cinema, cinema)
= (0.7 × 0.8) + (0.3 × 0.6) = 0.56 + 0.18
= **0.74**

6 a)

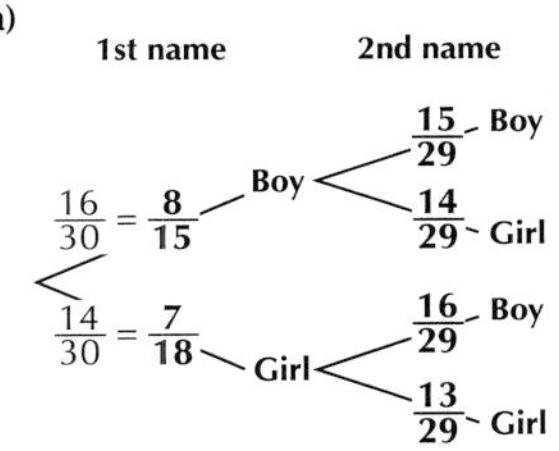

The probabilities change for the 2nd name because one name has been removed and not replaced.

b) P(boy, boy) = $\frac{8}{15} \times \frac{15}{29} = \mathbf{\frac{8}{29}}$

c) P(one girl) = P(boy, girl) + P(girl, boy)
$= \left(\frac{8}{15} \times \frac{14}{29}\right) + \left(\frac{7}{15} \times \frac{16}{29}\right)$
$= \frac{112}{435} + \frac{112}{435} = \mathbf{\frac{224}{435}}$

27.7 Sets and Venn Diagrams

Page 489 Exercise 1

1 a) (i) A = **{12, 14, 16, 18, 20, 22, 24}**
(ii) B = **{2, 3, 5, 7, 11, 13, 17, 19, 23, 29}**
(iii) C = **{1, 4, 9, 16, 25, 36, 49, 64, 81, 100, 121, 144, 169, 196}**
(iv) D = **{1, 2, 3, 5, 6, 10, 15, 30}**

b) n(A) means the number of elements in set A, so:
(i) **7** (ii) **10**
(iii) **14** (iv) **8**

2 a) A ∪ B is the union of set A and set B. All the elements in either set A or set B are **{1, 2, 4, 5, 6, 7, 8}**.

b) A ∩ B is the intersection of set A and set B. The elements that are in both sets are **{2, 4, 8}**.
Make sure you're happy with all the bits of set notation — the questions aren't too tricky as long as you know what each symbol means.

3 a) n(A) = **4** b) n(B) = **6**

c) A′ is the complement of set A, i.e. all the elements of the universal set that are not in set A. So A′ = **{2, 4, 5, 6, 8}**

d) A ∪ B = **{1, 2, 3, 5, 6, 7, 8, 9}**

e) B′ = **{1, 4, 7}**

f) A ∩ B = **{3, 9}**

g) n(A ∪ B) = **8**

h) A′ ∪ B is the union of the complement of set A and set B. This is all the elements in either the complement of set A or set B.
A′ ∪ B = **{2, 3, 4, 5, 6, 8, 9}**

Page 490 Exercise 2

1 a) A = {1, 3, 5}, B = {1, 3, 7}. 1 and 3 are in both sets, so these go in the overlap between the circles. 5 goes only in set A, 7 goes only in set B, and the rest of the positive integers ≤ 10 go around the outside of the circles.

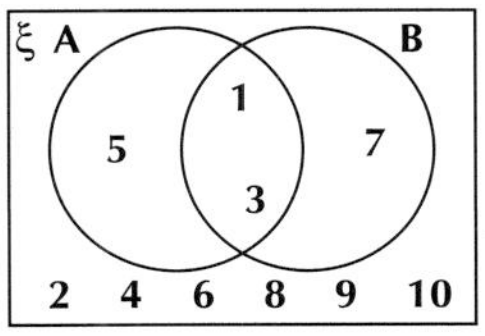

b) A = {2, 3, 4, 5}, B = {1, 3, 5, 7, 9}. 3 and 5 are in both sets, so these go in the overlap between the circles. 2 and 4 only go in set A, 1, 7 and 9 only go in set B, and the rest of the positive integers ≤ 10 go around the outside of the circles.

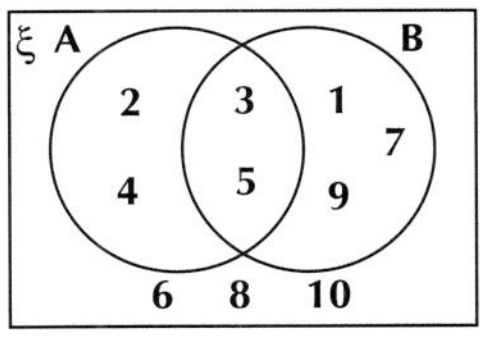

Using the same method for c)-d):

c)

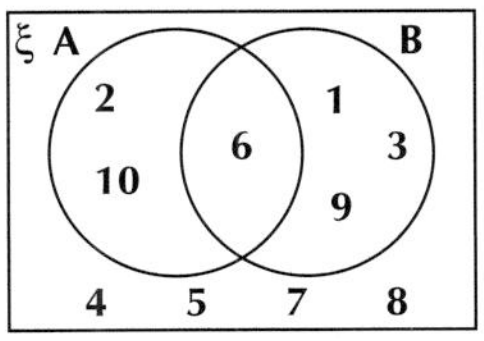

d)

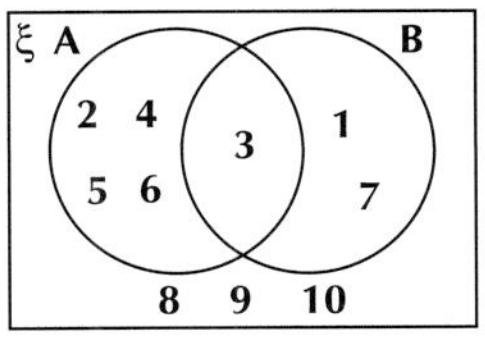

2 a) People who only had salad are in the part of the salad circle that's not overlapping the fries circle.
So P(only salad) = $\frac{76}{200} = \mathbf{\frac{19}{50}}$ = **0.38**.

b) People who only had fries are in the part of the fries circle that's not overlapping the salad circle.
So P(only fries) = $\mathbf{\frac{87}{200}}$ = **0.435**.

c) People who had no side are not part of any circle so are outside both circles.
So P(no side) = $\frac{10}{200} = \mathbf{\frac{1}{20}}$ = **0.05**.

d) All the customers in the salad circle: 76 + 27 = 103, so P(salad) = $\mathbf{\frac{103}{200}}$ = **0.515**.

e) All the customers in the fries circle:
87 + 27 = 114, so
P(fries) = $\frac{114}{200} = \mathbf{\frac{57}{100}}$ = **0.57**.

f) People who had fries and salad are in the intersection of the two circles,
so P(fries and salad) = $\mathbf{\frac{27}{200}}$ = **0.135**.

3 a) A = {1, 3, 5, 7, 9}, B = {2, 3, 5, 7}. 3, 5 and 7 are in both sets, so these go in the overlap. The positive integers ≤ 10 that aren't in A or B go outside the circles.

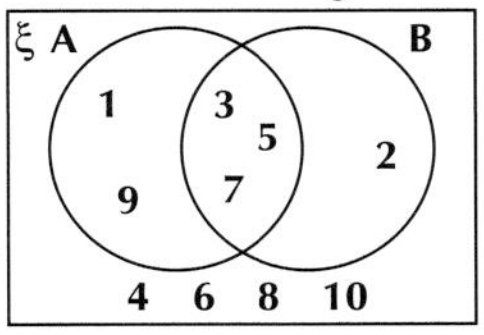

b) (i) n(B) = 4, so P(B) = $\frac{4}{10} = \mathbf{\frac{2}{5}}$ = **0.4**
(ii) A ∩ B = {3, 5, 7}, so n(A ∩ B) = 3 and P(A ∩ B) = $\mathbf{\frac{3}{10}}$ = **0.3**
(iii) A ∩ B′ means the numbers which are in A and not in B so are the numbers ≤ 10 which are odd and not prime (the ones in the part of circle A that's not overlapping circle B).
A ∩ B′ = {1, 9}, so n(A ∩ B′) = 2 and P(A ∩ B′) = $\frac{2}{10} = \mathbf{\frac{1}{5}}$ = **0.2**

4 a) 15 pupils like Maths and English, so 15 goes in the intersection. 23 – 15 = 8 pupils just like Maths and 18 – 15 = 3 pupils just like English, so 8 and 3 go in the other parts of the M and E circles respectively. That leaves 30 – 8 – 15 – 3 = 4 pupils who like neither Maths nor English, so 4 goes outside the circles.

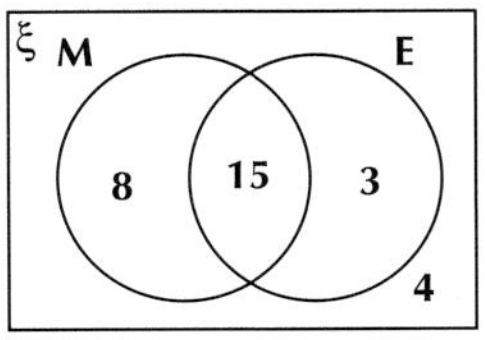

Always check that the numbers in your Venn diagram add up to the total from the question — here 8 + 15 + 3 + 4 = 30, which is the total number of pupils in the class.

b) (i) P(M and E) = $\frac{15}{30} = \mathbf{\frac{1}{2}}$
(ii) P(M only) = $\frac{8}{30} = \mathbf{\frac{4}{15}}$
(iii) P(E only) = $\frac{3}{30} = \mathbf{\frac{1}{10}}$
(iv) P(not M and not E) = $\frac{4}{30} = \mathbf{\frac{2}{15}}$

Page 491 Review Exercise

1 There are 4 + 6 + 5 + 5 = 20 pieces of fruit in total.

a) There are 4 apples and 6 bananas, so 4 + 6 = 10 pieces in total. So the likelihood is **evens**.

b) There are 5 oranges, so there are 20 – 5 = 15 pieces of fruit which are not oranges. So Bianca is **likely** to pick a piece of fruit which isn't an orange.

c) There are no lemons so it is **impossible** to pick one.

On the scale the arrows should be:

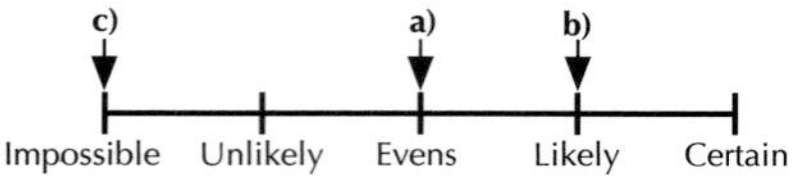

2 There are 6 possible outcomes — 1, 2, 3, 4, 5 and 6.

a) P(even) = P(2, 4 or 6) = $\frac{3}{6} = \mathbf{\frac{1}{2}}$

b) P(multiple of 3) = P(3 or 6) = $\frac{2}{6} = \mathbf{\frac{1}{3}}$

c) The answers to part a) and b) don't add up to 1 because these are not mutually exclusive outcomes that cover all the possibilities. (The two events aren't mutually exclusive because both events happen if a 6 is rolled, and the outcomes 1 and 5 are not included in either event.)

3 a) P(Derek wins) = $\mathbf{\frac{8}{15}}$

b) P(Eileen wins) = 1 – P(Derek wins)
$= 1 - \frac{8}{15} = \mathbf{\frac{7}{15}}$

4 a) **Yes**, these events are independent because the outcome of one does not affect the other.

b) 4 + 5 + 11 = 20 sweets in total.

(i) P(blue, blue) = P(blue) × P(blue)
$= \frac{4}{20} \times \frac{4}{20} = \frac{16}{400} = \mathbf{\frac{1}{25}}$

(ii) P(white then red)

= P(white) × P(red)

$= \frac{5}{20} \times \frac{11}{20} = \frac{55}{400} = \mathbf{\frac{11}{80}}$

5 a) After the first pick, there are only 9 marbles left, and 1 fewer of the colour that was picked, so:

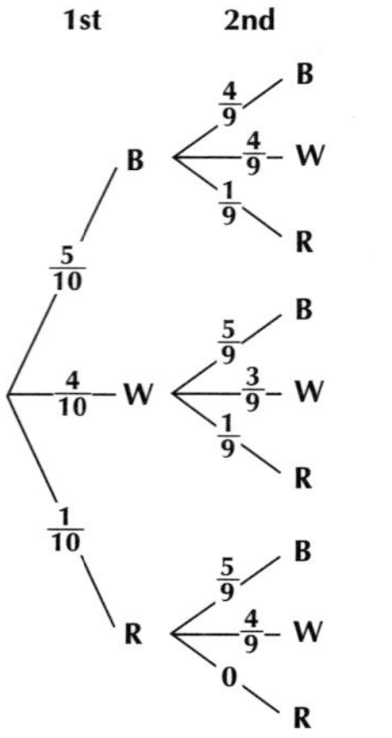

The probabilities on the second set of branches are different because the marbles are not replaced

b) (i) P(B, B) = $\frac{5}{10} \times \frac{4}{9} = \frac{20}{90} = \mathbf{\frac{2}{9}}$

(ii) P(at least one blue)
= 1 – P(not blue, not blue)
$= 1 - \left(\frac{5}{10} \times \frac{4}{9}\right) = 1 - \frac{2}{9} = \mathbf{\frac{7}{9}}$

You could also add up all of the probabilities that there is at least one blue marble.

(iii) P(different colours)
= 1 – P(same colours)
= 1 – P((B, B) or (W, W) or (R, R))
$= 1 - \left(\left(\frac{5}{10} \times \frac{4}{9}\right) + \left(\frac{4}{10} \times \frac{3}{9}\right) + \left(\frac{1}{10} \times 0\right)\right)$
$= 1 - \left(\frac{10}{45} + \frac{6}{45} + 0\right) = 1 - \frac{16}{45} = \mathbf{\frac{29}{45}}$

You could also add up all of the probabilities that the marbles are different colours.

6 a) 49 + 21 = **70 people**

b) 49 + 35 + 21 + 15 = 120 possible outcomes.
35 people bought only an ice lolly, so
P(only ice lolly) = $\frac{35}{120} = \mathbf{\frac{7}{24}}$

Page 492 Exam-Style Questions

1 The possible numbers she can make are 17, 18, 27, 28, 37, 38, 71, 72, 73, 81, 82, 83. So there are 12 possible outcomes in total *[1 mark]*. Five of these are even *[1 mark]*, so P(even) = $\mathbf{\frac{5}{12}}$ *[1 mark]*.

The question didn't ask you to list the outcomes, but it's a good idea to write them all down to make sure you don't miss any.

2 a)

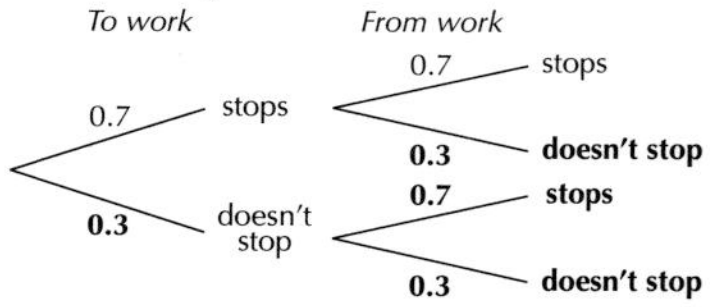

[2 marks available — 1 mark for calculating the probability that she doesn't have to stop, 1 mark for correctly completing the rest of the tree diagram]

b) P(stop only once) = P(stop, doesn't stop) + P(doesn't stop, stop)
= (0.7 × 0.3) + (0.3 × 0.7) *[1 mark]*
= 0.21 + 0.21 = **0.42** *[1 mark]*

3 a) Number of pupils asked = 7 + 9 = 16 *[1 mark]*.
Number of girls who went to the cinema last week = $\frac{2}{3} \times 9 = 6$ *[1 mark]*. 5 boys also went to the cinema, so probability
$= \frac{(6+5)}{16} = \mathbf{\frac{11}{16}}$ or **0.6875** *[1 mark]*.

b) E.g. His sample of Year 11 boys may not be representative of all Year 11 boys. *[1 mark]*
OR The sample size (of 7 boys) is too small to give a reliable answer *[1 mark]*.

4 a) 10 members do archery, and 6 members do both activities, so the number doing only archery is 10 – 6 = 4. 17 members do only basketball. That leaves 30 – (4 + 6 + 17) = 3 that do neither activity.
So the Venn diagram looks like this:

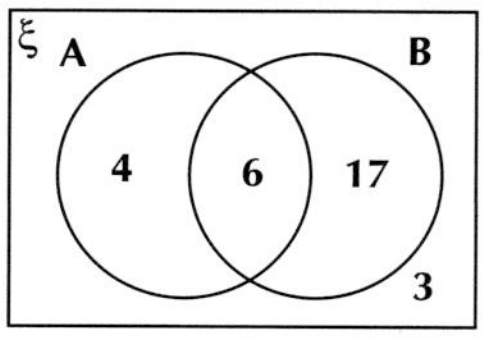

[2 marks available — 2 marks for a fully correct diagram, otherwise 1 mark for 2 correct numbers in the Venn diagram]

b) n(A ∪ B) is the number in the union of the sets = 4 + 6 + 17 = **27** *[1 mark]*

Mixed Exam-Style Questions

Page 493

1 a) Find the duration of the film using the start and finish times of the first showing:
+ 20 mins   + 1 hour   + 15 mins
18:40 ⟶ 19:00 ⟶ 20:00 ⟶ 20:15
So the film lasts 1 hour 35 minutes. *[1 mark]*
Add 1 hour 35 minutes to 20:30:
+ 1 hour   + 30 mins   + 5 mins
20:30 ⟶ 21:30 ⟶ 22:00 ⟶ 22:05
So the second showing will finish at **22:05** (or **10:05 pm**) *[1 mark]*.

b) The total cost was
£20 + (3 × £10) – £1.25 *[1 mark]* = £48.75
So one ticket costs £48.75 ÷ 5 *[1 mark]*
= **£9.75** *[1 mark]*.

2 a) **18** (2 × 9 = 18) *[1 mark]*

b) **25** (5^2 = 25) *[1 mark]*

c) **5** and **12** (5 × 12 = 60) *[1 mark]*

d) 3 + 12 + 18 + 25 = 58, so **5** was not included in the sum *[1 mark]*

e) For the difference between two numbers to be odd, one number must be even and one must be odd. You're looking for the biggest difference, so the two numbers must be either 18 and 3 or 25 and 12.
18 – 3 = 15 and 25 – 12 = 13. So the largest odd difference is **15** *[1 mark]*.

3 a) **A** and **C** are the same shape and size, so they are congruent *[1 mark]*.

b) E is the same shape as D, but half the size. So **D** and **E** are similar *[1 mark]*.

4 You can make 40p from 2p and 5p coins as follows:
8 × 5p = 40p
6 × 5p + 5 × 2p = 40p
4 × 5p + 10 × 2p = 40p
2 × 5p + 15 × 2p = 40p
20 × 2p = 40p
[2 marks available — 2 marks if all five correct ways have been found, otherwise 1 mark if three or four correct ways have been found]
The number of 5p's has to be even — the difference between 40p and an odd multiple of 5p would be odd, and so couldn't be a multiple of 2p.

5 The population of China is bigger, as it has the higher power of 10, so
larger population : smaller population
= $1.410 \times 10^9 : 5.936 \times 10^7$
To get this in the form n : 1, divide both sides by 5.936×10^7, to get 1 on the RHS.
On the LHS this gives
$\frac{1.410 \times 10^9}{5.936 \times 10^7}$ = 23.753... = 23.75 (to 4 s.f.)
So the ratio is **23.75 : 1**
[3 marks available — 1 mark for identifying China as having the larger population, 1 mark for attempting to divide by Italy's population, 1 mark for the correct final ratio]

6 a) On the first Friday, 36 students ate pizza and 14 ate pasta *[1 mark]*.
$\frac{14}{36+14} = \frac{14}{50} = \frac{28}{100}$ = **28%** chose pasta *[1 mark]*

b) On the second Friday, 26 students chose pizza and 12 chose pasta *[1 mark]*.
$\frac{26}{26+12} = \frac{26}{38} = \mathbf{\frac{13}{19}}$ chose pizza *[1 mark]*

c) E.g. The number of students having pizza has decreased each week.
or The number of students eating pasta has stayed roughly the same.
or The number of students eating in the canteen has decreased.
[1 mark for any suitable trend]

7 Uma worked for 4 × 7.5 + 7 = 37 hours from Monday to Friday.
Her pay for this work is 37 × £9.46 = £350.02 *[1 mark]*.
So for the Saturday overtime she was paid £406.78 – £350.02 = £56.76 *[1 mark]*
The hourly rate for overtime is £9.46 × 2 = £18.92, so she worked £56.76 ÷ £18.92 = **3 hours** overtime *[1 mark]*.

8 On the map, AB = 6 cm
1 cm = 5 km, so multiply by 5 to find the real-life distance: 6 × 5 = 30 km *[1 mark]*.
Time taken = distance ÷ speed
= 30 ÷ 12 *[1 mark]*
= **2.5 hours** or **2 hours 30 minutes** *[1 mark]*

9 a) $\frac{3}{4} = \frac{3 \times 2}{4 \times 2} = \frac{6}{\mathbf{8}}$ *[1 mark]*

b) $\frac{3}{4} = \frac{3 \times 11}{4 \times 11} = \frac{33}{44}$. $\frac{8}{11} = \frac{4 \times 8}{44} = \frac{32}{44}$ and $\frac{9}{11} = \frac{4 \times 9}{44} = \frac{36}{44}$, so $\frac{3}{4} < \frac{\mathbf{9}}{\mathbf{11}}$ *[1 mark]*

10 Find how much the special offer standard box contains: 25% = 25 ÷ 100 = 0.25, so multiplier = 1 + 0.25 = 1.25
500 g × 1.25 = 625 g *[1 mark]*
Now find the cost per gram for each box.
Family box:
225p ÷ 750 g = 0.3 pence per gram *[1 mark]*
Special offer standard box:
200p ÷ 625 g = 0.32 pence per gram *[1 mark]*
The cost per gram for the family box is less than that for the standard box, so the **family box** is better value *[1 mark]*.
You'd also get the marks if you divided the weight of each box by its cost to find the weight per penny. The more grams you get per penny, the better value.

11 Perimeter = 4x + 3 + 4x + 3 + 4x + 3
= 12x + 9 *[1 mark]*
So 12x + 9 > 480 *[1 mark]*
⇒ 12x > 480 – 9 ⇒ 12x > 471 *[1 mark]*
⇒ x > 471 ÷ 12 ⇒ **x > 39.25** *[1 mark]*

12 a) The shape can be divided into 7 small squares the same size as the shaded area:

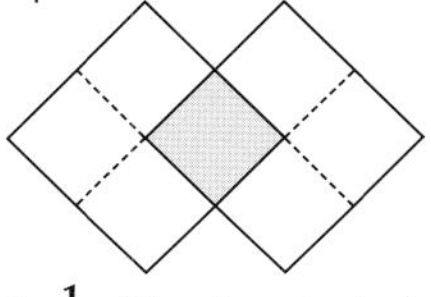

So $\frac{\mathbf{1}}{\mathbf{7}}$ of the shape is shaded *[1 mark]*

b) The shaded shape is a square, so it has side length $\sqrt{25}$ = **5 cm** *[1 mark]*.
Since the large squares overlap at the mid-point of their sides, side length of large square = 2 × 5 cm = 10 cm *[1 mark]*.

13 P(green grape) = $\frac{1}{3}$ and there are 8 green grapes, so 8 = $\frac{1}{3}$ of the total number of grapes. That means there are 8 × 3 = 24 grapes in the bowl *[1 mark]*. Let x be the number of black grapes in the bowl,
then 24 = 8 + 12 + x *[1 mark]* ⇒ 24 = 20 + x
⇒ x = 24 – 20 = 4
So there are **4** black grapes in the bowl *[1 mark]*.

14 Price of computer from US plus delivery
= \$432 + \$30 = \$462
20% = 20 ÷ 100 = 0.2, so multiplier for import tax is 1 + 0.2 = 1.2.
Cost of US computer including import tax:
\$462 × 1.2 = \$554.40 *[1 mark]*
Exchange rate is \$1 = £0.75, so multiply by 0.75 to find cost in £:
\$554.40 × 0.75 = £415.80 *[1 mark]*. So the **American seller** offers the better deal *[1 mark]*.
Alternatively, you could convert the UK price into US dollars and compare that to \$544.40.

15 Opposite sides of a parallelogram are equal:
5x + 2 = 20 – 10x *[1 mark]*
⇒ 5x + 10x = 20 – 2
⇒ 15x = 18 ⇒ x = 1.2 *[1 mark]*
So the shorter side of the parallelogram has length 3x = 3 × 1.2 = 3.6 cm and the longer side has length 5x + 2 = 5 × 1.2 + 2 = 8 cm *[1 mark]*.
So perimeter = 2 × 3.6 + 2 × 8 = **23.2 cm** *[1 mark]*
You could also find an expression for the perimeter in terms of x, and plug in your value of x. You'll get marks for any sensible method.

16 a) At A, y = 2x + 5 and y = 1 intersect, so 2x + 5 = 1 *[1 mark]*
⇒ 2x = –4 ⇒ x = –2
So A = (–2, 1) *[1 mark]*.
At B, y = 2x + 5 and x = –1 intersect, so y = 2 × (–1) + 5 *[1 mark]* = 3
So B = (–1, 3) *[1 mark]*.
At C, x = –1 and y = 1, so C = (–1, 1) *[1 mark]*.

b) ABC is a right-angled triangle with hypotenuse AB, so use Pythagoras.

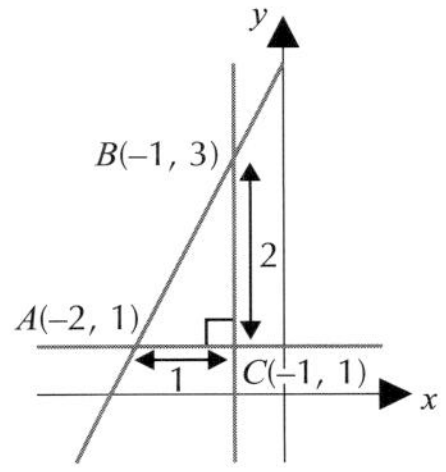

From a), A = (–2, 1) and C = (–1, 1), so length of AC is –1 – (–2) = 1 *[1 mark]*.
B = (–1, 3) so length of BC is 3 – 1 = 2 *[1 mark]*.
Length of AB = $\sqrt{(AC)^2 + (BC)^2}$ *[1 mark]*
= $\sqrt{1^2 + 2^2} = \sqrt{1 + 4}$
= $\sqrt{\mathbf{5}}$ *[1 mark]*
You don't need to have drawn the diagram in order to get all of the marks.

17 The sum of the interior angles in a polygon with n sides equals (n – 2) × 180°.
So the sum of the interior angles of a pentagon is (5 – 2) × 180° = 540° *[1 mark]*.
There are three right angles in the pentagon, so the sum of the other two angles is 540° – 3 × 90° = 270° *[1 mark]*.
These angles are in the ratio 4 : 11.
1 part = 270° ÷ (4 + 11) *[1 mark]* = 18°
So the largest angle is 11 x 18° = **198°** *[1 mark]*.

18 Volume of water = 1.14 × 1.14 × 1.75
= 2.2743 m^3 *[1 mark]*
When the tank is repositioned, the volume is the same so 1.14 × 2.25 × x = 2.2743 *[1 mark]*
⇒ x = 2.2743 ÷ (1.14 × 2.25) = 0.88666...,
so the depth is **0.887 m** (3 s.f.) *[1 mark]*.

19 Use the properties of the hexagon to work out the other lengths you need. The hexagon is regular, so all sides have length 10 cm. This also means the hexagon has a vertical line of symmetry, so an identical right-angled triangle can be drawn on the RHS.

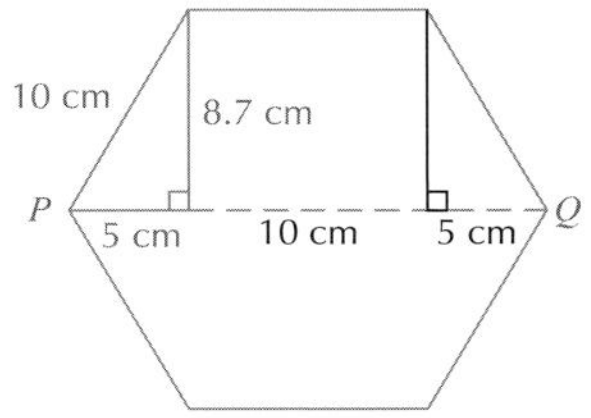

The area of a trapezium is given by
$A = \frac{1}{2}(a + b) \times h$
Considering the trapezium that makes up the top half of the hexagon: a = 10 cm, b = 5 + 10 + 5 = 20 cm and h = 8.7 cm.
So $A = \frac{1}{2}(10 + 20) \times 8.7 = 15 \times 8.7$
= 130.5 cm^2
So area of full hexagon = 2 × 130.5
= **261 cm^2**
[4 marks available — 1 mark for working out the missing lengths, 1 mark for attempting to use the formula for area, 1 mark for doubling to find the area of the hexagon, 1 mark for the correct final answer]
If you weren't sure about the formula for the area of a trapezium, you could instead work out the areas of the right-angled triangles and the rectangle, then add them together to get 130.5 cm^2.

Glossary

A

Acute-angled triangle A triangle where all angles are less than 90°.

Adjacent In a **right-angled triangle**, it's the side between the angle under consideration and the right angle.

Allied angles The pair of angles in a C- or U-shape formed when a straight line crosses two **parallel lines**. The angles add up to 180°.

Alternate angles The pair of equal angles in a Z-shape formed when a straight line crosses two **parallel lines**.

AND rule (for independent events) The **probability** of both A and B happening is equal to the probability of A happening multiplied by the probability of B happening.

Angle bisector The line that cuts an angle into two equal smaller angles.

Angle of depression/elevation The angle between a horizontal line and the line of sight of an observer at the same level looking down or up, respectively.

Arc A part of the **circumference** of a circle.

Area The space inside a 2D shape. It's measured in units squared.

Arithmetic sequence A **sequence** where the **terms** increase or decrease by the same amount each time (the **common difference**).

Average A way of representing a set of data with a central or typical value of the set. The three common averages used are the **mode**, **median** and **mean**.

B

Bar chart A chart to display **discrete** or categorical data. The height of bars shows the number (or frequency) of items in different categories.

Bearing A three-figure angle measured clockwise from the **north line** to tell you the position of one object in relation to another.

Bias (in outcomes) Applies to e.g. rolling dice, where the **outcomes** are not equally likely.

Biased sample One in which some members of a **population** are more likely to be included than others.

Bisect Split a line or angle exactly in half.

BODMAS The correct order to carry out mathematical operations — it stands for Brackets, Other, Division, Multiplication, Addition, Subtraction.

C

Centre of enlargement The point where an **enlargement** is measured from.

Chord A line between two points on the edge of a circle.

Circumference The distance around the outside of a circle.

Coefficient The number in front of a **variable**. E.g. in the term $2x$, 2 is the coefficient.

Common denominator The same bottom number in two or more **fractions**.

Common difference The number you add or subtract to get between **terms** in an **arithmetic sequence**.

Common factor A common **factor** of two or more numbers is a factor of both or all of those numbers.

Common multiple A common **multiple** of two or more numbers is a multiple of both or all of those numbers.

Common ratio The number you multiply by to get between **terms** in a **geometric sequence**.

Complement of a set All **elements** of the **universal set** that aren't in the set. The complement of a set A is written A′.

Composite bar chart A **bar chart** which has single bars split into different sections for each set of data displayed.

Composite shape A shape made up of two or more basic shapes.

Compound decay When a quantity gets smaller over time due to successive **percentage** decreases based on the decreasing value itself.

Compound growth When a quantity gets larger over time due to successive **percentage** increases based on the increasing value itself.

Compound inequality A combination of multiple **inequalities**, e.g. $a < x < b$.

Compound interest **Compound growth** applied to money.

Compound measure A measurement made up of two or more other measurements, e.g. speed = distance ÷ time.

Cone A 3D shape with a circular base and a curved sloping face that goes up to a point at the top.

Congruent Shapes which are exactly the same shape and size as each other.

Construction An accurate drawing made using a pair of compasses, a protractor and a ruler.

Continuous data Numerical data which can take any value in a given range.

Conversion factor The number you multiply or divide by to convert a measurement from one unit to another. E.g. A conversion factor of 100 is used to convert between centimetres and metres.

Coordinates Two numbers in a pair of brackets which describe the position of a point on a grid.

Correlation How two variables are related to each other. Positive correlation means the variables increase and decrease together. Negative correlation mean that as one variable increases, the other decreases.

Corresponding angles The pair of equal angles in an F-shape formed when a straight line crosses two **parallel lines**.

Cosine The cosine of an angle x in a **right-angled triangle** is the length of the **adjacent** side divided by the length of the **hypotenuse** side, i.e. $\cos x = \frac{\text{adjacent}}{\text{hypotenuse}}$.

Counter example An example which doesn't work, used to show that a statement is false.

Cross-section The face exposed when cutting through a 3D shape.

Cube A **cuboid** where all six faces are squares.

Cube (power) A number multiplied by itself twice — written as the **power** of 3, x^3.

Cubic graph The graph of a cubic **function**, which has x^3 as its highest **power**. Cubic graphs all have the same basic shape — a curve with a 'wiggle' in the middle.

Cuboid A 3D shape with six rectangular or square faces.

Cylinder A 3D shape with a constant circular **cross-section**.

D

Decimal place (d.p.) The position of a digit that comes after the decimal point.

Denominator The bottom number of a **fraction**.

Density The mass per unit **volume** of a substance.

Depreciation The loss of value over time due to **compound decay**.

Diameter The line from one side of a circle to the other through its centre. The diameter is twice the length of the **radius**.

Difference of two squares A **quadratic** expression with just two **square** terms separated by a minus sign, $a^2 - b^2$, which can be **factorised** as $(a + b)(a - b)$.

Direct proportion Two **variables** are in direct proportion when the **ratio** between them is always the same, i.e. $y = Ax$.

Discrete data Numerical data which can only take certain values.

Distance-time graph A graph with distance travelled on the vertical axis and time taken on the horizontal axis.

Dual bar chart A **bar chart** which shows two sets of data by having two bars per category.

E

Element An item contained in a **set**, also called a member of the set. $x \in A$ means 'x is an element of set A'.

Elevation The 2D view of a 3D object looking at it either from the front or side horizontally.

Enlargement A **transformation** where a shape is enlarged by a particular **scale factor**, sometimes in relation to a **centre of enlargement**.

Equation A way of showing that two **expressions** are equal to each other for a particular value or values of an unknown.

Equidistant When a set of points are the same distance from two points or a line.

Equilateral triangle A triangle with 3 equal sides and equal angles (of 60°).

Equivalent fraction A **fraction** that shows the same proportion as another fraction using a different **numerator** and **denominator**.

Error interval The range of values that a **rounded** number could actually be.

Estimate An approximation to the answer to a calculation or the size of an amount.

Event A set of one or more **outcomes** to which a **probability** is assigned.

Exact A value which is precise and completely accurate. Exact answers may include square roots or π, and are often given as **fractions** rather than decimals.

Exchange rate The **conversion factor** between two currencies.

Expanding brackets Removing brackets by multiplying everything inside the brackets by everything in front of the brackets.

Expected frequency The number of times an **event** is expected to happen — the **probability** of the event multiplied by the number of times an experiment is done.

Experimental probability The number of times a result has occurred divided by the number of times an experiment has been done. It's used to estimate **probabilities**. Also known as **relative frequency**.

Expression An algebraic expression is a combination of **terms** separated by + and – signs.

Exterior angles The angle between a side of a **polygon** and a line that extends out from a neighbouring side.

Extrapolation Using e.g. a **line of best fit** to predict values outside the range of data you have.

F

Factor The factors of a number are the numbers that divide into it exactly.

Factorising Finding a **common factor** in the terms of an **expression** and taking it outside a pair of brackets.

Fair (in probability) A situation where all **outcomes** are equally likely to happen.

Fair sample A sample selected at random so that every member in the population has an equal chance of being selected.

Fibonacci-type sequence A sequence where each term is found by adding together the two previous terms.

Formula The mathematical relationship between different quantities.

Fraction A value written as one number divided by another.

Frequency The number of times that a data value occurs.

Frequency table A table to record the frequency of a response or event.

Frequency tree A diagram made up of branches to show the different possible outcomes of multiple events. The number at the end of each branch shows how many times that event or combination of events happened.

Frustum The 3D shape left once you chop off the top bit of a cone parallel to its circular base.

Function A rule that turns one number (the input) into another number (the output).

G

Geometric sequence A sequence where the terms are found by multiplying by the same value each time (the common ratio).

Gradient The slope or steepness of a graph, which can be found by dividing the change in y by the change in x.

H

Highest common factor (HCF) The largest number that will divide exactly into both (or all) of a given pair (or set) of numbers.

Hypotenuse The longest side in a right-angled triangle, opposite the right angle.

I

Identity A way of showing that two expressions are always equal to each other, not just for a particular value or values. Identities use the sign '≡'.

Image The shape formed by carrying out a transformation.

Imperial units A set of non-metric units for measuring, e.g. inches, ounces, miles.

Improper fraction A fraction where the numerator (the top number) is bigger than the denominator (the bottom number).

Independent events Two (or more) events are independent if one of them happening has no effect on the probability of the other(s) happening.

Index (or power) A repeated multiplication of a number or variable. a^x means 'a to the power of x' or 'x lots of a multiplied together'.

Inequality A pair of expressions separated by one of the symbols $<$, $>$, $\leq$, $\geq$. Like an equation, but with a range of solutions.

Intercept The point where a graph crosses an axis.

Interior angles The angles inside each vertex (corner) of a polygon.

Interpolation Using e.g. a line of best fit to predict values within the range of data you have.

Intersection (of sets) The intersection of two sets ($A \cap B$) contains only the elements that are in both sets.

Inverse proportion Two variables are inversely proportional when one variable increases as the other decreases, i.e. $y = \frac{A}{x}$. The product of the two variables is constant.

Isometric drawing A 2D drawing of a 3D object on an isometric grid of dots or lines in a pattern of equilateral triangles.

Isosceles triangle A triangle with 2 equal sides and 2 equal angles.

K

Kite A quadrilateral with two pairs of equal sides and one pair of equal angles in opposite corners.

L

Like terms Algebraic terms that contain exactly the same combination of letters (in any order).

Line of best fit A straight line on a scatter graph to show the general trend of the data.

Line of symmetry A mirror line on a graph or 2D shape which divides it so that each half is a reflection of the other.

Locus A set of points which satisfy a particular condition, e.g. points that are a fixed distance away from a point or line. The plural of locus is loci.

Lowest common multiple (LCM) The smallest number that is a multiple of both (or all) of a given pair (or group) of numbers.

M

Magnitude (vectors) The size or length of a vector.

Mean The total of all the values in a set of data divided by the number of values.

Median The middle value in a set of data written in size order.

Member See element.

Metric units Units for measuring using a decimal-based system (so units are based on powers of 10), e.g. metres, kilograms.

Midpoint A point which is halfway between two other points.

Mirror line The line in which an object is reflected. **Image** points will be the same perpendicular distance from the mirror line as the corresponding points on the object, but on the opposite side.

Mixed number A number that has a whole number part and a **fraction** part.

Modal group/class The group with the highest **frequency** in a set of grouped data.

Mode The most common value in a set of data.

Multiple The multiples of a number are the numbers in its times table.

Mutually exclusive events **Events** that cannot happen at the same time. E.g. choosing a club and choosing a red card from a pack of cards are mutually exclusive events.

N

Net The 2D representation of a 3D object that can be folded up to make the object.

North line The line vertically upwards from a point, used as the start point for **bearings**.

***n*th term** A general **term** in a **sequence** in the position n, which can be used to find any term in the sequence.

Numerator The top number of a **fraction**.

O

Obtuse-angled triangle A triangle with one angle greater than 90°.

Opposite In a **right-angled triangle**, it's the side opposite the angle under consideration.

Order of rotational symmetry The number of positions you can **rotate** a shape into so that it still looks exactly the same.

Origin The point (0, 0) on a **coordinate** grid. This is where the x- and y-axis intersect.

OR rule (for mutually exclusive events) The **probability** of at least one of the **events** happening is the sum of the probabilities of each event happening. It's also called the addition rule.

OR rule (general) If two **events** are not **mutually exclusive**, the **probability** that at least one event happens is equal to the sum of the probabilities of each event happening, minus the probability that both events happen.

Outcome The result of an activity in **probability**. E.g. flipping a coin and getting tails.

Outlier An extreme value in a data set.

Overestimate An **estimated** value that's greater than the actual value.

P

Parabola The shape of a graph of a **quadratic** function.

Parallel Two lines are parallel if they have the same **gradient**. They are always at the same distance apart and never meet.

Parallelogram A **quadrilateral** with two pairs of equal, **parallel** sides.

Percentage A **proportion** of something compared to the whole, where the whole is taken to be 100.

Perimeter The distance around the outside of a shape. It's found by adding up the lengths of the edges.

Perpendicular bisector The perpendicular bisector of a line is at right angles to the line and cuts it in half.

Perpendicular height The height of a shape measured at a right angle to the base.

Perpendicular lines Two lines that cross at a right angle.

Pictogram A diagram which uses a symbol to represent a certain number of items.

Pie chart A circular chart showing the **proportion** of the data set in each category rather than the actual frequency.

Plan The 2D view of a 3D object looking vertically downwards on it.

Plane of symmetry A 2D shape that cuts a 3D solid into two identical halves.

Polygon A 2D shape with straight sides.

Population The whole group of people or things you want to find out about when collecting data.

Position to term rule The rule relating a number in a sequence to its position in the sequence. Often referred to as the ***n*th term**.

Power See **index**.

Pressure A force per unit area.

Primary data Data you have collected yourself.

Prime factorisation Breaking a number down into a unique string of **prime numbers** (its prime factors) multiplied together.

Prime number A number that has no **factors** except itself and 1.

Prism A 3D shape with a constant **cross-section**.

Probability How likely an **event** is to happen.

Projections **Plans** and **elevations** — 2D representations of a 3D object.

Proof A mathematical explanation to show that something is true.

Proportion How the size of one quantity relates to the size of another, or to the total.

Pyramid A 3D shape which has a **polygon** base and which rises to a point.

Pythagoras' theorem The rule connecting lengths of sides in **right-angled triangles**: $h^2 = a^2 + b^2$, where h is the **hypotenuse** of the triangle and a and b are the shorter sides.

Q

Quadrant One of the four sections of a **coordinate** grid separated by the x-axis and y-axis.

Quadratic An **expression**, **equation** or **function** where the highest **power** of the **variable** is 2. They take the form $ax^2 + bx + c$, where a, b and c are constants and x is a variable.

Quadratic graph The graph of a quadratic **function** — a u- or n-shaped symmetrical curve.

Quadratic sequence A **sequence** where the difference between terms changes by the same amount each time.

Quadrilateral A shape that has 4 sides.

Qualitative data Data which is descriptive, so it records words instead of numbers.

Quantitative data Data which is numerical. It can be **discrete** or **continuous**.

R

Radius The line from the centre to the edge of a circle.

Range The difference between the largest and smallest value in a set of data — a measure of spread.

Rate of flow How quickly a liquid is moving into, out of, or through a certain space.

Ratio A way of showing **proportion** between quantities in the form $a : b$.

Reciprocal The reciprocal of a number is 1 ÷ that number.

Reciprocal graph The graph of a **reciprocal** function, i.e. $y = \frac{A}{x}$.

Reflection A **transformation** where a shape is mirrored in a straight line.

Regular polygon A **polygon** where all its sides and angles are equal.

Relative frequency See **experimental probability**.

Remainder The amount left over after carrying out a division.

Resultant vector The sum of two or more **vectors**.

Rhombus A **parallelogram** where all sides are the same length.

Right-angled triangle A triangle with a right-angle (90°).

Root (of an equation) Another word for the solution of an **equation**, usually used when solving an equation that has zero on one side of the equals sign.

Root The inverse operation of **squaring** and **cubing** (and raising to other **powers**). Square roots are written as $\pm\sqrt{x}$, cube roots as $\sqrt[3]{x}$.

Rotation A **transformation** where a shape is turned about a particular point — the centre of rotation.

Rotational symmetry Where a shape looks exactly the same after it has been **rotated** by a certain number of degrees.

Rounding Where numbers are approximated to make them easier to work with, e.g. to• a certain number of **decimal places** or **significant figures**.

S

Sample A smaller group of the **population** used to represent the population and to collect data from.

Sample space diagram A list, grid or **two-way table** to show all the possible **outcomes** of an **event** in a systematic way. Also known as a possibility diagram.

Scalar A quantity with magnitude (size) but no direction.

Scale drawing A diagram where all lengths are related to their real-life lengths by a constant **scale factor**.

Scale factor The number that tells you how many times longer the sides of an **enlarged** shape are compared to the original shape. Or how many times bigger one quantity is in relation to another proportional quantity.

Scalene triangle A triangle whose sides and angles are all different.

Scatter graph A graph of two variables plotted against each other, which can show if these variables are related or not.

Seasonality A basic pattern in data that is repeated regularly over time.

Secondary data Data collected by someone else. You can get secondary data from. E.g. newspapers or the internet.

Sector An area of a circle from the centre to the edge, like a "slice of pie".

Segment An area of a circle between an **arc** and a **chord**.

Semicircle Half of a circle.

Sequence A list of numbers or shapes which follows a particular rule. Each number or shape is called a **term**.

Set A set is a group of items or numbers, written in a pair of curly brackets { }.

Set notation A collection of symbols used to help define the **elements** of (objects within) a **set**.

Significant figures (s.f.) The digits in a number after and including the first digit that is not a 0.

Similar Shapes that have the same size angles (and so are the same shape) but the lengths of corresponding sides are different, related by a **scale factor**.

Simple interest A **percentage** of an initial amount of money added on at regular intervals. The amount of interest added each time doesn't change.

Simplify Reduce an **expression**, **equation**, **fraction** or **ratio** to the form which is easiest to use.

Simultaneous equations A pair of **equations** which are both true for particular values of the unknowns.

Sine The sine of an angle x in a **right-angled triangle** is the length of the **opposite** side divided by the length of the **hypotenuse**, i.e. $\sin x = \frac{\text{opposite}}{\text{hypotenuse}}$.

Solid A 3D shape.

Speed The distance travelled per unit of time.

Sphere A 3D shape with one curved surface, no vertices (corners) and no edges.

Square (power) A number multiplied by itself — written as the **power** of 2, x^2.

Standard form A way to write very big or small numbers as $A \times 10^n$, where $1 \le A < 10$ and n is an integer.

Stem and leaf diagram A diagram that displays data values. They consist of 'stems' (the first digit(s)) and 'leaves' (the remaining digit) of the data values arranged in numerical order.

Subject The letter before the equals sign in a formula. E.g. in $s = \frac{d}{t}$, s is the subject.

Substituting Replacing the letters in a **formula**, **expression** or **equation** with actual values.

Surface area The total area of all the faces of a 3D shape.

T

Tally chart A way to record **qualitative** or **quantitative** data, where the number of marks (tallies) for each category gives the **frequency**.

Tangent (to a circle) A straight line that just touches the circle at a point.

Tangent The tangent of an angle x in a **right-angled triangle** is the length of the **opposite** side divided by the length of the **adjacent** side, i.e. $\tan x = \frac{\text{opposite}}{\text{adjacent}}$.

Term (algebra) An individual part of an **expression**, e.g. 3, $2x$, a^2b.

Term (sequence) A number in a **sequence**.

Term to term rule The rule which determines how to get from one number in a **sequence** to the next.

Tetrahedron A **pyramid** with a triangular base.

Theoretical probability **Probability** calculated without performing an experiment. It can be found when all possible **outcomes** are equally likely, by dividing the number of ways an **event** can occur by the total number of outcomes.

Time series graph A line graph that shows data collected at regular intervals over time.

Transformation The changing of the size and/or position of a shape.

Translation A **transformation** where a shape is moved horizontally and/or vertically but keeps its original shape and size.

Trapezium A **quadrilateral** with one pair of **parallel** sides.

Tree diagram A diagram made up of branches to show the **probabilities** of different possible **outcomes** of multiple **events**.

Trend A pattern in data.

Truncating Chopping off the digits of a decimal number after a certain number of **decimal places** without rounding.

Turning point The point on a curve where the **gradient** is zero, e.g. a maximum or minimum point.

Two-way table A data collection sheet which records two pieces of information about the same subject at once. It shows the **frequency** for two different variables.

U

Underestimate An **estimated** value which is less than the actual value.

Union The union of two **sets** ($A \cup B$) contains all the **elements** that are in either set.

Unique factorisation theorem This states that the **prime factorisation** of every number is unique to that number.

Unit fraction A **fraction** where the **numerator** is one.

Universal set The **set** of all things under consideration for a particular situation, denoted by ξ.

V

Variable Letters in **expressions**, **formulas** or **equations** which can take different numerical values.

Vector A quantity or straight line with **magnitude** (size) and direction.

Venn diagram A diagram with two or more circles used to represent **sets**, which may overlap.

Vertex A corner of a 2D or 3D shape (the plural is vertices).

Vertically opposite angles The pair of equal angles opposite each other when two lines intersect.

Volume The amount of space a 3D shape takes up.

Y

$y = mx + c$ The **equation** for a straight line where m is the **gradient** and c is the **y-intercept** (the point where the line crosses the y-axis).

Index

Published by CGP

Editors:
Adam Bartlett, Michael Bushell, Sarah George, Tom Miles, Ali Palin, Rosa Roberts, David Ryan, Caley Simpson, Michael Weynberg, Ruth Wilbourne.

Contributors:
Katharine Brown, Pamela Chatley, Eva Cowlishaw, Alastair Duncombe, Paul Garrett, Geoff Gibb, Stephen Green, Philip Hale, Phil Harvey, Judy Hornigold, Claire Jackson, Mark Moody, Charlotte O'Brien, Philip Potten, Rosemary Rogers, Manpreet Sambi, Neil Saunders, Jan Walker, Kieran Wardell, Jeanette Whiteman.

With thanks to Liam Dyer, Helen Kennedy, Sam Mann, Lauren McNaughten, Deanne Morris, Glenn Rogers and Emma Stubbs for the proofreading.

With thanks to Ana Pungartnik for the copyright research.

Cover design concept by emc design ltd.

Clipart from Corel®
Printed by Elanders Ltd, Newcastle upon Tyne.

GCSE Mathematics

Foundation Level

Q: What's the best way to get through GCSE Maths?
A: By taking the square route.

That was a joke, obviously. The real answer is: by using this CGP book. It's packed with study notes, examples and indispensable tips for every topic, plus enough practice questions to keep you happy throughout the course.

It even includes a free Online Edition to read on your PC, Mac or tablet!

CGP

How to get your free Online Edition

Go to **cgpbooks.co.uk/extras** and enter this code...

2414 2218 4222 2114

This code will only work once. If someone has used this book before you, they may [illegible] n.

Atmosphere, weather and climate

Atmosphere, weather and climate provides a thorough introduction to weather processes and climatic conditions. Since the last edition, the recognition of the reality and possible effects of human activities on the environment has revolutionized attitudes to the study of the atmosphere and of world climate.

Beginning with an extended treatment of atmospheric composition and energy, stressing the heat budget of the earth and the causes of the greenhouse effect, the authors turn to the manifestations and circulation of atmospheric moisture, including atmospheric stability and precipitation patterns in space and time. A consideration of atmospheric motion on small to large scales and modelling of general circulation leads to a discussion of the structure of air masses, frontal cyclones and weather forecasting on different time scales.

The treatment of weather and climate in temperate latitudes begins with studies of Europe and America, extending to the conditions of their polar and sub-tropical margins. Tropical weather and climate are also described through an analysis of the climatic mechanisms of monsoon Asia, Africa and Amazonia, together with the tropical margins of Africa and Australia and the effects of ocean movement and the El Niño-Southern Oscillation. Small scale climates – including urban climates – are considered from the perspective of energy budgets.

The final chapter presents a state of the art treatment of the nature and causes of climate change, stressing the possible effects of human activity and presenting possible scenarios of changes in atmospheric composition, temperature, precipitation and sea level over the next 100 years.

The sixth edition of *Atmosphere, weather and climate* will prove invaluable to students and teachers in secondary and tertiary education studying environmental and earth sciences, geography, ecology, agriculture, hydrology and related subjects.

Atmosphere, weather and climate

SIXTH EDITION

ROGER G. BARRY AND RICHARD J. CHORLEY

ROUTLEDGE

London and New York

First published in 1968 by
Methuen & Co. Ltd
Second edition 1971
Third edition 1976
Fourth edition 1982
Fifth edition 1987
Reprinted by Routledge 1989, 1990
11 New Fetter Lane, London EC4P 4EE
Sixth edition 1992
Reprinted 1995

Simultaneously published in the USA and Canada
by Routledge
29 West 35th Street, New York, NY 10001

Typeset by Solidus (Bristol) Limited
and printed in Great Britain by
Richard Clay (The Chaucer Press) Ltd, Bungay, Suffolk

British Library Cataloguing in Publication Data

Barry, R. G.
Atmosphere, weather and climate.—6th ed.
1. Meteorology
I. Title. II. Chorley, Richard J.
551.5 QC861.2

Library of Congress Cataloguing in Publication Data

Barry, Roger Graham
Atmosphere, weather and climate.

Bibliography: p.
Includes index.
1. Meteorology. 2. Atmospheric physics.
3. Climatology. I. Chorley, Richard J.
II. Title.
QC861.2.B36 1987 551.5 82–12337

ISBN 0–415–07761–3 (pbk)

Contents

Preface to the sixth edition

When the first edition of this book appeared in 1968, it was greeted as being 'remarkably up to date' (*Meteorological Magazine*). Since that time several new editions have extended and sharpened its description and analysis of atmospheric processes and global climates. Indeed, succeeding Prefaces provide a virtual commentary on recent advances in meteorology and climatology of relevance to students in these fields and to scholars in related disciplines.

In 1971 major additions were made regarding the energy budget of the earth, the spatial pattern of the heat budget components, atmospheric stability, orographic precipitation, oceanic circulation and associated climatic effects, vorticity, mesoscale systems in middle latitudes, rainfall variability and aspects of the climate of the sub-Arctic, the Mediterranean and eastern Asia.

The third edition of 1976 involved the substantial recasting of Chapters 2, 4 and 7 and the addition of a new Appendix on synoptic weather maps. New material was introduced on atmospheric composition and its variations with time, the radiation budget, adiabatic temperature changes, the effects of topographic barriers on winds, land and sea breezes, the southern hemisphere circulation and air masses, depression structure and the spatial distribution of precipitation, long-range forecasting, the climate of the Mediterranean, the intertropical confluence, tropical disturbances and sub-synoptic systems, and urban climates.

In 1982 Chapters 7 and 8 were substantially rewritten and the units were standardized throughout. Major additions and revisions involved the material on solar radiation, thunderstorm mechanisms, drought, mesoscale rainfall systems, tornado structure, disturbances within subtropical high-pressure belts, the energy balance of vegetated surfaces, urban climatology, atmospheric pollution and the nature and causes of climatic change. This fourth edition contained over 100 new or revised figures and many new plates, compared with the first edition.

The fifth edition of 1987 included new sections on modelling the atmospheric circulation, forecasting, the African monsoon and ENSO events, together with a new appendix on data sources. Additional material was provided on greenhouse gases and volcanic dust, sunspot activity, albedo, satellite measurements of the planetary energy budget, atmospheric stability, condensation nuclei and cloud processes, mechanisms of orographic precipitation, convergence and divergence, subtropical high-pressure behaviour, hemispheric circulation patterns, southern hemisphere air masses, frontal structures, mesoscale convective systems, super-cell thunderstorms, European blocking conditions, regional winds of Iberia, rainstorms of the south-west USA, the Intertropical Convergence Zone, tropical weather systems, cloud clusters, the Asian monsoon circulation, climate of the Sahara, climatic change in the Sahel Zone and Europe, the climatic effect of global vegetation changes, and the use of mathe-

matical models for climatic forecasting. Some seventy new or revised figures and plates were introduced.

Since the previous edition appeared, the recognition of the reality and possible effects of human activities on the environment has revolutionized attitudes to the study of the atmosphere and of world climate. This revolution has led us to make the most comprehensive of all our revisions of this book for the sixth edition, while retaining its well-tried format. A substantially changed Chapter 1 now includes extended treatments of greenhouse gases and their recent increases; the greenhouse effect; aerosols; the formation, destruction and distribution of ozone; the carbon cycle; global cloud cover; the thermal role of the oceans; and the exosphere and magnetosphere. Chapter 2 improves coverage of orographic precipitation and has an added section on acid precipitation; and Chapter 3 contains new material on aerodynamic roughness and airflow over topographic obstructions, a new section on the Walker circulation, and material on modelling the atmosphere/earth/ocean system. Chapter 4 treats medium-range and long-range forecasting more fully, and Chapter 5 has additional material on wind activity over Western Europe. Chapter 6 has been significantly revised with added coverage of hurricanes; Asian monsoon mechanisms; monsoon depressions; the timing, intensity and phases of the Asian summer monsoon; the Walker circulation; and the West African monsoon and associated circulations. Major features of Chapter 6 are new sections on the climates of Amazonia and southern Africa. Chapter 7 contains updates on tropical forest climatology, urban pollution, urban canyon climatology and urban heat islands, together with a new section on tropical urban climates. Chapter 8 has been extensively rewritten and expanded to contain new material on climatic change during the past 100 years; possible causes of climatic change, including forcing and feedback mechanisms of the atmosphere/earth/ocean system; Pleistocene climatic changes; and short-term changes with particular reference to anthropogenic causes of changes in atmospheric composition and global vegetation. The chapter ends with a preview of the changes in atmospheric composition, temperature and humidity, and of sea level projected to take place over the next 100 years, together with a cautionary postscript highlighting our present inadequate knowledge regarding the operations of the global climate system. SI units of energy flux density are now used exclusively throughout and this edition contains some ninety new or redrawn figures and tables, compared with its predecessor.

Wherever possible, the criticisms and suggestions of colleagues and reviewers have been taken into account in preparing this latest edition.

R. G. BARRY
Cooperative Institute for Research in Environmental Sciences and the Department of Geography, University of Colorado, Boulder

R. J. CHORLEY
Department of Geography and Sidney Sussex College, University of Cambridge

Acknowledgements

The authors are very much indebted to Mr A. J. Dunn for his considerable contribution to the first edition; also to Dr F. Kenneth Hare of Birkbeck College, London, now at the University of Toronto, Ontario, for his thorough and authoritative criticism of the preliminary text and his valuable suggestions for its improvement; also to Mr Alan Johnson of Barton Peveril School, Eastleigh, Hampshire, for helpful comments on Chapters 1–3; and to Dr C. Desmond Walshaw, formerly of the Cavendish Laboratory, Cambridge, and Mr R. H. A. Stewart of the Nautical College, Pangbourne, for offering valuable criticisms and suggestions at an early stage in the preparation of the original manuscript. Gratitude is also expressed to the following persons for their helpful comments with respect to the fourth edition: Dr Brian Knapp of Leighton Park School, Reading; Dr L. F. Musk of the University of Manchester; Dr A. H. Perry of University College, Swansea; Dr R. Reynolds of the University of Reading; and Dr P. Smithson of the University of Sheffield. Dr C. Ramage, of the University of Hawaii, made numerous helpful suggestions on the revision of Chapter 6 for the fifth edition. Dr Z. Toth and Dr D. Gilman of the National Meteorological Center, Washington, DC, kindly helped in the updating of Chapter 4, I. The authors accept complete responsibility for any remaining textual errors.

The figures were prepared by the cartographic and photographic staffs in the Geography Departments at Cambridge University (Mr I. Agnew, Mr R. Blackmore, Mr R. Coe, Mr I. Gulley, Mrs S. Gutteridge, Miss L. Judge, Miss R. King, Mr C. Lewis, Mrs P. Lucas, Miss G. Seymour, Mr A. Shelley, Miss J. Wyatt and, especially, Mr M. Young); at Southampton University (Mr A. C. Clarke, Miss B. Manning and Mr R. Smith); and at the University of Colorado, Boulder (Mr T. Wiselogel).

Our grateful thanks go to our families for their constant encouragement and forbearance.

The authors would like to thank the following learned societies, editors, publishers, organizations and individuals for permission to reproduce figures, tables and plates.

Learned societies

American Geographical Society for Figure 1.44 from the *Geographical Review*.
American Geophysical Union for Figures 1.25B, 4.11 and 4.12 from the *Review of Geophysics and Space Physics*; for Figures 1.35A–C and 1.39 from the *Journal of Geophysical Research*; and for Figure 7.6 from the *Transactions*.
American Meteorological Society for Figure 4.23 from the *Bulletin*; for Figures 3.30, 4.8, 7.1 and 8.11 from the *Journal of Applied Meteorology*; for Figures 4.2B and 4.4B from the *Meteorological Monographs*; for Figures 4.35 and 6.49 from the *Journal of Atmospheric Sciences*; and for Figures 2.25, 3.11, 3.25, 3.40, 3.41, 3.43, 4.6B, 4.25, 6.2B, 6.12, 6.13, 6.32, 6.44 and 6.45A from the *Monthly Weather Review*.

American Planning Association for Figure 7.35 from the *Journal*.

American Society for Testing and Materials for Figure 1.2.

Association of American Geographers for Figure 2.34 from the *Annals*; and for Figure 4.40 from *Resource Paper 11*.

European Space Agency, Darmstadt, for Plates 1 and 27.

Geographical Association for Figure 2.2 from *Geography*.

Indian National Science Academy, New Delhi, for Figure 6.28.

Institute of British Geographers for Figures 2.27, 2.28B, 7.22 and 7.36 from the *Transactions*; and for Figures 2.35 and 8.12 from the *Atlas of Drought in Britain 1975–76* by J. C. Doornkamp and K. J. Gregory (eds).

Institution of Civil Engineers for Figure 2.28A from the *Proceedings*.

Marine Technology Society, Washington, DC, for Figures 6.48C and D from the *Journal*.

National Geographic Society for Plate 22 from the *National Geographic Picture Atlas of Our Fifty States*.

Royal Geographical Society for Figure 3.33 from the *Journal*.

Royal Meteorological Society for Figures 2.21, 4.10, 5.7, 5.8, 6.3 and 7.15 from the *Quarterly Journal*; for Figure 8.4 from *World Climate* 8000–0 BC; for Figures 2.29, 5.9, 6.40 and 8.13 from the *Journal of Climatology*; and for Figures 1.23, 2.5, 2.30, 4.31, 5.5, 5.12, 5.27, 6.46 and 7.23B, and for Plates 14, 19 and 20 from *Weather*.

Royal Society of London for Figure 4.28 and Plate 15 from the *Proceedings, Section A*.

US National Academy of Sciences for Figure 8.3.

Editors

Advances in Space Research for Figures 1.21 and 2.20.

Climatic Change for Figure 8.9.

Endeavour for Figure 2.23.

Erdkunde for Figures 6.22, A.1.1B and A.1.2.

Geographical Magazine for Figure 6.37A.

Geographical Reports of Tokyo Metropolitan University for Figure 6.33.

Japanese Progress in Climatology for Figure 7.32.

Meteorological Magazine for Figures 3.37, 4.9 and 7.2, and Plate 25.

Meteorological Monographs for Figures 4.2B and 4.4B.

Meteorologische Rundschau for Figures 5.28 and 7.9.

New Scientist for Figures 4.26, 4.29 and 4.33A.

Progress in Physical Geography for Figures 7.27, 7.28 and 7.30.

Science for Figures 7.26C and 8.6.

Tellus for Figures 5.10, 5.11, 6.18 and 6.25.

Zeitschrift für Geomorphologie for Figure 7.5 from *Supplement 21*.

Publishers

Academic Press, New York, for Figures 3.44, 4.11, 4.12, 4.32, 6.11, 6.12 and 6.13 from *Advances in Geophysics*; for Figure 6.16 from *Monsoon Meteorology* by C. S. Ramage; and for Figure 6.37B from *Quaternary Research*.

Allen and Unwin, London, for Figures 1.26 and 1.28B from *Oceanography for Meteorologists* by H. V. Sverdrup.

Cambridge University Press for Figure 2.17 from *Clouds, Rain and Rainmaking* by B. J. Mason; for Figure 3.23 from *World Weather and Climate* by D. Riley and L. Spalton; for Figure 6.50 from *The Warm Desert Environment* by A. Goudie and J. Wilkinson; for Figure 7.20 from *The Tropical Rain Forest* by P. W. Richards; for Figure 7.24 from *Air: Composition and Chemistry* by P. Brimblecombe (ed.); for Figures 4.22 and 6.10 from *Weather Systems* by L. F. Musk; and for Figures 1.3, 1.7, 8.5, 8.10, 8.16, 8.17, 8.19 and 8.21 from *Climate Change: The IPCC Scientific Assessment*.

Cleaver-Hulme Press, London, for Figure 3.15 from *Realms of Water* by Ph. H. Kuenen.

The Controller, Her Majesty's Stationery Office

(Crown Copyright Reserved) for Figure 2.8 from *Geophysical Memoir 102* by J. K. Bannon and L. P. Steele; for Figure 6.30 from *Geophysical Memoir 115* by J. Findlater; for Figure 1.27 from *Meteorological Office Scientific Paper 6, M.O.685* by F. E. Lumb; for Figure 2.6 from *Ministry of Agriculture Technical Bulletin 4* by R. T. Pearl *et al.*; for Figures 3.37, 4.14 and 7.2, and for Plate 23 from the *Meteorological Magazine*; for Figures 4.9 and 4.13 from *A Course in Elementary Meteorology* by D. E. Pedgley; for Figure 4.15 from *British Weather in Maps* by J. A. Taylor and R. A. Yates (Macmillan, London); for Figure 4.30 from *Geophysical Memoirs* by D. E. Pedgley; for Figures 5.25 and 5.26 from *Weather in the Mediterranean 1*, 2nd edn (1962); and for the tephigram base of Figure 2.10 from *RAF Form 2810*.

Elsevier, Amsterdam, for Figure 6.56 from *Climates of Australia and New Zealand* by J. Gentilli (ed.); for Figure 6.35 from *Palaeogeography, Palaeoclimatology, Palaeoecology*; and for Figures 6.42 and 6.43 from *Climates of Central and South America* by W. Schwerdtfeger (ed.); and for Figure 8.8 from *Greenhouse-gas-induced Climatic Change* by M. E. Schlesinger (ed.).

Generalstabens Litografiska Anstalt, Stockholm, for Figure 4.16 from *Klimatologi* by G. H. Liljequist.

Harvard University Press, Cambridge, Mass., for Figures 1.28A, 1.33, 7.12, 7.13B and 7.14A from *The Climate near the Ground* (2nd edn) by R. Geiger.

Houghton Mifflin Company, Boston, for Plates 7, 12 and 18 from *A Field Guide to the Atmosphere* by V. T. Schaefer and J. A. Day.

Hutchinson, London, for Figures 7.23A and 7.31 from the *Climate of London* by T. J. Chandler; and for Figures 6.38 and 6.39 from *The Climatology of West Africa* by D. F. Hayward and J. S. Oguntoyinbo.

Justus Perthes, Gotha, for Figure 2.33 from *Petermanns Geographische Mitteilungen, Jahrgang 95*.

Kluwer Academic Publishers, Dordrecht, Holland for Figure 1.1B from *Air–Sea Exchange of Gases and Particles* by P. S. Liss and W. G. N. Slinn (eds); and for Figure 8.9 from *Climate Change* (Vol. 16) by G. A. Meehl and W. M. Washington.

Longman, London, for Figure 3.35 from *Contemporary Climatology* by A. Henderson-Sellers and P. J. Robinson.

Macmillan, London, for Figure 4.15 from *British Weather in Maps* by J. A. Taylor and R. A. Yates.

McGraw-Hill Book Company, New York, for Figure 2.26 from *Handbook of Meteorology* by F. A. Berry, E. Bollay and N. R. Beers (eds); for Figure 3.39 from *Dynamical and Physical Meteorology* by G. J. Haltiner and F. L. Martin; for Figures 7.13A and 7.14B from *Forest Influences* by J. Kittredge; for Figures 2.9, 2.22 and 3.24 from *Introduction to Meteorology* by S. Petterssen; for Figures 6.5 and 6.6 from *Tropical Meteorology* by H. Riehl; for Figures 5.18 and 6.22 from *The Earth's Problem Climates* by G. T. Trewartha; and for Figure 1.43 from *Handbook of Geophysics and Space Environments* by Shea L. Valley (ed.).

Methuen, London, for Figures 1.32, 2.32 and 6.41 from *Mountain Weather and Climate* by R. G. Barry; for Figures 2.1, 3.32 and 3.36 from *Models in Geography* by R. J. Chorley and P. Haggett (eds); and for Figures 7.6, 7.16, 7.21C and D, 7.25, 7.26, 7.29 and 7.33 from *Boundary Layer Climates* by T. R. Oke.

National Academy Press, Washington, DC, for Figure 8.3.

North-Holland Publishing Company, Amsterdam, for Figure 2.31 from the *Journal of Hydrology*.

Oliver and Boyd, Edinburgh, for Figure 7.11 from *Fundamentals of Forest Biogeocoenology* by V. Sukachev and N. Dylis.

Oxford University Press, Cape Town, for Figures 3.38, 6.53, 6.54 and 6.55 from *Climatic Change and Variability in Southern Africa* by P. D. Tyson.

Pitman, London, for Figure 3.18 from *Tropical and Equatorial Meteorology* by M. A. Garbel.

Princeton University Press for Figure 1.5 from *Design with Climate* by V. Olgyay.

D. Reidel, Dordrecht, for Figures 3.26 and 8.7 from *The Climate of Europe: Past, Present and Future* by H. Flohn and R. Fantechi (eds); for Figures 6.51 and 8.18 from *Climatic Change*; for Figure 7.30 from *Interactions of Energy and Climate* by W. Bach, J. Pankrath and J. Williams (eds).

Routledge, London, for Figures 6.38 and 6.39 from *The Climatology of West Africa* by D. F. Hayward and J. S. Oguntoyinbo.

Scientific American Inc., New York, for Figure 1.37 by R. E. Newell; for Figures 1.10 and 1.11 by M. R. Rapino and S. Self; for Figure 4.33B by J. Snow; for Figure 1.15 by P. V. Foukal; and for Figure 8.8 by P. D. Jones and T. M. L. Wigley.

Springer-Verlag, Vienna and New York, for Figure 1.46 from the *Meteorologische Rundschau*; and for Figures 2.30 and 3.10 from *Archiv für Meteorologie, Geophysik und Bioklimatologie.*

Time-Life Inc., Amsterdam, for Plate 5 from *The Grand Canyon* by R. Wallace.

University of California Press, Berkeley, for Figure 6.8 and Plate 32 from *Cloud Structure and Distributions over the Tropical Pacific Ocean* by J. S. Malkus and H. Riehl.

University of Chicago Press for Figures 1.14, 1.18, 1.33, 2.4, 7.7, 7.8 and 7.10 from *Physical Climatology* by W. D. Sellers.

University of Wisconsin Press, Madison, for Figures 5.18 and 6.33 from *The Earth's Problem Climates* by G. T. Trewartha.

Van Nostrand Reinhold Company, New York, for Figure 6.47 from *The Encyclopedia of Atmospheric Sciences and Astrogeology* by R. W. Fairbridge (ed.).

Walter De Gruyter, Berlin, for Figure 5.2 from *Allgemeine Klimageographie* by J. Blüthgen.

Weidenfeld and Nicolson, London, for Figure 4.21 from *Climate and Weather* by H. Flohn.

Westview Press, Boulder, for Figure 1.4 from *Climate Change and Society* by W. W. Kellogg and R. Schware.

John Wiley, Chichester, for Figures 5.9, 6.40 and 8.13 from the *Journal of Climatology*; for Figures 1.6, 1.9 and 8.20 from *The Greenhouse Effect, Climatic Change, and Ecosystems* by G. Bolin *et al.*; and Figure 7.34 from *Human Activity and Environmental Processes* by K. J. Gregory and D. E. Walling (eds); and for Figures 1.8 and 8.15 from *The Changing Atmosphere* by F. S. Roland and I. S. A. Isaksen (eds).

John Wiley, New York, for Figure 1.29A from *Physical Geography* (2nd edn) by A. N. Strahler; for Figures A.1.3, A.1.4 and Table A.1.1 from *Physical Geography* (3rd edn) by A. N. Strahler; for Figures 1.16E, 1.17 and 2.19 from *Introduction to Physical Geography* by A. N. Strahler; for Figure 1.19 from *Meteorology, Theoretical and Applied* by E. W. Hewson and R. W. Longley; for Figure 4.27 from *Weather and Climate Modification* by W. N. Hess (ed.); for Figure 4.37 from *Paleoclimate Analysis and Modeling* by A. D. Hecht; and for Figures 6.17, 6.29, 6.31 and 6.34 from *Monsoons* by J. S. Fein and P. L. Stephens (eds).

V. H. Winston and Sons, Silver Spring, Maryland, for Figure 5.3 from *Soviet Geography.*

Organizations

Center for Climatic Research, University of Delaware, Newark, Delaware, for Figure 5.23.

Department of Electronics (NERC Satellite Station), University of Dundee, for Plates 23 and 25.

Deutscher Wetterdienst, Zentralamt, Offenbach am Main, for Figure 6.27.

Environmental Science Service Administration (ESSA) for Plates 10 and 20.

Geographical Branch, Department of Energy, Mines and Resources, Ottawa, for Figure 5.14 from *Geographical Bulletin.*

Laboratory of Climatology, Centerton, New Jersey, for Figure 5.22.

National Academy of Sciences, Washington, DC, for Figure 8.14.

National Aeronautics and Space Administration

(NASA) for Plates 16, 29, 30, 34 and 38.

National Geophysical Data Center, Boulder, for Figure 1.15

National Hurricane Center, Miami, for Plate 35.

National Meteorological Center, Washington, DC, for Figure 4.36.

National Oceanic and Atmospheric Administration (NOAA), United States Department of Commerce, Washington, DC, for Figures 4.38, 4.39 and 4.41, and for Plates 2 and 3 from *Technical Memo. NESS 95*; and for Plates 31 and 37.

National Snow and Ice Data Center, Boulder, for Plates 11, 24, and 32.

Natural Environmental Research Council for Figure 1.5 from *Our Future World.*

National Space Development Agency of Japan for Plate 36.

Naval Weather Service Command, Washington, DC, for Figures 3.19 and 3.27.

New Zealand Meteorological Service, Wellington, New Zealand, for Figures 6.26 and 6.45 from the *Proceedings of the Symposium on Tropical Meteorology* by J. W. Hutchings (ed.).

Nigerian Meteorological Service for Figure 6.36 from *Technical Note 5.*

Press Association-Reuters Ltd, London, for Plate 6.

Quartermaster Research and Engineering Command, Natick, Mass., for Figure 5.17 by J. N. Rayner.

Risø National Laboratory, Roskilde, Denmark, for Table 3.1 and Figures 3.12 and 5.1 from *European Wind Atlas* by I. Troen and E. L. Petersen.

United Nations Food and Agriculture Organization, Rome, for Figure 7.19 from *Forest Influences.*

United States Department of Agriculture, Washington, DC, for Figures 7.17B and 7.18 from *Climate and Man.*

United States Department of Energy, Washington, DC, for Figure 1.25A.

United States Geological Survey, Washington, DC, for Figure 6.52 from *Professional Paper 1052.*

United States National Air Pollution Administration, Washington, DC, for Figures 7.21C and D and 7.25 from *Public Health Service Publication No. AP-63.*

United States Weather Bureau for Figures 2.25, 3.11, 3.25, 3.40, 3.41, 3.43, 4.6B, 4.25, 6.2B, 6.12 and 6.45A from the *Monthly Weather Review*; and for Figure 4.19 from *Research Paper 40.*

World Meteorological Organization for Figure 1.31 and Plates 4, 17 and 28 from *Technical Note 124*; and for Figures 6.48A and B from *The Global Climate System 1982–84.*

Individuals

Arizona State Climatologist, Phoenix, Arizona, for Figure 5.29.

Dr C. F. Armstrong and Dr C. K. Stidd, of the Desert Research Institute, University of Nevada, for Figure 2.31.

Dr B. W. Atkinson, of Queen Mary College, London, for Figure 7.34.

Dr August H. Auer, Jr., of the New Zealand Meteorological Service, for Plate 39.

Mr P. E. Baylis, of the University of Dundee, and Dr R. Reynolds, of the University of Reading, for Plate 21.

Dr R. P. Beckinsale, of Oxford University, for suggested modification to Figure 4.7.

Dr H. N. Bhalme, of Pune, India, for Figure 6.32.

Dr B. Bolin, of the University of Stockholm, for Figures 1.6, 1.9 and 8.20.

Dr P. Brimblecombe, of the University of East Anglia, for Figure 7.24.

Mr R. Bumpas, of the National Center for Atmospheric Research, Boulder, for Plate 8.

Dr G. C. Evans, of the University of Cambridge, for Figure 7.20A.

Dr H. Flohn, of the University of Bonn, for Figures 3.29 and 6.15.

Dr C. K. Folland, of the Meteorological Office, Bracknell, for Figures 8.5 and 8.10.

Dr S. Gregory, of the University of Sheffield, for Figure 6.14.

Mr Ernst Haas for Plate 5.

Dr S. L. Hastenrath, of the University of Wisconsin, for Figures 1.46 and 2.31.

Dr L. H. Horn and Dr R. A. Bryson, of the University of Wisconsin, for Figure 5.15.

Dr J. Houghton, of the Meteorological Office, Bracknell, for Figures 1.7, 8.17, 8.19 and 8.21.

Dr R. A. Houze, Jr., of the University of Washington, for Figures 4.11, 4.13, 4.32, 6.12 and 6.13.

Mr S. Jones for the description of Plate 36.

Dr V. E. Kousky, of São Paulo, for Figure 6.44.

Dr Y. Kurihara, of Princeton University, for Figure 6.11.

Mr E. Lantz for Plate 26.

Dr F. H. Ludlam, of Imperial College, London, for Plates 14 and 19.

Dr Kiuo Maejima, of Tokyo Metropolitan University, for Figure 6.33.

Dr J. Maley, of the Université des Sciences et des Techniques du Languedoc, for Figure 6.37B.

Dr Brooks Martner, of the University of Wyoming, for Plate 13.

Dr J. R. Mather, of the University of Delaware, for Figure 5.23.

Dr Yale Mintz, of the University of California, for Figure 3.35.

Dr L. F. Musk, of the University of Manchester, for Figures 4.22 and 6.10.

Dr T. R. Oke, of the University of British Columbia, for Figures 3.14 A and C, 7.3A and B, 7.6, 7.16, 7.21C and D, 7.25, 7.26B and C, 7.27, 7.28 and 7.33.

Dr L. R. Ratisbona, of the Servicio Meteorologico Nacional, Rio de Janeiro, for Figures 6.42 and 6.43.

Mr D. A. Richter, of Analysis and Forecast Division, National Meteorological Center, Washington, DC, for Figure 4.25.

Dr J. Roederer, of the University of Alaska, for Figure 1.45.

Dr J. C. Sadler, of the University of Hawaii, for Figure 6.20.

Dr B. Saltzman, of Yale University, for Figure 3.44.

Dr R. S. Scorer, of Imperial College, London, and Mrs Robert F. Symons, for Plate 9.

Dr Glenn E. Shaw, of the University of Alaska, for Figure 1.1A.

Dr K. P. Shine, of the University of Reading, for Figure 8.16.

Dr W. G. N. Slinn, for Figure 1.1B.

Dr R. S. Stolarski, of the Goddard Space Flight Center, for Figure 1.8.

Dr A. N. Strahler, of Santa Barbara, California, for Figures 1.16E, 1.17, 1.29A, 2.19, 8.2, A.1.3 and A.1.4; and for Table A.1.1.

Dr R. T. Watson, of NASA, Houston, for Figure 1.3.

Dr T. M.L. Wigley, of the Climatic Research Unit, University of East Anglia, for Figure 8.8.

Introduction

In this book we aim to provide a non-technical account of how the atmosphere works, thereby developing an understanding of weather phenomena and of global climates. The atmosphere, which is vital to terrestrial life, is a shallow envelope equivalent in thickness to less than 1 per cent of the earth's radius. Most weather systems form and decay within its lowest 10 km. It is thought that the earth's atmosphere evolved to its present form and composition at least 400 million years ago when extensive vegetation developed on land. Its presence provides an indispensable shield from harmful radiation from the sun and its gaseous contents sustain the plant and animal biosphere on which human life depends.

Over much of the globe, the state of the atmosphere is far from constant in response to varying *weather* processes. Weather extremes – gales, blizzards, tornadoes, floods – drastically affect human activities and frequently result in loss of life, even when anticipated. Hence, by seeking to understand atmospheric phenomena, we can hope to forecast their vagaries and in some instances control or modify them in a beneficial way. This broad endeavour constitutes the field of the atmospheric sciences. *Meteorology* is specifically concerned with the physics of weather processes. *Weather systems* – which produce the variety of instantaneous states of the atmosphere – differ in their size and life-span. Four scales are commonly recognized: *mesoscale* systems, such as thunderstorms, extend some 10 km horizontally with a lifetime of a few hours; *synoptic-scale* systems, like mid-latitude cyclones and tropical storms, have a diameter of a few thousand km and a lifetime of about 5 days; *planetary-scale* waves in the atmospheric circulation span 5000–10,000 km and usually persist for several weeks. In addition, small-scale wind eddies near the earth's surface and processes operating within vegetation canopies are the concern of *micro-meteorology*.

Climate introduces the longer time-scales operating in the atmosphere. It is sometimes loosely regarded as 'average weather', but it is more meaningful to define climate as the long-term state of the atmosphere encompassing the aggregate effect of weather phenomena – the extremes as well as the mean values. It is also usual to distinguish regional and global *macro-climate*, on the one hand, from *local* or *topo-climates* related to terrain features (valleys, hill slopes), on the other hand.

A recent view of climate, which has emerged over the past 10–15 years, envisages a *climate system* involving interactions between the atmosphere, oceans, land surface, snow and ice cover, and the biosphere (see Figures 3.43 and 8.14). Couplings between these components operating on short time-scales (hours to months), that may involve two-way interactions, are considered to represent processes *internal* to the system. Changes, usually over years to millennia, in the earth's surface, biosphere, atmospheric composition, and incoming solar radiation that may affect the whole climate

system without being affected by it, are considered to be *external* processes. This view of the climate system has resulted in a widening interest in climate processes and their effects by biologists, chemists, oceanographers, glaciologists, solar physicists and scientists in many other disciplines. This interest in climate processes has been the major theme of this book since its inception a quarter of a century ago; the concept of the world climate system was introduced in the edition of ten years ago and that of mathematical modelling in the fifth edition five years ago. In the sixth edition the concept of the world climate system is a central theme.

The structure of the book represents this viewpoint. We will look first at the composition and structure of the atmosphere and its role in the global exchange of energy, the moisture balance and wind systems. Then weather and climate in middle and low latitudes are discussed and, finally, small-scale climates and climatic change. The key to atmospheric processes is the radiant energy which the earth and its atmosphere receive from the sun. In order to study the receipt of this energy we need to begin by considering the nature of the atmosphere – its composition and basic properties.

1

Atmospheric composition and energy

A COMPOSITION OF THE ATMOSPHERE

1 Total atmosphere

Air is a mechanical mixture of gases, not a chemical compound. Table 1.1, illustrating the average composition of dry air, shows that four gases – nitrogen, oxygen, argon and carbon dioxide – account for 99.98 per cent of the air by volume. Moreover, rocket observations show that these gases are mixed in remarkably constant proportions up to about 80 km (50 miles).

Of especial significance, despite their relative scarcity, are the so-called *greenhouse gases* which play an important role in the thermodynamics of the atmosphere by trapping long-wave terrestrial re-radiation, producing the *greenhouse effect* (see E, this chapter). These gases, the concentrations of which are particularly susceptible to human (i.e. anthropogenic) activities, include:

1 Carbon dioxide (CO_2), which is involved in a complex global cycle (see A.4, this chapter). It is released from the interior of the earth

Table 1.1 Average composition of the dry atmosphere below 25 km.

Component	*Symbol*	*Volume % (dry air)*	*Molecular weight*
Nitrogen	N_2	78.08	28.02
Oxygen	O_2	20.98	32.00
*‡Argon	Ar	0.93	39.88
Carbon dioxide	CO_2	0.035	44.00
‡Neon	Ne	0.0018	20.18
*‡Helium	He	0.0005	4.00
†Ozone	O_3	0.00006	48.00
Hydrogen	H	0.00005	2.02
‡Krypton	Kr	0.0011	
‡Xenon	Xe	0.00009	
§Methane	CH_4	0.0017	

Notes: *Decay products of potassium and uranium.
†Recombination of oxygen.
‡Inert gases.
§At surface.

and produced by respiration, soil processes, combustion and oceanic evaporation. Conversely, it is dissolved in the ocean and consumed by the process of plant photosynthesis.

2 Methane (CH_4), which is produced primarily by anerobic (i.e. oxygen-deficient) processes in natural wetlands and rice paddies, by animal digestive processes, biomass burning and other human activities. It is destroyed in the troposphere by a reaction with the hydroxyl OH which is produced by the action of vegetation:

$$CH_4 + OH \rightarrow CH_3 + H_2O$$

3 Nitrous oxide (N_2O), which is produced by biological mechanisms in the oceans and soils, by industrial combustion, automobiles, aircraft, biomass burning, and as a result of the use of chemical fertilizers. It is destroyed by photochemical reactions in the stratosphere involving the production of nitrogen oxides (NO_x); its lifetime is 132 years.

4 Ozone (O_3), which is produced by the high-level breakup of oxygen molecules by solar ultraviolet radiation and destroyed by reactions involving nitrogen oxides and chlorine (Cl) (the latter generated by CFCs, volcanic eruptions and vegetation burning) in the middle and upper stratosphere.

5 Chlorofluorocarbons (CFCs: chiefly $CFCl_3$(F-11) and CF_2Cl_2(F-12)), which are entirely anthropogenically produced by aerosol propellants, refrigerator coolants (e.g. 'freon'), cleansers and air conditioners, and were not present in the atmosphere until the 1930s. CFC molecules rise slowly into the stratosphere and then move poleward, being decomposed by photochemical processes into chlorine after an estimated average lifetime of some 55–116 years.

In addition to these gases, water vapour (H_2O), the major greenhouse gas, is a vital atmospheric constituent. It averages about 1 per cent by volume but is very variable both in space and time, being involved in a complex global hydrological cycle (see Chapter 2).

In addition to the greenhouse gases, important *reactive gas species* are produced by the cycles of sulphur, nitrogen and chlorine halogens. These play key roles in acid precipitation and in ozone destruction. Sources of these species are as follows:

Nitrogen species. The reactive species of nitrogen are nitric oxide (NO) and nitrogen dioxide (NO_2). NO_x refers to these and other odd nitrogen species with oxygen. Fossil fuel combustion (approximately 2/3 for heating, 1/3 for cars and other transport) is the primary source of NO_x (mainly NO) accounting for 15–25 $\times$ 10^9 kg N/year. Biomass burning and lightning activity are other important sources. NO_x emissions increased some 200 per cent between 1940 and 1980. The total source of NO_x is about 40 $\times$ 10^9 kg N/year. About 25 per cent of this goes into the stratosphere where it undergoes photochemical dissociation. It is also removed as nitric acid (HNO_3) in snowfall. Odd nitrogen is also released as NH_x by ammonia oxidation in fertilizers and by domestic animals (6–10 $\times$ 10^9 kg N/year).

Sulphur species. Reactive species are sulphur dioxide (SO_2) and reduced sulphur (H_2S, DMS). Atmospheric sulphur is almost entirely anthropogenic in origin; 90 per cent from coal and oil combustion, and much of the remainder from copper smelting. The major sources are sulphur dioxide (80–100 $\times$ 10^9 kg S/year), hydrogen sulphide (H_2S) (20–40 $\times$ 10^9 g S/year) and dimethyl sulphide (DMS) (35–55 $\times$ 10^9 kg S/year). SO_2 emissions increased by about 50 per cent between 1940 and 1980. Volcanic activity releases approximately 10^9 kg S/year as sulphur dioxide. Because the lifetime of SO_2 and H_2S in the atmosphere is only about one day, atmospheric sulphur occurs largely as carbonyl sulphur (COS) which has a lifetime of about one year. The conversion of H_2S gas to sulphur particles is an important source of atmospheric aerosols.

Despite its short lifetime, sulphur dioxide is readily transported over long distances. It is removed from the atmosphere when condensation nuclei of SO_2 are precipitated as acid rain

containing sulphuric acid (H_2SO_4). The acidity of fog deposition can be more serious because up to 90 per cent of the fog droplets may be deposited. In Californian coastal fogs, pH values of only 2.0–2.5 are not uncommon. Peak pH readings in the eastern United States and Europe are ≤4.3 (pH = 7 is neutral, see Chapter 2, J). In these areas and central southern China, rainfall deposits > 1 g m^{-2} of SO_2 annually.

Halogens of chlorine, which are entirely anthropogenic gases, have increased sharply in the atmosphere over the last few decades. Trichloroethane ($C_2H_3Cl_3$), for example, which is used in dry cleaning and degreasing agents, increased fourfold in the 1980s and has a seven-year residence time in the atmosphere. 'Freons' (chlorofluorocarbons CF_2Cl_2 and $CFCl_3$) are even more of a problem, having residence times of about 55 and 116 years, respectively. Apart from their greenhouse effect, the chlorine is involved in ozone destruction. It is important in the chemistry of the lower stratosphere where OH radicals break it down.

There are also significant quantities of *Aerosols* in the atmosphere. These are suspended particles of sea salt, dust (particularly silicates), organic matter and smoke. Aerosols enter the atmosphere from a variety of natural and anthropogenic sources (see Table 1.2). Some originate as particles – soil grains and mineral dust from dry surfaces, carbon from fires, and volcanic dust. Others are converted

Table 1.2 Aerosol production estimates, less than 5 μm radius (10^9 kg/year) and typical concentrations near the surface ($\mu g\ m^{-3}$)

		Concentration	
	Production	*Remote*	*Urban*
Natural			
Primary production:			
Sea salt	500	5–10	
Mineral particles	250	0.5–5*	
Volcanic	25		
Forest fires	5		
Secondary production (gas → particle):			
Sulphates from H_2S	335	1–2	
Nitrates from NO_x	60		
Converted plant hydrocarbons	75		
Total Natural	1250		
Anthropogenic			
Primary production:			
Industrial processes	12		100–500† (Industrial processes to Miscellaneous)
Combustion	10		
Transportation	2		
Miscellaneous	6		
Biomass burning	35		
Secondary production (gas → particle):			
Sulphate from SO_2	200	0.5–1.5	10–20
Nitrates from NO_x	35	0.2	0.5
Converted hydrocarbons	25		0.5
Total Anthropogenic	325		

Source: After Bach 1976 and Bridgman 1990.
Notes: *10–60 $\mu g\ m^{-3}$ during dust episodes from the Sahara over the Atlantic.
†Total suspended particles.

into particles from inorganic gases (sulphur from anthropogenic SO_2 and natural H_2S; ammonium salts from NH_3; nitrogen from NO_x). Other sources are sea salts and organic matter (plant hydrocarbons and anthropogenically derived) (see Figure 1.1). Natural sources are probably 4–5 times larger than anthropogenic ones on a global scale, but the estimates are wide-ranging. Moreover, there is considerable spatial variability. For example, some 20,000 Tg (10^{12}g) of crustal material are picked up by the air annually, one-third from the Sahara. Most of this is deposited downwind over the Atlantic. Large particles rapidly sink back to the surface or are washed out (scavenged) by rain after a few days, but fine particles from volcanic eruptions may reside in the stratosphere above the level of weather processes for 1–3 years.

Figure 1.1 Atmosphere particles. A: Mass distribution, together with a depiction of the surface–atmosphere processes which create and modify atmospheric aerosols, illustrating the three size modes. Aitken nuclei are solid and liquid particles that act as condensation nuclei and capture ions, thus playing a role in cloud electrification. B: Area distribution.

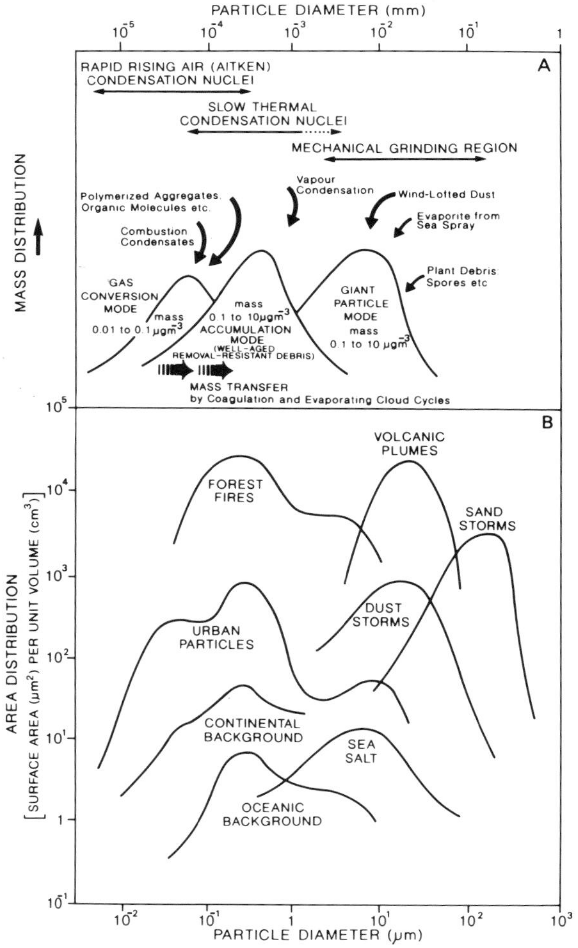

Sources: A: After Glenn E. Shaw, University of Alaska, Geophysical Institute. B: After Slinn (1983).

Having made these generalizations about the atmosphere, we now examine the variations which occur in composition with height, latitude and time.

2 Variations with height

The light gases (hydrogen and helium especially) might be expected to become more abundant in the upper atmosphere, but large-scale turbulent mixing of the atmosphere prevents such diffusive separation even at heights of many tens of km above the surface. The height variations which do occur are related to the source-locations of the two major non-permanent gases – water vapour and ozone. Since both absorb some solar and terrestrial radiation the heat budget and vertical temperature structure of the atmosphere are considerably affected by the distribution of these two gases.

Water vapour comprises up to 4 per cent of the atmosphere by volume (about 3 per cent by weight) near the surface, but only 3–6 ppm (parts per million by volume) above 10 to 12 km. It is supplied to the atmosphere by evaporation from surface water or by transpiration from plants and is transferred upwards by atmospheric turbulence. Turbulence is most effective below about 10 km and as the maximum possible water vapour density of cold air is anyway very low (see B.2, this chapter), there is little water vapour in the upper layers of the atmosphere.

Ozone (O_3) is concentrated mainly between 15 to 35 km. The upper layers of the atmosphere are irradiated by ultraviolet radiation from

the sun ($< 0.19\ \mu m$ wavelength; see C.1, this chapter) which causes the break-up of oxygen molecules at altitudes above 30 km (i.e. $O_2 \rightarrow O + O$). These separated atoms ($O + O$) may then individually combine with other oxygen molecules to create ozone, as illustrated by the simple photochemical scheme:

$$O_2 + O + M \rightarrow O_3 + M$$

where M represents the energy and momentum balance provided by collision with a third atom or molecule. Such three-body collisions are rare at 80 to 100 km because of the very low density of the atmosphere, while below about 35 km most of the incoming ultraviolet radiation has already been absorbed at higher levels. Therefore ozone is mainly formed between 30 and 60 km where collisions between O and O_2 are more likely. Ozone itself is unstable; its abundance is determined by three distinctly different photochemical interactions. Above 40 km odd oxygen is destroyed primarily by a cycle involving molecular oxygen; between 20 and 40 km NO_x cycles are dominant; while below 20 km hydrogen (HO_2) is responsible. An additional cycle involving chlorine (ClO) also contributes at around 40 km altitude. Collisions with monatomic oxygen may re-create oxygen (i.e. $O_3 + O \rightarrow O_2 + O_2$), but ozone is mainly destroyed through other photochemical cycles associated with longer wavelengths of ultraviolet radiation (i.e. 0.23–0.29 μm) involving catalytic reactions. The destruction of ozone involves a recombination with atomic oxygen, causing a net loss of the odd oxygen. This takes place through the catalytic effect of a radical such as OH (hydroxyl radical):

$$H + O_3 \rightarrow OH + O_2$$

$$OH + O \rightarrow H + O_2,$$

$$\text{thus} \quad O_3 + O \rightarrow 2\,O_2.$$

The odd hydrogen atoms and OH result from the dissociation of water vapour, molecular hydrogen and methane (CH_4).

Stratospheric ozone is similarly destroyed in the presence of nitrogen oxides (NO_x, i.e. NO_2 and NO) and chlorine radicals (Cl, ClO). The source gas of the NO_x is nitrous oxide (N_2O) which is produced by combustion and fertilizer use, while chloroflurocarbons (CFCI), manufactured for 'freon', give rise to the chlorines. These source gases are transported into the stratosphere from the surface and are converted by oxidation into NO_x, and by UV photodecomposition into chlorine radicals, respectively.

The chlorine chain involves:

$$2\,(Cl + O_3) \rightarrow ClO + O_2$$

$$CO + C1O \rightarrow C1\,O_2$$

$$\text{and} \quad C1 + O_3 \rightarrow C1O + O_2$$

$$OH + O_3 \rightarrow HO_2 + O_2.$$

Both reactions result in a conversion of O_3 to O_2 and the removal of all odd oxygens. Another cycle may involve an interaction of the oxides of chlorine and bromine. Nevertheless, it appears that the observed doubling of Cl_x species during the decade 1970–80 is quantitatively insufficient to explain the observed decrease of stratospheric ozone over Antarctica (see pp. 10–11). A mechanism that may enhance the catalytic process involves polar stratospheric clouds. These can form readily during the austral spring (October) when temperatures decrease to 190–195K permitting the formation of particles of nitric acid (HNO_3) ice and water ice. It is apparent, however, that anthropogenic sources of the trace gases are a primary factor in the ozone decline. Conditions in the Arctic are somewhat different as the stratosphere is warmer and there is more mixing of air from lower latitudes.

The constant metamorphosis of oxygen to ozone and from ozone back to oxygen involves a very complex set of photochemical processes which tend to maintain an approximate equilibrium above about 40 km. However, the ozone mixing ratio is at its maximum at about 35 km, whereas maximum ozone concentration (see Note 1) occurs lower down, between 20 and 25 km in low latitudes and between 10 and 20 km in high latitudes. This is the result of

some circulation mechanism transporting ozone downwards to levels where its destruction is less likely, allowing an accumulation of the gas to occur. Despite the importance of the ozone layer, it is essential to realize that if the atmosphere were compressed to sea level (at normal sea-level temperature and pressure) ozone would contribute only about 3 mm to the total atmospheric thickness of 8 km (see Figure 1.2).

3 Variations with latitude and season

Variations of atmospheric composition with latitude and season are particularly important in the case of water vapour and ozone.

Ozone content is low over the equator and high in subpolar latitudes in spring (see Figure 1.2). If the distribution were solely the result of photochemical processes the maximum would occur in June near the equator and the anomalous pattern must be due to a poleward transport of ozone. The movement is apparently from higher levels (30–40 km) in low latitudes towards lower levels (20–25 km) in high latitudes during winter months. Here the ozone is stored during the *polar night*, giving rise to an ozone-rich layer in early spring under natural conditions. It is this springtime feature which has been disrupted by the Antarctic ozone 'hole' (see pp. 11 and 327). The type of circulation responsible for this transfer is not yet known with certainty, although it does not seem to be a simple direct one.

The water-vapour content of the atmosphere is closely related to air temperature (see B.2, this chapter, and Chapter 2, A and B) and is therefore greatest in summer and in low latitudes. There are, however, obvious exceptions to this generalization, such as the tropical desert areas of the world.

The carbon dioxide content of the air (averaging about 353 parts per million (ppm)) has a large seasonal range in higher latitudes in the northern hemisphere associated with photosynthesis and decay of the biosphere. At 50°N the concentration ranges from 346 ppm in late summer to 360 ppm in spring. The low summer values are related to the assimilation of CO_2 by the cold polar seas. Over the year a small net transfer of CO_2 from low to high altitudes takes place to maintain an equilibrium content in the air.

4 Variations with time

The quantities of carbon dioxide, other greenhouse gases and particles in the atmosphere may be subject to long-term variations and these are of special significance because of their possible effect on the radiation budget.

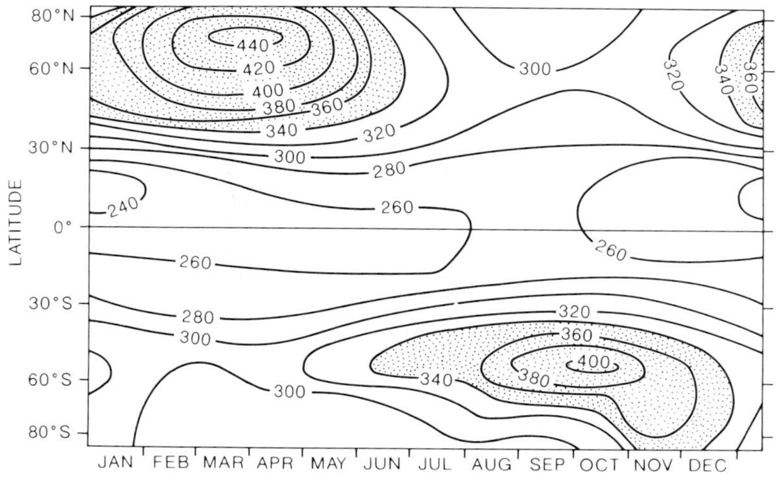

Figure 1.2 Average latitude–season cross-section of total ozone (1957–75). Values are in Dobson units (10^{-3} cm, at standard atmospheric temperature and pressure). Values in excess of 340 Dobson units are stippled.

Source: From London and Angell 1982 (*copyright American Society for Testing and Materials Reprinted with permission*).

Measurements of atmospheric trace gases show increases in nearly all of them since the Industrial Revolution began (see Table 1.3). The burning of fossil fuels is the primary source of these increasing trace gas concentrations. Heating, transportation and industrial activities generate almost 5×10^{20} J/year of energy. Oil and natural gas consumption account for 60 per cent of global energy and coal about 25 per cent. Natural gas is almost 90 per cent methane (CH_4) whereas the burning of coal and oil releases not only CO_2, but also odd nitrogen (NO_x), sulphur, and carbon monoxide (CO). Other factors relating to agricultural practices (land clearance, farming and cattle raising) also contribute to modifying the atmospheric composition. The concentrations and sources of the most important greenhouse gases are considered in turn.

Carbon dioxide (CO_2). The major reservoirs of carbon are in limestone sediments and fossil fuels on land and in the world oceans. The atmosphere contains about 750×10^{12} kg of carbon (C), corresponding to a CO_2 concentration of 350 ppm (see Figure 1.3). The major fluxes of atmospheric carbon dioxide are a result of solution/dissolution in the ocean and photosynthesis/respiration and decomposition by biota. The average time for a CO_2 molecule to be dissolved in the ocean or taken up by plants is about four years. Photosynthetic activity leading to primary production on land involves 50×10^{12} kg of carbon annually, representing 7 per cent of atmospheric carbon; this accounts for the 14 ppm annual oscillation in CO_2 observed in the northern hemisphere due to its extensive land biosphere.

The oceans play a key role in the global carbon cycle. Photosynthesis by phytoplankton generates organic compounds of aqueous carbon dioxide. Eventually, some of the biogenic matter sinks into deeper water where it undergoes decomposition and oxidation back into carbon dioxide. This process transfers carbon dioxide from the surface water and sequesters it in the ocean deep water. As a consequence, atmospheric concentrations of CO_2 can be maintained at a lower level than otherwise. This mechanism is known as a 'biologic pump'; long-term changes in its operation may have caused the rise in atmospheric CO_2 at the end of the last glaciation. Ocean biomass productivity is limited by the availability of nutrients and by light. Hence, unlike the land biosphere, increasing CO_2 levels will not necessarily affect ocean productivity; inputs of

Table 1.3 Anthropogenically-induced changes in concentration of atmospheric trace gases.

Gas	*Concentration*		*Increase (%)*	*Sources*
	*1850**	*1985*	*1975–85*	
Carbon dioxide	280 ppm	345 ppm	4.6	Fossil fuels
Methane	0.7–1.0 ppm	1.7 ppm	11.0	Rice paddies, cows, wetlands
Nitrous oxide	280 ppbv	305 ppbv	3.5	Microbiological activity, fertilizer, fossil fuel
CFC-11	0	0.24 ppbv	103	Freon†
CFC-12	0	0.4 ppbv	101	Freon†
Ozone (troposphere)	?	10–50 ppbv	?25	

Source: After Ramanathan and Rodhe.
Notes: *Pre-industrial levels are primarily derived from measurements in ice cores where air bubbles are trapped as snow accumulates on polar ice sheets.
†Production began in the 1930s.

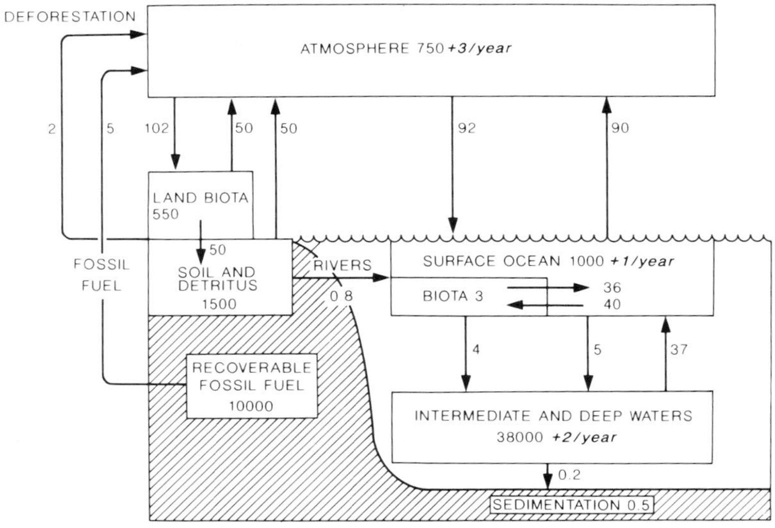

Figure 1.3 Global carbon reservoirs (gigatons of carbon GtC: where 1Gt = 10^9 metric tons = 10^{12} kg) and gross annual fluxes (GtC yr^{-1}). Numbers italicized in the reservoirs suggest the net annual accumulation due to anthropogenic causes. 1 Gt carbon = 37 Gt CO_2.

Source: After Sundquist, Trabalka, Bolin and Siegenthaler; from IPCC 1990.

fertilizers in river runoff may be a more significant factor. In the oceans the carbon dioxide ultimately goes to produce carbonate of lime, partly in the form of shells and the skeletons of marine creatures. On land the dead matter becomes humus which may subsequently form a fossil fuel. These transfers within the oceans and lithosphere involve very long time-scales compared with exchanges involving the atmosphere. As Figure 1.3 showed, the exchanges between the atmosphere and the other reservoirs are more or less balanced.

Yet this balance is not an absolute one; between AD1750 and 1990 the concentration of atmospheric CO_2 is estimated to have been increased by 25 per cent, from 280 to 353 ppm (see Figure 1.4). Half of this increase has taken place since the mid-1960s; currently, atmospheric CO_2 levels are increasing by 1.8 ppmv per year. The primary net source is fossil fuel combustion, accounting for 5×10^{12} kg C/year. Tropical deforestation and fires may contribute a further 2×10^{12} kg C/year; the figure is still uncertain. Fires destroy only above-ground biomass and a large fraction of the carbon is stored as charcoal in the soil. The consumption of fossil fuels should actually have produced an increase almost twice as great as observed. The difference is primarily accounted for by uptake and dissolution in the oceans and the terrestrial biosphere.

Carbon dioxide has a significant impact on global temperature by its absorption and re-emission of radiation from the earth and atmosphere (see Figure 1.14 and E, this chapter). Calculations suggest that the increase to 370 ppm (expected by AD2000 could raise the mean air temperature near the surface by 0.5°C compared with the 1960s (in the absence of other factors).

Research on deep ice cores taken from Antarctica has enabled changes in past atmospheric composition to be studied by extracting air bubbles trapped in the old ice. This shows large natural variations of CO_2 concentration over the last Ice Age cycle (see Figure 1.5). These variations of up to 100 ppm were contemporaneous with temperature changes estimated to be up to 10°C. These long-term variations in carbon dioxide and climate are discussed further in Chapter 8.

Methane (CH_4) is released almost equally by natural wetlands and rice paddies (together about 40 per cent of the total), as well as by enteric fermentation in animals, by termites, through coal and oil extraction, biomass

Figure 1.4 Observations of atmospheric CO_2 increase at Mauna Loa, Hawaii (1957–75), estimates from 1860–1960 based on early measurements, and projected trends into the twenty-first century.

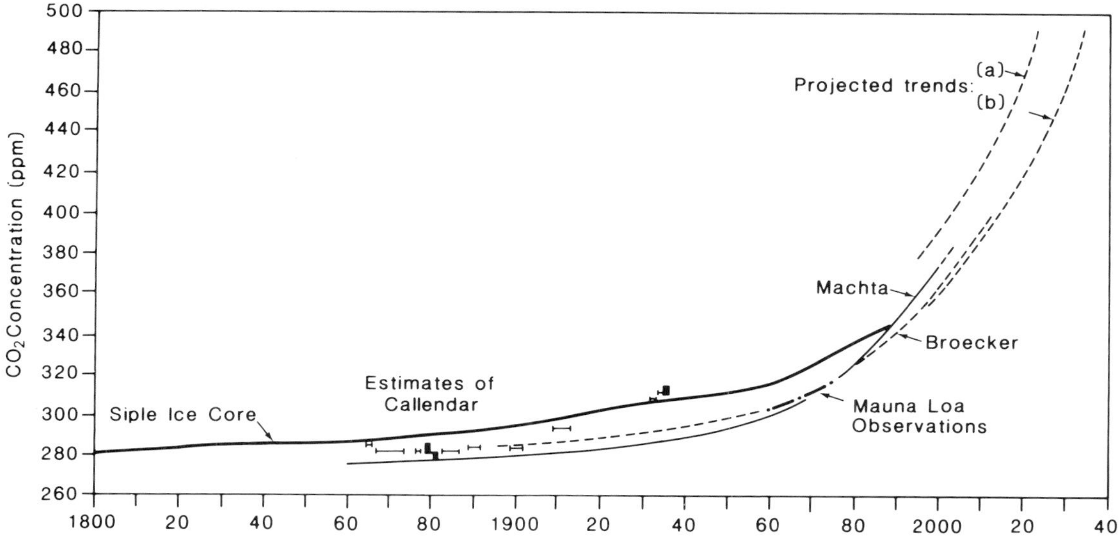

Source: After Keeling, Callendar, Machta, Broecker and others.
Note: (a) and (b) indicate different scenarios of global fossil fuel use (from Kellogg and Schware 1981). The thick line gives readings from air bubbles in the Siple Station ice core since 1800.

Figure 1.5 Changes in atmospheric CO_2 (ppmv: parts per million by volume) and estimates of the resulting global temperature deviations from the present value obtained from air trapped in ice bubbles in cores at Vostok, Antarctica.

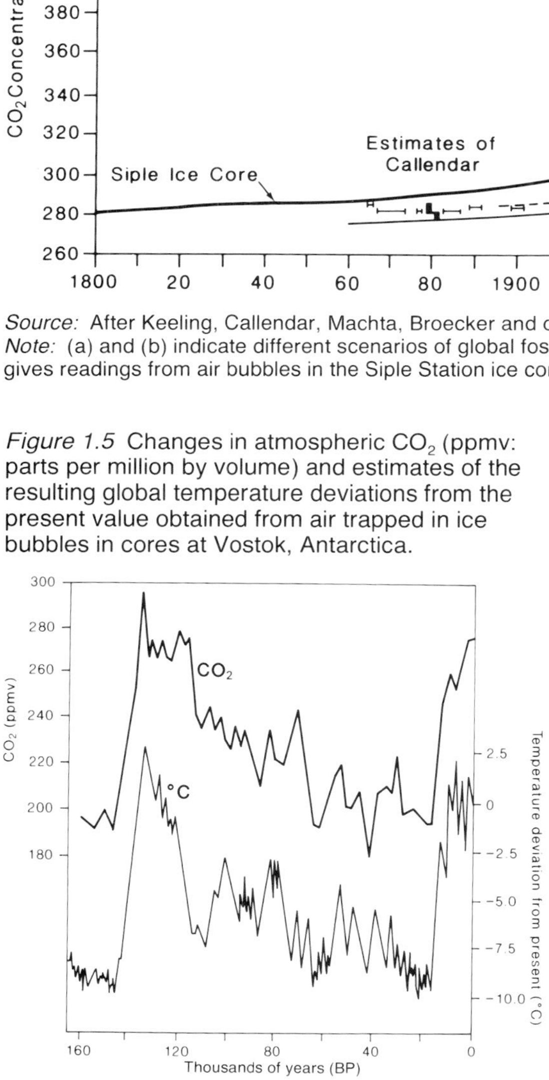

Source: Our Future World, Natural Environment Research Council (NERC) 1989.

burning, and from landfills. Almost two-thirds of the total is related to anthropogenic activity. Cattle populations have increased by 5 per cent/year over 30 years and paddy rice areas by 7 per cent/year, although it is uncertain whether these account quantitatively for the annual increase of 120 ppbv in methane over the last decade. Methane is destroyed by reaction with the hydroxyl (OH) radical and ultimately converted into CO_2 and H_2O. Present concentrations (1720 ppbv) are more than double the pre-industrial level (800 ppbv) and are increasing by 1 per cent annually (see Figure 1.6).

Nitrous oxide (N_2O), which is relatively inert, originates primarily from microbial activity (nitrification) in soils and in the ocean (4 to 8×10^9 kg N/year) with a further 0.2×10^9 kg N/year from fuel combustion. Other anthropogenic sources are nitrogen fertilizers and biomass burning. The concentration of N_2O has increased from a pre-industrial level of about 285 ppbv to 310 ppbv (in clean air). Its increase began around 1940 and is now about 0.8 ppbv/

Figure 1.6 Methane concentration (ppmv parts per million by volume) in air bubbles trapped in ice dating back to 10,000 years BP obtained from ice cores in Greenland and Antarctica.

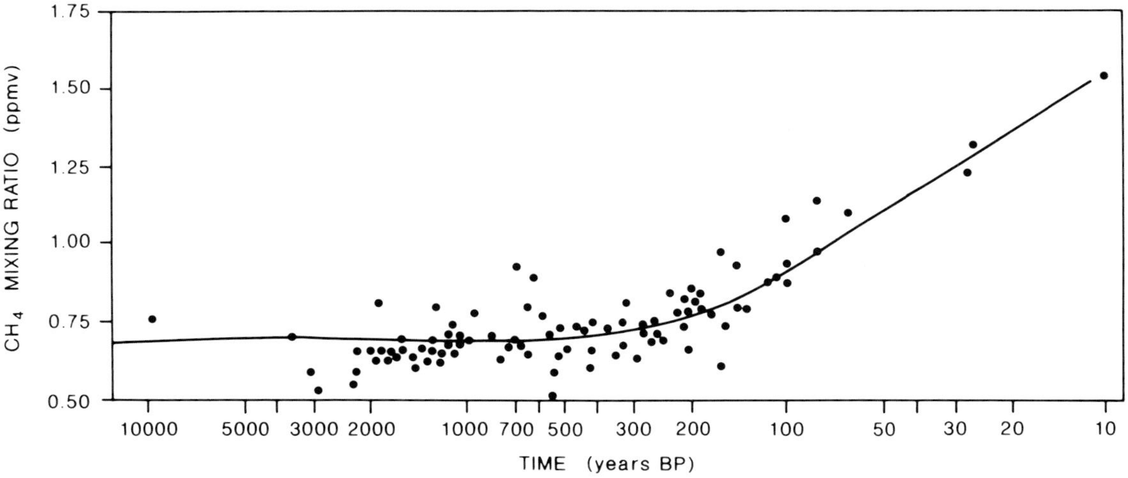

Source: Data from Rasmussen and Khalil, Craig and Chou, and Robbins; from Bolin *et al.* (eds) *The Greenhouse Effect, Climatic Change, and Ecosystems* (SCOPE 29). (*Copyright © 1986. Reprinted by permission of John Wiley & Sons, Inc.*)

year (see Figure 1.7A). The major sink of N_2O is in the stratosphere where it is oxidized into NO_x.

Chlorofluorocarbons (CF_2Cl_2 and $CFCl_3$), better known as 'freons' F-11 and F-12, respectively, were first produced in the 1930s and now have a total atmospheric burden of 10^{10} kg. They are increasing at 4–5 per cent per year (see Figure 1.7B). Although their concentration is < 1 ppbv, they are a significant contributor to the greenhouse effect. They have a residence time of 55–116 years in the atmosphere and recognition of this cumulative effect led to international agreements in 1989 and 1990 to curtail their production. However, while the replacement of CFCs by hydrohalocarbons (HCFCs) can significantly reduce the depletion of stratospheric ozone, HCFCs still have a large greenhouse potential.

Ozone (O_3) is distributed very unevenly with height and latitude (see Figure 1.2) as a result of the complex photochemistry involved in its production (A.2, this chapter). Since the late 1970s, dramatic declines in springtime total ozone have been detected over high latitudes. The normal increase in stratospheric ozone

Figure 1.7 Concentrations of (A) nitrous oxide which have increased since the mid-eighteenth century and especially since 1950; and of (B) CFC11 since 1950. Both in parts per billion by volume.

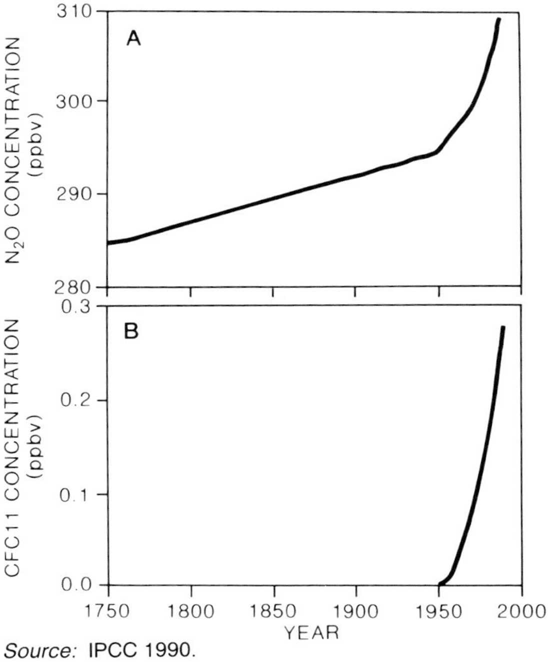

Source: IPCC 1990.

associated with increasing solar radiation in spring apparently failed to develop. Observations in Antarctica show a decrease in total ozone from 320 Dobson units (10^{-3} cm at standard atmospheric temperature and pressure) in October 1975 to about 200 in 1984 and 120 in 1987 (see Figure 1.8 for the results from one specific location). Similar reductions are also evident in the Arctic and at lower latitudes (see Figure 1.9). Between 1979 and 1986 there was a 30 per cent decrease in ozone at 30–40 km altitude between latitudes 20° and 50°N and S; along with this there has been an increase in ozone in the lowest 10 km as a result of anthropogenic activities. These changes in the vertical distribution of ozone concentration are likely to lead to changes in atmospheric heating (E, this chapter), with implications for future climate trends (see Chapter 8).

The effects of reduced stratospheric ozone are particularly important for their potential biological damage to living cells and human skin. It is estimated that a 1 per cent reduction in total ozone will increase ultraviolet-B radiation by 2 per cent, for example, and ultraviolet radiation at 0.30 μm is a thousand times more damaging for the skin than at 0.33 μm. The ozone

Figure 1.8 Mean October total ozone measurements over Halley Bay, Antarctica, since 1957. Readings after 1970 have been checked with those obtained from satellite remote sensing.

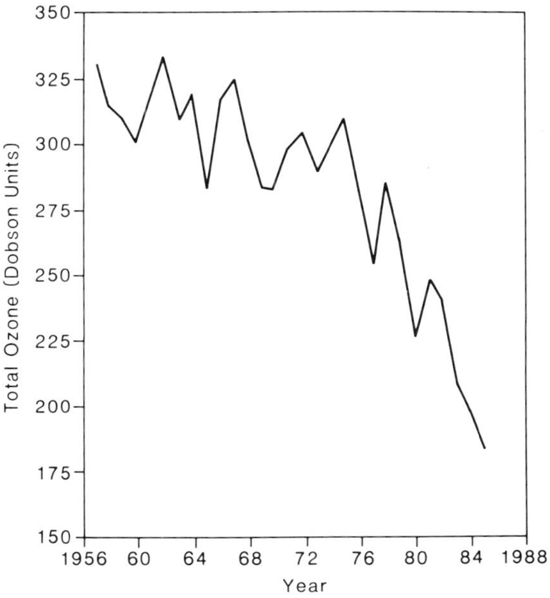

Source: From Stolarski, in Roland and Isaksen (eds), *The Changing Atmosphere*. (*Copyright © 1988. Reprinted by permission of John Wiley & Sons, Inc.*)

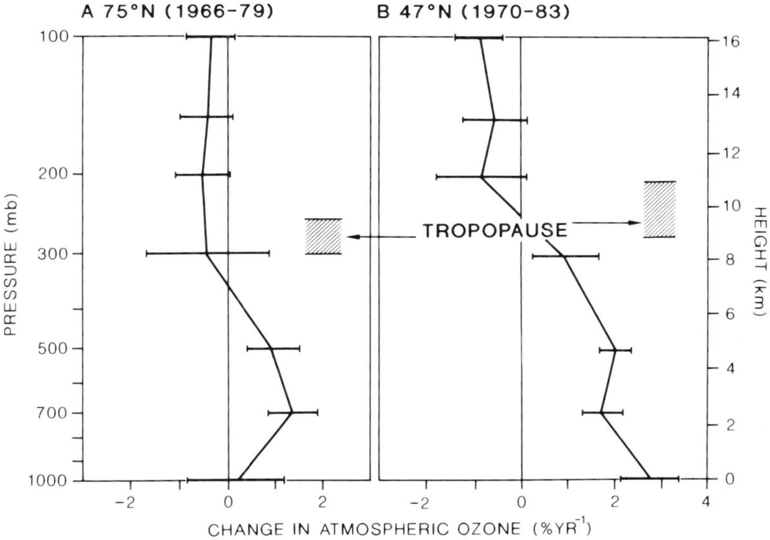

Figure 1.9 Change in atmospheric ozone with elevation (% per year) during 13-year periods at (A) Resolute, northern Canada and (B) Hohenpeissenberg, northern Germany. The horizontal bars indicate the 90 per cent confidence interval. The annual range in altitude of the tropopause is indicated.

Source: Data from Logan; from Bolin *et al.* (eds), *The Greenhouse Effect, Climatic Change, and Ecosystems* (SCOPE 29). (*Copyright © 1986. Reprinted by permission of John Wiley & Sons, Inc.*)

decrease would also be greater in higher latitudes. However, it should be noted that the mean latitudinal and altitudinal gradients of UV-B radiation imply that the effects of a 2 per cent UV-B increase in mid-latitudes could be offset by moving poleward 20 km or 100 m lower in altitude! Recent polar observations suggest more dramatic changes. Stratospheric ozone totals in 1990 over Palmer Station, Antarctica (65°S) were found to have maintained low levels from September until early December, instead of recovering in November. Hence, the altitude of the sun was higher and the incoming radiation much greater than in previous years, especially at wavelengths ≤ 0.30 μm. It remains, however, to determine the specific effects of increased UV radiation on marine biota, for instance.

Changes in atmospheric particle concentration derived from volcanic dust are extremely irregular (see Figure 1.10) but individual volcanic emissions are rapidly diffused geographically (see Figure 1.11). On the other hand, the contribution of man-made particles

Figure 1.10 The average opacity of the stratosphere of the northern hemisphere during the past century measured from the intensity of sunlight and starlight. Some general correlation with major volcanic eruptions is indicated.

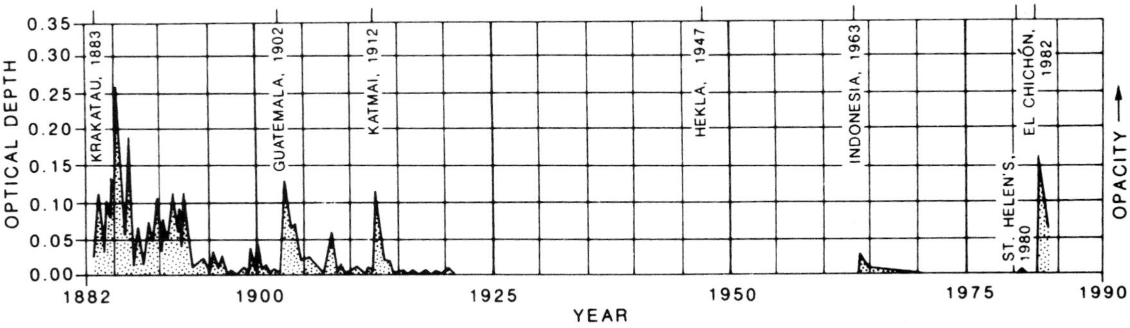

Source: From 'The atmospheric effects of El Chichón by M. R. Rampino and S. Self (*copyright © 1984 by Scientific American Inc. All rights reserved*).

Figure 1.11 The spread of the volcanic dust cloud following the main eruption of the El Chichón volcano in Mexico on 3 April 1982. A strong zonal wind circulation carried the dust cloud at an average speed of 20 m s^{-1} (45 mph) so that it encircled the globe in less than three weeks.

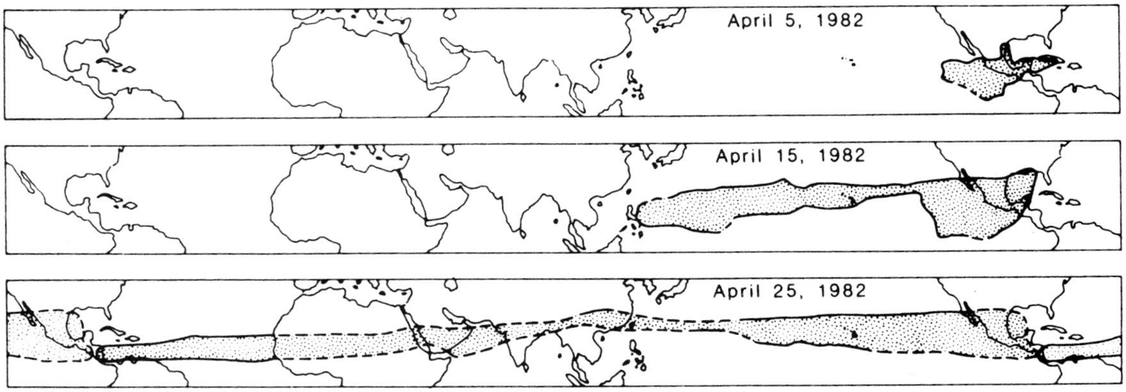

Source: From 'The atmospheric effects of El Chichón' by M. R. Rampino and S. Self (*copyright © 1984 by Scientific American Inc. All rights reserved*).

(particularly of sulphates and soil) has been progressively increasing, at present accounting for about 30 per cent of the total and this figure could double by AD2000. The overall effect on the lower atmosphere is uncertain; urban pollutants generally warm the atmosphere through absorption and reduce solar radiation reaching the surface. Aerosols may lower the planetary albedo above a high albedo desert or snow surface, but increase it over an ocean surface. Thus, the global role of tropospheric aerosols is difficult to evaluate, although some authorities consider it to be one of cooling. Volcanic eruptions, which inject dust and sulphur dioxide high into the stratosphere, are known to cause a small deficit in surface heating with a global effect of −0.1° to −0.2°C, but the effect is short-lived, lasting only a year or so after the event. Also, unless the eruption is in low latitudes, the dust and sulphate aerosols remain in one hemisphere and do not cross the equator.

B MASS OF THE ATMOSPHERE

It is necessary to examine some of the mechanical laws that the atmospheric gases obey. Two simple laws specify the main factors governing changes in pressure. The first, Boyle's Law, states that, at a constant temperature, the volume (V) of a mass of gas varies inversely as its pressure (P), i.e.

$$P = \frac{k_1}{V}$$

(k_1 is a constant); and the second, Charles's Law, that, at a constant pressure, volume varies directly with absolute temperature (T) measured in degrees Kelvin (see note 2):

$$V = k_2 T$$

These laws imply that the three qualities of pressure, temperature and volume are completely interdependent, such that any change in one of them will cause a compensating change to occur in one, or both, of the remainder. The gas laws may be combined to give the following relationship:

$$PV = RmT$$

where m = mass of air
R = a gas constant for dry air ($287\ \mathrm{J\ kg^{-1}\ K^{-1}}$) (see note 3).

If m and T are held fixed, we obtain Boyle's Law; if m and P are held fixed, we obtain Charles's Law. Since it is convenient to use density, ϱ (= mass/volume) rather than volume when studying the atmosphere, we can rewrite the equation in the form known as the equation of state

$$P = R\varrho T$$

1 Total pressure

Air is highly compressible, such that its lower layers are much more dense than those above. Fifty per cent of the total mass of air is found below 5 km (see Figure 1.12) and the average density decreases from about 1.2 kg m^{-3} at the surface to 0.7 kg m^{-3} at 5000 m (approximately 16,000 ft) close to the extreme limit of human habitation.

Figure 1.12 The percentage of the total mass of the atmosphere lying below elevations up to 80 km (50 miles). This illustrates the shallow character of the earth's atmosphere.

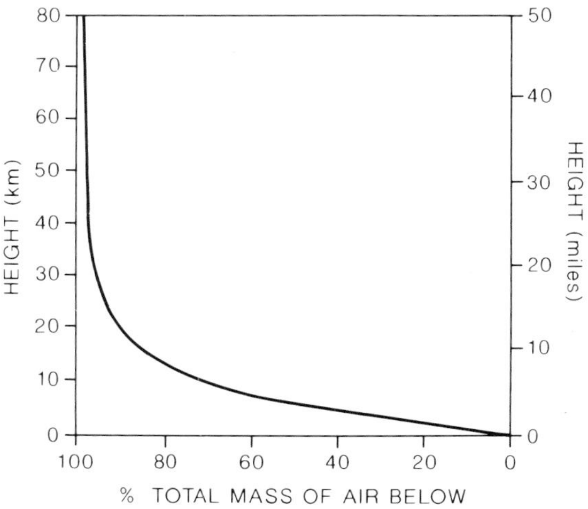

Pressure is measured as a force per unit area. The units used by meteorologists are called millibars (mb), 1 mb being equal to a force of 100 newtons acting on 1 m^2 (see Appendix 2). Pressure readings are made with a mercury barometer which in effect measures the weight of the column of mercury that the atmosphere is able to support in a vertical glass tube. The closed upper end of the tube has a vacuum space and its open lower end is immersed in a cistern of mercury. By exerting pressure downwards on the surface of mercury in the cistern, the atmosphere is able to support a mercury column in the tube of about 760 mm (29.9 in) height, or approximately 1013 mb.

Pressure data are standardized in three ways. The readings from a mercury barometer are adjusted to correspond to those for a standard temperature of 0°C (to allow for the thermal expansion of mercury); they are referred to a standard gravity value of 9.81 ms^{-2} at 45° latitude (to allow for the slight latitudinal variation in g from 9.78 ms^{-2} at the equator to 9.83 ms^{-2} at the poles); and they are calculated for mean sea level to eliminate the effect of station elevation. This third correction is the most significant because near sea level pressure decreases with height about 1 mb per 8 m. A fictitious temperature between the station and sea level has to be assumed and in mountain areas this commonly causes bias in the calculated mean sea-level pressure. (The method is summarized in note 4.)

Mean sea-level pressure (p_0) can be estimated by taking account of the total mass of the atmosphere (M), the mean acceleration due to gravity (g_0) and the mean earth radius (R_E):

$$p_0 = g_0 \, (M/4\pi R_E^2)$$

where the denominator is the surface area of a spherical earth. Substituting appropriate values into this expression: $M = 5.14 \times 10^{18}$ kg, $g_0 = 9.8$ m s^{-2}, $R_E = 6.37 \times 10^6$ m, we find $p_0 \simeq 10^5$ $kgm^{-1}s^{-2} = 10^5 Nm^{-2}$, or 10^5 pascals (Pa is the SI unit of pressure). Meteorologists still use the millibar (mb) unit; 1 millibar = 10^6 dyne $cm^{-2} = 10^2$ Pa (or hPa; h = hecto).

Hence the mean sea-level pressure is approximately 10^5 Pa or 1000 mb. The global mean value is 1013.25 mb (equivalent to 14.7 lb/in^2). On average, nitrogen contributes about 760 mb, oxygen 240 mb and water vapour 10 mb. In other words, each gas exerts a partial pressure independent of the others.

Atmospheric pressure, depending as it does on the weight of the overlying atmosphere, decreases logarithmically with height. This relationship is expressed by the *hydrostatic equation*:

$$\frac{\partial p}{\partial z} = -g\varrho$$

i.e. the rate of change of pressure (p) with height (z) is dependent on gravity (g) multiplied by the air density (ϱ). With increasing height, the drop in air density causes a decrease in this rate of pressure decrease. The temperature of the air also affects this rate, which is greater for cold dense air (see Chapter 3, C.1), although the relationship between pressure and height is so significant that meteorologists often express elevations in millibars: 1000 mb represents sea level, 500 mb about 5500 m and 300 mb about 9000 m. A conversion nomogram for an idealized (standard) atmosphere is given in Appendix 2.

2 Vapour pressure

At any given temperature there is a limit to the density of water vapour in the air, with a consequent upper limit to the vapour pressure. This is termed the *saturation vapour pressure* (e_s) and Figure 1.13 illustrates how it increases with temperature, reaching a maximum of 1013 mb (1 atmosphere) at boiling point. Attempts to introduce more vapour into the air when the vapour pressure is at saturation produce condensation of an equivalent amount of vapour. Figure 1.13 shows that, whereas the saturation vapour pressure has a single value at any temperature above freezing point, below 0°C the saturation vapour pressure above an ice

Figure 1.13 Plot (semi-logarithmic) of the saturation vapour pressure as a function of temperature (i.e. the dew-point curve). Below 0°C the atmospheric saturation vapour pressure is less with respect to an ice surface than with respect to a water drop. Thus, condensation may take place on an ice crystal at lower air humidity than is necessary for the growth of water drops.

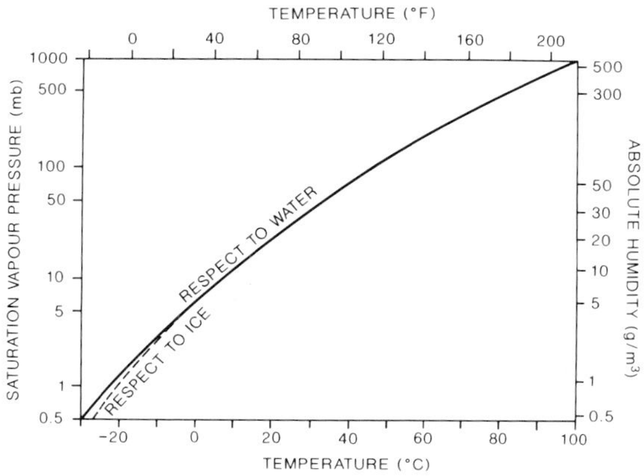

surface is lower than that above a super-cooled water surface. The significance of this will be discussed in Chapter 2, G.1.

Vapour pressure (*e*) varies with latitude and season from about 0.2 mb over northern Siberia in January to over 30 mb in the tropics in July, but this is not reflected in the pattern of surface pressure. Pressure decreases at the surface when some of the overlying air is displaced horizontally, and in fact the air in high-pressure areas is generally dry owing to dynamic factors, particularly vertical air motion (see Chapter 3, D.1), whereas air in low-pressure areas is usually moist.

C SOLAR RADIATION

The prime source of the energy injected into our atmosphere is the sun, which is continually shedding part of its mass by radiating waves of electro-magnetic energy and high energy particles into space. This constant emission is important because it represents in the long run almost all the energy available to the earth (except for a small amount emanating from the radioactive decay of earth minerals). The amount of energy received by the earth, assuming for the moment that there is no interference from the atmosphere, is affected by four factors: solar output, the sun–earth distance, altitude of the sun, and day length.

1 Solar output

Solar energy, which originates from nuclear reactions within the sun's hot core (16×10^6K), is transmitted to the sun's surface by radiation and hydrogen convection. Visible solar radiation (light) comes from a 'cool' (~6000 K) outer surface layer called the *photosphere*. Temperatures rise again in the outer chromosphere (10,000 K) and corona (10^6K) which is continually expanding into space. The outflowing hot gases (plasma) from the sun, referred to as the *solar wind* (with a speed of 1.5×10^6 km hr^{-1}), interact with the earth's magnetic field and upper atmosphere. The earth intercepts both the normal electro-magnetic radiation and energetic particles emitted by the sun during solar flares.

The sun behaves virtually as a *black body*, meaning that it both absorbs all energy received and in turn radiates energy at the maximum rate possible for a given temperature. The energy emitted at a particular wavelength by a perfect radiator of given temperature is described by a relationship due to Max Planck. The black body curves in Figure 1.14 illustrate this relationship. The total energy emitted by a black body is given by the area under each curve; its value is found by integration of Planck's equation, known as Stefan's Law:

$$F = \sigma T^4$$

where $\sigma = 5.67 \times 10^{-8}$ W m^{-2} K^{-4} (the Stefan–Boltzmann Constant), i.e. the energy emitted (F) is proportional to the fourth power of the absolute temperature of the body (T).

The total solar output to space, assuming a temperature of 5800 K for the sun, is 2.33 ×

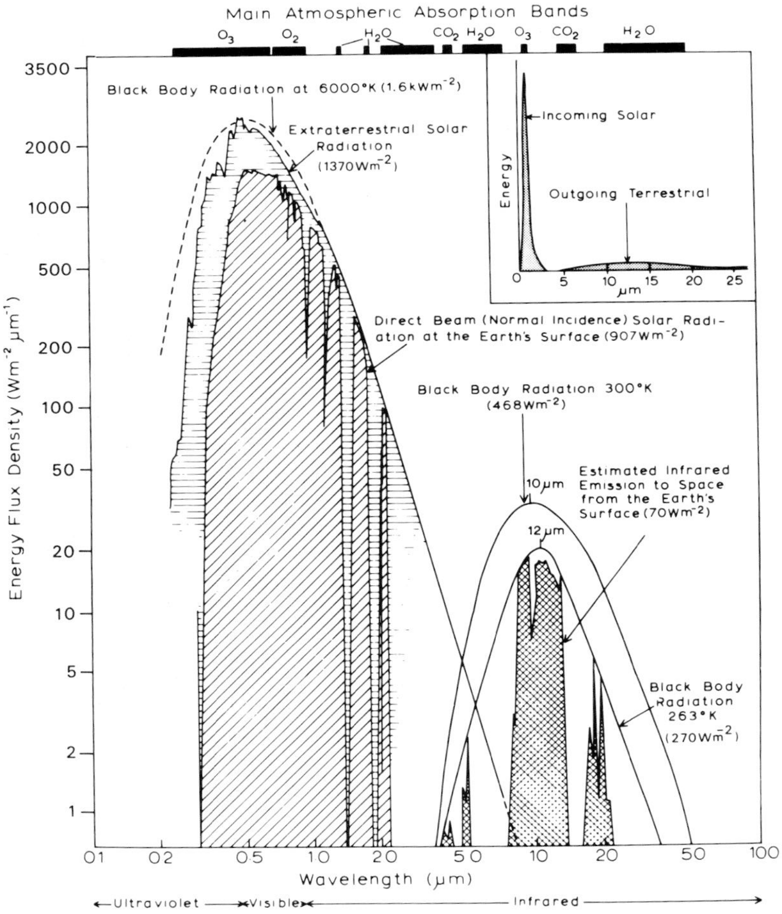

Figure 1.14 Spectral distribution of solar and terrestrial radiation, plotted logarithmically, together with the main atmospheric absorption bands. The cross-hatched areas in the infra-red spectrum indicate the 'atmospheric windows' where radiation escapes to space. The black-body radiation at 6000 K is that proportion of the flux which would be incident on the top of the atmosphere. The inset shows the same curves for incoming and outgoing radiation with the wavelength plotted arithmetically on an arbitrary vertical scale.

Source: Mostly after Sellers 1965.

10^{25} kJmin^{-1}, but only a tiny fraction of this is intercepted by the earth, because the energy received is inversely proportional to the square of the solar distance (150 million km) (see note 5).

The energy received at the top of the atmosphere on a surface perpendicular to the solar beam for mean solar distance is termed the *solar constant* (see note 5). The most recent satellite measurements indicate a value of about 1368 W m^{-2}, or 1.96 cal cm^{-2}min^{-1}. Figure 1.14 shows the wavelength range of solar (shortwave) radiation and the long-wave (infra-red) radiation emitted by the earth and atmosphere. For solar radiation, 8 per cent is ultraviolet and shorter wavelength emission, 39 per cent visible light (0.4–0.7 μm) and 53 per cent near-infra-red (> 0.7μm). The figure illustrates the black body radiation curves for 6000 K at the top of the atmosphere (which slightly exceeds the observed extraterrestrial radiation) for 300 K, and for 263 K. The mean temperature of the earth's surface is about 288 K (15°C) and of the atmosphere about 250 K (−23°C). Gases do not behave as black bodies, and Figure 1.14 shows the absorption bands in the atmosphere which cause its emission to be much less than that from an equivalent black body. The wavelength of maximum emission (λ_{max}) varies inversely with the absolute temperature of the radiating body:

$$\lambda_{max} = \frac{2897}{T} \times 10^{-6} \text{ m (Wien's Law).}$$

Thus solar radiation is very intense and is mainly short wave between about 0.2 and 4.0 μm, with a maximum (per unit wavelength) at 0.5 μm, whereas the much weaker terrestrial radiation has a peak intensity at about 10 μm and a range of about 4 to 100 μm (1 μm = 1 micrometer = 10^{-6} m).

There have been suggestions that the solar constant undergoes small periodic variations of about 0.1 per cent, perhaps related to sunspot activity, but since determinations of the solar constant are subject to errors of similar magnitude, the reality of such fluctuations has been doubted. *Sunspots* are dark (i.e. cooler) areas visible on the sun's surface. Their number and positions change in a regular manner known as the sunspot cycle ($\sim$ 11 years). Figure 1.15 shows the variation of sunspot activity over the last few centuries. Output within the ultraviolet part of the spectrum shows considerable variability, with up to twenty times more ultraviolet radiation emitted at certain wavelengths during a sunspot maximum than during a sunspot minimum. Satellite measurements during the 1980s, the latest solar cycle, show a small decrease in solar output as sunspot number approached its *minimum* and a subsequent

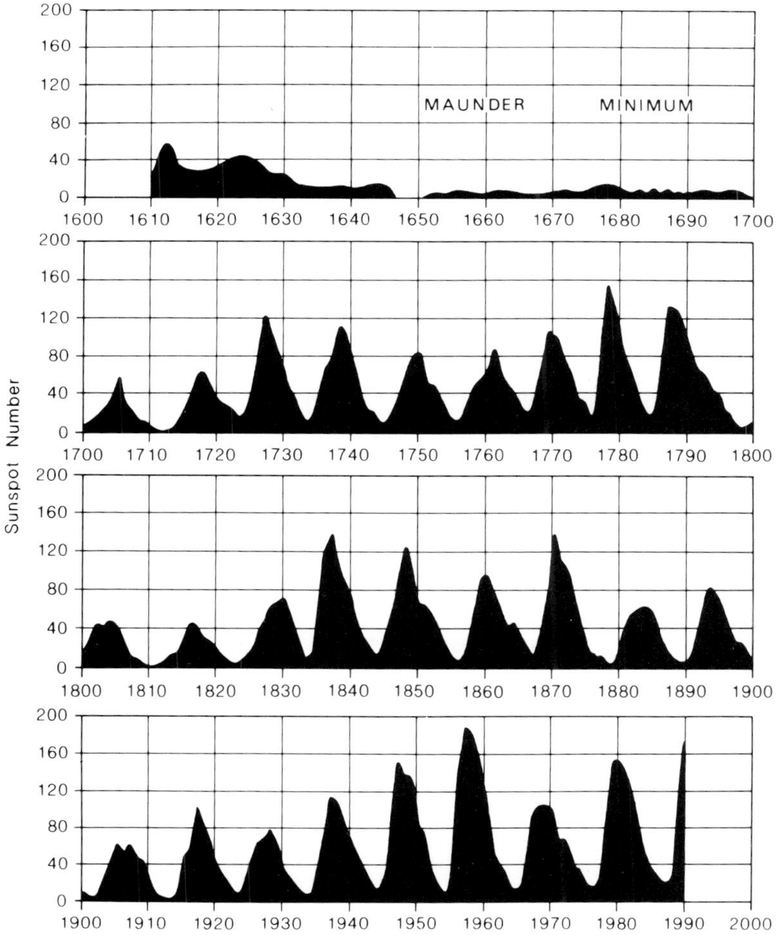

Figure 1.15 Yearly sunspot numbers for the sun's visible surface for the period 1610–1989. The pronounced Maunder Minimum (1645–1705) is believed by some to coincide with the most severe conditions of the Little Ice Age in Europe and North America.

Sources: Reproduced by courtesy of the National Geophysical Data Center, Boulder, Colorado. Data before 1700 courtesy of Foukal (1990) and *Scientific American*.

recovery. Although sunspot areas are cool spots, they are surrounded by bright areas of activity known as faculae that have higher temperatures; the net effect is for solar output to vary in parellel with the number of sunspots. Thus, the solar 'constant' decreased by about 1.5 W m^{-2} from sunspot maximum to minimum. In the long term, assuming that the earth behaves as a black body, a long-continued difference of 2 per cent in the solar constant could change the effective mean temperature of the earth's surface by as much as 1.2°C (2.2°F); however, the observed fluctuations of about 0.1 per cent would change the mean global temperature by ≤0.06°C.

2 Distance from the sun

The annually changing distance of the earth from the sun produces seasonal variations in our receipt of solar energy. Owing to the eccentricity of the earth's orbit round the sun, the receipt of solar energy on a surface normal to the beam is 7 per cent more on 3 January at the perihelion than on 4 July at the aphelion (see Figure 1.16). In theory (that is, discounting the interposition of the atmosphere and the difference in degree of conductivity between large land and sea masses) this difference should produce an increase in the effective January world surface temperatures of about 4°C (7°F), over those of July. It should also make northern hemisphere winters warmer than those in the southern, and southern hemisphere summers warmer than those in the northern. In practice, atmospheric heat circulation and the effects of continentality substantially mask this global tendency, and the actual seasonal contrast between the hemispheres is reversed. Moreover, the northern summer half-year (21 March–22 September) is 5 days longer than the southern hemisphere summer (22 September–21 March). This difference slowly changes; about 10,000 years ago the aphelion occurred in the northern hemisphere winter and northern summers received 3–4 per cent more radiation than today. This same pattern will return about 10,000 years from now (see Figure 1.16D).

Figure 1.17 graphically illustrates the seasonal variations of energy receipt, with latitude. Actual amounts of radiation received on a horizontal surface outside the atmosphere are given in Table 1.4. The intensity on a horizontal surface (I_h) is determined from

$$I_h = I_0 \sin d,$$

where I_0 = the solar constant and d = the angle between the surface and the solar beam.

3 Altitude of the sun

The altitude of the sun (i.e. the angle between its rays and a tangent to the earth's surface at the point of observation) also affects the amount of solar radiation received at the surface of the earth. The greater the sun's altitude the more concentrated is the radiation intensity per unit area at the earth's surface. There are, in addition, important variations with solar altitude of the proportion of radiation reflected by the surface, particularly in the case of a water surface (see D.5, this chapter). The principal factors that determine the sun's altitude are, of course, the latitude of the site, the time of day and the season (see Figure 1.16). At the June solstice the sun's altitude is a constant 23½° throughout the day at the North Pole and the sun is directly overhead at noon at the Tropic of Cancer (23½°N).

4 Length of day

The length of daylight also affects the amount of radiation that is received. Obviously the longer the time during which the sun shines the greater is the quantity of radiation which a given portion of the earth will be able to receive. At the equator, for example, the daylength is close to 12 hours in all months, whereas at the poles it varies between 0 and 24 hours from winter to summer (see Figure 1.16).

The combination of all these factors produces the pattern of receipt of solar energy at the top of the atmosphere shown by Figure 1.17. The polar regions receive their maximum amounts of

Figure 1.16 The astronomical (orbital) effects on the solar radiation reaching the earth and their time scales. A: Orbital stretch, or eccentricity (≈ 95,000-year period); B: Axial tilt (41,000-year); C: Wobble of the axial path (21,000-year) which causes a shift in the timing of perihelion (D). E illustrates the geometry of the present seasons.

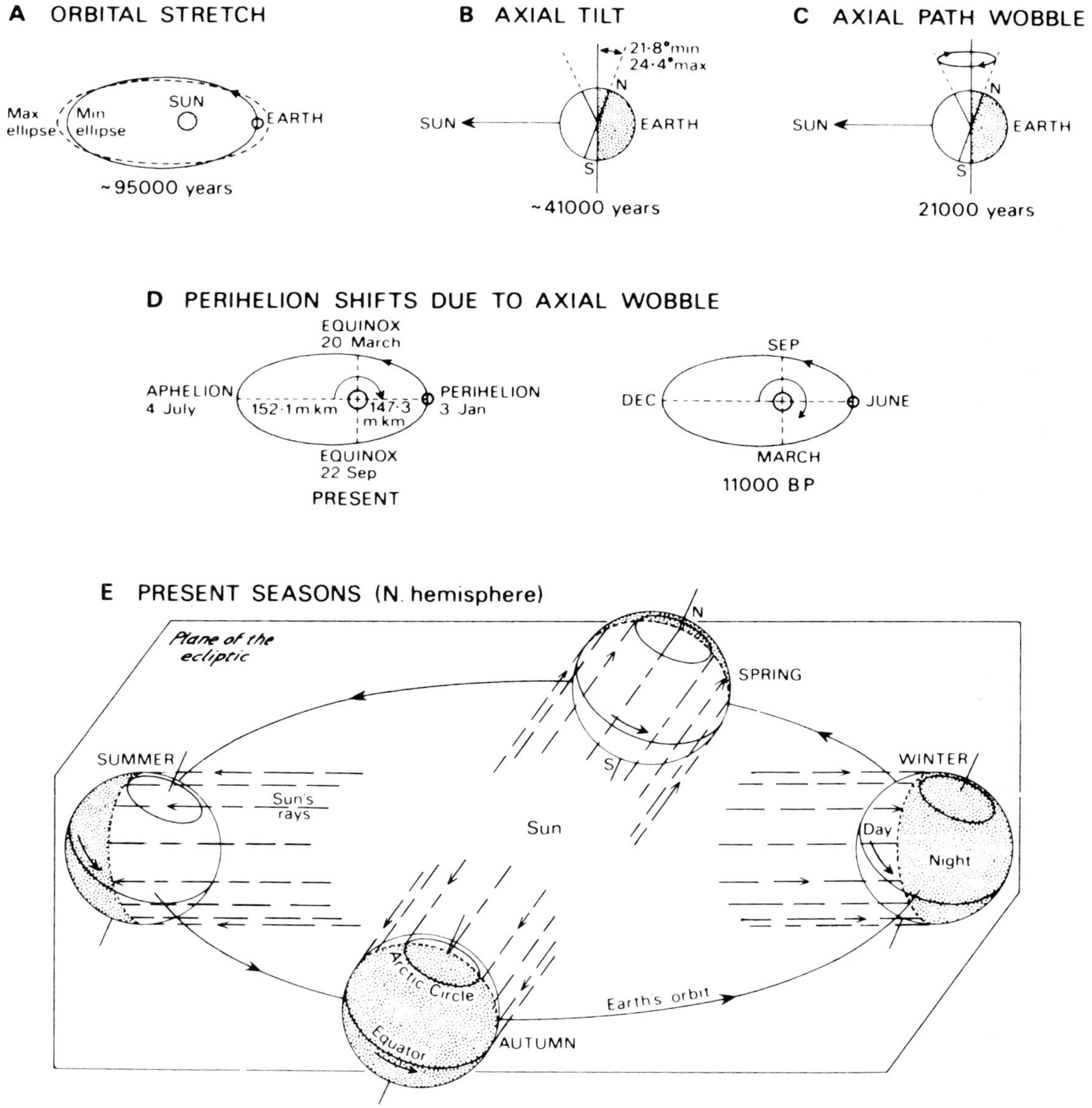

Source: E is after Strahler 1965.

Figure 1.17 The variations of solar radiation with latitude and season for the whole globe, assuming no atmosphere. This assumption explains the abnormally high amounts of radiation received at the poles in summer, when daylight lasts for 24 hours each day.

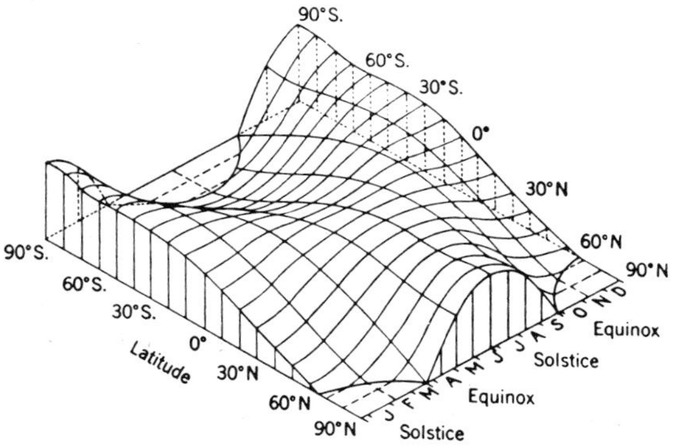

Source: After W. M. Davis; from Strahler 1965.

solar radiation during their summer solstices, which is the period of continuous day. The amount received during the December solstice in the southern hemisphere is theoretically greater than that received by the northern hemisphere during the June solstice, due to the previously mentioned elliptical path of the earth round the sun (see Table 1.4). The equator has two radiation maxima at the equinoxes and two minima at the solstices, due to the apparent passage of the sun during its double annual movement between the northern and southern hemispheres.

D SURFACE RECEIPT OF SOLAR RADIATION AND ITS EFFECTS

1 Energy transfer within the earth-atmosphere system

So far we have described the distribution of solar radiation as if it were all available at the earth's surface. This is of course an unreal view because of the effect of the atmosphere on energy transfer. Heat energy can be transferred by the three following mechanisms:

1 Radiation: Electromagnetic waves transfer energy (both heat and light) between two bodies without the necessary aid of an intervening material medium at a speed of 300×10^6 m s^{-1} (i.e. the speed of light). This is so with solar energy through space, whereas the earth's atmosphere only allows the passage of radiation at certain wavelengths and restricts that at others.

Radiation entering the atmosphere may be absorbed by atmospheric gases in certain wavelengths but, as shown in Figure 1.14, most short-wave radiation is transmitted without absorption. Scattering occurs if the direction of a photon of radiation is changed by interaction with atmospheric gases and aerosols. Two types of scattering are distinguished. For gas molecules smaller than the radiation wavelength (λ), *Rayleigh scattering* occurs in all directions and is proportional to $(1/\lambda^4)$. As a result, the scattering of blue light ($\lambda - 0.4\ \mu$m) is an order of magnitude (i.e.

Table 1.4 Daily solar radiation on a horizontal surface outside the atmosphere: W m^{-2}.

Date	*90°N*	*70*	*50*	*30*	*0*	*30*	*50*	*70*	*90°S*
Dec 22	0	0	88	233	421	520	528	540	574
Feb 4	0	12	144	284	438	486	454	392	404
Mar 21	0	153	287	387	447	387	287	153	0
May 6	386	350	433	464	418	271	138	12	0
June 22	538	505	494	487	394	218	82	0	0

Source: After Kondratyev.

$\times$ 10) greater than that of red light ($\lambda \sim$ 0.7 μm), thus creating the daytime blue sky. However, when water droplets or aerosol particles, with similar sizes (0.1–0.5 μm radius) to the radiation wavelength, are present, most of the light is scattered forward. This *Mie scattering* gives the grayish appearance of polluted atmospheres.

Within a cloud, or between low clouds and snow-covered surfaces, radiation undergoes multiple scattering. In the latter case, the 'white out' conditions typical of polar regions in summer (and mid-latitude snowstorms) are experienced, when surface features and the horizon become indistinguishable.

2 Conduction: Under this mechanism the heat passes through a substance from point to point by means of the transfer of adjacent molecular motions. Since air is a poor conductor this type of heat transfer can be virtually neglected in the atmosphere, but it is important in the ground.
3 Convection: This occurs in fluids (including gases) which are able to circulate internally and distribute heated parts of the mass. The low viscosity of air and its consequent ease of motion makes this the chief method of atmospheric heat transfer. It should be noted that *forced convection* (mechanical turbulence) occurs due to the development of eddies as air flows over uneven surfaces, even when there is no surface heating to set up *free* (thermal) *convection.*

Convection transfers energy in two forms. The first is the *sensible heat* content of the air (called enthalpy by physicists) which is transferred directly by the rising and mixing of warmed air. It is defined as c_pT where T is the temperature and c_p (= 1004 J kg^{-1}K^{-1}) is the specific heat at constant pressure (the heat absorbed by unit mass for unit temperature increase). Sensible heat is also transferred by conduction. The second form of energy transfer by convection is indirect, involving *latent heat.* Here, there is no temperature change. Whenever water is converted into water vapour by evaporation (or boiling) heat is required. This is referred to as the latent heat of vaporization (L). At 0°C, L is 2.50 $\times$ 10^6 J kg^{-1} of water, or 597 cal g^{-1}. More generally,

$$L\ (10^6\ \mathrm{J\ kg^{-1}}) \approx (2.5 - 0.00235T)$$

where T is in °C. When water condenses in the atmosphere (see Chapter 2, C) the same amount of latent heat is given off as is used for evaporation *at the same temperature.* Similarly, for melting ice at 0°C, the latent heat of melting is required which is 0.335 $\times$ 10^6 J kg^{-1} (80 cal g^{-1}). If ice evaporates, without melting, the latent heat of this sublimation process is 2.83 $\times$ 10^6 J kg^{-1} at 0°C (676 cal g^{-1}) (i.e. the sum of the latent heats of melting and vaporization). In all of these phase changes of water there is an energy transfer. We shall return to other aspects of these processes in Chapter 2.

2 Effect of the atmosphere

Solar radiation is virtually all in the short-wavelength range, less than 4 μm (see Figure 1.14). About 18 per cent of the incoming energy is absorbed directly by ozone and water vapour. Ozone absorption is concentrated in three solar spectral bands (0.20–0.31, 0.31–0.35 and 0.45–0.85 μm), while water vapour absorbs to a lesser degree in several bands between 0.9 and 2.1 μm (see Figure 1.14). Solar wavelengths shorter than 0.285 μm scarcely penetrate below 20 km altitude, whereas those > 0.295 μm, reach the surface. Thus, the 3 mm (equivalent) column of ozone attenuates ultraviolet radiation almost entirely, except for a partial window around 0.20 μm where radiation reaches the lower stratosphere. About 30 per cent is immediately reflected back into space from the atmosphere, clouds and the earth's surface, leaving approximately 70 per cent to heat the earth and its atmosphere. Of this, the greater part eventually heats the atmosphere, but much of this heat is received secondhand by the atmosphere via the earth's surface. The ultimate retention of this energy by the atmosphere is of prime importance, because if it did not occur the

average temperature of the earth's surface would fall by some 40°C, making most life obviously impossible. The surface absorbs almost half of the incoming energy available at the top of the atmosphere and re-radiates it outwards as long (infra-red) waves of greater than 3 μm (see Figure 1.14). Much of this re-radiated long-wave energy can be absorbed by the water vapour, carbon dioxide and ozone in the atmosphere, the rest escaping through atmospheric *windows* back into outer space, principally between 8 and 13 μm (see Figure 1.14). Figure 1.18 illustrates the relative roles of the atmosphere, clouds and the earth's surface in reflecting and absorbing solar radiation at different latitudes. (A more complete analysis of the total heat budget of the earth-atmosphere system is given in F, this chapter.)

Figure 1.18 The average annual latitudinal disposition of solar radiation in W m^{-2}. Of 100 per cent radiation entering the top of the atmosphere, 21 per cent is reflected back to space by clouds, 6 per cent by air (plus dust and water vapour), and 4 per cent by the earth's surface. Three per cent is absorbed by clouds, 18 per cent by the air, and 48 per cent by the earth.

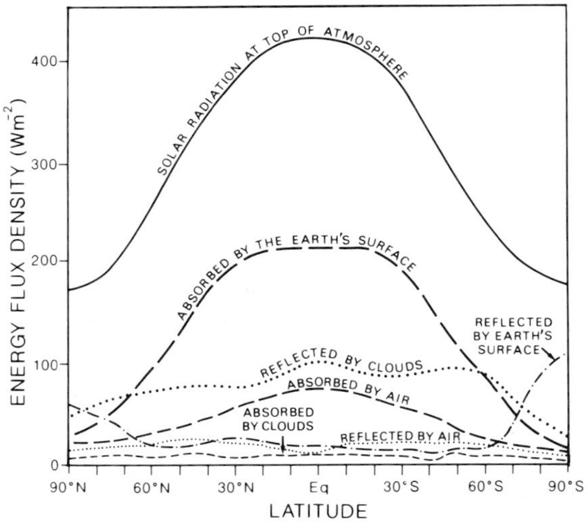

Source: After Sellers 1965.

3 Effect of cloud cover

Cloud cover can, if it is thick and complete enough, form a significant barrier to the penetration of radiation. The drop in surface temperature often experienced on a sunny day when a cloud temporarily cuts off the direct solar radiation illustrates our reliance upon the sun's radiant energy. How much radiation is actually reflected depends on the amount of cloud cover and thickness (see Figure 1.19). The proportion of incident radiation that is reflected is termed the *albedo*, or reflection coefficient (expressed as a fraction or percentage). Cloud type affects the albedo. Aircraft measurements show that the albedo of a complete overcast ranges from 44 to 50 per cent for cirrostratus to 90 per cent for cumulonimbus. Average albedos, as determined by satellites, aircraft and surface measurements, are summarized in Table 1.5 (see note 6).

The total (or global) solar radiation (direct, Q, and diffuse, q) received at the surface on cloudy days is

$$Q + q = (Q + q)_0[b + (1 - b)(1 - c)]$$

Figure 1.19 Percentage of reflection, absorption and transmission of solar radiation by cloud layers of different thickness.

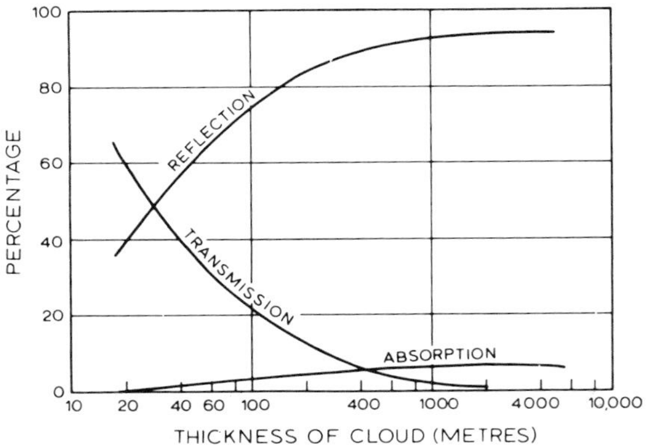

Source: From Hewson and Longley 1944.

Table 1.5 The average (integrated) albedo of various surfaces (0.3–4.0 μm).

Planet earth	0.31
Global surface	0.14–0.16
Global cloud	0.23
Cumulonimbus	0.9
Stratocumulus	0.6
Cirrus	0.4–0.5
Fresh snow	0.8–0.9
Melting snow	0.4–0.6
Sand	0.30–0.35
Grass, cereal crops	0.18–0.25
Deciduous forest	0.15–0.18
Coniferous forest	0.09–0.15
Tropical rainforest	0.07–0.15
Water bodies*	0.06–0.10

Note: *Increases sharply at low solar angles.

where $(Q+q)_0$ = global solar radiation for clear skies;
c = cloudiness (fraction of sky covered);
b = a coefficient depending on cloud type and thickness; and the depth of atmosphere through which the radiation must pass. For mean monthly values for the United States $b \approx 0.35$, so that

$$(Q+q) \approx (Q+q)_0[1-0.65c].$$

The effect of a cloud cover also operates in reverse, since it serves to retain much of the heat that would otherwise be lost from the earth by radiation throughout the day and night. This largely negative role of clouds means that their presence appreciably lessens the daily temperature range by preventing high maxima by day and low minima by night. As well as interfering with the transmission of radiation, clouds act as temporary thermal reservoirs for they absorb a certain proportion of the energy which they intercept. The global effect of cloud reflection and absorption of solar radiation is illustrated in Figures 1.19 and 1.20.

Global cloudiness is not yet accurately known. Ground-based observations are mostly at land stations and refer to a small (~ 250 km^2) area. Satellite estimates are derived from the reflected short-wave radiation and infra-red irradiance measurements, with various threshold assumptions for cloud presence/absence; typically they refer to a grid area of 2500 km^2 to 37,500 km^2. Surface-based observations tend to be about 10 per cent greater than satellite estimates due to the observer's perspective. Average winter and summer distributions of total cloud amount from surface observations are shown in Figure 1.21. The cloudiest areas are the Southern Ocean and the mid-high latitude North Pacific and North Atlantic storm tracks. Lowest amounts are over the Saharan–Arabian desert area. Total global cloud cover is just over 60 per cent in January and July.

Figure 1.20 The average receipt of solar radiation with latitude at the top of the atmosphere and at the earth's surface during the June solstice.

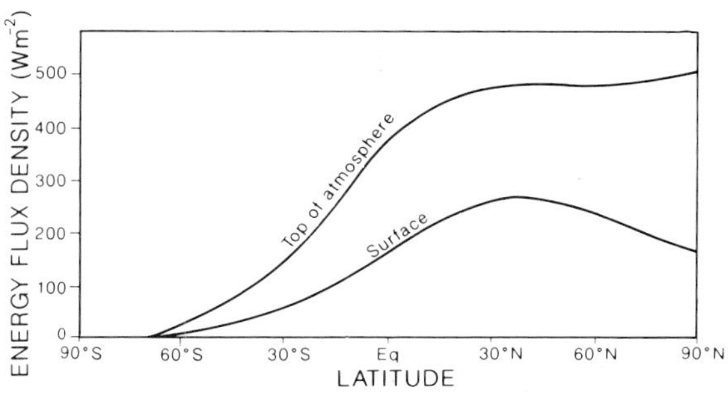

4 Effect of latitude

As Figure 1.17 has already shown, different parts of the earth's surface receive different amounts of solar radiation. The time of the year is one factor controlling this, more radiation being received in summer than in winter because of the higher altitude of the sun and the longer days. Latitude is a very important control because the geographical situation of a region will determine both the duration of daylight and the distance travelled through the atmosphere by the oblique rays from the sun. However, actual calculations show the effect of the latter to be negligible in the Arctic, apparently due to

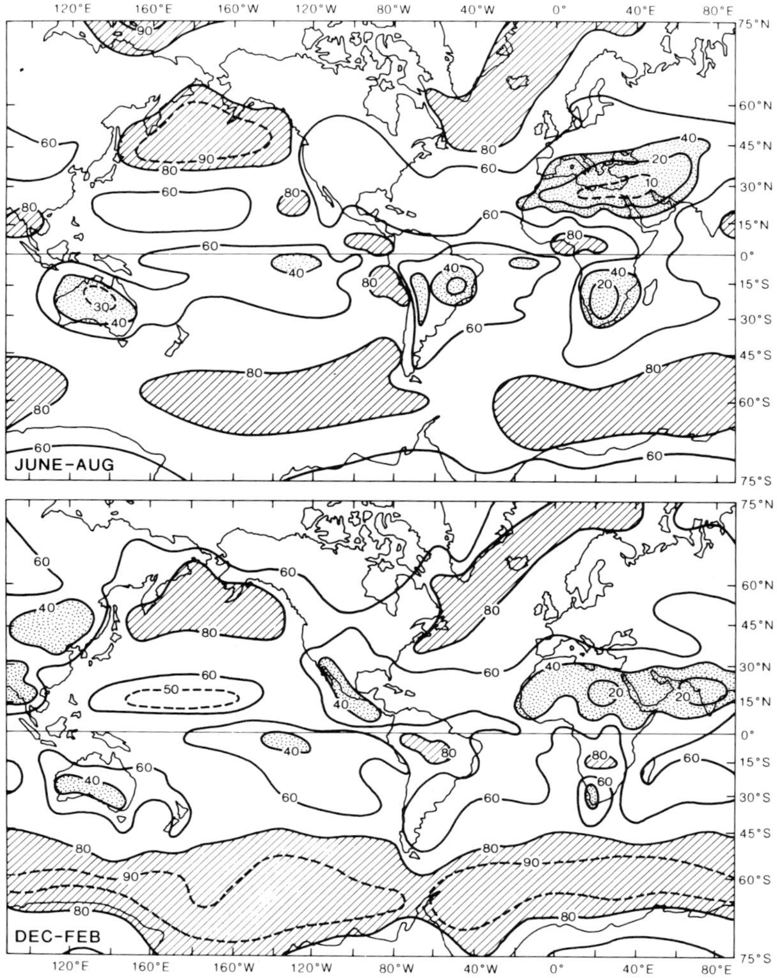

Figure 1.21 The global distribution of total cloud amount (per cent) derived from surface-based observations during the period 1971–81, averaged for the months June–August (*above*) and December–February (*below*). High percentages are shaded and low percentages stippled.

Source: From London *et al.* 1989.

the low vapour content of the air limiting the tropospheric absorption. Figure 1.20 shows that in the upper atmosphere over the North Pole there is a marked maximum of solar radiation at the June solstice yet only about 30 per cent is absorbed at the surface. This may be compared with the global average of 48 per cent of solar radiation being absorbed at the surface. The explanation lies in the high average cloudiness over the Arctic in summer and also in the high reflectivity of the snow and ice surfaces. This example illustrates the complexity of the radiation budget and the need to take into account the interaction of several factors.

A special feature of the latitudinal receipt of radiation is that the maximum temperatures experienced at the earth's surface do not occur at the equator, as one might expect, but at the tropics. A number of factors need to be taken into account. The apparent migration of the vertical sun is relatively rapid during its passage over the equator but its rate slows down as it reaches the tropics. Between 6°N and 6°S the sun's rays remain almost vertically overhead for only 30 days during each of the spring and autumn equinoxes, allowing little time for any large build-up of surface heat and high temperatures. On the other hand, between 17.5° and

23.5° latitude the sun's rays shine down almost vertically for 86 consecutive days during the period of the solstice. This longer sustained period, combined with the fact that the tropics experience longer days than at the equator, makes the maximum zones of heating occur nearer the tropics than the equator. In the northern hemisphere this poleward displacement of the zone of maximum heating is emphasized by the effect of *continentality* (see D.5, this chapter), while low cloudiness associated with the subtropical high-pressure belts is an additional factor. The clear skies are particularly effective in allowing large annual receipts of solar radiation in these areas. The net result of these influences is shown in Figure 1.22 in terms of the average annual solar radiation on a horizontal surface at ground level, and by Figure 1.23 in terms of the average daily maximum shade temperatures. Over the continents the highest values occur at about 23°N and 10°–15°S. In consequence the mean annual *thermal equator* (i.e. the zone of maximum temperature) is located at about 5°N. Nevertheless, the mean air temperatures, reduced to mean sea level, are very broadly related to latitude (see Figures 1.24A and B).

5 Effect of land and sea

Another important control on the effect of incoming solar radiation stems from the different ways in which land and sea are able to profit from it. Whereas water has a tendency to store the heat it receives, land, in contrast, quickly returns it to the atmosphere. There are several reasons for this.

A large proportion of the incoming solar radiation is reflected back into the atmosphere without heating the earth's surface at all. The proportion depends upon the type of surface (see Table 1.5). A sea surface reflects very little unless the angle of incidence of the sun's rays is large. The albedo for a calm water surface is only 2 to 3 per cent for a solar elevation angle exceeding 60°, but is more than 50 per cent

Figure 1.22 The average annual solar radiation on a horizontal surface at ground level in W m^{-2}. Maxima are found in the world's hot deserts, where as much as 80 per cent of the solar radiation annually incident on the top of the unusually clear atmosphere reaches the ground.

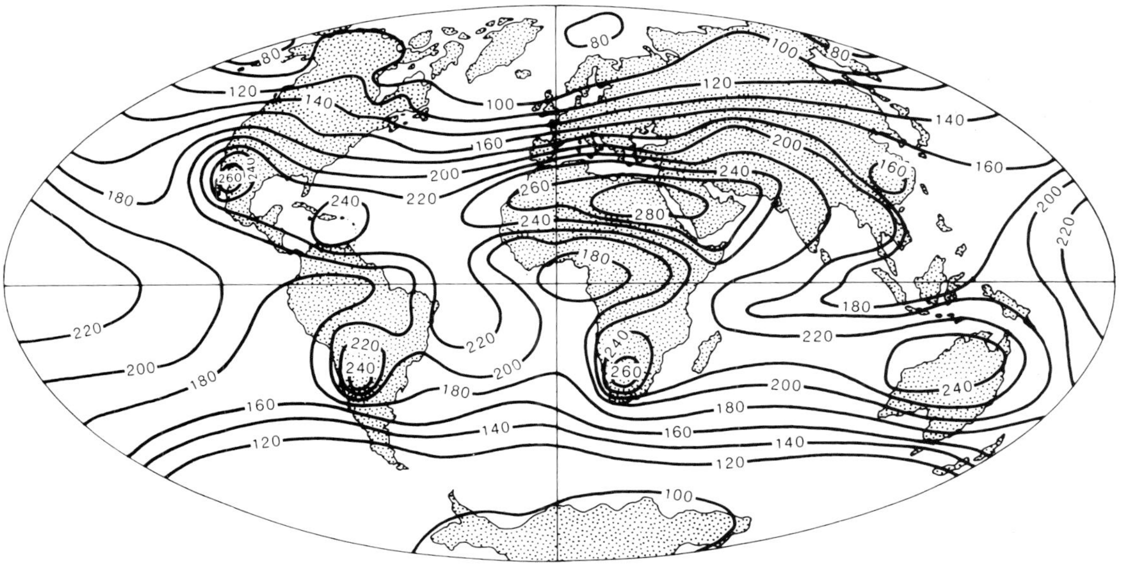

Source: After Budyko *et al.* 1962.

Figure 1.23 Mean daily maximum shade air temperatures (°C).

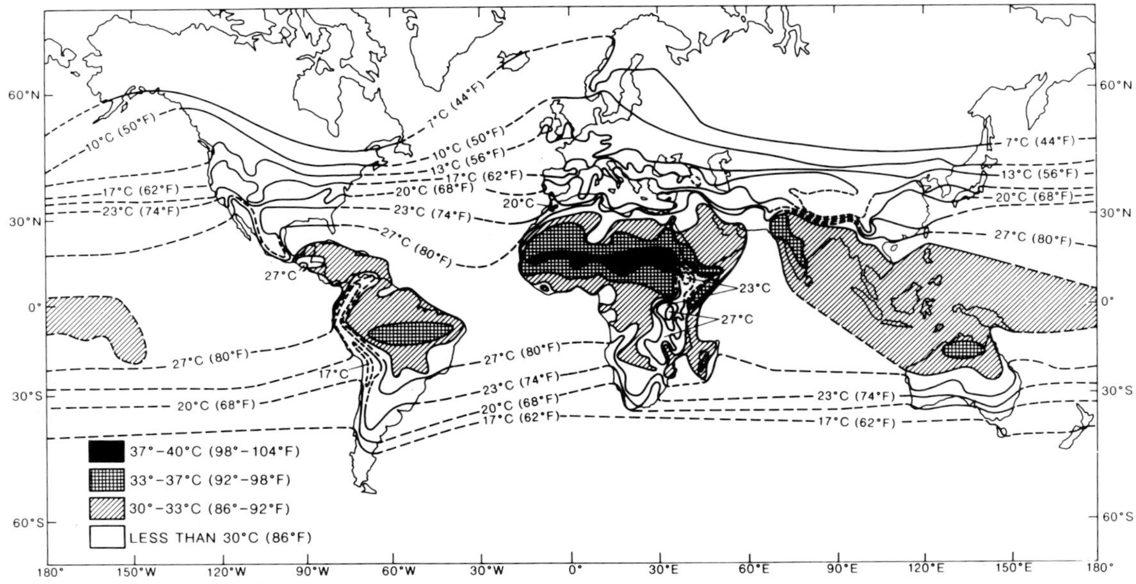

Source: After Ransom 1963.

when the angle is 15°. For land surfaces the albedo is generally between 8 and 40 per cent of the incoming radiation. The figure for forests is about 9 to 18 per cent according to the type of tree and density of foliage (see Chapter 7, C.2.a), for grass approximately 25 per cent, for cities 14 to 18 per cent, and desert sand 30 per cent. Fresh snow may reflect as much as 90 per cent of solar radiation, but snow cover on vegetated, especially forested, surfaces is much less reflective (30–50 per cent). The long duration of snow cover on the northern continents (see Figure 1.25A) causes much of the incoming radiation to be reflected in winter, although global distribution of annual average surface albedo (see Figure 1.25B) shows mainly the influence of the snow-covered Arctic sea ice and Antarctic ice sheet (see Figure 1.35A for planetary albedo).

The global solar radiation absorbed at the surface is determined from measurements of radiation incident on the surface and its albedo (a). It may be expressed as

$$(Q + q)(100 - a)$$

where the albedo is a percentage. A snow cover will absorb only about 15 per cent of the incident radiation, whereas for the sea the figure generally exceeds 90 per cent. The ability of the sea to absorb the heat received also depends upon its transparency. As much as 20 per cent of the radiation penetrates as far down as 9 m (30 ft). Figure 1.26 provides some indication of how much energy is absorbed by the sea at different depths. However, the heat absorbed by the sea is carried down to considerable depths by the turbulent mixing of water masses by the action of waves and currents. Figure 1.27, for example, illustrates the warming of the North Sea down to about 40 m in summer. In completely *still* water the annual heat penetration would only be apparent down to about 3–4 m.

A measure of the difference between the subsurfaces of land and sea is given in Figure 1.28, which shows ground temperatures at

Figure 1.24A Mean sea-level temperatures (°C) in January. The position of the thermal equator is shown approximately by the dashed line.

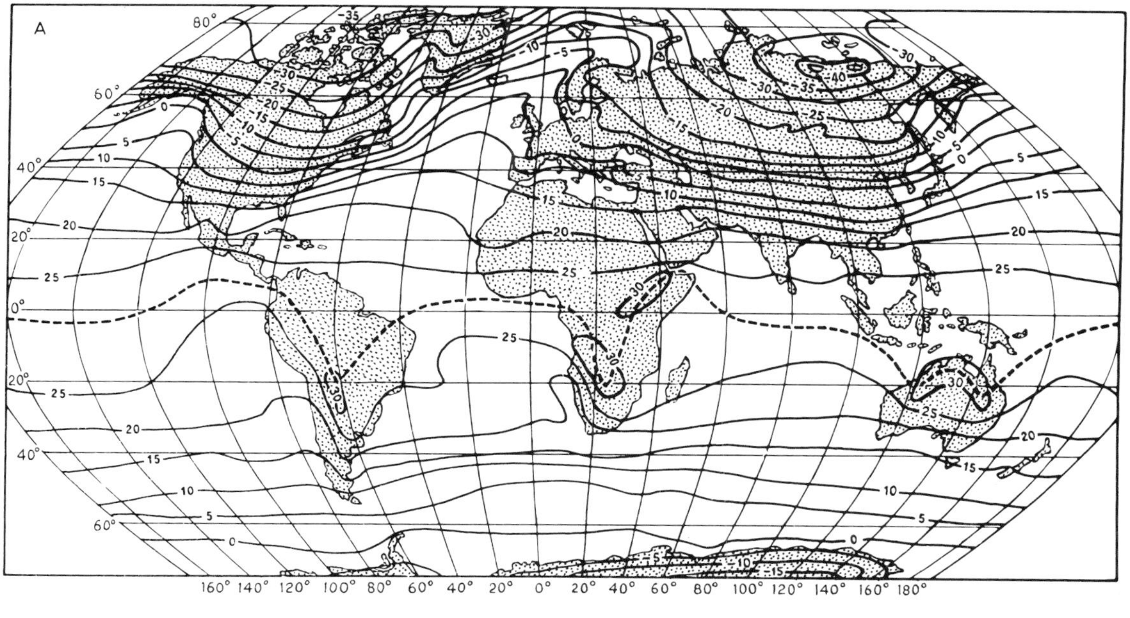

Figure 1.24B Mean sea-level temperatures (°C) in July. The position of the thermal equator is shown approximately by the dashed line.

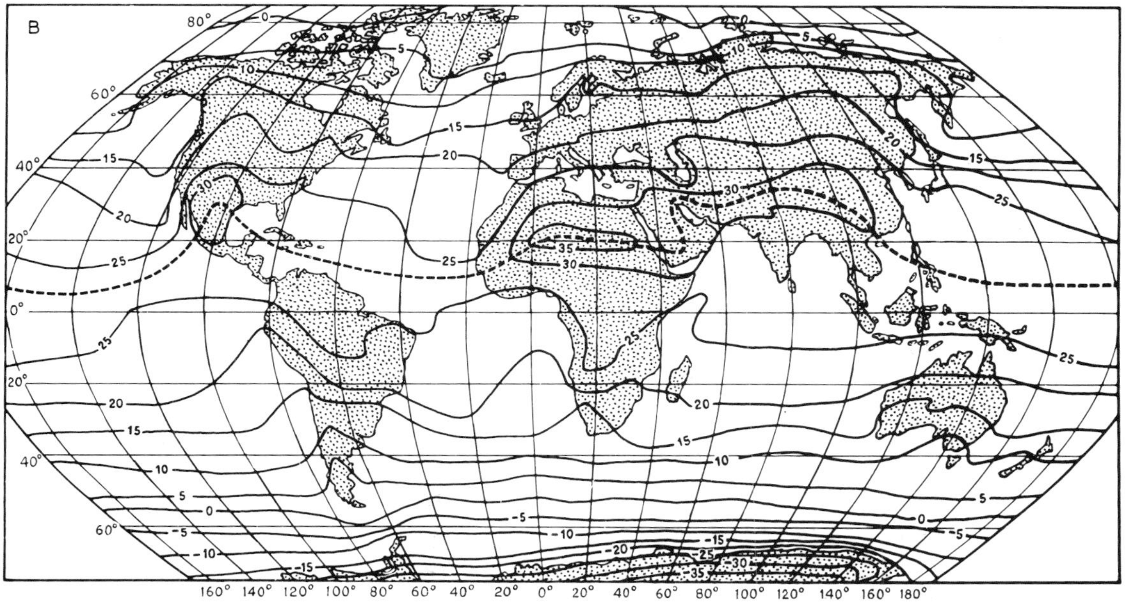

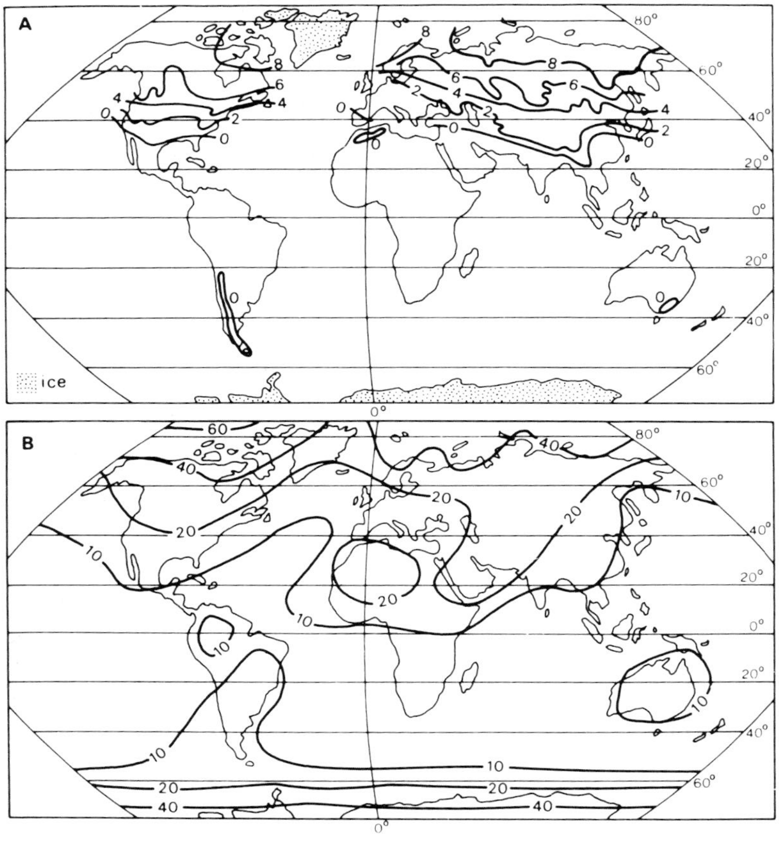

Figure 1.25 Surface albedo of the earth. A: Average snow-cover duration in months; B: Annual average surface albedo (per cent).

Source: After Hummel and Reck; from Henderson-Sellers and Wilson 1983.

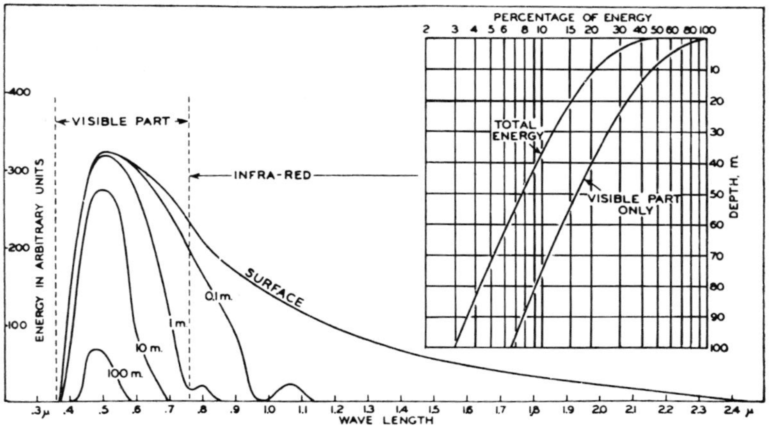

Figure 1.26 Schematic representation of the energy spectrum of the sun's radiation (in arbitrary units) which penetrates the sea surface to depths of 0.1, 1, 10 and 100 m. This illustrates the absorption of infra-red radiation by water, and also shows the depths to which visible (light) radiation penetrates.

Source: From Sverdrup 1945.

Figure 1.27 Mean temperatures for the upper 100 m of the North Sea for February, May, August and November.

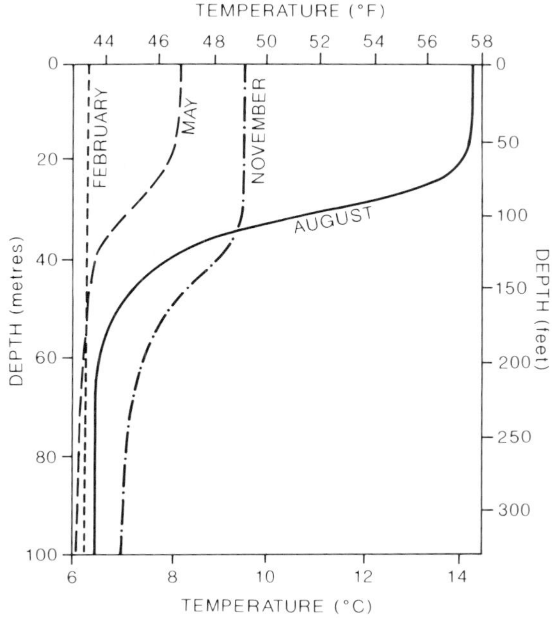

Source: From Lumb 1961 (*Crown Copyright Reserved*).

Figure 1.28 Annual variation of temperature at different depths in soil at Kaliningrad (*above*) and in the water of the Bay of Biscay (at approximately 47°N, 12°W) (*below*), illustrating the relatively deep penetration of solar energy into the oceans as distinct from that into land surfaces. The bottom figure shows the temperature deviations from the annual mean for each depth.

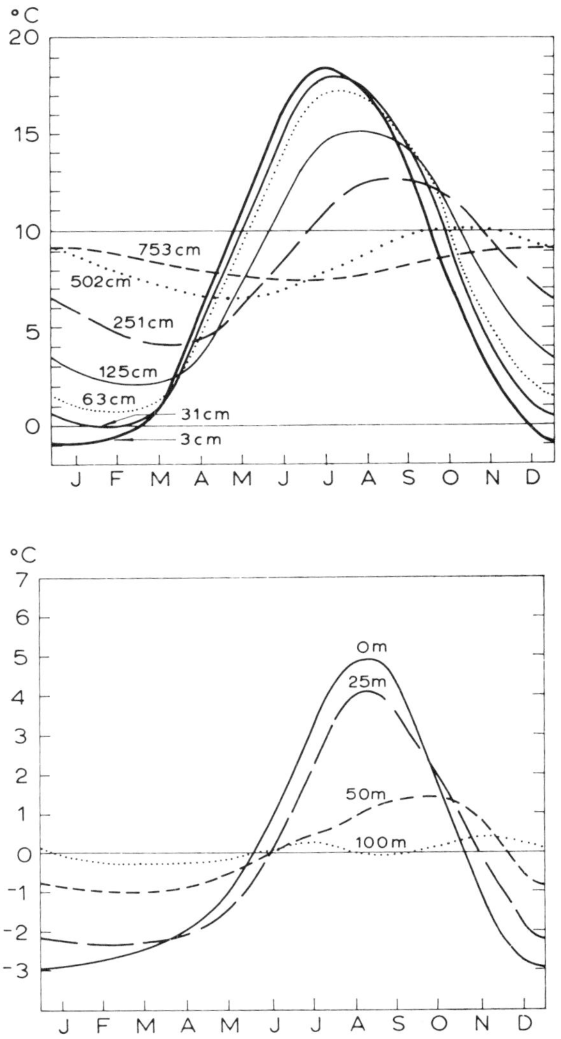

Sources: From Geiger 1965 and Sverdrup 1945.

Kaliningrad (Königsberg) and sea temperature deviations from the annual mean at various depths in the Bay of Biscay. Heat transmission in the soil is carried out almost wholly by conduction and the degree of conductivity varies with the moisture content and porosity of each particular soil.

Air is an extremely poor conductor and for this reason a loose, sandy soil surface heats up rapidly by day, as the heat is not conducted away. Increased soil moisture tends to raise the conductivity by filling the soil pores, but too much moisture increases the soil's heat capacity, thereby reducing the temperature response. The relative depths over which the annual and diurnal temperature variations are effective in wet and dry soils are roughly as follows:

	Diurnal variation	*Annual variation*
Wet soil	0.5 m	9 m
Dry sand	0.2 m	3 m

However, the *actual* temperature change is greater in dry soils. For example, the following values of diurnal temperature range have been observed during clear summer days at Sapporo, Japan:

	Sand	*Loam*	*Peat*	*Clay*
Surface	40°C	33°C	23°C	21°C
5 cm	20	19	14	14
15 cm	7	6	2	4

The different heating qualities of land and water are also partly accounted for by their different *specific heats*. The specific heat (c) of a substance can be represented by the number of thermal units required to raise a unit mass of it through 1°C. In cgs units the specific heat of water is conveniently 1.0 cal g^{-1} deg^{-1} (4184 J kg^{-1} K^{-1}). The specific heat of water is much greater than for most other common substances, and water must absorb five times as much heat energy to raise its temperature by the same amount as a comparable mass of dry soil. Thus for dry sand $c = 840$ J kg^{-1} K^{-1} (0.2 cal g^{-1} deg^{-1}).

If unit volumes of water and soil are considered the heat capacity, ρc, of the water, where ρ = density ($\rho c = 4.18 \times 10^6$ J m^{-3} K^{-1}, or 1.0 cal cm^{-3} deg^{-1}), exceeds that of the sand approximately threefold ($\rho c = 1.3 \times 10^6$ J m^{-3} K^{-1}, or 0.3 cal cm^{-3} deg^{-1}) if the sand is dry and twofold if it is wet. When this water is cooled the situation is reversed, for then a large quantity of heat is released. A metre-thick layer of sea water being cooled by as little as 0.1°C will release enough heat to raise the temperature of approximately a 30-m thick air layer by 10°C (18°F). In this way the oceans act as a very effective reservoir for much of the world's heat. Similarly, evaporation of sea water causes a large heat expenditure because a great amount of energy is needed to evaporate even a small quantity of water (see Chapter 2, A).

The thermal role of the ocean is an important and complex one. The ocean has three thermal layers:

1 The seasonal boundary, or upper mixed, layer, lying above the thermocline. This is less than 100 m deep in the tropics but is hundreds of metres deep in the sub-polar seas. It is subject to annual thermal mixing from the surface.
2 The warm water sphere or lower mixed layer. This underlies '1' and slowly exchanges heat with it down to many hundreds of metres.
3 The deep ocean. This contains some 80 per cent of the total oceanic water volume and exchanges heat with '1' in polar seas.

This vertical thermal circulation allows global heat to be conserved in the oceans, thus damping down the global effects of climatic change produced by thermal forcing (see Chapter 8). The time for heat energy to diffuse within the upper mixed layer is 2–7 months, within the lower mixed layer 7 years, and within the deep ocean upwards of 300 years. The comparative figure for the outer thermal layer of the solid earth is only 11 days.

These differences between land and sea help to produce what is termed *continentality*. Continentality implies, first, that a land surface heats and cools much quicker than that of an ocean. Over the land the lag between maximum and minimum periods of radiation and the maximum and minimum surface temperatures is only 1 month, but over the ocean and at coastal stations the lag is as much as 2 months (see Figure 1.29). Second, the annual and diurnal ranges of temperature are greater in continental than in coastal locations. Figure 1.29 illustrates the annual variation of temperature at Winnipeg and Stornoway, while Figure 1.36C shows the diurnal ranges experienced in continental and

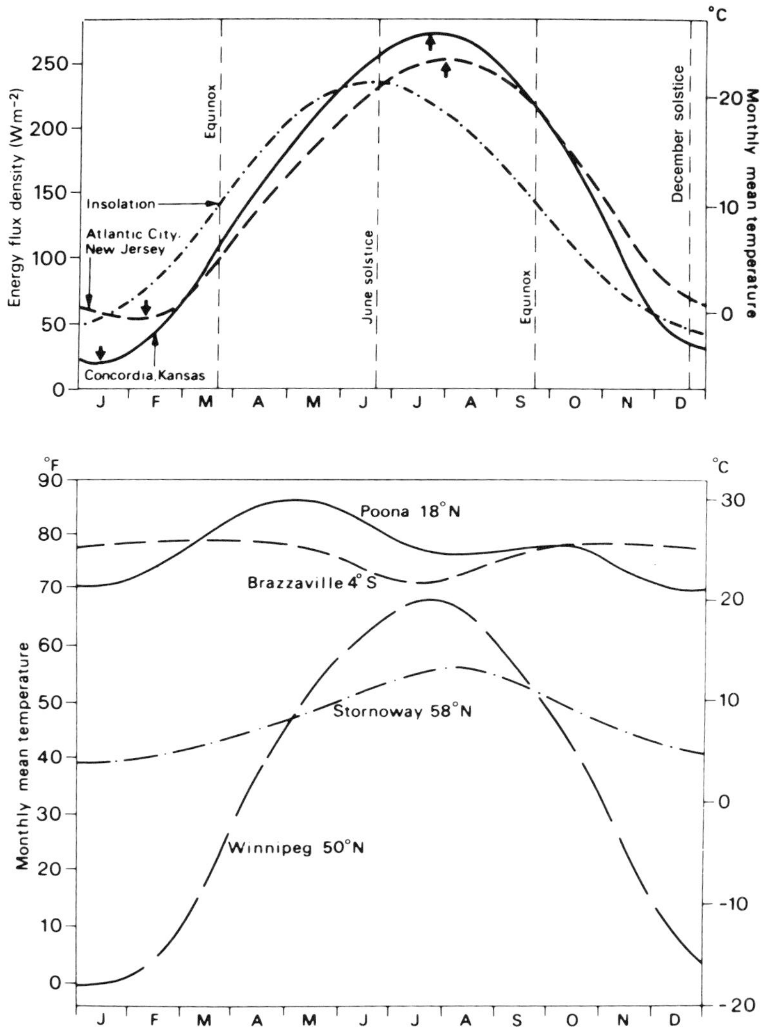

Figure 1.29 Mean annual temperature regimes in various climates and the relationships with solar radiation. *Above:* Temperatures at maritime (Atlantic City) and continental (Concordia, Kansas) locations in the middle latitudes. A curve of representative solar radiation is also given. Maximum and minimum points are indicated on the temperature curves, illustrating the respective time lags behind the radiation curve.† *Below:* Mean annual temperature regimes for Poona (Monsoon), Brazzaville (Equatorial), Stornoway (Temperate maritime) and Winnipeg (Temperate continental).

Source: †Data from Trewartha; after Strahler 1965.

maritime areas. This is described more fully below (see pp. 39–40). The third effect of continentality results from the global distribution of the land masses. The small sea area of the northern hemisphere causes the northern hemisphere summer to be warmer but its winters colder on the average than those of the southern hemisphere (summer, 22.4°C (72.3°F) versus 17.1°C (62.7°F); winter, 8.1°C (46.5°F) versus 9.7°C (49.5°F)). Heat storage in the oceans causes them to be warmer in winter and cooler in summer than land in the same latitude, although ocean currents give rise to some local departures from this rule. The distribution of temperature anomalies for the latitude in January and July (Figure 1.30) illustrates the significance of continentality and also the influence of the warm drift currents in the North Atlantic and the North Pacific in winter (compare Figure 3.42).

Sea temperatures can now be estimated by the use of infra-red satellite imagery (see E, this chapter). Plate 4 is an infra-red photograph taken at night of the south-east coast of the

Figure 1.30 World temperature anomalies (i.e. the difference between recorded temperatures (°C) and the mean for that latitude) for January and July. Solid lines indicate positive, and dashed lines negative, anomalies.

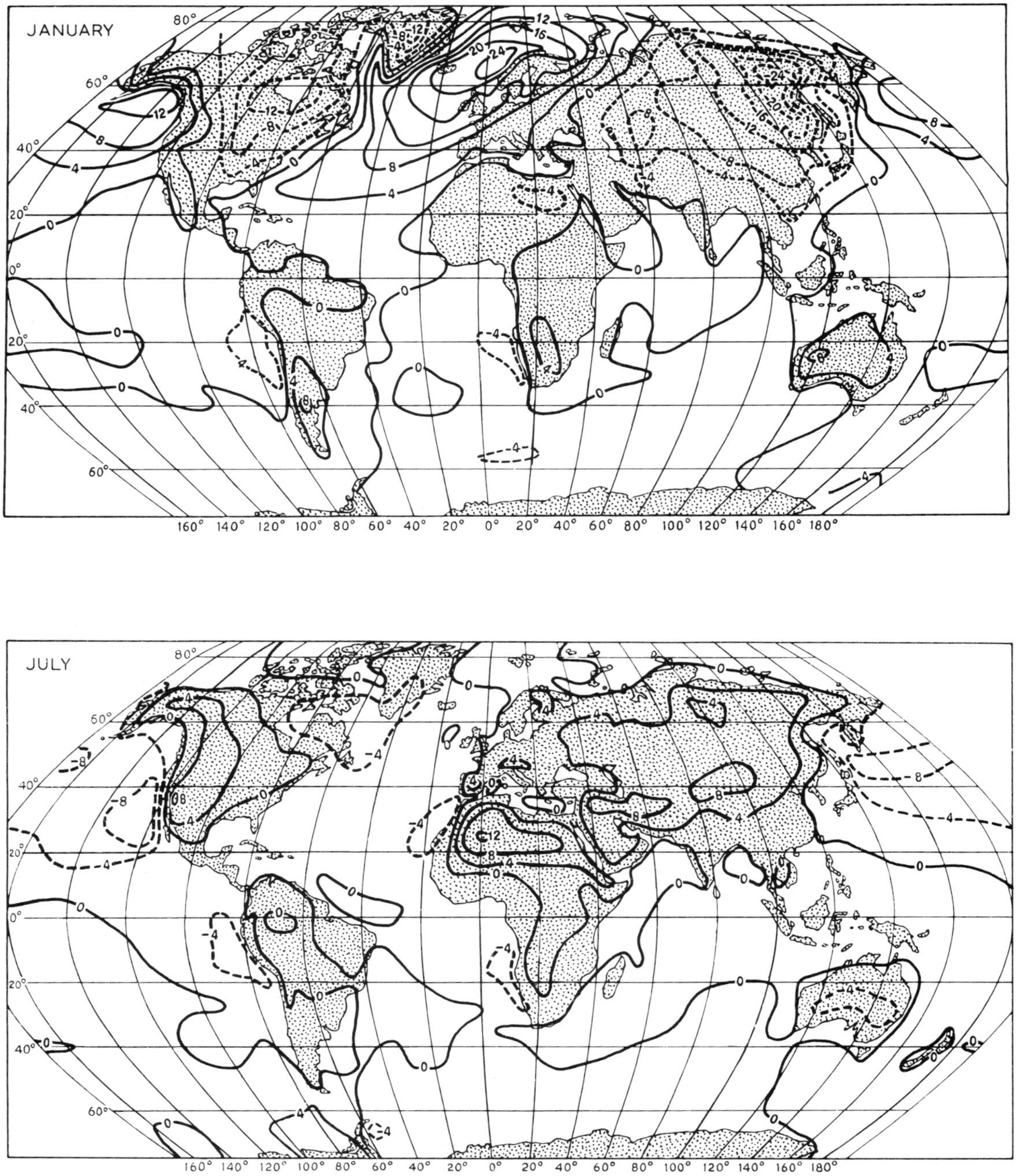

United States in which sea surface temperatures appear in various shades of grey with the darkest areas representing the relatively warm, meandering Gulf Stream. From such images, maps of sea-surface temperatures are now routinely constructed, as illustrated in Figure 1.31.

6 Effect of elevation and aspect

When we come down to the local scale, even differences in the elevation of the land and its *aspect* (that is the direction which the surface faces) will strikingly control the amount of solar radiation received.

Obviously some slopes are more exposed to the sun than others, whereas really high elevations that have a much smaller mass of air above them (see Figure 1.12) receive considerably more direct solar radiation under clear skies than locations near sea level, particularly below 2000–3000 m due to the concentration of water vapour in the lower troposphere (see Figure 1.32). On the average in middle latitudes the intensity of incident solar radiation increases by 5–15 per cent for each 1000 m increase in elevation in the lower troposphere. The difference between sites at 200 and 3000 m in the Alps, for instance, can amount to 70 W m^{-2} on cloudless summer days. However, there is also a correspondingly greater net loss of terrestrial radiation at higher elevations because the low density of the overlying air results in a smaller fraction of the outgoing radiation being absorbed. The overall effect is invariably complicated by the greater cloudiness associated with most mountain ranges and it is therefore impossible to generalize from the limited data at present available.

Figure 1.33 illustrates the effect of aspect and slope angle on theoretical maximum solar radiation receipts at two locations in the northern hemisphere. The general effect of latitude on insolation amounts is clearly shown, but it is also apparent that increasing latitude causes a relatively greater radiation loss for north-facing slopes, as distinct from south-facing ones. The radiation intensity on a sloping surface (I_s) is

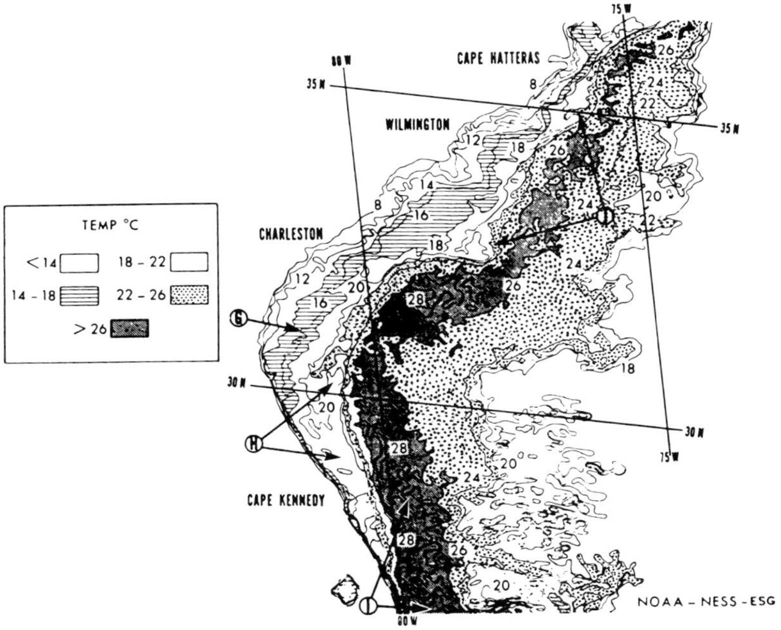

Figure 1.31 Sea-surface temperatures off the east coast of the United States at 0900 GMT on 15 February 1971, estimated from infra-red imagery (see Plate 4). Numbers show the point temperatures which were measured by scanning radiometer. The coldest shelf waters (8°–14°C) are indicated at G; the intermediate slope waters (H) have surface temperatures of 14°–22°C; the Gulf Stream surface (I) is at 26°–28°C and shows steep temperature gradients along certain of its margins (J).

Source: After Rao *et al.*, from WMO 1973.

Figure 1.32 Direct solar radiation as a function of altitude observed in the European Alps. The absorbing effects of water vapour and dust, particularly below about 3000 m, are shown by comparison with a theoretical curve for an ideal atmosphere.

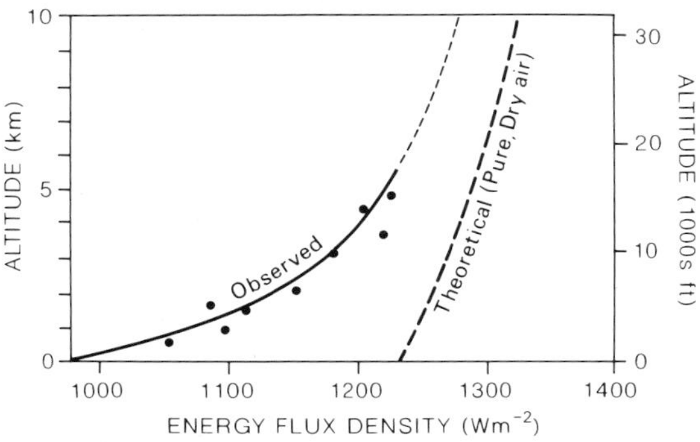

Source: After Albetti, Kastrov, Kimball and Pope; from Barry 1992.

$$I_s = I_0 \cos i$$

where$i$ = the angle between the solar beam and a beam normal to the sloping surface. Relief may also affect the quantity of insolation and the duration of direct sunlight when a mountain barrier screens the sun from valley floors and sides at certain times of day. In many alpine valleys settlement and cultivation are noticeably concentrated on southward-facing slopes (the adret or sunny side), whereas northward slopes (ubac or shaded side) remain forested.

E TERRESTRIAL INFRA-RED RADIATION AND THE GREENHOUSE EFFECT

Radiation from the sun is predominantly short wave, whereas that leaving the earth is long wave, or infra-red, radiation (see Figure 1.14).

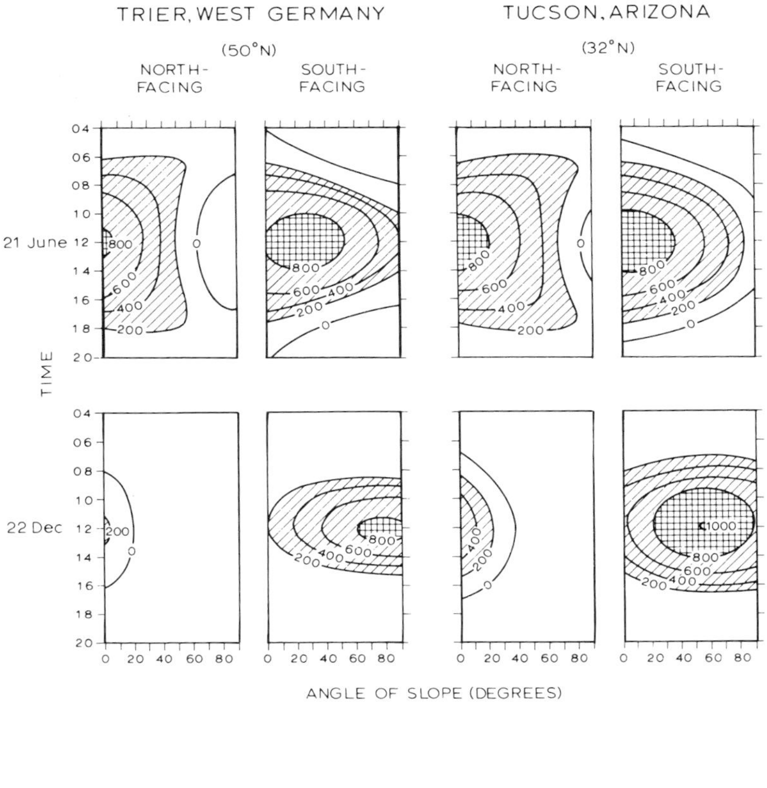

Figure 1.33 Average direct beam solar radiation (W m^{-2}) incident at the surface under cloudless skies at Trier, West Germany, and Tucson, Arizona, as a function of slope, aspect, time of day and season of year.

Source: After Geiger 1965 and Sellers 1965.

The infra-red emission from the surface is slightly less than that from a black body at the same temperature and, accordingly, Stefan's equation (see p. 15) is modified by an emissivity coefficient (ε) which is generally between 0.90 and 0.95, i.e. $F = \varepsilon\sigma T^4$. Figure 1.14 shows that the atmosphere is highly absorbent to infra-red radiation (due to the effects of water vapour, carbon dioxide and other trace gases), except between about 8.5 and 13.0 μm – the 'atmospheric window'. The opaqueness of the atmosphere to infra-red radiation, relative to its transparency to short-wave radiation, is commonly referred to as the greenhouse effect. However, in the case of an actual greenhouse the effect of the glass is probably as significant in reducing cooling by restricting the turbulent heat loss as it is in retaining the infra-red radiation.

The total 'greenhouse' effect results from the net infra-red absorption capacity of water vapour, carbon dioxide, and other trace gases – methane (CH_4), nitrous oxide (N_2O) and tropospheric ozone (O_3) – which absorb strongly at wavelengths within the atmospheric window region, in addition to their other absorbing bands (see Figure 1.14 and Table 1.6). Moreover, because concentrations of these trace gases are low, their radiative effects increase approximately linearly with concentration, whereas the effect of CO_2 is related to the logarithm of the concentration. In addition, because of the long atmospheric residence time of methane (10 years) and CFCs (55–116 years), the cumulative effects of human activities will be substantial. It is estimated that between 1765 and 1990, the radiative effect of increased CO_2 concentration was 1.5 W m^{-2}, and of all traces gases 2.5 W m^{-2} (cf. the solar constant value of 1370 W m^{-2}).

The net warming contribution of the natural (non-anthropogenic) greenhouse gases to the mean 'effective' planetary temperature of 255 K (corresponding to the emitted infra-red radiation) is approximately 33 K; of this, water vapour accounts for 21 K, carbon dioxide 7 K, ozone 2 K, and other trace gases (nitrous oxide, methane) about 3 K. The present global mean surface temperature is 288 K, but the surface was considerably warmer during the early evolution of the earth when the atmosphere contained large quantities of methane, water vapour and ammonia. The largely carbon dioxide atmosphere of Venus creates a 500 K greenhouse effect on that planet.

Stratospheric ozone absorbs significant amounts of both incoming ultraviolet radiation, harmful to life, and outgoing terrestrial long-

Table 1.6 Influence of greenhouse gases.

Gas	*Centres of main absorption bands (μm)*	*Temperature increase (K) for ×2 present concentration*	*Global warming potential on a weight basis (kg^{-1} of air)†*
Water vapour (H_2O)	6.3–8.0, > 15 (8.3–12.5)*		
Carbon dioxide (CO_2)	(5.2), (10), 14.7	3.0 ± 1.5	1
Methane (CH_4)	6.52, 7.66	0.3–0.4	11
Ozone (O_3)	4.7, 9.6, (14.3)	0.9	
Nitrous oxide (N_2O)	7.78, 8.56, 17.0	0.3	270
Chlorofluoromethanes			
($CF\,Cl_3$)	4.66, 9.22, 11.82	} 0.1	3400
($CF_2\,Cl_2$)	8.68, 9.13, 10.93		7100

Sources: After Campbell; Ramanathan; Lashof and Ahnja; Luther and Ellingson; IPCC 1992.
Notes: *Important in moist atmospheres.
† Refers to direct annual radiative forcing for the surface-troposphere system.

wave re-radiation, so that its overall thermal role is a complex one. Its net effect on earth surface temperatures depends on the elevation at which the absorption occurs, being to some extent a trade-off between short- and long-wave absorption in that:

1 An increase of ozone above about 30 km absorbs relatively more incoming short-wave radiation, causing a net *decrease* of surface temperatures.
2 An increase of ozone below about 25 km absorbs relatively more outgoing long-wave radiation, causing a net *increase* of surface temperatures.

It is worth emphasizing that long-wave radiation is not merely terrestrial in the narrow sense. The atmosphere radiates to space and clouds are particularly effective since they act as black bodies. For this reason cloudiness and cloud-top temperature can be mapped from satellites by day and night using infra-red sensors (see Plates 2 and 21). Radiative cooling of cloud layers averages about 1.5°C per day.

For the globe as a whole, satellite measurements for April 1985 show that in cloud-free conditions the mean reflected solar radiation is approximately 286 W m^{-2}, whereas the emitted terrestrial radiation is 266 W m^{-2}. Including cloud-covered areas, the corresponding global values are 241 W m^{-2} and 235 W m^{-2}, respectively. Clouds reduce the reflected solar radiation by 45 W m^{-2}, but reduce the emitted radiation only by 31 W m^{-2}. Hence, global cloud cover causes a net radiative loss of 14 W m^{-2}, due to the dominance of cloud albedo reducing short-wave radiation absorption. In lower latitudes, this effect is much larger (up to −50 to −100 W m^{-2}) whereas in high latitudes the two factors are close to balance, or the increased infra-red absorption by clouds may lead to a small positive value. These results are important in terms of changing concentrations of greenhouse gases since the net radiative forcing by cloud cover is four times that expected from CO_2 doubling (see Chapter 8).

F HEAT BUDGET OF THE EARTH

We can now summarize the net effect of the transfers of energy in the earth-atmosphere system averaged over the globe and over an annual period.

The incident solar radiation averaged over the globe is

$$\text{Solar constant} \times \pi r^2/4\pi r^2$$

where r = radius of the earth and $4\pi r^2$ is the surface area of a sphere. This figure is approximately 342 W m^{-2}, or 11 × 10^9 J m^{-2} yr^{-1} (10^9 J = GJ); for convenience we will regard it as 100 units. Referring to Figure 1.34, incoming radiation is absorbed in the stratosphere (3 units) by ozone mainly, and 18 units are absorbed in the troposphere by carbon dioxide (1), water vapour (12), dust (2) and water droplets in clouds (3). Twenty-one units are reflected back to space from clouds which cover about 60 per cent of the earth's surface, on average. A further 4 units are similarly reflected from the surface and 6 units are returned by atmospheric scattering. The total reflected radiation is the *planetary albedo* (31 per cent or 0.31). The remaining 48 units reach the earth either directly (27) or as diffuse radiation (21) transmitted via clouds or by downward scattering.

The pattern of outgoing terrestrial radiation is quite different (see Figure 1.34). The black-body radiation, assuming a mean surface temperature of 287 K is equivalent to 113 units of infra-red (long-wave) radiation. This is possible because most of the outgoing radiation is re-absorbed by the atmosphere; the *net* loss of infra-red radiation is only 16 units. These exchanges represent a time-averaged state for the whole globe. Recall that solar radiation affects only the sunlit hemisphere, where the incoming exceeds 342 W m^{-2}. Conversely, on the night-time hemisphere no solar radiation is received. Infra-red exchanges continue, however, due to the accumulated heat in the ground. Only about 6 units escape through the atmospheric window directly from the surface, but the atmosphere radiates 63 units to space

Figure 1.34 The balance of the atmospheric energy budget. The transfers are explained in the text. Solid lines indicate energy gains by the atmosphere and surface in the left-hand diagram and the troposphere in the right-hand diagram. The exchanges are referred to 100 units of incoming solar radiation at the top of the atmosphere (equal to 342 W m^{-2} or 0.5 cal cm^{-2} min^{-1}).

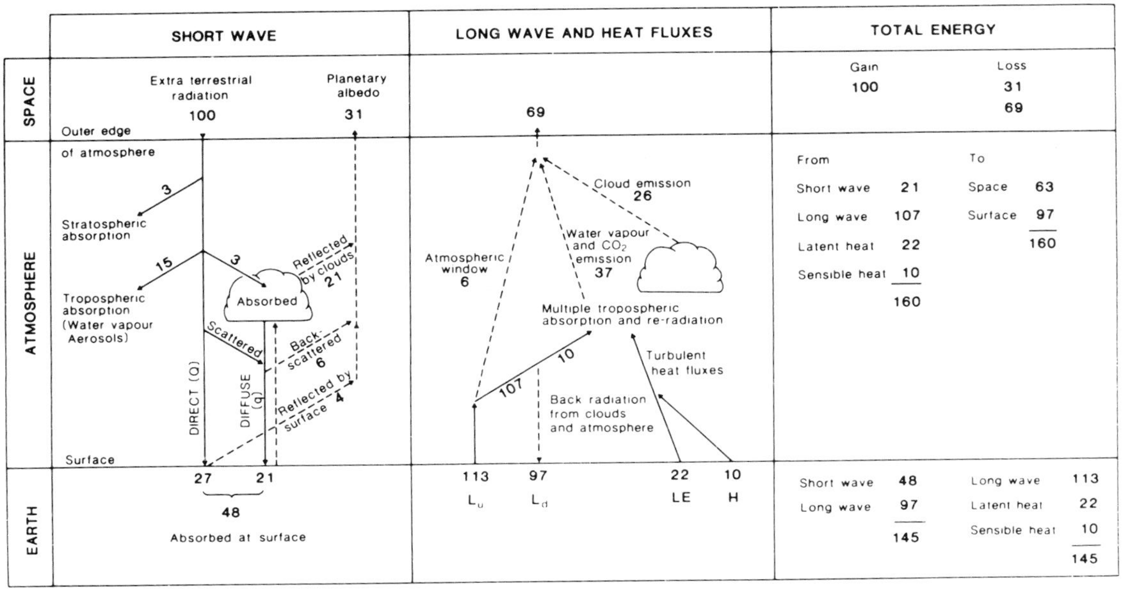

Source: Data after Fröhlich and London 1985.

(37 from the emission by water vapour and CO_2 in the atmosphere and 26 from cloud emission), giving a total of 69 units, as well as re-radiating 97 units back to the surface (L_d); $L_u + L_d = L_n$ is negative.

These radiation transfers can be expressed symbolically:

$$R_n = (Q + q)(1 - a) + L_n$$

where R_n = net radiation, $(Q + q)$ = global solar radiation, a = albedo and L_n = net long-wave radiation. At the surface R_n = 32 units. This surplus is conveyed to the atmosphere by the turbulent transfer of sensible heat, or enthalpy (10 units) and latent heat (22 units).

$$R_n = LE + H$$

where H = sensible heat transfer and LE = latent heat transfer. There is also a flux of heat into the ground (see D.5, this chapter), but for annual averages this is approximately zero.

Figure 1.34 summarizes the total balances at the surface (± 145 units) and for the atmosphere (± 160 units). The total absorbed solar radiation and emitted radiation for the entire earth-atmosphere system is estimated to be ± 7 GJ m^{-2} yr^{-1} (± 69 units). These estimates are still rather crude, however, and various uncertainties are still to be resolved.

Satellite measurements now provide global views of the energy balance at the top of the atmosphere. The incident solar radiation is almost symmetrical about the equator in the annual mean (cf. Table 1.4). The mean annual totals on a horizontal surface at the top of the atmosphere are approximately 420 W m^{-2} at the equator and 180 W m^{-2} at the poles. The distribution of the planetary albedo (see Figure 1.35A) shows the lowest values over the low-latitude oceans compared with the more persistent areas of cloud cover over the continents. The highest values are over the polar

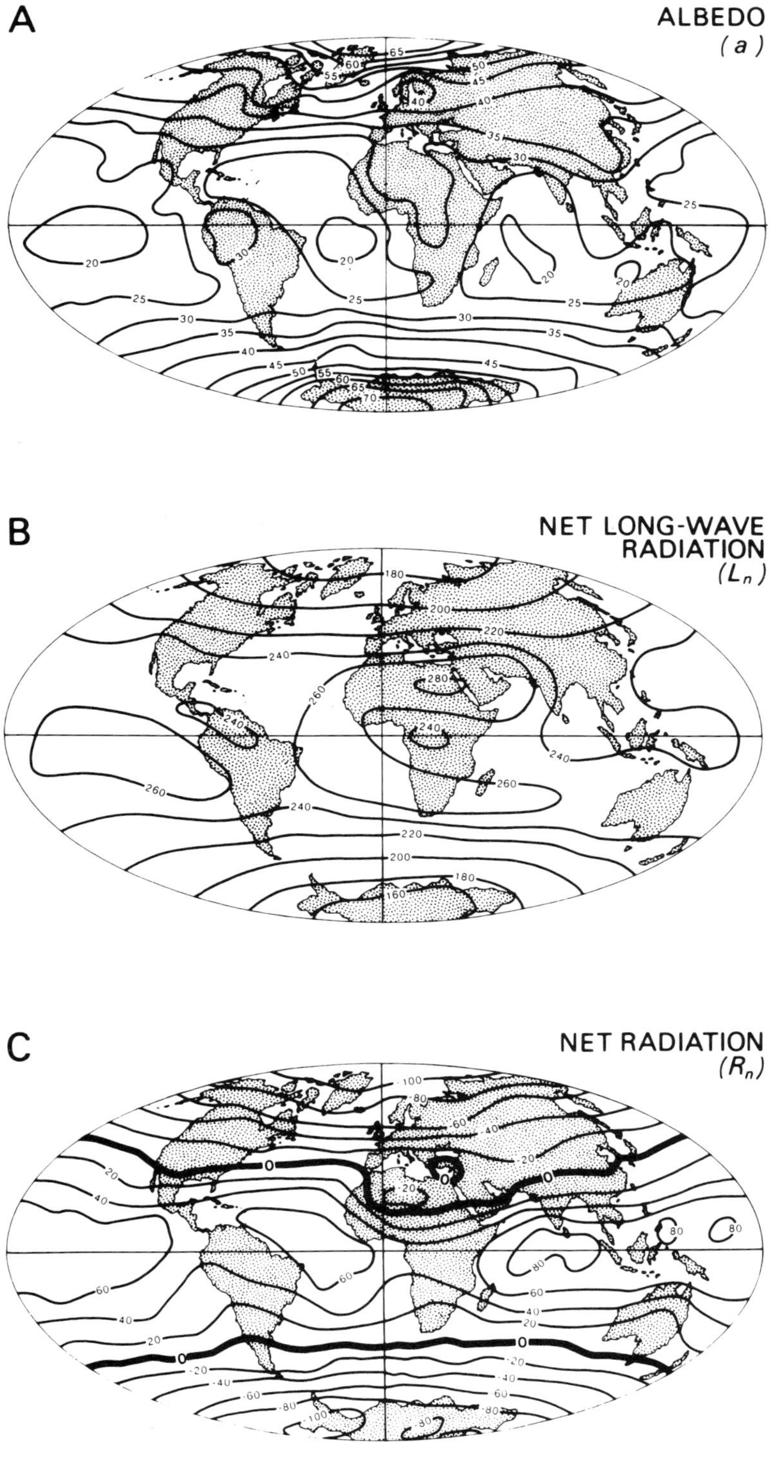

Figure 1.35 Average annual planetary albedo (per cent) (A), net long-wave radiation (W m^{-2}) (B) and net radiation (W m^{-2}) (C) on a horizontal surface at the top of the atmosphere.

Source: From satellite data analysed by Stephens *et al.* 1981.

icecaps. Correspondingly, the net (outgoing) long-wave radiation (see Figure 1.35B) shows the smallest losses where the temperatures are lowest and largest losses over the largely clear skies of the Saharan desert surface and over low-latitude oceans. The net radiation (see Figure 1.35C) shows that the earth-atmosphere system achieves balance about latitude 30°N. The consequences of a low-latitude energy surplus and a high-latitude deficit are examined in section G (this chapter).

The annual and diurnal variations of temperature are directly related to the local energy budget. Under clear skies, in middle and lower latitudes, the diurnal regime of radiative exchanges generally shows a midday maximum of absorbed solar radiation (see Figure 1.36A). A maximum of infra-red (long-wave) radiation (see Figure 1.14) is also emitted by the heated ground surface at midday when it is warmest. The atmosphere re-radiates infra-red radiation downward but there is a net loss at the surface

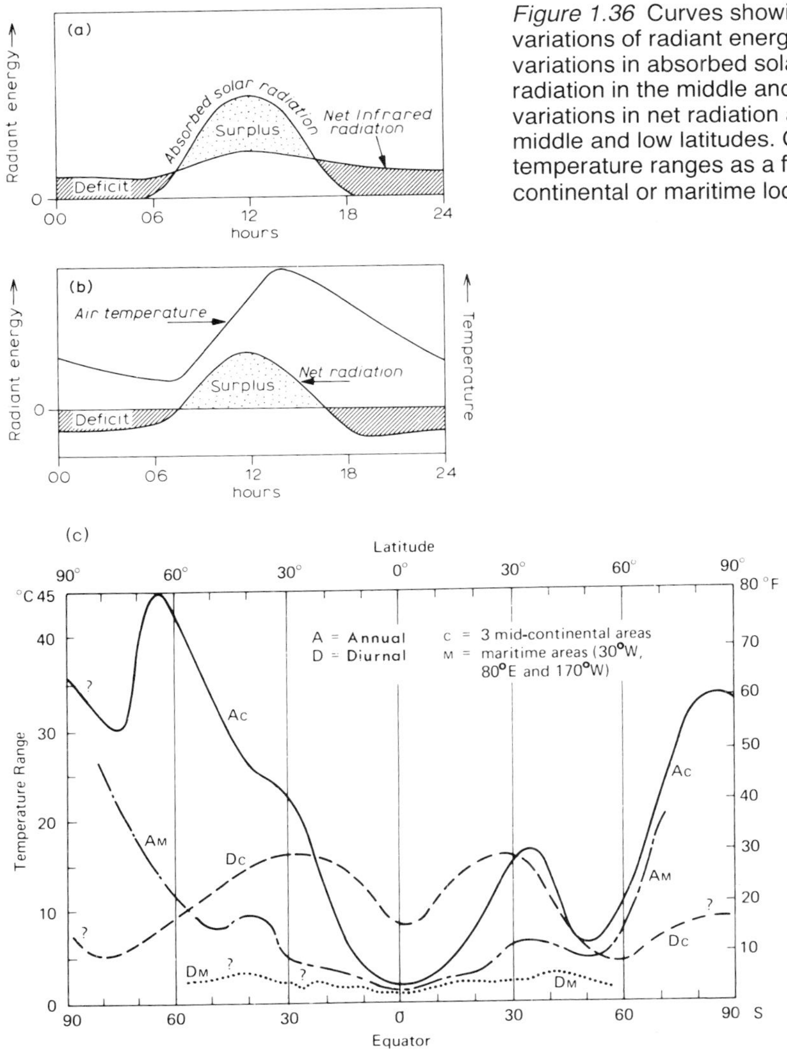

Figure 1.36 Curves showing diurnal and annual variations of radiant energy and temperature. A: Diurnal variations in absorbed solar radiation and infra-red radiation in the middle and low latitudes. B: Diurnal variations in net radiation and air temperature in the middle and low latitudes. C: Annual and diurnal temperature ranges as a function of latitude and of continental or maritime location.

Source: From Paffen 1967.

(L_n). The difference between the absorbed solar radiation and L_n is the net radiation, R_n; this is generally positive between about an hour after sunrise and an hour or so before sunset with a midday maximum. The delay in the occurrence of the maximum air temperature until about 1400 hours local time (see Figure 1.36B) is caused by the gradual heating of the air by convective transfer from the ground. Minimum R_n occurs in the early evening when the ground is still warm; there is a slight increase thereafter. The temperature decrease after midday is slowed by heat supplied from the ground. Minimum air temperature occurs shortly after sunrise due to the lag in the transfer of heat from the surface to the air. The annual pattern of the net radiation budget and temperature regime is closely analogous to the diurnal one.

There are marked latitudinal variations in the diurnal and annual ranges of temperature. Broadly, the annual range is a maximum in higher latitudes, with extreme values about 65°N related to the effects of continentality in Asia and North America. The diurnal range reaches a maximum at the tropics over land areas, but it is in the equatorial zone that the diurnal variation of heating and cooling exceeds the annual one (see Figure 1.36C). This is of course related to the small seasonal change in solar elevation angle at the equator. From the point of view of the total energy budget, the atmosphere, oceans and upper crustal skin of the earth form a complex system of storages and transfers. The global climate as a whole is dominated by the relatively small energy storages provided by the crustal skin and the atmosphere, compared with that of the oceans (see Chapter 3).

G ATMOSPHERIC ENERGY AND HORIZONTAL HEAT TRANSPORT

So far, we have described the gases and other constituents that make up our atmosphere, and have given some account of the earth's heat budget. We have already referred to two forms of energy: internal (or heat) energy due to the motion of individual air molecules and latent energy which is released by condensation of water vapour. Two other forms of energy are important: geopotential energy due to gravity and height above the surface and kinetic energy associated with air motion.

Geopotential and internal energy are interrelated, since the addition of heat to an air column not only increases its internal energy, but adds to its geopotential as a result of the vertical expansion of the air column. In a column extending to the top of the atmosphere the geopotential is approximately 40 per cent of the internal energy. These two are therefore usually considered together and termed the total potential energy (*PE*). For the whole atmosphere

potential energy ≈ 10^{24} J (23.4×10^{22} cal)
kinetic energy ≈ 10^{20} J

In a later section (Chapter 3, F) we shall see how energy is transferred from one form to another, but here we need only be concerned with heat energy. It is apparent that the receipt of heat energy is very unequal geographically and that this must lead to great lateral transfers of energy across the surface of the earth. Much present-day meteorological research is focused on these transfers, since undoubtedly they give rise, at least indirectly, to the observed patterns of global weather and climate.

The amounts of energy received at different latitudes vary substantially, the equator on the average receiving 2.5 times as much annual solar energy as the poles. Clearly if this process were not modified in some way the variations in receipt would cause a massive accumulation of heat within the tropics (associated with gradual increases of temperature) and a corresponding deficiency at the poles. Yet this does not seem to happen, and the earth as a whole is roughly in a state of thermal equilibrium in so far as no one region is obviously gaining heat at the expense of another. Some authors believe the Ice Ages to have been an exception to this rule. One explanation of this equilibrium could be that for each region of the world there is an equalization

between the amount of incoming and outgoing radiation. Observation shows that this is not so (see Figure 1.37), however, for, whereas incoming radiation varies appreciably with changes in latitude, being highest at the equator and declining to a minimum at the poles, outgoing radiation has a more even latitudinal distribution owing to the rather small variations in atmospheric temperature. Some other explanation therefore becomes necessary.

1 The horizontal transport of heat

If the net radiation for the whole earth-atmosphere system is calculated, it is found that there is a positive budget between 35°S and 40°N as shown in Figures 1.35C and 1.38. The latitudinal belts in each hemisphere separating the zones of positive and negative net radiation budgets oscillate dramatically with season (see Figure 1.39). As the tropics do not get progressively hotter or the high latitudes colder, a redistribution of world heat energy must constantly occur, taking the form of a continuous movement of energy from the tropics to the poles. In this way the tropics shed their excess heat and the poles are not allowed to reach extremes of cold. If there were no meridional interchange of heat, a radiation balance at each latitude would only be achieved if the equator were 14°C warmer and the North Pole 25°C colder than now. This poleward heat transport takes place within the atmosphere and oceans, and it is estimated that the former accounts for approx-

Figure 1.37 A meridional illustration of the balance between incoming solar radiation and outgoing radiation from the earth and atmosphere* in which the zones of permanent surplus and deficit are maintained in equilibrium by a poleward energy transfer.†

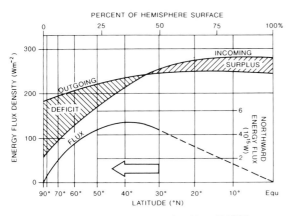

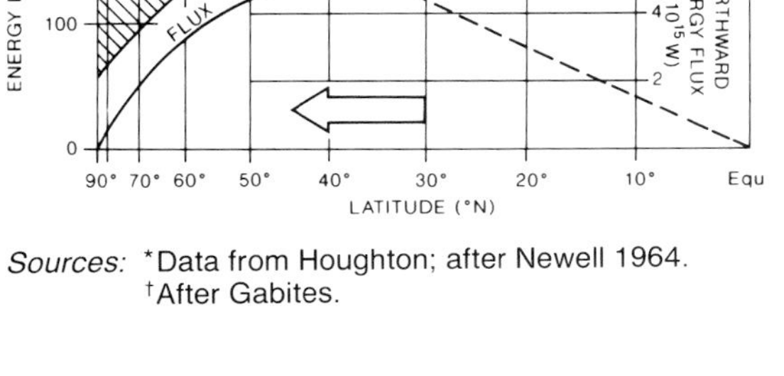

Sources: *Data from Houghton; after Newell 1964.
†After Gabites.

Figure 1.38 A: Net radiation balance for the earth's surface of 101 W m^{-2} (incoming solar radiation of 156 W m^{-2}, minus outgoing long-wave energy to the atmosphere of 55 W m^{-2}); for the atmosphere of −101 W m^{-2} (incoming solar radiation of 84 W m^{-2}, minus outgoing long-wave energy to space of 185 W m^{-2}); and for the whole earth-atmosphere system of zero. B: The average annual latitudinal distribution of the components of the poleward energy transfer (in 10^{15} W) in the earth-atmosphere system.

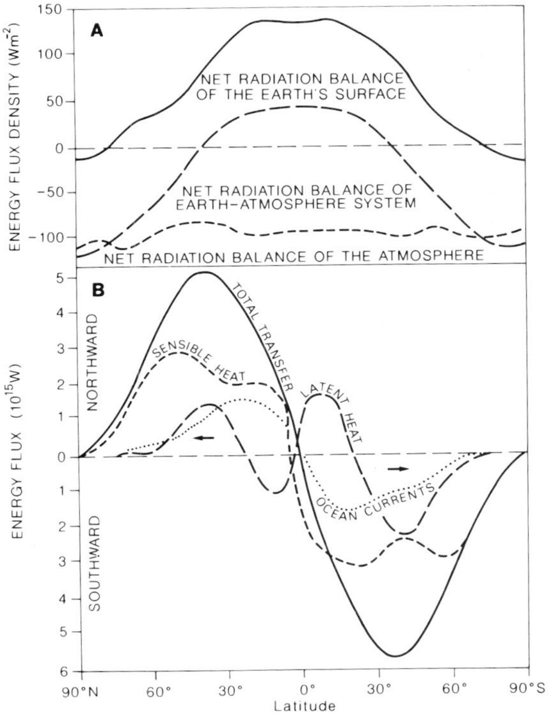

Source: From Sellers 1965.

Figure 1.39 Values of average net radiation (R_n) ($W\ m^{-2}$) for the earth-atmosphere system in January and July; the heavy zero lines indicate equality of incoming and outgoing radiation.

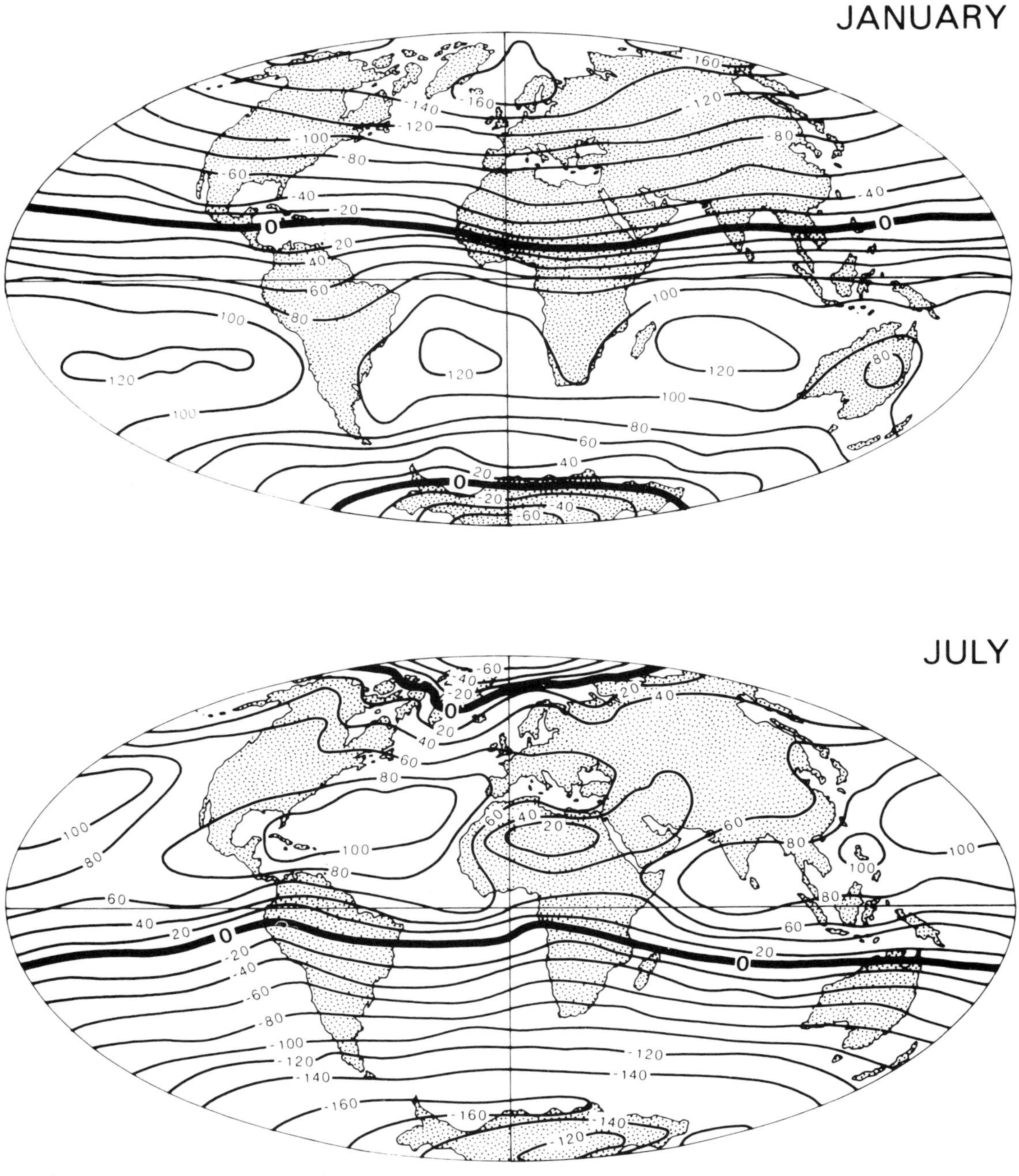

Source: From Stephens *et al.* 1981.

imately two-thirds of the required total. The horizontal transport (*advection* of heat) occurs in the form of both latent heat (that is water vapour which subsequently condenses) and sensible heat (that is warm air masses) (see Figure 1.38B). It varies in intensity according to the latitude and the season. Figure 1.38B shows the mean annual pattern of energy transfer by the three mechanisms. The latitudinal zone of maximum total transfer rate is found between latitudes 35° and 45° in both hemispheres, although the patterns for the individual components are quite different from one another. The latent heat transport, which occurs almost wholly in the lowest 2 or 3 km, reflects the global wind belts on either side of the subtropical high-pressure zones (see Chapter 3, E). The more important meridional transfer of sensible heat has a double maximum not only latitudinally but also in the vertical plane, where there are maxima near the surface and at about 200 mb. The high-level transport is particularly significant over the subtropics, whereas the primary latitudinal maximum about 50° to 60°N is related to the travelling low-pressure systems of the westerlies.

The intensity of the poleward energy flow is closely related to the meridional (that is, north–south) temperature gradient. In winter this temperature gradient is at a maximum and in consequence the hemispheric air circulation is most intense. The nature of the complex transport mechanisms will be discussed in Chapter 3, F.

As shown in Figure 1.38B, ocean currents account for a significant proportion of the poleward heat transfer in low latitudes. Indeed, recent satellite estimates of the required total poleward energy transport indicate that the previous figures are too low. The ocean transport may be 47 per cent of the total at 30°–35°N and as much as 74 per cent at 20°N; the Gulf Stream and Kuro Shio currents are particularly important. As a result of this factor, the energy budget equation for an ocean area must be expressed as

$$R_n = LE + H + G + \Delta A,$$

where ΔA = horizontal advection of heat by currents and G = the heat transferred into or out of storage in the water. The latter is more or less zero for annual averages.

2 Spatial pattern of the heat budget components

The mean latitudinal values of the heat budget components discussed above conceal great spatial variations. Figure 1.40 shows the global distribution of the annual net radiation at the surface. Broadly, its magnitude decreases poleward from about 25° latitude, although as a result of the considerable absorption of solar radiation by the sea, the net radiation is greater over the oceans – exceeding 160 W m^{-2} in latitudes 15°–20° – than over land areas, where it is about 80–105 W m^{-2} in the same latitudes. Net radiation is also rather lower in arid continental areas than in humid ones, because in spite of the increased insolation receipts under clear skies there is at the same time greater net loss of terrestrial radiation.

Figures 1.41 and 1.42 show the annual vertical transfers of latent and sensible heat to the atmosphere. Both maps show that the fluxes are distributed very differently over land and sea. Heat expenditure for evaporation is at a maximum in tropical and subtropical ocean areas, where it exceeds 160 W m^{-2}. It is less near the equator where wind speeds are somewhat lower and the air has a vapour pressure close to the saturation value (see Chapter 2, A). It is clear from Figure 1.41 that the major warm currents considerably augment the evaporation rate. On land the latent heat transfer is greatest in hot, humid regions. It is least in arid areas due to the low precipitation and in high latitudes where there is little available energy.

The largest exchange of sensible heat occurs in the tropical deserts where more than 80 W m^{-2} is transferred to the atmosphere. In contrast to latent heat, the sensible heat flux is generally small over the oceans, only reaching 25–40 W

Figure 1.40 Global distribution of the annual net radiation at the surface, in W m^{-2}.

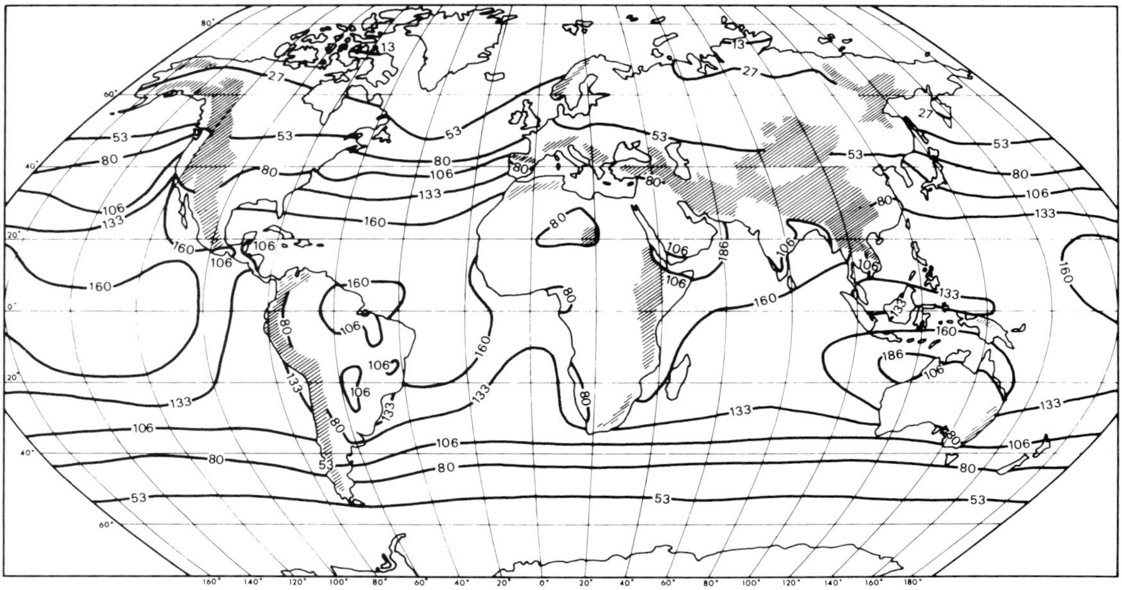

Source: After Budyko *et al.* 1962.

Figure 1.41 Global distribution of the vertical transfer of latent heat, in W m^{-2}.

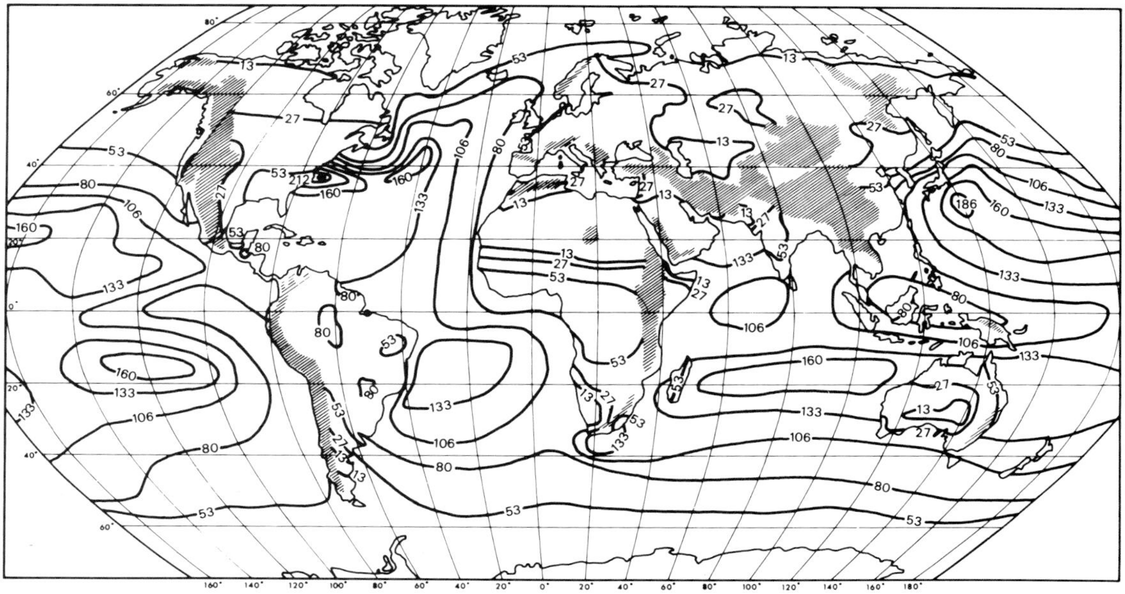

Source: After Budyko *et al.* 1962.

Figure 1.42 Global distribution of the vertical transfer of sensible heat, in W m^{-2}.

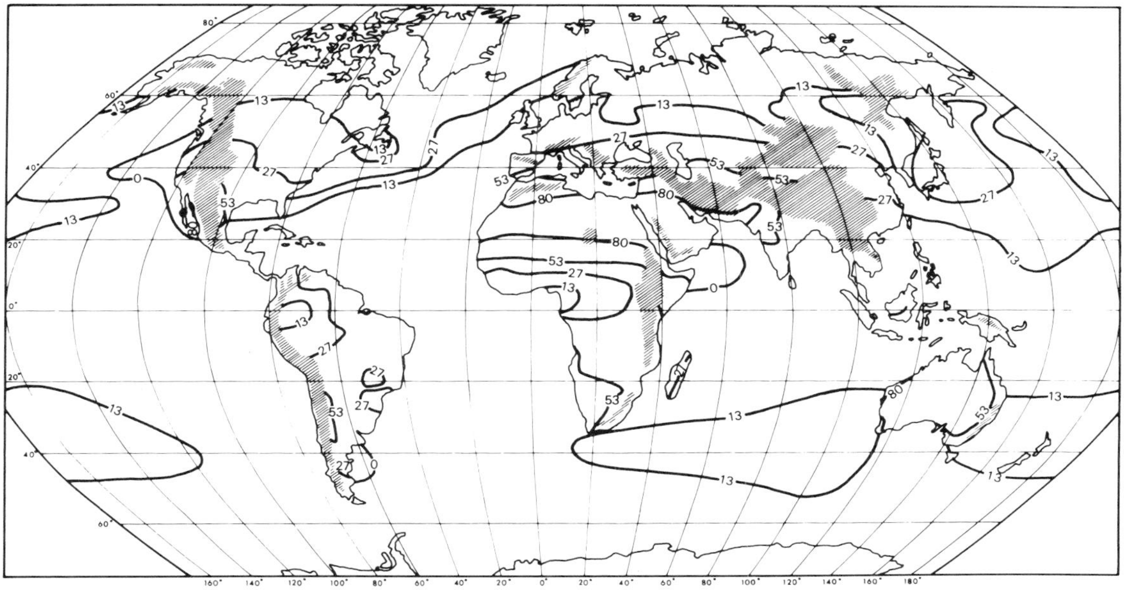

Source: After Budyko *et al.* 1962.

m^{-2} in areas of warm currents. Indeed, negative values occur (transfer *to* the ocean) where warm continental air masses move offshore over cold currents.

H THE LAYERING OF THE ATMOSPHERE

The atmosphere can be divided conveniently into a number of rather well-marked horizontal layers, mainly on the basis of temperature. The evidence for this structure comes from regular RAWINSONDE (radar wind-sounding) balloons, radio-wave investigations, and, more recently, from rocket flights and satellite sounding systems (Plates 2 and 3). Broadly, the pattern (see Figure 1.43) consists of three relatively warm layers (near the surface; between 50 and 60 km; and above about 120 km) separated by two relatively cold layers (between 10 and 30 km; and about 80 km). Mean January and July temperature sections illustrate the considerable latitudinal variations and seasonal trends that complicate the scheme (see Figure 1.44).

1 Troposphere

The lowest layer of the atmosphere is called the troposphere. It is the zone where weather phenomena and atmospheric turbulence are most marked, and contains 75 per cent of the total molecular or gaseous mass of the atmosphere and virtually all the water vapour and aerosols. Throughout this layer there is a general decrease of temperature with height at a mean rate of about 6.5°C/km (or 3.6°F/1000 ft), and the whole zone is capped in most places by a temperature inversion level (i.e. a layer of relatively warm air above a colder one) and in others by a zone that is isothermal with height. The troposphere thus remains to a large extent self-contained because the inversion acts as a 'lid' which effectively limits convection (see Chapter 2, E). This inversion level or weather ceiling is

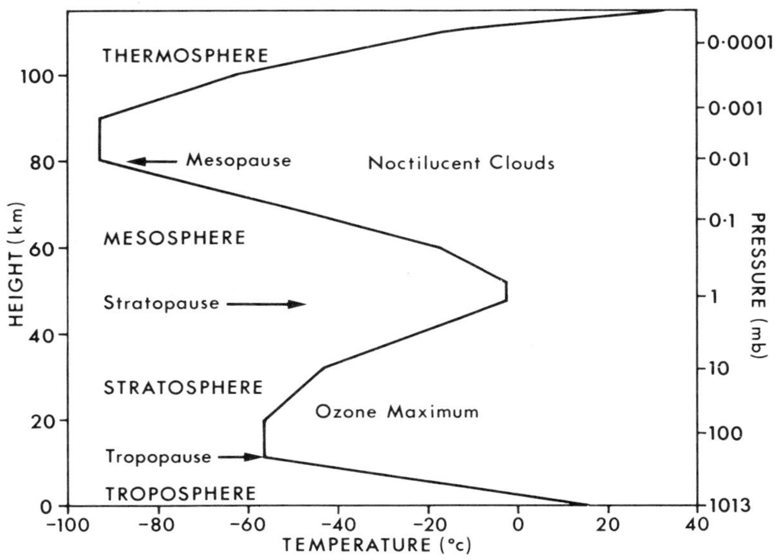

Figure 1.43 The generalized vertical distribution of temperature and pressure up to about 110 km. Note particularly the tropopause and the zone of maximum ozone concentration with the warm layer above.

Source: Based on data in Valley 1965.

called the *tropopause* (see note 7). Its height is not constant, either in space or time. It seems that the height of the tropopause at any point is correlated with sea-level temperature and pressure, which are in turn related to the factors of latitude, season and daily changes in surface pressure. There are marked variations in the altitude of the tropopause with latitude (see Figure 1.44), from about 16 km (10 miles) at the equator where there is great heating and vertical convective turbulence to only 8 km (5 miles) at the poles.

The meridional temperature gradients in the troposphere in summer and winter are roughly parallel, as are the tropopauses (see Figure 1.44), and the strong lower mid-latitude temperature gradient in the troposphere is reflected in the tropopause breaks (see also Figure 3.22). In these zones important interchanges can occur between the troposphere and stratosphere, or vice versa. Traces of water vapour probably penetrate into the stratosphere by this means, while dry, ozone-rich stratosphere air may be brought down into the mid-latitude troposphere. For example, above-average concentrations of ozone are observed in the rear of mid-latitude low-pressure systems where the tropopause elevation tends to be low. Both facts are probably the result of stratospheric subsidence, which warms the lower stratosphere and causes downward transfer of the ozone.

2 Stratosphere

The second major atmospheric layer is the stratosphere which extends upwards from the tropopause to about 50 km (30 miles). Although the stratosphere contains much of the total atmospheric ozone (it reaches a peak density at approximately 22 km), the maximum temperatures associated with the absorption of the sun's ultraviolet radiation by ozone occur at the *stratopause*, where temperatures may exceed 0°C (see Figure 1.44). The air density is much less here so that even limited absorption produces a large temperature increase. Temperatures increase fairly generally with height in summer, with the coldest air at the equatorial tropopause. In winter the structure is more complex with very low temperatures, averaging −80°C, at the equatorial tropopause which is highest at this season. Similar low temperatures

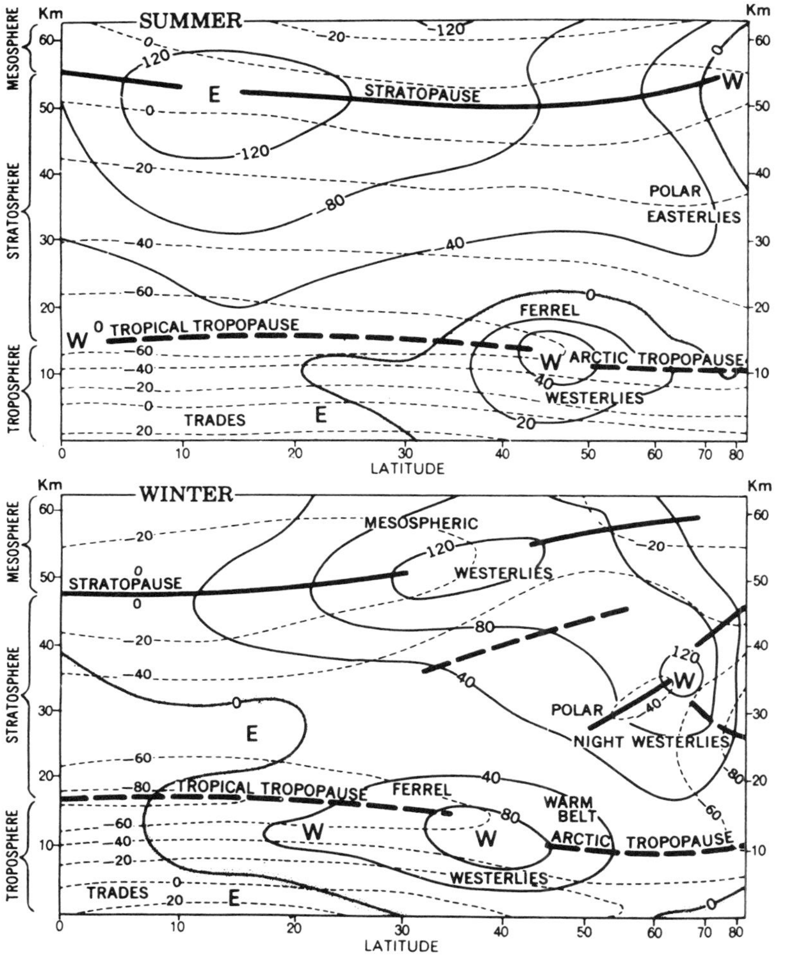

Figure 1.44 Mean zonal (westerly) winds (solid isolines, in knots; negative values from the east) and temperatures (in °C, dashed isolines), showing the broken tropopause near the mean Ferrel jet stream. The heavy black lines denote reversals of the vertical temperature gradient of the tropopause and stratopause. Summer and winter refer to the northern hemisphere.

Source: After Boville (from Hare 1962).
Notes: The term 'Ferrel Westerlies' was proposed by F. K. Hare in honour of W. Ferrel (see p. 124).

are found in the middle stratosphere at high latitudes, whereas over 50°–60°N there is a marked warm region with nearly isothermal conditions at about −45°C to −50°C. Marked seasonal changes of temperature affect the stratosphere. The cold 'polar night' winter stratosphere often undergoes dramatic *sudden warmings* associated with subsidence due to circulation changes in late winter or early spring when temperatures at about 25 km may jump from −80°C to −40°C over a 2-day period. The autumn cooling is a more gradual process. In the tropical stratosphere there is a quasi-biennial (26-month) wind regime, with easterlies in the layer 18 to 30 km for 12 to 13 months, followed by westerlies for a similar period. The reversal begins first at high levels and takes approximately 12 months to descend from 30 to 18 km (10 to 60 mb).

How far these events in the stratosphere are linked with temperature and circulation changes in the troposphere remains a topic of meteorological research. Any interactions that do exist, however, are likely to be complex, otherwise they would already have become evident.

3 Mesosphere

Above the stratopause average temperatures decrease to a minimum of about −90°C (183 K) around 80 km. This layer is commonly termed the mesosphere, although it must be noted that as yet there is no universal acceptance of terminology for the upper atmospheric layers. The layers between the tropopause and the lower thermosphere are now commonly referred to as the *Middle Atmosphere*, with the Upper Atmosphere designating the regions above about 100 km altitude. Above 80 km temperatures again begin rising with height and this inversion is referred to as the 'mesopause'. Molecular oxygen and ozone absorption bands contribute to heating around 85 km altitude. It is in this region that 'noctilucent clouds' are observed over high latitudes in summer. Their presence appears to be due to meteoric dust particles which act as ice crystal nuclei when traces of water vapour are carried upwards by high-level convection caused by the vertical decrease of temperature in the mesosphere. However, their formation is also thought to be related to increasing concentrations of atmospheric methane since they were not apparently observed prior to the Industrial Revolution.

Pressure is very low in the mesosphere, decreasing from about 1 mb at 50 km to 0.01 mb at 90 km.

4 Thermosphere

Above the mesopause atmospheric densities are extremely low, although the tenuous atmosphere still effects drag on space vehicles above 250 km. The lower portion of the thermosphere is composed mainly of nitrogen (N_2) and oxygen in molecular (O_2) and atomic (O) forms, whereas above 200 km atomic oxygen predominates over nitrogen (N_2 and N). Temperatures rise with height, owing to the absorption of extreme ultraviolet radiation (0.125–0.205 μm) by molecular and atomic oxygen, probably approaching 1200 K at 350 km, but these temperatures are essentially theoretical. For example, artificial satellites do not acquire such temperatures because of the rarefied air. 'Temperatures' in the upper thermosphere and exosphere undergo wide diurnal and seasonal variations. They are higher by day and are also higher during a sunspot maximum, although the changes are only represented in varying velocities of the sparse air molecules.

Above 100 km the atmosphere is increasingly affected by cosmic radiation, solar X-rays and ultraviolet radiation which cause *ionization*, or electrical charging, by separating negatively charged electrons from neutral oxygen atoms and nitrogen molecules, leaving the atom or molecule with a net positive charge (an *ion*). The Aurora Borealis and Aurora Australis are produced by the penetration of ionizing particles through the atmosphere from about 300 to 80 km, particularly in zones about 10°–20° latitude from the earth's magnetic poles. On occasion, however, the aurorae may appear at heights up to 1000 km, demonstrating the immense extension of a rarefied atmosphere. The term *ionosphere* is commonly applied to the layers above 80 km, although sometimes it is used only for the region of high electron density between about 100 and 300 km. In view of these different designations it seems preferable to avoid confusion by using the terminology adopted here.

5 Exosphere and magnetosphere

The base of the exosphere is between about 500 and 750 km. Here atoms of oxygen, hydrogen and helium (about 1 per cent of which are ionized) form the tenuous atmosphere and the gas laws (see B, this chapter) cease to be valid. Neutral helium and hydrogen atoms, which have low atomic weights, can escape into space since the chance of molecular collisions deflecting them downwards becomes less with increasing height. Hydrogen is replaced by the breakdown of water vapour and methane (CH_4) near the mesopause, while helium is produced by the action of cosmic radiation on nitrogen

and from the slow, but steady, breakdown of radioactive elements in the earth's crust.

Ionized particles increase in frequency through the exosphere and, beyond about 200 km, in the magnetosphere there are only electrons (negative) and protons (positive) derived from the solar wind – a plasma of electrically-conducting gas. Detailed investigation of these regions made possible by satellites, involves the study of magnetohydrodynamics – the behaviour of plasmas or gases of electrically-charged particles. Charged particles are concentrated in two bands centred at altitudes of about 3000 and 16,000 km (the Van Allen 'radiation' belts, or plasmasphere) as a result of trapping by the earth's magnetic field – the geomagnetic storm ring current. On the nightside of the earth (away from the sun), the magnetosphere has an extended tail, whereas on the dayside it is compressed by the solar wind, creating a 'bow shock' to about ten earth radii (57,000 km) (see Figure 1.45). Low energy plasma enters via the dayside polar cusp – an opening in the magnetopause – and becomes concentrated in the auroral oval. On the nightside, plasma particles attain higher energies and 'precipitate' into the earth's atmosphere producing ionization. Above about 80,000 km the earth's atmosphere merges with that of the sun (the Heliosphere).

I VARIATION OF TEMPERATURE WITH HEIGHT

The last section described the gross characteristics of the vertical temperature profile in the atmosphere. Now we examine in more detail some of the features of the temperature gradient at low levels.

Vertical temperature gradients are determined in part by energy transfers and in part by vertical motion of the air. The various factors interact in a highly complex manner. The energy terms are the release of latent heat by condensation (see D, this chapter), radiational cooling of the air and sensible heat transfer from the

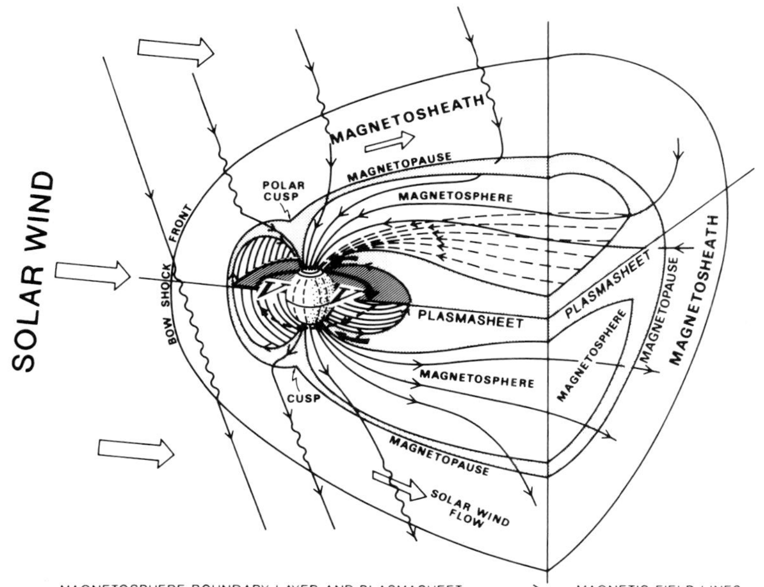

MAGNETOSPHERE BOUNDARY LAYER AND PLASMASHEET
V VAN ALLEN BELTS (TRAPPED RADIATION)
---→ GEOMAGNETIC STORM RING CURRENT
→ MAGNETIC FIELD LINES
▬► AURORAL PRECIPITATION

Figure 1.45 A schematic diagram of the structure of the magnetosphere and its interactions with the solar wind. The supersonic solar wind plasma forms a shock wave as it encounters the earth's magnetic field (the 'bow shock'). Auroral 'precipitation' causes excitation of neutral atoms and molecules by collisions with accelerating electrons. The trapped radiation in the plasmasphere is also termed the Van Allen 'radiation belts'.

Source: Mostly after. Roederer 1981.

ground. Horizontal temperature advection may also be important. Vertical motion is dependent on the type of pressure system. High-pressure areas are generally associated with descent and warming of deep layers of air, hence decreasing the temperature gradient and frequently causing temperature inversions in the lower troposphere. In contrast, low-pressure systems are associated with rising air which cools upon expansion and increases the vertical temperature gradient. This is only part of the story since moisture is an additional complicating factor (see Chapter 2, E). It remains true, however, that the middle and upper troposphere is relatively cold above a surface low-pressure area, leading to a steeper temperature gradient.

The overall vertical decrease of temperature, or *lapse rate*, in the troposphere is, as has been stated, about 6.5°C/km. However, this is by no means constant with height, season or location. Average global values calculated by C. E. P. Brooks for July show increasing lapse rate with height: 5°C/km in the lowest 2 km, 6°C/km between 4 and 6 km, and 7°C/km between 6 and 8 km. Winter values are generally smaller and in continental areas, such as central Canada or eastern Siberia, may even be negative (i.e. temperatures increase with height in the lower layer) as a result of excessive radiational cooling over a snow surface (see Figure 1.46). A similar situation occurs when dense, cold air accumulates in mountain basins on calm, clear nights. On such occasions the mountain tops may be many degrees warmer than the valley floor below (see Chapter 3, C.1). For this reason the adjustment of average temperatures of upland stations to mean sea level may produce misleading results. Observations in Colorado at Pike's Peak (4301 m or 14,111 ft) and Colorado Springs (1859 m or 6099 ft) show the mean lapse rate to be 4.1°C/km (2.2°F/1000 ft) in winter and 6.2°C/km (3.4°F/1000 ft) in summer! It should be noted that such topographic lapse rates may bear little relation to free-air lapse rates in nocturnal radiation conditions, and the two must be carefully distinguished.

Table 1.7 summarizes the seasonal characteristics of lapse rates in six major climatic zones and examples of five of these are illustrated in Figure 1.46. The seasonal regime is very pronounced in continental areas with cold winters whereas inversions persist for much of the year in the Arctic. In winter the Arctic inversion is due to intense radiational cooling but in summer it is the result of the surface cooling of advected warmer air. The winter lapse rate is only greater than the summer one in Mediterranean climates. In these regions there is more likelihood of rising air associated with low-pressure areas in winter. In contrast, subsidence is predominant in the desert zones in winter. The tropical and subtropic deserts have very steep lapse rates in summer when there is

Table 1.7 Temperature lapse rates in the lowest 1000–1500 metres.

Climate	*Season of maximum*	*Rate °C/km*	*Season of minimum*	*Rate °C/km*
Tropical rainy	Dry season	>5	Rainy season	>4.5
Tropical and subtropical deserts	Summer	>8	Winter	>5
Mediterranean	Winter	>5	Summer	<5
Mid-latitudes (cold winter)	Summer	>6	Winter	0–5
Boreal continental	Summer	>5	Winter	<0
Arctic	Summer	≤0	Winter	<0

Source: After Lautensach and Bögel 1956.

Figure 1.46 The annual variation of lapse rate in five climatic zones: 1 Tropical rainy climate (Togo); 2 Tropical desert (Arizona); 3 Mediterranean (Sicily); 4 Mid-latitude, cold winter climate (North Germany); 5 Boreal continental (Eastern Siberia).

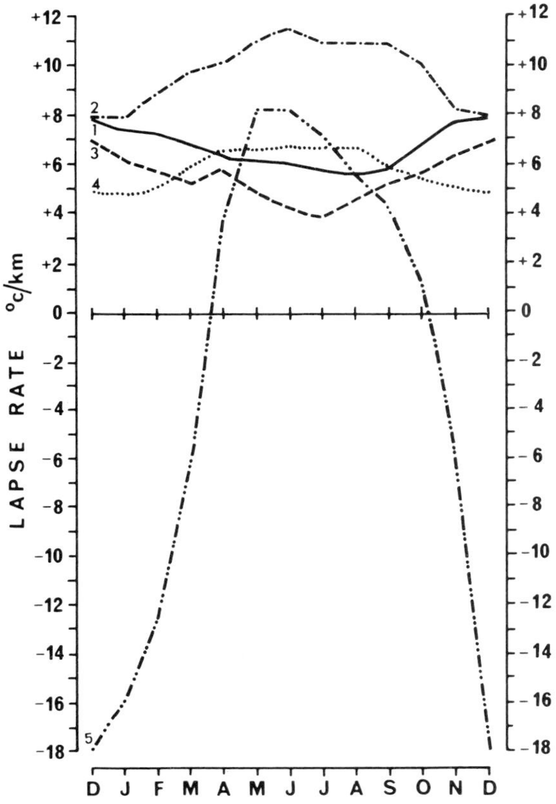

Source: Hastenrath 1968.

considerable heat transfer from the surface and generally ascending motion.

SUMMARY

The atmosphere is a mixture of gases with constant proportions up to 80 km or more. The exceptions are ozone, which is concentrated in the lower stratosphere, and water vapour in the lower troposphere. Carbon dioxide, methane and other trace gases are increasing in this century due to the burning of fossil fuels and other anthropogenic effects, but large natural fluctuations also have occurred in the past. Air is highly compressible, so that half of its mass occurs in the lowest 5 km and pressure decreases logarithmically with height from an average sea-level value of 1013 mb.

Almost all energy affecting the earth is derived from solar radiation, which is short wavelength (< 4 μm) due to the high temperature of the sun (~ 6000 K) (i.e. Wien's Law). The solar constant has a value of approximately 1368 W m^{-2}. The sun and the earth radiate almost as black bodies (Stefan's law, $F = \sigma T^4$) whereas the atmospheric gases do not. Terrestrial radiation, from an equivalent black body, amounts only to about 270 W m^{-2} due to its low radiating temperature (263 K) and its infra-red (long-wave) radiation between 4 and 100 μm. Water vapour and carbon dioxide are the major absorbing gases for infra-red radiation, whereas the atmosphere is largely transparent to solar radiation (the greenhouse effect). Trace gas increases are now augmenting the 'natural' greenhouse effect (33K). Solar radiation is lost by reflection, mainly from clouds, and by absorption (largely by water vapour). The planetary albedo is 31 per cent; 48 per cent of the extraterrestrial radiation reaches the surface. The atmosphere is heated primarily from the surface by the absorption of terrestrial infra-red radiation and by turbulent heat transfer. Temperature usually decreases with height at an average rate of about 6.5°C/km in the troposphere. In the stratosphere and thermosphere it increases with height due to the presence of radiation-absorbing gases.

The excess of net radiation in lower latitudes leads to a poleward energy transport from tropical latitudes by ocean currents and

continued

by the atmosphere. This is in the form of sensible heat (warm air masses/ocean water) and latent heat (atmospheric water vapour). Air temperature at any point is affected by the incoming solar radiation and other vertical energy exchanges, surface properties (slope, albedo, heat capacity), land and sea distribution and elevation, and also by horizontal advection due to air mass movements and ocean currents.

2

Atmospheric moisture

Terrestrial moisture is in a constant state of transformation, termed the *hydrologic cycle*, in which the three most important stages are evaporation, condensation and precipitation. Figure 2.1 indicates the relative average annual amounts of water involved in each phase of the cycle. It shows that the atmosphere holds only a very small amount of water although the exchanges with the land and oceans are very considerable. This is further emphasized by Table 2.1. The average storage of water vapour in the atmosphere (about 2.5 cm, or 1 in) is only sufficient for some 10 days' supply of rainfall over the earth as a whole. However, intense (horizontal) influx of moisture into the air over a given region makes possible short-term rainfall totals greatly in excess of 3 cm. The phenomenal record total of 187 cm (73.6 in) fell on the island of Réunion, off Madagascar, during 24 hours in March 1952, and much greater intensities have been observed over shorter periods (see I.1, this chapter).

The atmosphere acquires moisture by evaporation from oceans, lakes, rivers and damp soil or from moisture transpired from plants. Taken together, these are often referred to as *evapotranspiration* and the mechanisms involved will now be discussed in detail.

A EVAPORATION

Evaporation occurs whenever energy is transported to an evaporating surface if the vapour pressure in the air is below the saturated value (e_s). As illustrated in Figure 1.13, the saturation vapour pressure increases with temperature. The change in state from liquid to vapour requires energy to be expended in overcoming the intermolecular attractions of the water particles. This energy is generally provided by the removal of heat from the immediate surroundings causing an apparent heat loss (*latent heat*), as discussed on p. 21, and a consequent drop in temperature. The latent heat of vaporization to evaporate 1 kg of water at 0° C is 2.5×10^6 J (or 600 cal g^{-1}). Conversely, condensation releases this heat, and the temperature of an air mass in which condensation is occurring is increased as the water vapour reverts to the liquid state. The diurnal range of temperature if often moderated by damp air conditions, when evaporation takes place during the day and condensation at night.

Viewed another way, evaporation implies an addition of kinetic energy to individual water molecules and, as their velocity increases, so the chance of individual surface molecules escaping into the atmosphere becomes greater. As the faster molecules will generally be the first to escape, so the average energy (and therefore temperature) of those composing the remaining liquid will decrease and the quantities of energy required for their continued release become correspondingly greater. In this way evaporation decreases the temperature of the remaining liquid by an amount proportional to the latent heat of vaporization.

The rate of evaporation depends on a number

Figure 2.1 The hydrological cycle and water storage of the globe. The exchanges in the cycle are referred to 100 units which equal the mean annual global precipitation of 85.7 cm (33.8 in). The percentage storage figures for atmospheric and continental water are percentages of all *fresh* water. The saline ocean waters make up 97 per cent of *all* water. The horizontal advection of water vapour indicates the *net* transfer.

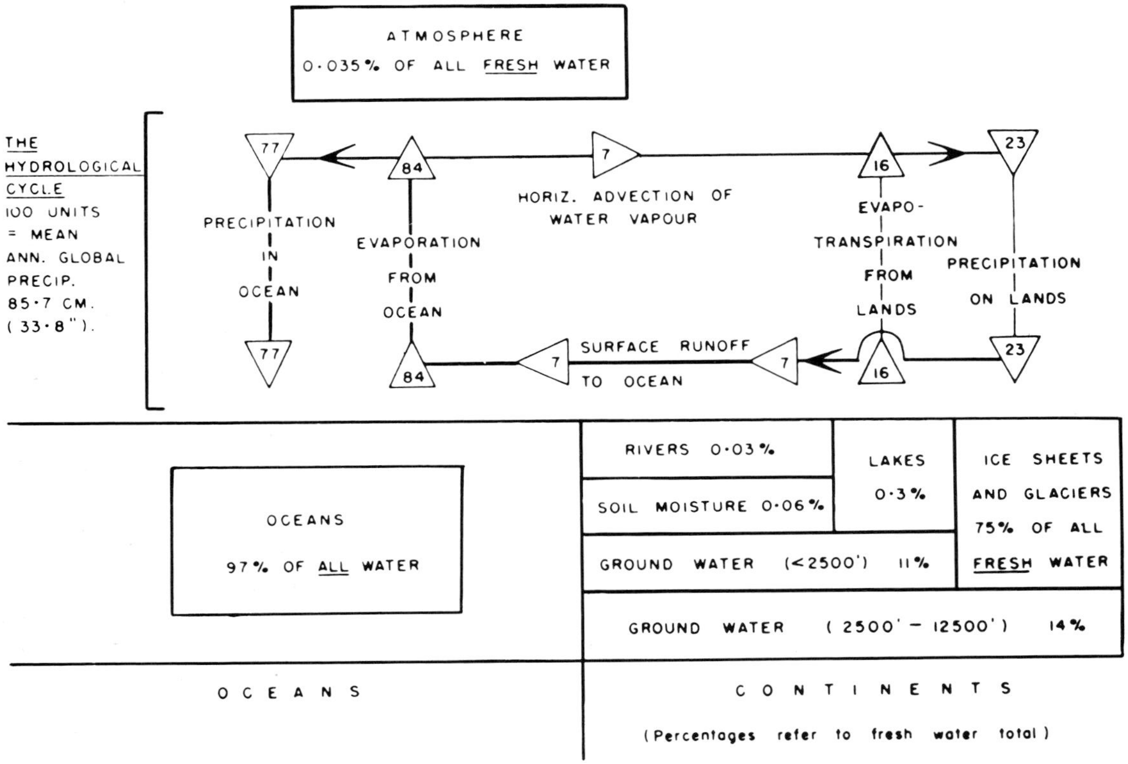

Source: From More 1967.

Table 2.1 Mean water content of the atmosphere (in cm of rainfall equivalent).

	Northern hemisphere	*Southern hemisphere*	*World*
January	1.9 (0.8 in)	2.5 (1.0 in)	2.2 (0.9 in)
July	3.4 (1.3 in)	2.0 (0.8 in)	2.7 (1.1 in)

Source: After Sutcliffe 1956.

of factors. The two most important are the difference between the saturation vapour pressure at the water surface and the vapour pressure of the air, and the existence of a continual supply of energy to the surface. Wind velocity also affects the evaporation rate because the wind is generally associated with the importation of fresh, unsaturated air which will absorb the available moisture.

Water loss from plant surfaces, chiefly leaves, is a complex process termed *transpiration.* It occurs when the vapour pressure in the leaf cells is greater than the atmospheric vapour pressure, and is vital as a life function in that it causes a rise of plant nutrients from the soil and cools the

leaves. The cells of the plant roots can exert an osmotic tension of up to about 15 atmospheres upon the water films between the adjacent soil particles. As these soil water films shrink, however, the tension within them increases. If the tension of the soil films exceeds the osmotic root tension the continuity of the plant's water supply is broken and wilting occurs. Transpiration is controlled by the atmospheric factors which determine evaporation as well as by plant factors such as the stage of plant growth, leaf area and leaf temperature, and also by the amount of soil moisture (see Chapter 7, C). It occurs mainly during the day when the *stomata* (i.e. small pores in the leaves) through which transpiration takes place are open. This opening is determined primarily by light intensity. Transpiration naturally varies greatly with season, and during the winter months in mid-latitudes conifers lose only 10–18 per cent of their total annual transpiration losses and deciduous trees less than 4 per cent.

In practice it is difficult to separate water evaporated from the soil, *intercepted moisture* remaining on vegetation surfaces after precipitation and subsequently evaporated, and transpiration. For this reason evaporation is sometimes applied as a general term for all these, or, more correctly, the composite term *evapotranspiration* may be used.

Evapotranspiration losses from natural surfaces cannot be measured directly. There are, however, various indirect methods of assessment, as well as theoretical formulae. One approximate means of indirect measurement is based on the moisture balance equation:

$$\text{precipitation} = \text{runoff} + \text{evapotranspiration} + \text{soil moisture storage change}$$

Essentially the method is to measure the percolation through an enclosed block of soil with a vegetation cover (usually grass) and to record the rainfall upon it. The block, termed a *lysimeter*, is weighed regularly so that weight changes unaccounted for by rainfall or runoff can be ascribed to evapotranspiration losses, provided the grass is kept short! The technique allows the determination of daily evapotranspiration amounts.

If the soil block is regularly 'irrigated' so that the vegetation cover is always yielding the maximum possible evapotranspiration the water loss is called the *potential evapotranspiration* (or PE). More generally, PE can be defined as the water loss corresponding to the available energy. Assuming a constant soil moisture storage, potential evapotranspiration is calculated as the difference between precipitation and percolation. A simple evapotranspirometer installation is shown in Figure 2.2; the double-

Figure 2.2 An evapotranspirometer installation for calculating potential evapotranspiration losses. The double installation allows an average of the two results to be determined, giving a more reliable estimate.

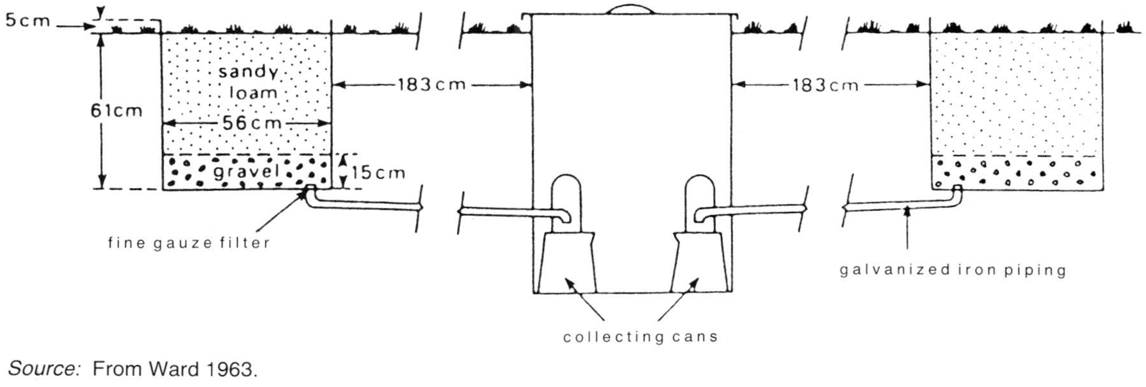

Source: From Ward 1963.

tank installation ensures that representative readings are obtained. Potential evapotranspiration forms the basis for one system of climate classification developed by C. W. Thornthwaite (see Appendix 1).

In regions where snow cover is long lasting, evaporation/sublimation from the snowpack can be estimated by lysimeters sunk into the snow that are regularly weighed. Except in dry, sunny and windy environments, however, evaporation from snow is of minor importance compared to melt.

A meteorological solution to the calculation of evaporation uses sensitive instruments to measure the net effect of eddies of air transporting moisture upward and downward near the surface. In this 'eddy correlation' technique the vertical component of wind and the atmospheric moisture content are measured simultaneously at the same level (say, 1.5 m) every few seconds. The product of each pair of measurements is then averaged over some time interval to determine the evaporation (or condensation). This method requires delicate rapid-response instruments and so it cannot be used in very windy conditions.

Theoretical methods for determining evaporation rates have followed two lines of approach. The first relates average monthly evaporation (E) from large water bodies to the mean wind speed (u) and the mean vapour pressure difference between the water surface and the air ($e_w - e_d$) in the form:

$$E = Ku(e_w - e_d)$$

where K is an empirical constant. This is termed the aerodynamic approach because it takes account of the factors responsible for removing vapour from the water surface. The second method is based on the energy budget. The *net balance* of solar and terrestrial radiation at the surface (R_n) is used for evaporation (E) and the transfer of heat to the atmosphere (H). A small proportion also heats the soil by day, but since nearly all of this is lost at night it can be disregarded. Thus:

$$R_n = LE + H,$$

where L is the latent heat of evaporation (2.5×10^6 J kg^{-1}). R_n can be measured with a net radiometer and the ratio $H/LE = \beta$, referred to as Bowen's ratio, can be estimated from measurements of temperature and vapour content at two levels near the surface. β ranges from < 0.1 for water to $\geqslant 10$ for a desert surface. The use of this ratio assumes that the vertical transfers of heat and water vapour by turbulence take place with equal efficiency. Evaporation is then determined from an expression of the form:

$$E = \frac{R_n}{L(1 + \beta)}$$

The most satisfactory climatological method so far devised combines the energy budget and aerodynamic approaches. In this way H. L. Penman succeeded in expressing evaporation losses in terms of four meteorological elements which are regularly measured, at least in Europe and North America. These are duration of sunshine (related to radiation amounts), mean air temperature, mean air humidity and mean wind speed (which limit the losses of heat and vapour from the surface). Direct measurements of net radiation can also be used if these are available.

The relative roles of the factors which have been mentioned are illustrated by the global pattern of evaporation (see Figure 2.3). Losses decrease sharply in high latitudes where there is little available energy. In middle and lower latitudes there are appreciable differences between land and sea (see Figure 2.4B). Rates are naturally high over the oceans in view of the unlimited availability of water, and on a seasonal basis the maximum rates occur in winter over the western Pacific and Atlantic, where cold continental air blows across warm ocean currents. On an annual basis maximum oceanic losses occur about 15°–20°N and 10°–20°S in the belts of the constant trade winds. The highest annual losses, estimated to be about 200 cm (80 in), are in the western Pacific and central Indian Ocean near

Figure 2.3 Mean evaporation (cm) for January and July.

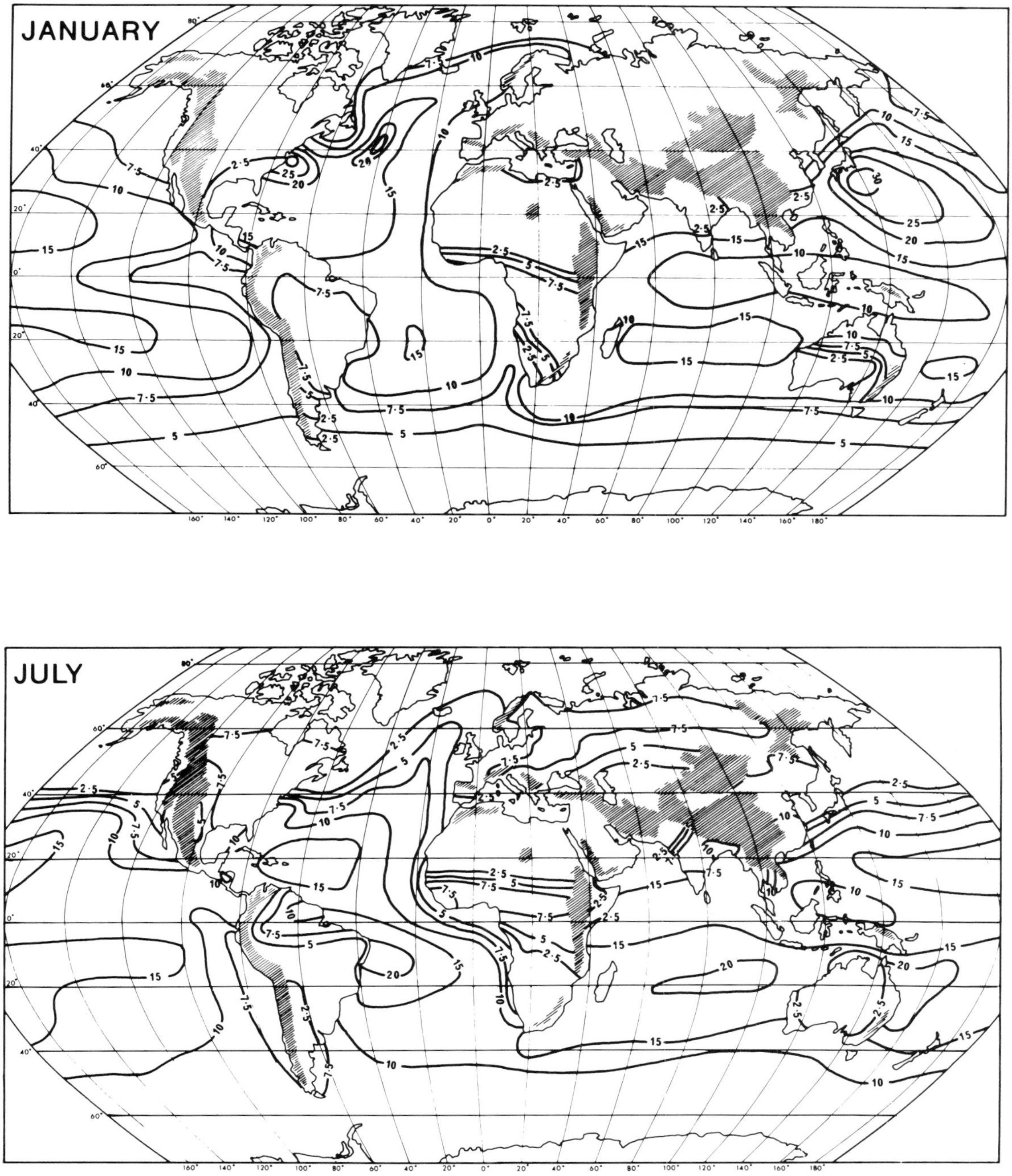

Figure 2.4 The average annual latitudinal distribution of (A) precipitation (in mm); (B) evaporation (in mm); and (C) meridional transfer of water vapour (in 10^{15} kg).

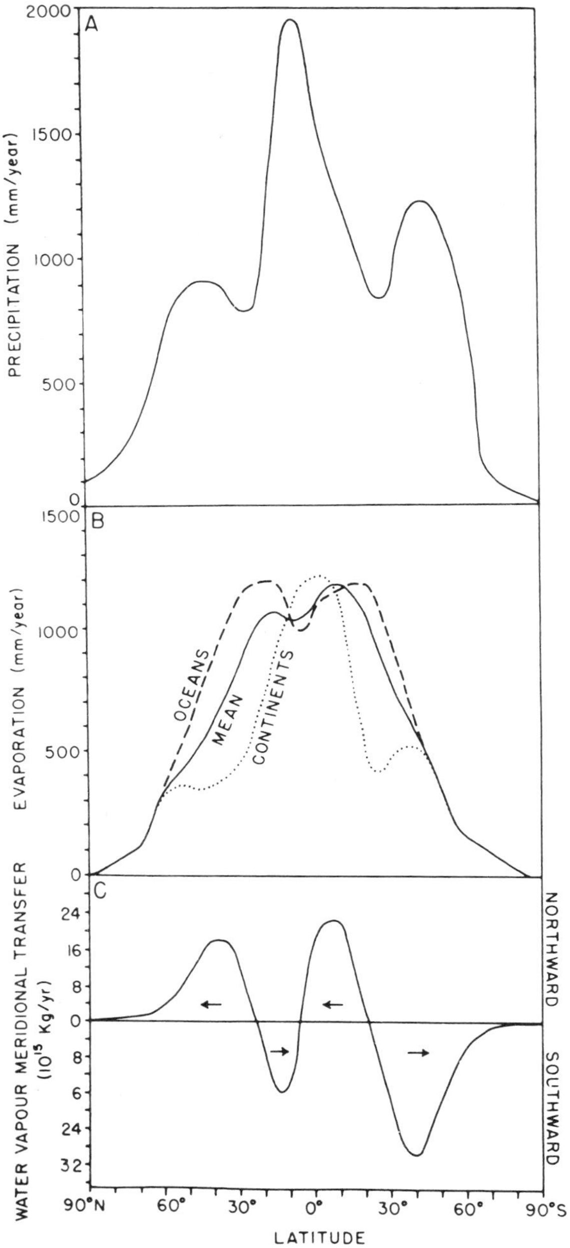

Source: Mostly from Sellers 1965.

15°S (cf. Figure 1.41; 2,460 MJ m^{-2} yr^{-1} (78 W m^{-2} over the year) is equivalent to an evaporation of 100 cm of water/cm^2). There is a subsidiary equatorial minimum over the oceans mainly as a result of the lower wind speeds in the doldrum belt and the proximity of the vapour pressure in the air to its saturation value, but the land maximum occurs more or less at the equator because of the relatively high solar radiation receipts and the large transpiration losses from the luxuriant vegetation of this region. The secondary maximum over land in mid-latitudes is related to the strong prevailing westerly winds. The other parts of Figure 2.4, incorporated here for convenient comparison, are discussed in later sections.

The annual evaporation over Britain, calculated by Penman's formula, ranges from about 38 cm (15 in) in Scotland to about 50 cm (20 in) in parts of south and south-east England. The annual potential evapotranspiration determined by Thornthwaite's method (based on mean temperature) is over 64 cm (25 in) in most of south-eastern England. Since this loss is concentrated in the period May–September there may be seasonal water deficits of 12–15 cm (5–6 in) in these parts of the country (as shown in Figure 2.5 for Southend), necessitating considerable use of irrigation water by farmers. Figure 2.6 indicates that in southern and south-eastern England it is necessary to irrigate in about nine years out of ten during the summer six months (April–September), assuming that the crop can extract 6.4 cm (2.5 in) of moisture from the soil.

In the United States, monthly moisture conditions are commonly evaluated on the basis of the Palmer Drought Severity Index (PDSI). This is determined from accumulated weighted differences between actual precipitation and the calculated amount required for evapotranspiration, soil recharge and runoff. Accordingly, it takes account of the persistence effects of drought situations. The PDSI has a range from $\geqslant 4$, extremely wet, to $\leqslant -4$, extreme drought.

Figure 2.5 The average annual moisture budget for stations in western, central and eastern Britain determined by Thornthwaite's method. When potential evaporation exceeds precipitation, soil moisture is used; at Berkhamsted in central England and Southend on the east coast, this is depleted by July–August. Autumn precipitation excess over potential evaporation goes into replenishing the soil moisture until field capacity is reached.

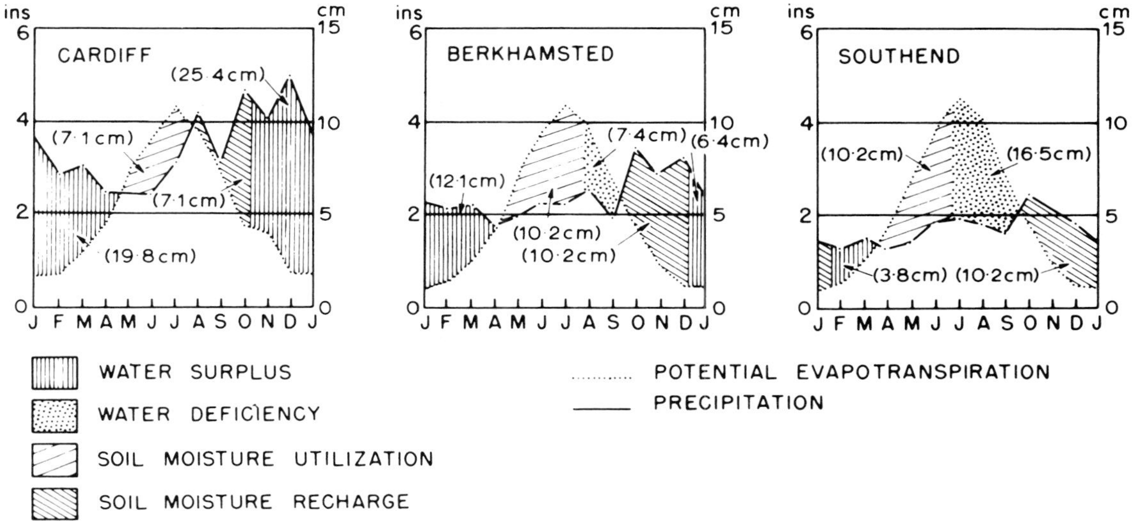

Source: From Howe 1956.

B HUMIDITY

1 Moisture content

The moisture content of the atmosphere can be expressed in several ways, apart from the vapour pressure, depending on which aspect the user wishes to emphasize. The total mass of water in a given volume of air, i.e. the density of the water vapour, is one such measure. This is termed the *absolute humidity* (ρ_w) and is measured in grams per cubic metre (g m^{-3}). Volumetric measurements are not greatly used in meteorology and more convenient is the *mass mixing ratio* (x). This is the mass of water vapour in grams per kilogram of dry air. For most practical purposes the *specific humidity* (q) is identical, being the mass of vapour per kilogram of air including its moisture.

The bulk of the atmosphere's moisture content is contained below 500 mb (5574 m), as Figure 2.7 clearly shows. It is also apparent that the seasonal effect is most marked in the lowest 3000 m (10,000 ft) or so, that is below about 700 mb. The global distribution of atmospheric vapour content in January and July is illustrated in Figure 2.8. Over southern Asia during the summer monsoon an air column holds 5–6 cm of precipitable water, compared with less than 1 cm in tropical desert areas. Minimum values of 0.1–0.2 cm occur over high latitudes and continental interiors of the northern hemisphere in winter.

Another important measure is *relative humidity* (r), which expresses the actual moisture content of a sample of air as a percentage of that contained in the same volume of saturated air at the same temperature. The relative humidity is defined with reference to the mixing ratio, but it can be determined approximately in several ways:

$$r = \frac{x}{x_s} \times 100 \approx \frac{q}{q_s} \times 100 \approx \frac{e}{e_s} \times 100$$

where the subscript s refers to the respective saturation values at the same temperature; e denotes vapour pressure.

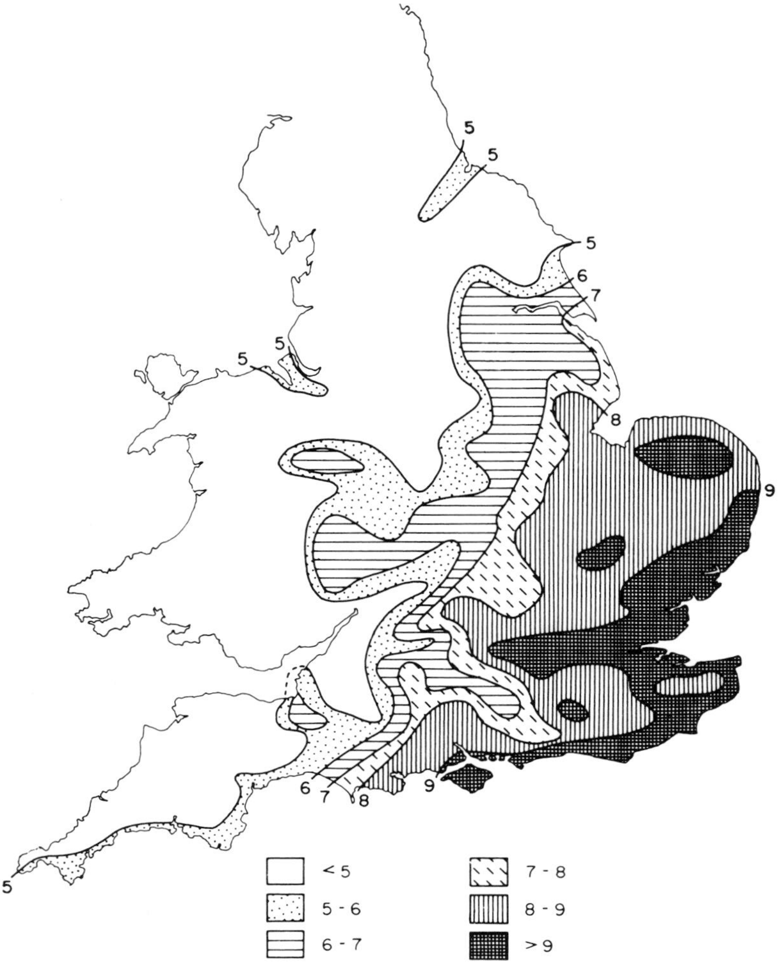

Figure 2.6 The average number of years in ten when irrigation is theoretically necessary for crops in England and Wales, based on Penman's formula.

Source: From Pearl *et al.*, 1954 (*Crown Copyright Reserved*).

A further index of humidity is the dew-point temperature. This is the temperature at which saturation occurs if air is cooled at constant pressure without addition or removal of vapour. When the air temperature and dew point are equal the relative humidity is 100 per cent and it is evident that relative humidity can also be determined from

$$\frac{e_s \text{ at dew point}}{e_s \text{ at air temperature}} \times 100$$

The relative humidity of a parcel of air will obviously change if either its temperature or its mixing ratio is changed. In general the relative humidity varies inversely with temperature during the day, tending to be lower in the early afternoon and higher at night.

Atmospheric moisture can be measured by at least five types of instrument. The most common for routine measurement is the *wet-bulb thermometer* installed in a louvred instrument shelter (Stevenson screen). The bulb of a standard thermometer is wrapped in muslin which is kept moist by a wick from a reservoir of pure water. The evaporative cooling of this wet bulb gives a reading which can be used in conjunction with a simultaneous dry

Figure 2.7 The average vertical variation of atmospheric vapour content at Portland, Maine, between 1946 and 1955.

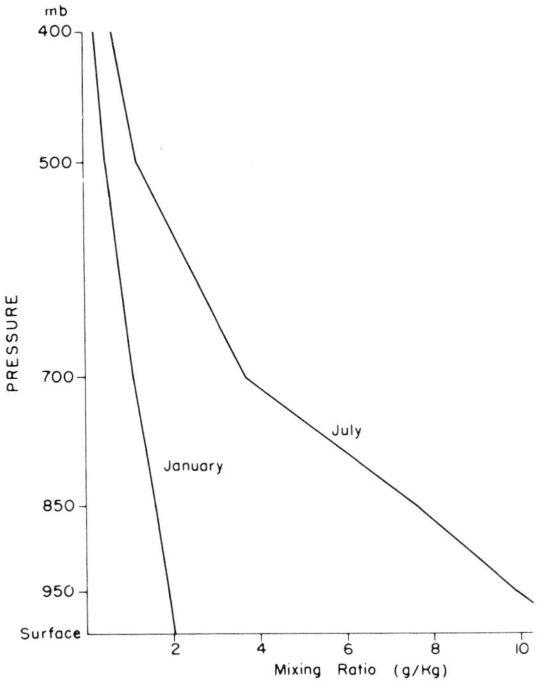

Source: Data from Reitan 1960.

bulb temperature reading to calculate the dew-point temperature. A similar portable device, called an aspirated *psychrometer*, uses a forced flow of air at a fixed rate over the dry and wet bulbs. A sophisticated instrument for determining dew point, based on a different principle, is the *dew-point hygrometer*. This detects when condensation first occurs on a cooled surface. Two other types of instrument are used to determine relative humidity. the *hygrograph* utilizes the expansion/contraction of a bundle of human hair, in response to humidity, to record relative humidity continuously by a mechanical coupling to a pen arm marking on a rotating drum. This has an accuracy of about ± 5–10 per cent. For upper air measurements, a *lithium chloride* element is used to detect changes in its electrical resistance to vapour pressure differences. Relative humidity changes are accurate to about ± 3 per cent.

2 Moisture transport

It is sometimes overlooked that the atmosphere transports moisture horizontally as well as vertically. Figure 2.1 shows a net transport from oceans to land areas, while Figure 2.4C illustrates the quantities which must be transported meridionally in order to maintain the required moisture balance at a given latitude (i.e. precipitation – evaporation = net horizontal transport of moisture into the air column). A prominent feature is the equatorward transport into low latitudes and the poleward transport in middle latitudes. The reader should inspect this diagram again in the light of the discussion of wind belts in Chapter 3, E.

At this point it is necessary to stress emphatically the fact that local evaporation is, in general, not the major source of local precipitation. For example, only 6 per cent of the annual precipitation of Arizona and 10 per cent of that over the Mississippi River basin is of local origin, the remainder being transported into these areas (i.e. moisture advection). Even when moisture is available in the atmosphere over a region only a small portion of it is usually precipitated. This depends on the efficiency of the condensation and precipitation mechanisms, both microphysical and large-scale, which we shall now consider.

C CONDENSATION

Condensation, the direct cause of all the various forms of precipitation, occurs under varying conditions which in one way or another are associated with change in one of the linked parameters of air volume, temperature, pressure or humidity. Thus, condensation takes place (i) when the temperature of the air is reduced but its volume remains constant and the air is cooled to dew point; (ii) if the volume of the air is increased without the addition of heat; this cooling

Figure 2.8 Mean atmospheric water vapour content in January and July, 1951–5, in cm of precipitable water.

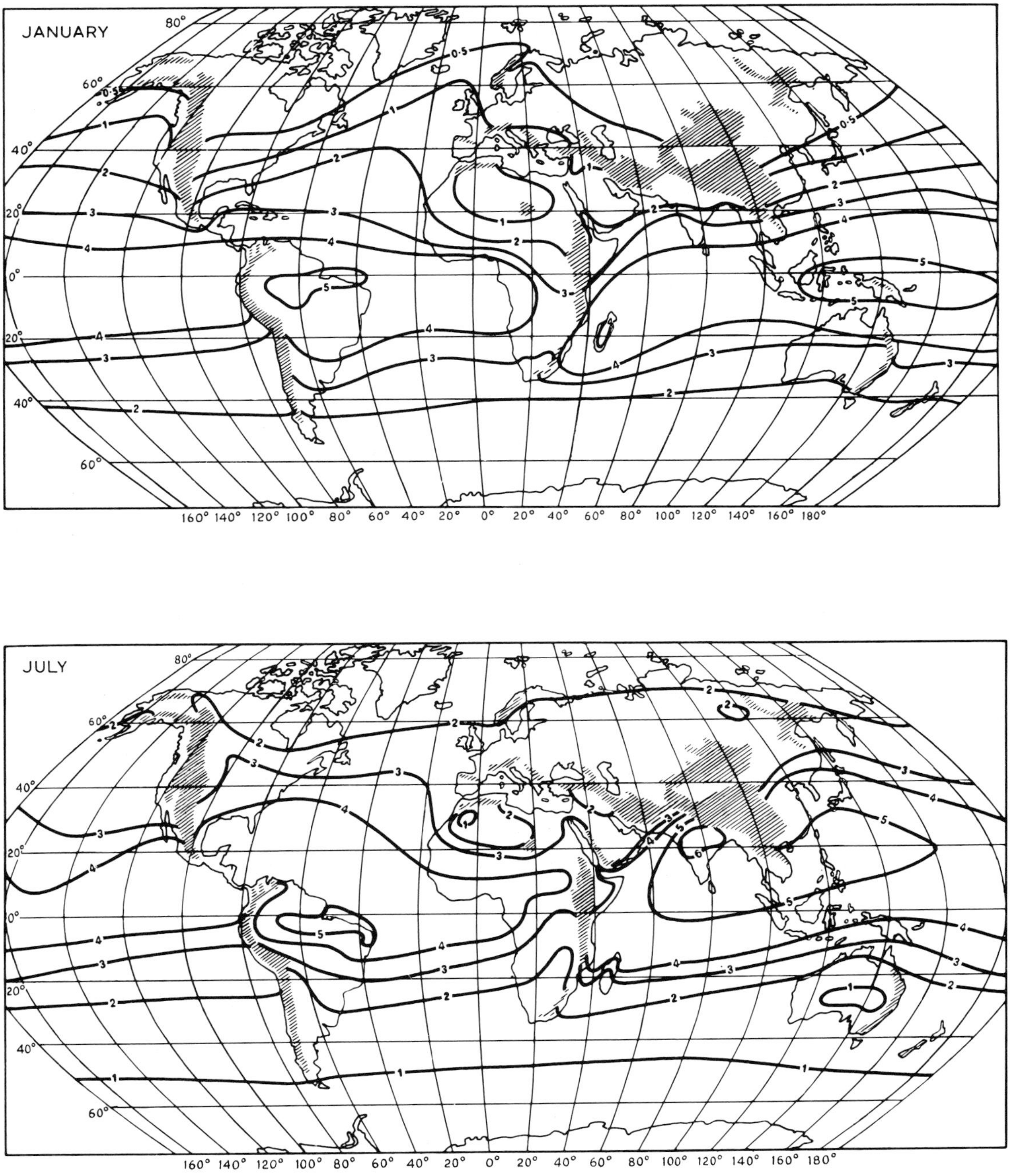

Source: After Bannon and Steele 1960 (*Crown Copyright Reserved*).

takes place because adiabatic expansion causes energy to be consumed through work (see D, this chapter); (iii) when a joint change of temperature and volume reduces the moisture-holding capacity of the air below its existing moisture content; or (iv) by evaporation adding moisture to the air. The key to the understanding of condensation clearly lies in the fine balance that exists between these variables. Whenever the balance between two or more of them is disturbed beyond a certain limit condensation may result.

The most common circumstances favourable to the production of condensation are those producing a drop in air temperature; namely contact cooling, mixing of air masses of different temperatures and dynamic cooling of the atmosphere. Contact cooling is produced, for example, within warm, moist air passing over a cold land surface (see Plate 5). On a clear winter's night strong radiation will cool the surface very quickly and this surface cooling will gradually extend to the moist lower air, reducing the temperature to a point where condensation occurs in the form of dew, fog or frost, depending on the amount of moisture involved, the thickness of the cooling air layer and the dew-point value. When the latter is below 0° C it is referred to as the hoar frost-point if the air is saturated with respect to ice.

The mixing of the differing layers within a single air mass or of two differing air masses can also produce condensation. Figure 2.9 indicates how the horizontal mixing of two air masses (A and B), of given temperature and moisture characteristics, may produce an air mass (C) which is oversaturated at the intermediate temperature and consequently forms cloud. Vertical mixing of an air layer, which is discussed below (see Figure 2.16) can have the same effect. Fog, or low stratus, with drizzle – known as 'crachin' – which is common along the coasts of South China and the Gulf of Tonkin in February–April, can develop as a result of either air mass mixing or warm advection over a colder surface.

The addition of moisture into the air near the surface by evaporation occurs when cold air

Figure 2.9 The effect of air-mass mixing. The horizontal mixing of two unsaturated air masses A and B will produce one supersaturated air mass C. The saturation vapour pressure curve is shown (cf. Figure 1.13 which is a semi-logarithmic plot).

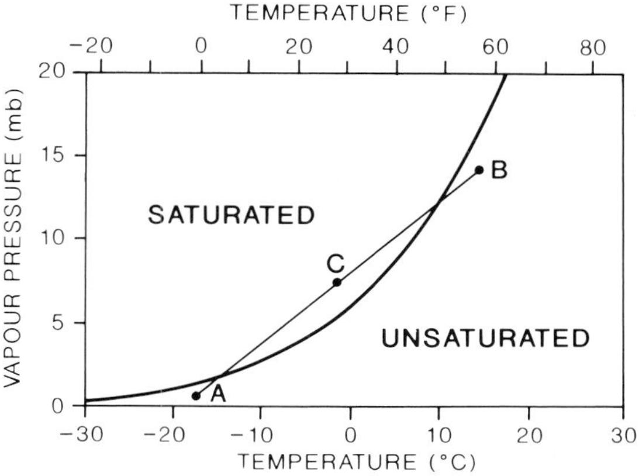

Source: From Petterssen 1941 [1969].

moves out over a warm water surface. This can cause steam fog to form, which is common in arctic regions. Attempts at fog dispersal are one area where some progress has been made in local weather modification. Cold fogs can be locally dissipated by the use of dry ice (frozen CO_2) or the release of propane gas through expansion nozzles to produce freezing and the subsequent fallout of ice crystals (cf. p. 76). Warm fogs (i.e. having drops above freezing temperatures) present bigger problems, but attempts at dissipation have shown some limited success in evaporating droplets by artificial heating, the use of large fans to draw down dry air from above, the sweeping out of fog particles by jets of water, and the injection of electrical charges into the fog to produce coagulation.

Undoubtedly the most effective cause of condensation, however, is the dynamic process of adiabatic cooling. This is considered in some detail in the next section.

D ADIABATIC TEMPERATURE CHANGES

The displacement of an air parcel to an environment of lower pressure (without heat exchange

with surrounding air) causes an increase in its volume and a consequent lowering of its temperature. A volume increase involves work and the consumption of energy, thus reducing the heat available per unit volume and hence the temperature. Such a temperature change, involving no subtraction or addition of heat, is termed *adiabatic*. Vertical displacements of air are obviously a major cause of adiabatic temperature changes.

Near the earth's surface most processes of change are non-adiabatic (sometimes termed *diabatic*) because of the tendency of air to mix and modify its characteristics by lateral movement, turbulence and related physical processes. When a parcel of air moves vertically the changes that take place often follow an adiabatic pattern because air is fundamentally a poor thermal conductor, and the air parcel as a whole tends to retain its own thermal identity which distinguishes it from the surrounding air masses. In some circumstances, on the other hand, mixing of air with its surroundings must be taken into account.

Let us consider the changes that occur when an air parcel rises and a decrease of pressure is accompanied by volume increase and temperature decrease (see Chapter 1,B). The rate at which temperature decreases in a rising, expanding air parcel is called the *adiabatic lapse rate*. If the upward movement of air does not produce condensation then the energy expended by expansion will cause the temperature of the mass to fall at what is called the *dry adiabatic lapse rate* or DALR (9.8°C/km or 5.4°F/1000 ft). However, prolonged reduction of the temperature invariably produces condensation, and when this happens latent heat is liberated, counteracting the dry adiabatic temperature decrease to a certain extent. It is therefore a distinguishing feature of rising and saturated (or precipitating) air that it cools at a slower rate (i.e. the *saturated adiabatic lapse rate* or SALR) than air which is unsaturated. Another difference between the dry and saturated adiabatic rates is that whereas the former remains constant the latter varies with temperature. This is because air masses at higher temperatures are able to hold more moisture and on condensation therefore release a greater quantity of latent heat. For high temperatures the saturated adiabatic lapse rate may be as low as 4°C/km (or 2.2°F/1000 ft), but this rate increases with decreasing temperatures, approaching 9°C/km (5°F/1000 ft) at −40°C (−40°F).

In all, three different lapse rates can be differentiated, two dynamic and one static. There is the environmental (or static) rate, which is the actual temperature decrease with height on any occasion, such as an observer ascending with a balloon would record. This is not an adiabatic rate therefore and may assume any form depending on local air temperature conditions. There are also the dynamic adiabatic dry and saturated lapse rates (or cooling rates) which apply to rising parcels of air moving through their environment. Close to the surface the vertical temperature gradient sometimes greatly exceeds the dry adiabatic lapse rate, that is, it is superadiabatic. This is particularly common in arid areas in summer (see Table 1.7). Over most ordinary dry surfaces the lapse rate approaches the dry adiabatic value at an elevation of 100 m or so.

The changing properties of moving air parcels can be conveniently expressed by plotting them as *path curves* on suitably constructed graphs. One such adiabatic diagram in common use is the tephigram (see Figure 2.10). This displays five sets of lines representing properties of the atmosphere:

1 Isotherms – i.e. lines of constant temperature (parallel lines from bottom left to top right).
2 Dry adiabats (parallel lines from bottom right to top left).
3 Isobars – i.e. lines of constant pressure (slightly curved nearly horizontal lines).
4 Saturated adiabats (curved lines sloping up from right to left).
5 Saturation mixing ratio lines (those at a slight angle to the isotherms).

The dry adiabats are also lines of constant potential temperature, θ (or isentropes). Potential temperature is the temperature of an air

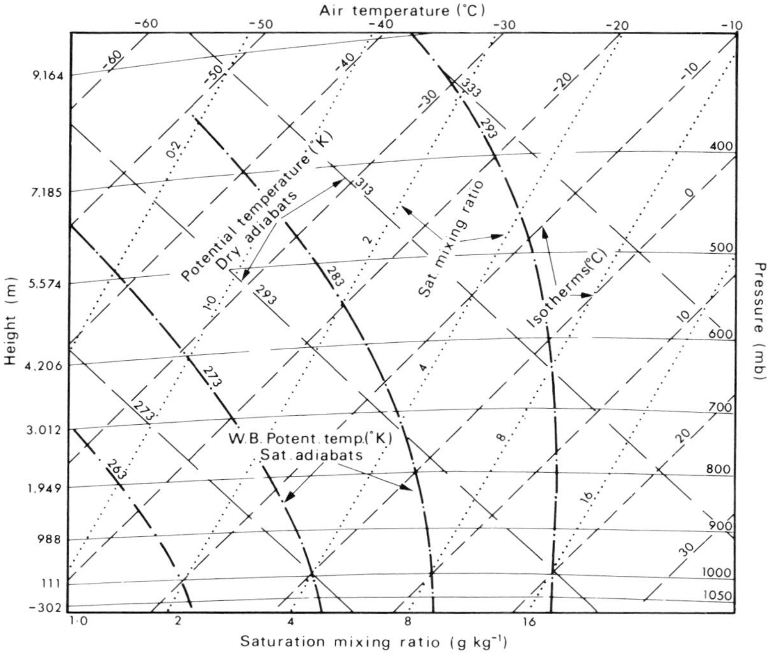

Figure 2.10 The tephigram, allowing the following properties of the atmosphere to be displayed: temperature, pressure, potential temperature, wet-bulb potential temperature and saturation (humidity) mixing ratio.

parcel brought dry-adiabatically to a pressure of 1000 mb. Mathematically,

$$\theta = T\left(\frac{1000}{p}\right)^{0.286}$$

where θ and T are in degK;

p = pressure (mb).

Figure 2.11 shows schematically the relationship between T and θ; also between T and θ_w, the wet-bulb potential temperature (where the air parcel is brought to a pressure of 1000 mb by a saturated adiabatic process).

Figure 2.11 illustrates that on an aerological diagram (such as the tephigram or adiabatic chart) a line along a dry adiabat (θ) through the dry-bulb temperature of the surface air (T_A), an isopleth of saturation mixing ratio (x_s) through the dew point (T_d) and a saturated adiabat (θ_w) through the wet-bulb temperature (T_w) all intersect at a point corresponding to saturation for the air mass. This relationship, known as Normand's theorem, is used to estimate the

Figure 2.11 Graph showing the relationships between temperature (T), potential temperature θ, wet-bulb potential temperature (θ_w) and saturation mixing ratio (X_s). T_d = dew point, T_w = wet-bulb temperature and T_A = air temperature.

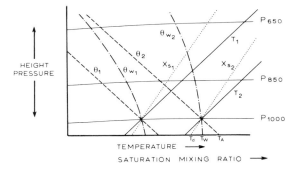

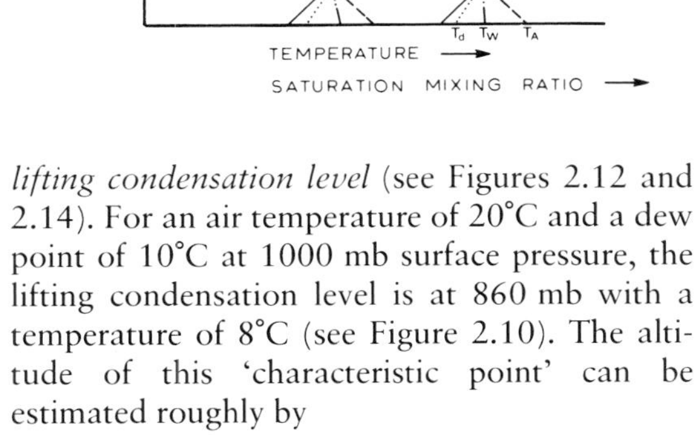

lifting condensation level (see Figures 2.12 and 2.14). For an air temperature of 20°C and a dew point of 10°C at 1000 mb surface pressure, the lifting condensation level is at 860 mb with a temperature of 8°C (see Figure 2.10). The altitude of this 'characteristic point' can be estimated roughly by

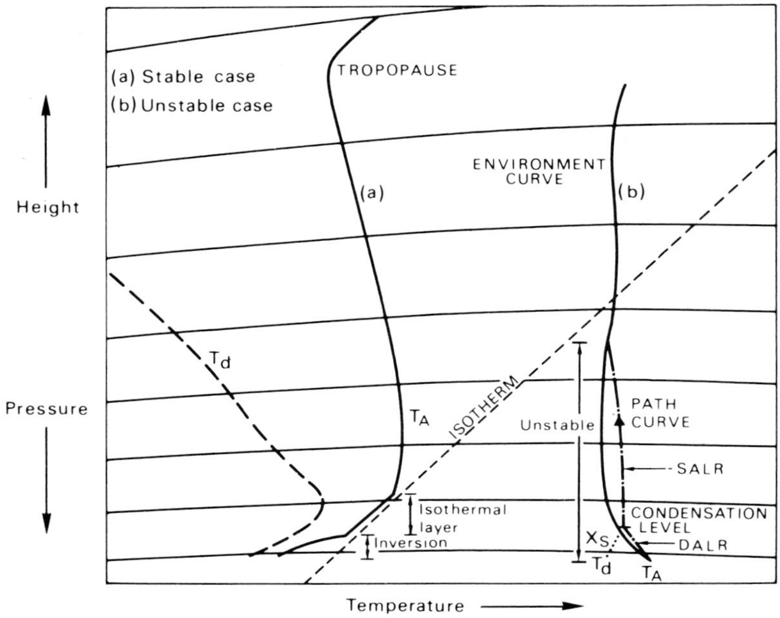

Figure 2.12 Tephigram showing (a) stable air case – T_A is the air temperature and T_d the dewpoint; and (b) unstable air case. The lifting condensation level is shown, together with the path curve (arrowed) of a rising air parcel. X_S is the saturation humidity mixing ratio line through the dewpoint temperature (see p. 65).

$$h\,(\mathrm{m}) = 120(T - T_d)$$

where T = air temperature and T_d = dew-point temperature at the surface in °C).

The lifting condensation level (LCL) formulation does not take account of vertical mixing. A modified calculation defines a *convective condensation level* (CCL). In the near-ground layer, surface heating may establish a superadiabatic lapse rate, but convection modifies this to the DALR profile. Daytime heating steadily raises the surface air temperature from T_0 to T_1, T_2 and T_3 (see Figure 2.13). Convection also equalizes the humidity mixing ratio, assumed equal to the value for the initial temperature. The CCL is located at the intersection of the environment temperature curve with a saturation mixing ratio line corresponding to the average mixing ratio in the surface layer (1000–1500 m). Expressed in another way, the surface air temperature is the minimum which enables cloud to form as a result of free convection. Because the air near the surface is often well mixed, the CCL and LCL, in practice, are commonly nearly identical.

Experimentation with a tephigram shows that both the convective and lifting condensation levels rise as the surface temperature increases with little change of dew point. This is commonly observed in the early afternoon when the base of cumulus clouds tends to be at higher levels.

Potential temperature provides an important yardstick for air mass characteristics, since if the air is only affected by dry adiabatic processes the potential temperature remains constant. This helps to identify different air masses and indicates when latent heat has been released through saturation of the air mass or when non-adiabatic temperature changes have occurred.

E AIR STABILITY AND INSTABILITY

The important characteristic of stable air is that if it is forced up or down it has a tendency to return to its former position once the motivating force ceases. Figure 2.12 shows the reason for this, in that the environmental temperature curve (*a*) lies to the right of any *path curve* repre-

senting the lapse rate of an unsaturated air parcel cooling dry adiabatically when forced to rise. At any level the rising parcel is cooler and more dense than its surroundings and therefore tends to revert to its former level. Similarly, if the air is forced downwards it will gain in temperature at the dry adiabatic rate, always be warmer and less dense than the surrounding air, and tend to return to its former position (unless prevented from doing so). However, if local surface heating causes the environmental lapse rate near the surface to exceed the dry adiabatic lapse rate (*b*), then the adiabatic cooling of a convective air parcel allows it to remain warmer and less dense than the surrounding air so that it continues to rise through buoyancy. Similarly, if an air parcel is impelled downwards under these same conditions from a higher level it will become colder than its surroundings and there will be no check on its downward progress until it reaches the surface. The characteristic of unstable air is a tendency to continue moving away from its original level when set in motion.

A further possibility is illustrated in Figure 2.14. The air is stable in the lower layers, but if the air is forced to rise, for example by passage over a mountain range or by local surface heating, until the path curve crosses the environment curve (the level of free convection) then the air, being warmer than its surroundings, is free to rise. This is termed *conditional instability*, as the development of instability is dependent on the air mass becoming saturated. Since the environmental lapse rate is frequently between the dry and saturated adiabatic rates the state of conditional instability is a common one.

The environment curve (*b*) in Figure 2.13 intersects the path curve at a higher level. Above the level of this intersection the atmosphere is stable, but the buoyant energy gained by the rising parcel enables it to move some distance into this region. The theoretical upper limit of cloud development can be estimated from the tephigram by determining an area (B) above the intersection of the environment and path curves equal to that between the two curves from the

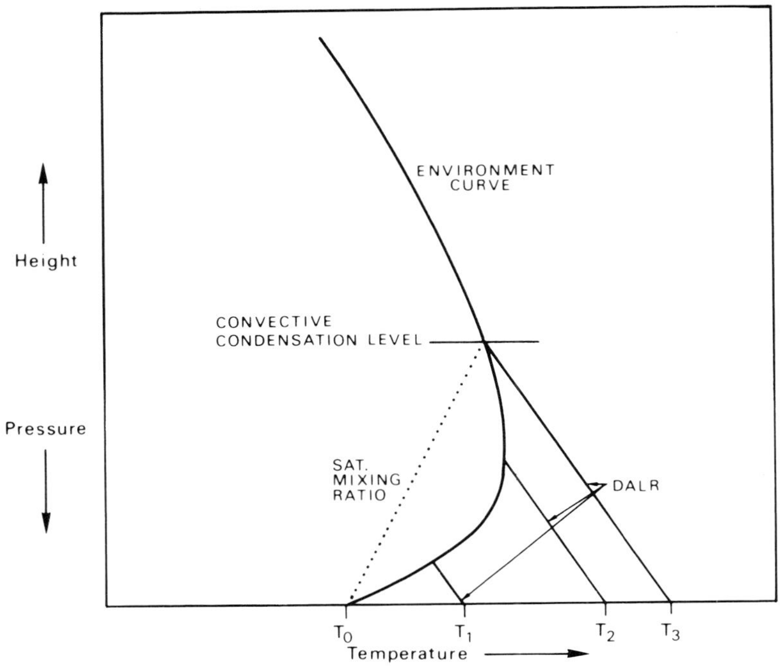

Figure 2.13 Schematic adiabatic chart used to determine the convective condensation level (see p. 66). T_0 represents the early-morning temperature: T_1, T_2 and T_3 illustrate daytime heating of the surface air.

level of free convection to the intersection (A in Figure 2.14); the tephigram is so constructed that equal areas represent equal energy.

These examples assume that a small air parcel is being displaced without any compensating air motion or mixing of the parcel with its surroundings. These assumptions are rather unrealistic since dilution of an ascending air parcel by mixing of the surrounding air with it through *entrainment* will reduce its buoyant energy. However, the method is generally satisfactory for routine forecasting, especially perhaps since the assumptions approximate conditions in the updraught of cumulonimbus clouds.

A further consideration is that a deep layer of air may be displaced by vertical motion over an extensive topographic barrier. Figure 2.15 shows a case where the air in the upper levels is less moist than that at lower levels. If the whole layer is forced bodily upwards the drier air at B follows the dry adiabatic rate, and so, for a while, will the air about A, but there eventually comes a time when the condensation level is reached, after which the lower layers of the rising air mass cool at the saturated adiabatic rate. This has the final effect of increasing the actual lapse rate of the total thickness of the raised layer, and, if this new rate is greater than that of the saturated adiabatic, the air layer becomes unstable, and may overturn. This is termed *convective (or potential) instability.*

Vertical mixing of air was referred to earlier as a possible cause of condensation. This is best illustrated by use of a tephigram. Figure 2.16 shows an initial distribution of temperature and dew point. Vertical mixing has the effect of averaging these conditions through the layer

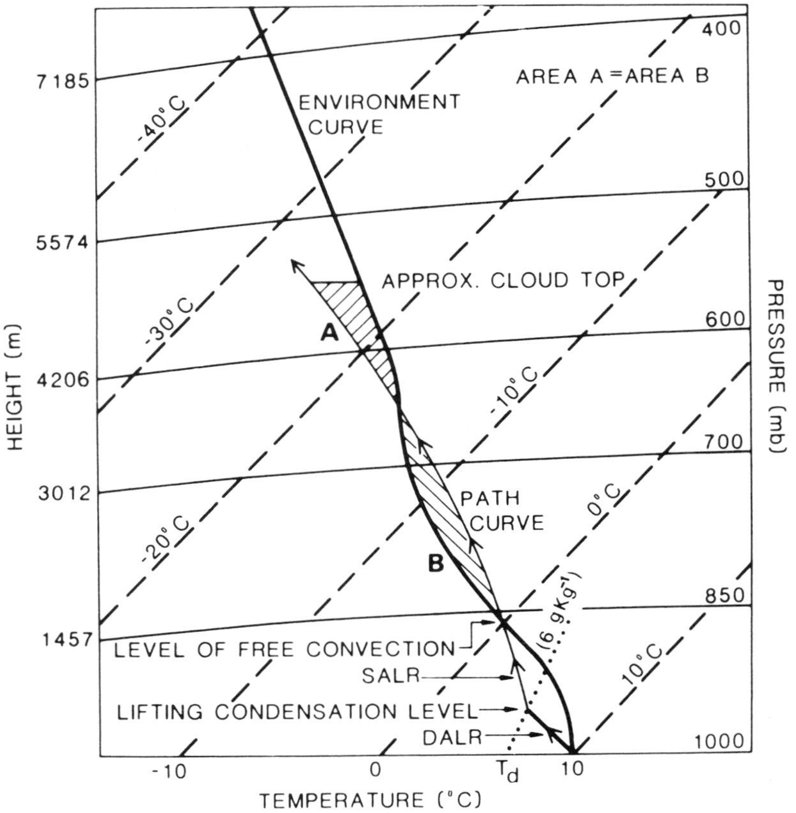

Figure 2.14 Schematic tephigram illustrating the conditions associated with the conditional instability of an air mass that is forced to rise. The saturation mixing ratio is a broken line and the lifting condensation level (cloud base) is below the level of free convection.

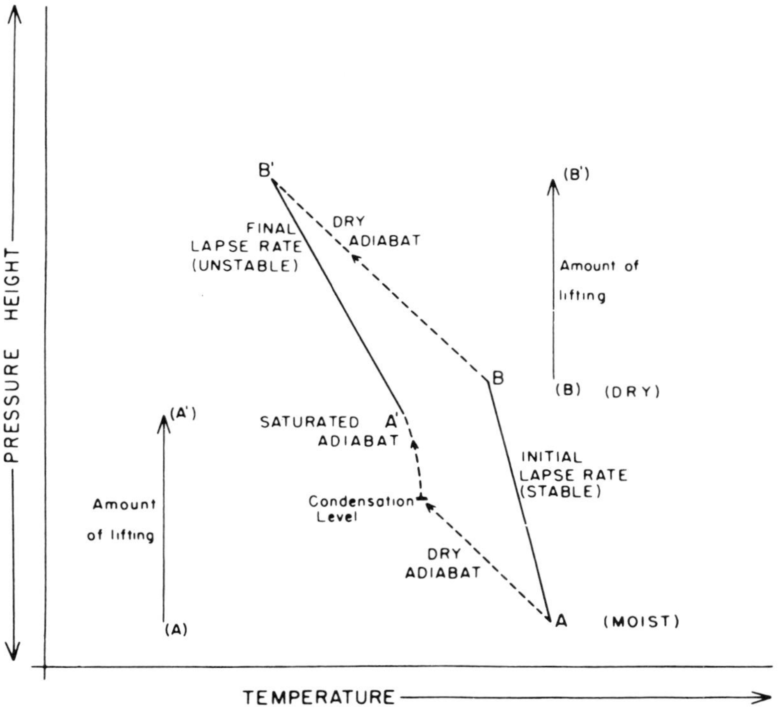

Figure 2.15 Convective instability. AB represents the initial state of an air column; moist at A, dry at B. After uplift of the whole air column the temperature gradient A′ B′ exceeds the saturated adiabatic lapse rate so that the air column is unstable.

Figure 2.16 Graph illustrating the effects of vertical mixing in an air mass. The horizontal lines are pressure surfaces (P_2, P_1). The initial temperature (T_1) and dew point temperature (T_{d_1}) gradients are modified by turbulent mixing to T_2 and T_{d_2}. The condensation level occurs where the dry adiabat (θ) through T_1 intersects the saturation humidity mixing ratio line (X_s) through T_{d_2}.

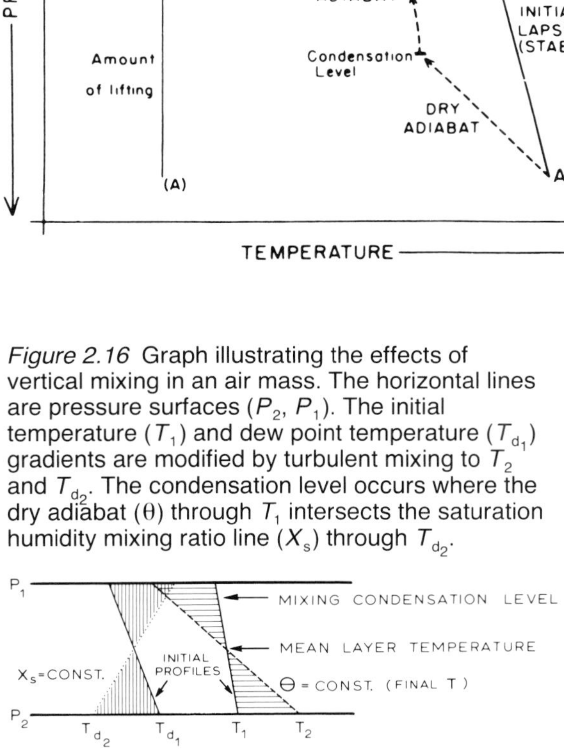

affected. Thus, the *mixing condensation level* is determined from the intersection of the average values of saturation humidity mixing ratio and potential temperature. The areas above and below the points where these average value lines cross the initial environment curves are equal.

Subsidence usually results from either radiational cooling or an excess of horizontal convergence of air in the upper troposphere. Subsiding air generally moves with a vertical velocity of only 1–10 cm s^{-1} (about 100–1000 ft hr^{-1}), although convectional conditions provide an exception (see H, this chapter). Subsidence can produce substantial changes in the atmosphere, and, for instance, if an air mass subsides about 300 m (1000 ft) all average-size cloud droplets will usually be evaporated.

F CLOUD FORMATION

The formation of clouds is dependent on atmospheric instability and vertical motion but it is also controlled by microscale processes. These are now discussed before we examine larger-scale aspects of cloud development and type.

1 Condensation nuclei

It is very important to note that condensation occurs with the utmost difficulty in *clean* air; moisture must generally find a suitable surface upon which it can condense. If pure air is reduced in temperature below its dew point it becomes *supersaturated* (i.e. relative humidity exceeding 100 per cent). To maintain a pure water drop of radius 10^{-7} cm (0.001 μm) requires a relative humidity of 320 per cent, and for one of 10^{-5} cm (0.1 μm) radius 101 per cent.

Usually condensation occurs on a foreign surface which can be a land or plant surface, as is the case for dew or frost, while in the free air condensation begins around so-called *hygroscopic nuclei*. These are microscopic particles – aerosols – the surfaces of which (like the weather enthusiast's seaweed!) have the property of *wettability*. Aerosols include dust, smoke, salts and chemical compounds. Sea salts, which are particularly hygroscopic, enter the atmosphere principally by the bursting of air bubbles in foam and are a major component of the aerosol load near the surface of the oceans. Other large contributions are from fine soil particles and various natural, industrial and domestic combustion products raised by the wind. A further source is the conversion of atmospheric trace gas to particles through photochemical reactions, particularly over urban areas. Nuclei range in size from those with a radius of 0.001 μm, which are ineffective because of the high supersaturations required for their activation, to *giants* of over 10 μm which do not remain airborne for very long (see pp. 3–4). On average, oceanic air contains 1 million condensation nuclei per litre (i.e. cm^3) and land air some 5 or 6 million. Near the surface marine aerosols are predominantly sea salts, but these tend to be rapidly removed due to their size. In the marine troposphere there are fine particles (mainly ammonium sulphate). A photochemical origin associated with anthropogenic activities accounts for about half of these in the northern hemisphere. Dimethyl sulphide (DMS) associated with algal decomposition also undergoes oxidation to sulphate. Over the tropical continents aerosols are produced biogenically, by forest vegetation and surface litter, and through biomass burning during the dry season; particulate organic carbon predominates. In mid-latitudes, remote from anthropogenic sources, coarse particles are mostly of crustal origin (calcium, iron, potassium and aluminium) whereas crustal, organic and sulphate particles are almost equally represented in the fine aerosol load.

Hygroscopic aerosols are soluble. This is very important since the saturation vapour pressure is less over a solution droplet (for example, sodium chloride or sulphuric acid) than over a pure water drop of the same size and temperature (see Figure 2.17). Indeed, condensation begins on hygroscopic particles before the air is saturated; in the case of sodium chloride nuclei at 78 per cent relative humidity. Figure 2.17 illustrates droplet radii for three sets of solution droplets of sodium chloride (a common sea salt) in relation to their equilibrium relative humidity, called Kohler curves. Droplets in an environment where values are below/above the appropriate line will evaporate/grow. Each curve has a maximum beyond which the droplet can grow in air with less supersaturation.

Once they are initially formed, the process of growth of water droplets is far from simple and much remains to be explained. In the early stages the solution effect is predominant and small drops grow more quickly than large ones, but, as the size of a droplet increases its growth rate by condensation decreases, as shown in Figure 2.18. The radial growth rate obviously slows down as the drop size increases because there is an increasingly greater surface area to add to with every increment of radius. However, the condensation rate is limited by the speed with which the released latent heat can be lost from the drop by conduction to the air and this heat reduces the vapour gradient. Moreover competition between droplets for the available moisture increasingly tends to reduce the degree of supersaturation.

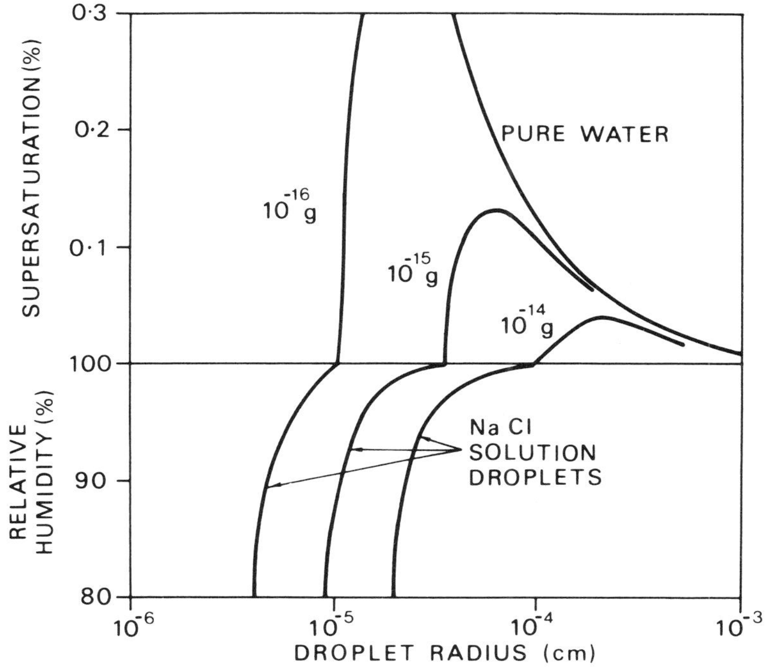

Figure 2.17 Kohler curves showing the variation of equillibrium relative humidity or supersaturation (%) with droplet radius for pure water and NaCl solution droplets. The numbers show the mass of sodium chloride (a similar family of curves is obtained for sulphate solutions). The pure water droplet line illustrates the curvature effect.

Source: After Mason 1975.

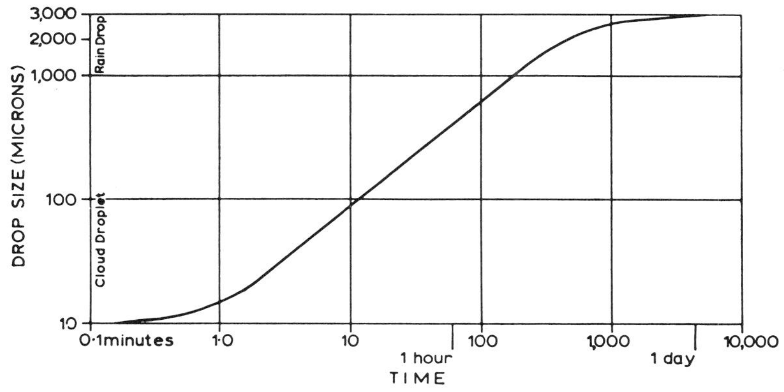

Figure 2.18 Droplet growth by condensation (note the logarithmic scale).

Supersaturation in clouds very rarely exceeds 1 per cent and, because the saturation vapour pressure is greater over a curved droplet surface than over a plane water surface, very small droplets ($< 0.1\ \mu m$ radius) are readily evaporated (see Figure 2.17). In the early stages the nucleus size is important; for supersaturation of 0.05 per cent a droplet of 1 μm radius with a salt nucleus of mass 10^{-13} g reaches 10 μm in 30 min, whereas one with a salt nucleus of 10^{-14} g would take 45 min. Later, when the dissolved salt has ceased to have significant effect, the radial growth rate becomes slow as a result of decreasing supersaturation.

Figure 2.18 illustrates not only the slow growth of droplets but also the immense size difference between cloud droplets (< 1 to 50 μm radius) and raindrops (exceeding 1 mm

diameter). These facts strongly suggest that the gradual process of condensation is inadequate to explain the rates of formation of raindrops which are often observed. For example, in most clouds precipitation develops within an hour. It must be remembered too that falling raindrops undergo evaporation in the unsaturated air below the cloud base. A droplet of 0.1 mm radius evaporates after falling only 150 m at a temperature of 5°C and 90 per cent relative humidity, but a drop of 1 mm radius would fall 42 km before evaporating. It seems likely then that cloud droplets are not necessarily the immediate source of raindrops. This point is taken up again in Section G.

2 Cloud types

The great variety of cloud forms necessitates a classification for purposes of weather reporting. The internationally adopted system is based upon (*a*) the general shape, structure and vertical extent of the clouds, and (*b*) their altitude.

These primary characteristics are used to define the ten basic groups (or genera) as shown in Figure 2.19. High cirriform cloud is composed of ice crystals giving a generally fibrous appearance. Stratiform clouds are layer-shaped, while cumuliform ones have a heaped appearance and usually show progressive vertical development. Other prefixes are *alto-* for middle level (medium) clouds and *nimbo-* for low clouds of considerable thickness, which appear dark grey and from which continuous rain is falling.

The height of the cloud base may show a considerable range for any of these types and varies with latitude. The approximate limits in thousands of metres for different latitudes are shown in Table 2.2.

Following taxonomic practice, the classification subdivides the major groups into species and varieties with Latin names according to details of their appearance. The World Meteorological Organization has produced an *International Cloud Atlas* illustrating all of these types.

Figure 2.19 The ten basic cloud groups classified according to height and form.

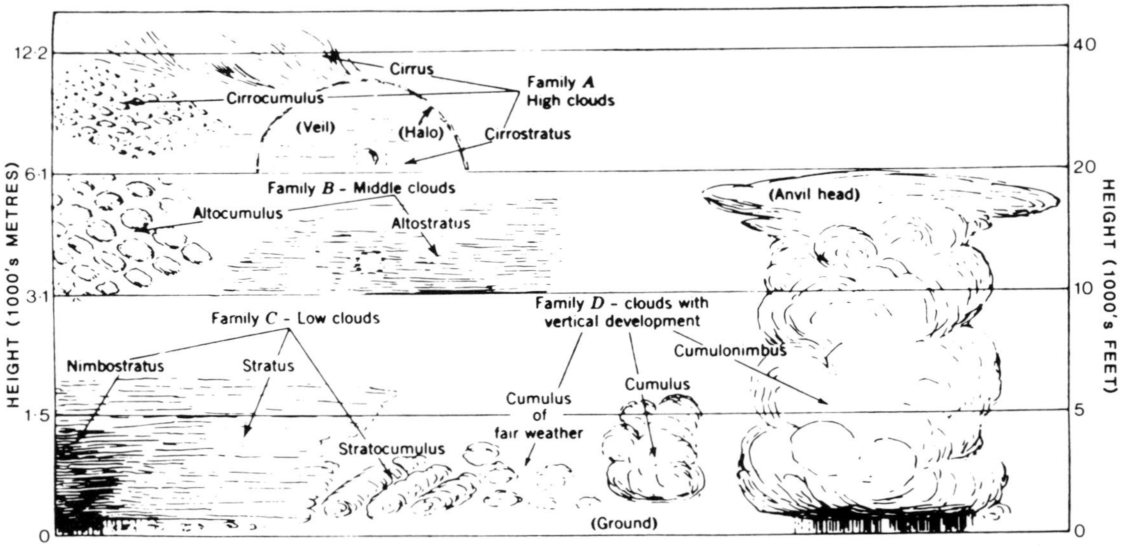

Source: From Strahler 1965.

Table 2.2 Cloud base height (in 000s m).

	Tropics	*Middle latitudes*	*High latitudes*
High cloud	Above 6	Above 5	Above 3
Medium cloud	2–7.5	2–7	2–4
Low cloud	Below 2	Below 2	Below 2

Other possible classifications of clouds take into account their mode of origin. For instance, a broad genetic grouping can be made according to the mechanism of vertical motion which produces condensation. Four categories are:

1 gradual uplift of air over a wide area in association with a low-pressure system;
2 thermal convection (on the local cumulus-scale);
3 uplift by mechanical turbulence (*forced convection*);
4 ascent over an orographic barrier.

Group 1 includes a wide range of cloud types and will be discussed more fully in Chapter 4, D.2. In connection with thermal convection, which forms cumuliform clouds (group 2), it is worth noting that upward convection currents (thermals) form plumes of warm air which, as they rise, expand and are carried downwind. Towers in cumulus and other clouds are caused not by thermals of surface origin, but by ones set up *within* the cloud as a result of the release of latent heat by condensation. Thermals gradually lose their impetus as mixing of cooler, drier air from the surroundings dilutes the more buoyant warm air. Cumulus towers also tend to evaporate as updraughts diminish, leaving a shallow oval-shaped 'shelf' cloud (*strato-cumulus cumulogenitus*) which may amalgamate with others to produce a high over-cast. Group 3 includes fog, stratus or stratocumulus and is important whenever air near the surface is cooled to dew point by conduction or night-time radiation and the air is stirred by irregularities of the ground. The final group (4) could include stratiform or cumulus clouds produced by forced uplift of air over mountains (see Plates 6 and 7). Hill fog is simply stratiform cloud enveloping high ground. A special and important category is the wave (lenticular) cloud which develops when air flows over hills setting up a wave motion in the air current downwind of the ridge (see Chapter 3, C.2). Clouds form in the crest of these waves if the air reaches its condensation level (see Plates 8 and 9).

A great deal of information on cloud amounts, especially in remote areas, and on cloud patterns in relation to weather systems is now being provided by operational weather satellites. These supply direct readout imagery and data that cannot be obtained by ground observations. Special classifications of cloud elements and patterns have been devised in order to analyse satellite imagery. The most common patterns seen on satellite photographs are cellular, or honeycomb-like, with a typical diameter of 30 km. These develop from the movement of cold air over a warmer sea surface. An open cellular pattern, where cumulus clouds are along the cell sides, forms where there is a large air–sea temperature difference, whereas closed polygonal cells occur if this difference is small. In both cases there is subsidence above the cloud layer. Open (closed) cells are more common over warm (cool) ocean currents to the east (west) of the continents. The honeycomb pattern has been attributed to mesoscale con-vective mixing, but the cells have a width–depth ratio of about 30:1, whereas laboratory thermal convection cells have a corresponding ratio of only 3:1. Thus the true explanation may be more complicated. Occasionally, subsidence over the tropical oceans leads to a modification of the cellular pattern referred to as 'actiniform'

(or radiating), illustrated in Plate 10. Another common pattern over oceans and uniform terrain is provided by linear cumulus cloud 'streets'. Helical motion in these two-dimensional cloud cells develops with surface heating, particularly when outbreaks of polar air move over warm seas (see Plates 11 and 29) and there is a capping inversion. Cloud patterns related to cyclonic systems are discussed in Chapter 4, D.

3 Global cloud cover

Clouds represent an important sink for radiative energy in the earth–atmosphere system by absorption, as well as a source by reflection and re-radiation (see Chapter 1, D.2 and D.3). In recent years satellites have provided large quantities of cloud data but, as noted earlier, surface-based estimates of cloud amounts are some 10 per cent greater than those derived from satellites, the greatest discrepancies occurring in summer in the subtropics and in the highest latitudes.

Total cloud percentage amounts show characteristic geographical, latitudinal and seasonal distributions (see Figures 1.21 and 2.20). During the northern summer there are high percentages over West Africa, north-west South America and South-East Asia; with minima over the southern hemisphere continents, southern Europe, North Africa and the Near East. During the northern winter there are high percentages over tropical land areas in the southern hemisphere, partly due to convection along the Inter–Tropical Convergence Zone, and in sub-polar ocean areas due to the advection of moist air. Minimal cloud cover is naturally associated with the sub-tropical high pressure regions throughout the year, whereas persistent maximum cloud cover occurs over the open ocean belt at 50°–70°S and over large parts of the northern hemisphere oceans north of about 45°N.

G FORMATION OF PRECIPITATION

The puzzle of raindrop formation has already been briefly mentioned. The simple growth of cloud droplets is apparently an inadequate mechanism by itself, but if this is the case then more complex processes have to be envisaged.

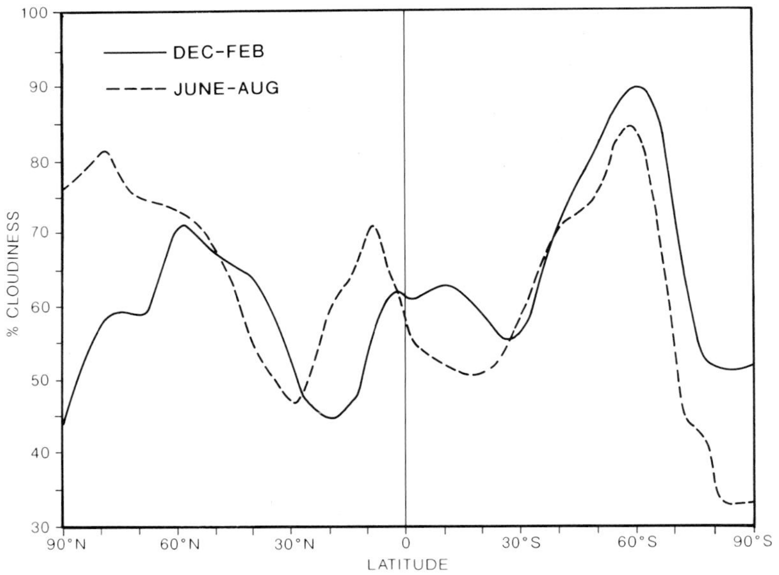

Figure 2.20 The average zonal distribution of total cloud amount (%) derived from surface observations over the total global surface (i.e. land plus water) for the months of December–February and June–August during the period 1971–81.

Source: From London *et al.* 1989.

Plate 1 Visible image of Africa, Europe and the Atlantic Ocean taken by METEOSAT on 19 August 1978 at 1155 hours GMT. An anticyclone is associated with clear skies over Europe and the Mediterranean, while frontal-wave cyclones are evident in the North Atlantic. Cloud clusters appear along the oceanic Intertropical Convergence Zone and there are extensive monsoon cloud masses over equatorial West Africa. Less-organized cloud cover is present over East Africa. The subtropical anticyclone areas are largely cloud-free but possess trade-wind cumulus, particularly in the South-East Trade Wind belt of the South Atlantic. The highly-reflective desert surfaces of the Sahara are prominent (*METEOSAT image supplied by the European Space Agency*).

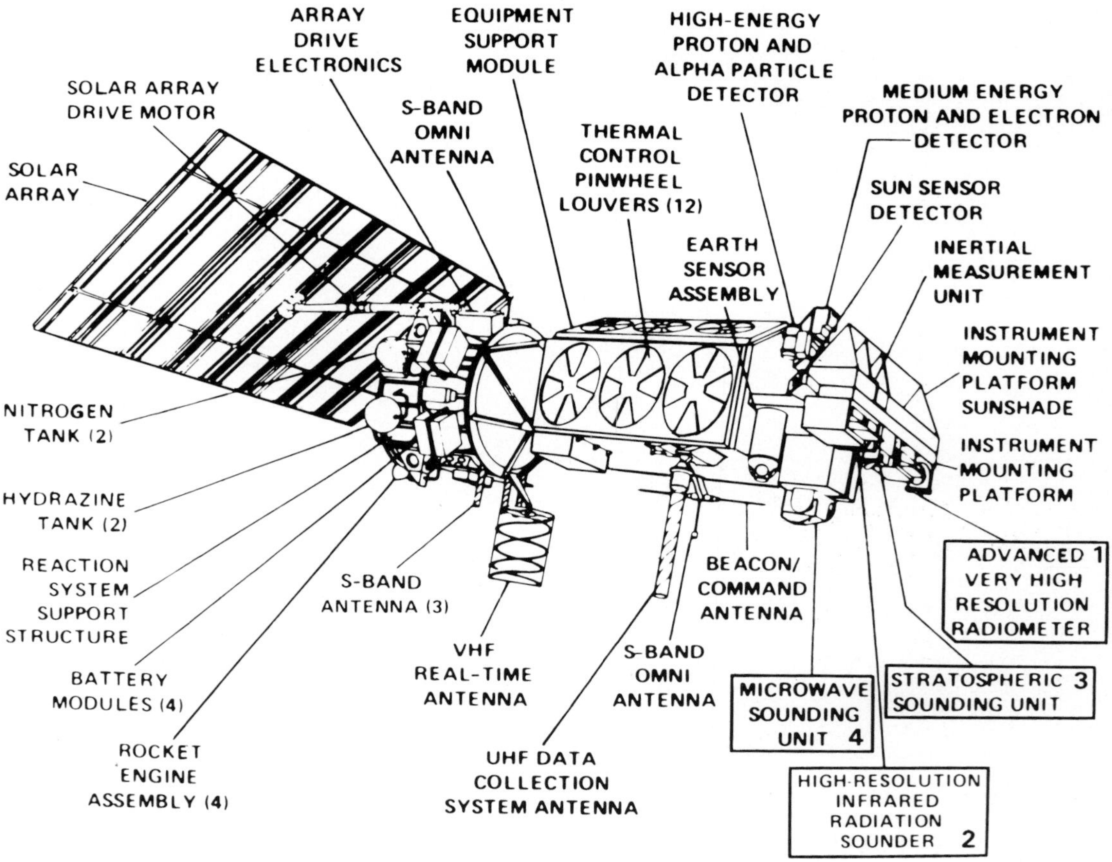

Plate 2 The TIROS-N spacecraft, having a length of 3.71 m and a weight of 1421 kg. The four instruments of particular meteorological importance are shown in the numbered boxes: 1 Visible and infra-red detector – discerns clouds, land–sea boundaries, snow and ice extent and temperatures of clouds, earth's surface and sea surface. 2 Infra-red detector – permits calculation of temperature profile from the surface to the 10-mb level, as well as the water vapour and ozone contents of the atmosphere. 3 Device for measuring radiation from the top of the atmosphere. 4 Device for measuring microwave radiation from the earth's surface (*NOAA: National Oceanic and Atmospheric Administration*).

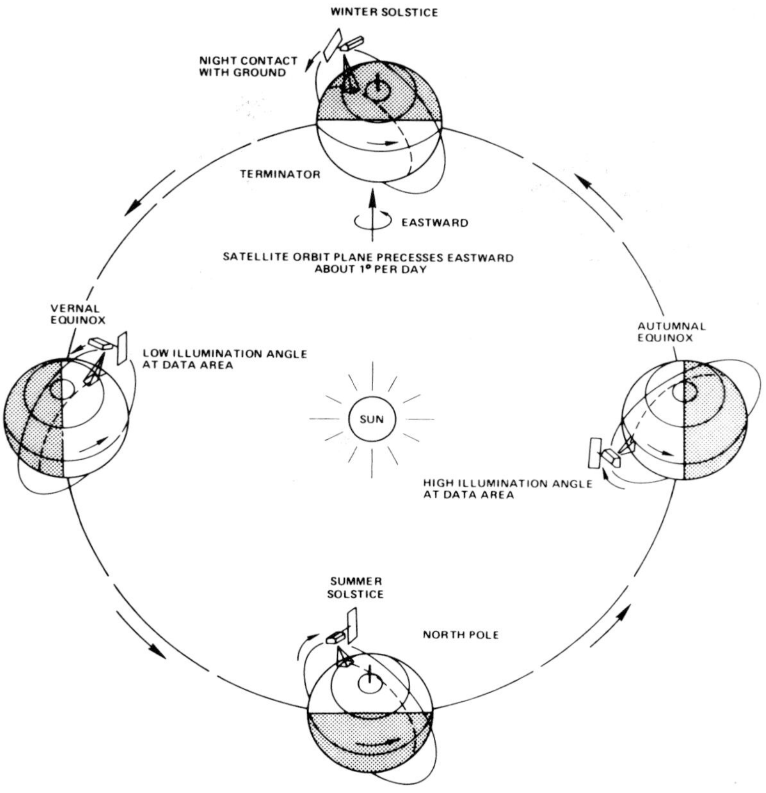

Plate 3 The TIROS-N satellite system consisting of two spacecraft in polar orbit at 833 and 870 km, respectively. The orbital plane of the second satellite lags 90° longitude behind that of the first and the orbital plane of each precesses eastward at about 1° longitude per day. Each satellite transmits data from a circular area of the earth's surface 6200 km in diameter. The satellites make 14.18 and 14.07 orbits of the earth per day, respectively, such that each point on the earth is sensed for 13–14 minutes at a time (*NOAA*).

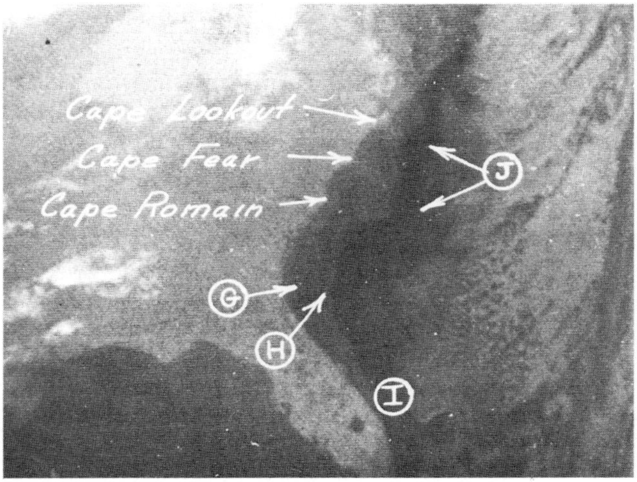

Plate 4 Night-time (0900 GMT) infra-red photograph showing sea-surface temperatures off the south-east coast of the United States on 15 February 1971 (see Figure 1.31). G: Coldest shelf water. H: Shelf water of intermediate temperature. I: Warm Gulf Stream, clearly showing meanders associated with cold water intrusions (J) (*World Meteorological Organization 1973*).

Plate 5 Cold, fog-laden air draining over the southern rim of the Grand Canyon, Arizona, elevation 2075 m (6800 ft) in the early morning (*photograph Ernst Haas; Courtesy Time/Life Publications*).

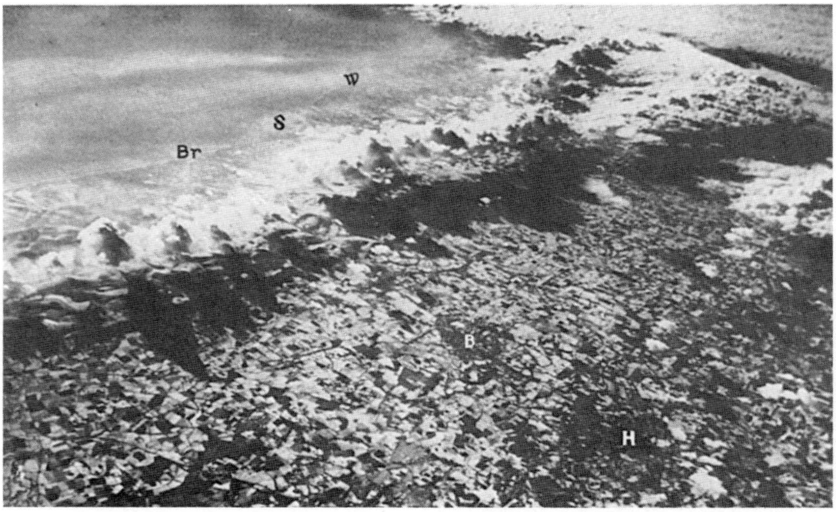

Plate 6 Cumulus orographic clouds developed over the dip-slope of the South Downs in Sussex, England. To the west (*right*), southern Hampshire is covered by stratiform clouds. The English Channel is in the upper left. This infra-red photograph was taken from an elevation of about 12,000 m (40,000 ft) (B = Burgess Hill; Br = Brighton; H = Haywards Heath; S = Shoreham; W = Worthing) (*P.A.–Reuters Ltd*).

Plate 7 Layered clouds formed as a moist, stable airflow is forced to rise over a mountain barrier (*from* A Field Guide to the Atmosphere *by Vincent J. Schaefer and John A. Day. Copyright © 1981 by Vincent J. Schaefer and John A. Day. Reprinted by permission of Houghton Mifflin Co.* All rights reserved.)

Plate 8 View north along the eastern front of the Colorado Rockies, showing lee-wave clouds (*NCAR photograph by Robert Bumpas*).

Plate 9 View looking south-south-east from about 9000 m (30,000 ft) along the Owen's Valley, California, showing a roll cloud developing in the lee of the Sierra Nevada mountains. The lee-wave crest is marked by the cloud layer, and the vertical turbulence is causing dust to rise high into the air (W = Mount Whitney, 4418 m (14,495 ft); I = Independence) (*photograph by Robert F. Symons; Courtesy R. S. Scorer*).

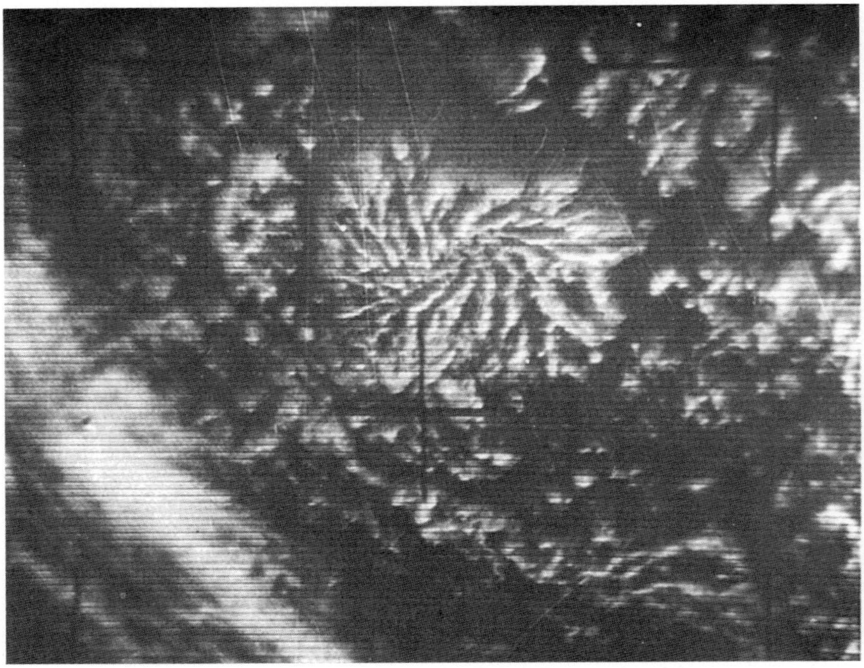

Plate 10 Radiating or dendritic cellular (actiniform) cloud pattern. These complex convective systems, some 150–250 km (90–150 miles) in diameter, were only discovered as a result of satellite photography. They usually occur in groups over areas of subsidence inversions intensified by cold ocean currents (e.g. in the low latitudes of the eastern Pacific) (*Environmental Science Service Administration*).

Plate 11 DMSP visible image of the coastal area off New England at 1433 hours GMT, 17 February 1979. A northerly airflow averaging 10 m s^{-1}, with surface air temperatures of about −15°C, moves offshore where sea-surface temperatures increase to 9°C within 250 km of the coast. Convective cloud streets are visible, also ice in James Bay (*upper left*), and in the Gulf of St Lawrence. (*Image courtesy of National Snow and Ice Data Center, University of Colorado, Boulder*). (see *Monthly Weather Review* **III**, 1983, p. 265).

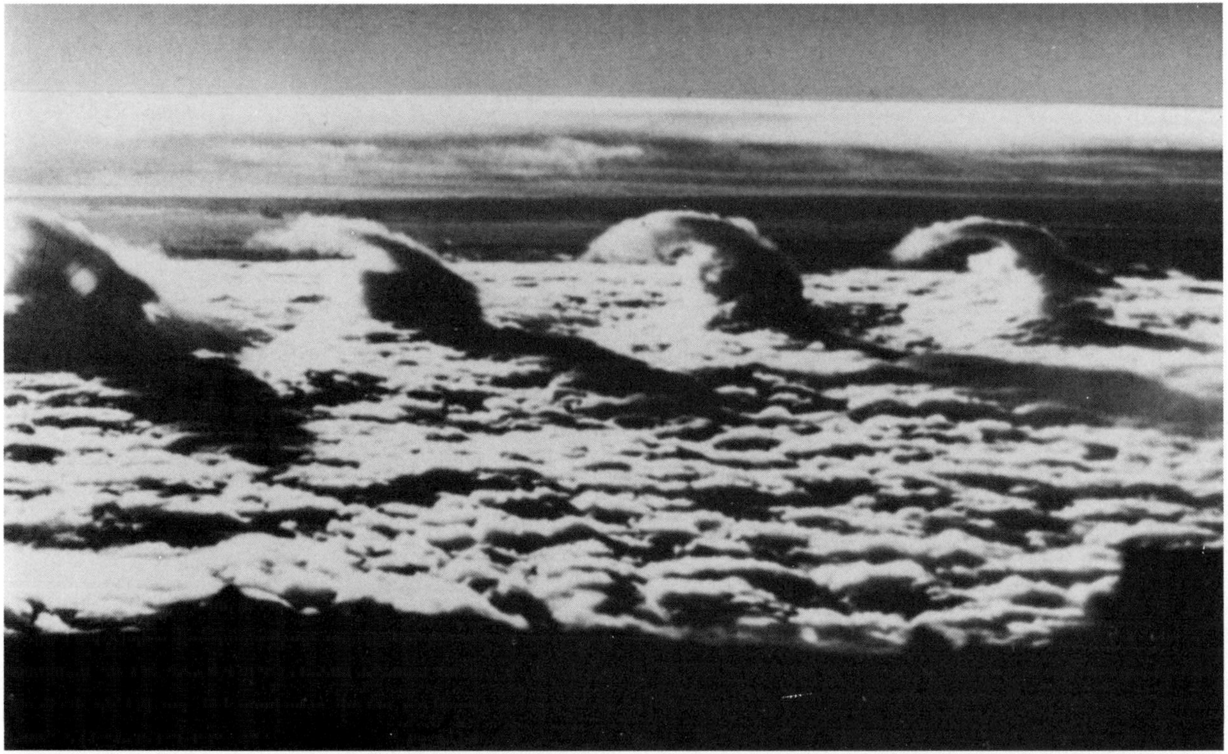

Plate 12 Four cumulus towers protruding through a stable altostratus cloud sheet over Montana. Tower tops are displaced by a 70 m s^{-1} (150 mph) jet stream (*from* A Field Guide to the Atmosphere *by Vincent J. Schaefer and John A. Day. Copyright © 1981 by Vincent J. Schaefer and John A. Day. Reprinted by permission of Houghton Mifflin Co*, All rights reserved.)

Plate 13 Isolated desert thunderstorm, 25 km (15 miles) west of Mt Lemmon, near Tucson, Arizona, on 31 July 1973 (*photograph courtesy Dr Brooks Martner, University of Wyoming*).

Plate 14 Thunderstorm approaching Östersund, Sweden, during late afternoon on 23 June 1955. Ahead of the region of intense precipitation there are rings of cloud formed over the squall-front (*copyright F. H. Ludlam; originally published in* Weather, *vol. XV, no. 2, 1960, p. 63*).

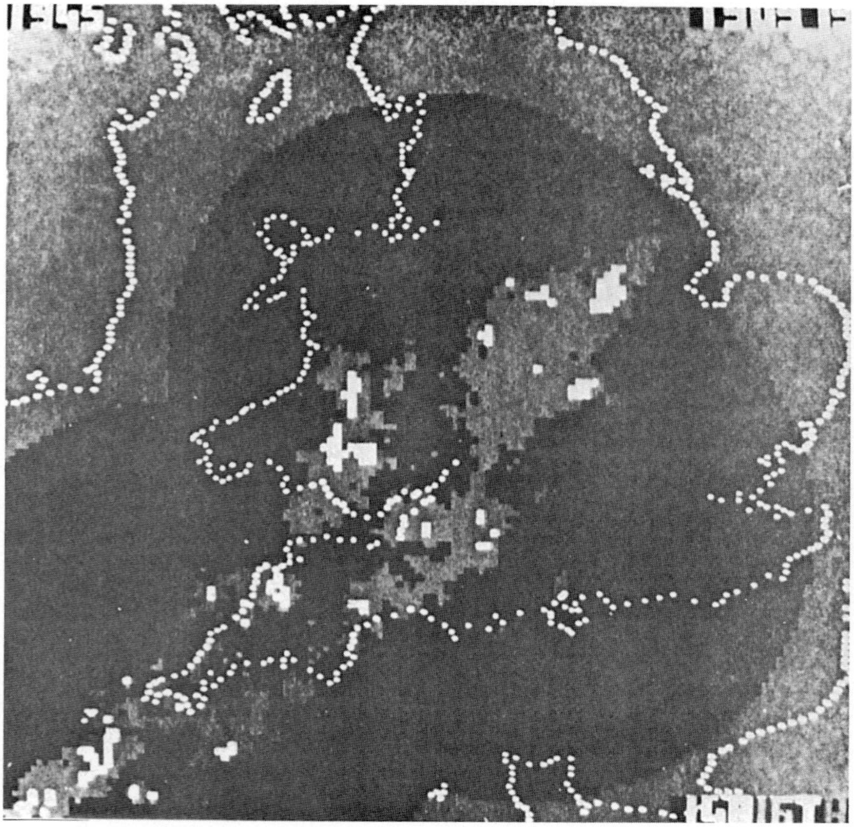

Plate 15 Photograph of a composite radar display from three radars (intersecting dark circles) showing the streaky rainfall distribution over southern Britain at 1945 hours on 19 September 1979. The display consists of a 128 × 128 matrix of 5-km squares, with light squares representing heavy rain and grey squares low to moderate rainfall intensity (*from Browning 1980*).

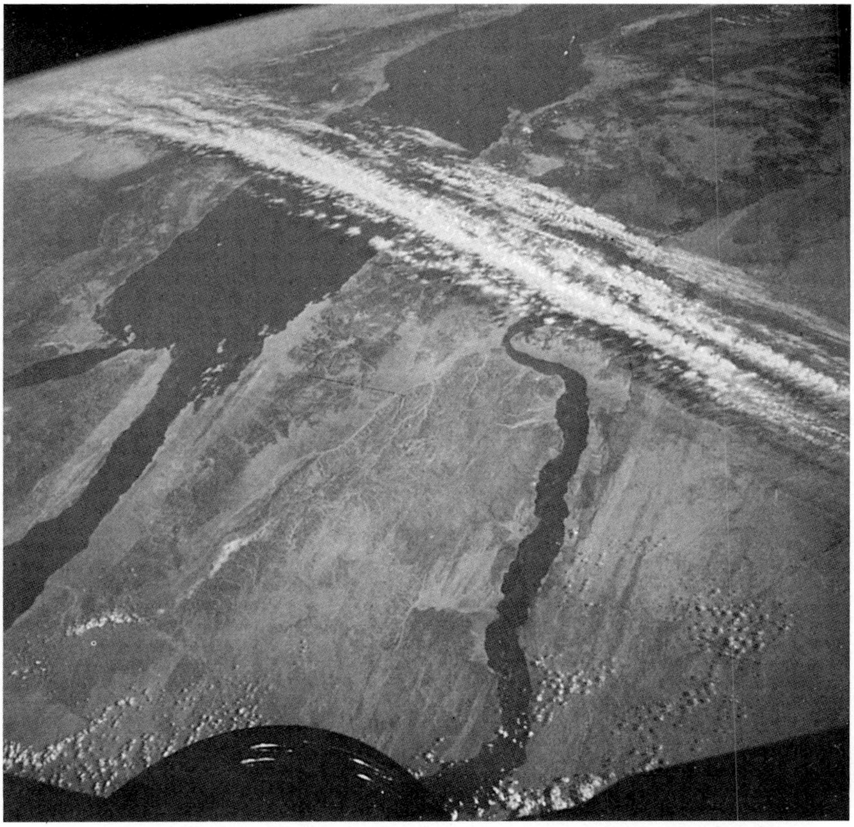

Plate 16 Photograph by an astronaut from Gemini XII manned spacecraft from an elevation of some 180 km (112 miles) looking south-east over Egypt and the Red Sea. The bank of cirrus clouds is associated with strong upper winds, possibly concentrated as a jet stream (*NASA photograph*).

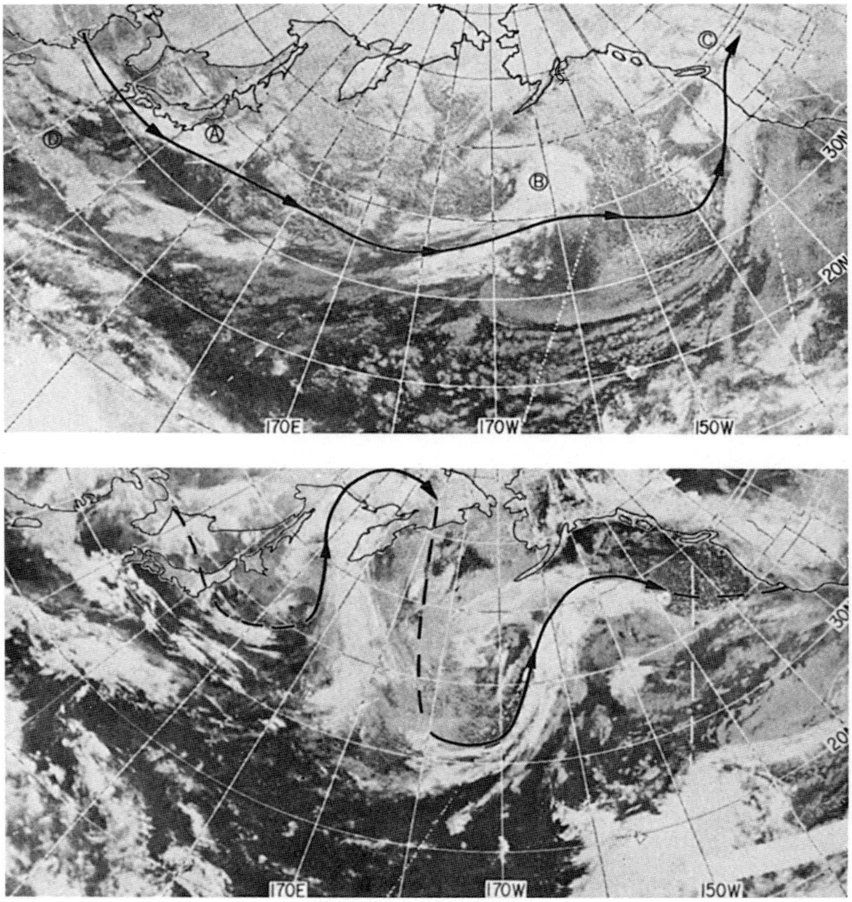

Plate 17 Infra-red photographs of the North Pacific, with the 200-mb jet stream inserted. *Above:* general zonal flow associated with a high zonal index, 12 March 1971; three major cloud systems (A, B, C) occur along the belt of zonal flow, and the large east–west belt of cloud (D) to the south of Japan is also characteristic of accentuated zonal flow. *Below:* large-amplitude flow regime associated with a lower zonal index, 23 April 1971 (*World Meteorological Organization 1973*).

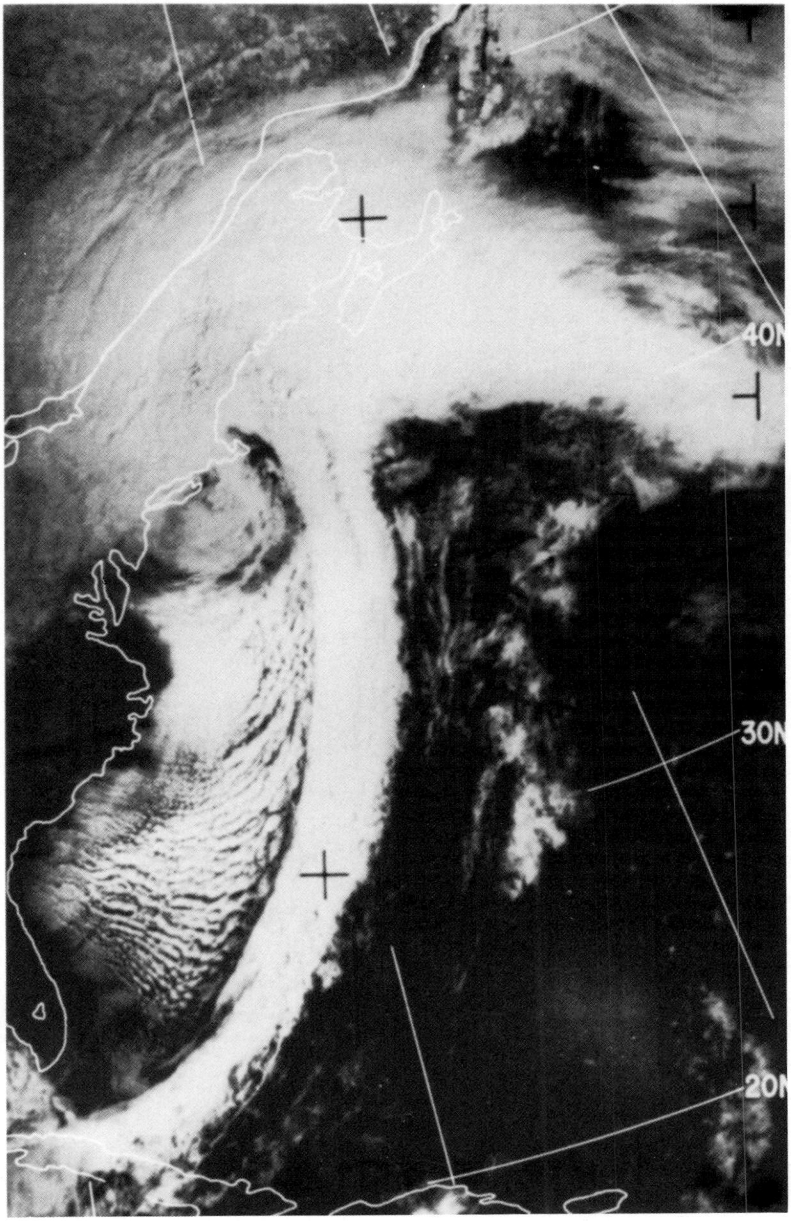

Plate 18 A well-developed, open-stage depression centred over the Maritime Provinces of Canada observed on 4 March 1971 from an ESSA 8 satellite. A broad warm front extends eastwards into the Atlantic and a narrow cold front parallels the east coast of the United States. The cooler north-westerly airflow behind the cold front is generating unstable cloud bands over the warmer ocean (*from* A Field Guide to the Atmosphere *by Vincent J. Schaefer and John A. Day. Copyright © 1981 by Vincent J. Schaefer and John A. Day. Reprinted by permission of Houghton Mifflin Co*, All rights reserved).

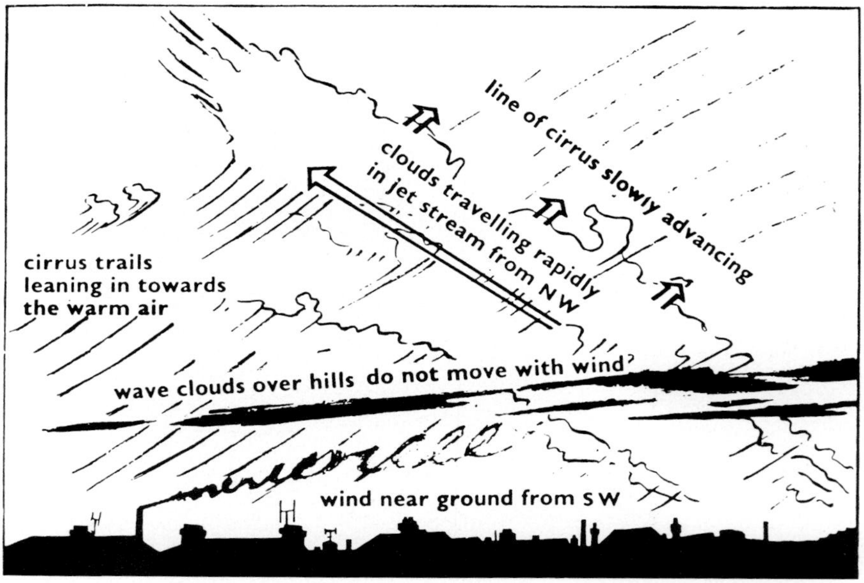

Plate 19 View looking westward towards an approaching warm front, with lines of jet-stream clouds extending from the north-west, from which trails of ice crystals are falling. In the middle levels are dark wave clouds formed in the lee of small hills by the south-westerly airflow, whereas the wind direction at the surface is more southerly – as indicated by the smoke from the chimney (*photograph copyright by F. H. Ludlam; diagram by R. S. Scorer; both published in* Weather, *vol. 18, no. 8, 1963, pp. 266–7*).

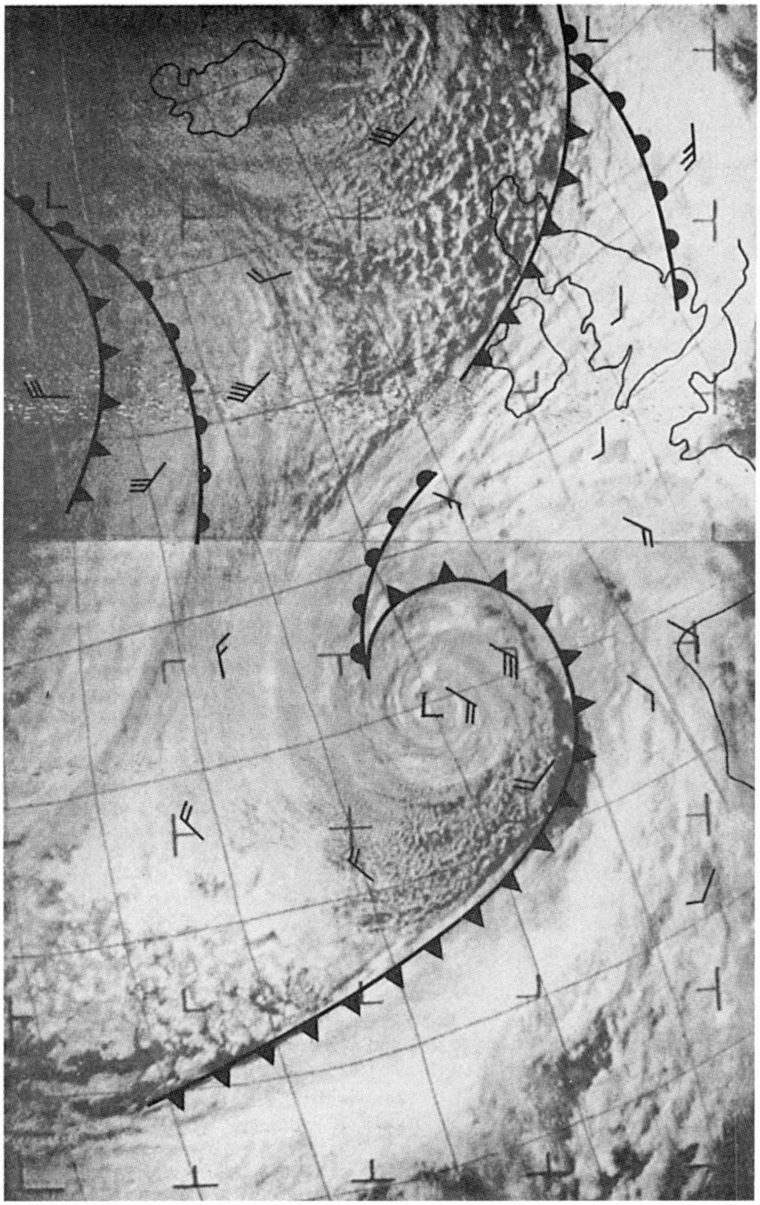

Plate 20 Photograph of a depression to the south-west of the British Isles taken by the ESSA 2 satellite at 1018 hours GMT on 12 November 1966, showing a well-marked spiral cloud structure and open cell cloud areas behind the cold fronts. The depression began as hurricane 'Lois' in the eastern Caribbean and moved north-eastwards until its winds fell below gale force on 10 November. Thereafter it began to deepen again and the pressure at the centre had dropped to 962 mb when this photograph was taken. The frontal zones are well developed, although the 1000–500-mb wind shear was unusually weak for such a well-defined system (*courtesy* Weather, *vol. XXIV, no. 6, 1969, p. 222; Crown Copyright Reserved*).

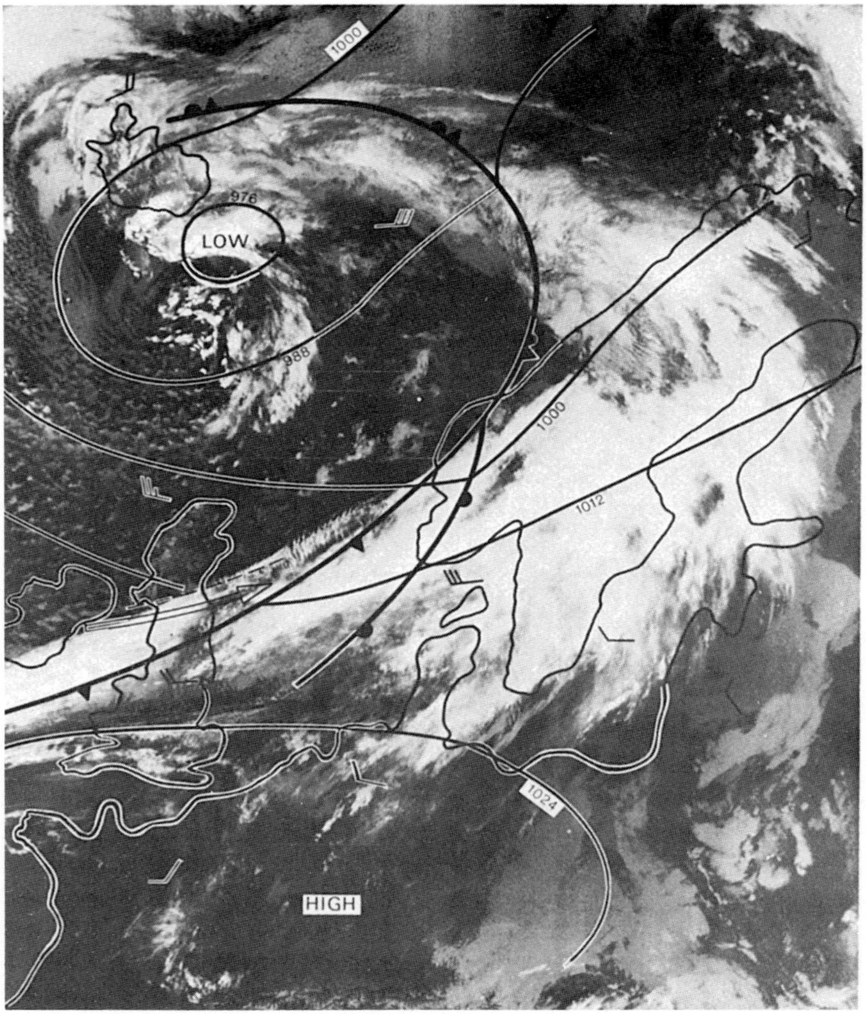

Plate 21 A partly-occluded frontal depression over north-west Europe at 1347 hours GMT on 19 October 1979. This TIROS infra-red image was processed to show the coldest surfaces (e.g. cloud tops) as white and the warmest (e.g. land surfaces) as black. Surface fronts, isobars and winds are shown, together with the upper tropospheric jet stream (arrowed), lying on the warm side of the cold front aloft (see Figure 3.23) (maximum velocity 65 m s^{-1} at about 250 mb; 11.3 km); also note the break (dashed) between the mid-latitude tropopause (200 mb; 11.8 km) and the polar tropopause (270 mb; 10.2 km). This added synoptic information relates to 1200 hours GMT (*image courtesy of P. E. Baylis, Department of Electrical Engineering and Electronics, University of Dundee; synoptic interpretation courtesy of Ross Reynolds, Department of Meteorology, University of Reading*).

Numerous early theories of raindrop growth have met with objections. For example, it was proposed that differently charged droplets could coalesce by electrical attraction, but it appears that distances between drops are too great and the difference between the electrical charges too small for this to happen. It was also suggested that large drops might grow at the expense of small ones, but observations show that the distribution of droplet size in a cloud tends to maintain a regular pattern, with the average radius between about 10 to 15 μm and a few larger than 40 μm. Another proposal was based on the variation of saturation vapour pressure with temperature such that if atmospheric turbulence brought warm and cold cloud droplets into close conjunction the supersaturation of the air with reference to the cold drop surfaces and the undersaturation with reference to the warm drop surfaces would cause the warm drops to evaporate and the cold ones to develop at their expense. However, except perhaps in some tropical clouds, the temperature of cloud droplets is too low for this differential mechanism to operate. Figure 2.9 shows that below about −10°C the slope angle of the saturation vapour pressure curve is low. Another theory was that raindrops grow around exceptionally large condensation nuclei (such as have been observed in some tropical storms). Large nuclei, it is known, do experience a more rapid rate of initial condensation, but after this stage they are subject to the same limiting rates of growth that apply to all other atmospheric water drops.

The two main groups of current theories which attempt to explain the rapid growth of raindrops involve the growth of ice crystals at the expense of water drops, and the coalescence of small water droplets by the sweeping action of falling drops.

1 Bergeron–Findeisen theory

This theory forms an important part of the presently accepted mechanism of raindrop growth, and is based on the fact that the relative humidity of air is greater with respect to an ice surface than with respect to a water surface. As the air temperature falls below 0°C the atmospheric vapour pressure decreases more rapidly over an ice surface than over water (see Figure 1.13). This results in the saturation vapour pressure over water becoming greater than that over ice, especially between temperatures of −5°C and −25°C where the difference exceeds 0.2 mb. If ice crystals and supercooled water droplets exist together in a cloud the latter tend to evaporate and direct deposition takes place from the vapour on to the ice crystals (this is often described by meteorologists as *sublimation*, which properly refers to direct evaporation from ice).

Just as the presence of condensation nuclei is necessary for the formation of water droplets, so *freezing nuclei* are necessary before ice particles can form – usually at very low temperatures (about −15°C to −25°C). Small water droplets can, in fact, be super-cooled in pure air to −40°C before spontaneous freezing occurs, but ice crystals generally predominate in clouds where temperatures are below about −22°C. Freezing nuclei are far less numerous than condensation nuclei; for example, there may be as few as 10 per litre at −30°C and probably rarely more than 1000. However, some become active at higher temperatures. Kaolinite, a common clay mineral, becomes initially active at −9°C and on subsequent occasions at −4°C. The origin of freezing nuclei has been the subject of much debate but it is generally considered that very fine soil particles are a major source. Another possibility is that meteoric dust provides the nuclei, although there seems to be no firm evidence of a relationship between meteorite showers and rainfall. Volcanic dust ejected into the upper stratosphere and troposphere during eruptions might be an additional terrestrial source. Recent work shows that common biogenic aerosols emitted by decaying plant litter, in the form of complex chemical compounds, serve as freezing nuclei. In the presence of certain associated bacteria, ice nucleation can take place at only −2°C to −5°C.

Once minute ice crystals have formed they grow readily by deposition from vapour, with different hexagonal forms of crystal developing at different temperature ranges. The number of ice crystals also tends to increase progressively because small splinters become detached during growth by air currents and act as fresh nuclei. The freezing of super-cooled water drops may also produce ice splinters (see H, this chapter). Ice crystals readily aggregate upon collision, due to their frequently branched (dendritic) shape, and tens of crystals may form a single snowflake. Temperatures between about 0°C and −5°C are particularly favourable to aggregation, because fine films of water on the crystals' surfaces freeze when two crystals touch, binding them together. When the fall-speed of the growing ice mass exceeds the existing velocities of the air upcurrents the snowflake falls, melting into a raindrop if it passes through a sufficient depth of air warmer than 0°C.

This theory seems to fit most of the observed facts, yet it is not completely satisfactory. Cumulus clouds over tropical oceans can give rain when they are only some 2000 m (6500 ft) deep and the cloud-top temperature is 5°C or more. Even in middle latitudes in summer precipitation may fall from cumuli which have no subfreezing layer (*warm clouds*). A suggested mechanism in such cases is that of 'droplet coalescence', which is discussed below (see pp. 76–7.

Practical rainmaking has been based on the Bergeron theory with some success, which at least supports its principal points. The basis of such experiments is the freezing nucleus. Super-cooled (water) clouds between −5°C and −15°C are *seeded* with especially effective materials, such as silver iodide or 'dry ice' (CO_2) from aircraft or ground-based generators in the case of silver iodide, promoting the growth of ice crystals and encouraging precipitation. The seeding of some cumulus clouds at these temperatures probably produces a mean increase of precipitation of 10–15 per cent from clouds which are already precipitating or are 'about to precipitate'. Increases of up to 10 per cent have resulted from seeding winter orographic storms. However, the seeding of depressions has produced no general precipitation increases, and it appears likely that clouds with an abundance of natural ice crystals or with above-freezing temperatures throughout are not susceptible to *rainmaking*. Premature release of precipitation may destroy the updraughts and cause dissipation of the cloud, however. This explains why some seeding experiments have actually *decreased* the rainfall! In other instances cloud growth and precipitation have been achieved, with some individually spectacular results having been obtained by such methods in Australia and the United States. Programmes aimed at increasing winter snowfall on the western slopes of the Sierra Nevada and Rocky Mountains by seeding cyclonic storms have been carried out for a number of years with mixed results. Their success depends in any case on the presence of suitable super-cooled clouds. When several cloud layers are present in the atmosphere, natural seeding may be important. For example, if ice crystals fall from high-level cirrostratus or altostratus (a 'releaser' cloud) into nimbostratus (a 'spender' cloud) composed of super-cooled water droplets, the latter can grow rapidly by the Bergeron process and such situations may lead to extensive and prolonged precipitation. This is a frequent occurrence in cyclonic systems in winter and is important in orographic precipitation (see I, this chapter).

Of all current attempts by man to control meteorological events none are more important than those relating to hurricanes. There is some indication that the seeding of the rising air in the cumulus eyewall may widen the ring of condensation and updraught, decrease the angular momentum of the storm, and thus the maximum speed of the winds. Such attempts are still in their infancy.

2 Collision theories

Alternative raindrop theories use collision, coalescence and 'sweeping' as the drop growth generator. It was originally thought that atmos-

pheric turbulence by making cloud particles collide would cause a significant proportion to coalesce. However, it was found that particles might just as easily break up if subjected to collisions and it was also observed that there is often no precipitation from highly turbulent clouds. Langmuir offered a variation of this simple collision theory by pointing out that falling drops have terminal velocities (typically 1–10 cm s^{-1}) directly related to their diameters such that the larger drops might overtake and absorb small droplets and that the latter might also be swept into the wake of the former and be absorbed by them. Figure 2.21 gives experimental results of the rate of growth of water drops by coalescence and also of ice particles by vapour deposition from an initial radius of 20 μm. Although coalescence is initially rather slow, the drop can reach 200 μm radius in 50 minutes; moreover, the growth rate is rapid for droplets with radii greater than 40 μm. Calculations show that drops must exceed 19 μm in radius before they can coalesce with other droplets, smaller droplets being swept aside without colliding. The initial presence of a few very large cloud droplets calls for the availability of giant nuclei (e.g. salt particles) if the cloud top does not reach above the freezing level. Observations show that maritime clouds do have relatively few large condensation nuclei (10–50 μm radius) and a high liquid water content, whereas continental air tends to contain many small nuclei (~ 1 μm) and less liquid water. Hence, rapid onset of showers is feasible by the coalescence mechanism in maritime clouds. Alternatively, if a few ice crystals are present at higher levels in the cloud (or if seeding occurs with ice crystals coming from higher cloud layers) they may eventually fall through the cloud as drops and the coalescence mechanism comes into action. Turbulence, especially in cumuliform clouds, serves to encourage collisions in the early stages and cloud electrification also increases the efficiency of coalescence. Thus, the coalescence process allows a more rapid growth than simple condensation can provide and is, in fact, common in 'warm' clouds in tropical maritime air masses, even in temperate latitudes.

Figure 2.21 Droplet growth by condensation and coalescence.

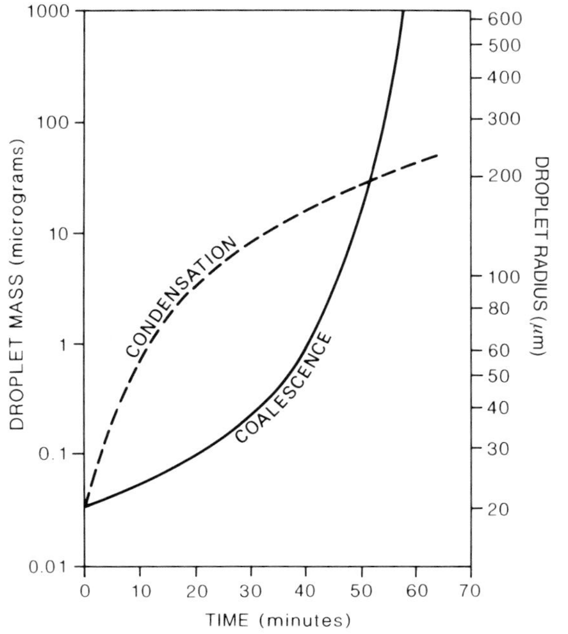

Source: From East and Marshall 1954.

3 Other types of precipitation

Rain has been discussed at length because it is the most common form of precipitation. Snow occurs when the freezing level is so near the surface that aggregations of ice crystals do not have time to melt before reaching the ground. Generally this means that the freezing level must be below 300 m (1000 ft). Mixed snow and rain ('sleet' in British usage) is especially likely when the air temperature at the surface is about 1.5°C (34° to 35°F). Snow rarely occurs with an air temperature exceeding 4°C (39°F).

Soft hail pellets (roughly spherical, opaque grains of ice with much enclosed air) occur when the Bergeron process operates in a cloud with a small liquid water content and ice particles grow mainly by deposition of water

vapour. Limited accretion of small super-cooled droplets forms an aggregate of soft, opaque ice particles 1 mm or so in radius. Showers of such pellets are quite common in winter and spring from cumulonimbus clouds.

Ice pellets may develop if the soft hail falls through a region of large liquid water content above the freezing level. Accretion forms a casing of clear ice around the pellet. Alternatively, an ice pellet consisting entirely of transparent ice may result from the freezing of a raindrop or the re-freezing of a melted snowflake.

True hailstones are roughly concentric accretions of clear and opaque ice. The embryo is a raindrop carried aloft in an updraught and frozen. Successive accretions of opaque ice (rime) occur due to impact of super-cooled droplets which freeze instantaneously, whereas the clear ice (glaze) represents a wet surface layer, developed as a result of very rapid collection of super-cooled drops in parts of the cloud with large liquid water content, which has subsequently frozen. A major difficulty in early theories was the necessity to postulate violently fluctuating upcurrents to give the observed banded hailstone structure, but a new thunderstorm model successfully accounts for this by demonstrating that the growing hailstones are re-cycled by the moving storm (see Chapter 4, H). On occasions hailstones may reach giant size, weighing up to 0.76 kg each (recorded in September 1970 at Coffeyville, Kansas). In view of their rapid fall speeds, hailstones may fall considerable distances with little melting. Hailstorms are a cause of severe damage to crops and property when large hail falls.

H THUNDERSTORMS

In temperate latitudes probably the most spectacular example of moisture changes and associated energy releases in the atmosphere is the thunderstorm. Unusually great upward and downward movements of air are both the principal ingredients and motivating machinery of such storms. They occur: (a) as rising cells of excessively heated moist air; (b) in association with the triggering-off of conditional instability by uplist over mountains; or (c) along a *squall line* in association with an air-mass discontinuity (see p. 169).

The life cycle of a local storm lasts only 1–2 hours, and begins when a parcel of air is either warmer than the air surrounding it or is actively undercut by colder encroaching air. In both instances the air begins to rise and the embryo thunder cell forms as an unstable updraught of warm air (see Figure 2.22). As condensation begins to form cloud droplets, latent heat is released and the initial upward impetus of the air parcel is augmented by an expansion and a decrease in density until the whole mass becomes completely out of thermal equilibrium with the surrounding air. At this stage updraughts may increase from 3 to 5 ms^{-1} at the cloud base to 8–10 ms^{-1} some 2–3 km higher and can exceed 30 ms^{-1}. The constant release of latent heat continuously injects fresh supplies of energy which accelerate the updraught. The rise of the air mass will continue as long as its temperature remains greater (or, in other words, its density less) than that of the surrounding air. Cumulonimbus clouds tend to form where the air is already moist, as a result of previous penetrating towers from a cluster of clouds, and there is persistent ascent.

Raindrops begin to develop rapidly when the ice stage (or freezing stage) is reached by the vertical build-up of the cell, allowing the Bergeron process to operate. They do not immediately fall to the ground because the updraughts are able to support them. The minimum cumulus depth for showers over ocean areas seems to be between 1 and 2 km, but 4–5 km is more typical inland. The corresponding minimum time intervals needed for showers to fall from growing cumulus are about 15 minutes over ocean areas and $\geqslant$ 30 minutes inland. Falls of hail require the special cloud processes, described in the last section, involving phases of ‘dry’ (rime accretion) and ‘wet’ growth on hail pellets. The mature stage of a storm (see

Figure 2.22 The cycle of a local thunderstorm. The arrows indicate the direction and speed of air currents; A The developing stage of the initial updraught; B The mature stage with updraughts and downdraughts; C The dissipating stage dominated by cool downdraughts.

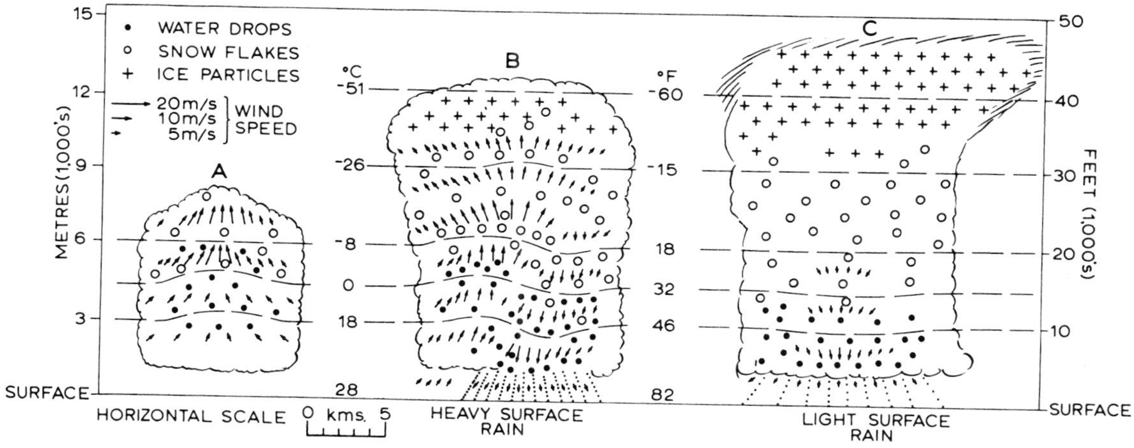

Source: After Byers and Braham; adapted from Petterssen 1958 [1969].

Figure 2.22B) is usually associated with precipitation downpours and lightning (see Plate 13). The precipitation causes frictional downdraughts of cold air. As these gather momentum cold air may eventually spread out below the thunder cell in a wedge. Gradually, as the moisture of the cell is expended, the supply of released latent heat energy diminishes, the downdraughts progressively gain in power over the warm updraughts, and the cell dissipates.

To simplify the explanation, a thunderstorm with only one cell was illustrated. Usually storms are far more complex in structure and consist of several cells arranged in clusters of 2–8 km across, 100 km or so in length and extending up to 10 km or more (see Plate 14). Such systems are known as squall lines (see Chapter 4, H).

Two general hypotheses have been developed to account for thunderstorm electrification. One involves the induction mechanism, the other non-inductive charge transfer. As an example of the first category, since the ionosphere is positively charged (owing to the action of cosmic and solar ultraviolet radiation in ionization) and the earth's surface is negatively charged during fine weather, cloud droplets can acquire an induced positive charge on their lower side and negative charge on their upper side. Non-inductive charge transfer requires contact between cloud or precipitation particles.

The typically observed distribution of charges in a thundercloud is shown in Figure 2.23. The separation of electrical charges of opposite sign may involve several mechanisms: raindrop breakup (large droplets retaining positive charge, the surface spray carrying negative ions); or the selective capture of negative atmospheric ions by falling cloud particles are possible factors, but do not appear to create sufficiently large charges. A third mechanism is the splintering of ice crystals during the freezing of cloud droplets. This operates as follows. A super-cooled droplet freezes inwards from its surface and this leads to a negatively-charged warmer core (OH^- ions) and a positively-charged colder surface due to the migration of H^+ ions outwards down the temperature gradient. When this soft hailstone ruptures during freezing small ice splinters carrying a positive charge are ejected by the ice shell and preferentially lifted to the top of the convection cell in updraughts. This theory can explain the charge distribution in Figure 2.23, which shows

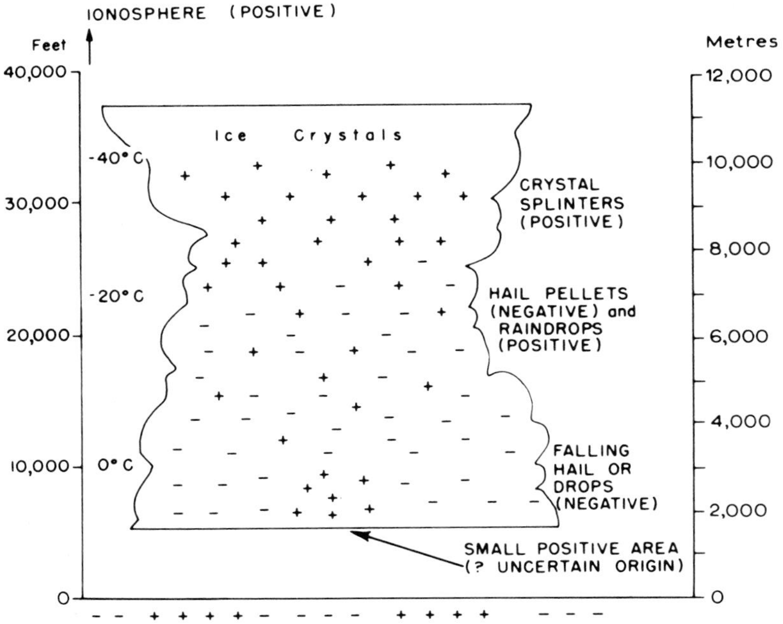

Figure 2.23 The distribution of electrostatic charges in a thunder-cloud.

Source: After Mason 1962.

that the upper part of the cloud (above about the −20°C isotherm) is positively charged. Equally, the negatively-charged hail pellets fall towards the cloud base. However, the ice-splintering mechanism appears only to work for a narrow range of temperature conditions and the charge transfer is small.

The processes discussed so far probably play a contributing role in cloud electrification, but according to J. Latham the major factor is non-inductive charge transfer. This involves collisions between splintered ice crystals and warmer pellets of soft hail. Previous accretion of super-cooled droplets on the hail pellets produces an irregular surface which is warmed as the droplets release latent heat on freezing. The impacts of ice crystals on this irregular surface generate negative charge whereas the crystals acquire positive charge. Negative charge is usually concentrated between about −10°C and −20°C in a thundercloud where ice crystal concentrations are large, due to splintering at about the −5°C level and then ascent of the crystals in upcurrents. Radar studies show that lightning is associated with both ice particles in clouds and with rising air currents causing upward motion of small hail. The origin of small positive areas near the cloud base (see Figure 2.23) is still under discussion. They could arise through the action of convective updraughts carrying positive charge. It is likely that the very varied electrical properties of thunderclouds (from cloud to cloud and within individual clouds as they develop) cannot be explained by any single theory of charge generation.

Lightning commonly begins more or less simultaneously with precipitation downpours. It may occur between the lower part of the cloud and the ground (which locally has an induced positive charge). The first (leader) stage of the flash bringing down negative charge from the cloud is met near the ground by a return stroke which rapidly takes positive charge upward along the already-formed channel of ionized air. Just as the leader is neutralized by the return stroke, so the latter is neutralized in turn within the clouds. Subsequent leaders and return strokes drain higher regions of the cloud until its supply

of negative charge is temporarily exhausted. The total flash typically lasts only about 0.2 seconds. Other more frequent flashes occur within a cloud or between clouds. The extreme heating and explosive expansion of air immediately round the path of the lightning sets up intense sound waves causing thunder to be heard. The sound travels at about 300 m s^{-1}.

Lightning is only one aspect of the atmospheric electricity cycle. During fine weather the earth's surface is negatively charged, the ionosphere positively charged. The potential gradient of this vertical electrical field in fine weather is about 100 V m^{-1} near the surface, decreasing to about 1 V m^{-1} at 25 km, whereas beneath a thundercloud it may exceed 1000 V m^{-1}. The 'breakdown potential' for lightning to occur in dry air is 3×10^6 V m^{-1}, but this is ten times the largest observed potential in thunderclouds. Hence, the necessity for localized cloud droplet/ice crystal charging processes, as already described, to initiate flash leaders.

Atmospheric ions conduct electrically from the ionosphere down to the earth, and hence a return supply must be forthcoming to maintain the observed electrical field. One source is the slow *point discharge*, from objects such as buildings and trees, of ions carrying positive charge induced by the negative thundercloud base. Similar upward currents occur above thunderstorm clouds. The other source (estimated to be smaller in its effect over the earth as a whole) is the instantaneous upward transfer of positive charge by lightning strokes, leaving the earth negatively charged. The joint operation of these supply currents, in approximately 1800 thunderstorms over the globe at any instant, is thought to be sufficient to balance the air–earth leakage, and this number seems to agree with observations. Thunderstorms globally are most frequent overland, occurring between 1200 and 2100 local time, with a minimum around 0300. Lightning is a significant environmental hazard. In the United States alone there are nearly 100 deaths per year, on average, as a result of lightning accidents.

The Russians have claimed some success in dissipating damaging hailstorms by the use of radar-directed artillery shells and rockets to inject silver iodide into high-liquid-water-content portions of clouds which freezes the available super-cooled water, so preventing it from accreting as shells on growing ice crystals. Attempts to 'drain off' lightning charges by seeding clouds with silver iodide or with millions of metallic needles have produced even less certain results.

I PRECIPITATION CHARACTERISTICS AND TYPES

Strictly, *precipitation* refers to all liquid and frozen forms of water (p. 78) – rain, snow, hail, dew, hoar-frost, fog-drip and rime (ice accretion on objects through the freezing on impact of super-cooled fog droplets) – but, in general, only rain and snow make significant contributions to precipitation totals. In many parts of the world the term rainfall can be used interchangeably with precipitation. The data in the following section refer to rainfall, since snowfall is less easily measured with the same degree of accuracy.

It must be emphasized that precipitation records are only *estimates*. Factors of site location, gauge height, large-and small-scale turbulence in the air flow, splash-in and evaporation all introduce errors in the catch. Also, there are more than fifty types of rain gauge in use by national meteorological services; design differences affect the airflow over the gauge aperture and evaporation losses from the container and inner side of the funnel. Falling snow is particularly subject to wind effects which can result in under-representation of the true amount by 50 per cent or more. Hence, corrections to gauge data need to take account of the proportion of precipitation falling in liquid and solid form, wind speeds during precipitation events, and precipitation intensity. Studies in Switzerland suggest that observed totals underestimate the true amounts by 7 per cent in summer and 11 per cent in winter below

2000 m, but by as much as 15 per cent in summer and 35 per cent in winter in the Alps between 2000 and 3000 m.

Precipitation data are also limited by the density of gauge networks. On a national basis, per 10,000 km^2 area, this ranges from 245 gauges in Britain, to ten in the United States and only three in Canada and Asia. In mountain areas the coverage is usually much worse.

1 Precipitation characteristics

There are many measures by which the various attributes of precipitation can be described. Long-term measures such as mean annual precipitation, annual variability and year-to-year trends have been of great interest to the geographer, and these statistical measures are treated in the concluding chapter (see Chapter 8, A). However, particularly in terms of hydrological considerations, the characteristics and relationships of individual rainstorms are being studied increasingly, and it is possible here to point to some of their commonly observed features. Weather observations usually indicate the amount, duration and frequency of precipitation and these enable other derived characteristics to be determined. Three of these are discussed below.

a Rainfall intensity

The intensity (= amount/duration) of rainfall during an individual storm, or a still shorter period, is of vital interest to hydrologists and water engineers concerned with flood forecasting and prevention, as well as to conservationists dealing with soil erosion. Chart records of the rate of rainfall (*hyetograms*) are necessary to assess intensity, which varies markedly with the time interval selected. Average intensities for short periods (thunderstorm-type downpours) are much greater than those for longer time intervals as Figure 2.24 illustrates for Washington, DC. In the case of extreme rates at different points over the earth (see Figure 2.25) the record intensity over 10 minutes is approximately three times that for

Figure 2.24 Generalized relationship between precipitation intensity and duration for Washington, DC.

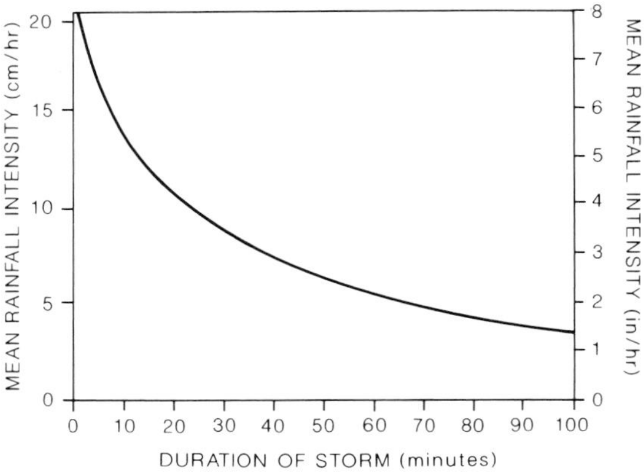

Source: After Yarnell 1935.

100 minutes, and the latter exceeds by as much again the record intensity over 1000 minutes (i.e. 16.5 hours). High-intensity rain is associated with increased drop size rather than an increased number of drops. For example, with precipitation intensities of 0.1, 1.3 and 10.2 cm/hr (i.e. 0.05, 0.5 and 4.0 in/hr), the most frequent raindrop diameters are 0.1, 0.2 and 0.3 cm, respectively. The occurrence of daily amounts exceeding 1.3 cm (0.5 in) is considered to be important for gully erosion in North America. Such falls account for 90 per cent of the annual rainfall on the Gulf Coast compared with only 20 per cent in the Great Basin.

b Areal extent of a rainstorm

The rainfall totals received in a given time interval vary according to the size of the area which is considered, showing a relationship analogous to that of rainfall duration and intensity. The maximum 24-hour rainfalls over areas of different extent in the United States (up to 1960) are shown in Table 2.3.

Figure 2.26, based on data of the type shown in Table 2.3, illustrates the maximum rainfall to be expected for a given storm area and given duration in the United States.

Figure 2.25 World record rainfalls and the 'envelope' of expected extremes at any place. The equation of the envelope line is given, together with the state or country where each record was established.

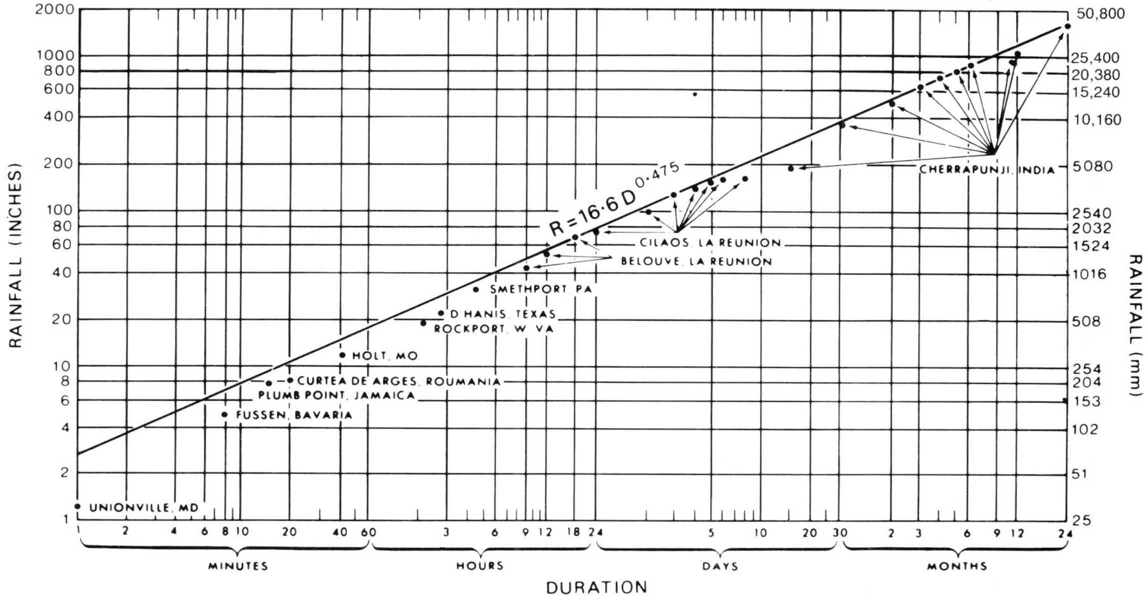

Source: From Paulhus 1965.

Figure 2.26 Enveloping depth/duration curves for maximum rainfall for areas of under 500 square miles 1295 km^2 in the United States.

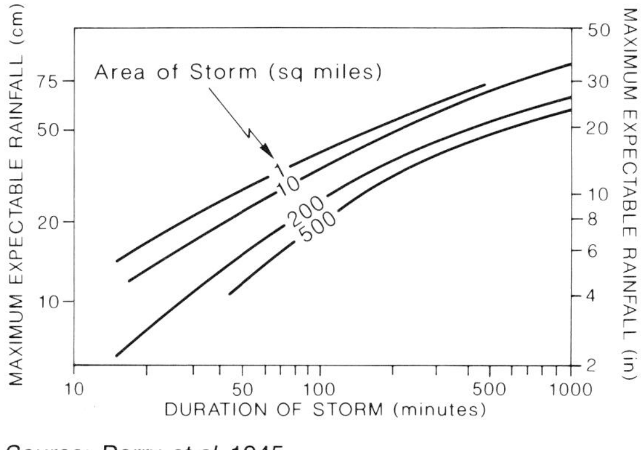

Source: Berry *et al.* 1945.

Table 2.3 Maximum 24-hour rainfalls in the United States (up to 1960).

Sq. miles	*cm*	*in*
10 (25.9 km^2)	98.3	38.7
10^2	89.4	35.2
10^3	76.7	30.2
10^4	30.7	12.1
10^5	10.9	4.3

Source: After Gilman 1964.

c Frequency of rainstorms

Another useful item is the average time-period within which a rainfall of specified amount or intensity can be expected to occur once. This is known as the *recurrence interval* or *return period*. Figure 2.27 gives this type of information on rainfall amount and duration for six contrasting stations. From this, for example, it would appear that on average, each 20 years, a

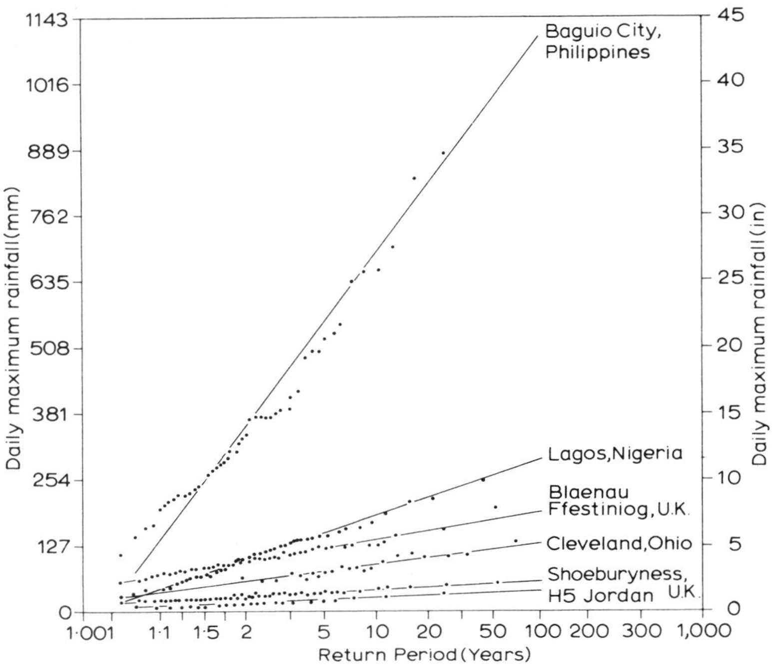

Figure 2.27 Rainfall/duration/frequency plots for daily maximum rainfalls in respect of a range of stations from the Jordan desert to an elevation of 1482 m in the monsoonal Philippines.

Source: After Rodda 1970; Linsley and Franzini 1964; and Ayoade 1976.

24-hour rainfall of at least 95 mm (3.75 in) is likely to occur at Cleveland and 216 mm (8.5 in) at Lagos. However, this *average* return period does not mean that such falls necessarily occur in the twentieth year of a selected period. Indeed, they might occur in the first! These estimates require long periods of observational data, but the approximately linear relationships shown by such graphs are of great practical significance for the design of flood-control systems.

Numerous case-studies of rainstorm events have been carried out in different climatic areas. Two examples are shown for south-west England and China in Figure 2.28A and B. The former was a 24-hour storm with an estimated 150–200 year return period, the latter was a 100-year storm and the figure shows the 1-hour rainfall. Despite less orographic assistance and the shorter recurrence interval, the Hong Kong storm produced three times the maximum hourly intensity and ten times the core area of the English storm.

2 Precipitation types

The above material can now be related to that of the previous sections in a discussion of precipitation types. A convenient starting point is the usual division into three main types – convective, cyclonic and orographic precipitation – according to the primary mode of uplift of the air. Essential to this analysis is some knowledge of storm systems. These are treated in later chapters and the newcomer to the subject may prefer to read the following in conjunction with them.

a 'Convective type' precipitation

This is associated with towering cumulus (*cumulus congestus*) and cumulonimbus clouds. Three subcategories can be distinguished according to their degree of spatial organization.

1 Scattered convective cells develop through strong heating of the land surface in summer, especially when low upper tropospheric temperatures facilitate the release of con-

Figure 2.28 Distribution of rainfall in mm (in parentheses in A, where inches are also given). A: Over Exmoor, England, during a 24-hour period on 15 August 1952. 75 per cent of this rainfall occurred during a 7-hour period and 18 per cent in one hour;* B: Over Hong Kong during 0630 to 0730 hours on 12 June 1966.†

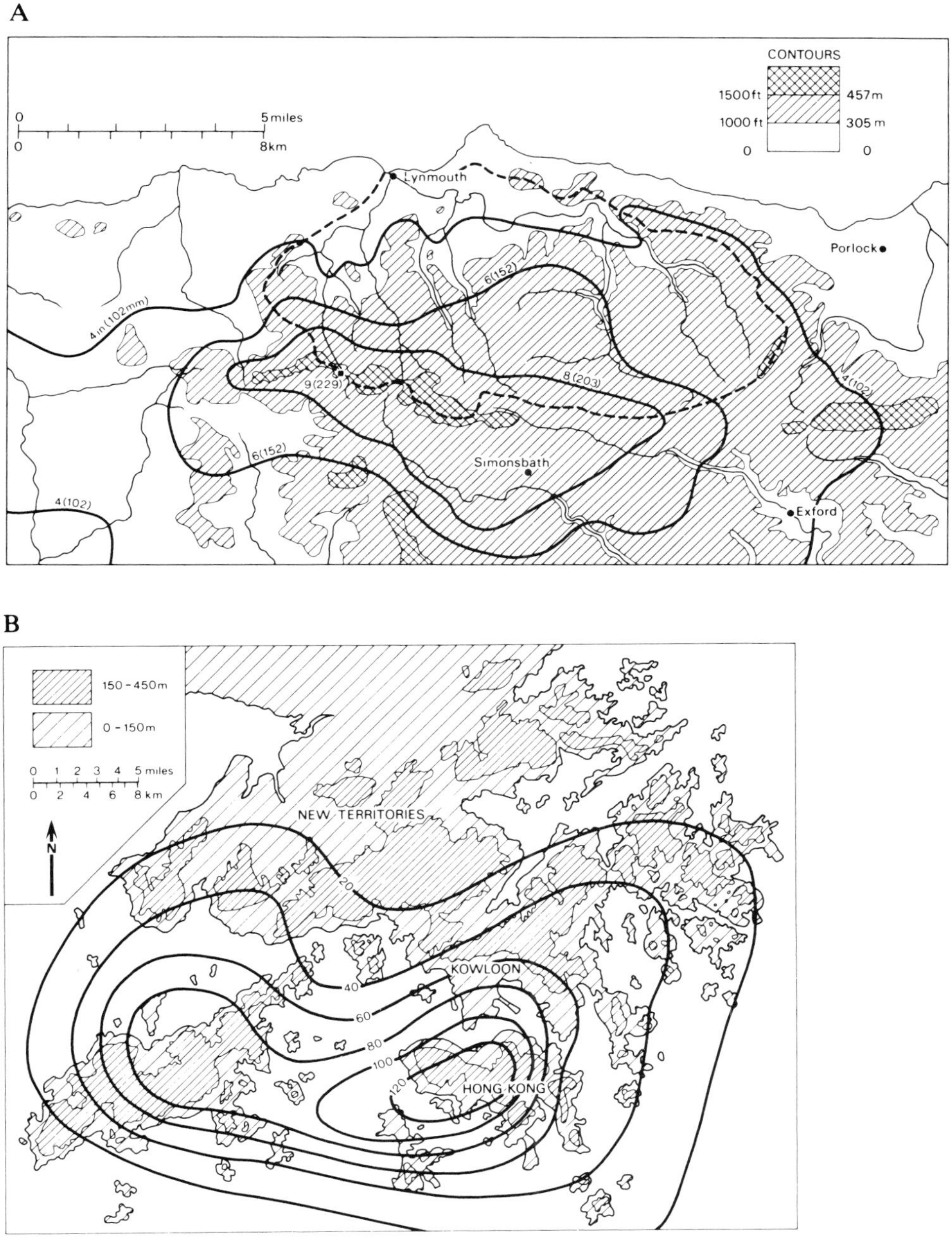

Sources: * From Dobbie and Wolf 1953.
†From So 1971.

ditional or convective instability (see E, this chapter). Precipitation, often including hail, is of the thunderstorm type, although thunder and lightning do not necessarily occur. Small areas (20 to 50 km^2) are affected by the individual heavy downpours, which generally last for about 30 minutes to 1 hour.

2 Showers of rain, snow or soft hail pellets may form in cold, moist unstable air passing over a warmer surface. Convective cells moving with the wind can produce a streaky distribution of precipitation parallel to the wind direction (see Plate 15), although over a period of several days the variable paths and intensities of the showers tend to obscure this pattern. Two locations in which these cells may occur are parallel to a surface cold front in the warm sector of a depression (sometimes as a squall line) or parallel to and ahead of the warm front (see Chapter 4, D). Hence the precipitation is widespread, though of brief duration at any locality.

3 In tropical cyclones cumulonimbus cells become organized about the centre in spiralling bands (see Chapter 6, C.2). Particularly in the decaying stages of such cyclones, typically over land, the rainfall can be very heavy and prolonged, affecting areas of thousands of square kilometres.

b 'Cyclonic type' precipitation

Precipitation characteristics vary according to the type of low-pressure system and its stage of development, but the essential mechanism is ascent of air through horizontal convergence of air-streams in an area of low pressure (see Chapter 3, B.1). In extra-tropical depressions this is reinforced by uplift of warm, less-dense air along an air-mass boundary (see Chapter 4, D.2). Such depressions give moderate and generally continuous precipitation over very extensive areas as they move usually eastward in the westerly wind belts between about 40° and 65° latitude. The precipitation belt in the forward sector of the storm can affect a locality in its path for 6 to 12 hours, whereas the belt in the rear gives a shorter period of thunderstorm-type precipitation. These sectors are, therefore, sometimes distinguished in precipitation classifications, and a more detailed breakdown is illustrated in Table 5.2. Polar lows (see Chapter 4, G.3) combine the effects of airstream convergence and convective activity of category *a*(2), above, whereas troughs in the equatorial low-pressure area give convective precipitation as a result of airstream convergence in the tropical easterlies (see Chapter 6, C.1).

c Orographic precipitation

Orographic precipitation is commonly regarded as a distinct type, but this requires careful qualification. Mountains are not especially efficient in causing moisture to be removed from airstreams crossing them, yet because precipitation falls repeatedly in more or less the same locations the cumulative totals are large. Orography, dependent on the alignment and size of the barrier, may cause (a) forced ascent on a smooth mountain slope, producing adiabatic cooling, condensation and precipitation; (b) triggering of conditional or convective instability by blocking of the airflow and upstream lifting; (c) triggering of convection by diurnal heating of slopes and upslope winds; (d) precipitation from low-level cloud over the mountains through 'seeding' of ice crystals or droplets from an upper-level feeder cloud (see Figure 2.30); (e) increased frontal precipitation by retarding the movement of cyclonic systems and fronts. West coast mountains with onshore flow, such as the Western Ghats, India, during the south-west summer monsoon; the west coasts of Canada, Washington and Oregon; or coastal Norway, in winter months, supposedly illustrate smooth forced ascent, yet most of the other processes seem to be involved. The limited width of the coastal ranges, with average wind speeds, generally allow insufficient time for the basic mechanisms of precipitation growth to operate (see G.1, this chapter). In view of the complexity of processes involved, Tor Bergeron proposed using the term 'orogeinic', rather than

orographic, precipitation, i.e. an origin related to various orographically-produced effects.

In mid-latitude areas where precipitation is predominantly of cyclonic origin, orographic effects tend to increase both frequency and intensity of winter precipitation, whereas during summer and in continental climates with a higher condensation level the main effect of relief is the occasional triggering of intense thunderstorm-type precipitation. The orographic influence occurs only in the proximity of high ground in the case of a stable atmosphere. Recent radar studies show that the main effect in this case is one of redistribution, whereas in the case of an unstable atmosphere precipitation appears to be increased, or at least redistributed on a larger scale, since the orographic effects may extend well downwind due to the activation of the mesoscale rain bands (see Figure 4.11).

In tropical highland areas there is a clearer distinction between orographic and convective contributions to total rainfall than in the mid-latitude cyclonic belt, although elements of tropical orographic rainfall are of convective origin due to the orographic release of potential instability. Figure 2.29 shows that in the mountains of Costa Rica the temporal character of convective and orographic rainfalls and their seasonal occurrences are quite distinguishable. Convective rain occurs mainly in the May–November period when 60 per cent of the rain falls in the afternoons between 1200 and 1800 hours; orographic rain predominates between December and April with a secondary maximum in June and July coinciding with an intensification of the Trades.

Even low hills may have an orographic effect. Research in Sweden shows that wooded hills rising only 30–50 m above the surrounding lowlands increase precipitation amounts locally by 50–80 per cent during cyclonic spells. Until Doppler radar studies of the motion of falling raindrops became feasible, the processes responsible for such effects were unknown. A principal cause is the 'seeder–feeder' ('releaser–spender') cloud mechanism, proposed by T. Bergeron (described above). This is illustrated in Figure

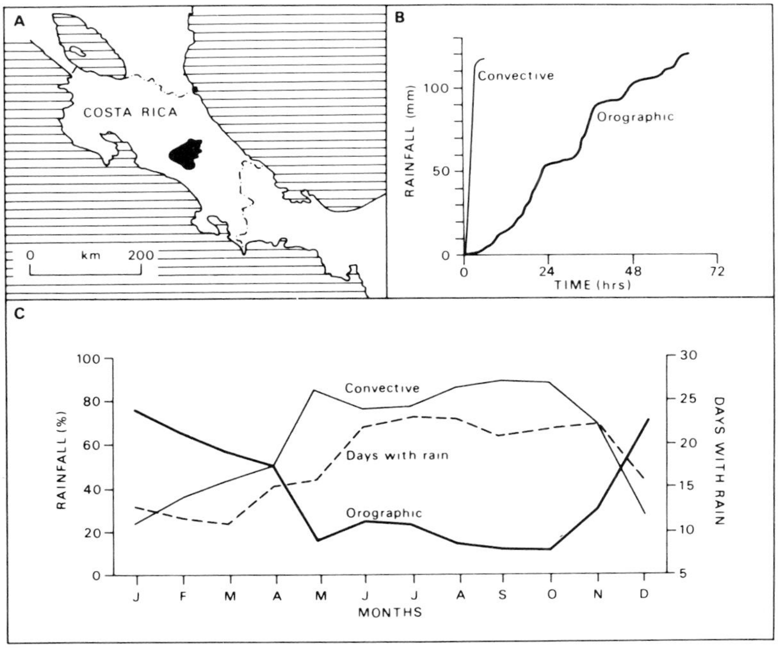

Figure 2.29 Orographic and convective rainfall in the Cachi region of Costa Rica for the period 1977–80. A: The Cachi region, elevation 500–3000 m. B: Typical accumulated rainfall distributions for individual convective (duration 1–6 hours, high intensity) and orographic (1–5 days, lower intensity except during convective bursts) rainstorms. C: Monthly rainfall divided into percentages of convective and orographic, plus days with rain, for Cachi (1018 m)

Source: From Chacon and Fernandez 1985 (*by permission of the Royal Meteorological Society*).

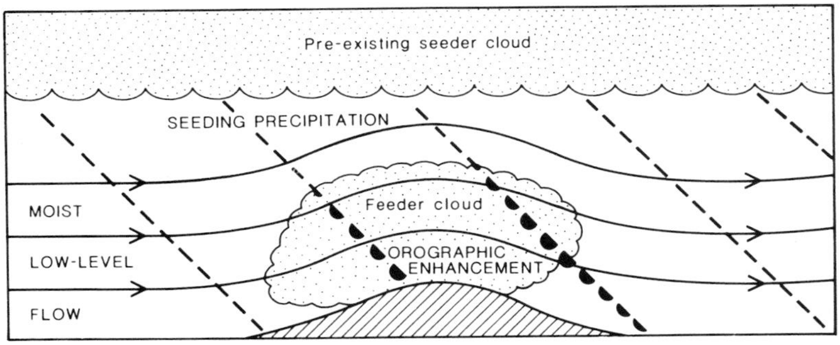

Figure 2.30 Schematic diagram of T. Bergeron's 'seeder–feeder' cloud model of orographic precipitation over hills.

Source: After Browning and Hill 1981 (*by permission of the Royal Meteorological Society*).
Note: This process may also operate in deep nimbostratus layers.

2.30. In moist stable airflows, shallow-cap clouds form over hilltops. Precipitation falling from an upper layer of altostratus (the seeder cloud) grows rapidly by the washout of droplets in the lower (feeder) cloud. The seeding cloud may release ice crystals, that subsequently melt, or droplets. Precipitation from the upper cloud layer alone would not give significant amounts at the ground as the droplets would have insufficient time to grow in the airflow which may traverse the hills in only 15–30 minutes. Most of the precipitation intensification takes place in the lowest kilometre of the atmosphere in moist, fast-moving airflows.

Two special cases of orographic effects may be mentioned. One is the general influence of surface friction which by turbulent uplift may assist the formation of stratus or stratocumulus layers when other conditions are suitable (see F.2, this chapter). Only light precipitation (drizzle, light rain or snow grains) is to be expected under these circumstances. The other case arises through frictional slowing down of an airstream moving inland over the coast. A particular instance of the convergence and uplift which this may initiate has been reported by Bergeron. During a 24-hour period in October 1945, a west-south-westerly airstream over Holland produced a belt of precipitation (3 cm or more) – a result of frictional convergence and uplift – in crossing the narrow zone of coastal sand dunes only a few metres high. Over the remainder of that virtually flat country a series of lee waves developed in the tropospheric airflow downwind from the coast (see Figure 3.11, for example), and these gave a series of transverse (north–south) bands of precipitation up to 2 cm in amount. On the following day the surface flow had changed little, but a temperature decrease from −20°C to −28°C at the 500-mb level altered the vertical stability so that the lee waves broke down and the precipitation distribution showed convective streaks, up to 4 cm per day, parallel to the surface wind direction.

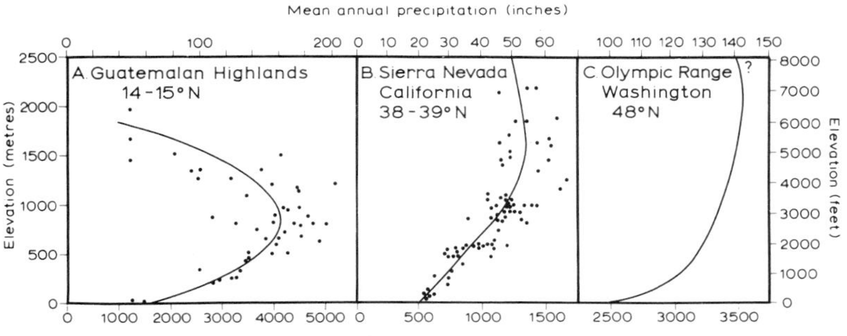

Figure 2.31 Generalized curves showing the relationship between elevation and mean annual precipitation for west-facing mountain slopes in Central and North America. The dots give some indication of the wide scatter of individual precipitation readings.

Sources: Adapted from Hastenrath 1967, and Armstrong and Stidd 1967.

3 Regional variations in the altitudinal maximum of precipitation

The increase of precipitation with height on mountain slopes is a world-wide characteristic, although actual profiles of precipitation differ regionally and seasonally. In middle latitudes an increase may be observed up to at least 3000–4000 m, as is the case in the Colorado Rocky Mountains and the Alps. In western Britain with mountains of about 1000 m, the maximum falls are recorded to leeward of the summits. This probably reflects the general tendency of air to go on rising for a while after it has crossed the crest line and the time-lag involved in the precipitation process after condensation. Over narrow uplands the horizontal distance may allow insufficient time for maximum cloud build-up and the occurrence of precipitation. However, a further factor may be the effect of eddies, set up in the airflow by the mountains, on the catch of rain gauges. Studies in Bavaria at the Hohenpeissenberg observatory show that standard rain gauges may overestimate amounts by about 10 per cent on the lee slopes and underestimate them by 14 per cent on windward slopes.

In the tropics and subtropics the maximum precipitation occurs below the higher mountain summits, from which level it decreases upwards towards the crests. Observations are generally sparse in the tropics, but numerous records from Java show that the average elevation of greatest precipitation is approximately 1200 m (4000 ft). Above about 2000 m the decrease in amounts becomes quite marked. Similar features are reported from Hawaii and, at a rather higher elevation, on mountains in East Africa (see Chapter 6, G.2). Figure 2.31A shows that, despite the wide range of records for individual stations, this effect is clearly apparent along the Pacific flank of the Guatemalan Highlands. Further north along the coast, the occurrence of a precipitation maximum below the mountain crest is observed in the Sierra Nevadas, despite some complication introduced by the shielding effect of the Coast Ranges (Figure 2.31B), but in the Olympic Mountains of Washington precipitation increases right up to the summits (Figure 2.31C). As has been previously mentioned, precipitation catches on mountain crests may underestimate the actual precipitation due to the effect of eddies, and this is particularly true where much of the precipitation falls in the form of snow which is very susceptible to blowing by the wind.

One explanation of this orographic difference between tropical and temperate rainfall is based on the concentration of moisture in a fairly shallow layer of air near the surface in the tropics (see Chapter 6, A). Much of the orographic precipitation seems to derive from warm clouds (particularly *cumulus congestus*), composed of water droplets, which commonly have an upper limit at about 3000 m. It is probable that the height of the maximum precipitation zone is close to the mean cloud base since the maximum size and number of falling drops will occur at that level. Thus, stations located above the level of mean cloud base will receive only a proportion of the orographic increment. In temperate latitudes much of the precipitation, especially in winter, falls from stratiform cloud which commonly extends through a considerable depth of the troposphere. In this case there tends to be a smaller fraction of the total cloud depth below the station level. These differences according to cloud type and depth are apparent even on a day-to-day basis in middle latitudes, as has been shown by detailed studies in the Bavarian Alps. Seasonal variations in the altitude of the mean condensation level and zone of maximum precipitation are similarly observed. In the mountains of central Asia (the Pamirs and Tian Shan), for instance, the maximum is reported to occur at about 1500 m (5000 ft) in winter and at 3000 m (9900 ft) or more in summer. A further difference between orographic effects on precipitation in the tropics and the mid-latitudes relates to the great instability of many tropical air masses. Where high mountains obstruct the flow of very moist maritime tropical air masses, the upwind turbulence may be sufficient to

trigger off sufficient precipitation to give a rainfall maximum at low elevations on the windward side of the range. By contrast, in more stable mid-latitude airflows the rainfall maximum is more closely related to the topography (see Figure 2.32).

4 The world pattern of precipitation

A glance at the maps of precipitation amount for December–February and June–August (Figure 2.33) indicates that the distributions are considerably more complex than those, for example, of mean temperature (see Figure 1.24B). Comparison of Figure 2.33 with the meridional profile of average precipitation for each latitude (see Figure 2.4A) brings out the marked longitudinal variations that are superimposed on the zonal pattern. The latter has three main features: an equatorial maximum which, like the thermal equator, is slightly displaced into the northern hemisphere; very low totals in high latitudes; and secondary minima in subtropical latitudes. Figure 2.33 demonstrates why the subtropics do not appear as particularly dry on the meridional profile in spite of the known aridity of the subtropical high pressure areas (see Chapters 3, D.1 and 5, D.2). In these latitudes the eastern sides of the continents receive considerable rainfall in summer.

In view of the complex controls involved, no brief explanation of these precipitation distributions can be very satisfactory. Various aspects of selected precipitation regimes are examined in Chapters 5 and 6, after consideration of the fundamental ideas about atmospheric motion, air masses and frontal zones. A classification of wind belts and precipitation characteristics is outlined in Appendix 1.C. It must suffice at this stage simply to note the factors which have to be taken into account in studying Figure 2.33.

1 The limit imposed on the maximum moisture content of the atmosphere by air temperature. This is important in high latitudes and in winter in continental interiors.
2 The major latitudinal zones of moisture influx due to atmospheric advection. This in itself is a reflection of the global wind systems

Figure 2.32 The relationship between precipitation (broken line) and relief in the tropics and the mid-latitudes. A: The highly saturated air masses over the Central Highlands of Papua New Guinea give seasonal maximum precipitations on the windward slopes of the mountains with changes in the monsoonal circulation;* B: Across the Jungfrau Massif in the Swiss Alps the precipitation is much less than in A and is closely correlated with the topography on the windward side of the mountains.† The arrows show the prevailing airflow directions.

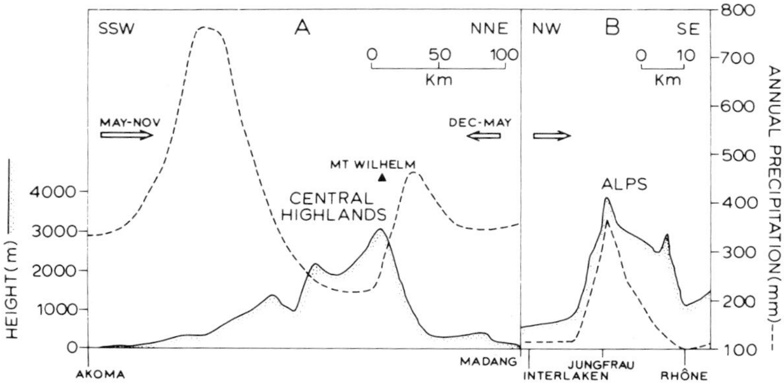

Sources: *After Barry 1992.
†After Maurer and Lütschg (from Barry).

Figure 2.33 Mean global precipitation (in mm) for the periods December–February and June–August.

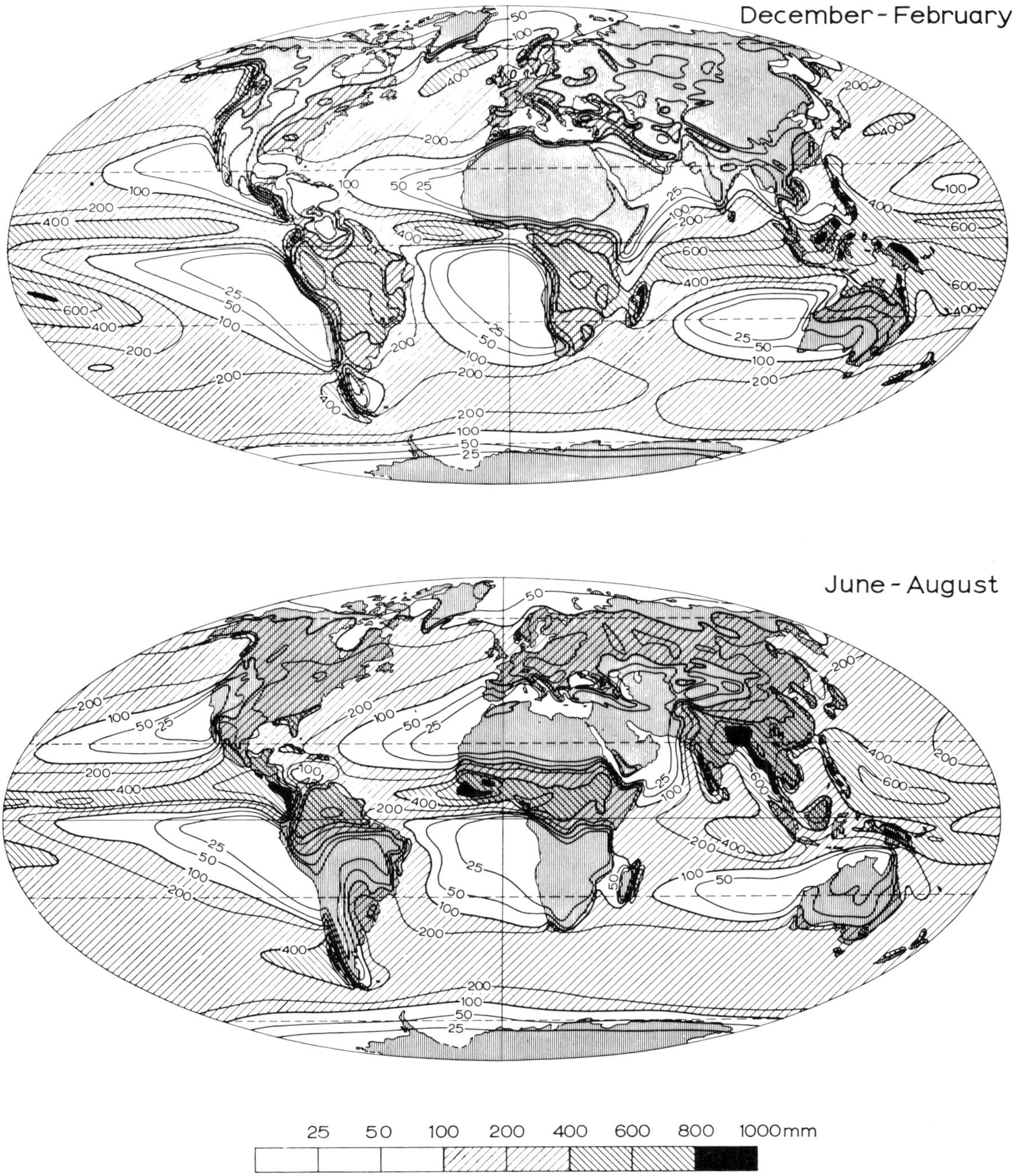

Source: From Möller 1951.

and their disturbances (i.e. the converging trade wind systems and the cyclonic westerlies, in particular).

3 The distribution of the land masses. It is noteworthy that the southern hemisphere lacks the vast, arid, mid-latitude continental interiors of the northern. The oceanic expanses of the southern hemisphere allow the mid-latitude storms to increase the zonal precipitation average for 45°S by about one-third compared with that of the northern hemisphere for 50°N (see Figure 2.4A). Another major non-zonal feature is the occurrence of the monsoon regimes, especially in Asia.

4 The orientation of mountain ranges with respect to the prevailing winds.

5 Drought

The term drought implies an absence of significant precipitation for a period which is long enough to cause moisture deficits in the soil due to evapotranspiration and decreases in stream flow, thereby disrupting the normal biological and human activities. Thus a drought condition may obtain after only three weeks without rain in parts of Britain, whereas some areas of the tropics regularly experience many successive dry months. There is then no precise, universally applicable definition of drought. At least 150 different definitions can be found in the literature developed by specialists in meteorology, agriculture, hydrology, and socio-economic studies having differing perspectives! All regions suffer the temporary but irregularly recurring condition of drought, but particularly those with marginal climates alternately influenced by differing climatic mechanisms. Drought conditions are especially associated with:

1 Increases in area and persistence of the subtropical high-pressure cells. Drought in southern Israel has been shown to be significantly related to this mechanism. The major droughts in the African Sahel (see Figure 8.11) have also been attributed to an eastward and southward expansion of the Azores anticyclone.

2 Changes in the summer monsoonal circulation, causing a postponement or failure of incursions of maritime tropical air, such as may occur in Nigeria or the Punjab of India.

3 Lower ocean surface temperatures produced by changes in currents or increased upwelling of cold waters. Rainfall in California and Chile may be affected by such mechanisms (see p. 266), and adequate rainfall in the drought-prone region of north-east Brazil appears to be strongly dependent on high sea-surface temperatures in the 0°–15°S belt of the South Atlantic.

4 Displacement of mid-latitude storm tracks associated with either an expansion of the circum-polar westerlies into lower latitudes, or with the development of persistent blocking patterns of circulation in middle latitudes (see Figure 3.39). It has been suggested that droughts on the Great Plains east of the Rockies in the 1890s and 1930s were due to such changes in the general circulation. However, the droughts of the 1910s and 1950s in this area were caused by persistent high pressure in the south-east and the northward displacement of storm tracks (Figure 2.34).

Clearly, most severe and prolonged droughts involve combinations of several mechanisms. The prolonged drought in the Sahel – a 3000 × 700 km zone stretching along the southern edge of the Sahara from Mauretania to Chad – which began in 1969 and has continued with interruptions up to the present (see Figure 8.11), has been attributed to several factors. These include an expansion of the circumpolar westerly vortex, shifting of the subtropical high-pressure belt towards the equator, lower sea-surface temperatures in the eastern North Atlantic, and 'desertification' due to overgrazing. The removal of vegetation, increasing the surface albedo, is thought to result in a reduction of precipitation. There is no evidence that the subtropical high pressure was further south, but

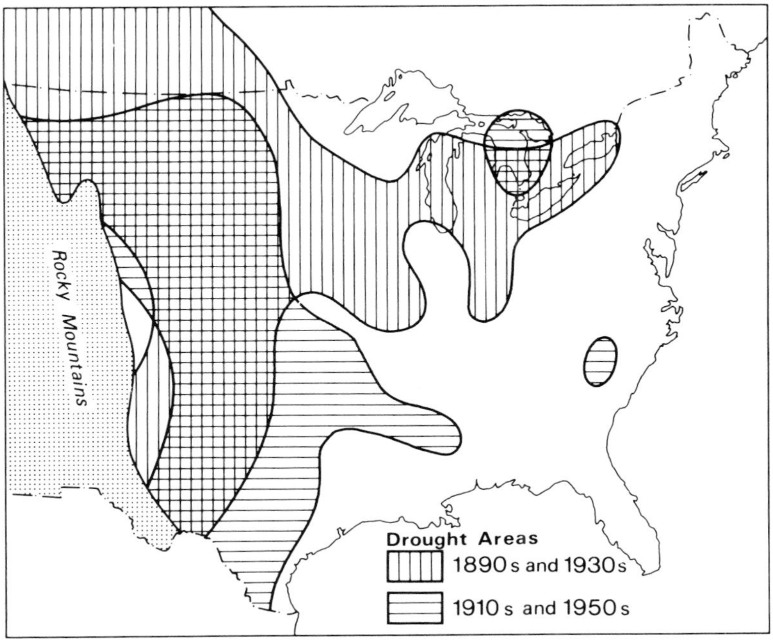

Figure 2.34 Drought areas for the central USA based on areas receiving less than 80 per cent of the normal July–August precipitation.

Source: After Borchert 1971.

dry easterly airflow was stronger across Africa during the drought years.

From May 1975 to August 1976 parts of north-west Europe from Sweden to western France experienced severe drought conditions. Southern England received less than 50 per cent of average rainfall, the most severe and prolonged drought since records began in 1727 (see Figure 2.35). The immediate causes of this regime were the establishment of a persistent blocking ridge of high pressure over the area, displacing depression tracks 5°–10° latitude northward over the eastern North Atlantic. Farther afield, the circulation over the North Pacific had changed earlier, with the development of a stronger high-pressure cell and stronger upper-level westerlies, perhaps associated with a cooler-than-average sea surface. The westerlies were displaced northward over both the Atlantic and the Pacific. Over Europe, the dry conditions at the surface increased the stability of the atmosphere, further lessening the possibility of precipitation.

J ACID PRECIPITATION

Acid precipitation includes both acid rain and snow (wet deposition) and dry deposition. The acidity of precipitation represents an excess of positive hydrogen ions $[H^+]$ in a water solution. Acidity is measured on the pH scale $(1/-\log[H^+])$ ranging from 0 to 14, where 7 is neutral, i.e. the hydrogen cations are balanced by anions of sulphate, nitrate and chloride. Over the oceans, the main anion is SO_4^{2-} from sea salt. The background level of acidity in rainfall is about pH 4.8 to 5.6, because atmospheric CO_2 reacts with water to form carbonic acid. Acid solutions in rainwater are enhanced by reactions involving both gas-phase and aqueous-phase chemistry with sulphur dioxide and nitrogen dioxide. For sulphur dioxide, rapid pathways are provided by:

$$HOSO_2 + O_2 \rightarrow HO_2 + SO_3$$
$$H_2O + SO_3 \rightarrow H_2SO_4 \text{ (gas phase)}$$
and
$$H_2O + HSO_3 \rightarrow H^+ + SO_4^{2-} + H_2O \text{ (aqueous phase)}$$

Figure 2.35: The drought of north-west Europe during May 1975 to August 1976. A: Conditions of a blocking high pressure over Britain, jet stream bifurcation and low sea-surface temperatures; B: Rainfall over Western Europe between May 1975 and August 1976 expressed as a percentage of a 30-year average.

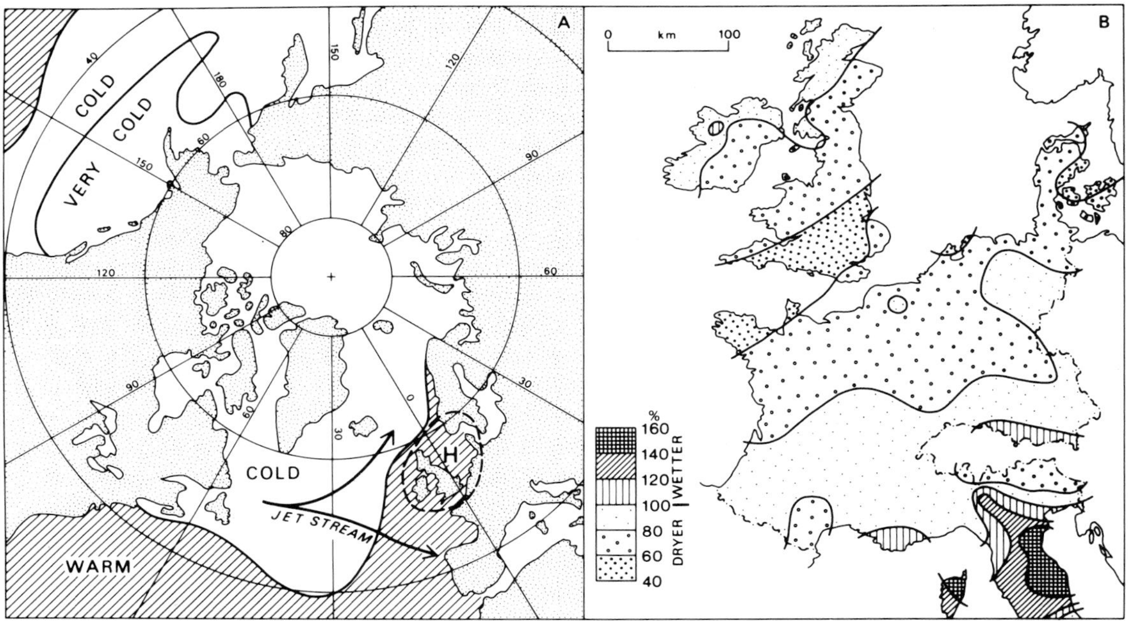

Source: From Doornkamp and Gregory 1980.

The OH radical is an important catalyst in gas phase reactions and hydrogen peroxide (H_2O_2) in the aqueous phase.

Acid deposition depends on the emission concentrations, atmospheric transport and chemical activity, the cloud types and cloud microphysical processes, and the type of precipitation. Observations in northern Europe and eastern North America in the mid-1970s, compared with the mid-1950s, showed a twofold to threefold increase in hydrogen ion deposition and rainfall acidity. Sulphate concentrations in rain water in Europe increased over this 20-year period by 50 per cent in southern Europe and 100 per cent in Scandinavia, although there has been a subsequent decrease apparently associated with reduced sulphur emissions in both Europe and North America. The emissions from coal and fuel oil in these regions have high sulphur contents (2–3 per cent) and, since major SO_2 emissions occur from elevated stacks, SO_2 is readily transported by the boundary layer winds. NO_x emissions, in contrast, are primarily from automobiles and thus NO_3^- is mainly deposited locally. SO_2 and NO_x have atmospheric residence times of 1–3 days. SO_2 is not readily dissolved in cloud or rain drops unless oxidized by OH or H_2O, but dry deposition is quite rapid. NO_3^- is insoluble in water, but it is oxidized to NO_2 by reaction with ozone.

In the western United States, where there are few major sources of emission, H^+ ion concentrations in rainwater are only 15–20 per cent of levels in the east, while sulphate and nitrate anion concentrations are one-third to one-half of those in the east. In China, where high-sulphur content coal is the main energy source and there are few automobiles, rainwater sulphate concentrations are high; observations in south-west China show levels six times those

in New York City. In winter, in Canada, snow has been found to have more nitrate and less sulphate than rain, apparently because falling snow scavenges nitrate faster and more effectively. Consequently, nitrate accounts for about half of the snowpack acidity. In spring, snow melt runoff causes an acid flush that may be harmful to fish populations in rivers and lakes especially at the egg or larval stages.

In areas subject to frequent fog, or hill cloud, acidity may be greater than with rainfall; North American data indicate pH values averaging 3.4 in fog. This is a result of several factors. Small fog or cloud droplets have a large surface area, higher levels of pollutants provide more time for aqueous phase chemical reactions, and the pollutants may act as nuclei for fog droplet condensation. In Los Angeles, fog water usually has high nitrate concentrations due to automobile traffic during morning rush hour and pH values as low as 2.2 to 4.0 are observed.

The impact of acid precipitation depends on the vegetation cover, soil and bedrock type. Neutralization may occur by addition of cations in the vegetation canopy or on the surface. Such buffering is greatest if there are carbonate rocks (Ca; Mg cations); otherwise the increased acidity augments normal leaching of bases from the soil.

SUMMARY

Atmospheric humidity can be described by the absolute mass of moisture in unit mass (or volume) of air, as a proportion of the saturation value, or in terms of the water vapour pressure. When cooled at constant pressure, air becomes saturated at the dew-point temperature.

The components of the surface moisture budget are total precipitation (including condensation on the surface), evaporation, storage change of water in the soil or in a snow cover, and runoff (on the surface or in the ground). Evaporation rate is determined by the available energy, the surface–air difference in vapour pressure, and the wind speed, assuming moisture supply is unlimited. If moisture supply is limited, the rate is affected by soil water tension, and plant factors. Evapotranspiration is best determined with a lysimeter. Otherwise, it may be calculated by formulae based on the energy budget, or on the aerodynamic profile method using the measured gradients of wind speed, temperature and moisture content near the ground.

Condensation in the atmosphere may occur by continued evaporation into the air; by mixing of air of different temperature and vapour pressure, such that the saturation point is reached; or by adiabatic cooling of the air through lifting until the condensation level is reached.

Air may be lifted through instability due to surface heating or mechanical turbulence, ascent of air at a frontal zone, or forced ascent over an orographic barrier. Instability is determined by the actual rate of temperature decrease with height in the atmosphere relative to the appropriate adiabatic rate. The dry adiabatic lapse rate is 9.8°C/km; the saturated adiabatic rate is less than the DALR due to latent heat released by condensation. It is least (around 5°C/km) at high temperatures, but approaches the DALR at subzero temperatures.

Condensation requires the presence of hygroscopic nuclei such as salt particles in the air. Otherwise, supersaturation occurs. Similarly, ice crystals only form naturally in clouds containing freezing nuclei (clay mineral particles). Otherwise, water droplets may super-cool to −39°C. Both super-cooled droplets and ice crystals may be present at cloud temperatures of −10° to −20°C.

Clouds are classified in ten basic types, according to altitude and cloud form. Satellites are providing new information on spatial patterns of cloudiness, revealing cellular

continued

(honeycomb) areas and linear cloud streets, as well as large-scale storm patterns.

Precipitation drops do not form directly by growth of cloud droplets through condensation. Two processes may be involved – coalescence of falling drops of differing sizes, and the growth of ice crystals by vapour deposition (the Bergeron–Findeisen process). Low-level cloud may be seeded naturally by ice crystals from upper cloud layers, or by introducing artificial nuclei. There is no single cause of the orographic enhancement of precipitation totals, and at least four contributing processes can be distinguished.

The freezing process appears to be a major element of cloud electrification in thunderstorms. Lightning plays a key role in maintaining the electrical field between the surface and the ionosphere.

Rainfall is described statistically by the intensity, areal extent and frequency (or recurrence interval) of rainstorms. Convective and cyclonic types of precipitation are commonly distinguished; orography generally intensifies the precipitation on windward slopes, but there are geographical differences in this altitudinal effect. Droughts may occur in many different climatic regions due to various causal factors.

Acid precipitation (by wet or dry deposition) results from the reaction of cloud droplets or catalysts with emissions of SO_2 and NO_x. There are large geographical variations in acid deposition.

3

Atmospheric motion

The atmosphere acts rather like a gigantic heat engine in which the constantly maintained difference in temperature existing between the poles and the equator provides the energy supply necessary to drive the planetary atmospheric circulation. The conversion of the heat energy into kinetic energy to produce motion must involve rising and descending air, but vertical movements are generally much less in evidence than horizontal ones, which may cover vast areas and persist for periods of a few days to several months.

Before considering these global aspects, however, it is important to look at the immediate controls on air motion. The downward-acting gravitational field of the earth sets up the observed decrease of pressure away from the earth's surface that is represented in the vertical distribution of atmospheric mass (see Figure 1.12). This mutual balance between the force of gravity and the vertical pressure gradient is referred to as *hydrostatic equilibrium* (p. 14). This state of balance, together with the general stability of the atmosphere and its shallow depth, greatly limits vertical air motion. Average horizontal wind speeds are of the order of one hundred times greater than average vertical movements, though individual exceptions occur – particularly in convective storms.

A LAWS OF HORIZONTAL MOTION

There are four controls on the horizontal movement of air near the earth's surface: pressure-gradient force, Coriolis force, centripetal acceleration and frictional forces. The primary cause of air movement is the development of a horizontal pressure gradient and the fact that such a gradient can persist (rather than being destroyed by air motion towards the low pressure) results from the effect of the earth's rotation in giving rise to the Coriolis force.

1 The pressure-gradient force

The pressure-gradient force has vertical and horizontal components but, as already noted, the vertical component is more or less in balance with the force of gravity. Horizontal differences in pressure can be due to thermal or mechanical causes (often not easily distinguishable), and these differences control the horizontal movement of an air mass. In effect the pressure gradient serves as the motivating force that causes the movement of air away from areas of high pressure and towards areas where it is lower, although other forces prevent air from moving directly across the isobars (lines of equal pressure). The pressure-gradient force per unit mass is expressed mathematically as

$$-\frac{1}{\rho}\frac{\mathrm{d}p}{\mathrm{d}n}$$

where ρ = air density and dp/dn = the horizontal gradient of pressure. Hence the closer the isobar spacing the more intense is the pressure gradient and the greater the wind speed. The pressure-gradient force is also inversely proportional to air density, and this relationship is of particular importance in understanding the behaviour of upper winds.

2 The earth's rotational deflective (Coriolis) force

The Coriolis force arises from the fact that the movement of masses over the earth's surface is usually referred to a moving co-ordinate system (i.e. the latitude and longitude grid which 'rotates' with the earth). The simplest way to begin to visualize the manner in which this deflecting force operates is to picture a rotating disc on which moving objects are deflected. Figure 3.1 shows the effect of such a deflective force operating on a mass moving outward from the centre of a spinning disc. The body follows a straight path in relation to a fixed frame of reference (for instance, a box which contains a spinning disc), but viewed relative to co-ordinates rotating with the disc the body swings to the right of its initial line of motion. This effect is readily demonstrated if a pencil line is drawn across a white disc on a rotating turntable. Figure 3.2 illustrates a case where the movement is not from the centre of the turntable and the object possesses an initial momentum in relation to its distance from the axis of rotation. In the analogous case of the rotating earth (with rotating reference co-ordinates of latitude and longitude), there is apparent deflection of moving objects to the right of their line of motion in the northern hemisphere and to the left in the southern hemisphere, as viewed by observers on the earth. The deflective force (per unit mass) is expressed by:

$$-2\omega V \sin \phi$$

where ω = the angular velocity of spin ($15° \text{ hr}^{-1}$ or $2\pi/24 \text{ rad hr}^{-1}$ for the earth = $7.29 \times 10^{-5} \text{ rad s}^{-1}$); ϕ = the latitude and V = the velocity of the mass. $2\omega \sin \phi$ is referred to as the Coriolis parameter (f).

The magnitude of the deflection is directly proportional to: (*a*) the horizontal velocity of the air (i.e. air moving at 11 m s^{-1} (25 mph) having half the deflective force operating on it as that moving at 22 m s^{-1} (50 mph)); and (*b*) the sine of the latitude ($\sin 0° = 0$; $\sin 90° = 1$). The effect is thus a maximum at the poles (i.e. where the plane of the deflecting force is parallel with the earth's surface) and decreases with the sine of the latitude, becoming zero at the equator (i.e. where there is no component of the deflection in a plane parallel to the surface). Values of f vary with latitude as follows:

Latitude	0°	10°	20°	43°	90°N
$f(10^{-4} \text{ s}^{-1})$	0	0.25	0.50	1.00	1.458

The Coriolis force always acts at right angles to the direction of the air motion, to the right in the northern hemisphere (f positive) and to the left in the southern hemisphere (f negative).

The earth's rotation also produces a vertical component of rotation about a horizontal axis. This is a maximum at the equator (zero at the poles), but it is much less important to atmospheric motions due to the existence of hydrostatic equilibrium.

Figure 3.1 The Coriolis deflecting force operating on an object moving outward from the centre of a rotating turntable.

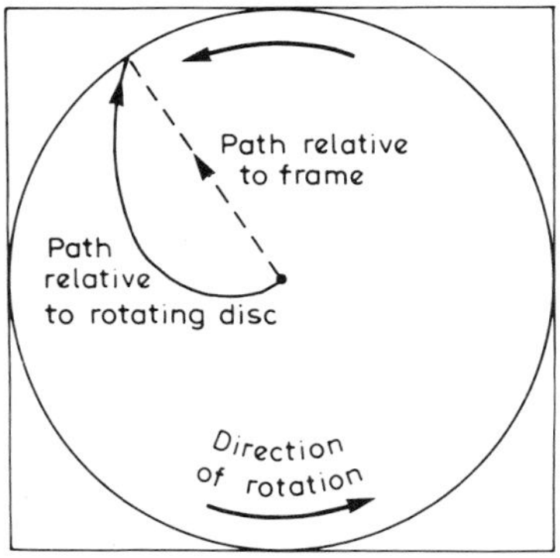

Figure 3.2 The Coriolis deflecting force on a rotating turntable. A: A man at X sees the object P and attempts to throw a ball towards it. Both locations are rotating anticlockwise. B: The man's position is now X′ and the object is at P′. To the man, the ball appears to follow a curved path and lands at Q. The man overlooked the fact that P was moving to his left and that the path of the ball would be affected by the initial impetus due to the man's own rotation.

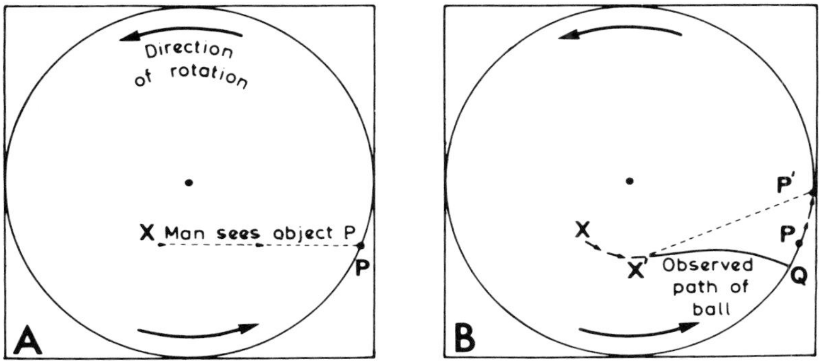

3 The geostrophic wind

Observations in the *free atmosphere* (above the level affected by surface friction at about 500 to 1000 m) show that the wind blows more or less at right angles to the pressure gradient (i.e. parallel to the isobars) with, for the northern hemisphere, the high-pressure core on the right and the low-pressure on the left when viewed downwind. This implies that for steady motion the pressure gradient force is exactly balanced by the Coriolis deflection acting in the diametrically opposite direction (see Figure 3.3).

Figure 3.3 The geostrophic wind case of balanced motion (northern hemisphere).

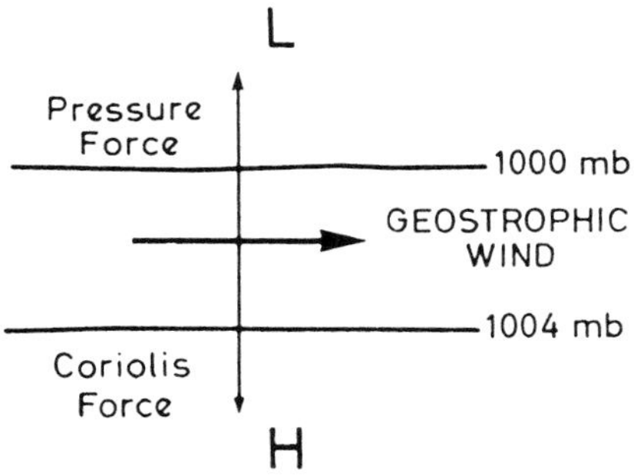

The wind in this idealized case is called a *geostrophic wind*, the velocity (V_g) of which is given by the following formula:

$$V_g = \frac{1}{2\omega \sin \phi \, \rho} \cdot \frac{dp}{dn}$$

where dp/dn = the pressure gradient. The velocity is thus inversely dependent on latitude, such that the same pressure gradient which will be associated with geostrophic wind speeds of 15 m s^{-1} (34 mph) at latitude 43° will produce a velocity of only 10 m s^{-1} (23 mph) at latitude 90°. Except in low latitudes, where the Coriolis deflection approaches zero, the geostrophic wind is a close approximation to the observed air motion in the free atmosphere. Since pressure systems are rarely stationary this fact implies that air motion must continually change towards a new balance. In other words mutual adjustments of the wind and pressure fields are constantly taking place. The common 'cause-and-effect' argument that a pressure gradient is formed and air begins to move towards low pressure before coming into geostrophic balance is an unfortunate oversimplification of reality.

4 The centripetal acceleration

For a body to follow a curved path there must be an inward acceleration towards the centre of rotation. This acceleration (c) is expressed by:

$$c = -\frac{mV^2}{r}$$

where m = the moving mass, V = its velocity and r = the radius of curvature. This factor is sometimes regarded for convenience as a centrifugal force operating radially outward (see note 1). In the case of the earth itself this is valid. The centrifugal effect due to rotation has in fact resulted in a slight bulging of the earth's mass in low latitudes and a flattening near the poles. The small decrease in apparent gravity towards the equator (see note 2) reflects the effect of the centrifugal force working against the gravitational attraction directed towards the earth's centre. It is only necessary therefore to consider the forces involved in the rotation of the air about a local axis of high or low pressure. Here the curved path of the air (parallel to the isobars) is maintained by an inward-acting, or centripetal, acceleration.

Figure 3.4 shows (for the northern hemisphere) that in a low-pressure system balanced flow is maintained in a curved path (referred to as the *gradient wind*) by the Coriolis force being weaker than the pressure force. The difference between the two gives the net centripetal acceleration inward. In the high-pressure case the inward acceleration is provided by the Coriolis force exceeding the pressure force. Since the pressure gradients are assumed to be equal, the different contributions of the Coriolis force in each case imply that the wind speed around the low-pressure must be less than the geostrophic value (*subgeostrophic*), whereas in the high-pressure case it is *supergeostrophic*. In reality this effect is obscured by the fact that the pressure gradient in a high is usually much weaker than in a low. Moreover, the fact that the earth's rotation is cyclonic imposes a limit on the speed of anticyclonic flow. The maximum occurs when the angular velocity is $f/2$ (= ω sin ϕ, at which value the absolute rotation of the air (viewed from space) is just cyclonic. Beyond this point anticyclonic flow breaks down ('dynamic instability'). There is no maximum speed in the case of cyclonic rotation.

The magnitude of the centripetal acceleration is generally small, but it becomes important where high-velocity winds are moving in very curved paths (i.e. about an intense low-pressure system). Two cases are of meteorological significance: first, in intense cyclones near the equator where the Coriolis force is negligible; and, second, in a narrow vortex such as a tornado. Under these conditions, when the large pressure gradient force provides the necessary centripetal acceleration for balanced flow parallel to the isobars, the motion is called *cyclostrophic*.

The above arguments all assume steady conditions of balanced flow. This simplification is useful, but it must be noticed that two factors prevent a continuous state of balance. Latitudinal motion changes the Coriolis parameter and the movement or changing intensity of a pressure system leads to acceleration or deceleration of the air, causing some degree of

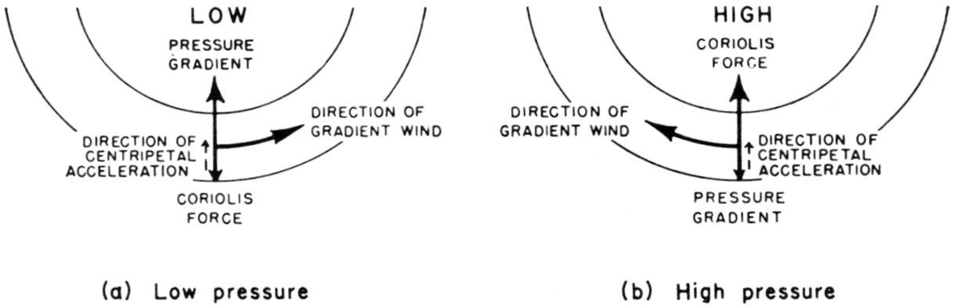

Figure 3.4 The gradient wind case of balanced motion around a low pressure (a) and a high pressure (b) in the northern hemisphere.

cross-isobaric flow. Pressure change itself depends on air displacement through the breakdown of the balanced state. If air movement were purely geostrophic there would be no growth or decay of pressure systems. The acceleration of air in moving at upper levels from a region of cyclonic isobaric curvature (subgeostrophic wind) to one of anticyclonic curvature (supergeostrophic wind) causes a fall of pressure at lower levels in the atmosphere through the removal of air aloft. The significance of this fact will be discussed in Chapter 4, F. The interaction of horizontal and vertical air motions is outlined in B.2, this chapter.

5 Frictional forces

The last force which has an important effect on air movement is that due to friction with the earth's surface. If we follow our study of geostrophic wind a little further we find that towards the surface (i.e. below about 500 m for flat terrain) friction begins to decrease the wind velocity below its geostrophic value. This has an effect on the deflective force (which is dependent on velocity) causing it also to decrease. As these two tendencies continue the wind consequently blows more and more obliquely across the isobars in the direction of the pressure gradient. The angle of obliqueness increases with the growing effect of frictional drag due to the earth's surface and it averages about 10°–20° at the surface over the sea and 25°–35° over land. The result is to produce a wind spiral with height (see Figure 3.5), analogous to the turning of ocean currents as the effect of wind stress diminishes with increasing depth. Both are referred to as 'Ekman spirals', after Ekman who investigated the variation of ocean currents with depth (see F.3, this chapter).

Wind velocity decreases exponentially close to the earth's surface due to the frictional effects produced by surface roughness. The aerodynamic roughness of terrain is characterized by the *roughness length* (z_0), defined by Oke as 'the height at which the neutral wind profile extrapolates to zero wind speed'. The concept of roughness length involves the lifting of airflow over a rough surface close to which the airflow is of low velocity. Typical roughness lengths are given in Table 3.1.

In summary, the surface wind (neglecting any curvature effects) represents a balance between the pressure gradient force and friction parallel to the air motion and between the pressure gradient force and the Coriolis force perpendicular to the air motion.

B DIVERGENCE, VERTICAL MOTION AND VORTICITY

These three terms essentially hold the key to a proper understanding of modern meteorological studies of wind and pressure systems on a synoptic and global scale. Mass uplift or descent of air occurs primarily in response to dynamic factors related to horizontal airflow and is only secondarily affected by air-mass stability. Hence the significance of these factors for weather processes.

1 Divergence

Different types of horizontal flow are shown in Figure 3.6A. The first panel shows that air may

Table 3.1 Typical roughness lengths (metres) associated with terrain surface characteristics.

Terrain surface characteristics	*Roughness length (m)*
Groups of high buildings	1–10
Temperate forest	0.8
Groups of medium buildings	0.7
Suburbs	0.5
Trees and bushes	0.2
Farmland	0.05–0.1
Grass	0.008
Bare soil	0.005
Snow	0.001
Smooth sand	0.0003
Water	0.0001

Source: After Troen and Petersen 1989.

Figure 3.5 The Ekman spiral of wind with height, in the northern hemisphere. The wind attains the geostrophic velocity at between 500 and 1000 m in the middle and higher latitudes as frictional drag effects become negligible. This is a theoretical profile of wind velocity under conditions of mechanical turbulence.

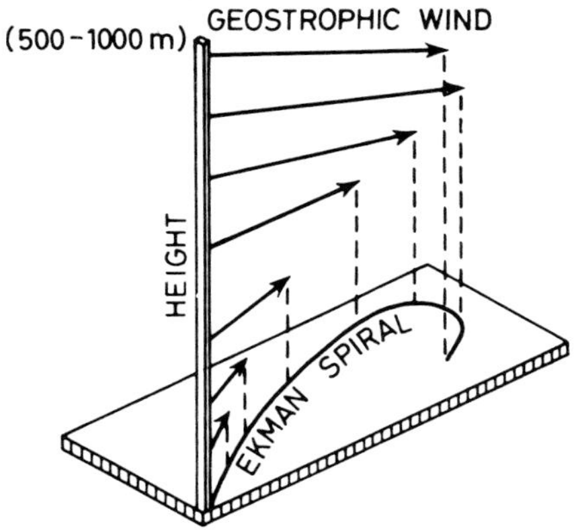

accelerate (decelerate) leading to velocity divergence (convergence). When streamlines (lines of instantaneous air motion) spread out or squeeze together, this is termed diffluence or confluence respectively. If the streamline pattern is strengthened by that of the isotachs (lines of equal wind speed), as shown in the third panels of Figure 3.6A, then there may be mass divergence or convergence at a point (see Figure 3.6B). In this case the compressibility of the air causes the density to decrease, or increase, respectively. Usually, however, confluence is associated with an increase in air velocity and diffluence with a decrease. In the intermediate case confluence is balanced by an increase in wind velocity and diffluence by a decrease in velocity. Hence, convergence (divergence) may give rise to vertical stretching (shrinking) as illustrated in Figure 3.6C. It is important to note that if all winds were geostrophic, there could be no convergence or divergence and hence no weather!

Other ways in which convergence or divergence can occur are the result of surface friction effects. Onshore winds undergo convergence at low levels when the air slows down on crossing the coastline owing to the greater friction overland, whereas offshore winds accelerate and become divergent. Frictional differences can also set up coastal convergence (or divergence) if the geostrophic wind is parallel to the coastline with, for the northern hemisphere, land to the right (or left) of the air current, viewed downwind.

2 Vertical motion

Horizontal inflow or outflow near the surface has to be compensated by vertical motion, as illustrated in Figure 3.7, if the low- or high-pressure systems are to persist and there is to be no continuous density increase or decrease. Air rises above a low-pressure cell and subsides over a high-pressure, with compensating divergence and convergence, respectively, in the upper troposphere. In the middle troposphere there must clearly be some level at which horizontal divergence or convergence is effectively zero; the mean 'level of non-divergence' is generally at about 600 mb. Large-scale vertical motion is extremely slow compared with convective and downdraught currents in cumulus, for example. Typical rates in large depressions and anticyclones are of the order of 5–10 cm s^{-1}, whereas updraughts in cumulus may exceed 10 m s^{-1}.

3 Vorticity

Vorticity implies the rotation or angular velocity of minute (imaginary) particles in any fluid system. The air within a depression can be regarded as comprising an infinite number of small air parcels, each rotating cyclonically about an axis vertical to the earth's surface (see Figure 3.8). Vorticity has three elements – magnitude (defined as *twice* the angular velocity, ω) (see note 3), direction (the horizontal or vertical axis about which the rotation occurs) and the sense of rotation. Rotation in the same sense as the earth's rotation – cyclonic in the northern hemisphere – is defined as

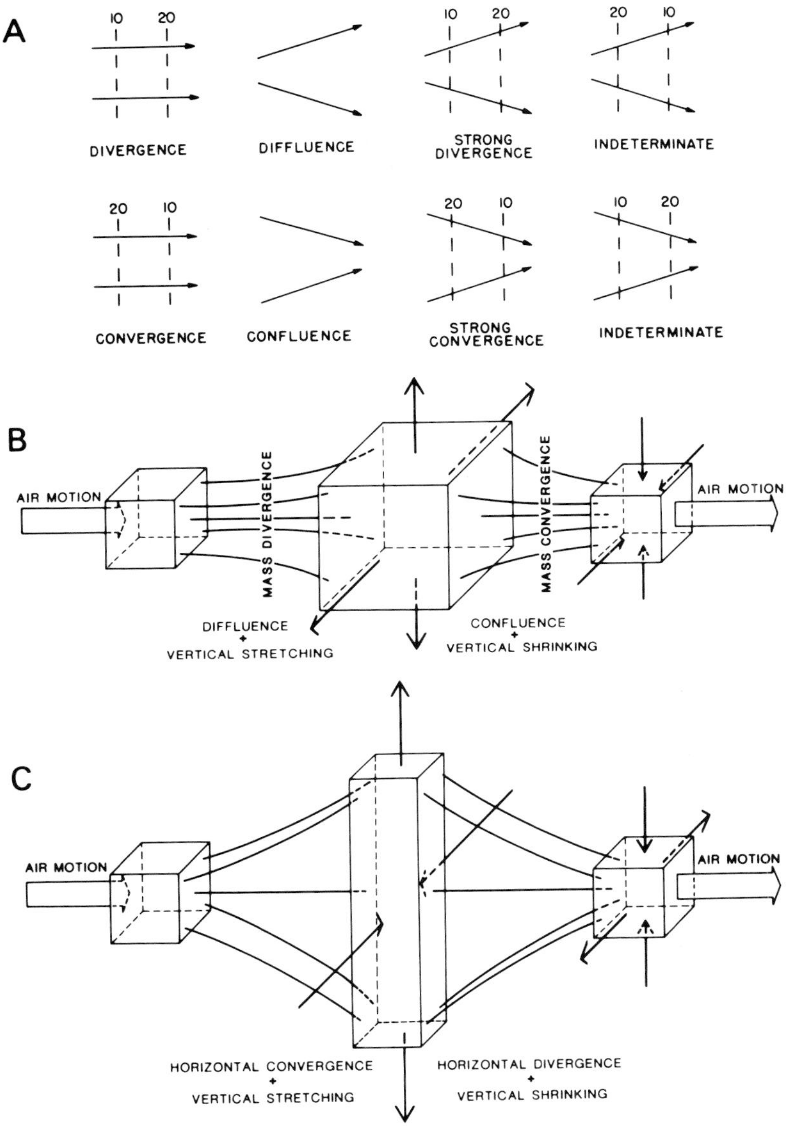

Figure 3.6 Convergence and divergence. A: Plan view of horizontal flow patterns producing divergence and convergence – the broken lines are schematic isopleths of wind speed (isotachs). B: Schematic illustration of local mass divergence and convergence, assuming density changes. C: Typical convergence-stretching and divergence-shrinking relationships in atmospheric flow.

positive. Cyclonic vorticity may result from cyclonic curvature of the streamlines, from cyclonic shear (stronger winds on the right side of the current, viewed downwind in the northern hemisphere), or a combination of the two (see Figure 3.9). Lateral shear (see Figure 3.9b) results from changes in isobar spacing. Anticyclonic vorticity occurs with the corresponding anticyclonic situation. The component of vorticity about a vertical axis is referred to as the vertical vorticity. This is generally the most important, but near the ground surface frictional shear causes vorticity about an axis parallel to the surface and normal to the wind direction.

Vorticity is related not only to air motion about a cyclone or anticyclone (*relative*

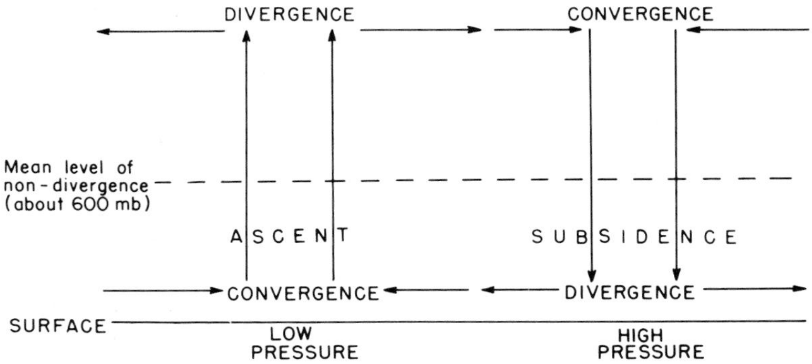

Figure 3.7 Cross-section of the patterns of vertical motion associated with (mass) divergence and convergence in the troposphere, illustrating mass continuity.

Figure 3.8 Sketch of the relative vertical vorticity (ζ) about a cyclone and an anticyclone in the northern hemisphere. The component of the earth's vorticity about its axis of rotation (or, the Coriolis parameter, f), is equal to twice the angular velocity (ω) times the sine of the latitude (ϕ). At the pole $f = 2\omega$, diminishing to 0 at the equator. Cyclonic vorticity is in the same sense as the earth's rotation about its own axis, viewed from above, in the northern hemisphere: this cyclonic vorticity is defined as positive ($\zeta > 0$).

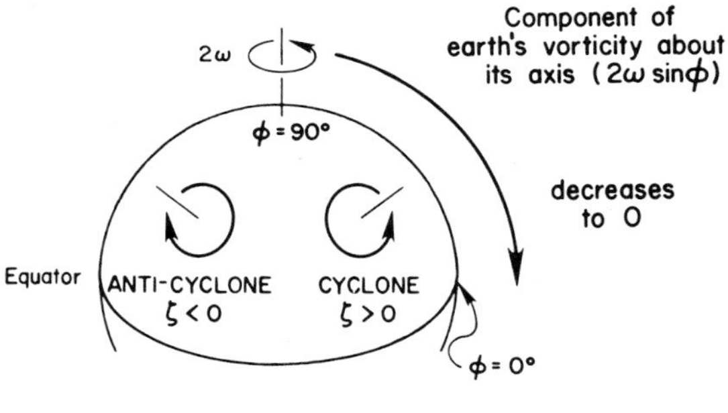

vorticity), but also to the location of that system on the rotating earth. The vertical component of *absolute vorticity* consists of the relative vorticity (ζ) and the latitudinal value of the Coriolis parameter, $f = 2\omega \sin \phi$ (see A.2, this chapter). At the equator the local vertical is at right angles to the earth's axis so that $f = 0$, but at the North Pole cyclonic relative vorticity and the earth's rotation act in the same sense (see Figure 3.8).

C LOCAL WINDS

To the practising meteorologist, local controls over air movement often provide more problems than the effects of the major planetary forces just discussed. Diurnal tendencies are superimposed upon both the large- and small-scale patterns of wind velocity. These are particularly noticeable in the case of local winds and therefore are examined before we consider the major types of local wind regime.

In normal conditions there is a general tendency for wind velocities to be least about dawn, at which time there is little vertical thermal mixing and the lower air does not therefore partake of the velocity of the more freely moving upper air (see D, this chapter). Conversely, velocities of some local winds are greatest between 1300 and 1400 hours, for this is the time when the air suffers its greatest tendency to move vertically due to terrestrial heating, allowing it, subject to surface frictional effects, to join in the freer upper-air movement. Upper air always moves more freely than air at surface levels because it is not subject to the retarding effects of friction and obstruction.

1 Mountain and valley winds

Terrain irregularities give rise to their own special meteorological conditions. On warm sunny days, the heated air in a valley is laterally

Figure 3.9 Streamline models illustrating in plan view the flow patterns with cyclonic and anticyclonic vorticity in the northern hemisphere. In c and d the effects of curvature (a_1 and a_2) and lateral shear (b_1 and b_2) are additive, whereas in e and f they more or less cancel out. Dashed lines are schematic isopleths of wind speed.

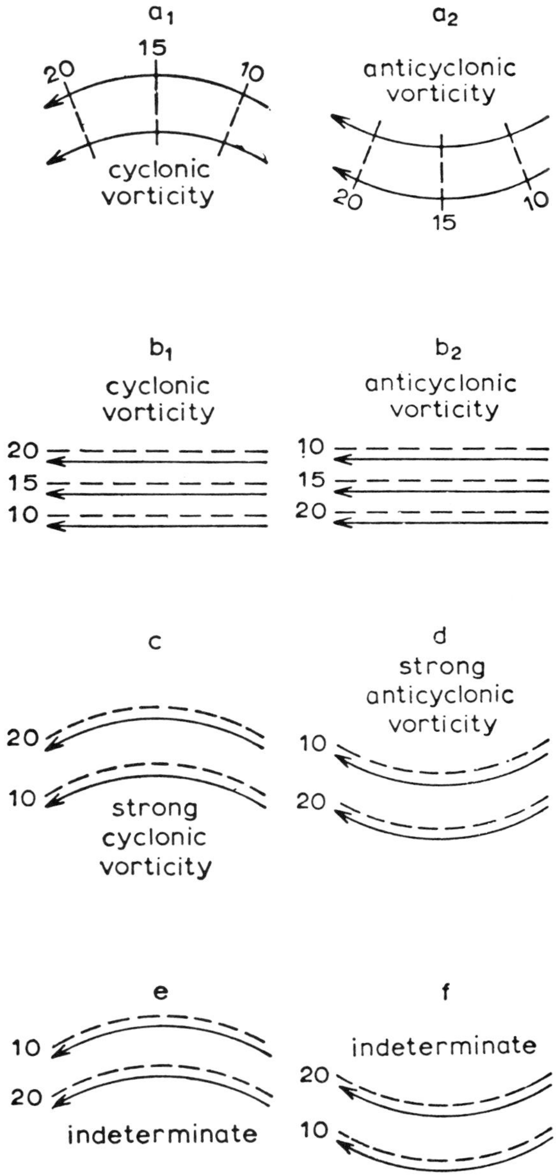

Source: After Riehl *et al.* 1954.

constricted, compared with that over an equivalent area of lowland, and so tends to expand vertically. The volume ratio of lowland: valley air is typically about 2 or 3:1 and this difference in heating sets up a pressure differential which causes air to flow from the lowland up the axis of the valley. This valley wind is generally very light and requires a weak regional pressure gradient in order to develop. This flow along the main valley develops more or less simultaneously with *anabatic* (upslope) winds which result from greater heating of the valley sides compared with the valley floor. These slope winds rise above the ridge lines and feed an upper return current along the line of the valley to compensate for the valley wind. This feature may be obscured, however, by the regional airflow. Speeds reach a maximum around 1400 hours. At night there is a reverse process as the cold denser air at higher elevations drains into depressions and valleys; this is known as a *katabatic* wind (see Figure 3.10).

If the air drains downslope into an open valley, a 'mountain wind' develops more or less simultaneously along the axis of the valley. This flows towards the plain where it replaces warmer, less dense air. The maximum velocity occurs just before sunrise at the time of maximum diurnal cooling. Like the valley wind, the mountain wind is also overlain by an upper return current, in this case up-valley.

Katabatic drainage is usually cited as the cause of frost pockets in hilly and mountainous areas. It is argued that greater radiational cooling on the slopes, especially if they are snow-covered, leads to a gravity flow of cold, dense air into the valley bottoms. Observations in California and elsewhere, however, suggest that the valley air remains colder than the slope air from the onset of nocturnal cooling, so that the air moving downslope slides over the denser air in the valley bottom. Moderate drainage winds will also act to raise the valley temperatures through turbulent mixing. One suggestion is that cold air pockets in valley bottoms and hollows may result from the cessation of turbulent heat transfer to the

Figure 3.10 Valley winds in an ideal V-shaped valley. a: Section across the valley. The valley wind and anti-valley wind are directed at right angles to the plane of the paper. The arrows show the slope and ridge wind in the plane of the paper, the latter diverging (div.) into the anti-valley wind system. b: Section running along the centre of the valley and out on to the adjacent plain, illustrating the valley wind (*below*) and the anti-valley wind (*above*).

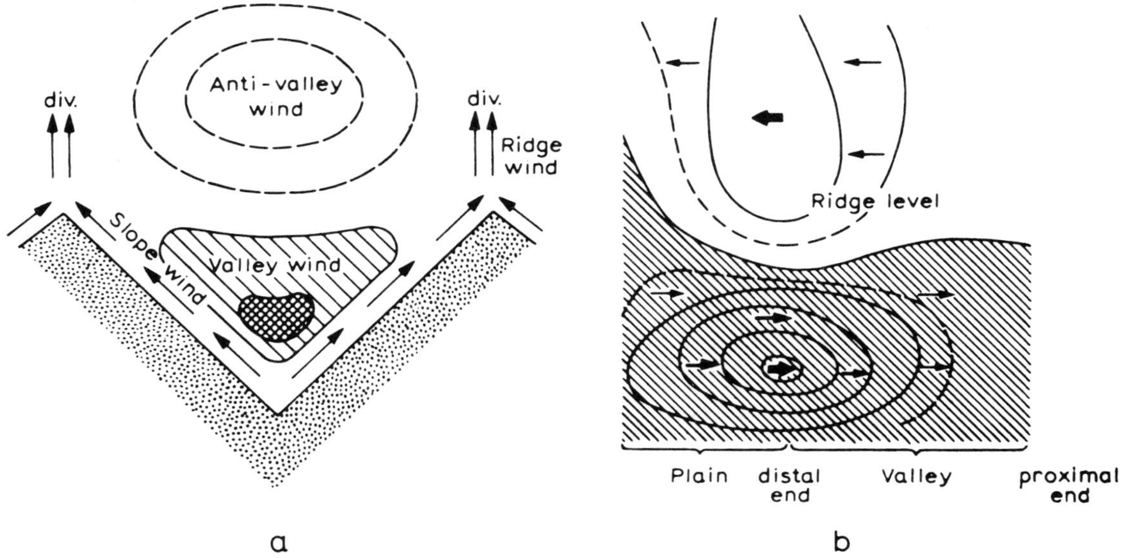

Source: After Buettner and Thyer 1965.

surface in sheltered locations rather than by cold air drainage. Clearly, the problem requires further study.

2 Winds due to topographic barriers

Mountain ranges have important effects on the airflow across them. The displacement of air upwards over the obstacle may trigger instability if the air is conditionally unstable (see Chapter 2, E), whereas stable air returns to its original level in the lee of the barrier and this descent often forms the first of a series of *lee waves* (or *standing waves*) downwind, as shown in Figure 3.11. The wave form remains more or less stationary relative to the barrier with the air moving quite rapidly through it. Below the crest of the waves there may be circular air motion in a vertical plane which is termed a *rotor*. The formation of such features is naturally of vital interest to pilots. The development of lee waves is commonly disclosed by the presence of lenticular clouds (see Plate 8), and on occasion a rotor causes reversal of the surface wind direction in the lee of high mountains (see Plate 9).

Airflow over a topographic obstruction causes convergence by vertical shrinking and an increase in near-surface velocities particularly at and near the crest line. Figure 3.12 shows instantaneous airflow conditions across Askervein Hill (relief c.120m) on the Isle of South Uist in the Scottish Hebrides where the near-crest wind speed at a height of 10 m above the surface approaches 80 per cent more than the undisturbed upstream velocity. In contrast, there was a 20 per cent decrease on the initial run-up to the hill and a 40 per cent decrease on the lee side, probably due to horizontal divergence.

Winds on summits are usually strong, at least in middle and higher latitudes. Average speeds on summits in the Rocky Mountains in winter months are around 12–15 m s^{-1}, for example, and on Mt Washington, New Hampshire, an extreme value of 103 m s^{-1} has been recorded.

Figure 3.11 Lee waves and rotors are produced by air flow across a long mountain range. The first wave crest usually forms less than one wavelength downwind of the ridge. There is a strong surface wind down the lee slope. Wave characteristics are determined by the wind speed and temperature relationships, shown schematically on the left of the diagram. The existence of an upper stable layer is particularly important.

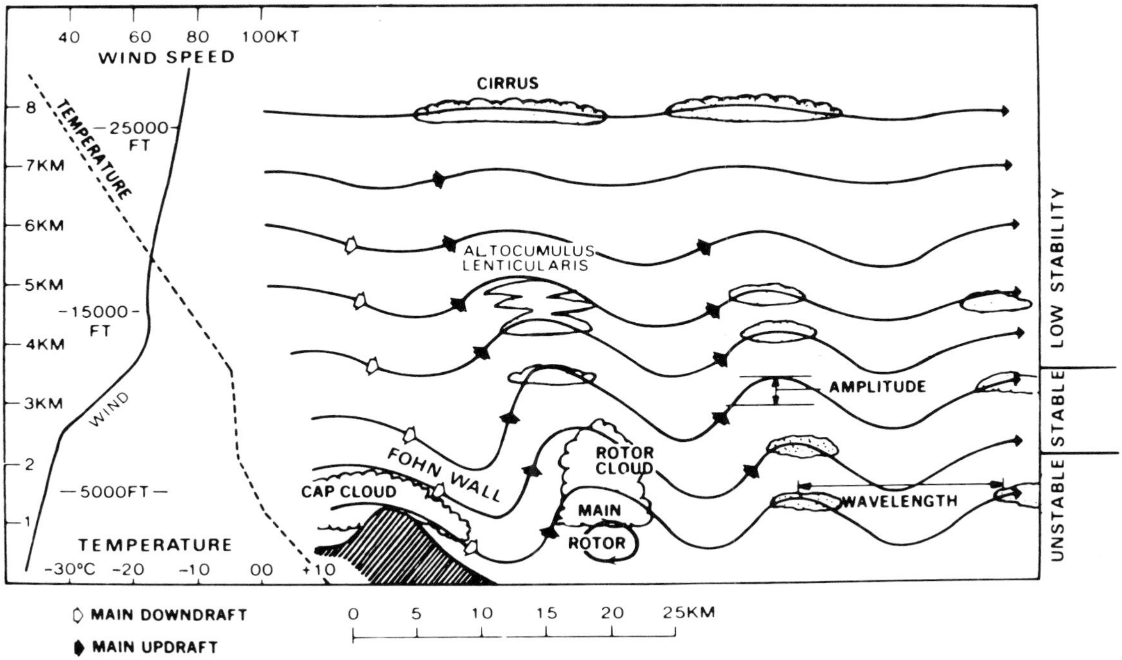

Source: After Ernst 1976.

Peak monthly speeds in excess of 40–50 m s^{-1} are typical in both these areas in winter. The air is constricted and thus accelerated over mountain barriers (the Venturi effect), but friction with the ground also retards the flow, compared with the free air at the same level. Many European summits record lower wind speeds than over adjacent lowlands at the same level.

A wind of local importance in mountain areas is the föhn or chinook. It is a strong, gusty, dry and warm wind which develops on the lee side of a mountain range when stable air is forced to flow over the barrier by the regional pressure gradient. Sometimes, there is a loss of moisture by precipitation on the mountains (see Figure 3.13) and the air, having cooled at the saturated adiabatic lapse rate above the condensation level, subsequently warms at the greater dry adiabatic lapse rate as it descends on the lee side with a consequent lowering of both the relative and absolute humidity. Other investigations show that in many instances there is no loss of moisture over the mountains. In such cases the föhn effect is the result of the blocking of air to windward of the mountains by a summit-level temperature inversion. This forces air from higher levels to descend and warm adiabatically. Föhn winds are common along the northern flanks of the Alps and the mountains of the Caucasus and central Asia in winter and spring, when the accompanying rapid temperature rise may help to trigger off avalanches on the snow-covered slopes. At Tashkent in central Asia, where the mean winter temperature is about the freezing point, temperatures may rise to more than 21°C during a föhn. In the same way the

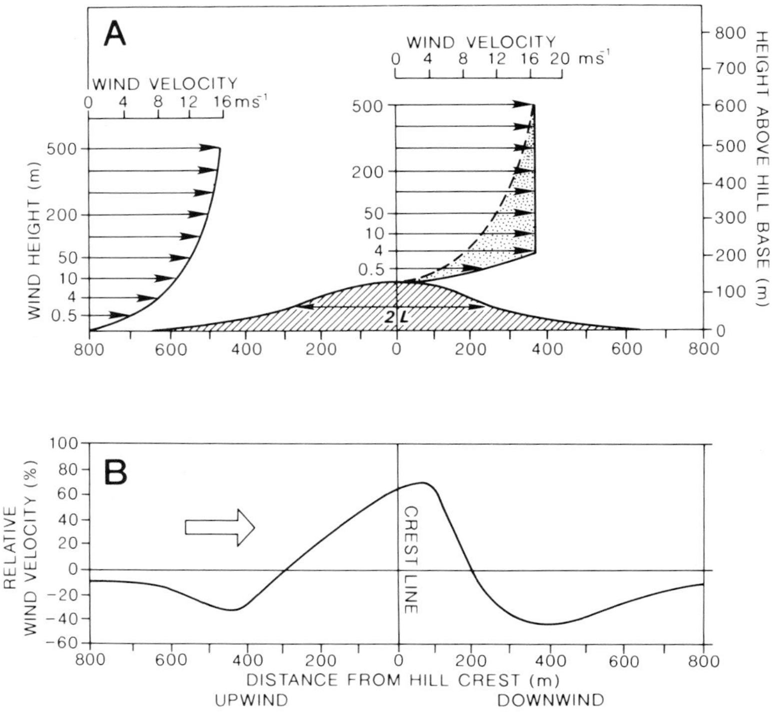

Figure 3.12 Airflow over Askervein Hill, South Uist, off the west coast of Scotland. A: Vertical airflow profiles (not true to scale) measured simultaneously 800 m upwind of the crest line and at the crest line. *L* is the *characteristic length* of the obstruction (i.e. one-half the hill width at mid elevation). *2L* (i.e. 500 m) is also the height above ground level to which the flow is increased by the topographic obstruction (stippled). The maximum speed-up of the airflow due to vertical convergence over the crest is to about 16.5 m s^{-1} at a height of 4 m. B: The relative speed-up (%) of airflow upwind and downwind of the crest line measured 14 m above ground level.

Source: After Taylor, Teunissen and Salmon *et al.* From Troen and Petersen 1989.

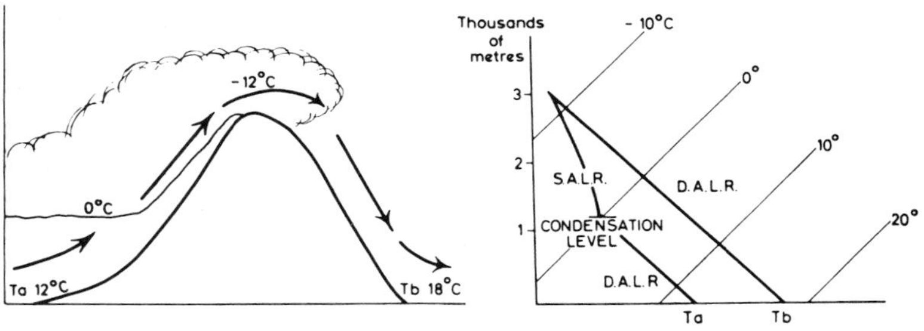

Figure 3.13 The föhn effect when an air parcel is forced to cross a mountain range. Ta refers to the temperature at the windward foot of the range and Tb to that at the leeward foot.

chinook is a significant feature at the eastern foot of the New Zealand Alps and of the Rocky Mountains. At Pincher Creek, Alberta, a temperature rise of 21°C (38°F) occurred in four minutes with the onset of a chinook on 6 January 1966. Less spectacular effects are also noticeable in the lee of the Welsh mountains, the Pennines and the Grampians, where the importance of föhn winds lies mainly in the dispersal of cloud by the subsiding dry air (see also Chapter 5, pp. 197 and 208). This is an important component of so-called 'rain shadow' effects.

In some parts of the world, winds descending on the lee slope of a mountain range are cold. The type example of such 'fall-winds' is the *bora* of the northern Adriatic, although similar winds occur on the northern Black Sea coast, in northern Scandinavia, Novaya Zemlya and in Japan. These winds occur when cold continental air masses are forced across a mountain range by the pressure gradient and, despite adiabatic warming, displace warmer air. They are therefore primarily a winter phenomenon.

On the eastern slope of the Rocky Mountains

in Colorado (and probably also in other similar continental locations), winds of either *bora* or chinook type can occur. Locally, at the foot of the mountains, such winds may attain hurricane force with gusts exceeding 45 m s^{-1} (100 mph). A few downslope storms of this type have caused millions of dollars of property damage in Boulder, Colorado and the immediate vicinity. These windstorms develop when a stable layer close to the mountain-crest level prevents airflow to windward from crossing over the mountains. Extreme amplification of a lee wave (see Figure 3.11) drags air from above the summit level (about 4000 m) down to the plains (1700 m) in a very short distance, so that the high velocities occur. However, the flow is not simply 'downslope'; the storm may affect the mountain slopes but not the foot of the slope, or vice versa, depending on the location of the lee wave trough. The high winds are caused by the acceleration of the air towards this pressure minimum.

3 Land and sea breezes

Another thermally-induced type of air movement is the land and sea breeze (see Figure 3.14). The vertical expansion of the air column which occurs daily during the hours of heating over the more rapidly heated land (see Chapter 1, D.5), tilts the isobaric surfaces downwards at the coast, causing onshore winds at the surface and a compensating offshore movement aloft. Typical land–sea pressure differences are of the order of 2 mb. At night the air over the sea is warmer and the situation is reversed, although this reversal is also the effect of downslope winds blowing off the land. Figure 3.15 shows that these local winds can have a decisive effect on coastal temperature and humidity. The advancing cool sea air may form a distinct line (or *front*, see Chapter 4, C) marked by cumulus cloud development, behind which there is a distinct wind velocity maximum. This often develops in summer, for example, along the Gulf Coast of Texas. On a smaller scale such features can also be observed in Britain, particularly along the south and east coasts. The sea breeze has a depth of about 1 km (3300 ft), although it thins towards the advancing edge, and may penetrate 50 km (30 miles) or more inland by 2100 hours. Typical wind speeds in such sea breezes are 4–7 m s^{-1} (about 10–15 mph), although these may be greatly increased where a well-marked low-level temperature inversion produces a 'Venturi effect' in constricting and accelerating the flow. The much shallower land breezes are usually only about 2 m s^{-1} (about 5 mph). The counter currents aloft are generally less evident and may be obscured by the regional airflow, but recent work along the Oregon coast has suggested that under certain conditions this upper return flow may be very closely related to the lower sea breeze conditions, even to the

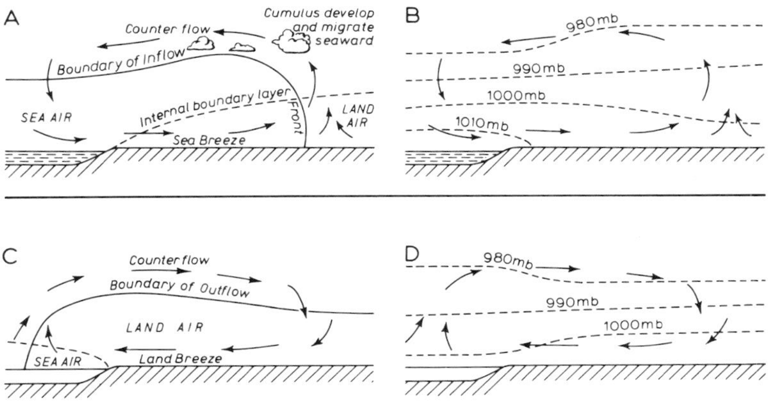

Figure 3.14 Diurnal land and sea breezes. A and B: Sea breeze circulation and pressure distribution in the early afternoon during anticyclonic weather. C and D: Land breeze circulation and pressure distribution at night during anticyclonic weather.

Source: A and C after Oke 1978.

Figure 3.15 The effect of the afternoon sea breeze on the temperature (°C) and relative humidity (%) at Joal on the Senegal coast, 8–10 February 1893.

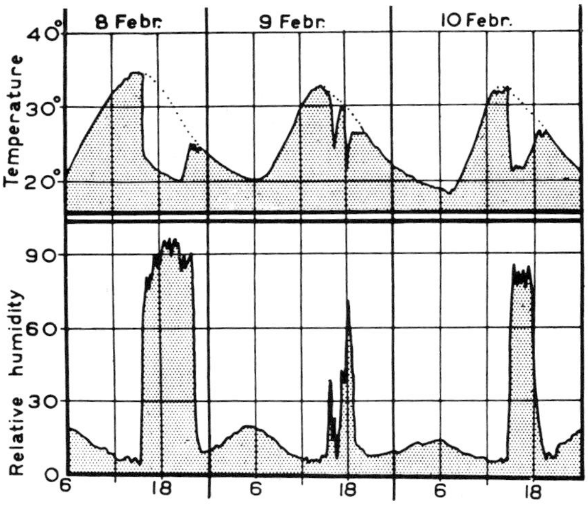

Source: After Angot and De Martonne; from Kuenen 1955.

extent of mirroring the surges in the latter. It is worth noting that in middle latitudes the Coriolis deflection causes turning of a well-developed onshore sea breeze (clockwise in the northern hemisphere) so that eventually it may blow more or less parallel to the shore. Analogous 'lake breeze' systems develop adjacent to large inland water bodies such as the Great Lakes.

D VARIATION OF PRESSURE AND WIND VELOCITY WITH HEIGHT

Changes of height reveal variations both of pressure and of wind characteristics. Above the level of surface frictional effects (about 500–1000 m) the wind increases in speed and becomes more or less geostrophic. With further height increase the reduction of air density leads to a general increase in wind speed (see A.1, this chapter). At 45°N a geostrophic wind of 14 m s^{-1} at 3 km is equivalent to one of 10 m s^{-1} at the surface for the same pressure gradient. There is also a seasonal variation in wind speeds aloft, these being much greater during winter months when the meridional temperature gradients are at a maximum. In addition, the persistence of these gradients causes the upper winds to be more constant in direction.

1 The vertical variation of pressure systems

The general relationships between surface and tropospheric pressure conditions are illustrated by the models of Figure 3.16. A low-pressure cell at sea level with a cold core will intensify with elevation, whereas one with a warm core tends to weaken and may be replaced by high pressure. A warm air column of relatively low density causes the pressure surfaces to bulge upwards and conversely a cold, more dense air column leads to downward contraction of the pressure surfaces. Thus, a surface high-pressure cell with a cold core (a *cold anticyclone*), such as the Siberian winter anticyclone, weakens with increasing elevation and is replaced by low pressure aloft. Cold anticyclones are shallow and rarely extend their influence above about 2500 m (8000 ft). By contrast, a surface high with a warm core (a *warm anticyclone*) intensifies with height (Figure 3.16D). This is characteristic of the large subtropical cells which maintain their warmth through dynamic subsidence. The warm low (Figure 3.16C) and cold high (Figure 3.16B) are consistent with the vertical motion schemes illustrated in Figure 3.7, whereas the other two types are primarily produced by dynamic processes. The high surface pressure in a warm anticyclone is linked hydrostatically with cold, relatively dense air in the lower stratosphere. Conversely a cold depression (Figure 3.16A) is associated with a warm lower stratosphere.

Mid-latitude low-pressure cells have cold air in the rear and in consequence the axis of low pressure slopes with height towards the colder air to the west. High-pressure cells slope towards the warmest air (see Figure 3.17) and in this manner the northern hemisphere subtropical high-pressure cells are displaced 10–15° south in latitude at the 3000-m level, as well as towards the west (see Figure 3.18). Even so, this slope of the high-pressure axes is not constant

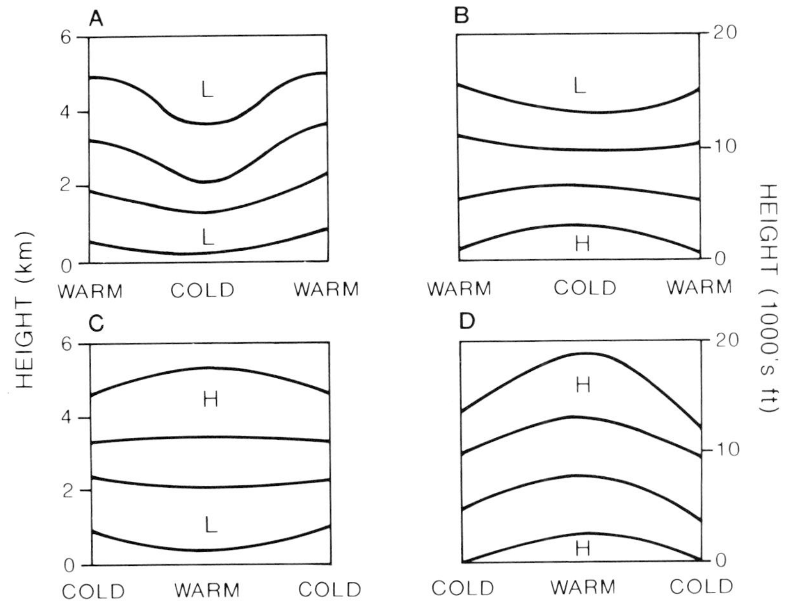

Figure 3.16 Models of the vertical pressure distribution in cold and warm air columns. A: a surface low pressure intensifies aloft in a cold air column. B: a surface high pressure weakens aloft and may become a low pressure in a cold air column. C: a surface low pressure weakens aloft and may become a high pressure in a warm air column. D: a surface high pressure intensifies aloft in a warm air column.

Figure 3.17 The characteristic slope of the axes of low- and high-pressure cells with height in the northern hemisphere.

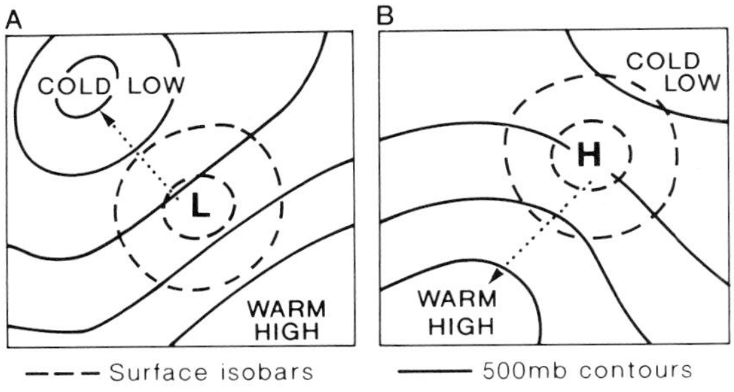

through time and stations located between the cells may experience widely fluctuating upper winds associated with variations in the inclination of the axes.

2 Mean upper-air patterns

It is helpful to begin by considering the patterns of pressure and wind in the middle troposphere. These are less complicated in appearance than surface maps as a result of the diminished effects of the land masses. Rather than using pressure maps at a particular height it is convenient to depict the height of a selected pressure surface; this is termed *contour chart* by analogy with topographic relief maps (see note 4). Figure 3.19 shows that in the middle troposphere of the southern hemisphere there is a vast circumpolar cyclonic vortex poleward of latitude 30°S in summer and winter. The vortex is more or less symmetrical about the pole, although the low centre is towards the Ross Sea sector. Corresponding charts for the northern hemisphere (see Figure 3.20) also show an extensive cyclonic vortex, but one which is markedly more asymmetric with a primary centre over the eastern Canadian Arctic and a secondary one over eastern Siberia. The major troughs and ridges form what are referred to as *long waves* (or *Rossby waves*) in the upper flow (see Chapter 4, F). The two major troughs at about 70°W and 150°E are thought to be induced by the combined influence on upper-air pressure and winds of large orographic barriers, like the Rockies and the Tibetan Plateau, and heat

Figure 3.18 Schematic horizontal and vertical structure of the subtropical high-pressure cells. Note particularly the convergence along the belts between the cells, the slope of the axes with height westward and equatorward, and the inclined spiral of air motion in the middle troposphere – up on the west sides (dynamically unstable air) and down on the east sides (dynamically stable air).

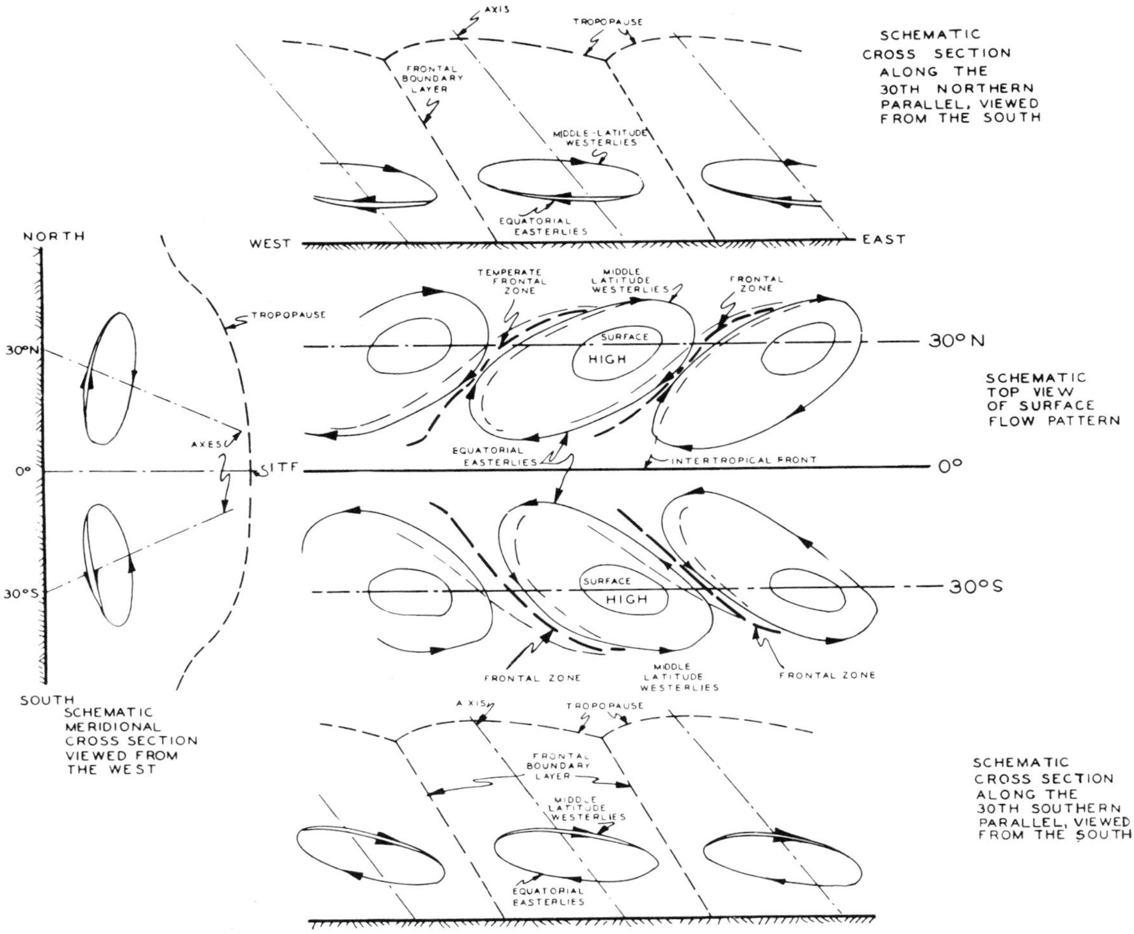

Source: From Garbell 1947.

sources such as warm ocean currents (in winter) or land masses (in summer). It is worth noting that land surfaces occupy over 50 per cent of the northern hemisphere between latitudes 40°N and 70°N. The subtropical high-pressure belt has only one clearly distinct cell in January over the eastern Caribbean, whereas in July cells are well developed over the Atlantic and the Pacific. In addition, the July map shows greater prominence of the subtropical high over the Sahara and southern North America. The northern hemisphere shows a marked summer to winter intensification of the mean circulation which is explained below.

In the southern hemisphere, the predominance of ocean surface (which comprises 81 per cent of the hemisphere) considerably reduces the development of long waves in the upper westerlies. Nevertheless, asymmetries are initiated by the effects on the atmosphere of such geo-

Figure 3.19 The mean contours (g.p.dkm) of the 700-mb pressure surface in January (A) and July (B) for the southern hemisphere, 1949–60.

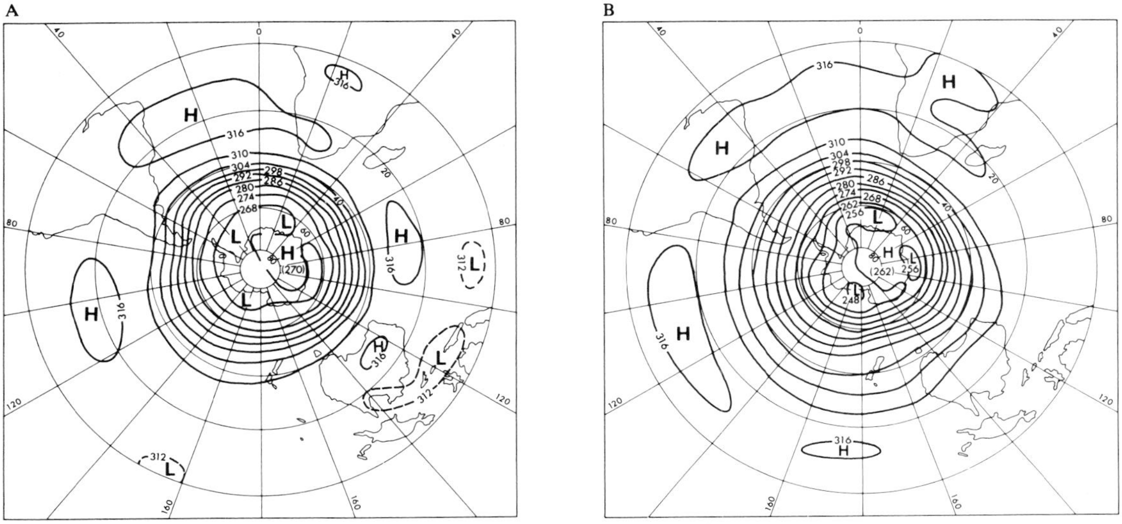

Source: After Taljaard *et al.* 1969.

Figure 3.20 The mean contours (g.p.dkm) of the 700-mb pressure surface in January (A) and July (B) for the northern hemisphere, 1950–9.

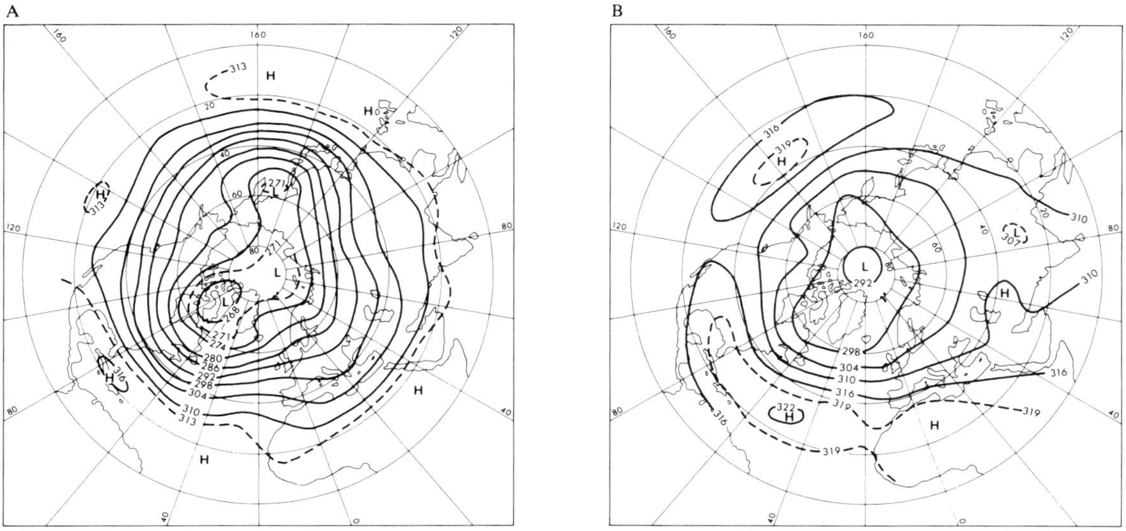

Source: Adapted from O'Connor 1961.

graphical features as the Andes, the elevated and extensive dome of eastern Antarctica, and ocean currents, particularly the Humboldt and Benguela Currents (see Figure 3.42) and the associated cold coastal upwellings.

3 Upper winds

It is a common observation that clouds at different levels move in different directions. The wind speeds at these levels may also be markedly different, although this is not so evident to the casual observer. The gradient of wind velocity with height is referred to as the (vertical) *wind shear*, and in the free air, above the friction level, the amount of shear depends upon the temperature structure of the air. This important relationship is illustrated in Figure 3.21. The diagram shows hypothetical contours of the 1000 and 500 mb pressure surfaces. The *thickness* of the 1000–500 mb layer is proportional to its mean temperature – low thickness values correspond to cold air, high thickness values to warm air. This relationship is apparent in the vertical sections of Figure 3.16. The theoretical wind vector (V_T), blowing parallel to the thickness lines with a velocity proportional to their gradient, is termed the *thermal wind*. The geostrophic wind velocity at 500 mb (G_{500}) is the vector sum of the 1 000-mb geostrophic wind (G_{1000}) and the thermal wind (V_T), as shown in Figure 3.21.

Since the thermal wind component blows with cold air (low thickness) to the left in the northern hemisphere, when viewed downwind, it is readily apparent that in the troposphere the poleward decrease of temperature should be associated with a large westerly component in the upper winds. Furthermore, since the meridional temperature gradient is steepest in winter (in the northern hemisphere) the zonal westerlies are most intense at this time.

The total result of the above influences is that in both hemispheres the mean upper geostrophic winds are dominantly westerly between the subtropical high-pressure cells (centred aloft about 15° latitude) and the polar low-pressure centre aloft. Between the subtropical high-pressure cells and the equator they are easterly. This dominant, westerly circulation reaches maximum speeds of 45–67 m s^{-1} (100–150 mph), which even increase to 135 m s^{-1} (300 mph) in winter. These maximum speeds are concentrated in a narrow band often situated

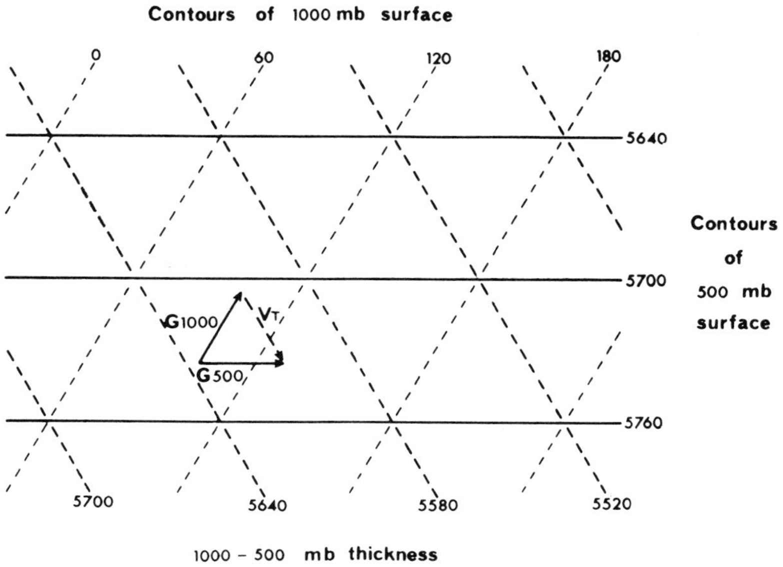

Figure 3.21 Schematic map of superimposed contours of isobaric height and thickness of the 1000–500 mb layer (in metres). G_{1000} is the geostrophic velocity at 1000 mb, G_{500} that at 500 mb, V_T is the resultant 'thermal wind' blowing parallel to the thickness lines.

at about 30° latitude, between 9000 and 15,000 m, called the *jet stream* (see note 5). Plate 16 shows bands of cirrus cloud which may have been related to jet-stream systems.

This stream, which is essentially a fast-moving mass of laterally concentrated air, is connected with the zone of maximum slope, folding, or fragmentation of the tropopause, which in turn coincides with the latitude of maximum poleward temperature gradient and energy transfer. The thermal wind, as described above, is a major component of the jet stream, but the basic reason for the concentration of the meridional temperature gradient in a narrow zone (or zones) is still uncertain. One theory is that the temperature gradient becomes accentuated when the upper wind pattern is confluent (see B.1, this chapter). Figure 3.22, giving a generalized view of the wind and temperature distribution in the northern hemisphere troposphere in winter, shows that there are two westerly jet streams (see Figure 1.44). The more northerly one, termed the *Polar Front Jet Stream* (see Chapter 4, E), is associated with the steep temperature gradient where polar and tropical air interact (see Figure 3.23), but the *Subtropical Jet Stream* is related to a temperature gradient confined to the upper troposphere. The Polar

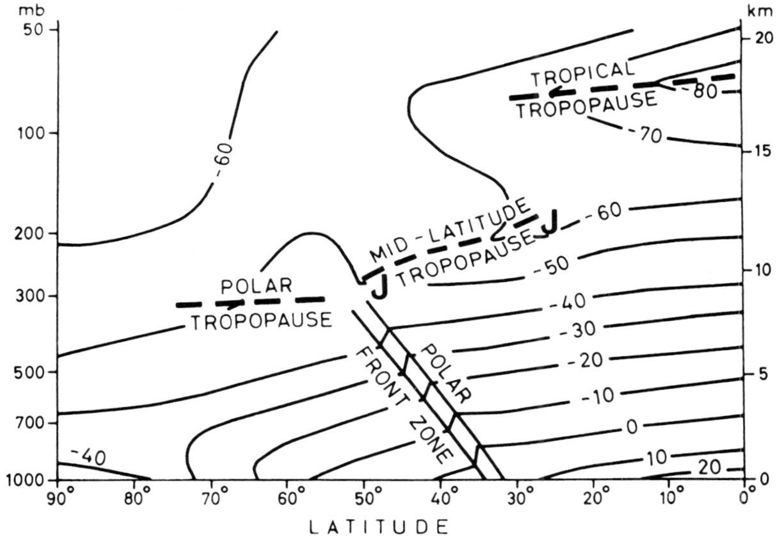

Figure 3.22 The typical distribution of temperature and the location of the westerly jet streams (J) in the northern hemisphere in winter.

Source: Partly after Defant and Taba 1957.

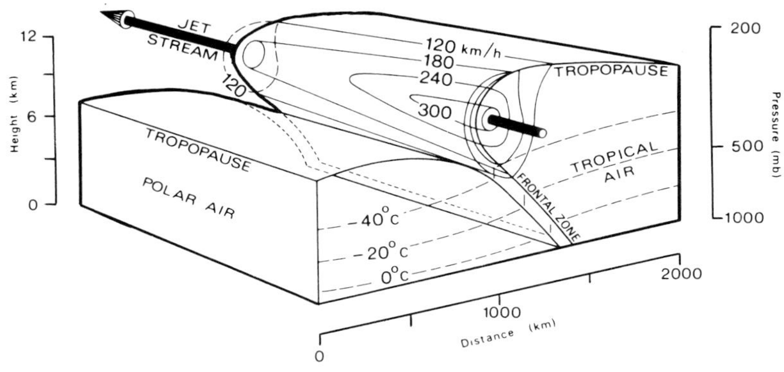

Figure 3.23 Structure of the mid-latitude frontal zone and associated jet stream showing generalized distribution of temperature, pressure and wind velocity.

Source: After Riley and Spalton 1981.

A

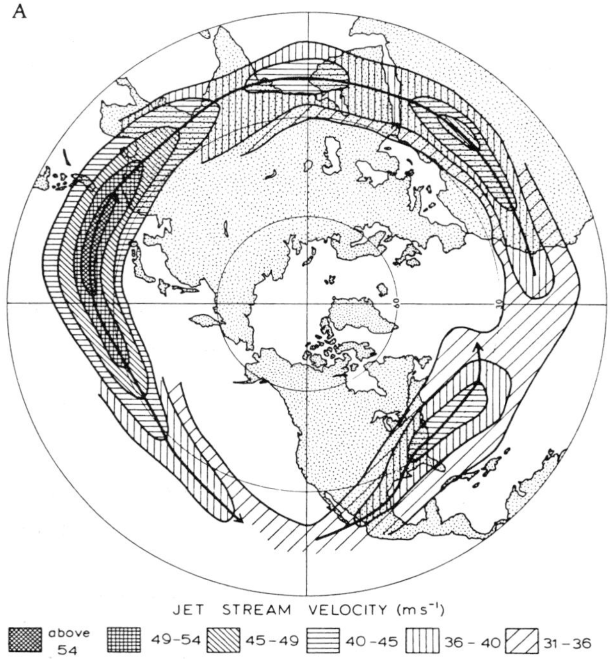

Figure 3.24A The mean location and velocities (m s^{-1}) of the westerly jet stream in the northern hemisphere in January.

B

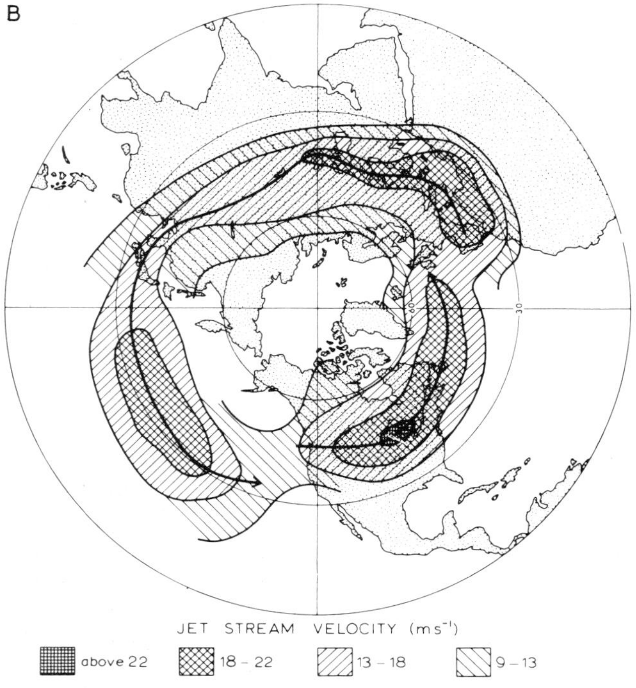

Figure 3.24B The mean location and velocities (m s^{-1}) of the westerly jet stream in the northern hemisphere in July. Note the different scale of wind intensity.

Source: After Namias and Clapp (adapted from Petterssen 1958).

Front Jet Stream is very irregular in its longitudinal location and is commonly discontinuous, whereas the Sub-tropical Jet Stream is much more persistent. For these reasons the location of the mean jet stream (see Figure 3.24) primarily reflects the position of the Subtropical Jet Stream. The synoptic pattern of jet-stream occurrence may be further complicated in some sectors by the presence of additional frontal zones (see Chapter 4, E), each associated with a jet stream. This situation is common in winter over North America. Comparison of Figures 3.20 and 3.24 indicates that the main jet-stream cores are associated with the principal troughs of the Rossby long waves. In summer, an *Easterly Tropical Jet Stream* forms in the upper troposphere over India and Africa due to regional reversal of the S–N temperature gradient (p. 247). The relationships between these upper tropospheric wind systems and surface weather and climate will be considered in Chapters 4, 5 and 6.

In the southern hemisphere, the mean jet stream in winter is similar in strength to its northern hemisphere winter counterpart and it weakens less in summer, because the meridional temperature gradient between 30° and 50°S is reinforced by heating over the southern continents. Upper air temperature gradients in the northern hemisphere, in contrast, are much stronger in winter, as noted above.

4 Surface pressure conditions

The most permanent features of the mean sea-level pressure maps are the oceanic subtropical high-pressure cells (see Figures 3.25 and 3.27). These anticyclones are located at about 30° latitude, suggestively situated below the mean Subtropical Jet Stream. They move a few degrees equatorward in winter and poleward in summer in response to the seasonal expansion and contraction of the two circumpolar vortices. In the northern hemisphere the subtropical ridges of high pressure weaken over the heated continents in summer but are thermally intensified over them in winter. The principal subtropical high-pressure cells are located: (a) over the Bermuda–Azores ocean region (aloft the centre of this cell lies over the east Caribbean); (b) over the south and south-west United States (the Great Basin or Sonoran cell) – this continental cell is naturally prone to seasonal decline, being replaced by a thermal surface low in summer; (c) over the east and north Pacific – a large and powerful cell (sometimes dividing into two, especially during the summer); and (d) over the Sahara – this, like other continental source areas, is seasonally variable both in intensity and extent, being most prominent in winter. In the southern hemisphere the subtropical anticyclones are oceanic, except over southern Australia in winter.

The latitude of the subtropical high-pressure belt depends on the meridional temperature difference between the equator and the pole and on the vertical temperature lapse rate (i.e. vertical stability). The greater the meridional temperature difference the more equatorward is the location of the subtropical high-pressure belt (see Figure 3.26).

Equatorward of the subtropical anticyclones there is an equatorial trough of low pressure, associated broadly with the zone of maximum insolation and tending to migrate with it, especially towards the heated continental interiors of the summer hemisphere. Poleward of the subtropical anticyclones lies a general zone of subpolar low pressure. In the southern hemisphere this sub-Antarctic Trough is virtually circumpolar (see Figure 3.27) whereas in the northern hemisphere the major centres are near Iceland and the Aleutians in winter and primarily over continental areas in summer. It is commonly stated that in high latitudes there is a surface anticyclone due to the cold polar air, but in the Arctic this is true only in spring near the Canadian Arctic Archipelago. In winter the Polar Basin is affected by high and low pressure cells with the major semi-permanent cold air anticyclones over Siberia and, to a lesser extent, north-western Canada. The shallow Siberian high is in part a result of the exclusion of tropical air masses from the interior by the

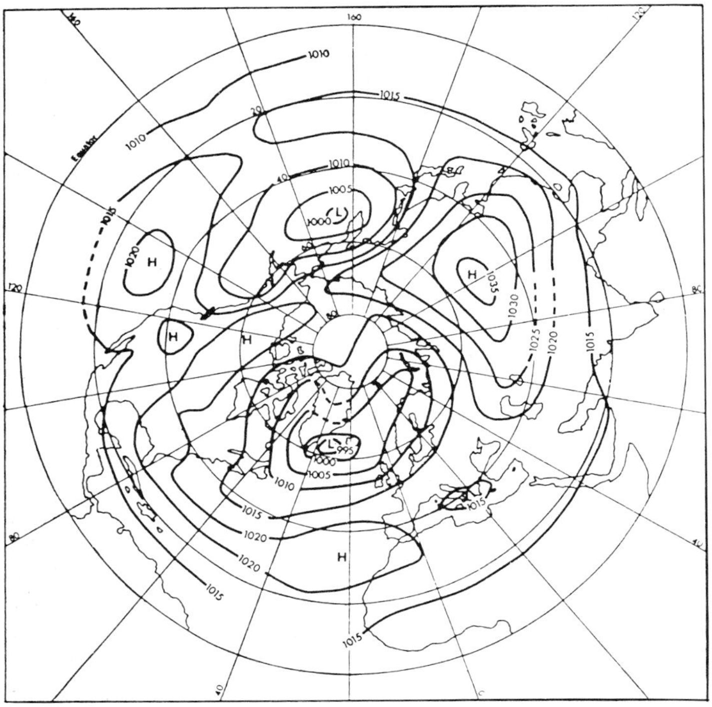

Figure 3.25A The mean sea-level pressure distribution (mb) in January for the northern hemisphere, 1950–59.

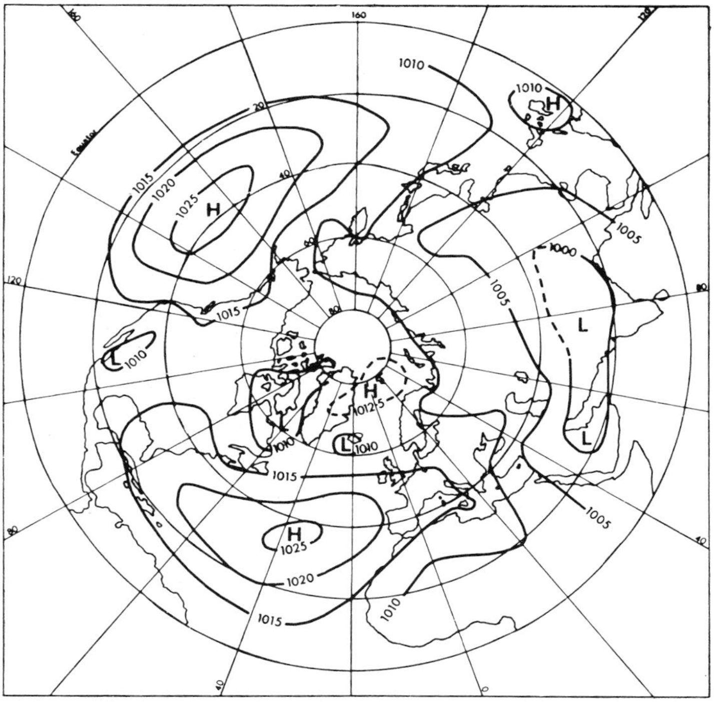

Figure 3.25B The mean sea-level pressure distribution (mb) in July for the northern hemisphere, 1950–59.

Source: After O'Connor 1961.

Figure 3.26 A plot of the meridional temperature difference at the 300–700 mb level in the previous month against the latitude of the centre of the subtropical high-pressure belt, assuming a constant vertical tropospheric lapse rate.

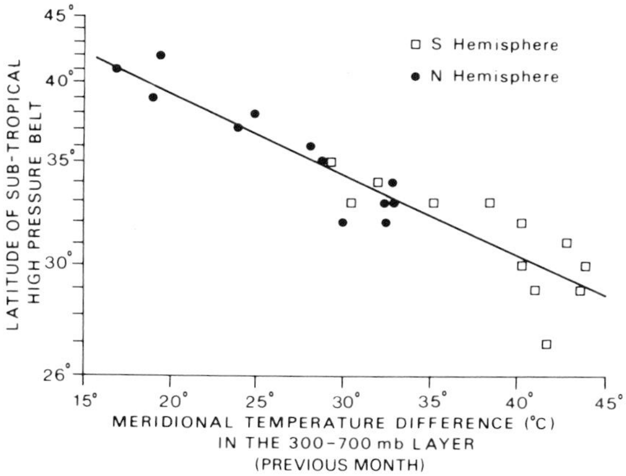

Source: After Flohn, in Flohn and Fantechi 1984 (*copyright © 1980/1982 by D. Reidel Publishing Company. Reprinted by permission*).

Tibetan massif and the Himalayas. Over Antarctica it is meaningless to speak of sea-level pressure but, on average, there is high pressure over the 3–4-km-high eastern Antarctic plateau.

The main circulation in the southern hemisphere is much more zonal at both 700 mb and sea level than in the northern hemisphere, due to the limited area and effect of the southern land masses. There is also little difference between summer and winter circulation intensity (see Figures 3.19 and 3.20). Indeed, it is important at this point to differentiate between mean pressure patterns and the highs and lows shown on synoptic weather maps. Thus, in the southern hemisphere, the zonality of the mean circulation conceals a high degree of day-to-day variability. The synoptic map is one which shows the principal pressure systems over a very large area at a given time; local wind features, for example, being ignored. The subpolar lows over Iceland and the Aleutians (see Figure 3.25A and B) shown on recurrent mean pressure maps represent the passage of

Figure 3.27 The mean sea-level pressure distribution (mb) in January (left) and July (right) for the southern hemisphere.

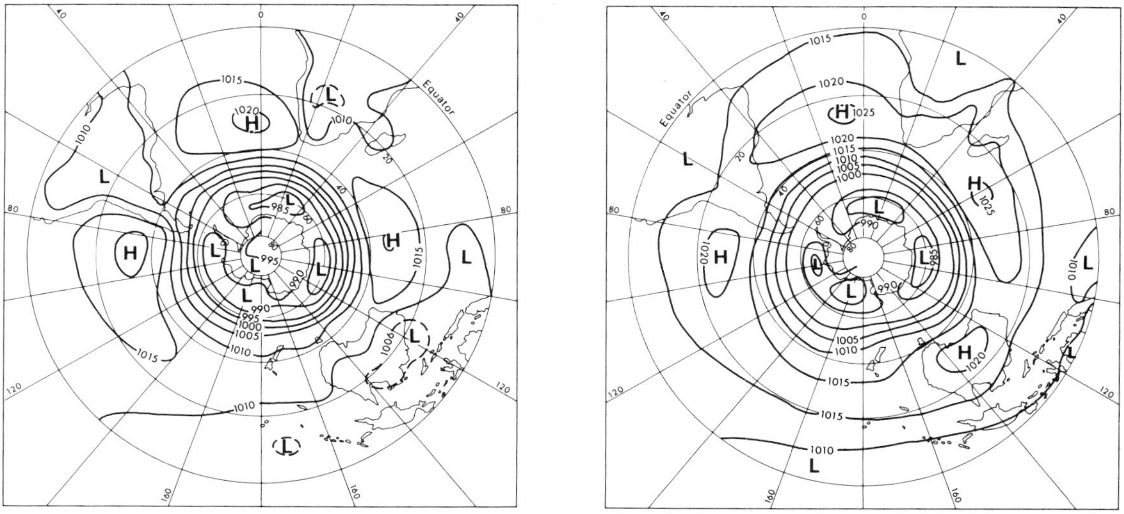

Source: From Taljaard *et al.* 1969.

deep depressions across these areas downstream of the upper long-wave troughs. The mean high-pressure areas, however, relate to more or less permanent highs. The intermediate areas, such as the zones about 50°–55°N and 40°–60°S, affected by travelling depressions and ridges of high pressure, appear on the mean maps as being of neither markedly high nor markedly low pressure. The movement of depressions is considered in Chapter 4, F.

On comparing the surface and tropospheric pressure distributions for January (see Figures 3.20 and 3.25A) it will be noticed that only the subtropical high-pressure cells extend to high levels. The reasons for this are evident from Figure 3.16B and D. In summer the equatorial low-pressure belt is also evident aloft over southern Asia. The subtropical cells are still discernible at 300 mb, showing them to be a fundamental feature of the global circulation and not merely a response to surface conditions.

E THE GLOBAL WIND BELTS

One fact that emerges from the preceding discussion is the importance of the subtropical high-pressure cells. Dynamic, rather than immediately thermal, in origin, and situated between 20° and 30° latitude, they seem to provide the key to the world's surface wind circulation. In the northern hemisphere the pressure gradients surrounding these cells are strongest between October and April. In terms of actual pressure, however, oceanic cells experience their highest pressure in summer, the belt being counterbalanced at low levels by thermal low-pressure conditions over the continents. Their strength and persistence clearly mark them as the dominating factor which controls the position and activities of both the trades and westerlies.

1 The trade winds

The trades (or tropical easterlies) are important because of the great extent of their activity; they blow over nearly half the globe (see Figure 3.28). They originate at low latitudes on the margins of the subtropical high-pressure cells, and their constancy of direction and speed (about 7 m s^{-1}) is remarkable. Trade winds, like the westerlies, are strongest during the winter half-year, which suggests they are both controlled by the same fundamental mechanism.

The two trade-wind systems tend to converge in the *Equatorial Trough* (of low-pressure). Over the oceans, particularly the central Pacific, the convergence of these air streams is often pronounced and in this sector the term *Inter-Tropical Convergence Zone* (ITCZ) is applicable. Generally, however, the convergence is discontinuous in space and time (see Plate 28). Equatorward of the main *root zones* of the trades over the eastern Pacific and eastern Atlantic are regions of light, variable winds, known traditionally as the *doldrums* and much feared in past centuries by the crews of sailing ships. Their seasonal extent varies considerably; from July to September they spread westward into the central Pacific while in the Atlantic they extend to the coast of Brazil. A third major doldrum zone is located in the Indian Ocean and western Pacific. In March–April it stretches 16,000 km from east Africa to 180° longitude and it is again very extensive during October–December.

2 The equatorial westerlies

In the summer hemisphere, and over continental areas especially, there is a zone of generally westerly winds intervening between the two trade-wind belts (see Figure 3.29). This westerly system is well marked over Africa and southern Asia in the northern hemisphere summer, when thermal heating over the continents assists the northward displacement of the Equatorial Trough (see Figure 3.28). Over Africa the westerlies reach to 2–3 km and over the Indian Ocean to 5–6 km. In Asia these winds are known as the 'Summer Monsoon' but this is now recognized to be a complex phenomenon the cause of which is partly global and partly

Figure 3.28 Map of the trade wind belts and the doldrums. The limits of the trades – enclosing the area within which 50 per cent of all winds are from the predominant quadrant – are shown by the solid (January) and the dashed (July) lines. The stippled area is affected by trade-wind currents in both months. Schematic streamlines are indicated by the arrows – dashed (July) and solid (January, or both months).

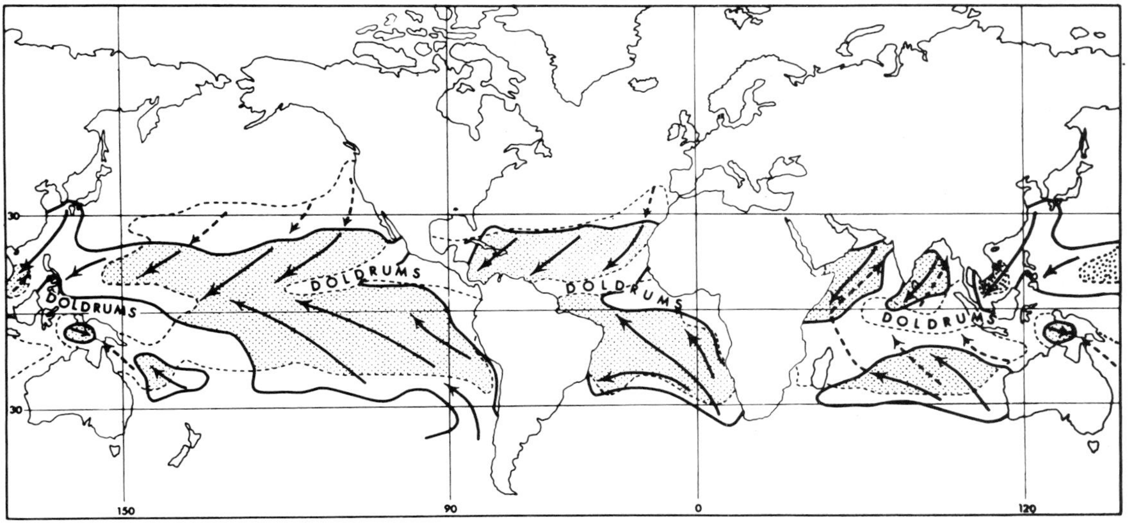

Source: Based on Crowe 1949 and 1950.

Figure 3.29 Distribution of the equatorial westerlies in any layer below 3 km (about 10,000 ft) for January and July.

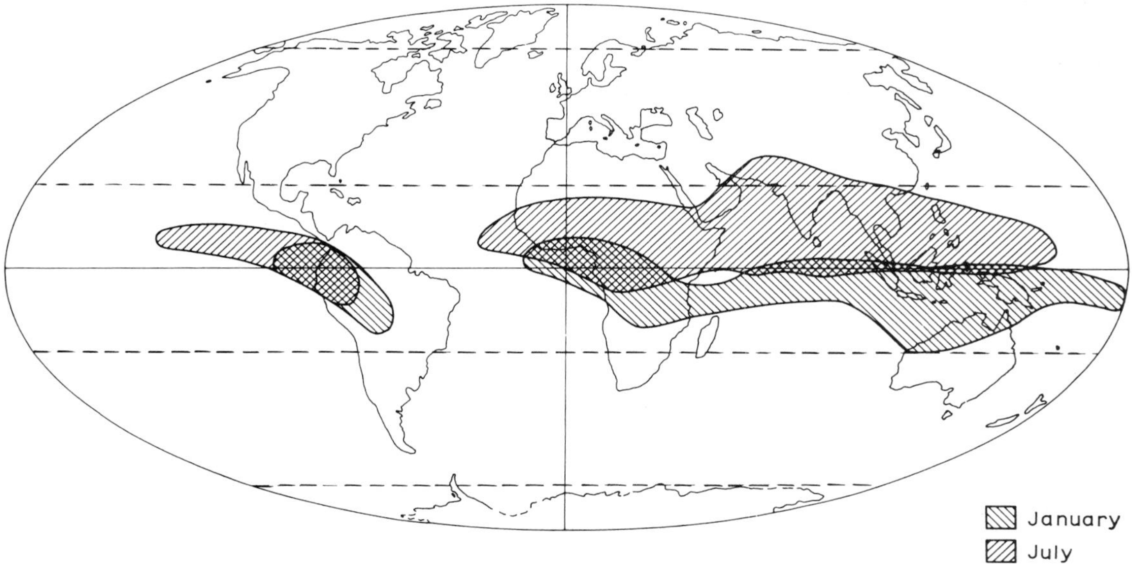

Source: After Flohn, in Indian Meteorological Department 1960.

regional in origin (see Chapter 6, D). The equatorial westerlies are not simply trades of the opposite hemisphere which recurve (due to the changed direction of the Coriolis deflection) on crossing the equator, since there is *on average* a westerly component in the Indian Ocean at 2°–3°S in June and July and at 2°–3°N in December and January. Over the Pacific and Atlantic Oceans the ITCZ does not shift sufficiently far from the equator to permit the development of this westerly wind belt.

3 The mid-latitude (Ferrel) westerlies

These are the winds of the mid-latitudes emanating from the poleward sides of the subtropical high-pressure cells. They are far more variable than the trades both in direction and intensity, for in these regions the path of air movement is frequently affected by cells of low and high pressure which travel generally eastwards within the basic flow (see Plate 1). Also in the northern hemisphere the preponderance of land areas with their irregular relief and changing seasonal pressure patterns tend to obscure the generally westerly airflow. The Scilly Islands, lying in the south-westerlies, record 46 per cent of winds from between south-west and north-west, but fully 29 per cent from the opposite sector between north-east and south-east.

The westerlies of the southern hemisphere are stronger and more constant in direction than those of the northern hemisphere because the broad expanses of ocean rule out the development of stationary pressure systems (see Figure 3.30). Kerguelen Island (49°S, 70°E) has an annual frequency of 81 per cent of winds from between south-west and north-west and the comparable figure of 75 per cent for Macquarie Island (54°S, 159°E) shows that this predominance is widespread over the southern oceans. However, the apparent zonality of the southern circumpolar vortex (see Figure 3.27) conceals considerable synoptic variability of wind velocity.

Figure 3.30 Profiles of the average west-wind component (m s^{-1}) at sea level in the northern and southern hemispheres during their respective summer (A) and winter (B) seasons.

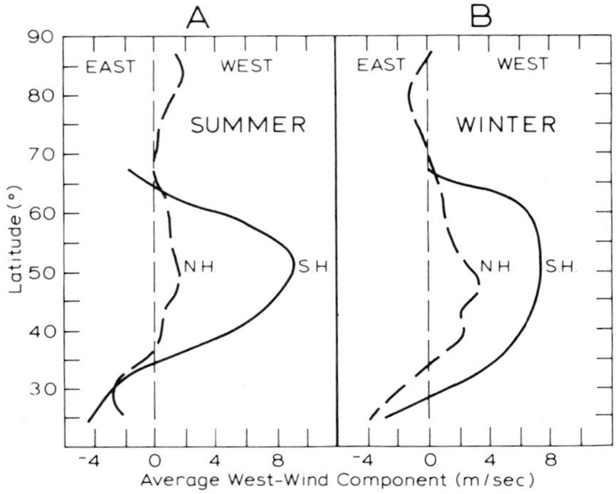

Source: After van Loon 1964.

4 The polar easterlies

This term is applied to winds that are supposed to occur between a polar high pressure and the belt of low pressure of the higher mid-latitudes. The polar high, as has already been pointed out, is by no means a quasi-permanent feature of the arctic circulation. Easterly winds occur mainly on the poleward sides of depressions over the North Atlantic and North Pacific and if average wind directions are calculated for entire latitude belts in high latitudes there is found to be little sign of a coherent system of polar easterlies. The situation in high latitudes of the southern hemisphere is complicated by the presence of Antarctica, but anticyclones appear to be frequent over the high plateau of eastern Antarctica and easterly winds prevail over the Indian Ocean sector of the Antarctic coastline. For example, in 1902–3 the expedition ship *Gauss* at 66°S, 90°E observed winds between north-east and south-east for 70 per cent of the time, and at many coastal stations the constancy of easterlies may be compared with that of the trades. However, westerly components predominate over the sea areas off west Antarctica.

F THE GENERAL CIRCULATION

The observed patterns of wind and pressure prompt consideration of the mechanisms maintaining the *general circulation* of the atmosphere – the large-scale patterns of wind and pressure which persist throughout the year or recur seasonally. Reference has already been made to one of the primary driving forces, the imbalance of radiation between lower and higher latitudes (see Chapter 1, G.1), but it is important also to appreciate the significance of energy transfers in the atmosphere. Energy is continually undergoing changes of form as shown schematically in Figure 3.31. Unequal heating of the earth and its atmosphere by solar radiation generates potential energy, some of which is converted into kinetic energy by the rising of warm air and the sinking of cold air. Ultimately, the kinetic energy of atmospheric motion on all scales is dissipated by friction and small-scale turbulent eddies (i.e. internal viscosity). In order to maintain the general circulation, the rate of generation of kinetic energy must obviously balance its rate of dissipation. These rates are estimated to be about 2 W m^{-2}, which amounts only to some 1 per cent of the average global solar radiation absorbed at the surface and in the atmosphere. In other words the atmosphere is a highly inefficient heat engine (see Chapter 1, G).

A second controlling factor is the angular momentum of the earth and its atmosphere. This is the tendency for the earth's atmosphere to move, with the earth, around the axis of rotation. Angular momentum is proportional to the rate of spin (that is the angular velocity) and the square of the distance of the air parcel from the axis of rotation. With a uniformly rotating earth and atmosphere, the total angular momentum must remain constant (in other words, there is a *conservation of angular momentum*). If, therefore, a large mass of air changes its position on the earth's surface such that its distance from the axis of rotation also changes, then its angular velocity must change in a manner so as to allow the angular momentum to remain constant. Naturally absolute angular momentum is high at the equator (see note 6) and decreases with latitude to become zero at the pole (that is, the axis of rotation), so air moving poleward tends to acquire progressively higher eastward velocities. For example, air travelling from 42° to 46° latitude and conserving its angular momentum, would increase its speed relative to the earth's surface by 29 m s^{-1}. This is the same principle which causes an ice skater to spin more violently when her arms are progressively drawn into the body. In practice this increase of air-mass velocity is countered or masked by the other forces affecting air movement (particularly friction), but there is no doubt that many of the important features of the general atmospheric circulation result from this poleward transfer of angular momentum.

The necessity for a poleward momentum transport is readily appreciated in terms of the maintenance of the mid-latitude westerlies. These winds continually impart westerly (east-

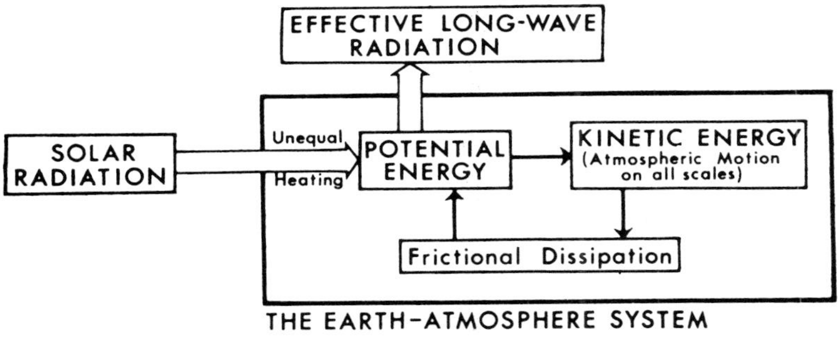

Figure 3.31 Schematic changes of energy involving the earth-atmosphere system.

ward) relative momentum to the earth by friction and it has been estimated that they would cease altogether due to this frictional dissipation of energy in little over a week if their momentum were not continually replenished from elsewhere. In low latitudes the extensive tropical easterlies are gaining westerly relative momentum by friction, as a result of the earth rotating in a direction opposite to their flow (see note 7), and this excess is transferred polewards with the maximum poleward transport occurring, significantly, in the vicinity of the mean subtropical jet stream at about 250 mb at 30°N and 30°S.

1 Circulations in the vertical and horizontal planes

There are two possible ways in which the atmosphere can transport heat and momentum. One is by circulation in the vertical plane as indicated in Figure 3.32 which shows three meridional cells. The low-latitude (or Hadley) cell and its counterpart in the southern hemisphere were considered to be analogous to the convective circulations set-up when a pan of water is heated over a flame and are referred to as *thermally direct* cells. Warm air near the equator was thought to rise and generate a low-level flow towards the equator, the earth's rotation deflecting these currents which thus form the north-east and south-east trades. This explanation was put forward by G. Hadley in 1735,

Figure 3.32 Three-cell model of the northern hemisphere meridional circulation.

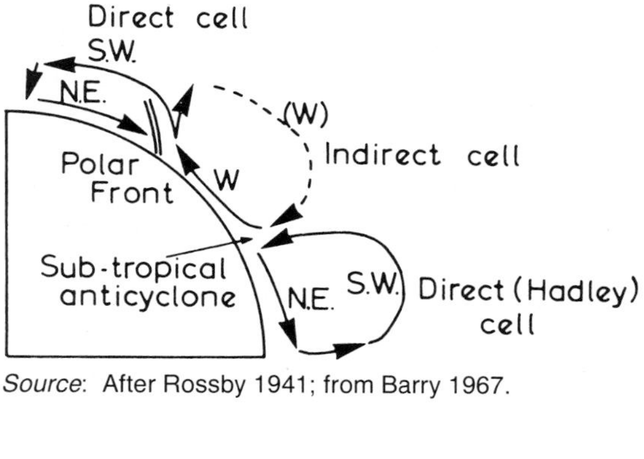

Source: After Rossby 1941; from Barry 1967.

although in 1856 W. Ferrel pointed out that the conservation of angular momentum would be a more effective factor in causing easterlies because the Coriolis force is small in low latitudes. The low-latitude cell, according to the above scheme, would be completed by poleward counter-currents aloft with the air sinking at about 30° latitude as it is cooled by radiation. However, this scheme is not entirely correct since the atmosphere does not have a simple heat source at the equator, the trades are not continuous around the globe (see Figure 3.28) and poleward upper flow is restricted mainly to the western ends of the subtropical high-pressure cells aloft (see Figures 3.20 and 3.27).

Figure 3.32 shows another thermally direct (polar) cell in high latitudes with cold dense air flowing out from a polar high pressure. The reality of this is doubtful, but it is in any case of limited importance to the general circulation in view of the small mass involved. It is worth noting at this point that a single direct cell in each hemisphere is not possible, because the easterly winds near the surface would slow down the earth's rotation. On average the atmosphere must rotate with the earth, requiring a balance between easterly and westerly winds over the globe.

The mid-latitude (Ferrel) cell in Figure 3.32 is thermally indirect and it would need to be driven by the other two. Momentum considerations indicate the necessity for upper easterlies in such a scheme, yet observations with upper-air balloons during the 1930s and 1940s demonstrated the existence of strong westerlies in the upper troposphere (see D.3, this chapter). Rossby modified the three-cell model to incorporate this fact, proposing that westerly momentum was transferred to middle latitudes from the upper branches of the cells in high and low latitudes. Such horizontal mixing could, for example, be accomplished by troughs and ridges in the upper flow. Figure 3.33 shows an idealized meridional circulation at 90°E containing northern and southern Hadley, Ferrel and Polar cells separated aloft by an easterly and four westerly jet streams.

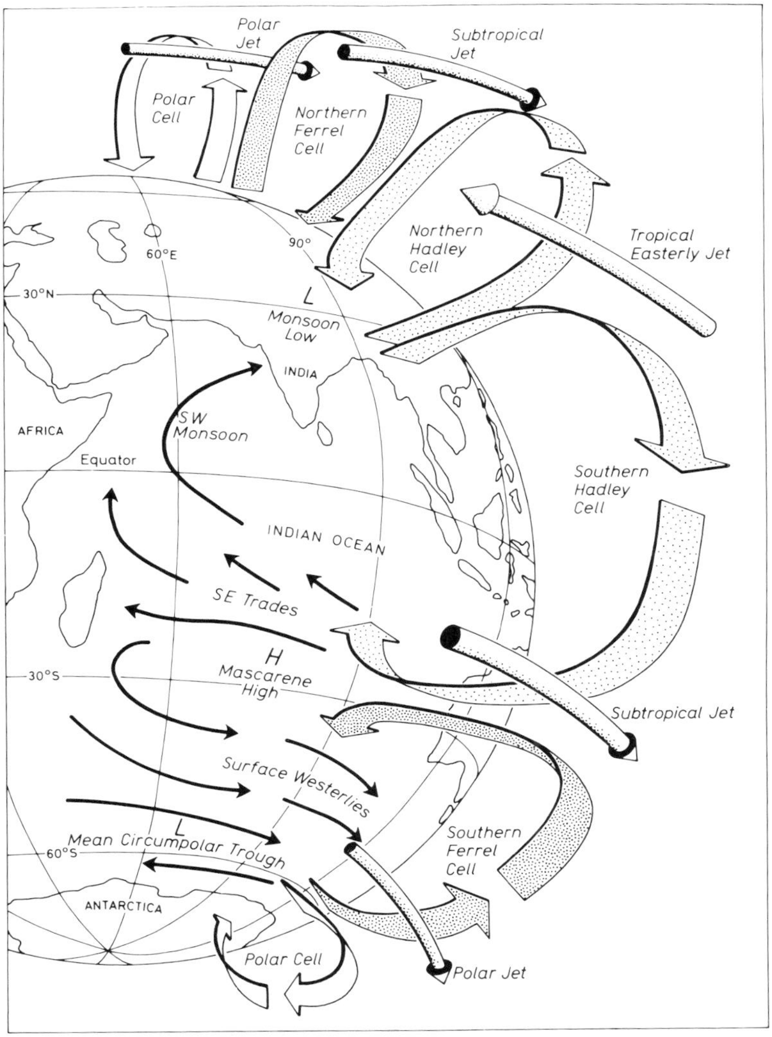

Figure 3.33 Meridional distribution of zonally-averaged circulation features at 90°E at the time of the Indian summer monsoon. The Hadley, Ferrel and Polar cells are shown, together with jet streams and surface winds over the Indian Ocean.

Source: After Meehl 1987b.

These views underwent radical amendment from about 1948 onwards. The alternative means of transporting heat and momentum – by horizontal circulations – had been suggested in the 1920s by A. Defant and H. Jeffreys but could not be tested until adequate upper-air data became available. Calculations for the northern hemisphere by V. P. Starr and R. M. White at Massachusetts Institute of Technology showed that in middle latitudes horizontal cells transport most of the required heat and momentum polewards. This operates through the mechanism of the quasi-stationary highs and the travelling highs and lows near the surface acting in conjunction with their related wave patterns aloft. The importance of such horizontal eddies for energy transport is shown in Figure 3.34 (see also Figure 1.38B). The modern concept of the general circulation therefore views the energy of the zonal winds as being derived from travelling waves, not from meridional circulations (see Figure 3.35). In lower latitudes, however, this

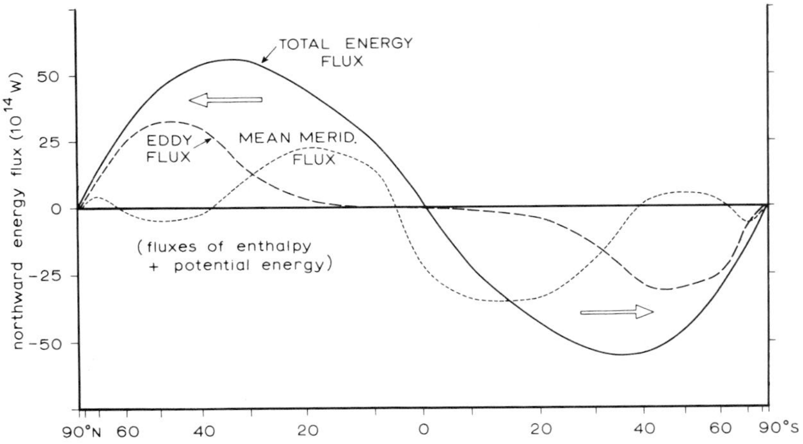

Figure 3.34 The poleward transport of energy, showing the importance of horizontal eddies.

mechanism may be insufficient by itself to account for the total energy transport estimated to be necessary for energy balance. For such reasons the mean Hadley cell still features in current representations of the general circulation, as Figure 3.36 shows, but the low-latitude circulation is recognized as being complex. In particular, the vertical heat transport in the Hadley cell is effected by giant cumulonimbus clouds in disturbance systems associated with the Equatorial Trough (of low pressure), which is located on average at 5°S in January and at 10°N in July (see Chapter 6, B). The Hadley cell of the winter hemisphere is by far the most important and it gives rise to low-level transequatorial flow into the summer hemisphere. The traditional model of global circulation with twin cells, symmetrical about the equator, is found only in spring/autumn. Longitudinally, the Hadley cells are linked with the monsoon regimes of the summer hemisphere. Rising air over southern Asia (and also South America and Indonesia) is associated with east–west (zonal) outflow and these systems are known as 'Walker circulations' (see pp. 129–30). The poleward return transport of the meridional Hadley cells takes place in troughs which extend into low latitudes from the mid-latitude westerlies. This tends to occur at the western ends of the upper tropospheric subtropical high pressure cells (see Figure 3.18). Horizontal mixing predominates in middle and high latitudes although it is also thought that there is a weak indirect mid-latitude cell in much reduced form (see Figure 3.36). The relationship of the jet streams to regions of steep meridional temperature gradient has already been noted (see Figure 3.22). A complete explanation of the two wind maxima and their role in the general circulation is still lacking, but they undoubtedly form an essential part of the story.

In the light of these theories, the origin of the subtropical anticyclones which play such an important role in the world's climates may be re-examined. Their existence has been variously ascribed to the piling-up of poleward-moving air as it is increasingly deflected eastwards through the earth's rotation and the conservation of angular momentum; to the sinking of poleward currents aloft by radiational cooling; to the general necessity for high pressure near 30° latitude separating approximately equal zones of east and west winds; or to combinations of such mechanisms. An adequate theory must account not only for their permanence but also for their cellular nature and the vertical inclination of the axes. The preceding discussion shows that ideas of a simplified Hadley cell and momentum conservation are only partially correct. Moreover, recent studies rather surprisingly show no

Figure 3.35 Mean zonal wind speeds (m s^{-1}) calculated for each latitude and for elevations up to more than 20 km. Note the weak mean easterly flow at all levels in low latitudes dominated by the Hadley cells and the strong upper westerly flow in the mid-latitudes localized into the subtropical jet streams.

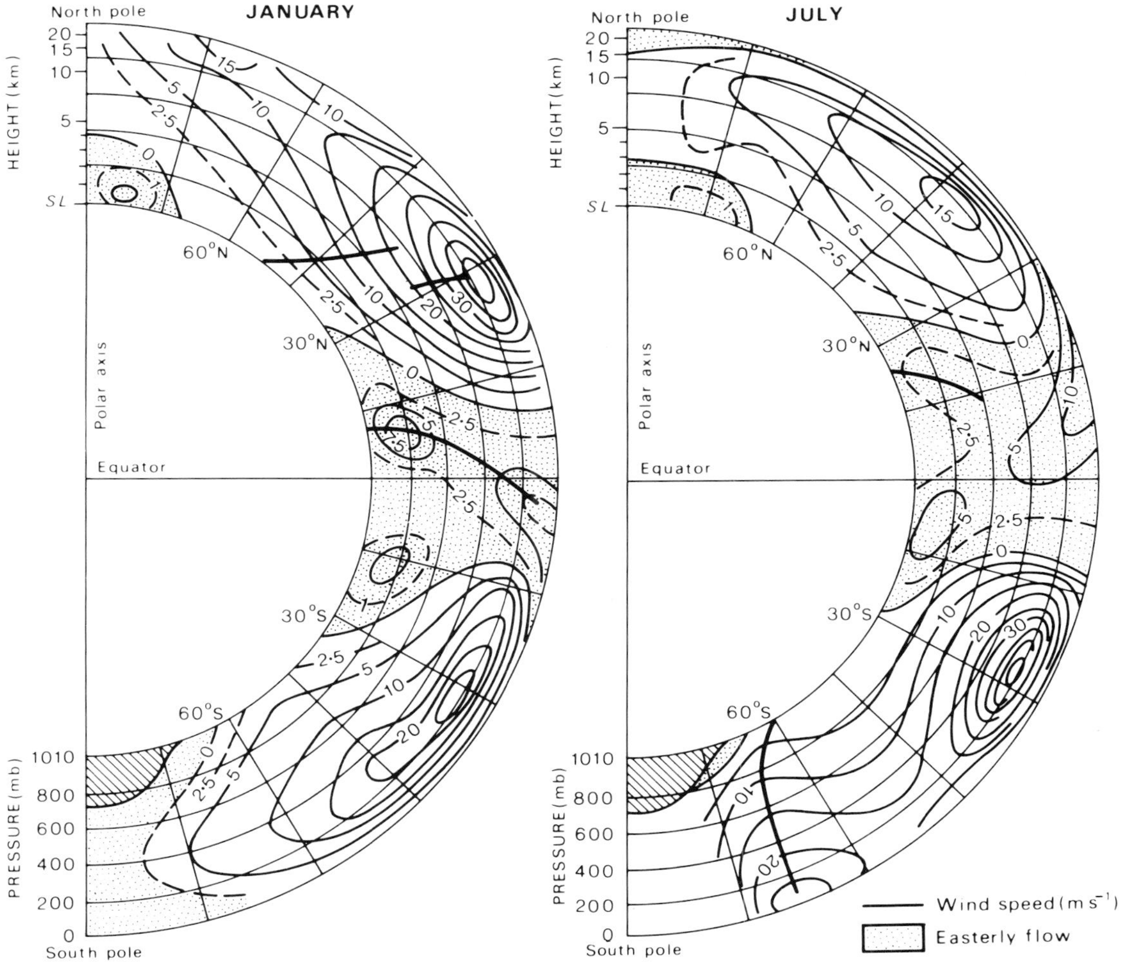

Source: After Mintz; from Henderson-Sellers and Robinson 1986.

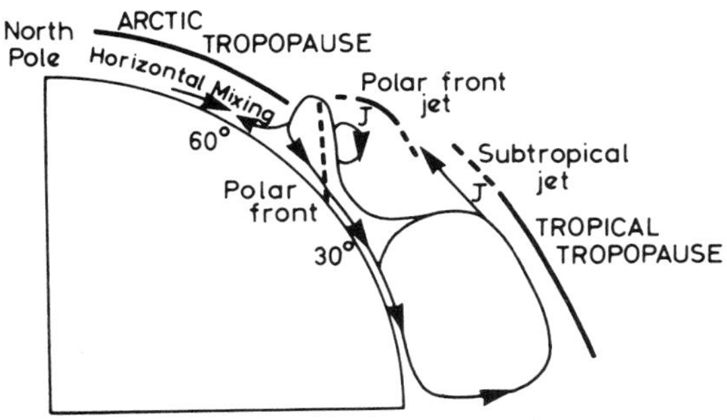

Figure 3.36 General meridional circulation model for the northern hemisphere in winter.

Source: After Palmén 1951; from Barry 1967.

relationship, on a seasonal basis, between the intensity of the Hadley cell and that of the subtropical highs.

It is probable that the high-level anticyclonic cells which are evident on *synoptic* charts (these tend to merge on mean maps) are related to anticyclonic eddies on the equatorward side of jet streams. Theoretical and observational studies show that, as a result of the latitudinal variation of the Coriolis parameter, cyclones in the westerlies tend to move poleward and anticyclonic cells equatorward. Hence the subtropical anticyclones are constantly regenerated. There is a statistical relationship between the latitude of the subtropical highs and the mean meridional temperature gradient (see Figure 3.26); a stronger gradient causes an equatorward shift of the high pressure, and vice versa. This shift is evident on a seasonal basis. The cellular pattern at the surface clearly reflects the influence of heat sources. The cells are stationary and elongated north–south over the northern hemisphere oceans in summer when continental heating creates low pressure and also the meridional temperature gradient is weak. In winter, on the other hand, the zonal flow is stronger in response to a greater meridional temperature gradient and continental cooling produces east–west elongation of the cells. Undoubtedly surface and high-level factors reinforce one another in some sectors and tend to cancel out in others. Indeed, it has

Figure 3.37 Tentative flow model relating summer convection, the easterly jet stream and high-pressure subsidence over northern Africa and the eastern North Atlantic.

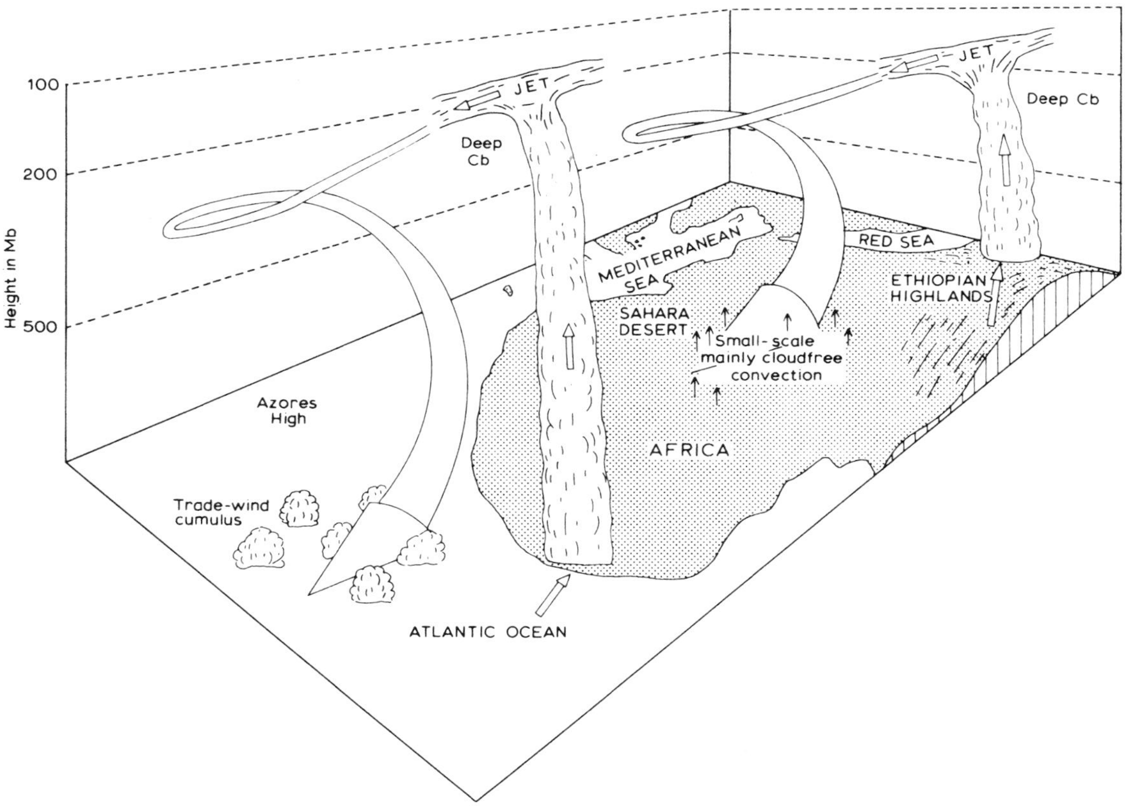

Source: Adapted from Walker 1972 (*Crown Copyright Reserved*).

been suggested that the Azores high-pressure cell, in particular, owes part of its summer intensification and its tendency to extend eastward to air masses which rise locally in areas of high monsoonal rainfall over Africa, enter the tropical easterly jet stream circulation (see Chapter 6, p. 256) and then subside over the western Sahara and the eastern North Atlantic (see Figure 3.37).

Just as Hadley circulations represent major meridional (i.e. north–south) components of the atmospheric circulation, so Walker circulations represent the large-scale zonal (i.e. east–west) components of tropical airflow. These latter circulations are driven by major east–west pressure gradients set up by differences between, on the one hand, air rising over heated continents and the warmer parts of the oceans and, on the other, air subsiding over cooler parts of the oceans, over continental areas where deep high-pressure systems have become established, and in association with subtropical high-pressure cells. The Walker circulations were first identified by Sir Gilbert Walker in 1922–3 as the result of an inverse correlation which he observed between pressure over the eastern Pacific Ocean and rainfall over Java and India. These Walker circulations are subject to fluctuations in which an oscillation (El Niño–Southern Oscillation: ENSO) between high phases (i.e. non-ENSO events) and low phases (i.e. ENSO events) is the most striking (see Figure 3.38 and Chapter 6, G.4):

Figure 3.38 The Walker circulation during high (A) and low (B) phases of the Southern Oscillation. Low phases correspond to ENSO events (see text).

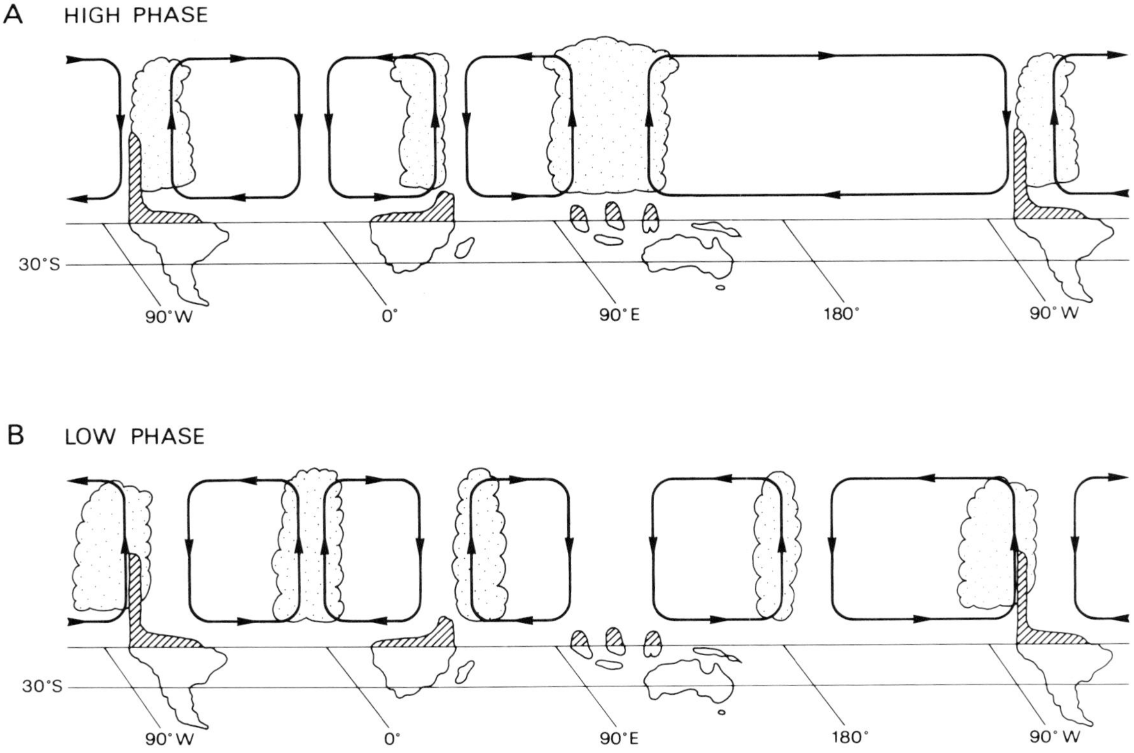

Source: After J. A. Lindesay; from Tyson 1986.

1 *High phase.* This phase features four major zonal cells involving rising low-pressure limbs and accentuated precipitation over Amazonia, central Africa and Indonesia/India; and subsiding high-pressure limbs and decreased precipitation over the eastern Pacific, South Atlantic and western Indian Ocean. During this phase low-level easterlies strengthen over the Pacific and subtropical westerly jet streams in both hemispheres weaken, as does the Pacific Hadley cell.
2 *Low phase.* This phase features five major zonal cells involving air rising in low-pressure limbs and accentuated precipitation over the South Atlantic, the western Indian Ocean, the western Pacific and the eastern Pacific; and subsiding high-pressure limbs and decreased precipitation over Amazonia, central Africa, Indonesia/India and the central Pacific. During this phase low-level westerlies and high-level easterlies dominate over the Pacific and subtropical westerly jet streams in both hemispheres intensify, as does the Pacific Hadley cell.

2 Variations in the circulation of the northern hemisphere

The pressure and contour patterns during certain periods of the year may be radically different from those indicated by the mean maps (see Figures 3.20, 3.40 and 3.41). These variations, of 3 to 8 weeks' duration, occur irregularly but are rather more noticeable in the winter months when the general circulation is strongest. The nature of the changes is illustrated schematically in Figure 3.39. The zonal westerlies over middle latitudes develop waves and the troughs and ridges become accentuated, ultimately splitting up into a cellular pattern with pronounced meridional flow at certain longitudes. The strength of the westerlies between 35° and 55°N is termed the *zonal index*; strong zonal westerlies are representative of high index and marked cellular patterns occur with low index (see Plate 17). A relatively low index may also occur if the westerlies are well south of their usual latitudes

Figure 3.39 The index cycle. A schematic illustration of the development of cellular patterns in the upper westerlies, usually occupying 3–8 weeks and being especially active in February and March in the northern hemisphere. Statistical studies indicate no regular periodicity in this sequence. A: High zonal index. The jet stream and the westerlies lie north of their mean position. The westerlies are strong, pressure systems have a dominantly east–west orientation, and there is little north–south air-mass exchange. B and C: The jet expands and increases in velocity, undulating with increasingly larger oscillations. D: Low zonal index. Complete break-up and cellular fragmentation of the zonal westerlies. Formation of stationary deep occluding cold depressions in lower mid-latitudes and deep warm blocking anticyclones at higher latitudes. This fragmentation commonly begins in the east and extends westward at a rate of about 60° of longitude per week.

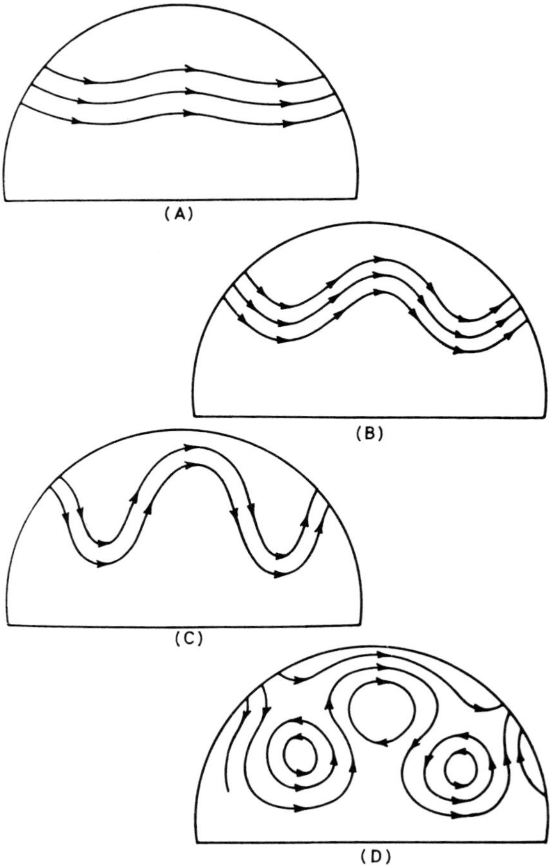

Source: After Namias; from Haltiner and Martin 1957.

and, paradoxically, such expansion of the zonal circulation pattern is associated with strong westerlies in lower latitudes than usual. Figures 3.40 and 3.41 illustrate the mean 700-mb contour patterns and zonal wind speed profiles for two contrasting months. In December 1957 the westerlies were stronger than normal northerlies of 40°N and the troughs and ridges were weakly developed, whereas in February 1958 there was a low zonal index and an expanded circumpolar vortex, giving rise to strong low-latitude westerlies. The 700-mb pattern shows very weak subtropical highs, deep meridional troughs and a blocking anticyclone off Alaska (see Figure 3.39D). The cause of these variations is still uncertain although it would appear that fast zonal flow is unstable and tends to break down. This tendency is certainly increased in the northern hemisphere by the arrangement of the continents and oceans.

Detailed studies are now beginning to show that the irregular index fluctuations, together with secondary circulation features, such as cells of low and high pressure at the surface or long waves aloft, play a major role in redistributing momentum and energy. Laboratory experiments with rotating 'dishpans' of water to simulate the atmosphere, and computer studies using numer-

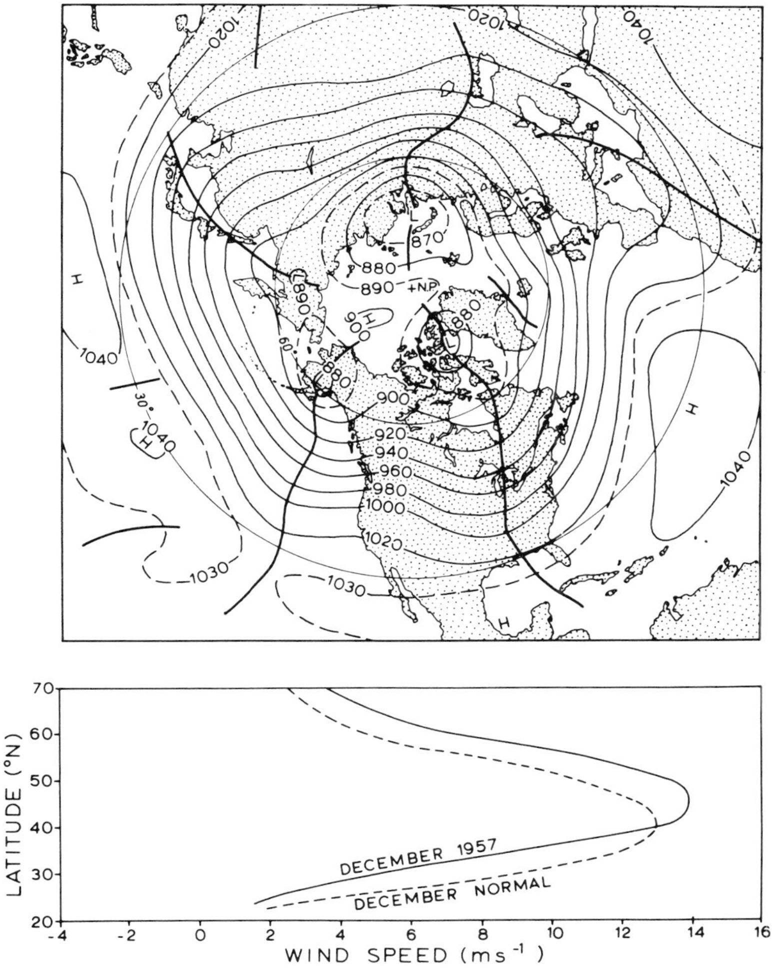

Figure 3.40 Left: Mean 700-mb contours (in tens of feet) for December 1957, showing a fast, westerly, small-amplitude flow typical of a high zonal index. *Below:* Mean 700-mb zonal wind speed profiles (m s^{-1}) in the western hemisphere for December 1957, compared with those of a normal December. The westerly winds were stronger than normal and displaced to the north.

Source: After *Monthly Weather Review* 85, 1957, pp. 410–11.

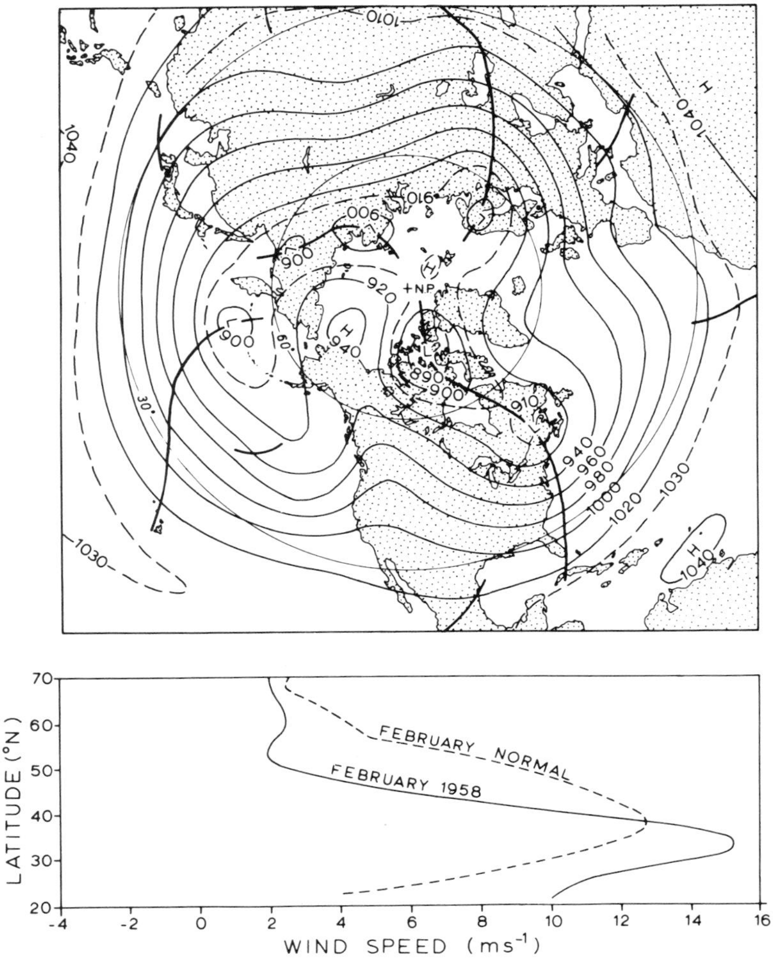

Figure 3.41 *Left:* Mean 700-mb contours (in tens of feet) for February 1958. *Below:* Mean 700-mb zonal wind speed profiles (m s^{-1}) in the western hemisphere for February 1958, compared with those of a normal February. The westerly winds were stronger than normal at low latitudes, with a peak at about 33°N.

Source: After *Monthly Weather Review* 86, 1958, pp. 62–3.

ical models of the atmosphere's behaviour, demonstrate that a Hadley circulation cannot provide an adequate mechanism for transporting heat poleward. In consequence, the meridional temperature gradient increases and eventually the flow becomes unstable in the Hadley mode, breaking down into a number of cyclonic and anticyclonic eddies. This phenomenon is referred to as *baroclinic instability*. In energy terms, the potential energy in the zonal flow is converted into potential and kinetic energy of the eddies. It is also now known that the kinetic energy of the zonal flow is derived *from* the eddies, the reverse of the classical picture which viewed the disturbances within the global wind belts as superimposed detail. The significance of atmospheric disturbances and the variations of the circulation are becoming increasingly evident. The mechanisms of the circulation are, however, greatly complicated by numerous interactions and *feedback* processes, one of the most important of which involves the oceanic circulation as outlined in F.3, this chapter. The significance of interactions between the oceanic and atmospheric heat and moisture budgets has already been discussed in Chapter 1, G, and Chapter 2, A.

3 The circulation of the ocean

The oceans occupy 71 per cent of the earth's surface with over 60 per cent of the global ocean area in the southern hemisphere. Three-quarters of the ocean area is between 3000 and 6000 m deep, whereas only 11 per cent of the land area exceeds 2000 m altitude.

The most obvious feature of the surface oceanic circulation is the control exercised over it by the low-level planetary wind circulation, especially by the subtropical oceanic high-pressure circulations and the westerlies. The oceanic circulation even partakes of the seasonal reversals of flow in the monsoonal regions of the

Figure 3.42 The general ocean current circulation of the globe, showing the mean anomalies of surface ocean temperatures (oC).

1 Gulf Stream
2 North Atlantic Drift
3 East Greenland Current
4 West Greenland Current
5 Labrador Current
6 Canary Current
7 North Equatorial Current
8 Caribbean Current
9 Antilles Current
10 South Equatorial Current
11 Brazil Current
12 Falkland Current
13 West Wind Drift
14 Benguela Current
15 Guinea Current
16 South-west and North-east Monsoon Drift
17 South Equatorial Current
18 Equatorial Counter Current
19 Mozambique Current
20 Agulhas Current
21 West Australian Current
22 Kuro Shio Current
23 North Pacific Drift
24 California Current
25 North Equatorial Current
26 Equatorial Counter Current
27 Alaska Current
28 Kamchatka Current
29 South Equatorial Current
30 East Australian Current
31 Peru or Humboldt Current
32 Equatorial Counter Current

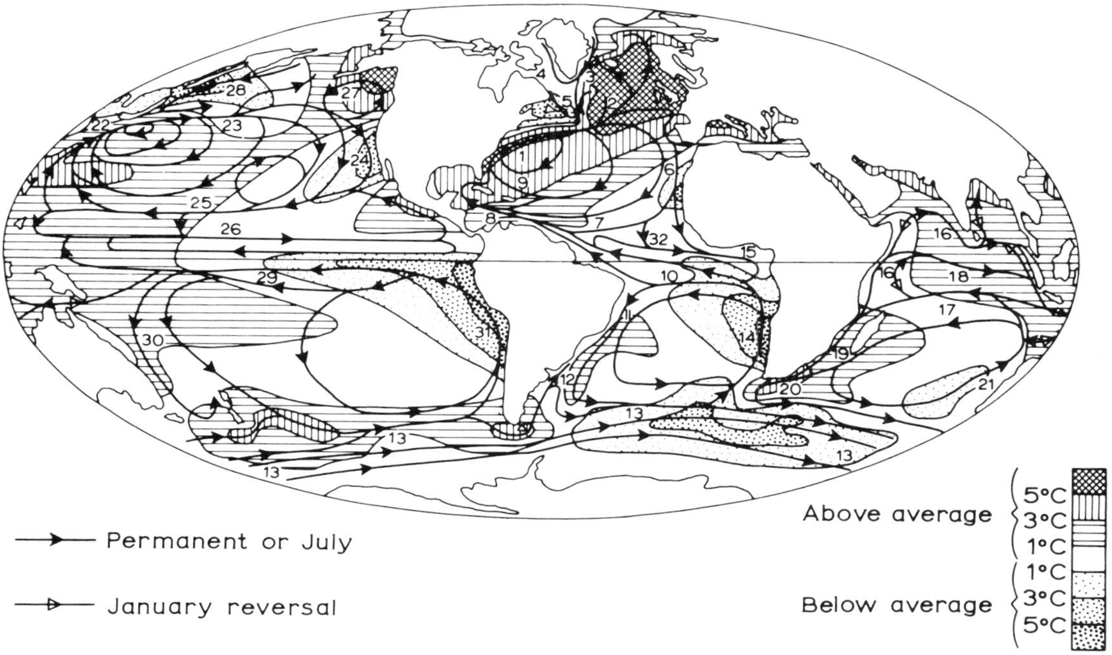

northern Indian Ocean, off east Africa and off northern Australia (see Figure 3.42). The Ekman effect (see A.5, this chapter) causes the flow to be increasingly deflected to the right (in the northern hemisphere) and to decrease in velocity as the influence of the wind stress diminishes with depth. However, the rate of change of flow direction with depth increases with latitude, such that near the equator there are no flow reversals at depth which are characteristic of higher latitudes. The depth at which this reversal occurs decreases poleward, but averages about 50 m over large areas of the ocean. In addition, as water moves meridionally the conservation of angular momentum implies changes in relative vorticity (see pp. 104) and 123), with poleward-moving currents acquiring anticyclonic vorticity and equatorward-moving currents acquiring cyclonic vorticity.

Equatorward of the subtropical high-pressure cells the persistent trade winds generate the broad North and South Equatorial Currents (see Figure 3.42). On the western sides of the oceans most of this water swings poleward with the airflow and thereafter increasingly comes under the influence of the Ekman deflection and of the anticyclonic vorticity effect. However, some water tends to pile up near the equator on the western sides of oceans, partly because here the Ekman effect is virtually absent with little poleward deflection and no reverse current at depth. To this is added some of the water which is displaced northward into the equatorial zone by the especially active subtropical high-pressure circulations of the southern hemisphere. This accumulated water flows back eastward down the hydraulic gradient as compensating narrow surface Equatorial Counter Currents, unimpeded by the weak surface winds. As the circulations swing poleward round the western margins of the oceanic subtropical high-pressure cells there is the tendency for water to pile up against the continents giving, for example, an appreciably higher sea level in the Gulf of Mexico than that along the Atlantic coast of the United States. This accumulated water cannot escape by sinking because of its relatively high temperature and resulting vertical stability, and it consequently continues poleward in the dominant direction of surface airflow. As a result of this movement the current gains anticyclonic vorticity which reinforces the similar tendency imparted by the winds, leading to relatively narrow currents of high velocity (for example, the Kuro Shio, Brazil, Mozambique–Agulhas and, to a less-marked extent, the East Australian Current). In the North Atlantic the configuration of the Caribbean Sea and Gulf of Mexico especially favours this pile-up of water, which is released poleward through the Florida Straits as the particularly narrow and fast Gulf Stream.

These poleward currents are opposed both by their friction with the nearby continental margins and by energy losses due to turbulent diffusion, such as those accompanying the formation and cutting off of meanders in the Gulf Stream. On the poleward sides of the subtropical high-pressure cells westerly currents dominate, and where they are unimpeded by land masses in the southern hemisphere they form the broad and swift West Wind Drift. In the northern hemisphere a great deal of the eastward-moving current in the Atlantic swings northward, leading to very anomalously high sea temperatures, and is compensated for by a southward flow of cold arctic water at depth. However, more than half of the water mass comprising the North Atlantic Drift, and almost all that of the North Pacific Drift, swings south round the east sides of the subtropical high-pressure cells, forming the Canary and California Currents. Their southern-hemisphere equivalents are the Benguela, Humboldt or Peru, and West Australian Currents. In contrast with the currents on the west sides of the oceans, these currents acquire cyclonic vorticity which is in opposition to the anticyclonic wind tendency, leading to relatively broad flows of low velocity. In addition the deflection due to the Ekman effect causes the surface water to move westward away from the coasts, leading to up-welling of cold water from depths of 100–300 m. Although the band of up-welling

may be quite narrow (about 200 km wide for the Benguela Current) the Ekman effect spreads this cold water westward. On the poleward margins of these cold-water coasts the meridional swing of the wind belts imparts a strong seasonality to the up-welling, the California Current up-welling, for example, being particularly well marked during the period March–July.

Whereas above the thermocline the ocean circulation is mainly wind driven, below it is powered by density gradients, predominantly due to salinity differences related to river runoff, the balance of precipitation minus evaporation, and sea-ice formation and melting. Exchanges between surface and deep ocean waters occur in regions of subsidence of high-density surface waters or in restricted belts of coastal up-welling associated with the Ekman effect, convergence at depth and the sweeping aside of surface waters by offshore winds (see Chapter 6, G.3). There are two main regions of subsidence:

1 The Weddell Sea, Antarctica, where sea-ice growth in winter releases brine into a cold surface water layer.
2 In the northern North Atlantic in winter where northward-flowing, high-salinity water in the lower mixed layer at intermediate depths (*c.* 800 m) rises to the surface as the strong westerly winds displace the surface water. The risen water cools, releasing heat to the atmosphere equivalent to about 30 per cent of the annual receipt of solar radiation in the region, becomes even more dense and sinks to the ocean bottom. This results in a southward-flowing density current on the Atlantic floor (completing the 'Atlantic conveyor' system) which joins the eastward-flowing circum-Antarctic bottom current and circulates into the less-saline Indian and Pacific Oceans with a global mixing time of several hundred years.

The atmosphere and the surface ocean waters are closely connected both in temperature and in CO_2 concentrations. The atmosphere contains less than 1.7 per cent of the CO_2 held by the oceans and the amount absorbed by surface ocean water rapidly regulates the concentration in the atmosphere. The absorption of CO_2 by the oceans is greatest where the water is rich in organic matter or where it is cold. Thus the oceans are capable of regulating atmospheric CO_2, of changing the greenhouse effect, and of making a contribution to global glacial conditions (during the last glaciation the atmospheric CO_2 concentration was only two-thirds of its interglacial value – see Figure 1.5).

Sea surface temperature anomalies in the Atlantic appear to have marked effects on the climate in Europe, Africa, and South America. For example, warmer sea surfaces off north-west Africa augment west African summer monsoon rainfall; dry conditions in the Sahel have been linked to a cooler North Atlantic. There are similar links between tropical sea surface temperatures and droughts in north-eastern Brazil. A circulation anomaly pattern known as the North Atlantic Oscillation (NAO) also shows strong air–sea interactions. The NAO is an oscillation in the pressure field between Iceland (65°N) and a zone about 40°N across the Atlantic. Fluctuations in the NAO give rise to alternating mild/severe winters in western Greenland–Labrador and north-west Europe. Severe winters in the Greenland area also have cold northerly airflow and more extensive sea ice in the Labrador Sea. On a longer time scale the NAO index (of S–N pressure difference) was generally low from 1925–70 when air temperatures in the northern hemisphere were above normal and cyclones over the east coast of North America tended to be located over the ocean thus causing longer, drier east coast summers. Prior to 1925 a regime of colder climatic conditions was associated with a higher NAO index.

G MODELLING THE ATMOSPHERIC CIRCULATION AND CLIMATE

Better understanding of the complex behaviour of the atmosphere and climate processes has been obtained in recent years through numerical

modelling studies. Here we can only sketch the essential features of this approach. There are three basic categories of model: the general circulation model (GCM), the energy budget model (EBM) and the radiative-convective model (RCM).

In the GCM, all dynamic and thermodynamic processes and the radiative and mass exchanges that have been treated in Chapters 1–3 are modelled using five basic sets of equations. These account for the horizontal and vertical components of motion (see A, B, this chapter), the conservation of energy (F, this chapter) and of water substance (Chapter 2), the continuity of mass (p. 102), and the equation of state (p. 13). In addition, radiative processes, including a diurnal and seasonal cycle, surface friction and cloud formation are usually represented. These are coupled in the manner shown schematically in Figure 3.43. Beginning with a set of initial atmospheric conditions, the equations are solved repeatedly for time steps of a few minutes at a large number of grid points over the earth and at several levels in the atmosphere. The horizontal grid is usually of the order of 5° latitude and longitude (except near the poles). Another computationally faster approach is to represent the horizontal fields by a series of two-dimensional sine and cosine functions (a spectral model). In the vertical, 2–15 levels may be used. Coastlines and mountains as well as essential elements of the surface vegetation (albedo, roughness) and soil (moisture content) may be incorporated. Sea-ice extent and sea-surface temperatures are usually specified for each month (unless the GCM is coupled to an interactive ocean model where the ocean circulation and heat transports are calculated). Snow cover on the ground is often computed as part of a hydrological cycle.

Similar models are used to simulate climatic change by representing inputs (i.e. forcings), storages and transfers (see Figure 3.43 and Chapter 8). The periods of time shown in Figure 3.44 refer to:

1 *Forcing times.* The characteristic time-spans over which solar and anthropogenic changes of input occur. In the case of the former these are the periods of solar radiation cycles and in the case of the latter the average time-span over which significant changes of such anthropogenic effects as increasing atmospheric CO_2 occur.
2 *Storage times.* For each compartment of the atmosphere and ocean subsystems these are the average times taken for an input of thermal energy to diffuse and mix within the compartment. For the earth subsystem the average times are those required for inputs of water to move through each compartment.

The complexity of the circulation of the

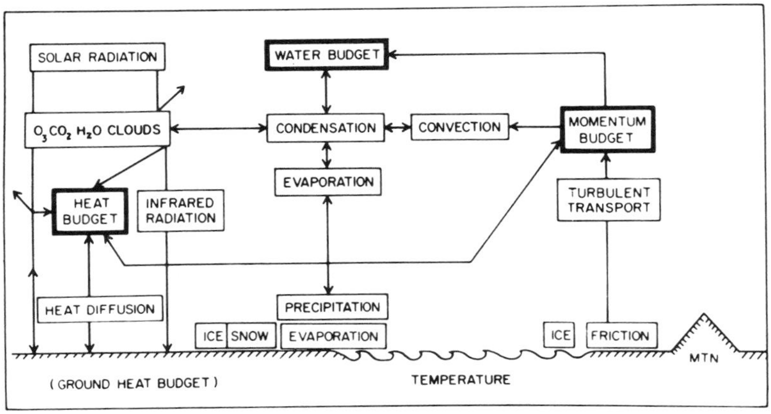

Figure 3.43 Schematic diagram of the interactions among physical processes in a general circulation model.

Source: From Druyan *et al.* 1975.

Figure 3.44 The earth/atmosphere/ocean system showing estimated equilibrium times, together with the wide time variations involving the external solar, tectonic, geothermal and anthropogenic forcing mechanisms.

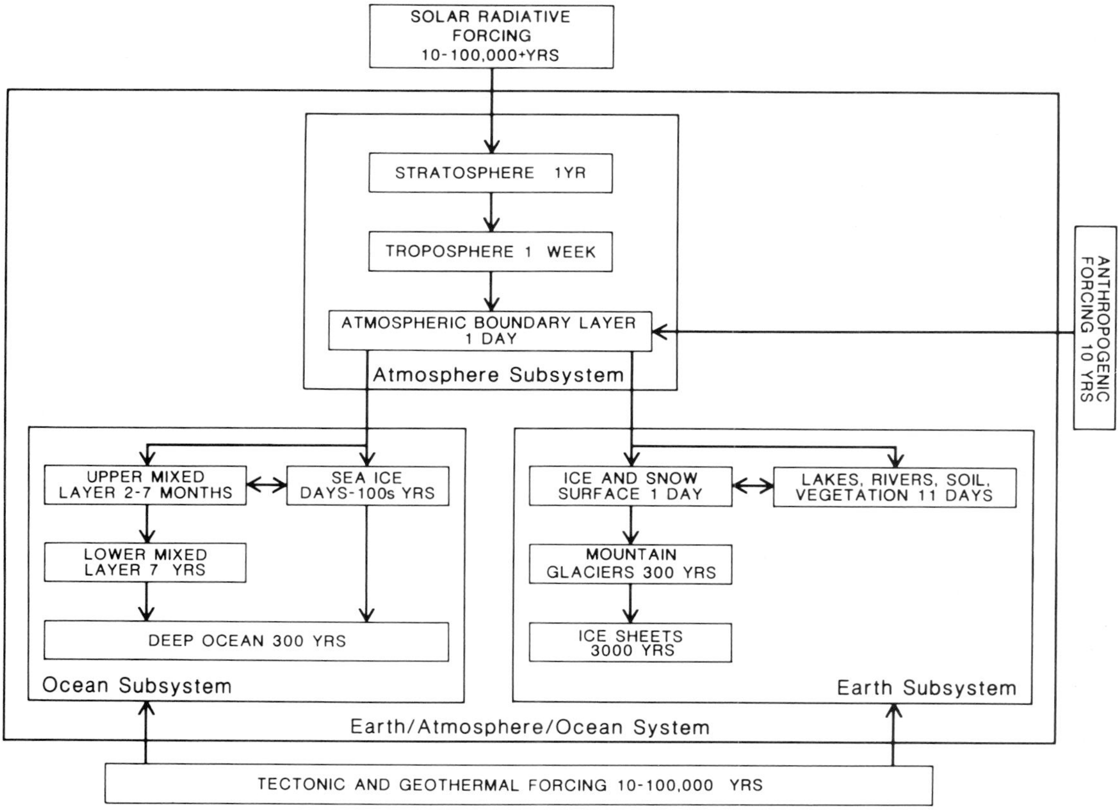

Source: After Saltzman 1983.

atmosphere/ocean/earth/ice system makes its modelling very difficult. Major uncertainties stem from a lack of understanding of the operation of the important feedback mechanisms which either amplify or damp down changes produced by the forcing agents (see Chapter 8) and also from a lack of knowledge regarding the mechanism of deep ocean circulations, particularly those involving up-welling.

Model experiments depend upon the nature of the problem to be studied. These include weather forecasting (see Chapter 4, I), the sensitivity of the climate to different conditions, short-term effects such as ocean temperature anomalies and volcanic eruptions, as well as long-term factors like changes in atmospheric composition or solar radiation. Figure 8.14 illustrates the components of the earth's climate system that are candidates for such climate experiments. The usual procedure is first to develop a model so that it adequately represents present-day conditions. This major task normally involves a group of scientists and computer programmers for several years and necessitates powerful computers. When satisfactory 'control cases' of present conditions are available for a series of model years of data, it is then feasible to examine the effect of changing

individual variables. This is important because in the real atmosphere changes will seldom if ever occur in isolation. In this way, the relative importance of different climatic variables can be determined quantitatively.

Because GCMs require massive computer resources, other approaches to modelling climate have developed. A variant of the GCM is the statistical-dynamical model (SDM) in which only zonally-averaged features are analysed and north–south energy and momentum exchanges are not treated explicitly, but are represented statistically through parameterization. Simpler still are the energy balance model (EBM) and the radiative convective model (RCM). The EBM assumes a global radiation balance and describes the integrated north–south transports of energy in terms of the poleward temperature gradients. EBMs can be one-dimensional (latitude variations only), two-dimensional (latitude–longitude, with simple land/ocean weightings or simplified geography) and even zero-dimensional (averaged for the globe). They are used particularly in climate change studies. The RCMs represent a single, globally-averaged vertical column. The vertical temperature structure is analysed in terms of radiative and convective exchanges. These less-complete models complement the GCMs because, for example, the RCM allows study of complex cloud-radiation interactions and atmospheric composition on lapse rates in the absence of large-scale dynamical processes.

SUMMARY

Air motion is described by its horizontal and vertical components; the latter are much smaller than the horizontal velocities. Horizontal motions compensate for vertical imbalances between gravitational acceleration and the vertical pressure gradient.

Horizontal wind velocity is determined by the horizontal pressure gradient, the earth's rotational effect (Coriolis force), and the curvature of the isobars (centripetal acceleration). All three factors are accounted for in the gradient wind equation, but this can be approximated in large-scale flow by the geostrophic wind relationship. Below 1500 m, the wind speed and direction are affected by surface friction.

Air ascends (descends) in association with surface convergence (divergence) of air. Air motion is also subject to relative vertical vorticity as a result of curvature of the streamlines and/or lateral shear; this, together with the earth's rotational effect, makes up the absolute vertical vorticity.

Local winds occur as a result of diurnally varying thermal differences setting-up local pressure gradients (mountain–valley winds and land–sea breezes) or due to the effect of a topographic barrier on airflow crossing it (examples are the lee-side föhn and bora winds).

The vertical change of pressure with height depends on the temperature structure. High (low) pressure systems intensify with altitude in a warm (cold) air column; thus warm lows and cold highs are shallow features. This 'thickness' relationship is illustrated by the upper level subtropical anticyclones and polar vortex in both hemispheres. The intermediate mid-latitude westerly winds thus have a large 'thermal wind' component. They become concentrated into upper tropospheric jet streams above sharp thermal gradients, such as fronts.

The upper flow displays a large-scale long-wave pattern, especially in the northern hemisphere, related to the influence of mountain barriers and land/sea differences. The surface pressure field is dominated by semi-permanent subtropical highs, subpolar lows and, in winter, shallow cold continental highs in Siberia and north-western Canada. The equatorial zone is predominantly low pressure. The associated global wind belts are

the easterly trade winds and the mid-latitude westerlies. There are more variable polar easterlies and over land areas in summer a band of equatorial westerlies representing the monsoon systems. This mean zonal (west–east) circulation is intermittently interrupted by 'blocking' highs; an idealized sequence is known as the index cycle.

The atmospheric general circulation which transfers heat and momentum polewards, is predominantly in a vertical meridional plane in low latitudes (the Hadley cell), but there are also important east–west circulations (Walker cells) between the major regions of subsidence and convective activity. Heat and momentum exchanges in middle and high latitudes are accomplished by horizontal waves and eddies (cyclones/anticyclones). Substantial energy is also carried poleward by ocean current systems. Various types of numerical models are now used to study the mechanisms of the atmospheric circulation and climate processes. These include vertical column models of radiative and convective processes, one- and two-dimensional energy balance models and complete three-dimensional general circulation models.

4

Air masses, fronts and depressions

An air mass may be defined as a large body of air whose physical properties, especially temperature, moisture content and lapse rate, are more or less uniform horizontally for hundreds of kilometres. The theoretical ideal is an atmosphere where surfaces of constant pressure are not intersected by isosteric (constant-density) surfaces, so that in any vertical cross-section, as shown in Figure 4.1, isobars and isotherms are parallel. Such an atmosphere is referred to as *barotropic*.

Three main factors tend to determine the nature and degree of uniformity of air-mass characteristics. They are: (a) the nature of the source area (from which the air mass obtains its original qualities); (b) the direction of movement and changes that occur in the constitution of an air mass as it moves over long distances; and (c) the age of the air mass. The physical properties of all air masses are classified according to the way in which they compare with the corresponding properties of the underlying surface region or with those of adjacent air masses.

Figure 4.1 A schematic temperature section for the northern hemisphere showing barotropic air masses and a baroclinic frontal zone (assuming that density decreases with height only).

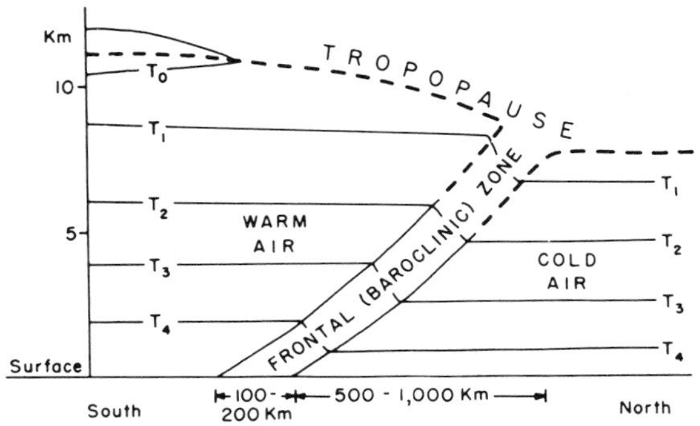

Study of the contrasting properties of different air masses leads naturally on to a consideration of air-mass boundaries or *fronts*. Their relationship to low-pressure centres and to the patterns of airflow aloft are also discussed in this chapter, and this is followed by a brief examination of the various approaches adopted in weather forecasting.

A NATURE OF THE SOURCE AREA

We have already observed how most of the physical processes of our atmosphere result from self-regulating attempts to equalize the major differences that arise from inequalities in the world distribution of heat, moisture and pressure. On the world scale the heat and momentum balances refer only to long-term average conditions. However, on a smaller scale, radiation and vertical mixing can produce some measure of equilibrium between the surface conditions and the properties of the overlying air mass if air remains over a given geographical region for a period of about 3 to 7 days. Naturally the chief source regions of air masses are areas of extensive, uniform surface type which are normally overlain by quasi-stationary

pressure systems. These requirements are fulfilled where there is slow divergent flow from the major thermal and dynamic high-pressure cells, whereas low-pressure regions are zones of convergence into which air masses move (see E, this chapter).

Air masses are classified on the basis of two primary factors. The first is the temperature, giving arctic, polar and tropical air, and the second is the type of surface in their region of origin, giving maritime and continental categories. The major cold- and warm-air masses will now be discussed.

1 Cold-air masses

The principal sources of cold air in the northern hemisphere are: (a) the continental anticyclones of Siberia and northern Canada, which originate continental polar (cP) air masses, and (b) the Arctic Basin, when it is dominated by high pressure (see Figure 4.2). Some classifications designate air of the latter category as continental arctic (cA), but the differences between cP and cA air masses are limited mainly to the middle and upper troposphere, where temperatures are lower in the cA air (see Table 4.1).

The snow-covered source regions of these two air masses lead to marked cooling of the lower layers (see Figure 4.3) and, since the vapour content of cold air is very limited, the air masses generally have a mixing ratio of only 0.1–0.5 g/kg near the surface. The stability produced by the effect of surface cooling prevents vertical mixing so that further cooling occurs more slowly by radiation losses only. The effect of this radiative cooling and the tendency for air-mass subsidence in high-pressure regions combine to produce a prominent temperature inversion from the surface up to about 850 mb in typical cA and cP air. In view of their extreme dryness these air masses are characterized by small cloud amounts and only occasional light snowfalls. In summer continental heating over northern Canada and Siberia causes the virtual disappearance of their sources of cold air. The Arctic Basin source remains (see Figure 4.4A), but the cold air here is very limited in depth at this time of year. In the southern hemisphere, the Antarctic continent and the ice shelves are a source of cA air in all seasons (see Figures 4.2B and 4.4B). There are no sources of cP air, however, due to the dominance of ocean areas in middle latitudes. At all seasons cA or cP air is greatly modified by a passage over the ocean. Secondary types of air mass are produced by such means and these will be considered in B, this chapter.

2 Warm-air masses

These have their origins in the subtropical high-pressure cells and, during the summer season, in the great accumulations of warm surface air which characterize the heart of large land areas.

The tropical (T) sources are either maritime (mT), originating in the oceanic subtropical high-pressure cells, or continental (cT), originating either from the continental parts of these subtropical cells (e.g. as does the North African *Harmattan*) or simply associated with regions of generally light variable winds, assisted by upper tropospheric subsidence, over the major continents in summer (e.g. Central Asia). In the southern hemisphere the source area of mT air covers about half of the hemisphere. There is no zone of significant temperature gradient between the equator and the oceanic Subtropical Convergence about 40°S.

The maritime type is characterized by high temperatures (accentuated by the warming action to which the descending air is subjected), high humidity of the lower layers over the oceans, and stable stratification. Since the air is warm and moist near the surface, stratiform cloud commonly develops as the air moves polewards from its source. The continental type in winter is restricted mainly to North Africa (see Figure 4.2, Table 4.1), where it is a warm, dry and stable air mass. In summer, warming of the lower layers by the heated land generates a steep lapse rate, but despite its instability the low relative and specific humidity prevent the development of cloud and precipitation. In the

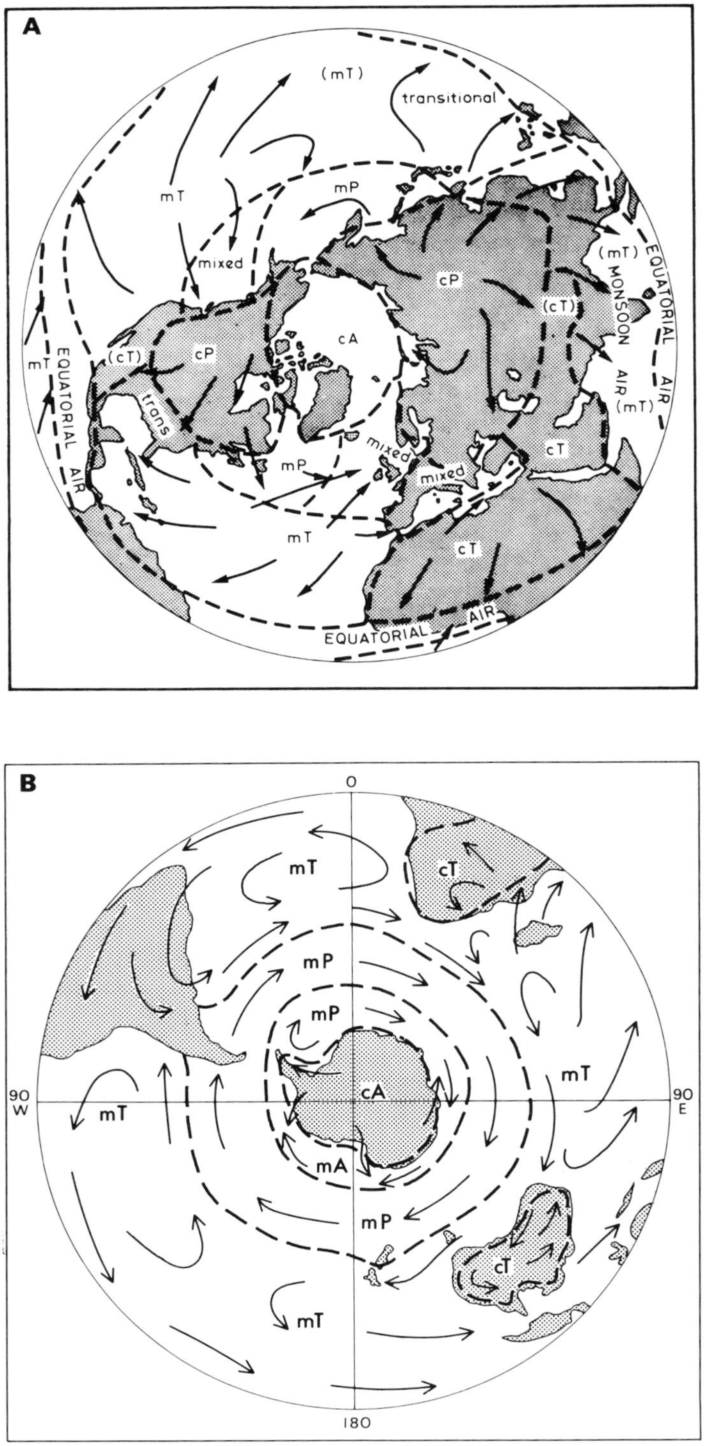

Figure 4.2 Air masses in winter. A: Northern hemisphere; B: Southern hemisphere.

Sources: A: After Petterssen 1950 and Crowe 1965. B: After Taljaard *et al.* 1969 and Newton 1972.

Table 4.1 Air-mass characteristics in winter

Air mass		*Level (mb)* *1000*	*850*	*700*	*500*
cA	(1) *T*	–	−31	−33	−42
	(3) *T*	1	−8	−21	−36
	(3) x	2.4	1.7	0.4	0.2
	(5) *T*	(0.33)††	−28	−30	−42
	(5) x	(0.2)††	0.3	0.2	0.1
mA	(1) *T*	–	−10	21	−38
†	(2) *T*	1	−9	−20	−40
	(2) x	3.1	1.7	0.7	0.6
	(3) *T*	4	−6	−14	−33
	(3) x	4.6	2.2	1.3	0.3
	(6) *T*	0	−10	−20	−35
	(6) x	3.0	1.6	0.8	0.2
cP	(1) *T*	–	−18	−20	−33
*	(2) *T*	−2	−12	−22	−41
	(2) x	2.6	1.5	0.6	0.1
	(3) *T*	7	−2	−13	−24
	(3) x	4.5	2.6	1.3	0.4
mPw	(1) *T*	–	5	−4	−23
**	(2) *T*	8	1	−9	−27
	(2) x	5.8	4.0	2.1	0.6
	(3) *T*	12	2	−7	−23
	(3) x	7.8	4.0	1.6	0.4
	(4) *T*	10	2	−7	−25
	(4) x	5.5	3.4	1.8	0.4
mT	(1) *T*	–	10	0	−17
‡	(2) *T*	11	6	−2	−17
	(2) x	6.8	5.6	3.5	1.2
	(3) *T*	–	10	2	−14
	(3) x	–	6.0	2.5	1.0
	(4) *T*	14	6	−2	−18
	(4) x	7.8	5.3	2.5	0.9
cT	(3) *T*	–	19	5	−17
	(3) x	–	1.8	1.3	0.6
Med	(3) *T*	14	3	−3	−19
	(3) x	7.0	3.7	2.5	0.9

Key and sources: 1 Typical values in North America, 45°–50°N (after Godson 1950).
2 Monthly means over the British Isles, using surface data at Kew in place of the 1000 mb values (after Belasco 1952).
3 Typical values in the Mediterranean (after *Weather in the Mediterranean*, MO 391, 1962).
4 Typical values in Australia, 33°S
5 Typical values in Antarctica, 75°S
6 Typical values in the Southern Oceans, 50°S
(4–6 after Taljaard *et al.* 1969).

Notes: *T* = air temperature (°C); x = humidity mixing ratio (g/kg)
Belasco's classification: †P_1, *A_1, **P_7, ‡T_1, ††950 mb level.

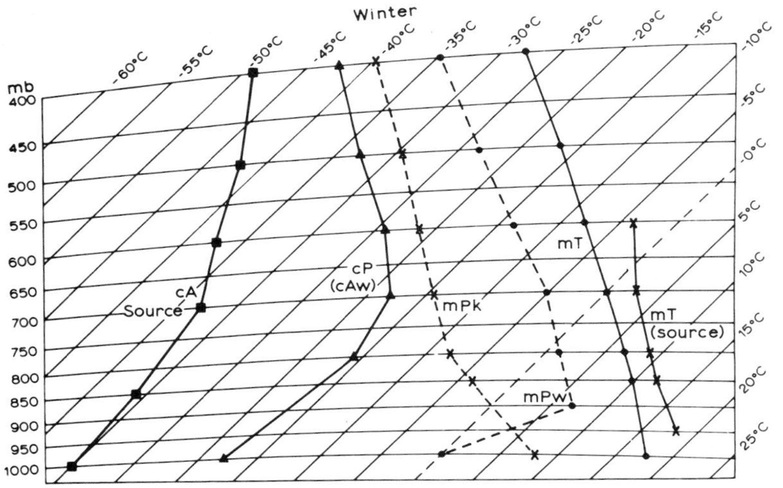

Figure 4.3 The average vertical temperature structure for selected air masses affecting North America at about 45°–50°N, recorded over their source areas or over North America in winter.

Source: After Godson 1950, Showalter 1939, and Willett.

southern hemisphere, cT air is rather more prevalent in winter over the subtropical continents except South America. In summer, much of southern Africa and northern Australia is affected by mT air, while there is a small source of cT air over Argentina (see Figure 4.4B).

The characteristics of the primary air masses are illustrated in Figures 4.3 and 4.5 and Tables 4.1 and 4.2. In some cases their properties have been considerably affected by movement away from the source region, and this question is discussed below (see pp. 146–50).

A different view of source regions can be obtained from analysis of airstreams. Streamlines of the mean resultant winds (see note 1) in individual months can be used to analyse areas of divergence representing air-mass source regions, downstream airflow and the confluence zones between different airstreams. Figure 4.6A shows the sources in the northern hemisphere and their annual duration. Four sources are dominant: the subtropical North Pacific and North Atlantic anticyclones, and their southern hemisphere counterparts. For the entire year air from these sources covers at least 25 per cent of the northern hemisphere; for 6 months of each year they affect almost three-fifths of the hemisphere. In the southern hemisphere, in contrast, the airstream climatology is much simpler as a result of the dominance of ocean surfaces (see Figure 4.6B). The source areas are associated with the three oceans, and especially their subtropical anticyclones. Antarctica is the major continental source, with another mainly in winter over Australia.

B AIR-MASS MODIFICATION

As air masses move away from their source region they are affected by different heat and moisture exchanges with the ground surface and by dynamic processes in the atmosphere. Thus an initially barotropic air mass is gradually changed into a moderately *baroclinic* airstream in which isosteric and isobaric surfaces intersect one another. The presence of horizontal temperature gradients means that air cannot travel as a solid block maintaining an unchanging internal structure. The trajectory (i.e. actual path) followed by an air parcel in the middle or upper troposphere will normally be quite different from that of a parcel nearer the surface, due to the increase of westerly wind velocity with height in the troposphere. The actual structure of an airstream at a given instant is determined to a large extent by the past history of air-mass modification processes. In spite of these quali-

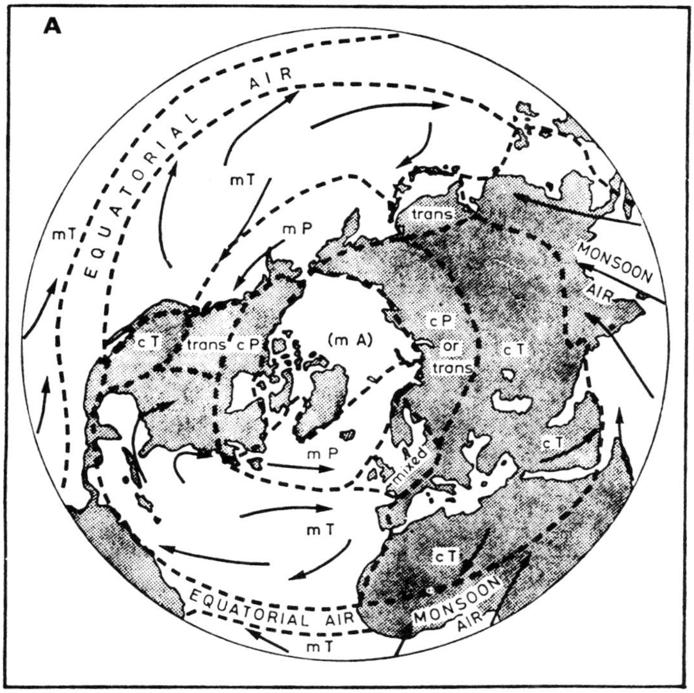

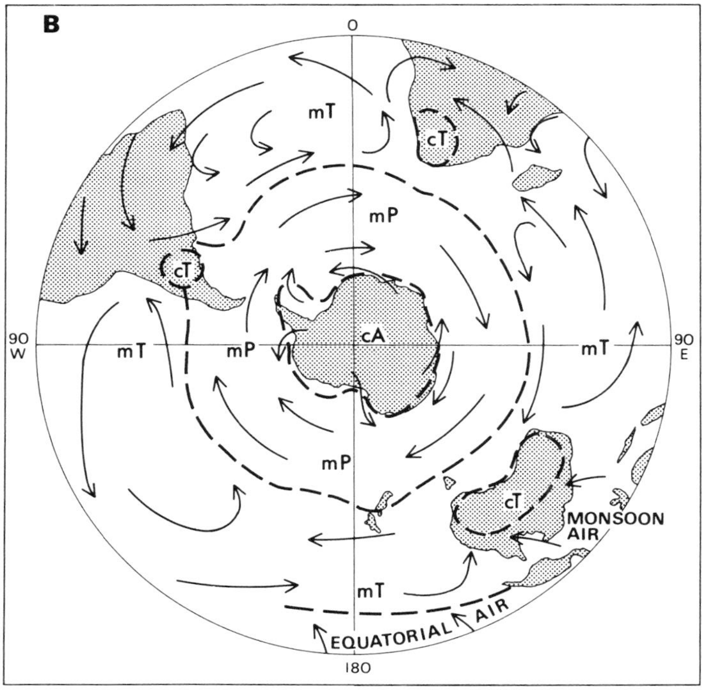

Figure 4.4 Air masses in summer. A: Northern hemisphere; B: Southern hemisphere.

Sources: A: After Petterssen 1950 and Crowe 1965. B: After Taljaard *et al.* 1969 and Newton 1972.

Table 4.2 Air-mass characteristics in summer

Air mass		*Level (mb)* 1000	*850*	*700*	*500*
cA	(5) *T*	(−9)‡	−13	−20	−33
	(5) *x*	(1.8)‡	1.1	0.7	0.7
mA	(1) *T*	–	−4	−14	−33
	(2) *T*	14	2	−7	−25
	(2) *x*	6.3	4.3	2.5	0.1
mP	(1) *T*	–	11	0	−19
*	(2) *T*	16	4	−6	−24
	(2) *x*	8.4	3.9	2.2	0.4
	(3) *T*	–	18	−2	−19
	(3) *x*	–	6.0	2.5	0.8
	(4) *T*	17	8	0	−14
	(4) *x*	8.0	6.0	3.1	1.0
cP	(3) *T*	26	13	4	−14
	(3) *x*	16.1	6.7	3.4	0.9
mT	(1) *T*	–	18	8	−8
	(2) *T*	19	12	4	−11
	(2) *x*	10.8	8.1	4.5	2.4
	(4) *T*	22	16	5	−11
	(4) *x*	13.4	8.0	4.8	1.7
cT	(1) *T*	–	22	10	−11
	(2) *T*	21	16	6	−11
	(2) *x*	12.1	3.9	3.4	1.1
†	(3) *T*	–	26	13	−10
	(3) *x*	–	4.5	2.5	0.5
	(4) *T*	27	20	7	−12
	(4) *x*	8.0	4.7	3.6	1.2
Med	(3) *T*	29	19	12	−6
	(3) *x*	14.1	7.4	3.0	0.9

Key and sources: See Table 4.1.
Notes: T = air temperature (°C); x = humidity mixing ratio (g/kg).
Belasco's classification: *P_3, †cT originating over Africa, ‡950 mb.

fications the air-mass concept nevertheless remains of considerable practical value.

1 Mechanisms of modification

The mechanisms by which air masses are modified are, for convenience, treated separately, although this rigid distinction is often not justified in practice.

a Thermodynamic changes

An air mass may be heated from below either by passing from a cold to a warm surface or by solar heating of the ground over which the air is located. Similarly, but in reverse, it can be cooled from below. Heating from below acts to increase air-mass instability so that the effect may be spread rapidly through a considerable thickness of air, whereas surface cooling

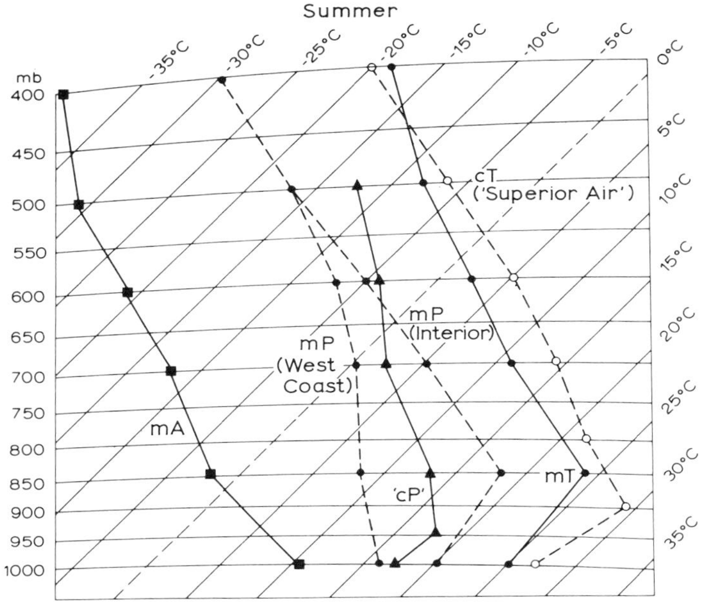

Figure 4.5 The average vertical temperature structure for selected air masses affecting North America in summer.

Source: After Godson 1950, Showalter 1939, and Willett.

Figure 4.6 Air-mass source regions in the northern hemisphere (A) and the southern hemisphere (B). Numbers show the areas affected by each air mass in months per year.

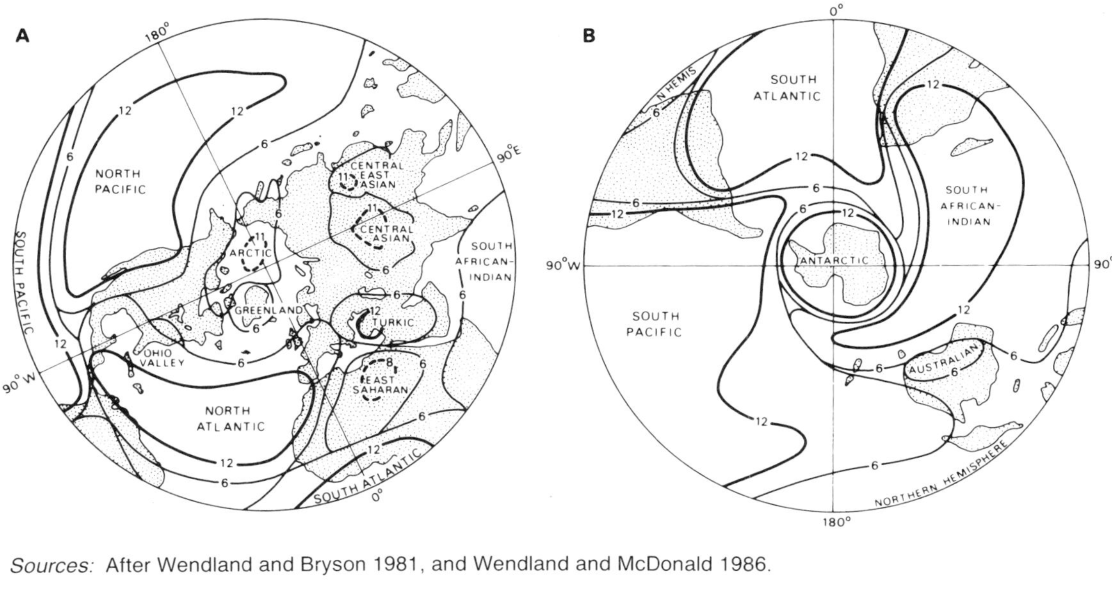

Sources: After Wendland and Bryson 1981, and Wendland and McDonald 1986.

produces a temperature inversion which greatly limits the vertical extent of the cooling. For this reason cooling mainly occurs through radiative heat loss by the air, a process which takes place only very gradually requiring up to two weeks.

Changes can also occur through increased evaporation, the moisture being supplied either from the underlying surface or by precipitation from an overlying air-mass layer. In reverse, the abstraction of moisture by condensation or precipitation can also cause changes. A parallel, and most important, change is the respective addition or loss of latent heat accompanying this condensation or evaporation. The distribution of latent and sensible heat transfers to the atmosphere are illustrated in Figures 1.41 and 1.42, although it must be noticed that these refer to annual values.

b Dynamic changes

Dynamic (or mechanical) changes are, superficially at any rate, different from thermodynamic changes because they involve mixing or pressure changes associated with the actual movement of the air mass. The distribution of the physical properties of air masses has been shown to be considerably modified, for example by a prolonged period of turbulent mixing (see Figure 2.16). This process is particularly important at low levels where surface friction intensifies the natural turbulence of airflow, providing a ready mechanism for the upward transfer of the effects of thermodynamic processes.

The radiative and advective exchanges discussed previously are non-adiabatic, but the ascent or descent of air causes adiabatic changes of temperature. Large-scale lifting may result from forced ascent by a mountain barrier or from airstream convergence. Conversely, sinking may occur when high-level convergence sets up subsidence or when stable air, having been forced up over high ground by the pressure gradient, descends in its lee. Dynamic processes in the middle and upper troposphere are in fact a major cause of air-mass modification. The decrease in stability aloft, as air moves away from the areas of subsidence, is a common example of this type of mechanism.

2 The results of modification: secondary air masses

The study of the ways in which air masses change in character tells us a great deal about our weather, for many common meteorological phenomena are the product of such modification.

a Cold air

Continental polar air frequently streams out from Canada over the western Atlantic in winter, where it undergoes rapid transformation. Heating over the Gulf Stream Drift rapidly makes the lower layers unstable and evaporation into the air leads to sharp increases of moisture content (see Figure 2.3) and cloud formation (see Plate 11). The turbulence associated with the convective instability is marked by gusty conditions. By the time the air has reached the central Atlantic it has become a cool, moist, maritime polar (mP) air mass. Analogous processes occur with outflow from Asia over the North Pacific (see Figure 4.2). Over middle latitudes of the southern hemisphere the circumpolar ocean gives rise to a continuous zone of mP air which, in summer, extends to the margin of Antarctica. In this season, however, there is a considerable gradient of ocean temperatures associated with the Antarctic Convergence that makes the zone far from uniform in its physical properties. The weather in cP airstreams is typically that of bright periods and squally showers, with a variable cloud cover of cumulus and cumulonimbus. As the air moves eastward towards Europe the cooler sea surface may produce a neutral or even stable stratification near the surface, especially in summer, but subsequent heating over land will again regenerate unstable conditions. Similar conditions, but with lower temperatures (see Table 4.1), result if cA air crosses sea areas in high latitudes producing maritime arctic (mA) air.

When cP air moves southward over land in winter, in central North America for example, it acquires higher temperatures and a greater tendency towards instability, but there is little gain in moisture content. The cloud type is scattered shallow cumulus which only rarely gives showers even in the afternoon when convectional instability is at a maximum. Exceptions occur in early winter around the eastern and southern shores of Hudson Bay and the Great Lakes. Until these water bodies freeze over, cold airstreams which cross them are rapidly warmed and supplied with moisture leading to locally heavy snowfalls.

Over Eurasia and North America cP air may move southwards and later recurve northwards. Some schemes of air-mass classification cater for such possibilities by specifying whether the air is colder (k), or warmer (w), than the surface over which it is passing. For example, cPk refers to cold, dry continental polar air which is moving over a warmer surface and is thereby likely to become unstable. Likewise, mPw indicates moist, maritime polar air which is being progressively cooled near the surface and, hence, becoming more stable.

In general, a 'k' air mass has gusty, turbulent winds which make for good visibility as smoke and haze are dispersed. The instability leads to cumuliform clouds. A 'w' air mass typically has stable or inversion conditions with stratiform cloud. Limited vertical mixing allows the concentration of smoke, haze and fog at low levels. Clearly, these and similar symbols provide a convenient shorthand description of the important parameters which characterize different air masses.

Many parts of the globe must be regarded as transitional regions where the surface and air circulation produce air masses with intermediate characteristics. Northern Asia and northern Canada fall into this category in summer. In a general sense the air has affinities with continental polar air masses but these land areas, particularly the Canadian shield, have extensive bog and water surfaces so that the air is moist and cloud amounts are quite high. In a similar manner, melt-water pools and leads in the arctic pack-ice make it more appropriate to regard the area as a source of maritime arctic (mA) air in summer (see Figure 4.4A). This designation is also applied to air over the antarctic pack-ice in winter that is much less cold in its lower levels than the air over the continent itself.

b Warm air

The modification of warm air masses is usually a gradual process. Air moving poleward over progressively cooler surfaces becomes increasingly stable in the lower layers. In the case of mT air with high moisture content, surface cooling may produce advection fog and this is particularly common, for example, in the south-western approaches to the English Channel during spring and early summer when the sea is still cool. Similar development of advection fog in mT air occurs along the South China coast in February–April, and also off Newfoundland and over the coast of northern California in spring and summer. If the wind velocity is sufficient for vertical mixing, low stratus cloud forms in the place of fog and drizzle may result. In addition, forced ascent of the air by high ground, or by overriding of an adjacent air mass, can produce heavy rainfall.

The cT air originating in those parts of the subtropical anticyclones situated over the arid subtropics in summer is extremely hot and dry (see Table 4.2). It is typically unstable at low levels and dust storms may occur, but the dryness and the subsidence of the upper air limit cloud development. In the case of North Africa this cT air may move out over the Mediterranean rapidly acquiring moisture, with the consequent release of potential instability triggering off showers and thunderstorm activity.

The air masses in low latitudes present considerable problems of interpretation. The temperature contrasts found in middle and high latitudes are virtually absent and what differences do exist are due principally to moisture content and, more particularly, to the presence or absence of subsidence. *Equatorial air* is usually cooler than that subsiding in the

subtropical anticyclones, for example. *Tropical air* masses can only be differentiated meaningfully in terms of moisture content and the effects of subsidence on the lapse rate. On the equatorial sides of the subtropical anticyclones in summer the air is moving westward from areas with cool sea surfaces (e.g. off north-west Africa and California) towards those of higher sea-surface temperatures. Moreover the south-western parts of the high-pressure cells are affected only by weak subsidence due to the vertical structure of the cells (see Figure 3.18). As a result of these circumstances the mT air moving westwards around the equatorward sides of the subtropical highs becomes much less stable than that on the north-eastern margin of the cells. Eventually such air forms the very warm, moist, unstable 'equatorial air' of the Inter-Tropical Convergence Zone (see Figures 4.2 and 4.4). *Monsoon air* is indicated separately in these figures, although there is no basic difference between it and mT air. The special difficulties of treating tropical climatology in terms of air masses are discussed in Chapter 6.

3 The age of the air mass

Eventually the mixing and modification necessarily accompanying the movement of an air mass away from its source will cause the rate of energy exchange with the surroundings to diminish, and the various weather phenomena associated with these changes will dissipate. This process will be associated with the loss of its original identity until, finally, its features merge with those of surrounding airstreams and the air may again become subject to the influence of a new source region.

North-west Europe is shown as an area of 'mixed' air masses in Figures 4.2 and 4.4. This is intended to refer to the variety of sources and directions from which air may invade the region, since weather processes associated with air-mass modification and with the frontal zones separating air masses are very much in evidence. The same is also true of the Mediterranean in winter, although the area does impart its own particular characteristics to polar and other air masses which stagnate over it. Such air is termed *Mediterranean*; typical temperature and humidity values are listed in Tables 4.1 and 4.2. In winter it is convectively unstable (see Figure 2.15) as a result of the moisture picked up over the Mediterranean Sea.

The length of time during which an air mass retains its original characteristics depends very much on the extent of the source area and the type of pressure pattern affecting the area. In general the lower air is changed much more rapidly than that at higher levels, although dynamic modifications aloft, which are sometimes overlooked by climatologists, are no less significant in terms of weather processes. Modern air-mass concepts must, therefore, be flexible from the point of view of both synoptic and climatological studies.

C FRONTOGENESIS

The first real advance in our detailed understanding of mid-latitude weather variations was made with the discovery that many of the day-to-day changes are associated with the formation and movement of boundaries, or *fronts* between different air masses. Observations of the temperature, wind directions, humidity and other physical phenomena during unsettled periods showed that discontinuities often persist between impinging air masses of differing characteristics. The term 'front' for these surfaces of air-mass conflict was a logical one, proposed during the First World War by a group of meteorologists (including V. and J. Bjerknes, H. Solberg and T. Bergeron) working in Norway, and their ideas are still an integral part of most weather analysis and forecasting particularly in middle and high latitudes.

1 Frontal waves

It was observed that the typical geometry of the air-mass interface, or front, resembles a wave form (see Figure 4.7). Similar wave patterns are,

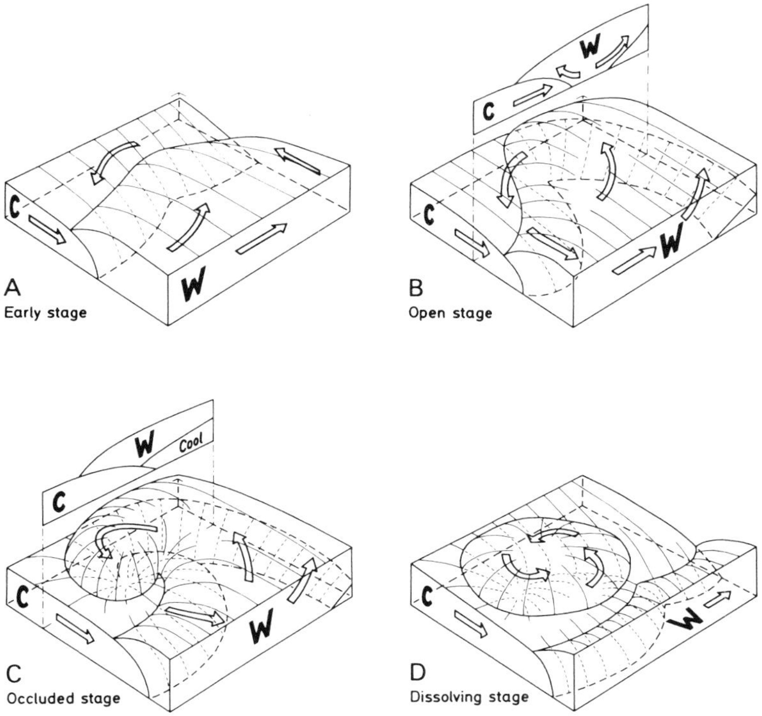

Figure 4.7 Four stages in the typical development of a mid-latitude depression. Satellite views of the cloud systems corresponding to these stages are shown in Figure 4.8.

Source: Mostly after Strahler 1965, modified after Beckinsale.
Notes: C = cold air; W = warm air.

in fact, found to occur on the interfaces between many different media, for example, waves on the sea surface, ripples on beach sand, aeolian sand dunes, etc. Unlike these wave forms, however, the frontal waves in the atmosphere are commonly unstable; that is, they suddenly originate, increase in size, and then gradually dissipate. Numerical model calculations show that in middle latitudes waves in a baroclinic atmosphere are unstable if their wavelength exceeds a few thousand kilometres. Frontal wave cyclones are typically 1500–3000 km in wavelength. The initially attractive analogy between atmospheric wave systems and waves formed on the interface of other media is, therefore, an insufficient basis on which to develop explanations of frontal waves. In particular, the circulation of the upper troposphere plays a key role in providing appropriate conditions for their development and growth, as will be shown below.

2 The frontal wave depression

A depression (also termed a low or cyclone) (see note 2) is an area of relatively low pressure, with a more or less circular isobaric pattern. It covers an area 1500–3000 km in diameter and usually has a life-span of 4–7 days. Systems with these characteristics, which are prominent on daily weather maps, are referred to as *synoptic scale* features. The depression, in mid-latitudes at least, is usually associated with a convergence of contrasting air masses. The interface between these air masses develops into a wave form with its apex located at the centre of the low-pressure area. The wave encloses a mass of warm air between modified cold air in front and fresh cold air in the rear. The formation of the wave also creates a distinction between the two sections of the original air-mass discontinuity for, although each section still marks the boundary between cold and warm air, the

weather characteristics found in the neighbourhood of each section are very different. The two sections of the frontal surface are distinguished by the names *warm front* for the leading edge of the wave and *cold front* for that of the cold air to the rear (see Figure 4.7).

The boundary between two adjacent air masses is marked by a strongly baroclininc zone of large temperature gradient 100–200 km wide (see B, this chapter, and Figure 4.1). Sharp discontinuities of temperature, moisture and wind properties at fronts, especially the warm front, are rather uncommon. Such discontinuities are usually the result of a pronounced surge of fresh, cold air in the rear sector of a depression, but in the middle and upper troposphere they are often caused by subsidence and may not coincide with the location of the baroclinic zone.

On satellite imagery, active cold fronts in a strong baroclinic zone commonly show marked spiral cloud bands formed as a result of the thermal advection (see Figure 4.8B, C). Warm fronts, however, are typically covered by a cirrus shield. As Figure 3.22 shows, an upper tropospheric jet stream is closely associated with the baroclinic zone, blowing roughly parallel to the line of the upper front (see Plate 19). This relationship is examined further in F, this chapter.

Air behind the cold front, away from the low centre, commonly has an anticyclonic trajectory and hence moves at a greater than geostrophic speed (see Chapter 3, A.4) impelling the cold front to acquire a supergeostrophic speed also. The wedge of warm air is pinched out at the surface and lifted bodily off the ground. This stage of *occlusion* eliminates the wave form at the surface (Figure 4.7). The occlusion gradually works outward from the centre of the depression along the warm front. Sometimes, the cold air wedge advances so rapidly that, in the friction layer close to the surface, cold air overruns the warm air and generates a *squall line* (see H (p. 169), this chapter).

The depression usually achieves its maximum intensity 12–24 hours after the beginning of occlusion.

By no means all frontal lows follow the idealized life cycle discussed above (cf. the caption for Plate 20). It is generally characteristic of oceanic cyclogenesis, but over North America many lows forming east of the Rocky Mountains in the lee pressure trough develop occluded fronts almost immediately. In winter months, the absence of moisture sources in this region greatly reduces the intensity of frontogenesis until the system moves eastward and draws in warm moist air from the south.

D FRONTAL CHARACTERISTICS

The activity of a front in terms of weather depends upon the vertical motion in the air

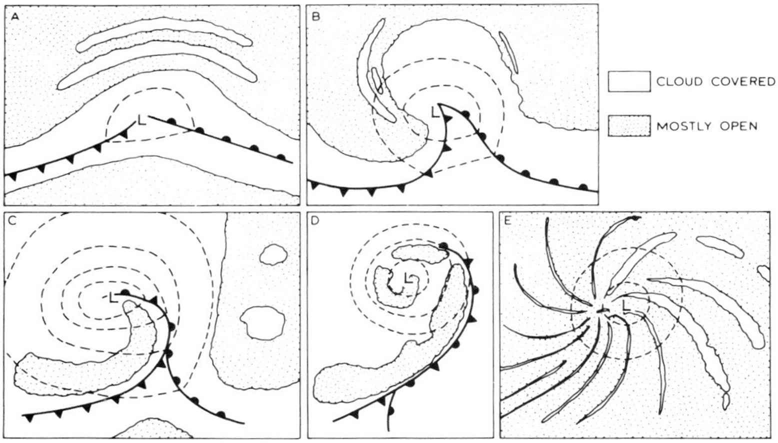

Figure 4.8 Schematic patterns of cloud cover (white) observed from satellites, in relation to surface fronts and generalized isobars. A, B, C and D correspond to the four stages in Figure 4.7.

Source: After Boucher and Newcomb 1962.

masses. If the air in the warm sector is rising relative to the frontal zone the fronts are usually very active and are termed *ana-fronts*, whereas sinking of the warm air relative to the cold air masses gives rise to less intense *kata-fronts* (see Figure 4.9).

1 The warm front

The warm front represents the leading edge of the warm sector in the wave. The frontal zone here has a very gentle slope, of the order ½°–1°, so that the cloud systems associated with the upper portion of the front herald its approach some 12 hours or more before the arrival of the surface front (see Plate 19). The ana-warm front, with rising warm air, has multi-layered cloud which steadily thickens and lowers towards the surface position of the front. The first clouds are thin, wispy cirrus, followed by sheets of cirrus and cirrostratus, and altostratus (see Figure 4.9A). The sun is obscured as the altostratus layer thickens and drizzle or rain begins to fall. The cloud often extends through most of the troposphere and with continuous precipitation occurring is generally designated as nimbostratus. Patches of stratus may also form in the cold air as rain falling through this air undergoes evaporation and quickly saturates it.

The descending warm air of the kata-warm front greatly restricts the development of medium- and high-level clouds. The frontal cloud is mainly stratocumulus, with a limited depth as a result of the subsidence inversions in both air masses (see Figure 4.9B). Precipitation is usually light rain or drizzle formed by coalescence since the freezing level tends to be above the inversion level, particularly in summer.

At the passage of the warm front the wind veers, the temperature rises and the fall of pressure is checked. The rain becomes intermittent or ceases in the warm air and the thin stratocumulus cloud sheet may break up.

Forecasting the extent of rain belts associated

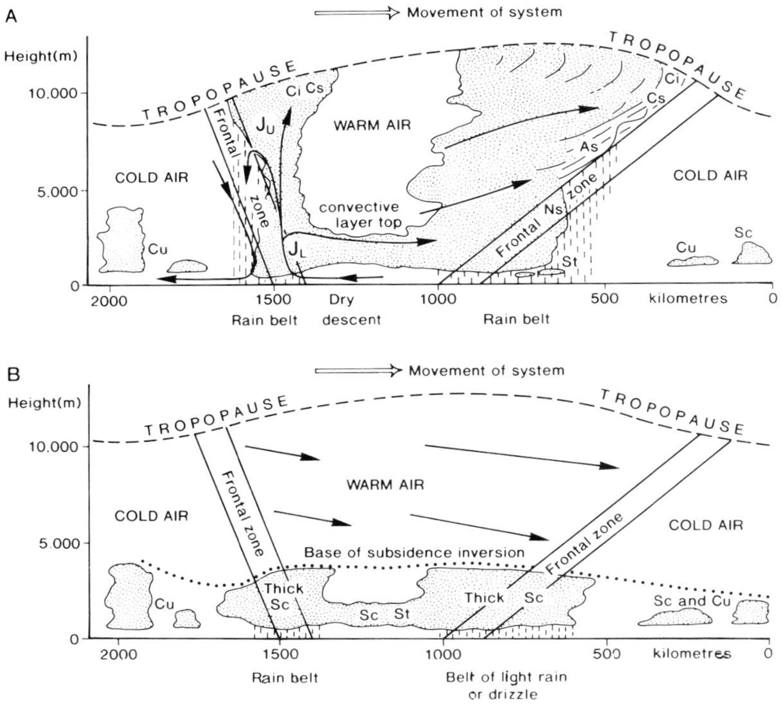

Figure 4.9 A: Cross-sectional model of a depression with ana-fronts where the air is rising relative to each frontal surface. Note that an ana-warm front may occur with a kata-cold front and vice versa. J_U and J_L show the locations of the upper and lower jet streams, respectively. B: Model of a depression with kata-fronts where the air is sinking relative to each frontal surface.

Sources: After Pedgley, 1962 *A Course in Elementary Meteorology*, and Bennetts *et al.* 1988 (*reproduced by permission of the Controller of Her Majesty's Stationery Office.*)

with the warm front is complicated by the fact that most fronts are not ana- or kata-fronts throughout their length or even at all levels in the troposphere. For this reason, radar is being used increasingly to determine by direct means the precise extent of rain belts and to detect differences in rainfall intensity.

Such studies have shown that most of the production and distribution of precipitation is controlled by a broad airflow a few hundred kilometres across and several kilometres deep, which flows parallel to and ahead of the surface cold front (see Figure 4.10).

Just ahead of the cold front the flow occurs as a low-level jet with winds up to 25–30 m s^{-1} at about 1 km above the surface. The air, which is warm and moist, rises over the warm front and turns south-eastward ahead of it as it merges with the mid-tropospheric flow (B in Figure 4.10). This flow has been termed a 'conveyor belt' (for large-scale heat and momentum transfer in mid-latitudes). Broad-scale convective (potential) instability is generated by the over-running of this low-level flow by potentially colder, drier air in the middle troposphere. Instability is released mainly in small-scale convection cells that are organized into clusters, known as mesoscale precipitation areas (MPAs). These MPAs are further arranged in bands, 50–100 km wide (see Figure 4.11). Ahead of the warm front, the bands are broadly parallel to the airflow in the rising section of the conveyor belt, whereas in the warm sector they parallel the cold front and the low-level jet. In some cases, cells and clusters are further arranged in bands within the warm sector and ahead of the warm front (see Figures 4.11 and 4.12). Precipitation from warm front rainbands often involves 'seeding' by ice particles falling from the upper cloud layers. It has been estimated that 20–35 per cent of the precipitation originates in the 'seeder' zone and the remainder in the lower clouds (see also Figure 2.30). Some of the cells and clusters are undoubtedly set up through orographic effects and these influences may extend well down-wind when the atmosphere is unstable.

Figure 4.10 Model of the large-scale flow and mesoscale precipitation structure of a partially-occluded depression typical of those affecting the British Isles. It shows the 'conveyor belt' (A) rising from 900 mb ahead of the cold front over the warm front. This is overlaid by a mid-tropospheric flow (B) of potentially colder air from behind the cold front. Most of the precipitation occurs in the well-defined region shown, within which it exhibits a cellular and banded structure.

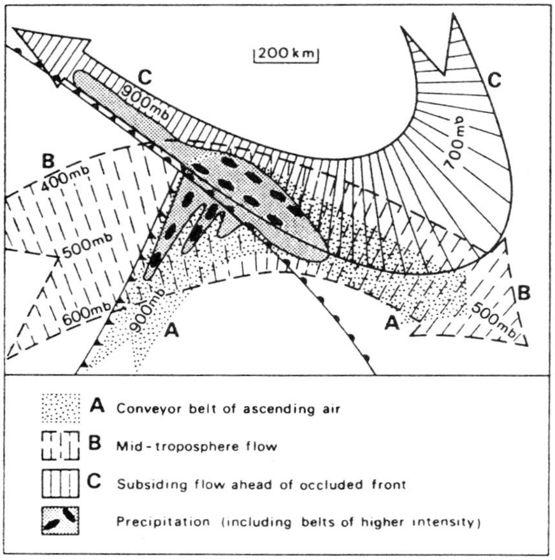

Source: After Harrold 1973.

2 The cold front

The weather conditions observed at cold fronts are equally variable, depending upon the stability of the warm sector air and the vertical motion relative to the frontal zone. The classical cold-front model is of the ana-type, and the cloud is usually cumulonimbus. Over the British Isles air in the warm sector is rarely unstable, so that nimbostratus occurs more frequently at the cold front (see Figure 4.9A). With the kata-cold front the cloud is generally stratocumulus, (see Figure 4.9B) and precipitation is light. With ana-cold fronts there are usually brief, heavy downpours sometimes accompanied by thunder. The steep slope of the cold front, roughly 2°, means that the bad weather is of shorter

Figure 4.11. Fronts and associated rain bands typical of a mature depression. The broken line X–Y shows the location of the cross-section given in Figure 4.12.

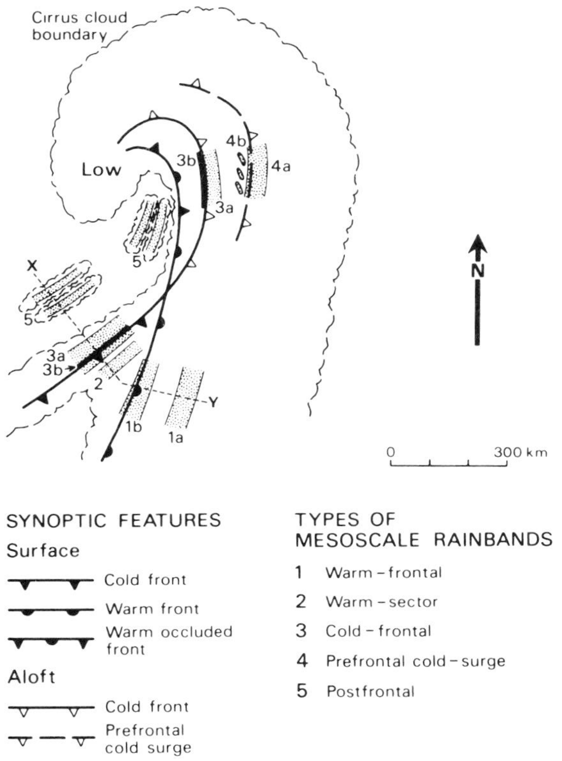

Source: After Hobbs; from Houze and Hobbs 1982.

duration than at the warm front. With the passage of the cold front the wind veers sharply, pressure begins to rise and temperature falls. The sky may clear very abruptly, even before the passage of the surface cold front in some cases, although with kata-cold fronts the changes are altogether more gradual.

3 The occlusion

Occlusions are classified as either *cold* or *warm*, the difference depending on the relative states of the cold-air masses lying in front and to the rear of the warm sector (see Figure 4.13). If the air is colder than the air following it then the occlusion is warm, but if the reverse is so (which is more likely over the British Isles) it is termed a cold occlusion. The air in advance of the depression is most likely to be coldest when depressions occlude over Europe in winter and very cold cP air is affecting the continent.

The line of the warm air wedge aloft is associated with a zone of layered cloud (similar to that found with a warm front) and often of precipitation. Hence its position is indicated separately on some weather maps and it is referred to by Canadian meteorologists as a *trowal* (trough of warm air aloft). The passage of an occluded front and trowal brings a change back to polar air-mass weather.

Figure 4.12 Cross-section along the line X–Y in Figure 4.11 showing cloud structures and rain bands. The vertical hatching represents rainfall location and intensity. Raindrop and ice-particle regions are shown, as are ice-particle concentrations and cloud liquid water content. Numbered belts refer to those shown in Figure 4.11. Scales are approximate.

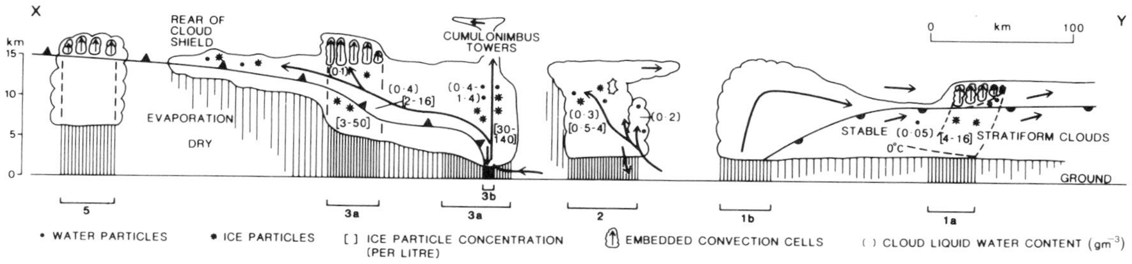

Source: After Hobbs and Matejka *et al.*; from Houze and Hobbs 1982.

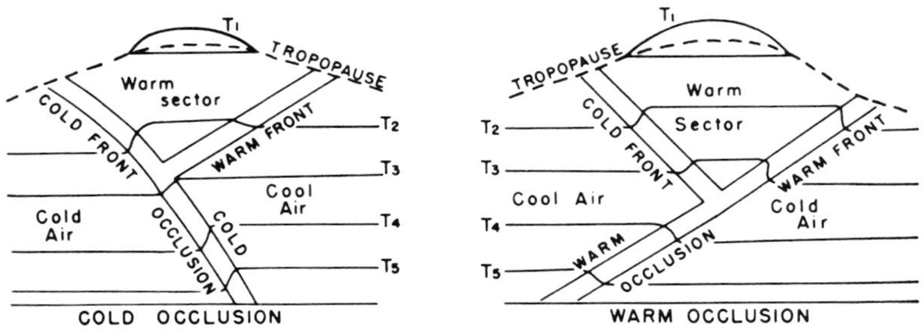

Figure 4.13 Schematic cross-sections of a cold and a warm occlusion.

Source: After Pedgley *A Course in Elementary Meteorology* (*reproduced by permission of the Controller of Her Majesty's Stationery Office*).

A different process occurs when there is interaction between a polar trough and the main polar front, giving rise to an instant occlusion. A warm conveyor belt on the polar front ascends as an upper tropospheric jet forming a stratiform cloud band (see Figure 4.14), while a low-level polar trough conveyor belt at right angles to it produces a convective cloud band and precipitation area poleward of the main polar front (see Plate 23) on the leading edge of the cold pool.

The occurrence of *frontolysis* (frontal decay) is not necessarily linked with occlusion, although it represents the final phase of a front's existence. Decay occurs when differences no longer exist between adjacent air masses. This may arise in four ways: through their mutual stagnation over a similar surface, as a result of both air masses moving on parallel tracks at the same speed, as a result of their movement in succession along the same track at the same speed, or by the system incorporating into itself air of the same temperature.

4 Frontal wave families

Observation has shown that frontal waves, or depressions, do not generally occur as separate units but in *families* of three or four (see Figure 4.15; Plate 33) with the depressions which succeed the original one forming as *secondaries* along the trailing edge of an extended cold front. Each new member follows a course which is south of its progenitor as the polar air pushes farther south to the rear of each depression in the series. Eventually the front trails far to the south and the cold polar air forms an extensive meridional wedge of high pressure terminating the sequence.

Another pattern of development may take place on the warm front, particularly at the point of occlusion, as a separate wave forms and runs ahead of the parent depression. This type of secondary is more likely with very cold (cA, mA or cP) air ahead of the warm front, and its formation is encouraged when the eastward

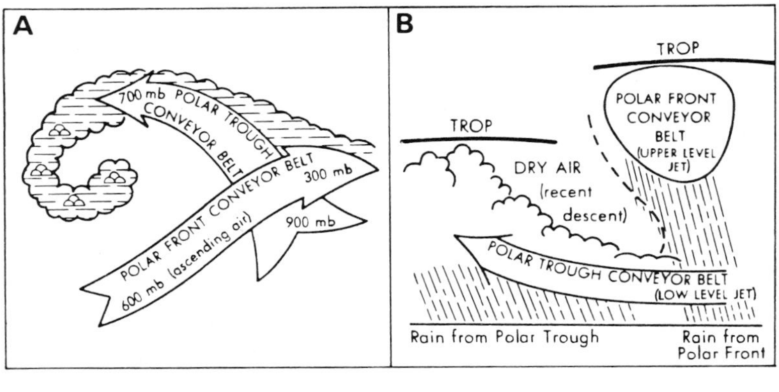

Figure 4.14 Schematic diagram showing the interaction between a polar trough and the polar front: (A) plan view and (B) vertical section along the polar trough axis.

Source: From Browning 1985 (*reproduced by permission of the Controller of Her Majesty's Stationery Office*).

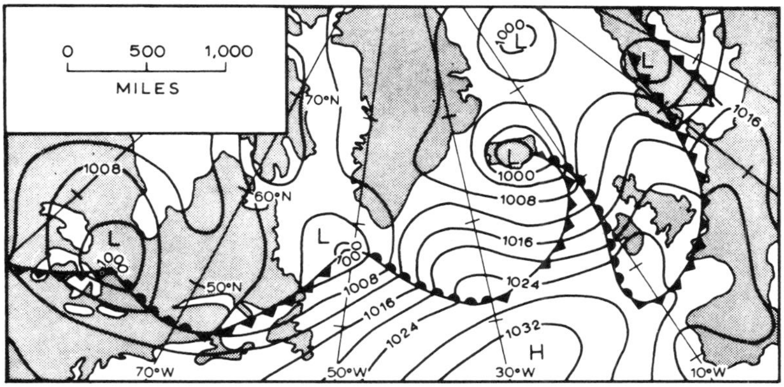

Figure 4.15 A depression family in the North Atlantic, 22 June 1954.

Source: After Taylor and Yates 1967 (*Crown Copyright Reserved*).

movement of the occlusion is barred by mountains. This situation often occurs when a primary depression is situated in the Davis Strait and a break-away wave forms south of Cape Farewell (the southern tip of Greenland), moving away eastwards. Analogous developments take place in the Skagerrak–Kattegat area when the occlusion is held up by the Scandinavian mountains.

E ZONES OF WAVE DEVELOPMENT AND FRONTOGENESIS

Fronts and associated depressions do not form everywhere and their development is restricted to well defined areas. Frontal formation in the temperate latitudes has been studied intensively for some years and knowledge of the general weather conditions to be expected is reasonably accurate. Not nearly so much is known about the nature of tropical fronts but the conditions of their formation and development are unlike those normally associated with higher-latitude fronts. World-wide air travel and the need of accurate forecasts for tropical routes is fast filling this gap. At the moment it seems that arctic and polar fronts are caused primarily by gross differences in air-mass characteristics, whereas tropical discontinuities within and between somewhat similar air masses are produced mainly by the nature of the large-scale air motion and especially by confluence within an airstream or between two air currents of different humidity.

The major zones of frontal wave development are naturally those areas which are most frequently baroclinic as a result of airstream confluence (see Figure 4.16). This is the case, for instance, off eastern Asia and eastern North America, especially in winter when there is a sharp temperature gradient between the snow-covered land and warm offshore currents. These zones are referred to respectively as the Pacific Polar and Atlantic Polar Fronts (see Figure 4.17). Their position is quite variable, but they show a general equatorward shift in winter, when the Atlantic Frontal Zone may extend into the Gulf of Mexico. In this area there is convergence of air masses of different stability between adjacent subtropical high-pressure cells (this frontal zone is sometimes misleadingly termed 'temperate'). Depressions developing here commonly move north-eastwards, sometimes following or amalgamating with others of the northern part of the Polar Front proper or of the Canadian Arctic Front. Frontal frequency remains high across the North Atlantic, but it decreases eastward in the North Pacific, perhaps owing to a less pronounced gradient of sea-surface temperature. Frontal activity is most common in the central North Pacific when the subtropical high is split into two cells with converging air currents between them.

Another section of the Polar Front, often referred to as the *Mediterranean Front*, is located

Figure 4.16 Mean pressure (mb) and surface winds for the world in January and July. The major frontal and convergence zones are shown as follows: Inter-Tropical Convergence Zone (ITCZ), South Pacific Convergence Zone (SPCZ), Monsoon Trough (MT), Zaïre Air Boundary (ZAB), Mediterranean Front (MF), northern and southern hemisphere Polar Fronts (PF), Arctic Fronts (AF) and Antarctic Fronts (AAF).

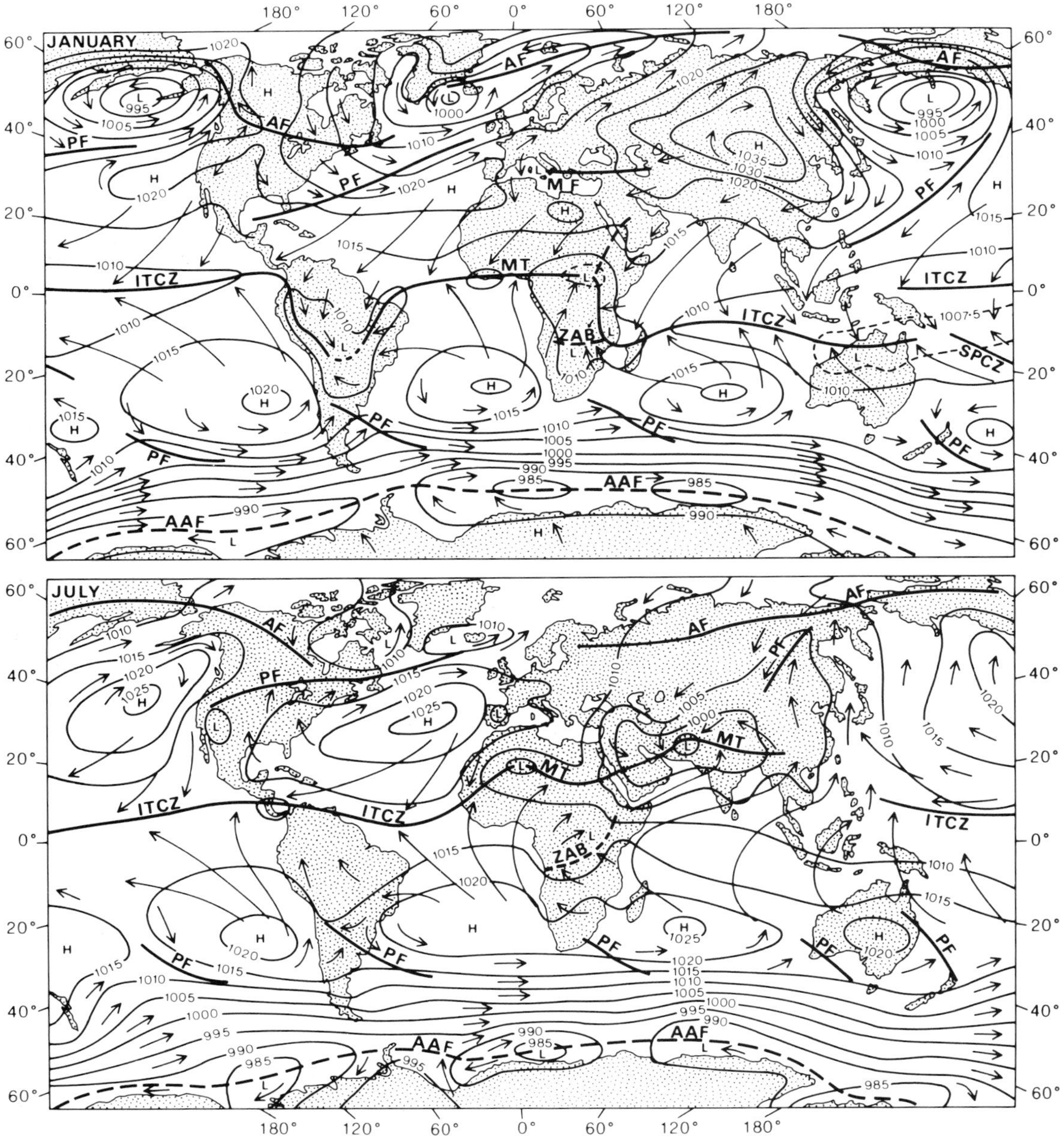

Source: Partly from Liljequist 1970.

Figure 4.17 The major northern hemisphere frontal zones in winter (A) and summer (B).

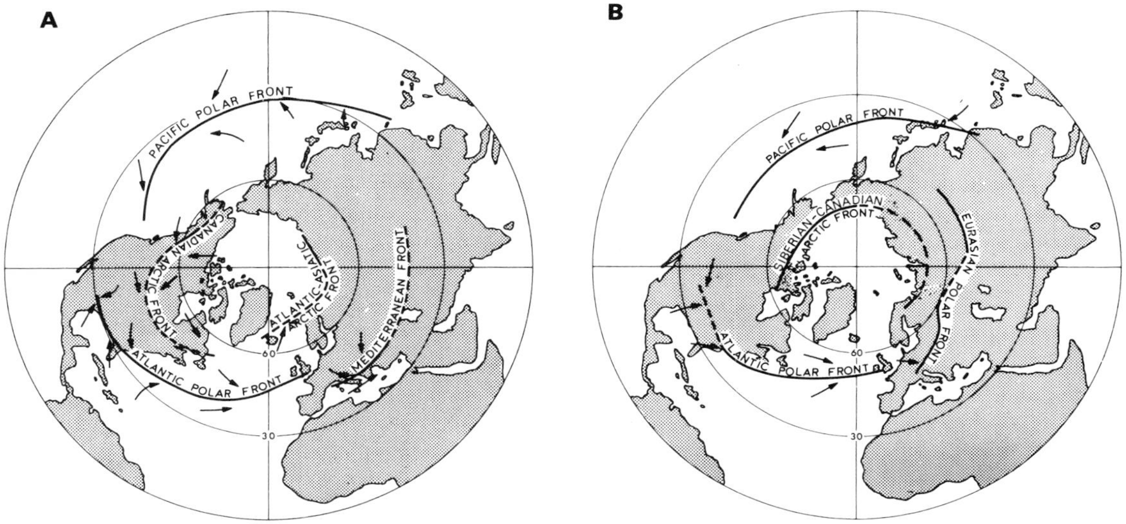

over the Mediterranean–Caspian Sea areas in winter. At intervals, fresh Atlantic mP air, or cool cP air from south-east Europe, converges with warmer air masses often of North African origin, over the Mediterranean Basin and initiates frontogenesis. In summer the area lies under the influence of the Azores subtropical high-pressure cell and the frontal zone is absent.

The summer locations of the Polar Front over the western Atlantic and Pacific are some 10° further north than in winter (see Figure 4.17) although the frontal zone is rather weak at this time of year. There is now a frontal zone over Eurasia and a corresponding one over middle North America. These reflect the general meridional temperature gradient and probably also the large-scale influence of orography on the general circulation (see F, this chapter).

In the southern hemisphere the Polar Front is on average about 45°S in January (summer) with branches spiralling poleward towards it from about 32°S off eastern South America and from 30°S, 150°W in the South Pacific (see Figure 4.18). In July (winter) there are two Polar Frontal Zones spiralling towards Antarctica from about 20°S; one starts over South America and the other at 170°W. They terminate some 4°–5° latitude further poleward than in summer. It is noteworthy that the southern hemisphere has more cyclonic activity in summer than does the northern hemisphere in its summer. This appears to be related to the seasonal strengthening of the meridional temperature gradient noted earlier (see p. 117).

The second major frontal zone is the Arctic Front, associated with the snow and ice margins of high latitudes (see Figure 4.17). In summer this zone is developed along the tundra margin in Siberia and North America. In winter over North America it is formed between cA (or cP) air and Pacific maritime air modified by crossing the Coast Ranges and Rockies. There is also a less pronounced arctic frontal zone in the North Atlantic–Norwegian Sea area, extending along the Siberian coast. A similar weak frontal zone is found in winter in the southern hemisphere. It is located at 65°–70°S near the edge of the antarctic pack-ice in the Pacific sector (see Figure 4.18), although rather few cyclones form there. Zones of airstream confluence in the southern hemisphere (cf. Figures 4.2 and 4.4) are less numerous and more persistent, particularly in coastal regions, than in the northern hemisphere.

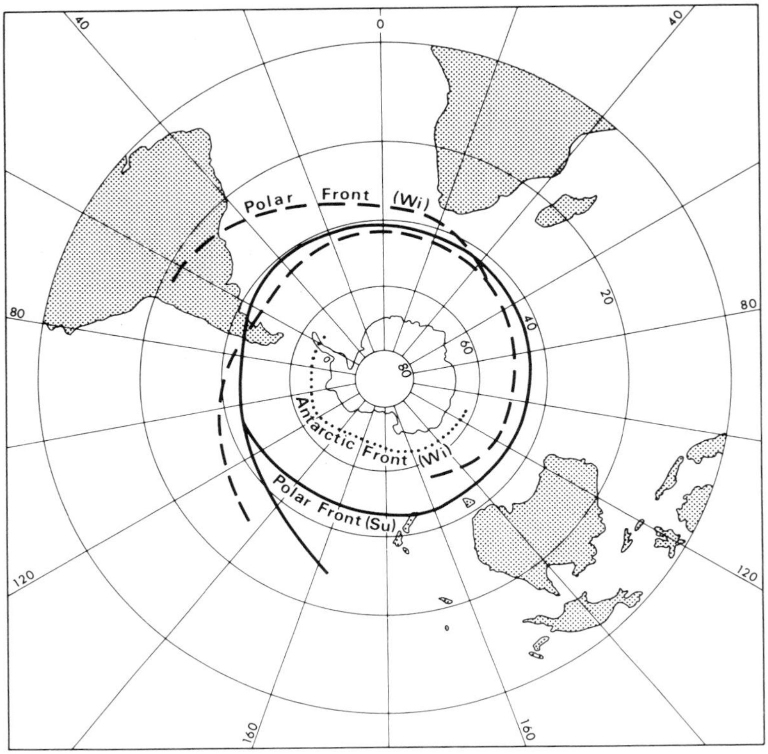

Figure 4.18 The major southern hemisphere frontal zones in winter (Wi) and summer (Su).

The principal tracks of depressions in the northern hemisphere in January are shown in Figure 4.19. The major ones reflect the primary frontal zones already discussed. In summer the Mediterranean route is absent and lows move across Siberia, but the other tracks are similar although generally more zonal at this season and located in higher latitudes (around 60°N).

Between the two hemispherical belts of subtropical high pressure there is a further major world convergence zone, the Inter-Tropical Convergence Zone (or ITCZ). Formerly this was referred to as the Inter-Tropical Front (ITF), but air-mass contrasts only occur in limited sectors. This zone moves seasonally north and south away from the equator, as the subtropical high-pressure cell activity alternates in opposite hemispheres. The contrast between the converging air masses obviously increases with the distance of the ITCZ from the equator, and the degree of difference in their characteristics is naturally associated with considerable variation in activity along the convergence zone. Activity is most intense in June–July over southern Asia and West Africa, when the contrast between the maritime and continental air masses which are involved is at a maximum. In these sectors the zone merits the term Inter-Tropical Front, although this does not imply that it behaves like a mid-latitude frontal zone. The nature of the ITCZ and its role in tropical weather are discussed in Chapter 6.

F SURFACE/UPPER-AIR RELATIONSHIPS AND THE FORMATION OF DEPRESSIONS

It has already been pointed out that a wave depression is associated with air-mass convergence, yet the barometric pressure at the centre of the low may decrease by 10–20 mb in 12–24 hours as the system intensifies. The

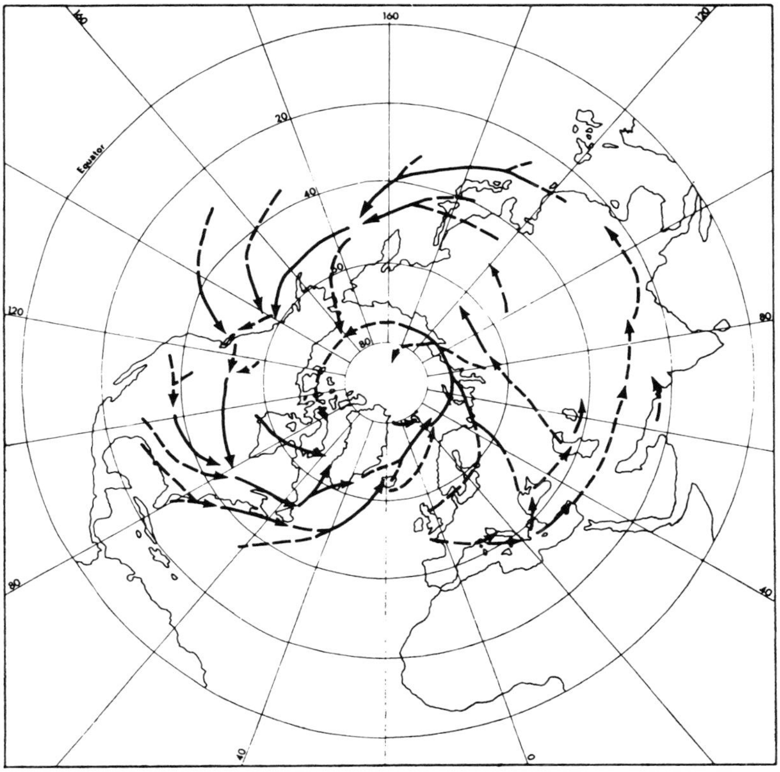

Figure 4.19 The principal northern hemisphere depression tracks in January. The full lines show major tracks, the dashed lines secondary ones which are less frequent and less well defined. The frequency of lows is a local maximum where arrow-heads end. An area of frequent cyclogenesis is indicated where a secondary track changes to a primary one or where two secondary tracks merge to form a primary.

Source: After Klein 1957.

explanation of this apparent discrepancy is that upper air divergence removes rising air more quickly than convergence at lower levels replaces it. The superimposition of a region of upper divergence over a frontal zone is the prime motivating force of *cyclogenesis* (i.e. depression formation).

The long (or *Rossby*) waves in the middle and upper troposphere, which were mentioned in Chapter 3, D.2, are particularly important in this respect, and it is worth considering first the reason why the hemispheric westerlies show this large-scale wave motion. The key to this problem lies in the rotation of the earth and the latitudinal variation of the Coriolis parameter (see Chapter 3, A.2). It can be shown that for large-scale motion the absolute vorticity about a vertical axis $(f + \zeta)$ tends to be conserved, i.e.

$$\frac{d(f+\zeta)}{dt} = 0$$

The symbol d/dt denotes a rate of change following the motion (a total differential). Consequently, if air moves poleward so that f increases, the cyclonic vorticity tends to decrease. The curvature thus becomes anticyclonic and the current returns towards lower latitudes. If the air moves equatorward of its original latitude f tends to decrease (see Figure 4.20), requiring ζ to increase, and the resulting cyclonic curvature again deflects the current polewards. In this manner large-scale flow tends to oscillate in a wave pattern.

Rossby related the motion of these waves to their wavelength (L) and the speed of the zonal current (u). The speed of the wave (or phase speed, c), is:

$$c = u - \beta\left(\frac{L}{2\pi}\right)^2$$

where $\beta = \partial f/\partial y$, i.e. the variation of the

Figure 4.20 A schematic illustration of the mechanism of long-wave development in the tropospheric westerlies.

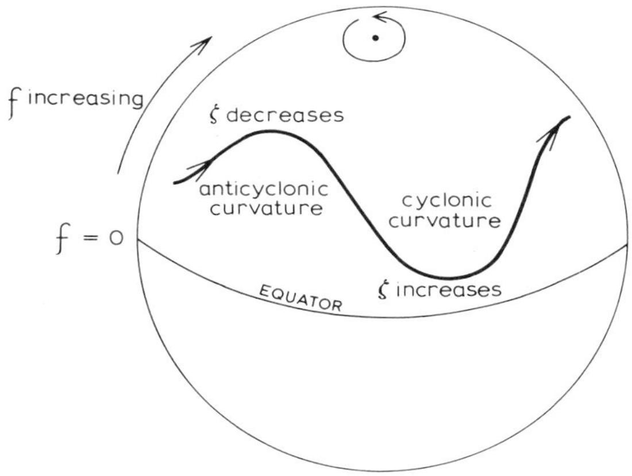

Coriolis parameter with latitude (a local, partial differential). For stationary waves, where $c = 0$, $L = 2\pi\sqrt{\mu/\beta}$. At 45° latitude this stationary wavelength is 3120 km for a zonal velocity of 4 m s^{-1}, increasing to 5400 km at 12 m s^{-1}. The wavelengths, at 60° latitude for zonal currents of 4 and 12 m s^{-1} are, respectively, 3170 and 6430 km. Long waves tend to remain stationary, or even to move westward against the current, so that $c \leqslant 0$. Shorter waves travel eastward with a speed close to that of the zonal current and tend to be steered by the quasi-stationary long waves.

The latitudinal circumference limits the circumpolar westerly flow to between three and six major Rossby waves, and these affect the formation and movement of surface depressions. It has been pointed out that the main stationary waves tend to be located about 70°W and 150°E in response to the influence on the atmospheric circulation of orographic barriers, such as the Rocky Mountains and the Tibetan plateau, and of heat sources. On the eastern limb of troughs in the upper westerlies of the northern hemisphere the flow is normally divergent, since the gradient wind is subgeostrophic in the trough but supergeostrophic in the ridge (see Chapter 3, A.4). Thus the sector ahead of an upper trough is a very favourable location for a surface depression to form or deepen, and it will be noted that the mean upper troughs are significantly positioned just west of the Atlantic and Pacific Polar Front Zones in winter.

With these ideas in mind we may now consider further the three-dimensional nature of depression development and the important links existing between upper and lower tropospheric flow. The basic theory relates to the vorticity equation which states that, for frictionless horizontal motion, the rate of change of the vertical component of absolute vorticity (dQ/dt or $d(f + \zeta)/dt$) is proportional to air-mass convergence ($-D$, i.e. negative divergence):

$$\frac{dQ}{dt} = -DQ \quad \text{or} \quad D = -\frac{1}{Q}\frac{dQ}{dt}$$

The relationship implies that a converging (diverging) air column has increasing (decreasing) absolute vorticity. The conservation of vorticity equation, which we have already discussed, is in fact a special case of this relationship.

In the sector ahead of an upper trough the decreasing cyclonic vorticity causes divergence (i.e. D positive), since the change in ζ outweighs that in f, thereby favouring surface convergence and low-level cyclonic vorticity. Once the surface cyclonic circulation has become established vorticity production is increased through the effects of thermal advection. Poleward transport of warm air in the warm sector and the eastward advance of the cold upper trough act to sharpen the baroclinic zone, strengthening the upper jet stream through the thermal wind mechanism (see p. 114). The vertical relationship between jet stream and front has already been shown (see Figure 3.22); a model depression sequence is demonstrated in plan view in Figure 4.21 and in three dimensions in Figure 4.22. The actual relationship may depart from this idealized case, although the jet is commonly located in the cold air as illustrated by the synoptic example of Figure 4.23. Velocity maxima (core zones) occur along the jet stream,

Figure 4.21 Stages in the development of an occluding depression. *Above:* Upper winds and Polar Jet Stream in relation to surface fronts, precipitation areas (dark stipple), and cloud (lighter stipple). *Below:* Cross-sections along the lines marked X–Y.

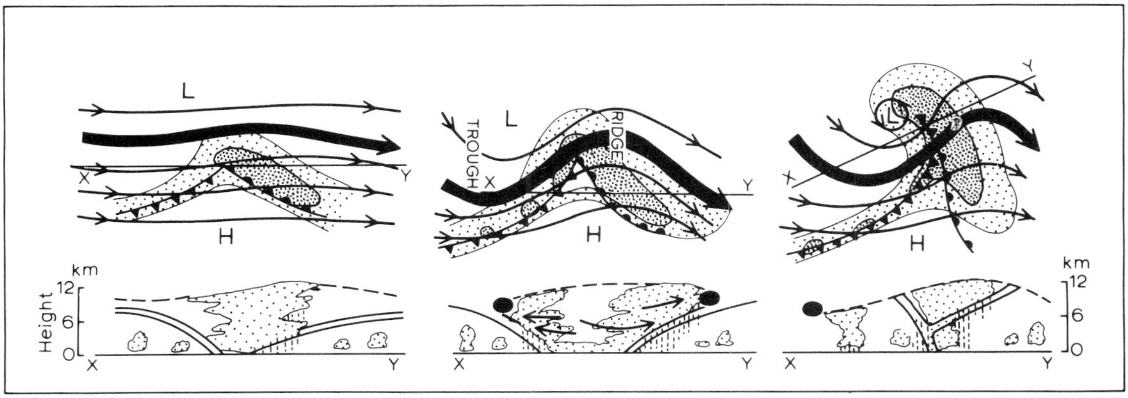

Source: After Flohn.

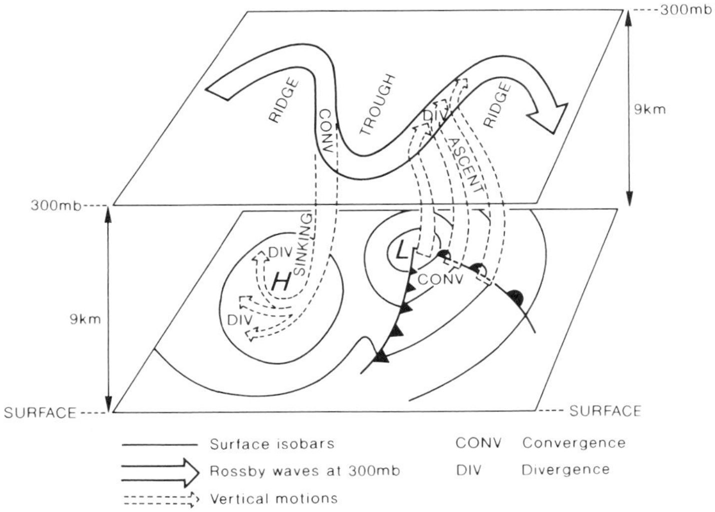

Figure 4.22 Schematic representation of the relationship between surface pressure and frontal systems and the location of troughs and ridges in the Rossby waves in the upper airflow.

Source: From Musk 1988.

as shown schematically in Figure 4.24, and the distribution of vertical motion upstream and downstream of these cores is quite different. For the jet entrance (i.e. upstream of the core) in the ridge divergence causes lower-level air to rise on the equatorward (i.e. right) side of the jet, whereas in the exit zone (downstream of the core) in the trough ascent is on the poleward side. The second depression in Figure 4.24 is moving eastward towards the area of maximum cyclogenetic tendency (upper divergence).

Figure 4.25 shows how precipitation is often more related to the position of the jet stream than to that of surface fronts; maximum precipitation areas are in the right entrance sector of the jet core. This vertical motion pattern is also of basic importance in the initial deepening stage of the depression. If the upper-air pattern

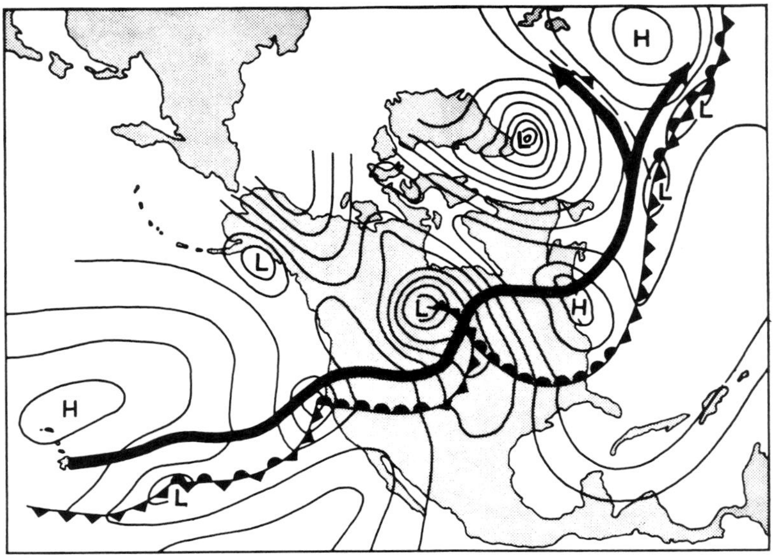

Figure 4.23 A typical depression family and its relationship with the jet stream. The thin lines are sea-level isobars.

Source: After Vederman 1954.

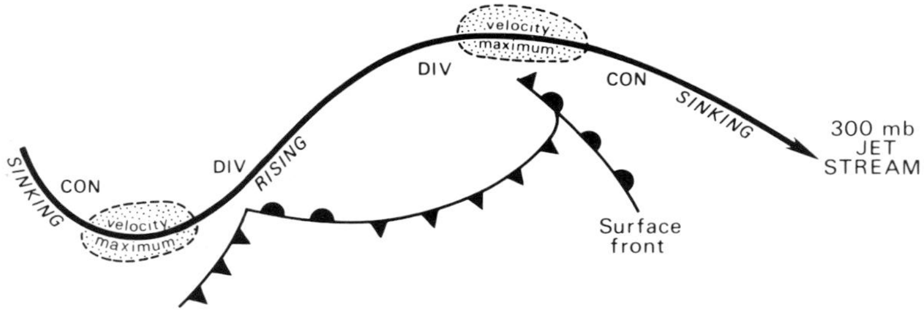

Figure 4.24 Model of the jet stream and surface fronts, showing zones of upper tropospheric divergence and convergence. Divergence covergence in the unlabelled quadrants of the jet cores where shear and curvature effects cancel out.

is unfavourable (e.g. beneath left entrance and right exit zones, where there is convergence) the depression will fill. Note that to the rear of the second depression in Figure 4.24 upper convergence will encourage a polar outbreak through subsidence.

The development of a depression can also be considered in terms of energy transfers. A cyclone requires the conversion of potential into kinetic energy and this is achieved by the upward (and poleward) motion of *warm* air. The rising warm air is driven by the vertical wind shear and by the superimposition of upper tropospheric divergence over a baroclinic zone. Intensification of this zone further strengthens the upper winds. The upper divergence allows surface convergence and pressure fall to occur simultaneously. Modern theory relegates the fronts to a quite subordinate role. They develop within depressions as narrow zones of intensified ascent, probably through the effects of cloud formation.

Recent research has identified a category of mid-latitude cyclones that develop and intensify rapidly, acquiring characteristics that resemble tropical hurricanes. These have been termed bombs in view of their explosive rate of deepening; pressure falls of at least 24 mb/24 hr are observed. For example, the *Q.E. II* storm, which battered the ocean liner *Queen Elizabeth II* off New York on 10 September 1978, developed a central pressure below 950 mb with hurricane-force winds and an eye-like storm centre within 24 hours (see Chapter 6, C.2).

Figure 4.25 The relations between surface fronts and isobars, surface precipitation (⩽ 25 mm vertical hatching; > 25 mm cross-hatching), and jet streams (wind speeds in excess of about 100 mph (45 m s^{-1}) occur within the dashed lines) over the United States on (A) 20 September 1958 and (B) 21 September 1958. This illustrates how the surface precipitation area is related more to the position of the jets than to that of the surface fronts. The air over the south-central United States was close to saturation, whereas that associated with the northern jet and the maritime front was much less moist.

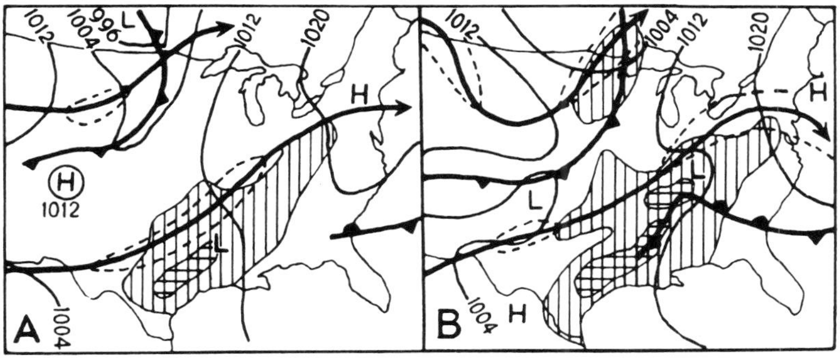

Source: After Richter and Dahl 1958.

These systems are observed mainly during the cold season off the east coast of the United States, off Japan, and over parts of the central and north-eastern North Pacific, in association with major baroclinic zones and close to strong gradients of sea-surface temperature. Synoptic analyses indicate that explosive cyclogenesis is favoured by an unstable lower troposphere and is often located downstream of a travelling 500-mb-level trough. They are characterized by strong vertical motion, associated with a sharply defined level of non-divergence near 500 mb, and large-scale release of latent heat. Wind maxima in the upper troposphere, organized as jet streaks, serve to amplify the lower-level instability and upward motion. Further studies have revealed that *average* cyclonic deepening rates over the North Atlantic and North Pacific are about 10 mb/24 hr, or three times greater than over the continental United States (3 mb/24 hr). Hence, it is suggested that explosive cyclogenesis may only represent a more intense version of typical maritime cyclone development.

The movement of depressions is determined essentially by the upper westerlies and, as a rule of thumb, a depression centre travels at about 70 per cent of the surface geostrophic wind speed in the warm sector. Records for the United States indicate that the average speed of depressions is 32 km h^{-1} (20 mph) in summer and 48 km h^{-1} (30 mph) in winter. The higher speed in winter reflects the stronger westerly flow in response to a greater meridional temperature gradient. Shallow depressions are mainly steered by the direction of the thermal wind in the warm sector and hence their path closely follows that of the upper jet stream (see Chapter 3, D.3). Deep depressions may greatly distort the thermal pattern, however, as a result of northward transport of warm air and the southward transport of cold air. In such cases the depression usually becomes slow-moving. The movement of a depression may be additionally guided by energy sources such as a warm sea surface, which generates cyclonic vorticity, or by mountain barriers. The depression may cross obstacles, such as the Rocky Mountains or the

Greenland Ice Sheet, as an upper low or trough and subsequently redevelop, aided by the lee-effects of the barrier or by fresh injections or contrasting air masses.

The location and intensity of storm tracks appears to be critically influenced by ocean surface temperatures. Investigations for the North Pacific and the North Atlantic demonstrate interactions with the atmospheric circulation on a near-hemispheric scale. Figure 4.26B suggests, for example, that an extensive relatively warm surface in the north-central Pacific in the winter of 1971–2 caused a northward displacement of the westerly jet stream together with a compensating southward displacement over the western United States, bringing in cold air there. This pattern contrasts with that observed during the 1960s (see Figure 4.26A) when a persistent cold anomaly in the central Pacific, with warmer water to the east, led to frequent storm development in the intervening zone of strong temperature gradient. The associated upper airflow produced a ridge over western North America with warm winters in California and Oregon. Models of the global atmospheric circulation support the view that persistent anomalies of sea-surface temperature exert an important control on local and large-scale weather conditions.

G NON-FRONTAL DEPRESSIONS

Not all depressions originate as frontal waves. Tropical depressions are indeed mainly non-frontal and these are considered in Chapter 6. In middle and high latitudes four types which develop in distinctly different situations are of particular importance and interest: the lee depression, the thermal low, the polar air depression, and the cold low.

1 The lee depression

Westerly airflow that is forced over a north–south mountain barrier undergoes vertical contraction over the ridge and expansion on the lee side. This vertical movement creates compensating lateral expansion and contraction, respectively. Hence there is a tendency for divergence and anticyclonic curvature over the crest, convergence and cyclonic curvature in the lee of the barrier. Wave troughs may be set up in this way on the lee side of low hills (see Figure 3.11) as well as major mountain chains. The airflow characteristics and the size of the barrier determine whether or not a closed low-pressure system actually develops. Such depressions, which at least initially tend to remain 'anchored' by the barrier, are frequent in winter to the south of the Alps when the mountains block the low-level flow of north-westerly airstreams. Fronts may occur in these depressions but it is important to recognize that the low does not form as a wave along a frontal zone.

2 The thermal low

These lows occur almost exclusively in summer, resulting from intense daytime heating of continental areas. Figure 3.16C illustrates their vertical structure. The most impressive examples are the summer low-pressure cells over Arabia, the northern part of the Indian sub-continent and Arizona. The Iberian peninsula is another region commonly affected by such lows. The weather accompanying them is usually hot and dry, but if sufficient moisture is present the instability caused by heating may lead to showers and thunderstorms. Thermal lows normally disappear at night when the heat source is cut off, but in fact those of India and Arizona persist.

3 Polar air depressions

Polar air depressions are a loosely defined class of mesoscale to subsynoptic-scale systems (a few hundred kilometres across) with a lifetime of 1–2 days. On satellite imagery they appear as a cloud spiral with one or several cloud bands, as a comma cloud (see Plate 24), or as a swirl in cumulus cloud streets. They develop mainly in winter months when unstable mP or mA air

Figure 4.26 Generalized relationships between ocean-surface temperatures, jet-stream tracks, storm-development zones and land temperatures over the North Pacific and North America during (A) average winter conditions in the 1960s, and (B) the winter of 1971–2, as determined by J. Namias.

A

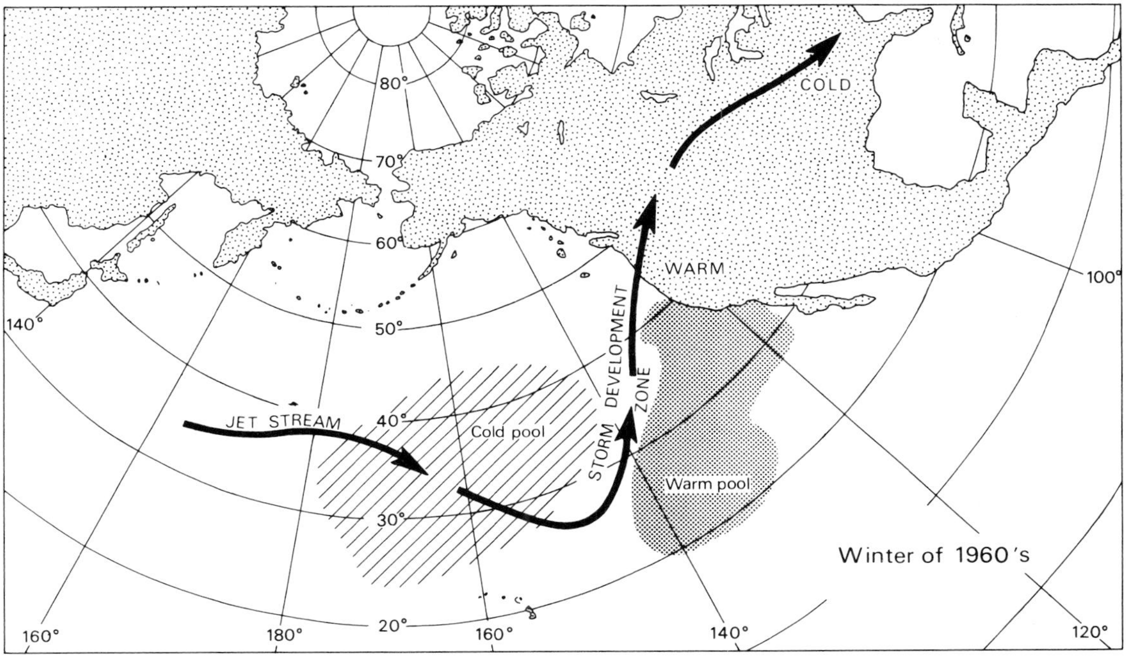

B

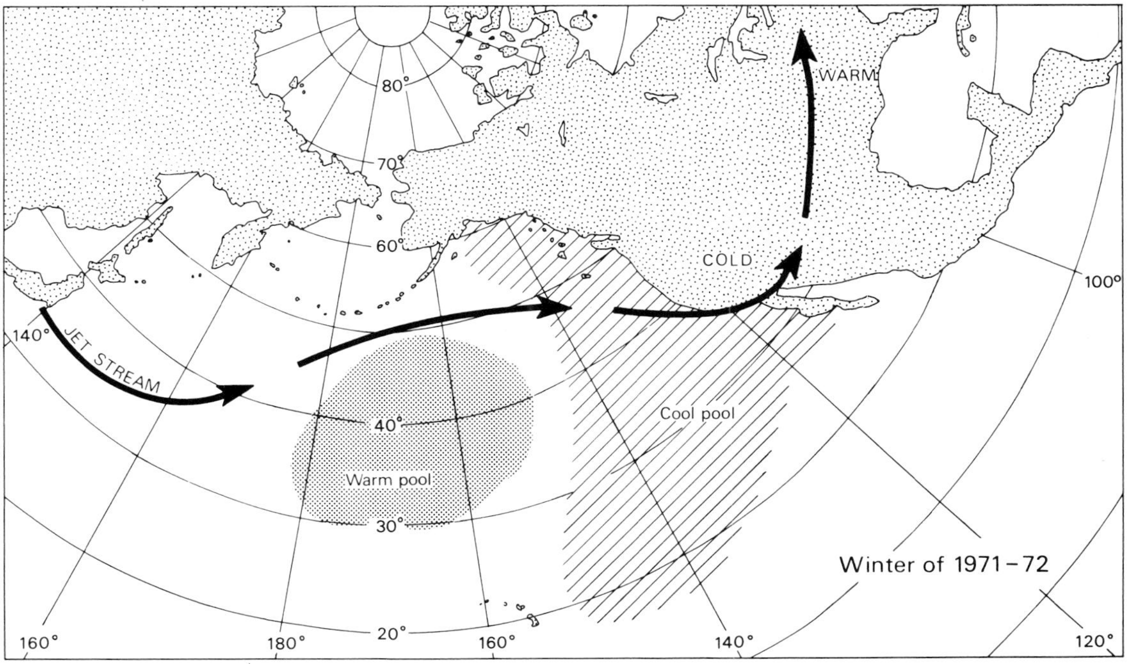

Source: After Wick 1973.

currents stream equatorward along the eastern side of a north–south ridge of high pressure, commonly in the rear of an occluding primary depression. They usually form within a baroclinic zone, e.g. near sea-ice margins where there are strong sea-surface temperature gradients, and their development may be stimulated by an initial upper-level disturbance.

In the northern hemisphere, the comma cloud type (which is mainly a cold core disturbance of the middle troposphere) is more common over the North Pacific, while the spiral-form polar low occurs more often in the Norwegian Sea. The latter is a low-level warm core disturbance which may have a closed cyclonic circulation up to about 800 mb or may consist simply of one or more troughs embedded in the polar airflow. A key feature is the presence of an ascending, moist, south-westerly flow *relative* to the low centre. This organization accentuates the general instability of the cold airstream to give considerable precipitation, often as snow. Heat input to the cold air from the sea continues by night and day so that in exposed coastal districts showers may occur at any time.

4 The cold low

The cold low (or *cold pool*) is usually most evident in the circulation and temperature fields of the middle troposphere. Characteristically it displays symmetrical isotherms about the low centre. Surface charts may show little or no sign of these persistent systems which are frequent over north-eastern North America and north-eastern Siberia. They probably form as the result of strong vertical motion and adiabatic cooling in occluding baroclinic lows along the arctic coastal margins. Such lows are especially important during the arctic winter in that they bring large amounts of medium and high cloud which offset radiational cooling of the surface. Otherwise they usually cause no 'weather' in the Arctic during this season. It is important to emphasize that tropospheric cold lows may be linked with either low- or high-pressure cells at the surface.

In middle latitudes cold lows may also form during periods of low-index circulation pattern (see Figure 3.41) by the cutting-off of polar air from the main body of cold air to the north (these are sometimes referred to as *cut-off lows*). This gives rise to weather of polar air-mass type, although rather weak fronts may also be present. Such lows are commonly slow-moving and give persistent unsettled weather with thunder in summer. Heavy precipitation over Colorado in spring and autumn is often associated with cold lows.

H MESOSCALE CONVECTIVE SYSTEMS

Mesoscale convective systems are intermediate in size and life-span between synoptic disturbances and individual cumulonimbus cells (see Figure 4.27). They occur in both middle lati-

Figure 4.27 The spatial scale and life-span of mesoscale and other meteorological systems.

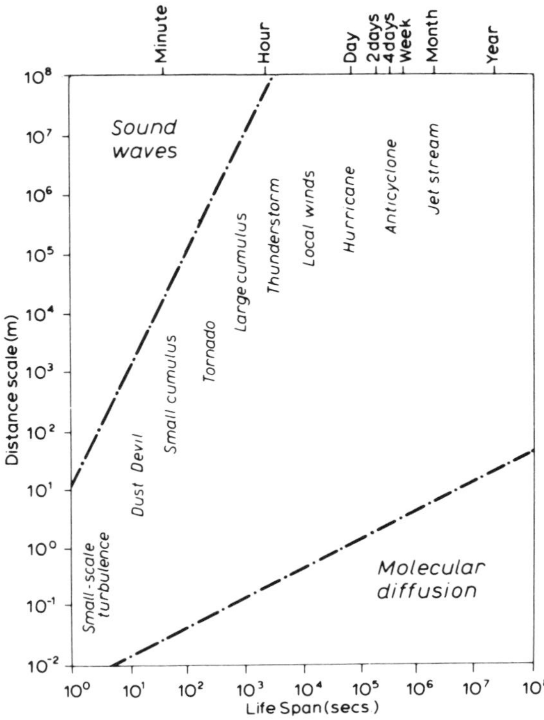

Source: After Smagorinsky 1974.

tudes and the tropics as either nearly circular clusters of convective cells or linear squall lines. The *squall line* consists of a narrow line of thunderstorm cells which may extend for hundreds of kilometres. It is marked by a sharp veer of wind direction and very gusty conditions. The squall line often occurs ahead of a kata-cold front maintained either as a self-propagating disturbance or by thunderstorm downdraughts. It may form a pseudo-cold front between rain-cooled air and a rainless zone within the same air mass. In frontal cyclones, cold air in the rear of the depression may overrun air in the warm sector. The intrusion of this nose of cold air sets up great instability and the subsiding cold wedge tends to act as a scoop forcing up the slower-moving warm air (see Plate 14).

Figure 4.28 shows the movement of clusters of convective cells, each cell about 1 km in diameter, as they crossed southern Britain with a cold front. Each individual cell may be short-lived, but cell clusters may persist for hours, strengthening or weakening due to orographic and other factors.

Figure 4.29 shows that the *relative* motion of the warm air is towards the squall line. Such conditions generate severe frontal thunderstorms like that which struck Wokingham, England, in September 1959. This moved from the south-west at about 20 m s^{-1}, steered by strong south-westerly flow aloft. The cold air subsided from high levels as a violent squall and the updraught ahead of this produced an intense hailstorm. The hailstones grow by accretion in the upper part of the updraught, where speeds in excess of 50 m s^{-1} are not uncommon, are blown ahead of the storm by strong upper winds, and begin to fall. This causes surface melting but the stone is caught up again by the advancing squall line and re-ascends. The melted surface freezes, giving glazed ice as the stone is carried above the freezing level and further growth occurs by the collection of super-cooled droplets (see also Chapter 2, pp. 75 and 00).

The *mesoscale convective complex* is common over the central United States in spring and summer, where it brings widespread severe weather. It develops from initially isolated cumulonimbus cells. The sequence, illustrated in Figure 4.30 appears to be as follows. As rain falls from the thunderstorm clouds, evaporative cooling of the air beneath the cloud bases sets up cold downdraughts (see Figure 2.22) and when these become sufficiently extensive they create a local high pressure of a few millibars' intensity. The downdraughts trigger the ascent of displaced warm air and a general warming of the middle troposphere results from latent heat

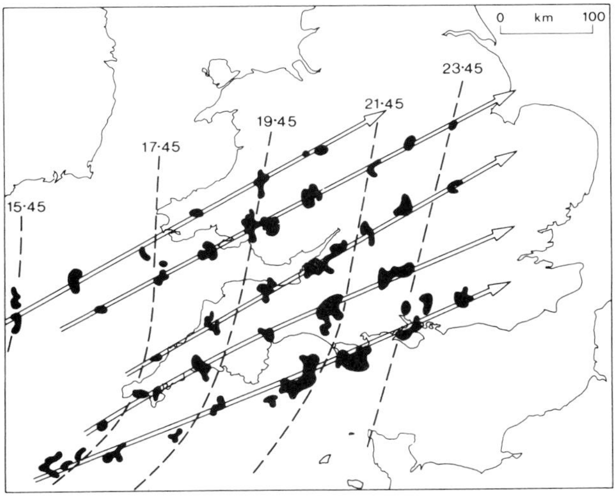

Figure 4.28 Successive positions of individual clusters of middle-tropospheric convective cells moving across southern Britain at about 50 km h^{-1} with a cold front. Cell location and intensity were determined by radar (see Plate 15).

Source: After Browning 1980.

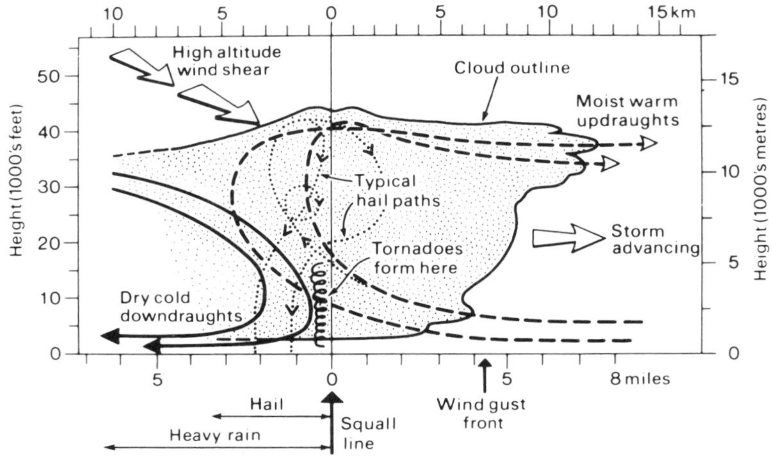

Figure 4.29 Thunder cell structure with hail and tornado formation.

Source: After Hindley 1977.

Figure 4.30 Stages in the development of a mesoscale thunderstorm system: growth (A), maturity (B), and decay (C). The large arrows show the directions of movement of parts of the system, and the small arrows the surface winds. Areas of heavy and light precipitation are delimited.

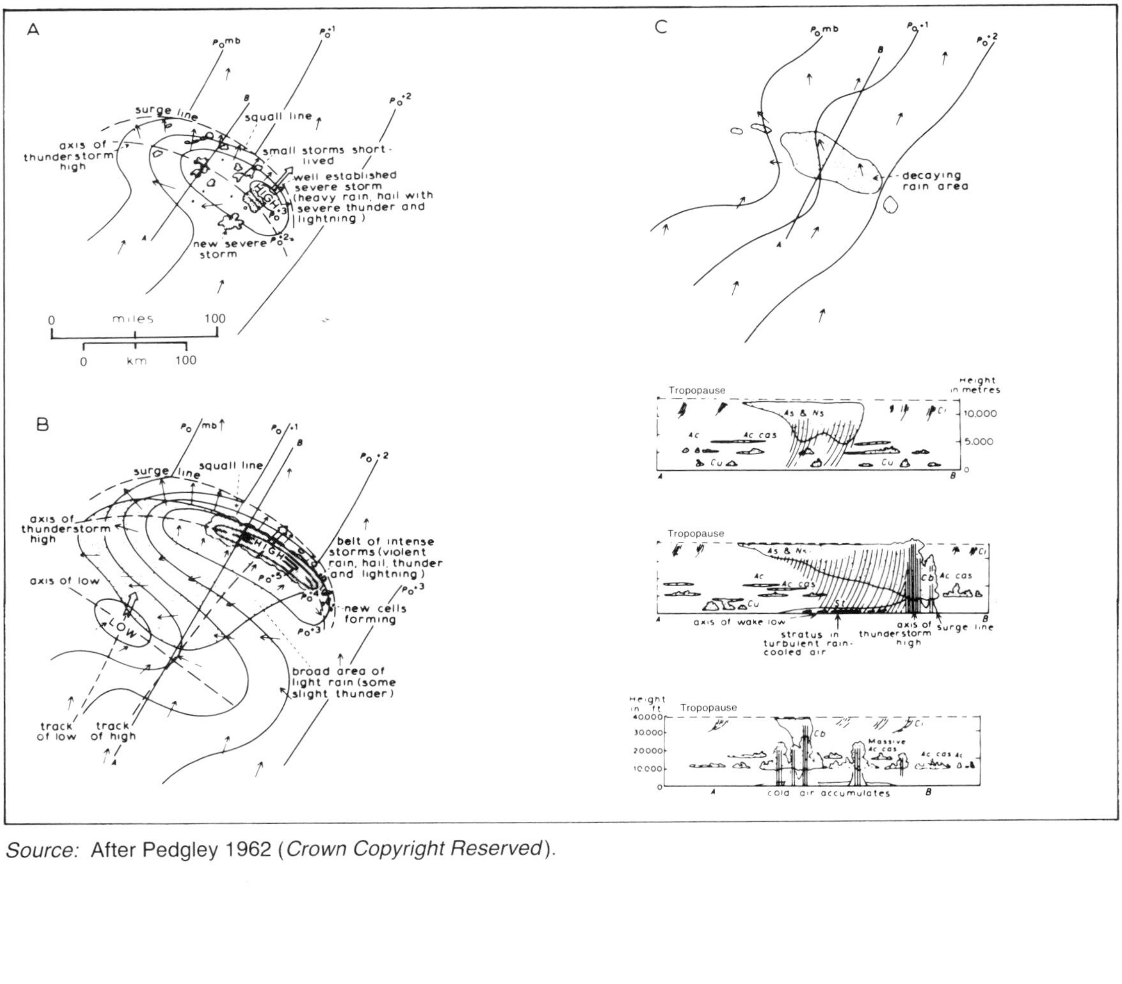

Source: After Pedgley 1962 (*Crown Copyright Reserved*).

release. Inflow develops towards this warm region, above the cold outflow, causing additional convergence of moist unstable air. In at least some cases, this inflow is provided by a low-level jet (see Figure 4.34). As individual cells become organized in a cluster along the leading edge of the surface high, new cells tend to form on the right flank (see Figure 4.30) through interaction of cold downdraughts with the surrounding air. Through this process and the decay of older cells on the left flank, the storm system tends to move 10°–20° to the right of the mid-tropospheric wind direction. As the thunderstorm high intensifies, a 'wake low', associated with clearing weather, develops to the rear of it. The system is now producing violent winds, and intense downpours of rain and hail accompanied by thunder. During the triggering of new cells tornadoes may form, as discussed later in this section. As the complex reaches maturity, during the evening and night hours over the Great Plains of the United States, the mesoscale circulation is capped by an extensive (> 100,000 km^2) cold upper-cloud shield, readily identified on infra-red satellite images. Statistics for forty-three systems over the Great Plains in 1978 showed that the systems lasted on average 12 hours, with initial mesoscale organization occurring in the early evening (1800–1900 LST) and maximum extent 7 hours later. During their life cycle, systems may travel from the Colorado–Kansas border to the Mississippi River or the Great Lakes, or from the Missouri–Mississippi river valley to the east coast. The complex usually decays when synoptic-scale features inhibit its self-propagation. The production of cold air is shut off when new convection ceases, so that the meso-high and -low are steadily weakened and the rainfall becomes light and sporadic, eventu-

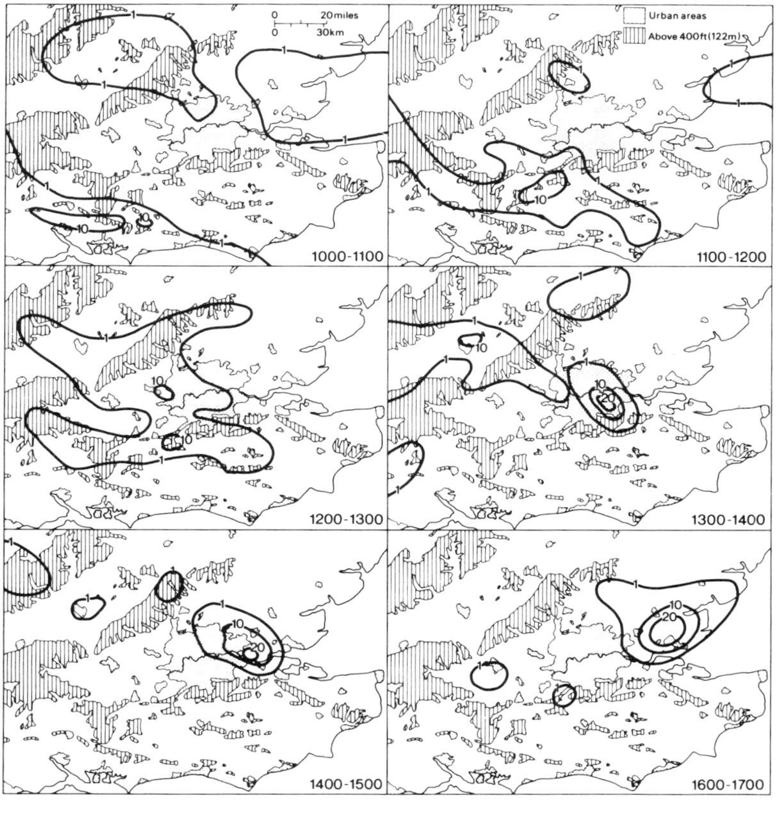

Figure 4.31 Hourly rainfalls (mm) over south-east England during the periods 1000–1500 hours and 1600–1700 hours on 14 September 1968.

Source: After Jackson 1977.

ally stopping altogether.

On 14 September 1968 a belt of intense thunderstorms moved northwards into south-east England. Strong convergence of air from the south and north-east occurred along this trough in association with jet streams aloft, allowing intense surface convection of hot moist air. The intense rainstorms during the period 1000–1700 hours (see Figure 4.31) were the first phase of a 2-day period of instability, yielding 200 mm (7.87 in) of rainfall at Tilbury just east of London, which proved to be the outstanding rainfall event of the century in south-east England.

In regions of great potential instability and with strong vertical wind shear supercell thunderstorms may develop (see Figure 4.32). These are about the same size as thundercell clusters but are dominated by one giant updraught and localized strong downdraughts. They are often associated with the production of large hailstones and tornadoes.

Tornadoes, which may develop from such squall-line thunderstorms, are common over the Great Plains of the United States, especially in spring and early summer (see Figure 4.33 and Plate 26). During this period cold, dry air from the high plateaux may override maritime tropical air (see note 3). Subsidence beneath the upper troposphere westerly jet (see Figure 4.34) caps the low-level moist air forming an inversion at about 1500–2000 m. The moist air is extended northward by a low-level southerly jet (cf. p. 154) and through continuing advection the air beneath the inversion becomes progressively more warm and moist. Eventually the general convergence and ascent in the depression trigger the potential instability of the air generating large cumulus clouds which penetrate the inversion. The convective trigger is sometimes provided by the approach of the cold front towards the western edge of the moist tongue although tornadoes also occur in association with tropical cyclones (see p. 237) and in other synoptic situations if the necessary vertical contrast is present in the temperature, moisture and wind fields.

The exact tornado mechanism is still not fully understood because of the observational problems involved. Tornadoes generally develop in a meso-low on the periphery of severe rotating thunderstorm systems where horizontal convergence increases the vorticity and rising air is replenished by moist air from progressively

Figure 4.32 A supercell thunderstorm.

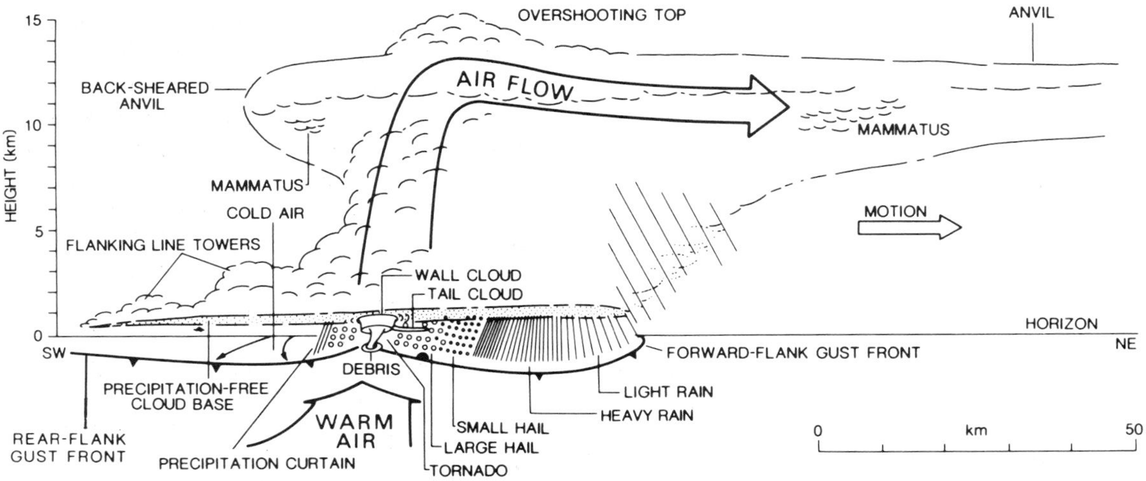

Source: After the National Severe Storms Laboratory, USA, and H. Bluestein; from Houze and Hobbs 1982 (*copyright ©Academic Press. Reproduced by permission*).

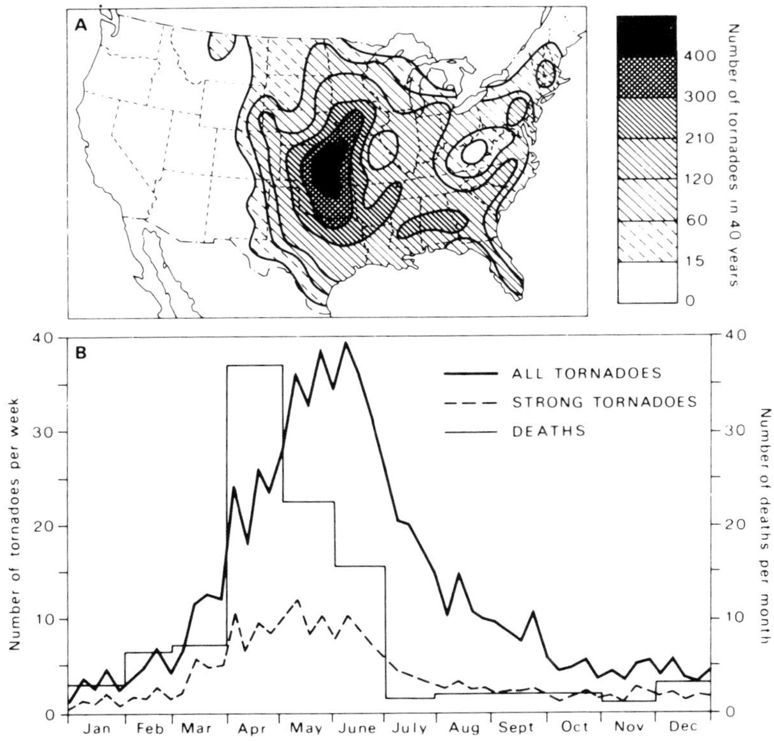

Figure 4.33 Incidence of tornadoes in the United States. A: Number of tornadoes reported over a 40-year period. B: Weekly average number of tornadoes (1950–82) and monthly averages of resulting deaths (1953–82).

Sources: A: From Hindley 1977 (this first appeared in *New Scientist*, London, the weekly review of science and technology). B: From 'The tornado' by J. T. Snow (*copyright © 1984 by Scientific American Inc. All rights reserved*).

Figure 4.34 The synoptic conditions favouring severe storms and tornadoes over the Great Plains.

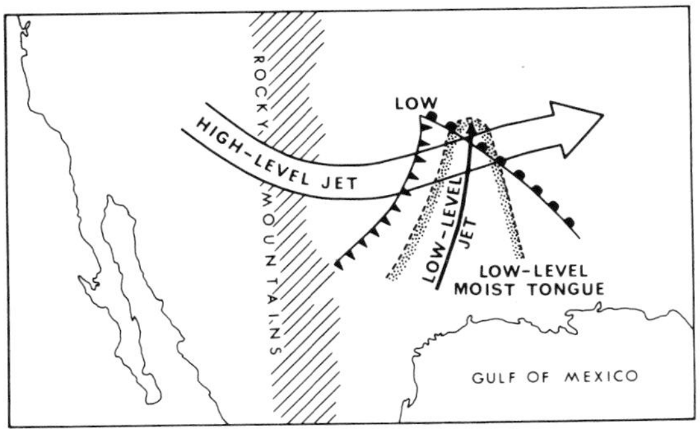

lower levels as the vortex descends and intensifies (see Figure 4.29). Such generating thunderstorms are identifiable, when seen in plan view on a radar display, by a 'hook echo' pattern representing spiral cloud bands about a small central eye. The pressures in the mesoscale thunderstorm low are only 2–5 mb less than in the surrounding environment. The tornado funnel has been observed to originate in the cloud base and extend towards the surface, and one idea is that convergence beneath the base of cumulonimbus clouds, aided by the interaction between cold precipitation downdraughts and neighbouring updraughts, may initiate rotation. Other observations suggest that the funnel forms simultaneously throughout a considerable depth of cloud, usually a towering cumulus. It appears that the upper portion of the tornado spire in this cloud may become linked to the main updraught of a neighbouring cumulonimbus, thereby causing rapid removal of air from the spire and allowing a violent pressure decrease at the surface. The pressure drop is estimated to exceed 200–250 mb in some cases and it is this which makes the funnel visible by causing air entering the vortex to reach saturation. The vortex is usually only a few

hundred metres in diameter (see Plate 26) and in an even more restricted band around the core the winds can attain speeds of 50–100 m s^{-1}. The fastest tornadoes often split into multiple vortices rotating anticlockwise with respect to the main tornado axis, each following a cycloidal path and the whole tornado system giving a complex pattern of destruction (see Figure 4.35), with maximum wind speeds being experienced on the right-side boundary (in the northern hemisphere) where the translational and rotational speeds are combined. Destruction results not only from the high winds, for buildings near the path of the vortex may explode outwards owing to the pressure reduction outside. Tornadoes commonly occur in families and move in rather straight paths (typically between 10 and 100 km long and 100 m to 2 km wide) at velocities dictated by the low-level jet. Thirty-year averages indicate some 750 tornadoes per year in the United States, with 60 per cent of these in April, May and June (see Figure 4.33B). They cause 100 fatalities each year, on average, although most of the deaths and destruction result from a few enduring mature tornadoes, making up only 1.5 per cent of the total reported. For example, the most severe recorded tornado travelled 200 km in 3 hours across Missouri, Illinois and Indiana on 18 March 1925, killing 689 people. These really intense tornadoes present problems as to their energy supply and it has recently been suggested that the release of heat energy by lightning, and other electrical discharges, may be a necessary additional source of energy.

Tornadoes are not unknown even in the British Isles. During 1960–82 there were 14 days per year with tornado occurrences. Most are minor outbreaks, but on 23 November 1981, 102 were reported during south-westerly flow ahead of a cold front. They are most common in autumn when cold air moves over relatively warm seas.

I METEOROLOGICAL FORECASTS

National meteorological services perform a variety of activities in order to provide weather forecasts. The principal ones are data collection, the preparation of basic analyses and prognostic charts of atmosphere conditions for use by local weather offices, the preparation of short- and long-term forecasts for the public as well as special services for aviation, shipping, agricultural and other commercial and industrial users, and the issuance of severe weather warnings.

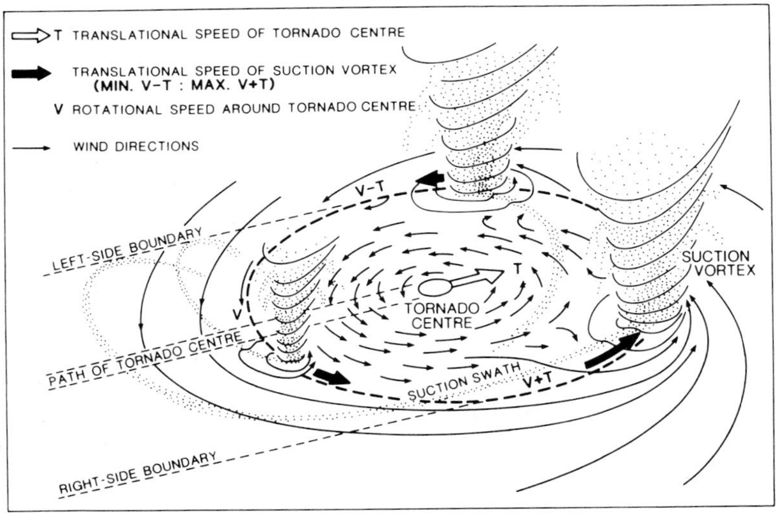

Figure 4.35 Schematic diagram of a complex tornado with multiple suction vortices.

Source: After Fujita, *J. Atmos. Sci.* 38, 1981, Fig. 15, p. 1251).

1 Data sources

The data required for forecasting and other services are provided by world-wide standard synoptic reports at 00, 06, 12, and 18 GMT (see Appendix 3), similar observations made hourly, particularly in support of national aviation requirements, upper-air soundings (at 00 and 12 GMT) satellite data and other specialized networks such as radar stations for severe weather. Under the World Weather Watch programme, synoptic reports are made at some 4000 land stations and by 7000 ships (see Figure 4.36). There are about 700 stations making upper-air soundings (temperature, pressure, humidity and wind) (see Figure 4.37). These data are transmitted in code via teletype and radio links to regional or national centres and into the high-speed Global Telecommunications System (GTS) connecting World Weather Centres in Melbourne, Moscow and Washington and eleven Regional Meteorological Centres for redistribution. Some 157 states and territories co-operate in this activity under the aegis of the World Meteorological Organization.

Meteorological information has been collected operationally by satellites of the United States and USSR since 1965 and, more recently, by the European Space Agency, India and Japan. There are two general categories of weather satellites: polar orbiters providing global coverage twice per 24 hours in orbital strips over the poles (such as the United States NOAA and TIROS series (see Plates 2 and 3), and the Russian Meteor); and geosynchronous satellites (such as the Geostationary Operational Environmental Satellites (GOES) and Meteosat), giving repetitive (30-minute) coverage of almost one-third of the earth's surface in low middle latitudes (see Figure 4.38). Information on the atmosphere is collected as digital data or direct readout visible and infra-red images of cloud cover and sea-surface temperature, but also includes global temperature and moisture profiles through the atmosphere obtained from multi-channel infra-red and microwave sensors

Figure 4.36 Synoptic reports from surface land stations and ships available over the Global Telecommunications System at the National Meteorological Center, Washington, DC.

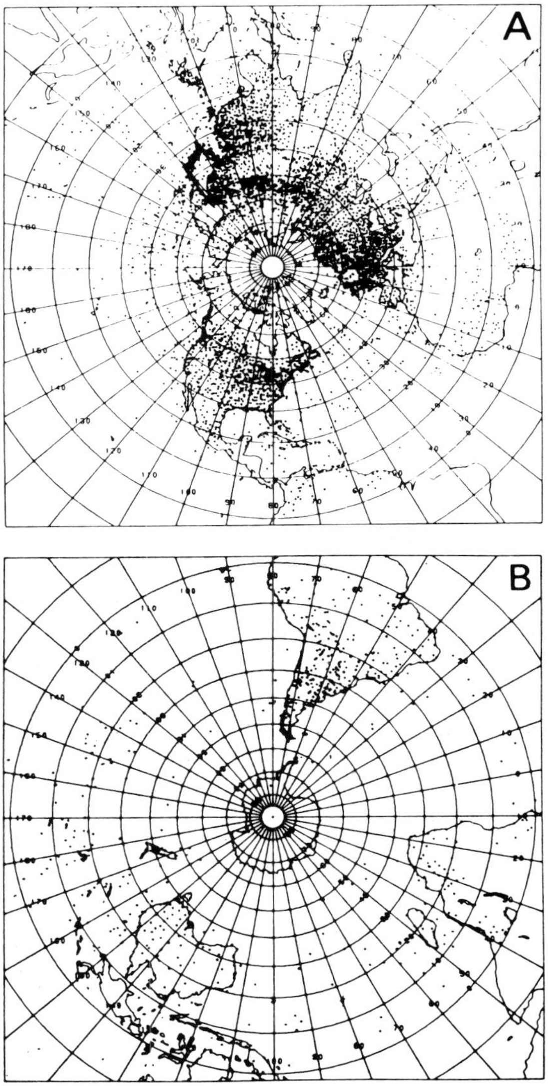

which receive radiation emitted from particular levels in the atmosphere. Additionally, satellites have a data collection system (DCS) that relays data on numerous environmental variables from ground platforms or ocean buoys to processing centres; GOES can also transmit processed

Figure 4.37 Synoptic reports from upper-air sounding stations available over the Global Telecommunications System at the National Meteorological Center, Washington, DC.

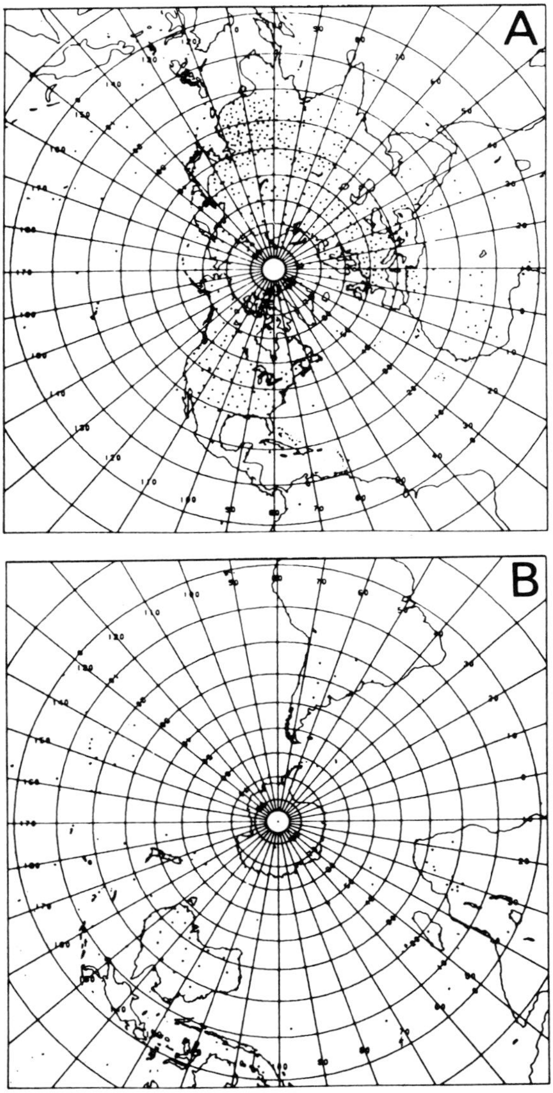

Source: From A. D. Hecht, *Paleoclimate Analysis and Modeling*, Wiley, New York, 1985) (*copyright © 1985 by John Wiley and Sons, Inc. Reproduced by permission*).

satellite images in fascimile and the NOAA polar orbiters have an automatic picture transmission (APT) system that is utilized at 900 stations world-wide.

2 Forecasting

Modern forecasting did not become possible until weather information could be rapidly collected, assembled and processed. The first development came in the middle of the last century with the invention of telegraphy, which permitted immediate analysis of weather data by the drawing of synoptic charts. These were first displayed in Britain at the Great Exhibition of 1851. Sequences of weather change were correlated with barometric pressure patterns both in space and time by such workers as Fitzroy and Abercromby, but it was not until later that theoretical models of weather systems were devised – notably the Bjerknes' depression model described earlier (see Figure 4.7).

Forecasts are usually referred to as short-range (1–2 days), extended range (3–5 days), medium range (up to 6–10 days) and long-range (monthly or seasonal) outlooks. The first two can for present purposes be considered together.

a Short- and extended-range forecasting

Forecasting procedures developed up to the 1950s were based on synoptic principles but, since the 1960s, practices have been revolutionized by numerical forecasting methods and the adoption of 'nowcasting' techniques.

During the first half of the century, short-range forecasts were based on synoptic principles, empirical rules and extrapolation of pressure changes. The Bjerknes' model of cyclone development for middle latitudes and simple concepts of tropical weather (see Chapter 6) served as the basic tools of the forecaster. The relationship between the development of surface lows and highs and the upper-air circulation was worked out during the 1940s and 1950s by C.-G. Rossby, R. C. Sutcliffe and others, providing the theoretical basis of synoptic forecasting. In this way, the position and intensities of low- and

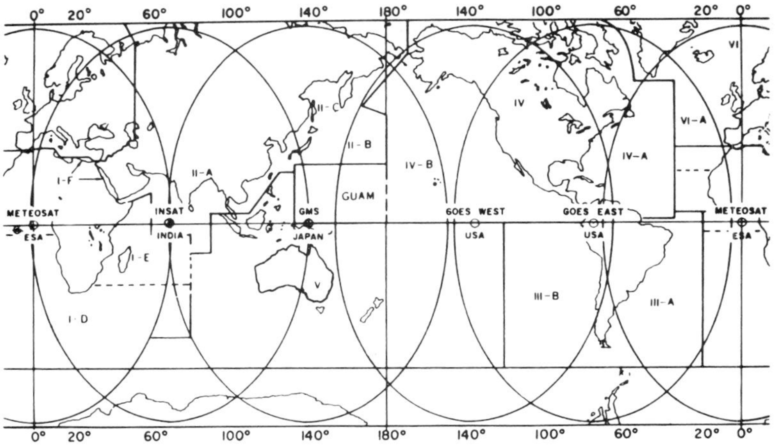

Figure 4.38 Coverage of geostationary satellites and WMO data collection areas (rectangular areas and numbers). The optimum coverage shown is not always available if satellite systems fail.

Source: Reproduced by courtesy of NOAA.

high-pressure cells and frontal systems were predicted.

Since 1955 in the United States – and 1965 in the United Kingdom – routine forecasts have been based on numerical models. These predict the evolution of physical processes in the atmosphere by determinations of the conservation of mass, energy and momentum. The basic principle is that the rise or fall of surface pressure is related to mass convergence or divergence, respectively, in the overlying air column. This prediction method was proposed by L. F. Richardson who, in 1922, made a laborious test calculation that gave very unsatisfactory results. The major reason for this lack of success is the fact that the *net* convergence or divergence in an air column is a small residual term compared with the large values of convergence and divergence at different levels in the atmosphere (see Figure 3.7). Small errors arising from observational limitations may, therefore, have a considerable effect on the correctness of the analysis.

Numerical weather prediction (NWP) methods developed in the 1950s use a less direct approach. The first developments assumed a one-level barotropic atmosphere with geostrophic winds and hence no convergence or divergence. The movement of systems could be predicted, but not changes in intensity. Despite the great simplifications involved in the barotropic model, it has been used for forecasting 500-mb contour patterns. The latest techniques employ multi-level baroclinic models and include frictional and other effects; hence the basic mechanisms of cyclogenesis are provided for. It is noteworthy that *fields* of continuous variables, such as pressure, wind and temperature, are handled and that fronts are regarded as secondary, derived features. The vast increase in the number of calculations which these models necessitate has required a new generation of super computers to allow the preparation of a forecast map to keep sufficiently ahead of the weather changes!

Forecast practices in the major national centres are basically similar. However, the specific information relates to the National Meteorological Center (NMC) in Washington, DC. Two types of global forecast are prepared each day at NMC; these are 3-day aviation forecasts and 6 to 10-day medium-range forecasts (MRF). A Global Data Assimilation System is used to prepare an analysis of the 00 GMT observations, incorporating as a first guess the prior 6-hour forecast from the MRF model. Different interpolation methods are used to obtain smoothed, gridded data on temperature, moisture, wind and geopotential height for the surface at standard pressure levels (850, 700, 500, 400, 300, 250, 200 and 100 mb) over the globe. The NMC currently has two basic pre-

diction models: a spectral model with eighteen levels (from the boundary layer into the upper stratosphere) and 160 km horizontal resolution, which is integrated for up to 10 days, and a regionally-applicable nested grid model with finer horizontal resolution. It should be noted that typically the computer time required increases several-fold when the grid spacing is halved. The essential forecast products are MSL pressure, temperature and wind velocity for standard pressure levels, 1 000–500-mb thickness, vertical motion and moisture content in the lower troposphere, and precipitation amounts.

Actual weather conditions are now commonly predicted using the Model Output Statistics (MOS) technique developed by the US National Weather Service. Rather than relating weather variables to the predicted pressure/height patterns and taking account of frontal models, for example, a series of regression equations are developed for specific locations between the variable of interest and up to ten predictors calculated by the numerical models. Weather elements so predicted for numerous locations include daily maximum/minimum temperature, 12-hour probability of precipitation occurrences and precipitation amount, probability of frozen precipitation, thunderstorm occurrence, cloud cover and surface winds. These forecasts are distributed as fascimile maps and tables to weather offices for local use.

In the United States, there are also separate centres with responsibilities for forecasting tropical storms (the National Hurricane Center in Miami for the Atlantic area, with warning centres at San Francisco for the eastern North Pacific and Honolulu for the central North Pacific) and also the National Severe Storms Forecasting Center in Kansas City, Missouri. The latter depends greatly on satellite and radar data to issue severe weather alerts.

The British Meteorological Office's 1982 global model has a 1.5° latitude by 1.875° longitude horizontal grid and fifteen levels between the surface and approximately 25 mb. All essential physical processes (diurnal radiation cycle, surface energy exchanges, cloud development and precipitation) are represented. Certain surface conditions are fixed, such as snow cover and sea-ice extent, albedo, soil moisture and thermal capacity, while sea-surface temperatures are entered from daily observations. By repeated calculations for time steps of a few minutes, atmospheric motion, temperature, cloud cover, precipitation, etc., are forecast up to six days ahead. For a more limited area (80°W–40°E, 30°–80°N) a 'fine mesh' model (0.75° latitude, 0.9375° longitude grid) is used to forecast up to 36 hours ahead.

Errors in numerical forecasts arise from several sources. One of the most serious is the limited accuracy of the initial analyses due to data deficiencies. The coverage over the oceans is sparse and only a quarter of the possible ship reports may be received within 12 hours; even over land more than one-third of the synoptic reports may be delayed beyond 6 hours. However, satellite-derived information and aircraft reports can help fill some gaps for the upper air. Another limitation is imposed by the horizontal and vertical resolution of the models and the need to parameterize subgrid processes such as cumulus convection. The small-scale nature of the turbulent motion of the atmosphere means that some weather phenomena are basically unpredictable, for example l the specific locations of shower cells in an unstable air mass. Greater precision than the 'showers and bright periods' or 'scattered showers' of the forecast language is impossible with present techniques. The procedure for preparing a forecast is becoming much less subjective, although in complex weather situations the skill of the experienced forecaster still makes the technique almost as much an art as a science. Detailed regional or local predictions can only be made within the framework of the general forecast situation for the country and demand thorough knowledge of possible topographic or other local effects by the forecaster.

b 'Nowcasting'

Severe weather is typically short-lived (< 2 hr) and, due to its mesoscale character (< 100 km), it affects local/regional areas necessitating site-specific forecasts. Included in this category are thunderstorms, gust fronts, tornadoes, high winds especially along coasts, over lakes and mountains, heavy snow and freezing precipitation. The development of radar networks, new instruments and high-speed communication links has provided a means of issuing warnings of such phenomena. Several countries have recently developed integrated satellite and radar systems to provide information on the horizontal and vertical extent of thunderstorms, for example. Such data are supplemented by networks of automatic weather stations (including buoys) that measure wind, temperature and humidity. In addition, for detailed boundary layer and lower troposphere data, there is now an array of vertical sounders – acoustic sounders (measuring wind speed and direction from echoes created by thermal eddies), specialized (Doppler) radar measuring winds in clear air by returns either from insects (3.5 cm wavelength radar) or from variations in the air's refractive index (10 cm wavelength radar). Nowcasting techniques use highly automated computers and image analysis systems to integrate data from a variety of sources rapidly. Interpretation of the data displays requires skilled personnel and/or extensive software to provide appropriate information. The prompt forecasting of wind shear and downburst hazards at airports is one example of the importance of nowcasting procedures.

Overall, the greatest benefits from improved forecasting can be expected in aviation and the electric power industry for forecasts less than 6 hours ahead, in transportation, construction and manufacturing for 12–24-hour forecasts and in agriculture for 2–5-day forecasts. In terms of economic losses, the last category could benefit the most from more reliable and more precise forecasts.

c Medium-range forecasting

The current 6–10 day forecasts prepared by the Climate Analysis Center (CAC) in Washington, DC are a 'blend' of the results of two numerical models: those produced by NMC's Medium-Range Forecast (MRF) model and by the European Centre for Medium-Range Weather Forecasts (ECMWF) model. First, a composite mean 500-mb height and height anomaly prediction is produced because the MRF model has a cold bias with heights in the middle and upper troposphere that are too low, whereas the ECMWF model has an opposite warm bias. Teleconnection patterns at 700 mb are examined to ensure internal consistency. Second, temperature and precipitation are predicted statistically using relationships of the kind illustrated in Figure 4.39. Temperature patterns over the contiguous United States are forecast from correlations with the predicted 6–10 day 700-mb height anomalies. Precipitation amounts derived from the MRF model are shown in three probability categories (above normal, near normal, below normal) (see Figure 4.40).

The methods of medium-range forecasting discussed above are unsuitable for predicting the probable trend of the weather for periods of a month or more, because they are concerned with individual synoptic disturbances with a life cycle of about 3 to 7 days. Theoretical considerations indicate that the limit of predictability using numerical techniques is less than 15 days. However, for large-scale features of the circulation, and for certain regimes such as blocking patterns, deterministic predictability may extend to 3–4 weeks. It is known, for example, that extended spells of particular types of weather are characteristic of mid-latitudes generally (see Chapter 5, A.4).

d Long-range outlooks

There are three approaches to long-range forecasting – dynamical, analogue and statistical. These are now outlined.

1 Dynamical methods. Dynamical 30-day forecasts are being attempted at various centres

Figure 4.39 Forecast of North American weather for December 1985 made one month ahead. A: Predicted 700-mb contours (gpdm). Solid arrows indicate main tracks of cyclones, open arrows of anticyclones, at sea level. The forecasting of such tracks has recently been discontinued. B and C: Forecast average temperature (B) and average precipitation (C) probabilities (in per cent). There are three classes of temperature, above normal, normal and below normal, and similarly heavy, near normal and light for precipitation. Each of the below and above normal classes is defined to occur 30 per cent of the time in the long run; near-normal temperature or moderate precipitation occur 40 per cent of the time. The 30 per cent heavy lines indicate indifference (for any departure from average), but near-normal values are most likely in unshaded areas.

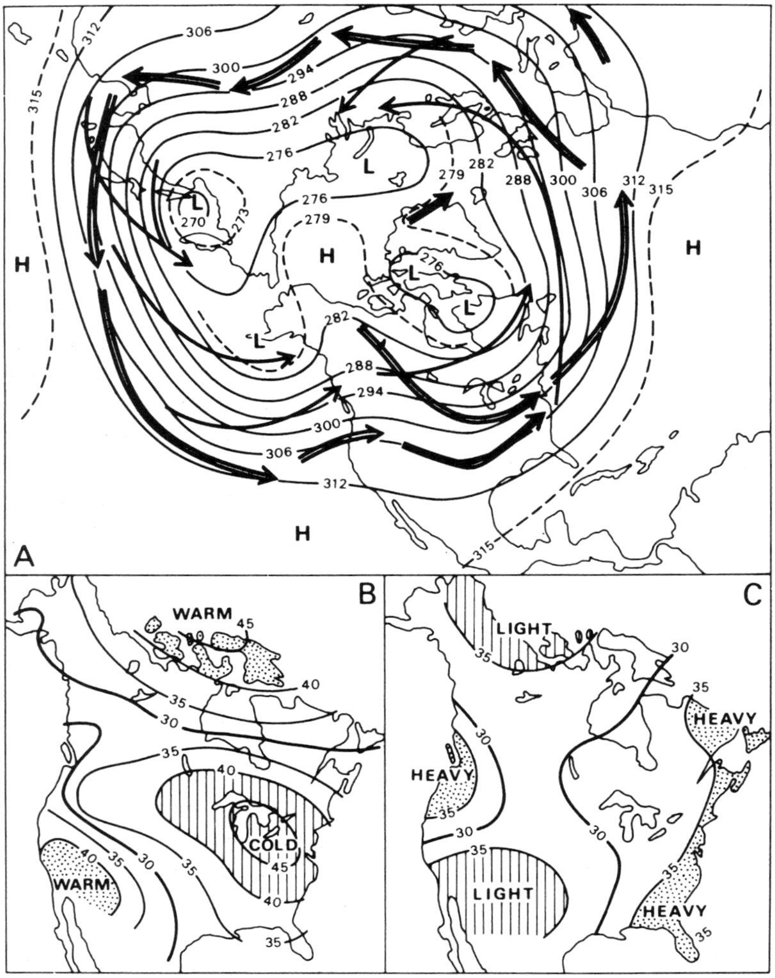

Source: From *Monthly and Seasonal Weather Outlook* 39 (23), 28 November 1985, Climate Analysis Center, NOAA, Washington, DC.

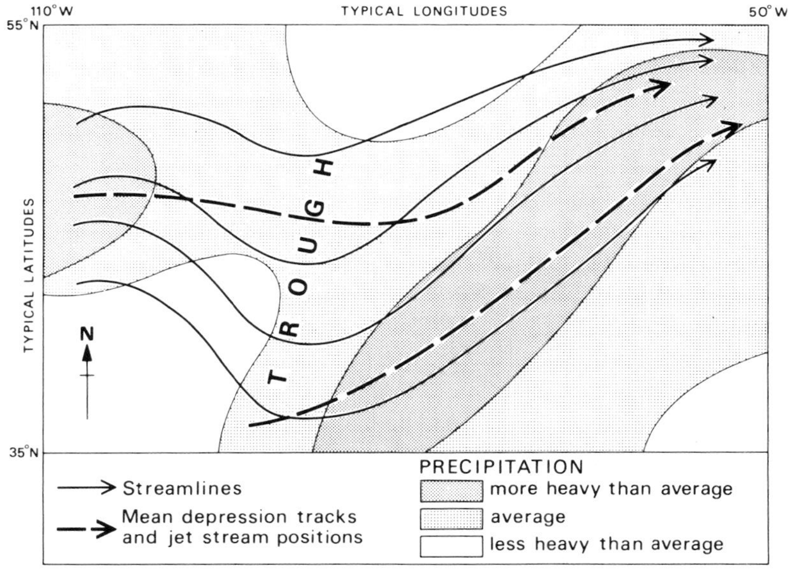

Figure 4.40 Generalized mean relationships between the westerly airflow pattern, depression tracks, jet-stream positions and precipitation zones in mid-latitudes in the northern hemisphere.

Source: After W. Klein; from Harman 1971.

using longer integrations of the medium-range models. For example, NMC prepared 108 contiguous 30-day forecasts for 14 December 1986 to 31 March 1987. It was found that the model generally performed better than a persistence forecast (of existing conditions), but the 30-day mean 500-mb circulation of the northern hemisphere is estimated best from the mean of the first 7 to 10 days. In particular, the unexpected development of blocking situations reduced the skill on several occasions. The sensitivity of model forecasts to external forcings such as anomalies of sea-surface temperature, soil moisture, snow cover and sea ice are being investigated.

2 Analogue methods. A statistical approach to long-term forecasting developed in Germany and Britain is based on the principle that sequences of weather events may tend to follow a similar course if the initial conditions are almost identical. The problem is then to find a period with weather conditions as closely analogous as possible to the present one and to use the past sequence of events as a guide to the future. The analogues are matched from a record of patterns of monthly temperature and pressure anomalies, and of sequences of *weather types*. The latter are actually types of pressure pattern or airflow over the country (see Chapter 5, A.3). Each category tends to be associated with a particular type of weather. The difficulty in analogue prediction arises from the fact that no two patterns or sequences of weather are ever identical. There may, for example, be five reasonable analogues for a particular month, but examination of the succeeding weather sequences might show mild, rainy weather in two cases and cold spells in the other three. In the preparation of the forecast, therefore, many factors which can affect the weather trends, such as sea temperatures and the extent of snow cover, have to be taken into account.

3 Statistical methods. The United States National Weather Service currently issues 30-day forecasts twice monthly using a method that has two principal steps. First, individual mean 500-mb height and anomaly forecasts for days 1–5 and 6–10 are prepared by NMC and ECMWF models. The likely utility of these forecasts is assessed from statistical studies; for example, in winter and summer the best correlations for the following 30-day mean period are given by the

previous 10–28 days, whereas in the autumn and spring shorter intervals of a week or less give the highest correlations with the following month. Thus, the forecast takes account of the degree of persistence in the circulation pattern expected from one month to the next, the recognition of probable changes of regime in the large-scale atmospheric circulation, as well as the possible effects of features such as snow cover and sea-surface temperature anomalies, and statistical records of the typical locations of troughs and ridges at that season. The second step involves deriving predicted surface temperature and precipitation values from the predicted 700-mb height anomalies for the month. Statistical analyses for the United States by W. H. Klein are the cornerstone of this procedure. These are partly based on analogues where the weather experienced during past cases with similar 700-mb patterns occurring in the same season serves as a guide to the conditions expected to recur. Figure 4.40 shows that more than average precipitation is probable ahead of the trough where there is maximum vorticity advection, especially in association with an active jet stream, whereas less than average precipitation is usual to the rear of the trough axis. With a low zonal index in the westerlies, long waves may persist in more or less stationary positions for periods of several weeks at a time, or even longer, and their mean positions are, therefore, reflected in surface patterns of temperature and precipitation (see also Chapter 5, B.1).

The 30-day outlook for the United States is presented as maps of the probability of three simple classes each of temperature (near-normal, above/below normal) and precipitation (near-average, above/below the median) (see Figure 4.39), together with tables for many cities.

The 90-day seasonal outlooks prepared by the Climate Analysis Center of the US National Weather Service are based primarily on persistence statistics that take account of trends and tendencies linking the circulation and weather patterns of the preceding several seasons. Overall, winter is more easily predictable and spring is the most variable season. Recurrent, large-scale abnormalities in climate, such as the El Niño phenomenon (see Chapter 6, G.4), may prove useful in seasonal forecasting in view of their world-wide *teleconnections* (i.e. distant areas of the globe where the climatic anomalies show either the same, or almost opposite, tendencies).

Figure 4.41A illustrates the observed height field corresponding to Figure 4.39A for December 1985, showing that the pattern is well represented on the forecast chart. Figure 4.41B and C shows that in this case, as is usual, the temperature forecasts are more reliable than those for precipitation.

Long-range forecasts prepared at the UK Meteorological Office have not been published since 1981. However, newer statistical procedures were developed in the 1980s. They involve a series of steps. The sea level pressure anomalies over the North Atlantic and Europe have been classified for 1899–1978 into six circulation types for each two-month interval (January–February, etc.) throughout the year. From these data, the probability of the different circulation types succeeding one another has been calculated. The likelihood of each outcome is determined according to an optimum combination of climatic states described by sea-surface-temperature anomalies in fifteen tropical and mid-latitude areas, and patterns of sea level pressure and 1 000–500-mb thickness anomalies over mid-latitudes of the northern hemisphere. On these bases, monthly forecasts of temperature and precipitation are made for ten areas of the British Isles. Statistical techniques are a useful complement to dynamical methods since they are constrained to be realistic by the past data.

The problems facing the long-range forecasters need to be recognized before criticisms are levelled at them. The complexity of the atmosphere's behaviour makes tentative wording of their predictions necessary and occasional failures inevitable at present.

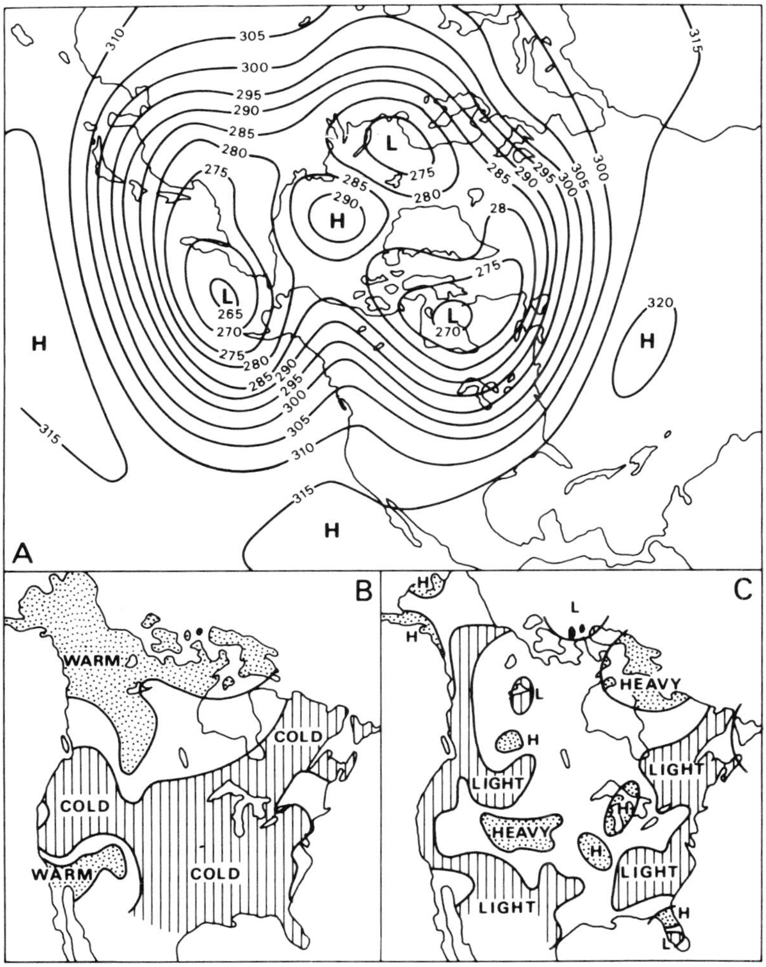

Figure 4.41 Actual North American weather for December 1985 (cf. Figure 4.39). A: Observed 700-mb contours (5 gpdm) corresponding to Figure 4.39A (from Climate Analysis Center, NOAA, Washington, DC.) B and C: Observed temperature (B) and precipitation (C), corresponding to Figure 4.39B and C.

Source: From *Monthly and Seasonal Weather Outlook* 40 (1), 1986, Climate Analysis Center, NOAA, Washington, DC.

SUMMARY

Ideal air masses are defined in terms of barotropic conditions, where isobars and isotherms are assumed to be parallel to each other and to the surface. The character of an air mass is determined by the nature of the source area, changes due to air-mass movement, and its age. On a regional scale, energy exchanges and vertical mixing lead to a measure of equilibrium between surface conditions and those of the overlying air, particularly in quasi-stationary high-pressure systems. Air masses are conventionally identified in terms of temperature characteristics (arctic, polar, tropical) and source region (maritime, continental). Primary air masses originate in regions of semi-permanent anticyclonic subsidence over extensive surfaces of like properties. Cold-air masses originate either in winter continental anticyclones (Siberia and Canada), where snow cover promotes low temperatures and stable strati-

continued

fication, or over high-latitude sea ice. Some sources are seasonal, like Siberia, others are permanent, such as Antarctica. Warm-air masses originate either as shallow tropical continental sources in summer or as deep, moist layers over tropical oceans. Air-mass movement causes stability changes by thermodynamic processes (heating/cooling from below and moisture exchanges) and by dynamic processes (mixing, lifting/subsidence), producing secondary air masses (e.g. mP air). The age of an air mass determines the degree to which it has lost its identity as the result of mixing with other air masses and of vertical exchanges with the underlying surface.

Air-mass boundaries give rise to fronts or baroclinic zones a few hundred kilometres wide. The classical (Norwegian) theory of mid-latitude cyclones considers that fronts are a key feature of their formation and life cycle. Depressions tend to form along major frontal zones – the polar fronts of the North Atlantic and North Pacific regions and of the southern oceans. Less-well-defined arctic fronts lie poleward and there are other seasonal frontal zones as in the Mediterranean. The Inter-Tropical Convergence Zone, between air masses from the opposing subtropical anticyclones, is different in character from frontal zones of high latitudes (see Chapter 6). Air masses and frontal zones move poleward/equatorward in summer/winter.

Newer cyclone theories regard fronts as rather incidental. Divergence of air in the upper troposphere is essential for large-scale uplift and low-level convergence. Surface cyclogenesis is therefore favoured on the eastern limb of an upper wave trough. 'Explosive' cyclogenesis appears to be associated with strong wintertime gradients of sea-surface temperature. Cyclones are basically steered by the quasi-stationary long (Rossby) waves in the hemispheric westerlies, the positions of which are strongly influenced by surface features (major mountain barriers and land/sea-surface temperature contrasts). Upper baroclinic zones are associated with jet streams at 300–200 mb which also follow the long-wave pattern.

The idealized weather sequence in an eastward-moving frontal depression involves increasing cloudiness and precipitation with an approaching warm front; the degree of activity depends on whether the warm sector air is rising or not (ana- or kata-fronts, respectively). The following cold front is often marked by a narrow band of convective precipitation, but rain both ahead of the warm front and in the warm sector may also be organized into locally intense mesoscale cells and bands due to the 'conveyor belt' of air in the warm sector. Associated with this airflow organization, there is often a well-developed squall line ahead of the cold front. In the central United States, especially, thunderstorm cells along such squall lines give severe thunder and hail conditions, sometimes with tornadoes. The updraughts and downdraughts in such cells set up clusters of developing and decaying storms.

Some low-pressure systems are essentially non-frontal. These include the lee depression encountered in the lee of mountain ranges; thermal lows due to summer heating; the polar air depression commonly formed in an outbreak of maritime Arctic air over oceans; and the upper cold low which is often a cut-off system in upper wave development or an occluded mid-latitude cyclone in the Arctic.

Analysis and forecasting of surface and upper-air weather maps is now highly automated in many national weather services. Short- and extended-range forecasting is based primarily on numerical prediction models. Automated integration of satellite and radar data is being widely developed for 'nowcasting' of severe weather events. Long-range outlooks are primarily statistical, taking account of possible effects of surface conditions on the large-scale circulation structure.

5

Weather and climate in temperate latitudes

In the two preceding chapters the general structure of the circulation and air-mass characteristics in middle latitudes have been outlined and the behaviour and origin of extra-tropical depressions examined. The direct contribution of pressure systems to the daily and seasonal variability of weather in the westerly wind belt is quite apparent to inhabitants of the temperate lands. Nevertheless there are equally prominent contrasts of regional climate in mid-latitudes which reflect the interaction of geographical and meterological factors. This chapter gives a selective synthesis of weather and climate in Europe, the Mediterranean and North America, drawing mainly on the principles already presented. The climatic conditions of the polar and subtropical margins of the westerly wind belt are examined in the final sections of the chapter. As far as possible different themes are used to illustrate some of the more significant aspects of the climate in each area.

A EUROPE

1 Pressure and wind conditions

The principal features of the mean North Atlantic pressure pattern are the Icelandic Low and the Azores High. These are present at all seasons (see Figure 3.25), although their location and relative intensity change considerably. The upper flow in this sector undergoes little seasonal change in pattern but the westerlies decrease in strength by over half from winter to summer. The other major pressure system influencing European climates is the Siberian winter anticyclone, the occurrence of which is intensified by the extensive winter snow cover and the marked continentality of Eurasia. Atlantic depressions frequently move towards the Norwegian or Mediterranean seas in winter, but if they travel due east they occlude and fill long before they can penetrate into the heart of Siberia. Thus the Siberian high pressure is quasi-permanent at this season and when it extends westward severe conditions affect much of Europe. In summer, pressure is low over all of Asia and depressions from the Atlantic tend to follow a more zonal path. Although the depression tracks over Europe do not shift poleward in summer (as a result of the local *southward* displacement of the Atlantic Arctic Front), the depressions at this season are rather less intense and the diminished air-mass contrasts produce weaker fronts.

Wind velocities over Western Europe bear strong relationships to the occurrence and movement of depressions. The strongest winds occur on coasts exposed to the north-west airflow which follows the passage of frontal systems or at constricted topographic locations which guide the movement of depressions or which funnel airflow into them (see Figure 5.1). In the latter regard, the Carcassonne Gap in south-west France provides a preferred southern route for depressions moving eastward from the Atlantic, and the Rhône and Ebro Valleys are funnels for strong airflow into the rear of

Figure 5.1 Average wind velocities (m s^{-1}) over Western Europe measured 50 m above ground level for sheltered terrain, open plains, sea coast, open sea and hilltops. Frequencies (per cent) of wind velocities for twelve locations are shown.

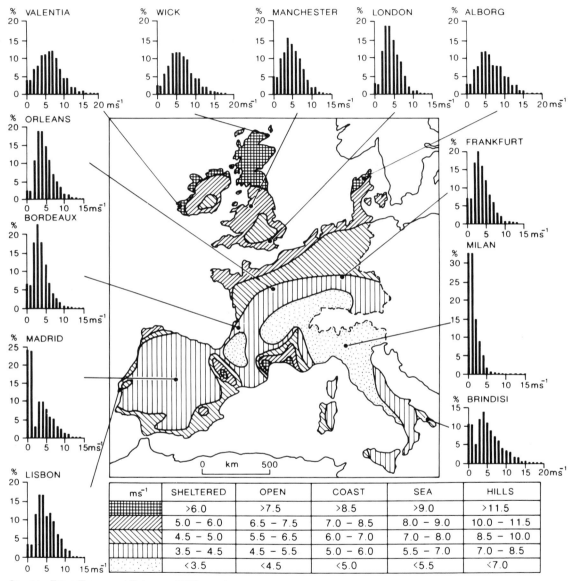

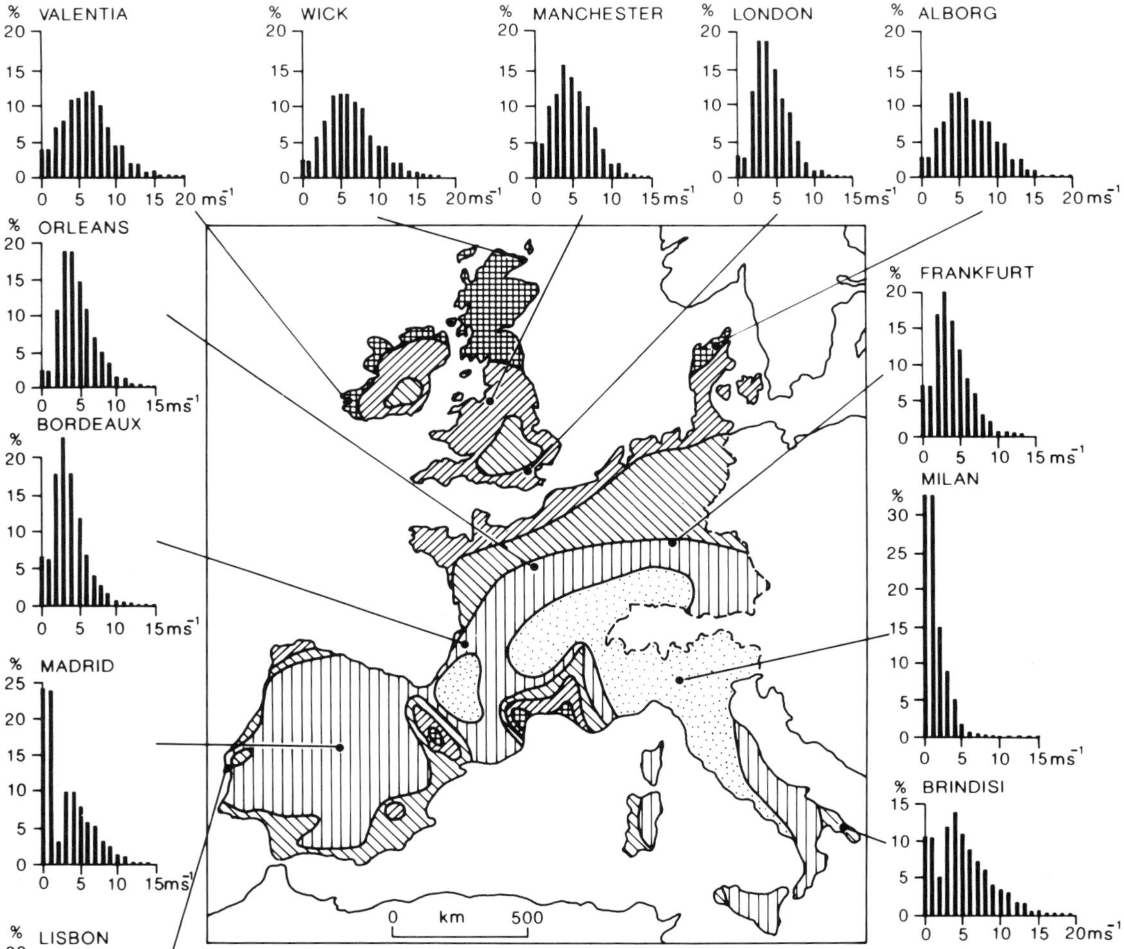

ms⁻¹	SHELTERED	OPEN	COAST	SEA	HILLS
(cross-hatched)	>6.0	>7.5	>8.5	>9.0	>11.5
(dense diagonal)	5.0 – 6.0	6.5 – 7.5	7.0 – 8.5	8.0 – 9.0	10.0 – 11.5
(diagonal)	4.5 – 5.0	5.5 – 6.5	6.0 – 7.0	7.0 – 8.0	8.5 – 10.0
(vertical)	3.5 – 4.5	4.5 – 5.5	5.0 – 6.0	5.5 – 7.0	7.0 – 8.5
(dotted)	<3.5	<4.5	<5.0	<5.5	<7.0

Source: From Troen and Petersen 1989.

depressions lying in the western Mediterranean, generating the Mistral and Cierzo winds, respectively, in winter (see D.1, this chapter). In all locations of Western Europe the mean velocity of winds on hilltops is at least 100 per cent greater than that in more sheltered locations; winds in open terrain are 25–30 per cent more intense on average than those in sheltered locations; and coastal wind velocities are at least 10–20 per cent less than those over

the adjacent sea areas (see Figure 5.1). Although mean wind-velocity figures are of some use, for example in making estimates of potential wind-power generation (on the assumption that a wind turbine can take advantage of some 20–30 per cent of available wind power), for many human activities it is the occurrence of infrequent strong winds, which may occur with very differing frequency at different locations, which is of especial significance.

2 Oceanicity and continentality

Winter temperatures in north-west Europe are some 11°C (20°F) or more above the latitudinal average (see Figure 1.30), a fact usually attributed solely to the presence of the North Atlantic Drift. There is, however, a complex interaction between the ocean and atmosphere. The Drift, which originates from the Gulf Stream off Florida strengthened by the Antilles current, is primarily a wind-driven current initiated by the prevailing south-westerlies. It flows at a velocity of 16 to 32 km (10–20 miles) per day and thus, from Florida, the water takes about eight or nine months to reach Ireland and about a year to reach Norway (see Chapter 3, F.3). The south-westerly winds transport both sensible and latent heat acquired over the western Atlantic towards Europe, and although they continue to gain heat supplies over the north-eastern Atlantic this local warming arises in the first place through the drag effect of the winds on the warm surface waters. Warming of air masses over the north-eastern Atlantic is mainly of significance when polar or arctic air flows south-eastwards from Iceland. For example, the temperature in such airstreams in winter may rise by 9°C (17°F) between Iceland and northern Scotland. By contrast, maritime tropical air cools on average about 4°C (8°F) between the Azores and south-west England in winter and summer. One very evident effect of the North Atlantic Drift is the absence of ice around the Norwegian coastline. However, as far as the climate of north-western Europe is concerned the primary factor is the occurrence of prevailing *onshore* winds transferring heat into the area.

The influence of maritime air masses can extend deep into Europe because there are few major topographic barriers to airflow and because of the presence of the Mediterranean Sea. Hence the change to a more continental climatic regime is relatively gradual except in Scandinavia where the mountain spine produces a sharp contrast between western Norway and Sweden. There are numerous indices expressing this continentality, but most are based on the annual range of temperature. Gorczynski's continentality index (K) is:

$$K = 1.7 \frac{A}{\sin \theta} - 20.4$$

where A is the annual temperature range (°C) and θ is the latitude angle. K is scaled to range from 0 at extreme oceanic stations to 100 at extreme continental stations. However, station values occasionally fall outside these limits. Some values in Europe are London 10, Berlin 21, Moscow 42. Figure 5.2 shows the variation of this index over Europe and Figure 5.3 plots its variation for Moscow between 1880 and 1980, showing its considerable oscillation with different large-scale atmospheric circulation regimes.

A very different approach by Berg relates the frequency of continental air masses (C) to that of all air masses (N) as an index of continentality, i.e. $K = C/N$ (per cent). Figure 5.2 shows that non-continental air occurs at least half the time over Europe west of 15°E as well as over Sweden and most of Finland.

A further illustration of maritime and continental regimes is provided by Figure 5.4. *Hythergraphs* for Valentia (Eire), Bergen and Berlin demonstrate the seasonal changes of mean temperature and precipitation in different locations. Valentia has a winter rainfall maximum and equable temperatures as a result of oceanic situation, whereas Berlin has a considerable temperature range and a summer maximum of rainfall. Bergen receives larger

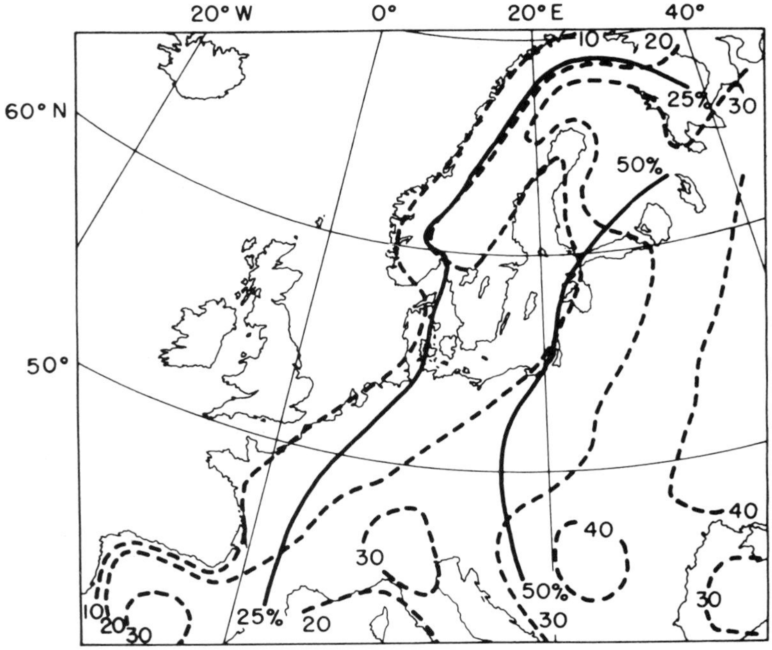

Figure 5.2 Continentality in Europe. The indices of Gorczynski (dashed) and Berg (solid) are explained in the text.

Source: Partly after Blüthgen 1966.

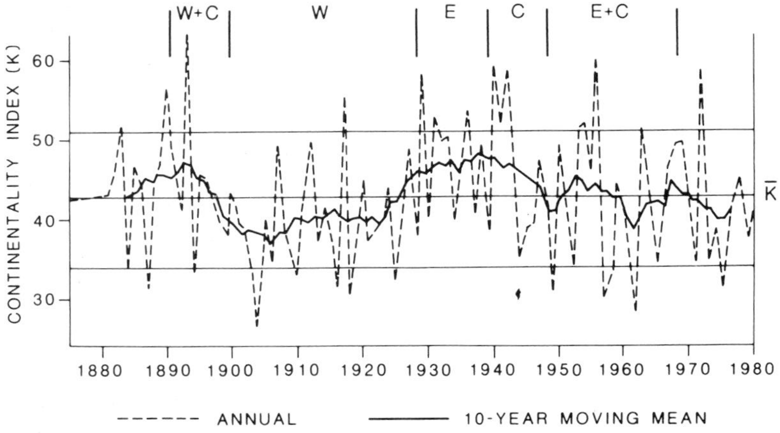

Figure 5.3 Variation of Gorczynski's continentality index (K) for Moscow (1880–1980) associated with dominant westerly (W), easterly (E) and meridional (C) hemispheric circulation patterns. Annual figures, the mean value ($\bar{K}$), the 10-year moving mean, and one standard deviation are shown.

Source: From Poltaraus and Staviskiy 1986.

rainfall totals due to orographic intensification and maximum in autumn and winter, its temperature range being intermediate to the other two. Such averages convey only a very general impression of climatic characteristics and therefore British weather patterns will now be examined in more detail.

3 British airflow patterns and their climatic characteristics

The daily weather maps for the British Isles sector (50°–60°N, 2°E–10°W) from 1873 to the present day have been classified by H. H. Lamb according to the airflow direction or isobaric

Figure 5.4 Hythergraphs for Valentia (Eire), Bergen and Berlin. Mean temperature and precipitation values for each month are plotted.

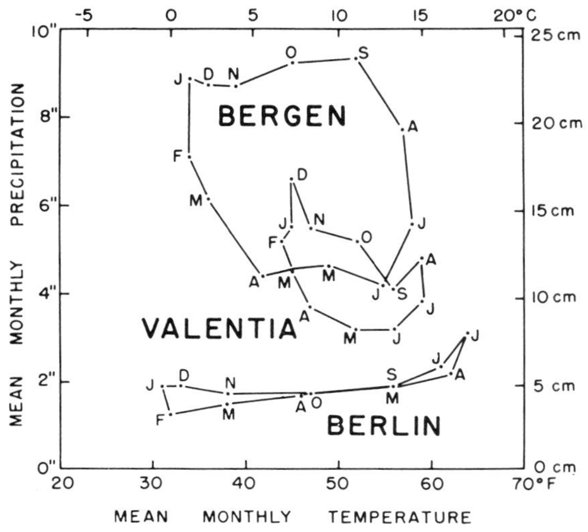

pattern. He identifies seven major categories: Westerly (W), North-westerly (NW), Northerly (N), Easterly (E) and Southerly (S) types – referring to the compass directions from which the airflow and weather systems are moving – and Cyclonic (C) or Anticyclonic (A) types when depressions or a high-pressure cell, respectively, dominate the weather map.

In theory, each category should produce a characteristic type of weather, depending on the season, and many writers use the term *weather type* to convey this idea. A few studies have been made of the *actual* weather conditions occurring in different localities with specific isobaric patterns – a field of study known as *synoptic climatology* – but the term *airflow type* seems preferable for Lamb's categories. The general weather conditions which are likely to be associated with a particular airflow type over the British Isles are summarized in Table 5.1.

On an annual basis, the most frequent airflow type is Westerly; including cyclonic and anticyclonic subtypes, it has a 35 per cent frequency in December–January and is almost as frequent in July–September (see Figure 5.5). The

Table 5.1 General weather characteristics and air masses associated with Lamb's 'Airflow Types' over the British Isles.

Type	*Weather conditions*
Westerly	Unsettled weather with variable wind directions as depressions cross the country. Mild and stormy in winter, generally cool and cloudy in summer (mP, mPw, mT).
North-westerly	Cool, changeable conditions. Strong winds and showers affect windward coasts especially, but the southern part of Britain may have dry, bright weather (mP, mA).
Northerly	Cold weather at all seasons, often associated with polar lows. Snow and sleet showers in winter, especially in the north and east (mA).
Easterly	Cold in the winter half-year, sometimes very severe weather in the south and east with snow or sleet. Warm in summer with dry weather in the west. Occasionally thundery (cA, cP).
Southerly	Warm and thundery in summer. In winter it may be associated with a low in the Atlantic giving mild, damp weather especially in the south-west or with a high over central Europe, in which case it is cold and dry (mT or cT, summer; mT or cP, winter).
Cyclonic	Rainy, unsettled conditions often accompanied by gales and thunderstorms. This type may refer either to the rapid passage of depressions across the country or to the persistence of a deep depression (mP, mPw, mT).
Anticyclonic	Warm and dry in summer, occasional thunderstorms (mT, cT). Cold and frosty in winter with fog, especially in autumn (cP).

Figure 5.5 Average climatic conditions associated with Lamb's circulation types for January, April, July and September. *Top:* mean daily temperature (°C) in central England for the straight (S) airflow types; at the right side are the quintiles of mean monthly temperature (i.e. 01/02 = 20 per cent, 04/05 = 80 per cent). *Middle:* mean daily rainfall (in millimetres) over England and Wales for the straight (S) and cyclonic (C) subdivisions of each type and terciles of the mean values (i.e. 01/02 = 33 per cent, 02/03 = 67 per cent). *Bottom:* mean frequency (per cent) for each circulation type, including Anticyclonic (A) and Cyclonic (C).

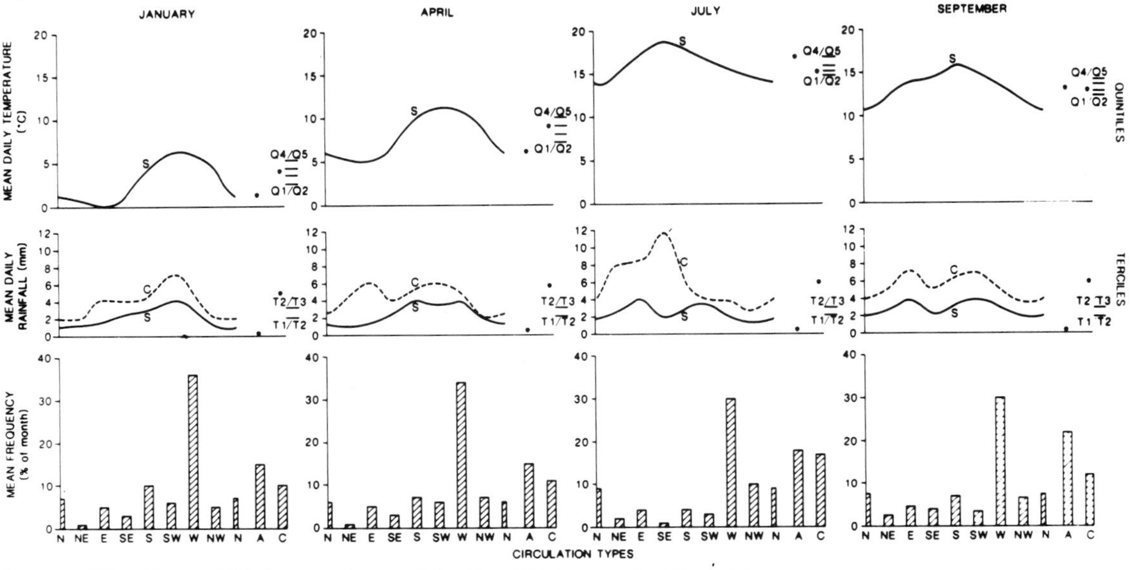

Source: After Storey 1982 (*by permission of the Royal Meteorological Society*).

minimum occurs in May (15 per cent) when Northerly and Easterly types reach their maxima (about 10 per cent each). Pure Cyclonic patterns are most frequent (13–17 per cent) in July–August and Anticyclonic patterns in June and September (20 per cent); Cyclonic patterns have ⩾ 10 per cent frequency in all months and Anticyclonic patterns ⩾ 13 per cent. Figure 5.5 illustrates the mean daily temperature in central England and the mean daily precipitation over England and Wales for each type in the mid-season months.

The general properties of air masses have been examined in Chapter 4, but certain aspects are of particular interest in respect of the British climate. The frequency of air masses in January, based on a study by Belasco for 1938–49, is illustrated for Kew in Figure 5.6. There is a clear predominance of polar maritime (mP and mPw) air which has a frequency of 30 per cent or more in all months except March. The maximum frequency of mP air at Kew is 33 per cent (with a further 10 per cent mPw) in July. The proportion is even greater in western coastal districts with mP and mPw occurring in the Hebrides, for example, on at least 38 per cent of days throughout the year.

North-westerly mP airstreams produce cool, showery weather at all seasons, especially on windward coasts. The air is unstable with cumuliform cloud, although inland in winter and at night the clouds disperse, giving low night temperatures. Over the sea, however, heating of the lower air continues by day and night in winter months so that showers and squalls can occur at any time, and these may affect windward coastal areas. The average daily mean temperatures with mP air (see Tables 4.1 and 4.2) are within about ± 1°C of the seasonal means in winter and summer, depending on the precise track of the air. More extreme conditions occur with mA air, the temperature

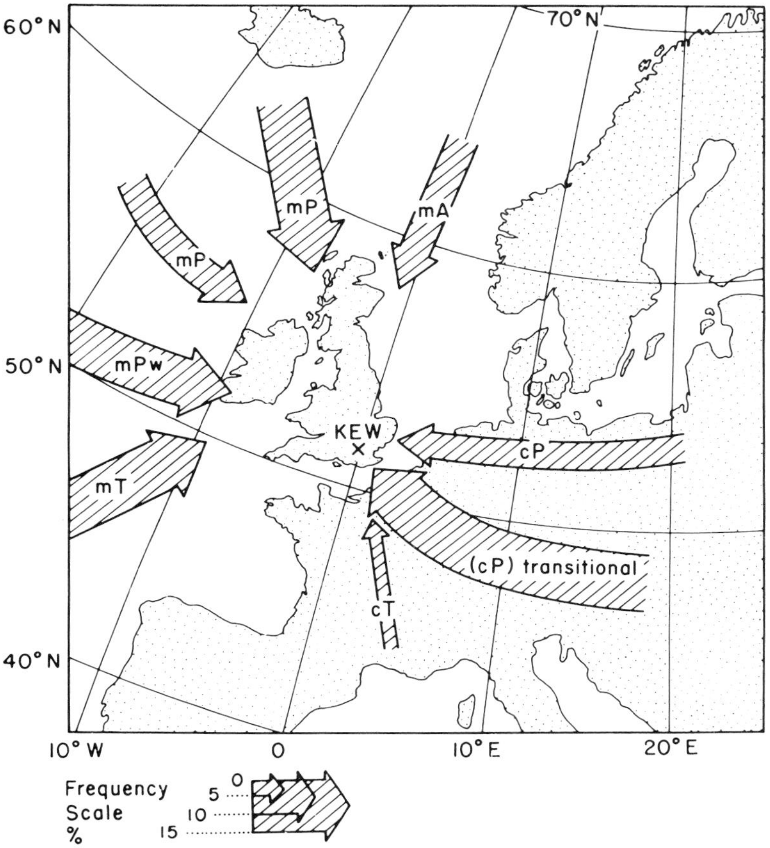

Figure 5.6 Average air-mass frequencies for Kew (London) in January. Anticyclonic types are included according to their direction of origin.

Source: Based on Belasco 1952.

Table 5.2 Percentage of the annual rainfall (1956–60) occurring with different synoptic situations.

Station	*Synoptic categories*								
	Warm front	*Warm sector*	*Cold front*	*Occlusion*	*Polar low*	*mP*	*cP*	*Arctic*	*Thunderstorm*
Cwm Dyli (99 m or 324 ft)*	18	30	13	10	5	22	0.1	0.8	0.8
Squires Gate (10 m or 33 ft)†	23	16	14	15	7	22	0.2	0.7	3
Rotherham (21 m or 70 ft)‡	26	9	11	20	14	15	1.5	1.1	3

Sources: After Shaw 1962, and R. P. Mathews (unpublished).
Notes: *Snowdonia.
†On the Lancashire coast (Blackpool).
‡In the Don Valley, Yorkshire.

departures at Kew being approximately −4°C in summer and winter. The visibility in mA air is usually very good. The contribution of mP and mA air masses to the mean annual rainfall over a 5-year period at three stations in northern England is given in Table 5.2, although it should be noted that both air masses may also be involved in non-frontal polar lows. Over much of southern England, and in areas to the lee of high ground, northerly and north-westerly airstreams usually give clear sunny weather with few showers. There is some indication of this in Table 5.2, for at Rotherham, in the lee of the Pennines, the percentage of the rainfall occurring with mP air is much lower than over Lancashire.

Maritime polar air which has followed a more southerly, cyclonic track over the Atlantic, approaching Britain from the south-west, or air which is moving northward ahead of a depression, is shown as mPw in Figure 5.6. This air has surface properties intermediate with mT air.

Maritime tropical air commonly forms the warm sector of depressions moving from between west and south towards Britain, but Figure 5.6 excludes cases of fronts and depressions (which have a frequency of about 10–12 per cent throughout the year). Hence the characteristic weather conditions of mT air occur rather more often than the percentage frequency might suggest (in January 11 per cent mT air and 4 per cent Anticyclonic air originating from south-west of Britain). The weather is unseasonably mild and damp with mT air in winter (see Table 4.1). There is generally a complete cover of stratus or stratocumulus cloud and drizzle or light rain (formed by coalescence) may occur, especially over high ground where low cloud produces hill fog. The clearance of cloud on nights with light winds readily cools the moist air to its dew point, forming mist and fog. Table 5.2 shows that a large proportion of the annual rainfall is associated with warm-front and warm-sector situations and therefore is largely attributable to convergence and frontal uplift within mT air. In summer the cloud cover with this air mass keeps temperatures closer to average than in winter (see Tables 4.1 and 4.2); night temperatures tend to be high, but daytime maxima remain rather low.

True continental polar air only affects the British Isles between December and February and even then it is relatively infrequent. Mean daily temperatures are well below average and maxima rise to only a degree or so above freezing point. The air mass is basically very dry and stable but a track over the central part of the North Sea supplies sufficient heat and moisture to cause showers, often in the form of snow, over eastern England and Scotland. *In toto* this provides only a very small contribution to the annual precipitation, as Table 5.2 shows, and on the west coast the weather is generally clear. A transitional cP–cT type of air mass reaches Britain from south-east Europe in all seasons, though less frequently in summer. Such airstreams are dry and stable.

Continental tropical air occurs on average about one day per month in summer, which accounts for the rarity of summer heat-waves since these south or south-east winds bring hot, settled weather. The lower layers are stable and the air is commonly hazy, but the upper layers tend to be unstable and occasionally intense surface heating may trigger off a thunderstorm. In winter such modified cT air sometimes reaches Britain from the Mediterranean bringing fine, hazy, mild weather.

4 Singularities and natural seasons

Most popular weather lore expresses the belief that each season has its own weather (for example, in England, 'February fill-dyke', 'April showers'), and ancient adages suggest that even the sequence of weather may be determined by the conditions established on a given date (e.g. 40 days of wet or fine weather following St Swithin's day, 15 July, in England). Some of these ideas are quite fallacious, but others contain more than a grain of truth if properly interpreted.

The tendency for a certain type of weather to recur with reasonable regularity around the same date is termed a *singularity*. Many calendars of singularities have been compiled, particularly in Europe, but the early ones (which concentrated upon anomalies of temperature or rainfall) did not prove very reliable. Greater success has been achieved by studying singularities of circulation pattern and catalogues have been prepared for the British Isles by Lamb and for central Europe by Flohn and Hess. Lamb's results are based on calculations of the daily frequency of the airflow categories between 1898 and 1947, some examples of which are shown in Figure 5.7. A noticeable feature is the infrequency of the westerly type in spring, the driest season of the year in the British Isles and also in northern France, northern Germany and in the countries bordering the North Sea. The catalogue of Flohn and Hess is based on a classification of large-scale patterns of airflow in the lower troposphere (*Grosswetterlage*) over central Europe originally proposed by F. Baur.

Some of the European singularities that occur most regularly are as follows:

Figure 5.7 The percentage frequency of anticyclonic, westerly and cyclonic conditions over Britain, 1898–1947.

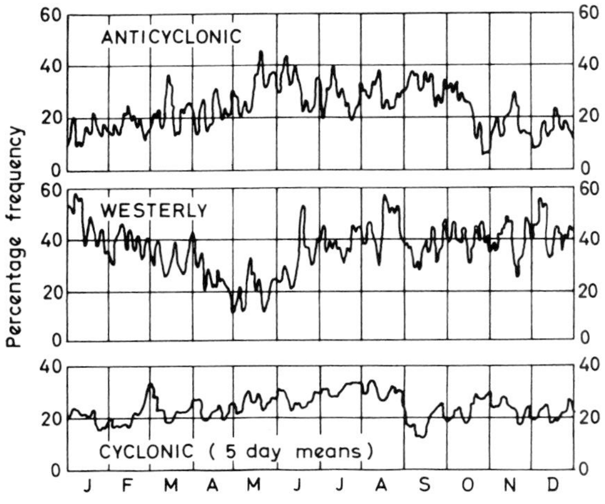

Source: After Lamb 1950.

1 A sharp increase in the frequency of Westerly and North-westerly type over Britain takes place about the middle of June. These invasions of maritime air also affect central Europe and this period is sometimes referred to as the beginning of the European 'summer monsoon'.
2 About the second week in September Europe and Britain are affected by a spell of Anticyclonic weather. This may be interrupted by Atlantic depressions giving stormy weather over Britain in late September, though anticyclonic conditions again affect central Europe at the end of the month and Britain during early October.
3 A marked period of wet weather often affects western Europe and also the western half of the Mediterranean at the end of October, whereas the weather in eastern Europe generally remains fine.
4 Anticyclonic conditions return to Britain and affect much of Europe about mid-November, giving rise to fog and frost.
5 In early December Atlantic depressions push eastwards to give mild, wet weather over most of Europe.

In addition to these singularities, major seasonal trends are recognizable and for the British Isles Lamb identified five *natural seasons* on the basis of spells of a particular type lasting for 25 days or more during the period 1898–1947 (see Figure 5.8). In as far as is possible to think in terms of a 'normal year', the seasons are:

1 *Spring–early summer* (the beginning of April to mid-June). This is a period of variable weather conditions during which long spells are least likely. Northerly spells in the first half of May are the most significant feature, although there is a marked tendency for anticyclones to occur in late May–early June.
2 *High summer* (mid-June to early September). Long spells of various types may occur in different years. Westerly and North-westerly types are the most common and they may be combined with either Cyclonic or Anti-

Figure 5.8 The frequency of long spells (25 days or more) of a given airflow type over Britain, 1898–1947. The diagram showing all long spells also indicates a division of the year into 'natural seasons'.

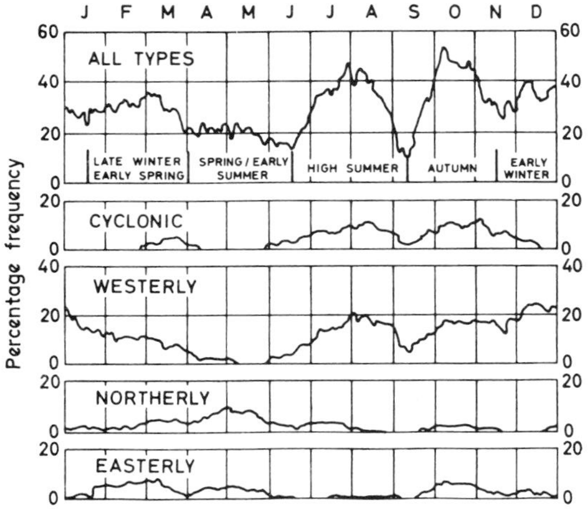

Source: After Lamb 1950.

cyclonic types. Persistent sequences of Cyclonic type occur more frequently than Anticyclonic ones.

3 *Autumn* (the second week in September to mid-November). Long spells are again present in most years. Anticyclonic ones mainly in the first half, Cyclonic and other stormy ones generally in October–November.
4 *Early winter* (from about the third week in November to mid-January). Long spells are less frequent than in summer and autumn. They are usually of Westerly type giving mild, stormy weather.
5 *Late winter and early spring* (from about the third week in January to the end of March). The long spells at this time of year can be of very different types, so that in some years it is mid-winter weather while in other years there is an early spring from about late February.

5 Synoptic anomalies

The mean climatic features of pressure, wind and the seasonal airflow regime provide an incomplete picture of climatic conditions. Some patterns of circulation occur irregularly and yet, because of their tendency to persist for weeks or even months, form an essential element of the climate.

Blocking patterns are an important example. It was noted in Chapter 3, F.2 that the zonal circulation in middle latitudes sometimes breaks down into a cellular pattern. This is commonly associated with a split of the jet stream into two branches over higher and lower middle latitudes and the formation of a cut-off low (see Chapter 4, G.4) south of a high-pressure cell. The latter is referred to as a *blocking anticyclone* since it prevents the normal eastward motion of depressions in the zonal flow. Figure 5.9 gives numerical values expressing the frequency of occurrence of blocking for part of the northern hemisphere with five major blocking centres shown (H). A major area of blocking is Scandinavia, particularly in spring. Depressions are diverted north-eastwards towards the Norwegian Sea or south-eastwards into southern Europe. This pattern, with easterly flow around the southern margins of the anticyclone, produces severe winter weather over much of northern Europe. In January–February 1947, for example, easterly flow across Britain as a result of blocking over Scandinavia led to extreme cold and frequent snowfall. Winds were almost continuously from the east between 22 January and 22 February and even daytime temperatures rose little above freezing point. Snow fell in some part of Britain every day from 22 January to 17 March 1947, and major snowstorms occurred as occluded Atlantic depressions moved slowly across the country. Other notably severe winter months – January 1881, February 1895 and January 1940 – were the result of similar pressure anomalies with pressure much above average to the north of the British Isles and below average to the south.

The average effects of a number of winter blocking situations over north-west Europe are shown in Figures 5.10 and 5.11. Precipitation amounts are above normal mainly over Iceland and the western Mediterranean as depressions

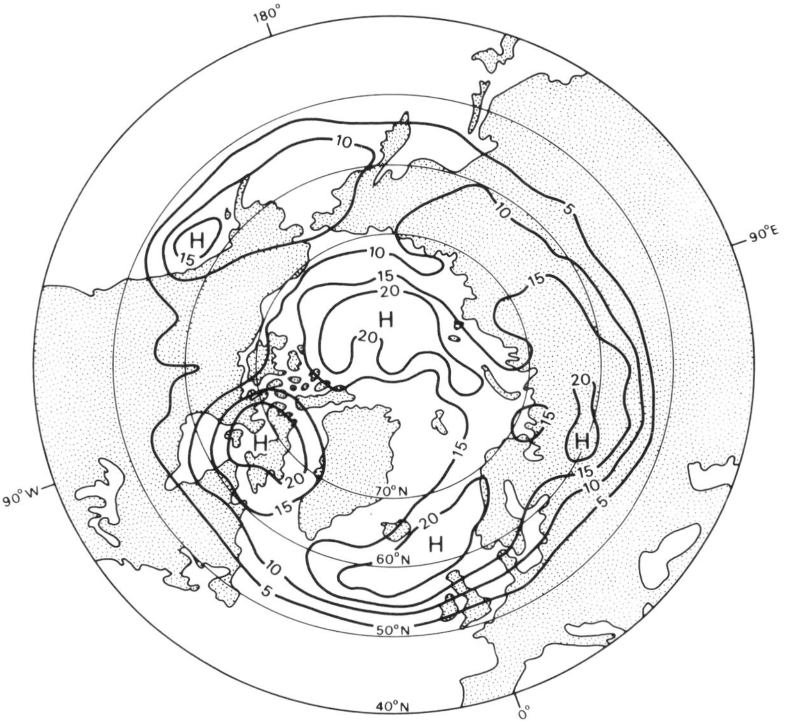

Figure 5.9 Frequency of occurrence of blocking conditions for the 500-mb level for all seasons. Values were calculated as 5-day means for 381 × 381 km squares for the period 1946–78.

Source: From Knox and Hay 1985 (*by permission of the Royal Meteorological Society*).

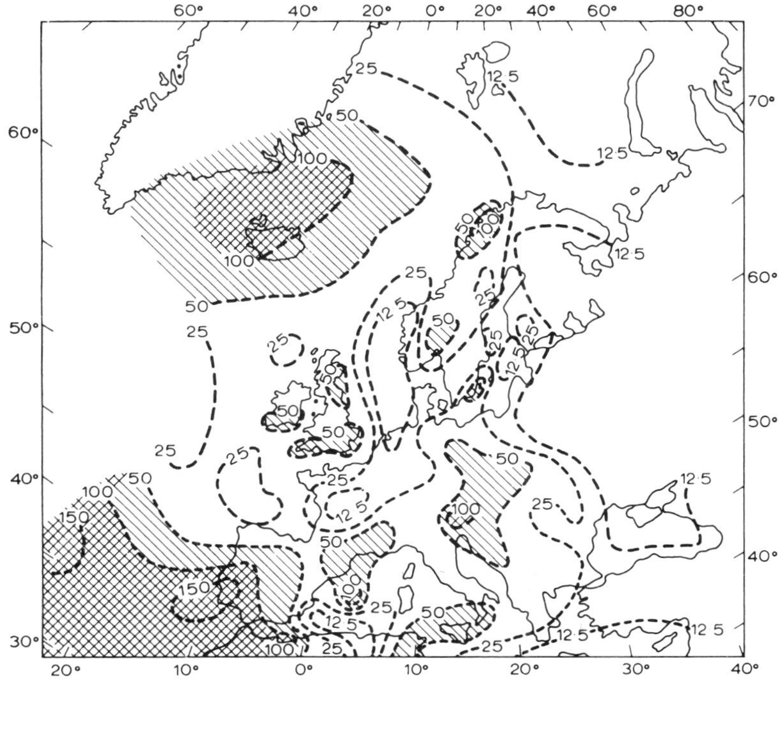

Figure 5.10 The mean precipitation anomaly, as a percentage of the average, during anticyclonic blocking in winter over Scandinavia. Areas above normal are cross-hatched, areas recording precipitation between 50 and 100 per cent of normal have oblique hatching.

Source: After Rex 1950.

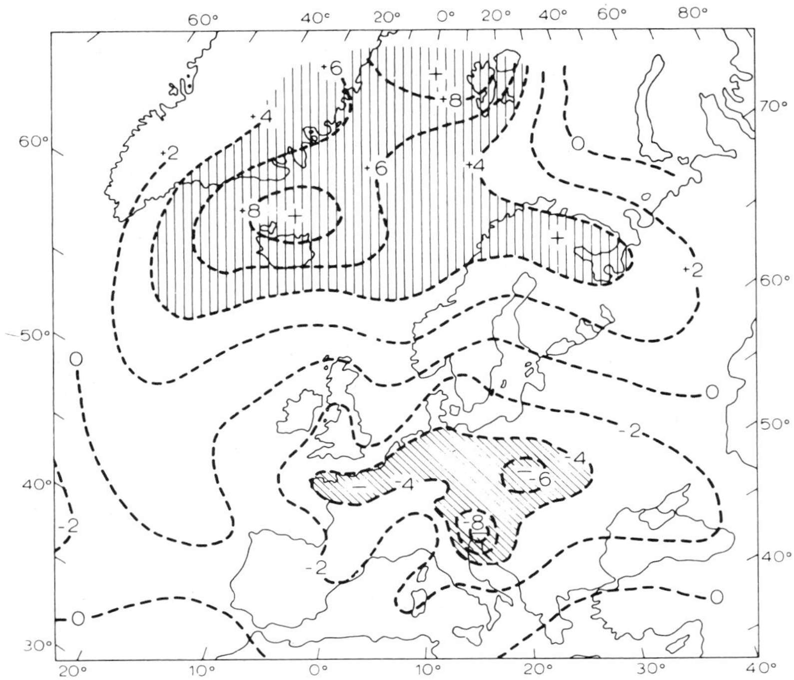

Figure 5.11 The mean surface temperature anomaly (°C) during anticyclonic blocking in winter over Scandinavia. Areas more than 4°C above normal have vertical hatching, those more than 4°C below normal have oblique hatching.

Source: After Rex 1950.

are steered around the blocking high following the path of the upper jet streams. Over most of Europe precipitation remains below average and this pattern is repeated with summer blocking. Winter temperatures are above average over the north-eastern Atlantic and adjoining land areas, but below average over central and eastern Europe and the Mediterranean due to the northerly outbreaks of cP air (see Figure 5.11). The negative temperature anomalies associated with cool northerly airflow in summer cover most of Europe, and only northern Scandinavia and the north-eastern Atlantic have above-average values.

Despite these generalizations the location of the block is of the utmost importance. For instance, in the summer of 1954 a blocking anticyclone across eastern Europe and Scandinavia allowed depressions to stagnate over the British Isles giving a dull, wet August, whereas in 1955 the blocking was located over the North Sea and a fine, warm summer resulted. The 1975–6 drought in Britain and the continent was caused by persistent blocking over north-western Europe. Another less common location of blocking is Iceland. A notable example was the 1962–3 winter when persistent high pressure south-east of Iceland led to northerly and north-easterly airflow over Britain. Temperatures in central England at that time were the lowest since 1740, with a mean of 0°C for December 1962 to February 1963. Central Europe was affected by easterly airstreams and mean January temperatures there were 6°C below average.

6 Topographic effects

In various parts of Europe topography has a marked effect on the climate, not only of the uplands themselves but also on that of adjacent areas. Apart from the more obvious effects upon temperatures, precipitation amounts and winds, the major mountain masses affect the movement of frontal systems too. Surface friction over mountain barriers tends to steepen the slope of

cold fronts and decrease the slope of warm fronts so that the latter are slowed down and the former accelerated. The cyclogenetic effect of mountain barriers in producing lee depressions has already been discussed (see Chapter 4, G.1).

The Scandinavian mountains are one of the most significant climatic barriers in Europe as a result of their orientation with regard to westerly airflow. Maritime air masses are forced to rise over the highland zone giving annual precipitation totals of over 250 cm (100 in) on the mountains of western Norway, whereas descent in their lee produces a sharp decrease in the amounts. The upper Gudbrandsdalen and Osterdalen in the lee of the Jotunheim and Dovre Mountains receive on average less than 50 cm (20 in), and similar low values are recorded in the Jämtland province of central Sweden around Östersund.

The mountains can equally function in the opposite sense. For example, Arctic air from the Barents Sea may move southwards in winter over the Gulf of Bothnia, usually when there is a depression over northern Russia, giving very low temperatures in Sweden and Finland. Western Norway is rarely affected, since the cold wave is contained to the east of the mountains. In consequence there is a sharp climatic gradient across the Scandinavian Highlands in the winter months.

The Alps provide a quite different illustration of topographic effects. Together with the Pyrenees and the mountains of the Balkans, the Alps effectively separate the Mediterranean climatic region from that of Europe. The penetration of warm air masses north of these barriers is comparatively rare and short-lived. However, with certain pressure patterns, air from the Mediterranean and north Italy is forced to cross the Alps, losing its moisture through precipitation on the southern slopes. Dry adiabatic warming on the northern side of the mountains can readily raise temperatures by 5°–6°C in the upper valleys of the Aar, Rhine and Inn. At Innsbruck there are approximately 50 days per year with föhn winds, with a maximum in spring, and these occurrences are often responsible for rapid melting of the snow, creating a risk of avalanches. When the airflow across the Alps has a northerly component föhns may occur in northern Italy, but their effects are generally less pronounced.

Features of upland climates in Britain will serve to illustrate some of the diverse effects of altitude. The mean annual rainfall on the west coasts near sea level is about 114 cm (45 in) but on the western mountains of Scotland, the Lake District and Wales averages exceed 380 cm (150 in) per year. The annual record is 653 cm (257 in) in 1954 at Sprinkling Tarn, Cumberland, and 145 cm (57 in) fell in a single month (October 1909) just east of the summit of Snowdon. The annual number of rain-days (days with at least 0.25 mm (0.01 in) of precipitation) increases from about 165 days in south-eastern England and the south coast to over 230 days in north-west Britain. There is little additional increase in the frequency of rainfall with height on the mountains of the north-west, so that the mean rainfall per rain-day rises sharply from 0.5 cm (0.2 in) near sea level in the west and north-west to over 1.3 cm (0.5 in) in the Western Highlands, the Lake District and Snowdonia. This demonstrates that 'orographic rainfall' here is primarily due to an intensification of the normal precipitation processes and is not a special type. It is more appropriate, therefore, to recognize an orographic component which increases the amounts of rain associated with frontal depressions and unstable airstreams (see Chapter 2, I.2).

Even quite low hills such as the Chilterns and South Downs cause a rise in rainfall, receiving about 12–13 cm (5 in) per year more than the surrounding lowlands. In South Wales, mean annual precipitation increases from 120 cm at the coast to 250 cm on the 500-m high Glamorgan Hills, 20 km inland. Studies using radar and a dense network of rain gauges indicate that orographic intensification is pronounced during strong low-level south-westerly airflows in frontal situations. Most of the enhancement of precipitation rate occurs in

the lowest 1500 m. Figure 5.12 shows the mean enhancement according to wind direction over England and Wales, averaged for several days with fairly constant wind velocities of about 20 m s^{-1} and nearly saturated low-level flow, attributable to a single frontal system on each day. Differences are apparent in Wales and southern England between winds from SSW and from WSW, whereas for SSE airflows the mountains of North Wales and the Pennines have little effect. Note also the areas of negative enhancement on the lee side of mountains.

The sheltering effects of the uplands produce low annual totals on the lee side (with respect to the prevailing winds). Thus, the lower Dee valley in the lee of the mountains of North Wales receives less than 75 cm (30 in) per year compared with over 250 cm (100 in) on Snowdonia.

The complexity of the various factors affecting rainfall in Britain is shown by the fact that a close correlation exists between annual totals in north-west Scotland, the Lake District and western Norway, which are directly affected by Atlantic depressions. At the same time there is an inverse relationship between annual amounts in the Western Highlands and lowland Aberdeenshire less than 240 km (150 miles) away. Annual precipitation in the latter area is more closely correlated with that in lowland eastern England. Essentially, the British Isles comprise two major climatic units for rainfall – first, an 'Atlantic' one with a winter season maximum, and, second, those central and

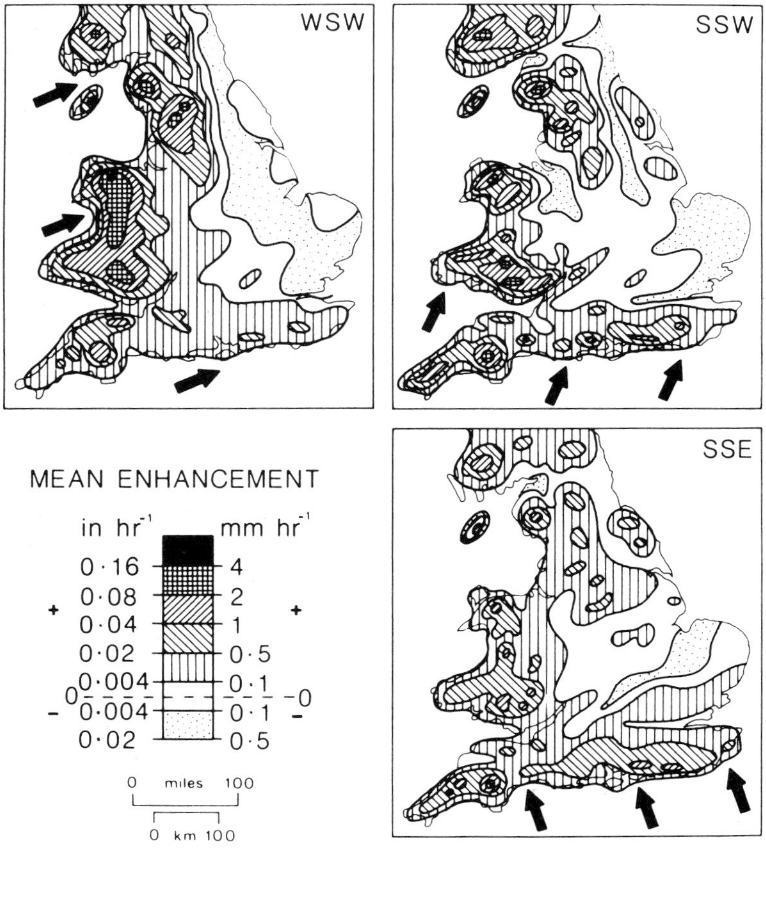

Figure 5.12 Mean orographic enhancement of precipitation over England and Wales, averaged for several days of fairly constant wind direction of about 20 m s^{-1} and nearly saturated low-level airflow.

Source: After Browning and Hill 1981 (*by permission of the Royal Meteorological Society*).

eastern districts with 'continental' affinities in the form of a weak summer maximum in most years. Other areas (eastern Ireland, eastern Scotland, north-east England and most of midland England and the Welsh border counties) generally have a wet second half of the year.

The occurrence of snow is another measure of altitude effects. Near sea level there are on average approximately 5 days per year with snow falling in south-west England, 15 days in the south-east and 35 days in northern Scotland. Between 60 and 300 m the frequency increases by about 1 day per 15 m of elevation and even more rapidly on higher ground. Approximate figures for northern Britain are 60 days at 600 m and 90 days at 900 m. The number of mornings with snow lying on the ground (more than half the ground covered) is closely related to mean temperature and hence altitude. Average figures range from about 5 days per year or less in much of southern England and Ireland, to between 30 and 50 days on the Pennines and over 100 days on the Grampian Mountains. In the last area (on the Cairngorms) and on Ben Nevis there are several semi-permanent snowbeds at about 1 160 m and it is estimated that the theoretical climatic snow-line – above which there would be *net* accumulation of snow – is at 1 620 m (5 300 ft) over Scotland.

The seasonal variability of lapse rates in mountain areas was mentioned in Chapter 1, I. There also exist marked geographical variations even within the British Isles. One measure of these variations is the length of the 'growing season'. Meteorological data can be used to determine an index of growth opportunity by counting the number of days on which the mean daily temperature exceeds an arbitrary threshold value – commonly 6°C (43°F). Along the south-west coasts of England the 'growing season', as calculated on this basis, is nearly 365 days per year and in this area it decreases by about 9 days per 30 m of elevation, but in northern England and Scotland the decrease is only about 5 days per 30 m from between 250–270 days near sea level. In continental climates the altitudinal decrease may be even more gradual; in central Europe and New England, for example, it is about 2 days per 30 m.

B NORTH AMERICA

The North American continent spans nearly 60° of latitude and, not surprisingly, exhibits a wide range of climatic conditions. Unlike Europe, the west coast is backed by the Pacific coast ranges rising to over 2750 m, which lie across the path of depressions in the mid-latitude westerlies and prevent the extension of maritime influences inland. In the interior of the continent there are no significant obstructions to air movement, and the absence of any east–west barrier allows air masses from the Arctic or the Gulf of Mexico to sweep across the interior lowlands, causing wide extremes of weather and climate. Maritime influences in eastern North America are greatly limited by the fact that the prevailing winds are westerly, so that the temperature regime is continental. Nevertheless, the Gulf of Mexico is a major source of moisture supply for precipitation over the eastern half of the United States and, as a result, the precipitation regimes are different from those found in eastern Asia.

We will look first at the broad characteristics of the atmospheric circulation over the continent.

1 Pressure systems

The mean pressure pattern for the middle troposphere displays a prominent trough over eastern North America in both summer and winter (see Figure 3.20). One theory is that this is a lee trough caused by the effect of the western mountain ranges on the upper westerlies, but at least in winter the strong baroclinic zone along the east coast of the continent is undoubtedly a major contributory factor. The implications of this mean wave pattern are that cyclones tend to move south-eastward over the Midwest carrying continental polar air southward, while the

cyclone paths are north-eastward along the Atlantic coast.

In individual months there may of course be considerable deviations from this average pattern, with important consequences for the weather in different parts of the continent, and, in fact, this relationship provides the basis for the monthly forecasts of the United States Weather Service. For example, if the trough is more pronounced than usual, temperatures may be much below average in the central, southern and eastern United States, whereas if the trough is weak the westerly flow is stronger with correspondingly less opportunity for cold outbreaks of polar air masses. Sometimes the trough is displaced to the western half of the continent causing a reversal of the usual weather pattern, since upper north-westerly airflow can bring cold, dry weather to the west while in the east there are very mild conditions associated with upper south-westerly flow. Precipitation amounts also depend on the depression tracks; if the upper trough is far to the west, depressions form ahead of it (see Chapter 4, E) over the south central United States and move north-eastwards towards the lower St Lawrence, giving more precipitation than usual in these areas and less along the Atlantic coast.

The major features of the surface pressure map in January (see Figure 3.25A) are the extension of the subtropical high over the south-western United States (called the Great Basin high) and the separate polar anticyclone of the Mackenzie district of Canada. Mean pressure is low off both the east and west coasts of higher middle latitudes, where oceanic heat sources indirectly give rise to the (mean) Icelandic and Aleutian lows. It is interesting to note that, on average, in December, of any region in the northern hemisphere for any month of the year, the Great Basin region has the most frequent occurrence of highs, whereas the Gulf of Alaska has the maximum frequency of lows. The Pacific coast as a whole has its most frequent cyclonic activity in winter as does the Great Lakes area, whereas over the Great Plains the maximum is in spring and early summer. Remarkably, the Great Basin in June has the most frequent cyclogenesis of any part of the northern hemisphere in any month of the year. Heating over this area in summer helps to maintain a shallow, quasi-permanent low-pressure cell, in marked contrast with the almost continuous subtropical high-pressure belt in the middle troposphere (see Figure 3.20). Continental heating also indirectly assists in the splitting of the Icelandic low to create a secondary centre over north-eastern Canada. The west-coast summer circulation is dominated by the Pacific anticyclone, while the south-eastern United States is affected by the Atlantic subtropical anticyclone cell (see Figure 3.25 B).

Broadly, there are three prominent depression tracks across the continent in winter (see Figure 4.19). One group moves from the west along a more or less zonal path about 45°–50°N, whereas a second loops southward over the central United States and then turns north-eastward towards New England and the Gulf of St Lawrence. Some of these depressions originate over the Pacific, cross the western ranges as an upper trough and redevelop in the lee of the mountains. Alberta is a noted area for this process and also for primary cyclogenesis since the arctic frontal zone is over north-west Canada in winter. This frontal zone involves much-modified mA air from the Gulf of Alaska and cold dry cA (or cP) air. Depressions of the third group form along the main polar frontal zone, which in winter is off the east coast of the United States, and move north-eastward towards Newfoundland. Sometimes this frontal zone is present over the continent about 35°N with mT air from the Gulf and cP air from the north or modified mP air from the Pacific. Polar front depressions forming over Colorado move north-eastward towards the Great Lakes and others developing over Texas follow a more or less parallel path, further to the south and east, towards New England.

Between the Arctic and Polar Fronts a third frontal zone is distinguished by Canadian meteorologists. This *maritime* (arctic) frontal zone is present when mA and mP (or mPc and

mPw) air masses interact along their common boundary. The three-front (i.e. four air-mass) model allows a detailed analysis to be made of the baroclinic structure of depressions over the North American continent using synoptic weather maps and cross-sections of the atmosphere. Figure 5.13 illustrates the three frontal zones and associated depressions on 29 May 1963. Along 95°W, from 60°N to 40°N, the following dew-point temperatures were reported in the four air masses: −8°C (17°F), 1°C (33°F), 4°C (40°F) and 13°C (55°F).

In summer, the east-coast depressions are less frequent and the tracks across the continent are displaced northwards with the main ones moving over Hudson Bay and Labrador–Ungava, or along the line of the St Lawrence. These are associated mainly with a rather poorly defined Maritime frontal zone. The Arctic Front is usually located along the north coast of Alaska, where there is a strong temperature gradient between the bare land and the cold Polar Sea and pack-ice. East from here the front is very variable in location from day to day and year to year. Broadly, it occurs most often in the vicinity of northern Keewatin and Hudson Strait, although one study of air-mass temperatures and airstream confluence regions suggests that an arctic frontal zone occurs further south over Keewatin in July and that its mean position (see Figure 5.14) is closely related to the boreal forest–tundra boundary. This relationship undoubtedly reflects the importance of arctic air-mass dominance for summer temperatures and consequently for tree-growth possibilities, but the precise nature of the interrelationships between atmospheric systems and vegetation boundaries requires more extensive investigation.

Several circulation singularities have been recognized in North America, as in Europe (see A.4, this chapter). Three which have received considerable attention in view of their prominence are: (a) the advent of spring in late March; (b) the midsummer high-pressure jump at the end of June; (c) the Indian summer in late September (and late October).

The arrival of spring is marked by different

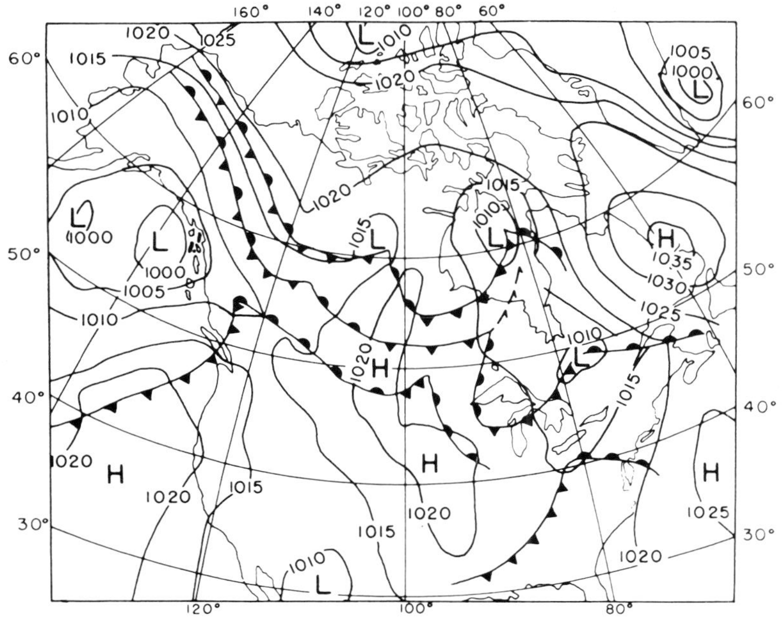

Figure 5.13 A synoptic example of depressions associated with three-frontal zones on 29 May 1963 over North America.

Source: Based on charts of the Edmonton Analysis Office and the Daily Weather Report.

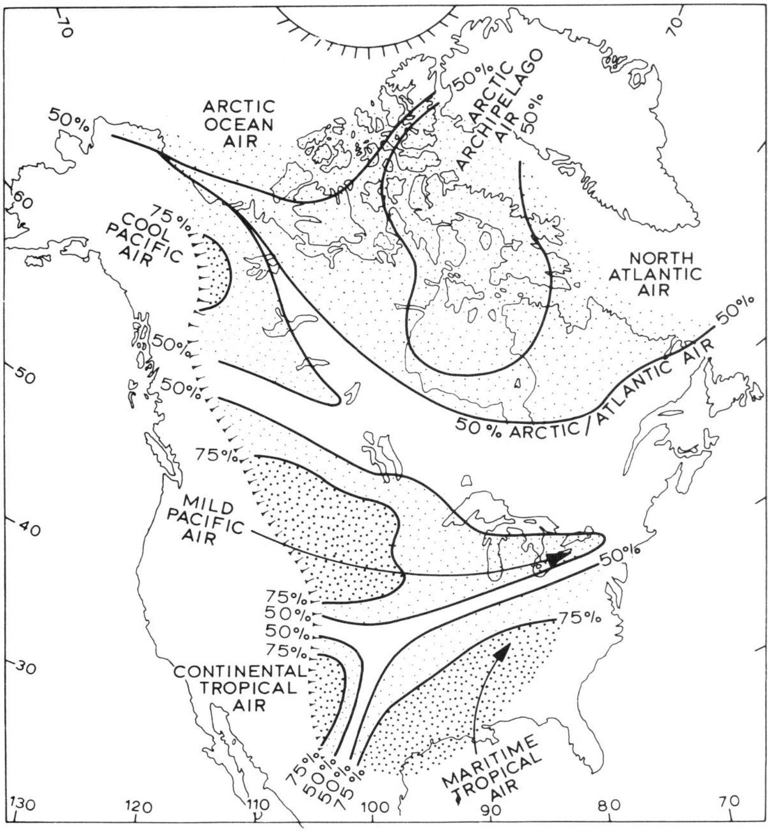

Figure 5.14 Regions in North America east of the Rocky Mountains dominated by the various air-mass types in July for more than 50 per cent and 75 per cent of the time. The 50 per cent frequency lines correspond to mean frontal positions.

Source: After Bryson 1966.

climatic responses in different parts of the continent. For example, there is a sharp decrease in March to April precipitation in California, due to the extension of the Pacific high, whereas precipitation intensity increases in the Midwest (see Figure 5.15) as a result of more frequent cyclogenesis in Alberta and Colorado and a northward extension of maritime tropical air over the Midwest from the Gulf of Mexico. These changes are part of a hemispheric readjustment of the circulation since at the beginning of April the Aleutian low-pressure cell, which from September to March is located about 55°N, 165°W, splits into two with one centre in the Gulf of Alaska and the other over northern Manchuria. This represents a decrease in the zonal index (see Chapter 3, F.2).

In late June there is a rapid northward displacement of the Bermuda and North Pacific subtropical high-pressure cells. In North America this pushes the depression tracks northward also with the result that precipitation decreases from June to July over the northern Great Plains, parts of Idaho and eastern Oregon (see Figure 5.15). Conversely, the southwesterly anticyclonic flow which affects Arizona in June is replaced by air from the Gulf of California and this causes the onset of the summer rains (see D.2, this chapter). It has been suggested by Bryson and Lahey that these circulation changes at the end of June may be connected with the disappearance of snow cover from the arctic tundra. This leads to a sudden decrease of surface albedo from about 75 to 15 per cent with consequent changes in the heat budget components and hence in the atmospheric circulation.

Frontal wave activity makes the first half of

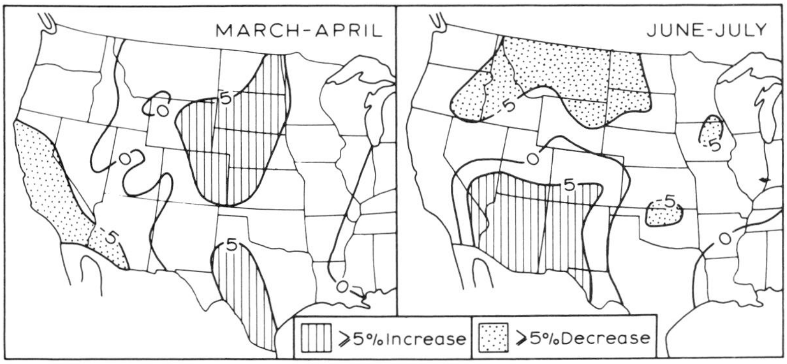

Figure 5.15 The precipitation changes between March–April (*left*) and June–July (*right*), as a percentage of the mean annual total, for the central and western United States.

Source: After Bryson and Lahey 1958.

September a rainy period in the northern Midwest states of Iowa, Minnesota and Wisconsin, but after about the 20th of the month anticyclonic conditions return with warm airflow from the dry south-west giving fine weather – the so-called Indian summer. Significantly, the hemispheric zonal index value rises in late September. This anticyclonic weather type has a second phase in the latter half of October, but at this time there are polar outbreaks. The weather is generally cold and dry, although if precipitation does occur there is a high probability of it being in the form of snow.

2 The temperate west coast and cordillera

The oceanic circulation of the North Pacific closely resembles that of the North Atlantic. The drift from the Kuro Shio current off Japan is propelled by the westerlies towards the western coast of North America and acts as a warm current between 40° and 60°N. Sea-surface temperatures are several degrees lower than in comparable latitudes off western Europe, however, due to the smaller volume of warm water involved. Also, in contrast to the Norwegian Sea, the shape of the Alaskan coastline prevents the extension of the drift to high latitudes (see Figure 3.42).

The Pacific coast ranges greatly restrict the inland extent of oceanic influences, and hence there is no extensive maritime temperate climate such as we have described for western Europe. The major climatic features duplicate those of the coastal mountains of Norway and those of New Zealand and southern Chile in the belt of Southern Westerlies. Topographic factors make the weather and climate of such areas very variable over short distances both vertically and horizontally, and, therefore, only a few salient characteristics are selected for consideration.

There is a regular pattern of rainy windward and drier lee slopes across the successive north-west to south-east ranges with a more general decrease towards the interior. The Coast Range in British Columbia has mean annual totals of precipitation exceeding 250 cm (100 in) with 500 cm (200 in) in the wettest places compared with 125 cm (50 in) or less on the summits of the Rockies, yet even on the leeward side of Vancouver Island the average figure at Victoria is only 70 cm (27 in). Analogous to the 'Westerlies-oceanic' regime of north-west Europe, there is a winter precipitation maximum along the littoral which also extends beyond the Cascades (in Washington) and the Coast Range (in British Columbia), but summers are drier due to the strong North Pacific anticyclone. The regime in the interior of British Columbia is transitional between that of the coastal region and the distinct summer maximum of central North America (see Figure 5.16), although at Kamloops in the Thompson valley (annual average 25 cm or 10 in) there is a slight summer maximum associated with thunderstorm type of rainfall. In general, the sheltered interior valleys receive less than 50 cm

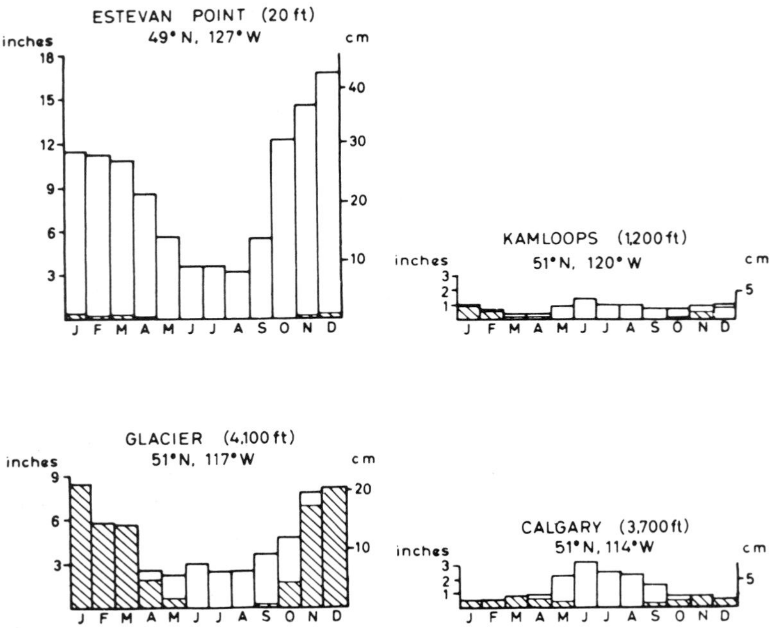

Figure 5.16 Precipitation graphs for stations in western Canada. The shaded portions represent snowfall, expressed as water equivalent. Elevations are 6 m, 366 m, 1250 m and 1128 m, respectively from top left to bottom right.

(20 in) per year and in the driest years certain localities have recorded only 15 cm (6 in). Above 1000 m much of the precipitation falls as snow (see Figure 5.16) and some of the greatest snow-depths in the world are reported from British Columbia, Washington and Oregon. For example, between 1000 and 1500 cm (400 and 600 in) falls on the Cascade Range at heights of about 1500 m (5000 ft) and even as far inland as the Selkirk Mountains the totals are considerable. The mean snowfall is 990 cm (390 in) at Glacier, British Columbia (elevation 1250 m) and this accounts for almost 70 per cent of the annual precipitation (see Figure 5.16). Near sea level on the outer coast, in contrast, very little precipitation falls as snow (for example, Estevan Point). It is estimated that the climatic snowline rises from about 1600 m (5250 ft) on the west side of Vancouver Island to 2900 m (9500 ft) in the eastern Coast Range. Inland its elevation increases from 2300 m (7550 ft) on the west slopes of the Columbia Mountains to 3100 m (10,170 ft) on the east side of the Rockies. This trend reflects the precipitation pattern referred to above.

Finally, mention must be made of the large diurnal variations which affect the cordilleran valleys. Strong diurnal rhythms of temperature (especially in summer) and wind direction are a feature of mountain climates and their effect is superimposed upon the general climatic characteristics of the area. Cold air drainage produces many remarkably low minima in the mountain valleys and basins. At Princeton, British Columbia (elevation 695 m), where the mean daily minimum in January is −14°C (7°F), there is on record an absolute low of −45°C (−49°F), for example. This leads in some cases to reversal of the normal lapse rate. Golden in the Rocky Mountain Trench has a January mean of −12°C (11°F), whereas 460 m (1500 ft) higher at Glacier (1250 m) it is −10°C (14°F).

3 Interior and eastern North America

Central North America has the typical climate of a continental interior with hot summers and cold winters (see Figures 5.17 and 5.18), yet the weather in winter is subject to marked variability. This is determined by the steep

temperature gradient between the Gulf of Mexico and the snow-covered northern plains; also by shifts of the upper wave patterns and jet stream. Cyclonic activity in winter is much more pronounced over central and eastern North America than in Asia, which is dominated by the Siberian anticyclone (see Figure 4.19), and consequently there is no climatic type with a winter minimum of precipitation in eastern North America.

The general temperature conditions in winter and summer are illustrated in Figure 5.17, showing the frequency with which hourly temperature readings exceed or fall below certain limits. The two chief features of all four maps are: (a) the dominance of the meridional temperature gradient, away from coasts, and (b) the continentality of the interior and east compared with the 'maritimeness' of the west coast. On the July maps additional influences are evident and these are referred to below.

a Continental and oceanic influences

The Labrador coast is fringed by the waters of a cold current, analogous to the Oya Shio off eastern Asia, but in both cases the prevailing westerlies greatly limit their climatic significance. The Labrador Current maintains drift ice off Labrador and Newfoundland until June and gives very low summer temperatures along the Labrador coast (see Figure 5.17C). The lower incidence of freezing temperatures in this area in January is related to the movement of some depressions into the Davis Strait, carrying Atlantic air northwards. A major role of the Labrador Current is in the formation of fog. Advection fogs are very frequent between May and August off Newfoundland, where the Gulf Stream and Labrador Current meet. Warm, moist, southerly airstreams are cooled rapidly over the cold waters of the Labrador Current and with steady, light winds such fogs may persist for several days, creating hazardous conditions for shipping. Southward-facing coasts are particularly affected and at Cape Race (Newfoundland), for example, there are on average 158 days per year with fog (visibility less than 1 km) at some time of day. The summer concentration is shown by the figures for Cape Race during individual months: May – 18 (days), June – 18, July – 24, August – 21, and September – 15.

Oceanic influence along the Atlantic coasts of the United States is very limited, and although there is some moderating effect on minimum temperatures at coastal stations this is scarcely evident on generalized maps such as Figure 5.17. More significant climatic effects are in fact found in the neighbourhood of Hudson Bay and the Great Lakes. Hudson Bay remains very cool in summer with water temperatures of about 7°–9°C and this depresses temperatures along its shore, especially in the east (see Figure 5.17C and D). Mean July temperatures are 12°C (54°F) at Churchill (59°N) and 8°C (47°F) at Inukjuak (58°N), on the west and east shores respectively, compared for instance with 13°C (56°F) at Aklavik (68°N) on the Mackenzie delta. The influence of Hudson Bay is even more striking in early winter when the land is snow-covered. Westerly airstreams crossing the open water are warmed on average by 11°C (20°F) in November, and moisture added to the air leads to considerable snowfall in western Ungava (see the graph for Inukjuak, Figure 5.21). By the beginning of January the Bay is frozen over almost entirely and no effects are evident. The Great Lakes influence their surroundings in much the same way. Heavy winter snowfalls are a notable feature of the southern and eastern shores of the Great Lakes. In addition to contributing moisture to north-westerly streams of cold cA and cP air, the heat source of the open water in early winter produces a low-pressure trough which increases the snowfall as a result of convergence. Yet a further factor is frictional convergence and orographic uplift at the shoreline. Mean annual snowfall exceeds 250 cm (100 in) along much of the eastern shore of Lake Huron and Georgian Bay, the south-eastern shore of Lake Ontario, the north-eastern shore of Lake Superior and its southern shore east of about 90° 30′W. Extremes include 114 cm (45 in) in one day at Watertown, New York,

Figure 5.17 The percentage frequency of hourly temperatures above or below certain limits for North America. A: January temperatures < 0°C; B: January temperatures > 10°C; C: July temperatures < 10°C; D: July temperatures > 21°C.

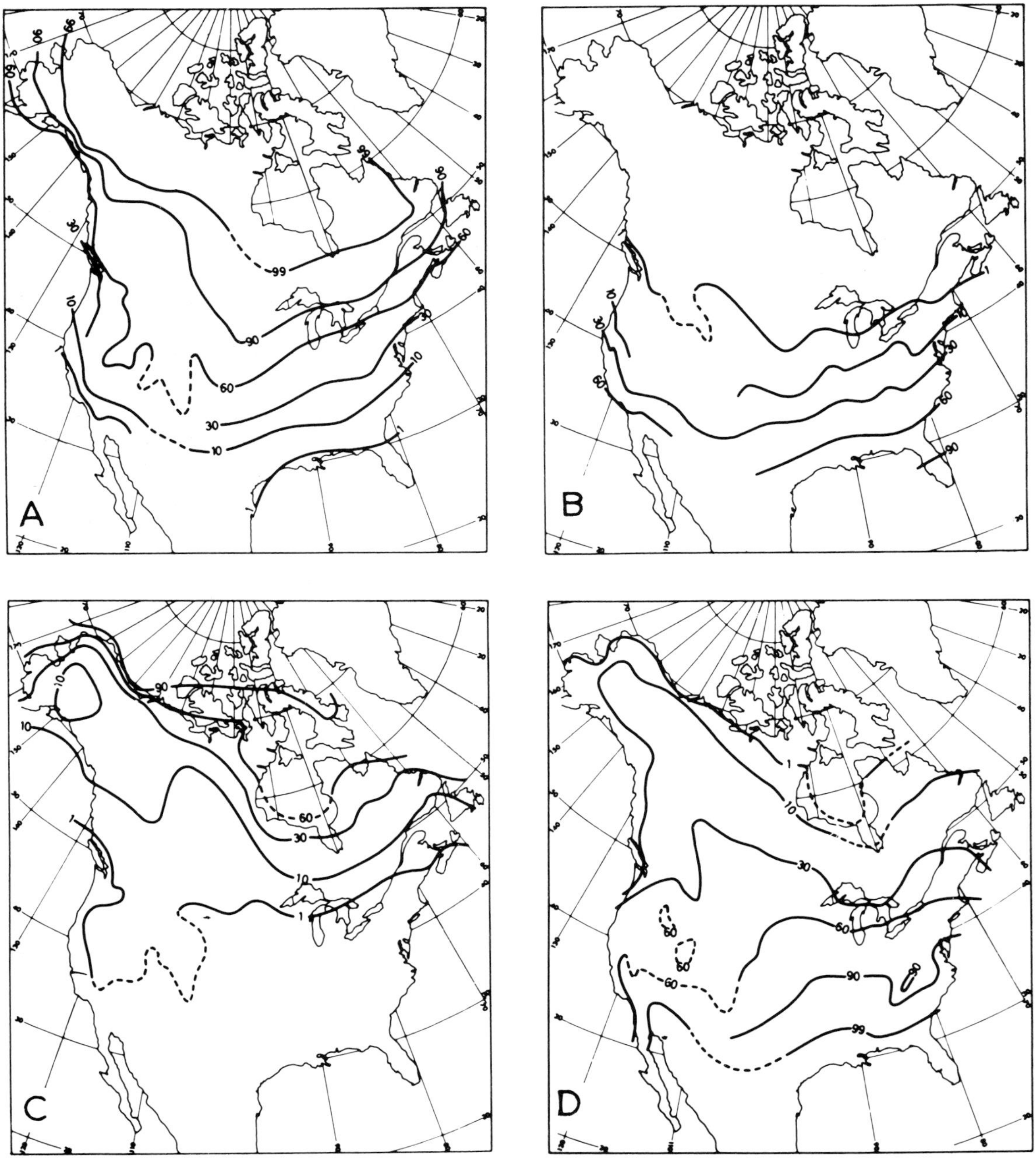

Source: After Rayner 1961.

and 894 cm (352 in) during the 1946–7 winter season at nearby Bennetts Bridge, both of which are close to the eastern end of Lake Ontario.

Transport in cities in these snow belts is quite frequently disrupted during winter snowstorms. The Great Lakes also provide an important tempering influence during winter months by raising average daily minimum temperatures at lakeshore stations by some 2°–4°C above those at inland locations. In mid-December the upper 60 m of Lake Erie has a uniform temperature of 5°C.

An indication of the seasonal range of temperature is provided by Figure 5.18, showing continentality (k) based on Conrad's formula:

$$k = \frac{1.7\,A}{\sin(\phi + 10)} - 14$$

where A is the average annual temperature range in °C and ϕ is the latitude angle. The results in middle and high latitudes are similar to those obtained by Gorczynski's method (see A.2, this chapter); with either of these empirical expressions it is only the relative magnitude of k that is of interest. The highest values form a tongue along the 100°W meridian with subsidiary areas on the 'Lake Plateau' of central Labrador–Ungava and on the high plateaux of Colorado and Utah. The 'maritimeness' of the Pacific coast, though of very limited inland extent, is pronounced, whereas on the east coast there is relatively high continentality. The map also illustrates the ameliorating effect of the Great Lakes.

b Warm and cold spells

Two types of synoptic condition are of particular significance for temperatures in the interior of North America. One is the 'cold wave' caused by a northerly outbreak of cP air, which in winter regularly penetrates deep into the central and eastern United States and occasionally affects even the Gulf coast, injuring frost-sensitive crops. Cold waves are arbitrarily defined as a temperature drop of at least 11°C (20°F) in 24 hours over most of the United States, and at least 9°C (16°F) in California, Florida and the Gulf Coast, to below a specified minimum depending on location and season. The winter criterion decreases from 0°C in California, Florida and the Gulf Coast to −18°C

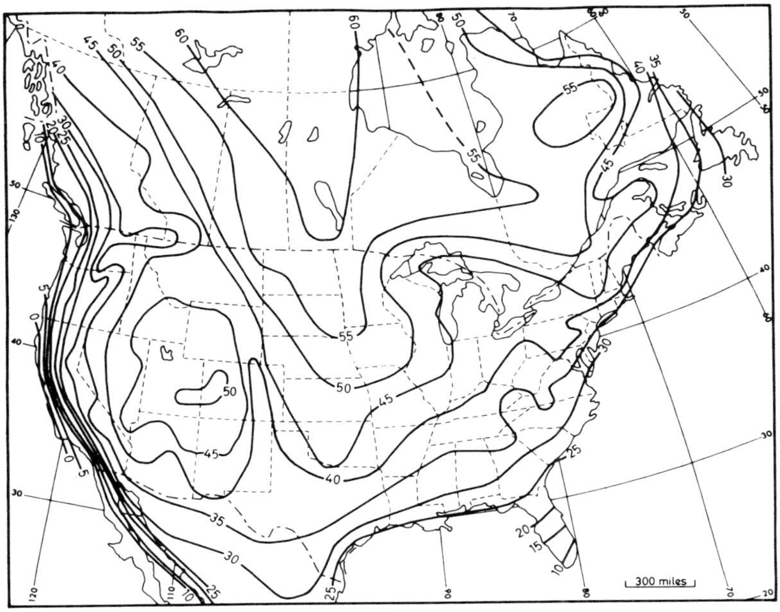

Figure 5.18 Continentality in North America according to Conrad's index.

Source: Modified after Trewartha 1981.

(0°F) over the northern Great Plains and the north-eastern States. The cold spells commonly occur with the build-up of a north–south anticyclone in the rear of a cold front. The polar air gives clearer, dry weather with strong, cold winds, although if the winds follow snowfall, fine, powdery snow may be whipped up by the wind, creating blizzard conditions. These are quite common over the northern plains.

Another type of temperature fluctuation is associated with the *Chinook* winds in the lee of the Rockies (see Chapter 3, C.2). The Chinook is particularly warm and dry as air of Pacific origin, after losing its moisture over the mountains, descends the eastern slopes and warms at the dry adiabatic lapse rate. The onset of the Chinook produces temperatures well above the seasonal normals so that snow is often thawed rapidly, and in fact the Indian word Chinook means snow-eater. Temperature rises of as much as 22°C (40°F) have been observed in 5 minutes and the occurrence of such warm spells is reflected by the high extreme maxima in winter months at Medicine Hat (see Figure 5.19). In Canada the Chinook effect may be observed a considerable distance from the Rockies into south-west Saskatchewan, but in Colorado its influence is rarely felt more than about 50 km from the foothills. No adequate definition of a Chinook has yet been established, but using an arbitrary criterion of winter days with a maximum temperature of at least 4.4°C (40°F), R. W. Longley has shown that in the Lethbridge area Chinooks occur on 40 per cent of days during December–February. However, since the phenomenon is the result of a particular type of airflow, it is evident that some wind characteristic should be included in future definitions.

Chinook conditions commonly develop in a Pacific airstream which is replacing a winter high-pressure cell over the high western plains. Sometimes the cold, stagnant cP air of the anticyclone is not dislodged by the descending Chinook and a marked inversion is formed, but on other occasions the boundary between the two air masses may reach ground level locally and, for example, the western suburbs of

Figure 5.19 Mean and extreme temperatures at Medicine Hat, Alberta.

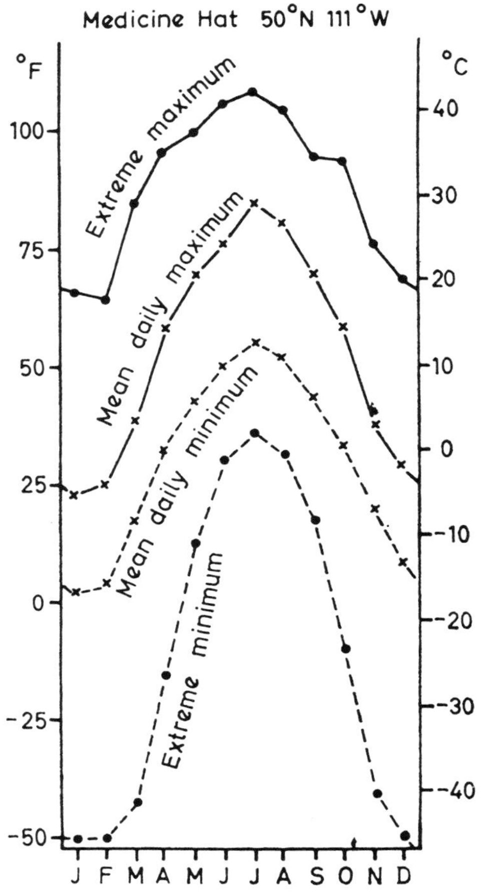

Calgary may record temperatures above 0°C while those to the east of the city remain below −15°C.

c Precipitation and the moisture balance

Longitudinal influences are apparent in the distribution of annual precipitation, although this is in large measure a reflection of the topography. The 60 cm annual isohyet in the United States approximately follows the 100°W meridian (see Figure 5.20), and westwards to the Rockies is an extensive dry belt in the rain shadow of the western mountain ranges. In the south-east totals exceed 125 cm, and 100 cm or more is received along the Atlantic coast as far

as New Brunswick and Newfoundland.

The major sources of moisture for precipitation over North America are the Pacific Ocean and the Gulf of Mexico. The former need not concern us here since comparatively little of the precipitation falling over the interior appears to be derived from that source. The Gulf source is extremely important in providing moisture for precipitation over central and eastern North America, but the predominance of south-westerly airflow means that little precipitation falls over the western Great Plains (see Figure 5.20). Over the south-eastern United States there is considerable evapotranspiration and this helps to maintain moderate annual totals northwards and eastwards from the Gulf by providing additional water vapour for the atmosphere. Along the east coast the Atlantic Ocean is an additional significant source of moisture for winter precipitation.

There are at least eight major types of seasonal precipitation regime in North America (see Figure 5.21); the winter maximum of the west coast and the transition type of the intermontane region in middle latitudes have already been mentioned and the subtropical types are discussed in the next section. Four primarily mid-latitude regimes are distinguished east of the Rocky Mountains:

1 A warm season maximum is found over much of the continental interior (e.g. Rapid City). In an extensive belt from New Mexico to the Prairie Provinces more than 40 per cent of the annual precipitation falls in summer. In New Mexico, the rain occurs mainly with late summer thunderstorms, but May–June is the wettest time over the central and northern Great Plains due to more frequent cyclonic activity. Winters are quite dry over the Plains, but the mechanism of the occasional heavy snowfalls is of interest. They occur over the north-western Plains during easterly upslope flow, usually in a ridge of high pressure.

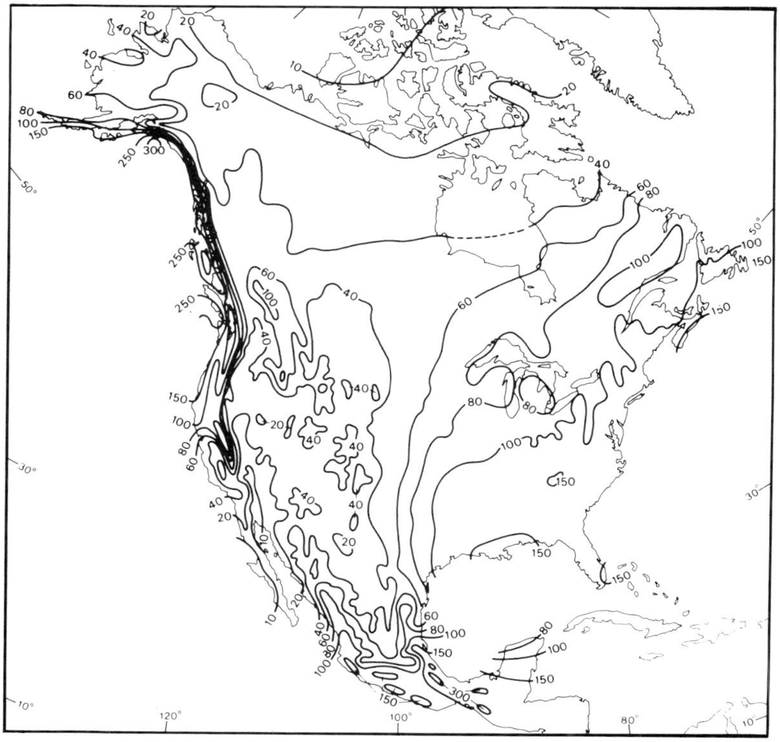

Figure 5.20 Mean annual precipitation (in centimetres) over North America for the period 1931–60. Isohyets in the Arctic underestimate the true totals by 30–50 per cent due to problems in recording snowfall accurately with precipitation gauges.

Source: Based on Hare and Hay, Court, and Mosiño Aleuán and Garcia).

Figure 5.21 North American rainfall regime regions and histograms showing mean monthly precipitations for each region (January, June and December are indicated). Note that the jet stream is anchored by the Rockies in more or less the same position at all seasons.

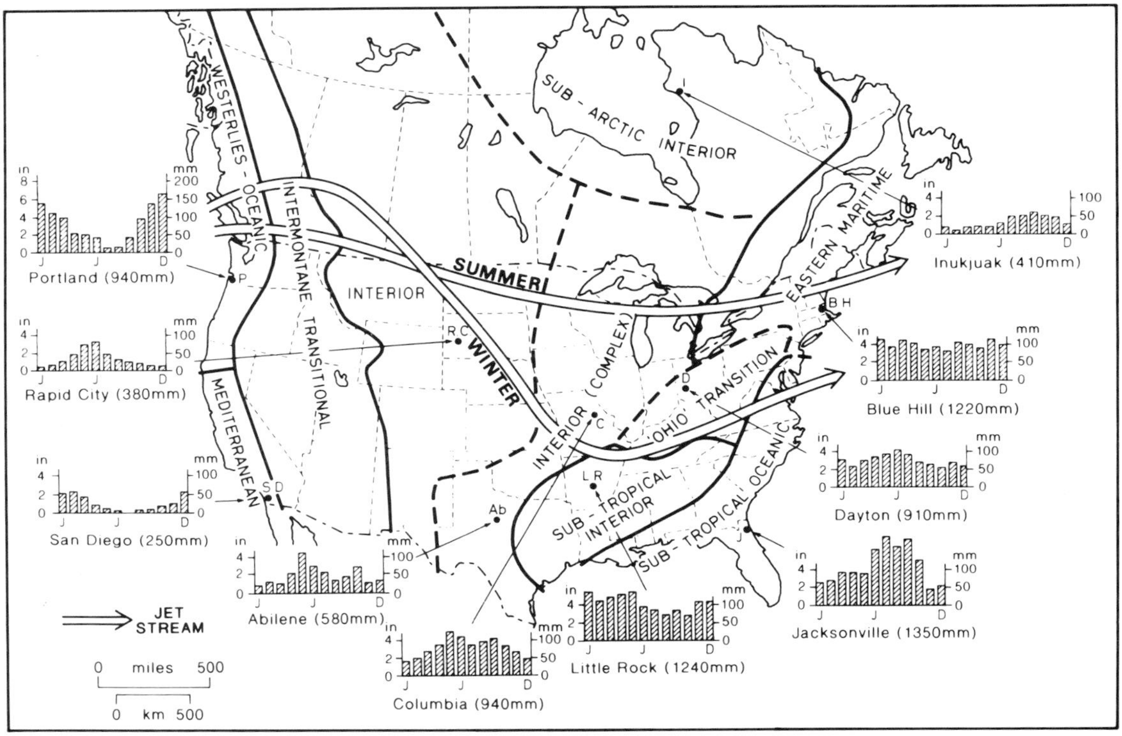

Source: Mostly after Trewartha 1981; additions by Henderson-Sellers and Robinson 1986.

Further north in Canada the maximum is commonly in late summer or autumn when depression tracks are in higher middle latitudes. There is a local maximum in autumn on the eastern shores of Hudson Bay (e.g. Inukjuak) due to the effect of open water.

2 Eastward and southward of the first zone there is a double maximum in May and September. In the upper Mississippi region (e.g. Columbia) there is a secondary minimum, paradoxically in July–August when the air is especially warm and moist, and a similar profile occurs in northern Texas (e.g. Abilene). An upper-level ridge of high pressure over the Mississippi valley seems to be responsible for reduced thunderstorm rainfall in midsummer, and a tongue of subsiding dry air extends southwards from this ridge towards Texas. In September, renewed cyclonic activity associated with the seasonal southward shift of the polar front, at a time when mT air from the Gulf is still warm and moist, causes a resumption of rainfall. Subsequently, however, drier westerly airstreams affect the continental interior as the general airflow becomes more zonal.

The diurnal occurrence of precipitation in the central United States is rather unusual for a continental interior. Sixty per cent or more of the summer precipitation falls during nocturnal thunderstorms (2000–0800 hr True Solar Time) in central Kansas, parts of Nebraska, Oklahoma and Texas. Hypotheses suggest that the nocturnal thunderstorm rain-

fall which occurs, especially with extensive mesoscale convective systems (see p. 169) may be linked to a tendency for nocturnal convergence and rising air over the plains east of the Rocky Mountains. The terrain profile appears to play a role here, as a large-scale inversion layer forms at night over the mountains setting up a low-level jet east of the mountains just above the boundary layer. This southerly flow, at 500–1000 m above the surface, can supply the necessary low-level moisture influx and convergence for the storms (see Figure 4.34).

3 East of the upper Mississippi, in the Ohio valley and south of the lower Lakes, there is a transitional regime between that of the interior and the east-coast type. Precipitation is reasonably abundant in all seasons but the summer maximum is still in evidence (e.g. Dayton).

4 In eastern North America (New England, the Maritimes, Quebec and south-east Ontario) precipitation is fairly evenly distributed throughout the year (e.g. Blue Hill). In Nova Scotia and locally around Georgian Bay there is a winter maximum, due in the latter case to the influence of open water. In the Maritimes it is related to winter (and also autumn) storm tracks.

It is worth comparing the eastern regime with the summer maximum which is found over eastern Asia. There the Siberian anticyclone excludes cyclonic precipitation in winter and monsoonal influences are felt in the summer months.

The seasonal distribution of precipitation is of vital interest for agricultural purposes. Rain falling in summer, for instance, when evaporation losses are high is less effective than an equal amount in the cool season. Figure 5.22 illustrates the effect of different regimes in terms of the moisture balance, calculated according to Thornthwaite's method. At Halifax (Nova Scotia) there is sufficient moisture stored in the soil to maintain evaporation at its maximum rate (i.e. actual evaporation = potential evaporation), whereas at Berkeley (California) there is a computed moisture deficit of nearly 5 cm in August. This is a guide to the amount of irrigation water which may be required by crops, although in dry regimes the Thornthwaite method generally underestimates the real moisture deficit.

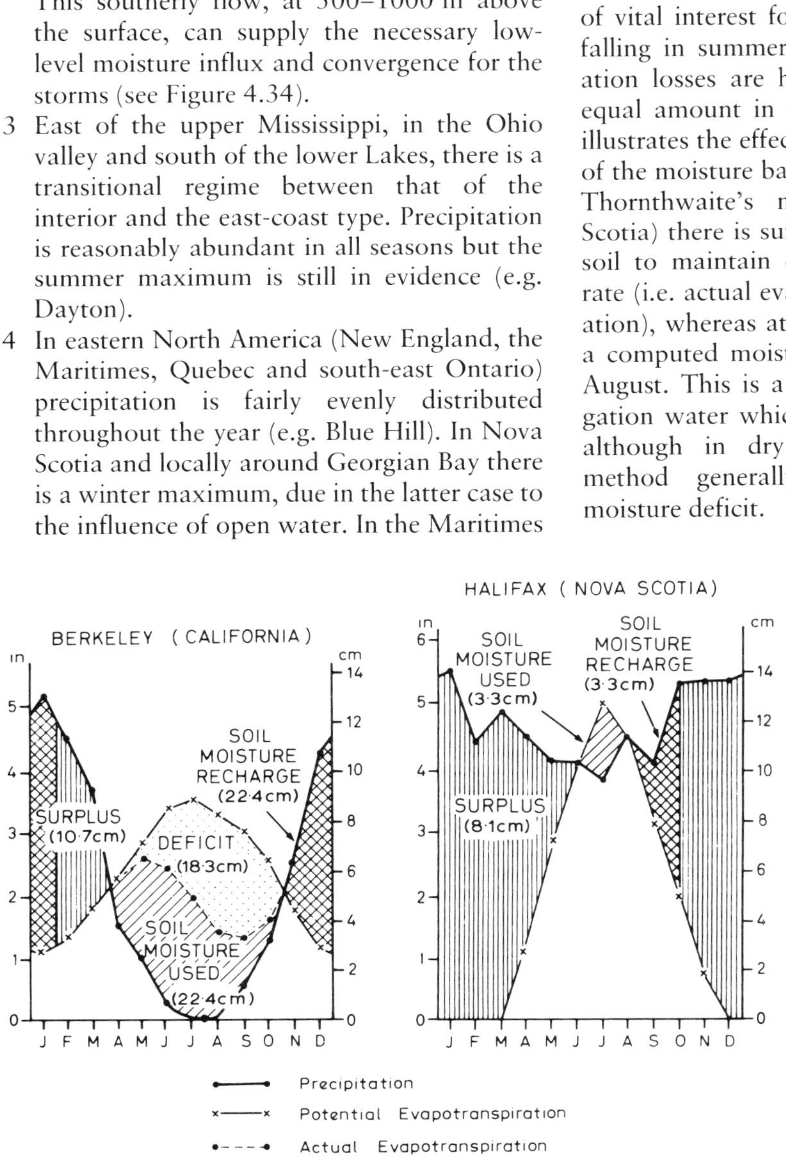

Figure 5.22 The moisture balances at Berkeley, California and Halifax, Nova Scotia.

Source: After Thornthwaite and Mather 1955.

The mean monthly potential evaporation (*PE*) is calculated in the method developed by C. W. Thornthwaite from tables based on a complex equation relating *PE* to air temperature. This only gives a general guide to the true *PE* in view of the different factors which affect evaporation (see Chapter 2, A), but the results are reasonably satisfactory in temperate latitudes. By determining the annual moisture surplus and the annual moisture deficit from graphs such as those in Figure 5.22, or from a monthly 'balance sheet', Thornthwaite obtained an index of aridity and humidity. The humidity index is 100 × water surplus/*PE* and the aridity index is 100 × water deficit/*PE*; there is generally a surplus in one season and a deficit in another. These may then be combined into a single moisture index (*Im*):

$$Im = \frac{100(\text{water surplus} - \text{water deficit})}{PE}$$

Here the deficit is given equal weight as the surplus held in the soil; any deficit means that the actual evaporation rate falls below the potential value after the soil water is used. The moisture index (see Table 5.3) is used to define certain climatic types (see also Appendix 1.B).

Figure 5.23 illustrates the distribution of these moisture regions over North America. The boundary separating the moist climates of the east from the dry climates of the west (apart from the west coast) follows the 95th meridian. The major humid areas are along the Appalachians, in the north-east and along the Pacific coast, while the most extensive arid areas are in the intermontane basins, the High Plains, the American Southwest and parts of northern Mexico. Some aspects of the precipitation climatology of this arid area are examined in D.2, this chapter.

Table 5.3 The moisture index.

Im	*climate*	*symbol*
> 100	perhumid	A
20 to 100	humid	B (with 4 subdivisions)
0 to 20	moist subhumid	C_2
−33 to 0	dry subhumid	C_1
−67 to −33	semi-arid	D
−100 to −67	arid	E

C THE POLAR MARGINS

The longitudinal differences in mid-latitude climates persist into the polar margins, giving rise to maritime and continental sub-types, modified by the extreme radiation conditions in winter and summer. For example, radiation receipts in summer along the arctic coast of Siberia compare favourably, by virtue of the long daylight, with those in lower middle latitudes. The maritime type is found in coastal Alaska, Iceland, northern Norway and adjoining parts of Russia. Winters are cold and stormy, with very short days. Summers are cloudy but mild with mean temperatures about 10°C. For example, Vardø in north Norway (70°N, 31°E) has monthly mean temperatures of −6°C (21°F) in January and 9°C (48°F) in July, while Anchorage, Alaska (61°N, 150°W) records −11°C (12°F) and 14°C (57°F), respectively. Annual precipitation is generally between 60 and 125 cm (25 and 50 in), with a cool season maximum and about 6 months of snow cover.

The weather is mainly controlled by depressions which are weakly developed in summer. In winter the Alaskan area is north of the main depression tracks and occluded fronts and upper troughs (trowals) are prominent, whereas northern Norway is affected by frontal depressions moving into the Barents Sea. Iceland is similar to Alaska, though depressions often move slowly over the area and occlude, whereas others moving north-eastwards along the Denmark Strait bring mild, rainy weather.

The interior, cold-continental climates have much more severe winters although precipitation amounts are less. At Yellowknife (62°N, 114°W), for instance, the mean January temper-

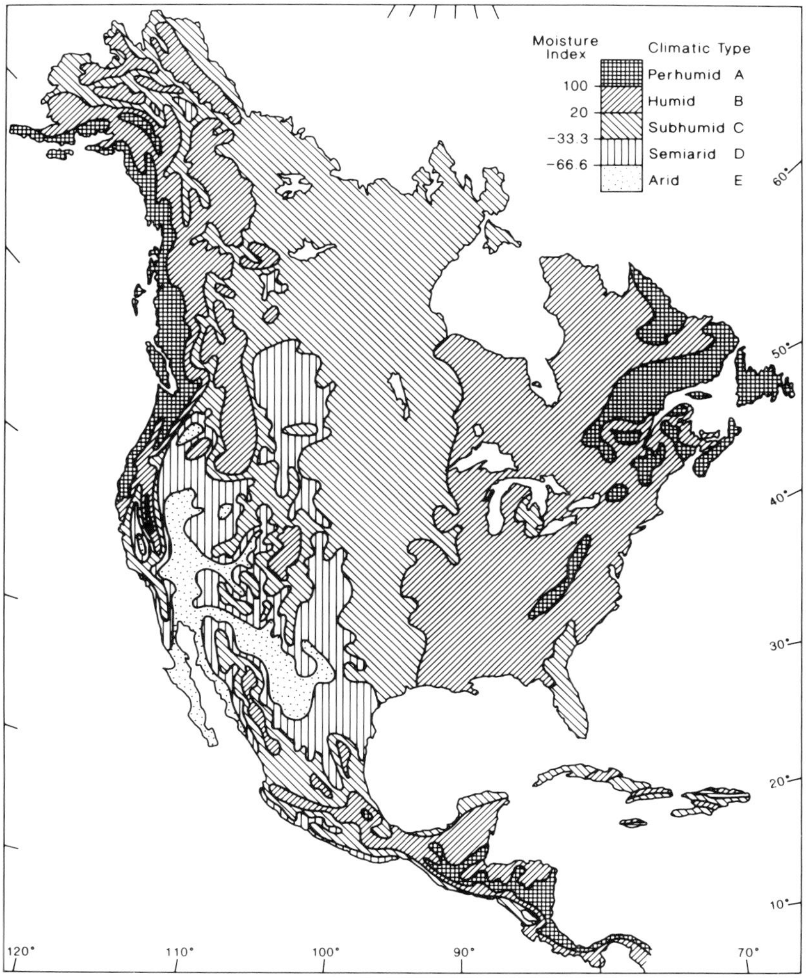

Figure 5.23 Moisture regions of North and Central America employing Thornthwaite's moisture index.

Source: From Mather 1985.

ature is only −28°C (−18°F). In these regions *permafrost* (permanently frozen ground) is widespread and often of great depth. In summer only the top 1–2 m of ground thaw and as the water cannot readily drain away this 'active layer' often remains waterlogged. Although frost may occur in any month, the long summer days usually give 3 months with mean temperatures above 10°C, and at many stations extreme maxima are 32°C (90°F) or more (see Figure 5.17D). The Barren Grounds of Keewatin, however, are much cooler in summer due to the extensive areas of lake and muskeg, and only July has a mean daily temperature of 10°C. Labrador–Ungava to the east is rather similar, with very high cloud amounts and maximum precipitation in June–September (see Figure 5.24). In winter, conditions fluctuate between periods of very cold, dry, high-pressure weather and spells of dull, bleak, snowy weather as depressions move eastwards or occasionally northwards over the area. In spite of the very low mean temperatures in winter, there have been occasions when maxima have exceeded 4°C (40°F) during incursions of maritime Atlantic air. Such variability is not found in

Figure 5.24 Selected climatological data for McGill Sub-Arctic Research Laboratory, Schefferville, 1955–62. The shaded portions of the precipitation represent snowfall, expressed as water equivalent.

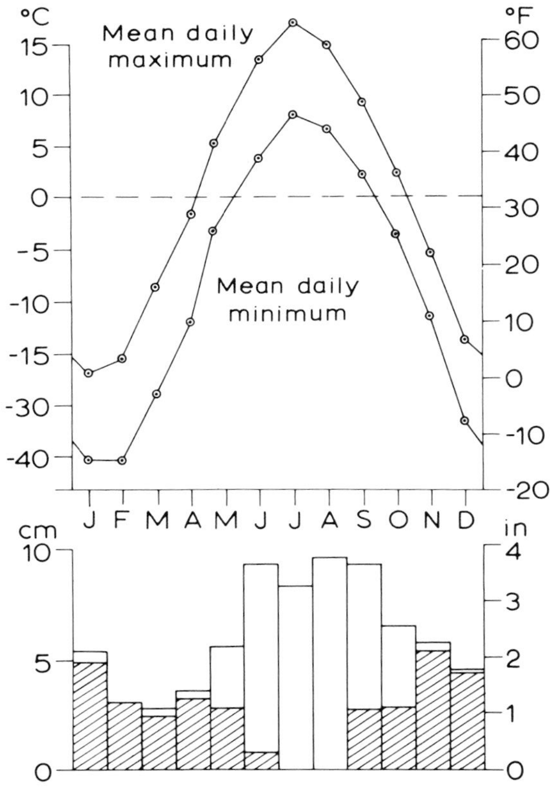

Source: Data from J. B. Shaw and D. G. Tout.

eastern Siberia which is intensely continental, apart from the Kamchatka peninsula, with the northern hemisphere's *cold pole* located in the remote north-east (see Figure 1.24A). Verkhoyansk and Oimyakon have a January mean of −50°C and both have recorded an absolute minimum of −67.7°C.

D THE SUBTROPICAL MARGINS

1 The Mediterranean

The characteristic west coast climate of the subtropics is the Mediterranean type with hot, dry summers and mild, relatively wet winters. It is interposed between the temperate maritime type and the arid subtropical desert climate but the Mediterranean regime is transitional in a special way for it is controlled by the westerlies in winter and by the subtropical anticyclone in summer. The seasonal change in position of the subtropical high and the associated subtropical westerly jet stream in the upper troposphere is evident in Figure 5.25. The type region is peculiarly distinctive, extending more than 3000 km into the Eurasian continent. Additionally, the configuration of seas and peninsulas produces great regional variety of weather and climate. The Californian region with similar conditions (see Figure 5.21) is of very limited extent, and attention is therefore concentrated on the Mediterranean basin itself.

The winter season sets in quite suddenly in the Mediterranean as the summer eastward extension of the Azores high-pressure cell collapses. This phenomenon can be observed on barographs throughout the region, but particularly in the western Mediterranean where a sudden drop in pressure occurs about 20 October and is accompanied by a marked increase in the probability of precipitation. The probability of receiving rain in any 5-day period increases dramatically from 50–70 per cent in early October to 90 per cent in late October. This change is associated with the first invasions by cold fronts, although thunder-shower rain has been common since August. The pronounced winter precipitation over the Mediterranean largely results from the relatively high sea-surface temperatures at that season; in January the sea temperature being about 2°C higher than the mean air temperature. Incursions of colder air into the region lead to convective instability along the cold front producing frontal and orographic rain. Incursions of arctic air are relatively infrequent (there being, on average, 6–9 invasions by cA and mA air each year), but the penetration by unstable mP air is much more common. It typically gives rise to deep cumulus development and is critical in the formation of Mediterranean depressions.

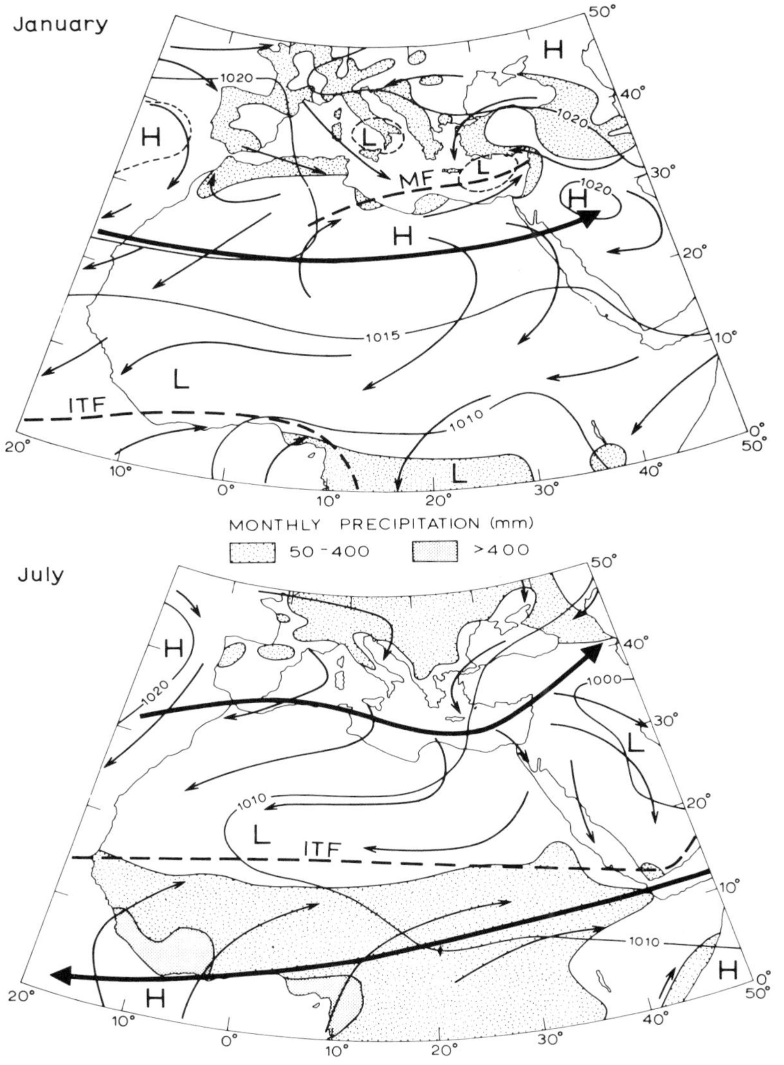

Figure 5.25 The distribution of surface pressure, winds and precipitation for the Mediterranean and North Africa during January and July. The average positions of the Subtropical Westerly and Tropical Easterly Jet Streams, together with the Intertropical (ITF) and Mediterranean (MF) Fronts, are also shown. The ITF is now interpreted as a Monsoon Tough.

Source: Partly after Weather in *the Mediterranean*, HMSO, 1962 (*Crown Copyright Reserved*).

The initiation and movement of these depressions (see Figure 5.26) is associated with a branch of the Polar Front Jet Stream located about 35°N. This jet develops during low index phases when the westerlies over the eastern Atlantic are distorted by a blocking anticyclone at about 20°W, leading to a deep stream of arctic air flowing southward over the British Isles and France.

Atlantic depressions entering the western Mediterranean as surface lows make up only 9 per cent of those affecting the region (see Figure 5.26); 17 per cent form as baroclinic waves south of the Atlas Mountains (the so-called *Saharan depressions*, which are most important sources of rainfall in late winter and spring) and fully 74 per cent develop in the western Mediterranean to the lee of the Alps and Pyrenees (see Chapter 4, G.1). The combination of the lee effect and that of unstable surface air over the western Mediterranean explains the frequent formation of these *Genoa-type depressions* when

Figure 5.26 Tracks of Mediterranean depressions, showing average annual frequencies, together with air-mass sources.

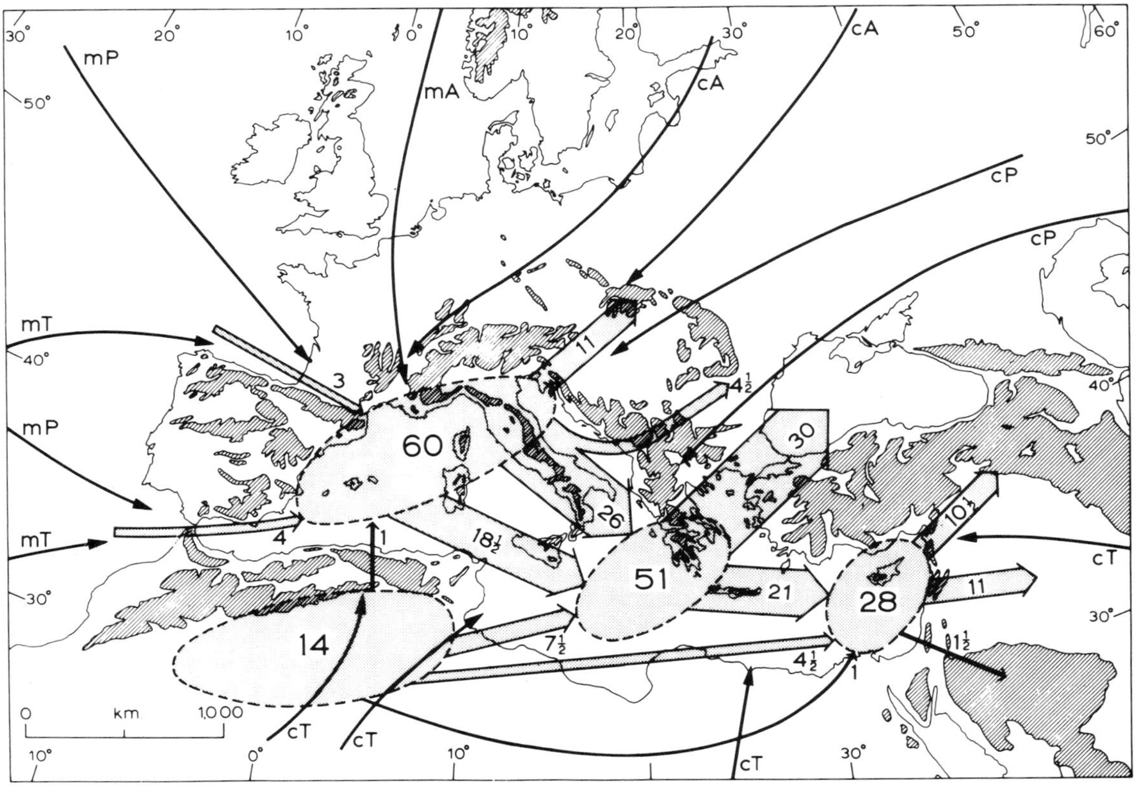

Source: After *Weather in the Mediterranean* HMSO, 1962 (*Crown Copyright Reserved*).

conditionally unstable mP air invades the region. These depressions are exceptional in that the instability of the local air in the warm sector gives unusually intense precipitation along the warm front, and the unstable mP air produces heavy showers and thunderstorm rainfall to the rear of the cold front, especially between 5° and 25°E. This warming of mP (or mA) air is so characteristic as to produce air designated as *Mediterranean* (see Table 4.1). The mean boundary between this Mediterranean air mass and cT air flowing north-eastward from the Sahara is referred to as the Mediterranean front (see Figure 5.25). There may be a temperature discontinuity as great as 12°–16°C (30°–40°F) across it in late winter. Sahara depressions and those from the western Mediterranean move eastward forming a belt of low pressure associated with this frontal zone and frequently draw cT air northward ahead of the cold front as the warm, dust-laden *scirocco* (especially in spring and autumn when Saharan air may spread into Europe). The movement of Mediterranean depressions is greatly complicated both by relief effects and by their regeneration in the eastern Mediterranean by fresh cP air from Russia or south-east Europe. Although many depressions travel eastward over Asia, there is a strong tendency for low-pressure centres to move north-eastward over the Black Sea and Balkans, especially as spring advances. Winter weather in the Mediterranean presents considerable variation, however, particularly as the Subtropical Westerly Jet Stream is highly mobile

and may occasionally even coalesce with the southerly-displaced Polar Front Jet Stream.

With high index zonal circulation over the Atlantic and Europe, depressions may pass far enough to the north so that their cold-sector air does not reach the Mediterranean, and then the weather there is generally settled and fine. Between October and April, anticyclones are the dominant circulation type for at least 25 per cent of the time over the whole Mediterranean area and in the western basin for 48 per cent of the time. This is reflected in the high mean pressure over the latter area in January (see Figure 5.25). Consequently, although the winter half-year is the rainy period, there are rather few rain-days. On average rain falls on only 6 days per month during winter in northern Libya and south-east Spain, yet there are 12 rain-days per month in western Italy, the western Balkan peninsula and the Cyprus area. The higher frequencies (and totals) are related to the areas of cyclogenesis and to the windward side of peninsulas.

Regional winds are also related to the meteorological and topographic factors. The familiar cold, northerly winds of the Gulf of Lions (the *mistral*), which are associated with northly mP airflows, are best developed when a depression is forming in the Gulf of Genoa east of a high-pressure ridge from the Azores anticyclone. Katabatic and funnelling effects strengthen the flow in the Rhône valley and similar localities so that violent winds are sometimes recorded. The mistral may last for several days until the outbreak of polar or continental air ceases. The frequency of these winds depends on their definition. The average frequency of a strong mistral in the south of France is shown in Table 5.4 (based on occurrence at one or more stations from Perpignan to the Rhône in 1924–7). Similar winds may occur along the Catalan coast of Spain (the *tramontana*, see Figure 5.27) and also in the northern Adriatic (the *bora*) and the northern Aegean seas when polar air flows southward in the rear of an eastward-moving depression and is forced over the mountains (see Chapter 3, C.2). In Spain, cold, dry, northerly winds occur in several different regions in winter. Figure 5.27 shows the *galerna* of the north coast and the *cierzo* of the Ebro valley.

The generally wet, windy and mild winter season in the Mediterranean is succeeded by a long indecisive spring lasting from March to May, with many false starts of summer weather.

The spring period, like that of early autumn, is especially unpredictable. In March 1966 a trough moving across the eastern Mediterranean, preceded by a warm southerly *khamsin* and followed by a northerly airstream, brought up to 70 mm (2.8 in) of rain in only 4 hours to an area of the southern Negev desert. Although April is normally a dry month in the eastern Mediterranean, Cyprus having an average of only 3 days with 1 mm of rainfall or more, high rainfalls can occur as in April 1971 when four depressions affected the region. Two of these were Saharan depressions moving eastward beneath the zone of diffluence on the cold side of a westerly jet and the other two were intensified in the lee of Cyprus. The rather rapid collapse of the Eurasian high-pressure cell in April, together with the discontinuous northward and eastward extension of the Azores anticyclone, encourages the northward

Table 5.4 Number of days with strong mistral in the south of France

Speed	*J*	*F*	*M*	*A*	*M*	*J*	*J*	*A*	*S*	*O*	*N*	*D*	*Year*
⩾11 m s^{-1} (21 kt)	10	9	13	11	8	9	9	7	5	5	7	10	103
⩾17 m s^{-1} (33 kt)	4	4	6	5	3	2	0.6	1	0.6	0	0	4	30

Source: After *Weather in the Mediterranean*, HMSO, 1962.

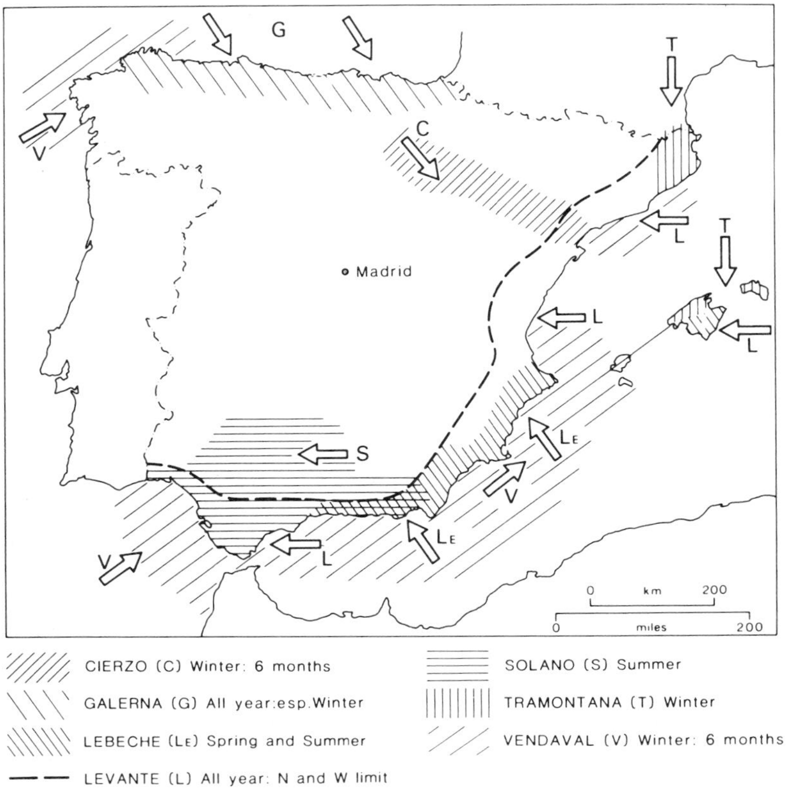

Figure 5.27 Areas affected by the major regional winds in Spain as a function of season.

Source: From Tout and Kemp 1985 (*by permission of the Royal Meteorological Society*).

displacement of depressions, and, even if higher latitude air does penetrate south to the Mediterranean, the sea surface there is relatively cooler and the air is more stable than during the winter.

By mid-June the Mediterranean basin is dominated by the expanded Azores anticyclone to the west, while to the south the mean pressure field shows a low-pressure trough extending across the Sahara from southern Asia (see Figure 5.25). The winds are predominantly northerly (e.g. the *etesians* of the Aegean) and represent an eastward continuation of the north-easterly trades. Locally, sea breezes reinforce these winds, but on the Levant coast they cause surface south-westerlies. The day-to-day weather of many parts of the North African coast is largely conditioned by land and sea breezes, involving air up to 1500 m (5000 ft) deep. Depressions are by no means absent in the summer months, but they are usually weak since the anticyclonic character of the large-scale circulation encourages subsidence, and air-mass contrasts are much reduced compared with winter (see Tables 4.1 and 4.2). Thermal lows form from time to time over Iberia and Anatolia, though thundery outbreaks are infrequent due to the low relative humidity.

The most important regional winds in summer are of continental tropical origin. There is a variety of local names for these usually hot, dry and dusty airstreams – *scirocco* (Algeria and the Levant), *lebeche* (south-east Spain) and *khamsin* (Egypt) – which move northwards ahead of eastward-moving depressions. In the Negev the onset of an easterly *khamsin* may cause the relative humidity to fall suddenly to less than 10 per cent and temperatures to rise to as much as 48°C (118°F). In southern Spain the easterly *solano* brings hot, humid weather to Andalucia in the summer half-year, whereas the

coastal *levante* – which has a long fetch over the Mediterranean – is moist and somewhat cooler (see Figure 5.27). Such regional winds occur when the Azores high extends over western Europe with a low-pressure system to the south.

Many stations in the Mediterranean receive only a few millimetres of rainfall in at least one summer month, yet it is important to realize that the seasonal distribution does not conform to the pattern of simple winter maximum over the whole of the Mediterranean basin. Figure 5.28 shows that this is found in the eastern and central Mediterranean, whereas Spain, southern France, northern Italy and the northern Balkans have more complicated profiles with a maximum in autumn or peaks in both spring and autumn. This double maximum can be interpreted generally as a transition between the continental interior type with summer maximum and the Mediterranean type with winter maximum. A similar transition region occurs in the south-western United States (see Figure 5.21), but local topography in this intermontane zone introduces further irregularities into the regimes.

2 The semi-arid south-western United States

Both the mechanisms and patterns of the climates of areas dominated by the subtropical high-pressure cells are rather obscure at present. The inhospitable nature of these arid regions inhibits data collection, and yet the proper interpretation of infrequent meteorological events requires a close network of stations maintaining continuous records over long periods. This difficulty is especially apparent in the interpretation of desert precipitation data, because much of the rain falls in very local storms irregularly scattered both in space and time. It is convenient to treat aspects of this climatic type here, in that most of the reliable data relate to the less arid regions marginal to the subtropical cell centres, and in particular to the south-western United States.

A series of observations at Tucson, Arizona, 730 m (2400 ft) between 1895 and 1957 showed a mean annual precipitation of 27.7 cm (10.91 in) falling on an average of about 45 days per year, with extreme annual figures of 61.4 cm (24.7 in) and 14.5 cm (5.72 in). Two moister periods in late November to March (receiving 30 per cent of the mean annual precipitation) and late June to September (50 per cent) are separated by more arid seasons from April to June (8 per cent) and October to November (12 per cent). The winter rains are generally prolonged and of low intensity (more than half the falls have an intensity of less than 0.5 cm (0.2 in) per hour), falling from alto-

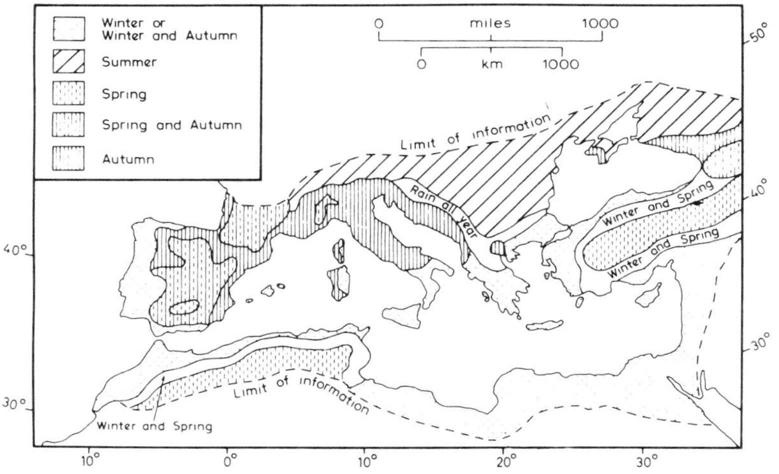

Figure 5.28 Rainfall regimes in the Mediterranean area.

Source: After Huttary 1950.

stratus clouds associated with the cold fronts of depressions which are forced to take southerly routes by strong blocking to the north. This occurs during phases of equatorial displacement of the Pacific subtropical high-pressure cell. The re-establishment of the cell in spring, before the main period of intense surface heating and convectional showers, is associated with the most persistent drought periods. Dry westerly to south-westerly flow from the eastern edge of the Pacific subtropical anticyclone is responsible for the low rainfall at this season. During one 29-year period in Tucson there were eight spells of more than 100 consecutive days of complete drought and twenty-four periods of more than 70 days. The dry conditions occasionally lead to dust storms. Yuma records nine per year, on average, associated with winds averaging 10–15 m s^{-1}. They occur both with cyclonic systems in the cool season and with summer convective activity. Phoenix experiences 6–7 per year, mainly in summer, with visibility reduced below 1 km in nearly half of these events.

The period of summer precipitation (locally known as the summer 'monsoon') is quite sharply defined, beginning in the last week in June and lasting until the middle of September. Precipitation mainly occurs from convective cells initiated by surface heating, convergence, or, less commonly, orographic lifting when the atmosphere is destabilized by upper-level troughs in the westerlies. These summer convective storms form in mesoscale clusters, the individual storm cells covering together less than 3 per cent of the surface area at any one time, and persisting for less than an hour on average. The storm clusters move across the country in the direction of the upper-air motion, and often seem to be controlled in movement by the existence of low-level jet streams at elevations of about 2500 m. The airflow associated with these storms is generally southerly along the southern and western margins of the Atlantic (or Bermudan) subtropical high, so that in contrast to the winter months the moisture is derived mainly from the Gulf of California, during 'surges' associated with the south-southwesterly low-level Sonoran jet (850–700 mb), and secondarily from the Gulf of Mexico during south-easterly flows. The southerly airflow regime at the surface and 700 mb (see Figures 3.20 and 3.25) over the Southwest often sets in abruptly around 1 July and it is therefore recognized as a singularity (see this chapter, A.4, and Figure 5.15).

Precipitation from these cells is extremely local (see Plate 13), and commonly concentrated in the mid-afternoon and evening. Intensities are much higher than in winter, half the summer rain falling at more than 0.4 in per hour. A particularly well-documented storm occurred over Phoenix, Arizona, at 0600 hours on 22 June 1972. Moving north-east, it produced up to 127 mm (5 in) of rainfall in the following 6 hours including hailstones measuring 19 mm (0.75 in) in diameter (see Figure 5.29). During the 29-year period about one quarter of the mean annual precipitation fell in storms giving 2.5 cm (1 in) or more per day. These intensities are much less than those associated with rainstorms in the humid tropics, but the sparsity of vegetation in the drier regions allows the rain to produce considerable surface erosion. Thus the highest measurements of surface erosion in the United States are from areas having 30–40 cm (12–15 in) of rain per year.

3 The interior and east coast of the United States

The climate of the subtropical south-eastern part of the United States has no exact counterpart in Asia, which is affected by the summer and winter monsoon systems. These are discussed in the next chapter and only the distinctive features of the North American subtropics are examined here. Seasonal wind changes are experienced in Florida, which is within the westerlies in winter and lies on the northern margin of the tropical easterlies in summer, but this is not comparable with the regime in southern and south-eastern Asia. Nevertheless, the summer season rainfall maximum (see Figure 5.21 for Jacksonville) is a

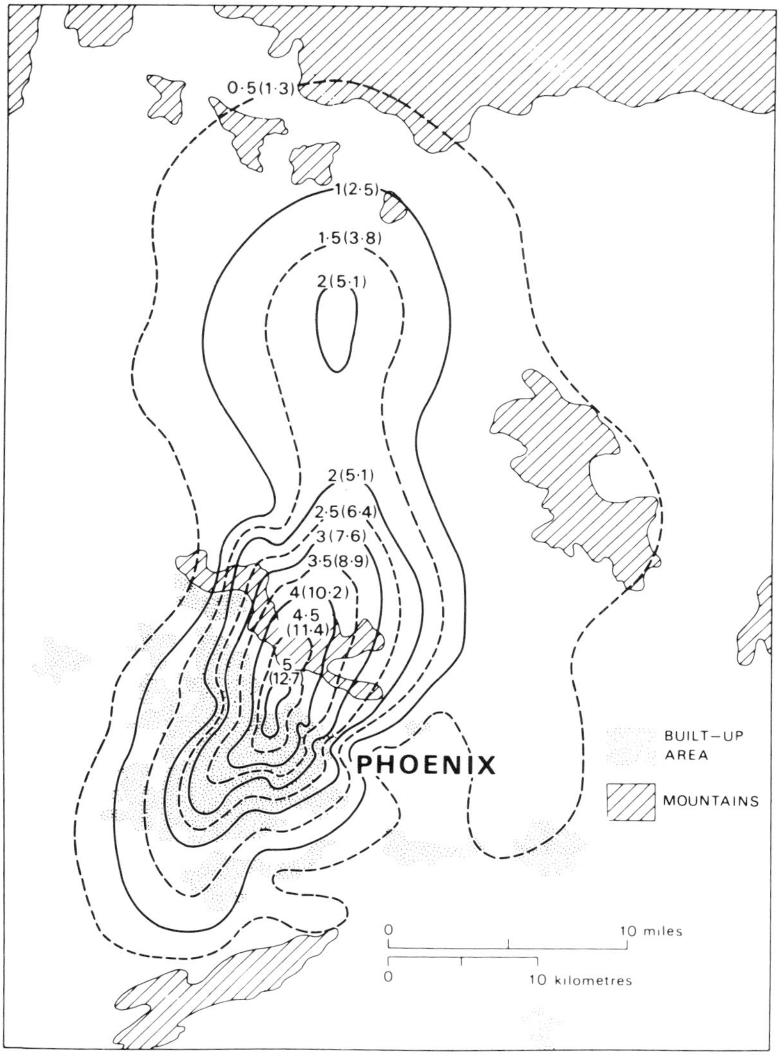

Figure 5.29 A rainstorm over Phoenix, Arizona, on 22 June 1972. The rainfall between 0600 and 1200 hours is shown in inches and in centimetres (in parentheses).

Source: After Durrenberger and Ingram 1978 (*courtesy State Climatologist, Arizona*).

result of this change-over. In June the upper flow over the Florida peninsula changes from northwesterly to southerly as a trough moves westward and becomes established in the Gulf of Mexico. This deep, moist southerly airflow provides appropriate conditions for convection, and indeed Florida probably ranks as the area with the highest annual number of days with thunderstorms – ninety or more, on average, in the vicinity of Tampa. These often occur in late afternoon although two factors apart from diurnal heating are thought to be important. One is the effect of sea-breezes converging from both sides of the peninsula, and the other is the northward penetration of disturbances in the easterlies (see Chapter 6). The latter may of course affect the area at any time of day. The westerlies resume control in September–October, although Florida remains under the easterlies during September when Caribbean tropical storms are most frequent and consequently the rainy season is prolonged. Tropical

cyclones contribute an average of 10–15 per cent of the annual rainfall near the Gulf Coast and in Florida.

The region of the Mississippi lowlands and the southern Appalachians to the west and north is not simply transitional to the 'Interior type' in terms at least of rainfall regime (see Figure 5.21). The profile shows a winter–spring maximum and a secondary summer maximum. The cool season peak is related to westerly depressions moving north-eastward from the Gulf coast area, and it is significant that the wettest month is commonly March when the mean jet stream is farthest south. The summer rains are associated with convection in humid air from the Gulf, though this convection becomes less effective inland as a result of the subsidence created by the anticyclonic circulation in the middle troposphere referred to previously (see B.3.c, this chapter).

SUMMARY

Seasonal changes in the Icelandic Low and the Azores High, together with variations in cyclone activity, control the climate of western Europe. The eastward penetration of maritime influences related to these atmospheric processes and to the warm waters of the North Atlantic Drift is illustrated by mild winter temperatures, the seasonality of precipitation regimes and indices of continentality in western Europe. Topographic effects on precipitation, snowfall, length of growing seasons and local winds are particularly marked over the Scandinavian Mountains, the Scottish Highlands and the Alps. Weather types in the British Isles can be described in terms of seven basic airflow patterns, the frequency and effects of which vary considerably with season. Recurrent weather spells about a particular date (singularities), such as the tendency for anticyclonic weather in mid-September, have been recognized in Britain and major seasonal trends in occurrence of airflow regimes can be used to define five natural seasons. Abnormal weather conditions (synoptic anomalies) are associated particularly with blocking anticyclones which are especially prevalent over Scandinavia, and may give rise to cold dry winters and warm dry summers.

The climate of North America is similarly affected by pressure systems which generate air masses of varying seasonal frequency. In winter, the subtropical high-pressure cell extends north over the Great Basin with anticyclonic cP air to the north over Hudson Bay. Major depression belts occur at about 45°–50°N, from the central USA to the St Lawrence, and along the east coast of Newfoundland. The arctic front is located over north-west Canada, the polar front lies along the north-east coast of the United States, and between the two a maritime (arctic) front may occur over Canada. In summer the frontal zones move north, the arctic front lying along the north coast of Alaska, Hudson Bay and the St Lawrence being the main locations of depressions tracks. Three major North American singularities concern the advent of spring in early March, the midsummer northward displacement of the subtropical high-pressure cell, and the Indian summer of September–October. In western North America the Coast Ranges inhibit the eastward spread of precipitation which may vary greatly locally (e.g. in British Columbia), especially as regards snowfall. The strongly continental interior and east of the continent experiences some moderating effects of Hudson Bay and the Great Lakes in early winter, but with locally significant snow belts. The climate of the east coast is dominated by continental pressure influences. Cold spells are produced by winter outbreaks of high-latitude cA/cP air in the rear of cold fronts. Zonal westerly airflows give rise to Chinook winds in the lee

of the Rockies. The major moisture sources of the Gulf of Mexico and the North Pacific produce regions of differing seasonal regime: the winter maximum of the west coast is separated by a transitional intermontane region from the interior with a general warm-season maximum; the north-east has a relatively even seasonal distribution. Moisture gradients which strongly influence vegetation and soil types are predominantly east–west in central North America, in contrast to the north–south isotherm pattern.

The polar margins have extensive areas of permanently frozen ground (permafrost) in the continental interiors, whereas the maritime regions of northern Europe and northern Canada–Alaska have cold stormy winters and cloudy milder summers influenced by the passage of depressions.

The subtropical margin of Europe consists of the Mediterranean region, lying between the belts dominated by the westerlies and the Saharan–Azores high pressure cells. The collapse of the Azores high-pressure cell in October allows depressions to move and form over the relatively warm Mediterranean Sea giving well-marked orographic winds (e.g. *mistral*) and stormy rainy winters. Spring is an unpredictable season marked by the collapse of the Eurasian high-pressure cell to the north and the strengthening of the Saharan–Azores anticyclone. In summer, the latter gives dry, hot conditions with strong local southerly airstreams (e.g. *scirocco*). The simple winter rainfall maximum is most characteristic of the eastern and southern Mediterranean, whereas in the north and west autumn and spring rains become more important.

The semi-arid south-western United States comes under the complex influence of the Pacific and Bermuda high-pressure cells, having extreme rainfall variations with winter and summer maxima mainly due to depressions and local thunderstorms, respectively. The interior and east coast of the United States is dominated by westerlies in winter and southerly thundery airflows in summer.

6

Tropical weather and climate

Fifty per cent of the surface of the globe lies between latitudes 30°N and 30°S and over 75 per cent of the world's population inhabits climatically tropical lands. Tropical climates are, therefore, of especial geographical interest.

The latitudinal limits of tropical climates vary greatly with longitude and season, and tropical weather conditions may reach well beyond the Tropics of Cancer and Capricorn. For example, the summer monsoon extends to 30°N in southern Asia, but only 20°N in West Africa, while in late summer and autumn tropical hurricanes may affect 'extra-tropical' areas of eastern Asia and eastern North America. Not only do the tropical margins extend seasonally polewards, but in the zone between the major subtropical high-pressure cells there is frequent interaction between temperate and tropical disturbances. Plate 22 illustrates a situation where there is such interaction between low and middle latitudes, whereas Plate 31, in contrast, shows distinct tropical and mid-latitude storms. In general, the tropical atmosphere is far from being a discrete entity and any meteorological or climatological boundaries must be arbitrary. There are, nevertheless, a number of distinctive features of tropical weather and these are discussed below.

A THE ASSUMED SIMPLICITY OF TROPICAL WEATHER

The study of tropical weather has passed broadly through three stages. First, for a long period which only ended some years before the Second World War, tropical weather conditions, patterns and mechanisms were assumed to be much more simple and obvious than those in higher latitudes. This belief was partly due to the paucity of pre-war meterological records, particularly over the vast tropical oceans, and partly due to certain theoretical and practical considerations. One reason was that temperature and hence air-mass contrasts seemed small compared with middle latitudes. Air masses were nevertheless identified on the basis of their moisture content, temperature and stability, although frontal activity was believed to be weak and weather systems correspondingly less evident. Obvious exceptions were tropical cyclones which were thought to result from special conditions of thermal convection. Another reason was the large extent of ocean surface which, it was assumed, would simplify the patterns of weather and climate.

Thus the following simple picture of *trade wind weather* evolved. Maritime tropical air masses, originating by subsidence in the subtropical high-pressure cells over the eastern halves of the oceans (see Figure 3.18), move steadily westwards and equatorwards in the easterly trade winds with quite constant speed and direction. Beneath the temperature inversion, formed between about 1 and 2 km altitude by subsidence, the air is moist with a layer of broken cumulus cloud. Conditions are invariably warm and dry, except where islands cause orographic clouds to form. Over the equa-

torial oceans the surface wind is light and variable (the doldrums) and the air is universally hot, humid and sultry (see Chapter 3, E.1).

A further element of simplicity was thought to be provided by the insolation regime. The persistently high altitude of the sun in low latitudes and the equality of length between the days and nights means that seasonal variations of radiation are minimized. Hence the annual rhythm was thought to produce simple rainfall regimes with a single maximum following the summer solstice at the tropics, and a double peak at the equator in response to the passage of the overhead sun at the equinoxes. Diurnal patterns of land and sea breezes giving an afternoon build-up of convective activity and thundery showers were regarded as characteristic features of nearly all tropical climates.

Following the evolution of this simple picture of tropical weather processes, attempts were made between 1920 and 1940 to introduce mid-latitude frontal concepts. Little real progress was made, however, as a result of the apparent general absence of significant air-mass contrasts. Furthermore, the small surface pressure gradients of most tropical disturbances (other than the hurricane variety) tend to go unnoticed due to the large semi-diurnal pressure oscillation in the tropics. Pressure varies by about 2–3 mb with maxima around 1000 and 2200 hours and minima at 0400 and 1600 hours. It must also be recalled that the wind direction provides no guide to the pressure pattern in low latitudes. The small Coriolis force prevents the wind from attaining geostrophic balance and consequently the techniques used for analysing mid-latitude weather maps have to be abandoned.

Weather observations made during the Second World War revealed the inadequacy of the earlier views. It became evident that weather changes are frequent and complex with quite distinct types of weather systems in different tropical lands and with considerable climatic differences even over the ocean areas. It also became apparent that much smaller trigger mechanisms are necessary to produce disturbances in the flow of high-energy, tropical air masses than those associated with mid-latitude depressions, and yet, paradoxically, tropical cyclones are infrequent. The systems responsible for these contrasts are examined in the following sections.

B THE INTERTROPICAL CONVERGENCE

The tendency for the trade wind systems of the two hemispheres to converge in the Equatorial (low-pressure) Trough has already been noted (see Chapter 3, E). Views on the exact nature of this feature have been subject to continual revision. From the 1920s to the 1940s the frontal concepts developed in mid-latitudes were applied in the tropics, and the streamline confluence of the North-East and South-East Trades was identified as the Intertropical Front (ITF). Over continental areas such as West Africa and southern Asia, where in summer hot, dry continental tropical air meets cooler, humid equatorial air, this term has some limited applicability (see Figure 6.1). Sharp temperature and moisture gradients can occur, but the front is seldom a weather-producing mechanism of the mid-latitude type. Elsewhere in low latitudes true fronts (with a marked density contrast) are rare.

Recognition in the 1940s–1950s of the significance of wind field convergence in tropical weather-production led to the designation of the trade wind convergence as the Intertropical Convergence Zone (ITCZ). This feature is apparent on a *mean* streamline map, but areas of convergence grow and decay, either *in situ* or within disturbances moving westward (see Plate 28), over periods of a few days. Moreover, convergence is infrequent even as a climatic feature in the doldrum zones (see Figure 3.28). Satellite photography has shown that over the oceans the position and intensity of the ITCZ varies greatly, even from day to day.

As climatic features the Equatorial Trough and ITC appear to move seasonally away from the equator (See Figure 6.1) in association with the *Thermal Equator* (zone of seasonal maximum temperature). The location of the

Figure 6.1 The position of the Equatorial Trough (Intertropical Convergence Zone or Intertropical Front in some sectors) in February and August. The cloud band in the southwest Pacific in February is known as the South Pacific Convergence Zone; over southern Asia and West Africa the term Monsoon Trough is used.

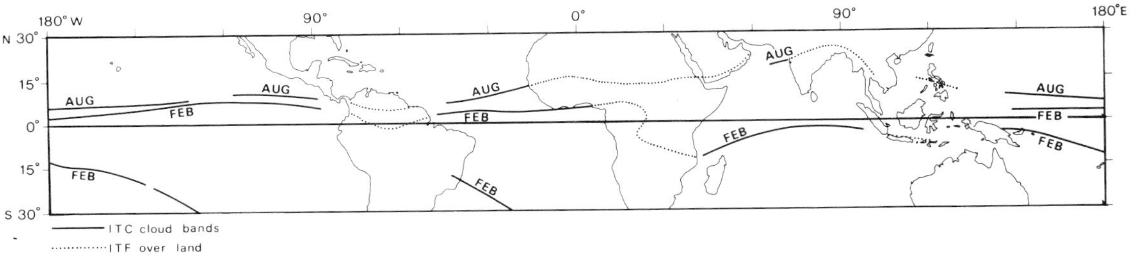

Sources: After Saha 1973, Riehl 1954, and Yoshino 1969.

Thermal Equator is directly related to solar heating (see Figures 1.22 and 1.24) and there is an obvious link between this and the Equatorial Trough in terms of thermal lows. However, if the ITC were to coincide with the Equatorial Trough then this zone of cloudiness would decrease incoming solar radiation, reducing the surface heating needed to maintain the low-pressure trough. In fact, this does not happen. Solar energy is available to heat the surface because the maximum surface wind convergence, uplift and cloud cover is commonly located several degrees equatorward of the trough. Figure 6.2 illustrates two patterns of Equatorial Trough and ITCZ. Cases involving convergence of two trade wind systems occur over the central North Atlantic in August and the eastern North Pacific in February, for example, while the situation where the Equatorial Trough is defined by easterlies on its poleward side and westerlies on its equatorward side is typical of West Africa in August and the western South Pacific in February. In this case the cloudiness maximum is distinct from the Equatorial Trough.

The dynamics of low-latitude atmosphere–ocean circulations are also involved. The convergence zone in the central equatorial Pacific moves seasonally between about 4°N in March–April and 8°N in September. This appears to be a response to the relative strengths of the north-east and south-east trades. The ratio of South Pacific/North Pacific trade wind strength exceeds 2 in September, but falls to 0.6 in April. Interestingly, the ratio varies in phase with the ratio of Antarctic–Arctic sea-ice areas; Antarctic ice is a maximum in September when Arctic ice is at its minimum. The convergence axis is often aligned close to the zone of maximum sea-surface temperatures, but is not anchored to it. Indeed, the SST maximum located within the Equatorial Counter Current (see Figure 3.42) is a result of the interactions between the trade winds and horizontal and vertical motions in the ocean surface layer.

Aircraft studies have revealed the complex structure of the central Pacific ITCZ. When moderately strong trades provide horizontal moisture convergence, convective cloud bands form, but the convergent lifting may be insufficient for rainfall in the absence of upper-level divergence. Moreover, although the south-east trades cross the equator, the mean monthly resultant winds between 115° and 180°W have, throughout the year, a more southerly component north of the equator and a more northerly one south of it, giving a zone of divergence (due to the sign change in the Coriolis parameter) along the equator.

In the south-western sectors of the Pacific and Atlantic oceans, satellite cloudiness studies indicate the presence of two semi-permanent confluence zones (see Figure 6.1). These do not occur in the eastern South Atlantic and South

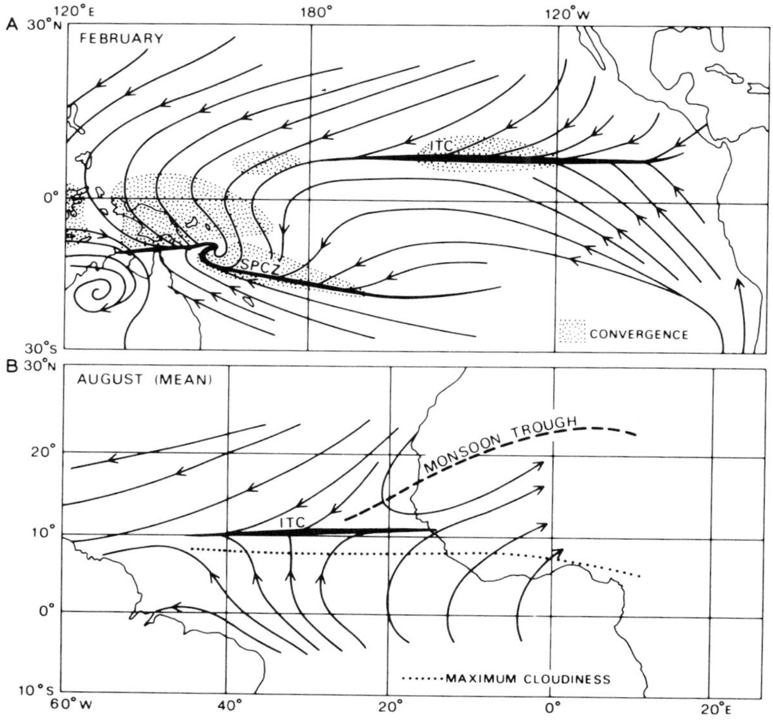

Figure 6.2 Illustrations of (A) streamline convergence forming an Intertropical Convergence (ITC) and South Pacific Convergence Zone (SPCZ) in February, and (B) the contrasting patterns of monsoon trough over West Africa, streamline convergence over the central tropical North Atlantic, and axis of maximum cloudiness to the south for August.

Sources: A: C. S. Ramage, personal communication, 1986. B: From Sadler 1975a.

Pacific where there are cold ocean currents. The convergence shown in the western South Pacific in February (summer) is now recognized as an important discontinuity and zone of maximum cloudiness (see Plate 28) termed the South Pacific Convergence Zone (SPCZ); it extends from the eastern tip of Papua New Guinea to about 30°S, 120°W. At sea level moist north-easterlies, west of the South Pacific subtropical anticyclone, converge with south-easterlies ahead of high-pressure systems moving eastward from Australia/New Zealand. The low-latitude section west of 180° longitude is part of the ITCZ system, related to warm surface waters. However, the maximum precipitation is south of the axis of maximum sea-surface temperature, and the surface convergence is south of the precipitation maximum in the central South Pacific. The south-eastward orientation of the SPCZ is caused by interactions with the mid-latitude westerlies. Its south-eastern end is associated with wave disturbances and jet stream clouds on the South Pacific polar front. The link across the subtropics appears to reflect upper-level tropical mid-latitude transfers of moisture and energy, especially during subtropical storm situations. Hence, the SPCZ shows substantial short-term and interannual variability in its location and development. The interannual variability is strongly associated with the phase of the Southern Oscillation (see p. 129).

C TROPICAL DISTURBANCES

It was not until the 1940s that detailed accounts were given of types of tropical disturbances other than the long-recognized tropical cyclone. However, our view of tropical weather systems has been radically revised following the advent of operational meteorological satellites in the 1960s. Special programmes of meteorological measurements at the surface and in the upper air, together with aircraft and ship observations,

have been carried out in the Pacific and Indian Oceans, the Caribbean and in the tropical eastern Atlantic.

It appears that five categories of weather system can be distinguished according to their space and time scales (see Figure 6.3). The smallest, with a life span of a few hours, is the *individual cumulus*, 1–10 km in diameter, which is generated by dynamically-induced convergence in the Trade Wind boundary layer. In fair weather, cumulus clouds are generally aligned in 'cloud streets', more or less parallel to the wind direction (see Plate 29), or form polygonal honeycomb-pattern cells, rather than scattered at random. This seems to be related to the boundary-layer structure and wind speed (see p. 73). There is little interaction between the air layers above and below the cloud base under these conditions, but in disturbed weather conditions updraughts and downdraughts cause interaction between the two layers which intensifies the convection. Individual cumulus towers, associated with violent thunderstorms, develop particularly in the Intertropical Convergence Zone sometimes reaching to more than 20,000 m (65,000 ft) in height and having updraughts of 10–14 m s^{-1} (25–30 mph) (see Plate 30). In this way the smallest scale of system can aid the development of larger disturbances. Convection is most active over sea surfaces with temperatures exceeding 26°C.

Figure 6.3 The mesoscale and synoptic structure of the equatorial trough zone (ITCZ), showing a model of the spatial distribution (*above*) and of the vertical structure (*below*) of convective elements which form the cloud clusters.

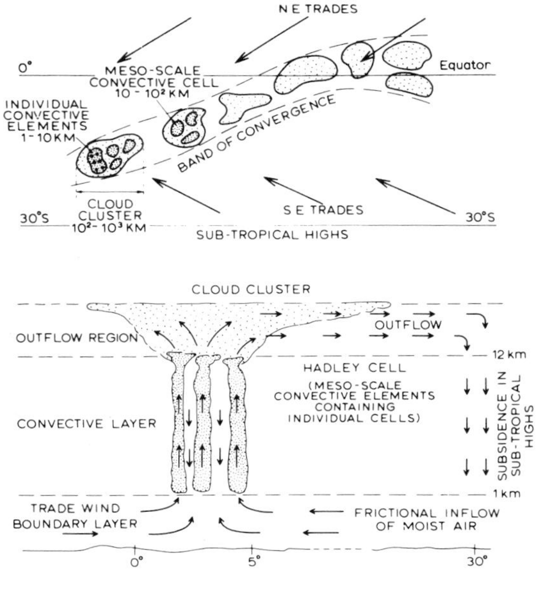

Source: From Mason 1970.

The second category of system develops through cumulus clouds becoming grouped into *mesoscale convective areas* (MCAs) up to 100 km across (see Figure 6.3). Third, several MCAs may comprise a *cloud cluster*, 100–1000 km in diameter. These subsynoptic scale systems are not very precisely defined as they were initially identified as amorphous cloud areas on satellite images (see Plate 37). They have been studied primarily from images of the tropical Atlantic and Pacific oceans. Their definition is rather arbitrary, but they may extend over an area 2° square up to 12° square. It is important to note that the peak convective activity has passed when cloud cover is most extensive through the spreading of cirrus canopies. Clusters in the Atlantic, defined as more than 50 per cent cloud cover extending over an area of 3° square, show maximum frequencies of 10–15 clusters per month near the ITC and also at 15°–20°N in the western Atlantic over zones of high sea-surface temperature. They consist of a cluster of mesoscale convective cells with the system having a deep layer of convergent airflow (see Figure 6.3). Some persist for only 1–2 days but others develop within synoptic-scale waves. Many aspects of their development and role remain to be determined (see also C.3, this chapter).

The fourth category of tropical weather system includes the synoptic scale waves and cyclonic vortices (discussed more fully below) and the fifth group is represented by the planetary-scale waves. The planetary waves (with a wavelength from 10,000 to 40,000 km) need not concern us in detail. Two types occur in the equatorial stratosphere and another in the

equatorial upper troposphere. While they may interact with lower tropospheric systems, they do not appear to be direct weather mechanisms. The synoptic-scale systems which determine much of the 'disturbed weather' of the tropics are sufficiently important and varied as to be discussed under the headings of wave disturbances and cyclonic storms.

1 Wave disturbances

There are several types of wave travelling westward in the equatorial and tropical tropospheric easterlies; the differences between them probably result from regional and seasonal variations in the structure of the tropical atmosphere. Their wavelength is between about 2000–4000 km and they have a life-span of 1–2 weeks, travelling some 6°–7° longitude per day.

The first wave type to be described in the Tropics was the *Easterly Wave* of the Caribbean area. This system is quite unlike a mid-latitude depression. There is a weak pressure trough which usually slopes eastward with height (see Figure 6.4) and typically the main development of cumulonimbus cloud and thundery showers *behind* the trough line. This pattern is associated with horizontal and vertical motion in the easterlies. Behind the trough low-level air undergoes convergence, while ahead of it there is divergence (see Chapter 3, B.1). This follows from the equation for the conservation of potential vorticity (compare Chapter 4, F), which assumes that the air travelling at a given level does not change its potential temperature (i.e. dry-adiabatic motion; see Chapter 2, D):

$$\frac{f+\zeta}{\Delta p} = k$$

f = the Coriolis parameter, ζ = relative vorticity (cyclonic positive) and Δp = the depth of the tropospheric air column. Air overtaking the trough line is moving both polewards (f increasing) and towards a zone of cyclonic curvature (ζ increasing), so that if the left-hand side of the equation is to remain constant Δp

Figure 6.4 A model of the areal (*above*) and vertical (*below*) structure of an easterly wave. Cloud is stippled and the precipitation area is shown in the vertical section. The streamline symbols refer to the areal structure, and the arrows on the vertical section indicate the horizontal and vertical motions.

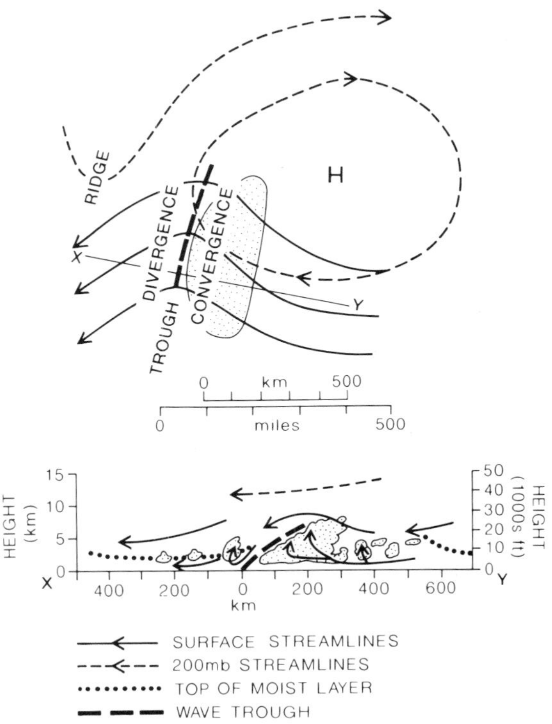

Source: Partly after Riehl and Malkus.

must increase. This vertical expansion of the air column necessitates horizontal contraction (convergence). Conversely there is divergence in the air moving southward ahead of the trough and curving anticyclonically. The divergent zone is characterized by descending, drying air with only a shallow moist layer near the surface, while in the vicinity of the trough and behind it the moist layer may be 4500 m (15,000 ft) or more deep. When the easterly air flow is *slower* than the speed of the wave, the reverse pattern of low-level convergence ahead of the trough and divergence behind it is observed as a consequence of the potential vorticity equation. Often

this is the case in the middle troposphere so that the pattern of vertical motion shown in Figure 6.4 is augmented.

The passage of such a transverse wave in the trades commonly produces the following weather sequence:

1 In the ridge ahead of the trough: fine weather, scattered cumulus cloud, some haze.
2 Close to the trough line: well-developed cumulus, occasional showers, improving visibility.
3 Behind the trough: veer of wind direction, heavy cumulus and cumulonimbus, moderate or heavy thundery showers and a decrease of temperature.

Satellite photography indicates that the simple easterly wave is rather less common than has been supposed. Many Atlantic disturbances show an 'inverted V' waveform in the low-level wind field and associated cloud, or a 'comma' cloud related to a vortex. They are often apparently linked with a wave pattern on the ITC further south. West African disturbances that move out over the eastern tropical Atlantic usually exhibit low-level confluence and upper-level diffluence *ahead* of the trough, giving maximum precipitation rates in this same sector. Many disturbances in the easterlies have a closed cyclonic wind circulation at about the 600-mb level.

It is obviously difficult to trace the growth of wave disturbances over the oceans and in continental areas with sparse data coverage. However, some generalizations can be made. At least 8 out of 10 disturbances develop some 2°–4° latitude poleward of the equatorial trough. Convection is probably set off by convergence of moisture in the airflow, accentuated by friction, and then maintained by entrainment into the thermal convective plumes (see Figure 6.3). About 90 tropical disturbances develop during the June–November hurricane season in the tropical Atlantic, approximately one system every 3 to 5 days. More than half of those disturbances originate over Africa. According to N. Frank, a high ratio of African depressions in the storm total in a given season indicates tropical characteristics, whereas a low ratio suggests storms originating from cold lows and the baroclinic zone between Saharan air and cooler, moist monsoon air. Many of them can be tracked westward into the eastern North Pacific. A quarter of the disturbances intensify into tropical depressions and 10 per cent become 'named' storms.

Developments in the Atlantic are closely related to the structure of the trades. In the eastern sectors of the subtropical anticyclones active subsidence maintains a pronounced inversion at 450 to 600 m (1500 to 2000 ft) (see Figure 6.5). Thus, the cool eastern tropical oceans are characterized by extensive, but shallow, marine stratocumulus which gives little rainfall. Downstream the height of the inversion base rises (see Figure 6.6) because the subsidence decreases away from the eastern part of the anticyclone and cumulus towers penetrate through the inversion from time to time, spreading moisture into the dry air above. Easterly waves tend to develop in the Caribbean when the trade wind inversion is weak or even absent during summer and autumn, whereas in winter and spring accentuated subsidence aloft inhibits their growth although disturbances may move westward above the inversion. Another factor which may initiate waves in the easterlies is the penetration of cold fronts into low latitudes. This is common in the sector between two subtropical high-pressure cells where the equatorward portion of the front tends to fracture, generating a westward-moving wave.

The influence of these features on regional climate is illustrated by the rainfall regime. For example, there is a late summer maximum at Martinique (see Figure 6.7) in the Windward Islands (15°N) when subsidence is weak, although some of the autumn rainfall is associated with tropical storms. In many trade wind areas the rainfall occurs in a few rainstorms associated with some form of disturbance. Over a 10-year period Oahu (Hawaii) had an average of twenty-four rainstorms per year of which ten accounted for more than two-thirds of the

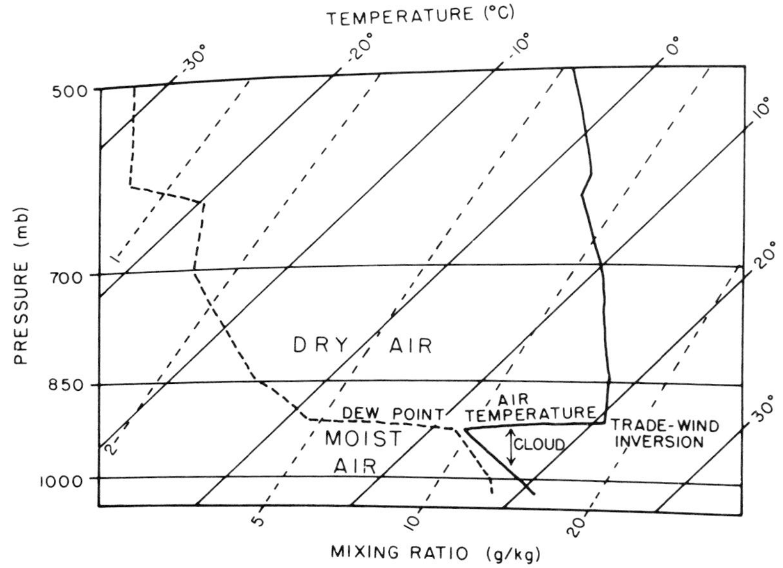

Figure 6.5 The vertical structure of trade wind air at 30°N, 140°W at 0300 GMT on 10 July 1949. The mixing ratio is the saturation value.

Source: Based on Riehl 1954.

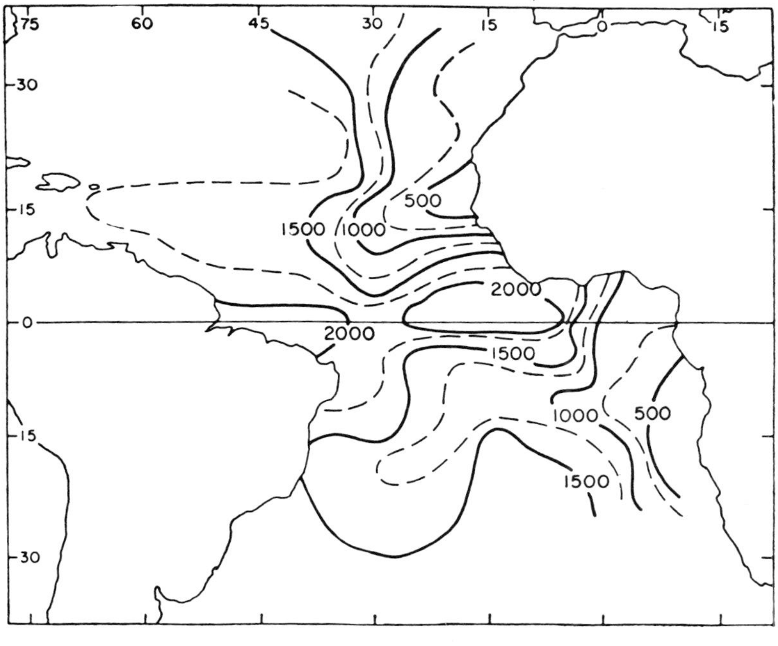

Figure 6.6 The height (in metres) of the base of the trade wind inversion over the tropical Atlantic.

Source: From Riehl 1954.

Figure 6.7 Average monthly rainfall at Fort de France, Martinique (based on 'CLIMAT' normals of the World Meteorological Organization for 1931–60). The mean annual rainfall is 184 cm (72.45 in).

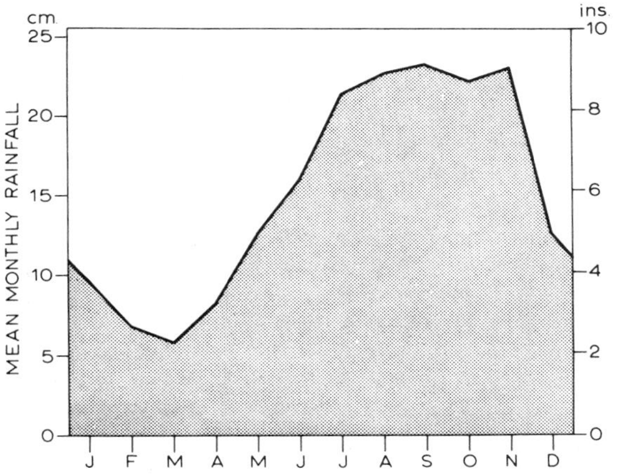

annual precipitation. There is quite high variability of rainfall from year to year in such areas, since a small reduction in the frequency of disturbances can have a large effect on rainfall totals.

In the central equatorial Pacific the trade wind systems of the two hemispheres converge in the Equatorial Trough and wave disturbances may be generated if the trough is sufficiently removed from the equator (usually to the north) to provide a small Coriolis force to begin cyclone motion. These disturbances quite often become unstable forming a cyclonic vortex as they travel westwards towards the Philippines, but the winds do not necessarily attain hurricane strength. The synoptic chart for part of the north-west Pacific on 17 August 1957 (see Figure 6.8), shows three developmental stages of tropical low-pressure systems. An incipient easterly wave has formed west of Hawaii which, however, filled and dissipated during the next 24 hours. A well-developed wave is evident near Wake Island, having spectacular cumulus towers extending up to over 9100 m (30,000 ft) along the convergence zone some 480 km (300 miles) east of it (see Plate 33). This wave developed within 48 hours into a circular tropical storm with winds up to 20 m s^{-1} (46 mph), but not into a full hurricane. A strong, closed, north-

Figure 6.8 The surface synoptic chart for part of the north-west Pacific on 17 August 1957. The movements of the central wave trough and of the closed circulation during the following 24 hours are shown by the dashed line and arrow, respectively. The dashed L just east of Saipan indicates the location in which another low-pressure system subsequently developed. Plate 33 shows the cloud formation along the convergence zone just east of Wake Island.

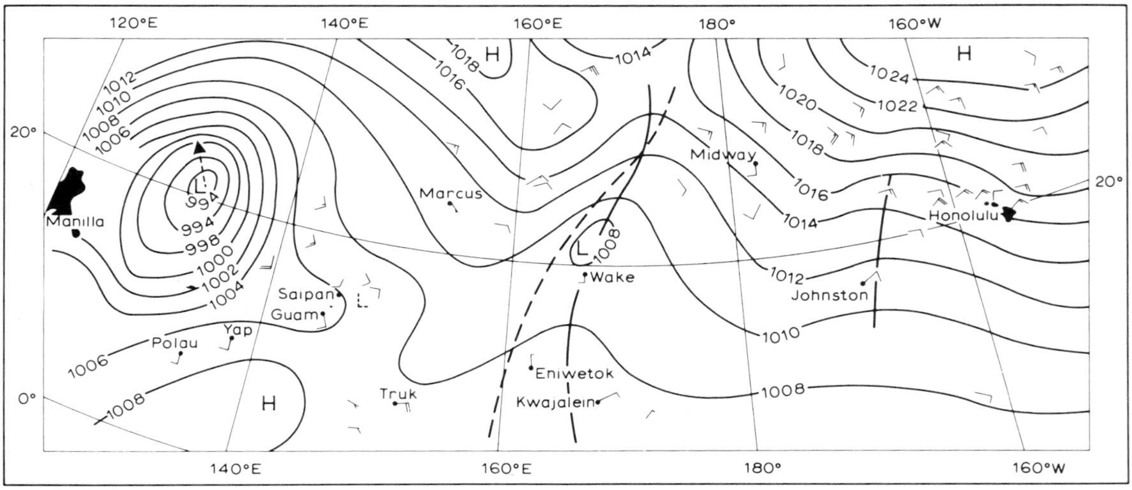

Source: From Malkus and Riehl 1964.

west moving circulation is situated east of the Philippines. Equatorial waves may form on both sides of the equator in an easterly current located between about 5°N and S. In such cases divergence ahead of a trough in the northern hemisphere is paired with convergence behind a trough line located further to the west in the southern hemisphere. The reader may confirm that this should be so by applying the equation for the conservation of potential vorticity, remembering that both f and ζ operate in the reverse sense in the southern hemisphere.

2 Cyclones

a Hurricanes

The most notorious type of cyclone is the tropical hurricane (or typhoon). Some 80 or so cyclones each year are responsible, on average, for 20,000 fatalities, as well as causing immense damage to property and a serious shipping hazard, due to the combined effects of high winds, high seas, flooding from the heavy rainfall and coastal storm surges. As a result, considerable attention has been given to forecasting their development and movement so that their origin and structure are beginning to be understood. Naturally the catastrophic force of a hurricane makes it a very difficult phenomenon to investigate, but some assistance is now obtained from aircraft reconnaissance flights sent out during the 'hurricane season', from radar observations of cloud and precipitation structure, and from satellite photography (see Plate 34).

The typical hurricane system has a diameter of about 650 km (400 miles), less than half that of a mid-latitude depression, although typhoons in the Western Pacific are often much larger. The central pressure at sea level is commonly 950 mb and exceptionally falls below 900 mb. Cyclonic-intensity storms are defined as having maximum sustained surface winds exceeding 33 m s^{-1} (74 mph) and in many storms they exceed 50 m s^{-1} (120 mph). The great vertical development of cumulonimbus clouds with tops at over 12,000 m (40,000 ft) reflects the immense convective activity concentrated in such systems. Radar and satellite studies show that the convective cells are normally organized in bands which spiral inward towards the centre.

Although the largest hurricanes are characteristic of the Pacific, the record is held by the Caribbean hurricane 'Gilbert' which was generated 320 km (200 miles) east of Barbados on 9 September 1988 and moved westward at an average speed of 24–27 km hr^{-1} (15–17 mph), dissipating off the east coast of Mexico. Aided by an upper tropospheric high-pressure cell north of Cuba, hurricane Gilbert intensified very rapidly, the pressure at its centre dropped to 888 mb (the lowest ever recorded in the western hemisphere) and maximum windspeeds near the core were in excess of 55 m s^{-1} (125 mph). More than 500 mm of rain fell on the highest parts of Jamaica in only 9 hours. However, the most striking feature of this record storm was its size, being some three times that of average Caribbean hurricanes. At its maximum extent the hurricane had a diameter of 3500 km, disrupting the ITCZ along more than one-sixth of the earth's equatorial circumference and drawing in air from as far away as Florida and the Galapagos Islands.

The main tropical cyclone activity in both hemispheres is in late summer and autumn during times of maximum northward and southward displacements of the Equatorial Trough (see Table 6.1). A small number of storms affect both the western North Atlantic and North Pacific areas as early as May and as late as December, and have occurred in every month in the latter area. In the Bay of Bengal, there is also a secondary early summer maximum. The annual frequency of cyclones shown in Table 6.1 are only approximate since in some cases it may be uncertain as to whether or not the winds actually exceeded hurricane force. In addition, storms in the more remote parts of the South Pacific and Indian Oceans frequently escaped detection prior to the use of weather satellites.

Table 6.1 Annual frequencies and usual seasonal occurrence of tropical cyclones (maximum sustained winds exceeding 25 m s^{-1}), 1958–77.

Location	*Annual frequency*	*Main occurrence*
Western North Pacific	26.3	July–October
Eastern North Pacific	13.4	August–September
Western North Atlantic	8.8	August–October
Northern Indian Ocean	6.4	May–June; October–November
Northern hemisphere total	54.6	
South-west Indian Ocean	8.4	January–March
South-east Indian Ocean	10.3	January–March
Western South Pacific	5.9	January–March
Southern hemisphere total	24.5	
Global total	79.1	

Source: After Gray 1979.
Note: Area totals are rounded.

A number of conditions are necessary, even if not always sufficient, for hurricane formation. One requirement as shown by Figure 6.9 is an extensive ocean area with a surface temperature greater than 27°C (80°F). Cyclones rarely form near the equator, where the Coriolis parameter is close to zero, or in zones of strong vertical wind shear (i.e. beneath a jet stream), as both factors inhibit the development of an organized vortex. There is also a definite connection between the seasonal position of the Equatorial Trough and zones of hurricane formation, which is borne out by the fact that no hurricanes occur in the South Atlantic (where the trough never lies south of 5°S) or in the south-east Pacific (where the trough remains north of the equator). On the other hand, satellite photographs over the north-east Pacific show an unexpected number of cyclonic vortices in summer, many of which move westwards near

Figure 6.9 Frequency of hurricane genesis (numbered isopleths) for a 20-year period. The principal hurricane tracks and the areas of sea surface having water temperatures greater than 27°C in the warmest month are also shown.

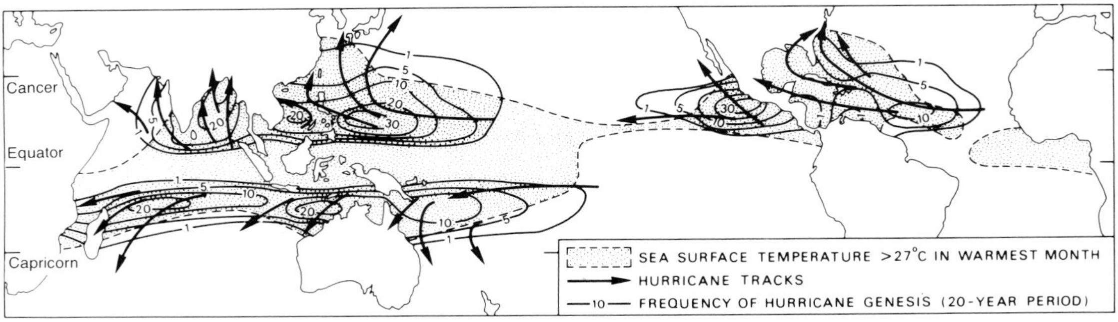

Source: After Palmén 1948 and Gray 1979.

the trough line about 10°–15°N. About 60 per cent of tropical cyclones seem to originate 5°–10° latitude poleward of the Equatorial Trough in the doldrum sectors, where the trough is at least 5° latitude from the equator. The development regions of hurricanes mainly lie over the western sections of the Atlantic, Pacific and Indian Oceans where the subtropical high-pressure cells do not cause subsidence and stability and the upper flow is divergent. About twice per season, in the western equatorial Pacific, tropical cyclones form almost simultaneously in each hemisphere near 5° latitude and along the same longitude. The cloud and wind patterns in these cyclone 'twins' are roughly symmetrical with respect to the equator (see Plate 36).

Early theories of hurricane development held that convection cells generated a sudden and massive release of latent heat to provide energy for the storm. Although convection cells were regarded as an integral part of the hurricane system, their scale was thought to be too small for them to account for the growth of a storm hundreds of kilometres in diameter. Recent research, however, is modifying this picture considerably. Energy is apparently transferred from the cumulus-scale to the large-scale circulation of the storm through the organization of the clouds into spiral bands (see Figure 6.10 and Plate 34), although the nature of the process is still being investigated. There is now ample evidence to show that hurricanes form from pre-existing disturbances, but while many of these disturbances develop as closed low-pressure cells few attain full hurricane intensity. The key to this problem is high level outflow (see Figure 6.11). This does not require an upper tropospheric anticyclone, but can occur on the eastern limb of an upper trough in the westerlies. This outflow in turn allows the development of very low pressure and high wind speeds near the surface. A distinctive feature of the hurricane is the warm vortex, since other tropical depressions and incipient storms have a cold core area of shower activity. The warm core develops through the action of 100–200 cumulonimbus towers releasing latent heat of

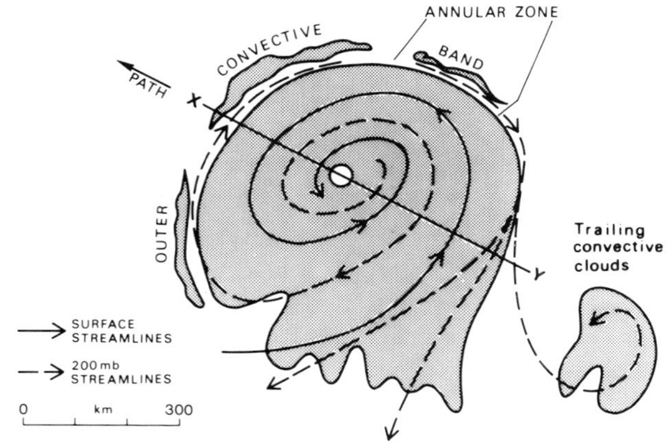

Figure 6.10 A model of the areal (*above*) and vertical (*below*) structure of a hurricane. Cloud (stippled), streamlines, convective features and path are shown.

Source: From Musk 1988.

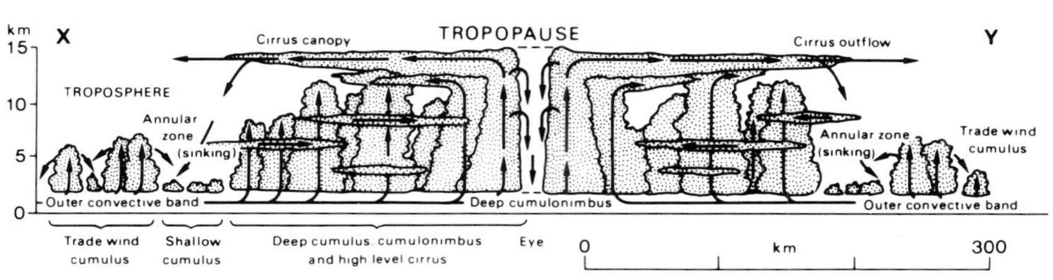

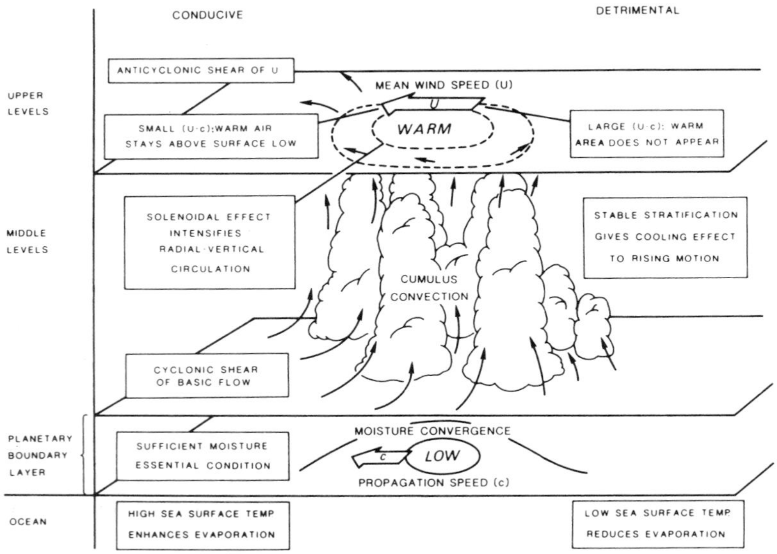

Figure 6.11 A schematic model of the conditions conducive (*left*) or detrimental (*right*) to the growth of a tropical storm in an easterly wave; *U* is the mean upper-level wind speed and *c* is the rate of propagation of the system. The warm vortex creates a thermal gradient that intensifies both the radial motion around it and the ascending air currents, termed the solenoidal effect.

Source: From Kurihara 1985 (*copyright © Academic Press. Reproduced by permission*).

condensation; about 15 per cent of the area of cloud bands is giving rain at any one time. Observations show that although these 'hot towers' form less than 1 per cent of the storm area within a radius of about 400 km (230 miles), their effect is sufficient to change the environment. The warm core is vital to hurricane growth because it intensifies the upper anticyclone, leading to a 'feedback' effect by stimulating the low-level influx of heat and moisture which further intensifies convective activity, latent heat release and therefore the upper-level high pressure. This enhancement of a storm system by cumulus convection is termed Conditional Instability of the Second Kind, or CISK (cf. the basic parcel instability described on p. 67). The thermally direct circulation converts the heat increment into potential energy and a small fraction of this – about 3 per cent – is transformed into kinetic energy. The remainder is exported by the anticyclonic circulation at about the 12 km (200-mb) level.

In the *eye*, or innermost region of the storm (see Figure 6.10), adiabatic warming of descending air accentuates the high temperatures, although since high temperatures are also observed in the eye-wall cloud masses, subsiding air can only be one contributory factor. Without this sinking air in the eye, the central pressure could not fall below about 1000 mb. The eye has a diameter of some 30–50 km (20–30 miles), within which the air is virtually calm and the cloud cover may be broken. The mechanics of the eye's inception are still largely unknown. If the rotating air conserved absolute angular momentum, wind speeds would become infinite at the centre and clearly this is not the case. The strong winds surrounding the eye are more or less in cyclostrophic balance, with the small radial distance providing a large centripetal acceleration (see p. 100). The air rises when the pressure gradient can no longer force it further inward. It is possible that the cumulonimbus anvils play a vital role in the complex link between the horizontal and vertical circulations around the eye by redistributing angular momentum in such a way as to set up a concentration of rotation near the centre.

The supply of heat and moisture combined with low frictional drag at the sea surface, the release of latent heat through condensation and the removal of the air aloft are essential conditions for the maintenance of hurricane

intensity. As soon as one of these ingredients diminishes the storm decays. This can occur quite rapidly if the track (determined by the general upper tropospheric flow) takes the vortex over a cool sea-surface or over land. In the latter case the increased friction causes greater cross-isobar air motion, temporally increasing the convergence and ascent. At this stage, increased vertical wind shear in thunderstorm cells may generate tornadoes, especially in the north-east quadrant of the storm (in the northern hemisphere). However, the most important effect of a land track is that cutting off of the moisture supply removes one of the major sources of heat. Rapid decay also occurs when cold air is drawn into the circulation or when the upper-level divergence pattern moves away from the storm.

Hurricanes usually move at some 16–24 kmph (10–15 mph), controlled primarily by the rate of movement of the upper warm core. Commonly they recurve poleward around the western margins of the subtropical high-pressure cells, entering the circulation of the westerlies, where they die out or degenerate into extra-tropical depressions (see Plate 20). Some of these systems retain an intense circulation and the high winds and waves can still wreak havoc. This is not uncommon along the Atlantic coast of the United States and occasionally eastern Canada. Similarly, in the western North Pacific, recurved typhoons are a major element in the climate of Japan (see D.4, this chapter) and may occur in any month. There is an average frequency of twelve typhoons per year over southern Japan and neighbouring sea areas.

To sum up. The hurricane develops from an initial disturbance which, under favourable environmental conditions, grows first into a tropical depression and then a tropical storm (with wind speeds of 17–33 m s^{-1} or 39–73 mph). The tropical storm stage may persist 4–5 days, whereas the hurricane stage usually lasts for only 2–3 days (4–5 days in the western Pacific). The main energy source is latent heat derived from condensed water vapour, and for this reason hurricanes are generated and continue to gather strength only within the confines of warm oceans. The cold-cored tropical storm is transformed into a warm-cored hurricane in association with the release of latent heat in cumulonimbus towers, and this establishes or intensifies an upper tropospheric anticyclonic cell. Thus high-level outflow maintains the ascent and low-level inflow in order to provide a continual generation of potential energy (from latent heat) and the transformation of this into kinetic energy. The inner eye which forms by sinking air is an essential element in the life cycle.

b Other tropical disturbances

Not all cyclonic systems in the tropics are of the intense hurricane variety. There are two other major types of cyclonic vortex. One is the *monsoon depression* which affects southern Asia during the summer. This disturbance is somewhat unusual in that the flow is westerly at low levels and easterly in the upper troposphere (see Figure 6.27). It is more fully described in D.4, this chapter.

The second type of system is usually relatively weak near the surface, but well-developed in the middle troposphere. In the eastern North Pacific and Indian Ocean such lows are referred to as *subtropical cyclones*. Some develop from the cutting-off in low latitudes of a cold upper-level wave in the westerlies (compare Chapter 4, G.4). They possess a broad eye of some 150 km (100 miles) in radius with little cloud, surrounded by a belt of cloud and precipitation about 300 km (200 miles) wide. In late winter and spring a few such storms make a major contribution to the rainfall of the Hawaiian Islands. These cyclones are very persistent and tend eventually to be re-absorbed by a trough in the upper westerlies. Other subtropical cyclones occur over the Arabian Sea in summer and make a major contribution to summer ('monsoon') rains in north-western India. These systems show upward motion mainly in the upper troposphere. Their development may be linked to air export at upper levels of cyclonic vorticity from the persistent heat low over Southern Asia, especially Arabia.

An infrequent and distinctly different weather system, known as a Temporal, occurs along the Pacific coasts of Central America in autumn and early summer. Its main feature is an extensive layer of altostratus, fed by individual convective cells, producing sustained moderate rainfall. These systems originate in the ITCZ over the eastern tropical North Pacific Ocean and are maintained by large-scale lower tropospheric convergence, localized convection and orographic uplift.

3 Tropical cloud clusters

These systems fall into two categories: non-squall and squall line. The former contain one or more mesoscale precipitation areas. They occur diurnally, for example, off the north coast of Borneo in winter where they are initiated by convergence of a nocturnal land breeze and the north-east monsoon flow (see Figure 6.12). By morning (0800 LST), cumulonimbus cells give precipitation. The cells are linked by an upper-level cloud shield which persists when the convection dies out around noon as a sea-breeze system replaces the nocturnal convergent flow.

Squall-line systems in the tropics (see Figure 6.13) form the leading edge of a line of cumulonimbus cells. The squall line and gust front move forward in the low-level flow and by the formation of new cells. These mature and eventually dissipate to the rear of the main line. The process is analogous to that of mid-latitude squall lines (see Figure 4.25) but the tropical cells are weaker. Squall-line systems cross Malaya from the west during the south-west monsoon season. They appear to be initiated by the convergent effects of land breezes in the Malacca Straits. These particular linear systems known as *sumatras*, give heavy rain and often thunder.

In West Africa, systems known as *disturbance lines* are an important feature of the climate in the summer half-year, when low-level south-westerly monsoon air is overrun by dry, warm

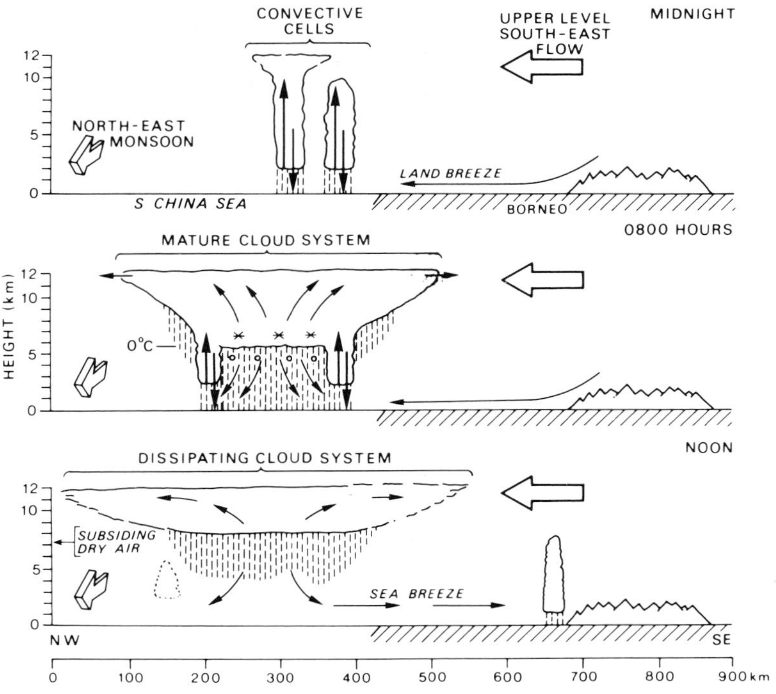

Figure 6.12 Schematic development of a non-squall cloud cluster off the north coast of Borneo: large arrows indicate the major circulation; small arrows, the local circulation; vertical shading, the zones of rain; stars, ice crystals; and circles, melted raindrops.

Source: After Houze *et al.* 1981.

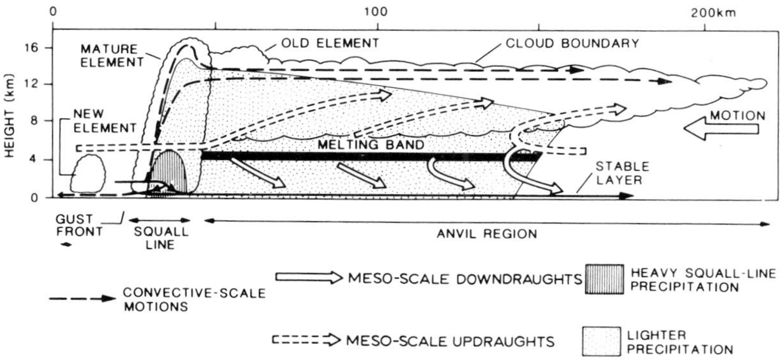

Figure 6.13 Cross-section of a tropical squall-line cloud cluster showing locations of precipitation and ice particle melting. Dashed arrows show the air motion generated by the squall-line convection and the broad arrows the mesoscale circulation.

Source: After Houze; from Houze and Hobbs 1982.

Saharan air. The meridional air mass contrast helps set up the lower-tropospheric African easterly jet shown in Figure 6.35 which illustrates the zonation of climate and associated weather systems. Disturbance lines tend to form when there is divergence in the upper troposphere north of the Tropical Easterly Jet (see Figure 6.37 also). They are several hundred kilometres long and travel westward associated with upper easterlies at about 50 km h^{-1} (30 mph) giving squalls and thundery showers before dissipating over cold-water areas of the North Atlantic. Spring and autumn rainfall in West Africa is derived in large part from these disturbances. Figure 6.14 for Kortright (Freetown, Sierra Leone) illustrates the daily rainfall amounts in 1960–1 associated with disturbance lines at 8°N. The summer monsoon rains make up the greater part of the total here, but their contribution diminishes northward.

D THE ASIAN MONSOON

The name *monsoon* is derived from the Arabic word (*mausim*) which means season, so explaining its application to large-scale seasonal reversals of the wind regime. The Asiatic seasonal wind reversal is notable for its immense extent and the penetration of its influence beyond tropical latitudes. For example, the surface circulation over China reflects this seasonal change (see Table 6.2).

However, such seasonal wind shifts at the

Figure 6.14 Daily rainfall at Kortright, Freetown, Sierra Leone, October 1960–September 1961.

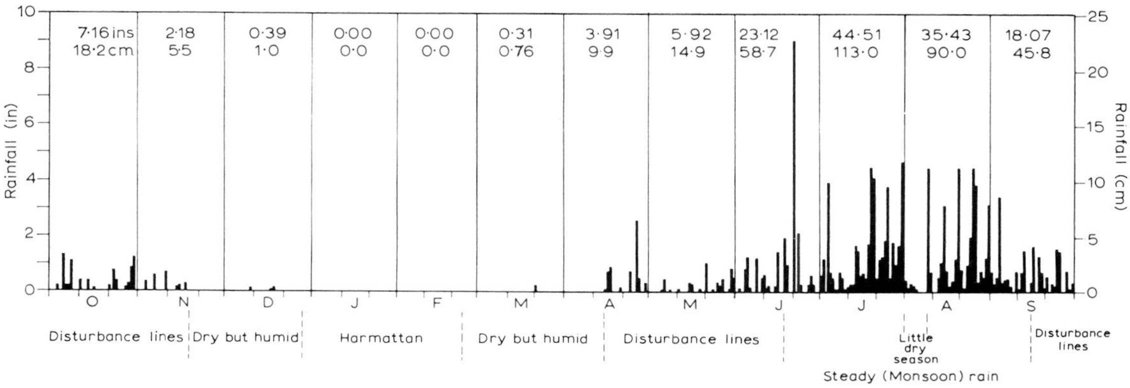

Source: After Gregory 1965.

Table 6.2 Surface circulation over China.

	January	*July*
North China	60 per cent of winds from W, NW and N	57 per cent of winds from SE, S and SW
South-east China	88 per cent of winds from N, NE and E	56 per cent of winds from SE, S and SW

surface are quite widespread and occur in many regions which would not traditionally be considered as monsoonal (see Figure 6.15). Although there is a rough accordance between these traditional regions and those experiencing over 60 per cent frequency of the prevailing octant, it is obvious that a variety of unconnected mechanisms can produce significant seasonal wind shifts. Nor is it possible to establish a simple relationship between seasonality of rainfall (see Figure 6.16) and seasonal wind shift. Areas traditionally designated as 'monsoonal' include some of the tropical and near-tropical regions experiencing a summer rainfall maximum and most of those having a double rainfall maximum. It is clear that a combination of criteria (for example, from Figures 6.15 and 6.16) is necessary to approach an adequate definition of monsoonal areas.

In summer the Equatorial Trough and the subtropical anticyclones are everywhere displaced northwards in response to the changing pattern of solar heating of the earth, and in southern Asia this movement is magnified by the effects of the land mass. However, the attractive simplicity of the traditional explanation, which envisages a monsoonal 'sea breeze' directed towards a summer thermal low pressure over the continent, unfortunately provides an inadequate basis for understanding the workings of the system. The Asiatic monsoon regime is a consequence of the interaction of both planetary and regional factors, both at the surface and in the upper troposphere. It is convenient to look at each season in turn and Figure 6.17 shows the generalized meridional circulations at 90°E over India and the Indian Ocean in winter (December–February), spring (April) and autumn (September), together with those associated with active and break periods during the June–August summer monsoon.

1 Winter

Near the surface this is the season of the outblowing 'winter monsoon', but aloft westerly airflow dominates. This, as we have seen, reflects the hemispheric pressure distribution. A shallow layer of cold, high-pressure air is

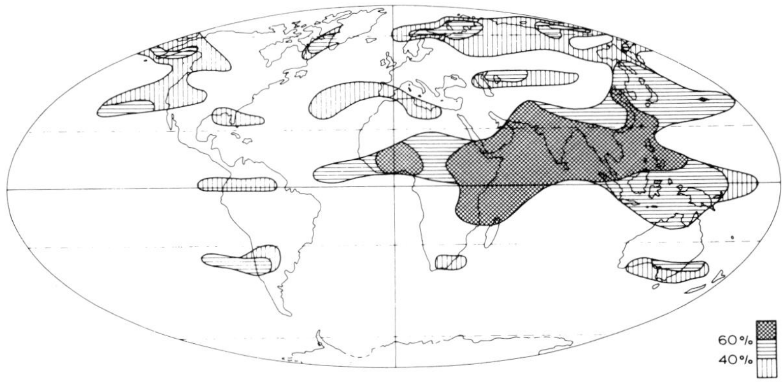

Figure 6.15 Regions experiencing a seasonal surface wind shift of at least 120°, showing the frequency of the prevailing octant.

Source: After Khromov.

Figure 6.16 The annual distribution of tropical rainfall. The shaded areas refer to periods during which more than 75 per cent of the mean annual rainfall occurs. Areas with less than 250 mm yr^{-1} (10 in yr^{-1}) are classed as deserts, and the unshaded areas are those needing at least 7 months to accumulate 75 per cent of the annual rainfall and are thus considered to exhibit no seasonal maximum.

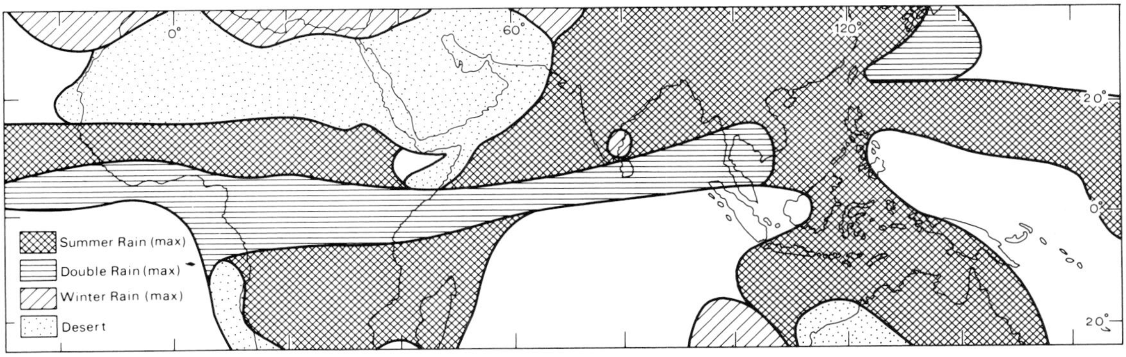

Source: After Ramage 1971.

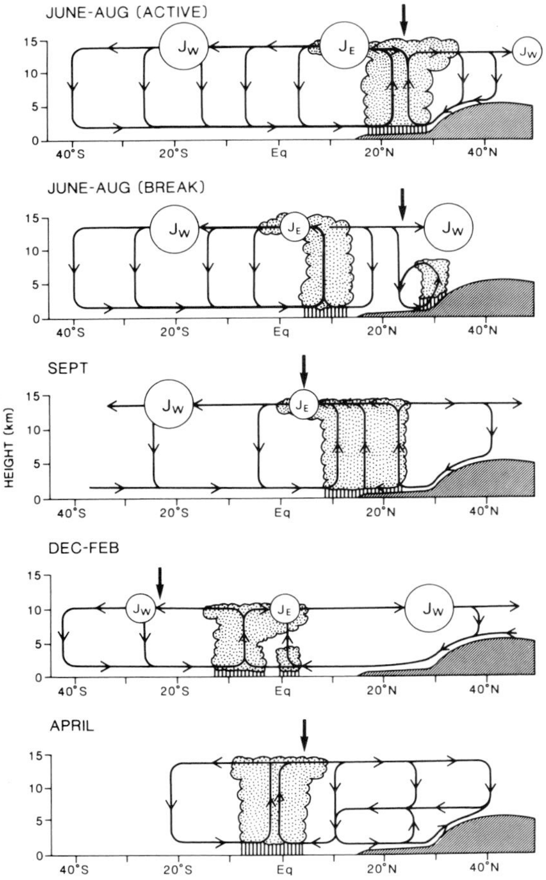

Figure 6.17 Schematic representation of the meridional circulation over India at 90°E at five characteristic periods of the year; the active summer monsoon (June–August); a break in the summer monsoon (June–August); and winter monsoon (December–February); approach of the monsoon season (April); the retreat of the summer monsoon (September). Easterly (J_E) and westerly (J_W) jet streams are shown at sizes depending on their strength; the arrows mark the positions of the overhead sun; and zones of maximum precipitation are indicated.

Source: After Webster 1987.

centred over the continental interior, but this has disappeared even at 700 mb (see Figure 3.20) where there is a trough over eastern Asia and zonal circulation over the continent. The upper westerlies split into two currents to the north and south of the high Tibetan (Qinghai-Xizang) plateau (see Figure 6.18), to reunite again off the east coast of China (see Figure 6.19). The plateau, which exceeds 4000 m (13,000 ft) over a vast area, is a tropospheric cold source in winter, particularly over its western part, although the strength of this source depends on the extent and duration of snow cover (snow-free ground acts as a heat source for the atmosphere in *all* months). Below 600 mb, the tropospheric heat sink gives rise to a shallow, cold plateau anticyclone, which is best developed in December and January. The two jet-stream branches have been attributed to the disruptive effect of the topographic barrier on the airflow, but this is limited to altitudes below about 4 km. In fact, the northern jet is highly mobile and may be located far from the Tibetan plateau. Two currents are also found to occur farther west where there is no obstacle to the flow. The branch over northern India corresponds with a strong latitudinal thermal gradient (from November to April) and it is probable that this factor, combined with the thermal effect of the barrier to the north, is responsible for the anchoring of the southerly jet. This southern branch is the stronger, with an average speed of more than 40 m s^{-1} (90 mph) at 200 mb, compared with about 20–25 m s^{-1} (45–55 mph) in the northern one. Where the two unite over north China and south Japan the average velocity exceeds 66 m s^{-1} (148 mph) (see Figure 6.20).

Air subsiding beneath this upper westerly current gives dry outblowing northerly winds from the subtropical anticyclone over north-western India and Pakistan. The surface wind direction is north-westerly over most of northern India, becoming north-easterly over Burma and Bangladesh and easterly over peninsular India. Equally important is the steering of winter depressions over northern India by the upper jet. The lows which are not usually frontal, appear to penetrate across the Middle East from the Mediterranean and are important sources of rainfall for northern India and Pakistan (e.g. Kalat, Figure 6.21), especially as it falls when evaporation is at a minimum. The equatorial trough of convergence and precipitation lies between the equator and about latitude 15°S (see Figure 6.17).

Some of these westerly depressions continue eastwards, redeveloping in the zone of jet-stream confluence about 30°N, 105°E over China, beyond the area of subsidence in the immediate lee of Tibet (see Figure 6.19), and it is significant that the mean axis of the winter jet stream over China shows a close correlation with the distribution of winter rainfall (see Figure 6.22). Other depressions affecting central and north China travel within the westerlies north of Tibet or are initiated by outbreaks of fresh cP air. In the rear of these depressions there are invasions of very cold air (e.g. the *buran* blizzards of Mongolia and Manchuria). The effect of such cold waves, comparable with the *northers* in the central and southern United States, is greatly to reduce mean temperatures. Winter mean temperatures in less-protected southern China are considerably below those at equivalent latitudes in India and, for example, temperatures at Calcutta and Hong Kong (both approximately 22½°N) are, respectively, 19°C (67°F) and 16°C (60°F) in January and 22°C (71°F) and 15°C (59°F) in February.

2 Spring

The key to change during this transition season is again found in the pattern of the upper airflow. In March the upper westerlies begin their seasonal migration northwards, but whereas the northerly jet strengthens and begins to extend across central China and into Japan, the southerly branch remains positioned south of Tibet, although weakening in intensity.

In April there is weak convection over India, where the circulation is dominated by subsiding air originating along the convective ITCZ

Figure 6.18 The mean zonal geostrophic wind (in ms^{-1}; solid lines) and temperatures (in °C; dashed lines) along longitude 105°E for January–March 1956. J denotes the jet stream cores.

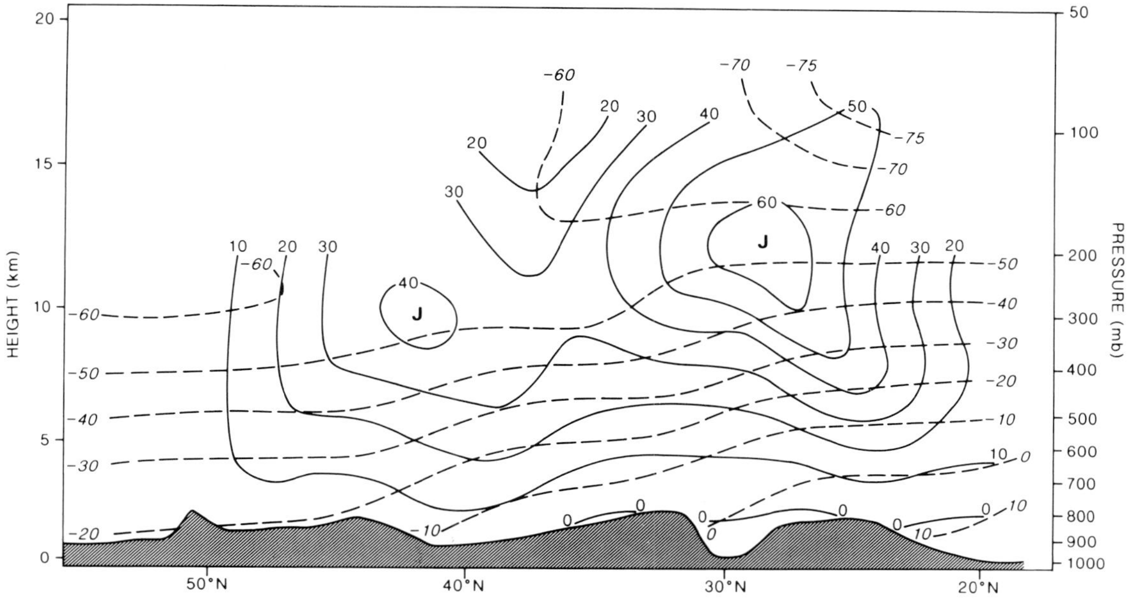

Source: From Academica Sinica 1958.

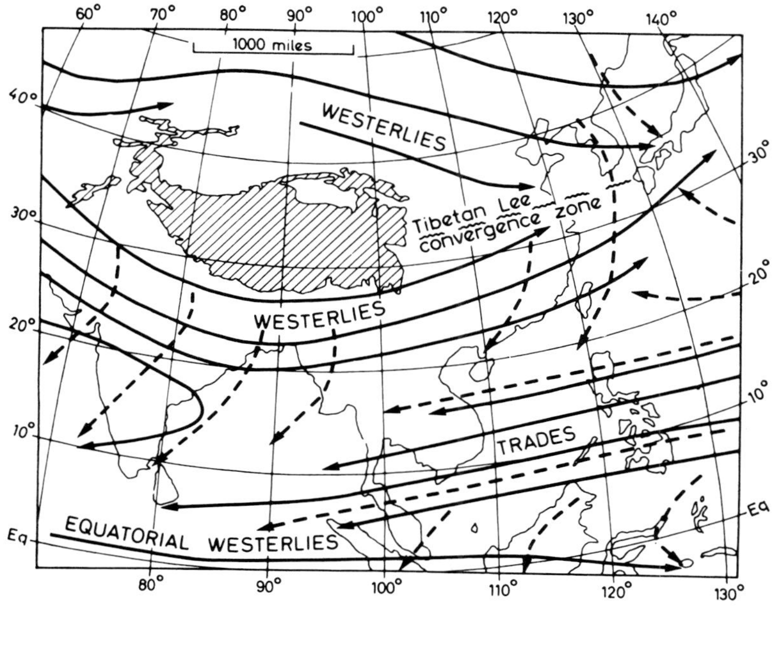

Figure 6.19 The characteristic air circulation over southern and eastern Asia in winter. Solid lines indicate airflow at about 3000 m (10,000 ft), and dashed lines that at about 600 m (2000 ft). The names refer to the wind systems aloft.

Sources: After Thompson 1951. Flohn 1968, Frost and Stephenson 1965, and others.

Figure 6.20 Mean 200-mb streamlines (→) and isotachs (- - -) in knots over south-east Asia (A) January (B) July, based on aircraft reports and sounding data.

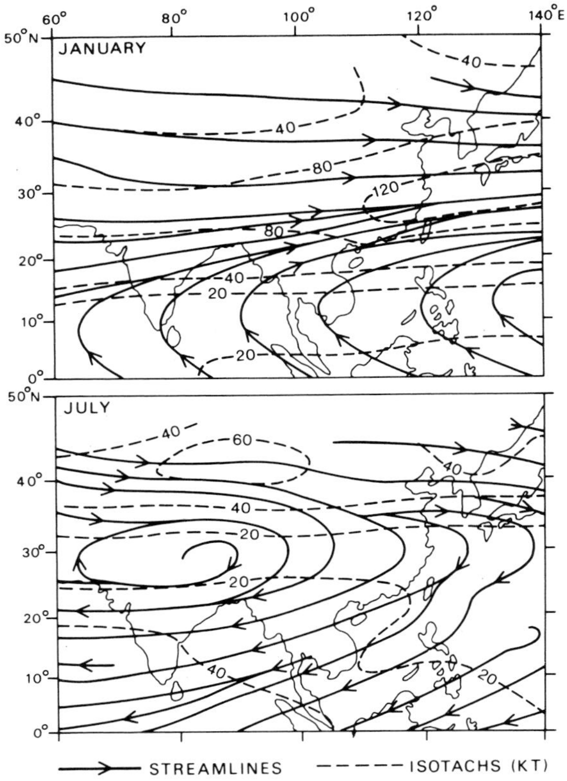

Source: From Sadler 1975b (*courtesy Dr. J. C. Sadler, University of Hawaii*).

Figure 6.21 Average monthly rainfall at six stations in the Indian region. The annual total in inches is given after the station name (see Appendix 2 for the metric conversion).

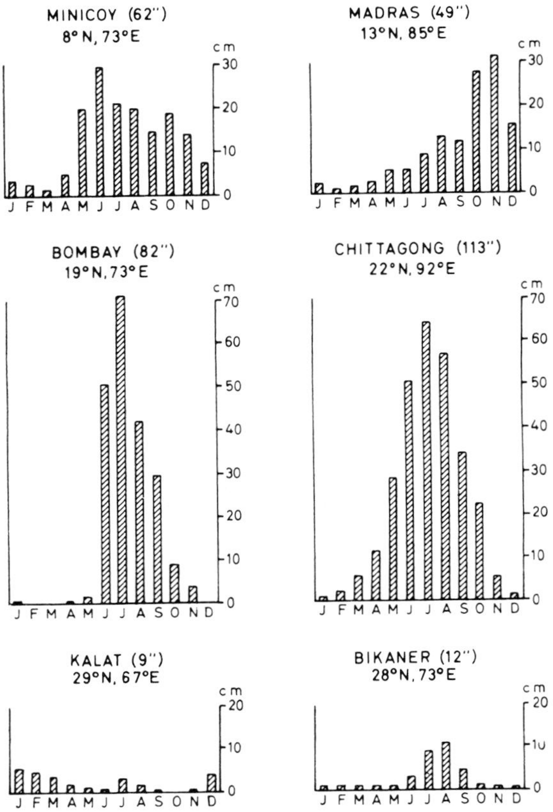

Source: Based on 'CLIMAT' normals of the World Meteorological organization for 1931–60.

trough centred over the equator and following the overhead sun northward over the warm Indian Ocean (see Figure 6.17). The weather over northern India becomes hot, dry and squally in response to the greater solar radiation heating. Mean temperatures at Delhi rise from 23°C (74°F) in March to 33°C (92°F) in May. The thermal low-pressure cell (see Chapter 4, G.2) reaches its maximum intensity at this time, but although onshore coastal winds develop the onset of the monsoon is still a month away and other mechanisms cause only limited precipitation. Some precipitation occurs in the north with 'westerly disturbances', particularly towards the Ganges delta where the low-level inflow of warm, humid air is overrun by dry, potentially cold air, triggering squall lines known as *nor'westers*. In the north-west, where less moisture is available the convection generates violent squalls and dust-storms termed *andhis*. The mechanism of these storms is not yet fully known, though high-level divergence in the waves of the subtropical westerly jet stream appears to be essential. The early onset of summer rains in Bengal, Bangladesh, Assam and Burma (e.g. Chittagong, Figure 6.21) is

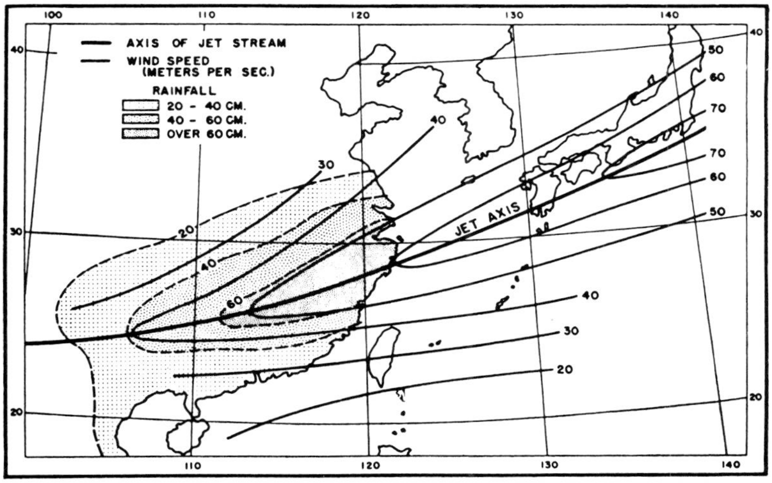

Figure 6.22 The mean winter jet stream axis at 12 km over the Far East and the mean winter precipitation over China in cm.

Source: After Mohri and Yeh; from Trewartha 1958.

favoured by an orographically produced trough in the westerlies at 300 mb, which is located at about 85°–90°E in May. Low-level convergence of maritime air from the Bay of Bengal, combined with the upper-level divergence ahead of the 300-mb trough, generates thunder squalls. Tropical disturbances in the Bay of Bengal are another source of these early rains. Rain also falls during this season over Sri Lanka and south India (e.g. Minicoy, Figure 6.21) in response to the northward movement of the Equatorial Trough.

China has no equivalent of India's hot, pre-monsoon season. The low-level, north-easterly winter monsoon (reinforced by subsiding air from the upper westerlies) persists in north China and even in the south it only begins to be replaced by maritime tropical air in April–May. Thus, at Canton mean temperatures rise from only 17°C (63°F) in March to 27°C (80°F) in May, some 6°C (12°F) less than the mean values over northern India.

Westerly depressions are most frequent over China in spring. They form more readily over central Asia at this season as the continental anticyclone begins to weaken; also many develop in the jet-stream confluence zone in the lee of the plateau. The average number crossing China per month during 1921–31 was as follows:

J	*F*	*M*	*A*	*M*	*J*	*J*	*A*	*S*	*O*	*N*	*D*	*Year*
7	8	9	11	10	8	5	3	3	6	7	7	86

Hence spring is wetter than winter and, over most of central and southern China, the three months March–May contribute between a quarter and a third of the annual rainfall.

3 Early summer

Generally during the last week in May the southern branch of the high-level jet begins to break down, becoming intermittent and then gradually shifting northward over the Tibetan plateau. At 500 mb and below, however, the plateau exerts a blocking effect on the flow and the jet axis there jumps from the south to north side of the plateau from May to June. Over India, the Equatorial Trough pushes northwards with each weakening of the upper westerlies south of Tibet, but the final *burst* of the monsoon, with the arrival of the humid, low-level south-westerlies, is not accomplished until the upper-air circulation has switched to its summer pattern (see Figures 6.20 and 6.23). Increased continental convection overcomes the spring subsidence and the return upper-level flow to the south is deflected by the Coriolis

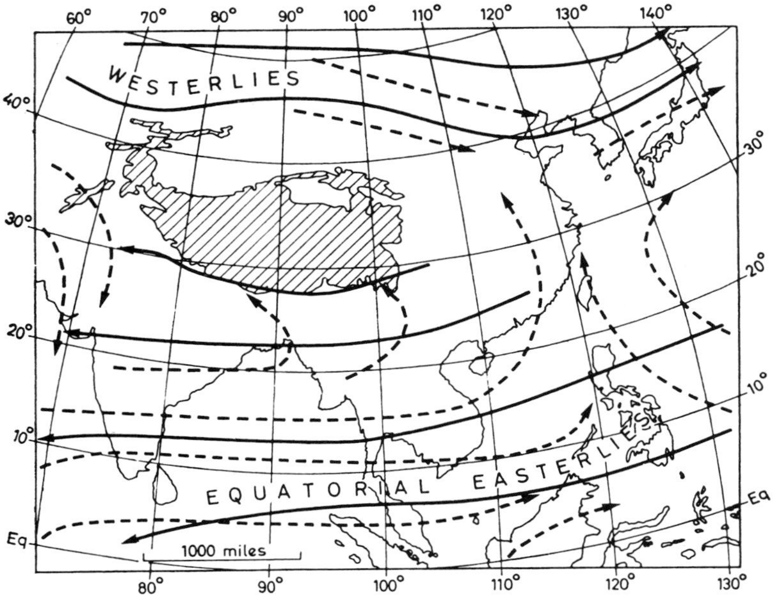

Figure 6.23 The characteristic air circulation over southern and eastern Asia in summer. Solid lines indicate air flow at about 6000 m (20,000 ft) and dashed lines at about 600 m (2000 ft). Note that the low-level flow is very uniform between about 600 and 3000 m.

Sources: After Thompson 1951, Flohn 1968, Frost and Stephenson 1965, and others.

force to produce a strengthening easterly jet located at about 10°–15°N and a westerly jet at about 20°S (see Figure 6.17). One theory suggests that the monsoon onset takes place in June as the col between the subtropical anticyclone cells of the west Pacific and the Arabian Sea at the 300-mb level is displaced northwestwards from a position about 15°N, 95°E in May towards central India. The north-westward movement of the monsoon (see Figure 6.24) is apparently related to the extension over India of the upper tropospheric easterlies.

The reorganization of the upper airflow has widespread effects in southern Asia. It is directly linked with the *Mai-yu* rains of China (which reach a peak about 10–15 June), the onset of the south-west Indian monsoon and the northerly retreat of the upper westerlies over the whole of the Middle East.

It must nevertheless be emphasized that it is still uncertain how far these changes are caused by events in the upper air or indeed whether the onset of the monsoon initiates a readjustment in the upper-air circulation. The presence of the Tibetan plateau is certainly of importance even if there is no significant barrier effect on the upper airflow. The plateau surface is strongly heated in spring and early summer (R_n is about 180 W m^{-2} in May) and nearly all of this is transferred via sensible heat to the atmosphere. This results in the formation of a shallow heat low on the plateau overlain, about 450 mb, by a warm anticyclone (see Figure 3.15). The plateau atmospheric boundary layer now extends over an area about twice the size of the plateau surface itself. Easterly airflow on the southern side of the upper anticyclone undoubtedly assists in the northward shift of the subtropical westerly jet stream. At the same time, the premonsoonal convective activity over the south-eastern rim of the plateau provides a further heat source, by latent heat release, for the upper anticyclone. The seasonal wind reversals over and around the Tibetan plateau have led Chinese meteorologists to distinguish a 'Plateau Monsoon' system, distinct from that over India.

Over China, the zonal westerlies retreat northwards in May–June and the westerly flow becomes concentrated north of the Tibetan plateau. The equatorial westerlies spread across south-east Asia from the Indian Ocean, giving a

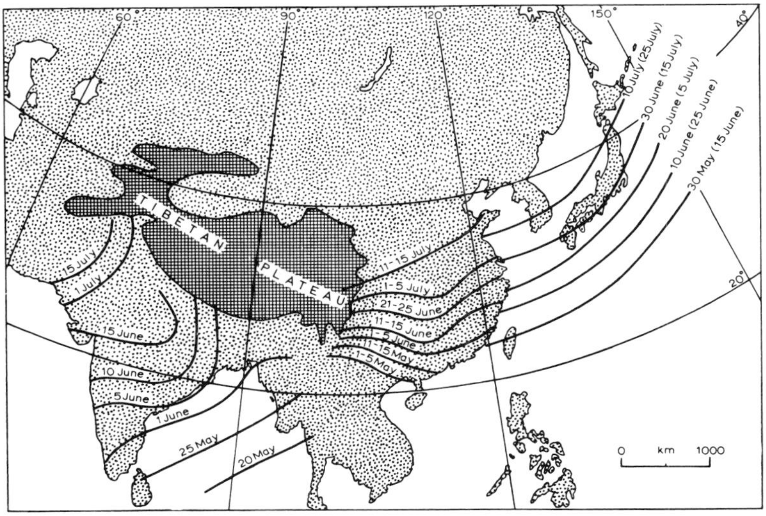

Figure 6.24 The advance of the summer monsoon over south and east Asia, based on: for India, the beginning of the rainy season;* for China, the northward shift of the 5-day mean wet-bulb temperature of 24°C;† for Japan, the onset of the 'bai-u rains'.‡

Sources: *After Chatterjee; †after Tu and Hwang; ‡after Takahashi and, in parentheses, after Kurashima 1968.

warm, humid air mass at least 3000 m deep, but the summer monsoon over southern China is apparently influenced less by the westerly flow over India than by southerly airflow over Indonesia near 100°E. Also, contrary to earlier views, the Pacific is only a moisture source when tropical south-easterlies extend westwards to affect the east coast.

The Mai-yu 'front' involves both the monsoon trough and the East Asian–West Pacific Polar Front, with weak disturbances moving eastward along the Yangtze valley and occasional cold fronts from the north-west. Its location shifts northward in three stages, from south of the Yangtze River in early May to north of it by the end of the month and into northern China in mid-July (see Figure 6.24), where it remains until late September.

4 Summer

By mid-July monsoon air covers most of southern and south-eastern Asia (see Figure 6.23) and in India the Equatorial Trough is located about 25°N. North of the Tibetan plateau there is a rather weak upper westerly current with a (subtropical) high-pressure cell over the plateau. The south-west monsoon in southern Asia is overlain by strong upper easterlies (see Figure 6.20) with a pronounced jet at 150 mb (about 15 km or 50,000 ft) which extends westwards across South Arabia and Africa (see Figure 6.25). No easterly jets have so far been observed over the tropical Atlantic or Pacific. The jet is related to a steep lateral temperature gradient with the upper air getting progressively colder to the south.

An important characteristic of the tropical easterly jet is the location of the main belt of summer rainfall on the right (i.e. north) side of the axis upstream of the wind maximum and on the left side downstream, except for areas where the orographic effect is predominant (see Figure 6.25). The mean jet maximum is located about 15°N, 50°–80°E.

The monsoon current does not give rise to a simple pattern of weather over India, despite the fact that much of the country receives 80 per cent or more of its annual precipitation during the monsoon season (see Figure 6.26). In the north-west a thin wedge of monsoon air is overlain by subsiding continental air. The inversion prevents convection and consequently little or no rain falls in the summer months in the arid north-west of the subcontinent (e.g. Bikaner and Kalat, Figure 6.21). This is similar to the Sahel zone in West Africa, discussed below.

Around the head of the Bay of Bengal and

Figure 6.25 The easterly tropical jet stream. *Above:* The location of the easterly jet streams at 200 mb on 25 July 1955. Streamlines are shown in solid lines and isotachs (wind speed) dashed. Wind speeds are given in knots (westerly components positive, easterly negative). *Below:* The average July rainfall (shaded areas receive more than 10 in or 25 cm) in relation to the location of the easterly jet streams.

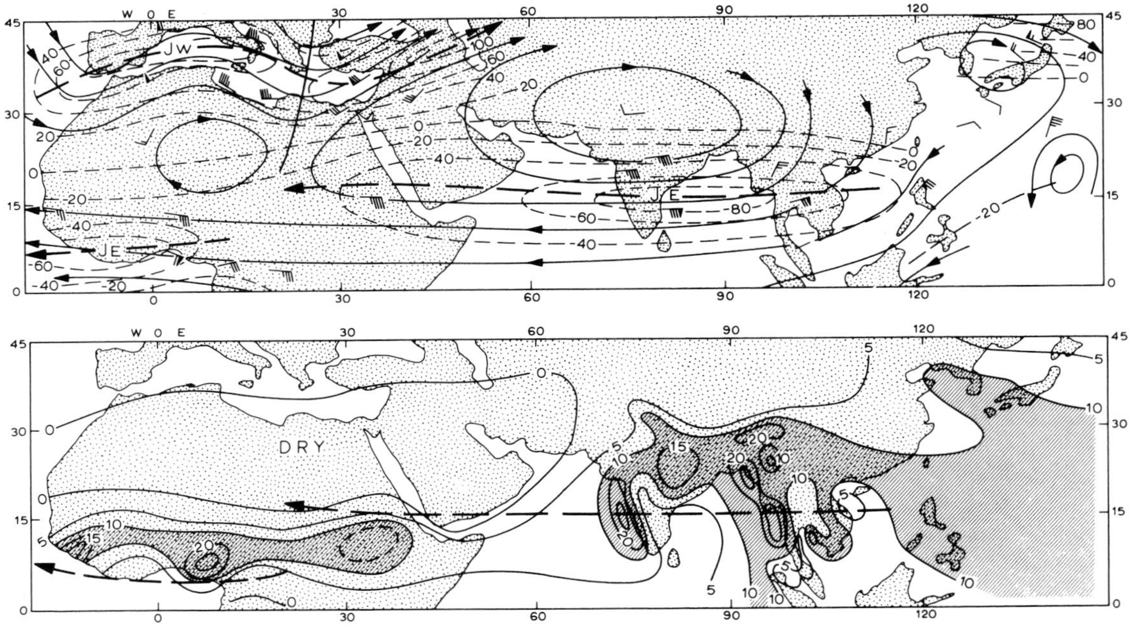

Source: From Koteswaram 1958.

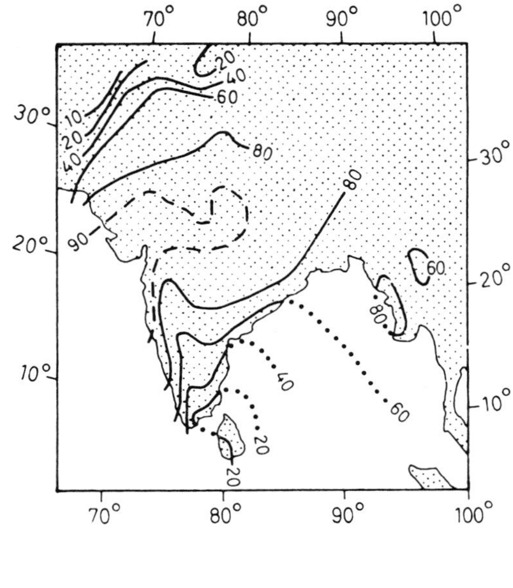

Figure 6.26 The percentage contribution of the monsoon rainfall (June to September) to the annual total.

Source: After Rao and Ramamoorthy, in Indian Meteorological Department 1960; and Ananthakrishnan and Rajagopalachari, in Hutchings 1964.

Figure 6.27 Monsoon depressions of 1200 GMT, 4 July 1957. The upper diagram shows the height (in tens of metres) of the 500-mb surface, the lower one the sea-level isobars. The broken line in the lower diagram represents the Equatorial Trough, and precipitation areas are shown by the oblique shading.

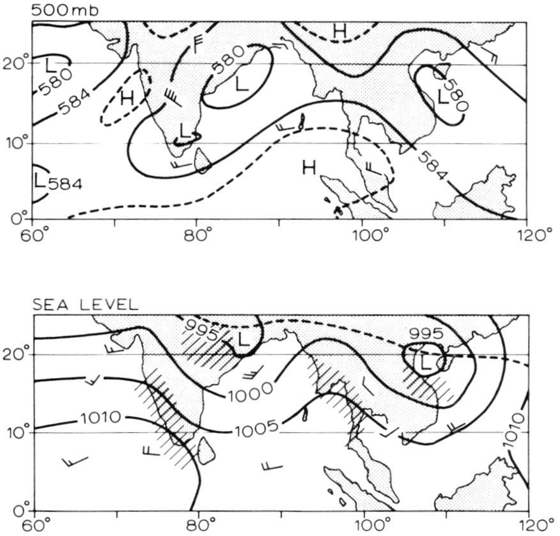

Source: Based on the IGY charts of the Deutscher Wetterdienst.

Figure 6.28 The normal track of monsoon depressions, together with a typical depression pressure distribution.

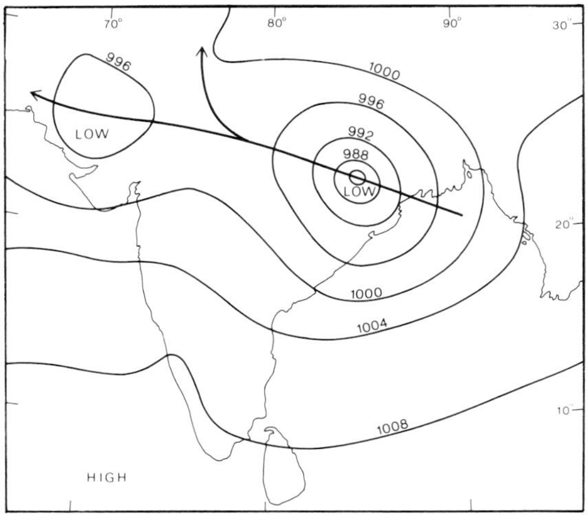

Source: After Das 1987.

Figure 6.29 The location of the Monsoon Trough in its normal position during an active summer monsoon phase (solid) and during breaks in the monsoon dashed.* Areas 1–4 indicate four successive daily areas of heavy rain (> 50 mm/day) during the period 7–10 July 1973 as a monsoon depression moved west along the Ganges Valley. Areas of lighter rainfall were much more extensive.†

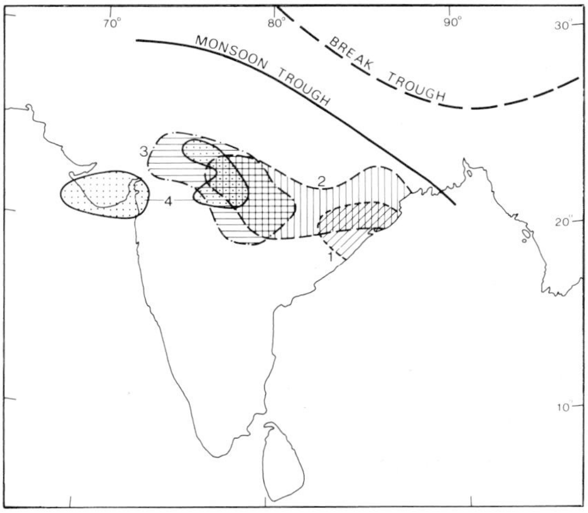

Source: *After Das 1987.
†From Webster 1987b.

along the Ganges valley the main weather mechanisms in summer are the 'monsoon depressions' (see Figure 6.27) which usually move westwards or north-westwards across India, steered by the upper easterlies (see Figure 6.28), mainly in July and August. On average, they occur about twice a month, apparently when an upper trough becomes superimposed over a surface disturbance in the Bay of Bengal. Monsoon depressions have cold cores, are generally without fronts and are some 1000–1250 km across, with a cyclonic circulation up to about 8 km, and a typical life time of 2–5 days. They produce average daily rainfalls of 10–20 cm, occurring mainly as convective rains in the south-west quadrant of the depression. The main rain areas typically lie south of the Equatorial or Monsoon Trough (see Figure 6.29) (in the south-west quadrant of the

monsoon depressions, resembling an inverted mid-latitude depression), and also tend to occur on windward coasts and mountains of India, Burma, and Malaya. Without such disturbances the distribution of monsoon rains would be controlled to a much larger degree by orography.

It has recently been discovered that part of the south-west monsoonal flow occurs in the form of a 15–45 m s^{-1} (30–100 mph) jet stream at a level of only 1000–1500 m (3250–5000 ft). This jet, strongest during active periods of the Indian monsoon, flows north-westward from Madagascar (see Figure 6.30), and crosses the equator from the south over East Africa where its core is often marked by a streak of cloud (similar to that in Plate 16) and where it may bring excessive local rainfall. The jet is displaced northward and strengthens from February to July and by May it becomes constricted against the Ethiopian Highlands, accelerates still more and is deflected eastward across the Arabian Sea towards the west coast of the Indian peninsula. This low-level jet, unique in the Trade Wind belt, flows offshore from the Horn of Africa bringing up cool waters and contributing to a temperature inversion which is also produced by dry upper air originating over Arabia or East Africa and by subsidence due to the convergent upper easterlies. The flow from the south-west over the Indian Ocean is relatively dry near the equator and near shore, apart from a shallow moist layer near the base. Downwind towards India, however, there is a strong temperature and moisture interaction between the ocean surface and the low-level jet flow such that deep convection builds up and convective instability is released, especially as the airflow slows down and converges near the west coast of India and as it is forced up over the Western Ghats. A portion of this south-west monsoon airflow is deflected by the Western Ghats to form 100-km-diameter offshore vortices lasting 2–3 days and capable of bringing 100 mm of rain in 24 hours along the western coastal belt of the peninsula. At Mangalore (13°N) there are on average 25 rain-days per month in June, 28 in July and 25 in August. The monthly rainfall averages are respectively 98 cm (38.6 in), 106 cm (41.7 in) and 58 cm (22.8 in), accounting for 75 per cent of the annual total. In the lee of the Ghats amounts are much reduced and there are semi-arid areas receiving less than 64 cm (25 in) per year.

In southern India, excluding the south-east, there is a marked tendency for less rainfall when the Equatorial Trough is farthest north. Figure 6.21 shows a maximum at Minicoy in June with a secondary peak in October as the Equatorial Trough and its associated disturbances withdraw southward. This double peak occurs in much of interior peninsular India south of about 20°N and in western Sri Lanka, although autumn is the wettest period.

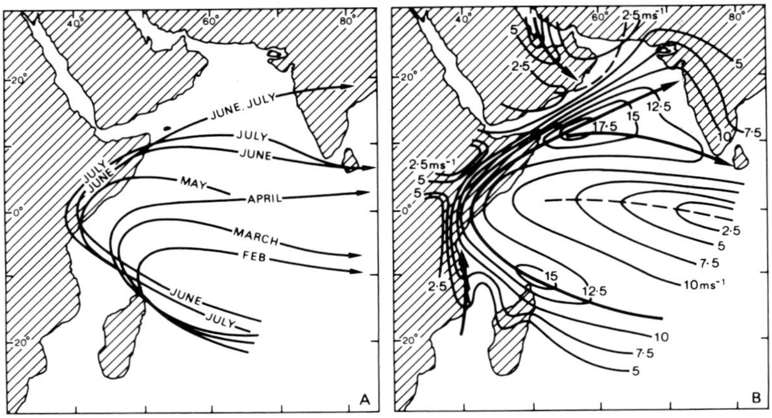

Figure 6.30 The mean monthly positions (A) and the mean July velocity (m s^{-1}) (B) of the low-level (1 km) Somali jet stream over the Indian Ocean.

Source: After Findlater 1971 (*reproduced by permission of the Controller of Her Majesty's Stationery Office*).

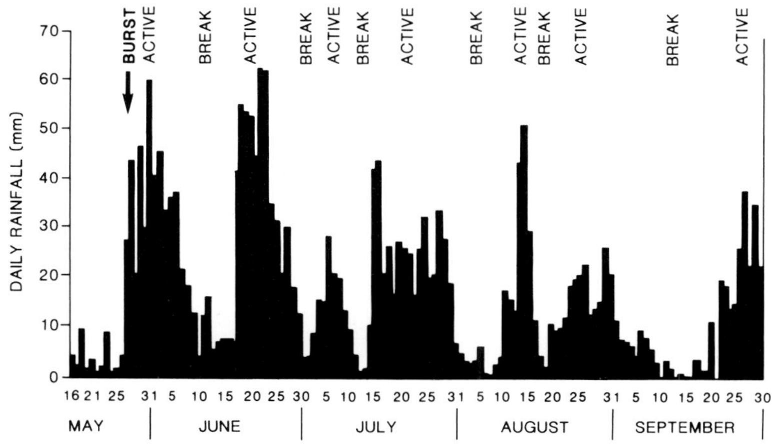

Figure 6.31 Mean daily rainfall (mm) along the west coast of India during the period 16 May to 30 September 1971 showing a pronounced burst of the monsoon followed by active periods and breaks of a periodic nature. All years do not exhibit these features as clearly.

Source: Ater Webster 1987b.

There is a variable pulse alternating between *active* and *break* periods in the May–September summer monsoon flow (see Figure 6.17) which, particularly at times of its strongest expression (e.g. 1971), produces periodic rainfall (see Figure 6.31). During active periods the convective Monsoon Trough is located in a southerly position giving heavy rain over north and central India and the west coast (see Figure 6.17). There is a consequently strong upper-level outflow to the south which strengthens both the easterly jet north of the equator and the westerly jet to the south over the Indian Ocean. The other upper-air outflow to the north fuels the weaker westerly jet there. Convective waves move east from the Indian Ocean to the cooler eastern Pacific with an irregular periodicity (on average 40–50 days for the most marked waves), finding maximum expression at the 850-mb level and being clearly connected with the Walker circulation. After the passage of an active convective wave there is a more stable break in the summer monsoon when the trough shifts to the south, the easterly jet weakens and subsiding air is forced to rise by the Himalayas along a *break trough* located above the foothills (see Figure 6.17), which replaces the Monsoon Trough during break periods. This circulation brings rain to the foothills of the Himalayas and the Brahmaputra Valley at a time of generally low rainfall elsewhere. The shift of the trough to the south of the subcontinent is associated with a similar movement and strengthening of the westerly jet to the north, weakening the Tibetan anticyclone or displacing it north-eastwards. The lack of rain over much of the subcontinent during break periods may be due in part to the eastward extension across India of the subtropical high-pressure cell centred over Arabia at this time.

It is important to realize that the monsoon rains are highly variable from year to year, emphasizing the role played by disturbances in generating rainfall within the generally moist south-westerly airflow. Droughts occur with some regularity in the Indian subcontinent and between 1891 and 1988 there were 18 years of extreme drought (see Figure 6.32) and at least five others of significant drought. These droughts are brought about by a combination of a late burst of the summer monsoon and an increase in the number and length of the break periods. Although breaks are most common in August–September, and last on average five days, they may occur at any time during the summer and can last as long as three weeks. It is thought that the mechanisms of burst and breaks are connected.

The strong surface heat source over the Tibetan plateau, which is most effective during the day, gives rise to a 50–85 per cent frequency of cumulonimbus clouds over central and

Figure 6.32 The yearly drought area index for the Indian subcontinent for the period 1891–1988, based upon the percentage of the total area experiencing moderate, extreme or severe drought. Years of extreme drought are dated. The dashed line indicates the lower limit of extreme droughts.

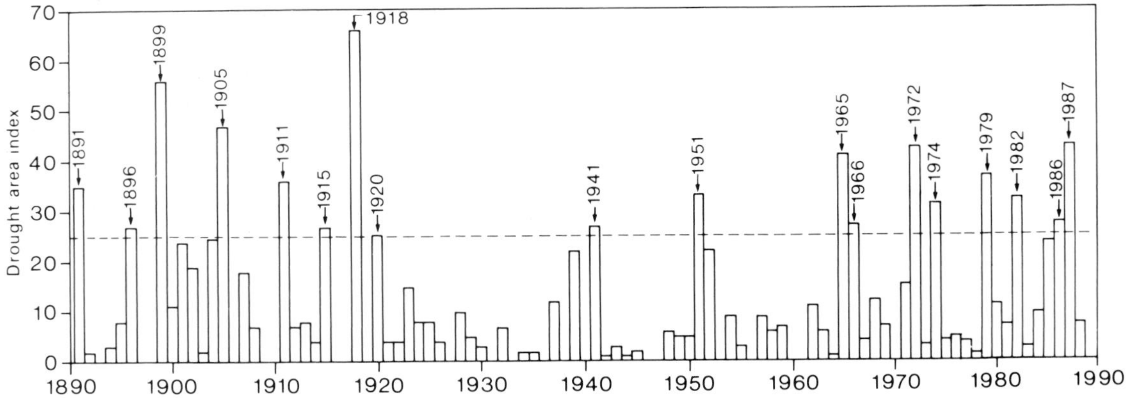

Source: After Bhalme and Mooley, 1980. Updated by courtesy of H. N. Bhalme *reproduced by permission of the American Meteorological Society.*

eastern Tibet in July. Late afternoon rain or hail showers are generally accompanied by thunder, but half or more of the precipitation falls at night, accounting for 70–80 per cent of the total in south-central and south-eastern Tibet. This may be related to large-scale plateau-induced local wind systems. However, the central and eastern plateau also has a frequency maximum of shear lines and associated weak lows at 500 mb during May through September. These plateau systems are shallow (2–2.5 km) and only 400–1000 km in diameter, but they are associated with cloud clusters on satellite imagery.

To the east over China, the surface airflow is south-westerly and the upper winds are weak with only a diffuse easterly current over southern China. According to traditional views the monsoon current reaches northern China by July. The annual rainfall regime shows a distinct summer maximum with, for example, 64 per cent of the annual total occurring at Tientsin (39°N) in July and August. Nevertheless, much of the rain falls during thunderstorms associated with shallow lows and the existence of the ITCZ in this region is doubtful (see Figure 6.1). The southerly winds, referred to above, which predominate over northern China in summer are not necessarily linked with the monsoon current farther south. Indeed this idea is the result of incorrect interpretation of streamline maps (of instantaneous airflow direction) as ones showing air trajectories (or the actual paths followed by air parcels). The depiction of the monsoon over China on Figure 6.22 is, in fact, based on a wet-bulb temperature value of 24°C. Cyclonic activity in northern China is attributable to the West Pacific Polar Front, forming between cP air and much-modified mT air.

In central and southern China the three summer months account for about 40–50 per cent of the annual average precipitation, with another 30 per cent or so being received in spring. In south-east China there is a rainfall singularity in the first half of July; a secondary minimum in the profile seems to result from the westward extension of the Pacific subtropical anticyclone over the coast of China.

A similar pattern of rainfall maxima occurs over southern and central Japan (see Figure 6.33), comprising two of the six natural seasons which have been recognized there. The main rains occur during the *Bai-u* season of the south-east monsoon resulting from waves, convergence zones and closed circulations moving mainly in the tropical airstream round

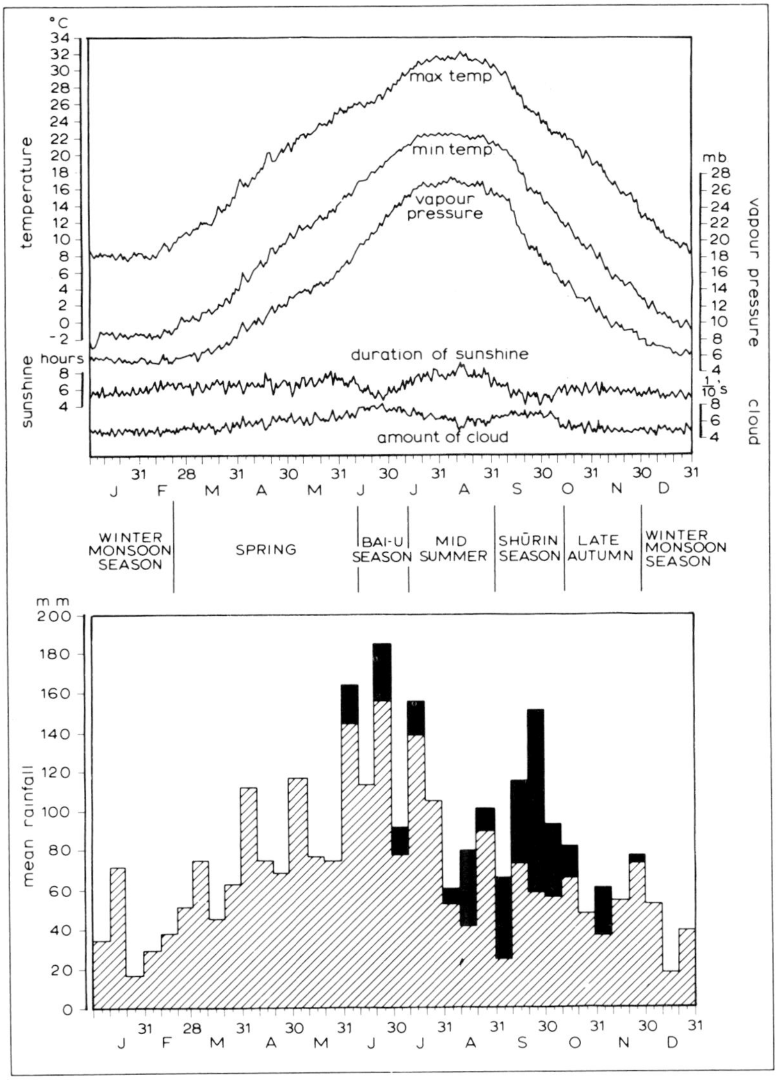

Figure 6.33 Above: Seasonal variation of daily normals at Nagoya, southern Japan, suggesting six natural sesons.* *Below:* Average 10-day precipitation amounts for a station in southern Japan, indicating in black the proportion of rainfall produced by typhoon circulations. The latter reaches a maximum during the Shūrin season.†

Sources: *From Maejima 1967. †After Saito 1959, from Trewartha 1981.

the Pacific subtropical anticyclone, but partly originating in a south-westerly stream which is the extension of the monsoon circulation of south Asia. The south-east circulation is displaced westwards from Japan by a zonal expansion of the subtropical anticyclone during late July and August giving a period of more settled sunny weather. The secondary precipitation maximum of the *Shūrin* season during September and early October coincides with an eastward contraction of the Pacific subtropical anticyclone allowing low-pressure systems and typhoons from the Pacific to swing north towards Japan. Although much of the Shūrin rainfall is believed to be of typhoon origin (see Figure 6.33). Some of it is undoubtedly associated with the southern sides of depressions, moving along the southward-migrating Pacific Polar Front to the north, because there is a marked tendency for the autumn rains to begin first in the north of Japan and to spread southwards.

5 Autumn

Autumn sees the southward swing of the Equatorial Trough and the zone of maximum convection which lies just to the north of the weakening easterly jet (see Figure 6.17). The break-up of the summer circulation systems is associated with the withdrawal of the monsoon rains, which is much less clearly defined than their onset (see Figure 6.34). By October the easterly trades of the Pacific affect the Bay of Bengal at the 500-mb level and generate disturbances at their confluence with the equatorial westerlies. This is the major season for Bay of Bengal cyclones and it is these disturbances, rather than the onshore northeasterly monsoon, which cause the October/November maximum of rainfall in south-east India (e.g. Madras, Figure 6.21).

During October the westerly jet re-establishes itself south of the Tibetan plateau, often within a few days, and cool season conditions are restored over most of southern and eastern Asia.

Figure 6.34 Normal dates for the withdrawal of monsoon rains over India.

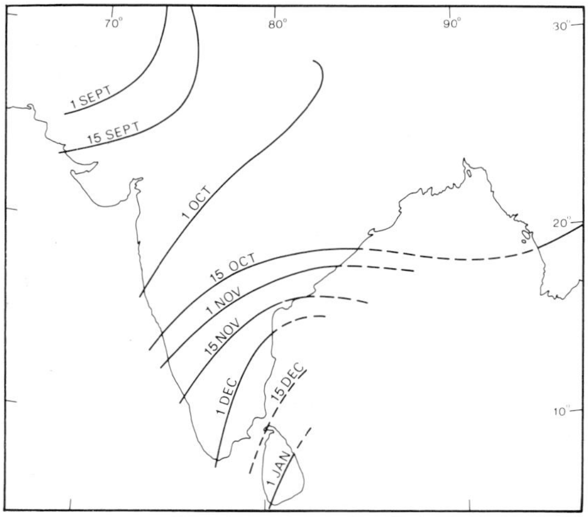

Source: After Das 1987.

E THE AFRICAN MONSOON

The annual climatic regime over West Africa has many similarities to that over southern Asia, the surface airflow being determined by the position of the leading edge of a Monsoon Trough (see Figure 6.2). This air flow is south-westerly to the south of the Trough and easterly to north-easterly to its north (see Figure 6.35). The major difference between the circulations of the two regions is largely due to the differing geography of the land–sea distribution and, particularly, to the topography. The lack of a large mountain range to the north of West Africa allows the Monsoon Trough to migrate regularly and continuously with the seasons without the pronounced northward surge associated with the early summer burst of the Indian monsoon. In general, the West African Monsoon Trough oscillates between annual extreme locations of about 2°N and 25°N (see Figure 6.36). In 1956, for example, these extreme positions were 5°N on 1 January and 23°N in August, and in 1958 6°N on 19 February and 23°N in August. The leading edge of the Monsoon Trough is complex in structure (see Figure 6.37B) and its position may oscillate greatly from day to day through several degrees of latitude.

In winter the south-westerly monsoon airflow over the coasts of West Africa is very shallow (i.e. 1000 m) with 3000 m of overriding easterly winds, which are themselves overlain by strong (> 20 m s^{-1}) winds (see Figure 6.38). North of the Monsoon Trough the surface north-easterlies (i.e. the 2000-m-deep Harmattan flow) blow clockwise outwards from the subtropical high-pressure centre, being compensated above 5000 m by an anticlockwise westerly airflow which, at about 12,000 m and 20°–30°N, is concentrated into a subtropical westerly jet stream of average speed 45 m s^{-1}. Mean January surface temperatures decrease from about 26°C along the southern coast to 14°C in southern Algeria.

With the approach of the northern summer, the strengthening of the South Atlantic subtropical high-pressure cell, combined with the

Figure 6.35 The major circulation in Africa in (A) June–August and (B) December–February. H: subtropical high-pressure cells; EW: equatorial westerlies (moist, unstable but containing the Congo high-pressure ridge); NW: the north-westerlies (summer extension of EW in the southern hemisphere); TE: tropical easterlies (trades); SW: south-westerly monsoonal flow in the northern hemisphere; W: extratropical westerlies; J: subtropical westerly jet stream; J_A and J_E: the (easterly) African jet streams; and MT: Monsoon Trough.

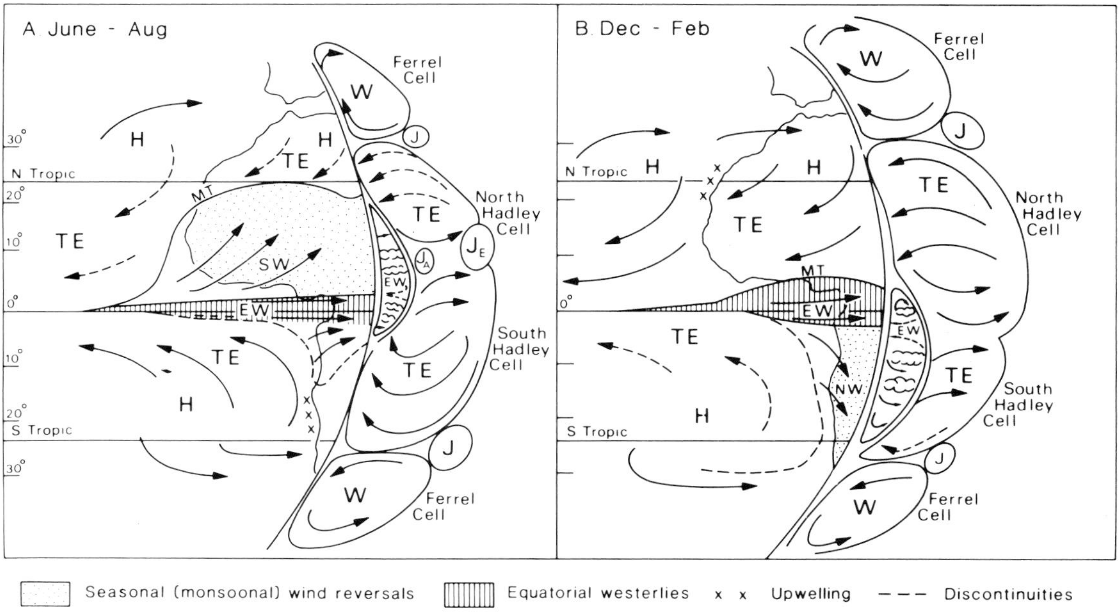

Source: From Rossignol-Strick 1985 (*by permission Elsevier Science Publishers B.V., Amsterdam*).

Figure 6.36 The daily position of the Monsoon Trough at longitude 3°E during 1957. This year experienced an exceptionally wide swing over West Africa, with the Trough reaching 2°N in January and 25°N on 1 August. Within a few days after the latter date the intensely oscillating Trough had swung southwards through 8° of latitude.

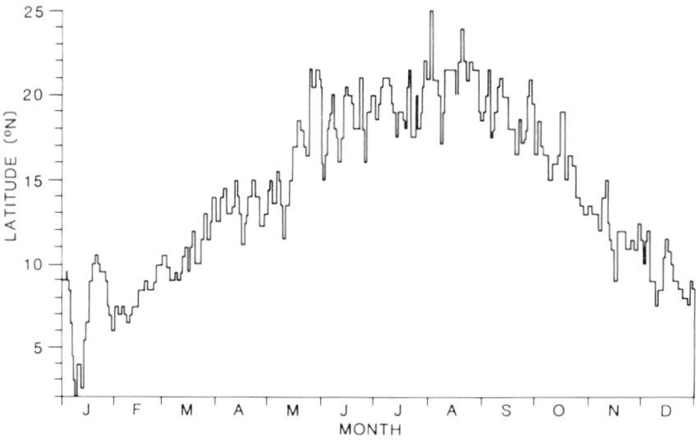

Source: After Clackson 1957, from Hayward and Oguntoyinbo 1987.

increased continental tempertures, establishes a strong south-westerly airflow at the surface which spreads northward behind the Monsoon Trough, lagging about six weeks behind the progress of the overhead sun. The northward migration of the trough oscillates diurnally with a northward progress of up to 200 km in the afternoons following a smaller southward retreat in the mornings. The northward spread of moist, unstable and relatively cool south-westerly airflow from the Gulf of Guinea brings rain in differing amounts to extensive areas of West Africa. Aloft, easterly winds spiral clockwise outwards from the subtropical high-pressure centre (see Figure 6.38) and are concentrated between June and August into two tropical easterly jet streams; the stronger (> 20 m s^{-1}) at about 15,000–20,000 m and the weaker (> 10 m s^{-1}) at about 4000–5000 m (see

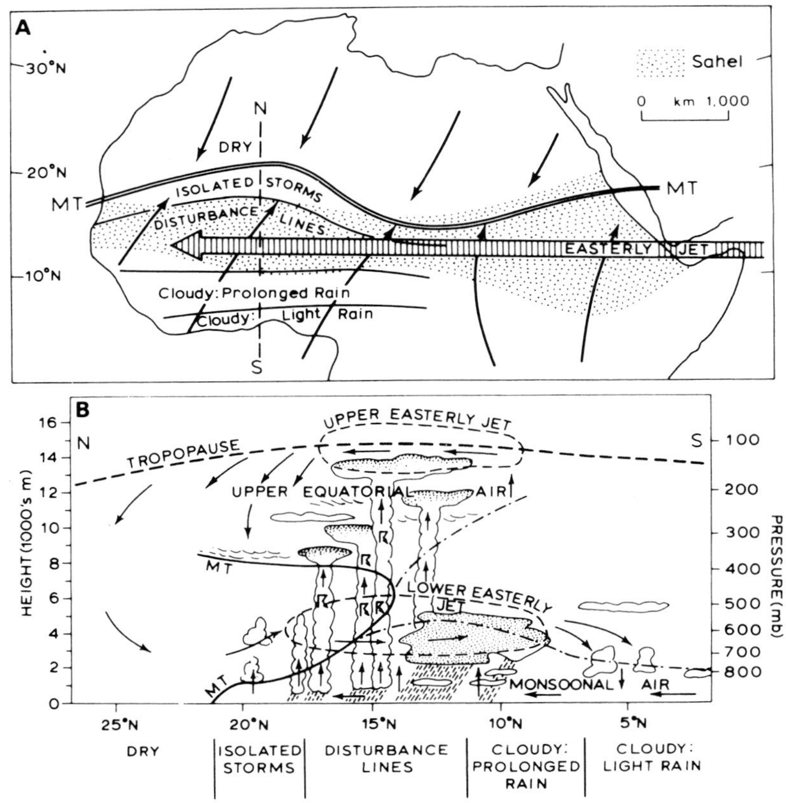

Figure 6.37 The structure of the circulation over northern Africa in August. A: Surface airflow and Easterly Tropical Jet. B: Vertical structure and resulting precipitation zones over West Africa.

Sources: A: Reproduction from the *Geographical Magazine*, London. B: From Maley 1982 (*copyright © Academic Press; reproduced by permission*), and Musk 1983.
Notes: ᴿ = thunderstorm activity; MT = Monsoon Trough.

Figure 6.37B). The lower jet occupies a broad band from 13°N to 20°N on the underside of which oscillations produce easterly waves which may develop into line squalls. By July the south-westerly monsoon airflow has spread far to the north, the leading trough reaching its extreme northern location, about 20°N, in August. At this time four major climatic belts can be identified over West Africa (see Figure 6.37A):

1 A coastal belt of cloud and light rain related to frictional convergence within the monsoon flow, overlain by subsiding easterlies.
2 A quasi-stationary zone of disturbances associated with deep stratiform cloud yielding prolonged light rains. Low-level convergence south of the easterly jet axes, apparently associated with easterly wave disturbances from east central Africa, causes instability in the monsoon air.
3 A broad zone underlying the easterly jet streams which help activate disturbance lines and thunderstorms. North–south lines of deep cumulonimbus cells may move westward steered by the jets. The southern, wetter part of this zone is termed the Soudan, the northern part the Sahel, but popular usage assigns the name Sahel to the whole belt.
4 Just south of the Monsoon Trough, the shallow tongue of humid air is overlain by drier subsiding air. Here there are only isolated storms, scattered showers, and occasional thunderstorms.

In contrast to winter conditions, August temperatures are lowest (i.e. 24°C–25°C) along the cloudy southern coasts and increase towards the north where they average 30°C in southern Algeria.

Both the summer airflows, the south-

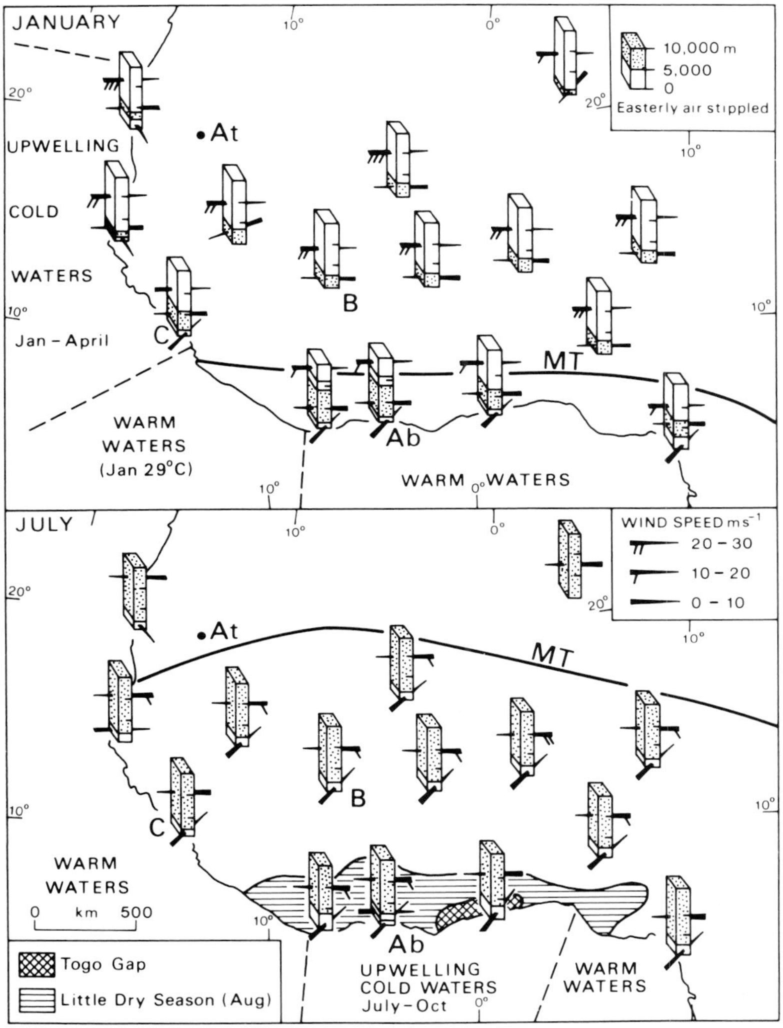

Figure 6.38 Mean wind speeds (m s^{-1}) and directions in January and July over West Africa up to about 15,000 m. Ocean water temperatures and the positions of the Monsoon Trough are also shown, as are the area affected by the August Little Dry Season and the location of the anomalous Togo Gap. The locations of Abidjan (Ab), Atar (At), Bamako (B) and Conakry (C) are given (see precipitation graphs in Figure 6.39).

Source: From Hayward and Oguntoyinbo 1987.

westerlies below and the easterlies aloft, are subject to perturbations which contribute significantly to the rainfall during this season. Three types of perturbations are particularly prevalent:

1 Waves in the south-westerlies. These are northward surges of the humid airflow, having periodicities of 4–6 days, producing bands of summer monsoon rain some 160 km broad and 50–80 km in north–south extent which have the most marked effect 1100–1400 km south of the surface Monsoon Trough, the position of which oscillates with the surges.
2 Waves in the easterlies. These develop on the interface between the lower south-westerly and the upper easterly airflows. These waves are from 1500 to 4000 km long from north to south and move across West Africa towards the west between mid-June and October with a periodicity of 3–5 days at speeds of about 5°–10° of longitude per day (i.e. 18–35 km hr^{-1}), sometimes developing

closed cyclonic circulations. At the height of the summer monsoon they produce most rainfall at around latitude 14°N, between 300 and 1100 km south of the Monsoon Trough. On average, some fifty easterly waves per year cross Dakar, some of these carry on in the general circulation across the Atlantic and it has been estimated that 60 per cent of West Indian hurricanes originate in West Africa as easterly waves.

3 Line squalls. Easterly waves vary greatly in intensity. Some give rise to little cloud and rain, whereas others develop as line squalls when the wave extends down to the surface producing updraughts, heavy rain and thunder. Line-squall formation is assisted where surface topographic convergence of the easterly flow occurs (e.g. near Lake Chad and north-west of the Niger delta). These disturbance lines travel at up to 60 km hr^{-1} from east to west across southern West Africa for distances of up to 3000 km (but averaging

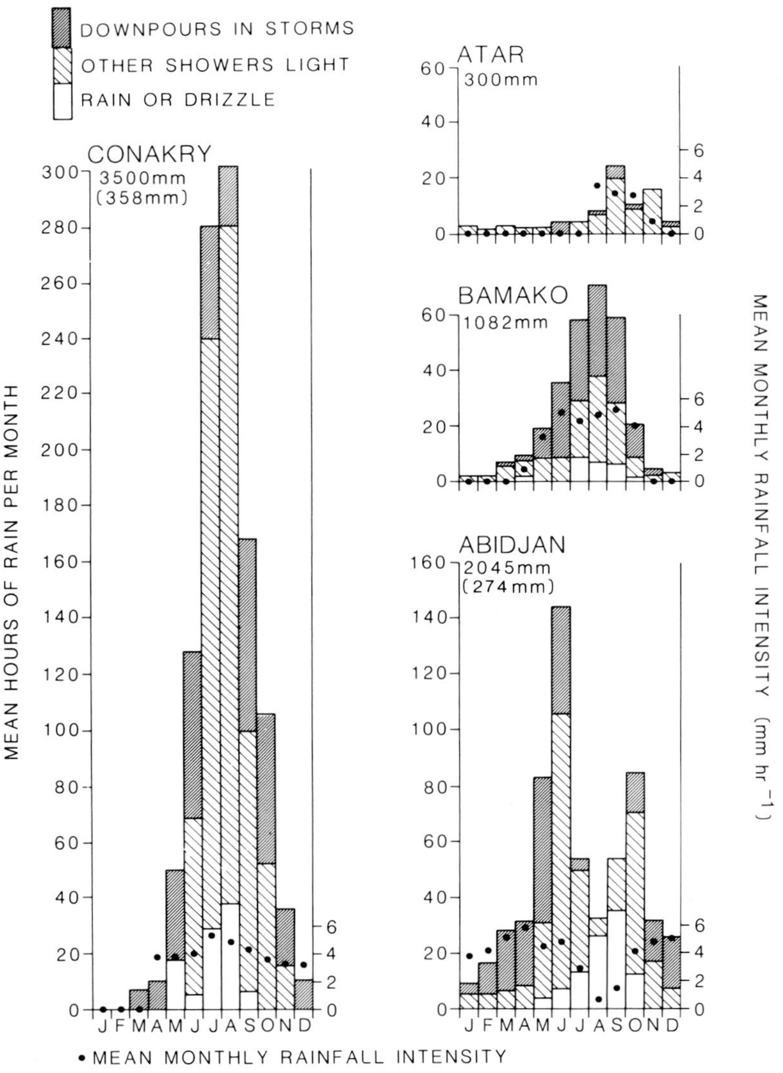

Figure 6.39 Mean number of hours of rain per month for four West African stations. Also shown are types of rainfall, mean annual totals (mm) and, in parentheses, maximum recorded daily rainfalls (mm) for Conakry (August) and Abidjan (June). Dots show the mean monthly rainfall intensities (mm hr^{-1}). Note the pronounced Little Dry Season at Abidjan. Station locations are marked on Figure 6.38.

Source: From Hayward and Oguntoyinbo 1987.

600 km) between June and September, yielding 40–90 mm of rain per day. Some coastal locations suffer about forty line squalls per year which account for more than 50 per cent of the annual rainfall (see Plate 37).

Annual rainfall decreases from 2000–3000 mm in the coastal belt (e.g. Conakry, Guinea) to about 100 mm at latitude 20°N (see Figure 6.39). Near the coast more than 300 mm per day of rain may fall during the rainy season but further north the variability increases due to the irregular extension and movement of the Monsoon Trough. Line squalls and other disturbances give a zone of maximum rainfall located 800–1000 km south of the surface position of the Monsoon Trough (see Figure 6.37B). Monsoon rains in the coastal zone of Nigeria (4°N) contribute 28 per cent of the annual total (about 2000 mm), thunderstorms 51 per cent and disturbance lines 21 per cent. At 10°N, 52 per cent of the total (about 1000 mm) is due to disturbance lines, 40 per cent to thunderstorms and only 9 per cent to the monsoon. Over most of the country, rainfall from disturbance lines has a double frequency maximum, thunderstorms a single one in summer (see Figure 6.40) for Minna, 9.5°N). In the northern parts of Nigeria and Ghana, rain falls in summer months, mostly from isolated storms or disturbance lines. The high variability of these rains from year to year characterizes the drought-prone Sahel environment.

Figure 6.40 The contributions of disturbance lines and thunderstorms to the average monthly precipitation at Minna, Nigeria (9.5°N).

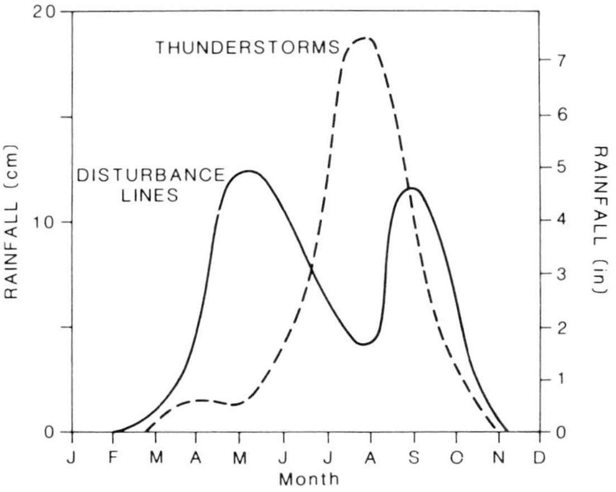

Source: After Omotosho 1985 (*by permission of the Royal Meteorological Society*).

The summer rainfall in the northern Soudano–Sahelian belts is determined partly by the northward penetration of the Monsoon Trough which may range up to 500–800 km beyond its average position (see Figure 6.41) and by the strength of the easterly jet streams. The latter affect the frequency of disturbance lines.

Anomalous climatic effects occur in a number of distinct West African localities at different times of the year. Although the temperatures of coastal waters always exceed 26°C and may reach 29°C in January, there are two areas of locally up-welling cold waters (see Figure 6.38). One lies north of Conakry along the coasts of Senegal and Mauretania where dominant offshore north-easterly winds in January–April skim off the surface waters causing cooler (20°C) water to rise, dramatically lowering the temperature of the afternoon onshore breezes (see Figure 3.15). The second area of cool ocean (19°–22°C) is located along the central southern coast west of Lagos during the period July–October, for a reason which is as yet unclear. From July to September an anomalously dry land area is located along the southern coastal belt (see Figure 6.38) during what is termed the *Little Dry Season*. The reason is that at this time the Monsoon Trough is in its most northerly position and this coastal zone, lying 1200–1500 km to the south of it and, more important, 400–500 km to the south of its major rain belt, has relatively stable air (see Figure 6.37B), a condition assisted by the relatively cool offshore coastal waters. Embedded within this relatively cloudy but dry belt is the smaller *Togo Gap*, lying between 0° and 3°E and having during the summer above average sunshine, subdued

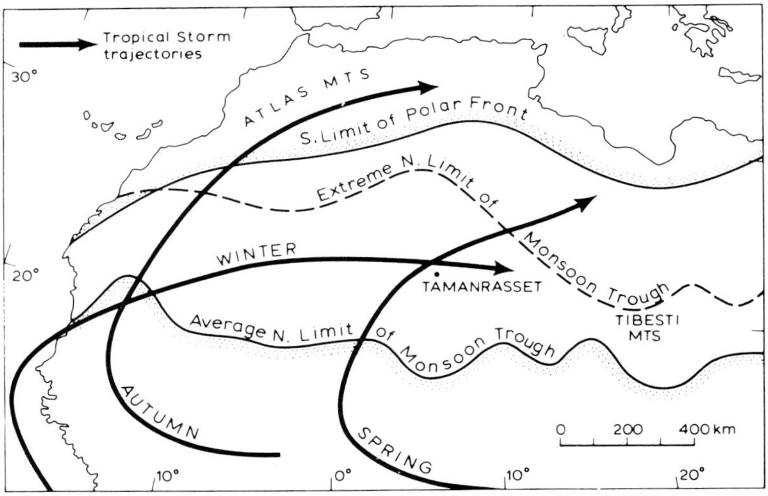

Figure 6.41 Extent of precipitation systems affecting western and central North Africa and typical tracks of Soudano–Sahelian depressions.

Source: Ater Dubief and Yacono; from Barry 1991.

convection, relatively low rainfall (i.e. less than 1000 mm) and low thunderstorm activity. It is considered that the trend of the coast here parallels the dominant low-level south-westerly winds, so limiting surface frictionally-induced convergence in an area where temperatures and convection are in any case inhibited by low coastal water temperatures.

F AMAZONIA

Amazonia lies athwart the equator (see Figure 6.42) and contains some 30 per cent of the total global biomass. The continuously high temperatures (24°–28°C) combine with the high transpiration to cause the region to behave at times as if it were a source of maritime equatorial air.

Important influences over the climate of Amazonia are the North and South Atlantic subtropical high-pressure cells. From these, stable easterly mT air invades Amazonia in a shallow (1000–2000 m), relatively cool and humid layer, overlain by warmer and drier air from which it is separated by a strong temperature inversion and humidity discontinuity. This shallow airflow gives some precipitation in coastal locations but produces drier conditions inland unless it is subjected to strong convection when a heat low is established over the continental interior. At such times the inversion rises to 3000–4000 m and may break down altogether associated with heavy precipitation, particularly in late afternoon or evening. The South Atlantic subtropical high-pressure cell expands westward over Amazonia in July producing drier conditions, as shown by the rainfall at inland stations such as Manaus (see Figure 6.42), but in September it begins to contract and the build-up of the continental heat low ushers in the October–April rainy season in central and southern Amazonia. The North Atlantic subtropical high-pressure cell is less mobile than its southern counterpart but varies in a more complex manner, having maximum westward extensions in July and February and minima in November and April. In northern Amazonia the rainy season is May to September. Rainfall over the region as a whole is mainly due to low-level convergence associated with convective activity, a poorly defined Equatorial Trough, instability lines, occasional incursions of cold fronts from the southern hemisphere, and relief effects.

Strong thermal convection over Amazonia can commonly produce more than 40 mm/day of rainfall over a period of a week and much

Figure 6.42 Mean annual precipitation (mm) (inches in parentheses) over the Amazon Basin, together with mean monthly precipitation amounts (mm) for eight stations.

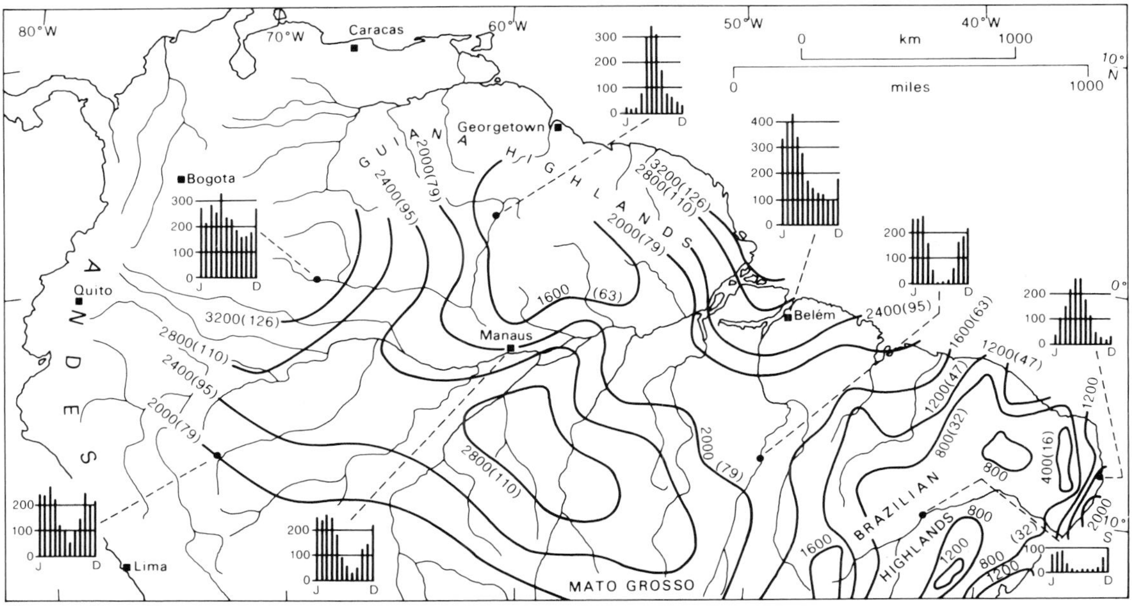

Source: From Ratisbona 1976.

higher average intensities over shorter periods. When it is recognized that 40 mm of rainfall in one day releases sufficient latent heat to warm the troposphere by 10°C, it is clear that sustained convection at this intensity is capable of fuelling the Walker circulation (see Figure 6.48). During high phases of ENSO air rises over Amazonia, whereas during the low phases the drought over north-east Brazil is intensified. In addition, convective air moving poleward may strengthen the Hadley circulation. This air tends to accelerate, due to the conservation of angular momentum, and strengthen the westerly jet streams such that correlations have been found between Amazonian convective activity and North American jet stream intensity and location.

The Intertropical Convergence Zone (ITCZ) does not exist in its characteristic form over the interior of South America and its passage only affects rainfall near the east coast. The intensity of this zone varies, being least when both the North and South Atlantic subtropical high-pressure cells are strongest (e.g. in July) giving a pressure increase which causes the equatorial trough to fill. The ITCZ swings to its most northerly position during July–October when invasions of more stable South Atlantic air are associated with drier conditions over central Amazonia, and to its most southerly in March–April (see Figure 6.43). At Manaus surface winds are predominantly south-easterly from May to August and north-easterly from September to April; whereas the upper tropospheric winds are north-westerly or westerly from May to September and southerly or south-easterly from December to April. This reflects the development in the austral summer of an upper tropospheric anticyclone that is located over the Peruvian–Bolivian Altiplano. This upper high is a result of sensible heating of the elevated plateau and the release of latent heat in frequent thunderstorms over the Altiplano, analogous to the situation over Tibet. Outflow from this high subsides in a broad area extending from eastern Brazil to West Africa.

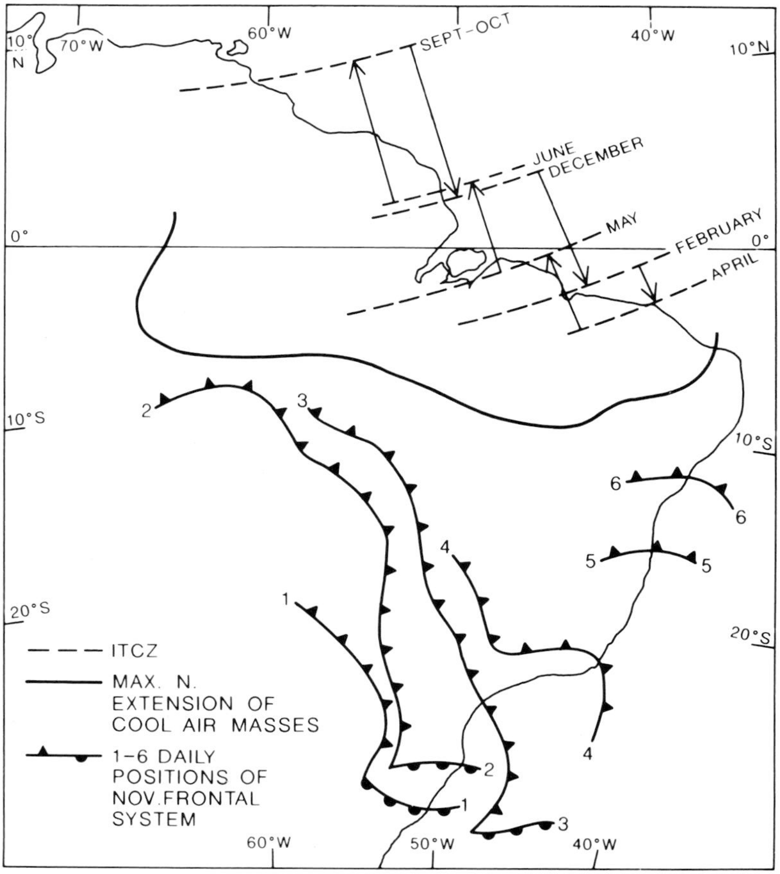

Figure 6.43 The synoptic elements of Brazil. The seasonal positions of the coastal Intertropical Convergence Zone; the maximum northerly extension of cool southerly mP air masses; and the positions of a typical frontal system during six successive days in November as the centre of the low pressure moves south-eastward into the South Atlantic.

Source: From Ratisbona 1976.

The drought-prone region of eastern Brazil is particularly moisture-deficient during periods when the ITCZ remains in a northerly position and relatively stable mT air from a cool South Atlantic surface is dominant (see Chapter 2, 1.5). Dry conditions may occur during January–May during strong ENSO events (see p. 129), when the descending branch of the Walker circulation covers most of Amazonia.

Significant Amazonian rainfall, particularly in the east, originates along mesoscale lines of instability which form near the coast due to converging trade winds and afternoon sea breezes, or to the interaction of nocturnal land breezes with onshore trade winds. These lines of instability move westward in the general airflow of speeds of about 50 km hr^{-1} (i.e. 30 mph), moving faster in January than in July and exhibiting a complex process of convective cell growth, decay, migration and regeneration. Many of these instability lines only reach 100 km or so inland, decaying after sunset (see Figure 6.44). However, the more persistent instabilities may produce a rainfall maximum about 500 km inland, and some remain active for up to 48 hours such that their precipitation effects reach as far west as the Andes. Other meso- to synoptic-scale disturbances form within Amazonia, especially between April and September. Precipitation also occurs with the penetration of cool mP air masses from the south, especially between September and November, which are heated from below and become unstable (see Figure 6.43).

Figure 6.44 Hourly rainfall fractions for Belém, Brazil, for January and July. The rain mostly results from convective cloud clusters developing offshore and moving inland, more rapidly in January.

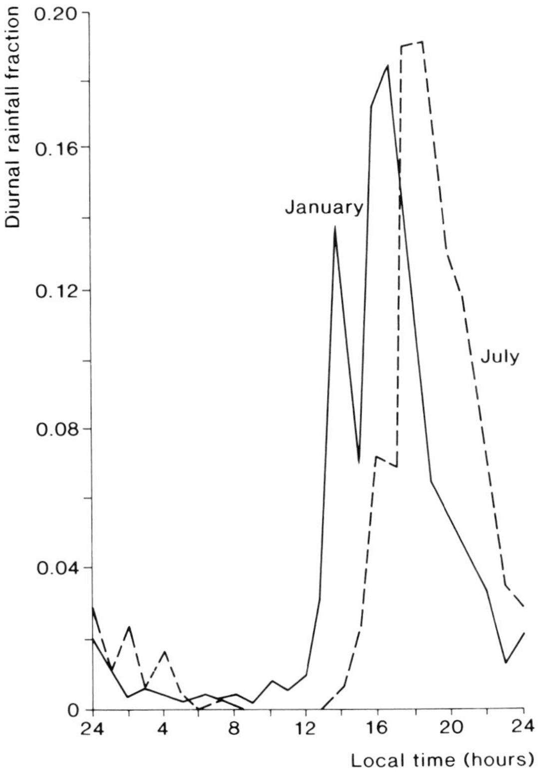

Source: After Kousky 1980.

The tropical easterlies over the margins of Amazonia are susceptible to the formation of easterly waves and closed vortices which move westward generating rain bands. Relief effects are naturally most noteworthy as airflow approaches the eastern slopes of the Andes where large-scale orographic convergence in a region of high evapotranspiration contributes to the high precipitation all through the year.

G OTHER SOURCES OF CLIMATIC VARIATIONS IN THE TROPICS

The major systems of tropical weather and climate have now been discussed, yet various other elements help to create contrasts in tropical weather both in space and time.

1 Diurnal variations

Diurnal weather variations are particularly evident at coastal locations in the trade wind belt and in the Indonesia–Malaysian Archipelago. Land and sea breeze regimes (see Chapter 3, C.3) are well developed, as the heating of tropical air over land can be up to five times that over adjacent water surfaces. The sea breeze normally sets in between 0800 and 1100, reaching a maximum velocity of 6–15 m s^{-1} (13–33 mph) about 1300 to 1600 and subsiding around 2000. It may be up to 1000–2000 m (about 3000–4000 ft) in height with a maximum velocity at an elevation of 200–400 m (650–1300 ft) and normally penetrates some 20–60 km (about 12–40 miles) inland.

In northern Australia, sea-breeze phenomena apparently extend up to 200 km inland from the Gulf of Carpentaria by late evening. During the August–November dry season, this may create suitable conditions for the bore-like 'Morning Glory' – a linear cloud roll and squall line which propagates, usually from the north-east, on the inversion created by the maritime air and nocturnal cooling. Sea breezes are usually associated with a heavy build-up of cumulus cloud and afternoon downpours. On large islands under calm conditions the sea breezes converge towards the centre so that an afternoon maximum of rainfall is observed. Under steady trade winds the pattern is displaced downwind so that descending air may be located over the centre of the island. A typical case of afternoon maximum is illustrated by Figure 6.45B for Nandi (Viti Levu, Fiji) in the south-west Pacific. The station has a lee exposure in both wet and dry seasons. This rainfall pattern is commonly believed to be widespread in the tropics, but

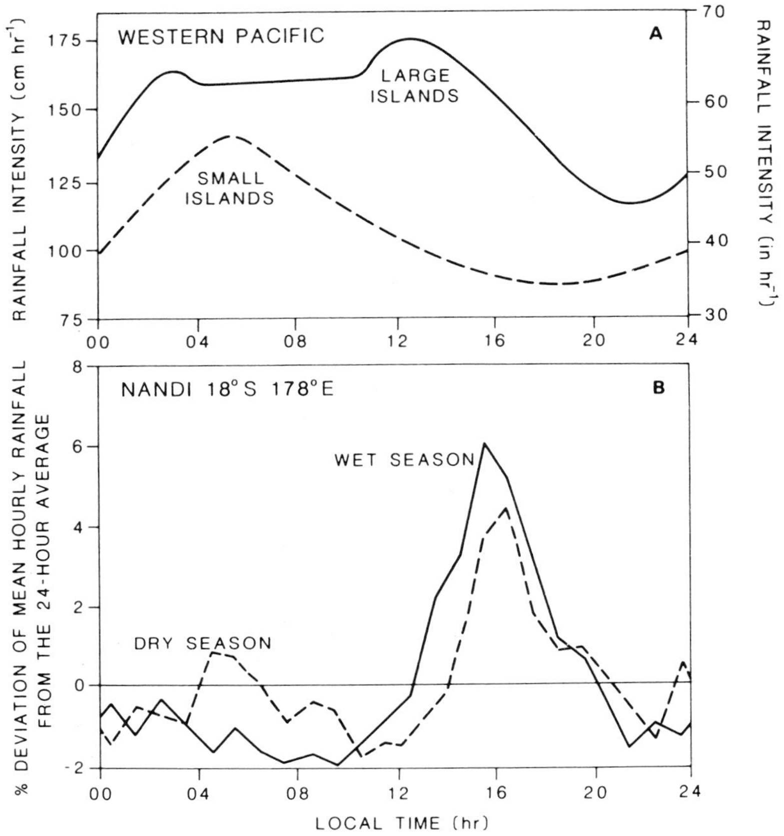

Figure 6.45 Diurnal variation of rainfall intensity for tropical islands in the Pacific. A: Large and small islands in the western Pacific. B: Wet and dry seasons for Nandi (Fiji) in the south-west Pacific (percentage deviation from the daily average).

Sources: A: After Gray and Jacobson 1977. B: After Finkelstein, in Hutchings 1964.

over the open sea and on small islands a nighttime maximum (often with a peak near dawn) seems to occur and even large islands can display this nocturnal regime when there is little synoptic activity. Figure 6.45A illustrates this nocturnal pattern at four small island locations in the western Pacific. Even large islands may show this effect, as well as the afternoon maximum associated with sea-breeze convergence and convection. There are several theories concerning the nocturnal rainfall peak. Recent studies point to a radiative effect, involving more effective nocturnal cooling of cloud-free areas around the mesoscale cloud systems. This favours subsidence which, in turn, enhances low-level convergence into the cloud systems and strengthens the ascending air currents. Strong cooling of cloud tops, relative to their surroundings, may also produce localized destabilization and encourage droplet growth by mixing of droplets at different temperatures (see Chapter 2, G). This effect would be at a maximum near dawn. Another factor is that the sea–air temperature difference, and consequently the oceanic heat supply to the atmosphere, is largest about 0300–0600 hours. Yet a further hypothesis suggests that the semi-diurnal pressure oscillation encourages convergence and therefore convective activity in the early morning and evening, but divergence and suppression of convection around midday.

The Malayan peninsula displays very varied diurnal rainfall regimes in summer. The effects of land and sea breezes, anabatic and katabatic winds and topography greatly complicate the rainfall pattern by their interactions with the

low-level south-westerly monsoon current. For example, there is a nocturnal maximum in the Malacca Straits region associated with the convection set off by the convergence of land breezes from Malaya and Sumatra (cf. p. 238), whereas on the east coast of Malaya the maximum occurs in the late afternoon to early evening when sea breezes extend about 30 km inland against the monsoon south-westerlies, and convective cloud develops in the deeper sea breeze current over the coast strip. On the interior mountains the summer rains have an afternoon maximum due to the unhindered convection process.

2 Topographic effects

Relief and surface configuration have a marked effect on rainfall amounts in tropical regions where hot, humid air masses are frequent. On the south-western slope of Mount Cameroon, Debundscha (9 m elevation) receives 1116 cm yr^{-1} on average (1960–80) from the south-westerly monsoon. In the Hawaiian Islands the mean annual total exceeds 760 cm (300 in) on the mountains, with one of the world's largest mean annual totals of 1199 cm (472 in) at 1569 elevation on Mt Waialeale (Kauai), but land on the lee side suffers correspondingly accentuated sheltering effects with less than 50 cm (20 in) over wide areas. On Hawaii itself the maximum falls on the eastern slopes at about 900 m, whereas the 4200 m summits of Mauna Loa and Mauna Kea, which rise above the trade wind inversion, receive only 25–50 cm (10–20 in). On the Hawaiian island of Oahu, the maximum precipitation occurs on the western slopes, just leeward of the 850 m (2785 ft) summit with respect to the easterly trade winds. The following measurements in the Koolau Mountains, Oahu, show that the orographic factor is pronounced during summer when precipitation is associated with the easterlies, but in winter when precipitation is from cyclonic disturbances it is more evenly distributed (see Table 6.3).

The Khasi Hills in Assam are an exceptional instance of the combined effect of relief and surface configuration. Part of the monsoon current from the head of the Bay of Bengal (see Figure 6.23) is channelled by the topography towards the high ground and the sharp ascent, which follows the convergence of the airstream in the funnel-shaped lowland to the south, results in some of the heaviest annual rainfall totals recorded anywhere. Mawsyuram (1400 m elevation), 16 km west of the more famous station of Cherrapunji, has a mean annual total (1941–69) of 1221 cm (481 in) and can claim to be the wettest spot in the world. Cherrapunji

Table 6.3 Precipitation in the Koolau Mountains, Oahu.

Location	*Elevation*		*Source of rainfall*		
	Metres	*Feet*	*Trade winds*	*Cyclonic disturbances*	
			28 May–3 Sept. 1957	2–28 Jan. 1957	5–6 March 1957
Summit	850	(2785)	71.3 cm (28.09 in)	49.9 cm (19.63 in)	32.9 cm (12.94 in)
760 m (2500 ft) west of summit	625	(2050)	121.0 cm (47.64 in)	54.4 cm (21.41 in)	37.0 cm (14.56 in)
7600 m (25,000 ft) west of summit	350	(1150)	32.9 cm (12.97 in)	46.7 cm (18.38 in)	33.4 cm (13.16 in)

Source: After Mink 1960 (original values in inches).

(1340 m) during the same period averaged 1102 cm; extremes recorded there include 569 cm (224 in) in July 1974 and 2440 cm (959 in) in the year (see Figure 2.25). However, throughout the monsoon area, topography plays a secondary role in determining rainfall distribution to the synoptic activity and large-scale dynamics.

Really high relief produces major changes in the main weather characteristics and is best treated as a special climatic type. The Kenya plateau, situated on the equator, has an average elevation of about 1500 m, above which rise the three volcanic peaks of Mt Kilimanjaro (5800 m), Mt Kenya (5200 m) and Ruwenzori (5200 m) nourishing permanent glaciers above 4270 m. Annual precipitation on the summit of Mt Kenya is about 114 cm (44 in), similar to amounts on the plateau to the south, but on the southern slopes between 2100 and 3000 m, and on the eastern slopes between about 1400 and 2400 m totals exceed 250 cm (100 in). Kabete (at an elevation of 1800 m near Nairobi) exhibits many of the features of tropical mountain climates, having a small annual temperature range (mean monthly temperatures are 19°C (67°F) for February and 16°C (60°F) for July), a high diurnal temperature range (averaging 9.5°C (17°F) in July and 13°C (24°F) in February) and a large average cloud cover (mean 7–8/10ths).

3 Cool ocean currents

Between the western coasts of the continents and the eastern rims of the subtropical high-pressure cells the ocean surface is relatively cold (see Figure 3.42). This is the result of the importation of water from higher latitudes by the dominant currents and the slow up-welling (sometimes at the rate of about 1 m in 24 hours) of water from intermediate depths due to the Ekman effect (see Chapter 3, F.3) and to the coastal divergence (see Chapter 3, B.1). This concentration of cold water gently cools the local air to dew point. As a result, dry warm air degenerates into a relatively cool, clammy, foggy atmosphere with a comparatively low temperature and little range along the west coast of North America off California (see Plate 38), off South America between latitudes 4°S and 31°S, and off south-west Africa (8°S and 32°S). Thus Callao, on the Peruvian coast, has a mean annual temperature of 19.4°C (67°F), whereas Bahia (at the same latitude on the Brazilian coast) has a corresponding figure of 25°C (77°F).

The temperature effect of offshore cold currents is not limited to coastal stations as it is carried inland during the day at all times of the year by a pronounced sea-breeze effect (see Chapter 3, C.3). Along the west coasts of South America and south-west Africa the sheltering effect from the dynamically-stable easterly trades aloft provided by the nearby Andes and Namib Escarpment, respectively, allows incursions of shallow tongues of cold air to roll in from the south-west. These tongues of air are capped by strong inversions at between 600 and 1500 m (2000 and 5000 ft) reinforcing the regionally low trade-wind inversions (see Figure 6.6) and thereby precluding the development of strong convective cells, except where there is orographically forced ascent. Thus, although the cool maritime air perpetually bathes the lower western slopes of the Andes in mist and low stratus cloud and Swakopmund (south-west Africa) has an average of 150 foggy days a year, little rain falls on the coastal lowlands. Lima (Peru) has a total mean annual precipitation of only 4.6 cm (1.8 in), although it receives frequent drizzle during the winter months of June to September, and Swakopmund in Namibia has a mean annual rainfall of 1.6 cm (0.65 in). Heavier rain occurs on the rare instances when large-scale pressure changes cause a cessation of the diurnal sea breeze or when modified air from the South Atlantic or South Indian Ocean is able to cross the continents at a time when the normal dynamic stability of the trade-wind is disturbed. In south-west Africa the inversion is most likely to break down during either October or April allowing convectional storms to form, and Swakopmund recorded some 5.1 cm (2 in) of rain on a single day in 1934. Under normal conditions,

however, the occurrence of precipitation is mainly limited to the higher seaward mountain slopes. Further north, tropical west coast locations in Angola and Gabon show that cold up-welling is a more variable phenomenon both in space and time; coastal rainfall varies strikingly with changing sea-surface temperatures (see Figure 6.46). In South America, from Colombia to northern Peru, the diurnal tide of cold air rolls inland for some 60 km (38 miles), rising up the seaward slopes of the Western Cordillera and overflowing into the longitudinal Andean valleys like water over a weir (see Figure 6.47). Plate 39 illustrates a similar overflow, giving rise to a downstream hydraulic jump where the flow decelerates, in the lee of mountains in Wyoming. On the west-facing slopes of the Andes of Colombia, air ascending or banked-up against the mountains, may under suitable conditions trigger off convectional instability in the overlying trades and produce thunderstorms. In south-west Africa, however, the 'tide' flows inland for some 130 km (80 miles) and rises up the 1 800-m (6000-ft) Namib Escarpment without producing much rain because convectional instability is not released and the adiabatic cooling of the air is more than offset by the radiational heating from the warm ground.

Figure 6.46 March rainfall along the south-western coast of Africa. (Gabon and Angola) associated with warm and cold sea-surface conditions for two 5-year composites.

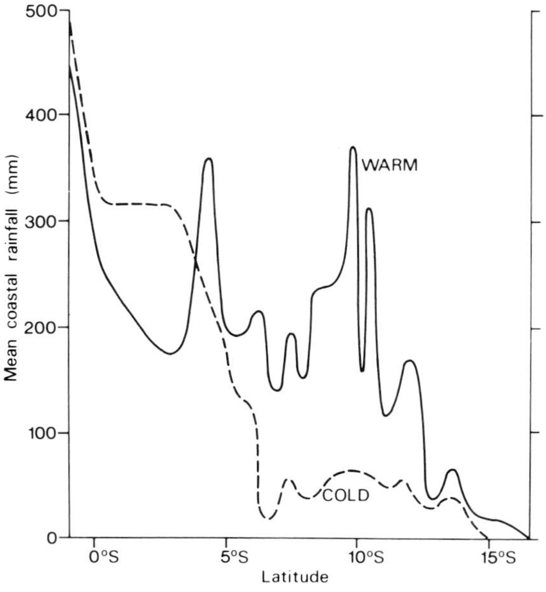

Source: After Nicholson and Entekhabi; from Nicholson 1989 (*by permission of the Royal Meteorological Society, redrawn*).

Figure 6.47 The structure of the sea breeze in western Colombia.

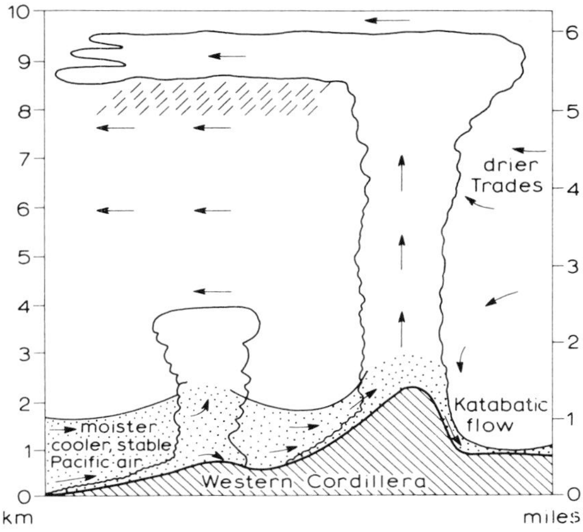

Source: After Lopez; from Fairbridge 1967.

4 El Niño–Southern Oscillation (ENSO) events

Each year, usually starting in December, a weak southward flow of warm water to about 6°S along the coast of Ecuador replaces the northward Peru Current and its associated cold up-welling (see Figure 3.42). The phenomenon is known as El Niño (the child), after the Christ child. At irregular intervals, of 2 to 10 years, this warm water becomes much more extensive and the coastal up-welling ceases entirely, with catastrophic consequences. The absence of the cold water and its nutrients leads to massive mortality of fish, and the birdlife which feeds on them, causing economic disaster for the fishing

and guano industries of Ecuador, Peru, and northern Chile. Major El Niños occurred in 1925–6, 1939–41, 1957–8, 1972–3 and 1982–3, but in between are less severe events giving an average recurrence interval of three years.

This phenomenon forms part of large-scale 1–4°C warming of the ocean surface in the eastern equatorial Pacific. It is associated with a decrease in the normal atmospheric pressure gradient between the subtropical high-pressure cell in the eastern South Pacific and low pressure over the Indonesia region (see pp. 129–30) that is referred to as the Southern Oscillation. The theory outlined in Figure 6.48 proposes that the normal (non-ENSO) trades push water offshore, allowing coastal up-welling, and raise sea level in the western South Pacific. As the South Pacific high pressure weakens and the south-east trades slacken, this weakens the wind-driven up-welling allowing the ITC and its associated rainfall to extend southward to Peru. In fact, strong trades did not precede the pronounced 1982–3 ENSO. Equatorial westerlies in the western Pacific that develop during the low

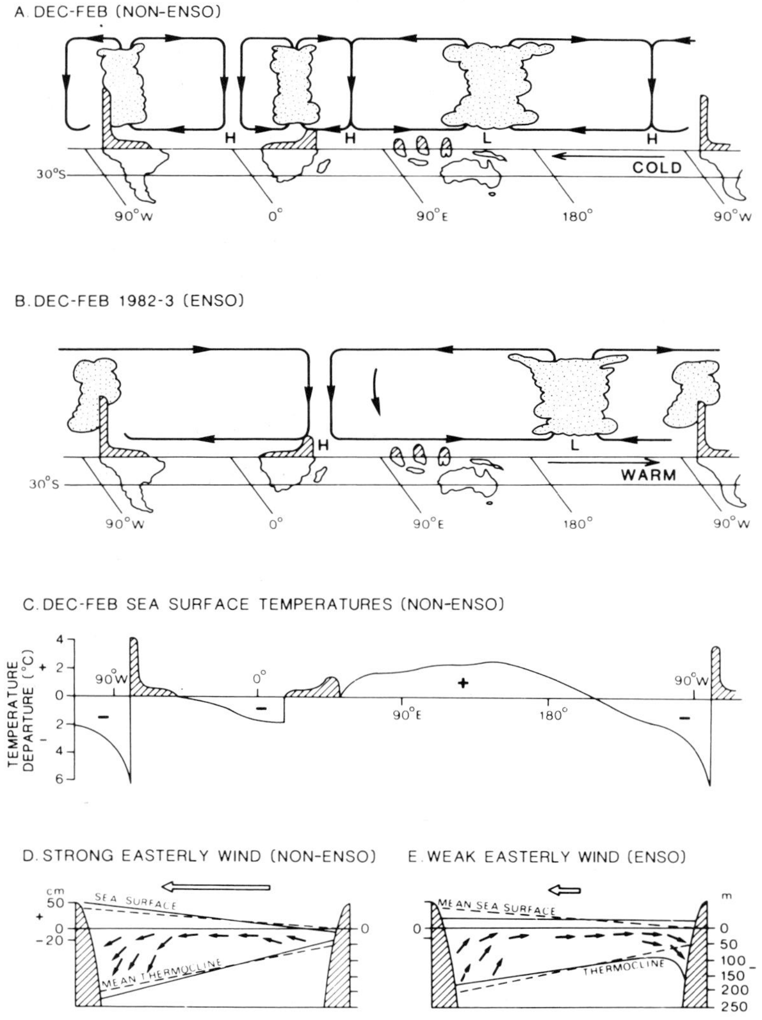

Figure 6.48 Schematic cross-sections of the Walker circulation along the equator based on computations of Y. Tourre (1984). A: Mean December–February regime (non-ENSO); rising air and heavy rains occur over the Amazon basin, central Africa and Indonesia–western Pacific, B: December–February 1982–3 ENSO pattern; the ascending Pacific branch is shifted east of the Date Line and suppressed convection occurs elsewhere due to subsidence. C: Departure of sea-surface temperature from its equatorial zonal mean, corresponding to non-ENSO case (A). D: Strong trades cause sea level to rise and the thermocline to deepen in the western Pacific for case (A). E: Winds relax, sea level rises in the eastern Pacific as water mass moves back eastward and the thermocline deepens off South America during ENSO events.

Source: Based on Wyrtki (*by permission World Meteorological Organization 1985*).

phase of the Southern Oscillation (perhaps in association with cyclone pairs north and south of the equator) create an eastward flow of warm water as large-scale, internal oceanic (Kelvin) waves. This depresses the thermocline off South America (see Figure 6.48E) preventing cold water from reaching the surface.

These closely related oceanic and atmospheric phenomena, together referred to as ENSO events, appear to give rise to anomalous weather regimes not only in the southern hemisphere, but also further afield. In the northern hemisphere winter, ENSO events with equatorial heating anomalies are associated with a strong trough and ridge teleconnection pattern, known as the Pacific–North American (PNA) pattern (see Figure 6.49), which may bring cloud and rain to the south-west United States and north-west Mexico.

Figure 6.49 Schematic Pacific–North America (PNA) circulation pattern in the upper troposphere during an ENSO event in December–February. The shading indicates a region of enhanced rainfall associated with anomalous westerly surface wind convergence in the equatorial western Pacific.

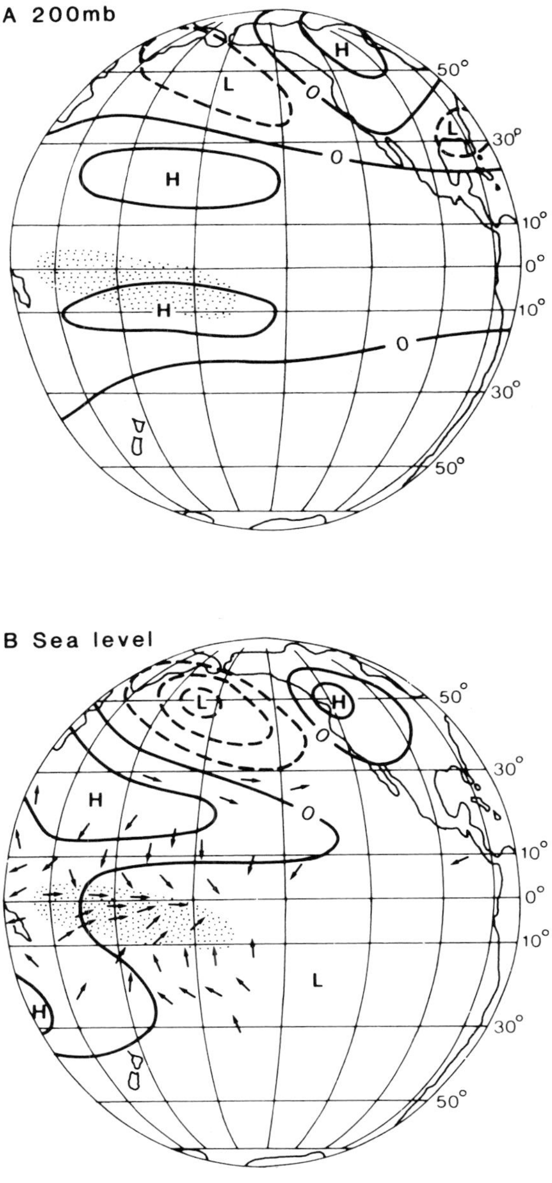

Source: After Shukla and Wallace 1983.

H CLIMATES OF THE TROPICAL MARGINS

1 Northern Africa

The dominance of high-pressure conditions in the Sahara is marked by the low average precipitation figures for this region. Over most of the central Sahara the mean annual precipitation is less than 25 mm, except for the high plateaus of the Ahaggar and Tibesti, which receive more than 100 mm. Parts of western Algeria have gone at least 2 years without more than 0.1 mm of rain in any 24-hour period, and most of south-west Egypt as much as 5 years. However, 24-hour storm rainfalls approaching 50 mm (more than 75 mm over the high plateaus) may be expected in scattered localities and, during a 35-year period of record, excessive short-period rainfall intensities occurred in the vicinity of west-facing slopes in Algeria, such as at Tamanrasset (46 mm in 63 minutes) (see Figure 6.50), El Golea (8.7 mm in 3 minutes) and Beni Abbes (38.5 mm in 25 minutes). During the summer, rainfall variability is introduced into the southern Sahara by the variable northward penetration of the Monsoon Trough (see Figures

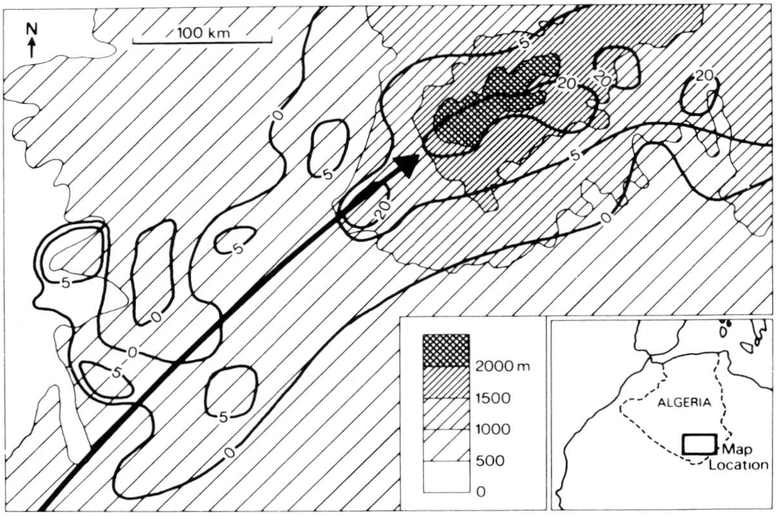

Figure 6.50 Track of a storm and the associated 3-hour rainfall (mm) during September 1950 around Tamanrasset in the vicinity of the Ahaggar Mountains, southern Algeria.

Source: Partly after Goudie and Wilkinson 1977.

6.2B and 6.41) which allows on occasion tongues of moist south-westerly air to penetrate far north and produce short-lived low-pressure centres. Study of these Saharan depressions has permitted a clearer picture to emerge of the region previously designated as the 'temperate front' (see Chapters 3, D.1 and 4, E). In the upper troposphere at about 200 mb (12 km), the westerlies overlie the poleward flanks of the subtropical high-pressure belt. Occasionally the individual high-pressure cells contract away from one another as meanders develop in the westerlies between them which may extend equatorward so as to interact with the low-level underlying tropical easterlies (see Figure 6.51). This interaction may lead to the development of lows which then move north-east along the meander trough associated with rain and

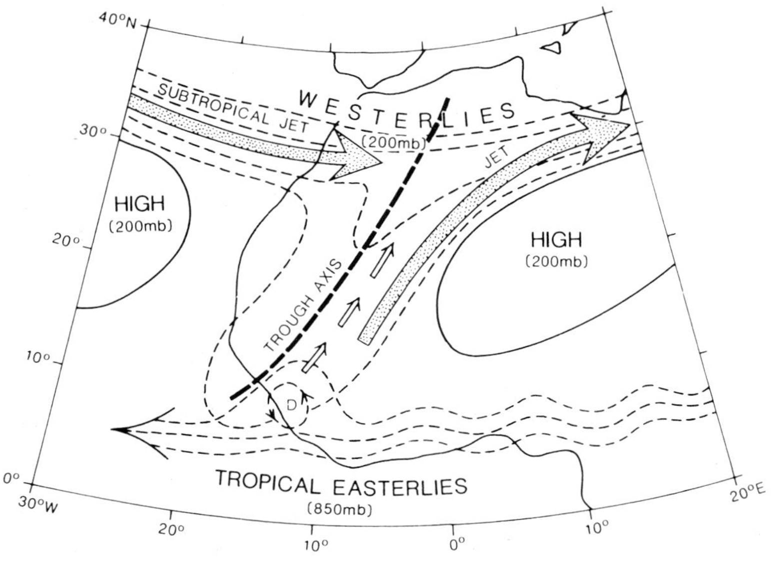

Figure 6.51 Interaction between the westerlies and the tropical easterlies leading to the production of Saharan depressions (D) which move north-eastward along a trough axis.

Source: After Nicholson and Flohn 1980 (*copyright © 1980/1982 by D. Reidel Publishing Company. Reprinted by permission*).

thunder. By the time they reach the central Sahara, they are frequently 'rained out' and give rise to duststorms, but they may be reactivated further north by the entrainment of moist Mediterranean air. The conditions of subtropical cell separation and the interaction of westerly and easterly circulations are most likely to occur around the equinoxes (see Figure 6.41), or sometimes in winter if the otherwise dominant Azores high-pressure cell contracts westwards. A less extreme case of the equatorward extension of the westerlies is the occasional penetration of cold fronts south from the Mediterranean bringing heavy rain to restricted desert areas (see Figure 6.52). In December 1976 such a depression produced up to 40 mm of rain during 2 days in southern Mauretania.

2 Southern Africa

Southern Africa lies between the South Atlantic and Indian Ocean subtropical high-pressure cells in a region subject to the interaction of tropical easterly and extra-tropical westerly airflows.

Figure 6.52 Southern extension of a cold front from a Mediterranean depression into the Sahara at 1200 hrs on 17 January 1972.

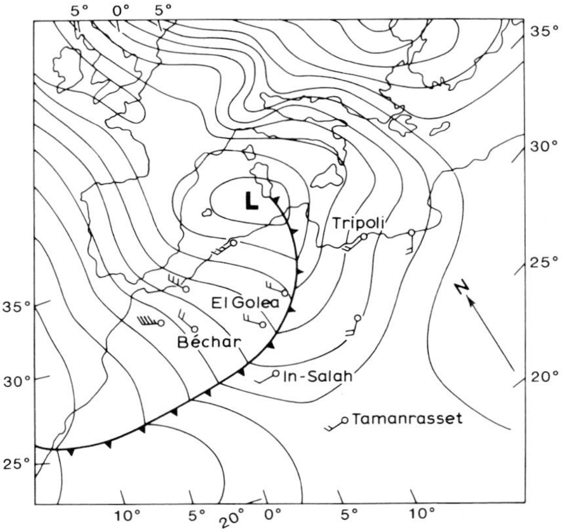

Source: From Breed *et al.* 1979.

Both these high-pressure cells shift west and intensify (see Figure 3.27) in the southern winter but, because the South Atlantic cell always extends 3° latitude further north than the Indian Ocean cell, it brings low-level westerlies to Angola and Zaïre at all seasons and high-level westerlies to central Angola in the southern summer. The seasonal longitudinal shifts of the subtropical high-pressure cells are especially significant to the climate of southern Africa in respect of the Indian Ocean cell. Whereas the 7°–13° longitudinal shift of the South Atlantic cell has relatively little effect, the westward movement of 24°–30° during the southern winter by the Indian Ocean cell brings an easterly flow at all levels to most of southern Africa. The seasonal airflows and convergence zones are shown in Figure 6.53.

In summer (e.g. January) low-level westerlies over Angola and Zaïre meet the north-east monsoon of East Africa along the Inter-tropical Convergence Zone (ITCZ) which extends east as the boundary between the recurved (westerly) winds from the Indian Ocean and the deep tropical easterlies further south. To the west these easterlies impinge on the Atlantic westerlies along the Zaïre Air Boundary (ZAB). The ZAB is subject to daily fluctuations and low-pressure systems form along it, either being stationary or moving slowly westward. When these are deep and associated with southward-extending troughs they may produce significant rainfall. It should be noted that the complex structure of the ITCZ and ZAB means that the major surface troughs and centres of low pressure do not coincide with them but are situated some distance upwind in the low-level airflow (see Figure 6.53), particularly in the easterlies. This low-level summer circulation is dominated by a combination of these frontal lows and convectional heat lows. By March a unified high-pressure system has been established giving a northerly flow of moist air which produces autumn rains in western regions. In winter (e.g. July) the ZAB separates the low-level westerly and easterly airflows from the Atlantic and Indian Oceans, although both are

Figure 6.53 Airflow over southern Africa during January and July, together with the locations of the Inter-Tropical Convergence Zone (ITCZ), the Zaïre Air Boundary (ZAB) and the major surface low pressure troughs.

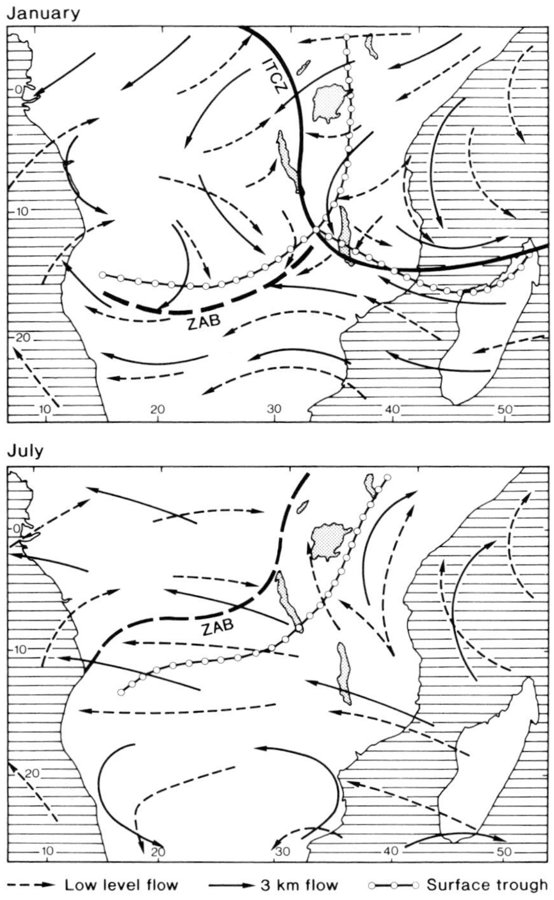

Source: After Taljaard; from Tyson 1986.

overlain by a high-level easterly flow. At this time the northerly displacement of the general circulation brings low- and high-level westerlies with rain to the southern Cape.

Thus tropical easterly airflows affect much of southern Africa throughout the year. A deep easterly flow dominates south of about 10°S in winter and south of 15°–18°S in summer. Over East Africa a north-easterly monsoonal flow occurs in summer, replaced by a south-easterly flow in winter. Easterly waves form in these airflows similar to, but less mobile than, those in other tropical easterlies. These waves form at the 850–700-mb level (i.e. 200–3000 m) in flows associated with easterly jets, often producing squall lines, belts of summer thunder cells and heavy rainfall. These waves are most common between December and February, when they may produce at least 40 mm of rain per day, but are rare between April and October. Tropical cyclones in the Indian Ocean occur particularly around February (see Figure 6.9 and Table 6.1) when the ITCZ lies at its extreme southerly position. These storms recurve south along the east coast of Tanzania and Mozambique but their influence is mainly limited to the coastal belt.

With few exceptions, deep westerly airflows are limited to the most southerly locations of southern Africa, especially in winter. As in northern mid-latitudes, disturbances in the westerlies involve:

1 Quasi-stationary Rossby waves.
2 Travelling waves, particularly marked at and above the 500-mb level, with axes tilted westward with height, divergence ahead and convergence in the rear, moving eastward at a speed of some 550 km/day, having a periodicity of 2–8 days and with associated cold fronts.
3 Cut-off low pressure centres. These are intense, cold-cored depressions most frequent during March–May and September–November, and rare during summer, December–February.

A feature of the climate of southern Africa is the prevalence of wet and dry spells, associated with broader features of the global circulation. Rainfall above normal, occurring as a north–south belt over the region, is associated with a high-phase Walker circulation (see p. 129) having an ascending limb over southern Africa; a strengthening of the ITCZ; an intensification of tropical lows and easterly waves, often in conjunction with a westerly wave aloft to the south; and a strengthening of the South Atlantic

subtropical high-pressure cell (see Figure 6.54). Such a wet spell may occur particularly during the spring to autumn period. Rainfall below normal is associated with a low-phase Walker circulation having a descending limb over southern Africa; a weakening of the ITCZ; a tendency to high pressure with a diminished occurrence of tropical lows and easterly waves; and a weakening of the South Atlantic subtropical high-pressure cell. At the same time there is a

Figure 6.54 Combination of A: a high-phase Walker circulation, continental low pressure and an upper westerly wave, and B: a strengthened South Atlantic subtropical high pressure cell, resulting in above-normal rainfall over southern Africa.

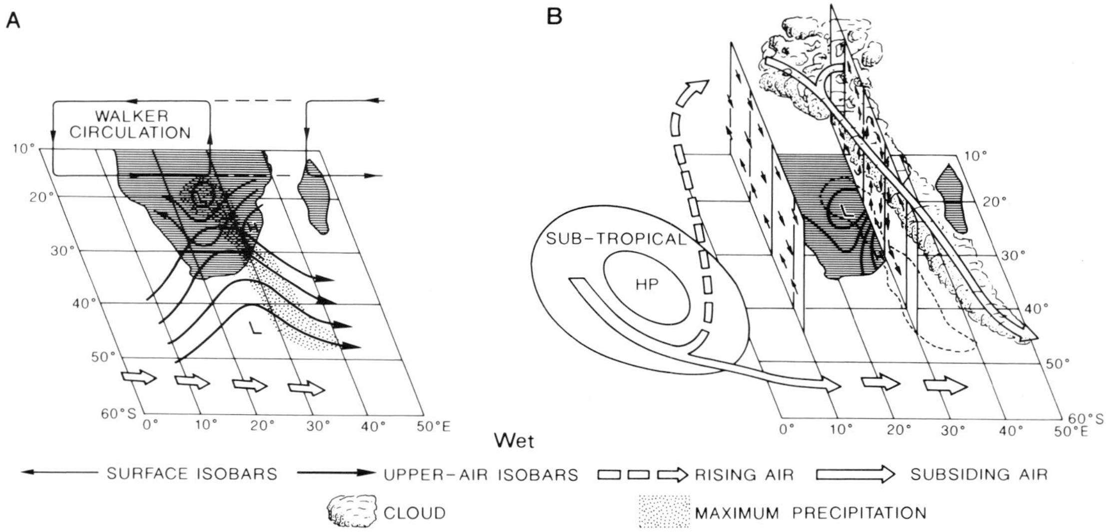

Source: After M. S. J. Harrison; from Tyson 1986.

Figure 6.55 Combination of A: a low-phase Walker circulation, continental high pressure, increased frequency of easterly waves and tropical lows over the western Indian Ocean and an upper westerly wave south of Madagascar and B: a weakened South Atlantic subtropical high pressure cell, resulting in below-normal rainfall over southern Africa and above-normal to the east.

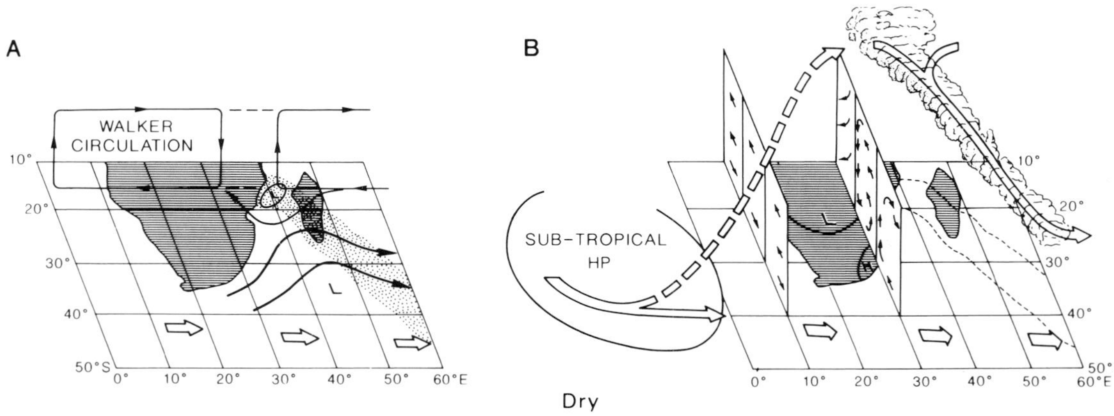

Source: After M. S. J. Harrison; from Tyson 1986.

belt of cloud and rain lying to the east in the western Indian Ocean associated with a rising limb of the Walker circulation and enhanced easterly disturbances in conjunction with a westerly wave aloft south of Madagascar (see Figure 6.55).

3 Australia

The continental high-pressure cell over central Australia inhibits average rainfall amounts, which total less than 250 mm per year over fully 37 per cent of the whole country. Indeed, one station exceeded 230 mm in only 1 out of 42 consecutive years. Detailed work on the variations in pressure of this cell has suggested that the apparently stable high-pressure cell is in reality caused by the local intensification of a constant progression of eastward-travelling anticyclones. Thus their seasonal and shorter-term variation in intensity allows the periodic inflow of the surrounding maritime tropical air masses from the Pacific (mTp) and Indian (mTi) Oceans and maritime polar air from the south. Occasional heavy rainfalls in the continental interior may result from northward incursions of troughs bringing mP air in January and eastward extension of mT air from the Indian Ocean in July (see Figure 6.56).

Figure 6.56 Air-mass frequencies, source areas, wind directions and dominance of the cT high-pressure cell over Australia in summer (*above*) and winter (*below*).

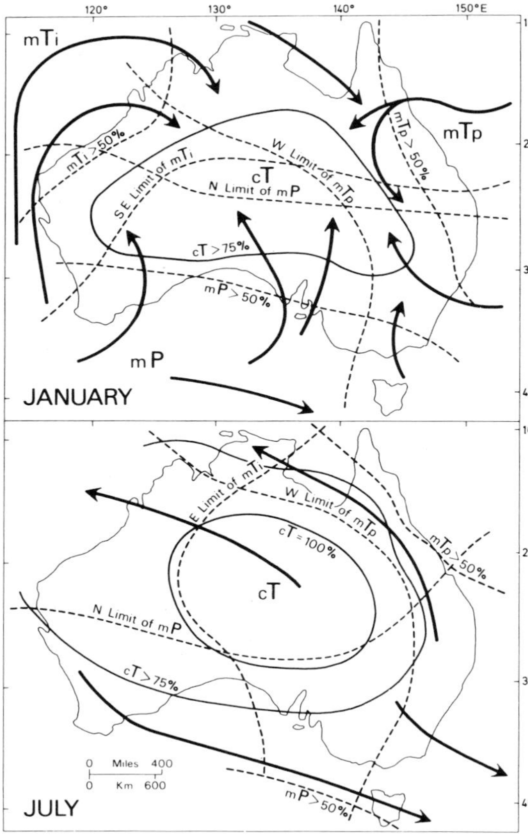

Source: After Gentilli 1971.

SUMMARY

The tropical atmosphere differs significantly from that in middle latitudes. Temperature gradients are generally weak and weather systems are mainly produced by airstream convergence triggering convection in the moist surface layer. Strong longitudinal differences in climate exist as a result of the zones of subsidence (ascent) on the eastern (western) margins of the subtropical high-pressure cells. In the eastern oceans, there is typically a strong trade wind inversion at about 1 km with dry subsiding air above, giving fine weather. Downstream this stable lid is gradually raised by convective clouds as the trades flow westward. Cloud masses are frequently organized into amorphous 'clusters' on a subsynoptic scale. The trade wind systems of the two hemispheres converge, but not in a spatially or temporally continuous manner. This Inter-tropical Convergence Zone also shifts poleward over the land sectors in summer, associated with the monsoon regimes of southern Asia, West Africa and northern Australia.

Wave disturbances in the tropical easterlies are variable in character. The 'classical'

easterly wave has maximum cloud build-up and precipitation behind (east of) the trough line. This distribution follows from the conservation of potential vorticity by the air. About 10 per cent of wave disturbances later intensify to become tropical storms or cyclones. This development requires a warm sea surface and low-level convergence to maintain the sensible and latent heat supply and upper-level divergence to maintain ascent. Cumulonimbus 'hot towers' nevertheless account for a small fraction of the spiral cloud bands. Tropical cyclones are most numerous in the western oceans of the northern hemisphere in the summer–autumn seasons.

The monsoon seasonal wind-reversal of southern Asia is the product of global and regional influences. The orographic barrier of the Himalayas and Tibetan Plateau plays an important role. In winter, the subtropical westerly jet stream is anchored south of the mountains. Subsidence occurs over northern India giving north-easterly surface (trade) winds. Occasional depressions from the Mediterranean penetrate to north-western India–Pakistan. The circulation reversal in summer is triggered by the development of an upper anticyclone over the elevated Tibet Plateau with upper easterly flow over India. This change is accompanied by the northward extension of low-level south-westerlies in the Indian Ocean which appear first in southern India and along the Burma coast and then extend north-westward. Rainfall is concentrated in spells associated with 'monsoon depressions' which travel westward steered by the upper easterlies. Monsoon rains fluctuate in intensity giving rise to 'active' and 'break' periods, in response to southward and northward displacements of the Monsoon Trough, respectively. There is also considerable year-to-year variability.

The West African monsoon has many similar features, but its northward advance is unhindered by a mountain barrier to the north. Four zonal climatic belts, related to the location of overlying easterly jet streams and east–west moving disturbances, are identified.

In Amazonia, where there are broad tropical easterlies but no well-defined ITCZ, precipitation is associated with convective activity triggering low-level convergence, with meso- to synoptic-scale disturbances forming *in situ*, and with instability lines generated by coastal winds that move inland.

Variability in tropical climates also occurs through diurnal effects, such as land–sea breezes, local topographic and coastal effects on airflow, and the penetration of extra-tropical weather systems and airflow into lower latitudes. At irregular intervals, the equatorial Pacific sector experiences an ENSO event with major repercussions both regionally and globally. This represents an alternation between strong easterlies in the equatorial Pacific (non-ENSO) and coastal up-welling off South America, with weak easterlies and warm offshore waters (ENSO). Convective activity normally over Indonesia –New Guinea is displaced eastward towards the central Pacific during an ENSO event.

The climates of the subtropical margins in Africa and Australia reflect the influence of the respective subtropical anticyclones, as well as seasonal influences from the extra-tropical westerlies in winter along polar margins and from the tropical easterlies in summer along equatorial margins. In southern Africa, however, there is also a complex interplay of low-level Indian Ocean easterlies and Atlantic westerlies in summer (January).

7

Small-scale climates

Meteorological phenomena encompass a wide range of space and time scales, from the instantaneous gusts of wind that swirl-up leaves and litter to the global-scale wind systems that shape the annual planetary climate. The weather systems discussed in Chapters 4 and 5 are conventionally designated as synoptic-scale systems, whereas tornadoes and thunderstorms (with a spatial scale of 1–50 km and a time scale of a few hours) are referred to as *mesoscale systems*. Other wind systems of comparable scale to the latter, like mountain and valley winds and land–sea breezes, can give rise to distinctive *local climates* (see Chapter 3,C). Small-scale turbulence, with wind eddies of a few metres dimension and lasting only a few seconds, represent the domain of *micrometeorology*. Atmospheric airflow disturbances in the boundary layer near the earth's surface can be classified into four main types, depending on their altitudinal and horizontal scales – i.e. synoptic and mesoscale, thermal, topographic and surface friction on 'flat' surfaces of varying roughness (see Figure 7.1). Their life-span and the kinetic energy which they involve are illustrated in Figure 7.2, in comparison with those for a range of human activities. For our purposes, we can consider such phenomena in relation to

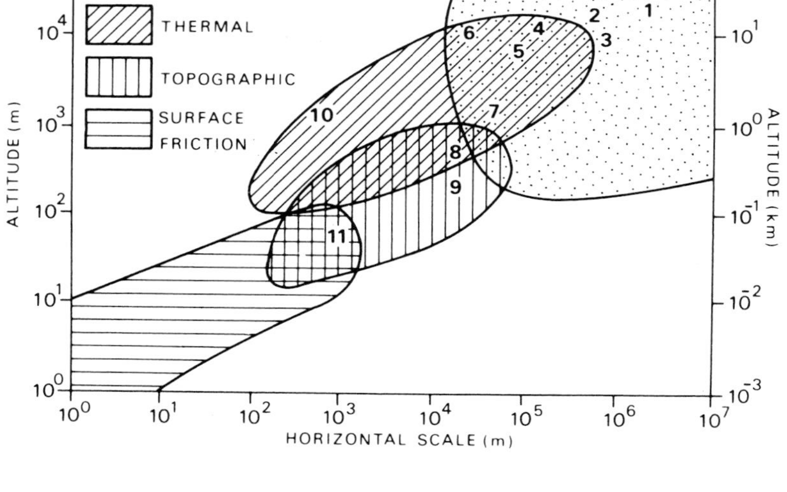

Figure 7.1 Types of airflow disturbances in the near-surface boundary layer of the atmosphere. Examples of specific phenomena are: 1 Depression (altitude 12 km; horizontal scale 1500–2000 km); 2 Hurricane (15 km; 700 km); 3 Easterly wave (10 km; 800 km); 4 Thunderstorm (13 km; 100 km); 5 Mountain lee wave (7 km; 75 km); 6 ITCZ convective cell (15 km; 10–100 km); 7 Sea breeze (1 km; 50 km); 8 Urban heat island (300 m+; 20 km); 9 Urban friction hump (of buildings); 10 Small cumulus cell (2000 m; 100–1000 m); 11 Forest (60 m; 1500 m).

Source: From Anderson 1971.

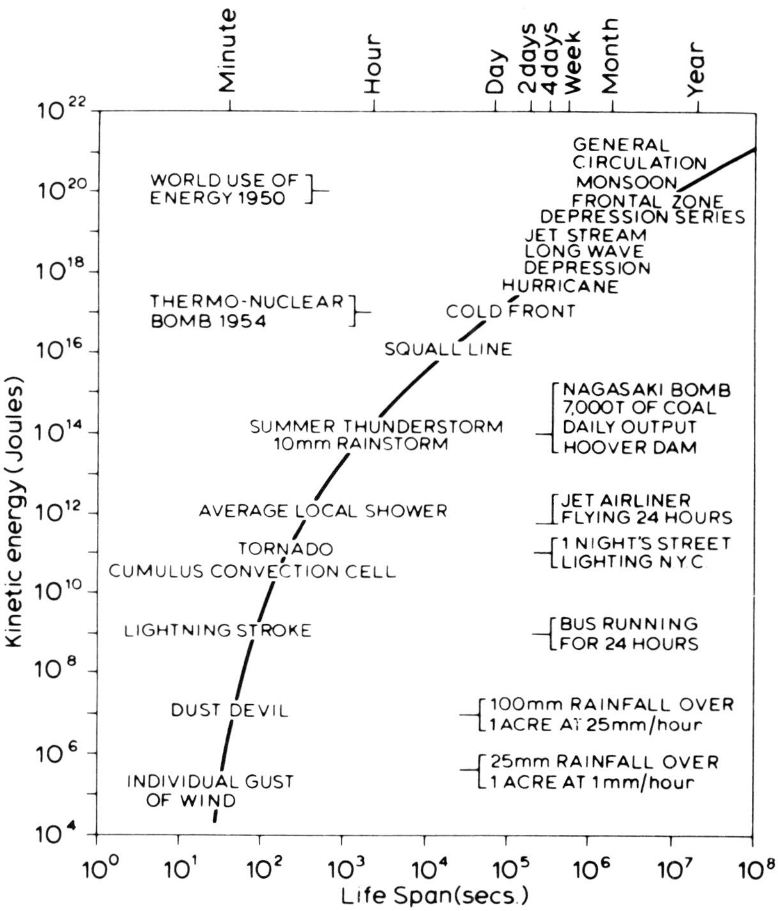

Figure 7.2 The relationship between the life-span (in seconds) of a range of meteorological phenomena and their kinetic energy (in joules). The kinetic energy of a number of human activities is also shown.

Source: After Koppány 1975.

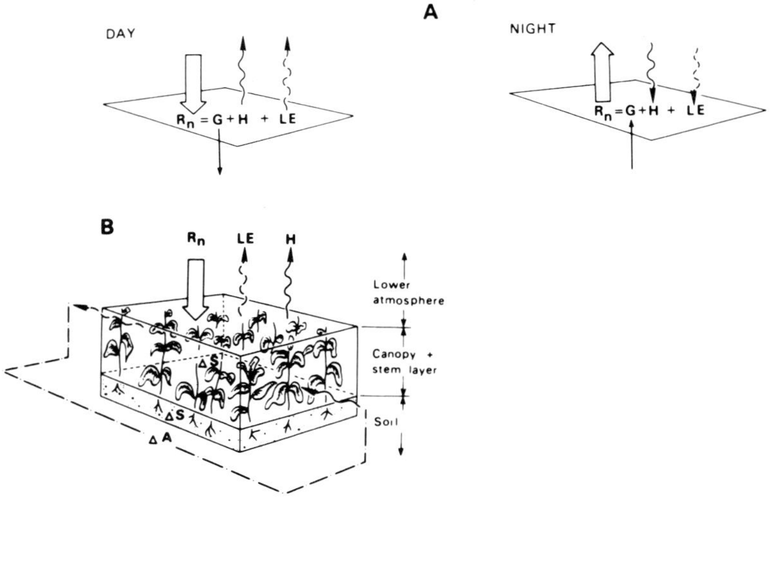

Figure 7.3 Energy flows involved in the energy balance of a simple surface during day and night (A) and a vegetated surface (B).

Source: After Oke 1978.

the climatic processes within a crop canopy, a forest stand or a cluster of city buildings.

A SURFACE-ENERGY BUDGETS

We will first review the processes of energy exchange between the atmosphere and an unvegetated surface. In Chapter 1, F, it was noted that the surface energy budget equation is usually written:

$$R_n = H + LE + G$$

where R_n, the net all-wavelength radiation
$= [S(1 - a)] + L_n$
S = incoming short-wave radiation,
a = fractional albedo of the surface, and
L_n = the net outgoing long-wave radiation.

R_n is usually positive by day since the absorbed solar radiation exceeds the net outgoing long-wave radiation; at night, when $S = 0$, R_n is determined by the negative magnitude of L_n.

The surface energy flux terms are

G = ground heat flux,
H = turbulent sensible heat flux to the atmosphere,
LE = turbulent latent heat flux to the atmosphere (E = evaporation; L = latent heat of vaporization).

Positive values denote a flux *away* from the surface interface. By day, the available net radiation is balanced by turbulent fluxes of sensible heat (H) and latent heat (LE) into the atmosphere and by conductive heat flux into the ground (G). At night, the negative R_n caused by net outgoing long-wave radiation is offset by the supply of conductive heat from the soil (G) and turbulent heat from the air (H) (see Figure 7.3A). Occasionally, condensation may contribute heat to the surface.

Commonly, there is a small residual heat storage (ΔS) in the soil in spring/summer and a return of heat to the surface in autumn/winter. Where a vegetation canopy is present there may also be a small additional biochemical heat storage, due to photosynthesis, as well as physical heat storage by leaves and stems (see Figure 7.3B).

An additional energy component to be considered in areas of mixed canopy cover (forest/grassland, desert/oasis), and in water bodies, is the horizontal transfer (*advection*) of heat by wind and currents (ΔA; see Figure 7.3B). The atmosphere transports both sensible and latent heat.

B NON-VEGETATED NATURAL SURFACES

The energy exchanges of dry desert surfaces are relatively simple and straightforward. Figure 7.4 illustrates the instantaneous noon and evening fluxes on a granite surface in July in California and the resulting large temperature range. Surface properties modify the heat penetration, as shown in Figure 7.5 from mid-August measurements in the Sahara. The maximum surface temperatures reached on bare dark-coloured basalt and light-coloured sandstone were almost identical, but the greater thermal conductivity of the former (3.1 W m^{-1} K^{-1} for basalt, versus 2.4 W m^{-1} K^{-1} for sandstone) gives a greater diurnal range and deeper penetration of the diurnal temperature wave, to about 1 m in the basalt. In sand (see Figure 7.5C), the temperature wave is negligible at 30 cm due to the low conductivity of inter-

Figure 7.4 Energy balance of a granite surface in California on 18 July at noon (solar altitude 70°) and 1800 hours (solar altitude 10°). Figures in W m^{-2}.

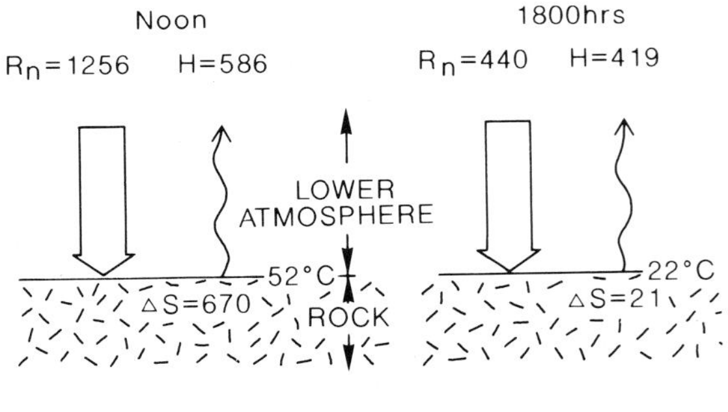

Source: Based on data from Miller 1965.

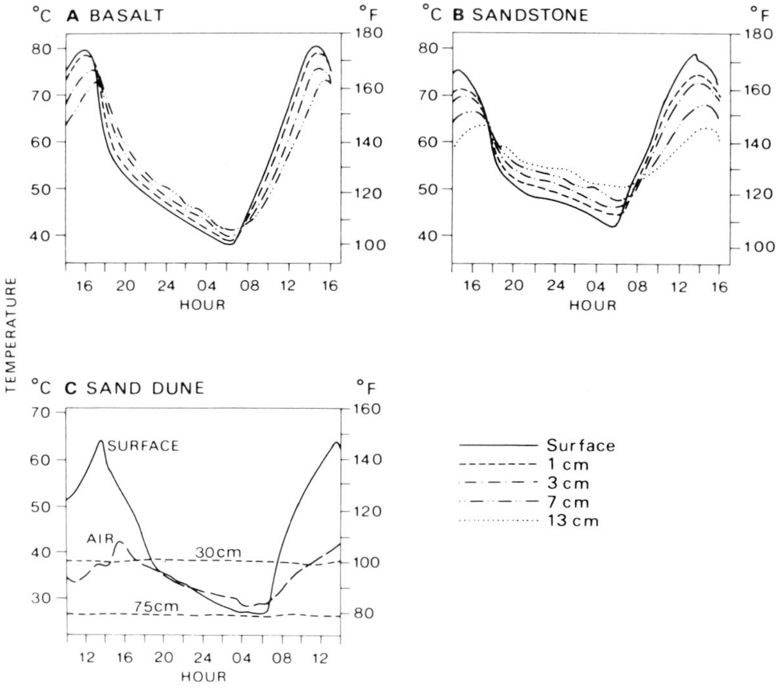

Figure 7.5 Diurnal temperatures near, at, and below the surface in the Tibesti region, central Sahara, in mid-August 1961. A: At the surface and at 1 cm, 3 cm and 7 cm below the surface of a basalt. B: At the surface and at 1 cm, 3 cm, 7 cm and 13 cm below the surface of a light-coloured sandstone. C: In the surface air layer, at the surface, and at 30 cm and 75 cm below the surface of a sand dune.

Source: After Peel 1974.

granular air. Note that the surface range of temperature is several times that in the air. Sand also has an albedo of about 0.35 compared with 0.2 for a rock surface.

A representative diurnal pattern of energy exchange over desert surfaces is shown in Figure 7.6. The 2-m air temperature varies between 17° and 29°C, although the surface of the dry lake bed reaches 57°C at midday. R_n reaches a maximum about 1300 hours. At this time most of the heat is transferred to the air by turbulent convection, while in the early morning the heating goes into the ground. At night, this soil heat is returned to the surface, offsetting radiational cooling. Over a 24-hour period, about 90 per cent of the net radiation goes into sensible heat, 10 per cent into ground flux.

For a water body, the energy fluxes are very differently apportioned. Figure 7.7 illustrates the seasonal regime for Lake Mead, Arizona, in 1952–3. The incoming short-wave radiation penetrates to about 10 m depth (see Chapter 1, D.5) and there is an important horizontal advective term (ΔA) due to the changing density stratification in the lake. Warmer water rises to the surface in winter (ΔA positive) whereas in summer there is a large loss as a result of turbulent mixing of the water. There is a strong annual cycle in the flux into and out of the water body (G), whereas the evaporative loss in excess of 200 cm per year occurs at all seasons. Wind effects in autumn cause LE to exceed the net radiation term.

C VEGETATED SURFACES

From the viewpoint of energetics, as well as of climate within the plant canopy, it is useful to consider short crops and forests separately.

1 Short green crops

Short green crops, up to a metre or so high, supplied with sufficient water and exposed to similar solar radiation conditions, all have a

Figure 7.6 Energy flows involved at a dry-lake surface at El Mirage, California (35°N), on 10–11 June 1950. Wind speed due to surface turbulence was measured at a height of 2 m.

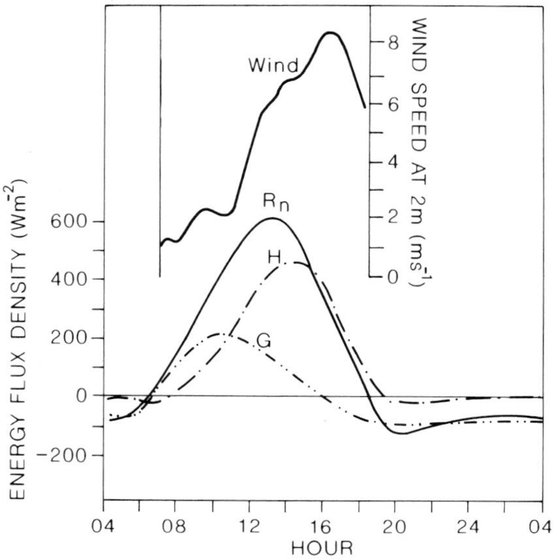

Source: After Vehrencamp 1953 and Oke 1978.

similar receipt of net radiation (R_n). This is largely because of the small range of albedos, 20–30 per cent for short green crops compared with 9–18 per cent for forests. Canopy structure appears to be the primary reason for this difference.

General figures for rates of energy dispersal at noon on a June day in a 20-cm high stand of grass in the higher mid-latitudes are shown in Table 7.1.

Table 7.1 Rates of energy dispersal at noon in a 20-cm stand of grass (in higher mid-latitudes on a June day).

	W m^{-2}
Net radiation at the top of the crop	550
Physical heat storage in leaves	6
Biochemical heat storage (i.e. growth processes)	22
Received at soil surface	200

Figure 7.8A and B shows the diurnal and annual energy balances of a field of short grass near Copenhagen (56°N). For an average 24-hour period in June about 58 per cent of the incoming radiation is involved in evapotranspiration, whereas in December the small amount of net outgoing radiation (i.e. negative R_n) is composed of 55 per cent heat supplied by the soil and 45 per cent sensible heat transfer from the air to the grass.

It is possible to generalize regarding the microclimates of short growing crops (see Figure 7.9, and Oke 1978):

1 Temperature. In early afternoon there is a temperature maximum just below the vegetation crown where the maximum energy absorption is occurring; the temperature is lower near the soil surface where heat flows into the soil. At night the crop cools mainly by long-wave emission and by some continued transpiration, producing a temperature minimum at about two-thirds the height of the crop. Under calm conditions, a temperature inversion may form just above the crop.
2 Wind speed. This is a minimum in the upper crop canopy where the foliage is most dense. Below there is a slight increase and a marked increase above.
3 Water vapour. The maximum diurnal evapotranspiration rate and supply of water vapour occurs at about two-thirds the crop height where the canopy is most dense.
4 Carbon dioxide. During the day CO_2 is absorbed by the photosynthesis of growing plants and emitted at night due to respiration. This maximum sink and source of CO_2 is at about two-thirds the crop height.

Finally, it is instructive to look at the conditions accompanying the growth of irrigated crops. Figure 7.10A and B shows the energy relationships in a 1-m high stand of irrigated Sudan Grass at Tempe, Arizona, on 20 July 1962. The air temperature varied between 25°C (77°F) and 45°C (113°F). During the day evapotranspiration in the dry air is near its potential

Figure 7.7 Energy flows involving the upper layers of water. Annual figures for Lake Mead, Arizona (36.1°N) during 1952–3.

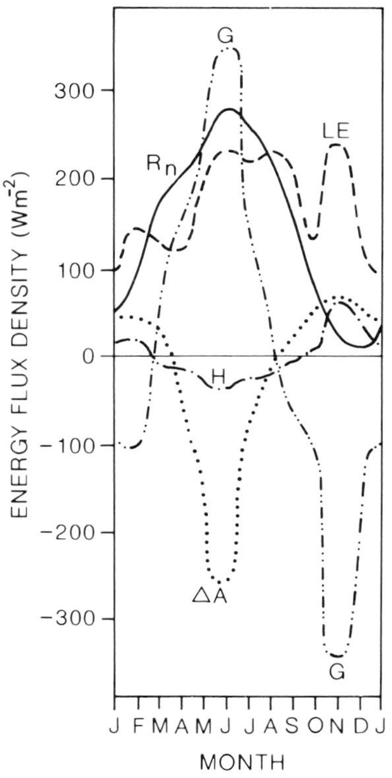

Source: After Sellers 1965.

and LE (anomalously high due to a local temperature inversion) exceeds R_n, the deficiency being made up by a transfer of sensible heat from the air (negative H). Evaporation continues during the night due to a quite high wind speed (7 m s^{-1}) sustained by the continued heat flow from the air. The evapotranspiration thus gives comparatively low diurnal temperatures and a shallow inversion within irrigated desert crops.

2 Forests

The vertical structure of a forest, which depends on the vegetational species, the ecological associations, the age of the stand and other botanical considerations, largely determines the forest microclimate. Much of the climatic influence of a forest can be explained in terms of the geometry of the forest, including morphological characteristics, size, coverage and stratification. Morphological characteristics include amount of branching (bifurcation), the periodicity of growth (i.e. evergreen or deciduous), together with the size, density and texture of the leaves. Tree size is obviously important. In temperate forests the sizes may be closely similar, whereas in tropical forests a great variety of sizes may be

Figure 7.8 Energy fluxes over short grass near Copenhagen (56°N). A: Totals for a day in June (17 hours daylight; maximum solar altitude 58°) and December (7 hours daylight; maximum solar altitude 11°). B: Seasonal curves of net radiation (R_n), latent heat (LE), sensible heat (H) and ground-heat flux (G).

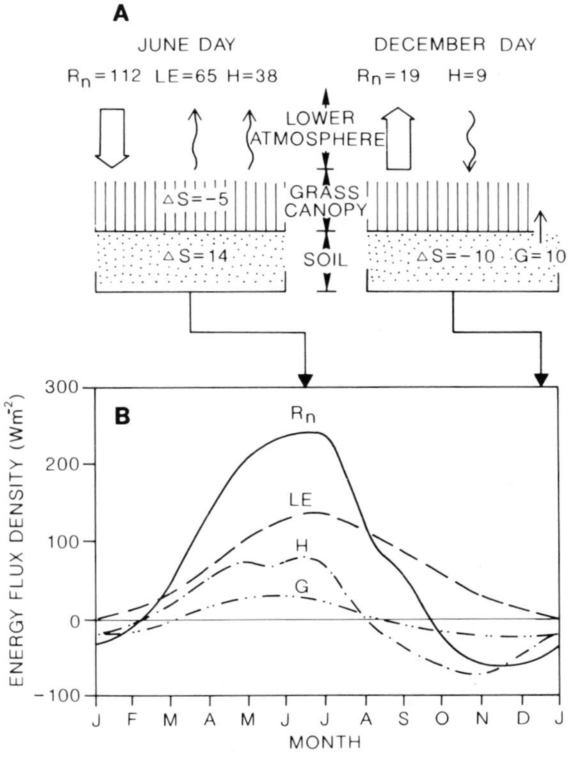

Source: Data from Miller 1965; and after Sellers 1965.

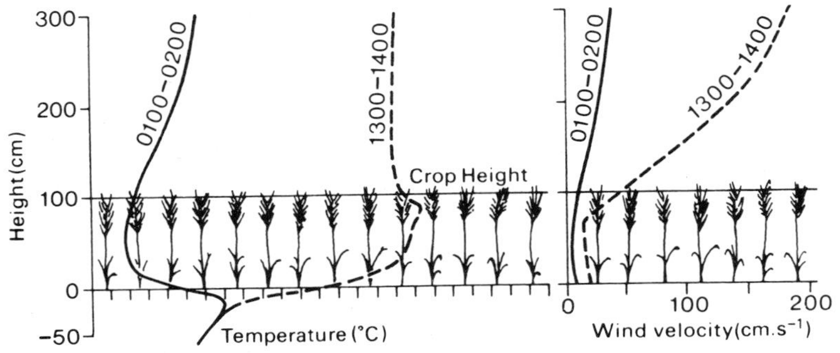

Figure 7.9 Temperature and wind-velocity profiles within and above a metre-high stand of barley at Rothamsted, southern England, on 23 July 1963 at 0100–0200 hours and 1300–1400 hours.

Source: After Long *et al.* 1964.

present locally. Crown coverage is important in terms of the physical obstruction presented by the canopy to radiation exchange and air movements.

An obvious example of the microclimatic effects of different spatial forest organizations can be gained by comparing the features of tropical rain forests and temperate forests. In tropical forests the average height of the taller trees is of the order of 46–55 m (150–180 ft), individuals rising to over 60 m (200 ft). The dominant height of temperate forests is up to 30 m, so that neither temperate, nor most tropical forests can compare in height with the western American redwoods (*Sequoia sempervirens*). Tropical forests commonly possess a great variety of species, there being seldom less than forty per hectare (100 hectares = 1 km^2) and sometimes over 100; comparing with less than twenty-five (occasionally only one) tree species with a trunk diameter greater than 10 cm (or 4 in) in Europe and North America. Many British woodlands, for example, have almost continuous canopy stratification from low shrubs to the tops of 36 m beeches, whereas tropical forests are strongly stratified with dense undergrowth, unbranching trunks, and commonly two upper strata of foliage. This stratification, the second or lower of which is usually the more dense, results in rather more complex microclimates in tropical forests than in temperate stands.

It is convenient to describe the climatic effects of forest stands in terms of their modification of energy transfers and of the airflow, their modification of the humidity environment, and their modification of the thermal environment.

a Modification of energy transfers

Forest canopies significantly change the pattern of incoming and outgoing radiation. The short-

Figure 7.10 Energy flows involved in the diurnal energy balance of irrigated Sudan Grass at Tempe, Arizona, on 20 July 1962 (Wm^{-2}).

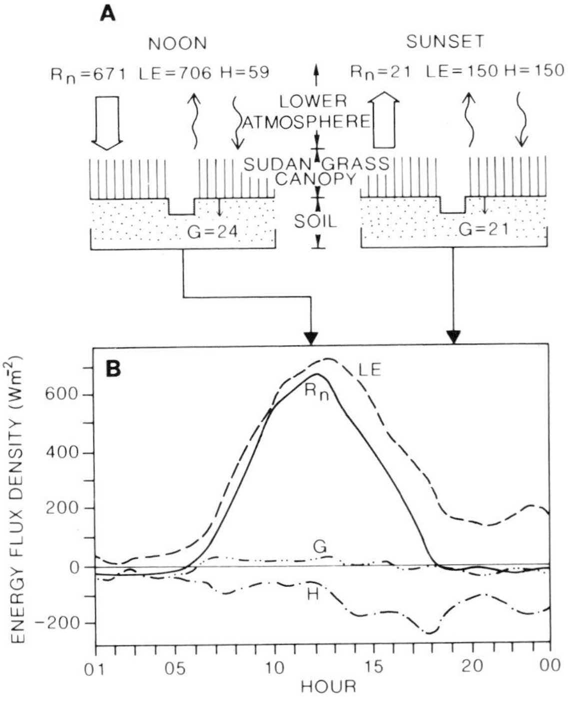

Source: After Sellers 1965.

Figure 7.11 Energy flows in a 30-year-old oak stand in the Tellerman Experimental Forest, Voronezh District, USSR, on an average summer day (June to August).

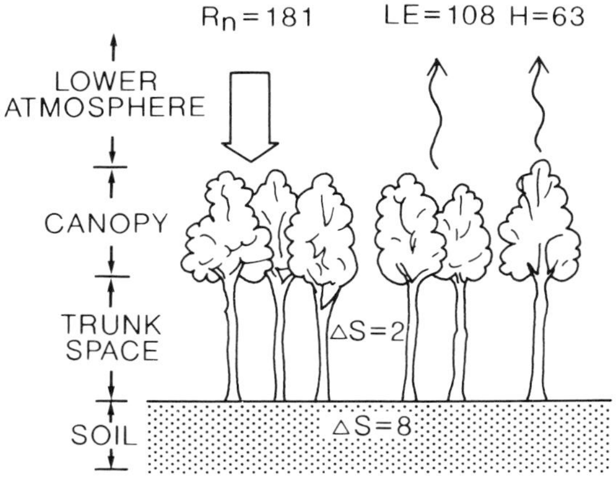

Source: After Sukachev and Dylis 1968.

wave reflectivity of forests depends somewhat on the characteristics of the trees and their density. Coniferous forest have albedos of about 8–14 per cent and values for deciduous range between 12 and 18 per cent, increasing as the canopy becomes more open. Values for semi-arid savannas and scrub woodland are much higher.

Besides reflecting energy, the forest canopy traps energy (see Figure 7.11), and it has been calculated that for dense red beeches (*Fagus sylvatica*) 80 per cent of the incoming radiation is intercepted by the tree-tops and less than 5 per cent reaches the forest floor. The greatest trapping occurs in sunny conditions, for when the sky is overcast the incoming diffuse radiation has greater possibility of penetration laterally to the trunk space (see Figure 7.12). Visible light, however, does not give an altogether accurate picture of total energy penetration for more ultraviolet than infra-red radiation is absorbed in the crowns. For example, only 7.6 per cent of short-wave radiation (less than 0.5 μm) reached a forest floor in Nigeria, as against 45.3 per cent greater than 0.6 μm. As far as light penetration is concerned there are obviously great variations depending on type of tree, tree spacing, time of year, age, crown density and height. About 50–75 per cent of the outside light intensity may penetrate to the floor of a birch–beech forest, 20–40 per cent for pine, 10–25 per cent for spruce and fir; but for tropical Congo forests the figure may be as low as 0.1 per cent and one of 0.01 per cent has been recorded for a dense elm stand in Germany. One of the most important effects of this is to reduce the length of daylight. For deciduous trees, more than 70 per cent of the light may penetrate when they are leafless. Tree age is also important in that this controls both crown cover and height. Figure 7.12 shows this

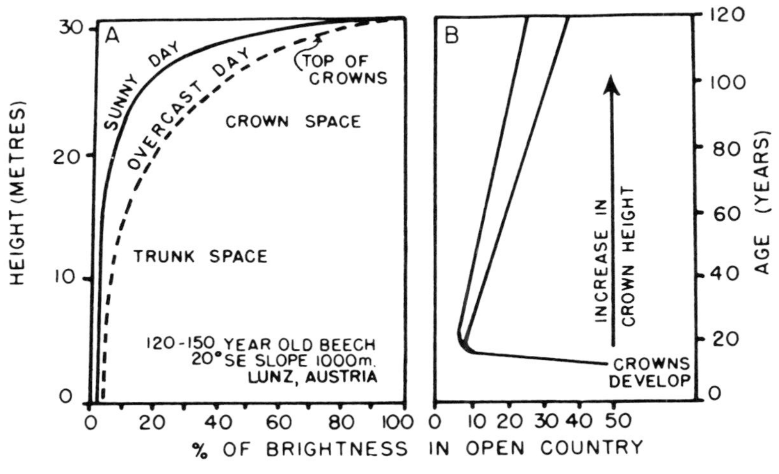

Figure 7.12 Amount of light as a function of height for a thick stand of red beeches (*Fagus sylvatica*) in Austria (A), and as a function of age for a Thuringian spruce forest (B).

Source: After Geiger 1965.

rather complicated effect for spruce in the Thuringian Forest, Germany. For a Scots pine (*Pinus sylvestris*) forest in Germany 50 per cent of the outside light intensity was recorded at 1.3 years, only 7 per cent at 20 years and 35 per cent at 130 years.

b Modification of airflow

Forests impede both the lateral and vertical movement of air, but it is more convenient to treat the latter in connection with thermal modifications. In general, air movement within forests is slight compared with that in the open, and quite large variations of outside wind velocity have little effect inside woods (see Figure 7.13). Measurements for European forests show that 30 m (100 ft) of penetration reduces wind velocities to 60–80 per cent, 60 m (200 ft) to 50 per cent and 120 m (400 ft) to only 7 per cent. A speed of 2.2 m s^{-1} (4.9 mph) outside a Brazilian evergreen forest was found to be reduced to 0.5 m s^{-1} (1.1 mph) at about 100 m (300 ft) within, and to be negligible at 1000 m. In the same location external storm winds of 28 m s^{-1} (62.7 mph) were reduced to 2 m s^{-1} (4.2 mph) some 11 km (7 miles) deep in the forest. Where there is a complex vertical structuring of the forest wind velocities become more complex. Thus whereas in the crowns (23 m; 75 ft) of a Panama rain forest the wind velocity was 75 per cent of that outside, it was only 20 per cent in the undergrowth (2 m; 6.5 ft). Other influences include the density of the stand and the season. For dense pine stands in Idaho simultaneous recordings showed the wind velocity to be 0.6 m s^{-1} (1.3 mph) in a cut area, 0.4 m s^{-1} (0.9 mph) half cut, and 0.1 m s^{-1} (0.2 mph) in the uncut stand. The effect of season on wind velocities in deciduous forests is shown by Figure 7.13. Observations in a Tennessee mixed-oak forest showed January forest-wind velocities to be 12 per cent of those in the open whereas those in August had dropped to 2 per cent.

Knowledge of the effect of forest barriers on winds has been utilized in the construction of wind breaks to protect crops and soil, and, for example, the cypress breaks of the southern Rhône valley and the Lombardy poplars (*Populus nigra*) of the Netherlands form distinctive features of the landscape. It has been found that the denser the obstruction the greater the

Figure 7.13 Influence on wind velocity profiles exercised by (A) dense 45-year-old ponderosa pine (*Pinus ponderosa*) stands in the Shasta Experimental Forest, California, the dashed lines indicating the wind profile over open country; and (B) an oak grove.

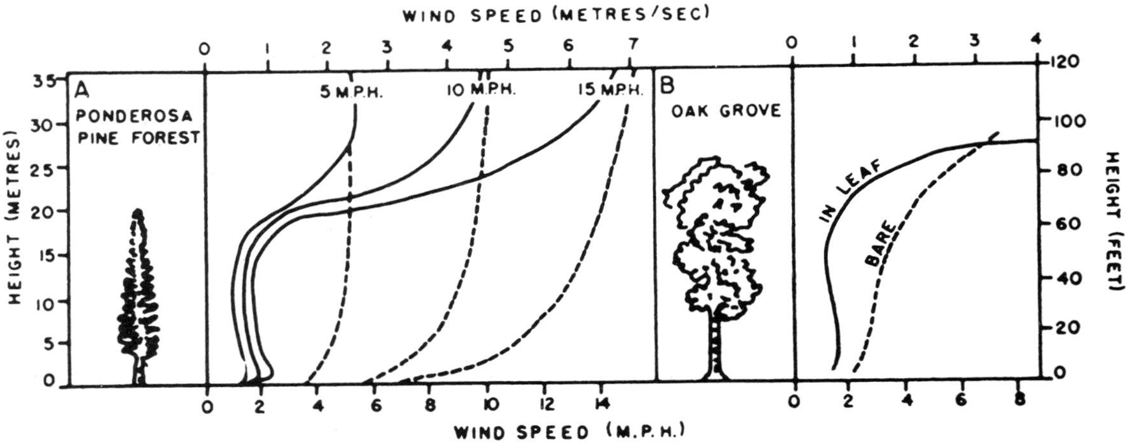

Sources: A: After Fons, and Kittredge 1948.
B: After R. Geiger and H. Amann, and Geiger 1965.

shelter immediately behind it, although the downwind extent of its effect is reduced by lee turbulence set up by the barrier. The maximum protection is given by the filtering mechanism produced by a break of about 40 per cent penetrability (see Figure 7.14). An obstruction begins to have an effect about eighteen times its own height upwind, and the downwind effect can be increased by the *back coupling* of more than one belt (see Figure 7.14A).

There are also some less obvious microclimatic effects of forest barriers. One of the most important is that the reduction of horizontal air movement in forest clearings increases the frost hazard on winter nights. Another aspect is the removal of dust and fog droplets from the air by the filtering action of forests. Measurements 1.5 km upwind, on the lee side and 1.5 km downwind of a kilometre-wide German forest gave dust counts (particles per litre) of 9000, less than 2000 and more than 4000, respectively. Measurements in association with a 2-m high and 13-m thick shelter belt on the south-east Hokkaido coast, Japan, in July 1952 showed this filtering effect on advection fog rolling in from the sea, in that 20 m downwind of the obstruction the humidity was only 0.1 g m^{-3} (mean wind velocity 2.55 m s^{-1}), compared with 0.3 g m^{-3} (mean wind velocity 3.4 m s^{-1}) a similar distance upwind. In extreme cases so much fog can be filtered from laterally moving air that *negative interception* can occur, where there is a higher precipitation catch within a forest than outside. The winter rainfall

Figure 7.14 The influence of shelterbelts on wind-velocity distributions (expressed as percentages of the velocity in the open). A: The effects of one shelter belt of three different densities, and of two back-coupled medium-dense shelter belts. B: The detailed effects of one half-solid shelter belt.

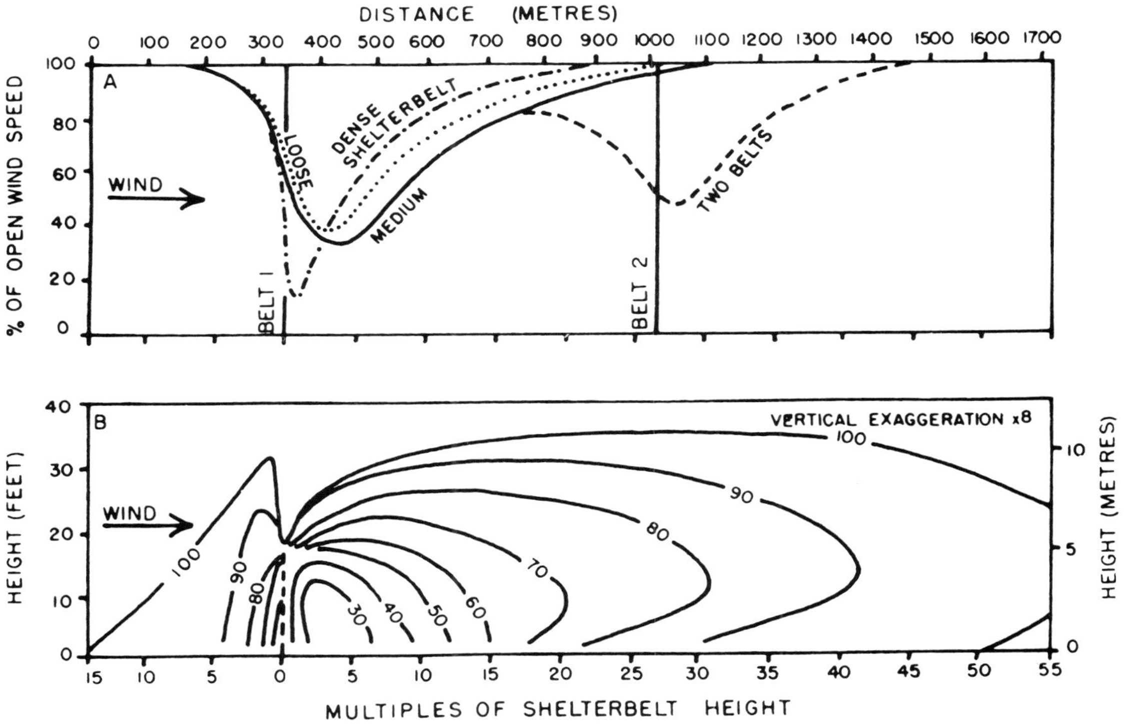

Sources: A: After W. Nägeli, and Geiger 1965. B: After Bates and Stoeckeler, and Kittredge 1948.

catch outside a eucalyptus forest near Melbourne, Australia was 50 cm (19.7 in), whereas inside the forest it was 60 cm (23.6 in).

c Modification of the humidity environment

The humidity conditions within forest stands contrast strikingly with those in the open. Evaporation from the forest floor is usually much less because of the decreased direct sunlight, lower wind velocities, lower maximum temperatures, and generally higher forest air humidity. Evaporation from the bare floors of pine forests is 70 per cent of that in the open for Arizona in summer and only 42 per cent for the Mediterranean region, although such measurements have little real significance in that water losses from vegetated surfaces are controlled by the plant evapotranspiration.

Unlike many cultivated crops, forest trees exhibit a wide range of physiological resistance to transpiration processes and, as a consequence, the proportions of forest energy flows involved in evapotranspiration (*LE*) are varied, as is the inversely-linked contribution by sensible heat exchange (*H*). In view of the global climatic significance of the Amazonian tropical broadleaved forest, it is especially interesting that estimates suggest that after rain up to 80 per cent of the net solar radiation (R_n) is involved in evapotranspiration (*LE*) (see Figure 7.15). Figure 7.16 compares diurnal energy flows during July in respect of a pine forest in

Figure 7.15 A computer simulation of energy flows involved in the diurnal energy balance of a primary tropical broadleaved forest in the Amazon during a high-sun period on the second dry day following a 22-mm daily rainfall.

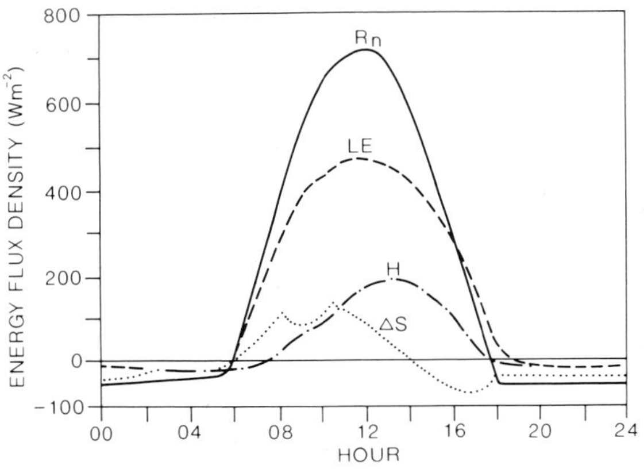

Source: After a BATS model from Dickinson and Henderson-Sellers 1988 (*by permission of the Royal Meteorological Society, redrawn*).

Figure 7.16 Energy components on a July day in two forest stands. A: Scots and Corsican pine at Thetford, England (52°N), on 7 July 1971. Cloud cover was present during the period 0000–0500 hours. B: Douglas fir stand at Haney, British Columbia (49°N), on 10 July 1970. Cloud cover was present during the period 1100–2000 hours.

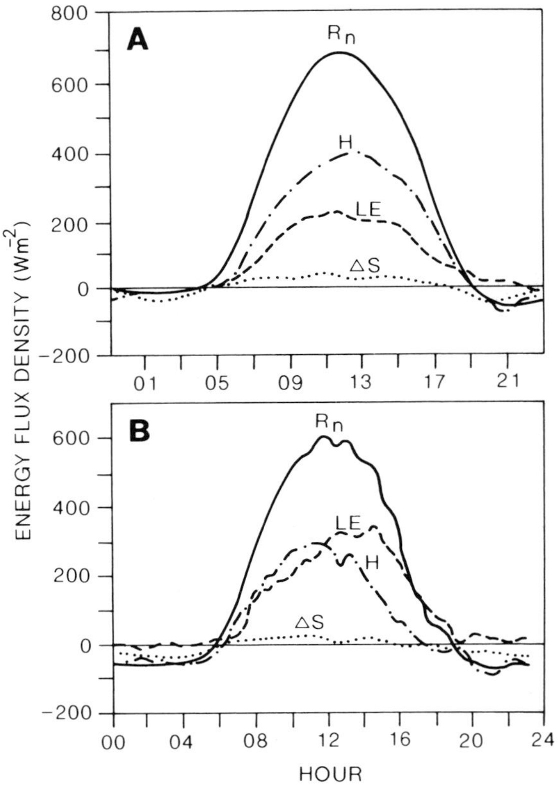

Sources: A: Data from Gay and Stewart 1974; after Oke 1978.
B: Data from McNaughton and Black 1973; after Oke 1978.

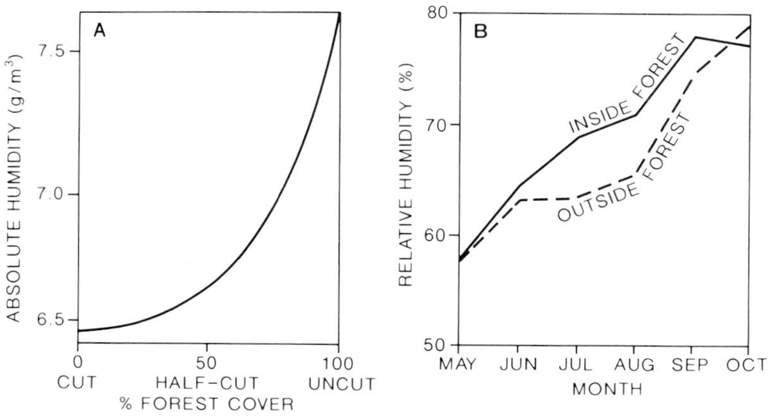

Figure 7.17 The effects (A) of percentage of white pine (*Pinus monticola*) cover on absolute humidity in northern Idaho in July and August 1931–3; and (B) of season on relative humidity in a Michigan birch–beech–maple forest at 1700 hours during May to October 1927–36.

Source: A: After Kittredge 1948. B: After US Dept. *Agriculture Yearbook*, 1941.

eastern England and a fir forest in British Columbia. In the former area only 0.33 R_n is employed for *LE* due to the high resistance of the pines to transpiration, whereas 0.66 R_n is similarly employed in the British Columbia fir forest, especially during the afternoon. Like short green crops, however, only a very small proportion of R_n is used ultimately for tree growth, an average figure being about 1.3 W m^{-2}, of which some 60 per cent produces wood tissue and 40 per cent forest litter.

During daylight leaves transpire water through open pores, or *stomata*, so that this loss is controlled by the length of day, the leaf temperature (modified by evaporational cooling), the leaf surface area, the tree species and its age, as well as by the meteorological factors of available radiant energy, atmospheric vapour pressure and wind speed (see Chapter 2.A). Total evaporation figures are therefore extremely varied; also the evaporation of water intercepted by the vegetation surfaces enters into the totals, in addition to direct transpiration. Calculations made for a catchment covered with Norway spruce (*Picea abies*) in the Harz Mountains of Germany showed an estimated annual evapotranspiration of 34 cm and additional interception losses of 24 cm.

The humidity of forest stands is very much linked to the amount of evapotranspiration and increases with the density of vegetation present (see Figure 7.17A). The increase of forest humidity over that outside averages some 3–10 per cent of relative humidity and is especially marked in summer (see Figure 7.17B). Mean annual relative humidity forest excesses for Germany and Switzerland are for beech 9.4 per cent, Norway spruce (*Picea abies*) 8.6 per cent, larch 7.9 per cent and Scots pine (*Pinus sylvestris*) 3.9 per cent. However, humidity comparisons in these terms are rather unsatisfactory in that forest temperatures differ strikingly from those in the open. Forest vapour pressures were found to be higher within an oak stand in Tennessee than outside for every month except December. Tropical forests exhibit almost complete night saturation irrespective of elevation in the trunk space, whereas during the day humidity is inversely related to elevation. Measurements in Amazonia show that in dry conditions daytime humidities in the lower trunk space (1.5 m) are near 20 g kg^{-1} compared with 18 g kg^{-1} at the top of the canopy (36 m); the deficit of absolute humidity (with respect to saturation) is reduced at the 1.5-m level to between one-third and one-half that at the top.

The influence of forest structures on precipitation is still very much an unresolved problem. This is partly due to the difficulties of comparing rain-gauge catches in the open with those near forests, within clearings or beneath trees. For example, on the windward side of a forest the dominance of low-level upcurrents decreases the

amount of precipitation actually caught in the rain gauge, whereas the reverse occurs where there are lee-side downdraughts. In small clearings the low wind velocities cause little turbulence around the opening of the gauge and catches are generally greater than outside the forest, although the actual precipitation amounts may be identical. On the other hand, it is sometimes found that the larger the clearing the more prevalent are downdraughts and consequently the precipitation catch increases. In a 25-m high pine and beech forest in Germany, catches in clearings of 12-m diameter were only 87 per cent of that upwind of the forest, but this catch rose to 105 per cent in clearings of 38-m diameter. An analysis of precipitation records for Letzlinger Heath (Germany) before and after afforestation suggested a mean annual increase of 6 per cent, with the greatest excesses occurring during drier years. It is generally agreed, however, that forests have little effect on cyclonic rain, but that they may have a marginal orographic effect in increasing lifting and turbulence, which is of the order of 1–3 per cent in temperate regions.

A far more important obstructional influence of forests on precipitation is in terms of the direct interception of rainfall by the canopy. This obviously varies with crown coverage, with season, and with the rainfall intensity. Measurements in German beech forests indicate that, on average, they intercept 43 per cent of precipitation in summer and 23 per cent in winter. Pine forests may intercept up to 94 per cent of low-intensity precipitation but as little as 15 per cent of high intensities, the average for temperate pines being about 30 per cent. The intercepted precipitation either evaporates on the canopy, runs down the trunk, or drips to the ground. Assessment of the total precipitation reaching the ground (the *throughfall*) requires very detailed measurements of the stem-flow and the contribution of drips from the canopy. Canopy evaporation is not necessarily a total loss of moisture from the woodland, since the solar energy used in the evaporating process is not available to remove soil moisture or transpiration water, but the vegetation does not derive the benefit of the cycling of the water through it via the soil. Evaporation from the canopy is very much a function of net radiation receipts (20 per cent of the total precipitation evaporates from the canopy of Brazilian evergreen forests), and of the type of species. Some Mediterranean oak forests yield virtually no stem-flow and their 35 per cent interception almost all evaporates from the canopy.

Recent investigations of the water balance of forests provide some evidence that evergreen forests may permit 10–50 per cent more evapotranspiration than grass in the same climatic conditions. Grass normally reflects 10–15 per cent more solar radiation than coniferous tree species and hence less energy is available for evaporation. In addition, trees have a greater surface roughness, which increases turbulent air motion and, therefore, the evaporation efficiency. Evergreens also allow transpiration to occur throughout the year. Nevertheless many more detailed and careful studies are required to check these results and to test the various hypotheses.

d Modification of the thermal environment

From what has been said it is apparent that forest vegetation has an important effect on

Figure 7.18 Seasonal regimes of mean daily maximum and minimum temperatures inside and outside a birch–beech–maple forest in Michigan.

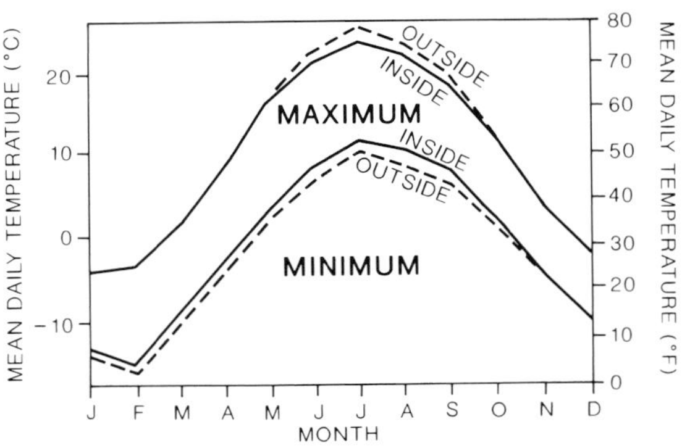

Source: After US Dept. *Agriculture Yearbook*, 1941.

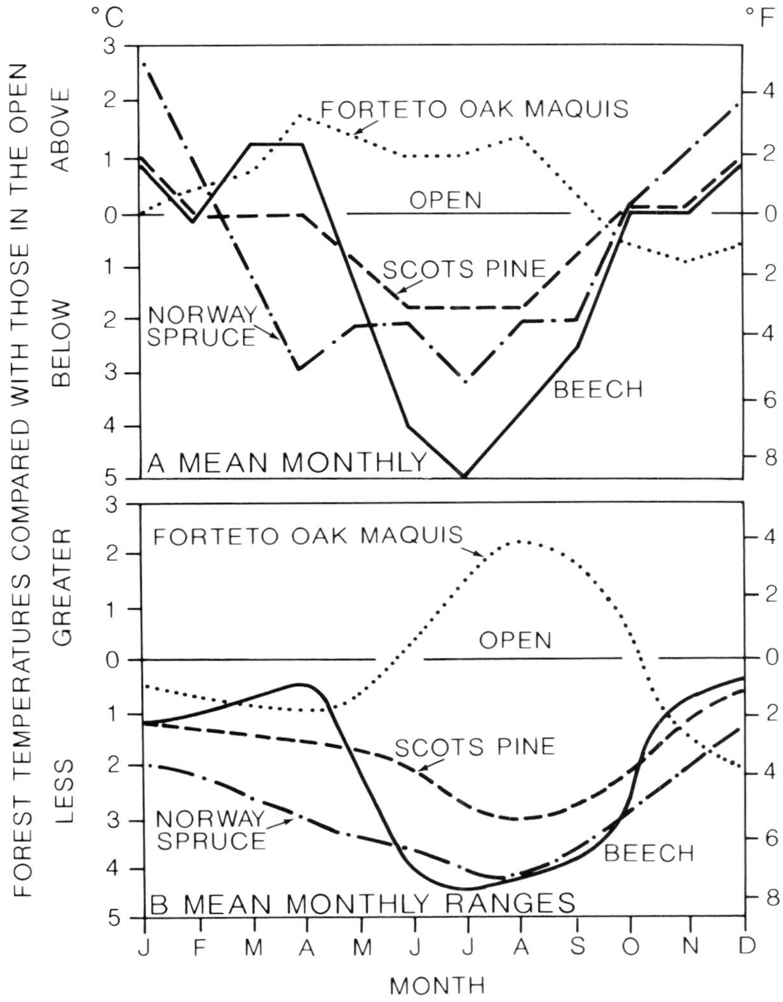

Figure 7.19 Seasonal regimes of (A) mean monthly temperatures and (B) mean monthly temperature ranges, compared with those in the open, for four types of Italian forest. Note the anomalous conditions associated with the *forteto* oak maquis, which transpires little.

Source: Food and Agriculture Organization of the United Nations 1962.

microscale temperature conditions. Shelter from the sun, blanketing at night, heat loss by evapotranspiration, reduction of wind speed, and the impeding of vertical air movement all influence the temperature environment. The most obvious effect of canopy blanketing is that inside the forest daily maximum temperatures are lower and minima are higher (see Figure 7.18). This is particularly apparent during periods of high summer evapotranspiration which depress daily maximum temperatures and cause mean monthly temperatures in tropical and temperate forests to fall well below those outside. In temperate forests at sea level the mean annual temperature may be about 0.6°C (1°F) less than that in surrounding open country, the mean monthly differences may reach 2.2°C (4°F) in summer but not exceed 0.1°C in winter, and on hot summer days the difference can be more than 2.8°C (5°F). Mean monthly temperatures and temperature ranges for temperate beech, spruce and pine forests are given in Figure 7.19, which also show that when trees do not transpire greatly in the summer (e.g. the *forteto* oak maquis of the Mediterranean) the high day temperatures reached in the sheltered woods

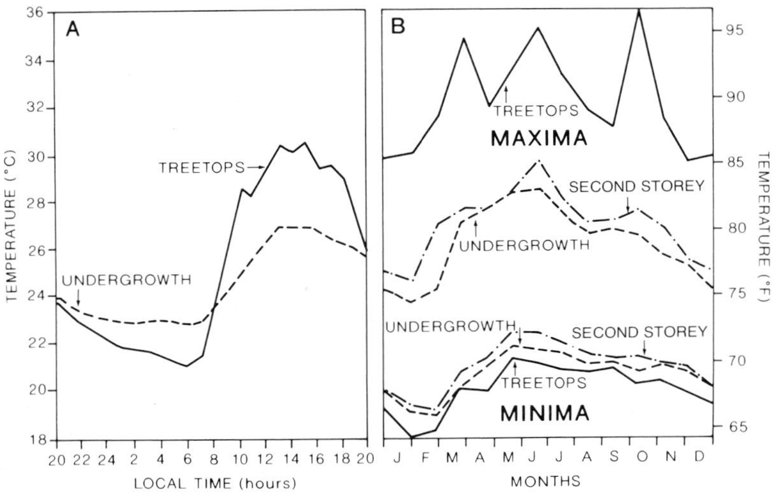

Figure 7.20 The effect of tropical rain-forest stratification on temperature.* A: Daily march of temperature (10–11 May 1936) in the tree tops (24 m) and in the undergrowth (0.7 m) during the wet season in primary rain forest at Shasha Reserve, Nigeria. B: Average weekly maximum and minimum temperatures in three layers of primary (Dipterocarp) forest, Mt Maquiling, Philippine Islands.

Sources: *After Richards 1952; A: after Evans; B: after Brown.

may cause mean monthly figures to reverse the trend exhibited by temperate forests. Even within individual climatic regions it is difficult to generalize, however, for at elevations of 1000 m the lowering of temperate forest mean temperatures below those in open country may be double that at sea level.

The complex vertical structure of forest stands is a further factor in complicating forest temperatures. Even in relatively simple stands vertical differences are very apparent. For example, in a ponderosa pine forest (*Pinus ponderosa*) in Arizona the recorded mean June–July maximum was increased by 0.8°C (1.5°F) simply by raising the thermometer from 1.5 to 2.4 m (5 to 8 ft) above the forest floor. In stratified tropical forests the thermal picture is much more complicated. The dense canopy heats up very much during the day and quickly loses its heat at night, showing a much greater diurnal temperature range than the undergrowth (see Figure 7.20A). Whereas daily maximum temperatures of the second storey are intermediate between those of the tree tops and the undergrowth, the nocturnal minima are higher than either tree tops or undergrowth because the second storey is insulated by trapped air both above and below (see Figure 7.20B). During dry conditions in the Amazonian rain forest there is a similar decoupling of the air in the lower storey from the upper two-thirds of the canopy, as reflected by the reduced amplitude of diurnal temperature range. At night, the pattern is reversed; temperatures respond to the radiative cooling of the lowest two-thirds of the vegetation canopy. Temperature variations within this deeper layer up to 25 m height are now decoupled from those in the tree tops and above.

D URBAN SURFACES

From a total of 4.4 billion in 1980, world population is forecast to increase to 8.2 billion in 2025, with the proportion of urban dwellers rising from 40 to 60 per cent during the same period. Thus, from about the turn of the century the majority of the human race will live and work in association with urban climatic influences.

The construction of every house, road or factory destroys existing microclimates and creates new ones of great complexity depending on the design, density and function of the building. Despite the great internal variation of urban climatic influences, it is possible to make certain generalizations regarding the effects of urban structures under three main headings:

1 Modification of atmospheric composition.
2 Modification of the heat budget.
3 Other effects of modifications of surface roughness and composition.

1 Modification of atmospheric composition

The city atmosphere is notoriously liable to pollution, particularly as the result of combustion, and this has effects that involve changes in the thermal properties of the atmosphere, cutting down the passage of sunlight, and providing abundant condensation nuclei. Although pollution poses a general problem for both city dwellers and planners, it is convenient to examine its sources under the following headings:

(a) *Aerosols.* The production of suspended particulate matter (measured in mg m^{-3} or μg m^{-3}) chiefly of carbon, lead and aluminium compounds and silica.

(b) *Gases.* The production of gases (measured in parts per million:ppm) can be viewed either from the traditional standpoint with its concentration on industrial and domestic coal burning and the production of such gases as sulphur dioxide (SO_2), or from the newer standpoint of petrol and oil combustion and the production of carbon monoxide (CO), hydrocarbons (Hc), nitrogen oxides (NO_x), ozone (O_3) and the like.

In dealing with atmospheric pollution it must be remembered, first, that the diffusion or concentration of pollutants is a function both of atmospheric stability (especially the presence of inversions) and of the features of horizontal air motion; second, that aerosols are removed from the atmosphere by settling out and by washing out; and, third, that certain gases are susceptible to complex chains of photochemical changes which may destroy some gases but produce others.

a Aerosols

As has already been pointed out (see Chapter 1, A.2 and A.4), the thermal economy of the globe is significantly affected by the natural production of aerosols which are deflated from deserts, erupted from volcanoes, produced by fires and so on. The overall thermal effect of low-level particulate aerosols is probably one of warming, due to increased absorption, and this may augment the warming associated with increasing amounts of carbon dioxide and certain trace gases (see Figure 1.4). Over the last century the average dust concentration has increased, particularly in Eurasia, due only in part to such eruptions as those of Mt Agung in Bali (1963) and El Chichen (1982). The proportion of atmospheric dust directly or indirectly attributable to human activity has been estimated at 30 per cent (see Chapter 1, A.4). As an example of the latter it is interesting that the North African tank battles of the Second World War disturbed the desert surface to such an extent that the material subsequently deflated was visible in clouds over the Caribbean.

The concentration of small Aitken nuclei (0.01–0.1 μm diameter) averages some 9500 cm^{-3} in the British countryside, but is typically 150,000 cm^{-3} and can reach 4,000,000 cm^{-3} in cities, as was measured near ground level in the industrial section of Vienna in 1946. Similarly, the concentration of larger particles (0.5–10 μm diameter) has been measured at 25–30 cm^{-3} in the city of Leipzig, as against 1–2 cm^{-3} in the rural environs. The greatest concentrations of smoke generally occur with low wind speeds, low vertical turbulence, temperature inversions, high relative humidities and air moving from the pollution sources of factory districts or areas of high density housing (see Plate 40). The character of domestic heating and power demands causes city smoke pollution to take on striking seasonal and diurnal cycles, with the greatest concentrations occurring at about 0800 in early winter (see Figure 7.21). The sudden morning increase is also partly a result of natural processes. Pollution trapped during the night beneath a stable layer a few hundred metres above the surface may be brought back to ground level when thermal

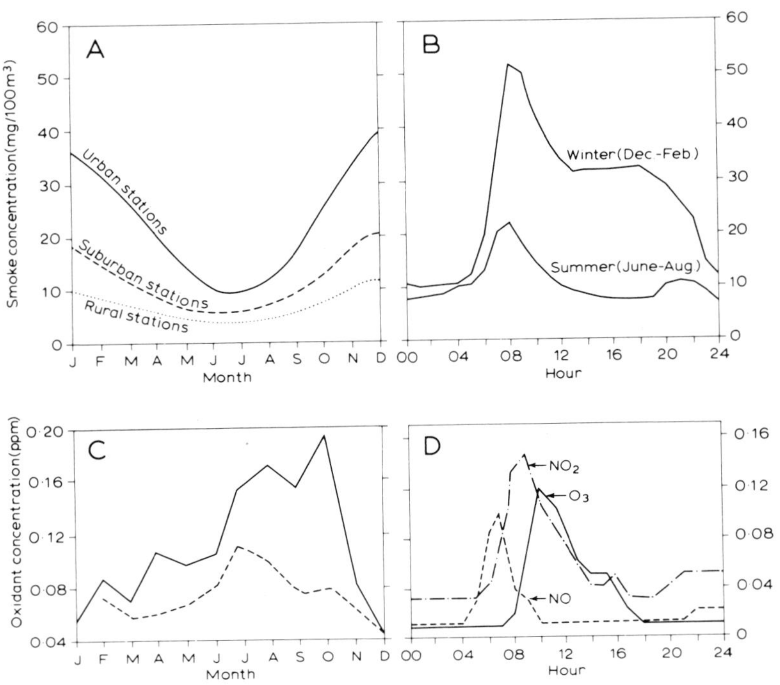

Figure 7.21 Annual and daily pollution cycles. A: Annual cycle of smoke pollution in and around Leicester, England, during the period 1937–9, before smoke abatement legislation was introduced. B: Diurnal cycle of smoke pollution at Leicester during summer and winter, 1937–9. C: Annual cycle of mean daily maximum 1-hour average oxidant concentrations for Los Angeles (1964–5) and Denver (1965) (dashed). D: Diurnal cycles of nitric oxide (NO), nitrogen dioxide (NO_2) and ozone (O_3) concentrations in Los Angeles on 19 July 1965.

Sources: A, B: After Meetham 1952 [*et al.* 1980]. C, D: After US DHEW 1970 and Oke 1978.

convection sets off vertical mixing. Figure 7.22 shows the striking results of the accumulation of air pollution that occurred over the British industrial city of Sheffield in mid-December 1964 during a period of cloudless skies, weak air flow, maximum long-wave radiation and the development of near-surface temperature inversions and radiation fog. These conditions were associated with a smoke concentration of 10 per cent above the monthly average on 14 December, which increased to 100 per cent above the average on 16 December.

The most direct effect of particulate pollution is to reduce incoming radiation and sunshine. Pollution, plus the associated fogs (termed *smog*), used to cause some British cities to lose 25–55 per cent of the incoming solar radiation during the period November to March (see Plate 40). In 1945 it was estimated that the city of Leicester lost 30 per cent of incoming radiation in winter, as against 6 per cent in summer. These losses are naturally greatest when the sun's rays strike the smog layer at low angle. Compared with the radiation received in the surrounding countryside, Vienna loses 15–21 per cent of radiation when the sun's altitude is 30°, but the loss rises to 29–36 per cent with an altitude of 10°. The effect of smoke pollution is dramatically illustrated by Figure 7.23 by comparing conditions before and after the enforcing of the Clean Air Act of 1956 in London. Before 1950 there was a striking difference of sunshine between the surrounding rural areas and the city centre (see Figure 7.23A) which could mean a loss of mean daily sunshine of 16 minutes in the outer suburbs, 25 minutes in the inner suburbs and 44 minutes in the city centre. It must be remembered, however, that smog layers also impeded the re-radiation of surface heat at night and that this blanketing effect contributed to higher night-time city temperatures. The use of smokeless fuels and other pollution controls cut London's total smoke emission from 1.4×10^8 kg (141,000 tons) in 1952 to 0.9×10^8 kg (89,000 tons) in 1960, and Figure 7.23B shows the increase in

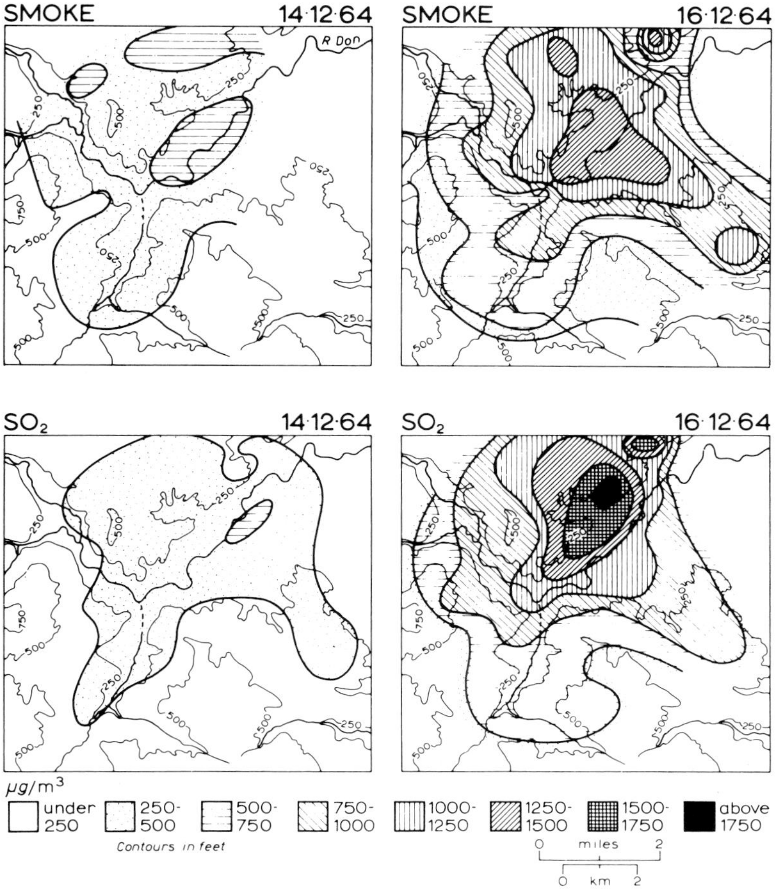

Figure 7.22 Average values of air pollution by smoke and sulphur dioxide for Sheffield, England, on 14 and 16 December 1964.

Source: From Garnett 1967.

average monthly sunshine figures for the period 1958–67 as compared with those of 1931–60.

The abundance of condensation nuclei in the city atmospheres, particularly those situated on low-lying land adjacent to large rivers, explains the former abundance of city fogs. Between August 1944 and December 1946, for example, suburban Greenwich had a monthly average of more than 20 days with good visibility at 0900 hours, whereas central London had less than 15. Occasionally very stable atmospheric conditions combine with excessive pollution production to give dense smog of a lethal character. During the period 5–9 December 1952 a temperature inversion over London caused a dense fog with visibility less than 10 m for 48 consecutive hours, resulting in 12,000 more deaths (mainly from chest complaints) during the period December 1952 to February 1953 compared with the same period the previous year. The close association of the incidence of fog with increasing industrialization and urbanization was well shown by the city of Prague, where the mean annual number of days with fog rose from 79 during the period 1860–80 to 217 during 1900–20.

b Gases

Along with the pollution by smoke and other particulate matter produced by the traditional urban and industrial activities involving the combustion of coal and coke, has been associated the generation of pollutant gases. Before the Clean Air Act it was estimated that, whereas

Figure 7.23 Sunshine in and around London. A: Mean monthly bright sunshine recorded in the city and suburbs for the years 1921–50, expressed as a percentage of that in adjacent rural areas. This shows clearly the effects of winter atmospheric pollution in the city. B: Mean monthly bright sunshine recorded in the city, suburbs and surrounding rural areas during the period 1958–67, expressed as a percentage of the averages for the period 1931–60. This shows the effect of the 1956 Clean Air Act in increasing the receipt of winter sunshine, in particular, in central London.

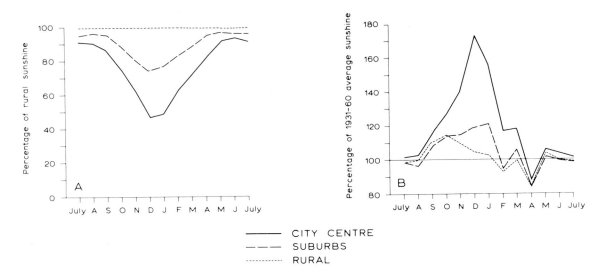

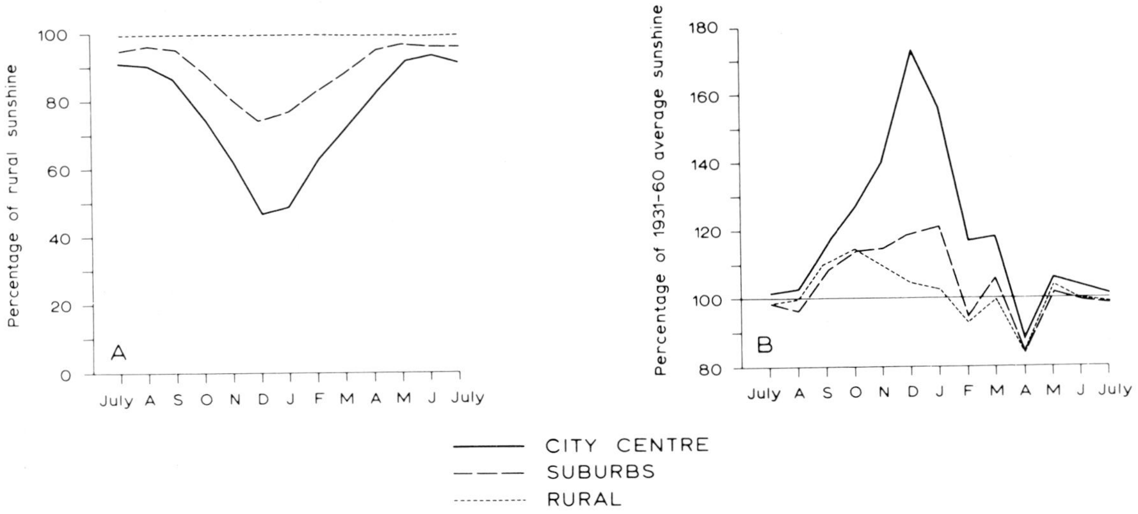

Sources: A: After Chandler 1965; B: after Jenkins 1969.

80–90 per cent of London's smoke was produced by domestic fires, these were responsible for only 30 per cent of the sulphur dioxide released in the atmosphere – the remainder being contributed by electricity power stations (41 per cent) and factories (29 per cent). Figure 7.22 illustrates the association between pollution by smoke and that by sulphur dioxide in Sheffield some 28 years ago; it is significant that on 16 December 1964 the concentration of sulphur dioxide in the city air had risen to three times the monthly average. After the early 1960s improved technology, the phasing out of coal fires, and anti-pollution regulations brought about a striking decline in sulphur dioxide pollution in many European and North American cities (see Figure 7.24). The effect of the regulations was not always clear, however, in that the decrease in London's atmospheric pollution was not apparent until some eight years after the introduction of the 1956 Clean Air Act, whereas in New York City the observed decrease began in the same year (1964) – *prior* to the air pollution control regulations there.

Urban complexes are being affected by a newer less obvious, but nevertheless equally serious, form of pollution resulting from the combustion of petrol and oil by cars, lorries and aircraft, as well as from petro-chemical industries. Los Angeles, lying in a topographically constricted basin and often subject to temperature inversions, is the prime example of such pollution, although this affects all modern cities to some extent. In Los Angeles seven million people use some four million private cars, consuming 30 million litres of gasoline per day and producing more than 12,000 tonnes of pollutants. To this is added the results of the consumption of 0.5 million litres per day of diesel fuel by 13,500 lorries and buses and 2.5 million litres of aviation fuel consumed in the vicinity of the city. Even with controls, 7 per

Figure 7.24 Annual mean concentration of sulphur dioxide ($\mu g\ m^{-3}$) measured in New York and London during a 25–30-year period. These show dramatic decreases of urban pollution by SO_2.

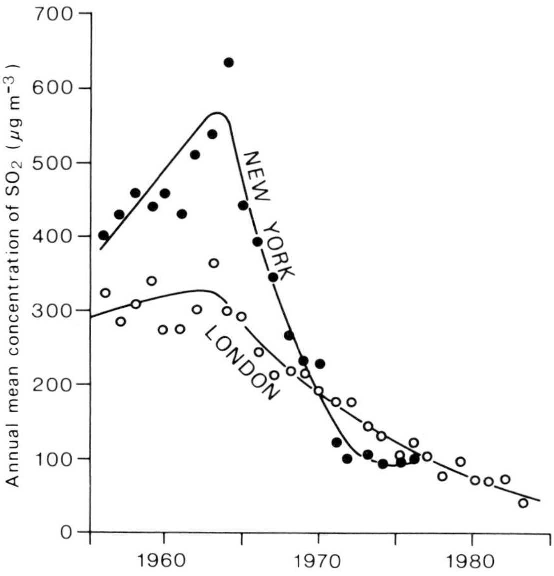

Source: From Brimblecombe 1986.

cent of the gasoline from private cars is emitted in an unburned or poorly-oxidized form, another 3.5 per cent as photochemical smog and 33–40 per cent as carbon monoxide. Smog involves at least four main components: carbon soot, particulate organic matter (POM), sulphate (SO_4^{2-}) and nitrate. Half of the aerosol mass is typically POM and sulphate. However, there are important regional differences. For example, the sulphur content of fuels used in California is lower than in the eastern United States and Europe and NO_2 emissions greatly exceed those of SO_2 in California. The production of the Los Angeles smog which, unlike traditional city smogs, occurs characteristically during the daytime in summer and autumn (see Figure 7.21C and D) is the result of a very complex chain of chemical reactions termed the disrupted photolytic cycle (see Figure 7.25). Ultraviolet radiation (0.37–0.42 μm) dissociates natural NO_2 into NO and O. Monatomic oxygen (O) may then combine with natural oxygen (O_2) to produce ozone (O_3). The ozone in turn reacts with the artificial NO to produce NO_2 (which goes back into the photochemical cycle forming

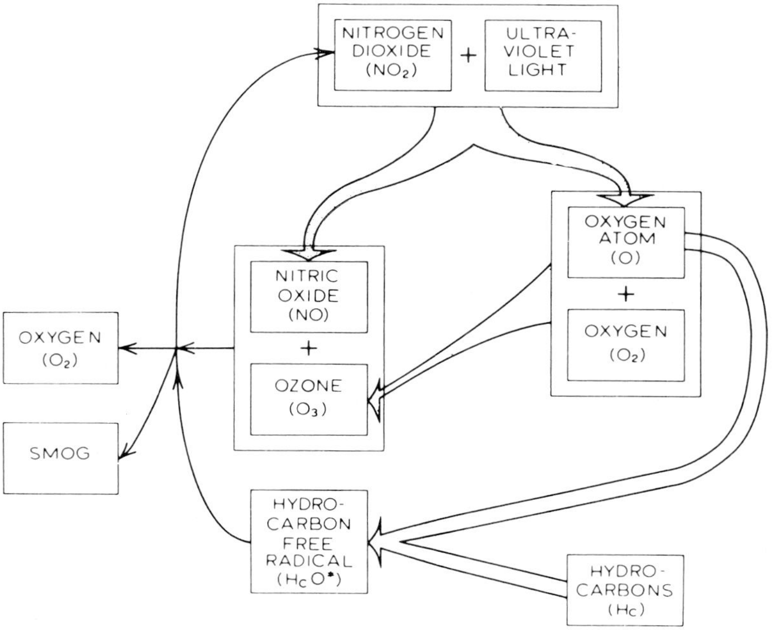

Figure 7.25 The NO_2 photolytic cycle disrupted by hydrocarbons to produce photochemical smog.

Sources: US DHEW 1970 and Oke 1978.

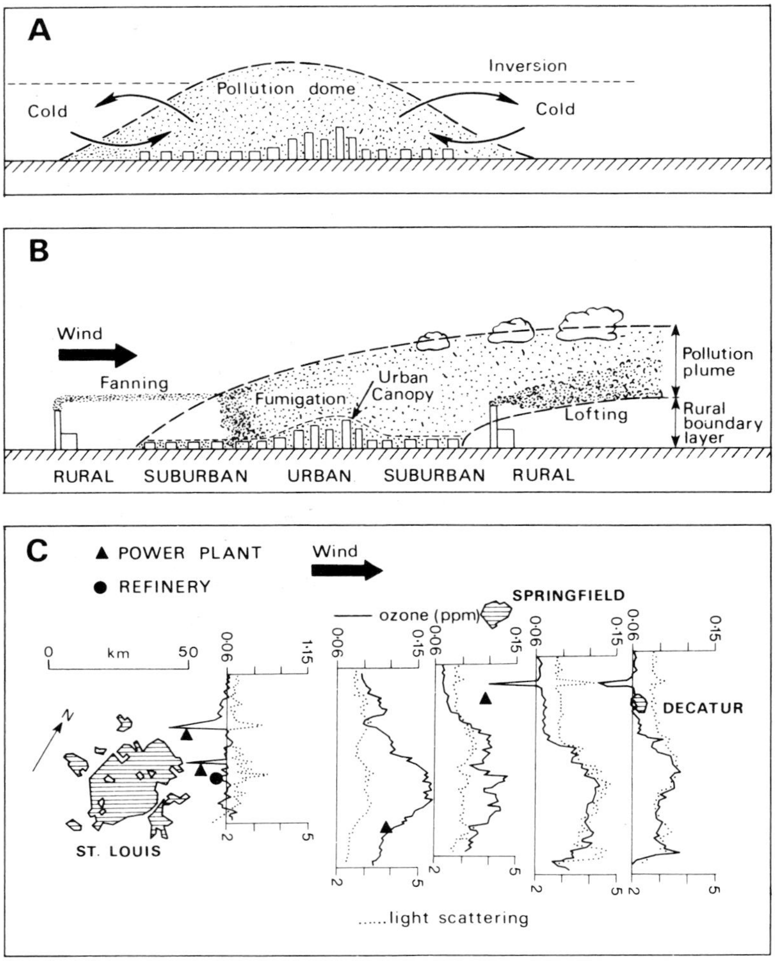

Figure 7.26 Configurations of urban pollution. A: Urban pollution dome. B: Urban pollution plume in a stable situation (i.e. early morning following a clear night). Fanning is indicative of vertical atmospheric stability. C: Pollution plume north-east of St Louis, Missouri, on 18 July 1975.

Sources: B: After Oke, 1978. C: After White *et al.* 1976 and Oke 1978.

a dangerous positive feedback loop) and oxygen. The hydrocarbons produced by the combustion of petrol combine with oxygen atoms to produce the hydrocarbon-free radical H_cO^*, and these react with the products of the O_3–NO reaction to generate oxygen and photochemical smog. This smog exhibits well-developed annual and diurnal cycles in the Los Angeles basin (see Figure 7.21C and D). Annual levels of photochemical smog pollution in Los Angeles (derived from averages of the daily highest hourly figures) are greatest in late summer and autumn when clear skies, light winds and temperature inversions combine with amounts of high solar radiation. The diurnal figures of variations in individual components of the disrupted photolytic cycle reflect the complex reactions with, for example, an early morning concentration of NO_2 due to the build-up of traffic, and a peak of O_3 when receipts of incoming radiation are high. The effect of this smog is not only to modify the radiation budget of cities but to produce a human health hazard: in Tokyo, for instance, citizens sometimes wear respiratory masks in the street in self-defence!

c Pollution distribution

Polluted atmospheres commonly assume well-

marked physical configurations around urban areas, which are very dependent upon environmental lapse rates, particularly the presence of temperature inversions and on wind speed. A pollution dome forms as a result of the collection of pollution below an inversion forming the urban boundary layer (see Figure 7.26A). A wind speed as low as 2 m s^{-1} (5 mph) is sufficient to displace the Cincinnati pollution dome downwind, and a wind speed of 3.5 m s^{-1} (8 mph) will disperse it into a plume. Figure 7.26B shows a section of an urban plume with the volume above the urban canopy of the building tops filled by buoyant mixing circulations. *Fumigation* is the term used when an inversion lid prevents upward dispersion, but lapse conditions due to morning heating of the surface air allow convective plumes and associated downdraughts to bring pollution back to the surface. Downwind, *lofting* occurs above the temperature inversion at the top of the rural boundary layer dispersing the pollution upwards. Figure 7.26C illustrates some of the features of a pollution plume up to 160 km (100 miles) downwind of St Louis on 18 July 1975. In view of the complexity of photochemical reactions, it is of note that ozone increases downwind due to photochemical reactions within the plume but decreases over power plants as the result of other reactions with the

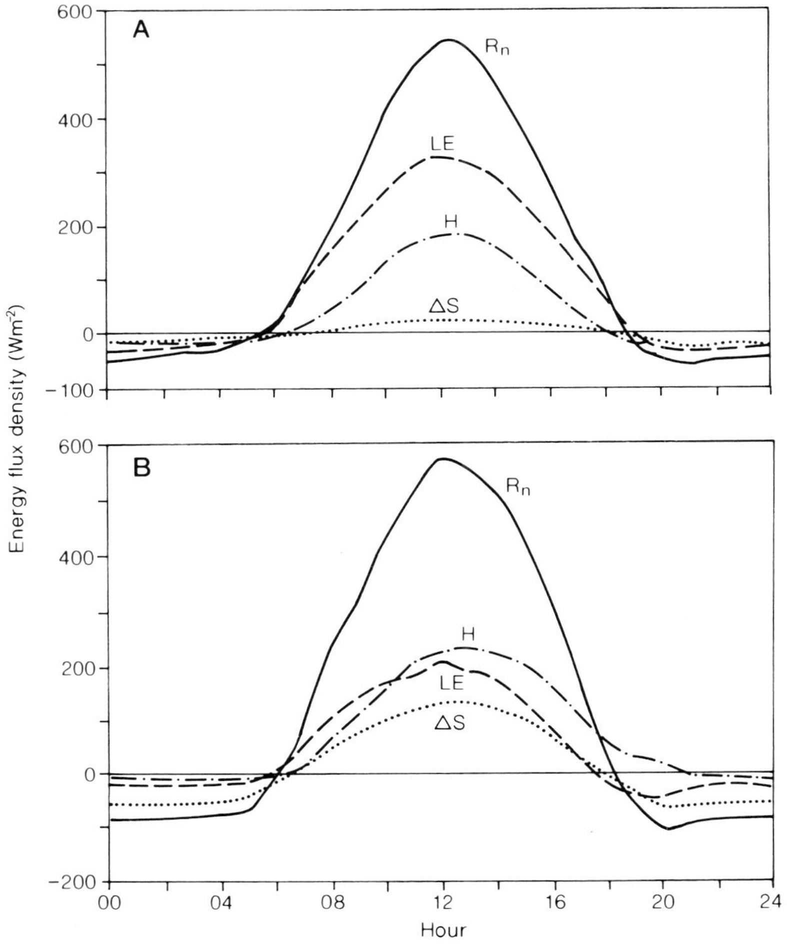

Figure 7.27 Average diurnal energy balances for (A) rural and (B) suburban locations in Greater Vancouver for 30 summer days.

Source: After Clough and Oke, from Oke 1988.

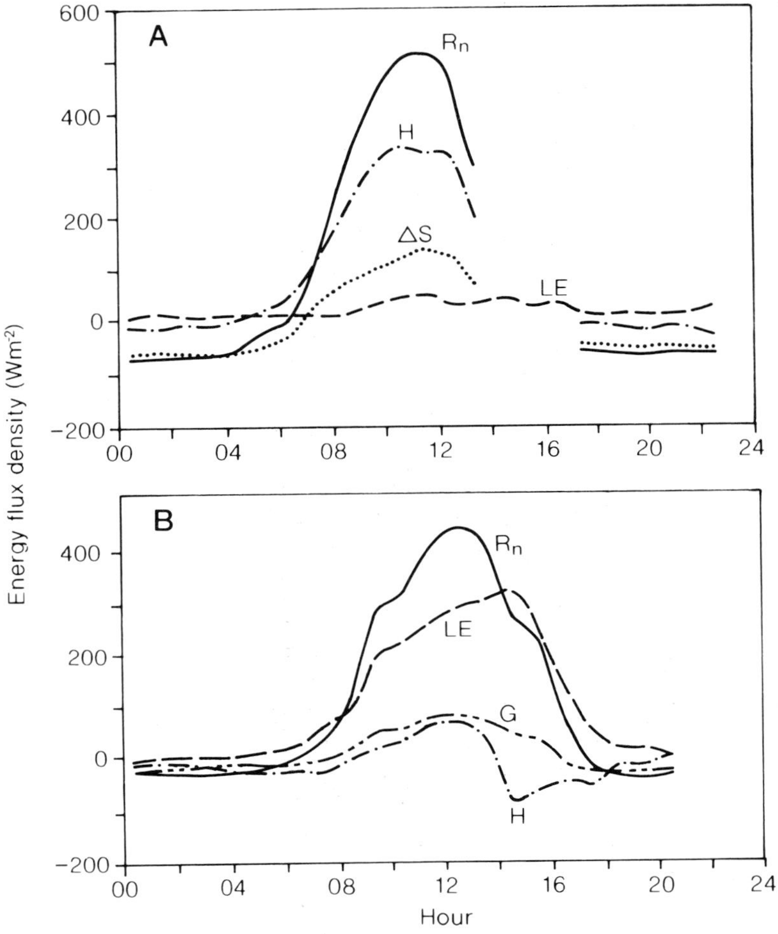

Figure 7.28 Diurnal energy balances for (A) the top of an urban canyon and (B) an irrigated suburban lawn in Vancouver, British Columbia, for 9–11 September 1973 and 6 August 1978, respectively.

Source: After Oke and Nunez, from Oke 1988.

emissions. This plume was observed to stretch for a total distance of 240 km (150 miles), but under conditions of an intense pollution source, steady large-scale surface airflow and vertical atmospheric stability pollution plumes may extend downwind for hundreds of kilometres. Plumes originating in the Chicago–Gary conurbation have been observed from high-flying aircraft to extend almost to Washington, DC, 950 km (600 miles) away.

2 Modification of the heat budget

The energy balance of the built surface is similar to those described earlier in this chapter except for the heat production resulting from human energy consumption by combustion, which may even exceed R_n during the winter in some cities. Although R_n may not be greatly different from that in nearby rural areas (except during times of significant pollution) heat storage by surfaces is greater (20–30 per cent of R_n by day) leading to greater nocturnal values of H and, in particular, LE is much less in city centres. After long dry periods evapotranspiration may be zero in city centres, except for certain industrial operations, and in the case of irrigated parks and gardens where LE may exceed R_n. This lack of LE means that by day 70–80 per cent of R_n may be transferred to the atmosphere as sensible heat (H).

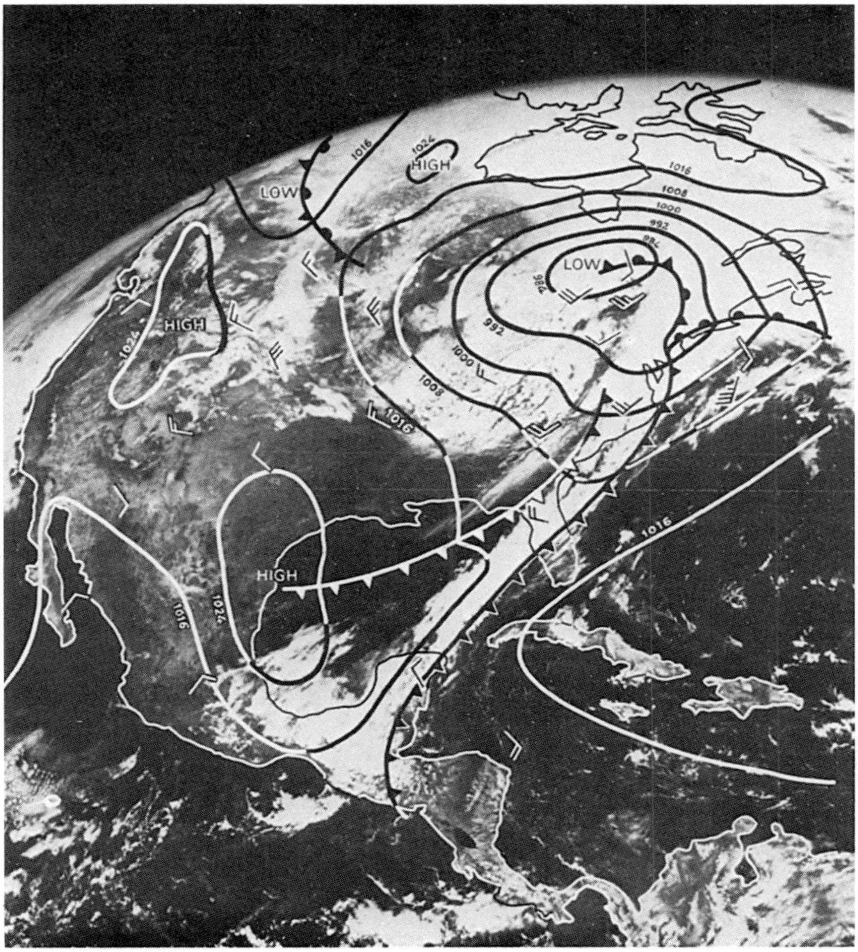

Plate 22 Satellite photograph of North and Central America taken from 37,000 km (23,000 miles) on an April day. A depression lies to the south-east of Hudson Bay and a belt of rain or drizzle marks the cold front from Yucatan to New England. This cloud band suggests high-level tropical/mid-latitude interaction. The Great Lakes are snowy and windy but elsewhere conditions are mostly clear with light winds (*NOAA photograph; Courtesy National Geographic Society*).

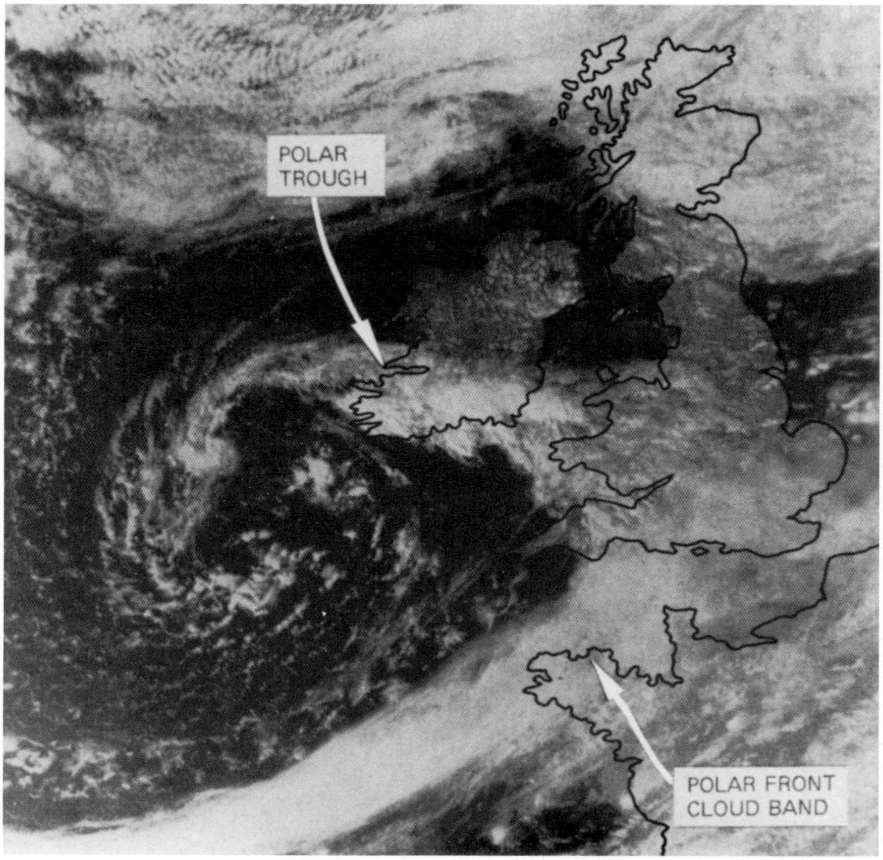

Plate 23 Infra-red photograph from a NOAA satellite taken at 1502 hours GMT on 9 September 1983 showing a convective cloud band associated with a polar trough conveyor belt (low-level jet) wrapped around the leading edge of a cold pool lying to the north-west of a stratiform cloud band marking a polar front conveyor belt (upper-level jet) (*from Browning 1985, fig. 21b, p. 314; see also Figure 4.14, this volume*). (*Permission from Dept of Electrical Engineering and Electronics, University of Dundee.*) (*Reproduced by permission of the Controller of Her Majesty's Stationery Office.*)

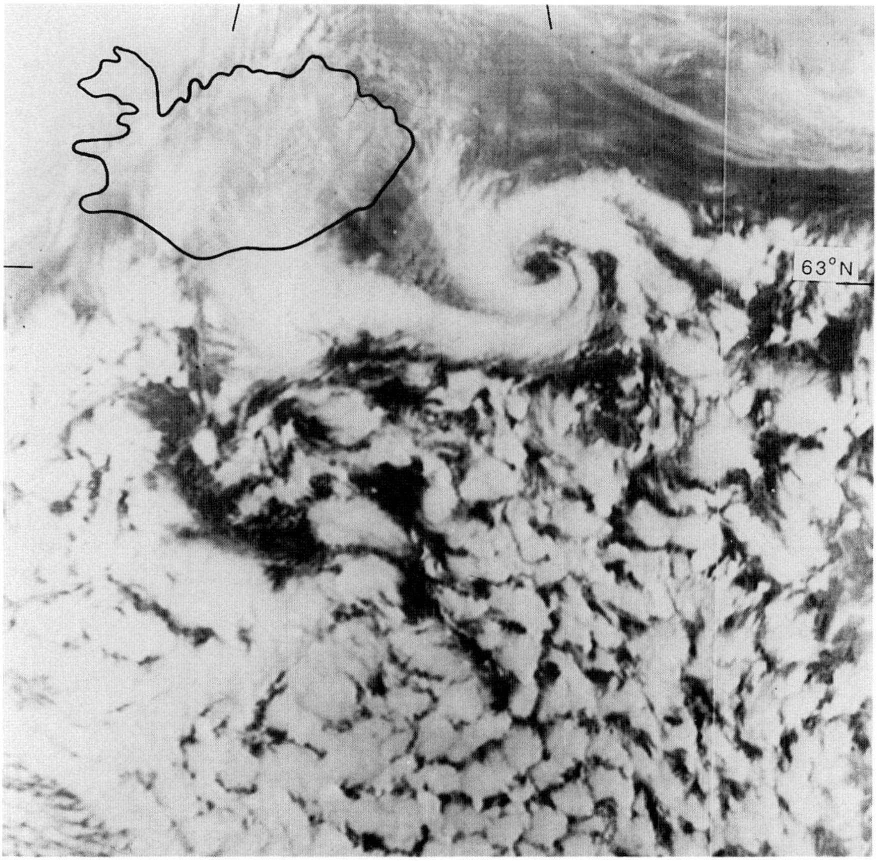

Plate 24 A polar low near Iceland, 14 January 1984, as seen on a visible band DMSP satellite image. This mesoscale low and the closed cellular cloud patterns to the south developed in a northerly airflow behind an occluded depression situated over the coast of Norway (*courtesy National Snow and Ice Data Center, University of Colorado, Boulder*).

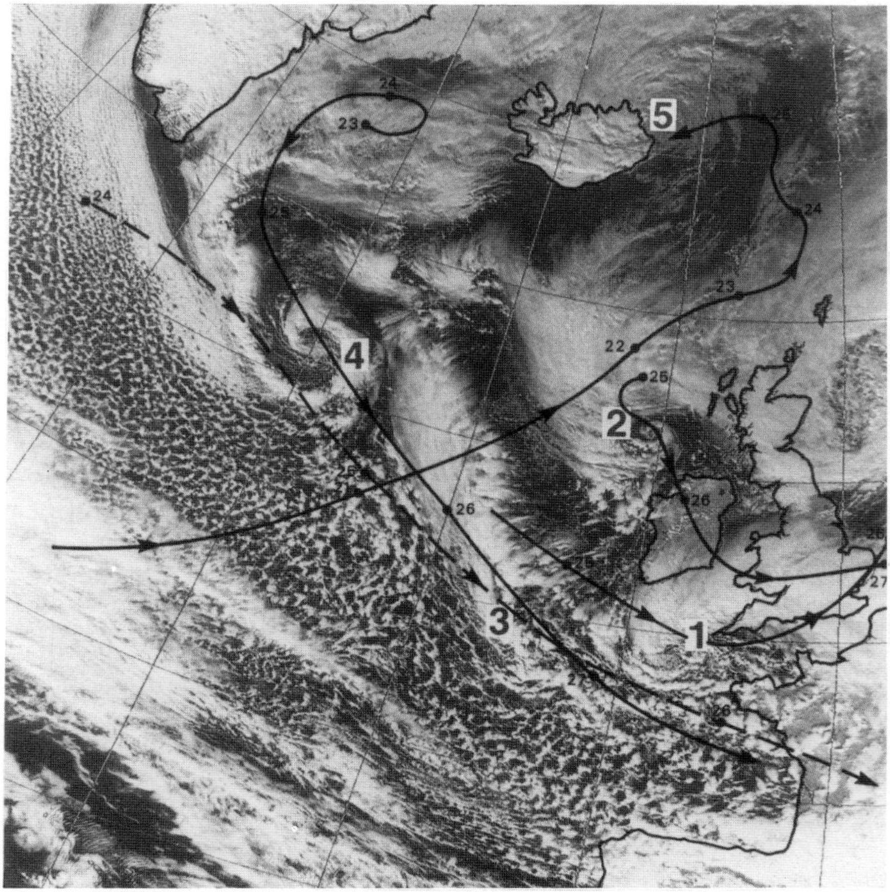

Plate 25 Visible image from NOAA-11 at 1418 GMT on 25 February 1989 showing a cold, north-westerly airstream over the North Atlantic with open cellular convection, bounded on the south by a weak surface front. To the north-east are several low pressure systems: 5 is the filling remains of a frontal depression which crossed the area in a north-easterly direction between the 21 and 25 February, behind which the cold north-westerly outbreak occurred. The passage of this depression assisted the formation of four polar low vortices (1–4) which formed between 24 and 25 February and moved in a south-easterly direction, suppressing convection in their vicinity; 3 moved the fastest but 1 had a central pressure of only 950 mb and brought strong winds and damage to the south of France. The dates marked on the trajectories of the low-pressure centres show their positions in the early morning of those days (*photograph courtesy of NERC Satellite Station, University of Dundee; Description from the* Meteorological Magazine*, 1989, p. 116, by permission of the Controller of Her Majesty's Stationery Office*).

Plate 26 View from 5 km (3 miles) distance of a tornado north-east of Tracy, Minnesota, on 13 June 1968 (*photograph courtesy of Eric Lantz and Associated Press*). A convectively-unstable tongue of warm air extended north from Texas and by mid-afternoon its temperature had risen to 32°C (90°F), and severe thunderstorms had set in ahead of a cold front lying to its west. Surface pressure continued to drop in this belt, which was supported by a trough at 500 mb and surmounted by a jet of over 45 m s^{-1} (100 mph) at the tropopause extending from Oregon to eastern Canada. Thunderstorm activity reached a maximum at about 1800 hours as the unstable belt moved into Minnesota, and individual cells were shown by radar to have built up to over 15,240 m (50,000 ft). This combination of conditions was ideal for tornado inception and on that afternoon thirty-four funnels were sighted within 480 km (300 miles) of Minneapolis. The Tracy tornado appeared 13 km (8 miles) south-west of the town at 1900 hours and moved north-east at 13 m s^{-1} (35 mph) for 21 km (13 miles), cutting a 90 to 150-m (300 to 500-ft) wide belt of total destruction through Tracy, killing nine people, injuring 125 and causing $3 million worth of damage. Unlike most tornadoes, it did not lift off the ground on encountering the rough urban surface but 'dug in' for its whole course until it suddenly dissolved a few seconds after the photograph was taken (*description courtesy of the Director, National Severe Storms Forecast Center, Kansas City*).

Plate 27 Visible image taken by METEOSAT-2, geostationary some 35,900 km (22,300 miles) above the equator, at 1130 hours GMT on 25 September 1983. A broad ITCZ lies across West and East Africa exhibiting centres of varied convective activity including cumulus towers over Central and East Africa. The subtropical high-pressure belt is particularly well developed in the northern hemisphere, as are depressions in the belts of the northern and southern westerlies. A broad wedge of cloud extends west over the South Atlantic from Angola and Gabon in the tropical easterly flow and there is a narrow belt of low cloud in the zone of coastal up-welling off Namibia (*METEOSAT image supplied by the European Space Agency*).

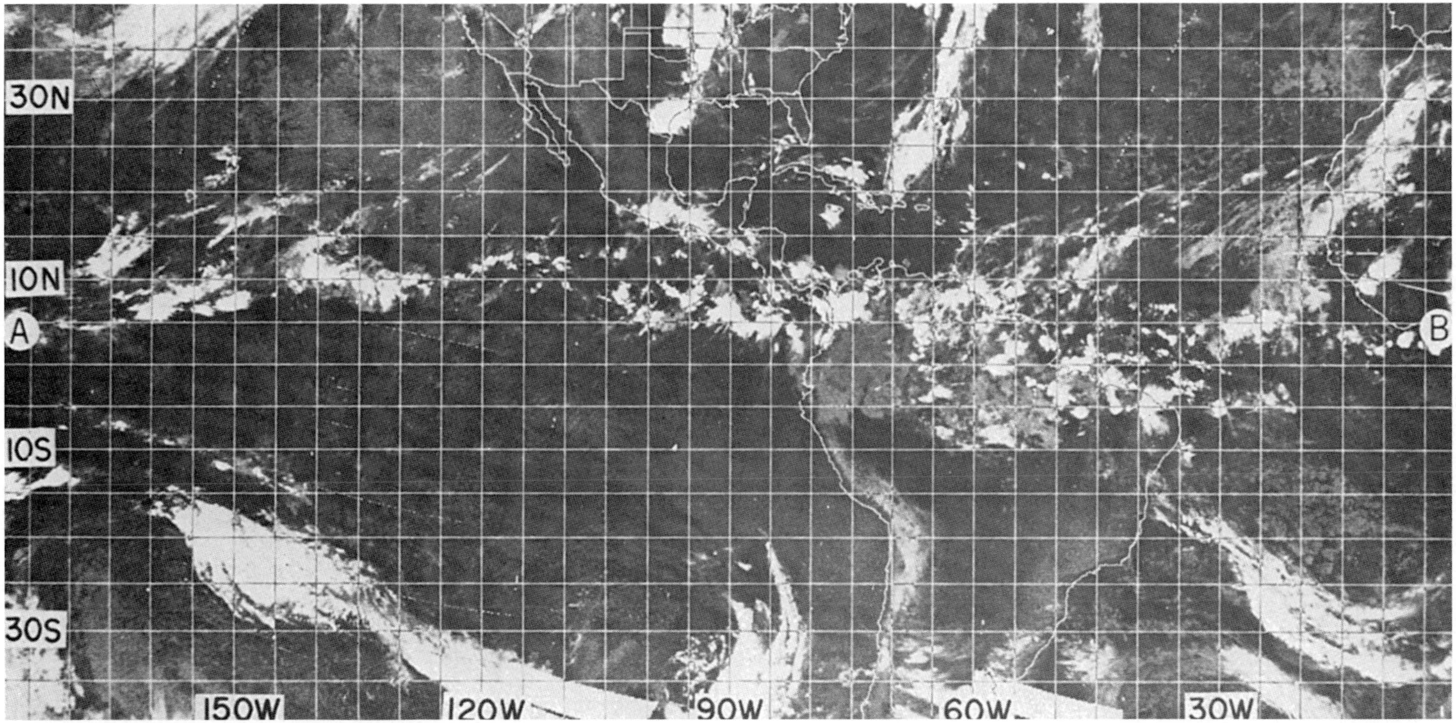

Plate 28 Infra-red photograph showing the Intertropical Convergence Zone on 1 June 1971, along which active thunderstorms appear as bright spots. There is also a cyclonic cloud band in the western South Pacific representing the South Pacific Convergence Zone (*NOAA-1 photograph, World Meteorological Organization 1973*).

Plate 29 View south over Florida from the Gemini V manned spacecraft at an elevation of 180 km (112 miles) on 22 August 1965, with Cape Kennedy launching site in the foreground. Cumulus clouds have formed over the warmer land, with a tendency to align in east–west 'streets', and are notably absent over Lake Okeechobee. In the south, thunder-head anvils can be seen (*NASA photograph*).

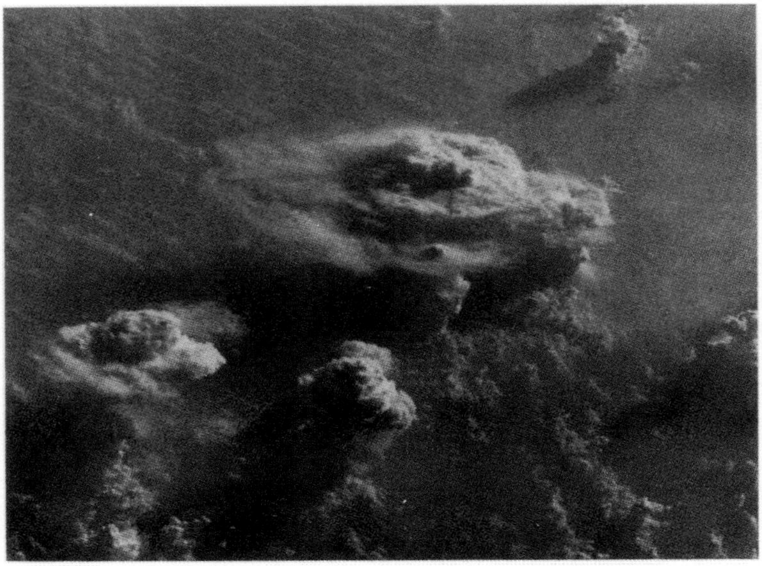

Plate 30 Cumulus towers with powerful thunderstorms along the ITCZ over Zaire photographed in April 1983 from the Space Shuttle at an elevation of 280 km (175 miles). The largest tower shows a double mushroom cap reaching to more than 15,240 m (50,000 ft) and the symmetrical form of the caps indicates a lack of pronounced airflow at high levels (*Courtesy NASA*).

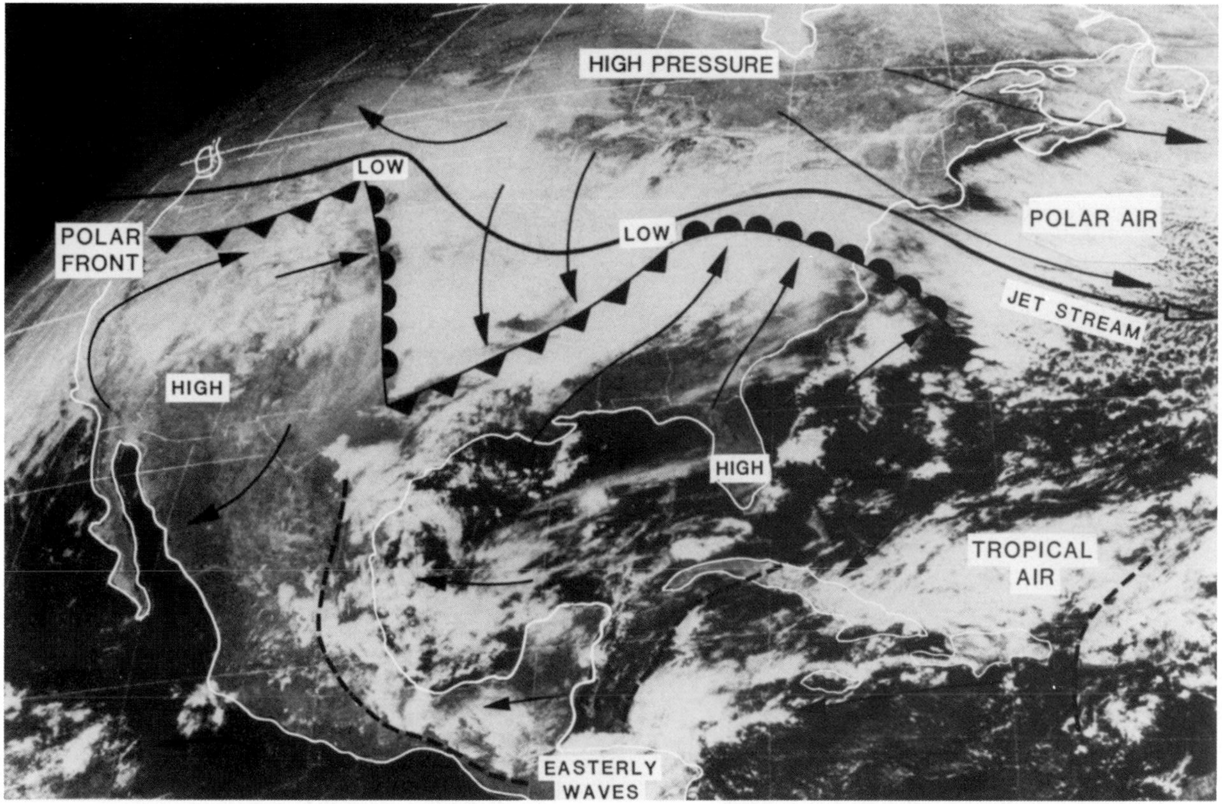

Plate 31 Satellite photograph of North America taken at 1700 hours GMT on 12 February 1979 from a GOES weather satellite located some 35,800 km (22,250 miles) above the equator. Two depressions are located below an upper jet stream located between polar and subtropical high-pressure cells. The more westerly depression is forming in the lee of the Rocky Mountains; cold polar air streaming off New England is becoming cloudy over the warmer sea; weak easterly waves (dashed) have developed equatorward of the subtropical high-pressure cells over the Caribbean and Central America (*courtesy NOAA*).

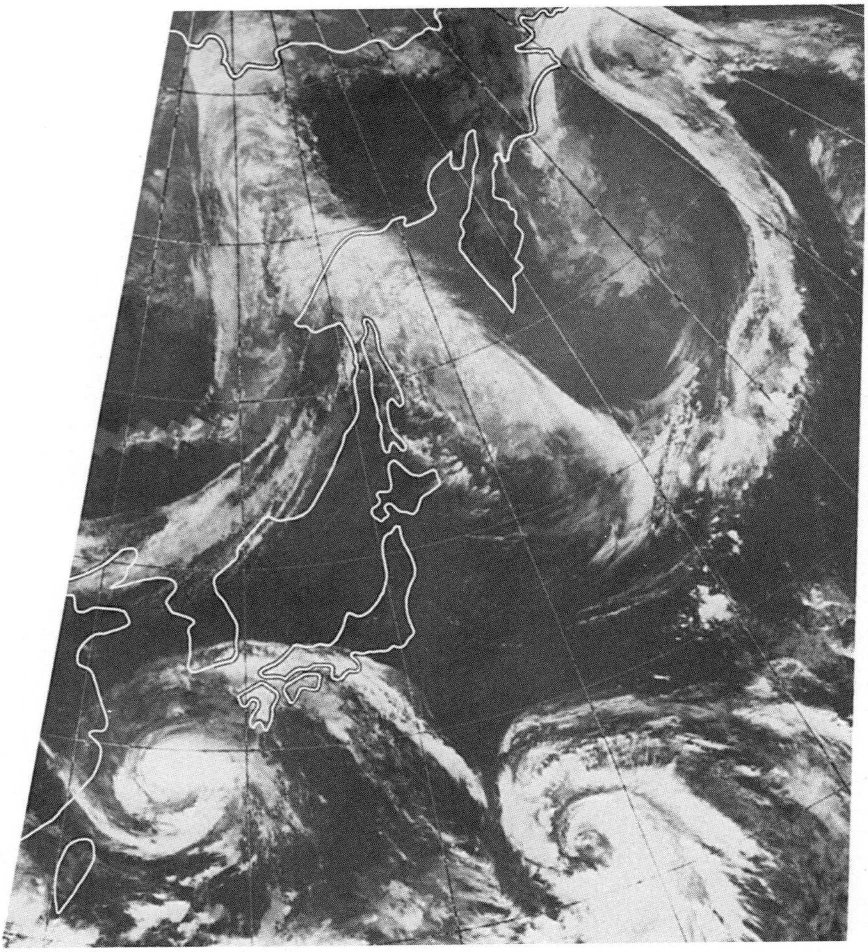

Plate 32 A satellite infra-red mosaic of eastern Asia and the western North Pacific showing two mid-latitude depression systems and typhoons Wendy (28°N, 126°E) and Virginia (22°N, 147°E) on 29 July 1978, about 0900 local time (Tokyo). The typhoons had maximum winds of about 36 m s^{-1} (80 mph) and sea-level pressure minima of about 965 mb (Wendy) and 975 mb (Virginia). A subtropical high-pressure ridge about 35°N separates the tropical and mid-latitude storms minimizing any interaction (*Defense Meteorological Satellite Program imagery, National Snow and Ice Data Center, University of Colorado, Boulder*).

Plate 33 An air view looking south-eastwards towards the line of high cumulus towers marking the convergence zone near the Wake Island wave trough shown in Figure 6.8 (*from Malkus and Riehl 1964*).

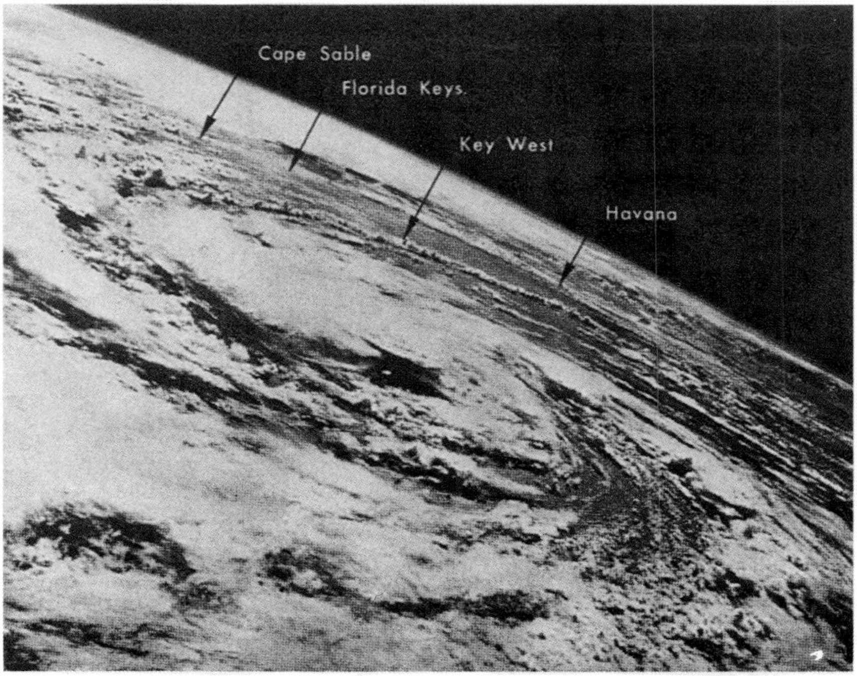

Plate 34 Hurricane 'Gladys' just west of Florida, photographed towards the south from Apollo 7 manned spacecraft at an elevation of 179 km (111 miles) on 17 October 1968. At this time, wind speeds up to 40 m s^{-1} (90 mph) were reported (*NASA photograph*).

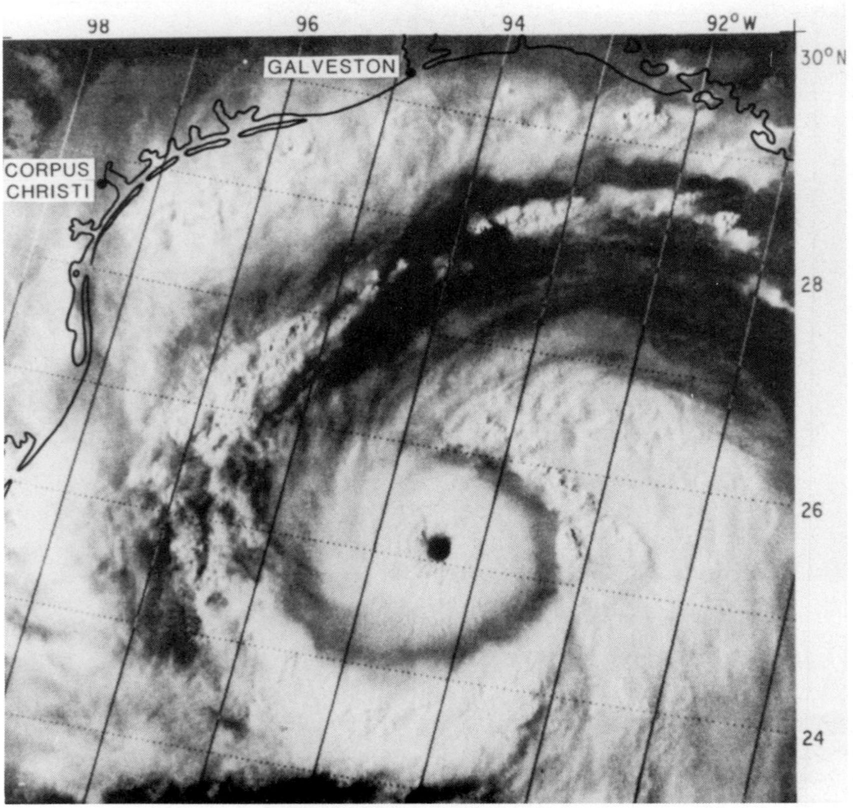

Plate 35 Visible satellite image of Hurricane Allen, located south of the Texas coast, taken at 2123 hours GMT on 8 August 1980 showing the clear eye (*photograph courtesy National Hurricane Center, Miami, Florida, USA*).

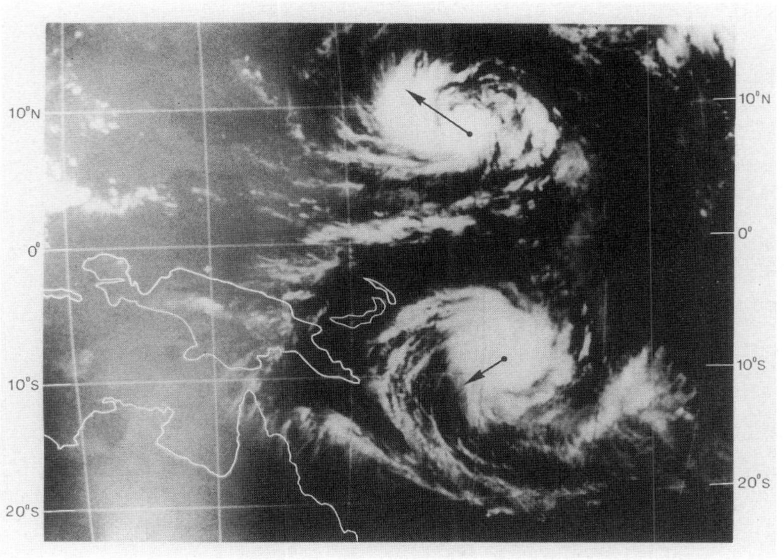

Plate 36 Two hurricanes with opposite rotary circulations formed simultaneously on either side of the equator just east of New Guinea. Cyclone Namn is in the southern hemisphere. The GMS-3 image was taken at 2100 hours LST (1000 hours GMT) on 18 May 1986. The arrows show the directions and distances of the movement of the hurricane eyes during the subsequent 15 hours (*courtesy National Space Development Agency of Japan; description courtesy of Mr S. Jones*).

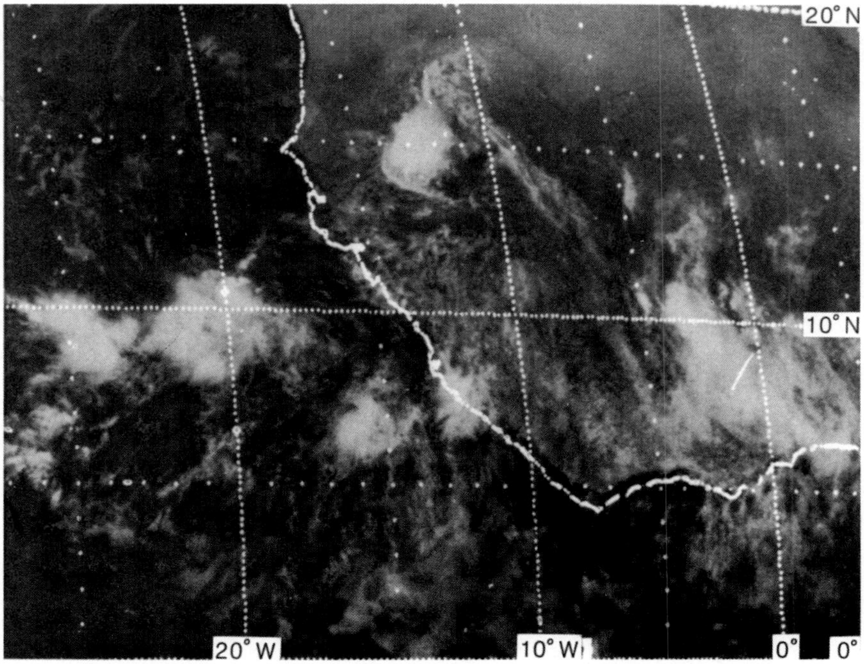

Plate 37 Visible satellite image showing five large tropical cloud clusters topped by cirrus shields situated between latitudes 5° and 10°N in the vicinity of West Africa, together with one squall-line cloud cluster at 15°N having a well-defined arc cloud squall line on its leading (south-west) edge. Taken by SMS-1 satellite at 1130 hours GMT on 5 September 1974 (*courtesy NOAA*).

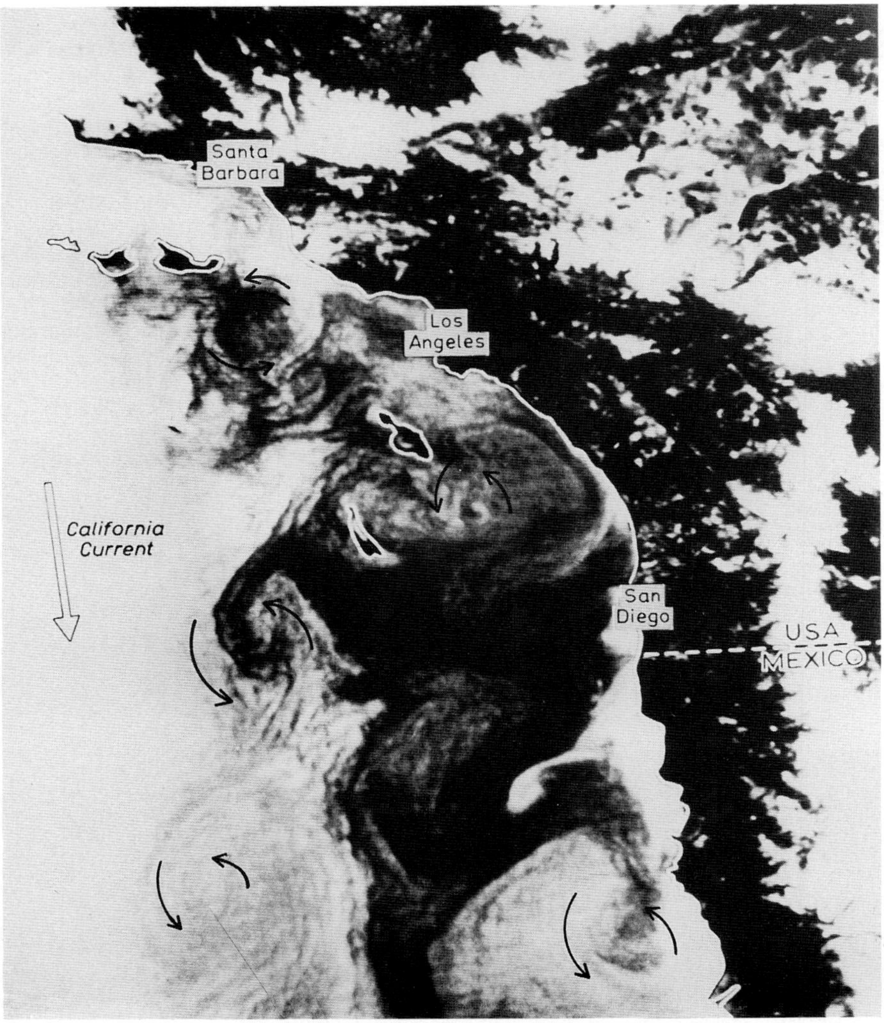

Plate 38 Infra-red photograph showing large vortices of cold water (*light*) up-welling in the warmer surface coastal waters (*dark*) off southern California. The colder offshore California Current is clearly shown (*NASA photograph*).

Plate 39 Hydraulic jump in the lee (i.e. eastern side) of Elk Mountain, Wyoming, at 1345 hours on 1 February 1973, looking south-west from the north-eastern edge of the Elk Mountain (*photograph courtesy Dr August H. Auer, Jr, University of Wyoming*).

Plate 40 View westwards over Edinburgh showing Arthur's Seat in the foreground and the Castle rising above urban smog in the middle distance. The smog thickens to the north (*right*) over the industrial suburb of Leith, merging with sea mist over the Firth of Forth at the far right. This photograph was taken some considerable time ago (*courtesy Aerofilms Ltd*).

Beneath the urban canopy, the microclimates of the streets and 'urban canyons' are dominated by the effects of elevation and aspect on the energy balance, which may vary strikingly within even one street.

The complex nature of the urban modification of the heat budget is well illustrated by observations made in and around the city of Vancouver. Figure 7.27 compares the summer diurnal energy balances for rural and suburban locations, with the former showing considerable consumption of net radiation (R_n) by evapotranspiration (*LE*) during the day, giving lower temperatures than in the suburbs, whereas the suburban gain of net radiation is greater by day but the loss greater during evening and night due to release of turbulent sensible heat from the suburban fabric (i.e. ΔS negative). The diurnal energy balance for a regularly-watered suburban lawn (see Figure 7.28B) shows features similar to those of irrigated crops (see Figure 7.10) in that some three-quarters of the net radiation is consumed by evaporative processes (*LE*) which exceed R_n after about 1400 hours, the deficiency being made up by the conduction of sensible heat (*H*) from the air to the ground after that time. The symmetrical energy balance for the dry top of an urban canyon (see Figures 7.28A and 7.29C) shows that two-thirds of the net radiation is transferred into atmospheric sensible heat and one-third into heat storage in the building material (ΔS). Figure 7.29A–C rationalizes this canyon-top energy balance symmetry in terms of the behaviour of its components (i.e. canyon floor, east-facing wall and west-facing wall) making up a white, windowless urban canyon in early September aligned north–south and with a canyon height equal to its width. The east-facing wall receives the first radiation in the early morning, reaching a maximum at 1000 hours, but being totally in shadow after 1200 hours. The low total R_n is because the east-facing wall is often in shadow. The street level (i.e. canyon floor) is sunlit only in the middle of the day and R_n and *H* dispositions are symmetrical. The third component of the urban canyon total energy balance is the west-facing wall which is a mirror-image (centred on noon) of that of the east-facing wall. Consequently, the symmetry of the street-level energy balance and the mirror images of the east- and west-facing walls produces the symmetrical diurnal energy balance of R_n, *H* and ΔS observed at the canyon top.

The thermal characteristics of urban areas are in marked contrast to those of the surrounding

Figure 7.29 Diurnal variation of energy balance components for a N–S oriented urban canyon in Vancouver, British Columbia, having white concrete walls, no windows, and a width/height ratio of 1:1, during the period 9–11 September 1973. A: The average for the E-facing wall. B: The average for the floor. C: Averages of fluxes through the canyon top.

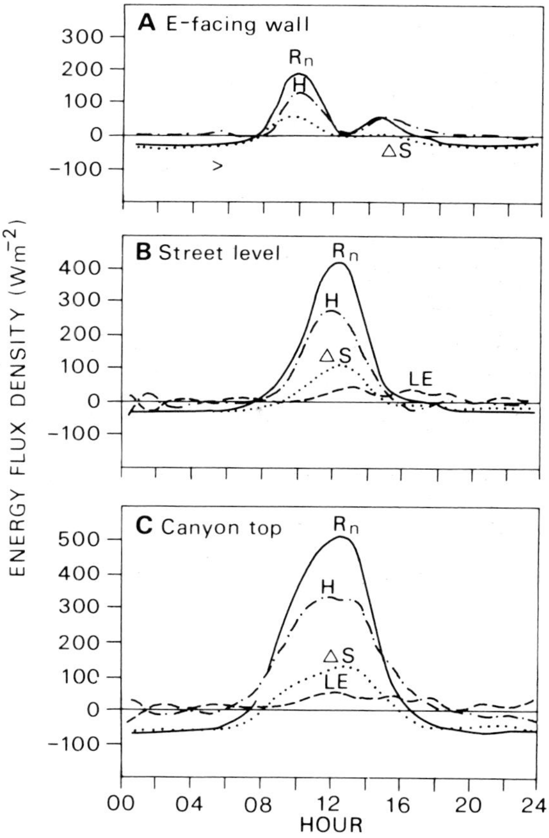

Source: After Nunez and Oke, from Oke 1978.

countryside, and the generally higher urban temperatures are the result of the interaction of the following factors:

1 Changes in the radiation balance due to atmospheric composition.
2 Changes in the radiation balance due to the albedo and thermal capacity of urban surface materials, and to canyon geometry.
3 The production of heat by buildings, traffic and industry.
4 The reduction of heat diffusion due to changes in airflow patterns as the result of urban surface roughness.
5 The reduction in thermal energy required for evaporation and evapotranspiration due to the surface character, rapid drainage and generally lower wind speeds of urban areas.

Consideration of the last two factors will be left to D.3, this chapter.

a Atmospheric composition

Air pollution makes the transmissivity of urban atmospheres significantly less than that of nearby rural areas. For example, during the period 1960–9 the atmospheric transmissivity over Detroit averaged 9 per cent less than that for nearby areas, and reached 25 per cent less under calm conditions. The increased absorption of solar radiation by aerosols plays a role in daytime heating of the boundary-layer pollution dome (see Figure 7.26A), but is less important within the urban canopy layer, which extends to mean roof-top height (see Figure 7.26B). Table 7.2 gives comparable urban and rural energy budget figures for the Cincinnati region during the summer of 1968 under anticyclonic conditions with < 3/10 cloud and wind speed of < 2 m s^{-1} (5 mph). The data show that pollution reduces the incoming short-wave radiation, but this is counter-balanced by a lower albedo and the greater surface area within urban canyons. The increased urban L_n at 1200 and 2000 LST is largely offset by anthropogenic heating (see below).

b Urban surfaces

Primary controls over a city's thermal climate are the character and density of urban surfaces, that is, the total *surface* area of buildings and roads, as well as the building geometry. Table 7.2 shows the relatively high heat absorption of the city surface. A problem of measurement is that the stronger the urban thermal influence, the weaker is the heat absorption *at street level*, and, consequently, observations made only in streets may lead to erroneous results. The geometry of urban canyons is particularly important, since it involves an increase in effective surface area and the trapping by multiple reflection of short-wave radiation, as well as a reduced 'sky view' (proportional to the area of

Table 7.2 Energy budget figures for the Cincinnati region during the summer of 1968 (W m^{-2}).

	Central business district			*Surrounding country*		
	0800	*1300*	*2000*	*0800*	*1300*	*2000*
Short wave, incoming $(Q+q)$	288*	763	–	306	813	–
Short wave, reflected $[(Q+q)a]$	42†	120†	–	80	159	–
Net long wave radiation (L_n)	−61	−100	−98	−61	−67	−67
Net radiation (R_n)	184	543	−98	165	587	−67
Heat produced by human activity	36	29	26‡	0	0	0

Source: From Bach and Patterson 1969.
Notes: *Pollution peak.
†An urban surface reflects less than agricultural land, and a rough skyscraper complex can absorb up to 6 times more incoming radiation.
‡Replaces more than 25 per cent of the long-wave radiation loss in the evening.

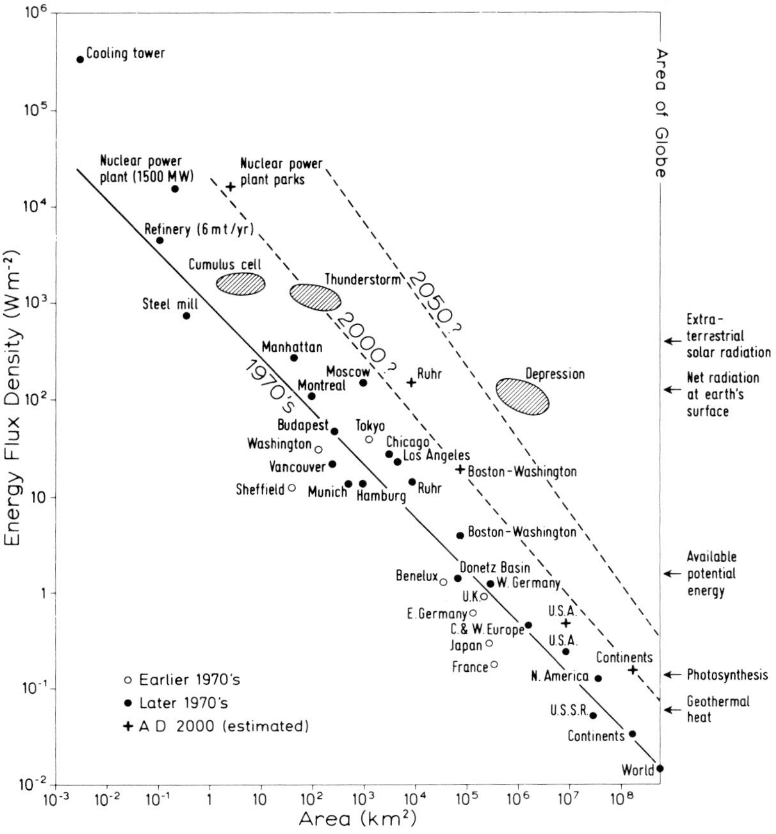

Figure 7.30 A comparison of natural and artificial heat sources in the global climate system on small, meso- and synoptic scales. Generalized regressions are given for artificial heat releases in the 1970s (early 1970s circles, late 1970s dots), together with predictions for the years 2000 (crosses) and 2050.

Sources: After Pankrath 1980 and Bach 1979.

the hemisphere open to the sky) which decreases the loss of infra-red radiation. From analyses by T. R. Oke, there appears to be an inverse linear relationship on calm, clear summer nights between the sky view factor (0–1.0) and the maximum urban–rural temperature difference. The difference is 10°–12°C for a sky view factor of 0.3, but only 3°C for a sky view factor of 0.8–0.9.

c Human heat production

Numerous studies show that large urban conurbations now produce energy through combustion at rates comparable with incoming solar radiation in winter. Winter receipts from solar radiation average around 25 W m^{-2} in Europe, compared with similar heat production from large cities. Figure 7.30 illustrates the magnitude and spatial scale of artificial and natural energy fluxes. In Cincinnati, a significant proportion of the energy budget is generated by human activity, even in summer (see Table 7.2). This heat production averaged 26 W m^{-2} or more, of which two-thirds was produced by industrial, commercial and domestic sources and one-third by cars. In terms of the future, it has been estimated that by AD 2000 the Boston–Washington megapolis may accommodate up to 56 million people in a continuous urban area of 30,000 km^2 (11,600 sq miles), and that this concentration of human activity would produce heat equivalent to 50 per cent of the total winter solar radiation recorded on a horizontal surface and 15 per cent of the total summer solar radiation. In the extreme situation of arctic settlements during

polar darkness, the energy balance during calm conditions depends only on net long-wave radiation and heat production by anthropogenic activities.

d Heat islands

The net effect of the urban thermal processes is to make city temperatures generally higher than those in the surrounding rural areas. This is most pronounced after sunset (for mid-latitude cities during calm, clear weather) when cooling rates in the rural areas greatly exceed those in the urban area. The energy balance differences that cause this effect depend on the radiation geometry and thermal properties of the surface. It is thought that the canyon geometry effect dominates in the urban canopy layer, whereas the sensible heat input from urban surfaces determines the boundary layer heating. By day the urban boundary layer is heated by increased absorption of short-wave radiation due to the pollution, as well as by sensible heat transferred from below and entrained by turbulence from above.

The *heat island* effect may result in minimum urban temperatures being 5°–6°C (9°–11°F) greater than those of the surrounding countryside, and these differences may be as much as 6°–8°C (11°–14°F) in the early hours of calm, clear nights in large cities, when the heat stored by urban surfaces during the day (augmented by combustion heating) is released. It should be noted that, because this is a *relative* phenomena, the heat island effect also depends on the rate of rural cooling, which is influenced by the magnitude of the regional environmental lapse rate.

For the period 1931–60, the centre of London had a mean annual temperature of 11.0°C (51.8°F), comparing with 10.3°C (50.5°F) for the suburbs, and 9.6°C (49.2°F) for the surrounding countryside. Calculations for London in the 1950s indicated that domestic fuel consumption gave rise to a 0.6°C warming in the city in winter and that accounted for one-third to one-half of the average city heat excess compared with adjacent rural areas. Differences are even more marked during still air conditions, especially at night under a regional inversion (see Figure 7.31). For this heat island effect to operate effectively there must be wind speeds of less than 5–6 m s^{-1} (12–14 mph), and it is especially in evidence on calm nights during summer and early autumn when it has steep cliff-like margins on the upwind edge of the city

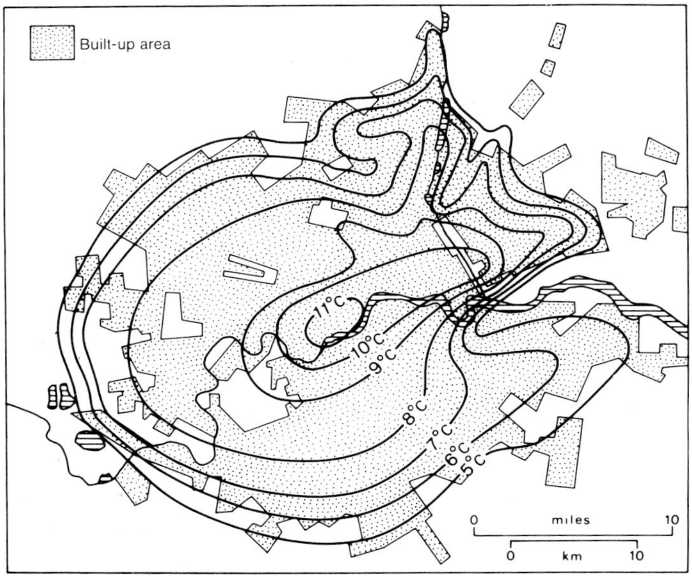

Figure 7.31 Distribution of minimum temperatures (°C) in London on 14 May 1959, showing the relationship between the 'urban heat island' and the built-up area.

Source: After Chandler 1965.

and the highest temperatures associated with the highest density of urban dwellings. In the absence of regional winds, a well-developed heat island may even generate its own inward local wind circulation at the surface. Thus the thermal contrasts of a city, like those of many other of its climatic features, depend on its topographic situation, and are greatest for sheltered sites with light winds. The fact that for London urban/rural temperature differences are greatest in summer, when direct heat combustion and atmospheric pollution are at a minimum, indicates that heat loss from buildings by radiation is the most important single factor contributing to the heat island effect. Seasonal differences are not necessarily the same, however, in other macroclimatic zones.

The effects on minimum temperatures are especially significant. Cologne, for example, has an average of 34 per cent less days with minima below 0°C (32°F) than its surrounding area, the corresponding figure for Basle being 25 per cent less. In London, Kew has an average of some 72 more days with frost-free screen temperatures than rural Wisley. Precipitation characteristics are also affected; in pre-1917 Berlin, 21 per cent of the incidences of rural snowfall were associated with either sleet or rain in the city centre.

Although it is difficult to isolate changes in temperatures that are due to urban controls from those due to other influences (see Chapter 8), it has been suggested that city growth is often accompanied by an increase in mean annual temperature, that of Osaka, Japan, rising by 2.6°C (4.5°F) in the last 100 years. Under calm conditions, the maximum difference in urban–rural temperatures is statistically related to population size, being nearly linear with the logarithm of the number of inhabitants. In North America, the maximum urban–rural temperature difference reaches 2.5°C for towns of 1,000, 8°C for cities of 100,000 and 12°C for cities of one million people. European cities show a smaller temperature difference for equivalent populations, perhaps as a result of the generally lower buildings and shallower urban canyons.

Figure 7.32 The built-up area of Tokyo in 1946 and 1975 (A), the mean January minimum temperatures (B), and number of days with sub-zero temperatures (C) between 1880 and 1975. During the Second World War the population of the city fell from 7.36 million to 3.49 million and then increased to 7.38 million in 1953 and 11.67 million in 1975.

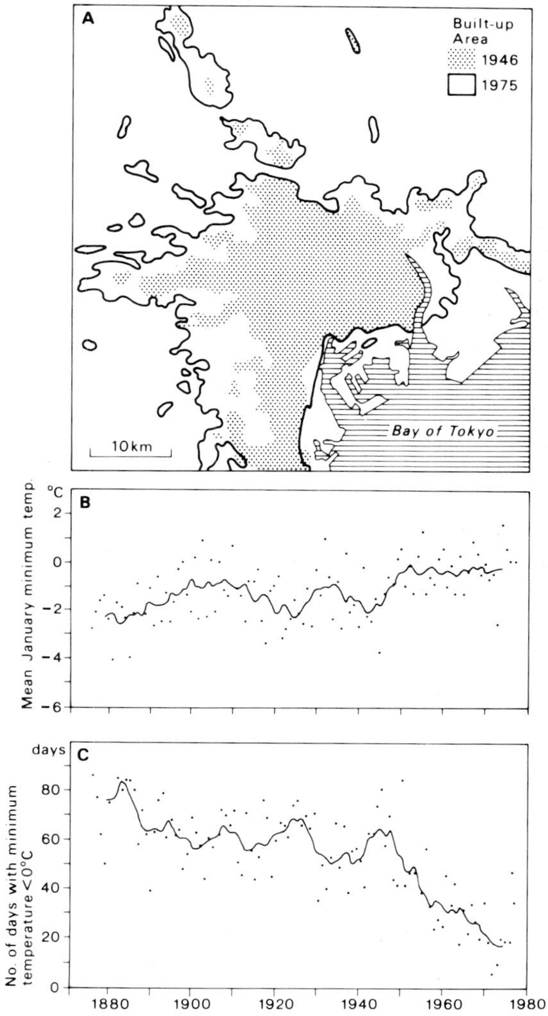

Source: After Maejima *et al.* 1982.

One of the most convincing examples of the possible relationship between urban growth and climate is that of the city of Tokyo, which expanded greatly after 1880 and particularly

after 1946 (see Figure 7.32A), when the population increased to 7.38 million in 1953 and to 11.67 million in 1975. During the period 1880–1975 there was a significant increase in mean January minimum temperatures and a decrease in the number of days with minimum temperatures below 0°C (see Figure 7.32B and C). Although these graphs suggest a reversal of these trends during the Second World War (1942–5), when evacuation almost halved Tokyo's population, it is clear that the basis of correlations of urban climate with population is complex. Urban density, industrial activity and the production of anthropogenic heat are all involved. Leicester, for example, when it had a population of 270,000, exhibited warming comparable in intensity with that of central London over smaller sectors. This suggests that the thermal influence of city size is not as important as that of urban density. The vertical extent of the heat island is little known, but is thought to exceed 100–300 m, especially early in the night. In the case of cities with skyscrapers, the vertical and horizontal patterns of wind and temperature are very complex (see Figure 7.33 for wind conditions).

3 Modification of surface characteristics

a Airflow

On the average, city wind speeds are lower than those recorded in the surrounding open country owing to the sheltering effect of the buildings, and central average wind speeds are usually at least 5 per cent less than those of the suburbs. In 1935, for example, winds exceeding 10.5 m s^{-1} (24 mph) were recorded on the relatively open Croydon Airport (London suburbs) for a total of 371 hours, whereas the corresponding figure was only 13 hours for the closely built-up South Kensington. However, the urban effect on air motion varies greatly depending on the time of day and season. During the day, city wind speeds are considerably less than those surrounding rural areas, but during the night the greater mechanical turbulence over the city means that the higher wind speeds aloft are transferred to the air at lower levels by turbulent mixing. During the day (1300 hours) the mean annual wind speed for London Airport (open country within the suburbs) was 2.9 m s^{-1} (6.4 mph), compared with a 2.1 m s^{-1} (4.7 mph) in central London for the period 1961–2. The comparative figures for night (0100 hours) were 2.2 m s^{-1} (4.9 mph) and 2.5 m s^{-1} (5.6 mph). Rural–urban wind speed differences are most marked with strong winds and the effects are therefore more evident during winter than during summer, when a higher proportion of low speeds is recorded in temperate latitudes.

Urban structures have considerable effects on

Figure 7.33 Details of urban airflow around two buildings of differing size and shape. Numbers give relative wind speeds; stippled areas are those of high wind velocity and turbulence at street level.

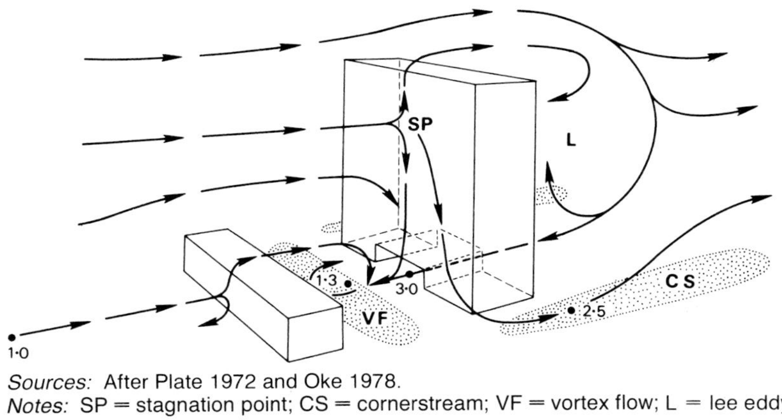

Sources: After Plate 1972 and Oke 1978.
Notes: SP = stagnation point; CS = cornerstream; VF = vortex flow; L = lee eddy.

Figure 7.34 The climatic effect of London. A: Surface potential temperature (°K) at noon on 21 August 1959 and the area receiving more than 12 mm total precipitation on that day. B: Divergence of the surface wind field (10^{-5} s^{-1}) at 0900 on 1 September 1960 over south-east England and the total precipitation (mm) on that day. The high frictional effect of the city surface caused airflow to converge over the city (i.e. negative divergence to occur).

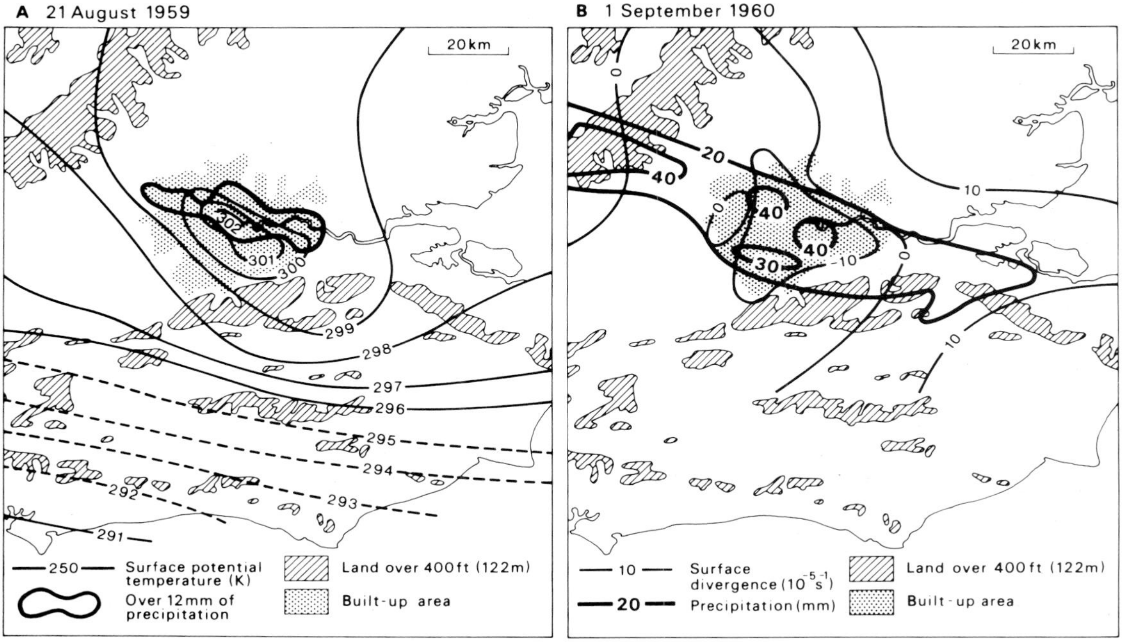

Source: From Atkinson 1987, in *Human Activity and Environmental Processes* by Gregory, K. J. and Walling, D. E. (eds). (*Copyright © 1987*; *reprinted by permission of John Wiley & Sons, Inc.*)

the movement of air both by producing turbulence as a result of their roughening the surface and by the channelling effects of the urban canyons. Some idea of the complexity of airflow round urban structures is given by Figure 7.33, illustrating the great differences in ground-level wind velocity and direction, the development of vortices and lee eddies, and the reverse flows which may occur. Structures play a major role in the diffusion of pollution within the urban canopy; for example, narrow streets often cannot be flushed by vortices. The formation of high-velocity streams and eddies in the usually dry and dusty urban atmosphere, where there is ample debris supply, leads to general urban airflows of only 5 m s^{-1} (11 mph) being annoying, and those of more than 20 m s^{-1} (45 mph) being dangerous.

b Moisture

The effect of urbanization on surface-moisture relationships is also important. The absence of large bodies of standing water and the rapid removal of surface run-off through drains decreases local evaporation. In addition, the lack of an extensive vegetational cover eliminates much evapotranspiration and this is an important source of augmenting urban heat. For these reasons, the air of mid-latitude cities has a tendency towards lower absolute

humidities than that of their surroundings, especially under conditions of light wind and cloudy skies. On other occasions of calm, clear weather the streets trap warm air, which retains its moisture because less dew is deposited on the warm surfaces of the city. Humidity contrasts between urban and rural areas are most noticeable in the case of relative humidity, which can be as much as 30 per cent less in the city by night as a result of the higher temperatures.

Urban influences on precipitation (excluding fog) are much more difficult to determine with any precision, partly because there are few rain gauges in cities and because air turbulence makes their 'catch' unreliable. It is now fairly certain, however, that urban areas in Europe and North America are responsible for local conditions which, in summer especially, can trigger off excesses of precipitation under marginal conditions. These triggers mainly involve the thermal effects and increased frictional convergence of built-up areas. These effects are well exemplified in the case of London for rainstorms on 21 August 1959 and 1 September 1960, respectively (see Figure 7.34). Recordings for Munich showed 11 per cent more days of light rain (0.1–0.5 mm or 0.004–0.12 in) than in the surrounding countryside, and Nürenberg has recorded 14 per cent more thunderstorms than its rural environs. European and North American cities apparently record 6–7 per cent more days with rain per year than their surrounding regions, giving 5–10 per cent increase in urban precipitation. The effect is generally more marked in the cold season in North America, although urban areas in the Midwest of the United States significantly increase summer convective activity with more frequent thunderstorms and hail downwind of industrial areas of St Louis (for a distance of 30–40 km) compared with rural areas (see Figure 7.35). The anomalies shown in Figure 7.35 are among the best documented of urban

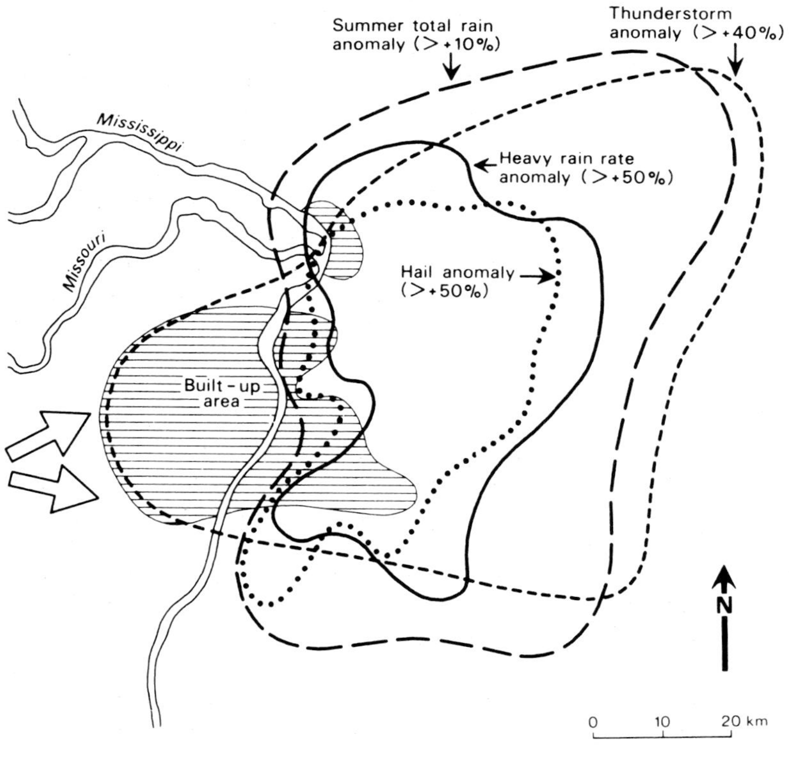

Figure 7.35 Anomalies of summer rainfall, rate of heavy rains, hail frequency and thunderstorm frequency downwind of the St Louis metropolitan area. Large arrows indicate the prevailing direction of motion of summer rain systems.

Source: After Changnon 1979.

Table 7.3 Average urban climatic conditions compared with those of surrounding rural areas.

Atmospheric composition	carbon dioxide	× 2
	sulphur dioxide	× 200
	nitrogen oxides	× 10
	carbon monoxide	× 200(+)
	total hydrocarbons	× 20
	particulate matter	× 3 to 7
Radiation	global solar	−15 to 20%
	ultraviolet (winter)	−30%
	sunshine duration	−5 to 15%
Temperature	winter minimum (average)	+1° to 2°C
	heating degree days	−10%
Wind speed	annual mean	−20 to 30%
	number of calms	+5 to 20%
Fog	winter	+100%
	summer	+30%
Cloud		+5 to 10%
Precipitation	total	+5 to 10%
	days with <5 mm (<0.2 in)	+10%

Source: Partly after World Meteorological Organization 1970.

effects. Over south-east England during 1951–60, summer thunderstorm rain (which comprised 5–15 per cent of the total precipitation) was especially concentrated in west, central and southern London (see Figure 7.36) and contrasted strikingly with the distribution of mean annual total rainfall. During this period London's thunderstorm rain was of the order of 20–25 cm (8–10 in) greater than that in rural south-east England.

Many of the results discussed in connection with urban influences are based on limited case studies, but a summary of average climatic differences between cities and their surrounding rural areas is presented in Table 7.3.

4 Tropical urban climates

A striking feature of recent and projected world population growth is the relative increase in the tropical and subtropical regions. At present there are thirty-four world cities with more than 5 million people, of which twenty-one are in the less-developed countries, but by the year 2025 it is predicted that, of the thirteen world cities which will have populations in the 20–30 million range, eleven will be in less-developed countries (i.e Mexico City, São Paulo, Lagos, Cairo, Karachi, Delhi, Bombay, Calcutta, Dacca, Shanghai and Jakarta).

Despite the difficulties in extrapolating knowledge of urban climatic effects from one region to another, it is clear that the ubiquitous high-technology architecture of most modern city centres and multi-storey residential areas imposes strikingly similar tendencies on their differing background climates. Nevertheless, most tropical urban-built land differs from that in higher latitudes in that it is commonly composed of high-density, single-storey buildings with few open spaces and poor drainage. In such a setting, as Oke points out, the composition of roofs is relatively more important than that of walls in terms of thermal energy exchanges, and the production of anthropogenic heat is more uniformly distributed spatially and of lower intensity than in European and North American cities. In the dry tropics buildings have a relatively high thermal mass to delay heat penetration and this, combined with the low soil moisture in the surrounding bare rural areas, makes the ratio of urban/rural thermal absorbtion greater than in temperate regions. However, it is not always possible to

Figure 7.36 The distribution of total thunderstorm rain in south-east England during the period 1951–60.

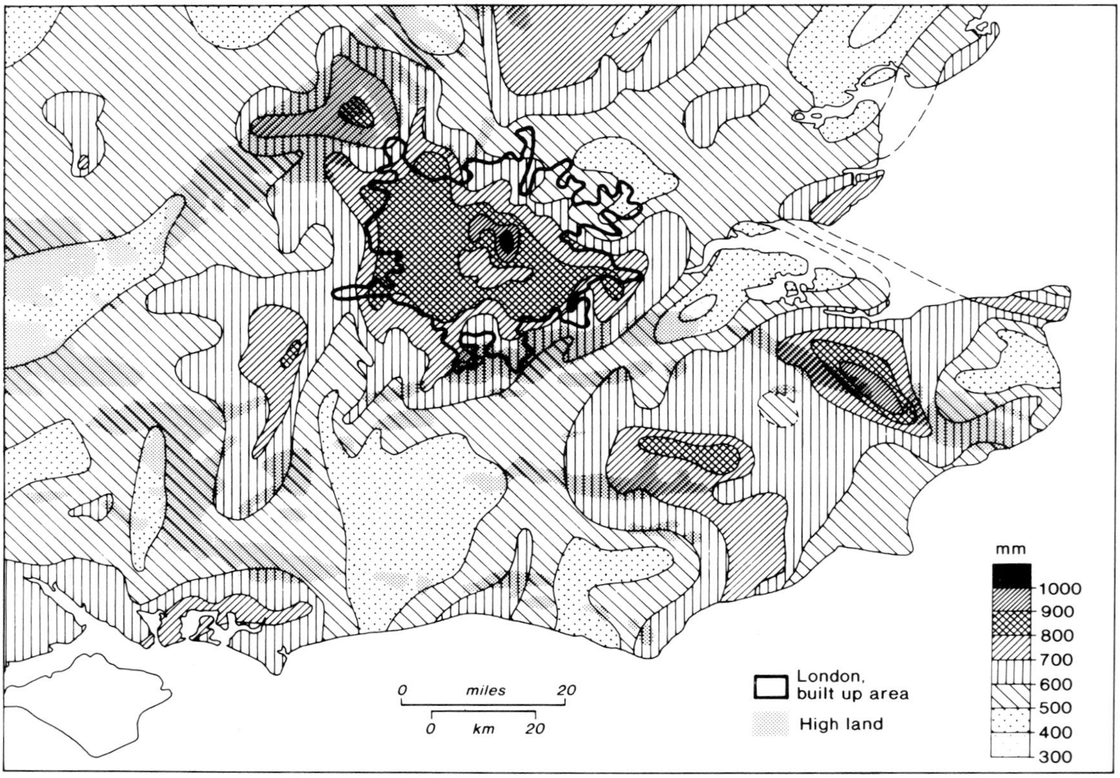

Source: After Atkinson 1968.

generalize regarding the thermal effects of cities in the dry tropics due to the differing 'oasis' effects of urban vegetation. Building construction in the humid tropics is characteristically lightweight to promote the much-needed ventilation. These cities differ from dry tropical ones in that the surrounding rural thermal absorbtion is greater than the urban because of the high rural soil moisture levels and the high urban albedos.

Tropical heat island tendencies are similar to those of temperate cities but are generally weaker, with somewhat different timings for temperature maxima, and with complications introduced by the effects of afternoon and evening convective rainstorms and by diurnal breezes. Tropical heat islands are naturally most marked at night during the dry season, although topographic effects are important. Quito, Ecuador (2851 m: 9350 ft), shows a maximum heat island effect by day (as much as 4°C) and weaker night-time effects, probably due to the nocturnal drainage of cold air from the nearby volcano Pichincha. Mexico City, on the other hand, despite its high elevation (2274 m: 7460 ft), exhibits classic heat island behaviour because of its plateau location. Despite a marked lack of data, there appears to be some urban precipitation enhancement in tropical regions, which is maintained for a larger part of the year than that associated with the short-term high-sun convective period in temperate latitudes.

SUMMARY

Small-scale climates are determined largely by the relative importance of the surface-energy-budget components, which vary in amount and sign depending on time of day and season. Bare land surfaces may have wide temperature variations controlled by H and G, whereas those of surface water bodies are strongly conditioned by LE and advective flows. Vegetated surfaces have more complex exchanges which are usually dominated by LE; this may account for more than 50 per cent of the incoming radiation, especially where there is an ample water supply (including irrigation). Forests have a lower albedo (< 0.10 for conifers) than most other vegetated surfaces (0.20–0.25). Their vertical structure produces a number of distinct microclimatic layers, particularly in tropical rain forests. Wind speeds are characteristically low in forests and trees form important shelter belts. Unlike short vegetation, the various types of trees exhibit a variety of rates of evapotranspiration and thereby differentially affect local temperatures and forest humidity. The effect of forests on precipitation has not yet been resolved but they may have a marginal topographic effect under convective conditions in temperate regions. The disposition of forest moisture is very much affected by canopy interception and evaporation, but forested catchments appear to have greater evapotranspiration losses than ones with a grass cover. Another major feature of forest microclimates is their lower temperatures and smaller diurnal ranges compared with surrounding areas.

Urban climates are dominated by the geometry and composition of built-up surfaces and by the effects of human urban activities. The composition of the urban atmosphere is modified by the addition of aerosols, producing smoke pollution and fogs, by industrial gases such as sulphur dioxide, and by a chain of chemical reactions, initiated by automobile exhaust fumes, which causes smog and inhibits both incoming and outgoing radiation. Pollution domes and plumes are produced around cities under appropriate conditions of vertical temperature structure and wind velocity. The urban heat budget is dominated by H and G, except in city parks, and as much as 70–80 per cent of incoming radiation may become sensible heat which is very variably distributed between the complex urban-built forms. Urban influences combine to give generally higher temperatures than in the surrounding countryside, not least because of the growing importance of heat production by human activities. These factors give rise to the urban heat island which may be 6°–8°C warmer than surrounding areas in the early hours of calm, clear nights when heat stored by urban surfaces is being released. The urban–rural temperature difference under calm conditions is statistically related to the city population size; the urban canyon geometry and sky view factor are major controlling factors. The heat island may be a few hundred metres deep, depending on the building configuration. Urban wind speeds are generally less than for rural areas by day, but the wind flow is extremely complex, depending on the geometry of city-built forms. Cities naturally tend to be less humid than rural areas but their topography, roughness and thermal qualities tend to intensify the effects of summer convective activity over and downwind of the urban areas, giving more thunderstorms and heavier storm rainfall. The increasing importance of tropical cities is stressed, together with their climatic comparison with temperate cities.

8

Climatic change

Probably the aspect of climate which most interests the layman is speculation regarding its possible trends. Unfortunately, as well as being the most interesting, it is also the most uncertain aspect of meteorological research. Realization that climate has changed radically with time came only during the 1840s when indisputable evidence of former ice ages was obtained, yet in many parts of the world the climate has altered sufficiently, even within the last few thousand years, to affect the possibilities for agriculture and settlement. Reliable weather records have only been kept during the last hundred years or so and therefore it is only the recent climatic fluctuations which can be investigated adequately. The discussion in this chapter is mainly limited to these events, but first it is worth while to consider the methods of handling the meteorological records which are available.

A CLIMATIC DATA

1 Averages

The climate of a place is often regarded simply as its 'average weather', but vital climatic information is overlooked if the range and frequency of extremes are neglected. Averages can be markedly affected by extreme values and this is particularly true of the arithmetic mean (which is obtained by totalling the individual values and dividing by the number of occurrences). For this reason a 30- or 35-year period is normally required for the determination of climatic averages. Even so, certain types of data are very inadequately summarized by the arithmetic mean, especially when small values are frequent but very large ones occur occasionally. This situation is illustrated by Figure 8.1, where the *histograms* or frequency-distribution graphs of annual rainfall at Helwan (Egypt) and Aden are clearly dissimilar to those at Greenwich (England) and Padua (Italy). The profile for Padua approximates a symmetrical 'normal' distribution, where half the values lie above and half below the mean, and where the most frequent (or *modal*) category is equal to the mean (see note 1). For 1725–1924, the annual rainfall at Padua has a mean of 859 mm, and a mode of about 884 mm. The *median* for the same period at Padua is 847 mm. This is another useful measure of central tendency, which has exactly half the number of items in the data series above it and half below it. Consequently it is not biased by the occurrence of a few extreme figures. Modal or median values would be much more meaningful indicators of annual precipitation at Aden and Helwan where the graphs exhibit strong 'positive' skewness, i.e. 'tail' of the distribution is towards values greater than the mean. Positive skewness is especially marked when the mean of the distribution approaches zero, when a large number of records includes some infrequent events of high magnitude, and when the length of time to which the event is related is short (for example, rainfall figures referring to any one month are usually more skewed than those relating to the

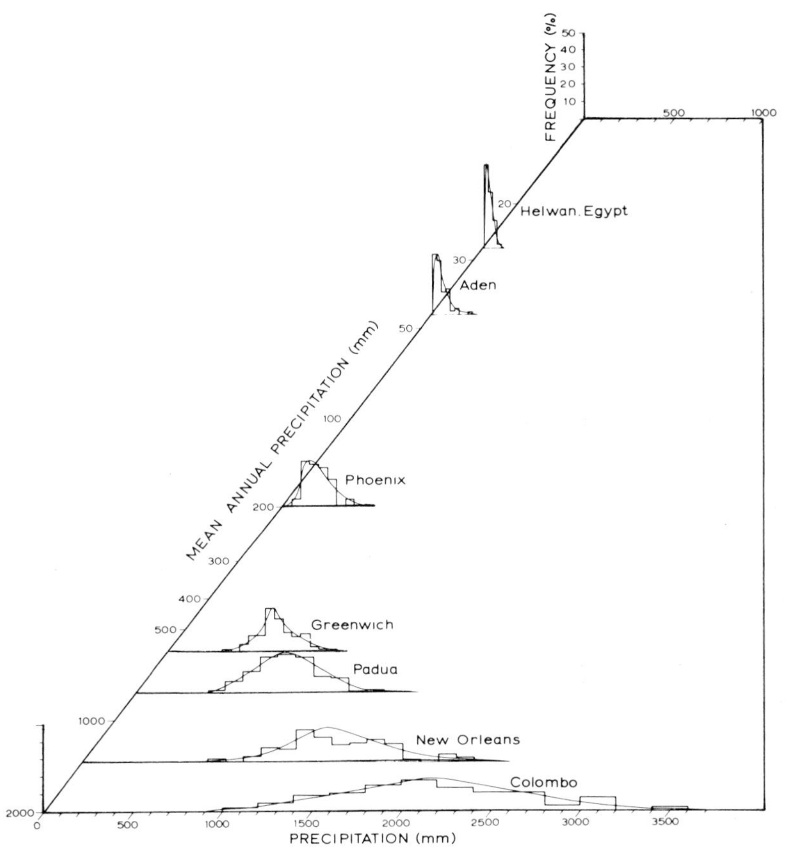

Figure 8.1 Frequency distribution curves of annual rainfall for: Helwan (Cairo), Egypt (averaged over 37 years); Aden (55 years); Phoenix, Arizona (45 years); Greenwich (London), England (100 years); Padua, Italy (200 years); New Orleans (88 years); and Colombo, Ceylon (70 years).

whole year). Another problem of frequency distributions is that they may have more than one frequency peak, as shown in Figure 8.1 for New Orleans, which is bimodal.

2 Variability

Variability about the average can be expressed in several ways. When the median is used it is common also to determine the upper and lower quartiles (Q_1 and Q_2), which are the central values between the median and the upper and lower extremes, respectively, i.e. the 25 and 75 per cent points in the frequency distribution. The average deviation from the median is given by $(Q_1 - Q_2)/2$. A more widely used measure of variability is the standard deviation (σ, pronounced *sigma*) which is calculated by summing the square of the deviation of each value from the mean, dividing by the number of cases and then taking the square root.

$$\sigma = \sqrt{\frac{\Sigma (x_i - \bar{x})^2}{n}}$$

where:

x_i = an individual value

$\bar{x} = \Sigma \frac{x_i}{n}$ (the mean)

n = number of cases

Σ = sum of all values for $i = 1$ to n.

It is therefore a measure of average deviation, where the difficulty created by positive and

negative departures (i.e. values greater and less than the mean) is removed by squaring each deviation and rectifying this by finally calculating the square root. Variability of rainfall may be compared between stations if the standard deviation is expressed as a percentage of the mean (the *coefficient of variation, CV*):

$$CV = \frac{\sigma}{\bar{x}} \times 100(\%)$$

Where the standard deviation is not available, use is sometimes made of the *mean deviation, MD*.

$$MD = \frac{\Sigma | x_i - \bar{x} |}{n}$$

where | — | denotes the absolute difference, disregarding sign.

This measure of variability, standardized against the mean, ranges for annual precipitation from about 10–20 per cent in western Europe and parts of monsoon India to over 50 per cent in arid areas of the world (see Figure 8.2). It is in such areas that a small change in the frequency of rainstorms can markedly affect the 'average' rainfall over a given period of years. It should be noted, however, that detailed examination of the precipitation in many diverse climatic regions shows that the apparent inverse relationship between annual total and variability is only very approximate. Moreover, a coefficient of variation $\geqslant 50$ per cent in fact violates the statistical assumption of a normal frequency distribution on which this statistic is based.

3 Trends

It is obvious that the great year-to-year variability of climatic conditions may conceal gradual trends from one type of regime towards another. The effect of short-term irregularities can be removed by various statistical techniques, of which the simplest is the *running mean* (or

Figure 8.2 Distribution of rainfall variability over the continents based on the mean deviation.

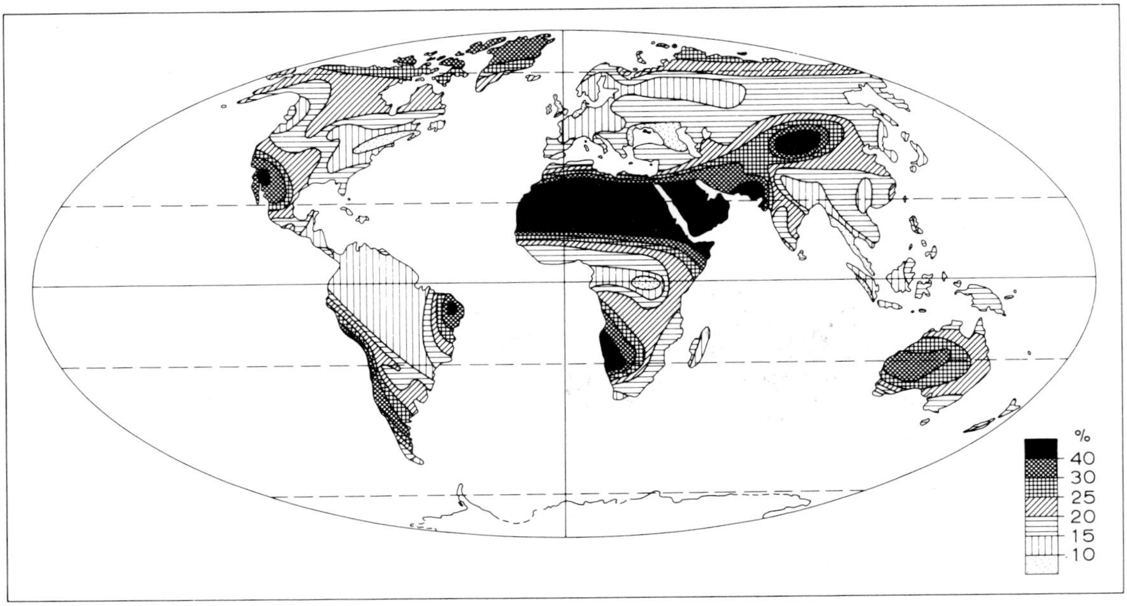

Source: After Erwin Biel.

moving average). The method is to calculate mean values for successive, overlapping periods of perhaps 5, 10 or 30 years, e.g.

$$\frac{\text{Year 1} + \text{Year 2} + \text{Year 3} + \text{Year 4} + \text{Year 5}}{5} = \text{Mean for Year 3}$$

$$\frac{\text{Year 2} + \text{Year 3} + \text{Year 4} + \text{Year 5} + \text{Year 6}}{5} = \text{Mean for Year 4, etc.}$$

This device smooths out the short-term fluctuations if periods of 20 or 30 years are used, thereby emphasizing the long-term trends. However, running means can also generate apparent regular periodic fluctuations where none exist. This can be shown by calculating running means for a series of random numbers.

B THE CLIMATIC RECORD

1 Evidence of climatic change

To understand the significance of climatic trends over the last hundred years they need to be viewed against the background of our general knowledge about past conditions.

Land and ocean sediments record numerous alternations between glacial and interglacial conditions during the last several million years. At least eight such cycles have occurred within the last million years, each averaging 125,000 years, with only about 10 per cent of each cycle as warm as the present day (see Figure 8.3 D and E). Only four or five glaciations are identified in the land evidence due to the absence of continuous sedimentary records, but it is likely that in each of these glacial intervals large ice sheets covered northern North America and northern Europe. Sea levels were also lowered by about 100–150 m due to the large volume of water locked up in the ice. Records from tropical lake basins show that these regions were generally arid at these times. The last such glacial maximum climaxed about 18,000 years ago, but 'modern' climatic conditions only became established during post-glacial time – conventionally dated to 10,000 years Before Present (BP). This time scale is used whenever dates are

Figure 8.3 Main trends in global climate during the past million years or so. A: Northern hemisphere, average land air temperatures. B: Eastern Europe, winter temperatures. C: Northern hemisphere, average land air temperatures. D: Northern hemisphere, average air temperatures based partly on sea-surface temperatures. E: Global average temperatures derived from deep-sea cores.

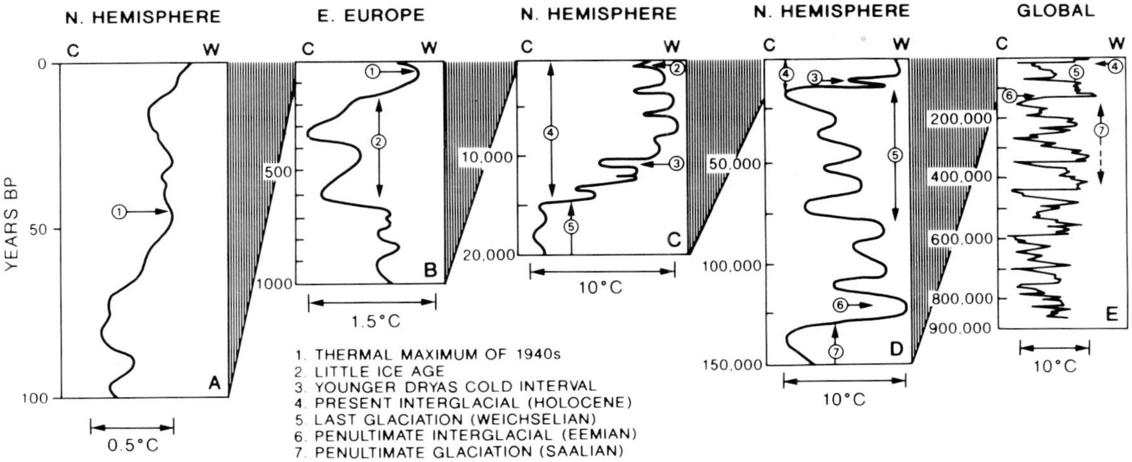

Source: Data taken from *Understanding Climatic Change: A Program for Action* (1975), published by National Academy Press, Washington, DC.

based on radiocarbon (carbon-14, ^{14}C) dating or other radiometric methods involving isotopic decay processes, such as potassium–argon (K–Ar).

For the last glacial cycle and post-glacial time, information on climatic conditions is obtained indirectly from *proxy* records. For example, the advance and retreat of glaciers represents a response to winter snowfall and summer melt. The history of vegetation, which indicates temperature and moisture conditions, can be traced from pollen types preserved in lake sediments and peat bogs. Former lake shorelines indicate changes in moisture balance. Estimates of seasonal climatic elements can be made from studies of annual snow/ice layers in cores taken from polar ice sheets where no melt occurs. These layers also record past volcanic events through the inclusion of microparticles and chemical compounds in the ice. In forested areas where trees form an annual growth layer, the ring-width can be interpreted through dendroclimatological studies in terms of moisture availability (in semi-arid regions) and summer warmth (near the polar and alpine tree-lines). Pollen sequences, lake-level records and ice cores usually span the last 10,000–20,000 years (and exceptionally back to 125,000 BP), while tree rings rarely extend more than 1500 years ago. For more recent time, historical documents often record crop harvests or extreme weather events (droughts, river and lake freezing, etc.).

2 Post-glacial conditions

Following the final retreat of the continental ice sheets from Europe and North America between 10,000 and 7000 years ago the climate rapidly ameliorated in middle and higher latitudes. In the subtropics this interval was also generally wetter, with high lake levels in Africa and the Middle East. A *thermal maximum* was reached in the mid-latitudes about 5000 years ago, when summer temperatures are known to have been 1–2°C higher than today (see Figure 8.3C) and the arctic tree-line was several hundred kilo-

Figure 8.4 Air temperatures in the lowlands of central England. Trends of the supposed 1000-year and 100-year averages since 10,000 BC (the latter calculated for the last millennium). Shaded ovals indicate the approximate ranges within which the temperature estimates lie and error margins of the radiocarbon dates.

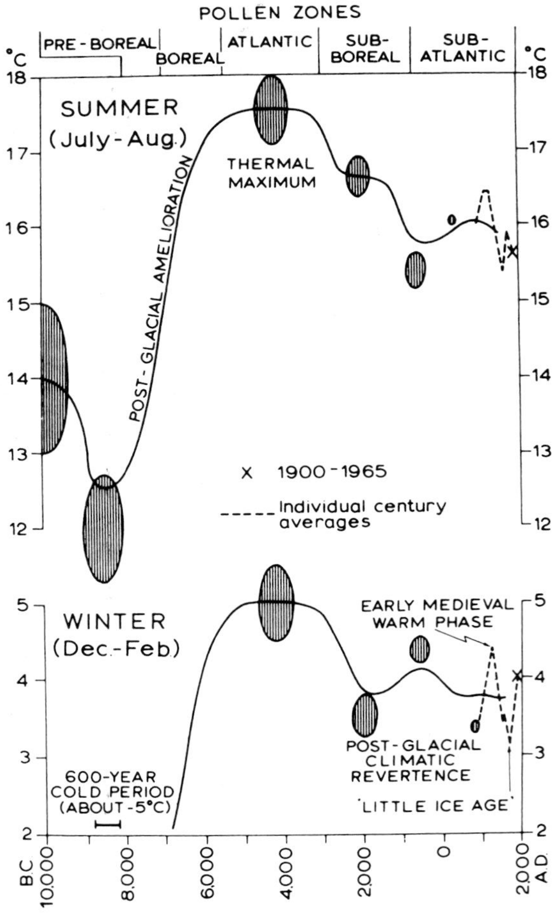

Source: After Lamb 1966.

metres further north in Eurasia and North America. By this time, subtropical desert regions were again very dry and were largely abandoned by primitive man. A temperature decline set in around 2000 years ago with colder, wetter conditions in Europe and North America. Although temperatures have not since equalled those of the thermal maximum there was

certainly a warmer period in many parts of the world between about AD 1000 and 1250 (see Figure 8.3B), and this phase was marked by the Viking colonization of Greenland and the occupation of Ellesmere Island in the Canadian Arctic by Eskimos. A further deterioration followed and severe winters between AD 1550 and 1850 gave a 'Little Ice Age' with extensive arctic pack-ice and glacier advances in some areas to maximum positions since the end of the Ice Age. These advances occurred at dates ranging from the mid-seventeenth to the late nineteenth century in Europe, as a result of the lag in glacier response and minor climatic fluctuations. Figure 8.4 attempts to summarize seasonal trends in central England, but it must be stressed that at present only the gross features are represented, as we know little or nothing about short-term fluctuations before the medieval period, for example, and even the relative magnitudes of the changes before about AD 1700 can only be indicated in a very general way.

3 The last 100 years or so

Long instrumental records for stations in Europe and the eastern United States indicate that the warming trend which ended the 'Little Ice Age' began early in the nineteenth century. Global

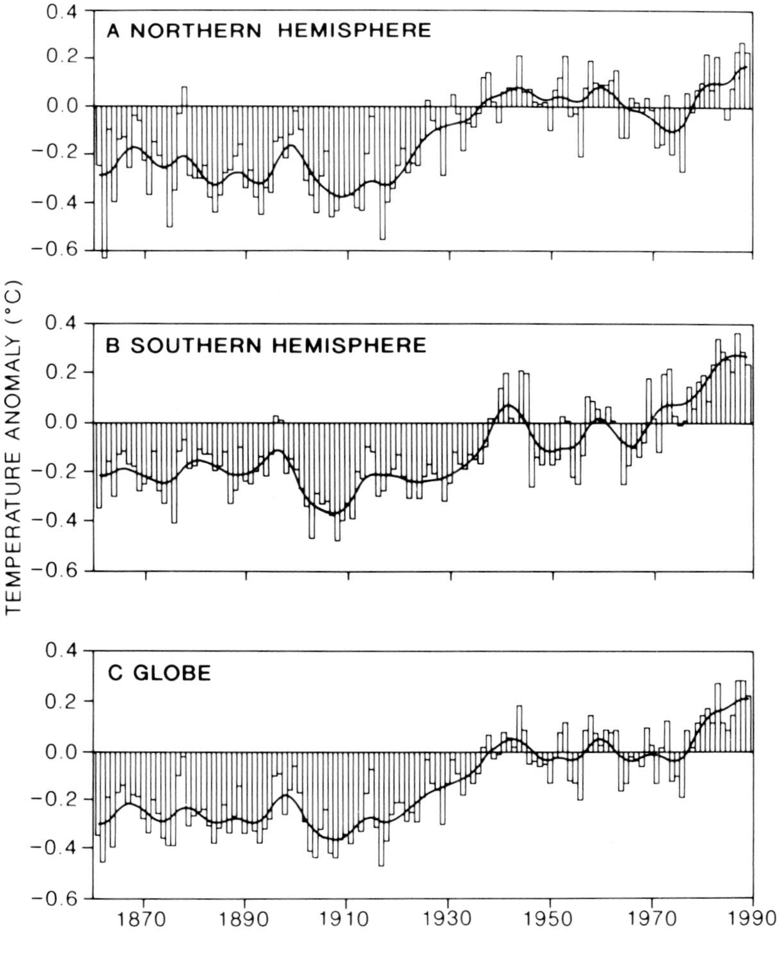

Figure 8.5 Annual combined land-air and sea-surface temperature anomalies between 1861 and 1989, relative to the 1951–80 average (designated zero). A: Northern hemisphere; B: Southern hemisphere; and C: Globe.

Source: After IPCC 1990.

Figure 8.6 Observed surface air temperature trends (5-year running means) for three latitudinal bands and the entire globe. Scales for low latitudes and the global mean are on the right.

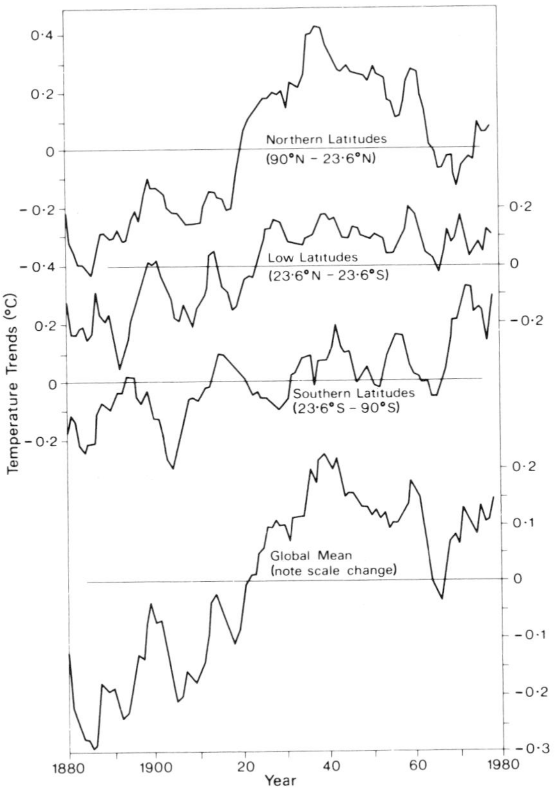

Source: After Hansen *et al.* 1981.

records since 1861 (see Figure 8.5) show a significant, but irregular, temperature rise of between 0.3° and 0.6°C, probably closer to the upper estimate. This trend was least in the tropics and greatest in cloudy, maritime regions of high latitude (see Figure 8.6). Winter temperatures were most affected (see Figure 8.7), and on Svalbard (77°N) the 1920–39 mean January temperature was 7.8°C (14°F) greater than the 1900–19 mean.

The general temperature rise has not been continuous, however, and four phases can be identified:

1 1861–1920, during which there were annual temperature oscillations within extreme limits of 0.4°C but no consistent trend.
2 1920–mid-1940s, during which there was considerable warming averaging 0.4°C.
3 Mid-1940s–early 1970s, during which there were oscillations within extreme limits of less than 0.4°C, with the northern hemisphere cooling slightly on average and the southern hemisphere remaining fairly constant in temperature. Regionally, northern Siberia, the eastern Canadian Arctic and Alaska, experienced a mean temperature lowering of 2°–3°C (4°–6°F) of winter temperatures between 1940–9 and 1950–9; this was partly compensated by a slight warming in the western United States, eastern Europe and Japan.
4 Early 1970s–1991, during which there was a marked overall warming of about 0.2°C, except for areas of the North and South Pacific, North Atlantic, Europe, Amazonia, and Antarctica (see Figure 8.8).

Global temperatures reached their highest observed levels during the 1980s which included *many* of the warmest years on record (see Figure 8.5). In the southern hemisphere, where the twentieth-century warming was delayed, there has been a more or less continuous warming of the order of 0.4°C. The twentieth-century global warming is approximately 0.5°C, which appears to exceed natural trends (estimated from statistical modelling) of 0.3°C/100 years. It is now widely considered likely that this global warming may be a result of increasing greenhouse gas concentrations, but the causal linkage is not yet definitively established. For example, experiments with general circulation models for doubled CO_2 concentrations, such as Figure 8.9, indicate significant amplification of warming in high latitudes in winter. This signal, attributable to feedback effects between surface albedo and temperature, associated with reduced snow cover and sea ice, was apparent in northern high latitudes between 1910 and 1940, but has been absent during the 1970–80s warming (see Figure 8.8).

The effects of the temperature rise are

Figure 8.7 Long-term records of winter and summer 30-year running mean temperatures at European stations. Station mean values are shown at each side.

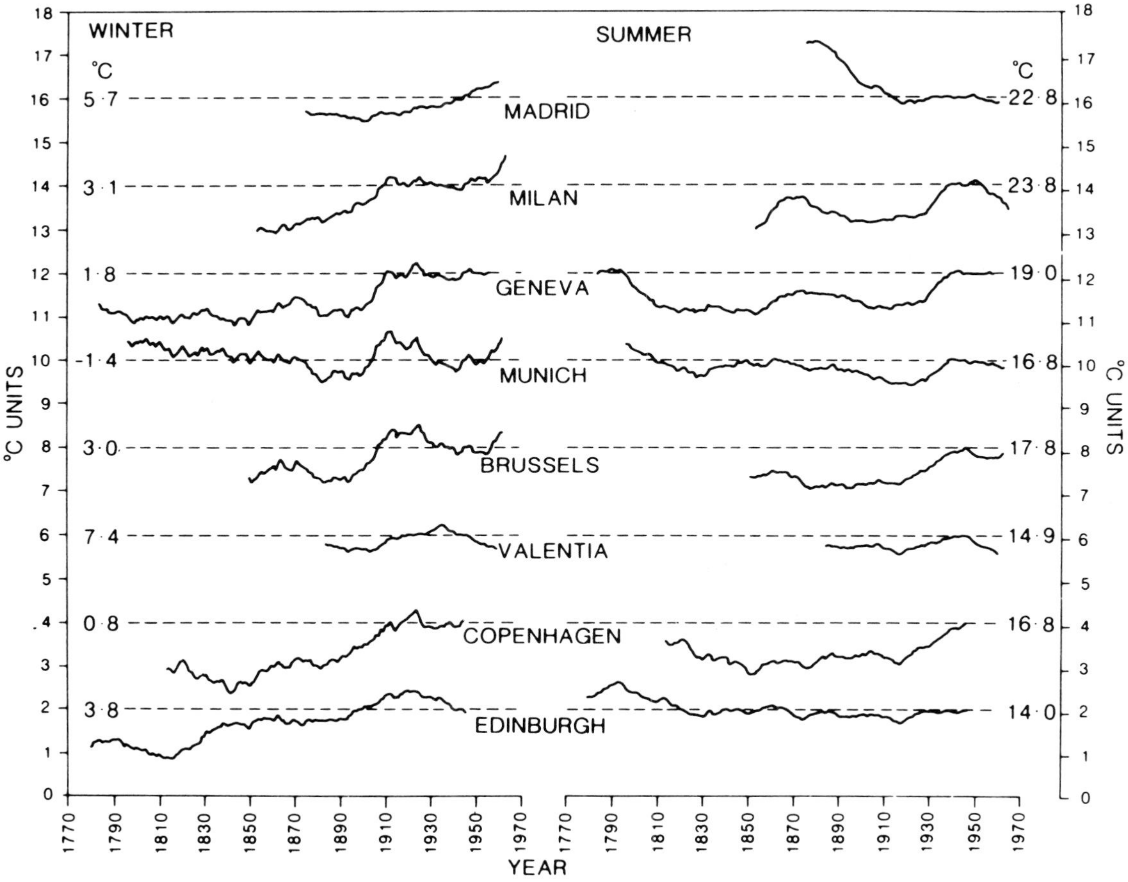

Source: From Schuurmans 1984 (*copyright © 1980/1982 by D. Reidel Publishing Company. Reprinted by permission*).

apparent in many ways. There has been, for example, a rapid retreat of most of the world's glaciers. Glaciers in the North Atlantic area retreated extensively between about 1920 and the present, due largely to temperature increases that have the effect of lengthening the ablation season with a corresponding raising of the snowline. Another tendency illustrating world warming was the retreat of arctic sea ice. Ports in the Arctic remained free of ice for longer periods during the 1920s–50s and cod extended their feeding northwards off west Greenland by 9° of latitude between 1919 and 1948.

Precipitation records are much more difficult to characterize. In East Africa and monsoon India there is no evidence of long-term trends since the late nineteenth century, although large fluctuations have occurred (see Figure 8.10C, D). In contrast, a steady increase in annual amounts appears to have taken place in the former USSR (see Figure 8.10A).

The West African records for this century (see Figure 8.11) show a tendency for both wet and dry years to occur in runs of up to 10–18 years.

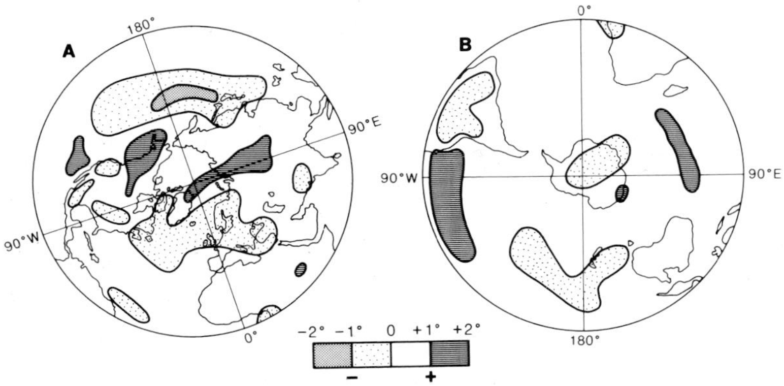

Figure 8.8 Global temperature trends (°C) observed between 1967 and 1986.

Sources: After Jones and Wigley 1990, and Jones, Wigley and Farmer 1991. (*Reproduced by permission of the* Scientific American *and Elsevier Science Publishers, Amsterdam.*)

Precipitation minima were experienced in the 1910s, 1940s and post-1968, with intervening wet years, in all of sub-Saharan West Africa. Throughout the two northern zones outlined in Figure 8.11, means for 1970–84 were generally < 50 per cent of those for 1950–9 with deficits during 1981–4 equal to or exceeding those of the disastrous early 1970s' drought. It has been suggested that this is related to weakening of the tropical easterly jet stream and limited northward penetration of the West African south-westerly monsoon flow. However, S.E. Nicholson attributes the precipitation fluctuations to contraction/expansion of the Saharan arid core, rather than to north–south shifts of the desert margin. In Australia rainfall changes have been related to changes in location and intensity of the subtropical anticyclones. The arid area appears to have increased in extent from the turn of the century to the 1930s and in central Australia annual amounts appear to have declined between 1910 and 1970.

In the middle latitudes, precipitation changes are usually less pronounced. Figure 8.12 illustrates long-term fluctuations for England and Wales and for individual stations. For the country as a whole, decadal departures are only about ± 10 per cent. The individual station graphs show that even over relatively short distances there may be considerable differences in the magnitude of anomalies (e.g. Manchester and Oxford). This is further illustrated by the pattern of fluctuations in central Europe (see Figure 8.13). The rising trend in the Bauer series, representing primarily West Germany (the former Federal Republic), is distinctly different from the other series. Poland in the early part of its record also differs from the regions to the south. For these countries the decadal range is approximately ± 7 per cent about the longer-term means.

Growing concern for possible anthropogenic causes of the greenhouse effect (see Chapter 1, A.4 and C.3, this chapter) has generated a strong body of opinion that the general temperature increase since the middle of the last century, and in particular the temperature increase and supposed unusual climatic fluctuations during the past twenty years, is due to changes in atmospheric composition induced by human activities. For example, during the past two decades Britain has experienced two major droughts (1976 and 1983), seven severe winter cold spells between 1978 and 1987 (compared with only three in the preceding forty years) and two major windstorms (1987 and 1989). World-wide, 1990 and 1991 were the two warmest years on record, 1983 was the year of the most intense El Niño event for a century, and hurricane Gilbert (1988) was the most severe storm on record.

In the face of these changes and speculations, however, it must be remembered that non-anthropogenic temperature changes of considerable magnitude have been common since the end of the last glaciation, and it is

Figure 8.9 Surface air temperature departures (°C) from the present predicted for the 990 mb-level by the NCAR Community Climate Model coupled with a mixed-layer ocean model, assuming a doubled CO_2 concentration.

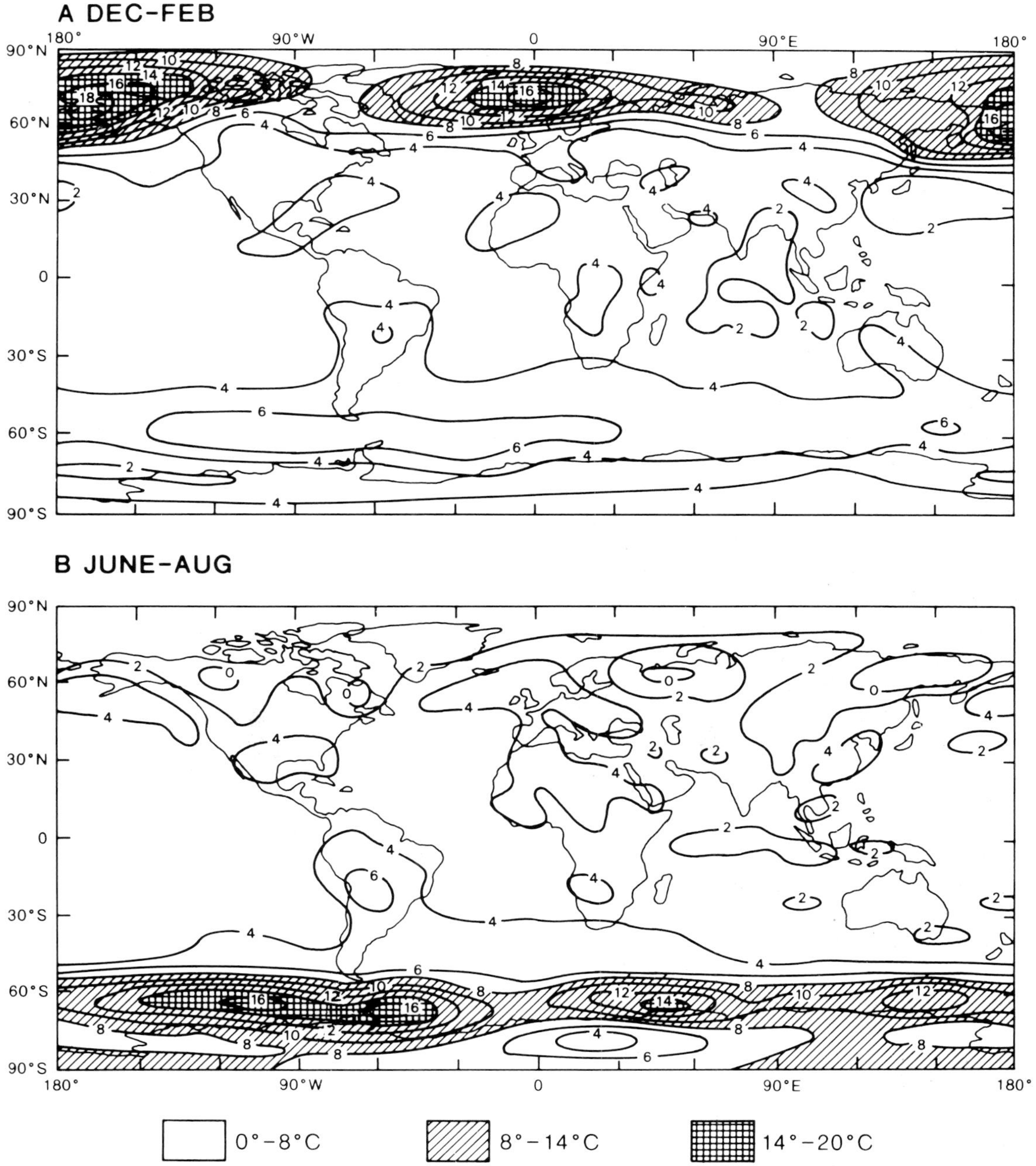

Source: From Meehl and Washington 1990.

Figure 8.10 Standardized regional annual precipitation anomalies. A: Former USSR, B: Sahel, C: All-India monsoon, D: East Africa. Note that Sahel values are annual averages of standardized *station* values restandardized to reflect the standard deviation of these annual averages.

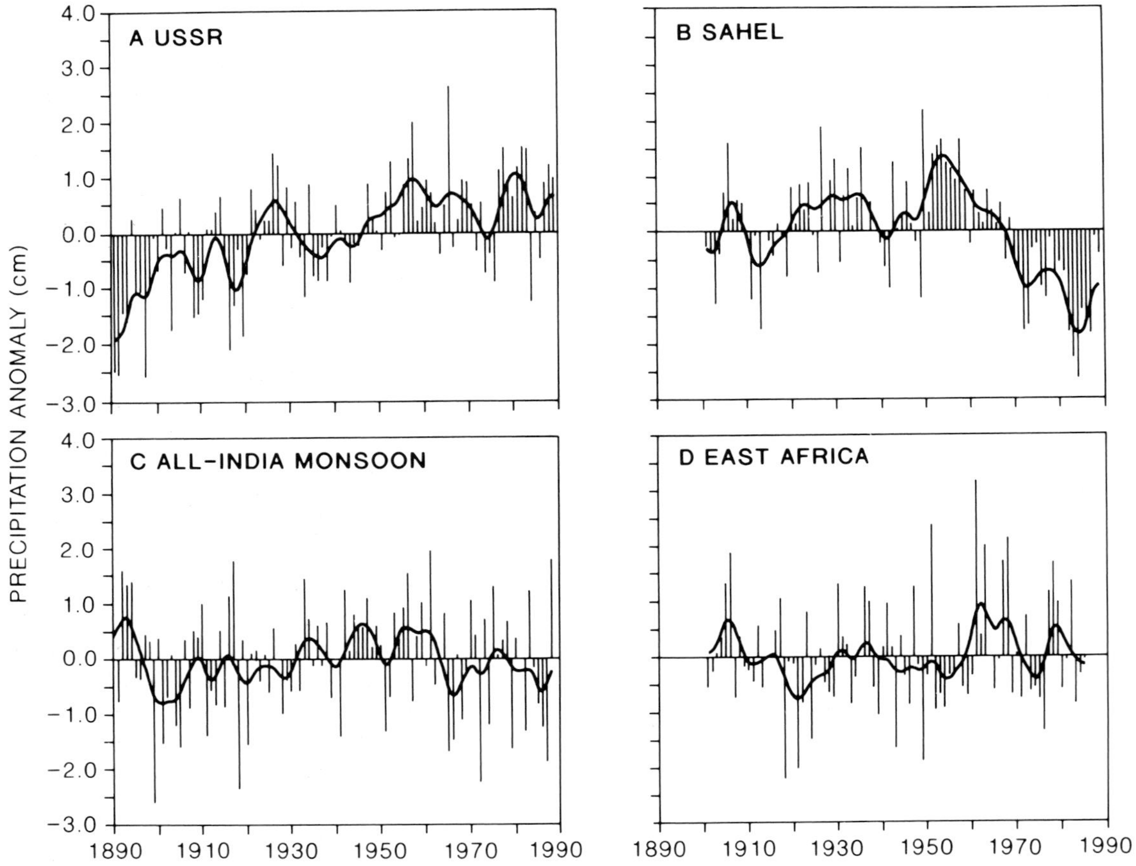

Source: From IPCC 1990.

possible that much of the temperature increase during the past 100 years or so may be part of the general temperature upturn following the Little Ice Age. It is therefore to be expected that global or hemispheric temperatures may vary through several tenths of a degree Celsius within a few years. Neither is it easy to explain the existence of periods of relatively stable global temperatures (e.g. 1861–1920 and mid-1940s–early 1970s) at times of continuing increase of greenhouse gas concentration. Nevertheless, the general warming since the middle of the last century and the most recent climatic events seem to be generally in harmony with the changes in atmospheric composition noted in Chapter 1, A.4 (rather than with other forcing mechanisms such as solar variations). In addition, it is recognized that the course of climatic change is undoubtedly complicated by little-known global mechanisms such as the inertial effects of the oceans and the possibly periodic effects of deep ocean currents. Overall, a strong and growing impression remains that natural climatic variabilities may from time to time be merely offsetting an inexorable and increasingly menacing anthropogenic global temperature enhancement.

Figure 8.11 Rainfall variations (percentage of standard departure) since 1900 for the Sahelo-Sahara, Sahel and Soudan zones of West Africa. The average positions of the Monsoon Trough (MT) in northern Nigeria during 1952–8 and in 1972 are shown.

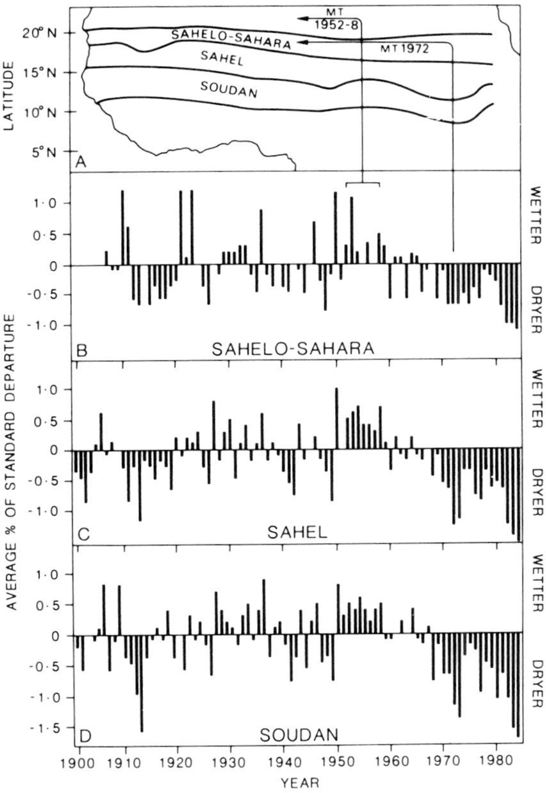

Source: From Nicholson 1985.

C POSSIBLE CAUSES OF CLIMATIC CHANGE

1 Forcing and feedback

It has become increasingly necessary to view global climate as operating within a complex atmosphere/earth/ocean/ice system (see Figures 8.14 and 3.44). Changes to this system are produced by the so-called *forcing agents*, which may be either *non-radiative* or *radiative*. The non-radiative forcing agents are mainly those which affect climate through their influence over the geometry of the earth's surface, such as location and size of mountain ranges, position and size of the ocean basins, etc. Changes of these kinds usually take place on time-scales in excess of a million years and, although they have had immense palaeoclimatic significance, are of less immediate concern to contemporary climatologists than the radiative forcing agents. Radiative forcing agents affect the supply and disposition of solar radiation. Solar radiation changes, like the non-radiative forcing agents, are external inputs into the atmosphere/earth/ocean/ice system but, unlike them, occur at a range of time-scales varying from ten to hundreds of thousands and, probably, millions of years. Thus solar radiation is both a long-term and a short-term external forcing agent.

Internal radiative forcing agents are mainly those which involve changes in atmospheric composition, cloud cover, aerosols and surface albedo. Although subject to long-term changes, it is their susceptibility to short-term anthropogenic changes which makes them of particular interest to contemporary climatologists. The interactive relations between short-term external solar radiative forcing and these internal radiative forcing agents lie at the heart of the understanding and prediction of short-term global climatic changes, through a complex set of *feedback mechanisms* which can be either *positive* (i.e. self-enhancing) or *negative* (i.e. self-regulating or damping).

Positive feedback mechanisms affecting global climate appear to be widespread and to be particularly effective in response to temperature changes, which is a matter of especial current concern. Increases in global temperature produce increases in atmospheric water vapour, increases in plant respiration, decreases of CO_2 dissolved in the oceans, and an increase in methane emissions from wetlands. All of these, in turn, tend to increase the global concentration of greenhouse gases and, hence, to increase global temperature further. Ice and snow cover is involved in especially important positive feedback effects in that a more extensive cover creates higher albedo and lower temperatures, which, in turn, will further extend the ice and

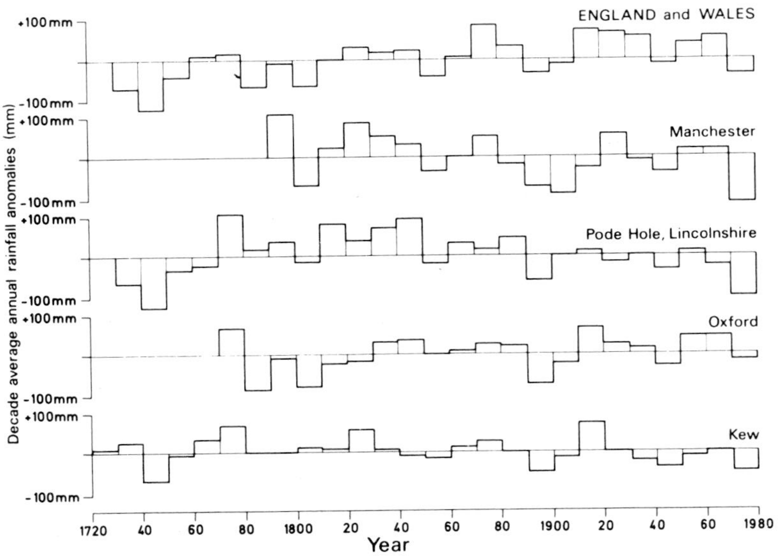

Figure 8.12 Decadal anomalies of average annual rainfall (mm) for England and Wales and for four individual stations in England (the 'last decade' includes data up to September 1978).

Source: After Kelly 1980.

snow cover, producing additional cooling. Conversely, a warming effect which melts ice and snow cover decreases the surface albedo, allows the absorption of more incoming solar radiation and leads to increased surface heating and higher temperatures.

Negative feedback mechanisms appear to be, unfortunately, much less important in the face of short-term radiative forcing and it is important to understand that, for example, they can only *reduce* the rate of warming but cannot, of themselves, cause global cooling. Cloud cover is a particularly complex global feedback mechanism, producing both positive and negative effects. For example, negative feedback may operate when increased global heating leads to greater evaporation and greater amounts of high-altitude cloud cover which will reflect more incoming solar radiation and thus damp down the global temperature rise.

2 Long-term changes

Several ice ages have been identified in the geologic past at intervals of 250 million years or more. These seem to coincide with the occurrence of continents in high northern or southern latitudes through continental drift or, in its modern interpretation, 'plate tectonics'. Nevertheless, it appears that although this locational effect is a necessary factor, it may not be a sufficient cause for ice ages or very warm climates of the geological past. The Cretaceous period apparently experienced an anomalously warm global climate and, even allowing for the effects of a clustering of the continents in low latitudes, it seems that additional factors must be sought. The possibility that atmospheric CO_2 concentrations were several times greater than now – giving an amplified 'greenhouse' effect – is one suggested explanation.

A more certain link has been established between the astronomical variations in the earth's orbit around the sun and the earth's climate, building on the theory of M. Milankovitch. Three interacting variations occur that involve regular changes in (a) the shape of the elliptical orbit (with a time scale of about 95,000 years); (b) the tilt of the earth's axis of rotation (approximately 42,000 years); and (c) the time of year when the earth is closest to the sun, or perihelion (about 21,000 years). The first affects the total annual radiation received by the earth, whereas the amount of axial tilt (21.8°–24.4°) modifies the summer–winter contrast. The time of perihelion – at present on

Figure 8.13 Area-averaged annual precipitation data, smoothed by 11-year running means, for central Europe, 1881–1980. The Bauer series for West Germany is an arithmetic mean of 12 stations in the Federal Republic and 2 in the German Democratic Republic (GDR); the GDR series is a mean of 18 stations, the Hungarian series a mean of 10 stations. The series for Poland and Slovakia made double use of weighted arithmetic means of station data, first, for station heights and, second, for the geographical area of height intervals. The Bohemian series was derived by planimetry of monthly isohyetal maps.

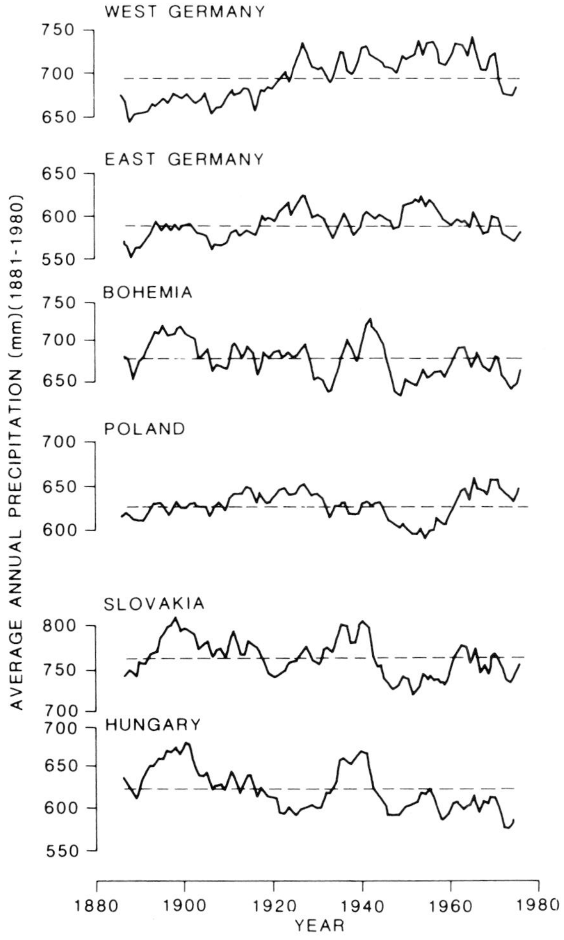

Source: From Brádzil *et al.* 1985 (by permission of the *Royal Meteorological Society*).

3 January (see Figure 1.16) – determines the relative warmth of summer/winter in the respective hemisphere. It has been shown that the timing of glacial conditions is favoured by a diminished seasonal contrast with mild winters and cool summers, i.e. small axial tilt and perihelion in the northern winter. Compared with the beginning of post-glacial time, when perihelion occurred in June, astronomical conditions today are closer to those of the last glaciation. However, a future (minor) glaciation is not expected for at least 3000–4000 years. These orbital effects have also been suggested as the basic cause of global fluctuations in the levels of tropical lakes (9°–27°N). These were high around 9000 years ago, when incoming solar radiation was about 25 W m^{-2} greater (less) than now in July (January) as a result of increased axial tilt. These lakes were very low around 15,000 years ago and again 5000 years ago.

The controls of interglacial–glacial transitions are still a subject of debate. Statistical analysis of various palaeoenvironmental variables (oceanic foraminifera, isotope records in marine shells and in ice cores) shows that climatic variability is pronounced at periods around 95,000, 41,000 and 23,000 years. The Milankovich orbital frequencies show closely identical maxima and phasing. However, the environmental records demonstrate maximum variability at the longest period (95,000), representing orbital eccentricity, despite the fact that this causes only small variations in global solar radiation compared with the other two frequencies. It appears that various terrestrial variables serve to reinforce the initial external forcing of the climatic shift. For example, the atmospheric concentration of carbon dioxide decreased from 280 ppm (interglacial) to 200 ppm (glacial) and that of methane from 650 to 350 ppbv. These reductions in greenhouse gases apparently followed global cooling by a few thousand years although CO_2 and temperature changes are broadly in phase over the last 150,000 years. Model experiments suggest that feedback effects of snow and ice cover on

Figure 8.14 The 'climate system'. External processes are shown by solid arrows; internal processes by open arrows.

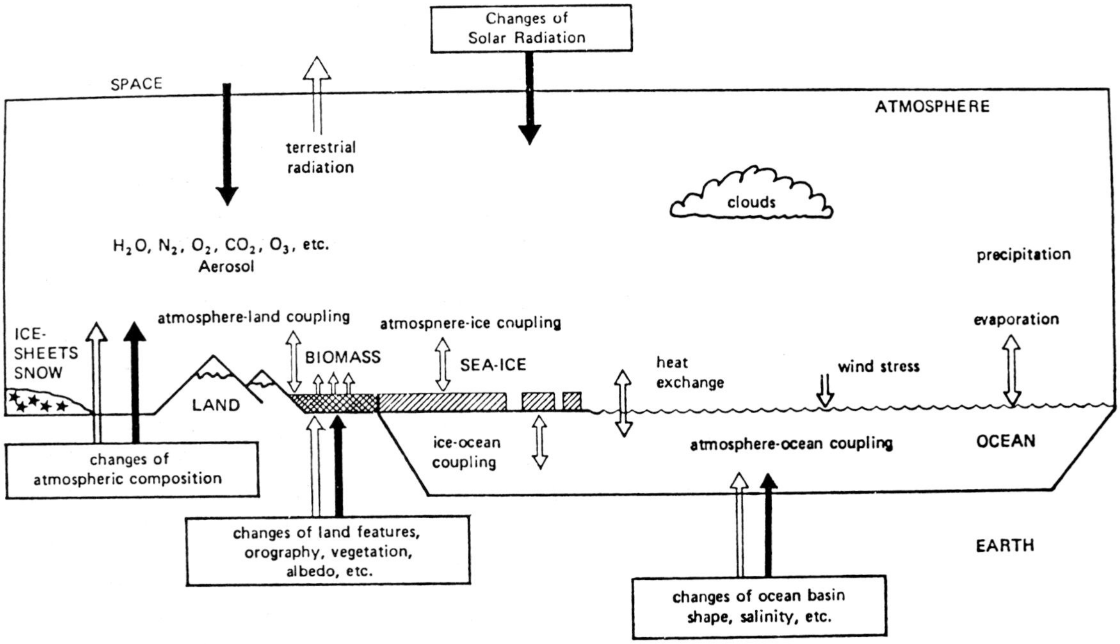

Source: After US GARP Committee.

albedo, and consequently temperature, may have played a major role.

Although the major glacial periods are now recognized as primarily due to astronomical causes, regional climatic changes of shorter duration may have been effected by many other mechanisms. The warming following the end of the Last Glaciation (*c.* 14,000 BP) was punctuated by a regional cooling (of as much as 6°C) in Europe and North America during the Younger Dryas (11,000–10,200 BP), which has been suggested as resulting from a breakdown of heat release in the North Atlantic due to a sudden massive pulse of fresh pro-glacial melt water draining along the St Lawrence outlet, so lowering the salinity of the North Atlantic and cutting off the 'Atlantic conveyor' circulation (see Chapter 3, F.3). The southward spread of sea-ice could also impede this conveyor-belt circulation, cutting off heat release to the atmosphere. This may have been one of the possible causes of the Little Ice Age (1550–1850), along with the occurrences of the Maunder sunspot minimum (see Figure 1.15) and a persistence of volcanic aerosols. It is noteworthy than an impeding of the conveyor belt due to an anomalous export of Arctic sea-ice may have produced the 'Great Salinity Anomaly' of 1968–82 in the northern North Atlantic.

Another possible variable over long time-scales is solar output. Sunspot cycles show no undisputed evidence of a link with weather conditions, although certain other short-term indicators of solar activity do suggest some possible relationships. Despite the absence of any observational evidence for changes of even 1–2 per cent in total output (see p. 17), however, astronomical theories of star birth, evolution and eventual 'death', spanning many billion years, suggest the likelihood of a variable sun.

3 Short-term changes

Remarkably, the causes of the observed climatic changes during the last few centuries are less well understood than those of the last glaciation. A multiplicity of possible explanations exists and, indeed, more than one factor is likely to be involved. It is useful to try to distinguish between the natural causes that have existed throughout the earth's history and those that are attributable to human activities in modifying the global environment. However, in many instances – such as the production of tropospheric aerosols – the two operate together.

a Circulation changes

The immediate cause of the recent climatic fluctuations appears to be the strength of the global wind circulation. The first 30 years of this century saw a pronounced increase in the vigour of the westerlies over the North Atlantic, the north-east trades, the summer monsoon of southern Asia and the southern hemisphere westerlies (in summer). Over the North Atlantic, these changes consisted of an increased pressure gradient between the Azores high and the Icelandic low, as the latter deepened, and also between the Icelandic low and the Siberian high which spread westward. These changes were accompanied by more northerly depression tracks, and this resulted in reduced continentality in Russia (see Figure 5.3) and a significant increase in the frequency of mild south-westerly airflow over the British Isles between about 1900 and 1930, as reflected by the average annual frequency of Lamb's Westerly airflow type (see Chapter 5, A.3). For 1873–97, 1898–1937, 1938–61 and 1962–91 the figures are 27, 38, 30 and 20 per cent respectively.

The recent decrease of westerly airflow, especially in winter, is linked with greater continentality in Europe and Russia (see Figure 5.3) and an increase of northerly airflow, giving cold spells and snowfalls in Britain. The southward shift of the main depression tracks produced a number of cool, wet summers in Britain (notably 1954, 1956, 1958 and 1960). These regional indicators reflect a general decline in the overall strength of the mid-latitude circumpolar westerlies, accompanying an apparent expansion of the polar vortex.

As has already been pointed out (see Chapter 3, D.4), global climate is closely related to the position and strength of the subtropical high-pressure cells. It has been estimated that a warming of the Arctic tropopause (winter +10°C; summer +3°C; annual +7°C), without changing equatorial or antarctic temperatures, would cause an annual shift of the subtropical high-pressure belt from its present average position of 37°N to 41°–43°N (i.e. some 100–200 km in summer but as much as 800 km in winter). This would bring drought to the Mediterranean, California, the Middle East, Turkestan and the Punjab; as well as displacing the thermal equator from 6°N to 9°–10°N, increasing the desertification in the belt 0°–20°S.

The key to these atmospheric variations must be linked to the heat balance of the earth/atmosphere system and this forces us to return to the fundamental energy considerations with which we began this book. The evidence for fluctuations in the 'solar constant' is inconclusive, although variations apparently do occur in the emission of high-energy particles and ultraviolet radiation during brief solar flares. All solar activity follows the well-known cycle of approximately 11 years, which is usually measured with reference to the period between sunspot maximum and minimum (see Figure 1.15) but numerous attempts to establish secure correlations between sunspot activity and terrestrial climates have produced mostly negative results. A statistical relationship has, nevertheless, been found between the occurrence of drought in the western United States over the last 300 years and the approximately 22-year double (Hale) cycle of the reversal of the solar magnetic polarity. Drought areas are most extensive in the 2–5 years following a Hale sunspot minimum (i.e. alternate 11-year sunspot minima).

Changes in atmospheric composition may

also have modified the atmospheric heat budget. The presence of increased amounts of volcanic dust and sulphate aerosols in the stratosphere is one suggested cause of the 'Little Ice Age'. Major eruptions can result in a surface cooling of perhaps 0.2°C for a few years after the event. Hence, frequent volcanic activity would be required for persistently cooler conditions. Conversely, it is suggested that reduced volcanic activity after 1914 may have contributed in part to the early twentieth-century warming. New interest in this question has been aroused by eruptions of El Chichón (March 1982) (see Chapter 1, A.4) and Mount Pinatub, Luzon (June 1991). It has been estimated that a huge volcanic eruption such as that of El Chichón, can, during a given decade, produce a forcing effect on global temperature about one-third as great as that exerted by greenhouse gases – *but in the opposite direction* (i.e. to produce surface cooling). The role of low-level aerosols is also complex. These originate naturally, from wind-blown soil and silt for example, as well as from atmospheric pollution due to human activities (industry, domestic heating and modern transportation).

b Anthropogenic factors

The growing influence of human activities on the environment is being increasingly recognized and concern over the potential for global warming caused by such anthropogenic effects is growing. Four categories of climatic variable are subject to change (see Table 8.1), and will now be considered in turn.

Changes in atmospheric composition associated with the explosive growth of world population, industry and technology have been described in Chapter 1, A.4, and it is clear that these have led to drastic increases in the concentration of greenhouse gases. The tendency of these increases is to increase radiative forcing and global temperatures; the percentage apportionment of radiative forcing by these greenhouse gas increases are summarized in Table 8.2. The radiative forcing effect of the minor trace gases is projected to increase steadily. Up to 1960, the cumulative CO_2 contribution since AD 1750 was about 67 per cent of the calculated 1.2 W m^{-2} forcing, whereas for 1980–90 the CO_2 contribution decreased to 56 per cent with CFCs contributing 24 per cent and methane 11 per cent. For the entire period from AD 1765 to 2050 the CO_2 contribution is projected to range from 4.15 W m^{-2}, out of a 6.5 W m^{-2} total (64 per cent), for a 'business-as-usual' scenario to 2.6 W m^{-2}, out of a 4.0 W m^{-2} total (65 per cent), if emission control policies are implemented rapidly.

The recent increase of global temperature forcing by the release of CFCs is particularly worrying. Ozone, which at high altitudes absorbs incoming short-wave radiation, is being dramatically destroyed above 25 km in the stratosphere (see Chapter 1, A.4) by emissions of H_2O and NO_x by jet aircraft and by surface emissions of N_2O by combustion and, especially, of CFCs. It is estimated that CFCs are now accumulating in the atmosphere five times

Table 8.1 The four categories of climatic variable subject to anthropogenic change.

Variable changed	*Scale of effect*	*Sources of change*
Atmospheric composition	Local–global	Release of aerosols and trace gases
Surface properties; energy budgets	Regional	Deforestation; desertification; urbanization
Wind regime	Local–regional	Deforestation; urbanization
Hydrological cycle components	Local–regional	Deforestation; desertification; irrigation; urbanization

Table 8.2 Percentage apportionment of radiative forcing due to greenhouse gas increases.

	CO_2	CH_4	CFCs	Stratospheric H_2O*	N_2O
Since 1795	61	17	12	6	4
1980–90	56	11	24	4	6

Note: *From the breakdown of CH_4.

faster than they can be destroyed by ultraviolet radiation. Ozone circulates in the stratosphere from low to high latitudes and thus the occurrence of ozone in polar regions is particularly diagnostic of its global concentration. In October 1984 an area of marked ozone depletion (the so-called 'ozone hole') was observed in the lower stratosphere (i.e. 12–24 km) centred on, but extending far beyond, the Antarctic continent. Ozone depletion is always greatest in the Antarctic spring but in this year the ozone concentration was more than 40 per cent less than that of October 1977. During the mid-1980s, Antarctic ozone concentrations fell to less than 200 Dobson units in September–October (see Figure 8.15), compared with some 400 units in the 1970s. It has been estimated that, because of the slowness of the global circulation of CFCs and of its reaction with ozone, even a cut in CFC emissions to the level of that in 1970 would not eliminate the Antarctic ozone hole for at least 50 years. However, this ozone reduction is also estimated to be sufficient to offset the radiative forcing effect of CFCs during the 1980s.

Figure 8.15 Mean ozone levels (Dobson units) over the southern hemisphere on 5 October 1987 obtained from the Total Ozone Mapping Spectrometer on NASA's Nimbus 7 satellite.

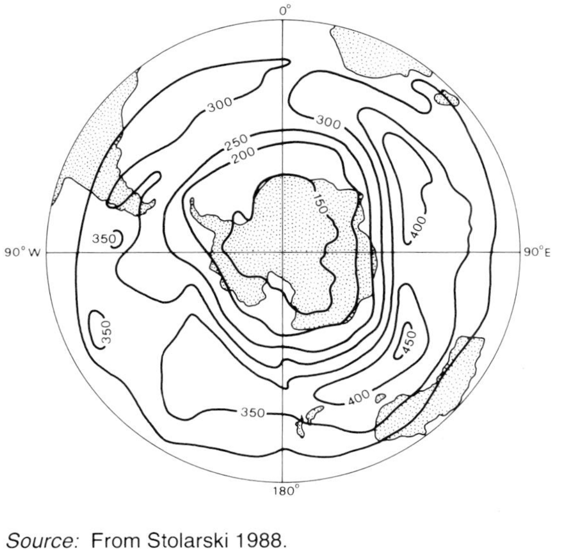

Source: From Stolarski 1988.

Indirect anthropogenic factors, such as increasing population pressures leading to overgrazing and forest clearance, may increase desertification which also contributes to the increase of wind-blown soil. The 'dust-bowl' years of the 1930s in the United States and the African Sahel drought since 1972 illustrate this. Evidence from central Eurasia shows a sharp rise in dust-fall on mountain snowfields since 1930, and atmospheric turbidity has increased by 57 per cent over Washington, DC, over the period 1905–64, and by 85 per cent over Davos, Switzerland (1920–58). The presence of particles in the atmosphere increases the backscatter of short-wave radiation, thereby increasing the planetary albedo and causing cooling, but the effect on infra-red radiation is one of surface warming. The net result is complicated by the surface albedo. Man-made aerosols cause net warming over snow and ice and most land surfaces, but cooling over the oceans which have a low albedo. Natural aerosols probably cause general cooling. The overall effect on global surface temperature appears to be one of cooling.

Changes in surface albedo occur naturally with season, but climatic forcing is also caused by anthropogenic vegetation changes. Human effects on vegetation cover have a long history. Deliberate burning of vegetation by Aborigines in Australia has been practised for perhaps 40,000 years. However, significant deforestation began in Eurasia during Neolithic times (*c.* 5000 BP), as evidenced by the appearance of

agricultural species and weeds. Deforestation expanded in these areas between about AD 700 and 1700 as populations slowly grew, but it did not take place in North America until the westward movement of settlement in the eighteenth and nineteenth centuries. During the last half-century extensive deforestation has occurred in the tropical rain forests of south-east Asia, Africa, and South America. Estimates of current tropical deforestation suggest losses of 10^5 km^2/year, out of a total tropical forest area of 9×10^6 km^2. This annual figure is more than half the total land surface at present under irrigation and twice the annual loss of marginal land to desertification. Forest destruction causes an increase in albedo of perhaps 10 per cent locally, with consequences for surface energy and moisture budgets. However, the large-scale effect of deforestation in temperate and tropical latitudes on global surface albedo is estimated to be < 0.001. It should also be noted that deforestation is difficult to define and monitor; it can refer to loss of forest cover with complete clearance and conversion to a different land use, or species' impoverishment without major changes in physical structure. The term desertification, applied in semi-arid regions, creates similar difficulties. The process of vegetation change and associated soil degradation are not solely attributable to human-induced changes, but are triggered by natural rainfall fluctuations.

Deforestation affects world climate in two main ways – first, by altering the atmospheric composition and, second, by affecting the hydrological cycle and local soil conditions:

1 Forests store great amounts of carbon dioxide, so buffering the carbon dioxide cycle in the atmosphere. The carbon retained in the vegetation of the Amazon basin is equivalent to at least 20 per cent of the entire atmospheric CO_2. Destruction of the vegetation would release about four-fifths of this to the atmosphere, of which about one-half would dissolve in the oceans but the other half would be added to the 16 per cent increase of atmospheric CO_2 already observed this century. The effect of this would be to accelerate the increase of world temperatures. A further effect of tropical forest destruction would be to reduce the natural production of nitrous oxide. Tropical forests and their soils produce up to one-half of the world's nitrous oxide which helps to destroy stratospheric ozone. Any increase in ozone would warm the stratosphere, but lower global surface temperatures.

2 Dense tropical forests have a great effect on the hydrological cycle through their high evapotranspiration and their reduction of surface runoff (about one-third of the rain never reaches the ground, being intercepted and evaporating off the leaves). Forest destruction decreases evapotranspiration, atmospheric humidity, local rainfall amounts, interception, effective soil depth, the height of the water table and surface roughness (and thereby atmospheric turbulence and heat transfer). Conversely, deforestation increases

Figure 8.16 Comparison of different radiative forcing mechanisms for a 50-year period in the future. The greenhouse gas forcings are for the period 2000–2050, using the four policy scenarios. Forcings due to changes in solar variability and sulphur emissions could be either positive or negative over the two periods.

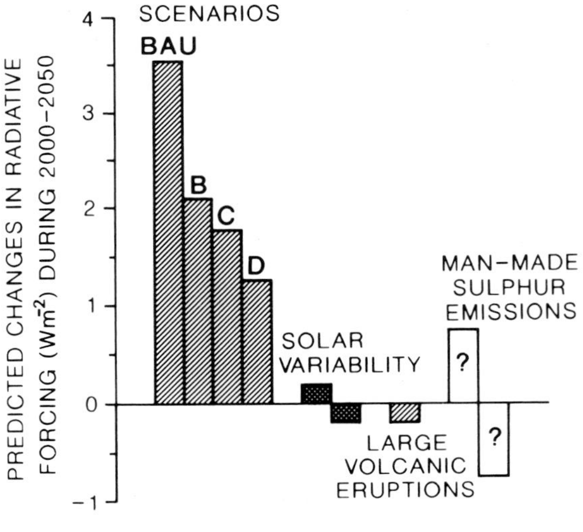

Source: From IPCC 1990.

the seasonality of rainfall, surface runoff, soil erosion, soil temperatures and surface albedo (and therefore near-surface air temperatures). All these tendencies operate to degrade existing primary and secondary tropical forests into savanna. Models designed to simulate the operation of Amazonian forests having a 27°C air temperature and a mean monthly rainfall of 220 mm (falling in four showers every third day each lasting 30 minutes at an intensity of 0.003 mm s^{-1}) predict that their degradation to savanna conditions would lead to a decrease of evapotranspiration by up to 40 per cent, an increase of runoff from 14 per cent of rainfall to 43 per cent, and an average increase of soil temperature from 27°C to 32°C.

D PREDICTED CHANGES OVER THE NEXT 100 YEARS

The so-called long-range forecasting which has been described earlier (see Chapter 4, I.2a), based on assumptions of persistence in general circulation patterns or with reference to analogous weather systems, looked forward for only a month or so. However, concern regarding the longer-term results of possible anthropogenic effects has prompted much recent speculation and research directed towards global climatic change during the coming century. For this purpose global mathematical models (see Chapter 3, G) have been employed, assuming radiative forcing by changes in atmospheric composition and based on assumptions of feedback within the atmosphere/earth/ocean/ice system (see C.1, this chapter).

Figure 8.17 Predicted changes of CO_2, CH_4 and CFC 11 between 1980 and 2100 with 'business as usual' (BAU) and with three other scenarios (B–D).

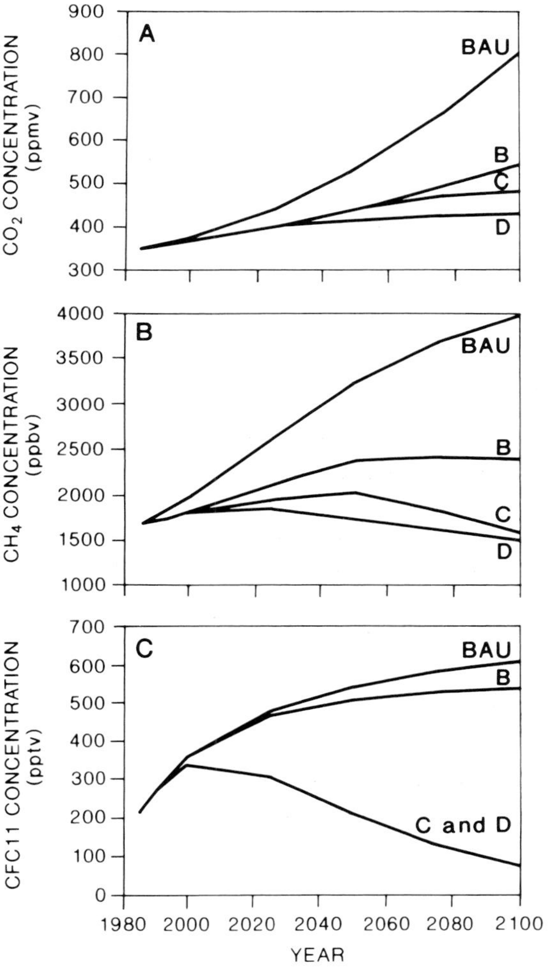

Source: From IPCC 1990.
Note: Units are in parts per million by volume (ppmv), parts per billion (ppbv), and parts per trillion (pptv), for A, B, and C respectively.

1 Atmospheric composition

The striking increases in the atmospheric concentration of greenhouse gases (especially CO_2) since the growth of industry and population, especially of most recent years (see Chapter 1, A.4), has attracted scientific attention as potentially the most potent climatic forcing agent. The current addition of CO_2 to the atmosphere by combustion is three to five times more important than that resulting from deforestation, and it has been estimated that with 'business as usual' (BAU) by the year 2070 its atmospheric concentration will be more than twice the level which existed during the pre-industrial era. Such estimates are, of course, very

speculative because of inexact knowledge regarding significant parts of the global carbon cycle. Although the oceans are the biggest store of global carbon, very little is in the form of CO_2 and the total marine biomass contains only half of one per cent of the carbon stored in terrestrial vegetation. The net flux of CO_2 into the oceans is not known but possibly up to 50 per cent of CO_2 produced from fossil fuel burning is taken up by the oceans. It is ignorance of such important mechanisms as this which causes the large bands of error which are associated with all current predictions resulting from global modelling.

It is important to view the potential effects of radiative forcing mechanisms on global climate in an integrated manner. Reference has been made to the estimated contributions of greenhouse gases; for AD 2000–2050, these total about 1.3 to 3.5 W m^{-2} according to the global energy scenarios adopted. In contrast, the contributions from possible large volcanic eruptions, from solar variability and from anthropogenic emissions of sulphur are considerably smaller and even of uncertain sign (see Figure 8.16). These may be ± 0.1 W m^{-2} for solar variations, -0.2 W m^{-2} in a decade with a major eruption and ± 0.75 W m^{-2} for the effects of SO_2 emissions on cloud condensation nuclei and therefore cloud albedo. These cloud effects will act regionally where SO_2 emissions are concentrated, greatly complicating their estimation. Tropospheric aerosols have an important, but poorly known, effect on global climate. Anthropogenic aerosols account for at least 25 per cent of all aerosols, and possibly as much as 50 per cent. Their radiative forcing may amount to between 0.5 and 1.5 W m^{-2}, or between 50–60 per cent of the greenhouse forcing, but of opposite sign (i.e. cooling). The effect of recent changes in surface characteristics on land albedo are very minor ($\leqslant 0.03$ W m^{-2}), although their implications for changes in regional water budgets are certainly of concern. For example, summer soil moisture levels are projected to decrease by 15 to 20 per cent in central North America and southern Europe by AD 2030 for the business-as-usual emissions scenario.

It has been suggested that to stabilize the concentration of greenhouse gases at present levels would require the following *percentage reductions* in emissions resulting from human activities: CO_2 > 60 per cent; CH_4 15–20 per cent; N_2O 70–80 per cent; CFCs 70–85 per cent. Such immediate huge reductions are clearly economically and politically impracticable and therefore the predictive models have been used by the Intergovernmental Panel on Climate Change (1990) to explore the possible effects of a range of scenarios. With 'business as usual' the concentration of CO_2 by the year 2100 may be expected to be 236 per cent of that in 1980; of CH_4, 242 per cent; and of CFC 11, 350 per cent (see Figure 8.17). Assuming the most severe cutback scenario (D), namely with CO_2 reduced to 50 per cent of its present level by the year 2050 and those of other major greenhouse gases proportionately, the resulting percentage changes of the concentration of greenhouse gases by the year 2100 (compared with 1980) might be expected to be: CO_2, 129 per cent; CH_4, 91 per cent; CFC 11, 57 per cent.

2 Temperature and humidity

An early use of mathematical models to attempt an understanding of global temperature changes involved the superimposition of the assumed effects of a number of factors in an attempt to simulate known historical changes and thereby to evaluate their relative influence, which could then be projected into forecasts. One such model for the northern hemisphere has employed estimates of temperature variations associated with the observed increase of CO_2, with the 76-, 22- and 12.4-year oscillations of solar output, together with the observed 100-year effects of volcanic aerosols to produce a composite temperature curve (see Figure 8.18) which compares well with the observed temperature record. However, not all workers would agree with the particular temperature response functions adopted for this simulation, or with

Figure 8.18 Mathematical model of temperature variations for the northern hemisphere associated with observed CO_2 and volcanic dust measurements and with three solar energy cycles over the past century. The composite curve for these five causes compares well with the observed temperature record (dots) which has been smoothed with a 1.7-year filter.

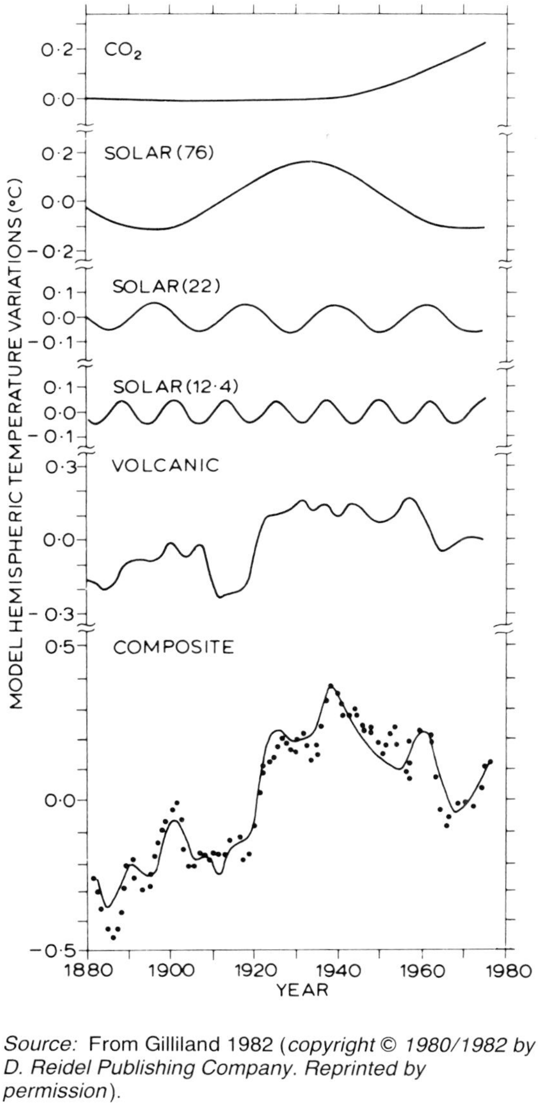

Source: From Gilliland 1982 (*copyright © 1980/1982 by D. Reidel Publishing Company. Reprinted by permission*).

their relative weighting. Further, although reasonable predictions of CO_2 levels and of regular solar oscillations are possible, that of concentrations of volcanic aerosols is obviously not.

More recent global temperature predictions (see Figure 8.19) are centred mainly on estimates of projected emissions of greenhouse gases, based on 'business as usual' with emissions continuing to grow at present rates, through intermediate scenarios (B and C) to scenario D which was introduced in D.1, this chapter.

More sophisticated spatially differentiated global simulation models have been employed by the Intergovernmental Panel on Climate Change (1990) who ran some twenty numerical simulation models to equilibrium (reached at *c.* 2030) for the period 1850–2050, assuming a doubling of CO_2 during that period. The major predictions were as follows:

1 A general warming of the earth's surface and troposphere (by up to more than 12°C in some high southern latitudes), and a cooling of the stratosphere and upper atmosphere as a result of the necessity for radiative equilibrium between the incoming solar radiation and the outgoing terrestrial radiation. The stratosphere and mesosphere would cool by about 10°C and the thermosphere by 50°C, for CO_2 doubling, perhaps causing more frequent noctilucent clouds in the polar mesosphere.
2 A consequential increase in the overall temperature lapse rate making the atmosphere more turbulent, increasing the frequency and intensity of depressions, tropical storms and hurricanes, and making mid-latitude winter storm tracks more stable in their positions.
3 Stronger warming of the earth's surface and troposphere in higher latitudes (especially at about 60° latitude) in late autumn and winter (by 4°–8°C), compare Figure 8.9 for example.
4 A tropical warming less than the global mean (i.e. 2°–3°C).

Figure 8.19 Simulated increase of global mean temperature above that for 1765 based on observed increases of greenhouse gases 1850–1990 and predictions of rises 1990–2100 assuming 'business as usual' (BAU) and three control scenarios (B–D).

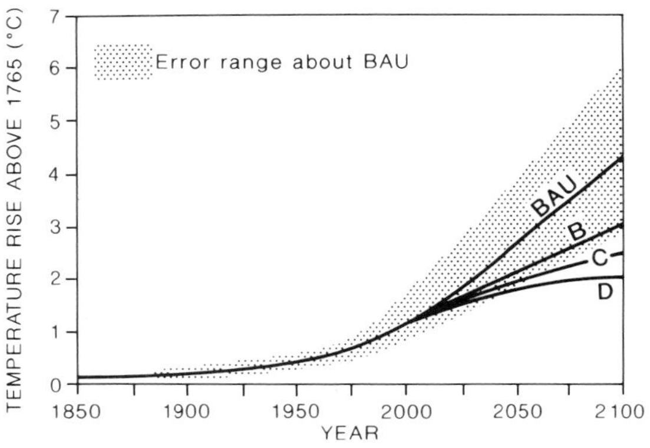

Source: From IPCC 1990.
Note: D represents accelerated implementation of emission controls.

5 Amplified warming over northern mid-latitude continents in summer (by 4°–6°C).
6 Increased precipitation in high latitudes and in the tropics throughout the year (by +10–20 per cent), and in mid-latitudes in winter. This increase is mostly associated with zones of lower-level convergence (i.e. mid-latitude storm tracks and the ITCZ).
7 Small changes only in the dry subtropics.
8 Soil moisture increases in winter and decreases in summer in northern mid-latitude continents.
9 A reduced extent of sea-ice extent and thickness.

3 Sea level

The mechanisms influencing sea level over the globe are extremely complex. Present sea level is not easy to define, estimates of sea levels over the past 100 years are difficult to make, and predictions over the next 100 years highly speculative. Sea-level changes are influenced by the following mechanisms (those of short time-scales – i.e. tens of years – italicized):

1 *Changes in ocean water mass – e.g. exchanges with glaciers; changes in the atmosphere/earth/ocean/ice/water distribution.*
2 *Changes in ocean water volume – e.g. thermal expansion and contraction; salinity changes; changes in atmospheric pressure.*
3 Changes in earth crustal levels.
  (a) Tectonic – e.g. rise of ocean ridges; sea-floor subsidence; plate movements.
  (b) Isostatic – e.g. tectonic loading; *ice and water loading.*
4 Changes in the global distribution of water.
  (a) Terrestrial rotation effects.
  (b) Global axis changes.
  (c) Terrestrial gravity variations.
  (d) Changes in the attraction of sun and moon.
  (e) *Changes in the velocity of ocean currents.*

Over the past 100 years the general global sea level has risen by 10–15 cm or more (see Figure 8.20). This rise has been attributed proportionately to the following causes:

1 Thermal expansion of ocean waters (50 per cent or more). This is difficult to estimate due to lack of knowledge regarding oceanic circulations, such that estimates vary from 30 to 60 per cent.
2 Glacier and small ice-cap melting (30 per cent i.e. 4.6 cm ± 2.6 cm). Estimates of this contribution to sea-level rise go as high as 48 per cent.
3 Greenland ice-cap melting (15 per cent – i.e. 2.3 cm ± 1.6 cm). This could be as great as 25 per cent and as little as 5 per cent.
4 Antarctic ice-sheet melting (3 per cent). The Antarctic ice sheet is a large and complex system with its own internal mechanisms and a mass balance which changes slowly. This source has probably not yet contributed greatly to the global sea-level rise but may do so very significantly in the future.

Forecasts of sea level changes during the next 100 years exhibit wide regions of possible error around a probable general rise (see Figure 8.21). Table 8.3 gives predictions for some more specific locations:

Table 8.3 Predictions of seasonal changes at specific locations (1850–2050).

Location	*Season*	*Temperature*	*Rainfall*	*Soil Moisture*
Central N. America	Winter	+2°–4°C	+0–15%	–
	Summer	+2°–3°C	−5–10%	−15–20%
India	Winter	+1°–2°C	–	–
	Summer	annually	+5–15%	+5–10%
Sahel	Winter	+1°–3°C	Changes marginal and	
	Summer	annually	regularly variable	
Southern Europe	Winter	+2°C	Some increase –	
	Summer	+2°–3°C	−5–15%	−15–25%
Australia	Winter	+2°C	–	–
	Summer	+1°–2°C	+10%	–

This area of considerable uncertainty derives largely from a lack of knowledge regarding the structure of the West Antarctic ice sheet and regarding the extent to which it is keyed to bedrock. If it is not well grounded, it may float and melt bodily as sea level rises, carrying a strong positive feedback effect. The time-scale for this to happen is 300 years or so. A further uncertainty is connected with the possibility that global warming may cause an increase in high-latitude snowfall leading to an *increase* in the volume of global ice and a lowering of sea level.

4 A cautionary postscript

Although there is considerable scientific consensus for the projected responses of the climate system to increasing trace gas concentrations in the atmosphere, various caveats must be recognized. First, cloud cover exerts a major influence on the radiation balance. Small increases of low cloud amount of only 15–20 per cent could essentially cancel out the greenhouse warming effect. Second, the role of the oceans in heat transport and in CO_2 uptake is

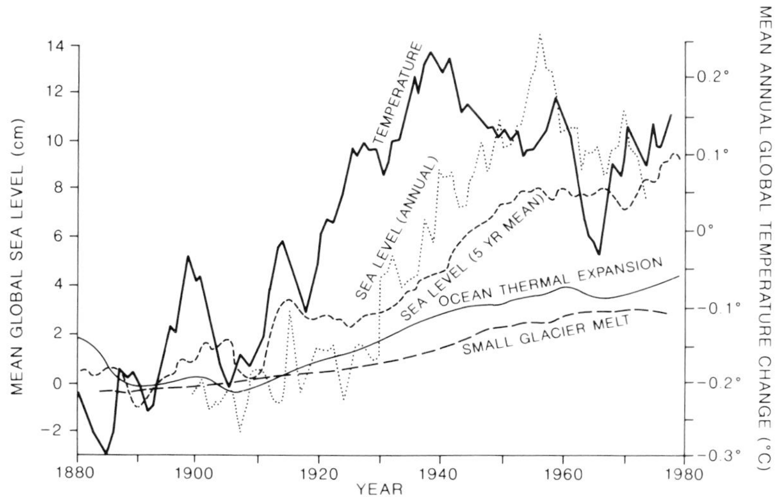

Figure 8.20 Conservative estimates of annual and 5-year running mean global sea-level changes (1880–1980), together with estimates of the contributions to these changes made by the thermal expansion of ocean waters and by the melting of small glaciers and ice caps (all values in cm). Mean annual global temperature changes (°C) are also given.

Source: From G. de Q. Robin in *The Greenhouse Effect, Climatic Change, and Ecosystems* (SCOPE 29) by Bolin, B., Döös, B. R., Jäger, J. and Warrick, R. A. (eds). (*Copyright © 1986; reprinted by permission of John Wiley & Sons, Inc.*)

Figure 8.21 Predicted sea-level rises based on 'business as usual' emissions (BAU), together with the possible error range between high and low estimates. Also illustrated are estimates associated with three other scenarios (B–D).

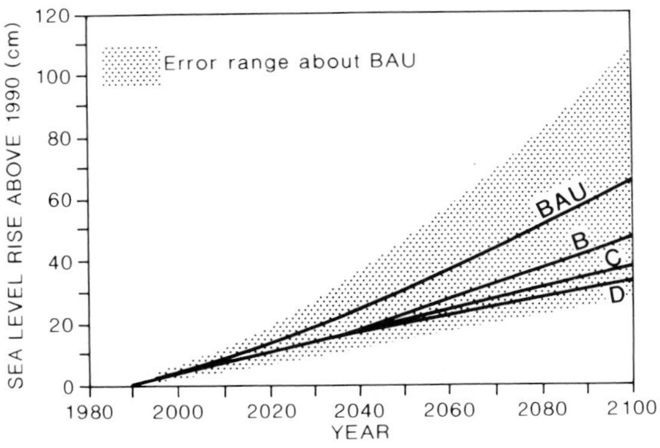

Source: From IPCC 1990.

complex and poorly known. At the least, the oceans may cause a delay of several decades in the climate system reaching equilibrium with greenhouse gas forcing. Third, the inherent variability of the climate system, and especially its response to *rapid* changes in radiative forcings, is unknown. Model experiments indicate the possibility of significant *random* shifts in the earth's climate in either direction. Revised predictions for the period 1990–2100, based on new estimates of greenhouse gas emissions, feedback effects, atmospheric albedo and absorption of CO_2 by the biosphere, suggest that the IPCC (1990) predictions of rises in global mean temperature and sea level should be reduced by at least 20 per cent, according to a 1992 study.

SUMMARY

Climatic data are described in terms of an average value (mean, median, mode) and a measure of the variability about the average (standard deviation, range, and so on). For skewed distributions, such as daily rainfall amounts, it is essential to know the frequency distribution.

Changes in climate involve factors both external to and within the climate system. External ones include solar variability, astronomical effects of the earth's orbit, and tectonic activity. Internal factors include variability within the atmosphere and ocean, and their feedbacks. During the last century, man-induced climatic change on local and global scales has become a reality primarily through changes in atmospheric composition and surface properties.

Climatic changes on geological time-scales involve continental drift, volcanic activity and possible changes in solar output. Within the last few million years glacial–interglacial cycles appear to be strongly controlled by astronomical variations in the earth's orbit, although atmosphere–ocean–cryosphere feedbacks must also be involved.

During the twentieth century there has been a significant average global temperature increase of some 0.5°C, greatest in the high-latitude maritime regions, particularly during the periods 1920–1940s and since the early 1970s. Precipitation trends are less clear, particularly in the mid-latitudes, but the precipitation of dry subtropical regions has tended to oscillate widely. Climatic behaviour during the past 20 years has tended to support a growing impression that the anthropogenically induced increase of greenhouse gases is permanently affecting global climate.

Possible causes of climatic change are examined from the point of view of the atmosphere/earth/ocean/ice global system and with respect to forcing and feedback mechanisms. Whereas longer-term changes are probably due to astronomical forcing mechanisms, short-term changes (i.e. last 100 years) appear to be more obviously linked to anthropogenic-factors. These are mainly changes in atmospheric composition, including depletion of ozone, and destruction of forest vegetation.

Predictions covering the next 100 years

point to a 236 per cent increase of CO_2 with 'business as usual', which might produce global temperature increases of 2°C and 4°C by the years 2030 and 2100, respectively, together with sea-level rises of 20 cm and 60 cm. The magnitude of such predictions, based on computer modelling, are very speculative however and are subject to very large error bands due to our restricted knowledge of the operations of the atmosphere/earth/ocean/ice global system.

APPENDIX 1
Climatic classification

The purpose of any classification system is to obtain an efficient arrangement of information in a simplied and generalized form. Thus, climatic statistics can be organized in order to describe and delimit the major types of climate in quantitative terms. Obviously no single classification can serve more than a limited number of purposes satisfactorily and many different schemes have therefore been developed. Some schemes merely provide a convenient nomenclature system, whereas others are an essential preliminary to further study. Many climatic classifications, for instance, are concerned with the relationships between climate and vegetation or soils, but surprisingly few attempts have been made to base a classification on the direct effects of climate on man.

Only the basic principles of the four groups of the most widely known classification systems are summarized here. Further information may be found in the listed references.

A GENERIC CLASSIFICATIONS RELATED TO PLANT GROWTH OR VEGETATION

The numerous schemes that have been suggested for relating climatic limits to plant growth or vegetation groups rely on two basic criteria – the degree of aridity and of warmth.

Aridity is not simply a matter of low precipitation, but of the 'effective precipitation' (i.e. precipitation minus evaporation). The ratio of rainfall/temperature has been used as such an index of precipitation effectiveness, on the grounds that higher temperatures increase evaporation. The ratio r/t was proposed by R. Lang in 1915 (where r = mean annual rainfall in mm, and t = mean annual temperature in °C), such that $r/t < 40$ is considered arid and $r/t > 160$ perhumid.

The work of W. Köppen is the prime example of this type of classification. Between 1900 and 1936 he published several classification schemes involving considerable complexity in their full detail. Nevertheless, the system has been used extensively in geographical teaching. The key features of Köppen's final classification are temperature criteria and aridity criteria.

Temperature criteria

Five of the six major climatic types are recognized on the basis of monthly mean temperature.

A Tropical rainy climate: coldest month > 18°C (64.4°F).
B Dry climates.
C Warm temperate rainy climates: coldest month between −3° and 18°C, warmest month > 10°C (50°F).
D Cold boreal forest climates: coldest month < −3° (26.6°F), warmest month > 10°C (see note 1).
E Tundra climate: warmest month 0°–10°C.
F Perpetual frost climate: warmest month < 0°C.

The arbitrary temperature limits stem from a variety of criteria, the supposed significance of the selected values being as follows: the 10°C summer isotherm correlates with the poleward limit of tree growth; the 18°C winter isotherm is critical for certain tropical plants; and the −3°C isotherm indicates a few weeks of snow cover. However, these correlations are far from precise! The criteria were determined from the study of vegetation groups defined on a physiological basis (i.e. according to the internal functions of plant organs) by De Candolle in 1874.

Aridity criteria:

Precipitation	*Steppe (BS)/ desert (BW) boundary*	*Forest/steppe boundary*
Winter precipitation maximum	$r/t = 1$	$r/t = 2$
Precipitation evenly distributed	$r/(t+7) = 1$	$r/(t+7) = 2$
Summer precipitation maximum	$r/(t+14) = 1$	$r/(t+14) = 2$

where: r = annual precipitation (cm)
t = mean annual temperature (°C)

The criteria imply that, with winter precipitation, arid (desert) conditions occur where $r/t < 1$, semi-arid conditions where $1 < r/t < 2$. If the rain falls in summer, a larger amount is required to offset evaporation and maintain an equivalent total of effective precipitation.

Subdivisions of each major category are made with reference, first, to the seasonal distribution of precipitation (the most common of which are: f = no dry season; m = monsoonal, with a short dry season and heavy rains during the rest of the year; s = summer dry season; w = winter dry season) and, second, to additional temperature characteristics. (For the B climates: h = mean annual temperature > 18°C; k = mean annual temperature < 18°C (warmest month > 18°C); k' = mean annual temperature < 18°C (warmest month < 18°C). Figure A.1.1A illustrates the distribution of the major Köppen climatic types on a hypothetical continent of low and uniform elevation.

A somewhat similar scheme has been proposed by A. A. Miller (1951), using the following criteria:

Boundary of arid conditions: $r/t = 1/5$.
Boundary of semi-arid conditions: $r/t = 1/3$.
Where: r = mean annual rainfall (in)
t = mean annual temperature (°F).

The thermal units relate to the *accumulated temperature*, which Miller estimated approximately by using 'month-degrees' – the excess of mean monthly temperatures above 43°F (6°C) – rather than the usual day-degrees based on daily mean temperatures above this limit.

C. W. Thornthwaite introduced a complex, empirical classification in 1931. An expression for *precipitation efficiency* was obtained by relating measurements of pan evaporation to temperature and precipitation. For each month the ratio

$$115(r/t - 10)^{10/9}$$

where: r = mean monthly rainfall (in)
t = mean monthly temperature (°F)

is calculated. The sum of the twelve monthly ratios gives the *precipitation efficiency* (*P-E*) index. By determining boundary values for the major vegetation regions the following humidity provinces were defined (with *P-E* index in parentheses): A: Rain forest (> 127); B: Forest (64–127); C: Grasslands (32–63); D: Steppe (16–31); E: Desert (< 16).

The second element of the classification is an index of *thermal efficiency* (*T-E*), expressed by the positive departure of monthly mean temperatures from freezing point. The index is thus the annual sum of $(t - 32)/4$ for each month. On this scale zero is 'frost climate' and over 127 is 'tropical'. Unlike Köppen, Thornthwaite makes moisture the primary classificatory factor for a *T-E* index of over 31 (the taiga/cool temperate boundary). Maps of the distribution of these

Figure A.1.1 A: The distribution of the major Köppen climatic types on a hypothetical continent of low and uniform elevation. Tw = mean temperature of warmest month; Tc = mean temperature of coldest month. B: The distribution of Flohn's climatic types on a hypothetical continent of low and uniform elevation (see Note 1).

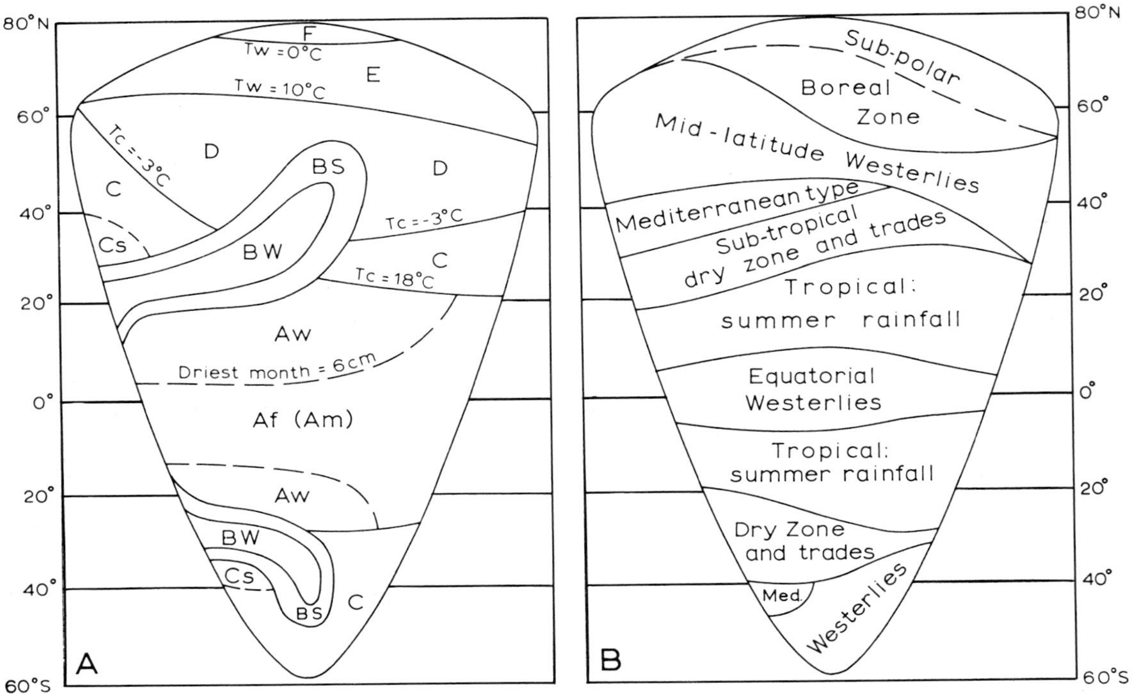

Source: From Flohn 1950.

climatic provinces in North America and over the world have been published, but the classification is now largely of historical interest.

B ENERGY AND MOISTURE BUDGET CLASSIFICATIONS

Thornthwaite's most important contribution was his second (1948) classification. It is based on the concept of potential evapotranspiration and the moisture budget (see Chapters 2, A and 5, B.3.c). The potential evapotranspiration (*PE*) is calculated from the mean monthly temperature (in °C), with corrections for day length. For a 30-day month (12-hour days):

$$PE \text{ (in cm)} = 1.6(10t/I)^a$$

where: I = the sum for 12 months of $(t/5)^{1.514}$
a = a further complex function of I.

Tables have been prepared for the easy computation of these factors.

The monthly water surplus (S) or deficit (D) is determined from a moisture budget assessment, taking into account stored soil moisture. A moisture index (Im) is given by:

$$Im = (100S - 60D)/PE$$

The weighting of a deficit by 0.6 is supposed to allow for the beneficial action of a surplus in one season, when moisture is stored in the subsoil, to be drawn on during subsequent droughts by deep-rooted perennials. In 1955 this weighting factor was omitted since it was recognized that a deficit can begin as soon as any moisture is

removed from the soil by evaporation. The later revision also allows for a variable soil moisture storage according to vegetation cover and soil type, and permits the evaporation rate to vary with the actual soil moisture content.

Thus, $Im = 100(S - D)/PE$ (Thornthwaite and Mather 1955; Mather 1985). The average water balance is calculated through a book-keeping procedure. For each month, the mean values of the following variables are determined in turn: PE, potential evapotranspiration (from the relationship given above); precipitation minus PE; Ws, soil water storage, a value assumed appropriate for that soil type at field capacity. Ws is decreased as the soil dries (ΔWs). AE, actual evapotranspiration. There are two cases: $AE = PE$, when Ws is at field capacity, or $(P - PE) > 0$; otherwise $AE = P + \Delta Ws$. The monthly moisture deficit, D, or surplus (S), is determined from $D = (PE - AE)$, or $S = (P - PE) > 0$ when $Ws \leq$ field capacity. The monthly deficits or surpluses are carried forward to the subsequent month.

A novel feature of the system is that the thermal efficiency is derived from the PE value because this itself is a function of temperature. The climatic types defined by these two factors are shown in Table A.1.1, both elements being subdivided according to the season of moisture deficit or surplus and the seasonal concentration of thermal efficiency.

The system has been applied to many regions, although no world map has yet been published. In tropical and semi-arid areas the method is not very satisfactory, but in eastern North America, for example, vegetation boundaries have been shown to coincide reasonably closely with particular PE values. This classification, unlike that of Köppen (1931) and many others, does not use vegetation boundaries to determine climatic ones.

M. I. Budyko in the Soviet Union developed a similar, but more fundamental approach using net radiation rather than temperature (see Chapter 2, A). He related the net radiation available for evaporation from a wet surface (R_o) to the heat required to evaporate the mean annual precipitation (Lr). This ratio R_o/Lr (where L = latent heat of evaporation) is called the *radiational index of dryness*. It has a value less than unity in humid areas and greater than unity in dry areas. Boundary values, with R_o/Lr in parentheses, are: Desert (> 3.0); Semi-desert (2.0–3.0); Steppe (1.0–2.0); Forest (0.33–1.0); Tundra (< 0.33). By way of comparison with the revised Thornthwaite index ($Im = 100(r/$

Table A.1.1 Thornthwaite's climatic classification

*Im (1955 system)**		*PE* *cm*	*PE* *in*	*Climatic type*
> 100	Perhumid (A)	> 114	> 44.9	Megathermal (A')
20 to 100	Humid (B_1 to B_4)	57 to 114	22.4 to 44.9	Mesothermal (B'_1 to B'_4)
0 to 20	Moist subhumid (C_2)	28.5 to 57	11.2 to 22.4	Microthermal (C'_1 to C'_2)
−33 to 0	Dry subhumid (C_1)	14.2 to 28.5	5.6 to 11.2	Tundra (D')
−67 to −33	Semi-arid (D)	< 14.2	< 5.6	Frost (E')
−100 to −67	Arid (E)			

Notes: * $Im = 100(S - D)/PE$ is equivalent to $100(r/PE - 1)$, where r = annual precipitation.

$PE - 1)$) it may be noted that $Im = 100(Lr/R_o - 1)$ if all the net radiation is used for evaporation from the wet surface (i.e. none is transferred into the ground by conduction or into the air as sensible heat). A general world map of R_o/Lr has appeared but over large parts of the earth there are as yet no measurements of net radiation.

Energy fluxes have also been used by Terjung and Louie (1972) to categorize the magnitude of energy input (net radiation and advection) and outputs (sensible heat and latent heat), and their seasonal range. On this basis, sixty-two climatic types are distinguished (in six broad groups), and a world map is presented.

C GENETIC CLASSIFICATIONS

The genetic basis of large-scale (or macro-) climates is the atmospheric circulation, and this can be related to regional climatology in terms of wind regimes or air masses. One attempt, made by A. Hettner in 1931, incorporated the wind system, rainfall amount and duration, position relative to the sea, and elevation. A very generalized scheme using air masses, according to their seasonal dominance, was put forward by B. P. Alissov in 1936.

A more satisfactory system, however, was proposed in 1950 by H. Flohn. His major categories, which are based on the global wind belts and the precipitation characteristics, are as follows:

1 Equatorial westerly zone: constantly wet.
2 Tropical zone, winter trades: summer rainfall.
3 Subtropical dry zone (trades or subtropical high pressure): dry conditions prevail.

Figure A.1.2 A genetic classification of world climates by E. Neef.

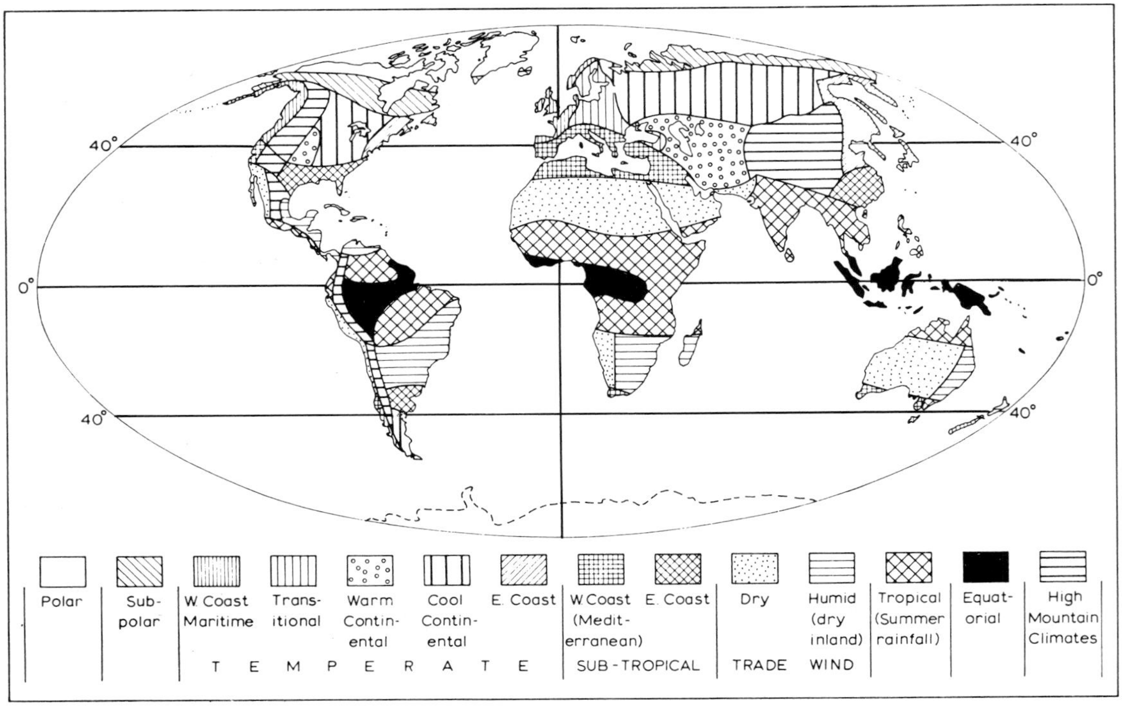

Source: From Flohn 1957.

4 Subtropical winter-rain zone (Mediterranean type): winter rainfall.
5 Extratropical westerly zone: precipitation throughout the year.
6 Subpolar zone: limited precipitation throughout the year.
6a Boreal, continental subtype: summer rainfall; limited winter snowfall.
7 High polar zone: meagre precipitation; summer rainfall, early winter snowfall.

It will be noted that temperature does not appear explicitly in the scheme. Figure A.1.1B shows the distribution of these types on a hypothetical continent. Rough general agreement between these types and those of Köppen's scheme is apparent. Note that the boreal subtype is restricted to the northern hemisphere and that the subtropical zones do not occur on the east side of a land mass. Flohn's approach has much to commend it as an introductory teaching outline. Although no world map of the distribution of these zones has been published, two maps prepared along similar lines by E. Neef and E. Kupfer were presented by Flohn in 1957. Neef's map is reproduced in Figure A.1.2.

Another simple, but extremely effective, genetic classification of world climates has been proposed by Strahler. He makes a major tripartite diversion into:

1 Low-latitude climates, controlled by equatorial and tropical air masses.
2 Middle-latitude climates, controlled by both tropical and polar air masses.
3 High-latitude climates, controlled by polar and arctic air masses.

These are subdivided into fourteen climatic regions, to which is added that of Highland Climates (see Table A.1.2). Figure A.1.3 shows the world distribution of these fifteen regions, and Figure A.1.4 gives mean monthly climatic data for representatives of thirteen of them.

Figure A.1.3 Simplified world map showing the distribution of Strahler's genetic climatic regions.

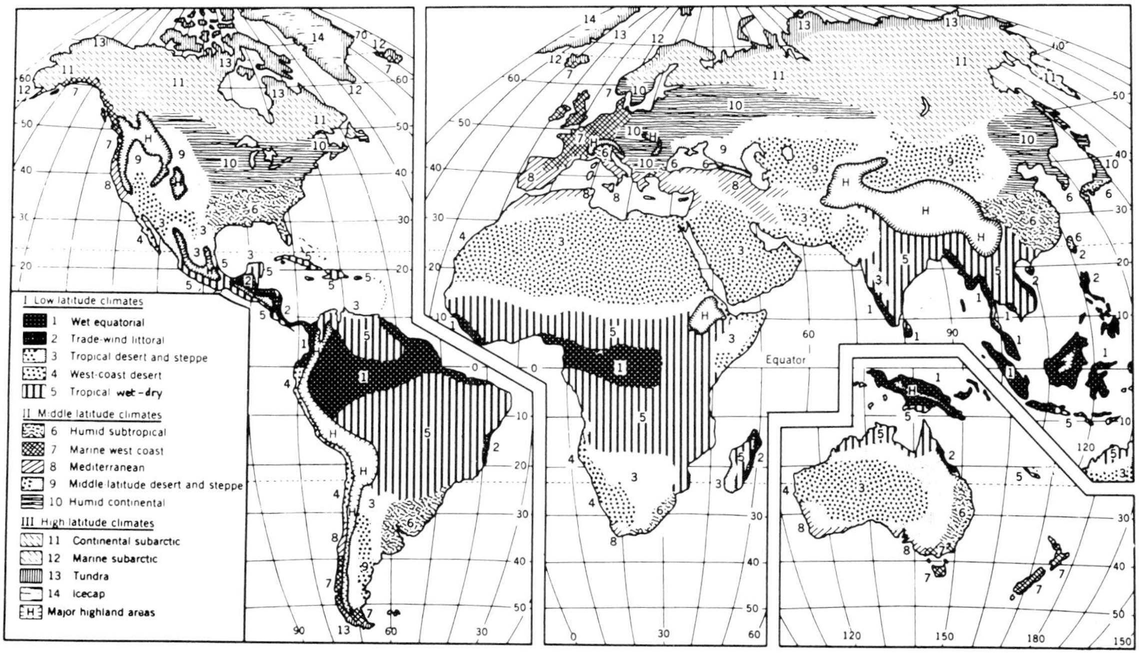

Source: From Strahler 1969.

Table A.1.2 Strahler's climatic classification.

Climate name	*Köppen symbol*		*Air mass source regions and frontal zones, general climate characteristics*
Group 1: Low-latitude climates (controlled by equatorial and tropical air masses)			
1 Wet equatorial climate 10°N–10°S lat. (Asia 10°–20°N)	*Af* *Am*	Tropical rain forest climate, and Tropical rain forest climate, monsoon type	Equatorial trough (convergence zone) climates are dominated by warm, moist tropical maritime (mT) and equatorial (mE) air masses yielding heavy rainfall through convectional storms. Remarkably uniform temperatures prevail throughout the year.
2 Trade wind littoral climate 10°–25°N and S lat.	*Af-Am*	Included in climate 1	Tropical easterlies (trades) bring maritime tropical (mT) air masses from moist western sides of oceanic subtropical high-pressure cells to give narrow east-coast zones of heavy rainfall and uniformly high temperatures. Rainfall shows strong seasonal variation.
3 Tropical desert and steppe climates 15°–35°N and S lat.	*BWh* *BSh*	Desert climate, hot, and Steppe climate, hot	Source regions of continental-tropical (cT_s) air masses in high-pressure cells at high level over lands astride the Tropics of Cancer and Capricorn give arid to semi-arid climate with very high maximum temperatures and moderate annual range.
4 West-coast desert climate 15°–30°N and S lat.	*BWk* *BWh*	Desert climate, cool, and Desert climate, hot (*BWn* in earlier versions, *n* meaning frequent fog)	On west coasts bordering the oceanic subtropical high-pressure cells, subsiding maritime tropical (mT_s) air masses are stable and dry. Extremely dry, but relatively cool, foggy desert climates prevail in narrow coastal belts. Annual temperature range is small.
5 Tropical wet-dry climate 5°–25°N and S lat.	*Aw*	Tropical rainy climate, savanna; also	Seasonal alternation of moist mT or mE air masses with dry cT air masses gives climate with wet season at time of high sun, dry season at time of low sun.
Group 2: Middle-latitude climates (controlled by both tropical and polar air masses)			
6 Humid subtropical climate 20°–35°N and S lat.	*Cfa* *Cwa*	Temperate rainy (humid mesothermal) climate, hot summers Temperate rainy (humid mesothermal) climate, dry winter, hot summer	Subtropical, eastern continental margins dominated by moist maritime (mT) air masses flowing from the western sides of oceanic high-pressure cells. In high-sun season, rainfall is copious and temperatures high. Winters are cool with frequent continental polar (cP) air mass invasions. Frequent cyclonic storms.
7 Marine west-coast climate 40°–60°N and S lat.	*Cfb* *Cfc*	Temperate rainy (humid mesothermal) climate, warm summers, and same but cool, short summers	Windward, middle-latitude west coasts receive frequent cyclonic storms with cool, moist maritime polar (mP) air masses. These bring much cloudiness and well-distributed precipitation, but with winter maximum. Annual temperature range is small for middle latitudes.
8 Mediterranean climate 30°–45°N and S lat.	*Csa* *Csb*	Temperate rainy (humid mesothermal) climate, dry, hot summer, and same, but dry, warm summer	This wet-winter, dry-summer climate results from seasonal alternation of conditions causing climates 4 and 7; mP air masses dominate in winter with cyclonic storms and ample rainfall, mT air masses dominate in summer and extreme drought. Moderate annual temperature range.

9 Middle-latitude desert and steppe-climates 35°–50°N and S lat.	*BWk* *BWk'* *BSk* *BSk'*	Desert climate, cool same, but cold; and Steppe climate, cool same, but cold	Interior, middle-latitude deserts and steppes of regions shut off by mountains from invasions of maritime air masses (mT or mP), but dominated by continental tropical (cT) air masses in summer and continental polar (cP) air masses in winter. Great annual temperature range; hot summers, cold winters.
10 Humid continental climate 35°–60°N lat.	*Dfa* *Dfb* *Dwa* *Dwb*	Cold, snowy forest (humid microthermal) climate, moist all year, hot summers; and same, but warm summers; also Cold, snowy forest (humid microthermal) climate, dry winters, hot summers; and same, but warm summers	Located in central and eastern parts of continents of middle latitudes, these climates are in the polar front zone, the 'battle ground' of polar and tropical air masses. Seasonal contrasts are strong and weather highly variable. Ample precipitation throughout the year is increased in summer by invading maritime tropical (mT) air masses. Cold winters are dominated by continental polar (cP) air masses invading frequently from northern source regions.
Group 3: High-latitude climates (controlled by polar and arctic air masses)			
11 Continental sub-arctic climate 50°–70°N lat.	*Dfc* *Dfd* *Dwc* *Dwd*	Cold, snowy forest (humid microthermal) climate, moist all year, cool summers; and same, but very cold winters; also Cold, snowy forest (humid microthermal) climate, dry winter, cool summer; and same, but very cold winter	This climate lies in source region of continental polar (cP) air masses, which in winter are stable and very cold. Summers are short and cool. Annual temperature range is enormous. Cyclonic storms, into which maritime polar (mP) air is drawn, supply light precipitation, but evaporation is small and the climate is therefore effectively moist.
12 Marine sub-arctic climate 50°–60°N and 45°–60°S	*ET*	Polar, tundra climate	Located in the arctic frontal zones of the winter season, these windward coasts and islands of subarctic latitudes are dominated by cool mP air masses. Precipitation is relatively large and annual temperature range small for so high a latitude.
13 Tundra climate north of 55°N south of 50°S		Polar, tundra climate	The arctic coastal fringes lie along a frontal zone, in which polar (mP, cP) air masses interact with arctic (*A*) air masses in cyclonic storms. Climate is humid and severely cold with no warm season or summer. Moderating influence of ocean water prevents extreme winter severity as in climate 11.
14 Ice-cap climate (Greenland, Antarctica)	*EF*	Polar climate, perpetual frost	Source regions of arctic (*A*) and antarctic (*AA*) air masses situated upon the great continental ice-caps have climate with annual temperature average far below all other climates and no above-freezing monthly average. High altitudes of ice plateaux intensify air mass cold.
Highland climates			Cool to cold moist climates, occupying high-altitude zones of the world's mountain ranges, are localized in extent and not included in classification system.

Source: From Strahler 1969.

Figure A.1.4 Climatic data for representative stations in thirteen of Strahler's climatic regions.

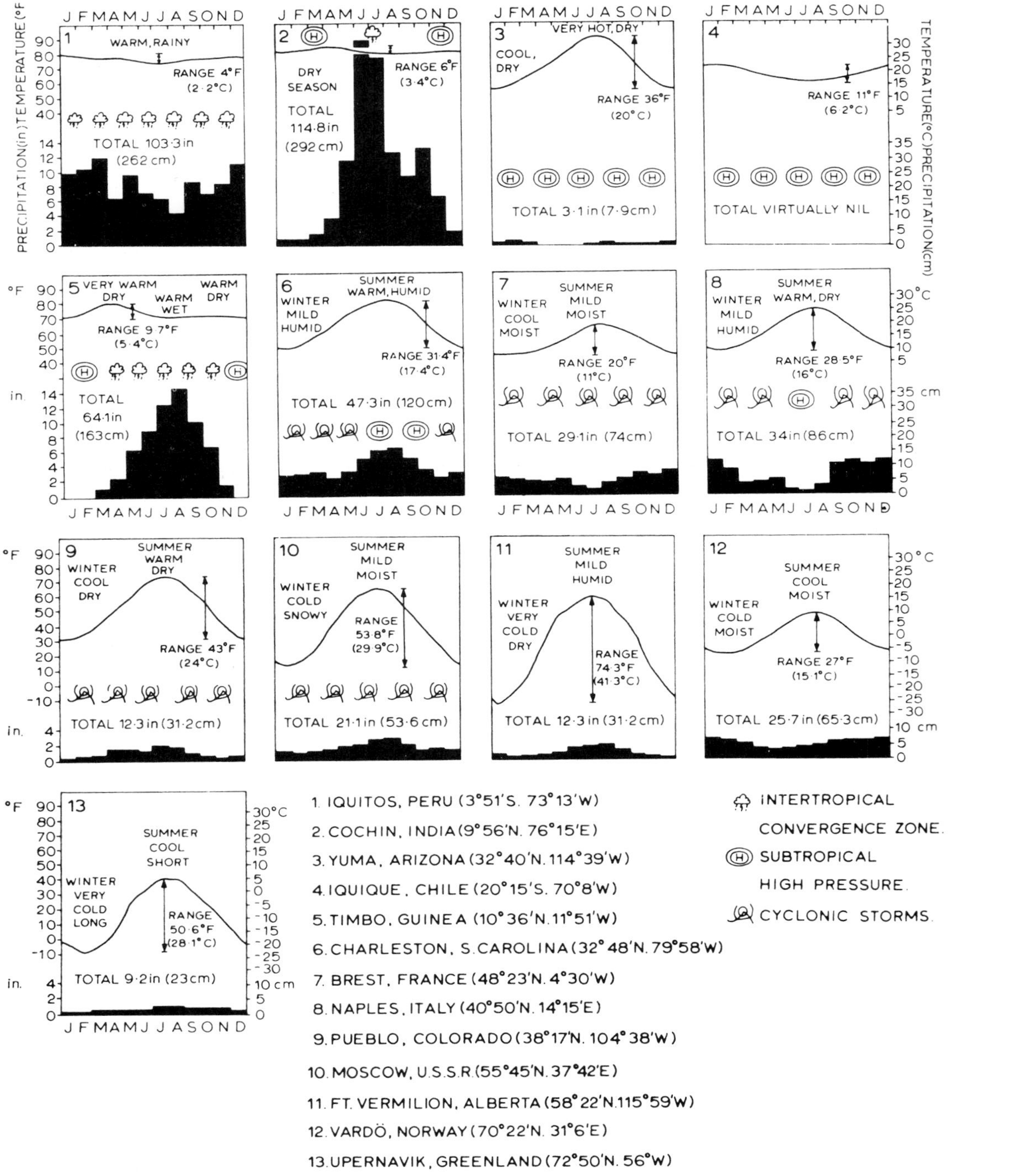

Source: Mostly after Strahler 1969.

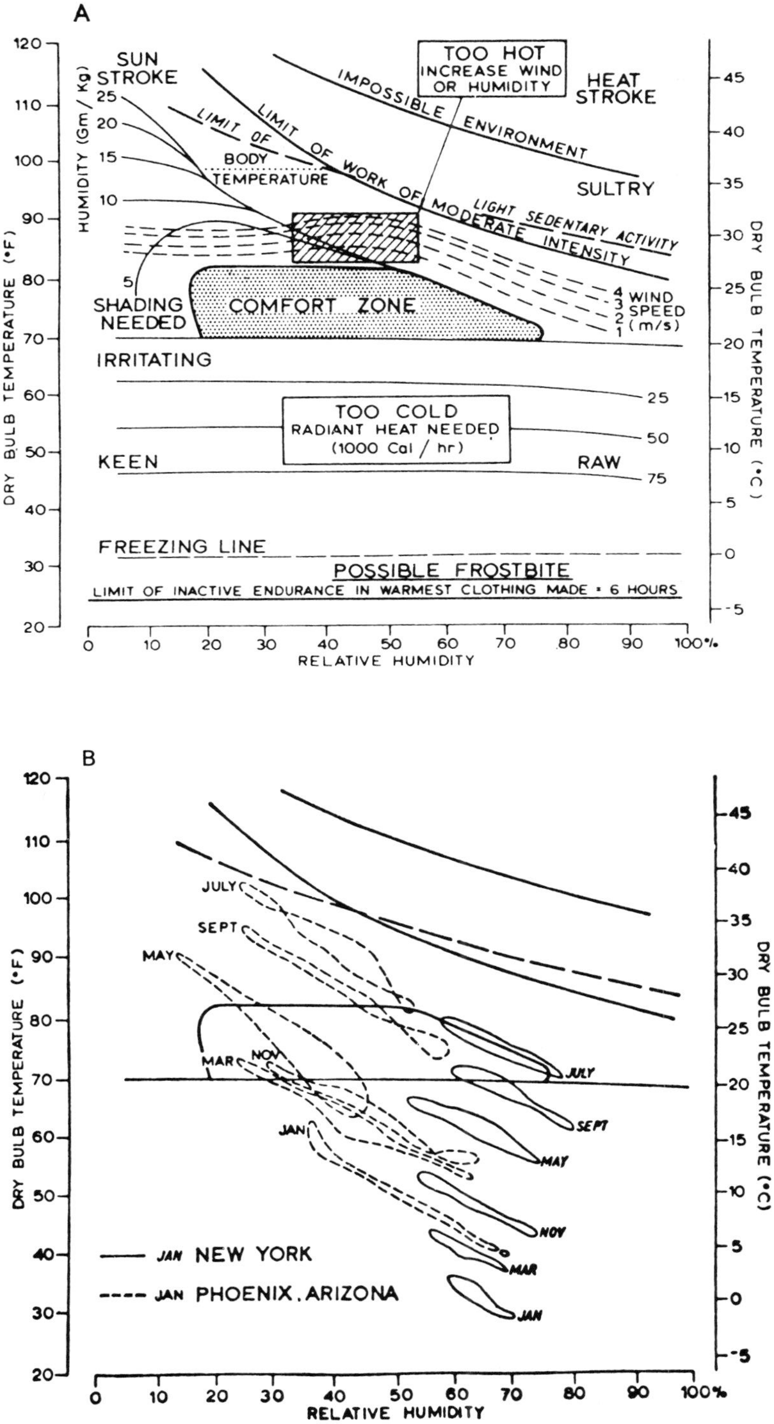

Figure A.1.5 The human relevance of climatic ranges. A: Atmospheric comfort, discomfort and danger for inhabitants of temperate climatic zones. Within the stippled limits of dry-bulb temperature and relative humidity the body feels comfortable at elevations of less than 300 m with customary indoor clothing, and performing sedentary or light work. Outside these limits corrective measures are necessary to restore the feeling of comfort. Below the 'comfort zone' radiant heat is needed, above it the skin temperature can be lowered by increasing the evaporation and heat convection either by raising the wind speed (for humid air) or by an increase of atmospheric moisture (for dry air). B: Mean daily dry-bulb temperatures and relative humidities during alternate months for New York City and Phoenix, Arizona, plotted on the 'comfort chart', indicating the need for both central heating and air conditioning.

Source: After Olgyay 1963.

D CLASSIFICATIONS OF CLIMATIC COMFORT

Climatic comfort indices have been established mainly by physiological experiments in test chambers. The most significant bioclimatic elements are air temperature, relative humidity, and wind speed (Buettner 1962). Figure A.1.5A illustrates the ranges of human comfort, discomfort and danger (heat stroke, frostbite) for one type of bioclimatic chart developed in the United States by V. Olgyay. The frequency of mean climatic conditions outside the *comfort range* in New York and Phoenix is shown in Figure A.1.5B, indicating the need for supplementary heat or clothing and air conditioning or evaporative cooling systems. Body heat is lost primarily by net radiation (60 per cent) and evaporation from the lungs and skin (25 per cent) in indoor conditions, whereas outdoors additional heat is lost convectively due to the wind. At subfreezing temperatures the effective *wind-chill* temperature may be substantially below the air temperature. Thus, a 15 m s^{-1} wind with an air temperature of −10°C gives a wind-chill equivalent temperature of −25°C.

A bioclimatic classification which incorporates estimates of comfort using temperature, relative humidity, sunshine and wind-speed data has been proposed for the United States by W. H. Terjung (1966).

BIBLIOGRAPHY

Bailey, H. P. (1960) A method for determining the temperateness of climate, *Geografiska Annaler* **42**, 1–16.

Basile, R. M. and Corbin, S. W. (1969) A graphical method for determining Thornthwaite climatic classification, *Ann. Assn. Amer. Geog.* **59**, 561–72.

Budyko, M. I. (1956) *The Heat Balance of the Earth's Surface* (trans. by N. I. Stepanova), US Weather Bureau, Washington, DC.

Budyko, M. I. (1974) *Climate and Life* (trans. by D. H. Miller), Academic Press, New York (508 pp).

Buettner, K. J. (1962) Human aspects of bioclimatological classification; in Tromp, S. W. and Weihe, W. H. (eds) *Biometeorology*, Pergamon, Oxford and London, pp. 128–40.

Carter, D. B. (1954) Climates of Africa and India according to Thornthwaite's 1948 classification, *Publications in Climatology* 7(4), Laboratory of Climatology, Centerton, NJ.

Chang, J.-H. (1959) An evaluation of the 1948 Thornthwaite classification, *Ann. Assn. Amer. Geog.* **49**, 24–30.

Crowe, P. R. (1957) Some further thoughts on evapotranspiration: a new estimate, *Geographical Studies* **4**, 56–75.

Flohn, H. (1950) Neue Anschauungen über die allgemeine Zirkulation der Atmosphäre und ihre klimatische Bedeutung, *Erdkunde* **4**, 141–62.

Flohn, H. (1957) Zur Frage der Einteilung der Klimazonen, *Erdkunde* **11**, 161–75.

Gentilli, J. (1958) *A Geography of Climate*, University of Western Australia Press, pp. 120–66.

Gregory, S. (1954) Climatic classification and climatic change, *Erdkunde* **8**, 246–52.

Hare, F. K. (1951) Climatic classification; in Stamp, L. D. and Wooldridge, S. W. (eds) *London Essays in Geography*, Longmans Green, London, pp. 111–34.

Mather, J. R. (1985) The water budget and the distribution of climates, vegetation and soils, *Publications in Climatology* **38**(2), Center for Climatic Research, University of Delaware, Newark (36 pp.).

Miller, A. A. (1951) Three new climatic maps, *Trans. Inst. Brit. Geog.* **17**, 13–20.

Olgyay, V. (1963) *Design with Climate: Bioclimatic approach to architectural regionalism*, Princeton University Press, Princeton, NJ (190 pp.).

Oliver, J. E. (1970) A genetic approach to climatic classification, *Ann. Assn. Amer. Geog.* **60**, 615–37. (Commentary, see **61**, 815–20.)

Papadakis, J. (1975) *Climates of the World and their Potentialities*, Buenos Aires (200 pp.).

Shear, J. A. (1966) A set-theoretic view of the Köppen dry climates, *Ann. Assn. Amer. Geog.* **56**, 508–15.

Sibbons, J. L. H. (1962) A contribution to the study of potential evapotranspiration, *Geografiska Annaler* **44**, 279–92.

Strahler, A. N. (1965) *Introduction to Physical Geography*, Wiley, New York (455 pp.).

Strahler, A. N. (1969) *Physical Geography* (3rd edn), Wiley, New York (733 pp.).

Terjung, W. H. (1966) Physiologic climates of the conterminous United States: a bioclimatological classification based on man, *Ann. Assn. Amer. Geog.* **56**, 141–79.

Terjung, W. H. and Louie, S. S.-F. (1972) Energy input–output climates of the world, *Arch. Met.*

Geophys. Biokl. B, **20**, 127–66.
Thornthwaite, C. W. (1933) The climates of the earth, *Geog. Rev.* **23**, 433–40.
Thornthwaite, C. W. (1943) Problems in the classification of climates, *Geog. Rev.* **33**, 233–55.
Thornthwaite, C. W. (1948) An approach towards a rational classification of climate, *Geog. Rev.* **38**, 55–94.
Thornthwaite, C. W. and Hare, F. K. (1955) Climatic classification in forestry, *Unasylva* **9**, 50–9.
Thornthwaite, C. W. and Mather, J. R. (1955) The water balance, *Publications in Climatology* **8**(1), Laboratory of Climatology, Centerton, NJ (104 pp.).
Thornthwaite, C. W. and Mather, J. R. (1957) Instructions and tables for computing potential evapotranspiration and the water balance, *Publications in Climatology* **10**(3), Laboratory of Climatology, Centerton, NJ (127 pp.).
Troll, C. (1958) Climatic seasons and climatic classification, *Oriental Geographer* **2**, 141–65.

APPENDIX 2
Système International (SI) units

Quantity	*Dimensions*	*SI*	*cgs metric*	*British*
length	L	m	10^2 cm	3.2808 ft
area	L^2	m^2	10^4 cm^2	10.7640 ft^2
volume	L^3	m^3	10^6 cm^3	35.3140 ft^3
mass	M	kg	10^3 g	2.2050 lb
density	ML^{-3}	kg m^{-3}	10^{-3} g cm^{-3}	
time	T	s	s	
velocity	LT^{-1}	m s^{-1}	10^2 cm s^{-1}	2.24 mi hr^{-1}
acceleration	LT^{-2}	m s^{-2}	10^2 cm s^{-2}	
force	MLT^{-2}	newton (kg m s^{-2})	10^5 dynes (10^5g cm^{-1} s^{-2})	
pressure	$ML^{-1}T^{-2}$	N m^{-2} (pascal)	10^{-2} mb	
energy, work	ML^2T^{-2}	joule (kg m^2 s^{-2})	10^7 ergs (10^7g cm^2 s^{-2})	
power	ML^2T^{-3}	watt (kg m^2 s^{-3})	10^7 ergs s^{-1}	1.340×10^{-3} hp
temperature	θ	kelvin (K)	°C	1.8°F
heat energy	ML^2T^{-2} (or H)	joule (J)	0.2388 cal	9.470×10^{-4} BTU
heat/radiation flux	HT^{-1}	watt (W) or J s^{-1}	0.2388 cal s^{-1}	3.412 BTU hr^{-1}
heat flux density	$HL^{-2}T^{-1}$	W m^{-2}	2.388×10^{-5} cal cm^{-2} s^{-1}	

The *basic SI units* are metre, kilogram, second (m, kg, s):

1 m	= 3.2808 feet	1 ft	= 0.3048 m
1 km	= 0.6214 miles	1 mi	= 1.6090 km
1 kg	= 2.2046 lb	1 lb	= 0.4536 kg
1 m s^{-1}	= 2.2400 mi hr^{-1}	1 mi hr^{-1}	= 0.4460 m s^{-1}

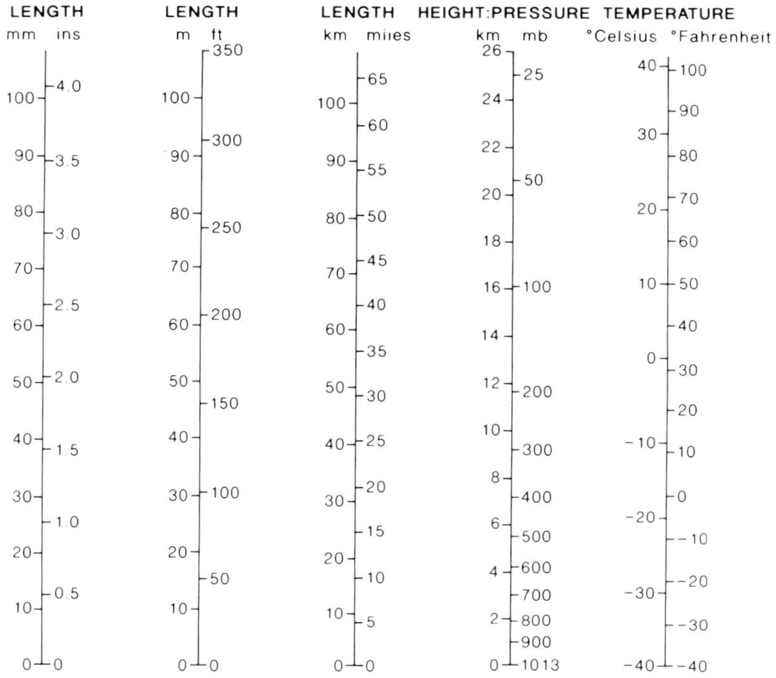

Figure A.2.1 Nomograms of height, pressure, length and temperature.

1 m^2	= 10.7640 ft^2	1 ft^2	= 0.0929 m^2
1 km^2	= 0.3861 mi^2	1 mi^2	= 2.5900 km^2
1°C	= 1.8°F	1°F	= 0.555°C

Temperature conversions can be determined by noting that:

$$\frac{T\,(°C)}{5} = \frac{T\,(°F) - 32}{9}$$

Energy conversion factors:

4.1868 J	= 1 calorie
J cm^{-2}	= 0.2388 cal cm^{-2}
Watt	= J s^{-1}
W m^{-2}	= 1.433×10^{-3} cal^{-2} min^{-1}
697.8 W m^{-2}	= 1 cal cm^{-2} min^{-1}

For time sums:

Day: 1 W m^{-2} = 8.64 J cm^{-2} dy^{-1}
= 2.064 cal cm^{-2} dy^{-1}

Day: 1 W m^{-2} = 8.64×10^4 J m^{-2} dy^{-1}
Month: 1 W m^{-2} = 2.592 M J m^{-2} $(30\ dy)^{-1}$
= 61.91 cal cm^{-2} $(30\ dy)^{-1}$
Year: 1 W m^{-2} = 31.536 M J m^{-2} yr^{-1}
= 753.4 cal cm^{-2} yr^{-1}

Gravitational acceleration (g) = 9.81 m s^{-2}

Latent heat of vaporization (288 K) = 2.47×10^6 J kg^{-1}

Latent heat of fusion (273 K) = 3.33×10^5 J kg^{-1}

APPENDIX 3
Synoptic weather maps

The synoptic weather map provides a generalized view of weather conditions over a large area at a given time. The map analysis smooths out local pressure and wind departures from the broad pattern. Such maps are usually prepared at 6- or 12-hourly intervals. Maps are generally prepared for mean sea-level pressure (or of height contours for the 1000-mb pressure surface) and at standard isobaric surfaces – 850, 700, 500, 300 mb, etc. The MSL pressure map typically shows isobars at 4 or 5 mb intervals, surface fronts and weather information.

Weather phenomena shown on the map are as follows:

temperature	type and height of cloud base
dew point	present weather
wind direction	past weather (last 6 hours)
wind speed	pressure tendency
pressure	pressure change (last 3 hours)
cloud amount	visibility

These data are presented in coded or symbolic form for each weather station. The plotting convention ('station model') is illustrated in Figure A.3.1. The basic weather symbols are illustrated in Figure A.3.2, and the synoptic code is given in Table A.3.1.

Note: Meteorological Office Leaflet No. 12. HMSO, London, gives titles and prices of meteorological reports, maps, reporting forms and diagrams.

REFERENCE

Stubbs, M. W. (1981) New code for reporting surface observations – an introduction, *Weather* **36**, 357–66.

Table A.3.1 Synoptic code (World Meteorological Organization, January 1982).

Symbol	*Key*	*Example*	*Comments*
yy	Day of the month (GMT)	05	All groups are in
GG	Time (GMT) to nearest hour	06	blocks of 5 digits
i_w	Indicator for type of wind speed observation and units	4	Measured by anemometer (knots)
IIiii	International index number of station		
i_R	Indicator: precipitation data included/omitted (code)	3	Data omitted
i_X	Indicator: station type + ww W_1 W_2 included/omitted (code)	1	Manned station with ww W_1 W_2 included
h	Height of lowest cloud (code)	3	
vv	Visibility (code)	66	
N	Total cloud amount (oktas)	7	
dd	Wind direction (tens of degrees)	32	
ff	Wind speed (knots, or m s^{-1})	20	Knots
1	Header	1	
s_n	Sign of temperature (code)	0	Positive value
TTT	Temperature (0.1°C), plotted rounded to nearest 1°C	203	(1 = negative value)
2	Header	2	
s_n	Sign of temperature (code)	0	
$T_dT_dT_d$	Dewpoint temperature (as TTT)	138	
4	Header	4	
PPPP	Mean sea-level pressure (tenths of mb, omitting thousands)	0105	
5	Header	5	
a	Characteristic of pressure tendency (coded symbol)	3	
ppp	3-hour pressure tendency (tenths of mb)	005	
7	Header	7	
ww	Present weather (coded symbol)	80	
W_1	Past weather (coded symbol)	9	(W_1 must be greater
W_2	Past weather (coded symbol)	8	than W_2)
8	Header	8	
N_h	Amount of low cloud (oktas)	4	
C_L	Low cloud type (coded symbol)	2	
C_M	Medium cloud type (coded symbol)	5	
C_H	High cloud type (coded symbol)	2	

Note: Group 3 is for a report of surface pressure and group 6 for precipitation data.

Basic station model for plotting weather data

MODEL (enlarged)

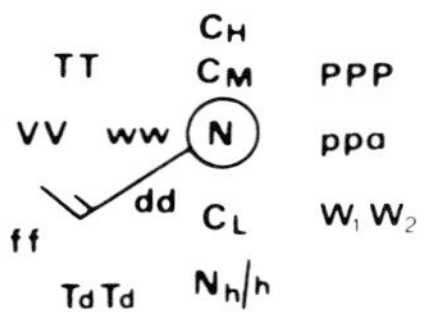

EXAMPLE

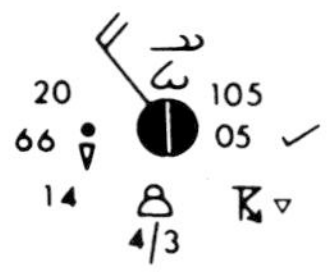

	KEY	EXAMPLE
N	Total cloud (oktas)[1]	7
dd	Wind direction (tens of degrees)	32
ff	Wind speed (knots)	20
VV	Visibility (code)	66
ww	Present weather (coded symbol)	80
W_1	Past weather (coded symbol)	9
W_2	" " " "	8
PPP	Sea-level pressure (mb)[2]	105
TT	Temperature (°C)[4]	20
N_h	Low cloud (oktas)	4
C_L	Low cloud type (coded symbol)	2
h	Height of C_L (code)	3
C_M	Medium cloud type (coded symbol)	5
C_H	High cloud type (coded symbol)	2
Td Td	Dew-point temperature (°C)[4]	14
a	Barograph trace (coded symbol)	3
pp	3-hour pressure change (mb)[3]	05

[1] okta = eighth
[2] Pressure in tens units and tenths mb: omitting initial 9 or 10 i.e. 105 = 1010·5
[3] Pressure change in units and tenths mb
[4] Rounded to nearest °C

Figure A.3.1 Basic station model for plotting weather data. The key and example are tabulated in the internationally agreed sequence for teletype messages. These data would be preceded by an identifying station number, date and time.

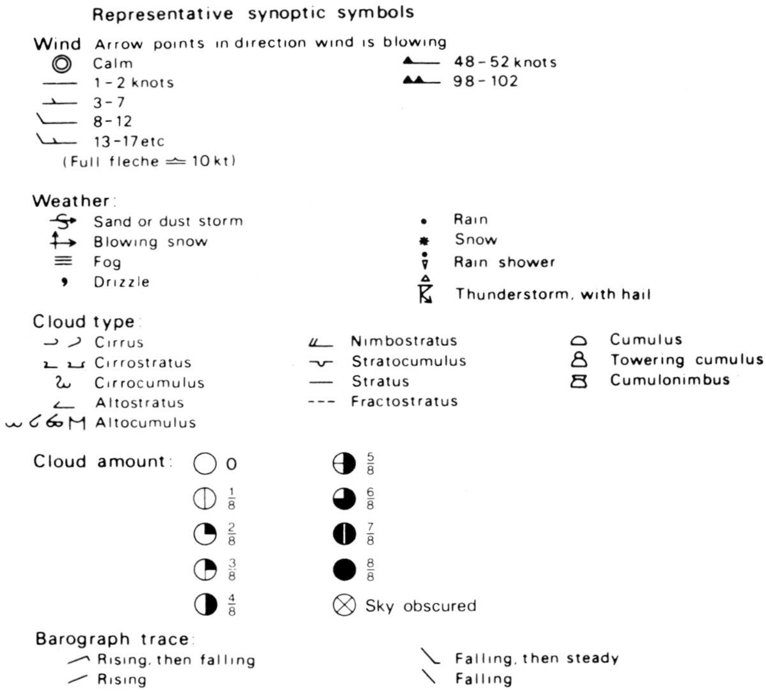

Figure A.3.2 Representative synoptic symbols.

APPENDIX 4
Data sources

A DAILY WEATHER MAPS AND DATA

Western Europe/North Atlantic: *Daily Weather Summary* (synoptic chart, data for the UK). London Weather Centre, 284 High Holborn, London WC1V 7HX, England.

Western Europe/North Atlantic: *Monthly Weather Report* (published about 15 months in arrears; tables for approximately 600 stations in the UK). London Weather Centre.

Europe – eastern/North Atlantic: *European Daily Weather Report* (synoptic chart). Deutsche Wetterdienst, Zentralamt D6050 Offenbach, Germany.

Europe – eastern/North Atlantic: *Weather Log* (daily synoptic chart, supplement to *Weather* magazine). Royal Meteorological Society, Bracknell, Berkshire, England.

North America: *Daily Weather Reports* (weekly publication). National Environmental Satellite Data and Information Service, NOAA, US Government Printing Office, Washington, DC 20402, USA.

B SATELLITE DATA

NOAA operational satellites (imagery, digital data). Satellite Data Services Division, NOAA/NESDIS, World Weather Building, Washington, DC 20023, USA.

Defense Meteorological Satellite Program (imagery). National Snow and Ice Data Center, University of Colorado, Boulder, CO 80309-0449, USA.

Metsat (imagery, digital data). ESOC, Robert-Bosil Str. 5, D-6100 Darmstadt, Federal Republic of Germany.

NASA research satellites (digital data). National Space Science Data Center, Goddard Space Flight Center, Greenbelt, MD 20771, USA.

UK Direct readout data from NOAA and Meteosat satellites are received at Dundee, Scotland. Dr. P. E. Baylis, Department of Electrical Engineering, University of Dundee, Scotland, UK.

C CLIMATIC DATA

Canadian Climate Center: *Climatic Perspectives* (197 weekly and monthly summary charts). Atmospheric Environment Service, 4905 Dufferin Street, Downsview, Ontario, Canada M3H 5T4.

Carbon Dioxide Information Center (CDIC): Data holdings and publications on climate – related variables and indices. Carbon Dioxide Information Center, Oak Ridge National Laboratory, Oak Ridge, TN 37931, USA.

National Center for Atmospheric Research, Boulder, CO 80307-3000, USA. Archives most global analyses and many global climate records.

National Climatic Center, NOAA/NESDIS: *Local Climatological Data* (1948–) (monthly tabulations, charts); *Monthly Climatic Data for the World* (May 1948–). National Climatic Data Center, Federal Building, Asheville, NC 28801, USA.

Climate Analysis Center, NOAA/NESDIS: *Climatic Diagnostics Bulletin* (1983–) (monthly summaries of selected diagnostic product from NMC analyses). Climate Analysis Center, NMC NOAA/NWS, World Weather Building, Washington, DC 20233, USA.

Climatic Research Unit, University of East Anglia: *Climate Monitor* (1976–) (monthly summaries, global and UK). Climatic Research Unit, University of East Anglia, Norwich NR4 7TJ, England.

World Climate Data Programme: *Climate System Monitoring Bulletin* (1984–) (monthly). World Climate Data Programme, WMO Secretariat, CP5, Geneva 20 CH-1211, Switzerland.

World Meteorological Centre, Melbourne: *Climate Monitoring Bulletin for the Southern Hemisphere* (1986–). Bureau of Meteorology, GPO Box 1289 K, Melbourne, Victoria 3001, Australia.

BIBLIOGRAPHY

Carleton, A. M. (1991) *Satellite Remote Sensing in Climatology*, Belhaven Press, London and CRC Press, Boca Raton, Fla. (291 pp.).

European Space Agency (1978) *Introduction to the Metsat System*, European Space Operations Center, Darmstadt (54 pp.).

European Space Agency (1981) *Atlas of Meteosat Imagery, Atlas Meteosat*, ESA-SP-1030, ESTEC, Nordwijk, Netherlands (494 pp.).

Finger, F. G., Laver, J. D., Bergman, K. H. and Patterson, V. L. (1985) The Climate Analysis Center's user information service, *Bull. Amer. Met. Soc.* **66**, 413–20.

Hastings, D. A., Emery, W. J., Weaver, R. L., Fischer, W. J. and Ramsay, J. W. (1987) *Proceedings North American NOAA Polar Orbiter Users Group First Meeting*, NOAA/NESDIS, US Dept of Commerce, Boulder, CO, National Geophysical Data Center (273 pp.).

Hattemer-Frey, H. A., Karl, T. R. and Quinlan, F. T. (1986) *An Annotated Inventory of Climatic Indices and Data Sets*, DOE/NBB-0080, Office of Energy Research, US Dept of Energy, Washington, DC (195 pp.).

Jenne, R. L. and McKee, T. B. (1985) Data; in Houghton, D. D. (ed.) *Handbook of Applied Meteorology*, Wiley, New York, pp. 1175–281.

Meteorological Office (1958) *Tables of Temperature, Relative Humidity and Precipitation for the World*, HMSO, London.

Singleton, F. (1985) Weather data for schools, *Weather* **40**, 310–13.

US Department of Commerce (1983) *NOAA Satellite Programs Briefing*, National Oceanic and Atmospheric Administration, Washington, DC (203 pp.).

US Department of Commerce (1984) *North American Climate Data Catalog. Part 1*, National Environmental Data Referral Service, Publication NEDRES-1, National Oceanic and Atmospheric Administration, Washington, DC (614 pp.).

World Meteorological Organization (1965) *Catalogue of Meteorological Data for Research*, WMO No. 174. TP-86, World Meteorological Organization, Geneva.

Problems

CHAPTER 1

1 The solar energy received at the top of the atmosphere (S) is proportional to $1/D^2$ where D is the solar distance. For mean solar distance (149.5×10^6 km), S = 1.35 k W m^{-2}. What are the amounts for maximum distance (152×10^6 km) and minimum distance (147×10^6 km)?

2. Calculate the noontime solar radiation received on a horizontal surface at the top of the atmosphere (ignoring variations in solar distance) at latitudes 0°, 23½°, 45°, 66½°, 90°N for the following dates: 22 December, 20 March and 21 June, and graph your results. The noontime solar elevation angles are:

	22 December	*20March*	*21 June*
0°	66½°	90°	66½°
23½°	43°	66½°	90°
45°	21½°	45°	68½°
66½°	0°	23½°	47°
90°N	–	0°	23½°

3 How does terrestrial radiation differ from solar radiation? Explain the physical basis of this difference.

4 Determine the radiation emitted from black bodies with temperatures of 6000 K and 300 K, respectively. The Stefan–Boltzmann constant is, $\sigma = 5.67 \times 10^{-8}$ W m^{-2} K^{-4}.

5 Show that the effective planetary temperature is approximately 255 K using Stefan's law and the estimated emitted infra-red radiation (from Figure 1.34).

6 What is the role of (a) ozone, (b) carbon dioxide, (c) dust particles, (d) water vapour, in the earth's radiation budget?

7 Assess the importance of cloud cover as a factor determining climate conditions at the earth's surface. (Note the typical range of cloud albedo, cloud amounts and surface albedos.)

8 What sections of the electromagnetic spectrum would be suitable for determining the following from a satellite?
(a) surface and cloud-top temperature;
(b) night-time cloud cover;
(c) planetary albedo;
(d) stratospheric temperatures.
(Refer to Figure 1.14.)

9 What is the basis for the division of the vertical structure of the atmosphere? Explain the zones of temperature increase in the upper atmosphere.

CHAPTER 2

1 Why does cooling eventually cause a mass of air to reach its saturation point? Describe the cooling processes which may result in cloud formation.

2 Mean daily evaporation over the globe is about 2.5 mm. What percentage of average incoming solar radiation absorbed at the surface does this represent?

3 Air at 10°C with a relative humidity of 50

per cent is cooled at constant pressure. At approximately what temperature will it reach its dew point? (Use Figure 1.13.)

4 If the temperature of an air parcel at 1000 mb is 20°C, what is its temperature at 700 mb following (a) unsaturated ascent, (b) saturated ascent? (Use Figure 2.10.)

5 By how much is an air parcel warmed if it ascends, saturated, from the 1 000-mb level at 10°C to 800 mb and descends unsaturated to 1000 mb?

6 Discuss the physical conditions that make 'cloud seeding' possible.

7 Explain the different types of cloud pattern visible on satellite imagery. How are the various patterns related to weather systems?

8 What methods would you use to describe the areal variation of magnitude and frequency of rainstorms?

9 How do fogs form and in what geographical regions are they common?

10 Discuss the definition of drought. Which regions of the world are most susceptible to drought?

11 Determine the annual moisture regime for a station in your area by the Penman and Thornthwaite methods (see the references to Chapter 2: Pearl *et al.* 1954, and Appendix 1: Thornthwaite and Mather 1957).

12 Consider what physical processes serve to limit extreme rainfall amounts? Some new records are described in *Weather* **39** (1984), p. 12; compare these with the graph of expected extremes. (Use Figure 2.25).

CHAPTER 3

1 Determine the balance of forces for cyclonic and anticyclonic gradient wind flow in the southern hemisphere. Compare these with the geostrophic wind case.

2 Calculate the geostrophic wind speed ($m\ s^{-1}$) at latitudes 20° and 43°N for a pressure gradient of 1.5 mb/100 km. Assume air density $\rho = 1.2\ kg\ m^{-3}$. (Note $1\ mb = 100\ kg\ m^{-1}\ s^{-2}$.)

3 Using Figures 1.30 and 3.25 plot a graph for 40°N of latitudinal temperature departure and the sign of the meridional MSL wind component in January. (Take winds of SE-SW as positive, NW-NE as negative; W or E winds = zero.) Note the correlation between the sign of the temperature anomaly and the meridional component of the wind.

4 Explain the locations of the major centres of low and high pressure in Figure 3.25.

5 Using Figures 1.30 and 3.25 determine the direction of the thermal wind component in July over (a) 35°N, 20°W, (b) 30°N, 100°W.

6 Explain the location of the major desert areas of the world.

7 Under what circumstances may local climatic influences be more important than large-scale controls?

8 Outline the characteristics of jet streams in the upper troposphere. What are their relationships with surface weather and climate?

9 Describe the role of the tropics in the general circulation of the atmosphere.

10 What is the zonal index and how is it related to characteristics of the circulation in middle latitudes?

CHAPTER 4

1 What are the three primary factors determining air-mass weather?

2 Determine the source regions from which air masses are likely to affect your home area in summer and winter. Outline the weather conditions likely to be associated with them following air-mass modifications *en route*.

3 What weather conditions are typically associated with a noontime tropical air mass moving northward (a) over sea (b) over land?

4 Explain the relationship between frontal zones and upper tropospheric jet streams.

(Consider a vertical section and a plan view of a frontal cyclone.)

5 Explain how a low pressure system may deepen or fill.

6 Compare Figures 4.19 and 3.20A and comment on the features identified.

7 Describe the types of *non*-frontal low pressure system and explain their occurrence.

8 What are the bases of methods of short-range and long-range weather forecasts?

9 Maintain a daily log of forecast weather conditions for your location (based on newspaper, TV or radio broadcast information) and tabulate the actual temperature, wind and weather conditions that occurred. In what synoptic situations and seasons are the forecasts more/less reliable?

CHAPTER 5

1 Examine Figures 1.30, 3.25 and 3.42 in relation to winter and summer temperature conditions in north-western Europe, eastern Siberia and north-eastern Canada. What principal factors are operative in each region?

2 What is meant by 'continentality' and what factors determine it?

3 List the major influences of the large-scale orography on climate in western North America.

4 Discuss the circumstances that give rise to long spells of a particular type of weather over Europe.

5 Using daily weather maps and daily precipitation records for your locality, determine the proportion of the precipitation occurring over a winter and summer season with frontal and non-frontal situations and different air masses.

6 Select cases of strong zonal flow and of blocking for Scandinavia or Alaska from daily weather maps and analyse the patterns of temperature and precipitation that result. If upper air charts are available, compare also the jet-stream patterns.

CHAPTER 6

1 What are the major differences between weather systems in the tropics and in middle latitudes?

2 Why are hurricanes absent from the South Atlantic and the eastern South Pacific?

3 What are the equatorial westerlies and what is their climatological significance?

4 What effects does the trade wind inversion have on tropical weather and climate?

5 Examine the role of synoptic systems in shaping the character of the monsoon regime of southern Asia.

6 In what respects is the monsoon regime of West Africa similar to that of southern Asia?

7 What are the most important local climatic influences in tropical regions?

CHAPTER 7

1 What are the main determinants of urban heat islands? Which factor is considered to be most important in mid-latitude cities in (a) winter; (b) summer?

2 What effects do differences in surface type (urban area, forest, lake, etc.) have on climatic parameters?

3 Enumerate meteorological considerations which could mitigate some of the undesirable features of urban climate if taken into account in city planning.

4 Discuss the view that a city has many small-scale climates.

5 What are the principal microclimatic effects of (a) shelter belts (b) forest clearings?

6 Topics for group topoclimatic investigations:

Spatial and temporal features of fog occurrence; climatic measurements (wind speed,

temperature, light intensity, etc.) inside and outside a forested area; comparison of the duration of sunshine and of the diurnal course of temperature on slopes of northerly and southerly aspect.

7 Analyse the monthly frequency of light winds (less than 2 m s^{-1}) in your area. Use hourly measurements if possible. Compile comparative data on fog frequency and air quality if appropriate. (A source of data on air quality is the *Journal of the Air Pollution Control Association.*)

CHAPTER 8

1 Plot frequency distribution of annual precipitation totals using data for a station in your locality and a contrasting climatic regime. (Use at least 30 years of data and not more than eight classes.) Compare with Figure 8.1. Determine appropriate averages and measures of variability.
2 For corresponding 30-year series of temperature data (such as mean daily temperature for January and July) determine arithmetic means and standard deviations.
3 Enumerate the various terrestrial and extraterrestrial factors which may be involved in global climatic change and consider the time-scales over which each is likely to be significant. Discuss the processes which are involved in their climatic effects.
4 Using some long-term temperature and precipitation records, compare the trends since the late nineteenth century in high, middle and low latitudes. Graph the values by individual years and as 10-year running means.

SOLUTIONS TO PROBLEMS

Chapter 1

1 1.40 and 1.31 k W m^{-2}

	Lat.	*22 December*	*20 March*	*21 June*
2	0°	1.26	1.37	1.26 k W m^{-2}
	23½°	0.93	1.26	1.37
	45°	0.50	0.97	1.27
	66½°	0.0	0.55	1.00
	90°N	0.0	0.0	0.55

4 73×10^3 k W m^{-2}; 459.3 k W m^{-2}

Chapter 2

2 20.6 per cent
3 0°C
4 (a) −8.5°C, (b) 6°C
5 8°C

Chapter 3

2 (a) 25 m s^{-1}
(b) 12.5 m s^{-1}
5 (a) Westerly
(b) Southerly

Notes

1 ATMOSPHERIC COMPOSITION AND ENERGY

1 Mixing ratio = ratio of number of molecules of ozone to molecules of air (parts per million by volume, ppm(v)). Concentration = mass per unit volume of air (molecules/cubic metre).

2 K = degrees Kelvin (or Absolute). The degree symbol is omitted.
°C = degrees Celsius
°C = K − 273
Conversions for °C and °F are given in Appendix 2.

3 Joule = 0.2388 cal. The units of the International Metric System are given in Appendix 2. At present the data in many references are still in calories; a calorie is the heat required to raise the temperature of 1 g of water from 14.5°C to 15.5°C. In the United States, another unit in common use is the Langley (ly) (ly min^{-1} = 1 cal cm^{-2} min^{-1}).

4 The equation for the so-called 'reduction' (actually the adjusted value is normally greater!) of station pressure (p_h) to sea level pressure (p_0) is written:

$$p_0 = p_h \exp \left[\frac{g_0}{R_d \bar{T}_v} Z_h \right]$$

where R_d = gas content for dry air; g_0 = global average of gravitational acceleration (9.8 ms^{-2}); Z_h = geopotential height of the station (≃ geometric height in the lowest kilometre or so); $\bar{T}_v$ = mean virtual temperature. This is a fictitious temperature used in the ideal gas equation to compensate for the fact that the gas constant of moist air exceeds that of dry air. Even for hot moist air, $\bar{T}_v$ is only a few degrees greater than the air temperature.

5 The radiation flux (per unit area) received normal to the beam at the top of the earth's atmosphere is calculated from the total solar output weighted by $1/(4\pi D^2)$, where the solar distance D = 1.5×10^{11} m, since the surface area of a sphere of radius r (here equivalent to D) is $4\pi r^2$ – i.e. the radiation flux is $2.33 \times 10^{25}/4\pi$ (2.25×10^{22}) = 82.4 kJ m^{-2} min^{-1} (or 1373 W m^{-2}).

6 The albedos refer to the solar radiation received on each given surface; thus the incident radiation is different for planet earth, the global surface and global cloud cover, as well as between any of these and the individual cloud types or surfaces.

7 The official definition is the lowest level at which the lapse rate decreases to less than, or equal to, 2°C/km (provided that the average lapse rate of the 2-km layer does not exceed 2°C/km).

3 ATMOSPHERIC MOTION

1 The centrifugal force is equal in magnitude and opposite in sign to the centripetal acceleration.

2 Apparent gravity, g = 9.78 m s^{-2} at the equator, 9.83 m s^{-2} at the poles.

3 The vorticity, or circulation, about a rotating circular fluid disc is given by the product of the rotation on its boundary (ωR) and the circumference ($2\pi R$) where R = radius of the disc. The vorticity is then $2\omega\pi R^2$, or 2ω per unit area.

4 The geostrophic wind concept is equally applicable to contour charts. Heights on these charts are given in geopotential metres (g.p.m.) or dekametres (g.p.dkm).

5 The World Meteorological Organization recommends an arbitrary lower limit of 30 m s^{-1}.

6 Equatorial speed of rotation is 465 m s^{-1}.

7 Note that, at the equator, an east/west wind of 5 m s^{-1} represents an absolute motion of 460/470 m s^{-1} towards the east.

4 AIR MASSES, FRONTS AND DEPRESSIONS

1 Resultant wind is the vector average of all wind directions and speeds.
2 This latter term is tending to be restricted to the tropical (hurricane) variety.
3 It is significant that some storms also occur downwind of other and plateau regions in Mexico, the Iberian peninsula and West Africa.

8 CLIMATIC CHANGE

1 Fuller details of elementary statistical procedures may be found in S. Gregory (1973) *Statistical Methods and the Geographer* (Longman), R. Hammond and P. McCullogh (1974) *Quantitative Techniques in Geography* (Oxford University Press) or J. Silk (1979) *Statistical Concepts in Geography* (George Allen and Unwin).

APPENDIX I

1 Note that many American workers use a modified version with O°C as the C/D boundary.

Bibliography

GENERAL

Anthes, R. A., Panofsky, H. A., Cahir, J. J. and Rango, A. (1981) *The Atmosphere* (3rd edn), C. E. Merrill, Columbus, Ohio (531 pp.).

Atkinson, B. W. (1981a) *Meso-Scale Atmospheric Circulations*, Academic Press, London (496 pp.).

*Atkinson, B. W. (ed.) (1981b) *Dynamical Meteorology*, Methuen, London (250 pp.).

Barrett, E. C. and Martin, D. W. (1981) *The Use of Satellite Data in Rainfall Monitoring*, Academic Press, London (340 pp.).

Barry, R. G. (1992) *Mountain Weather and Climate* (2nd edn), Routledge, London and New York (402 pp.).

Barry, R. G. and Perry, A. H. (1973) *Synoptic Climatology: Methods and Applications*, Methuen, London (555 pp.).

Battan, L. J. (1984) *Fundamentals of Meteorology* (2nd edn), Prentice-Hall, Englewood Cliffs, New Jersey (304 pp.).

*Berry, F. A., Bollay, E. and Beers, N. R. (eds) (1945) *Handbook of Meteorology*, McGraw-Hill, New York (1068 pp.).

Blüthgen, J. (1966) *Allgemeine Klimageographie* (2nd edn), W. de Gruyter, Berlin (720 pp.).

Bruce, J. P. and Clark, R. H. (1966) *Introduction to Hydrometeorology*, Pergamon, Oxford (319 pp.).

*Byers, H. R. (1974) *General Meteorology* (4th edn), McGraw-Hill, New York (480 pp.).

Carleton, A.M. (1991) *Satelite Remote Sensing in Climatology*, Belhaven Press, London (291pp.).

Chang, J. H. (1972) *Atmospheric Circulation Systems and Climates*, Oriental Publishing Co., Honolulu (326 pp.).

Cole, F. W. (1980) *Introduction to Meteorology* (3rd edn), Wiley, New York (505 pp.).

Crowe, P. R. (1971) *Concepts in Climatology*, Longman, London (589 pp.).

Fairbridge, R. W. (ed.) (1967) *The Encyclopaedia of Atmospheric Sciences and Astrotgeology*, Reinhold, New York (1200 pp.).

Fleagle, R. G. and Businger, J. A. (1980) *Introduction to Atmospheric Physics* (2nd edn), Academic Press, New York (432 pp.).

*Flohn, H. (ed.) (1969) *General Climatology* **2**, World Survey of Climatology **2**, Elsevier, Amsterdam (266 pp.).

*Flohn, H. (ed.) (1981) *General Climatology* **3**, World Survey of Climatology **3**, Elsevier, Amsterdam (408 pp.).

Gedzelman, S. D. (1980) *The Science and Wonders of the Atmosphere*, Wiley, New York (535 pp.).

Geiger, R. (1965) *The Climate Near the Ground* (2nd edn), Harvard University Press, Cambridge, Mass. (611 pp.).

*Haltiner, G. J. and Martin, F. L. (1957) *Dynamical and Physical Meteorology*, McGraw-Hill, New York (470 pp.).

Hecht, A. D. (ed.) (1985) *Paleoclimate Analysis and Modeling*, Wiley, New York (445 pp.).

Henderson-Sellers, A. and Robinson, P. J. (1986) *Contemporary Climatology*, Longman, London (439 pp.).

Herman, J. R. and Goldberg, R. A. (1985) *Sun, Weather and Climate*, Dover, New York (360 pp.).

*Hess, S. L. (1959) *Introduction to Theoretical Meteorology*, Henry Holt, New York (362 pp.).

Hewson, E. W. and Longley, R. W. (1944) *Meteorology, Theoretical and Applied*, Wiley, New York (468 pp.).

Hobbs, J. (1980) *Applied Climatology*, Westview Press, Boulder, Colo. (250 pp.).

Houghton, D. D. (1985) *Handbook of Applied Meteorology*, Wiley, New York (1461 pp.).

*Indicates more advanced texts with reference particularly to the physical bases of meteorology, often including mathematical material.

Houghton, H. G. (1985) *Physical Meteorology*, MIT Press, Cambridge, Mass. (442 pp.).
Houghton, J. T. (ed.) (1984) *The Global Climate*, Cambridge University Press, Cambridge (233 pp.).
*Humphreys, W. J. (1929) *Physics of the Air*, McGraw-Hill, New York (654 pp.).
Huschke, R. E. (ed.) (1959) *Glossary of Meteorology*, American Meteorological Society, Boston (638 pp.).
Hutchinson, G. E. (1957) *A Treatise on Limnology* **1**, Wiley, New York (1015 pp.).
Kendrew, W. G. (1961) *The Climates of the Continents* (5th edn), Oxford University Press, London (608 pp.).
Lamb, H. H. (1972) *Climate: Present, Past and Future* **1**: *Fundamentals and Climate Now*, Methuen, London (613 pp.).
Liljequist, G. H. (1970) *Klimatologi*, Generalstabens Litografiska Anstalt, Stockholm.
List, R. J. (1951) *Smithsonian Meteorological Tables* (6th edn), Smithsonian Institution, Washington (527 pp.).
Lockwood, J. G. (1974) *World Climatology: An environmental approach*, Arnold, London (330 pp.).
Lockwood, J. G. (1979) *Causes of Climate*, Arnold, London (260 pp.).
Lutgens, F. K. and Tarbuck, E. J. (1986) *The Atmosphere: An introduction to meteorology* (2nd edn), Prentice-Hall, Englewood Cliffs, New Jersey (492 pp.).
McBoyle, G. (ed.) (1973) *Climate in Review*, Houghton Mifflin, Boston (314 pp.).
McIlveen, R. (1986) *Basic Meteorology: A physical outline*, Van Nostrand Reinhold, UK (457 pp.).
McIntosh, D. H. and Thom A. S. (1972) *Essentials of Meteorology*, Wykeham Publications, London (239 pp.).
*Malone, T. F. (ed.) (1951) *Compendium of Meteorology*, American Meteorological Society, Boston (1334 pp.).
Manley, G. (1952) *Climate and the British Scene*, Collins, London (314 pp.).
Miller, A. and Anthes, R. A. (1980) *Meteorology* (4th edn), C. E. Merrill, Columbus, Ohio (170 pp.).
Monteith, J. L. (1973) *Principles of Environmental Physics*, Arnold, London (241 pp.).
Monteith, J. L. (ed.) (1975) *Vegetation and the Atmosphere*, Vol. 1: *Principles*, Academic Press, London (278 pp.).
Munn, R. E. (1966) *Descriptive Micrometeorology*, Academic Press, New York (245 pp.).
Musk, L. F. (1988) *Weather Systems*, Cambridge University Press, Cambridge (160 pp.).
Neiburger, M., Edinger, J. G. and Bonner, W. D. (1982) *Understanding Our Atmospheric Environment* (2nd edn), W. H. Freeman, San Francisco (453 pp.).
Oliver, J. E. and Fairbridge, R. W. (eds) (1987) *The Encyclopedia of Climatology*, Van Nostrand Reinhold, New York (986 pp.).
*Palmén, E. and Newton, C. W. (1969) *Atmosphere Circulation Systems: Their structure and physical interpretation*, Academic Press, New York (603 pp.).
Pedgley, D. E. (1962) *A Course of Elementary Meteorology*, HMSO, London (189 pp.).
Perry, A. H. and Walker, J. M. (1977) *The Ocean–Atmosphere System*, Longman, London (160 pp.).
*Petterssen, S. (1956) *Weather Analysis and Forecasting* (2 vols), McGraw-Hill, New York (428 pp. and 266 pp.).
Petterssen, S. (1969) *Introduction to Meteorology* (3rd edn), McGraw-Hill, New York (333 pp.) (previous editions 1941 and 1958).
Pickard, G. L. and Emery, W. J. (1990) *Descriptive Physical Oceanography* (5th edn), Pergamon Press, Oxford.
*Reiter, E. R. (1963) *Jet Stream Meteorology*, University of Chicago Press, Chicago, Ill. (515 pp.).
Rex, D. F. (ed.) (1969) *Climate of the Free Atmosphere*, World Survey of Climatology **4**, Elsevier, Amsterdam (450 pp.).
Schaefer, V. J. and Day, J. A. (1981) *A Field Guide to the Atmosphere*, Houghton Mifflin, Boston, Mass. (359 pp.).
Schwerdtfeger, W. (1984) *Weather and Climate of the Antarctic*, Elsevier, Amsterdam (261 pp.).
*Sellers, W. D. (1965) *Physical Climatology*, University of Chicago Press, Chicago, Ill. (272 pp.).
Smith, W. L., Bishop, W. P., Dvorak, V.F., Hayden, C. M., McElroy, J. H., Mosher, F. R., Oliver, V. J., Purdom, J. F. and Wark, D.Q. (1986) The meteorological satellite: overview of 25 years of operations, *Science* **231**, 455–62.
Strahler, A. N. (1965) *Introduction to Physical Geography*, Wiley, New York (455 pp.).
Strahler, A. N. and Strahler, A. H. (1987) *Modern Physical Geography* (3rd edn), John Wiley & Sons, New York (332 pp.).
Stringer, E. T. (1972a) *Foundations of Climatology*, W. H. Freeman, San Francisco (586 pp.).
Stringer, E. T. (1972b) *Techniques of Climatology*, W. H. Freeman, San Francisco (539 pp.).
Sverdrup, H. V. (1945) *Oceanography for Meteorologists*, Allen & Unwin, London (235 pp.).
Sverdrup, H. V., Johnson, M. W. and Fleming, R. H. (1942) *The Oceans; Their physics, chemistry and general biology*, Prentice-Hall, New York (1087 pp.).
Taljaard, J. J., van Loon, H., Crutcher, H. L. and

Jenne, R. L. (1969) *Climate of the Upper Air, Part 1. Southern Hemisphere* 1, Naval Weather Service Command, Washington, DC, NAVAIR 50-1C-55.

Taylor, J. A. and Yates, R. A. (1967) *British Weather in Maps* (2nd edn), Macmillan, London (315 pp.).

Trewartha, G. T. (1981) *The Earth's Problem Climates* (2nd edn), University of Wisconsin Press, Madison (371 pp.).

Trewartha, G. T. and Horne, L. H. (1980) *An Introduction to Climate* (5th edn), McGraw-Hill, New York (416 pp.).

van Loon, H. (ed.) (1984) *Climates of the Oceans: World Survey of Climatology* **15**, Elsevier, Amsterdam (716 pp.).

Wallace, J. M. and Hobbs, P. V. (1977) *Atmospheric Science: An introductory survey*, Academic Press, New York (467 pp.).

Willett, H. C. and Sanders, F. (1959) *Descriptive Meteorology* (2nd edn), Academic Press, New York (355 pp.).

World Meteorological Organization (1962) *Climatological Normals (CLINO) for CLIMAT and CLIMAT SHIP stations for the period 1931–60*, World Meteorological Organization, Geneva.

World Meteorological Organization (1973) The use of satellite pictures in weather analysis and forecasting, *WMO Technical Note* 124, Geneva (275 pp.).

CHAPTER 1 ATMOSPHERIC COMPOSITION AND ENERGY

Ahmad, S. A. and Lockwood, J. G. (1979) Albedo, *Prog. Phys. Geog.* **3**, 520–43.

Andreae, M. O. and Schimel, D. S. (1989) *Exchange of Trace Gases Between Terrestrial Ecosystems and the Atmosphere*, J. Wiley & Sons, Chichester (347 pp.).

Bach, W. (1976) Global air pollution and climatic change, *Rev. Geophys. Space Phys.* **14**, 429–74.

Barry, R. G. (1985) The cryosphere and climatic change; in MacCracken, M. C. and Luther, F. M. (eds) *Detecting the Climatic Effects of Increasing Carbon Dioxide*, DOE/ER-0235, US Department of Energy, Washington, DC, pp. 109–48.

Barry, R. G. and Chambers, R. E. (1966) A preliminary map of summer albedo over England and Wales, *Quart. J. Roy. Met. Soc.* **92**, 543–8.

Beckinsale, R. P. (1945) The altitude of the zenithal sun: a geographical approach, *Geog. Rev.* **35**, 596–600.

Bolin, B., Degens, E. T., Kempe, S. and Ketner, P. (eds) (1979) *The Global Carbon Cycle* (SCOPE 13), J. Wiley & Sons, Chichester (528 pp.).

Bolin, B., Döös, B. R., Jäger, J. and Warrick, R. A. (eds) (1986) *The Greenhouse Effect, Climatic Change, and Ecosystems* (SCOPE 29), J. Wiley & Sons, Chichester (541 pp.).

Bolle, H.-J., Seiler, W. and Bolin, B. (1986) Other greenhouse gases and aerosols; in Bolin, B. *et al.* (eds) *The Greenhouse Effect, Climatic Change, and Ecosystems*, J. Wiley & Sons, Chichester, pp. 157–203.

Bridgman, H. A. (1990) *Global Air Pollution: Problems for the 1990s*, Belhaven Press, London (201 pp.).

Brimblecombe, P. (1986) *Air: Composition and Chemistry*, Cambridge University Press, Cambridge (224 pp.).

Budyko, J. I., Nayefimova, N. A., Zubenok, L. I. and Strokhina, L. A. (1962) The heat balance of the surface of the earth, *Soviet Geography* **3**(5), 3–16.

Campbell, I. M. (1986) *Energy and the Atmosphere. A Physical-Chemical Approach* (2nd edn), John Wiley & Sons, Chichester (337 pp.).

Craig, R. A. (1965) *The Upper Atmosphere; meteorology and physics*, Academic Press, New York (509 pp.).

Defant, F. R. and Taba, H. (1957) The threefold structure of the atmosphere and the characteristics of the tropopause, *Tellus* **9**, 259–74.

Foukal, P. V. (1990) The variable sun, *Sci. American* **262**, 26–33.

Fröhlich, C. and London, J. (1985) *Radiation Manual*, World Meteorological Organization, Geneva.

Garnett, A. (1937) Insolation and relief, *Trans. Inst. Brit. Geog.* **5** (71 pp.).

Hare, F. K. (1962) The stratosphere, *Geog. Rev.* **52**, 525–47.

Hastenrath, S. L. (1968) Der regionale und jahrzeitliche Wandel des vertikalen Temperaturgradienten und seine Behandlung als Wärmhaushaltsproblem, *Meteorologische Rundschau* **1**, 46–51.

Henderson-Sellers, A. and Wilson, M. F. (1983) Surface albedo data for climate modeling, *Rev. Geophys. Space Phys.* **21**, 1743–78.

Intergovernmental Panel on Climate Change (IPCC) (eds: Houghton, J. T., Jenkins, G. J. and Ephraums, J. J.) (1990) *Climate Change: The IPCC Scientific Assessment*, Cambridge University Press, Cambridge (365 pp.).

Keeling, C. D. (1973) The carbon dioxide cycle: reservoir models to depict the exchange of atmospheric carbon dioxide with the oceans, and land plants; in Rasool, S. I. (ed.) *Chemistry of the Lower Atmosphere*, Plenum Press, New York, pp. 251–329.

Kellogg, W. W. and Schware, R. (1981) *Climate Change and Society*, Westview Press, Boulder, Colo. (178 pp.).

Kondratyev, K. Ya. and Moskalenko, N. I. (1984) The role of carbon dioxide and other minor gaseous components and aerosols in the radiation budget; in Houghton, J. T. (ed.) *The Global Cli-*

mate, Cambridge University Press, Cambridge, pp. 225–33.

Kung, E. C., Bryson, R. A. and Lenschow, D. H. (1964) Study of a continental surface albedo on the basis of flight measurements and structure of the earth's surface cover over North America, *Monthly Weather Review* **92**, 543–64.

Lashof, D. A. and Ahnja, D. R. (1990) Relative contributions of greenhouse gas emissions to global warming, *Nature* **344**, 529–31.

Lautensach, H. and Bogel, R. (1956) Der Jahrsgang des mittleren geographischem Hohengradienten der Lufttemperatur in den verschiedenen Klimagebieten der Erde, *Erdkunde* **10**, 270–82.

London, J. (1985) The observed distribution of atmospheric ozone and its variations; in Whitten, R. C. and Prasad, T. S. (eds) *Ozone in the Free Atmosphere*, Van Nostrand Reinhold, New York, pp. 11–80.

London, J. and Angell, J. K. (1982) The observed distribution of ozone and its variations; in Bower, F. A. and Ward, R. B. (eds) *Stratospheric Ozone and Man*, Chemical Rubber Company, CRC Press, Boca Raton, Fla., pp. 7–12.

London, J. and Sasamori, T. (1971) Radiative energy budget of the atmosphere, *Space Research* **11**, 639–49.

London, J., Warren, S. G. and Hahn, C. J. (1989) The global distribution of observed cloudiness – a contribution to the ISCCP, *Adv. Space Res.* **9**, 161–5.

Lumb, F. E. (1961) *Seasonal variation of the sea surface temperature in coastal waters of the British Isles*, Sci. Paper No. 6, Meteorological Office, HMSO, London (21 pp.).

McElroy, M. B. and Salawitch, R. J. (1989) Changing composition of the global stratosphere, *Science* **243**, 763–70.

McFadden, J. D. and Ragotzkie, R. A. (1967) Climatological significance of albedo in central Canada, *Jour. Geophys. Res.* **72**, 1135–43.

Machta, L. (1972) The role of the oceans and biosphere in the carbon dioxide cycle; in Dyrssen, D. and Jagner, D. (eds) *The Changing Chemistry of the Oceans*, Nobel Symposium 20, Wiley, New York, pp. 121–45.

Miller, D. H. (1968) A Survey Course: the energy and mass budget at the surface of the earth, *Pub. No. 7, Assn. Amer. Geog.*, Washington, DC (142 pp.).

NASA (n.d.) *From Pattern to Process: The Strategy of the Earth Observing System* (Vol. II), Eos Science Steering Committee Report, NASA, Houston.

NERC (1989) *Our Future World: Global Environmental Research*, NERC, London (28 pp.).

Newell, R. E. (1964) The circulation of the upper atmosphere, *Sci. American* **210**, 62–74.

Newton, H. W. (1958) *The Face of the Sun*, Pelican, London (208 pp.).

Paffen, K. (1967) Das Verhältniss der Tages-zur Jahreszeitlichen Temperaturschwankung, *Erdkunde* **21**, 94–111.

Pearce, F. (1989) Methane: the hidden greenhouse gas, *New Scientist* **122**, 37–41.

Plass, G. M. (1959) Carbon dioxide and climate, *Sci. American* **201**, 41–7.

Ramanathan, V., Cicerone, R. J., Singh, H. B. and Kiehl, J. T. (1985) Trace gas trends and their potential role in climatic change, *J. Geophys. Res.* **90**(D3), 5547–66.

Ramanathan, V., Cess, R. D., Harrison, E. F., Minnis, P., Barkstrom, B. R., Ahmad, E. and Hartmann, D. (1989) Cloud-radiative forcing and climate: results from the Earth Radiation Budget Experiment, *Science* **243**, 57–63.

Ramanathan, V., Barkstrom, B. R. and Harrison, E. F. (1990) Climate and the earth's radiation budget, *Physics Today* **42**, 22–32.

Rampino, M. R. and Self, S. (1984) The atmospheric effects of El Chichón, *Sci. American* **250**(1), 34–43.

Ransom, W. H. (1963) Solar radiation and temperature, *Weather* **8**, 18–23.

Raval, A. and Ramanathan, V. (1989) Observational determination of the greenhouse effect, *Nature* **342**, 758-61.

Rodhe, H. (1990) A comparison of the contribution of various gases to the greenhouse effect, *Science* **244**, 763–70.

Roederer, J. (1981) The perception of sound in association with auroras, *J. Acoustical Soc. Amer.* **69**.

Roland, F. S. and Isaksen, I. S. A. (eds) (1988) *The Changing Atmosphere*, J. Wiley & Sons, Chichester.

Sellers, W. D. (1980) A comment on the cause of the diurnal and annual temperature cycles, *Bull. Amer. Met. Soc.* **61**, 741–55.

Shine, K. (1990) Effects of CFC substitutes, *Nature* **344**, 492–3.

Slinn, W. G. N. (1983) Air-to-sea transfer of particles; in Liss, P. S. and Slinn, W. G. N. (eds) *Air-Sea Exchange of Gases and Particles*, D. Reidel, Dordrecht, pp. 299–407.

Solomon, S. (1988) The mystery of the Antarctic ozone hole, *Rev. Geophys.* **26**, 131–48.

Stephens, G. L., Campbell, G. G. and Vonder Haar, T. H. (1981) Earth radiation budgets, *J. Geophys. Res* **86**(C10), 9739–60.

Stone, R. (1955) Solar heating of land and sea, *Geography* **40**, 288.

Valley, S. L. (ed.) (1965) *Handbook of Geophysics and Space Environments*, McGraw-Hill, New York.

World Meteorological Organization (1964) Regional basic networks. *WMO Bulletin* **13**, 146–7.

CHAPTER 2 ATMOSPHERIC MOISTURE

Acreman, M. (1989) Extreme rainfall in Calderdale, 19 May 1989, *Weather* **44**, 438–46.
Andersson, T. (1980) Bergeron and the oreigenic (orographic) maxima of precipitation, *Pure Appl. Geophys.* **119**, 558–76.
Armstrong, C. F. and Stidd, C. K. (1967) A moisture-balance profile in the Sierra Nevada, *J. of Hydrology* **5**, 258–68.
Atlas, D., Chou, S.-H. and Byerly, W. P. (1983) The influence of coastal shape on winter mesoscale air–sea interactions, *Monthly Weather Review* **111**, 245–52.
Ayoade, J. A. (1976) A preliminary study of the magnitude, frequency and distribution of intense rainfall in Nigeria, *Hydro. Sci. Bull.* **21**(3), 419–29.
Bannon, J. K. and Steele, L. P. (1960) Average water-vapour content of the air, *Geophysical Memoirs 102*, Meteorological Office (38 pp.).
Baumgartner, A. and Reichel, E. (1975) *The World Water Balance: Mean annual global, continental and maritime precipitation, evaporation and runoff*, Elsevier, Amsterdam (179 pp.).
Bennetts, D. A., McCallum, E. and Grant, J. R. (1986) Cumulonimbus clouds: an introductory review, *Met. Mag.* **115**, 242–56.
Bergeron, T. (1960) Problems and methods of rainfall investigation; in *The Physics of Precipitation*, Geophysical Monograph 5, Amer. Geophys. Union, Washington, DC, 5–30.
Biswas, M. R. and Biswas, A. K. (eds) (1980) *Desertification*, Pergamon, Oxford (523 pp.).
Borchert, J. R. (1971) The dust bowl in the 1970s, *Ann. Assn. Amer. Geogr.* **61**, 1–22.
Braham, R. R. (1959) How does a raindrop grow? *Science* **129**, 123–9.
Bridgman, H. A. (1990) *Global Air Pollution*, Belhaven Press, London (261 pp.).
Browning, K. A. (1980) Local weather forecasting, *Pro. Roy. Soc. Lond. Sect. A* **371**, 179–211.
Browning, K. A. (1985) Conceptual models of precipitation systems, *Met. Mag.* **114**, 293–319.
Browning, K. A. and Hill, F. F. (1981) Orographic rain, *Weather* **36**, 326–9.
Bryson, R. A. (1973) Drought in the Sahel: who or what is to blame? *The Ecologist* **3**(10), 366–71.
Byers, H. R. and Braham, R. R. (1949) *The Thunderstorm*, US Weather Bureau (286 pp.).
Chacon, R. E. and Fernandez, W. (1985) Temporal and spatial rainfall variability in the mountainous region of the Reventazon River Basin, Costa Rica, *J. Climatology* **5**, 175–88.
Dalby, D. and Harrison Church, R. J. (eds) (1973) *Drought in Africa*, Centre for African Studies, School of Oriental and African Studies, University of London (124 pp.).
Deacon, E. L. (1969) Physical processes near the surface of the earth; in Flohn, H. (ed.) *General Climatology*, World Survey of Climatology 2, Elsevier, Amsterdam, pp. 39–104.
Dobbie, C. H. and Wolf, P. O. (1953) The Lynmouth flood of August 1952, *Pro. Inst. Civ. Eng.* Part III, 522–88.
Doornkamp, J. C. and Gregory, K. J. (eds) (1980) *Atlas of Drought in Britain 1975–6*, Institute of British Geographers, London (82 pp.).
Dorman, C. E. and Bourke, R. H. (1981) Precipitation over the Atlantic Ocean, 30°S to 70°N, *Monthly Weather Review* **109**, 554–63.
Durbin, W. G. (1961) An introduction to cloud physics, *Weather* **16**, 71–82 and 113–25.
East, T. W. R. and Marshall, J. S. (1954) Turbulence in clouds as a factor in precipitation, *Quart. J. Roy. Met. Soc.* **80**, 26–47.
Garcia-Prieto, P. R., Ludlam, F. H. and Saunders, P. M. (1960) The possibility of artificially increasing rainfall on Tenerife in the Canary Islands, *Weather* **15**, 39–51.
Gilman, C. S. (1964) Rainfall; in Chow, V. T. (ed.) *Handbook of Applied Hydrology*, McGraw-Hill, New York, section 9.
Harrold, T. W. (1966) The measurement of rainfall using radar, *Weather* **21**, 247–9 and 256–8.
Hastenrath, S. L. (1967) Rainfall distribution and regime in Central America, *Archiv. Met. Geophys. Biokl.* B., **15**(3), 201–41.
Hershfield, D. M. (1961) Rainfall frequency atlas of the United States for durations from 30 minutes to 24 hours and return periods of 1 to 100 years, *US Weather Bureau, Tech. Rept.* 40.
Hopkins, M. M., Jr (1967) An approach to the classification of meteorological satellite data, *J. Appl. Met.* **6**, 164–78.
Houze, R. A., Jr and Hobbs, P. V. (1982) Organization and structure of precipitating cloud systems, *Adv. Geophys.* **24**, 225–315.
Howe, G. M. (1956) The moisture balance in England and Wales, *Weather* **11**, 74–82.
Howells, G. (1990) *Acid Rain and Acid Waters*, Ellis Horwood, New York (215 pp.).
Ilesanmi, O. O. (1971) An empirical formulation of an ITD rainfall model for the tropics: a case study for Nigeria, *J. App. Met.* **10**(5), 882–91.
Jaeger, L. (1976) Monatskarten des Niederschlags für die ganze Erde, *Berichte des Deutsches*

Wetterdienstes **18**(139), Offenbach am Main (38 pp. + plates).

Jiusto, J. E. and Weickmann, H. K. (1973) Types of snowfall, *Bull. Amer. Met. Soc.* **54**, 148–62.

Kelly, P. M. and Wright, P. B. (1978) The European drought of 1975–6 and its climatic context, *Prog. Phys. Geog.* **2**, 237–63.

Landsberg, H. E. (1974) Drought, a recurring element of climate, *Graduate Program in Meteorology, University of Maryland*, Contribution no. 100 (47 pp.).

Latham, J. (1966) Some electrical processes in the atmosphere, *Weather* **21**, 120–7.

Likens, G. E., Wright, R. F., Galloway, J. N. and Butler, T. J. (1979) Acid rain, *Sci. American* **241**, 39–47.

Linsley, R. K. and Franzini, J. B. (1964) *Water-Resources Engineering*, McGraw-Hill, New York, (654 pp.).

London, J., Warren, S. G. and Hahn, C. J. (1989) The global distribution of observed cloudiness – a contribution to the ISCPP, *Adv. Space Res.* **9**(7), 161–5.

Ludlam, F. H. (1980) *Clouds and Storms. The behavior and effect of water in the atmosphere*, Pennsylvania State University, University Park and London (405 pp.).

MacDonald, J. E. (1962) The evaporation-precipitation fallacy, *Weather* **17**, 168–77.

Markham, C. G. and McLain, D. R. (1977) Sea-surface temperature related to rain in Ceará, north-eastern Brazil, *Nature* **265**, 320–3.

Mason, B. J. (1959) Recent developments in the physics of rain and rainmaking, *Weather* **14**, 81–97.

Mason, B. J. (1962) Charge generation in thunderstorms, *Endeavour* **21**, 156–63.

Mason, B. J. (1975) *Clouds, Rain and Rainmaking* (2nd edn), Cambridge University Press, Cambridge and New York (189 pp.).

Mather, J. R. (1985) The water budget and the distribution of climates, vegetation and soils, *Publications in Climatology* **38**(2), University of Delaware, Center for Climatic Research, Newark, Del. (36 pp.).

Miller, D. H. (1977) *Water at the Surface of the Earth*, Academic Press, New York (557 pp.).

Möller, F. (1951) Vierteljahrkarten des Niederschlags für die ganze Erde, *Petermanns Geographische Mitteilungen*, 95 Jahrgang, 1–7.

More, R. J. (1967) Hydrological models and geography; in Chorley, R. J. and Haggett, P. (eds) *Models in Geography*, Methuen, London, pp. 145–85.

Palmer, W. C. (1965) Meteorological drought, *Research Paper No. 45*, US Weather Bureau, Washington, DC.

Paulhus, J. L. H. (1965) Indian Ocean and Taiwan rainfall set new records, *Monthly Weather Review* **93**, 331–5.

Pearl, R. T. *et al.* (1954) *The calculation of irrigation need*, Tech. Bull. No. 4, Min. Agric., Fish and Food, HMSO, London (35 pp.).

Penman, H. L. (1963) *Vegetation and Hydrology*, Tech. Comm. No. 53, Commonwealth Bureau of Soils, Harpenden (124 pp.).

Ratcliffe, R. A. S. (1978) Meteorological aspects of the 1975–6 drought, *Pro. Roy. Soc. Lond. Sect. A* **363**, 3–20.

Reitan, C. H. (1960) Mean monthly values of precipitable water over the United States, 1946–56. *Monthly Weather Review* **88**, 25–35.

Rodda, J. C. (1970) Rainfall excesses in the United Kingdom, *Trans. Inst. Brit. Geog.* **49**, 49–60.

Rodhe, H. (1989) Acidication in a global perspective, *Ambio* **18**, 155–60.

Sawyer, J. S. (1956) The physical and dynamical problems of orographic rain, *Weather* **11**, 375–81.

Schermerhorn, V. P. (1967) Relations between topography and annual precipitation in western Oregon and Washington, *Water Resources Research* **3**, 707–11.

Schwartz, S. E. (1989) Acid deposition: unravelling a regional phenomenon, *Science* **243**, 753–63.

Sevruk, B. (ed.) (1985) *Correction of Precipitation Measurements*, Zürcher Geogr. Schriften no. 23 (also appears as WMO Rep. no. 24, Instruments and Observing Methods, WMO, Geneva) (288 pp.).

Smith, R. B. (1989) Mechanisms of orographic precipitation, *Met. Mag.* **118**, 85–8.

So, C. L. (1971) Mass movements associated with the rainstorm of June 1966 in Hong Kong, *Trans. Inst. Brit. Geog.* **53**, 55–65.

Sutcliffe, R. C. (1956) Water balance and the general circulation of the atmosphere, *Quart. J. Roy. Met. Soc.* **82**, 385–95.

Thompson, B. W. (1986) Small-scale katabatics and cold hollows, *Weather* **41**, 146–53.

Twort, A. C., Hoather, R. C. and Law, F. M. (1985) *Water Supply*, Arnold, London.

Ward, R. C. (1963) Measuring potential evapotranspiration, *Geography* **47**, 49–55.

Weischet, W. (1965) Der tropische-konvective und der ausser tropischeadvektive Typ der vertikalen Niederschlagsverteilung, *Erdkunde* **19**, 6–14.

Weston, K. J. (1977) Cellular cloud patterns, *Weather* **32**, 446–50.

Wilhite, D. A. and Glantz, M. H. (1982) Understanding the drought phenomenon: the role of definitions, *Water Internat.* **10**, 111–30.

World Meteorological Organization (1956) *Inter-*

national Cloud Atlas, Geneva.

World Meteorological Organization (1972) *Distribution of precipitation in mountainous areas* (2 vols), WMO no. 326, Geneva (228 and 587 pp.).

Yarnell, D. L. (1935) Rainfall intensity–frequency data, US Dept. Agr., Misc. Pub. no. 204.

CHAPTER 3 ATMOSPHERIC MOTION

Barry, R. G. (1967) Models in meteorology and climatology; in Chorley, R. J. and Haggett, P. (eds) *Models in Geography*, Methuen, London, pp. 97–144.

Barry, R. G. (1979) Recent advances in climate theory based on simple climate models, *Prog. Phys. Geog.* **3**, 259–86.

Beran, W. D. (1967) Large amplitude lee waves and chinook winds, *J. Appl. Met.* **6**, 865–77.

Borchert, J. R. (1953) Regional differences in world atmospheric circulation, *Ann. Assn. Amer. Geog.* **43**, 14–26.

Brinkmann, W. A. R. (1971) What is a foehn? *Weather* **26**, 230–9.

Brinkmann, W. A. R. (1974) Strong downslope winds at Boulder, Colorado, *Monthly Weather Review* **102**, 592–602.

Broecker, W. S., Peteet, D. M. and Rind, D. (1985) Does the ocean-atmosphere system have more than one stable mode of operation?, *Nature* **315**, 21–6.

Broecker, W. S. and Denton, G. H. (1990) What drives glacial cycles?, *Sci. American* **262**(1), 43–50.

Buettner, K. J. and Thyer, N. (1965) Valley winds in the Mount Rainier area, *Archiv. Met. Geophys. Biokl.* B **14**, 125–47.

Corby, G. A. (ed.) (1970) *The global circulation of the atmosphere*, Roy. Met. Soc., London (257 pp.).

Crowe, P. R. (1949) The trade wind circulation of the world, *Trans. Inst. Brit. Geog.* **15**, 38–56.

Crowe, P. R. (1950) The seasonal variation in the strength of the trades, *Trans. Inst. Brit. Geog.* **16**, 23–47.

Defant, F. and Taba, H. (1957) The threefold structure of the atmosphere and the characteristics of the tropopause, *Tellus* **9**, 259–74.

Dietrich, G. (1963) *General Oceanography: An introduction*, Wiley, New York (588 pp.).

Druyan, L. M., Somerville, R. J. C. and Quirk, W. J. (1975) Extended-range forecasts with the GISS model of the global atmosphere, *Monthly Weather Review* **103**, 779–95.

Eddy, A. (1966) The Texas coast sea-breeze: a pilot study, *Weather* **21**, 162–50.

Ernst, J. A. (1976) SMS-1 Night-time infrared imagery of low-level mountain waves, *Monthly Weather Review* **104**, 207–9.

Flohn, H. (1969) Local wind systems; in Flohn, H. (ed.) *General Climatology*, World Survey of Climatology 2, Elsevier, Amsterdam, pp. 139–71.

Flohn, H. and Fantechi, R. (eds) (1984) *The Climate of Europe: Past, Present and Future*, D. Reidel, Dordrecht (356 pp.).

Garbell, M. A. (1947) *Tropical and Equatorial Meteorology*, Pitman, London (237 pp.).

Geiger, R. (1969) Topoclimates; in Flohn, H. (ed.) *General Climatology*, World Survey of Climatology 2, Elsevier, Amsterdam, pp. 105–38.

Glenn, C. L. (1961) The chinook, *Weatherwise* **14**, 175–82.

Hare, F. K. (1965) Energy exchanges and the general circulation, *Geography* **50**, 229–41.

Houghton, J. (ed.) (1984) *The Global Climate*, Cambridge University Press, Cambridge (233 pp.).

Indian Meteorological Department (1960) *Monsoons of the World*, Delhi (270 pp.).

Johnson, A. and O'Brien, J. J. (1973) A study of an Oregon sea breeze event, *J. Appl. Met.* **12**, 1267–83.

Kuenen, Ph. H. (1955) *Realms of Water*, Cleaver-Hulme Press, London (327 pp.).

Lamb, H. H. (1960) Representation of the general atmospheric circulation, *Met. Mag.* **89**, 319–30.

LeMarshall, J. F., Kelly, G. A. M. and Karoly, D. J. (1985) An atmospheric climatology of the southern hemisphere based on 10 years of daily numerical analyses (1972–1982): I, Overview, *Austral. Met. Mag.* **33**, 65–86.

Levitus, S. (1982) Climatological atlas of the world ocean, *NOAA Professional Paper No. 13*, Rockville, Md. (173 pp.).

Lockwood, J. G. (1962) Occurrence of föhn winds in the British Isles, *Met. Mag.* **91**, 57–65.

Lorenz, E. N. (1967) *The nature and theory of the general circulation of the atmosphere*, World Meteorological Organization, Geneva (161 pp.).

McDonald, J. E. (1952) The Coriolis effect, *Sci. American* **186**, 72–8.

Meehl, G. A. (1984) Modelling the earth's climate, *Climatic Change* **6**, 259–86.

Meehl, G. A. (1987a) The annual cycle and interannual variability in the tropical Pacific and Indian Ocean regions, *Monthly Weather Review* **115**, 51–74.

Meehl, G. A. (1987b) The tropics and their role in the global climate system, *Geographical Journal* **153**, 21–36.

Namias, J. (1972) Large-scale and long-term fluctuations in some atmospheric and ocean variables; in Dyrssen, D. and Jagner, D. (eds) *The Changing*

Chemistry of the Oceans, Nobel Symposium 20, Wiley, New York, pp. 27–48.

O'Connor, J. F. (1961) Mean circulation patterns based on 12 years of recent northern hemispheric data, *Monthly Weather Review* **89**, 211–28.

Oke, T. R. (1978) *Boundary Layer Climates*, Methuen, London (372 pp.) (2nd edn 1987, 435 pp.).

Palmén, E. (1951) The role of atmospheric disturbances in the general circulation, *Quart. J. Roy. Met. Soc.* **77**, 337–54.

Pfeffer, R. L. (1964) The global atmospheric circulation, *Trans. New York Acad. Sci.*, ser. 11, **26**, 984–97.

Riehl, H. (1962a) General atmospheric circulation of the tropics, *Science* **135**, 13–22.

Riehl, H. (1962b) *Jet streams of the atmosphere*, Tech. Paper No. 32, Colorado State University (117 pp.).

Riehl, H. (1969) On the role of the tropics in the general circulation of the atmosphere, *Weather* **24**, 288–308.

Riehl, H. *et al.* (1954) The jet stream, *Met. Monogr.* **2**(7). American Meteorological Society, Boston, Mass. (100 pp.).

Riley, D. and Spalton, L. (1981) *World Weather and Climate* (2nd edn), Cambridge University Press, London (128 pp.).

Rossby, C.-G. (1941) The scientific basis of modern meteorology, US Dept. of Agriculture Yearbook *Climate and Man*, pp. 599–655.

Rossby, C.-G. (1949) On the nature of the general circulation of the lower atmosphere; in Kuiper, G. P. (ed.) *The Atmosphere of the Earth and Planets*, University of Chicago Press, Chicago, Ill., pp. 16–48.

Saltzman, B. (1983) Climatic systems analysis, *Adv. Geophys.* **25**, 173–233.

Sawyer, J. S. (1957) Jet stream features of the earth's atmosphere, *Weather* **12**, 333–4.

Scorer, R. S. (1958) *Natural Aerodynamics*, Pergamon Press, Oxford (312 pp.).

Scorer, R. S. (1961) Lee waves in the atmosphere, *Sci. American* **204**, 124–34.

Starr, V. P. (1956) The general circulation of the atmosphere, *Sci. American* **195**, 40–5.

Steinacker, R. (1984) Area–height distribution of a valley and its relation to the valley wind, *Contrib. Atmos. Phys.* **57**, 64–74.

Streten, N. A. (1980) Some synoptic indices of the southern hemisphere mean sea level circulation 1972–77, *Monthly Weather Review* **108**, 18–36.

Taljaard, J. J., van Loon, H., Crutcher, H. L. and Jenne, R. L. (1969) *Climate of the Upper Air, Part 1. Southern Hemisphere* 1, Naval Weather Service Command, Washington, DC, NAVAIR 50-1C-55.

Troen, I. and Petersen, E. L. (1989) *European Wind Atlas*, Commission of the Economic Community, Risø National Laboratory, Roskilde, Denmark (656 pp.).

Tucker, G. B. (1962) The general circulation of the atmosphere, *Weather* **17**, 320–40.

Tyson, P. D. (1986) *Climatic Change and Variability in Southern Africa*, Oxford University Press, Cape Town (220 pp.).

van Arx, W. S. (1962) *Introduction to Physical Oceanography*, Addison-Wesley, Reading, Mass. (422 pp.).

van Loon, H. (1964) Mid-season average zonal winds at sea level and at 500 mb south of 25°S and a brief comparison with the northern hemisphere, *J. Appl. Met.* **3**, 554–63.

van Loon, H. (ed.) (1984) *Climates of the Oceans*; Vol. 15 in Landsberg, H. E. (ed.) *World Survey of Climatology*, Elsevier, Amsterdam (716 pp.).

Waco, D. E. (1968) Frost pockets in the Santa Monica Mountains of southern California, *Weather* **23**, 456–61.

Walker, J. M. (1972) Monsoons and the global circulation, *Met. Mag.* **101**, 349–55.

Wallington, C. E. (1960) An introduction to lee waves in the atmosphere, *Weather* **15**, 269–76.

Wallington, C. E. (1969) Depressions as moving vortices, *Weather* **24**, 42–51.

Wickham, P. G. (1966) Weather for gliding over Britain, *Weather* **21**, 154–61.

CHAPTER 4 AIR MASSES, FRONTS AND DEPRESSIONS

Barrett, E. C. (1964) Satellite meteorology and the geographer, *Geography* **49**, 377–86.

Bates, F. C. (1962) Tornadoes in the central United States, *Trans. Kansas Acad. Sci.* **65**, 215–46.

Belasco, J. E. (1952) Characteristics of air masses over the British Isles, Meteorological Office, *Geophysical Memoirs* **11**(87) (34 pp.).

Bennetts, D. A., Grant, J. R. and McCallum, E. (1988) An introductory review of fronts: Pt. I Theory and observations, *Met. Mag.* **117**, 357–70.

Bosart, L. (1985) Weather forecasting; in Houghton, D. D. (ed.) *Handbook of Applied Meteorology*, Wiley, New York, pp. 205–79.

Boucher, R. J. and Newcomb, R. J. (1962) Synoptic interpretation of some TIROS vortex patterns: a preliminary cyclone model, *J. Appl. Met.* **1**, 122–36.

Boyden, C. J. (1960) The use of upper air charts in forecasting, *The Marine Observer* **30**, 27–31.

Boyden, C. J. (1963) Development of the jet stream

and cut-off circulations, *Met. Mag.* **92**, 287–99.
Browning, K. A. (1968) The organization of severe local storms, *Weather* **23**, 429–34.
Browning, K. A. (1980) Local weather forecasting, *Pro. Roy. Soc. Lond. Sect. A* **371**, 179–211.
Browning, K. A. (ed.) (1983) *Nowcasting*, Academic Press, New York (256 pp.).
Browning, K. A. (1985) Conceptual models of precipitation systems, *Met. Mag.* **114**, 293–319.
Browning, K. A. (1986) Weather radar and FRONTIERS, *Weather* **41**, 9–16.
Browning, K. A. and Hill, F. F. (1981) Orographic rain, *Weather* **36**, 326–9.
Browning, K. A., Bader, M. J., Waters, A. J., Young, M. V. and Monk, G. A. (1987) Application of satellite imagery to nowcasting and very short range forecasting, *Met. Mag.* **116**, 161–79.
Businger, S. (1985) The synoptic climatology of polar low outbreaks, *Tellus* **37A**, 419–32.
Carleton, A. M. (1985) Satellite climatological aspects of the 'polar low' and 'instant occlusion', *Tellus* **37A**, 433–50.
Crowe, P. R. (1949) The trade wind circulation of the world, *Trans. Inst. Brit. Geog.* **15**, 38–56.
Crowe, P. R. (1965) The geographer and the atmosphere, *Trans. Inst. Brit. Geog.* **36**, 1–19.
Freeman, M. H. (1961) Fronts investigated by the Meteorological Research Flight, *Met. Mag.* **90**, 189–203.
Gadd, A. J. (1985) The 15-level weather prediction model, *Met. Mag.* **114**, 222–6.
Galloway, J. L. (1958a) The three-front model: its philosophy, nature, construction and use, *Weather* **13**, 3–10.
Galloway, J. L. (1958b) The three-front model, the tropopause and the jet stream, *Weather* **13**, 395–403.
Galloway, J. L. (1960) The three-front model, the developing depression and the occluding process, *Weather* **15**, 293–309.
Gilchrist, A. (1986) Long-range forecasting, *Quart. J. Roy. Met. Soc.* **112**, 567–92.
Godson, W. L. (1950) The structure of North American weather systems, *Cent. Proc. Roy. Met. Soc.* London, pp. 89–106.
Gyakum, J. R. (1983) On the evolution of the *QE II* storm, I: Synoptic aspects, *Monthly Weather Review* **111**, 1137–55.
Hare, F. K. (1960) The westerlies, *Geog. Rev.* **50**, 345–67.
Harman, J. R. (1971) Tropical waves, jet streams, and the United States weather patterns, Association of American Geographers, Commission on College Geography, *Resource Paper No. 11* (37 pp.).
Harrold, T. W. (1973) Mechanisms influencing the distribution of precipitation within baroclinic disturbances, *Quart. J. Roy. Met. Soc.* **99**, 232–51.
Hindley, K. (1977) Learning to live with twisters, *New Scientist* **70**, 280–2.
Hobbs, P. V. (1978) Organization and structure of clouds and precipitation on the meso-scale and micro-scale of cyclonic storms, *Rev. Geophys. and Space Phys.* **16**, 1410–11.
Houze, R. A. and Hobbs, P. V. (1982) Organization and structure of precipitating cloud systems, *Adv. Geophys.* **24**, 225–315.
Hunt, R. D. (1985) The models in action, *Met. Mag.* **114**, 261–72.
Jackson, M. C. (1977) Meso-scale and small-scale motions as revealed by hourly rainfall maps of an outstanding rainfall event: 14–16 September 1968, *Weather* **32**, 2–16.
Kalnay, E., Kanamitsu, M. and Baker, W. E. (1990) Global numerical weather prediction at the National Meteorological Center, *Bull. Amer. Met. Soc.* **71**, 1410–28.
Kessler, E. (ed.) (1983) *The Thunderstorm in Human Affairs* (2nd edn), University of Oklahoma Press, Norman, Okla (250 pp.).
Klein, W. H. (1948) Winter precipitation as related to the 700 mb circulation, *Bull. Amer. Met. Soc.* **29**, 439–53.
Klein, W. H. (1957) Principal tracks and mean frequencies of cyclones and anticyclones in the Northern Hemisphere, *Research Paper No. 40*, Weather Bureau, Washington, DC (60 pp.).
Klein, W. H. (1982) Statistical weather forecasting on different time scales, *Bull. Amer. Met. Soc.* **63**, 170–7.
Lamb, H. H. (1951) Essay on frontogenesis and frontolysis, *Met. Mag.* **80**, 35–6, 65–71 and 97–106.
Liljequist, G. H. (1970) *Klimatologi*, Generalstabens Litografiska Anstalt, Stockholm.
Ludlam, F. H. (1961) The hailstorm, *Weather* **16**, 152–62.
Lyall, I. T. (1972) The polar low over Britain, *Weather* **27**, 378–90.
Maddox, R. A. (1980) Mesoscale convective complexes, *Bull. Amer. Met. Soc.* **61**, 1374–87.
Mason, B. J. (1974) The contribution of satellites to the exploration of the global atmosphere and to the improvement of weather forecasting, *Met. Mag.* **103**, 181–201.
Miles, M. K. (1962) Wind, temperature and humidity distribution at some cold fronts over SE England, *Quart. J. Roy. Met. Soc.* **88**, 286–300.
Miller, R. C. (1959) Tornado-producing synoptic patterns, *Bull. Amer. Met. Soc.* **40**, 465–72.
Miller, R. C. and Starrett, L. G. (1962) Thunder-

storms in Great Britain, *Met. Mag.* **91**, 247–55.

Newton, C. W. (1966) Severe convective storms, *Adv. Geophys.* **12**, 257–308.

Newton, C. W. (ed.) (1972) Meteorology of the Southern Hemisphere, *Met. Monogr. No. 13* (35), American Meteorological Society, Boston, Mass. (263 pp.).

Pedgley, D. E. (1962) A meso-synoptic analysis of the thunderstorms on 28 August 1958, *Geophysical Memoirs Meteorological Office* **14**(1) (30 pp.).

Penner, C. M. (1955) A three-front model for synoptic analyses, *Quart. J. Roy. Met. Soc.* **81**, 89–91.

Petterssen, S. (1950) Some aspects of the general circulation of the atmosphere, *Cent. Proc. Roy. Met. Soc.*, London, pp. 120–55.

Pothecary, I. J. W. (1956) Recent research on fronts, *Weather* **12**, 147–50.

Reed, R J. (1960) Principal frontal zones of the northern hemisphere in winter and summer, *Bull. Am. Met. Soc.* **41**, 591–8.

Richter, D. A. and Dahl, R. A. (1958) Relationship of heavy precipitation to the jet maximum in the eastern United States, *Monthly Weather Review* **86**, 368–76.

Riley, D. and Spalton, L. (1974) *World Weather and Climate*, Cambridge University Press, Cambridge (120 pp.).

Roebber, P. J. (1989) On the statistical analysis of cyclone deepening rates, *Monthly Weather Review* **117**, 2293–8.

Sanders, F. and Gyakum, J. R. (1980) Synoptic-dynamic climatology of the 'bomb', *Monthly Weather Review* **108**, 1589–1606.

Showalter, A. K. (1939) Further studies of American air mass properties, *Monthly Weather Review* **67**, 204–18.

Slater, P. M. and Richards, C. J. (1974) A memorable rainfall event over southern England, *Met. Mag.* **103**, 255–68 and 288–300.

Smagorinsky, J. (1974) Global atmospheric modeling and the numerical simulation of climate; in Hess, W. N. (ed.) *Weather and Climate Modification*, Wiley, New York, pp. 633–86.

Smith, W. L. (1985) Satellites; in Houghton, D. D. (ed.) *Handbook of Applied Meteorology*, Wiley, New York, pp. 380–472.

Snow, J. T. (1984) The tornado, *Sci. American* **250**(4), 56–66.

Sutcliffe, R. C. and Forsdyke, A. G. (1950) The theory and use of upper air thickness patterns in forecasting, *Quart. J. Roy. Met. Soc.* **76**, 189–217.

United States Department of Commerce (1981) *Operations of the National Weather Service*, National Oceanic and Atmospheric Administration, Silver Springs, Md. (249 pp.).

Vederman, J. (1954) The life cycles of jet streams and extratropical cyclones, *Bull. Amer. Met. Soc.* **35**, 239–44.

Wagner, A. J. (1989) Medium- and long-range weather forecasting, *Weather and Forecasting* **4**, 413–26.

Wallington, C. E. (1963) Meso-scale patterns of frontal rainfall and cloud, *Weather* **18**, 171–81.

Weatherwise (1983) A conversation with Donald Gilman, Predicting the weather for the long term, *Weatherwise* **36**, 290–7.

Wendland, W. M. and Bryson, R. A. (1981) Northern hemisphere airstream regions, *Monthly Weather Review* **109**, 255–70.

Wendland, W. M. and McDonald, N. S. (1986) Southern hemisphere airstream climatology, *Monthly Weather Review* **114**, 88–94.

Wick, G. (1973) Where Poseidon courts Aeolus, *New Scientist*, 18 January, pp. 123–6.

Yoshino, M. M. (1967) Maps of the occurrence frequencies of fronts in the rainy season in early summer over east Asia, *Science Reports of the Tokyo University of Education* **89**, 211–45.

CHAPTER 5 WEATHER AND CLIMATE IN TEMPERATE LATITUDES

Bailey, H. P. (1964) Toward a unified concept of the temperate climate, *Geog. Rev.* **54**(4), 516–45.

Balling, R. C., Jr (1985) Warm seasonal nocturnal precipitation in the Great Plains of the United States, *J. Climate Applied Met.* **24**, 1383–7.

Barry, R. G. (1963) Aspects of the synoptic climatology of central south England, *Met. Mag.* **92**, 300–8.

Barry, R. G. (1967a) Seasonal locational of the arctic front over North America, *Geog. Bull.* **9**, 79–95.

Barry, R. G. (1967b) The prospect for synoptic climatology: a case study; in Steel, R. W. and Lawton, R. (eds) *Liverpool Essays in Geography*, Longman, London, pp. 85–106.

Barry, R. G. (1973) A climatological transect on the east slope of the Front Range, Colorado, *Arct. Alp. Res.* **5**, 89–110.

Barry, R. G. and Hare, F. K. (1974) Arctic climate; in Ives, J. D. and Barry, R. G. (eds) *Arctic and Alpine Environments*, Methuen, London, pp. 17–54.

Belasco, J. E. (1948) The incidence of anticyclonic days and spells over the British Isles, *Weather* **3**, 233–42.

Belasco, J. E. (1952) Characteristics of air masses over the British Isles, Meteorological Office, *Geophysical Memoirs* **11**(87) (34 pp.).

Boast, R. and McQuingle, J. B. (1972) Extreme weather conditions over Cyprus during April 1971, *Met. Mag.* **101**, 137–53.

Borchert, J. (1950) The climate of the central North American grassland, *Ann. Assn. Amer. Geog.* **40**, 1–39.

Browning, K. A. and Hill, F. F. (1981) Orographic rain, *Weather* **36**, 326–9.

Bryson, R. A. (1966) Air masses, streamlines and the boreal forest, *Geog. Bull.* **8**, 228–69.

Bryson, R. A. and Hare, F. K. (eds) (1974) *Climates of North America*, World Survey of Climatology 11, Elsevier, Amsterdam (420 pp.).

Bryson, R. A. and Lahey, J. F. (1958) *The March of the Seasons*, Meteorological Department, University of Wisconsin (41 pp.).

Bryson, R. A. and Lowry, W. P. (1955) Synoptic climatology of the Arizona summer precipitation singularity, *Bull. Amer. Met. Soc.* **36**, 329–39.

Burbridge, F. E. (1951) The modification of continental polar air over Hudson Bay, *Quart. J. Met. Soc.* 77, 365–74.

Butzer, K. W. (1960) Dynamic climatology of large-scale circulation patterns in the Mediterranean area, *Meteorologische Rundschau* **13**, 97–105.

Carleton, A. M. (1986) Synoptic-dynamic character of 'bursts' and 'breaks' in the southwest US summer precipitation singularity, *J. Climatol.* **6**, 605–23.

Chandler, T. J. and Gregory, S. (eds) (1976) *The Climate of the British Isles*, Longman, London (390 pp.).

Derecki, J. A. (1976) Heat storage and advection in Lake Erie, *Water Resources Research* **12**(6), 1144–50.

Durrenberger, R. W. and Ingram, R. S. (1978) Major storms and floods in Arizona 1862–1977, *State of Arizona, Office of the State Climatologist, Climatological Publications, Precipitation Series* No. 4 (44 pp.).

Easterling, D. R. and Robinson, P. J. (1985) The diurnal variation of thunderstorm activity in the United States, *J. Climate Appl. Met.* **24**, 1048–58.

Elsom, D. M. and Meaden, G. T. (1984) Spatial and temporal distribution of tornadoes in the United Kingdom 1960–1982, *Weather* **39**, 317–23.

Environmental Science Services Administration (1965) *APT Users' Guide*, US Department of Commerce, Washington, DC (80 pp.).

Environmental Science Services Administration (1968) *Climatic Atlas of the United States*, US Department of Commerce, Washington, DC (80 pp.).

Evenari, M., Shanan, L. and Tadmor, N. (1971) *The Negev*, Harvard University Press, Cambridge, Mass. (345 pp.).

Ferguson, E. W., Ostby, F. P., Leftwich, P. W., Jr, and Hales, J. E., Jr (1986) The tornado season of 1984, *Monthly Weather Review* **114**, 624–35.

Flohn, H. (1954) *Witterung und Klima in Mitteleuropa*, Zurich (218 pp.).

Gentilli, J. (ed.) (1971) *Climates of Australia and New Zealand*, World Survey of Climatology **13**, Elsevier, Amsterdam (405 pp.).

Gorczynski, W. (1920) Sur le calcul du degré du continentalisme et son application dans la climatologie, *Geografiska Annaler* **2**, 324–31.

Green, C. R. and Sellers, W. D. (1964) *Arizona Climate*, University of Arizona Press, Tucson (503 pp.).

Hales, J. E., Jr (1974) South-western United States summer monsoon source – Gulf of Mexico or Pacific Ocean, *J. Appl. Met.* **13**, 331–42.

Hare, F. K. (1968) The Arctic, *Quart. J. Roy. Met. Soc.* **74**, 439–59.

Hare, F. K. and Thomas, M. K. (1979) *Climate Canada* (2nd edn) Wiley, Canada (230 pp.).

Hawke, E. L. (1933) Extreme diurnal range of air temperature in the British Isles, *Quart. J. Roy. Met. Soc.* **59**, 261–5.

Hill, F. F., Browning, K. A. and Bader, M. J. (1981) Radar and rain gauge observations of orographic rain over South Wales, *Quart. J. Roy. Met. Soc.* **107**, 643–70.

Horn, L. H. and Bryson, R. A. (1960) Harmonic analysis of the annual march of precipitation over the United States, *Ann. Assn. Am. Geog.* **50**, 157–71.

Huttary, J. (1950) Die Verteilung der Niederschläge auf die Jahreszeiten im Mittelmeergebeit, *Meteorologische Rundschau* **3**, 111–19.

Klein, W. H. (1963) Specification of precipitation from the 700-mb circulation, *Monthly Weather Review* **91**, 527–36.

Knox, J. L. and Hay, J. E. (1985) Blocking signatures in the northern hemisphere: frequency distribution and interpretation, *J. Climatology* **5**, 1–16.

Lamb, H. H. (1950) Types and spells of weather around the year in the British Isles: annual trends, seasonal structure of the year, singularities, *Quart. J. Roy. Met. Soc.* **76**, 393–438.

Linacre, W. and Hobbs, J. (1977) *The Australian Climatic Environment*, Wiley, Brisbane (354 pp.).

Longley, R. W. (1967) The frequency of Chinooks in Alberta, *The Albertan Geographer* **3**, 20–2.

Lumb, F. E. (1961) Seasonal variations of the sea surface temperature in coastal waters of the British Isles, *Met. Office Sci. Paper No. 6*, MO 685 (21 pp.).

Manley, G. (1944) Topographical features and the climate of Britain, *Geog. Jour.* **103**, 241–58.

Manley, G. (1945) The effective rate of altitude change in temperate Atlantic climates, *Geog. Rev.* **35**, 408–17.

Mather, J. R. (1985) The water budget and the

distribution of climates, vegetation and soils, *Publications in Climatology* **38**(2), Center for Climatic Research, University of Delaware, Newark (36 pp.).
Meteorological Office (1952) *Climatological Atlas of the British Isles*, MO 488, HMSO, London (139 pp.).
Meteorological Office (1962) *Weather in the Mediterranean I, General Meteorology* (2nd edn) MO 391, HMSO, London (362 pp.).
Meteorological Office (1964a) *Weather in the Mediterranean II* (2nd edn) MO 391b, HMSO, London (372 pp.).
Meteorological Office (1964b) *Weather in Home Fleet Waters I, The Northern Seas*, Part 1, MO 732a, HMSO, London (265 pp.).
Namias, J. (1964) Seasonal persistence and recurrence of European blocking during 1958–60, *Tellus* **16**, 394–407.
Nickling, W. G. and Brazel, A. J. (1984) Temporal and spatial characteristics of Arizona dust storms (1965–1980), *Climatology* **4**, 645–60.
Poltaraus, B. V. and Staviskiy, D. B. (1986) The changing continentality of climate in central Russia, *Soviet Geography* **27**, 51–8.
Rayner, J. N. (1961) *Atlas of Surface Temperature Frequencies for North America and Greenland*, Arctic Meteorological Research Group, McGill University, Montreal.
Rex, D. F. (1950–1) The effect of Atlantic blocking action upon European climate, *Tellus* **2**, 196–211 and 275–301; **3**, 100–11.
Schick, A. P. (1971) A desert flood, *Jerusalem Studies in Geography* **2**, 91–155.
Shaw, E. M. (1962) An analysis of the origins of precipitation in Northern England, 1956–60, *Quart. J. Roy. Met. Soc.* **88**, 539–47.
Sivall, T. (1957) Sirocco in the Levant, *Geografiska Annaler* **39**, 114–42.
Stone, J. (1983) Circulation type and the spatial distribution of precipitation over central, eastern and southern England, *Weather* **38**, 173–7, 200–5.
Storey, A. M. (1982) A study of the relationship between isobaric patterns over the UK and central England temperature and England–Wales rainfall, *Weather* **37**, 2–11, 46, 88–9, 122, 151, 170, 208, 244, 260, 294, 327, 360.
Sumner, E. J. (1959) Blocking anticyclones in the Atlantic-European sector of the northern hemisphere, *Met. Mag.* **88**, 300–11.
Thomas, M. K. (1964) *A Survey of Great Lakes Snowfall*, Great Lakes Research Division, University of Michigan, Publication No. 11, pp. 294–310.
Thornthwaite, C. W. and Mather, J. R. (1955) The moisture balance, *Publications in Climatology* **8**(1), Laboratory of Climatology, Centerton, NJ (104 pp.).
Tout, D. G. and Kemp, V. (1985) The named winds of Spain, *Weather* **40**, 322–9.
Troen, I. and Petersen, E. L. (1989) *European Wind Atlas*, Commission of the Economic Community, Risø National Laboratory, Roskilde, Denmark (656 pp.).
United States Weather Bureau (1947) *Thunderstorm Rainfall*, Vicksburg, Mississippi (331 pp.).
Villmow, J. R. (1956) The nature and origin of the Canadian dry belt, *Ann. Assn. Amer. Geog.* **46**, 221–32.
Visher, S. S. (1954) *Climatic Atlas of the United States*, Harvard University Press, Cambridge, Mass. (403 pp.).
Wallace, J. M. (1975) Diurnal variations in precipitation and thunderstorm frequency over the coterminous United States, *Monthly Weather Review* **103**, 406–19.
Wallén, C. C. (1960) Climate; in Somme, A. (ed.) *The Geography of Norden*, Cappelens Forlag, Oslo, pp. 41–53.
Wallén, C. C. (ed.) (1970) *Climates of Northern and Western Europe*, World Survey of Climatology 5, Elsevier, Amsterdam (253 pp.).
Woodroffe, A. (1988) Summary of the weather pattern developments of the storm of 15/16 October 1987, *Met. Mag.* **117**, 99–103.

CHAPTER 6 TROPICAL WEATHER AND CLIMATE

Academica Sinica (1957–8) On the general circulation over eastern Asia, *Tellus* **9**, 432–46; **10**, 58–75 and 299–312.
Anthes, R. A. (1982) Tropical cyclones: their evolution, structure, and effects, *Met. Monogr.* Amer. Met. Soc., Boston, Mass. **19**(41) (208 pp.).
Arakawa, H. (ed.) (1969) *Climates of Northern and Eastern Asia*, World Survey of Climatology 8, Elsevier, Amsterdam (248 pp.).
Atkinson, G. D. (1971) *Forecasters Guide to Tropical Meteorology*, Headquarters Air Weather Service, US Air Force. Tech. Rep. 240 (360 pp.).
Avila, L. A. (1990) Atlantic tropical systems of 1989, *Monthly Weather Review* **118**, 1178–85.
Barry, R. G. (1978) Aspects of the precipitation characteristics of the New Guinea mountains, *J. Trop. Geog.* **47**, 13–30.
Beckinsale, R. P. (1957) The nature of tropical rainfall, *Tropical Agriculture* **34**, 76–98.
Bhalme, H. N. and Mooley, D. A. (1980) Large-scale droughts/floods and monsoon circulation, *Monthly Weather Review* **108**, 1197–1211.

Blumenstock, D. I. (1958) Distribution and characteristics of tropical climates, *Proc. 9th Pacific Sci. Congr.* **20**, 3–23.

Breed, C. S. *et al.* (1979) Regional studies of sand seas, using Landsat (ERTS) imagery, *US Geological Survey Professional Paper No. 1052*, 305–97.

Chang, J.-H. (1962) Comparative climatology of the tropical western margins of the northern oceans, *Ann. Assn. Amer. Geog.* **52**, 221–7.

Chang, J.-H. (1967) The Indian summer monsoon,*Geog. Rev.* **57**, 373–96.

Chang, J.-H. (1971) The Chinese monsoon, *Geog. Rev.* **61**, 370–95.

Chopra, K. P. (1973) Atmospheric and oceanic flow problems introduced by islands, *Adv. Geophys.* **16**, 297–421.

Clackson, J. R. (1957) The seasonal movement of the boundary of northern air, Nigerian Meteorological Service, *Technical Note 5* (see Addendum 1958).

Crowe, P. R. (1949) The trade wind circulation of the world, *Trans. Inst. Brit. Geog.* **15**, 37–56.

Crowe, P. R. (1951) Wind and weather in the equatorial zone, *Trans. Inst. Brit. Geog.* **17**, 23–76.

Cry, G. W. (1965) Tropical cyclones of the North Atlantic Ocean, *Tech. Paper No. 55*, Weather Bureau, Washington, DC (148 pp.).

Curry, L. and Armstrong, R. W. (1959) Atmospheric circulation of the tropical Pacific ocean, *Geografiska Annaler* **41**, 245–55.

Das, P. K. (1987) Short- and long-range monsoon prediction in India; in Fein, J. S. and Stephens, P. L. (eds) *Monsoons*, John Wiley & Sons, New York, pp. 549–78.

Dickinson, R. E. (ed.) (1987) *The Geophysiology of Amazonia: Vegetation and Climate Interactions*, John Wiley & Sons, New York (526 pp.).

Dubief, J. (1963) Le climat du Sahara. *Memoire de l'Institut de Recherches Sahariennes, Université d'Alger, Algiers* (275 pp.).

Dunn, G. E. and Miller, B. I. (1960) *Atlantic Hurricanes*, Louisiana State University Press, Baton Rouge, La. (326 pp.).

Eldridge, R. H. (1957) A synoptic study of West African disturbance lines, *Quart. J. Roy. Met. Soc.* **83**, 303–14.

Fett, R. W. (1964) Aspects of hurricane structure: new model considerations suggested by TIROS and Project Mercury observations, *Monthly Weather Review* **92**, 43–59.

Findlater, J. (1971) Mean monthly airflow at low levels over the Western Indian Ocean, *Geophysical Memoirs 115, Meteorological Office* (53 pp.).

Findlater, J. (1974) An extreme wind speed in the low-level jet-stream system of the western Indian Ocean, *Met. Mag.* **103**, 201–5.

Flohn, H. (1968) *Contributions to a Meteorology of the Tibetan Highlands.* Atmos. Sci. Paper No. 130, Colorado State University, Fort Collins (120 pp.).

Flohn, H. (1971) Tropical circulation patterns, *Bonn. Geogr. Abhandl* **15** (55 pp.).

Fosberg, F. R., Garnier, B. J. and Küchler, A. W. (1961) Delimitation of the humid tropics, *Geog. Rev.* **51**, 333–47.

Frank, N. L. and Hubert, P. J. (1974) Atlantic tropical systems of 1973, *Monthly Weather Review* **102**, 290–5.

Frost, R. and Stephenson, P. H. (1965) Mean streamlines and isotachs at standard pressure levels over the Indian and west Pacific Oceans and adjacent land areas, *Geophys. Mem.* **14**(109), HMSO, London (24 pp.).

Gao, Y.-X. and Li, C. (1981) Influence of Qinghai-Xizang plateau on seasonal variation of general atmospheric circulation; in *Geoecological and Ecological Studies of Qinghai-Xizang Plateau*, Vol. 2, Science Press, Beijing, pp. 1477–84.

Garnier, B. J. (1967) Weather conditions in Nigeria, *Climatological Research Series No. 2*, McGill University Press, Montreal (163 pp.).

Gentilli, J. (ed.) (1971) *Climates of Australia and New Zealand*, World Survey of Climatology 13, Elsevier, Amsterdam (405 pp.).

Goudie, A. and Wilkinson, J. (1977) *The Warm Desert Environment*, Cambridge University Press, Cambridge (88 pp.).

Gray, W. M. (1968) Global view of the origin of tropical disturbances and hurricanes, *Monthly Weather Review* **96**, 669–700.

Gray, W. M. (1979) Hurricanes: their formation, structure and likely role in the tropical circulation; in Shaw, D. B. (ed.) *Meteorology over the Tropical Oceans*, Royal Meteorological Society, Bracknell, pp. 155–218.

Gray, W. M. and Jacobson, R. W. (1977) Diurnal variation of deep cumulus convection, *Monthly Weather Review* **105**, 1171–88.

Gregory, S. (1965) *Rainfall over Sierra Leone*, Geography Department, University of Liverpool, Research Paper No. 2 (58 pp.).

Griffiths, J. F. (ed.) (1972) *Climates of Africa*, World Survey of Climatology 10, Elsevier, Amsterdam (604 pp.).

Hamilton, M. G. (1979) *The South Asian Summer Monsoon*, Arnold, Australia (72 pp.).

Hastenrath, S. (1991) *Climate Dynamics of the Tropics*, Kluwer Academic Publishers, Dordrecht (488 pp.).

Hayward, D. F. and Oguntoyinbo, J. S. (1987) *The Climatology of West Africa*, Hutchinson, London (271 pp.).

Houze, R. A., Goetis, S. G., Marks, F. D. and West, A. K. (1981) Winter monsoon convection in the

vicinity of North Borneo, *Monthly Weather Review* **109**, 1595–614.

Houze, R. A. and Hobbs, P. V. (1982) Organization and structure of precipitating cloud systems, *Adv. Geophys.* **24**, 225–315.

Hutchings, J. W. (ed.) (1964) *Proceedings of the Symposium on Tropical Meteorology*, New Zealand Meteorological Service, Wellington (737 pp.).

Indian Meteorological Department (1960) *Monsoons of the World*, Delhi (270 pp.).

Jackson, I. J. (1977) *Climate, Water and Agriculture in the Tropics*, Longman, London (248 pp.).

Jalu, R. (1960) Etude de la situation météorologique au Sahara en Janvier 1958, *Ann. de Géog.* **69**(371), 288–96.

Jordan, C. L. (1955) Some features of the rainfall at Guam, *Bull. Amer. Met. Soc.* **36**, 446–55.

Kamara, S. I. (1986) The origins and types of rainfall in West Africa, *Weather* **41**, 48–56.

Kiladis, G. N. and von Storch, H. (1989) Origin of the South Pacific convergence zone, *J. Climate.* **2**, 1185–95.

Knox, R. A. (1987) The Indian Ocean: interaction with the monsoon; in Fein, J. S. and Stephens, P. L. (eds) *Monsoons*, John Wiley & Sons, New York, pp. 365–97.

Koteswaram, P. (1958) The easterly jet stream in the tropics, *Tellus* **10**, 43–57.

Kousky, V. E. (1980) Diurnal rainfall variation in northeast Brazil, *Monthly Weather Review* **108**, 488–98.

Kreuels, R., Fraedrich, K. and Ruprecht, E. (1975) An aerological climatology of South America, *Met. Rundsch.* **28**, 17–24.

Krishnamurti, T. N. (ed.) (1977) Monsoon meteorology, *Pure Appl. Geophys.* **115**, 1087–529.

Kurashima, A. (1968) Studies on the winter and summer monsoons in east Asia based on dynamic concept, *Geophys. Mag.* (Tokyo) **34**, 145–236.

Kurihara, Y. (1985) Numerical modeling of tropical cyclones; in Manabe, S. (ed.) *Issues in Atmospheric and Oceanic Modeling. Part B, Weather Dynamics. Advances in Geophysics*, Academic Press, New York, pp. 255–87.

Lander, M. A. (1990) Evolution of the cloud pattern during the formation of tropical cyclone twins symmetrical with respect to the Equator, *Monthly Weather Review* **118**, 1194–1202.

Lau, K.-M. and Li, M.-T. (1984) The monsoon of East Asia and its global associations – a survey, *Bull. Amer. Met. Soc.* **65**, 114–25.

Le Borgue, J. (1979) Polar invasion into Mauretania and Senegal, *Ann. de Géog.* **88**(485), 521–48.

Lighthill, J. and Pearce, R. P. (eds) (1979) *Monsoon Dynamics*, Cambridge University Press, Cambridge (735 pp.).

Lockwood, J. G. (1965) The Indian monsoon – a review, *Weather* **20**, 2–8.

Logan, R. F. (1960) *The Central Namib Desert, South-west Africa*, National Academy of Sciences, National Research Council, Publication 758, Washington, DC (162 pp.).

Lowell, W. E. (1954) Local weather of the Chicama Valley, Peru, *Archiv Met. Geophys. Biokl.* B **5**, 41–51.

Lydolph, P. E. (1957) A comparative analysis of the dry western littorals, *Ann. Assn. Amer. Geog.* **47**, 213–30.

Maejima, I. (1967) Natural seasons and weather singularities in Japan, *Geog. Report No. 2.* Tokyo Metropolitan University, pp. 77–103.

Maley, J. (1982) Dust, clouds, rain types, and climatic variations in tropical North Africa, *Quaternary Res.* **18**, 1–16.

Malkus, J. S. (1955–6) The effects of a large island upon the trade-wind air stream, *Quart. J. Roy. Met. Soc.* **81**, 538–50; **82**, 235–8.

Malkus, J. S. (1958) Tropical weather disturbances: why do so few become hurricanes?, *Weather* **13**, 75–89.

Malkus, J. S. and Riehl, H. (1964) *Cloud Structure and Distributions over the Tropical Pacific Ocean*, University of California Press, Berkeley and Los Angeles (229 pp.).

Mason, B. J. (1970) Future developments in meteorology: an outlook to the year 2000, *Quart. J. Roy. Met. Soc.* **96**, 349–68.

Meehl, G. A. (1987) The tropics and their role in the global climate system, *Geog. Jour.* **153**, 21–36.

Mink, J. F. (1960) Distribution pattern of rainfall in the leeward Koolau Mountains, Oahu, Hawaii, *J. Geophys. Res.* **65**, 2869–76.

Molion, L. C. B. (1987) On the dynamic climatology of the Amazon Basin and associated rain-producing mechanisms; in Dickinson, R. E. (ed.) *The Geophysiology of Amazonia*, John Wiley & Sons, New York, pp. 391–405.

Musk, L. (1983) Outlook – changeable, *Geog. Mag.* **55**, 532–3.

Neal, A. B., Butterworth, L. J. and Murphy, K. M. (1977) The morning glory, *Weather* **32**, 176–83.

Nicholson, S. E. (1989) Long-term changes in African rainfall, *Weather* **44**, 46–56.

Nicholson, S. E. and Flohn, H. (1980) African environmental and climatic changes and the general atmospheric circulation in late Pleistocene and Holocene, *Climatic Change* **2**, 313–48.

Nieuwolt, S. (1977) *Tropical Climatology*, John Wiley & Sons, London (207 pp.).

Omotosho, J. B. (1985) The separate contributions of line squalls, thunderstorms and the monsoon to

the total rainfall in Nigeria, *J. Climatology* **5**, 543–52.

Palmén, E. (1948) On the formation and structure of tropical hurricanes, *Geophysica* **3**, 26–38.

Palmer, C. E. (1951) Tropical meteorology; in Malone, T. F. (ed.) *Compendium of Meteorology*, American Meteorological Society, Boston, Mass., pp.859–80.

Physik, W. L. and Smith, R. K. (1985) Observations and dynamics of sea breezes in northern Australia, *Austral. Met. Mag.* **33**, 51–63.

Raghavan, K. (1967) Influence of tropical storms on monsoon rainfall in India, *Weather* **22**, 250–5.

Ramage, C. S. (1952) Relationships of general circulation to normal weather over southern Asia and the western Pacific during the cool season, *J. Met.* **9**, 403–8.

Ramage, C. S. (1964) Diurnal variation of summer rainfall in Malaya, *J. Trop. Geog.* **19**, 62–8.

Ramage, C. S. (1968) Problems of a monsoon ocean, *Weather* **23**, 28–36.

Ramage, C. S. (1971) *Monsoon Meteorology*, Academic Press, New York and London (296 pp.).

Ramage, C. S. (1986) El Niño, *Sci. American* **254**, 76–83.

Ramage, C. S., Khalsa, S. J. S. and Meisner, B. N. (1980) The central Pacific near-equatorial convergence zone, *J. Geophys. Res.* **86**(7), 6580–98.

Ramaswamy, C. (1956) On the sub-tropical jet stream and its role in the development of large-scale convection, *Tellus* **8**, 26–60.

Ramaswamy, C. (1962) Breaks in the Indian summer monsoon as a phenomenon of interaction between the easterly and the sub-tropical westerly jet streams, *Tellus* **14**, 337–49.

Ratisbona, L. R. (1976) The climate of Brazil; in Schwerdtfeger, W. (ed.) *Climates of Central and South America*, World Survey of Climatology 12, Elsevier, Amsterdam, pp. 219–93.

Reynolds, R. (1985) Tropical meteorology, *Prog. Phys. Geog.* **9**, 157–86.

Riehl, H. (1954) *Tropical Meteorology*, McGraw-Hill, New York (392 pp.).

Riehl, H. (1963) On the origin and possible modification of hurricanes, *Science* **141**, 1001–10.

Riehl, H. (1979) *Climate and Weather in the Tropics*, Academic Press, New York (611 pp.).

Rossignol-Strick, M. (1985) Mediterranean Quaternary sapropels, an immediate response of the African monsoon to variation in isolation, *Palaeogeography, Palaeoclimatology, Palaeoecology* **49**, 237–63.

Sadler, J. C. (1975a) The monsoon circulation and cloudiness over the GATE area, *Monthly Weather Review* **103**, 369–87.

Sadler, J. C. (1975b) *The Upper Tropospheric Circulation over the Global Tropics*, UHMET-75-05, Department of Meteorology, University of Hawaii (35 pp.).

Saha, R. R. (1973) Global distribution of double cloud bands over the tropical oceans, *Quart. J. Roy. Met. Soc.* **99**, 551–5.

Saito, R. (1959) The climate of Japan and her meteorological disasters, *Proceedings of the International Geophysical Union*, Regional Conference in Japan, Tokyo, pp. 173–83.

Sawyer, J. S. (1970) Large-scale disturbance of the equatorial atmosphere, *Met. Mag.* **99**, 1–9.

Schwerdtfeger, W. (ed.) (1976) *Climates of Central and South America*, World Survey of Climatology 12, Elsevier, Amsterdam (532 pp.).

Shaw, D. B. (ed.) (1978) *Meteorology over the Tropical Oceans*, Royal Meteorological Society, Bracknell (278 pp.).

Shukla, J. and Wallace, J. (1983) Numerical simulation of the atmospheric response to the equatorial Pacific sea surface temperature anomalies, *J. Atmos. Sci.* **40**, 1613–40.

Sikka, D. R. (1977) Some aspects of the life history, structure and movement of monsoon depressions, *Pure and Applied Geophysics* **115**, 1501–29.

Thompson, B. W. (1951) An essay on the general circulation over South-East Asia and the West Pacific, *Quart. J. Roy. Met. Soc.* 569–97.

Trenberth, K. E. (1976) Spatial and temporal oscillations in the Southern Oscillation, *Quart. J. Roy. Met. Soc.* **102**, 639–53.

Trewartha, G. T. (1958) Climate as related to the jet stream in the Orient, *Erdkunde* **12**, 205–14.

Trewartha, G. T. (1981) *The Earth's Problem Climates* (2nd edn), University of Wisconsin Press, Madison (371 pp.).

Tyson, P. D. (1986) *Climatic Change and Variability in Southern Africa*, Oxford University Press, Cape Town (220 pp.).

Watts, I. E. M. (1955) *Equatorial Weather, with particular reference to South-east Asia*, Oxford University Press, London (186 pp.).

Webster, P. J. (1987a) The elementary monsoon; in Fein, J. S. and Stephens, P. L. (eds) *Monsoons*, John Wiley & Sons, New York, pp. 3–32.

Webster, P. J. (1987b) The variable and interactive monsoon; in Fein, J. S. and Stephens, P. L. (eds) *Monsoons*, John Wiley & Sons, New York, pp. 269–330.

World Meteorological Organization (1972) Synoptic analysis and forecasting in the tropics of Asia and the south-west Pacific, *WMO No. 321*, Geneva (524 pp.).

World Meteorological Organization (n.d.) *The Global Climate System. A critical review of the*

climate system during 1982–1984, World Climate Data Programme, WMO, Geneva (52 pp.).

Wyrtki, K. (1982) The Southern Oscillation, ocean–atmosphere interaction and El Niño, *Marine Tech. Soc. J.* **16**, 3–10.

Yarnal, B. (1985) Extratropical teleconnections with El Niño/Southern Oscillation (ENSO) events, *Prog. Phys. Geog.* **9**, 315–52.

Ye, D. (1981) Some characteristics of the summer circulation over the Qinghai-Xizang (Tibet) Plateau and its neighbourhood, *Bull. Amer. Met. Soc.* **62**, 14–19.

Ye, D. and Gao, Y.-X. (1981) The seasonal variation of the heat source and sink over Qinghai-Xizang plateau and its role in the general circulation; in *Geoecological and Ecological Studies of Qinghai-Xizang Plateau*, Vol. 2, Science Press, Beijing, pp. 1453–61.

Yoshino, M. M. (1969) Climatological studies on the polar frontal zones and the intertropical convergence zones over South, South-east and East Asia, *Climatol. Notes* **1**, Hosei University (71 pp.).

Yoshino, M. M. (ed.) (1971) *Water Balance o, Monsoon Asia*, University of Tokyo Press (308 pp.).

Young, J.A. (co-ordinator) (1972) *Dynamics of the Tropical Atmosphere* (Notes from a Colloquium), National Center for Atmospheric Research, Boulder, Colo. (587 pp.).

CHAPTER 7 SMALL-SCALE CLIMATES

Anderson, G. E. (1971) Mesoscale influences on wind fields, *J. App. Met.* **10**, 377–86.

Atkinson, B. W. (1968) A preliminary examination of the possible effect of London's urban area on the distribution of thunder rainfall 1951–60, *Trans. Inst. Brit. Geog.* **44**, 97–118.

Atkinson, B. W. (1977) *Urban Effects on Precipitation: An Investigation of London's Influence on the Severe Storm of August 1975*, Dept. Geog. Queen Mary Coll. London, Occasional Paper 8 (31 pp.).

Atkinson, B. W. (1987) Precipitation; in Gregory, K. J. and Walling, D. E. (eds) *Human Activity and Environmental Processes*, John Wiley & Sons, Chichester, pp. 31–50.

Bach, W. (1971) Atmospheric turbidity and air pollution in Greater Cincinnati, *Geog. Rev.* **61**, 573–94.

Bach, W. (1979) Short-term climatic alterations caused by human activities, *Prog. Phys. Geog.* **3**(1), 55–83.

Bach, W. and Patterson, W. (1969) Heat budget studies in Greater Cincinnati, *Proc. Assn. Amer. Geog.* **1**, 7–16.

Brimblecombe, P. (1986) *Air: Composition and Chemistry*, Cambridge University Press, Cambridge (224 pp.).

Bryson, R. A. and Kutzbach, J. E. (1968) *Air Pollution*, Association of American Geographers, Commission on College Geography, Resource Paper 2 (42 pp.).

Caborn, J. M. (1955) The influence of shelter-belts on microclimate, *Quart. J. Roy. Met. Soc.* **81**, 112–15.

Chandler, T. J. (1965) *The Climate of London*, Hutchinson, London (292 pp.).

Chandler, T. J. (1967) Absolute and relative humidities in towns, *Bull. Amer. Met. Soc.* **48**, 394–9.

Changnon, S. A. (1969) Recent studies of urban effects on precipitation in the United States, *Bull. Amer. Met. Soc.* **50**, 411–21.

Changnon, S. A. (1979) What to do about urban-generated weather and climate changes, *J. Amer. Plan. Assn.* **45**(1), 36–48.

Committee on Air Pollution (1955) *Report*, Cmnd. 9322, HMSO, London.

Coutts, J. R. H. (1955) Soil temperatures in an afforested area in Aberdeenshire, *Quart. J. Roy. Met. Soc.* **81**, 72–9.

Dickinson, R. E. and Henderson-Sellers, A. (1988) Modelling tropical deforestation: a study of GCM land-surface parameterizations, *Quart. J. Roy. Met. Soc.***114**, 439–62.

Duckworth, F. S. and Sandberg, J. S. (1954) The effect of cities upon horizontal and vertical temperature gradients, *Bull. Amer. Met. Soc.* **35**, 198–207.

Food and Agriculture Organization of the United Nations (1962) *Forest Influences*, Forestry and Forest Products Studies No. 15, Rome (307 pp.).

Garnett, A. (1967) Some climatological problems in urgban geography with special reference to air pollution, *Trans. Inst. Brit. Geog.* **42**, 21–43.

Gay, L. W. and Stewart, J. B. (1974) Energy balance studies in coniferous forests, *Report No. 23*, Inst. Hydrol., Nat. Env. Res. Coun., Wallingford.

Goldreich, Y. (1984) Urban topo-climatology, *Prog. Phys. Geog.* **8**, 336–64.

Goldsmith, J. R. (1969) Los Angeles smog, *Science Jour.* **5**, 44–9.

Heintzenberg, J. (1989) Arctic haze: air pollution in polar regions, *Ambio* **18**, 50–5.

Hewson, E. W. (1951) Atmospheric pollution; in Malone, T. F. (ed.) *Compendium of Meteorology*, American Meteorological Society, Boston, Mass., pp. 1139–57.

Jäger, J. (1983) *Climate and Energy Systems. A*

review of their interactions, Wiley, New York (231 pp.).
Jauregi, E. (1987) Urban heat island development in medium and large urban areas in Mexico, *Erdkunde* **41**, 48–51.
Jenkins, I. (1969) Increases in averages of sunshine in Greater London, *Weather* **24**, 52–4.
Kessler, A. (1985) Heat balance climatology; in Essenwanger, B.M. (ed.) *World Survey of Climatology, Vol. 1A: General climatology*, Elsevier, Amsterdam (224 pp.).
Kittredge, J. (1948) *Forest Influences*, McGraw-Hill, New York (394 pp.).
Koppány, Gy. (1975) Estimation of the life span of atmospheric motion systems by means of atmospheric energetics, *Met. Mag.* **104**, 302–6.
Landsberg, H. E. (1981) City climate; in Landsberg, H. E. (ed.) *General Climatology 3*, World Survey of Climatology 3, Elsevier, Amsterdam, pp. 299–334.
Long, I. F., Monteith, J. L., Penman, H. L. and Szeicz, G. (1964) The plant and its environment, *Meteorologische Rundschau* **17**(4), 97–101.
Lowry, W. P. (1969) *Weather and Life*, Academic Press, New York (305 pp.).
McNaughton, K. and Black, T. A. (1973) A study of evapotranspiration from a Douglas fir forest using the energy balance approach, *Water Resources Research* **9**, 1579–90.
Maejima, I. *et al.* (1982) Recent climatic change and urban growth in Tokyo and its environs, *Japanese Prog. Climatology*, March 1983, 1–22.
Marshall, W. A. L. (1952) *A Century of London Weather*, Met. Office, Air Ministry, Rept MO 508. HMSO, London (103 pp.).
Meetham, A.R. (1955) Know your fog, *Weather* **10**, 103–5.
Meetham, A. R. *et al.* (1980) *Atmospheric Pollution* (4th edn) (1st edn 1952), Pergamon Press, Oxford and London.
Miess, M. (1979) The climate of cities; in Laurie, I. C. (ed.) *Nature in Cities*, Wiley, Chichester, pp. 91–104.
Miller, D. H. (1965) The heat and water budget of the earth's surface, *Adv. Geophys.* **11**, 175–302.
Nicholas, F. W. and Lewis, J.E. (1980) Relationships between aerodynamic roughness and land use and land cover in Baltimore, Maryland, *US Geol. Surv. Prof. Paper 1099-C* (36 pp.).
Oke, T. R. (1978) *Boundary Layer Climates*, Methuen, London (372 pp.), (2nd edn 1987, 435 pp.).
Oke, T. R. (1979) *Review of Urban Climatology 1973–76*, WMO Technical Note No. 169, Geneva, World Meteorological Organization (100 pp.).
Oke, T. R. (1980) Climatic impacts of urbanization; in Bach, W., Pankrath, J. and Williams, J. (eds) *Interactions of Energy and Climate*, D. Reidel, Dordrecht, Holland, pp. 339–56.
Oke, T. R. (1982) The energetic basis of the heat island, *Quart. J. Roy. Met. Soc.* **108**, 1–24.
Oke, T. R. (ed.) (1986) *Urban Climatology and its Applications with Special Regard to Tropical Areas*, World Meteorological Organization Publication No. 652, Geneva (534 pp.).
Oke, T. R. (1988) The urban energy balance, *Prog. Phys. Geog.* **12**(4), 471–508.
Oke, T. R. and East, C. (1971) The urban boundary layer in Montreal, *Boundary-Layer Met.* **1**, 411–37.
Pankrath, J. (1980) Impact of heat emissions in the Upper-Rhine region; in Bach, W., Pankrath, J. and Williams, J. (eds) *Interactions of Energy and Climate*, D. Reidel, Dordrecht, Holland, pp. 363–81.
Parry, M. (1966) The urban 'heat island'; in Tromp, S. W. and Weite, W. H. (eds) *Biometeorology* 2. Pergamon, Oxford and London, pp. 616–24.
Pease, R. W., Jenner, C. B. and Lewis, J. E. (1980) The influences of land use and land cover on climate analysis: an analysis of the Washington–Baltimore area, *US Geol. Surv. Prof. Paper 1099-A* (39 pp.).
Peel, R. F. (1974) Insolation and weathering: some measures of diurnal temperature changes in exposed rocks in the Tibesti region, central Sahara, *Zeit.f, Geomorph. Supp.* **21**, 19–28.
Peterson, J. T. (1971) Climate of the city; in Detwyler, T. R. (ed.) *Man's Impact on Environment*, McGraw-Hill, New York, pp. 131–54.
Plate, E. (1972) Berücksichtigung von Windströmungen in der Bauleitplanung; in *Seminarberichte Rahmenthema Unweltschutz*. Institut für Stätebau und Landesplanung, Selbstverlag, Karlsruhe, pp. 201–29.
Reynolds, E. R. C. and Leyton, L. (1963) Measurement and significance of throughfall in forest stands; in Whitehead, F. M. and Rutter, A. J. (eds) *The Water Relations of Plants*, Blackwell Scientific Publications, Oxford, pp. 127–41.
Richards, P. W. (1952) *The Tropical Rain Forest*, Cambridge University Press, Cambridge (450 pp.).
Rutter, A. J. (1967) Evaporation in forests, *Endeavour* **97**, 39–43.
Scorer, R. (1968) *Air Pollution*, Pergamon, Oxford and London (151 pp.).
Seinfeld, J. H. (1989) Urban air pollution: state of the science, *Science* **243**, 745–52.
Sellers, W. D. (1965) *Physical Climatology*, University of Chicago Press, Chicago (272 pp.).
Shuttleworth, W. J. *et al.* (1985) Daily variation of

temperature and humidity within and above the Amazonian forest, *Weather* **40**, 102–8.

Smagorinsky, J. (1974) Global atmospheric modelling and the numerical simulation of climate; in Hess, W. N. (ed.) *Weather and Climate Modification*, Wiley, New York, pp. 633–86.

Sopper, W. E. and Lull, H. W. (eds) (1967) *International Symposium on Forest Hydrology*, Pergamon, Oxford and London (813 pp.).

Stearn, A. C. (ed.) (1968) *Atmospheric Pollution* (3 vols), Academic Press, New York.

Sukachev, V. and Dylis, N. (1968) *Fundamentals of Forest Biogeocoenology*, Oliver and Boyd, Edinburgh (672 pp.).

Terjung, W. H. (1970) Urban energy balance climatology, *Geog. Rev.* **60**, 31–53.

Terjung, W. H. and Louis, S. S.-F. (1973) Solar radiation and urban heat islands, *Ann. Assn. Amer. Geog.* **63**, 181–207.

Terjung, W. H. and O'Rourke, P. A. (1980) Simulating the causal elements of urban heat islands, *Boundary-Layer Met.* **19**, 93–118.

Terjung, W. H. and O'Rourke, P. A. (1981) Energy input and resultant surface temperatures for individual urban interfaces, *Archiv. Met. Geophys. Biokl.* B, **29**, 1–22.

Turner, W. C. (1955) Atmospheric pollution, *Weather* **10**, 110–19.

Tyson, P. D., Garstang, M. and Emmitt, G. D. (1973) *The Structure of Heat Islands*, Occasional Paper No. 12, Dept. of Geography and Environmental Studies, University of the Witwatersrand, Johannesburg (71 pp.).

US DHEW (1970) *Air Quality Criteria for Photochemical Oxidants*, National Air Pollution Control Administration, US Public Health Service, Publication No. AP-63, Washington, DC.

Vehrencamp, J. E. (1953) Experimental investigation of heat transfer at an air–earth interface, *Trans. Amer. Geophys. Union* **34**, 22–30.

White, W. H., Anderson, J. A., Blumenthal, D. L., Husar, R. B., Gillani, N. V., Husar, J. D. and Wilson, W. E. (1976) Formation and transport of secondary air pollutants: ozone and aerosols in the St Louis urban plume, *Science* **194**, 187–9.

World Meteorological Organization (1970) *Urban Climates*, WMO Technical Note No. 108 (390 pp.).

Zon, R. (1941) Climate and the nation's forests; in US Dept. of Agriculture Yearbook, *Climate and Man*, pp. 477–98.

CHAPTER 8 CLIMATIC CHANGE

Bayce, A. (ed.) (1979) *Man's Influence on Climate*, D. Reidel, Dordrecht, Holland (113 pp.).

Beckinsale, R. P. (1965) Climatic change: a critique of modern theories; in Whittow, J. B. and Wood, P. D. (eds) *Essays in Geography for Austin Miller*, University of Reading Press, pp. 1–38.

Bolin, B., Döös, B. R., Jäger, J. and Warrick, R. A. (eds) (1986) *The Greenhouse Effect, Climatic Change, and Ecosystems*, John Wiley & Sons, Chichester (541 pp.).

Bradley, R. S. (1985) *Quaternary Paleoclimatology*, Allen & Unwin, Boston (472 pp.).

Brádzil, R., Samaj, F. and Valovič, S. (1985) Variation of spatial annual precipitation sums in central Europe in the period 1881–1980, *J. Climatology* **5**, 617–31.

Broecker, W. S. and Denton, G. S. (1990) What drives glacial cycles? *Sci. American* **262**, 48–56.

Bryson, R. A. (1968) All other factors being constant ..., *Weatherwise* **21**, 51–61.

Bryson, R. A. (1974) A perspective on climatic change, *Science* **184**, 753–60.

Callendar, G. S. (1961) Temperature fluctuations and trends over the earth, *Quart. J. Roy. Met. Soc.* **87**, 1–12.

Conrad, V. and Pollak, L. W. (1950) *Methods in Climatology*, Harvard University Press, Cambridge, Mass. (See ch. 2, Statistical analysis of climatic elements, pp. 17–60).

Dickinson, R. E. and Henderson-Sellers, A. (1988) Modelling tropical deforestation: a study of GCM land-surface parameterizations, *Quart, J. Roy. Met. Soc.* **114**, 439–62.

Frakes, L. A. (1979) *Climates throughout Geologic Time*, Elsevier, Amsterdam (310 pp.).

Gilliland, R. L. (1982) Solar, volcanic and CO_2 forcing of recent climatic change, *Climatic Change* **4**, 111–31.

Goudie, A. (1981) *The Human Impact: Man's role in environmental change*, Blackwell, Oxford (316 pp.).

Gregory, S. (1962 and subsequent editions) *Statistical Methods and the Geographer*, Longman, London (240 pp.).

Gregory, S. (1969) Rainfall reliability; in Thomas, M. F. and Whittington, G. W. (eds) *Environment and Land Use in Africa*, Methuen, London, pp. 57–82.

Gribbin, J. (ed.) (1978) *Climatic Change*, Cambridge University Press, Cambridge (280 pp.).

Gribbin, J. (1987) An atmosphere in convulsions, *New Scientist* **116**(1588), 30–1.

Grove, J. M. (1988) *The Little Ice Age*, Methuen, London (498 pp.).

Hansen, J., Johnson, D., Lacis, A., Lebedeff, S., Lee, R., Rind, D. and Russell, G. (1981) Climate impact of increasing carbon dioxide, *Science* **213**, 957–66.

Hansen, J., Fung, I., Lacis, A., Rind, D., Lebedeff, S., Ruedy, R. and Russell, G. (1988) Global climatic changes as forecast by Goddard Institute for Space Studies three-dimensional model, *J. Geophys. Res.* **93**(D8), 9341–64.

Hansen, J. and Lacis, A. (1990) Sun and dust versus greenhouse gases: an assessment of their relative roles in global climate change, *Nature* **346**(6286), 713–19.

Henderson-Sellers, A. and Wilson, M. F. (1983) Surface albedo data for climate modelling, *Rev. Geophys. Space Phys.* **21**, 1743–8.

Hughes, M. K., Kelly, P. M., Pilcher, J. R. and La Marche, V. (eds) (1981) *Climate from Tree Rings*, Cambridge University Press, Cambridge (400 pp.).

Imbrie, J. and Imbrie, K. P. (1979) *Ice Ages: Solving the mystery*, Macmillan, London (224 pp.).

Intergovernmental Panel on Climate Change (IPCC) (1900) *Climate Change: The IPCC Scientific Assessment*, (eds Houghton, J. T., Jenkins, G. J. and Ephraums, J. J.), Cambridge University Press, Cambridge (365 pp.).

Intergovernmental Panel on Climate Change (IPCC) (1992) Climate Change 1992: *The Supplementary Report to the IPCC Scientific Assessment*, Houghton, J. T., Callander, B. A. and Varney, S. K. (eds), Cambridge University Press, Cambridge (200 pp.).

Jäger, J. and Barry, R. G. (1991) 'Climate' in B. L. Turner II (ed.) *The Earth as Transformed by Human Actions*, Cambridge University Press, Cambridge, pp. 335–51.

Jones, P. D. and Wigley, T. M. L. (1990) Global warming trends, *Sci. American* **262**(8), 66–73.

Jones, P. D., Wigley, T. M. L. and Farmer, G. (1991) Marine and land temperature data sets: a comparison and a look at recent trends; in Schlesinger, M. E. (ed.) *Greenhouse-gas-induced Climatic Change*, Elsevier, Amsterdam, pp. 153–72.

Kelly, P. M. (1980) Climate: historical perspective and climatic trends; in Doornkamp, J. C. and Gregory, K. J. (eds) *Atlas of Drought in Britain, 1975–6*, Institute of British Geographers, London, pp. 9–11.

Kutzbach, J. E. and Street-Perrott, A. (1985) Milankovitch forcings of fluctuations in the level of tropical lakes from 18 to 0 kyr BP, *Nature* **317**, 130–9.

Lamb, H. H. (1965) Frequency of weather types, *Weather* **20**, 9–12.

Lamb, H. H. (1966) *The Changing Climate: Selected Papers*, Methuen, London (236 pp.).

Lamb, H. H. (1970) Volcanic dust in the atmosphere; with a chronology and an assessment of its meteorological significance, *Phil. Trans. Roy. Soc.*, A **266**, 425–533.

Lamb, H. H. (1977) *Climate: Present, Past and Future, 2: Climatic History and the Future*, Methuen, London (835 pp.).

Landsberg, H. E. (1970) Man-made climatic changes, *Science* **170**, 1265–74.

Leopold, L. B. (1951) Rainfall frequency: an aspect of climatic variation, *Trans. Amer. Geophys. Union.* **32**(3), 347–57.

Lewis, P. (1960) The use of moving averages in the analysis of time-series, *Weather* **15**, 121–6.

McCormac, B. M. and Seliga, T. A. (1979) *Solar–Terrestrial Influences on Weather and Climate*, D. Reidel, Dordrecht, Holland (340 pp.).

Macdonald, G. J. F. (1966) Weather and climate modification: problems and prospects, *Bull. Amer. Met. Soc.* **47**, 4–19.

Manley, G. (1958) Temperature trends in England, 1698–1957, *Archiv. Met. Geophys. Biokl.* (Vienna), B. **9**, 413–33.

Mather, J. R. and Sdasyuk, G. V. (eds) (1991) *Global Change: Geographical Approaches* (Sections 3.2.2, 3.2.3), Tucson, University of Arizona Press.

Meehl, G. A. and Washington, W. M. (1990) CO_2 climate sensitivity and snow–sea-ice albedo parameterization in an atmospheric GCM coupled to a mixed-layer ocean model, *Climatic Change* **16**, 283–306.

Mitchell, J. M. Jr (ed.) (1968) Causes of climatic change, *Amer. Met. Soc. Monogr.* **8**(30) (159 pp.).

Mitchell, J. M., Jr (1972) The natural breakdown of the present interglacial and its possible intervention by human activities, *Quat. Res.* **2**, 436–45.

Mitchell, J. M., Jr, Stockton, C. W. and Meko, D. M. (1979) Evidence of a 22-year rhythm of drought in the western United States related to the Hale solar cycle since the seventeenth century; in McCormac, B. M. and Seliga, T. A. (eds) *Solar–Terrestrial Influences on Weather and Climate*, D. Reidel, Dordrecht, Holland, pp. 125–43.

Nicholson, S. E. (1980) The nature of rainfall fluctuations in subtropical West Africa, *Mon. Wea. Rev.* **108**, 473–87.

Nicholson, S. E. (1985) Sub-Saharan rainfall 1981–84, *J. Applied Met.* **24**, 1388–91.

Pfister, C. (1985) Snow cover, snow lines and glaciers in central Europe since the 16th century; in Tooley, M. J. and Sheail, G. M. (eds) *The Climatic Scene*, Allen & Unwin, London, pp. 154–74.

Pittock, A. B., Frakes, L. A., Jenssen, D., Peterson, J. A. and Zillman, J. W. (eds) (1978) *Climatic Change and Variability: A southern perspective*, Cambridge University Press, Cambridge (455 pp.).

Schneider, S. H. and Kellogg, W. W. (1973) The chemical basis for climate change; in Rasool, S. I. (ed.) *Chemistry of the Lower Atmosphere*, Plenum, New York, pp. 203–49.

Schuurmans, C. J. E. (1984) Climate variability and its time changes in European countries, based on instrumental observations; in Flohn, H. and Fantechi, R. (eds) *The Climate of Europe: Past, Present and Future*, D. Reidel, Dordrecht, pp. 65–101.

Sewell, W. R. D. (ed.) (1966) *Human Dimensions of Weather Modification*, University of Chicago, Dept. of Geography, Research Paper No. 105 (423 pp).

Sioli, H. (1985) The effects of deforestation in Amazonia, *Geog. J.* **151**, 197–203.

Stolarski, R. S. (1988a) The Antarctic ozone hole, *Sci. American* **258**(1), 20–6.

Stolarski, R. S. (1988b) Changes in ozone over the Antarctic; in Roland, F. S. and Isaksen, I. S. A. (eds) *The Changing Atmosphere*, John Wiley & Sons, Chichester, pp. 105–99.

Street, F. A. (1981) Tropical palaeoenvironments, *Progr. Phys. Geog.* **5**, 157–85.

Study of Man's Impact on Climate (SMIC) (1971) *Inadvertent Climate Modification*, Massachusetts Institute of Technology Press, Cambridge, Mass. (308 pp.).

Thompson, R. D. (1989) Short-term climatic change: evidence, causes, environmental consequences and strategies for action, *Prog. Phys. Geog.* **13**(3), 315–47.

Tooley, M. J. and Shennan, I. (eds) (1987) Sea-level changes, *Trans. Inst. Brit. Geog. Sp. Pub. 20* (397 pp.).

Toon, O. B. and Pollack, J. B. (1980) Atmospheric aerosols and climate, *Amer. Scientist* **68**, 268–78.

Wigley, T. M. L., Ingram, M. J. and Farmer, G. (eds) (1981) *Climate and History*, Cambridge University Press, Cambridge (530 pp.).

Wigley, T. M. L. and Raper, S. C. B. (1992) 'Implications for climate and sea level of fixed IPCC emissions scenarios', *Nature*, **357**, pp. 293–300.

Williams, J. (ed.) (1978) *Carbon Dioxide, Climate and Society*, Pergamon, Oxford (332 pp.).

Index